D0080983

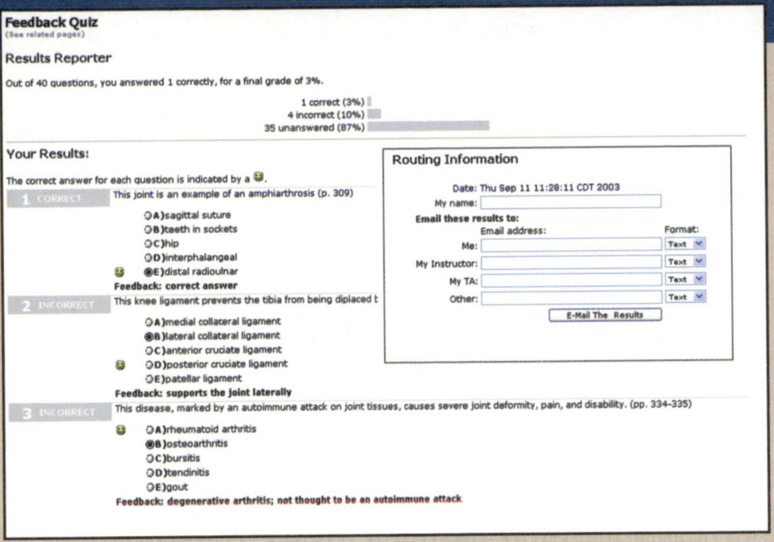

Test Yourself

Take a quiz at the *Human Anatomy* Online Learning Center to gauge your mastery of chapter content. Each chapter quiz is specially constructed to test your comprehension of key concepts. Immediate feedback on your responses explains why an answer is correct or incorrect. You can even e-mail your quiz results to your professor!

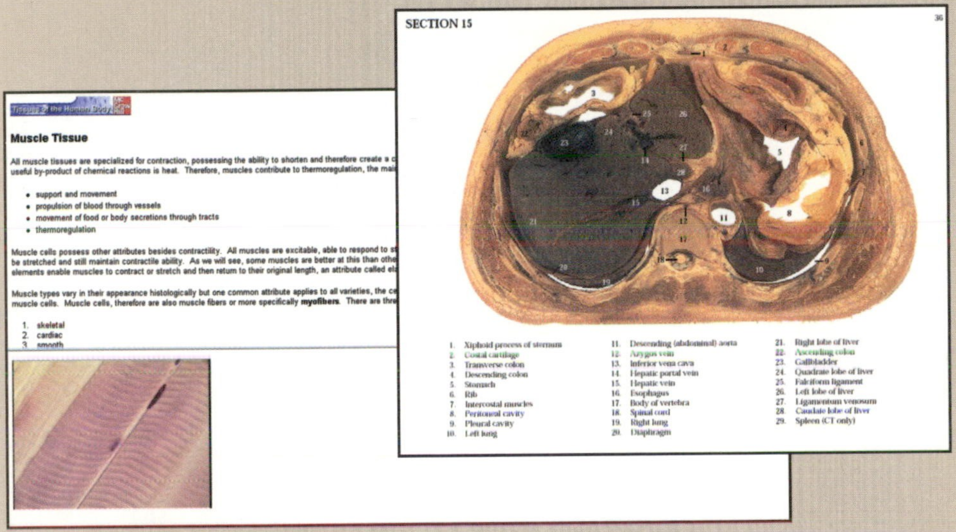

Course Tools

Here you'll find chapter-specific study outlines and a listing of relevant websites. The *Human Anatomy* Online Learning Center also features cutting-edge online histology and anatomy atlases plus general study tips and career information.

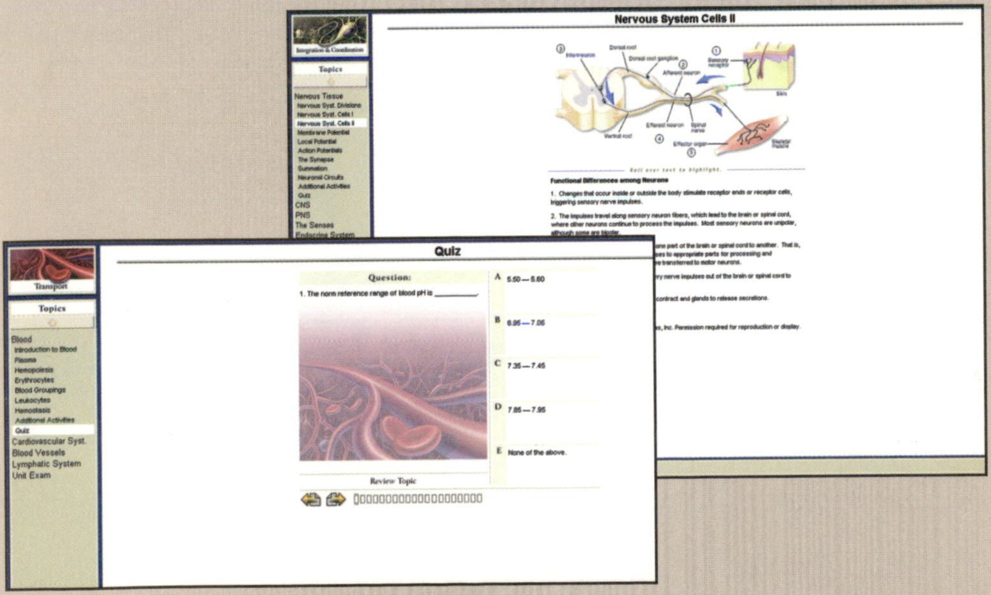

Access to Premium Learning Materials

The *Human Anatomy* Online Learning Center is your portal to exclusive interactive study tools like McGraw-Hill's **Essential Study Partner**.

Visit www.mhhe.com/saladinha1 today!

IMPORTANT:

HERE IS YOUR REGISTRATION CODE TO ACCESS YOUR PREMIUM McGRAW-HILL ONLINE RESOURCES.

For key premium online resources you need THIS CODE to gain access. Once the code is entered, you will be able to use the Web resources for the length of your course.

If your course is using **WebCT** or **Blackboard**, you'll be able to use this code to access the McGraw-Hill content within your instructor's online course.

Access is provided if you have purchased a new book. If the registration code is missing from this book, the registration screen on our Website, and within your WebCT or Blackboard course, will tell you how to obtain your new code.

Registering for McGraw-Hill Online Resources

To gain access to your McGraw-Hill web resources simply follow the steps below:

1. USE YOUR WEB BROWSER TO GO TO: **www.mhhe.com/saladinha1**
2. CLICK ON **FIRST TIME USER**.
3. ENTER THE REGISTRATION CODE* PRINTED ON THE TEAR-OFF BOOKMARK ON THE RIGHT.
4. AFTER YOU HAVE ENTERED YOUR REGISTRATION CODE, CLICK **REGISTER**.
5. FOLLOW THE INSTRUCTIONS TO SET-UP YOUR PERSONAL UserID AND PASSWORD.
6. WRITE YOUR UserID AND PASSWORD DOWN FOR FUTURE REFERENCE. KEEP IT IN A SAFE PLACE.

TO GAIN ACCESS to the McGraw-Hill content in your instructor's **WebCT** or **Blackboard** course simply log in to the course with the UserID and Password provided by your instructor. Enter the registration code exactly as it appears in the box to the right when prompted by the system. You will only need to use the code the first time you click on McGraw-Hill content.

Thank you, and welcome to your McGraw-Hill online resources!

Mc Graw Hill Higher Education

*YOUR REGISTRATION CODE CAN BE USED ONLY ONCE TO ESTABLISH ACCESS. IT IS NOT TRANSFERABLE.

0-07-292130-7 T/A SALADIN: HUMAN ANATOMY, 1/E

MCGRAW-HILL
ONLINE RESOURCES

REGISTRATION CODE

6VYK-ZZD7-D7SA-GL37-9K4W

Mc Graw Hill Higher Education

How's Your Math?

Do you have the math skills you need to succeed?

Why risk not succeeding because you struggle with your math skills?

Get access to a web-based, personal math tutor:

- Available 24/7, unlimited use
- Driven by artificial intelligence
- Self-paced
- An entire month's subscription **for much less** than the cost of one hour with a human tutor

ALEKS is an inexpensive, private, infinitely patient math tutor that's accessible any time, anywhere you log on.

ALEKS® McGraw Hill

Log On for a **FREE** 48-hour Trial

www.highedstudent.aleks.com

ALEKS is a registered trademark of ALEKS Corporation.

HUMAN ANATOMY

KENNETH S. SALADIN
Georgia College and State University

McGraw Hill Higher Education

Boston Burr Ridge, IL Dubuque, IA Madison, WI New York San Francisco St. Louis
Bangkok Bogotá Caracas Kuala Lumpur Lisbon London Madrid Mexico City
Milan Montreal New Delhi Santiago Seoul Singapore Sydney Taipei Toronto

The McGraw·Hill Companies

Higher Education

HUMAN ANATOMY

Published by McGraw-Hill, a business unit of The McGraw-Hill Companies, Inc., 1221 Avenue of the Americas, New York, NY 10020. Copyright © 2005 by The McGraw-Hill Companies, Inc. All rights reserved. No part of this publication may be reproduced or distributed in any form or by any means, or stored in a database or retrieval system, without the prior written consent of The McGraw-Hill Companies, Inc., including, but not limited to, in any network or other electronic storage or transmission, or broadcast for distance learning.

Some ancillaries, including electronic and print components, may not be available to customers outside the United States.

This book is printed on acid-free paper.

3 4 5 6 7 8 9 0 VNH/VNH 0 9 8 7 6 5

ISBN 0–07–039080–0

Publisher: *Martin J. Lange*
Sponsoring editor: *Michelle Watnick*
Director of development: *Kristine Tibbetts*
Marketing manager: *James F. Connely*
Senior project manager: *Mary E. Powers*
Senior production supervisor: *Laura Fuller*
Senior media project manager: *Tammy Juran*
Media technology producer: *Renee Russian*
Design manager: *K. Wayne Harms*
Cover/interior designer: *Kaye Farmer*
Senior photo research coordinator: *John C. Leland*
Photo research: *Mary Reeg*
Supplement producer: *Brenda A. Ernzen*
Compositor: *Carlisle Communications, Ltd.*
Typeface: *10/12 Minion*
Printer: *Von Hoffmann Corporation*

Cover images:
Left image: *Paolo Mascagni,* Anatomia universa, 1823–1832. *Image provided through the courtesy of The John Martin Rare Book Room, Hardin Library for the Health Sciences, University of Iowa.*
Center image: *Photo Researchers/© Science Photo Library*
Right image: *Getty Images/Suza Scalora*

The credits section for this book begins on page C-1 and is considered an extension of the copyright page.

Library of Congress Cataloging-in-Publication Data

Saladin, Kenneth S.
Human anatomy / Kenneth S. Saladin. — 1st ed.
p. cm.
Includes index.
ISBN 0–07–039080–0
1. Human anatomy. I. Title.

2005 2003067540
CIP

www.mhhe.com

ABOUT THE AUTHOR

KEN SALADIN is Distinguished Professor of Biology at Georgia College and State University in Milledgeville, Georgia. He received his B.S. from Michigan State University in zoology and his Ph.D. from Florida State University in parasitology. He has taught at GC&SU since 1977, where his course assignments have covered a very broad range including majors' and nonmajors' biology, general zoology, biological etymology, comparative animal behavior, sociobiology, parasitology, histology, neuroanatomy, human anatomy and physiology, and varied senior and graduate seminars on topics in evolution and biophilosophy. He also leads an annual Study Abroad class in the Galápagos Islands. Ken has been repeatedly recognized for superior teaching and scholarship, with six Phi Kappa Phi Honor Professor awards, a 1998 Excellence in Research and Publication Award, and selection as Distinguished Professor in 2001. He is an active member of the Human Anatomy and Physiology Society and the Society for Integrative and Comparative Biology. After several years of reviewing textbooks and writing instructor supplements for various publishers, Ken began his first textbook in 1993. His *Anatomy and Physiology: The Unity of Form and Function,* now in its third edition (McGraw-Hill, 2004), rose quickly to become one of the few most widely used books in that market. *Human Anatomy* is his second book. You can write to Ken at ken.saladin@gcsu.edu.

•••••

This book is dedicated to my father

ALBERT SALADIN

who taught me more than either of us
realized at the time.

•••••

BRIEF CONTENTS

Preface viii

Part One
Organization of the Body

1 The Study of Human Anatomy 1
Atlas A Survey of the Human Body 23
2 Cytology—The Study of Cells 47
3 Histology—The Study of Tissues 77
4 Human Development 105

Part Two
Support and Movement

5 The Integumentary System 129
6 Bone Tissue 151
7 The Axial Skeleton 171
8 The Appendicular Skeleton 207
9 Joints 227
10 The Muscular System—Introduction 255
11 The Axial Musculature 283
12 The Appendicular Musculature 313
Atlas B Surface Anatomy 349

Part Three
Integration and Control

13 Nervous Tissue 365
14 The Spinal Cord and Spinal Nerves 385
15 The Brain and Cranial Nerves 413
16 The Autonomic Nervous System and Visceral Reflexes 455
17 Sense Organs 475
18 The Endocrine System 513

Part Four
Maintenance

19 The Circulatory System I—Blood 541
20 The Circulatory System II—The Heart 561
21 The Circulatory System III—Blood Vessels 585
22 The Lymphatic System and Immunity 629
23 The Respiratory System 653
24 The Digestive System 675
25 The Urinary System 707

Part Five
Reproduction

26 The Reproductive System 727

Appendix: Answers to Chapter Review Questions A-1
Glossary G-1
Credits C-1
Index I-1

TABLE OF CONTENTS

Preface *viii*

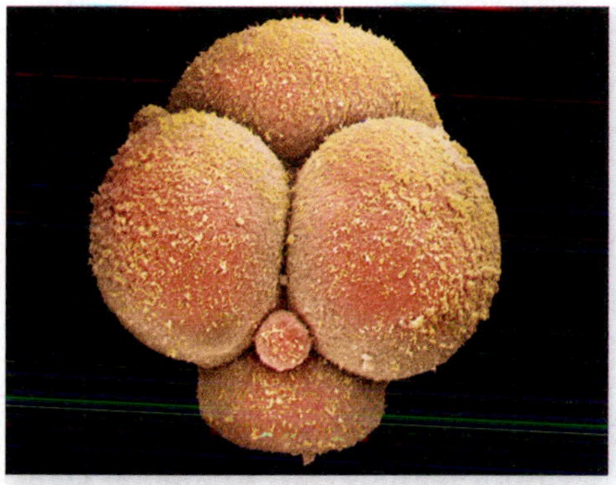

PART ONE
Organization of the Body *1*

CHAPTER 1
The Study of Human Anatomy **1**
The Scope of Human Anatomy 2
Early Anatomists 8
The Nature of Human Life 11
The Evolution of Human Structure 14
The Language of Anatomy 16
Chapter Review 20

ATLAS A
Survey of the Human Body **23**
General Anatomical Terminology 24
Body Regions 26
Body Cavities and Membranes 26
Organ Systems 30
A Visual Survey of the Body 31
Atlas Review 44

CHAPTER 2
Cytology—The Study of Cells **47**
The Study of Cells 48
The Cell Surface 52
The Cytoplasm 63
The Life Cycle of Cells 68
Chapter Review 73

CHAPTER 3
Histology—The Study of Tissues **77**
The Study of Tissues 78
Epithelial Tissue 80
Connective Tissue 86
Nervous and Muscular Tissue—Excitable Tissues 94
Glands and Membranes 96
Tissue Growth, Development, Death, and Repair 99
Chapter Review 101

CHAPTER 4
Human Development **105**
Gametogenesis and Fertilization 106
Stages of Prenatal Development 108
Clinical Perspectives 120
Chapter Review 125

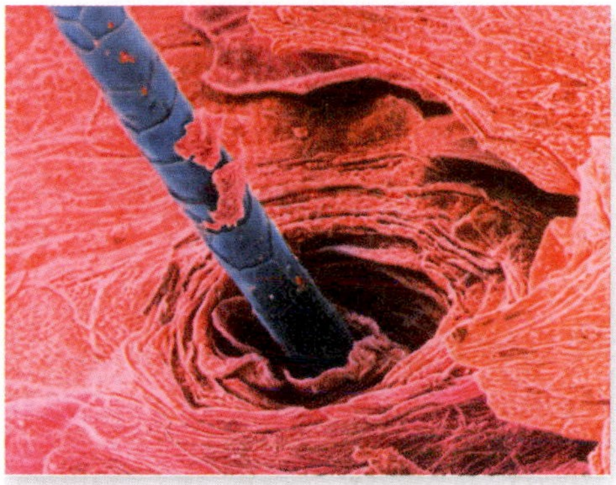

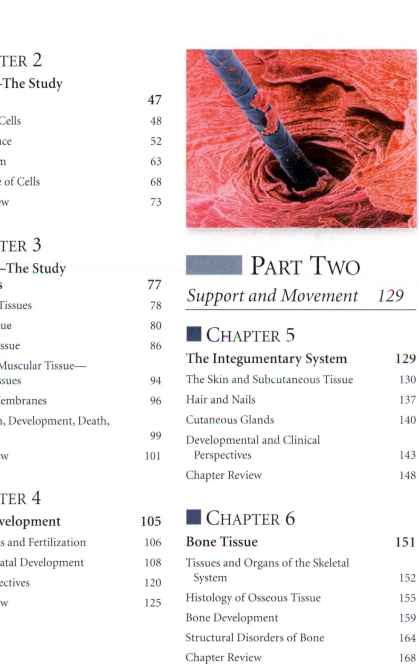

PART TWO
Support and Movement *129*

CHAPTER 5
The Integumentary System **129**
The Skin and Subcutaneous Tissue 130
Hair and Nails 137
Cutaneous Glands 140
Developmental and Clinical Perspectives 143
Chapter Review 148

CHAPTER 6
Bone Tissue **151**
Tissues and Organs of the Skeletal System 152
Histology of Osseous Tissue 155
Bone Development 159
Structural Disorders of Bone 164
Chapter Review 168

Chapter 7
The Axial Skeleton 171
Overview of the Skeleton 172
The Skull 175
The Vertebral Column and Thoracic Cage 188
Developmental and Clinical Perspectives 196
Chapter Review 203

Chapter 8
The Appendicular Skeleton 207
The Pectoral Girdle and Upper Limb 208
The Pelvic Girdle and Lower Limb 213
Developmental and Clinical Perspectives 221
Chapter Review 225

Chapter 9
Joints 227
Joints and Their Classification 228
Synovial Joints 231
Anatomy of Selected Synovial Joints 239
Clinical Perspectives 246
Chapter Review 252

Chapter 10
The Muscular System—Introduction 255
Muscle Types and Functions 256
General Anatomy of Muscles 257
Microscopic Anatomy 263
Functional Perspectives 269
Cardiac and Smooth Muscle 272
Developmental and Clinical Perspectives 275
Chapter Review 278

Chapter 11
The Axial Musculature 283
Learning Approaches 284
Muscles of the Head and Neck 287
Muscles of the Trunk 298
Chapter Review 309

Chapter 12
The Appendicular Musculature 313
Muscles Acting on the Shoulder and Upper Limb 314
Muscles Acting on the Hip and Lower Limb 329
Muscle Injuries 344
Chapter Review 346

Atlas B
Surface Anatomy 349
The Importance of External Anatomy 350
The Head and Neck 351
The Trunk 352
The Upper Limb 356
The Lower Limb 358
Muscle Self-Test 364

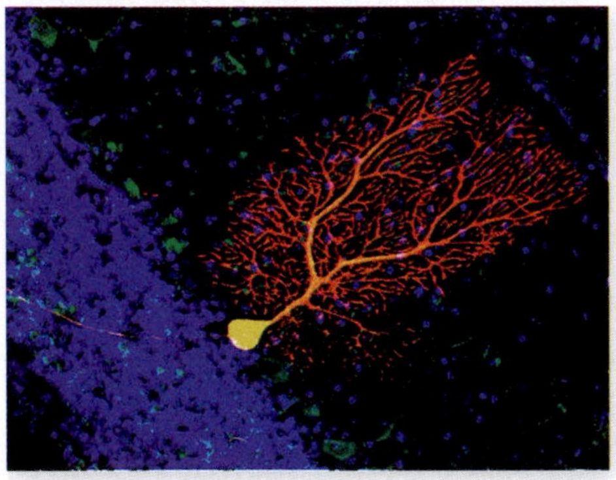

Part Three
Integration and Control 365

Chapter 13
Nervous Tissue 365
Overview of the Nervous System 366
Nerve Cells (Neurons) 367
Supportive Cells (Neuroglia) 371
Synapses and Neural Circuits 374
Developmental and Clinical Perspectives 378
Chapter Review 382

Chapter 14
The Spinal Cord and Spinal Nerves 385
The Spinal Cord 386
The Spinal Nerves 393
Somatic Reflexes 405
Clinical Perspectives 407
Chapter Review 409

Chapter 15
The Brain and Cranial Nerves 413
Overview of the Brain 414
The Hindbrain and Midbrain 421
The Forebrain 427
The Cranial Nerves 440
Developmental and Clinical Perspectives 449
Chapter Review 451

Chapter 16
The Autonomic Nervous System and Visceral Reflexes 455
General Properties of the Autonomic Nervous System 456
Anatomy of the Autonomic Nervous System 458
Autonomic Effects 465
Developmental and Clinical Perspectives 469
Chapter Review 471

Chapter 17
Sense Organs 475
Receptor Types and the General Senses 476
The Chemical Senses 481
The Ear 484
The Eye 494
Developmental and Clinical Perspectives 504
Chapter Review 508

Chapter 18
The Endocrine System 513
Overview of the Endocrine System 514
The Hypothalamus and Pituitary Gland 517
Other Endocrine Glands 523
Developmental and Clinical Perspectives 532
Chapter Review 537

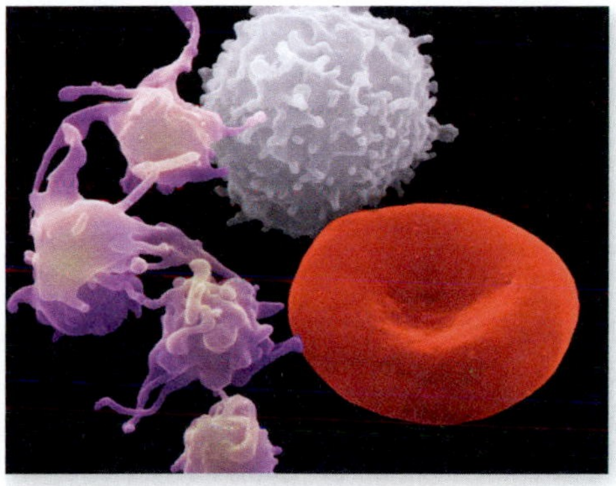

Part Four
Maintenance 541

Chapter 19
The Circulatory System I—Blood 541
Introduction 542
Erythrocytes 545
Leukocytes 548
Platelets 552
Clinical Perspectives 555
Chapter Review 558

Chapter 20
The Circulatory System II—The Heart 561
Overview of the Cardiovascular System 562
Gross Anatomy of the Heart 565
Blood Supply to the Cardiac Muscle 569
The Cardiac Conduction System and Cardiac Muscles 573
Developmental and Clinical Perspectives 576
Chapter Review 581

Chapter 21
The Circulatory System III—Blood Vessels 585
General Anatomy of the Blood Vessels 586
Anatomy of the Pulmonary Circuit 595
Anatomy of the Systemic Arteries 596
Anatomy of the Systemic Veins 610
Developmental and Clinical Perspectives 620
Chapter Review 625

Chapter 22
The Lymphatic System and Immunity 629
Lymph and Lymphatic Vessels 630
Lymphatic Cells, Tissues, and Organs 634
The Lymphatic System in Relation to Immunity 644
Developmental and Clinical Perspectives 646
Chapter Review 649

Chapter 23
The Respiratory System 653
Overview of the Respiratory System 654
The Upper Respiratory Tract 655
The Lower Respiratory Tract 659
Neuromuscular Aspects of Respiration 666
Developmental and Clinical Perspectives 668
Chapter Review 671

Chapter 24
The Digestive System 675
Digestive Processes and General Anatomy 676
The Mouth Through Esophagus 680
The Stomach 685
The Small Intestine 689
The Large Intestine 692
Accessory Glands of Digestion 694
Developmental and Clinical Perspectives 699
Chapter Review 703

Chapter 25
The Urinary System 707
Functions of the Urinary System 708
Anatomy of the Kidney 709
Anatomy of the Ureters, Urinary Bladder, and Urethra 717
Developmental and Clinical Perspectives 719
Chapter Review 723

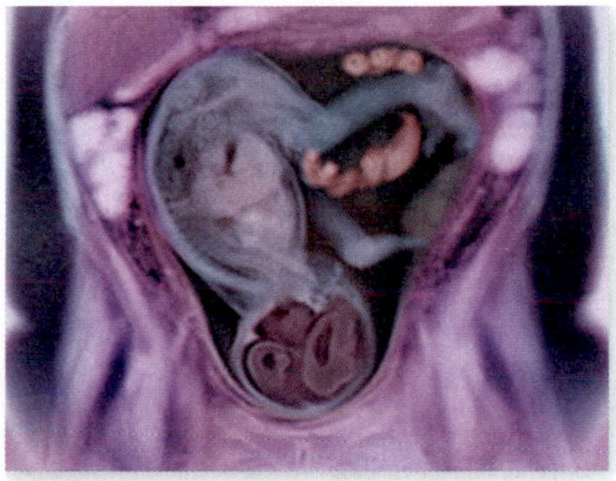

Part Five
Reproduction 727

Chapter 26
The Reproductive System 727
Sexual Reproduction 728
Male Reproductive Anatomy 729
Female Reproductive Anatomy 740
Developmental and Clinical Perspectives 751
Chapter Review 759

Appendix: Answers to Chapter Review Questions A-1
Glossary G-1
Credits C-1
Index I-1

PREFACE

Human Anatomy is designed primarily for a one-semester course, usually taken in the first or second year of college in preparation for admission to programs in nursing, therapy, health education, or pre-professional health programs. This book has evolved through extensive research on the needs and likes of anatomy students and instructors. In developing this first edition we commissioned detailed reviews from scores of instructors and held focus groups in which instructors discussed their course, challenges, text illustration programs, and the general content of anatomy textbooks. We created consultant panels of anatomy instructors to thoroughly analyze the entire book and its art program. These efforts have generated thousands of pages of reviews, all of which I read carefully in developing this book.

Audience

Human Anatomy is based on an assumption that most users are just beginning or returning to college. At this stage, many are still developing the study habits and skills necessary for success in a health science curriculum. The complexity of human anatomy can be a daunting subject, and I have tried to make it more manageable through a variety of learning aids described in this preface. Also mindful that English is not the primary language of many students who take human anatomy, I have tried to keep the prose free of unnecessary jargon and idioms, and as clear as any writing on this complex subject can be.

I also realize that many human anatomy students have taken no prior college biology or chemistry, since many institutions have no prerequisites for human anatomy. Other students, too, return to college to train for a health career after extended absences to raise families or try other careers. So even if the student has had college biology or chemistry, we cannot assume that he or she remembers it. Some chemistry is needed even for the study of anatomy, but chemistry is introduced infrequently and in relatively simple terminology in this book. All anatomy is based ultimately on cell biology, which is covered in chapter 2. This introduction provides all the background on cytology necessary for understanding the later chapters.

How We Met Your Needs

Reviewers and focus group members consistently tell us that the most important qualities of an acceptable textbook are accuracy, writing style, and quality of illustrations.

Accuracy

Textbook inaccuracies are an important source of frustration for instructors, students, and writers alike. We have taken several measures to avoid them in this book. The book itself was diligently reviewed by colleagues during its development—in the first and second draft manuscripts and the first and revised page proofs—to ensure that the content is accurate, concise, and clear. Page proofs were double-checked not only by me but also the editor against the manuscript to ensure the correction of any errors introduced during page composition (typesetting).

To produce an accurate and dependable textbook, I consider myself obligated, of course, to continue learning. It is not just an obligation but a pleasure to increase the depth of my own understanding, keep my knowledge updated, and arrive at better and clearer ways of explaining human form and function. As Isaac Asimov once said, "the greatest satisfaction for any conscientious and enthusiastic author comes from what one learns by writing." What stronger motivation than teaching and writing can there be for pursuing a life of perpetual scholarship? What better reward for knowledge can there be than these opportunities to share it?

My approaches to this life of scholarly inquiry and sharing include keeping up with the biological and medical journals that arrive in my mailbox almost daily; keeping my reference library updated with the newest editions of the most highly regarded biomedical text and reference books; enlightening discussions with colleagues on the HAPP-L listserv of the Human Anatomy and Physiology Society; attending annual conferences; and enrolling in continuing education courses in human anatomy and physiology, including courses I have taken during recent summers in neuroanatomy and neurophysiology, musculoskeletal anatomy and kinesiology, and cadaver dissection.

Writing Style

My writing style has also been shaped greatly by more than a decade of feedback from skillful editors and perceptive colleagues and students. The style that has drawn so many gratifying compliments to my previous book has, of course, been employed in this one as well. Students benefit most from a book they enjoy reading; a book that goes beyond presenting information to also telling an interesting story; and a book that steers a middle course between dry formality on one hand and a chatty condescending tone on the other. This has been my guiding principle in developing the right voice for my books. It is not the place of any writer to judge how successful he or she has been in achieving such stylistic ideals; that is for the reader to say. But I feel confident in inviting the reader to choose topics that students typically find most difficult, reading the presentation in this book alongside those of other books written for the same audience, and deciding which presentation will best serve his or her classes.

Quality of Illustrations

For the visual appeal and instructional value of this book, I am highly indebted to the professional medical illustrators and graphic artists who rendered the art in such beautiful and captivating style. The art program has benefited greatly from reviewers of my older textbook who, over the course of three editions, gave us valuable direction with respect to the desirable size, color palette and saturation, and amount of labeling appropriate to their esthetic tastes and teaching needs.

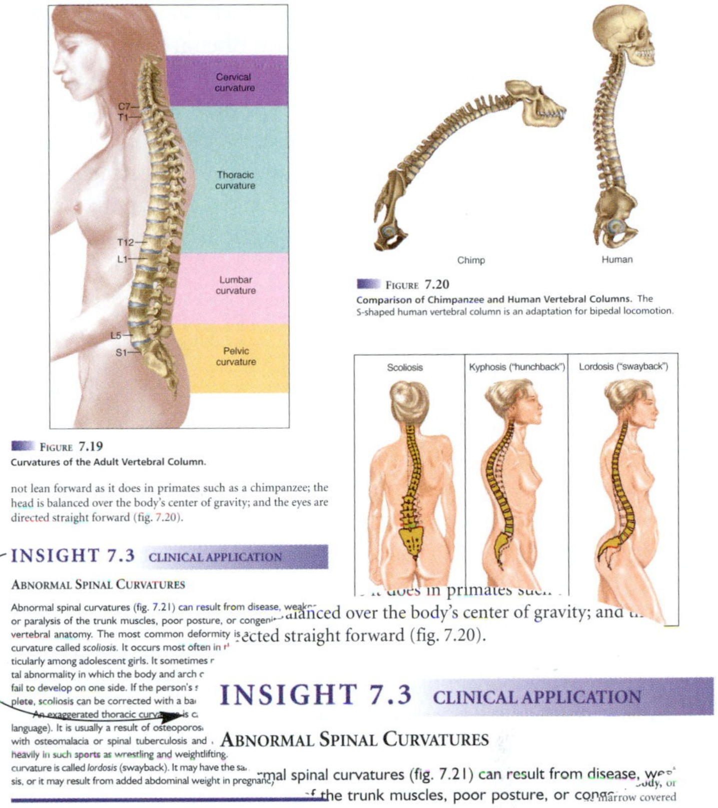
Cervical curvature
Thoracic curvature
Lumbar curvature
Pelvic curvature
C7
T1
T12
L1
L5
S1
FIGURE 7.19
Curvatures of the Adult Vertebral Column.
Chimp
Human
FIGURE 7.20
Comparison of Chimpanzee and Human Vertebral Columns. The S-shaped human vertebral column is an adaptation for bipedal locomotion.
Scoliosis
Kyphosis ("hunchback")
Lordosis ("swayback")
not lean forward as it does in primates such as a chimpanzee; the head is balanced over the body's center of gravity; and the eyes are directed straight forward (fig. 7.20).
INSIGHT 7.3 CLINICAL APPLICATION
ABNORMAL SPINAL CURVATURES
INSIGHT 7.3 CLINICAL APPLICATION
ABNORMAL SPINAL CURVATURES

What Sets This Book Apart

The following features are designed to serve the student's needs and adapt the book to the abilities of most beginning college students.

Anatomy Atlases

Basic anatomical terminology such as directional terms, body regions, and body cavities, as well as a broad overview of the 11 organ systems, are provided in atlas A following chapter 1. In many other books, this is included in chapter 1, but reviewers and users of my previous book have found it more useful to have these fundamental concepts covered in a module of their own.

Surface anatomy incorporates elements of integumentary, skeletal, and muscular anatomy and is therefore presented through a series of photographs in atlas B, following chapter 12. Photographic cadaver anatomy is presented in atlas A and in many individual chapters. For those especially interested in photographic anatomy of the cadaver, all such photos are listed in the index at "cadaveric anatomy."

"Insight" Sidebars

Each chapter has from two to five special topics set apart as sidebars called Insights, listed by title and page number on the chapter opener page. These are of three types: Clinical Applications, Evolutionary Medicine essays, and Medical History essays. Additional clinical, evolutionary, and historical remarks are interwoven with the main text of each chapter.

Clinical Applications

The Clinical Application insights are by no means meant to make this a clinical textbook. Rather, their importance and purpose is that most students who study from this book will be interested in clinical careers, and clinical insights show how the basic biology of the body is relevant to those interests. The importance of bone collagen, for example, may not be convincingly obvious to all readers, but it becomes more so when reading about osteogenesis imperfecta (brittle bone disease), a tragic result of defective collagen deposition (insight 3.3). Similarly, the warming and humidifying function of the nasal cavity becomes especially apparent when it is bypassed by tracheostomy (insight 23.1). Each organ system chapter also has a section on developmental and clinical perspectives at the end, including a table that briefly describes some of the most common or interesting dysfunctions of that system.

Evolutionary Medicine

No understanding of the human body can be complete without taking its evolutionary history into account; the human body today must be seen as reflecting adaptations to past environments. Since the mid-1990s, an increasing number of books on evolutionary (darwinian) medicine have appeared, along with many articles in medical journals exploring evolutionary interpretations of human structure, function, and disease. This trend shows no signs of abating. And indeed, the 38th edition of *Gray's Anatomy* (1995) is thoroughly evolutionary, with 11 pages of evolution in its first chapter, and pervasive evolutionary interpretations of human anatomy throughout that esteemed tome. Yet no other human anatomy textbook for this introductory undergraduate market has incorporated evolutionary medicine into its perspective. There is little room to delve very far into this subject, but the importance of evolution to human anatomy is introduced in a section of chapter 1, "The Evolution of Human Structure," and is reinforced by six of the insight essays and by evolutionary comments elsewhere in the main body of text. Insight 4.2, for example, provides an evolutionary interpretation of morning sickness as an adaptation for protecting the embryo from teratogens, and insight 25.2 clarifies the function of the nephron loop through an evolutionary comparison of humans to aquatic and desert mammals.

Medical History

Other books also say little if anything about the history and personalities behind the science of human anatomy. They seem to expect students to accept the information *ex cathedra* without asking, "Who says? How do they know that?" Again, introductory anatomy textbooks allow little room or luxury to discuss history at any great length, but I do provide a brief history of anatomy ("Early Anatomists") in chapter 1, and add historical remarks in Medical History insight essays, such as an insight into the function of the prefrontal cortex from the accident of Phineas Gage (insight 15.2). Historical comments are also found in the general text, such as Hippocrates' interpretation of brain function (chapter 15), Harvey's discoveries in blood circulation (chapter 21), and William Beaumont's experiments in gastric physiology (chapter 24). Such stories add considerable human interest to human anatomy, taking it beyond the realm of merely memorizing the facts.

Developmental Biology

My manuscript reviewers had widely disparate opinions of how much embryology this book should contain. Some said they have no time to teach embryology and wanted none at all, while others regarded chapter 4 (Human Development) to be the most important chapter in the book and wanted much more depth. The modal response was that there should be a moderate amount of embryology on each organ system, but not very much detail. I have aimed at this middle ground.

Chapter 4 presents basic embryology and lays a foundation for understanding the more specialized embryology of individual organ systems. For each organ system, there is a developmental section near the end of the chapter that goes briefly into its further development from the basic primordia described in chapter 4. These sections are not meant to be encyclopedic treatments of human embryology, but broad overviews and key examples. The eye and ear suffice for sense organ embryology, and the pituitary, thyroid, and adrenal glands for the endocrine system, for example. Neither space limitations nor, apparently, the interests of prospective users warrant greater detail or a comprehensive treatment of the development of every organ.

Aging

At the other end of the life span are the degenerative changes of old age. In view of the steadily increasing average age of the North American population, these are becoming increasingly important to health care providers. The effects of aging are presented for each organ system in a section following prenatal development ("The Aging Vascular System," for example).

Pedagogy

The following features are designed to serve the student's needs and adapt the book to the abilities of most beginning college students.

Brushing Up

Each chapter opener page (beginning with chapter 2) has a "Brushing Up" box which lists concepts from earlier chapters that the reader should understand before embarking on the new one. It helps to tie the organ systems together and show their relevance to each other. It also serves as an aid in courses that teach the systems in a different order from the one presented here, and for students returning after an absence from college who may need to refresh their memories of some concepts.

Objectives and "Before You Go On" Questions

Each chapter is broken down into typically three to six major sections, framed between a set of learning objectives at the beginning and a set of review questions ("Before You Go On") at the end of each section. Blocking the chapters out in this manner makes it easier for a student to plan a study session around concrete goals with a defined beginning and end. "Before You Go On" is an opportunity to test one's comprehension of the preceding material, or for instructors to test that comprehension, before moving on to a new section.

Vocabulary Aids

Among the greatest hurdles to studying human anatomy are its massive vocabulary and many students' unfamiliarity with biological word roots based heavily in Greek and Latin. Even as a graduate teaching assistant, I developed the habit of breaking words down into familiar roots in my lectures, and I have taught a course on biomedical etymology for many years. I am convinced that students find such terms as *pterygoid* and *extensor carpi radialis brevis* less forbidding, and easier to pronounce, spell, and remember, if they cultivate the habit of looking for familiar roots and affixes.

I have brought my etymological habit to *Human Anatomy.* Chapter 1 has a section, unique among human anatomy textbooks at this level, titled "The Language of Anatomy." It aims to instill the habit of breaking words down into familiar roots, intuiting the meaning of new terms from a familiarity with frequently used roots, and perceiving the relationship between singular and plural forms such as *corpus, corpora.* It explains the historic rationale for a medical language based in Greek and Latin, and the importance of precision and spelling in not confusing similar words such as *malleus* and *malleolus,* or *ileum* and *ilium.*

Following up on this, every chapter has footnotes identifying the roots and origins of new vocabulary terms, and easily understood "pro-NUN-see-AY-shun" guides for terms whose pronunciation is not intuitively obvious. The most frequently used roots, prefixes, and suffixes are listed with their meanings and biomedical examples inside the back cover of the book.

Terminology

The vocabulary in this book follows the *Terminologia Anatomica,* which has been the global standard for anatomical terms since 1998. My adherence to the TA is not absolute, however; I retain some traditional terms where TA would seem more confusing than helpful to the beginning student. Following the recommendations of the *AMA Manual of Style* and *Stedman's Medical Dictionary,* I also minimize the use of eponyms and substitute descriptive names, such as *tactile disc* for *Merkel disc* and *intestinal crypts* for *crypts of Lieberkühn.* I give the traditional eponyms in parentheses when first introducing the term. Some eponyms remain unavoidable (*Golgi complex* and the *Broca area,* for example). Also following the AMA's and Stedman's recommendations, when I do use eponyms, I use nonpossessive forms—thus, *Cushing syndrome* and *Alzheimer disease* rather than *Cushing's syndrome* and *Alzheimer's disease.*

Concept Reviews

Each chapter has a Review of Key Concepts at the end, a concise restatement of the chapter's main points for the purpose of study and review. Key vocabulary terms are italicized to make them stand out in this review activity.

Self-Testing Exercises

There are multiple types of self-testing questions in each chapter. At the end of the chapter are 10 multiple choice and 10 sentence completion questions on simple recall of information (Testing Your Recall); 10 True or False questions that call for more than just identifying (or guessing) which statements are true or false, but also for briefly explaining *why* the false statements are untrue; and 5 essay questions (Testing Your Comprehension) that call for deeper interpretive thought or application of the chapter's information to new clinical scenarios.

Within the body of each chapter there are an average of 17 "Before You Go On" questions and 3 "Think About It" questions. The latter are questions dispersed through the chapter calling for the student to apply what he or she has just read to a new situation, draw comparisons between concepts in different chapters, and so forth.

The questions in each chapter thus draw upon three levels of cognitive skill: (1) simple recall and recognition, as in Testing Your Recall; (2) ability to express concepts in one's own words, as in Before You Go On; and (3) analytical insight, as in Think About It, Testing Your Comprehension, and the explanation task in the True or False questions.

Suggestions Welcome!

Even though this book is now post-partum and dressed in hard covers, it is still very much a work in progress. It has benefited greatly from the many reviewers who provided critiques of the manuscript and art during its development. Undoubtedly it will improve still more as I hear from students and colleagues who use it, and who wish to point out its strong features or make suggestions for improvement. I welcome any user to send feedback to me at the following address, and will be grateful for your contribution to the quality and accuracy of future editions.

Ken Saladin
Department of Biology
Georgia College and State University
Milledgeville, GA 31061
USA
478-445-0816
ken.saladin@gcsu.edu

Teaching and Learning Supplements

McGraw-Hill offers various tools and technology products to support *Human Anatomy.* Students can order supplemental study materials by contacting their local bookstore or by calling 800-262-4729. Instructors can obtain teaching aids by calling the Customer Service Department at 800-338-3987, visiting our A&P website at www.mhhe.com/ap, or contacting your local McGraw-Hill sales representative.

For the Instructor:

DIGITAL CONTENT MANAGER CD-ROM

This multimedia collection of visual resources allows instructors to utilize artwork from the text in multiple formats to create customized classroom presentations, visually based tests and quizzes,

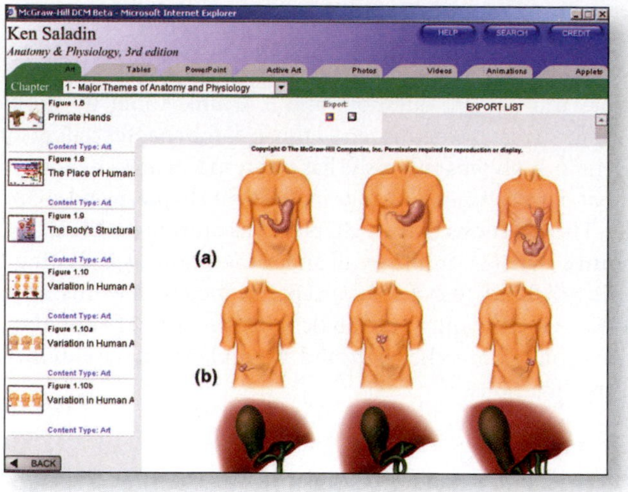

dynamic course website content, or attractive printed support material. The digital assets on this cross-platform CD-ROM include:

Art Library—Full-color digital files of all illustrations in the book, plus the same art saved in unlabeled and gray scale versions, can be readily incorporated into lecture presentations, exams, or custom-made classroom materials. These images are also pre-inserted into blank PowerPoint slides for ease of lecture preparation.

TextEdit Art Library—Every line art piece is placed in a PowerPoint presentation that allows the user to revise and/or move or delete labels as desired for creation of customized presentations and/or for testing purposes.

Active Art Library—Active Art consists of art files that have been converted to a format that allows the artwork to be edited inside of PowerPoint. Each piece can be broken down to its core elements, grouped or ungrouped, and edited to create customized illustrations.

Animations Library—Full-color presentations involving key figures in the book have been brought to life via animation. These animations offer total flexibility for instructors and were designed to be used in lecture. Instructors can pause, rewind, fast forward, and turn audio off and on to create dynamic lecture presentations.

Photo Library—Like the Art Library, digital files of all photographs from the book are available.

Table Library—Every table that appears in the book is provided in electronic form.

PowerPoint Lecture Outlines—Based on the information in the Instructor's Manual described below, it is possible to create ready-made presentations that combine art and lecture notes for each of the 26 chapters of the book. Written by Roger Gilchrist, University of Alabama-Birmingham, these lectures can be used as they are, or can be customized to reflect your preferred lecture topics and sequences.

INSTRUCTOR'S TESTING AND RESOURCE CD-ROM

The cross-platform CD-ROM contains the Instructor's Manual and Test Item File, written by Robin McFarland, Cabrillo College, are both available in Word and PDF formats. The manual follows the order of sections and subsections in the textbook and summarizes the main points in the text, figures, and tables. The Instructor's Manual also includes links to relevant websites, the answers to the problem sets at the end of each chapter, and a Test Bank of additional questions that can be used for homework assignments and/or the preparation of exams. These additional questions are found in a computerized test bank utilizing Brownstone Diploma testing software to quickly create customized exams. This user-friendly program allows instructors to search questions by topic, format, or difficulty level; edit existing questions or add new ones; and scramble questions and answer keys for multiple versions of the same test.

LABORATORY MANUAL

The *Human Anatomy Laboratory Manual* by Eric Wise, Santa Barbara City College, is expressly written to coincide with the chapters of *Human Anatomy.* This lab manual uses the same four-color art program as this book. It is accompanied by a separate Instructor's Manual, which contains solutions and keys for grading laboratory reports.

TRANSPARENCIES

A set of 600 transparency overheads includes nearly every piece of line art in the text. The images are printed with better visibility and contrast than ever before, and labels are large and bold for clear projection.

CLINICAL APPLICATIONS MANUAL

This manual written by Michael Hendrix, Southwest Missouri State University, expands on *Human Anatomy*'s clinical themes and introduces new clinical topics and case studies to develop the student's ability to apply his or her knowledge to realistic situations.

eINSTRUCTION

This Classroom Performance System (CPS) brings interactivity into the classroom/lecture hall. It is a wireless response system that gives the instructor and students immediate feedback from the entire class. The wireless response pads are essentially remotes that are easy to use and engage students. CPS allows you to motivate student preparation, interactivity and active learning so you can receive immediate feedback and know what students understand.

COURSE DELIVERY SYSTEMS

With help from our partners, WebCT, Blackboard, TopClass, eCollege, and other course management systems, instructors can take complete control over their course content. These course cartridges also provide online testing and powerful student tracking features. The Saladin Online Learning Center is available within all of these platforms.

For the Student:

ONLINE LEARNING CENTER (OLC)

The website offers an extensive array of learning and teaching tools. The site includes quizzes for each chapter, links to websites, clinical applications, interactive activities, art labeling exercises, and study outlines. Instructor resources at this site include lecture outlines and teaching tips.

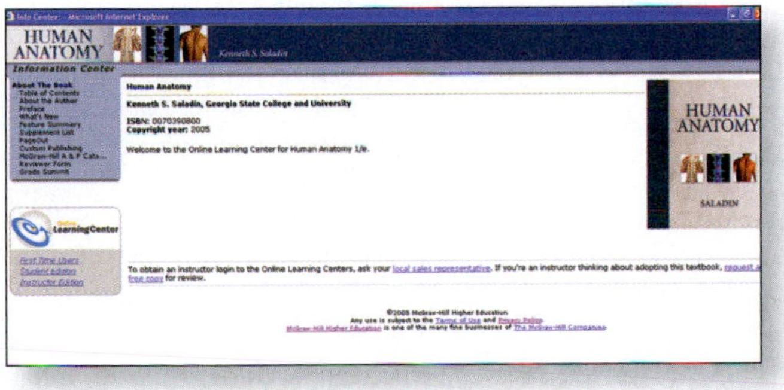

SALADIN INTERACTIVE CD-ROM

Set up in easy to use tabular format, this dual-platform CD-ROM is a fully interactive learning tool. The CD is organized chapter-by-chapter and provides a link directly to the text's Online Learning Center. Standard features include chapter-based quizzes, animations of complex processes, and PowerPoints of all the images found in the textbook. Saladin Interactive CD-ROM offers an indispensable resource for enhancing topics covered within the text.

INTERACTIVE CLINICAL RESOURCE CD-ROM

The Interactive Clinical Resource CD-ROM offers 150 3-D animations and 3-D models of human disease and disorders. It also contains 13 sections of clinical content (and nearly every body system) including Urinary, Skeletal, Reproductive, Nervous, Muscular, Immune, Digestive, Circulatory, and Endocrine.

VIRTUAL ANATOMY DISSECTION REVIEW CD-ROM

This multimedia program created by John Waters of Pennsylvania State University, and Melissa Janssen and Donna White of Collin County Community College, contains vivid, high-quality, labeled cat dissection photographs. The program helps students easily identify and compare the corresponding structures and functions between the cat and the human body.

STUDENT STUDY ART NOTEBOOK

This visual guide contains every illustration from the text to make it easier for students to learn anatomy. This collection of images provides a comprehensive resource for studying anatomical structures and a convenient place to write notes during lecture or lab.

Acknowledgments

As every textbook author knows, a book on this scale is never his or hers alone, but the product of many people's hard work and support. I am very indebted to Vice President and Editor-in-Chief Michael Lange and Publisher Marty Lange for their unfailing encouragement and material support; Sponsoring Editor Michelle Watnick and Marketing Manager Jim Connely for their infectious enthusiasm and promotion of the book; my editor Kristine Tibbetts, Director of Development, who has worked with me from year one of my textbook writing endeavors and spared no effort to make this book of highest quality; Mary E. Powers, Senior Project Manager, for keeping all parts of this project meshed like fine clockwork; Designer K. Wayne Harms for the esthetic appearance of the book; Photo Coordinator John Leland and Photo Researcher Mary Reeg for locating the great number of photographs between these covers; Copyeditor Cathy Conroy, for a sharp eye that spared me from innumerable embarrassments; Jack Haley and his team of medical illustrators and graphic artists at Imagineering STA Media Services in Toronto; and photographer Tim Vacula and illustrator Linda Chandler, who worked with me at Milledgeville on illustrative concepts for the book. And as always, I owe a thousand thanks to Colin Wheatley for talking me into textbook writing in the first place.

For factual accuracy and effective presentation, I also greatly appreciate the reviewers listed on the following pages, who provided a great deal of very helpful and detailed corrections, feedback, and encouragement during the writing process.

To my family—Diane, Emory, and Nicole—I thank you once again for your forbearance with the obsessions of an author and for helping me dispose of the royalties in ways that have been fun and enlightening.

Reviewers

No words could adequately convey my indebtedness and gratitude to the anatomy instructors and experts who have reviewed this book, and who have provided such a wealth of scientific information, corrections, suggestions for effective presentation, and encouragement. For making the book beautiful, I am indebted to the team described earlier. For making it *right,* I am thankful to the colleagues listed on the following pages.

Reviewers

Jane Horlings Aloi
Saddleback College
Jonathan Anning
Slippery Rock University
Amy Lynn Aulthouse
Ohio Northern University
Sharon R. Barnewall
Columbus State Community College
Fredric Bassett
Rose State College
David Bastedo
San Bernardino Valley College
Mark G. Birchette
Long Island University
Leann Blem
Virginia Commonwealth University
Ty W. Bryan
Bossier Parish Community College
Carolyn Williamson Burroughs
Bossier Parish Community College
Luis Cardenas
California State University-Northridge
Pamela J. Carlton
CUNY-College of Staten Island
Katherine P. Clark
Solano Community College
Harold Cleveland
University of Central Oklahoma
James E. Cordes
Louisiana State University-Eunice
Paul V. Cupp, Jr.
Eastern Kentucky University
Mary Beth Dawson
Kingsborough Community College
Robert Droual
Modesto Junior College
David K. Ferris
University of South Carolina-Spartanburg
Michael L. Foster
Eastern Kentucky University
Pamela B. Fouché
Walters State Community College
Carl D. Frailey
Johnson County Community College
Timothy J. Gaudin
University of Tennessee-Chattanooga
Todd Gordon
Kansas City Kansas Community College
Douglas J. Gould
University of Kentucky
David L. Hammerman
Long Island University
Steve Hardin
Ozarks Technical Community College
John P. Harley
Eastern Kentucky University
Ruth Heisler
University of Colorado-Boulder
Michael Hendrix
Southwest Missouri State University
Phyllis C. Hirsch
East Los Angeles College
Jon Jackson
University of North Dakota
Alice Khin
University of Alberta
Lyle W. Konigsberg
University of Tennessee
Karen M. Krabbenhoft
University of Wisconsin-Madison
Gavin R. Lawson
Bridgewater College
William I. Lutterschmidt
Sam Houston State University
Malinda B. McMurry
Morehead State University
Craig L. Mekow
Palmer College of Chiropractic
Michele Monlux
Modesto Junior College
John G. Osborne
East Tennessee State University
Wayne H. Preston
College of Sequoias
William Saltarelli
Central Michigan University
Beth Scaglione Sewell
Valparaiso University
Tom Sourisseau
Cabrillo College
Leeann S. Sticker
Northwestern State University of Louisiana
William J. Swartz
Louisiana State University
R. Brent Thomas
University of South Carolina-Spartanburg
Richard D. Torres
Ohio Northern University
Harry A. Tracy, Jr.
University of Texas-San Antonio
Laurie Walter
Chicago State University
Barbara Ann Walton
University of Tennessee-Chattanooga
Everett D. Wilson
Sam Houston State University
Mary Leslie Wilson
Gordon College
MaryJo A. Witz
Monroe Community College

Text Consultant Panel

Frank Baker
Golden West College
Leann Blem
Virginia Commonwealth University
Carolyn Williamson Burroughs
Bossier Parish Community College
Michael L. Foster
Eastern Kentucky University
Carl D. Frailey
Johnson County Community College
Douglas J. Gould
University of Kentucky
Phyllis C. Hirsch
East Los Angeles College
Glen Johnson
University of Nebraska-Lincoln
Malinda B. McMurry
Morehead State University
William Saltarelli
Central Michigan University
Carl F. Sievert
University of Wisconsin-Madison
Suzanne G. Strait
Marshall University
Harry A. Tracy, Jr.
University of Texas-San Antonio
Sandy Whisler
Central Texas College

Art Consultant Panel

Jane Aloi Horlings
Saddleback College
Lucy G. Andrews
University of Alabama-Birmingham
Frank Baker
Golden West College
Sharon R. Barnewall
Columbus State Community College
Mark G. Birchette
Long Island University
Jennifer K. Brueckner
University of Kentucky Community College System
Ty W. Bryan
Bossier Parish Community College
Carolyn Williamson Burroughs
Bossier Parish Community College
Paul V. Cupp, Jr.
Eastern Kentucky University
Luis Cardenas
California State University-Northridge
Alan Dietsche
University of Rochester
Christine Eckel
Salt Lake Community College
David K. Ferris
University of South Carolina-Spartanburg
Michael L. Foster
Eastern Kentucky University
Pamela B. Fouché
Walters State Community College
Carl D. Frailey
Johnson County Community College
Patrick B. Fulks
Bakersfield College
Ron Gaines
Cameron University
Timothy J. Gaudin
University of Tennessee-Chattanooga
Douglas J. Gould
University of Kentucky
Eric S. Hall
Rhode Island College
John P. Harley
Eastern Kentucky University
Cynthia Herbrandson
Kellogg Community College
Phyllis C. Hirsch
East Los Angeles College
Jon Jackson
University of North Dakota
W.R. Jones
Loyola University-Chicago
Jeanne Kowalczyk
University of South Carolina-Spartanburg
Gary Lange
Saginaw Valley State University
Gavin R. Lawson
Bridgewater College
Malinda B. McMurry
Morehead State University
Craig L. Mekow
Palmer College of Chiropractic
Michele Monlux
Modesto Junior College
Tim R. Mullican
Dakota Wesleyan University
Wayne H. Preston
College of Sequoias
Jacob A. Sapiro
Fullerton College
Carl F. Sievert
University of Wisconsin-Madison
Lewis M. Stein
Chatham College
Leeann S. Sticker
Northwestern State University of Louisiana
Suzanne G. Strait
Marshall University
Judy Sullivan
Antelope Valley College
Stuart Sumida
California State University-San Bernardino
R. Brent Thomas
University of South Carolina-Spartanburg
Mary Tracy
University of Detroit-Mercy
Laurie Walter
Chicago State University
Barbara Ann Walton
University of Tennessee-Chattanooga
Patricia Brady Wilhelm
Community College of Rhode Island
MaryJo A. Witz
Monroe Community College
Glenn Yoshida
Los Angeles Southwest College

Focus Group Participants

Kathleen Andersen
University of Iowa

Frank Baker
Golden West College
Mike Bodri
Northwestern State University
Christine Byrd
Western Michigan University
Marian Diamond
University of California-Berkeley
Martha Dixon
Diablo Valley College
Christine Eckel
Salt Lake Community College
Michael L. Foster
Eastern Kentucky University
Roger Gilchrist
University of Alabama-Birmingham
Douglas J. Gould
University of Kentucky
Leslie Hendon
University of Alabama-Birmingham
Michael Hendrix
Southwest Missouri State University
Phyllis C. Hirsch
East Los Angeles College
Jon Jackson
University of North Dakota
Lyle W. Konigsberg
University of Tennessee
Karen M. Krabbenhoft
University of Wisconsin-Madison
Dennis Landin
Louisiana State University-Baton Rouge
Robin McFarland
Cabrillo College
Michele Monlux
Modesto Junior College
John G. Osborne
East Tennessee State University
Roland Rodriguez
Irvine Valley College
Robert Seegmiller
Brigham Young University
Suzanne G. Strait
Marshall University
Stuart Sumida
California State University-San Bernadino
Mark Terrell
Indiana University Purdue University Indianapolis
Gail Turner
Virginia Commonwealth University
John Wilkins
Ball State University
David A. Woodman
University of Nebraska-Lincoln

Questionnaire Respondents

Mitch Albers
Minneapolis College
Jane Aloi Horlings
Saddleback College
Kathleen Andersen
University of Iowa
Deborah Anderson
St. Norbert College
Lucy G. Andrews
University of Alabama, Birmingham
Paul Arnold
Young Harris College
Josephine Arogyasami
Southern Virginia University
Michael Atkinson
Oakland City University
Amy Lynn Aulthouse
Ohio Northern University
Caryn Babaian
Manor College
Charlotte Bacon
University of Hartford
Frank Baker
Golden West College
Susan Baldi
Santa Rosa Junior College
Bobby Baldridge
Asbury College
Hirendra Banerjee
Elizabeth City State University
Mick Bondello
Allan Hancock College
Laurie Bonneau
Trinity College
Joan Bowden
Alfred University
Chad Brown
Capital University
Ty W. Bryan
Bossier Parish Community College
Stacey Buser
University of Akron
Craig Canby
Des Moines University
Marnie Chapman
University of Alaska
Bill Cockerham
Fresno Pacific University
Kimberly Coleman
Landmark College
Warren Darling
University of Iowa
Rosemary Davenport
Gulf Coast Communty College
Latonya Derrick
Allen University
Martha Dixon
Diablo Valley College
Leah Dvorak
Concordia University Wisconsin
Ruth Ebeling
Biola University
David K. Ferris
University of South Carolina, Spartanburg
Esther Fleischmann
University of Maryland Baltimore Campus
Pamela B. Fouché
Walters State Community College
Claire Fuller
Murray State University
Ron Gaines
Cameron University
Anne Geller
San Diego Mesa College
Joel Gober
Long Beach City College
Miriam Golbert
College of The Canyons
Douglas J. Gould
University of Kentucky College Of Medicine
Eric S. Hall
Rhode Island College
David L. Hammerman
Long Island University
Wesley Hanson
Southern Nazarene University
Rosemary Harkins
Langston University
John P. Harley
Eastern Kentucky University
Ron Harris
Marymount College
Lauraine Hawkins
Pennsylvania State University
Elizabeth Hayes
University of Wisconsin-Fond du Lac
Ruth Heisler
University of Colorado-Boulder
Maureen Helgren
Quinnipiac University
Andrew Hufford
Butte College
Jerry Johnson
Western Baptist College
Gary Lange
Saginaw Valley State University
Stephen Langjahr
Antelope Valley College
Al Martyn
Little Priest Tribal College
Christopher McNair
Hardin-Simmons University
Craig L. Mekow
Palmer College of Chiropractic
Michele Monlux
Modesto Junior College
Chris Nicolay
University of North Carolina-Asheville
Mark Nielsen
University of Utah
Susan Orsbon
Newman University
John G. Osborne
East Tennessee State University
Jeff Parmelee
Simpson College
James Paruk
Feather River College
Mary Perry
Allan Hancock College
Wendy Perryman
Oral Roberts University
Andrew Petto
University of the Arts
Tricia Reichert
Colby Community College
Deborah Rhoden
Snead State Community College
Lewis Royal
Community College of Rhode Island
Katherine Schmeidler
Irvine Valley College
Richard Search
Thomas University
Donald Serva
Wheeling Jesuit University
Mark A. Shoop
Tennessee Wesleyan College
Sharon Simpson
Broward Community College
Colleen Sinclair
Johns Hopkins University
Tracy Soltesz
Pikeville College
Tom Southworth
College of Alameda
Robert Stark
Depauw University
Lewis M. Stein
Indiana University of Pennsylvania-Indiana
Suzanne G. Strait
Marshall University
Stuart Sumida
California State University San Bernardino
William J. Swartz
Louisiana State University
Anthony Tolvo
Molloy College
Janice Toyoshima
Evergreen Valley College
Mary Tracy
University of Detroit-Mercy
Curt Walker
Dixie State College
Mark Waters
Ohio University
Deloris Wenzel
University of Georgia
Linda Winkler
University of Pittsburgh
Mark Womble
Youngstown State University

Vivid and Captivating Illustrations Contribute to Learning

Saladin's *Human Anatomy* brings key concepts to life with its unique style of biomedical illustration. The digitally rendered images have a vivid three-dimensional look that will not only stimulate your students' interest and enthusiasm, but also give them the clearest possible understanding of important concepts.

Unparalleled Art Program

Saladin's illustration program includes digital line art, numerous cadaver photographs, and light, TEM, and SEM photomicrographs. Larger images and brighter colors in the third edition will help draw your students into the subject.

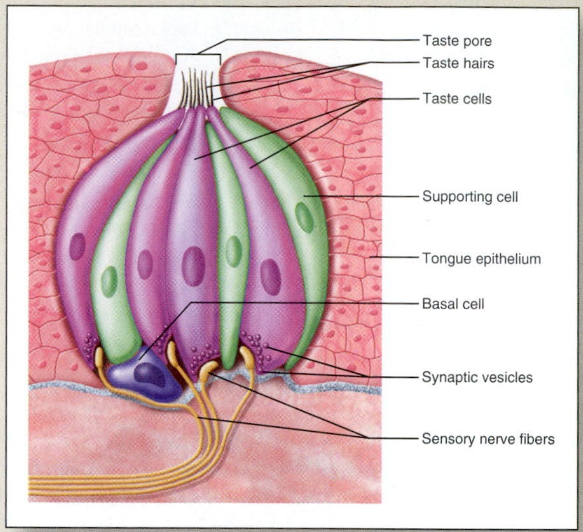

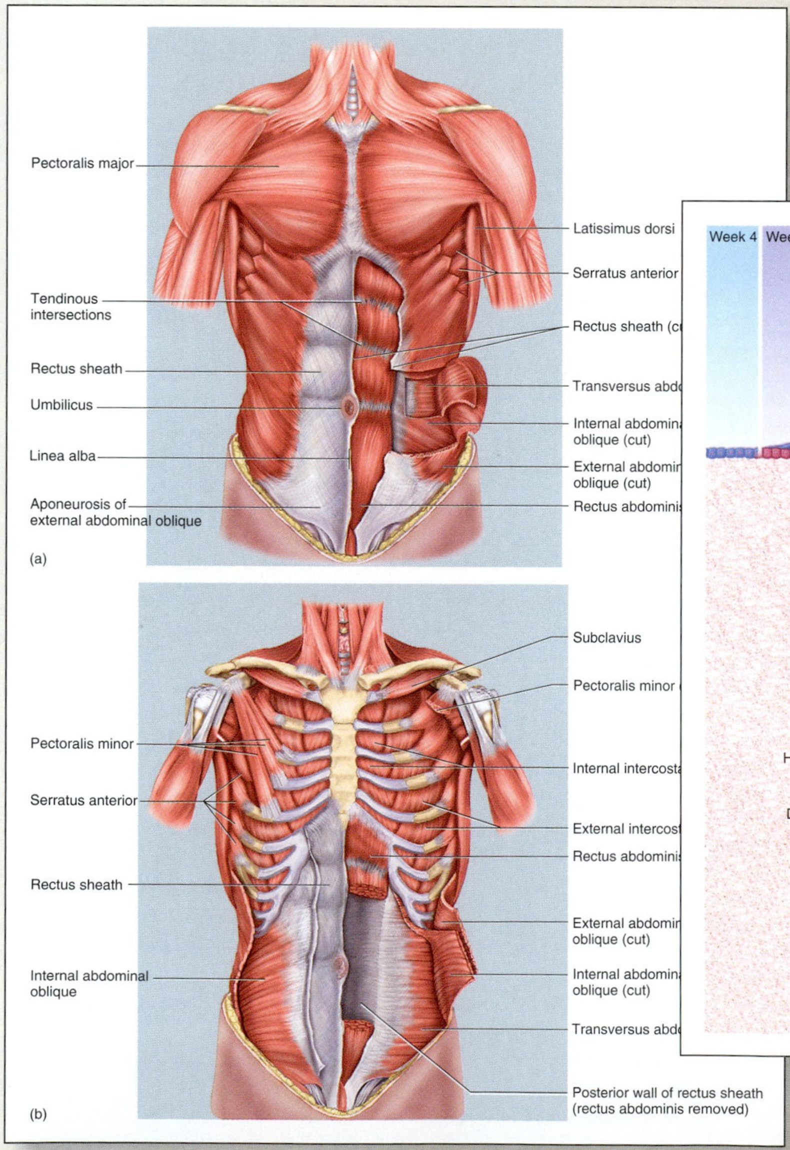

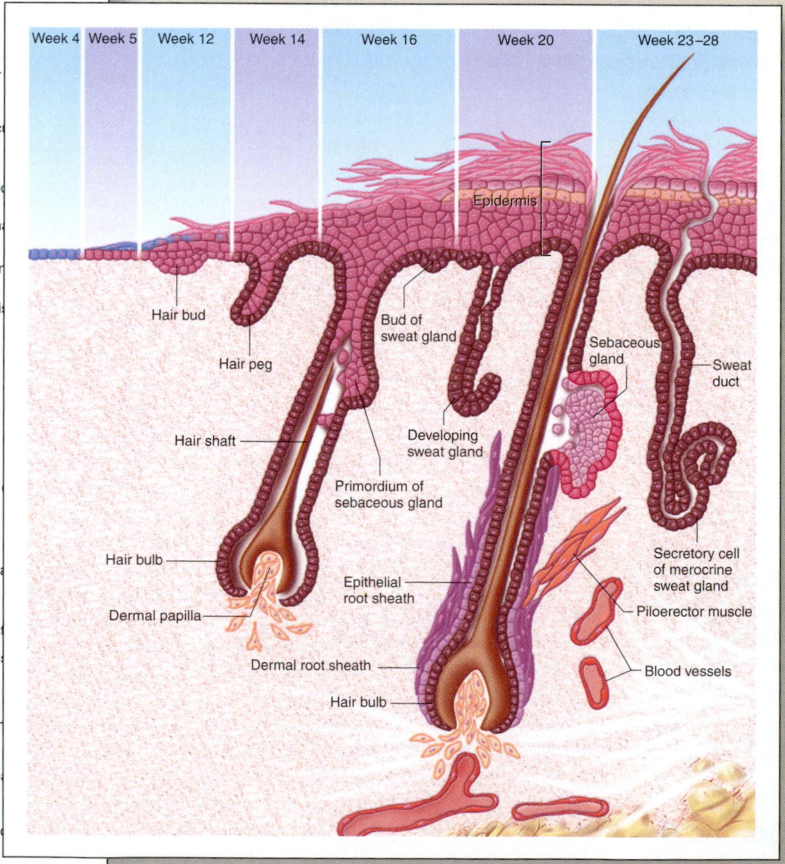

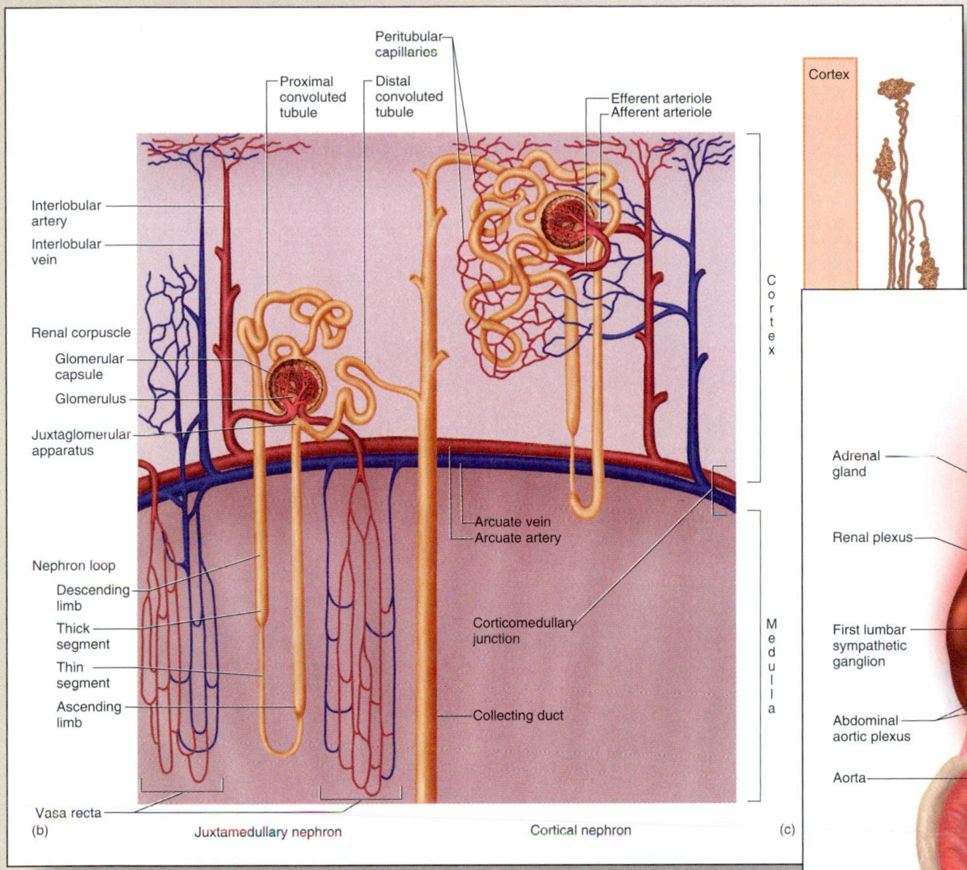
Peritubular capillaries
Proximal convoluted tubule
Distal convoluted tubule
Efferent arteriole
Afferent arteriole
Cortex
Interlobular artery
Interlobular vein
Renal corpuscle
Glomerular capsule
Glomerulus
Juxtaglomerular apparatus
C o r t e x
Arcuate vein
Arcuate artery
Nephron loop
Descending limb
Thick segment
Thin segment
Ascending limb
Corticomedullary junction
M e d u l l a
Collecting duct
Vasa recta
(b)
Juxtamedullary nephron
Cortical nephron
(c)

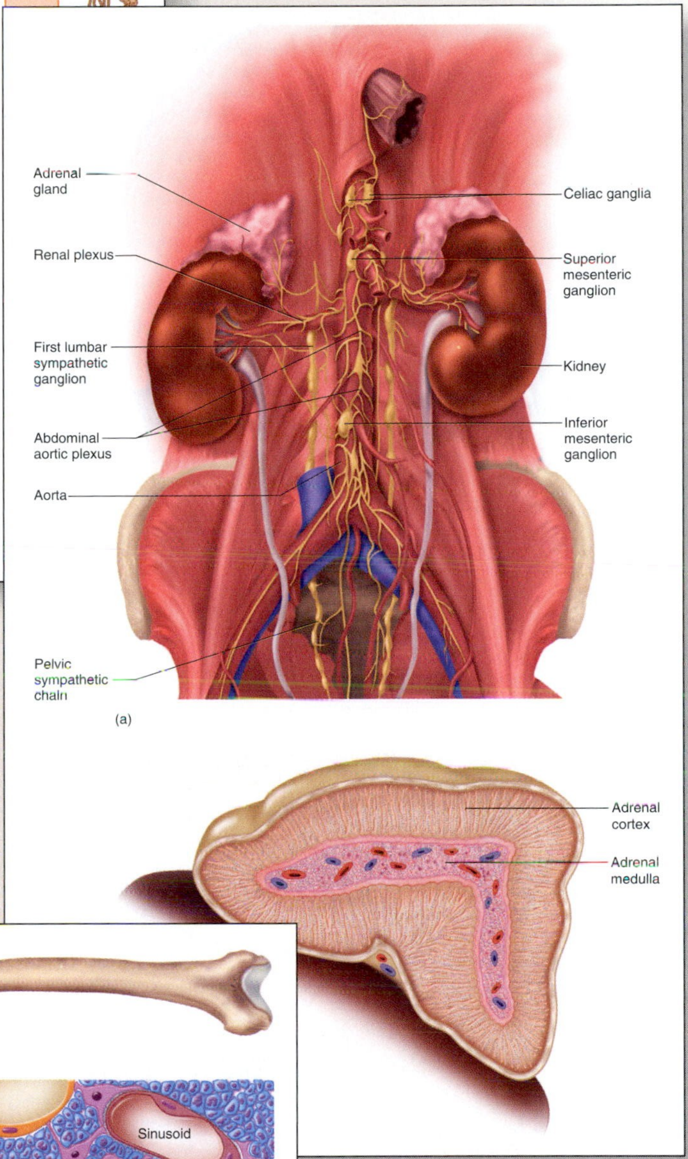
Adrenal gland
Celiac ganglia
Renal plexus
Superior mesenteric ganglion
First lumbar sympathetic ganglion
Kidney
Abdominal aortic plexus
Inferior mesenteric ganglion
Aorta
Pelvic sympathetic chain
(a)
Adrenal cortex
Adrenal medulla

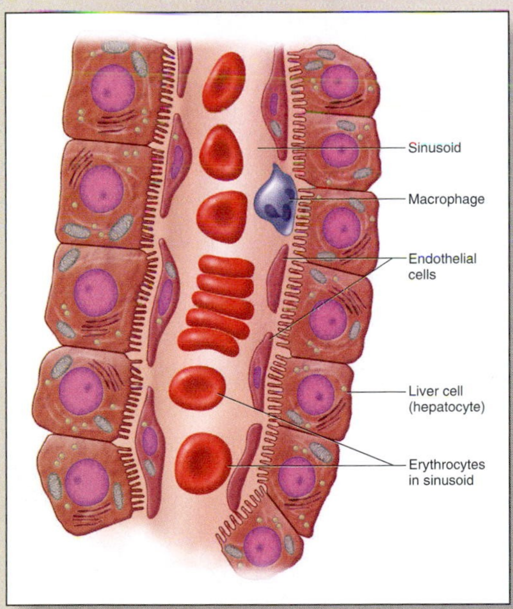
Sinusoid
Macrophage
Endothelial cells
Liver cell (hepatocyte)
Erythrocytes in sinusoid

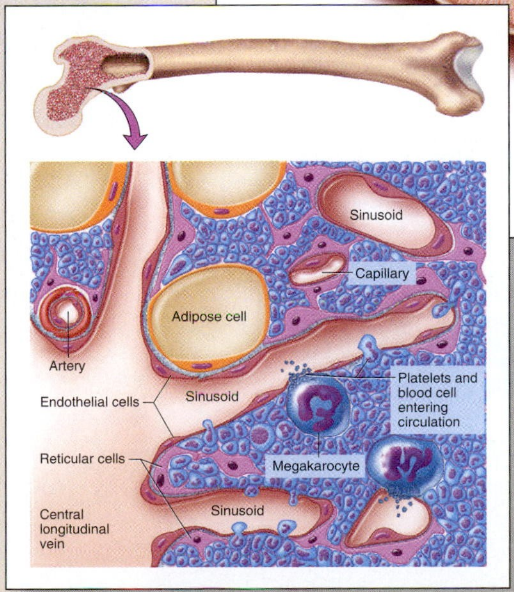
Sinusoid
Capillary
Adipose cell
Artery
Endothelial cells
Sinusoid
Platelets and blood cell entering circulation
Reticular cells
Megakarocyte
Sinusoid
Central longitudinal vein

Atlas Quality Cadaver Images

Color photographs of cadavers dissected specifically for this book allow students to see the real texture of organs and their relationships to each other. This anatomical realism combines with the simplified clarity of line art to give your students a holistic view of bodily structure.

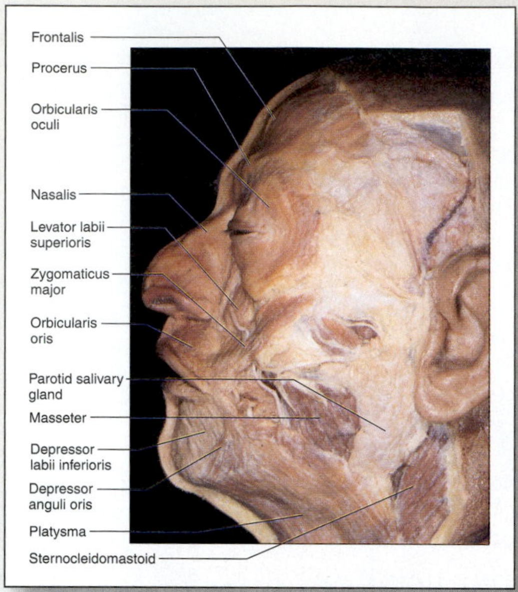
Frontalis
Procerus
Orbicularis oculi
Nasalis
Levator labii superioris
Zygomaticus major
Orbicularis oris
Parotid salivary gland
Masseter
Depressor labii inferioris
Depressor anguli oris
Platysma
Sternocleidomastoid

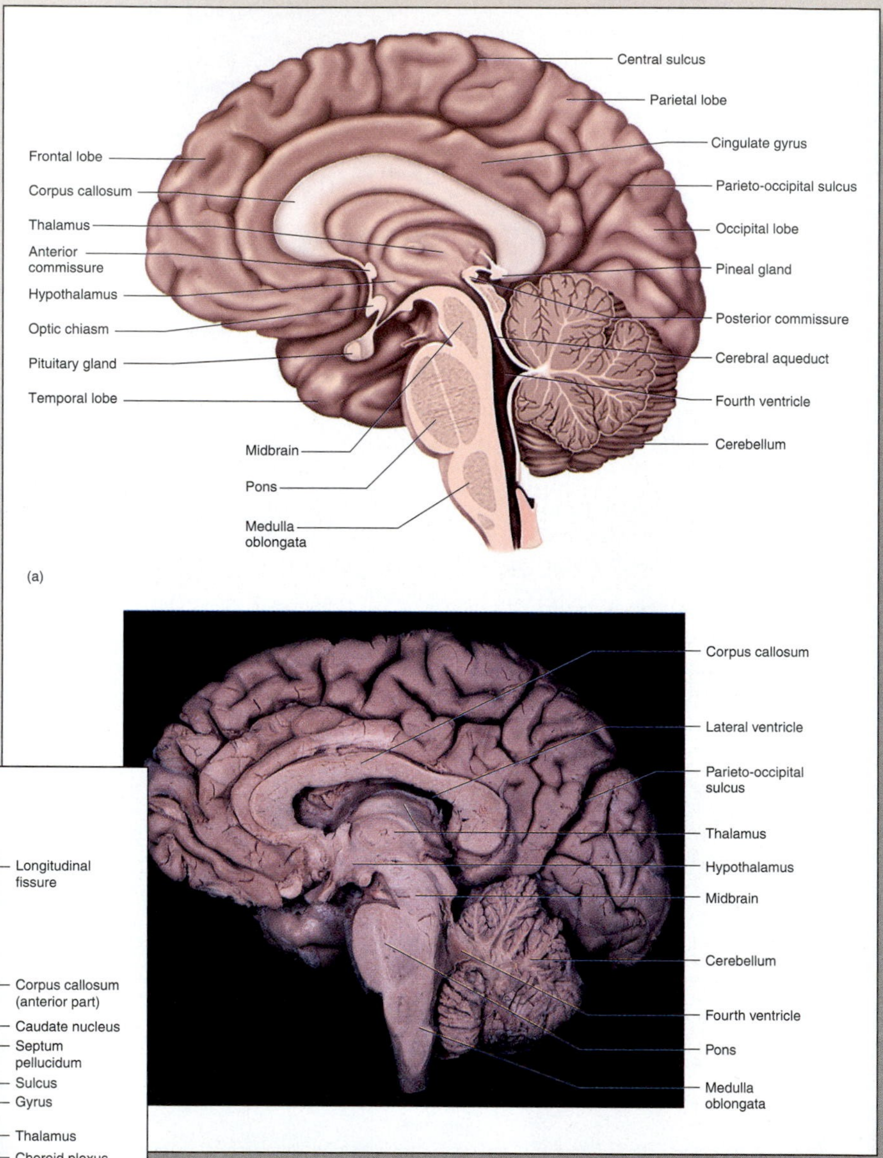
Central sulcus
Parietal lobe
Frontal lobe
Cingulate gyrus
Corpus callosum
Parieto-occipital sulcus
Thalamus
Occipital lobe
Anterior commissure
Pineal gland
Hypothalamus
Posterior commissure
Optic chiasm
Cerebral aqueduct
Pituitary gland
Temporal lobe
Fourth ventricle
Cerebellum
Midbrain
Pons
Medulla oblongata
(a)
Corpus callosum
Lateral ventricle
Parieto-occipital sulcus
Thalamus
Hypothalamus
Midbrain
Cerebellum
Fourth ventricle
Pons
Medulla oblongata

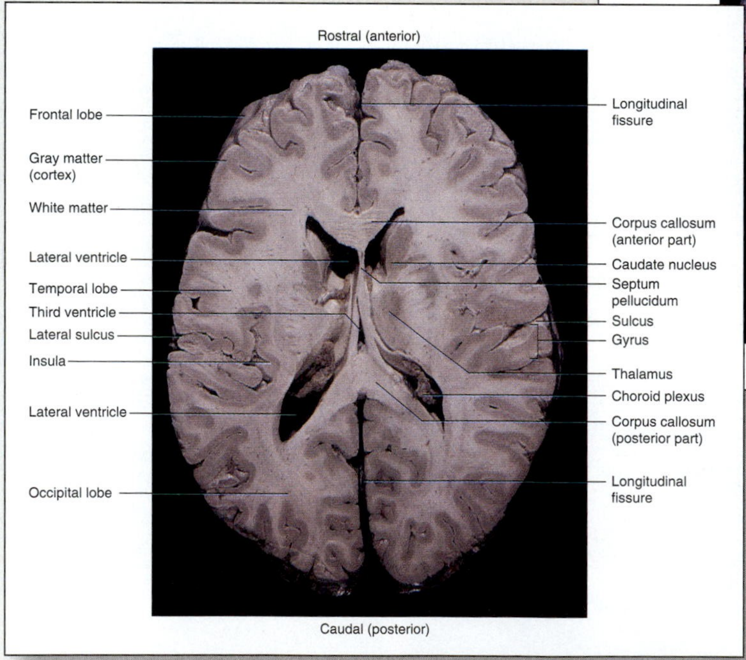
Rostral (anterior)
Frontal lobe
Longitudinal fissure
Gray matter (cortex)
White matter
Corpus callosum (anterior part)
Lateral ventricle
Caudate nucleus
Temporal lobe
Septum pellucidum
Third ventricle
Sulcus
Lateral sulcus
Gyrus
Insula
Thalamus
Choroid plexus
Lateral ventricle
Corpus callosum (posterior part)
Occipital lobe
Longitudinal fissure
Caudal (posterior)

Micrographs

All life processes are ultimately cellular processes. Saladin drives this point home with a variety of histological micrographs in LM, SEM, and TEM formats, including many colorized electron micrographs.

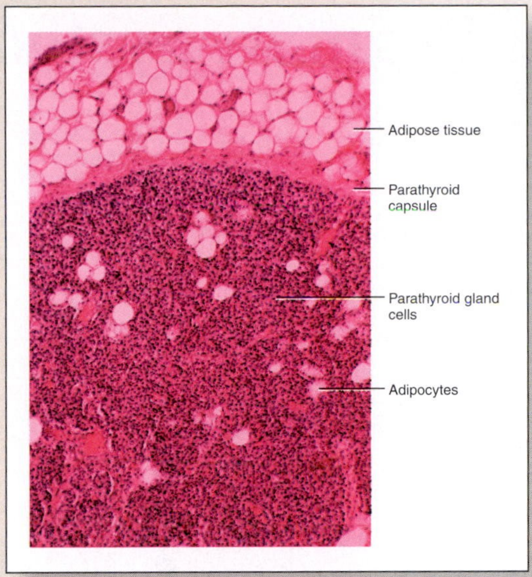
Adipose tissue
Parathyroid capsule
Parathyroid gland cells
Adipocytes

Photomicrographs Correlated with Line Art

Saladin juxtaposes histological photomicrographs with line art. Much like the combination of cadaver gross photographs and line art, this gives students the best of both perspectives: the realism of photos and the explanatory clarity of line drawings.

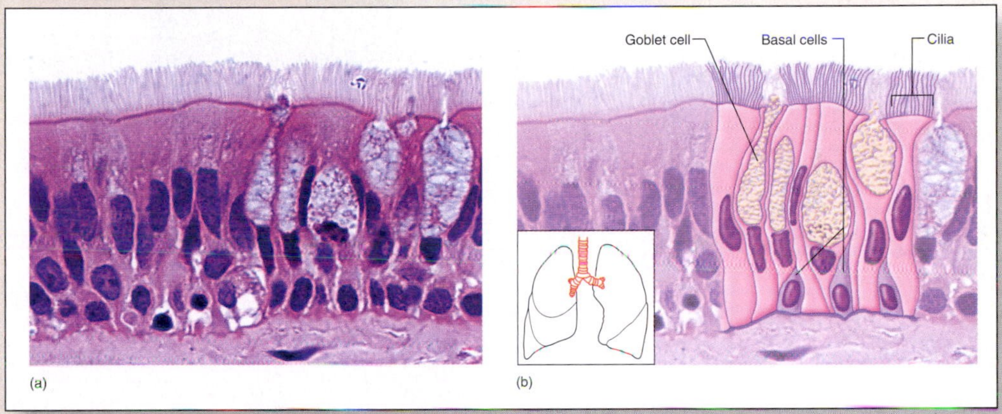
Goblet cell
Basal cells
Cilia
(a)
(b)

From Macroscopic to Microscopic

Saladin's line art guides students from the intuitive level of gross anatomy to the functional foundations revealed by microscopic anatomy.

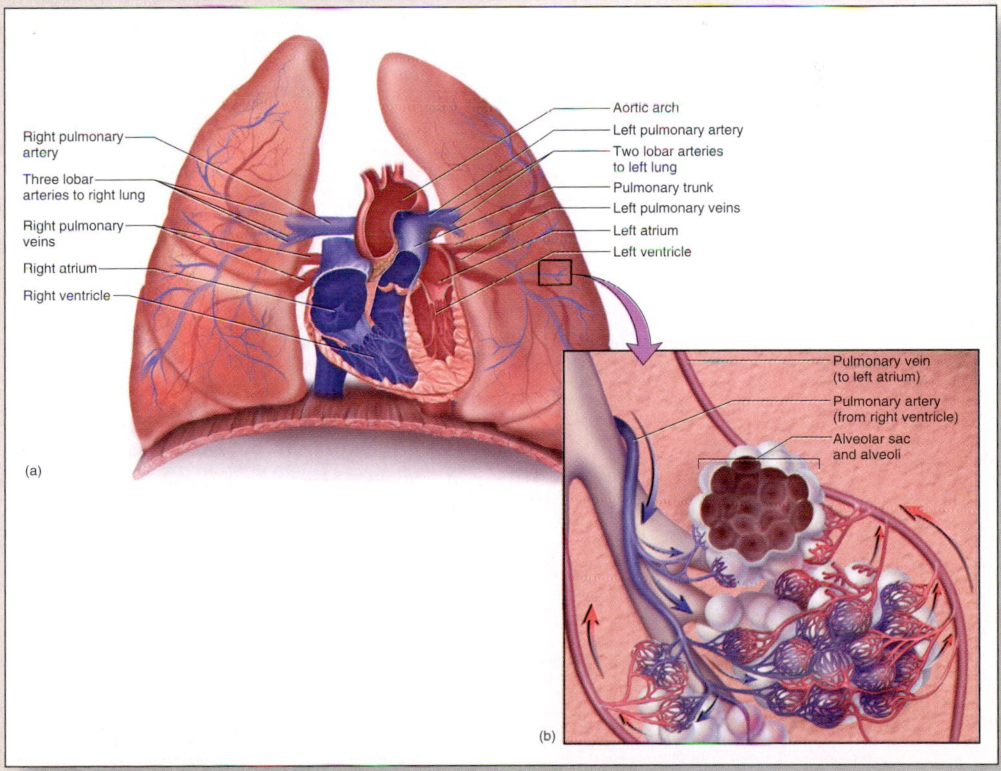
Right pulmonary artery
Three lobar arteries to right lung
Right pulmonary veins
Right atrium
Right ventricle
Aortic arch
Left pulmonary artery
Two lobar arteries to left lung
Pulmonary trunk
Left pulmonary veins
Left atrium
Left ventricle
(a)
Pulmonary vein (to left atrium)
Pulmonary artery (from right ventricle)
Alveolar sac and alveoli
(b)

Pedagogical Aids Promote Systematic Learning

Saladin structures each chapter around a consistent and unique framework of pedagogic devices. No matter what the subject matter of a chapter, this enables students to develop a consistent learning strategy, making *Human Anatomy* a superior learning tool.

15

CHAPTER FIFTEEN

The Brain and Cranial Nerves

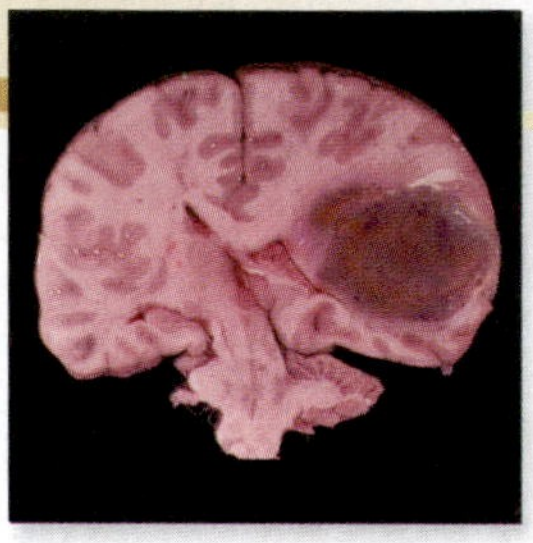

Frontal section of a brain with a large tumor (glioblastoma) in the left cerebral hemisphere

CHAPTER OUTLINE

Overview of the Brain 414
- Major Landmarks 414
- Gray and White Matter 414
- Meninges 414
- Ventricles and Cerebrospinal Fluid 417
- Blood Supply and the Brain Barrier System 418

The Hindbrain and Midbrain 421
- The Medulla Oblongata 421
- The Pons 421
- The Cerebellum 421
- The Midbrain 425
- The Reticular Formation 426

The Forebrain 427
- The Diencephalon 428
- The Cerebrum 428
- Higher Forebrain Functions 432

The Cranial Nerves 440
- Classification 440
- Nerve Pathways 440
- An Aid to Memory 440

Developmental and Clinical Perspectives 449
- The Aging Central Nervous System 449
- Two Neurodegenerative Diseases 449

Chapter Review 451

INSIGHTS

15.1 Clinical Application: Meningitis 417
15.2 Medical History: An Accidental Lobotomy 438
15.3 Clinical Application: Some Cranial Nerve Disorders 449

BRUSHING UP

To understand this chapter, you may find it helpful to review the following concepts:
- Anatomy of the cranium (pp. 175–183)
- Glial cells and their functions (p. 371)
- Embryonic development of the central nervous system (pp. 378–379)
- Meninges (p. 386)
- Gray and white matter (pp. 338–389)
- Tracts of the spinal cord (pp. 389–392)
- Structure of nerves and ganglia (pp. 393–395)

Brushing Up

A Brushing Up list at the beginning of the chapter ties chapters together and reminds students that all organ systems are conceptually related to each other. They discourage the habit of forgetting about a chapter after the exam is over. Brushing Up lists are also useful to instructors who present the subject in a different order from the textbook.

CHAPTER FIFTEEN The Brain and Cranial Nerves 417

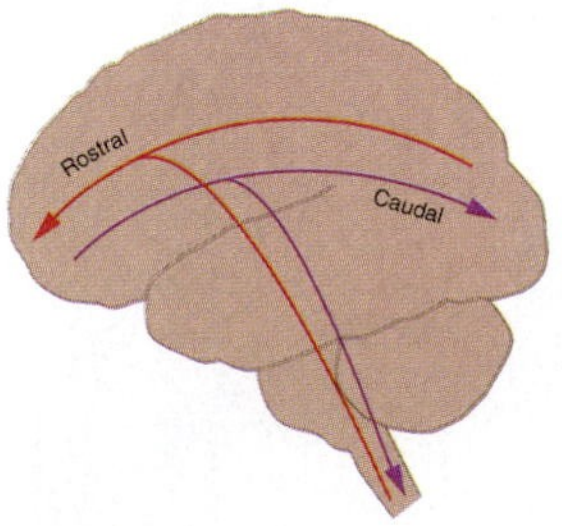

FIGURE 15.3
Directional Terms in CNS Anatomy. The *rostral* direction is from the rear of the head or from lower points in the brainstem or spinal cord toward the forehead. The *caudal* direction is from the forehead toward the rear of the head or toward lower points in the brainstem or spinal cord.

around the spinal cord. It is attached to the cranial bone only in limited places—around the foramen magnum, the sella turcica, the crista galli, and the sutural lines of the skull.

In some places, the two layers of dura are separated by **dural sinuses,** spaces that collect blood that has circulated through the brain. Two major dural sinuses are the **superior sagittal sinus,** found just under the cranium along the midsagit-...

INSIGHT 15.1 CLINICAL APPLICATION

MENINGITIS

Meningitis—inflammation of the meninges—is one of the most serious diseases of infancy and childhood. It occurs especially between 3 months and 2 years of age. Meningitis is caused by a variety of bacteria and viruses that invade the CNS by way of the nose and throat, often following respiratory, throat, or ear infections. The pia mater and arachnoid are most often affected, and from here the infection can spread to the adjacent nervous tissue. In bacterial meningitis, the brain swells, the ventricles enlarge, and the brainstem may exhibit hemorrhages. Signs include a high fever, stiff neck, drowsiness, and intense headache and may progress to vomiting, loss of sensory and motor functions, and coma. Death can occur within hours of the onset. Infants and toddlers with a high fever should therefore receive immediate medical attention.

Meningitis is diagnosed partly by examining the cerebrospinal fluid (CSF) for bacteria and white blood cells. The CSF is obtained by making a *lumbar puncture (spinal tap)* between two lumbar vertebrae and drawing fluid from the subarachnoid space (see insight 14.1).

Ventricles and Cerebrospinal Fluid

The brain has four internal chambers called **ventricles.** The largest are the two **lateral ventricles,** which form an arc in each cerebral hemisphere (fig. 15.5). Through a pore called the **interventricular foramen,** each lateral ventricle is connected to the **third ventricle,** a narrow medial space inferior to the corpus callosum. From here, ...

Insights

Each chapter has from two to five special topic Insight essays on the history behind the science, the evolution behind human form and function, and especially the clinical implications of the basic science. Insight sidebars lend the subject deeper meaning, intriguing perspectives, and career relevance to the student.

Objectives

Each new section of a chapter begins with a list of learning objectives. Students and instructors find this more useful than a single list of objectives at the beginning of a chapter, where few students ever refer back to them as they progress with their reading.

Think About It

Success in health professions requires far more than memorization. More important is your insight and ability to apply what you remember in new cases and problems. Think About It questions, which can be found strategically distributed throughout each chapter, encourage stopping and thinking more deeply about the meaning of broader significance.

386 PART THREE Integration and Control

We studied the nervous system at a cellular level in chapter 13. In these next four chapters, we move up the structural hierarchy to study the nervous system at the organ and system levels of organization. We begin with the *spinal cord*, an "information highway" between the brain and the trunk and limbs. It is about as thick as a finger, and extends through the vertebral canal as far as the first or second lumbar vertebra. At regular intervals, it gives off a pair of *spinal nerves*, which receive sensory input from the skin, muscles, bones, joints, and viscera, and which issue motor commands back to muscle and gland cells. The spinal cord is a component of the central nervous system and the spinal nerves are a component of the peripheral nervous system, but these central and peripheral components are so closely linked structurally and functionally that it is appropriate that we consider them together in this chapter. The brain and cranial nerves will be discussed in chapter 15.

The Spinal Cord

Objectives

When you have completed this section, you should be able to

- name the two types of tissue in the central nervous system and state their locations;
- describe the gross and microscopic anatomy of the spinal cord; and
- name the major conduction pathways of the spinal cord and state their functions.

Functions

The spinal cord serves three principal functions:

1. **Conduction.** The spinal cord contains bundles of nerve fibers that conduct information up and down the body, connecting different levels of the trunk with each other and with the brain. It enables sensory information to reach the brain, motor commands to reach the effectors, and input received at one level of the cord to affect output from another level.
2. **Locomotion.** Walking involves repetitive, coordinated contractions of several muscle groups in the limbs. Motor neurons in the brain initiate walking and determine its speed, distance, and direction, but the simple repetitive muscle contractions that put one foot in front of another, over and over, are coordinated by groups of neurons called **central pattern generators** in the cord. These neuronal circuits produce the sequence of outputs to the extensor and flexor muscles that cause alternating movements of the legs.
3. **Reflexes.** Reflexes are involuntary stereotyped responses to stimuli. They involve the brain, spinal cord, and peripheral nerves.

Surface Anatomy

The **spinal cord** (fig. 14.1) is a cylinder of nervous tissue that begins at the foramen magnum of the skull and passes through the vertebral canal as far as the inferior margin of the first lumbar vertebra (L1) or slightly beyond. In adults, it averages about 1.8 cm thick and 45 cm long. Thus, it occupies only the upper two-thirds of the vertebral canal; the lower one-third is described shortly. The cord gives rise to 31 pairs of spinal nerves. The first pair pass between the skull and vertebra C1, and the rest pass through the intervertebral foramina. Although the spinal cord is not visibly segmented, the part supplied by each pair of spinal nerves is called a *segment*. The cord exhibits longitudinal grooves on its ventral and dorsal sides—the *ventral median fissure* and *dorsal median sulcus*, respectively.

The spinal cord is divided into **cervical, thoracic, lumbar,** and **sacral regions.** It may seem odd that it has a sacral region when the cord itself ends well above the sacrum. These regions, however, are named for the level of the vertebral column from which the spinal nerves emerge, not for the vertebrae that contain the cord itself.

The cord widens at two points along its course: a **cervical enlargement** in the inferior cervical region, where it gives rise to nerves of the upper limbs; and a similar **lumbar enlargement** in the lumbosacral region, where it gives rise to nerves of the pelvic region and lower limbs. Inferior to the lumbar enlargement, the cord tapers to a point called the **medullary cone.** The lumbar enlargement and medullary cone give off a bundle of nerve roots that occupy the vertebral canal from L2 to S5. This bundle, named the **cauda equina**[1] (CAW-duh ee-KWY-nah) for its resemblance to a horse's tail, innervates the pelvic organs and lower limbs.

Think About It!

Spinal cord injuries commonly result from fractures of vertebrae C5 to C6, but never from fractures of L3 to L5. Explain both observations.

Meninges of the Spinal Cord

The spinal cord and brain are enclosed in three connective tissue membranes called **meninges** (meh-NIN-jeez)—singular, *meninx*[2] (MEN-inks). These membranes separate the soft tissue of the central nervous system from the bones of the vertebrae and skull. From superficial to deep, they are the dura mater, arachnoid mater, and pia mater.

The **dura mater**[3] (DOO-ruh MAH-tur) forms a loose-fitting sleeve called the **dural sheath** around the spinal cord. It is a tough collagenous membrane with a thickness and texture similar to a rubber kitchen glove. The space between the sheath and vertebral bone, called the **epidural space,** is occupied by blood vessels, adipose tissue, and loose connective tissue (fig. 14.2*a*). Anesthetics are sometimes introduced to this space to block pain signals during childbirth or surgery; this procedure is called *epidural anesthesia*.

[1] *cauda* = tail + *equin* = horse
[2] *menin* = membrane
[3] *dura* = tough + *mater* = mother, womb

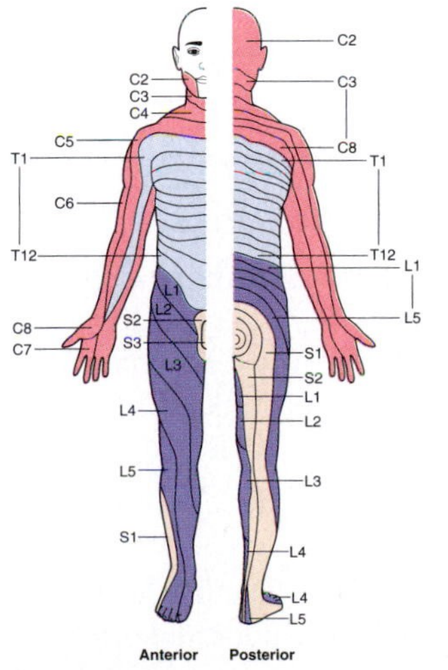
C2
C3
C4
C5
C6
C7
C8
T1
T12
L1
L2
L3
L4
L5
S1
S2
S3
Anterior
Posterior

Figure 14.18
A Dermatome Map of the Body. Anterior and posterior views. Each

CHAPTER FOURTEEN The Spinal Cord and Spinal Nerves 405

Before You Go On

Answer the following questions to test your understanding of the preceding section:

5. What is meant by the dorsal and ventral roots of a spinal nerve? Which of these is sensory and which is motor?
6. Where are the somas of the dorsal root located? Where are the somas of the ventral root?
7. List the five plexuses of spinal nerves and state where each one is located.
8. State which plexus gives rise to each of the following nerves: axillary, ilioinguinal, obturator, phrenic, pudendal, radial, and sciatic.

INSIGHT 14.4 CLINICAL APPLICATION

Spinal Nerve Injuries

The radial and sciatic nerves are especially vulnerable to injury. The radial nerve, which passes through the axilla, may be compressed against the humerus by improperly adjusted crutches, causing *crutch paralysis*. A similar injury often resulted from the discredited practice of trying to correct a dislocated shoulder by putting a foot in a person's armpit and pulling on the arm. One consequence of radial nerve injury is *wrist drop*—the fingers, hand, and wrist are chronically flexed because the extensor muscles supplied by the radial nerve are paralyzed.

Because of its position and length, the sciatic nerve of the hip and thigh is the most vulnerable nerve in the body. Trauma to this nerve produces *sciatica*, a sharp pain that travels from the gluteal region along the posterior side of the thigh and leg as far as the ankle. Ninety percent of cases result from a herniated intervertebral disc or osteoarthritis of the lower spine,

Before You Go On

Saladin divides each chapter into short "digestible" segments of about three to five pages each. Each segment ends with a few content review questions, so students can pause to evaluate their understanding of the previous few pages before going on.

Learning System

Chapter Review

Briefly restates the key points of the chapter.

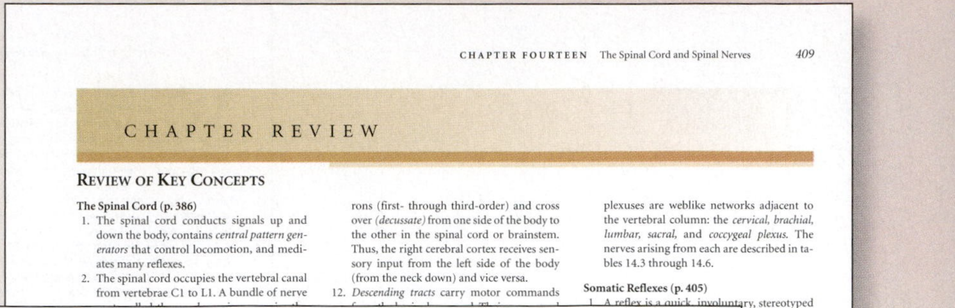
CHAPTER FOURTEEN The Spinal Cord and Spinal Nerves 409

CHAPTER REVIEW

REVIEW OF KEY CONCEPTS

The Spinal Cord (p. 386)

1. The spinal cord conducts signals up and down the body, contains *central pattern generators* that control locomotion, and mediates many reflexes.
2. The spinal cord occupies the vertebral canal from vertebrae C1 to L1. A bundle of nerve

rons (first- through third-order) and cross over (*decussate*) from one side of the body to the other in the spinal cord or brainstem. Thus, the right cerebral cortex receives sensory input from the left side of the body (from the neck down) and vice versa.

12. *Descending tracts* carry motor commands

plexuses are weblike networks adjacent to the vertebral column: the *cervical, brachial, lumbar, sacral,* and *coccygeal plexus.* The nerves arising from each are described in tables 14.3 through 14.6.

Somatic Reflexes (p. 405)

Testing Your Recall

Multiple choice and short answer questions allow students to check their knowledge.

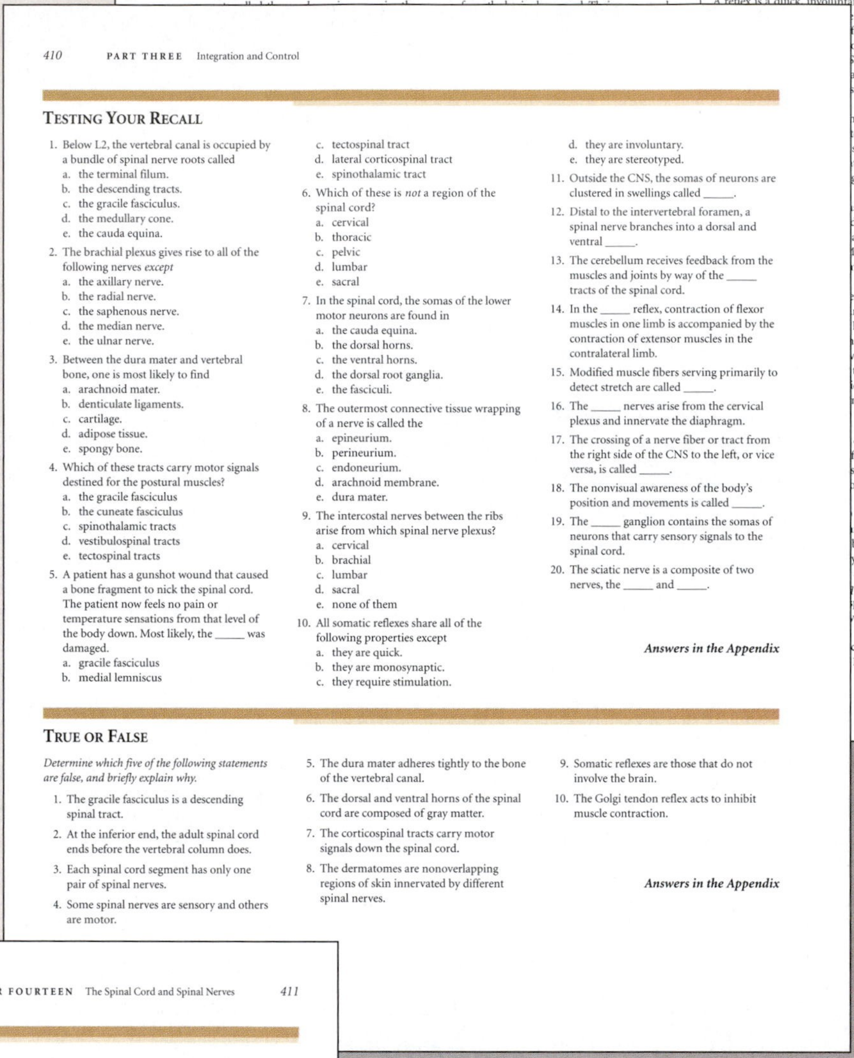
410 PART THREE Integration and Control

TESTING YOUR RECALL

1. Below L2, the vertebral canal is occupied by a bundle of spinal nerve roots called
   a. the terminal filum.
   b. the descending tracts.
   c. the gracile fasciculus.
   d. the medullary cone.
   e. the cauda equina.
2. The brachial plexus gives rise to all of the following nerves *except*
   a. the axillary nerve.
   b. the radial nerve.
   c. the saphenous nerve.
   d. the median nerve.
   e. the ulnar nerve.
3. Between the dura mater and vertebral bone, one is most likely to find
   a. arachnoid mater.
   b. denticulate ligaments.
   c. cartilage.
   d. adipose tissue.
   e. spongy bone.
4. Which of these tracts carry motor signals destined for the postural muscles?
   a. the gracile fasciculus
   b. the cuneate fasciculus
   c. spinothalamic tracts
   d. vestibulospinal tracts
   e. tectospinal tracts
5. A patient has a gunshot wound that caused a bone fragment to nick the spinal cord. The patient now feels no pain or temperature sensations from that level of the body down. Most likely, the _____ was damaged.
   a. gracile fasciculus
   b. medial lemniscus
   c. tectospinal tract
   d. lateral corticospinal tract
   e. spinothalamic tract
6. Which of these is *not* a region of the spinal cord?
   a. cervical
   b. thoracic
   c. pelvic
   d. lumbar
   e. sacral
7. In the spinal cord, the somas of the lower motor neurons are found in
   a. the cauda equina.
   b. the dorsal horns.
   c. the ventral horns.
   d. the dorsal root ganglia.
   e. the fasciculi.
8. The outermost connective tissue wrapping of a nerve is called the
   a. epineurium.
   b. perineurium.
   c. endoneurium.
   d. arachnoid membrane.
   e. dura mater.
9. The intercostal nerves between the ribs arise from which spinal nerve plexus?
   a. cervical
   b. brachial
   c. lumbar
   d. sacral
   e. none of them
10. All somatic reflexes share all of the following properties except
    a. they are quick.
    b. they are monosynaptic.
    c. they require stimulation.
    d. they are involuntary.
    e. they are stereotyped.
11. Outside the CNS, the somas of neurons are clustered in swellings called _____.
12. Distal to the intervertebral foramen, a spinal nerve branches into a dorsal and ventral _____.
13. The cerebellum receives feedback from the muscles and joints by way of the _____ tracts of the spinal cord.
14. In the _____ reflex, contraction of flexor muscles in one limb is accompanied by the contraction of extensor muscles in the contralateral limb.
15. Modified muscle fibers serving primarily to detect stretch are called _____.
16. The _____ nerves arise from the cervical plexus and innervate the diaphragm.
17. The crossing of a nerve fiber or tract from the right side of the CNS to the left, or vice versa, is called _____.
18. The nonvisual awareness of the body's position and movements is called _____.
19. The _____ ganglion contains the somas of neurons that carry sensory signals to the spinal cord.
20. The sciatic nerve is a composite of two nerves, the _____ and _____.

Answers in the Appendix

TRUE OR FALSE

Determine which five of the following statements are false, and briefly explain why.

1. The gracile fasciculus is a descending spinal tract.
2. At the inferior end, the adult spinal cord ends before the vertebral column does.
3. Each spinal cord segment has only one pair of spinal nerves.
4. Some spinal nerves are sensory and others are motor.
5. The dura mater adheres tightly to the bone of the vertebral canal.
6. The dorsal and ventral horns of the spinal cord are composed of gray matter.
7. The corticospinal tracts carry motor signals down the spinal cord.
8. The dermatomes are nonoverlapping regions of skin innervated by different spinal nerves.
9. Somatic reflexes are those that do not involve the brain.
10. The Golgi tendon reflex acts to inhibit muscle contraction.

Answers in the Appendix

True or False

Saladin's True or False questions are more than they appear. They also require the student to explain why the false statements are untrue, thus challenging the student to think more deeply into the material and to appreciate and express subtle points. Answers can be found in the Appendix.

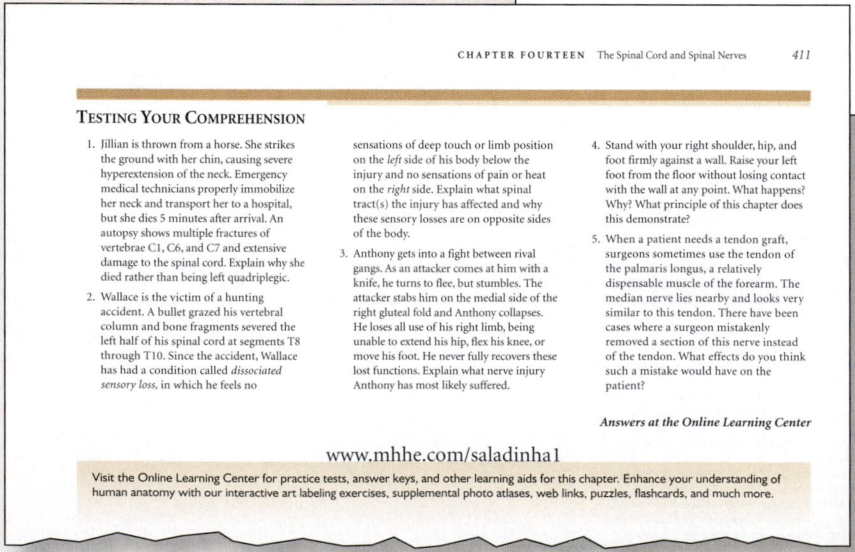
CHAPTER FOURTEEN The Spinal Cord and Spinal Nerves 411

TESTING YOUR COMPREHENSION

1. Jillian is thrown from a horse. She strikes the ground with her chin, causing severe hyperextension of the neck. Emergency medical technicians properly immobilize her neck and transport her to a hospital, but she dies 5 minutes after arrival. An autopsy shows multiple fractures of vertebrae C1, C6, and C7 and extensive damage to the spinal cord. Explain why she died rather than being left quadriplegic.
2. Wallace is the victim of a hunting accident. A bullet grazed his vertebral column and bone fragments severed the left half of his spinal cord at segments T8 through T10. Since the accident, Wallace has had a condition called *dissociated sensory loss,* in which he feels no sensations of deep touch or limb position on the *left* side of his body below the injury and no sensations of pain or heat on the *right* side. Explain what spinal tract(s) the injury has affected and why these sensory losses are on opposite sides of the body.
3. Anthony gets into a fight between rival gangs. As an attacker comes at him with a knife, he turns to flee, but stumbles. The attacker stabs him on the medial side of the right gluteal fold and Anthony collapses. He loses all use of his right limb, being unable to extend his hip, flex his knee, or move his foot. He never fully recovers these lost functions. Explain what nerve injury Anthony has most likely suffered.
4. Stand with your right shoulder, hip, and foot firmly against a wall. Raise your left foot from the floor without losing contact with the wall at any point. What happens? Why? What principle of this chapter does this demonstrate?
5. When a patient needs a tendon graft, surgeons sometimes use the tendon of the palmaris longus, a relatively dispensable muscle of the forearm. The median nerve lies nearby and looks very similar to this tendon. There have been cases where a surgeon mistakenly removed a section of this nerve instead of the tendon. What effects do you think such a mistake would have on the patient?

Answers at the Online Learning Center

www.mhhe.com/saladinha1

Visit the Online Learning Center for practice tests, answer keys, and other learning aids for this chapter. Enhance your understanding of human anatomy with our interactive art labeling exercises, supplemental photo atlases, web links, puzzles, flashcards, and much more.

Testing Your Comprehension

Questions that go beyond memorization to require a deeper level of analysis and clinical application. Scenarios from *Morbidity and Mortality Weekly Reports* and other sources prompt students to apply the chapter's basic science to real-life case histories.

Website

Located at the end of the Chapter Review is a reminder that additional study questions and other learning activities for anatomy appear on the Online Learning Center.

Online Learning Center

The Online Learning Center that accompanies Human Anatomy is found at **www.mhhe.com/saladinha** This online resource offers an extensive array of quizzing and learning tools that will help the student master the topics covered in the textbook. Each chapter offers a series of interactive art labeling exercises, vocabulary flashcards, and other engaging activities designed to reinforce learning. Access code included with every new copy of the text.

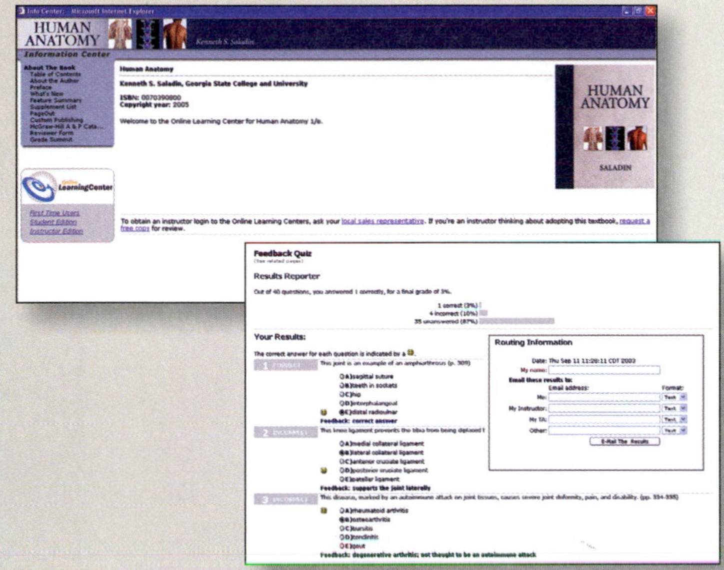

Interactive Clinical Resource CD-ROM

The Interactive Clinical Resource CD-ROM offers over 150 3-D animations and 3-D models of human disease and disorders. Students can play the 3-D animations, explore the 3-D models, print the associated text and view the slides with labels and definitions of key structures related to disease/disorder. Students will learn how the various diseases/disorders affect the human body system along with possible treatments. Contact your bookstore for ordering information, or ask your instructor to order this CD with the textbook.

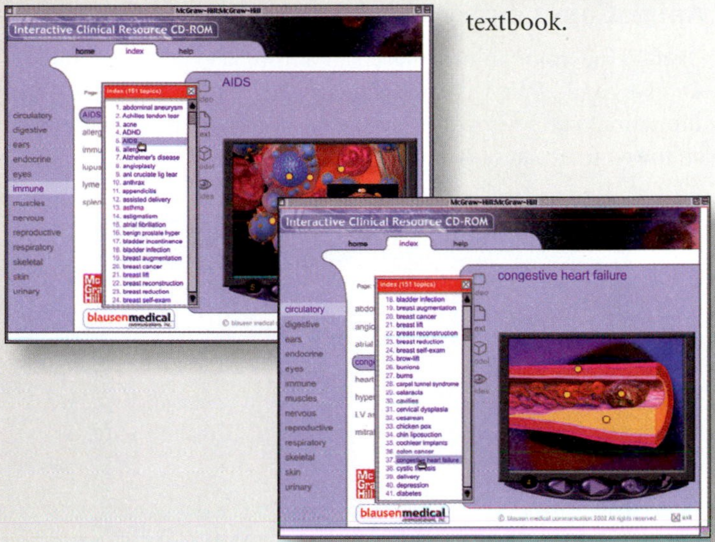

Virtual Anatomy Dissection Review CD-ROM

This CD-ROM contains over 100 high quality cat dissection photographs identifying over 600 anatomical structures. An audio feature offers pronunciations for each of the 600 terms and brief video clips demonstrate the various movements of the human body. Contact your bookstore for ordering information, or ask your instructor to order this CD with the textbook.

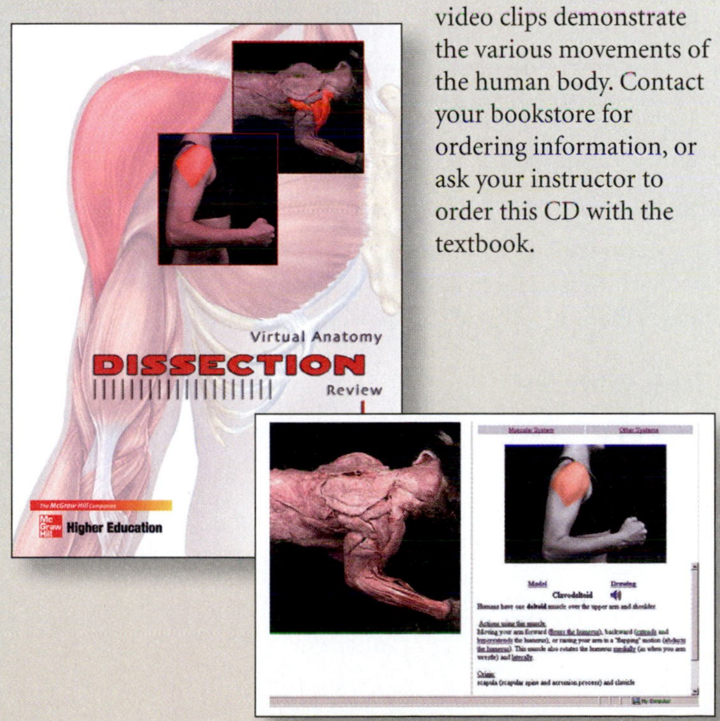

The Dynamic Human CD-ROM

The Dynamic Human CD-ROM illustrates the important relationships between anatomical structures and their functions in the human body. Realistic computer visualizations and three dimensional visualizations are the premier features of this CD-ROM. Contact your bookstore for ordering information, or ask your instructor to order this CD with the textbook.

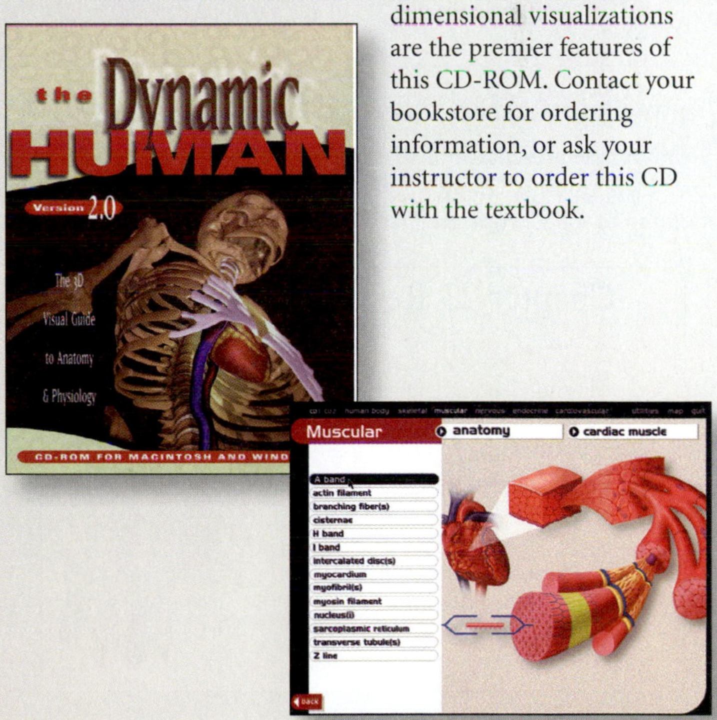

A Powerful Lecture Resource

The Digital Content Manager CD-ROM is a multimedia collection of visual resources that allows you to utilize artwork from the text in multiple formats to create your own customized course tools. Take a look at what this CD-ROM has to offer.

Animations Library

Over 25 full-color animations created from key figures in *Human Anatomy* and other McGraw-Hill anatomy textbooks. These animations can be imported into your classroom presentations or online course materials.

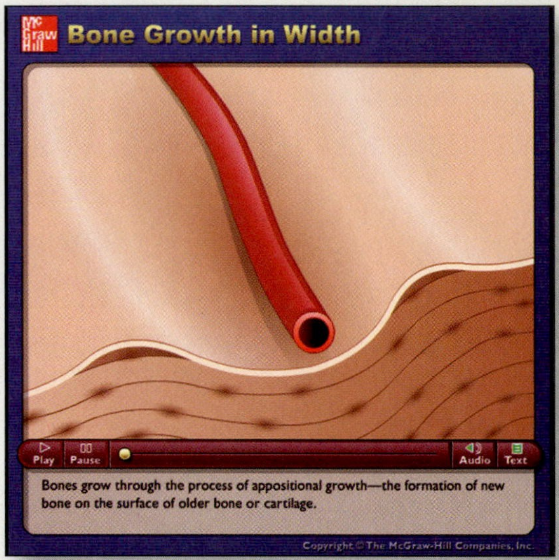

PowerPoint Lecture Outlines

Ready-made presentations that combine art and lecture notes are provided for every chapter in the text. These lectures can be tailored to reflect your preferred lecture topic or sequence.

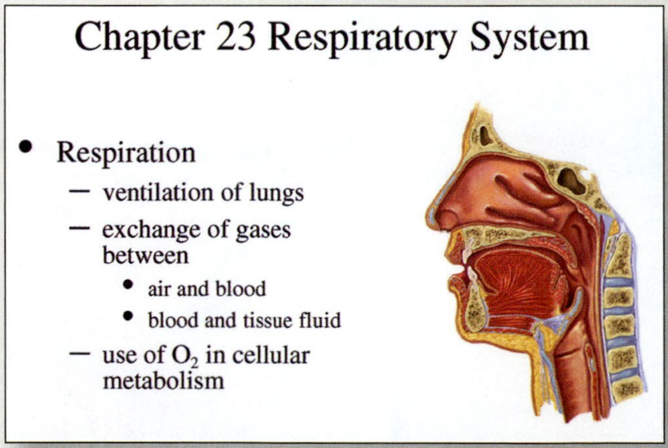

Art Library

Full-color and grayscale digital files of all the illustrations in *Human Anatomy* can be readily incorporated into lecture presentations, exams, or custom-made classroom materials.

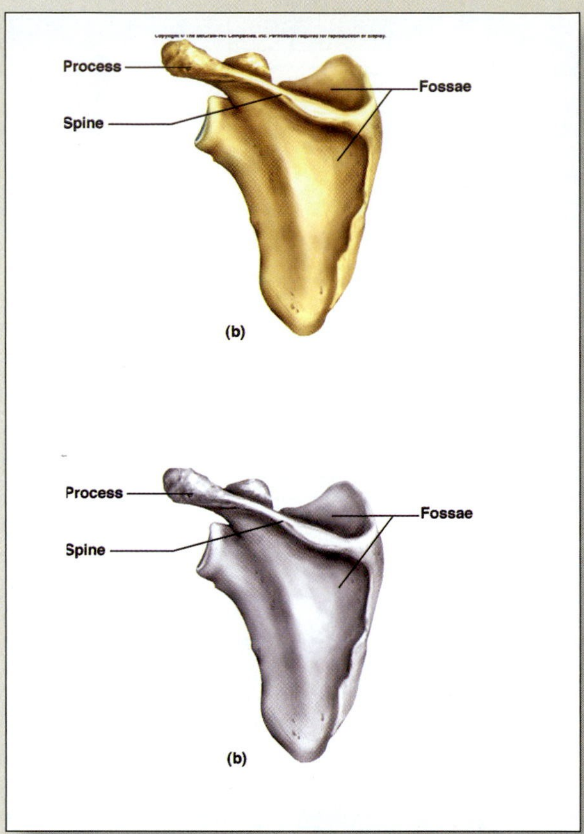

Photo Library

Digital files of instructionally significant photographs from the text, including cadaver, bone, histology, and surface anatomy images that can be reproduced for multiple classroom uses.

Table Library

Every table that appears in the text is provided in electronic form.

TextEdit Art

Every line art piece in *Human Anatomy* is placed into a PowerPoint presentation that allows labels to be revised, moved or deleted for creation of customized presentations and/or for testing purposes.

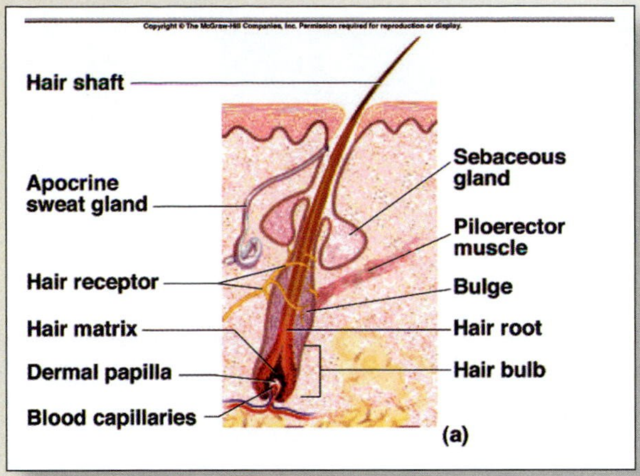

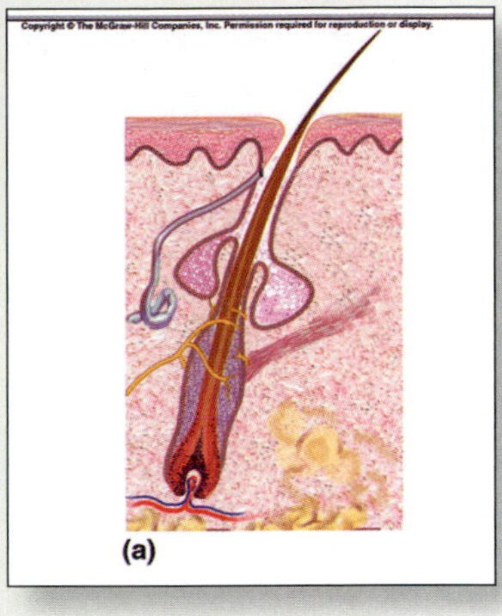

Active Art

Active Art presents key art pieces as a series of PowerPoint slides that illustrate difficult concepts in a step-by-step manner. Artwork is broken down into small, digestible frames, allowing the instructor to bring each piece into lecture in whatever sequence or format desired. These figures can also be customized in almost any way imaginable. Because every Active Art image is completely ungroupable, any part of an Active Art Slide can be used as "biological clipart" in any other PowerPoint presentation, or a component in your own rendition of a figure.

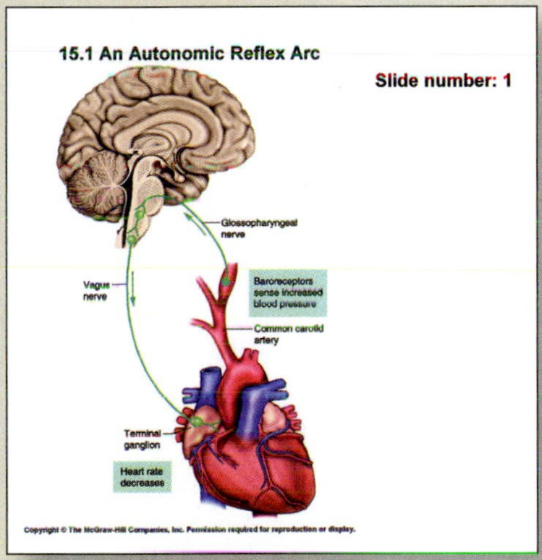

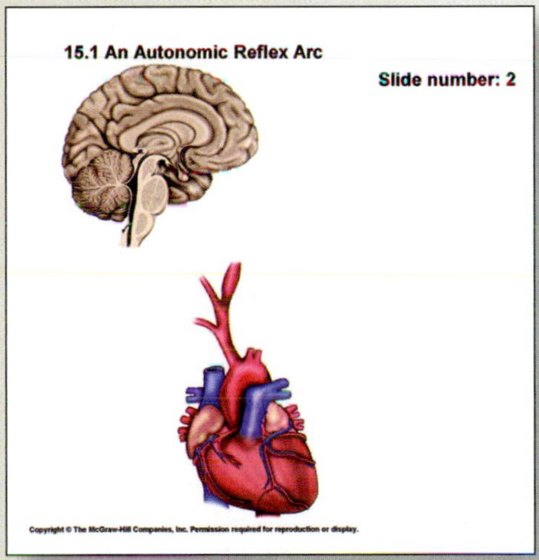

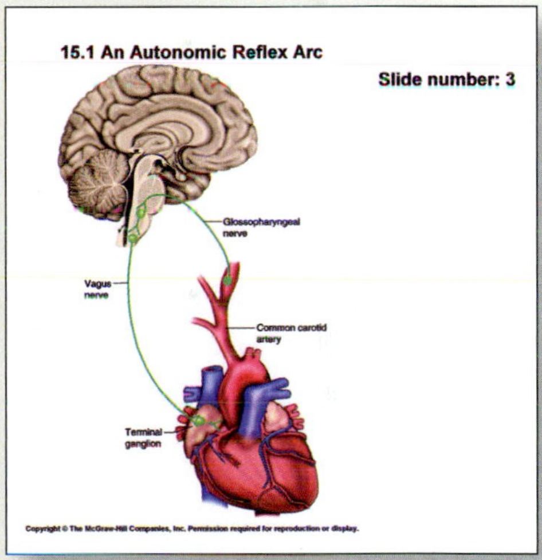

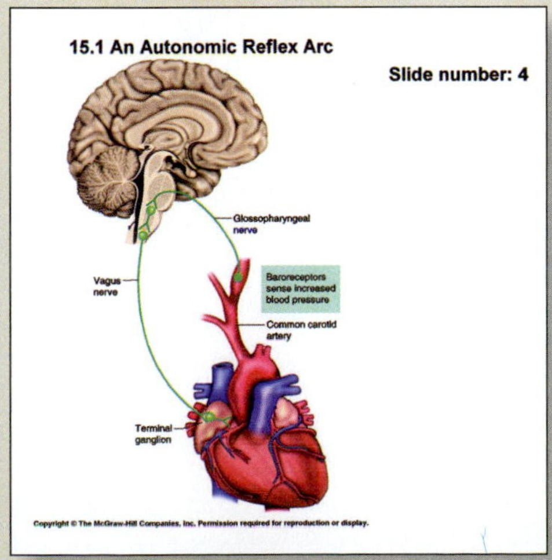

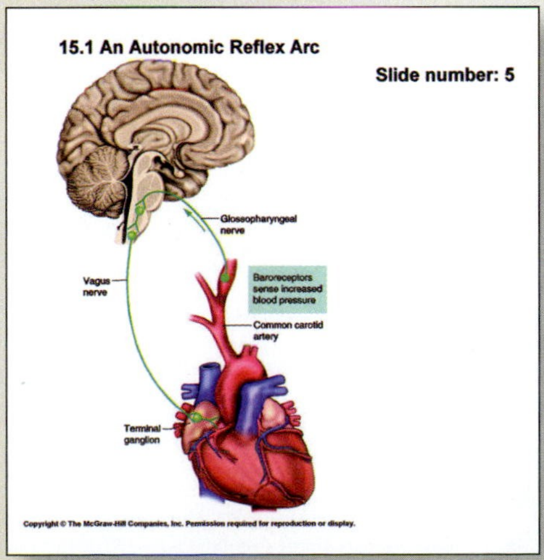

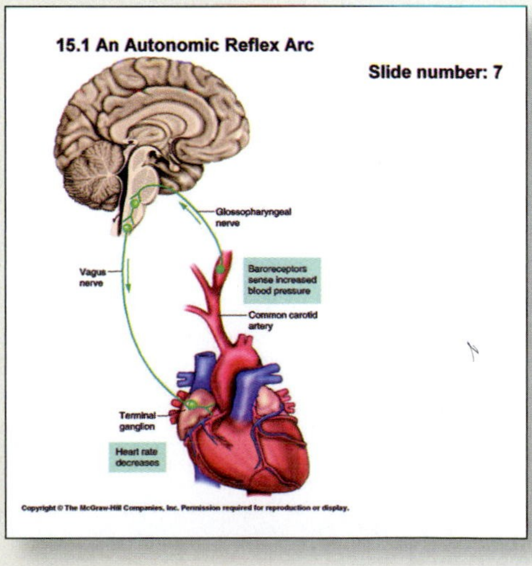

1

CHAPTER ONE

The Study of Human Anatomy

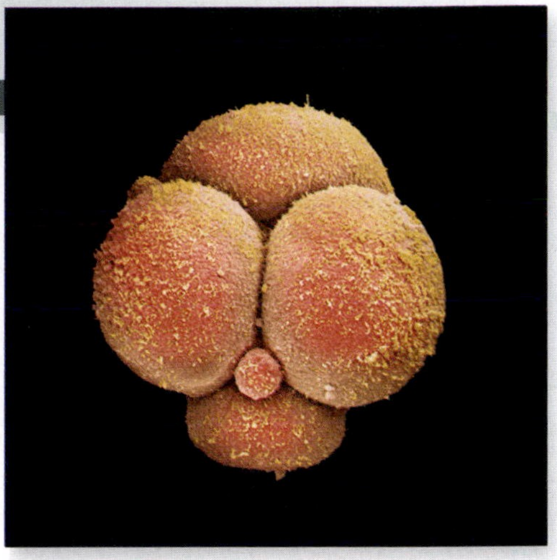

A new life begins—
a human embryo after the first two cell divisions

CHAPTER OUTLINE

The Scope of Human Anatomy *2*
- The Anatomical Sciences 2
- Methods of Study 2
- Variation in Human Structure 5
- Levels of Human Structure 6

Early Anatomists *8*
- The Greek and Roman Legacy 8
- Medieval Anatomy in Christendom and Islam 9
- The Birth of Modern Anatomy 9
- The Discovery of Microscopic Anatomy 10

The Nature of Human Life *11*
- What Is Life? 12
- What Is a Human? 12

The Evolution of Human Structure *14*
- Evolution, Selection, and Adaptation 14
- Life in the Trees 15
- Walking Upright 16

The Language of Anatomy *16*
- The History of Anatomical Terminology 17
- Analyzing Medical Terms 18
- Singular and Plural Noun Forms 18
- The Importance of Precision 18

Chapter Review *20*

INSIGHTS

1.1 Clinical Application: Situs Inversus and Other Unusual Anatomy 6

1.2 Evolutionary Medicine: Vestiges of Human Evolution 15

1.3 Medical History: Obscure Word Origins 18

PART ONE

This book is an introduction to the structure of the human body. It is meant primarily to provide a foundation for advanced study in fields related to health and fitness. Beyond that purpose, however, the study of anatomy can also provide a satisfying sense of self-understanding. Even as children, we are curious about what's inside the body. Dried skeletons, museum exhibits, and beautifully illustrated atlases of the body have long elicited special fascination.

This chapter lays a foundation for our study of anatomy by considering some broad themes—What does this science encompass? What methods are used to study anatomy? How did our understanding of human anatomy develop? What aspects of human anatomy differentiate us from other animals and define us as humans? How did the human body come to be as it is? How can a beginner more easily master the complex language of anatomy?

The Scope of Human Anatomy

Objectives

When you have completed this section, you should be able to:

- define *anatomy* and some of its subdisciplines;
- name and describe some approaches to studying anatomy;
- describe some methods of medical imaging;
- describe some variations in human anatomy; and
- state the levels of human structure from organismal to subatomic.

Human **anatomy** is the study of the structure of the human body. It provides an essential foundation for the understanding of **physiology,** the study of function; anatomy and physiology together are the bedrock of the health sciences. You can study human anatomy from an atlas; yet as beautiful, fascinating, and valuable as many anatomy atlases are, they teach almost nothing but the locations, shapes, and names of structures. This book is much different; it deals with what biologists call **functional morphology**[1]—not just the structure of organs, but the functional reasons behind that structure. Anatomy and physiology are complementary to each other; each makes sense of the other, and each molds the other in the course of human development and evolution. We cannot delve into the details of physiology in this book, but enough will be said of function to help you make sense of human structure and to more deeply appreciate the beauty of human form.

The Anatomical Sciences

There are several perspectives on human structure that form subdisciplines of anatomy. **Gross anatomy** is the study of structure visible to the naked eye, either by surface observation or dissection. **Surface anatomy** is the external structure of the body, and is especially important in conducting a physical examination of a patient.

[1]*morpho* = form, structure + *logy* = study of

Systemic anatomy is the study of one organ system at a time and is the approach taken by most introductory textbooks such as this one. **Regional anatomy** is the study of multiple organ systems at once in a given region of the body, such as the head or chest. Medical schools typically teach anatomy from this perspective, because it is more logical to dissect all structures of the head and neck, the chest, or a limb, than it would be to try to dissect the entire digestive system, then the cardiovascular system, and so forth. Dissecting one system almost invariably destroys organs of another system that stand in the way. Furthermore, as surgeons operate on a particular area of the body, they must think from a regional perspective and attend to the interrelationships of all structures in that area.

Ultimately, the structure and function of the body result from its individual cells. To see those, we usually take tissue specimens, thinly slice and stain them, and observe them under the microscope. This approach is called **microscopic anatomy (histology). Histopathology**[2] is the microscopic examination of tissues for signs of disease. **Cytology**[3] is the study of the structure and function of individual cells. **Ultrastructure** refers to fine detail, down to the molecular level, revealed by the electron microscope.

Anatomy, of course, is not limited to the study of humans, but extends to all living organisms. Even students of human structure benefit from **comparative anatomy**—the study of more than one species in order to learn generalizations and evolutionary trends. Anatomy students often begin by dissecting other animals with which we share a common ancestry and many structural similarities. Indeed, many of the reasons for human structure become apparent only when we look at the structure of other animals. In chapter 25, for example, you will see that physiologists had little idea of the purpose of certain tubular loops in the kidney *(nephron loops)* until they compared human kidneys with those of desert and aquatic animals, which have greater and lesser needs to conserve water. The greater an animal's need to conserve water (the drier its habitat), the longer these loops are. Thus, comparative anatomy hinted at the function of the nephron loop.

Methods of Study

The structure of the human body is studied in several ways. The simplest is **inspection**—simply looking at the body's appearance, as in performing a physical examination or making a clinical diagnosis from surface appearance. Physical examinations involve not only looking at the body for signs of normalcy or disease, but also touching and listening to it. **Palpation**[4] means feeling a structure with the hands, such as palpating a swollen lymph node or taking a pulse. **Auscultation**[5] (AWS-cul-TAY-shun) is listening to the natural sounds made by the body, such as heart and lung sounds. In **percussion,** the examiner taps on the body, feels for abnormal resistance, and listens to the emitted sound for signs of abnormalities such as pockets of fluid or air.

[2]*histo* = tissue + *patho* = disease + *logy* = study of
[3]*cyto* = cell + *logy* = study of
[4]*palp* = touch, feel
[5]*auscult* = listen

A deeper understanding of the body depends on **dissection**—the careful cutting and separation of tissues to reveal their relationships. In many schools of health science, one of the first steps in the training of students is dissection of the **cadaver,**[6] a dead human body (fig. 1.1). The very words *anatomy*[7] and *dissection*[8] both mean "cutting apart;" until the nineteenth century, dissection was called "anatomizing."

Dissection, of course, is not the method of choice when studying a living person! It was once common to diagnose disorders through **exploratory surgery**—opening the body and taking a look inside to see what was wrong and what could be done about it. Any breach of the body cavities is risky, however, and most exploratory surgery has now been replaced by **medical imaging** techniques—methods of viewing the inside of the body without surgery. The branch of medicine concerned with imaging is called **radiology.** Imaging methods are called *noninvasive* if they do not involve any penetration of the skin or body orifices. *Invasive* imaging techniques may entail inserting ultrasound probes into the esophagus, vagina, or rectum to get closer to the organ to be imaged, or injecting substances into the bloodstream or body passages to enhance image formation.

Any anatomy student today must be acquainted with the basic techniques of radiology and their respective advantages and limitations. Many of the images printed in this book have been produced by the following techniques.

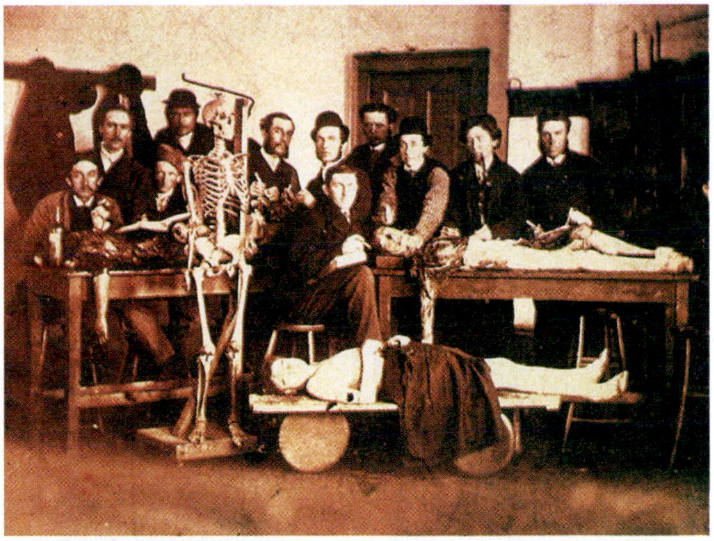

FIGURE 1.1
Early Medical Students in the Gross Anatomy Laboratory with Three Cadavers. Students of the health sciences have long begun their professional training by dissecting cadavers.

RADIOGRAPHY

Radiography is the process of photographing internal structures with X rays, a form of high-energy radiation. X rays penetrate soft tissues of the body and darken photographic film on the other side. They are absorbed, however, by dense tissues such as bone, teeth, tumors, and tuberculosis nodules, which leave the film lighter in these areas (fig. 1.2*a*). The term *X ray* also applies to a photograph *(radiograph)* made by this method. Radiography is commonly used in dentistry, mammography, diagnosis of fractures, and examination of the chest. Hollow organs can be visualized by filling them with a *radiopaque* substance that absorbs X rays. Barium sulfate, for example, is given orally for examination of the esophagus, stomach, and small intestine or by enema for examination of the large intestine. Other substances are given by injection for *angiography,* the examination of blood vessels (fig. 1.2*b*). Some disadvantages of radiography are that images of overlapping organs can be confusing and slight differences in tissue density are not easily detected. Nevertheless, radiography still accounts for over 50% of all clinical imaging. Until the 1960s, it was the only method widely available.

SONOGRAPHY

Sonography[9] is the second oldest and second most widely used method of imaging. It is an outgrowth of sonar technology developed in World War II. A hand-held device held firmly against the skin produces high-frequency ultrasound waves and receives the signals reflected back from internal organs. Sonography avoids the harmful effects of X rays, and the equipment is inexpensive and portable. Its primary disadvantage is that it does not produce a very sharp image (fig. 1.3). Although sonography was first used medically in the 1950s, images of significant clinical value had to wait until computer technology had developed enough to analyze differences in the way tissues reflect ultrasound. Sonography is not very useful for examining bones or lungs, but it is the method of choice in obstetrics, where the image *(sonogram)* can be used to locate the placenta and evaluate fetal age, position, and development. *Echocardiography* is the sonographic examination of the beating heart.

COMPUTED TOMOGRAPHY

Computed tomography (a **CT scan**), formerly called a *computerized axial tomographic*[10] *(CAT) scan,* is a more sophisticated application of X rays developed in 1972. The patient is moved through a ring-shaped machine that emits low-intensity X rays on one side and receives them with a detector on the opposite side. A computer analyzes signals from the detector and produces an image of a "slice" of the body about as thin as a coin (fig. 1.4). The computer can "stack" a series of these images to construct a three-dimensional image of the body. CT scanning has the advantage of imaging thin sections of the body, so there is little organ overlap and the image is much sharper than a conventional X ray. CT scanning is useful for identifying tumors, aneurysms, cerebral hemorrhages, kidney stones, and other abnormalities. It has replaced most exploratory surgery.

[6]from *cadere* = to fall down or die
[7]*ana* = apart + *tom* = cut
[8]*dis* = apart + *sect* = cut
[9]*sono* = sound + *graphy* = recording process
[10]*tomo* = section, cut, slice + *graphic* = pertaining to a recording

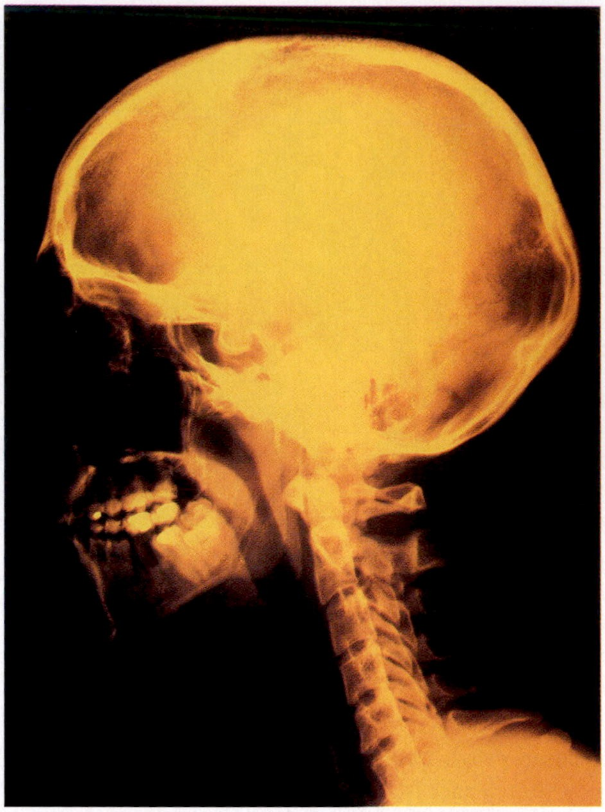

(a)

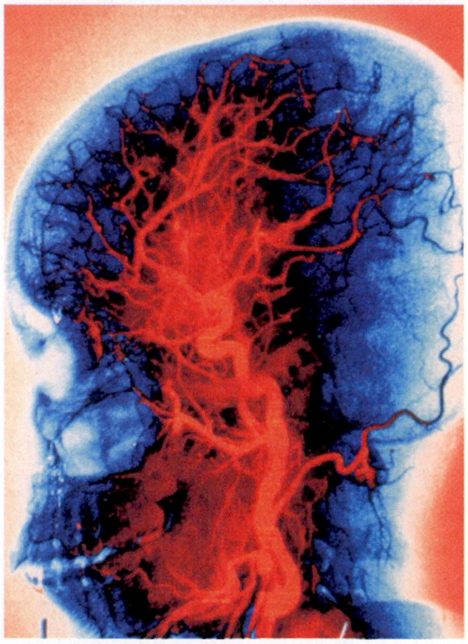

(b)

FIGURE 1.2

Radiography. (*a*) An X ray (radiograph) of the head and neck. (*b*) A cerebral angiogram, made by injecting a substance opaque to X rays into the circulation and then taking an X ray of the head to visualize the blood vessels. The vessels are enhanced with false color in this photograph.

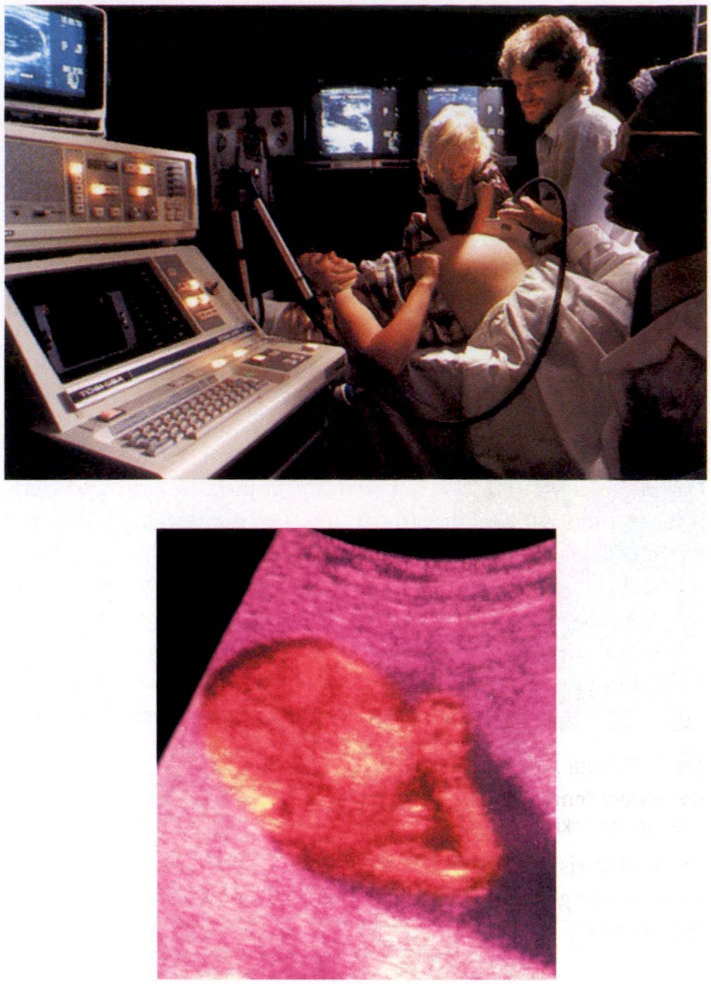

FIGURE 1.3

Fetal Sonogram. Shows the head and right arm of a 28-week-old fetus sucking its thumb.

MAGNETIC RESONANCE IMAGING

Magnetic resonance imaging (MRI) was conceived as a technique superior to CT for visualizing soft tissues. The patient lies within a chamber surrounded by a large electromagnet that creates a magnetic field 3,000 to 60,000 times as strong as the earth's. Hydrogen atoms in the tissues align themselves with the magnetic field. The patient is then bombarded with radio waves, which cause the hydrogen atoms to absorb additional energy and align in a different direction. When the radio waves are turned off, the hydrogen atoms abruptly realign to the magnetic field, giving off their excess energy at rates that depend on the type of tissue. A computer analyzes the emitted energy to produce an image of the body. MRI can "see" clearly through the skull and vertebral column to produce images of the nervous tissue. Moreover, it is better than CT for distinguishing between soft tissues such as the white and gray matter of the nervous system (fig. 1.5). *Functional MRI (fMRI)* is a form of MRI that visualizes moment-to-moment

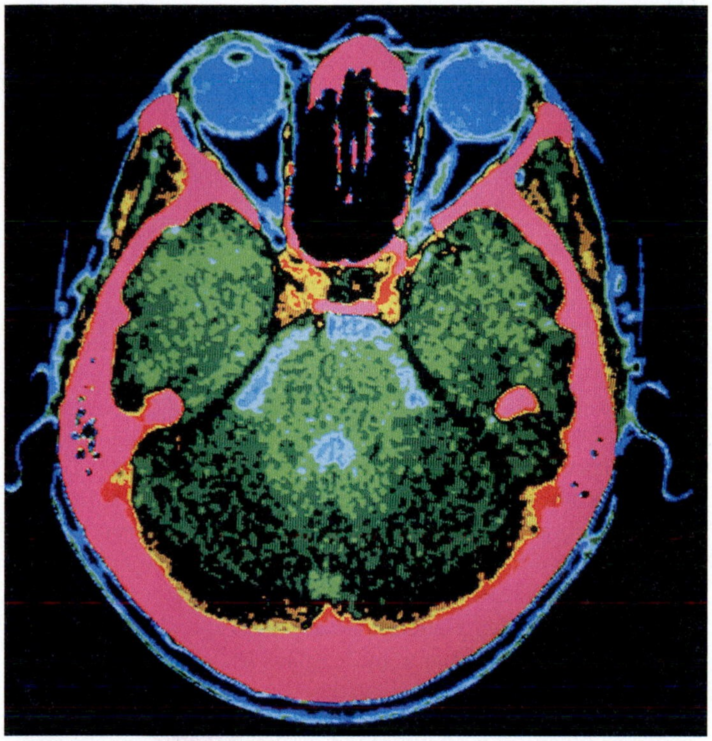

FIGURE 1.4
Computed Tomographic (CT) Scan of the Head at the Level of the Eyes. The eyes and skin are shown in blue, bone in red, and the brain in green.

changes in tissue function; fMRI scans of the brain, for example, show shifting patterns of activity as the brain applies itself to a specific task.

Think About It!

The concept of MRI was conceived in 1948 but was not put into clinical practice until the 1970s. Speculate on a possible reason for this delay.

POSITRON EMISSION TOMOGRAPHY

Positron emission tomography (a **PET scan**), developed in the 1970s, is used to assess the metabolic state of a tissue and to distinguish which tissues are most active at a given moment (fig. 1.6). The procedure begins with an injection of radioactively labeled glucose, which emits positrons (electron-like particles with a positive charge). When a positron and electron meet, they annihilate each other and give off gamma rays that can be detected by sensors and analyzed by computer. The result is a color image that shows which tissues were using the most glucose at the moment. In cardiology, PET scans can show the extent of damaged heart tissue. Since damaged tissue consumes little or no glucose, it appears dark. In neuroscience, PET scans are used, like fMRI, to show which regions of the brain are most active when a person performs a specific task. The PET scan is an example of **nuclear medicine**—the use of radioisotopes to treat disease or to form diagnostic images of the body.

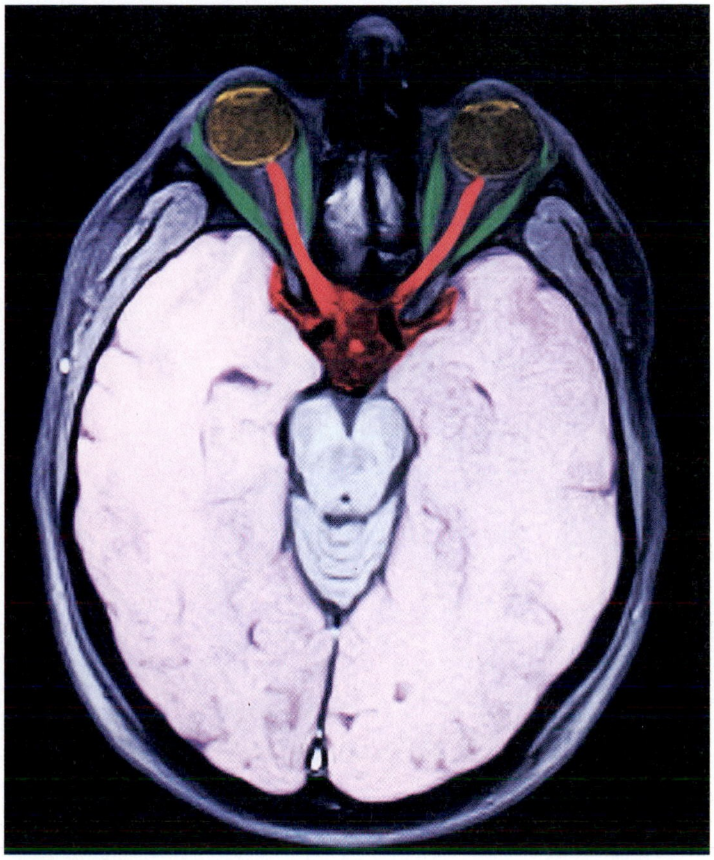

FIGURE 1.5
Magnetic Resonance Image (MRI) of the Head at the Level of the Eyes. The optic nerves appear in red and the muscles that move the eyes appear in green.

Variation in Human Structure

A quick look around any classroom is enough to show that no two humans look exactly alike; on close inspection, even identical twins exhibit differences. Anatomy atlases and textbooks can easily give you the impression that everyone's internal anatomy is the same, but this simply is not true. Books such as this one can only teach you the most common structure—the anatomy seen in approximately 70% or more of people. Someone who thinks that all human bodies are the same internally would make a very confused medical student or an incompetent surgeon.

Some people completely lack certain organs. For example, most of us have a *palmaris longus* muscle in the forearm and a *plantaris* muscle in the leg, but not everyone has them. Most of us have five lumbar vertebrae (bones of the lower spine), but some have four and some have six. Most of us have one spleen, but some people have two. Most have two kidneys, but some have only one. Most kidneys are supplied by a single *renal artery* and drained by one *ureter*, but in some people a single kidney has two renal arteries or ureters. Figure 1.7 shows some common variations in human anatomy, and Insight 1.1 describes a particularly dramatic variation.

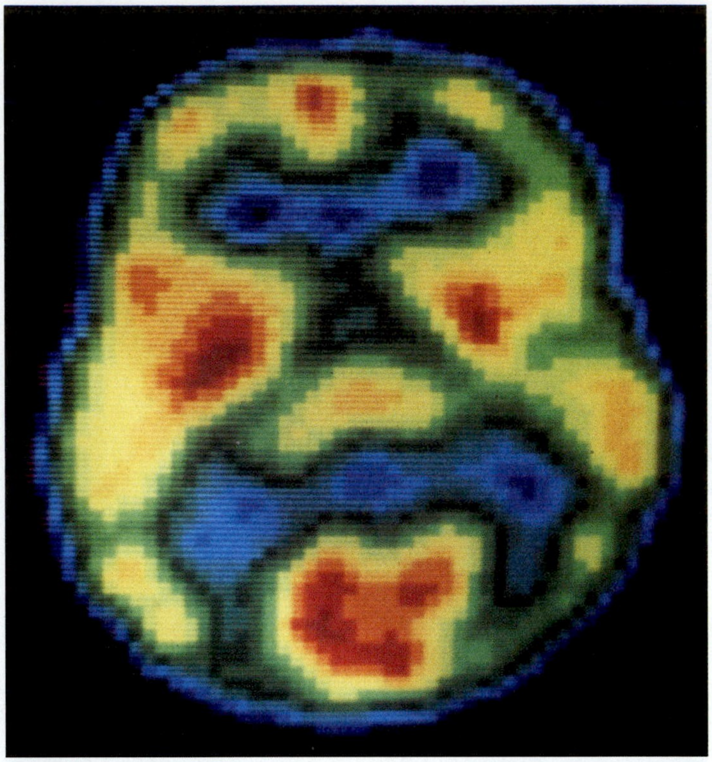

FIGURE 1.6

Positron Emission Tomographic (PET) Scan of the Brain of an Unmedicated Schizophrenic Patient. Red areas indicate high glucose consumption (high metabolism). In this patient, the visual center of the brain (rear of head, *bottom of photo*) was especially active when this scan was made.

INSIGHT 1.1 CLINICAL APPLICATION

Situs Inversus and Other Unusual Anatomy

In most people, the heart tilts toward the left, the spleen and sigmoid colon are on the left, the liver and gallbladder lie mainly on the right, the appendix is on the right, and so forth. This normal arrangement of the viscera is called *situs* (SITE-us) *solitus.* About 1 in 8,000 people are born, however, with a striking developmental abnormality called *situs inversus*—the organs of the thoracic and abdominal cavities are reversed between right and left. A selective right-left reversal of the heart is called *dextrocardia.* In *situs perversus,* a single organ occupies an atypical position, not necessarily a left-right reversal—for example, a kidney located low in the pelvic cavity instead of high in the abdominal cavity.

Some conditions, such as dextrocardia in the absence of complete situs inversus, can cause serious medical problems. Complete situs inversus, however, usually causes no functional problems because all of the viscera, though reversed, maintain their normal relationships to each other. Situs inversus is often diagnosed prenatally by sonography, but many people remain unaware of their condition for several decades until it is discovered by medical imaging, on physical examination, or in surgery. However, you can easily imagine the importance of such conditions in diagnosing appendicitis, performing gallbladder surgery, interpreting an X ray, or auscultating the heart valves.

Think About It!

People who are allergic to penicillin or aspirin often wear Medic Alert bracelets or necklaces that note this fact in case they need emergency medical treatment and are unable to communicate. Why would it be important for a person with situs inversus to have this noted on a Medic Alert bracelet?

Levels of Human Structure

Although this book is concerned mainly with gross anatomy, the study of human structure spans all levels from the organismal to the subatomic. Consider for a moment an analogy to human structure: The English language, like the human body, is very complex, yet an endless array of ideas can be conveyed with a limited number of words. All words in the English language are, in turn, composed of various combinations of just 26 letters. Between an essay and the alphabet are successively simpler levels of organization: paragraphs, sentences, words, and syllables. Language has a hierarchy of complexity, with letters, syllables, words, and so forth being successive levels of the hierarchy. Humans have an analogous hierarchy of complexity (fig. 1.8), as follows:

The organism is composed of organ systems,
organ systems are composed of organs,
organs are composed of tissues,
tissues are composed of cells,
cells are composed (in part) of organelles,
organelles are composed of molecules,
molecules are composed of atoms, and
atoms are composed of subatomic particles.

The **organism** is a single, complete individual.

An **organ system** is a group of organs that carry out a basic function of the organism such as circulation, respiration, or digestion. The human body has 11 organ systems, defined and illustrated in atlas A following this chapter: the integumentary, skeletal, muscular, nervous, endocrine, circulatory, lymphatic, respiratory, urinary, digestive, and reproductive systems. Usually, the organs of a system are physically interconnected, such as the kidneys, ureters, urinary bladder, and urethra that compose the urinary system. The endocrine system, however, is a group of hormone-secreting glands and tissues that, for the most part, have no physical connection to each other.

An **organ** is a structure composed of two or more tissue types that work together to carry out a particular function. Organs have definite anatomical boundaries and are visibly distinguishable from adjacent structures. Most organs and higher levels of structure are within the domain of gross anatomy. However, there are organs within organs—the large organs visible to the naked eye often contain smaller organs visible only with the microscope. The skin, for example, is the body's largest organ. Included within it are thousands of smaller organs: each hair follicle, nail, sweat gland, nerve, and blood vessel of the skin is an organ in itself.

A **tissue** is a mass of similar cells and cell products that forms a discrete region of an organ and performs a specific function. The

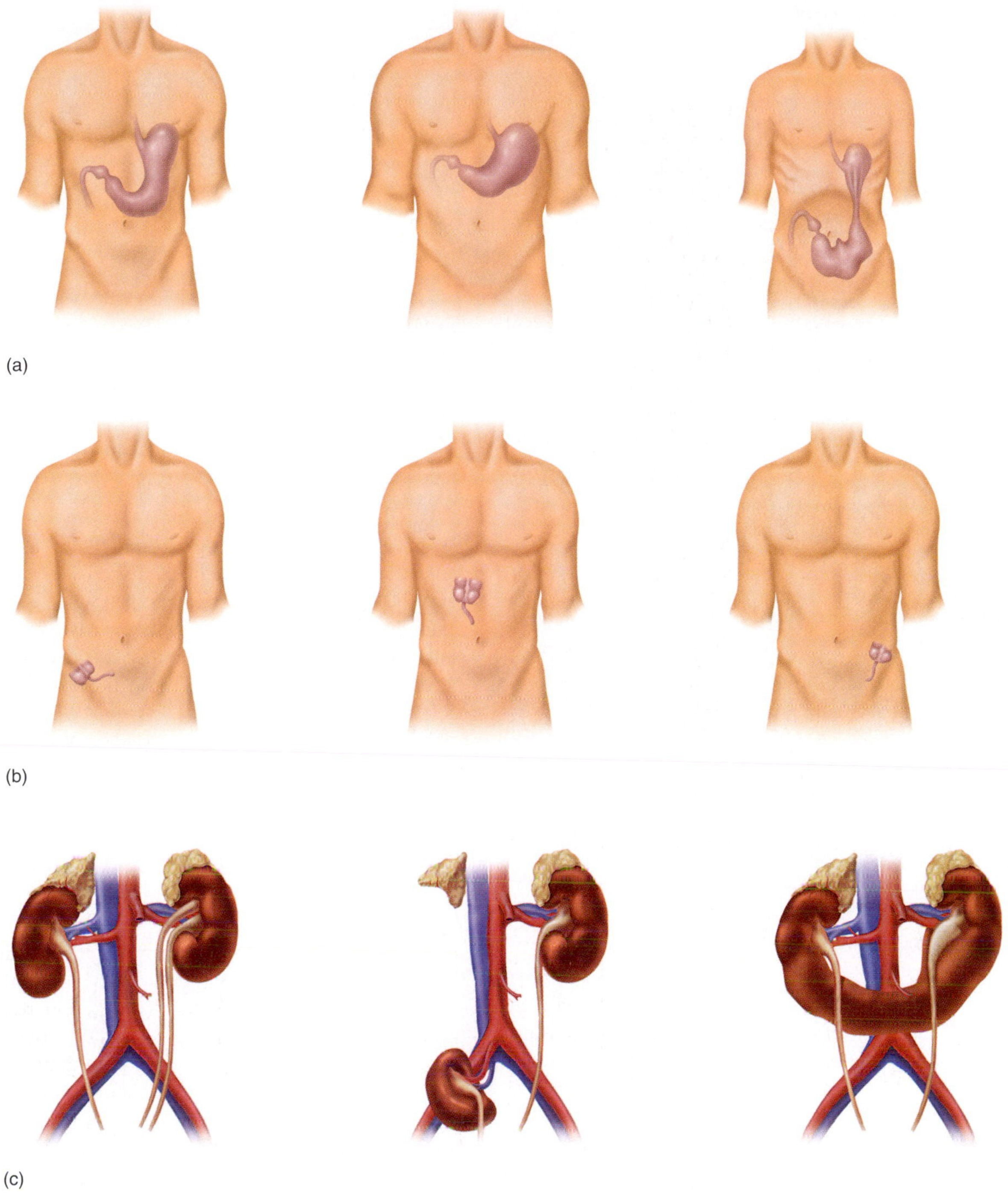

Figure 1.7

Variation in Human Anatomy. The left-hand figure in each case depicts the most common anatomy. (*a*) Variations in stomach shape correlated with body weight. (*b*) Variations in the position of the appendix. (*c*) Variations in the position of the kidneys (*left,* a normal kidney and one with two ureters, both kidneys in normal position; *center,* pelvic kidney; *right,* horseshoe kidney).

body is composed of only four primary classes of tissue—epithelial, connective, nervous, and muscular tissues. *Histology,* the study of tissues, is the subject of chapter 3.

Cells are the smallest units of an organism that carry out all the basic functions of life; nothing simpler than a cell is considered alive. A cell is enclosed in a *plasma membrane* composed of lipids and protein. Most cells have one *nucleus,* an organelle that contains most of its DNA. *Cytology,* the study of cells and organelles, is the subject of chapter 2.

Organelles[11] are microscopic structures in a cell that carry out its individual functions. Examples include mitochondria, centrioles, and lysosomes.

Organelles and other cellular components are composed of **molecules.** The largest molecules, such as proteins, fats, and DNA,

[11]*elle* = little

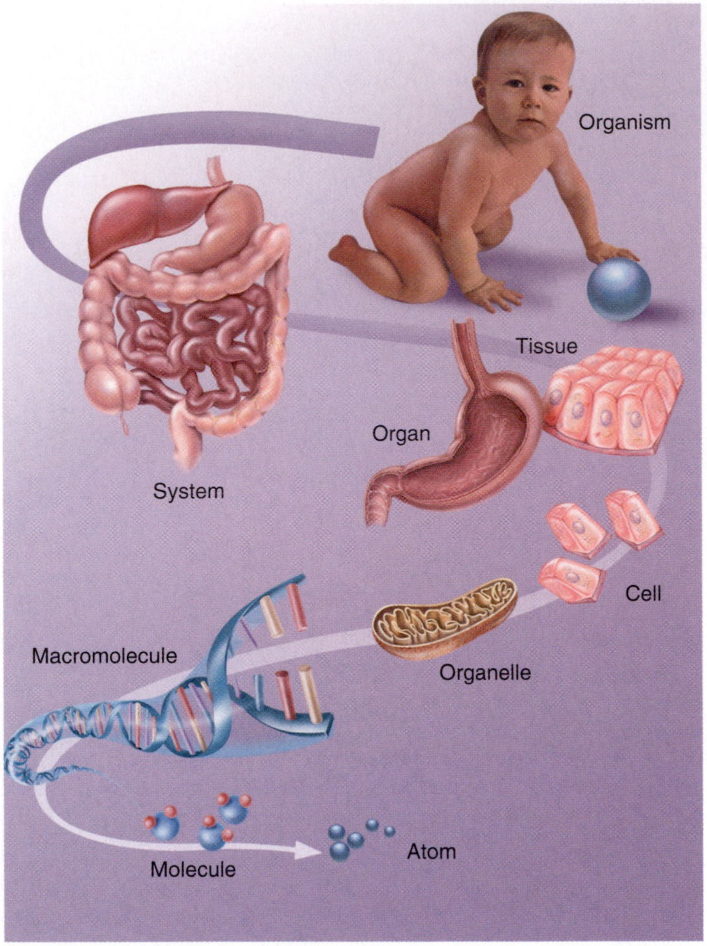

Figure 1.8
The Body's Structural Hierarchy.

are called *macromolecules.* A molecule is a particle composed of at least two **atoms,** and an atom is composed of **subatomic particles**—protons, neutrons, and electrons.

Think About It!

Architect Louis Henri Sullivan coined the phrase, "Form ever follows function." What do you think he meant by this? Discuss how this idea could be applied to the human body and cite a specific example of human anatomy to support it. Identify some exceptions to this rule.

Before You Go On

Answer the following questions to test your understanding of the preceding section:

1. How does functional morphology differ from the sort of anatomy taught by a photographic atlas of the body?
2. Why would regional anatomy be a better learning approach than systemic anatomy for a cadaver dissection course?
3. What are some reasons that sonography would be unsuitable for examining the size and location of a brain tumor?
4. Put the following alphabetical list in order from the largest and most complex to the smallest and least complex components of the human body: cells, molecules, organelles, organs, organ systems, tissues.

Early Anatomists

Objectives

When you have completed this section, you should be able to:

- describe the major contributions of ancient Greece and Rome to anatomy;
- identify some contributors to medicine in the Middle Ages;
- describe the change that anatomy underwent in the Renaissance; and
- describe the origin and importance of microscopic anatomy.

Any science is more enjoyable if we consider not just the current state of knowledge, but how it compares to past understandings of the subject and how our current knowledge was gained. Of all sciences, medicine has one of the most fascinating histories. Medical science has progressed far more in the last 25 years than in the 2,500 years before that, but the field did not spring up overnight. It is built upon centuries of thought and controversy, triumph and defeat. We cannot fully understand its present state without understanding people who had the curiosity to try new things, the vision to look at human form and function in new ways, and the courage to question authority.

The Greek and Roman Legacy

Anatomy is an ancient human interest, undoubtedly older than any written language we know. We can only guess when people began deliberately cutting into human bodies out of curiosity, simply to know what was inside. The Greek philosopher **Aristotle** (384–322 B.C.E.) was one of the earliest to write about anatomy. In his book, *Of the Parts of Animals,* he tried to identify unifying themes in anatomy and argued that complex structures are built from a smaller variety of simple components—as we have seen in our look at the hierarchy of human structure. Aristotle believed that diseases and other natural events could have either supernatural causes, which he called *theologi,* or natural ones, which he called *physici* or *physiologi.* We derive such terms as *physician* and *physiology* from the latter. Until the nineteenth century, physicians were called "doctors of physic."

Herophilus (c. 335–280 B.C.E.) was the most experienced anatomist of antiquity. A Greek working at Alexandria, Egypt, he dissected hundreds of cadavers and gave public demonstrations of anatomy. He named the duodenum; wrote good descriptions of the retina, optic nerve, ovaries, uterus, prostate gland, liver, and pancreas; and distinguished between cranial and spinal nerves, sensory and motor nerves, and arteries and veins. His work remained the highest achievement in human anatomy until the Renaissance.

But **Claudius Galen** (129–c. 199 C.E.), Greek-born physician to the Roman gladiators, was more influential. His medical textbook was worshipped to excess by European medical professors for centuries to follow. Yet cadaver dissection was banned in Galen's time because of some horrid excesses that preceded him, including public dissections of living slaves and prisoners. Aside from what he could learn by treating the gladiators' wounds, Galen was limited to dissecting pigs, monkeys, and other animals. Because he was not permitted to dissect cadavers, he had to guess at much of human anatomy and made some incorrect deductions from animal dissections. He described the human liver, for example, as having five fingerlike lobes, somewhat like a baseball glove, because that is what he had seen in baboons. But Galen saw science as a method of discovery, not as a body of fact to be taken on faith. He knew that he might be wrong, and he advised his followers to trust their own observations more than they trusted any book, including his own. Unfortunately, his advice was not heeded. For nearly 1,500 years, medical professors dogmatically taught what they read in Aristotle and Galen, and few dared to question the authority of these "ancient masters."

Medieval Anatomy in Christendom and Islam

Western medical science advanced very little during the Middle Ages. Even though many of the most famous medical schools of Europe were founded during this era, the professors taught medicine primarily as a dogmatic commentary on Galen and Aristotle. European culture was dominated by the Church, which saw little need for further research and discouraged science in general, on the argument that the masters had already discovered everything that needed to be known. Medieval medical illustrations were crude representations of the body, serving more to decorate a page than to give any meaningful instruction to the reader (fig. 1.9). Many were astrological charts that showed which sign of the zodiac was thought to influence each organ of the body. From such pseudoscience came the word *influenza,* Italian for *influence.*

Medieval medicine advanced more in the Muslim world than in Christendom, as Islam (the Muslim religion) placed less restriction on scientific inquiry. Probably the most influential Muslim physician was **Ibn-Sina,** known in Europe as **Avicenna** (980–1037), court physician to the vizier of southern Persia. He wrote 16 medical treatises, most famously the *Canon of Medicine*—a five-volume, million-word encyclopedia of medical knowledge up to that time. The *Canon* was widely translated and became a leading reference for over 500 years in European medical schools, where Avicenna was nicknamed "the Galen of Islam."

Ibn an-Nafis (1210–88) was the personal physician of the sultan of Egypt. He discovered the pulmonary circuit and coronary circulation and corrected many of the errors of Galen and Avicenna. His work remained largely unknown in Europe for 300 years, but after his books were rediscovered and translated to Latin in 1547, they strongly influenced leading European anatomists including Vesalius, discussed next.

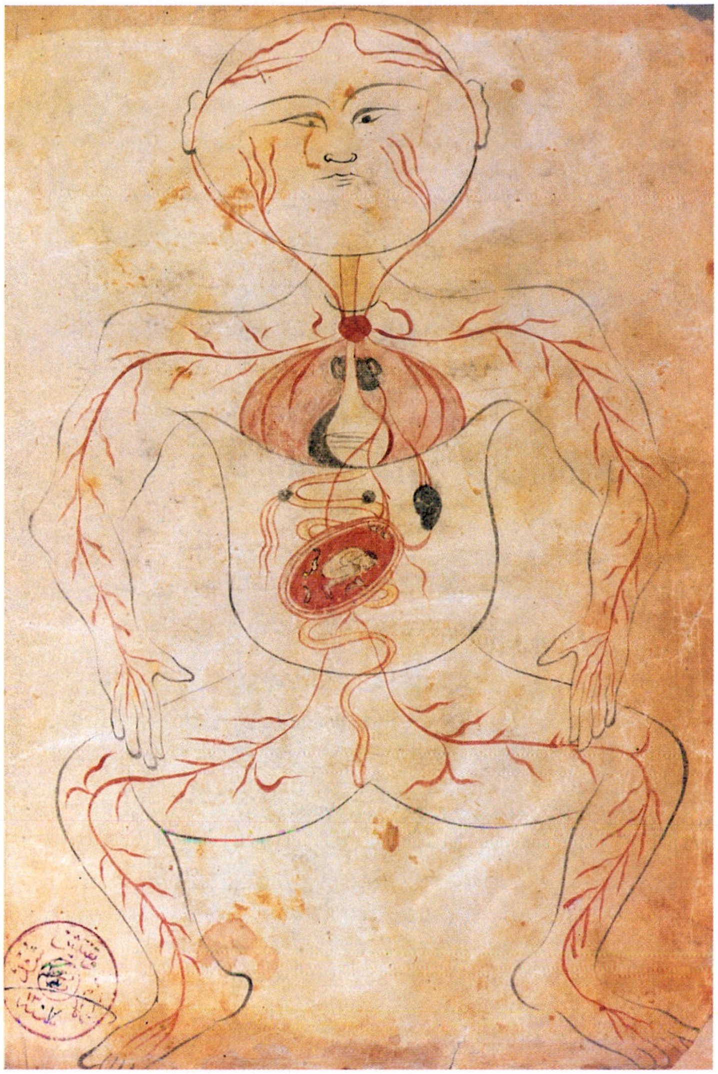

FIGURE 1.9
Medieval Medical Illustration. This figure depicts a woman showing the heart, lungs, arteries, digestive tract, and a fetus in the uterus.

The Birth of Modern Anatomy

Andreas Vesalius (1514–64) is commonly regarded as the pioneer of modern anatomy. He was a Flemish physician who taught in Italy. In his time, cadaver dissection had resumed for legal purposes (autopsies) and gradually found its way into the training of medical students throughout Europe. Furthermore, the Italian Renaissance created an environment more friendly to creative scholarship. Dissection was an unpleasant business, however, and most professors considered it beneath their dignity. In those days before refrigeration or embalming, the odor from the decaying cadaver was unbearable. Dissections were often conducted outdoors in a nonstop 4-day race against decay. Bleary medical students had to fight the urge to vomit, lest they offend their overbearing professor. The professor typically sat in an elevated chair, the *cathedra,* reading dryly from Galen or Aristotle while a lower-ranking *barber-surgeon* removed putrefying organs from the cadaver and held them up for the students to see. Barbering and surgery were considered to be "kindred arts of the knife"; today's barber poles date from this era, their red and white stripes symbolizing blood and bandages.

Vesalius broke with tradition by coming down from the cathedra and doing the dissections himself. He was quick to point out that Galen's books were wrong on many points, and he was the first to publish accurate anatomical illustrations (fig. 1.10). When others began to plagiarize them, Vesalius published the first atlas of anatomy, *De Humani Corporis Fabrica (On the Structure of the Human Body)*, in 1543. This book began a rich tradition of medical illustration that has been handed down through the vividly illustrated atlases and textbooks of today.

The Discovery of Microscopic Anatomy

Modern anatomy also owes an enormous debt to two early inventors—Leeuwenhoek and Hooke. **Antony van Leeuwenhoek** (an-TOE-nee vahn LAY-wen-hook) (1632–1723) was the first to invent a microscope capable of visualizing single cells. He was a Dutch textile merchant whose original motive in building a microscope was to examine the weave of fabrics more closely so that he could better judge their quality and price. He ground a beadlike lens and mounted it in a metal plate equipped with a movable specimen clip (fig. 1.11*a*). This *simple* (single-lens) *microscope* magnified objects 200 to 300 times.

Out of curiosity, Leeuwenhoek examined a drop of lake water and was astonished to find a variety of microorganisms—"little animalcules," he called them, "very prettily a-swimming." He went on to observe practically everything he could get his hands on, including blood cells, blood capillaries, sperm, and muscular tissue. Probably no one in history had looked at nature in such a revolutionary way. Leeuwenhoek opened the door to an entirely new understanding of human structure and the causes of disease. He was praised at first, and reports of his observations were eagerly received by scientific societies, but this enthusiasm did not last. By the end of the seventeenth century, the microscope was treated as a mere toy for the upper classes, as amusing and meaningless as a kaleidoscope. Leeuwenhoek had even become the brunt of satire.

Leeuwenhoek's most faithful admirer was the Englishman **Robert Hooke** (1635–1703), who developed the first practical *compound microscope*—a tube with a lens at each end, using the second lens to further magnify the image produced by the first (fig. 1.11*b*). Galileo had already made several compound microscopes by 1612, but Hooke invented many of the features found in microscopes used by scientists and students today: a stage to hold the specimen, an illuminator, and coarse and fine focus controls. His microscopes produced poor images with blurry edges *(spherical aberration)* and rainbow-colored distortions *(chromatic aberration),* but poor images were better than none. Although Leeuwenhoek was the first to see cells, Hooke named them. In 1663, he observed thin shavings of cork with his microscope and observed that they "consisted of a great many little boxes," which he called *cells* after the cubicles of a monastery. He published these observations in his book, *Micrographia,* in 1665.

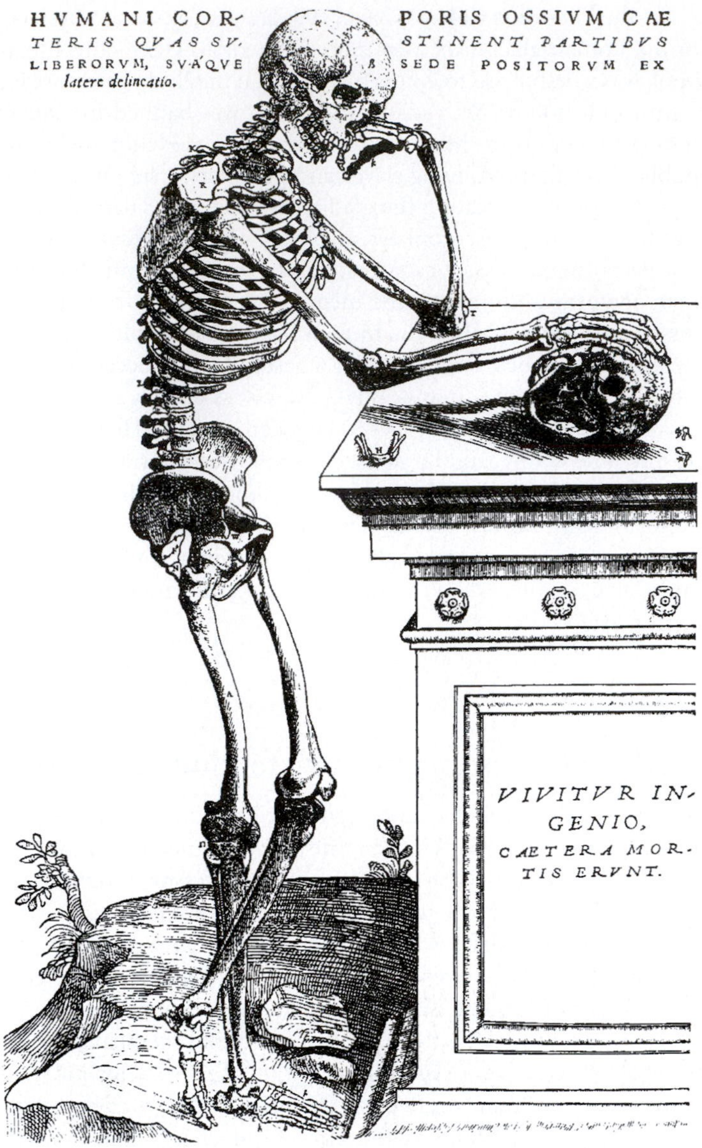

FIGURE 1.10
The Art of Vesalius. Andreas Vesalius revolutionized medical illustration with the realistic art commissioned for his 1543 book, *De Humani Corporis Fabrica.*

In nineteenth-century Germany, **Carl Zeiss** (1816–88) and his business partner, physicist **Ernst Abbe** (1840–1905), greatly improved the compound microscope, adding the condenser and developing superior optics that reduced chromatic and spherical aberration. Chapter 2 describes some more recently invented types of microscopes.

With improved microscopes, biologists began eagerly examining a wider variety of specimens. Botanist **Matthias Schleiden** (1804–81) and zoologist **Theodor Schwann** (1810–82) concluded by 1839 that all organisms were composed of cells. This was the first tenet of the **cell theory,** added to by later biologists. The cell theory was perhaps the most important breakthrough in biomedical history, because all functions of the body are now interpreted as the effects of cellular function.

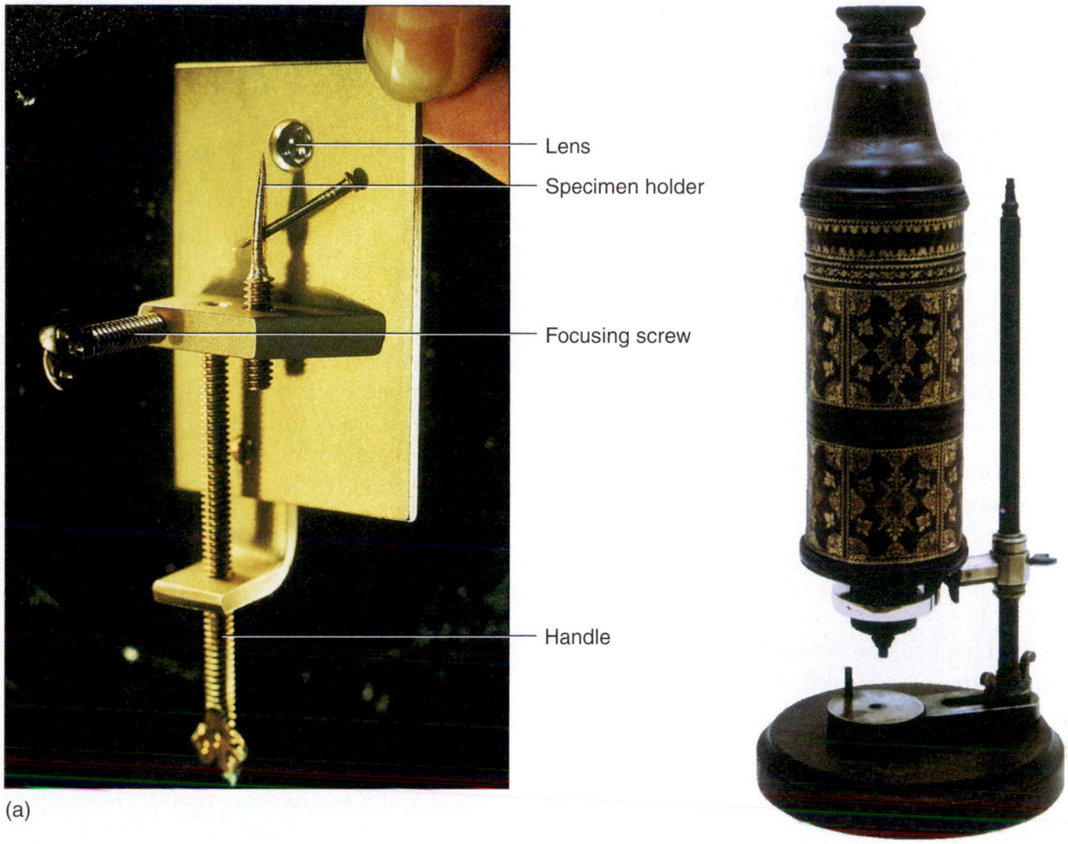
Lens
Specimen holder
Focusing screw
Handle
(a)
(b)

FIGURE 1.11
Early Microscopes. (*a*) A replica of Leeuwenhoek's simple microscope made by Dr. William Walter of Montana State University. (*b*) One of Hooke's compound microscopes, with a lens at each end of a tubular body.

Although the philosophical foundation for modern medicine was largely established by the time of Leeuwenhoek and Hooke, clinical practice was still in a dismal state. Only a few doctors attended medical school or received any formal education in basic science or human anatomy. Physicians tended to be ignorant, ineffective, and pompous. Their practice was heavily based on expelling imaginary toxins from the body by bleeding their patients or inducing vomiting, sweating, or diarrhea. They performed operations with filthy hands and instruments, spreading lethal infections from one patient to another. Fractured limbs often became gangrenous and had to be amputated, and there was no anesthesia to lessen the pain. Disease was still widely attributed to demons and witches, and many people felt they would be interfering with God's will if they tried to cure an illness.

Before You Go On

Answer the following questions to test your understanding of the preceding section:

5. In what way did the followers of Galen disregard his advice? How does Galen's advice apply to you?
6. Why did medical science develop more freely in Muslim countries than in Christian culture in the Middle Ages?
7. Describe two ways in which Vesalius improved medical education and set standards that remain relevant today.
8. How is our concept of human form and function today affected by the inventions of Leeuwenhoek and Hooke?

The Nature of Human Life

Objectives

When you have completed this section, you should be able to:

- state the characteristics that distinguish living organisms from nonliving objects;
- outline the classification of humans within the animal kingdom; and
- describe the anatomical features that define "human" and distinguish humans from other animals.

This book is a study of human life, so before we go much further, we should consider what we mean by the expression, "human life." This is really two questions: What is life? and What is a human?

What Is Life?

Why do we consider a growing child to be alive, but not a growing crystal? Is abortion the taking of a human life? If so, what about a contraceptive foam that kills only sperm? As a patient is dying, at what point does it become ethical to disconnect life-support equipment and remove organs for donation? If these organs are alive, as they must be to serve someone else, then why isn't the donor considered alive? Such questions have no easy answers, but they demand a concept of what life is—a concept that may differ with one's biological, medical, or legal perspective.

From a biological viewpoint, life is not a single property. It is a collection of properties that help to distinguish living from nonliving things:

- **Organization.** Living things exhibit a far higher level of organization than the nonliving world around them. They expend a great deal of energy to maintain order, and a breakdown in this order is accompanied by disease and often death.
- **Cellular composition.** Living matter is always compartmentalized into one or more cells.
- **Biochemical unity.** All living things have a universal chemical composition that includes DNA, proteins, lipids, and carbohydrates. Such compounds are not found in anything of nonbiological origin.
- **Metabolism.**[12] Living things take in molecules from the environment and chemically change them into molecules that form their own structures, control their physiology, or provide energy. **Metabolism** is the sum of all this internal chemical change. It consists of two classes of reactions: *anabolism,*[13] in which relatively complex molecules are synthesized from simpler ones (for example, protein synthesis), and *catabolism,*[14] in which relatively complex molecules are broken down into simpler ones (for example, protein digestion). There is a constant turnover of molecules in the body; although you sense a continuity of personality and experience from your childhood to the present, nearly every molecule of your body has been replaced within the past year.
- **Excitability.** The ability of organisms to sense and react to *stimuli* (changes in their environment) is called *excitability, irritability,* or *responsiveness.* It occurs at all levels from the single cell to the entire body, and it characterizes all living things from bacteria to humans. Responsiveness is especially obvious in animals because of nerve and muscle cells that exhibit high sensitivity to environmental stimuli, rapid transmission of information, and quick reactions.
- **Homeostasis.**[15] Homeostasis is the tendency for the internal conditions of the body to remain stable in spite of changes in the environment around the organism. Homeostasis is the purpose of nearly all normal physiology. There are hormonal and neural mechanisms, for example, for maintaining a stable body temperature, blood pressure, and blood glucose concentration. As one case in point, if the body temperature drops, blood vessels of the skin constrict to minimize heat loss and one may shiver to generate more heat. If the body becomes too warm, blood vessels of the skin dilate to enhance heat loss, and one may sweat.
- **Growth.** Some nonliving things grow, but not in the way your body does. When a saturated sugar solution evaporates, crystals grow from it, but not through a change in the composition of the sugar. They merely add more sugar molecules from the solution to the crystal surface. The growth of the body, by contrast, occurs through chemical change; for the most part, the body is not composed of the molecules one eats, but of molecules made by chemically altering the food.
- **Development.** Development is any change in form or function over the lifetime of the organism. It includes not only growth, but also *differentiation*—the transformation of cells with no specialized function to cells that are committed to a particular task. For example, a single embryonic, unspecialized tissue called mesoderm differentiates into muscle, bone, cartilage, blood, and several other specialized tissues.
- **Reproduction.** All living organisms can produce copies of themselves, thus passing their genes on to new, younger "containers"—their offspring.
- **Evolution.** All living species exhibit genetic change from generation to generation, and therefore evolve. This occurs because *mutations* (changes in the genes) are inevitable and environmental conditions favor some individuals over others, thus perpetuating some genes and eliminating others from the population. Unlike the other characteristics of life, evolution is a characteristic seen only in the population as a whole. No single individual evolves over the course of its life.

Clinical and legal criteria of life differ from these biological criteria. A person who has shown no brain waves for 24 hours, and has no reflexes, respiration, or heartbeat other than what is provided by artificial life support, can be declared legally dead. At such time, however, most of the body is still biologically alive and its organs may be useful for transplant.

What Is a Human?

Our second question was What is a human? We belong to the animal kingdom—as opposed to plants, fungi, protists, or bacteria—but what distinguishes us from other animals? To answer this, it helps to begin with an outline of our classification within the kingdom Animalia. We belong to each of the following progressively smaller groups:

[12]*metabol* = change + *ism* = process
[13]*ana* = up
[14]*cata* = down
[15]*homeo* = the same + *stasis* = stability

Phylum Chordata
Subphylum Vertebrata
Class Mammalia
Order Primates
Family Hominidae
Genus *Homo*
Species *Homo sapiens*

(A species name is always a two-word name that includes the genus.)

OUR CHORDATE CHARACTERISTICS

During the course of human embryonic development, we exhibit the following structures.

- A *notochord,* a dorsal, flexible rod found only in the embryo.
- *Pharyngeal arches,* a series of bulges that develop in the pharyngeal (throat) region (fig. 1.12). *Pharyngeal pouches* between these open and form gill slits in fish and amphibians, but not in humans.
- A *tail* that extends beyond the anus. The small bones of the *coccyx* ("tailbone") remain after birth as a remnant of this.
- A dorsal hollow *nerve cord,* a column of nervous tissue that passes along the dorsal (upper) side of the body and has a central canal filled with fluid.

The first three of these features are found only in the embryo and fetus; only the nerve cord persists through life, as the spinal cord.

These four features identify humans as members of the phylum *Chordata.* They distinguish us from nonchordates such as clams, worms, and insects, but not from fellow chordates such as fish, lizards, and birds. They only begin to narrow down our concept of what it means, anatomically, to be human.

OUR VERTEBRATE CHARACTERISTICS

Additional features of humans include:

- A well-developed brain and sense organs.
- An internal skeleton.
- A jointed *vertebral column* (spine).
- A *cranium,* the protective bony enclosure for the brain.

These features narrow our classification down a little more, to the subphylum *Vertebrata.*[16]

OUR MAMMALIAN CHARACTERISTICS

To narrow things down further, humans are members of the class *Mammalia.*[17] Mammals share the following characteristics:

- *Mammary glands* for nourishment of the young.
- Hair, which serves in most mammals to retain body heat.

[16]*vertebr* = backbone + *ata* = possessing
[17]*mamma* = breast + *alia* = possessing

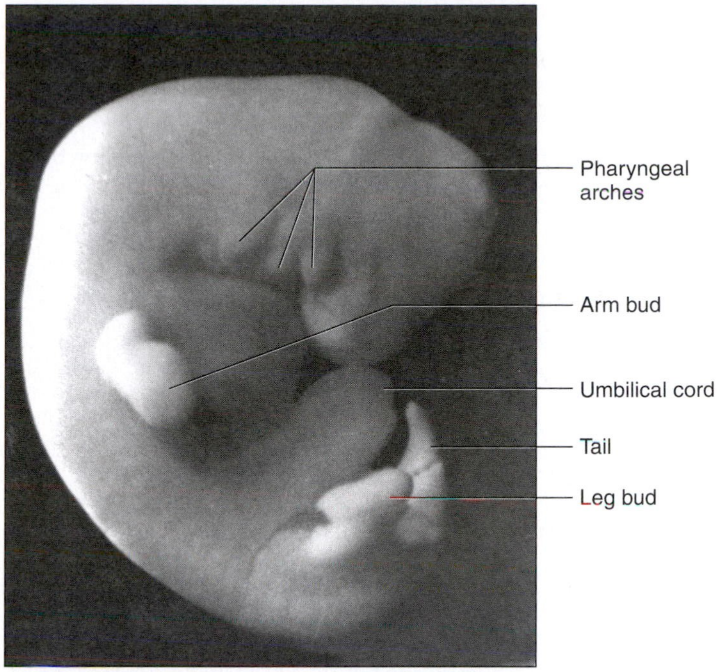

FIGURE 1.12
Primitive Chordate Characteristics in Humans. This 38-day human embryo shows the tail and pharyngeal arches that characterize all Chordata at some point in their development.

- *Endothermy,*[18] the ability to generate most of their body heat by metabolic means instead of having to warm up by basking in the sun or seeking other warm places.
- *Heterodonty,*[19] the possession of varied types of teeth (incisors, canines, premolars, and molars) specialized to puncture, cut, and grind food. These varied teeth break food into small pieces, making chemical digestion faster. Rapid digestion is necessary to support the high metabolic rate needed to maintain endothermic animals.
- A single lower jawbone (mandible).
- Three middle-ear bones (commonly called the *hammer, anvil,* and *stirrup*).

Less than 0.2% of the known animal species are mammals. Thus, we have narrowed down the classification of humans quite a lot, but still have not distinguished ourselves from rats, horses, dogs, or monkeys, which also have the characteristics listed above.

OUR PRIMATE CHARACTERISTICS

Mammals are divided into 19 orders. Humans belong to the *Primates,* an order that contains about 4% of the mammals including not just humans but also apes, monkeys, lemurs, and a few other species. We and most other primates have the following characteristics:

- Four upper and four lower incisors, the front cutting teeth.
- A pair of *clavicles* (collarbones).

[18]*endo* = within + *therm* = heat
[19]*hetero* = different, varied + *odont* = teeth

- Only two mammary glands.
- Forward-facing eyes with stereoscopic vision.
- Flat *nails* in place of claws.
- *Opposable thumbs* that can touch the fingertips, enabling the hand to encircle and grasp objects.

OUR HOMINID CHARACTERISTICS

Humans are in the family **Hominidae** (ho-MIN-ih-dee), with the following characteristics:

- Large brains.
- Complex speech.
- Tool making.
- *Bipedalism,*[20] the habit of walking on two legs, which is perhaps the clearest distinction between hominids and other primates. We could, therefore, define humans as bipedal primates.

All bipedal primates that have ever lived are classified either in the genus *Australopithecus* (aus-TRAL-oh-PITH-eh-cus) or the genus *Homo; Homo sapiens* is the only surviving species. There is no unanimously accepted definition of *human.* Some authorities treat all Hominidae as humans; some restrict the word *human* to the genus *Homo;* and some go even further and limit it to *Homo sapiens.*

Before You Go On

Answer the following questions to test your understanding of the preceding section:

9. List four biological criteria of life and one clinical criterion. Explain how a person could be considered clinically dead but biologically alive.
10. Explain why humans are classified as vertebrates, why they are classified as mammals, and why they are classified as primates.
11. As a matter of opinion, do you consider *Australopithecus* to be human? Why or why not?

The Evolution of Human Structure

Objectives

When you have completed this section, you should be able to:

- define *evolution, natural selection,* and *adaptation;*
- explain the relevance of evolution to modern anatomy and medicine;
- explain how the tree-dwelling habits of early primates account for certain aspects of modern human anatomy;
- describe some human characteristics that evolved in connection with upright walking; and
- name the major species of *Homo* and the relative times at which they existed.

If any two theories have the broadest implications for understanding the human body, they are probably the cell theory and the theory of natural selection. *Natural selection,* an explanation of how species originate and change through time, was the brainchild primarily of **Charles Darwin** (1809–82)—probably the most influential biologist who ever lived. His book, *On the Origin of Species by Means of Natural Selection* (1859), has been called "the book that shook the world." In presenting the first well-supported theory of evolution, the *Origin of Species* not only caused the restructuring of all of biology, but also profoundly changed the prevailing view of our origin, nature, and place in the universe.

While the *Origin of Species* scarcely touched upon human biology, its unmistakable implications for humans created an intense storm of controversy that continues even today. In *The Descent of Man* (1871), Darwin directly addressed the issue of human evolution and emphasized features of anatomy and behavior that reveal our relationship to other animals. No understanding of human form and function is complete without an understanding of our evolutionary history.

Evolution, Selection, and Adaptation

Evolution simply means change in the genetic composition of a population of organisms. Examples include the evolution of bacterial resistance to antibiotics, the appearance of new strains of the AIDS virus, and the emergence of new species of organisms. **Natural selection,** the prevailing theory of how evolution works, is essentially this: Some individuals within a species have hereditary advantages over their competitors—for example, better camouflage, disease resistance, or ability to attract mates—that enable them to produce more offspring. They pass these advantages on to their offspring, and such characteristics therefore become more and more common in successive generations. This brings about the genetic change in a population that constitutes evolution.

Natural forces that favor some individuals over others are called **selection pressures.** They include such things as climate, predators, disease, competition, and the availability of food. **Adaptations** are features of a species' anatomy, physiology, and behavior that have evolved in response to selection pressures and enable the organism to cope with the challenges of its environment. We will consider shortly some selection pressures and adaptations that were important to human evolution.

Several aspects of our anatomy make little sense without an awareness that the human body has a history (see insight 1.2). Our evolutionary relationship to other species is also important in choosing animals for biomedical research. If there were no issues of cost, availability, or ethics, we might test drugs on our nearest living relatives, the chimpanzees, before approving them for human use. Their

[20]*bi* = two + *ped* = foot

genetics, anatomy, and physiology are most similar to ours, and their reactions to drugs therefore afford the best prediction of how the human body would react. On the other hand, if we had no kinship with any other species, the selection of a test species would be arbitrary; we might as well use frogs or snails. In reality, we compromise. Rats and mice are used extensively for research because they are fellow mammals with a physiology similar to ours, but present fewer of the aforementioned issues than chimpanzees or other mammals do. An animal species or strain selected for research on a particular problem is called a **model**—for example, a mouse model for leukemia.

Think About It!

The human species may yet undergo evolutionary change, but it is impossible for any one person to do so. Explain why.

INSIGHT 1.2 EVOLUTIONARY MEDICINE

Vestiges of Human Evolution

One of the classic lines of evidence for evolution, debated even before Darwin was born, is *vestigial organs.* These structures are the reduced remnants of organs that apparently were more functional in the ancestors of a species. They now serve little or no purpose or, in some cases, have been converted to new functions.

Our bodies, for example, are covered with millions of hairs, each equipped with a useless little muscle called a *piloerector.* In other mammals, these muscles fluff the hair and conserve heat. In humans, they merely produce goosebumps. Above each ear, we have three *auricularis muscles.* In other mammals, they move the ears to receive sounds better, but most people cannot contract them at all. As Darwin said, it makes no sense that humans would have such structures were it not for the fact that we came from nonhuman ancestors in which they were functional.

Life in the Trees

As we have already seen, humans belong to the order Primates, which also includes the monkeys and apes. Primates originated 55 to 60 million years ago, when certain squirrel-sized, insect-eating mammals (insectivores) took up life in the trees of Africa. This **arboreal**[21] (treetop) habitat probably afforded greater safety from predators, less competition, and a rich food supply of leaves, fruits, insects, and lizards. But the forest canopy is a challenging world, with dim and dappled sunlight, swaying branches, and prey darting about in the dense foliage. Any new feature that enabled arboreal animals to move about more easily in the treetops would have been strongly favored by natural selection. Thus, the shoulder became more mobile, enabling primates to reach out in any direction (even overhead, which other mammals cannot do). The thumbs became opposable and thus made the hands **prehensile**[22]—able to grasp objects by encircling them with the thumb and fingers (fig. 1.13). The thumb is so important to us that it receives highest priority in the repair of hand injuries. If the thumb can be saved, the hand can be reasonably functional; if it is lost, hand functions are severely diminished.

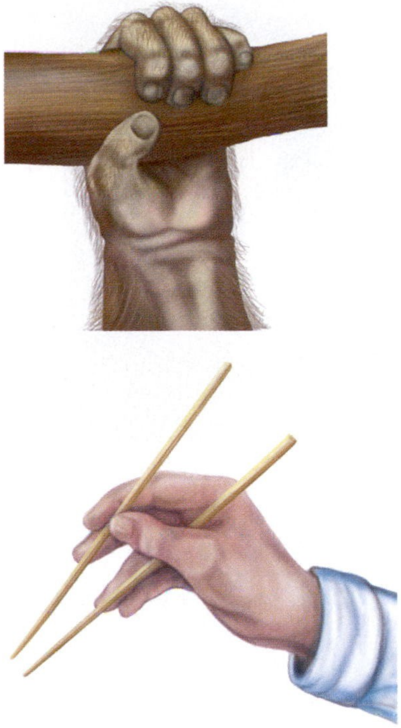

FIGURE 1.13
Primate Hands. The opposable thumb makes the primate hand prehensile, able to encircle and grasp objects.

The eyes of primates moved to a more forward-facing position (fig. 1.14), allowing for **stereoscopic**[23] **vision** (depth perception). This adaptation provided better hand-eye coordination in catching and manipulating prey, with the added advantage of making it easier to judge distances accurately in leaping from tree to tree. Color vision, rare among mammals, is also a primate hallmark. Primates eat mainly fruit and leaves. The ability to distinguish subtle shades of orange and red enables them to distinguish ripe, sugary fruits from unripe ones. Distinguishing subtle shades of green helps them to differentiate between tender young leaves and tough, more toxic older foliage.

Various fruits ripen at different times and widely separated places in the tropical forest. This requires a good memory of what will be available, when it will be available, and how to get to it. Larger brains may have evolved in response to the challenge of efficient food finding and, in turn, laid the foundation for more sophisticated social organization.

None of this is meant to imply that humans evolved from monkeys or apes—a common misconception about evolution that no biologist believes. Observations of monkeys and apes, however, provide insight into how primates have adapted to the arboreal habitat and how certain aspects of human anatomy may have originated.

[21]*arbor* = tree + *eal* = pertaining to
[22]*prehens* = to seize
[23]*stereo* = solid + *scop* = vision

FIGURE 1.14
Primitive Tool Use in a Chimpanzee. Chimpanzees exhibit the prehensile hands and forward-facing eyes typical of most primates. Such traits endow primates with stereoscopic vision and good hand-eye coordination, two supremely important factors in human evolution.

Walking Upright

About 4 to 5 million years ago, Africa became hotter and drier and much of the forest was replaced by savanna (grassland). Some primates adapted to living on the savanna, but this was a dangerous place with more predators and less protection. Just as squirrels stand briefly on their hind legs to look around for danger, so would these early ground-dwelling primates. Being able to stand up not only helps an animal stay alert, but also frees the forelimbs for purposes other than walking. Chimpanzees sometimes walk upright to carry food or weapons (sticks and rocks), and it is reasonable to suppose that our early ancestors did so too. They could also carry their infants. Footprints preserved in volcanic ash in Tanzania indicate that humans walked upright as early as 3.6 million years ago.

These advantages were so great that they favored skeletal modifications that made bipedalism easier. The anatomy of the human pelvis, femur, knee, great toe, foot arches, spinal column, skull, arms, and many muscles became adapted for bipedal locomotion, as did many aspects of human family life and society. As the skeleton and muscles became adapted for bipedalism, brain volume increased dramatically (table 1.1). It must have become increasingly difficult for a fully developed, large-brained infant to pass through the mother's pelvic outlet at birth. This may explain why humans are born in a relatively immature, helpless state compared to other mammals, before their nervous systems have matured and the bones of the skull have fused. The relative helplessness of human young may explain why human mates formed longer-lasting pair bonds and became more nearly monogamous than most other primates.

The oldest bipedal primates are classified in the genus *Australopithecus.* About 2.5 million years ago, *Australopithecus* gave rise to *Homo habilis,* the earliest member of our own genus. *Homo habilis* differed from *Australopithecus* in height, brain volume, some details of skull anatomy, and tool-making ability. It was probably the first primate able to speak. *Homo habilis* gave rise to *Homo erectus* about 1.1 million years ago, which in turn led to our own species, *Homo sapiens,* about 300,000 years ago (fig. 1.15). *Homo sapiens* includes the extinct Neanderthal and Cro-Magnon people as well as modern humans.

TABLE 1.1
Brain Volumes of the Hominidae

Genus or Species	Time of Origin (millions of years ago)	Brain Volume (cc)
Australopithecus	3–4	400
Homo habilis	2.5	650
Homo erectus	1.1	1,100
Homo sapiens	0.3	1,350

This brief account barely begins to explain how human anatomy, physiology, and behavior have been shaped by ancient selection pressures. Later chapters further demonstrate that the evolutionary perspective provides a meaningful understanding of why humans are the way we are. Evolution is the basis for comparative anatomy and physiology, which have been so fruitful for the understanding of human biology. The emerging science of **evolutionary (darwinian) medicine** traces some of our diseases and imperfections to our evolutionary past.

Before You Go On

Answer the following questions to test your understanding of the preceding section:

12. Define *adaptation* and *selection pressure.* Why are these concepts important in understanding human anatomy?
13. Select any two human characteristics and explain how they may have originated in primate adaptations to an arboreal habitat.
14. Select two other human characteristics and explain how they may have resulted from adaptation to a savanna habitat.

THE LANGUAGE OF ANATOMY

Objectives

When you have completed this section, you should be able to:

- explain why modern anatomical terminology is so heavily based on Greek and Latin;
- recognize eponyms when you see them;
- describe the efforts to achieve an internationally uniform anatomical terminology;
- discuss the Greek, Latin, or other derivations of medical terms;
- state some reasons why the literal meaning of a word may not lend insight into its definition;
- relate singular noun forms to their plural forms; and
- discuss why precise spelling is important to anatomical communication.

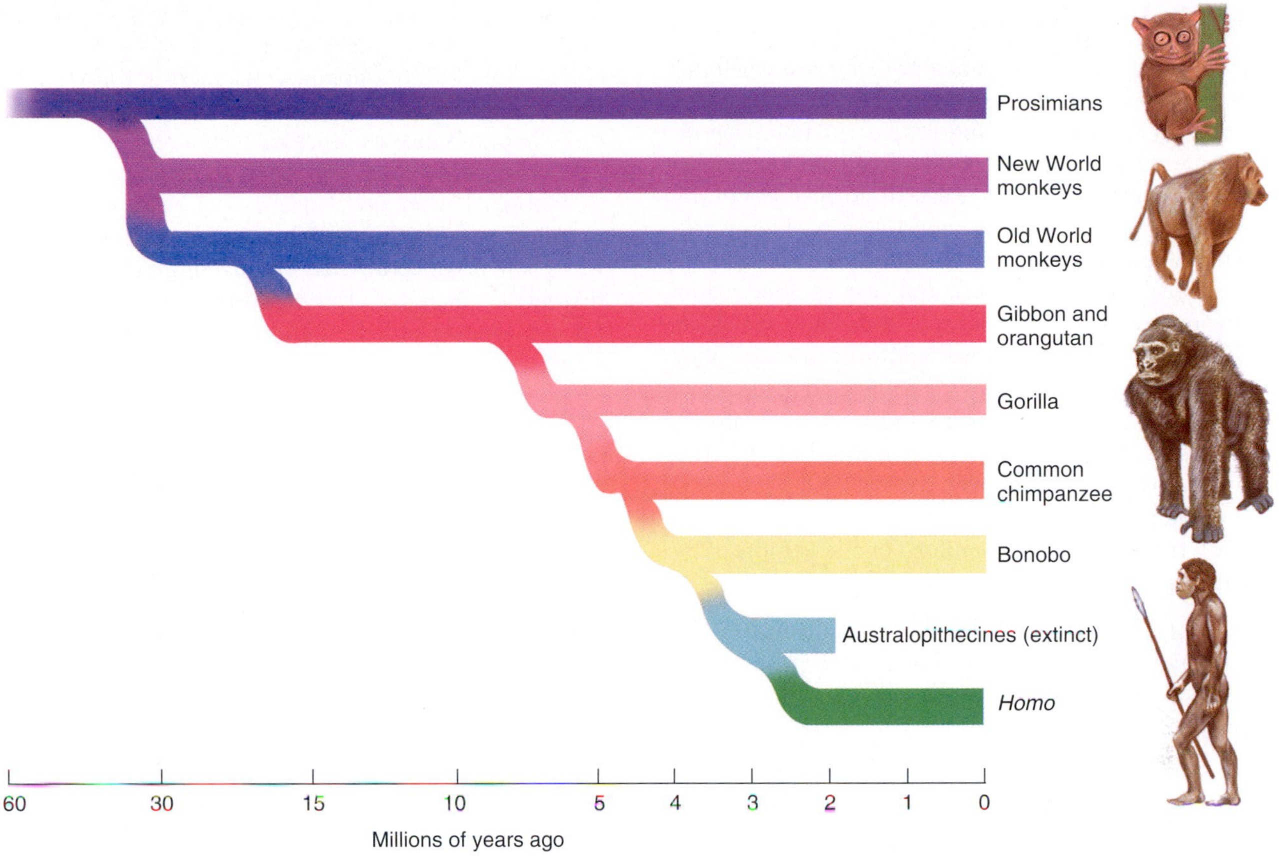
Prosimians
New World monkeys
Old World monkeys
Gibbon and orangutan
Gorilla
Common chimpanzee
Bonobo
Australopithecines (extinct)
Homo
60
30
15
10
5
4
3
2
1
0
Millions of years ago

FIGURE 1.15

Primate Phylogeny. Figures at *right* show representative primates alive today. The branch points in this "family tree" show approximate times that different lines diverged from a common ancestor. Note that the time scale is not uniform; recent events are expanded for clarity.

One of the greatest challenges faced by students of anatomy and physiology is the vocabulary. In this book, you will encounter such Latin terms as *corpus callosum* (a brain structure), *ligamentum arteriosum* (a small fibrous band near the heart), and *extensor carpi radialis longus* (a forearm muscle). You may wonder why structures aren't named in "just plain English," and how you will ever remember such formidable names. This section will give you some answers to these questions and some useful tips on mastering anatomical terminology.

The History of Anatomical Terminology

The major features of human gross anatomy have standard international names prescribed by a book titled the *Terminologia Anatomica* (TA). The TA was codified in 1998 by an international body of anatomists, the Federated Committee on Anatomical Terminology, and approved by professional associations of anatomists in more than 50 countries.

About 90% of today's medical terms are formed primarily from about 1,200 Greek and Latin roots. Scientific investigation began in ancient Greece and soon spread to Rome. The ancient Greeks and Romans coined many of the words still used in human anatomy today: *duodenum, uterus, prostate, cerebellum, diaphragm, sacrum, amnion,* and others. In the Renaissance, the fast pace of anatomical discovery required a profusion of new terms to describe things. Anatomists in different countries began giving different names to the same structures. Adding to the confusion, they often named new structures and diseases in honor of their esteemed teachers and predecessors, giving us such nondescriptive terms as the *fallopian tube* and *duct of Santorini.* Terms coined from the names of people, called **eponyms,**[24] afford little clue as to what a structure or condition is.

In hopes of resolving this growing confusion, anatomists began meeting as early as 1895 to try to devise a uniform international terminology. After several false starts, a list of terms titled the *Nomina Anatomica* (NA) was agreed upon. The NA rejected all eponyms as unofficial and gave each structure a unique Latin name to be used worldwide. Even if you were to look at an anatomy atlas in Japanese or Arabic, the illustrations would be labeled with the same Latin terms as in an English-language atlas. The NA served for many decades until recently replaced by the TA, which prescribes both Latin names and accepted English equivalents. The terminology in this book conforms to the TA except where undue confusion would result from abandoning widely used, yet unofficial terms.

[24]*epo* = after, related to + *nym* = name

Analyzing Medical Terms

The task of learning anatomical terminology seems overwhelming at first, but there is a simple trick to becoming more comfortable with the technical language of medicine. People who find scientific terms confusing and difficult to pronounce, spell, and remember usually feel more confident once they realize the logic of how terms are composed. A term such as *hyponatremia* is less forbidding once we recognize that it is composed of three common word elements: *hypo-* (below normal), *natr-* (sodium), and *-emia* (blood condition). Thus, hyponatremia is a deficiency of sodium in the blood. Those three word elements appear over and over in many other medical terms: *hypothermia, natriuretic, anemia,* and so on. Once you learn the meanings of *hypo-, natri-,* and *-emia,* you already have the tools at least to partially understand hundreds of other biomedical terms. Inside the back cover, you will find a lexicon of the 400 word elements most commonly footnoted in this book.

Scientific terms are typically composed of one or more of the following elements:

- At least one *root (stem)* that bears the core meaning of the word. In *cardiology,* for example, the root is *cardi-* (heart). Many words have two or more roots. In *adipocyte,* the roots are *adip-* (fat) and *cyte* (cell).
- *Combining vowels,* which are often inserted to join roots and make the word easier to pronounce. The letter *o* is the most common combining vowel (as in *adipocyte*), but all vowels are used in this way, such as *a* in *ligament, e* in *vitreous,* the first *i* in *spermicidal, u* in *ovulation,* and *y* in *tachycardia.* Some words have no combining vowels. A combination of a root and combining vowel is called a *combining form:* for example, *odont* (tooth) + *o* (the combining vowel) make the combining form *odonto-,* as in *odontoblast* (a cell that produces the dentin of a tooth).
- A *prefix* may be present to modify the core meaning of the word. For example, *gastric* (pertaining to the stomach or to the belly of a muscle) takes on a wide variety of new meanings when prefixes are added to it: *epigastric* (above the stomach), *hypogastric* (below the stomach), *endogastric* (within the stomach), and *digastric* (a muscle with two bellies).
- A *suffix* may be added to the end of a word to modify its core meaning. For example, *microscope, microscopy, microscopic,* and *microscopist* have different meanings because of their suffixes alone. Often two or more suffixes, or a root and suffix, occur together so often that they are treated jointly as a *compound suffix;* for example, *log* (study) + *y* (process) form the compound suffix *-logy* (the study of).

To summarize these basic principles, consider the word *gastroenterology,* a branch of medicine dealing with the stomach and small intestine. It breaks down into:

gastro/entero/logy

gastro = a combining form meaning "stomach"
entero = a combining form meaning "small intestine"
logy = a compound suffix meaning "the study of"

"Dissecting" words in this way and paying attention to the word-origin footnotes throughout this book will help make you more comfortable with the language of anatomy. Knowing how a word breaks down and knowing the meaning of its elements make it far easier to pronounce a word, spell it, and remember its definition.

There are a few unfortunate exceptions, however. The path from original meaning to current usage has often become obscured by history (see insight 1.3). The foregoing approach also is no help with eponyms or **acronyms**—words composed of the first letter, or first few letters, of a series of words. For example, *calmodulin,* a calcium-binding protein found in many cells, is cobbled together from a few letters of the three words, *cal*cium *modul*ating prote*in.*

INSIGHT 1.3 MEDICAL HISTORY

Obscure Word Origins

The literal translation of a word doesn't always provide great insight into its modern meaning. The history of language is full of twists and turns that are fascinating in their own right and say much about the history of the whole of human culture, but they can create confusion for students.

For example, the *amnion* is a transparent sac that forms around the developing fetus. The word is derived from *amnos,* from the Greek for "lamb." From this origin, *amnos* came to mean a bowl for catching the blood of sacrificial lambs, and from there the word found its way into biomedical usage for the membrane that emerges (quite bloody) as part of the afterbirth. The *acetabulum,* the socket of the hip joint, literally means "vinegar cup." Apparently the hip socket reminded an anatomist of the little cups used to serve vinegar as a condiment on dining tables in ancient Rome. The word *testicles* literally means "little witnesses." The history of medical language has several amusing conjectures as to why this word was chosen to name the male gonads.

Singular and Plural Noun Forms

A point of confusion for many beginning students is how to recognize the plural forms of medical terms. Few people would fail to recognize that *ovaries* is the plural of *ovary,* but the connection is harder to make in other cases: for example, the plural of *cortex* is *cortices* (COR-ti-sees), the plural of *corpus* is *corpora,* and the plural of *epididymis* is *epididymides* (EP-ih-DID-ih-MID-eze). Table 1.2 will help you make the connection between common singular and plural noun terminals.

The Importance of Precision

A final word of advice for your study of anatomy: Be precise in your use of anatomical terms. It may seem trivial if you misspell *trapezius* as *trapezium,* but in doing so, you would be changing the name of a back muscle to the name of a wrist bone. Similarly, changing *occipitalis* to *occipital* or *zygomaticus* to *zygomatic* changes other muscle names to bone names. A "little" error such as misspelling *ileum* as *ilium* changes the name of the final portion of the small intestine

TABLE 1.2

Singular and Plural Forms of Some Noun Terminals

Singular Ending	Plural Ending	Examples
-a	-ae	axilla, axillae
-ax	-aces	thorax, thoraces
-en	-ina	lumen, lumina
-ex	-ices	cortex, cortices
-is	-es	diagnosis, diagnoses
-is	-ides	epididymis, epididymides
-ix	-ices	appendix, appendices
-ma	-mata	carcinoma, carcinomata
-on	-a	ganglion, ganglia
-um	-a	septum, septa
-us	-era	viscus, viscera
-us	-i	villus, villi
-us	-ora	corpus, corpora
-x	-ges	phalanx, phalanges
-y	-ies	ovary, ovaries
-yx	-ices	calyx, calices

to the name of the hip bone. Changing *malleus* to *malleolus* changes the name of a middle-ear bone to the name of a bony protuberance of the ankle. *Elephantiasis* is a disease that produces an elephant-like thickening of the limbs and skin. Many people misspell this *elephantitis;* if such a word existed, it would mean inflammation of an elephant.

The health professions demand the utmost attention to detail and precision—people's lives may one day be in your hands. The habit of carefulness must extend to your use of language as well. Many patients die because of miscommunication in the hospital.

Before You Go On

Answer the following questions to test your understanding of the preceding section:

15. Explain why modern anatomical terminology is so heavily based on Greek and Latin.
16. Distinguish between an eponym and an acronym, and explain why both of these present difficulties for interpreting anatomical terms.
17. Break each of the following words down into its roots and affixes and state their meanings, following the example of *gastroenterology* analyzed earlier: *pericardium, appendectomy, subcutaneous, arteriosclerosis, hypercalcemia.* Consult the list of word elements inside the back cover of the book for help.
18. Write the singular form of each of the following words: *pleurae, gyri, nomina, ganglia, fissures.* Write the plural form of each of the following: *villus, tibia, encephalitis, cervix, stoma.*

CHAPTER REVIEW

Review of Key Concepts

The Scope of Human Anatomy (p. 2)

1. *Functional morphology* is the study of anatomy not merely from the standpoint of the appearances and names of structures, but how structure relates to function.
2. Approaches to the study of anatomy include *gross, surface, systemic, regional, and comparative anatomy. Microscopic anatomy (histology), cytology, histopathology,* and *ultrastructure* are studies of structure at the tissue to cellular level.
3. Methods of study include *inspection, palpation, auscultation, percussion,* and *dissection.*
4. The internal anatomy of a living person can be examined by such imaging methods as *radiography, sonography, computed tomography (CT), magnetic resonance imaging (MRI),* and *positron emission tomography (PET).*
5. Introductory textbooks teach the most typical anatomy of a given organ or system, but organs can vary not only in appearance but also in number and location from one person to another.
6. The human body exhibits a hierarchy of structural complexity. From the largest and most complex, to the smallest and simplest, the principal levels of human structure are the *organism, organ systems, organs, tissues, cells, organelles, molecules,* and *atoms.*

Early Anatomists (p. 8)

1. Some of the most innovative and influential anatomists of ancient Greece and Rome were *Aristotle, Herophilus,* and *Galen.*
2. In the middle ages, science and medicine developed little within Christian culture, but significant advances were made in Muslim culture by such physician-scientists as *Ibn-Sina (Avicenna)* and *Ibn an-Nafis.*
3. Gross anatomy was modernized by Renaissance physician and professor *Andreas Vesalius,* who commissioned the first accurate anatomical art.
4. In the seventeenth to eighteenth centuries, *Antony van Leeuwenhoek* and *Robert Hooke* developed microscopes that extended the study of anatomy to the cellular level. *Carl Zeiss* and *Ernst Abbe* greatly improved microscopes in the early nineteenth century.
5. With such instruments, *Theodor Schwann* and *Mathias Schleiden* examined a broad range of animal and plant tissues and concluded that all organisms are composed of cells. The *cell theory* is one of the major foundations of modern anatomy, physiology, and medicine.

The Nature of Human Life (p. 11)

1. Life is not a single, easily defined property. Rather, living organisms are distinguished from nonliving matter by a combination of characteristics: a high degree of organization, cellular composition, biochemical unity, metabolism, responsiveness, homeostasis, growth, development, reproduction, and evolution.
2. Clinical criteria of death typically include an absence of reflexes, respiration, heartbeat, and brain waves.
3. Within the animal kingdom, humans belong to the phylum *Chordata,* whose members have a notochord, pharyngeal arches, postanal tail, and dorsal hollow nerve cord.
4. Within the Chordata, humans belong to the subphylum *Vertebrata,* characterized by a well-developed brain and sense organs, internal skeleton, jointed vertebral column, and cranium.
5. Within the Vertebrata, humans belong to the class *Mammalia,* defined by mammary glands, hair, endothermy, heterodonty (varied types of teeth), a single mandible, and three middle-ear bones.
6. Within the Mammalia, humans belong to the order *Primates,* which also includes monkeys and apes. Primates are characterized by four upper and four lower incisors, a pair of clavicles, only two mammary glands, forward-facing eyes, flat nails, and opposable thumbs.
7. The family *Hominidae* contains the bipedal primates—those that habitually walk on two legs. These include the extinct genus *Australopithecus* and the living genus *Homo,* with the only living species being *Homo sapiens.* Hominids are characterized by their bipedalism as well as large brains, complex speech, and tool making.

The Evolution of Human Structure (p. 14)

1. Human anatomy is most fully understood from the standpoint of how it evolved.
2. *Natural selection* is a theory of evolution that says that populations evolve because some individuals have hereditary advantages over their competitors and pass those advantages to more offspring than their competitors do.
3. Environmental factors called *selection pressures*—such as climate, food, disease, and predation—shape the evolution of *adaptations* that promote survival and reproductive success.
4. The mobile shoulder, opposable thumbs, prehensile hands, stereoscopic vision, and color vision of humans are common to most primates, and probably first evolved as adaptations to selection pressures in the arboreal habitat of early primates.
5. Bipedalism, which entailed extensive remodeling of the skeleton, probably evolved later as an adaptation to the African savanna. Bipedalism in turn may have been a factor in evolution of early childbirth of humans, and long-term pair-bonding between mates.
6. Humans belong to the species *Homo sapiens.* The genus *Homo* evolved from early bipedal primates in the genus *Australopithecus.*
7. Evolutionary medicine is a branch of medical science that interprets some human dysfunctions and diseases in terms of the evolutionary history of *Homo sapiens.*

The Language of Anatomy (p. 16)

1. Anatomical and medical terminology, most of it derived from Latin and Greek, can be an obstacle to the beginning anatomy student. Insight into word derivations, however, can make it significantly easier to understand, remember, spell, and pronounce biomedical terms. This goal is supported by word derivation footnotes throughout this book.
2. Anatomists began by 1895 to standardize international anatomical terminology. Official international terms are now codified in the *Terminologia Anatomica* (TA) of 1998.
3. The TA rejects eponyms (anatomical terms based on the names of people) and provides preferred Latin and English terms for most human structures in gross anatomy.
4. Scientific words can typically be broken down into one or more word *roots (stems)* and *affixes* (prefixes and suffixes), often joined by *combining vowels.*
5. About 90% of medical terms are composed of various combinations of only 1,200 roots and affixes. Therefore a relatively modest vocabulary of roots and affixes can give one insight into the meanings of most medical terms.
6. It is sometimes difficult for a beginner to recognize that two words are merely the singular and plural forms of the same noun (*corpus* and *corpora,* for example). Table 1.2 should be an aid to becoming familiar with these singular/plural relationships.
7. Precision is extremely important in medical communication. Minor changes in spelling can radically change the meaning of a word—for example, a one-letter difference changing the name of part of the intestine to the name of a hip bone. Similarly "minor" errors can make life-and-death differences in the hospital, and it is therefore crucial to cultivate the habit of precision early in one's studies.

Testing Your Recall

1. Structure that can be observed with the naked eye is called
   a. gross anatomy.
   b. ultrastructure.
   c. microscopic anatomy.
   d. macroscopic anatomy.
   e. cytology.
2. Which of the following techniques requires an injection of radioisotopes into a patient's bloodstream?
   a. sonography
   b. a PET scan
   c. radiography
   d. a CT scan
   e. an MRI scan
3. The simplest structures considered to be alive are
   a. organs.
   b. tissues.
   c. cells.
   d. organelles.
   e. proteins.
4. Which of the following people revolutionized the teaching of gross anatomy?
   a. Vesalius
   b. Aristotle
   c. Hippocrates
   d. Leeuwenhoek
   e. Avicenna
5. Which of the following characteristics do humans *not* share with all other chordates?
   a. pharyngeal arches
   b. a hollow nerve cord
   c. a tail extending beyond the anus
   d. a notochord
   e. a vertebral column
6. Which of the following men argued that all living organisms are composed of cells?
   a. Hippocrates
   b. an-Nafis
   c. Schwann
   d. Leeuwenhoek
   e. Avicenna
7. When a person's blood sugar (glucose) level rises, insulin is secreted. This stimulates cells to absorb glucose and thus brings the glucose level back to normal. This is an example of
   a. homeostasis.
   b. differentiation.
   c. organization.
   d. anabolism.
   e. catabolism.
8. The word root *histo-* means
   a. visible.
   b. diseased.
   c. cellular.
   d. tissue.
   e. microscopic.
9. The word root *patho-* means
   a. doctor.
   b. medicine.
   c. disease.
   d. organ.
   e. health.
10. The prefix *hetero-* means
    a. same.
    b. different.
    c. both.
    d. solid.
    e. below.
11. Cutting and separating tissues to reveal their structural relationships is called _____.
12. _____ invented many components of the compound microscope and named the cell.
13. The term for all chemical change in the body is _____.
14. Most physiology serves the purpose of _____, maintaining a stable internal environment in the body.
15. _____ is a science that doesn't merely describe bodily structure but interprets structure in terms of its function.
16. When a doctor presses on the upper abdomen to feel the size and texture of the liver, he or she is using a technique of physical examination called _____.
17. _____ is a method of medical imaging that uses X rays and a computer to generate images of thin slices of the body.
18. A/an _____ is the simplest body structure to be composed of two or more types of tissue.
19. Depth perception, or the ability to form three-dimensional images, is called _____ vision.
20. Our hands are said to be _____ because they can encircle an object such as a branch or tool. The presence of a/an _____ thumb is important to this ability.

Answers in the Appendix

True or False

Determine which five of the following statements are false, and briefly explain why.

1. Regional anatomy is a variation of gross anatomy.
2. The inventions of Carl Zeiss and Ernst Abbe are necessary to the work of a modern histopathologist.
3. Abnormal skin color or dryness could be one piece of diagnostic information gained by auscultation.
4. Radiology refers only to those medical imaging methods that use radioisotopes.
5. It is more harmful to have only the heart reversed from left to right than to have all of the thoracic and abdominal organs reversed.
6. There are more cells than organelles in the body.
7. Leeuwenhoek was a biologist who invented the microscope in order to study cells.
8. All Vertebrata have a notochord but not all of them are endothermic.
9. Human stereoscopic vision probably evolved in response to the demands of the savanna habitat of the first hominids.
10. The word *scuba,* derived from the words *self-contained underwater breathing apparatus,* is an acronym.

Answers in the Appendix

Testing Your Comprehension

1. Classify each of the following radiologic techniques as invasive or noninvasive and explain your reasoning for each: angiography, sonography, CT, MRI, and PET.
2. Beginning medical students are always told to examine multiple cadavers and not confine their study to just one. Other than the obvious purpose of studying both male and female anatomy, why is this instruction so important to medical education?
3. Which characteristics of living things are possessed by an automobile? What bearing does this have on our definition of life?
4. Why is a monkey not classified as human? Why is a horse not classified as a primate? (Hint: What characteristics must an animal have to be a human or a primate, and which of these are monkeys or horses lacking?)
5. Why do you think the writers of the *Terminologia Anatomica* decided to reject eponyms? Do you agree with that decision? Why do you think they decided to name structures in Latin? Do you agree with that decision? Explain your reasons for agreeing or disagreeing with each.

Answers at the Online Learning Center

www.mhhe.com/saladinha1

Visit the Online Learning Center for practice tests, answer keys, and other learning aids for this chapter. Enhance your understanding of human anatomy with our interactive art labeling exercises, supplemental photo atlases, web links, puzzles, flashcards, and much more.

A

ATLAS A

Survey of the Human Body

General Anatomical Terminology *24*
- Anatomical Position 24
- Anatomical Planes 24
- Directional Terms 25

Body Regions *26*
- Axial Region 26
- Appendicular Region 26

Body Cavities and Membranes *26*
- Dorsal Body Cavity 29
- Ventral Body Cavity 29

Organ Systems *30*

A Visual Survey of the Body *31*

Atlas Review *44*

General Anatomical Terminology

Anatomical Position

Anatomical position is a stance in which a person stands erect with the feet flat on the floor, arms at the sides, and the palms, face, and eyes facing forward (fig. A.1). This position provides a precise and standard frame of reference for anatomical description and dissection. Without such a frame of reference, to say that a structure such as the sternum, thymus, or aorta is "above the heart" would be vague, since it would depend on whether the subject was standing, lying face down, or lying face up. From the perspective of anatomical position, however, we can describe the thymus as *superior* to the heart, the sternum as *anterior* or *ventral* to it, and the aorta as *posterior* or *dorsal* to it. These descriptions remain valid regardless of the subject's position.

Unless stated otherwise, assume that all anatomical descriptions refer to anatomical position. Bear in mind that if a subject is facing you in anatomical position, the subject's left will be on your right and vice versa. In most anatomical illustrations, for example, the left atrium of the heart appears toward the right side of the page, and while the appendix is located in the right lower quadrant of the abdomen, it appears on the left side of most illustrations.

The forearm is said to be **supine** when the palms face up or forward and **prone** when they face down or rearward (fig. A.2). The difference is particularly important to descriptions of anatomy of this region. In the supine position, the two forearm bones (radius and ulna) are parallel and the radius is lateral to the ulna. In the prone position, the radius and ulna cross; the radius is lateral to the ulna at the elbow but medial to it at the wrist. Descriptions of nerves, muscles, blood vessels, and other structures of the arm assume that the arm is supine.

Anatomical Planes

Many views of the body are based on real or imaginary "slices" called sections or planes. *Section* implies an actual cut or slice to reveal internal anatomy, whereas *plane* implies an imaginary flat surface passing through the body. The three major anatomical planes are *sagittal, frontal,* and *transverse* (fig. A.3).

Figure A.1
Anatomical Position. The feet are flat on the floor and close together, the arms are held downward and supine, and the face is directed forward.

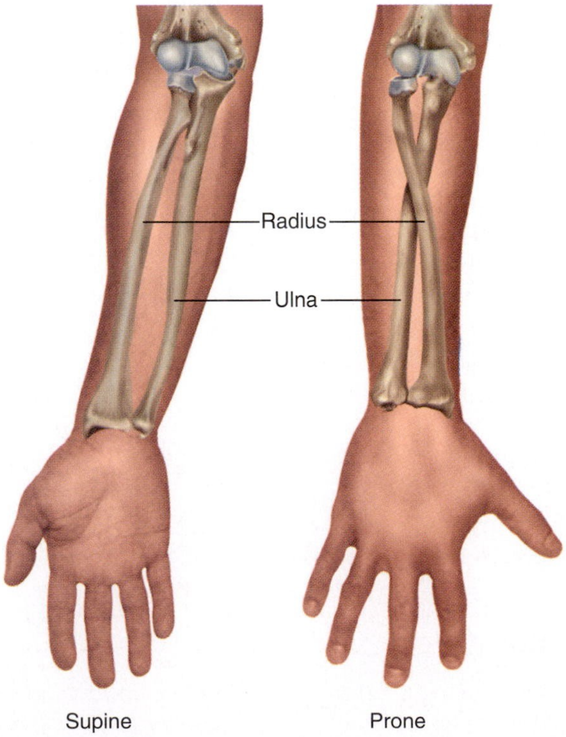

Figure A.2
Positions of the Forearm. When the forearm is supine, the palm faces forward; when prone, it faces rearward. Note the differences in the relationship of the radius to the ulna.

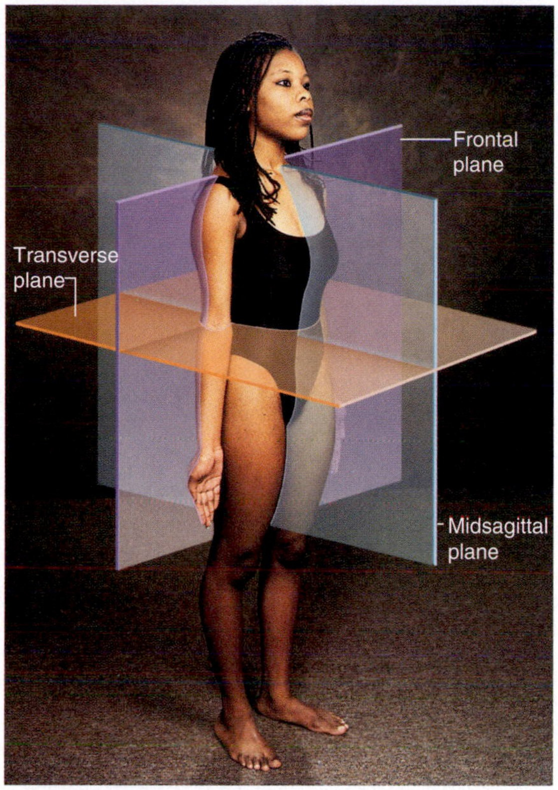

Figure A.3
Anatomical Planes of Reference.

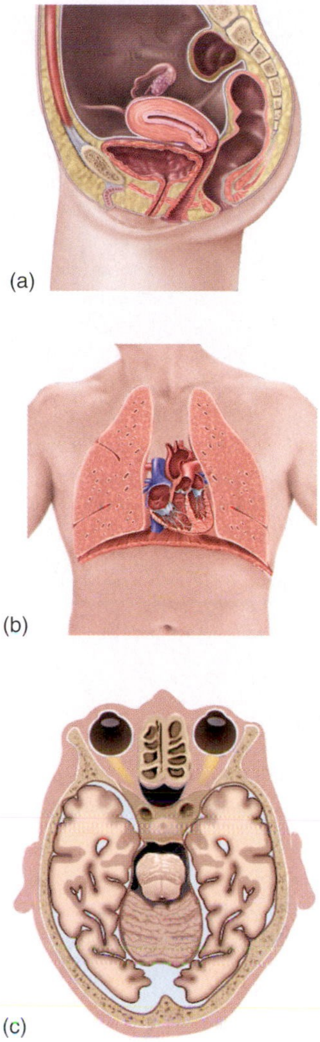

Figure A.4
Views of the Body in the Three Primary Anatomical Planes.
(*a*) Sagittal section of the pelvic region. (*b*) Frontal section of the thoracic region. (*c*) Transverse section of the head at the level of the eyes.

A **sagittal**[1] (SADJ-ih-tul) **plane** extends vertically and divides the body or an organ into right and left portions. The **median (midsagittal) plane** passes through the midline of the body and divides it into *equal* right and left halves. Other sagittal planes parallel to this (off center) divide the body into unequal right and left portions. The head and pelvic organs are commonly illustrated in midsagittal views (fig. A.4*a*).

A **frontal (coronal) plane** also extends vertically, but it is perpendicular to the sagittal plane and divides the body into anterior (front) and posterior (back) portions. A frontal section of the head, for example, would divide it into one portion bearing the face and another bearing the back of the head. Contents of the thoracic and abdominal cavities are most commonly shown in frontal section (fig. A.4*b*).

A **transverse (horizontal) plane** passes across the body or an organ perpendicular to its long axis (fig. A.4*c*); therefore, it divides the body or organ into superior (upper) and inferior (lower) portions. CT scans are typically transverse sections (see fig. 1.4, p. 5).

Directional Terms

Table A.1 summarizes frequently used terms that describe the position of one structure relative to another. Intermediate directions are often indicated by combinations of these terms. For example, one structure may be described as *dorsolateral* to another (toward the back and side). The dorsal surface of a structure is also called the **dorsum;** for example, the dorsum of your hand is what most people call the "back" of the hand. The dorsum of the foot, however, is its upper surface.

Because of the bipedal, upright stance of humans, some directional terms have different meanings for humans than they do for other animals. *Anterior,* for example, denotes the region of the body that leads the way in normal locomotion. For a four-legged animal such as a cat, this is the head end of the body; for a human, however, it is the front of the chest and abdomen. Thus, *anterior* has the same meaning as *ventral* for a human but not for a cat. *Posterior* denotes the region of the body that comes last in normal locomotion—the tail end of a cat but the dorsal side (back) of a human. These differences must be kept in mind when dissecting other animals for comparison to human anatomy.

[1] *sagitta* = arrow

TABLE A.1

Directional Terms in Human Anatomy

Term	Meaning	Examples of Usage
Ventral	Toward the front* or belly	The aorta is *ventral* to the vertebral column.
Dorsal	Toward the back or spine	The vertebral column is *dorsal* to the aorta.
Anterior	Toward the ventral side*	The sternum is *anterior* to the heart.
Posterior	Toward the dorsal side*	The esophagus is *posterior* to the trachea.
Cephalic	Toward the head or superior end	The cephalic end of the embryonic neural tube develops into the brain.
Rostral	Toward the forehead or nose	The forebrain is *rostral* to the brainstem.
Caudal	Toward the tail or inferior end	The spinal cord is *caudal* to the brain.
Superior	Above	The heart is *superior* to the diaphragm.
Inferior	Below	The liver is *inferior* to the diaphragm.
Medial	Toward the midsagittal plane	The heart is *medial* to the lungs.
Lateral	Away from the midsagittal plane	The eyes are *lateral* to the nose.
Proximal	Closer to the point of attachment or origin	The elbow is *proximal* to the wrist.
Distal	Farther from the point of attachment or origin	The fingernails are at the *distal* ends of the fingers.
Superficial	Closer to the body surface	The skin is *superficial* to the muscles.
Deep	Farther from the body surface	The bones are *deep* to the muscles.

*In humans only; definition differs for other animals.

BODY REGIONS

Knowledge of the external anatomy and landmarks of the body is important in performing a physical examination and many other clinical procedures. For purposes of study, the body is divided into two major regions called the *axial* and *appendicular regions.* Smaller areas within the major regions are described in the following paragraphs and illustrated in figure A.5.

Axial Region

The **axial region** consists of the **head, cervical**[2] region (neck), and **trunk.** The trunk is further divided into the **thoracic region** above the diaphragm and the **abdominal region** below it.

One way of referring to the locations of abdominal structures is to divide the region into quadrants. Two perpendicular lines intersecting at the umbilicus (navel) divide the abdomen into a **right upper quadrant (RUQ), right lower quadrant (RLQ), left upper quadrant (LUQ),** and **left lower quadrant (LLQ)** (fig. A.6*a, b*). The quadrant scheme is often used to describe the site of an abdominal pain or abnormality.

The abdomen also can be divided into nine regions defined by four lines that intersect like a tic-tac-toe grid (fig. A.6*c, d*). Each vertical line is called a *midclavicular line* because it passes through the midpoint of the clavicle (collarbone). The superior horizontal line is called the *subcostal*[3] *line* because it connects the inferior borders of the lowest costal cartilages (cartilage connecting the tenth rib on each side to the inferior end of the sternum). The inferior horizontal line is called the *intertubercular*[4] *line* because it passes from left to right between the tubercles *(anterior superior spines)* of the pelvis—two points of bone located about where the front pockets open on most pants. The three lateral regions of this grid, from upper to lower, are the **hypochondriac,**[5] **lateral (lumbar),** and **inguinal**[6] **(iliac) regions.** The three medial regions from upper to lower are the **epigastric,**[7] **umbilical,** and **hypogastric (pubic)** regions.

Appendicular Region

The **appendicular** (AP-en-DIC-you-lur) **region** of the body consists of the **upper limbs** and **lower limbs** (also called *appendages* or *extremities*). The upper limb includes the **brachium** (BRAY-kee-um) (arm), **antebrachium**[8] (AN-teh-BRAY-kee-um) (forearm), **carpus** (wrist), **manus** (hand), and **digits** (fingers). The lower limb includes the **thigh, crus** (leg), **tarsus** (ankle), **pes** (foot), and **digits** (toes).

In strict anatomical terms, *arm* refers only to that part of the upper limb between the shoulder and elbow. *Leg* refers only to that part of the lower limb between the knee and ankle.

BODY CAVITIES AND MEMBRANES

The body is internally divided into two major body cavities, the *dorsal* and *ventral body cavities* (fig. A.7). The organs within these cavities are called the **viscera** (VISS-er-uh) (singular, *viscus*[9]). Various membranes line the cavities, cover the viscera, and hold the viscera in place (table A.2).

[2] *cervic* = neck
[3] *sub* = below + *cost* = rib
[4] *inter* = between + *tubercul* = little swelling

[5] *hypo* = below + *chondr* = cartilage
[6] *inguin* = groin
[7] *epi* = above, over + *gastr* = stomach
[8] *ante* = fore, before + *brachi* = arm
[9] *viscus* = body organ

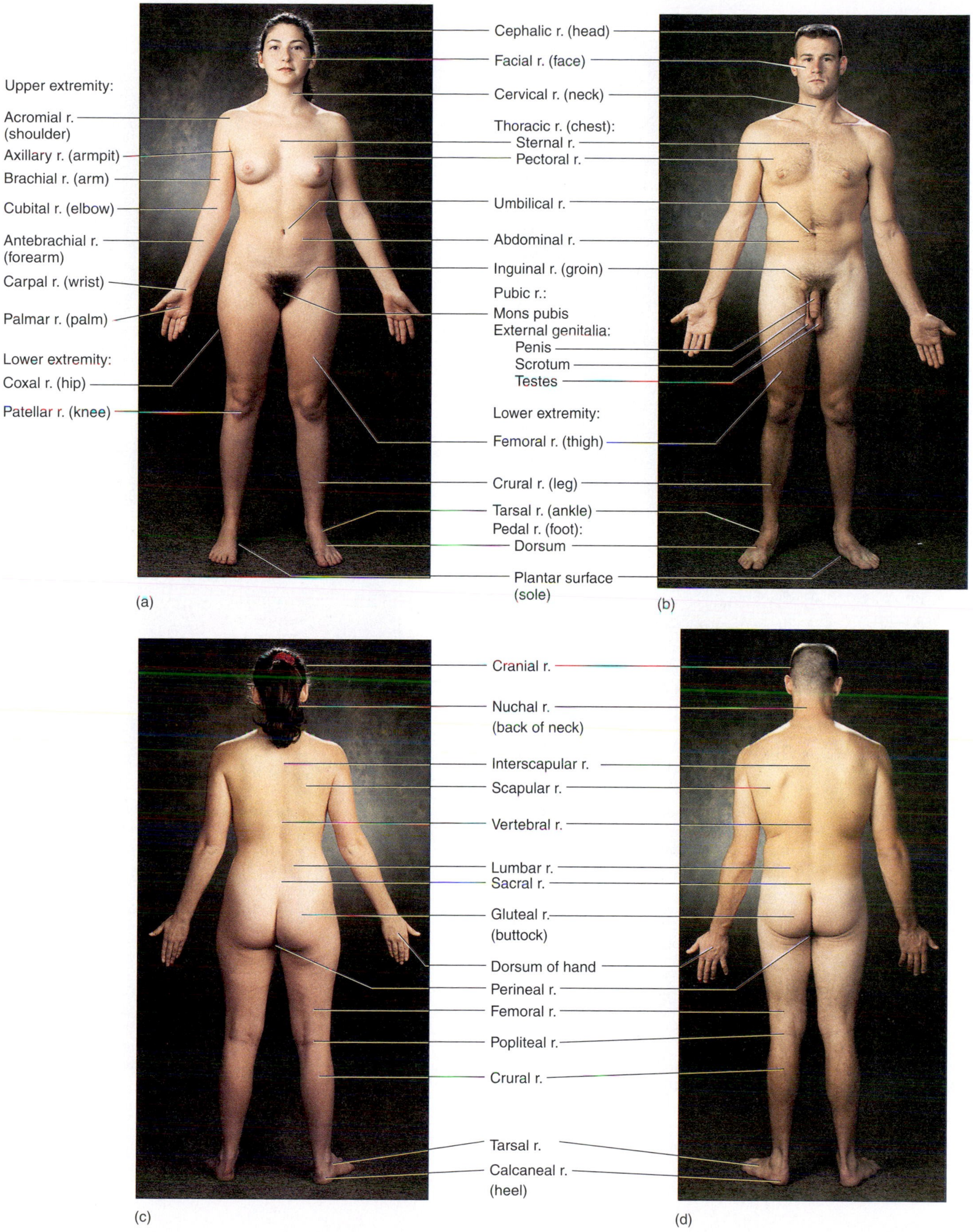

Upper extremity:
Acromial r. (shoulder)
Axillary r. (armpit)
Brachial r. (arm)
Cubital r. (elbow)
Antebrachial r. (forearm)
Carpal r. (wrist)
Palmar r. (palm)
Lower extremity:
Coxal r. (hip)
Patellar r. (knee)
Cephalic r. (head)
Facial r. (face)
Cervical r. (neck)
Thoracic r. (chest):
Sternal r.
Pectoral r.
Umbilical r.
Abdominal r.
Inguinal r. (groin)
Pubic r.:
Mons pubis
External genitalia:
Penis
Scrotum
Testes
Lower extremity:
Femoral r. (thigh)
Crural r. (leg)
Tarsal r. (ankle)
Pedal r. (foot):
Dorsum
Plantar surface (sole)
(a)
(b)
Cranial r.
Nuchal r. (back of neck)
Interscapular r.
Scapular r.
Vertebral r.
Lumbar r.
Sacral r.
Gluteal r. (buttock)
Dorsum of hand
Perineal r.
Femoral r.
Popliteal r.
Crural r.
Tarsal r.
Calcaneal r. (heel)
(c)
(d)

Figure A.5
The Adult Male and Female Bodies. (*a* and *b*) Ventral aspect; (*c* and *d*) dorsal aspect (*r.* = region).

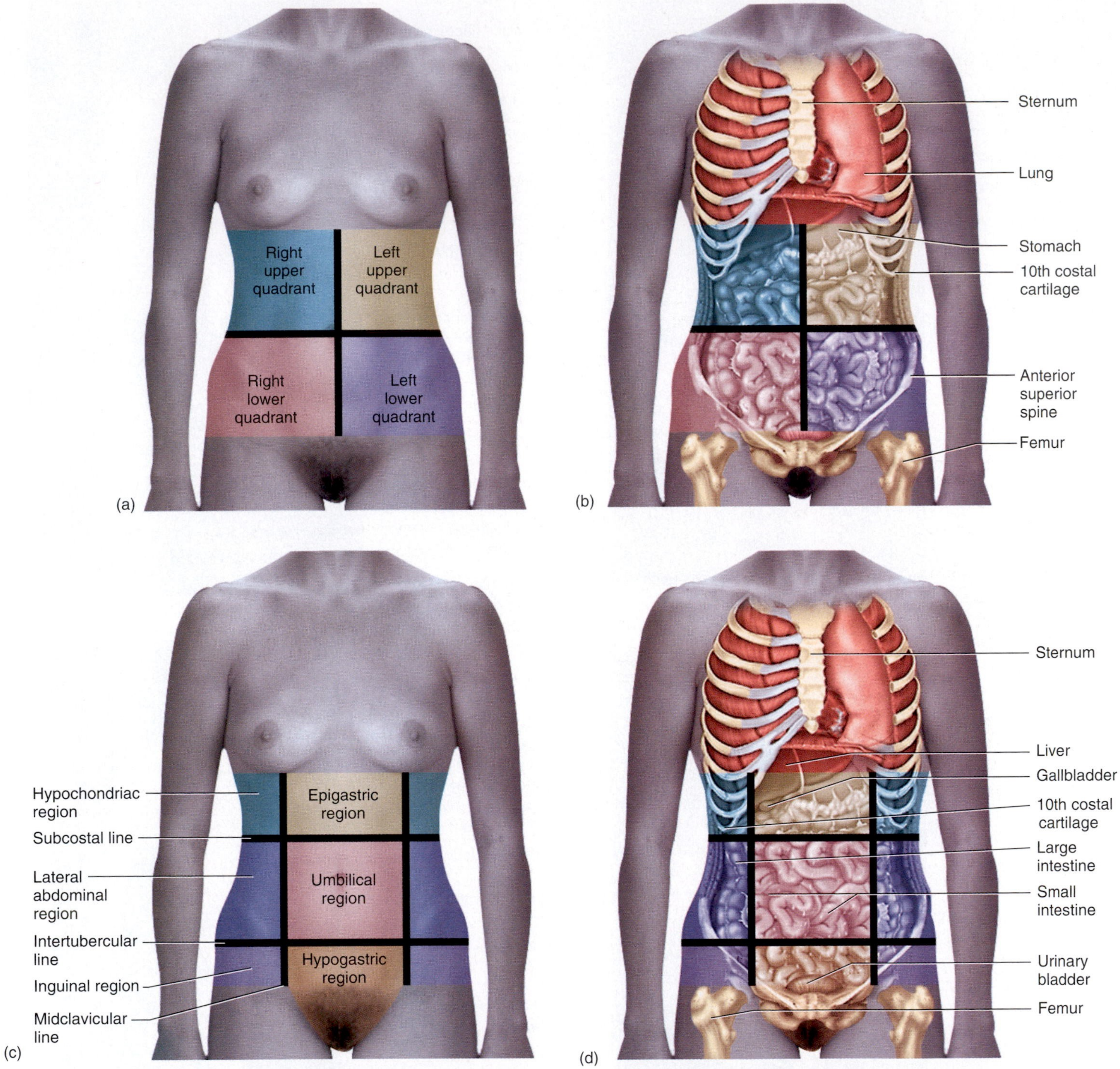

FIGURE A.6

Four Quadrants and Nine Regions of the Abdomen. (*a*) External division into four quadrants. (*b*) Internal anatomy correlated with the four quadrants. (*c*) External division into nine regions. (*d*) Internal anatomy correlated with the nine regions.

Dorsal Body Cavity

The **dorsal body cavity** has two subdivisions: (1) the **cranial** (CRAY-nee-ul) **cavity,** which is enclosed by the cranium (braincase) and contains the brain, and (2) the **vertebral canal,** which is enclosed by the vertebral column (spine) and contains the spinal cord. The dorsal body cavity is lined by three membrane layers called the **meninges** (meh-NIN-jeez). Among other functions, the meninges protect the delicate nervous tissue from the hard protective bone that encloses it.

Ventral Body Cavity

During embryonic development, a space called the *coelom* (SEE-loam) forms within the trunk and eventually gives rise to the **ventral body cavity.** This cavity later becomes partitioned by a muscular sheet, the **diaphragm,** into a superior **thoracic cavity** and an inferior **abdominopelvic cavity.** The thoracic and abdominopelvic cavities are lined with thin **serous membranes,** which secrete a lubricating film of moisture similar to blood serum (hence the name *serous*).

THORACIC CAVITY

The thoracic cavity is divided into right, left, and medial portions by a partition called the **mediastinum**[10] (ME-dee-ah-STY-num) (fig. A.7). The right and left sides contain the lungs and are lined by a two-layered membrane called the **pleura**[11] (PLOOR-uh) (fig. A.8*a*). The outer layer, or *parietal*[12] (pa-RY-eh-tul) *pleura,* lies against the inside of the rib cage; the inner layer, or *visceral* (VISS-er-ul) *pleura,* forms the external surface of the lung. The narrow, moist space between the visceral and parietal pleurae is called the **pleural cavity** (see fig. A.19). It is lubricated by a slippery **pleural fluid.**

The medial portion, or mediastinum, is occupied by the esophagus and trachea, a gland called the thymus, and the heart and major blood vessels connected to it. The heart is enclosed by a two-layered membrane called the **pericardium.**[13] The *visceral*

TABLE A.2

Body Cavities and Membranes

Name of Cavity	Principal Viscera	Membranous Lining
Dorsal Body Cavity		
Cranial cavity	Brain	Meninges
Vertebral canal	Spinal cord	Meninges
Ventral Body Cavity		
Thoracic Cavity		
Pleural cavities (2)	Lungs	Pleurae
Pericardial cavity	Heart	Pericardium
Abdominopelvic Cavity		
..Abdominal cavity	Digestive organs, spleen, kidneys	Peritoneum
Pelvic cavity	Bladder, rectum, reproductive organs	Peritoneum

[10] *mediastinum* = in the middle
[11] *pleur* = rib, side
[12] *pariet* = wall
[13] *peri* = around + *cardi* = heart

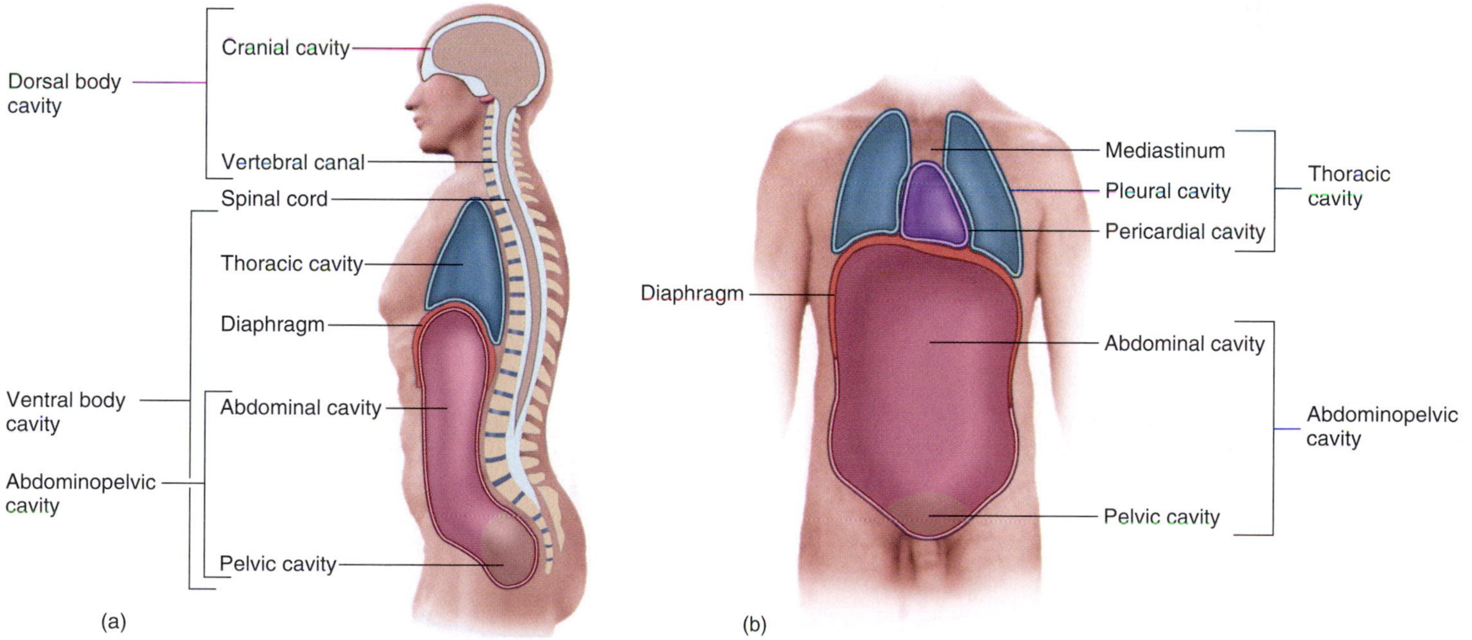

FIGURE A.7
The Major Body Cavities. (*a*) Left lateral view; (*b*) anterior view.

pericardium forms the heart surface while the *parietal pericardium* is separated from it by a space called the **pericardial cavity** (fig. A.8*b*). This space is lubricated by **pericardial fluid.**

ABDOMINOPELVIC CAVITY

The abdominopelvic cavity consists of the **abdominal cavity** above the brim of the pelvis and the **pelvic cavity** below the brim (see fig. A.16). The abdominal cavity contains most of the digestive organs as well as the spleen, kidneys, and ureters. The pelvic cavity is markedly narrower and its lower end tilts dorsally (see fig. A.7*a*). It contains the distal part of the large intestine, the urinary bladder and urethra, and the reproductive organs.

The abdominopelvic cavity contains a moist serous membrane called the **peritoneum**[14] (PERR-ih-toe-NEE-um). The *parietal peritoneum* lines the walls of the cavity, while the *visceral peritoneum* covers the external surfaces of most digestive organs. The **peritoneal cavity** is the space between the parietal and visceral layers. It is lubricated by **peritoneal fluid.**

Some organs of the abdominal cavity lie against the dorsal body wall and are covered by peritoneum only on the side facing the peritoneal cavity. They are said to have a **retroperitoneal**[15] position (fig. A.9). These include the kidneys, ureters, adrenal glands, most of the pancreas, and abdominal portions of two major blood vessels—the aorta and inferior vena cava (see fig. A.15). Organs that are encircled by peritoneum and connected to the dorsal body wall by peritoneal sheets are described as **intraperitoneal.**[16]

Strictly speaking, none of the viscera lie within the peritoneal cavity—that is, between the parietal and visceral peritoneum—just as, strictly speaking, the heart is not within the pericardial cavity and the lungs are not within the pleural cavities.

The intestines are suspended from the dorsal abdominal wall by a translucent peritoneal sheet called the **mesentery**[17] (MEZ-en-tare-ee). This is a continuation of the peritoneum that wraps around the intestines, forming a moist membrane called the **serosa** (seer-OH-sa) on their outer surfaces (fig. A.10). The mesentery of the large intestine is called the **mesocolon.** The visceral peritoneum consists of the mesenteries and serosae.

A fatty membrane called the **greater omentum**[18] hangs like an apron from the inferolateral margin of the stomach and overlies the intestines (figs. A.10 and A.13). It is unattached at its inferior border and can be lifted to reveal the intestines. A smaller **lesser omentum** extends from the superomedial border of the stomach to the liver.

ORGAN SYSTEMS

The human body has 11 organ systems (fig. A.11) and an immune system, which is better described as a population of cells than as an organ system. These systems are classified in the following list by their principal functions, but this is an unavoidably flawed classification. Some organs belong to two or more systems—for example, the male urethra is part of both the urinary and reproductive

[14]*peri* = around + *tone* = stretched
[15]*retro* = behind
[16]*intra* = within

[17]*mes* = in the middle + *enter* = intestine
[18]*omentum* = covering

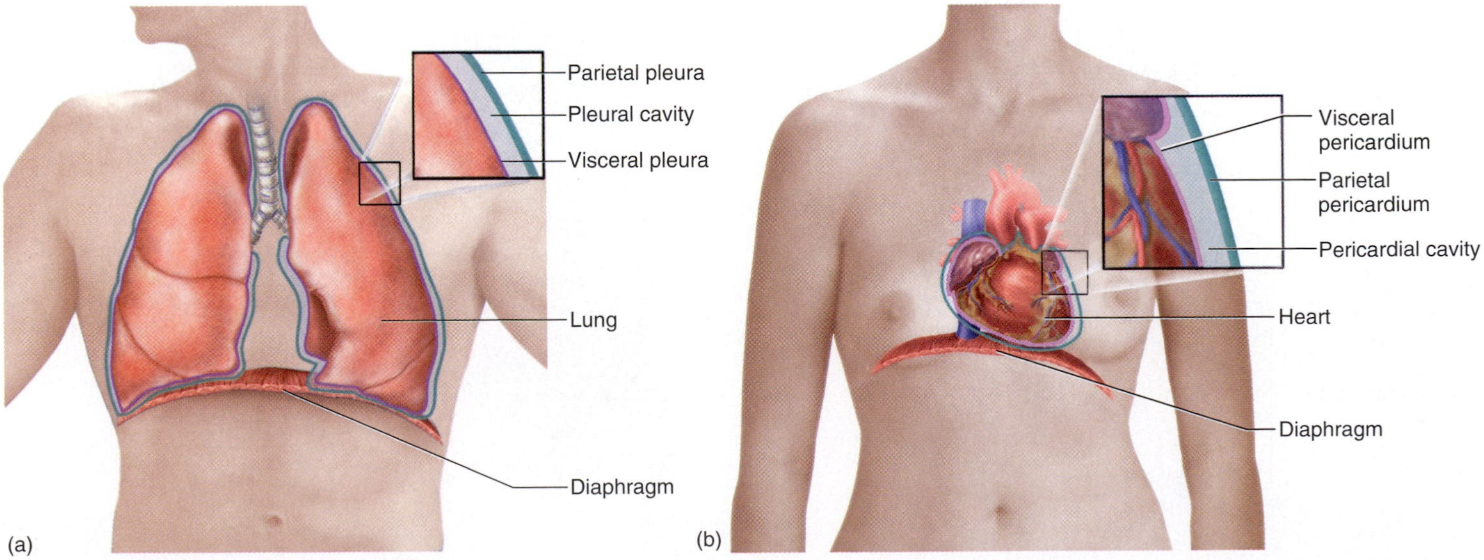

FIGURE A.8
Parietal and Visceral Layers of Double-Walled Membranes. (*a*) The pleura; (*b*) the pericardium.

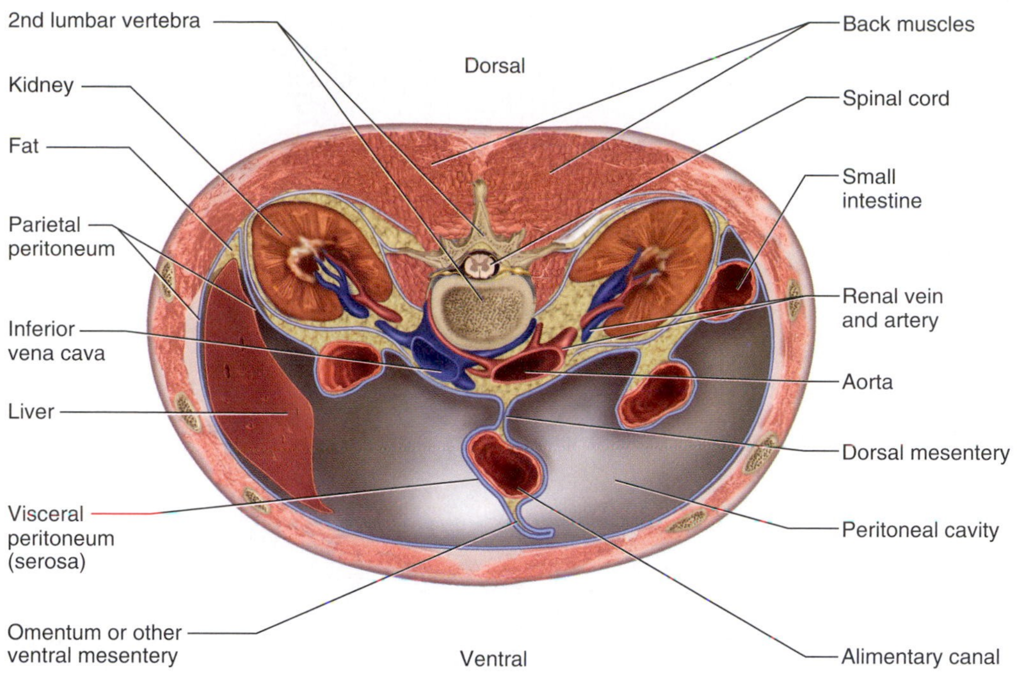

2nd lumbar vertebra
Kidney
Fat
Parietal peritoneum
Inferior vena cava
Liver
Visceral peritoneum (serosa)
Omentum or other ventral mesentery
Dorsal
Ventral
Back muscles
Spinal cord
Small intestine
Renal vein and artery
Aorta
Dorsal mesentery
Peritoneal cavity
Alimentary canal

FIGURE A.9

Transverse Section Through the Abdominal Cavity. Shows the peritoneum, peritoneal cavity (with most viscera omitted), and some retroperitoneal organs (the kidneys, aorta, and inferior vena cava).

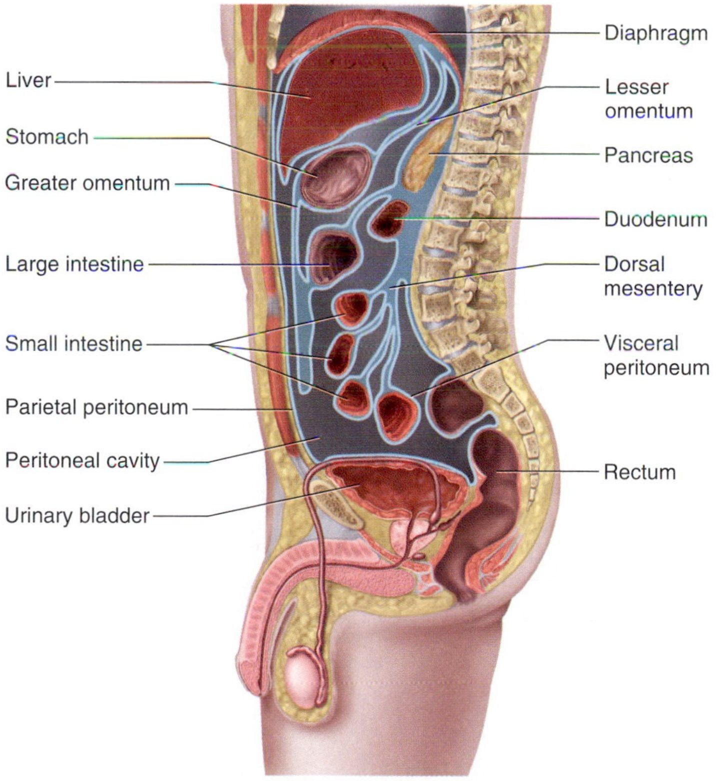

Liver
Stomach
Greater omentum
Large intestine
Small intestine
Parietal peritoneum
Peritoneal cavity
Urinary bladder
Diaphragm
Lesser omentum
Pancreas
Duodenum
Dorsal mesentery
Visceral peritoneum
Rectum

FIGURE A.10

Serous Membranes of the Abdominal Cavity. Sagittal section, left lateral view.

systems; the pharynx is part of the respiratory and digestive systems; and the mammary glands can be considered part of the integumentary and female reproductive systems.

Protection, Support, and Movement
- Integumentary system
- Skeletal system
- Muscular system

Internal Communication and Integration
- Nervous system
- Endocrine system

Fluid Transport
- Circulatory system
- Lymphatic system

Defense
- Immune system

Input and Output
- Respiratory system
- Urinary system
- Digestive system

Reproduction
- Reproductive system

A Visual Survey of the Body

Figures A.12 through A.16 provide an overview of the anatomy of the trunk and internal organs of the thoracic and abdominopelvic cavities. Figures A.17 through A.22 are photographs of the cadaver showing the major organs of the dorsal and ventral body cavities.

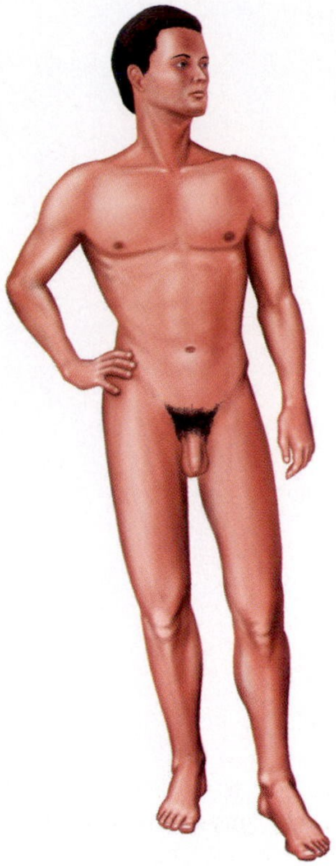

A.11*a* Integumentary system
Principal organs: Skin, hair, nails, cutaneous glands
Principal functions: Protection, water retention, thermoregulation, vitamin D synthesis, cutaneous sensation, nonverbal communication

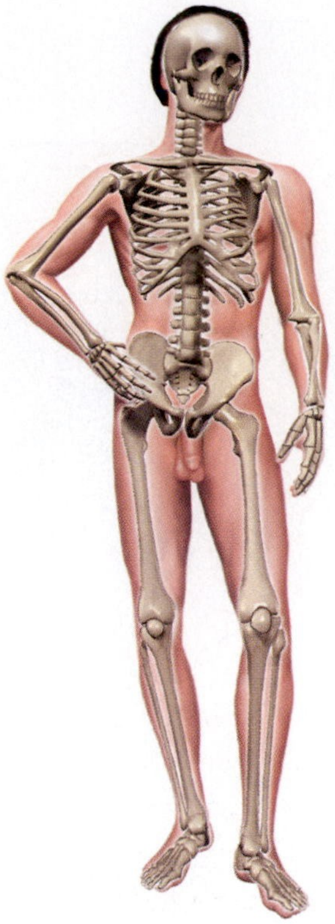

A.11*b* Skeletal system
Principal organs: Bones, cartilages, ligaments
Principal functions: Support, movement, protective enclosure of viscera, blood formation, electrolyte and acid-base balance

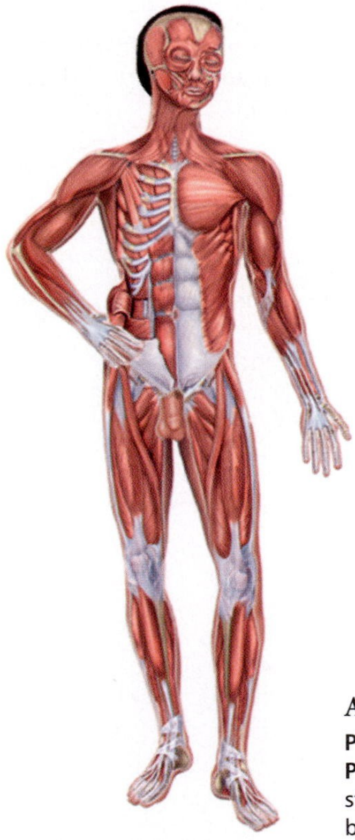

A.11*c* Muscular system
Principal organs: Skeletal muscles
Principal functions: Movement, stability, communication, control of body openings, heat production

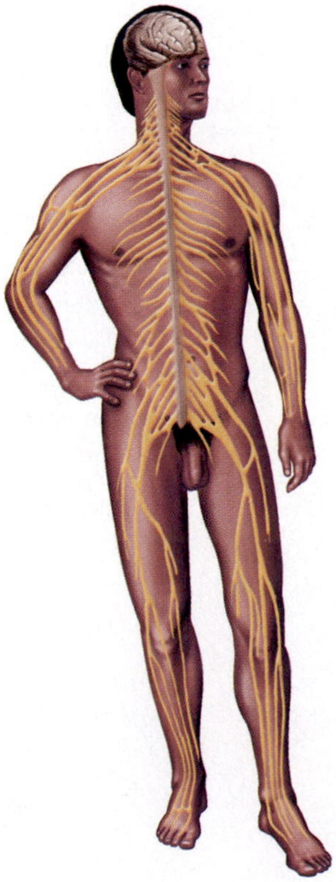

A.11*d* Nervous system
Principal organs: Brain, spinal cord, nerves, ganglia
Principal functions: Rapid internal communication and coordination, sensation

Figure A.11
The Human Organ Systems.

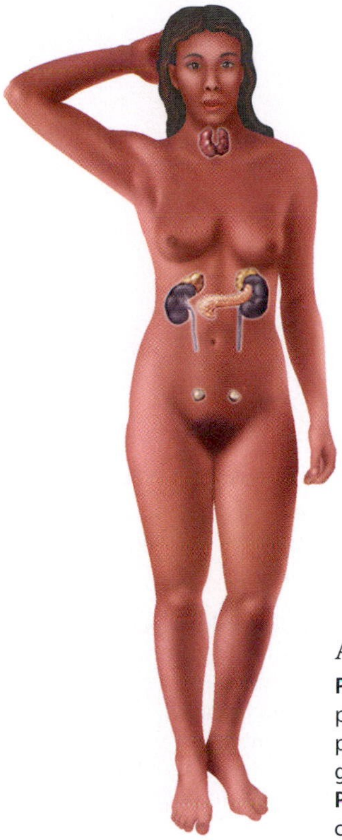

A.11*e* Endocrine system
Principal organs: Pituitary gland, pineal gland, thyroid gland, parathyroid glands, thymus, adrenal glands, pancreas, testes, ovaries
Principal functions: Internal chemical communication and coordination

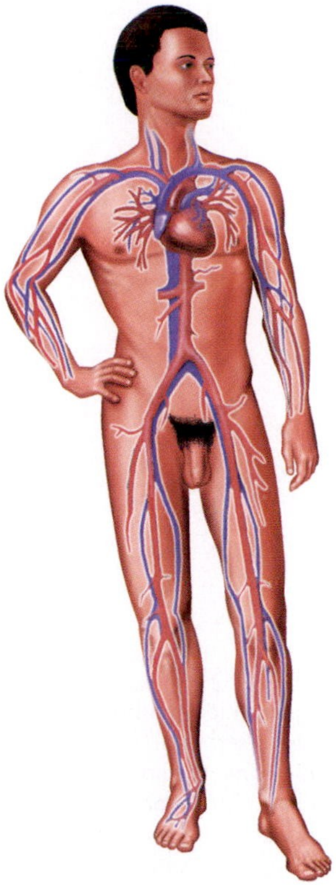

A.11*f* Circulatory system
Principal organs: Heart, blood vessels
Principal functions: Distribution of nutrients, oxygen, wastes, hormones, electrolytes, heat, immune cells, and antibodies; fluid, electrolyte, and acid-base balance

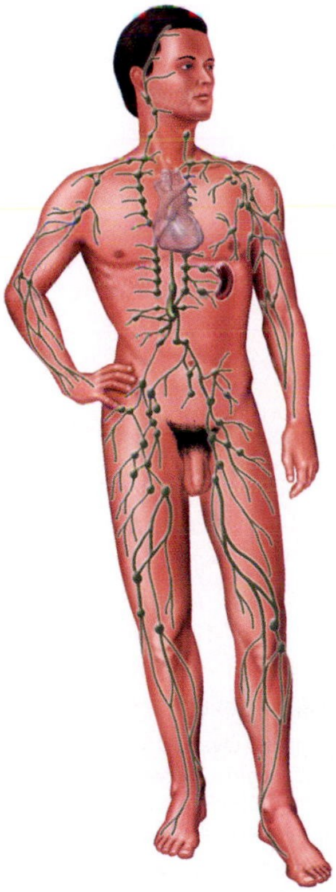

A.11*g* Lymphatic system
Principal organs: Lymph nodes, lymphatic vessels, thymus, spleen, tonsils
Principal functions: Recovery of excess tissue fluid, detection of pathogens, production of immune cells, defense

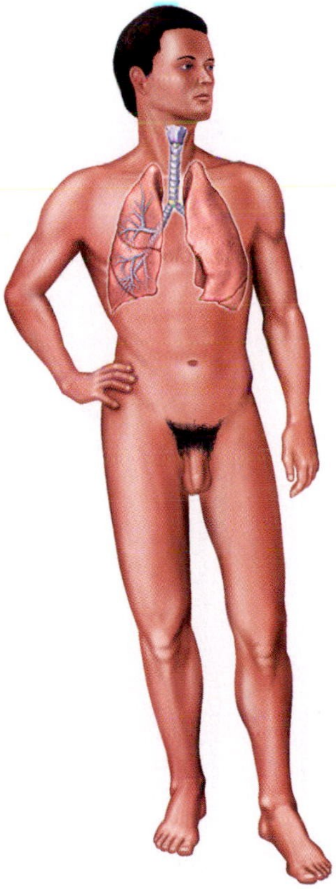

A.11*h* Respiratory system
Principal organs: Nose, pharynx, larynx, trachea, bronchi, lungs
Principal functions: Absorption of oxygen, discharge of carbon dioxide, acid-base balance, speech

Figure A.11
The Human Organ Systems *(continued)*.

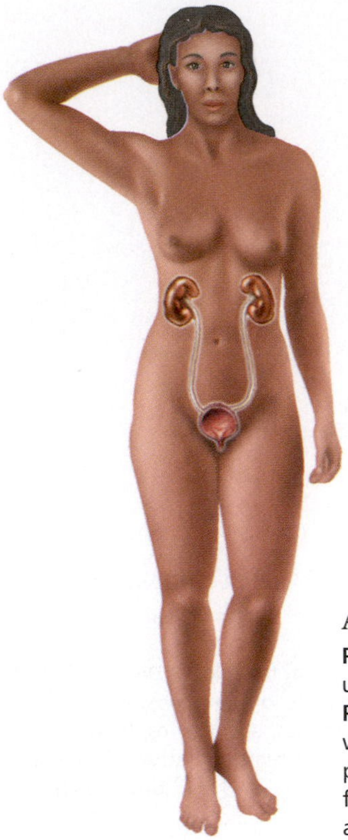

A.11*i* Urinary system

Principal organs: Kidneys, ureters, urinary bladder, urethra

Principal functions: Elimination of wastes; regulation of blood volume and pressure; stimulation of red blood cell formation; control of fluid, electrolyte, and acid-base balance; detoxification

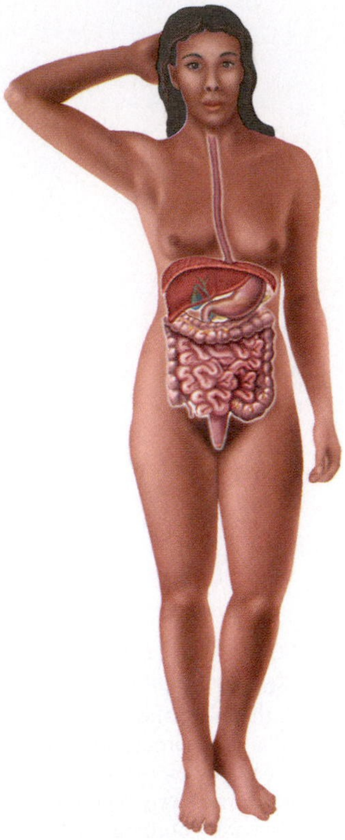

A.11*j* Digestive system

Principal organs: Teeth, tongue, salivary glands, esophagus, stomach, small and large intestines, liver, pancreas

Principal functions: Nutrient breakdown and absorption; liver functions including metabolism of carbohydrates, lipids, proteins, vitamins, and minerals, synthesis of plasma proteins, disposal of drugs, toxins, and hormones, and cleansing of blood

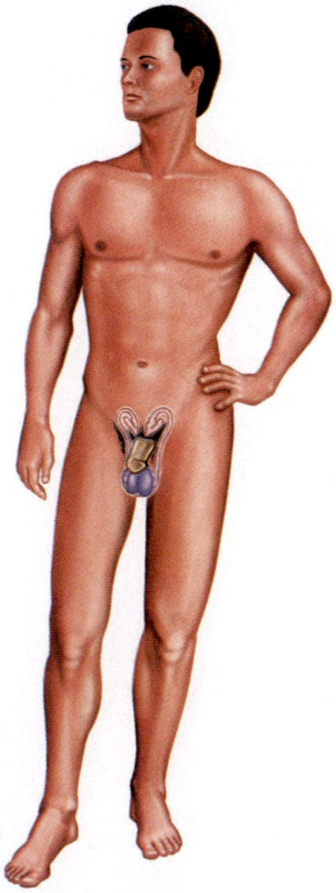

A.11*k* Male reproductive system

Principal organs: Testes, epididymides, spermatic ducts, seminal vesicles, prostate gland, bulbourethral glands, penis

Principal functions: Production and delivery of sperm

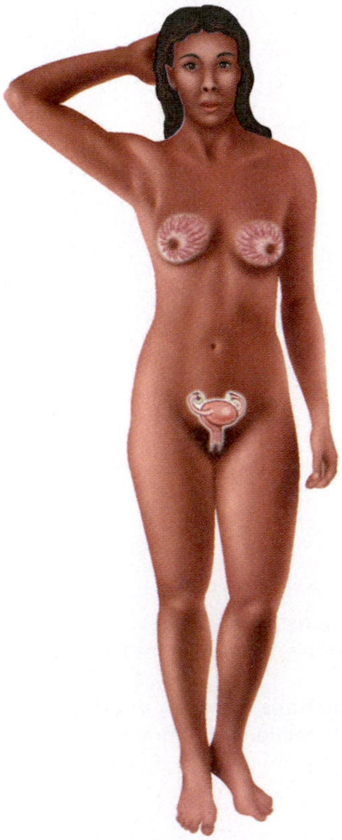

A.11*l* Female reproductive system

Principal organs: Ovaries, uterine tubes, uterus, vagina, vulva, mammary glands

Principal functions: Production of eggs, site of fertilization and fetal development, fetal nourishment, birth, lactation

Figure A.11

The Human Organ Systems *(continued)*.

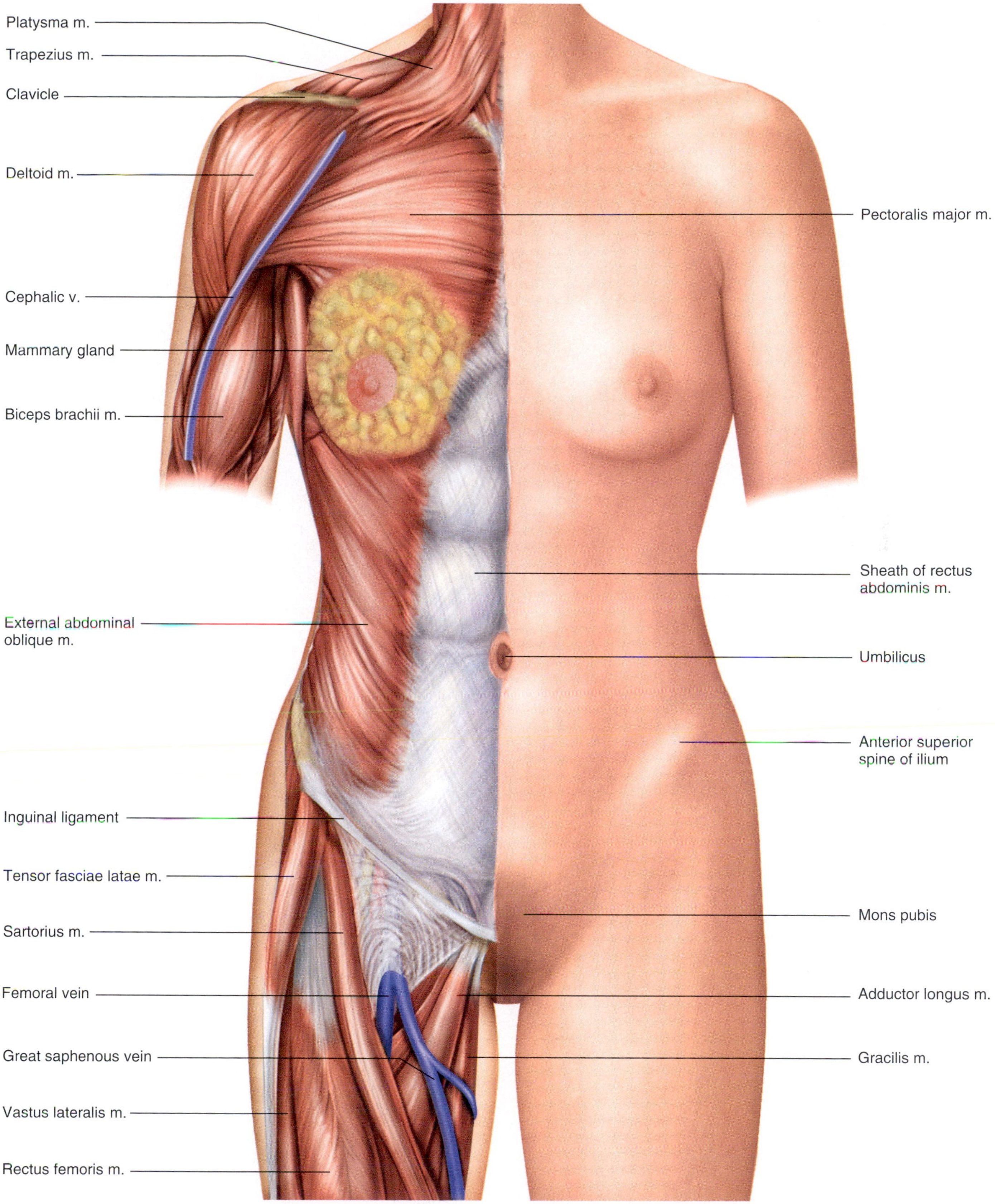
Platysma m.
Trapezius m.
Clavicle
Deltoid m.
Pectoralis major m.
Cephalic v.
Mammary gland
Biceps brachii m.
Sheath of rectus abdominis m.
External abdominal oblique m.
Umbilicus
Anterior superior spine of ilium
Inguinal ligament
Tensor fasciae latae m.
Sartorius m.
Mons pubis
Femoral vein
Adductor longus m.
Great saphenous vein
Gracilis m.
Vastus lateralis m.
Rectus femoris m.

FIGURE A.12

Superficial Anatomy of the Trunk (female). Surface anatomy is shown on the anatomical left and structures immediately deep to the skin on the right (*m.* = muscle; *v.* = vein).

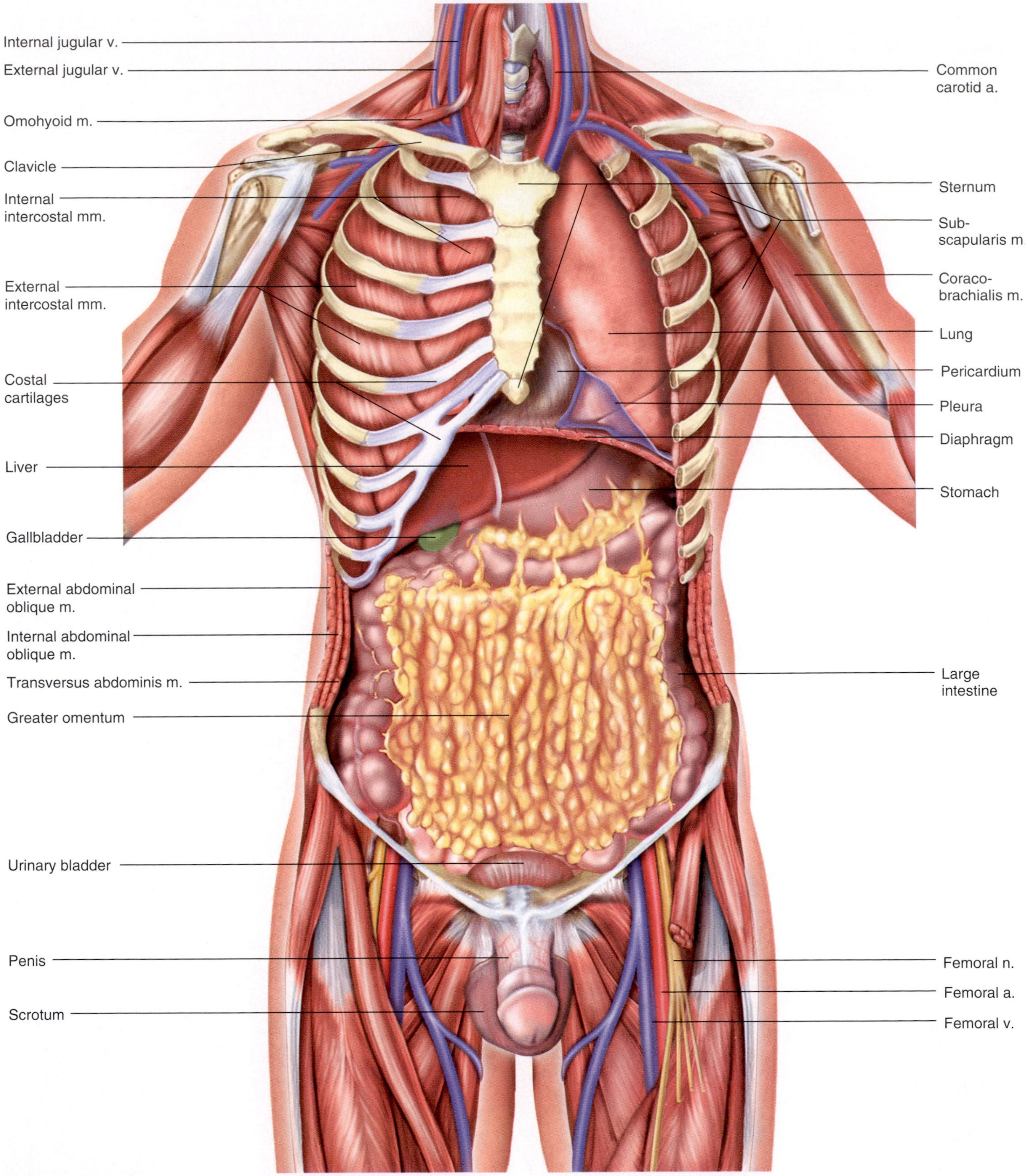
Internal jugular v.
External jugular v.
Omohyoid m.
Clavicle
Internal intercostal mm.
External intercostal mm.
Costal cartilages
Liver
Gallbladder
External abdominal oblique m.
Internal abdominal oblique m.
Transversus abdominis m.
Greater omentum
Urinary bladder
Penis
Scrotum
Common carotid a.
Sternum
Subscapularis m.
Coracobrachialis m.
Lung
Pericardium
Pleura
Diaphragm
Stomach
Large intestine
Femoral n.
Femoral a.
Femoral v.

FIGURE A.13

Anatomy at the Level of the Rib Cage and Greater Omentum (male). The anterior body wall is removed and the ribs, intercostal muscles, and pleura are removed from the anatomical left (*a.* = artery; *v.* = vein; *m.* = muscle; *mm.* = muscles; *n.* = nerve).

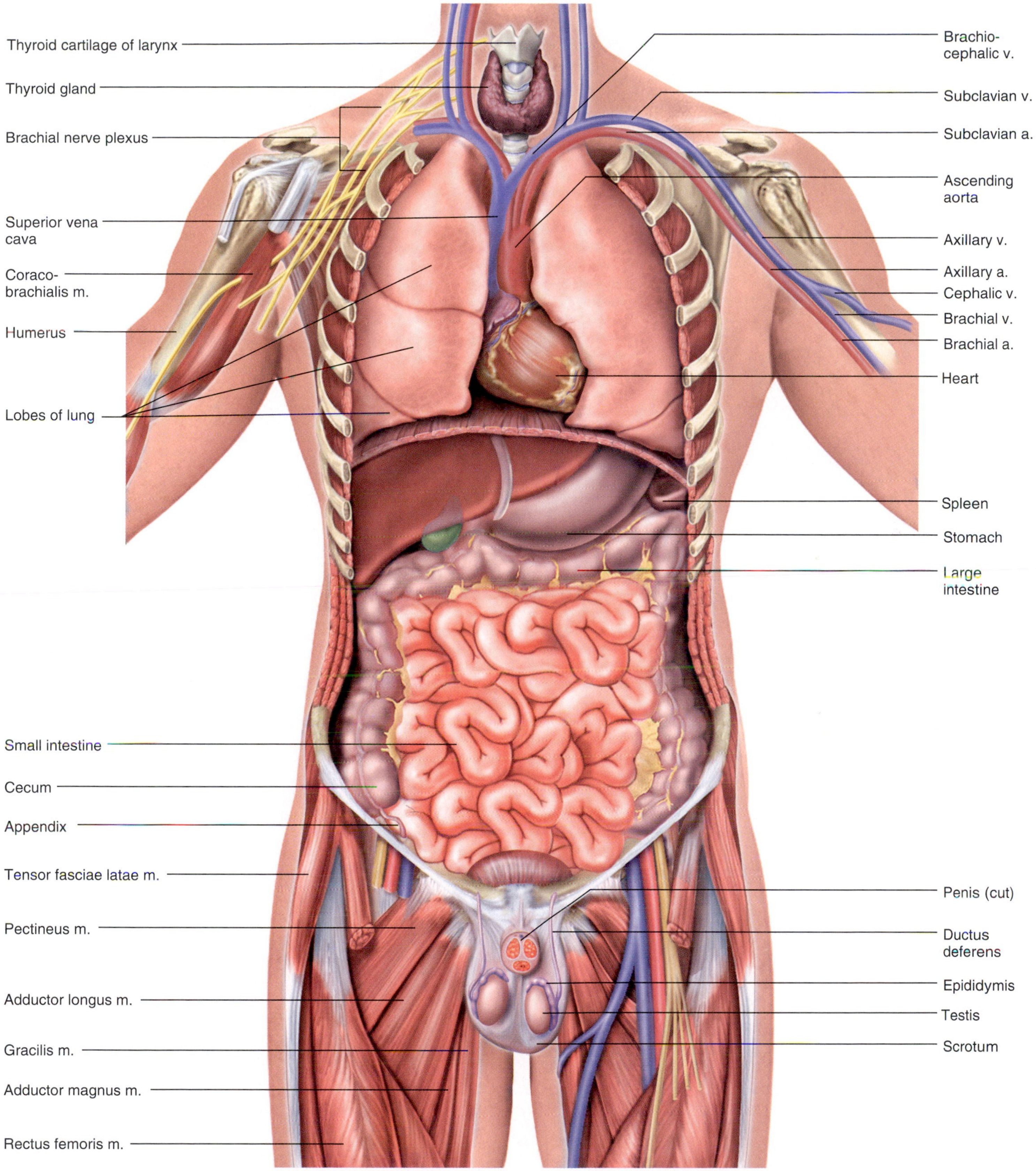

Thyroid cartilage of larynx
Thyroid gland
Brachial nerve plexus
Superior vena cava
Coraco-brachialis m.
Humerus
Lobes of lung
Small intestine
Cecum
Appendix
Tensor fasciae latae m.
Pectineus m.
Adductor longus m.
Gracilis m.
Adductor magnus m.
Rectus femoris m.
Brachio-cephalic v.
Subclavian v.
Subclavian a.
Ascending aorta
Axillary v.
Axillary a.
Cephalic v.
Brachial v.
Brachial a.
Heart
Spleen
Stomach
Large intestine
Penis (cut)
Ductus deferens
Epididymis
Testis
Scrotum

FIGURE A.14

Anatomy at the Level of the Lungs and Intestines (male). The sternum, ribs, and greater omentum are removed (*a.* = artery; *v.* = vein; *m.* = muscle).

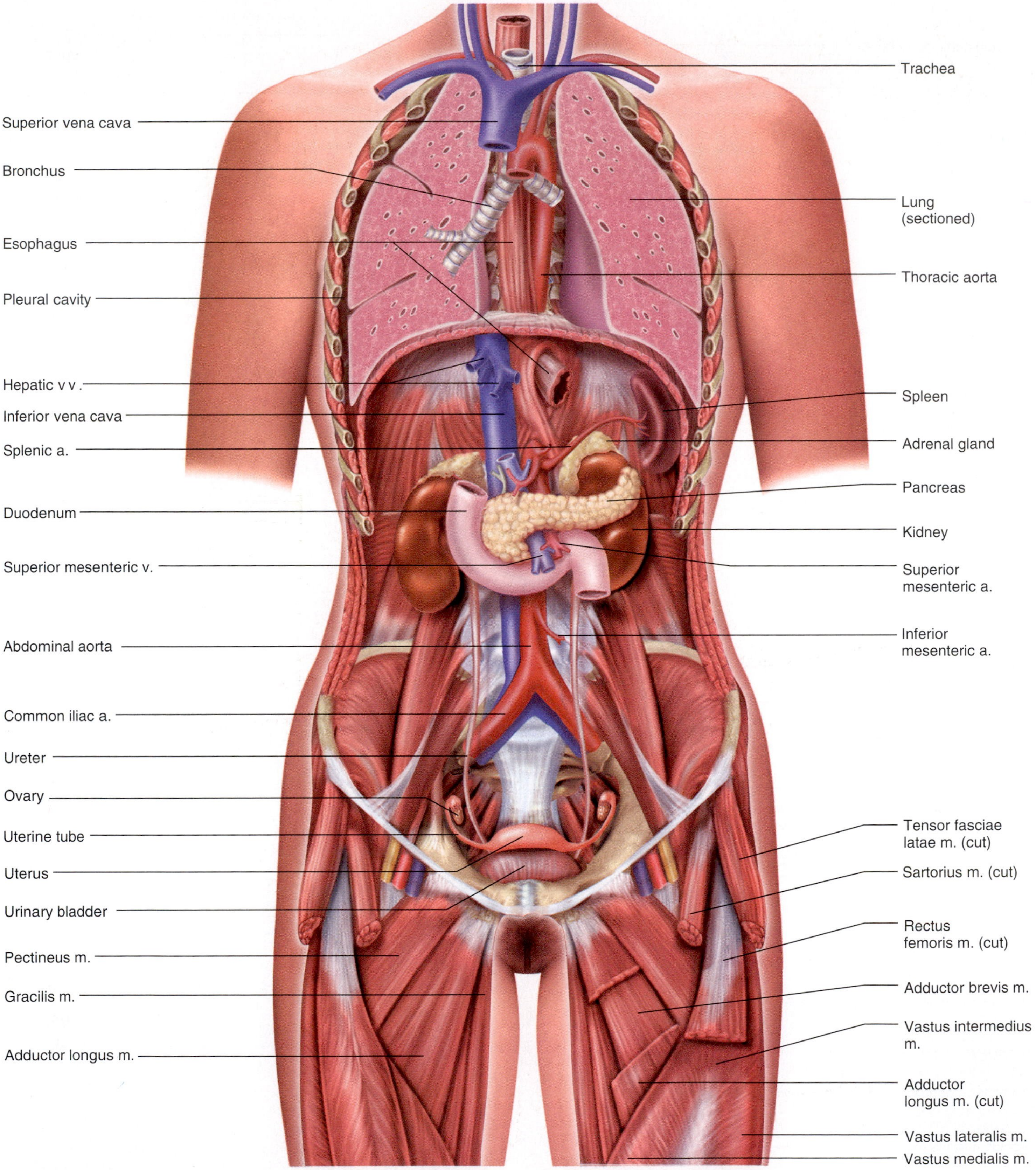

Trachea
Superior vena cava
Bronchus
Lung (sectioned)
Esophagus
Thoracic aorta
Pleural cavity
Hepatic v v .
Spleen
Inferior vena cava
Adrenal gland
Splenic a.
Pancreas
Duodenum
Kidney
Superior mesenteric v.
Superior mesenteric a.
Abdominal aorta
Inferior mesenteric a.
Common iliac a.
Ureter
Ovary
Tensor fasciae latae m. (cut)
Uterine tube
Uterus
Sartorius m. (cut)
Urinary bladder
Rectus femoris m. (cut)
Pectineus m.
Gracilis m.
Adductor brevis m.
Vastus intermedius m.
Adductor longus m.
Adductor longus m. (cut)
Vastus lateralis m.
Vastus medialis m.

FIGURE A.15

Anatomy at the Level of the Retroperitoneal Viscera (female). The heart is removed, the lungs are frontally sectioned, and the viscera of the peritoneal cavity and the peritoneum itself are removed (*a.* = artery; *v.* = vein; *vv.* = veins; *m.* = muscle).

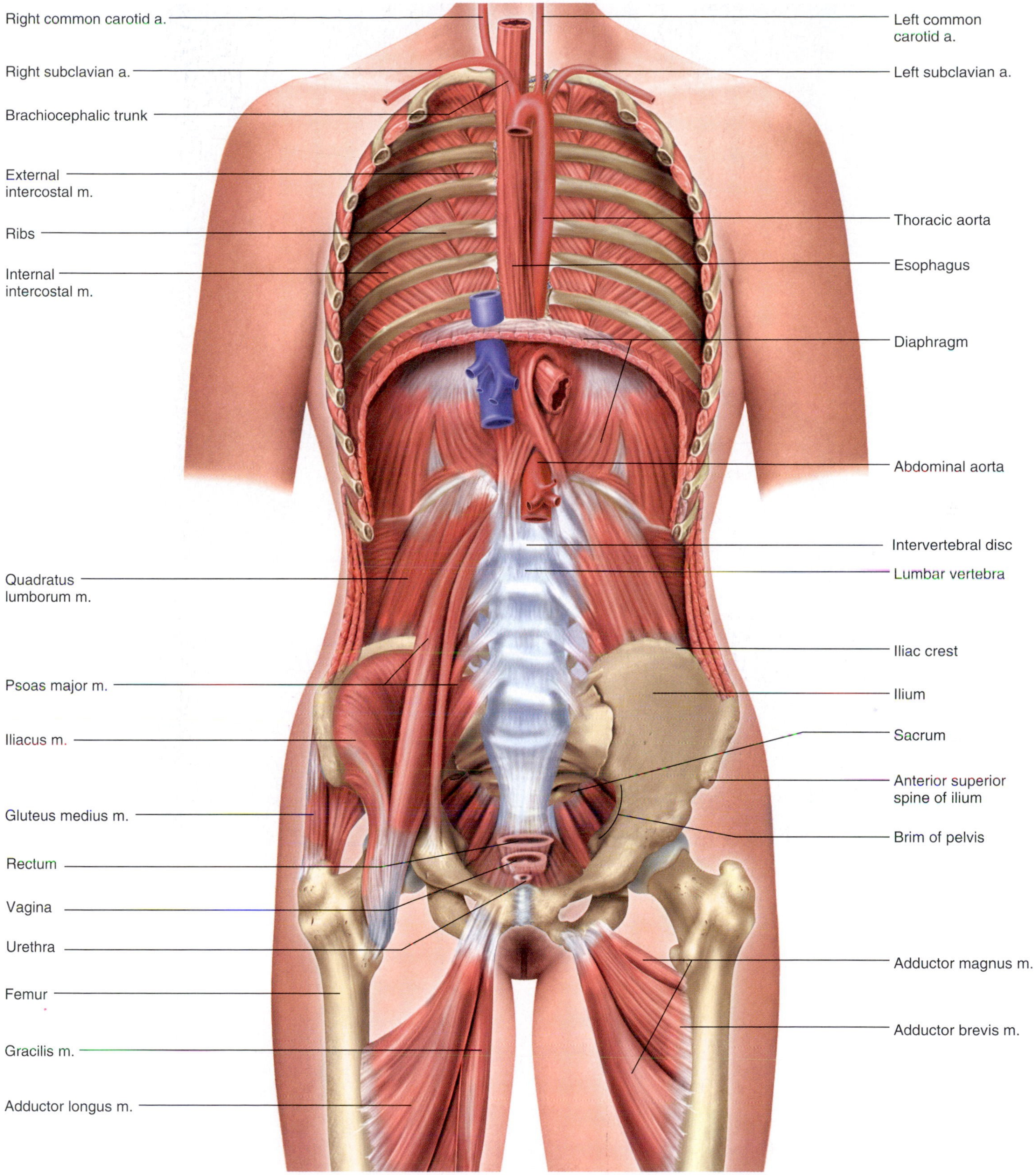

Right common carotid a.
Right subclavian a.
Brachiocephalic trunk
External intercostal m.
Ribs
Internal intercostal m.
Quadratus lumborum m.
Psoas major m.
Iliacus m.
Gluteus medius m.
Rectum
Vagina
Urethra
Femur
Gracilis m.
Adductor longus m.
Left common carotid a.
Left subclavian a.
Thoracic aorta
Esophagus
Diaphragm
Abdominal aorta
Intervertebral disc
Lumbar vertebra
Iliac crest
Ilium
Sacrum
Anterior superior spine of ilium
Brim of pelvis
Adductor magnus m.
Adductor brevis m.

FIGURE A.16

Anatomy at the Level of the Dorsal Body Wall (female). The lungs and retroperitoneal viscera are removed (*a.* = artery; *m.* = muscle).

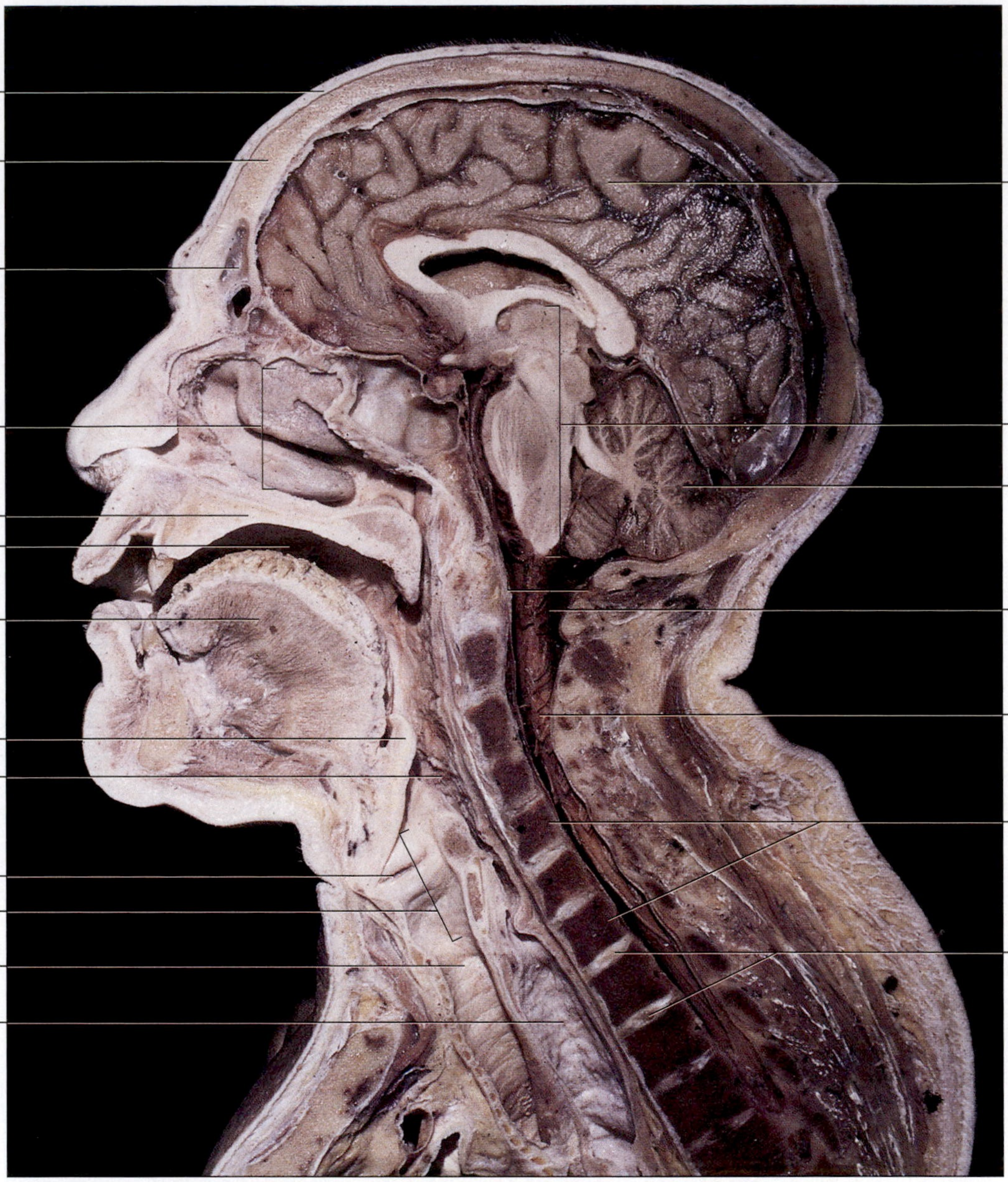

Scalp
Cranium
Frontal sinus
Nasal cavity
Palate
Oral cavity
Tongue
Epiglottis
Pharynx
Vocal cord
Larynx
Trachea
Esophagus
Cerebrum
Brainstem
Cerebellum
Foramen magnum of skull
Spinal cord
Vertebral column
Intervertebral discs

FIGURE A.17

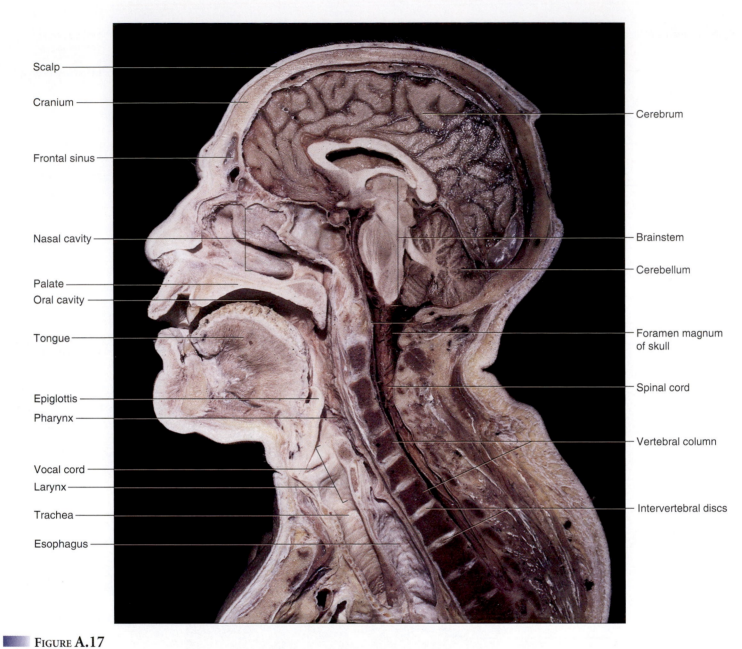

Sagittal Section of the Head. Shows contents of the cranial, nasal, and buccal cavities.

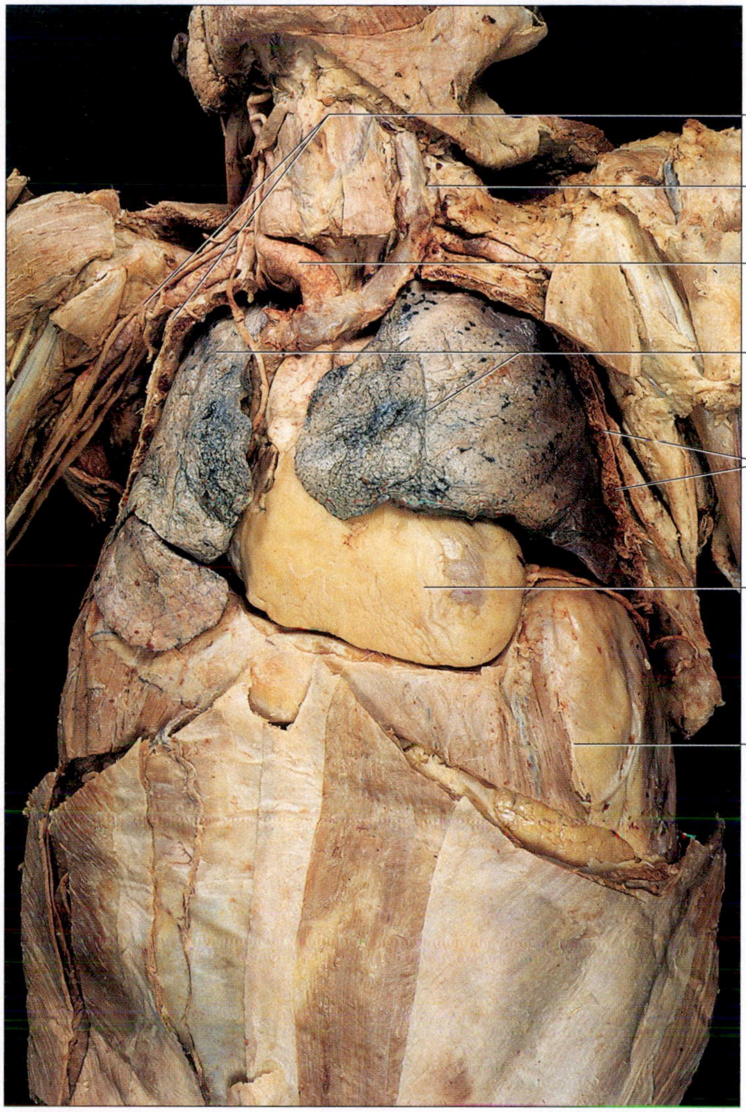
Nerves
Internal jugular vein
Subclavian vein
Lungs
Ribs
Heart
Diaphragm

FIGURE A.18
Frontal View of the Thoracic Cavity.

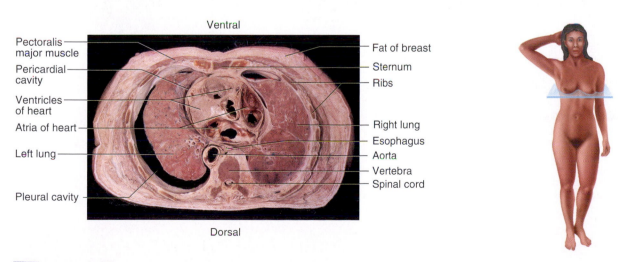

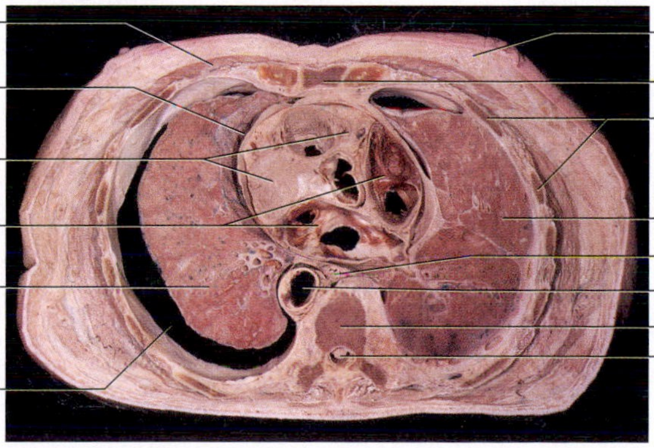
Ventral
Pectoralis major muscle
Pericardial cavity
Ventricles of heart
Atria of heart
Left lung
Pleural cavity
Fat of breast
Sternum
Ribs
Right lung
Esophagus
Aorta
Vertebra
Spinal cord
Dorsal

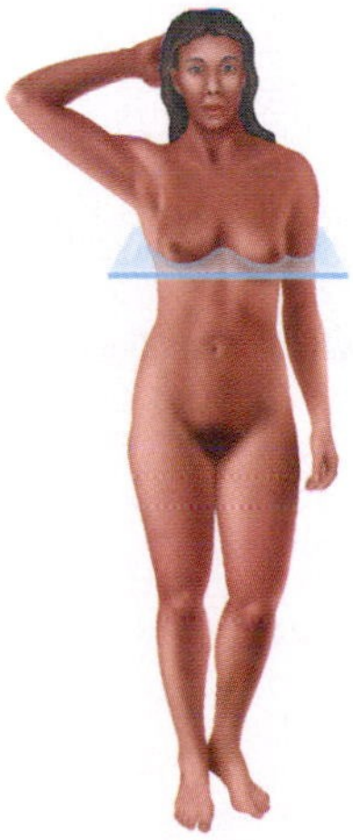

FIGURE A.19
Transverse Section of the Thoracic Cavity. Section taken at the level shown by the inset and oriented the same as the reader's body.

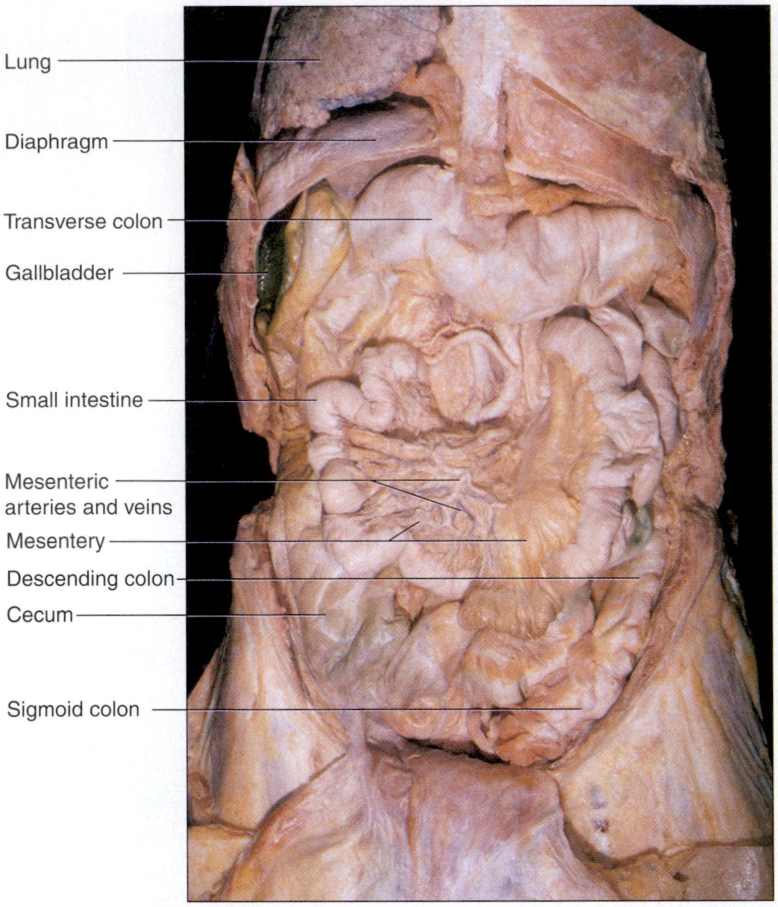
Lung
Diaphragm
Transverse colon
Gallbladder
Small intestine
Mesenteric
arteries and veins
Mesentery
Descending colon
Cecum
Sigmoid colon

FIGURE A.20
Frontal View of the Abdominal Cavity.

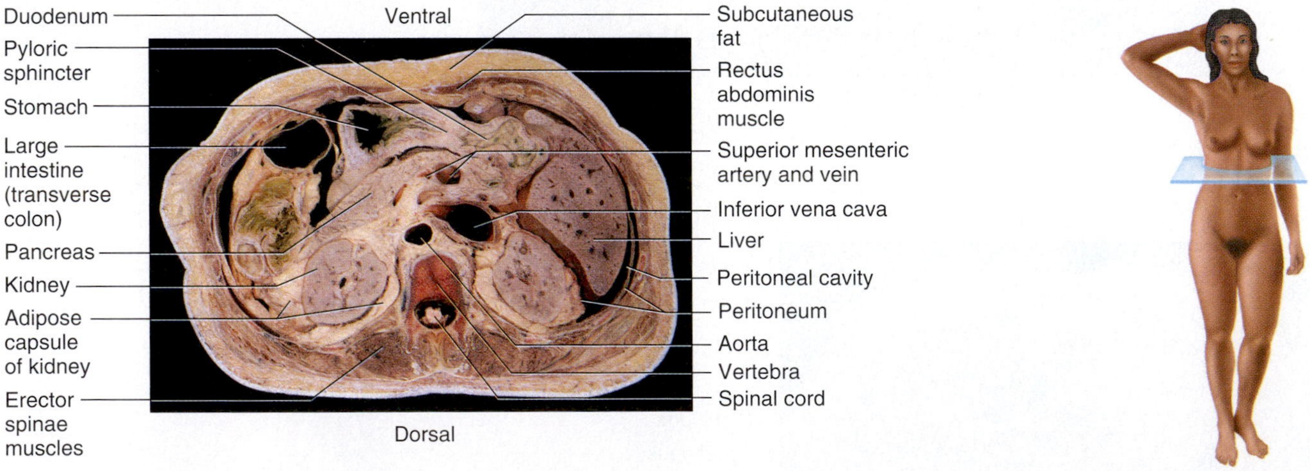
Duodenum
Pyloric
sphincter
Stomach
Large
intestine
(transverse
colon)
Pancreas
Kidney
Adipose
capsule
of kidney
Erector
spinae
muscles
Ventral
Dorsal
Subcutaneous
fat
Rectus
abdominis
muscle
Superior mesenteric
artery and vein
Inferior vena cava
Liver
Peritoneal cavity
Peritoneum
Aorta
Vertebra
Spinal cord

FIGURE A.21
Transverse Section of the Abdominal Cavity. Section taken at the level shown by the inset and oriented the same as the reader's body.

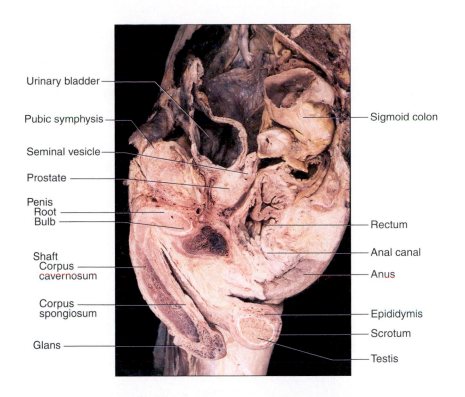

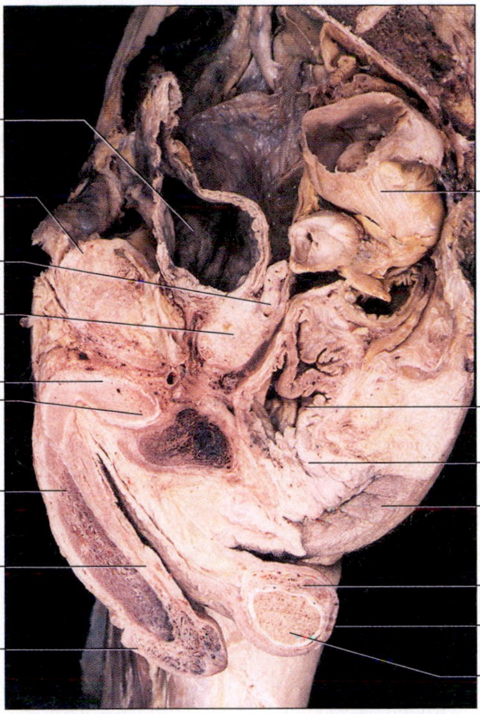
Urinary bladder
Pubic symphysis
Seminal vesicle
Prostate
Penis
Root
Bulb
Shaft
Corpus
cavernosum
Corpus
spongiosum
Glans
Sigmoid colon
Rectum
Anal canal
Anus
Epididymis
Scrotum
Testis

(a)

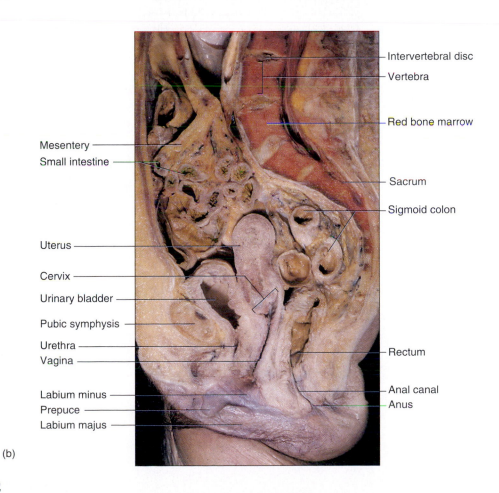

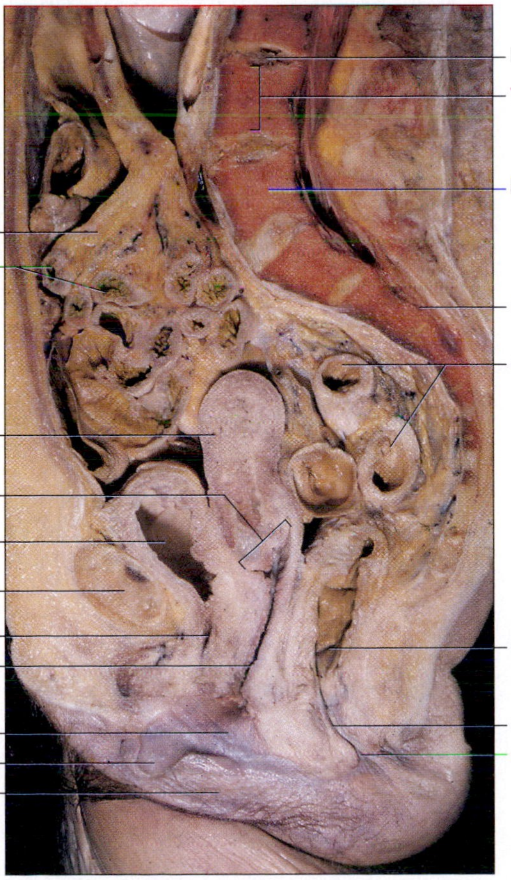
Intervertebral disc
Vertebra
Red bone marrow
Mesentery
Small intestine
Sacrum
Sigmoid colon
Uterus
Cervix
Urinary bladder
Pubic symphysis
Urethra
Vagina
Rectum
Labium minus
Prepuce
Labium majus
Anal canal
Anus

(b)

FIGURE A.22

Sagittal Section of the Pelvic Cavity. (*a*) Male. (*b*) Female. Both viewed from the left.

ATLAS REVIEW

REVIEW OF KEY CONCEPTS

General Anatomical Terminology (p. 24)

1. *Anatomical position* (fig. A.1) provides a standard frame of reference so that directional terminology remains consistent regardless of the orientation of the subject's body relative to the observer.
2. In anatomical position, the forearm is held *supine* (palms forward) rather than *prone* (fig. A.2). The feet are close together and flat on the floor, arms to the sides, and the head and eyes directed forward.
3. Three mutually perpendicular planes through the body are the *sagittal, frontal,* and *transverse planes* (figs. A.3 and A.4). The sagittal plane that divides the body or an organ into equal halves is called the *medial plane.*
4. The positions of structures relative to each other are described by standard directional terms defined in table A.1. Some of these definitions are different for a human than for four-legged animals. Such differences must be kept in mind when doing laboratory animal dissections for comparison to humans.

Body Regions (p. 26)

1. The *axial region* of the body consists of the head, cervical region, and trunk. The trunk is subdivided into the *thoracic* and *abdominal regions,* separated by the diaphragm.
2. The abdominal region can be divided into four quadrants (right and left, upper and lower) or nine smaller regions (the *hypochondriac, lateral,* and *inguinal regions* on each side, and the *epigastric, umbilical,* and *hypogastric regions* medially) (fig. A.6). These divisions are useful for anatomical and clinical descriptions of the locations of organs, pain, or other abnormalities.
3. The *appendicular regions* are the upper and lower limbs. The upper limb is divided into *brachium* (arm proper), *antebrachium* (forearm), *carpus* (wrist), *manus* (hand), and *digits* (fingers); the lower limb is divided into *thigh, crus* (leg proper), *tarsus* (ankle), *pes* (foot), and *digits* (toes).

Body Cavities and Membranes (p. 26)

1. The *dorsal body cavity* consists of the *cranial cavity* containing the brain, and the *vertebral canal* containing the spinal cord (fig. A.7). It is lined by three membranes called *meninges.*
2. The *ventral body cavity* consists of the *thoracic cavity* above the diaphragm and *abdominopelvic cavity* below (fig. A.7).
3. The *mediastinum* is a thick medial wall in the chest containing the heart, major blood vessels, esophagus, trachea, and thymus.
4. The *pleurae* are serous membranes that envelop the lungs (fig. A.8*a*). The *parietal pleura* lies against the rib cage and the *visceral pleura* forms the lung surface. The space between the parietal and visceral layers is called the *pleural cavity* and is moistened and lubricated by *pleural fluid.*
5. The *pericardium* is a serous membrane that envelops the heart (fig. A.8*b*). The *visceral pericardium* forms the heart surface and the *parietal pericardium* is separated from it by a space called the *pericardial cavity,* lubricated by *pericardial fluid.*
6. The *abdominopelvic cavity* is divided into the *abdominal cavity* above the pelvic brim and the *pelvic cavity* below (fig. A.7).
7. The *peritoneum* is a serous membrane that lines the abdominopelvic cavity. The *parietal peritoneum* lies against the body wall and the *visceral peritoneum* folds inward as the *mesentery,* encircles many of the abdominal viscera, and forms their outer layer, the *serosa* (fig. A.9). The visceral peritoneum continues beyond some viscera as a sheet, the *ventral mesentery.* The *greater* and *lesser omentum* are ventral mesenteries attached to the stomach (fig. A.10).
8. The space between the parietal and visceral peritoneum is the *peritoneal cavity* and is lubricated by *peritoneal fluid.*
9. Organs that lie against the dorsal body wall and have peritoneum covering only their ventral surfaces are *retroperitoneal*—for example, the kidneys, ureters, and adrenal glands (fig. A.9). Organs that are suspended in the abdominopelvic cavity and encircled by peritoneum are called *intraperitoneal*—for example, the liver, spleen, stomach, and small intestine.

Organ Systems (p. 30)

1. The body has 11 organ systems (fig. A.11). Some organs play roles in two or more of these systems.
2. The *integumentary, skeletal,* and *muscular systems* provide protection, support, and movement.
3. The *nervous* and *endocrine systems* provide internal communication and integration.
4. The *circulatory* and *lymphatic systems* provide fluid transport.
5. The *respiratory, urinary,* and *digestive systems* provide for the input of gases and nutrients and the output of metabolic wastes.
6. The *reproductive system* produces offspring and thus serves for continuity of the species.
7. The *immune system* is not an organ system but a population of cells that colonize many of the organ systems and provide defense against pathogens.

Testing Your Recall

1. Which of the following is *not* an essential part of anatomical position?
   a. eyes facing forward
   b. feet flat on the floor
   c. forearms supine
   d. mouth closed
   e. arms down to the sides
2. A ring-shaped section of the small intestine would be a _____ section.
   a. posterior
   b. midsagittal
   c. transverse
   d. frontal
   e. medial
3. The tarsal region is _____ to the popliteal region.
   a. medial
   b. superficial
   c. superior
   d. dorsal
   e. distal
4. The greater omentum is _____ to the small intestine.
   a. posterior
   b. parietal
   c. deep
   d. superficial
   e. proximal
5. A _____ line passes through the sternum, umbilicus, and mons pubis.
   a. central
   b. proximal
   c. midclavicular
   d. midsagittal
   e. intertubercular
6. The _____ region is immediately medial to the coxal region.
   a. inguinal
   b. hypochondriac
   c. umbilical
   d. popliteal
   e. antecubital
7. Which of the following regions is *not* part of the upper limb?
   a. plantar
   b. carpal
   c. antecubital
   d. brachial
   e. palmar
8. Which of these organs is intraperitoneal?
   a. urinary bladder
   b. kidneys
   c. heart
   d. small intestine
   e. brain
9. In which area do you think pain from the gallbladder would be felt?
   a. umbilical region
   b. right upper quadrant
   c. hypogastric region
   d. left hypochondriac region
   e. left lower quadrant
10. Which of the following is *not* an organ system?
    a. muscular system
    b. integumentary system
    c. endocrine system
    d. lymphatic system
    e. immune system
11. The forearm is said to be _____ when the palms are facing forward.
12. The more superficial layer of the pleura is called the _____ pleura.
13. The right and left pleural cavities are separated by a thick wall called the _____.
14. The back of the head is called the _____ region, and the back of the neck is the _____ region.
15. The manus is more commonly known as the _____ and the pes is more commonly known as the _____.
16. The dorsal body cavity is lined by membranes called the _____.
17. Abdominal organs that lie against the dorsal abdominal wall and are covered with peritoneum only on the anterior side are said to have a _____ position.
18. The sternal region is _____ to the pectoral region.
19. The pelvic cavity can be described as _____ to the abdominal cavity in position.
20. The anterior pit of the elbow is the _____ region, and the corresponding (but posterior) pit of the knee is the _____ fossa.

Answers in the Appendix

True or False

Determine which five of the following statements are false, and briefly explain why.

1. A single sagittal section of the body can pass through one lung but not through both.
2. It would be possible to see both eyes in one frontal section of the head.
3. The knee is both superior and proximal to the tarsal region.
4. The diaphragm is ventral to the lungs.
5. The esophagus is in the dorsal body cavity.
6. The liver is in the lateral abdominal region.
7. The heart is in the mediastinum.
8. Both kidneys could be shown in a single coronal section of the body.
9. The peritoneum lines the inside of the stomach and intestines.
10. The sigmoid colon is in the lower right quadrant of the abdomen.

Answers in the Appendix

Testing Your Comprehension

1. Identify which anatomical plane—sagittal, frontal, or transverse—is the only one that could *not* show (*a*) both the brain and tongue, (*b*) both eyes, (*c*) both the hypogastric and gluteal regions, (*d*) both kidneys, (*e*) both the sternum and vertebral column, and (*f*) both the heart and uterus.
2. Laypeople often misunderstand anatomical terminology. What do you think people really mean when they say they have "planter's warts"?
3. Name one structure or anatomical feature that could be found in each of the following locations relative to the ribs: medial, lateral, superior, inferior, deep, superficial, posterior, and anterior. Try not to use the same example twice.
4. Based on the illustrations in this atlas, identify an internal organ that is (*a*) in the upper left quadrant and retroperitoneal, (*b*) in the lower right quadrant of the peritoneal cavity, (*c*) in the hypogastric region, (*d*) in the right hypochondriac region, and (*e*) in the pectoral region.
5. Why do you think people with imaginary illnesses came to be called hypochondriacs?

Answers at the Online Learning Center

www.mhhe.com/saladinha1

Visit the Online Learning Center for practice tests, answer keys, and other learning aids for this chapter. Enhance your understanding of human anatomy with our interactive art labeling exercises, supplemental photo atlases, web links, puzzles, flashcards, and much more.

2

CHAPTER TWO

Cytology—The Study of Cells

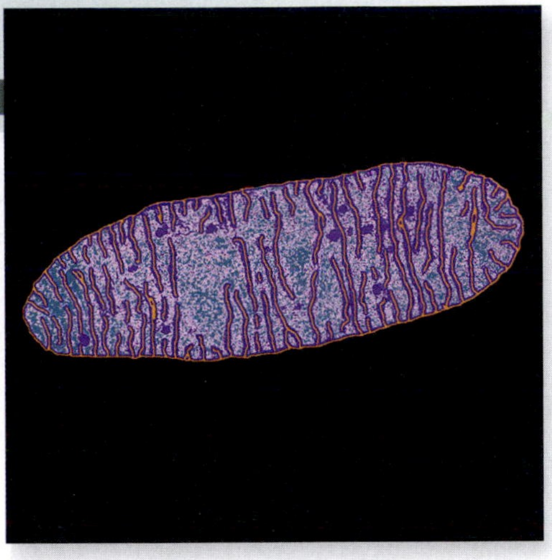

A mitochondrion photographed through a transmission electron microscope (TEM)

CHAPTER OUTLINE

The Study of Cells 48
- Microscopy 48
- Cell Shapes and Sizes 49
- The Components of a Cell 50

The Cell Surface 52
- The Plasma Membrane 52
- Membrane Transport 57
- The Glycocalyx 58
- Microvilli, Cilia, and Flagella 59
- Intercellular Junctions 60

The Cytoplasm 63
- The Cytoskeleton 63
- Organelles 63
- Inclusions 68

The Life Cycle of Cells 68
- The Cell Cycle 68
- Cell Division 69
- Stem Cells 72

Chapter Review 73

INSIGHTS

2.1 Clinical Application: When Desmosomes Fail 63
2.2 Evolutionary Medicine: The Origin of Mitochondria 68
2.3 Clinical Application: Cancer 71

BRUSHING UP

To understand this chapter, you may find it helpful to review the following concepts:

- The hierarchy of human organization (p. 6)
- History of the microscope (p. 10)

The most important revolution in the history of medicine was the realization that all bodily functions result from cellular activity. By extension, nearly every dysfunction of the body is now recognized as stemming from a dysfunction at the cellular level. The majority of new medical research articles published every week are on cellular function, and all drug development is based on an intimate knowledge of how cells work. **Cytology,**[1] the study of cellular structure and function, has thus become indispensable to any true understanding of the structure and function of the human body, the mechanisms of disease, and the rationale of therapy.

This chapter therefore begins our study of anatomy at the cellular level. We will see how continued developments in microscopy have deepened our insight into cell structure, examine the structural components of cells, and briefly survey two aspects of cellular function—transport through the plasma membrane and the cell life cycle. It is the derangement of that life cycle that gives rise to one of the most dreaded of human diseases, cancer.

The Study of Cells

Objectives

When you have completed this section, you should be able to:

- state the modern tenets of the cell theory;
- discuss the way that developments in microscopy have changed our view of cell structure;
- outline the major structural components of a cell;
- identify cell shapes from their descriptive terms; and
- state the size range of human cells and explain why cell size is limited.

The scientific study of cellular structure and function is called cytology. Some historians date the birth of this science to April 15, 1663, the day Robert Hooke named the cell when he observed the little boxes formed by the cell walls of cork. However, this arguably gives too little credit to Leeuwenhoek, who was not only the first to see and publish descriptions of living cells, but also the first to study human cells. Cytology was greatly advanced by refinements in microscope technology in the nineteenth century. By 1900, it was established beyond reasonable doubt that every living organism is made of cells; that cells now arise only through the division of preexisting cells rather than springing spontaneously from nonliving matter; and that all cells have the same basic chemical components, such as carbohydrates, lipids, proteins, and nucleic acids. These and other principles have been codified as the **cell theory** (table 2.1).

Table 2.1

Tenets of the Modern Cell Theory

1. All organisms are composed of cells and cell products.
2. The cell is the simplest structural and functional unit of life. There are no smaller subdivisions of a cell or organism that, in themselves, are alive.
3. An organism's structure and all of its functions are ultimately due to the activities of its cells.
4. Cells now come only from preexisting cells, not from nonliving matter. All life, therefore, traces its ancestry to the same original cells.
5. Because of this common ancestry, the cells of all species have many fundamental similarities in their chemical composition and metabolic mechanisms.

Microscopy

Cytology would not exist without the microscope. Throughout this book, you will find many *photomicrographs*—photographs of tissues and cells taken through the microscope. The microscopes used to produce these photographs fall into three basic categories: the light microscope, transmission electron microscope, and scanning electron microscope.

The **light microscope (LM)** uses visible light to produce its images. It is the least expensive type of microscope, the easiest to use, and the most often used, but it is also the most limited in the amount of useful magnification it can produce. Leeuwenhoek's single-lens microscopes magnified specimens about 200 times, and light microscopes today magnify up to 1,200 times. There are several varieties of light microscopes, including fluorescence microscopes (see fig. 2.15*a*).

Most of the structure we study in this chapter is not visible with the LM, not because the LM cannot magnify cells enough, but because it cannot reveal enough detail. The most important thing about a good microscope is not magnification but **resolution**—the ability to reveal detail. Any image can be photographed and enlarged as much as we wish, but if enlargement fails to reveal any more useful detail, it is *empty magnification.* A big fuzzy image is not nearly as informative as one that is small and sharp. For reasons of physics beyond the scope of this chapter, it is the wavelength of light that places a limit on resolution. Visible light has wavelengths ranging from about 400 to 700 nanometers (nm). At these wavelengths, the LM cannot distinguish between two objects any closer together than 200 nm (0.2 micrometers, or μm).

Resolution improves when objects are viewed with radiation of shorter wavelengths. *Electron microscopes* achieve higher resolution by using short-wavelength (0.005 nm) beams of electrons in place of light. The **transmission electron microscope (TEM),** invented in the mid-twentieth century, is usually used to study specimens that have been sliced ultrathin with diamond knives and stained with heavy metals such as osmium, which absorbs electrons. The TEM resolves details as small as 0.5 nm and attains useful magnifications of biological material up to 600,000 times. This is good enough to see proteins, nucleic acids, and other large molecules. Such fine detail is called cell *ultrastructure.* Even at the same magnifications as the LM, the TEM reveals far more detail (fig. 2.1). It usually produces two-dimensional black-and-white images, but electron photomicrographs are often colorized for instructional purposes.

The **scanning electron microscope (SEM)** uses a specimen coated with vaporized metal (usually gold). The electron beam

[1] *cyto* = cell + *logy* = study of

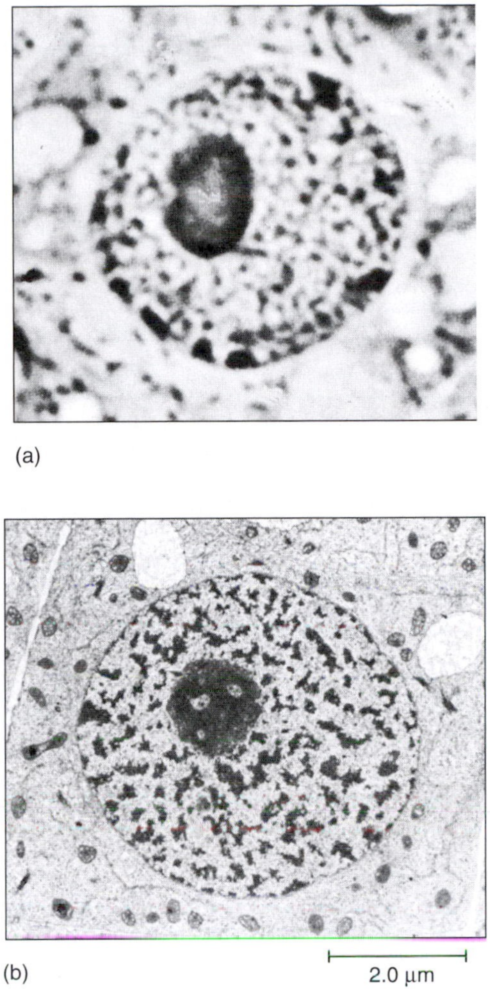
(a)
(b)
2.0 μm

FIGURE 2.1

Magnification Versus Resolution. These cell nuclei were photographed at the same magnification (about × 750) through (*a*) a light microscope (LM) and (*b*) a transmission electron microscope (TEM). Note the finer detail visible with the TEM.

strikes the specimen and discharges secondary electrons from the metal coating. These electrons then produce an image on a fluorescent screen. The SEM yields less resolution than the TEM and is used at lower magnification, but it produces dramatic three-dimensional images that are sometimes more informative than TEM images, and it does not require that the specimen be cut into thin slices. The SEM can only view the surfaces of specimens; it does not see through an object like the LM or TEM. Cell interiors can be viewed, however, by a *freeze-fracture* method in which a cell is frozen, cracked open, and coated with gold vapor, then viewed by either TEM or SEM. Figure 2.2 compares red blood cells photographed with the LM, TEM, and SEM.

THINK ABOUT IT!

Beyond figure 2.2, list all of the photomicrographs in this chapter that you believe were made with the LM, with the TEM, and with the SEM.

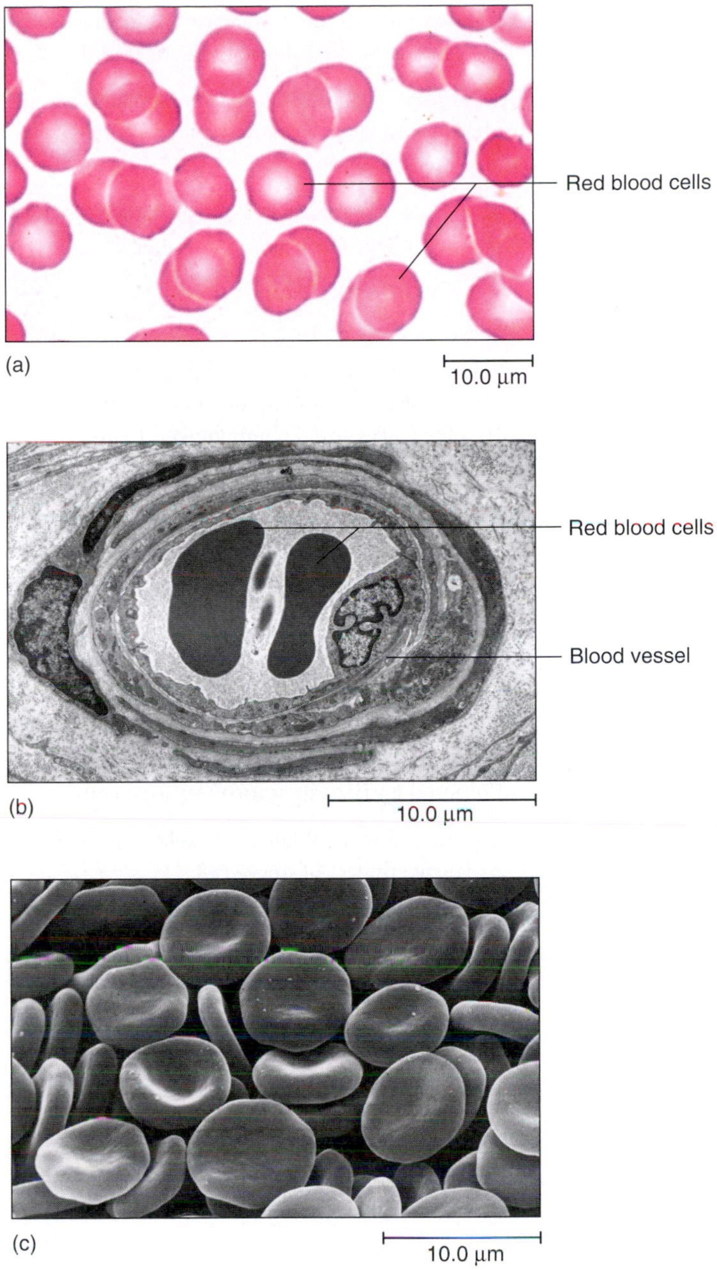
Red blood cells
(a)
10.0 μm
Red blood cells
Blood vessel
(b)
10.0 μm
(c)
10.0 μm

FIGURE 2.2

Images Produced by Three Kinds of Microscopes. Red blood cells as they appear with (*a*) the light microscope (LM), (*b*) the transmission electron microscope (TEM), and (*c*) the scanning electron microscope (SEM).

Cell Shapes and Sizes

We will shortly examine the structure of a generic cell, but the generalizations to be drawn should not blind you to the diversity of cellular form and function in humans. There are about 200 kinds of cells in the human body, with a variety of shapes, sizes, and functions.

Descriptions of organ and tissue structure often refer to the shapes of the constituent cells by the following terms (fig. 2.3):

- *Squamous*[2] (SQUAY-mus)—a thin, flat, scaly shape; squamous cells line the esophagus and form the surface layer (epidermis) of the skin.
- *Cuboidal*[3] (cue-BOY-dul)—squarish and about equal in height and width; liver cells are a good example.
- *Columnar*—distinctly taller than wide, such as the inner lining cells of the stomach and intestines.
- *Polygonal*[4] (pa-LIG-ah-nul)—having irregularly angular shapes with four, five, or more sides. Squamous, cuboidal, and columnar cells often look polygonal when viewed from above rather than from the side.
- *Stellate*[5]—having multiple pointed processes projecting from the body of a cell, giving it a somewhat starlike shape. The cell bodies of nerve cells are often stellate.
- *Spheroid* to *ovoid*—round to oval, as in egg cells and white blood cells.
- *Discoid*—disc-shaped, as in red blood cells.
- *Fusiform*[6] (FEW-zih-form)—spindle-shaped; elongated, with a thick middle and tapered ends, as in smooth muscle cells.
- *Fibrous*—long, slender, and threadlike, as in skeletal muscle cells and the axons (nerve fibers) of nerve cells.

In some cells, it is important to distinguish one surface from another, because the surfaces may differ in function and membrane composition. This is especially true in *epithelia,* cell layers that cover organ surfaces. An epithelial cell rests on a lower **basal surface** attached to an extracellular *basement membrane* (see chapter 3). The upper surface of the cell is called the **apical surface.** Its sides are **lateral surfaces.**

The most useful unit of measurement for designating cell sizes is the **micrometer (μm)**—one-millionth (10^{-6}) of a meter, one-thousandth (10^{-3}) of a millimeter. The smallest objects most people can see with the naked eye are about 100 μm, which is about one-quarter the size of the period at the end of this sentence. A few human cells fall within this range, such as the egg cell and some fat cells, but most human cells are about 10 to 15 μm wide. The longest human cells are nerve cells (sometimes over a meter long) and muscle cells (up to 30 cm long), but both are too slender to be seen with the naked eye.

There are several factors that limit the size of cells. If a cell swells to excessive size, it ruptures like an overfilled water balloon. Also, if a cell were too large, molecules could not diffuse from place to place fast enough to support its metabolism. The time required for diffusion is proportional to the square of distance, so if cell diameter doubled, the travel time for molecules within the cell would increase fourfold. A nucleus can therefore effectively control only a limited volume of cytoplasm.

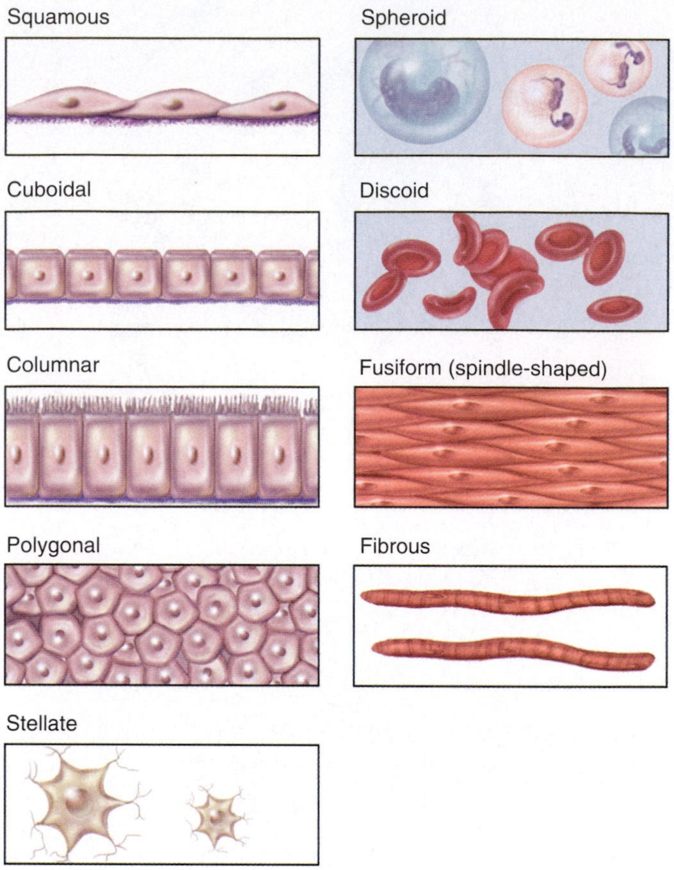
Squamous
Spheroid
Cuboidal
Discoid
Columnar
Fusiform (spindle-shaped)
Polygonal
Fibrous
Stellate

FIGURE 2.3
Common Cell Shapes.

In addition, cell size is limited by the relationship between its volume and surface area. The surface area of a cell is proportional to the square of its diameter, while volume is proportional to the cube of diameter. Thus, for a given increase in diameter, cell volume increases much faster than surface area. Picture a cuboidal cell 10 μm on each side (fig. 2.4). It would have a surface area of 600 μm^2 (10 μm × 10 μm × 6 sides) and a volume of 1,000 μm^3 (10 × 10 × 10 μm). Now, suppose it grew by another 10 μm on each side. Its new surface area would be 20 μm × 20 μm × 6 = 2,400 μm^2, and its volume would be 20 × 20 × 20 μm = 8,000 μm^3. The 20 μm cell has eight times as much cytoplasm needing nourishment and waste removal, but only four times as much membrane surface through which wastes and nutrients can be exchanged. In short, a cell that is too big cannot support itself.

The Components of a Cell

Before electron microscopy, little was known about structural cytology except that cells were enclosed in a membrane and contained a nucleus. The material between the nucleus and surface membrane was thought to be little more than a gelatinous mixture of chemi-

[2]*squam* = scale + *ous* = characterized by
[3]*cub* = cube + *oidal* = like, resembling
[4]*poly* = many + *gon* = angles
[5]*stell* = star + *ate* = characterized by
[6]*fusi* = spindle + *form* = shape

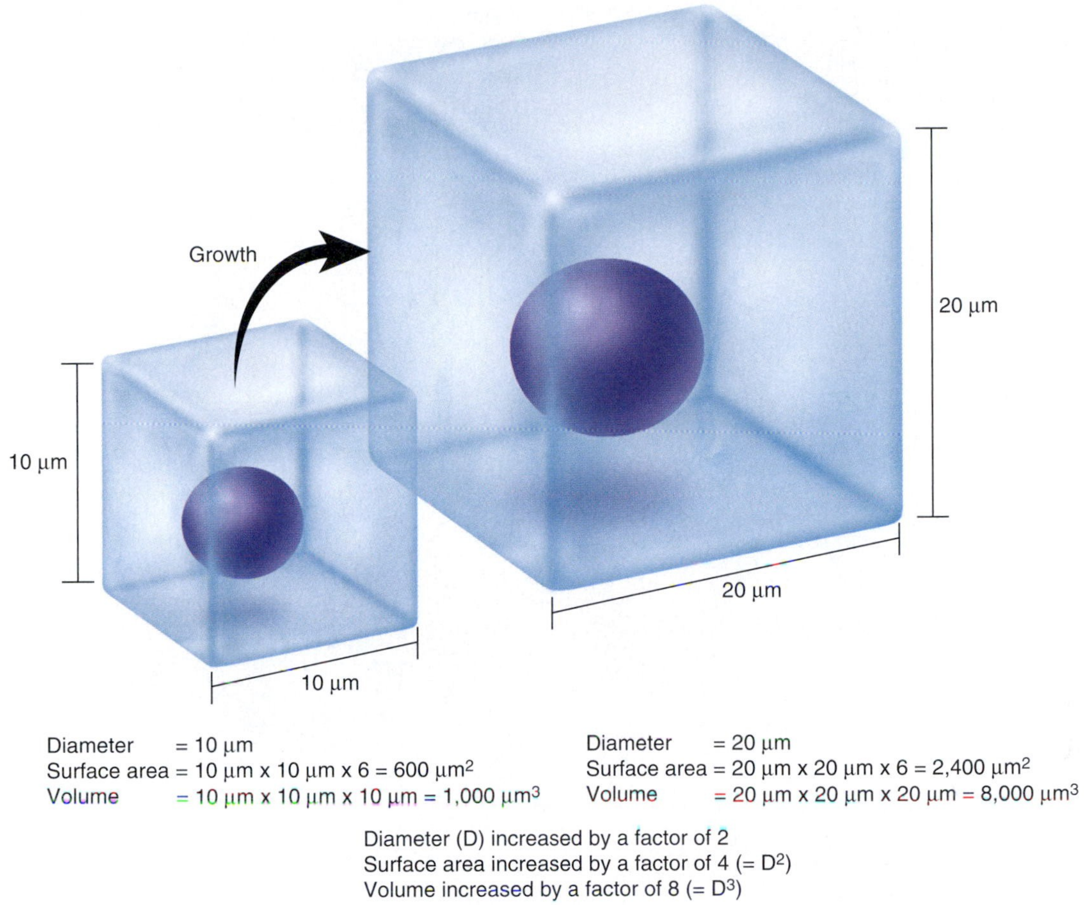
Growth
20 µm
10 µm
20 µm
10 µm
Diameter = 10 µm
Surface area = 10 µm x 10 µm x 6 = 600 µm2
Volume = 10 µm x 10 µm x 10 µm = 1,000 µm3
Diameter = 20 µm
Surface area = 20 µm x 20 µm x 6 = 2,400 µm2
Volume = 20 µm x 20 µm x 20 µm = 8,000 µm3
Diameter (D) increased by a factor of 2
Surface area increased by a factor of 4 (= D2)
Volume increased by a factor of 8 (= D3)

FIGURE 2.4

The Relationship Between Cell Surface Area and Volume. As a cell doubles in width, its volume increases eightfold, but its surface area increases only fourfold. A cell that is too large may have too little plasma membrane to support the metabolic needs of its volume of cytoplasm.

cals and vaguely defined particles. But electron microscopy has revealed that the cytoplasm is crowded with a maze of passages, compartments, and fibers (fig. 2.5). Earlier microscopists were little aware of this detail simply because most of these structures are too small to be visible with the LM (table 2.2).

We now view cell structure as having the following major components:

- Plasma membrane
- Cytoplasm
  - Cytoskeleton
  - Organelles
  - Inclusions
  - Cytosol
- Nucleoplasm

The **plasma membrane (cell membrane)** forms the surface boundary of the cell. The material between the plasma membrane and the nucleus is the **cytoplasm**[7] and the material within the nucleus is the **nucleoplasm.** The cytoplasm contains the *cytoskeleton,* a supportive framework of protein filaments and tubules; an abundance of *organelles,* diverse structures that perform various metabolic tasks for the cell; and *inclusions,* which are not metabolically active parts of the cell but include stored cell products such as lipids and pigments, and foreign bodies such as dust and bacteria. The cytoskeleton, organelles, and inclusions are embedded in a clear gel called the **cytosol.**

The fluid within the cell (the cytosol) is also called the **intracellular fluid (ICF).** All body fluids not contained in the cells are collectively called the **extracellular fluid (ECF).** The extracellular fluid located amid the cells is also called **tissue (interstitial) fluid.** Some other extracellular fluids include blood plasma, lymph, and cerebrospinal fluid.

Before You Go On

Answer the following questions to test your understanding of the preceding section:

1. What are the tenets of the cell theory?
2. What is the main advantage of an electron microscope over a light microscope?
3. Define *cytoplasm, cytosol,* and *organelle.*
4. Explain why cells cannot grow to unlimited size.

[7]*cyto* = cell + *plasm* = formed, molded

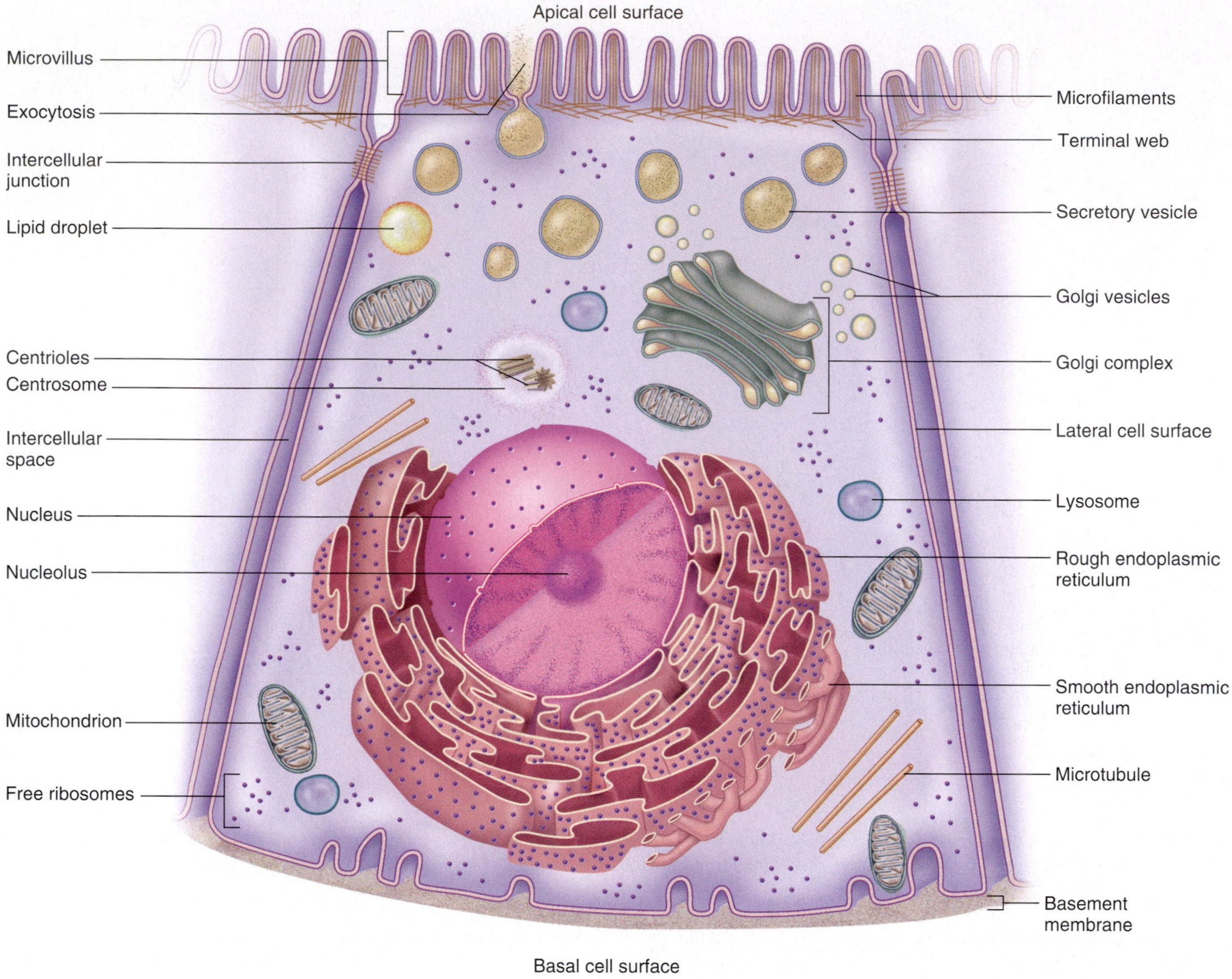
Apical cell surface
Microvillus
Exocytosis
Intercellular junction
Lipid droplet
Centrioles
Centrosome
Intercellular space
Nucleus
Nucleolus
Mitochondrion
Free ribosomes
Microfilaments
Terminal web
Secretory vesicle
Golgi vesicles
Golgi complex
Lateral cell surface
Lysosome
Rough endoplasmic reticulum
Smooth endoplasmic reticulum
Microtubule
Basement membrane
Basal cell surface

FIGURE 2.5
Structure of a Generalized Cell. The cytoplasm is usually more crowded with organelles than is shown here. The organelles are not all drawn to the same scale.

The Cell Surface

Objectives

When you have completed this section, you should be able to

- describe the structure of a plasma membrane;
- explain the functions of the lipid, protein, and carbohydrate components of the plasma membrane;
- describe the processes for moving material into and out of a cell; and
- describe the structure and functions of microvilli, cilia, flagella, and intercellular junctions.

A great deal of human physiology takes place at the cell surface—for example, the binding of signaling molecules such as hormones, the stimulation of cellular activity, the attachment of cells to each other, and the transport of materials into and out of cells. This, then, is where we begin our study of cellular structure and function. In this section, we examine the plasma membrane that defines the outer boundary of a cell; a carbohydrate coating called the *glycocalyx* on the membrane surface; hairlike extensions of the cell surface; the attachment of cells to each other and to extracellular materials; and the processes of membrane transport. We will examine the interior of the cell only after we have explored its boundary.

The Plasma Membrane

The plasma (cell) membrane defines the boundary of a cell, governs its interactions with other cells, maintains differences in chemical composition between the ECF and ICF, and controls the passage of materials into and out of the cell. With the TEM, it looks like two parallel dark lines (fig. 2.6*a*). The side that faces the cytoplasm is

TABLE 2.2

Sizes of Some Biological Structures in Relation to the Resolving Power of the Human Eye, Light Microscope (LM), and Transmission Electron Microscope (TEM)

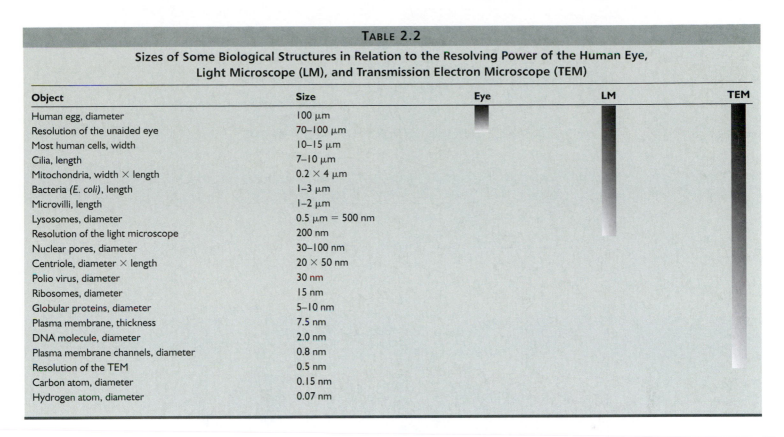

Object	Size	Eye	LM	TEM
Human egg, diameter	100 µm			
Resolution of the unaided eye	70–100 µm			
Most human cells, width	10–15 µm			
Cilia, length	7–10 µm			
Mitochondria, width × length	0.2 × 4 µm			
Bacteria (*E. coli*), length	1–3 µm			
Microvilli, length	1–2 µm			
Lysosomes, diameter	0.5 µm = 500 nm			
Resolution of the light microscope	200 nm			
Nuclear pores, diameter	30–100 nm			
Centriole, diameter × length	20 × 50 nm			
Polio virus, diameter	30 nm			
Ribosomes, diameter	15 nm			
Globular proteins, diameter	5–10 nm			
Plasma membrane, thickness	7.5 nm			
DNA molecule, diameter	2.0 nm			
Plasma membrane channels, diameter	0.8 nm			
Resolution of the TEM	0.5 nm			
Carbon atom, diameter	0.15 nm			
Hydrogen atom, diameter	0.07 nm			

called the *intracellular face* of the membrane and the side that faces outward is the *extracellular face.* The total thickness of the plasma membrane is about 7.5 nm. The term *unit membrane* refers to this as well as similar membranes that enclose most organelles, but the term *plasma membrane* refers exclusively to the unit membrane that forms the cell surface.

The **fluid-mosaic model** of the unit membrane depicts it as an oily, two-layered lipid film with proteins embedded in it (fig. 2.6*b*). Some of the proteins are anchored in place and some drift around. Some penetrate from one side of the lipid bilayer to the other, and some adhere only to the intracellular or extracellular face. By weight, the membrane is about half lipid and half protein. Since the lipid molecules are smaller and lighter, however, they constitute about 90% to 99% of the molecules in the membrane.

MEMBRANE LIPIDS

About 75% of the membrane lipid molecules are phospholipids. A **phospholipid** (fig. 2.7) consists of a three-carbon backbone called glycerol with fatty acid tails attached to two of the carbons and a phosphate-containing head attached to the third. The two fatty acid tails are *hydrophobic* (water-repellent), while the head is *hydrophilic* (attracted to water). Thus the molecule as a whole is *amphiphilic (amphipathic)*—partially attracted to water and partially repelled by it. The heads of the phospholipids face the ECF and ICF, while the tails orient away from the water, toward the middle of the membrane. The phospholipids drift laterally from place to place, spin on their axes, and flex their tails. These movements keep the membrane fluid.

Fat-soluble substances such as steroid hormones, oxygen, and carbon dioxide easily pass in and out of the cell through the phospholipid bilayer. However, the phospholipids severely restrict the movement of water-soluble substances such as glucose, salts, and water itself, which must pass through the membrane proteins discussed shortly.

About 20% of the lipid molecules are **cholesterol.** Cholesterol has an important impact on the fluidity of the membrane. If there is too little cholesterol, plasma membranes become excessively fragile. People with abnormally low cholesterol levels suffer an increased incidence of strokes because of the rupturing of fragile blood vessels. Excessively high concentrations of cholesterol in the membrane can inhibit the action of enzymes and receptor proteins in the membrane.

The remaining 5% of the membrane lipids are **glycolipids**—phospholipids with short carbohydrate chains bound to them. Glycolipids occur only on the extracellular face of the membrane. They contribute to the *glycocalyx,* a sugary cell coating discussed later.

An important quality of the plasma membrane is its capacity for self-repair. When a physiologist inserts a probe into a cell, it does not pop the cell like a balloon. The probe slips through the oily film and the membrane seals itself around it. When cells take in matter by endocytosis (described later), they pinch off bits of their own membrane, which form bubblelike vesicles in the cytoplasm. As these vesicles pull away from the membrane, they do not leave gaping holes; the lipids immediately flow together to seal the break.

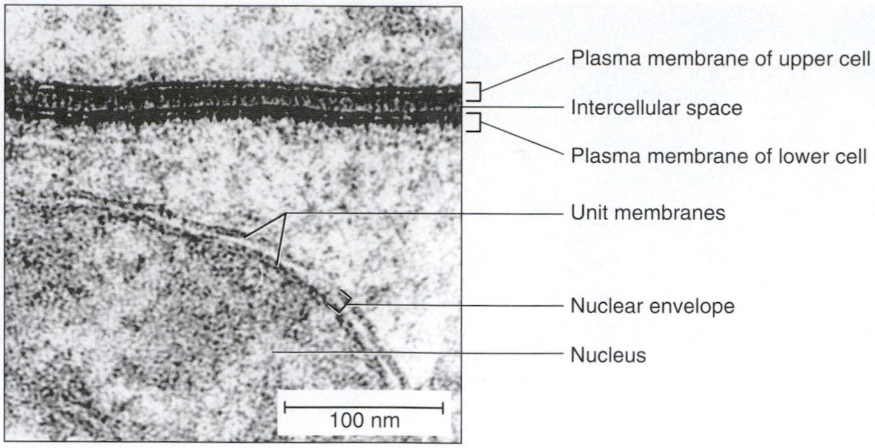

(a)

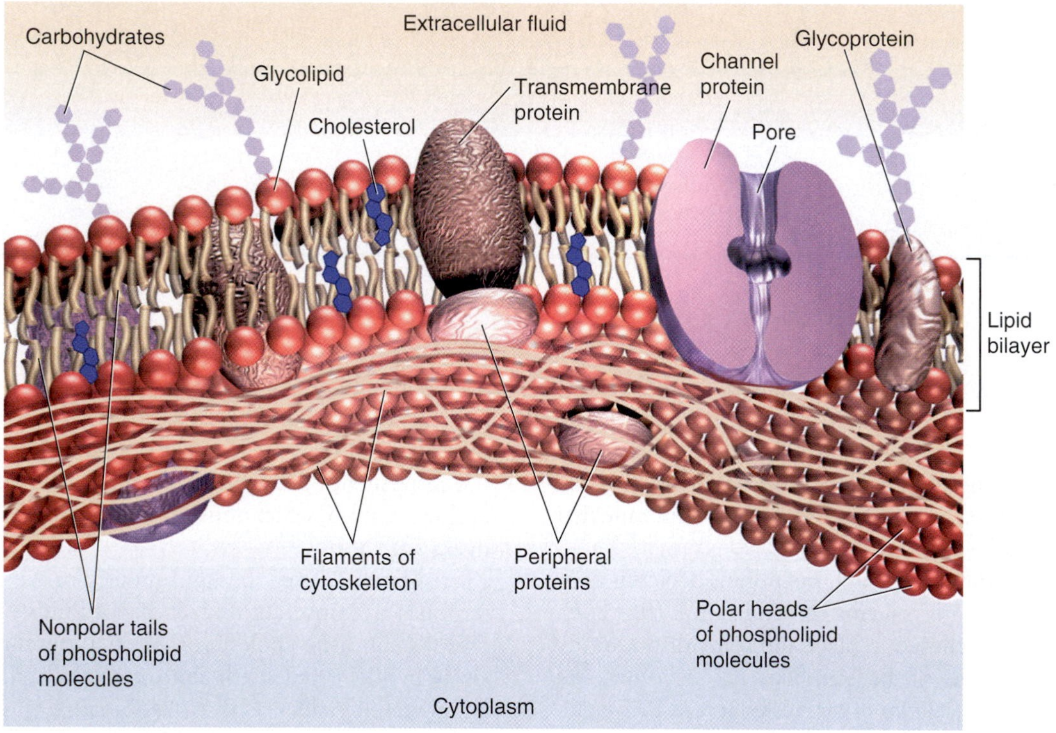

(b)

FIGURE 2.6

The Plasma Membrane. (*a*) Plasma membranes of two adjacent cells (TEM). Note also that the nuclear envelope is composed of two *unit membranes,* each of which is similar to a plasma membrane. (*b*) Molecular structure of the plasma membrane.

MEMBRANE PROTEINS

Proteins that pass all the way through a plasma membrane are called **integral (transmembrane) proteins.** They have hydrophilic regions in contact with the cytoplasm and extracellular fluid, and hydrophobic regions that pass back and forth through the lipid of the membrane (fig. 2.8). Most of the integral proteins are **glycoproteins,** which, like glycolipids, have carbohydrate chains linked to them and help form the glycocalyx. **Peripheral proteins** are those that do not protrude into the phospholipid layer but adhere to either face of the membrane, usually the intracellular face. Some integral proteins drift about freely in the plasma membrane, while others are anchored to the cytoskeleton and thus held in one place. Most peripheral proteins are anchored to the cytoskeleton and associated with integral proteins.

The functions of membrane proteins are very diverse and are among the most interesting aspects of cell physiology. These proteins serve in the following roles:

- **Receptors** (fig. 2.9*a*). Cells communicate with each other by chemical signals such as hormones and neurotransmitters. Some of these messengers (epinephrine, for example) cannot enter their target cells but can only "knock on the door" with their message. They bind to a membrane protein called a

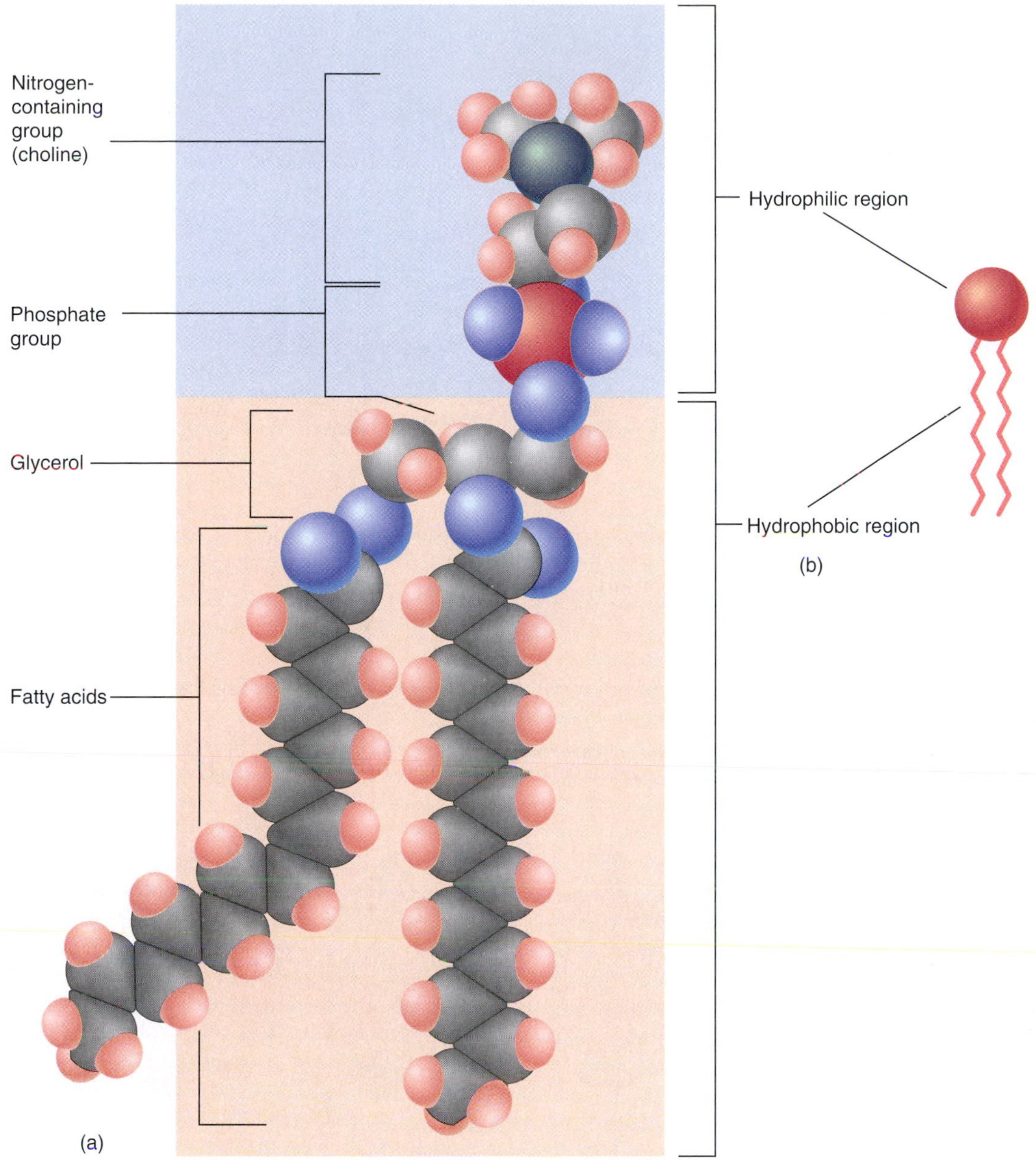

Nitrogen-containing group (choline)
Phosphate group
Glycerol
Fatty acids
(a)
Hydrophilic region
Hydrophobic region
(b)

FIGURE 2.7

Phospholipid Structure and Symbol. (*a*) Space-filling molecular model of a phospholipid showing its four major structural components. Carbon atoms are represented in *gray,* nitrogen in *black,* hydrogen in *pink,* oxygen in *blue,* and phosphorus in *red.* (*b*) Symbolic representation of the hydrophobic and hydrophilic parts of the phospholipid molecule commonly used in diagrams of plasma membranes.

receptor, and the receptor triggers physiological changes inside the cell.

- **Enzymes** (fig. 2.9*b*). Some membrane proteins are enzymes that break down chemical messengers after the message has been received. Enzymes in the plasma membranes of intestinal cells carry out the final stages of starch and protein digestion before the cell absorbs the digested nutrients.
- **Channel proteins** (fig. 2.9*c*). Some membrane proteins have tunnels through them that allow water and hydrophilic solutes to enter or leave a cell. These are called **channel proteins.** Some channels are always open, while others, called **gates** (fig. 2.9*d*), open or close when they are stimulated and thus allow things to enter or leave the cell only at appropriate times. Membrane gates are responsible for firing of the heart's pacemaker, muscle contraction, and most of our sensory processes, among other things.
- **Cell-identity markers** (fig. 2.9*e*). The glycoproteins and glycolipids of the membrane are like genetic identification tags, unique to an individual (or to identical twins). They enable the body to distinguish what belongs to it from what does not—especially from foreign invaders such as bacteria and parasites.

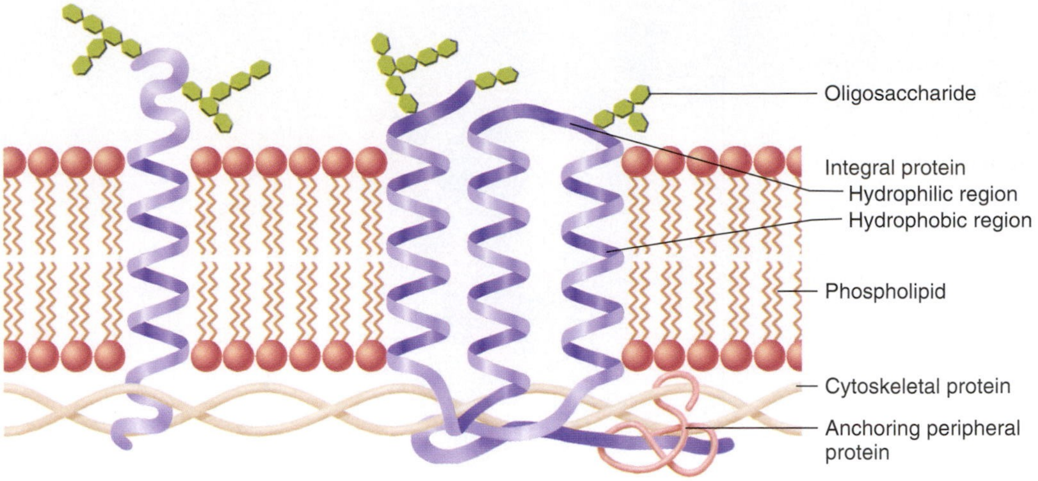

FIGURE 2.8

Transmembrane Proteins. A transmembrane protein has hydrophobic regions embedded in the phospholipid bilayer and hydrophilic regions projecting into the extracellular and intracellular fluids. The protein may cross the membrane once *(left)* or multiple times *(right)*. The intracellular "domain" of the protein is often anchored to the cytoskeleton by peripheral proteins.

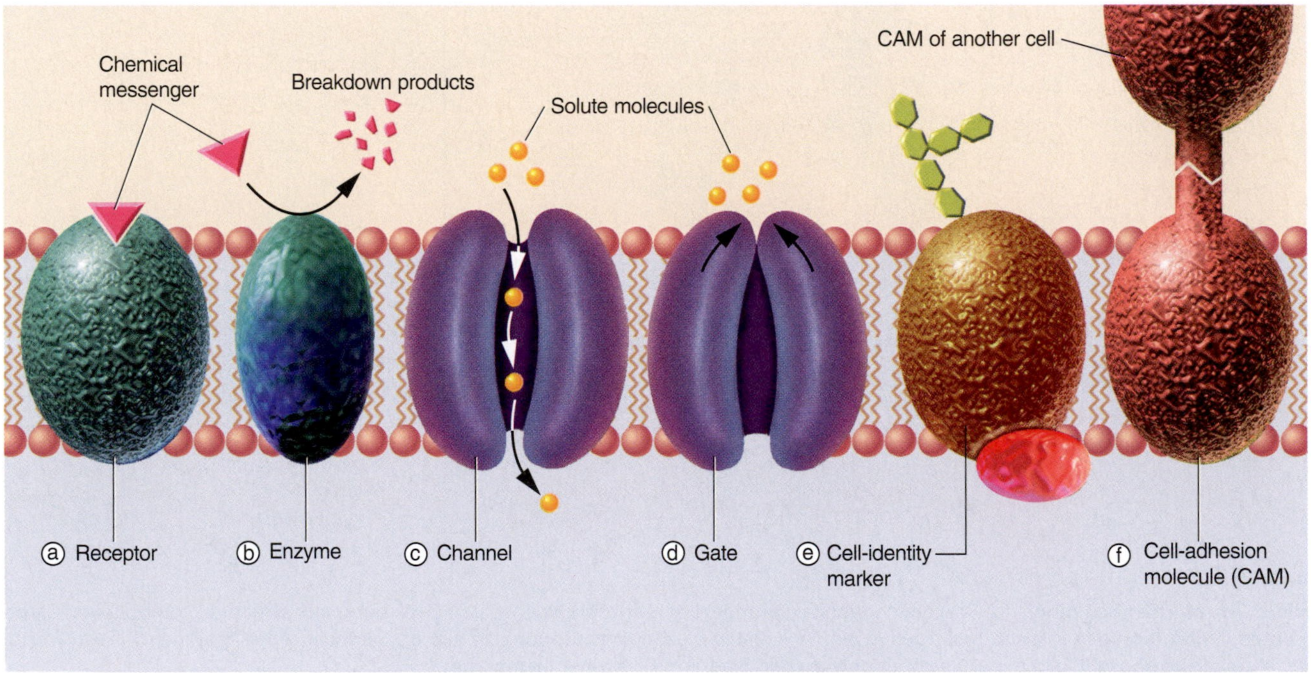

FIGURE 2.9

Some Functions of Plasma Membrane Proteins. (*a*) A receptor that binds a chemical messenger such as a hormone sent by another cell. (*b*) An enzyme that breaks down the chemical messenger to stop a signal that has served its purpose. (*c*) A channel protein that is constantly open and allows solutes to pass through the membrane. (*d*) A gated channel that opens or closes to allow solutes through at certain times (shown in the closed state). (*e*) A glycoprotein serving as a cell-identity marker. (*f*) A cell-adhesion molecule (CAM) that binds one cell to another.

- **Cell-adhesion molecules** (fig. 2.9*f*). Cells adhere to each other and to extracellular material through membrane proteins called **cell-adhesion molecules (CAMs).** With few exceptions (such as blood cells and metastasizing cancer cells), cells do not grow or survive normally unless they are mechanically linked to the extracellular material. Special events such as sperm-egg binding and the binding of an immune cell to a cancer cell also require CAMs.
- **Carriers** (fig. 2.10). Some membrane proteins, called **carriers,** don't merely open to allow substances through—they actively bind to a substance on one side of the membrane and release it on the other side. Carriers are responsible for transporting glucose, amino acids, sodium, potassium, calcium, and many other substances into and out of cells.

Membrane Transport

One of the most important functions of the plasma membrane is to control the passage of materials into and out of the cell. Figure 2.10 illustrates three methods of movement through plasma membranes, as well as filtration, an important mode of transport across the walls of blood capillaries.

FILTRATION

Filtration (fig. 2.10*a*) is a process in which a physical pressure forces material through a membrane, like the weight of water forcing it through the paper filter in a drip coffeemaker. In the body, the prime example of filtration is blood pressure forcing fluid to seep through the walls of the blood capillaries into the tissue fluid. This is how water, salts, organic nutrients, and other solutes pass from the bloodstream to the tissue fluid, where they can get to the cells surrounding a blood vessel, and it is how the kidneys filter wastes from the blood. The capillary wall holds back large particles such as blood cells and proteins.

SIMPLE DIFFUSION

Simple diffusion (fig. 2.10*b*) is the net movement of particles from a place of high concentration to a place of low concentration—in other words, *down a concentration gradient.* Diffusion is how oxygen and steroid hormones enter cells and potassium ions leave, for example. The cell does not have to expend any energy to achieve this; all molecules are in spontaneous random motion, and this alone provides the energy for their diffusion through space. Molecules diffuse through air, liquids, and solids; they can penetrate both living membranes (the plasma membrane) and nonliving ones (such as dialysis tubing and cellophane) if the membrane has large enough gaps or pores. We say that the plasma membrane is *selectively permeable* because it lets some particles through but holds back larger ones.

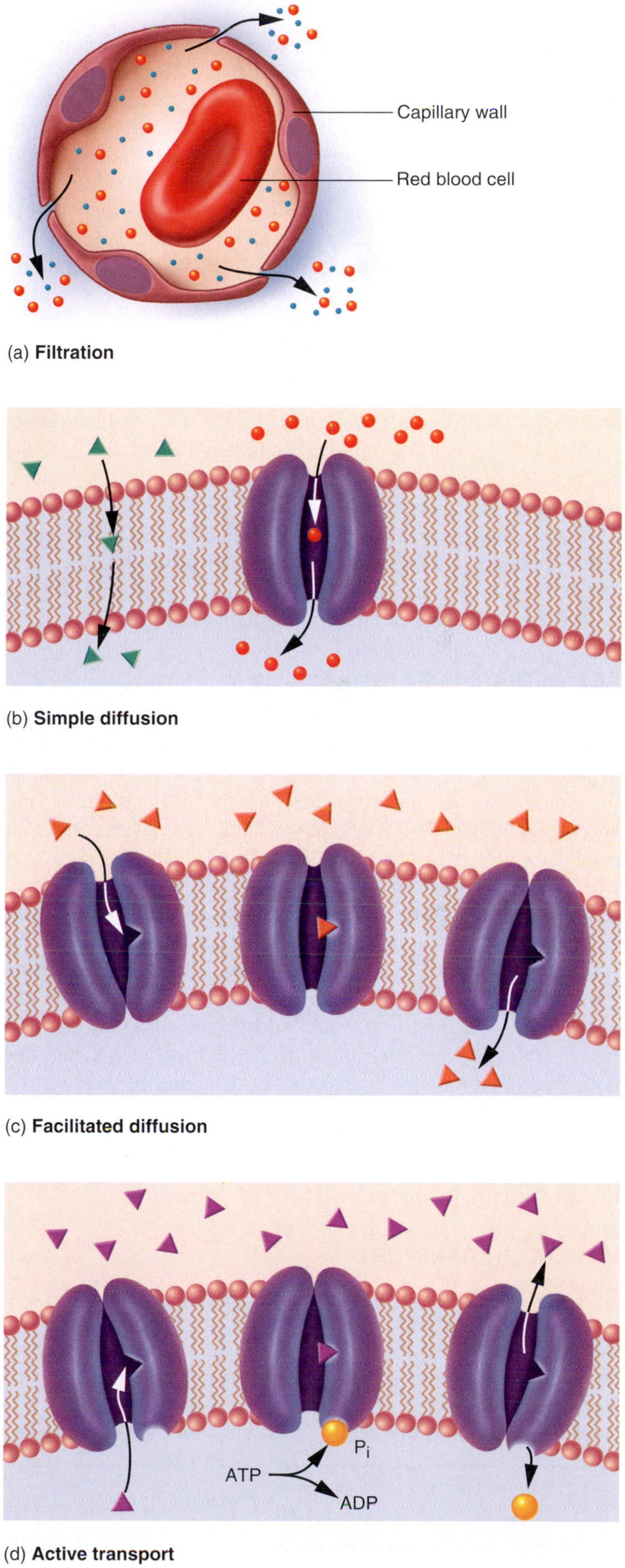

FIGURE 2.10

Some Modes of Membrane Transport. (*a*) Filtration. The blood pressure within a capillary forces water *(blue)* and small dissolved particles such as salts *(red)* through small clefts between the cells that line the capillary. These clefts hold back larger particles such as the red blood cell shown. (*b*) Simple diffusion. Lipid-soluble molecules *(triangles)* such as oxygen, carbon dioxide, and alcohol diffuse through the phospholipid regions of the membrane. Water-soluble particles *(circles)* such as salts and glucose diffuse through channel proteins. All diffusing substances pass from the side of the membrane where they are more concentrated to the side where they are less so (that is, down their concentration gradients). (*c*) Facilitated diffusion. Solutes bind to a receptor site *(shown as a notch)* on a transport protein. The protein then changes shape and releases the solute to the other side of the membrane. Solute movement is again down the concentration gradient. (*d*) Active transport. A solute particle binds to a receptor site on a transport protein; ATP breaks down into ADP and an inorganic phosphate group (P_i); P_i binds to a site of its own on the transport protein; the protein changes shape, releases the solute to the other side of the membrane, and releases the P_i. The solute moves up its concentration gradient.

OSMOSIS

Osmosis[8] (oz-MO-sis) is a special case of simple diffusion—the diffusion of water through a selectively permeable membrane from the side where the water is more concentrated to the side where it is less so. It is important to note that water molecules are *more* concentrated where dissolved matter is *less* so, because dissolved matter occupies less space there. Thus, if the fluids on two sides of a cell membrane differ in the concentration of dissolved matter (and these solutes cannot penetrate the membrane), water tends to pass by osmosis from the more dilute side to the less dilute side. Blood capillaries absorb fluid from the tissues by osmosis.

FACILITATED DIFFUSION

The next two processes, facilitated diffusion and active transport, are called *carrier-mediated transport* because they employ carrier proteins in the plasma membrane. **Facilitated**[9] **diffusion** (fig. 2.10*c*) can be defined as the movement of a solute through a unit membrane, down its concentration gradient, with the aid of a carrier. The carrier binds to a particle on one side of a membrane, where the solute is more concentrated, and releases it on the other side, where it is less concentrated. One use of facilitated diffusion is to absorb the sugars and amino acids from digested food.

ACTIVE TRANSPORT

Active transport (fig. 2.10*d*) is the carrier-mediated transport of a solute through a unit membrane *up its concentration gradient,* with the expenditure of *adenosine triphosphate (ATP).* ATP is essential to this process because moving particles up a gradient requires an energy input, like getting a wagon to roll uphill. If a cell dies and stops producing ATP, active transport ceases immediately. One use of active transport is to pump calcium out of cells. Calcium is more concentrated in the ECF than in the ICF, so this is an "uphill" movement.

An especially well-known active transport process is the **sodium-potassium (Na^+-K^+) pump,** which binds three sodium ions from the ICF and ejects them from the cell, then binds two potassium ions from the ECF and releases these into the cell. The Na^+-K^+ pump plays roles in controlling cell volume; generating body heat; maintaining the electrical excitability of your nerves, muscles, and heart; and providing energy for other transport pumps to draw upon in moving such solutes as glucose through the plasma membrane. About half of the calories that you "burn" every day are used just to operate your Na^+-K^+ pumps.

VESICULAR TRANSPORT

All of the processes discussed up to this point move molecules or ions individually through the plasma membrane. In **vesicular transport,** however, cells move much larger particles or droplets of fluid through the membrane in bubblelike *vesicles.* Vesicular processes that bring matter into a cell are called **endocytosis**[10] (EN-doe-sy-TOE-sis) and those that release material from a cell are called **exocytosis**[11] (EC-so-sy-TOE-sis). Like active transport, all forms of vesicular transport require ATP. Figure 2.11 illustrates the following modes of vesicular transport.

There are three forms of endocytosis: *phagocytosis, pinocytosis,* and *receptor-mediated endocytosis.* In **phagocytosis**[12] (FAG-oh-sy-TOE-sis), or "cell eating," a cell reaches out with **pseudopods** (footlike extensions) and surrounds a particle such as a bacterium or a bit of cell debris and engulfs it, taking it into a cytoplasmic vesicle called a *phagosome* to be digested (fig. 2.11*a*). Phagocytosis is carried out especially by white blood cells and *macrophages,* which are described in chapter 3. Some macrophages consume as much as 25% of their own volume in material per hour, thus playing a vital role in cleaning up the tissues.

Pinocytosis[13] (PIN-oh-sy-TOE-sis), or "cell drinking," occurs in all human cells. In this process, dimples form in the plasma membrane and progressively cave in until they pinch off as *pinocytotic vesicles* containing droplets of ECF (fig. 2.11*b*). Kidney tubule cells use this method to reclaim the small amount of protein that filters out of the blood, thus preventing the protein from being lost in the urine.

Receptor-mediated endocytosis (fig. 2.11*c*) is more selective. It enables a cell to take in specific molecules from the ECF with a minimum of unnecessary fluid. Molecules in the ECF bind to specific receptor proteins on the plasma membrane. The receptors then cluster together and the membrane sinks in at this point, creating a pit. The pit soon pinches off to form a vesicle in the cytoplasm. Cells use receptor-mediated endocytosis to absorb cholesterol and insulin from the blood. Hepatitis, polio, and AIDS viruses "trick" our cells into admitting them by receptor-mediated endocytosis.

Exocytosis (fig. 2.11*d*) is the process of discharging material from a cell. It is used, for example, by digestive glands to secrete enzymes, by breast cells to secrete milk, and by sperm cells to release enzymes for penetrating an egg. It resembles endocytosis in reverse. A *secretory vesicle* in the cell migrates to the surface and fuses with the plasma membrane. A pore opens up that releases the products from the cell, and the empty vesicle usually becomes part of the plasma membrane. In addition to releasing cell products, exocytosis is the cell's way of replacing the bits of membrane removed by endocytosis.

The Glycocalyx

The carbohydrate components of the glycoproteins and glycolipids of the plasma membrane form a fuzzy, sugary coating called the **glycocalyx**[14] (GLY-co-CAY-licks) on every cell surface (fig. 2.12). The glycocalyx has multiple functions. It cushions the plasma membrane and protects it from physical and chemical injury. It functions in cell identity and thus in the body's ability to distinguish its own healthy cells from diseased cells, invading organisms,

[8]*osm* = push, thrust + *osis* = condition, process
[9]*facil* = easy
[10]*endo* = into + *cyt* = cell + *osis* = process
[11]*exo* = out of + *cyt* = cell + *osis* = process
[12]*phago* = eating + *cyt* = cell + *osis* = process
[13]*pino* = drinking + *cyt* = cell + *osis* = process
[14]*glyco* = sugar + *calyx* = cup, vessel

and transplanted tissues. Human blood types and transfusion compatibility are determined by the glycocalyx. The glycocalyx also includes the cell-adhesion molecules described earlier, and thus helps to bind tissues together and enables a sperm to bind to an egg and fertilize it.

Microvilli, Cilia, and Flagella

Many cells have surface extensions called *microvilli, cilia,* and *flagella.* These aid in absorption, movement, and sensory processes.

MICROVILLI

Microvilli[15] (MY-cro-VIL-eye; singular, *microvillus*) are extensions of the plasma membrane that serve primarily to increase its surface area (fig. 2.12). Microvilli are best developed in cells specialized for absorption, such as the epithelial cells of the intestines and kidney tubules. They give such cells 15 to 40 times as much absorptive surface area as they would have if their apical surfaces were flat. On many cells, microvilli are little more than tiny bumps on the plasma membrane. On cells of the taste buds and inner ear, they are well developed but serve sensory rather than absorptive functions.

Individual microvilli cannot be distinguished very well with the light microscope because they are only 1 to 2 μm long. On some cells, they are very dense and appear as a fringe called the **brush border** at the apical cell surface. With the scanning electron microscope, they resemble a deep-pile carpet. With the transmission electron microscope, microvilli typically look like finger-shaped projections of the cell surface. They show little internal structure, but often have a bundle of stiff supportive filaments of a protein called *actin.* Actin filaments attach to the inside of the plasma membrane at the tip of the microvillus, and at its base they extend a little way into the cell and anchor the microvillus to a protein mesh called the *terminal web.* When tugged by another protein in the cytoplasm, actin can shorten a microvillus to "milk" its absorbed contents downward into the cell.

CILIA

Cilia (SIL-ee-uh; singular, *cilium*[16]) are hairlike processes about 7 to 10 μm long. Nearly every cell has a solitary, nonmotile *primary cilium.* Its function is not yet known in most cases, but

[15] *micro* = small + *villi* = hairs
[16] *cilium* = eyelash

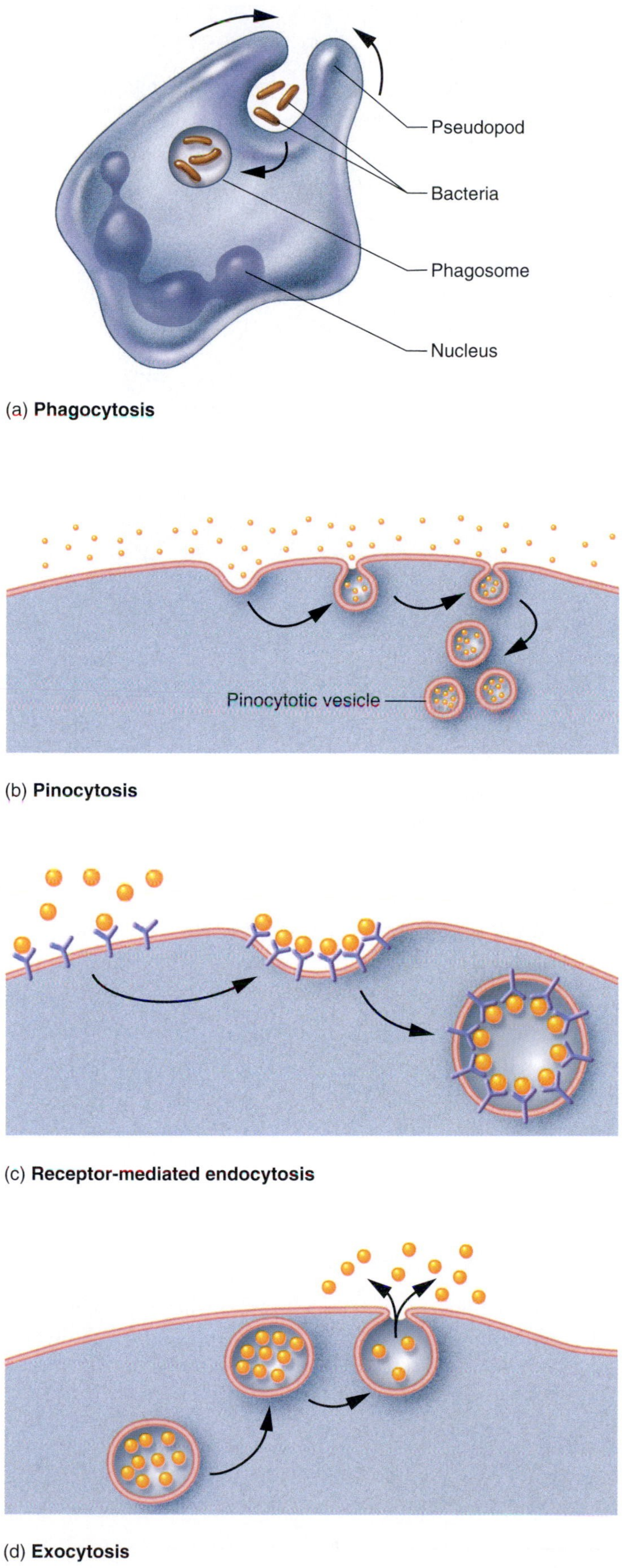

FIGURE 2.11

Modes of Vesicular Transport. (*a*) Phagocytosis. A white blood cell engulfing bacteria with its pseudopods. (*b*) Pinocytosis. A cell imbibing droplets of extracellular fluid. (*c*) Receptor-mediated endocytosis. The *Y*s in the plasma membrane represent membrane receptors. The receptors bind a solute in the extracellular fluid, then cluster together. The membrane caves in at that point until a vesicle pinches off into the cytoplasm bearing the receptors and bound solute. (*d*) Exocytosis. A cell releasing a secretion or waste product.

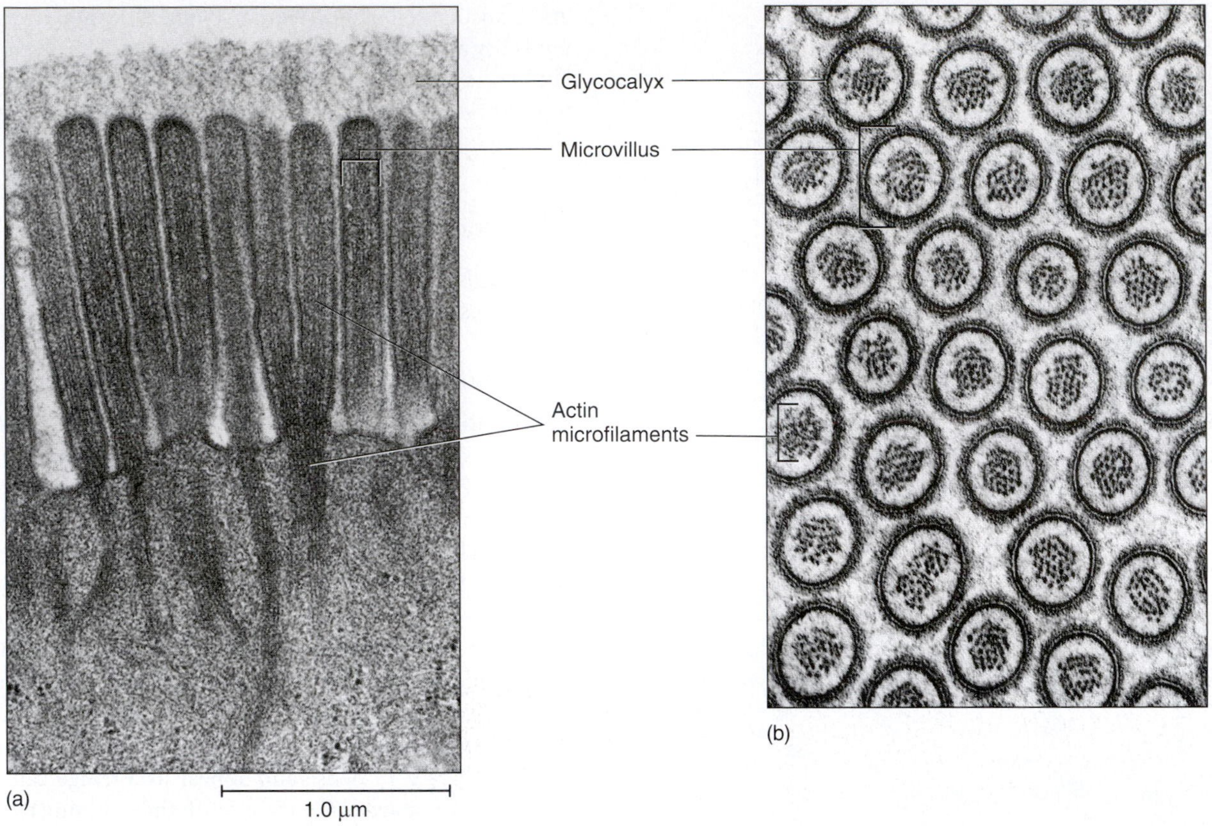
Glycocalyx
Microvillus
Actin
microfilaments
(a)
1.0 µm
(b)

FIGURE 2.12

Microvilli and the Glycocalyx. The microvilli are anchored by microfilaments of actin, which occupy the core of each microvillus and project into the cytoplasm. (*a*) Longitudinal sections, perpendicular to cell surface. (*b*) Cross sections.

some primary cilia are sensory. The light-absorbing parts of the retinal cells in the eye are modified primary cilia; in the inner ear, they play a role in the senses of motion and balance; and in kidney tubules, they are thought to monitor fluid flow. Odor molecules bind to nonmotile cilia on the sensory cells of the nose.

Motile cilia are less widespread, occurring mainly in the respiratory tract and uterine (fallopian) tubes. Cells here typically have 50 to 200 cilia each (fig. 2.13*a*). These cilia beat in waves that sweep across the surface of an epithelium, always in the same direction, moving substances such as mucus and egg cells.

Cilia possess a central core called the **axoneme**[17] (ACK-so-neem), an orderly array of thin protein cylinders called *microtubules.* There are two central microtubules surrounded by a pinwheel-like ring of nine microtubule pairs (fig. 2.13*b–d*). The central microtubules stop at the cell surface, but the peripheral microtubules continue a short distance into the cell as part of a **basal body** that anchors the cilium. In each pair of peripheral microtubules, one tubule has two little **dynein**[18] (DINE-een) **arms.** Dynein, a *motor protein,* uses energy from ATP to "crawl" up the adjacent pair of microtubules. When microtubules on the front of the cilium crawl up the microtubules behind them, the cilium bends toward the front.

[17]*axo* = axis + *neme* = thread
[18]*dyn* = power, energy + *in* = protein

FLAGELLA

A **flagellum**[19] (fla-JEL-um) is more or less like a long solitary cilium. Except for its great length, its structure is the same. The only functional flagellum in humans is the whiplike tail of a sperm cell.

THINK ABOUT IT!

Kartagener syndrome is a hereditary disease in which dynein is lacking from cilia and flagella. How do you think Kartagener syndrome will affect a man's ability to father a child? How might it affect his respiratory health? Explain your answers.

Intercellular Junctions

Also at the cell surface are certain arrangements of proteins called **intercellular junctions** that link cells together and attach them to the extracellular material. Such attachments enable cells to grow and divide normally, resist stress, and communicate with each other. Without them, cardiac muscle cells would pull apart when they contracted, and every swallow of food would scrape away the lining of the esophagus. We will examine three types of junctions—tight junctions, desmosomes, and gap junctions (fig. 2.14).

[19]*flagellum* = whip

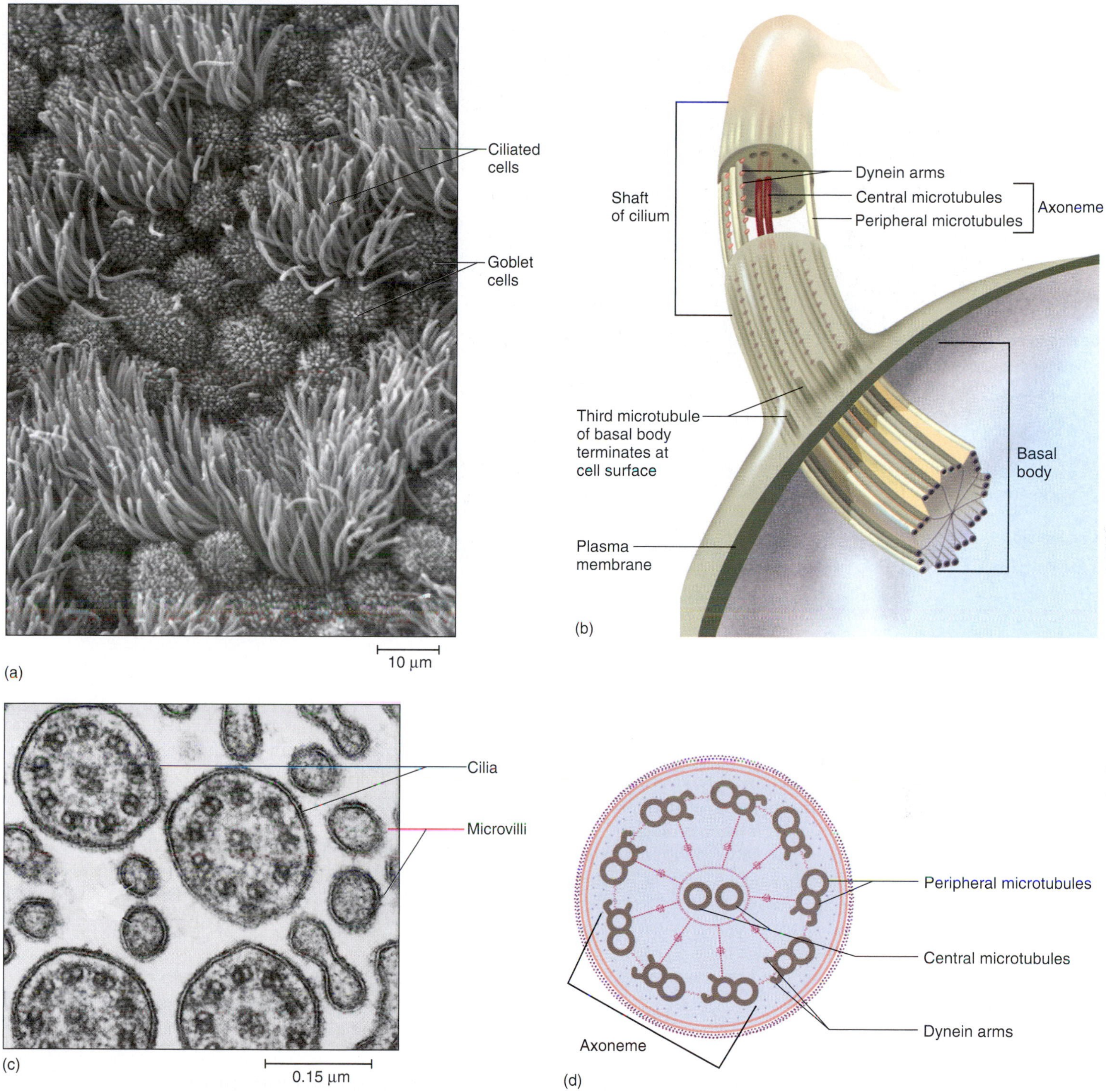
Ciliated cells
Goblet cells
10 µm
(a)
Dynein arms
Central microtubules
Peripheral microtubules
Axoneme
Shaft of cilium
Third microtubule of basal body terminates at cell surface
Basal body
Plasma membrane
(b)
Cilia
Microvilli
(c)
0.15 µm
Peripheral microtubules
Central microtubules
Dynein arms
Axoneme
(d)

FIGURE 2.13

The Structure of Cilia. (*a*) Cilia of the trachea. Several nonciliated, mucus-secreting goblet cells are visible among the ciliated cells. The goblet cells have short microvilli, giving them a shaggy surface. Note the relative sizes of cilia and microvilli. (*b*) Three-dimensional structure of a cilium and its basal body. (*c*) Cross section of several cilia and microvilli. Note the two central microtubules and nine pairs of peripheral microtubules in each cilium. (*d*) Details of the ciliary axoneme.

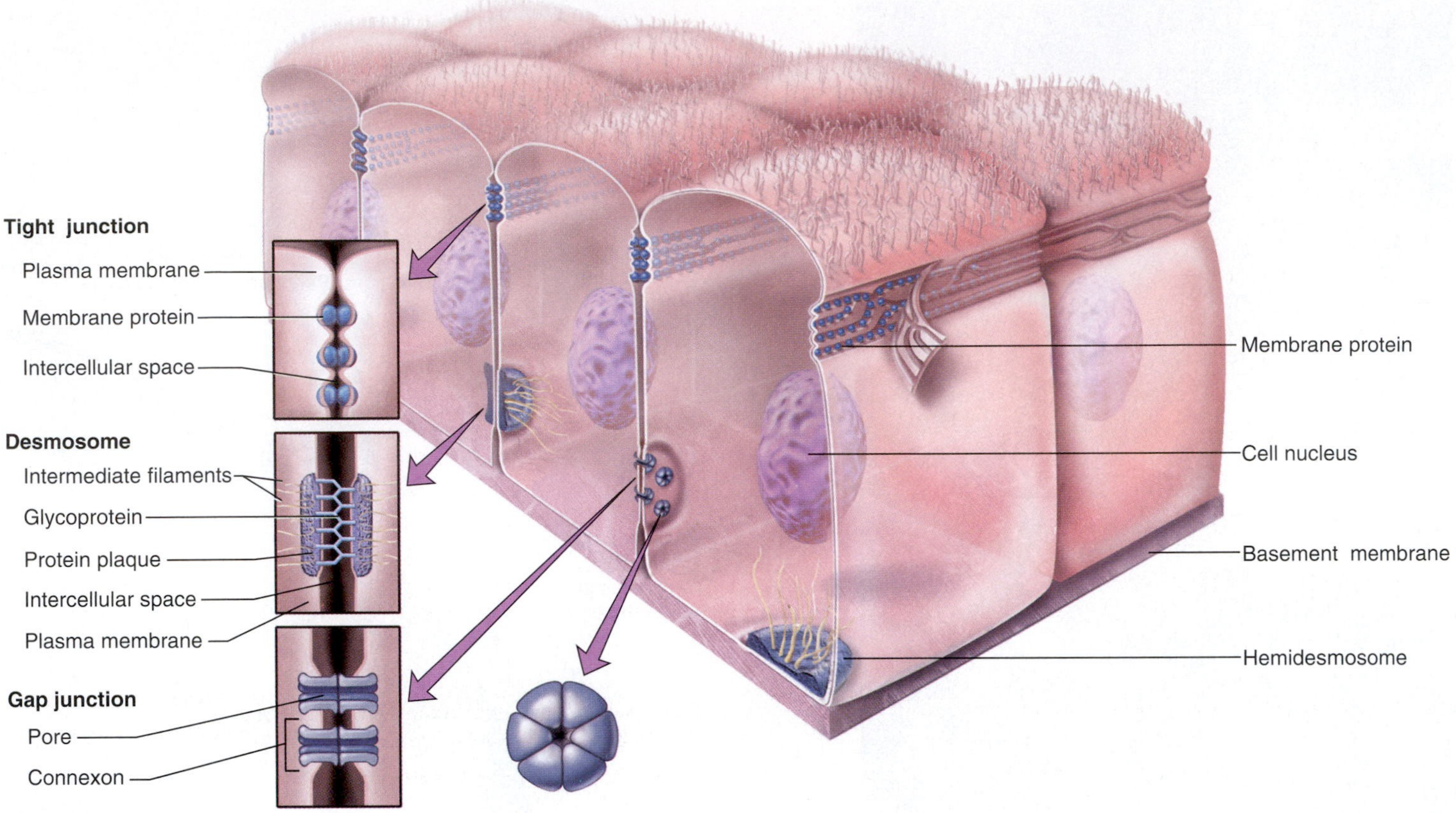
Tight junction
Plasma membrane
Membrane protein
Intercellular space
Desmosome
Intermediate filaments
Glycoprotein
Protein plaque
Intercellular space
Plasma membrane
Gap junction
Pore
Connexon
Membrane protein
Cell nucleus
Basement membrane
Hemidesmosome

FIGURE 2.14
Types of Intercellular Junctions.

TIGHT JUNCTIONS

A **tight junction** completely encircles an epithelial cell near its apical surface and joins it tightly to the neighboring cells. Proteins in the membranes of two adjacent cells form a zipperlike pattern of complementary grooves and ridges. This seals off the intercellular space and makes it difficult for substances to pass between the cells. In the stomach and intestines, for example, tight junctions prevent digestive juices from seeping between epithelial cells and digesting the underlying connective tissue. They also help to prevent intestinal bacteria from invading the tissues, and they ensure that most digested nutrients pass *through* the epithelial cells and not *between* them.

DESMOSOMES

A **desmosome**[20] (DEZ-mo-some) is a patch that holds cells together more like a snap than a zipper. Desmosomes are not continuous and therefore cannot prevent substances from passing between cells. They serve to keep cells from pulling apart and thus enable a tissue to resist mechanical stress. Desmosomes are common in the epidermis, cardiac muscle, and the cervix of the uterus. Hooklike J-shaped proteins arise from the cytoskeleton, approach the cell surface, penetrate into a thick protein plaque associated with the plasma membrane, and then the short arm of the J turns back into the cell. Proteins of the plaque are linked to a glycoprotein mesh between the two cells. This mesh is then linked to the plaque and cytoskeleton of the next cell. Thus, each cell mirrors the other and contributes half of the desmosome. The basal cells of epithelial tissue have *hemidesmosomes*—half-desmosomes that anchor them to the underlying basement membrane.

THINK ABOUT IT!

Why would desmosomes not be suitable as the sole intercellular junctions between epithelial cells of the stomach?

GAP (COMMUNICATING) JUNCTIONS

A **gap junction** is formed by a ringlike *connexon,* which consists of six transmembrane proteins arranged somewhat like the segments of an orange, surrounding a water-filled channel. Ions, glucose, amino acids, and other small solutes can diffuse through the channel directly from the cytoplasm of one cell into the next. In the human embryo, nutrients pass from cell to cell through gap junctions until the circulatory system forms and takes over the role of nutrient distribution. In cardiac muscle, gap junctions allow electrical excitation to pass directly from cell to cell so that the cells contract in near-unison.

[20]*desmo* = band, bond, ligament + *som* = body

INSIGHT 2.1 CLINICAL APPLICATION

When Desmosomes Fail

We often get our best insights into the importance of a structure from the dysfunctions that occur when it breaks down. Desmosomes are destroyed in a disease called *pemphigus vulgaris*[21] (PEM-fih-gus vul-GAIR-iss), in which misguided antibodies (defensive proteins) called *autoantibodies* attack the desmosome proteins, especially in the skin. The resulting breakdown of desmosomes between the epidermal cells leads to widespread blistering of the skin and oral mucosa, loss of tissue fluid, and sometimes death. The condition can be controlled with drugs that suppress the immune system, but such drugs compromise the body's ability to fight off infections.

[21] *pemphigus* = blistering + *vulgaris* = common

Before You Go On

Answer the following questions to test your understanding of the preceding section:

5. Generally speaking, what sort of substances can enter a cell by diffusing through its phospholipid membrane? What sort of substances can enter only through the channel proteins?
6. Compare the structure and function of integral proteins with peripheral proteins.
7. What membrane transport processes get all the necessary energy from the spontaneous movement of molecules? What ones require ATP as a source of energy? What membrane transport processes are carrier-mediated? What ones are not?
8. Identify several reasons why the glycocalyx is important to human survival.
9. How do microvilli and cilia differ in structure and function?
10. What are the functional differences between tight junctions, gap junctions, and desmosomes?

The Cytoplasm

Objectives

When you have completed this section, you should be able to

- describe the cytoskeleton and its functions;
- list the main organelles of a cell and explain their functions; and
- give some examples of cell inclusions and explain how inclusions differ from organelles.

We now probe more deeply into the cell to study the structures in the cytoplasm. These are classified into three groups—cytoskeleton, organelles, and inclusions—all embedded in the clear, gelatinous cytosol.

The Cytoskeleton

The **cytoskeleton** is a network of protein filaments and tubules that structurally support a cell, determine its shape, organize its contents, move substances through the cell, and contribute to movements of the cell as a whole. It can form a very dense supportive web in the cytoplasm (fig. 2.15). It is connected to transmembrane proteins of the plasma membrane and they in turn are connected to protein fibers external to the cell, so there is a strong structural continuity from extracellular material to the cytoplasm. Cytoskeletal elements may even connect to chromosomes in the nucleus, enabling physical tension on a cell to move nuclear contents and mechanically stimulate genetic function.

The cytoskeleton is composed of *microfilaments, intermediate filaments,* and *microtubules.* **Microfilaments** are about 6 nm thick and are made of the protein actin. They form a fibrous **terminal web (membrane skeleton)** on the cytoplasmic side of the plasma membrane. The phospholipids of the plasma membrane are spread out over the terminal web like butter on a slice of bread. It is thought that the phospholipids would break up into little droplets without this support. As described earlier, actin microfilaments also form the supportive cores of the microvilli and play a role in cell movement. Muscle cells are especially packed with actin, which is pulled upon by the motor protein myosin to make muscle contract.

Intermediate filaments (8–10 nm in diameter) are thicker and stiffer than microfilaments. The J-shaped filaments of desmosomes are in this category. So is the tough protein *keratin* that fills the cells of the epidermis and gives strength to the skin.

A **microtubule** (25 nm in diameter) is a hollow cylinder made of 13 parallel strands called *protofilaments.* Each protofilament is a long chain of globular proteins called *tubulin* (fig. 2.16). Microtubules radiate from the centrosome (see p. 68) and hold organelles in place, form bundles that maintain cell shape and rigidity, and act somewhat like railroad tracks to guide organelles and molecules to specific destinations in a cell. They form the ciliary and flagellar basal bodies and axonemes described earlier, and as discussed later under organelles, they form the centrioles and mitotic spindle involved in cell division. Microtubules are not permanent structures. They appear and disappear moment by moment as tubulin molecules assemble into a tubule and then suddenly break apart again to be used somewhere else in the cell. The double and triple sets of microtubules in cilia, flagella, basal bodies, and centrioles, however, are more stable.

Organelles

The minute, metabolically active structures within a cell are called **organelles** (literally "little organs") because they are to the cell what organs are to the body—structures that play individual roles in the survival of the whole (see fig. 2.5). A cell may have 10 billion protein molecules, some of which are potent enzymes with the potential to destroy the cell if they are not contained and isolated from other cellular components. You can imagine the enormous problem of keeping track of all this material, directing molecules to the correct destinations, and maintaining order against the incessant tug of disorder. Cells maintain order partly by compartmentalizing their contents in organelles. Figure 2.17 shows some major organelles.

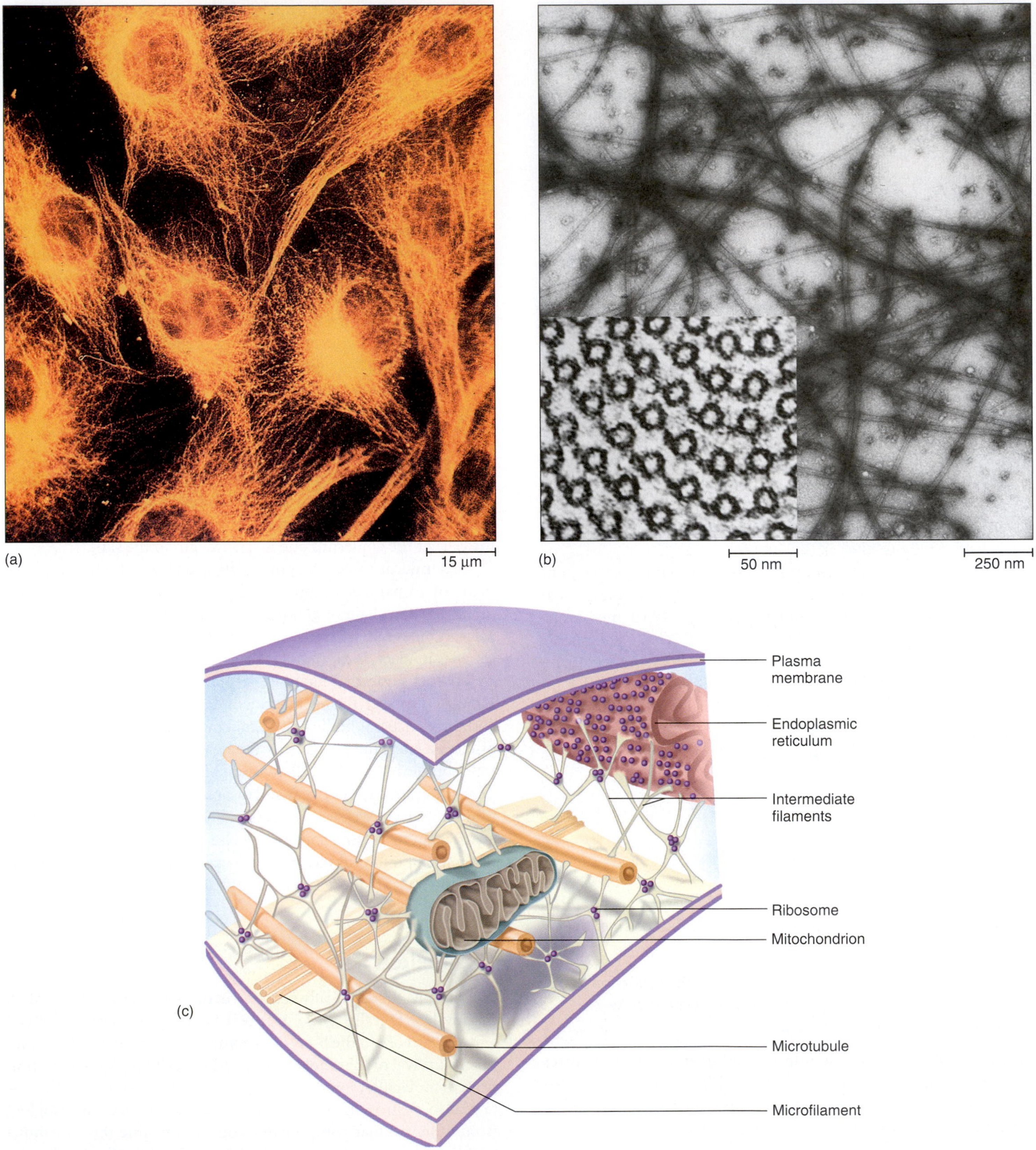
(a)
15 µm
(b)
50 nm
250 nm
Plasma membrane
Endoplasmic reticulum
Intermediate filaments
Ribosome
Mitochondrion
(c)
Microtubule
Microfilament

FIGURE 2.15

The Cytoskeleton. (*a*) The fibrous cytoskeleton made visible by labeling it with fluorescent antibodies and photographing it through a fluorescence microscope. (*b*) Electron micrograph of a cell of the testis showing numerous microtubules in longitudinal section and cross section *(inset)*. (*c*) Diagram of the cytoskeleton.

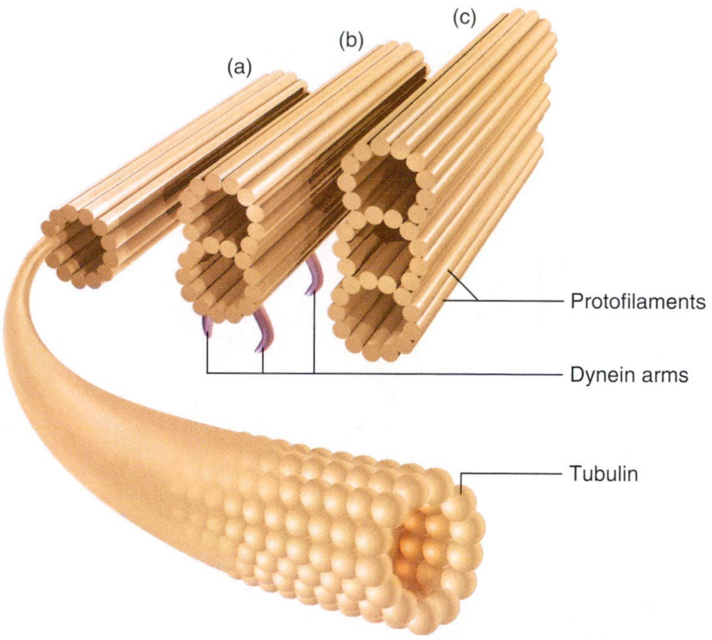
(a)
(b)
(c)
Protofilaments
Dynein arms
Tubulin

FIGURE 2.16

Microtubules. (*a*) A single microtubule is composed of 13 protofilaments. Each protofilament is a helical chain of globular proteins called tubulin. (*b*) One of the nine microtubule pairs of a cilium. (c) One of the nine microtubule triplets of a centriole.

THE NUCLEUS

The **nucleus** (fig. 2.17*a*) is the largest organelle and usually the only one visible with the light microscope. It contains the cell's chromosomes and is therefore the genetic control center of cellular activity. Organelles called ribosomes are produced here, and the early steps in protein synthesis occur here under the direction of the genes. Most cells have only one nucleus, but there are exceptions. Mature red blood cells have none; they are *anuclear.*[22] A few cell types are *multinucleate*—having 2 to 50 nuclei—including some liver cells, skeletal muscle cells, and certain bone-dissolving and platelet-producing cells.

The nucleus is usually spheroid to elliptical in shape and averages about 5 μm in diameter. It is surrounded by a **nuclear envelope** consisting of two parallel unit membranes. The envelope is perforated with **nuclear pores,** about 30 to 100 nm in diameter, formed by a ring-shaped complex of proteins. These proteins regulate molecular traffic into and out of the nucleus and bind the two unit membranes together.

The material within the nucleus is called the *nucleoplasm.* Suspended in this nucleoplasm, most human cells have 46 **chromosomes**[23]—long strands composed of DNA and proteins. In nondividing cells, the chromosomes are in the form of very fine filaments broadly dispersed throughout the nucleus, visible only with the TEM. Collectively, this material is called **chromatin** (CRO-muh-tin). When cells prepare to divide, the chromosomes condense into thick rodlike bodies visible with the LM, as described later in this chapter. The nuclei of nondividing cells also usually exhibit one or more dense masses called **nucleoli** (singular, *nucleolus*), where subunits of the ribosomes are made before they are transported out to the cytoplasm.

ENDOPLASMIC RETICULUM

The term *endoplasmic reticulum (ER)* literally means "little network within the cytoplasm." It is a system of interconnected channels called **cisternae**[24] (sis-TUR-nee) enclosed by a unit membrane (fig. 2.17*b*). In areas called **rough endoplasmic reticulum,** the network consists of parallel, flattened cisternae covered with ribosomes, which give it its rough or granular appearance. The rough ER is continuous with the outer membrane of the nuclear envelope, and adjacent cisternae are often connected by perpendicular bridges. In areas called **smooth endoplasmic reticulum,** the membrane lacks ribosomes, the cisternae are more tubular in shape, and they branch more extensively. The cisternae of the smooth ER are continuous with those of the rough ER, so the two are functionally different parts of the same cytoplasmic network.

The endoplasmic reticulum synthesizes steroids and other lipids, detoxifies alcohol and other drugs, and manufactures all of the membranes of the cell. The rough ER produces the phospholipids and proteins of the plasma membrane. It also synthesizes proteins that are either secreted from the cell or packaged in organelles called *lysosomes.* Rough ER is most abundant in cells that synthesize large amounts of protein, such as antibody-producing cells and cells of the digestive glands.

Most cells have only a scanty smooth ER, but it is relatively abundant in cells that engage extensively in detoxification, such as liver and kidney cells. Long-term abuse of alcohol, barbiturates, and other drugs leads to tolerance partly because the smooth ER proliferates and detoxifies the drugs more quickly. Smooth ER is also abundant in cells of the testes and ovaries that synthesize steroid hormones. Skeletal muscle and cardiac muscle contain extensive networks of smooth ER that release calcium to trigger muscle contraction and store the calcium between contractions.

RIBOSOMES

Ribosomes are small granules of protein and ribonucleic acid (RNA) found in the cytosol and on the outer surfaces of the rough ER and nuclear envelope. Ribosomes "read" coded genetic messages (messenger RNA) from the nucleus and assemble amino acids into proteins specified by the code. The unattached ribosomes found scattered throughout the cytoplasm make enzymes and other proteins for use in the cytosol. The ribosomes attached to the rough ER make proteins that will either be packaged in lysosomes or, as in digestive enzymes, secreted from the cell.

[22]*a* = without + *nucle* = nucleus
[23]*chromo* = color + *some* = body

[24]*cistern* = reservoir

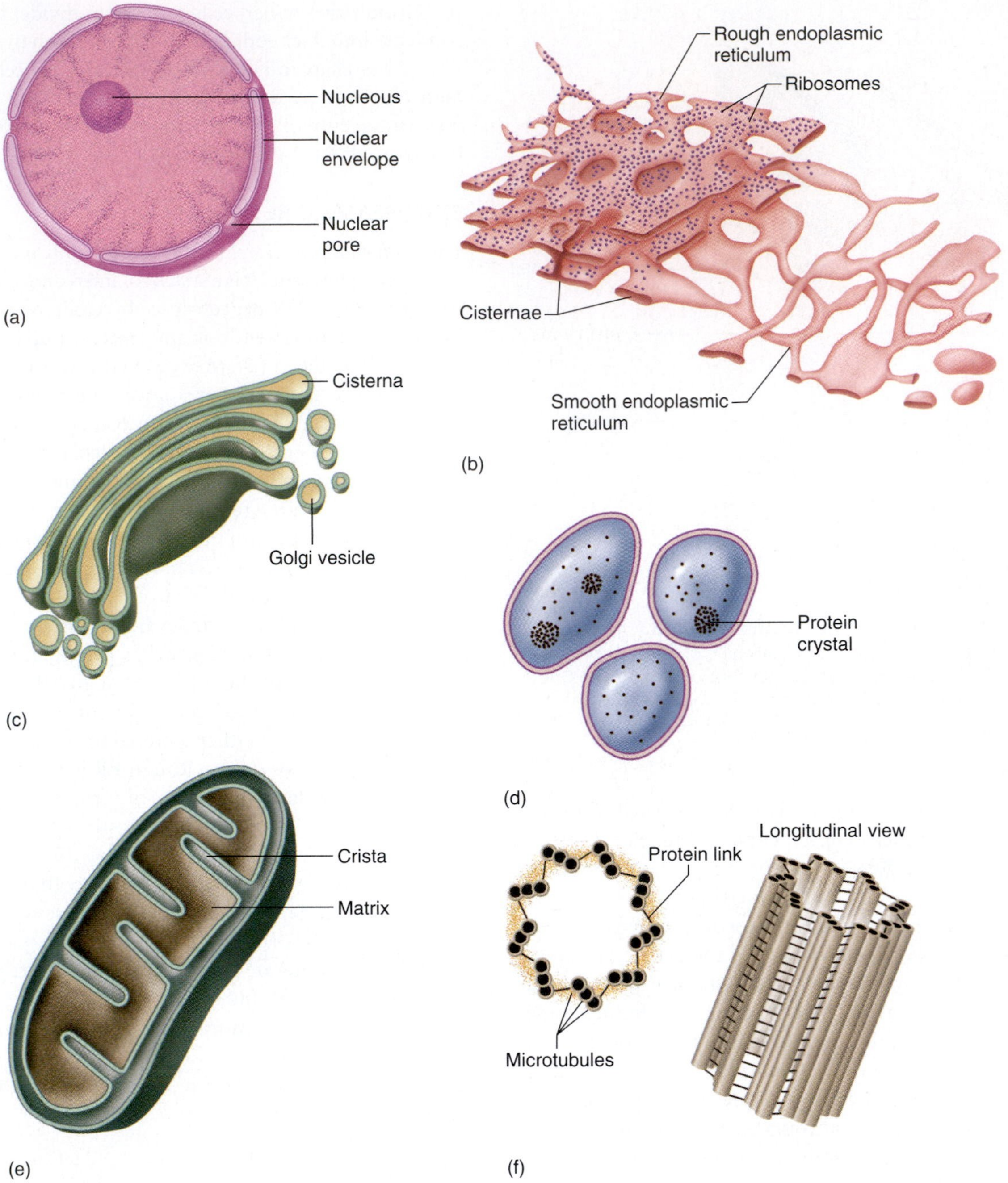
Nucleous
Nuclear envelope
Nuclear pore
(a)
Rough endoplasmic reticulum
Ribosomes
Cisternae
Smooth endoplasmic reticulum
(b)
Cisterna
Golgi vesicle
(c)
Protein crystal
(d)
Crista
Matrix
(e)
Longitudinal view
Protein link
Microtubules
(f)

FIGURE 2.17

Major Organelles. (*a*) Nucleus. (*b*) Endoplasmic reticulum, showing rough and smooth regions. (*c*) Golgi complex and Golgi vesicles. (*d*) Lysosomes. (*e*) Mitochondrion. (*f*) A pair of centrioles. Centrioles are typically found in pairs, perpendicular to each other so that an electron micrograph shows one in cross section and one in longitudinal section.

GOLGI COMPLEX

The **Golgi**[25] (GOAL-jee) **complex** is a small system of cisternae that synthesize carbohydrates and put the finishing touches on protein and glycoprotein synthesis (fig. 2.17*c*). The complex resembles a stack of pita bread. Typically, it consists of about six cisternae, slightly separated from each other, each of them a flattened, slightly curved sac with swollen edges.

Figure 2.18 shows the functional interaction between the ribosomes, endoplasmic reticulum, and Golgi complex. Ribosomes link amino acids together in a genetically specified order to make a particular protein. This new protein threads its way into the cisterna of the rough ER, where enzymes trim and modify it. The altered protein is then shuffled into a little **transport vesicle,** a small, spheroidal organelle that buds off the ER and carries the protein to the nearest cisterna of the Golgi complex. The Golgi complex sorts these proteins, passes them along from one cisterna to the next, cuts and splices

[25]Camillo Golgi (1843–1926), Italian histologist

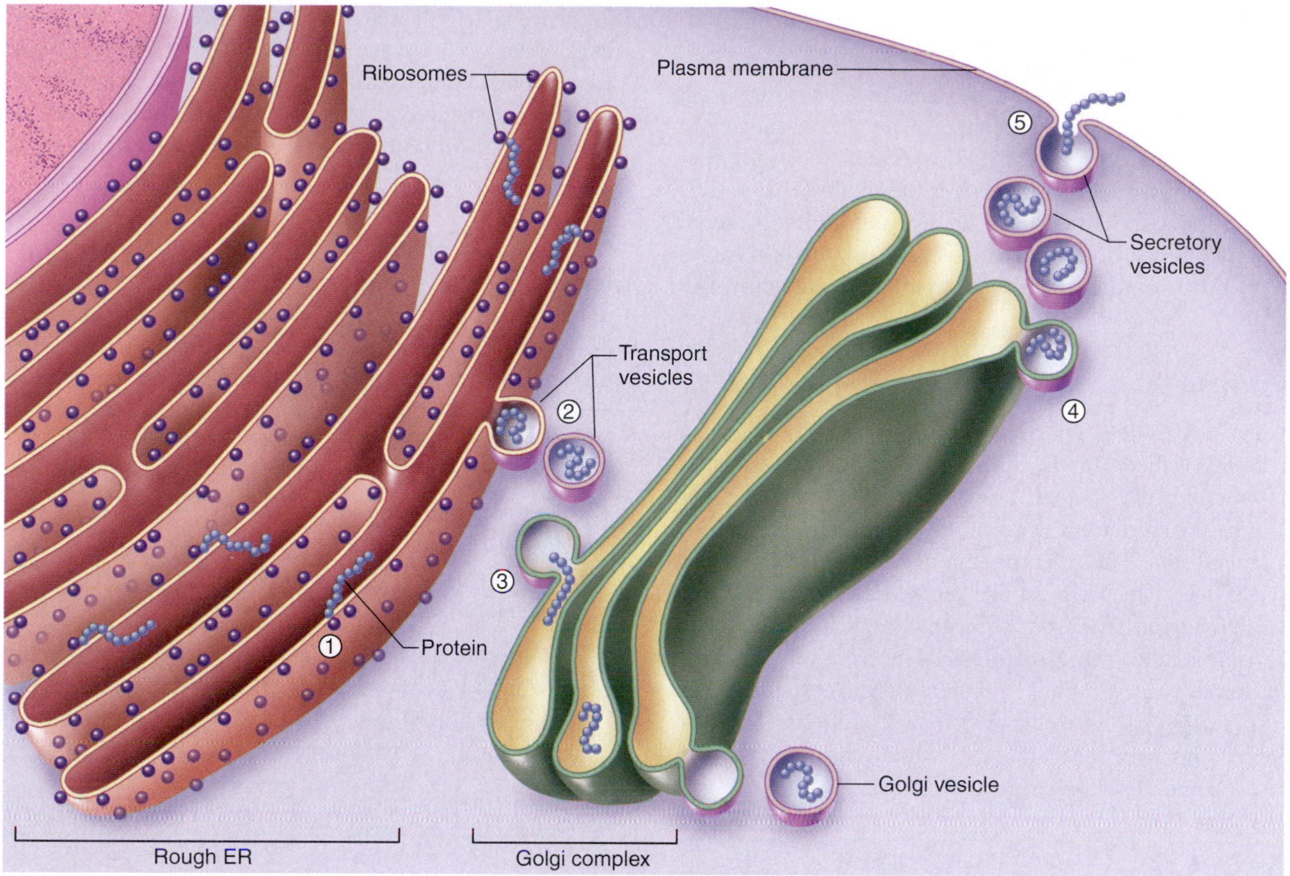

Ribosomes
Plasma membrane
⑤
Secretory vesicles
Transport vesicles
②
④
③
①
Protein
Golgi vesicle
Rough ER
Golgi complex

FIGURE 2.18

The Functional Relationship of Ribosomes, Rough Endoplasmic Reticulum, the Golgi Complex, and Secretory Vesicles in the Synthesis and Secretion of a Protein. The steps in protein synthesis and secretion are numbered *1* through *5*. (*1*) Acting on instructions from the nucleus, each ribosome assembles amino acids in the correct sequence to make a particular protein. The protein is threaded into the cisterna of the rough ER as it is synthesized. The rough ER cuts and splices proteins and may make other modifications. (*2*) The rough ER packages the modified protein into transport vesicles that carry it to the Golgi complex. (*3*) The transport vesicle fuses with a cisterna of the Golgi complex and unloads its protein. The Golgi complex may further modify the protein. (*4*) The Golgi complex buds off Golgi vesicles containing the finished protein. (*5*) Some Golgi vesicles become secretory vesicles, which travel to the plasma membrane and release the product by exocytosis.

some of them, adds carbohydrates to some of them, and finally packages the proteins in membrane-bounded **Golgi vesicles.** These vesicles bud off the swollen rim of the cisterna farthest from the ER. They are seen in abundance in the neighborhood of the Golgi complex.

Some Golgi vesicles become lysosomes, the organelles discussed next; some migrate to the plasma membrane and fuse with it, contributing fresh protein and phospholipid to the membrane; and some become **secretory vesicles** that store a cell product, such as breast milk, mucus, or digestive enzymes, for later release by exocytosis.

LYSOSOMES

A **lysosome**[26] (LY-so-some) (see fig. 2.17*d*) is a package of enzymes contained in a single unit membrane. Although often round or oval, lysosomes are extremely variable in shape. When viewed with the TEM, they often exhibit dark gray contents devoid of structure, but sometimes show crystals or parallel layers of protein. At least 50 lysosomal enzymes have been identified. They break down proteins, nucleic acids, carbohydrates, phospholipids, and other substances. White blood cells called *neutrophils* phagocytize bacteria and digest them with the enzymes of their lysosomes. Lysosomes also digest and dispose of worn-out mitochondria and other organelles; this process is called *autophagy*[27] (aw-TOFF-uh-jee). They are also responsible for a sort of "cell suicide" called *apoptosis (programmed cell death),* in which cells that are no longer needed undergo a prearranged death. The uterus, for example, weighs about 900 g at full-term pregnancy and shrinks to 60 g within 5 or 6 weeks after birth through apoptosis.

[26]*lyso* = loosen, dissolve + *some* = body

[27]*auto* = self + *phagy* = eating

PEROXISOMES

Peroxisomes resemble lysosomes (and are not illustrated) but contain different enzymes and are not produced by the Golgi complex. They are especially abundant in liver and kidney cells, where they neutralize free radicals and detoxify alcohol and other drugs. They are named for the hydrogen peroxide (H_2O_2) they produce in the course of detoxifying alcohol and killing bacteria. Peroxisomes also break down fatty acids into two-carbon molecules that the mitochondria can use as an energy source for ATP synthesis.

MITOCHONDRIA

Mitochondria[28] (MY-toe-CON-dree-uh) (singular, *mitochondrion*) are organelles specialized for ATP synthesis. They have a variety of shapes: spheroid, rod-shaped, bean-shaped, or threadlike (see fig. 2.17*e*). Like the nucleus, a mitochondrion is surrounded by a double unit membrane. The inner membrane usually has folds called **cristae**[29] (CRIS-tee), which project like shelves across the organelle. Cristae bear the enzymes that produce most of the ATP. The space between the cristae is called the **mitochondrial matrix.** It contains enzymes, ribosomes, and a small, circular DNA molecule called *mitochondrial DNA (mtDNA)*, which is genetically different from the DNA in the cell's nucleus (see insight 2.2). Mutations in mtDNA are responsible for some muscle, heart, and eye diseases.

INSIGHT 2.2 EVOLUTIONARY MEDICINE

THE ORIGIN OF MITOCHONDRIA

There is evidence that mitochondria evolved from bacteria that either invaded or were engulfed by other ancient cells, then evolved a mutually beneficial metabolic relationship with them. In size and physiology, mitochondria resemble certain bacteria that live within other cells in a state of symbiosis. Mitochondrial DNA (mtDNA) resembles bacterial DNA more than it does nuclear DNA, and it is replicated independently of nuclear DNA.

While nuclear DNA is reshuffled in each generation by the process of sexual reproduction, mtDNA remains unchanged from generation to generation except by the slow pace of random mutation. Biologists and anthropologists have used mtDNA as a "molecular clock" to trace evolutionary lineages in humans and other species. They have gained some evidence, although still controversial, that of all females who lived in Africa about 200,000 years ago, only one has any descendents still living today—everyone now on earth is descended from this "mitochondrial Eve."

CENTRIOLES

A **centriole** (SEN-tree-ole) is a short cylindrical assembly of microtubules arranged in nine groups of three microtubules each (fig. 2.17*f*). Near the nucleus, most cells have a small, clear patch of cytoplasm called the **centrosome**[30] containing a pair of mutually perpendicular centrioles (see fig. 2.5). These centrioles play a role in cell division described later. In ciliated cells, each cilium also has a **basal body** composed of a single centriole oriented perpendicular to the plasma membrane. Two microtubules of each triplet form the peripheral microtubules of the axoneme of the cilium.

Inclusions

Inclusions are of two kinds: stored cellular products such as pigments, fat droplets, and granules of glycogen (a starchlike carbohydrate), and foreign bodies such as dust particles, viruses, and intracellular bacteria. Inclusions are never enclosed in a unit membrane, and unlike the organelles and cytoskeleton, they are not essential to cell survival.

Before You Go On

Answer the following questions to test your understanding of the preceding section:

11. Describe at least three functions of the cytoskeleton.
12. Briefly state how each of the following cell components can be recognized in electron micrographs: the nucleus, a mitochondrion, a lysosome, and a centriole. What is the primary function of each?
13. Distinguish between organelles and inclusions. State two examples of each.
14. What three organelles are involved in protein synthesis?
15. Define *centriole, microtubule, cytoskeleton,* and *axoneme.* How are these structures related to each other?

THE LIFE CYCLE OF CELLS

Objectives

When you have completed this section, you should be able to

- describe the life cycle of a cell;
- name the stages of mitosis and describe the events that occur in each one; and
- discuss the types and clinical uses of stem cells.

This chapter concludes with an examination of the typical life cycle of human cells, including the process of cell division. Finally, we examine an issue of current controversy, the therapeutic use of embryonic stem cells.

The Cell Cycle

Most cells periodically divide into two daughter cells, so a cell has a life cycle extending from one division to the next. This **cell cycle** (fig. 2.19) is divided into four main phases: G_1, S, G_2, and M.

The **first gap (G_1) phase** is an interval between cell division and DNA replication. During this time, a cell synthesizes proteins,

[28] *mito* = thread + *chondr* = grain
[29] *crista* = crest
[30] *centro* = central + *some* = body

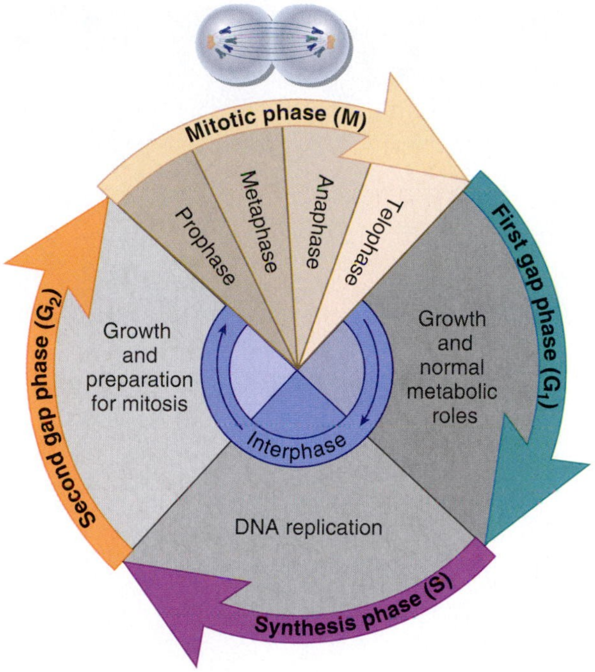

FIGURE 2.19
The Cell Cycle.

grows, and carries out its preordained tasks for the body. Almost all human physiology pertains to what cells do in the G_1 phase. Cells in G_1 also begin to replicate their centrioles in preparation for the next mitosis and they accumulate the materials needed in the next phase to replicate their DNA.

In the **synthesis (S) phase,** a cell carries out DNA replication. Each of its DNA molecules uncoils into two separate strands and each strand acts as a "template" for the synthesis of the missing strand. A cell begins the S phase with 46 molecules of DNA and ends this phase with 92. The cell then has two identical sets of DNA molecules, which are available to be divided up between daughter cells at the next cell division.

The **second gap (G_2) phase** is a relatively brief interval between DNA replication and cell division. In G_2, a cell finishes replicating its centrioles and synthesizes enzymes that control cell division.

The **mitotic (M) phase** is the period in which a cell replicates its nucleus, divides its DNA into two identical sets (one per nucleus), and pinches in two to form two genetically identical daughter cells. The details of this phase are considered in the next section. Phases G_1, S, and G_2 are collectively called **interphase**—the time between M phases.

The length of the cell cycle varies greatly from one cell type to another. Cultured connective tissue cells called fibroblasts divide about once a day and spend 8 to 10 hours in G_1, 6 to 8 hours in S, 4 to 6 hours in G_2, and 1 to 2 hours in M. Stomach and skin cells divide rapidly, bone and cartilage cells slowly, and skeletal muscle cells and nerve cells not at all. Some cells leave the cell cycle for a "rest" and cease to divide for days, years, or the rest of one's life. Such cells are said to be in the **G_0 (G-zero) phase.** The balance between cells that are actively cycling and those standing by in G_0 is an important factor in determining the number of cells in the body. An inability to stop cycling and enter G_0 is characteristic of cancer cells.

Cell Division

Cells divide by two mechanisms called mitosis and meiosis. Meiosis, however, is restricted to one purpose, the production of eggs and sperm, and is therefore treated in chapter 26 on reproduction. **Mitosis** serves all the other functions of cell division: the development of an individual, composed of some 40 trillion cells, from a one-celled fertilized egg; continued growth of all the organs after birth; the replacement of cells that die; and the repair of damaged tissues. Four phases of mitosis are recognizable—*prophase, metaphase, anaphase,* and *telophase* (fig. 2.20).

In **prophase,**[31] at the outset of mitosis, the chromosomes coil into short, dense rods that are easier to distribute to daughter cells than the long, delicate chromatin of interphase. At this stage, a chromosome consists of two genetically identical bodies called **chromatids,** joined together at a pinched spot called the **centromere** (fig. 2.21). There are 46 chromosomes, two chromatids per chromosome, and one molecule of DNA in each chromatid. The nuclear envelope disintegrates during prophase and releases the chromosomes into the cytosol. The centrioles begin to sprout elongated microtubules called **spindle fibers,** which push the centrioles apart as they grow. Eventually, a pair of centrioles come to lie at each pole of the cell. Microtubules grow toward the chromosomes, where some of them become attached to a platelike protein complex called the **kinetochore**[32] (kih-NEE-toe-core) on each side of the centromere. The spindle fibers then tug the chromosomes back and forth until they line up along the midline of the cell.

In **metaphase,**[33] the chromosomes are aligned on the cell equator, oscillating slightly and awaiting a signal that stimulates each chromosome to split in two at the centromere. The spindle fibers now form a football-shaped array called the **mitotic spindle,** with long microtubules reaching out from each centriole to the chromosomes, and shorter microtubules forming a starlike *aster*[34] that anchors the assembly to the inside of the plasma membrane at each end of the cell.

Anaphase[35] begins with activation of an enzyme that cleaves the two sister chromatids from each other at the centromere. Each chromatid is now regarded as a separate, single-stranded *daughter chromosome.* One daughter chromosome migrates to each pole of

[31]*pro* = first
[32]*kineto* = motion + *chore* = place
[33]*meta* = next in a series
[34]*aster* = star
[35]*ana* = apart

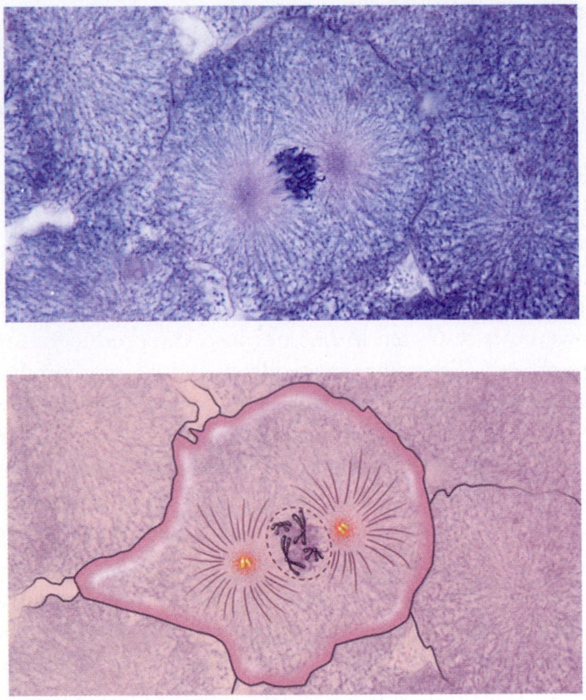

Prophase
Chromatin condenses.
Nucleoli and nuclear envelope break down.
Spindle fibers grow from centrioles.
Centrioles migrate to opposite poles of cell.

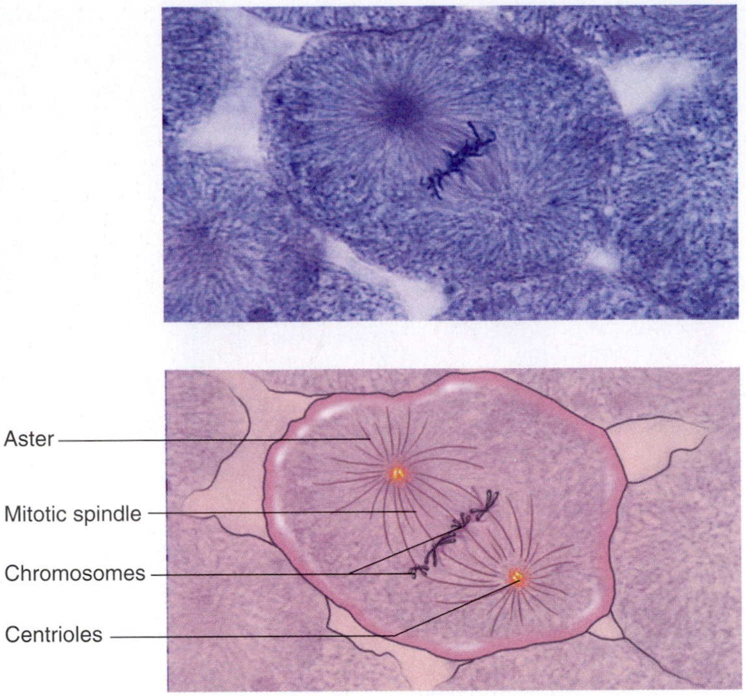
Aster
Mitotic spindle
Chromosomes
Centrioles

Metaphase
Chromosomes lie along midline of cell.
Some spindle fibers attach to kinetochores.
Fibers of aster attach to plasma membrane.

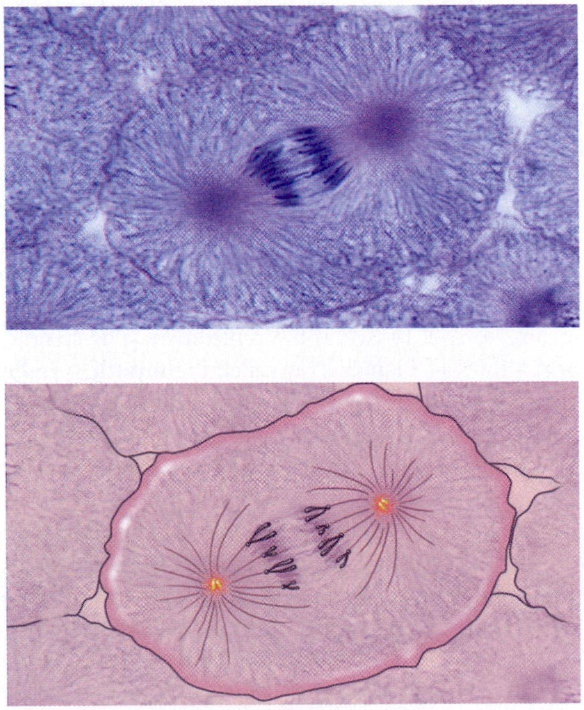

Anaphase
Centromeres divide in two.
Spindle fibers pull sister chromatids to opposite poles of cell.
Each pole (future daughter cell) now has an identical set of genes.

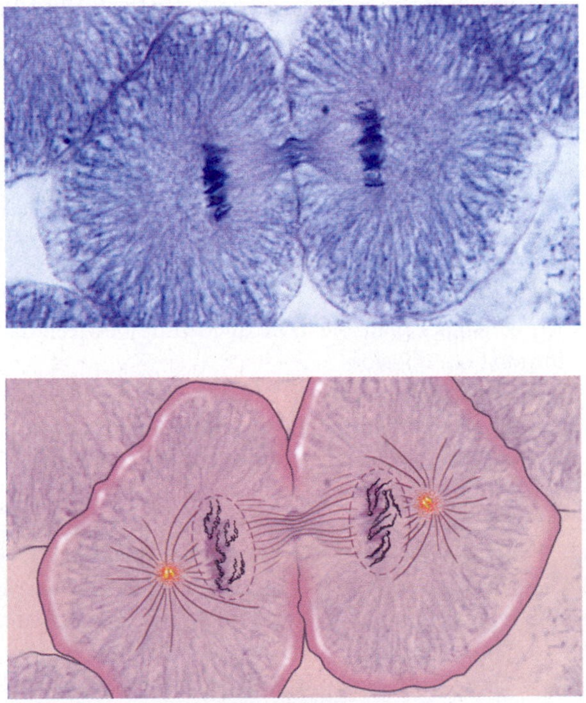

Telophase
Chromosomes gather at each pole of cell.
Chromatin decondenses.
New nuclear envelope appears at each pole.
New nucleoli appear in each nucleus.
Mitotic spindle vanishes.
(Above photo also shows cytokinesis.)

FIGURE 2.20

Mitosis. The photographs show mitosis in whitefish eggs, where chromosomes are relatively easy to observe. The drawings show a hypothetical cell with only two chromosome pairs. In humans, there are 23 pairs.

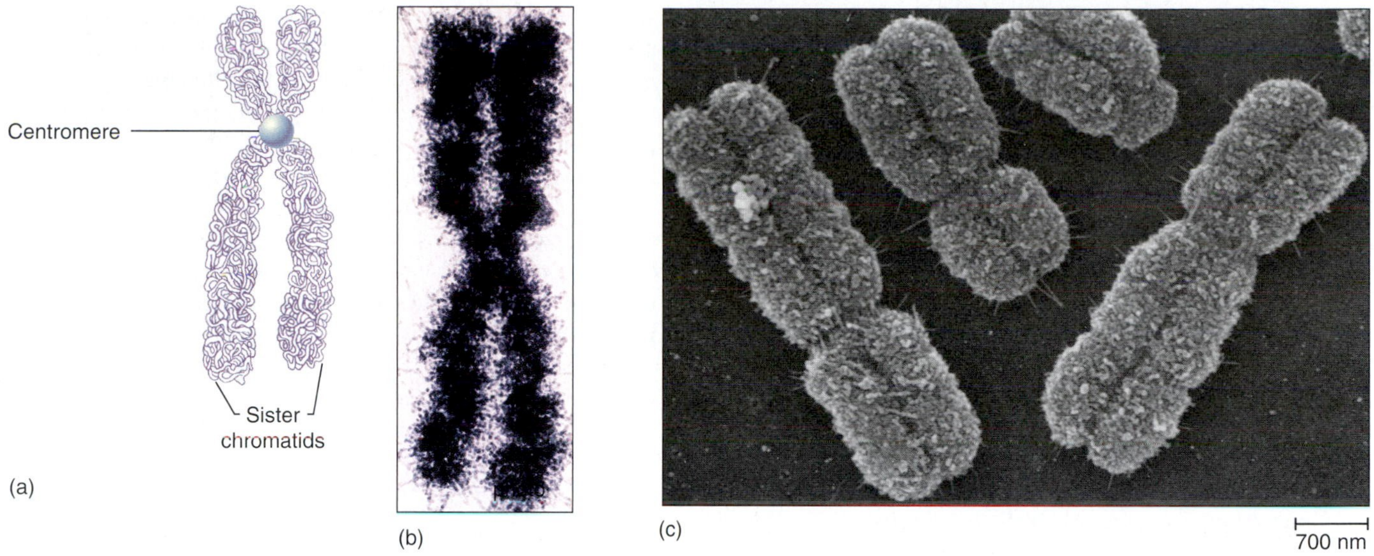
Centromere
Sister chromatids
(a)
(b)
(c)
700 nm

FIGURE 2.21

Chromosomes. (*a*) Diagram of a chromosome in metaphase. From the end of the S phase of the cell cycle to the beginning of anaphase in mitosis, a chromosome consists of two genetically identical chromatids. (*b*) TEM of a metaphase chromosome. (*c*) SEM of four metaphase chromosomes.

the cell, with its centromere leading the way and the arms trailing behind. Migration is achieved by means of motor proteins in the kinetochore crawling along the spindle fiber as the fiber itself is "chewed up" and disassembled at the chromosomal end. Since sister chromatids are genetically identical, and since each daughter cell receives one chromatid from each chromosome, the daughter cells of mitosis are genetically identical.

In **telophase,**[36] the chromatids cluster on each side of the cell. The rough ER produces a new nuclear envelope around each cluster, and the chromatids begin to uncoil and return to the thinly dispersed chromatin form. The mitotic spindle breaks up and vanishes. Each new nucleus forms nucleoli, indicating it has already begun making RNA and preparing for protein synthesis.

Telophase is the end of nuclear division but overlaps with **cytokinesis**[37] (SY-toe-kih-NEE-sis), division of the cytoplasm. Cytokinesis is achieved by the motor protein myosin pulling on microfilaments of actin in the membrane skeleton. This creates a crease called the *cleavage furrow* around the equator of the cell, and the cell eventually pinches in two. Interphase has now begun for these new cells.

[36]*telo* = end, final
[37]*cyto* = cell + *kinesis* = action, motion

INSIGHT 2.3 CLINICAL APPLICATION

CANCER

A *tumor* (*neoplasm*[38]) is a mass of tissue produced when the rate of cell division exceeds the rate of cell death in a tissue. When a tumor is especially fast-growing, lacks a confining fibrous capsule, and its cells are capable of breaking free and spreading to other organs (*metastasizing*[39]), the tumor is said to be *malignant*[40] and constitutes a *cancer.* Cancer was named by Hippocrates, who compared the distended veins in some breast tumors to the outstretched legs of a crab.[41]

All cancer is caused by mutations (changed in DNA or chromosome structure), which can be induced by chemicals, viruses, or radiation, or simply occur through errors in DNA replication in the cell cycle. Agents that cause mutation are called *mutagens,*[42] and those that induce cancer are also called *carcinogens.*[43] Many forms of cancer stem from mutations in two gene families, the oncogenes and tumor-suppressor genes. *Oncogenes*[44] are mutated genes that promote the synthesis of excessive amounts of growth factors (chemicals that stimulate cell division) or excessive sensitivity of target cells to growth factors. *Tumor-suppressor (TS) genes* inhibit the development of cancer by opposing oncogenes, promoting DNA repair, and other means. Cancer occurs when TS genes are unable to perform this function. Oncogenes are like an accelerator to the cell cycle, while TS genes are like a brake.

Untreated cancer is almost always fatal. Tumors destroy healthy tissue; they can grow to block major blood vessels or respiratory airways; they can damage blood vessels and cause hemorrhaging; they can compress and kill brain tissue; and they tend to drain the body of nutrients and energy as they "hungrily" consume a disproportionate share of the body's oxygen and nutrients.

[38]*neo* = new + *plasm* = growth, formation
[39]*meta* = beyond + *stas* = being stationary
[40]*mal* = bad, evil
[41]*cancer* = crab
[42]*muta* = change + *gen* = to produce
[43]*carcino* = cancer + *gen* = to produce
[44]*onco* = tumor

Stem Cells

One of the most controversial scientific issues in the last few years has been stem cell research. **Stem cells** are immature cells with the ability to develop into one or more types of mature, specialized cells. **Adult stem (AS) cells** exist in most of the body's organs. They multiply and replace older cells that are lost to damage or normal cellular turnover. Some stem cells are **unipotent,** able to develop into only one mature cell type, such as the cells that develop into sperm or epidermal squamous cells. Some are **multipotent,** able to differentiate into multiple mature cell types, such as certain bone marrow cells that give rise to multiple types of white blood cells.

Embryonic stem (ES) cells comprise human embryos (technically, pre-embryos; see chapter 4) of up to 150 cells. They are **pluripotent,** able to develop into any type of embryonic or adult cell. ES cells are easily obtained from the excess embryos created in fertility clinics when a couple attempts to conceive a child by *in vitro fertilization* (IVF). In IVF, eggs are fertilized in glassware and allowed to develop to about 8 to 16 cells. Some of these are then transplanted into the mother's uterus. Excess embryos are created to compensate for the low probability of success. Those that are not transplanted to the uterus are usually destroyed, but present a potential source of stem cells for research and therapy.

Skin and bone marrow adult stem cells have been used in therapy for many years. There is hope that stem cells can be manipulated to replace a broader range of tissues, such as cardiac muscle damaged by a heart attack; injured spinal cords; the brain cells lost in Parkinson and Alzheimer diseases; or the insulin-secreting cells needed by people with diabetes mellitus. Adult stem cells seem, however, to have limited developmental potential and to be unable to produce all cell types needed to treat a broad range of diseases. In addition, they are present in very small numbers and are difficult to isolate and culture in the quantities needed for therapy. Embryonic stem cells are easier to obtain and culture, and have more developmental flexibility, but their use remains embroiled in political, religious, and ethical debate. Some would argue that if the excess embryos from IVF are destined to be destroyed, it would seem sensible to use them for beneficial purposes. Others argue, however, that potential medical benefits cannot justify the destruction of a human embryo, or even a pre-embryo of scarcely more than 100 cells.

Before You Go On

Answer the following questions to test your understanding of the preceding section:

16. State what occurs in each of the four phases of the cell cycle.
17. State what occurs in each of the four phases of mitosis.
18. Explain how a cell ensures that each of its daughter cells gets an identical set of genes.
19. Define *unipotent, multipotent* and *pluripotent stem cells.* Give an example of each.
20. Discuss the advantages and disadvantages of adult and embryonic stem cells for therapy.

CHAPTER REVIEW

Review of Key Concepts

The Study of Cells (p. 48)

1. *Cytology* is the study of cellular structure and function.
2. Cytology employs various kinds of microscopes, including the *light microscope (LM), transmission electron microscope (TEM),* and *scanning electron microscope (SEM).* Electron microscopes produce images of higher *resolution* than the LM.
3. Some common cells shapes are *squamous, cuboidal, columnar, polygonal, stellate, spheroid, ovoid, discoid, fusiform,* and *fibrous* (fig. 2.3).
4. The *basal, apical,* and *lateral surfaces* of a cell may vary in molecular structure and function.
5. Cell sizes are usually expressed in *micrometers* (μm) ($1\ \mu m = 10^{-6}\ m = 10^{-3}\ mm$). Most human cells are about 10 to 15 μm wide, but a few types are about 100 μm wide and barely visible to the naked eye.
6. Cell sizes are limited by such factors as their physical strength and diffusion rates. As a cell increases in width, its volume increases by the cube of the width and its surface area by the square of the width. Above a certain size, there is not enough cell surface to metabolically serve its volume of cytoplasm.
7. A cell is enclosed by a *plasma membrane.* The material between the plasma membrane and nucleus is the *cytoplasm;* the material within the nucleus is the *nucleoplasm.* The cytoplasm contains a *cytoskeleton, organelles,* and *inclusions,* embedded in the gelatinous *cytosol.*
8. The fluid within a cell is called *intracellular fluid (ICF).* All body fluids not contained in cells are collectively called *extracellular fluid (ECF).* The ECF amid the cells of a tissue is called *tissue fluid.*

The Cell Surface (p. 52)

1. The boundary of a cell is defined by the *plasma membrane,* a double layer composed primarily of lipids and protein, about 7.5 nanometers (nm) thick. A similar *unit membrane* encloses many of the organelles.
2. The lipids of the plasma membrane are about 75% *phospholipids,* 20% *cholesterol,* and 5% *glycolipids* (in terms of number of molecules). Phospholipids form sheets with their hydrophilic heads facing the ECF and ICF and their hydrophobic fatty acid tails facing each other in the middle of the membrane.
3. The cholesterol content of a plasma membrane affects its fluidity and strength.
4. Glycolipids are phospholipids with carbohydrate chains attached. They and glycoproteins form the *glycocalyx* of the cell.
5. *Transmembrane proteins* penetrate from one side of the plasma membrane to the other. Most of these are glycoproteins, with carbohydrate chains attached. *Peripheral proteins* are attached to the intracellular or extracellular face of the membrane and do not penetrate into the phospholipid layer.
6. Membrane proteins serve a variety of functions: receptors for chemical signals, enzymes, channels, gates, carriers, cell-identity markers, and cell-adhesion molecules.
7. *Filtration* is a method of transport in which a physical force drives water and small solutes through the plasma membrane. It is especially important in the transfer of substances from the bloodstream through capillary walls to the tissues.
8. *Simple diffusion* is a process in which molecules move spontaneously *down a concentration gradient* from a point of high concentration to a point of lower concentration. Substances can diffuse through a plasma membrane if they are small enough to fit through channels in the membrane, or are soluble in its phospholipid.
9. *Osmosis* is the diffusion of water through a selectively permeable membrane from a side with less dissolved matter (where water is more concentrated) to the side with more dissolved matter (where water is less concentrated).
10. *Facilitated diffusion* is the transport of solutes through a membrane, down their concentration gradient, by carrier proteins. The carriers transport solutes that otherwise would not pass through the membrane, or would pass through less efficiently.
11. *Active transport* is the transport of solutes through a membrane, up their concentration gradient, by carrier proteins that consume ATP in the process. The *Na^+-K^1 pump* is an especially important example of this.
12. *Vesicular transport* is the transport of larger quantities of matter through the membrane by means of membrane-bounded vesicles. It is called *endocytosis* when solutes are transported into a cell and *exocytosis* when they are transported out.
13. The three forms of endocytosis are: (1) *phagocytosis,* in which a cell surrounds a particle with pseudopods and engulfs it; (2) *pinocytosis,* in which the plasma membrane sinks inward and nonselectively imbibes droplets of ECF; and (3) *receptor-mediated endocytosis,* a more selective process in which solutes of the ECF bind to membrane receptors, and the membrane then sinks inward to internalize the receptors and the material bound to them.
14. Exocytosis resembles endocytosis in reverse. It is a way for a cell to discharge wastes or to release its own secretions. Most gland cells release their secretions by this method. Exocytosis also replaces the plasma membrane that has been internalized by endocytosis.
15. The glycocalyx is a carbohydrate coating on every cell surface, formed by the carbohydrate components of glycolipids and glycoproteins. It functions in cell identity, in the body's ability to distinguish its own tissues from foreign invaders, and in cell adhesion.
16. *Microvilli* are cell surface extensions that increase a cell's surface area. They are especially abundant in cells heavily engaged in absorption, as in the intestines and kidneys. On some cells, they form a fringe called the *brush border.* The core of a microvillus often has supportive bundles of actin filaments.
17. *Cilia* are hairlike processes that usually have a core structure called the *axoneme,* composed of two central microtubules surrounded by nine microtubule pairs. In the respiratory tract and uterine tube, cilia are motile and serve to propel mucus and egg cells. In the inner ear, retina, nasal cavity, and kidney tubules, cilia serve sensory roles. Many cell types have a solitary *primary cilium* of unknown function.
18. A *flagellum* is similar to a cilium but much longer. It has an axoneme and is motile. The only functional flagellum in humans is a sperm tail.

19. Cells are linked to each other by *intercellular junctions* of three major types: tight junctions, desmosomes, and gap junctions.
20. *Tight junctions* form a zipperlike seal that encircles a cell and joins it tightly to neighboring cells. They prevent the nonselective passage of materials between epithelial cells, ensuring that most substances that do pass through must travel through the cytoplasm of the cells themselves.
21. *Desmosomes* are protein patches that mechanically link one cell to another and enable tissues to resist stress. *Hemidesmosomes,* which are like half a desmosome, bind epithelial cells to an underlying basement membrane.
22. *Gap junctions* are pores surrounded by a ringlike *connexon,* a circle of six membrane proteins. Solutes can pass directly from cell to cell through gap junctions.

The Cytoplasm (p. 63)

1. The cytoplasm consists of a clear gelatinous cytosol in which are embedded the cytoskeleton, organelles, and inclusions.
2. The *cytoskeleton* is a supportive framework for the cell composed of protein microfilaments, intermediate filaments, and microtubules.
3. *Microfilaments* are made of the protein *actin.* They form a supportive *terminal web* on the inner face of the plasma membrane, support the microvilli, and provide for cell movements such as muscle contraction.
4. *Intermediate filaments* are larger, stiffer protein filaments, such as the ones found in desmosomes and the keratin in epidermal cells.
5. *Microtubules* are hollow cylinders composed of the protein *tubulin.* They hold organelles in place, form bundles that maintain cell shape, form tracks that guide the movements of organelles and other materials within a cell, and form structures such as centrioles, basal bodies, axonemes, and mitotic spindles.
6. *Organelles* are structures in the cytoplasm that carry out metabolic functions of the cell.
7. The *nucleus* is the largest organelle. It contains most of the cell's DNA. It is bordered by a *nuclear envelope* composed of two unit membranes perforated with large *nuclear pores.* The *nucleoplasm,* or nuclear contents, contains 46 chromosomes and often one or more nucleoli.
8. The *endoplasmic reticulum* (ER) is a system of interconnected channels called *cisternae,* which often occupy most of the cytoplasm. Areas called *rough ER* have relatively flat cisternae and are studded with *ribosomes.* Areas called *smooth ER* have more tubular cisternae and lack ribosomes. The ER synthesizes phospholipids, steroids, and other lipids; produces all the membranes of the cell; and detoxifies some drugs. The rough ER is a major site of protein synthesis. Smooth ER is scanty in most cells, but abundant in cells that synthesize steroids or engage in detoxification. It functions as a calcium reservoir in muscle and some other cells.
9. *Ribosomes* are protein-synthesizing granules of RNA and enzymes, found either free in the cytosol or attached to the rough ER.
10. The *Golgi complex* is a group of cisternae that synthesize carbohydrates, modify proteins produced by the rough ER, and package cellular products into lysosomes and secretory vesicles.
11. *Lysosomes* are membrane-enclosed packets of enzymes that break down macromolecules, expired organelles, and phagocytized foreign matter, and assist in *programmed cell death (apoptosis).*
12. *Peroxisomes* also are membrane-enclosed packets of enzymes. They serve to detoxify alcohol and other drugs, oxidize fatty acids, and neutralize free radicals.
13. *Mitochondria* are ATP-synthesizing organelles. They are enclosed in a double unit membrane and usually have inward folds of the inner membrane called *cristae.* They contain their own DNA, a variety called *mitochondrial DNA,* similar to the DNA of bacteria.
14. A *centriole* is a short cylindrical array of nine triplets of microtubules. There are usually two centrioles in a clear patch of cytoplasm called the *centrosome.* Each cilium and flagellum also has a solitary basal centriole called a *basal body,* which gives rise to the axoneme.
15. *Inclusions* are either cell products such as fat droplets and pigment granules, or foreign matter such as viruses or dust particles.

The Life Cycle of Cells (p. 68)

1. The life cycle of a cell *(cell cycle)* consists of the *G_1 (first gap) phase* in which a cell grows and carries out its tasks for the body; an *S (synthesis) phase* in which it replicates its DNA; a *G_2 (second gap) phase* in which it prepares for mitosis; and an *M (mitotic) phase* in which it divides. G_1, S, and G_2 collectively constitute the *interphase* between cell divisions.
2. Cells that have left the cell cycle and stopped dividing, either temporarily or permanently, are in the *G_0* phase. Mature skeletal muscle cells, neurons, and some other cells are incapable of mitosis and stay in G_0 permanently.
3. Mitosis is the mode of cell division employed in embryonic development, growth, replacement of dead cells, and repair of injured tissues. It consists of four phases called prophase, metaphase, anaphase, and telophase.
4. In *prophase,* the chromosomes condense and become visible by LM as paired *sister chromatids* joined by a centromere. The nuclear envelope disintegrates and microtubules grow from the centrioles.
5. In *metaphase,* the chromosomes align on the equator of the cell, while microtubules attach to their centromeres and form a mitotic spindle.
6. In *anaphase,* the centromeres divide and the sister chromatids separate from each other, becoming single-stranded *daughter chromosomes.* These chromosomes migrate toward opposite poles of the cell.
7. In *telophase,* the daughter chromosomes cluster at each end of the cell, uncoil, and become finely dispersed chromatin, as a new nuclear envelope forms around each cluster.
8. *Cytokinesis* begins during telophase and consists of a division of the cytoplasm into two distinct cells.
9. Nearly all organs and tissues contain undifferentiated *stem cells* that multiply and differentiate into specialized mature cells.
10. The only *pluripotent* stem cells, capable of differentiating into any type of embryonic or adult cell, are *embryonic stem cells,* from pre-embryos composed of up to 150 cells. Adult stem cells can be *multipotent* (able to differentiate into multiple mature cell types) or *unipotent* (able to differentiate into only one mature cell type).
11. Stem cells are used in therapy for replacing damaged tissues. A major research effort is presently underway to manipulate either adult or embryonic stem cells into producing replacements for a wider range of lost or damaged cells and tissues.

TESTING YOUR RECALL

1. The clear, structureless gel in a cell is its
   a. nucleoplasm.
   b. endoplasm.
   c. cytoplasm.
   d. neoplasm.
   e. cytosol.
2. New nuclei form and a cell begins to pinch in two during
   a. prophase.
   b. metaphase.
   c. interphase.
   d. telophase.
   e. anaphase.
3. The amount of _____ in a plasma membrane affects its stiffness versus fluidity.
   a. phospholipid
   b. cholesterol
   c. glycolipid
   d. glycoprotein
   e. transmembrane protein
4. Cells specialized for absorption of matter from the ECF are likely to show an abundance of
   a. lysosomes.
   b. microvilli.
   c. mitochondria.
   d. secretory vesicles.
   e. ribosomes.
5. Osmosis is a special case of
   a. pinocytosis.
   b. carrier-mediated transport.
   c. active transport.
   d. facilitated diffusion.
   e. simple diffusion.
6. Embryonic stem cells are best described as
   a. pluripotent.
   b. multipotent.
   c. unipotent.
   d. more developmentally limited than adult stem cells.
   e. more difficult to culture and harvest than adult stem cells.
7. The amount of DNA in a cell doubles during
   a. prophase.
   b. metaphase.
   c. anaphase.
   d. the S phase.
   e. the G_2 phase.
8. Fusion of a secretory vesicle with the plasma membrane and release of the vesicle's contents is
   a. exocytosis.
   b. receptor-mediated endocytosis.
   c. active transport.
   d. pinocytosis.
   e. phagocytosis.
9. Most cellular membranes are made by
   a. the nucleus.
   b. the cytoskeleton.
   c. enzymes in the peroxisomes.
   d. the endoplasmic reticulum.
   e. replication of existing membranes.
10. Matter can leave a cell by any of the following means *except*
    a. active transport.
    b. pinocytosis.
    c. facilitated diffusion.
    d. simple diffusion.
    e. exocytosis.
11. Most human cells are 10 to 15 _____ wide.
12. When a hormone cannot enter a cell, it binds to a _____ at the cell surface.
13. _____ are channels in the plasma membrane that open or close in response to various stimuli.
14. An adult stem cell would be classified as _____ if it can develop into any of five types of specialized cells.
15. High-resolution photomicrographs with a three-dimesional appearance are most often produced with a _____ microscope.
16. Thin scaly cells are described by the term _____.
17. Two human organelles that are surrounded by a double unit membrane are the _____ and _____.
18. Liver cells can detoxify alcohol with two organelles, the _____ and _____.
19. Cells adhere to each other and to extracellular material by means of membrane proteins called _____.
20. A macrophage would use the process of _____ to engulf a dying tissue cell.

Answers in the Appendix

TRUE OR FALSE

Determine which five of the following statements are false, and briefly explain why.

1. The shape of a cell is determined mainly by its cytoskeleton.
2. The most important quality of a microscope is how much magnification it can produce.
3. A plasma membrane is too thin to be seen with the light microscope.
4. The hydrophilic heads of membrane phospholipids are in contact with both the ECF and ICF.
5. Water-soluble substances usually must pass through channel proteins to enter a cell.
6. Cells must use ATP to move substances down a concentration gradient.
7. Osmosis is a type of active transport involving water.
8. Cilia and flagella have an axoneme but microvilli do not.
9. Desmosomes enable substances to pass from cell to cell.
10. A nucleolus is an organelle within the nucleoplasm.

Answers in the Appendix

Testing Your Comprehension

1. What would probably happen to the plasma membrane of a cell if it were composed of hydrophilic molecules such as carbohydrates?
2. Since electron microscopes are capable of much more resolution than light microscopes, why do you think biologists go on using light microscopes? Why are students in introductory biology courses not provided with electron microscopes?
3. This chapter mentions that the polio virus enters cells by means of receptor-mediated endocytosis. Why do you think the viruses don't simply enter through the channels in the plasma membrane? Cite some specific facts from this chapter to support your conjecture.
4. A major tenet of the cell theory is that all bodily structure and function results from the function of cells. Yet the structural properties of bone are due more to its extracellular material than to its cells. Is this an exception to the cell theory? Why or why not?
5. If a cell were poisoned so its mitochondria ceased to function, what membrane transport processes would immediately stop? What ones could continue?

Answers at the Online Learning Center

www.mhhe.com/saladinha1

Visit the Online Learning Center for practice tests, answer keys, and other learning aids for this chapter. Enhance your understanding of human anatomy with our interactive art labeling exercises, supplemental photo atlases, web links, puzzles, flashcards, and much more.

3

CHAPTER THREE

Histology—The Study of Tissues

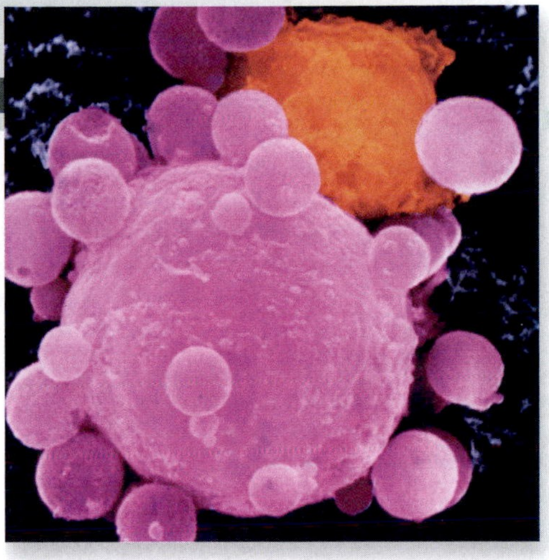

A cancer cell (mauve) undergoing apoptosis (cell suicide) under attack by an immune cell (orange) (SEM)

CHAPTER OUTLINE

The Study of Tissues 78
- The Primary Tissue Classes 78
- Interpreting Tissue Sections 78

Epithelial Tissue 80
- Simple Epithelia 80
- Stratified Epithelia 83

Connective Tissue 86
- Overview 86
- Fibrous Connective Tissue 86
- Cartilage 90
- Bone 92
- Blood 92

Nervous and Muscular Tissue—Excitable Tissues 94
- Nervous Tissue 94
- Muscular Tissue 94

Glands and Membranes 96
- Glands 96
- Membranes 97

Tissue Growth, Development, Death, and Repair 99
- Changes in Tissue Type 99
- Tissue Growth 99
- Tissue Shrinkage and Death 99
- Tissue Repair 100

Chapter Review 101

INSIGHTS

3.1 Clinical Application: Collagen Diseases 87
3.2 Clinical Application: The Consequences of Defective Elastin 88
3.3 Clinical Application: Brittle Bone Disease 93
3.4 Clinical Application: Keloids 100

BRUSHING UP

To understand this chapter, you may find it helpful to review the following concepts:

- The hierarchy of human structure (p. 6)
- Membranes of the body (pp. 29–30)
- Cell shapes (p. 50)
- Cell structure (pp. 50–68)

With its 50 trillion cells and thousands of organs, the human body may seem to be a structure of forbidding complexity. Fortunately for our health, longevity, and self-understanding, the biologists of past generations were not discouraged by this complexity, but discovered patterns that made it more understandable. One pattern is the fact that these trillions of cells belong to only 200 different types or so, and these cells are organized into tissues that fall into just 4 broad categories—*epithelial, connective, nervous,* and *muscular tissue.*

Histology[1] (microscopic anatomy) is the study of tissues and how they are arranged into organs. Histology bridges the gap between the *cytology* of the preceding chapter and the *organ system* approach of the chapters that follow. Here we study the four tissue classes; variations within each class; how to recognize tissue types microscopically and relate their microscopic anatomy to their function; and how tissues are arranged to form an organ. This chapter describes only mature tissue types. Embryonic tissues are discussed in chapter 4.

The Study of Tissues

Objectives

When you have completed this section, you should be able to

- name the four primary classes into which all adult tissues are classified; and
- visualize the three-dimensional shape of a structure from a two-dimensional tissue section.

The Primary Tissue Classes

A **tissue** is a mass of similar cells and cell products that forms a discrete region of an organ and performs a specific function. The four **primary tissues** are epithelial, connective, nervous, and muscular tissue (table 3.1). These tissues differ from each other in the types and functions of their cells, the characteristics of the **matrix (extracellular material)** that surrounds the cells, and the relative amount of space occupied by cells versus matrix. In epithelial and muscular tissue, the cells are so close together that the matrix is barely visible, while in connective tissue, the matrix usually occupies more space than the cells do.

The matrix is composed of fibrous proteins and **ground substance.** The latter is also variously known as the *extracellular fluid (ECF), interstitial[2] fluid, tissue fluid,* or *tissue gel,* although in cartilage and bone, the matrix is rubbery or stony in consistency.

In summary, a tissue is composed of cells and matrix, and the matrix is composed of fibers and ground substance.

[1] *histo* = tissue + *logy* = study of
[2] *inter* = between + *stit* = to stand

Interpreting Tissue Sections

In your study of histology, you may be presented with various types of tissue preparations mounted on microscope slides. Most such preparations are thin slices called **histological sections,** and are artificially colored to bring out detail. The best anatomical insight depends on an ability to deduce the three-dimensional structure of an organ from these two-dimensional sections. This ability, in turn, depends on an awareness of how tissues are prepared for study.

Histologists use a variety of techniques to preserve, section (slice), and stain tissues to show their structural details as clearly as possible. Tissue specimens are preserved in a **fixative**—a chemical such as formalin that prevents decay and makes the tissue more firm. After fixation, most tissues are sectioned by a machine called a *microtome,* which makes slices that are typically only one or two cells thick. This is necessary so the light of a microscope can pass through and so the image is not confused by too many superimposed layers of cells. The sections are then mounted on slides and artificially colored with histological **stains** to enhance detail. If they were not stained, most tissues would appear pale gray. With stains that bind to different components of a tissue, however, you may see pink cytoplasm, violet nuclei, and blue, green, or golden brown protein fibers, depending on the stain used.

When viewing such sections, you must try to translate the microscopic image into a mental image of the whole structure. Like the boiled egg and elbow macaroni in figure 3.1, an object may look quite different when it is cut at various levels, or *planes of section.* A coiled tube, such as a gland of the uterus (fig. 3.1*c*), is often broken up into multiple portions since it meanders in and out of the plane of section. An experienced viewer, however, would recognize that the separated pieces are parts of a single tube winding its way to the organ surface. Note that a grazing slice through a boiled egg might miss the yolk. Similarly, a grazing slice through a cell may miss the nucleus and give the false impression that the cell did not have one. In some tissue sections, you are likely to see many cells with nuclei and many others in which the nucleus did not fall in the plane of section, and is therefore absent.

Many anatomical structures are significantly longer in one direction than another—the humerus and esophagus, for example. A tissue cut in the long direction is called a **longitudinal section (l.s.),** and one cut perpendicular to this is a **cross section (c.s.** or **x.s.**), or **transverse section (t.s.).** A section cut at an angle between a longitudinal and cross section is an **oblique section.** Figure 3.2 shows how certain organs look when sectioned on each of these planes.

Not all histological preparations are sections. Liquid tissues such as blood and soft tissues such as spinal cord may be prepared as **smears,** in which the tissue is rubbed or spread across the slide rather than sliced. Some membranes and cobwebby tissues like the *areolar tissue* described later in this chapter are sometimes mounted as **spreads,** in which the tissue is laid out on the slide, like placing a small square of tissue paper or a tuft of lint on a sheet of glass.

TABLE 3.1

The Four Primary Tissue Classes

Type	Definition	Representative Locations
Epithelial	Tissue composed of layers of closely spaced cells that cover organ surfaces or form glands, and serve for protection, secretion, and absorption	Epidermis Inner lining of digestive tract Liver and other glands
Connective	Tissue with more matrix than cell volume, often specialized to support, bind, and protect organs	Tendons and ligaments Cartilage and bone Blood and lymph
Nervous	Tissue containing excitable cells specialized for rapid transmission of coded information to other cells	Brain Spinal cord Nerves
Muscular	Tissue composed of elongated, excitable cells specialized for contraction	Skeletal muscles Heart (cardiac muscle) Walls of viscera (smooth muscle)

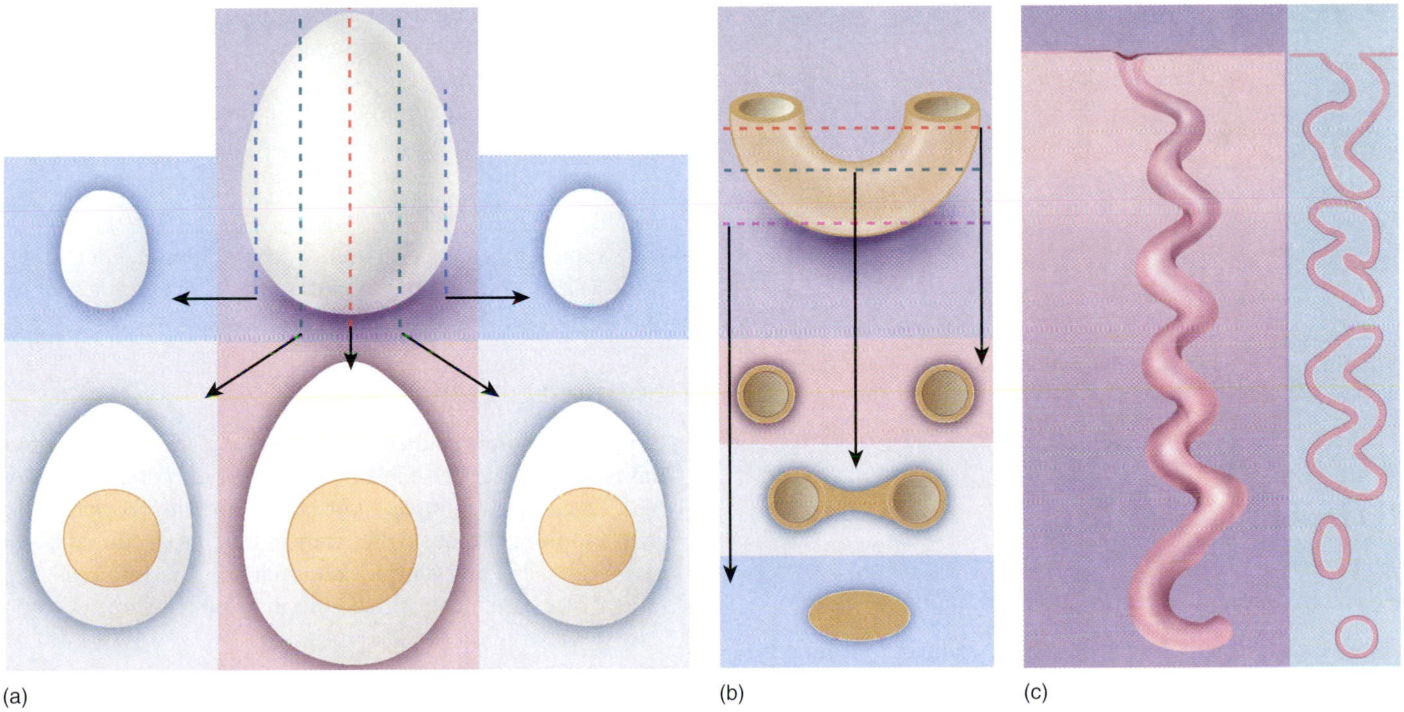

FIGURE 3.1

Three-Dimensional Interpretation of Two-Dimensional Images. (*a*) A boiled egg. Note that the grazing sections (*far left* and *right*) miss the yolk, just as a tissue section may miss a nucleus or other structure. (*b*) Elbow macaroni, which resembles many curved ducts and tubules. A section far from the bend would give the impression of two separate tubules; a section near the bend would show two interconnected lumina (cavities); and a section still farther down could miss the lumen completely. (*c*) A coiled gland in three dimensions and as it would look in a vertical tissue section.

Before You Go On

Answer the following questions to test your understanding of the preceding section:

1. Define *tissue* and distinguish a tissue from a cell and an organ.
2. Classify each of the following into one of the four primary tissue classes: the skin surface, fat, the spinal cord, most heart tissue, bones, tendons, blood, and the inner lining of the stomach.
3. What are tissues composed of in addition to cells?
4. What is the term for a thin, stained slice of tissue mounted on a microscope slide?
5. Sketch what a pencil would look like in a longitudinal section, cross section, and oblique section.

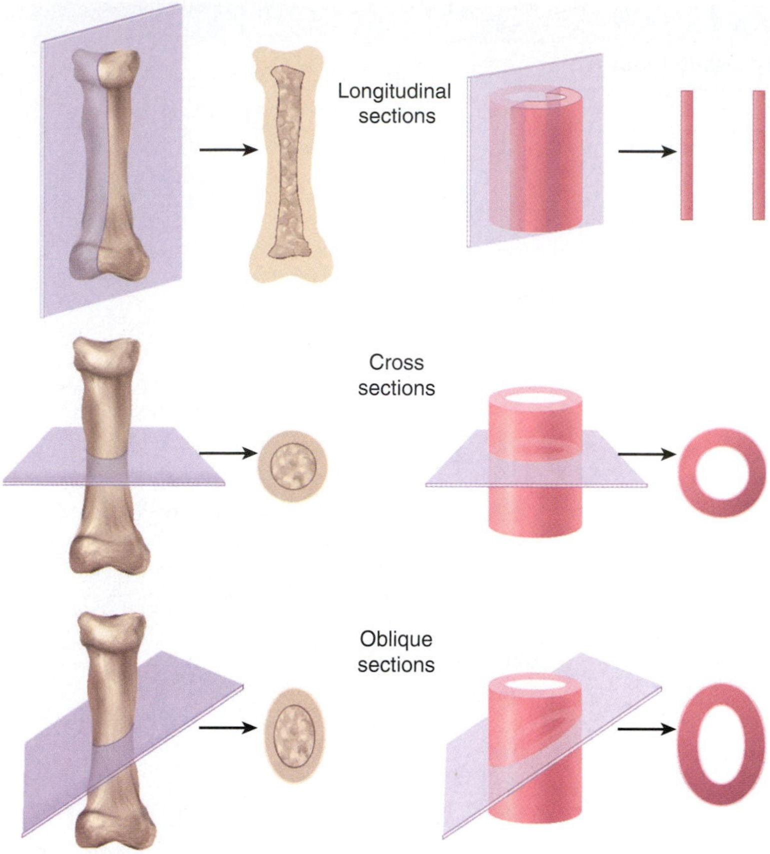

Figure 3.2
Longitudinal, Cross, and Oblique Sections. Note the effect of the plane of section on the two-dimensional appearance of elongated structures such as bones and blood vessels.

Epithelial Tissue

Objectives
When you have completed this section, you should be able to

- describe the properties that distinguish epithelium from other tissue classes;
- list and classify eight types of epithelium, distinguish them from each other, and state where each type can be found in the body;
- explain how the structural differences between epithelia relate to their functional differences; and
- visually recognize each epithelial type from specimens or photographs.

Epithelium[3] is a type of tissue composed of one or more layers of closely adhering cells, either covering an organ surface or forming the secretory tissue and ducts of a gland. Epithelia form the external and internal linings of many organs, line the body cavities, and form the epidermis of the skin. The extracellular material of an epithelium is so thin it is barely visible with the light microscope, and epithelia allow no room for blood vessels. They do, however, almost always lie on a layer of loose connective tissue and depend on its blood vessels for nourishment and waste removal.

Between an epithelium and the underlying connective tissue is a layer called the **basement membrane,** usually too thin to be visible with the light microscope. It contains collagen, adhesive glycoproteins called *laminin* and *fibronectin,* and a large protein-carbohydrate complex called *heparan sulfate.* It gradually blends with other protein fibers on the connective tissue side. The basement membrane serves to anchor an epithelium to the connective tissue below it. The surface of an epithelial cell that faces the basement membrane is its *basal surface,* and the one that faces away from the basement membrane is the *apical surface.*

Epithelia are classified into two broad categories, *simple* and *stratified,* with four types in each category:

- Simple epithelia
  - Simple squamous epithelium
  - Simple cuboidal epithelium
  - Simple columnar epithelium
  - Pseudostratified columnar epithelium
- Stratified epithelia
  - Stratified squamous epithelium
  - Stratified cuboidal epithelium
  - Stratified columnar epithelium
  - Transitional epithelium

In a simple epithelium, every cell touches the basement membrane, whereas in a stratified epithelium, some cells rest on top of others and do not contact the basement membrane (fig. 3.3).

Simple Epithelia

Generally, a simple epithelium has only one layer of cells, although this is a somewhat debatable point in the *pseudostratified columnar* type. Three types of simple epithelia are named for the shapes of their cells: **simple squamous**[4] (thin scaly cells), **simple cuboidal** (square or round cells), and **simple columnar** (tall narrow cells). In the fourth type, **pseudostratified columnar epithelium,** not all cells

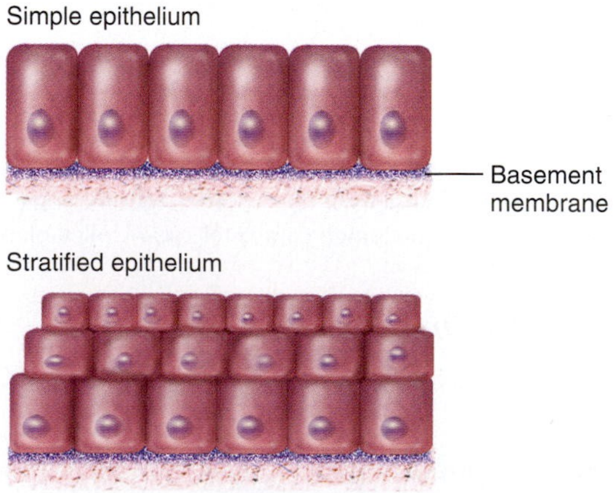

Figure 3.3
Comparison of Simple and Stratified Epithelia.

[3]*epi* = upon + *theli* = nipple, female

[4]*squam* = scale

reach the free surface; the taller cells cover the shorter ones. This epithelium looks multilayered (stratified) in most tissue sections, but careful examination, especially with the electron microscope, shows that every cell reaches the basement membrane. Simple columnar and pseudostratified columnar epithelia often have wineglass-shaped **goblet cells** that produce protective mucus over the mucous membranes.

Table 3.2 illustrates and summarizes the structural and functional differences among these four types.

TABLE 3.2

Simple Epithelia

Simple Squamous Epithelium

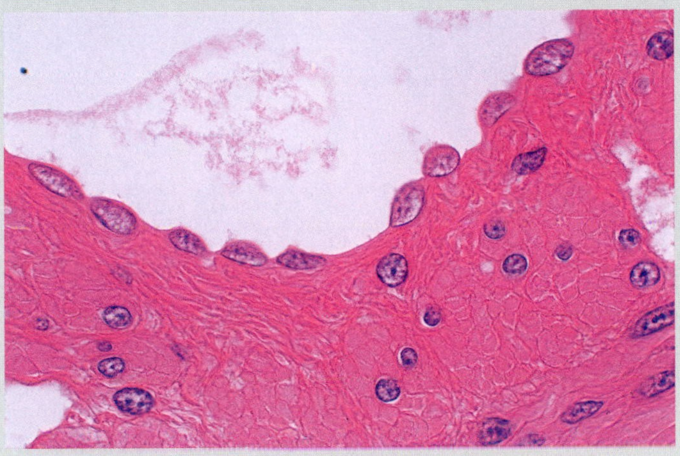

(a)

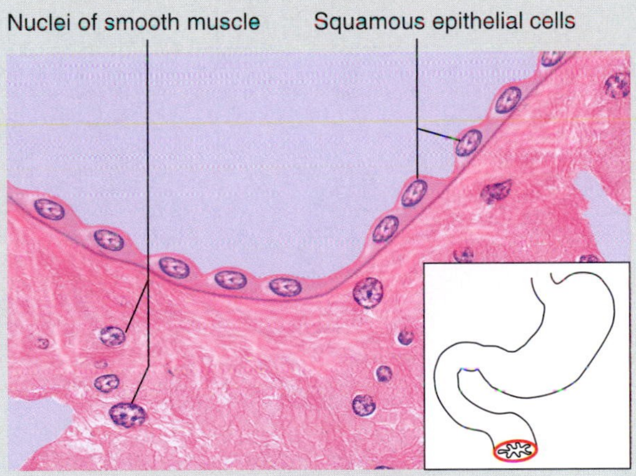

(b)

FIGURE 3.4

External Surface (serosa) of the Small Intestine.

Microscopic appearance: Single layer of thin cells, shaped like fried eggs with bulge where nucleus is located; nucleus flattened in the plane of the cell, like an egg yolk; cytoplasm may be so thin it is hard to see in tissue sections; in surface view, cells have angular contours and nuclei appear round

Representative locations: Air sacs (alveoli) of lungs; glomerular capsules of kidneys; some kidney tubules; inner lining (endothelium) of heart and blood vessels; serous membranes of stomach, intestines, and some other viscera; surface mesothelium of pleurae, pericardium, peritoneum, and mesenteries

Functions: Allows rapid diffusion or transport of substances through membrane; secretes lubricating serous fluid

Simple Cuboidal Epithelium

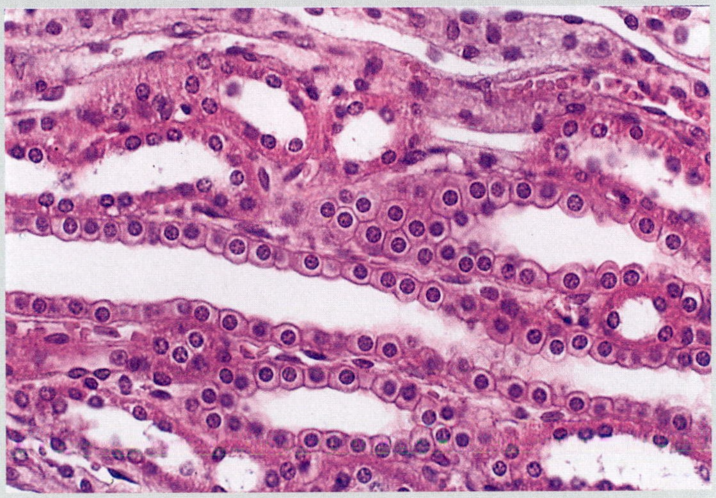

(a)

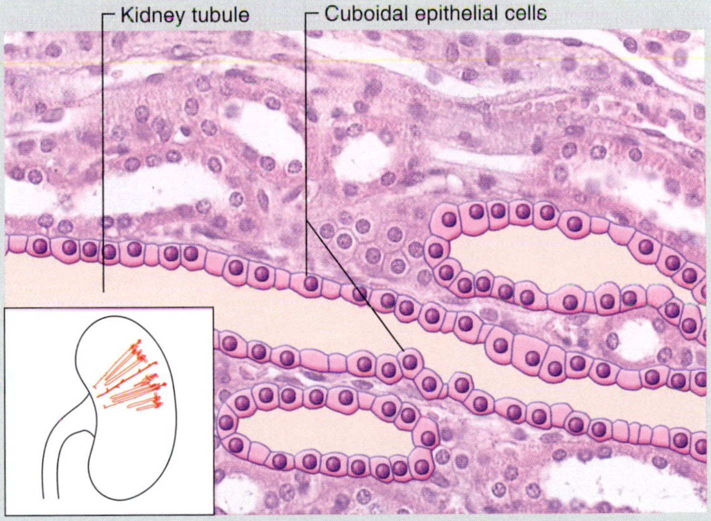

(b)

FIGURE 3.5

Kidney Tubules.

Microscopic appearance: Single layer of square or round cells; in glands, cells often pyramidal and arranged like segments of an orange around a central space; spherical, centrally placed nuclei; often with a brush border of microvilli in some kidney tubules; ciliated in bronchioles of lung

Representative locations: Liver, thyroid, mammary, salivary, and other glands; most kidney tubules; bronchioles

Functions: Absorption and secretion; production and movement of respiratory mucus

(continued)

Table 3.2
Simple Epithelia *(continued)*

Simple Columnar Epithelium

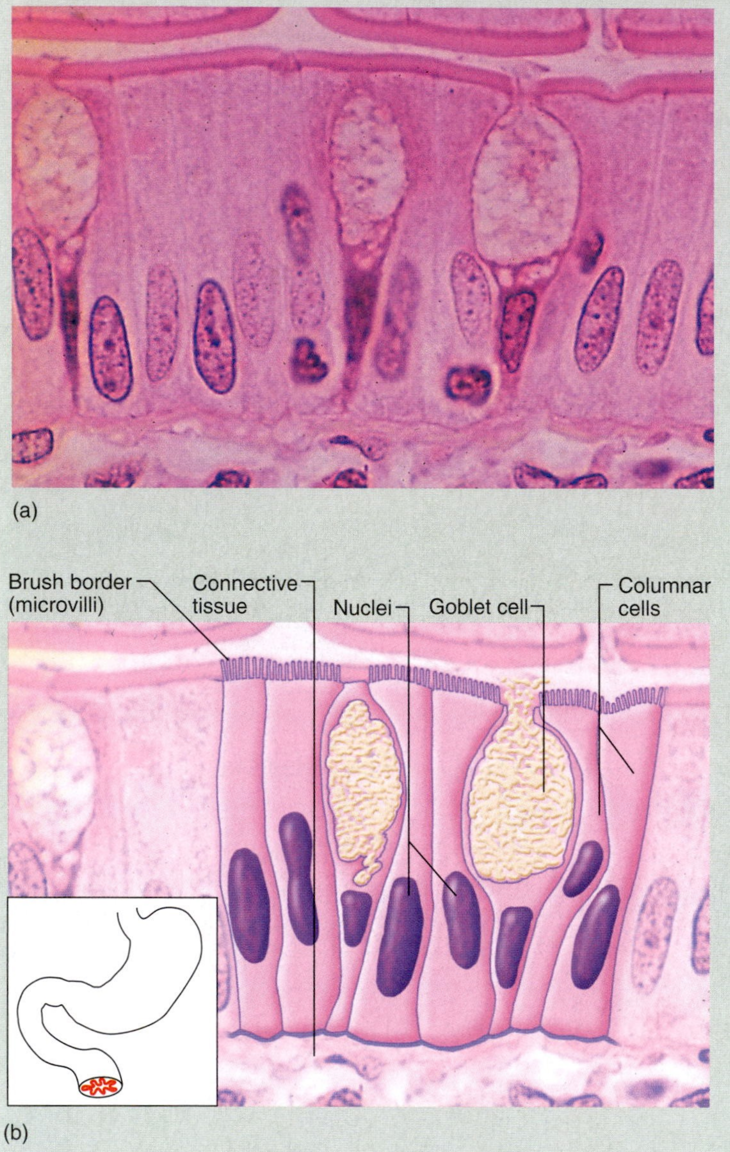
(a)
Brush border (microvilli)
Connective tissue
Nuclei
Goblet cell
Columnar cells
(b)

Figure 3.6

Internal Surface (mucosa) of the Small Intestine.

Microscopic appearance: Single layer of tall, narrow cells; oval or sausage-shaped nuclei, vertically oriented, usually in basal half of cell; apical portion of cell often shows secretory vesicles visible with TEM; often shows a brush border of microvilli; ciliated in some organs; may possess goblet cells

Representative locations: Inner lining of stomach, intestines, gallbladder, uterus, and uterine tubes; some kidney tubules

Functions: Absorption; secretion of mucus and other products; movement of egg and embryo in uterine tube

Pseudostratified Columnar Epithelium

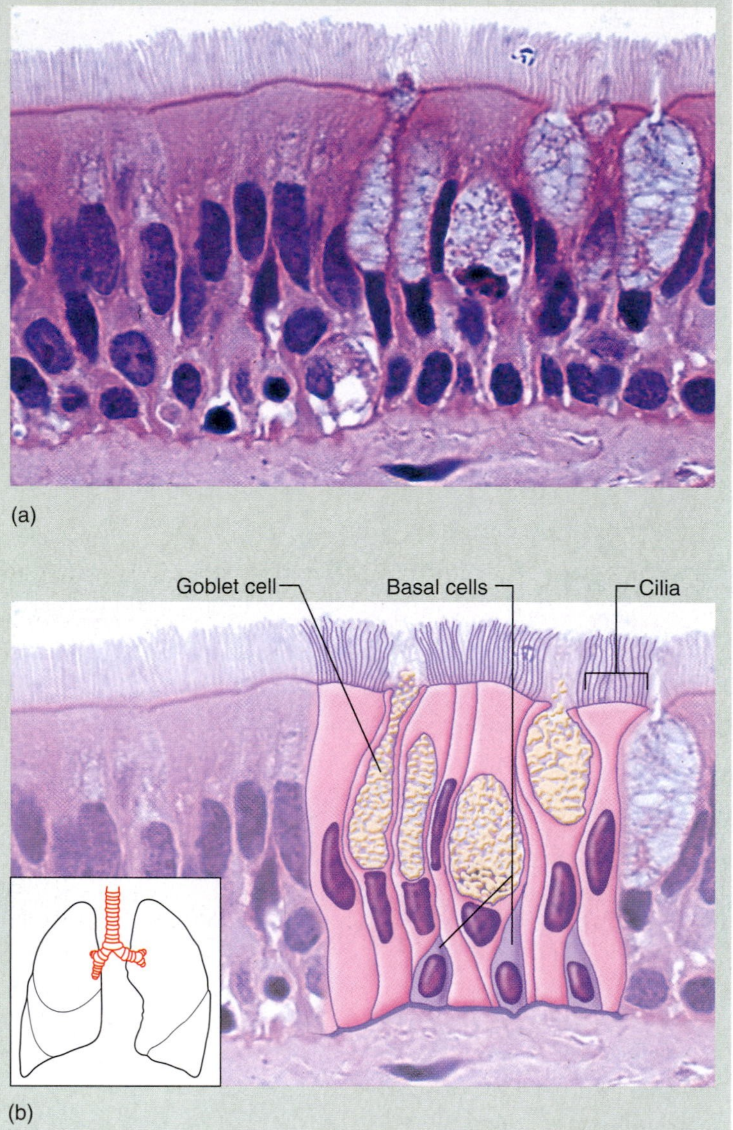
(a)
Goblet cell
Basal cells
Cilia
(b)

Figure 3.7

Mucosa of the Trachea.

Microscopic appearance: Looks multilayered; some cells do not reach free surface but all cells reach basement membrane; nuclei at several levels in deeper half of epithelium; often with goblet cells; often ciliated

Representative locations: Respiratory tract from nasal cavity to bronchi; portions of male urethra

Functions: Secretes and propels mucus

Stratified Epithelia

Stratified epithelia range from 2 to 20 or more layers of cells, with some cells resting directly on others and only the deepest layer resting on the basement membrane. Three of the stratified epithelia are named for the shapes of their surface cells: **stratified squamous, stratified cuboidal,** and **stratified columnar.** The deeper cells, however, may be of a different shape than the surface cells. The fourth type, **transitional epithelium,** was named when it was thought to represent a transitional stage between stratified squamous and stratified columnar epithelium. This is now known to be untrue, but the name has persisted.

Stratified columnar epithelium is rare—seen only in short stretches where two other epithelial types meet, as in limited regions of the pharynx, larynx, anal canal, and male urethra. We will not consider this type any further. The other three types are illustrated and summarized in table 3.3.

The most widespread epithelium in the body is stratified squamous epithelium, which warrants further discussion. Its deepest layer of cells are cuboidal to columnar, and undergo continual mitosis. Their daughter cells push toward the surface and become flatter (more squamous) as they migrate farther upward, until they finally die and flake off. Their separation from the surface is called **exfoliation (desquamation)** (fig. 3.8); the study of exfoliated cells is called *exfoliate cytology.* You can easily study exfoliated cells by scraping your gums with a toothpick, smearing this material on a slide, and staining it. A similar procedure is used in the *Pap smear,* an examination of exfoliated cells from the cervix for signs of uterine cancer (see fig. 26.29).

Stratified squamous epithelia are of two kinds—keratinized and nonkeratinized. A **keratinized (cornified)** epithelium, found on the skin surface (epidermis), is covered with a layer of compact, dead squamous cells. These cells are packed with the durable protein **keratin** and coated with a water-repellent glycolipid. The skin surface is therefore relatively dry, it retards water loss from the body, and it resists penetration by disease organisms. (Keratin is also the protein of which animal horns are made, hence its name.[5]) The tongue, oral mucosa, esophagus, vagina, and a few other internal membranes are covered with the **nonkeratinized** type, which lacks the surface layer of dead cells. This type provides a surface that is, again, abrasion-resistant, but also moist and slippery. These characteristics are well suited to resist stress produced by the chewing and swallowing of food and by sexual intercourse and childbirth.

[5]*kerat* = horn

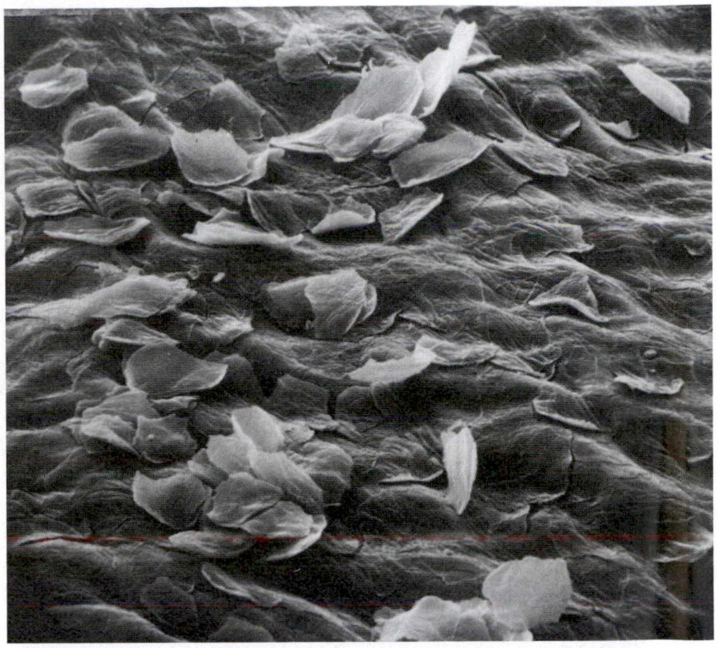

FIGURE 3.8
Exfoliation of Squamous Cells from the Vaginal Mucosa (SEM). A Pap smear is prepared from loose cells scraped from the epithelial surface. From R. G. Kessel and R. H. Kardon, *Tissues and Organs: A Text-Atlas of Scanning Electron Microscopy.* (W. H. Freeman, 1979).

Before You Go On

Answer the following questions to test your understanding of the preceding section:

6. Distinguish between simple and stratified epithelia, and explain why pseudostratified columnar epithelium belongs in the former category.
7. Explain how to distinguish a stratified squamous epithelium from a transitional epithelium.
8. What function do keratinized and nonkeratinized stratified squamous epithelia have in common? What is the structural difference between these two? How is this structural difference related to a functional difference between them?
9. How do the epithelia of the esophagus and stomach differ? How does this relate to their respective functions?

Table 3.3
Stratified Epithelia

Stratified Squamous Epithelium—Keratinized

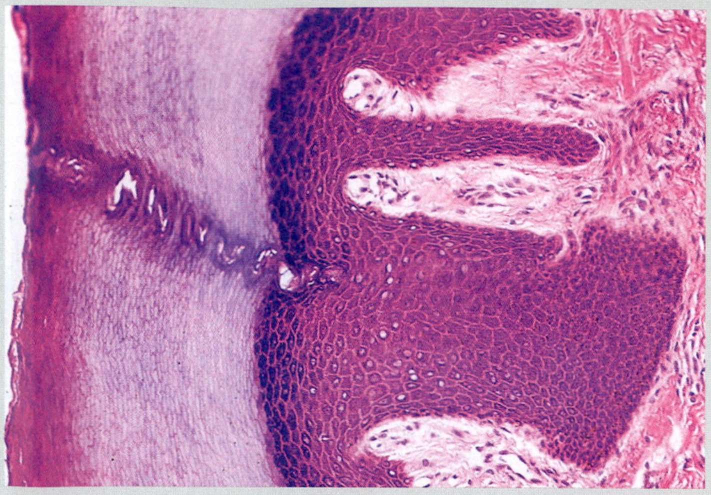

(a)

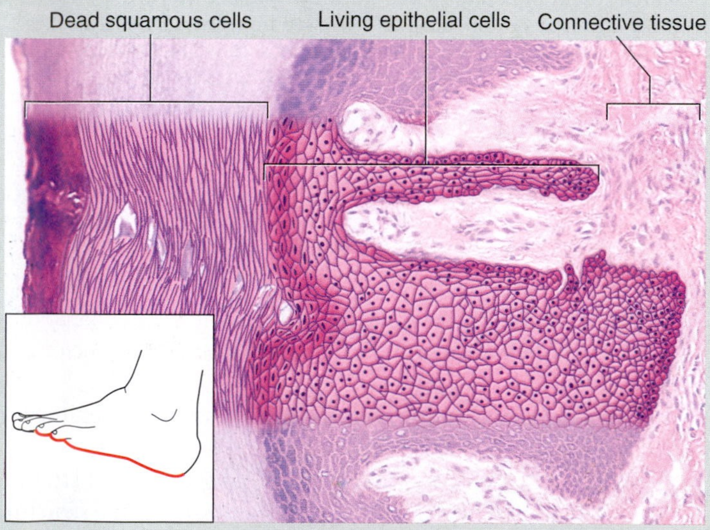
Dead squamous cells
Living epithelial cells
Connective tissue

(b)

Figure 3.9

Skin from the Sole of the Foot.

Microscopic appearance: Multiple cell layers with cells becoming increasingly flat and scaly toward surface; surface covered with a layer of compact dead cells without nuclei; basal cells may be cuboidal to columnar

Representative locations: Epidermis; palms and soles are especially heavily keratinized

Functions: Resists abrasion; retards water loss through skin; resists penetration by pathogenic organisms

Stratified Squamous Epithelium—Nonkeratinized

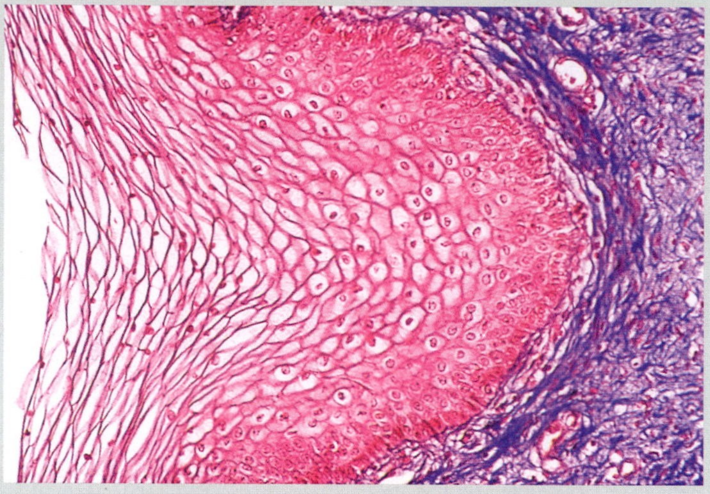

(a)

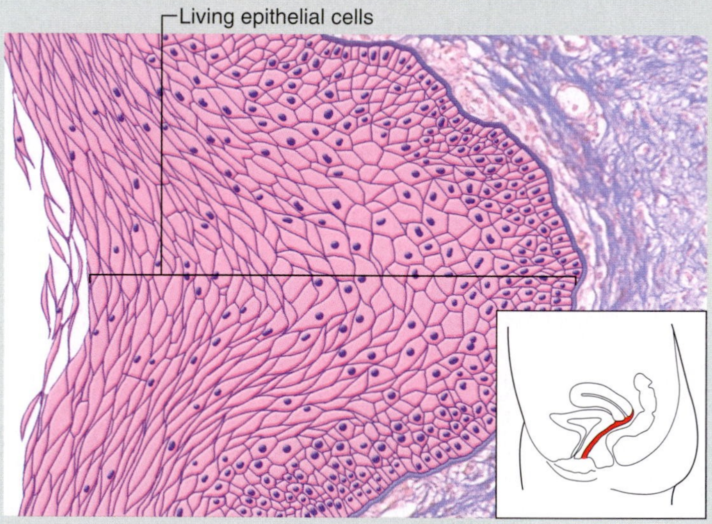
Living epithelial cells

(b)

Figure 3.10

Mucosa of the Vagina.

Microscopic appearance: Same as keratinized epithelium but without the surface layer of dead cells

Representative locations: Tongue, oral mucosa, esophagus, anal canal, vagina

Functions: Resists abrasion and penetration by pathogenic organisms

Stratified Cuboidal Epithelium

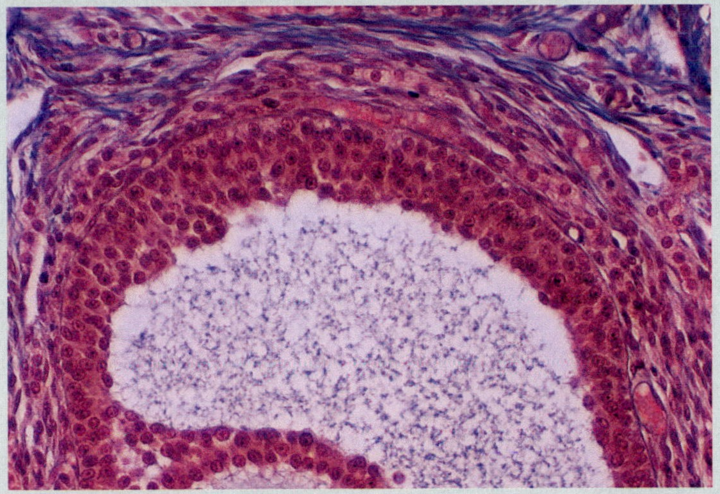

(a)

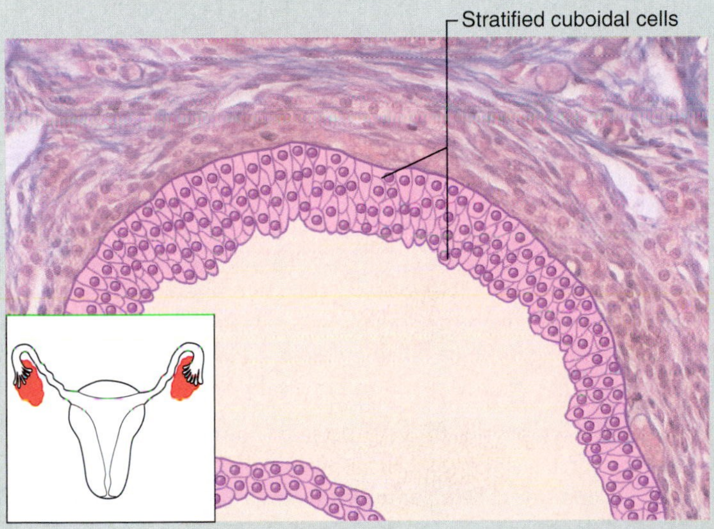

(b)

FIGURE 3.11

Wall of a Follicle in the Ovary.

Microscopic appearance: Two or more layers of cells; surface cells square or round

Representative locations: Sweat gland ducts; egg-producing vesicles (follicles) of ovaries; sperm-producing ducts (seminiferous tubules) of testis

Functions: Contributes to sweat secretion; secretes ovarian hormones; produces sperm

Transitional Epithelium

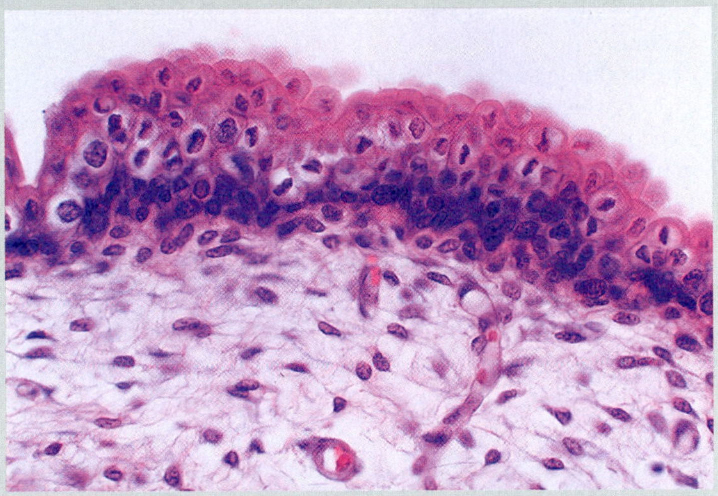

(a)

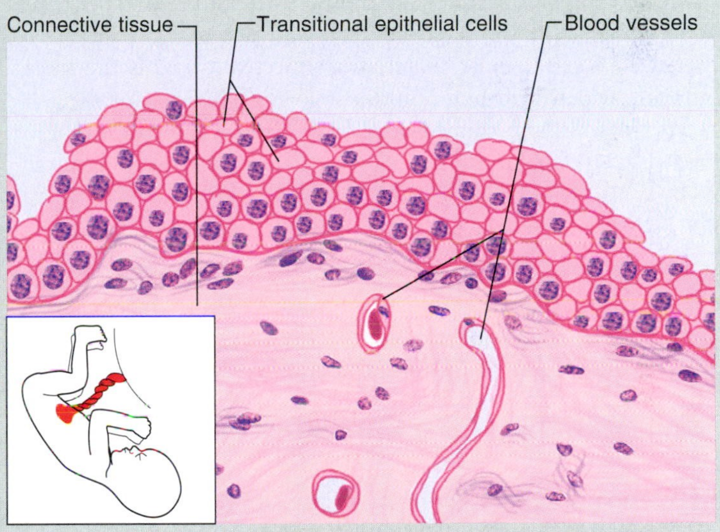

(b)

FIGURE 3.12

Allantoic Duct of the Umbilical Cord.

Microscopic appearance: Somewhat resembles stratified squamous epithelium, but surface cells are rounded, not flattened, and often bulge above surface; typically five or six cells thick when relaxed, two or three cells thick when stretched; cells may be flatter and thinner when epithelium is stretched (as in a distended bladder); some cells have two nuclei

Representative locations: Urinary tract—part of kidney, ureter, bladder, part of urethra; allantoic duct and external surface of umbilical cord

Functions: Stretches to allow filling of urinary tract

Connective Tissue

Objectives

When you have completed this section, you should be able to

- describe the properties that most connective tissues have in common;
- discuss the types of cells found in connective tissue;
- explain what the matrix of a connective tissue is and describe its components;
- name 10 types of connective tissue, describe their cellular components and matrix, and explain what distinguishes them from each other; and
- visually recognize each connective tissue type from specimens or photographs.

Overview

Connective tissue is a type of tissue in which cells usually occupy less space than the extracellular material, and which serves in most cases to bind organs to each other (for example, the way a tendon connects muscle to bone) or to support and protect organs. Most cells of a connective tissue are not in direct contact with each other, but are well separated by extracellular material. Connective tissue is the most abundant, widely distributed, and histologically variable of the primary tissues. Mature connective tissues fall into three broad categories: *fibrous connective tissue, supportive connective tissue* (cartilage and bone), and *fluid connective tissue* (blood).

The functions of connective tissue include the following:

- **Binding of organs.** Tendons bind muscle to bone, ligaments bind one bone to another, fat holds the kidneys and eyes in place, and fibrous tissue binds the skin to underlying muscle.
- **Support.** Bones support the body and cartilage supports the ears, nose, trachea, and bronchi.
- **Physical protection.** The cranium, ribs, and sternum protect delicate organs such as the brain, lungs, and heart; fatty cushions around the kidneys and eyes protect these organs.
- **Immune protection.** Connective tissue cells attack foreign invaders, and connective tissue fiber forms a "battlefield" under the skin and mucous membranes where immune cells can be quickly mobilized against disease agents.
- **Movement.** Bones provide the lever system for body movement, cartilages are involved in movement of the vocal cords, and cartilages on bone surfaces ease joint movements.
- **Storage.** Fat is the body's major energy reserve; bone is a reservoir of calcium and phosphorus that can be drawn upon when needed.
- **Heat production.** Metabolism of brown fat generates heat in infants and children.
- **Transport.** Blood transports gases, nutrients, wastes, hormones, and blood cells.

Fibrous Connective Tissue

Fibrous connective tissue is the most diverse type of connective tissue. It is also called *fibroconnective tissue* or *connective tissue proper*. Nearly all connective tissues contain fibers, but the tissues considered here are classified together because the fibers are so conspicuous. The tissue, of course, also includes cells and ground substance. Before examining specific types of fibrous connective tissue, we will examine these components.

COMPONENTS OF FIBROUS CONNECTIVE TISSUE

Cells The cells of fibrous connective tissue include the following types:

- **Fibroblasts.**[6] These are large, flat cells that often appear tapered at the ends and show slender, wispy branches. They produce the fibers and ground substance that form the matrix of the tissue.
- **Macrophages.**[7] These are large phagocytic cells that wander through the connective tissues. They phagocytize and destroy bacteria, other foreign matter, and dead or dying cells of our own body, and they activate the immune system when they sense foreign matter called *antigens.* They arise from certain white blood cells called *monocytes* and from the stem cells that produce monocytes.
- **Leukocytes,**[8] or **white blood cells (WBCs).** WBCs travel briefly in the bloodstream, then crawl out through the capillary walls and spend most of their time in the connective tissues. The two most common types are *neutrophils,* which wander about in search of bacteria, and *lymphocytes,* which react against bacteria, toxins, and other foreign agents. Lymphocytes often form dense patches in the mucous membranes.
- **Plasma cells.** Certain lymphocytes turn into plasma cells when they detect foreign agents. The plasma cells then synthesize disease-fighting proteins called *antibodies.* Plasma cells are rarely seen except in inflamed tissue and the walls of the intestines.
- **Mast cells.** These cells, found especially alongside blood vessels, secrete a chemical called *heparin* that inhibits blood clotting, and one called *histamine* that increases blood flow by dilating blood vessels.
- **Adipocytes** (AD-ih-po-sites), or **fat cells.** These appear in small clusters in some fibrous connective tissues. When they dominate an area, the tissue is called *adipose tissue.*

Fibers Three types of protein fibers are found in fibrous connective tissues:

- **Collagenous** (col-LADJ-eh-nus) **fibers.** These fibers, made of collagen, are tough and flexible and resist stretching. Collagen is about 25% of the body's protein, the most abundant type. It is the base of such animal products as gelatin, leather, and glue.[9] In fresh tissue, collagenous fibers have a glistening white appearance, as seen in tendons and some cuts of meat (fig. 3.13); thus, they are

[6] *fibro* = fiber + *blast* = producing
[7] *macro* = big + *phage* = eater
[8] *leuko* = white + *cyte* = cell
[9] *colla* = glue + *gen* = producing

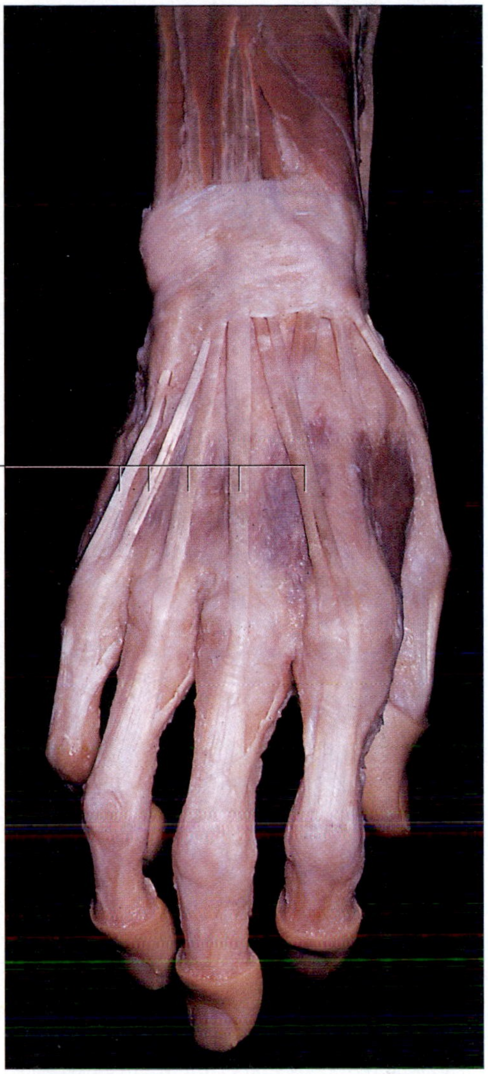

FIGURE 3.13

Tendons of the Hand. The white glistening appearance results from the collagen of which tendons are composed.

often called *white fibers.* In tissue sections, collagen forms coarse, wavy bundles, often dyed pink, blue, or green by the most common histological stains. Tendons, ligaments, and the deep layer of the skin (the dermis) are made mainly of collagen. Less visibly, collagen pervades the matrix of cartilage and bone.

- **Reticular**[10] **fibers.** These are thin collagen fibers coated with glycoprotein. They form a spongelike framework for such organs as the spleen and lymph nodes.
- **Elastic fibers.** These are thinner than collagenous fibers, and they branch and rejoin each other along their course. They are made of a protein called **elastin,** whose coiled structure allows it to stretch and recoil like a rubber band. Elastic fibers account for the ability of the skin, lungs, and arteries to spring back after they are stretched. (Elasticity is not the ability to stretch, but the tendency to recoil when tension is released.) Fresh elastic fibers are yellowish and therefore often called *yellow fibers.*

[10]*ret* = network + *icul* = little

INSIGHT 3.1 CLINICAL APPLICATION

COLLAGEN DISEASES

The gene for collagen is especially subject to mutation, and there are consequently several diseases that stem from hereditary defects in collagen synthesis. Since collagen is such a widespread protein in the body, the effects are very diverse. People with *Ehlers-Danlos*[11] syndrome have abnormally long, loose collagen fibers, which show their effects in unusually stretchy skin, loose joints, slow wound healing, and abnormalities in the blood vessels, intestines, and urinary bladder. Infants with this syndrome are often born with dislocated hips. *Osteogenesis imperfecta* is a hereditary collagen disease that affects bone development (see insight 3.3).

Not all collagen diseases are hereditary, however. *Scurvy,* for example, results from a dietary deficiency of vitamin C (ascorbic acid). Ascorbic acid is a cofactor needed for the metabolism of proline and lysine, two amino acids that are especially abundant in collagen. The signs of scurvy include bleeding gums, loose teeth, subcutaneous and intramuscular hemorrhages, and poor wound healing.

[11]Edward L. Ehlers (1863–1937), Danish dermatologist; Henri A. Danlos (1844–1912), French dermatologist

Ground Substance Amid the cells and fibers in some connective tissue sections, there appears to be a lot of empty space. In life, this space is occupied by the featureless ground substance. Ground substance usually has a gelatinous to rubbery consistency resulting from three classes of large molecules composed of protein and carbohydrate, called *glycosaminoglycans (GAGs), proteoglycans,* and *adhesive glycoproteins.* Some of these molecules are up to 20 μm long—larger than some cells. GAGs also form a very slippery lubricant in the joints and constitute much of the jellylike *vitreous humor* of the eyeball. In connective tissue, such molecules form a gel that slows down the spread of bacteria and other pathogens (disease-causing agents). Adhesive glycoproteins bind plasma membrane proteins to collagen and proteoglycans outside the cell. They bind all the components of a tissue together and mark pathways that guide migrating embryonic cells to their destinations in a tissue.

TYPES OF FIBROUS CONNECTIVE TISSUE

Fibrous connective tissue is divided into two broad categories according to the relative abundance of fiber: *loose* and *dense connective tissue.* In **loose connective tissue,** much of the space is occupied by ground substance, which is dissolved out of the tissue during histological fixation and leaves empty space in prepared tissue sections. The loose connective tissues we will discuss are *areolar, reticular,* and *adipose tissue.* In **dense connective tissue,** fiber occupies more space than the cells and ground substance, and appears closely packed in tissue sections. The two dense connective tissues we will discuss are *dense regular* and *dense irregular connective tissue.*

Areolar[12] (AIR-ee-OH-lur) **tissue** exhibits loosely organized fibers, abundant blood vessels, and a lot of seemingly empty space. It possesses all six of the aforementioned cell types. Its fibers run in random directions and are mostly collagenous, but elastic and reticular fibers are also present. Areolar tissue is highly variable in appearance. In many serous membranes, it looks like figure 3.14, but in the skin and mucous membranes, it is more compact (see fig. 3.9)

[12]*areola* = little space

and sometimes difficult to distinguish from dense irregular connective tissue. Some advice on how to tell them apart is given after the discussion of dense irregular connective tissue.

Areolar tissue is found in tissue sections from almost every part of the body. It surrounds blood vessels and nerves and penetrates with them even into the small spaces of muscles, tendons, and other tissues. Nearly every epithelium rests on a layer of areolar tissue, whose blood vessels provide the epithelium with nutrition, waste removal, and a ready supply of infection-fighting leukocytes in times of need. Because of the abundance of open, fluid-filled space, leukocytes can move about freely in areolar tissue and can easily find and destroy pathogens.

Reticular tissue is a mesh of reticular fibers and fibroblasts. It forms the structural framework (stroma) of such organs and tissues as the lymph nodes, spleen, thymus, and bone marrow. The space amid the fibers is filled with blood cells. Imagine a kitchen sponge soaked with blood; the sponge fibers are analogous to the reticular tissue stroma.

Adipose tissue or **fat,** is tissue in which adipocytes are the dominant cell type. Adipocytes may also occur singly or in small clusters in areolar tissue. Adipocytes usually range from 70 to 120 μm in diameter, but they may be five times as large in obese people. The space between adipocytes is occupied by areolar tissue, reticular tissue, and blood capillaries.

Fat is the body's primary energy reservoir. The quantity of stored fat and the number of adipocytes are quite stable in a person, but this doesn't mean stored fat is stagnant. New triglycerides (fat molecules) are constantly being synthesized and stored as others are broken down and released into circulation. Thus, there is a constant turnover of stored triglycerides, with an equilibrium between synthesis and breakdown, energy storage and energy use. Adipose tissue also provides thermal insulation, and it contributes to body contours such as the female breasts and hips.

Most adipose tissue is of a type called *white fat,* but fetuses, infants, and children also have a heat-generating tissue called *brown fat,* which accounts for up to 6% of an infant's weight. Brown fat gets its color from an unusual abundance of blood vessels and lysosomes. It stores fat in the form of multiple droplets rather than one large one. Brown fat has numerous mitochondria, but their oxidative metabolism is not linked to ATP synthesis. Therefore, when these cells oxidize fats, they release all of the energy as heat. Hibernating animals accumulate brown fat in preparation for winter.

Table 3.4 summarizes the three types of loose connective tissue.

Think About It!

Why would infants and children have more need for brown fat than adults do? (Hint: Smaller bodies have a higher ratio of surface area to volume than larger bodies do.)

Dense regular connective tissue is named for two properties: (1) the collagen fibers are closely packed and leave relatively little open space, and (2) the fibers are parallel to each other. It is found especially in tendons and ligaments. The parallel arrangement of fibers is an adaptation to the fact that tendons and ligaments are pulled in predictable directions. With some minor exceptions such as blood vessels and sensory nerve fibers, the only cells in this tissue are fibroblasts, visible by their slender, violet-staining nuclei squeezed between bundles of collagen. This type of tissue has few blood vessels and receives a meager supply of oxygen and nutrients, so injured tendons and ligaments are slow to heal.

The vocal cords, suspensory ligament of the penis, and some ligaments of the vertebral column are made of a type of dense regular connective tissue called **yellow elastic tissue.** In addition to the densely packed collagen fibers, it exhibits branching elastic fibers and more fibroblasts. The fibroblasts have larger, more conspicuous nuclei than seen in most dense regular connective tissue.

Elastic tissue also takes the form of wavy sheets in the walls of the large and medium arteries. When the heart pumps blood into the arteries, these sheets enable them to expand and relieve some of the pressure on smaller vessels downstream. When the heart relaxes, the arterial wall springs back and keeps the blood pressure from dropping too low between heartbeats. The importance of this elastic tissue becomes especially clear in diseases such as atherosclerosis, where it is stiffened by lipid and calcium deposits, and Marfan syndrome, a genetic defect in elastin synthesis (see insight 3.2).

Dense irregular connective tissue also has thick bundles of collagen and relatively little room for cells and ground substance, but the collagen bundles run in random directions. This arrangement enables the tissue to resist unpredictable stresses. Dense irregular connective tissue constitutes most of the dermis, where it binds the skin to the underlying muscle and connective tissue. It forms a protective capsule around organs such as the kidneys, testes, and spleen and a tough fibrous sheet around the bones, nerves, and most cartilages.

It is sometimes difficult to judge whether a tissue is areolar or dense irregular. In the dermis, for example, these tissues occur side by side, and the transition from one to the other is not at all obvious. A relatively large amount of clear space suggests areolar tissue, and thicker bundles of collagen with relatively little clear space suggests dense irregular tissue. The dense connective tissues are summarized in table 3.5.

INSIGHT 3.2 CLINICAL APPLICATION

The Consequences of Defective Elastin

Marfan[13] *syndrome* is a hereditary defect in elastin fibers, usually resulting from a mutation in the gene for *fibrillin,* a glycoprotein that forms the structural scaffold for elastin. Clinical signs of Marfan syndrome include hyperextensible joints, hernias of the groin, and vision problems resulting from abnormally elongated eyes and deformed lenses. People with Marfan syndrome typically show unusually tall stature, long limbs, spidery fingers, abnormal spinal curvature, and a protruding "pigeon breast." More serious problems are weakened heart valves and arterial walls. The aorta, where blood pressure is highest, is sometimes enormously dilated close to the heart and may rupture. Marfan syndrome is present in about 1 out of 20,000 live births, and most victims die by their mid-30s. Some authorities speculate that Abraham Lincoln's tall, gangly physique and spindly fingers were signs of Marfan syndrome, which may have ended his life prematurely had he not been assassinated. A number of star athletes have died at a young age of Marfan syndrome, including Olympic volleyball champion Flo Hyman, who died of a ruptured aorta during a game in Japan in 1986, at the age of 31.

[13]Antoine Bernard-Jean Marfan (1858–1942), French physician

TABLE 3.4
Loose Connective Tissues

Areolar Tissue

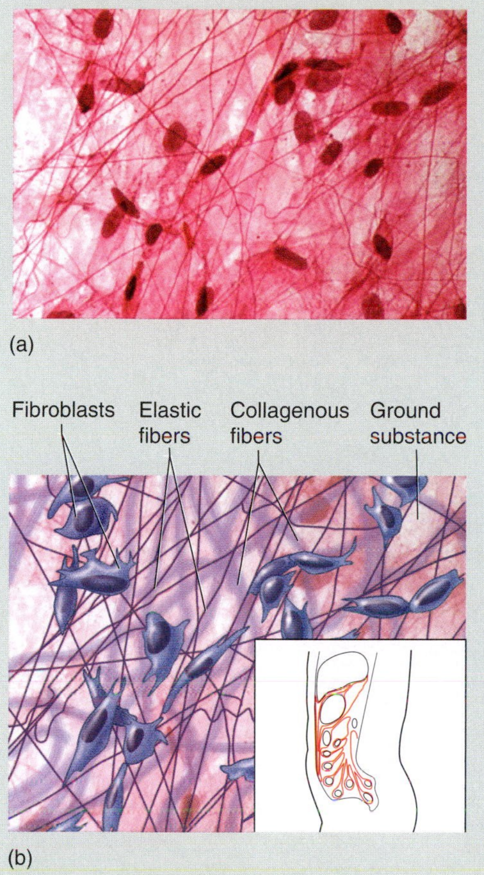
(a)
Fibroblasts
Elastic fibers
Collagenous fibers
Ground substance
(b)

FIGURE 3.14
Spread of the Mesentery.

Microscopic appearance: Loose arrangement of collagenous and elastic fibers; scattered cells of various types; abundant ground substance; numerous blood vessels

Representative locations: Underlying nearly all epithelia; surrounding blood vessels, nerves, esophagus, and trachea; fasciae between muscles; mesenteries; visceral layers of pericardium and pleura

Functions: Loosely binds epithelia to deeper tissues; allows passage of nerves and blood vessels through other tissues; provides an arena for immune defense; provides nutrients and waste removal for overlying epithelia

Reticular Tissue

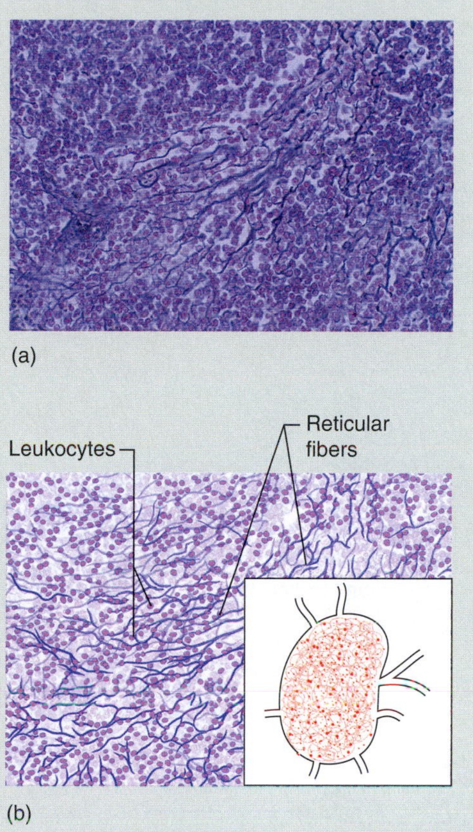
(a)
Leukocytes
Reticular fibers
(b)

FIGURE 3.15
Lymph Node.

Microscopic appearance: Loose network of reticular fibers and cells, infiltrated with numerous lymphocytes and other blood cells

Representative locations: Lymph nodes, spleen, thymus, bone marrow

Functions: Supportive stroma (framework) for lymphatic organs

Adipose Tissue

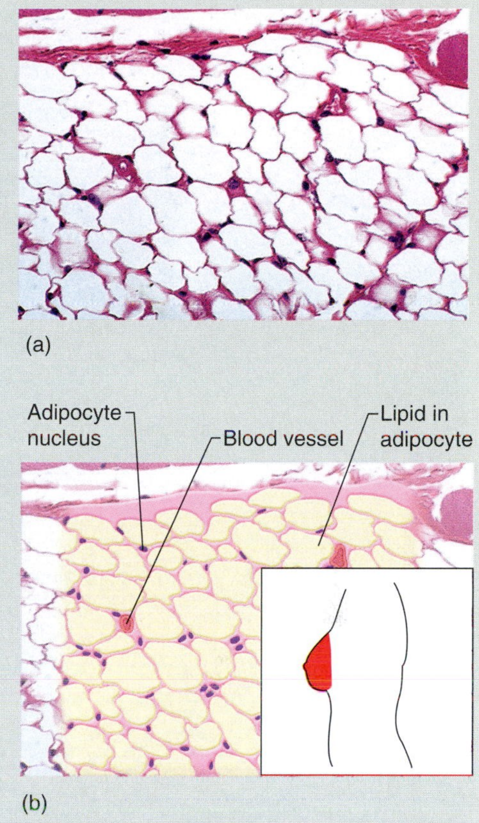
(a)
Adipocyte nucleus
Blood vessel
Lipid in adipocyte
(b)

FIGURE 3.16
Adipose Tissue.

Microscopic appearance: Dominated by adipocytes—large, empty-looking cells with thin margins; adipocytes usually shrunken by histological fixatives; nucleus pressed against plasma membrane; tissue sections often pale because of scarcity of stained cytoplasm; blood vessels often present

Representative locations: Subcutaneous fat beneath skin; breast; mesenteries; surrounding organs such as heart, kidneys, and eyes

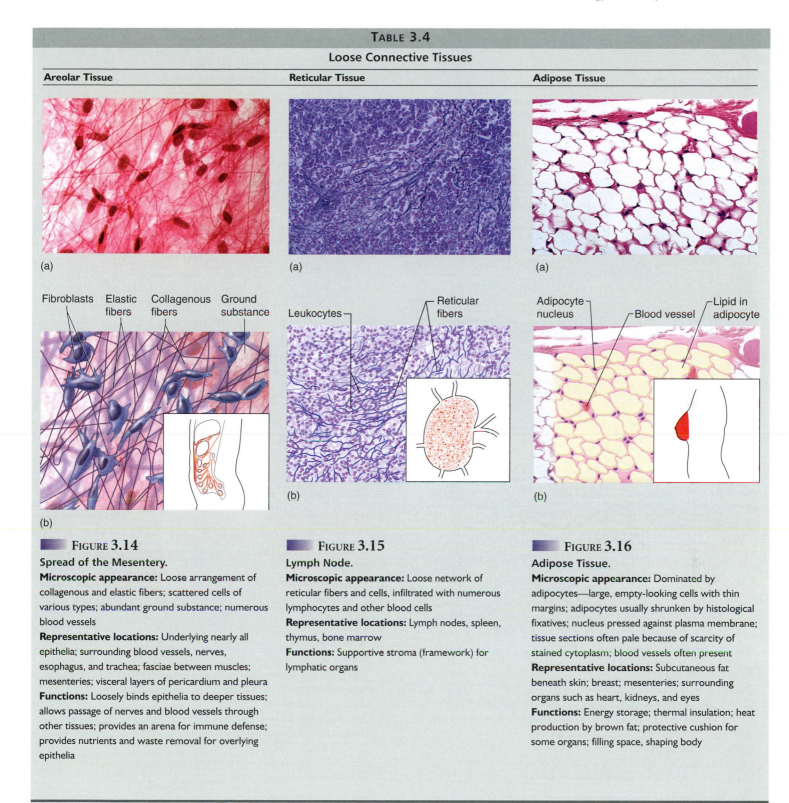

Functions: Energy storage; thermal insulation; heat production by brown fat; protective cushion for some organs; filling space, shaping body

TABLE 3.5

Dense Connective Tissues

Dense Regular Connective Tissue

(a)

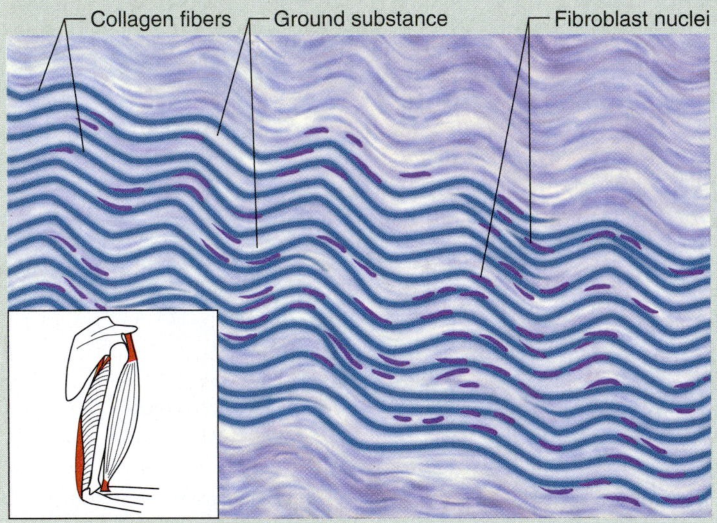
Collagen fibers
Ground substance
Fibroblast nuclei

(b)

FIGURE 3.17

Tendon.

Microscopic appearance: Densely packed, parallel, often wavy collagen fibers; slender fibroblast nuclei compressed between collagen bundles; scanty open space (ground substance); scarcity of blood vessels

Representative locations: Tendons and ligaments

Functions: Ligaments tightly bind bones together and resist stress; tendons attach muscle to bone and move the bones when the muscles contract.

Dense Irregular Connective Tissue

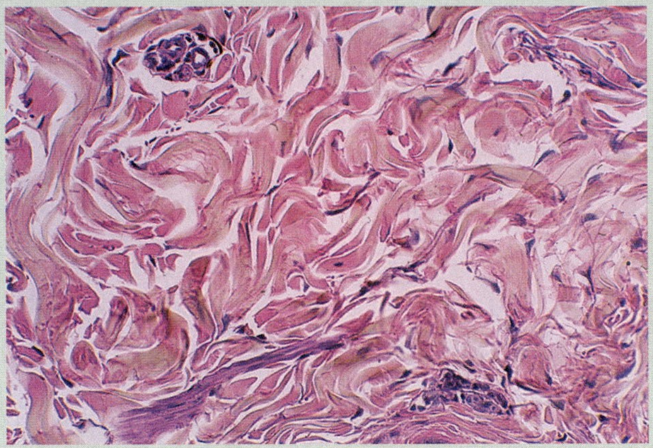

(a)

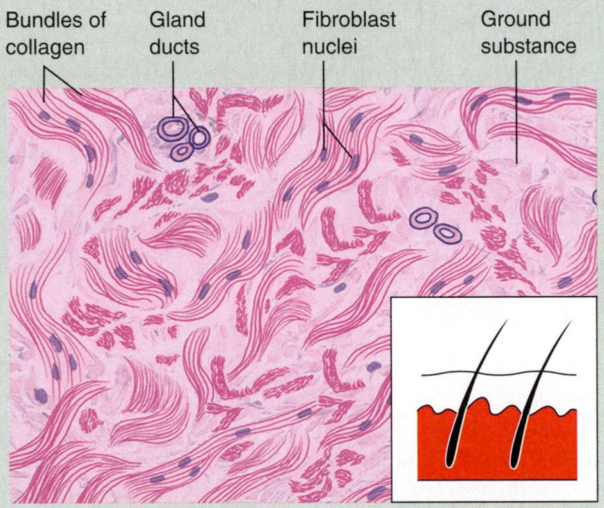
Bundles of collagen
Gland ducts
Fibroblast nuclei
Ground substance

(b)

FIGURE 3.18

Dermis of the Skin.

Microscopic appearance: Densely packed collagen fibers running in random directions; scanty open space (ground substance); few visible cells; scarcity of blood vessels

Representative locations: Deeper portion of dermis of skin; capsules around viscera such as liver, kidney, spleen; fibrous sheaths around muscles, nerves, cartilages, and bones

Functions: Durable, hard to tear; withstands stresses applied in unpredictable directions

Cartilage

Cartilage (table 3.6) is a supportive connective tissue with a flexible rubbery matrix. It gives shape to the external ear, the tip of the nose, and the larynx (voicebox)—the most easily palpated cartilages in the body. Cells called **chondroblasts**[14] (CON-dro-blasts) secrete the matrix and surround themselves with it until they become trapped in little cavities called **lacunae**[15] (la-CUE-nee). Once enclosed in lacunae, the cells are called **chondrocytes** (CON-dro-sites). Cartilage rarely exhibits blood vessels except when transforming into bone; thus nutrition and waste removal depend on solute diffusion through the stiff matrix. Because this is a slow process, chondrocytes have low rates of

[14] *chondro* = cartilage, gristle + *blast* = forming

[15] *lacuna* = lake, cavity

TABLE 3.6

Types of Cartilage

Hyaline Cartilage

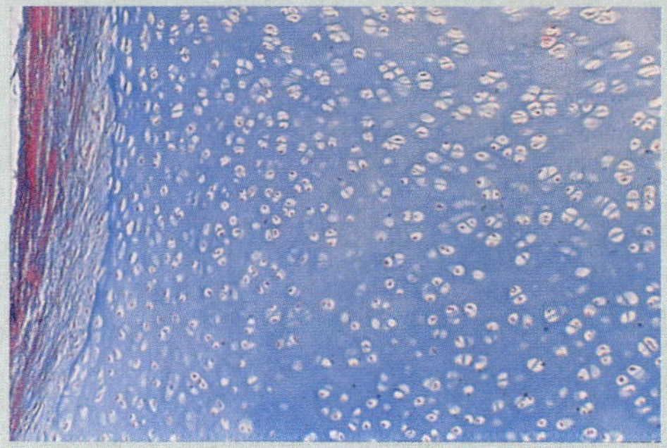
(a)

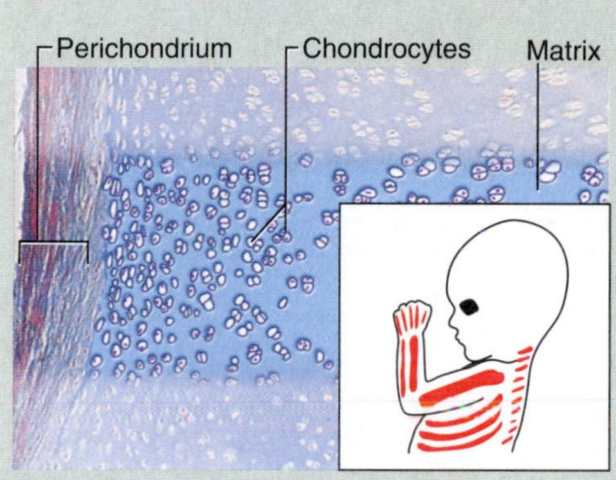
Perichondrium
Chondrocytes
Matrix
(b)

FIGURE 3.19
Fetal Skeleton.
Microscopic appearance: Clear, glassy matrix, often stained light blue or pink in tissue sections; fine, dispersed collagen fibers, not usually visible; chondrocytes often in small clusters of three or four cells (*cell nests*), enclosed in lacunae; usually covered by perichondrium
Representative locations: Forms a thin *articular cartilage*, lacking perichondrium, over the ends of bones at movable joints; a *costal cartilage* attaches the end of a rib to the breastbone; forms supportive rings and plates around trachea and bronchi; forms a boxlike enclosure around the larynx; forms much of the fetal skeleton
Functions: Eases joint movements; holds airway open during respiration; moves vocal cords during speech; a precursor of bone in the fetal skeleton and forms the growth zones of long bones of children

Elastic Cartilage

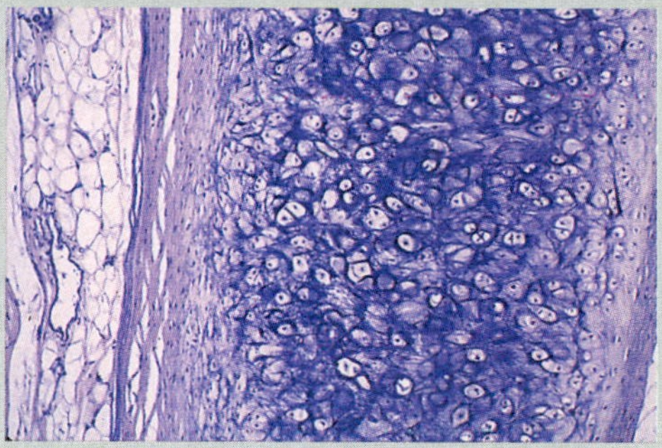
(a)

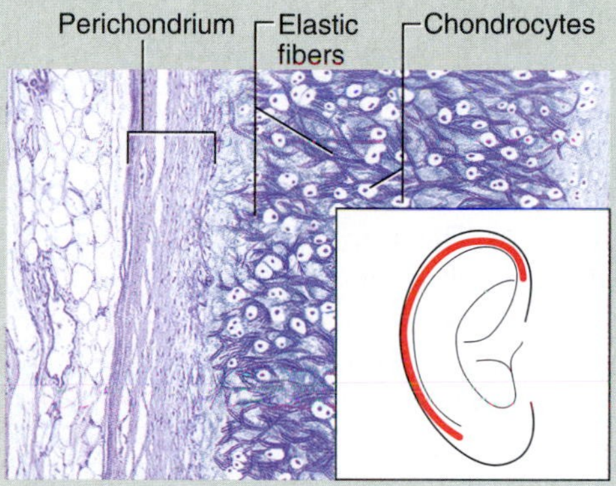
Perichondrium
Elastic fibers
Chondrocytes
(b)

FIGURE 3.20
External Ear.
Microscopic appearance: Elastic fibers form weblike mesh amid lacunae; always covered by perichondrium
Representative locations: External ear; epiglottis
Functions: Provides flexible, elastic support

Fibrocartilage

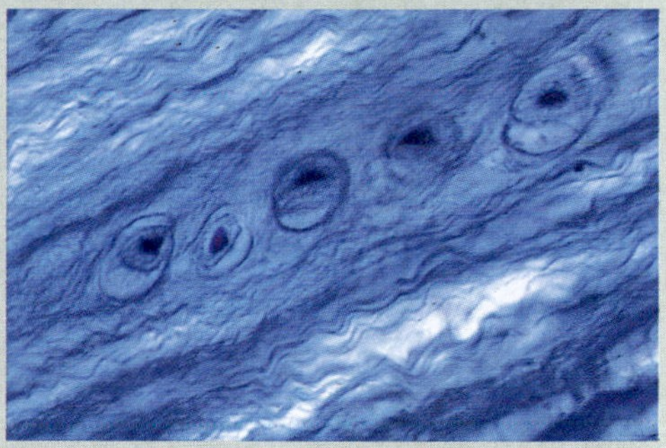
(a)

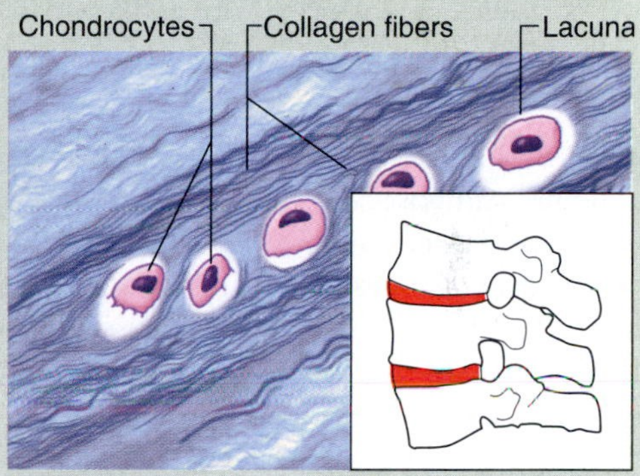
Chondrocytes
Collagen fibers
Lacuna
(b)

FIGURE 3.21
Intervertebral Disc.
Microscopic appearance: Parallel collagen fibers similar to those of tendon; rows of chondrocytes in lacunae between collagen fibers; never has a perichondrium
Representative locations: Pubic symphysis (anterior joint between two halves of pelvic girdle); intervertebral discs that separate bones of spinal column; menisci, or pads of shock-absorbing cartilage, in knee joint; at points where tendons insert on bones near articular hyaline cartilage
Functions: Resists compression and absorbs shock in some joints; often a transitional tissue between dense connective tissue and hyaline cartilage (for example, at some tendon-bone junctions)

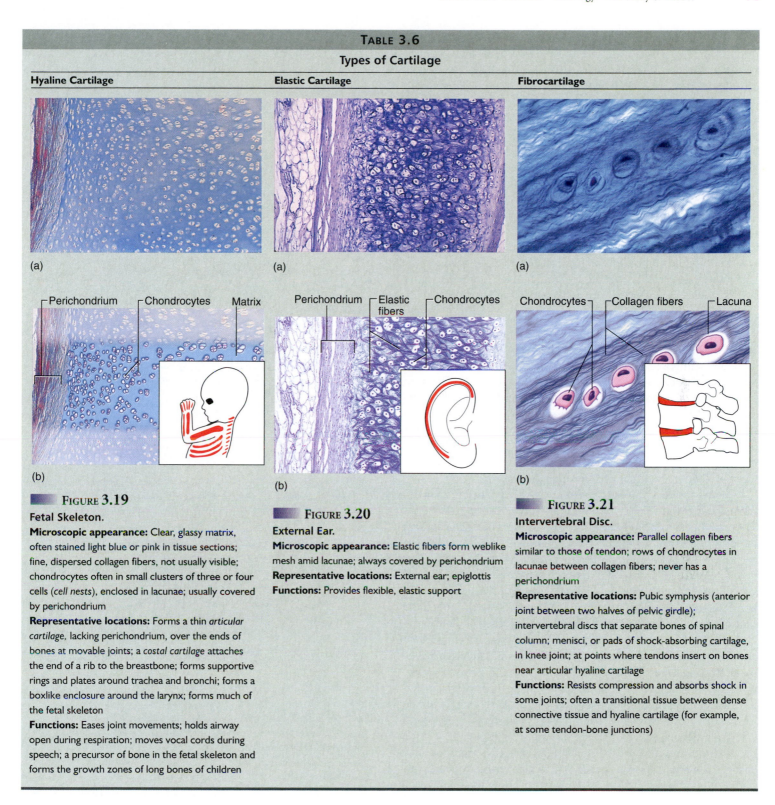

metabolism and cell division, and injured cartilage heals slowly. The matrix is rich in glycosaminoglycans and contains collagen fibers that range in thickness from invisibly fine to conspicuously coarse. Differences in the fibers provide a basis for classifying cartilage into three types: *hyaline cartilage, elastic cartilage,* and *fibrocartilage.*

THINK ABOUT IT!

When the following tissues are injured, which do you think is the fastest to heal and which do you think is the slowest—cartilages, adipose tissue, or tendons? Explain your reasoning.

Hyaline[16] (HY-uh-lin) **cartilage** is named for its clear, glassy microscopic appearance, which stems from the usually invisible fineness of its collagen fibers. **Elastic cartilage** is named for its conspicuous elastic fibers, and **fibrocartilage** for its coarse, readily visible bundles of collagen. Elastic cartilage and most hyaline cartilage are surrounded by a sheath of dense irregular connective tissue called the **perichondrium.**[17] A reserve population of chondroblasts between the perichondrium and cartilage contributes to cartilage growth throughout life. There is no perichondrium around fibrocartilage.

You can feel the texture of hyaline cartilage by palpating the tip of your nose, your "Adam's apple" at the front of the larynx (voicebox), and periodic rings of cartilage around the trachea (windpipe) just below the larynx. Hyaline cartilage is easily seen in many grocery items—it is the "gristle" at the ends of pork ribs, on chicken leg and breast bones, and at the joints of pigs' feet, for example. Elastic cartilage gives shape to the external ear. You can get some idea of its springy resilience by folding your ear down and releasing it.

Bone

The term *bone* refers both to organs of the body such as the femur and mandible, composed of multiple tissue types, and to the bone tissue, or **osseous tissue,** that makes up most of the mass of bones. There are two forms of osseous tissue: (1) **Spongy bone** fills the heads of the long bones (see fig. 6.4*a*). Although it is calcified and hard, its delicate slivers and plates give it a spongy appearance. (2) **Compact (dense) bone** is a more dense calcified tissue with no spaces visible to the naked eye. It forms the external surfaces of all bones, so spongy bone, when present, is always covered by compact bone.

Further differences between compact and spongy bone are described in chapter 6. Here, we examine only compact bone (table 3.7). Most specimens you study will probably be chips of dried compact bone ground to microscopic thinness. In such preparations, the cells are absent but spaces reveal their former locations. Most compact bone is arranged in cylinders of tissue that surround **central (haversian**[18]**)** canals, which run longitudinally through the shafts of long bones such as the femur. Blood vessels and nerves travel through the central canals in life. The bone matrix is deposited in concentric **lamellae,**[19] onionlike layers around each central canal. A central canal and its surrounding lamellae are called an **osteon.** Tiny lacunae between the lamellae are occupied in life by mature bone cells, or **osteocytes.**[20] Delicate canals called **canaliculi**[21] radiate from each lacuna to its neighbors and allow the osteocytes to contact each other. The bone as a whole is covered with a tough fibrous **periosteum**[22] (PERR-ee-OSS-tee-um) similar to the perichondrium of cartilage.

About one-third of the dry weight of bone is composed of collagen fibers and glycosaminoglycans, which enable bone to bend slightly under stress; two-thirds consists of minerals (mainly calcium salts) that enable bones to withstand compression by the weight of the body.

TABLE 3.7

Bone

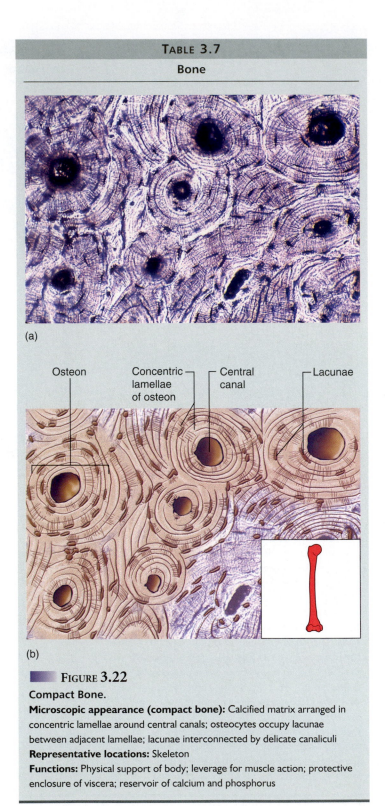

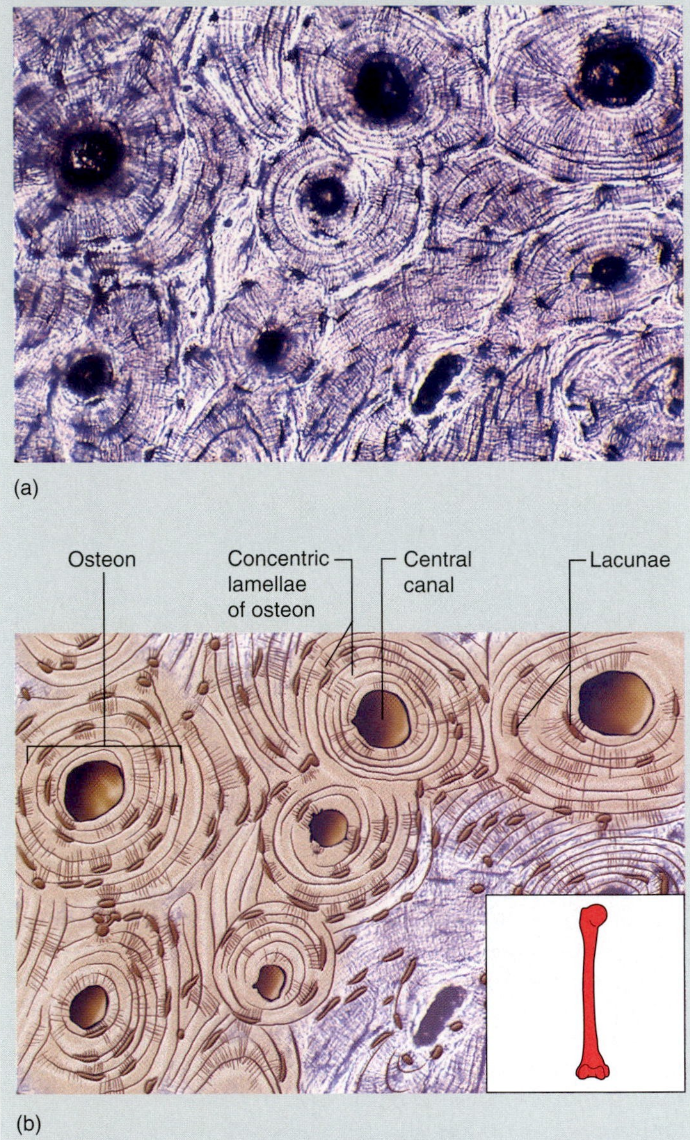
(a)
Osteon
Concentric lamellae of osteon
Central canal
Lacunae
(b)

FIGURE 3.22

Compact Bone.

Microscopic appearance (compact bone): Calcified matrix arranged in concentric lamellae around central canals; osteocytes occupy lacunae between adjacent lamellae; lacunae interconnected by delicate canaliculi

Representative locations: Skeleton

Functions: Physical support of body; leverage for muscle action; protective enclosure of viscera; reservoir of calcium and phosphorus

Blood

Blood (table 3.8) is a liquid connective tissue that travels through tubular vessels. Its primary function is to transport cells and dissolved matter from place to place. Blood consists of a ground substance called **plasma** and of cells and cell fragments collectively called **formed elements.** The most abundant formed elements are

[16]*hyal* = glass
[17]*peri* = around + *chondri* = cartilage
[18]Clopton Havers (1650–1702), English anatomist
[19]*lam* = plate + *ella* = little
[20]*osteo* = bone + *cyte* = cell
[21]*canal* = canal, channel + *icul* = little
[22]*peri* = around + *oste* = bone

INSIGHT 3.3 CLINICAL APPLICATION

Brittle Bone Disease

Osteogenesis[23] *imperfecta* is a hereditary defect of collagen deposition in the bones. Collagen-deficient bones are exceptionally brittle, and so this disorder is also called *brittle bone disease.* Bone fractures are often present at birth; children suffer from frequent spontaneous fractures, their teeth are often deformed, and they may have a hearing impairment due to deformity of the middle-ear bones. Children with osteogenesis imperfecta are sometimes mistaken for battered children before the disease is diagnosed. In severe cases, the child is stillborn or dies soon after birth. Little can be done for children with this disease except for very careful handling, prompt treatment of fractures, and orthopedic braces to minimize skeletal deformity.

[23]*osteo* = bone + *genesis* = formation

erythrocytes[24] (eh-RITH-ro-sites), or **red blood cells (RBCs).** In stained blood smears, they look like pink discs with thin, pale centers and no nuclei. Erythrocytes transport oxygen and carbon dioxides. **Leukocytes,** or **white blood cells (WBCs),** serve various roles in defense against infection and other diseases. They travel from one organ to another in the bloodstream and lymph but spend most of their lives in the connective tissues. Leukocytes are somewhat larger than erythrocytes and have conspicuous nuclei that usually appear violet in stained preparations. There are five kinds, distinguished partly by variations in nuclear shape: *neutrophils, eosinophils, basophils, lymphocytes,* and *monocytes.* Their individual characteristics are considered in detail in chapter 19. **Platelets** are small cell fragments scattered amid the blood cells. They are involved in clotting and other mechanisms for minimizing blood loss, and they secrete growth factors that promote blood vessel growth and maintenance.

[24]*erythro* = red + *cyte* = cell

Before You Go On

Answer the following questions to test your understanding of the preceding section:

10. What features do most or all connective tissues have in common to set this class apart from nervous, muscular, and epithelial tissue?
11. List the cell and fiber types found in fibrous connective tissues and state their functional differences.
12. What substances account for the gelatinous consistency of connective tissue ground substance?
13. What is areolar tissue? How can it be distinguished from any other kind of connective tissue?
14. Discuss the difference between dense regular and dense irregular connective tissue as an example of the relationship between form and function.
15. Describe some similarities, differences, and functional relationships between hyaline cartilage and bone.
16. What are the three basic kinds of formed elements in blood, and what are their respective functions?

Table 3.8
Blood

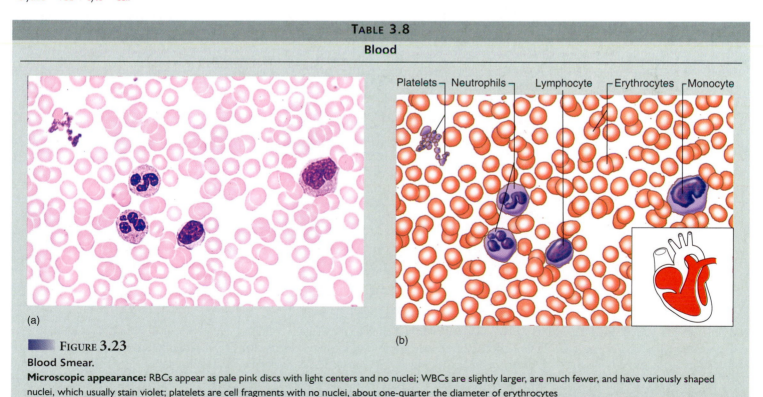

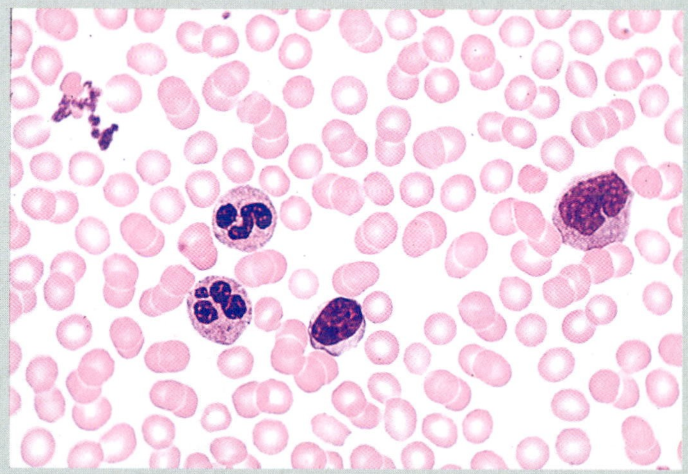

(a)

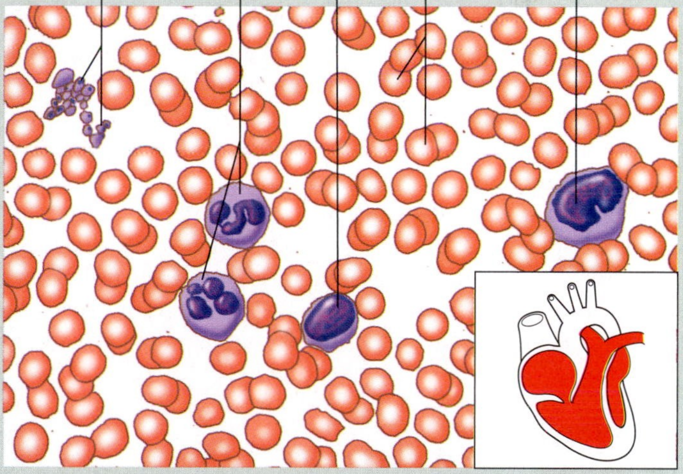

(b)

Figure 3.23
Blood Smear.

Microscopic appearance: RBCs appear as pale pink discs with light centers and no nuclei; WBCs are slightly larger, are much fewer, and have variously shaped nuclei, which usually stain violet; platelets are cell fragments with no nuclei, about one-quarter the diameter of erythrocytes

Representative locations: Contained in heart and blood vessels

Functions: Transports gases, nutrients, wastes, chemical signals, and heat throughout body; provides defensive WBCs; contains clotting agents to minimize bleeding; platelets secrete growth factors that promote tissue maintenance and repair

Nervous and Muscular Tissue—Excitable Tissues

Objectives

When you have completed this section, you should be able to

- explain what distinguishes excitable tissues from other tissues;
- name the cell types that compose nervous tissue;
- identify the major parts of a nerve cell;
- name the three kinds of muscular tissue and describe the differences between them; and
- visually identify nervous and muscular tissues from specimens or photographs.

Excitability is a characteristic of all living cells, but it is developed to its highest degree in nervous and muscular tissue, which are therefore described as **excitable tissues.** The basis for their excitation is an electrical charge difference (voltage) called the *membrane potential,* which occurs across the plasma membranes of all cells. Nervous and muscular tissues respond quickly to outside stimuli by means of rapid changes in membrane potential. In nerve cells, these changes result in the rapid transmission of signals to other cells. In muscle cells, they result in contraction, or shortening of the cell.

Nervous Tissue

Nervous tissue is specialized for rapid communication by means of electrical and chemical signals. It consists of **neurons** (NOOR-ons), or nerve cells, and a much greater number of supportive **neuroglia** (noo-ROG-lee-uh), or **glial** (GLEE-ul) **cells,** which protect and assist the neurons (table 3.9). Neurons are specialized to detect stimuli, respond quickly, and transmit coded information rapidly to other cells. Each neuron has a prominent **soma,** or cell body, that houses the nucleus and most other organelles. This is the cell's center of genetic control and protein synthesis. Somas are usually round, ovoid, or stellate in shape. Extending from the soma, there are usually multiple short, branched processes called **dendrites,**[25] which receive signals from other cells and transmit messages to the soma, and a single, much longer **axon (nerve fiber),** which sends outgoing signals to other cells. Some axons are more than a meter long and extend from the brainstem to the foot. Nervous tissue is found in the brain, spinal cord, nerves, and ganglia (aggregations of neuron cell bodies forming knotlike swellings in nerves). Local variations in the structure of nervous tissue are described in chapters 13 to 16.

Table 3.9

Nervous Tissue

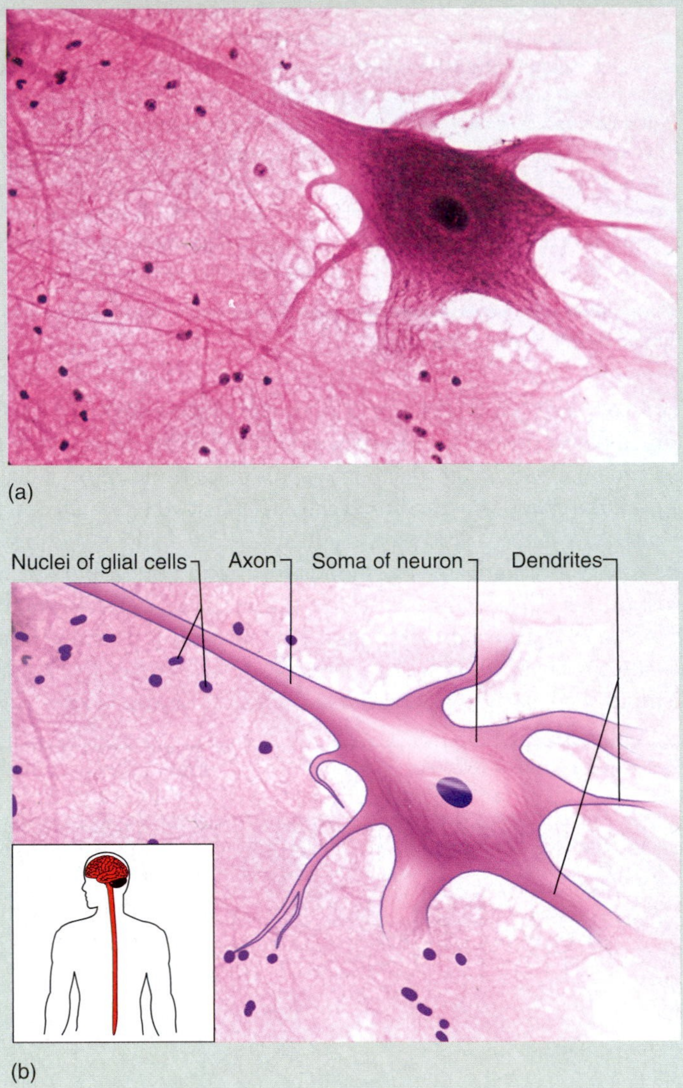
(a)
Nuclei of glial cells
Axon
Soma of neuron
Dendrites
(b)

Figure 3.24

Spinal Cord Smear.

Microscopic appearance: Most sections show a few large neurons, usually with rounded or stellate cell bodies (somas) and fibrous processes (axon and dendrites) extending from the somas; neurons are surrounded by a greater number of much smaller glial cells, which lack dendrites and axons

Representative locations: Brain, spinal cord, nerves, ganglia

Function: Internal communication

Muscular Tissue

Muscular tissue is specialized to contract when it is stimulated, and thus to exert a physical force on other tissues—for example, when a skeletal muscle pulls on a bone, the heart contracts and expels blood, or the bladder contracts and expels urine. Not only do movements of the body and its limbs depend on muscular tissue, but so do such processes as digestion, waste elimination, breathing, speech, and blood circulation. The muscles are also an important source of body heat. The word *muscle* means "little mouse," apparently referring to the appearance of muscles rippling under the skin.

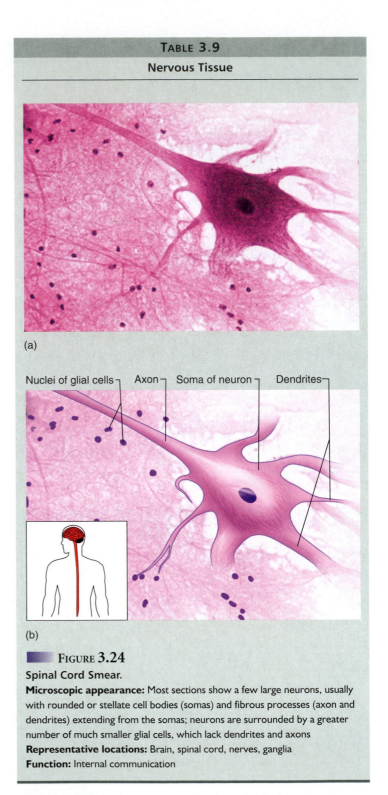

[25]*dendr* = tree + *ite* = little

Table 3.10

Muscular Tissue

Skeletal Muscle

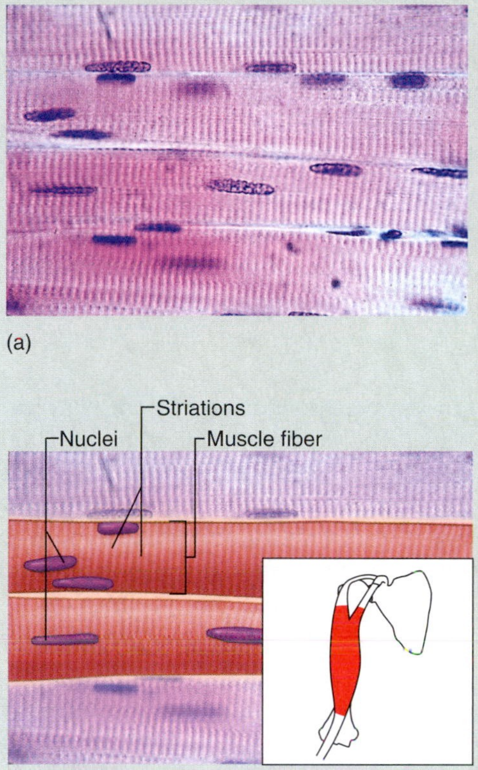

Figure 3.25

Skeletal Muscle.

Microscopic appearance: Long, cylindrical, unbranched cells (fibers), relatively parallel in longitudinal tissue sections; striations; multiple nuclei per cell, near plasma membrane

Representative locations: Skeletal muscles, mostly attached to bones but also including voluntary sphincters of the lips, eyelids, urethra, and anus; diaphragm; tongue; some muscles of esophagus

Functions: Body movements, facial expression, posture, breathing, speech, swallowing, control of urination and defecation, and childbirth; under voluntary control

Cardiac Muscle

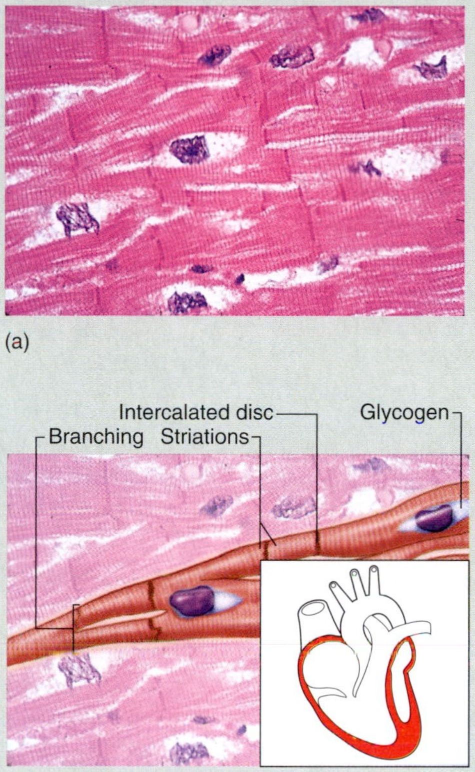

Figure 3.26

Cardiac Muscle.

Microscopic appearance: Short branched cells (myocytes); less parallel appearance in tissue sections; striations; intercalated discs; one nucleus per cell, centrally located and often surrounded by a light zone

Representative locations: Heart

Functions: Pumping of blood; under involuntary control

Smooth Muscle

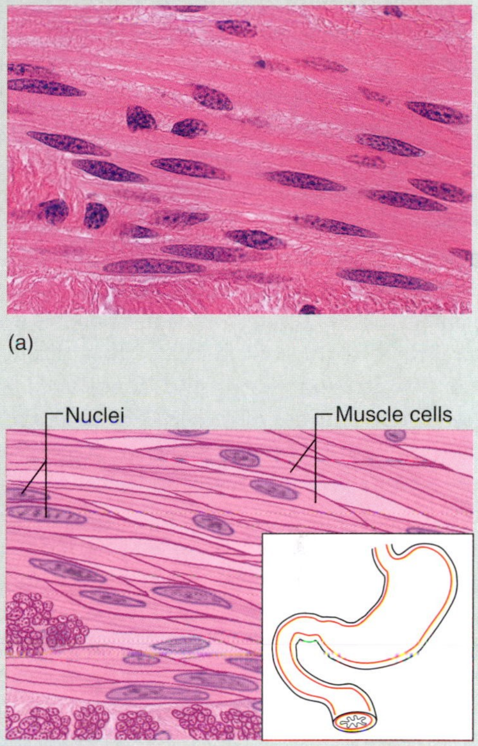

Figure 3.27

Smooth Muscle, Wall of the Small Intestine.

Microscopic appearance: Short fusiform cells overlapping each other; nonstriated; one nucleus per cell, centrally located

Representative locations: Usually found as sheets of tissue in walls of viscera; also in iris and associated with hair follicles; involuntary sphincters of urethra and anus

Functions: Swallowing; contractions of stomach and intestines; expulsion of feces and urine; labor contractions; control of blood pressure and flow; control of respiratory airflow; control of pupillary diameter; erection of hairs; under involuntary control

There are three types of muscular tissue—*skeletal, cardiac,* and *smooth*—which differ in appearance, physiology, and function (table 3.10). **Skeletal muscle** consists of long, cylindrical cells called **muscle fibers.** Most of it is attached to bones, but there are exceptions in the tongue, upper esophagus, some facial muscles, and some **sphincter**[26] (SFINK-tur) muscles (ringlike or cufflike muscles that open and close body passages). Each cell contains multiple nuclei adjacent to the plasma membrane. Skeletal muscle is described as *striated* and *voluntary.* The first term refers to alternating light and dark bands, or **striations** (stry-AY-shuns), created by the overlapping pattern of cytoplasmic protein filaments that cause muscle contraction. The second term, **voluntary,** refers to the fact that we usually have conscious control over skeletal muscles.

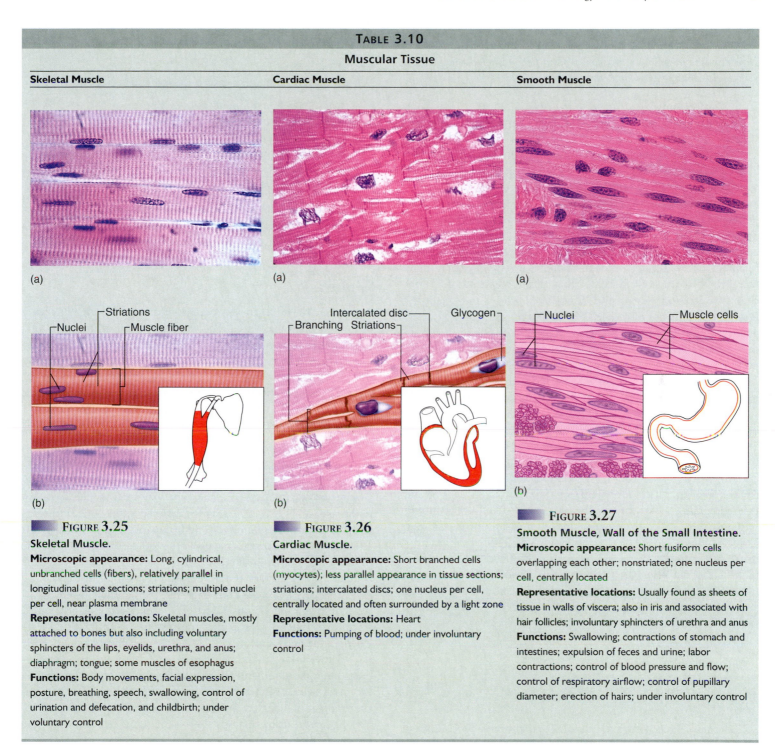

[26]*sphinc* = squeeze, bind tightly

THINK ABOUT IT!

How does the meaning of the word fiber *differ in the following uses: muscle fiber, nerve fiber, and connective tissue fiber?*

Cardiac muscle is limited to the heart. It, too, is striated, but it differs from skeletal muscle in its other features. Its cells are much shorter, so they are commonly called **myocytes**[27] rather than fibers. The myocytes are branched and contain only one nucleus, which is located near the center. A light-staining region of glycogen, a starchlike energy source, often surrounds the nucleus. Cardiac myocytes are joined end to end by junctions called **intercalated**[28] (in-TUR-ku-LAY-ted) **discs.** Intercalated discs appear as dark transverse lines separating each myocyte from the next. They may be only faintly visible, however, unless the tissue has been specially stained for them. Gap junctions in the discs enable a wave of excitation to travel rapidly from cell to cell, so that all the myocytes of a heart chamber are stimulated, and contract, almost simultaneously. Desmosomes in the discs keep the myocytes from pulling apart when the heart contracts. Cardiac muscle is considered **involuntary** because it is not usually under conscious control; it contracts even if all nerve connections to it are severed.

Smooth muscle lacks striations and is involuntary. Smooth muscle cells (also called myocytes) are fusiform and relatively short. They have a single centrally placed nucleus. Small amounts of smooth muscle are found in the iris of the eye and in the skin, but most of it, called **visceral muscle,** forms layers in the walls of the digestive, respiratory, and urinary tracts, uterus, blood vessels, and other organs. In locations such as the esophagus and small intestine, smooth muscle forms adjacent layers, with the cells of one layer encircling the organ and the cells of the other layer running longitudinally. When the circular smooth muscle contracts, it may propel contents such as food through the organ. When the longitudinal layer contracts, it makes the organ shorter and thicker. By regulating the diameter of blood vessels, smooth muscle is very important in controlling blood pressure and flow. Both smooth and skeletal muscle form sphincters that control the emptying of the bladder and rectum.

Before You Go On

Answer the following questions to test your understanding of the preceding section:

17. What do nervous and muscular tissue have in common? What is the primary function of each?
18. What kinds of cells compose nervous tissue, and how can they be distinguished from each other?
19. Name the three kinds of muscular tissue, describe how to distinguish them from each other in microscopic appearance, and state a location and function for each one.

[27]*myo* = muscle + *cyte* = cell
[28]*inter* = between + *calated* = inserted

GLANDS AND MEMBRANES

Objectives

When you have completed this section, you should be able to

- describe or define different types of glands;
- describe the typical anatomy of a gland;
- name and compare different modes of glandular secretion;
- describe the way tissues are organized to form the body's membranes; and
- name and describe the major types of membranes in the body.

We have surveyed all of the fundamental categories of human tissue, and we will now look at the way in which multiple tissue types are assembled to form the body's glands and membranes.

Glands

A **gland** is a cell or organ that secretes substances for use in the body or for elimination as waste. The gland product may be something synthesized by the gland cells (such as digestive enzymes) or something removed from the tissues and modified by the gland (such as urine). Glands are composed predominantly of epithelial tissue, but usually have a supporting connective tissue framework and capsule.

ENDOCRINE AND EXOCRINE GLANDS

Glands are broadly classified as endocrine or exocrine. Both types originate as invaginations of a surface epithelium. Multicellular **exocrine**[29] (EC-so-crin) **glands** maintain their contact with the surface by way of a **duct,** an epithelial tube that conveys their secretion to the surface. The secretion may be released to the body surface, as in the case of sweat, mammary, and tear glands, but more often it is released into the lumen (cavity) of another organ such as the mouth or intestine. **Endocrine**[30] (EN-doe-crin) **glands** lose their contact with the surface and have no ducts. They do, however, have a high density of blood capillaries and secrete their products directly into the blood. The secretions of endocrine glands, called **hormones,** function as chemical messengers to stimulate cells elsewhere in the body. Endocrine glands are the subject of chapter 18.

Unicellular glands are isolated gland cells found in an epithelium that is predominantly nonsecretory. For example, the respiratory tract, which is lined mainly by ciliated cells, also has a liberal scattering of nonciliated, mucus-secreting goblet cells (see fig. 3.7). The digestive tract has many scattered endocrine cells, which secrete hormones that coordinate digestive processes.

EXOCRINE GLAND STRUCTURE

Figure 3.28 shows a generalized multicellular exocrine gland—a structural arrangement found in such organs as the mammary gland, pancreas, and salivary glands. Most glands are enclosed in a fibrous **capsule.** The capsule often gives off extensions called

[29]*exo* = out + *crin* = to separate, secrete
[30]*endo* = in, into

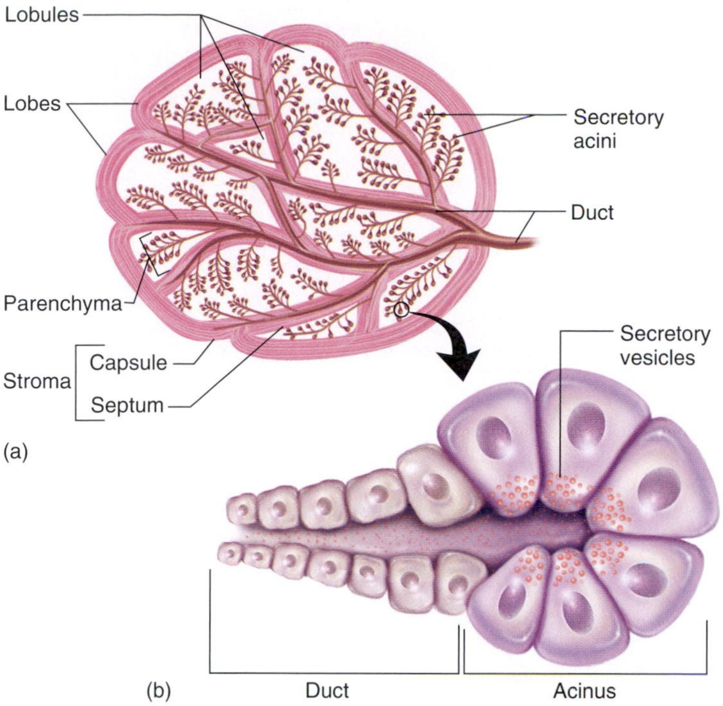
Lobules
Lobes
Secretory acini
Duct
Parenchyma
Stroma
Capsule
Septum
(a)
Secretory vesicles
(b)
Duct
Acinus

FIGURE 3.28

General Structure of an Exocrine Gland. (*a*) The gland duct branches repeatedly, following the connective tissue septa, until its finest divisions end in saccular acini of secretory cells. (*b*) Detail of an acinus and the beginning of a duct.

septa,[31] or **trabeculae**[32] (trah-BEC-you-lee), that divide the interior of the gland into compartments called **lobes,** which are visible to the naked eye. Finer connective tissue septa may further subdivide each lobe into microscopic **lobules** (LOB-yools). Blood vessels, nerves, and the gland's ducts generally travel through these septa. The connective tissue framework of the gland, called its **stroma,** supports and organizes the glandular tissue. The cells that perform the tasks of synthesis and secretion are collectively called the **parenchyma**[33] (pa-REN-kih-muh). This is typically simple cuboidal or simple columnar epithelium.

Exocrine glands are classified as **simple** if they have a single unbranched duct and **compound** (fig. 3.29) if they have a branched duct. If the duct and secretory portion are of uniform diameter, the gland is called **tubular.** If the secretory cells form a dilated sac, the gland is called **acinar** and the sac is an **acinus**[34] (ASS-ih-nus), or **alveolus**[35] (AL-vee-OH-lus). A gland in which both the acini and tubules secrete a product is called a **tubuloacinar gland.**

TYPES OF SECRETIONS

Glands are classified not only by their structure but also by the nature of their secretions. **Serous** (SEER-us) **glands** produce relatively thin, watery fluids such as perspiration, milk, tears, and digestive juices. **Mucous glands,** found in the tongue and roof of the mouth among other places, secrete a glycoprotein called *mucin* (MEW-sin). After it is secreted, mucin absorbs water and forms the sticky product *mucus.* (Note that *mucus,* the secretion, is spelled differently from *mucous,* the adjective form of the word.) **Mixed glands,** such as the two pairs of salivary glands in the chin, contain both serous and mucous cells and produce a mixture of the two types of secretions. **Cytogenic**[36] **glands** release whole cells. The only examples of these are the testes and ovaries, which produce sperm and egg cells.

METHODS OF SECRETION

Glands are classified as merocrine or holocrine depending on how they produce their secretions. **Merocrine**[37] (MERR-oh-crin) **glands,** also called **eccrine**[38] (EC-rin) **glands,** release their secretion by exocytosis, as described in chapter 2. These include the tear glands, pancreas, gastric glands, and most others. In **holocrine**[39] **glands,** cells accumulate a product and then the entire cell disintegrates, so the secretion is a mixture of cell fragments and the substance the cell had synthesized prior to its disintegration. Only a few glands use this mode of secretion, such as the oil-producing glands of the scalp and certain glands of the eyelid.

Some glands, such as the axillary (armpit) sweat glands and mammary glands, are named **apocrine**[40] **glands** from a former belief that the secretion was composed of bits of apical cytoplasm that broke away from the cell surface. Closer study showed this to be untrue; these glands are primarily merocrine in their mode of secretion. They are nevertheless different from other merocrine glands in function and histological appearance, so they are still referred to as apocrine glands.

Membranes

In Atlas A, the major cavities of the body were described, as well as some of the membranes that line them and cover their viscera. We now consider some histological aspects of the major body membranes.

The largest membrane of the body is the **cutaneous** (cue-TAY-nee-us) **membrane**—or more simply, the skin (detailed in chapter 5). It consists of a stratified squamous epithelium (epidermis) resting on a layer of connective tissue (dermis). The two principal kinds of internal membranes are mucous and serous membranes. **Mucous membranes** (fig. 3.30), also called **mucosae** (mew-CO-see), line passageways that open to the exterior: the digestive, respiratory, urinary, and reproductive tracts. A mucosa consists of two to three layers: (1) an epithelium, (2) an areolar connective tissue layer called the **lamina propria**[41] (LAM-ih-nuh PRO-pree-uh), and sometimes (3) a layer of smooth muscle called the **muscularis** (MUSK-you-LAIR-iss) **mucosae.** Mucous membranes

[31]*septum* = wall
[32]*trab* = plate + *cula* = little
[33]*par* = beside + *enchym* = pour in
[34]*acinus* = berry
[35]*alveol* = cavity, pit
[36]*cyto* = cell + *genic* = producing
[37]*mero* = part + *crin* = to separate, secrete
[38]*ec* = *ex* = out
[39]*holo* = whole, entire
[40]*apo* = from, off, away
[41]*lamina* = layer + *propria* = of one's own

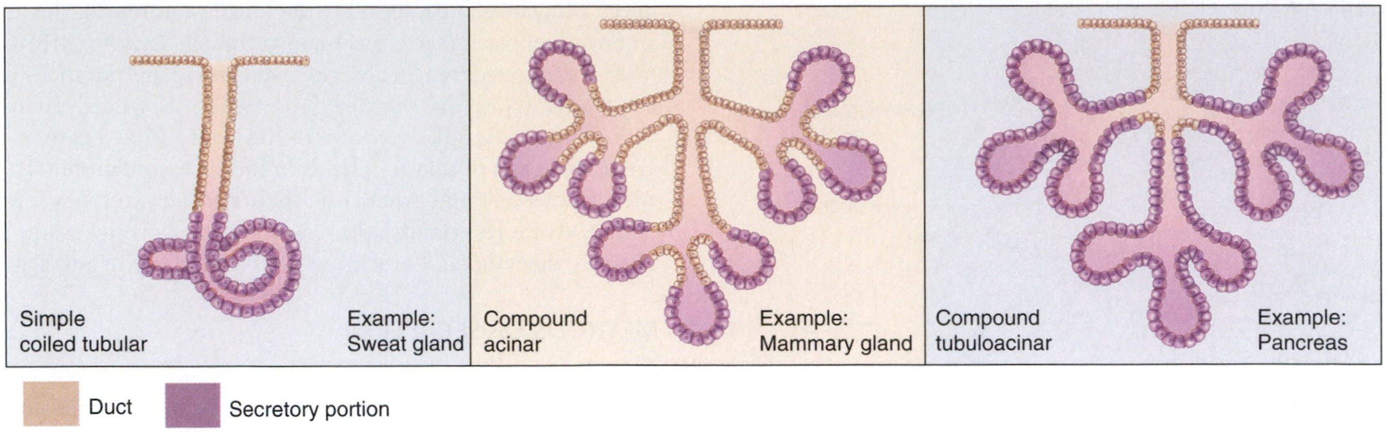
Simple
coiled tubular
Example:
Sweat gland
Compound
acinar
Example:
Mammary gland
Compound
tubuloacinar
Example:
Pancreas
Duct
Secretory portion

FIGURE 3.29

Some Types of Exocrine Glands. Glands are simple if their ducts do not branch and compound if they do. They are tubular if they have a uniform diameter, acinar if their secretory cells are limited to saccular acini, and tubuloacinar if they have secretory cells in both the acinar and tubular regions.

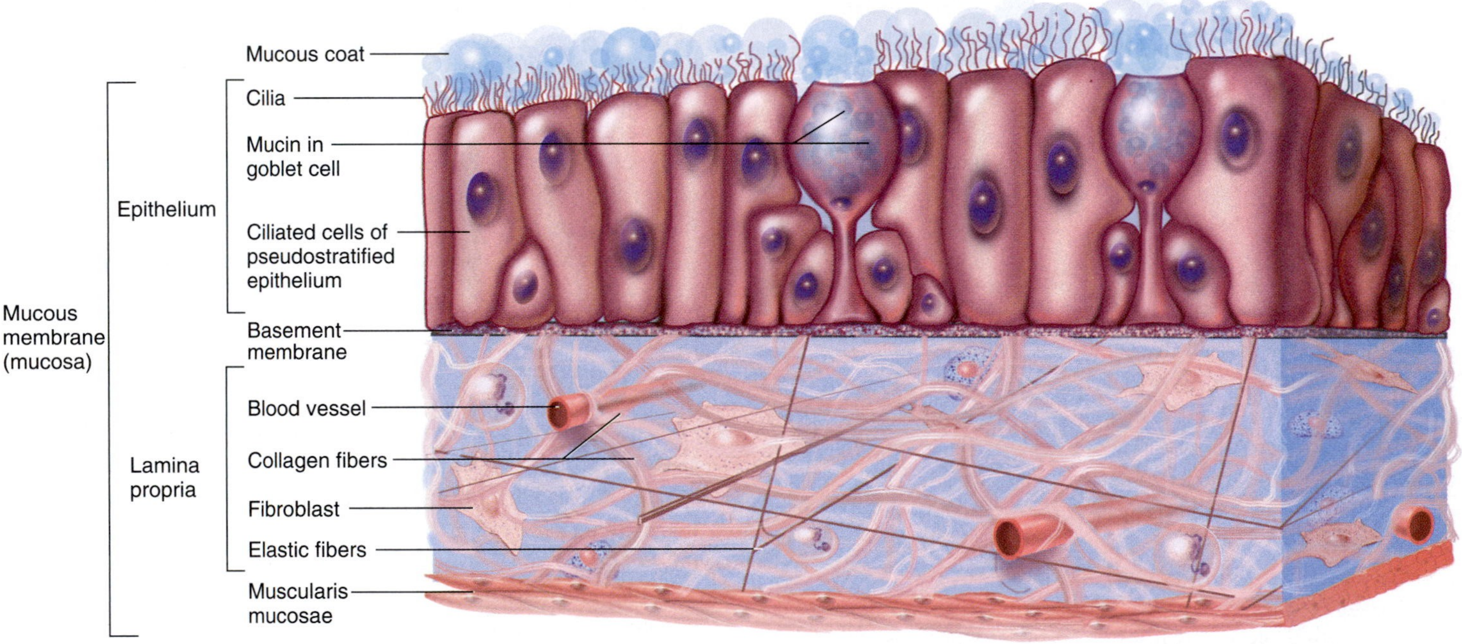
Mucous coat
Cilia
Mucin in
goblet cell
Epithelium
Ciliated cells of
pseudostratified
epithelium
Mucous
membrane
(mucosa)
Basement
membrane
Blood vessel
Lamina
propria
Collagen fibers
Fibroblast
Elastic fibers
Muscularis
mucosae

FIGURE 3.30

Histology of a Mucous Membrane.

have absorptive, secretory, and protective functions. They are often covered with mucus secreted by goblet cells, multicellular mucous glands, or both. The mucus traps bacteria and foreign particles, which keeps them from invading the tissues and aids in their removal from the body. The epithelium of a mucous membrane may also include absorptive, ciliated, and other types of cells.

A **serous membrane (serosa)** is composed of a simple squamous epithelium resting on a thin layer of areolar connective tissue. Serous membranes produce watery **serous** (SEER-us) **fluid,** which arises from the blood and derives its name from the fact that it is similar to blood serum in composition. Serous membranes line the insides of some body cavities and form a smooth surface on the outer surfaces of some of the viscera, such as the digestive tract. The pleurae, pericardium, and peritoneum described in Atlas A are serous membranes.

The circulatory system is lined with a simple squamous epithelium called **endothelium.** The endothelium rests on a thin layer of areolar tissue, which often rests in turn on an elastic sheet. Collectively, these tissues make up a membrane called the *tunica interna* of the blood vessels and *endocardium* of the heart. The simple squamous epithelium that lines the pleural, pericardial, and peritoneal cavities is called **mesothelium.**

Some joints of the skeletal system are enclosed by fibrous **synovial** (sih-NO-vee-ul) **membranes,** made only of connective tissue. These membranes span the gap from one bone to the next and secrete slippery *synovial fluid* into the joint.

Before You Go On

Answer the following questions to test your understanding of the preceding section:

20. Distinguish between a simple gland and a compound gland, and give an example of each. Distinguish between a tubular gland and an acinar gland, and give an example of each.
21. Contrast the merocrine and holocrine methods of secretion, and name a gland product produced by each method.
22. Describe the differences between a mucous and a serous membrane.
23. Name the layers of a mucous membrane, and state which of the four primary tissue classes composes each layer.

TISSUE GROWTH, DEVELOPMENT, DEATH, AND REPAIR

Objectives

When you have completed this section, you should be able to

- name and describe the ways in which a tissue can change from one type to another;
- name and describe the modes of tissue growth;
- name and describe the modes and causes of tissue shrinkage and death; and
- name and describe the ways the body repairs damaged tissues.

Changes in Tissue Type

You have studied the form and function of more than two dozen discrete types of human tissue in this chapter. You should not leave this subject, however, with the impression that once these tissue types are established in the adult, they never change. Tissues are, in fact, capable of changing from one type to another within certain limits. Most obviously, unspecialized tissues of the embryo develop into more diverse and specialized types of mature tissue—embryonic *mesenchyme* to muscle, for example. This development of a more specialized form and function is called **differentiation.**

Epithelia sometimes exhibit **metaplasia,**[42] a change from one type of mature tissue to another. For example, the vaginal epithelium changes from simple cuboidal in childhood to stratified squamous in adolescence and adulthood. The latter is better adapted to the demands of intercourse and childbirth. The nasal cavity is lined with pseudostratified columnar epithelium. However, if we block one nostril and breathe through the other one for several days, the epithelium in the unblocked passage changes to stratified squamous. In smokers, the ciliated pseudostratified columnar epithelium of the bronchi may transform into a stratified squamous epithelium.

THINK ABOUT IT!

What functions of a ciliated pseudostratified columnar epithelium could not be served by a stratified squamous epithelium? In light of this, what might be some consequences of bronchial metaplasia in heavy smokers?

Tissue Growth

Tissues grow either because their cells increase in number or because the existing cells grow larger. Most embryonic and childhood growth occurs by **hyperplasia**[43] (HY-pur-PLAY-zhuh), tissue growth through cell multiplication. Skeletal muscle and adipose tissue grow, however, through **hypertrophy**[44] (hy-PUR-truh-fee), the enlargement of preexisting cells. **Neoplasia**[45] (NEE-oh-PLAY-zhuh) is the development of a tumor (neoplasm) (see chapter 2).

Tissue Shrinkage and Death

The shrinkage of a tissue through a loss in cell size or number is called **atrophy**[46] (AT-ruh-fee). It results from aging or disuse. Muscles that are not exercised exhibit *disuse atrophy* as their cells become smaller. Limbs set in a cast for a broken bone shrivel for this reason.

Necrosis[47] (neh-CRO-sis) is the premature, pathological death of tissue due to trauma, toxins, infection, and so forth. **Gangrene** is any tissue necrosis resulting from an insufficient blood supply. **Infarction** is the sudden death of tissue, such as heart muscle *(myocardial infarction),* which occurs when its blood supply is cut off. A *decubitus ulcer* (bed sore) is tissue necrosis that occurs when immobilized persons, such as those confined to a hospital bed or wheelchair, are unable to move, and continual pressure on the skin cuts off blood flow to an area. Cells dying by necrosis usually swell and rupture, triggering inflammation in the surrounding tissue.

Apoptosis[48] (AP-oh-TOE-sis), or **programmed cell death,** is the normal death of cells that have completed their function and best serve the body by dying and getting out of the way. Cells undergoing apoptosis activate enzymes that degrade their DNA and proteins. The cells shrink and are quickly phagocytized by macrophages. The cell contents never escape, so there is no inflammatory response. Although billions of cells die every hour by apoptosis, they are engulfed so quickly that they are almost never seen except within

[42] *meta* = change + *plas* = form, growth

[43] *hyper* = excessive + *plas* = growth
[44] *trophy* = nourishment
[45] *neo* = new
[46] *a* = without
[47] *necr* = death + *osis* = process
[48] *apo* = away + *ptosis* = falling

macrophages. One example of apoptosis is that in embryonic development, we produce about twice as many neurons as we need. Those that make connections with target cells survive, while the excess 50% die. Apoptosis also "dissolves" the webbing between the fingers and toes during embryonic development, it frees the earlobe from the side of the head in people with detached earlobes, and it causes shrinkage of the uterus after pregnancy ends. Immune cells can stimulate cancer cells to "commit suicide" by apoptosis (see photo on p. 77).

Tissue Repair

Damaged tissues can be repaired by either *regeneration* or *fibrosis.* **Regeneration** is the replacement of dead or damaged cells by the same type of cells as before, thus restoring normal function to an organ. Most skin injuries (cuts, scrapes, and minor burns) heal by regeneration. **Fibrosis** is the replacement of damaged tissue with scar tissue, composed mainly of collagen. Scar tissue helps to hold an organ together, but it does not restore normal function. Examples include the healing of severe cuts and burns, the healing of muscle injuries, and scarring of the lungs in tuberculosis.

INSIGHT 3.4 CLINICAL APPLICATION

Keloids

In some people, especially dark-skinned adults, healing skin wounds exhibit excessive fibrosis, producing raised, shiny scars called *keloids* (fig. 3.31). Keloids extend beyond the boundaries of the original wound and tend to return even if they are surgically removed. Keloids may result from the excessive secretion of a fibroblast-stimulating growth factor by macrophages and platelets. They occur most often on the upper trunk and earlobes. Some tribespeople practice *scarification*—scratching or cutting the skin to induce keloid formation as a way of decorating the body.

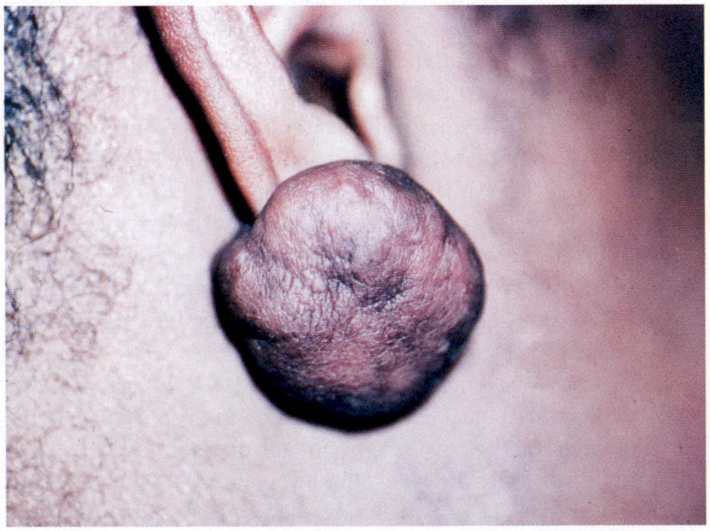

FIGURE 3.31
A Keloid. This scar is the result of ear piercing.

Before You Go On

Answer the following questions to test your understanding of the preceding section:

24. Distinguish between *differentiation* and *metaplasia.*
25. Tissues can grow through an increase in cell size or cell number. What are the respective terms for these two kinds of growth?
26. Distinguish between *atrophy, necrosis,* and *apoptosis,* and describe a circumstance under which each of these forms of tissue loss may occur.
27. Distinguish between regeneration and fibrosis. Which process restores normal cellular function? What good is the other process if it does not restore function?

CHAPTER REVIEW

Review of Key Concepts

The Study of Tissues (p. 78)

1. A *tissue* is a mass of similar cells and cell products that forms a discrete region of an organ and performs a specific function. The study of tissues is called *histology*.
2. The four primary tissues are epithelial, connective, nervous, and muscular tissue. These types differ in the types and functions of their cells, the characteristics of their *matrix*, and the relative volume of cells and matrix.
3. The matrix is composed of fibrous proteins and *ground substance*.
4. Most tissues are studied in the form of thin slices called *histological sections*, colored with *stains* to enhance their details.
5. Tissues are typically cut into *longitudinal*, *cross*, or *oblique sections*. Some are prepared as *smears* or *spreads*.

Epithelial Tissue (p. 80)

1. *Epithelium* is a type of tissue composed of one or more layers of closely adhering cells, either covering an organ surface or forming the secretory tissue and ducts of a gland. Epithelia have little intercellular material and no blood vessels.
2. An epithelium rests on a *basement membrane* of adhesive glycoproteins and other extracellular material, which separates it from and binds it to the underlying connective tissue.
3. Epithelia are classified as *simple* if all the cells contact the basement membrane. The four kinds of simple epithelia are *simple squamous*, *simple cuboidal*, *simple columnar*, and *pseudostratified columnar* (table 3.2).
4. Epithelia are classified as *stratified* if some cells rest atop others, without touching the basement membrane. The four kinds of stratified epithelia are *stratified squamous*, *stratified cuboidal*, *stratified columnar*, and *transitional* (table 3.3).
5. Stratified squamous epithelium is the most widespread of all eight types. It has *keratinized* and *nonkeratinized* forms.

Connective Tissue (p. 86)

1. *Connective tissue* is a type of tissue in which cells usually occupy less space than the extracellular material, and which serves in most cases to bind organs to each other or to support and protect organs. Other functions include immunity, movement, storage, heat production, and transport.
2. *Fibrous connective tissue* exhibits conspicuous fibers in its matrix. The fiber types are *collagenous*, *reticular*, and *elastic*.
3. The cell types common in this tissue are *fibroblasts*, *macrophages*, *leukocytes*, *plasma cells*, *mast cells*, and *adipocytes*.
4. The ground substance of this tissue acquires a gelatinous to rubbery consistency due to three types of large protein-carbohydrate complexes: *glycosaminoglycans*, *proteoglycans*, and *adhesive glycoproteins*.
5. The *loose fibrous connective tissues* are *areolar tissue*, *reticular tissue*, and *adipose tissue* (table 3.4).
6. The *dense fibrous connective tissues* are *dense regular* and *dense irregular connective tissue* (table 3.5).
7. *Cartilage* is a supportive connective tissue with a flexible rubbery matrix. It exhibits cells called *chondrocytes* occupying cavities called *lacunae* scattered throughout the matrix. The three types of cartilage are *hyaline cartilage*, *elastic cartilage*, and *fibrocartilage* (table 3.6).
8. *Bone (osseous tissue)* is a supportive connective tissue with a calcified matrix. It exhibits cells called *osteocytes*, which occupy lacunae between layers or *lamellae* of matrix. The lamellae are arranged in concentric cylinders around a *central canal* in each *osteon*. The two types of bone are *compact bone* (table 3.7) and *spongy bone*.
9. *Blood* is a liquid connective tissue that serves to transport cells and solutes from place to place. It consists of three kinds of *formed elements—erythrocytes*, *leukocytes*, and *platelets*—suspended in a liquid ground substance, the *plasma* (table 3.8).

Nervous and Muscular Tissue—Excitable Tissues (p. 94)

1. All living tissues are excitable, but nervous and muscular tissue have developed this property to the highest degree. Excitability means that a cell responds to stimulation with a change in the membrane potential (voltage) across the plasma membrane.
2. *Nervous tissue* is specialized for rapid communication by means of electrical and chemical signals. It consists of *neurons* (nerve cells) and *neuroglia* (supportive cells). Neurons carry out the communicative function.
3. A neuron has a *soma* (cell body), usually multiple *dendrites* which conduct incoming signals to the soma, and usually a single *axon* (nerve fiber) to carry signals away to another cell.
4. *Muscular tissue* is specialized to contract when stimulated, and thus to exert force on other tissues. Muscular tissue serves such purposes as body movements, movement of material through the digestive tract, waste elimination, breathing, speech, and blood circulation.
5. There are three kinds of muscular tissue: skeletal, cardiac, and smooth muscle.
6. *Skeletal muscle* consists of elongated, striated, multinucleated cells called *muscle fibers*. Most skeletal muscle is attached to bones and moves joints when it contracts. Skeletal muscle is under voluntary control.
7. *Cardiac muscle* consists of shorter cells called *myocytes*, with a single nucleus, striations, and *intercalated discs* where the cells meet end to end. Cardiac muscle is involuntary and is limited to the heart.
8. *Smooth muscle* consists of short fusiform myocytes with a single nucleus and no striations. It occurs in the iris, skin, blood vessels, and the walls of the digestive, respiratory, urinary, and reproductive tracts, among other locations.

Glands and Membranes (p. 96)

1. A *gland* is a cell or organ that secretes substances for use in the body or for elimination as waste.
2. *Exocrine glands* usually have ducts and release their secretions onto the body surface or into the lumen of another organ. *Endocrine glands* lack ducts and secrete their products, called *hormones*, into the bloodstream.
3. *Unicellular glands* are isolated gland cells found in predominantly nonglandular epithelia. They can be exocrine (such as *goblet cells*) or endocrine (such as hormone-secreting cells of the digestive tract).
4. Most glands are enclosed in a connective tissue *capsule*, which issues fibrous *septa (trabeculae)* into the interior of the gland, dividing it into *lobes* and microscopic *lobules*.

This supportive connective tissue framework is called the *stroma.* The secretory portions and ducts of a gland, called the *parenchyma,* are composed of epithelial cells.

5. *Simple glands* have a single unbranched duct and *compound glands* have branched ducts. In *tubular glands,* the duct and secretory portion form a tubule of uniform diameter. In an *acinar* gland, the secretory cells are limited to a saclike *acinus* at the end of a duct; in a *tubuloacinar* gland, the secretory cells form both acini and tubules leading away from them.
6. Glands are classified as *serous* if they secrete a relatively thin, watery fluid; *mucous* if they secrete mucus; *mixed* if they secrete both; and *cytogenic* if their product is intact cells (eggs or sperm).
7. Glands vary in their mode of secretion. *Merocrine (eccrine)* glands secrete their products by exocytosis. In *holocrine* glands, entire cells disintegrate and become the secretion. *Apocrine* glands are predominantly merocrine in their mode of secretion but have a distinctive histological appearance.
8. The body has numerous membranes of various types. The most extensive are cutaneous, serous, and mucous membranes. The *cutaneous membrane* is the skin.
9. *Mucous membranes* line passages that open to the exterior (digestive, respiratory, urinary, and reproductive tracts). Mucous membranes are composed of a secretory epithelium, a connective tissue *lamina propria,* and sometimes a *muscularis mucosae.*
10. *Serous membranes* consist of a simple squamous epithelium on a thin bed of areolar connective tissue. They secrete watery *serous fluid.* They include the *endothelium* of the blood vessels and *mesothelium* of the body cavities.
11. *Synovial membranes* are nonepithelial, connective tissue membranes enclosing joint cavities.

Tissue Growth, Development, Death, and Repair (p. 99)

1. Tissues can change from one type to another through *differentiation* (transformation of an unspecialized embryonic tissue into a specialized mature tissue) or *metaplasia* (transformation of one mature tissue type to another).
2. Tissues can grow by means of *hyperplasia* (cell multiplication), *hypertrophy* (enlargement of preexisting cells), or *neoplasia* (tumor growth).
3. Tissues can shrink or degenerate by means of *atrophy* (shrinkage through aging or lack of use), *necrosis* (pathological tissue death due to trauma, toxins, infections, etc.—as in *gangrene* and *infarction*), or *apoptosis (programmed cell death).*
4. Damaged tissues can be repaired by *regeneration,* which restores the preexisting tissue type and its functionality, or by *fibrosis,* the formation of scar tissue.

Testing Your Recall

1. Transitional epithelium is found in
   a. the urinary system.
   b. the respiratory system.
   c. the digestive system.
   d. the reproductive system.
   e. all of the above.
2. The external surface of the stomach is covered by
   a. a mucosa.
   b. a serosa.
   c. the parietal peritoneum.
   d. a lamina propria.
   e. a basement membrane.
3. The interior of the respiratory tract is lined with
   a. a serosa.
   b. mesothelium.
   c. a mucosa.
   d. endothelium.
   e. peritoneum.
4. A seminiferous tubule of the testis is lined with _____ epithelium.
   a. simple cuboidal
   b. pseudostratified columnar ciliated
   c. stratified squamous
   d. transitional
   e. stratified cuboidal
5. When the blood supply to a tissue is cut off, the tissue is most likely to undergo
   a. metaplasia.
   b. hyperplasia.
   c. apoptosis.
   d. necrosis.
   e. hypertrophy.
6. A fixative serves to
   a. stop tissue decay.
   b. improve contrast.
   c. repair a damaged tissue.
   d. bind epithelial cells together.
   e. bind cardiac myocytes together.
7. The collagen of areolar tissue is produced by
   a. macrophages.
   b. fibroblasts.
   c. mast cells.
   d. leukocytes.
   e. chondrocytes.
8. Tendons are composed of _____ connective tissue.
   a. skeletal
   b. areolar
   c. dense irregular
   d. yellow elastic
   e. dense regular
9. The shape of the external ear is due to
   a. skeletal muscle.
   b. elastic cartilage.
   c. fibrocartilage.
   d. articular cartilage.
   e. hyaline cartilage.
10. The most abundant formed element(s) of blood is/are
    a. plasma.
    b. erythrocytes.
    c. platelets.
    d. leukocytes.
    e. proteins.
11. The prearranged death of a cell that has completed its task is called _____.
12. The simple squamous epithelium that lines the peritoneal cavity is called _____.
13. Osteocytes and chondrocytes occupy little cavities called _____.
14. Muscle cells and axons are often called _____ because of their shape.
15. Tendons and ligaments are made mainly of the protein _____.
16. A rubbery tissue without a perichondrium is _____.
17. An epithelium rests on a layer called the _____ between its deepest cells and the underlying connective tissue.
18. Fibers and ground substance make up the _____ of a connective tissue.
19. In _____ glands, the secretion is formed by the complete disintegration of the gland cells.
20. Any epithelium in which every cell touches the basement membrane is called a/an _____ epithelium.

Answers in the Appendix

True or False

Determine which five of the following statements are false, and briefly explain why.

1. If we assume that the aorta is cylindrical, an oblique section of it would have an oval shape.
2. Everything in a tissue that is not a cell is classified as ground substance.
3. The colors seen in prepared histology slides are not the natural colors of those tissues.
4. The parenchyma of the liver is a simple cuboidal epithelium.
5. The tongue is covered with keratinized stratified squamous epithelium.
6. Macrophages are large phagocytic cells that develop from lymphocytes.
7. Most of the body's protein is collagen.
8. Brown fat produces more ATP than white fat.
9. After tissue differentiation is complete, a tissue cannot change type.
10. Erythrocytes have no nuclei.

Answers in the Appendix

Testing Your Comprehension

1. A woman in labor is often told to push. In doing so, is she consciously contracting her uterus to expel the baby? Justify your answer based on the muscular composition of the uterus.
2. The clinical application insights in this chapter describe some hereditary defects in collagen and elastin. Predict some pathological consequences that might result from a hereditary defect in keratin.
3. When cartilage is compressed, water is squeezed out of it, and when pressure is taken off, water flows back into the matrix. This being the case, why do you think cartilage at weight-bearing joints such as the knees can degenerate from lack of exercise?
4. The epithelium of the respiratory tract is mostly of the pseudostratified ciliated type, but in the alveoli—the tiny air sacs where oxygen and carbon dioxide are exchanged between the blood and inhaled air—the epithelium is simple squamous. Explain the functional significance of this histological difference. That is, why don't the alveoli have the same kind of epithelium as the rest of the respiratory tract?
5. Suppose you cut your finger on a broken bottle. How might mast cells promote healing of the injury?

Answers at the Online Learning Center

www.mhhe.com/saladinha1

Visit the Online Learning Center for practice tests, answer keys, and other learning aids for this chapter. Enhance your understanding of human anatomy with our interactive art labeling exercises, supplemental photo atlases, web links, puzzles, flashcards, and much more.

4

CHAPTER FOUR

Human Development

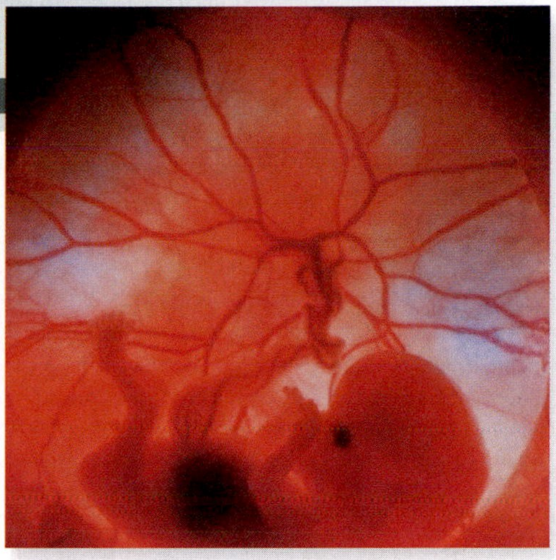

Human fetus at 11 weeks' gestation, with umbilical cord and circular placenta in background

CHAPTER OUTLINE

Gametogenesis and Fertilization 106
- Gametogenesis 106
- Sperm Migration and Capacitation 106
- Fertilization 107

Stages of Prenatal Development 108
- Preembryonic Stage 108
- Embryonic Stage 110
- Fetal Stage 118

Clinical Perspectives 120
- Spontaneous Abortion 120
- Birth Defects 120

Chapter Review 125

INSIGHTS

4.1 Clinical Application: Ectopic Pregnancy 111
4.2 Evolutionary Medicine: Morning Sickness 114
4.3 Clinical Application: Placental Disorders 118

BRUSHING UP

To understand this chapter, you may find it helpful to review the following concepts:

- Fetal sonography (p. 3)
- Chordate characteristics of humans (p. 13)
- Cell division (p. 69)

Perhaps the most dramatic, seemingly miraculous aspect of human life is the transformation of a one-celled fertilized egg into an independent, fully developed individual. From the beginning of recorded thought, people have pondered how a baby forms in the mother's body and how two parents can produce another human being who, although unique, possesses characteristics of each. Aristotle dissected the embryos of various birds, established their sequence of organ development, and speculated that the hereditary traits of a child resulted from the mixing of the male's semen with the female's menstrual blood. Misconceptions about human development persisted for many centuries. Scientists of the seventeenth century thought that all the features of the infant existed in a preformed state in the egg or the sperm, and that they simply unfolded and expanded as the embryo developed. Some thought that the head of the sperm had a miniature human curled up in it, while others thought the miniature person existed in the egg, and the sperm were merely parasites in the semen. The modern science of **embryology** was not born until the nineteenth century, largely because darwinism at last gave biologists a systematic framework for asking the right questions and discovering unifying themes in the development of diverse species of animals, including humans.

The rest of this book deals with individual organ systems. The anatomy of any organ system, however, is best understood from the standpoint of its prenatal development. Therefore, this chapter describes some of the earliest and most general developments of the human embryo, while the chapters following it briefly describe the further development of the respective systems.

Gametogenesis and Fertilization

Objectives

When you have completed this section, you should be able to

- describe the major features of sperm and egg production;
- explain how sperm migrate to the egg and acquire the capacity to fertilize it; and
- describe the fertilization process and how an egg prevents fertilization by more than one sperm.

Gametogenesis

Reproduction requires sperm and eggs, which are known as the sex cells or **gametes.** The production of these cells is called **gametogenesis.**[1] This process is detailed in chapter 26, because it is best understood relative to the anatomy of the testes and ovaries. However, a few basic facts are needed here in order to best understand fertilization and the beginning of human development.

One of the most important properties of the gametes is that they have only half as many chromosomes as other cells of the body. They are called **haploid**[2] for this reason, whereas the other cells of the body are **diploid.**[3] Haploid cells have a single set of 23 chromosomes. Diploid cells have two complete sets of 23 chromosomes each (46 in all), one set from the mother and one from the father. Haploid gametes are necessary to sexual reproduction because if the gametes were diploid, the fertilization of an egg with 46 chromosomes by a sperm with 46 would produce a *zygote* (fertilized egg) with 92 chromosomes. All cells descended from the zygote by mitosis would also have 92. Then in that generation, a sperm with 92 chromosomes would fertilize an egg with 92, and the next generation would have 184 chromosomes in each cell, and so on. Obviously, if sexual reproduction is to combine a cell from each parent in each generation, there must be a mechanism for maintaining the normal chromosome number. The solution is to reduce the chromosome number by one-half as the gametes are formed, and this function is achieved by a special form of cell division called **meiosis**[4] **(reduction division).**

Gametogenesis begins with diploid stem cells that sustain their numbers through mitosis. Some of these cells then set off on a path that leads to egg or sperm cells through meiosis. Meiosis consists of two cell divisions (meiosis I and II) that have the effect of creating new genetic variety in the chromosomes and halving the chromosome number.

In sperm production *(spermatogenesis),* meiosis I and II result in four equal-sized cells called *spermatids,* which then grow tails and develop into sperm without further division. In egg production *(oogenesis),* however, each meiotic division produces one large cell (the future gamete) and a small cell called a *polar body,* which soon dies. The polar body is merely a way of disposing of the excess chromosomes. This uneven division produces an egg with as much cytoplasm as possible—the raw material for early preembryonic development. In oogenesis, an unfertilized egg dies after meiosis I. Meiosis II never occurs unless the egg is fertilized.

Sperm Migration and Capacitation

The human ovary usually releases one egg (oocyte) per month, around day 14 of a typical 28-day ovarian cycle. This egg is swept into the *uterine (fallopian) tube* by the beating of cilia on the tube's epithelial cells, and begins a 3-day trip down the tube toward the uterus. If the egg is not fertilized, it dies within 24 hours and gets no more than one-third of the way to the uterus. Therefore, if a sperm is to fertilize an egg, it must migrate up the tube to meet it. The vast majority of sperm never make it. While a typical ejaculation may contain 300 million sperm, many of these are destroyed by vaginal acid or drain out of the vagina; others fail to get through the cervical canal into the uterus; still more are destroyed by leukocytes in the uterus; and half of the survivors of all these ordeals are likely to go up the wrong uterine tube. Finally, 2,000 to 3,000 spermatozoa (0.001%) reach the general vicinity of the egg.

Freshly ejaculated sperm cannot immediately fertilize an egg. They must undergo a process called **capacitation,** which takes about 10 hours and occurs during their migration in the female reproductive tract. In fresh sperm, the plasma membrane is toughened by cholesterol. During capacitation, cholesterol is

[1]*gameto* = marriage, union + *genesis* = production
[2]*haplo* = half
[3]*diplo* = double
[4]*meio* = less, fewer

leached from the membrane. As a result, the membrane becomes more fragile so it can break open more easily upon contact with the egg. It also becomes more permeable to calcium ions, which diffuse into the sperm and stimulate more powerful lashing of the tail.

The anterior tip of the sperm contains a specialized lysosome called the **acrosome,** a packet of enzymes used to penetrate the egg and certain barriers around it (see chapter 26). When the sperm contacts an egg, the acrosome undergoes exocytosis—the *acrosomal reaction*—releasing these enzymes. But the first sperm to reach an egg is not the one to fertilize it. The egg is surrounded by a gelatinous membrane called the *zona pellucida* and, outside this, a layer of small *granulosa cells.* It may require up to 100 sperm to clear a path through these barriers before one of them can penetrate into the egg itself.

Fertilization

When a sperm contacts the egg's plasma membrane, it digests a hole into the membrane and the sperm head and midpiece enter the egg (fig. 4.1). The *midpiece,* a short segment of the tail behind the head, contains sperm mitochondria, the "powerhouses" that synthesize ATP for sperm motility. The egg, however, destroys the sperm mitochondria, so only the mother's mitochondria (and mitochondrial DNA) pass to the offspring.

It is important that only one sperm be permitted to fertilize an egg. If two or more sperm did so—an event called **polyspermy**—the fertilized egg would have 69 or more chromosomes and would fail to develop. The egg has two mechanisms for preventing such a wasteful fate: (1) In the *fast block to polyspermy,* binding of the sperm to the egg opens sodium channels in the egg membrane. The rapid inflow of sodium ions depolarizes the membrane and inhibits the binding of any more sperm. (2) In the *slow block to polyspermy,* sperm penetration triggers an inflow of calcium ions. Calcium stimulates a *cortical reaction*—the exocytosis of secretory vesicles called **cortical granules** just beneath the egg membrane. The secretion from these granules swells with water, pushes all remaining sperm away from the egg, and creates an impenetrable **fertilization membrane** between the egg and zona pellucida.

Upon fertilization, the egg completes meiosis II and discards a second polar body. The sperm and egg nuclei swell and become **pronuclei.** A mitotic spindle forms between them, each pronucleus ruptures, and the chromosomes of the two gametes mix into a single diploid set. This mingling of the maternal and paternal chromosomes is called **amphimixis.**[5] The cell, now called a zygote, is ready for its first mitotic division. Pregnancy has begun.

[5] *amphi* = both + *mixis* = mingling

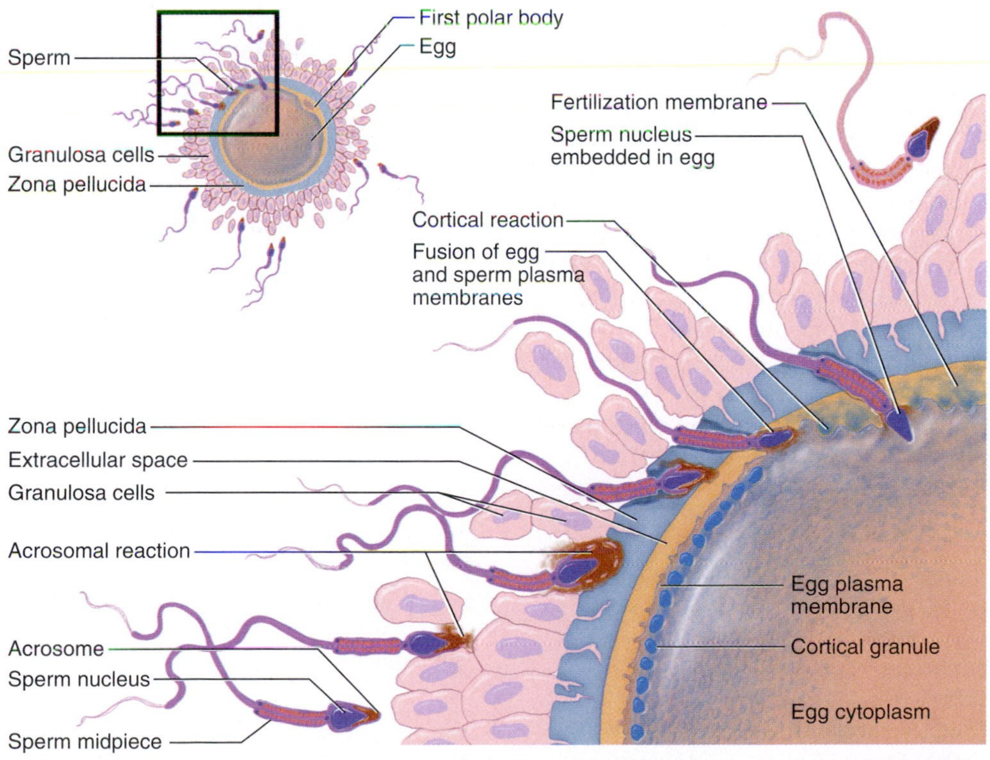

FIGURE 4.1

Fertilization. The sequence of events proceeds from *lower left* to *upper right.* The fertilization membrane formed at the *right* side of the figure is the slow block to polyspermy.

Before You Go On

Answer the following questions to test your understanding of the preceding section:

1. Why is it necessary for gametogenesis to reduce the chromosome number of the sex cells by one-half?
2. Explain why sperm cannot fertilize an egg immediately after ejaculation.
3. Explain why all of your mitochondria originate only from your mother.
4. Describe two ways a fertilized egg prevents the entry of excess sperm.

Stages of Prenatal Development

Objectives

When you have completed this section, you should be able to

- name and define the three basic stages of prenatal development;
- describe the implantation of a conceptus in the uterine wall;
- describe the major events that transform a fertilized egg into an embryo.
- define and describe the amnion, chorion, allantois, and yolk sac associated with the embryo;
- describe three ways in which the conceptus is nourished during its development;
- describe the formation and functions of the placenta; and
- describe some major developments in the fetal stage.

Human **gestation** (pregnancy) lasts an average of 266 days (38 weeks), from **conception** (fertilization) to **parturition** (childbirth). Since the date of conception is seldom known with certainty, the gestational calendar is usually measured from the first day of a woman's last menstrual period (LMP), and birth is predicted to occur about 280 days (40 weeks) thereafter. Time periods in this chapter, however, are measured from the date of conception.

All the products of conception are collectively called the **conceptus.** This includes all developmental stages from zygote through fetus, and the associated structures such as the umbilical cord, placenta, and amniotic sac.

Clinically, the course of a pregnancy is divided into 3-month intervals called **trimesters:**

1. The **first trimester** (first 12 weeks) extends from fertilization through the first month of fetal life. This is the most precarious stage of development; more than half of all embryos die in the first trimester. Stress, drugs, and nutritional deficiencies are most threatening to the conceptus during this time.
2. The **second trimester** (weeks 13 through 24) is a period in which the organs complete most of their development. It becomes possible with sonography to see good anatomical detail in the fetus. By the end of this trimester, the fetus looks distinctly human, and with intensive clinical care, infants born at the end of the second trimester have a chance of survival.
3. In the **third trimester** (weeks 25 to birth), the fetus grows rapidly and the organs achieve enough cellular differentiation to support life outside the womb. Some organs, such as the brain, liver, and kidneys, however, require further differentiation after birth to become fully functional. At 35 weeks from fertilization, the fetus typically weighs about 2.5 kg (5.5 lb). It is considered mature at this weight, and usually survives if born early. Most twins are born at about 35 weeks' gestation.

From a more biological than clinical standpoint, human development is divided into three stages called the preembryonic, embryonic, and fetal stages (table 4.1).

1. The **preembryonic stage** begins with the **zygote**[6] (fertilized egg) and lasts about 16 days. It involves three main processes: (1) *cleavage,* or cell division; (2) *implantation,* in which the conceptus becomes embedded in the mucosal lining *(endometrium)* of the uterus; and (3) *embryogenesis,* in which the embryonic cells migrate and differentiate into three tissue layers called the *ectoderm, mesoderm,* and *endoderm*—collectively known as the **primary germ layers.** Once these layers exist, the individual is called an *embryo.*
2. The **embryonic stage** extends from day 17 until the end of week 8. It is a stage in which the primary germ layers develop into the rudiments of all the organ systems. When all of the organ systems are represented (even though not yet functional), the individual is considered a *fetus.*
3. The **fetal stage** of development extends from the beginning of week 9 until birth. This is a stage in which the organs grow, differentiate, and become capable of functioning outside the mother's body.

Preembryonic Stage

The three major events of the preembryonic stage are cleavage, implantation, and embryogenesis.

CLEAVAGE

Cleavage consists of mitotic divisions that occur in the first 3 days after fertilization, dividing the zygote into smaller and smaller cells called **blastomeres.**[7] It begins as the conceptus migrates down the uterine tube (fig. 4.2). The first cleavage occurs in about 30 hours. Blastomeres divide again at shorter and shorter time intervals, doubling the number of cells each time. In the early divisions, the blastomeres divide simultaneously, but as cleavage progresses they become less synchronized.

By the time the conceptus arrives in the uterus, about 72 hours after ovulation, it consists of 16 or more cells and has a bumpy surface similar to a mulberry—hence it is called a **morula.**[8] The morula is no

---

[6] *zygo* = union
[7] *blast* = bud, precursor + *mer* = segment, part
[8] *mor* = mulberry + *ula* = little

TABLE 4.1

The Stages of Prenatal Development

Stage	Age*	Major Developments and Defining Characteristics
Preembryonic Stage		
Zygote	0–30 hours	A single diploid cell formed by the union of egg and sperm
Cleavage	30–72 hours	Mitotic division of the zygote into smaller, identical blastomeres
Morula	3–4 days	A spherical stage consisting of 16 or more blastomeres
Blastocyst	4–16 days	A fluid-filled, spherical stage with an outer mass of trophoblast cells and inner mass of embryoblast cells; becomes implanted in the endometrium; inner cell mass forms an embryonic disc and differentiates into the three primary germ layers
Embryonic Stage	16 days–8 weeks	A stage in which the primary germ layers differentiate into organs and organ systems; ends when all organ systems are present
Fetal Stage	8–40 weeks	A stage in which organs grow and mature at a cellular level to the point of being capable of supporting life independently of the mother

* From the time of fertilization

larger than the zygote; cleavage merely produces smaller and smaller blastomeres. This increases the ratio of cell surface area to volume, which favors efficient nutrient uptake and waste removal, and it produces a larger number of cells from which to form different embryonic tissues.

The morula lies free in the uterine cavity for 4 or 5 days and divides into 100 cells or so. It becomes a hollow sphere called the **blastocyst,** with an internal cavity called the **blastocoel** (BLAST-oh-seal). The wall of the blastocyst is a layer of squamous cells called the **trophoblast,**[9] which is destined to form part of the placenta and play an important role in nourishing the embryo. On one side of the blastocoel, adhering to the inside of the trophoblast, is an inner cell mass called the **embryoblast,** which is destined to become the embryo itself.

[9]*troph* = food, nourishment + *blast* = to produce

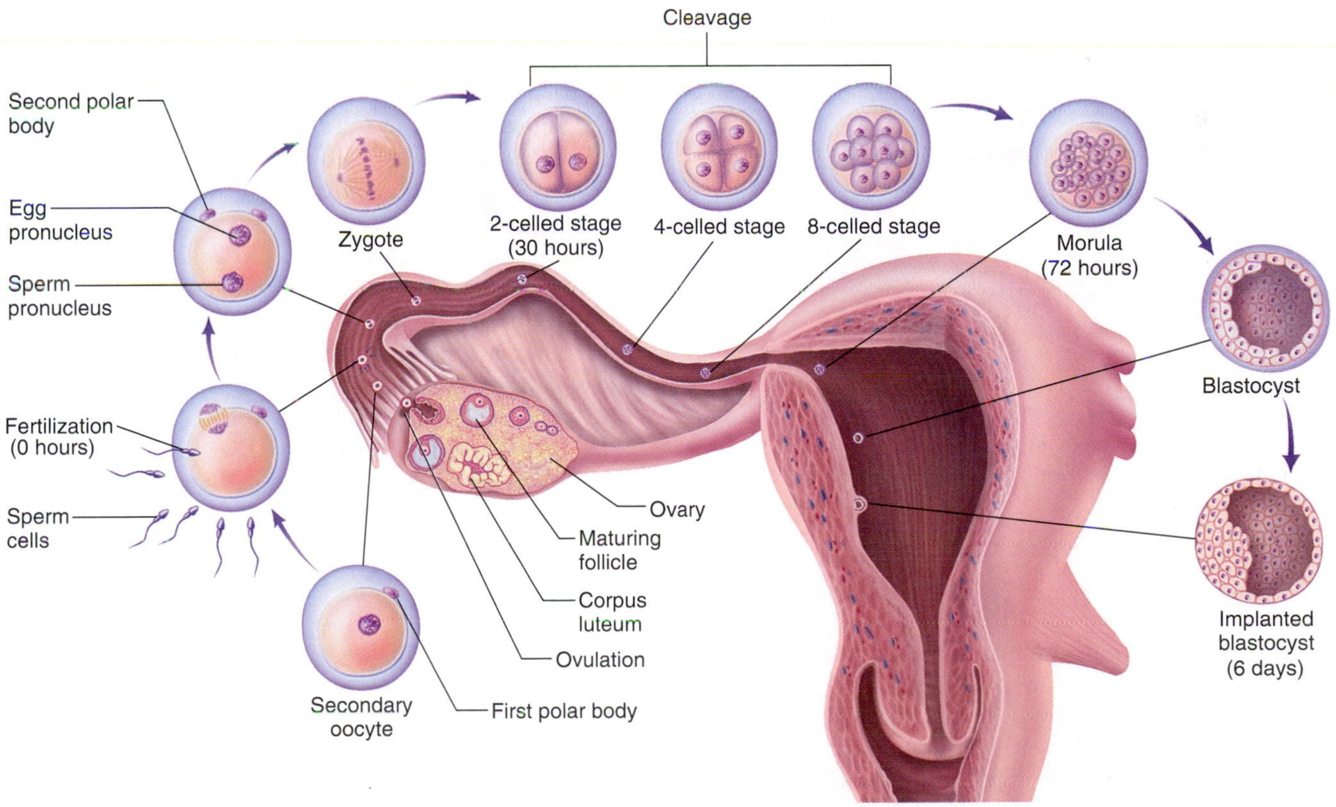

FIGURE 4.2

Migration of the Conceptus. The egg is fertilized in the distal end of the fallopian tube, and the preembryo begins cleavage as it migrates to the uterus.

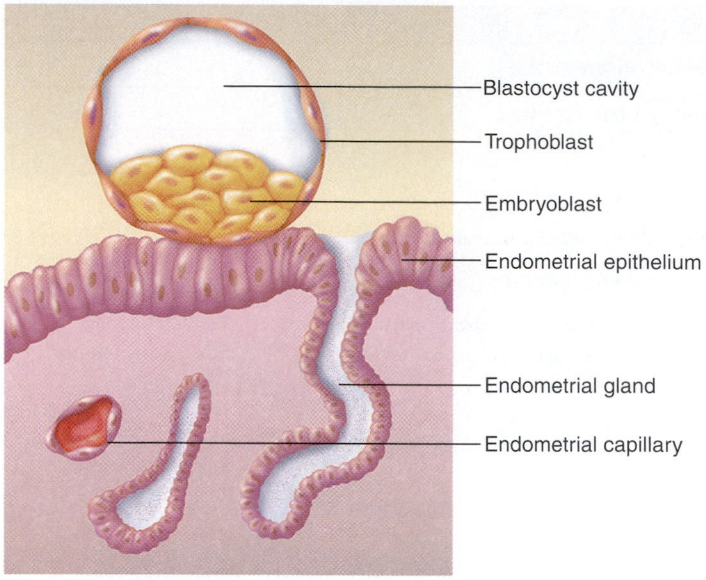

(a)

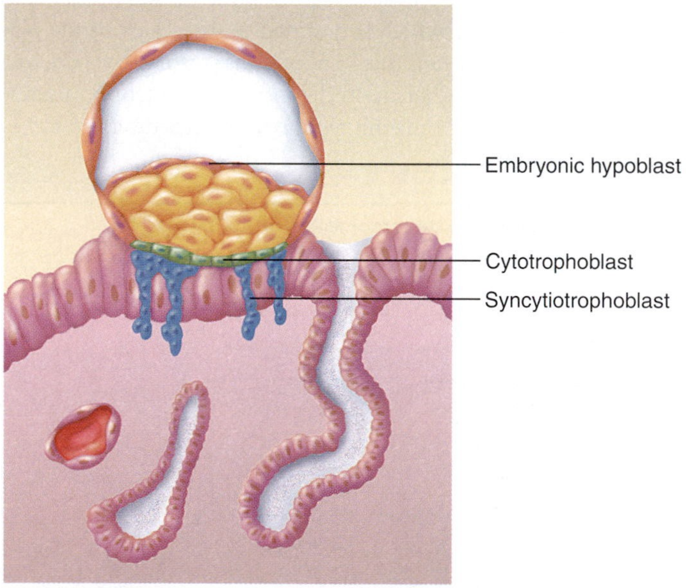

(b)

FIGURE 4.3

Implantation. (*a*) Structure of the blastocyst about 6 to 7 days after ovulation. (*b*) The progress of implantation about 1 day later. The syncytiotrophoblast has begun growing rootlets, which penetrate the endometrium.

IMPLANTATION

About 6 days after ovulation, the blastocyst attaches to the endometrium, usually on the "ceiling" or on the dorsal wall of the uterus. The process of attachment, called **implantation,** begins when the blastocyst adheres to the endometrium (fig. 4.3). The trophoblast cells on this side separate into two layers. In the superficial layer, in contact with the endometrium, the plasma membranes break down and the trophoblast cells fuse into a multinucleate mass called the **syncytiotrophoblast**[10] (sin-SISH-ee-oh-TRO-foe-blast). (A *syncytium* is any body of protoplasm containing multiple nuclei.) The deep layer, close to the embryoblast, is called the **cytotrophoblast**[11] because it retains individual cells divided by membranes.

The syncytiotrophoblast grows into the uterus like little roots, digesting endometrial cells along the way. The endometrium reacts to this injury by growing over the trophoblast and eventually covering it, so the conceptus becomes completely buried in endometrial tissue. Implantation takes about a week and is completed about the time the next menstrual period would have occurred if the woman had not become pregnant.

EMBRYOGENESIS

During implantation, the embryoblast undergoes **embryogenesis**—arrangement of the blastomeres into the three primary germ layers. At the beginning of this phase, the embryoblast separates slightly from the trophoblast, creating a narrow space between them called the **amniotic cavity.** The embryoblast flattens into an **embryonic disc (blastodisc)** composed of two cell layers: the **epiblast** facing the amniotic cavity and the **hypoblast** facing away. Some hypoblast cells multiply and form a membrane called the *yolk sac* enclosing the blastocoel. Now the embryonic disc is flanked by two spaces: the amniotic cavity on one side and the yolk sac on the other (fig. 4.4).

Meanwhile, the embryonic disc elongates and, around day 15, a groove called the **primitive streak** forms along the midline of the epiblast. These events make the embryo bilaterally symmetric and define its future right and left sides, dorsal and ventral surfaces, and **cephalic**[12] and **caudal**[13] ends.

The next step is **gastrulation**—multiplying epiblast cells migrate medially toward the primitive streak and down into it. These cells replace the original hypoblast with a layer now called the **endoderm,** which will become the inner lining of the digestive tract, among other things. A day later, migrating epiblast cells form a third layer between the first two, called the **mesoderm** (fig. 4.5). Once this is formed, the epiblast is called **ectoderm.** Thus, the three primary germ layers all arise from the original epiblast. The ectoderm and endoderm are epithelia composed of tightly joined cells, but the mesoderm is a more loosely organized tissue.

The mesoderm later differentiates into a loose fetal connective tissue called **mesenchyme,**[14] from which such tissues as muscle, bone, and blood develop. Mesenchyme is composed of fibroblasts and fine, wispy collagen fibers embedded in a gelatinous ground substance.

Once the three primary germ layers are formed, embryogenesis is complete and the individual is considered an **embryo.** It is about 2 mm long and 16 days old at this point.

Embryonic Stage

The embryonic stage of development begins around day 16 and extends to the end of week 8. During this time, the placenta and other

[10] *syn* = together + *cyt* = cell
[11] *cyto* = cell
[12] *cephal* = head
[13] *caud* = tail
[14] *mes* = middle + *enchym* = poured into

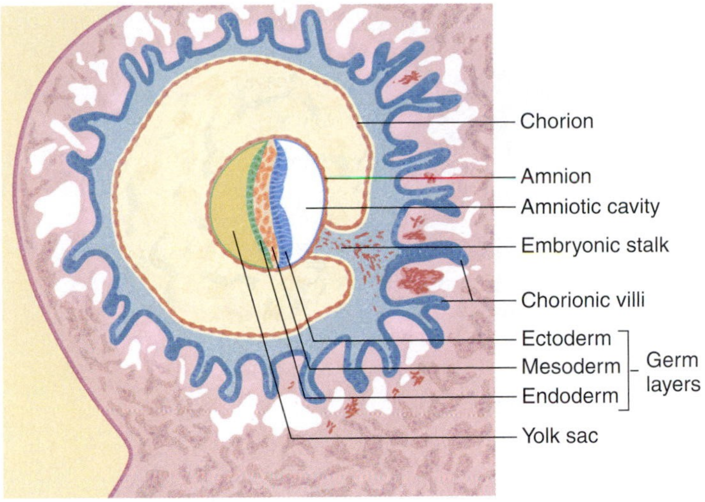

FIGURE 4.4

The Implanted Conceptus at 16 Days. The primary germ layers and extraembryonic membranes have formed, and the conceptus is now an embryo.

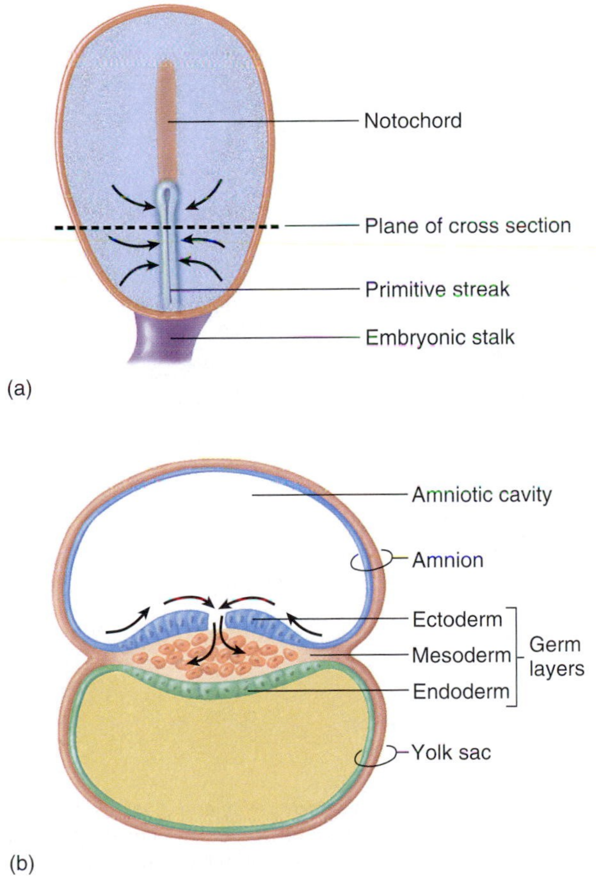

FIGURE 4.5

Gastrulation. (*a*) Dorsal view of the embryonic disc at 16 days; epiblast cell migration indicated by *arrows*. (*b*) Cross section of the embryonic disc at the level indicated in *a*. Mesoderm arises from epiblast cells that migrate into the primitive streak; the remaining epiblast cells become ectoderm.

INSIGHT 4.1 CLINICAL APPLICATION

Ectopic Pregnancy

In about 1 out of 300 pregnancies, the blastocyst implants somewhere other than the uterus, producing an *ectopic*[15] *pregnancy.* Most cases begin as *tubal pregnancies,* implantation in the uterine tube. This usually occurs because the conceptus encounters an obstruction such as a constriction resulting from earlier pelvic inflammatory disease, tubal surgery, previous ectopic pregnancies, or repeated miscarriages. The uterine tube cannot expand enough to accommodate the growing conceptus for long; if the situation is not detected and treated early, the tube usually ruptures within 12 weeks. This can kill the mother, or the conceptus can reimplant in the abdominopelvic cavity, producing an *abdominal pregnancy.* The conceptus can grow anywhere it finds an adequate blood supply—for example, on the outside of the uterus, colon, or bladder. About 1 pregnancy in 7,000 is abdominal. Abdominal pregnancy is a serious threat to the mother's life and usually requires abortion, but about 9% of abdominal pregnancies end in live birth by cesarean section.

[15]*ec* = outside + *top* = place

accessory structures develop, the embryo begins receiving nutrition primarily from the placenta, and the germ layers differentiate into organs and organ systems. Although these organs are still far from functional, it is their presence at 8 weeks that marks the transition from the embryonic stage to the fetal stage.

EMBRYONIC FOLDING AND ORGANOGENESIS

In week 4, the embryo grows rapidly and folds around the yolk sac, converting the flat embryonic disc into a somewhat cylindrical form. As the cephalic and caudal ends curve around the ends of the yolk sac, the embryo becomes C-shaped, with the head and tail almost touching (fig. 4.6). The lateral margins of the disc fold around the sides of the yolk sac to form the ventral surface of the embryo. This lateral folding encloses a longitudinal channel, the *primitive gut,* which later becomes the digestive tract.

As a result of embryonic folding, the entire surface is covered with ectoderm, which later produces the epidermis of the skin. In the meantime, the mesoderm splits into two layers. One of them adheres to the ectoderm and the other to the endoderm, thus opening a space called the **coelom** (SEE-loam) between them. The coelom becomes divided into the thoracic cavity and peritoneal cavity by a wall, the diaphragm. By the end of week 5, the thoracic cavity further subdivides into pleural and pericardial cavities.

The formation of organs and organ systems during this time is called **organogenesis.** Table 4.2 lists the major tissues and organs that arise from each primary germ layer.

Think About It!

List the four primary tissue types of the adult body (see chapter 3) and identify which of the three primary germ layers of the embryo predominantly gives rise to each.

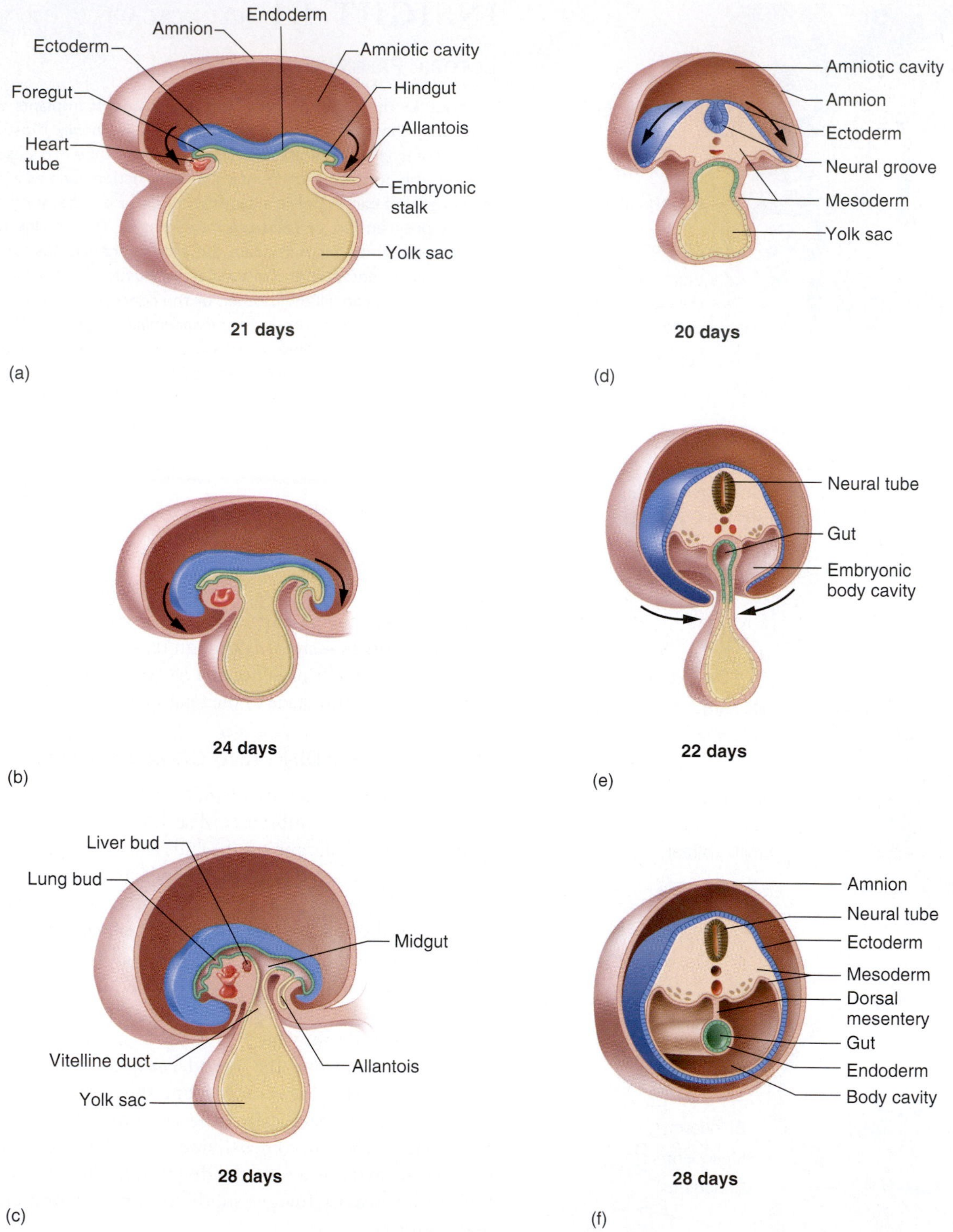

FIGURE 4.6

Embryonic Folding. Transformation from a flat embryonic disc to a more rounded and curled body, enclosing a body cavity and digestive tract (gut). Figures *a–c* are longitudinal sections and figures *d–f* are cross sections. *Arrows* indicate the directions of embryonic folding at the cephalic and caudal ends in *a–b* and the lateral margins in *d–e.* Note the progressive narrowing of the connection between the yolk sac and primitive gut as the embryo folds and the gut becomes regionally differentiated.

TABLE 4.2

Derivatives of the Three Primary Germ Layers

Layer	Major Derivatives
Ectoderm	Epidermis; hair follicles and piloerector muscles; cutaneous glands; nervous system; adrenal medulla; pineal and pituitary glands; lens, cornea, and intrinsic muscles of the eye; internal and external ear; salivary glands; epithelia of the nasal cavity, oral cavity, and anal canal
Mesoderm	Dermis; skeleton; skeletal, cardiac, and most smooth muscle; cartilage; adrenal cortex; middle ear; blood and lymphatic vessels; blood; bone marrow; lymphoid tissue; epithelium of kidneys, ureters, gonads, and genital ducts; mesothelium of ventral body cavity
Endoderm	Most of the mucosal epithelium of the digestive and respiratory tracts; mucosal epithelium of urinary bladder and parts of urethra; epithelial components of accessory reproductive and digestive glands (except salivary glands); thyroid and parathyroid glands; thymus

Three major events of organogenesis are especially important for understanding organ development in later chapters: development of the *neural tube,* outpocketings of the throat region called *pharyngeal pouches,* and body segments called *somites.*

The formation of the **neural tube** is called **neurulation.** This process is detailed in chapter 13, but a few essential points are needed here. By week 3, a thick ridge of ectoderm called the *neural plate* appears along the midline of the embryonic disc. This is the source of the entire nervous system. As development progresses, the neural plate sinks and becomes a *neural groove,* with a raised edge called the *neural fold* on each side. The edges of the neural fold meet and close, somewhat like a zipper, beginning in the middle and progressing toward both ends (fig. 4.7). By 4 weeks, this process creates an enclosed channel, the neural tube. The cephalic end of the tube develops bulges or *vesicles* that develop into different regions of the brain, and the more caudal part becomes the spinal cord. Neurulation is one of the most sensitive periods of prenatal development. Abnormal developments called *neural tube defects* are among the most common and devastating birth defects (see p. 380).

Pharyngeal (branchial) pouches are five pairs of pockets that form in the walls of the future throat of the embryo around 4 to 5 weeks' gestation (fig. 4.8). They are separated by **pharyngeal arches,** which appear as external bulges in the neck region (fig. 4.9).

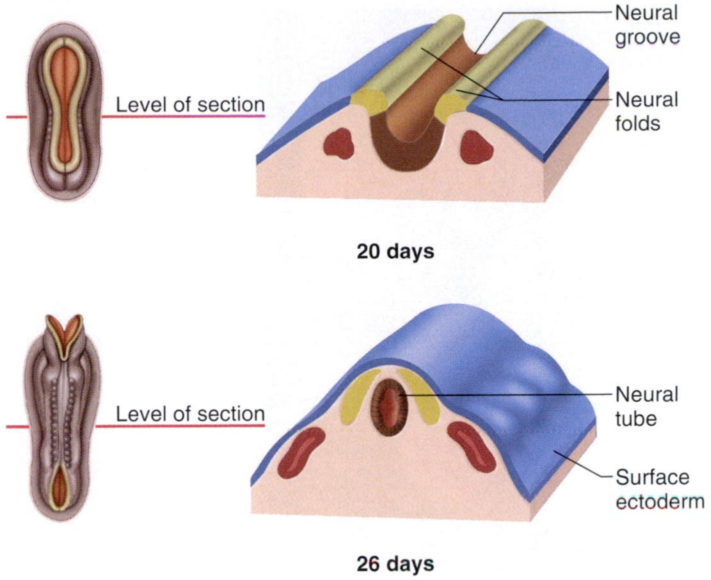

FIGURE 4.7

Neurulation. (*a*) The neural groove at 20 days. (*b*) The neural tube at 26 days.

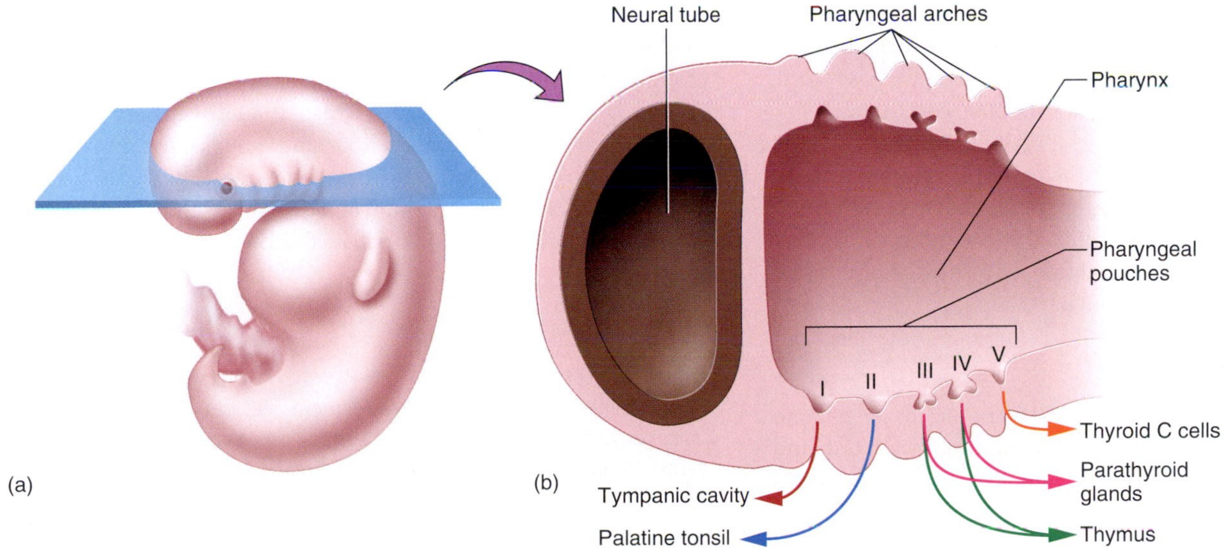

FIGURE 4.8

The Pharyngeal Pouches. (*a*) Level at which the section in figure *b* is taken. (*b*) Superior view of the pharyngeal region showing the five pairs of pharyngeal pouches (I–V) and their developmental fates. Compare figure 4.9.

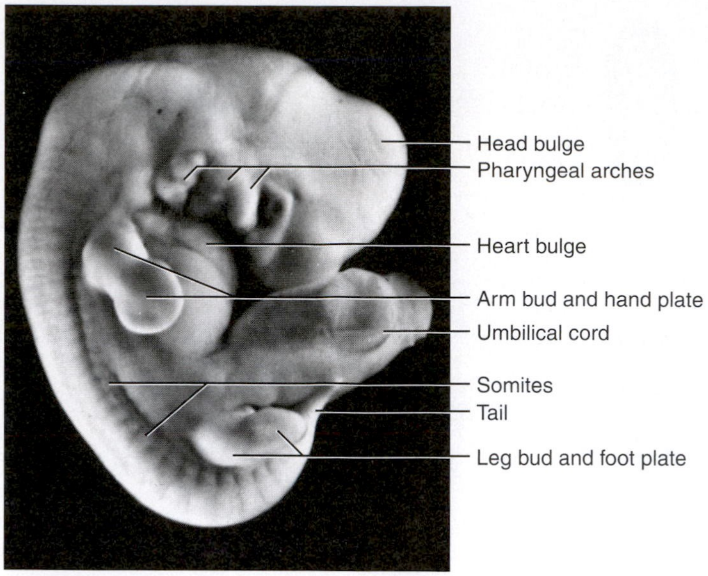
Head bulge
Pharyngeal arches
Heart bulge
Arm bud and hand plate
Umbilical cord
Somites
Tail
Leg bud and foot plate

FIGURE 4.9
A 5-week Embryo (SEM).

INSIGHT 4.2 EVOLUTIONARY MEDICINE

Morning Sickness

The nausea called *morning sickness* is often a woman's earliest sign that she may be pregnant. It sometimes progresses to vomiting. Severe and prolonged vomiting, called *hyperemesis gravidarum,*[19] can necessitate hospitalization for fluid therapy to restore electrolyte and acid-base balance. The physiological cause of morning sickness is unknown; it may result from the steroids of pregnancy inhibiting intestinal motility. It is also uncertain whether it is merely an undesirable effect of pregnancy or whether it has a biological purpose. An evolutionary hypothesis is that morning sickness is an adaptation to protect the embryo from toxins. The embryo is most vulnerable to toxins at the same time that morning sickness peaks, and women with morning sickness tend to prefer bland foods and to avoid spicy and pungent foods, which are highest in toxic compounds. Pregnant women tend also to be especially sensitive to flavors and odors that suggest spoiled food. Women who do not experience morning sickness are more likely to miscarry or bear children with birth defects.

[19]*hyper* = excessive + *emesis* = vomiting + *gravida* = pregnant woman + *arum* = of

Pharyngeal pouches are among the basic defining characteristics of all Chordata (see chapter 1). In humans, they give rise to such structures as the middle-ear cavity, palatine tonsil, thymus, parathyroid glands, and part of the thyroid gland.

Somites are bilaterally paired blocks of mesoderm that give the embryo a segmented appearance (fig. 4.9). They represent a primitive vertebrate segmentation that is more distinctly visible in fish, snakes, and other lower vertebrates than in mammals. Humans, however, show traces of this segmentation in the linear series of vertebrae, ribs, spinal nerves, and trunk muscles. Somites begin to appear by day 20, and number 42 to 44 pairs by day 35. Beginning in week 4, each somite subdivides into three tissue masses: a **sclerotome,**[16] which surrounds the neural tube and gives rise to bone tissue of the vertebral column; a **myotome,**[17] which gives rise to muscles of the trunk; and a **dermatome,**[18] which gives rise to the dermis of the skin and to its associated subcutaneous tissue.

At 5 weeks, the embryo exhibits a prominent **head bulge** at the cephalic end and a pair of optic vesicles destined to become the eyes. A large **heart bulge** contains a heart, which has been beating since day 22. The **arm buds** and **leg buds,** the future limbs, are present at 24 and 28 days, respectively. Figure 4.10 shows the external appearance of embryos from 37 to 56 days.

EMBRYONIC MEMBRANES

The conceptus develops a number of accessory organs external to the embryo itself. These include the placenta, umbilical cord, and four **embryonic membranes**—the *amnion, yolk sac, allantois,* and *chorion* (fig. 4.11). To understand these membranes, it helps to realize that all mammals evolved from egg-laying reptiles. Within the shelled, self-contained egg of a reptile, the embryo rests atop a yolk, which is enclosed in the yolk sac; it is suspended in a pool of liquid contained in the amnion; it stores toxic wastes in another sac, the allantois; and to breathe, it has a chorion permeable to gases. All of these membranes persist in mammals, including humans, but are modified in their functions.

The **amnion** is a transparent sac that develops from epiblast cells of the embryonic disc. It grows to completely enclose the embryo and is penetrated only by the umbilical cord. The amnion becomes filled with **amniotic fluid,** which enables the embryo to develop symmetrically; keeps its surface tissues from adhering to each other; protects it from trauma, infection, and temperature fluctuations; allows the freedom of movement important to muscle development; and plays a role in normal lung development. At first, the amniotic fluid forms by filtration of the mother's blood plasma, but beginning at 8 to 9 weeks, the fetus urinates into the amniotic cavity about once an hour and contributes substantially to the fluid volume. The volume remains stable, however, because the fetus swallows amniotic fluid at a comparable rate. At term, the amnion contains 700 to 1,000 mL of fluid.

Think About It!

Oligohydramnios[20] *is an abnormally low volume of amniotic fluid. Renal agenesis*[21] *is a failure of the fetal kidneys to develop. Which of these do you think is most likely to cause the other one? Explain why. What could be some consequences of oligohydramnios to fetal development?*

[16]*sclero* = hard + *tom* = segment
[17]*myo* = muscle + *tom* = segment
[18]*derma* = skin + *tom* = segment

[20]*oligo* = few, little + *hydr* = water, fluid + *amnios* = amniotic
[21]*a* = without + *genesis* = formation, development

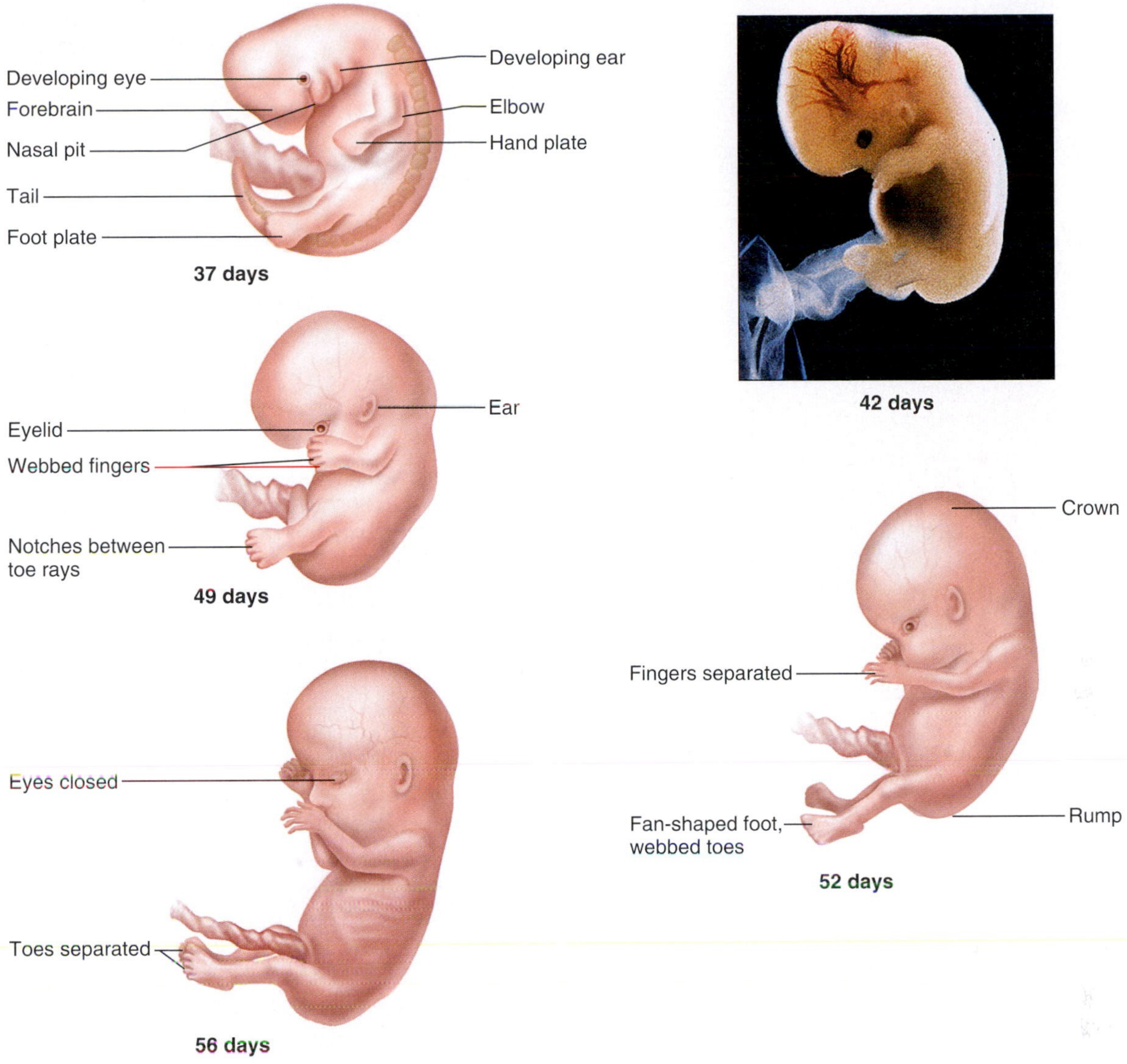
Developing eye
Forebrain
Nasal pit
Tail
Foot plate
Developing ear
Elbow
Hand plate
37 days
42 days
Eyelid
Webbed fingers
Notches between toe rays
Ear
49 days
Crown
Fingers separated
Fan-shaped foot, webbed toes
Rump
52 days
Eyes closed
Toes separated
56 days

FIGURE 4.10
Development of the Embryo from 37 to 56 Days. At 56 days (8 weeks), all organ systems are present and the individual begins the fetal phase.

The **yolk sac,** as we have already seen, arises from cells of the embryonic hypoblast opposite the amnion. Initially it is larger than the embryo and is broadly connected to almost the entire length of the primitive gut. During embryonic folding, however, its connection to the gut becomes constricted and reduced to a narrow passage called the **vitelline**[22] **duct.** Since the embryo continues growing long after the yolk sac stops, the yolk sac becomes a relatively small pouch suspended from the ventral side of the embryo. It produces the first blood cells and the stem cells of gametogenesis. These cells migrate by ameboid movement into the embryo, where the blood cells colonize the bone marrow and other tissues, and the gametogenic stem cells colonize the future gonads.

The **allantois** (ah-LON-toe-iss) is initially an outpocketing of the yolk sac; eventually, as the embryo grows, it becomes an outgrowth of the caudal end of the gut connected to it by the **allantoic duct.** It forms a foundation for growth of the umbilical cord and becomes part of the urinary bladder. The allantoic duct can be seen in histological cross sections of the umbilical cord if they are cut close enough to the fetal end.

The **chorion** (CORE-ee-on) is the outermost membrane, enclosing all the rest of the membranes and the embryo. Initially it has shaggy processes called **chorionic villi** around its entire surface. As the pregnancy advances, the villi on the placental side grow and branch, and this surface is then called the *villous chorion.* The villous chorion forms the fetal portion of the placenta. The villi degenerate over the rest of the surface, which is then called the *smooth chorion.*

[22]*vitell* = yolk

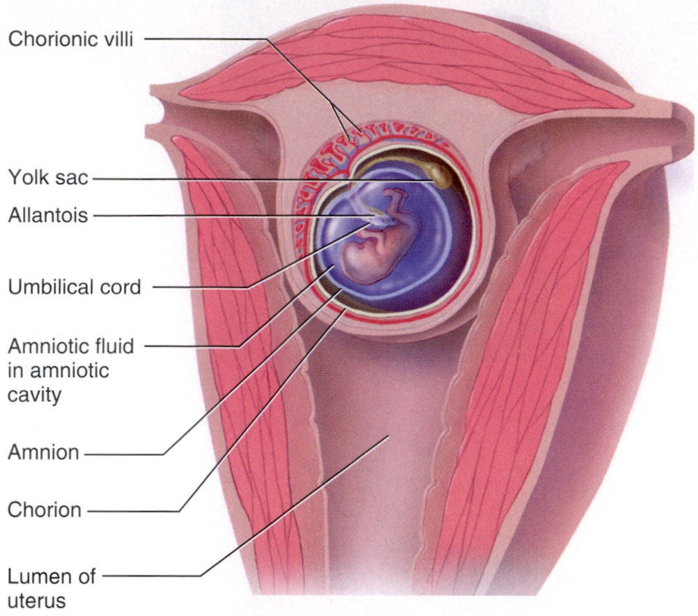
Chorionic villi
Yolk sac
Allantois
Umbilical cord
Amniotic fluid in amniotic cavity
Amnion
Chorion
Lumen of uterus

FIGURE 4.11

The Embryonic Membranes. Frontal section of the uterus with an 8-week fetus and accessory organs.

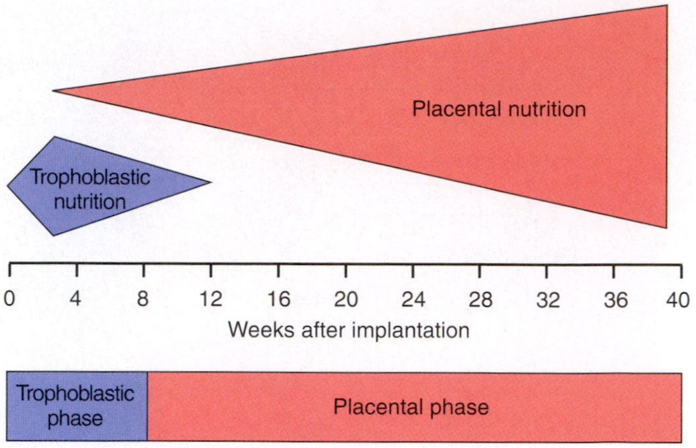
Placental nutrition
Trophoblastic nutrition
0
4
8
12
16
20
24
28
32
36
40
Weeks after implantation
Trophoblastic phase
Placental phase

FIGURE 4.12

The Phases of Intrauterine Nutrition. Trophoblastic nutrition peaks at 2 weeks and ends by 12 weeks. Placental nutrition begins at 2 weeks and becomes increasingly important until birth, 39 weeks after implantation.

PRENATAL NUTRITION

Over the course of gestation, the conceptus is nourished in three different, overlapping ways. As it travels down the uterine tube and lies free in the uterine cavity before implantation, it absorbs a glycogen-rich secretion of the uterine glands called **uterine milk.** It is the accumulation of this fluid that forms the blastocyst cavity.

As it implants, the conceptus makes a transition to **trophoblastic (deciduous) nutrition,** in which the trophoblast digests cells of the endometrium called **decidual**[23] **cells.** Under the influence of progesterone, these cells proliferate and accumulate a rich store of glycogen, proteins, and lipids. As the conceptus burrows into the endometrium, the syncytiotrophoblast digests them and supplies the nutrients to the embryoblast. Trophoblastic nutrition is the only mode of nutrition for the first week after implantation. It remains the dominant source of nutrients through the end of week 8; the period from implantation through week 8 is therefore called the **trophoblastic phase** of the pregnancy. Trophoblastic nutrition wanes as placental nutrition takes over, and ceases entirely by the end of week 12 (fig. 4.12).

In **placental nutrition,** nutrients from the mother's blood diffuse through the placenta into the fetal blood. The **placenta**[24] is a vascular organ attached to the uterine wall on one side and, on the other side, connected to the fetus by way of the **umbilical cord.** It begins to develop about 11 days after conception, becomes the dominant mode of nutrition around the beginning of week 9, and is the sole mode of nutrition from the end of week 12 until birth. The period from week 9 until birth is called the **placental phase** of the pregnancy.

Placentation, the development of a placenta, begins when extensions of the syncytiotrophoblast, the first chorionic villi, penetrate more and more deeply into the endometrium, like the roots of a tree penetrating into the nourishing "soil" of the uterus (fig. 4.13). As they digest their way through uterine blood vessels, the villi become surrounded by pools of free blood. These pools eventually merge to form a blood-filled cavity, the **placental sinus.** Exposure to maternal blood stimulates increasingly rapid growth of the villi, which become branched and treelike. Embryonic mesenchyme grows into the villi and gives rise to the blood vessels that connect to the embryo by way of the umbilical cord.

The fully developed placenta is a disc of tissue about 20 cm in diameter and 3 cm thick (fig. 4.14). At birth, it weighs about one-sixth as much as the baby. The surface facing the fetus is smooth and gives rise to the umbilical cord. The surface facing the uterine wall is rougher. It consists of the chorionic villi, which are contributed by the fetus, and a region of the mother's endometrium called the *decidua basalis.*

The umbilical cord contains two **umbilical arteries** and one **umbilical vein.** Pumped by the fetal heart, blood flows into the placenta by way of the umbilical arteries and then returns to the fetus by way of the umbilical vein. The placental villi are *filled with* fetal blood and *surrounded by* maternal blood; the two bloodstreams do not mix unless there is damage to the placental barrier. The barrier, however, is only 3.5 μm thick—half the diameter of a single red blood cell. Early in development, the chorionic villi have thick membranes that are not very permeable to nutrients and wastes, and their total surface area is relatively small. As the villi grow and branch, their surface area increases and the membranes become thinner and more permeable. Thus there is a dramatic increase in *placental conductivity,* the rate at which substances diffuse through the membrane. Oxygen and nutrients pass from the maternal

[23]*decid* = falling off
[24]*placenta* = flat cake

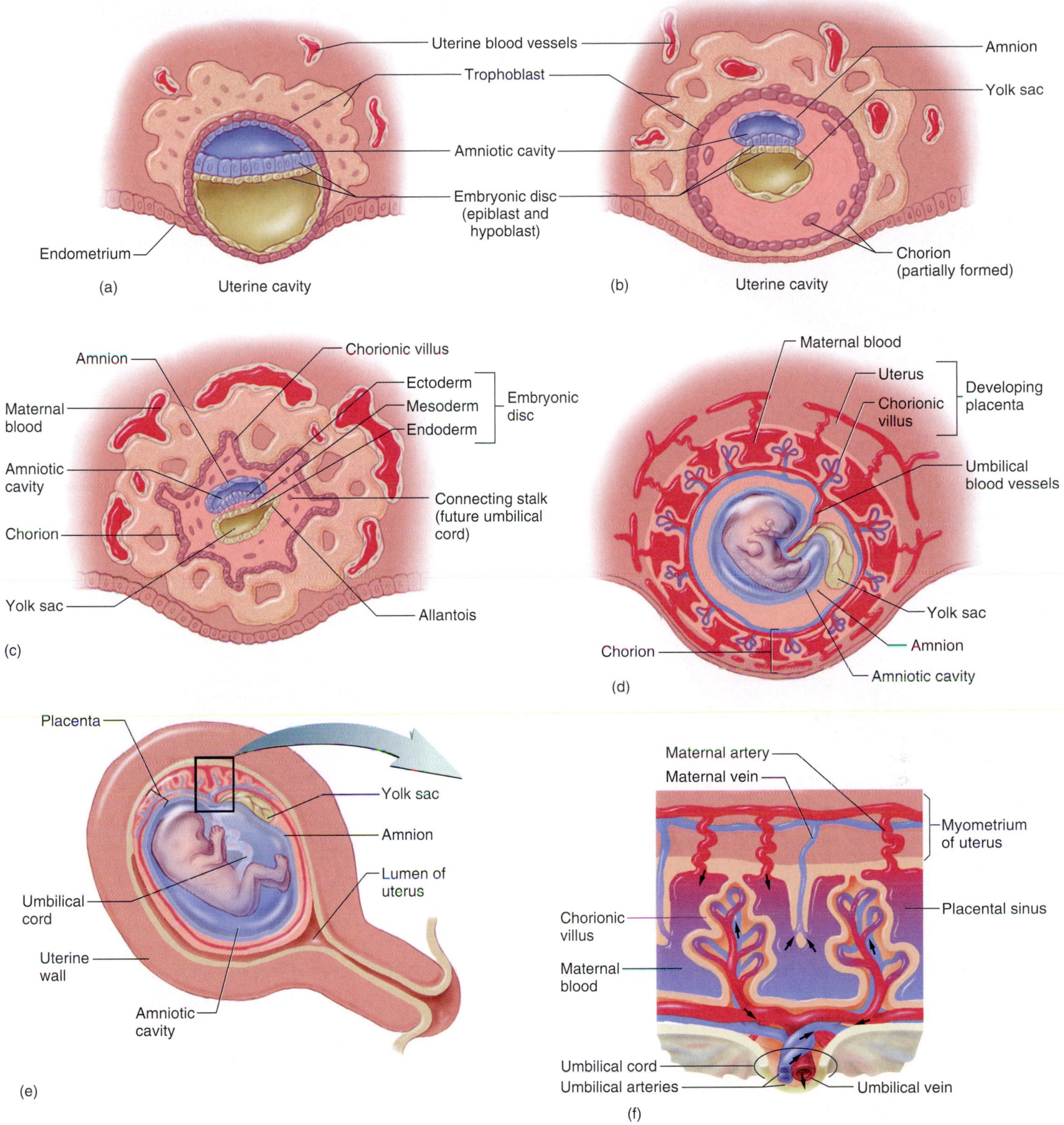

FIGURE 4.13

Development of the Placenta and Fetal Membranes. (*a*) Implantation nearly complete; an amniotic cavity has formed within the embryonic disc, lined by epiblast cells that will become the amnion. (*b*) Conceptus about 9 days after fertilization; hypoblast cells have now formed the yolk sac. (*c*) Conceptus at 16 days; the allantois is beginning to form, and the chorion is forming from the trophoblast and embryonic mesoderm. (*d*) Embryo at 4.5 weeks, enclosed in the amnion and chorion. (*e*) Embryo at 13.5 weeks; placentation is complete. (*f*) A portion of the mature placenta and umbilical cord. *Arrows* indicate blood flow.

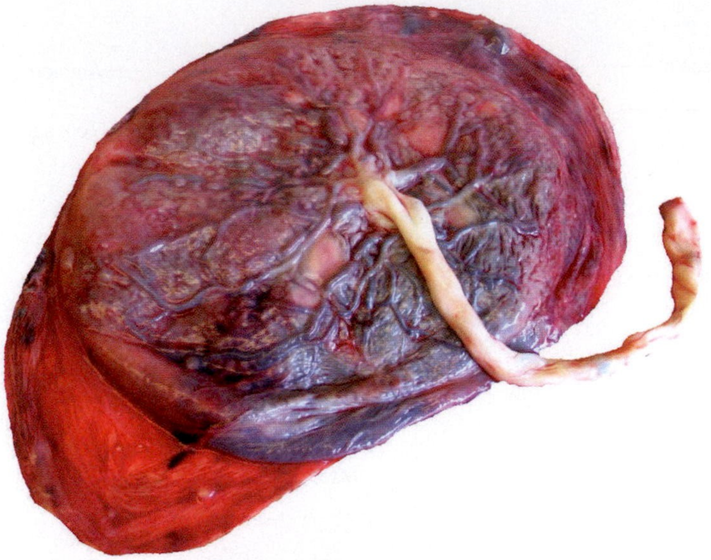

(a)

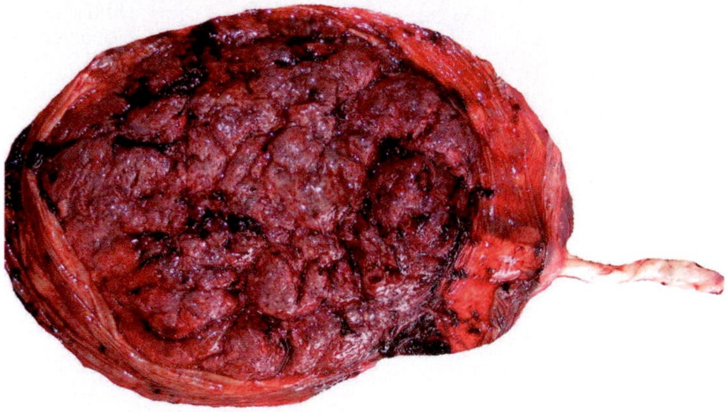

(b)

FIGURE 4.14

The Placenta. (*a*) The fetal side, showing blood vessels, the umbilical cord, and some of the amniotic sac attached to the lower left margin of the placenta. (*b*) The maternal (uterine) side.

blood to the fetal blood, while fetal wastes pass the other way to be eliminated by the mother. Unfortunately, the placenta is also permeable to nicotine, alcohol, and most other drugs in the maternal bloodstream. Nutrition, excretion, and other functions of the placenta are summarized in table 4.3.

Fetal Stage

At the end of 8 weeks, all of the organ systems are present, the individual is about 3 cm long, and it is now considered a **fetus** (fig. 4.15). Its bones have just begun to calcify and the skeletal muscles exhibit spontaneous contractions, although these are too weak to be felt by the mother. The heart, which has been beating since the fourth week, now circulates blood. The heart and liver are very large and form the prominent ventral bulges seen in figure 4.9. The head is nearly half the body length.

INSIGHT 4.3 CLINICAL APPLICATION

Placental Disorders

The two primary causes of third-trimester bleeding are placental disorders called *placenta previa* and *abruptio placentae*. The two disorders are similar and easily mistaken for each other. A suspicion of either condition calls for a sonogram to differentiate the two and decide on a course of action.

The conceptus usually implants high on the body of the uterus or on its ceiling. In about 0.5% of births, however, the placenta is so low on the uterine wall that it partially or completely blocks the cervical canal. This condition, called *placenta previa,* makes it impossible for the infant to be born without the placenta separating from the uterine wall first. Thus, there is a possibility of life-threatening hemorrhaging during pregnancy or birth. If placenta previa is detected by sonography, the infant is delivered by cesarean (C) section.

Abruptio placentae (ah-BRUP-she-oh pla-SEN-tee) is the premature partial or total separation of the placenta from the uterine wall. It occurs in 0.4% to 3.5% of pregnancies. Slight separations may require no more than bed rest and observation, but more severe cases can threaten the life of the mother, fetus, or both. Such cases require early delivery, usually by C section.

TABLE 4.3

Functions of the Placenta

Nutritional Roles	Transports nutrients such as glucose, amino acids, fatty acids, minerals, and vitamins from the maternal blood to the fetal blood; stores nutrients such as carbohydrates, protein, iron, and calcium in early pregnancy and releases them to the fetus later, when fetal demand is greater than the mother can absorb from the diet
Excretory Roles	Transports nitrogenous wastes such as ammonia, urea, uric acid, and creatinine from the fetal blood to the maternal blood
Respiratory Roles	Transports O_2 from mother to fetus, and CO_2 from fetus to mother
Endocrine Roles	Secretes hormones (estrogen, progesterone, relaxin, human chorionic gonadotropin, and human chorionic somatomammotropin); allows other hormones synthesized by the conceptus to pass into the mother's blood and maternal hormones to pass into the fetal blood
Immune Roles	Transports maternal antibodies into fetal blood to confer immunity on fetus

The primary changes in the fetal period are that the organ systems become functional and the fetus rapidly gains weight and becomes more human looking (fig. 4.16). Full-term fetuses average about 36 cm from the crown of the head to the curve of the buttock in a sitting position (*crown-to-rump length, CRL*). Most neonates (newborn infants) weigh between 3.0 and 3.4 kg (6.6–7.5 lb). About 50% of this weight is gained in the last 10 weeks. Most neonates weighing 1.5 to 2.5 kg are viable, but with difficulty. Neonates weighing under 500 g rarely survive.

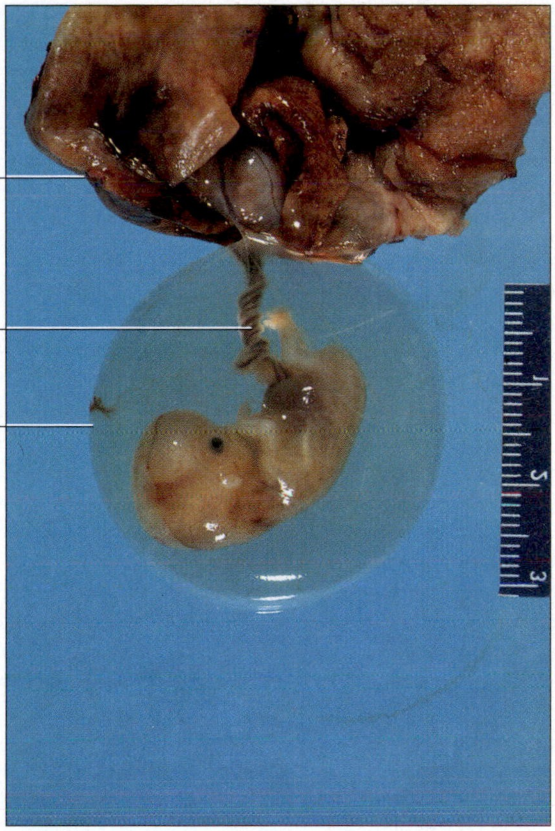

FIGURE 4.15
An 8-week Fetus. This is the age of transition from embryo to fetus.

The face acquires a more distinctly human appearance during the last trimester. The head grows more slowly than the rest of the body, so its relative length drops from one-half of the CRL at 8 weeks to one-fourth at birth. The skull has the largest circumference of any body region at term (about 10 cm), and passage of the head is therefore the most difficult part of labor. The limbs grow more rapidly than the trunk during the fetal stage, and achieve their final relative proportions to the trunk by 20 weeks. Other highlights of fetal development are described in table 4.4, and the development of individual organ systems is detailed in the chapters that follow.

Before You Go On

Answer the following questions to test your understanding of the preceding section:

5. What is the criterion for classifying a developing individual as an embryo? What is the criterion for classifying it as a fetus? At what gestational ages are these stages reached?
6. In the blastocyst, what are the cells called that eventually give rise to the embryo? What are the cells that carry out implantation?
7. Name and define the three principal processes that occur in the preembryonic stage.
8. Name the three primary germ layers and explain how they develop in the embryonic disc.
9. Distinguish between trophoblastic and placental nutrition.
10. State the functions of the placenta, amnion, chorion, yolk sac, and allantois.
11. Define and describe the neural tube, primitive gut, somites, and pharyngeal pouches.

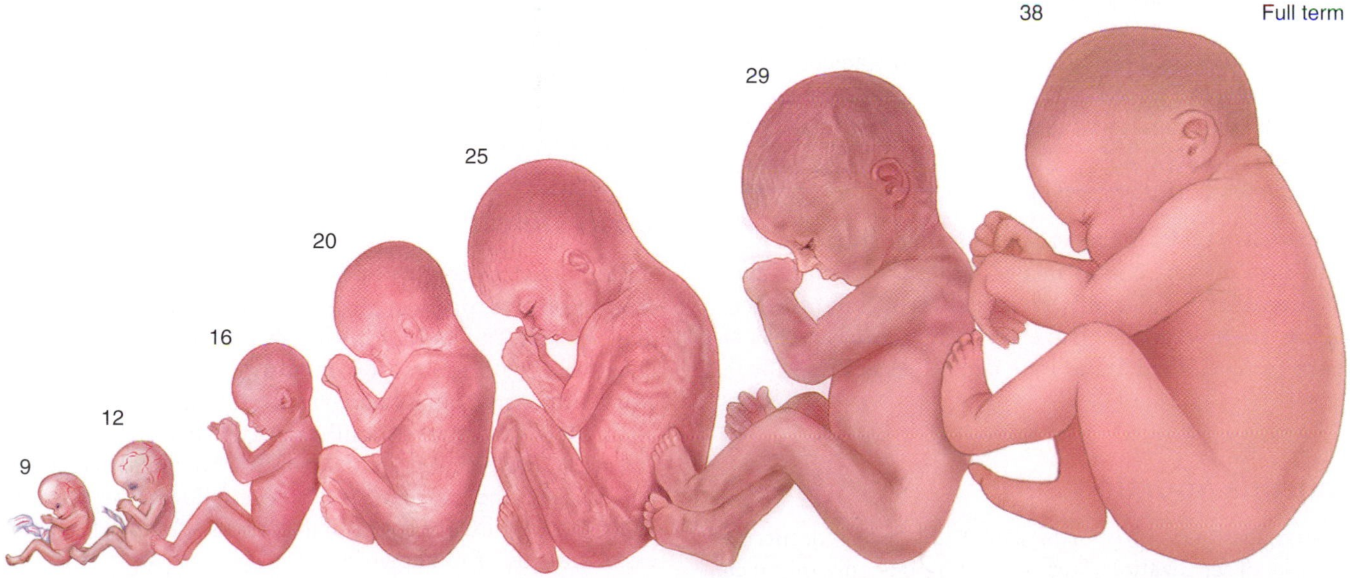

FIGURE 4.16
Growth of the Fetus. Number by each fetus is the age in weeks.

Table 4.4

Major Events of Prenatal Development, with Emphasis on the Fetal Stage

End of Week	Crown-to-Rump Length; Weight	Developmental Events
4	0.6 cm; <1 g	Vertebral column and central nervous system begin to form; limbs represented by small limb buds; heart begins beating around day 22; no visible eyes, nose, or ears
8	3 cm; 1 g	Eyes form, eyelids fused shut; nose flat, nostrils evident but plugged with mucus; head nearly as large as the rest of the body; brain waves detectable; bone calcification begins; limb buds form paddlelike hands and feet with ridges called **digital rays,** which then separate into distinct fingers and toes; blood cells and major blood vessels form; genitals present but sexes not yet distinguishable
12	9 cm; 45 g	Eyes well developed, facing laterally; eyelids still fused; nose develops bridge; external ears present; limbs well formed, digits exhibit nails; fetus swallows amniotic fluid and produces urine; fetus moves, but too weakly for mother to feel it; liver is prominent and produces bile; palate is fusing; sexes can be distinguished
16	14 cm; 200 g	Eyes face anteriorly, external ears stand out from head, face looks more distinctly human; body larger in proportion to head; skin is bright pink, scalp has hair; joints forming; lips exhibit sucking movements; kidneys well formed; digestive glands forming and **meconium**[25] (fetal feces) accumulating in intestine; heartbeat can be heard with a stethoscope
20	19 cm; 460 g	Body covered with fine hair called **lanugo**[26] and cheeselike sebaceous secretion called **vernix caseosa,**[27] which protects it from amniotic fluid; skin bright pink; brown fat forms and will be used for postpartum heat production; fetus is now bent forward into "fetal position" because of crowding; **quickening** occurs—mother can feel fetal movements
24	23 cm; 820 g	Eyes partially open; skin wrinkled, pink, and translucent; lungs begin producing surfactant; rapid weight gain
28	27 cm; 1,300 g	Eyes fully open; skin wrinkled and red; full head of hair present; eyelashes formed; fetus turns into upside-down **vertex position;** testes begin to descend into scrotum; marginally viable if born at 28 weeks
32	30 cm; 2,100 g	Subcutaneous fat deposition gives fetus a more plump, babyish appearance, with lighter, less wrinkled skin; testes descending; twins usually born at this stage
36	34 cm; 2,900 g	More subcutaneous fat deposited, body plump; lanugo is shed; nails extend to fingertips; limbs flexed; firm hand grip
38	36 cm; 3,400 g	Prominent chest, protruding breasts; testes in inguinal canal or scrotum; fingernails extend beyond fingertips

[25] *mecon* = poppy juice, opium
[26] *lan* = down, wool
[27] *vernix* = varnish + *caseo* = cheese

Clinical Perspectives

Objectives

When you have completed this section, you should be able to:

- discuss the frequency and causes of early spontaneous abortion;
- discuss the frequency of birth defects and major categories of their causes;
- describe some syndromes that result from chromosomal nondisjunction; and
- explain what teratogens are and describe some of their effects.

Expectant parents worry a great deal about the possibilities of miscarriage or birth defects. It is estimated that, indeed, more than half of all pregnancies end in miscarriage, often without the parents realizing that a pregnancy had even begun, and 2% to 3% of infants born in the United States have clinically significant birth defects.

Spontaneous Abortion

Most miscarriages are *early spontaneous abortions,* occurring within 3 weeks of fertilization. Such abortions are easily mistaken for a late and unusually heavy menstrual period. One investigator estimated that 25% to 30% of blastocysts fail to implant; 42% of implanted blastocysts die by the end of the second week; and 16% of those that make it through 2 weeks are seriously abnormal and abort within the next week. Another study found that 61% of early spontaneous abortions were due to chromosomal abnormalities.

Even later in development, spontaneously aborted fetuses show a significantly higher incidence of neural tube defects, cleft lip, and cleft palate than do newborns or induced abortions. Spontaneous abortion may in fact be a natural mechanism for preventing the development of nonviable fetuses or the birth of severely deformed infants.

Birth Defects

A birth defect, or **congenital anomaly,**[28] is the abnormal structure or position of an organ at birth, resulting from a defect in prenatal development. The study of birth defects is called **teratology.**[29] Birth defects are the single most common cause of infant mortality in North America. Not all congenital anomalies are noticeable at birth; some are detected months to years later. Thus, by the age of 2 years, 6% of children are diagnosed with congenital anomalies, and by age 5 the incidence is 8%. The following sections discuss some known causes of congenital anomalies, but in 50% to 60% of cases, the cause is unknown.

[28] *con* = with + *gen* = born
[29] *terato* = monster + *logy* = study of

MUTAGENS AND GENETIC ANOMALIES

Genetic anomalies are the most common known cause of birth defects, accounting for an estimated one-third of all cases and 85% of those with an identifiable cause. One cause of genetic defects is **mutations,** or changes in DNA structure. Among other things, mutations cause achondroplastic dwarfism (see insight 6.3, p. 163), microcephaly (abnormal smallness of the head), stillbirth, and childhood cancer. Mutations can occur through errors in DNA replication during the cell cycle or under the influence of environmental agents called **mutagens,** including some chemicals, viruses, and radiation.

Some of the most common genetic disorders result not from mutagens, however, but from **aneuploidy**[30] (AN-you-ploy-dee), an abnormal number of chromosomes in the zygote. Aneuploidy results from **nondisjunction,** a failure of one of the 23 pairs of chromosomes to separate during meiosis I, so that both members of the pair go to the same daughter cell. For example, suppose nondisjunction resulted in an egg with 24 chromosomes instead of the normal 23. If this egg were fertilized by a normal sperm, the zygote would have 47 chromosomes instead of the usual 46.

Figure 4.17 compares normal disjunction of the X chromosomes with some effects of nondisjunction. In nondisjunction, an egg cell may receive both X chromosomes. If it is fertilized by an X-bearing sperm, the result is an XXX zygote and a suite of defects called the **triplo-X syndrome.** Triplo-X females are sometimes infertile and may have mild intellectual impairments. If an XX egg is fertilized by a Y-bearing sperm, the result is an XXY combination,

[30]*an* = not + *eu* = good, normal + *ploid* = form

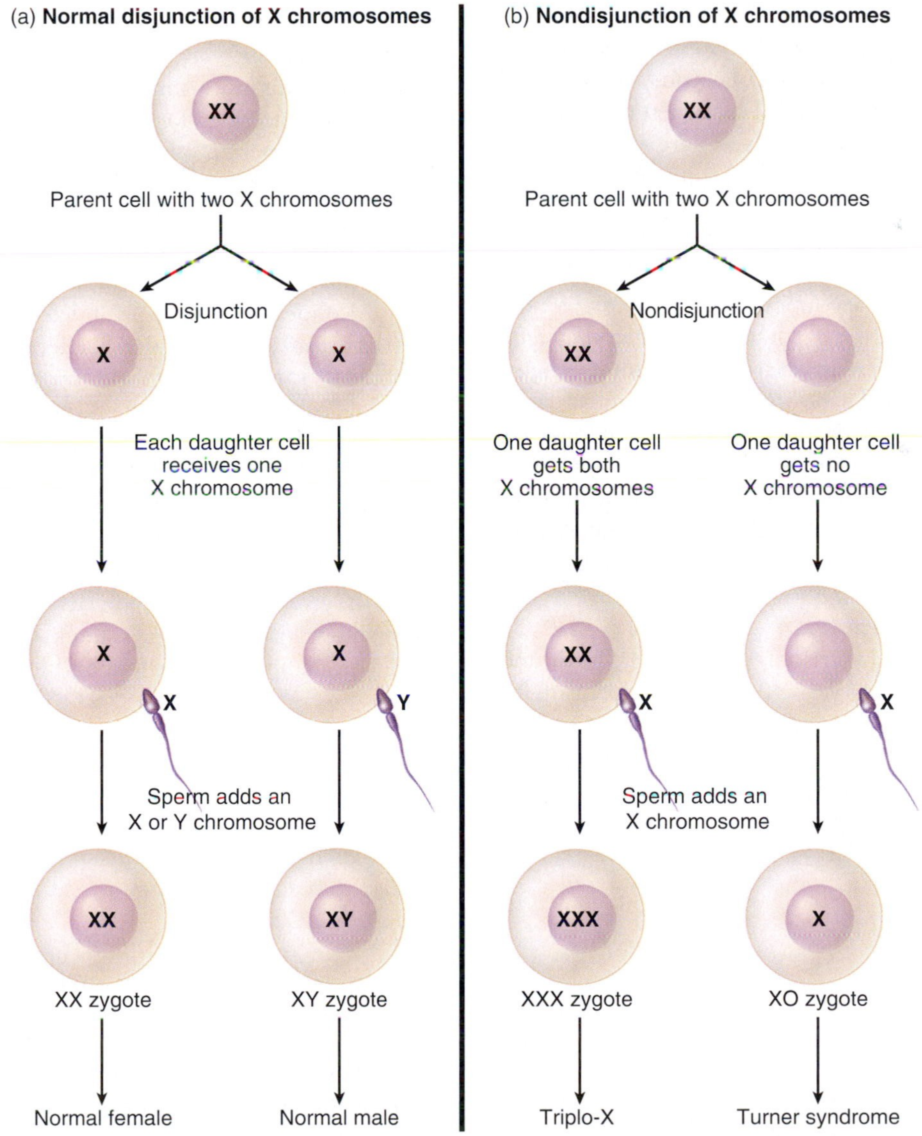
(a) Normal disjunction of X chromosomes
XX
Parent cell with two X chromosomes
Disjunction
X
X
Each daughter cell receives one X chromosome
X
X
X
Y
Sperm adds an X or Y chromosome
XX
XY
XX zygote
XY zygote
Normal female
Normal male
(b) Nondisjunction of X chromosomes
XX
Parent cell with two X chromosomes
Nondisjunction
XX
One daughter cell gets both X chromosomes
One daughter cell gets no X chromosome
XX
X
X
Sperm adds an X chromosome
XXX
X
XXX zygote
XO zygote
Triplo-X
Turner syndrome

FIGURE 4.17

Disjunction and Nondisjunction. (*a*) The outcome of normal disjunction and fertilization by X- or Y-bearing spermatozoa. (*b*) Two of the possible outcomes of nondisjunction followed by fertilization with an X-bearing spermatozoon.

causing **Klinefelter**[31] **syndrome.** People with Klinefelter syndrome are sterile males, usually of average intelligence, but with undeveloped testes, sparse body hair, unusually long arms and legs, and enlarged breasts (*gynecomastia*[32]). This syndrome often goes undetected until puberty, when failure to develop the secondary sex characteristics may prompt genetic testing.

The other possible outcome of X chromosome nondisjunction is that an egg cell may receive no X chromosome (both X chromosomes are discarded in the first polar body). If fertilized by a Y-bearing sperm, such an egg dies for lack of the indispensable genes on the X chromosome. If it is fertilized by an X-bearing sperm, however, the result is a female with **Turner**[33] **syndrome,** with an XO combination (*O* represents the absence of one sex chromosome). Only 3% of fetuses with Turner syndrome survive to birth. Girls who survive show no serious impairments as children, but tend to have a webbed neck and widely spaced nipples. At puberty, the secondary sex characteristics fail to develop. The ovaries are nearly absent, the girl remains sterile, and she usually has a short stature.

The other 22 pairs of chromosomes (the *autosomes*) are also subject to nondisjunction. Nondisjunction of chromosomes 13 and 18 results in **Edward syndrome (trisomy-13)** and **Patau syndrome (trisomy-18),** respectively. Affected individuals have three copies of the respective chromosome. Nearly all fetuses with these trisomies die before birth. Live-born infants with these syndromes are severely deformed, and fewer than 5% survive for one year. The most common autosomal anomaly is **Down**[34] **syndrome (trisomy-21).** The signs include retarded physical development; short stature; a relatively flat face with a flat nasal bridge; low-set ears; *epicanthal folds* at the medial corners of the eyes; an enlarged, protruding tongue; stubby fingers; and a short broad hand with only one palmar crease (fig. 4.18). People with Down syndrome tend to have outgoing, affectionate personalities. Mental retardation is common and sometimes severe, but is not inevitable. Down syndrome occurs in about 1 out of 700 to 800 live births in the United States and increases in proportion to the age of the mother. The chance of having a child with Down syndrome is about 1 in 3,000 for a woman under 30, 1 in 365 by age 35, and 1 in 9 by age 48.

About 75% of the victims of trisomy-21 die before birth, and about 20% die before the age of 10 years. Typical causes of death include immune deficiency and abnormalities of the heart or kidneys. For those who survive beyond 10 years, modern medical care has extended life expectancy to about 60. After the age of 40, however, many of these people develop early-onset Alzheimer disease, linked to a gene on chromosome 21.

[31]Harry F. Klinefelter, Jr. (1912–), American physician
[32]*gyneco* = female + *mast* = breast + *ia* = condition
[33]Henry H. Turner (1892–1970), American endocrinologist
[34]John Langdon H. Down (1828–96), British physician

TERATOGENS

Teratogens[35] are agents that cause anatomical deformities in the fetus; they include viruses, drugs, other chemicals, infectious diseases, and radiation such as X rays. The effect of a teratogen depends on the genetic susceptibility of the embryo, the dosage of the teratogen, and the time of exposure. Teratogen exposure during the first two weeks usually does not cause birth defects, but may cause spontaneous abortion. Teratogens can exert destructive effects at any stage of development, but the period of greatest vulnerability is weeks 3 through 8. Different organs have different critical periods. For example, limb abnormalities are most likely to result from teratogen exposure at 24 to 36 days, and brain abnormalities from exposure at 3 to 16 weeks.

Perhaps the most notorious teratogenic drug is thalidomide, a sedative first marketed in 1957. Thalidomide was taken by women in early pregnancy, often before they knew they were pregnant; it caused over 5,000 babies to be born with unformed arms or legs (fig. 4.19) and often with defects of the ears, heart, and intestines. It was taken off the market in 1961, but has recently been reintroduced for more limited purposes. Many teratogens produce less obvious effects, including physical or mental retardation, hyperirritability, inattention, strokes, seizures, respiratory arrest, crib death, and cancer. A general lesson to be learned from the thalidomide tragedy and other cases is that pregnant women should avoid all sedatives, barbiturates, and opiates. Even the acne medicine Acutane has caused severe birth defects.

Alcohol causes more birth defects than any other teratogen. Even one drink a day has adverse effects on fetal and childhood development, some of which are not noticed until a child begins school. Alcohol abuse during pregnancy can cause **fetal alcohol syndrome (FAS),** characterized by a small head, malformed facial features, cardiac and central nervous system defects, stunted growth, and behavioral symptoms such as hyperactivity, nervousness, and a poor attention span. Cigarette smoking also contributes to fetal and infant mortality, ectopic pregnancy, anencephaly (failure of the cerebrum to develop), cleft lip and palate, and cardiac abnormalities. Diagnostic X rays should be avoided during pregnancy because radiation can have teratogenic effects.

THINK ABOUT IT!

Martha is showing a sonogram of her unborn baby to her coworkers. Her friend Betty tells her she shouldn't have sonograms made because X rays can cause birth defects. Is Betty's concern well founded? Explain.

Infectious diseases are largely beyond the scope of this book, but it must be noted at least briefly that several microorganisms can cross the placenta and cause serious congenital anomalies, stillbirth, or neonatal death. Common viral infections of the fetus and newborn include herpes simplex, rubella, cytomegalovirus, and human immunodeficiency virus (HIV). Congenital bacterial infections include gonorrhea and syphilis.

[35]*terato* = monster + *gen* = producing

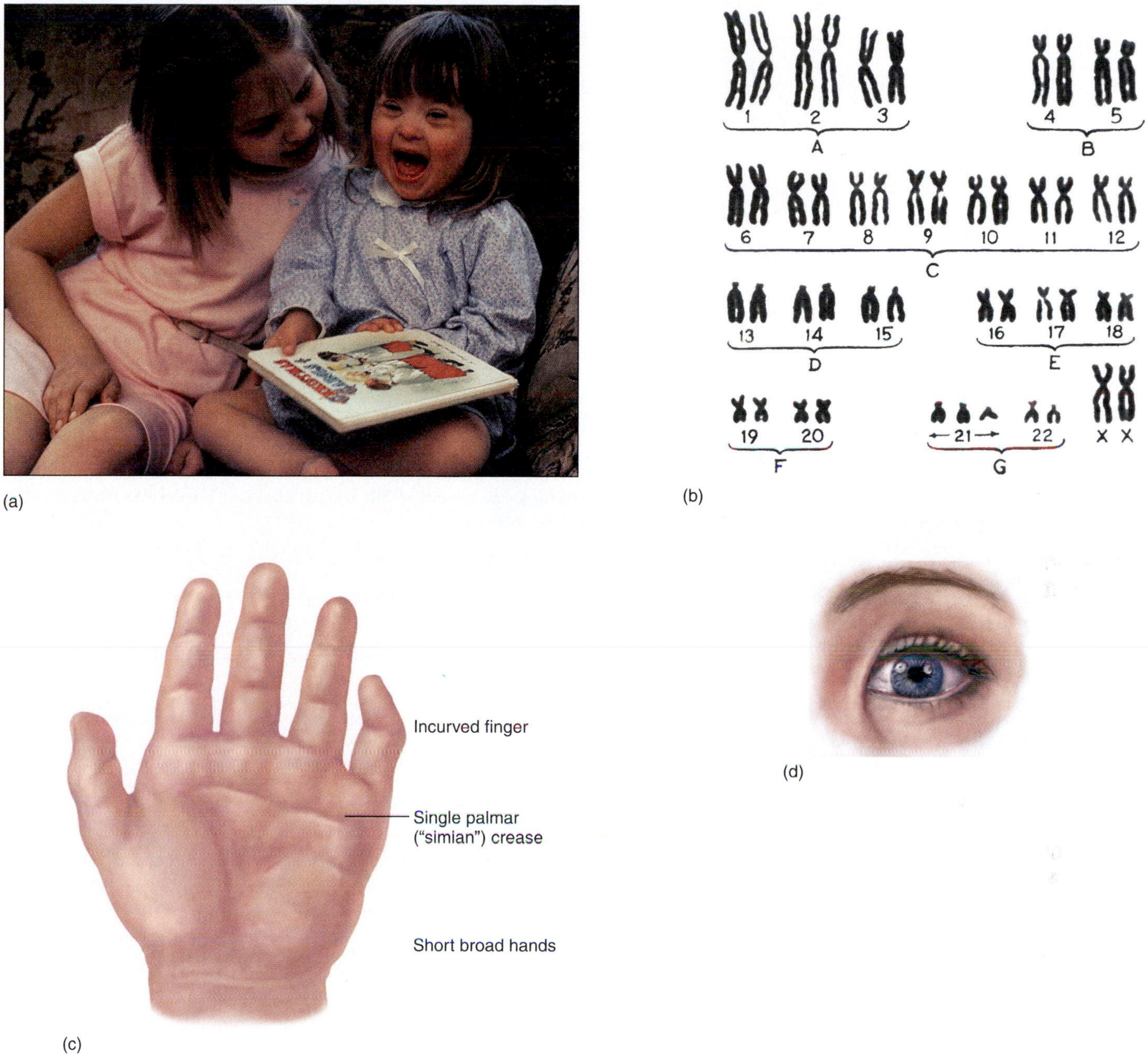
(a)
1 2 3
A
4 5
B
6 7 8 9 10 11 12
C
13 14 15
D
16 17 18
E
19 20
F
←21→ 22
G
X X
(b)
Incurved finger
Single palmar ("simian") crease
Short broad hands
(c)
(d)

FIGURE 4.18

Down syndrome. (*a*) A child with Down syndrome *(right)* plays with her big sister. (*b*) The karyotype (chart of the chromosomes) in Down syndrome, showing the trisomy of chromosome 21. (*c*) Characteristics of the hand seen in Down syndrome. (*d*) The epicanthal fold over the medial commissure (canthus) of the left eye.

Toxoplasma, a protozoan contracted from meat, unpasteurized milk, and housecats, is another common cause of fetal deformity. Some of these pathogens have relatively mild effects on adults, but because of its immature immune system, the fetus is vulnerable to devastating effects such as blindness, hydrocephalus, cerebral palsy, seizures, and profound physical and mental retardation. Infections of the fetus and newborn are treated in greater detail in microbiology textbooks.

Some terata and other developmental disorders are described in table 4.5.

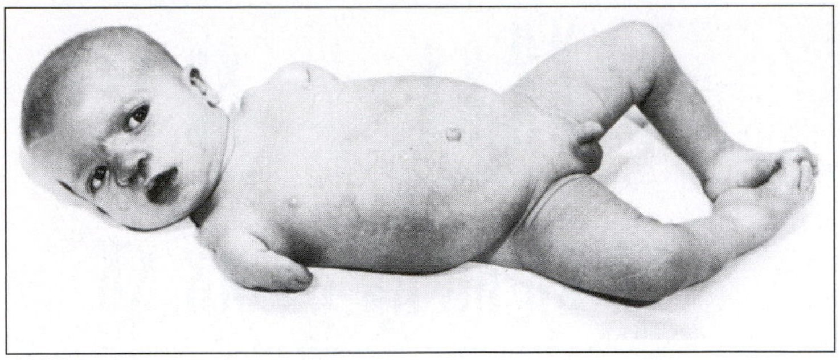

Figure 4.19

Effects of Thalidomide. Taken as a sedative in early pregnancy, thalidomide proved to be a teratogen with severe effects on embryonic limb development.

Before You Go On

Answer the following questions to test your understanding of the preceding section:

12. In what sense can spontaneous abortion be considered a protective mechanism?
13. What is the difference between mutation and nondisjunction?
14. Name and describe two birth defects resulting from nondisjunction of autosomes and two from nondisjunction of sex chromosomes.
15. Name three distinctly different classes of teratogens and give one example from each class.

Table 4.5

Some Disorders of Human Development

Anencephaly	Lack of a forebrain due to failure of the cranial roof to form, leaving the forebrain exposed. The exposed tissue dies, and the fetus is born (or stillborn) with only a brainstem. Live-born anencephalic infants are very short-lived.
Cleft Lip and Palate	Failure of the right and left sides of the lip or palate to fuse medially, resulting in abnormal facial appearance, defective speech, and difficulty suckling.
Clubfoot (talipes)	Deformity of the foot involving an ankle bone, the talus. The sole of the foot is commonly turned medially, and as a child grows, he or she may walk on the ankles rather than on the soles.
Cri du Chat[36]	A congenital anomaly due to deletion of a portion of chromosome 5. Infants with cri du chat have microcephaly, congenital heart disease, profound mental retardation, and a weak catlike cry.
Hydrocephalus	Abnormal accumulation of cerebrospinal fluid in the brain. When it occurs in the fetus, the cranial bones separate, the head becomes abnormally large, and the face looks disproportionately small. May reduce the cerebrum to a thin shell of nervous tissue. Fatal for about half of patients but can be treated by inserting a shunt that drains fluid from the brain to a vein in the neck.
Meromelia	Partial absence of limbs, such as the lack of some digits, a hand, or a forearm. Complete absence of a limb is *amelia.*

Disorders Described Elsewhere

Abruptio placentae 118
Achondroplastic dwarfism 163
Birthmarks 137
Congenital defects of the kidney 721
Cryptorchidism 739, 756
Dextrocardia 6
Down syndrome 122
Ectopic pregnancy 111
Fetal alcohol syndrome 122
Hypospadias 739, 756
Klinefelter syndrome 122
Osteogenesis imperfecta 93
Patent ductus arteriosus 579
Placenta previa 118
Premature birth 669
Respiratory distress syndrome 669
Situs inversus 6
Situs perversus 6
Spina bifida 380
Spontaneous abortion 120
Triplo-X syndrome 121
Turner syndrome 122

[36] *cri du chat* = cry of the cat (French)

CHAPTER REVIEW

Review of Key Concepts

Gametogenesis and Fertilization (p. 106)

1. *Gametogenesis* is the production of sperm and eggs, which are the *gametes* (sex cells).
2. Gametes are *haploid,* with 23 chromosomes—half the number found in most cells of the body. When two gametes unite, they form a *zygote* with 46 chromosomes, the normal human *diploid* number. The haploid number is produced by a form of cell division called *meiosis.*
3. In *spermatogenesis* (sperm production), meiosis produces four equal-sized sperm. In egg production, it produces three *polar bodies* and only one egg.
4. The ovaries usually release one egg per month. The egg travels down the uterine tube to the uterus, a trip that requires 3 days, although the egg lives only 1 day if it is not fertilized. Fertilization therefore must occur long before the egg reaches the uterus, and requires that the sperm migrate up the female reproductive tract to meet the egg in the uterine tube.
5. Freshly ejaculated sperm cannot fertilize an egg. During migration, sperm undergo *capacitation,* acquiring the ability to fertilize the egg.
6. Upon contact, a sperm releases enzymes from its *acrosome* that digest a path through the barriers around the egg or into the egg itself. The sperm head and midpiece enter the egg cytoplasm. The egg deploys a *fast block* and a *slow block* to prevent *polyspermy,* or fertilization by more than one sperm.
7. Upon fertilization, the egg and sperm nuclei swell to become *pronuclei;* the egg forms a mitotic spindle; and the pronuclei release their chromosomes, which mingle in *amphimixis* into a single diploid set. The cell is now a zygote.

Stages of Prenatal Development (p. 108)

1. Gestation, or pregnancy, lasts an average of 266 days (38 weeks) from *conception* to *parturition.* Birth is predicted to occur about 280 days after the start of the last menstrual period.
2. All products of conception—the embryo or fetus, and the placenta, amnion, and other accessory organs—are called the *conceptus.*
3. The period of gestation can be divided from a clinical perspective into three *trimesters* (3 months each), or from a biological perspective into the *preembryonic, embryonic,* and *fetal stages.* These stages are not equivalent to trimesters; the preembryonic and embryonic stage and the first month of the fetal stage all occur in the first trimester.
4. The first 16 days of development, called the *preembryonic stage,* consists of three major events—cleavage, implantation, and embryogenesis—resulting in an embryo.
5. *Cleavage* is the mitotic division of the zygote into cells called *blastomeres.* The stage that arrives at the uterus is a *morula* of about 16 blastomeres. It develops into a hollow ball called the *blastocyst,* with an outer cell mass called the *trophoblast* and inner cell mass called the *embryoblast.*
6. *Implantation* is the attachment of the blastocyst to the uterine wall. The trophoblast differentiates into a cellular mass called the *cytotrophoblast* next to the embryo, and a multinucleate mass called the *syncytiotrophoblast,* which grows rootlets into the endometrium. The endometrium grows over the blastocyst and soon completely covers it.
7. *Embryogenesis* occurs during implantation, and consists of the arrangement of the blastomeres into three *primary germ layers*—the *ectoderm, mesoderm,* and *endoderm.*
8. An *amniotic cavity* forms between the embryoblast and trophoblast, and the embryoblast flattens into an *embryonic disc,* with two cell layers called *epiblast* and *hypoblast.* The embryonic disc soon elongates and forms a median *primitive streak* in the epiblast.
9. In *gastrulation,* epiblast cells migrate into the primitive streak and replace the hypoblast, then form a middle layer of cells called mesoderm. The three cell layers are now called the ectoderm, mesoderm, and endoderm, and the individual is considered to be an embryo.
10. The next 6 weeks of development are the *embryonic stage,* marked by formation of the embryonic membranes, placental nutrition, and appearance of the organ systems.
11. The embryonic disc folds at the cephalic and caudal ends and along both lateral margins, acquiring a body that is C-shaped longitudinally and rounded in cross section. This embryonic folding encloses a ventral passage, the *primitive gut.*
12. A coelom, or body cavity, appears within the mesoderm and then becomes partitioned into thoracic and peritoneal cavities; the thoracic cavity subdivides into pleural and pericardial cavities.
13. Development of the organs from the primary germ layers is called *organogenesis.* Three major events in this stage are *neurulation,* the formation of a neural tube in the ectoderm; the appearance of *pharyngeal pouches;* and segmentation of the body into a series of *somites.*
14. Each somite divides into three cell masses—the *sclerotome, myotome,* and *dermatome*—which are the forerunners, respectively, of bone, muscle, and the dermis of the skin.
15. At the end of 8 weeks, all organ systems are present in rudimentary form, and the individual is considered a fetus.
16. Four membranes are associated with the embryo and fetus: the amnion, yolk sac, allantois, and chorion.
17. The *amnion* is a transparent sac that encloses the embryo in a pool of *amniotic fluid.* This fluid protects the embryo from trauma and temperature fluctuations and allows freedom of movement and symmetric development.
18. The *yolk sac* contributes to development of the digestive tract and produces the first blood and germ stem cells of the embryo.
19. The *allantois* is an outgrowth of the yolk sac that forms a structural foundation for umbilical cord development and becomes part of the urinary bladder.
20. The *chorion* encloses all of the other membranes and forms the fetal part of the placenta.
21. After implantation, the conceptus is fed by *trophoblastic nutrition,* in which the trophoblast digests *decidual cells* of the endometrium. This is the dominant mode of nutrition for 8 weeks.
22. The *placenta* begins to form 11 days after conception as chorionic villi of the trophoblast invade uterine blood vessels, eventually creating a blood-filled cavity called the *placental sinus.* The chorionic villi grow into branched treelike structures surrounded by the maternal blood in the sinus. Nutrients diffuse from the maternal blood into embryonic blood vessels in the villi, and embryonic wastes diffuse the other way to be disposed of by the mother. *Placental nutrition* becomes dominant at the start of week 9 and continues until birth.

23. The placenta communicates with the embryo and fetus by way of two arteries and a vein contained in the umbilical cord.
24. The fetal stage is marked especially by rapid weight gain and by the organs attaining functionality.
25. The individual weighs about 1 g at the start of the fetal stage and 3,400 g at birth. About 50% of this weight gain occurs in the last 10 weeks.
26. The trunk grows faster than the head, and the limbs grow faster than the trunk during the fetal stage, so the body acquires normal proportions. Other major developments in the fetal stage are summarized in table 4.4.

Clinical Perspectives (p. 120)

1. More than half of all conceptions end in miscarriage and 2% to 3% of live births show significant birth defects. By age 5, 8% of children exhibit defects traced to abnormal intrauterine development.
2. Most miscarriages are *early spontaneous abortions* (within 3 weeks of fertilization), easily confused with a late and heavy menstrual period. Most early spontaneous abortions are due to chromosomal abnormalities.
3. A *birth defect,* or *congenital anomaly,* is the presence of one or more organs that are abnormal in structure or position at birth. The study of birth defects is *teratology.*
4. The most common cause of birth defects is genetic anomalies. These can be mutations (changes in DNA or chromosome structure) or *aneuploidies* (abnormalities of chromosome count).
5. Aneuploidy results from *nondisjunction* of a chromosome pair at meiosis I. Nondisjunction of sex chromosomes causes *triplo-X, Klinefelter,* and *Turner syndromes.* Nondisjunction of autosomes causes *Edward syndrome (trisomy-13), Patau syndrome (trisomy-18),* and *Down syndrome (trisomy-21).*
6. *Teratogens* are agents that cause anatomical deformities in the fetus. The greatest sensitivity to teratogens is from weeks 3 through 8. Examples of teratogens include thalidomide, alcohol, cigarette smoke, X rays, and several viruses, bacteria, and protozoans.

Testing Your Recall

1. When a conceptus arrives in the uterus, it is at what stage of development?
   a. zygote
   b. morula
   c. blastomere
   d. blastocyst
   e. embryo
2. The entry of a sperm nucleus into an egg must be preceded by
   a. the cortical reaction.
   b. the acrosomal reaction.
   c. the fast block.
   d. implantation.
   e. cleavage.
3. The primitive gut develops as a result of
   a. gastrulation.
   b. cleavage.
   c. embryogenesis.
   d. embryonic folding.
   e. aneuploidy.
4. Chorionic villi develop from
   a. the zona pellucida.
   b. the endometrium.
   c. the syncytiotrophoblast.
   d. the embryoblast.
   e. the epiblast.
5. Which of these results from aneuploidy?
   a. Turner syndrome
   b. fetal alcohol syndrome
   c. nondisjunction
   d. mutation
   e. rubella
6. Fetal urine accumulates in the _____ and contributes to the fluid there.
   a. placental sinus
   b. yolk sac
   c. allantois
   d. chorion
   e. amnion
7. A preembryo has
   a. ectoderm.
   b. a heart bulge.
   c. a cytotrophoblast.
   d. a coelom.
   e. decidual cells.
8. Lanugo is
   a. fetal hair.
   b. fetal feces.
   c. a teratogen.
   d. an embryonic membrane.
   e. the head-down position at full term.
9. The first blood and germ stem cells come from
   a. the mesoderm.
   b. the hypoblast.
   c. the syncytiotrophoblast.
   d. the placenta.
   e. the yolk sac.
10. For the first 8 weeks of gestation, a conceptus is nourished mainly by
    a. the placenta.
    b. amniotic fluid.
    c. colostrum.
    d. decidual cells.
    e. yolk cytoplasm.
11. Viruses and chemicals that cause congenital anatomical deformities are called _____.
12. Aneuploidy is caused by _____, the failure of a pair of chromosomes to separate in meiosis.
13. The brain and spinal cord develop from a longitudinal ectodermal channel called the _____.
14. Attachment of the conceptus to the uterine wall is called _____.
15. Fetal blood flows through growths called _____, which project into the placental sinus.
16. The enzymes with which a sperm penetrates an egg are contained in an organelle called the _____.
17. Fertilization occurs in a part of the female reproductive tract called the _____.
18. Bone, muscle, and dermis arise from segments of mesoderm called _____.
19. The egg cell has fast and slow blocks to _____, or fertilization by more than one sperm.
20. A developing individual is first classified as a/an _____ when the three primary germ layers have formed.

Answers in the Appendix

True or False

Determine which five of the following statements are false, and briefly explain why.

1. Freshly ejaculated sperm are more capable of fertilizing an egg than are sperm several hours old.
2. Fertilization normally occurs in the uterus.
3. An egg is usually fertilized by the first sperm that contacts it.
4. By the time a conceptus reaches the uterus, it has already undergone several cell divisions and consists of 16 cells or more.
5. The individual is first considered a fetus when all of the organ systems are present.
6. The placenta becomes increasingly permeable as it develops.
7. During cleavage, the preembryo acquires a greater number of cells but does not increase in size.
8. In oogenesis, a germ cell divides into four equal-sized egg cells.
9. The stage of the conceptus that implants in the uterine wall is the blastocyst.
10. The energy for sperm motility comes from its acrosome.

Answers in the Appendix

Testing Your Comprehension

1. Only one sperm is needed to fertilize an egg, yet a man who ejaculates fewer than 10 million sperm is usually infertile. Explain this apparent contradiction. Supposing 10 million sperm were ejaculated, predict how many would come within close range of the egg. How likely is it that any one of these sperm would fertilize it?
2. What is the difference between embryology and teratology?
3. At what point in the timeline of table 4.4 do you think thalidomide exerts its teratogenic effect?
4. A teratologist is studying the cytology of a fetus that aborted spontaneously at 12 weeks. She concludes that the fetus was triploid. What do you think this term means? How many chromosomes do you think she found in each of its cells? To produce this state, what normal process of human development apparently failed?
5. A young woman finds out she is about 4 weeks pregnant. She tells her doctor that she drank heavily at a party three weeks earlier, and she is worried about the possible effects of this on her baby. If you were the doctor, would you tell her that there is serious cause for concern, or not? Why?

Answers at the Online Learning Center

www.mhhe.com/saladinha1

Visit the Online Learning Center for practice tests, answer keys, and other learning aids for this chapter. Enhance your understanding of human anatomy with our interactive art labeling exercises, supplemental photo atlases, web links, puzzles, flashcards, and much more.

5

CHAPTER FIVE

The Integumentary System

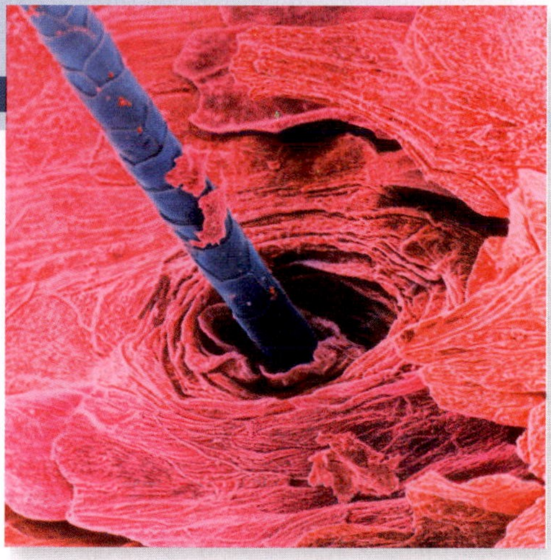

A human hair emerging from its follicle

CHAPTER OUTLINE

The Skin and Subcutaneous Tissue 130
- Functions of the Skin 130
- The Epidermis 130
- The Dermis 133
- The Hypodermis 135
- Skin Color 135
- Skin Markings 137

Hair and Nails 137
- Hair 137
- Nails 140

Cutaneous Glands 140
- Sweat Glands 140
- Sebaceous Glands 142
- Ceruminous Glands 142
- Mammary Glands 142

Developmental and Clinical Perspectives 143
- Prenatal Development of the Integumentary System 143
- The Aging Integumentary System 145
- Skin Disorders 145

Chapter Review 148

INSIGHTS

5.1 Clinical Application: Tension Lines and Surgery 134
5.2 Clinical Application: Birthmarks 137
5.3 Clinical Application: Extra Nipples 142
5.4 Clinical Application: Skin Grafts and Artificial Skin 147

BRUSHING UP

To understand this chapter, you may find it helpful to review the following concepts:

- Cancer (p. 71)
- Keratinized Stratified Squamous Epithelium (p. 83)
- Areolar and Dense Irregular Connective Tissues (p. 87–88)
- Merocrine, Holocrine, and Apocrine Gland Types (p. 97).

PART TWO

The skin is also known as the **integument,**[1] while the **integumentary system** consists of the skin and its accessory organs—the hair, nails, and cutaneous glands. We pay more attention to this organ system than to any other. It is, after all, the most visible one, and its appearance strongly affects our social interactions. Few people venture out of the house without first looking in a mirror to see if their skin and hair are presentable. Social considerations aside, the integumentary system is important to one's self-image, and a positive self-image is important to the attitudes that promote overall health. Care of the integumentary system is thus a particularly important part of the total plan of patient care.

The scientific study and medical treatment of the integumentary system is called **dermatology.**[2] Inspection of the skin, hair, and nails is a significant part of a physical examination. The integumentary system provides clues not only to its own health, but also to deeper disorders such as liver cancer, anemia, and heart failure. The skin also is the most vulnerable of our organs, exposed to radiation, trauma, infection, and injurious chemicals. Consequently, it needs and receives more medical attention than any other organ system.

The Skin and Subcutaneous Tissue

Objectives

When you have completed this section, you should be able to

- list the functions of the skin and relate them to its structure;
- describe the histological structure of the epidermis, dermis, and subcutaneous tissue;
- describe the normal and pathological colors that the skin can have and explain their causes; and
- describe the common markings of the skin.

The **skin** is the body's largest organ. In adults, it covers an area of 1.5 to 2.0 m^2 and accounts for about 15% of the body weight. The skin consists of two layers: a stratified squamous epithelium called the *epidermis* and a deeper connective tissue layer called the *dermis* (fig. 5.1). Below the dermis is another connective tissue layer, the *hypodermis,* which is not part of the skin but is customarily studied in conjunction with it.

Most of the skin is 1 to 2 mm thick, but it ranges from less than 0.5 mm on the eyelids to 6 mm between the shoulder blades. The difference is due mainly to variation in the thickness of the dermis, although skin is classified as thick or thin based on the relative thickness of the epidermis alone. **Thick skin** covers the palms, soles, and corresponding surfaces of the fingers and toes. It has an epidermis that is 400 to 600 μm thick, due to a very thick surface layer of dead cells called the *stratum corneum* (fig. 5.2). Thick skin has sweat glands but no hair follicles or sebaceous (oil) glands. The rest of the body is covered with **thin skin,** which has an epidermis 75 to 150 μm thick with a thin stratum corneum (see fig. 5.4*a*). It possesses hair follicles, sebaceous glands, and sweat glands.

[1]*integument* = covering
[2]*dermat* = skin + *logy* = study of

Functions of the Skin

The skin is much more than a container for the body. It has a variety of important functions that go well beyond appearance, as we shall see here.

1. **Resistance to trauma and infection.** The skin bears the brunt of most physical injuries to the body, but it resists and recovers from trauma better than other organs do. The epidermal cells are packed with the tough protein **keratin** and linked by strong desmosomes that give this epithelium its durability. Few infectious organisms can penetrate the intact skin. Bacteria and fungi colonize the skin surface, but their numbers are kept in check by the relative dryness and slight acidity (pH 4–6) of the surface. This protective acidic film is called the *acid mantle.*
2. **Water retention.** The skin is important as a barrier to water. It prevents the body from absorbing excess water when you are swimming or bathing, but even more importantly, it prevents the body from losing excess water.
3. **Vitamin D synthesis.** The skin carries out the first step in the synthesis of vitamin D, which is needed for bone development and maintenance. The liver and kidneys complete the process.
4. **Sensation.** The skin is our most extensive sense organ. It is equipped with a variety of nerve endings that react to heat, cold, touch, texture, pressure, vibration, and tissue injury (see chapter 16). These sensory receptors are especially abundant on the face, palms, fingers, soles, nipples, and genitals. There are relatively few on the back and in skin overlying joints such as the knees and elbows.
5. **Thermoregulation.** In response to chilling, the skin helps to retain heat. The dermis has nerve endings called **thermoreceptors** that transmit signals to the brain, and the brain sends signals back to the dermal blood vessels. **Vasoconstriction,** or narrowing of these blood vessels, reduces the flow of blood close to the skin surface and thus reduces heat loss. When one is overheated, **vasodilation** or widening of the dermal blood vessels increases cutaneous blood flow and increases heat loss. If this is not enough to restore normal temperature, the brain also triggers sweating.
6. **Nonverbal communication.** The skin is an important means of communication. Humans, like other primates, have much more expressive faces than most mammals. Complex skeletal muscles insert on dermal collagen fibers and pull on the skin to create subtle and varied facial expressions (fig. 5.3).

The Epidermis

The **epidermis**[3] is a keratinized stratified squamous epithelium, as discussed in chapter 3. That is, its surface consists of dead cells packed with keratin. Like other epithelia, the epidermis lacks

[3]*epi* = above, upon + *derm* = skin

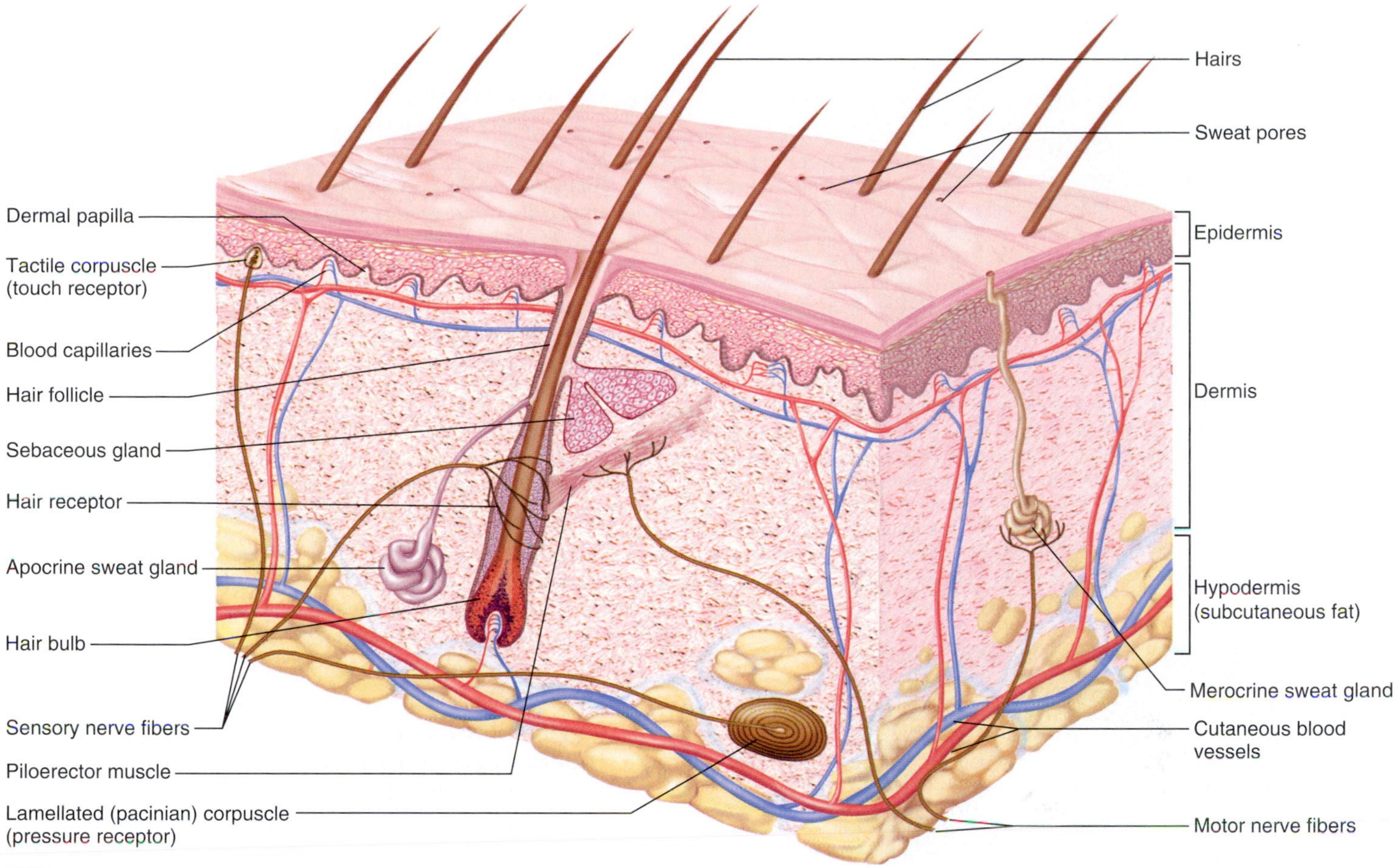

Hairs
Sweat pores
Epidermis
Dermis
Hypodermis
(subcutaneous fat)
Merocrine sweat gland
Cutaneous blood
vessels
Motor nerve fibers
Dermal papilla
Tactile corpuscle
(touch receptor)
Blood capillaries
Hair follicle
Sebaceous gland
Hair receptor
Apocrine sweat gland
Hair bulb
Sensory nerve fibers
Piloerector muscle
Lamellated (pacinian) corpuscle
(pressure receptor)

FIGURE 5.1

Structure of the Skin and Subcutaneous Tissue.

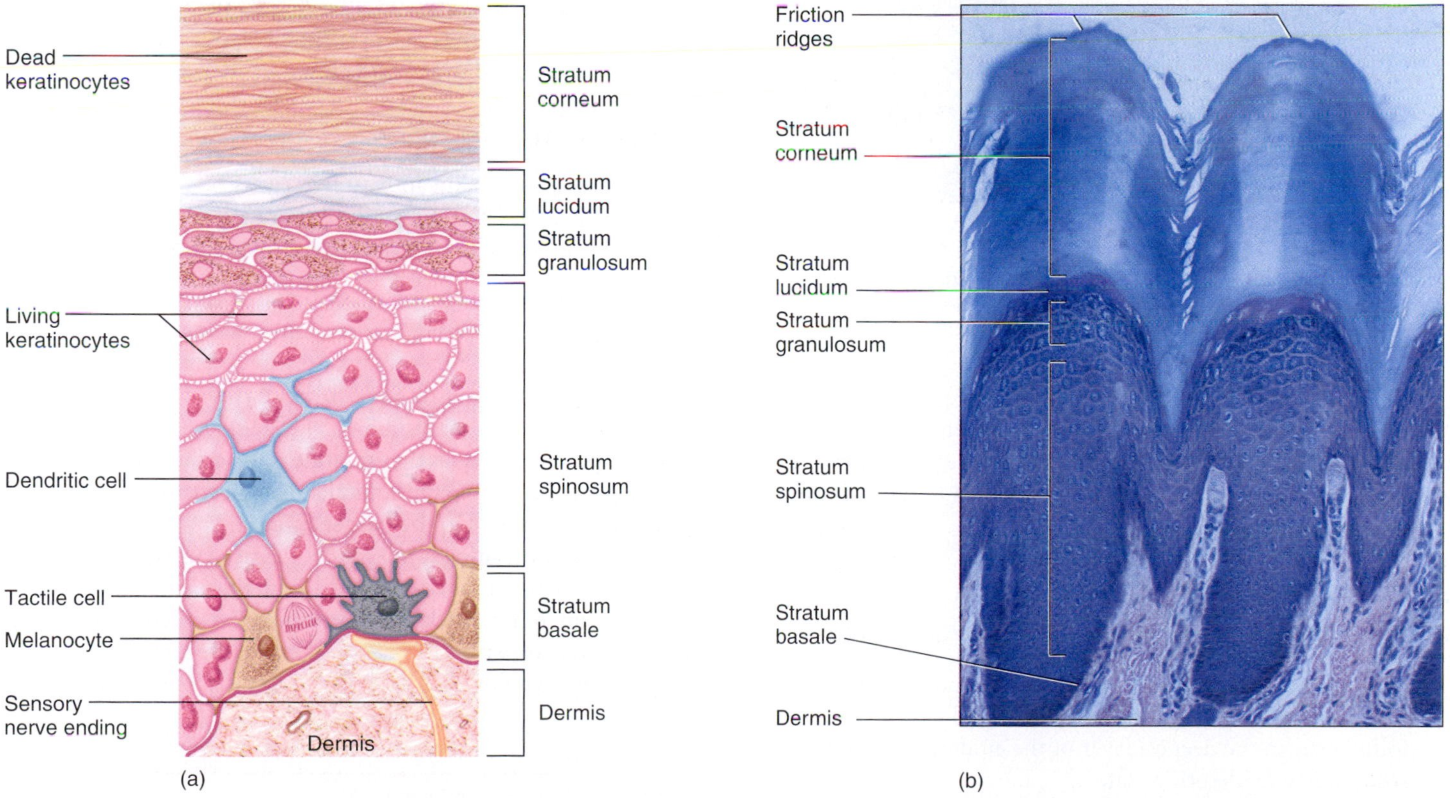

Dead
keratinocytes
Living
keratinocytes
Dendritic cell
Tactile cell
Melanocyte
Sensory
nerve ending
Dermis
Stratum
corneum
Stratum
lucidum
Stratum
granulosum
Stratum
spinosum
Stratum
basale
Dermis
(a)
Friction
ridges
Stratum
corneum
Stratum
lucidum
Stratum
granulosum
Stratum
spinosum
Stratum
basale
Dermis
(b)

FIGURE 5.2

The Epidermis. (*a*) Drawing of the epidermal layers and cell types. (*b*) Photograph of thick skin from the fingertip.

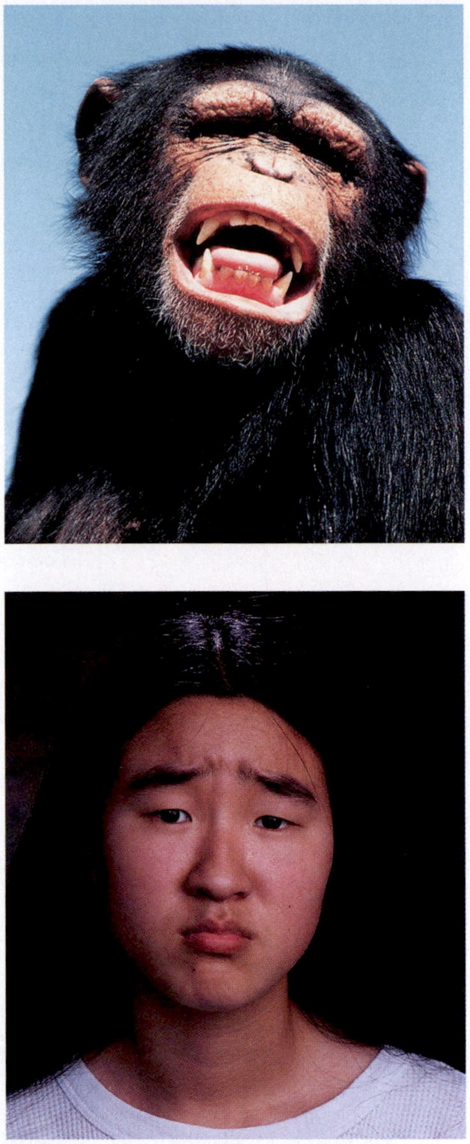

FIGURE 5.3
Importance of the Skin in Nonverbal Expression. Primates differ from other mammals in having very expressive faces due to facial muscles that insert on collagen fibers of the dermis and move the skin.

blood vessels and depends on the diffusion of nutrients from the underlying connective tissue. It has sparse nerve endings for touch and pain, but most sensations of the skin are due to nerve endings in the dermis.

CELLS OF THE EPIDERMIS

The epidermis is composed of five types of cells (see fig. 5.2):

1. **Stem cells** are undifferentiated cells that undergo mitosis and give rise to the keratinocytes described next. They are found only in the deepest layer of the epidermis, called the *stratum basale* (described later).
2. **Keratinocytes** (keh-RAT-ih-no-sites) are the great majority of epidermal cells. They are named for their role in synthesizing keratin. In ordinary histological specimens, nearly all of the epidermal cells you can see are keratinocytes.
3. **Melanocytes** also occur only in the stratum basale, amid the stem cells and deepest keratinocytes. They synthesize the brown to black pigment *melanin.* They have long branching processes that spread among the keratinocytes and continually shed melanin-containing fragments from their tips. The keratinocytes phagocytize these fragments and accumulate melanin granules on the "sunny side" of the nucleus. Like a parasol, the pigment shields the DNA from ultraviolet radiation. People of all races have about equal numbers of melanocytes. Differences in skin color result from differences in the rate of melanin synthesis and how clumped or spread-out the melanin is. In light skin, the melanin is less abundant and is relatively clumped near the keratinocyte nucleus, imparting less color to the cells.
4. **Tactile (Merkel[4]) cells,** relatively few in number, are receptors for the sense of touch. They, too, are found in the basal layer of the epidermis and are associated with an underlying dermal nerve fiber. The tactile cell and its nerve fiber are collectively called a *tactile (Merkel) disc.*
5. **Dendritic[5] (Langerhans[6]) cells** are found in two layers of the epidermis called the *stratum spinosum* and *stratum granulosum* (described in the next section). They are macrophages that originate in the bone marrow but migrate to the epidermis and epithelia of the oral cavity, esophagus, and vagina. The epidermis has as many as 800 dendritic cells per square millimeter. They "stand guard" against toxins, microbes, and other pathogens that penetrate into the skin. When they detect such invaders, they alert the immune system so the body can defend itself.

LAYERS OF THE EPIDERMIS

The epidermis consists of four to five layers of cells (five in thick skin) (see fig. 5.2). This description progresses from deep to superficial, and from the youngest to the oldest keratinocytes.

1. The **stratum basale** (bah-SAY-lee) consists mainly of a single layer of cuboidal to low columnar stem cells and keratinocytes resting on the basement membrane. Scattered among these are also found melanocytes and tactile cells. As stem cells of the stratum basale undergo mitosis, they give rise to keratinocytes that migrate toward the skin surface and replace lost epidermal cells. The life history of these cells is described in the next section.
2. The **stratum spinosum** (spy-NO-sum) consists of several layers of keratinocytes; in most skin, this is the thickest stratum, but in thick skin it is usually exceeded by the stratum corneum. The deepest cells of the stratum spinosum retain the capability of mitosis, but as they are

[4] F. S. Merkel (1845–1919), German anatomist
[5] *dendr* = tree, branch
[6] Paul Langerhans (1847–88), German anatomist

pushed farther upward, they cease dividing. Instead, they produce more and more keratin filaments, which cause the cells to flatten. Therefore, the higher up you look in the stratum spinosum, the flatter the cells appear. Dendritic cells are also found throughout the stratum spinosum, but are not usually visible in tissue sections.

The stratum spinosum is named for an artificial appearance *(artifact)* created by the histological fixation of tissue specimens. Keratinocytes are firmly attached to each other by numerous desmosomes, which partly account for the toughness of the epidermis. Histological fixatives shrink the keratinocytes and cause them to pull away from each other, but they remain attached to each other by the desmosomes—like two people holding hands while they step farther away from each other. The desmosomes thus create bridges from cell to cell, giving each cell a spiny appearance from which we derive the word *spinosum.*

3. The **stratum granulosum** consists of three to five layers of flat keratinocytes—more in thick skin than in thin skin—and some dendritic cells. The keratinocytes of this layer contain coarse, dark-staining *keratohyalin granules* that give the layer its name. The functional significance of these granules will be explained shortly.
4. The **stratum lucidum**[7] (LOO-sih-dum) is a thin translucent zone superficial to the stratum granulosum, seen only in thick skin. Here, the keratinocytes are densely packed with *eleidin* (ee-LEE-ih-din), an intermediate product in the production of keratin. The cells have no nuclei or other organelles. Because organelles are absent and eleidin does not stain well, this zone has a pale, featureless appearance with indistinct cell boundaries.
5. The **stratum corneum** consists of up to 30 layers of dead, scaly, keratinized cells that form a durable water-resistant surface layer.

THE LIFE HISTORY OF A KERATINOCYTE

Dead cells constantly flake off the skin surface. They float around as tiny white specks in the air, settling on household surfaces and forming much of the house dust that accumulates there. Because we constantly lose these epidermal cells, they must be continually replaced.

Keratinocytes are produced deep in the epidermis by the mitosis of stem cells in the stratum basale. Some of the deepest keratinocytes in the stratum spinosum also remain mitotic and thus increase their number. Mitosis requires an abundant supply of oxygen and nutrients, which these deep epidermal cells can acquire from the blood vessels in the nearby dermis. Once the epidermal cells migrate more than two or three cell layers away from the dermis, their mitosis ceases. Mitosis is seldom seen in prepared slides of the skin, because it occurs mainly at night and most histological specimens are taken during the day.

As new keratinocytes are formed, they push the older ones toward the surface. Over the course of 30 to 40 days, a keratinocyte makes its way to the skin surface and then flakes off. This migration is slower in old age, and faster in skin that has been injured or stressed. Injured epidermis regenerates more rapidly than any other tissue in the body. Mechanical stress from manual labor or tight shoes accelerates keratinocyte multiplication and results in *calluses* or *corns,* thick accumulations of dead keratinocytes on the hands or feet.

As keratinocytes are shoved upward by the dividing cells below, their cytoskeleton proliferates, the cells grow flatter, and they produce lipid-filled **membrane-coating vesicles.** In the stratum granulosum, three important developments occur. (1) The keratinocytes undergo apoptosis (programmed cell death). (2) The keratohyalin granules release a substance that binds to the intermediate filaments of the cytoskeleton and converts them to keratin. (3) The membrane-coating vesicles release a lipid mixture that spreads out over the cell surface and waterproofs it.

An *epidermal water barrier* forms between the stratum granulosum and the stratum spinosum. It consists of the lipids secreted by the keratinocytes, tight junctions between the keratinocytes, and a thick layer of insoluble protein on the inner surfaces of the keratinocyte plasma membranes. The epidermal water barrier is crucial to retaining water in the body and preventing dehydration. Cells above the barrier quickly die because the barrier cuts them off from the supply of nutrients below. Thus, the stratum corneum consists of compact layers of dead keratinocytes and keratinocyte fragments. Dead keratinocytes soon *exfoliate* (fall away) from the epidermal surface as tiny specks called **dander.** *Dandruff* is composed of clumps of dander stuck together by sebum (oil).

The Dermis

Beneath the epidermis is a connective tissue layer, the **dermis.** It ranges from 0.2 mm thick in the eyelids to about 4 mm thick in the palms and soles. It is composed mainly of collagen but also contains elastic and reticular fibers, fibroblasts, and the other cells typical of fibrous connective tissue (described in chapter 3). It is well supplied with blood vessels, sweat glands, sebaceous glands, and nerve endings. The hair follicles and nail roots are embedded in the dermis. Smooth muscles (piloerector muscles) associated with hair follicles contract in response to such stimuli as cold, fear, and touch. This makes the hairs stand on end, causes "goose bumps," and wrinkles the skin in areas such as the scrotum and areola. In the face, skeletal muscles attach to dermal collagen fibers and produce such expressions as a smile, a wrinkle of the forehead, and the lifting of an eyebrow.

The boundary between the epidermis and dermis is histologically conspicuous and usually wavy. The upward waves are fingerlike extensions of the dermis called **dermal papillae,**[8] and the downward waves are extensions of the epidermis called **epidermal ridges.** The dermal and epidermal boundaries thus interlock like corrugated cardboard, an arrangement that resists slippage of the epidermis across the dermis. If you look closely at your hand and wrist, you will see delicate furrows that divide the skin into tiny rectangular to rhomboidal areas. The dermal papillae produce the raised areas between the furrows (see fig. 5.2*b*). On the fingertips,

[7] *lucid* = light, clear

[8] *pap* = nipple + *illa* = little

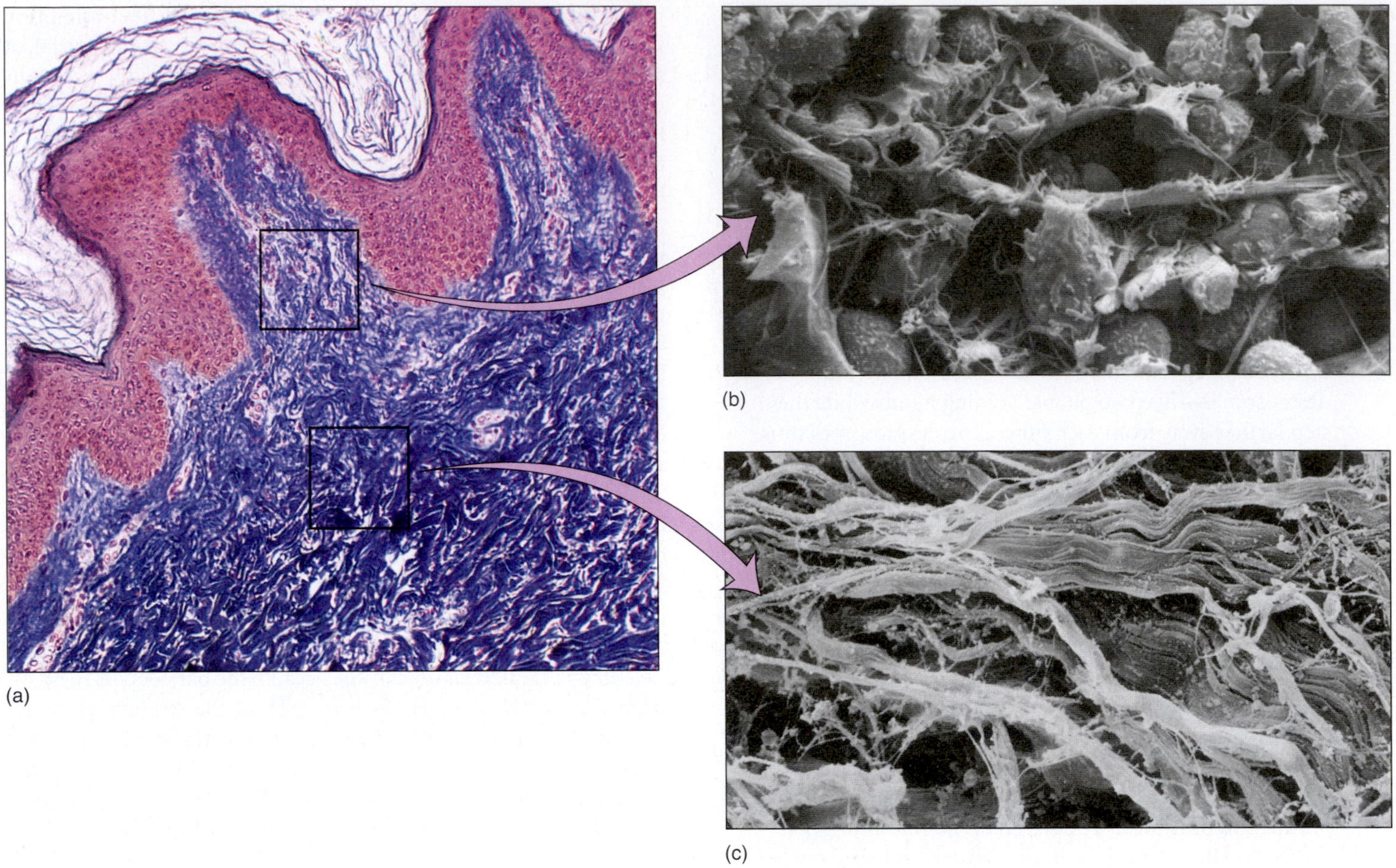

FIGURE 5.4

Layers of the Dermis. (*a*) Light micrograph of axillary skin, with the collagen stained blue. (*b*) The papillary layer, made of loose (areolar) tissue, forms the dermal papillae. (*c*) The reticular layer, made of dense irregular connective tissue, forms the deeper four-fifths of the dermis. Figures *b* and *c* from R. G. Kessel and R. H. Kardon, *Tissues and Organs: A Text-Atlas of Scanning Electron Microscopy* (W. H. Freeman, 1979).

this wavy boundary forms the *friction ridges* that leave fingerprints on the things we touch. In highly sensitive areas such as the lips and genitals, tall dermal papillae allow nerve fibers and blood capillaries to reach close to the skin surface.

Think About It!

Dermal papillae are relatively high and numerous in palmar and plantar skin but low and few in number in the face and abdomen. What do you think is the functional significance of this difference?

There are two zones of dermis called the papillary and reticular layers (fig. 5.4). The **papillary** (PAP-ih-lerr-ee) **layer** is a thin zone of areolar tissue in and near the dermal papillae. The loosely organized tissue of the papillary layer allows for mobility of leukocytes and other defenses against organisms introduced through breaks in the epidermis.

The **reticular layer** of the dermis is deeper and much thicker. It consists of dense irregular connective tissue. Leather is composed of the reticular layer of animal skin. The boundary between the

INSIGHT 5.1 CLINICAL APPLICATION

Tension Lines and Surgery

The collagen bundles in the dermis are arranged mostly in parallel rows that run longitudinally to obliquely in the limbs and encircle the neck, trunk, wrists, and a few other areas. They keep the skin under constant tension and are thus called *tension lines (Langer[9] lines)*. If an incision is made in the skin, especially if it is perpendicular to the tension lines, the wound gapes because the collagen bundles pull the edges of the incision apart. Even if the skin is punctured with a circular object such as an ice pick, the wound gapes with a lemon-shaped opening, the direction of the wound axis being perpendicular to the tension lines. Such gaping wounds are relatively difficult to close and tend to heal with excessive scarring. Some surgeons make incisions parallel to the tension lines—for example, making a transverse incision when delivering a baby by cesarean section—so that the incisions will gape less and heal with less scarring.

[9] Karl Langer (1819–87), Austrian physician

papillary and reticular layers is often vague. In the reticular layer, the collagen forms thicker bundles with less room for ground substance, and there are often small clusters of adipocytes. Stretching of the skin in obesity and pregnancy can tear the collagen fibers and produce *striae* (STRY-ee), or stretch marks. These occur especially in areas most stretched by weight gain: the thighs, buttocks, abdomen, and breasts.

The Hypodermis

Beneath the skin is a layer called the **hypodermis,**[10] **subcutaneous tissue,** or **superficial fascia**[11] (FASH-ee-uh). The boundary between the dermis and hypodermis is indistinct, but the hypodermis generally has more areolar and adipose tissue. The hypodermis binds the skin to the underlying tissues and pads the body. Drugs are introduced here by hypodermic injection because the subcutaneous tissue is highly vascular and absorbs them quickly.

Subcutaneous fat is hypodermis composed predominantly of adipose tissue. This fat serves as an energy reservoir and thermal insulation. It is not uniformly distributed; for example, it is virtually absent from the scalp but relatively abundant in the breasts, abdomen, hips, and thighs. The subcutaneous fat averages about 8% thicker in women than in men, and is different in distribution (fig. 5.5). It also varies with age. Infants and elderly people have less subcutaneous fat than other people and are therefore more sensitive to cold.

Table 5.1 summarizes the layers of the skin and hypodermis.

Skin Color

The most significant factor in skin color is **melanin,**[12] which is produced by the melanocytes but which accumulates in the keratinocytes of the stratum basale and stratum spinosum (fig. 5.6).

[10]*hypo* = below + *derm* = skin
[11]*fasc* = band
[12]*melan* = black

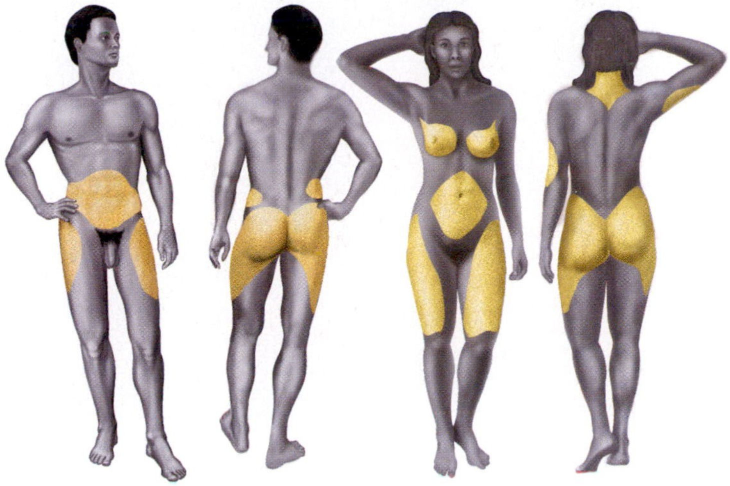

FIGURE 5.5
Distribution of Subcutaneous Fat in Men and Women.

There are two forms of melanin—a brownish black **eumelanin** and a reddish yellow sulfur-containing pigment, **pheomelanin.**[13] People of different races have essentially the same number of melanocytes, but in dark-skinned people, the melanocytes produce greater quantities of melanin, and the melanin in the keratinocytes breaks down more slowly. Thus, melanized cells may be seen throughout the epidermis, from stratum basale to stratum corneum. In light-skinned people, the melanin breaks down more rapidly and little of it is seen beyond the stratum basale, if even there.

The amount of melanin in the skin also varies with exposure to the ultraviolet (UV) rays of sunlight, which stimulate melanin synthesis and darken the skin. A suntan fades as melanin is degraded in older keratinocytes and as the keratinocytes migrate to

[13]*pheo* = dusky + *melan* = black

TABLE 5.1
Stratification of the Skin and Hypodermis

Layer	Description
Epidermis	Keratinized stratified squamous epithelium
Stratum corneum	Dead, keratinized cells of the skin surface
Stratum lucidum	Clear, featureless, narrow zone seen only in thick skin
Stratum granulosum	Two to five layers of cells with dark-staining keratohyalin granules; scanty in thin skin
Stratum spinosum	Many layers of keratinocytes, typically shrunken in fixed tissues but attached to each other by desmosomes, which give them a spiny look; progressively flattened the farther they are from the dermis. Dendritic cells occur here but are not visible in routinely stained preparations.
Stratum basale	Single layer of cuboidal to columnar cells resting on basal lamina; site of most mitosis; consists of stem cells, keratinocytes, melanocytes, and tactile cells, but these are not all distinguishable with routine stains. Melanin is conspicuous in keratinocytes of this layer in black to brown skin.
Dermis	Fibrous connective tissue, richly endowed with blood vessels and nerve endings. Sweat glands and hair follicles originate here and in hypodermis.
Papillary layer	Superficial one-fifth of dermis; composed of areolar tissue; often extends upward as dermal papillae.
Reticular layer	Deeper four-fifths of dermis; dense irregular connective tissue
Hypodermis	Areolar or adipose tissue between skin and muscle

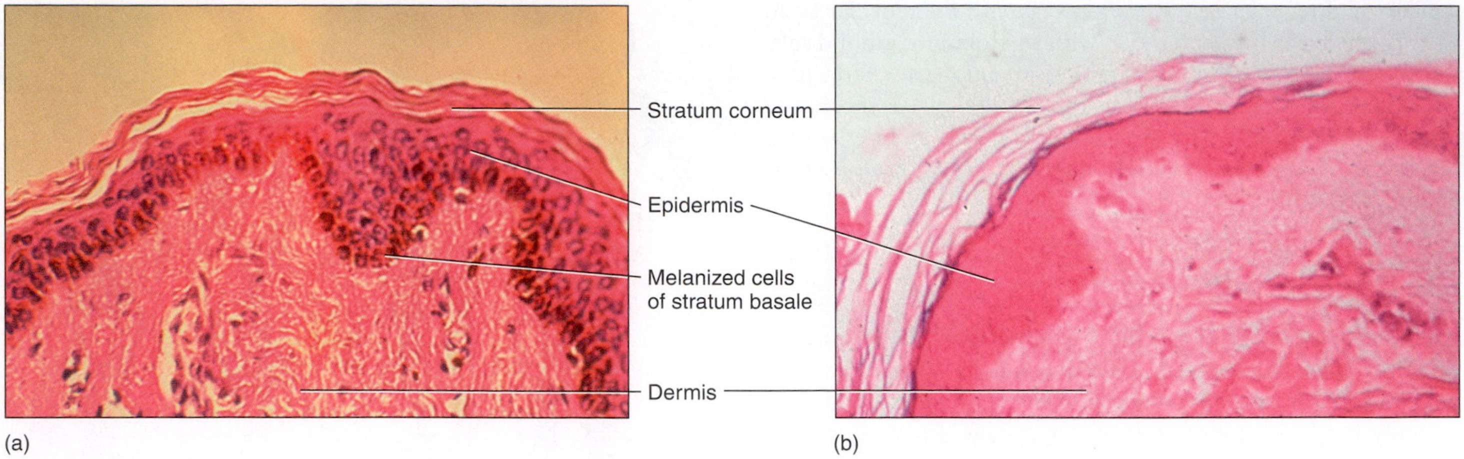

FIGURE 5.6

Variations in Skin Pigmentation. (*a*) Keratinocytes in and near the stratum basale have heavy deposits of melanin in dark skin. (*b*) Light skin shows little or no visible melanin in the keratinocytes.

the surface and exfoliate. The amount of melanin also varies substantially from place to place on the body. It is relatively concentrated in freckles and moles, on the dorsal surfaces of the hands and feet as compared to the palms and soles, in the nipple and surrounding area (areola) of the breast, around the anus, in the scrotum and penis, and on the lateral surface of the female genital folds *(labia majora)*. The contrast between heavily melanized and lightly melanized regions of the skin is more pronounced in some races than in others, but it exists to some extent in nearly everyone.

Other factors in skin color are hemoglobin and carotene. **Hemoglobin,** the red pigment of blood, imparts reddish to pinkish hues to the skin. Its color is lightened by the white of the dermal collagen. The skin is redder in places such as the lips, where blood capillaries come closer to the surface and the hemoglobin shows through more vividly. **Carotene**[14] is a yellow pigment acquired from egg yolks and yellow and orange vegetables. Depending on the diet, it can become concentrated to various degrees in the stratum corneum and subcutaneous fat. It is often most conspicuous in skin of the heel and in "corns" or calluses of the feet because this is where the stratum corneum is thickest.

The skin may also exhibit abnormal colors of diagnostic value:

- **Cyanosis**[15] is blueness of the skin resulting from a deficiency of oxygen in the circulating blood. Oxygen deficiency turns the hemoglobin a reddish violet color. It can result from conditions that prevent the blood from picking up a normal load of oxygen in the lungs, such as airway obstructions in drowning and choking, lung diseases such as emphysema, or respiratory arrest. Cyanosis also occurs in situations such as cold weather and cardiac arrest, when blood flows so slowly through the skin that most of its oxygen is extracted faster than freshly oxygenated blood arrives.
- **Erythema**[16] (ERR-ih-THEE-muh) is abnormal redness of the skin. It occurs in such situations as exercise, hot weather, sunburns, anger, and embarrassment. Erythema is caused by increased blood flow in dilated cutaneous blood vessels or by dermal pooling of red blood cells that have escaped from abnormally permeable capillaries.
- **Pallor** is a pale or ashen color that occurs when there is so little blood flow through the skin that the white color of the dermal collagen shows through. It can result from emotional stress, low blood pressure, circulatory shock, cold temperatures, or severe anemia.
- **Albinism**[17] is a genetic lack of melanin that results in white hair, pale skin, and pink eyes. Melanin is synthesized from the amino acid tyrosine by the enzyme tyrosinase. People with albinism have inherited a recessive, nonfunctional tyrosinase gene from both parents.
- **Jaundice**[18] is a yellowing of the skin and whites of the eyes resulting from high levels of bilirubin in the blood. Bilirubin is a hemoglobin breakdown product. When erythrocytes get old, they disintegrate and release their hemoglobin. The liver converts hemoglobin to bilirubin and other pigments, which are excreted in the bile. Bilirubin can accumulate enough to discolor the skin, however, in such situations as a rapid rate of erythrocyte destruction; when diseases such as cancer, hepatitis, and cirrhosis interfere with liver function; and in premature infants, where the liver is not well enough developed to dispose of bilirubin efficiently.
- **Bronzing** is a golden-brown skin color that results from Addison disease, a deficiency of glucocorticoid hormones from the adrenal cortex. John F. Kennedy had Addison disease and bronzing of the skin, which many people mistook for a suntan.

---

[14]*carot* = carrot
[15]*cyan* = blue + *osis* = condition
[16]*eryth* = red + *em* = blood
[17]*alb* = white + *ism* = state, condition
[18]*jaun* = yellow

INSIGHT 5.2 CLINICAL APPLICATION

Birthmarks

Birthmarks, or *hemangiomas,*[19] are patches of discolored skin caused by benign tumors of the blood capillaries. *Capillary hemangiomas* (strawberry birthmarks) usually develop about a month after birth. They become bright red to deep purple and develop small capillary-dense elevations that give them a strawberry-like appearance. About 90% of capillary hemangiomas disappear by the age of 5 or 6 years. *Cavernous hemangiomas* are flatter and duller in color. They are present at birth, enlarge up to 1 year of age, and then regress. About 90% disappear by the age of 9 years. A port-wine stain *(nevus flammeus)* is flat and pinkish to dark purple in color. It can be quite large, and remains for life.

[19]*hem* = blood + *angi* = vessel + *oma* = tumor

- A **hematoma,**[20] or bruise, is a mass of clotted blood showing through the skin. It is usually due to accidental trauma (blows to the skin), but it may indicate hemophilia, other metabolic or nutritional disorders, or physical abuse.

Think About It!

An infant brought to a clinic shows abnormally yellow skin. What sign could you look for to help decide whether this was due to jaundice or to a large amount of carotene from strained vegetables in the diet?

Skin Markings

The skin is marked by many lines, creases, and ridges. **Friction ridges** are the markings on the fingertips that leave distinctive oily fingerprints on surfaces we touch (see fig. 5.2*b*). They are characteristic of most primates. They help prevent monkeys, for example, from slipping off a branch as they walk across it, and they enable us to manipulate small objects more easily. Friction ridges form during fetal development and remain essentially unchanged for life. Everyone has a unique pattern of friction ridges; not even identical twins have identical fingerprints.

Flexion lines (flexion creases) are lines on the flexor surfaces of the digits, palms, wrists, elbows, and other places (see fig. B.8 in atlas B). They mark sites where the skin folds during flexion of the joints. The skin is tightly bound to the deep fascia along these lines.

Freckles and moles are tan to black aggregations of melanocytes. **Freckles** are flat melanized patches that vary with heredity and exposure to the sun. A **mole (nevus)** is an elevated patch of melanized skin, often with hair. Moles are harmless and sometimes even regarded as "beauty marks," but they should be watched for changes in color, diameter, or contour that may suggest malignancy (skin cancer).

[20]*hemat* = blood + *oma* = mass

Before You Go On

Answer the following questions to test your understanding of the preceding section:

1. What is the major histological difference between thick and thin skin? Where on the body is each type of skin found?
2. How does the skin help control body temperature?
3. List the five cell types of the epidermis. Describe their locations and functions.
4. List the five layers of epidermis from deep to superficial. What are the distinctive features of each layer?
5. What are the two layers of the dermis? What type of tissue composes each layer?
6. Name the pigments responsible for normal skin colors, and explain how certain conditions can produce discolorations of the skin.

Hair and Nails

Objectives

When you have completed this section, you should be able to

- distinguish between three types of hair;
- describe the histology of a hair and its follicle;
- discuss some theories of the purposes served by various kinds of hair; and
- describe the structure and function of nails.

The hair, nails, and cutaneous glands are the **accessory organs (appendages)** of the skin. Hair and nails are composed mostly of dead, keratinized cells. While the stratum corneum of the skin is made of pliable **soft keratin,** the hair and nails are composed mostly of **hard keratin.** Hard keratin is more compact than soft keratin and is toughened by numerous cross-linkages between the keratin molecules.

Hair

A hair is also known as a **pilus** (PY-lus); in the plural, *pili* (PY-lye). It is a slender filament of keratinized cells that grows from an oblique tube in the skin called a **hair follicle** (fig. 5.7).

DISTRIBUTION AND TYPES

Hair is found almost everywhere on the body except the lips, nipples, parts of the genitals, palms and soles, ventral and lateral surfaces of the fingers and toes, and distal segment of the fingers. Hairless skin is sometimes called *glabrous*[21] *skin.* The extremities and trunk have about 55 to 70 hairs per square centimeter, and the face has about 10 times as many. There are about 30,000 hairs in a man's beard and about 100,000 hairs on the average person's scalp. The number of hairs in a given area does not differ much from one person to another or even between the sexes. Differences in apparent hairiness are due mainly to differences in the texture and pigmentation of the hair.

[21]*glab* = smooth

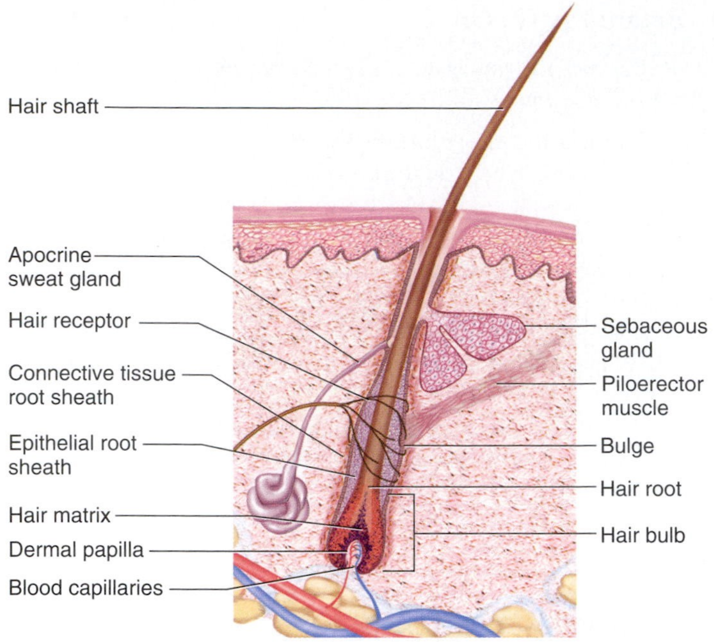
Hair shaft
Apocrine sweat gland
Hair receptor
Connective tissue root sheath
Epithelial root sheath
Hair matrix
Dermal papilla
Blood capillaries
Sebaceous gland
Piloerector muscle
Bulge
Hair root
Hair bulb

(a)

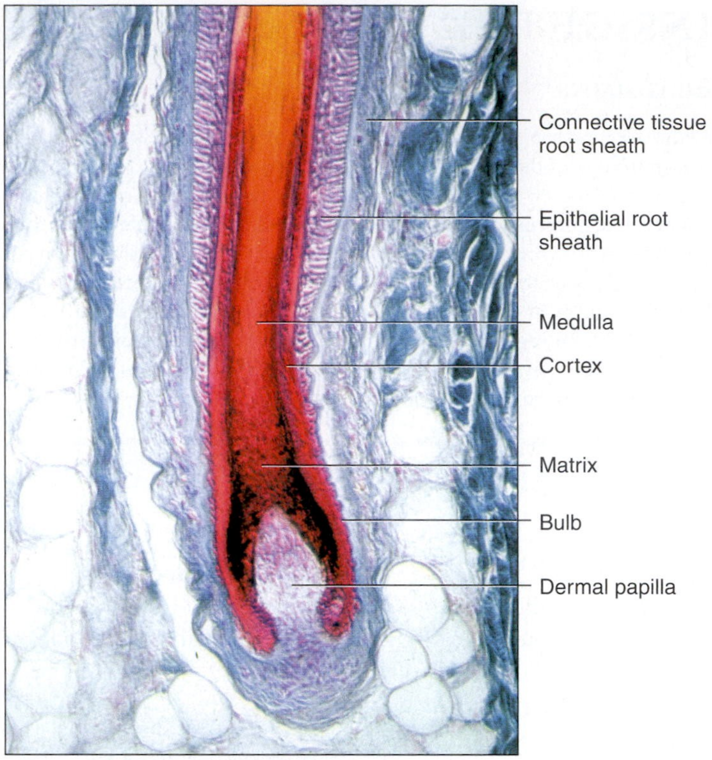
Connective tissue root sheath
Epithelial root sheath
Medulla
Cortex
Matrix
Bulb
Dermal papilla

(b)

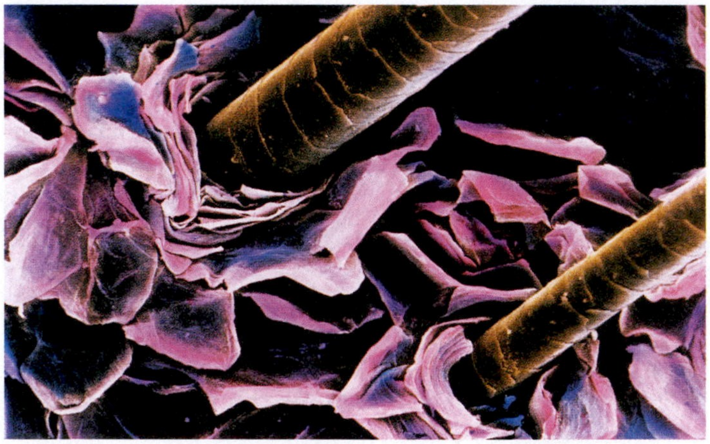

(c)

FIGURE 5.7

Structure of a Hair and Its Follicle. (*a*) Anatomy of the follicle and associated structures. (*b*) Light micrograph of the base of a hair follicle. (*c*) Electron micrograph of two hairs emerging from their follicles. Note the exfoliating epidermal cells encircling the follicles like rose petals.

Not all hair is alike, even on one person. Over the course of our lives, we grow three kinds of hair: lanugo, vellus, and terminal hair. **Lanugo**[22] is fine, downy, unpigmented hair of the fetus. By the time of birth, it is replaced by **vellus,**[23] a similarly fine, unpigmented hair. Vellus constitutes about two-thirds of the hair of women, one-tenth of the hair of men, and all of the hair of children except for the eyebrows, eyelashes, and hair of the scalp. **Terminal hair** is longer, coarser, and pigmented. It forms the eyebrows and eyelashes, covers the scalp, and after puberty, it forms the axillary and pubic hair, the male facial hair, and some of the hair on the trunk and limbs.

STRUCTURE OF THE HAIR AND FOLLICLE

A hair is divisible into three zones along its length: (1) the **bulb,** a swelling at the base where the hair originates in the dermis; (2) the **root,** which is the remainder of the hair within the follicle; and (3) the **shaft,** which is the portion above the skin surface. Except near the bulb, all the tissue is dead. The hair bulb grows around a bud of vascular connective tissue called the **dermal**

[22] *lan* = down, wool
[23] *vellus* = fleece

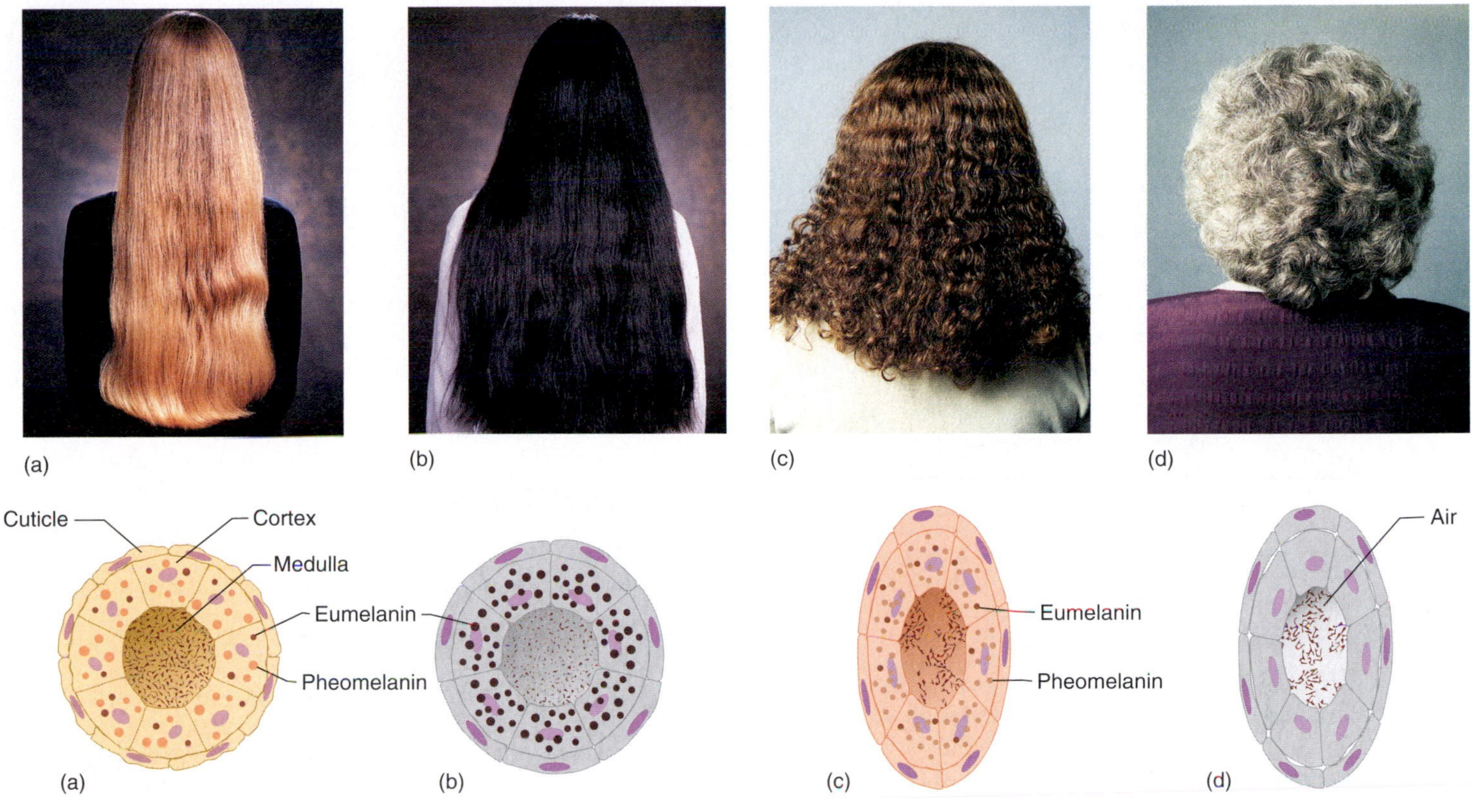

FIGURE 5.8

Basis of Hair Color and Texture. Straight hair (*a* and *b*) is round in cross section while curly hair (*c* and *d*) is flatter. Blonde hair (*a*) has scanty eumelanin and a moderate amount of pheomelanin. Eumelanin predominates in black and brown hair (*b*). Red hair (*c*) derives its color predominantly from pheomelanin. Gray and white hair (*d*) lack pigment and have air in the medulla.

papilla, which provides the hair with its sole source of nutrition. Immediately above the papilla is a region of mitotically active cells, the **hair matrix,** which is the hair's growth center. All cells higher up are dead.

In cross section, a hair reveals three layers: (1) the **medulla,** a core of loosely arranged cells and air spaces found in thick hairs, but absent from thin ones; (2) the **cortex,** a layer of keratinized cuboidal cells; and (3) the **cuticle,** a surface layer of scaly cells that overlap each other like roof shingles, with their free edges directed upward (fig. 5.7*c*). Cells lining the follicle are like shingles facing in the opposite direction. They interlock with the scales of the hair cuticle and resist pulling on the hair. When a hair is pulled out, this layer of follicle cells comes with it.

The follicle is a diagonal tube that dips deeply into the dermis and sometimes extends as far as the hypodermis. It has two principal layers: an **epithelial root sheath** and a **connective tissue root sheath.** The epithelial root sheath, which is an extension of the epidermis, lies immediately adjacent to the hair root. The connective tissue root sheath, derived from the dermis, surrounds the epithelial sheath and is somewhat denser than the adjacent dermal connective tissue.

Associated with the follicle are nerve and muscle fibers. Nerve fibers called **hair receptors** entwine each follicle and respond to hair movements. You can feel their effect by carefully moving a single hair with a pin or by lightly running your finger over the hairs of your arm without touching the skin. Also associated with each hair is a **piloerector muscle (arrector pili**[24]**)**, a bundle of smooth muscle cells extending from dermal collagen fibers to the connective tissue root sheath of the follicle (see figs. 5.1 and 5.7). In response to cold, fear, or other stimuli, the sympathetic nervous system stimulates these muscles to contract and thereby makes the hair stand on end. In other mammals, this traps an insulating layer of warm air next to the skin or makes the animal appear larger and less vulnerable to a potential enemy. In humans, it pulls the follicles into a vertical position and causes "goose bumps" but serves no useful purpose.

HAIR TEXTURE AND COLOR

The texture of hair is related to differences in cross-sectional shape (fig. 5.8)—straight hair is round, wavy hair is oval, and tightly curly hair is relatively flat. Hair color is due to pigment granules in the cells of the cortex. Brown and black hair are rich in eumelanin. Red hair has less eumelanin but a high concentration of pheomelanin. Blond hair has an intermediate amount of pheomelanin but very little eumelanin. Gray and white hair results from a scarcity or absence of melanins in the cortex and the presence of air in the medulla.

[24]*arrect* = erect + *pil* = of hair

Table 5.2

Functions of Hair

Hair of the Torso and Limbs	Vestigial, but serves a sensory purpose as in detection of small insects crawling on the skin
Scalp Hair	Heat retention, protection from sun
Beard, Pubic, and Axillary (armpit) Hair	Advertises sexual maturity; associated with apocrine scent glands in these areas and modulates the dispersal of sexual scents (pheromones) from these glands
Guard Hairs (vibrissae)	Help keep foreign objects out of nostrils and auditory canal; eyelashes help keep debris from eyes
Eyebrows	Enhance facial expression, may reduce glare of sun and help keep forehead perspiration from eyes

FUNCTIONS OF HAIR

In most mammals, hair serves to retain body heat. Humans have too little hair to serve this purpose except on the scalp, where there is no insulating fat. Hair elsewhere on the body serves a variety of functions that are somewhat speculative, but probably best inferred by comparison to the specialized types and patches of hair in other mammals (table 5.2).

Nails

Fingernails and toenails are clear, hard derivatives of the stratum corneum. They are composed of very thin, dead, scaly cells, densely packed together and filled with parallel fibers of hard keratin. Most mammals have claws, whereas flat nails are one of the distinguishing characteristics of primates (see chapter 1). Flat nails allow for more fleshy and sensitive fingertips, while they also serve as strong keratinized "tools" that can be used for digging, grooming, picking apart food, and other manipulations.

The anatomical features of a nail are shown in figure 5.9. The most important features are the **nail matrix,** a growth zone concealed beneath the skin at the proximal edge of the nail, and the **nail plate,** which is the visible portion covering the tip of the finger or toe. The nail groove and space under the free edge accumulate dirt and bacteria and require special attention when scrubbing for duty in an operating room or nursery.

Fingernails grow about 1 mm per week and toenails somewhat more slowly. New cells are added to the nail plate by mitosis in the nail matrix. Contrary to some advertising claims, adding gelatin to the diet has no effect on the growth or hardness of the nails.

The appearance of the nails can be valuable to medical diagnosis. An iron deficiency, for example, may cause the nails to become flat or concave (spoonlike) rather than convex. The fingertips become swollen or *clubbed* in response to long-term hypoxemia stemming from conditions such as congenital heart defects and emphysema.

Before You Go On

Answer the following questions to test your understanding of the preceding section:

7. What is the difference between vellus and terminal hair?
8. Describe the three regions of a hair from its base to its tip, and the three layers of a hair seen in cross section.
9. State the function of the hair papilla, hair receptors, and piloerector muscle.
10. State a reasonable theory for the different functions of hair of the eyebrows, eyelashes, scalp, nostrils, and axilla.
11. Describe some similarities between a nail and a hair.

Cutaneous Glands

Objectives

When you have completed this section, you should be able to

- name two types of sweat glands, and describe the structure and function of each;
- describe the location, structure, and function of sebaceous and ceruminous glands; and
- discuss the distinction between breasts and mammary glands, and explain their respective functions.

The skin has five types of glands: *merocrine sweat glands, apocrine sweat glands, sebaceous glands, ceruminous glands,* and *mammary glands.*

Sweat Glands

Sweat glands, or **sudoriferous**[25] (soo-dor-IF-er-us) **glands,** are of two kinds, merocrine and apocrine (fig. 5.10). **Merocrine (eccrine) sweat glands,** the most numerous type, produce watery perspiration that serves primarily to cool the body. There are 3 to 4 million merocrine sweat glands in the adult skin, with a total weight about equal to that of a kidney. They are especially abundant on the palms, soles, and forehead, but they are widely distributed over the rest of the body as well. Each is a simple tubular gland with a twisted coil in the dermis or hypodermis and an undulating or coiled duct leading to a sweat pore on the skin surface. This duct is lined by a stratified cuboidal epithelium in the dermis and by keratinocytes in the epidermis. Amid the secretory cells at the deep end of the gland, there are specialized **myoepithelial**[26] **cells** with properties similar to smooth muscle. They contract in response to stimuli from the sympathetic nervous system and squeeze perspiration up the duct.

Apocrine sweat glands occur in the groin, anal region, axilla, and areola, and in mature males, they also occur in the beard area. They are absent from the axillary region of Koreans and are very sparse in the Japanese. Their ducts lead into nearby hair follicles rather than opening directly onto the skin surface. They produce their secretion in the same way that merocrine glands do—that is,

[25] *sudor* = sweat + *fer* = carry, bear
[26] *myo* = muscle

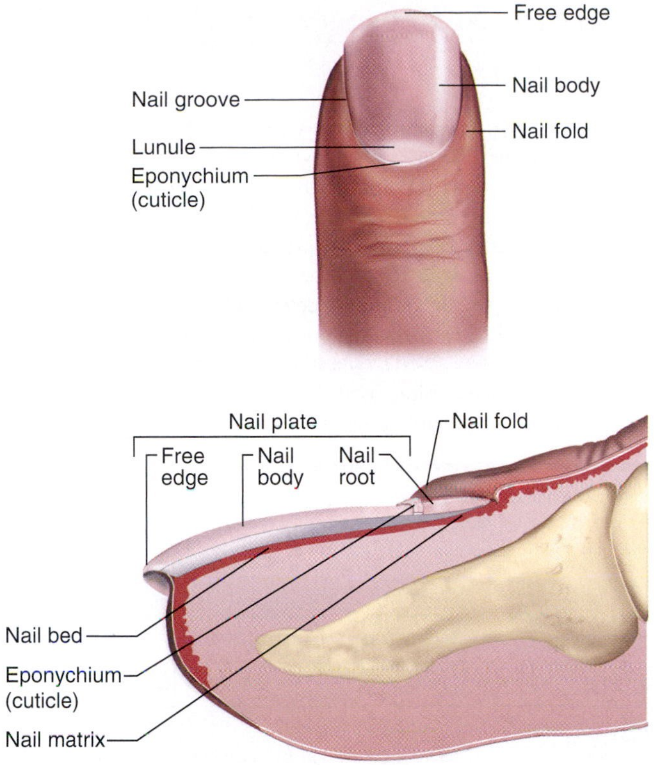

FIGURE 5.9
Anatomy of a Fingernail.

Nail Bed	The skin on which the nail plate rests
Nail Plate	The clear, keratinized portion of a nail
Root	The proximal end of a nail, underlying the nail bed
Body	The major portion of the nail plate, overlying the nail bed
Free Edge	The portion of the nail plate that extends beyond the end of the digit
Hyponychium[a]	The epithelium of the nail bed
Nail Fold	The fold of skin around the margins of the nail plate
Nail Groove	The groove where the nail fold meets the nail plate
Eponychium[b]	Dead epidermis that covers the proximal end of the nail; commonly called the cuticle
Nail Matrix	The growth zone (mitotic tissue) at the proximal end of the nail, corresponding to the stratum basale of the epidermis
Lunule[c]	The region at the base of the nail that appears as a small white crescent because it overlies a thick stratum basale that obscures dermal blood vessels from view

[a] *hypo* = under + *onych* = nail
[b] *ep* = above + *onych* = nail
[c] *lun* = moon + *ule* = little

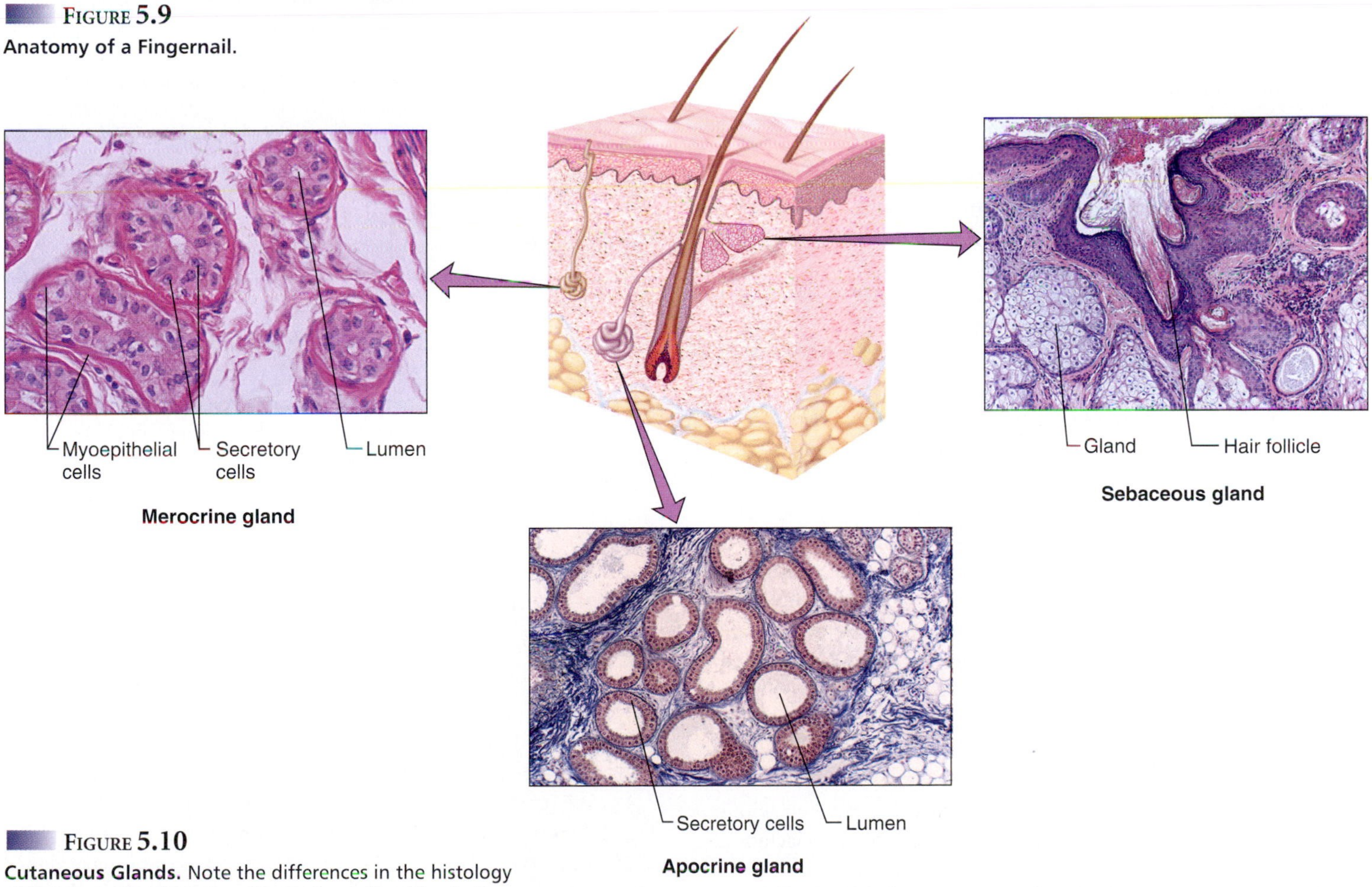

FIGURE 5.10
Cutaneous Glands. Note the differences in the histology of the three gland types and in their relationships to the hair follicle.

by exocytosis. The secretory part of an apocrine gland, however, has a much larger lumen than that of a merocrine gland, so these glands have continued to be referred to as apocrine glands to distinguish them functionally and histologically from the merocrine type. Apocrine sweat is thicker and more milky than merocrine sweat because it has more fatty acids in it.

Apocrine sweat glands are scent glands that respond especially to stress and sexual stimulation. They do not develop until puberty, and in women, they enlarge and shrink in phase with the menstrual cycle. These facts, as well as experimental evidence, suggest that their function is to secrete chemicals called *sex pheromones,* which exert subtle effects on the sexual behavior and physiology of other people. They apparently correspond to the scent glands that develop in other mammals on attainment of sexual maturity. Fresh apocrine sweat does not have a disagreeable odor, and indeed it is considered attractive or arousing in some cultures. Stale apocrine sweat acquires a rancid odor from the action of bacteria on the lipids in the perspiration. Disagreeable body odor is called *bromhidrosis.*[27] It occasionally indicates a metabolic disorder, but more often it reflects poor hygiene.

Many mammals have apocrine scent glands associated with specialized tufts of hair. In humans, apocrine glands are found almost exclusively in regions covered by the pubic hair, axillary hair, and beard, suggesting that they are similar to other mammalian scent glands in function. The hair serves to retain the aromatic secretion and regulate its rate of evaporation from the skin. Thus, it seems no mere coincidence that women's faces lack both apocrine scent glands and a beard.

Sebaceous Glands

Sebaceous[28] (seh-BAY-shus) **glands** produce an oily secretion called **sebum** (SEE-bum). They are flask-shaped, with short ducts that usually open into a hair follicle (see fig. 5.10), although some of them open directly onto the skin surface. These are holocrine glands with little visible lumen. Their secretion consists of broken-down cells that are replaced by mitosis at the base of the gland. Sebum keeps the skin and hair from becoming dry, brittle, and cracked. The sheen of well-brushed hair is due to sebum distributed by the hairbrush.

Ceruminous Glands

Ceruminous (seh-ROO-mih-nus) **glands** are found only in the auditory (external ear) canal, where their secretion combines with sebum and dead epidermal cells to form earwax, or **cerumen.**[29] They are simple, coiled, tubular glands with ducts leading to the skin surface. Cerumen keeps the eardrum pliable, waterproofs the canal, and kills bacteria.

Mammary Glands

Mammary glands are milk-producing glands that develop within the breasts *(mammae)* under conditions of pregnancy and lactation. They are not synonymous with the breasts, which are present in both sexes and which, even in females, usually contain only small traces of mammary gland. Mammary glands are modified apocrine sweat glands that produce a richer secretion than other apocrine glands and channel it through ducts to a nipple for more efficient conveyance to the offspring. The mammary glands are discussed in more detail in chapter 26. Table 5.3 summarizes the cutaneous glands.

[27]*brom* = stench + *hidros* = sweat
[28]*seb* = fat, tallow + *aceous* = possessing
[29]*cer* = wax

INSIGHT 5.3 CLINICAL APPLICATION

Extra Nipples

In most mammals, two rows of mammary glands develop along lines called the *mammary ridges* or *milk lines,* which extend from the axillary to the inguinal region. Primates have dispensed with all but the most anterior pair of mammary glands. Some women, however, develop additional nipples along the milk line, so a breast may have two nipples or there may be an additional nipple just inferior or superior to the breast. This condition is called *polythelia.*[30] The extra nipple often is so little developed that it is mistaken for a mole. In a few cases, fully formed additional breasts develop inferior to the primary ones—a condition called *polymastia.*[31] In the Middle Ages and colonial America, polythelia was often used to incriminate women as supposed witches, and women were sometimes put to death because of it.

[30]*poly* = many, multiple + *theli* = nipples + *ia* = condition
[31]*poly* = many, multiple + *mast* = breasts + *ia* = condition

Table 5.3

Cutaneous Glands

Gland Type	Definition
Sudoriferous glands	Sweat glands
Merocrine	Sweat glands that function in evaporative cooling; widely distributed over the body surface; open by ducts onto the skin surface
Apocrine	Sweat glands that function as scent glands; found in the regions covered by the pubic, axillary, and male facial hair; open by ducts into hair follicles
Sebaceous glands	Holocrine oil-producing glands associated with hair follicles
Ceruminous glands	Glands of the ear canal that produce cerumen (earwax)
Mammary glands	Milk-producing glands located in the breasts

Before You Go On

Answer the following questions to test your understanding of the preceding section:

12. How do merocrine and apocrine sweat glands differ in structure and function?
13. What types of hair are associated with apocrine glands? Why?
14. What other type of gland is associated with hair follicles? How does its mode of secretion differ from that of sweat glands?
15. What is the difference between a breast and mammary gland? What other type of cutaneous gland is most closely related to mammary glands?

Developmental and Clinical Perspectives

Objectives

When you have completed this section, you should be able to

- describe the prenatal development of the skin, hair, and nails;
- describe the three most common forms of skin cancer; and
- describe the three classes of burns and the priorities in burn treatment.

Prenatal Development of the Integumentary System

SKIN

The epidermis develops from the embryonic ectoderm, and the dermis from the mesoderm. In week 4 of embryonic development, ectodermal cells multiply and organize into two layers—a superficial *periderm* of squamous cells and a deeper *basal layer* (fig. 5.11). In week 11, the basal layer gives rise to a new *intermediate layer* of cells between these two. From then until birth, the basal layer is known as the *germinative layer.* Its cells remain for life as the stem cells of the stratum basale. Cells of the intermediate layer synthesize keratin and become the first keratinocytes. The intermediate layer becomes stratified into three layers of keratinocytes—the stratum spinosum, granulosum, and corneum—as the periderm is sloughed off into the amniotic fluid. By week 21, the periderm is gone and the stratum corneum is the outermost layer of the fetal integument.

Beneath the developing epidermis, the embryonic mesoderm differentiates into a gelatinous connective tissue called mesenchyme. Mesenchymal cells begin producing collagenous and elastic fibers by week 11, and the mesenchyme takes on the characteristics of typical fibrous connective tissue. Dermal papillae appear along the dermal-epidermal boundary in the third month. Blood vessels appear in the dermis late in week 6. At birth, the skin has 20 times as many blood vessels as it needs to support its metabolism. The excess may serve to regulate the body temperature of the newborn.

HAIR AND NAILS

The first hair follicles appear around the end of the second month on the eyebrows, eyelids, upper lip, and chin; follicles do not appear elsewhere until the fourth month. At birth, there are about 5 million hair follicles in both sexes; no additional follicles form after one is born.

A hair follicle begins as a cluster of ectodermal cells called a *hair bud,* which pushes down into the dermis and elongates into a rodlike *hair peg* (fig. 5.12). The lower end of the peg expands into a *hair bulb.* The dermal papilla first appears as a small mound of dermal tissue just below the bulb, and then expands into the bulb itself. Ectodermal cells overlying the papilla form the *germinal matrix,* a mass of mitotically active cells that produce the hair shaft. The first hair to develop in the fetus is lanugo, which appears in week 12 and is abundant by week 20. By the time of birth, most lanugo is replaced by vellus.

The first indications of nail development are epidermal thickenings that appear on the ventral surfaces of the fingers around 10 weeks, and on the toes around 14 weeks. They soon migrate to the dorsal surfaces of the digits, where they form a shallow depression called the *primary nail field.* The margins of the nail

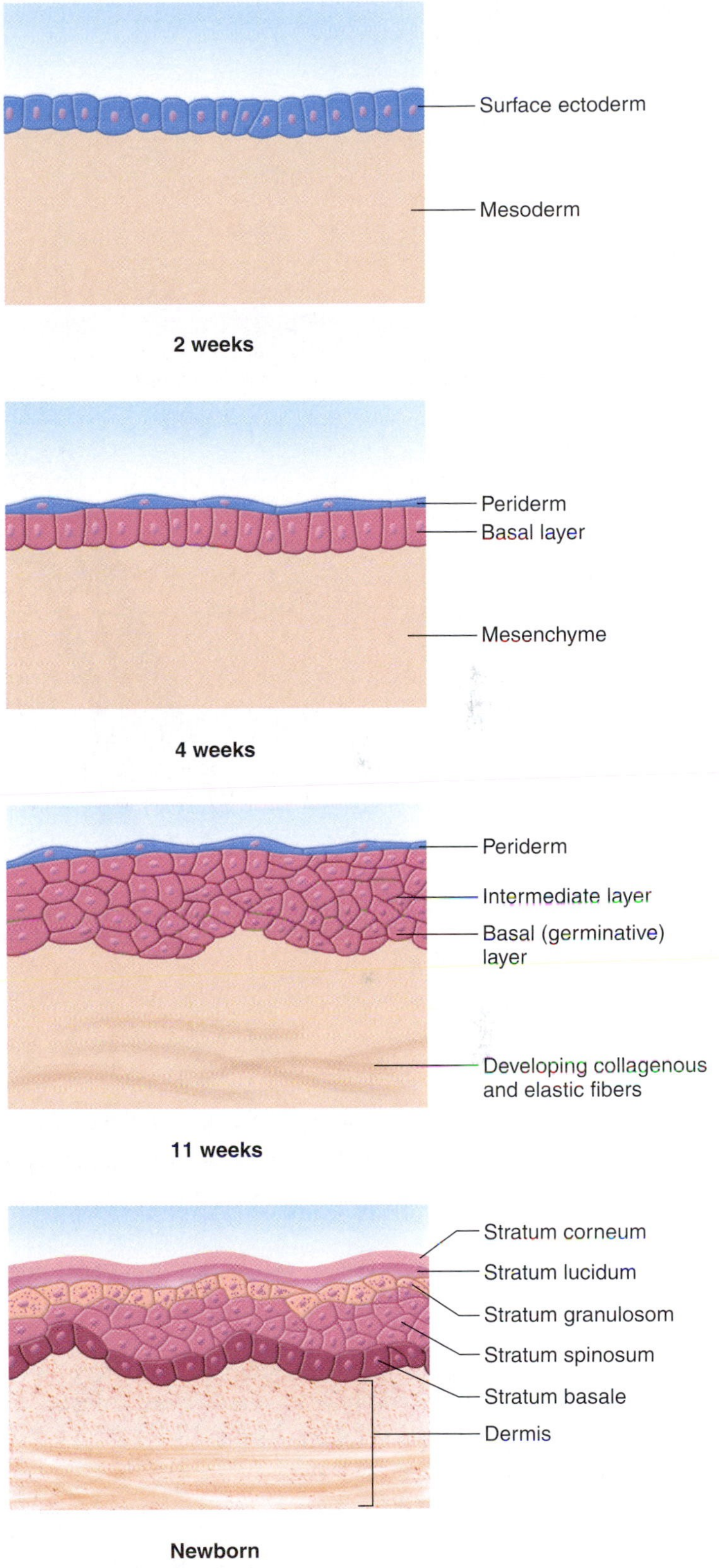
Surface ectoderm
Mesoderm
2 weeks
Periderm
Basal layer
Mesenchyme
4 weeks
Periderm
Intermediate layer
Basal (germinative) layer
Developing collagenous and elastic fibers
11 weeks
Stratum corneum
Stratum lucidum
Stratum granulosom
Stratum spinosum
Stratum basale
Dermis
Newborn

FIGURE 5.11
Prenatal Development of the Epidermis and Dermis.

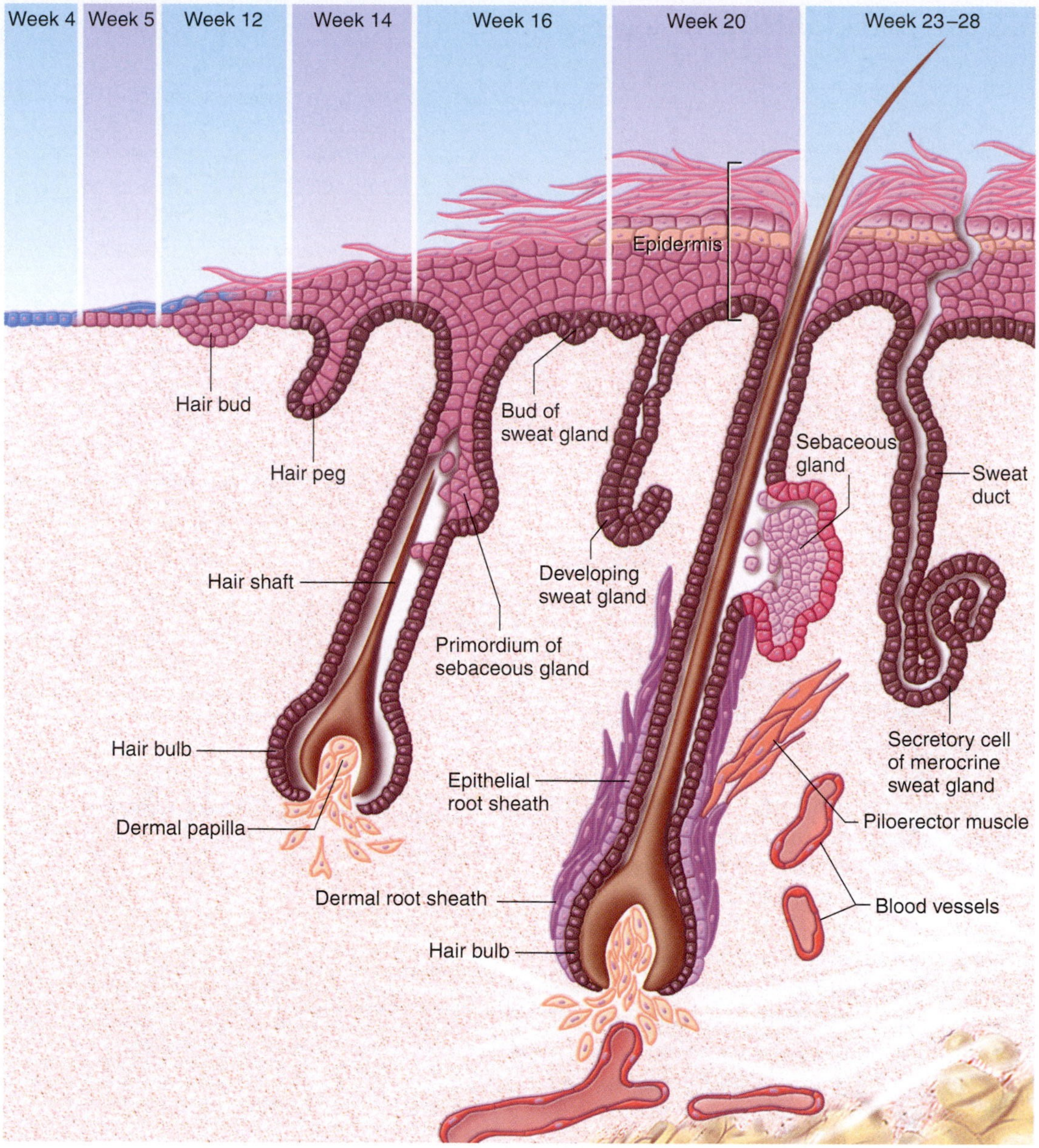
Week 4
Week 5
Week 12
Week 14
Week 16
Week 20
Week 23–28
Epidermis
Hair bud
Hair peg
Bud of sweat gland
Sebaceous gland
Sweat duct
Hair shaft
Developing sweat gland
Primordium of sebaceous gland
Secretory cell of merocrine sweat gland
Hair bulb
Epithelial root sheath
Dermal papilla
Piloerector muscle
Dermal root sheath
Blood vessels
Hair bulb

FIGURE 5.12
Prenatal Development of a Hair Follicle, Merocrine Sweat Gland, and Sebaceous Gland.

field are called the *nail folds.* In the proximal nail fold of each digit, the germinal layer of epidermis develops into the *nail root.* Mitosis in the root produces the keratinocytes that become compressed into the hard nail plate. The nail plate reaches the fingertips by 8 months and the toe tips by birth.

GLANDS

Sebaceous glands begin to bud from the sides of a hair follicle about 4 weeks after the hair germ begins to elongate (see fig. 5.12). Mature sebaceous glands are present on the face by 6 months, and secrete very actively before birth. Their sebum mixes with epidermal and peridermal cells to form a white, greasy skin coating called the *vernix caseosa.*[32] The vernix protects the skin from abrasions and from the amniotic fluid, which can otherwise cause the fetal skin to chap and harden. Its slipperiness also aids in the birth passage through the vagina. The vernix is anchored to the skin by the lanugo and later by the vellus. Sebaceous glands become largely dormant by the time of birth, and are reactivated at puberty under the influence of the sex hormones.

Apocrine sweat glands also develop as outgrowths from the hair follicles. They appear over most of the body at first, but then degenerate except in the limited areas described earlier—especially in the axillary and genital regions. Like the sebaceous glands, they become active at puberty.

[32]*vernix* = varnish + *case* = cheese + *osa* = having the qualities of

Merocrine sweat glands develop as buds of the embryonic germinative layer which grow and push their way down into the dermis. These buds develop at first into solid cords of epithelial tissue, but cells in the center of the cord later degenerate to form the lumen of the sweat duct, while cells at the lower end differentiate into secretory and myoepithelial cells.

The Aging Integumentary System

Senescence (age-related degeneration) of the integumentary system often becomes noticeable by the late 40s. The hair turns grayer and thinner as melanocytes die out, mitosis slows down, and dead hairs are not replaced. Atrophy of the sebaceous glands leaves the skin and hair drier. As epidermal mitosis declines and collagen is lost from the dermis, the skin becomes almost paper-thin and translucent. It becomes looser because of a loss of elastic fibers and flattening of the dermal papillae. If you pinch a fold of skin on the back of a child's hand, it quickly springs back when you let go; do the same on an older person, and the skin fold remains longer. Because of its loss of elasticity, aged skin sags to various degrees and may hang loosely from the arm and other places.

Aged skin has fewer blood vessels than younger skin, and those that remain are more fragile. The skin can become reddened as broken vessels leak into the connective tissue. Many older people exhibit *rosacea*—patchy networks of tiny, dilated blood vessels visible especially on the nose and cheeks. Because of the fragility of the dermal blood vessels, aged skin bruises more easily. Injured skin heals slowly in old age because of poorer circulation and a relative scarcity of immune cells and fibroblasts. Antigen-presenting dendritic cells decline by as much as 40% in the aged epidermis, leaving the skin more susceptible to recurring infections.

Thermoregulation can be a serious problem in old age because of the atrophy of cutaneous blood vessels, sweat glands, and subcutaneous fat. Older people are more vulnerable to hypothermia in cold weather and heatstroke in hot weather. Heat waves and cold spells take an especially heavy toll among the elderly poor, who suffer from a combination of reduced homeostasis and inadequate housing.

Degeneration of the skin is accelerated by excessive exposure to the ultraviolet radiation of sunlight. This *photoaging* accounts for more than 90% of the changes that people find medically troubling or cosmetically disagreeable: skin cancer; yellowing and mottling of the skin; age spots, which resemble enlarged freckles on the back of the hand and other sun-exposed areas; and wrinkling, which especially affects the most exposed areas of skin (face, hands, and arms). Sun-damaged skin shows many malignant and premalignant cells, extensive damage to the dermal blood vessels, and dense masses of coarse, frayed elastic fibers underlying the surface wrinkles and creases.

Skin Disorders

Because it is the most exposed of all our organs, skin is not only the most vulnerable to injury and disease, but is also the one place where we are most likely to notice anything out of the ordinary. We focus here on two particularly common and serious disorders, skin cancer and burns. Other skin diseases are briefly summarized in table 5.4.

TABLE 5.4

Some Disorders of the Integumentary System

Acne	Inflammation of the sebaceous glands, especially beginning at puberty; follicle becomes blocked with keratinocytes and sebum and develops into a blackhead *(comedo)* composed of these and bacteria; continued inflammation of follicle results in pus production and pimples.
Dermatitis	Any inflammation of the skin, typically marked by itching and redness; often *contact dermatitis*, caused by exposure to toxins such as poison ivy.
Eczema (ECK-zeh-mah)	Itchy, red, "weeping" skin lesions cause by an allergy, usually beginning before age 5; may progress to thickened, leathery, darkly pigmented patches of skin.
Psoriasis (so-RY-ah-sis)	Recurring, reddened plaques covered with silvery scale; sometimes disfiguring; possibly caused by an autoimmune response; runs in families.
Ringworm	A fungal infection of the skin (not a worm) that sometimes grows in a circular pattern; common in moist areas such as the axilla, groin, and foot *(athlete's foot)*.
Rosacea (ro-ZAY-she-ah)	A red rashlike area, often in the area of the nose and cheeks, marked by fine networks of dilated blood vessels; worsened by hot drinks, alcohol, and spicy food.
Warts	Benign, elevated, rough lesions caused by human papillomaviruses (HPV). *Common warts* are most common in late childhood on the fingers, elbows, and other areas of skin subject to stress. *Plantar warts* occur on the soles and *venereal warts* on the genitals. Warts can be treated by freezing with liquid nitrogen, electric cauterization (burning), laser vaporization, surgical excision, and some medicines such as salicylic acid.

Disorders Described Elsewhere

Abnormal skin coloration 136	Pemphigus vulgaris 63
Birthmarks 137	Polythelia and polymastia 142
Burns 146	Skin cancer 146
Keloids 100	

SKIN CANCER

Skin cancer is induced by the ultraviolet rays of the sun. It occurs most often on the head and neck, where exposure is greatest. It is most common in fair-skinned people and the elderly, who have had the longest lifetime UV exposure and have less melanin to shield the keratinocyte DNA from radiation. The popularity of sun tanning, however, has caused an alarming increase in skin cancer among younger people. While sunscreens protect against sunburn, there is no evidence that they afford protection from skin cancer. Skin cancer is one of the most common cancers, but it is also one of the easiest to treat and has one of the highest survival rates when it is detected and treated early.

There are three types of skin cancer named for the epidermal cells in which they originate: basal cell carcinoma, squamous cell carcinoma, and malignant melanoma. The three types are also distinguished from each other by the appearance of their **lesions**[33] (zones of tissue injury).

Basal cell carcinoma[34] is the most common type, but it is also the least dangerous because it seldom metastasizes. It arises from cells of the stratum basale and eventually invades the dermis. On the surface, the lesion first appears as a small, shiny bump. As the bump enlarges, it often develops a central depression and a beaded "pearly" edge (fig. 5.13*a*).

Squamous cell carcinoma arises from keratinocytes of the stratum spinosum. The lesions have a raised, reddened, scaly appearance, later forming a concave ulcer with raised edges (fig. 5.13*b*). The chance of recovery is good with early detection and surgical removal, but if it goes unnoticed or is neglected, this cancer tends to metastasize to the lymph nodes and can be lethal.

Malignant melanoma is the most deadly skin cancer but accounts for only 5% of all cases. It often arises from the melanocytes of a preexisting mole. It metastasizes quickly and is often fatal if not treated immediately. The risk for malignant melanoma is greatest in people who experienced severe sunburns as children, especially redheads. It is important to distinguish a mole from malignant melanoma. A mole usually has a uniform color and even contour, and it is no larger in diameter than the end of a pencil eraser (about 6 mm). If it becomes malignant, however, it forms a large, flat, spreading lesion with a scalloped border (fig. 5.13*c*). The American Cancer Society suggests an "ABCD rule" for recognizing malignant melanoma: *A* for asymmetry (one side of the lesion looks different from the other); *B* for border irregularity (the contour is not uniform but wavy or scalloped); *C* for color (often a mixture of brown, black, tan, and sometimes red and blue); and *D* for diameter (greater than 6 mm).

Skin cancer is treated by surgical excision, radiation therapy, or destruction of the lesion by heat (electrodesiccation) or cold (cryosurgery).

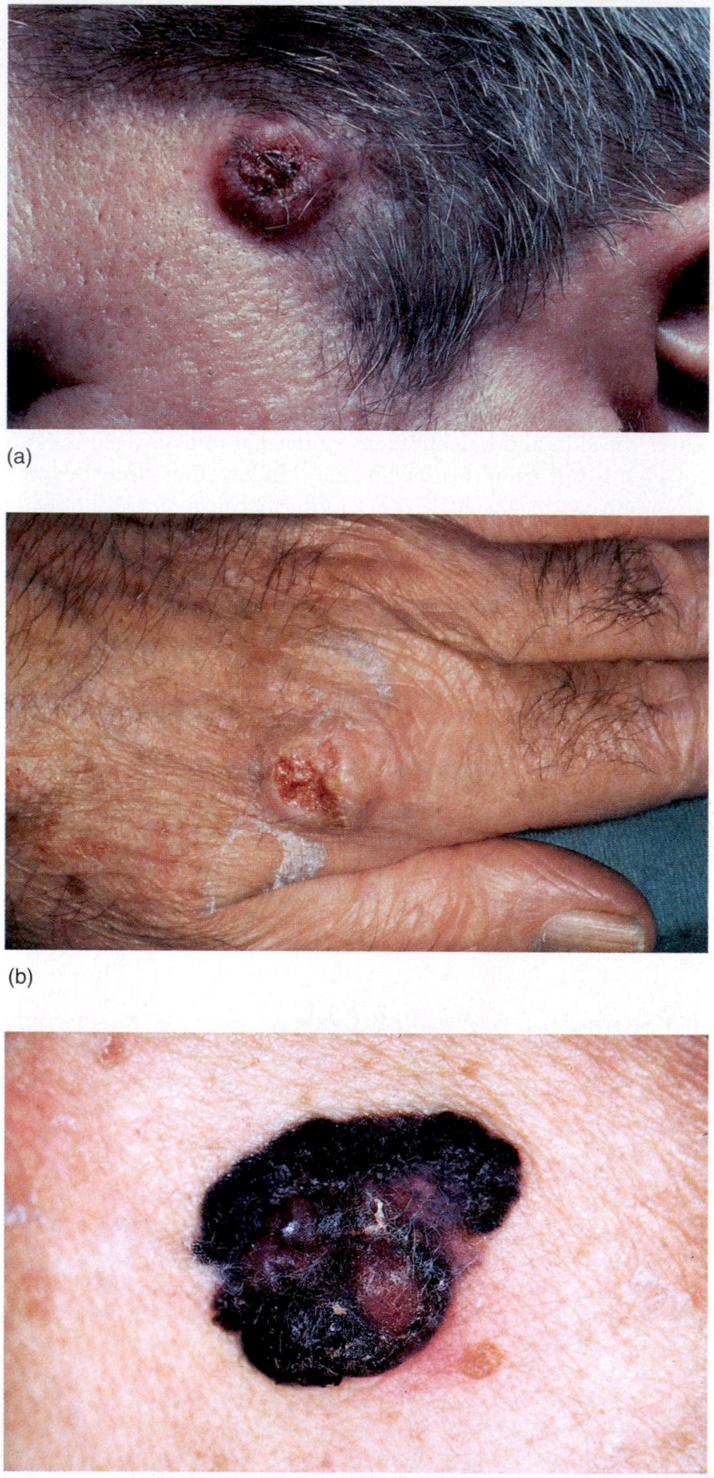

(a)

(b)

(c)

FIGURE 5.13

Skin Cancer Lesions. (*a*) Basal cell carcinoma. (*b*) Squamous cell carcinoma. (*c*) Malignant melanoma.

BURNS

Burns are the leading cause of accidental death. They are usually caused by UV radiation, fires, kitchen spills, or excessively hot bath water, but they also can be caused by other forms of radiation, strong acids and bases, or electrical shock. Burn deaths result primarily from fluid loss, infection, and the toxic effects of **eschar** (ESS-car)—the burned, dead tissue.

[33] *lesio* = injure
[34] *carcin* = cancer + *oma* = tumor

Burns are classified according to the depth of tissue involvement (fig. 5.14). **First-degree burns** involve only the epidermis and are marked by redness, slight edema, and pain. They heal in a few days and seldom leave scars. Most sunburns are first-degree burns.

Second-degree burns involve the epidermis and part of the dermis but leave at least some of the dermis intact. First- and second-degree burns are therefore also known as **partial-thickness burns.** A second-degree burn may be red, tan, or white and is blistered and very painful. It may take from 2 weeks to several months to heal and may leave scars. The epidermis regenerates by division of epithelial cells in the hair follicles and sweat glands and those around the edges of the lesion. Some sunburns and many scalds are second-degree burns.

Third-degree burns are also called **full-thickness burns** because the epidermis and dermis are completely destroyed. Sometimes even deeper tissue is damaged (hypodermis, muscle, and bone). Since no dermis remains, the skin can regenerate only from the edges of the wound. Third-degree burns often require skin grafts (see insight 5.4). If a third-degree burn is left to itself to heal, contracture (abnormal connective tissue fibrosis) and severe disfigurement may result.

Think About It!

A third-degree burn may be surrounded by painful areas of first- and second-degree burns, but the region of the third-degree burn is painless. Explain the reason for this lack of pain.

Before You Go On

Answer the following questions to test your understanding of the preceding section:

16. What types of cells are involved in each type of skin cancer?
17. Which type of skin cancer is most dangerous? What are its early warning signs?
18. What is the difference between a first-, second-, and third-degree burn?
19. What are the two most urgent priorities in treating a burn victim? How are these needs dealt with?

INSIGHT 5.4 CLINICAL APPLICATION

Skin Grafts and Artificial Skin

Third-degree burns leave no dermis to regenerate what was lost, and therefore require skin grafts. Ideally, these should come from elsewhere on the same patient's body *(autografts)* so there is no problem with immune rejection, but this may not be feasible in patients with extensive burns. A skin graft from another person *(allograft)* or even skin from another species *(xenograft)*, such as pig skin, may be used, but they present problems with immune rejection. At least two bioengineering companies produce artificial skin as a temporary burn covering. One such product is made by culturing fibroblasts on a collagen gel to produce a dermis, then culturing keratinocytes on this substrate to produce an epidermis. This is being used to treat not only burn patients but also patients with leg and foot ulcers caused by diabetes mellitus.

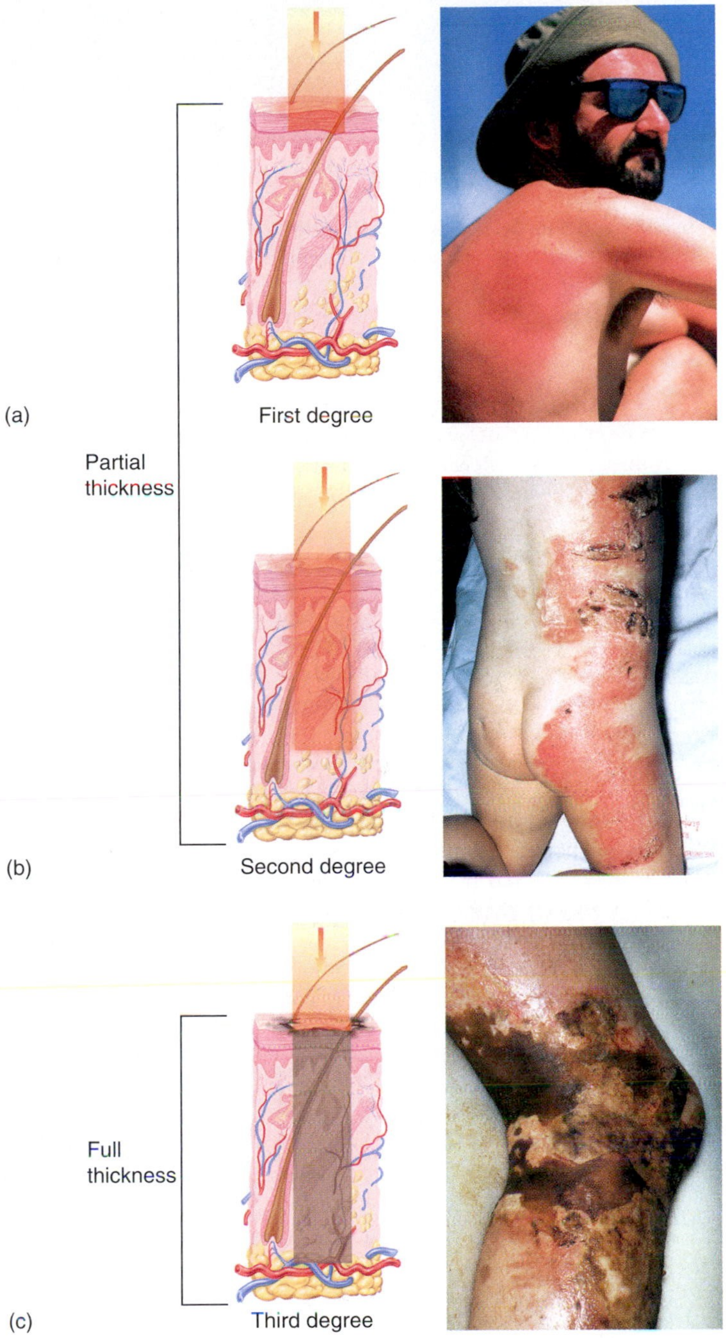
(a)
First degree
Partial thickness
(b)
Second degree
Full thickness
(c)
Third degree

FIGURE 5.14

Three Degrees of Burns. *(a)* First-degree burn, involving only the epidermis. *(b)* Second-degree burn, involving the epidermis and part of the dermis. *(c)* Third-degree burn, extending through the entire dermis and often involving even deeper tissue.

CHAPTER REVIEW

Review of Key Concepts

The Skin and Subcutaneous Tissue (p. 130)

1. *Dermatology* is the study of the *integumentary system,* a system that includes the skin *(integument),* hair, nails, and cutaneous glands.
2. The skin is composed of a superficial *epidermis* of keratinized stratified squamous epithelium, and a deeper *dermis* of fibrous connective tissue. Beneath the skin is a connective tissue *hypodermis.*
3. Skin ranges from less than 0.5 mm to 6 mm thick. *Thick skin* is named for a thick, heavily keratinized epidermis, not necessarily for total thickness; it is found on the palms, soles, and corresponding surfaces of the digits, and it is hairless. The rest of the body is covered with *thin skin,* in which the epidermis is more lightly keratinized.
4. Functions of the skin include resistance to trauma and infection, water retention, vitamin D synthesis, sensation, thermoregulation, and nonverbal communication.
5. The epidermis has five types of cells: *stem cells, keratinocytes, melanocytes, tactile cells,* and *dendritic cells.*
6. Layers of the epidermis from deep to superficial are the *stratum basale, stratum spinosum, stratum granulosum, stratum lucidum* (in thick skin only), and *stratum corneum.*
7. Keratinocytes are the majority of epidermal cells. They originate by mitosis of stem cells in the stratum basale and push the older keratinocytes upward. Keratinocytes flatten and produce *membrane-coating vesicles* and cytoskeletal filaments as they migrate upward. In the stratum granulosum, the cytoskeletal filaments are transformed to keratin, the membrane-coating vesicles release lipids that help to render the cells water-resistant, and the cells undergo apoptosis. Above the stratum granulosum, dead keratinocytes become compacted into the stratum corneum. Thirty to 40 days after its mitotic birth, the average keratinocyte flakes off the epidermal surface. This loss of the dead cells is called *exfoliation.*
8. The dermis is 0.2 to 4 mm thick. It is composed mainly of collagen but includes elastic and reticular fibers, fibroblasts, and other cell types. It contains blood vessels, sweat glands, sebaceous glands, nerve endings, hair follicles, nail roots, smooth muscle, and in the face, skeletal muscle.
9. In most places, upward projections of the dermis called *dermal papillae* interdigitate with downward *epidermal ridges* to form a wavy boundary. The papillae form the friction ridges of the fingertips and irregular ridges, separated by furrows, elsewhere.
10. The dermis is composed of a superficial *papillary layer,* which is composed of areolar tissue and forms the dermal papillae; and a thicker, deeper *reticular layer* composed of dense irregular connective tissue. The reticular layer provides toughness to the dermis, while the papillary layer forms an arena for the mobilization of defenses against pathogens that breach the epidermis.
11. The *hypodermis (subcutaneous tissue)* is composed of more areolar and adipose tissue than the reticular layer of dermis. It pads the body and binds the skin to underlying muscle or other tissues. In areas composed mainly of adipocytes, it is called *subcutaneous fat.*
12. Normal skin colors result from various proportions of *eumelanin, pheomelanin,* the hemoglobin of the blood, the white collagen of the dermis, and dietary *carotene.* Pathological conditions with abnormal skin coloration include *cyanosis, erythema, pallor, albinism, jaundice, bronzing, hematomas,* and *hemangiomas (birthmarks).*
13. Skin markings include *friction ridges* of the fingertips (the source of oily fingerprints), *flexion creases* of the palms, *flexion lines* of the wrist and other places, *freckles,* and *moles.*

Hair and Nails (p. 137)

1. Hair and nails are composed of compact, highly cross-linked *hard keratin.*
2. A hair *(pilus)* is a slender filament of keratinized cells growing from an oblique *hair follicle.*
3. The three types of hair are *lanugo,* present only prenatally; *vellus,* a fine unpigmented body hair; and the coarser, pigmented *terminal hair* of the eyebrows, scalp, beard, and other areas.
4. Deep in the follicle, a hair begins with a dilated *bulb,* continues as a narrower *root* below the skin surface, and extends above the skin as the *shaft.* The bulb contains a *papilla* of vascularized connective tissue. The *hair matrix* just above the papilla is the site of hair growth by mitosis of the matrix cells. In cross section, a hair exhibits a thin outer *cuticle,* a thicker layer of keratinized cells forming the hair *cortex,* and a core called the *medulla.*
5. Differences in hair texture are attributable to differences in cross-sectional shape—straight hair is round, wavy hair is oval, and tightly curly hair is relatively flat.
6. Variations in hair color arise from the relative amounts of eumelanin and pheomelanin.
7. A hair follicle consists of an inner *epithelial root sheath* (an extension of the epidermis) and an outer *connective tissue root sheath.* It is supplied by nerve endings called *hair receptors* that detect hair movements, and a bundle of smooth muscle called the *piloerector muscle,* which erects the hair.
8. Table 5.2 lists the functions of hair of various types and locations, including thermal insulation, protection from the sun and from foreign objects, sensation, facial expression, signaling sexual maturity, and regulating the dispersal of pheromones.
9. Fingernails and toenails are hard plates of densely packed, dead, keratinized cells. They arise from a growth zone called the *nail matrix.*

Cutaneous Glands (p. 140)

1. The most abundant and widespread sweat glands are *merocrine sweat glands,* which produce a watery secretion that cools the body. Merocrine glands release their product by exocytosis.
2. *Apocrine sweat glands* are associated with hair follicles in the groin, anal region, axilla, areola, and beard. They develop at puberty along with the appearance of hair in these regions, and apparently function to secrete sex pheromones. Apocrine sweat glands also release their secretion by exocytosis.
3. *Sebaceous glands,* also usually associated with hair follicles, produce an oily secretion called *sebum,* which keeps the skin and hair pliable. These are holocrine glands; their cells break down in entirety to form the secretion.
4. *Ceruminous glands* are found in the auditory canal. *Cerumen,* or earwax, is a mixture of ceruminous gland secretion, sebum, and dead epidermal cells. It keeps the eardrum pliable, waterproofs the auditory canal, and kills bacteria.

5. *Mammary glands* are modified apocrine sweat glands that develop in the breasts during pregnancy and lactation, and produce milk.

Developmental and Clinical Perspectives (p. 143)

1. The epidermis develops from ectoderm through a process that involves the formation of a superficial, temporary *periderm,* then an *intermediate layer* of cells that differentiate into keratinocytes, and then loss of the periderm. The original ectodermal layer becomes a *germinative layer* of stem cells while the intermediate layer gives rise to the stratum spinosum, granulosum, and corneum.
2. The dermis develops from mesoderm, which differentiates into embryonic mesenchyme. As mesenchymal cells produce collagenous and elastic fibers, the mesenchyme differentiates into mature fibrous connective tissue.
3. A hair follicle begins as an ectodermal thickening called the *hair germ,* which elongates into a *hair peg* with a dilated *hair bulb* at its lower end. A dermal papilla forms just below the hair bulb and then grows into its center. Ectodermal cells just above the papilla become the *germinal matrix,* where mitosis produces the cells of the *hair shaft.* The fetus develops a temporary hair called *lanugo,* which falls out before birth.
4. A fingernail or toenail begins as a ventral epidermal thickening which migrates to the dorsal side of the digit and forms a *primary nail field.* The germinal layer of the proximal nail fold becomes the *nail root.* Mitosis here produces the cells that become keratinized and densely compressed to form the hard nail plate.
5. Sebaceous glands bud from the sides of developing hair follicles. They produce *vernix caseosa* before birth, become dormant by the time of birth, and are reactivated at puberty.
6. Apocrine sweat glands also bud from the hair follicles, closer to the epidermis than do the sebaceous glands. They initially develop over most of the body, then degenerate except in limited areas such as the axillary and genital regions, and become active at puberty.
7. Merocrine sweat glands arise as cords of tissue that grow downward from the germinative layer of the epidermis. Cells in the center of the cord degenerate to produce the gland lumen, and cells at the lower end differentiate into secretory and myoepithelial cells.
8. Senescence of the integumentary system is marked by thinning and graying of the hair, dryness of the skin and hair due to atrophy of the sebaceous glands, and thinning and loss of elasticity in the skin. Aged skin is more vulnerable than younger skin to trauma and infection, and it heals more slowly. The loss of subcutaneous fat reduces the capability for thermoregulation. UV radiation accelerates the aging of the skin, promoting wrinkling, age spots, and skin cancer.
9. Skin cancer is of three forms distinguished by the cells of origin and the appearance of the lesions: *basal cell carcinoma, squamous cell carcinoma,* and *malignant melanoma.* Malignant melanoma is the least common form, but is the most dangerous because of its tendency to metastasize quickly.
10. Burns are classified as first-, second- and third-degree. *First-degree burns* involve epidermis only; *second-degree burns* extend through part of the dermis; and *third-degree burns* extend all the way through the dermis and often into deeper tissues.

Testing Your Recall

1. Cells of the ___ are keratinized and dead.
   a. papillary layer
   b. stratum spinosum
   c. stratum basale
   d. stratum corneum
   e. stratum granulosum
2. Which of the following terms is *least* related to the rest?
   a. subcutaneous fat
   b. superficial fascia
   c. reticular layer
   d. hypodermis
   e. subcutaneous tissue
3. Which of the following skin conditions or appearances would most likely result from liver failure?
   a. pallor
   b. erythema
   c. pemphigus vulgaris
   d. jaundice
   e. melanization
4. All of the following interfere with microbial invasion of the skin *except*
   a. the acid mantle.
   b. melanin.
   c. cerumen.
   d. keratin.
   e. sebum.
5. The hair on a 6-year-old's arms is
   a. vellus.
   b. lanugo.
   c. trichosiderin.
   d. terminal hair.
   e. rosacea.
6. Which of the following terms is *least* related to the rest?
   a. lunule
   b. nail plate
   c. hyponychium
   d. free edge
   e. cortex
7. Which of the following is a scent gland?
   a. an eccrine gland
   b. a sebaceous gland
   c. an apocrine gland
   d. a ceruminous gland
   e. a merocrine gland
8. _____ are skin cells with a sensory role.
   a. Tactile cells
   b. Dendritic cells
   c. Prickle cells
   d. Melanocytes
   e. Keratinocytes
9. The embryonic periderm becomes part of
   a. the vernix caseosa.
   b. the lanugo.
   c. the stratum corneum.
   d. the stratum basale.
   e. the dermis.
10. Which of the following skin cells alert the immune system to pathogens?
    a. fibroblasts
    b. melanocytes
    c. keratinocytes
    d. dendritic cells
    e. Merkel cells
11. Two common word roots that refer to the skin in medical terminology are _____ and _____.
12. A muscle that causes a hair to stand on end is called a/an _____.
13. The most abundant protein of the epidermis is _____, while the most abundant protein of the dermis is _____.

14. Blueness of the skin due to low oxygen concentration in the blood is called _____.
15. Projections of the dermis toward the epidermis are called _____.
16. Cerumen is more commonly known as _____.
17. The holocrine glands that secrete into a hair follicle are called _____.
18. The scaly outermost layer of a hair is called the _____.
19. A hair is nourished by blood vessels in a connective tissue projection called the _____.
20. A _____ burn destroys part of the dermis, but not all of it.

Answers in the Appendix

True or False

Determine which five of the following statements are false, and briefly explain why.

1. Dander consists of dead keratinocytes.
2. The term *integument* means only the skin, but *integumentary system* refers also to the hair, nails, and cutaneous glands.
3. The dermis is composed mainly of keratin.
4. Vitamin D is synthesized by certain cutaneous glands.
5. Cells of the stratum granulosum cannot undergo mitosis.
6. Dermal papillae are better developed in skin that is subject to a lot of mechanical stress than in skin that is subject to less stress.
7. The three layers of the skin are the epidermis, dermis, and hypodermis.
8. People of African descent have a much higher density of epidermal melanocytes than do people of northern European descent.
9. Pallor indicates a genetic lack of melanin.
10. Apocrine scent glands develop at the same time in life as the pubic and axillary hair.

Answers in the Appendix

Testing Your Comprehension

1. Many organs of the body contain numerous smaller organs, perhaps even thousands. Describe an example of this in the integumentary system.
2. Certain aspects of human form and function are easier to understand when viewed from the perspective of comparative anatomy and evolution. Discuss examples of this in the integumentary system.
3. Explain how the complementarity of form and function is reflected in the fact that the dermis has two histological layers and not just one.
4. Cold weather does not normally interfere with oxygen uptake by the blood, but it can cause cyanosis anyway. Why?
5. Why is it important for the epidermis to be effective, but not *too* effective, in screening out UV radiation?

Answers at the Online Learning Center

www.mhhe.com/saladinha1

Visit the Online Learning Center for practice tests, answer keys, and other learning aids for this chapter. Enhance your understanding of human anatomy with our interactive art labeling exercises, supplemental photo atlases, web links, puzzles, flashcards, and much more.

6

CHAPTER SIX

Bone Tissue

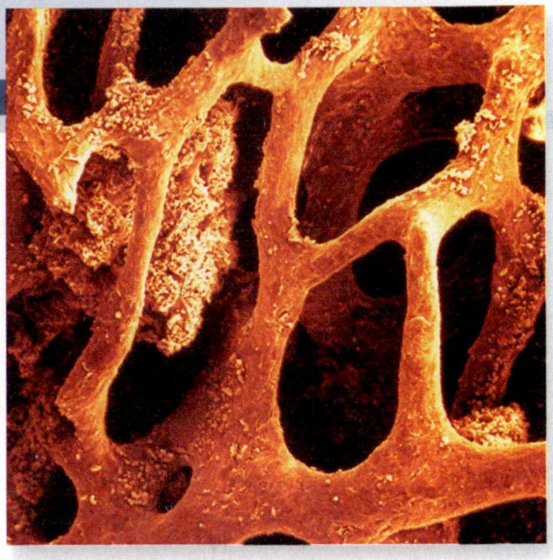

Spongy bone of the human femur

CHAPTER OUTLINE

Tissues and Organs of the Skeletal System 152
- Functions of the Skeleton 152
- Bones and Osseous Tissue 152
- The Shapes of Bones 153
- General Features of Bones 153

Histology of Osseous Tissue 155
- Cells 155
- Matrix 155
- Compact Bone 156
- Spongy Bone 158
- Bone Marrow 158

Bone Development 159
- Endochondral Ossification 159
- Intramembranous Ossification 161
- Bone Growth and Remodeling 161
- Nutritional and Hormonal Factors 163
- The Aging Skeletal System 163

Structural Disorders of Bone 164
- Fractures 164
- Osteoporosis 165
- Other Structural Disorders 167

Chapter Review 168

INSIGHTS

6.1 Medical History: Radioactivity and Bone Cancer 152

6.2 Clinical Application: Polymers, Ceramics, and Bones 156

6.3 Clinical Application: Achondroplastic Dwarfism 163

6.4 Clinical Application: When *Not* to Eat Your Spinach 165

BRUSHING UP

To understand this chapter, you may find it helpful to review the following concepts:

- Hyaline Cartilage (p. 92)
- Introduction to Bone Histology (p. 92)

In art and history, nothing has symbolized death so much as a skull or skeleton.[1] Bones and teeth are the most durable remains of a once-living body and the most vivid reminder of the impermanence of life. The dry bones presented for laboratory study may wrongly suggest that the skeleton is a nonliving scaffold for the body, like the steel girders of a building. Seeing it in such a sanitized form makes it easy to forget that the living skeleton is made of dynamic tissues, full of cells—that it continually remodels itself and interacts physiologically with all of the other organ systems. The skeleton is permeated with nerves and blood vessels, evidence of its sensitivity and metabolic activity.

Osteology,[2] the study of bone, is the subject of chapters 6 through 8. In this chapter, we study bone as a tissue—its composition, development, and growth. This will provide a basis for understanding the skeleton, joints, and muscles in the chapters that follow.

INSIGHT 6.1 MEDICAL HISTORY

Radioactivity and Bone Cancer

Radioactivity captured the public imagination when Marie and Pierre Curie and Henri Becquerel shared the 1903 Nobel Prize for its discovery. Not for several decades, however, did anyone realize its dangers. Factories employed women to paint luminous numbers on watch and clock dials with radium paint. The women moistened the paint brushes with their tongues to keep them finely pointed and ingested radium in the process. Their bones readily absorbed the radium, and many of the women developed *osteosarcoma,* the most common and deadly form of bone cancer.

Even more horrific, in the wisdom of hindsight, was a deadly health fad in which people drank "tonics" made of radium-enriched water. One famous enthusiast was the champion golfer and millionaire playboy Eben Byers, who drank several bottles of radium tonic each day and praised its virtues as a wonder drug and aphrodisiac. Like the factory women, Byers contracted osteosarcoma. By the time of his death, holes had formed in his skull and doctors had removed his entire upper jaw and most of his mandible in an effort to halt the spreading cancer. Byers's bones and teeth were so radioactive they could expose photographic film in complete darkness. Brain damage left him unable to speak, but he remained mentally alert to the bitter end. His tragic decline and death in 1932 shocked the world and put an end to the radium tonic fad.

Tissues and Organs of the Skeletal System

Objectives

When you have completed this section, you should be able to

- name the tissues and organs that compose the skeletal system;
- state several functions of the skeletal system;
- distinguish between bone as a tissue and as an organ.
- describe how bones are classified by shape; and
- describe the general features of a long bone.

The **skeletal system** is composed of bones, cartilages, and ligaments tightly joined to form a strong, flexible framework for the body. Cartilage, the embryonic forerunner of most bones, covers many joint surfaces in the mature skeleton. Ligaments hold bones together at the joints and are discussed in chapter 9. Tendons are structurally similar to ligaments but attach muscles to bones; they are discussed with the muscular system in chapters 11 and 12.

Functions of the Skeleton

The skeleton obviously provides the body with physical support, but it plays many other roles that go unnoticed by most people. Its functions include:

- **Support.** Bones of the legs, pelvis, and vertebral column hold up the body; the jaw bones support the teeth; and nearly all bones provide support for muscles.
- **Movement.** Skeletal muscles would serve little purpose if not for their attachment to the bones and ability to move them.
- **Protection.** Bones enclose and protect such delicate organs and tissues as the brain, spinal cord, lungs, heart, pelvic viscera, and bone marrow.
- **Blood formation.** Red bone marrow is the major producer of blood cells, including most cells of the immune system.
- **Electrolyte balance.** The skeleton is the body's main mineral reservoir. It stores calcium and phosphate and releases them when needed for other purposes.
- **Acid-base balance.** Bone buffers the blood against excessive pH changes by absorbing or releasing alkaline salts such as calcium phosphate.
- **Detoxification.** Bone tissue removes heavy metals and other foreign elements from the blood and thus reduces their toxic effects on other tissues. It can later release these contaminants more slowly for excretion. The tendency of bone to absorb foreign elements can, however, have terrible consequences (see insight 6.1).

Bones and Osseous Tissue

Bone, or **osseous**[3] **tissue,** is a connective tissue in which the matrix is hardened by the deposition of calcium phosphate and other minerals. The hardening process is called **mineralization** or **calcification.** Osseous tissue, however, is only one of the tissues that make up a bone. Also present are blood, bone marrow, cartilage, adipose tissue, nervous tissue, and fibrous connective tissue. The word *bone* can denote an organ composed of all these tissues, or it can denote just the osseous tissue.

[1]*skelet* = dried up
[2]*osteo* = bone + *logy* = study of
[3]*os, osse, oste* = bone

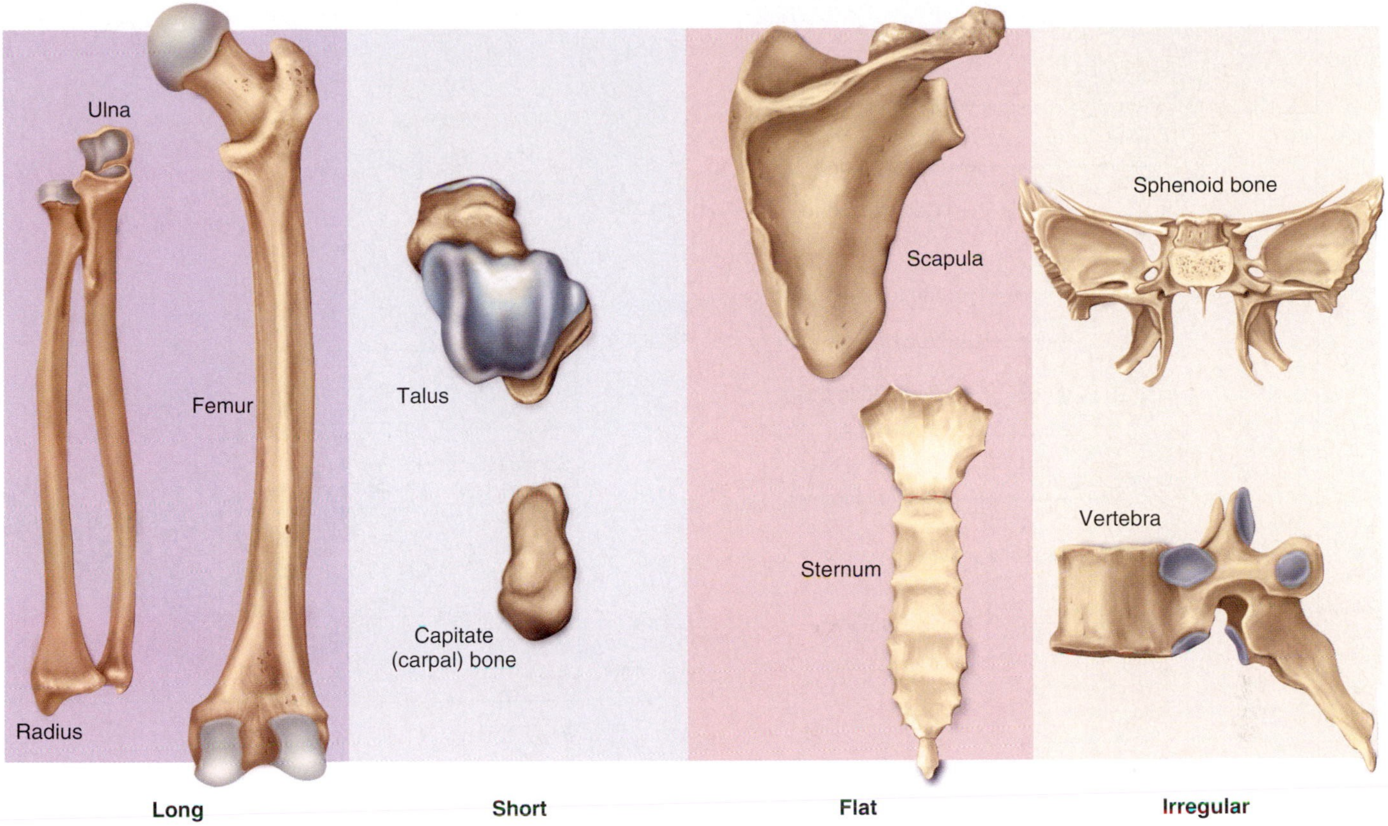
Ulna
Femur
Radius
Talus
Capitate
(carpal) bone
Scapula
Sternum
Sphenoid bone
Vertebra
Long
Short
Flat
Irregular

FIGURE 6.1
Classification of Bones by Shape.

The Shapes of Bones

Bones are classified into four groups according to their shapes and corresponding functions (fig. 6.1):

1. **Long bones** are roughly cylindrical in shape and significantly longer than wide. Like crowbars, they serve as rigid levers that are acted upon by the skeletal muscles to produce body movements. Long bones include the humerus of the arm, the radius and ulna of the forearm, the metacarpals and phalanges of the hand, the femur of the thigh, the tibia and fibula of the leg, and the metatarsals and phalanges of the feet.
2. **Short bones** are more nearly equal in length and width. They include the carpal (wrist) and tarsal (ankle) bones. They have limited motion and merely glide across one another, enabling the ankles and wrists to bend in multiple directions.
3. **Flat bones** enclose and protect soft organs and provide broad surfaces for muscle attachment. They include most cranial bones, the ribs, the sternum (breastbone), the scapula (shoulder blade), and the ossa coxae (hipbones).
4. **Irregular bones** have elaborate shapes that do not fit into any of the preceding categories. They include the vertebrae and some skull bones, such as the sphenoid and ethmoid bones.

General Features of Bones

Bones have an outer shell of dense white osseous tissue called **compact (dense) bone,** usually enclosing a more loosely organized form of osseous tissue called **spongy (cancellous) bone.** The skeleton is about three-quarters compact bone and one-quarter spongy bone by weight. Compact and spongy bone are described later in more detail.

Figure 6.2*a* shows a longitudinal section through a long bone. The principal features of a long bone are its shaft, called the **diaphysis**[4] (dy-AF-ih-sis), and an expanded head at each end, called the **epiphysis**[5] (eh-PIF-ih-sis). The diaphysis consists largely of a cylinder of compact bone enclosing a space called the **medullary**[6] (MED-you-lerr-ee) **cavity.** The epiphysis is filled with spongy bone. Bone marrow occupies the medullary cavity and the spaces amid the spongy bone of the epiphysis. The diaphysis of a long bone provides leverage, while the epiphysis is enlarged to strengthen the joint and provide added surface area for the attachment of tendons and ligaments.

In children and adolescents, an **epiphyseal** (EP-ih-FIZZ-ee-ul) **plate** of hyaline cartilage separates the marrow spaces of the epiphysis and diaphysis (see fig. 6.2*a*). On X rays, it appears as a transparent

[4]*dia* = across + *physis* = growth; originally named for a ridge on the shaft of the tibia
[5]*epi* = upon, above + *physis* = growth
[6]*medulla* = marrow

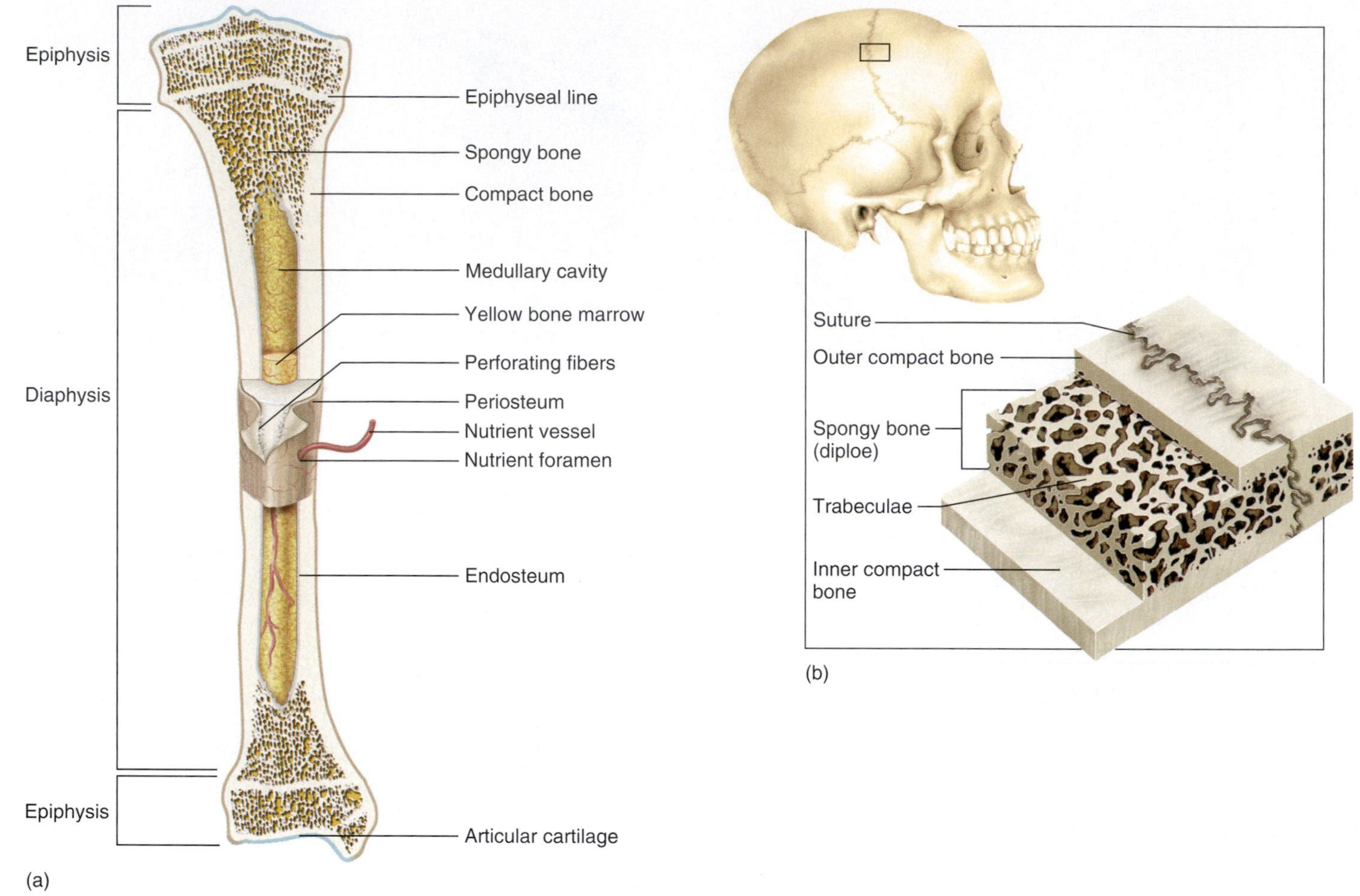
Epiphysis
Diaphysis
Epiphysis
Epiphyseal line
Spongy bone
Compact bone
Medullary cavity
Yellow bone marrow
Perforating fibers
Periosteum
Nutrient vessel
Nutrient foramen
Endosteum
Articular cartilage
(a)
Suture
Outer compact bone
Spongy bone (diploe)
Trabeculae
Inner compact bone
(b)

FIGURE 6.2
General Anatomy of Long and Flat Bones. (*a*) A long bone, the tibia. (*b*) Two flat bones of the cranium, joined at a suture.

line at the end of a long bone (see fig. 6.9). The epiphyseal plate is a zone where the bones grow in length. In adults, the epiphyseal plate is depleted and the bones can grow no longer, but an *epiphyseal line* on the bone surface marks the former location of the plate.

Externally, most of the bone is covered with a sheath called the **periosteum.**[7] This has a tough, outer *fibrous layer* of collagen and an inner *osteogenic layer* of bone-forming cells. Some collagen fibers of the outer layer are continuous with the tendons that bind muscle to bone, and some penetrate into the bone matrix as **perforating (Sharpey**[8]**) fibers.** The periosteum thus provides strong attachment and continuity from muscle to tendon to bone. The osteogenic layer is important to the growth of bone and healing of fractures. Blood vessels of the periosteum penetrate into the bone through minute holes called **nutrient foramina** (for-AM-ih-nuh); we will trace where they go when we consider bone histology. The internal surface of a bone is lined with **endosteum,**[9] a thin layer of reticular connective tissue and cells that deposit and dissolve osseous tissue.

At most joints, the ends of the adjoining bones have no periosteum but rather a thin layer of hyaline cartilage, the **articular**[10] **cartilage.** Together with a lubricating fluid secreted between the bones, this cartilage enables a joint to move far more easily than it would if one bone rubbed directly against the other.

Flat bones have a sandwichlike construction, with two layers of compact bone enclosing a middle layer of spongy bone (fig. 6.2*b*). In the skull, the spongy layer is called the **diploe**[11] (DIP-lo-ee). A moderate blow to the skull can fracture the outer layer of compact bone, but the diploe can sometimes absorb the impact and leave the inner layer of compact bone unharmed.

Before You Go On

Answer the following questions to test your understanding of the preceding section:

1. Name five tissues found in a bone.

[7] *peri* = around + *oste* = bone
[8] William Sharpey (1802–80), Scottish histologist
[9] *endo* = within + *oste* = bone
[10] *artic* = joint
[11] *diplo* = double

2. List three or more functions of the skeletal system other than supporting the body and protecting some of the internal organs.
3. Name the four bone shapes and give an example of each.
4. Explain the difference between compact and spongy bone, and describe their spatial relationship to each other.
5. State the anatomical terms for the shaft, head, growth zone, and fibrous covering of a long bone.

Histology of Osseous Tissue

Objectives

When you have completed this section, you should be able to

- list and describe the cells, fibers, and ground substance of bone tissue;
- state the functional importance of each constituent of bone tissue;
- compare the histology of the two types of bone tissue; and
- distinguish between two types of bone marrow.

Cells

Like any other connective tissue, bone consists of cells, fibers, and ground substance. There are four types of bone cells (fig. 6.3):

1. **Osteogenic**[12] **(osteoprogenitor) cells** are stem cells found in the endosteum, the inner layer of the periosteum, and within the central canals of the osteons. They arise from embryonic fibroblasts. Osteogenic cells multiply continually, and some of them differentiate into the *osteoblasts* described next.
2. **Osteoblasts**[13] are bone-forming cells that synthesize the organic matter of the matrix and help to mineralize the bone. They line up in rows in the endosteum and inner layer of periosteum, and resemble a cuboidal epithelium on the bone surface (see fig. 6.11). Osteoblasts are nonmitotic, so the only source of new osteoblasts is mitosis and differentiation of the osteogenic cells. Stress and fractures stimulate accelerated mitosis of osteogenic cells, and therefore a rapid rise in the number of osteoblasts.
3. **Osteocytes** are former osteoblasts that have become trapped in the matrix they deposited. They live in tiny cavities called **lacunae,**[14] which are connected to each other by slender channels called **canaliculi**[15] (CAN-uh-LIC-you-lye). Each osteocyte has delicate cytoplasmic processes that reach into the canaliculi to meet the processes of neighboring osteocytes. The processes of neighboring osteocytes are joined by gap junctions, which allow osteocytes to pass nutrients and chemical signals to each other and to transfer wastes to the nearest blood vessels for disposal. Osteocytes also communicate by gap junctions with the osteoblasts on the bone surface. Osteocytes play no major role in depositing or resorbing bone. Rather, they are strain detectors. When they detect strain in a bone, they communicate with osteoblasts at the surface. The osteoblasts then deposit bone where needed—for example, building up bone in response to weight-bearing exercise. Osteoblasts also chemically signal *osteoclasts* to remove bone elsewhere.
4. **Osteoclasts**[16] are bone-dissolving macrophages found on bone surfaces. They develop from the same marrow cells that produce monocytes of the blood. Several of these marrow cells fuse with each other to form an osteoclast; thus, osteoclasts are unusually large (up to 150 μm in diameter) and typically have 3 or 4 nuclei, but sometimes up to 50. The side of the osteoclast facing the bone has a *ruffled border* with many deep infoldings of the plasma membrane, increasing its surface area. Hydrogen pumps in the ruffled border secrete hydrogen ions (H^+) into the extracellular fluid, and chloride ions (Cl^-) follow by electrical attraction; thus, the space between the osteoclast and the bone becomes filled with hydrochloric acid (HCl) with a pH of about 4. HCl dissolves the minerals of the adjacent bone, then lysosomes of the osteoclast release enzymes that digest the organic component. Osteoclasts often reside in little pits called *resorption bays (Howship*[17] *lacunae)* that they have etched into the bone surface.

Think About It!

Considering the function of osteoblasts, what organelles do you think are especially abundant in their cytoplasm?

Matrix

The matrix of osseous tissue is, by dry weight, about one-third organic and two-thirds inorganic matter. The organic matter includes collagen and various large protein-carbohydrate complexes called glycosaminoglycans, proteoglycans, and glycoproteins. The inorganic matter is about 85% **hydroxyapatite,** a crystallized calcium phosphate salt [$Ca_{10}(PO_4)_6(OH)_2$], 10% calcium carbonate ($CaCO_3$), and lesser amounts of magnesium, sodium, potassium, fluoride, sulfate, carbonate, and hydroxide ions.

The collagen and minerals form a composite that gives bones a combination of flexibility and strength similar to fiberglass (see insight 6.2). The minerals resist compression (crumbling or sagging when weight is applied). When bones are deficient in calcium salts, they become soft and bend easily. This is the central problem in a childhood disease called **rickets.** Rickets occurs when a child is deficient in vitamin D and therefore cannot absorb enough dietary calcium. For lack of calcium, the bones are soft and the legs become bowed outward by the weight of the body.

[12]*osteo* = bone + *genic* = producing
[13]*osteo* = bone + *blast* = form, produce
[14]*lac* = lake, hollow + *una* = little
[15]*canal* = canal, channel + *icul* = little
[16]*osteo* = bone + *clast* = destroy, break down
[17]J. Howship (1781–1841), English surgeon

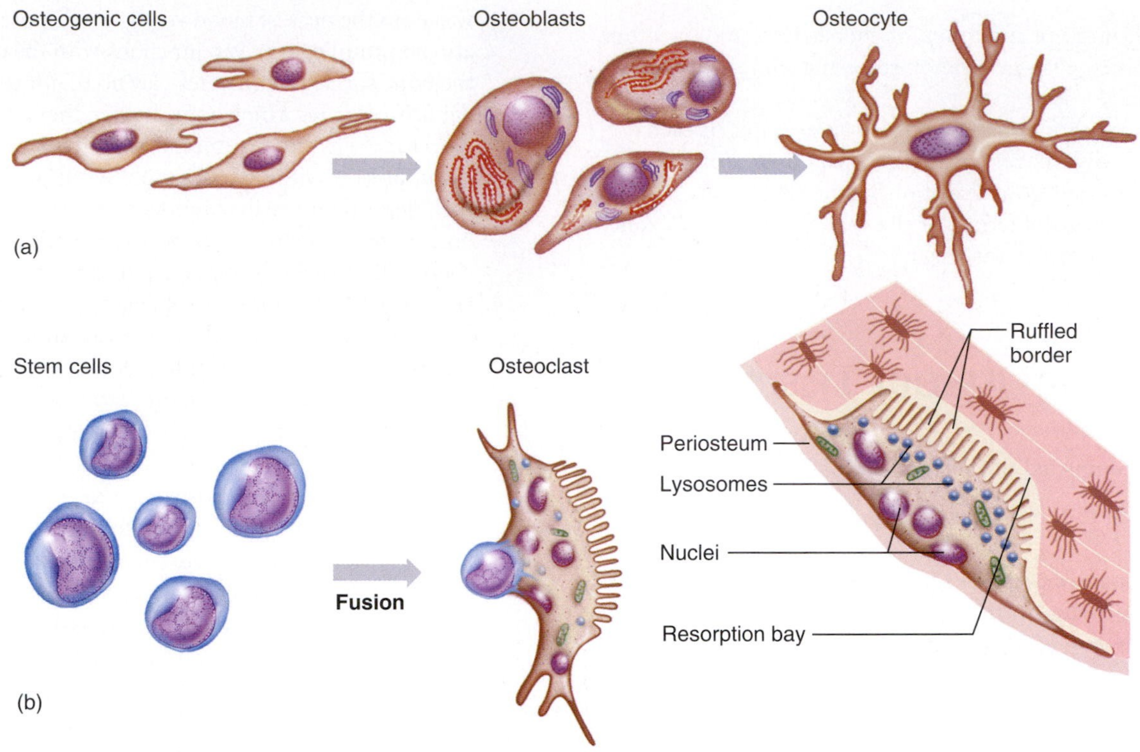
Osteogenic cells
Osteoblasts
Osteocyte
(a)
Stem cells
Osteoclast
Ruffled border
Periosteum
Lysosomes
Nuclei
Resorption bay
Fusion
(b)

Figure 6.3

Bone Cells and Their Development. (*a*) Osteogenic cells give rise to osteoblasts, which deposit matrix around themselves and transform into osteocytes. (*b*) Bone marrow cells fuse to form osteoclasts.

The collagen fibers of bone give it the ability to resist tension, so that the bone can bend slightly without snapping. Without collagen, the bones become very brittle, as in brittle bone disease (see insight 3.3, p. 93). Without collagen, a jogger's bones would shatter under the impact of running.

INSIGHT 6.2 CLINICAL APPLICATION

Polymers, Ceramics, and Bones

The physical properties of bone can be understood by analogy to some principles of engineering. Engineers use four kinds of construction materials: metals, ceramics (stone, glass, cement), polymers (rubber, plastic, cellulose), and composites (mixtures of two or more of the other classes). Bone is a composite of polymer (protein) and ceramic (mineral). The protein gives it flexibility and resistance to tension, while the mineral gives it resistance to compression. Owing to the mineral component, bone can support the weight of the body without sagging, and owing to the protein component, it can bend a little when subjected to stress.

Bone is somewhat like a fiberglass fishing rod, which is made of a ceramic (glass fibers) embedded in a polymer (resin). The polymer alone would be too flexible and limp to serve the purpose of a fishing rod, while the ceramic alone would be too brittle and would easily break. The combination of the two, however, gives the rod strength and flexibility. Unlike fiberglass, however, the ratio of ceramic to polymer in a bone varies from one location to another, adapting osseous tissue to different amounts of tension and compression exerted on different parts of the skeleton.

Compact Bone

The histological study of compact bone usually uses slices that have been dried, cut with a saw, and ground to translucent thinness. This procedure destroys the cells and much of the other organic content but reveals fine details of the inorganic matrix (fig. 6.4*a*). Such sections show onionlike **concentric lamellae**—layers of matrix concentrically arranged around a **central (haversian**[18] or **osteonic) canal.** A central canal and its lamellae constitute an **osteon (haversian system)**—the basic structural unit of compact bone. Along their length, central canals are joined by transverse or diagonal passages. The central canals contain blood vessels and nerves. Lacunae lie between adjacent layers of matrix and are connected with each other by canaliculi. Canaliculi of the innermost lacunae open into the central canal.

In longitudinal views and three-dimensional reconstructions, we find that an osteon is a cylinder of tissue surrounding a central canal. In each lamella, the collagen fibers are laid down in a helical pattern like the threads of a screw. In areas where the bone must resist tension (bending), the helix is loosely coiled like the threads on a wood screw and the fibers are more nearly longitudinal. In weight-bearing areas, where the bone must resist compression, the helix is more tightly coiled like the closely spaced threads on a bolt and the fibers are more nearly transverse. Often, the helices coil in one direction in one lamella and in the opposite

[18]Clopton Havers (1650–1702), English anatomist

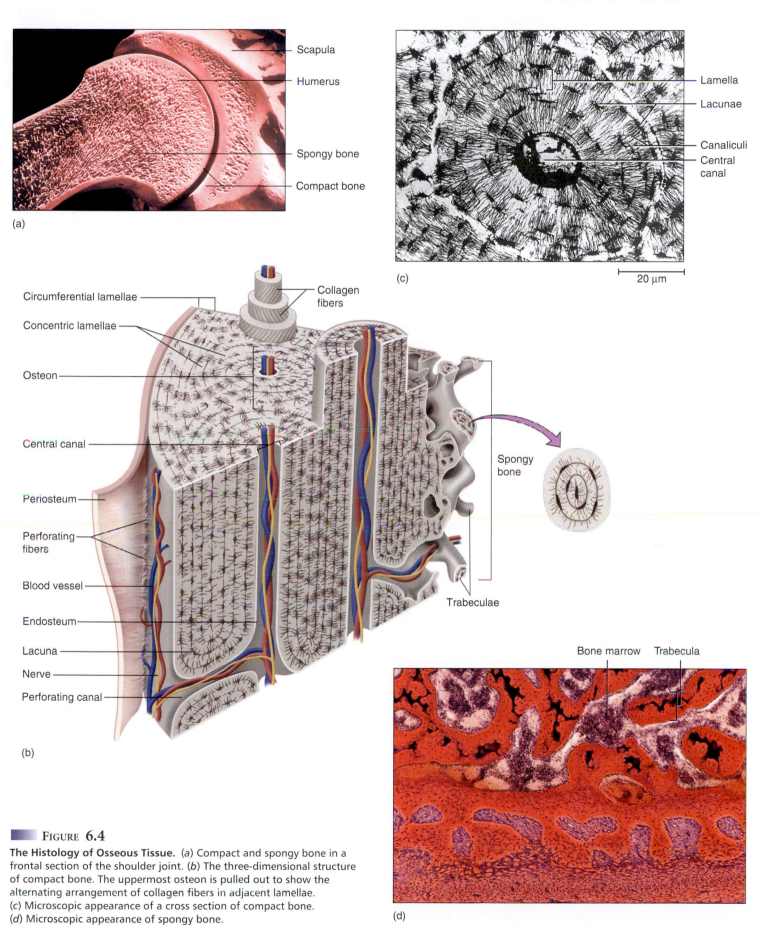

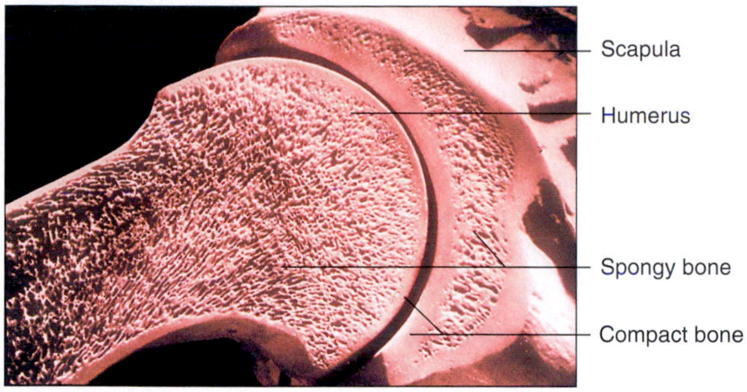
Scapula
Humerus
Spongy bone
Compact bone

(a)

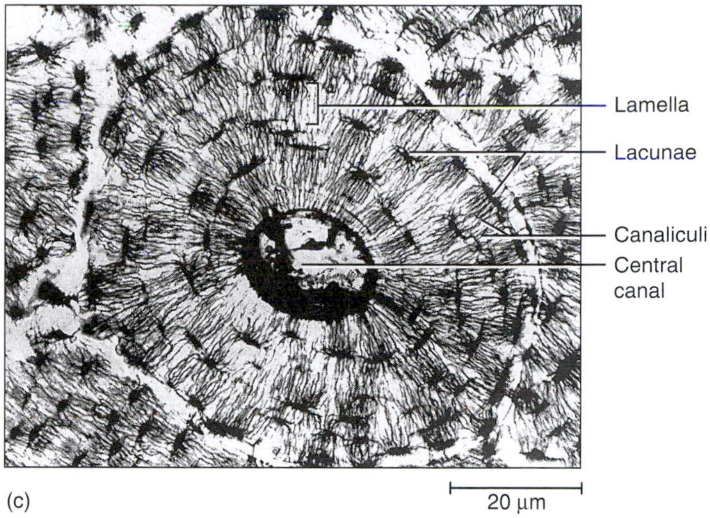
Lamella
Lacunae
Canaliculi
Central canal
20 µm

(c)

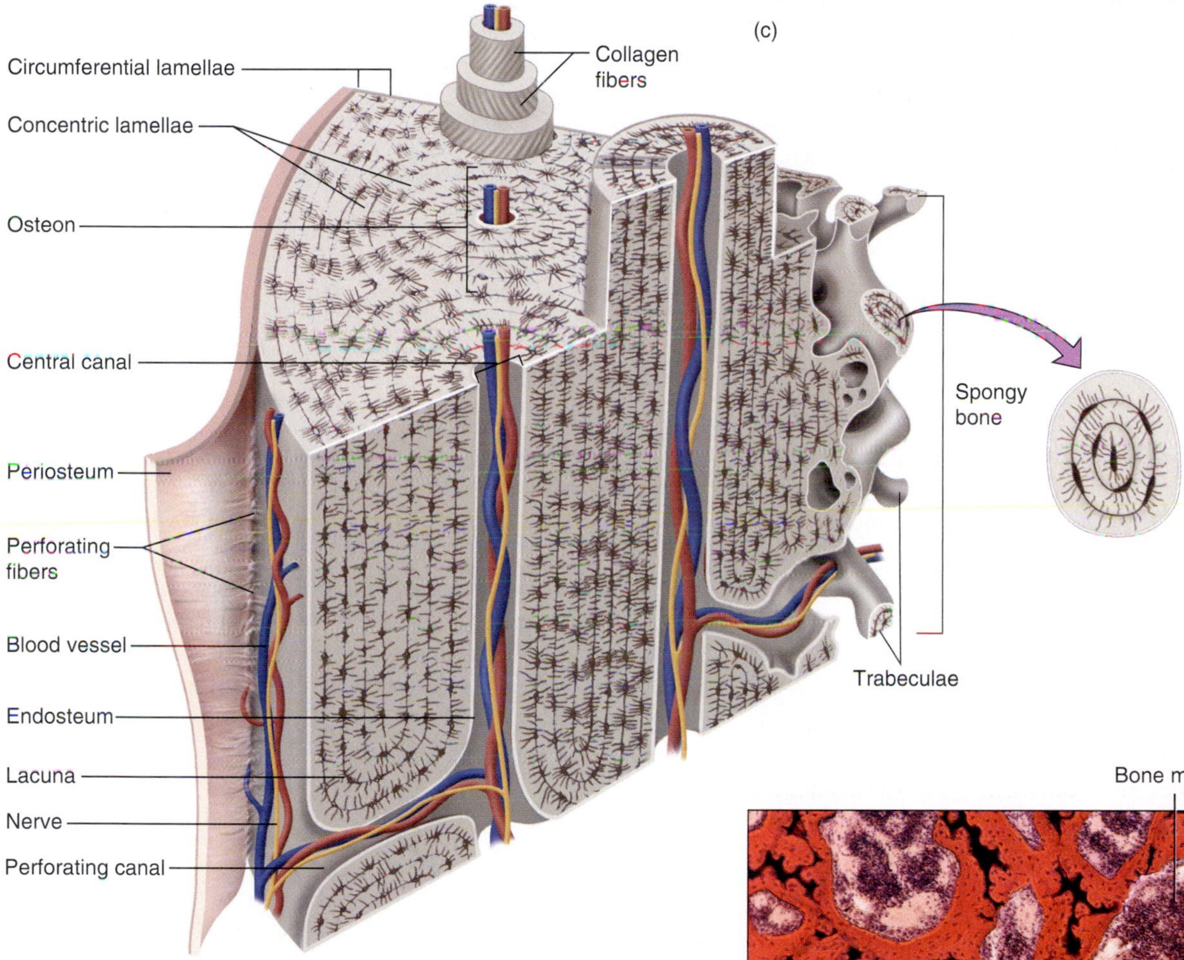
Circumferential lamellae
Concentric lamellae
Osteon
Central canal
Periosteum
Perforating fibers
Blood vessel
Endosteum
Lacuna
Nerve
Perforating canal
Collagen fibers
Spongy bone
Trabeculae

(b)

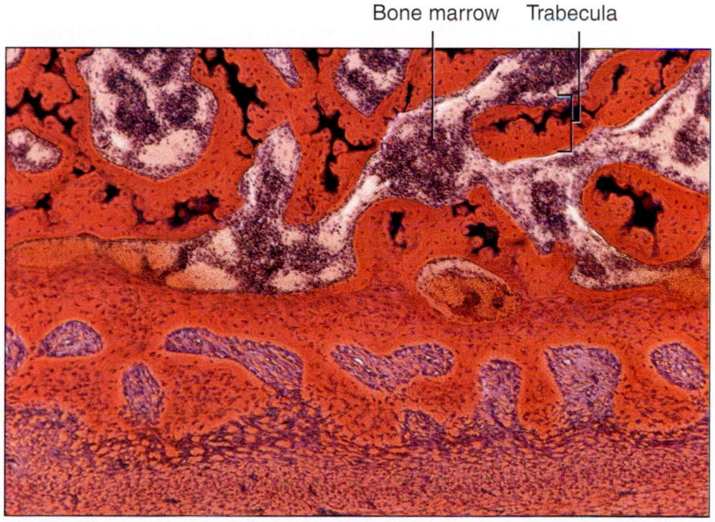
Bone marrow
Trabecula

(d)

FIGURE 6.4

The Histology of Osseous Tissue. (*a*) Compact and spongy bone in a frontal section of the shoulder joint. (*b*) The three-dimensional structure of compact bone. The uppermost osteon is pulled out to show the alternating arrangement of collagen fibers in adjacent lamellae. (*c*) Microscopic appearance of a cross section of compact bone. (*d*) Microscopic appearance of spongy bone.

direction in the next lamella. Like alternating layers of a sheet of plywood, this makes the bone stronger and enables it to resist tension in multiple directions.

The skeleton receives about half a liter of blood per minute. Blood vessels, along with nerves, enter the bone tissue through nutrient foramina on the surface. These open into narrow **perforating (Volkmann[19]) canals** that cross the matrix and lead to the central canals. The innermost osteocytes around each central canal receive nutrients from these blood vessels and pass them along through their gap junctions to neighboring osteocytes. They also receive wastes from their neighbors and convey them to the central canal for removal by the bloodstream. Thus, the cytoplasmic processes of the osteocytes maintain a two-way flow of nutrients and wastes between the central canal and the outermost cells of the osteon.

Not all of the matrix is organized into osteons. The inner and outer boundaries of dense bone are arranged in *circumferential lamellae* that run parallel to the bone surface. Between osteons, we can find irregular patches of *interstitial lamellae,* the remains of old osteons that broke down as the bone grew and remodeled itself.

Spongy Bone

Spongy bone consists of a lattice of thin plates called **trabeculae**[20] and rods and spines called **spicules**[21] (fig. 6.4*d;* see also p. 151). Although calcified and hard, spongy bone is named for its spongelike appearance; it is permeated by spaces filled with bone marrow. The matrix is arranged in lamellae like those of compact bone, but there are few osteons. Central canals are not needed here because no osteocyte is very far from the blood supply in the marrow. Spongy bone is well designed to impart strength to a bone with a minimum of weight. Its trabeculae are not randomly arranged as they might seem at a glance, but develop along the bone's lines of stress (fig. 6.5).

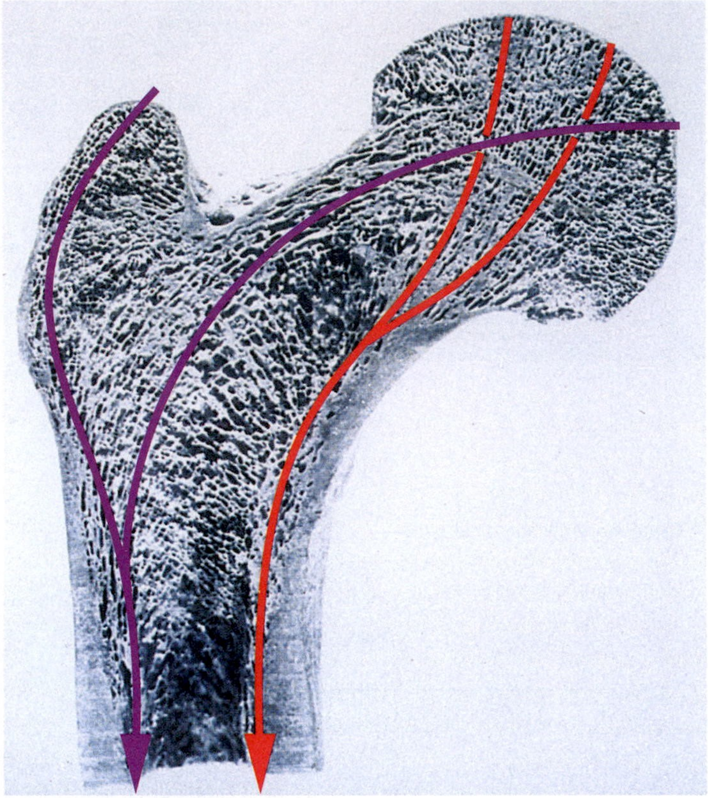

FIGURE 6.5

Spongy Bone Structure in Relation to Mechanical Stress. In this frontal section of the femur (thighbone), the trabeculae of spongy bone can be seen oriented along lines of mechanical stress applied by the weight of the body.

Bone Marrow

Bone marrow is a general term for soft tissue that occupies the medullary cavity of a long bone, the spaces amid the trabeculae of spongy bone, and the larger central canals. In a child, the medullary cavity of nearly every bone is filled with **red bone marrow (myeloid tissue).** This is a *hemopoietic*[22] (HE-mo-poy-ET-ic) tissue—that is, it produces blood cells. Red bone marrow looks like blood but with a thicker consistency. It consists of a delicate mesh of reticular tissue saturated with immature blood cells and scattered adipocytes.

With age, the red bone marrow is gradually replaced by fatty **yellow bone marrow,** like the fat at the center of a ham bone. By early adulthood, red bone marrow is limited to the vertebrae, sternum, ribs, pectoral (shoulder) and pelvic (hip) girdles, and the proximal heads of the humerus and femur, while the rest of the skeleton contains yellow marrow (fig. 6.6). Yellow bone marrow no longer produces blood, although in the event of severe or chronic anemia, it can transform back into red marrow and resume that role.

Before You Go On

Answer the following questions to test your understanding of the preceding section:

6. Suppose you had unlabeled electron micrographs of the four kinds of bone cells and their neighboring tissues. Name each of the four cells and explain how you could visually distinguish each one from the other three.
7. Name three organic components of the bone matrix.
8. What are the mineral crystals of bone called, and what are they made of?
9. Sketch a cross section of an osteon and label its major parts.
10. What are the three kinds of bone marrow? What does *hemopoietic tissue* mean? Which type of bone marrow fits this description?

[19]Alfred Volkmann (1800–77), German physiologist
[20]*trabe* = plate + *cul* = little
[21]*spicul* = dart, little point
[22]*hemo* = blood + *poietic* = forming

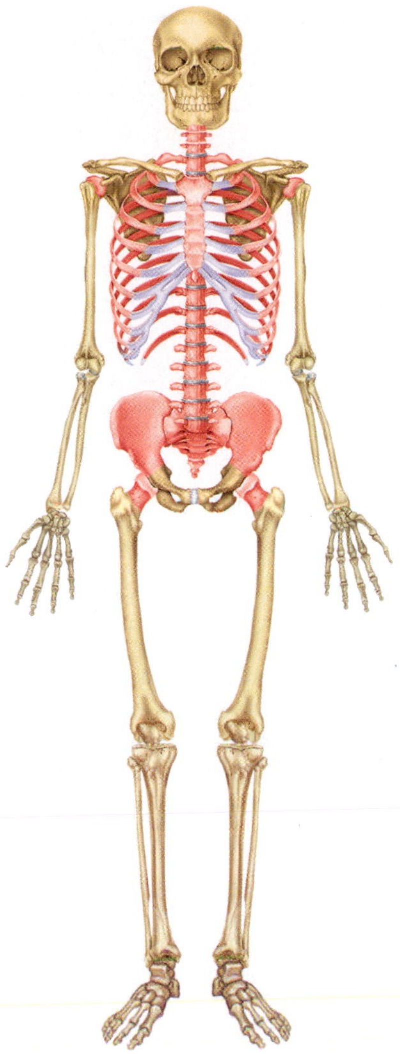

FIGURE 6.6

Distribution of Red and Yellow Bone Marrow. In an adult, red bone marrow occupies the medullary cavities of the vertebrae, sternum, and ribs, and proximal heads of the humerus and femur. Yellow bone marrow occurs in the long bones of the limbs.

Bone Development

Objectives

When you have completed this section, you should be able to

- describe two mechanisms of bone formation;
- explain how a child grows in height; and
- explain how mature bone continues to grow and remodel itself.

The formation of bone is called **ossification** (OSS-ih-fih-CAY-shun), or **osteogenesis.** There are two methods of ossification—*endochondral* and *intramembranous.*

Endochondral Ossification

Endochondral[23] (EN-doe-CON-drul) **ossification** is a process in which a bone develops from hyaline cartilage. Most bones form by this method, including the vertebrae, pelvic bones, and limb bones. In endochondral ossification, embryonic mesenchyme condenses into a hyaline cartilage *model* that resembles the shape of the bone to come. The cartilage is then broken down, reorganized, and calcified to form a bone (fig. 6.7).

THE PRIMARY OSSIFICATION CENTER

In the cartilage model, the first sign of endochondral ossification is the multiplication and swelling of chondrocytes near the center, forming a **primary ossification center.** As the lacunae enlarge, the matrix between them is reduced to thin walls and the model becomes weak at this point. It soon gets reinforcement, however. Some cells of the perichondrium become osteoblasts, which produce a bony collar around the model. This collar acts like a splint to provide temporary support for the model, and it cuts off the diffusion of nutrients to the chondrocytes, hastening their death. Once the collar has formed, the fibrous sheath around it is considered periosteum rather than perichondrium.

Buds of connective tissue grow from this periosteum into the cartilage and penetrate the thin walls between the enlarged lacunae. They break down the lacunae and transform the primary ossification center into a cavity called the **primary marrow space.** Osteogenic cells invade the cartilage model by way of the connective tissue buds, transform into osteoblasts, and line the marrow space. The osteoblasts deposit an organic matrix called **osteoid**[24] **tissue**—soft collagenous tissue similar to bone except for a lack of minerals—and then calcify it to form a temporary framework of bony trabeculae. As ossification progresses, osteoclasts break down these trabeculae and enlarge the primary marrow space. The ends of the bone are still composed of hyaline cartilage at this stage.

THE METAPHYSIS

At the boundary between the marrow space and each cartilaginous head of a developing long bone, there is a transitional zone called the **metaphysis** (meh-TAF-ih-sis). It exhibits five histological zones of transformation from cartilage to bone (fig. 6.8):

1. **Zone of reserve cartilage.** In this zone, farthest from the marrow space, the resting cartilage as yet shows no sign of transforming into bone.
2. **Zone of cell proliferation.** A little closer to the marrow space, chondrocytes multiply and become arranged into longitudinal columns of flattened lacunae.
3. **Zone of cell hypertrophy.** Next, the chondrocytes cease to divide and begin to hypertrophy, just as they did in the primary ossification center. The cartilage walls between lacunae become very thin. Cell multiplication in zone 2 and

[23] *endo* = within + *chondr* = cartilage
[24] *oste* = bone + *oid* = like, resembling

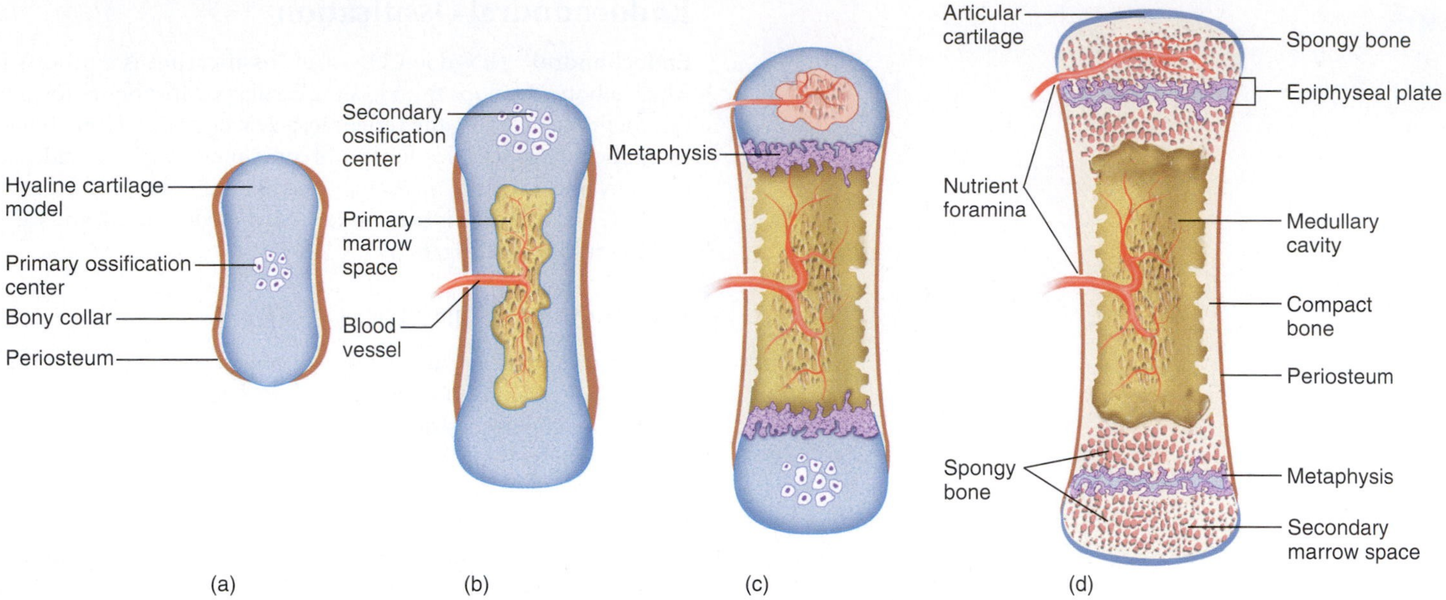

Hyaline cartilage model
Primary ossification center
Bony collar
Periosteum
Secondary ossification center
Primary marrow space
Blood vessel
Metaphysis
Articular cartilage
Nutrient foramina
Spongy bone
Spongy bone
Epiphyseal plate
Medullary cavity
Compact bone
Periosteum
Metaphysis
Secondary marrow space
(a)
(b)
(c)
(d)

FIGURE 6.7

Stages of Endochondral Ossification. (*a*) Chondrocyte hypertrophy at the center of the cartilage model and formation of a supportive bony collar. (*b*) Invasion of the model by blood vessels and creation of a primary marrow space. (*c*) Typical state of a long bone at the time of birth, with blood vessels growing into the secondary marrow space and well-defined metaphyses at each end of the primary marrow space. (*d*) Appearance of a long bone in childhood. By adulthood, the epiphyseal plates will be depleted and the primary and secondary marrow spaces will be united.

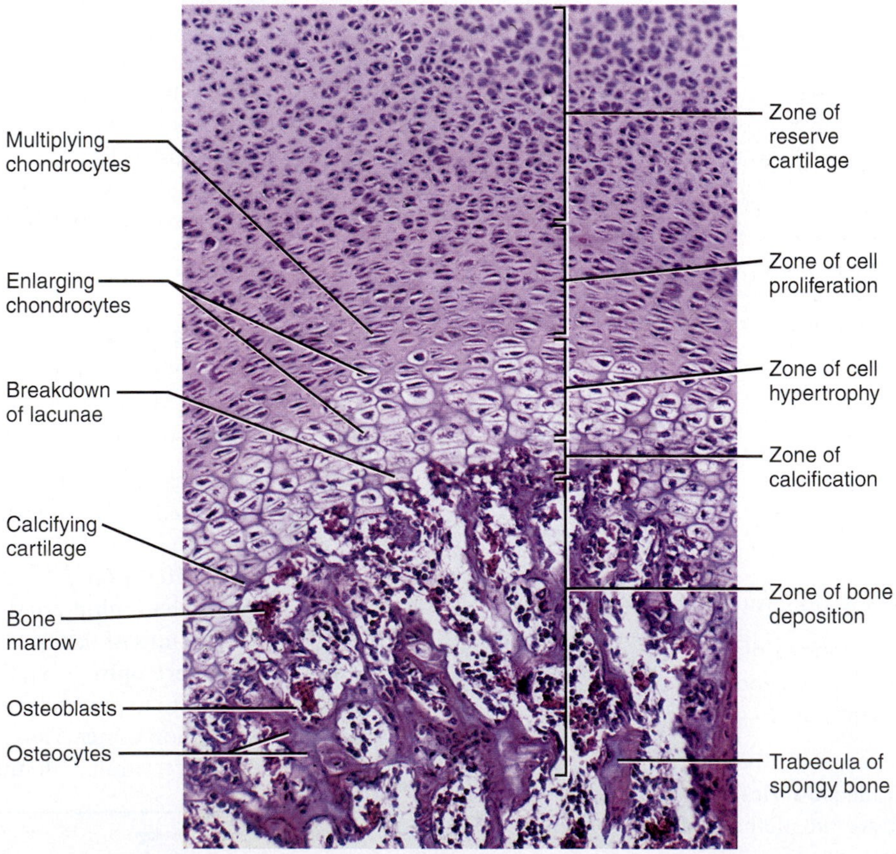

Multiplying chondrocytes
Enlarging chondrocytes
Breakdown of lacunae
Calcifying cartilage
Bone marrow
Osteoblasts
Osteocytes
Zone of reserve cartilage
Zone of cell proliferation
Zone of cell hypertrophy
Zone of calcification
Zone of bone deposition
Trabecula of spongy bone

FIGURE 6.8

Zones of the Metaphysis. This micrograph shows the transition from cartilage to bone in the growth zone of a long bone.

hypertrophy in zone 3 continually push the zone of reserve cartilage toward the ends of the bone and make the bone grow longer.

4. **Zone of calcification.** Minerals are deposited in the matrix between columns of lacunae and calcify the cartilage for temporary support.
5. **Zone of bone deposition.** Within each column, the walls between lacunae break down and the chondrocytes die. This converts each column into a longitudinal channel, which is quickly invaded by marrow and blood vessels from the primary marrow space. Osteoclasts dissolve the calcified cartilage while osteoblasts line up along the walls of these channels and begin depositing concentric layers of bone matrix. The channel therefore grows smaller and smaller as one layer after another is laid down, until only a narrow channel remains in the middle—now a central canal. Osteoblasts trapped in their own matrix become osteocytes and stop producing matrix.

Think About It!

In a given osteon, which lamellae are the oldest—those immediately adjacent to the central canal or those around the perimeter of the osteon? Explain your answer.

The primary ossification centers of a 12-week-old fetus are shown in figure 7.30. The joints are translucent because they have not yet ossified. They are still cartilaginous even at the time of birth, which is one reason human newborns cannot walk.

THE SECONDARY OSSIFICATION CENTER

Around the time of birth, **secondary ossification centers** begin to form in the epiphyses (see fig. 6.7*b*). Here, too, chondrocytes enlarge, the walls of matrix between them dissolve, and the chondrocytes die. Vascular buds arise from the perichondrium and grow into the cartilage, bringing osteogenic cells and osteoclasts with them. The cartilage is eroded from the center of the epiphysis outward in all directions. Thin trabeculae of cartilage matrix calcify to form spongy bone. Hyaline cartilage persists in two places—on the epiphyseal surfaces as the articular cartilages and at the junction of the diaphysis and epiphysis, where it forms the epiphyseal plate (fig. 6.9). Each side of the epiphyseal plate has a metaphysis, where the transformation of cartilage to bone occurs.

Intramembranous Ossification

Intramembranous[25] (IN-tra-MEM-bruh-nus) **ossification** produces the flat bones of the skull and most of the clavicle (collarbone). It begins when some of the embryonic connective tissue (mesenchyme) condenses into a sheet of soft tissue with a dense supply of blood capillaries (fig. 6.10). The cells of this sheet enlarge

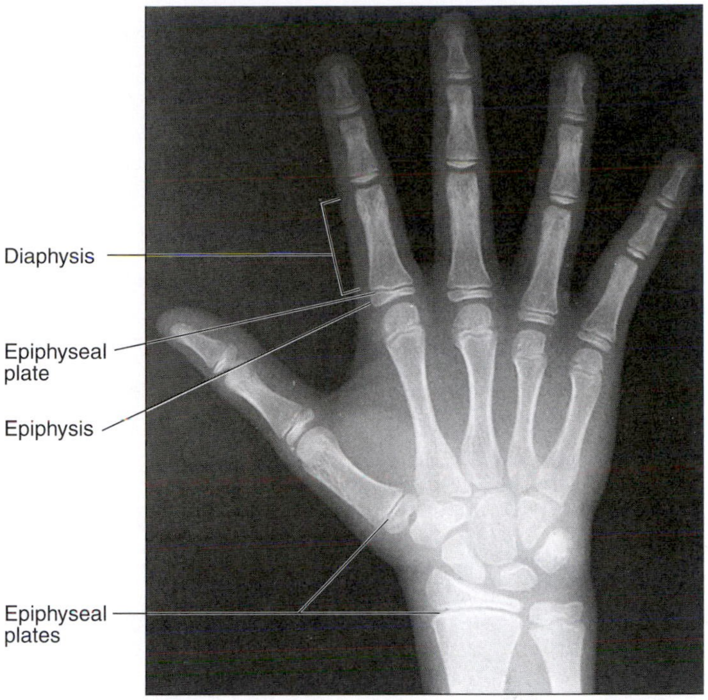
Diaphysis
Epiphyseal plate
Epiphysis
Epiphyseal plates

FIGURE 6.9
X Ray of a Child's Hand. The cartilaginous epiphyseal plates are evident at the ends of the long bones. These will disappear, and the epiphyses will fuse with the diaphyses, by adulthood. Compare the X ray on page 207.

and differentiate into osteogenic cells, and some of the mesenchyme transforms into a network of soft trabeculae. Osteogenic cells gather on the trabeculae, become osteoblasts, and deposit osteoid tissue (fig. 6.11). As the trabeculae grow thicker, calcium phosphate is deposited in the matrix and some osteoblasts become trapped in lacunae. Once trapped, they differentiate into osteocytes. Some of the now-calcified trabeculae form permanent spongy bone. Osteoclasts soon appear on these trabeculae, resorbing and remodeling bone and creating a marrow space. Trabeculae at the surface continue to calcify until the spaces between them are filled in, thereby converting the spongy bone to compact bone. This process gives rise to the typical structure of a flat cranial bone—a sandwichlike arrangement of spongy bone between two surface layers of compact bone. Mesenchyme at the surface of the developing bone remains uncalcified, but becomes increasingly fibrous and eventually gives rise to the periosteum.

Bone Growth and Remodeling

Bones continue to grow and remodel themselves throughout life, changing size and shape to accommodate the changing forces applied to the skeleton. For example, in children the femurs grow longer, the curvature of the cranium increases to accommodate a growing brain, and many bones develop surface bumps, spines, and ridges (described in chapter 7) as a child begins to walk and the muscles exert tension on the bones. The prominence of these

[25]*intra* = within + *membran* = membrane

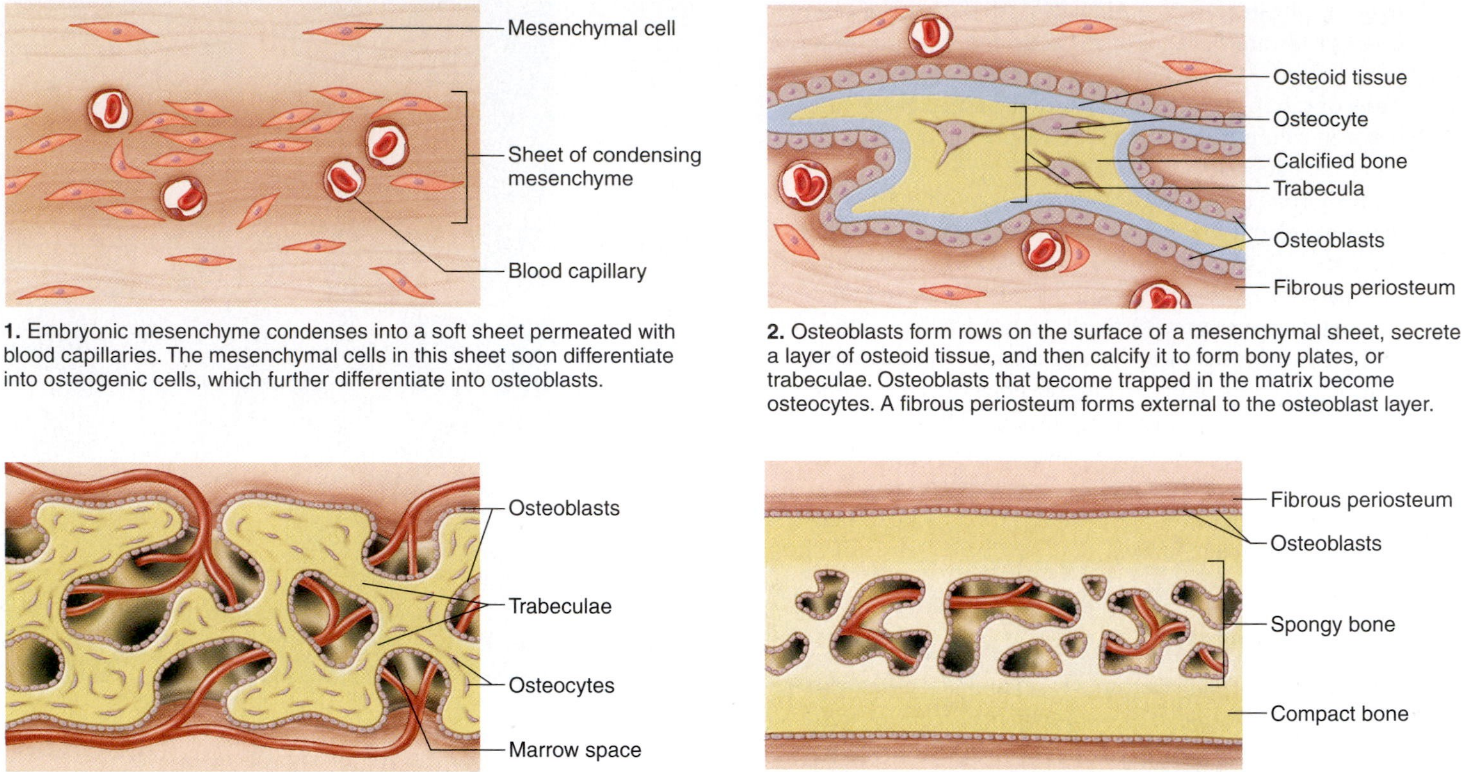
Mesenchymal cell
Sheet of condensing mesenchyme
Blood capillary
Osteoid tissue
Osteocyte
Calcified bone
Trabecula
Osteoblasts
Fibrous periosteum
Osteoblasts
Trabeculae
Osteocytes
Marrow space
Fibrous periosteum
Osteoblasts
Spongy bone
Compact bone

1. Embryonic mesenchyme condenses into a soft sheet permeated with blood capillaries. The mesenchymal cells in this sheet soon differentiate into osteogenic cells, which further differentiate into osteoblasts.

2. Osteoblasts form rows on the surface of a mesenchymal sheet, secrete a layer of osteoid tissue, and then calcify it to form bony plates, or trabeculae. Osteoblasts that become trapped in the matrix become osteocytes. A fibrous periosteum forms external to the osteoblast layer.

3. Continued bone deposition forms a honeycomb of bony trabeculae enclosing marrow spaces with blood vessels.

4. Further ossification at the surface of the bone fills in the spaces and produces surface plates of compact bone. Spongy bone remains in the center of the plate, forming the typical sandwichlike arrangement of a flat bone. In the skull, this middle layer of spongy bone is called the diploe.

FIGURE 6.10
Stages of Intramembranous Ossification.

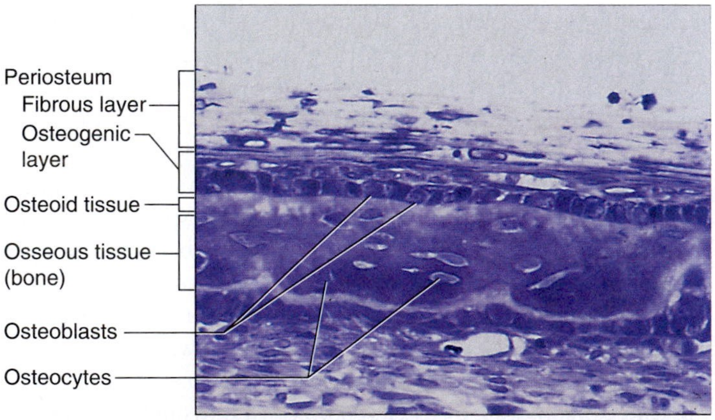
Periosteum
Fibrous layer
Osteogenic layer
Osteoid tissue
Osseous tissue (bone)
Osteoblasts
Osteocytes

FIGURE 6.11
A Developing Flat Bone (fetal cranium). Note the layers of osteoid tissue, osteoblasts, and fibrous periosteum on both sides of the bone.

surface features and the density of bone depend on the amount of stress to which a bone has been subjected. On average, the bones have greater density and mass in athletes and people engaged in heavy manual labor than they do in sedentary people. Anthropologists who study skeletal remains can distinguish between members of different social classes by the degree of bone development—a reflection of the individual's nutritional status and history of manual labor. The skeleton also yields information about sex, race, height, weight, and medical history that can be useful to *forensic pathology*—study of the body to determine the identity of a person, cause and time of death, and so forth.

Cartilage grows by two mechanisms—**interstitial**[26] **growth** (adding more matrix internally) and **appositional**[27] **growth** (adding more to the surface). Interstitial growth in the epiphyseal plate adds to the length of a bone. A mature bone, however, grows only by the appositional mechanism. Osteocytes have little room as it is and none to spare for the deposition of more matrix. The only way an adult bone can grow, therefore, is by adding more osseous tissue to the surface.

Appositional growth is similar to intramembranous ossification. The osteogenic cells in the inner layer of periosteum differentiate into osteoblasts. These deposit osteoid tissue on the bone surface, calcify it, and become trapped in it as osteocytes. At the bone surface, matrix is laid down in layers parallel to the surface, not in cylindrical osteons like those deeper in the bone. While deposition occurs at the outer surface of a bone, osteoclasts dissolve bone on the inner surface and thus enlarge the marrow cavity as the bone grows. There is a critical balance between bone deposition and removal. If one process outpaces the other, or if both of them occur too rapidly, various bone deformities can occur (see table 6.2, especially *osteitis deformans*).

[26]*inter* = between + *stit* = to place, stand
[27]*ap* = *ad* = to, near + *posit* = to place

INSIGHT 6.3 CLINICAL APPLICATION

Achondroplastic Dwarfism

Achondroplastic[28] (ah-con-dro-PLAS-tic) *dwarfism* is a condition in which the long bones of the limbs stop growing in childhood, while the growth of other bones is unaffected. As a result, a person has a short stature but a normal-sized head and trunk (fig. 6.12). As its name implies, achondroplastic dwarfism results from a failure of cartilage growth—specifically, failure of the chondrocytes in zones 2 and 3 of the metaphysis to multiply and enlarge. This is different from *pituitary dwarfism,* in which a deficiency of growth hormone stunts the growth of all of the bones and a person has short stature but normal proportions throughout the skeletal system.

Achondroplastic dwarfism results from a spontaneous mutation that can arise any time DNA is replicated. Two people of normal height with no family history of dwarfism can therefore have a child with achondroplastic dwarfism. The mutant allele is dominant, so the children of a heterozygous achondroplastic dwarf have at least a 50% chance of exhibiting dwarfism, depending on the genotype of the other parent.

[28]*a* = without + *chondro* = cartilage + *plast* = growth

Figure 6.12
Achondroplastic Dwarfism. The student on the *right,* pictured with her roommate of normal height, is an achondroplastic dwarf with a height of about 122 cm (48 in.). Her parents were of normal height. Note the normal proportion of head to trunk but shortening of the limbs.

Nutritional and Hormonal Factors

The balance between bone deposition and resorption is influenced by nearly two dozen nutrients, hormones, and growth factors. The most important factors that promote bone deposition are as follows.

- **Calcium** and **phosphate** are needed as raw materials for the calcified ground substance of bone.
- **Vitamin A** promotes synthesis of the glycosaminoglycans (GAGs) of the bone matrix.
- **Vitamin C (ascorbic acid)** promotes the cross-linking of collagen molecules in bone and other connective tissues.
- **Vitamin D (calcitriol)** is necessary for calcium absorption by the small intestine, and it reduces the urinary loss of calcium and phosphate. Vitamin D is synthesized by one's own body. The process begins when the ultraviolet radiation in sunlight acts on a cholesterol derivative (7-dehydrocholesterol) in the keratinocytes of the epidermis. The product produced here is picked up by the blood stream, and the liver and kidneys complete its conversion to vitamin D.
- **Calcitonin,** a hormone secreted by the thyroid gland, stimulates osteoblast activity. It functions chiefly in children and pregnant women; it seems to be of little significance in nonpregnant adults.
- **Growth hormone** promotes intestinal absorption of calcium, the proliferation of cartilage at the epiphyseal plates, and the elongation of bones.
- **Sex steroids** (estrogen and testosterone) stimulate osteoblasts and promote the growth of long bones, especially in adolescence.

Bone deposition is also promoted by thyroid hormone, insulin, and local *growth factors* produced within the bone itself. Bone resorption is stimulated mainly by one hormone:

- **Parathyroid hormone (PTH)** is produced by four small *parathyroid glands,* which adhere to the back of the thyroid gland in the neck. The parathyroid glands secrete PTH in response to a drop in blood calcium level. PTH stimulates osteoblasts, which then secrete an *osteoclast-stimulating factor* that promotes bone resorption by the osteoclasts. The principal purpose of this response is not to maintain bone composition but to maintain an appropriate level of blood calcium, without which a person can suffer fatal muscle spasms. PTH also reduces urinary calcium losses and promotes calcitriol synthesis.

The Aging Skeletal System

The predominant effect of aging on the skeleton is a loss of bone mass and strength. After age 30, osteoblasts become less active than osteoclasts. The imbalance between deposition and resorption leads to **osteopenia,**[29] the loss of bone; when the loss is severe enough to compromise physical activity and health, it is called *osteoporosis* (discussed in the next section). After age 40, women lose about 8% of their bone mass per decade and men lose about 3%. Bone loss from the jaws is a contributing factor in tooth loss. Not only does bone density decline with age, but the bones become more brittle as the osteoblasts synthesize less protein. Fractures occur more easily and heal more slowly. Arthritis, a family of joint disorders associated with aging, is discussed in chapter 9.

[29]*osteo* = bone + *penia* = lack

Before You Go On

Answer the following questions to test your understanding of the preceding section:

11. Describe the five zones of a metaphysis and the major distinctions between them.
12. Describe the stages of intramembranous ossification. Name a bone that is formed in this way.
13. Identify the nutrients most important to bone growth.
14. Identify the principal hormones that stimulate bone growth.

Structural Disorders of Bone

Objectives

When you have completed this section, you should be able to

- name and describe the types of fractures;
- explain how a fracture is repaired;
- discuss the causes and effects of osteoporosis; and
- briefly describe a few other structural defects of the skeleton.

Fractures are probably the most familiar disorder of the skeletal system, although the most common structural defect is *osteoporosis.* This section describes both of these defects and briefly defines a few others.

Fractures

There are multiple ways of classifying bone fractures. A **stress fracture** is a break caused by abnormal trauma to a bone, such as fractures incurred in falls, athletics, and military combat. A **pathologic fracture** is a break in a bone weakened by some other disease, such as bone cancer or osteoporosis, usually caused by a stress that would not normally fracture a bone. Fractures are also classified according to the direction of the fracture line, whether or not the skin is broken, and whether a bone is merely cracked or is broken into separate pieces (table 6.1; fig. 6.13).

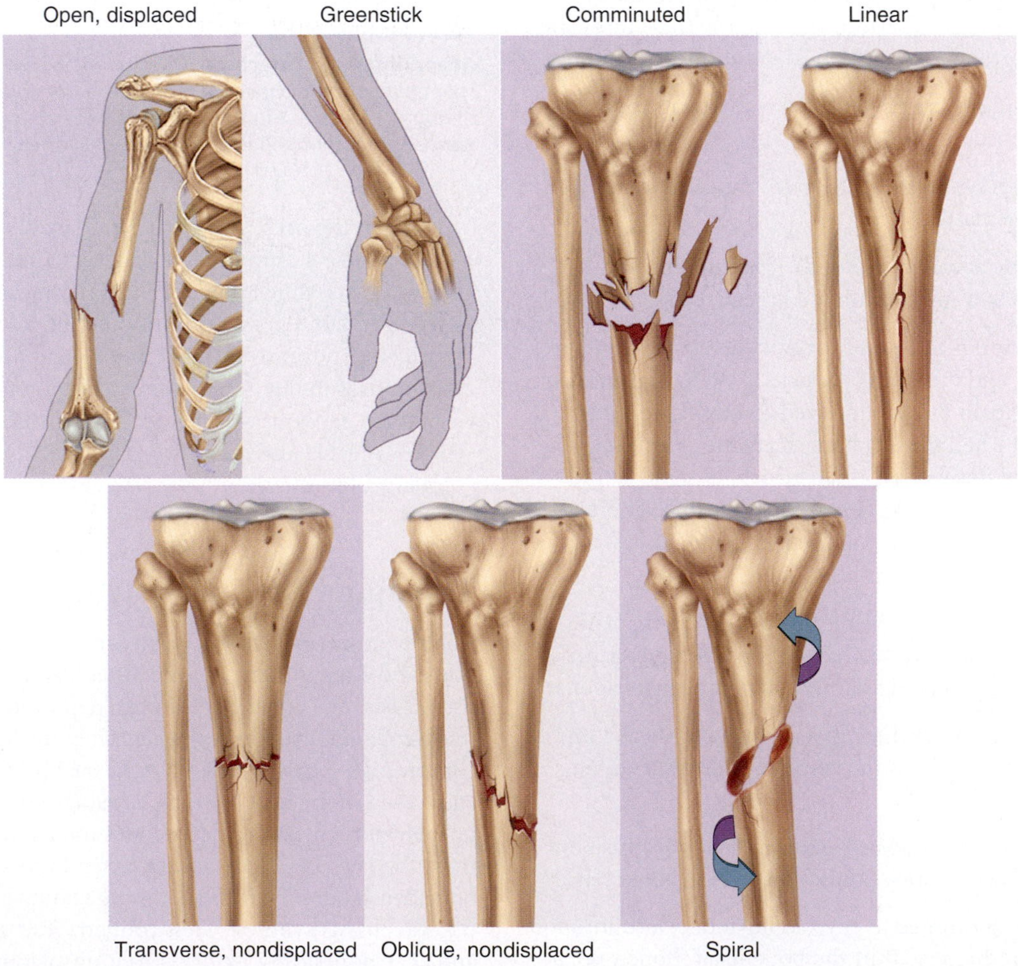
Open, displaced
Greenstick
Comminuted
Linear
Transverse, nondisplaced
Oblique, nondisplaced
Spiral

FIGURE 6.13
Some Types of Bone Fractures. Compare table 6.1.

TABLE 6.1

Classification of Fractures

Type	Description
Closed	Skin is not broken (formerly called a *simple* fracture)
Open	Skin is broken; bone protrudes through skin or wound extends to fractured bone (formerly called a *compound* fracture)
Complete	Bone is broken into two or more pieces
Incomplete	Partial fracture that extends only partway across bone; pieces remain joined
Greenstick	Bone is bent on one side and has incomplete fracture on opposite side
Hairline	Fine crack in which sections of bone remain aligned; common in skull
Comminuted	Bone is broken into three or more pieces
Displaced	The portions of a fractured bone are out of anatomical alignment
Nondisplaced	The portions of bone are still in correct anatomical alignment
Impacted	One bone fragment is driven into the medullary space or spongy bone of the other
Depressed	Broken portion of bone forms a concavity, as in skull fractures
Linear	Fracture parallel to long axis of bone
Transverse	Fracture perpendicular to long axis of bone
Oblique	Diagonal fracture, between linear and transverse
Spiral	Fracture spirals around axis of long bone, the result of a twisting stress, often produced when an abusive adult roughly picks a child up by the arm

Most fractures are set by *closed reduction,* a procedure in which the bone fragments are manipulated into their normal positions without surgery. *Open reduction* involves the surgical exposure of the bone and the use of plates, screws, or pins to realign the fragments (fig. 6.14*b*). To stabilize the bone during healing, fractures are often set in fiberglass casts. *Traction* is used to treat fractures of the femur in children. It aids in the alignment of the bone fragments by overriding the force of the strong thigh muscles. Traction is rarely used for elderly patients, however, because the risks from long-term confinement to bed outweigh the benefits. Hip fractures are usually pinned, and early ambulation (walking) is encouraged because it promotes blood circulation and healing.

An uncomplicated fracture heals in 8 to 12 weeks, but complex fractures take longer and all fractures heal more slowly in older people. Figure 6.15 shows the healing process. Usually, a healed fracture leaves a slight thickening of the bone visible by X ray, but in some cases healing is so complete that no trace of the fracture can be found.

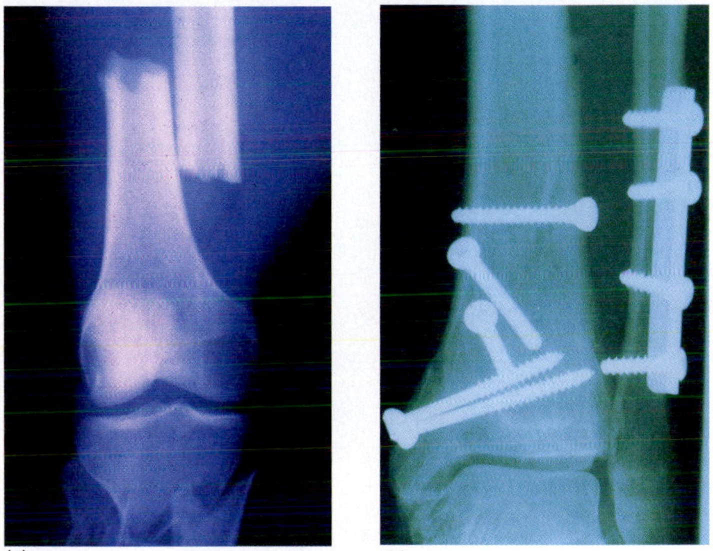

(a) (b)

FIGURE 6.14

X Rays of Bone Fractures. (*a*) A displaced fracture of the femur. (*b*) An ankle fracture involving both the tibia and fibula. This fracture has been set by open reduction, a process of surgically exposing the bone and realigning the fragments with plates and screws.

INSIGHT 6.4 CLINICAL APPLICATION

WHEN *NOT* TO EAT YOUR SPINACH

Many a child has been exhorted to "Eat your spinach! It's good for you." There is one time, however, when it may not be healthy. People with healing bone fractures are sometimes advised not to eat it. Why? Spinach is rich in oxalate, an organic compound that binds calcium and magnesium in the digestive tract and interferes with their absorption. Consequently, the oxalate can deprive a fractured bone of the free calcium that it needs in order to heal. There are about 571 milligrams of oxalate per 100 grams of spinach. Some other foods high in oxalate are cocoa (623 mg), rhubarb (447 mg), and beets (109 mg).

Osteoporosis

Osteoporosis[30] (OSS-tee-oh-pore-OH-sis)—literally, "porous bones"—is a disease in which the bones lose mass and become increasingly brittle and subject to fractures. It involves loss of proportionate amounts of organic matrix and minerals, and it affects spongy bone in particular, since this is the most metabolically active type (fig. 6.16). The bone that remains is histologically normal but insufficient in quantity to support the body's weight.

[30]*osteo* = bone + *por* = porous + *osis* = condition

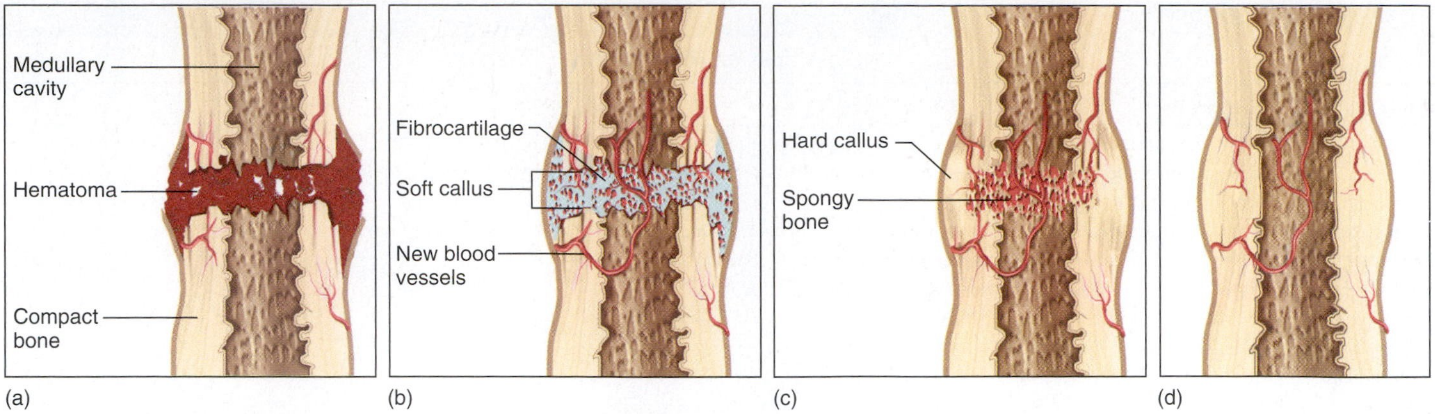

FIGURE 6.15

The Healing of a Bone Fracture. (*a*) Blood vessels are broken at the fracture line; blood clots and forms a fracture hematoma. (*b*) Blood vessels grow into the clot and a soft callus of fibrocartilage forms. (*c*) Mineral deposition hardens the soft callus and converts it to a hard callus of spongy bone. (*d*) Osteoclasts remove excess tissue from the hard callus and the bone eventually resembles its original appearance.

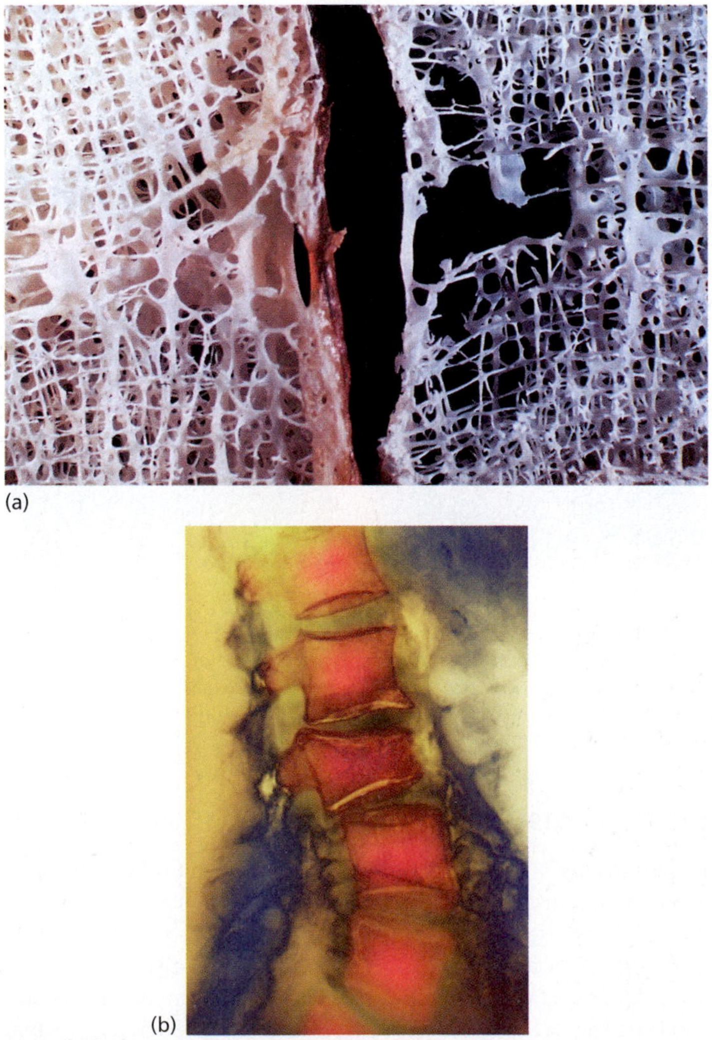

FIGURE 6.16

Osteoporosis. (*a*) Spongy bone of a healthy lumbar vertebra (*left*) and a lumbar vertebra with osteoporosis (*right*). (*b*) Colorized X ray of lumbar vertebrae severely damaged by osteoporosis.

The most serious consequence of osteoporosis is pathologic fractures, which occur especially in the hip, wrist, and vertebral column and under stresses as slight as sitting down too quickly. Among the elderly, hip fractures often lead to fatal complications such as pneumonia. For half of those who survive, a hip fracture involves a long, costly recovery. As the weight-bearing bodies of the vertebrae lose spongy bone, they become compressed like marshmallows. Consequently, many people lose height after middle age, and some develop a spinal deformity called **kyphosis,**[31] an exaggerated thoracic curvature ("widow's hump" or "dowager's hump") (fig. 6.17).

Postmenopausal white women are at greatest risk for osteoporosis for multiple reasons: (1) women have less bone mass than men to begin with, (2) they begin losing it earlier (starting around age 35), (3) they lose it faster than men do, and (4) after menopause, the ovaries no longer produce estrogen, an important stimulus to bone deposition. By age 70, the average white woman has lost 30% of her bone mass, and some have lost as much as 50%. Young black women develop more bone mass than whites. Although they, too, lose bone after menopause, the loss usually does not reach the threshold for osteoporosis and pathologic fractures. Men of both races suffer osteoporosis less than white women but more than black women. In men, bone loss begins around age 60 and seldom exceeds 25%. Osteoporosis also occurs among young female runners and dancers in spite of their vigorous exercise. Their percentage of body fat is so low that their ovaries secrete unusually low levels of estrogen and the women may stop ovulating.

Estrogen replacement therapy cannot reverse osteoporosis, but it can slow its progress. Furthermore, in some women, estrogen therapy increases the risk of breast cancer. Alternatives to estrogen therapy are becoming available, but each has its own undesirable side effects. Some patients are now treated with a calcitonin nasal spray. Milk and other calcium sources and moderate exercise can also slow the progress of osteoporosis, but only slightly.

[31] *kypho* = bent, humpbacked + *osis* = condition

FIGURE 6.17

Woman with Osteoporosis and Kyphosis. The pronounced curvature of the thoracic spine results from compression fractures of the weakened vertebrae.

The risk of osteoporosis is best minimized by exercise and ample calcium intake (850–1,000 mg/day) early in life, especially between the ages of 25 and 40, when the skeleton is building to its maximum mass. The risk factors for osteoporosis include not only sex, race, inadequate exercise, and inadequate calcium intake, but also smoking, vitamin C deficiency, and diabetes mellitus.

Other Structural Disorders

Several additional bone disorders are summarized in table 6.2. **Orthopedics**[33] is the branch of medicine that deals with the prevention and correction of injuries and disorders of the bones, joints, and muscles. As the word suggests, this field originated as the treatment of skeletal deformities in children, but it is now much more extensive. It includes the design of artificial joints and limbs and the treatment of athletic injuries.

Before You Go On

Answer the following questions to test your understanding of the preceding section:

15. Name and describe any five types of bone fractures.
16. What is a callus? How does it contribute to fracture repair?
17. List the major risk factors for osteoporosis and concisely describe some ways of preventing it.

[33]*ortho* = straight + *ped* = child, foot

TABLE 6.2

Structural Disorders of Bone

Rickets	Defective mineralization of bone in children, usually as a result of insufficient sunlight or vitamin D, sometimes due to a dietary deficiency of calcium or phosphate or to liver or kidney diseases that interfere with calcitriol synthesis. Causes bone softening and deformity, especially in the weight-bearing bones of the lower limbs.
Osteomalacia	Adult form of rickets, most common in poorly nourished women who have had multiple pregnancies. Bones become softened, deformed, and more susceptible to fractures.
Osteitis deformans (Paget[32] disease)	Excessive osteoclast proliferation and bone resorption, with osteoblasts attempting to compensate by depositing extra bone. This results in rapid, disorderly bone remodeling and weak, deformed bones. Osteitis deformans usually passes unnoticed, but in some cases it causes pain, disfiguration, and fractures. It is most common in males over the age of 50.

Disorders Described Elsewhere

Achondroplastic dwarfism 163	Fractures 164
Brittle bone disease 93	Osteopenia 163
	Osteoporosis 165

[32]Sir James Paget (1814–99), English surgeon

CHAPTER REVIEW

Review of Key Concepts

Tissues and Organs of the Skeletal System (p. 152)

1. The *skeletal system* is a framework composed of bones, cartilages, and ligaments. The study of this system is called *osteology.*
2. The functions of this system include support, movement, protection of soft tissues, blood formation, electrolyte balance, acid-base balance, and detoxification.
3. *Osseous tissue* (bone) is a connective tissue in which the matrix is hardened by calcium phosphate and other minerals. Other tissues found in a bone include blood, bone marrow, cartilage, adipose tissue, nervous tissue, and fibrous connective tissue.
4. Bones are classified into four categories by shape: *long, short, flat,* and *irregular* bones.
5. A bone has an outer shell of *compact bone* which usually encloses more loosely organized *spongy bone.*
6. A long bone has a relatively narrow, long *diaphysis* (shaft) with an expanded *epiphysis* (head) at each end. The epiphysis is filled with spongy bone. Bone marrow occupies the diaphysis and the spaces amid the spongy bone of the epiphysis.
7. A cartilaginous *epiphyseal plate* separates the marrow spaces of the epiphysis and diaphysis in children and adolescents. It is the site of bone elongation.
8. A bone is externally covered with a fibrous *periosteum,* which is bound to the bone by collagenous *perforating fibers.* The medullary cavity is lined with a fibrous *endosteum.*
9. At most joints, the ends of a bone have no periosteum but are covered with hyaline *articular cartilage.*
10. Flat bones consist of a sandwichlike arrangement of spongy bone enclosed between two layers of compact bone. The spongy bone layer of the skull is called the *diploe.*

Histology of Osseous Tissue (p. 155)

1. Osseous tissue has four kinds of cells: osteogenic cells, osteoblasts, osteocytes, and osteoclasts.
2. *Osteogenic cells* are stem cells found in the endosteum, periosteum, and central canals. They give rise to osteoblasts.
3. *Osteoblasts* are bone-depositing cells found on the bone surfaces. They produce the organic components of the bone matrix and promote its mineralization.
4. *Osteocytes* are bone cells found within the lacunae and surrounded by bone matrix. They communicate with each other and with surface osteoblasts by way of cytoplasmic processes in the *canaliculi* of the matrix. They function as strain detectors and stimulate bone deposition by osteoblasts.
5. *Osteoclasts* are bone-dissolving cells found on the bone surfaces. They secrete hydrochloric acid, which dissolves the inorganic salts of the bone matrix, and they produce enzymes that digest the organic components.
6. The matrix of bone is about one-third organic and two-thirds inorganic matter by dry weight.
7. The inorganic part of the matrix is about 85% hydroxyapatite (crystalline calcium phosphate), 10% calcium carbonate, and 5% other minerals.
8. The organic part of the matrix consists of collagen and large protein-carbohydrate complexes called glycosaminoglycans, proteoglycans, and glycoproteins.
9. The mineral component of the bone renders it resistant to compression, so that it does not crumble under the body's weight, while the protein component renders it resistant to tension, so that it can bend slightly without breaking.
10. Compact bone is composed largely of cylindrical units called *osteons,* in which the matrix is arranged in concentric *lamellae* around a central canal. *Lacunae,* occupied by osteocytes, lie between the lamellae of matrix and are connected to each other by *canaliculi.*
11. Collagen fibers wind helically along the length of each lamella, with the helices coiling in alternating directions in adjacent lamellae to give the matrix added strength.
12. Blood vessels enter the bone matrix through *nutrient foramina* on the surface and pass by way of *perforating canals* to reach the central canals.
13. In addition to the concentric lamellae of osteons, compact bone exhibits circumferential lamellae that travel parallel to the inner and outer bone surfaces, and *interstitial lamellae* located between osteons, representing the remains of older osteons that have partially broken down.
14. Spongy bone consists of thin *trabeculae* of osseous tissue, with spaces between the trabeculae occupied by bone marrow. The matrix is arranged in lamellae but shows few osteons. Spongy bone provides a bone with maximal strength in proportion to its light weight.
15. There are two kinds of bone marrow: blood-forming (hemopoietic) *red marrow* and fatty *yellow marrow.* Red marrow occupies the medullary spaces of nearly all bones in children and adolescents. By adulthood, red marrow is limited to the vertebrae, ribs, sternum, pectoral and pelvic girdles, and proximal heads of the humerus and femur; it is replaced by yellow marrow elsewhere.

Bone Development (p. 159)

1. *Endochondral ossification* is a process in which bone develops from hyaline cartilage. Most bones form this way, including the vertebrae, pelvic bones, and limb bones.
2. Endochondral ossification begins in a *primary ossification center* of the cartilage model. Chondrocytes and their lacunae enlarge, while some cells of the perichondrium become osteoblasts and produce a supportive bony collar around the middle of the cartilage model. The breakdown of cartilage lacunae in the primary ossification center creates a cavity, the *primary marrow space,* which grows toward the ends of the bone.
3. Osteogenic cells invade the primary marrow space by way of blood vessels, and differentiate into osteoblasts. The osteoblasts deposit *osteoid tissue* and then calcify it to form temporary trabeculae of bone. Osteoclasts later enlarge the marrow space by breaking down these trabeculae.
4. At the boundary between the primary marrow space and cartilaginous head of the bone, there is a zone called the *metaphysis,* where cartilage is replaced by bone. The metaphysis has five zones: the *zone of reserve cartilage* farthest from the marrow space; the *zone of cell proliferation,* where chondrocytes multiply and form longitudinal columns of cells; the *zone of cell hypertrophy,* where these chondrocytes enlarge; the *zone of calcification,* where the matrix becomes temporarily calcified; and nearest the marrow space, the *zone of bone deposition,* where lacunae break down, chondrocytes die, and bone is deposited.
5. Near the time of birth, a *secondary ossification center* appears in the middle of the epi-

physis. Ossification proceeds from here outward, toward the epiphyseal plate on one side, and leaving a layer of articular cartilage over the end of the bone.

6. *Intramembranous ossification* produces the flat bones of the skull and most of the clavicle. It is a mode of bone formation that does not pass through a cartilage model.
7. Intramembranous ossification begins when mesenchyme condenses into a sheet of soft tissue populated with osteogenic cells. Osteogenic cells gather along soft trabeculae of mesenchyme and differentiate into osteoblasts. The osteoblasts deposit soft osteoid tissue and then calcify it. Calcified trabeculae become spongy bone, while compact surface bone is formed by filling in the spaces between trabeculae with osseous tissue.
8. Bones are remodeled throughout life to accommodate bodily growth and changes in force applied to the skeleton. In childhood, bones increase in length by means of the *interstitial growth* of cartilage in the epiphyseal plates. In adulthood, long bones no longer grow in length, but can increase in thickness by *appositional growth,* the addition of osseous tissue to the surface.
9. Some nutrients required for bone development include calcium, phosphate, and vitamins A, C, and D. Hormones that stimulate bone growth include calcitonin, growth hormone, estrogen, testosterone, thyroid hormone, and insulin. Parathyroid hormone promotes bone resorption by osteoclasts.

Structural Disorders of Bone (p. 164)

1. The prevention and treatment of bone, joint, and muscle disorders is called *orthopedics.*
2. Bones can break because of trauma *(stress fracture)* or diseases that weaken a bone and make it unable to withstand normal levels of stress *(pathologic fracture).* Table 6.1 defines the various types of fractures. Fractures are set by either *closed reduction,* which does not involve surgical exposure of the bone, or *open reduction,* which involves the surgical use of plates, screws, or pins to align bone fragments.
3. The most common bone disease is *osteoporosis,* a loss of bone mass (especially spongy bone) causing increasing susceptibility to pathologic fractures. Fractures of the vertebrae, wrist, and hip, and a spinal deformity called *kyphosis* (exaggerated thoracic curvature) commonly result from osteoporosis.
4. Osteoporosis can occur in either sex, any race, and a wide range of ages, but risk factors that increase its incidence include being of female sex, white race, light build, and postmenopausal age, as well as inadequate exercise, low calcium intake, smoking, vitamin C deficiency, and diabetes mellitus.

Testing Your Recall

1. Which cells have a ruffled border and secrete hydrochloric acid?
   a. chondrocytes
   b. osteocytes
   c. osteogenic cells
   d. osteoblasts
   e. osteoclasts
2. The medullary cavity of a child's bone may contain
   a. red bone marrow.
   b. hyaline cartilage.
   c. periosteum.
   d. osteocytes.
   e. articular cartilages.
3. The long bones of the limbs grow in length by cell proliferation and hypertrophy in
   a. the epiphysis.
   b. the epiphyseal line.
   c. the dense bone.
   d. the epiphyseal plate.
   e. the spongy bone.
4. Osteoclasts are most closely related, by common descent, to
   a. osteocytes.
   b. osteogenic cells.
   c. monocytes.
   d. fibroblasts.
   e. osteoblasts.
5. The walls between cartilage lacunae break down in the zone of
   a. cell proliferation.
   b. calcification.
   c. reserve cartilage.
   d. bone deposition.
   e. cell hypertrophy.
6. Which of these does *not* promote bone deposition?
   a. dietary calcium
   b. vitamin D
   c. parathyroid hormone
   d. calcitonin
   e. testosterone
7. A child jumps to the ground from the top of a playground "jungle gym." His leg bones do not shatter mainly because they contain
   a. an abundance of glycosaminoglycans.
   b. young, resilient osteocytes.
   c. an abundance of calcium phosphate.
   d. collagen fibers.
   e. hydroxyapatite crystals.
8. One long bone meets another at its
   a. diaphysis.
   b. epiphyseal plate.
   c. periosteum.
   d. metaphysis.
   e. epiphysis.
9. Calcitriol is made from
   a. calcitonin.
   b. 7-dehydrocholesterol.
   c. hydroxyapatite.
   d. estrogen.
   e. PTH.
10. One sign of osteoporosis is
    a. osteitis deformans.
    b. osteomalacia.
    c. a stress fracture.
    d. kyphosis
    e. a calcium deficiency.
11. Calcium phosphate crystallizes in bone as a mineral called _____.
12. Osteocytes contact each other through channels called _____ in the bone matrix.
13. A bone increases in diameter only by _____ growth, the addition of new surface osteons.
14. Most compact bone is organized in cylindrical units called _____, composed of lamellae encircling a central canal.
15. The _____ glands secrete a hormone that stimulates cells to resorb bone and return its minerals to the blood.
16. The ends of a bone are covered with a layer of hyaline cartilage called the _____.
17. The cells that deposit new bone matrix are called _____.
18. The most common bone disease is _____.
19. The transitional region between epiphyseal cartilage and the primary marrow cavity of a young bone is called the _____.
20. The cranial bones develop from a flat sheet of condensed mesenchyme in a process called _____.

Answers in the Appendix

True or False

Determine which five of the following statements are false, and briefly explain why.

1. Spongy bone is normally covered by compact bone.
2. Most bones develop from hyaline cartilage.
3. Fractures are the most common bone disorder.
4. The growth zone of the long bones of adolescents is the articular cartilage.
5. Osteoclasts develop from osteoblasts.
6. Osteocytes develop from osteoblasts.
7. The protein of the bone matrix is called hydroxyapatite.
8. Blood vessels travel through the central canals of compact bone.
9. Yellow bone marrow has a hemopoietic function.
10. Parathyroid hormone promotes bone resorption and raises blood calcium concentration.

Answers in the Appendix

Testing Your Comprehension

1. Most osteocytes of an osteon are far removed from blood vessels, but still receive blood-borne oxygen and nutrients. Explain how this is possible.
2. Predict what symptoms a person might experience if he or she suffered a degenerative disease in which the articular cartilages were worn away and the fluid between the bones dried up.
3. One of the more common fractures in children and adolescents is an *epiphyseal fracture,* in which the epiphysis of a long bone separates from the diaphysis. Explain why this would be more common in children than in adults.
4. Describe how the arrangement of trabeculae in spongy bone demonstrates the complementarity of form and function.
5. Identify two bone diseases you would expect to see if the epidermis were a completely effective barrier to UV radiation and a person took no dietary supplements to compensate for this. Explain your answer.

Answers at the Online Learning Center

www.mhhe.com/saladinha1

Visit the Online Learning Center for practice tests, answer keys, and other learning aids for this chapter. Enhance your understanding of human anatomy with our interactive art labeling exercises, supplemental photo atlases, web links, puzzles, flashcards, and much more.

7

CHAPTER SEVEN

The Axial Skeleton

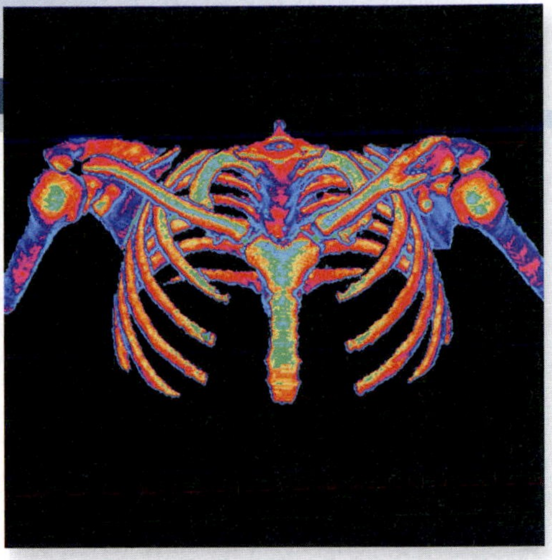

Superior view of the thoracic cage and pectoral girdle (colorized CT scan)

CHAPTER OUTLINE

Overview of the Skeleton 172
- Bones of the Skeletal System 172
- Surface Features of Bones 174

The Skull 175
- Cranial Bones 175
- Facial Bones 184
- Bones Associated with the Skull 186
- Adaptations of the Skull for Bipedalism 186

The Vertebral Column and Thoracic Cage 188
- General Features of the Vertebral Column 188
- General Structure of a Vertebra 189
- Intervertebral Discs 190
- Regional Characteristics of Vertebrae 190
- The Thoracic Cage 194

Developmental and Clinical Perspectives 196
- Development of the Axial Skeleton 196
- Pathology of the Axial Skeleton 200

Chapter Review 203

INSIGHTS

7.1 Clinical Application: Injury to the Ethmoid Bone 183

7.2 Evolutionary Medicine: Evolutionary Significance of the Palate 184

7.3 Clinical Application: Abnormal Spinal Curvatures 189

BRUSHING UP

To understand this chapter, you may find it helpful to review the following concepts:

- Directional Terms *Rostral* and *Caudal* (p. 26)
- The Axial and Appendicular Body Regions (p. 26)
- General Features of Bones (p. 153)
- Endochondral and Intramembranous Ossification (p. 159)

A knowledge of skeletal anatomy will be useful as you study later chapters. It provides a point of reference for studying the gross anatomy of other organ systems because many organs are named for their relationships to nearby bones. The subclavian artery and vein, for example, are located beneath the clavicles; the temporalis muscle is attached to the temporal bone; the ulnar nerve and radial artery travel beside the ulna and radius of the forearm; and the frontal, parietal, temporal, and occipital lobes of the brain are named for adjacent bones of the cranium. An understanding of how the muscles produce body movements also depends on knowledge of skeletal anatomy. In addition, the positions, shapes, and processes of bones can serve as landmarks for a clinician in determining where to give an injection or record a pulse, what to look for in an X ray, or how to perform physical therapy and other medical procedures.

Overview of the Skeleton

Objectives

When you have completed this section, you should be able to

- state the approximate number of bones in the adult body;
- explain why this number varies with age and from one person to another; and
- define several terms that denote surface features of bones.

The skeleton (fig. 7.1) is divided into two regions: the **axial skeleton** and the **appendicular skeleton.** The axial skeleton, studied in this chapter, forms the central supporting axis of the body and includes the skull, vertebral column, and thoracic cage (ribs and sternum). The appendicular skeleton, studied in chapter 8, includes the bones of the upper limb and pectoral girdle, and bones of the lower limb and pelvic girdle.

Bones of the Skeletal System

It is often stated that there are 206 bones in the skeleton, but this is only a typical adult count. At birth there are about 270, and even more bones form during childhood. With age, however, the number decreases as separate bones fuse. For example, each half of the adult pelvis is a single bone called the *os coxae* (oss COC-see), which results from the fusion of three childhood bones—the ilium, ischium, and pubis. The fusion of several bones, completed by late adolescence to the mid-20s, brings about the average adult number of 206. These bones are listed in table 7.1

This number varies even among adults. One reason is the development of **sesamoid**[1] **bones**—bones that form within some tendons in response to stress. The patella (kneecap) is the largest of these; most of the others are small, rounded bones in such locations as the knuckles. Another reason for adult variation is that some people have extra bones in the skull called **sutural** (SOO-chure-ul), or **wormian,**[2] **bones** (see fig. 7.6).

Table 7.1
Bones of the Adult Skeletal System

Axial Skeleton	
Skull	*Total 22*
Cranial bones	
Frontal bone (1)	
Parietal bones (2)	
Occipital bone (1)	
Temporal bones (2)	
Sphenoid bone (1)	
Ethmoid bone (1)	
Facial bones	
Maxillae (2)	
Palatine bones (2)	
Zygomatic bones (2)	
Lacrimal bones (2)	
Nasal bones (2)	
Vomer (1)	
Inferior nasal conchae (2)	
Mandible (1)	
Auditory Ossicles	*Total 6*
Malleus (2)	
Incus (2)	
Stapes (2)	
Hyoid Bone (1)	*Total 1*
Vertebral Column	*Total 26*
Cervical vertebrae (7)	
Thoracic vertebrae (12)	
Lumbar vertebrae (5)	
Sacrum (1)	
Coccyx (1)	
Thoracic Cage	*Total 25*
Ribs (24)	
Sternum (1)	
Appendicular Skeleton	
Pectoral Girdle	*Total 4*
Scapulae (2)	
Clavicles (2)	
Upper Limbs	*Total 60*
Humerus (2)	
Radius (2)	
Ulna (2)	
Carpals (16)	
Metacarpals (10)	
Phalanges (28)	
Pelvic Girdle	*Total 2*
Ossa coxae (2)	
Lower Limbs	*Total 60*
Femur (2)	
Patella (2)	
Tibia (2)	
Fibula (2)	
Tarsals (14)	
Metatarsals (10)	
Phalanges (28)	
	Grand Total: 206

[1] *sesam* = sesame seed + *oid* = resembling
[2] Ole Worm (1588–1654), Danish physician

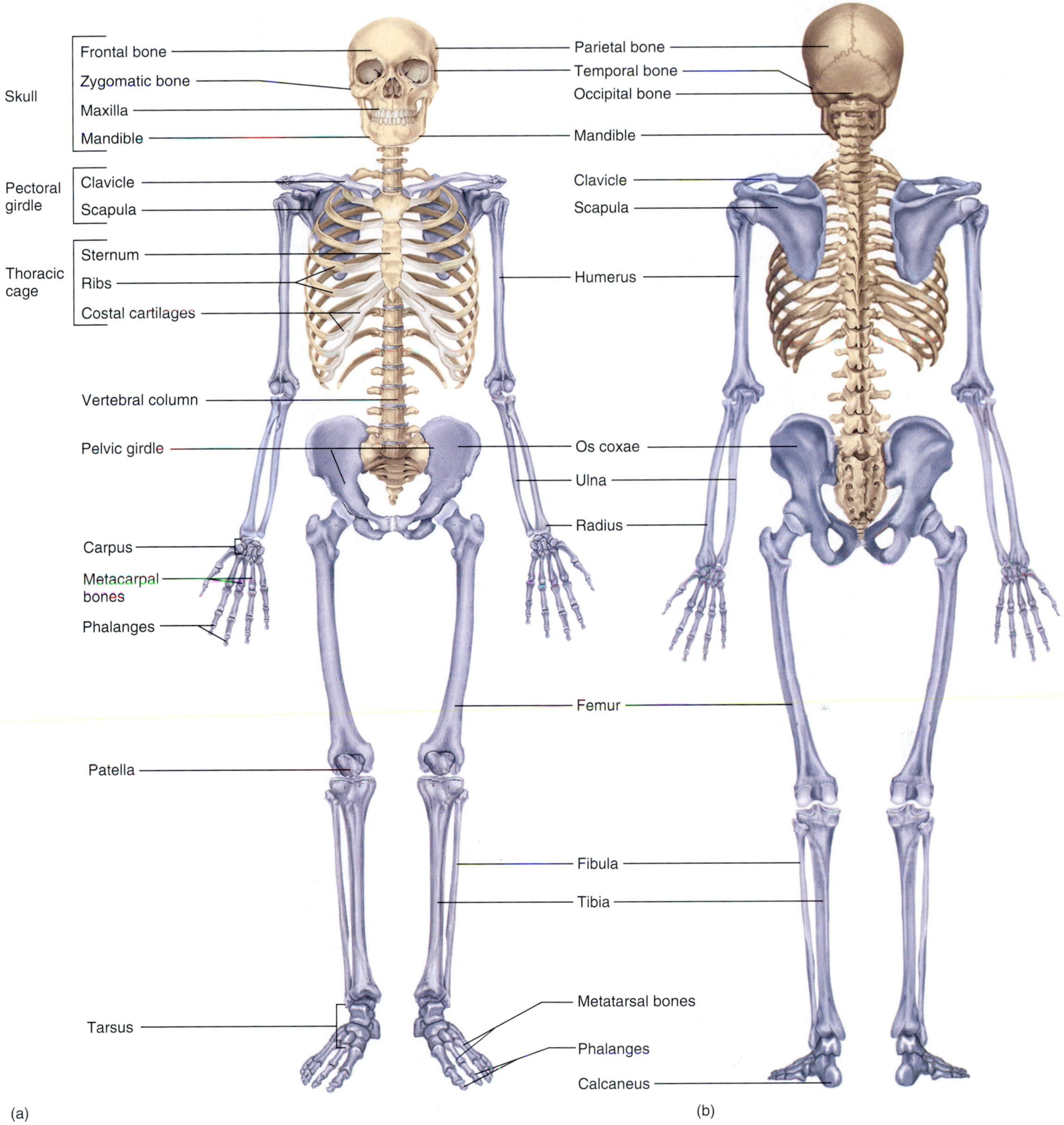
Skull
Frontal bone
Zygomatic bone
Maxilla
Mandible
Pectoral girdle
Clavicle
Scapula
Thoracic cage
Sternum
Ribs
Costal cartilages
Vertebral column
Pelvic girdle
Carpus
Metacarpal bones
Phalanges
Patella
Tarsus
(a)
Parietal bone
Temporal bone
Occipital bone
Mandible
Clavicle
Scapula
Humerus
Os coxae
Ulna
Radius
Femur
Fibula
Tibia
Metatarsal bones
Phalanges
Calcaneus
(b)

FIGURE 7.1

The Adult Skeleton. (*a*) Ventral view. (*b*) Dorsal view. The appendicular skeleton is colored *blue,* and the rest is axial skeleton.

TABLE 7.2

Surface Features (markings) of Bones

Term	Description and Example
Articulations	
Condyle	A rounded knob (occipital condyles of the skull)
Facet	A smooth, flat, slightly concave or convex articular surface (articular facets of the vertebrae)
Head	The prominent expanded end of a bone, sometimes rounded (head of the femur)
Extensions and Projections	
Crest	A narrow ridge (iliac crest of the pelvis)
Epicondyle	A projection superior to a condyle (medial epicondyle of the femur)
Line	A slightly raised, elongated ridge (nuchal lines of the skull)
Process	Any bony prominence (mastoid process of the skull)
Protuberance	A bony outgrowth or protruding part (mental protuberance of the chin)
Spine	A sharp, slender, or narrow process (spine of the scapula)
Trochanter	Two massive processes unique to the femur
Tubercle	A small, rounded process (greater tubercle of the humerus)
Tuberosity	A rough surface (tibial tuberosity)
Depressions	
Alveolus	A pit or socket (tooth socket)
Fossa	A shallow, broad, or elongated basin (mandibular fossa)
Fovea	A small pit (fovea capitis of the femur)
Sulcus	A groove for a tendon, nerve, or blood vessel (intertubercular sulcus of the humerus)
Passages	
Canal	A tubular passage or tunnel in a bone (condylar canal of the skull)
Fissure	A slit through a bone (orbital fissures behind the eye)
Foramen	A hole through a bone, usually round (foramen magnum of the skull)
Meatus	An opening into a canal (acoustic meatus of the ear)

Surface Features of Bones

The surface of a bone may exhibit a variety of ridges, spines, bumps, depressions, canals, pores, slits, and articular surfaces, often called *surface markings.* It is important to know the names of these features because later descriptions of joints, muscle attachments, and the routes traveled by nerves and blood vessels are based on this terminology. The terms for the most common of these features are listed in table 7.2, and several of them are illustrated in figure 7.2.

As you study the skeleton, use yourself as a model. You can easily palpate (feel) many of the bones and some of their details through the skin. Rotate your forearm, cross your legs, palpate your skull and wrist, and think about what is happening beneath the surface or what you can feel through the skin. You will gain the most from this chapter (and indeed, the entire book) if you are conscious of your own body in relation to what you are studying.

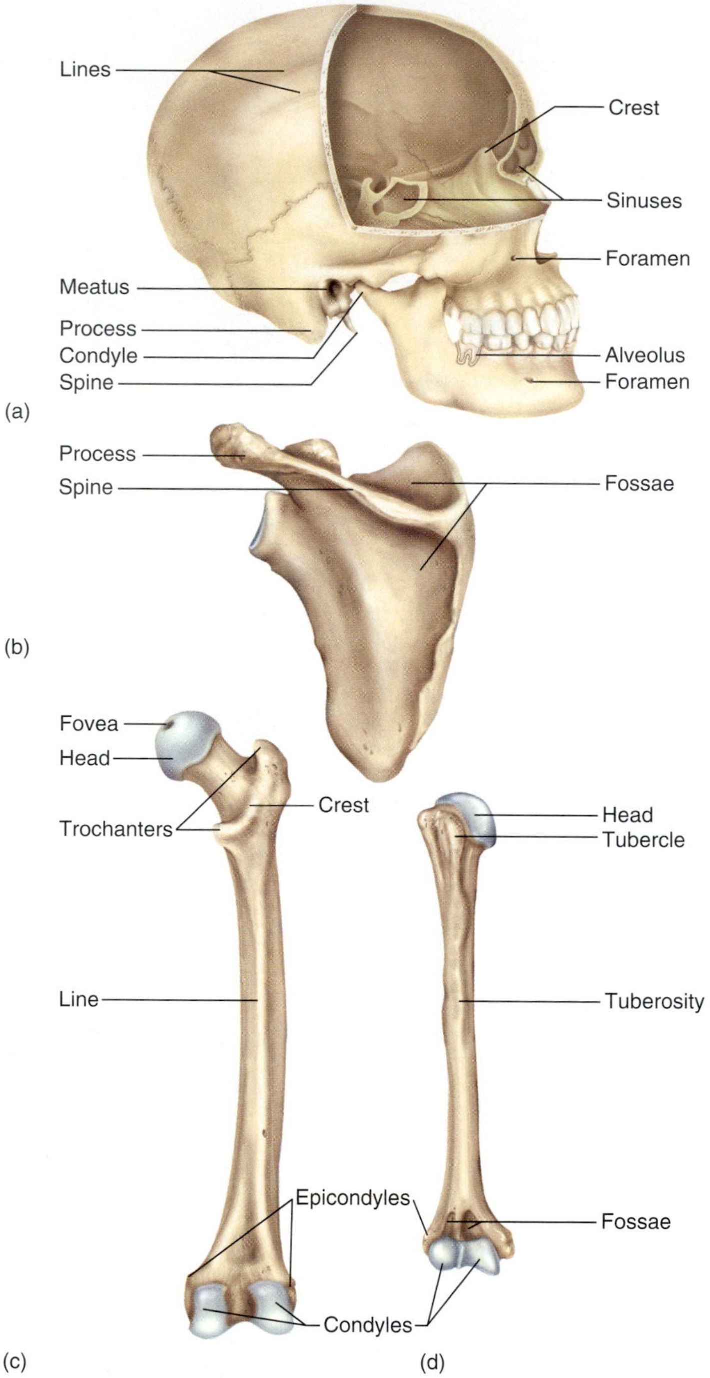

FIGURE 7.2
Surface Features of Bones. (*a*) Skull. (*b*) Scapula. (*c*) Femur. (*d*) Humerus.

Before You Go On

Answer the following questions to test your understanding of the preceding section:

1. Name the major components of the axial skeleton. Name those of the appendicular skeleton.

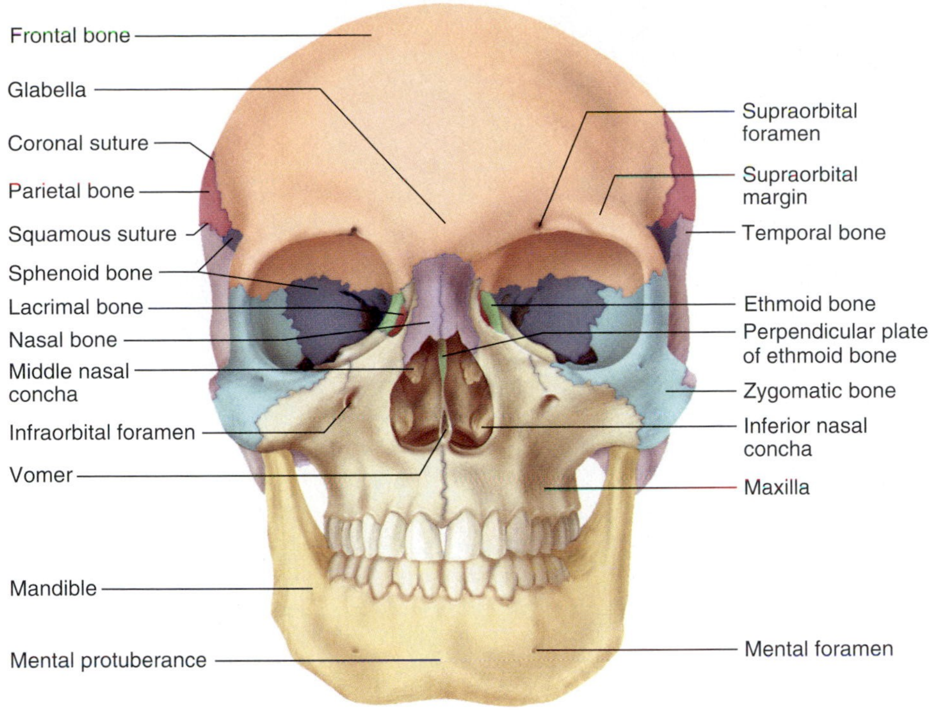
Frontal bone
Glabella
Coronal suture
Parietal bone
Squamous suture
Sphenoid bone
Lacrimal bone
Nasal bone
Middle nasal concha
Infraorbital foramen
Vomer
Mandible
Mental protuberance
Supraorbital foramen
Supraorbital margin
Temporal bone
Ethmoid bone
Perpendicular plate of ethmoid bone
Zygomatic bone
Inferior nasal concha
Maxilla
Mental foramen

FIGURE 7.3
The Skull, Anterior View.

2. Explain why an adult does not have as many bones as a child does. Explain why one adult may have more bones than another adult of the same age.
3. Briefly describe each of the following bone features: condyle, epicondyle, process, tubercle, fossa, sulcus, and foramen.

THE SKULL

Objectives

When you have completed this section, you should be able to

- name the bones of the skull and their anatomical features;
- identify the cavities within the skull and in some of its individual bones;
- identify the sutures that join bones of the skull; and
- describe some adaptations of the skull for upright locomotion.

The skull is the most complex part of the skeleton. Figures 7.3 to 7.6 present an overview of its general anatomy. Although the skull may seem to consist only of the mandible (lower jaw) and "the rest," it is composed of 22 bones and sometimes more. Most of them are rigidly joined by **sutures** (SOO-chures), joints that appear as seams on the cranial surface (fig. 7.4). These are important landmarks in the descriptions that follow.

The skull contains several prominent cavities (fig. 7.7). The largest, with an adult volume of about 1,300 mL, is the **cranial cavity,** which encloses the brain. Other cavities include the **orbits** (eye sockets), **nasal cavity, buccal** (BUCK-ul) **cavity** (mouth), **middle-** and **inner-ear cavities,** and **paranasal sinuses.** The paranasal sinuses are named for the bones in which they occur (fig. 7.8)—the **frontal, ethmoid, sphenoid,** and **maxillary sinuses.** These cavities are connected with the nasal cavity, lined by a mucous membrane, and filled with air. They lighten the anterior portion of the skull and act as chambers that add resonance to the voice.

Bones of the skull have especially conspicuous **foramina**—singular, *foramen* (fo-RAY-men)—holes that allow passage for nerves and blood vessels. The major foramina are summarized in table 7.3. The details of this table will mean more to you when you study cranial nerves and blood vessels in later chapters.

Cranial Bones

The cranial cavity is enclosed by the **cranium**[3] (braincase), which protects the brain and associated sensory organs. The cranium is composed of eight bones called the **cranial bones:**

1 frontal bone	1 occipital bone
2 parietal bones	1 sphenoid bone
2 temporal bones	1 ethmoid bone

[3]*crani* = helmet

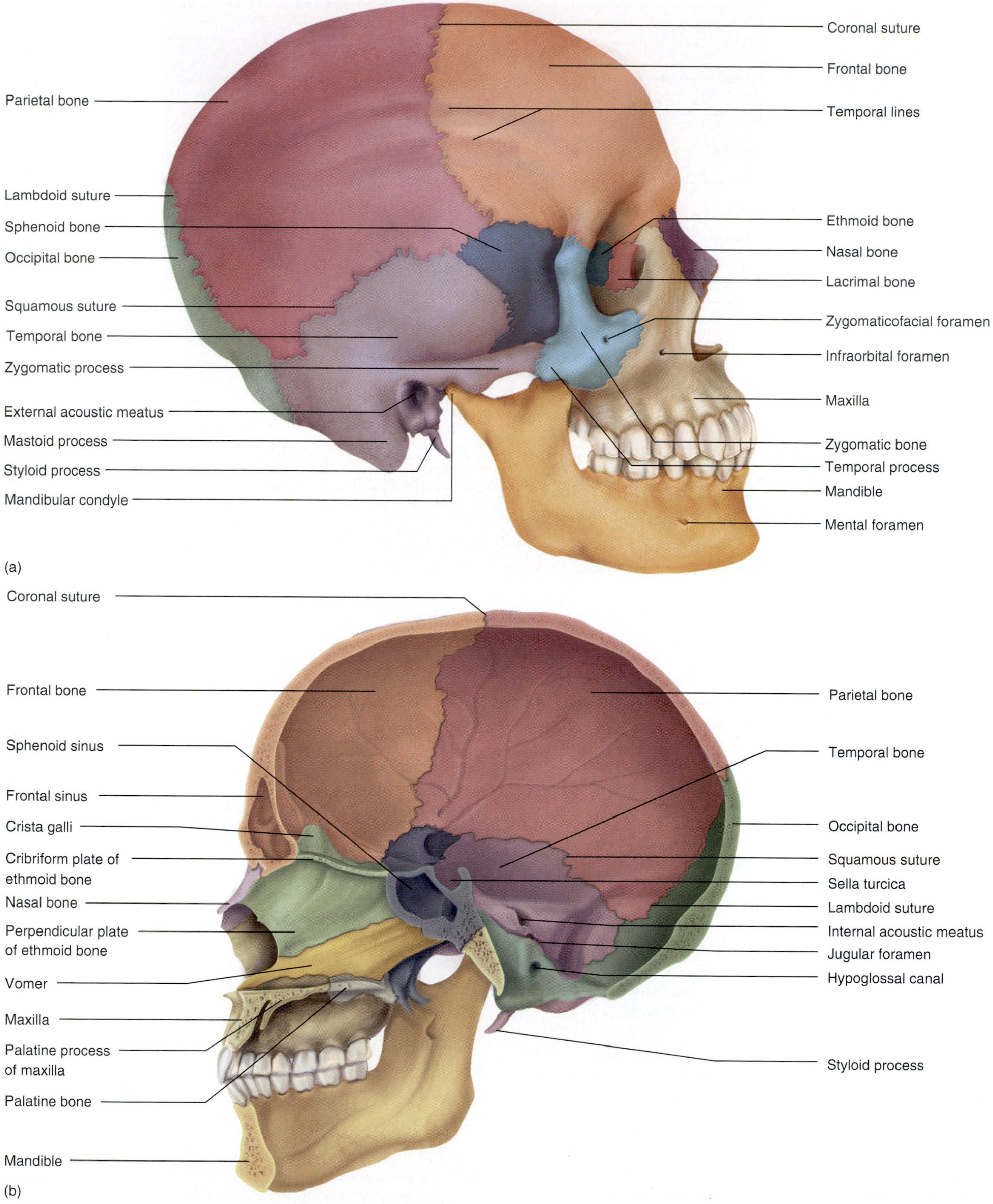
Coronal suture
Frontal bone
Parietal bone
Temporal lines
Lambdoid suture
Sphenoid bone
Occipital bone
Ethmoid bone
Nasal bone
Lacrimal bone
Squamous suture
Temporal bone
Zygomaticofacial foramen
Infraorbital foramen
Zygomatic process
External acoustic meatus
Maxilla
Mastoid process
Zygomatic bone
Styloid process
Temporal process
Mandible
Mandibular condyle
Mental foramen
(a)
Coronal suture
Frontal bone
Parietal bone
Sphenoid sinus
Temporal bone
Frontal sinus
Crista galli
Occipital bone
Cribriform plate of
ethmoid bone
Squamous suture
Sella turcica
Nasal bone
Lambdoid suture
Internal acoustic meatus
Perpendicular plate
of ethmoid bone
Jugular foramen
Hypoglossal canal
Vomer
Maxilla
Palatine process
of maxilla
Styloid process
Palatine bone
Mandible
(b)

FIGURE 7.4

The Skull. (*a*) Right lateral view. (*b*) Interior of the right half.

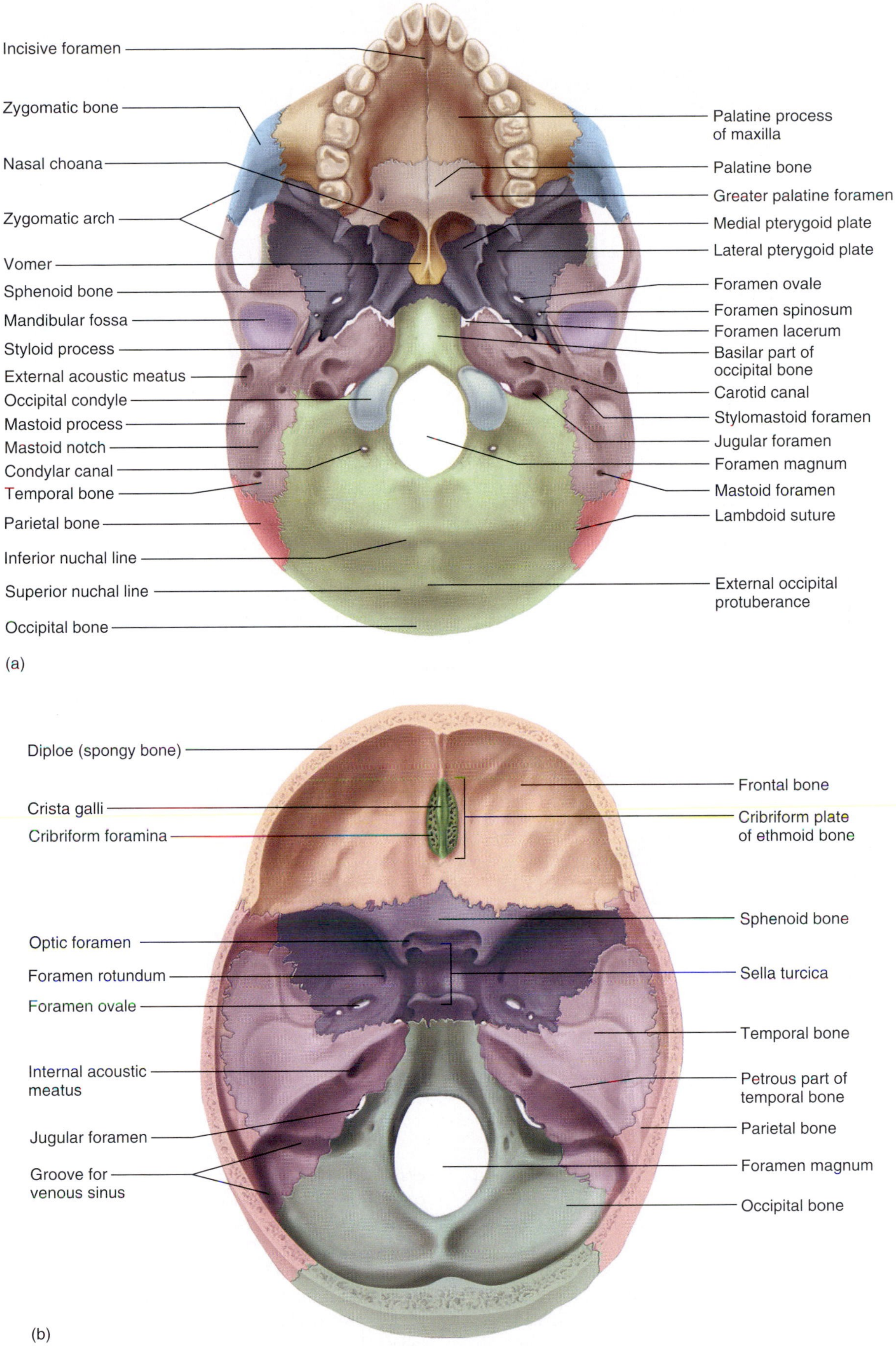

Incisive foramen
Zygomatic bone
Nasal choana
Zygomatic arch
Vomer
Sphenoid bone
Mandibular fossa
Styloid process
External acoustic meatus
Occipital condyle
Mastoid process
Mastoid notch
Condylar canal
Temporal bone
Parietal bone
Inferior nuchal line
Superior nuchal line
Occipital bone
Palatine process of maxilla
Palatine bone
Greater palatine foramen
Medial pterygoid plate
Lateral pterygoid plate
Foramen ovale
Foramen spinosum
Foramen lacerum
Basilar part of occipital bone
Carotid canal
Stylomastoid foramen
Jugular foramen
Foramen magnum
Mastoid foramen
Lambdoid suture
External occipital protuberance
(a)
Diploe (spongy bone)
Crista galli
Cribriform foramina
Optic foramen
Foramen rotundum
Foramen ovale
Internal acoustic meatus
Jugular foramen
Groove for venous sinus
Frontal bone
Cribriform plate of ethmoid bone
Sphenoid bone
Sella turcica
Temporal bone
Petrous part of temporal bone
Parietal bone
Foramen magnum
Occipital bone
(b)

Figure 7.5

Base of the Skull. (*a*) Inferior view. (*b*) Internal view of the cranial floor.

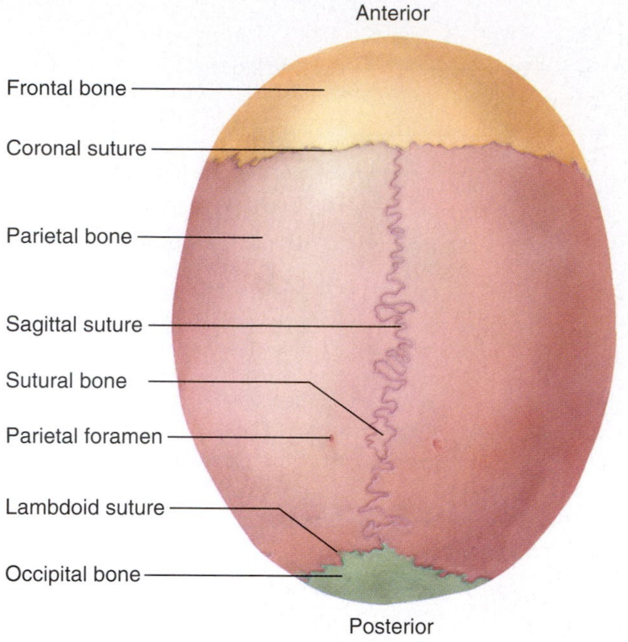

FIGURE 7.6
The Calvaria (skullcap), Superior View.

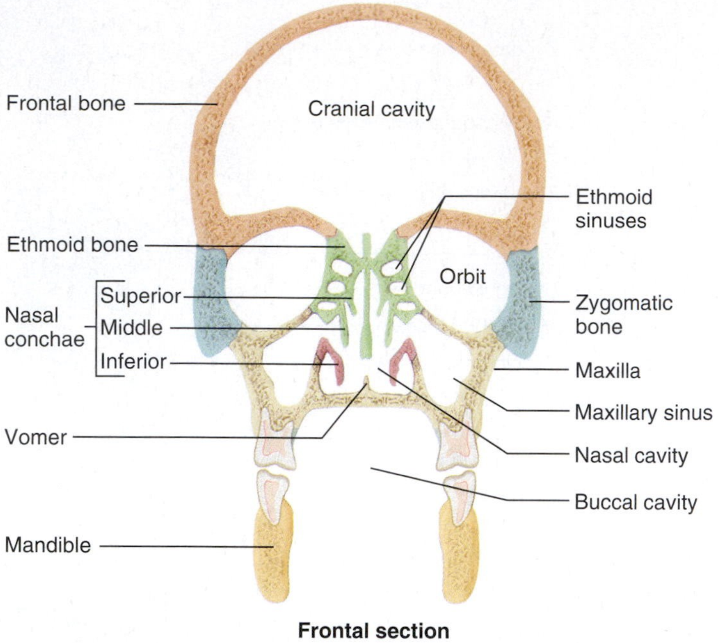

FIGURE 7.7
Major Cavities of the Skull, Frontal Section.

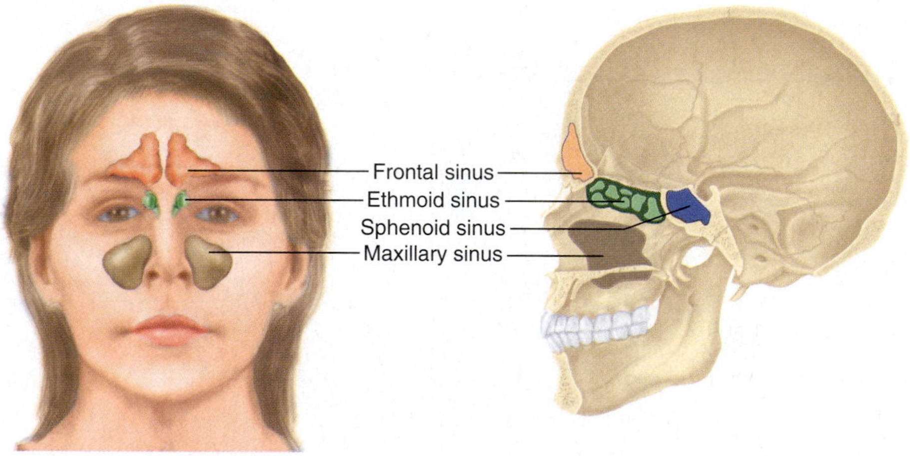

FIGURE 7.8
The Paranasal Sinuses.

The delicate brain tissue does not directly touch the cranial bones but is separated from them by three membranes, the *meninges* (meh-NIN-jeez) (see chapter 13). The thickest and toughest of these, the *dura mater* (DUE-rah MAH-tur), is essentially the periosteum of the cranial bones. It lies loosely against the cranium in most places but is attached to it at a few points.

The cranium consists of two major parts—the calvaria and the base. The **calvaria**[4] (skullcap) forms the roof and walls (see fig. 7.6). In study skulls it is often sawed so that part of it can be lifted off for examination of the interior. This reveals the **base** (floor) of the cranial cavity (see fig. 7.5*b*), which is divided into three basins called **cranial fossae.** The fossae correspond to the contour of the inferior surface of the brain (fig. 7.9). The relatively shallow *anterior cranial fossa* is crescent-shaped and accommodates the frontal lobes of the brain. The *middle cranial fossa,* which drops abruptly deeper, is shaped like a pair of outstretched bird's wings and accommodates the temporal lobes. The *posterior cranial fossa* is deepest and houses a large posterior division of the brain called the cerebellum.

We now consider the eight cranial bones and their distinguishing features.

[4] *calvar* = bald, skull

TABLE 7.3

Foramina of the Skull and the Nerves and Blood Vessels Transmitted Through Them

Bones and Their Foramina*	Structures Transmitted
Frontal Bone	
Supraorbital foramen or notch	Supraorbital nerve, artery, and vein; ophthalmic nerve
Parietal Bone	
Parietal foramen	Emissary vein of superior longitudinal sinus
Temporal Bone	
Carotid canal	Internal carotid artery
External acoustic meatus	Sound waves to eardrum
Internal acoustic meatus	Vestibulocochlear nerve; internal auditory vessels
Stylomastoid foramen	Facial nerve
Mastoid foramen	Meningeal artery; vein from sigmoid sinus
Temporal–Occipital Region	
Jugular foramen	Internal jugular vein; glossopharyngeal, vagus, and accessory nerves
Temporal–Occipital–Sphenoid Region	
Foramen lacerum	No major nerves or vessels; closed by cartilage
Occipital Bone	
Foramen magnum	Spinal cord; accessory nerve; vertebral arteries
Hypoglossal canal	Hypoglossal nerve to muscles of tongue
Condylar canal	Vein from transverse sinus
Sphenoid Bone	
Foramen ovale	Mandibular division of trigeminal nerve; accessory meningeal artery
Foramen rotundum	Maxillary division of trigeminal nerve
Foramen spinosum	Middle meningeal artery; spinosal nerve; part of trigeminal nerve
Optic foramen	Optic nerve; ophthalmic artery
Superior orbital fissure	Oculomotor, trochlear, and abducens nerves; ophthalmic division of trigeminal nerve; ophthalmic veins
Ethmoid Bone	
Olfactory foramina	Olfactory nerves
Maxilla	
Infraorbital foramen	Infraorbital nerve and vessels; maxillary division of trigeminal nerve
Incisive foramen	Nasopalatine nerves
Maxilla–Sphenoid Region	
Inferior orbital fissure	Infraorbital nerve; zygomatic nerve; infraorbital vessels
Lacrimal Bone	
Lacrimal foramen	Tear duct leading to nasal cavity
Palatine Bone	
Greater palatine foramen	Palatine nerves
Zygomatic Bone	
Zygomaticofacial foramen	Zygomaticofacial nerve
Zygomaticotemporal foramen	Zygomaticotemporal nerve
Mandible	
Mental foramen	Mental nerve and vessels
Mandibular foramen	Inferior alveolar nerves and vessels to the lower teeth

* When two or more bones are listed together (for example, temporal–occipital), it indicates that the foramen passes between them.

FRONTAL BONE

The **frontal bone** extends from the forehead back to a prominent *coronal suture,* which crosses the crown of the head from right to left and joins the frontal bone to the parietal bones (see figs. 7.3 and 7.4). The frontal bone forms the anterior wall and about one-third of the roof of the cranial cavity, and it turns inward to form nearly all of the anterior cranial fossa and the roof of the orbit. Deep to the eyebrows it has a ridge called the **supraorbital margin.** The center of each margin is perforated by a single **supraorbital foramen** (see figs. 7.3 and 7.14), which provides passage for a nerve, artery, and vein. In some people, the edge of this foramen breaks through the margin of the orbit and forms a *supraorbital notch.* The frontal

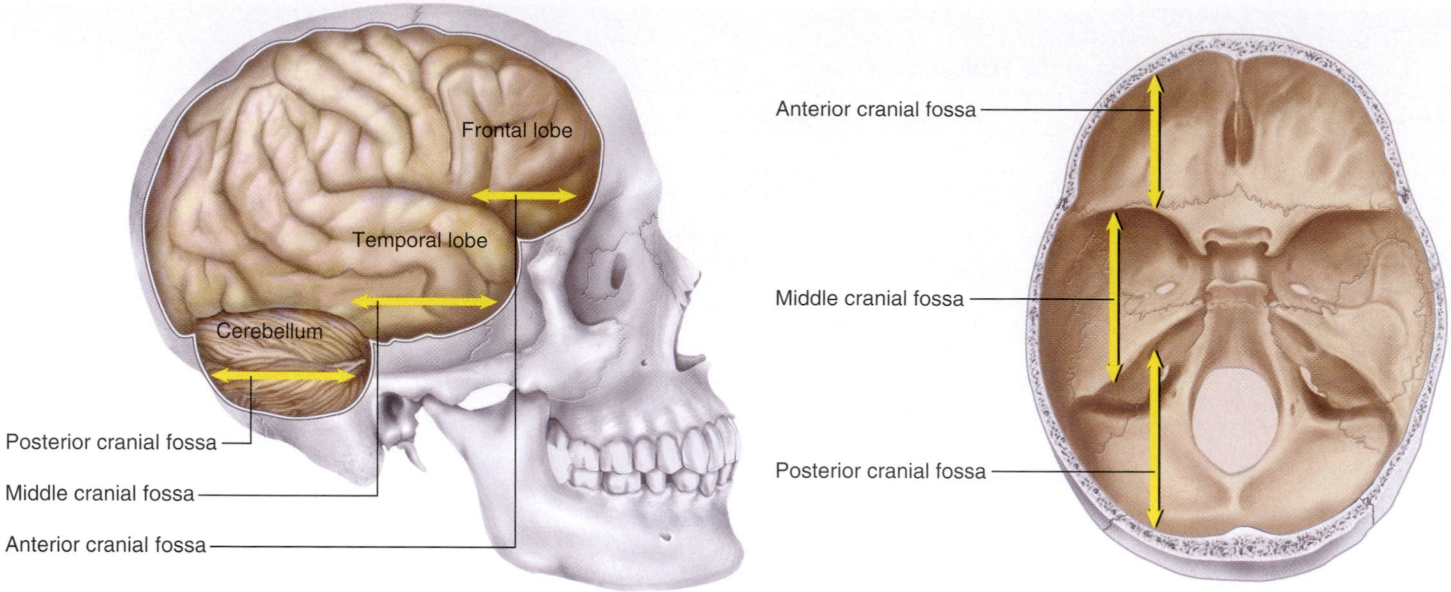

FIGURE 7.9
Cranial Fossae. The three fossae conform to the contours of the base of the brain.

bone also contains the frontal sinus. You may not see this on some study skulls. It is absent from some people, and on some skulls the calvaria is cut too high to show the sinus. Along the cut edge of the calvaria, you can see the diploe—the layer of spongy bone in the middle of the cranial bones (see fig. 7.5*b*).

PARIETAL BONES

The right and left **parietal** (pa-RYE-eh-tul) **bones** form most of the cranial roof and part of its walls (see figs. 7.4 and 7.6). Each is bordered by four sutures that join it to the neighboring bones: (1) the **sagittal suture** between the parietal bones; (2) the **coronal**[5] **suture** at the anterior margin; the **lambdoid**[6] (LAM-doyd) **suture** at the posterior margin; and (4) the **squamous suture** laterally. Small sutural (wormian) bones are often seen along the sagittal and lambdoid sutures, like little islands of bone with the suture lines passing around them. Internally, the parietal and frontal bones have markings that look a bit like aerial photographs of river tributaries (see fig. 7.4*b*). These represent places where the bone has been molded around blood vessels of the meninges.

Externally, the parietal bones have few features. A **parietal foramen** sometimes occurs near the corner of the lambdoid and sagittal sutures (see fig. 7.6). A pair of slight thickenings, the superior and inferior **temporal lines,** form an arc across the parietal and frontal bones (see fig. 7.4*a*). They mark the attachment of the large, fan-shaped *temporalis* muscle, a chewing muscle that passes between the zygomatic arch and temporal bone and inserts on the mandible.

TEMPORAL BONES

If you palpate your skull just above and anterior to the ear—that is, the temporal region—you can feel the **temporal bone,** which forms the lower wall and part of the floor of the cranial cavity (fig. 7.10). The temporal bone derives its name from the fact that people often develop their first gray hairs on the temples with the passage of time.[7] The relatively complex shape of the temporal bone is best understood by dividing it into four parts:

1. The **squamous**[8] **part** (which you just palpated) is relatively flat and vertical. It is encircled by the squamous suture. It bears two prominent features: (1) the **zygomatic process,** which extends anteriorly to form part of the **zygomatic arch** (cheekbone), and (2) the **mandibular fossa,** a depression where the mandible articulates with the cranium.
2. The **tympanic**[9] **part** is a small ring of bone that borders the **external acoustic meatus** (me-AY-tus), the opening into the ear canal. It has a pointed spine on its inferior surface, the **styloid process,** named for its resemblance to the stylus used by ancient Greeks and Romans to write on wax tablets. The styloid process provides attachment for muscles of the tongue, pharynx, and hyoid bone.
3. The **mastoid**[10] **part** lies posterior to the tympanic part. It bears a heavy **mastoid process,** which you can palpate as a prominent lump behind the earlobe. It is filled with small air sinuses that communicate with the middle-ear cavity. These sinuses are subject to infection and inflammation

[5]*corona* = crown
[6]Shaped like Greek letter lambda (λ)

[7]*tempor* = time
[8]*squam* = flat + *ous* = characterized by
[9]*tympan* = drum (eardrum) + *ic* = pertaining to
[10]*mast* = breast + *oid* = resembling

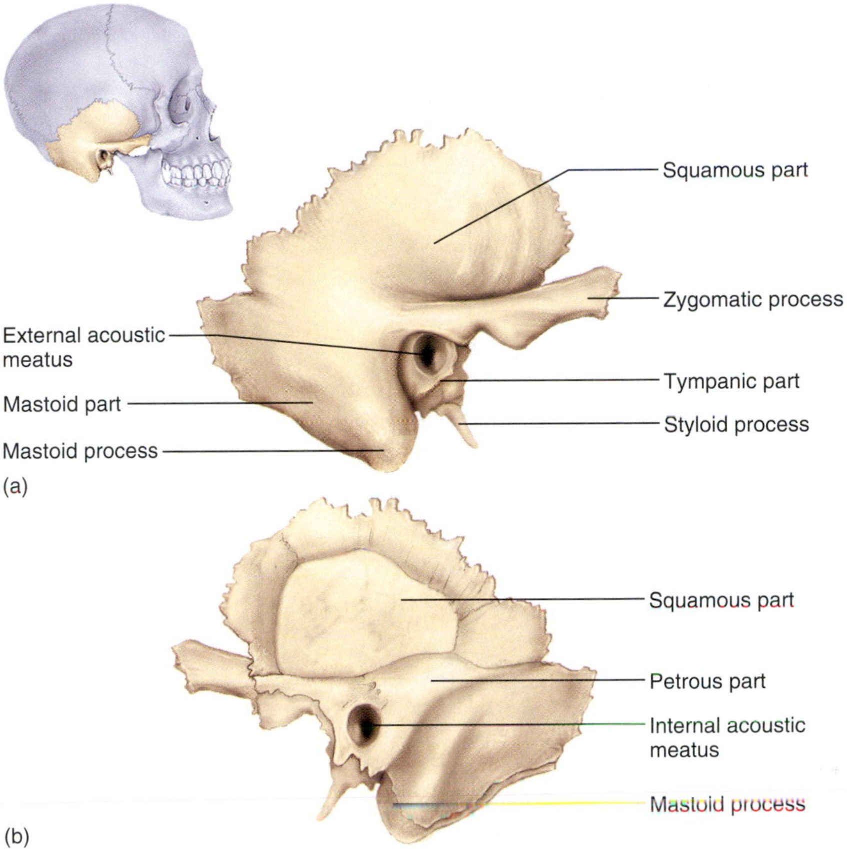

Squamous part
Zygomatic process
External acoustic meatus
Tympanic part
Mastoid part
Styloid process
Mastoid process
(a)
Squamous part
Petrous part
Internal acoustic meatus
Mastoid process
(b)

FIGURE 7.10
The Right Temporal Bone. (*a*) Lateral (external) view. (*b*) Medial (internal) view.

(mastoiditis), which can erode the bone and spread to the brain. Ventrally, there is a groove called the **mastoid notch** medial to the mastoid process (see fig. 7.5*a*). It is the origin of the *digastric muscle,* a muscle that opens the mouth. The notch is perforated by the **stylomastoid foramen** at its anterior end and the **mastoid foramen** at its posterior end.

4. The **petrous**[11] **part** can be seen in the cranial floor, where it resembles a little mountain range separating the middle cranial fossa from the posterior fossa (fig. 7.10*b*). It houses the middle- and inner-ear cavities. The **internal acoustic meatus,** an opening on its posteromedial surface, allows passage of the vestibulocochlear (vess-TIB-you-lo-COC-lee-ur) nerve, which carries sensations of hearing and balance from the inner ear to the brain. On the ventral surface of the petrous part are two prominent foramina named for the major blood vessels that pass through them (see fig. 7.5*a*): (1) The **carotid canal** is a passage for the internal carotid artery, a major blood supply to the brain. This artery is so close to the inner ear that you can sometimes hear the pulsing of its blood when your ear is resting on a pillow or your heart is beating hard. (2) The **jugular foramen** is a large, irregular opening just medial to the styloid process, between the temporal and occipital bones. Blood from the brain drains through this foramen into the internal jugular vein of the neck. Three cranial nerves also pass through this foramen.

[11] *petr* = stone, rock + *ous* = like

OCCIPITAL BONE

The **occipital** (oc-SIP-ih-tul) **bone** forms the rear of the skull *(occiput)* and much of its base (see fig. 7.5). Its most conspicuous feature is a large opening, the **foramen magnum** (literally "big hole"), which admits the spinal cord to the cranial cavity and provides a point of attachment for the dura mater. An important consideration in treatment of head injuries is swelling of the brain. Since the cranium cannot enlarge, swelling puts pressure on the brain and results in even more tissue damage. Severe swelling may force the brainstem out through the foramen magnum, usually with fatal consequences.

The occipital bone continues anterior to the foramen magnum as a thick medial plate, the **basilar part.** On each side of the foramen magnum is a smooth knob called the **occipital condyle** (CON-dile), where the skull rests on the vertebral column. At the

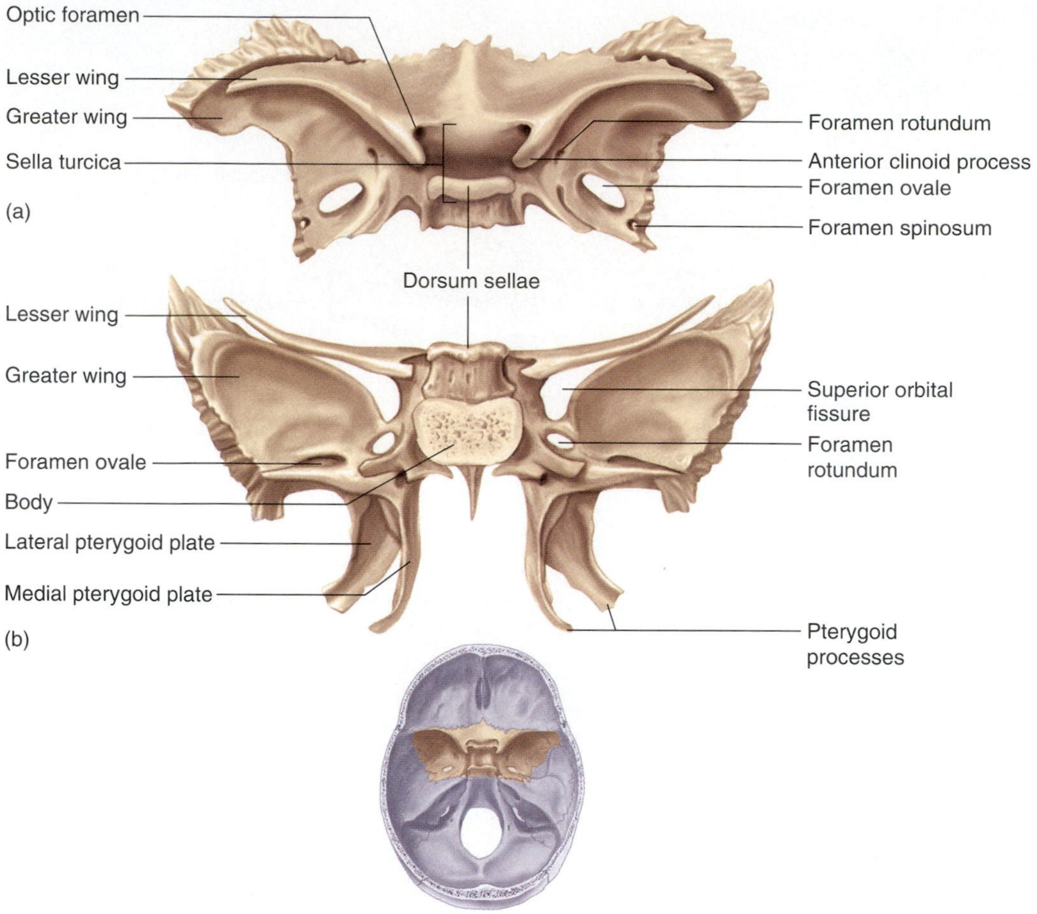
Optic foramen
Lesser wing
Greater wing
Sella turcica
(a)
Foramen rotundum
Anterior clinoid process
Foramen ovale
Foramen spinosum
Dorsum sellae
Lesser wing
Greater wing
Foramen ovale
Body
Lateral pterygoid plate
Medial pterygoid plate
(b)
Superior orbital fissure
Foramen rotundum
Pterygoid processes

FIGURE 7.11
The Sphenoid Bone. (*a*) Superior view. (*b*) Posterior (caudal) view.

anterolateral edge of each condyle is a **hypoglossal**[12] **canal,** named for the hypoglossal nerve that passes through it to supply the muscles of the tongue. In some people, a **condylar** (CON-dih-lur) **canal** occurs posterior to each occipital condyle.

Internally, the occipital bone displays impressions left by large venous sinuses that drain blood from the brain (see fig. 7.5*b*). One of these grooves travels along the midsagittal line. Just before reaching the foramen magnum, it branches into right and left grooves that wrap around the occipital bone like outstretched arms before terminating at the jugular foramina.

Other features of the occipital bone can be palpated on the back of your head. One is a prominent medial bump called the **external occipital protuberance**—the attachment for the **nuchal**[13] (NEW-kul) **ligament,** which binds the skull to the vertebral column. A ridge, the **superior nuchal line,** can be traced horizontally from the external occipital protuberance toward the mastoid process (see fig. 7.5*a*). It defines the superior limit of the neck and provides attachment for several neck and back muscles to the skull. By pulling down on the occipital bone, some of these muscles help to keep the head erect. The **inferior nuchal line** provides attachment for some of the deep neck muscles. This inconspicuous ridge cannot be palpated on the living body but is visible on an isolated skull.

SPHENOID BONE

The **sphenoid**[14] (SFEE-noyd) **bone** has a complex shape with a thick medial **body** and outstretched **greater** and **lesser wings,** which give the bone as a whole a somewhat ragged mothlike shape. Most of it is best seen from the superior perspective (fig. 7.11*a*). In this view, the lesser wings form the posterior margin of the anterior cranial fossa and end at a sharp bony crest, where the sphenoid drops abruptly to the greater wings. These form about half of the middle cranial fossa (the temporal bone forming the rest) and are perforated by several foramina to be discussed shortly.

The greater wing forms part of the lateral surface of the cranium just anterior to the temporal bone (see fig. 7.4*a*). The lesser wing forms the posterior wall of the orbit and contains the **optic foramen,** which permits passage of the optic nerve and ophthalmic artery (see fig. 7.14). Superiorly, a pair of bony spines of the lesser wing called the **anterior clinoid processes** appear to guard the op-

[12]*hypo* = below + *gloss* = tongue
[13]*nucha* = back of the neck
[14]*sphen* = wedge + *oid* = resembling

tic foramina. A gash in the posterior wall of the orbit, the **superior orbital fissure,** angles upward lateral to the optic foramen. It serves as a passage for nerves that supply some of the muscles that move the eyes.

The body of the sphenoid has a saddlelike prominence named the **sella turcica**[15] (SEL-la TUR-sih-ca). It consists of a deep pit called the *hypophyseal fossa,* which houses the pituitary gland, a raised anterior margin called the *tuberculum sellae* (too-BUR-cu-lum SEL-lee), and a posterior margin called the *dorsum sellae.* In life, a fibrous membrane is stretched over the sella turcica. A stalk penetrates the membrane to connect the pituitary gland to the floor of the brain.

Lateral to the sella turcica, the sphenoid is perforated by several foramina (see fig. 7.5). The **foramen rotundum** and **foramen ovale** (oh-VAY-lee) are passages for two branches of the trigeminal nerve. The **foramen spinosum,** about the diameter of a pencil lead, provides passage for an artery of the meninges. An irregular gash called the **foramen lacerum**[16] (LASS-eh-rum) occurs at the junction of the sphenoid, temporal, and occipital bones. It is filled with cartilage in life and transmits no major vessels or nerves.

In an inferior view, the sphenoid can be seen just anterior to the basilar part of the occipital bone. The internal openings of the nasal cavity seen here are called the **nasal choanae**[17] (co-AH-nee), or **internal nares.** Lateral to each choana, the sphenoid bone exhibits a pair of parallel plates—the **medial pterygoid**[18] (TERR-ih-goyd) **plate** and **lateral pterygoid plate** (see fig. 7.5*a*). These provide attachment for some of the jaw muscles. The sphenoid sinus occurs within the body of the sphenoid bone.

ETHMOID BONE

The **ethmoid**[19] (ETH-moyd) **bone** is located between the orbital cavities and forms the roof of the nasal cavity (fig. 7.12). An inferior projection of the ethmoid, called the **perpendicular plate,** forms the superior part of the **nasal septum,** which divides the nasal cavity into right and left **nasal fossae** (FOSS-ee). Three curled, scroll-like **nasal conchae**[20] (CON-kee), or **turbinate**[21] **bones,** project into each fossa from the lateral wall (see figs. 7.7 and 7.13). The superior and middle conchae are extensions of the ethmoid bone. The inferior concha—a separate bone—is included in the discussion of facial bones in the next section. The conchae are covered with the mucous membrane of the nasal cavity. The superior concha and the adjacent region of the nasal septum also bear the receptor cells for the sense of smell (olfactory sense). The ethmoid bone also includes a large, delicate mass on each side of the perpendicular plate, honeycombed with chambers called **ethmoid air cells;** collectively, these constitute the ethmoid sinus.

From the interior of the skull, one can see only a small superior part of the ethmoid bone. It exhibits a medial crest called the **crista galli**[22] (GAL-eye), a point of attachment for the meninges (see figs. 7.4*b* and 7.5*b*). On each side of the crista is a horizontal **cribriform**[23] (CRIB-rih-form) **plate** marked by numerous perforations, the **cribriform (olfactory) foramina.** These foramina allow nerve fibers for the sense of smell to pass from the nasal cavity to the brain.

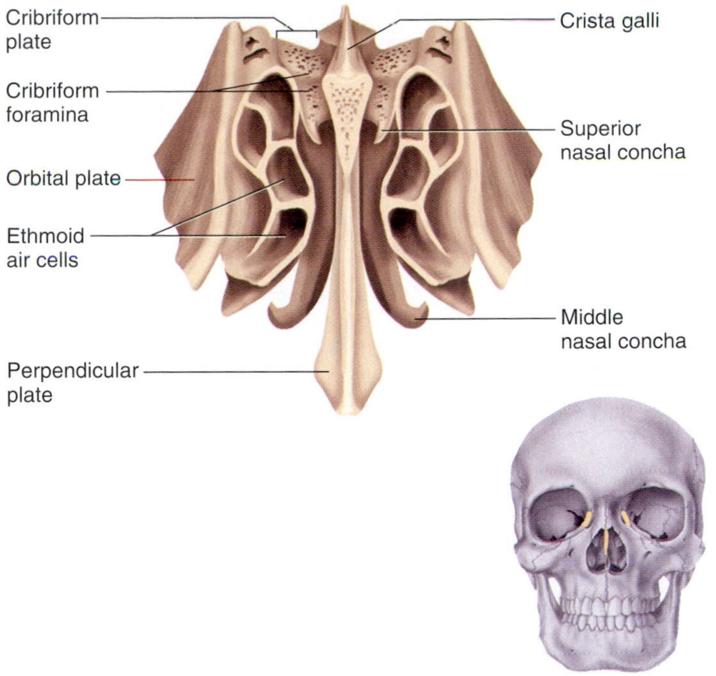

FIGURE 7.12
The Ethmoid Bone, Anterior View.

INSIGHT 7.1 CLINICAL APPLICATION

INJURY TO THE ETHMOID BONE

The ethmoid bone is very delicate and is easily injured by a sharp upward blow to the nose, such as a person might suffer by striking an automobile dashboard in a collision. The force of a blow can drive bone fragments through the cribriform plate into the meninges or brain tissue. Such injuries are often evidenced by leakage of cerebrospinal fluid into the nasal cavity, and may be followed by the spread of infection from the nasal cavity to the brain. Blows to the head can also shear off the olfactory nerves that pass through the ethmoid bone and cause *anosmia,* an irreversible loss of the sense of smell and a great reduction in the sense of taste (most of which depends on smell). This not only deprives life of some of its pleasures, but can also be dangerous, as when a person fails to smell smoke, gas, or spoiled food.

---

[15] *sella* = saddle + *turcica* = Turkish
[16] *lacerum* = torn, lacerated
[17] *choana* = funnel
[18] *pterygo* = wing
[19] *ethmo* = sieve, strainer + *oid* = resembling
[20] *conchae* = conchs (large marine snails)
[21] *turbin* = whirling, turning
[22] *crista* = crest + *galli* = of a rooster
[23] *cribri* = sieve + *form* = in the shape of

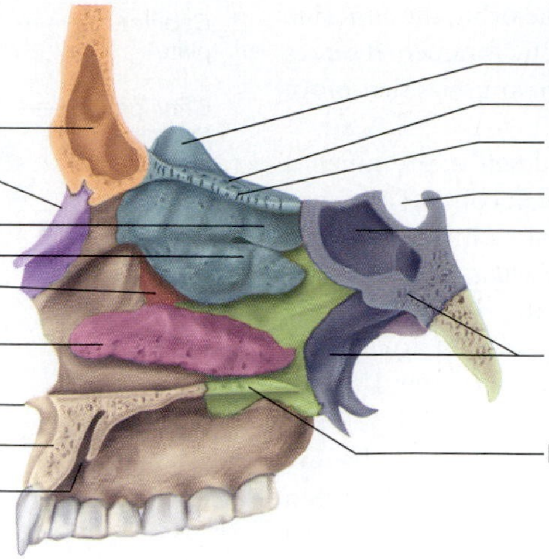
Frontal sinus
Nasal bone
Superior nasal concha
Middle nasal concha
Lacrimal bone
Inferior nasal concha
Anterior nasal spine
Maxilla
Incisive foramen
Crista galli
Cribriform plate
Cribriform foramina
Sella turcica
Sphenoid sinus
Sphenoid bone
Palatine bone

FIGURE 7.13
The Right Nasal Cavity, Sagittal Section.

Facial Bones

The **facial bones** are those that have no direct contact with the brain or meninges. They support the teeth, give shape and individuality to the face, form part of the orbital and nasal cavities, and provide attachment for the muscles of facial expression and mastication. There are 14 facial bones:

2 maxillae	2 nasal bones
2 palatine bones	2 inferior nasal conchae
2 zygomatic bones	1 vomer
2 lacrimal bones	1 mandible

MAXILLAE

The **maxillae** (mac-SILL-ee) are the largest facial bones. They form the upper jaw and meet each other at a medial suture (see figs. 7.3 and 7.4*a*). Small points of maxillary bone called **alveolar processes** grow into the spaces between the bases of the teeth. The root of each tooth is inserted into a deep socket, or **alveolus.** If a tooth is lost or extracted so that chewing no longer puts stress on the maxilla, the alveolar processes are resorbed and the alveolus fills in with new bone, leaving a smooth area on the maxilla. Even though the teeth are preserved with the skull, they are not bones. The teeth are discussed in detail in chapter 24.

THINK ABOUT IT!

Suppose you were studying a skull with some teeth missing. How could you tell whether the teeth had been lost after the person's death or years before it?

Each maxilla extends from the teeth to the inferomedial wall of the orbit. Just below the orbit, it exhibits an **infraorbital foramen,** which provides passage for a blood vessel to the face and a nerve that receives sensations from the nasal region and cheek. This nerve emerges through the foramen rotundum into the cranial cavity. The maxilla forms part of the floor of the orbit, where it exhibits a gash called the **inferior orbital fissure** that angles downward and medially (fig. 7.14). The inferior and superior orbital fissures form a sideways V whose apex lies near the optic foramen. The inferior orbital fissure is a passage for blood vessels and a nerve that supply some of the muscles that control eye movements.

The **palate** forms the roof of the mouth and floor of the nasal cavity. It consists of a bony **hard palate** in front and a fleshy **soft palate** in the rear. Most of the hard palate is formed by horizontal extensions of the maxilla called **palatine** (PAL-uh-tine) **processes** (see fig. 7.5*a*). Near the anterior margin of each palatine process, just behind the incisors, is an **incisive foramen.** The palatine processes normally meet at a median *intermaxillary suture* at about 12 weeks of fetal development. Failure to join causes cleft palate (see table 7.7, p.201).

PALATINE BONES

The **palatine bones** form the rest of the hard palate, part of the wall of the nasal cavity, and part of the floor of the orbit (see figs. 7.5*a* and 7.13). At the posterolateral corners of the hard palate are the two large **greater palatine foramina.**

INSIGHT 7.2 EVOLUTIONARY MEDICINE

EVOLUTIONARY SIGNIFICANCE OF THE PALATE

In most vertebrates, the nasal passages open into the oral cavity. Mammals, by contrast, have a palate that separates the nasal cavity from the oral cavity. In order to maintain our high metabolic rate, we must digest our food rapidly; in order to do this, we chew it thoroughly to break it up into small, easily digested particles before swallowing it. We would be unable to breathe freely during this prolonged chewing if we lacked a palate to separate the airflow from the oral cavity.

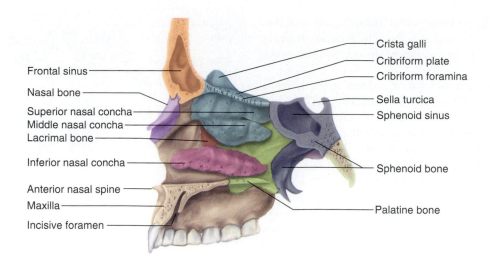

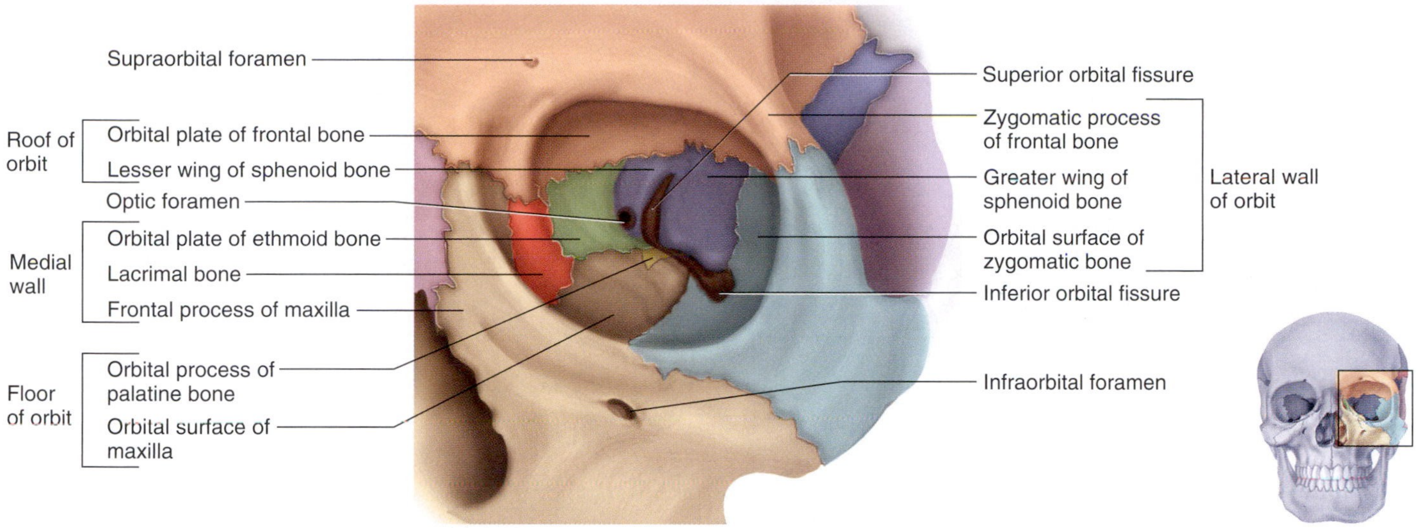
Supraorbital foramen
Roof of orbit
Orbital plate of frontal bone
Lesser wing of sphenoid bone
Optic foramen
Medial wall
Orbital plate of ethmoid bone
Lacrimal bone
Frontal process of maxilla
Floor of orbit
Orbital process of palatine bone
Orbital surface of maxilla
Superior orbital fissure
Zygomatic process of frontal bone
Greater wing of sphenoid bone
Orbital surface of zygomatic bone
Lateral wall of orbit
Inferior orbital fissure
Infraorbital foramen

FIGURE 7.14
The Left Orbit, Anterior View.

ZYGOMATIC BONES

The **zygomatic**[24] **bones** form the angles of the cheeks inferolateral to the eyes, and form part of the lateral wall of each orbit; they extend about halfway to the ear (see figs. 7.4*a* and 7.5*a*). Each zygomatic bone has an inverted T shape and usually a small **zygomaticofacial** (ZY-go-MAT-ih-co-FAY-shul) **foramen** near the intersection of the stem and crossbar of the T. The prominent zygomatic arch that flares from each side of the skull is formed by the union of the zygomatic and temporal bones.

LACRIMAL BONES

The **lacrimal**[25] (LACK-rih-mul) **bones** form part of the medial wall of each orbit (fig. 7.14). A depression called the **lacrimal fossa** houses a membranous *lacrimal sac* in life. Tears from the eye collect in this sac and drain into the nasal cavity.

NASAL BONES

Two small rectangular **nasal bones** form the bridge of the nose (see fig. 7.3) and support cartilages that shape the lower portion of the nose. If you palpate the bridge, you can easily feel where the nasal bones end and the cartilages begin. The nasal bones are often fractured by blows to the nose.

INFERIOR NASAL CONCHAE

There are three conchae in the nasal cavity. The superior and middle conchae, as discussed earlier, are parts of the ethmoid bone. The **inferior nasal concha (inferior turbinate bone)**—the largest of the three—is a separate bone (see fig. 7.13).

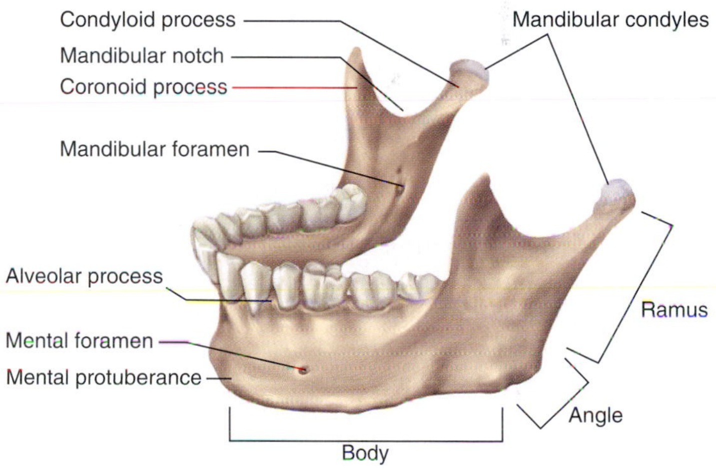
Condyloid process
Mandibular notch
Coronoid process
Mandibular foramen
Alveolar process
Mental foramen
Mental protuberance
Mandibular condyles
Ramus
Angle
Body

FIGURE 7.15
The Mandible.

VOMER

The **vomer** forms the inferior portion of the nasal septum (see figs. 7.3 and 7.4*b*). Its name literally means "plowshare," which refers to its resemblance to the blade of a plow. The superior half of the nasal septum is formed by the perpendicular plate of the ethmoid bone, as mentioned earlier. The vomer and perpendicular plate support a wall of *septal cartilage* that forms most of the anterior part of the nasal septum.

MANDIBLE

The **mandible** (fig. 7.15) is the strongest bone of the skull and the only one that can move. It supports the lower teeth and provides attachment for muscles of mastication and facial expression. The horizontal portion is called the **body;** the vertical-to-oblique posterior portion is the **ramus** (RAY-mus)—plural, *rami* (RAY-my); and these

[24]*zygo* = to join, unite
[25]*lacrim* = tear, to cry

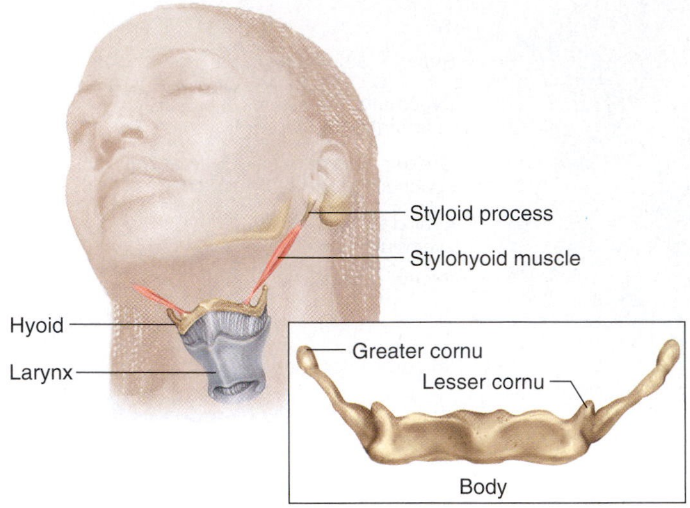

FIGURE 7.16
The Hyoid Bone.

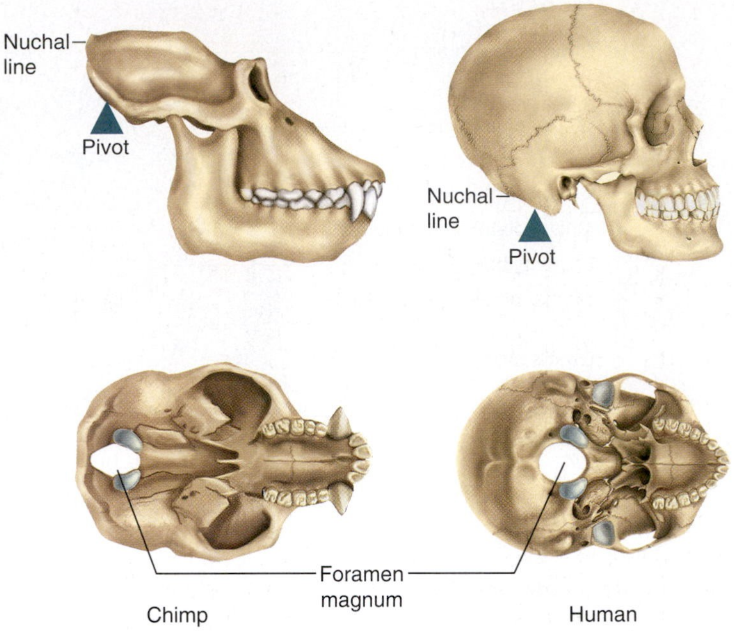

FIGURE 7.17
Adaptations of the Skull for Bipedalism. Comparison of chimpanzee and human skulls. The foramen magnum is shifted rostrally and the face is flatter in humans. Thus the skull is balanced on the vertebral column and the gaze is directed forward when a person is standing.

two portions meet at a corner called the **angle.** The point of the chin is the **mental protuberance.** On the anterolateral surface of the body, the **mental foramen** permits the passage of nerves and blood vessels of the chin. The inner surface of the body has a number of shallow depressions and ridges to accommodate muscles and salivary glands. The angle of the mandible has a rough lateral surface for insertion of the *masseter,* a muscle of mastication. Like the maxilla, the mandible has pointed alveolar processes between the teeth.

The ramus is somewhat Y-shaped. Its posterior branch, called the **condyloid** (CON-dih-loyd) **process,** bears the **mandibular condyle**—an oval knob that articulates with the mandibular fossa of the temporal bone. The hinge of the mandible is the **temporomandibular joint (TMJ).** The anterior branch of the ramus, called the **coronoid process,** is the point of insertion for the temporalis muscle, which pulls the mandible upward when you bite. The U-shaped arch between the two processes is called the **mandibular notch.** Just below the notch, on the medial surface of the ramus, is the **mandibular foramen,** a passage for the nerve and blood vessels that supply the lower teeth. Dentists inject anesthetic near here to deaden sensation from the lower teeth, and near the foramen rotundum to deaden sensation from the upper teeth.

Bones Associated with the Skull

Seven bones are closely associated with the skull but not considered part of it. These are the three auditory ossicles in each middle-ear cavity and the hyoid bone beneath the chin. The **auditory ossicles**[26]—named the **malleus** (hammer), **incus** (anvil), and **stapes** (STAY-peez) (stirrup)—are discussed in connection with hearing in chapter 17. The **hyoid**[27] **bone** is a slender bone between the chin and larynx (fig. 7.16). It is one of the few bones that does not articulate with any other. The hyoid is suspended from the styloid processes of the skull, somewhat like a hammock, by the small *stylohyoid muscles* and *stylohyoid ligaments.* The medial **body** of the hyoid is flanked on either side by hornlike projections called the **greater** and **lesser cornua**[28] (CORN-you-uh)—singular, *cornu* (COR-new). The hyoid bone serves for attachment of several muscles that control the mandible, tongue, and larynx. Forensic pathologists look for a fractured hyoid as evidence of strangulation.

Adaptations of the Skull for Bipedalism

Some mammals can stand, hop, or walk briefly on their hind legs, but humans are the only mammals that are habitually bipedal. Chapter 1 explored some possible reasons why bipedal locomotion originally evolved in the Hominidae. Efficient bipedal locomotion is possible only because of several adaptations of the feet, legs, vertebral column, and skull.

The human head is balanced on the vertebral column with the gaze directed forward. This was made possible in part by an evolutionary remodeling of the skull. The foramen magnum moved to a more inferior location in the course of human evolution, and the face is much flatter than an ape's face, so there is less weight anterior to the occipital condyles (fig. 7.17). Being balanced on the spine, the head does not require strong muscles to hold it erect. Apes have prominent supraorbital ridges for the attachment of muscles that pull back on the skull. In humans, these ridges are much lighter and the muscles of the forehead serve only for facial expression, not to hold the head up.

Table 7.4 summarizes the bones of the skull.

[26]*os* = bone + *icle* = little
[27]*hy* = the letter *U* + *oid* = resembling
[28]*cornu* = horn

TABLE 7.4
Anatomical Checklist for the Skull and Associated Bones

Cranial Bones

Frontal Bone (figs. 7.3 to 7.7)
- Supraorbital margin
- Supraorbital foramen or notch
- Frontal sinus

Parietal Bones (figs. 7.4 and 7.6)
- Temporal lines
- Parietal foramen

Temporal Bones (figs. 7.4, 7.5, and 7.10)
- Squamous part
  - Zygomatic process
  - Mandibular fossa
- Tympanic part
  - External acoustic meatus
  - Styloid process
- Mastoid part
  - Mastoid process
  - Mastoid notch
  - Mastoid foramen
  - Stylomastoid foramen
- Petrous part
  - Internal acoustic meatus
  - Carotid canal
  - Jugular foramen

Occipital Bone (figs. 7.4, 7.5, and 7.6)
- Foramen magnum
- Basilar part
- Occipital condyles
- Hypoglossal canal

Occipital Bone (figs. 7.4, 7.5, and 7.6)—(Cont.)
- Condylar canal
- External occipital protuberance
- Superior nuchal line
- Inferior nuchal line

Sphenoid Bone (figs. 7.4, 7.5, and 7.11)
- Body
- Lesser wing
  - Optic foramen
  - Anterior clinoid process
  - Superior orbital fissure
- Greater wing
  - Foramen ovale
  - Foramen rotundum
  - Foramen spinosum
  - Foramen lacerum
  - Medial and lateral pterygoid plates
- Nasal choanae
- Sphenoid sinus
- Sella turcica
- Dorsum sellae

Ethmoid Bone (figs. 7.4, 7.7, and 7.12)
- Perpendicular plate
- Superior nasal concha (superior turbinate bone)
- Middle nasal concha (middle turbinate bone)
- Ethmoid sinus (air cells)
- Crista galli
- Cribriform plate

Facial Bones

Maxilla (figs. 7.3, 7.4, and 7.5a)
- Alveoli
- Alveolar processes
- Infraorbital foramen
- Inferior orbital fissure
- Palatine processes
- Incisive foramen
- Maxillary sinus

Palatine Bones (figs. 7.4b, 7.5a and 7.13)
- Greater palatine foramen

Zygomatic Bones (figs. 7.4a and 7.5a)
- Zygomaticofacial foramen

Lacrimal Bones (figs. 7.3 and 7.14)
- Lacrimal fossa

Nasal Bones (figs. 7.3 and 7.13)

Inferior Nasal Concha (fig. 7.13)

Vomer (figs. 7.3 and 7.4b)

Mandible (figs. 7.3 and 7.15)
- Body
  - Mental protuberance
  - Mental foramen
- Angle
- Ramus
  - Condyloid process
  - Mandibular condyle
  - Coronoid process
  - Mandibular notch
  - Mandibular foramen

Associated with the Skull

Auditory Ossicles
- Malleus (hammer)
- Incus (anvil)
- Stapes (stirrup)

Hyoid Bone (fig. 7.16)
- Body
- Greater cornu
- Lesser cornu

Before You Go On

Answer the following questions to test your understanding of the preceding section:

4. Name the paranasal sinuses and state their locations. Name any four other cavities in the skull.
5. Explain the difference between a cranial bone and a facial bone. Give four examples of each.
6. Draw an oval representing a superior view of the calvaria. Draw lines representing the coronal, lambdoid, and sagittal sutures. Label the four bones separated by these sutures.
7. State which bone has each of these features: a squamous part, hypoglossal foramen, greater cornu, greater wing, condyloid process, and cribriform plate.
8. Determine which of the following structures cannot normally be palpated on a living person: the mastoid process, crista galli, superior orbital fissure, palatine processes, zygomatic bone, mental protuberance, and stapes. You may find it useful to palpate some of these on your own skull as you try to answer.

The Vertebral Column and Thoracic Cage

Objectives

When you have completed this section, you should be able to

- describe the general features of the vertebral column and those of a typical vertebra;
- describe the special features of vertebrae in different regions of the vertebral column, and discuss the functional significance of the regional differences;
- relate the shape of the vertebral column to upright locomotion; and
- describe the anatomy of the sternum and ribs and how the ribs articulate with the thoracic vertebrae.

General Features of the Vertebral Column

The **vertebral column** physically supports the skull and trunk, allows for their movement, protects the spinal cord, and absorbs stresses produced by walking, running, and lifting. It also provides attachment for the limbs, thoracic cage, and postural muscles. Although commonly called the backbone, it does not consist of a single bone but a chain of 33 **vertebrae** with **intervertebral discs** of fibrocartilage between most of them. The adult vertebral column averages about 71 cm (28 in.) long, with the 23 intervertebral discs accounting for about one-quarter of the length.

As shown in figure 7.18, the vertebrae are divided into five groups: 7 *cervical* (SUR-vih-cul) *vertebrae* in the neck, 12 *thoracic vertebrae* in the chest, 5 *lumbar vertebrae* in the lower back, 5 *sacral vertebrae* at the base of the spine, and 4 tiny *coccygeal* (coc-SIDJ-ee-ul) *vertebrae*. To help remember the numbers of cervical, thoracic, and lumbar vertebrae—7, 12, and 5—you might think of a typical work day: go to work at 7, have lunch at 12, and go home at 5.

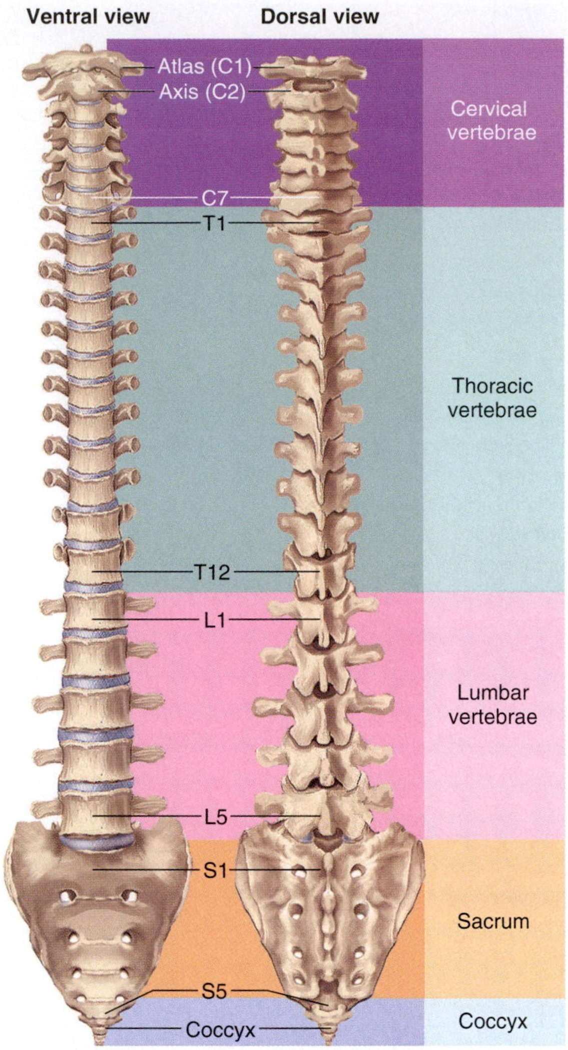
Ventral view
Dorsal view
Atlas (C1)
Axis (C2)
C7
T1
T12
L1
L5
S1
S5
Coccyx
Cervical vertebrae
Thoracic vertebrae
Lumbar vertebrae
Sacrum
Coccyx

Figure 7.18
The Vertebral Column, Ventral and Dorsal Views.

Variations in this arrangement occur in about 1 person in 20. For example, the last lumbar vertebra is sometimes incorporated into the sacrum, producing 4 lumbar and 6 sacral vertebrae. In other cases, the first sacral vertebra fails to fuse with the second, producing 6 lumbar and 4 sacral vertebrae. The cervical and thoracic vertebrae are more constant in number.

Beyond the age of 3 years, the vertebral column is slightly S-shaped, with four bends called the **cervical, thoracic, lumbar,** and **pelvic curvatures** (fig. 7.19). The thoracic and pelvic curvatures are called *primary curvatures* because they are present at birth, when the spine has a single C-shaped curvature. The cervical and lumbar curvatures are called *secondary curvatures* because they develop later, in the child's first few years of crawling and walking, as described later in this chapter. The resulting S shape makes sustained bipedal walking possible because the trunk of the body does

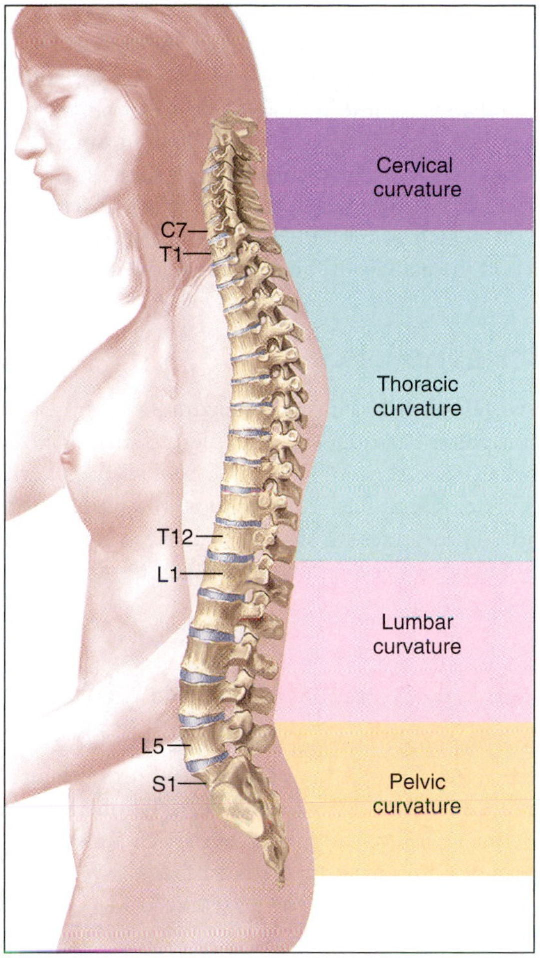

FIGURE 7.19
Curvatures of the Adult Vertebral Column.

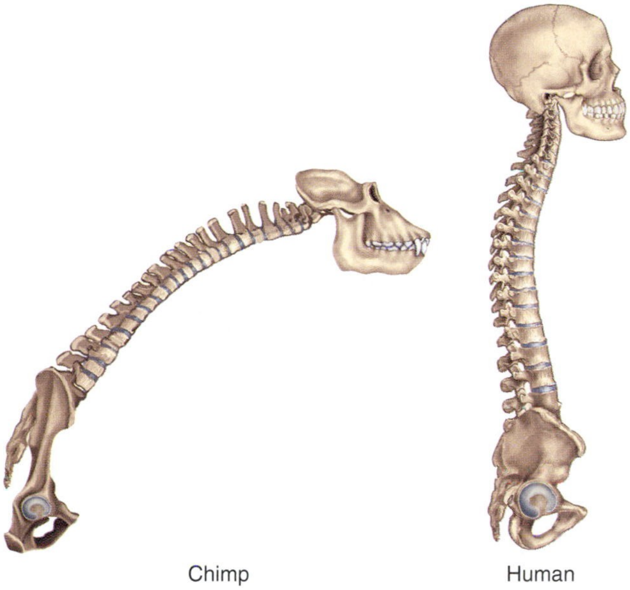

FIGURE 7.20
Comparison of Chimpanzee and Human Vertebral Columns. The S-shaped human vertebral column is an adaptation for bipedal locomotion.

not lean forward as it does in primates such as a chimpanzee; the head is balanced over the body's center of gravity; and the eyes are directed straight forward (fig. 7.20).

INSIGHT 7.3 CLINICAL APPLICATION

ABNORMAL SPINAL CURVATURES

Abnormal spinal curvatures (fig. 7.21) can result from disease, weakness, or paralysis of the trunk muscles, poor posture, or congenital defects in vertebral anatomy. The most common deformity is an abnormal lateral curvature called *scoliosis.* It occurs most often in the thoracic region, particularly among adolescent girls. It sometimes results from a developmental abnormality in which the body and arch of a vertebra (described later) fail to develop on one side. If the person's skeletal growth is not yet complete, scoliosis can be corrected with a back brace.

An exaggerated thoracic curvature is called *kyphosis* (hunchback, in lay language). It is usually a result of osteoporosis, but it also occurs in people with osteomalacia or spinal tuberculosis and in adolescents who engage heavily in such sports as wrestling and weightlifting. An exaggerated lumbar curvature is called *lordosis* (swayback). It may have the same causes as kyphosis, or it may result from added abdominal weight in pregnancy or obesity.

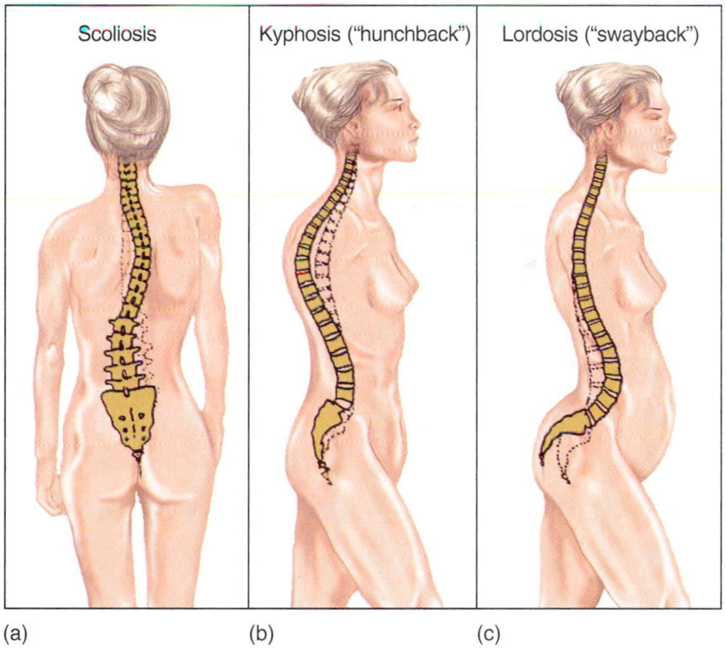

FIGURE 7.21
Abnormal Spinal Curvatures. (*a*) Scoliosis, an abnormal lateral deviation. (*b*) Kyphosis, an exaggerated thoracic curvature. (*c*) Lordosis, an exaggerated lumbar curvature.

General Structure of a Vertebra

A representative vertebra and intervertebral disc are shown in figure 7.22. The most obvious feature of a vertebra is the **body,** or **centrum**—a mass of spongy bone and red bone marrow covered

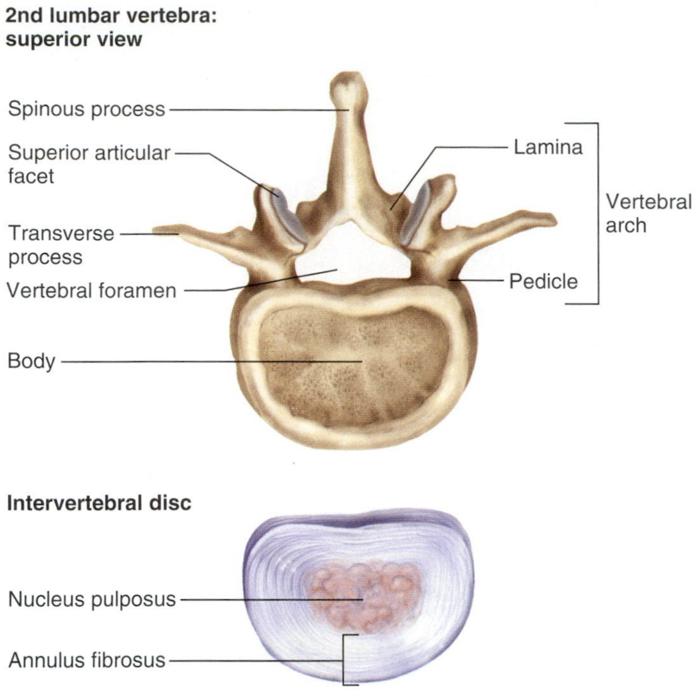
2nd lumbar vertebra:
superior view
Spinous process
Superior articular facet
Transverse process
Vertebral foramen
Body
Lamina
Vertebral arch
Pedicle
Intervertebral disc
Nucleus pulposus
Annulus fibrosus

FIGURE 7.22

A Representative Vertebra and Intervertebral Disc, Superior Views.

with a thin layer of compact bone. This is the weight-bearing portion of the vertebra. Its rough superior and inferior surfaces provide firm attachment to the intervertebral discs.

THINK ABOUT IT!

The lower we look on the vertebral column, the larger the vertebral bodies and intervertebral discs are. What is the functional significance of this?

Dorsal to the body of each vertebra is an ovoid to triangular canal called the **vertebral foramen.** Collectively, these foramina form the **vertebral canal,** a passage for the spinal cord. The foramen is bordered by a bony **vertebral arch** composed of two parts, a pillarlike **pedicle**[29] and platelike **lamina,**[30] on each side. Extending from the apex of the arch, a projection called the **spinous process** is directed toward the rear and downward. You can see and feel the spinous processes as a row of bumps along the spine. A **transverse process** extends laterally from the point where the pedicle and lamina meet. The spinous and transverse processes provide points of attachment for spinal muscles and ligaments.

A pair of **superior articular processes** project upward from one vertebra and meet a similar pair of **inferior articular processes** that project downward from the vertebra just above (fig. 7.23*a*). Each process has a flat articular surface (facet) facing that of the adjacent vertebra. These processes restrict twisting of the vertebral column, which could otherwise severely damage the spinal cord.

When two vertebrae are joined, they exhibit an opening between their pedicles called the **intervertebral foramen.** This allows passage for spinal nerves that connect with the spinal cord at regular intervals. Each foramen is formed by an **inferior vertebral notch** in the pedicle of the superior vertebra and a **superior vertebral notch** in the pedicle of the one just below it (fig. 7.23*b*).

Intervertebral Discs

An **intervertebral disc** is a pad consisting of an inner gelatinous **nucleus pulposus** surrounded by a ring of fibrocartilage, the **annulus fibrosus** (see fig. 7.22). The discs bind adjacent vertebrae together, enhance spinal flexibility, support the weight of the body, and absorb shock. Under stress—for example, when you lift a heavy weight—the discs bulge laterally. Excessive stress can cause a *herniated disc* (see p. 202).

Regional Characteristics of Vertebrae

We are now prepared to consider how vertebrae differ from one region of the vertebral column to another and from the generalized anatomy just described. Knowing these variations will enable you to identify the region of the spine from which an isolated vertebra was taken. More importantly, these modifications in form reflect functional differences among the vertebrae.

CERVICAL VERTEBRAE

The **cervical vertebrae** (C1–C7) are the smallest and lightest. The first two (C1 and C2) have unique structures that allow for head movements (fig. 7.24). Vertebra C1 is called the **atlas** because it supports the head in a manner reminiscent of the Titan of Greek mythology who was condemned by Zeus to carry the world on his shoulders. It scarcely resembles the typical vertebra; it is little more than a delicate ring surrounding a large vertebral foramen. On each side is a **lateral mass** with a deeply concave **superior articular facet** that articulates with the occipital condyle of the skull. In nodding motion of the skull, as in gesturing "yes," the occipital condyles rock back and forth on these facets. The **inferior articular facets,** which are comparatively flat or only slightly concave, articulate with C2. The lateral masses are connected by an **anterior arch** and a **posterior arch,** which bear slight protuberances called the **anterior** and **posterior tubercle,** respectively.

Vertebra C2, the **axis,** allows rotation of the head as in gesturing "no." Its most distinctive feature is a prominent knob called the **dens** (denz), or **odontoid**[31] **process,** on its anterosuperior side. No other vertebra has a dens. It begins to form as an independent ossification center during the first year of life and fuses with the axis by the age of 3 to 6 years. It projects into the vertebral foramen of the atlas, where it is nestled in a facet and held in place by a **trans-**

[29] *ped* = foot + *icle* = little
[30] *lamina* = layer, plate
[31] *dens* = *odont* = tooth + *oid* = resembling

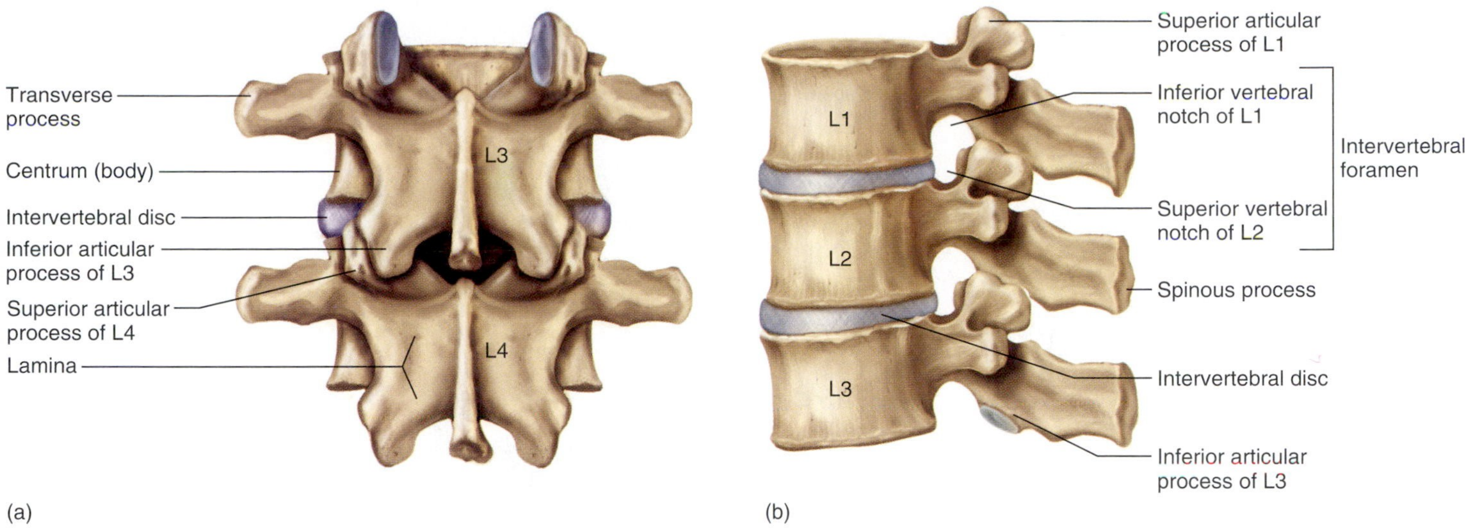

Transverse process
Centrum (body)
Intervertebral disc
Inferior articular process of L3
Superior articular process of L4
Lamina
L3
L4
(a)
Superior articular process of L1
Inferior vertebral notch of L1
Intervertebral foramen
Superior vertebral notch of L2
Spinous process
Intervertebral disc
Inferior articular process of L3
L1
L2
L3
(b)

FIGURE 7.23

Articulated Vertebrae. (*a*) Dorsal view of vertebrae L3 to L4. (*b*) Left lateral view of vertebrae L1 to L3.

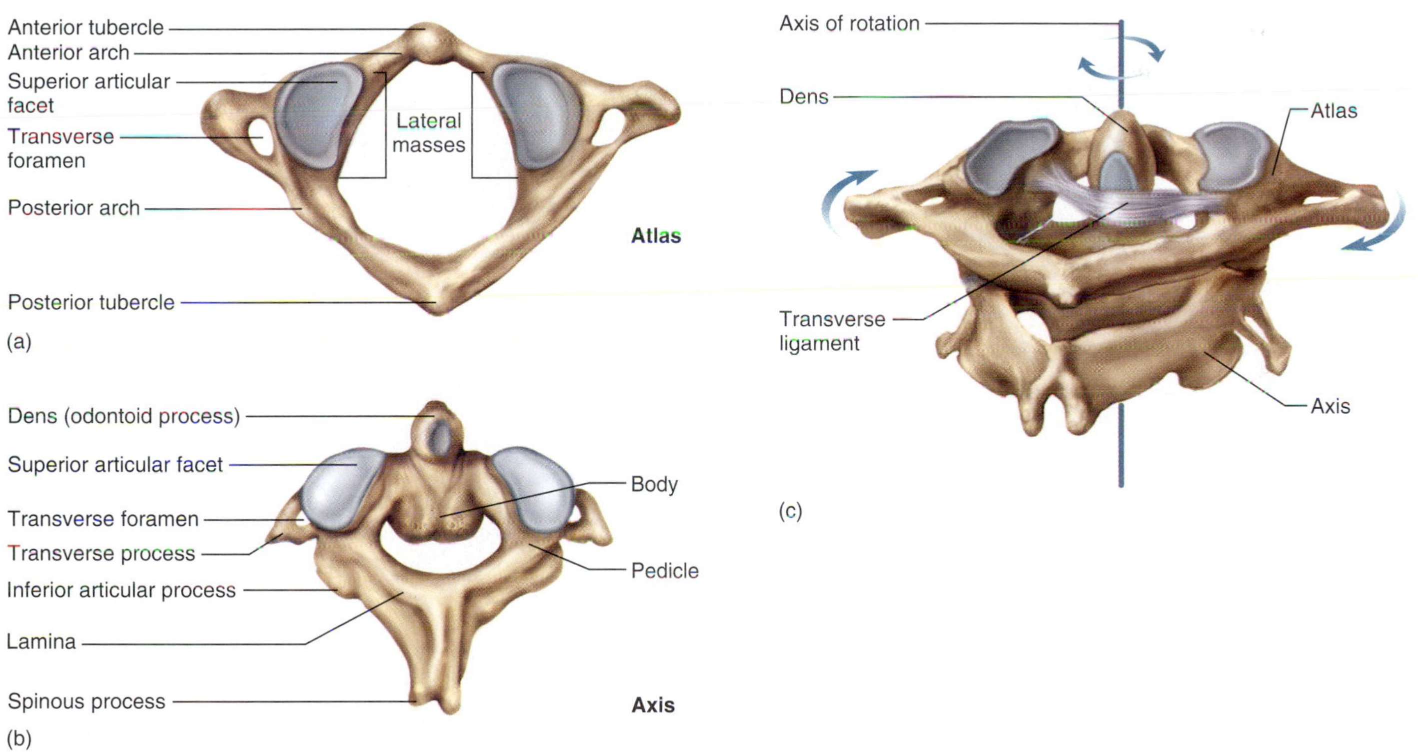

Anterior tubercle
Anterior arch
Superior articular facet
Transverse foramen
Posterior arch
Posterior tubercle
Lateral masses
Atlas
(a)
Dens (odontoid process)
Superior articular facet
Transverse foramen
Transverse process
Inferior articular process
Lamina
Spinous process
Body
Pedicle
Axis
(b)
Axis of rotation
Dens
Atlas
Transverse ligament
Axis
(c)

FIGURE 7.24

The Atlas and Axis, Cervical Vertebrae C1 and C2. (*a*) The atlas, superior view. (*b*) The axis, superodorsal view. (*c*) Articulation of the atlas and axis and rotation of the atlas. This movement turns the head from side to side, as in gesturing "no." Note the transverse ligament holding the dens of the axis in place.

verse ligament (fig. 7.24*c*). A heavy blow to the top of the head can cause a fatal injury in which the dens is driven through the foramen magnum into the brainstem. The articulation between the atlas and the cranium is called the **atlanto-occipital joint;** the one between the atlas and axis is called the **atlantoaxial joint.**

The axis is the first vertebra that exhibits a spinous process. In vertebrae C2 to C6, the process is forked, or *bifid,*[32] at its tip (fig. 7.25*a*). This fork provides attachment for the *nuchal ligament* of the

[32]*bifid* = cleft into two parts

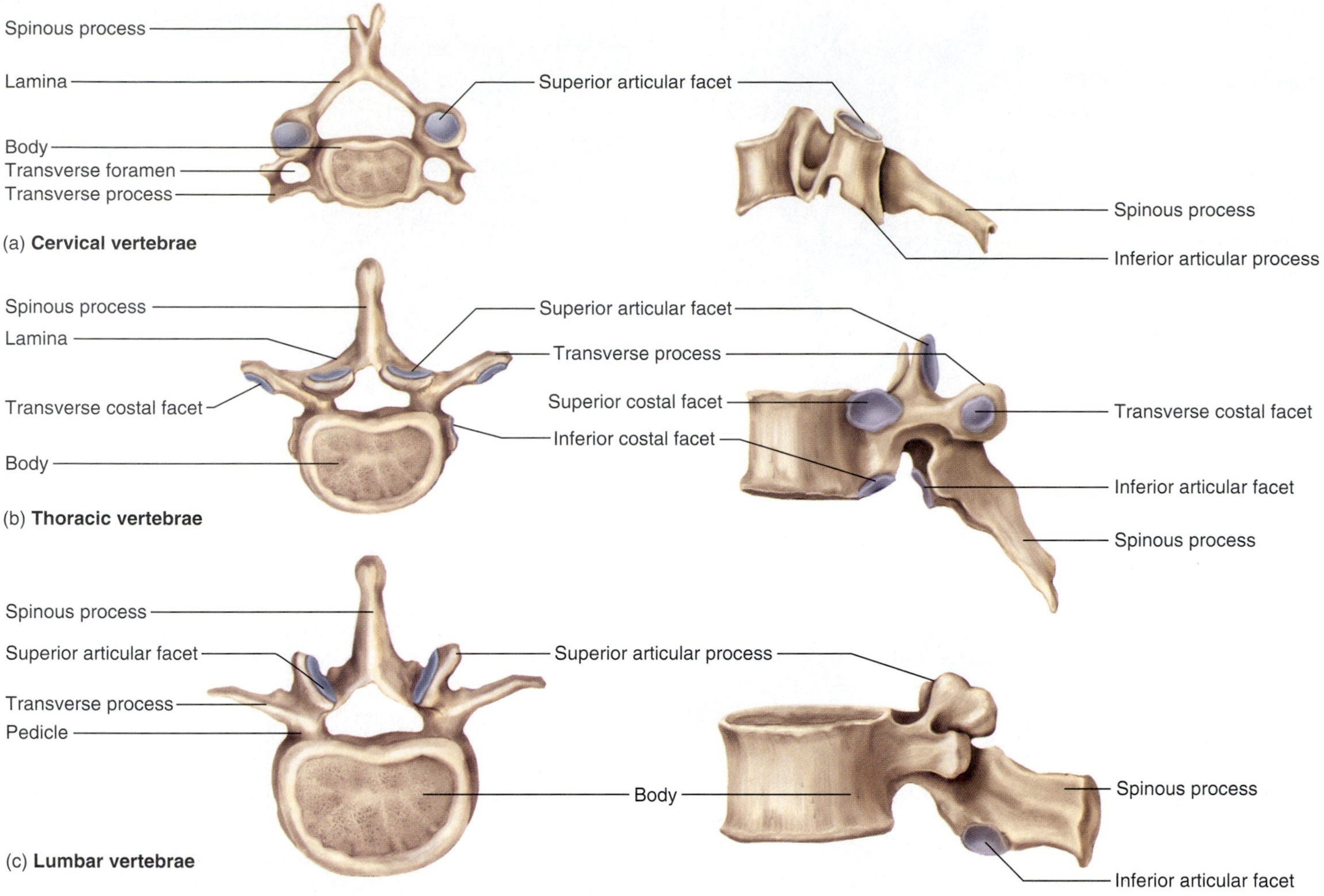

FIGURE 7.25
Typical Cervical, Thoracic, and Lumbar Vertebrae. The *left-hand* figures are superior views, and the *right-hand* figures are left lateral views.

back of the neck. All seven cervical vertebrae have a prominent round **transverse foramen** in each transverse process. In vertebrae C1 through C6, these foramina provide passage and protection for the *vertebral arteries,* which supply blood to the brain. The transverse foramen of C7 allows passage for the *vertebral veins.* Transverse foramina occur in no other vertebrae and thus provide an easy means of recognizing a cervical vertebra.

THINK ABOUT IT!

How would head movements be affected if vertebrae C1 and C2 had the same structure as C3? What is the functional advantage of the lack of a spinous process in C1?

Cervical vertebrae C3 to C6 are similar to the typical vertebra described earlier, with the addition of the transverse foramina and bifid spinous processes. Vertebra C7 is a little different—its spinous process is not bifid, but it is especially long and forms a prominent bump on the lower back of the neck. C7 is sometimes called the *vertebra prominens* because of this especially conspicuous spinous process. This feature is a convenient landmark for counting vertebrae.

THORACIC VERTEBRAE

There are 12 **thoracic vertebrae** (T1–T12), corresponding to the 12 pairs of ribs attached to them. They lack the transverse foramina and bifid processes that distinguish the cervicals, but possess the following distinctive features of their own (fig. 7.25*b*):

- The spinous processes are relatively pointed and angle sharply downward.
- The body is somewhat heart-shaped and more massive than in the cervical vertebrae but less than in the lumbar vertebrae.
- The body has small, smooth, slightly concave spots called *costal facets* (to be described shortly) for attachment of the ribs.

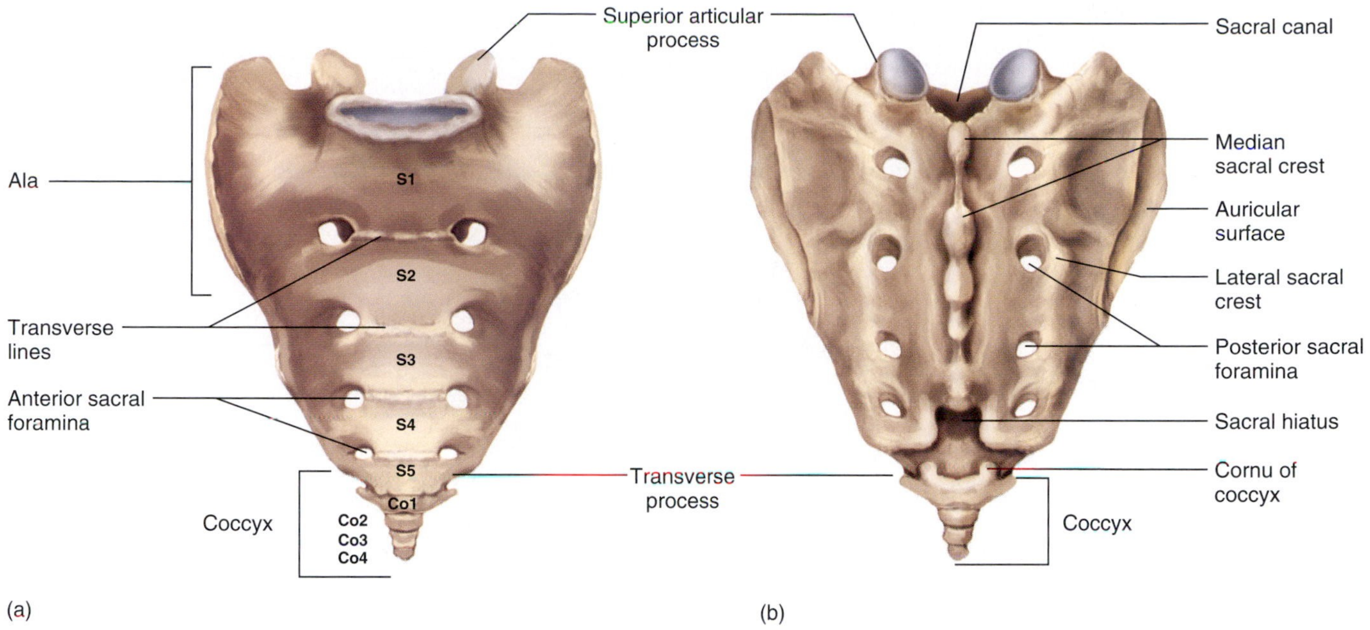

FIGURE 7.26

The Sacrum and Coccyx. (*a*) Anterior surface, which faces the viscera of the pelvic cavity. (*b*) Posterior surface. The processes of this surface can be palpated in the sacral region.

- Vertebrae T1 to T10 have a shallow, cuplike **transverse costal**[33] **facet** at the end of each transverse process. These provide a second point of articulation for ribs 1 to 10. There are no transverse costal facets on T11 and T12 because ribs 11 and 12 attach only to the bodies of the vertebrae.

No other vertebrae have ribs articulating with them. Thoracic vertebrae vary among themselves mainly in the way the ribs articulate with them. In most cases, a rib inserts between two vertebrae, so each vertebra contributes one-half of the articular surface—the rib articulates with the **inferior costal facet** of the upper vertebra and the **superior costal facet** of the vertebra below that. This terminology may be a little confusing, but note that the facets are named for their position on the vertebral body, not for which part of the rib's articulation they provide. Vertebrae T1 and T10 to T12, however, have complete costal facets on the bodies for ribs 1 and 10 to 12, which articulate on the vertebral body instead of between vertebrae. Vertebrae T11 and T12, as noted, have no transverse costal facets. These variations will be more functionally understandable after you have studied the anatomy of the ribs, so we will return then to the details of these articular surfaces.

Vertebra T12 differs in its articular processes from those above it. Its superior articular processes face dorsally to meet the ventrally-facing inferior processes of T11, while the inferior articular processes of T12 face laterally like those of the lumbar vertebrae, described next. T12 thus represents a transition between the thoracic and lumbar pattern.

LUMBAR VERTEBRAE

There are five **lumbar vertebrae** (L1–L5). Their most distinctive features are a thick, stout body and a blunt, squarish spinous process (fig. 7.25*c*). In addition, their articular processes are oriented differently than those of other vertebrae. In thoracic vertebrae, the superior processes face dorsally and the inferior processes face ventrally. In lumbar vertebrae, the superior processes face medially (like the palms of your hands about to clap), and the inferior processes face laterally, toward the superior processes of the next vertebra. This arrangement makes the lumbar region of the spine especially resistant to twisting. These differences are best observed on an articulated (assembled) skeleton.

SACRUM

The **sacrum** is a bony plate that forms the dorsal wall of the pelvic cavity (fig. 7.26). It is named for the fact that it was once considered the seat of the soul.[34] In children, there are five separate **sacral vertebrae** (S1–S5). They begin to fuse around age 16 and are fully fused by age 26.

The anterior surface of the sacrum is relatively smooth and concave and has four transverse lines that indicate where the five vertebrae have fused. This surface exhibits four pairs of large **anterior sacral (pelvic) foramina,** which allow for the passage of nerves and arteries to the pelvic organs. The dorsal surface of the sacrum is very rough. The spinous processes of the vertebrae fuse into a dorsal ridge called the **median sacral crest.** The transverse processes fuse into a

[33]*costa* = rib + *al* = pertaining to

[34]*sacr* = sacred

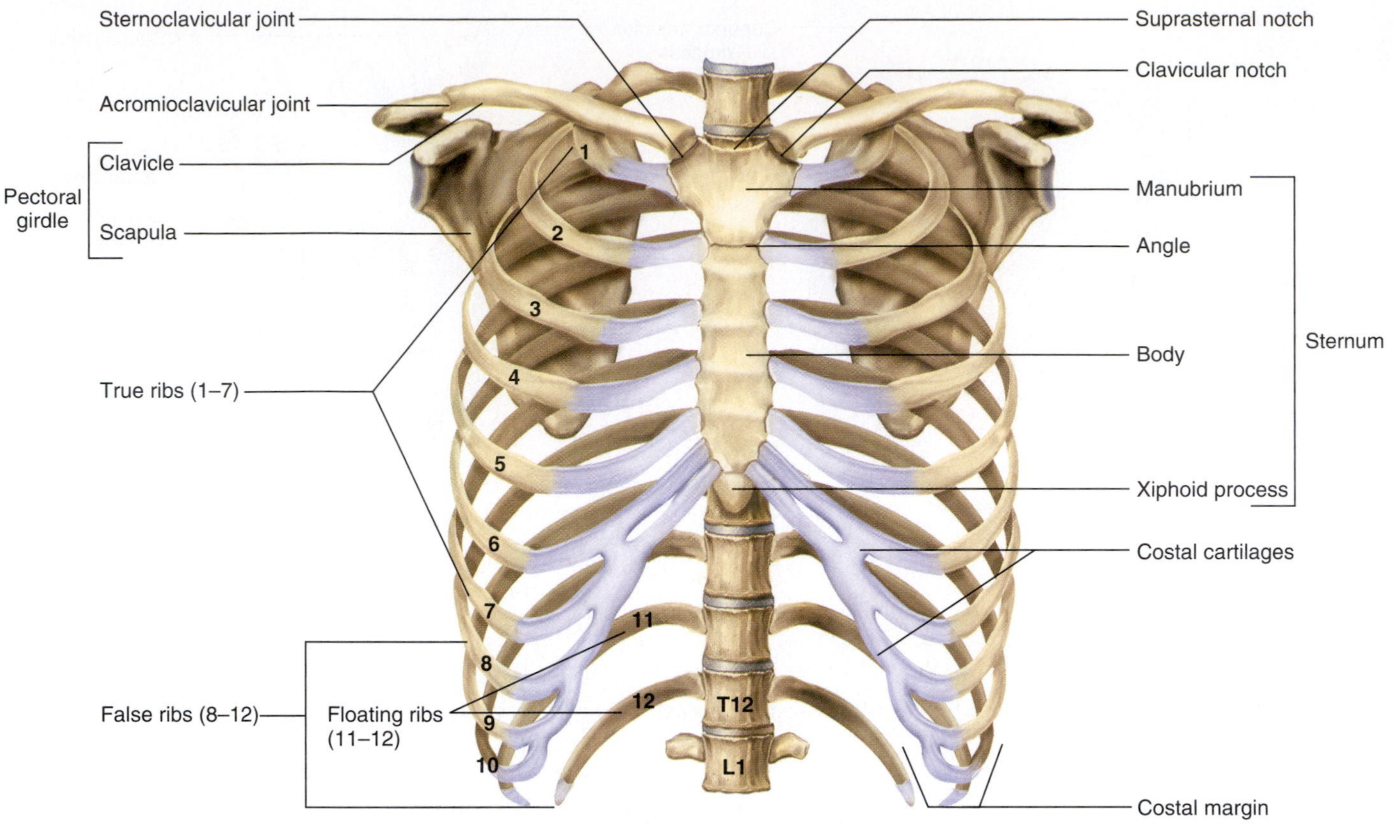

Sternoclavicular joint
Acromioclavicular joint
Pectoral girdle
Clavicle
Scapula
True ribs (1–7)
False ribs (8–12)
Floating ribs (11–12)
1
2
3
4
5
6
7
8
9
10
11
12
T12
L1
Suprasternal notch
Clavicular notch
Manubrium
Angle
Body
Sternum
Xiphoid process
Costal cartilages
Costal margin

FIGURE 7.27
The Thoracic Cage and Pectoral Girdle, Ventral View.

less prominent **lateral sacral crest** on each side of the median crest. Again on the dorsal side of the sacrum, there are four pairs of openings for spinal nerves, the **posterior sacral foramina.** The nerves that emerge here supply the gluteal region and lower limb.

A **sacral canal** runs through the sacrum and ends in an inferior opening called the **sacral hiatus** (hy-AY-tus). This canal contains spinal nerve roots in life. On each side of the sacrum is an ear-shaped region called the **auricular**[35] (aw-RIC-you-lur) **surface.** This articulates with a similarly shaped surface on the os coxae and forms the strong, nearly immovable **sacroiliac** (SAY-cro-ILL-ee-ac) **(SI) joint.** At the superior end of the sacrum, lateral to the median crest, are a pair of **superior articular processes** that articulate with vertebra L5. Lateral to these are a pair of large, rough, winglike extensions called the **alae**[36] (AIL-ee).

COCCYX

The **coccyx**[37] (fig. 7.26) usually consists of four (sometimes five) small vertebrae, Co1 to Co4, which fuse by the age of 20 to 30 into a single triangular bone. Vertebra Co1 has a pair of hornlike projections, the **cornua,** which serve as attachment points for ligaments that bind the coccyx to the sacrum. The coccyx can be fractured by a difficult childbirth or a hard fall on the buttocks. Although it is the vestige of a tail, it is not entirely useless; it provides attachment for muscles of the pelvic floor.

The Thoracic Cage

The **thoracic cage** (fig. 7.27) consists of the thoracic vertebrae, sternum, and ribs. It forms a more or less conical enclosure for the lungs and heart and provides attachment for the pectoral girdle and upper limb. It has a broad base and a somewhat narrower superior apex; it is rhythmically expanded by the respiratory muscles to create a vacuum that draws air into the lungs. The inferior border of the thoracic cage is formed by a downward arc of the ribs called the **costal margin.** The ribs protect not only the thoracic organs but also the spleen, most of the liver, and to some extent the kidneys.

STERNUM

The **sternum** (breastbone) is a bony plate anterior to the heart (fig. 7.27). It is subdivided into three regions: the manubrium, body, and xiphoid process. The **manubrium**[38] (ma-NOO-bree-um) is the broad superior portion. It has a medial **suprasternal notch (jugular**

[35]*auri* = ear + *cul* = little + *ar* = pertaining to
[36]*alae* = wings
[37]*coccyx* = cuckoo (named for resemblance to a cuckoo's beak)
[38]*manubrium* = handle

notch), which you can easily palpate between your clavicles (collarbones), and right and left **clavicular notches,** where it articulates with the clavicles. The **body,** or **gladiolus,**[39] is the longest part of the sternum. It joins the manubrium at the **sternal angle,** which can be palpated as a transverse ridge at the point where the sternum projects farthest forward. In some people, however, the angle is rounded or concave. The second rib attaches here, making the sternal angle a useful landmark for counting ribs in a physical examination. The manubrium and body have scalloped lateral margins where cartilages of the ribs are attached. At the inferior end of the sternum is a small, pointed **xiphoid**[40] (ZIF-oyd) **process** that provides attachment for some of the abdominal muscles. In cardiopulmonary resuscitation, improperly performed chest compression can drive the xiphoid process into the liver and cause a fatal hemorrhage.

RIBS

There are 12 pairs of **ribs,** with no difference between the sexes. Each is attached at its posterior (proximal) end to the vertebral column. A strip of hyaline cartilage called the **costal cartilage** extends from the anterior (distal) ends of ribs 1 to 7 to the sternum. Ribs 1 to 7 are thus called **true ribs.** The costal cartilages of ribs 8, 9, and 10 attach to the costal cartilage of rib 7, and ribs 11 and 12 do not attach to anything at the distal end but are embedded in muscle. Ribs 8 to 12 are therefore called **false ribs,** and ribs 11 and 12 are also called **floating ribs** for lack of any connection to the sternum.

Ribs 1 to 10 each have a proximal **head** and **tubercle,** connected by a narrow **neck;** ribs 11 and 12 have a head only (fig. 7.28). Ribs 2 to 9 have beveled heads that come to a point between a **superior articular facet** above and an **inferior articular facet** below. Rib 1, unlike the others, is a flat horizontal plate. Ribs 2 to 10 have a sharp turn called the **angle,** distal to the tubercle, and the remainder consists of a flat blade called the **shaft.** Along the inferior margin of the shaft is a **costal groove** that marks the path of the intercostal blood vessels and nerve.

Variations in rib anatomy relate to the way different ribs articulate with the vertebrae. Once you observe these articulations on an intact skeleton, you will be better able to understand the anatomy of isolated ribs and vertebrae. Vertebra T1 has a complete superior costal facet on the body that articulates with rib 1, as well as a small inferior costal facet that provides half of the articulation with rib 2. Ribs 2 through 9 all articulate between two vertebrae, so these vertebrae have both superior and inferior costal facets on the respective margins of the body. The inferior costal facet of each vertebra articulates with the superior articular facet of the rib, and the superior costal facet of the next vertebra articulates with the inferior articular facet of the same rib (fig. 7.29*a*). Ribs 10 through 12 each articulate with a single costal facet on the body of the respective vertebra.

Ribs 1 to 10 each have a second point of attachment to the vertebrae: the tubercle of the rib articulates with the transverse costal facet of the same-numbered vertebra (fig. 7.29*b*). Ribs 11 and 12 articulate only with the vertebral bodies; they do not have tubercles and vertebrae T11 and T12 do not have transverse costal facets.

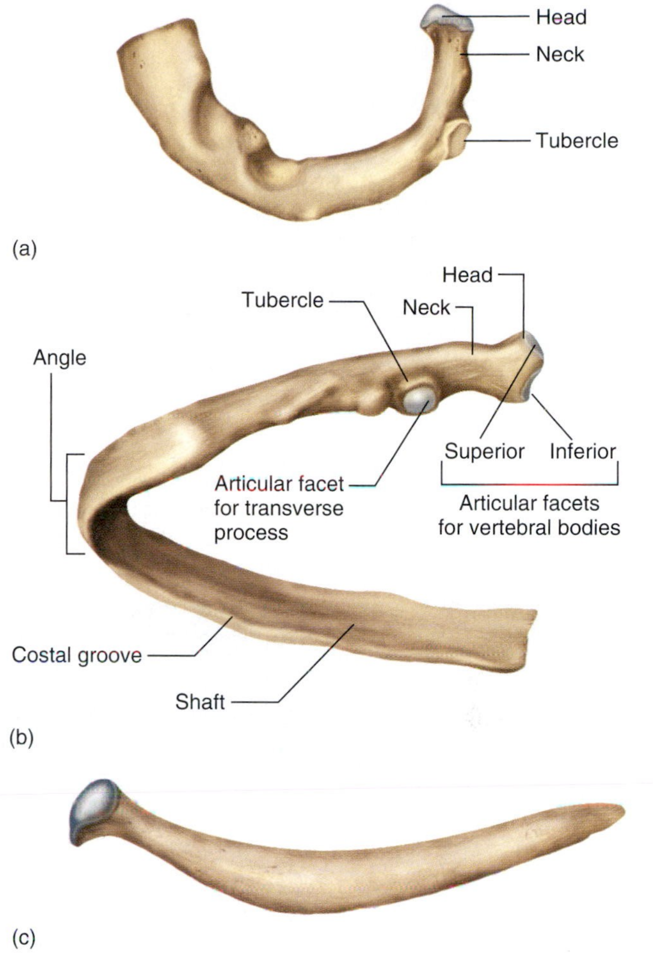

FIGURE 7.28
Anatomy of the Ribs. (*a*) Rib 1 is an atypical flat plate. (*b*) Typical features of ribs 2 to 10. (*c*) Appearance of the floating ribs, 11 and 12.

Table 7.5 summarizes these variations. Table 7.6 provides a checklist that you can use to review your knowledge of the vertebral column and thoracic cage.

Before You Go On

Answer the following questions to test your understanding of the preceding section:

9. Make a table with three columns headed "cervical," "thoracic," and "lumbar." In each column, list the identifying characteristics of each type of vertebra.
10. Describe how rib 5 articulates with the spine. How do ribs 1 and 12 differ from this and from each other in their modes of articulation?
11. Distinguish between true, false, and floating ribs. State which ribs fall into each category.
12. Name the three divisions of the sternum and list the sternal features that can be palpated on a living person.

[39]*gladiolus* = sword
[40]*xipho* = sword + *oid* = resembling

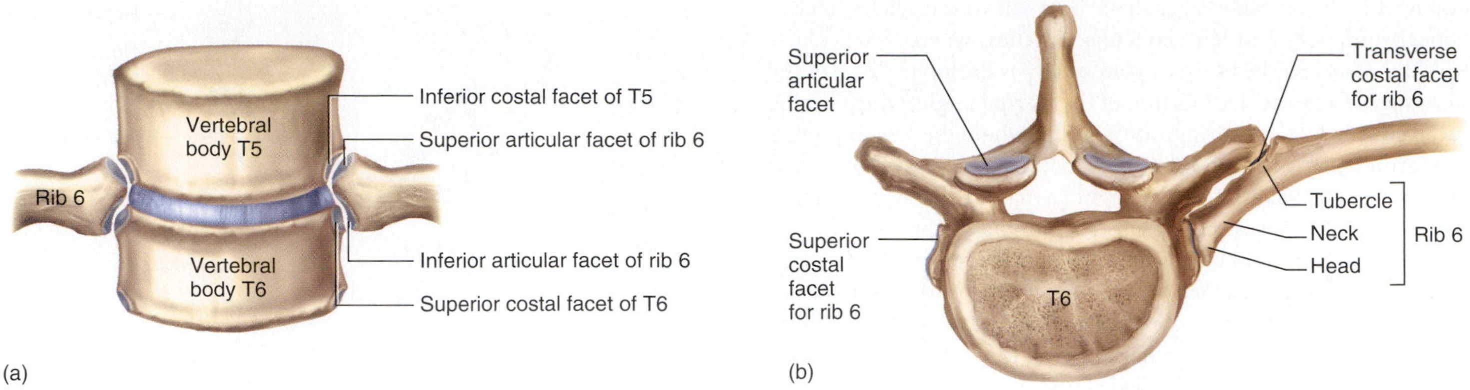

FIGURE 7.29
Articulation of Rib 6 with Vertebrae T5 and T6. (*a*) Ventral view. Note the relationship of the articular facets of the rib with the costal facets of the two vertebrae. (*b*) Superior view. Note that the rib articulates with a vertebra at two points: the costal facet of the vertebral body and the transverse costal facet on the transverse process.

TABLE 7.5
Articulations of the Ribs

Rib	Type	Costal Cartilage	Articulating Vertebral Bodies	Articulating with a Costal Facet?	Rib Tubercle
1	True	Individual	T1	Yes	Present
2	True	Individual	T1 and T2	Yes	Present
3	True	Individual	T2 and T3	Yes	Present
4	True	Individual	T3 and T4	Yes	Present
5	True	Individual	T4 and T5	Yes	Present
6	True	Individual	T5 and T6	Yes	Present
7	True	Individual	T6 and T7	Yes	Present
8	False	Shared with rib 7	T7 and T8	Yes	Present
9	False	Shared with rib 7	T8 and T9	Yes	Present
10	False	Shared with rib 7	T10	Yes	Present
11	False, floating	None	T11	No	Absent
12	False, floating	None	T12	No	Absent

DEVELOPMENTAL AND CLINICAL PERSPECTIVES

Objectives

When you have completed this section, you should be able to

- describe the prenatal development of the axial skeleton; and
- describe some common disorders of the axial skeleton.

Development of the Axial Skeleton

The axial skeleton develops primarily by endochondral ossification. This is a two-step process: (1) **chondrification,** in which embryonic mesenchyme condenses and differentiates into hyaline cartilage, and (2) **ossification,** in which the cartilage is replaced by bone, as described in chapter 6. Significant parts of the skull develop by intramembranous ossification, with no cartilage precursor. Bones that form by this mode are called *membranous bones.*

THE SKULL

Development of the skull is extremely complex, and we will take only a broad overview of the process here. We can view the skull as developing in three major parts: the base, the calvaria, and the facial bones. The base and calvaria are collectively called the **neurocranium** because they enclose the brain; the facial skeleton is called the **viscerocranium** because it develops from the pharyngeal (visceral) arches. (These arches are described in chapter 4.) Both the neurocranium and viscerocranium have regions of cartilaginous and membranous origin. The cartilaginous neurocranium is also called the **chondrocranuim.**

The base of the cranium develops from several pairs of cartilaginous plates inferior to the brain. These plates undergo endochondral ossification and give rise to most parts of the sphenoid, ethmoid, temporal, and occipital bones. The flat bones of the calvaria form, in contrast, by the intramembranous method. They begin to ossify in week 9, slightly later than the cranial base. As a membranous bone ossifies, trabeculae and spicules of osseous tissue first appear in the center and then spread toward the edges (fig. 7.30).

TABLE 7.6

Anatomical Checklist for the Vertebral Column and Thoracic Cage

Vertebral Column

Spinal Curvatures (fig. 7.19)
- Cervical curvature
- Thoracic curvature
- Lumbar curvature
- Pelvic curvature

General Vertebral Structure (figs. 7.22 and 7.23)
- Body (centrum)
- Vertebral foramen
- Vertebral canal
- Vertebral arch
  - Pedicle
  - Lamina
- Spinous process
- Transverse process
- Superior articular process
- Inferior articular process
- Intervertebral foramen
  - Inferior vertebral notch
  - Superior vertebral notch

Intervertebral Discs (fig. 7.22)
- Annulus fibrosus
- Nucleus pulposus

Cervical Vertebrae (figs. 7.24 and 7.25a)
- Transverse foramina
- Bifid spinous process
- Atlas
  - Anterior arch
  - Anterior tubercle

Cervical Vertebrae (figs. 7.24 and 7.25a)—(Cont.)
  - Posterior arch
  - Posterior tubercle
  - Lateral mass
  - Superior articular facet
  - Inferior articular facet
  - Transverse ligament
- Axis
  - Dens (odontoid process)

Thoracic Vertebrae (fig. 7.25b)
- Superior costal facet
- Inferior costal facet
- Transverse costal facet

Lumbar Vertebrae (figs. 7.23 and 7.25c)

Sacral Vertebrae (fig. 7.26)
- Sacrum
- Anterior sacral foramina
- Dorsal sacral foramina
- Median sacral crest
- Lateral sacral crest
- Sacral canal
- Sacral hiatus
- Auricular surface
- Superior articular process
- Ala

Coccygeal Vertebrae (fig. 7.26)
- Coccyx
- Cornu

Thoracic Cage

Sternum (fig. 7.27)
- Manubrium
  - Suprasternal notch
  - Clavicular notch
  - Sternal angle
- Body (gladiolus)
- Xiphoid process

Ribs (figs. 7.27 to 7.29)
- True ribs (ribs 1–7)
- False ribs (ribs 8–12)
- Floating ribs (ribs 11–12)

Ribs (figs. 7.27 to 7.29)—(Cont.)
- Head
- Superior articular facet
- Inferior articular facet
- Neck
- Tubercle
- Angle
- Shaft
- Costal groove
- Costal cartilage

Facial bones develop mainly from the first two pharyngeal arches. Although these arches are initially supported by cartilage, the cartilages do not transform into bone. They become surrounded by developing membranous bone, and while some of the cartilages become middle-ear bones and part of the hyoid bone, some simply degenerate and disappear. Thus, the facial bones are built *around* cartilages but develop by the intramembranous process.

The skull therefore develops from a multitude of separate pieces. These pieces undergo considerable fusion by the time of birth, but their fusion is by no means complete then. At birth, both the mandible and frontal bone, for example, are still paired. The right and left halves of the mandible are joined at the chin by a fibrous **mental symphysis** (SIM-fih-sis). The symphysis ossifies during the first year after birth, uniting the two mandibular bones into one. The frontal bones usually fuse by the age of 5 or 6 years, but in some people a *metopic*[41] *suture* persists between them. Traces of this suture are evident in some adult skulls.

[41]*met* = beyond + *op* = the eyes

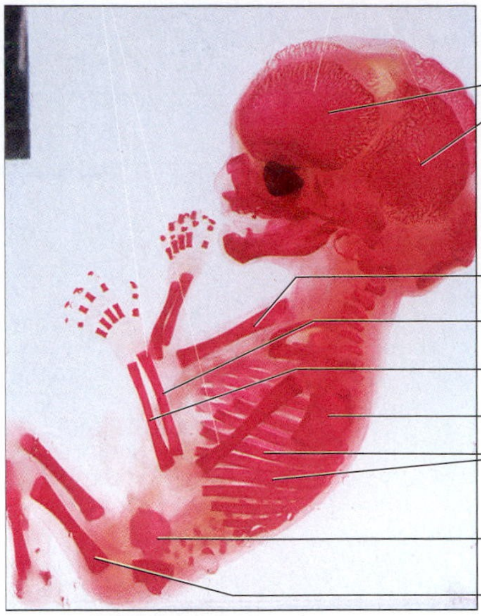

FIGURE 7.30

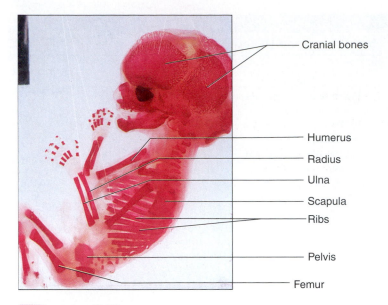

The Fetal Skeleton at 12 Weeks. The red-stained regions are ossified at this age, whereas the elbow, wrist, knee, and ankle joints appear translucent because they are still cartilaginous. The cranial bones are still widely separated.

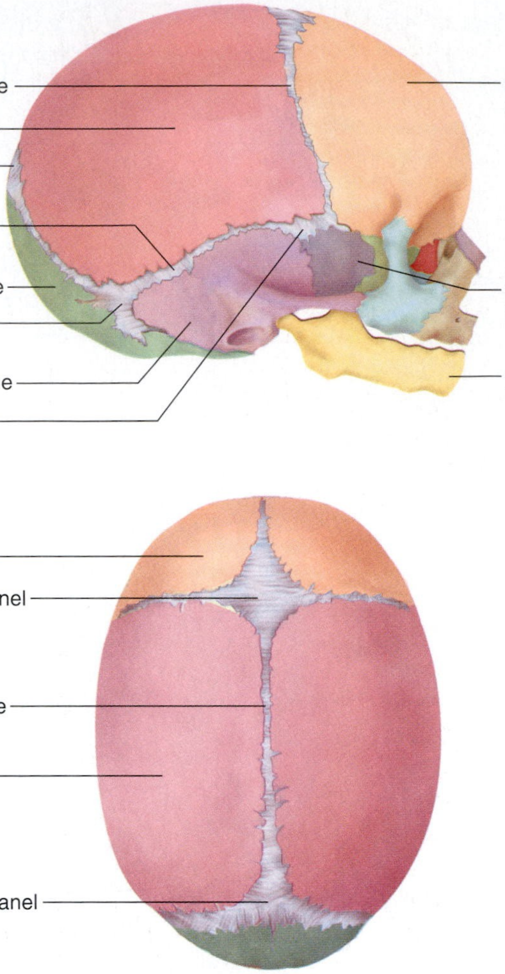

FIGURE 7.31

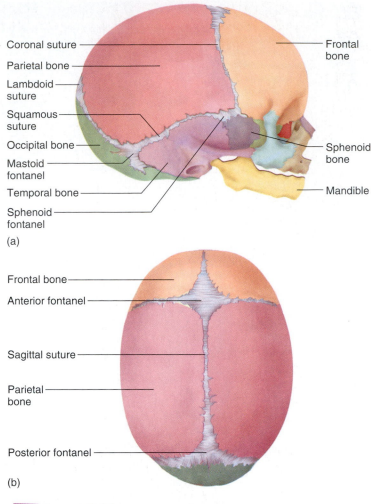

The Fetal Skull Near the Time of Birth. (*a*) Right lateral view. (*b*) Superior view. The cranial bones are separated by fibrous sutures and fontanels.

The cranial bones are still separated at birth by gaps called **fontanels,**[42] bridged by fibrous membranes (fig. 7.31). The term refers to the fact that pulsation of the infant's blood can be felt there. Fontanels permit the bones to shift and the skull to deform during the birth passage. This shifting may deform the infant's head, but the head usually assumes a normal shape within a few days after birth. Four of the fontanels are especially prominent and regular in location: the **anterior, posterior, sphenoid,** and **mastoid fontanels.** The fontanels close by intramembranous ossification. Most are fully ossified by 12 months of age, but the largest one, the anterior fontanel, does not close for 18 to 24 months. A "soft spot" can still be palpated in the corner formed by the frontal and parietal bones up to that age.

The face of a newborn is flat and small compared to the large cranium. It enlarges as the teeth and paranasal sinuses develop. To accommodate the growing brain, a child's skull grows more rapidly than the rest of the skeleton. It reaches about half its adult size by 9 months of age, three-quarters by age 2, and nearly final size by 8 or 9 years. The heads of babies and young children are therefore much larger in proportion to the trunk than the heads of adults—an attribute thoroughly exploited by cartoonists and advertisers who draw big-headed characters to give them a more endearing or immature appearance. In humans and other animals, the large rounded heads of the young are thought to promote survival by stimulating parental caregiving instincts.

THE VERTEBRAL COLUMN

One of the universal characteristics of all chordate animals, including humans, is the **notochord,** a flexible, middorsal rod of mesodermal tissue. In humans, the notochord is evident inferior to the neural tube in the third week of development. In the fourth week, part of each embryonic somite becomes a sclerotome flanking the notochord—so-named because it is destined to give rise to bone. The sclerotomes are temporarily separated by zones of looser mesenchyme (fig. 7.32*a*). As shown in figure 7.32*b*, each vertebral body arises from portions of two adjacent sclerotomes and the loose mesenchyme between them. The midportion of each sclerotome gives rise to the annulus fibrosus of the intervertebral disc. The notochord degenerates and disappears in the regions of the developing vertebral bodies, but persists and expands between the vertebrae to form the nucleus pulposus of the intervertebral discs.

Meanwhile, mesenchyme surrounding the neural tube condenses and forms the vertebral arches of the vertebrae. Approaching the end of the embryonic phase, the mesenchyme of the sclerotomes forms the cartilaginous forerunners of the vertebral bodies. The two halves of the vertebral arch fuse with each other and with the body, and the spinous and transverse processes grow outward from the arch. Thus, a complete cartilaginous vertebral column is established.

[42]*fontan* = fountain + *el* = little

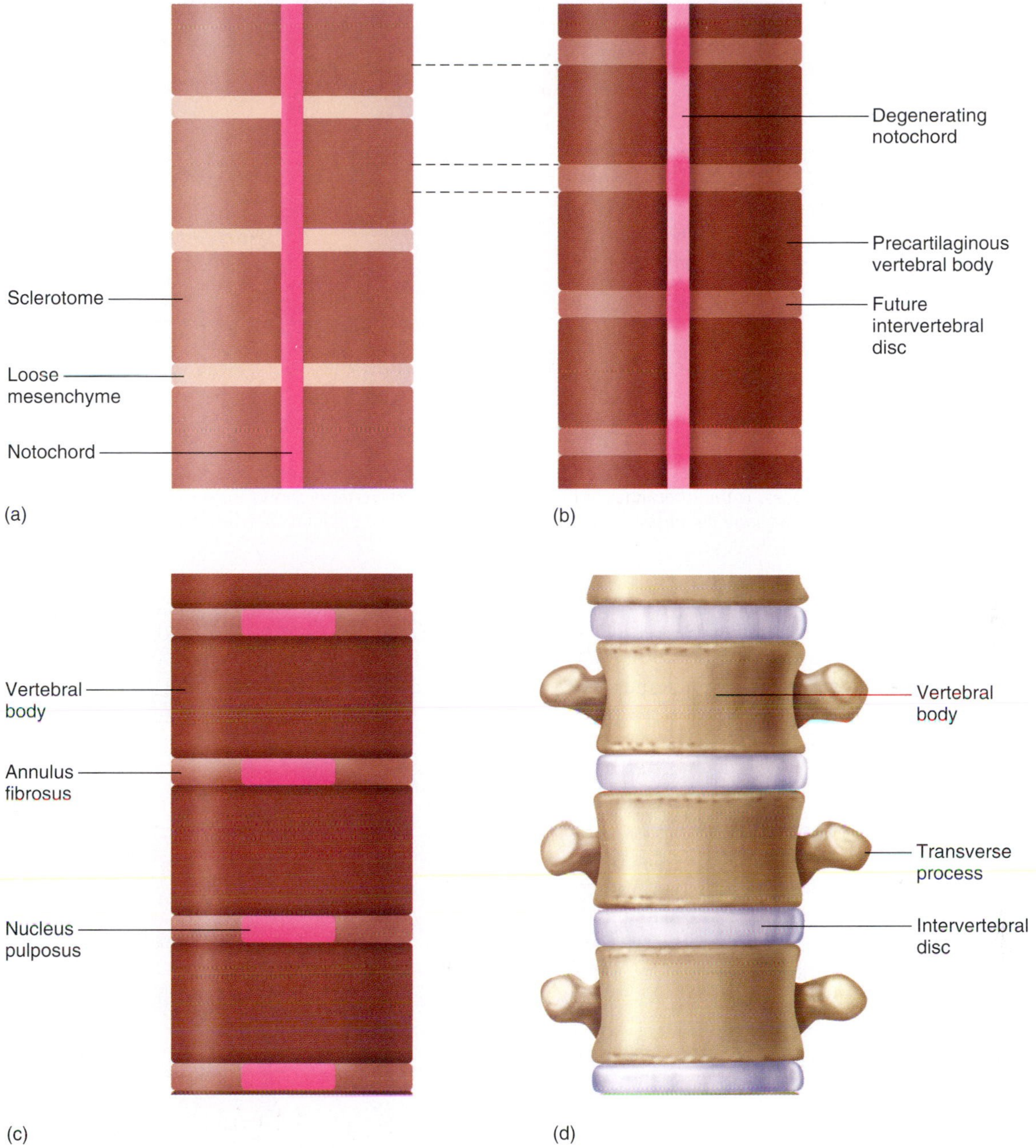
Degenerating notochord
Precartilaginous vertebral body
Future intervertebral disc
Sclerotome
Loose mesenchyme
Notochord
(a)
(b)
Vertebral body
Annulus fibrosus
Nucleus pulposus
(c)
Vertebral body
Transverse process
Intervertebral disc
(d)

FIGURE 7.32

Development of the Vertebrae and Intervertebral Discs. (*a*) The notochord is flanked by sclerotomes, which are separated by zones of loose mesenchyme. (*b*) Each vertebral body forms by the condensation of parts of two sclerotomes and the loose mesenchyme between them. The midregion of each sclerotome remains less condensed and forms the annulus fibrosus of the intervertebral disc. The notochord degenerates in the regions of condensing mesenchyme but persists between vertebral bodies as the nucleus pulposus. Dashed lines indicate which regions of the sclerotomes in figure *a* give rise to the vertebral body and intervertebral disc in figure *b*. (*c*) Further condensation of the vertebral bodies. The notochord has now disappeared except at the nucleus pulposus of each disc. (*d*) Chondrification and ossification give rise to the fully developed vertebral bodies.

Ossification of the vertebrae begins during the embryonic period and is not completed until age 25. Each vertebra develops three primary ossification centers: one in the body and one in each half of the vertebral arch. At birth, these three bony parts of the vertebra are still connected by hyaline cartilage (fig. 7.33). The bony halves of the vertebral arch finish ossifying and fuse around 3 to 5 years of age, beginning in the lumbar region and progressing rostrally. The attachments of the arch to the body remain cartilaginous for a time in order to allow for growth of the spinal cord. These attachments ossify at an age of 3 to 6 years. Secondary ossification centers form in puberty at the tips of the spinous and transverse processes and in a ring encircling the body. They unite with the rest of the vertebra by age 25.

At birth, the vertebral column exhibits one continuous C-shaped curve (fig. 7.34), as it does in monkeys, apes, and most other four-legged animals. As an infant begins to crawl and lift its head, the cervical curvature forms, enabling an infant on its belly to look forward. The lumbar curvature begins to develop as a toddler begins walking.

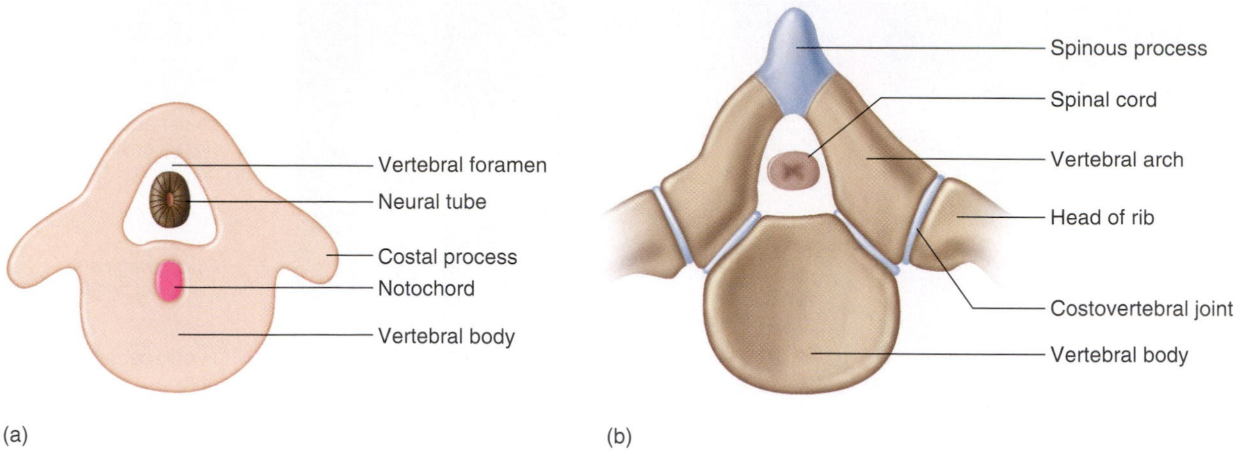

FIGURE 7.33

Development of a Thoracic Vertebra. (*a*) At 5 weeks. (*b*) At birth. In figure *a,* the vertebra is composed of mesenchyme surrounding the neural tube. The notochord is still present. The costal process is the forerunner of the rib. In figure *b,* the vertebra shows three centers of ossification at birth—the body and the two vertebral arches. Hyaline cartilage *(blue)* still composes the spinous process, the joints between the vertebral arches and the body, and the joints between the ribs and the vertebra.

THE RIBS AND STERNUM

At 5 weeks, a developing thoracic vertebra consists of a body of mesenchyme with a vertebral body, vertebral foramen, and a pair of winglike lateral extensions called the **costal processes** (see fig. 7.33*a*), which soon give rise to the ribs. At 6 weeks, a chondrification center develops at the base of each costal process. At 7 weeks, these centers begin to undergo endochondral ossification. A *costovertebral joint* now appears at the base of the costal process, separating it from the vertebral body (see fig. 7.33*b*). By this time, the first seven ribs (true ribs) connect to the sternum by way of costal cartilages. An ossification center soon appears at the angle of the rib, and endochondral ossification proceeds from there to the distal end of the shaft. Secondary ossification centers appear in the tubercle and head of the rib during adolescence.

The sternum begins as a pair of longitudinal strips of condensed mesenchyme called the **sternal bars.** These form initially in the ventrolateral body wall and migrate medially during chondrification. The right and left sternal bars begin to fuse in week 7 as the most cranial pairs of ribs contact them. Fusion of the sternal bars progresses caudally, ending with the formation of the xiphoid process in week 9. The sternal bones form by endochondral ossification beginning rostrally and progressing caudally. Ossification begins in month 5 and is completed shortly after birth. In some cases, the sternal bars fail to fuse completely at the caudal end, so the infant xiphoid process is forked or perforated.

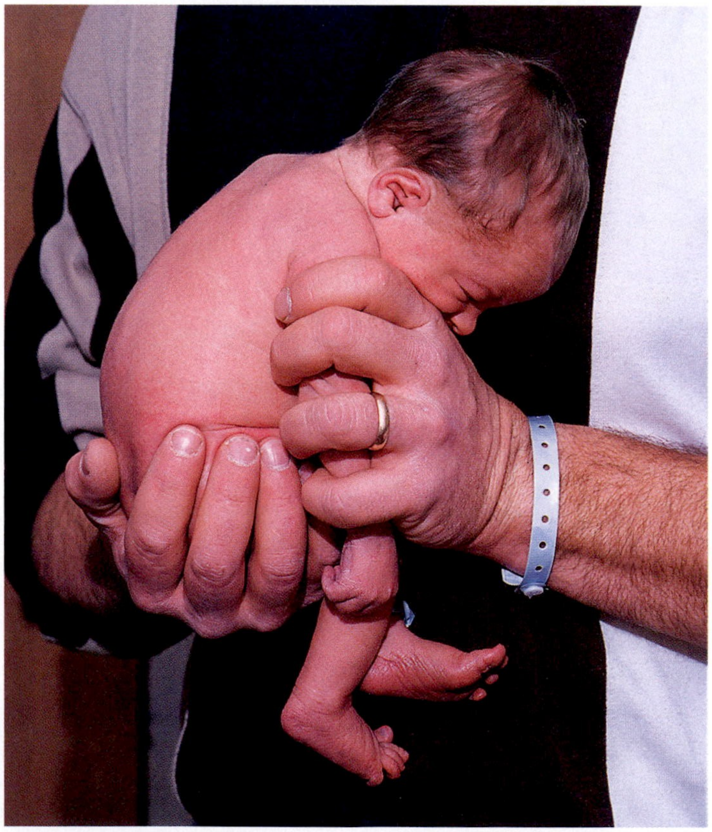

FIGURE 7.34

Spinal Curvature of the Newborn Infant. At this age, the spine forms a single C-shaped curve.

Pathology of the Axial Skeleton

Disorders that affect all parts of the skeleton are discussed in chapter 6, especially fractures and osteoporosis. Table 7.7 lists some disorders that affect especially the axial skeleton. We will consider in slightly more depth skull fractures, vertebral fractures and dislocations, and herniated intervertebral discs.

TABLE 7.7

Disorders of the Axial Skeleton

Cleft Palate	Failure of the palatine processes of the maxilla to fuse during fetal development, resulting in a fissure connecting the oral and nasal cavities, and often accompanied by cleft lip. Causes difficulty for an infant in nursing. Can be surgically corrected with good cosmetic results.
Craniosynostosis	Premature closure of the cranial sutures within the first two years after birth, resulting in skull asymmetry, deformity, and sometimes mental retardation. Cause is unknown. Surgery can limit brain damage and improve appearance.
Spinal Stenosis	Abnormal narrowing of the vertebral canal or intervertebral foramina caused by hypertrophy of the vertebral bone. Most common in middle-aged and older people. May compress spinal nerves and cause low back pain or muscle weakness.
Spondylosis	A defect of the laminae of the lumbar vertebrae. Defective vertebrae may shift anteriorly, especially at the L5 to S1 level. Stress on the bone may cause microfractures in the laminae and eventual dissolution of the laminae. May be treated by nonsurgical manipulation or by surgery, depending on severity.

Disorders Described Elsewhere

Ethmoid bone fractures 183	Lordosis 189	Spina bifida 380
Herniated discs 202	Scoliosis 189	Vertebral fractures 202
Kyphosis 189	Skull fractures 201	

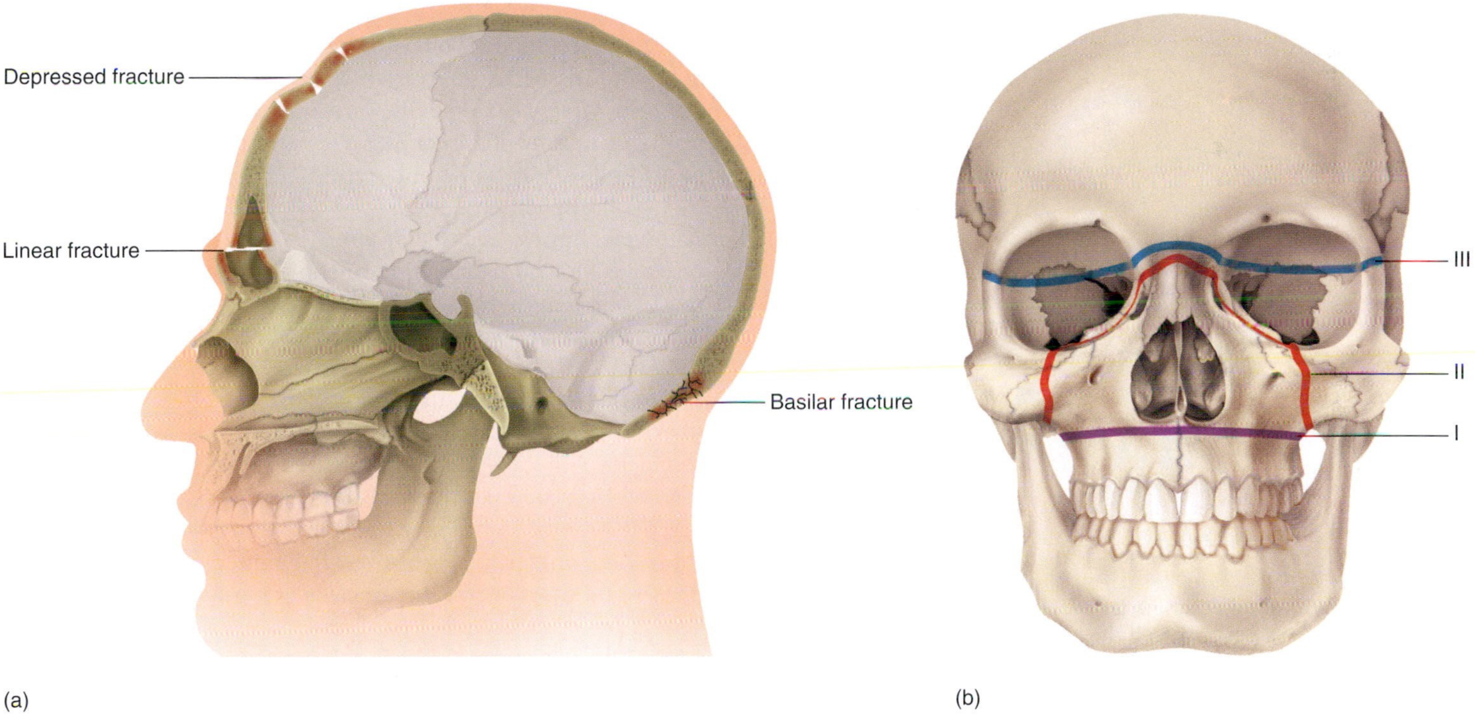

FIGURE 7.35

Skull Fractures. (*a*) Medial view showing linear and depressed fractures of the frontal bone, and basilar fracture of the occipital bone. (*b*) The three types of Le Fort fractures of the facial bones.

SKULL FRACTURES

The domelike shape of the skull distributes the force of most blows and tends to minimize their effects. Hard blows can nevertheless fracture the calvaria (fig. 7.35*a*). Most cranial fractures are *linear fractures* (elongated cracks), which can radiate away from the point of impact. In a *depressed fracture,* the cranium caves inward and may compress and damage underlying brain tissue. If a blow occurs in an area where the calvaria is especially thick, as in the occipital region, the bone may bend inward at the point of impact without breaking, but as the force is distributed through the cranium it can fracture it some distance away, even on the opposite side of the skull (a *contrafissura* fracture). In addition to damaging brain tissue, skull fractures can damage cranial nerves and meningeal blood vessels. A break in a blood vessel may cause a hematoma (mass of clotted blood) that compresses the brain tissue, potentially leading to death within a few hours.

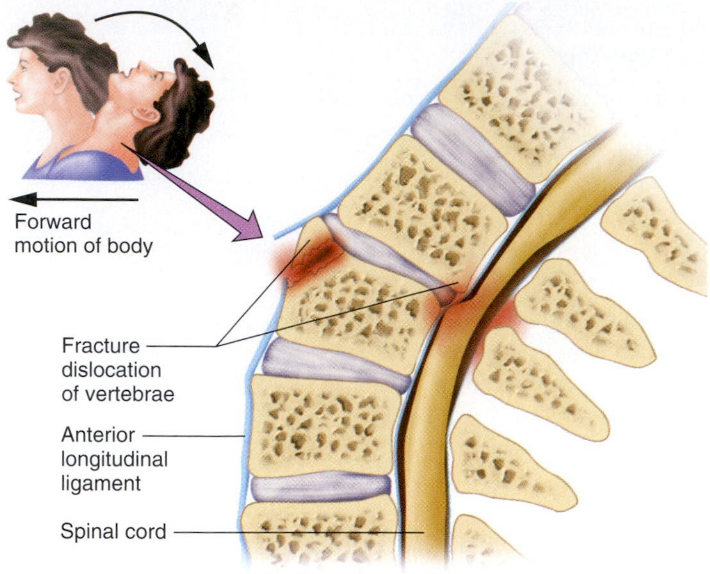
Forward
motion of body
Fracture
dislocation
of vertebrae
Anterior
longitudinal
ligament
Spinal cord

(a)

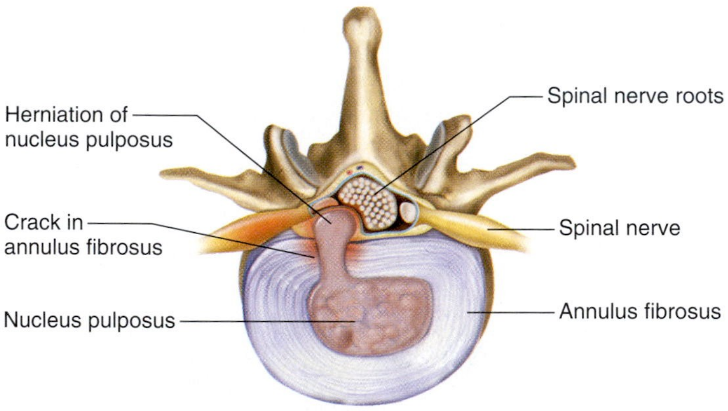
Herniation of
nucleus pulposus
Crack in
annulus fibrosus
Nucleus pulposus
Spinal nerve roots
Spinal nerve
Annulus fibrosus

(b)

FIGURE 7.36

Injuries to the Vertebral Column. (*a*) Whiplash injury. Violent hyperextension of the neck has torn the anterior longitudinal ligament and fractured the vertebral body. (*b*) Herniated intervertebral disc. The nucleus pulposus is oozing into the vertebral canal and compressing a bundle of spinal nerve roots that passes through the lumbar vertebrae.

Blows to the face often produce linear *Le Fort*[43] *fractures,* which predictably follow lines of weakness in the facial bones. The three typical Le Fort fractures are shown in figure 7.35*b*. The type II Le Fort fracture separates the entire central region of the face from the rest of the skull.

VERTEBRAL FRACTURES AND DISLOCATIONS

Injury to the cervical vertebrae (a "broken neck") often results from violet blows to the head, as in diving, motorcycle, and equestrian accidents, and sudden flexion or extension of the neck, as in automobile accidents. Such injuries often crush the body or arches of a vertebra or cause one vertebra to slip forward relative to the one below it. The dislocation of one vertebra relative to the next can cause irreparable damage to the spinal cord. "Whiplash" often results from rear-end automobile collisions causing violent hyperextension of the neck (backward jerking of the head). This stretches or tears the *anterior longitudinal ligament* that courses anteriorly along the vertebral bodies, and it may fracture the vertebral body (fig. 7.36*a*). Dislocations are relatively rare in the thoracic and lumbar regions because of the way the vertebrae are tightly interlocked. When fractures occur in these regions ("broken back"), they most often involve vertebra T11 or T12, at the transition from the thoracic to lumbar spine.

HERNIATED DISCS

A **herniated** ("slipped" or "ruptured") **disc** is cracking of the annulus fibrosis of an intervertebral disc under strain, sometimes caused by violent flexion of the vertebral column or by lifting heavy weights. Cracking of the annulus allows the gelatinous nucleus pulposus to ooze out, sometimes putting pressure on a spinal nerve root or the spinal cord (fig. 7.36*b*). Pressure on the spinal cord sometimes must be relieved by *laminectomy,* a surgery in which the vertebrae laminae are cut and sometimes the laminae and spinous process are removed. Back pain results from both pressure on the nervous tissue and inflammation stimulated by the nucleus pulposus. About 95% of disc herniations occur at levels L4/L5 and L5/S1. The nucleus pulposus usually escapes in a posterolateral direction, where the annulus is thinnest. Herniated discs rarely occur in young people because their discs are well hydrated and absorb pressure well. As people get older, the discs become dehydrated and they degenerate and grow thinner, becoming more susceptible to herniation. After middle age, however, the annulus fibrosus becomes thicker and tougher, and the nucleus pulposus is smaller, so disc herniations again become less common.

Before You Go On

Answer the following questions to test your understanding of the preceding section:

13. Define *chondrocranium* and *viscerocranium* and explain why each of them is named that.
14. What is the functional significance of fontanels? When does the last fontanel close?
15. What structure in the adult is a remnant of the embryonic notochord?
16. What is a Le Fort fracture? What is whiplash?
17. Explain why a herniated disc can cause nerve pain (neuralgia).

[43]Léon C. Le Fort (1829–93), French surgeon and gynecologist

CHAPTER REVIEW

Review of Key Concepts

Overview of the Skeleton (p. 172)

1. The skeleton is divided into the *axial skeleton* (skull, vertebral column, and thoracic cage) and *appendicular skeleton* (limbs, pectoral girdle, and pelvic girdle).
2. There are typically about 270 bones at birth. Additional bones form during childhood, but then certain bones begin to fuse and the number drops to an average of 206 by adulthood.
3. The number of adult bones varies, especially the number of *sesamoid* and *sutural bones*.
4. Bones have a variety of surface features that provide muscle attachments, articular surfaces, and passages for nerves and blood vessels. The terminology of these features (table 7.2) is important in skeletal anatomy.

The Skull (p. 175)

1. The skull consists of 22 bones (and sometimes additional sutural bones), most of which are rigidly joined by *sutures*.
2. Cavities in the skull include the cranial cavity, orbits, nasal cavity, buccal cavity, middle- and inner-ear cavities, and the frontal, ethmoid, sphenoid, and maxillary sinuses.
3. The skull has numerous foramina that provide passage for nerves and blood vessels (table 7.3).
4. Eight of the skull bones are called *cranial bones* because they form the cranial cavity: the frontal, parietal, temporal, occipital, sphenoid, and ethmoid bones. The parietal and temporal bones are paired and the others are single.
5. The brain is separated from these bones by membranes called meninges. One of these, the *dura mater,* is the periosteum of the cranial bones.
6. The roof of the cranial cavity is called the *calvaria* and its floor is the *base.* The base exhibits *anterior, middle,* and *posterior cranial fossae* that conform to the contours of the base of the brain.
7. Fourteen of the skull bones are called facial bones; they do not contribute to the cranial cavity but are located anteriorly on the skull and shape the face. These are the maxillae, inferior nasal conchae, vomer, mandible, and the palatine, zygomatic, lacrimal, and nasal bones. The vomer and mandible are single and the others are paired.
8. The anatomical features of the 22 skull bones are summarized in table 7.4.
9. Seven other bones are closely associated with the skull: three *auditory ossicles* (*malleus, incus,* and *stapes*) in each ear and a single *hyoid bone* just below the chin. The auditory ossicles transfer sound to the inner ear, and the hyoid bone provides attachment for muscles of the mandible, tongue, and larynx.
10. The infant skull exhibits gaps *(fontanels)* between the cranial bones where the bones have not yet fused; the *anterior* and *posterior fontanels* are located at the anterior and posterior ends of the sagittal suture. The *sphenoid* and *mastoid fontanels* are paired and laterally situated anterior and posterior to the temporal bone on each side.
11. The frontal bone and mandible are each represented by separate right and left bones in the newborn infant. The halves fuse in early childhood. The skull reaches nearly adult size by 8 or 9 years of age.

The Vertebral Column and Thoracic Cage (p. 188)

1. The vertebral column normally consists of 33 *vertebrae* and 23 cartilaginous *intervertebral discs.* The vertebrae are divided into five groups: 7 *cervical vertebrae* in the neck; 12 *thoracic vertebrae* in the chest, with ribs attached to them; 5 *lumbar vertebrae* in the lower back; 5 *sacral vertebrae* fused in a single adult bone, the *sacrum;* and a "tailbone" *(coccyx)* of 4 fused *coccygeal vertebrae.* The numbers differ in about 5% of adults, especially in the lumbar to sacral area.
2. The vertebral column is C-shaped at birth. Beyond the age of 3 years, it has four bends called the *cervical, thoracic, lumbar,* and *pelvic curvatures.*
3. Some major features of a vertebra are the *body* or *centrum;* a *vertebral arch* composed of two *pedicles* and two *laminae;* a dorsal *spinous process* arising where the two laminae meet; and a pair of lateral *transverse processes.*
4. The vertebral arch encloses a space called the *vertebral foramen;* collectively, the vertebral foramina form the *vertebral canal,* which houses the spinal cord.
5. Each vertebra joins the one above it through its *superior articular processes* and the one below it through its *inferior articular processes.*
6. Between two adjacent vertebrae, there is a gap called the *intervertebral foramen,* which allows for the passage of a spinal nerve.
7. An intervertebral disc consists of a fibrous ring, the *annulus fibrosus,* enclosing a gelatinous center, the *nucleus pulposus.* The discs bind adjacent vertebrae together, add flexibility to the vertebral column, support the body weight, and absorb shock.
8. Cervical vertebrae (C1–C7) are relatively small. All are characterized by a *transverse foramen* in each transverse process, through which the vertebral arteries travel. C2 through C6 typically exhibit a forked spinous process. C1, the *atlas,* is a relatively simple ring of bone with a pair of *lateral masses* joined by an *anterior* and *posterior arch,* but with no centrum. C2, the *axis,* has a unique superior process called the *dens.*
9. Thoracic vertebrae (T1–T12) are specialized for rib attachment. They all exhibit *costal facets* on the centrum, and T1 through T10 also have *transverse costal facets* at the ends of the transverse processes. Ribs 1 through 10 attach to their respective vertebrae at two points, the vertebral body and transverse process; ribs 11 and 12 attach to the body only.
10. The lumbar vertebrae (L1–L5) have no unique features, but have especially heavy bodies and stout, squarish spinous processes. Their articular facets meet each other in a lateral-to-medial direction (instead of dorsoventrally like those of other vertebrae), except at the T12-L1 joint and the L5-S1 joint.
11. The adult sacrum is a triangular plate of bone formed by the fusion of five sacral vertebrae (S1–S5). Fusion of their spinous processes forms a dorsal *median sacral crest,* and fusion of their transverse processes produces a *lateral sacral crest* on each side. The intervertebral foramina are represented by the *anterior* and *posterior sacral foramina.* The sacrum and os coxae (hip bone) have complementary *auricular surfaces* where they meet at the *sacroiliac joint.*
12. The coccyx is a small pointed "tailbone" formed by the fusion of usually four *coccygeal vertebrae* (Co1–Co4).
13. The *thoracic cage* is a bony enclosure for the lungs and heart, and is composed of the thoracic vertebrae, ribs, and sternum.
14. The sternum (breastbone) consists of a superior *manubrium,* a long middle *body,* and a small, pointed *xiphoid process* at the inferior end. Its margins are scalloped where

they receive the *costal cartilages* of the ribs, and the superolateral corners of the manubrium exhibit *clavicular notches* for articulation with the clavicles.

15. There are 12 pairs of ribs. Most of them exhibit a *head* where they articulate with the body of a vertebra; a narrow *neck;* a rough *tubercle* where they articulate with the transverse process of a vertebra; and a flat bladelike *shaft.* In most, the shaft has a squared end where it meets a *costal cartilage,* which connects the rib to the sternum.
16. Ribs 1 through 7 are called *true ribs* because they each connect to the sternum by their own costal cartilage. Ribs 8 through 12 are called *false ribs* because they do not have independent attachments to the sternum. The costal cartilages of ribs 8 through 10 connect to the cartilage of rib 7. Ribs 11 and 12 have no costal cartilages and do not attach to the sternum at all; thus they are also called *floating ribs.* They attach only to the bodies of vertebrae T11 and T12, and have no tubercles.

Developmental and Clinical Perspectives (p. 196)

1. Parts of the skull develop by intramembranous ossification. The rest of the skull and most of the rest of the skeleton develop by endochondral ossification. In the latter case, bone development begins when mesenchyme condenses and differentiates into hyaline cartilage, a process called *chondrification.* The cartilage is then replaced by bone in the process of endochondral ossification.
2. The base of the skull develops primarily by the endochondral ossification of several cartilaginous plates inferior to the brain. The calvaria develops mainly by intramembranous ossification. The base and calvaria are collectively called the *neurocranium* because they enclose the cranial cavity and surround the brain. The portion arising by endochondral ossification is also called the *chondrocranium.*
3. The facial skeleton is called the *viscerocranium* because it develops from the pharyngeal (visceral) arches. The facial bones form primarily by intramembranous ossification.
4. Ossification of the skull continues for years after birth as the two frontal bones and two mandibular bones unite medially and the fontanels close. The face of the newborn is small and flat compared to the cranium, but grows to its mature proportions as the teeth and paranasal sinuses develop.
5. The vertebral column is preceded by a dorsal rod, the *notochord,* formed around day 22 to 24. Most of the notochord later degenerates as the *sclerotomes* of mesoderm produce the vertebral bodies and intervertebral discs, but notochordal tissue persists as the nucleus pulposus of the intervertebral discs.
6. The vertebral column develops at first by chondrification of the mesenchyme of the sclerotomes, creating a cartilaginous fetal vertebral column. Vertebral ossification begins during the embryonic period but is not completed until 25 years after birth. The vertebral column is C-shaped at birth but acquires its cervical and lumbar curvatures in infancy and early childhood, in association with lifting of the head and walking.
7. The ribs develop as lateral extensions of the vertebrae called *costal processes.* The processes chondrify and then ossify and separate from the vertebral body. Secondary ossification centers do not appear in the ribs until adolescence.
8. The sternum begins as a pair of longitudinal *sternal bars* of mesenchyme. These migrate medially and fuse as the ribs attach to them. Ossification is by the endochondral method, beginning in month 5 and concluding soon after birth.
9. Skull fractures may be *linear* (elongated cracks) or *depressed* (indentations of the bone). They can damage cranial nerves, meningeal blood vessels, and brain tissue, and may result in hematomas (blood clots) that fatally compress the brain. Blows to the face may produce Le Fort fractures along lines of weakness that separate regions of the face from the rest of the skull.
10. The cervical vertebrae are often fractured, displaced, or both by violent blows to the head or extreme flexion or extension of the neck. Such injuries can cause irreparable damage to the spinal cord. Fractures lower in the vertebral column are most often at vertebra T11 or T12.
11. A herniated disc is the cracking of the annulus fibrosus and oozing of the nucleus pulposus through the crack. This triggers inflammation and puts pressure on the spinal cord and spinal nerves, causing back pain. Some cases require a *laminectomy* to relieve pressure on the spinal cord.

Testing Your Recall

1. Which of these is *not* a paranasal sinus?
   a. frontal
   b. temporal
   c. sphenoid
   d. ethmoid
   e. maxillary
2. Which of these is a facial bone?
   a. frontal
   b. ethmoid
   c. occipital
   d. temporal
   e. lacrimal
3. Which of these *cannot* be palpated on a living person?
   a. the crista galli
   b. the mastoid process
   c. the zygomatic arch
   d. the superior nuchal line
   e. the hyoid bone
4. All of the following are groups of vertebrae *except* for _____, which is a spinal curvature.
   a. thoracic
   b. cervical
   c. lumbar
   d. pelvic
   e. sacral
5. Thoracic vertebrae do *not* have
   a. transverse foramina.
   b. costal facets.
   c. transverse costal facets.
   d. transverse processes.
   e. pedicles.
6. Which of these bones forms by intramembranous ossification?
   a. a vertebra
   b. a parietal bone
   c. the occipital bone
   d. the sternum
   e. a rib
7. The viscerocranium includes
   a. the maxilla.
   b. the parietal bones.
   c. the occipital bone.
   d. the temporal bone.
   e. the atlas.
8. Which of these is *not* a suture?
   a. parietal
   b. coronal

c. lambdoid
d. sagittal
e. squamous

9. The word root *ethmo-* means
   a. wedge.
   b. funnel.
   c. sieve.
   d. wing.
   e. crest.
10. The nasal septum is composed partly of the same bone as
    a. the zygomatic arch.
    b. the hard palate.
    c. the cribriform plate.
    d. the nasal concha.
    e. the centrum.
11. Gaps between the cranial bones of an infant are called _____.
12. The external acoustic meatus is an opening in the _____ bone.
13. Bones of the skull are joined along lines called _____.
14. The _____ bone has greater and lesser wings and protects the pituitary gland.
15. A herniated disc occurs when a ring called the _____ cracks.
16. The transverse ligament of the atlas holds the _____ of the axis in place.
17. The sacroiliac joint is formed where the _____ surface of the sacrum articulates with that of the ilium.
18. We have five pairs of _____ ribs and two pairs of _____ ribs.
19. Ribs 1 to 10 are joined to the sternum by way of strips of connective tissue called _____.
20. The point at the inferior end of the sternum is the _____.

Answers in the Appendix

True or False

Determine which five of the following statements are false, and briefly explain why.

1. The bodies of the vertebrae are derived from the notochord of the embryo.
2. Adults have more bones than children do.
3. A smooth round knob on a bone is called a condyle.
4. The zygomatic arch consists entirely of the zygomatic bone.
5. The dura mater adheres tightly to the entire inner surface of the cranial cavity.
6. The sphenoid bone forms part of the orbit.
7. The nasal septum is not entirely bony.
8. Not everyone has a frontal sinus.
9. The anterior surface of the sacrum is smoother than the posterior surface.
10. The lumbar vertebrae do not articulate with any ribs and therefore do not have transverse processes.

Answers in the Appendix

Testing Your Comprehension

1. A child was involved in an automobile collision. She was not wearing a safety restraint, and her chin struck the dashboard hard. When the physician looked into her auditory canal, he could see into her throat. What do you infer from this about the nature of her injury?
2. Chapter 1 noted that there are significant variations in the internal anatomy of different people (p. 5). Give some examples from this chapter other than pathological cases (such as cleft palate) and normal age-related differences.
3. Vertebrae T12 and L1 look superficially similar and are easily confused. Explain how to tell the two apart.
4. What effect would you predict if an ossification disorder completely closed off the superior and inferior orbital fissures?
5. For each of the following bones, name all the other bones with which it articulates: parietal, zygomatic, temporal, and ethmoid bones.

Answers at the Online Learning Center

www.mhhe.com/saladinhal

Visit the Online Learning Center for practice tests, answer keys, and other learning aids for this chapter. Enhance your understanding of human anatomy with our interactive art labeling exercises, supplemental photo atlases, web links, puzzles, flashcards, and much more.

8

CHAPTER EIGHT

The Appendicular Skeleton

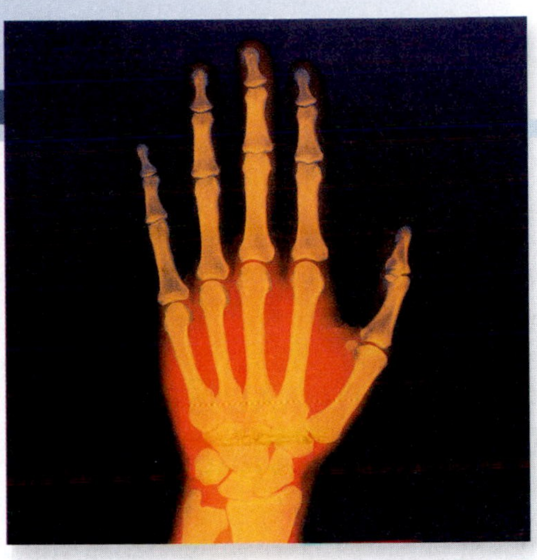

X ray of an adult hand

CHAPTER OUTLINE

The Pectoral Girdle and Upper Limb 208
- Pectoral Girdle 208
- Upper Limb 209

The Pelvic Girdle and Lower Limb 213
- Pelvic Girdle 213
- Lower Limb 215

Developmental and Clinical Perspectives 221
- Development of the Appendicular Skeleton 221
- Pathology of the Appendicular Skeleton 223

Chapter Review 225

INSIGHTS

8.1 Clinical Application: Fractured Clavicle 208
8.2 Clinical Application: Femoral Fractures 218
8.3 Medical History: Anatomical Position—Clinical and Biological Perspectives 223

BRUSHING UP

To understand this chapter, you may find it helpful to review the following concepts:

- Evolution of Bipedal Locomotion in Humans (p. 16)
- Terminology of the Appendicular Region (p. 26)
- General Features of Bones (p. 153)
- Endochondral and Intramembranous Ossification (p. 159)

In this chapter, we turn our attention to the *appendicular skeleton*—the bones of the upper and lower limbs and of the pectoral and pelvic girdles that attach them to the axial skeleton. We depend so heavily on the limbs for mobility and the ability to manipulate objects, that deformities and injuries to the appendicular skeleton are more disabling than most disorders of the axial skeleton. Hand injuries, especially, can disable a person far more than a comparable amount of tissue injury elsewhere on the body. Injuries to the appendicular skeleton are especially common in athletics, recreation, and the workplace. A knowledge of the anatomy of the appendicular skeleton is therefore especially important.

The Pectoral Girdle and Upper Limb

Objectives

When you have completed this section, you should be able to

- identify and describe the features of the clavicle, scapula, humerus, radius, ulna, and bones of the wrist and hand; and
- describe the evolutionary innovations of the human forelimb.

Pectoral Girdle

The **pectoral girdle** (shoulder girdle) supports the arm. It consists of two bones on each side of the body: the *clavicle* (collarbone) and the *scapula* (shoulder blade). The medial end of the clavicle articulates with the sternum at the **sternoclavicular joint,** and its lateral end articulates with the scapula at the **acromioclavicular joint** (see fig. 7.27, p. 194). The scapula also articulates with the humerus at the **humeroscapular**(shoulder) **joint.** These are loose attachments that result in a shoulder far more flexible than that of most other mammals, but they also make the shoulder joint easy to dislocate.

Think About It!

How is the unusual flexibility of the human shoulder joint related to the habitat of our primate ancestors?

CLAVICLE

The **clavicle**[1] (fig. 8.1) is a slightly S-shaped bone, somewhat flattened dorsoventrally and easily seen and palpated on the upper thorax (see fig. B.1*b*, p. 351). The superior surface is relatively smooth, whereas the inferior surface is marked by grooves and ridges for muscle attachment. The medial **sternal end** has a rounded, hammerlike head, and the lateral **acromial end** is markedly flattened. Near the acromial end is a rough tuberosity called the **conoid tubercle**—a ligament attachment that faces dorsally and slightly downward. The clavicle braces the shoulder. It is thickened in people who do heavy manual labor, and in most people the right clavicle is stronger and shorter than the left. Without the clavicles, the pectoralis major muscles would pull the shoulders forward and medially, as occurs when a clavicle is fractured. Indeed, the clavicle is the most commonly fractured bone in the body because it is so close to the surface and because people often reach out with their arms to break a fall (see insight 8.1).

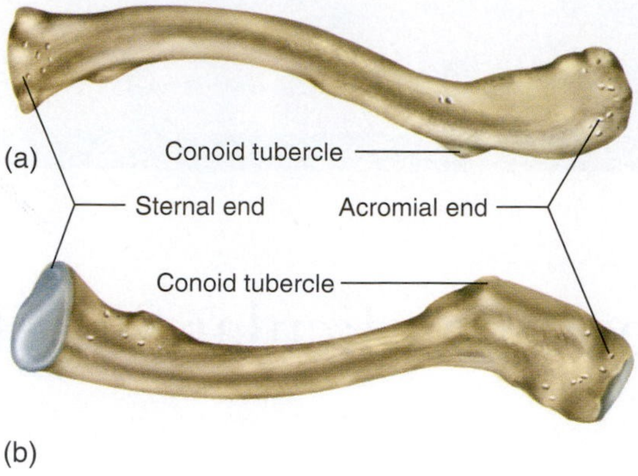
(a)
Conoid tubercle
Sternal end
Acromial end
Conoid tubercle
(b)

Figure 8.1
The Right Clavicle (collarbone). (*a*) Superior view. (*b*) Inferior view.

INSIGHT 8.1 CLINICAL APPLICATION

Fractured Clavicle

The clavicle is the most frequently broken bone of the body. It can break when one falls directly on the shoulder, or when one thrusts out an arm to break a fall and the force of the fall is transmitted through the limb bones to the pectoral girdle. Fractures most often occur at a weak point about one-third of the length of the bone from the lateral end. When the clavicle is broken, the shoulder tends to drop, while the sternocleidomastoid muscle of the neck elevates the medial fragment and the pectoralis major muscle of the chest may pull the lateral fragment toward the sternum. The clavicle is sometimes fractured during birth in wide-shouldered infants, but these neonatal fractures heal quickly. In children, clavicular fractures are often of the greenstick type (see fig. 6.14).

SCAPULA

The **scapula** (fig. 8.2) is a triangular plate that dorsally overlies ribs 2 to 7. The three sides of the triangle are called the **superior, medial** (vertebral), and **lateral** (axillary) **borders,** and its three angles are the **superior, inferior,** and **lateral angles.** A conspicuous **suprascapular notch** in the superior border provides passage for a nerve. The broad anterior surface of the scapula, called the **subscapular fossa,** is slightly concave and relatively featureless. The posterior surface has a transverse ridge called the **spine,** a deep indentation superior to the spine called the **supraspinous fossa,** and a broad surface inferior to it called the **infraspinous fossa.**[2] The scapula is held in place by numerous muscles attached to these three fossae.

[1] *clav* = hammer, club + *icle* = little

[2] *supra* = above; *infra* = below

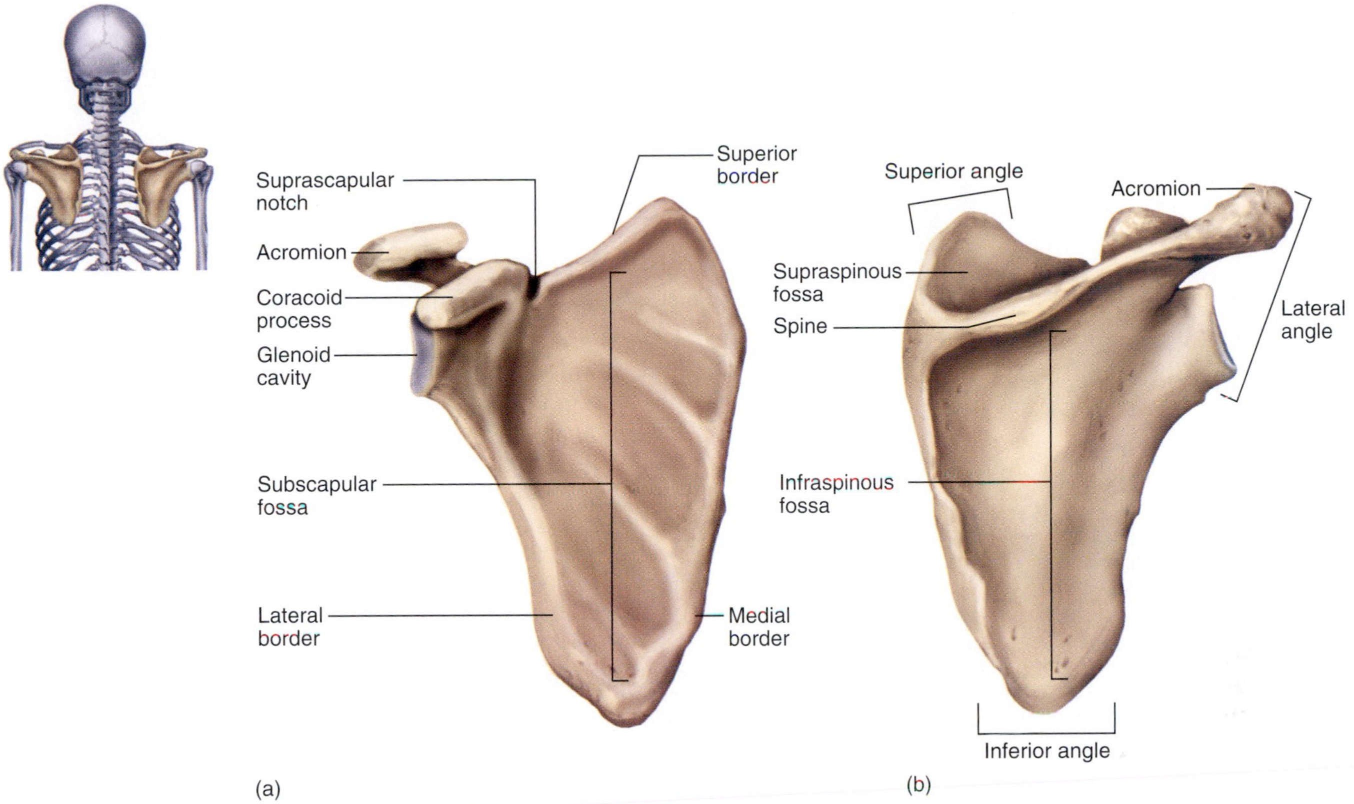

FIGURE 8.2
The Right Scapula. (*a*) Anterior view. (*b*) Posterior view.

The most complex region of the scapula is its lateral angle, which has three main features:

1. The **acromion**[3] (ah-CRO-me-on) is a platelike extension of the scapular spine that forms the apex of the shoulder. It articulates with the clavicle—the sole point of attachment of the arm and scapula to the axial skeleton.
2. The **coracoid**[4] (COR-uh-coyd) **process** is shaped like a finger but named for a vague resemblance to a crow's beak; it provides attachment for the biceps brachii and other muscles of the arm.
3. The **glenoid**[5] (GLEN-oyd) **cavity** is a shallow socket that articulates with the head of the humerus.

THINK ABOUT IT!

What part of the scapula do you think is most commonly fractured? Why?

Upper Limb

The upper limb is divided into 4 regions containing a total of 30 bones per limb:

1. The **brachium**[6] (BRAY-kee-um), or arm proper, extends from shoulder to elbow. It contains only one bone, the *humerus.*
2. The **antebrachium,**[7] or forearm, extends from elbow to wrist and contains two bones—the *radius* and *ulna.* In anatomical position, these bones are parallel and the radius is lateral to the ulna.
3. The **carpus,**[8] or wrist, contains eight small bones arranged in two rows.
4. The **manus,**[9] or hand, contains 19 bones in two groups—5 *metacarpals* in the palm and 14 *phalanges* in the digits.

HUMERUS

The **humerus** has a hemispherical **head** that articulates with the glenoid cavity of the scapula (fig. 8.3). The smooth surface of the head (covered with articular cartilage in life) is bordered by a

[3]*acr* = extremity, point + *omi* = shoulder
[4]*corac* = crow + *oid* = resembling
[5]*glen* = pit, socket

[6]*brachi* = arm
[7]*ante* = before
[8]*carp* = wrist
[9]*man* = hand

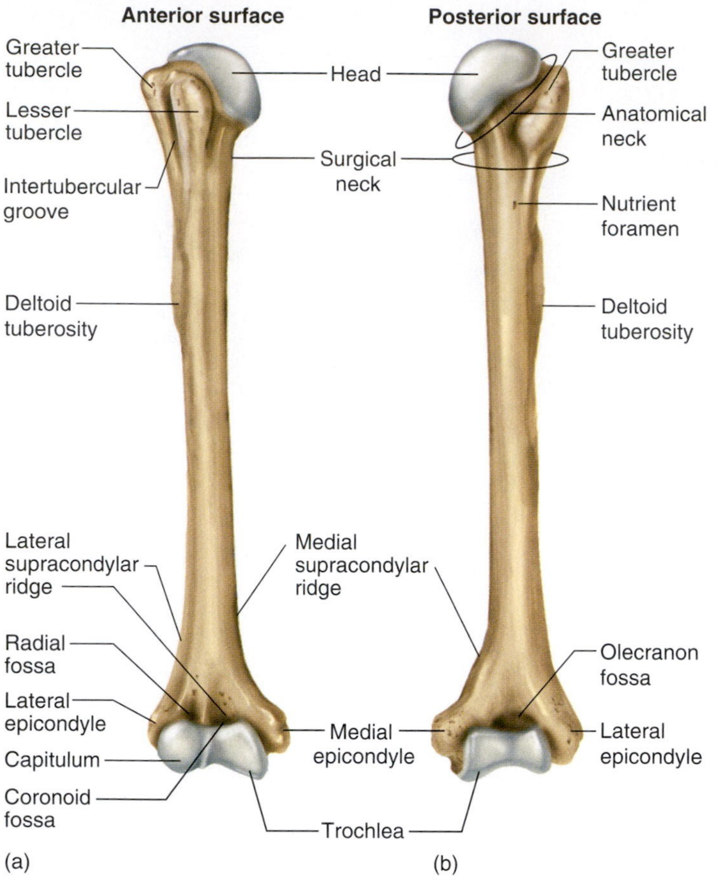

FIGURE 8.3
The Right Humerus. (*a*) Anterior view. (*b*) Posterior view.

groove called the **anatomical neck.** Other prominent features of the proximal end are muscle attachments called the **greater** and **lesser tubercles** and an **intertubercular groove** between them that accommodates a tendon of the biceps muscle. The **surgical neck,** a common fracture site, is a narrowing of the bone just distal to the tubercles, at the transition from the head to the shaft.

The shaft has a rough area called the **deltoid tuberosity** on its lateral surface. This is an insertion for the deltoid muscle of the shoulder. The distal end of the humerus has two smooth condyles. The lateral one, called the **capitulum**[10] (ca-PIT-you-lum), is shaped somewhat like a fat tire and articulates with the radius. The medial one, called the **trochlea**[11] (TROCK-lee-uh), is pulleylike and articulates with the ulna. Immediately proximal to these condyles, the humerus flares out to form two bony processes, the **lateral** and **medial epicondyles.** The medial epicondyle protects the ulnar nerve, which passes close to the surface across the back of the elbow. This epicondyle is popularly known as the "funny bone" because a sharp blow to the elbow at this point stimulates the ulnar nerve and produces an intense tingling sensation. Proximal to the epicondyles, the margins of the humerus are sharply angular and form muscle attachments called the **lateral** and **medial supracondylar ridges.**

The distal end of the humerus also shows three deep pits—two anterior and one posterior. The posterior pit, called the **olecranon** (oh-LEC-ruh-non) **fossa,** accommodates the olecranon of the ulna when the elbow is extended. On the anterior surface, a medial pit called the **coronoid fossa** accommodates the coronoid process of the ulna when the elbow is flexed. The lateral pit is the **radial fossa,** named for the nearby head of the radius.

RADIUS

The proximal head of the **radius** (fig. 8.4) is a distinctive disc that rotates freely on the humerus when the palm is turned forward and back. It articulates with the capitulum of the humerus and radial notch of the ulna. On the shaft, immediately distal to the head, is a medial rough **tuberosity,** which is the insertion of the biceps muscle. The distal end of the radius has the following features, from lateral to medial:

1. a bony point, the **styloid process,** which can be palpated proximal to the thumb;
2. two shallow depressions (articular facets) that articulate with the scaphoid and lunate bones of the wrist; and
3. the **ulnar notch,** which articulates with the end of the ulna.

ULNA

At the proximal end of the **ulna** (fig. 8.4) is a deep, C-shaped **trochlear notch** that wraps around the trochlea of the humerus. The posterior side of this notch is formed by a prominent **olecranon**—the bony point where you rest your elbow on a table. The anterior side is formed by a less prominent **coronoid process.** Laterally, the head of the ulna has a less conspicuous **radial notch,** which accommodates the head of the radius.

At the distal end of the ulna is a medial **styloid process.** The bony lumps you can palpate on each side of your wrist are the styloid processes of the radius and ulna. The radius and ulna are attached along their shafts by a ligament called the **interosseous** (IN-tur-OSS-ee-us) **membrane,** which is attached to an angular ridge called the **interosseous margin** on the medial side of each bone.

CARPAL BONES

The **carpal bones,** which form the wrist, are arranged in two rows of four bones each (fig. 8.5). These short bones allow movements of the wrist from side to side and up and down. The carpal bones of the proximal row, starting at the lateral (thumb) side, are the **scaphoid (navicular), lunate, triquetral** (tri-QUEE-trul), and **pisiform** (PY-sih-form). Translating from the Latin, these words mean boat-, moon-, triangle-, and pea-shaped, respectively. Unlike the other carpal bones, the pisiform is a sesamoid bone; it develops within the tendon of the *flexor carpi ulnaris muscle.*

The bones of the distal row, again starting on the lateral side, are the **trapezium,**[12] **trapezoid, capitate,**[13] and **hamate.**[14] The hamate can be recognized by a prominent hook, or **hamulus,** on the palmar side.

[10]*capit* = head + *ulum* = little
[11]*troch* = wheel, pulley
[12]*trapez* = table, grinding surface
[13]*capit* = head + *ate* = possessing
[14]*ham* = hook + *ate* = possessing

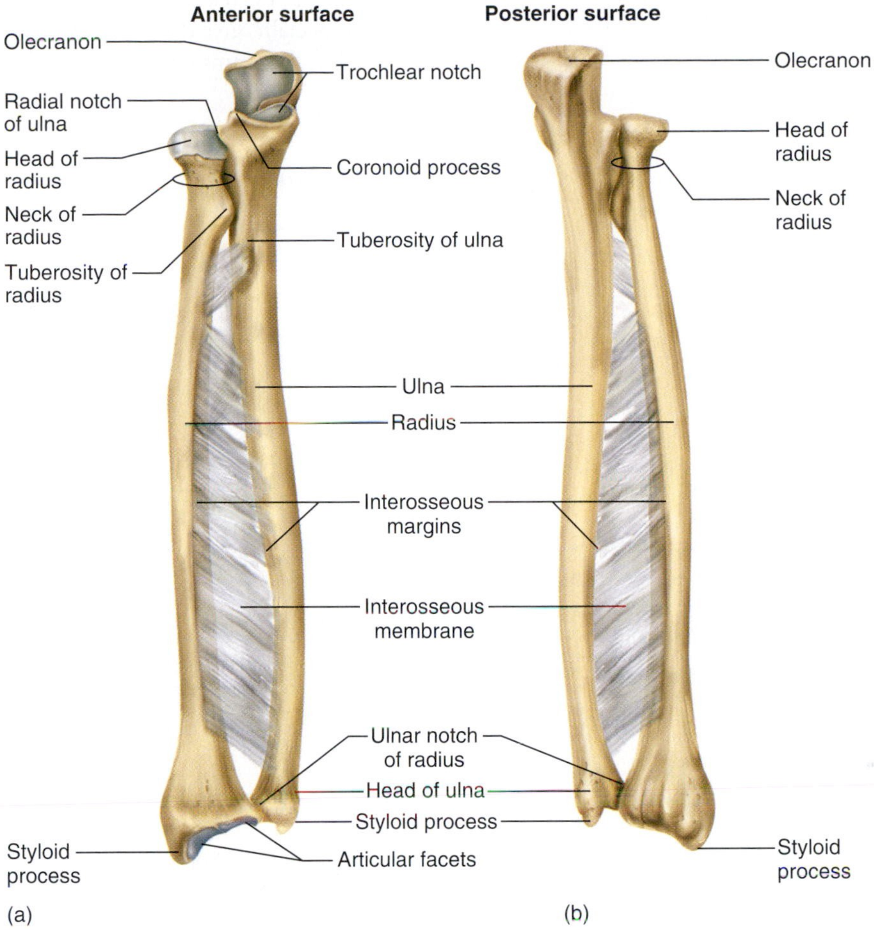

FIGURE 8.4
The Right Radius and Ulna. (*a*) Anterior view. (*b*) Posterior view.

METACARPAL BONES

Bones of the palm are called **metacarpals.**[15] Metacarpal I is located at the base of the thumb and metacarpal V at the base of the little finger (fig. 8.5). On a skeleton, the metacarpals look like extensions of the fingers, so that the fingers seem much longer than they really are. The proximal end of a metacarpal bone is called the **base,** the shaft is called the **body,** and the distal end is called the **head.** The heads of the metacarpals form the knuckles when you clench your fist.

PHALANGES

The bones of the fingers are called **phalanges** (fah-LAN-jeez); in the singular, *phalanx* (FAY-lanks). There are two phalanges in the **pollex** (thumb) and three in each of the other digits (fig. 8.5). Phalanges are identified by Roman numerals preceded by *proximal, middle,* and *distal.* For example, proximal phalanx I is in the basal segment of the thumb (the first segment beyond the web between the thumb and palm); the left proximal phalanx IV is where people usually wear wedding rings; and distal phalanx V forms the tip of the little finger. The three parts of a phalanx are the same as in a metacarpal: *base, body,* and *head.* The ventral surface of a phalanx is slightly concave from end to end and flattened from side to side; the dorsal surface is rounder and slightly convex from end to end.

EVOLUTION OF THE FORELIMB

Elsewhere in chapters 7 and 8, we examine how the evolution of bipedal locomotion in humans has affected the skull, vertebral column, and lower limb. The effects of bipedalism on the upper limb are less immediately obvious, but nevertheless substantial. In apes, all four limbs are adapted primarily for walking and climbing, and the forelimbs are longer than the hindlimbs. Thus, the shoulders are higher than the hips when the animal walks. When some apes such as orangutans and gibbons walk bipedally, they typically hold their long forelimbs over their heads to prevent them from dragging on the ground. By contrast, the human forelimbs are adapted primarily for reaching out, exploring the environment, and manipulating objects. They are shorter than the hindlimbs and far less muscular than the forelimbs of apes. No longer needed for locomotion, our forelimbs, especially the hands, have become better adapted for carrying objects, holding things closer to the eyes, and manipulating them

[15] *meta* = beyond + *carp* = wrist

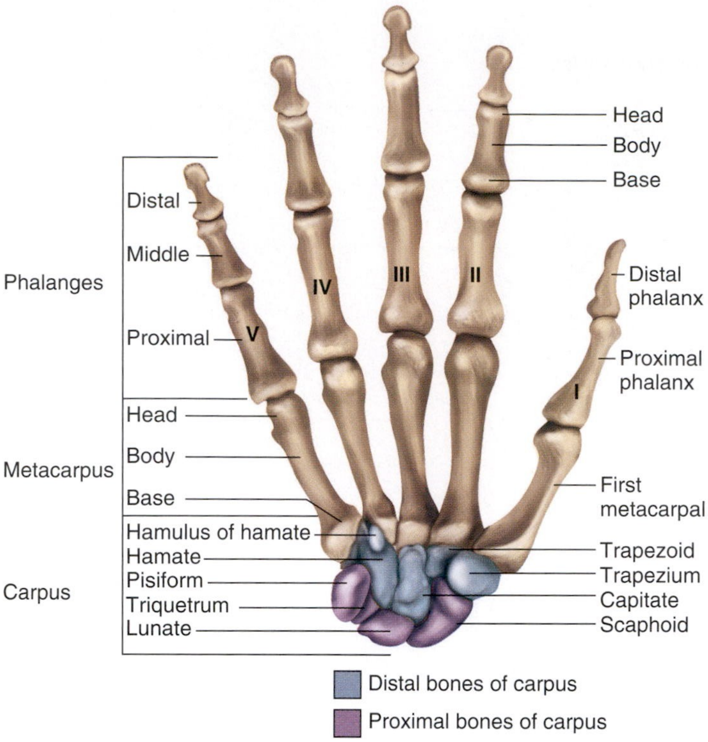
Head
Body
Base
Distal
Middle
Phalanges
Proximal
V
IV
III
II
Distal phalanx
Proximal phalanx
I
Head
Body
Metacarpus
Base
First metacarpal
Hamulus of hamate
Hamate
Pisiform
Carpus
Triquetrum
Lunate
Trapezoid
Trapezium
Capitate
Scaphoid
Distal bones of carpus
Proximal bones of carpus

FIGURE 8.5

The Right Wrist and Hand. Anterior (palmar) View. Carpal bones are color-coded to distinguish the proximal and distal rows.

more precisely. Although the forelimbs have the same basic bone and muscle pattern as the hindlimbs, the joints of the shoulder and hands, especially, give the forelimbs far greater mobility.

Table 8.1 summarizes the bones of the pectoral girdle and upper limb.

Before You Go On

Answer the following questions to test your understanding of the preceding section:

1. Describe how to distinguish the medial and lateral ends of the clavicle from each other, and how to distinguish its superior and inferior surfaces.
2. Name the three fossae of the scapula and describe the location of each.
3. What three bones meet at the elbow? Identify the fossae, articular surfaces, and processes of this joint and state to which bone each of these features belongs.
4. Name the four bones of the proximal row of the carpus from lateral to medial, and then the four bones of the distal row in the same order.
5. Name the four bones from the tip of the little finger to the base of the hand on that side.

TABLE 8.1

Anatomical Checklist for the Pectoral Girdle and Upper Limb

Pectoral Girdle

Clavicle (fig. 8.1)	*Scapula (fig. 8.2) (Cont.)*	*Scapula (fig. 8.2) (Cont.)*
Sternal end	Lateral (axillary) border	Fossae
Acromial end	Angles	Subscapular fossa
Conoid tubercle	Superior angle	Supraspinous fossa
Scapula (fig. 8.2)	Inferior angle	Infraspinous fossa
Borders	Lateral angle	Acromion
Superior border	Suprascapular notch	Coracoid process
Medial (vertebral) border	Spine	Glenoid cavity

Upper Limb

Humerus (fig. 8.3)	*Radius (fig. 8.4)*	*Carpal Bones (fig. 8.5) (Cont.)*
Proximal end	Head	Triquetral
Head	Tuberosity	Pisiform
Anatomical neck	Styloid process	Distal group
Surgical neck	Articular facets	Trapezium
Greater tubercle	Ulnar notch	Trapezoid
Lesser tubercle	*Ulna (fig. 8.4)*	Capitate
Intertubercular groove	Trochlear notch	Hamate
Shaft	Olecranon	Hamulus
Deltoid tuberosity	Coronoid process	*Bones of the Hand (fig. 8.5)*
Distal end	Radial notch	Metacarpal bones I–V
Capitulum	Styloid process	Base
Trochlea	Interosseous margin	Body
Lateral epicondyle	Interosseous membrane	Head
Medial epicondyle	*Carpal Bones (fig. 8.5)*	Phalanges I–V
Olecranon fossa	Proximal group	Proximal phalanx
Coronoid fossa	Scaphoid	Middle phalanx
Radial fossa	Lunate	Distal phalanx

The Pelvic Girdle and Lower Limb

Objectives

When you have completed this section, you should be able to

- identify and describe the features of the pelvic girdle, femur, patella, tibia, fibula, and bones of the foot;
- compare the anatomy of the male and female pelvis and explain the functional significance of the differences; and
- describe the evolutionary adaptations of the pelvis and hindlimb for bipedal locomotion.

Pelvic Girdle

The adult **pelvic**[16] **girdle** is composed of three bones: a right and left **os coxae** (plural, ossa coxae) and the sacrum (fig. 8.6). Another term for the os coxae—arguably the most self-contradictory term in anatomy—is the *innominate*[17] (ih-NOM-ih-nate) *bone*, "the bone with no name." The pelvic girdle supports the trunk on the legs and encloses and protects the viscera of the pelvic cavity—mainly the lower colon, urinary bladder, and reproductive organs.

Each os coxae is joined to the vertebral column at one point, the sacroiliac joint, where its **auricular surface** matches the auricular

[16] *pelv* = basin, bowl
[17] *in* = without + *nomin* = name + *ate* = having

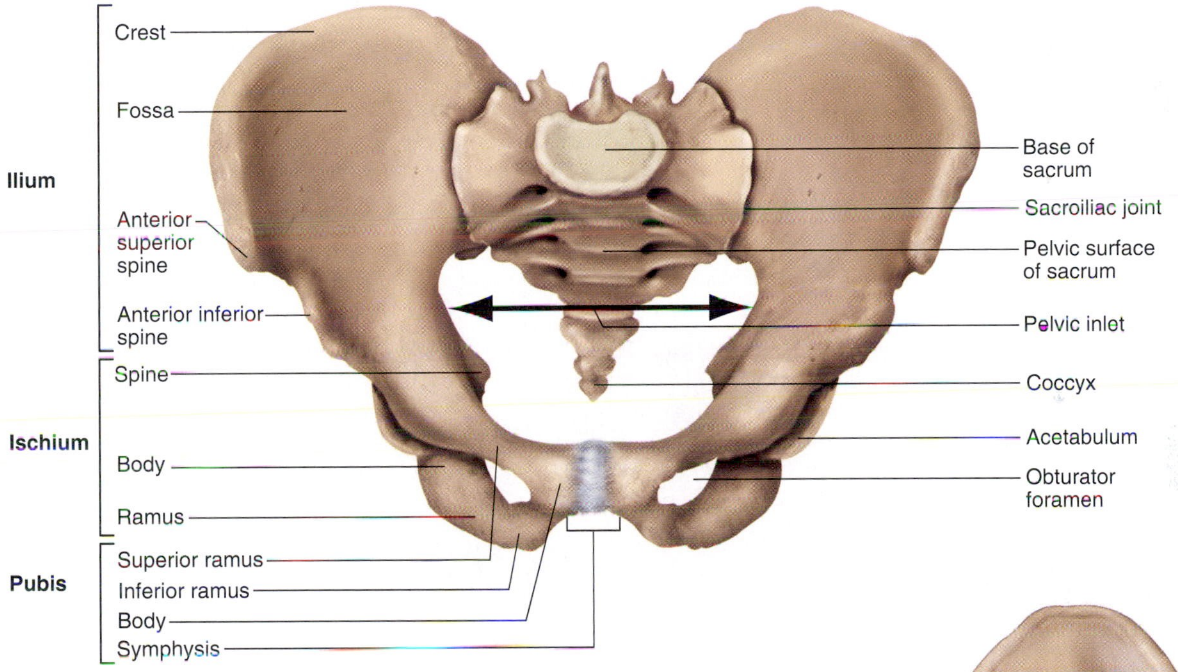
Crest
Fossa
Ilium
Anterior superior spine
Anterior inferior spine
Spine
Ischium
Body
Ramus
Pubis
Superior ramus
Inferior ramus
Body
Symphysis
Base of sacrum
Sacroiliac joint
Pelvic surface of sacrum
Pelvic inlet
Coccyx
Acetabulum
Obturator foramen

(a)

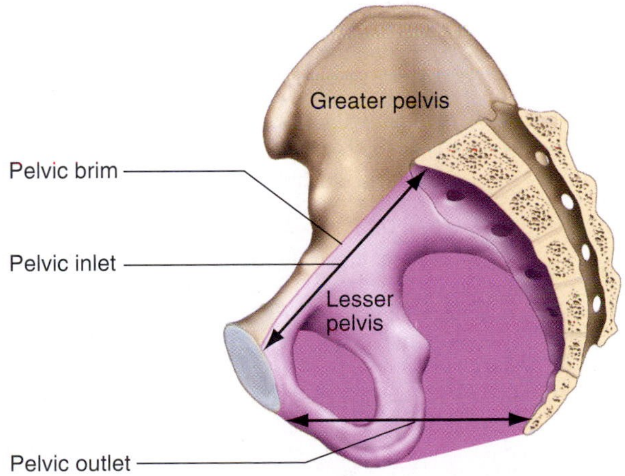
Greater pelvis
Pelvic brim
Pelvic inlet
Lesser pelvis
Pelvic outlet

(b)

Figure 8.6

The Pelvic Girdle. (*a*) Anterosuperior view of the female pelvis. The pelvic girdle consists of the ossa coxae, sacrum, and coccyx. (*b*) Medial section of the pelvic girdle showing the greater (false) pelvis *(brown)* and lesser (true) pelvis *(violet)*, separated by the pelvic brim.

surface of the sacrum. On the anterior side of the pelvis is the **pubic symphysis,**[18] the point where the right and left pubic bones are joined by a pad of fibrocartilage, the **interpubic disc.** The symphysis can be palpated immediately above the genitalia.

The pelvic girdle has a bowl-like shape with the broad **greater (false) pelvis** between the flare of the hips and the narrow **lesser (true) pelvis** below (fig. 8.6*b*). The two are separated by a somewhat round margin called the **pelvic brim.** The opening circumscribed by the brim is called the **pelvic inlet**—an entry into the lesser pelvis through which an infant's head passes during birth. The lower margin of the lesser pelvis is called the **pelvic outlet.**

The os coxae has three distinctive features that will serve as landmarks for further description. These are the **iliac**[19] **crest** (superior crest of the hip); **acetabulum**[20] (ASS-eh-TAB-you-lum) (the hip socket—named for its resemblance to vinegar cups used in ancient Rome); and **obturator**[21] **foramen** (a large round-to-triangular hole below the acetabulum, closed in life by a ligament called the *obturator membrane*).

The adult os coxae forms by the fusion of three childhood bones called the *ilium* (ILL-ee-um), *ischium* (ISS-kee-um), and *pubis* (PEW-biss), identified by color in figure 8.7. The largest of these is the **ilium,** which extends from the iliac crest to the superior wall of the acetabulum. The iliac crest extends from a point or angle on the anterior side, called the **anterior superior spine,** to a sharp posterior angle, called the **posterior superior spine.** In a lean person, the anterior superior spines form visible anterior protrusions, and the posterior superior spines are sometimes marked by dimples above the buttocks where connective tissue attached to the spines pulls inward on the skin (see fig. B.4, p. 354).

Below the superior spines are the **anterior** and **posterior inferior spines.** Below the posterior inferior spine is a deep **greater sciatic** (sy-AT-ic) **notch,** named for the sciatic nerve that passes through it and continues down the posterior side of the thigh.

The posterolateral surface of the ilium is relatively rough-textured because it serves for attachment of several muscles of the buttocks and thighs. The anteromedial surface, by contrast, is the smooth, slightly concave **iliac fossa,** covered in life by the broad *iliacus* muscle. Medially, the ilium exhibits an auricular surface that matches the one on the sacrum, so that the two bones form the sacroiliac joint.

The **ischium** forms the inferoposterior portion of the os coxae. Its heavy **body** is marked with a prominent **spine.** Inferior to the spine is a slight indentation, the **lesser sciatic notch,** and then the thick, rough-surfaced **ischial tuberosity,** which supports your

[18] *sym* = together + *physis* = growth
[19] *ili* = flank, loin + *ac* = pertaining to
[20] *acetabulum* = vinegar cup
[21] *obtur* = to close, stop up + *ator* = that which

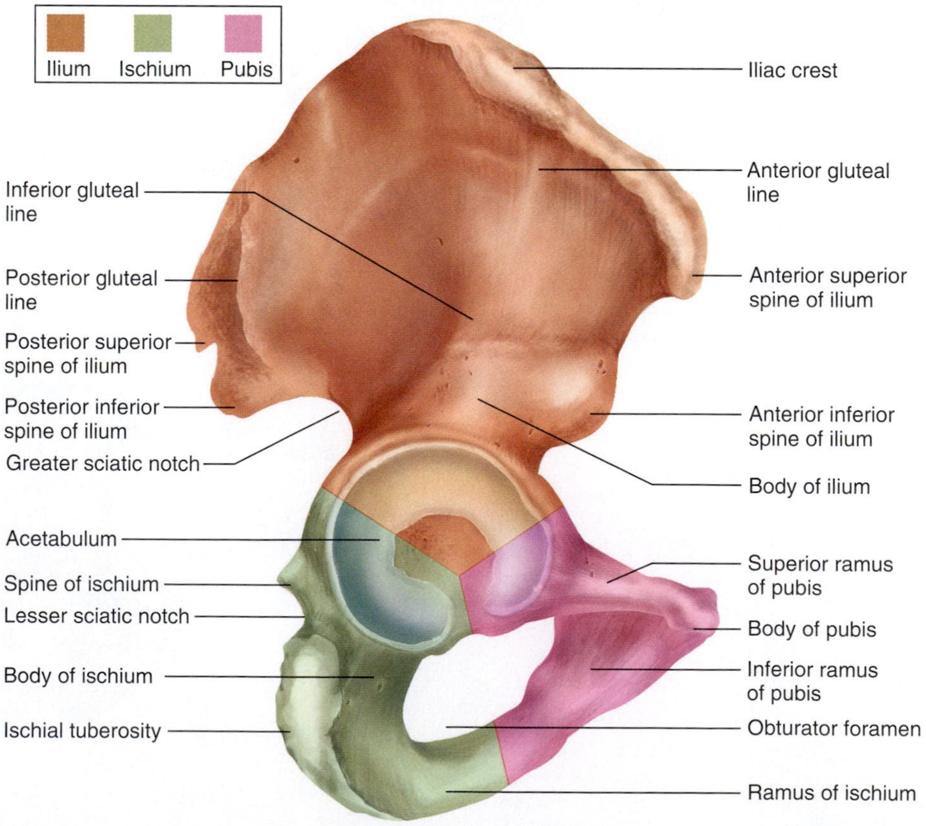

FIGURE 8.7
The Right Os Coxae. Lateral View. The three childhood bones that fuse to form the adult os coxae are identified by color.

body when you are sitting. The tuberosity can be palpated by sitting on your fingers. The **ramus** of the ischium joins the inferior ramus of the pubis anteriorly.

The **pubis** (pubic bone) is the most anterior portion of the os coxae. It has a **superior** and **inferior ramus** and a triangular **body.** The body of one pubis meets the body of the other at the pubic symphysis. The pubis and ischium encircle the obturator foramen. The pubis is often fractured when the pelvis is subjected to violent anteroposterior compression, as in seat-belt injuries.

The pelvic girdle is anatomically adapted to two requirements—bipedalism and childbirth. In apes and other quadrupedal (four-legged) mammals, the abdominal viscera are supported by the muscular wall of the abdomen. In humans, the viscera bear down on the floor of the pelvic cavity, and a bowl-shaped pelvis is necessary to support their weight. This has resulted in a narrower pelvic outlet—a condition quite incompatible with the fact that we, including our infants, are such a large-brained species. The pain of childbirth is unique to humans and, one might say, a price we must pay for having both a large brain and a bipedal stance. Narrowing of the pelvic outlet is thought to be the reason why human infants are born in such an immature state compared to those of other primates. They must be born before the cranial bones fuse so that the head can squeeze through the pelvic outlet.

The largest muscle of the buttock, the *gluteus maximus,* serves in apes primarily as an abductor of the thigh—that is, it moves the leg laterally. In humans, however, the ilium has expanded posteriorly, so the gluteus maximus originates behind the hip joint. This changes the function of the muscle—instead of abducting the thigh, it pulls the thigh back in the second half of a stride (pulling back on your right thigh, for example, when your left foot is off the ground and swinging forward). This action accounts for the smooth, efficient stride of a human as compared to the awkward, shuffling gait of a chimpanzee or gorilla when it is walking upright. The posterior growth of the ilium is the reason the greater sciatic notch is so deeply concave (fig. 8.8).

The pelvis is the most *sexually dimorphic* part of the skeleton—that is, the one whose anatomy most differs between the sexes. In identification of skeletal remains to sex, attention is focused especially on the pelvis. The average male pelvis is more robust (heavier and thicker) than the female's owing to the forces exerted on the bone by stronger muscles. The female pelvis is adapted to the needs of pregnancy and childbirth. It is wider and shallower, and has a larger pelvic inlet and outlet for passage of the infant's head. Table 8.2 and figure 8.9 summarize the most useful features of the pelvis in sex identification.

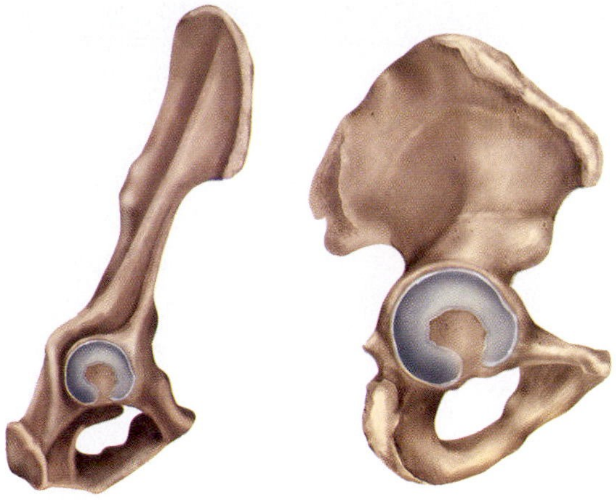

FIGURE 8.8

Chimpanzee and Human Os Coxae. The human ilium forms a more bowl-like greater pelvis and is expanded dorsally *(page left)* so that the gluteus maximus muscle produces an effective backswing of the thigh during the stride.

Lower Limb

The number and arrangement of bones in the lower limb are similar to those of the upper limb. In the lower limb, however, they are adapted for weight-bearing and locomotion and are therefore shaped and articulated differently. The lower limb is divided into four regions containing a total of 30 bones per limb:

1. The **femoral region,** or thigh, extends from hip to knee and contains the *femur* (the longest bone in the body). The *patella* (kneecap) is a sesamoid bone at the junction of the femoral and crural regions.
2. The **crural** (CROO-rul) **region,** or leg proper, extends from knee to ankle and contains two bones, the medial *tibia* and lateral *fibula.*
3. The **tarsal region (tarsus),** or ankle, is the union of the crural region with the foot. The tarsal bones are treated as part of the foot.
4. The **pedal region (pes),** or foot, is composed of 7 *tarsal bones,* 5 *metatarsals,* and 14 *phalanges* in the toes.

FEMUR

The **femur** (FEE-mur) (fig. 8.10) has a nearly spherical head that articulates with the acetabulum of the pelvis, forming a quintessential *ball-and-socket joint.* A ligament extends from the acetabulum to a pit, the **fovea capitis**[22] (FOE-vee-uh CAP-ih-tiss), in the head of the femur. Distal to the head is a constricted **neck** and then two massive, rough processes called the **greater** and **lesser trochanters** (tro-CAN-turs), which are insertions for the powerful muscles of the hip. They are connected on the posterior side by a thick oblique ridge of bone, the **intertrochanteric crest,** and on the anterior side by a more delicate **intertrochanteric line.**

The primary feature of the shaft is a posterior ridge called the **linea aspera**[23] (LIN-ee-uh ASS-peh-ruh) at its midpoint. It branches into less conspicuous lateral and medial ridges at its inferior and superior ends.

The distal end of the femur flares into **medial** and **lateral epicondyles,** which serve as sites of muscle and ligament attachment. Distal to these are two smooth round surfaces of the knee joint, the **medial** and **lateral condyles,** separated by a groove called the **intercondylar** (IN-tur-CON-dih-lur) **fossa.** On the anterior side of the femur, a smooth medial depression called the **patellar surface** articulates with the patella.

[22]*fovea* = pit + *capitis* = of the head
[23]*linea* = line + *asper* = rough

TABLE 8.2

Comparison of the Male and Female Pelvic Girdles

	Male	Female
General Appearance	More massive; rougher; heavier processes	Less massive; smoother; more delicate processes
Tilt	Upper end of pelvis relatively vertical	Upper end of pelvis tilted forward
Ilium	Deeper; projects farther above sacroiliac joint	Shallower; does not project as far above sacroiliac joint
Sacrum	Narrower and deeper	Wider and shallower
Coccyx	Less movable; more vertical	More movable; tilted dorsally
Width of Greater Pelvis	Anterior superior spines closer together; hips less flared	Anterior superior spines farther apart; hips more flared
Pelvic Inlet	Heart-shaped	Round or oval
Pelvic Outlet	Smaller	Larger
Pubic Symphysis	Taller	Shorter
Greater Sciatic Notch	Narrower	Wider
Obturator Foramen	Round	Triangular to oval
Acetabulum	Larger, faces more laterally	Smaller, faces slightly ventrally
Pubic Arch	Usually 90° or less	Usually greater than 100°

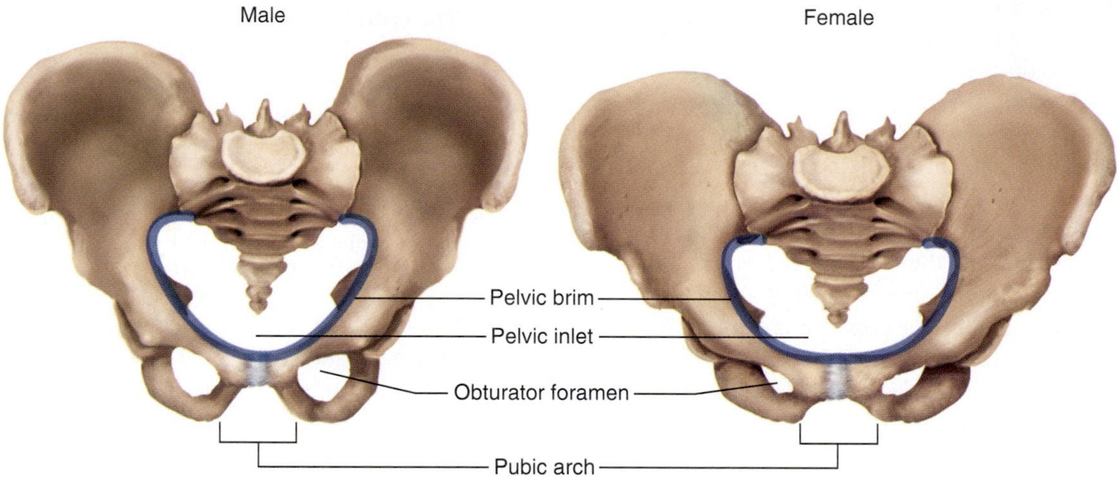
Male
Female
Pelvic brim
Pelvic inlet
Obturator foramen
Pubic arch

FIGURE 8.9
Comparison of the Male and Female Pelvic Girdles. Compare table 8.2.

While the femurs of apes are nearly vertical, in humans they angle medially from the hip to the knee (fig. 8.11). This places our knees closer together, beneath the body's center of gravity. We lock our knees when standing (see chapter 12), allowing us to maintain an erect posture with little muscular effort. Apes cannot do this, and they cannot stand on two legs for very long without tiring—much as you would if you tried to maintain an erect posture with your knees slightly bent.

PATELLA

The **patella,** or kneecap (see fig. 8.10), is a roughly triangular sesamoid bone that forms within the tendon of the knee as a child begins to walk. It has a broad superior **base,** a pointed inferior **apex,** and a pair of shallow **articular facets** on its posterior surface where it articulates with the femur. The lateral facet is usually larger than the medial. The *quadriceps femoris tendon* extends from the anterior muscle of the thigh (the *quadriceps femoris*) to the patella, and it continues as the *patellar ligament* from the patella to the tibia.

THINK ABOUT IT!

A hard fall on the knee can shatter the patella into multiple pieces and require its surgical removal. How would you classify such a fracture (see table 6.1)? How would the removal of the patella affect the action of the quadriceps femoris muscle on the tibia? How might this affect a person's gait (walk)?

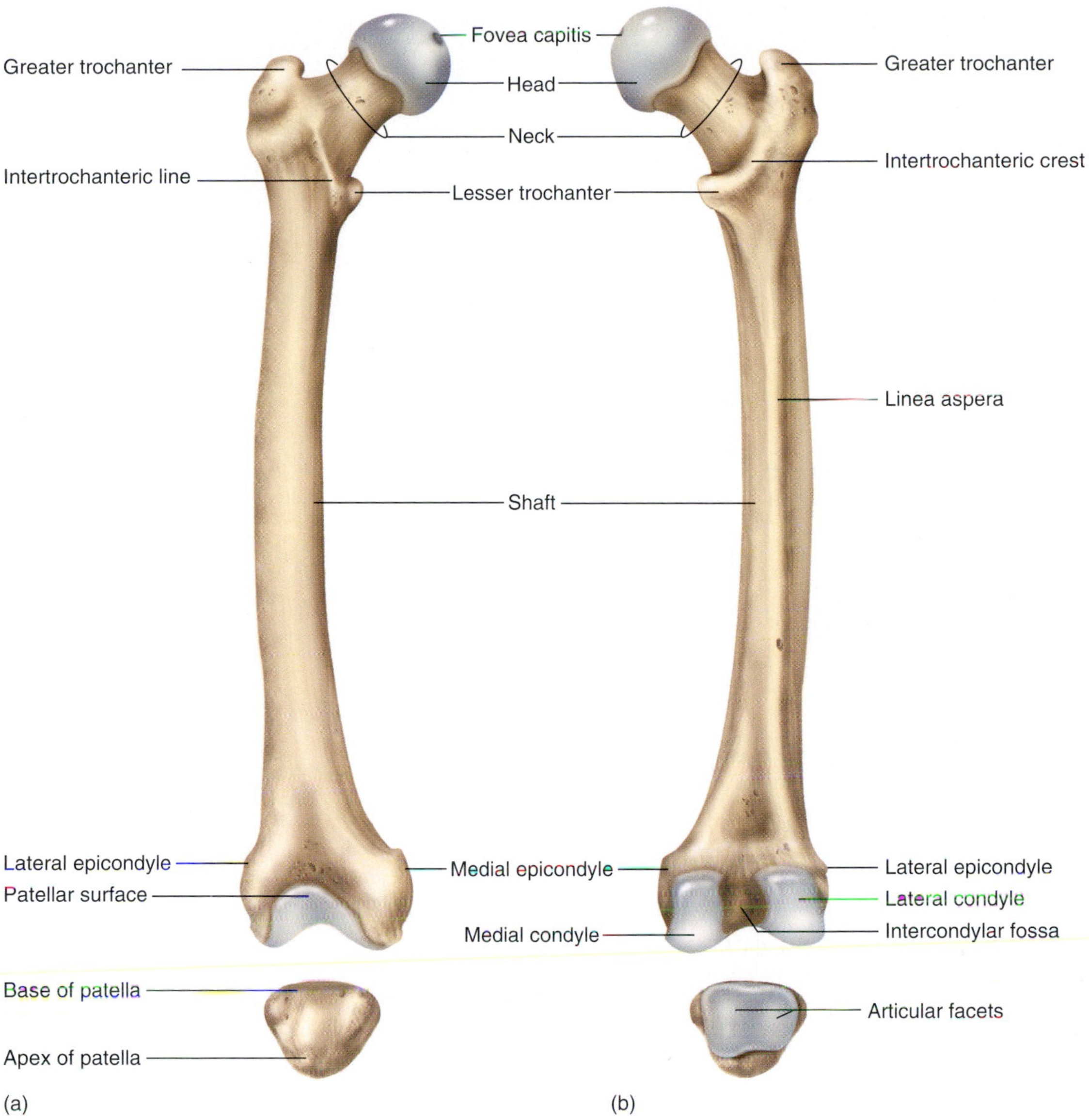
Greater trochanter
Intertrochanteric line
Fovea capitis
Head
Neck
Lesser trochanter
Greater trochanter
Intertrochanteric crest
Linea aspera
Shaft
Lateral epicondyle
Patellar surface
Medial epicondyle
Medial condyle
Lateral epicondyle
Lateral condyle
Intercondylar fossa
Base of patella
Apex of patella
Articular facets
(a)
(b)

FIGURE 8.10

The Right Femur and Patella. (*a*) Anterior view. (*b*) Posterior view.

TIBIA

The leg has two bones—a thick strong tibia (TIB-ee-uh) on the medial side and a slender fibula (FIB-you-luh) on the lateral side (fig. 8.13). The **tibia** is the only weight-bearing bone of the crural region. Its broad superior head has two fairly flat articular surfaces, the **medial** and **lateral condyles,** separated by a ridge called the **intercondylar eminence.** The condyles of the tibia articulate with those of the femur. The rough anterior surface of the tibia, the **tibial tuberosity,** can be palpated just below the patella. This is where the patellar ligament inserts and the thigh muscles exert their pull when they extend the leg. Distal to this, the shaft has a sharply angular **anterior crest,** which can be palpated in the shin region. At the ankle, just above the rim of a standard dress shoe, you can palpate a prominent bony knob on each side. These are the **medial** and **lateral malleoli**[24] (MAL-ee-OH-lie). The medial malleolus is part of the tibia, and the lateral malleolus is part of the fibula.

[24]*malle* = hammer + *olus* = little

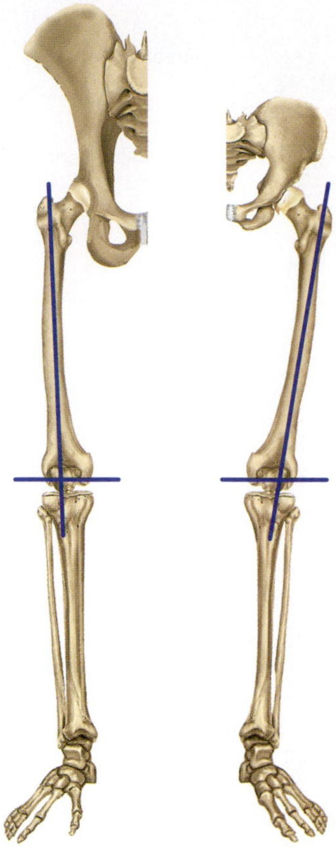

Figure 8.11

Adaptation of the Lower Limb for Bipedalism. In contrast to chimpanzees, which are quadrupedal, humans have the femurs angled medially so that the knees are more nearly directly below the body's center of gravity.

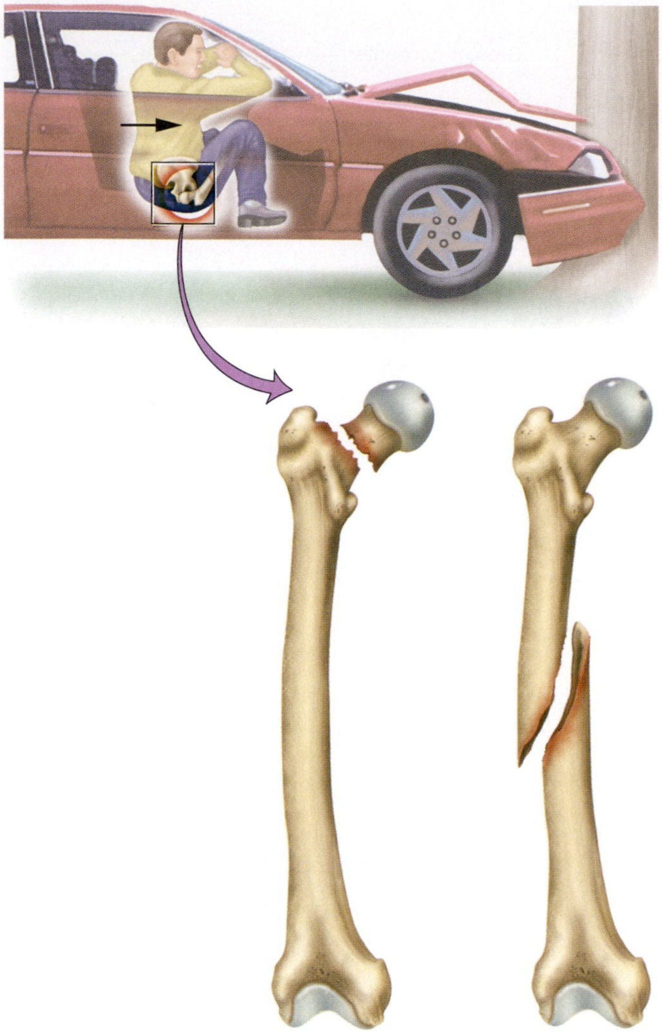

Figure 8.12

Fractures of the Femur. The femoral neck often fractures in elderly people as a result of falls. Violent trauma, as in automobile accidents, may cause spiral fractures of the femoral shaft.

INSIGHT 8.2 CLINICAL APPLICATION

Femoral Fractures

The femur is a very strong bone, well guarded by the thigh muscles, and it is not often fractured. Nevertheless, it can break in high-impact trauma suffered in automobile and equestrian accidents, figure skating falls, and so forth. If a person in an automobile collision has the feet braced against the floor or brake pedal with the knees locked, the force of impact is transmitted up the shaft and may fracture the shaft or neck of the femur (fig. 8.12). Comminuted and spiral fractures of the shaft can take up to a year to heal.

A "broken hip" is usually a fracture of the femoral neck, the weakest part of the femur. Elderly people often break the femoral neck when they stumble or are knocked down—especially women whose femurs are weakened by osteoporosis. Fractures of the femoral neck heal poorly because this is an anatomically unstable site and it has an especially thin periosteum with limited potential for ossification. In addition, fractures in this site often break blood vessels and cut off blood flow, resulting in degeneration *(posttraumatic avascular necrosis)* of the head.

FIBULA

The **fibula** (fig. 8.13) is a slender lateral strut that helps to stabilize the ankle. It does not bear any of the body's weight. The fibula is somewhat thicker and broader at its proximal end, the **head,** than at the distal end. The point of the head is called the **apex,** or **styloid process.** The distal expansion is the lateral malleolus.

THE ANKLE AND FOOT

The **tarsal bones** of the ankle are arranged in proximal and distal groups somewhat like the carpal bones of the wrist (fig. 8.14). Because of the load-bearing role of the ankle, however, their shapes and arrangement are conspicuously different from those of the

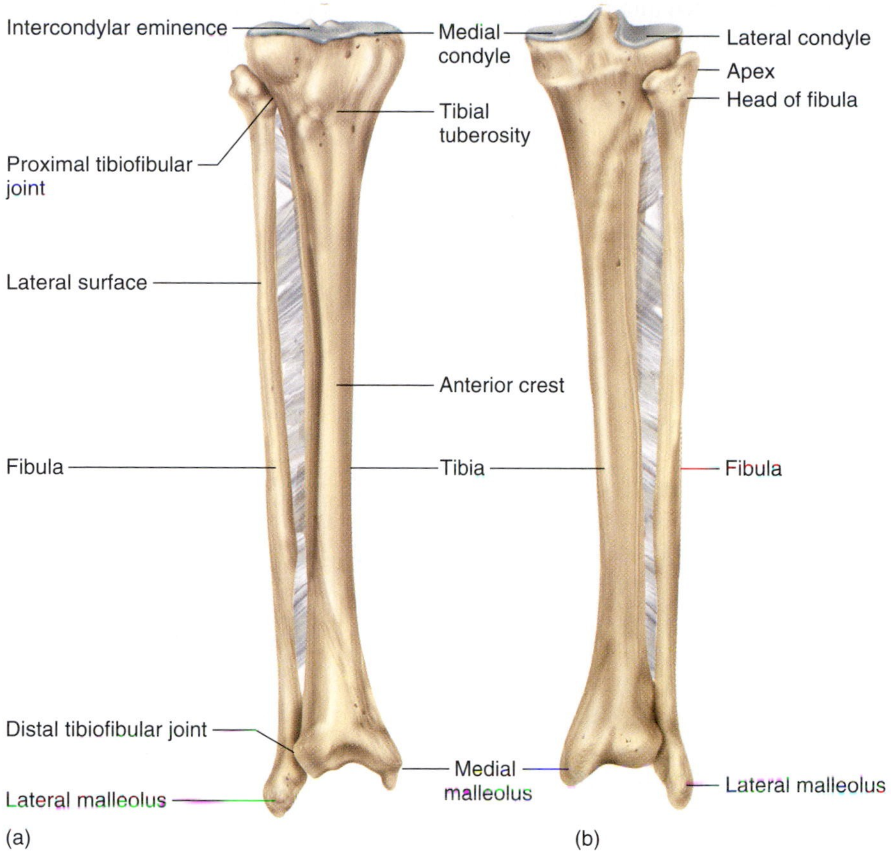

Intercondylar eminence
Medial condyle
Proximal tibiofibular joint
Tibial tuberosity
Lateral surface
Anterior crest
Fibula
Tibia
Distal tibiofibular joint
Medial malleolus
Lateral malleolus
(a)
Lateral condyle
Apex
Head of fibula
Fibula
Lateral malleolus
(b)

FIGURE 8.13
The Right Tibia and Fibula. (*a*) Anterior view. (*b*) Posterior view.

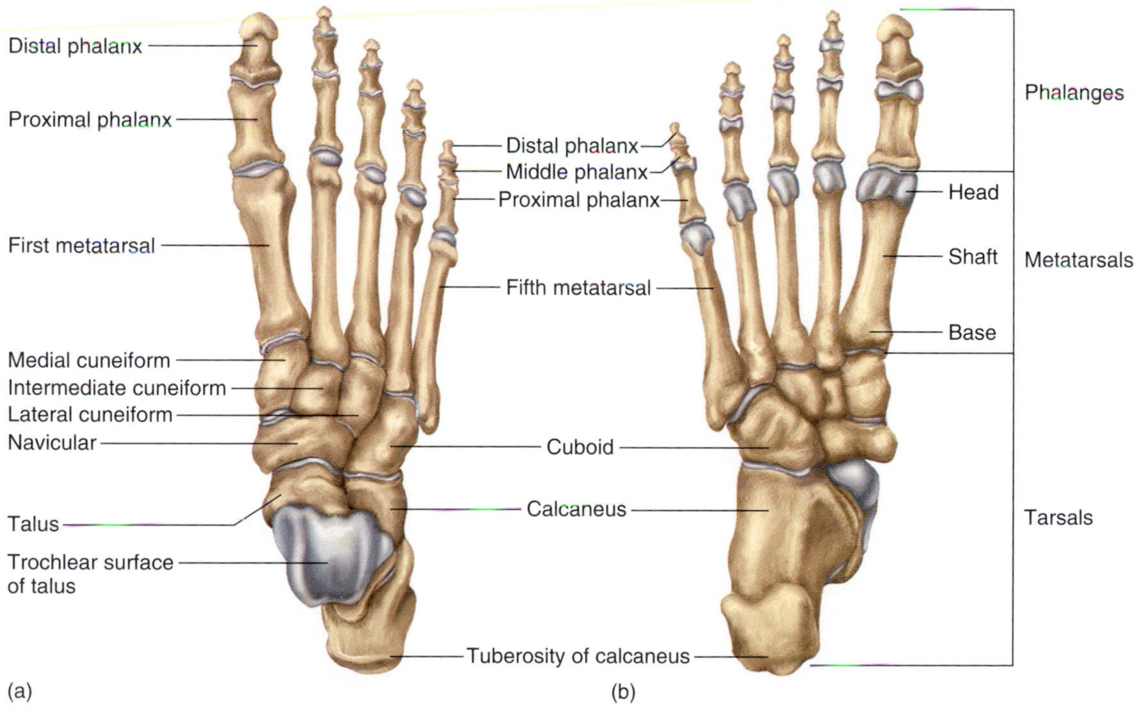

Distal phalanx
Proximal phalanx
First metatarsal
Medial cuneiform
Intermediate cuneiform
Lateral cuneiform
Navicular
Talus
Trochlear surface of talus
Distal phalanx
Middle phalanx
Proximal phalanx
Fifth metatarsal
Cuboid
Calcaneus
Tuberosity of calcaneus
(a)
(b)
Phalanges
Head
Shaft
Base
Metatarsals
Tarsals

FIGURE 8.14
The Right Foot. (*a*) Superior (dorsal) view. (*b*) Inferior (plantar) view.

carpal bones, and they are thoroughly integrated into the structure of the foot. The largest tarsal bone is the **calcaneus**[25] (cal-CAY-nee-us), which forms the heel. Its posterior end is the point of attachment for the **calcaneal (Achilles) tendon** from the calf muscles. The second-largest tarsal bone, and the most superior, is the **talus.** It has three articular surfaces: an inferoposterior one that articulates with the calcaneus, a superior **trochlear surface** that articulates with the tibia, and an anterior surface that articulates with a short, wide tarsal bone called the **navicular.** The talus, calcaneus, and navicular are considered the proximal row of tarsal bones. ("Navicular" is also used as a synonym for the scaphoid bone of the wrist.)

The distal group forms a row of four bones. Proceeding from the medial side to the lateral, these are the **first, second,** and **third cuneiforms**[26] (cue-NEE-ih-forms) and the **cuboid.** The cuboid is the largest.

The remaining bones of the foot are similar in arrangement and name to those of the hand. The proximal **metatarsals**[27] are similar to the metacarpals. They are **metatarsals I** to **V** from medial to lateral, metatarsal I being proximal to the great toe. (Note that Roman numeral I represents the *medial* group of bones in the foot but the *lateral* group in the hand. In both cases, however, Roman numeral I refers to the largest digit of the limb [see insight 8.3]). Metatarsals I to III articulate with the first through third cuneiforms; metatarsals IV and V both articulate with the cuboid.

Bones of the toes, like those of the fingers, are called phalanges. The great toe is the **hallux** and contains only two bones, the proximal and distal phalanx I. The other toes each contain a proximal, middle, and distal phalanx. The metatarsal and phalangeal bones each have a base, body, and head, like the bones of the hand. All of them, especially the phalanges, are slightly concave on the ventral side.

As important as the hand has been to human evolution, the foot may be an even more significant adaptation. Unlike other mammals, humans support their entire body weight on two feet. The tarsal bones are tightly articulated with each other, and the calcaneus is strongly developed. The hallux (great toe) is not opposable as it is in most Old World monkeys and apes (fig. 8.15), but it is highly developed so that it provides the "toe-off" that pushes the body forward in the last phase of the stride. For this reason, loss of the hallux has a more crippling effect than the loss of any other toe.

While apes are flat-footed, humans have strong, springy foot arches that absorb shock as the body jostles up and down during walking and running (fig. 8.16). The **medial longitudinal arch,** which essentially extends from heel to hallux, is formed from the calcaneus, talus, navicular, cuneiforms, and metatarsals I to III. The **lateral longitudinal arch** extends from heel to little toe and includes the calcaneus, cuboid, and metatarsals IV and V. The **transverse arch** includes the cuboid, cuneiforms, and proximal heads of the metatarsals. These arches are held together by short, strong ligaments. Excessive weight, repetitious stress, or congenital weakness of these ligaments can stretch them, resulting in *pes planis* (commonly called flat feet or fallen arches). This condition makes a person less tolerant of prolonged standing and walking.

Table 8.3 summarizes the pelvic girdle and lower limb.

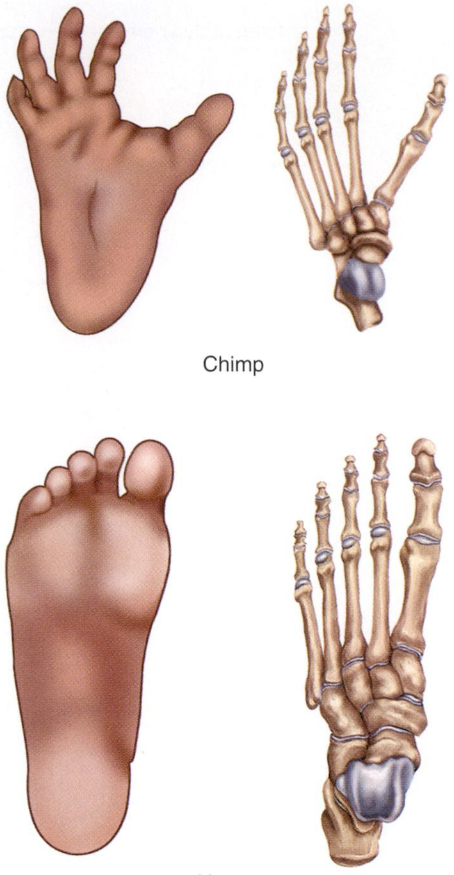

FIGURE 8.15

Some Adaptations of the Foot for Bipedalism. In contrast to the prehensile great toe (hallux) of the chimp, the human great toe is nonprehensile but is more robust and is adapted for the toe-off part of the stride.

Before You Go On

Answer the following questions to test your understanding of the preceding section:

6. Name the bones of the adult pelvic girdle. What three bones of a child fuse to form the os coxae of an adult?
7. Name any four structures of the pelvis that you can palpate, and describe where to palpate them.
8. Describe several ways in which the male and female pelvis differ.
9. What parts of the femur are involved in the hip joint? What parts are involved in the knee joint?

[25]*calc* = stone, chalk
[26]*cunei* = wedge + *form* = in the shape of
[27]*meta* = beyond + *tars* = ankle

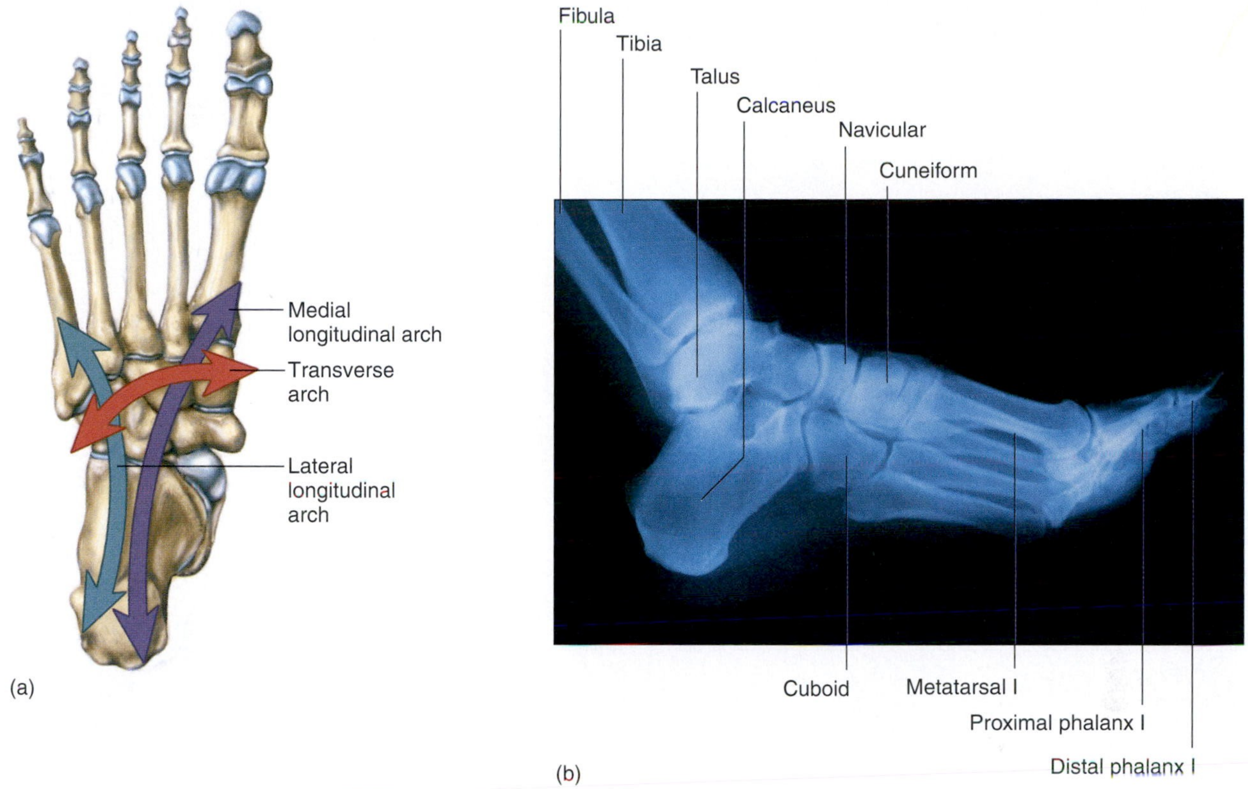
Medial longitudinal arch
Transverse arch
Lateral longitudinal arch
(a)
Fibula
Tibia
Talus
Calcaneus
Navicular
Cuneiform
Cuboid
Metatarsal I
Proximal phalanx I
Distal phalanx I
(b)

FIGURE 8.16
Arches of the Foot. (*a*) Inferior view of the right foot. (*b*) X ray of the right foot, lateral view, showing the lateral longitudinal arch.

10. Name the prominent knobs on each side of your ankle. What bones contribute to these structures?
11. Name all the bones that articulate with the talus and describe the location of each.
12. Describe several ways in which the human and ape pelvis and hindlimb differ, and the functional reason for the differences.

Developmental and Clinical Perspectives

Objectives

When you have completed this section, you should be able to

- describe the pre- and postnatal development of the appendicular skeleton; and
- describe some common disorders of the appendicular skeleton.

Development of the Appendicular Skeleton

With one exception, the bones of the limbs and their girdles form by endochondral ossification. The process begins when mesenchyme condenses and differentiates into hyaline cartilage (chondrification) and continues as the cartilage is replaced by osseous tissue (ossification). The exception is the clavicle, which forms primarily by intramembranous ossification. Although limb and limb girdle ossification is well under way at the time of birth, it is not completed until a person is in his or her twenties.

The first sign of limb development is the appearance of upper **limb buds** around day 26 to 27 and lower limb buds 1 or 2 days later. A limb bud consists of a core of mesenchyme covered with ectoderm. The limb buds elongate as the mesenchyme proliferates. The distal ends of the limb buds flatten into paddlelike **hand** and **foot plates.** By day 38 in the hand and day 44 in the foot, these plates show parallel ridges called **digital rays,** the future fingers and toes. The mesenchyme between the digital rays breaks down by apoptosis, forming notches between the rays which deepen until, at the end of week 8, the digits are well separated.

Condensed mesenchymal models of the future limb bones begin to appear during week 5. Chondrification is apparent by the end of that week, and by the end of week 6, a complete cartilaginous limb skeleton is present. The long bones begin to ossify in the following week, in the manner described in chapter 6 (see fig. 6.7). The humerus, radius, ulna, femur, and tibia develop primary ossification centers in weeks 7 to 8; the scapula and ilium in week 9; the metacarpals, metatarsals, and phalanges over the next 3 weeks; and the ischium and pubis in weeks 15 and 20, respectively. The clavicle ossifies intramembranously beginning early in week 7. During ossification, the upper limbs rotate laterally about 90° and the lower limbs rotate about 90° medially, so the elbows face dorsally and the knees face ventrally.

TABLE 8.3

Anatomical Checklist for the Pelvic Girdle and Lower Limb

Pelvic Girdle	
Os Coxae (figs. 8.6–8.9)	*Os Coxae (figs. 8.6–8.9) (Cont.)*
Pubic symphysis	Greater sciatic notch
Interpubic disc	Iliac pillar
Greater (false) pelvis	Iliac fossa
Lesser (true) pelvis	Auricular surface
Pelvic brim	Ischium
Pelvic inlet	Body
Pelvic outlet	Ischial spine
Acetabulum	Lesser sciatic notch
Obturator foramen	Ischial tuberosity
Ilium	Ramus
Iliac crest	Pubis
Anterior superior spine	Superior ramus
Anterior inferior spine	Inferior ramus
Posterior superior spine	Body
Posterior inferior spine	
Lower Limb	
Femur (fig. 8.10)	*Fibula (fig. 8.13)*
Proximal end	Head
Head	Apex (styloid process)
Fovea capitis	Lateral malleolus
Neck	*Tarsal Bones (fig. 8.14)*
Greater trochanter	Proximal group
Lesser trochanter	Calcaneus
Intertrochanteric crest	Talus
Intertrochanteric line	Navicular
Shaft	Distal group
Linea aspera	First cuneiform
Distal end	Second cuneiform
Medial condyle	Third cuneiform
Lateral condyle	Cuboid
Intercondylar fossa	*Bones of the Foot (figs. 8.14–8.16)*
Medial epicondyle	Metatarsal bones I–V
Lateral epicondyle	Phalanges
Patellar surface	Proximal phalanx
Patella (fig. 8.10)	Middle phalanx
Base	Distal phalanx
Apex	Arches of the foot
Articular facets	Medial longitudinal arch
Tibia (fig. 8.13)	Lateral longitudinal arch
Medial condyle	Transverse arch
Lateral condyle	
Intercondylar eminence	
Tibial tuberosity	
Anterior crest	
Medial malleolus	

The carpal bones are still cartilaginous at birth. Some of them ossify as early as 2 months of age (the capitate) and some as late as 9 years (the pisiform). Among the tarsal bones, the calcaneus and talus begin to ossify prenatally at 3 and 6 months, respectively; the cuboid begins to ossify just before or after birth; and the cuneiforms do not ossify until the first to third years after birth. The patella forms at 3 to 6 years of age. The epiphyses of the long bones are cartilaginous, and their secondary ossification centers just beginning to form, at birth. The epiphyseal plates persist until about age 20, at which time the epiphysis and diaphysis fuse and bone elongation ceases. The ilium, ischium, and pubis are not fully fused into a single os coxae until the age of 25.

INSIGHT 8.3 MEDICAL HISTORY

Anatomical Position—Clinical and Biological Perspectives

It may seem puzzling that we count metacarpal bones I to V progressing from lateral to medial, but count metatarsal bones I to V progressing from medial to lateral. This minor point of confusion is the legacy of a committee of anatomists who met in the early 1900s to define anatomical position. A controversy arose as to whether the arms should be presented with the palms forward or facing the rear in anatomical position. Veterinary anatomists argued that palms to the rear (forearms pronated) would be a more natural position comparable to forelimb orientation in other animals. It is more comfortable to stand with the forearms pronated, and if you watch a child crawl on all fours, you will see that he or she does so with the palms on the floor, in the pronated position. In this animal-like stance, the largest digits (the thumbs and great toes) are medial on all four limbs. Human clinical anatomists, however, argued that if you ask patients to "show me your arms" or "show me your hands," most will present the palms forward or upward—that is, supinated. The clinical anatomists won the debate, forcing us, the heirs to this terminology, to number the hand and foot bones in a biologically less rational order.

A few other anatomical terms also reflect less than perfect logic. The *dorsum* of the foot is its superior surface—it does not face dorsally—and the dorsal artery and nerve of the penis (see p. 739) lie along the surface that faces anteriorly (ventrally). In a cat, dog, or other quadrupedal mammal, however, the dorsum of the foot and the dorsal artery and nerve of the penis do face dorsally (upward). However illogical some of these terms may seem, we inherit them from comparative anatomy and the habit of naming human structures after the corresponding structures in other species.

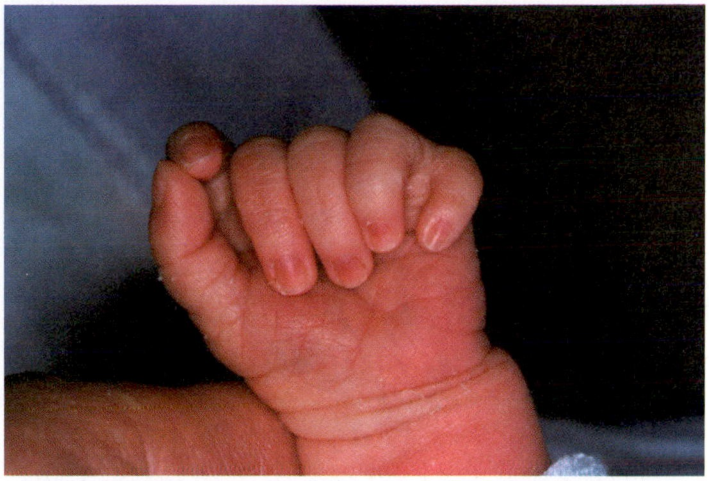

(a)

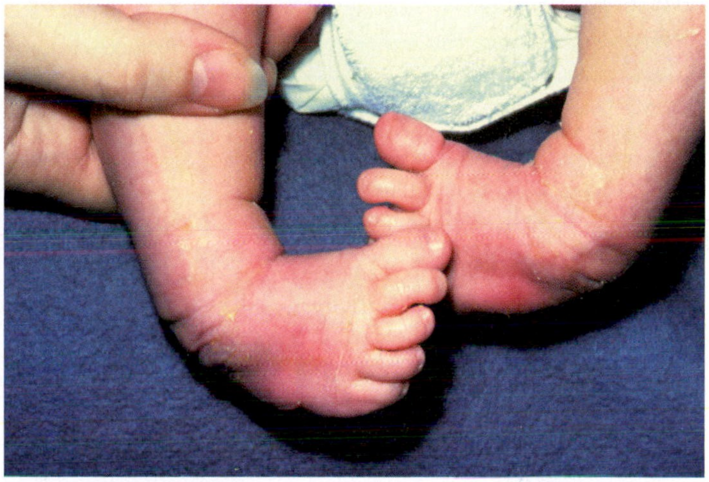

(b)

Figure 8.17

Congenital Deformities of the Hands and Feet. (*a*) Polydactyly. (*b*) Talipes (clubfoot).

Pathology of the Appendicular Skeleton

The appendicular skeleton is subject to several developmental abnormalities, occurring in as many as 2 out of 1,000 live births. The most striking is **amelia,** a complete absence of one or more limbs. The partial absence of a limb is called **meromelia.**[28] Meromelia typically entails an absence of the long bones, with rudimentary hands or feet attached directly to the trunk. Such defects are often accompanied by deformities of the heart, urogenital system, or craniofacial skeleton. These abnormalities are usually hereditary, but they can be induced by teratogenic chemicals such as thalidomide (see fig. 4.19). The limbs are most vulnerable to teratogens in the fourth and fifth weeks of development.

Another class of developmental limb disorders includes **polydactyly,**[29] the presence of extra fingers or toes (fig. 8.17*a*), and **syndactly,**[30] the fusion of two or more digits. The latter results from a failure of the digital rays to separate. *Cutaneous syndactyly,* most common in the foot, is the persistence of a skin web between the digits, and is relatively easy to correct surgically. *Osseous syndactyly* is fusion of the bones of the digits, owing to failure of the notches to form between the embryonic digital rays. Polydactyly and syndactyly are usually hereditary, but can also be induced by teratogens.

Clubfoot, or **talipes**[31] (TAL-ih-peez) is a congenital deformity in which the feet are adducted and plantar flexed (defined in chapter 9), with the soles turned medially (fig. 8.17*b*). This is a relatively common birth defect, present in about 1 out of 1,000 live births, but the cause remains obscure. It is sometimes hereditary, and some think it may also result from a malpositioning of the fetus in the uterus, but the latter hypothesis remains unproven. Children with talipes cannot support their weight on their feet and tend to walk on their ankles. In some cases, talipes requires that the foot be manipulated and set in a new cast every week beginning in the neonatal nursery, and lasting 4 to 6 months. Some cases require surgery at 6 to 9 months to release tight ligaments and tendons and realign the foot.

[28] *mero* = part + *melia* = limb
[29] *poly* = many + *dactyl* = finger
[30] *syn* = together + *dactyl* = finger
[31] *tali* = heel + *pes* = foot

TABLE 8.4

Disorders of the Appendicular Skeleton

Avulsion	A fracture in which a body part, such as a finger, is completely torn from the body, as in many accidents with farm and factory machinery. The term can also refer to non-osseous structures such as the avulsion of an ear.
Calcaneal (heel) Spurs	Abnormal outgrowths of the calcaneus. Often results from high-impact exercise such as aerobics and running, especially if done with inappropriate footwear. Stress on the plantar aponeurosis (a connective tissue sheet in the sole of the foot) stimulates exostosis, or growth of the bony spur, and can cause severe foot pain.
Colles[32] Fracture	Pathologic fracture at the distal end of the radius and ulna, often occurring when stress is placed on the wrist (as in pushing oneself up from an armchair) and the bones have been weakened by osteoporosis.
Epiphyseal Fracture	Separation of the epiphysis from the diaphysis of a long bone. Common in children and adolescents because of their cartilaginous epiphyseal plates. May present a threat to normal completion of bone growth.
Pes Planis[33]	"Flat feet" or "fallen arches" (absence of visible arches) in adolescents and adults. Caused by stretching of plantar ligaments due to prolonged standing or excess weight.
Pott[34] Fracture	Fracture of the distal end of the tibia, fibula, or both; a sports injury common in football, soccer, and snow skiing.
Disorders Described Elsewhere	
Fracture of the clavicle 208	Fracture of the femur 218

[32]Abraham Colles (1773–1843), Irish surgeon
[33]*pes* = foot + *planis* = flat
[34]Sir Percivall Pott (1713–88), British surgeon

The most common noncongenital disorders of the appendicular skeleton are fractures, dislocations, and arthritis. Even though these disorders can affect the axial skeleton as well, they are more common and more often disabling in the appendicular skeleton. The general classification of bone fractures is discussed in chapter 6, and some fractures specific to the appendicular skeleton are discussed in insights 8.1 and 8.2 of this chapter. Arthritis, dislocation, and other joint disorders are described in chapter 9. Table 8.4 describes some other disorders of the appendicular skeleton.

Before You Go On

Answer the following questions to test your understanding of the preceding section:

13. Describe the progression from a limb bud to a hand with fully formed and separated fingers.
14. Name some appendicular bones that do not ossify until a person is at least a few years old.
15. Distinguish between amelia and meromelia, and between polydactyly and syndactyly.

CHAPTER REVIEW

Review of Key Concepts

The Pectoral Girdle and Upper Limb (p. 208)

1. The *pectoral girdle* supports the upper limb. It consists of a clavicle and scapula on each side of the body. The *clavicle* articulates with the sternum medially and the scapula laterally. The *scapula* also articulates with the humerus of the arm at the *humeroscapular joint*. The anatomical features of the clavicle and scapula are summarized in table 8.1.
2. The upper limb has 30 bones divided into 4 regions: the *brachium* (arm), containing the *humerus;* the *antebrachium* (forearm) containing the *radius* and *ulna;* the *carpus* (wrist), containing 2 rows of 4 *carpal bones* each; and the *manus* (hand) containing 5 *metacarpal bones* in the palmar region and 14 *phalanges* in the digits. The anatomical features of these bones are summarized in table 8.1.
3. The evolution of bipedal locomotion is reflected in the anatomy of the upper limb, whose function has changed from locomotion to reaching and grasping. Its adaptations include a shorter length, less heavy musculature, and increased mobility of the joints of the shoulder and hand.

The Pelvic Girdle and Lower Limb (p. 213)

1. The adult *pelvic girdle* supports the trunk on the lower limbs and encloses and protects viscera of the pelvic cavity. It consists of three bones: an *os coxae* on each side and the sacrum of the vertebral column.
2. Each os coxae articulates with the vertebral column at the *sacroiliac joint*. The ossa coxae articulate anteriorly with each other at the *pubic symphysis*, where a fibrocartilage *interpubic disc* separates the two bones. Each os coxae has a deep socket, the *acetabulum*, where it articulates with the femur.
3. The pelvic girdle is bowl-like, with a broad *greater pelvis* superiorly, a narrow *lesser pelvis* inferiorly, and a *pelvic brim* marking the boundary between them. In childbirth, an infant's head descends from the greater pelvis, through the *pelvic inlet* surrounded by the brim, into the lesser pelvis, and then out the *pelvic outlet*.
4. The adult os coxae forms by fusion of three childhood bones, the *ilium*, *ischium*, and *pubis*.
5. The anatomy of the pelvic girdle exhibits adaptations to bipedalism in both sexes and to childbirth in women. This is the most sexually dimorphic region of human skeletal anatomy (table 8.2).
6. The anatomical features of the pelvic girdle are summarized in table 8.3.
7. The lower limb has 30 bones divided into 4 regions: the *femoral region* (thigh) containing the *femur;* the *patella;* the *crural region* (leg) containing the *tibia* and *fibula;* the *tarsal region* (ankle), whose bones are regarded as part of the foot; and the *pes* (foot) containing 7 *tarsal bones*, 5 *metatarsal bones*, and 14 *phalanges*. Anatomical features of the lower limb bones are summarized in table 8.3.
8. Adaptations of the lower limb to bipedalism include medially angled femurs, locking knees, a great toe strengthened for the toe-off part of the stride, and foot arches.

Developmental and Clinical Perspectives (p. 221)

1. The clavicle forms mainly by intramembranous ossification and the other limb and girdle bones by endochondral ossification. The latter process proceeds through chondrification (condensation of mesenchyme into cartilage) and then ossification (replacement of the cartilage with bone).
2. The limbs begin as *limb buds*, which elongate and form paddlelike *hand* and *foot plates*. Ridges called *digital rays* appear in each plate, and the tissue between the rays breaks down by apoptosis, separating the fingers and toes by the end of the embryonic stage (8 weeks). The limb bones begin to ossify at various times from 7 weeks of embryonic development to 9 years after birth, and some bones do not complete their ossification until a person is 25 years old.
3. Developmental abnormalities of the appendicular skeleton include the complete absence of one or more limbs *(amelia)*, partial absence *(meromelia)*, excess fingers or toes *(polydactyly)*, failure of toes or fingers to separate *(syndactyly)*, and clubfoot *(talipes)*.
4. The most common disorders appearing after birth are fractures, joint dislocations, and arthritis.

Testing Your Recall

1. The os coxae is attached to the axial skeleton through its
   a. auricular surface.
   b. articular surface.
   c. pubic symphysis.
   d. conoid tubercle.
   e. coronoid process.
2. Which of these bones supports the most body weight?
   a. ilium
   b. pubis
   c. femur
   d. tibia
   e. talus
3. Which of these structures can be most easily palpated on a living person?
   a. the deltoid tuberosity
   b. the greater sciatic notch
   c. the medial malleolus
   d. the spine of the scapula
   e. the glenoid cavity
4. Compared to the male pelvic girdle, the pelvic girdle of a female
   a. has a less movable coccyx.
   b. has a rounder pelvic inlet.
   c. is narrower between the iliac crests.
   d. has a narrower pubic arch
   e. has a narrower sacrum.
5. The lateral and medial malleoli are most similar to
   a. the radial and ulnar styloid processes.
   b. the humeral capitulum and trochlea.
   c. the acromion and coracoid process.
   d. the base and head of a metacarpal bone.
   e. the anterior and posterior superior spines.
6. When you rest your hands on your hips, you are resting them on
   a. the pelvic inlet.
   b. the pelvic outlet.
   c. the pelvic brim.
   d. the iliac crests.
   e. the auricular surfaces.

7. The disc-shaped head of the radius articulates with the _____ of the humerus.
   a. radial tuberosity
   b. trochlea
   c. capitulum
   d. olecranon process
   e. glenoid cavity
8. All of the following are carpal bones, *except* the _____, which is a tarsal bone.
   a. trapezium
   b. cuboid
   c. trapezoid
   d. triquetral
   e. pisiform
9. The bone that supports your body weight when you are sitting down is
   a. the acetabulum.
   b. the pubis.
   c. the ilium.
   d. the coccyx.
   e. the ischium.
10. Which of these is the bone of the heel?
    a. cuboid
    b. calcaneus
    c. navicular
    d. trochlear
    e. talus
11. The Latin anatomical name for the thumb is _____ and the name for the great toe is _____.
12. The acromion and coracoid process are parts of what bone?
13. How many phalanges, total, does the human body have?
14. The bony prominences on each side of your elbow are the lateral and medial _____ of the humerus.
15. One of the wrist bones, the _____, is characterized by a prominent hook.
16. The fibrocartilage pad that holds the pelvic girdle together anteriorly is called the _____.
17. The leg proper, between the knee and ankle, is called the _____ region.
18. The _____ processes of the radius and ulna form bony protuberances on each side of the wrist.
19. Two massive protuberances unique to the proximal end of the femur are the greater and lesser _____.
20. The _____ arch of the foot extends from the heel to the great toe.

Answers in the Appendix

True or False

Determine which five of the following statements are false, and briefly explain why.

1. There are more carpal bones than tarsal bones.
2. The hands have more phalanges than the feet.
3. The upper limb is attached to the axial skeleton at only one point, the acromioclavicular joint.
4. On a living person, it would be possible to palpate the muscles in the infraspinous fossa but not those of the subscapular fossa.
5. In strict anatomical terminology, the words *arm* and *leg* both refer to regions with only one bone.
6. If you rest your chin on your hands with your elbows on a table, the olecranon of the ulna rests on the table.
7. The most frequently broken bone in humans is the humerus.
8. The proximal end of the radius articulates with both the humerus and ulna.
9. The pisiform bone and patella are both sesamoid bones.
10. The pelvic outlet is the opening in the floor of the greater pelvis leading into the lesser pelvis.

Answers in the Appendix

Testing Your Comprehension

1. In adolescents, trauma sometimes separates the head of the femur from the neck. Why do you think this is more common in adolescents than in adults?
2. By palpating the hind leg of a cat or dog or examining a laboratory skeleton, you can see that cats and dogs stand on the heads of their metatarsal bones; the calcaneus does not touch the ground. How is this similar to the stance of a woman wearing high-heeled shoes? How is it different?
3. Contrast the tarsal bones with the carpal bones. Which ones are similar in name, location, or both? Which ones are different?
4. A surgeon has removed 8 cm of Joan's radius because of osteosarcoma, a bone cancer, and replaced it with a graft taken from one of the bones of Joan's lower limb. What bone do you think would most likely be used as the source of the graft? Explain your answer.
5. Andy, a 55-year-old, 75-kg (165-lb) roofer, is shingling the steeply pitched roof of a new house when he loses his footing and slides down the roof and over the edge, feet first. He braces himself for the fall, and when he hits the ground he cries out and doubles up in excruciating pain. Emergency medical technicians called to the scene tell him he has broken his hips. Describe, more specifically, where his fractures most likely occurred. On the way to the hospital, Andy says, "You know it's funny, when I was a kid, I used to jump off roofs that high, and I never got hurt." Why do you think Andy was more at risk of a fracture as an adult than he was as a boy?

Answers at the Online Learning Center

www.mhhe.com/saladinha1

Visit the Online Learning Center for practice tests, answer keys, and other learning aids for this chapter. Enhance your understanding of human anatomy with our interactive art labeling exercises, supplemental photo atlases, web links, puzzles, flashcards, and much more.

9

CHAPTER NINE

Joints

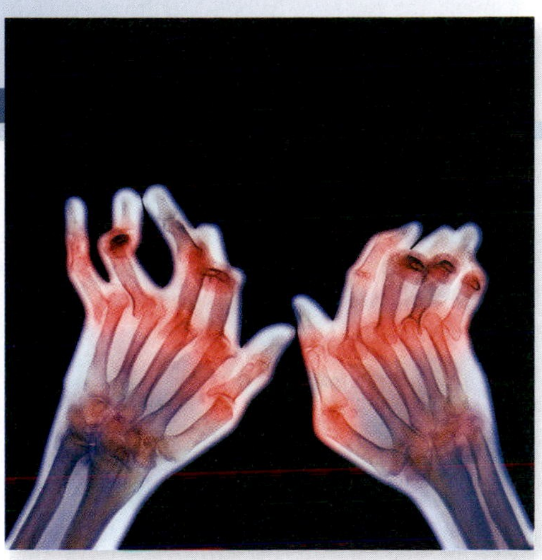

X ray of hands with severe rheumatoid arthritis

CHAPTER OUTLINE

Joints and Their Classification 228
- Bony Joints 228
- Fibrous Joints 228
- Cartilaginous Joints 230

Synovial Joints 231
- General Anatomy 231
- Types of Synovial Joints 233
- Movements of Synovial Joints 234
- Range of Motion 239

Anatomy of Selected Synovial Joints 239
- The Jaw Joint 239
- The Shoulder Joint 240
- The Elbow Joint 241
- The Hip Joint 242
- The Knee Joint 243
- The Ankle Joint 246

Clinical Perspectives 246
- Arthritis 247
- Joint Prostheses 250

Chapter Review 252

INSIGHTS

9.1 Clinical Application: Exercise and Articular Cartilage 232
9.2 Clinical Application: TMJ Syndrome 239
9.3 Clinical Application: Pulled Elbow 243
9.4 Clinical Application: Knee Injuries and Arthroscopic Surgery 247

BRUSHING UP

To understand this chapter, you may find it helpful to review the following concepts:
- The Distinction Between Hyaline Cartilage and Fibrocartilage (p. 92)
- Anatomy of the Skeletal System (p. 171–224)

In order for the skeleton to serve the purposes of protection and movement, the bones must be joined together. A **joint,** or **articulation,** is any point at which two bones meet, regardless of whether they are movable at that point. Your shoulder, for example, is a very movable joint, whereas the skull sutures described in chapter 6 are immovable joints.

The science of joint structure, function, and dysfunction is called **arthrology.** The study of musculoskeletal movement is **kinesiology** (kih-NEE-see-OL-oh-jee). Kinesiology is a branch of **biomechanics,** which deals with a broad range of motions and mechanical processes in the body, including the physics of blood circulation, respiration, and hearing.

This chapter describes the joints of the skeleton and the types of joint movements relevant to the actions of skeletal muscles described in chapters 11 and 12.

Joints and Their Classification

Objectives

When you have completed this section, you should be able to

- explain what joints are, how they are named, and what functions they serve;
- name and describe the four major classes of joints;
- describe the three types of fibrous joints and give an example of each;
- distinguish between the three types of sutures;
- describe the two types of cartilaginous joints and give an example of each; and
- name some joints that become synostoses as they age.

Joints such as the shoulder, elbow, and knee are remarkable specimens of biological design—self-lubricating, almost frictionless, and able to bear heavy loads and withstand compression while executing smooth and precise movements. Yet, it is equally important that other joints be less movable or even immovable. Such joints are better able to support the body and provide protection for delicate organs. The vertebral column, for example, must provide a combination of support and flexibility; thus its joints are only moderately movable. The immovable joints between the cranial bones afford the best possible protection for the brain and sense organs.

The name of a joint is typically derived from the names of the bones involved. For example, the *atlanto-occipital joint* is where the occipital condyles meet the atlas, the *humeroscapular joint* is where the humerus meets the scapula, and the *coxal joint* is where the femur meets the os coxae.

Joints are classified according to the manner in which the adjacent bones are bound to each other, with corresponding differences in how freely the bones can move. Authorities differ in their classification schemes, but one common view places the joints in four major categories. From the least to the greatest freedom of movement, they are called *bony, fibrous, cartilaginous,* and *synovial joints.* This section will describe the first three of these and the subclasses of each. The remainder of the chapter will then be concerned primarily with synovial joints.

Bony Joints

A **bony joint,** or **synostosis**[1] (SIN-oss-TOE-sis), is an immovable joint formed when the gap between two bones ossifies and they become, in effect, a single bone. Bony joints can form by ossification of either fibrous or cartilaginous joints. An infant is born with right and left frontal and mandibular bones, for example, but these soon fuse seamlessly into a single frontal bone and mandible. In old age, some cranial sutures become obliterated by ossification and the adjacent cranial bones, such as the parietal bones, fuse. The epiphyses and diaphyses of the long bones are joined by cartilaginous joints in childhood and adolescence, and these become synostoses in early adulthood. The attachment of the first rib to the sternum also becomes a synostosis with age.

Fibrous Joints

A **fibrous joint** is also called a **synarthrosis**[2] (SIN-ar-THRO-sis) or **synarthrodial joint.** It is a point at which adjacent bones are bound by collagen fibers that emerge from the matrix of one bone, cross the space between them, and penetrate into the matrix of the other (fig. 9.1). There are three kinds of fibrous joints: *sutures, gomphoses,* and *syndesmoses.* In sutures and gomphoses, the fibers are very short and allow for little or no movement. In syndesmoses, the fibers are longer and the attached bones are more movable.

SUTURES

Sutures are immovable fibrous joints that closely bind the bones of the skull to each other; they occur nowhere else. In chapter 7, we did not take much notice of the differences between one suture and another, but some differences may have caught your attention as you studied the diagrams in that chapter or examined laboratory specimens. Sutures can be classified as *serrate, lap,* and *plane sutures.* Readers with some knowledge of woodworking may recognize that the structures and functional properties of these sutures have something in common with basic types of carpentry joints (fig. 9.2).

Serrate sutures appear as wavy lines along which the adjoining bones firmly interlock with each other by their serrated margins, like pieces of a jigsaw puzzle. Serrate sutures are analogous to a dovetail wood joint. Examples include the coronal, sagittal, and lambdoid sutures that border the parietal bones.

Lap (squamous) sutures occur where two bones have overlapping beveled edges, like a miter joint in carpentry. On the surface, a lap suture appears as a relatively smooth (nonserrated) line. An example is the squamous suture between the temporal and parietal bones.

Plane (butt) sutures occur where two bones have straight, nonoverlapping edges. The two bones merely border on each other, like two boards glued together in a butt joint. This type of suture is seen between the palatine processes of the maxillae in the roof of the mouth.

[1]*syn* = together + *ost* = bone + *osis* = condition
[2]*syn* = together + *arthr* = joined + *osis* = condition

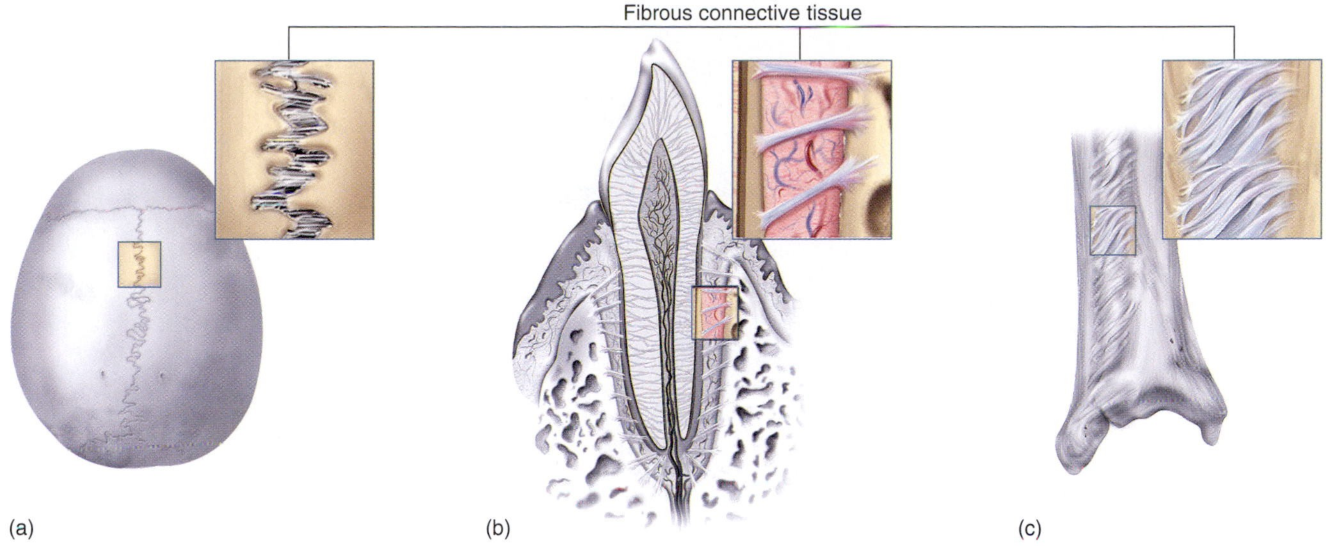
Fibrous connective tissue
(a)
(b)
(c)

FIGURE 9.1

Types of Fibrous Joints. (*a*) A suture between the parietal bones. (*b*) A gomphosis between a tooth and the jaw. (*c*) A syndesmosis between the tibia and fibula.

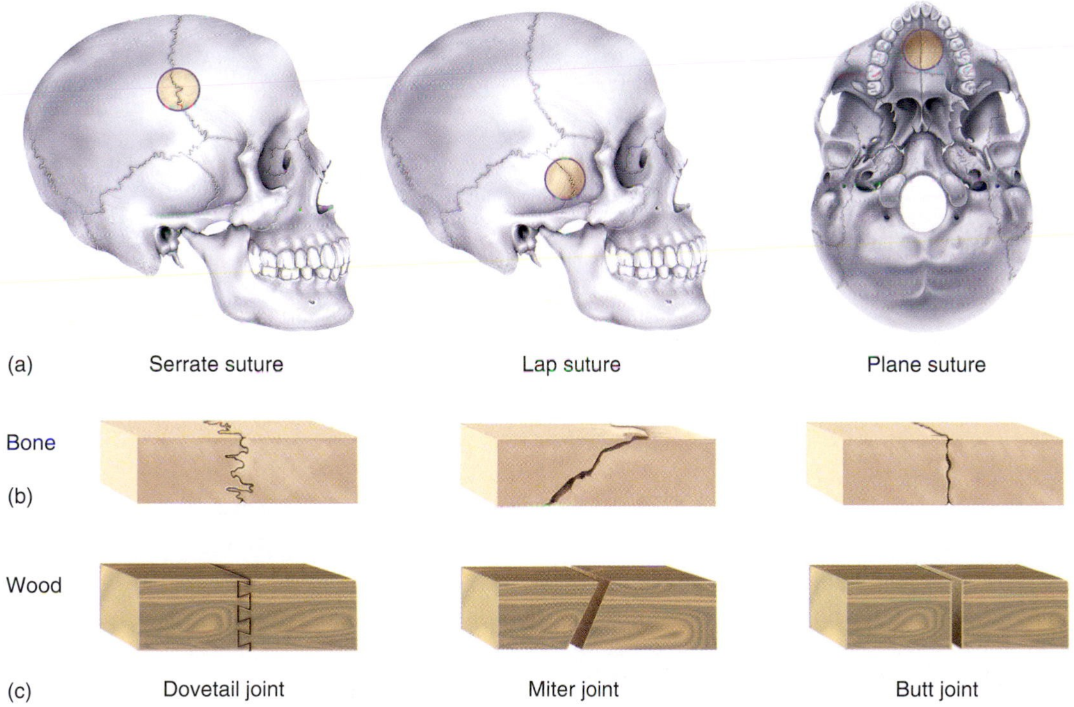
(a)
Serrate suture
Lap suture
Plane suture
Bone
(b)
Wood
(c)
Dovetail joint
Miter joint
Butt joint

FIGURE 9.2

Types of Sutures. (*a*) Locations. (*b*) Structure of the adjoining bones. (*c*) Functional analogies to some common wood joints.

GOMPHOSES

Even though the teeth are not bones, the attachment of a tooth to its socket is classified as a joint called a **gomphosis** (gom-FOE-sis). The term refers to its similarity to a nail hammered into wood.[3] The tooth is held firmly in place by a fibrous **periodontal ligament,** which consists of collagen fibers that extend from the bone matrix of the jaw into the dental tissue (see fig. 9.1*b*). The periodontal ligament allows the tooth to move or "give" a little under the stress of chewing. This allows us to sense how hard we are biting or to sense a particle of food stuck between the teeth.

[3]*gomph* = nail, bolt + *osis* = condition

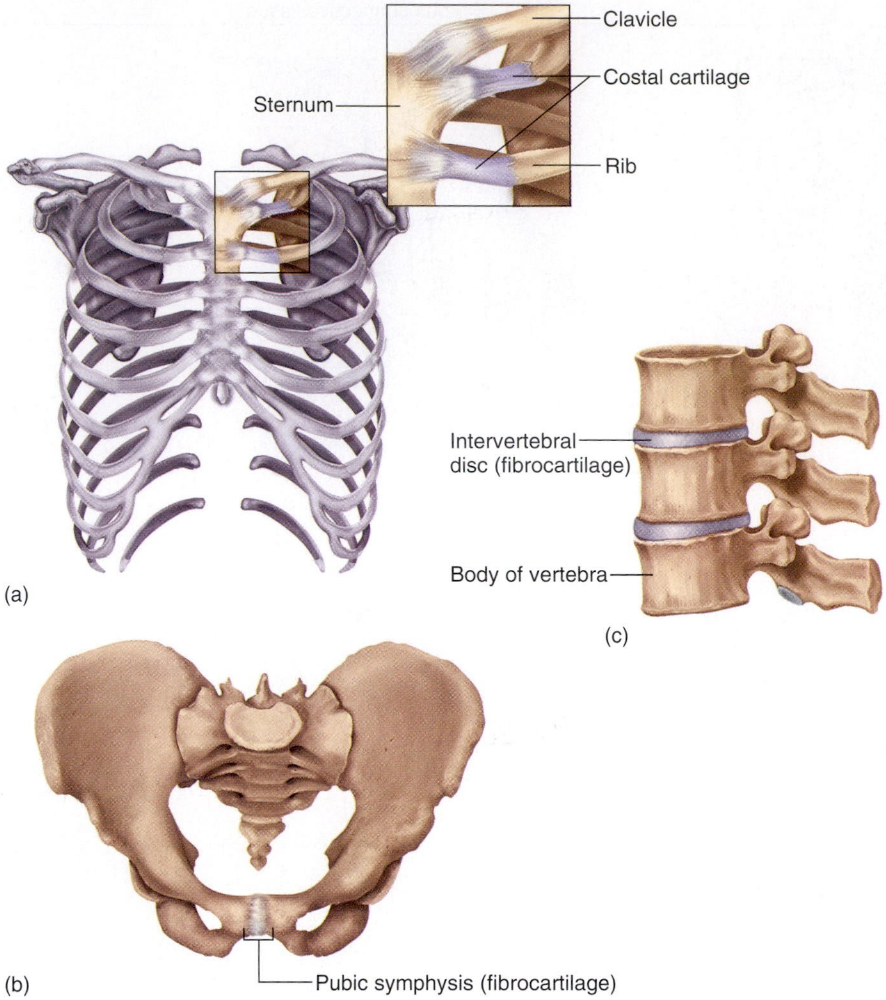
Clavicle
Costal cartilage
Sternum
Rib
(a)
Intervertebral disc (fibrocartilage)
Body of vertebra
(c)
(b)
Pubic symphysis (fibrocartilage)

FIGURE 9.3

Cartilaginous Joints. (*a*) Synchondroses, represented by costal cartilages joining the ribs to the sternum. (*b*) The pubic symphysis. (*c*) Intervertebral discs, which join adjacent vertebrae to each other by symphyses.

SYNDESMOSES

A **syndesmosis**[4] (SIN-dez-MO-sis) is a fibrous joint at which two bones are bound by longer collagenous fibers than in a suture or gomphosis, giving the bones more mobility. While the range of motion differs greatly among syndesmoses, all of them are more mobile than sutures or gomphoses. One of the less movable syndesmoses is the joint that binds the distal ends of the tibia and fibula together, side by side. A more movable one exists between the shafts of the radius and ulna, which are joined by a broad fibrous sheet called an **interosseous membrane** that allows for movement such as pronation and supination of the forearm (see fig. 9.1*c*).

[4]*syn* = together + *desm* = band + *osis* = condition

Cartilaginous Joints

A **cartilaginous joint** is also called an **amphiarthrosis**[5] (AM-fee-ar-THRO-sis) or **amphiarthrodial joint.** In these joints, two bones are linked by cartilage (fig. 9.3). The two types of cartilaginous joints are *synchondroses* and *symphyses.*

SYNCHONDROSES

A **synchondrosis**[6] (SIN-con-DRO-sis) is a joint in which the bones are bound by hyaline cartilage. An example is the temporary joint between the epiphysis and diaphysis of a long bone in a child, formed by the cartilage of the epiphyseal plate. Another is the attachment of a rib to the sternum by a hyaline costal cartilage (fig. 9.3*a*).

[5]*amphi* = on all sides + *arthr* = joined + *osis* = condition
[6]*syn* = together + *chondr* = cartilage + *osis* = condition

SYMPHYSES

In a **symphysis**[7] (SIM-fih-sis), two bones are joined by fibrocartilage (fig. 9.3*b, c*). One example is the pubic symphysis, in which the right and left pubic bones are joined by the cartilaginous interpubic disc. Another is the joint between the bodies of two vertebrae, united by an intervertebral disc. The surface of each vertebral body is covered with hyaline cartilage. Between the vertebrae, this cartilage becomes infiltrated with collagen bundles to form fibrocartilage. Each intervertebral disc permits only slight movement between adjacent vertebrae, but the collective effect of all 23 discs gives the spine considerable flexibility.

Think About It!

The intervertebral joints are symphyses only in the cervical through the lumbar region. How would you classify the intervertebral joints of the sacrum and coccyx in a middle-aged adult?

Before You Go On

Answer the following questions to test your understanding of the preceding section:

1. What is the difference between arthrology and kinesiology?
2. Explain the distinction between a diarthrosis, amphiarthrosis, and synarthrosis.
3. Define *suture, gomphosis,* and *syndesmosis,* and explain what these three joints have in common.
4. Name the three types of sutures and describe how they differ.
5. Name two synchondroses and two symphyses.
6. Give some examples of joints that become synostoses with age.

Synovial Joints

Objectives

When you have completed this section, you should be able to

- describe the anatomy of a synovial joint and its associated structures;
- describe the six types of synovial joints;
- list and demonstrate the types of movements that occur at diarthroses; and
- discuss the factors that affect the range of motion of a joint.

The most familiar type of joint is the **synovial** (sih-NO-vee-ul) **joint,** also called a **diarthrosis**[8] (DY-ar-THRO-sis) or **diarthrodial joint.** Ask most people to point out any joint in the body, and they are likely to point to a synovial joint such as the elbow, knee, or knuckles. Many synovial joints, like these examples, are freely movable. Others, such as the joints between the wrist and ankle bones and between the articular processes of the vertebrae, have a more limited range of motion. Synovial joints are the most structurally complex type of joint, and are the most likely to develop uncomfortable and crippling dysfunctions.

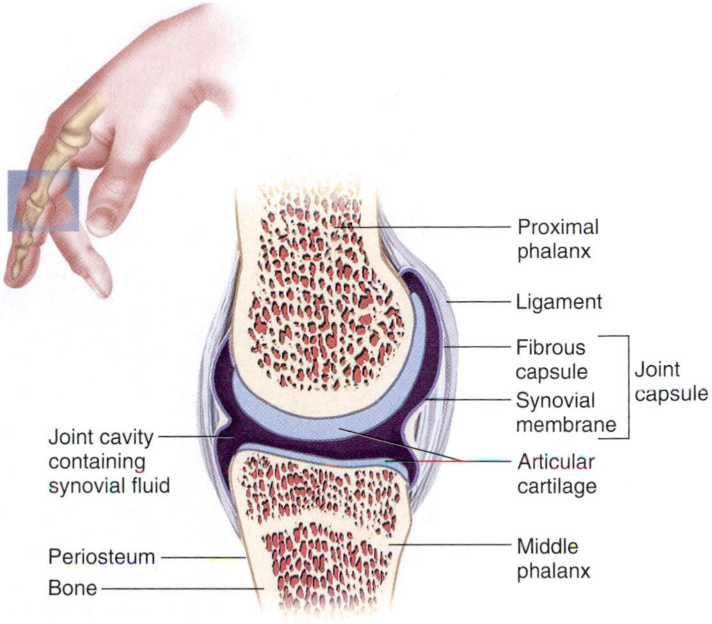

Figure 9.4
Structure of a Simple Synovial Joint.

General Anatomy

In synovial joints, the facing surfaces of the two bones are covered with **articular cartilage,** a layer of hyaline cartilage about 2 mm thick. These surfaces are separated by a narrow space, the **joint (articular) cavity,** containing a slippery lubricant called **synovial fluid** (fig. 9.4). This fluid, for which the joint is named, is rich in albumin and hyaluronic acid, which give it a viscous, slippery texture similar to raw egg white.[9] It nourishes the articular cartilages, removes their wastes, and makes movements at synovial joints almost friction-free. A connective tissue **joint (articular) capsule** encloses the cavity and retains the fluid. It has an outer **fibrous capsule** continuous with the periosteum of the adjoining bones, and an inner, cellular **synovial membrane.** The synovial membrane is composed of fibroblast-like cells that secrete the fluid and macrophages that remove debris from the joint cavity.

In several synovial joints, fibrocartilage grows inward from the joint capsule and forms a pad between the articulating bones. In the jaw (temporomandibular) and distal radioulnar joints, and at both ends of the clavicle (sternoclavicular and acromioclavicular joints), the pad crosses the entire joint capsule and is called an **articular disc** (see fig. 9.14*c*). In the knee, two cartilages extend inward from the left and right but do not entirely cross the joint (see fig. 9.19). Each is called a **meniscus**[10] because of its crescent shape.

[7]*sym* = together + *physis* = growth
[8]*dia* = separate, apart + *arthr* = joint + *osis* = condition
[9]*ovi* = egg
[10]*men* = moon, crescent + *iscus* = little

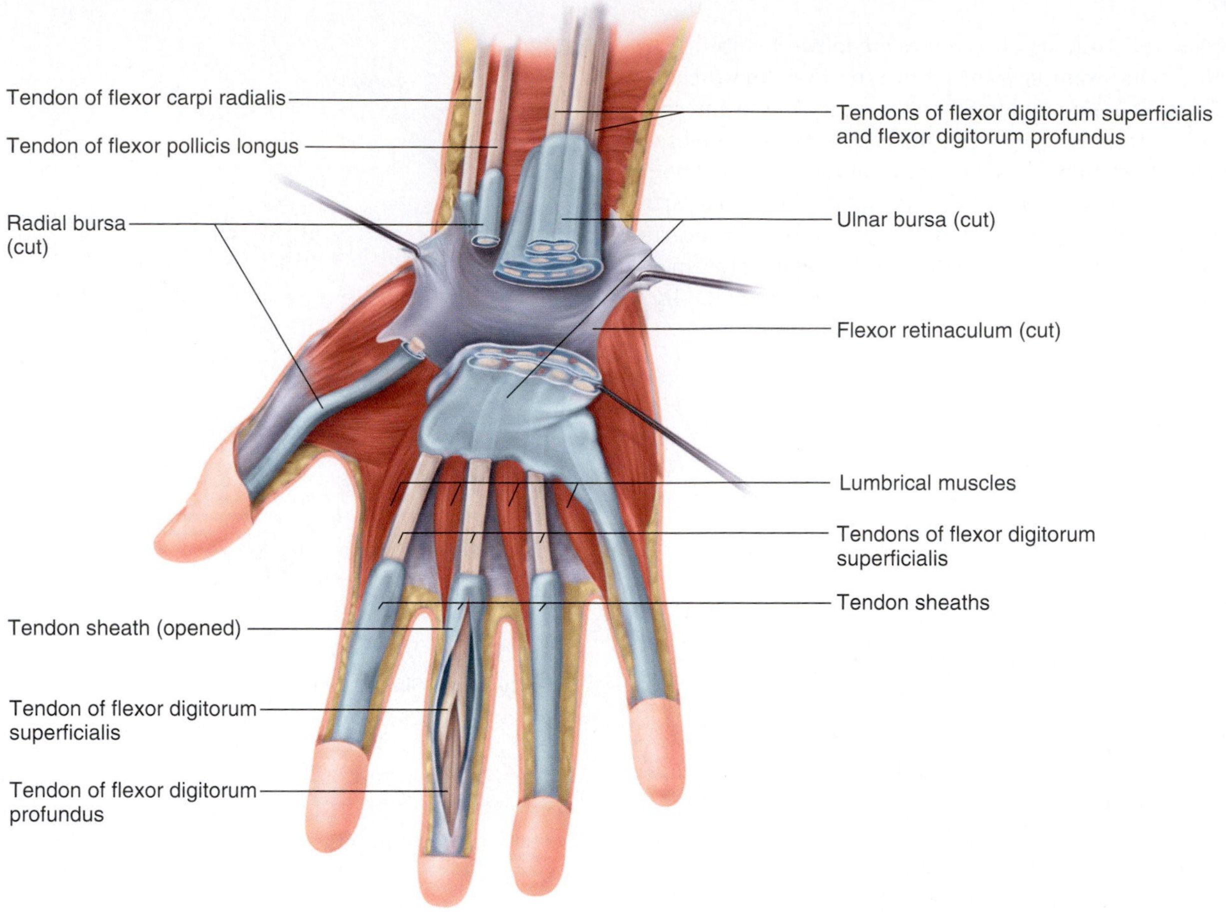
Tendon of flexor carpi radialis
Tendon of flexor pollicis longus
Radial bursa (cut)
Tendons of flexor digitorum superficialis and flexor digitorum profundus
Ulnar bursa (cut)
Flexor retinaculum (cut)
Lumbrical muscles
Tendons of flexor digitorum superficialis
Tendon sheaths
Tendon sheath (opened)
Tendon of flexor digitorum superficialis
Tendon of flexor digitorum profundus

FIGURE 9.5
Tendon Sheaths and Other Bursae in the Hand and Wrist.

These cartilages absorb shock and pressure, guide the bones across each other, improve the fit between the bones, and stabilize the joint, reducing the chance of dislocation.

Accessory structures associated with a synovial joint include tendons, ligaments, and bursae. A **tendon** is a strip or sheet of tough, collagenous connective tissue that attaches a muscle to a bone. Tendons are often the most important structures in stabilizing a joint. A **ligament** is a similar tissue that attaches one bone to another. Several ligaments are named and illustrated in our later discussion of individual joints, and tendons are more fully considered in chapters 10 through 12 along with the gross anatomy of muscles.

A **bursa**[11] is a fibrous sac filled with synovial fluid, located between adjacent muscles or where a tendon passes over a bone (see fig. 9.15*a*). Bursae cushion muscles, help tendons slide more easily over the joints, and sometimes enhance the mechanical effect of a muscle by modifying the direction in which its tendon pulls. Bursae called **tendon sheaths** are elongated cylinders wrapped around a tendon. These are especially numerous in the hand and foot (fig. 9.5). **Bursitis** is inflammation of a bursa, usually due to overexertion of a joint. **Tendinitis** is a form of bursitis in which a tendon sheath is inflamed.

[11]*burs* = purse

INSIGHT 9.1 CLINICAL APPLICATION

EXERCISE AND ARTICULAR CARTILAGE

When synovial fluid is warmed by exercise, it becomes thinner (less viscous) and more easily absorbed by the articular cartilage. The cartilage then swells and provides a more effective cushion against compression. For this reason, a warm-up period before vigorous exercise helps protect the articular cartilage from undue wear and tear.

Because cartilage is nonvascular, its repetitive compression during exercise is important to its nutrition and waste removal. Each time a cartilage is compressed, fluid and metabolic wastes are squeezed out of it. When weight is taken off the joint, the cartilage absorbs synovial fluid like a sponge, and the fluid carries oxygen and nutrients to the chondrocytes. Lack of exercise causes the articular cartilages to deteriorate more rapidly from lack of nutrition, oxygenation, and waste removal.

Weight-bearing exercise builds bone mass and strengthens the muscles that stabilize many of the joints, thus reducing the risk of joint dislocations. Excessive joint stress, however, can hasten the progression of osteoarthritis (p. 247) by damaging the articular cartilage. Swimming is a good way of exercising the joints with minimal damage.

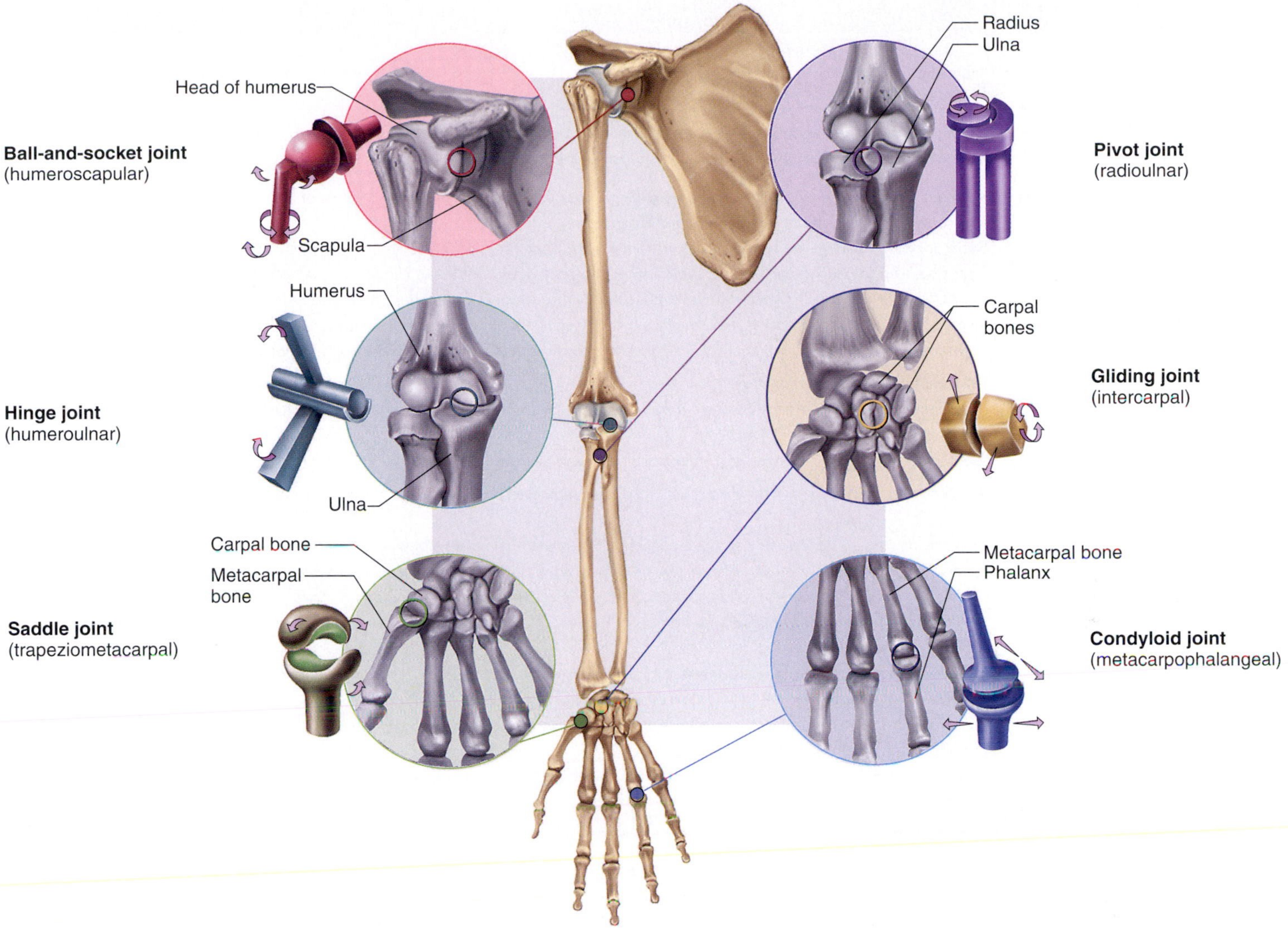

FIGURE 9.6
The Six Types of Synovial Joints. All six have representatives in the forelimb. Mechanical models show the types of motion possible at each joint.

Types of Synovial Joints

There are six types of synovial joints, distinguished by patterns of motion determined by the shapes of the articular surfaces of the bones (fig. 9.6; table 9.1). A bone's movement at a joint can be described with reference to three mutually perpendicular planes in space (*x*, *y*, and *z*). If the bone can move in only one plane, the joint is said to be **monaxial;** if it can move in two planes, the joint is **biaxial;** and if three, it is **multiaxial.**

1. **Hinge joints.** At a hinge joint, one bone has a convex surface that fits into a concave depression of the other one. Hinge joints are monaxial—like a door hinge, they can move in only one plane. Examples include the elbow, knee, and interphalangeal (finger and toe) joints.
2. **Gliding (plane) joints.** Here, the articular surfaces are flat or only slightly concave and convex. The adjacent bones slide over each other and have rather limited monaxial movement. Gliding joints occur between the carpal and tarsal bones, between the articular processes of the vertebrae, and at the sternoclavicular joint. To feel a gliding joint in motion, palpate your sternoclavicular joint as you raise your arm above your head.
3. **Pivot joints.** These are monaxial joints in which one bone has a projection that fits into a ringlike ligament of another, and the first bone rotates on its longitudinal axis relative to the other. One example is the atlantoaxial joint between the first two vertebrae—the dens of the axis projects into the vertebral foramen of the atlas, where it is held against the arch of the atlas by a ligament (see fig. 7.24). This joint pivots when you rotate your head as in gesturing "no." Another example is the proximal radioulnar joint, where the *annular ligament* on the ulna encircles the head of the radius (see fig. 9.16*b*) and permits the radius to rotate during pronation and supination of the forearm.
4. **Saddle joint.** The trapeziometacarpal joint at the base of the thumb (between the trapezium and metacarpal I) and

TABLE 9.1

Anatomical Classification of the Joints

Joint	Characteristics and Examples
Bony Joint (synostosis)	Former fibrous or cartilaginous joint in which adjacent bones have become fused by ossification. Examples: midsagittal line of frontal bone; fusion of epiphysis and diaphysis of an adult long bone; and fusion of ilium, ischium, and pubis to form os coxae
Fibrous Joint (synarthrosis)	Adjacent bones bound by collagen fibers extending from the matrix of one into the matrix of the other
Suture (figs. 9.1*a*, 9.2)	Immovable fibrous joint between cranial and facial bones
Serrate suture	Bones joined by a wavy line formed by interlocking teeth along the margins. Examples: coronal, sagittal, and lambdoid sutures
Lap suture	Bones beveled to overlap each other; superficial appearance is a smooth line. Example: squamous suture around temporal bone
Plane suture	Bones butted against each other without overlapping or interlocking. Example: palatine suture
Gomphosis (fig. 9.1*b*)	Insertion of a tooth into a socket, held in place by collagen fibers of periodontal ligament
Syndesmosis (fig. 9.1*c*)	Slightly movable joint held together by ligaments or interosseous membranes. Examples: tibiofibular joint and radioulnar joint
Cartilaginous Joint (amphiarthrosis)	Adjacent bones bound by cartilage
Synchondrosis (fig. 9.3*a*)	Bones held together by hyaline cartilage. Examples: articulation of ribs with sternum, and epiphyseal plate uniting the epiphysis and diaphysis of a long bone of a child
Symphysis (fig. 9.3*b*, *c*)	Slightly movable joint held together by fibrocartilage. Examples: intervertebral discs and pubic symphysis
Synovial Joint (diarthrosis) (figs. 9.4 and 9.6)	Adjacent bones covered with hyaline articular cartilage, separated by lubricating synovial fluid and enclosed in a fibrous joint capsule
Hinge	Monaxial diarthrosis, able to flex and extend in only one plane. Examples: elbow, knee, and interphalangeal joints
Gliding	Synovial amphiarthrosis with slightly concave or convex bone surfaces that slide across each other. Examples: intercarpal, intertarsal, and sternoclavicular joints; joints between the articular processes of the vertebrae
Ball-and-socket	Multiaxial diarthrosis in which a smooth hemispherical head of one bone fits into a cuplike depression of another. Examples: shoulder and hip joints
Pivot	Joint in which a projection of one bone fits into a ringlike ligament of another, allowing one bone to rotate on its longitudinal axis. Examples: atlantoaxial joint and proximal radioulnar joint
Saddle	Joint in which each bone surface is saddle-shaped (concave on one axis and convex on the perpendicular axis). Examples: trapeziometa carpal and sternoclavicular joints
Condyloid (ellipsoid)	Biaxial diarthrosis in which an oval convex surface of one bone articulates with an elliptical depression of another. Examples: radiocarpal and metacarpophalangeal joints

sternoclavicular joint between the clavicle and sternum are saddle joints. The articular surface of each bone is shaped like a saddle—concave in one direction and convex in the other. These are biaxial joints. If you compare the range of motion of your thumb with that of your fingers, you can see that a saddle joint is more movable than a condyloid or hinge joint. This is the joint responsible for that hallmark of primate anatomy, the opposable thumb.

5. **Condyloid (ellipsoid) joints.** These joints exhibit an oval convex surface on one bone that fits into a similarly shaped depression on the next. The radiocarpal joint of the wrist and the metacarpophalangeal (MET-uh-CAR-po-fuh-LAN-jee-ul) joints at the bases of the fingers are examples. These are considered biaxial joints because they can move in two directions, for example up and down and side to side. To demonstrate, hold your hand with your palm facing you. Flex your index finger back and forth as if gesturing to someone, "come here," and then move the finger from side to side toward the thumb and away. This shows the biaxial motion of the condyloid joint.
6. **Ball-and-socket joints.** These occur at the shoulder and hip, where one bone has a smooth hemispherical head that fits within a cuplike depression on the other. The head of the humerus fits into the glenoid cavity of the scapula, and the head of the femur fits into the acetabulum of the os coxae. These are the only multiaxial joints of the skeleton.

In table 9.1 the joints are classified by structural criteria. Some joints are difficult to classify, however, because they have elements of more than one type. The jaw joint, for example, has some aspects of condyloid, hinge, and gliding joints for reasons that will be apparent later.

Movements of Synovial Joints

In physical therapy, kinesiology, and other medical and scientific fields, specific terms are used to describe the movements of diarthroses. You will need a command of these terms to understand the muscle actions in chapters 11 and 12. In the following discussion, many of them are grouped to describe opposite or contrasting movements.

FLEXION, EXTENSION, AND HYPEREXTENSION

Flexion (fig. 9.7*a, e, f, g*) is movement that decreases the angle of a joint, usually in a sagittal plane. Examples are bending the elbow or knee and bending the neck to look down at the floor. Bending at the

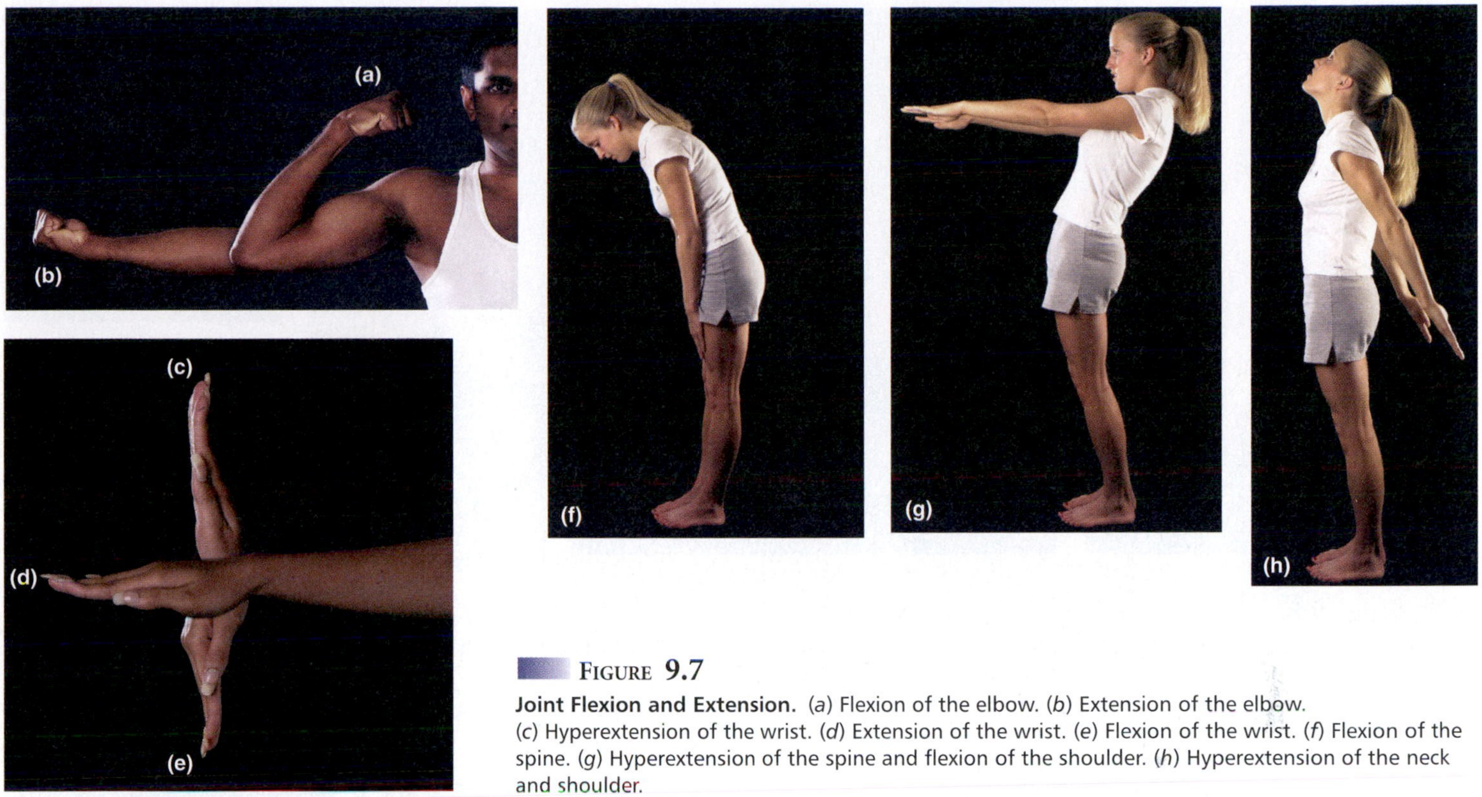

FIGURE 9.7

Joint Flexion and Extension. (*a*) Flexion of the elbow. (*b*) Extension of the elbow. (*c*) Hyperextension of the wrist. (*d*) Extension of the wrist. (*e*) Flexion of the wrist. (*f*) Flexion of the spine. (*g*) Hyperextension of the spine and flexion of the shoulder. (*h*) Hyperextension of the neck and shoulder.

waist, as if taking a bow, is flexion of the spine. Flexion of the shoulder consists of raising the arm from anatomical position in a sagittal plane, as if to point in front of you or toward the ceiling. Flexion of the hip entails raising the thigh, as in a high-stepping marching stance.

Extension (fig. 9.7*b, d*) is movement that straightens a joint and generally returns a body part to anatomical position—for example, straightening the elbow or knee, raising the head to look directly forward, straightening the waist, or moving the arm back to a position parallel to the trunk.

Hyperextension (fig. 9.7*c, g, h*) is the extension of a joint beyond 180°. For example, raising the back of your hand, as if admiring a new ring, hyperextends the wrist. If you look up toward the ceiling, you are hyperextending your neck. If you move your arm to a position posterior to the shoulder, you are hyperextending your shoulder.

THINK ABOUT IT!

Some synovial joints have articular surfaces or ligaments that prevent them from being hyperextended. Try hyperextending some of your synovial joints and list a few for which this is impossible.

ABDUCTION AND ADDUCTION

Abduction[12] (ab-DUC-shun) (fig. 9.8*a, c, d*) is movement of a body part away from the median plane—for example, raising the arm to one side of the body or standing spread-legged. To abduct the fingers is to spread them apart. **Adduction**[13] (ah-DUC-shun) (fig. 9.8*b, e*) is movement toward the median plane, returning an abducted body part to anatomical position. Some movements are open to alternative interpretations. Bending the head to one side or bending sideways at the waist may be regarded as abduction or lateral flexion.

ELEVATION AND DEPRESSION

Elevation (fig. 9.9*a*) is movement that raises a bone vertically. The mandible is elevated when biting off a piece of food, and the clavicles and scapulae are elevated when shrugging the shoulders as if to gesture, "I don't know." The opposite of elevation is **depression**—lowering the mandible to open the mouth or lowering the shoulders, for example (fig. 9.9*b*).

PROTRACTION AND RETRACTION

Protraction[14] is movement of a bone anteriorly (forward) on a horizontal plane, and **retraction**[15] is movement posteriorly (fig. 9.10*a, b*). Jutting the jaw outward, rounding the shoulders forward, or thrusting the pelvis forward are examples of protraction. The clavicles are retracted when standing at military attention. Most people have some degree of overbite and so must protract the mandible to make the incisors meet when taking a bite of fruit, for example. The mandible is then retracted to make the molars meet and grind food between them.

[12] *ab* = away + *duc* = to carry, lead

[13] *ad* = toward + *duc* = to carry, lead

[14] *pro* = forward + *trac* = pull, draw

[15] *re* = back + *tract* = pull, draw

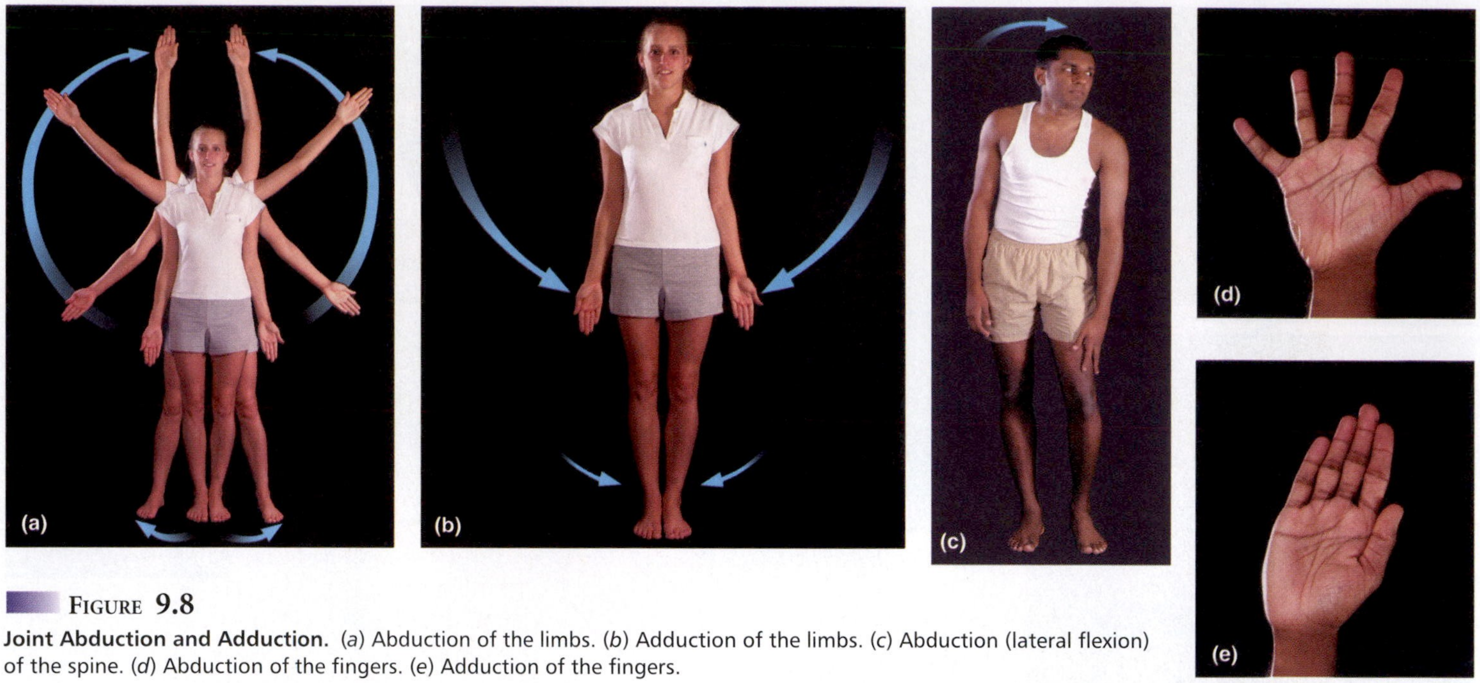

FIGURE 9.8

Joint Abduction and Adduction. (*a*) Abduction of the limbs. (*b*) Adduction of the limbs. (*c*) Abduction (lateral flexion) of the spine. (*d*) Abduction of the fingers. (*e*) Adduction of the fingers.

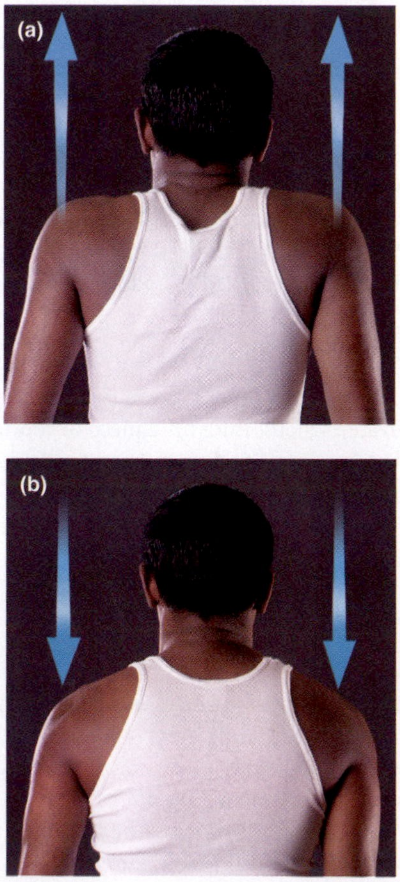

FIGURE 9.9

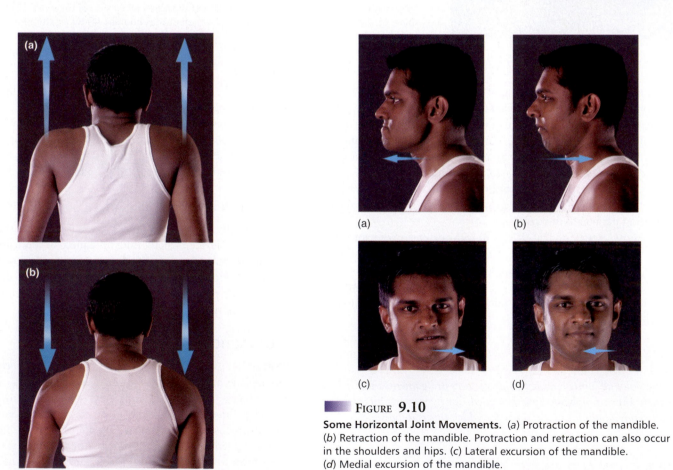

Elevation and Depression. (*a*) Elevation of the shoulders. (*b*) Depression of the shoulders.

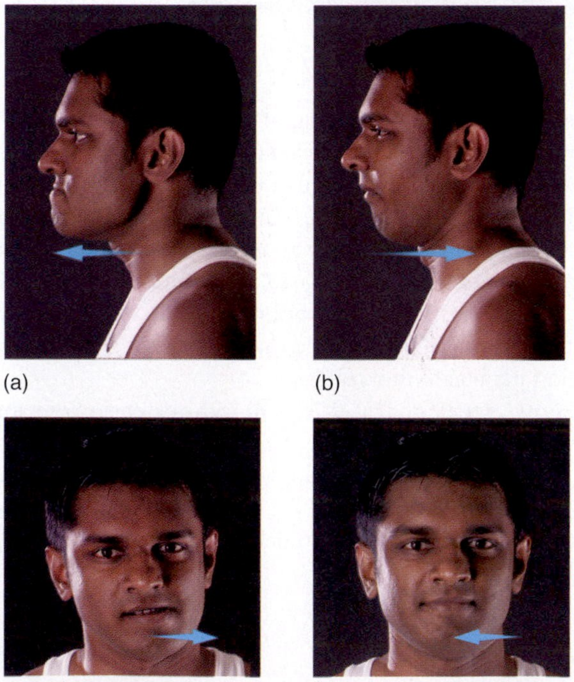

(a) (b) (c) (d)

FIGURE 9.10

Some Horizontal Joint Movements. (*a*) Protraction of the mandible. (*b*) Retraction of the mandible. Protraction and retraction can also occur in the shoulders and hips. (*c*) Lateral excursion of the mandible. (*d*) Medial excursion of the mandible.

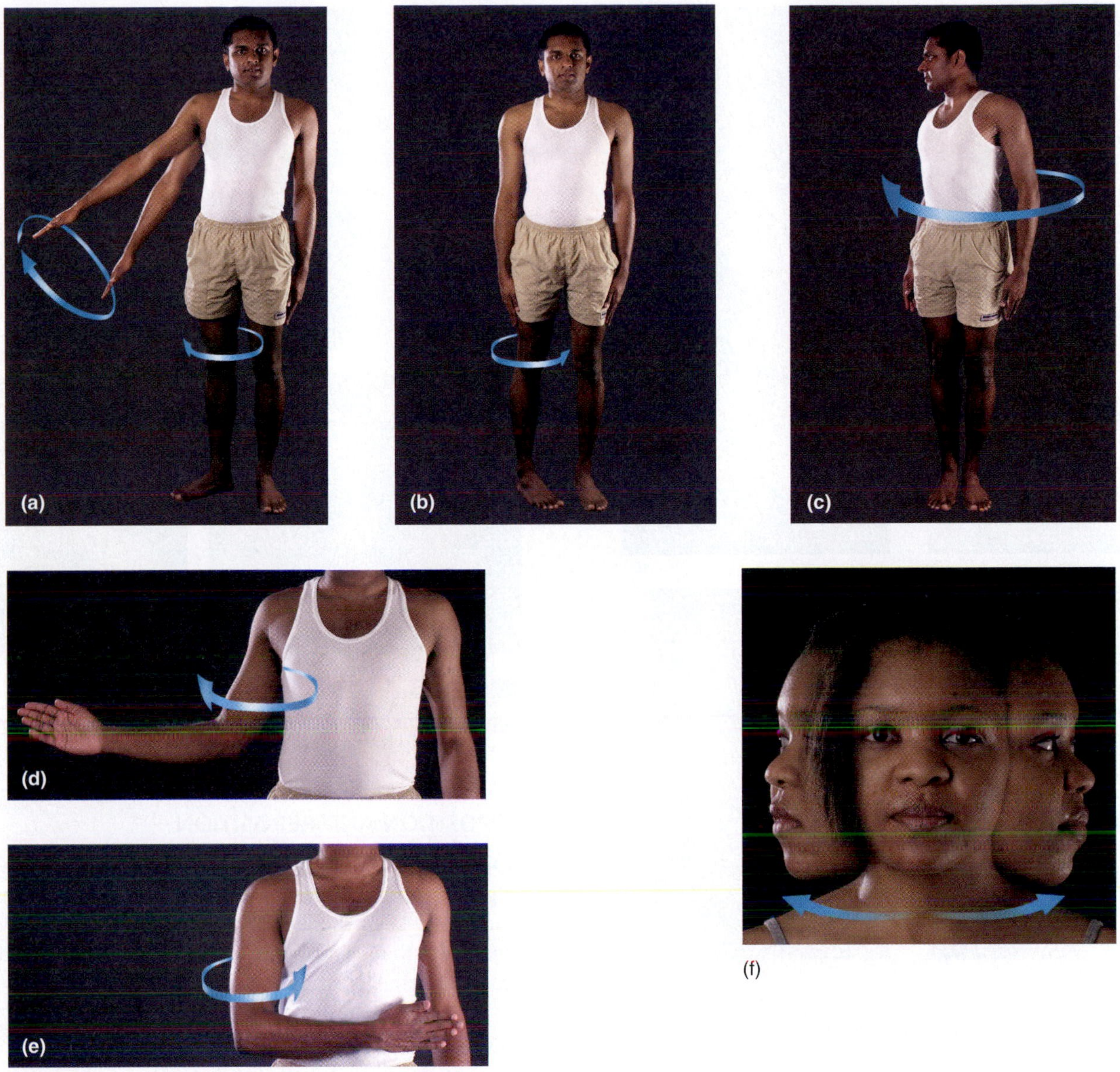

FIGURE 9.11

Circumduction and Rotation. (*a*) Circumduction of the forelimb and lateral rotation of the femur. (*b*) Medial rotation of the femur. (*c*) Right rotation of the trunk. (*d*) Lateral rotation of the humerus. (*e*) Medial rotation of the humerus. (*f*) Right and left rotation of the head.

LATERAL AND MEDIAL EXCURSION

Biting and chewing food require several movements of the jaw: up and down (elevation-depression), forward and back (protraction-retraction), and side-to-side grinding movements. The last of these are called **lateral excursion** (sideways movement to the right or left) and **medial excursion** (movement back to the midline) (fig. 9.10*c, d*).

CIRCUMDUCTION

Circumduction[16] (fig. 9.11*a*) is movement in which one end of an appendage remains relatively stationary while the other end makes a circular motion. The appendage as a whole thus describes a conical space. For example, if an artist standing at an easel reaches out and draws a circle on the canvas, the shoulder remains stationary while the hand makes a circle. The extremity as a whole thus exhibits circumduction. A baseball player winding up for the pitch circumducts the arm in a more extreme "windmill" fashion. Circumduction is actually a sequence of flexion, abduction, extension, and adduction.

ROTATION

Rotation is a movement in which a bone turns on its longitudinal axis. Figure 9.11*b, d,* and *e* show the limb movements that occur in **lateral (external)** and **medial (internal) rotation** of the femur and humerus. Twisting at the waist and turning the head from side to side are called **right** and **left rotation** (fig. 9.11*c, f*).

[16]*circum* = around + *duc* = to carry, lead

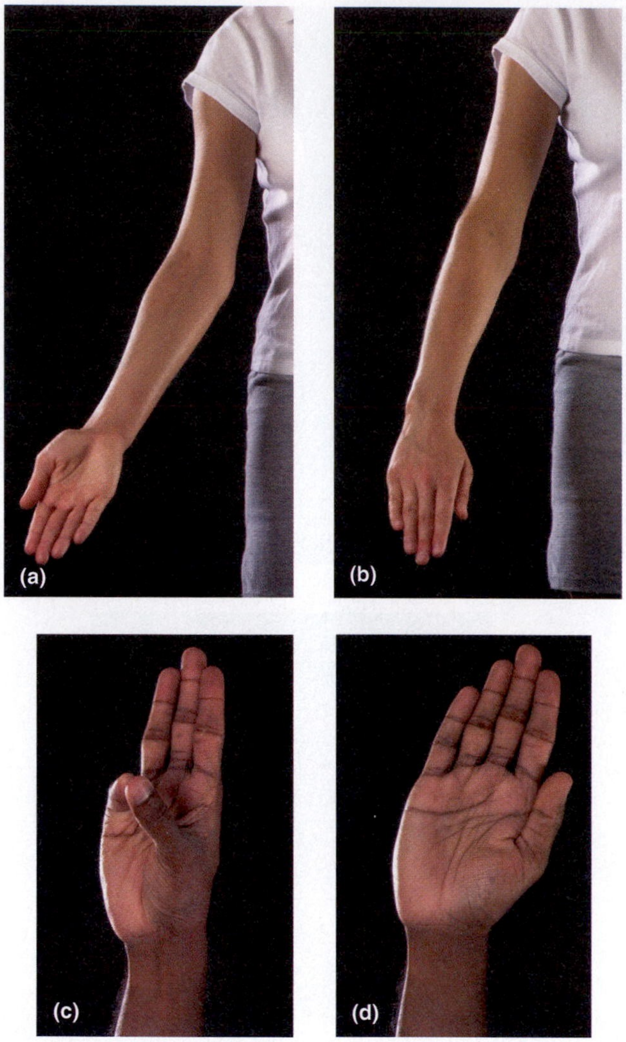

FIGURE 9.12

Joint Movements of the Forearm and Thumb. (*a*) Supination of the forearm. (*b*) Pronation of the forearm. (*c*) Opposition of the thumb. (*d*) Reposition of the thumb.

SUPINATION AND PRONATION

These movements are limited to the forearm. **Supination**[17] (SOO-pih-NAY-shun) (fig. 9.12*a*) is rotation of the forearm so that the palm faces forward or upward; in anatomical position, the forearm is supine. **Pronation**[18] (fig. 9.12*b*) is rotation of the forearm so that the palm faces toward the rear or downward. As an aid to memory, think of it this way: you are *prone* to stand in the most comfortable position, which is with the palm *pronated.* If you were holding a bowl of *soup* in your hand, your forearm would have to be *supinated.* These movements are achieved with muscles discussed in chapter 12. The *supinator* muscle is the most powerful, and supination is the sort of movement you would usually make with the right hand to turn a doorknob clockwise or drive a screw into a piece of wood.

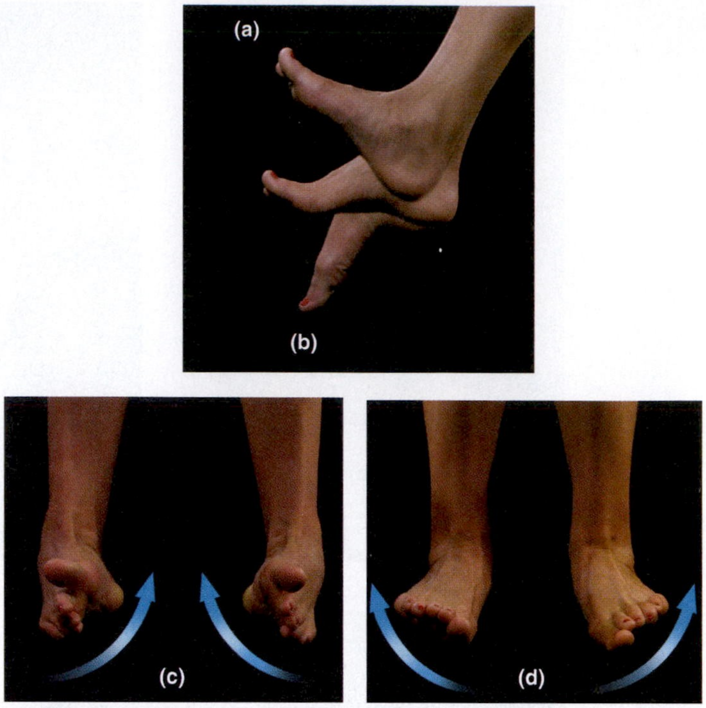

FIGURE 9.13

Joint Movement of the Foot. (*a*) Dorsiflexion. (*b*) Plantar flexion. (*c*) Inversion. (*d*) Eversion.

OPPOSITION AND REPOSITION

Opposition[19] is movement of the thumb to approach or touch the fingertips, and **reposition**[20] is its movement back to anatomical position, parallel to the index finger (fig. 9.12*c, d*). Opposition is the movement that enables the hand to grasp objects and is the single most important hand function.

DORSIFLEXION AND PLANTAR FLEXION

Special names are given to vertical movements at the ankle. **Dorsiflexion** (DOR-sih-FLEC-shun) is a movement in which the toes are raised (as one might do to apply toenail polish) (fig. 9.13*a*). Dorsiflexion occurs in each step you take as your foot comes forward. It prevents your toes from scraping on the ground and results in a "heel strike" when that foot touches down in front of you. **Plantar flexion** is a movement that points the toes downward, as in standing on tiptoe or pressing the gas pedal of a car (fig. 9.13*b*). This motion also produces the "toe-off" in each step you take, as the heel of the foot behind you lifts off the ground.

INVERSION AND EVERSION

These terms also apply exclusively to the feet. **Inversion**[21] is a movement that lifts the medial border of the foot so the soles turn medially and face each other; **eversion**[22] is a movement that lifts the

---

[17]*supin* = to lay back
[18]*pron* = to bend forward

[19]*op* = against + *posit* = to place
[20]*re* = back + *posit* = to place
[21]*in* = inward + *version* = turning
[22]*e* = outward + *version* = turning

lateral border of the foot so the soles face away from each other (fig. 9.13*c, d*). While dorsiflexion and plantar flexion are movements at the tibiotalar joint, inversion and eversion result from gliding movements of the tarsal bones. Inversion and eversion are common in fast sports such as tennis and football and often result in ankle sprains. These terms also refer to congenital deformities of the feet, which are often corrected by orthopedic shoes or braces.

Range of Motion

We can see from the movements just described that the **range of motion (ROM)** of a joint varies greatly from one type to another. ROM obviously affects a person's functional independence and quality of life. It is also an important consideration in training for athletics or dance, in clinical diagnosis, and in monitoring the progress of rehabilitation. Several factors affect the ROM and stability of a joint:

- **Structure and action of the muscles.** The two most important factors in stabilizing a joint are tendons and muscle tone (a state of partial contraction of a "resting" muscle). Tendons, ligaments, and muscles have sensory nerve endings called *proprioceptors* (PRO-pree-oh-SEP-turs) that continually monitor joint angle and muscle tension. Upon receiving this information, the spinal cord sends nerve signals back to the muscles to increase or decrease their state of contraction and adjust the position of the joint and tautness of the tendons.
- **Structure of the articular surfaces of the bones.** You cannot hyperextend your elbow, for example, because the olecranon of the ulna fits into the olecranon fossa of the humerus and prevents further movement in that direction.
- **Strength and tautness of ligaments, tendons, and the joint capsule.** You cannot hyperextend your knee because its *cruciate ligament* is pulled tight when the knee is extended, and thus prevents further motion. Gymnasts and acrobats increase the ROM of their joints by gradually stretching their ligaments during training. "Double-jointed" people have unusually large ROMs at some joints, not because the joint is actually double or fundamentally different from normal in its anatomy, but because the ligaments are unusually long and slack.

Before You Go On

Answer the following questions to test your understanding of the preceding section:

7. What are the two components of a joint capsule? What is the function of each?
8. Give at least one example each of a monaxial, biaxial, and multiaxial joint, and explain the reason for its classification.
9. Name the joints that would be involved if you reached directly overhead and screwed a light bulb into a ceiling fixture. Describe the joint actions that would occur.

Anatomy of Selected Synovial Joints

Objectives

When you have completed this section, you should be able to

- identify the major anatomical features of the jaw, shoulder, elbow, hip, knee, and ankle joints; and
- explain how the anatomical differences between these joints are related to differences in function.

We now examine the gross anatomy of certain diarthroses. It is beyond the scope of this book to discuss all of them, but the ones selected here most often require medical attention and many of them have a strong bearing on athletic performance.

The Jaw Joint

The **temporomandibular joint** (TMJ) is the articulation of the condyle of the mandible with the mandibular fossa of the temporal bone (fig. 9.14). You can feel its action by pressing your fingertips against the jaw immediately anterior to the ear while opening and closing your mouth. This joint combines elements of condyloid, hinge, and gliding joints. It functions in a hingelike fashion when the mandible is elevated and depressed, it glides slightly forward when the jaw is protracted to take a bite, and it glides from side to side to grind food between the molars.

The synovial cavity of the TMJ is divided into superior and inferior chambers by an articular disc, which permits lateral and medial excursion of the mandible. Two ligaments support the joint. The **temporomandibular ligament** on the lateral side prevents posterior displacement of the mandible. If the jaw receives a hard blow, this ligament normally prevents the condyloid process from being driven upward and fracturing the base of the skull. The **sphenomandibular ligament** on the medial side of the joint extends from the sphenoid bone to the ramus of the mandible. A *stylomandibular ligament* extends from the styloid process to the angle of the mandible but is not part of the TMJ proper.

INSIGHT 9.2 CLINICAL APPLICATION

TMJ Syndrome

TMJ syndrome has received medical recognition only recently, although it may affect as many as 75 million Americans. It can cause moderate intermittent facial pain, clicking sounds in the jaw, limitation of jaw movement, and in some people, more serious symptoms—severe headaches, vertigo (dizziness), tinnitus (ringing in the ears), and pain radiating from the jaw down the neck, shoulders, and back. It seems to be caused by a combination of psychological tension and malocclusion (misalignment of the teeth). Treatment may involve psychological management, physical therapy, analgesic and anti-inflammatory drugs, and sometimes corrective dental appliances to align the teeth properly.

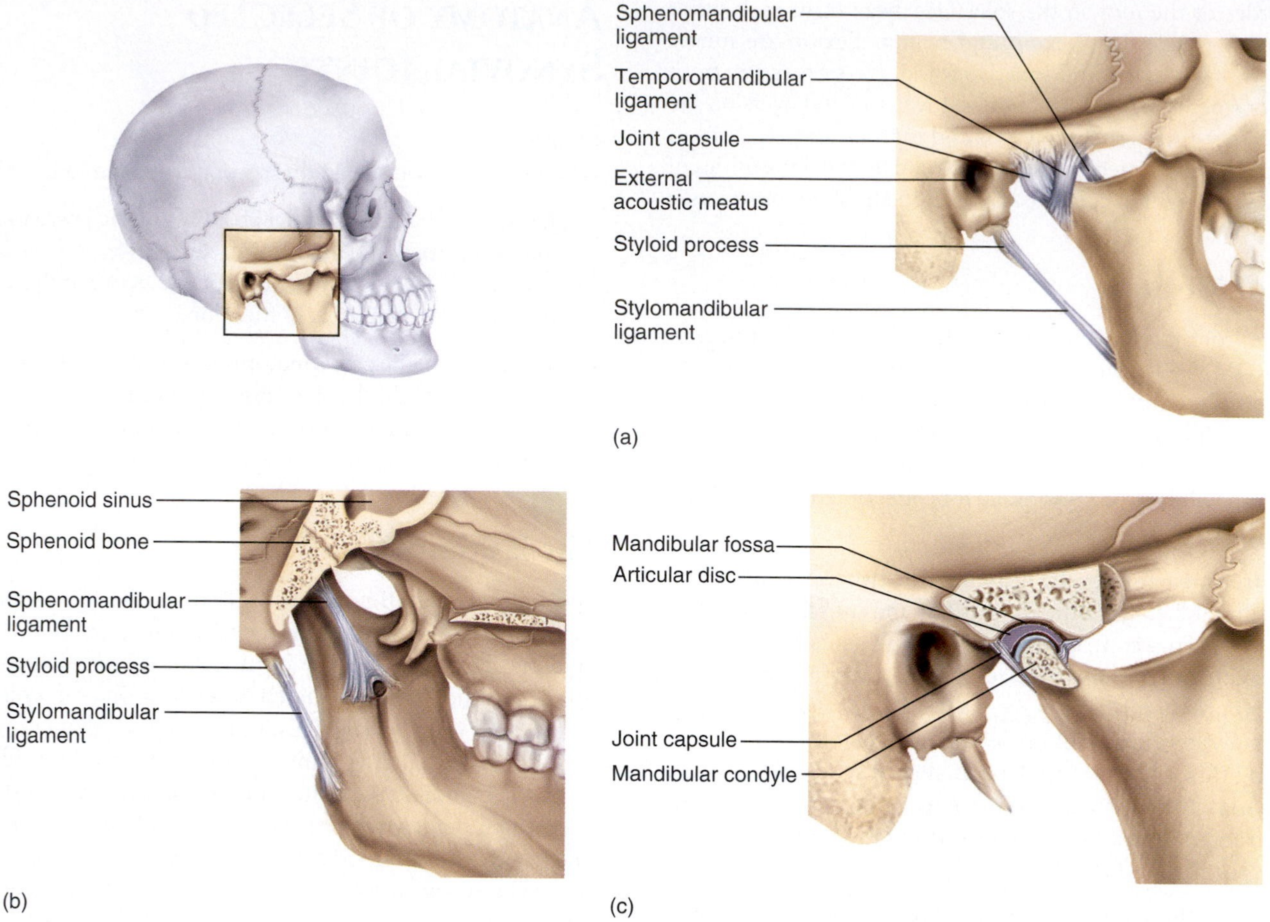
Sphenomandibular ligament
Temporomandibular ligament
Joint capsule
External acoustic meatus
Styloid process
Stylomandibular ligament
(a)
Sphenoid sinus
Sphenoid bone
Sphenomandibular ligament
Styloid process
Stylomandibular ligament
(b)
Mandibular fossa
Articular disc
Joint capsule
Mandibular condyle
(c)

FIGURE 9.14

The Temporomandibular Joint (TMJ). (*a*) Lateral view. (*b*) Medial view. (*c*) Sagittal section through the joint cavity.

A deep yawn or other strenuous depression of the mandible can dislocate the TMJ by making the condyle pop out of the fossa and slip forward. The joint can be relocated by pressing down on the molars while pushing the jaw backward.

The Shoulder Joint

The shoulder joint is called the **humeroscapular (glenohumeral) joint** (fig. 9.15). It is the most freely movable joint of the body but also one of the most commonly injured. The shallowness of the glenoid cavity and looseness of the joint capsule sacrifice joint stability for freedom of movement. The cavity, however, has a ring of fibrocartilage called the **glenoid labrum**[23] around its margin, which makes it somewhat deeper than it appears on a dried skeleton.

Five principal ligaments support this joint. The **coracohumeral ligament** extends from the coracoid process of the scapula to the greater tubercle of the humerus, and the **transverse humeral ligament,** which extends from the greater to the lesser tubercle of the humerus, creating a tunnel through which a tendon of the biceps brachii passes. The other three ligaments, called **glenohumeral ligaments,** are relatively weak and sometimes absent.

The tendon of the biceps brachii muscle is the most important stabilizer of the shoulder. It originates on the margin of the glenoid cavity, passes through the joint capsule, and emerges into the intertubercular groove of the humerus, where it is held by the transverse humeral ligament. Inferior to this groove, it merges into the biceps brachii. Thus, the tendon functions as a taut, adjustable strap that holds the humerus against the glenoid cavity.

In addition to the biceps brachii, four muscles important in stabilizing the humeroscapular joint are the *subscapularis, supraspinatus, infraspinatus,* and *teres minor.* The tendons of these four muscles form the *rotator cuff,* which is fused to the joint capsule on all sides except ventrally. The rotator cuff is discussed more fully in chapter 12.

Shoulder dislocations are very painful and can result in permanent damage. The most common dislocation is downward displacement of the humerus, because (1) the rotator cuff protects the joint in all directions except ventrally, and (2) the joint is protected from above by the coracoid process, acromion process, and clavicle. Dislocations most often occur when the arm is abducted and then receives a blow from above—for example, when the outstretched

[23]*labrum* = lip

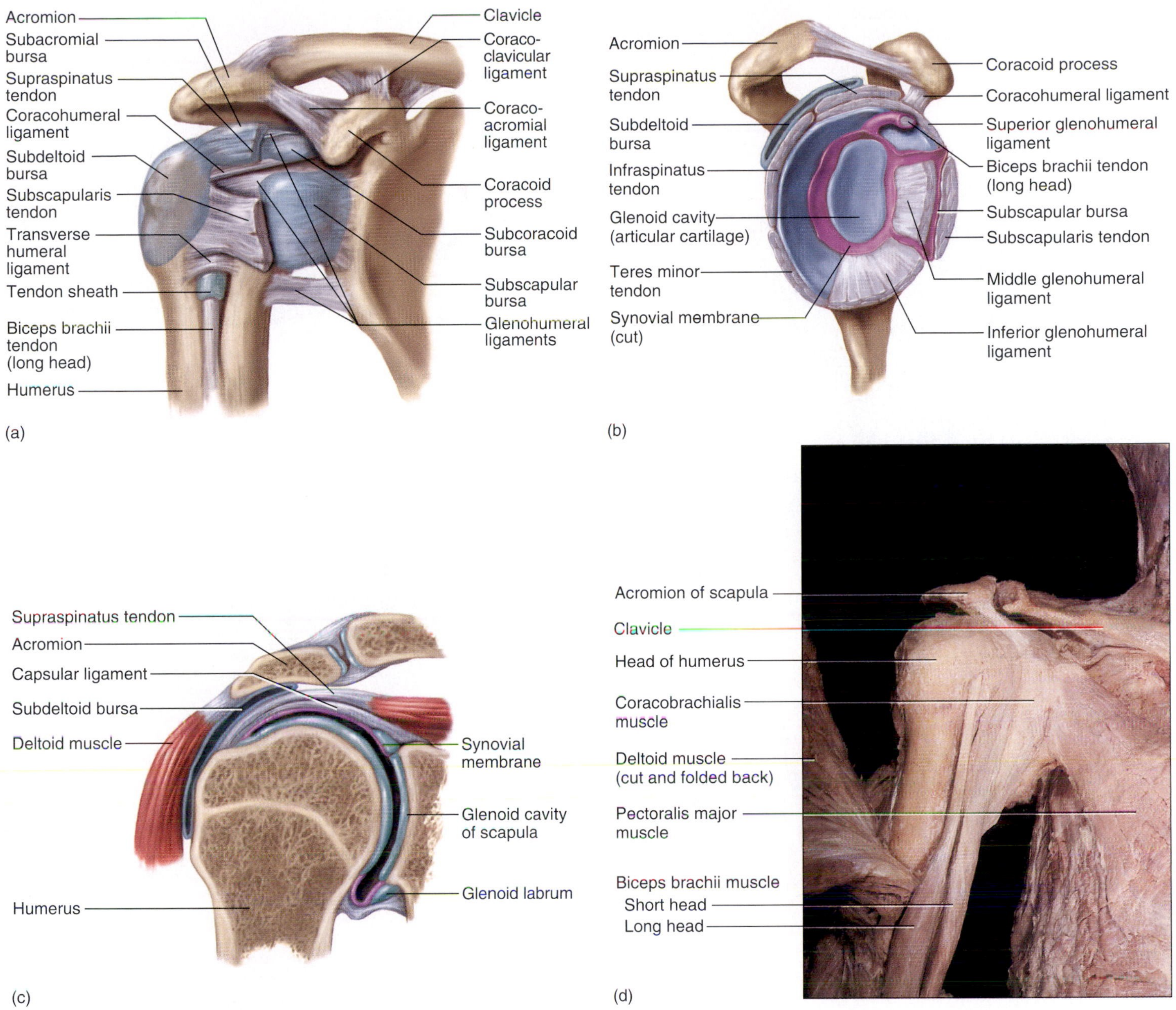

FIGURE 9.15

The Shoulder (humeroscapular) Joint. (*a*) Anterior view. (*b*) Lateral view of the glenoid cavity and labrum with the humerus removed. (*c*) Frontal section of the right shoulder joint, anterior view. (*d*) Anterior dissection of the joint.

arm is struck by heavy objects falling off a shelf. They also occur in children who are jerked off the ground by one arm or forced to follow by a hard tug on the arm. Children are especially prone to such injury not only because of the inherent stress caused by such abuse, but also because a child's shoulder is not fully ossified and the rotator cuff is not strong enough to withstand such stress. Because this joint is so easily dislocated, you should never attempt to move an immobilized person by pulling on his or her arm.

Four bursae are associated with the shoulder joint. Their names describe their locations—the **subdeltoid, subacromial, subcoracoid,** and **subscapular bursae** (fig. 9.15).

The Elbow Joint

The elbow is a hinge joint composed of two articulations—the **humeroulnar joint,** where the trochlea of the humerus joins the trochlear notch of the ulna, and the **humeroradial joint,** where the capitulum of the humerus meets the head of the radius (fig. 9.16). Both are enclosed in a single joint capsule. On the posterior side of the elbow, there is a prominent **olecranon bursa** to ease the movement of tendons over the elbow. Side-to-side motions of the elbow joint are restricted by a pair of ligaments, the **radial (lateral) collateral ligament** and **ulnar (medial) collateral ligament.**

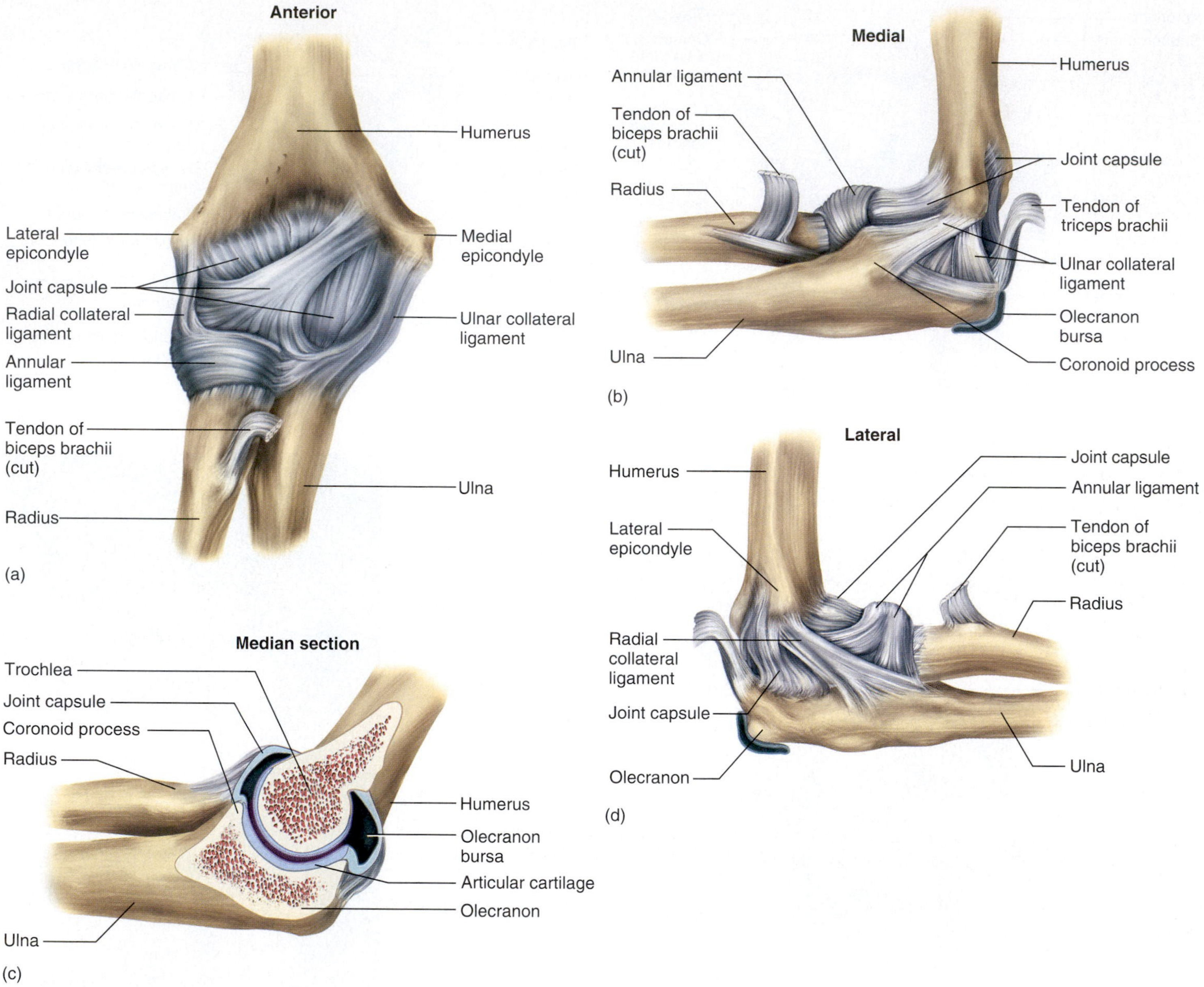
Anterior
Humerus
Lateral epicondyle
Medial epicondyle
Joint capsule
Radial collateral ligament
Ulnar collateral ligament
Annular ligament
Tendon of biceps brachii (cut)
Ulna
Radius
(a)
Medial
Humerus
Annular ligament
Tendon of biceps brachii (cut)
Joint capsule
Radius
Tendon of triceps brachii
Ulnar collateral ligament
Olecranon bursa
Ulna
Coronoid process
(b)
Median section
Trochlea
Joint capsule
Coronoid process
Radius
Humerus
Olecranon bursa
Articular cartilage
Olecranon
Ulna
(c)
Lateral
Joint capsule
Humerus
Annular ligament
Lateral epicondyle
Tendon of biceps brachii (cut)
Radius
Radial collateral ligament
Joint capsule
Olecranon
Ulna
(d)

FIGURE 9.16
The Elbow Joint. (*a*) Anterior view. (*b*) Medial view. (*c*) Median section. (*d*) Lateral view.

Another joint occurs in the elbow region, the **proximal radioulnar joint,** but it is not involved in the hinge. At this joint, the disclike head of the radius fits into the radial notch of the ulna and is held in place by the **annular ligament,** which encircles the head of the radius and attaches at each end to the ulna. The radial head rotates like a wheel against the ulna as the forearm is rotated.

The Hip Joint

The **coxal** (hip) **joint** is the point where the head of the femur inserts into the acetabulum of the os coxae (fig. 9.18). Because the coxal joints bear much of the body's weight, they have deep sockets and are much more stable than the shoulder joint. The depth of the socket is somewhat greater than you see on dried bones because a horseshoe-shaped ring of fibrocartilage, the **acetabular labrum,** is attached to its rim. Dislocations of the hip are rare, but some infants suffer congenital dislocations because the acetabulum is not deep enough to hold the head of the femur in place. This condition can be treated by placing the infant in traction until the acetabulum develops enough strength to support the body's weight.

THINK ABOUT IT!

Where else in the body is there a structure similar to the acetabular labrum? What do those two locations have in common?

INSIGHT 9.3 CLINICAL APPLICATION

Pulled Elbow

The immature skeletons of children and adolescents are especially vulnerable to injury. Pulled elbow (dislocation of the radius), a common injury in preschool children (especially girls), typically occurs when an adult lifts or jerks a child up by one arm when the arm was pronated, as in lifting a child into a high chair or shopping cart (fig. 9.17). This tears the annular ligament from the head of the radius, and the radius pulls partially or entirely out of the ligament. The proximal part of the torn ligament is then painfully pinched between the radial head and the capitulum of the humerus. Radial dislocation is treated by supinating the forearm with the elbow flexed and then putting the arm in a sling for about 2 weeks—time enough for the annular ligament to heal.

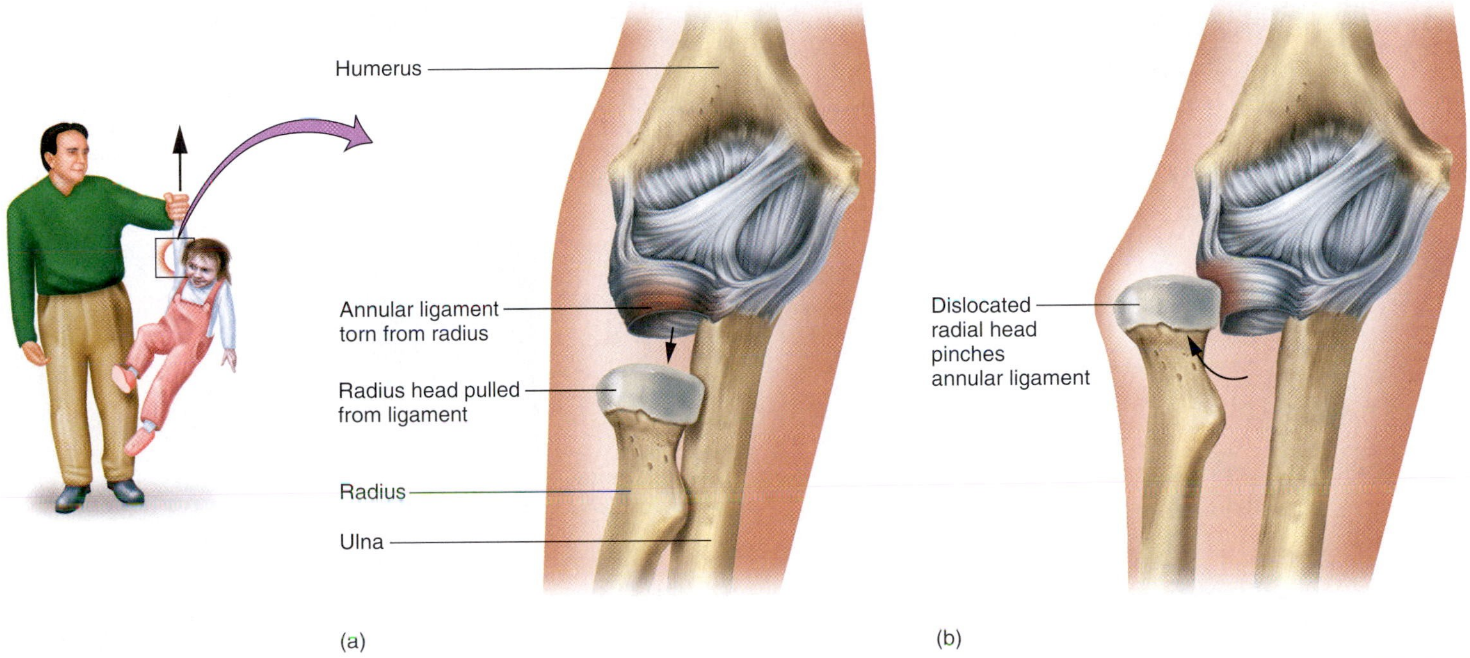
Humerus
Annular ligament torn from radius
Radius head pulled from ligament
Radius
Ulna
(a)
Dislocated radial head pinches annular ligament
(b)

Figure 9.17
Pulled Elbow. Lifting a child by the arm can dislocate the radius. (*a*) The annular ligament tears and the radial head is pulled from the ligament. (*b*) Muscle contraction pulls the radius upward. The head of the radius produces a lump on the lateral side of the elbow and may painfully pinch the annular ligament.

Ligaments that support the coxal joint include the **iliofemoral** (ILL-ee-oh-FEM-oh-rul) and **pubofemoral** (PYU-bo-FEM-or-ul) **ligaments** on the anterior side and the **ischiofemoral** (ISS-kee-oh-FEM-or-ul) **ligament** on the posterior side. The name of each ligament refers to the bones to which it attaches—the femur and the ilium, pubis, or ischium. When you stand up, these ligaments become twisted and pull the head of the femur tightly into the acetabulum. The head of the femur has a conspicuous pit called the *fovea capitis.* The **round ligament,** or **ligamentum teres**[24] (TERR-eez), arises here and attaches to the lower margin of the acetabulum. This is a relatively slack ligament, so it is questionable whether it plays a significant role in holding the femur in its socket. It does, however, contain an artery that supplies blood to the head of the femur. A **transverse acetabular ligament** bridges a gap in the inferior margin of the acetabular labrum.

[24]*teres* = round

The Knee Joint

The **tibiofemoral** (knee) **joint** is the largest and most complex diarthrosis of the body (figs. 9.19 and 9.20). It is primarily a hinge joint, but when the knee is flexed it is also capable of slight rotation and lateral gliding. The patella and patellar ligament also form a gliding **patellofemoral joint** with the femur.

The joint capsule encloses only the lateral and posterior aspects of the knee joint, not the anterior. The anterior aspect is covered by the patellar ligament and the **lateral** and **medial patellar retinacula** (not illustrated). These are extensions of the tendon of the *quadriceps femoris* muscle, the large anterior muscle of the thigh. The knee is stabilized mainly by the quadriceps tendon in front and the tendon of the *semimembranosus* muscle on the rear of the thigh. Developing strength in these muscles therefore reduces the risk of knee injury.

The joint cavity contains two cartilages called the **lateral meniscus** and **medial meniscus,** joined by a **transverse ligament.**

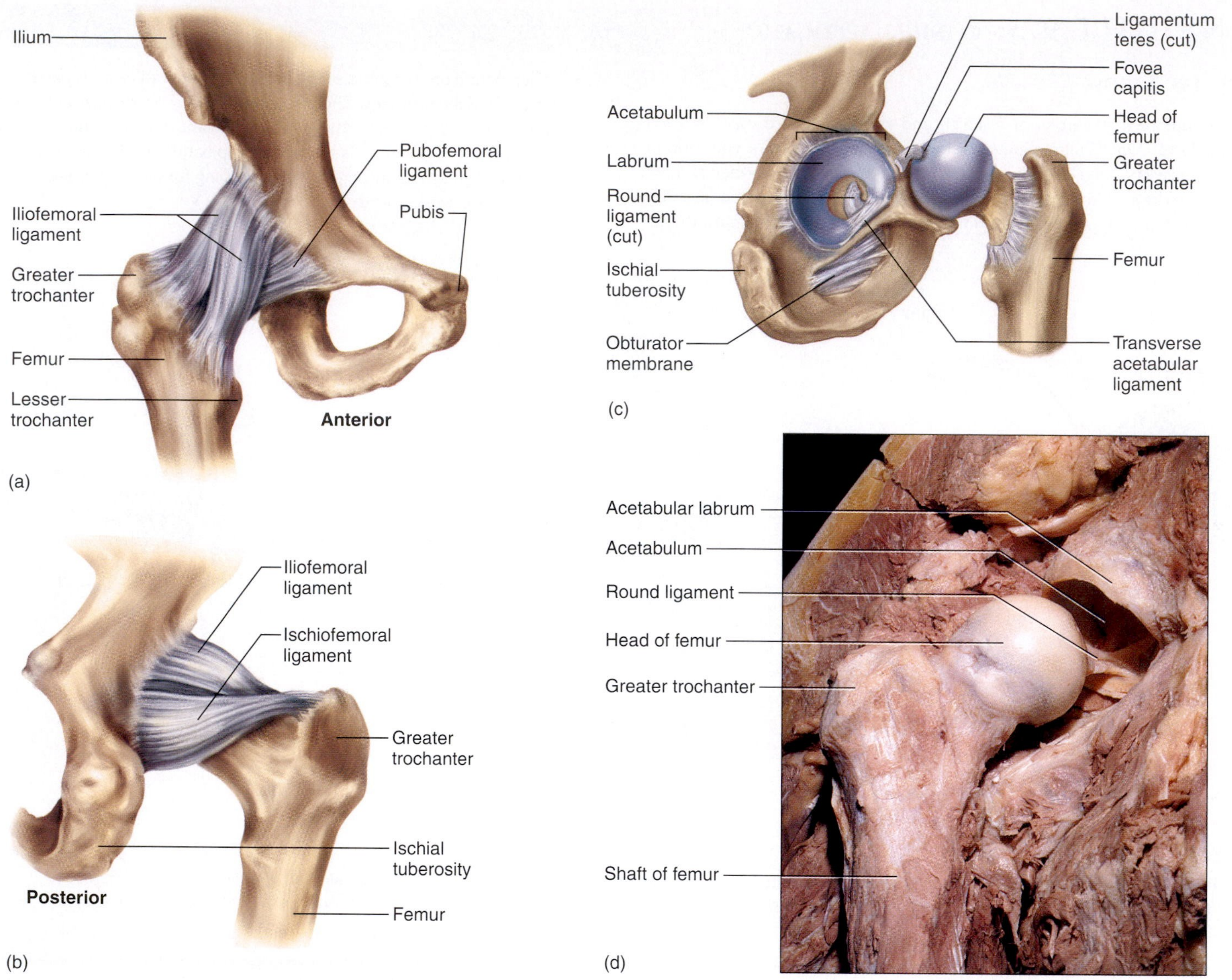

FIGURE 9.18

The Coxal (hip) Joint. (*a*) Anterior view. (*b*) Posterior view. (*c*) The acetabulum with the femoral head retracted. (*d*) Photograph of the right hip with the femoral head retracted, anterior view.

These menisci absorb the shock of the body weight jostling up and down on the knee and prevent the femur from rocking from side to side on the tibia. The posterior "pit" of the knee, the **popliteal** (pop-LIT-ee-ul) **region,** is supported by a complex array of **intracapsular ligaments** within the joint capsule and **extracapsular ligaments** external to it. The extracapsular ligaments are the **oblique popliteal ligament** (an extension of the semimembranosus tendon), **arcuate** (AR-cue-et) **popliteal ligament, fibular (lateral) collateral ligament,** and **tibial (medial) collateral ligament.** The two collateral ligaments prevent the knee from rotating when the joint is extended.

There are two intracapsular ligaments deep within the joint cavity. The synovial membrane folds around them, however, so that they are excluded from the fluid-filled synovial cavity. These ligaments cross each other in the form of an X; hence, they are called the **anterior cruciate**[25] (CROO-she-ate) **ligament (ACL)** and **posterior cruciate ligament (PCL).** These are named according to whether they attach to the anterior or posterior side of the tibia, not for their attachments to the femur. When the knee is extended, the ACL is pulled tight and prevents hyperextension. The PCL prevents the femur from sliding off the front of the tibia and prevents the tibia from being displaced backward.

An important aspect of human bipedalism is the ability to "lock" the knees and stand erect without tiring the extensor muscles of the leg. When the knee is extended to the fullest degree allowed by the ACL, the femur rotates medially on the tibia. This

[25]*cruci* = cross + *ate* = characterized by

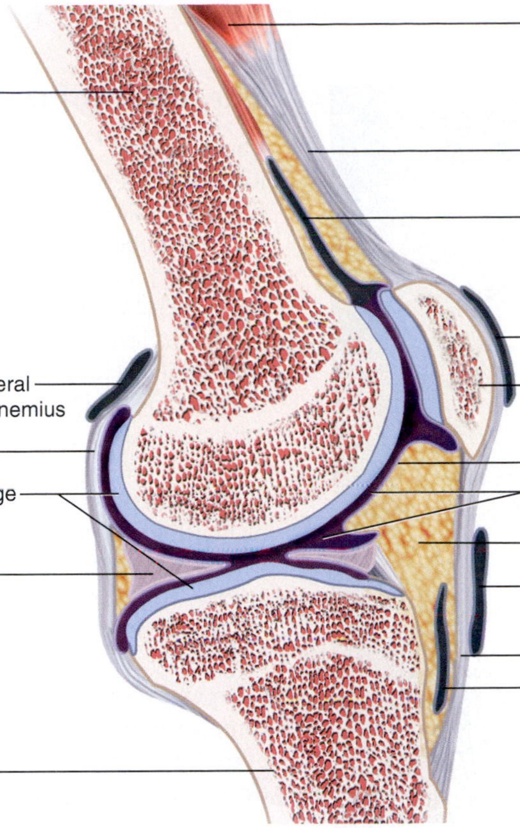

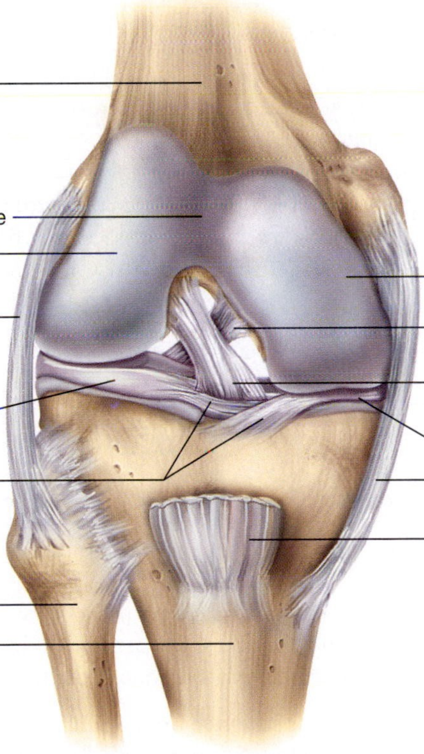

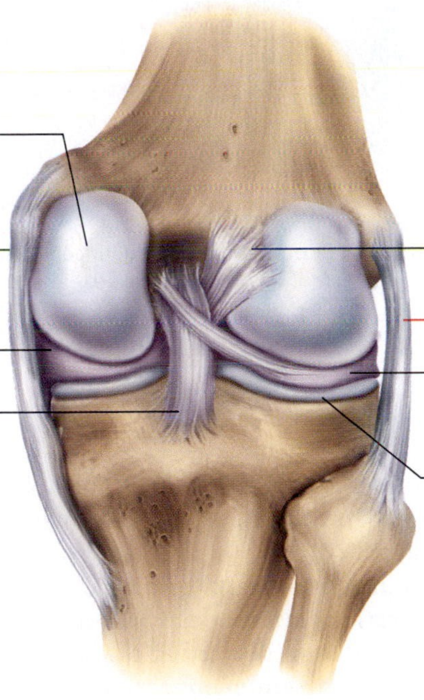

FIGURE 9.19

The Knee Joint. (*a*) Diagram of a midsagittal section. (*b*) Anterior view of structures in the joint cavity of the right knee. (*c*) Posterior view of the right knee.

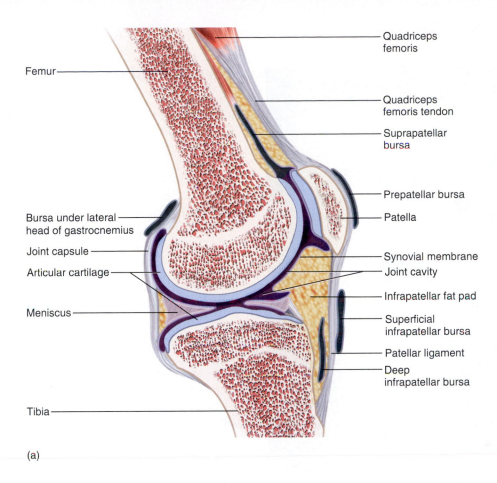

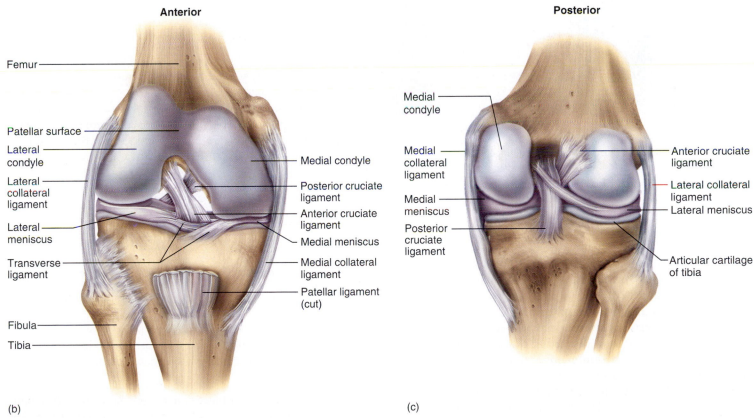

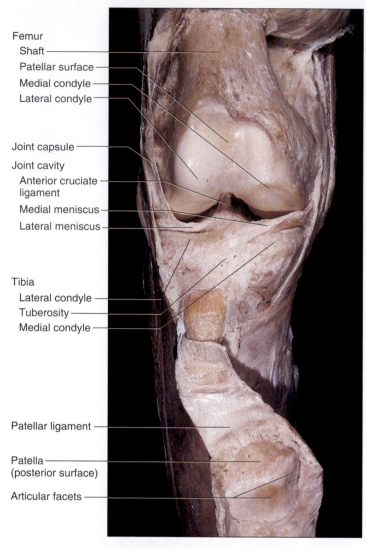

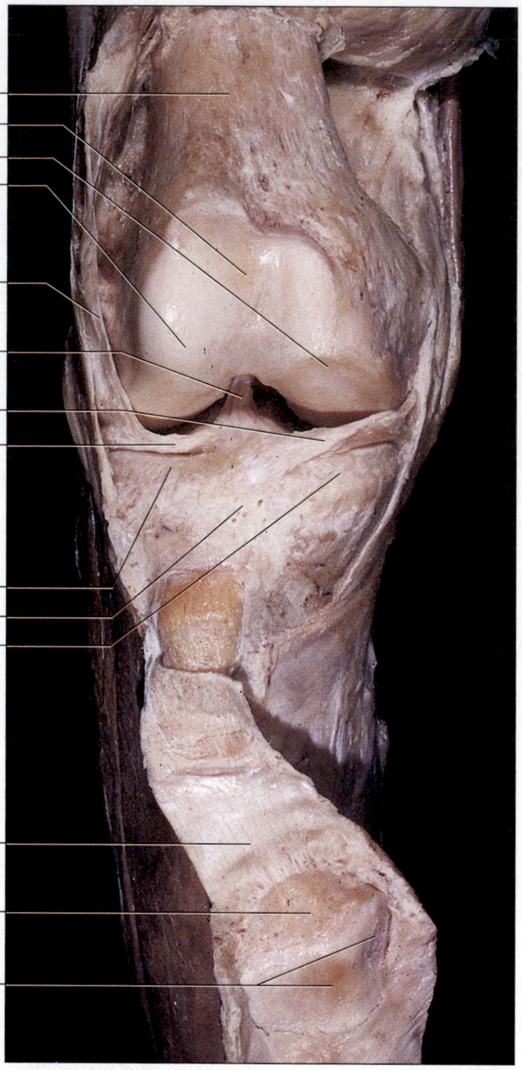
Femur
Shaft
Patellar surface
Medial condyle
Lateral condyle
Joint capsule
Joint cavity
Anterior cruciate ligament
Medial meniscus
Lateral meniscus
Tibia
Lateral condyle
Tuberosity
Medial condyle
Patellar ligament
Patella (posterior surface)
Articular facets

FIGURE 9.20

Anterior Dissection of the Knee Joint. The quadriceps tendon has been cut and folded (reflected) downward to expose the joint cavity and the posterior surface of the patella.

action locks the knee, and in this state all the major knee ligaments are twisted and taut. To unlock the knee, the *popliteus* muscle rotates the femur laterally, causing the ligaments to untwist.

The knee joint has at least 13 bursae. Four of these are anterior—the **superficial infrapatellar, suprapatellar, prepatellar,** and **deep infrapatellar.** Located in the popliteal region are the **popliteal bursa** and **semimembranosus bursa** (not illustrated). At least seven more bursae are found on the lateral and medial sides of the knee joint. From figure 9.19*a,* your knowledge of the relevant word elements *(infra-, supra-, pre-),* and the terms *superficial* and *deep,* you should be able to work out the reasoning behind most of these names and develop a system for remembering the locations of these bursae.

The Ankle Joint

The **talocrural**[26] (ankle) **joint** includes two articulations—a medial joint between the tibia and talus and a lateral joint between the fibula and talus, both enclosed in one joint capsule (fig. 9.22). The malleoli of the tibia and fibula overhang the talus on each side like a cap and prevent most side-to-side motion (fig. 9.23). The ankle therefore has a more restricted range of motion than the wrist.

The ligaments of the ankle include (1) **anterior** and **posterior tibiofibular ligaments,** which bind the tibia to the fibula; (2) a multipart **deltoid ligament,** which binds the tibia to the foot on the medial side; and (3) a multipart **lateral collateral ligament,** which binds the fibula to the foot on the lateral side. The **calcaneal (Achilles) tendon** extends from the calf muscles to the calcaneus. It plantarflexes the foot and limits dorsiflexion. Plantar flexion is limited by extensor tendons on the anterior side of the ankle and by the anterior part of the joint capsule.

Sprains (torn ligaments and tendons) occur especially often at the ankle, especially when the foot is suddenly inverted or everted to an excessive extent. These sprains are painful and usually accompanied by immediate swelling. They are best treated by immobilizing the joint and reducing swelling with an ice pack, but in extreme cases they may require a cast or surgery.

The synovial joints described in this section are summarized in table 9.2.

Before You Go On

Answer the following questions to test your understanding of the preceding section:

10. What keeps the mandibular condyle from slipping out of its fossa in a posterior direction?
11. Explain how the biceps tendon braces the shoulder joint.
12. What keeps the femur from slipping backward off the tibia?
13. What keeps the tibia from slipping sideways off the talus?

CLINICAL PERSPECTIVES

Objectives

When you have completed this section, you should be able to

- define *rheumatism* and describe the scope of the profession of rheumatology;
- define arthritis and describe its forms and causes;
- discuss the design and application of artificial joints; and
- identify several joint diseases other than arthritis.

Our quality of life depends so much on mobility, and mobility depends so much on proper functioning of the diarthroses, that joint dysfunctions are common medical complaints. **Rheumatism** is a

[26]*talo* = ankle + *crural* = pertaining to the leg

INSIGHT 9.4 CLINICAL APPLICATION

Knee Injuries and Arthroscopic Surgery

Although the knee can bear a lot of weight, it is highly vulnerable to rotational and horizontal stress, especially when the knee is flexed (as in skiing or running) and receives a blow from behind or from the lateral side (fig. 9.21). The most common injuries are to a meniscus or the anterior cruciate ligament (ACL). Knee injuries heal slowly because ligaments and tendons have a very scanty blood supply and cartilage has no blood vessels at all.

The diagnosis and surgical treatment of knee injuries has been greatly improved by *arthroscopy,* a procedure in which the interior of a joint is viewed with a pencil-thin instrument, the *arthroscope,* inserted through a small incision. The arthroscope has a light source, a lens, and fiber optics that allow a viewer to see into the cavity, take photographs or videotapes of the joint, and withdraw samples of synovial fluid. Saline is often introduced through one incision to expand the joint and provide a clearer view of its structures. If surgery is required, additional small incisions can be made for the surgical instruments and the procedures can be observed through the arthroscope or on a monitor. Arthroscopic surgery produces much less tissue damage than conventional surgery and enables patients to recover more quickly.

Orthopedic surgeons now often replace a damaged ACL with a graft from the patellar ligament. The surgeon "harvests" a strip from the middle one-third of the patient's patellar ligament, drills a hole into the femur and tibia within the joint cavity, threads the ligament through the holes, and fastens it with screws. The grafted ligament is more taut and "competent" than the damaged ACL. It becomes ingrown with blood vessels and serves as a substrate for the deposition of more collagen, which further strengthens it in time. Following arthroscopic ACL reconstruction, a patient typically must use crutches for 7 to 10 days and undergo supervised physical therapy for 6 to 10 weeks, followed by self-directed exercise therapy. Healing is completed in about 9 months.

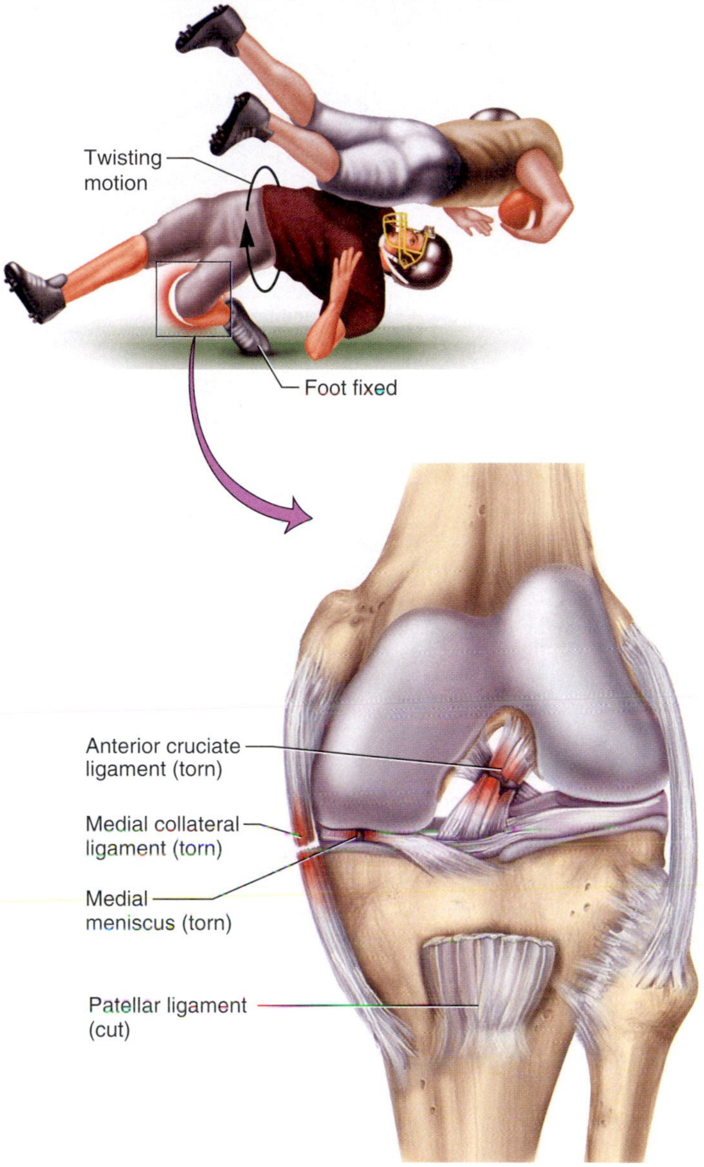
Twisting motion
Foot fixed
Anterior cruciate ligament (torn)
Medial collateral ligament (torn)
Medial meniscus (torn)
Patellar ligament (cut)

Figure 9.21
Knee Injuries.

broad term for any pain in the supportive and locomotory organs of the body, including bones, ligaments, tendons, and muscles. Physicians who deal with the study, diagnosis, and treatment of joint disorders are called **rheumatologists.**

Arthritis

The most common crippling disorder in the United States is **arthritis,**[27] a broad term that embraces more than a hundred diseases of largely obscure or unknown causes. In general, arthritis means inflammation of a joint. Nearly everyone develops arthritis to some degree after middle age.

The most common form of arthritis is **osteoarthritis (OA),** also called "wear-and-tear arthritis" because it is apparently a normal consequence of years of wear on the joints. As joints age, the articular cartilage softens and degenerates. As the cartilage becomes roughened by wear, joint movement may be accompanied by crunching or crackling sounds called *crepitus.* OA affects especially the fingers, intervertebral joints, hips, and knees. As the articular cartilage wears away, exposed bone tissue often develops spurs that grow into the joint cavity, restrict movement, and cause pain.

[27]*arthr* = joint + *itis* = inflammation

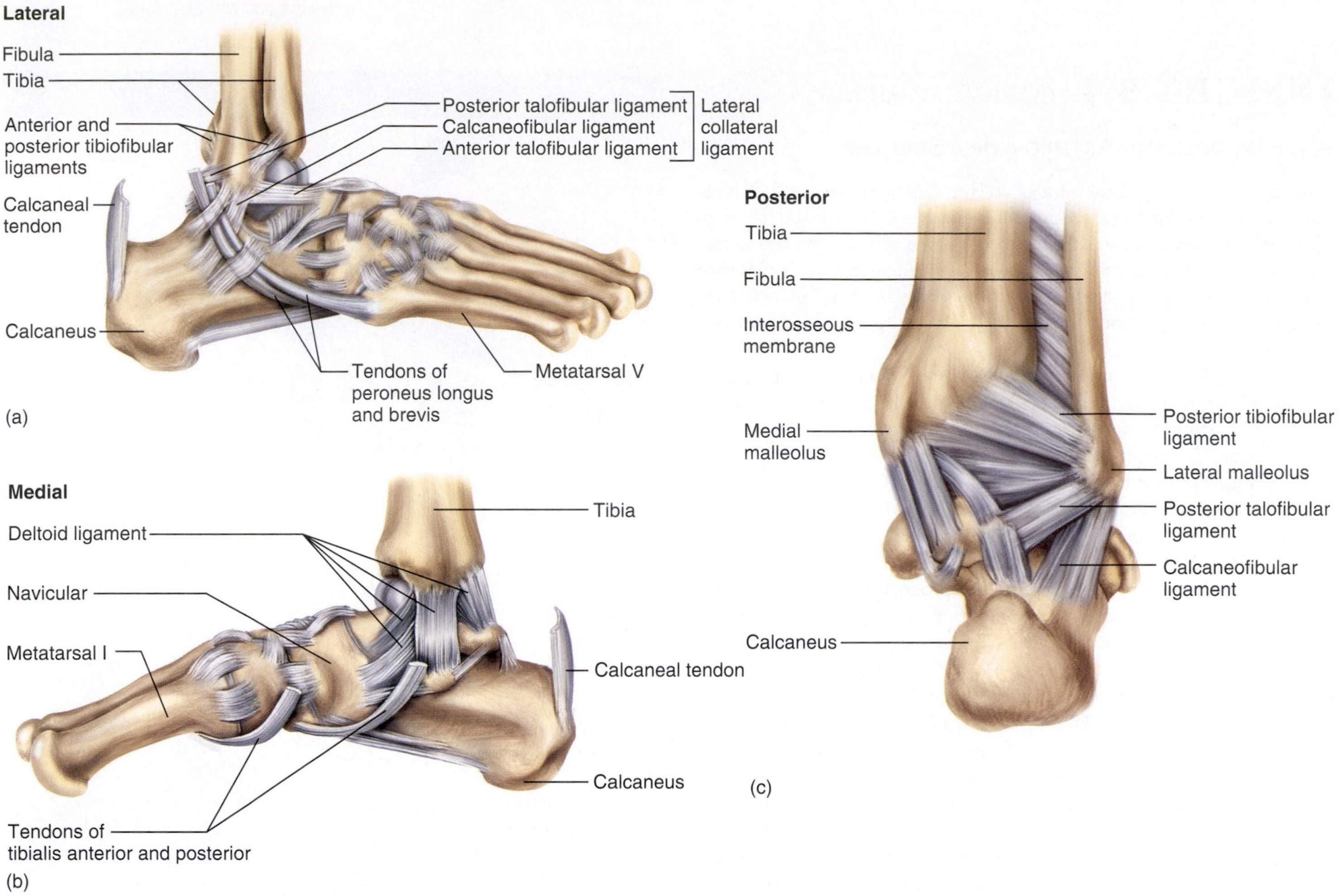
Lateral
Fibula
Tibia
Anterior and posterior tibiofibular ligaments
Calcaneal tendon
Calcaneus
Posterior talofibular ligament
Calcaneofibular ligament
Anterior talofibular ligament
Lateral collateral ligament
Tendons of peroneus longus and brevis
Metatarsal V
(a)
Medial
Deltoid ligament
Navicular
Metatarsal I
Tibia
Calcaneal tendon
Calcaneus
Tendons of tibialis anterior and posterior
(b)
Posterior
Tibia
Fibula
Interosseous membrane
Medial malleolus
Posterior tibiofibular ligament
Lateral malleolus
Posterior talofibular ligament
Calcaneofibular ligament
Calcaneus
(c)

FIGURE 9.22

The Talocrural (ankle) Joint and Ligaments of the Right Foot. (*a*) Lateral view. (*b*) Medial view. (*c*) Posterior view.

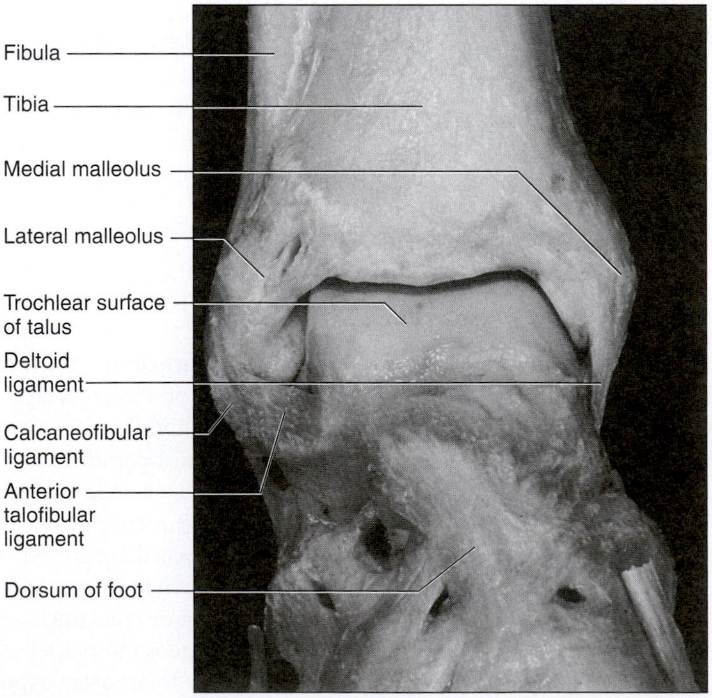
Fibula
Tibia
Medial malleolus
Lateral malleolus
Trochlear surface of talus
Deltoid ligament
Calcaneofibular ligament
Anterior talofibular ligament
Dorsum of foot

FIGURE 9.23

Anterior Dissection of the Talocrural Joint.

TABLE 9.2

Review of the Principal Diarthroses

Joint	Major Anatomical Features and Actions
Temporomandibular Joint (fig. 9.14)	*Type:* condyloid, hinge, and gliding *Movements:* elevation, depression, protraction, retraction, lateral and medial excursion *Articulation:* condyle of mandible, mandibular fossa of temporal bone *Ligaments:* temporomandibular, sphenomandibular *Cartilage:* articular disc
Humeroscapular Joint (fig. 9.15)	*Type:* ball-and-socket *Movements:* adduction, abduction, flexion, extension, circumduction, medial and lateral rotation *Articulation:* head of humerus, glenoid fossa of scapula *Ligaments:* coracohumeral, transverse humeral, three glenohumerals *Tendons:* rotator cuff (tendons of subscapularis, supraspinatus, infraspinatus, teres minor), tendon of biceps brachii *Bursae:* subdeltoid, subacromial, subcoracoid, subscapular *Cartilage:* glenoid labrum
Elbow (fig. 9.16)	*Type:* hinge and pivot *Movements:* flexion, extension, pronation, supination, rotation *Articulations:* humeroulnar—trochlea of humerus, trochlear notch of ulna; humeroradial—capitulum of humerus, head of radius; radioulnar—head of radius, radial notch of ulna *Ligaments:* radial collateral, ulnar collateral, annular *Bursa:* olecranon
Coxal Joint (fig. 9.18)	*Type:* ball-and-socket *Movements:* adduction, abduction, flexion, extension, circumduction, medial and lateral rotation *Articulations:* head of femur, acetabulum of os coxae *Ligaments:* iliofemoral, pubofemoral, ischiofemoral, ligamentum teres, transverse acetabular *Cartilage:* acetabular labrum
Knee Joint (fig. 9.19)	*Type:* primarily hinge *Movements:* flexion, extension, slight rotation *Articulations:* tibiofemoral, patellofemoral *Ligaments:* anterior—lateral patellar retinaculum, medial patellar retinaculum; popliteal intracapsular—anterior cruciate, posterior cruciate; popliteal extracapsular—oblique popliteal, arcuate popliteal, lateral collateral, medial collateral *Bursae:* anterior—superficial infrapatellar, suprapatellar, prepatellar, deep infrapatellar; popliteal—popliteal, semimembranosus; medial and lateral—seven other bursae not named in this chapter *Cartilages:* lateral meniscus, medial meniscus (connected by transverse ligament)
Ankle Joint (fig. 9.22)	*Type:* hinge *Movements:* dorsiflexion, plantar flexion, extension *Articulations:* tibia-talus, fibula-talus, tibia-fibula *Ligaments:* anterior and posterior tibiofibular, deltoid, lateral collateral *Tendon:* calcaneal (Achilles)

OA rarely occurs before age 40, but it affects about 85% of people older than 70. It usually does not cripple, but in severe cases it can immobilize the hip.

Rheumatoid arthritis (RA), which is far more severe, results from an autoimmune attack against the joint tissues. RA stems from an autoantibody called *rheumatoid factor.* Autoantibodies are misguided antibodies that attack the body's own tissues instead of limiting their attack to foreign matter. Rheumatoid factor attacks the synovial membranes. Inflammatory cells accumulate in the synovial fluid and produce enzymes that degrade the articular cartilage. The synovial membrane thickens and adheres to the articular cartilage, fluid accumulates in the joint capsule, and the capsule is invaded by fibrous connective tissue. As articular cartilage degenerates, the joint begins to ossify, and sometimes the bones become solidly fused and immobilized, a condition called **ankylosis**[28] (fig. 9.24). The disease tends to develop symmetrically—if the right wrist or hip develops RA, so does the left.

RA tends to flare up and subside (go into remission) periodically.[29] It affects women far more than men, and typically begins between the ages of 30 and 40. There is no cure, but joint damage can be slowed with hydrocortisone or other steroids. Because long-term

[28]*ankyl* = bent, crooked + *osis* = condition
[29]*rheumat* = tending to change

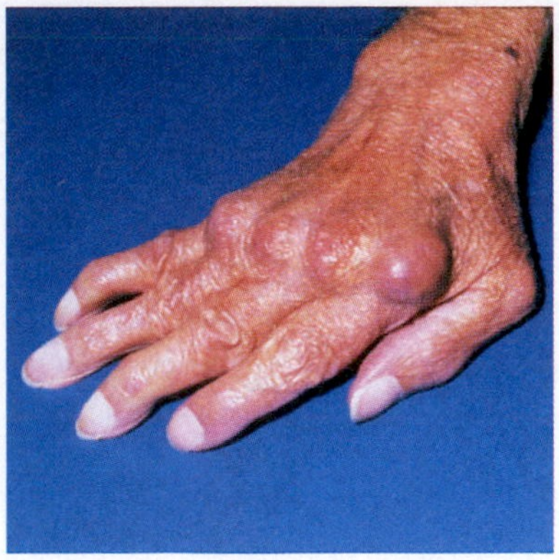

FIGURE 9.24
Rheumatoid Arthritis (RA). A severe case with ankylosis of the joints. Compare the X ray on page 227.

use of steroids weakens the bone, however, aspirin is the treatment of first choice to control the inflammation. Physical therapy is also used to preserve the joint's range of motion and the patient's functional ability.

Several common pathologies of the joints are briefly described in table 9.3.

TABLE 9.3

Disorders of the Joints

Dislocation (luxation)	Displacement of a bone from its normal position at a joint, usually accompanied by a sprain of the adjoining connective tissues. Most common at the fingers, thumb, shoulder, and knee.
Gout	A hereditary disease, most common in men, in which uric acid crystals accumulate in the joints and irritate the articular cartilage and synovial membrane. Causes *gouty arthritis,* with swelling, pain, tissue degeneration, and sometimes fusion of the joint. Most commonly affects the great toe.
Strain	Painful overstretching of a tendon or muscle without serious tissue damage. Often results from inadequate warm-up before exercise.
Subluxation	Partial dislocation in which two bones maintain contact between their articular surfaces.
Synovitis	Inflammation of a joint capsule, often as a complication of a sprain.

Disorders Described Elsewhere

Ankle sprains 246	Dislocation of the shoulder 240	Rheumatoid arthritis 249
Bursitis 232	Knee injuries 247	Rotator cuff injury 319
Congenital hip dislocation 242	Osteoarthritis 247	Tendinitis 232
Dislocation of the elbow 243		TMJ syndrome 239

Joint Prostheses

Arthroplasty,[30] a treatment of last resort, is the replacement of a diseased joint with an artificial device called a **joint prosthesis.**[31] Joint prostheses were first developed to treat war injuries in World War II and the Korean War. Total hip replacement (THR), first performed in 1963 by English orthopedic surgeon Sir John Charnley, is now the most common orthopedic procedure for the elderly. The first knee replacements were performed in the 1970s. Joint prostheses are now available for finger, shoulder, and elbow joints, as well as for hip and knee joints. Arthroplasty is performed on over 250,000 patients per year in the United States, primarily to relieve pain and restore function in elderly people with OA or RA.

Arthroplasty presents ongoing challenges for biomedical engineering. An effective prosthesis must be strong, nontoxic, and corrosion-resistant. In addition, it must bond firmly to the patient's bones and enable a normal range of motion with a minimum of friction. The heads of long bones are usually replaced with prostheses made of a metal alloy such as cobalt-chrome, titanium alloy, or stainless steel. Joint sockets are made of polyethylene (fig. 9.25). Prostheses are bonded to the patient's bone with screws or bone cement.

About 80% to 90% of hip replacements and at least 60% of ankle replacements remain functional for 2 to 10 years. The most common form of failure is detachment of the prosthesis from the bone. This problem has been reduced by using *porous-coated prostheses,* which become infiltrated by the patient's own bone and create a firmer bond. A prosthesis is not as strong as a natural joint, however, and is not an option for many young, active patients.

Before You Go On

Answer the following questions to test your understanding of the preceding section:

14. Define *arthritis.* How do the causes of osteoarthritis and rheumatoid arthritis differ? Which type is more common?
15. What are the major engineering problems in the design of joint prostheses? What is the most common cause of failure of a prosthesis?

[30]*arthro* = joint + *plasty* = surgical repair
[31]*prosthe* = something added

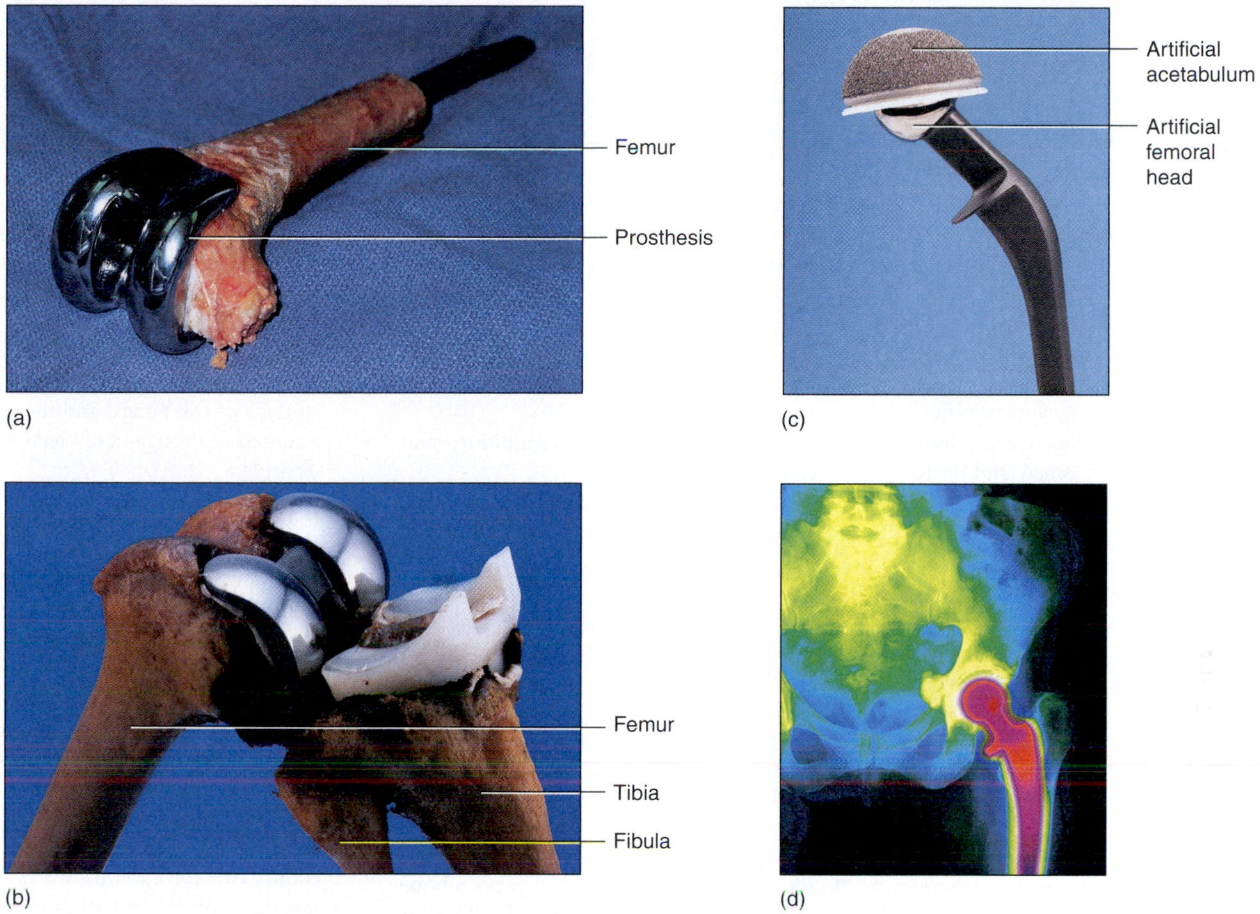
Femur
Prosthesis
(a)
Artificial acetabulum
Artificial femoral head
(c)
Femur
Tibia
Fibula
(b)
(d)

FIGURE 9.25

Joint Prostheses. (*a*) An artificial femoral head inserted into the femur. (*b*) An artificial knee joint bonded to a natural femur and tibia. (*c*) A porous-coated hip prosthesis. The caplike portion replaces the acetabulum of the os coxae, and the ball and shaft below it are bonded to the proximal end of the femur. (*d*) X ray of a patient with a total hip replacement.

CHAPTER REVIEW

Review of Key Concepts

Joints and Their Classification (p. 228)

1. A *joint (articulation)* is any point at which two bones meet. Not all joints are movable.
2. The sciences dealing with joints include arthrology, kinesiology, and biomechanics.
3. Joints are typically named after the bones involved, such as the humeroscapular joint.
4. Joints are classified according to the manner in which the bones are joined and corresponding differences in how freely the bones can move.
5. *Bony joints (synostoses)* are joints at which the original gap between two bones becomes ossified and adjacent bones become, in effect, a single bone—for example, union of the two frontal bones into a single bone.
6. *Fibrous joints (synarthroses)* are joints at which two bones are united by collagenous fibers. The three types of fibrous joints are *sutures* (which are of *serrate, lap,* and *plane* types), *gomphoses* (teeth in their sockets), and *syndesmoses* (exemplified by long bones joined along their shafts by interosseous membranes).
7. *Cartilaginous joints (amphiarthroses)* are joints at which two bones are united by cartilage. In *synchondroses,* the cartilage is hyaline (as in the epiphyseal plates of juvenile bones), and in *symphyses,* it is fibrocartilage (as in the intervertebral discs and pubic symphysis).

Synovial Joints (p. 231)

1. A *synovial joint (diarthrosis)* is a joint at which two bones are separated by a *joint cavity* which contains lubricating *synovial fluid.* The ends of the adjacent bones are covered with hyaline *articular cartilages.* The cavity is enclosed by a *joint capsule,* which is composed of an outer *fibrous capsule* and inner *synovial membrane.* The synovial membrane secretes the synovial fluid. Most synovial joints are highly movable.
2. Some synovial joints contain a fibrocartilage *articular disc* or (in the knee) a pair of *menisci,* which absorb shock and pressure, guide bone movements, improve the fit between the bones, and stabilize the joint.
3. Accessory structures of a synovial joint include *tendons* (from muscle to bone), *ligaments* (from bone to bone), and *bursae.* Bursae are fibrous sacs continuous with the joint cavity and filled with synovial fluid. Some bursae are elongated cylinders called *tendon sheaths.*
4. Synovial joints are described as *monaxial, biaxial,* or *multiaxial* based on the number of geometric planes (one to three) along which a bone can move.
5. The six categories of synovial joints are *hinge, gliding, pivot, saddle, condyloid,* and *ball-and-socket joints* (fig. 9.6).
6. Table 9.1 summarizes the definitions and classifications of joints.
7. *Flexion* is a movement that decreases a joint angle, usually in a sagittal plane, such as flexing the elbow. *Extension* is the opposite movement; it increases a joint angle, as in straightening the elbow. Some joints such as the wrist are also capable of *hyperextension,* which increases the joint angle beyond 180 degrees.
8. *Abduction* is the movement of a body part away from the median plane, as in spreading the fingers or raising the arm to one's side. *Adduction* is the opposite movement, moving a body part toward the median plane.
9. *Elevation* is a movement that raises a bone vertically, as in shrugging the shoulders or biting, and *depression* is the lowering of a bone, as in dropping the shoulders or opening the mouth.
10. *Protraction* is movement of a bone anteriorly, as in rounding the shoulders or jutting the mandible, and *retraction* is the movement of a bone posteriorly, as in squaring the shoulders or pulling the mandible inward.
11. *Lateral* and *medial excursion* are movements of the mandible to either side and back to the midline, respectively.
12. *Circumduction* is a movement in which the attached end of an appendage remains relatively stationary while the free end describes a circle.
13. *Rotation* is the turning of a bone such as the humerus on its longitudinal axis, twisting at the waist, or turning the head from side to side.
14. *Supination* is a forearm movement that turns the palm forward or up, and *pronation* is a forearm movement that turns the palm rearward or down.
15. *Opposition* is movement of the thumb toward the fingertips, and *reposition* is returning the thumb to its normal resting position.
16. *Dorsiflexion* is an ankle movement that raises the toes and *plantar flexion* is an ankle movement that lowers them.
17. *Inversion* of the feet turns the soles medially, facing each other; *eversion* turns them laterally, away from each other.
18. The *range of motion (ROM)* of a joint depends on the structure and action of the associated muscles; structure of the articular surface of the bones; and the strength and tautness of the ligaments, tendons, and joint capsule.

Anatomy of Selected Synovial Joints (p. 239)

1. This chapter describes six synovial joints or joint groups: the temporomandibular joint (TMJ), humeroradial (shoulder) joint, the elbow (a complex of three joints), the coxal (hip) joint, the tibiofemoral (knee) joint, and the talocrural (ankle) joint. The principal structures and types of movement at each joint are summarized in table 9.2.
2. The *temporomandibular joint* is involved in biting and chewing, and has an articular disc to absorb the pressure produced in such actions. Two common disorders of this joint are dislocation and TMJ syndrome.
3. The *humeroscapular joint* is notable for its great mobility and shallow socket (glenoid cavity), making it very susceptible to dislocation. Rotator cuff injuries (chapter 12) are also common at this joint.
4. The elbow contains three joints—*humeroulnar, humeroradial,* and *proximal radioulnar.* It allows for hingelike movements of the forearm and for rotation of the radius on the ulna when the forearm is pronated and supinated.
5. The *coxal joint* is an important weight-bearing ball-and-socket joint and therefore has an especially deep socket, the acetabulum of the os coxae. When a person stands, some of the ligaments at this joint twist and pull the head of the femur more tightly into the acetabulum.
6. The *tibiofemoral joint* is the most complex diarthrosis of the body. It has numerous ligaments and bursae. The most important stabilizing structures within the joint cavity are the *anterior* and *posterior cruciate ligaments* and the *lateral* and *medial menisci.* Injuries to these ligaments and cartilages are common.
7. At the *talocrural joint,* the tibia and fibula articulate with each other, and each articulates with the talus. Numerous ligaments support this joint. Ankle sprains are tearing of these ligaments and adjacent tendons.

Clinical Perspectives (p. 246)

1. *Rheumatism* is a broad term for pain in the bones, joints, ligaments, tendons, or muscles. Physicians who specialize in joint disorders are called *rheumatologists.*
2. *Arthritis* is a general term for more than 100 inflammatory joint diseases. The most common form of arthritis is *osteoarthritis (OA),* which occurs to some degree in almost everyone as a result of years of wear and tear on the joints. It is marked especially by erosion of the articular cartilages.
3. *Rheumatoid arthritis (RA)* is a more severe autoimmune joint disease caused by an antibody called *rheumatoid factor* that damages synovial membranes. RA can be severely crippling, sometimes causing a fusion of the bones called *ankylosis.*
4. *Arthroplasty* is the replacement of a diseased joint with an artificial joint, or *joint prosthesis.* It was first developed for hip and knee joints but is now also performed at finger, shoulder, and elbow joints.

Testing Your Recall

1. Which of the following is unique to the thumb?
   a. gliding joint
   b. hinge joint
   c. saddle joint
   d. condyloid joint
   e. pivot joint
2. Which of the following is the least movable?
   a. diarthrosis
   b. synostosis
   c. symphysis
   d. syndesmosis
   e. condyloid joint
3. Which of the following movements are unique to the foot?
   a. dorsiflexion and inversion
   b. elevation and depression
   c. circumduction and rotation
   d. abduction and adduction
   e. opposition and reposition
4. Which of the following joints cannot be circumducted?
   a. carpometacarpal
   b. metacarpophalangeal
   c. humeroscapular
   d. coxal
   e. interphalangeal
5. Which of the following terms denotes a general condition that includes the other four?
   a. gout
   b. arthritis
   c. rheumatism
   d. osteoarthritis
   e. rheumatoid arthritis
6. In the adult, the ischium and pubis are united by
   a. a synchondrosis.
   b. a diarthrosis.
   c. a synostosis.
   d. an amphiarthrosis.
   e. a symphysis.
7. Articular discs are found only in certain
   a. synostoses.
   b. symphyses.
   c. diarthroses.
   d. synchondroses.
   e. amphiarthroses.
8. Which of the following joints has anterior and posterior cruciate ligaments?
   a. the shoulder
   b. the elbow
   c. the hip
   d. the knee
   e. the ankle
9. To bend backward at the waist involves _____ of the vertebral column.
   a. rotation
   b. hyperextension
   c. dorsiflexion
   d. abduction
   e. flexion
10. If you sit on a sofa and then raise your left arm to rest it on the back of the sofa, your left shoulder joint undergoes
    a. lateral excursion.
    b. abduction.
    c. elevation.
    d. adduction.
    e. extension.
11. The lubricant of a diarthrosis is _____.
12. A fluid-filled sac that eases the movement of a tendon over a bone is called a/an _____.
13. A _____ joint allows one bone to swivel on another.
14. _____ is the science of movement.
15. The joint between a tooth and the mandible is called a/an _____.
16. In a _____ suture, the articulating bones have interlocking wavy margins, somewhat like a dovetail joint in carpentry.
17. In kicking a football, what type of action does the knee joint exhibit?
18. The angle through which a joint can move is called its _____.
19. A person with a degenerative joint disorder would most likely be treated by a physician called a _____.
20. The femur is prevented from slipping sideways off the tibia in part by a pair of cartilages called the lateral and medial _____.

Answers in the Appendix

True or False

Determine which five of the following statements are false, and briefly explain why.

1. More people get rheumatoid arthritis than osteoarthritis.
2. A doctor who treats arthritis is called a kinesiologist.
3. Synovial joints are also known as synarthroses.
4. Most ligaments, but not all, connect one bone to another.
5. Reaching behind you to take something out of your hip pocket involves hyperextension of the elbow.
6. The anterior cruciate ligament normally prevents hyperextension of the knee.
7. There is no meniscus in the elbow joint.
8. The knuckles are diarthroses.
9. Synovial fluid is secreted by the bursae.
10. Unlike most ligaments, the periodontal ligaments do not attach one bone to another.

Answers in the Appendix

Testing Your Comprehension

1. Why is there a pair of menisci in the knee joint but not in the elbow, the corresponding joint of the upper limb? Why is there an articular disc in the temporomandibular joint?
2. What ligaments would most likely be torn if you slipped and your foot were suddenly forced into an excessively inverted position: (*a*) the posterior talofibular and calcaneofibular ligaments, or (*b*) the deltoid ligament? Explain. What would the resulting condition of the ankle be called?
3. In order of occurrence, list the joint actions (flexion, pronation, etc.) and the joints where they would occur as you (*a*) sit down at a table, (*b*) reach out and pick up an apple, (*c*) take a bite, and (*d*) chew it. Assume that you start in anatomical position.
4. What structure in the elbow joint serves the same purpose as the anterior cruciate ligament (ACL) of the knee?
5. List the six types of synovial joints and for each one, if possible, identify a joint in the upper limb and a joint in the lower limb that falls into each category. Which of these six joints have no examples in the lower limb?

Answers at the Online Learning Center

www.mhhe.com/saladinha1

Visit the Online Learning Center for practice tests, answer keys, and other learning aids for this chapter. Enhance your understanding of human anatomy with our interactive art labeling exercises, supplemental photo atlases, web links, puzzles, flashcards, and much more.

10

CHAPTER TEN

The Muscular System—Introduction

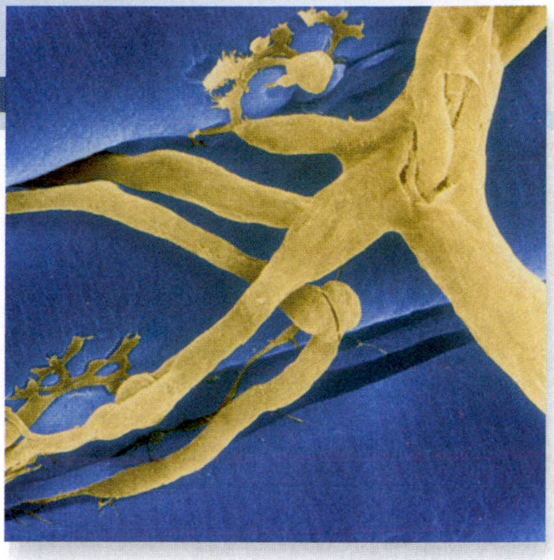

Neuromuscular junctions. Muscle fibers are shown in *blue* and nerve fibers in *yellow* (SEM).

CHAPTER OUTLINE

Muscle Types and Functions 256
- Types of Muscle 256
- Functions of Muscle 256
- Properties of Muscle 257

General Anatomy of Muscles 257
- Connective Tissues and Fascicles 257
- Fascicles and Muscle Shapes 258
- Muscle Attachments 259
- Functional Groups of Muscles 260
- Intrinsic and Extrinsic Muscles 260
- Muscles, Bones, and Levers 260

Microscopic Anatomy 263
- Ultrastructure of Muscle Fibers 263
- The Nerve-Muscle Relationship 266
- The Blood Supply 268

Functional Perspectives 269
- Contraction and Relaxation 269
- Muscle Growth and Atrophy 269
- Physiological Classes of Muscle Fibers 270

Cardiac and Smooth Muscle 272
- Cardiac Muscle 273
- Smooth Muscle 274

Developmental and Clinical Perspectives 275
- Embryonic Development of Muscle 275
- The Aging Muscular System 275
- Diseases of the Muscular System 276

Chapter Review 278

INSIGHTS

10.1 Clinical Application: Muscle-Bound 260

10.2 Clinical Application: Neuromuscular Toxins and Paralysis 267

BRUSHING UP

To understand this chapter, you may find it helpful to review the following concepts:

- Components of a neuron (p. 94)
- The three types of muscle (p. 95)
- Embryonic mesoderm, somites, and myotomes (p. 114)
- Skeletal anatomy (pp. 171–246)

The structure and function of the muscular system occupy a place of central importance in several fields of health care and fitness. Physical and occupational therapists must be well acquainted with the muscular system to design and carry out rehabilitation programs. To give intramuscular injections safely requires some understanding of the skeletal muscles and their associated nerves and blood vessels. Even safely and effectively moving a patient who is physically incapacitated requires a knowledge of the joints and muscles. Coaching, kinesiology, sports medicine, and dance also benefit from knowledge of skeletomuscular anatomy and mechanics.

The term **muscular system** refers only to the skeletal muscles. The study of this system is called **myology.**[1] The subject is closely related to what we have covered in the preceding chapters. It relates muscle attachments to the bone structures described in chapters 7 and 8, and muscle function to the joint movements described in chapter 9. This chapter describes general aspects of muscle gross anatomy, followed by a description of the ultrastructure of muscle cells, how this relates to the functional properties of muscle, and how cardiac and smooth muscle compare with skeletal muscle. Chapters 11 and 12 describe the anatomy and function of specific skeletal muscles of the axial and appendicular regions, respectively.

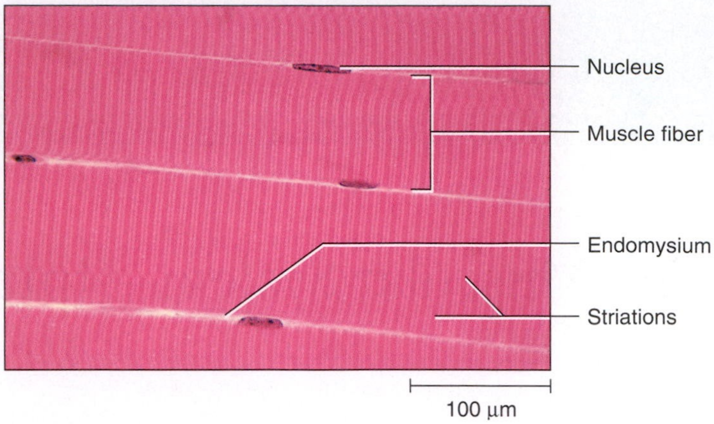
Nucleus
Muscle fiber
Endomysium
Striations
100 µm

FIGURE 10.1
Skeletal Muscle Fibers.

Muscle Types and Functions

Objectives

When you have completed this section, you should be able to

- describe the distinctions between the three types of muscular tissue; and
- list the functions of muscles and the properties that muscular tissue must have to carry out these functions.

Muscle is a tissue specialized to produce movement of a body part or an organ's contents, such as blood and food. Its cells convert the chemical energy of ATP into the mechanical energy of motion and exert a useful pull on other cells or tissues.

Types of Muscle

Skeletal muscle may be defined as voluntary striated muscle that is usually attached to one or more bones. A typical skeletal muscle cell is about 100 µm in diameter and 3 cm long; some are as thick as 500 µm and as long as 30 cm. Because of their extraordinary length, skeletal muscle cells are usually called **muscle fibers** or **myofibers.** A skeletal muscle fiber is packed with protein microfilaments that overlap each other in such a way as to produce alternating light and dark bands, or **striations** (fig. 10.1). Skeletal muscle is called **voluntary** because it is usually subject to conscious control.

Cardiac muscle is also striated, but it is **involuntary**—not normally under conscious control. Its cells are not fibrous in shape, and are therefore called **myocytes** or **cardiocytes.**

Smooth muscle contains the same contractile proteins as skeletal and cardiac muscle, but they are not arranged in a regularly overlapping way, so there are no striations in smooth muscle. Its cells, also called myocytes, are relatively short and fusiform in shape—that is, thick in the middle and tapered at the ends. Smooth muscle, like cardiac, is involuntary.

Functions of Muscle

The functions of muscular tissue are as follows:

- **Movement.** Most obviously, the muscles enable us to move from place to place and to move individual body parts. Muscular contractions also move body contents in the course of respiration, circulation, digestion, defecation, urination, and childbirth.
- **Stability.** Muscles maintain posture by resisting the pull of gravity and preventing unwanted movements. They also hold some articulating bones in place by maintaining tension on the tendons.
- **Communication.** Muscles are used for facial expression, other body language, writing, and speech.
- **Control of body openings and passages.** Ringlike *sphincter muscles* around the eyelids, pupils, and mouth control the admission of light, food, and drink into the body; others that encircle the urethral and anal orifices control elimination of waste; and other sphincters control the movement of food, bile, and other materials through the body.
- **Heat production.** The skeletal muscles produce as much as 85% of our body heat, which is vital to the functioning of enzymes and therefore to all of our metabolism.

[1] *myo* = muscle + *logy* = study of

Properties of Muscle

To carry out the foregoing functions, muscle cells must have the following properties:

- **Excitability (responsiveness).** Excitability is a property of all living cells, but it is developed to the highest degree in muscle and nerve cells. When stimulated by chemical signals, stretch, and other stimuli, muscle cells respond with electrical changes across the plasma membrane.
- **Conductivity.** The local electrical excitation produced at the point of muscle stimulation is conducted throughout the entire plasma membrane, initiating the events that lead to contraction.
- **Contractility.** Muscle fibers are unique in their ability to shorten substantially when stimulated. This enables them to pull on bones and other tissues and organs to create movement.
- **Extensibility.** In order to contract, a cell must also be extensible—able to stretch again between contractions. Most cells rupture if they are stretched even a little, but skeletal muscle fibers can stretch to as much as three times their contracted length.
- **Elasticity.** When a muscle cell is stretched and then the tension is released, it recoils to its original length. (Elasticity refers to the tendency to recoil, not the ability to stretch.)

From this point on, this chapter concerns skeletal muscles unless otherwise stated.

Before You Go On

Answer the following questions to test your understanding of the preceding section:

1. What general function of muscular tissue distinguishes it from other tissue types?
2. What are the basic structural differences between skeletal, cardiac, and smooth muscle?
3. State five functions of the muscular system.
4. State five special properties of muscular tissue that enable it to perform its functions.

General Anatomy of Muscles

Objectives

When you have completed this section, you should be able to

- describe the connective tissues and associated structural organization of a muscle;
- describe types of muscles defined by the arrangement of their fiber bundles (fascicles);
- describe the parts of a typical muscle;
- describe the types of muscle-bone attachments;
- describe the way that muscles are arranged in groups with complementary actions at a joint;
- explain what intrinsic and extrinsic muscles are; and
- describe three types of musculoskeletal levers and their respective advantages.

Connective Tissues and Fascicles

A skeletal muscle is more than muscular tissue. It also contains connective tissue, nervous tissue, and blood vessels. In this section, we will examine the connective tissue components of a skeletal muscle. From the smallest to largest, and from deep to superficial, these are (fig. 10.2):

- **Endomysium**[2] (EN-doe-MIZ-ee-um), a thin sleeve of areolar connective tissue that surrounds each muscle fiber. This allows room for blood capillaries and nerve fibers to reach every muscle fiber.
- **Perimysium,**[3] a thicker connective tissue sheath that wraps muscle fibers together in bundles called **fascicles**[4] (FASS-ih-culs). Fascicles are visible to the naked eye as parallel strands—the "grain" in a cut of meat; tender roast beef is easily pulled apart along its fascicles.
- **Epimysium,**[5] a fibrous sheath that surrounds the entire muscle. The epimysium extends beyond the ends of many muscles as a fibrous band, the **tendon,** connecting it to the periosteum of a bone.
- **Deep fascia** (FASH-ee-uh), sheets of connective tissue that separate neighboring muscles from each other.
- **Superficial fascia** (= hypodermis; see chapter 5), a layer of connective tissue that separates the muscles from the overlying skin (fig. 10.2*b*). In places such as the abdomen and buttocks, the superficial fascia is very fatty, while in areas such as the forehead and dorsum of the hand, fat is scanty or absent.

The **series-elastic components** of a muscle are the connective tissue elements from endomysium to tendon, linking muscle fibers to bones. They form a strong collagenous continuity from muscle to bone: endomysium → perimysium → epimysium → tendon → periosteum → bone matrix. These tissues are extensible and elastic—they stretch under tension and recoil when released. Elastic recoil of the tendons adds significantly to the power output and efficiency of the muscles. When you are running, for example, recoil of the calcaneal (Achilles) tendon helps to lift the heel and produce some of the thrust as your toes push off from the ground. (This recoil is also responsible for the long, energy-efficient leaping of kangaroos.)

[2]*endo* = within + *mys* = muscle
[3]*peri* = around
[4]*fasc* = bundle + *icle* = little
[5]*epi* = upon, above

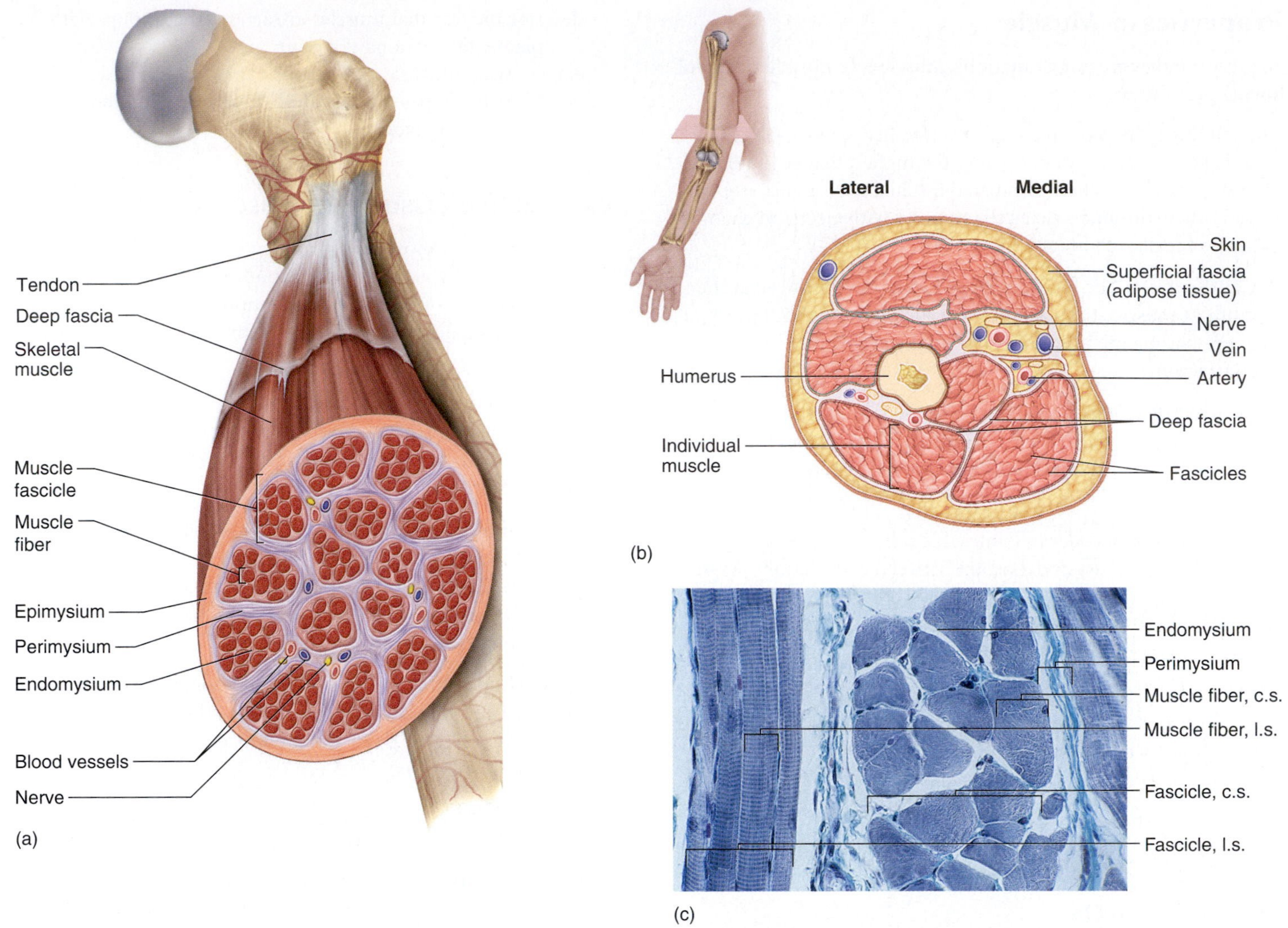

FIGURE 10.2

The Connective Tissues of a Skeletal Muscle. (*a*) The muscle-bone attachment. (*b*) A cross section of the arm showing the relationship of neighboring muscles to fascia and bone. (*c*) Muscle fascicles in the tongue. Vertical fascicles, cut in longitudinal section (l.s.), pass between the dorsal and ventral surfaces of the tongue and alternate with horizontal fascicles, cut in cross section (c.s.), that pass from the root to the tip of the tongue. A fibrous perimysium can be seen between the fascicles, and endomysium can be seen between individual muscle fibers within a fascicle.

Fascicles and Muscle Shapes

The strength of a muscle and the direction of its pull are determined partly by the orientation of its fascicles. Differences in fascicle orientation are the basis for classifying muscles into five types (fig. 10.3):

1. **Fusiform[6] muscles** are thick in the middle and tapered at each end. Their contractions are moderately strong. The *biceps brachii* of the arm and *gastrocnemius* of the calf are examples of this type.
2. **Parallel muscles** are long, straplike muscles of uniform width and parallel fascicles. They can span a great distance and shorten more than other muscle types, but they are weaker than fusiform muscles. Examples include the *rectus abdominis* of the abdomen, *sartorius* of the thigh, and *zygomaticus major* of the face.
3. **Convergent muscles** are fan-shaped—broad at the origin and converging toward a narrower insertion. These muscles are relatively strong because all of their fascicles exert their tension on a relatively small insertion. The *pectoralis major* in the chest is a muscle of this type.
4. **Pennate[7] muscles** are feather-shaped. Their fascicles insert obliquely on a tendon that runs the length of the muscle,

[6] *fusi* = spindle + *form* = shape

[7] *penna* = feather

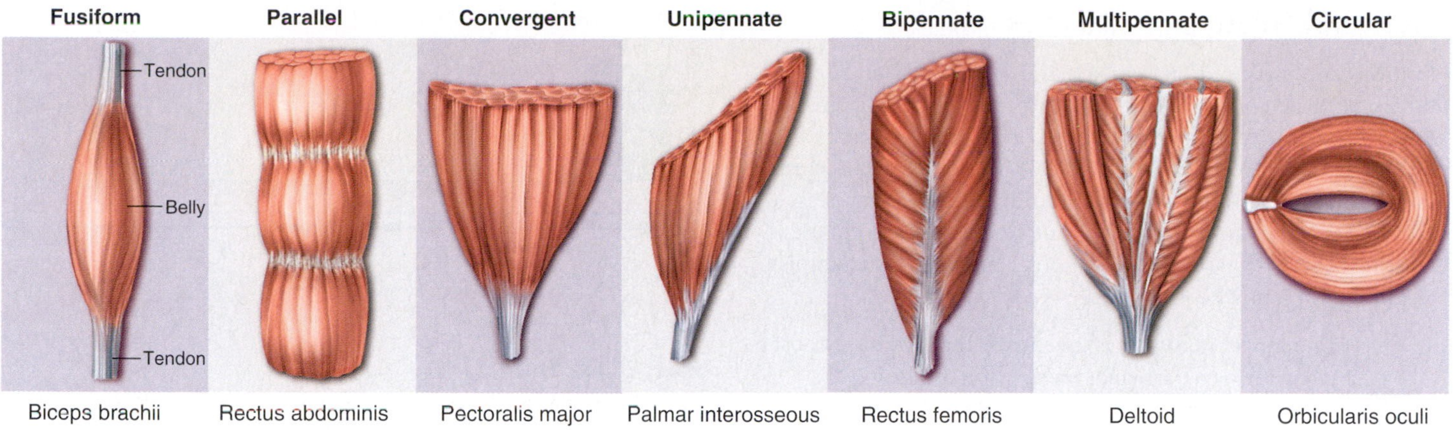

Fusiform
Parallel
Convergent
Unipennate
Bipennate
Multipennate
Circular
Tendon
Belly
Tendon
Biceps brachii
Rectus abdominis
Pectoralis major
Palmar interosseous
Rectus femoris
Deltoid
Orbicularis oculi

FIGURE 10.3
Classification of Muscles According to Fascicle Orientation. The fascicles are the "grain" visible in each muscle illustration.

like the shaft of a feather. There are three types of pennate muscles: *unipennate,* in which all fascicles approach the tendon from one side (for example, the *palmar interosseous muscles* of the hand and *semimembranosus* of the thigh); *bipennate,* in which fascicles approach the tendon from both sides (for example, the *rectus femoris* of the thigh); and *multipennate,* shaped like a bunch of feathers with their quills converging on a single point (for example, the *deltoid* of the shoulder).

5. **Circular muscles (sphincters)** form rings around body openings. These include the *orbicularis oris* of the lips and *orbicularis oculi* of the eyelids.

Muscle Attachments

Most skeletal muscles are attached to a different bone at each end, so either the muscle or its tendon spans at least one joint. When the muscle contracts, it moves one bone relative to the other. The muscle attachment at the relatively stationary end is called its **origin,** or **head.** Its attachment at the more mobile end is called its **insertion.** For the biceps brachii, for example, the origin is the scapula and the insertion is the radius (fig. 10.4). The middle region between the origin and insertion is called the **belly.**

There are two ways a muscle can attach to a bone. In a **direct (fleshy) attachment,** collagen fibers of the epimysium are continuous with the periosteum, the fibrous sheath around a bone. The red muscle tissue appears to emerge directly from the bone, as we see along the margin of the brachialis muscle in figure 10.4. In an **indirect attachment,** a tendon emerges from the connective tissue of the muscle and merges into the periosteum of the bone, as we see at both ends of the biceps brachii in figure 10.4. Some collagen fibers of the periosteum continue into the bone matrix as *perforating fibers* (see chapter 5), so there is a strong structural continuity of tendon to periosteum to bone matrix. Excessive stress is more likely to tear a tendon than to pull it loose from the muscle or bone.

In some cases, the epimysium of one muscle attaches to the fascia or tendon of another or to collagen fibers of the dermis. The

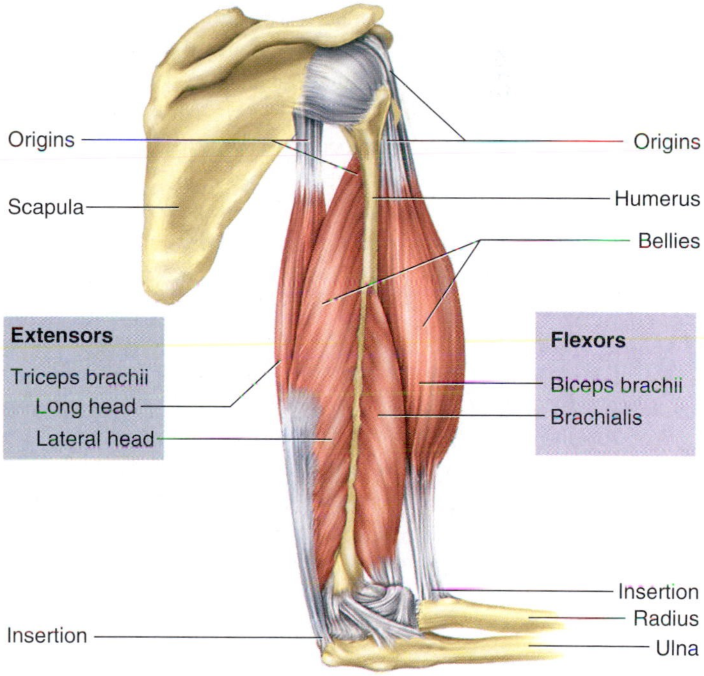

Origins
Scapula
Origins
Humerus
Bellies
Extensors
Triceps brachii
Long head
Lateral head
Flexors
Biceps brachii
Brachialis
Insertion
Radius
Ulna
Insertion

FIGURE 10.4
A Muscle Group Acting on the Elbow. The biceps brachii and brachialis are synergists in elbow flexion. The biceps is the prime mover in flexion. The triceps brachii is an antagonist of these two muscles and is the prime mover in elbow extension.

ability of a muscle to produce facial expressions depends on the latter type of attachment. Some muscles are connected to a broad sheetlike tendon called an **aponeurosis**[8] (AP-oh-new-RO-sis). This term originally referred to the tendon located beneath the scalp, but now it also refers to similar tendons associated with certain abdominal, lumbar, hand, and foot muscles.

[8] *apo* = upon, above + *neuro* = nerve

Functional Groups of Muscles

The movement produced by a muscle is called its **action.** Skeletal muscles seldom act independently; instead, they function in groups whose combined actions produce the coordinated motion of a joint. Muscles can be classified into at least four categories according to their actions, but it must be stressed that a particular muscle can act in a certain way during one joint action and in a different way during other actions of the same joint. The following examples are illustrated in figure 10.4:

1. The **prime mover (agonist)** is the muscle that produces most of the force during a particular joint action. In flexing the elbow, for example, the prime mover is the biceps brachii.
2. A **synergist**[9] (SIN-ur-jist) is a muscle that aids the prime mover. Several synergists acting on a joint can produce more power than a single larger muscle. The *brachialis,* for example, lies deep to the biceps brachii and works with it as a synergist to flex the elbow. The actions of a prime mover and its synergist are not necessarily identical and redundant. If the prime mover worked alone at a joint, it might cause rotation or other undesirable movements of a bone. A synergist may stabilize a joint and restrict these movements, or modify the direction of a movement, so that the action of the prime mover is more coordinated and specific.
3. An **antagonist**[10] is a muscle that opposes the prime mover. In some cases, it relaxes to give the prime mover almost complete control over an action. More often, however, the antagonist maintains some tension on a joint and thus limits the speed or range of the agonist, preventing excessive movement and joint injury. If you extend your arm to reach out and pick up a cup of tea, your *triceps brachii* on the posterior side of the humerus is the prime mover of elbow extension, while your biceps brachii acts as an antagonist to slow the extension and stop it at the appropriate point. If you extend your arm rapidly to throw a dart, however, the biceps must be more relaxed. The biceps and triceps brachii represent an **antagonistic pair** of muscles that act on opposite sides of a joint. We need antagonistic pairs at a joint because a muscle can only pull, not push—a single muscle cannot flex *and* extend the elbow, for example. Which member of the pair acts as the agonist depends on the motion under consideration. In flexion of the elbow, the biceps is the agonist and the triceps is the antagonist; when the elbow is extended, their roles are reversed.
4. A **fixator** is a muscle that prevents a bone from moving. To *fix* a bone means to hold it steady, allowing another muscle attached to it to pull on something else. For example, consider again the flexion of the elbow by the biceps brachii. The biceps originates on the scapula and inserts on the radius. The scapula is loosely attached to the axial skeleton, so when the biceps contracts, it seems that it would pull the scapula laterally. However, there are fixator muscles that attach the scapula to the vertebral column. They contract at the same time as the biceps, holding the scapula firmly in place and ensuring that the force generated by the biceps moves the radius rather than the scapula.

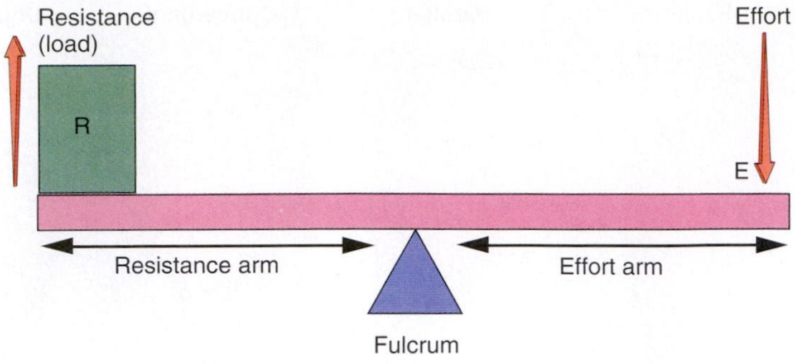

FIGURE 10.5
Basic Components of a Lever. This example is a first-class lever.

Intrinsic and Extrinsic Muscles

In places such as the tongue, larynx, back, hand, and foot, anatomists distinguish between intrinsic and extrinsic muscles. An **intrinsic muscle** is entirely contained within a particular region, having both its origin and insertion there. An **extrinsic muscle** acts upon a designated region but has its origin elsewhere. For example, some movements of the fingers are produced by extrinsic muscles in the forearm, whose long tendons reach to the phalanges; other finger movements are produced by the intrinsic muscles located between the metacarpal bones of the hand.

Muscles, Bones, and Levers

Many bones, especially the long bones, act as levers on which the muscles exert their force. A **lever** is any elongated, rigid object that rotates around a fixed point called the **fulcrum** (fig. 10.5). Familiar examples include a seesaw and a crowbar. Rotation occurs when an **effort** applied to one point on the lever overcomes a **resistance**

INSIGHT 10.1 CLINICAL APPLICATION

Muscle-Bound

Any well-planned program of resistance (strength) training or bodybuilding must include exercises aimed at proportional development of the different members of a muscle group, such as flexors and extensors of the arm. Otherwise, the muscles on one side of a joint may develop out of proportion to their antagonists and restrict the joint's range of motion (ROM). If the biceps brachii is heavily developed without proportionate attention to the triceps brachii, for example, the stronger biceps will cause the elbow to be somewhat flexed constantly, and the ROM of the elbow will be restricted. The joint is then said to be "muscle-bound." People with muscle-bound joints move awkwardly and are poor at activities that require agility, such as dance and ball games.

[9] *syn* = together + *erg* = work
[10] *ant* = against + *agonist* = actor, competitor

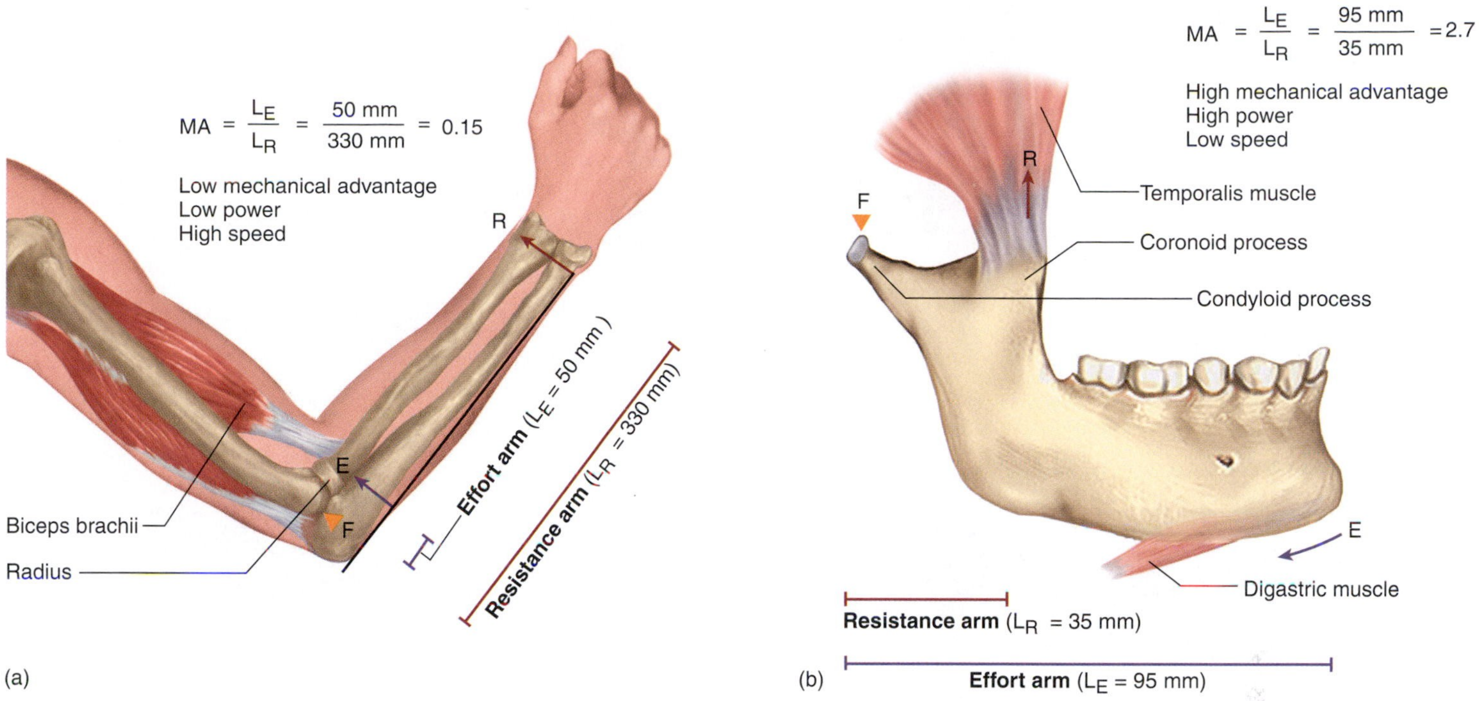
MA = LE / LR = 50 mm / 330 mm = 0.15
Low mechanical advantage
Low power
High speed
R
E
F
Biceps brachii
Radius
Effort arm (LE = 50 mm)
Resistance arm (LR = 330 mm)
(a)
MA = LE / LR = 95 mm / 35 mm = 2.7
High mechanical advantage
High power
Low speed
R
F
Temporalis muscle
Coronoid process
Condyloid process
E
Digastric muscle
Resistance arm (LR = 35 mm)
Effort arm (LE = 95 mm)
(b)

FIGURE 10.6

Mechanical Advantage (MA). MA is calculated as the length of the effort arm divided by the length of the resistance arm. (*a*) The forearm acts as a third-class lever during flexion of the elbow. (*b*) The mandible acts as a second-class lever when the jaw is forcibly opened. The digastric muscle and others provide the effort, while tension in the temporalis and other muscles provide the resistance.

(**load**) located at some other point. The part of a lever from the fulcrum to the point of effort is called the **effort arm,** and the part from the fulcrum to the point of resistance is the **resistance arm.** In the body, a long bone acts as a lever, a joint serves as the fulcrum, and the effort is generated by a muscle attached to the bone.

The function of a lever is to produce a gain in the speed, distance, or force of a motion—either to exert more force against a resisting object than the force applied to the lever (for example, in moving a heavy boulder with a crowbar), or to move the resisting object farther or faster than the effort arm is moved (as in swinging a baseball bat). A single lever cannot confer both advantages. There is a trade-off between force on one hand and speed or distance on the other—as one increases, the other decreases.

The **mechanical advantage (MA)** of a lever is the ratio of its output force to its input force. It is equal to the length of the effort arm, L_E, divided by the length of the resistance arm, L_R; that is, $MA = L_E/L_R$. If MA is greater than 1.0, the lever produces more force, but less speed or distance, than the force exerted on it. If MA is less than 1.0, the lever produces more speed or distance, but less force, than the input. Consider the elbow joint, for example (fig. 10.6*a*). The resistance arm of the ulna is longer than the effort arm, so we know from the preceding formula that the mechanical advantage is less than 1.0. The figure shows some representative values for L_E and L_R that yield $MA = 0.15$. The biceps brachii muscle puts more power into the lever than we get out of it, but the hand moves farther and faster than the insertion of the biceps tendon. By contrast, when the digastric muscle depresses the mandible, the MA is about 2.7. The coronoid process of the mandible moves with greater force, but a shorter distance, than the insertion of the digastric (fig. 10.6*b*).

As we have already seen, some joints have two or more muscles acting on them that seemingly produce the same effect, such as elbow flexion. At first, you might consider this arrangement redundant, but it makes sense if the tendinous insertions of the muscles are at slightly different places and produce different mechanical advantages. A runner taking off from the starting line, for example, uses "low-gear" (high-MA) muscles that do not generate much speed but have the power to overcome the inertia of the body. A runner then "shifts into high gear" by using muscles with different insertions that have a lower mechanical advantage but produce more speed at the feet. This is analogous to the way an automobile transmission works to get a car to move and then cruise at high speed.

There are three classes of levers that differ with respect to which component—the fulcrum (F), effort (E), or resistance (R)—is in the middle. A **first-class lever** (fig. 10.7*a*) is one with the fulcrum in the middle (EFR), such as a seesaw. An anatomical example is the atlanto-occipital joint of the neck, where the muscles of the back of the neck pull down on the occipital bone of the skull and oppose the tendency of the head to tip forward. Loss of muscle tone here can be embarrassing if you nod off in class.

A **second-class lever** (fig. 10.7*b*) is one in which the resistance is in the middle (FRE). Lifting the handles of a wheelbarrow, for example, makes it pivot on its wheel at the opposite end and lift

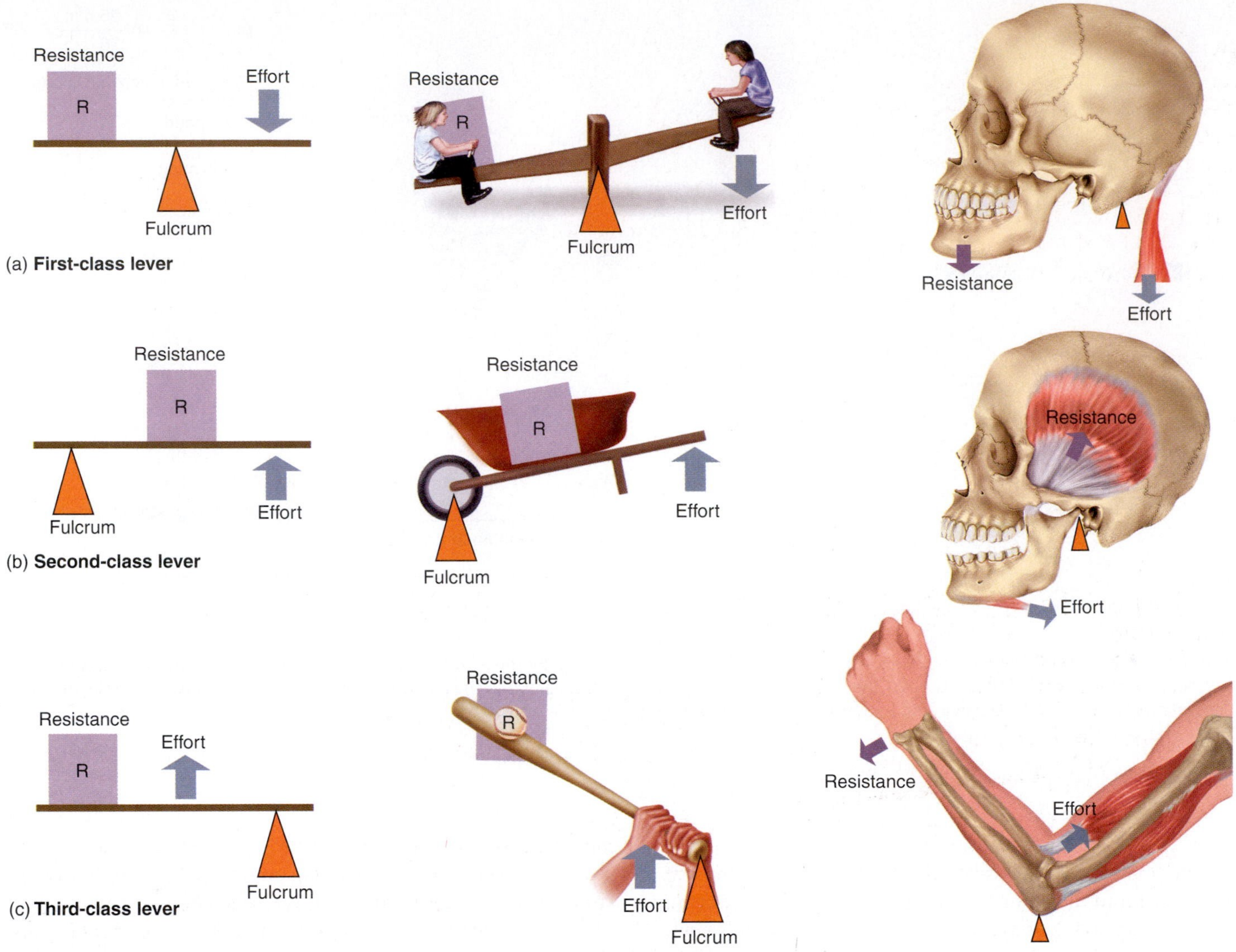

Figure 10.7

The Three Classes of Levers. *Left:* The lever classes defined by the relative positions of the resistance (load), fulcrum, and effort. *Center:* Mechanical examples. *Right:* Anatomical examples. (*a*) Muscles at the back of the neck pull down on the occipital bone to oppose the tendency of the head to tip forward. The fulcrum is the occipital condyle. (*b*) To open the mouth, the digastric muscle pulls down on the chin. It is resisted by the temporalis muscle on the side of the head. The fulcrum is the temporomandibular joint. (*c*) In flexing the elbow, the biceps brachii exerts an effort on the radius. Resistance is provided by the weight of the forearm or anything held in the hand. The fulcrum is the elbow joint.

a load in the middle. The mandible acts as a second-class lever when the *digastric muscle* pulls down on the chin to open the mouth. The fulcrum is the temporomandibular joint, the effort is applied to the chin by the digastric muscle, and the resistance is the tension of muscles such as the *temporalis,* which is used to bite and to hold the mouth closed. (This arrangement is upside down relative to a wheelbarrow, but the mechanics remain the same.)

In a **third-class lever** (fig. 10.7*c*), the effort is applied between the fulcrum and resistance (FER). A baseball bat, for example, acts as a third-class lever. For a right-handed batter, the left hand near the knob of the bat acts as the fulcrum, the right hand on the handle produces the force, and the baseball is the resistance. Most levers in the human body are third-class levers. At the elbow, the fulcrum is the joint between the ulna and humerus; the effort is applied by the biceps brachii muscle, and the resistance can be provided by any weight in the hand or the weight of the forearm itself. The mandible acts as a third-class lever when you close your mouth to bite off a piece of food. Again, the temporomandibular joint is the fulcrum, but now the temporalis muscle exerts the effort, while the resistance is supplied by the item of food being bitten.

Think About It!

Sit on the edge of a desk with your feet off the floor. Plantarflex your foot. Where is the effort? Where is the fulcrum? (Name the specific joint, based on chapter 9.) Where is the resistance? Which class of lever does the foot represent in plantar flexion?

Before You Go On

Answer the following questions to test your understanding of the preceding section:

5. Name the connective tissue layers of a muscle beginning with the individual muscle fiber and ending with the tissue that separates the muscles from the skin.
6. Sketch the fascicle arrangements that define a fusiform, parallel, convergent, pennate, and circular muscle.
7. Define the *origin, insertion,* and *action* of a muscle.
8. Distinguish between a prime mover, synergist, antagonist, and fixator.
9. What is the difference between intrinsic and extrinsic muscles that control the fingers?
10. Define a first-, second-, and third-class lever and give an example of each in the musculoskeletal system.
11. What is the principal advantage of a joint action with a mechanical advantage less than 1.0? What is the principal advantage of a joint action with a mechanical advantage greater than 1.0?

Microscopic Anatomy

Objectives

When you have completed this section, you should be able to

- describe the ultrastructure of a muscle fiber and its myofilaments;
- explain what accounts for the striations of skeletal muscle;
- describe the relationship of a nerve fiber to a muscle fiber;
- define a *motor unit* and discuss its functional significance; and
- describe the blood vessels of a skeletal muscle.

Ultrastructure of Muscle Fibers

In order to understand muscle function, you must know how the organelles and protein microfilaments of a muscle fiber are arranged. A skeletal muscle fiber (fig. 10.8) has multiple flattened or sausage-shaped nuclei pressed against the inside of the plasma membrane. This unusual condition results from their embryonic development—several unspecialized cells called **myoblasts**[11] fuse to produce each muscle fiber, with each myoblast contributing a nucleus to the mature fiber. Some myoblasts remain as unspecialized **satellite cells** between the muscle fiber and endomysium. When a muscle is injured, satellite cells can multiply and produce new muscle fibers to some degree. Most muscle repair, however, is by fibrosis rather than regeneration of functional muscle.

The plasma membrane, called the **sarcolemma,**[12] has tunnel-like infoldings called **transverse (T) tubules** that penetrate through the fiber and emerge on the other side. The function of a T tubule is to carry an electrical current from the surface of the cell to the interior when the cell is stimulated. The cytoplasm, called **sarcoplasm,** is occupied mainly by long protein bundles called **myofibrils** about 1 μm in diameter. Most other organelles of the cell, such as mitochondria and smooth endoplasmic reticulum (ER), are located between adjacent myofibrils. The sarcoplasm also contains an abundance of **glycogen,** which provides stored energy for the muscle to use during exercise, and a red pigment called **myoglobin,** which binds oxygen until it is needed for muscular activity.

The smooth ER of a muscle fiber is called **sarcoplasmic reticulum (SR).** It forms a network around each myofibril, and alongside the T tubules it exhibits dilated sacs called **terminal cisternae.** The SR is a reservoir for calcium ions; it has gated channels in its membrane that can release a flood of calcium into the cytosol, where the calcium activates the muscle contraction process.

MYOFILAMENTS

Most of the muscle fiber is filled with myofibrils. Each myofibril is composed of parallel protein microfilaments called **myofilaments.** The key to muscle contraction lies in the arrangement and action of these myofilaments, so we must examine these at a molecular level. There are three kinds of myofilaments:

1. **Thick filaments** (fig. 10.9*a, b*) are about 15 nm in diameter. Each is made of several hundred molecules of a protein called **myosin.** A myosin molecule is shaped like a golf club, with two chains that intertwine to form a shaftlike *tail* and a double globular *head (cross-bridge)* projecting from it at an angle. A thick filament may be likened to a bundle of 200 to 500 such "golf clubs," with their heads directed outward in a spiral array around the bundle. The heads on one half of the thick filament angle to the left, and the heads on the other half angle to the right; in the middle is a *bare zone* with no heads.
2. **Thin filaments** (fig. 10.9*c, d*), 7 nm in diameter, are composed primarily of two intertwined strands of a protein called **fibrous (F) actin.** Each F actin is like a bead necklace—a string of subunits called **globular (G) actin.** Each G actin has an **active site** that can bind to the head of a myosin molecule (see fig. 10.15).

   A thin filament also has 40 to 60 molecules of yet another protein called **tropomyosin.** When a muscle fiber is relaxed, tropomyosin blocks the active sites of actin and prevents myosin from binding to it. Each tropomyosin molecule, in turn, has a smaller calcium-binding protein called **troponin** bound to it.
3. **Elastic filaments** (fig. 10.10*b, c*), 1 nm in diameter, are made of a huge springy protein called **titin**[13] **(connectin).** They run through the core of a thick filament, emerge from the end of it, and connect it to a structure called the *Z disc,* explained shortly. They help to keep thick and thin filaments aligned with each other, resist overstretching of a muscle, and help the cell recoil to resting length after it is stretched.

[11] *myo* = muscle + *blast* = precursor
[12] *sarco* = flesh, muscle + *lemma* = husk
[13] *tit* = giant + *in* = protein

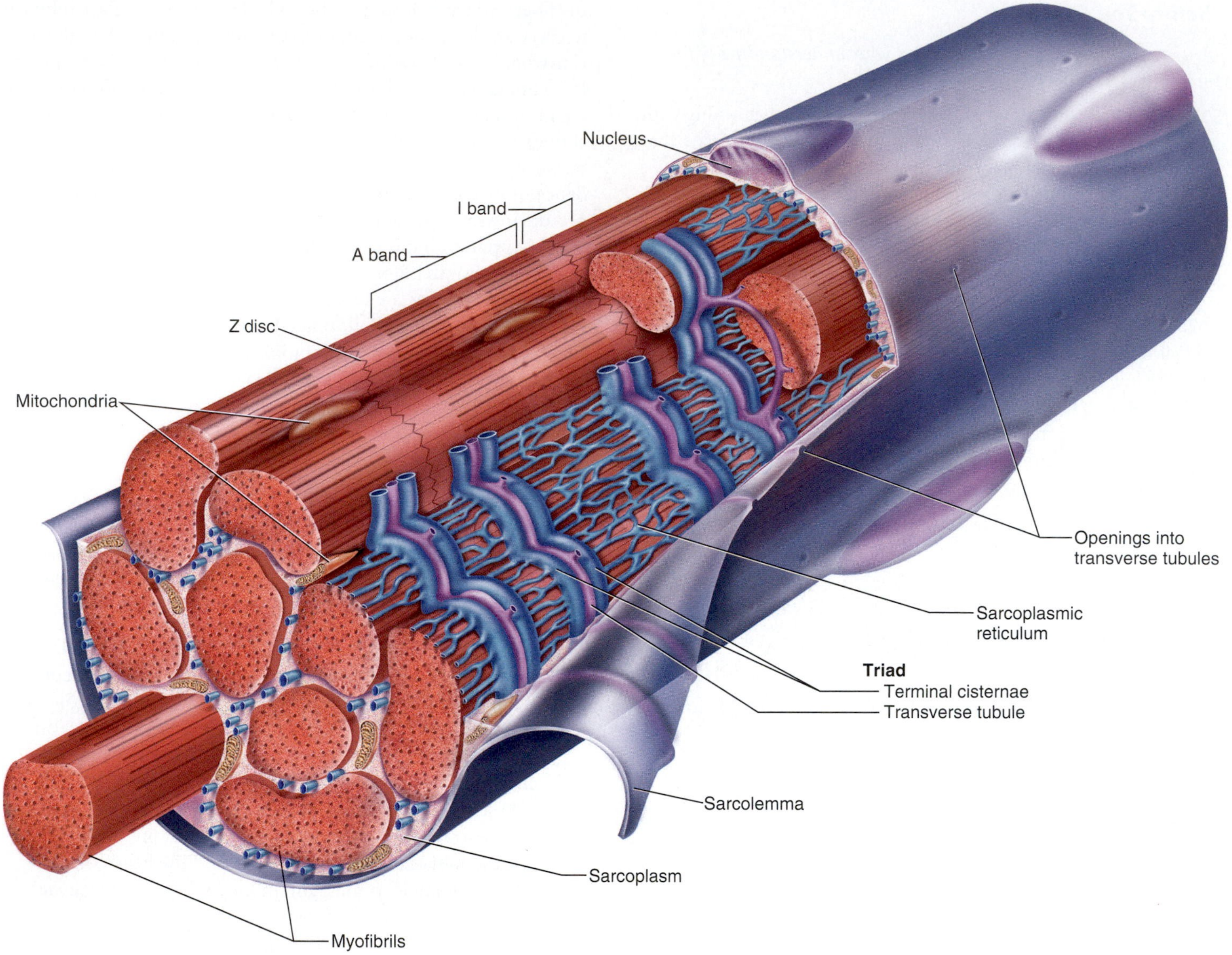

Nucleus
I band
A band
Z disc
Mitochondria
Openings into transverse tubules
Sarcoplasmic reticulum
Triad
Terminal cisternae
Transverse tubule
Sarcolemma
Sarcoplasm
Myofibrils

FIGURE 10.8

Structure of a Skeletal Muscle Fiber. This is a single cell containing 11 myofibrils (9 shown at the *left* and 2 cut off at midfiber). A muscle fiber can contain thousands of myofibrils.

Myosin and actin are called the **contractile proteins** of muscle because they do the work of shortening the muscle fiber. Tropomyosin and troponin are called the **regulatory proteins** because they act like a switch to determine when it can contract and when it cannot.

STRIATIONS AND SARCOMERES

Myosin and actin are not unique to muscle; these proteins occur in all cells, where they function in cellular motility, mitosis, and transport of intracellular materials. In skeletal and cardiac muscle they are especially abundant, however, and are organized into a precise array that accounts for the striations of these two muscle types (fig. 10.10).

Striated muscle has dark **A bands** alternating with lighter **I bands.** (*A* stands for *anisotropic* and *I* for *isotropic,* which refers to the way these bands affect polarized light. To help remember which band is which, think "d**A**rk" and "l**I**ght.") Each A band consists of thick filaments lying side by side. Part of the A band, where thick and thin filaments overlap, is especially dark. In this region, each thick filament is surrounded by thin filaments. In the middle of the A band, there is a lighter region called the **H band,**[14] into which the thin filaments do not reach.

Each light I band is bisected by a dark narrow **Z disc**[15] (Z line) composed of titin. The Z disc provides anchorage for the thin

[14]H = *helle* = bright (German)
[15]Z = *Zwichenscheibe* = between disc (German)

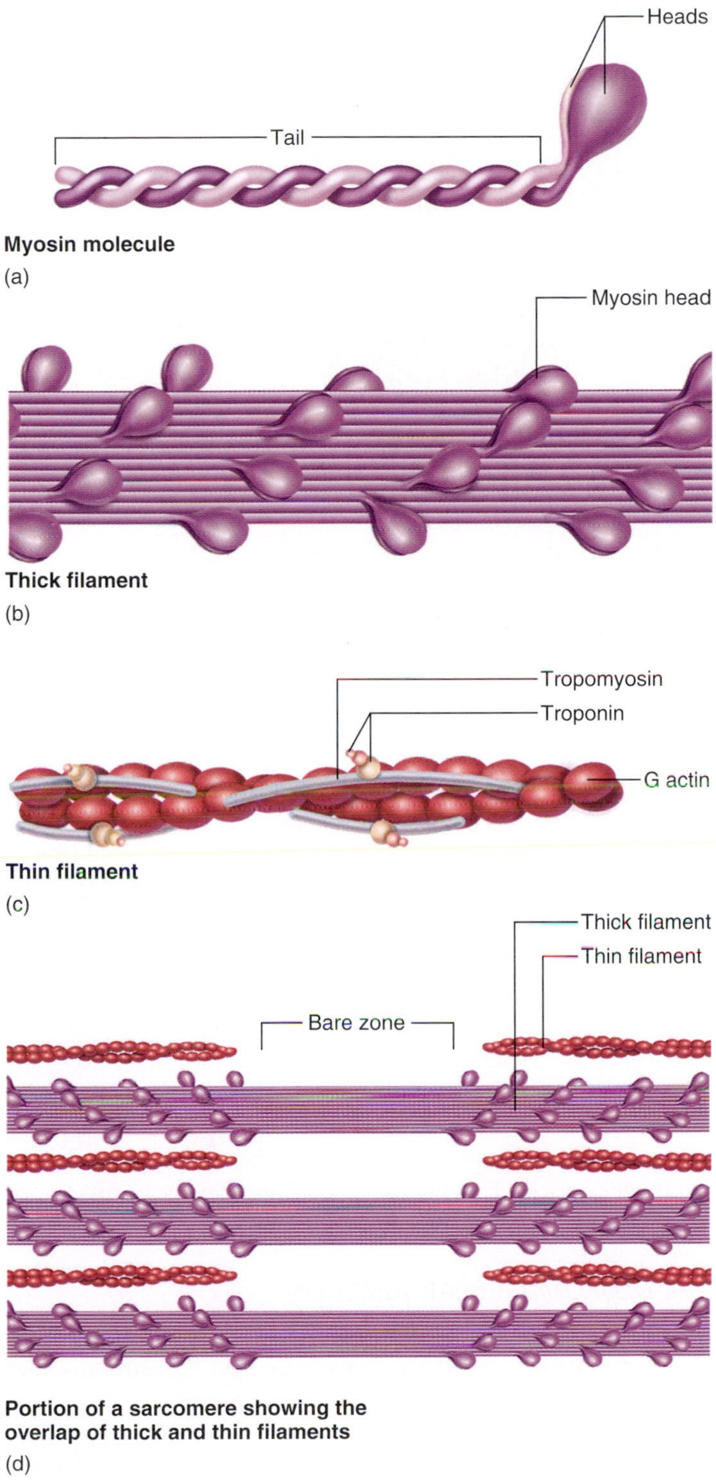
Heads
Tail
Myosin molecule
(a)
Myosin head
Thick filament
(b)
Tropomyosin
Troponin
G actin
Thin filament
(c)
Thick filament
Thin filament
Bare zone
Portion of a sarcomere showing the overlap of thick and thin filaments
(d)

FIGURE 10.9

Molecular Structure of Thick and Thin Filaments. (*a*) A single myosin molecule consists of two intertwined proteins forming a filamentous tail and a double globular head. (*b*) A thick filament consists of 200 to 500 myosin molecules bundled together with the heads projecting outward in a spiral array. (*c*) A thin filament consists of two intertwined chains of G actin molecules, smaller filamentous tropomyosin molecules, and a three-part protein called troponin associated with the tropomyosin. (*d*) A region of overlap between the thick and thin myofilaments.

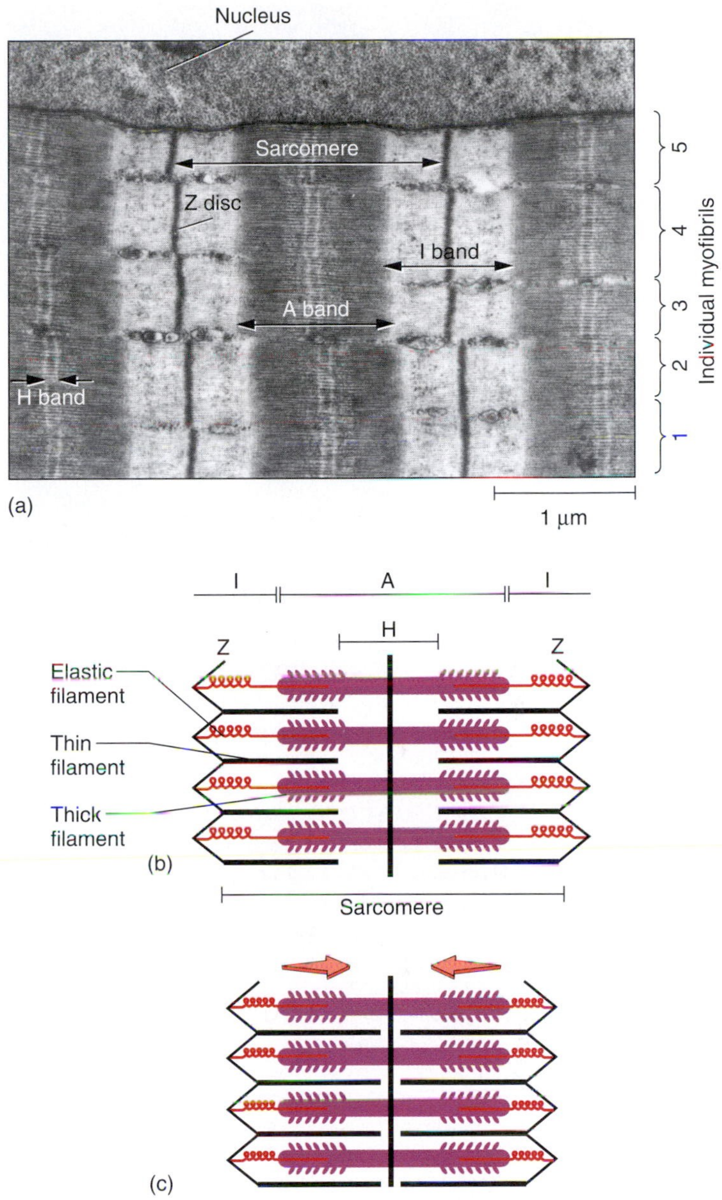
Nucleus
Sarcomere
Z disc
I band
A band
H band
Individual myofibrils
1
2
3
4
5
(a)
1 µm
I
A
I
H
Z
Z
Elastic filament
Thin filament
Thick filament
(b)
Sarcomere
(c)

FIGURE 10.10

Muscle Striations and Their Molecular Basis. (*a*) Five myofibrils of a single muscle fiber, showing the striations in the relaxed state (TEM). (*b*) The overlapping pattern of thick and thin myofilaments that accounts for the striations seen in figure *a.* (*c*) The pattern of myofilaments in a contracting muscle fiber. Note that all myofilaments in figure *c* are the same length as before, but they overlap to a greater extent.

filaments and elastic filaments. Each segment of a myofibril from one Z disc to the next is called a **sarcomere**[16] (SAR-co-meer), the functional contractile unit of the muscle fiber.

The terminology of muscle fiber structure is reviewed in table 10.1.

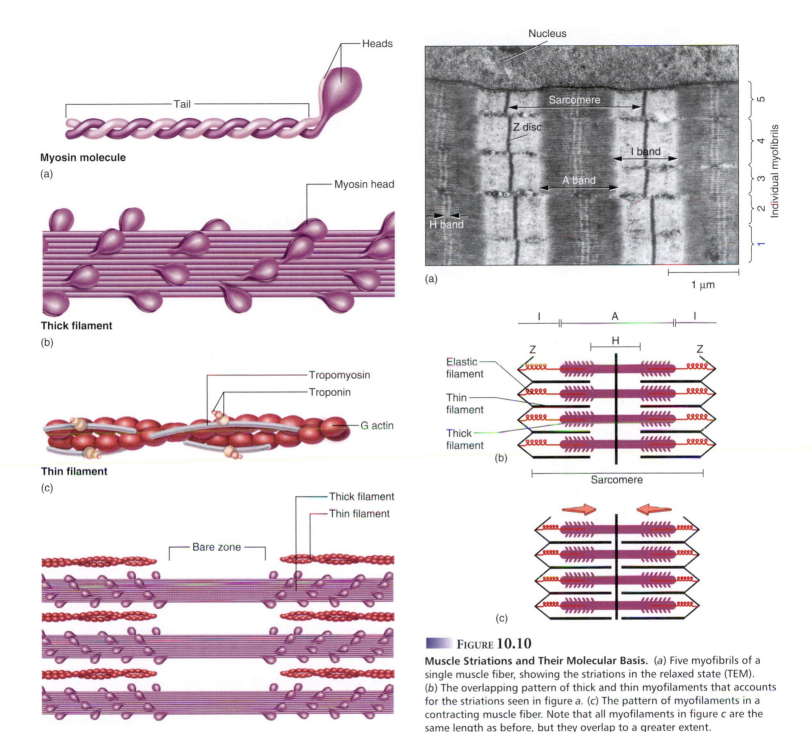

[16]*sarco* = muscle + *mere* = part, segment

Table 10.1

Structural Components of a Muscle Fiber

Term	Definition
General Structure and Contents of the Muscle Fiber	
Sarcolemma	The plasma membrane of a muscle fiber
Sarcoplasm	The cytoplasm of a muscle fiber
Glycogen	An energy-storage polysaccharide abundant in muscle
Myoglobin	An oxygen-storing red pigment of muscle
T tubule	A tunnel-like extension of the sarcolemma extending from one side of the muscle fiber to the other; conveys electrical signals from the cell surface to its interior
Sarcoplasmic reticulum	The smooth ER of a muscle fiber; a Ca^{2+} reservoir
Terminal cisternae	The dilated ends of sarcoplasmic reticulum adjacent to a T tubule
Myofibrils	
Myofibril	A bundle of protein microfilaments (myofilaments)
Myofilament	A threadlike complex of several hundred contractile protein molecules
Thick filament	A myofilament about 11 nm in diameter composed of bundled myosin molecules
Elastic filament	A myofilament about 1 nm in diameter composed of a giant protein, titin, that emerges from the core of a thick filament and links it to a Z disc; aids in the recoil of a relaxing muscle fiber.
Thin filament	A myofilament about 5 to 6 nm in diameter composed of actin, troponin, and tropomyosin
Myosin	A protein with a long shaftlike tail and a globular head; constitutes the thick myofilament
F actin	A fibrous protein made of a long chain of G actin molecules twisted into a helix; main protein of the thin myofilament
G actin	A globular subunit of F actin with an active site for binding a myosin head
Regulatory proteins	Troponin and tropomyosin, proteins that do not directly engage in the sliding filament process of muscle contraction but regulate myosin-actin binding
Tropomyosin	A regulatory protein that lies in the groove of F actin and, in relaxed muscle, blocks the myosin-binding active sites
Troponin	A regulatory protein associated with tropomyosin that acts as a calcium receptor
Titin	A springy protein that forms the elastic filaments and Z discs
Striations and Sarcomeres	
Striations	Alternating light and dark transverse bands across a myofibril
A band	Dark band formed by parallel thick filaments that partly overlap the thin filaments
H band	A lighter region in the middle of an A band that contains thick filaments only; thin filaments do not reach this far into the A band in relaxed muscle
I band	A light band composed of thin filaments only
Z disc	A disc of titin to which thin filaments and elastic filaments are anchored at each end of a sarcomere; appears as a narrow dark line in the middle of the I band
Sarcomere	The distance from one Z disc to the next; the contractile unit of a muscle fiber

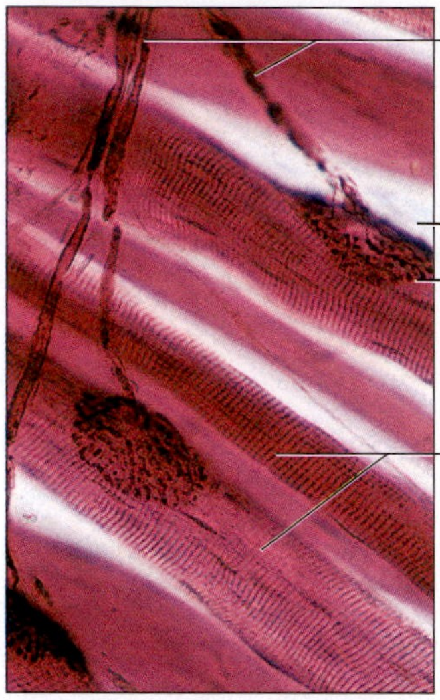
Motor nerve fibers
Neuromuscular junction
Muscle fibers
100 µm

Figure 10.11
Innervation of Skeletal Muscle (LM). Compare page 255.

The Nerve-Muscle Relationship

Skeletal muscle never contracts unless it is stimulated by a nerve (or artificially with electrodes). If its nerve connections are severed or poisoned, a muscle is paralyzed. Thus, muscle contraction cannot be understood without first understanding the relationship between nerve and muscle cells.

MOTOR NEURONS

Skeletal muscles are innervated by *somatic motor neurons* (see fig. 3.24). These are nerve cells whose cell bodies lie in the brainstem and spinal cord and whose axons, called **somatic motor fibers,** lead to the skeletal muscles. At its distal end, each somatic motor fiber branches about 200 times, with each branch leading to a different muscle fiber (fig. 10.11). Each muscle fiber is innervated by only one motor neuron.

THE NEUROMUSCULAR JUNCTION

The functional connection between a nerve fiber and any target cell that it stimulates is called a **synapse** (SIN-aps). A **neuromuscular junction** is a synapse between a nerve fiber and a muscle cell (fig. 10.12). Each branch of a motor nerve fiber ends in a bulbous swelling called a **synaptic** (sih-NAP-tic) **knob,** which is nestled in a depression on the sarcolemma called the **motor end plate.** The two cells do not actually touch each other but are separated by a tiny gap, 60 to 100 nm wide, called the **synaptic cleft.** A third cell, called a *Schwann cell,* envelops the entire junction and isolates it from the surrounding tissue fluid.

The synaptic knob contains spheroid organelles called **synaptic vesicles,** which are filled with a chemical called **acetylcholine** (ASS-eh-till-CO-leen) **(ACh).** When a nerve signal arrives

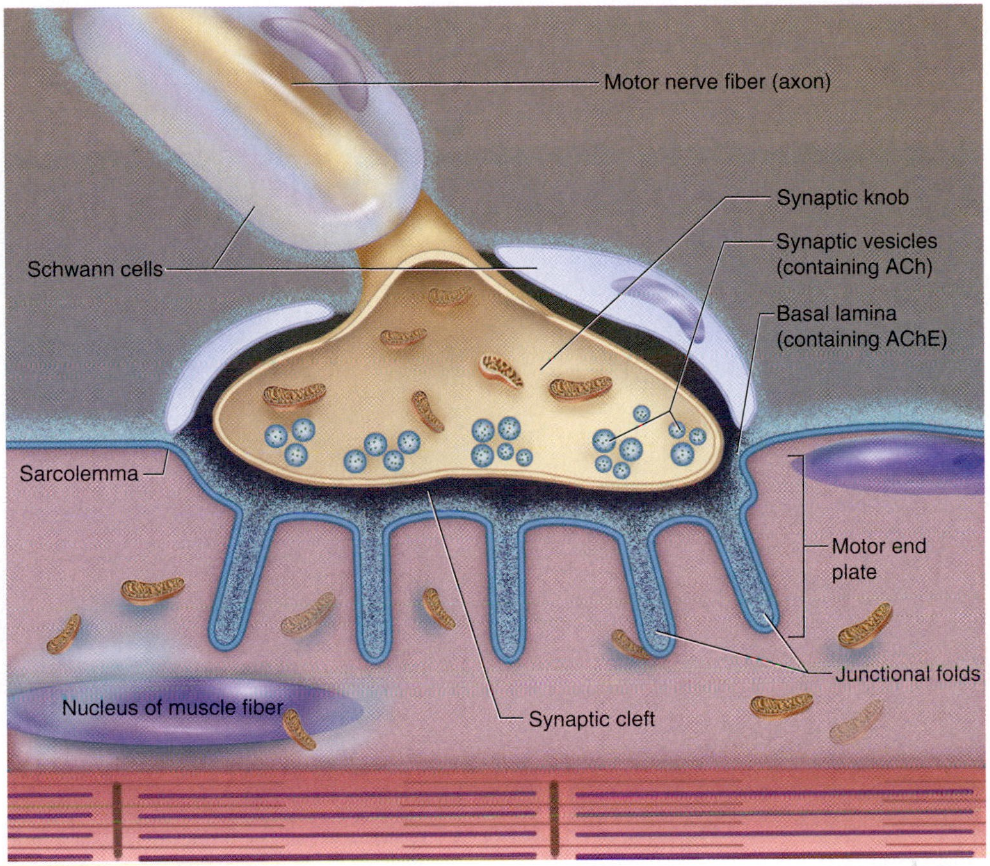
Motor nerve fiber (axon)
Synaptic knob
Synaptic vesicles (containing ACh)
Basal lamina (containing AChE)
Schwann cells
Sarcolemma
Motor end plate
Junctional folds
Nucleus of muscle fiber
Synaptic cleft

FIGURE 10.12
A Neuromuscular Junction.

at the synaptic knob, some of these vesicles release their ACh by exocytosis. ACh diffuses across the synaptic cleft and binds to **ACh receptors,** membrane proteins of the sarcolemma. These receptors respond to ACh by initiating events that lead to muscle contraction. The sarcolemma has infoldings called **junctional folds** that increase the membrane surface area and allow for more ACh receptors, and thus more sensitivity of the muscle fiber to nervous stimulation. When ACh has completed its task of stimulating the muscle, it is broken down by an enzyme called **acetylcholinesterase** (ASS-eh-till-CO-lin-ESS-ter-ase) **(AChE),** found in the sarcolemma and the synaptic cleft.

THE MOTOR UNIT

When a nerve signal approaches the end of an axon, it spreads out over all of its terminal branches and stimulates all the muscle fibers supplied by them. Thus, these muscle fibers contract in unison. Since they behave as a single functional unit, one nerve fiber and all the muscle fibers innervated by it are called a **motor unit.** The muscle fibers of a single motor unit are not all clustered together but are dispersed throughout a muscle (fig. 10.13). Thus, when they are stimulated, they cause a weak contraction over a wide area—not just a localized twitch in one small region.

INSIGHT 10.2 CLINICAL APPLICATION

NEUROMUSCULAR TOXINS AND PARALYSIS

Toxins that interfere with synaptic function can paralyze the muscles. Some pesticides, for example, contain *cholinesterase inhibitors* that bind to AChE and prevent it from degrading ACh. This causes *spastic paralysis,* a state of continual contraction of the muscle, which poses the danger of suffocation if it affects the laryngeal and respiratory muscles.

Tetanus (lockjaw) is a form of spastic paralysis caused by the toxin of a bacterium, *Clostridium tetani.* In the spinal cord, a chemical called glycine normally stops motor neurons from producing unwanted muscle contractions. The tetanus toxin blocks glycine release and thus causes overstimulation and spastic paralysis of the muscles.

Flaccid paralysis is a state in which the muscles are limp and cannot contract. This too can cause respiratory arrest if it affects the thoracic muscles. Flaccid paralysis can be caused by poisons such as curare (cue-RAH-ree) that compete with ACh for receptor sites but do not stimulate the muscle. Curare is extracted from certain plants and used by some South American natives to poison blowgun darts. It has been used to treat muscle spasms in some neurological disorders and to relax abdominal muscles for surgery, but other muscle relaxants have now replaced curare for most purposes.

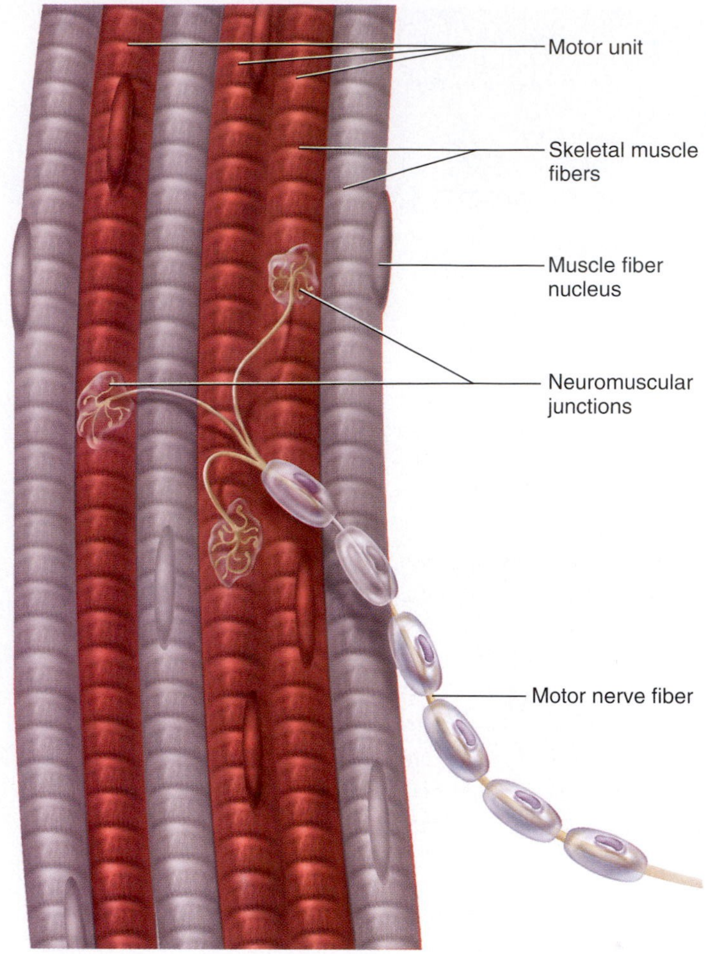
Motor unit
Skeletal muscle fibers
Muscle fiber nucleus
Neuromuscular junctions
Motor nerve fiber

FIGURE 10.13
A Motor Unit. The motor nerve fiber shown here branches to supply those muscle fibers shown in color. The other muscle fibers *(gray)* belong to other motor units.

Earlier it was stated that a motor nerve fiber supplies about 200 muscle fibers, but this is just a representative number. Where fine control is needed, we have *small motor units.* In the muscles of eye movement, for example, there are about 3 to 6 muscle fibers per nerve fiber. Small motor units are not very strong, but they provide the fine degree of control needed for subtle movements. They also have small neurons that are easily stimulated. Where strength is more important than fine control, we have large motor units. The gastrocnemius muscle of the calf, for example, has about 1,000 muscle fibers per nerve fiber. Large motor units are much stronger, but have larger neurons that are harder to stimulate and they do not produce such fine control.

One advantage of having multiple motor units in a muscle is that they are able to "work in shifts." Muscle fibers fatigue when subjected to continual stimulation. If all of the fibers in one of your postural muscles fatigued at once, for example, you might collapse. To prevent this, other motor units take over while the fatigued ones rest, and the muscle as a whole can sustain long-term contraction. The role of motor units in muscular strength is discussed later in the chapter.

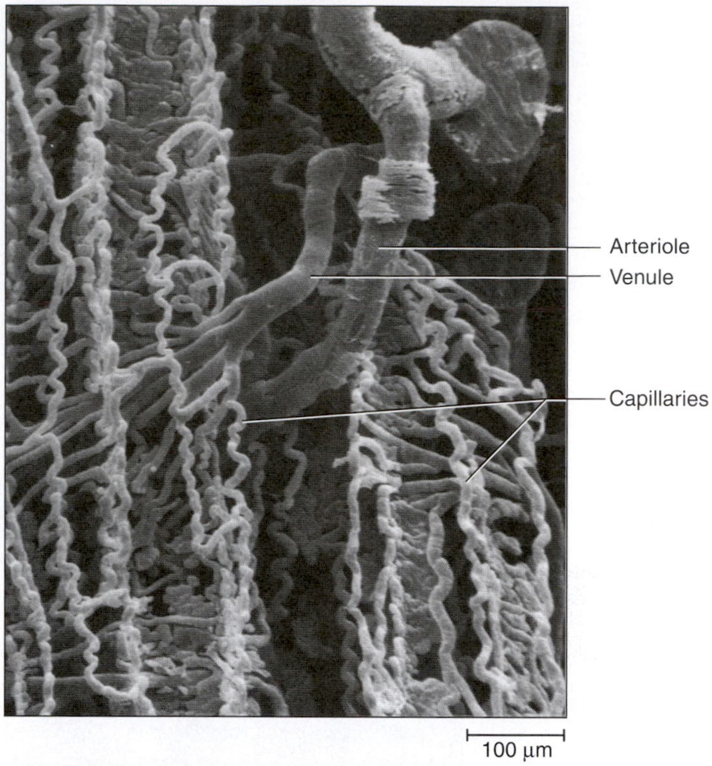
Arteriole
Venule
Capillaries
100 µm

FIGURE 10.14
A Vascular Cast of the Blood Vessels in a Contracted Skeletal Muscle. This was prepared by injecting the blood vessels with a polymer, digesting away the tissue to leave a replica of the vessels, and photographing the cast through the SEM. From R. G. Kessell and R. H. Kardon, *Tissues and Organs: A Text-Atlas of Scanning Electron Microscopy* (W. H. Freeman & Co., 1979).

The Blood Supply

The muscular system as a whole receives about 1.25 L of blood per minute at rest—which is about one-quarter of the blood pumped by the heart. During heavy exercise, total cardiac output rises and the muscular system's share of it is more than three-quarters, or 11.6 L/min. Working muscle has a great demand for glucose and oxygen. Blood capillaries ramify through the endomysium to reach every muscle fiber, sometimes so intimately associated with the muscle fibers that the muscle fibers have surface indentations to accommodate them. The capillaries of skeletal muscle undulate or coil when the muscle is contracted (fig. 10.14), allowing them enough slack to stretch out straight, without breaking, when the muscle lengthens.

Before You Go On

Answer the following questions to test your understanding of the preceding section:

12. What special terms are given to the plasma membrane, cytoplasm, and smooth ER of a muscle cell?
13. Name the proteins that compose the thick and thin filaments of a muscle fiber and describe their structural arrangement.

14. Define *sarcomere.* Describe the striations of a sarcomere and sketch the arrangement of thick and thin filaments that accounts for the striations.
15. Describe the role of a synaptic knob, synaptic vesicles, synaptic cleft, and acetylcholine in neuromuscular function.
16. What is a motor unit? How do large and small motor units differ functionally?
17. Why is it important that the blood capillaries of a contracted muscle have an undulating or coiled arrangement?

Functional Perspectives

Objectives

When you have completed this section, you should be able to

- explain how a muscle fiber contracts and relaxes, and relate this to its ultrastructure;
- describe how muscle grows and shrinks with use and disuse; and
- discuss two physiological categories of muscle fibers and their respective advantages.

The fine structure of a muscle fiber means little without some understanding of what purpose it serves. In this section, we will relate the foregoing ultrastructure to the mechanism of contraction and relaxation, see how exercise and disuse affect muscle, and examine some physiological types of muscle fibers that relate to functional differences between various muscles of the body.

Contraction and Relaxation

Muscle contraction and relaxation occurs in four stages (fig. 10.15):

1. **Excitation.** Excitation begins when a signal in a nerve fiber triggers the release of acetylcholine (ACh). ACh diffuses across the synaptic cleft and binds to receptors in the sarcolemma. These receptors are gated sodium-potassium channels that open as long as ACh is bound to them. The flow of Na^+ and K^+ through the gated channels produces a change in the voltage across the sarcolemma. This change, in turn, sets off a chain reaction of electrical excitation that spreads in all directions along the muscle fiber, like ripples spreading across the surface of a pond after you drop a stone into the water. The excitation spreads even down the T tubules to the interior of the muscle fiber.
2. **Excitation-contraction coupling.** This stage links excitation to the initiation of muscle contraction. The T tubules, as we have seen, are closely associated with the terminal cisternae of the sarcoplasmic reticulum (SR). Electrical signals spreading down the T tubules indirectly open calcium channels in the SR. The SR releases a flood of Ca^{2+} into the cytosol. Ca^{2+} binds to the troponin of the thin filaments, and the activated troponin causes tropomyosin to shift position so that it no longer blocks the active sites on actin. Now that these active sites are exposed, the heads of the myosin filaments can bind to them.
3. **Contraction.** A myosin head, when activated by ATP, swings forward into a high-energy "cocked" position and binds to an active site of actin. The myosin head then flexes and tugs on the actin, pulling it a short distance—like flexing your elbow to pull in the rope of a boat anchor. This movement is called the *power stroke* of the myosin head. Myosin then binds a new ATP, recocks, and repeats the process. Whenever one myosin head lets go of the thin filament, other myosin heads hold on, so the thin filament is never entirely released as long as the muscle is contracting. The myosin heads "take turns" pulling on the actin filament and letting go. The overall effect is that the thin filament slides smoothly alongside the thick filament; this model of muscle contraction is therefore called the *sliding filament theory.* No myofilaments get shorter during contraction—they merely slide across each other.

   As the thin filaments slide across the thick filaments, they pull on the Z discs, bringing them closer together. As shown in figure 10.10*c*, this shortens each sarcomere. The Z discs are connected to the sarcolemma by way of the cytoskeleton, so as the Z discs move closer together, they pull on the sarcolemma and the entire cell shortens.
4. **Relaxation.** Relaxation begins when nerve signals stop arriving at the neuromuscular junction and the nerve fiber stops releasing ACh. An enzyme in the synapse breaks down the ACh that is already present, thus halting the stimulation of the muscle fiber. When electrical excitation of the sarcolemma ceases, the sarcoplasmic reticulum begins pumping Ca^{2+} back into its cisternae for storage. As the Ca^{2+} level in the cytosol declines, tropomyosin moves back into its resting position, blocking the active sites of the thin filaments. Myosin-actin cross-bridges can no longer form, and the muscle relaxes.

Think About It!

During muscle contraction, which band(s) of the muscle striations would you expect to become narrower or disappear? Which would remain the same width as in relaxed muscle? Explain.

Muscle Growth and Atrophy

It is common knowledge that muscles grow larger when exercised and shrink when they are not used. This is the basis for *resistance exercises* such as weight lifting. And yet, skeletal muscle fibers are incapable of mitosis. We have about the same number of muscle fibers in adulthood as we do in late childhood. How, then, does a muscle grow? Exercise stimulates the muscle fiber to produce more protein myofilaments. As a result, the myofibrils grow thicker. At a certain point, a large myofibril splits longitudinally, so a well-conditioned muscle has more myofibrils per muscle fiber than does a weakly conditioned one. The entire muscle grows in bulk (thickness), not by the mitosis of existing cells (hyperplasia), but by the enlargement of cells that have existed since childhood (hypertrophy). Some authorities, however, think that entire muscle fibers (not just their myofibrils) may split longitudinally when they reach a certain size, thus giving rise to an increase in the number of

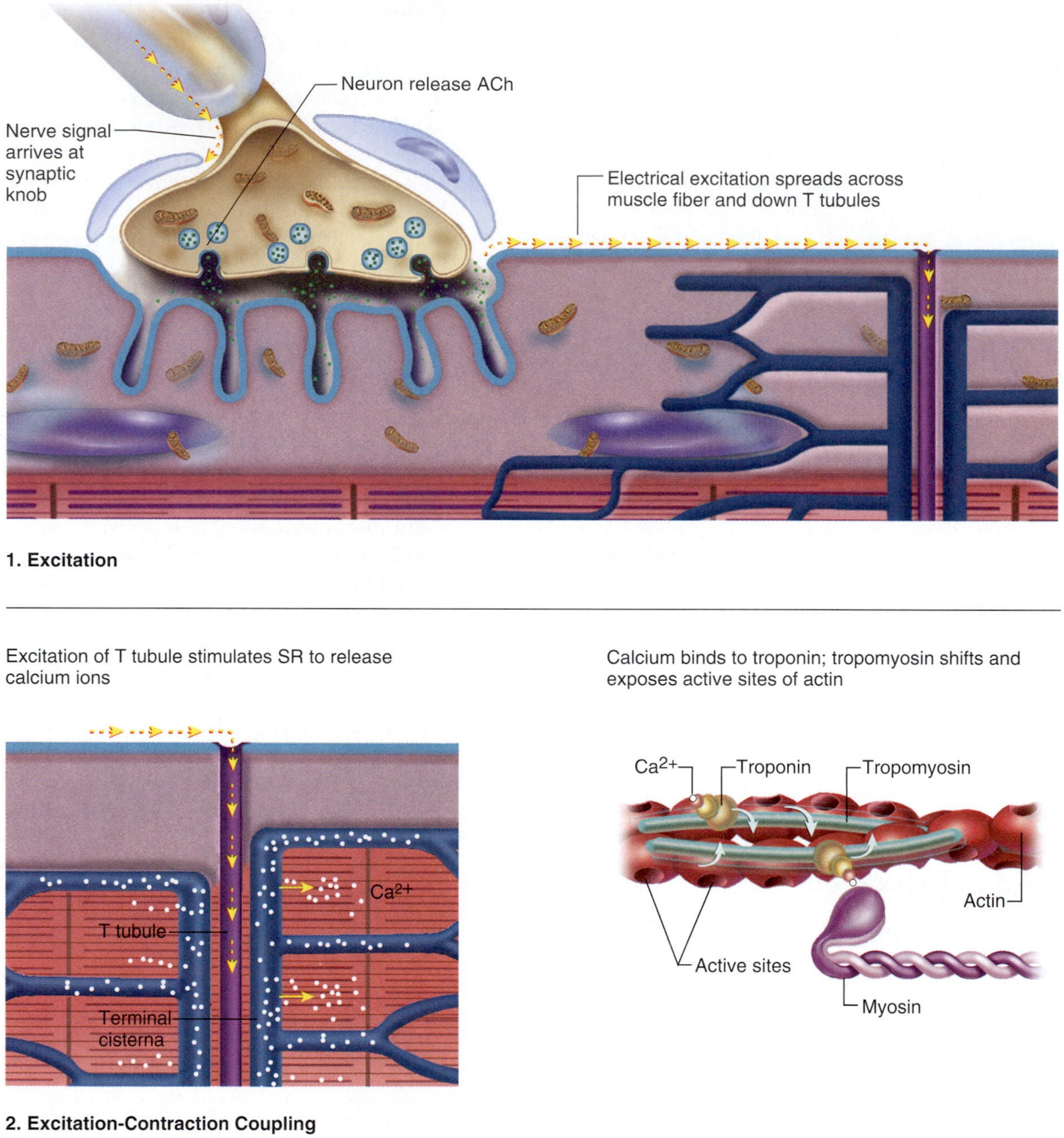
Neuron release ACh
Nerve signal arrives at synaptic knob
Electrical excitation spreads across muscle fiber and down T tubules
1. Excitation
Excitation of T tubule stimulates SR to release calcium ions
Calcium binds to troponin; tropomyosin shifts and exposes active sites of actin
Ca2+
T tubule
Terminal cisterna
Ca2+
Troponin
Tropomyosin
Actin
Active sites
Myosin
2. Excitation-Contraction Coupling

Figure 10.15

The Principal Events in Muscle Contraction and Relaxation.

fibers—not by mitosis but by a process more akin to tearing. Well-exercised muscles also develop more mitochondria, more myoglobin and glycogen, and a greater density of blood capillaries.

When a muscle is not used, it shrinks (atrophies). This can result from spinal cord injuries or other injuries that sever the nerve connections to a muscle *(denervation atrophy)*, from lack of exercise *(disuse atrophy)*, or from aging *(senescence atrophy)*. The shrinkage of a limb that has been in a cast for several weeks is a good example of disuse atrophy. Muscle quickly regrows when exercise resumes, but if the atrophy becomes too advanced, muscle fibers die and are not replaced. Physical therapy is therefore important for maintaining muscle mass in people who are unable to use the muscles voluntarily.

Physiological Classes of Muscle Fibers

Not all muscle fibers are metabolically alike or adapted to perform the same task. Some respond slowly but are relatively resistant to fatigue, while others respond more quickly but also fatigue quickly. Each primary type of fiber goes by several names:

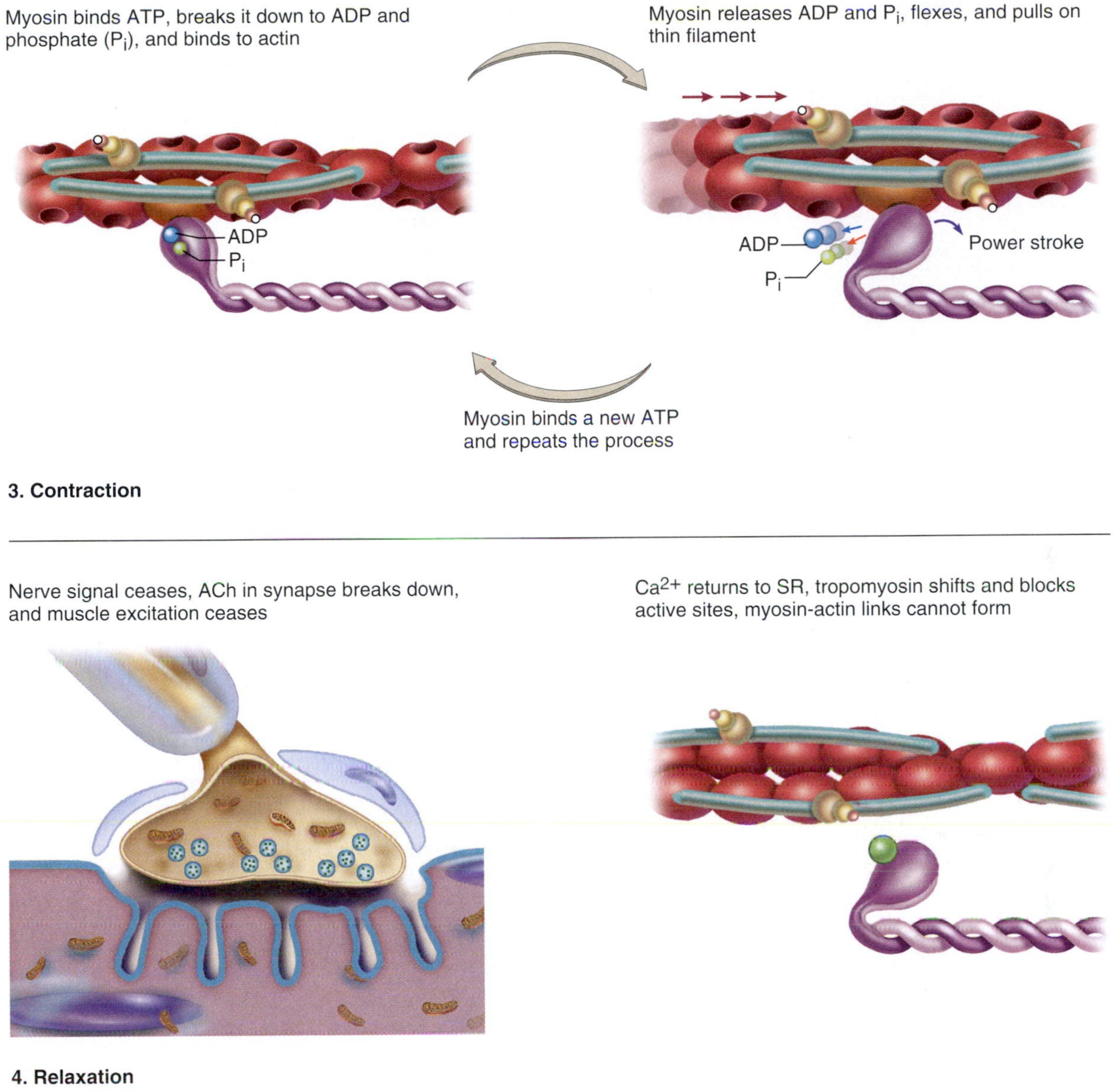
Myosin binds ATP, breaks it down to ADP and phosphate (P$_i$), and binds to actin
ADP
P$_i$
Myosin releases ADP and P$_i$, flexes, and pulls on thin filament
ADP
P$_i$
Power stroke
Myosin binds a new ATP and repeats the process
3. Contraction
Nerve signal ceases, ACh in synapse breaks down, and muscle excitation ceases
Ca^{2+} returns to SR, tropomyosin shifts and blocks active sites, myosin-actin links cannot form
4. Relaxation

FIGURE 10.15
The Principal Events in Muscle Contraction and Relaxation. *(continued)*

- **Slow oxidative (SO), slow-twitch, red,** or **type I fibers.** These fibers have relatively abundant mitochondria, myoglobin, and blood capillaries, and therefore a relatively deep red color. They are well adapted to aerobic respiration, a means for making ATP that does not generate lactic acid, a major contributor to muscle fatigue. Thus, these fibers do not fatigue easily. However, in response to a single stimulus, they exhibit a relatively long *twitch,* or contraction, lasting about 100 milliseconds (msec). The soleus muscle of the calf and the postural muscles of the back are composed mainly of these slow-twitch, fatigue-resistant fibers.
- **Fast glycolytic (FG), fast-twitch, white,** or **type II fibers.** These fibers are rich in enzymes for anaerobic fermentation, a process that produces lactic acid. They respond quickly, with twitches as short as 7.5 msec, but because of the lactic acid they generate, they fatigue more easily than SO fibers. They are poorer in mitochondria, myoglobin, and blood capillaries than SO fibers, so they are relatively pale (hence the expression *white* fibers). They are well adapted for quick responses but not for endurance. Thus, they are especially important in sports such as basketball that require stop-and-go activity and frequent changes of pace. The gastrocnemius muscle of the calf, biceps brachii of the arm, and the muscles of eye movement consist mainly of FG fibers.

Some authorities recognize two subtypes of FG fibers called types IIA and IIB. Type IIB is the common type just described, while IIA, or **intermediate fibers,** combine fast-twitch responses with aerobic fatigue-resistant metabolism. Type IIA fibers,

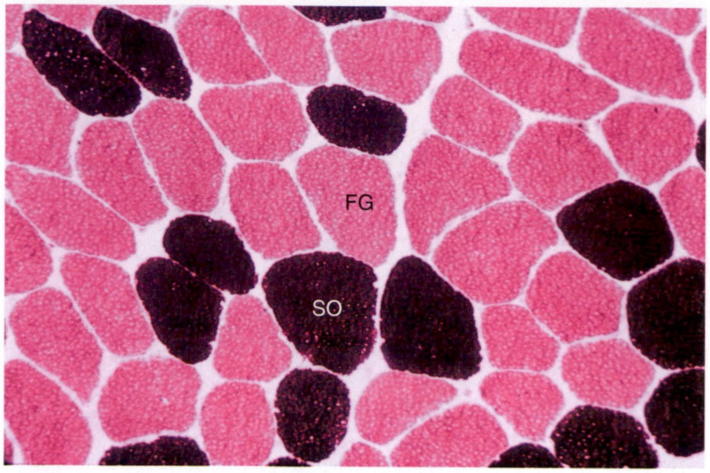

FIGURE 10.16

Types of Muscle Fibers in a Skeletal Muscle. *FG,* fast glycolytic fiber. *SO,* slow oxidative fiber.

TABLE 10.2

Classification of Skeletal Muscle Fibers

	Fiber Type	
Properties	**Slow Oxidative**	**Fast Glycolytic**
Relative diameter	Smaller	Larger
ATP synthesis	Aerobic	Anaerobic
Fatigue resistance	Good	Poor
ATP hydrolysis	Slow	Fast
Glycolysis	Moderate	Fast
Myoglobin content	Abundant	Low
Glycogen content	Low	Abundant
Mitochondria	Abundant and large	Fewer and smaller
Capillaries	Abundant	Fewer
Color	Red	White, pale
Representative Muscles in Which Fiber Type Is Predominant		
	Soleus	Gastrocnemius
	Erector spinae	Biceps brachii
	Quadratus lumborum	Muscles of eye movement

however, are relatively rare except in some endurance-trained athletes. The three fiber types can be differentiated histologically by using stains for certain mitochondrial enzymes and other cellular components (fig. 10.16). Table 10.2 summarizes the difference between SO and FG fibers.

All muscles are composed of both SO and FG fibers, but the proportions of these fiber types differ from one muscle to another. Muscles composed mainly of SO fibers are called *red muscles* and those composed mainly of FG fibers are called *white muscles.* People with different types and levels of physical activity differ in the proportion of one fiber type to another even in the same muscle, such as the *quadriceps femoris* of the anterior thigh (table 10.3). It is thought that people are born with a genetic predisposition for a certain ratio of fiber types. Those who go into competitive sports discover the sports at which they can excel and gravitate toward those for which heredity has best equipped them. One person might be a "born sprinter" and another a "born marathoner."

TABLE 10.3

Proportion of Slow- and Fast-Twitch Fibers in the Quadriceps Femoris of Male Athletes

Sample Population	Slow-Twitch (SO)	Fast-Twitch (FG)
Marathon runners	82%	18%
Swimmers	74	26
Average males	45	55
Sprinters and jumpers	37	63

We noted earlier that sometimes two or more muscles act across the same joint and superficially seem to have the same function. We have already seen some reasons why such muscles are not as redundant as they seem. Another reason is that they may differ in the proportion of SO to FG fibers. For example, the gastrocnemius and soleus muscles of the calf both insert on the calcaneus through the same tendon, the calcaneal tendon, so they exert the same pull on the heel. The gastrocnemius, however, is a predominantly fast glycolytic muscle adapted for quick, powerful movements such as jumping, whereas the soleus is a predominantly slow oxidative muscle that does most of the work in endurance exercises such as jogging and skiing.

Before You Go On

Answer the following questions to test your understanding of the preceding section:

18. What role does the sarcoplasmic reticulum play in muscle contraction? What role does it play in muscle relaxation?
19. Why does tropomyosin have to move before a muscle fiber can contract? What makes it move?
20. What role does ATP play in muscle contraction?
21. What is the mechanism of muscle growth? Describe the growth process in muscle and distinguish it from hyperplasia.
22. What are the basic functional differences between slow oxidative and fast glycolytic muscle fibers?

CARDIAC AND SMOOTH MUSCLE

Objectives

When you have completed this section, you should be able to

- describe cardiac muscle tissue and compare its structure and physiology to the other types; and
- describe smooth muscle tissue and compare its structure and physiology to the other types.

In this section, we will compare cardiac muscle and smooth muscle to skeletal muscle. Cardiac and smooth muscle have special structural and physiological properties related to their distinctive functions.

TABLE 10.4

Comparison of Skeletal, Cardiac, and Smooth Muscle

Feature	Skeletal Muscle	Cardiac Muscle	Smooth Muscle
Location	Associated with skeletal system	Heart	Walls of viscera and blood vessels, iris of eye, piloerectors of hair follicles
Cell shape	Long cylindrical fibers	Short branched cells	Fusiform cells
Cell length	100 μm–30 cm	50–100 μm	50–200 μm
Cell width	10–100 μm	10–20 μm	2–10 μm
Striations	Present	Present	Absent
Nuclei	Multiple nuclei, adjacent to sarcolemma	Usually one nucleus, near middle of cell	One nucleus, near middle of cell
Connective tissues	Endomysium, perimysium, epimysium	Endomysium only	Endomysium only
Sarcoplasmic reticulum	Abundant	Present	Scanty
T tubules	Present, narrow	Present, wide	Absent
Gap junctions	Absent	Present in intercalated discs	Present in single-unit smooth muscle
Ca^{2+} source	Sarcoplasmic reticulum	Sarcoplasmic reticulum and extracellular fluid	Mainly extracellular fluid
Innervation and control	Somatic motor fibers (voluntary)	Autonomic fibers (involuntary)	Autonomic fibers (involuntary)
Nervous stimulation required?	Yes	No	No
Mode of tissue repair	Limited regeneration, mostly fibrosis	Limited regeneration, mostly fibrosis	Relatively good capacity for regeneration

Cardiac Muscle

Cardiac muscle constitutes most of the heart. Its form and function are discussed extensively in chapter 20 so that you will be able to relate these to the actions of the heart. Here, we only briefly compare it to skeletal and smooth muscle (table 10.4).

Cardiac muscle is striated like skeletal muscle, but otherwise exhibits several differences. Its *cardiocytes (myocytes)* are not long multinucleate fibers but short, stumpy, slightly branched cells. Microscopically, cardiac muscle exhibits characteristic dark lines called **intercalated** (in-TUR-kuh-LAY-ted) **discs** where the cells meet. An intercalated disc is a steplike region containing electrical gap junctions that allow the cells to communicate with each other, and various mechanical junctions that prevent the cells from pulling apart when they contract (see details in chapter 20). Each myocyte can join several others at its intercalated discs.

Cardiocytes usually have only one, centrally placed nucleus (occasionally two) (fig. 10.17*a*), often surrounded by glycogen. Cardiac muscle is very rich in glycogen and myoglobin, and it has especially large mitochondria that fill about 25% of the cell, compared to smaller mitochondria occupying about 2% of a skeletal muscle fiber. Cardiac muscle is therefore very well adapted to aerobic respiration and very resistant to fatigue, although it is highly vulnerable to interruptions in its oxygen supply. The sarcoplasmic reticulum is less developed than in skeletal muscle, but the T tubules are larger and admit supplemental Ca^{2+} from the extracellular fluid. Cardiac myocytes have little capacity for mitosis. Furthermore, cardiac muscle has no satellite cells, so the repair of damaged cardiac muscle is primarily by fibrosis (scarring).

Cardiac muscle is innervated by the *autonomic nervous system (ANS)* rather than by somatic motor neurons. The ANS is a division of the nervous system that usually operates without one's

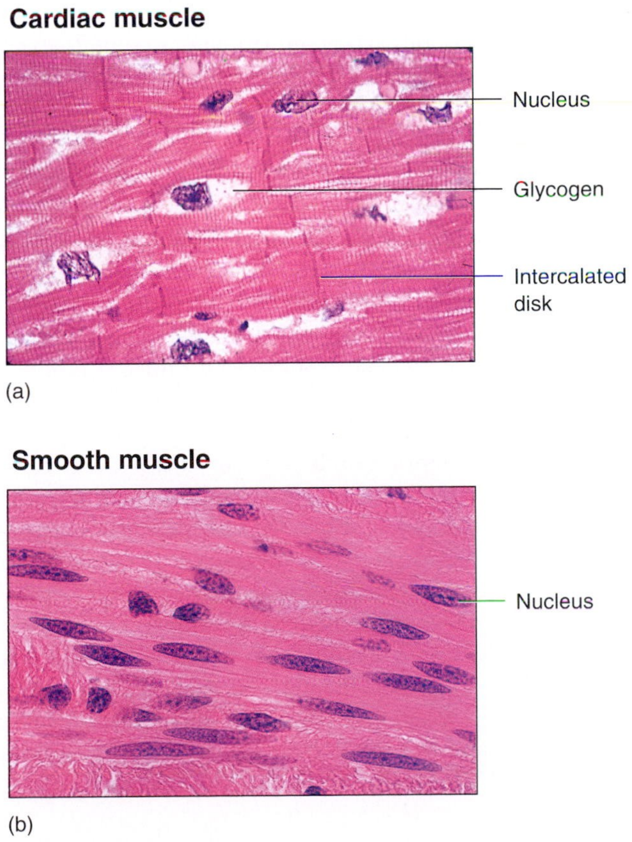
Cardiac muscle
Nucleus
Glycogen
Intercalated disk
(a)
Smooth muscle
Nucleus
(b)

FIGURE 10.17
Cardiac and Smooth Muscle.

conscious awareness or control. It does not generate the heartbeat, but it modulates the heart rate and contraction strength. Cardiocytes pulsate rhythmically even without nervous stimulation; this property is called *autorhythmicity.* In an intact heart, their beating is triggered by a pacemaker called the *sinoatrial node.*

Smooth Muscle

Smooth muscle is composed of myocytes with a fusiform shape, about 30 to 200 μm long, 5 to 10 μm wide at the middle, and tapering to a point at each end (fig. 10.17*b*). There is only one nucleus, located near the middle of the cell. Thick and thin filaments are both present, but they are not aligned with each other and produce no visible striations or sarcomeres; this is the reason for the name *smooth* muscle. Z discs are absent; instead, the thin filaments are attached by way of the cytoskeleton to **dense bodies,** little masses of protein scattered throughout the sarcoplasm and on the inner face of the sarcolemma.

The sarcoplasmic reticulum is scanty, and there are no T tubules. The calcium needed to activate smooth muscle contraction comes mainly from the extracellular fluid by way of calcium channels in the sarcolemma. During relaxation, calcium is pumped back out of the cell. Some smooth muscle has no nerve supply, but when nerve fibers are present, they come from the autonomic nervous system, like those of the heart.

Unlike skeletal and cardiac muscle, smooth muscle is capable of mitosis and hyperplasia. Thus, an organ such as the pregnant uterus can grow by adding more myocytes, and injured smooth muscle regenerates well.

There are two functional categories of smooth muscle called *multiunit* and *single-unit* types (fig. 10.18). **Multiunit smooth muscle** occurs in some of the largest arteries and pulmonary air passages, in the piloerector muscles of the hair follicles, and in the iris of the eye. Its innervation, although autonomic, is otherwise similar to that of skeletal muscle—the terminal branches of a nerve fiber synapse with individual myocytes and form a motor unit. Each motor unit contracts independently of the others, hence the name of this muscle type.

Single-unit smooth muscle is more widespread. It occurs in most blood vessels and in the digestive, respiratory, urinary, and reproductive tracts—thus, it is also called **visceral muscle.** The nerve fibers in this type of muscle do not synapse with individual muscle cells, but pass through the tissue and exhibit swellings called **varicosities** at which they release neurotransmitters. Neurotransmitters nonselectively stimulate multiple muscle cells in the vicinity of a varicosity. The muscle cells themselves are electrically coupled to each other by gap junctions. Thus, they directly stimulate each other and a large number of cells contract as a unit, almost as if they were a single cell. This is the reason that this muscle type is called *single-unit* smooth muscle.

In many of the hollow internal organs, visceral muscle forms two or more layers—typically an inner *circular layer,* in which the fibers encircle the organ, and an outer *longitudinal layer,* in which the fibers run lengthwise along the organ (fig. 10.19). When the cir-

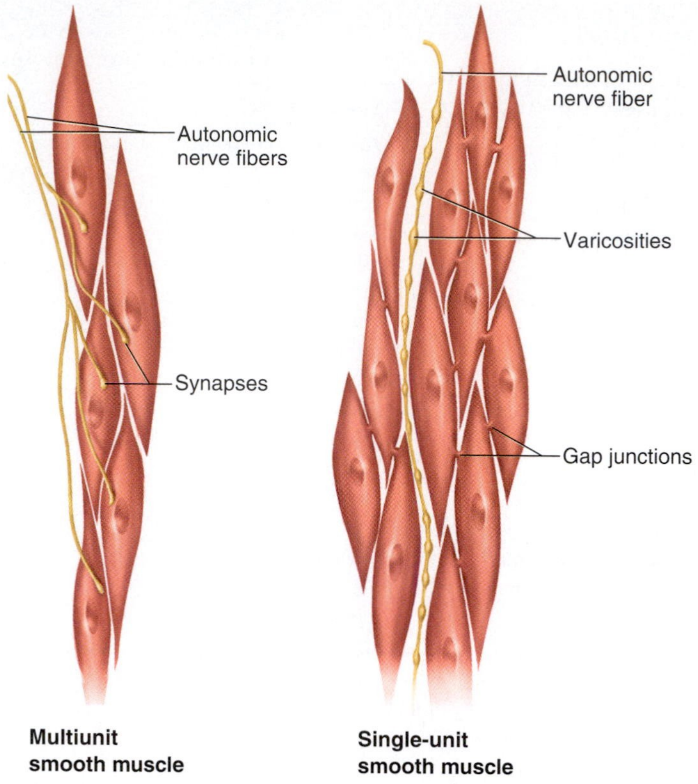

FIGURE 10.18

Smooth Muscle Types. In multiunit smooth muscle, each muscle cell receives its own nerve supply. In single-unit smooth muscle, a nerve fiber passes through the tissue without synapsing with any specific muscle cell, and the muscle cells are connected to each other by gap junctions.

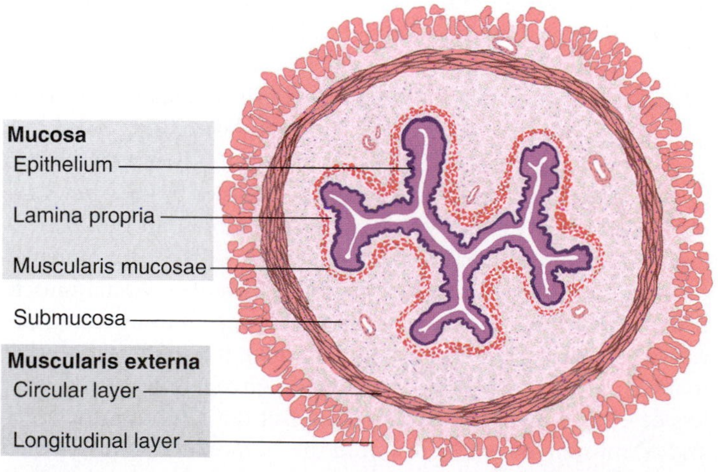

FIGURE 10.19

Layers of Visceral Muscle in the Wall of the Esophagus. Many hollow organs have alternating circular and longitudinal layers of smooth muscle.

cular layer of muscle contracts, it narrows the organ and may make it longer (like a roll of dough squeezed in your hands); when the longitudinal muscle contracts, it makes the organ shorter and thicker.

Smooth muscle contracts and relaxes slowly, and in response not only to nervous stimulation but also to chemicals and stretch. Its metabolism is mostly aerobic, but it has a very low energy requirement compared to skeletal and cardiac muscle, so it is highly fatigue-resistant. This enables smooth muscle to maintain a state of continual, partial contraction called **smooth muscle tone.** Smooth muscle tone maintains blood pressure by keeping the blood vessels partially constricted and it prevents such organs as the stomach, intestines, urinary bladder, and uterus from becoming flaccid. In the digestive tract and some other locations, smooth muscle is responsible for waves of contraction called **peristalsis** that propel the contents through an organ (food in the esophagus and urine in the ureters, for example).

Table 10.4 compares some properties of skeletal, cardiac, and smooth muscle.

Before You Go On

Answer the following questions to test your understanding of the preceding section:

23. What organelles are more abundant and larger in cardiac muscle than in skeletal muscle? What is the functional significance of this?
24. What organelle is less developed in cardiac muscle than in skeletal muscle? How does this affect the activation of muscle contraction in the heart?
25. What factors make cardiac muscle more resistant to fatigue than skeletal muscle? What accounts for the relative fatigue resistance of smooth muscle?
26. How are single-unit and multiunit smooth muscle different? Which type is more abundant?

Developmental and Clinical Perspectives

Objectives

When you have completed this section, you should be able to

- describe how the three types of muscle develop in the embryo;
- describe the changes that occur in the muscular system in old age;
- discuss two muscle diseases, muscular dystrophy and myasthenia gravis; and
- briefly define and discuss several other disorders of the muscular system.

Embryonic Development of Muscle

Muscular tissue arises from embryonic mesoderm, with the exception of the piloerector muscles of the skin and the muscles within the eye. As described in chapter 4, the mesoderm of the trunk forms segmentally arranged blocks of tissue called somites, which then divide into regions called the dermatomes, sclerotomes, and myotomes. The myotomes give rise to the skeletal muscles of the trunk. Beginning in week 4, mesodermal cells of the myotomes begin to elongate and assume a fusiform shape. The cells, now called **myoblasts,**[17] multiply rapidly (fig. 10.20). Even as myoblasts continue to proliferate, some of them fuse to form multinucleated **myotubes.** By week 8, the future muscles are differentiated and in their proper positions relative to the developing skeleton. Their growth is influenced by the cartilaginous models of the bones. Muscles of the head and limbs develop from less sharply defined masses of mesoderm, but otherwise by the same general process.

By week 9, myofilaments begin to appear in the myotubes. By week 17, the muscle fibers exhibit striations, and fetal muscle contractions are strong enough to be felt by the mother. It was once thought that the fetus first comes alive at this time, so this stage of development is called *quickening.*[18]

Myoblasts continue to fuse with the fetal muscle fibers and contribute to their growth until close to the time of birth. Some myoblasts persist as *satellite cells* in skeletal muscle. These cells contribute to new muscle growth in childhood and may regenerate a limited amount of damaged skeletal muscle even in adults. New muscle fibers are added up to about 1 year after birth, but there is no further mitosis in skeletal muscle thereafter.

Cardiac muscle develops in association with an embryonic *heart tube* described in chapter 20. Mesenchymal cells near the heart tube differentiate into myoblasts and these proliferate mitotically as they do in the development of skeletal muscle. But in contrast to skeletal muscle development, the myoblasts do not fuse. They remain joined to each other and develop intercalated discs at their points of adhesion. The heart begins beating in week 3. Mitosis in cardiac myocytes continues after birth and is active until about age 9, although there is now evidence of limited mitotic capability even in adults. There is understandable interest in being able to stimulate this process in hopes of promoting regeneration of cardiac muscle damaged by heart attacks.

Smooth muscle develops similarly from myoblasts associated with the embryonic gut, blood vessels, and other organs. As in cardiac muscle, these myoblasts never fuse with each other, but in single-unit smooth muscle, they do become interconnected through gap junctions.

The Aging Muscular System

One of the most noticeable changes we experience with age is the replacement of lean body mass (muscle) with fat. The change is dramatically exemplified by CT scans of the thigh. In a young well-

[17]*myo* = muscle + *blast* = precursor
[18]*quick* = alive

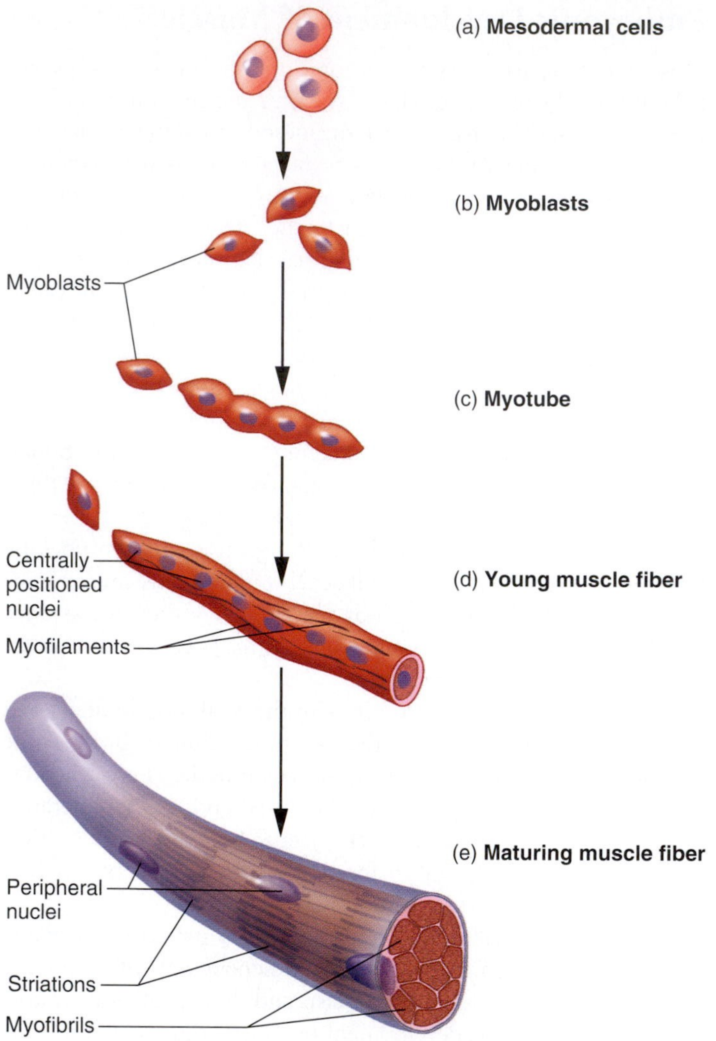
(a) Mesodermal cells
(b) Myoblasts
Myoblasts
(c) Myotube
Centrally positioned nuclei
(d) Young muscle fiber
Myofilaments
(e) Maturing muscle fiber
Peripheral nuclei
Striations
Myofibrils

FIGURE 10.20

Embryonic Development of Skeletal Muscle Fibers. (*a*) Nearly all muscle arises from mesodermal cells. (*b*) Some mesodermal cells differentiate into fusiform myoblasts. (c) Myoblasts fuse to form a multinucleate myotube. (*d*) Myofilaments begin to appear in the young muscle fiber, while additional myoblasts join the fiber and increase its length. (e) Myofilaments become organized into myofibrils with striations, and muscle contractions begin.

conditioned male, muscle accounts for 90% of the cross-sectional area of the midthigh, whereas in a frail 90-year-old woman, it is only 30%. Muscular strength and mass peak in the 20s; by the age of 80, most people have only half as much strength and endurance. A large percentage of people over age 75 cannot lift a 4.5 kg (10 lb) weight with their arms; such simple tasks as carrying a sack of groceries into the house may become impossible. The loss of strength is a major contributor to falls, fractures, and dependence on others for the routine activities of daily living. Fast glycolytic (fast-twitch) fibers exhibit the earliest and most severe atrophy, increasing reaction time and reducing coordination.

There are multiple reasons for the loss of strength. Aged muscle fibers have fewer myofibrils, so they are smaller and weaker. The sarcomeres are increasingly disorganized, and muscle mitochondria are smaller and have reduced quantities of oxidative enzymes. Aged muscle has less ATP, glycogen, and myoglobin; consequently, it fatigues quickly. Muscles also exhibit more fat and fibrous tissue with age, which limits their movement and blood circulation. With reduced circulation, muscle injuries heal more slowly and with more scar tissue.

But the weakness and easy fatigue of aged muscle also stems from the aging of other organ systems. There are fewer motor neurons in the spinal cord, and some muscle shrinkage may represent denervation atrophy. The remaining neurons produce less acetylcholine and show less efficient synaptic transmission, which makes the muscles slower to respond to stimulation. As muscle atrophies, motor units have fewer muscle fibers per motor neuron, and more motor units must be recruited to perform a given task. Tasks that used to be easy, such as buttoning the clothes or eating a meal, take more time and effort. The sympathetic nervous system is also less efficient in old age, and less effective in increasing the blood flow to the muscles during exercise. This contributes to reduced endurance.

Diseases of the Muscular System

Diseases of muscular tissue are called **myopathies.** The muscular system suffers fewer diseases than any other organ system, but two of particular importance are muscular dystrophy and myasthenia gravis.

Muscular dystrophy[19] is a collective term for several hereditary diseases in which the skeletal muscles degenerate, lose strength, and are gradually replaced by fat and scar tissue. This new connective tissue impedes blood circulation, which in turn accelerates muscle degeneration, creating a fatal spiral of positive feedback. The most common form of the disease is *Duchenne*[20] *muscular dystrophy (DMD),* a sex-linked trait that occurs especially in males (about 1 in 3,500 male live births). It results from a defective gene for *dystrophin,* a large protein that links actin filaments to membrane glycoproteins. In muscle fibers lacking dystrophin, the sarcolemma becomes torn and the muscle fiber dies. DMD is not evident at birth, but difficulties appear as a child begins to walk. The child falls frequently and has difficulty standing up again. The disease affects the hips first, then the legs, and progresses to the abdominal and spinal muscles. The muscles shorten as they atrophy, causing postural abnormalities such as scoliosis. DMD is incurable but is treated with exercise to slow the atrophy and with braces to reinforce the weakened hips and correct the posture. Patients are usually confined to a wheelchair by early adolescence and rarely live beyond the age of 20.

Myasthenia gravis[21] (MY-ass-THEE-nee-uh GRAV-is) **(MG)** is most prevalent in women from 20 to 40 years old. It is an autoimmune disease in which antibodies attack the neuromuscular junctions and trigger the destruction of ACh receptors. As a result, the muscle fibers become less and less sensitive to ACh. The effects often appear first in the facial muscles and include drooping eyelids

[19] *dys* = bad, abnormal + *trophy* = growth
[20] Guillaume B. A. Duchenne (1806–75), French physician
[21] *my* = muscle + *asthen* = weakness + *grav* = severe

TABLE 10.5

Disorders of the Muscular System

Charley Horse	Slang for any painful tear, stiffness, and blood clotting in a muscle caused by *contusion* (a blow to the muscle causing hemorrhaging).
Contracture	Abnormal muscle shortening not caused by nervous stimulation. Can result from a persistence of calcium in the sarcoplasm after stimulation, or from contraction of scar tissue.
Crush Syndrome	A shocklike state following the massive crushing of muscles, associated with a high and potentially fatal fever, cardiac irregularities caused by K^+ released from the injured muscles, and kidney failure caused by blockage of the renal tubules with myoglobin released by the traumatized muscle. Myoglobin in the urine *(myoglobinuria)* is a common sign.
Delayed Onset Muscle Soreness	Pain and stiffness felt from several hours to a day after strenuous exercise. Associated with microtrauma to the muscles, with disrupted Z discs, myofibrils, and plasma membranes, and with elevated blood levels of myoglobin and enzymes released by damaged muscle fibers.
Rhabdomyoma	A rare, benign muscle tumor, usually occurring in the tongue, neck, larynx, nasal cavity, throat, heart, or vulva. Treated by surgical removal.
Rhabdomyosarcoma	A malignant muscle tumor; the most common form of pediatric soft-tissue sarcoma, although accounting for <3% of childhood cancers and rarely seen in adults. Results from abnormal proliferation of myoblasts. Begins as a painless mass in a muscle but metastasizes rapidly. Diagnosed by biopsy and treated with surgery, chemotherapy, or radiation therapy.

Disorders Described Elsewhere

Atrophy 270	Compartment syndrome 342	Muscular dystrophy 276	Paralysis 267
Back injuries 306	Hamstring injuries 336	Myasthenia gravis 276	Sports injuries 345

and double vision (due to weakness of the eye muscles). These signs are often followed by difficulty swallowing, weakness of the limbs, and poor physical endurance. Some people with MG die quickly as a result of respiratory failure, while others have normal life spans. The symptoms can be controlled with cholinesterase inhibitors, which retard the breakdown of ACh and prolong its action on the muscles, and with drugs that suppress the immune system and thus slow the attack on ACh receptors.

Some other disorders of the muscular system in general are briefly described in table 10.5, whereas disorders more specific to the axial or appendicular musculature are described in chapters 11 and 12.

Before You Go On

Answer the following questions to test your understanding of the preceding section:

27. What cells come between mesodermal cells and the muscle fiber in the stages of skeletal muscle development? Describe how several uninucleated cells transform into a multinucleated muscle fiber.
28. What is the principal difference between the way cardiac and smooth muscle form and the way skeletal muscle forms?
29. Describe the major changes seen in the muscular system in old age.
30. What is the root cause of Duchenne muscular dystrophy? What is the normal function of dystrophin?
31. How is synaptic function altered in myasthenia gravis? How does this synaptic dysfunction affect a person with MG?
32. In a game of baseball, the pitcher hits a man in the thigh with the ball. Which of the following conditions would this most likely cause: atrophy, a charley horse, contracture, crush syndrome, or a rhabdomyoma. Explain why the accident causes the condition you select.

CHAPTER REVIEW

Review of Key Concepts

Muscle Types and Functions (p. 256)

1. *Muscular system* refers only to skeletal muscles. The study of the muscular system is *myology.*
2. Skeletal muscle is voluntary striated muscle that is usually attached to bones. Its long slender cells are called *muscle fibers.* Cardiac muscle is involuntary striated muscle. Its cells are not fibrous in shape and are called *myocytes* or *cardiocytes.* Smooth muscle is involuntary nonstriated muscle. Its cells, also called myocytes, are fusiform in shape.
3. Muscular tissue serves for movement, stability of the body, communication, control of body passages and openings, and heat production.
4. Muscle cells have five properties that enable them to carry out their functions: excitability, conductivity, contractility, extensibility, and elasticity.

General Anatomy of Muscles (p. 257)

1. A skeletal muscle is composed of muscular tissue, connective tissue, nervous tissue, and blood vessels.
2. The connective tissues of a muscle include a thin *endomysium* around each muscle fiber, a thicker *perimysium* that binds fibers into bundles called *fascicles,* and an *epimysium* that surrounds the entire muscle. Deep fascia separate neighboring muscles from each other, and superficial fascia separate muscles from the skin (fig. 10.2).
3. The *series elastic components* of a muscle are the connective tissue elements from endomysium to tendon. They stretch when a muscle contracts and recoil when tension is released, and they enhance the power output of a muscle.
4. The strength of a muscle and direction of its pull are determined partly by the orientation of its fascicles. Muscles are classified into five categories according to fascicle orientation: *fusiform, parallel, convergent, pennate,* and *circular muscles* (fig. 10.3).
5. Most skeletal muscles are attached to a different bone at each end, span at least one joint, and move one bone relative to another when they contract. The muscle attachment at the stationary end is its *origin,* the attachment at the moving end is the *insertion,* and the middle region of a muscle is its *belly.*
6. Some muscles have a *direct attachment* to bone in which the muscle fibers extend nearly to the bone surface and the epimysium is continuous with the periosteum; others have an *indirect attachment* in which a tendon spans the distance between the muscular tissue and bone.
7. Some muscles attach to the fascia or tendon of another muscle or to collagen fibers of the dermis. Some have broad sheetlike tendons called *aponeuroses.*
8. The movement produced by a muscle is called its *action.* Muscles work in functional groups that act on a single joint with different effects. In a given joint movement, a *prime mover* is the muscle that produces most of the force; a *synergist* is a muscle that aids the prime mover by adding power, stabilizing a joint, or modifying the direction of joint movement; an *antagonist* is a muscle that opposes the prime mover (such as an extensor that opposes a flexor); and a *fixator* is a muscle that holds a bone still.
9. *Intrinsic muscles* have both their origin and insertion within a specified region such as the head or hand; *extrinsic muscles* have their origin outside of a specified region, and they or their tendons extend into that region, such as muscles of the forearm whose tendons extend into the hand.
10. Many bones, especially long bones, act as levers when moved by the muscles. An *effort* is applied at one point on a lever to overcome a *load (resistance)* located at some other point; the lever rotates around a fixed *fulcrum.* The part of the lever from the fulcrum to the point of effort is the *effort arm,* and the part from the fulcrum to the load or point of resistance is the *resistance arm.*
11. Levers produce a gain in the speed, distance, or force of a movement. The ratio of the length of the effort arm to the length of the resistance arm is called the *mechanical advantage (MA)* of a lever. If MA > 1, a lever produces more force, but less speed or distance, than the force applied to it. If MA < 1, it produces a gain in speed or distance but exerts less force than the effort applied to it. When two or more muscles span and move the same joint, they may differ in MA and therefore in the qualities of the movement they produce.
12. A *first-class lever* has the effort at one end, resistance at the other end, and the fulcrum between the effort and resistance, as in a crowbar or the atlanto-occipital joint of the neck. A *second-class lever* has the fulcrum at one end, effort applied at the other end, and the load or resistance between the fulcrum and effort, as in a wheelbarrow or in the way the mandible behaves as the mouth is opened. A *third-class lever* is one with the fulcrum at one end, the resistance at the other end, and the effort applied between the fulcrum and resistance, as in a baseball bat or the action of the biceps brachii muscle on the forearm.

Microscopic Anatomy (p. 263)

1. The key to understanding muscle contraction lies in the microscopic structure of individual skeletal muscle cells (muscle fibers).
2. A muscle fiber is a long, slender cell with multiple nuclei just inside the plasma membrane *(sarcolemma).* The sarcolemma extends inward as tunnel-like *transverse (T) tubules* that cross the cell and open to the surface on both sides. The cytoplasm *(sarcoplasm)* is occupied mainly by *myofibrils,* which are threadlike bundles of protein filaments. Between the myofibrils, the muscle fiber has an abundance of mitochondria and smooth endoplasmic reticulum (ER). The cytoplasm also contains an abundance of *glycogen* (an energy-storage carbohydrate) and *myoglobin* (an oxygen-binding protein).
3. The smooth ER, or *sarcoplasmic reticulum,* forms an extensive branching network amid the myofibrils and has dilated *terminal cisternae* flanking each T tubule. It is a reservoir of calcium ions and has gated channels that can release a flood of Ca^{2+} into the cytosol.
4. Each myofibril is a bundle of protein *myofilaments.* There are three kinds of myofilaments: *thick filaments* composed of a motor protein called *myosin; thin filaments* composed mainly of *actin,* but also containing the regulatory proteins *tropomyosin* and *troponin;* and *elastic filaments* composed of the protein *titin.*
5. Elastic filaments keep the thick and thin filaments aligned with each other, resist overstretching of a muscle, and aid in recoil of a

muscle to its resting length. The work of contraction is carried out by the thick and thin filaments.

6. In striated (skeletal and cardiac) muscle, myosin and actin are organized in such a way that they overlap and produce alternating dark *A bands* and light *I bands* that repeat at regular intervals along the length of the cell. These bands are called *striations*. The dark A bands consist of a midregion called the *H band* where only thick filaments occur, flanked by even darker regions where the thick and thin filaments overlap. The light A bands are bisected by a thin dark line called a *Z disc*, composed of titin. The thin filaments and elastic filaments are anchored to the Z discs.
7. The region from one Z disc to the next is called a *sarcomere*. This is the functional unit of muscle contraction. When a muscle fiber contracts, the sarcomeres become shorter and the Z discs are pulled closer together.
8. Skeletal muscle contracts only when stimulated by a *somatic motor neuron*. The axon *(somatic motor fiber)* of one neuron branches at its tip and leads to typically about 200 muscle fibers, but each muscle fiber receives only one nerve fiber.
9. The nerve and muscle fiber meet at a synapse called a *neuromuscular junction*. Each tip of the nerve fiber ends in a dilated bulb, the *synaptic knob*, nestled in a depression of the muscle fiber sarcolemma called the *motor end plate*. A narrow gap, the *synaptic cleft*, separates the synaptic bulb from the sarcolemma.
10. The synaptic knob contains *synaptic vesicles* filled with the chemical *acetylcholine (ACh)*. The sarcolemma across from the knob has proteins that act as ACh receptors. An enzyme called *acetylcholinesterase (AChE)*, found in the synaptic cleft and as part of the sarcolemma of the motor end plate, breaks down ACh to terminate stimulation of the muscle fiber.
11. One nerve fiber and all the muscle fibers innervated by it are called a *motor unit*, because stimulation by that nerve fiber causes all these muscle fibers to contract in unison. Small motor units have as few as 3 to 6 muscle fibers per nerve fiber, and produce precise, finely controlled movements, as in the muscles of eye movement. Large motor units may have up to 1,000 muscle fibers per nerve fiber, and produce strong but not finely controlled movements, as in the muscles of the thigh and leg. Motor units can "work in shifts" so that fresh units take over the contraction of a muscle when other units fatigue.
12. The muscular system receives from one-quarter of the blood pumped by the heart at rest, to three-quarters of it during exercise. Blood capillaries penetrate into the endomysium to reach every muscle fiber.

Functional Perspectives (p. 269)

1. Muscle contraction and relaxation occur in four steps: excitation, excitation-contraction coupling, contraction, and relaxation.
2. In *excitation*, a signal in the motor nerve fiber triggers the release of acetylcholine (ACh) from the synaptic vesicles. ACh crosses the synaptic cleft and binds with receptors on the muscle fiber. These receptors are gated Na^+/K^+ channels that open to allow flow of these ions through the sarcolemma, producing a voltage change. This sets off a chain reaction of electrical excitation that spreads along the fibers and down the T tubules to the interior of the muscle fiber.
3. In *excitation-contraction coupling*, electrical signals in the T tubules indirectly open the Ca^{2+} channels of the sarcoplasmic reticulum, releasing calcium ions into the cytosol. Ca^{2+} binds to the troponin molecules of the thin filaments. This induces tropomyosin to move away from the active sites on the actin, so these sites are exposed to the action of myosin.
4. In *contraction*, the heads of the myosin molecules are activated by ATP, swing forward into an extended or "cocked" position, bind to the active sites of actin, then flex and pull the actin filament a short distance. Each myosin head then binds a new ATP, recocks, and repeats the process. By repetition, the thin filament slides across the thick filament, pulling the Z discs closer together. The Z discs are linked to the sarcolemma, so their movement shortens the cell as a whole.
5. In *relaxation*, the nerve signal stops, ACh is no longer released, and the existing ACh in the synaptic cleft is degraded by AChE. Stimulation of the muscle fiber therefore ceases. The sarcoplasmic reticulum pumps Ca^{2+} back into its cisternae. As the Ca^{2+} level in the cytosol falls, tropomyosin moves back into its resting position, blocking the active sites of actin. Myosin can no longer bind actin, and the muscle relaxes.
6. Muscles grow in response to *resistance exercise*, not by the mitotic production of more muscle fibers but by the production of more myofilaments and thickening of the fibers that already exist. Well-exercised muscles also develop more mitochondria, myoglobin, glycogen, and blood capillaries.
7. Muscle shrinkage, or *atrophy*, occurs when the nerve connection to a muscle is severed *(denervation atrophy)*, a muscle is not exercised *(disuse atrophy)*, or simply as a result of aging *(senescence atrophy)*.
8. *Slow oxidative (SO) muscle fibers* employ aerobic respiration and are relatively fatigue-resistant, but have relatively long, slow twitches (contractions). Postural muscles of the back and the soleus muscle of the calf are composed predominantly of SO fibers.
9. Fast glycolytic (FG) muscle fibers employ anaerobic fermentation and fatigue relatively quickly, but produce quick twitches. The gastrocnemius muscle of the calf and the muscles of eye movement are composed predominantly of FG fibers.
10. *Intermediate fibers* are a type of FG fibers that combine fast twitches with aerobic fatigue-resistant metabolism. These are relatively rare except in some endurance-trained athletes.

Cardiac and Smooth Muscle (p. 272)

1. Cardiac muscle consists of short, thick, branched *myocytes* connected to each other through electrical and mechanical junctions in its *intercalated discs*.
2. Cardiac muscle contracts spontaneously without need of nervous stimulation, although the nervous system does modify the heart rate and contraction strength. The contractions of cardiac muscle are very prolonged compared to those of skeletal muscle, allowing time for the heart to eject blood.
3. Cardiac muscle has an abundance of myoglobin and glycogen, and has numerous large mitochondria; thus it is highly resistant to fatigue.
4. Smooth muscle cells contain myosin and actin like skeletal and cardiac muscle, but the myofilaments of smooth muscle are not regularly aligned with each other, so there are no striations. Smooth muscle has no T tubules and has very little sarcoplasmic reticulum; the calcium needed to activate its contraction comes mainly from the extracellular fluid. Unlike skeletal and cardiac muscle, smooth muscle cells are capable of mitosis.
5. *Multiunit smooth muscle* is found in some blood vessels and pulmonary air passages, the iris, and piloerector muscles of the skin. In this type of muscle, each cell is innervated by a nerve fiber and contracts independently of other muscle cells.
6. Most smooth muscle is *single-unit smooth muscle (visceral muscle)*, found in most blood vessels and in the digestive, respiratory, urinary, and reproductive tracts.
7. In single-unit smooth muscle, nerve fibers do not synapse with individual muscle cells. *Varicosities* of the nerve fiber release neurotransmitters, which diffuse to nearby muscle cells and may stimulate them to contract. The muscle cells are connected through electrical gap junctions and contract in unison.

8. Smooth muscle contracts and relaxes slowly and is very fatigue-resistant. It maintains muscle tone in organs such as the uterus, bladder, and blood vessels, and produces waves of contraction called *peristalsis* in the digestive tract and other tubular organs.

Developmental and Clinical Perspectives (p. 275)

1. Most skeletal muscle develops from embryonic mesoderm. Mesenchymal cells differentiate into *myoblasts,* which fuse to form multinucleated *myotubes.* Myofilaments and striations appear in the myotubes as they mature into functional muscle fibers. New muscle fibers are added until about 1 year of age; after that, the growth of muscles is primarily by cellular enlargement (hypertrophy) rather than an increase in cell number.
2. Cardiac and smooth muscle also develop from myoblasts, but the myoblasts do not fuse as they do in skeletal muscle. In cardiac muscle, myoblasts remain attached to each other and develop intercalated discs at their points of adhesion. In single-unit smooth muscle, they form gap junctions.
3. In old age, the skeletal muscles exhibit substantial atrophy and replacement of muscular tissue with fat and fibrous tissue. Aged muscle fibers exhibit fewer myofibrils and mitochondria, less glycogen and myoglobin, and more disorganized sarcomeres. Fewer motor neurons innervate the muscles and those that do are less efficient at stimulating them. Reduced blood circulation to aged muscles also contributes to reduced endurance.
4. *Muscular dystrophy* is a family of hereditary *myopathies* (muscle diseases) in which the skeletal muscles degenerate and are replaced by fat and scar tissue. *Duchenne muscular dystrophy,* the most common form, is a sex-linked trait affecting mostly boys. It results from the lack of a protein, *dystrophin,* that links actin to the sarcolemma. It is a crippling and incurable disease that usually claims the victim's life by the age of 20.
5. *Myasthenia gravis* is an autoimmune disease most commonly affecting young women. It is caused by autoantibodies that destroy ACh receptors and render muscle unresponsive to ACh. The result is muscular weakness, often first seen in facial muscles. Some victims die of respiratory failure, whereas some people live a normal life span with the aid of cholinesterase inhibitors and drugs to suppress the immune response.
6. Several other muscular system disorders are described in table 10.5.

Testing Your Recall

1. A fascicle is bounded and defined by the
   a. endomysium.
   b. deep fascia.
   c. superficial fascia.
   d. epimysium.
   e. perimysium.
2. Muscle cells must have all of the following properties *except* _____ to carry out their function.
   a. extensibility
   b. elasticity
   c. autorhythmicity
   d. contractility
   e. conductivity
3. If a tendon runs longitudinally throughout a muscle and fascicles insert obliquely on it along both sides, the muscle is classified as
   a. parallel.
   b. oblique.
   c. bipennate.
   d. convergent.
   e. multipennate.
4. A muscle that holds a bone still during a particular action is called
   a. a fixator.
   b. an antagonist.
   c. an agonist.
   d. a synergist.
   e. an intrinsic muscle.
5. Skeletal muscle fibers have _____, whereas smooth muscle cells do not.
   a. T tubules
   b. ACh receptors
   c. thick myofilaments
   d. thin myofilaments
   e. dense bodies
6. Smooth muscle cells have _____, whereas skeletal muscle fibers do not.
   a. T tubules
   b. ACh receptors
   c. thick myofilaments
   d. thin myofilaments
   e. dense bodies
7. ACh receptors are found in
   a. synaptic vesicles.
   b. terminal cisternae.
   c. thick filaments.
   d. thin filaments.
   e. junctional folds.
8. Single-unit smooth muscle cells can stimulate each other because they have
   a. a pacemaker.
   b. diffuse junctions.
   c. gap junctions.
   d. tight junctions.
   e. calcium pumps.
9. A second-class lever always has
   a. the fulcrum in the middle.
   b. the effort applied between the fulcrum and resistance.
   c. a mechanical advantage less than 1.
   d. a mechanical advantage greater than 1.
   e. the resistance at one end.
10. Slow oxidative muscle fibers have all of the following *except*
    a. an abundance of myoglobin.
    b. an abundance of glycogen.
    c. high fatigue resistance.
    d. a red color.
    e. a high capacity to synthesize ATP aerobically.
11. Acetylcholine is released from organelles called _____.
12. The _____ is a depression in the sarcolemma that receives a motor nerve ending.
13. Parts of the sarcoplasmic reticulum called _____ lie on each side of a T tubule.
14. Thick myofilaments consist mainly of the protein _____.
15. The tissue between skin and muscle is called _____.
16. Muscle contains an oxygen-storage pigment called _____.
17. The _____ of skeletal muscle play the same role as dense bodies in smooth muscle.
18. A circular muscle that controls a body opening or passage is called a/an _____.
19. Skeletal muscle fibers develop by the fusion of embryonic cells called _____.
20. A wave of contraction passing along the esophagus or small intestine is called _____.

Answers in the Appendix

True or False

Determine which five of the following statements are false, and briefly explain why.

1. Fusiform muscles are stronger than parallel muscles.
2. A given muscle may be an agonist in one joint movement and an antagonist in a different movement of that joint.
3. Extrinsic muscles are not located entirely within the body region that they control.
4. Cardiac myocytes are of the fatigue-resistant, slow oxidative type.
5. One motor neuron can supply only one muscle fiber.
6. To initiate muscle contraction, calcium ions must bind to the myosin heads.
7. A first-class lever can have a mechanical advantage that is either greater than or less than 1.
8. Slow oxidative fibers are more fatigue-resistant than fast glycolytic fibers.
9. The blood vessels of a skeletal muscle are more wavy or coiled when a muscle is relaxed than when it contracts.
10. Thick, well-developed muscles consist of more muscle fibers than thinner, less-developed ones.

Answers in the Appendix

Testing Your Comprehension

1. Give three distinctly different reasons why two muscles that act across the same side of the same joint are not necessarily redundant in function.
2. What would be the consequences for muscular system function if muscle fibers were not elastic?
3. For each of the following muscle pairs, state which muscle you think would have the higher percentage of fast glycolytic fibers: (*a*) Muscles that move the eyes or muscles of the upper throat than initiate swallowing? (*b*) The abdominal muscles employed in doing sit-ups or the muscles employed in handwriting? (*c*) Muscles of the tongue or the skeletal muscle sphincter of the anus? Explain each answer.
4. Discuss some reasons why the heart could not function effectively if it were composed of the fast glycolytic type of muscle fiber.
5. Botulism is a form of food poisoning that occurs when a bacterium, *Clostridium botulinum*, releases a neurotoxin that prevents motor neurons from releasing ACh. In view of this, what early signs of botulism would you predict? Explain why a person with botulism could die of suffocation.

Answers at the Online Learning Center

www.mhhe.com/saladinha1

Visit the Online Learning Center for practice tests, answer keys, and other learning aids for this chapter. Enhance your understanding of human anatomy with our interactive art labeling exercises, supplemental photo atlases, web links, puzzles, flashcards, and much more.

11

CHAPTER ELEVEN

The Axial Musculature

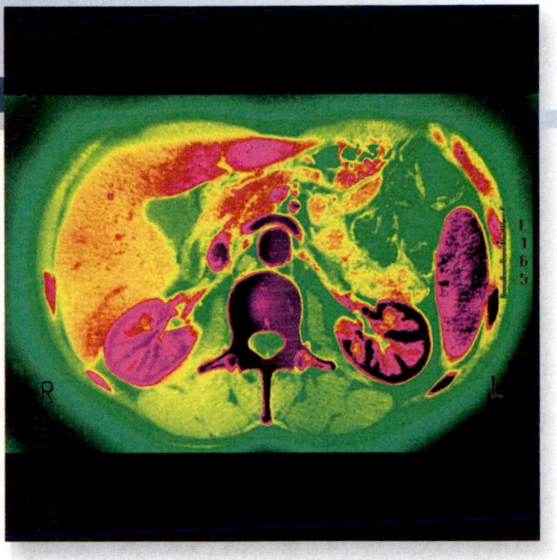

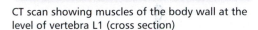

CT scan showing muscles of the body wall at the level of vertebra L1 (cross section)

CHAPTER OUTLINE

Learning Approaches 284
- How Muscles Are Named 284
- Muscle Innervation 284
- A Learning Strategy 287

Muscles of the Head and Neck 287
- Muscles of Facial Expression 288
- Muscles of Chewing and Swallowing 292
- Muscles Acting on the Head 296

Muscles of the Trunk 298
- Muscles of Respiration 299
- Muscles of the Abdomen 300
- Muscles of the Back 303
- Muscles of the Pelvic Floor 307

Chapter Review 309

INSIGHTS

11.1 Clinical Application: Difficulty Breathing 299
11.2 Clinical Application: Heavy Lifting and Back Injuries 306
11.3 Clinical Application: Hernias 306

BRUSHING UP

To understand this chapter, you may find it helpful to review the following concepts:
- Anatomy of the axial skeleton (chapter 7)
- Terminology of joint actions (pp. 234–239)
- Shapes of muscles (fusiform, pennate, circular, etc.) (p. 258)
- Direct and indirect muscle attachments (p. 259)
- Prime movers, synergists, antagonists, and fixators (p. 260)
- Intrinsic and extrinsic muscles (p. 260)

There are about 600 skeletal muscles in the human body. Chapters 11 and 12 describe fewer than one-third of these. This chapter deals with the muscles that act on the axial division of the body—that is, on the head and trunk. Muscles that act on the limbs and limb girdles (the appendicular division) are described in chapter 12. This chapter opens with some tips to help you study the muscles more insightfully.

Learning Approaches

Objectives

When you have completed this section, you should be able to

- translate several Latin words commonly used in the naming of muscles;
- define the *origin, insertion, action,* and *innervation* of a muscle;
- describe the sources of the nerves to the head-neck and trunk muscles and explain the numbering system for the cranial and spinal nerves; and
- describe and practice some methods that will help in the learning of the skeletal muscles.

How Muscles Are Named

Figure 11.1 shows an overview of the major superficial muscles. Learning the names of these and other muscles may seem a forbidding task at first, especially when some of them have such long Latin names as *depressor labii inferioris* and *flexor digiti minimi brevis.* Such names, however, typically describe some distinctive aspects of the structure, location, or action of a muscle, and become very helpful once we grow familiar with a few common Latin words. For example, the depressor labii inferioris is a muscle that lowers (depresses) the bottom (inferior) lip, and the flexor digiti minimi brevis is a short (brevis) muscle that flexes the smallest (minimi) finger (digit). Several of the most common words in muscle names are interpreted in table 11.1, and others are explained in footnotes throughout the chapter. Familiarity with these terms will help you translate muscle names and remember the location, appearance, and action of the muscles.

Muscle Innervation

The **innervation** of a muscle refers to the identity of the nerve that stimulates it. Knowing the innervation to each muscle enables clinicians to diagnose nerve and spinal cord injuries from their effects on muscle function, and to set realistic goals for rehabilitation. The innervations described in this chapter will be more meaningful after you have studied the peripheral nervous system (chapters 14 and 15), but a brief orientation will be helpful here. The muscles are innervated by two groups of nerves:

- **Spinal nerves,** which arise from the spinal cord, emerge through the intervertebral foramina and innervate muscles below the neck. Spinal nerves are identified by letters and numbers that refer to the vertebrae—for example, T6 for the sixth thoracic nerve and S2 for the second sacral nerve.

TABLE 11.1

Words Commonly Used to Name Muscles

Criterion	Term a'nd Meaning	Examples of Usage
Size	Major (large)	Pectoralis major
	Maximus (largest)	Gluteus maximus
	Minor (small)	Pectoralis minor
	Minimus (smallest)	Gluteus minimus
	Longus (long)	Abductor pollicis longus
	Longissimus (longest)	Longissimus thoracis
	Brevis (short)	Extensor pollicis brevis
Shape	Rhomboideus (rhomboidal)	Rhomboideus major
	Trapezius (trapezoidal)	Trapezius
	Teres (round, cylindrical)	Pronator teres
	Deltoid (triangular)	Deltoid
Location	Capitis (of the head)	Splenius capitis
	Cervicis (of the neck)	Semispinalis cervicis
	Pectoralis (of the chest)	Pectoralis major
	Thoracis (of the thorax)	Spinalis thoracis
	Intercostal (between the ribs)	External intercostals
	Abdominis (of the abdomen)	Rectus abdominis
	Lumborum (of the lower back)	Quadratus lumborum
	Femoris (of the femur, or thigh)	Quadriceps femoris
	Peroneus (of the fibula)	Peroneus longus
	Brachii (of the arm)	Biceps brachii
	Carpi (of the wrist)	Flexor carpi ulnaris
	Digiti (of a finger or toe, singular)	Extensor digiti minimi
	Digitorum (of the fingers or toes, plural)	Flexor digitorum profundus
	Pollicis (of the thumb)	Opponens pollicis
	Indicis (of the index finger)	Extensor indicis
	Hallucis (of the great toe)	Abductor hallucis
	Superficialis (superficial)	Flexor digitorum superficialis
	Profundus (deep)	Flexor digitorum profundus
Number of Heads	Biceps (two heads)	Biceps femoris
	Triceps (three heads)	Triceps brachii
	Quadriceps (four heads)	Quadriceps femoris
Orientation	Rectus (straight)	Rectus abdominis
	Transversus (transverse)	Transversus abdominis
	Oblique (slanted)	External abdominal oblique
Action	Adductor	Adductor pollicis
	Abductor	Abductor digiti minimi
	Flexor	Flexor carpi radialis
	Extensor	Extensor carpi radialis
	Pronator	Pronator teres
	Supinator	Supinator
	Levator	Levator scapulae
	Depressor	Depressor anguli oris

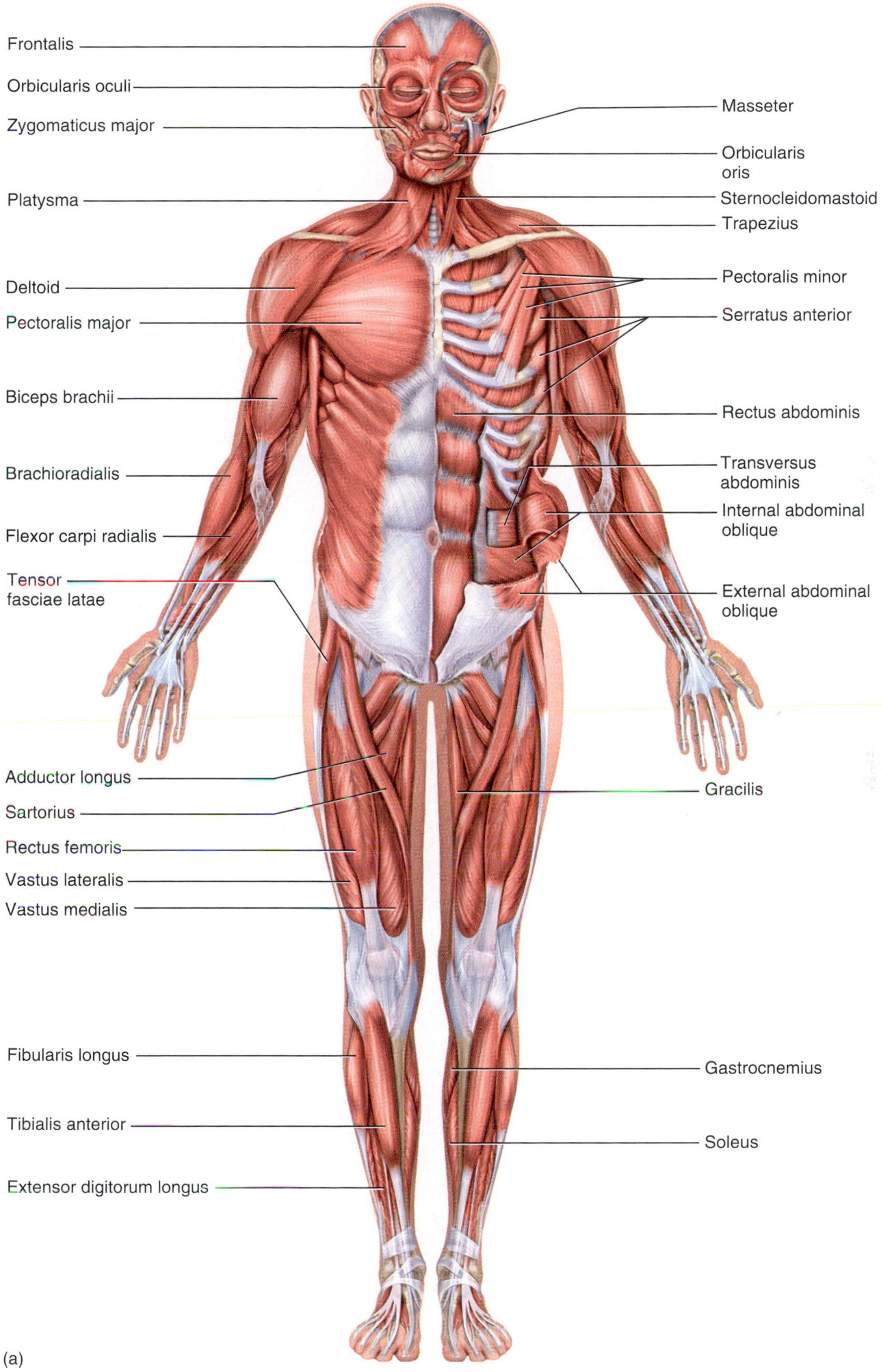
Frontalis
Orbicularis oculi
Zygomaticus major
Platysma
Deltoid
Pectoralis major
Biceps brachii
Brachioradialis
Flexor carpi radialis
Tensor fasciae latae
Adductor longus
Sartorius
Rectus femoris
Vastus lateralis
Vastus medialis
Fibularis longus
Tibialis anterior
Extensor digitorum longus
Masseter
Orbicularis oris
Sternocleidomastoid
Trapezius
Pectoralis minor
Serratus anterior
Rectus abdominis
Transversus abdominis
Internal abdominal oblique
External abdominal oblique
Gracilis
Gastrocnemius
Soleus
(a)

FIGURE 11.1

The Muscular System. (*a*) Anterior aspect. In each figure, major superficial muscles are shown on the anatomical right, and some of the deeper muscles of the trunk are shown on the left. Muscles not labeled here are shown in more detail in later figures.

(continued)

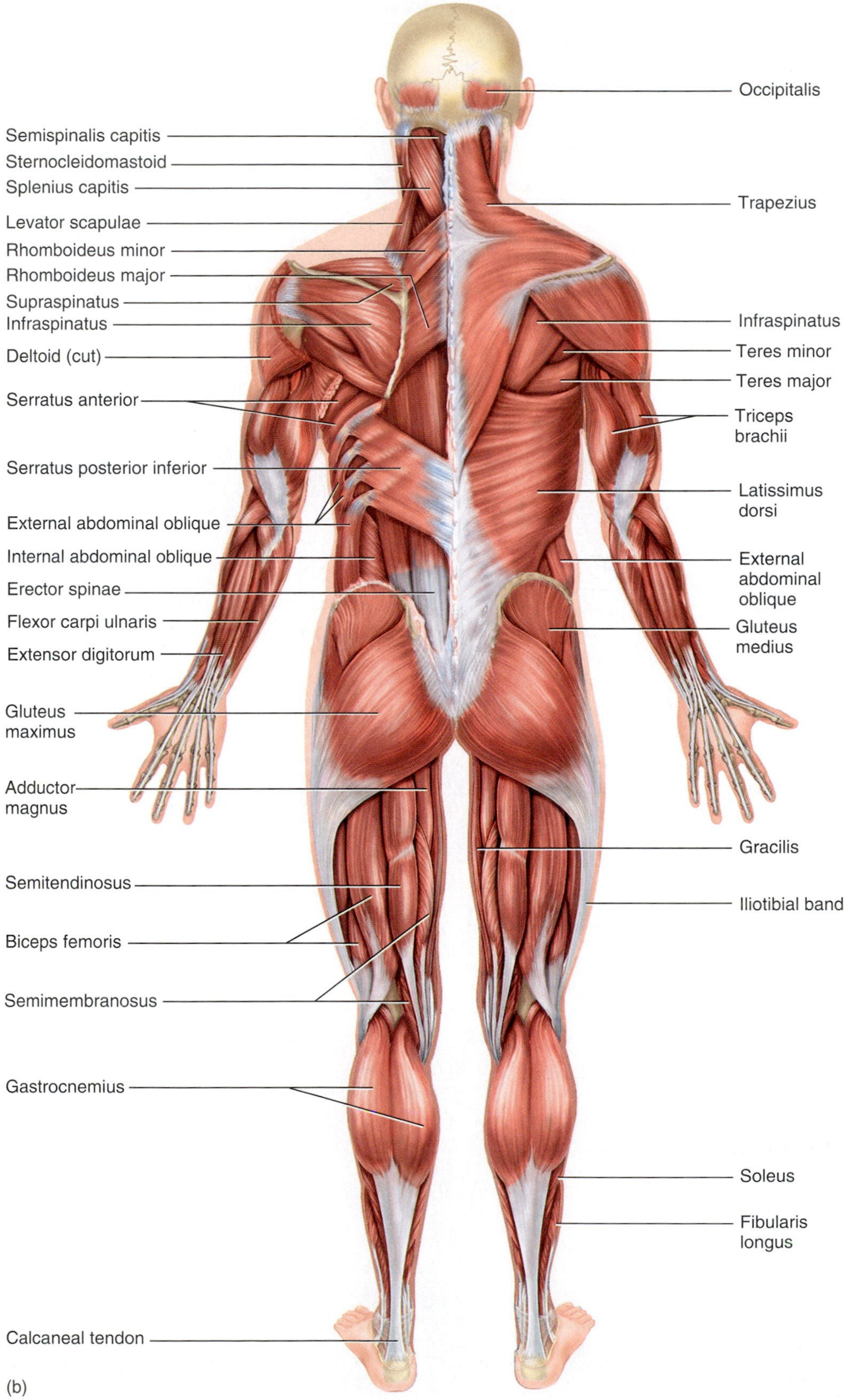
Occipitalis
Semispinalis capitis
Sternocleidomastoid
Splenius capitis
Trapezius
Levator scapulae
Rhomboideus minor
Rhomboideus major
Supraspinatus
Infraspinatus
Infraspinatus
Teres minor
Deltoid (cut)
Teres major
Serratus anterior
Triceps brachii
Serratus posterior inferior
Latissimus dorsi
External abdominal oblique
Internal abdominal oblique
External abdominal oblique
Erector spinae
Gluteus medius
Flexor carpi ulnaris
Extensor digitorum
Gluteus maximus
Adductor magnus
Gracilis
Semitendinosus
Iliotibial band
Biceps femoris
Semimembranosus
Gastrocnemius
Soleus
Fibularis longus
Calcaneal tendon
(b)

FIGURE 11.1
The Muscular System *(continued)*. (*b*) Posterior aspect.

Immediately after emerging from an intervertebral foramen, each spinal nerve branches into a *dorsal* and *ventral ramus*.[1] You will note references to nerve numbers and rami in many of the muscle tables. The term *plexus* in some of the tables refers to weblike networks of spinal nerves adjacent to the vertebral column. All of the nerves named here are illustrated, and most are also discussed, in chapter 14.

- **Cranial nerves,** which arise from the base of the brain, emerge through the skull foramina and innervate muscles of the head and neck. Cranial nerves are identified by numerals I to XII and by names given in chapter 15, although not all 12 of them innervate skeletal muscles.

A Learning Strategy

The following suggestions can help you develop a rational strategy for learning the skeletal muscles as you first encounter them in the textbook and laboratory:

- Examine models, cadavers, dissected animals, or an anatomical atlas as you read about the muscles. Visual images are often easier to remember than words, and direct observation of a muscle may stick in your memory better than descriptive text or two-dimensional drawings.
- When studying a particular muscle, palpate it on yourself if possible. Contract the muscle to feel it bulge and sense its action. Doing so will make muscle locations and actions less abstract. Atlas B following chapter 12 shows where you can see and palpate several muscles on the living body.
- Locate the origins and insertions of muscles on an articulated skeleton. Some study skeletons are painted and labeled to show these. This will help you visualize the locations of muscles and understand how they produce particular joint actions.
- Study the derivations of the muscle names; look for descriptive meaning in their names.
- Say the names *aloud* to yourself or a study partner. It is harder to remember and spell terms you cannot pronounce, and silent pronunciation is not nearly as effective as speaking and hearing the names. Pronunciation guides are provided in the muscle tables for all but the most obvious cases.

Before You Go On

Answer the following questions to test your understanding of the preceding section:

1. What is meant by the innervation of a muscle? Why is it important to know this? What two major groups of nerves innervate the skeletal muscles?
2. In table 11.1, pick a muscle name from the right column that you think meets each of the following descriptions: (*a*) lies beside the radius and straightens the wrist; (*b*) pulls down the corners of your mouth when you frown; (*c*) raises your shoulder blades; (*d*) moves your little finger laterally, away from the fourth digit; (*e*) the largest muscle deep to the breast.

[1]*ramus* = branch

Muscles of the Head and Neck

Objectives

When you have completed this section, you should be able to

- name and locate the muscles that produce facial expressions;
- name and locate the muscles used for chewing and swallowing;
- name and locate the neck muscles that move the head; and
- identify the origin, insertion, action, and innervation of any of these muscles.

Muscles of the head and neck will be treated here from a regional and functional perspective, thus placing them in the following groups: muscles of facial expression, muscles of chewing and swallowing, and muscles that move the head as a whole (tables 11.2–11.4). These tables and the ones following them provide the names and pronunciations of the major muscles and identify their origin (stationary attachment), insertion (movable attachment), action (motion produced), and innervation (nerve supply).

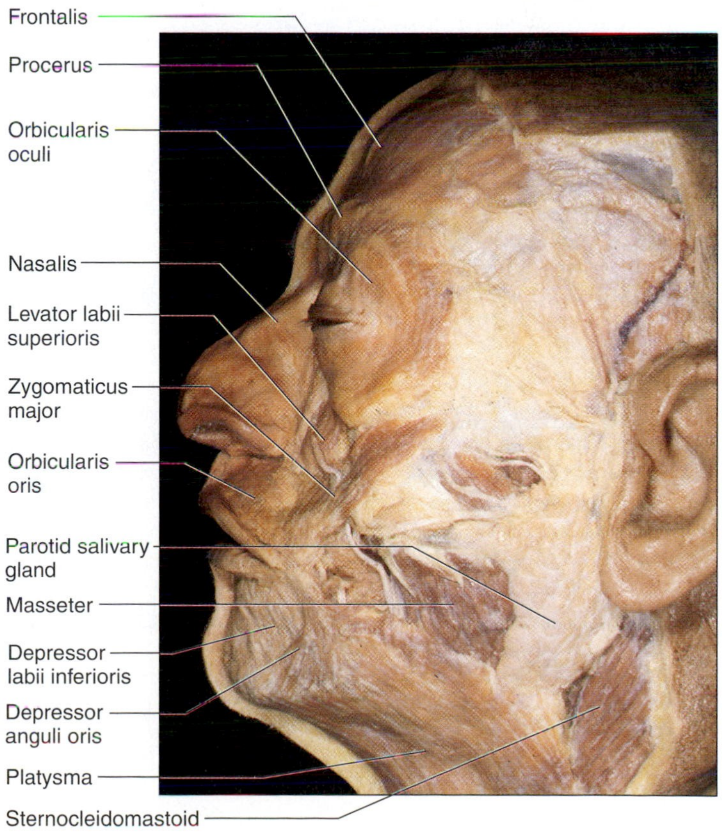
Frontalis
Procerus
Orbicularis oculi
Nasalis
Levator labii superioris
Zygomaticus major
Orbicularis oris
Parotid salivary gland
Masseter
Depressor labii inferioris
Depressor anguli oris
Platysma
Sternocleidomastoid

FIGURE 11.2
Some Muscles of Facial Expression in the Cadaver.

Table 11.2

Muscles of Facial Expression (figs. 11.2–11.4)

O = origin, I = insertion, N = innervation (n. = nerve)

Humans have much more expressive faces than most mammals because of a complex array of muscles that insert in the dermis of the skin. These muscles tense the skin when they contract and produce such expressions as a smile, frown, or wink, as well as contributing to speech through movements of the lips. All of these muscles but one are innervated by the facial nerve (cranial nerve VII). This nerve is especially vulnerable to injury from lacerations and skull fractures, which can paralyze the muscles and cause parts of the face to sag.

The Scalp. The *occipitofrontalis* is a broad sheetlike muscle underlying the scalp. It is divided into the *frontalis* of the forehead and *occipitalis* at the rear of the head, connected to each other by a broad aponeurosis, the **galea aponeurotica**[2] (gay-LEE-UH AP-oh-new-ROT-ih-cuh).

Occipitofrontalis (oc-SIP-ih-toe-frun-TAY-lis)

Occipitalis

Retracts scalp; fixes galea aponeurotica

O: superior nuchal line	I: galea aponeurotica	N: facial n. (VII)

Frontalis

Raises eyebrows and creates wrinkles in forehead when occipitalis is contracted; draws scalp forward when occipitalis is relaxed

O: galea aponeurotica	I: skin of forehead	N: facial n. (VII)

The Ocular and Nasal Regions. The *orbicularis oculi* is a sphincter muscle in the eyelid that encircles and closes the eye. The *levator palpebrae superioris* lies deep to the orbicularis oculi, in the eyelid and roof of the orbit, and opens the eye. Other muscles in this group move the eyelids and skin of the forehead, and compress and flare the nostrils. Muscles within the orbit that move the eyeball itself are discussed in chapter 17.

Orbicularis Oculi[3] (or-BIC-you-LERR-iss OC-you-lye)

Closes eye; compresses lacrimal gland to promote flow of tears

O: medial wall of orbit	I: eyelid	N: facial n. (VII)

Levator Palpebrae Superioris[4] (leh-VAY-tur pal-PEE-bree)

Opens eye; raises upper eyelid

O: roof of orbit	I: upper eyelid	N: oculomotor n. (III)

Corrugator Supercilii[5] (COR-oo-GAY-tur SOO-per-SIL-ee-eye)

Medially depresses eyebrows and draws them closer together; wrinkles skin between eyebrows

O: superciliary ridge	I: skin of eyebrow	N: facial n. (VII)

Procerus[6] (pro-SEE-rus)

Wrinkles skin between eyebrows; draws skin of forehead down

O: skin on bridge of nose	I: skin of forehead	N: facial n. (VII)

Nasalis[7] (nay-SAY-liss)

One part widens nostrils; another part depresses nasal cartilages and compresses nostrils

O: maxilla and nasal cartilages	I: bridge and alae of nose	N: facial n. (VII)

The Oral Region. The mouth is the most expressive part of the face, so it is not surprising that the muscles here are especially diverse. The *orbicularis oris* is a sphincter muscle in the lips that encircles the mouth like the orbicularis oculi does the eye. Other muscles in this region approach the orbicularis oris from all directions. The *levator labii superioris, zygomaticus minor* and *major, levator anguli oris,* and *risorius* insert on the upper lip or corners of the mouth and draw the lip upward and laterally, in expressions such as smiling, laughing, and grimacing. Inserting on the lower lip are the *depressor anguli oris (triangularis)* and *depressor labii inferioris,* which draw the lower lip downward. All of these muscles are very important in speech.

Medially, a pair of tiny *mentalis* muscles originate on the mandible and insert in the dermis of the chin. Unlike the foregoing muscles, they do not act directly on the lips. They pull the soft tissues of the chin upward, which wrinkles the chin and pushes the lower lip out, as in a pouting expression. People with especially thick mentalis muscles have a groove between them, the *mental cleft,* externally visible as a dimple of the chin.

(continued)

[2] *galea* = helmet + *apo* = above + *neuro* = nerves, the brain
[3] *orb* = circle + *ocul* = eye
[4] *levat* = to raise + *palpebr* = eyelid + *superior* = upper
[5] *corrug* = wrinkle + *supercilii* = of the eyebrow
[6] *procer* = long, slender
[7] *nasalis* = of the nose

TABLE 11.2

Muscles of Facial Expression *(continued)*

Orbicularis Oris[8] (or-BIC-you-LERR-iss OR-iss)

Closes lips; protrudes lips as in kissing; aids in speech		
O: muscle fibers around mouth	I: mucous membrane of lips	N: facial n. (VII)

Levator Labii Superioris[9]

Elevates upper lip		
O: zygomatic bone, maxilla	I: upper lip	N: facial n. (VII)

Levator Anguli Oris (ANG-you-lye)

Elevates angle (corner) of mouth, as in smiling and laughing		
O: maxilla	I: superior corner of mouth	N: facial n. (VII)

Zygomaticus[10] Major (ZY-go-MAT-ih-cus) and Zygomaticus Minor

Draws corners of mouth laterally and upward, as in smiling and laughing		
O: zygomatic bone	I: superolateral corner of mouth	N: facial n. (VII)

Risorius[11] (rih-SOR-ee-us)

Draws corner of mouth laterally, as in grimacing		
O: fascia near ear	I: corner of mouth	N: facial n. (VII)

Depressor Anguli Oris[12]

Depresses angle (corner) of mouth, as in frowning		
O: mandible	I: inferolateral corner of mouth	N: facial n. (VII)

Depressor Labii Inferioris

Depresses lower lip		
O: near mental protuberance	I: lower lip	N: facial n. (VII)

Mentalis[13] (men-TAY-lis)

Pulls skin of chin upward; elevates and protrudes lower lip, as in pouting		
O: near mental protuberance	I: skin of chin	N: facial n. (VII)

The Buccal Region. The muscle of the cheek is the *buccinator,* which has multiple functions in blowing, sucking, and chewing. If the cheek is inflated with air, compression of the buccinator blows it out. Sucking is achieved by contracting the buccinators to draw the cheeks inward, and then relaxing them. This action is especially important to nursing infants. To appreciate this action, hold your fingertips lightly on your cheeks as you make a kissing noise. You will feel the relaxation of the buccinators at the moment air is sharply drawn in through the pursed lips. The buccinators also aid chewing by pushing and retaining food between the teeth.

Buccinator[14] (BUCK-sin-AY-tur)

Compresses cheek; pushes food between teeth; expels air or liquid from mouth; creates suction		
O: lateral aspects of maxilla and mandible	I: orbicularis oris	N: facial n. (VII)

The Cervical and Mental Region. The *platysma* is a thin superficial muscle that arises from the shoulder and upper chest and inserts broadly along the mandible and overlying skin.

Platysma[15] (plah-TIZ-muh)

Depresses mandible, opens and widens mouth, tenses skin of neck		
O: fasciae of deltoid and pectoralis major muscles	I: mandible, skin of lower face, muscles at corners of mouth	N: facial n. (VII)

[8] *orb* = circle + *oris* = of the mouth
[9] *levat* = to raise + *labi* = lip + *superior* = upper
[10] refers to the zygomatic arch
[11] *risor* = laughter
[12] *depress* = to lower + *angul* = angle + *oris* = of the mouth
[13] *mentalis* = of the chin
[14] *bucc* = cheek; *buccinator* = trumpeter
[15] *platy* = flat

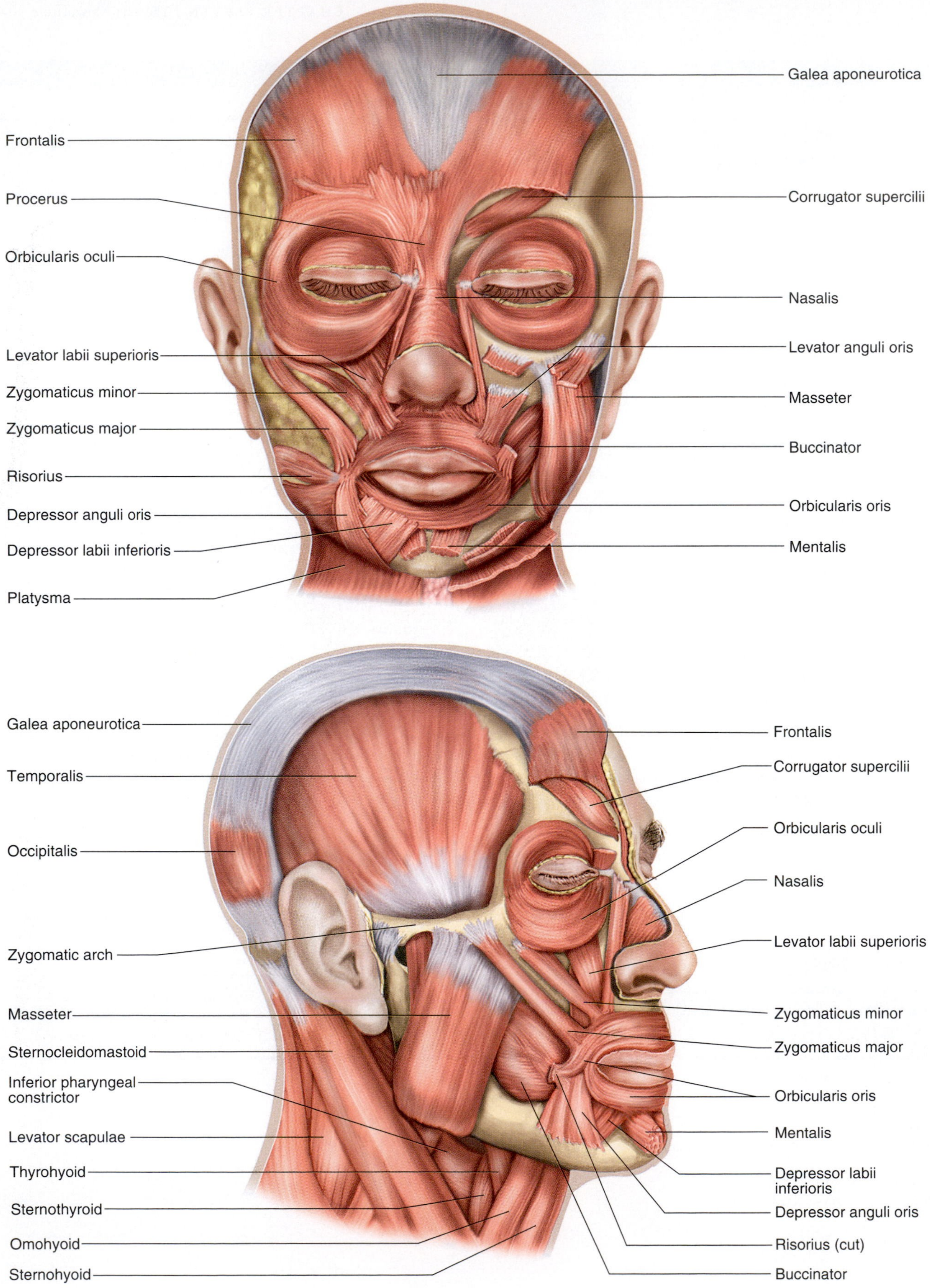

Galea aponeurotica
Frontalis
Procerus
Corrugator supercilii
Orbicularis oculi
Nasalis
Levator labii superioris
Levator anguli oris
Zygomaticus minor
Masseter
Zygomaticus major
Buccinator
Risorius
Orbicularis oris
Depressor anguli oris
Mentalis
Depressor labii inferioris
Platysma
Galea aponeurotica
Frontalis
Temporalis
Corrugator supercilii
Occipitalis
Orbicularis oculi
Nasalis
Zygomatic arch
Levator labii superioris
Masseter
Zygomaticus minor
Sternocleidomastoid
Zygomaticus major
Inferior pharyngeal constrictor
Orbicularis oris
Levator scapulae
Mentalis
Thyrohyoid
Depressor labii inferioris
Sternothyroid
Depressor anguli oris
Omohyoid
Risorius (cut)
Sternohyoid
Buccinator

FIGURE 11.3
Muscles of Facial Expression.

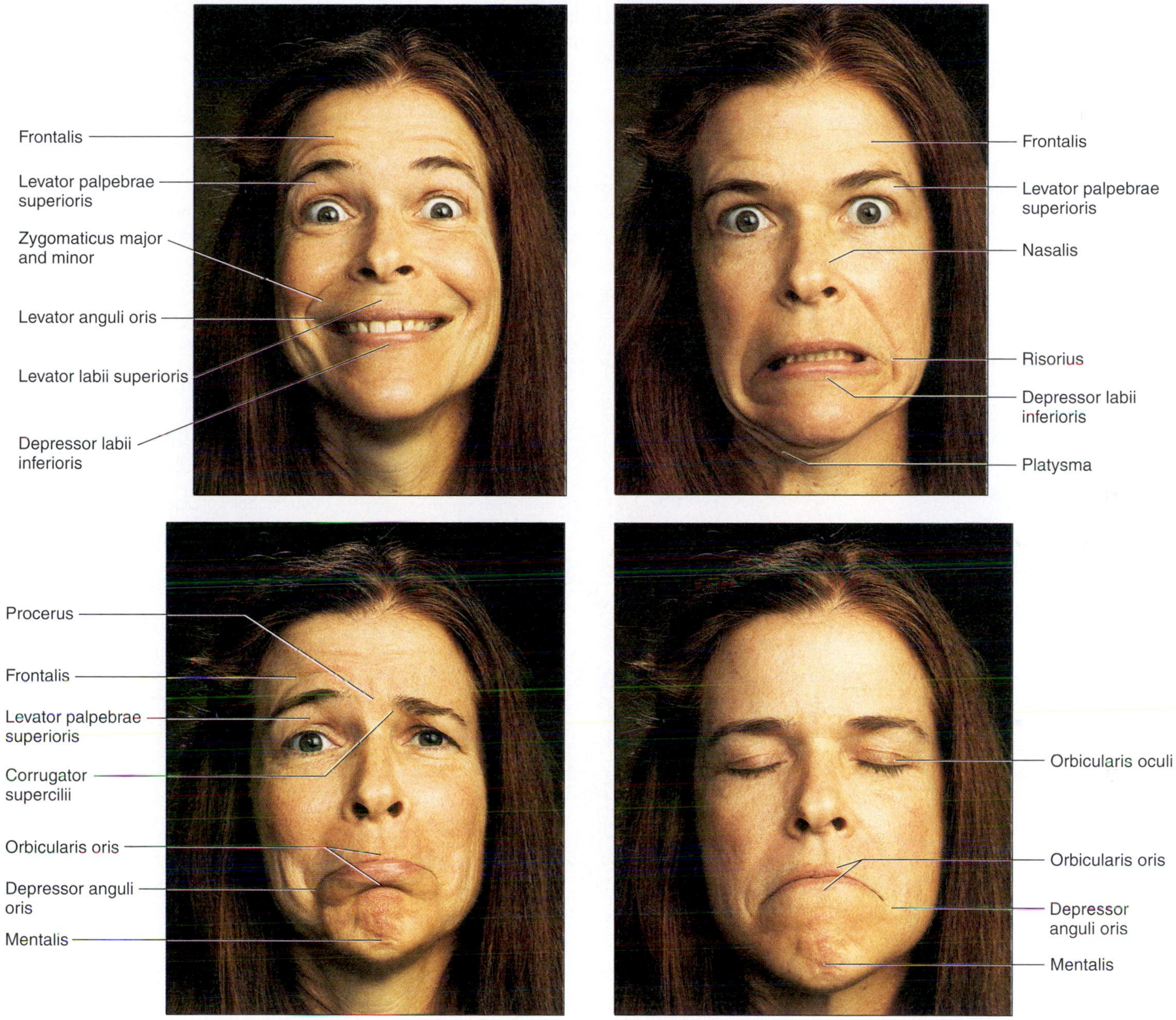
Frontalis
Levator palpebrae superioris
Zygomaticus major and minor
Levator anguli oris
Levator labii superioris
Depressor labii inferioris
Frontalis
Levator palpebrae superioris
Nasalis
Risorius
Depressor labii inferioris
Platysma
Procerus
Frontalis
Levator palpebrae superioris
Corrugator supercilii
Orbicularis oris
Depressor anguli oris
Mentalis
Orbicularis oculi
Orbicularis oris
Depressor anguli oris
Mentalis

FIGURE 11.4

Expressions Produced by Several of the Facial Muscles. The ordinary actions of these muscles are usually more subtle than these demonstrations.

TABLE 11.3

Muscles of Chewing and Swallowing (figs. 11.5–11.7)

O = origin, I = insertion, N = innervation (n. = nerve)

The following muscles contribute to facial expression and speech but are primarily concerned with manipulation of food, including tongue movements, chewing, and swallowing.

Extrinsic Muscles of the Tongue. The tongue is a very agile organ. It pushes food between the molars for chewing (mastication) and later forces the chewed food into the pharynx for swallowing. Both intrinsic and extrinsic muscles are responsible for its complex movements. The intrinsic muscles, which have no specific names, consist of a variable number of vertical fascicles that extend from the superior to inferior side of the tongue, transverse fascicles that extend from left to right, and longitudinal fascicles that extend from root to tip (see fig. 10.2c). The extrinsic muscles connect the tongue to other structures in the head (fig. 11.5).

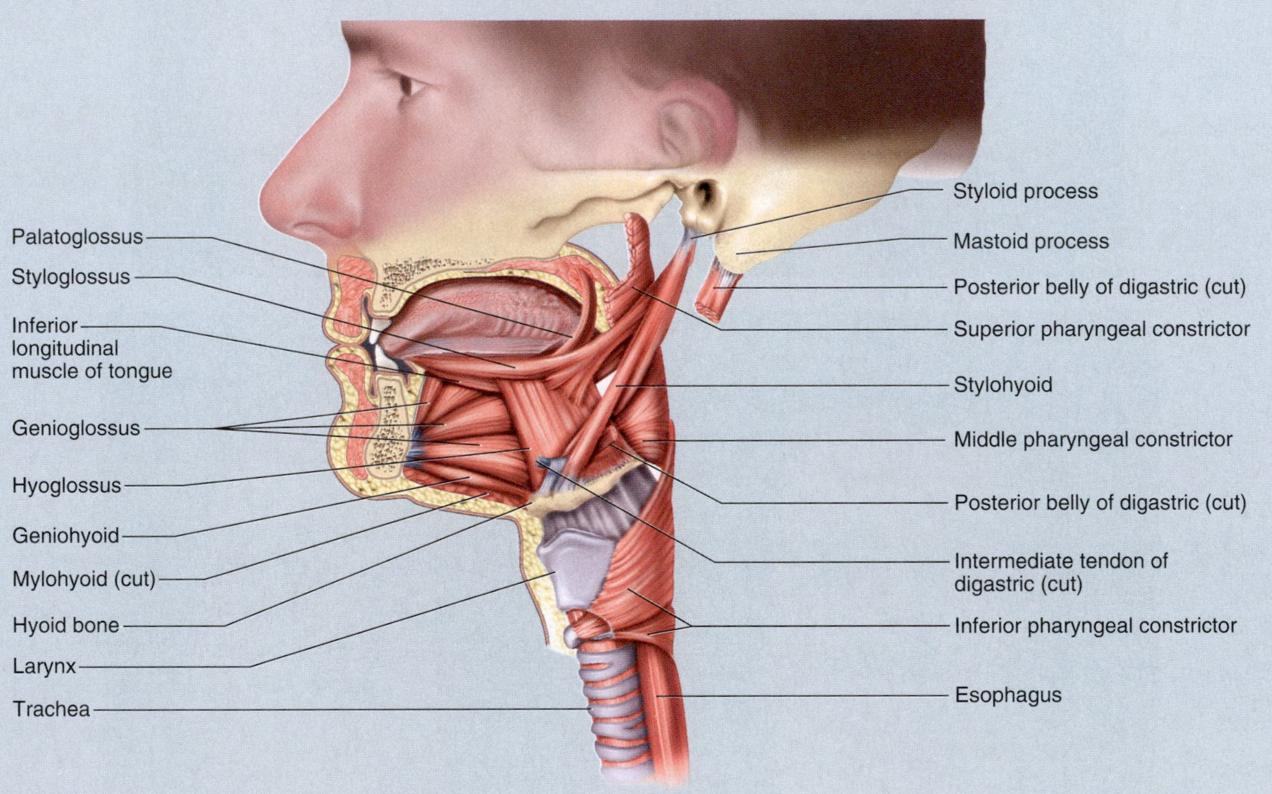

FIGURE 11.5
Muscles of the Tongue and Pharynx. Left lateral view.

Genioglossus[16] **(JEE-nee-oh-GLOSS-us)**

Depresses and protrudes tongue; creates dorsal groove in tongue that enables infants to grasp nipple and channel milk to pharynx

O: mental spines of mandible	I: hyoid bone, lateral aspect of tongue	N: hypoglossal n. (XII)

Hyoglossus[17]

Depresses sides of tongue

O: hyoid bone	I: lateral aspect of tongue	N: hypoglossal n. (XII)

Styloglossus[18]

Elevates and retracts tongue

O: styloid process	I: lateral aspect of tongue	N: hypoglossal n. (XII)

Palatoglossus[19]

Elevates posterior part of tongue; constricts fauces (entry to pharynx)

O: soft palate	I: lateral aspect of tongue	N: accessory n. (XI)

(continued)

[16]*genio* = chin + *gloss* = tongue
[17]*hyo* = hyoid bone + *gloss* = tongue
[18]*stylo* = styloid process + *gloss* = tongue
[19]*palato* = palate + *gloss* = tongue

TABLE 11.3

Muscles of Chewing and Swallowing *(continued)*

Muscles of Mastication. There are four paired muscles of mastication—the temporalis, masseter, and medial and lateral pterygoids. The *temporalis* is a broad, fan-shaped muscle that arises from the temporal lines of the skull, passes behind the zygomatic arch, and inserts on the coronoid process of the mandible (fig. 11.6*a*). The *masseter* is shorter and superficial to the temporalis, arising from the zygomatic arch and inserting on the lateral surface of the angle of the mandible. It is a thick muscle easily palpated on the side of the jaw. The temporalis and masseter elevate the mandible to bite and chew food; they are two of the most powerful muscles in the body. Similar action is provided by the *medial* and *lateral pterygoids.* They arise from the pterygoid processes of the sphenoid bone and insert on the medial surface of the mandible (fig. 11.6*b*). The pterygoids elevate and protract the mandible and produce the lateral excursions that grind food between the molars.

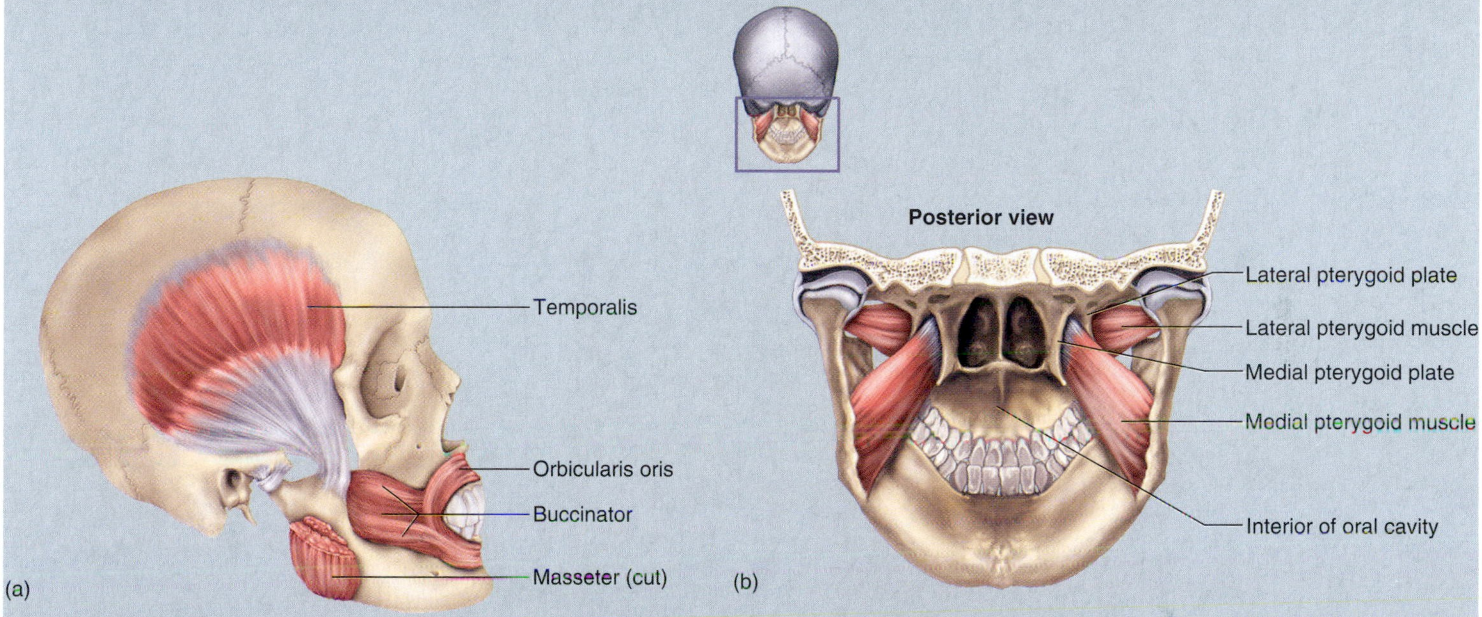
Temporalis
Orbicularis oris
Buccinator
Masseter (cut)
(a)
Posterior view
Lateral pterygoid plate
Lateral pterygoid muscle
Medial pterygoid plate
Medial pterygoid muscle
Interior of oral cavity
(b)

FIGURE 11.6

Muscles of Chewing. (*a*) Right lateral view. In order to expose the insertion of the temporalis muscle on the mandible, part of the zygomatic arch and masseter muscle are removed. (*b*) View of the pterygoid muscles looking into the oral cavity from behind the skull.

Temporalis[20] (TEM-po-RAY-liss)

Elevates mandible for biting and chewing; retracts mandible

O: temporal lines	I: coronoid process	N: trigeminal n. (V)

Masseter[21] (ma-SEE-tur)

Elevates mandible for biting and chewing; causes some lateral excursion of mandible

O: zygomatic arch	I: lateral aspect of mandibular ramus and angle	N: trigeminal n. (V)

Medial Pterygoid[22] (TERR-ih-goyd)

Elevates mandible; produces lateral excursion

O: pterygoid process of sphenoid bone	I: medial aspect of mandibular angle	N: trigeminal n. (V)

Lateral Pterygoid

Protracts mandible; produces lateral excursion

O: pterygoid process of sphenoid bone	I: slightly anterior to mandibular condyle	N: trigeminal n. (V)

(continued)

[20]temporal region
[21]*masset* = chew
[22]*pteryg* = wing + *oid* = like

TABLE 11.3

Muscles of Chewing and Swallowing *(continued)*

Hyoid Muscles—Suprahyoid Group. Several of the actions of chewing and swallowing are aided by eight pairs of *hyoid muscles* associated with the hyoid bone. Four of them, superior to the hyoid, form the *suprahyoid group*—the digastric, geniohyoid, mylohyoid, and stylohyoid. (See fig. 11.5 for the geniohyoid and fig. 11.7 for the others.) Those inferior to the hyoid form the *infrahyoid group.* Several hyoid muscles receive their innervation from the *ansa cervicalis,* a loop of nerve at the side of the neck formed by certain fibers of the first to third cervical nerves.

The *digastric* muscle arises from the mastoid process and thickens into a *posterior belly* beneath the margin of the mandible. It then narrows, passes through a connective tissue loop *(fascial sling)* attached to the hyoid bone, widens into an *anterior belly,* and attaches to the mandible near the mental protuberance. When it contracts, it pulls on the sling and elevates the hyoid bone, but if the hyoid is fixed by the infrahyoid muscles, the digastric muscle opens the mouth. The mouth normally drops open by itself when the temporalis and masseter muscles are relaxed, but the digastric, platysma, and mylohyoid can open it more widely, as when ingesting food or yawning.

The *geniohyoid* protracts the hyoid to widen the pharynx when food is swallowed. The *mylohyoid* muscles fuse at the midline, form the floor of the mouth, and work synergistically with the digastric to forcibly open the mouth. The *stylohyoid* elevates the hyoid bone.

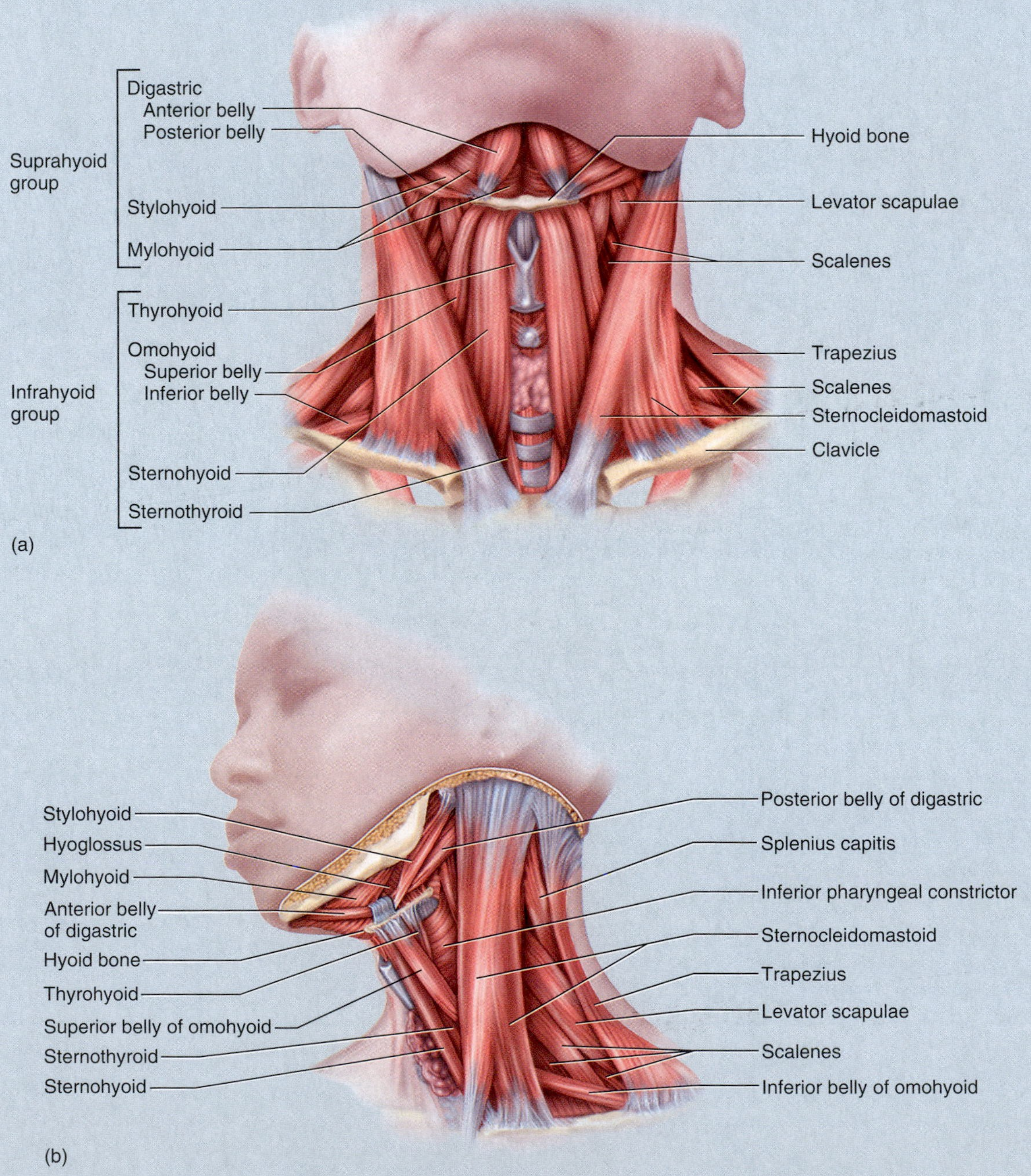

FIGURE 11.7

Muscles of the Neck. (*a*) The hyoid muscles, anterior view. (*b*) Left lateral view. The geniohyoid is deep to the mylohyoid and can be seen in figure 11.5.

(continued)

TABLE 11.3

Muscles of Chewing and Swallowing *(continued)*

Digastric[23]

Retracts mandible; elevates and fixes hyoid; opens mouth when hyoid is fixed by other muscles

O: mastoid notch and inner aspect of mandible near protuberance	I: hyoid, via fascial sling	N: trigeminal n. (V), facial n. (VII)

Geniohyoid[24] **(JEE-nee-oh-HY-oyd)**

Elevates and protracts hyoid; dilates pharynx to receive food; opens mouth when hyoid is fixed by other muscles

O: inner aspect of mental protuberance	I: hyoid	N: hypoglossal n. (XII)

Mylohyoid[25]

Forms floor of mouth; elevates hyoid; opens mouth when hyoid is fixed by other muscles

O: inferior margin of mandible	I: hyoid	N: trigeminal n. (V)

Stylohyoid

Elevates hyoid

O: styloid process	I: hyoid	N: facial n. (VII)

Hyoid Muscles—Infrahyoid Group. The next four muscles are located inferior to the hyoid and are therefore called the *infrahyoid group*. By fixing the hyoid from below, they enable the suprahyoid muscles to open the mouth. The thyrohyoid, named for the hyoid bone and large *thyroid cartilage* of the larynx, helps prevent choking. It elevates the thyroid cartilage so that the larynx becomes sealed by a flap of tissue, the epiglottis. You can feel this effect by placing your fingers on your "Adam's apple" (a prominence of the thyroid cartilage) and feeling it bob up as you swallow. The sternothyroid then pulls the larynx down again. These infrahyoid muscles that act on the larynx are regarded as the extrinsic muscles of the larynx. The larynx also has intrinsic muscles, which are concerned with control of the vocal cords and laryngeal opening (see chapter 23).

Thyrohyoid

Depresses hyoid; elevates larynx; fixes hyoid during opening of mouth

O: thyroid cartilage of larynx cartilage 1	I: hyoid	N: hypoglossal n. (XII)

Omohyoid[26]

Depresses hyoid; fixes hyoid during opening of mouth

O: superior border of scapula	I: hyoid	N: ansa cervicalis

Sternohyoid

Depresses hyoid; fixes hyoid during opening of mouth

O: manubrium, costal cartilage 1	I: hyoid	N: ansa cervicalis

Sternothyroid

Depresses larynx; fixes hyoid during opening of mouth

O: manubrium, costal cartilage 1 or 2	I: thyroid cartilage of larynx	N: ansa cervicalis

Pharyngeal Constrictors. During swallowing, the *superior, middle,* and *inferior pharyngeal constrictors* of the throat (see fig. 11.5) constrict in that order to force chewed food into the esophagus.

Pharyngeal Constrictors (three muscles)

Constrict pharynx to force food into esophagus

O: mandible, medial pterygoid plate, hyoid bone, larynx	I: posterior median raphe (fibrous seam) of pharynx	N: glossopharyngeal n. (IX), vagus n. (X)

[23]*di* = two + *gastr* = belly
[24]*genio* = chin
[25]*mylo* = mill, molar teeth
[26]*omo* = shoulder

TABLE 11.4

Muscles Acting on the Head (figs. 11.7–11.9)

O = origin, I = insertion, N = innervation (n. = nerve, nn. = nerves)

Flexors of the Neck. Muscles that move the head originate on the vertebral column, thoracic cage, and pectoral girdle and insert on the cranial bones. The principal flexors of the neck are the sternocleidomastoid and three scalenes on each side (see fig. 11.7). *The superior, middle,* and *inferior scalenes* flex the neck laterally and aid in respiration.

The prime mover of neck flexion, however, is the *sternocleidomastoid,* a thick cordlike muscle that extends from the sternum and clavicle to the mastoid process behind the ear. It is most easily seen and palpated when the head is turned to one side and slightly extended. When both sternocleidomastoids contract, the neck flexes forward, as when you look at something between your feet. When only the left one contracts, the head tilts down and to the right. To visualize this action, hold the index finger of your left hand on your left mastoid process and the index finger of your right hand on your sternal notch. Now contract the left sternocleidomastoid in a way that brings the two fingertips as close together as possible. You will note that this action causes you to look downward and to the right.

As the sternocleidomastoid passes obliquely across the neck, it divides the neck into *anterior* and *posterior triangles.* Other muscles and landmarks subdivide each of these into smaller triangles of surgical importance (fig. 11.8).

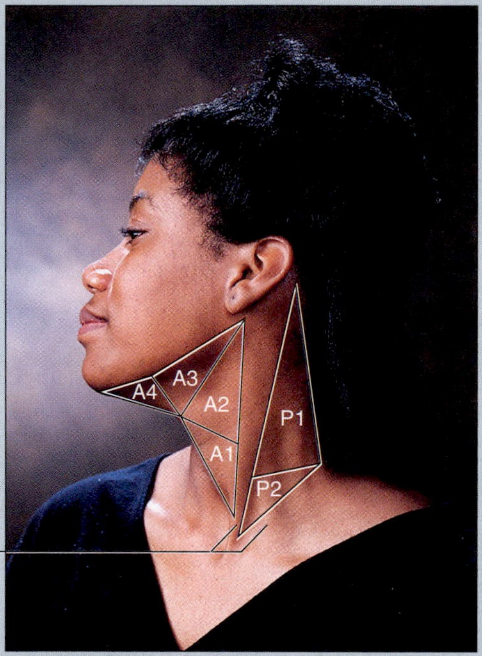
Anterior triangles
A1. Muscular
A2. Carotid
A3. Submandibular
A4. Suprahyoid
Posterior triangles
P1. Occipital
P2. Omoclavicular
Sternocleidomastoid
A1
A2
A3
A4
P1
P2

FIGURE 11.8
Triangles of the Neck. The sternocleidomastoid muscle separates the anterior triangles from the posterior triangles.

Sternocleidomastoid[27] (STIR-no-CLY-doe-MASS-toyd)

Contraction of either one draws head down and toward the side opposite the contracting muscle; contraction of both draws head forward and down, as in looking between the feet

O: clavicle, manubrium | I: mastoid process | N: accessory n. (XI)

Scalenes[28] (SCAY-leens) (three muscles)

Flex neck laterally; elevate ribs 1 and 2 in inspiration

O: vertebrae C2–C6 | I: ribs 1–2 | N: C5–C8

Extensors of the Neck. The extensors are located in the nuchal region (the back of the neck). Their actions include extension (holding the head erect), hyperextension (as in looking upward toward the sky), abduction (tilting the head to one side), and rotation (as in looking to the left and right). Extension and hyperextension involve equal action of the right and left members of a pair; the other actions require the muscle on one side to contract more strongly than the opposite muscle. Many head movements result from a combination of these actions—for example, looking up over the shoulder involves a combination of rotation and extension.

[27]*sterno* = sternum + *cleido* = clavicle + *mastoid* = mastoid process of skull
[28]*scal* = staircase

(continued)

TABLE 11.4

Muscles Acting on the Head *(continued)*

Three major extensors are the trapezius, splenius capitis, and semispinalis capitis (figs. 11.9, 11.14, and 11.15). The *trapezius* is a vast triangular muscle of the upper back and neck; together, the right and left trapezius muscles form a trapezoid. The long origin of the trapezius extends from the occipital protuberance of the skull to vertebra T12. The trapezius converges to an insertion on the shoulder. The *splenius capitis*, which lies just deep to the trapezius on the neck, has oblique fascicles that diverge from the vertebral column toward the ears. It is nicknamed the "bandage muscle" because of the way it tightly binds deeper neck muscles. The *semispinalis capitis* is slightly deeper, and its fascicles travel vertically up the back of the neck to insert on the occipital bone. It is named for the fact that it is attached in part to the spinous processes of the vertebrae. A complex array of smaller, deeper extensors are synergists of these prime movers; they extend the head, rotate it, or both.

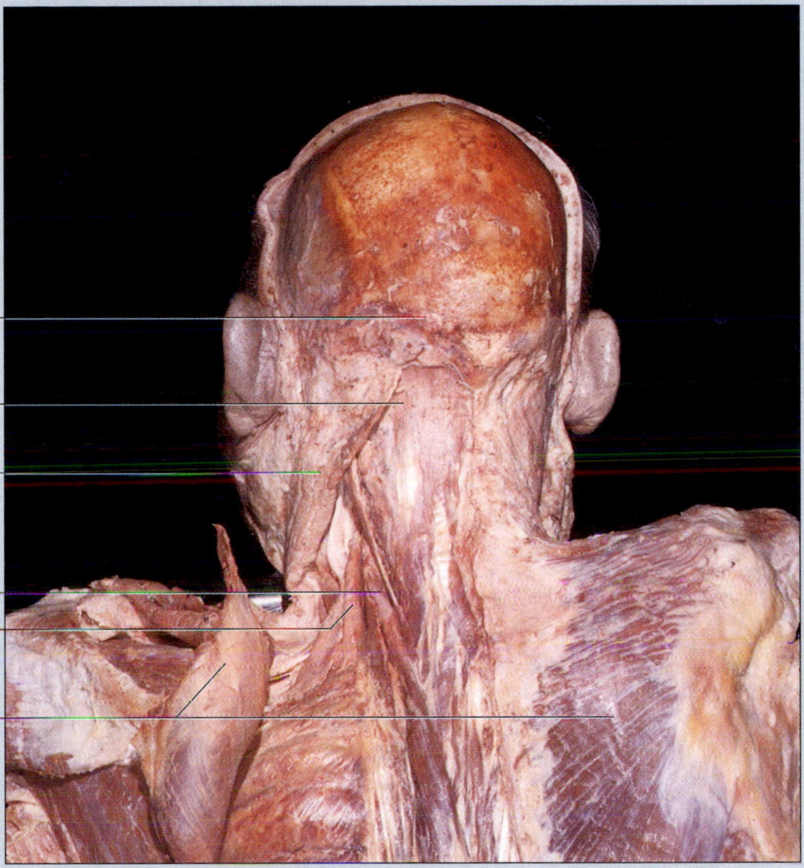
Superior nuchal line
Semispinalis capitis
Sternocleidomastoid
Longissimus capitis
Longissimus cervicis
Trapezius

FIGURE 11.9
Muscles of the Shoulder and Nuchal Regions.

Trapezius[29] (tra-PEE-zee-us)

Abducts head, extends neck (see other functions in chapter 12)

O: external occipital protuberance, nuchal ligament, spinous processes of vertebrae C7–T12	I: clavicle, acromion, scapular spine	N: accessory n. (XI), C3–C4

Splenius Capitis[30] (SPLEE-nee-us CAP-ih-tis) and Splenius Cervicis[31] (SIR-vih-sis)

Rotate head, extend neck

O: *capitis*—spinous processes of vertebrae C7 to T3 or T4; *cervicis*—spinous processes of T3–T6	I: *capitis*—mastoid process, superior nuchal line; *cervicis*—transverse processes of C1 to C2 or C3	N: dorsal rami of middle and lower cervical nn.

Semispinalis Capitis[32] (SEM-ee-spy-NAY-liss)

Rotates head, extends neck (see other parts of semispinalis in table 11.7)

O: transverse processes of vertebrae T1–T6, articular processes of C4–C7	I: occipital bone	N: dorsal rami of cervical nn.

[29] *trapez* = table, trapezoid
[30] *spleni* = bandage + *capitis* = of the head
[31] *cervicis* = of the neck
[32] *semi* = half + *spin* = spinous processes of vertebrae

Think About It!

Of the muscles you have studied so far, name three that you would consider intrinsic muscles of the head and three that you would classify as extrinsic. Explain your reason for each.

Before You Go On

Answer the following questions to test your understanding of the preceding section:

3. Name two muscles that elevate the upper lip and two that depress the lower lip.
4. Name the four paired muscles of mastication and state where they insert on the mandible.
5. Distinguish between the functions of the suprahyoid and infrahyoid muscles.
6. List the prime movers of neck extension and flexion.

Muscles of the Trunk

Objectives

When you have completed this section, you should be able to

- name and locate the muscles of respiration and explain how they affect abdominal pressure;
- name and locate the muscles of the abdominal wall, back, and pelvic floor; and
- identify the origin, insertion, action, and innervation of any of these muscles.

In this section, we will examine muscles of the trunk of the body in three functional groups concerned with respiration, support of the abdominal wall and pelvic floor, and movement of the vertebral column (tables 11.5–11.8). In the illustrations, you will note some major muscles that are not discussed in the associated tables—for example, the pectoralis major and serratus anterior. Although they are *located in* the trunk, they *act upon* the limbs and limb girdles, and are therefore discussed in chapter 12.

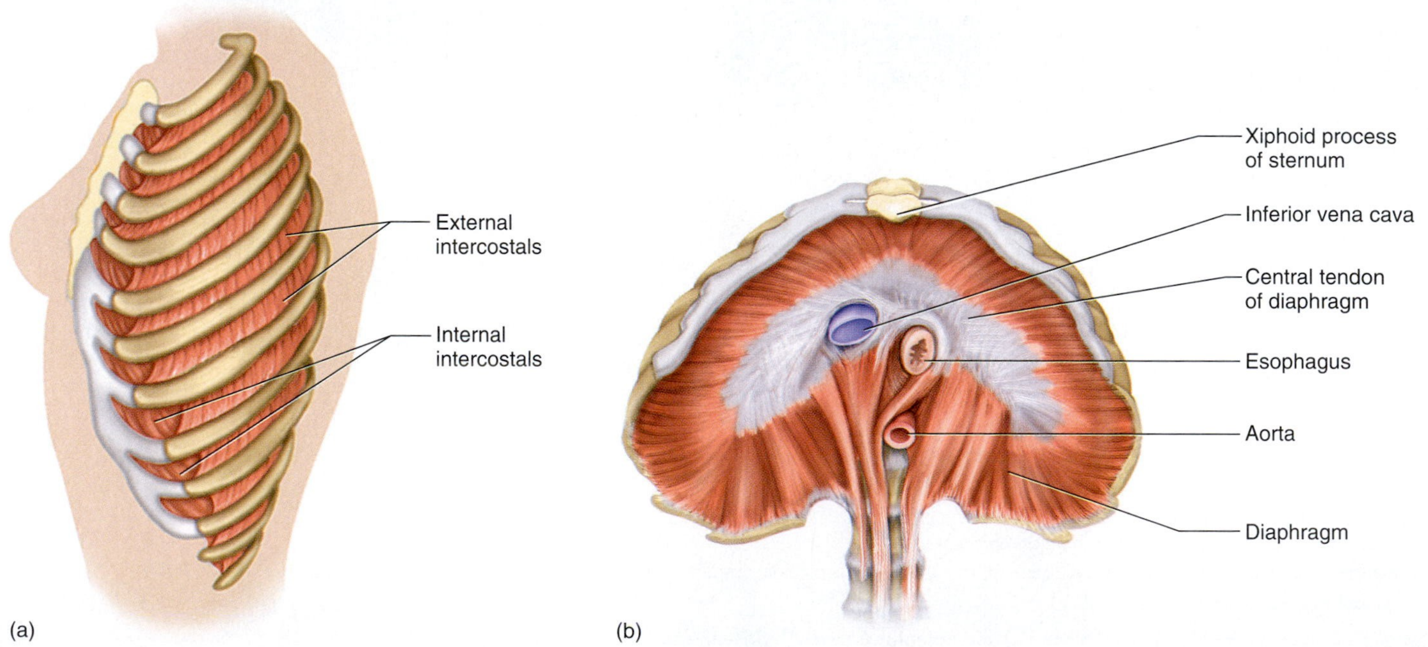

Figure 11.10

Muscles of Respiration. (*a*) The intercostal muscles, viewed from the left. (*b*) The diaphragm, viewed from below.

TABLE 11.5

Muscles of Respiration (fig. 11.10)

O = origin, I = insertion, N = innervation (n. = nerve, nn. = nerves)

We breathe primarily by means of muscles that enclose the thoracic cavity—the diaphragm, which forms its floor; 11 pairs of external intercostal muscles, which lie superficially between the ribs; and 11 pairs of internal intercostal muscles, which lie between the ribs deep to the external intercostals (fig. 11.10). The lungs themselves contain no skeletal muscle; they do not play an active part in their own ventilation.

The *diaphragm* is a muscular dome between the abdominal and thoracic cavities. It has openings that allow passage of the esophagus and major blood vessels. Its fascicles converge from the margins toward a fibrous **central tendon.** When the diaphragm contracts, it flattens slightly, increasing the volume of the thoracic cage and creating a partial vacuum that draws air into the lungs. Its contraction also raises pressure in the abdominal cavity below, thus helping to expel the contents of the bladder and rectum and facilitating childbirth—which is why people tend to take a deep breath and hold it during these functions.

The *external intercostals* extend obliquely downward and anteriorly from each rib to the rib below it. When the scalenes fix the first rib, the external intercostals lift the others, pulling them up somewhat like bucket handles. This action pulls the ribs closer together and draws the entire rib cage upward and outward, expanding the thoracic cage and promoting inhalation.

No muscular effort is required to exhale—when the diaphragm and external intercostals relax, the thoracic cage springs back to its prior size and expels the air. However, forced expiration—exhaling more than the usual amount of air or exhaling quickly as in blowing out a candle—is achieved mainly by the *internal intercostals.* These also extend from one rib to the next, but they lie deep to the external intercostals and have fascicles at right angles to them. The abdominal muscles also aid in forced expiration by pushing the viscera up against the diaphragm.

Diaphragm[33] (DY-uh-fram)

Prime mover of inspiration; compresses abdominal viscera to aid in such processes as defecation, urination, and childbirth

O: xiphoid process, ribs 10–12, costal cartilages 5–9, lumbar vertebrae	I: central tendon	N: phrenic n.

External Intercostals[34] (IN-tur-COSS-tulz)

When scalenes fix rib 1, external intercostals draw ribs 2–12 upward and outward to expand thoracic cavity and inflate lungs

O: inferior margins of ribs 1–11	I: superior margins of ribs 2–12	N: intercostal nn.

Internal Intercostals

When quadratus lumborum and other muscles fix rib 12, internal intercostals draw ribs downward and inward to compress thoracic cavity and force air from lungs; not needed for relaxed expiration

O: inferior margins of ribs 1–11	I: superior margins of ribs 2–12	N: intercostal nn.

[33] *dia* = across + *phragm* = partition
[34] *inter* = between + *costa* = rib

INSIGHT 11.1 CLINICAL APPLICATION

DIFFICULTY BREATHING

Asthma, emphysema, heart failure, and other conditions can cause *dyspnea,* difficulty catching one's breath. People with dyspnea make increased use of accessory muscles to aid the diaphragm and intercostals in breathing, and often lean on a table or chair back to breathe more deeply. This action fixes the clavicles and scapulae so that the accessory muscles—such as the *pectoralis major* and *serratus anterior* (see chapter 12)—move the ribs instead of the bones of the pectoral girdle.

THINK ABOUT IT!

What muscles are eaten as "spare ribs"? What is the tough fibrous membrane between the meat and the bone?

TABLE 11.6

Muscles of the Abdomen (figs. 11.11–11.13)

O = origin, I = insertion, N = innervation (n. = nerve, nn. = nerves)

The anterior and lateral walls of the abdomen are reinforced by four pairs of sheetlike muscles that support the viscera, stabilize the vertebral column during heavy lifting, and aid in respiration, urination, defecation, vomiting, and childbirth. They are the *rectus abdominis, external abdominal oblique, internal abdominal oblique,* and *transversus abdominis.*

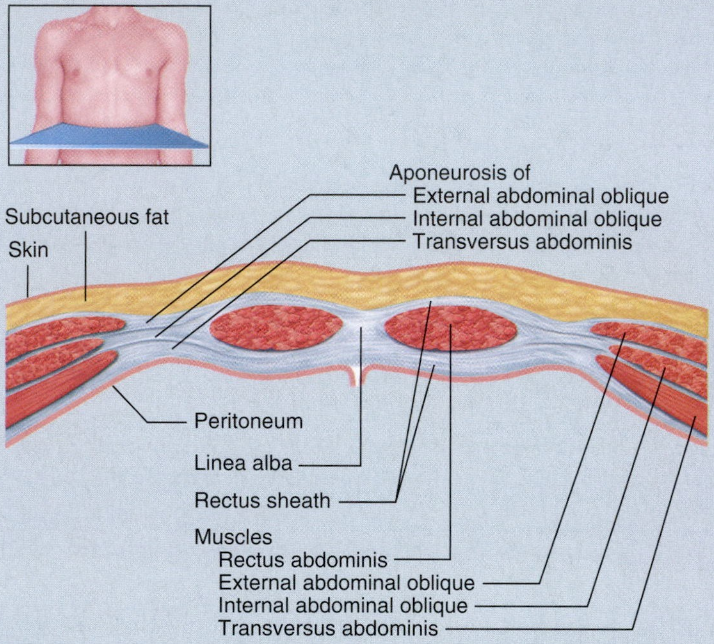

FIGURE 11.11
Cross Section of the Anterior Abdominal Wall.

The *rectus abdominis* is a medial straplike muscle extending vertically from the pubis to the sternum. It is separated into four segments by fibrous **tendinous intersections** that give the abdomen a segmented appearance in well-muscled individuals. The rectus abdominis is enclosed in a fibrous sleeve called the **rectus sheath,** and the right and left muscles are separated by a vertical fibrous strip called the **linea alba.**[35]

The *external abdominal oblique* is the most superficial muscle of the lateral abdominal wall. Its fascicles run anteriorly and downward. Deep to it is the *internal abdominal oblique,* whose fascicles run anteriorly and upward. Deepest of all is the *transversus abdominis,* whose fascicles run horizontally across the abdomen. Unlike the thoracic cavity, the abdominal cavity lacks a protective bony enclosure. However, the wall formed by these three muscle layers is strengthened by the way their fascicles run in different directions like layers of plywood.

The tendons of the abdominal muscles are aponeuroses. They continue medially to form the rectus sheath and terminate at the linea alba. At its inferior margin, the aponeurosis of the external oblique forms a strong, cordlike **inguinal ligament** that extends from the pubis to the anterior superior spine of the ilium.

Rectus Abdominis[36] (ab-DOM-ih-niss)

Supports abdominal viscera; flexes vertebral column as in sit-ups; depresses ribs; stabilizes pelvis during walking; increases intra-abdominal pressure to aid in urination, defecation, and childbirth

O: pubis   I: xiphoid process, costal cartilages 5–7   N: intercostal nn. 7–12

[35] *linea* = line + *alb* = white
[36] *rect* = straight + *abdominis* = of the abdomen

(continued)

TABLE 11.6

Muscles of the Abdomen *(continued)*

External Abdominal Oblique

Flexes abdomen as in sit-ups; flexes and rotates vertebral column

O: ribs 5–12	I: xiphoid process, linea alba	N: intercostal nn. 8–12, iliohypogastric n., ilioinguinal n.

Internal Abdominal Oblique

Similar to external oblique

O: inguinal ligament, iliac crest, thoracolumbar fascia	I: xiphoid process, linea alba, pubis, ribs 10–12	N: same as external oblique

Transversus Abdominis

Compresses abdomen, increases intra-abdominal pressure, flexes vertebral column

O: inguinal ligament, iliac crest, thoracolumbar fascia, costal cartilages 7–12	I: xiphoid process, linea alba, pubis, inguinal ligament	N: intercostal nn. 8–12, iliohypogastric n., ilioinguinal n.

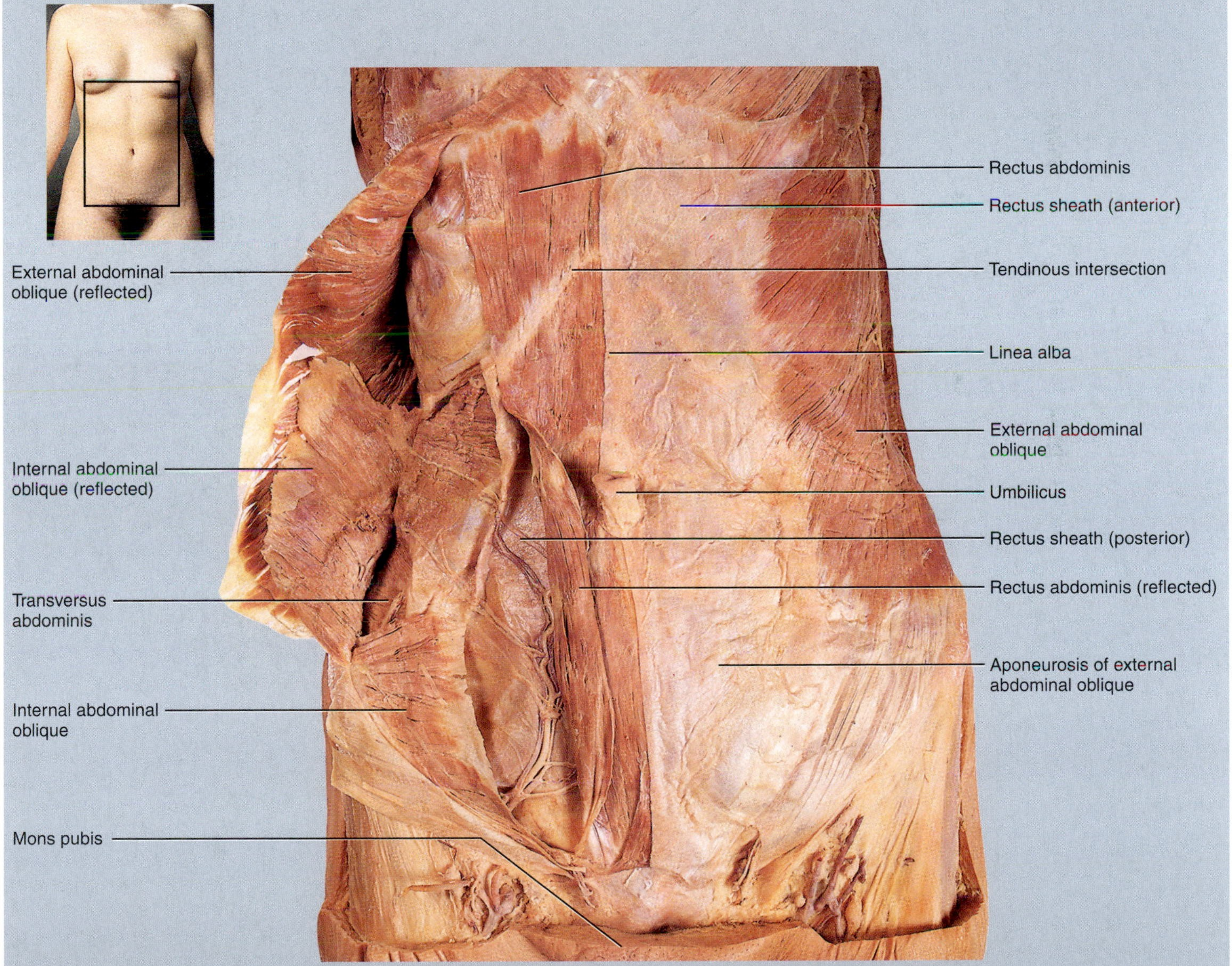

FIGURE 11.12

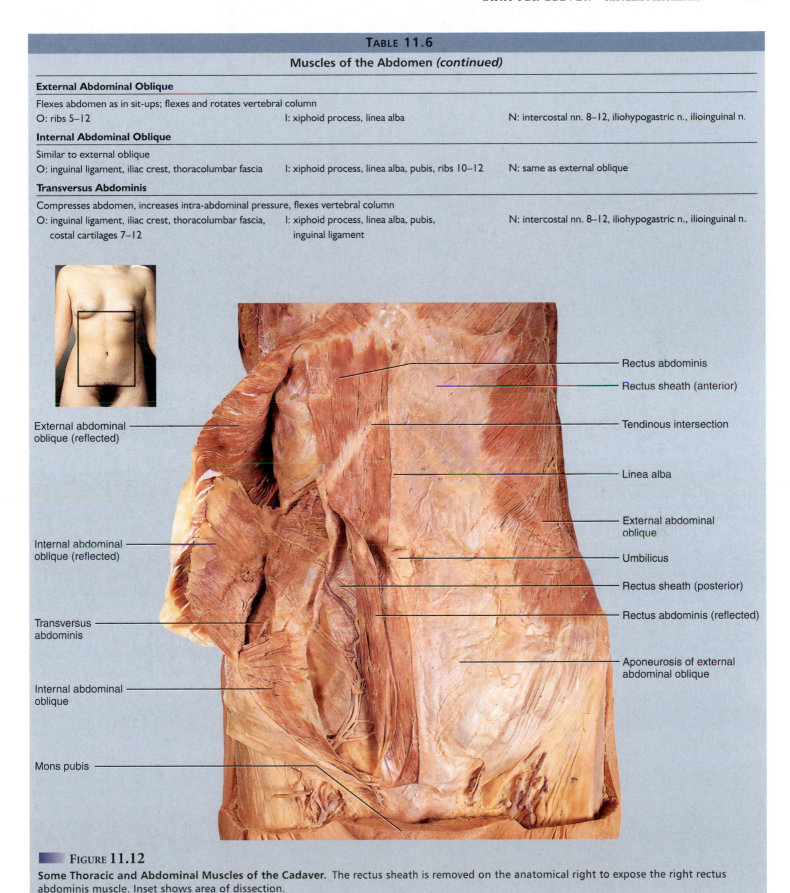

Some Thoracic and Abdominal Muscles of the Cadaver. The rectus sheath is removed on the anatomical right to expose the right rectus abdominis muscle. Inset shows area of dissection.

(continued)

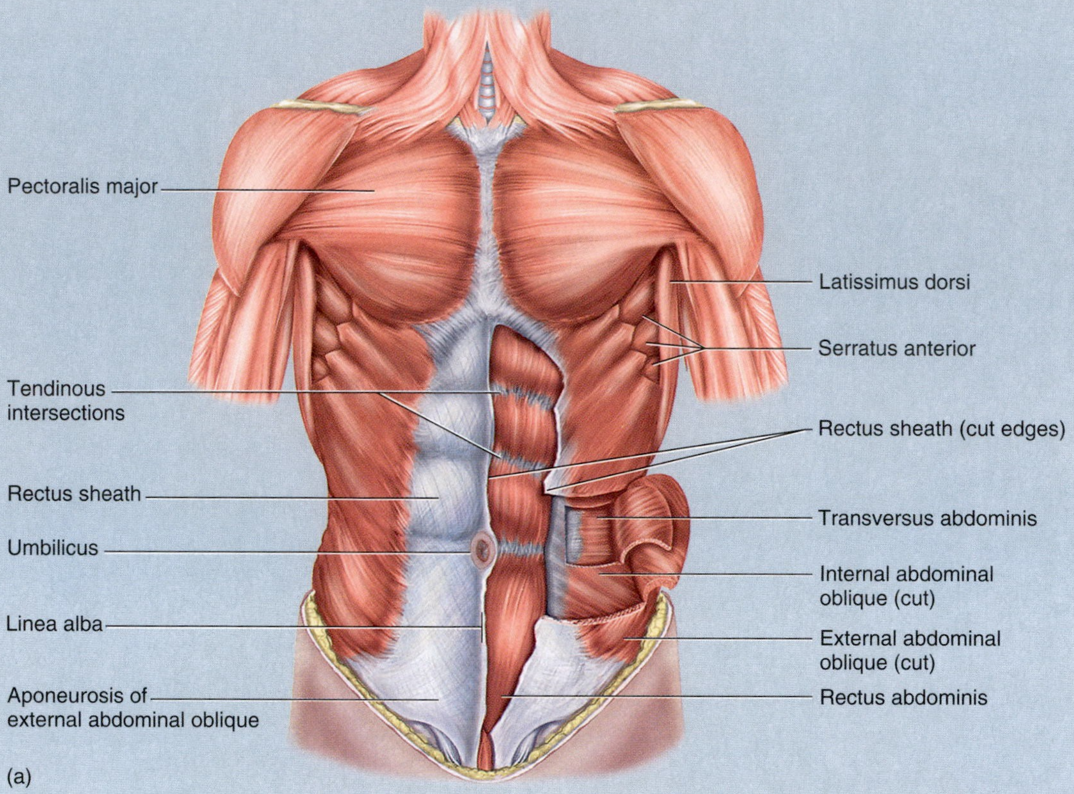
Pectoralis major
Latissimus dorsi
Serratus anterior
Tendinous intersections
Rectus sheath (cut edges)
Rectus sheath
Transversus abdominis
Umbilicus
Internal abdominal oblique (cut)
Linea alba
External abdominal oblique (cut)
Aponeurosis of external abdominal oblique
Rectus abdominis
(a)

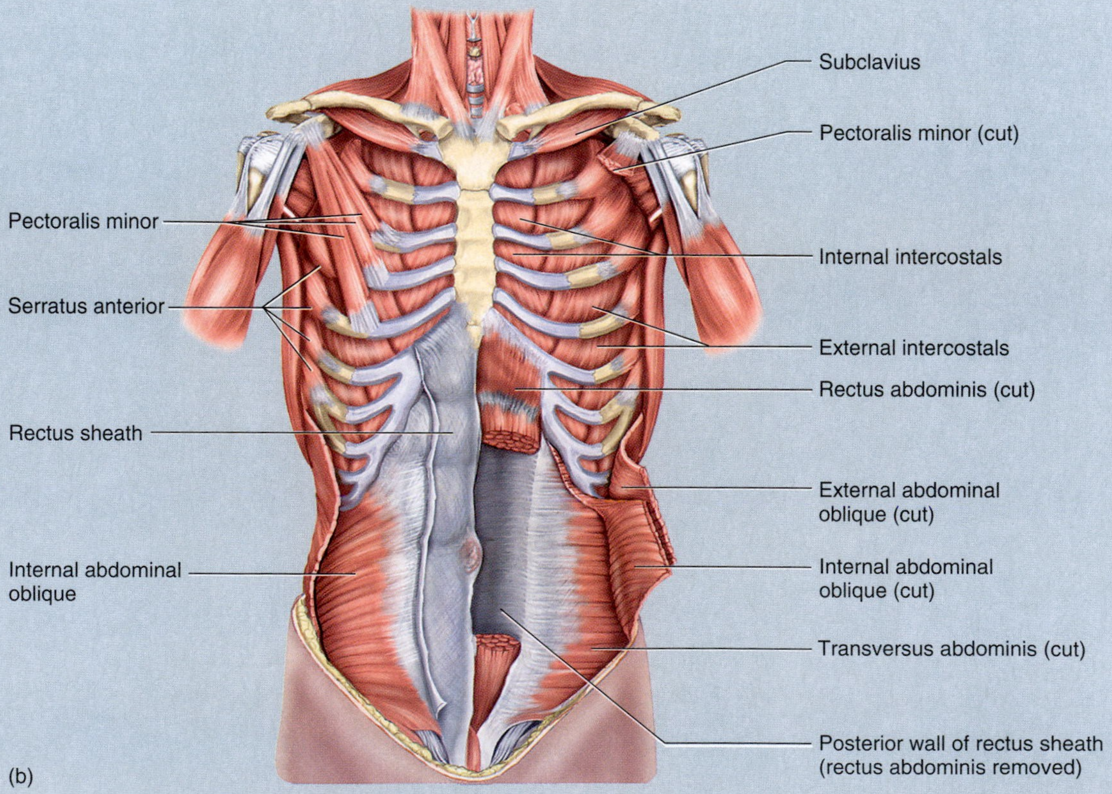
Subclavius
Pectoralis minor (cut)
Pectoralis minor
Internal intercostals
Serratus anterior
External intercostals
Rectus abdominis (cut)
Rectus sheath
External abdominal oblique (cut)
Internal abdominal oblique
Internal abdominal oblique (cut)
Transversus abdominis (cut)
Posterior wall of rectus sheath (rectus abdominis removed)
(b)

FIGURE 11.13

Thoracic and Abdominal Muscles. (*a*) Superficial muscles. The left rectus sheath is cut away to expose the rectus abdominis muscle. (*b*) Deep muscles. On the anatomical right, the external abdominal oblique has been removed to expose the internal abdominal oblique, and the pectoralis major removed to expose the pectoralis minor. On the anatomical left, the internal abdominal oblique has been cut to expose the transversus abdominis, and the rectus abdominis has been cut to expose the posterior rectus sheath.

TABLE 11.7

Muscles of the Back (figs. 11.14–11.16)

O = origin, I = insertion, N = innervation (n. = nerve, nn. = nerves)

The back muscles considered here extend, rotate, and abduct the vertebral column. Back muscles that act on the pectoral girdle and arm are considered in chapter 12. The muscles associated with the vertebral column moderate your motion when you bend forward and contract to return the trunk to the erect position. They are classified into two groups—a *superficial group* which extends from the vertebrae to the ribs and a *deep group* which connects the vertebrae to each other.

Superficial Back Muscles. In the superficial group, the prime mover of spinal extension is the *erector spinae* (ee-RECK-tur SPY-nee). You use this muscle to maintain your posture and to stand up straight after bending at the waist. It is divided into three "columns"—the *iliocostalis, longissimus,* and *spinalis.* These are complex, multipart muscles with cervical, thoracic, and lumbar portions. Some of their portions move the head and have already been discussed, while those that act on cervical and lower parts of the vertebral column are described in this table. Most of the lower back (lumbar) muscles are in the longissimus group. Two *serratus posterior* muscles—superior and inferior—overlie the erector spinae and aid breathing by moving the ribs.

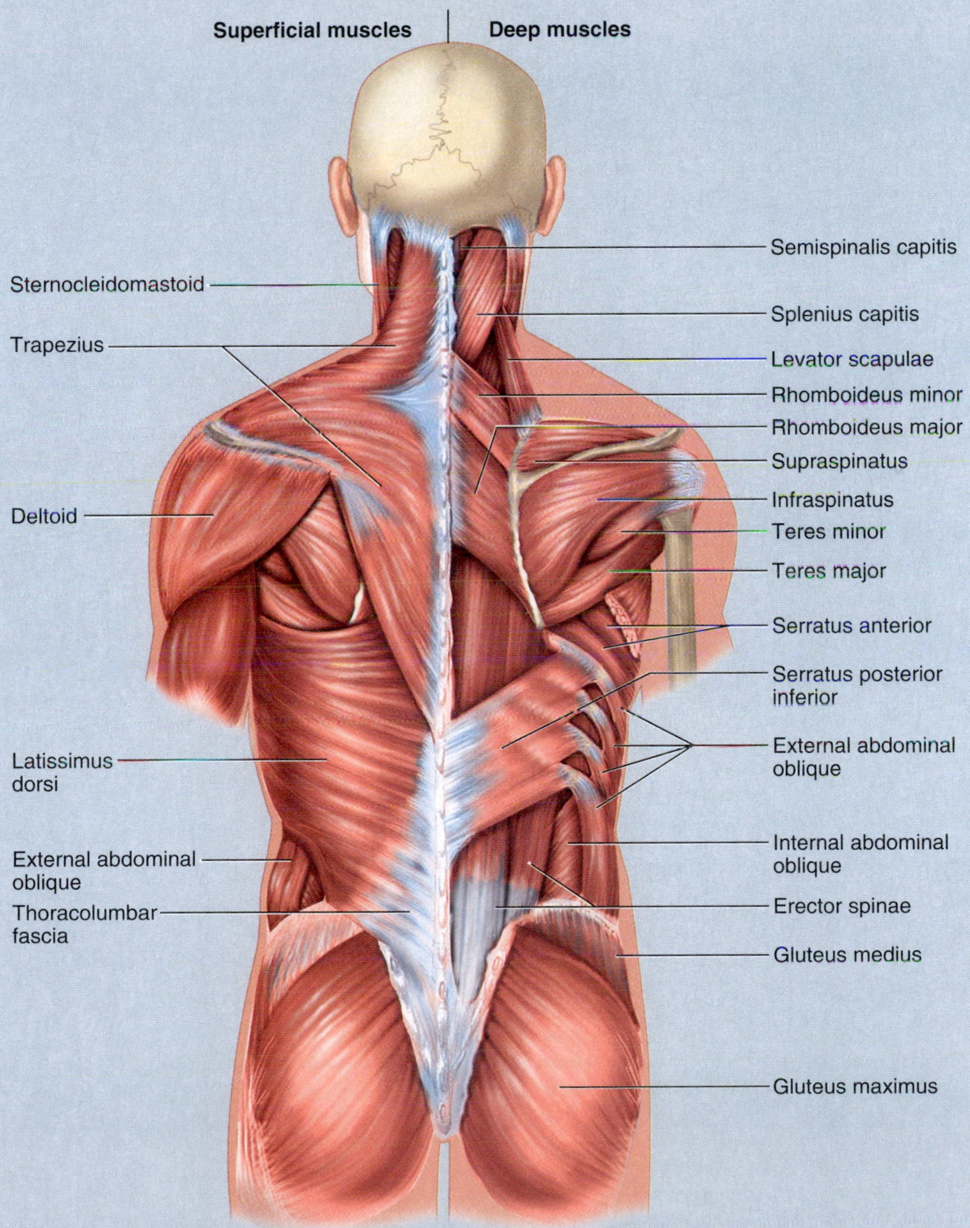

Superficial muscles
Deep muscles
Sternocleidomastoid
Trapezius
Deltoid
Latissimus dorsi
External abdominal oblique
Thoracolumbar fascia
Semispinalis capitis
Splenius capitis
Levator scapulae
Rhomboideus minor
Rhomboideus major
Supraspinatus
Infraspinatus
Teres minor
Teres major
Serratus anterior
Serratus posterior inferior
External abdominal oblique
Internal abdominal oblique
Erector spinae
Gluteus medius
Gluteus maximus

FIGURE 11.14
Neck, Back, and Gluteal Muscles. The most superficial muscles are shown on the *left,* and the next deeper layer on the *right.*

(continued)

TABLE 11.7

Muscles of the Back *(continued)*

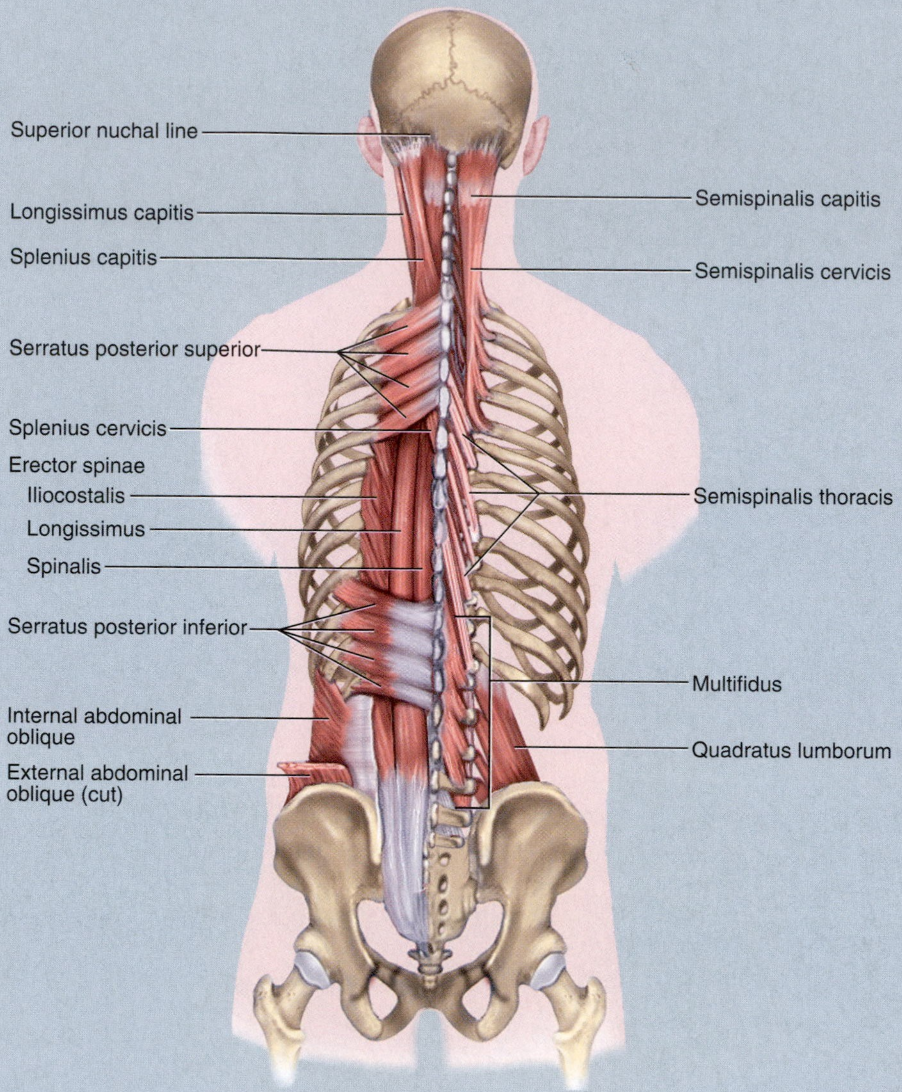
Superior nuchal line
Longissimus capitis
Splenius capitis
Serratus posterior superior
Splenius cervicis
Erector spinae
Iliocostalis
Longissimus
Spinalis
Serratus posterior inferior
Internal abdominal oblique
External abdominal oblique (cut)
Semispinalis capitis
Semispinalis cervicis
Semispinalis thoracis
Multifidus
Quadratus lumborum

FIGURE 11.15

Muscles Acting on the Vertebral Column. Those on the *right* are deeper than those on the *left*.

Iliocostalis[37] Cervicis (ILL-ee-oh-coss-TAH-liss SIR-vih-sis), Iliocostalis Thoracis (tho-RA-sis), and Iliocostalis Lumborum[38] (lum-BORE-um)

Extend and laterally flex vertebral column; thoracis and lumborum rotate ribs during forceful inspiration

O: angles of ribs, sacrum, iliac crest	I: *cervicis*—vertebrae C4–C6; *thoracis*—vertebra C7, angles ribs 1–6; *lumborum*—angles of ribs 7–12	N: dorsal rami of spinal nn.

Longissimus[39] Cervicis (lawn-JISS-ih-muss) and Longissimus Thoracis

Extend and laterally flex vertebral column

O: *cervicis*—vertebrae T1 to T4 or T5; *thoracis*—sacrum, iliac crest, vertebrae T1–L5	I: *cervicis*—vertebrae C2–C6; *thoracis*—vertebrae T1–T12, ribs 3 or 4 to 12	N: dorsal rami of spinal nn.

[37]*ilio* = ilium + *cost* = ribs
[38]*lumborum* = of the lower back
[39]*longissimus* = longest

(continued)

TABLE 11.7

Muscles of the Back *(continued)*

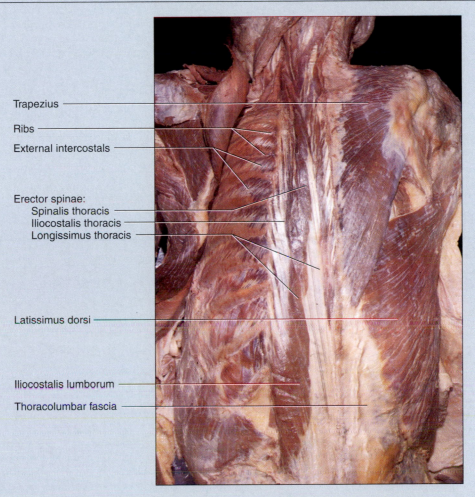

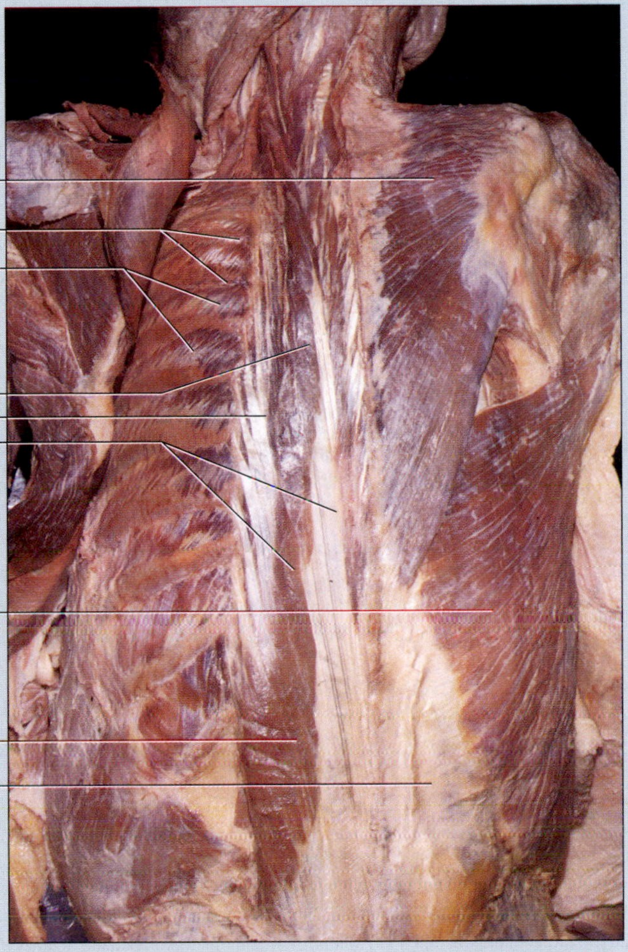
Trapezius
Ribs
External intercostals
Erector spinae:
Spinalis thoracis
Iliocostalis thoracis
Longissimus thoracis
Latissimus dorsi
Iliocostalis lumborum
Thoracolumbar fascia

FIGURE 11.16
Some Deep Back Muscles of the Cadaver.

Spinalis[40] Cervicis (spy-NAY-liss) and Spinalis Thoracis

Extend vertebral column

O: *cervicis*—nuchal ligament, spinous process of vertebra C7; *thoracis*—spinous processes of T11–L2	I: *cervicis*—spinous process of axis; *thoracis*—spinous processes of upper thoracic vertebrae	N: dorsal rami of spinal nn.

Serratus[41] Posterior Superior

Elevates ribs

O: spinous processes of C7–T2	I: angles of ribs 2–5	N: ventral rami of T1–T4

Serratus Posterior Inferior

Draws ribs back and downward

O: spinous processes of T11–L2	I: ribs 9–12	N: ventral rami of T9–T12

(continued)

[40] *spinalis* = of the spinous processes
[41] *serra* = saw

TABLE 11.7

Muscles of the Back *(continued)*

Deep Back Muscles. The major deep thoracic muscle is the *semispinalis* (see fig. 11.15). This is divided into three parts, the *semispinalis capitis* which we have already studied (see table 11.4), the *semispinalis cervicis,* and *semispinalis thoracis,* in that order from superior to inferior. In the lumbar region, the major deep muscle is the *quadratus lumborum.* The erector spinae and quadratus lumborum are enclosed in a fibrous sheath called the **thoracolumbar fascia,** which is the origin of some of the abdominal and lumbar muscles. The *multifidus* muscle deep to this connects the vertebrae to each other from the cervical to the lumbar region, and acts to extend and rotate the vertebral column.

Semispinalis Cervicis[42] (SEM-ee-spy-NAY-liss SUR-vih-sis) and Semispinalis Thoracis[43] (tho-RA-sis)		
Extend neck; extend and rotate vertebral column		
O: transverse processes of vertebrae T1–T10	I: spinous processes of vertebrae C2–T5	N: dorsal rami of spinal nn.
Quadratus Lumborum[44] (quad-RAY-tus lum-BORE-um)		
Laterally flexes vertebral column, depresses rib 12		
O: iliac crest, lower lumbar vertebrae, thoracolumbar fascia	I: upper lumbar vertebrae, rib 12	N: ventral rami of L1–L3
Multifidus[45] (mul-TIFF-ih-dus)		
Extends and rotates vertebral column		
O: sacrum, iliac crest, vertebrae C4–L5	I: laminae and spinous processes of vertebrae above origins	N: dorsal rami of spinal nn.

[42]*cervicis* = of the neck
[43]*thoracis* = of the thorax
[44]*quadrat* = four-sided + *lumborum* = of the lower back
[45]*multi* = many + *fid* = split, sectioned

INSIGHT 11.2 CLINICAL APPLICATION

Heavy Lifting and Back Injuries

When a skeletal muscle is excessively stretched, its sarcomeres are so stretched that its thick and thin myofilaments have little or no overlap. When such a muscle is stimulated to contract, few of the myosin heads are able to attach to the actin filaments (see chapter 10), the contraction is very weak, and the muscle and connective tissues are subject to injury.

When you are fully bent over forward, as in touching your toes, the erector spinae is extremely stretched. Standing up from such a position is therefore initiated by the hamstring muscles on the back of the thigh and the gluteus maximus of the buttocks. The erector spinae joins in the action when it is partially contracted. Standing too suddenly or improperly lifting a heavy weight, however, can strain the erector spinae, cause painful muscle spasms, tear tendons and ligaments of the lower back, and rupture intervertebral discs. The lumbar muscles are adapted for maintaining posture, not for lifting. This is why it is important, in heavy lifting, to kneel and use the powerful extensor muscles of the thighs and buttocks to lift the load.

INSIGHT 11.3 CLINICAL APPLICATION

Hernias

A hernia is any condition in which the viscera protrude through a weak point in the muscular wall of the abdominopelvic cavity. The most common type to require treatment is an *inguinal hernia.* In the male fetus, each testis descends from the pelvic cavity into the scrotum by way of a passage called the *inguinal canal* through the muscles of the groin. This canal remains a weak point in the pelvic floor, especially in infants and children. When pressure rises in the abdominal cavity, it can force part of the intestine or bladder into this canal or even into the scrotum. This also sometimes occurs in men who hold their breath while lifting heavy weights. When the diaphragm and abdominal muscles contract, pressure in the abdominal cavity can soar to 1,500 pounds per square inch—more than 100 times the normal pressure and quite sufficient to produce an inguinal hernia, or "rupture." Inguinal hernias rarely occur in women.

Two other sites of hernia are the diaphragm and navel. A *hiatus hernia* is a condition in which part of the stomach protrudes through the diaphragm into the thoracic cavity. This is most common in overweight people over 40. It may cause heartburn due to the regurgitation of stomach acid into the esophagus, but most cases go undetected. In an *umbilical hernia,* abdominal viscera protrude through the navel.

TABLE 11.8

Muscles of the Pelvic Floor (fig. 11.17)

O = origin, I = insertion, N = innervation (n. = nerve)

The floor of the pelvic cavity is formed by three layers of muscles and fasciae that span the pelvic outlet and support the viscera. It is penetrated by the anal canal, urethra, and vagina, which open into a diamond-shaped region between the thighs called the **perineum** (PERR-ih-NEE-um). The perineum is bordered by four bony landmarks—the pubic symphysis anteriorly, the coccyx posteriorly, and the ischial tuberosities laterally. The anterior half of the perineum is the **urogenital triangle** and the posterior half is the **anal triangle** (fig. 11.17*b*). These are especially important landmarks in obstetrics.

Superficial Perineal Space. The pelvic floor is divided into three layers or "compartments." The one just deep to the skin, called the **superficial perineal space** (fig. 11.17*a, b*), contains three muscles: the ischiocavernosus, bulbospongiosus, and superficial transverse perineus. The *ischiocavernosus* muscles converge like a V from the ischial tuberosities toward the penis or clitoris and assist in erection. In males, the *bulbospongiosus (bulbocavernosus)* forms a sheath around the base (bulb) of the penis; it expels semen during ejaculation. In females, it encloses the vagina like a pair of parentheses and tightens on the penis during intercourse. Voluntary contractions of this muscle in both sexes also help void the last few milliliters of urine. The **superficial transverse perineus** extends from the ischial tuberosities to a strong median fibromuscular anchorage, the **perineal body.**

Ischiocavernosus[46] (ISS-kee-oh-CAV-er-NO-sus)

Aids in erection of penis and clitoris

O: ischial and pubic rami, ischial tuberosity	I: penis, clitoris	N: pudendal n.

Bulbospongiosus[47] (BUL-bo-SPUN-jee-OH-sus)

Male: compresses urethra to expel semen or urine   Female: constricts vaginal orifice

O: central tendon of perineum, bulb of penis	I: fasciae of perineum, penis or clitoris	N: pudendal n.

Superficial Transverse Perineus (PERR-ih-NEE-us)

Fixes central tendon of perineum, supports pelvic floor

O: ischial ramus	I: central tendon	N: pudendal n.

Middle Compartment. In the middle compartment, the urogenital triangle is spanned by a thin triangular sheet called the **urogenital diaphragm.** This is composed of a fibrous membrane and two muscles—the **deep transverse perineus** and the **external urethral sphincter** (fig. 11.17*c, d*). The anal triangle has one muscle at this level, the **external anal sphincter.**

Deep Transverse Perineus

Fixes perineal body; supports pelvic floor; expels last drops of urine in both sexes and semen in male

O: ischial ramus	I: central tendon	N: pudendal n.

External Urethral Sphincter

Compresses urethra to voluntarily inhibit urination

O: ischial and pubic rami	I: perineal raphe of male, vaginal wall of female	N: pudendal n.

External Anal Sphincter

Compresses anal canal to voluntarily inhibit defecation

O: anococcygeal raphe	I: central tendon	N: pudendal n., S4

Pelvic Diaphragm. The deepest compartment, the **pelvic diaphragm,** is similar in both sexes. It consists of two muscle pairs shown in figure 11.17*e*—the **levator ani** and **coccygeus.**

Levator Ani (leh-VAY-turAY-nye)

Supports viscera; resists pressure surges in abdominal cavity; elevates anus during defecation; forms vaginal and anorectal sphincters

O: os coxae from pubis to ischial spine	I: coccyx, anal canal, anococcygeal raphe	N: pudendal n., S3–S4

Coccygeus (coc-SIDJ-ee-us)

Draws coccyx anteriorly after defecation or childbirth; supports and elevates pelvic floor; resists abdominal pressure surges

O: ischial spine	I: lower sacrum to upper coccyx	N: S3 or S4

Before You Go On

Answer the following questions to test your understanding of the preceding section:

7. Which muscles are used more often, the external intercostals or internal intercostals? Explain.
8. Explain how pulmonary ventilation affects abdominal pressure and vice versa.
9. Name a major superficial muscle and two major deep muscles of the back.
10. Define *perineum, urogenital triangle,* and *anal triangle.*
11. Name one muscle in the superficial perineal space, one in the urogenital diaphragm, and one in the pelvic diaphragm. State the function of each.

[46] *ischio* = ischium + *cavernosus* = corpus cavernosum of the penis or clitoris
[47] *bulbo* = bulb of penis + *spongiosus* = corpus spongiosum of penis

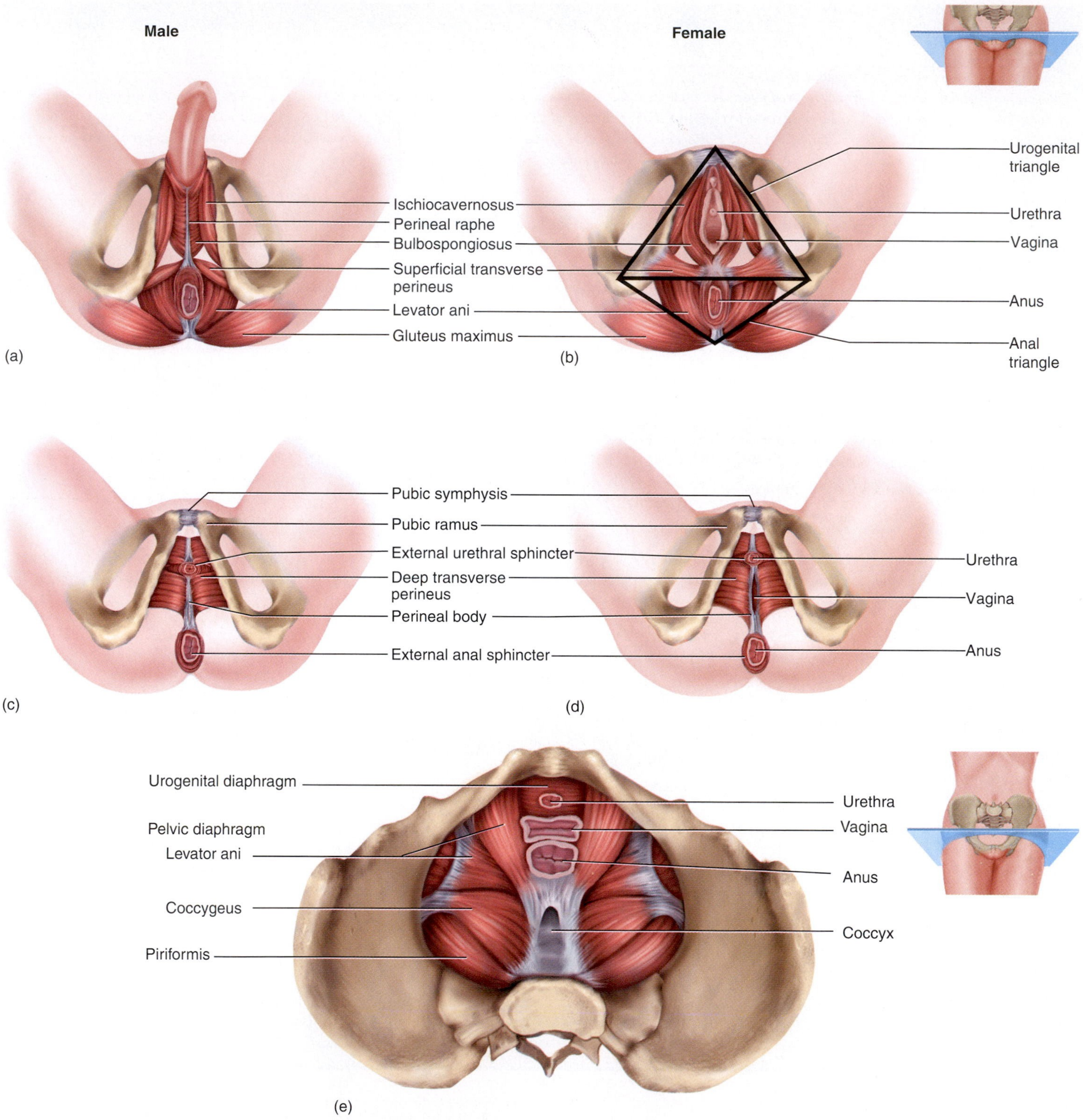

Male
Female
Ischiocavernosus
Perineal raphe
Bulbospongiosus
Superficial transverse perineus
Levator ani
Gluteus maximus
Urogenital triangle
Urethra
Vagina
Anus
Anal triangle
(a)
(b)
Pubic symphysis
Pubic ramus
External urethral sphincter
Deep transverse perineus
Perineal body
External anal sphincter
Urethra
Vagina
Anus
(c)
(d)
Urogenital diaphragm
Pelvic diaphragm
Levator ani
Coccygeus
Piriformis
Urethra
Vagina
Anus
Coccyx
(e)

FIGURE 11.17

Muscles of the Pelvic Floor. (*a, b*) The superficial perineal space, inferior view. Triangles of the perineum are marked in *b*. (*c, d*) The urogenital diaphragm, inferior view; this is the next deeper layer after the muscles in *a* and *b*. (*e*) The pelvic diaphragm, the deepest layer, superior view (seen from within the pelvic cavity).

CHAPTER REVIEW

Review of Key Concepts

Learning Approaches (p. 284)

1. Learning muscle anatomy requires that one become familiar with a few Latin words that are used in naming muscles. These words describe such characteristics as the size, shape, location, number of heads, orientation, and action of a muscle (table 11.1).
2. Four cardinal facts about a given skeletal muscle are its *origin* (its stationary bone attachment), *insertion* (the bone attachment that moves when the muscle contracts), *action* (the motion it produces), and *innervation* (the identity of the nerve that stimulates a given skeletal muscle).
3. Muscles below the neck are innervated by *spinal nerves,* which arise from the spinal cord and emerge through the intervertebral foramina. Spinal nerves are identified by a letter and number that refers to the vertebrae, such as spinal nerve T6 for the sixth thoracic nerve.
4. Muscles of the head and neck are innervated by *cranial nerves,* which arise from the brainstem and emerge through the skull foramina. Cranial nerves are identified by names (see chapter 15) and by Roman numerals I through XII.

Muscles of the Head and Neck (p. 287)

1. Humans and other primates have much more expressive faces than other animals, and they have correspondingly complex facial muscles.
2. The *occipitofrontalis* moves the scalp, eyebrows, and forehead (table 11.2).
3. The eyelid and other tissues around the eye are moved by the *orbicularis oculi, levator palpebrae superioris, corrugator supercilii,* and *procerus* (table 11.2).
4. The *nasalis* muscle flares and compresses the nostrils (table 11.2).
5. The lips are acted upon by the *orbicularis oris, levator labii superioris, levator anguli oris, zygomaticus major* and *minor, risorius, depressor anguli oris, depressor labii inferioris,* and *mentalis* (table 11.2).
6. The cheeks are acted upon by the *buccinator* muscles (table 11.2).
7. The *platysma* acts upon the mandible and the skin of the neck (table 11.2).
8. The tongue is controlled by a set of unnamed *intrinsic muscles* and several *extrinsic muscles:* the *genioglossus, hyoglossus, styloglossus,* and *palatoglossus* (table 11.3).
9. Biting and chewing are achieved by the actions of the *temporalis, masseter, medial pterygoid,* and *lateral pterygoid* muscles on the mandible (table 11.3).
10. Four muscles are associated with the hyoid bone and located superior to it, and are thus called the *suprahyoid group:* the *digastric, geniohyoid, mylohyoid,* and *stylohyoid* (table 11.3). These muscles act on the mandible and hyoid bone to forcibly open the mouth and to aid in swallowing.
11. Another four muscles associated with the hyoid bone are inferior to it and therefore called the *infrahyoid group:* the *thyrohyoid, omohyoid, sternohyoid,* and *sternothyroid* (table 11.3). These muscles depress or fix the hyoid and elevate or depress the larynx, especially in association with swallowing.
12. The *superior, middle,* and *inferior pharyngeal constrictors* contract in sequence to force food down and into the esophagus (table 11.3).
13. The *sternocleidomastoid* and three *scalene* muscles flex the neck. The *trapezius, splenius capitis, splenius cervicis* and *semispinalis capitis* are the major extensors of the neck. Some of these are also employed in rotation of the head (table 11.4).

Muscles of the Trunk (p. 298)

1. Breathing is achieved by the muscles of respiration, especially the *diaphragm, external intercostals,* and *internal intercostals* (table 11.5).
2. The abdominal wall is supported by the sheetlike *rectus abdominis, external abdominal oblique, internal abdominal oblique,* and *transversus abdominis* muscles, which support the abdominal viscera, stabilize the vertebral column during lifting, and aid in respiration, urination, defecation, vomiting, and childbirth (table 11.6).
3. The back has numerous complex muscles that extend, rotate, abduct the vertebral column and aid in breathing. The superficial back muscles include the *erector spinae* (which is subdivided into the *iliocostalis, longissimus,* and *spinalis* muscle columns) and the *serratus posterior superior* and *serratus posterior inferior* muscles. The deep back muscles include the *semispinalis* (subdivided into the *semispinalis capitis, semispinalis cervicis,* and *semispinalis thoracis*); the *quadratus lumborum;* and the *multifidus* (table 11.7).
4. The pelvic floor is spanned by three layers of muscles and fasciae (table 11.8). The anal canal, urethra, and vagina penetrate the pelvic floor muscles and open into the *perineum,* a diamond-shaped space between the thighs bordered by the pubic symphysis, coccyx, and ischial tuberosities. The anterior half of the perineum is the *urogenital triangle* and the posterior half is the *anal triangle.*
5. The most superficial compartment of the pelvic floor is the *superficial perineal space.* It contains three muscles: the *ischiocavernosus, bulbospongiosus,* and *superficial transverse perineus.*
6. The middle compartment of the pelvic floor, in the urogenital triangle, consists of the *urogenital diaphragm,* which is composed of a fibrous membrane and two muscles, the *deep transverse perineus* and *external urethral sphincter.* In the anal triangle, the middle compartment has one muscle, the *external anal sphincter.*
7. The deepest compartment of the pelvic floor is the *pelvic diaphragm.* It consists of two muscles, the *levator ani* and *coccygeus.*

Testing Your Recall

1. Which of the following muscles is the prime mover in spitting out a mouthful of liquid?
   a. platysma
   b. buccinator
   c. risorius
   d. masseter
   e. palatoglossus
2. The word _____ in a muscle name indicates a function related to the head.
   a. cervicis
   b. carpi
   c. capitis
   d. hallucis
   e. teres
3. Which of these is *not* a suprahyoid muscle?
   a. genioglossus
   b. geniohyoid
   c. stylohyoid
   d. mylohyoid
   e. digastric
4. Which of these muscles is an extensor of the neck?
   a. external oblique
   b. sternocleidomastoid
   c. splenius capitis
   d. iliocostalis
   e. latissimus dorsi
5. Which of these muscles of the pelvic floor is the deepest?
   a. superficial transverse perineus
   b. bulbospongiosus
   c. ischiocavernosus
   d. deep transverse perineus
   e. levator ani
6. The facial nerve supplies all of the following muscles *except*
   a. the frontalis.
   b. the orbicularis oculi.
   c. the orbicularis oris.
   d. the depressor labii inferioris.
   e. the mylohyoid.
7. The _____ produce(s) lateral grinding movements of the jaw.
   a. pterygoids
   b. temporalis
   c. hyoglossus
   d. zygomaticus major and minor
   e. risorius
8. All of the following muscles act on the vertebral column *except*
   a. the serratus posterior superior.
   b. the iliocostalis thoracis.
   c. the longissimus thoracis.
   d. the spinalis thoracis.
   e. the multifidus.
9. A muscle that aids in chewing without moving the mandible is
   a. the temporalis.
   b. the mentalis.
   c. the buccinator.
   d. the levator anguli oris.
   e. the splenius cervicis.
10. Which of the following muscles raises the upper lip?
    a. levator palpebrae superioris
    b. orbicularis oris
    c. masseter
    d. zygomaticus minor
    e. mentalis
11. The prime mover of spinal extension is the _____.
12. Ejaculation results from contraction of the _____ muscle.
13. The muscle that opens your eyes is the _____.
14. As its name implies, the _____ nerve controls several muscles of the tongue.
15. The _____ muscle, named for its two bellies, opens the mouth.
16. The anterior half of the perineum is a region called the _____.
17. The abdominal aponeuroses converge on a midsagittal fibrous band on the abdomen called the _____.
18. The thyrohyoid muscle inserts on the thyroid cartilage of the _____.
19. The _____ muscles diverge like a V from the middle of the upper thorax to insertions behind the ears.
20. The largest muscle of the upper back is the _____.

Answers in the Appendix

True or False

Determine which five of the following statements are false, and briefly explain why.

1. The origin of the sternocleidomastoid is the mastoid process.
2. The largest deep muscle of the lower back is the quadratus lumborum.
3. The muscle used to stick out your tongue is the genioglossus.
4. The abdominal oblique muscles rotate the vertebral column.
5. Exhaling requires contraction of the internal intercostal muscles.
6. The digastric muscles form the floor of the mouth.
7. The scalenes are superficial to the trapezius.
8. Cutting the phrenic nerves would paralyze the prime mover of respiration.
9. The orbicularis oculi and orbicularis oris are sphincters.
10. All of the cranial nerves innervate muscles of the head and neck.

Answers in the Appendix

Testing Your Comprehension

1. Name one antagonist of each of the following muscles: (*a*) orbicularis oculi, (*b*) genioglossus, (*c*) masseter, (*d*) sternocleidomastoid, (*e*) rectus abdominis.
2. Name one synergist of each of the following muscles: (*a*) temporalis, (*b*) procerus, (*c*) platysma, (*d*) semispinalis capitis, (*e*) bulbospongiosus.
3. Dental procedures, vaccination, HIV infection, and some other infections occasionally injure branches of the facial nerve and weaken or paralyze the affected

muscles. Predict the problems that a person would have if the orbicularis oris and buccinator muscles were paralyzed by such a nerve lesion.

4. Removal of cancerous lymph nodes from the neck sometimes requires removal of the sternocleidomastoid on that side. How would this affect a patient's range of head movement?

5. In a disease called tick paralysis, the saliva from a tick bite paralyzes skeletal muscles beginning with the lower limbs and progressing superiorly. What would be the most urgent threat to the life of a tick paralysis patient?

Answers at the Online Learning Center

www.mhhe.com/saladinha1

Visit the Online Learning Center for practice tests, answer keys, and other learning aids for this chapter. Enhance your understanding of human anatomy with our interactive art labeling exercises, supplemental photo atlases, web links, puzzles, flashcards, and much more.

12

CHAPTER TWELVE

The Appendicular Musculature

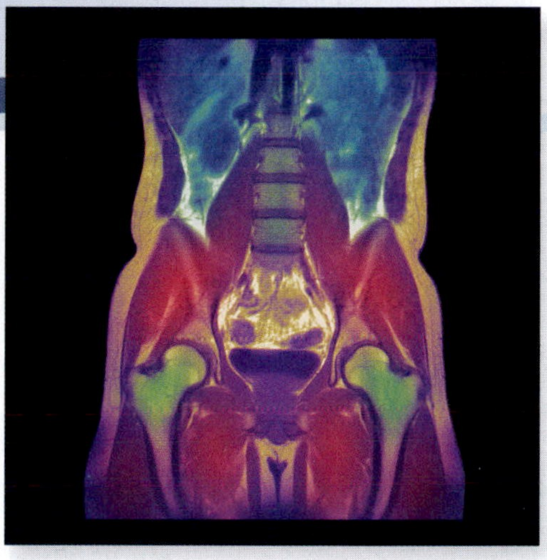

MRI scan showing muscles of the lumbar, pelvic, and upper femoral regions

CHAPTER OUTLINE

Muscles Acting on the Shoulder and Upper Limb 314
- Muscles Acting on the Scapula 314
- Muscles Acting on the Humerus 316
- Muscles Acting on the Forearm 320
- Muscles Acting on the Wrist and Hand 322
- Intrinsic Muscles of the Hand 327

Muscles Acting on the Hip and Lower Limb 329
- Muscles Acting on the Hip and Femur 329
- Muscles Acting on the Knee and Leg 334
- Muscles Acting on the Foot 336
- Intrinsic Muscles of the Foot 342

Muscle Injuries 344

Chapter Review 346

INSIGHTS

12.1 Clinical Application: Carpal Tunnel Syndrome 326
12.2 Clinical Application: Intramuscular Injections 333
12.3 Clinical Application: Hamstring Injuries 336
12.4 Clinical Application: Compartment Syndrome 342

BRUSHING UP

To understand this chapter, you may find it helpful to review the following concepts:
- Terminology of the limb regions (p. 26)
- Anatomy of the appendicular skeleton (chapter 8)
- Terminology of joint actions (pp. 234–239)
- Shapes of muscles (fusiform, pennate, circular, etc.) (p. 258)
- The meaning of muscle origin, insertion, and action (pp. 259–260)
- Prime movers, synergists, antagonists, and fixators (p. 260)
- Intrinsic and extrinsic muscles (p. 260)
- Greek and Latin words commonly used to name muscles (p. 284, table 11.1)
- Muscle innervation (pp. 284, 287)

In this chapter we continue our study of the muscular system with muscles of the pectoral girdle, upper limb, pelvic girdle, and lower limb. This chapter also describes several muscle injuries, which are more common in the appendicular region than in the axial region. The terminology of muscles outlined in table 11.1 will be of further help in interpreting the names of the appendicular muscles in this chapter.

Muscles Acting on the Shoulder and Upper Limb

Objectives

When you have completed this section, you should be able to

- name and locate the muscles that act on the pectoral girdle, shoulder, elbow, wrist, and hand;
- relate the actions of these muscles to the joint movements described in chapter 9; and
- describe the origin, insertion, and innervation of each muscle.

The upper limb is used for a broad range of both powerful and subtle actions, ranging from climbing, grasping, and throwing to writing, playing musical instruments, and manipulating small objects. It therefore has an especially complex array of muscles, but the muscles fall into logical groups that make their functional relationships and names easier to understand. Tables 12.1 through 12.7 group these into muscles that act on the scapula, those that act on the humerus and shoulder joint, those that act on the forearm and elbow joint, extrinsic (forearm) muscles that act on the wrist and hand, and intrinsic (hand) muscles that act on the fingers.

Table 12.1

Muscles Acting on the Scapula (fig. 12.1)

O = origin, I = insertion, N = innervation (n. = nerve, nn. = nerves)

Muscles that act on the pectoral girdle originate on the axial skeleton and insert on the clavicle and scapula. The scapula is loosely attached to the thoracic cage and is capable of considerable movement—rotation (as in raising and lowering the apex of the shoulder), elevation and depression (as in shrugging and lowering the shoulders), and protraction and retraction (pulling the shoulders forward or back). The clavicle braces the shoulder and moderates these movements.

Anterior Group. Muscles of the pectoral girdle fall into anterior and posterior groups (see fig. 12.2*b* and *d*). The major muscles of the anterior group are the *pectoralis minor* and *serratus anterior* (see fig. 11.12*b*).

Pectoralis Minor (PECK-toe-RAY-liss)

Protracts and depresses scapula when ribs are fixed; elevates ribs when scapula is fixed

O: ribs 3–5	I: coracoid process	N: medial and lateral pectoral nn.

Serratus Anterior (serr-AY-tus)

Holds scapula against rib cage; elevates ribs; protracts and rotates scapula to tilt glenoid cavity upward; forcefully depresses scapula; abducts and elevates arm; prime mover in forward thrusting, throwing, and pushing ("boxer's muscle")

O: ribs 1–9	I: medial border of scapula	N: long thoracic n.

Posterior Group. In the posterior group are the large, superficial *trapezius* and three deep muscles, the *levator scapulae, rhomboideus major,* and *rhomboideus minor* (see fig. 11.14). We studied the trapezius in chapter 11 for its actions on the neck, but we now focus on its actions on the scapula. Its actions depend on whether its superior, middle, or inferior parts contract and whether it acts alone or with other muscles. The levator scapulae and superior part of the trapezius act together to elevate the scapula, as when you shrug your shoulders; but if these muscles act alone, the levator scapulae rotates the scapula clockwise *(medial rotation)* and the trapezius rotates it counterclockwise *(lateral rotation)*, as viewed from the rear (fig. 12.1). Depression of the scapula occurs mainly by gravitational pull, but the trapezius and serratus anterior can cause faster, more forcible depression, as in swimming, hammering, and rowing. The rhomboideus muscles contribute to elevation, medial rotation, and retraction of the scapula.

Trapezius (tra-PEE-zee-us)

Superior fibers elevate scapula or rotate it to tilt glenoid cavity upward; middle fibers retract scapula; inferior fibers depress scapula. When scapula is fixed, one trapezius acting alone flexes neck laterally and both trapezius muscles working together extend neck.

O: external occipital protuberance, superior nuchal line, spinous processes of C7–T12	I: clavicle, acromion, scapular spine	N: accessory n. (XI), C3–C4

Levator Scapulae (leh-VAY-tur SCAP-you-lee)

Rotates scapula to tilt glenoid cavity downward; flexes neck when scapula is fixed; elevates scapula when acting with superior fibers of trapezius

O: transverse processes of vertebrae C1–C4	I: superior angle to medial border of scapula	N: C3–C4, dorsal scapular n.

(continued)

TABLE 12.1
Muscles Acting on the Scapula *(continued)*

Rhomboideus Major (rom-BOY-dee-us) and Rhomboideus Minor

Retract and elevate scapula; rhomboideus major also fixes scapula and rotates it to tilt glenoid cavity downward

O: spinous processes of vertebrae C7–T1 (r. minor) and T2–T5 (r. major)	I: medial border of scapula	N: dorsal scapular n.

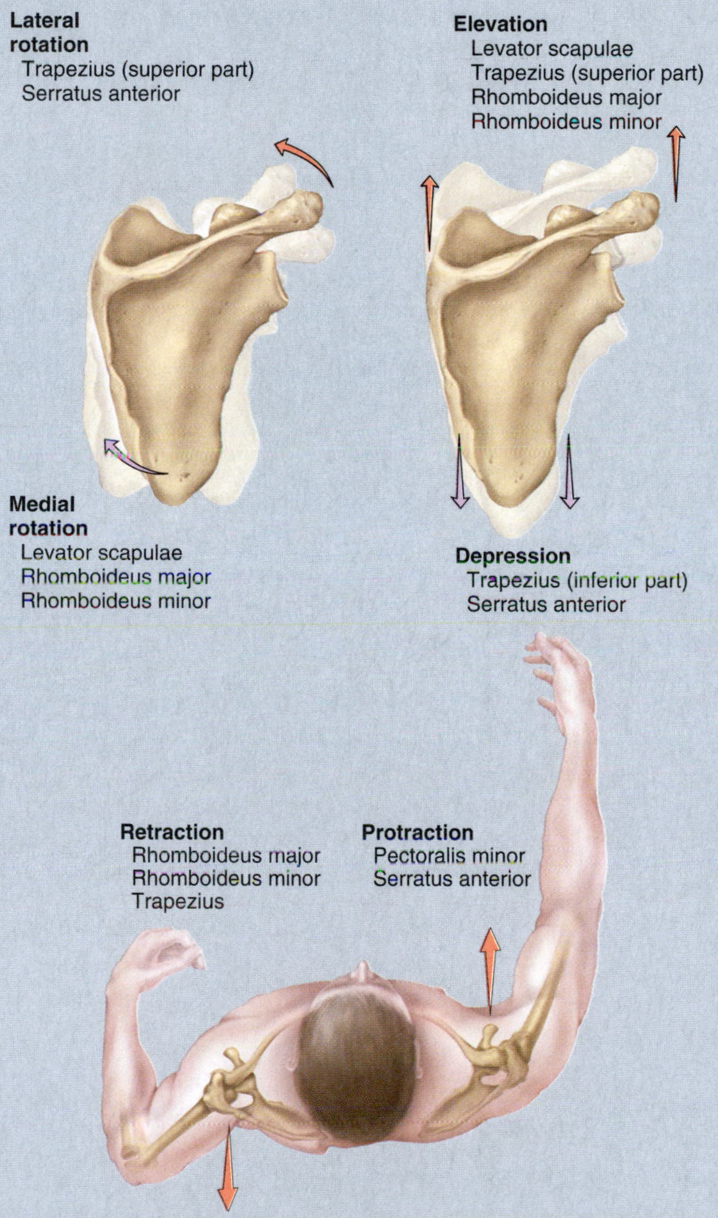
Lateral rotation
Trapezius (superior part)
Serratus anterior
Elevation
Levator scapulae
Trapezius (superior part)
Rhomboideus major
Rhomboideus minor
Medial rotation
Levator scapulae
Rhomboideus major
Rhomboideus minor
Depression
Trapezius (inferior part)
Serratus anterior
Retraction
Rhomboideus major
Rhomboideus minor
Trapezius
Protraction
Pectoralis minor
Serratus anterior

FIGURE 12.1

Actions of Some Thoracic Muscles on the Scapula. Note that an individual muscle can contribute to multiple actions, depending on which fibers contract and what synergists act with it.

Table 12.2

Muscles Acting on the Humerus (figs. 12.2–12.4)

O = origin, I = insertion, N = innervation (n. = nerve, nn. = nerves)

Nine muscles cross the humeroscapular (shoulder) joint and insert on the humerus. The *pectoralis major* and *latissimus dorsi* bear the primary responsibility for attachment of the arm to the trunk and are prime movers of this joint (figs. 12.2*a, b;* 12.3). The pectoralis major is the thick, fleshy muscle of the mammary region, and the latissimus dorsi is a broad muscle of the back that extends from the waist to the axilla. The axilla (armpit) is bordered by the *axillary folds,* formed by the latissimus dorsi posteriorly and the pectoralis major anteriorly (see fig. B.5 in the atlas following this chapter). The pectoralis major flexes the shoulder as in pointing at something in front of you or hugging someone, and the latissimus dorsi extends it as in pointing at something behind you—thus, they are antagonists.

The other seven muscles of the shoulder originate on the scapula (fig. 12.2*b, d*). In this group, the prime mover is the *deltoid*—the thick muscle that caps the shoulder. It acts like three different muscles. Its anterior fibers flex the shoulder, its posterior fibers extend it, and its lateral fibers abduct it. Abduction by the deltoid is antagonized by the combined action of the pectoralis major and latissimus dorsi. The *teres major* assists in extension of the shoulder and the *coracobrachialis* assists in flexion and adduction. The other four muscles form the rotator cuff, discussed shortly.

Pectoralis Major

Prime mover of shoulder flexion; adducts and medially rotates humerus; depresses pectoral girdle; elevates ribs; aids in climbing, pushing, and throwing

O: clavicle, sternum, costal cartilages 1–6, aponeurosis of external abdominal oblique	I: intertubercular groove of humerus	N: medial and lateral pectoral nn.

Latissimus Dorsi[1] (la-TISS-ih-mussDOR-sye)

Adducts and medially rotates humerus; extends shoulder joint; produces strong downward strokes of arm, as in hammering or swimming (“swimmer’s muscle”); pulls body upward in climbing

O: vertebrae T7–L5, lower three or four ribs, thoracolumbar fascia, iliac crest, inferior angle of scapula	I: intertubercular groove of humerus	N: thoracodorsal n.

Deltoid

Lateral fibers abduct humerus; anterior fibers flex and medially rotate it; posterior fibers extend and laterally rotate it

O: clavicle, scapular spine, acromion	I: deltoid tuberosity of humerus	N: axillary n.

Teres Major (TERR-eez)

Adducts and medially rotates humerus; extends shoulder joint

O: from inferior angle to lateral border of scapula	I: intertubercular groove of humerus	N: subscapular n.

Coracobrachialis (COR-uh-co-BRAY-kee-AL-iss)

Adducts arm; flexes shoulder joint

O: coracoid process	I: medial aspect of shaft of humerus	N: musculocutaneous n.

[1]*latissimus* = broadest + *dorsi* = of the back

(continued)

TABLE 12.2

Muscles Acting on the Humerus *(continued)*

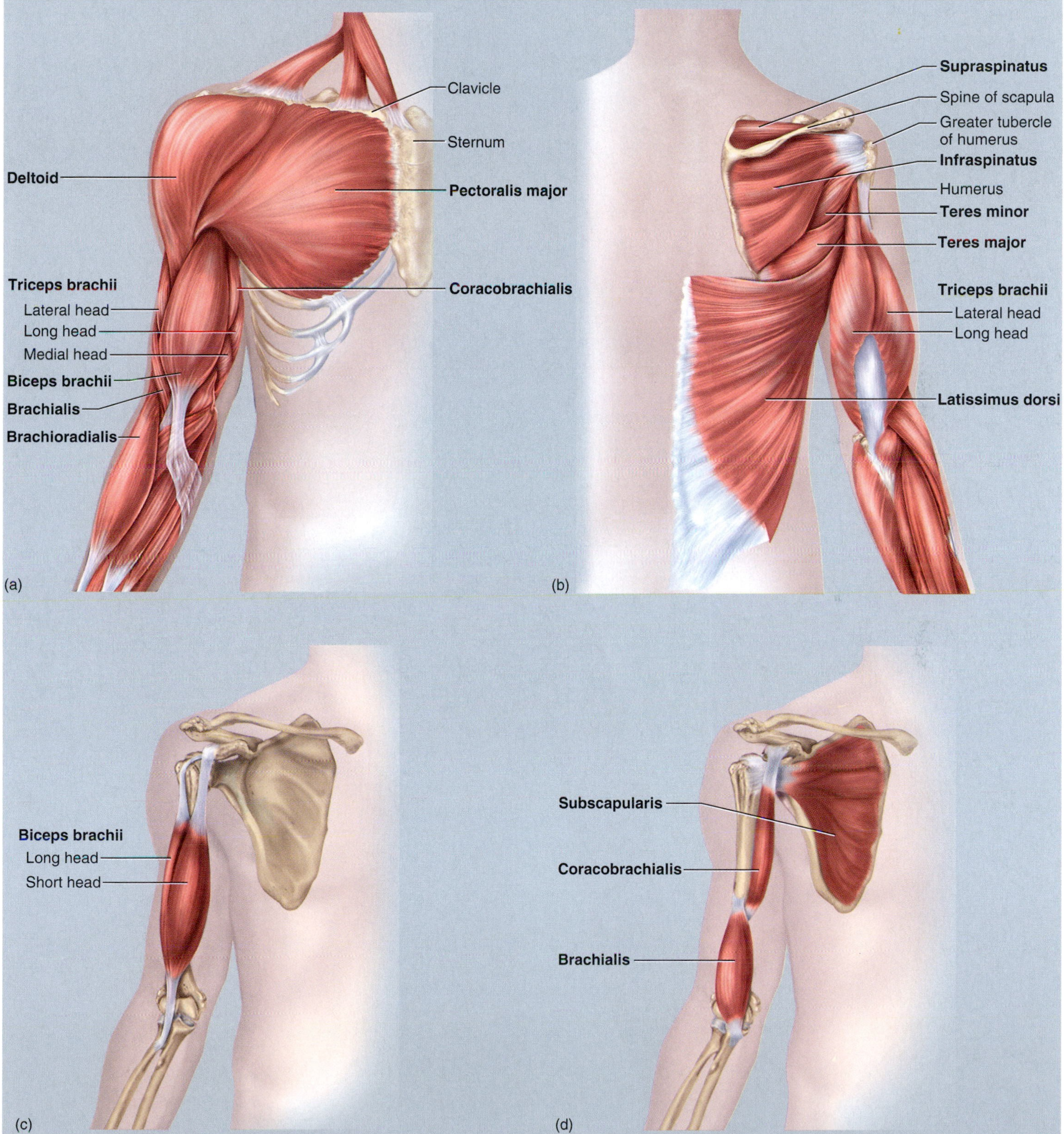

FIGURE 12.2

Pectoral and Brachial Muscles. (*a*) Anterior view. (*b*) Posterior view. (*c*) The biceps brachii, the superficial flexor of the elbow. (*d*) The brachialis, the deep flexor of the elbow, and the coracobrachialis and subscapularis, which act on the humerus.

(continued)

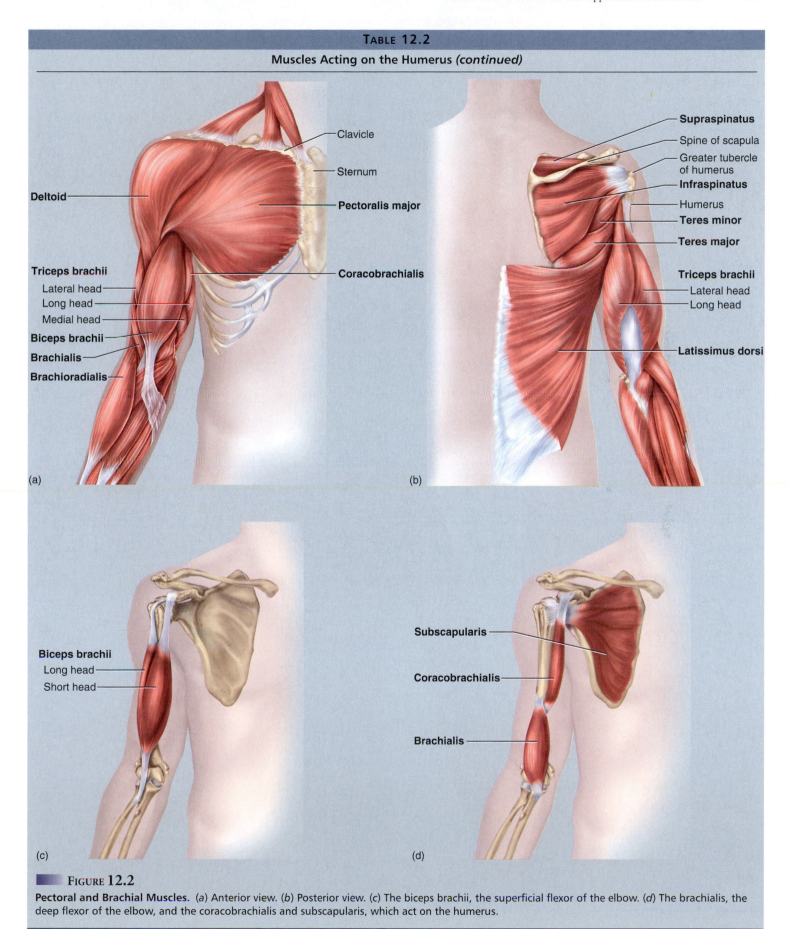

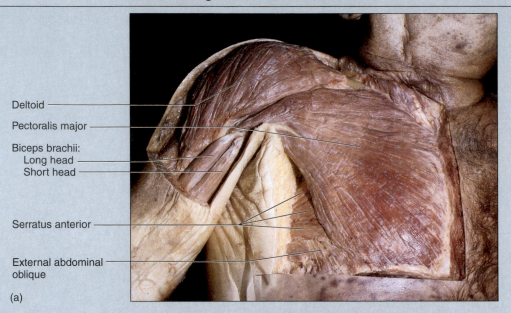

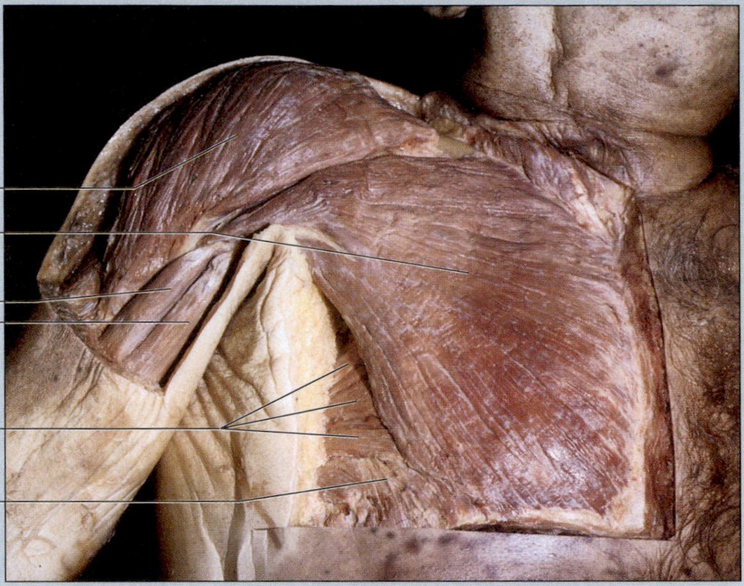
Deltoid
Pectoralis major
Biceps brachii:
Long head
Short head
Serratus anterior
External abdominal oblique

(a)

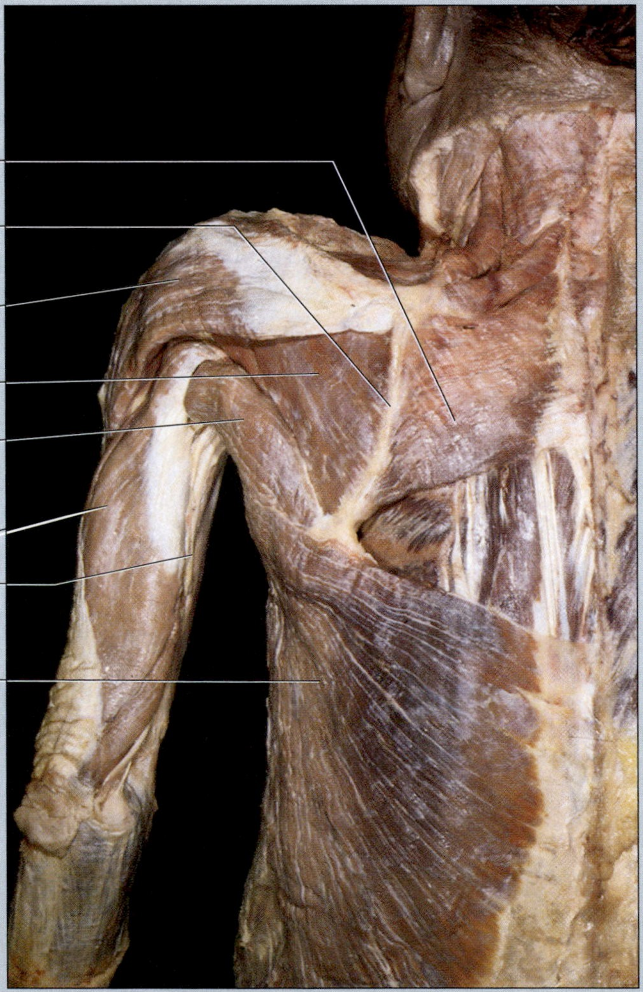
Rhomboideus major
Medial border of scapula
Deltoid
Infraspinatus
Teres major
Triceps brachii:
Lateral head
Long head
Latissimus dorsi

(b)

FIGURE 12.3
Muscles of the Chest and Arm of the Cadaver. (*a*) Anterior view. (*b*) Posterior view.

(continued)

TABLE 12.2

Muscles Acting on the Humerus *(continued)*

Rotator Cuff. The *rotator cuff* is composed of the tendons of four scapular muscles: the *supraspinatus, infraspinatus, teres minor,* and *subscapularis* (the "SITS muscles," taking the first letter of each) (figs. 12.2*b, d* and 12.4). The first three originate on the posterior surface and are listed from superior to inferior. The supraspinatus lies above the scapular spine in the supraspinous fossa, the infraspinatus occupies most of the infraspinous fossa below the spine, and the teres minor is the first muscle inferior to the infraspinatus. The subscapularis lies on the anterior surface of the scapula, occupying the subscapular fossa. The tendons of these muscles merge with the joint capsule of the shoulder as they pass it en route to the humerus. They insert on the proximal end of the humerus, forming a partial sleeve around it. The rotator cuff reinforces the joint capsule and holds the head of the humerus in the glenoid cavity.

Rotator cuff injuries are common in sports and recreation. The tendon of the supraspinatus, especially, is easily damaged by strenuous circumduction (as in baseball pitching), falls (as in skiing), or hard blows from the side (as when a hockey player is slammed against the boards).

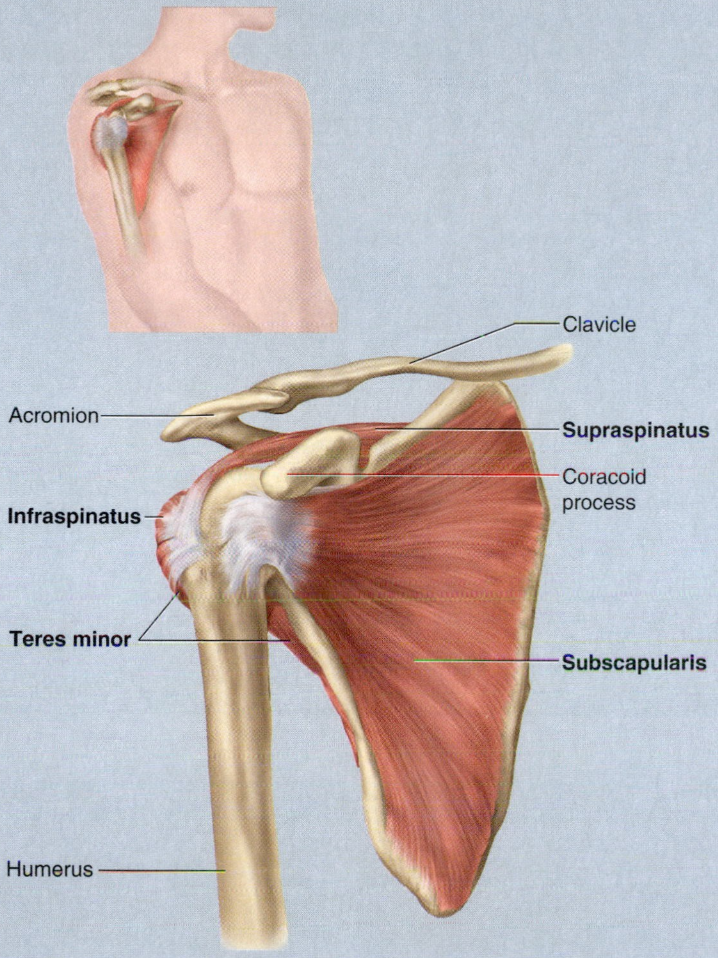

FIGURE 12.4

The Rotator Cuff. Anterolateral view of the right shoulder. The rotator cuff muscles, labeled in boldface, are sometimes nicknamed the SITS muscles for the first letters of their formal names.

Supraspinatus (SOO-pra-spy-NAY-tus)

Abducts humerus; resists downward displacement when carrying heavy weight

O: supraspinous fossa of scapula	I: greater tubercle of humerus	N: suprascapular n.

Infraspinatus (IN-fra-spy-NAY-tus)

Extends and laterally rotates humerus

O: infraspinous fossa of scapula	I: greater tubercle of humerus	N: suprascapular n.

Teres Minor

Adducts and laterally rotates humerus

O: lateral border of scapula	I: greater tubercle of humerus	N: axillary n.

Subscapularis (SUB-SCAP-you-LERR-iss)

Medially rotates humerus

O: subscapular fossa of scapula	I: lesser tubercle of humerus	N: subscapular n.

Think About It!

Since a muscle can only pull on a bone, and not push, antagonistic muscles are needed to produce opposite actions at a joint. Reconcile this fact with the observation that the deltoid muscle both flexes and extends the shoulder.

Since the humeroscapular joint is capable of such a wide range of movements and is acted upon by so many muscles, its actions are summarized in table 12.3.

Muscles acting on the forearm are classified by their actions in table 12.5.

Table 12.3

Actions of the Shoulder (humeroscapular) Joint

Italics indicate prime movers; others are synergists. Parentheses indicate only a slight effect.

Flexion	Extension
Anterior deltoid	*Posterior deltoid*
Pectoralis major	*Latissimus dorsi*
Coracobrachialis	Teres major
Biceps brachii	
Abduction	**Adduction**
Lateral deltoid	*Pectoralis major*
Supraspinatus	*Latissimus dorsi*
	Coracobrachialis
	Triceps brachii
	Teres major
	(Teres minor)
Medial Rotation	**Lateral Rotation**
Subscapularis	*Infraspinatus*
Teres major	*Teres minor*
Latissimus dorsi	Deltoid
Deltoid	
Pectoralis major	

Table 12.4

Muscles Acting on the Forearm (figs. 12.2, 12.3, 12.5, and 12.6)

O = origin, I = insertion, N = innervation (n. = nerve, nn. = nerves)

Muscles with Bellies in the Arm (brachium). The elbow and forearm are capable of four motions: flexion, extension, pronation, and supination. The prime movers of flexion are on the anterior side of the humerus and include the superficial *biceps brachii* and deeper *brachialis* (see fig. 12.2*c, d*). In flexion of the elbow, the biceps elevates the radius while the brachialis elevates the ulna. The biceps is named for its two heads, which arise from separate tendons at the scapula. The tendon of the long head is important in holding the humerus in the glenoid cavity and stabilizing the shoulder joint. The two heads converge close to the elbow on a single distal tendon. The prime mover of extension is the *triceps brachii* on the posterior side of the humerus (see figs. 12.2*b* and 12.3*b*).

Biceps Brachii[2] (BY-seps BRAY-kee-eye)

Flexes elbow; abducts arm; supinates forearm; holds head of humerus in glenoid cavity

O: *long head*—supraglenoid tubercle; *short head*—coracoid process	I: tuberosity of radius, fascia of forearm	N: musculocutaneous n.

Brachialis (BRAY-kee-AL-iss)

Flexes elbow

O: anterior distal shaft of humerus	I: coronoid process and tuberosity of ulna	N: musculocutaneous n., radial n.

Triceps Brachii (TRI-seps BRAY-kee-eye)

Extends elbow; long head adducts humerus

O: *long head*—infraglenoid tubercle of scapula; *lateral head*—proximal posterior shaft of humerus; *medial head*—posterior shaft of humerus	I: olecranon of ulna	N: radial n.

[2]*bi* = two + *ceps* = head + *brachi* = arm. Note that *biceps* is singular; there is no such word as *bicep*. The plural form is *bicipites* (by-SIP-ih-teez).

(continued)

TABLE 12.4

Muscles Acting on the Forearm *(continued)*

Muscles with Bellies in the Forearm (antebrachium). The *anconeus* is a weak synergist of extension that crosses the posterior side of the elbow (see fig. 12.6*d*, *e*). The *brachioradialis* is a synergist in elbow flexion. Its belly lies in the antebrachium (forearm) beside the radius, rather than in the brachium with the other two flexors (see figs. 12.2*a* and 12.6*a*). It forms the thick, fleshy mass on the lateral side of the forearm just distal to the elbow. Its origin is on the distal end of the humerus, and its insertion is on the distal end of the radius. Since its insertion is so far from the fulcrum, the brachioradialis does not generate as much power as the prime movers; it is effective mainly when the prime movers have partially flexed the elbow.

Pronation is achieved by two anterior muscles in the forearm—the *pronator teres* near the elbow and *pronator quadratus* near the wrist. Supination is achieved by the biceps brachii and the *supinator* of the posterior forearm (fig. 12.5).

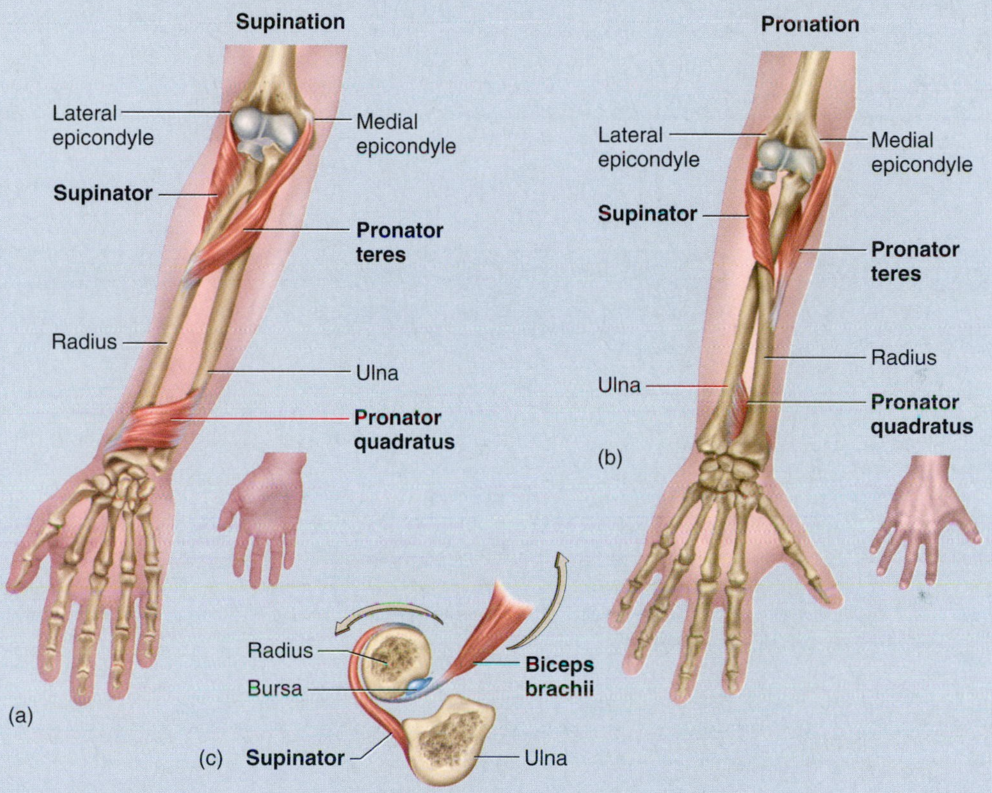

FIGURE 12.5

Actions of the Rotator Muscles on the Forearm. (*a*) Supination. (*b*) Pronation. (*c*) Cross section just distal to the elbow, showing how the biceps brachii aids in supination.

Anconeus[3] (an-CO-nee-us)

Extends elbow		
O: lateral epicondyle of humerus	I: olecranon and posterior aspect of ulna	N: radial n.

Brachioradialis (BRAY-kee-oh-RAY-dee-AL-iss)

Flexes elbow		
O: lateral supracondylar ridge of humerus	I: styloid process of radius	N: radial n.

Pronator Teres (PRO-nay-tur TERR-eez)

Pronates forearm, flexes elbow		
O: medial epicondyle of humerus, coronoid process of ulna	I: lateral midshaft of radius	N: median n.

Pronator Quadratus (PRO-nay-tur quad-RAY-tus)

Pronates forearm		
O: anterior distal shaft of ulna	I: anterior distal shaft of radius	N: median n.

Supinator (SOO-pih-NAY-tur)

Supinates forearm		
O: lateral epicondyle of humerus, proximal shaft of ulna	I: proximal shaft of radius	N: radial n.

[3]*ancon* = elbow

Table 12.5

Actions of the Forearm

Italics indicate prime movers; others are synergists. Parentheses indicate only a slight effect.

Flexion	Extension
Biceps brachii	*Triceps brachii*
Brachialis	Anconeus
Brachioradialis	
Flexor carpi radialis	
(Pronator teres)	

Pronation	Supination
Pronator teres	*Supinator*
Pronator quadratus	Biceps brachii

Table 12.6

Muscles Acting on the Wrist and Hand (fig. 12.6)

O = origin, I = insertion, N = innervation (n. = nerve, nn. = nerves)

The hand is acted upon by extrinsic muscles in the forearm and intrinsic muscles in the hand itself. The bellies of the extrinsic muscles form the fleshy roundness of the proximal forearm; their tendons extend into the wrist and hand. Their actions are mainly flexion and extension, but the wrist and fingers can be abducted and adducted and the thumb and fingers can be opposed.

Several of these muscles originate on the humerus; therefore, they cross the elbow joint and weakly contribute to flexion and extension of the elbow. This action is relatively negligible, however, and we focus on their action at the wrist and fingers. Although these muscles are numerous and complex, most of their names suggest their actions, and from their actions, their approximate locations in the forearm can generally be deduced.

The deep fasciae divide the muscles of the forearm into **anterior** and **posterior compartments** and each compartment into superficial and deep layers. The muscles are classified here by compartment and layer. Some muscles of the forearm were considered earlier (the pronator quadratus, pronator teres, supinator, anconeus, and brachioradialis) because they act on the radius and ulna rather than on the hand.

Anterior Compartment—Superficial Layer. Most muscles of the anterior compartment are flexors of the wrist and fingers that arise from a common tendon on the humerus (fig. 12.6*a, b*). The two prominent tendons that you can palpate at the wrist belong to the *palmaris longus* on the medial side and the *flexor carpi radialis* on the lateral side. The latter is an important landmark for finding the radial artery, where the pulse is usually taken. At the distal end, the tendon of the palmaris longus passes over the flexor retinaculum while the other tendons pass beneath it. The palmaris longus is absent on one or both sides (most often the left) in about 14% of people. To see if you have one, flex your wrist and touch the tips of your thumb and little finger together. If present, the palmaris longus tendon will stand up prominently on the wrist.

Flexor Carpi Radialis (CAR-pie RAY-dee-AY-liss)

Powerful wrist flexor; abducts hand; synergist in elbow flexion

O: medial epicondyle of humerus	I: base of metacarpals II and III	N: median n.

Flexor Carpi Ulnaris (ul-NAY-riss)

Flexes and adducts wrist; fixes wrist during extension of fingers

O: medial epicondyle of humerus	I: pisiform, hamate, metacarpal V	N: ulnar n.

Flexor Digitorum Superficialis (DIDJ-ih-TOE-rum SOO-per-FISH-ee-AY-liss)

Flexes fingers II–V at proximal interphalangeal joints; aids in flexion of wrist and metacarpophalangeal joints

O: medial epicondyle of humerus, radius, coronoid process of ulna	I: four tendons leading to middle phalanges II–V	N: median n.

Palmaris Longus (pall-MERR-iss)

Weakly flexes wrist; tenses palmar aponeurosis; often absent

O: medial epicondyle of humerus	I: palmar aponeurosis, flexor retinaculum	N: median n.

Anterior Compartment—Deep Layer. The anterior compartment has two deep flexors, the *flexor pollicis longus,* which flexes the thumb, and the *flexor digitorum profundus,* which flexes the other four digits (fig. 12.6*c*). Both of them also contribute to wrist flexion.

Flexor Pollicis Longus (PAHL-ih-sis)

Flexes interphalangeal joint of thumb

O: radius, interosseous membrane	I: distal phalanx I	N: median n.

(continued)

Table 12.6

Muscles Acting on the Wrist and Hand *(continued)*

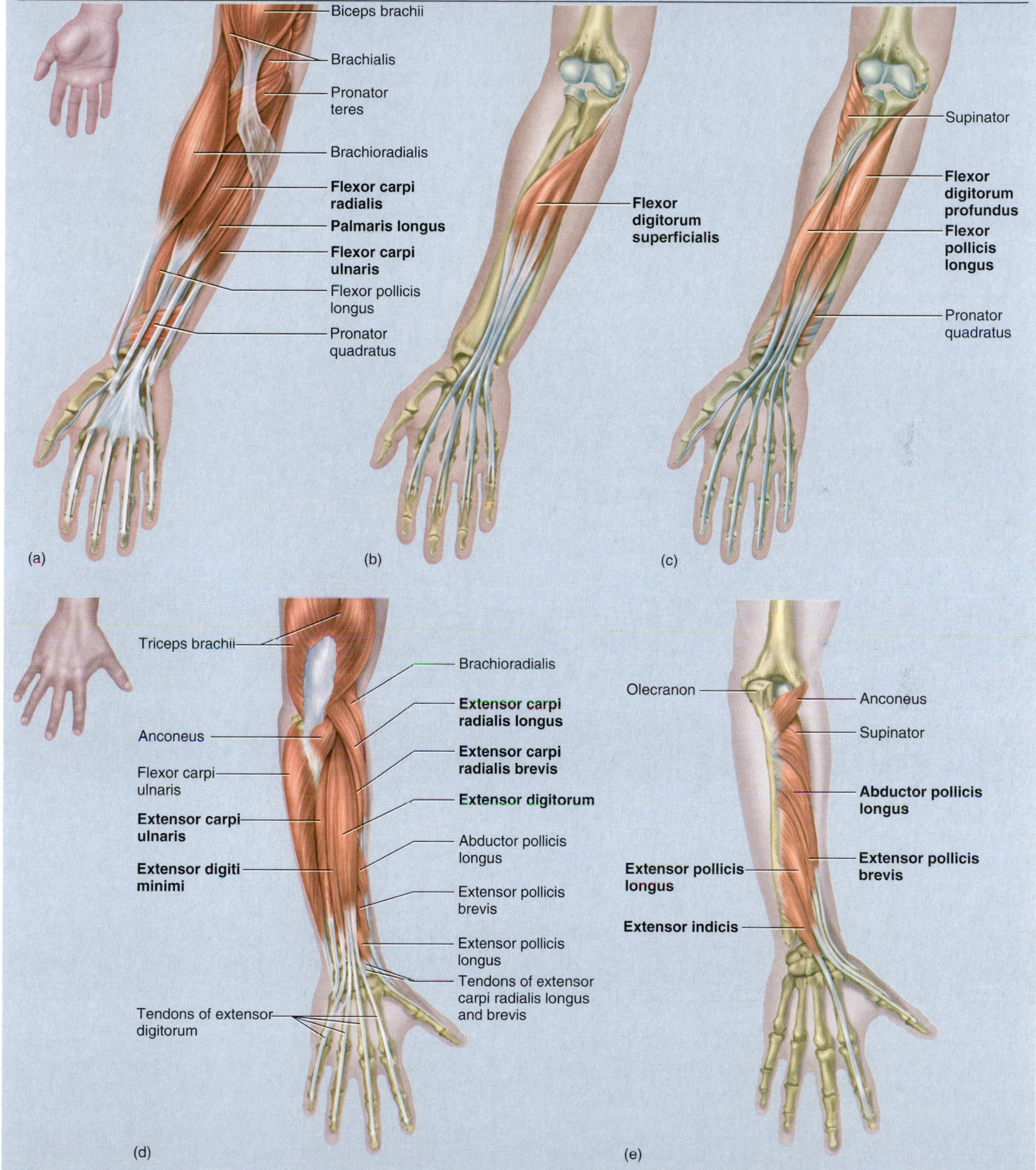

Figure 12.6

Muscles of the Forearm. Figures *a–c* are anterior views and figures *d–e* are posterior. Muscles labeled in boldface are: (*a*) superficial flexors; (*b*) the flexor digitorum superficialis, deep to the muscles in *a* but also classified as a superficial flexor; (*c*) deep flexors; (*d*) superficial extensors; and (*e*) deep extensors.

(continued)

Table 12.6

Muscles Acting on the Wrist and Hand *(continued)*

Flexor Digitorum Profundus

Flexes wrist and distal interphalangeal joints of digits II–V

O: shaft of ulna, interosseous membrane	I: four tendons to distal phalanges II–V	N: median and ulnar nn.

Posterior Compartment—Superficial Layer. Muscles of the posterior compartment are mostly wrist and finger extensors that share a single proximal tendon arising from the humerus. One of the superficial muscles on this side, the *extensor digitorum,* has four distal tendons that can easily be seen and palpated on the back of the hand when the fingers are strongly hyperextended (fig. 12.6*d,* and see figs. B.7 and B.8). By strongly abducting and extending the thumb into a hitchhiker's position, you should also be able to see a deep dorsolateral pit at the base of the thumb, with a taut tendon on each side of it. This depression is called the *anatomical snuffbox* because it was once fashionable to place a pinch of snuff here and inhale it (see fig. B.8*b*). It is bordered laterally by the tendons of the *abductor pollicis longus* and *extensor pollicis brevis* and medially by the tendon of the *extensor pollicis longus.*

Extensor Carpi Radialis Longus

Extends and abducts wrist

O: lateral supracondylar ridge of humerus	I: base of metacarpal II	N: radial n.

Extensor Carpi Radialis Brevis

Extends and abducts wrist; fixes wrist during finger flexion

O: lateral epicondyle of humerus	I: base of metacarpal III	N: radial n.

Extensor Carpi Ulnaris

Extends and adducts wrist

O: lateral epicondyle of humerus, posterior shaft of ulna	I: base of metacarpal V	N: radial n.

Extensor Digitorum (DIDJ-ih-TOE-rum)

Extends fingers II–V at metacarpophalangeal joints; extends wrist

O: lateral epicondyle of humerus	I: dorsal aspect of phalanges II–V	N: radial n.

Extensor Digiti Minimi (DIDJ-ih-ty MIN-ih-my)

Extends metacarpophalangeal joint of little finger; sometimes considered to be a detached portion of extensor digitorum

O: lateral epicondyle of humerus	I: distal and middle phalanges V	N: radial n.

Posterior Compartment—Deep Layer. The posterior compartment contains one muscle that extends the index finger, the *extensor indicis,* and three that act on the thumb, the *abductor pollicis longus, extensor pollicis longus,* and *extensor pollicis brevis* (fig. 12.6*e*).

Extensor Indicis (IN-dih-sis)

Extends index finger at metacarpophalangeal joint

O: shaft of ulna, interosseous membrane	I: middle and distal phalanges II	N: radial n.

Abductor Pollicis Longus

Abducts and extends thumb; abducts wrist

O: posterior aspect of radius and ulna, interosseous membrane	I: trapezium, base of metacarpal I	N: radial n.

Extensor Pollicis Longus

Extends thumb at metacarpophalangeal joint

O: shaft of ulna, interosseous membrane	I: distal phalanx I	N: radial n.

Extensor Pollicis Brevis

Extends thumb at metacarpophalangeal joint

O: shaft of radius, interosseous membrane	I: proximal phalanx I	N: radial n.

Figure 12.7 shows cross sections through the proximal and distal arm and proximal forearm, showing the grouping of the foregoing muscles into the superficial and deep flexor and extensor groups.

It may seem as if the tendons of the forearm muscles would stand up at the wrist like taut bowstrings when these muscles were contracted, but this is prevented by the fact that most of them pass under a **flexor retinaculum (transverse carpal ligament)** on the anterior side of the wrist and an **extensor retinaculum (dorsal carpal ligament)** on the posterior side (see figs. 3.13 and 12.9*c*). The **carpal tunnel** is a tight space between the carpal bones and flexor retinaculum (fig. 12.8). The flexor tendons passing through the tunnel are enclosed in tendon sheaths that enable them to slide back and forth quite easily, although this region can become painfully inflamed by repetitive motion (see insight 12.1).

Think About It!

Why are the prime movers of finger extension and flexion located in the forearm rather than in the hand, closer to the fingers?

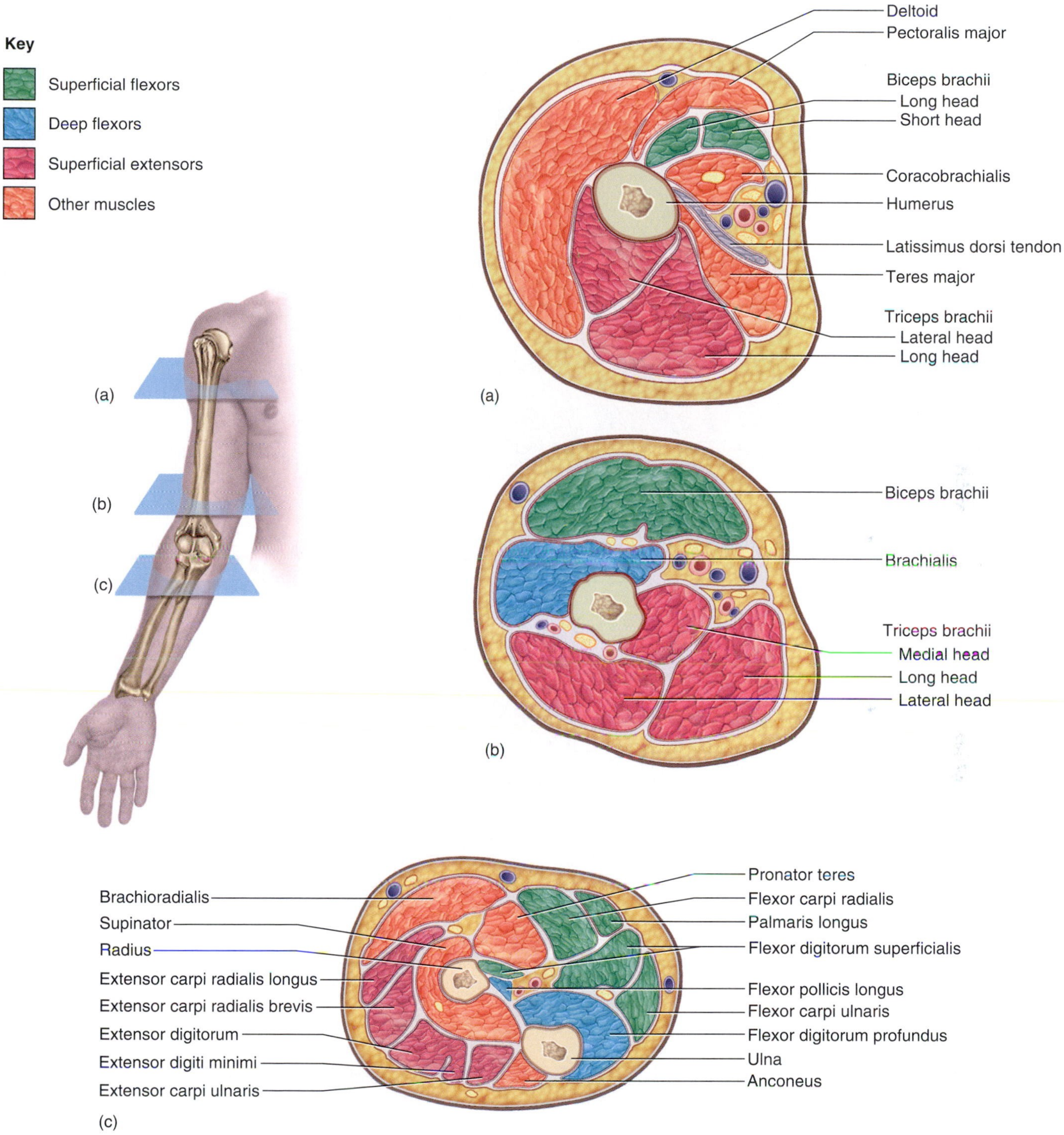
Key
Superficial flexors
Deep flexors
Superficial extensors
Other muscles
(a)
(b)
(c)
Deltoid
Pectoralis major
Biceps brachii
Long head
Short head
Coracobrachialis
Humerus
Latissimus dorsi tendon
Teres major
Triceps brachii
Lateral head
Long head
(a)
Biceps brachii
Brachialis
Triceps brachii
Medial head
Long head
Lateral head
(b)
Brachioradialis
Supinator
Radius
Extensor carpi radialis longus
Extensor carpi radialis brevis
Extensor digitorum
Extensor digiti minimi
Extensor carpi ulnaris
Pronator teres
Flexor carpi radialis
Palmaris longus
Flexor digitorum superficialis
Flexor pollicis longus
Flexor carpi ulnaris
Flexor digitorum profundus
Ulna
Anconeus
(c)

FIGURE 12.7

Serial Cross Sections Through the Upper Limb. Each section is taken at the correspondingly lettered level in the figure at the *left* and is pictured with the posterior muscle compartment facing the *bottom* of the page, as if you were viewing the right upper limb of a person facing you with the limb extended and the palm up.

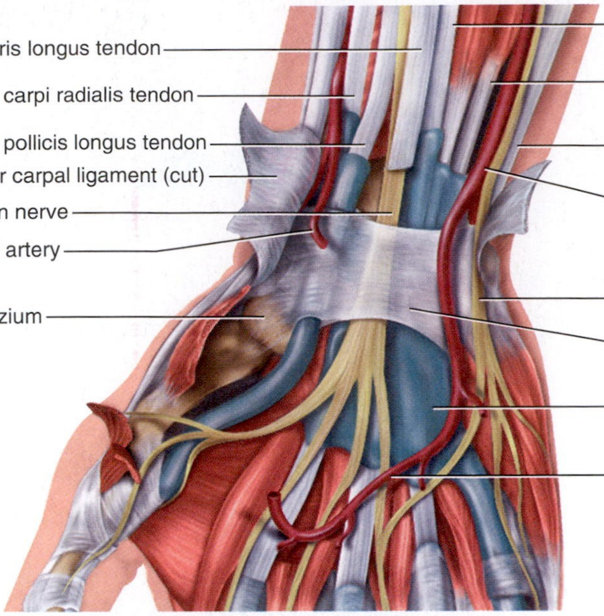

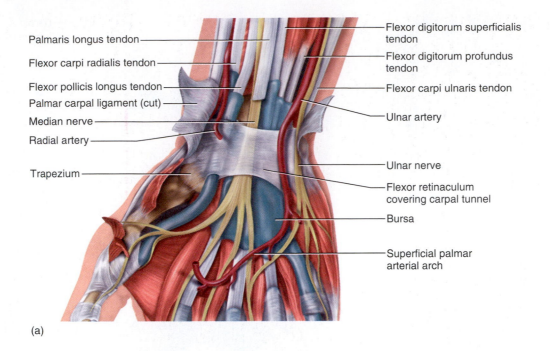

Palmaris longus tendon
Flexor carpi radialis tendon
Flexor pollicis longus tendon
Palmar carpal ligament (cut)
Median nerve
Radial artery
Trapezium
Flexor digitorum superficialis tendon
Flexor digitorum profundus tendon
Flexor carpi ulnaris tendon
Ulnar artery
Ulnar nerve
Flexor retinaculum covering carpal tunnel
Bursa
Superficial palmar arterial arch
(a)

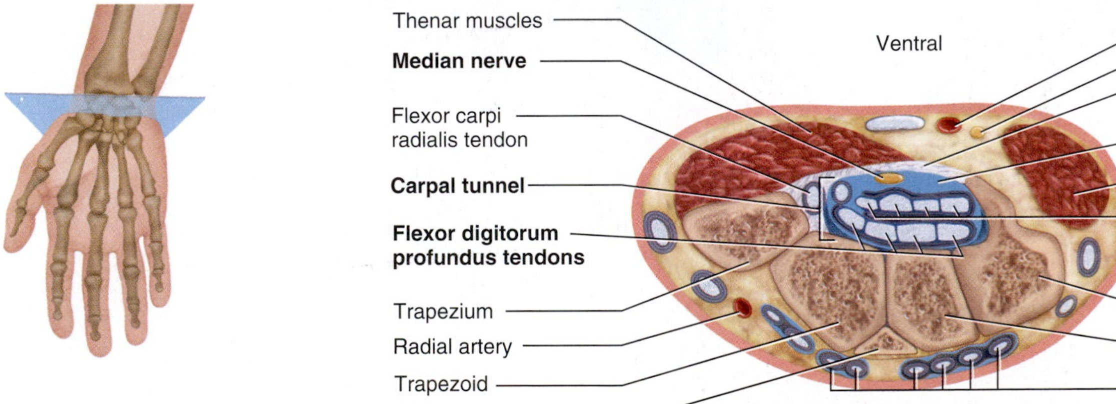

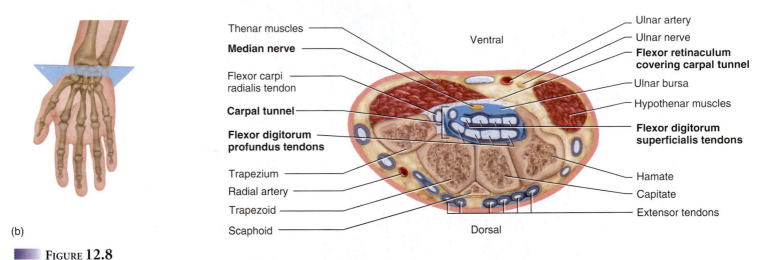

Thenar muscles
Median nerve
Flexor carpi radialis tendon
Carpal tunnel
Flexor digitorum profundus tendons
Trapezium
Radial artery
Trapezoid
Scaphoid
Ventral
Dorsal
Ulnar artery
Ulnar nerve
Flexor retinaculum covering carpal tunnel
Ulnar bursa
Hypothenar muscles
Flexor digitorum superficialis tendons
Hamate
Capitate
Extensor tendons
(b)

FIGURE 12.8

The Carpal Tunnel. (*a*) Dissection of the wrist (anterior aspect) showing the tendons, nerve, and bursae that pass under the flexor retinaculum. (*b*) Cross section of the wrist, ventral (anterior side) up. Note how the flexor tendons and median nerve are confined in the tight space between the carpal bones and flexor retinaculum.

INSIGHT 12.1 CLINICAL APPLICATION

Carpal Tunnel Syndrome

Prolonged, repetitive motions of the wrist and fingers can cause tissues in the carpal tunnel to become inflamed, swollen, or fibrotic. Since the carpal tunnel cannot expand, swelling puts pressure on the median nerve of the wrist, which passes through the carpal tunnel with the flexor tendons. This pressure causes tingling and muscular weakness in the palm and lateral side of the hand and pain that may radiate to the arm and shoulder. This condition, called *carpal tunnel syndrome,* is common among pianists, meat cutters, and others who spend long hours making repetitive wrist motions. It can also be caused by other factors that reduce the size of the carpal tunnel, including tumors, infections, and bone fractures. Carpal tunnel syndrome is treated with aspirin and other anti-inflammatory drugs, immobilization of the wrist, and sometimes surgical removal of part or all of the flexor retinaculum to relieve pressure on the nerve.

Table 12.7

Intrinsic Muscles of the Hand (fig. 12.9)

O = origin, I = insertion, N = innervation (n. = nerve, nn. = nerves)

The intrinsic muscles of the hand assist the flexors and extensors of the forearm and make finger movements more precise. They are divided into three groups—the *thenar group* at the base of the thumb, the *hypothenar group* at the base of the little finger, and the *midpalmar group* in between.

Thenar Group. The thenar group of muscles forms the thick fleshy mass *(thenar eminence)* at the base of the thumb, except for the *adductor pollicis,* which forms the web between the thumb and palm. All are concerned with thumb movements.

Abductor Pollicis Brevis (PAHL-ih-sis)

Abducts thumb

O: scaphoid, trapezium, flexor retinaculum	I: lateral aspect of proximal phalanx I	N: median n.

Adductor Pollicis

Adducts thumb and opposes it to the fingers

O: trapezium, trapezoid, capitate, metacarpals II–III	I: medial aspect of proximal phalanx I	N: ulnar n.

Flexor Pollicis Brevis

Flexes thumb at metacarpophalangeal joint

O: scaphoid, trapezium, flexor retinaculum	I: proximal phalanx I	N: median and ulnar nn.

Opponens Pollicis (op-OH-nens)

Opposes thumb to fingers; medially rotates metacarpal I

O: trapezium, flexor retinaculum	I: metacarpal I	N: median n.

Hypothenar Group. The hypothenar group of muscles forms the fleshy mass *(hypothenar eminence)* at the base of the little finger. All are concerned with the movements of that digit.

Abductor Digiti Minimi

Abducts little finger

O: pisiform, tendon of flexor carpi ulnaris	I: medial aspect of proximal phalanx V	N: ulnar n.

Flexor Digiti Minimi Brevis

Flexes little finger at metacarpophalangeal joint

O: hamulus of hamate, flexor retinaculum	I: medial aspect of proximal phalanx V	N: ulnar n.

Opponens Digiti Minimi

Opposes little finger to thumb; laterally rotates metacarpal V; deepens pit of palm

O: hamulus of hamate, flexor retinaculum	I: medial aspect of metacarpal V	N: ulnar n.

Midpalmar Group. The midpalmar group of muscles spans the hollow of the palm. This group has 11 muscles divided into three subgroups:

1. Four *dorsal interosseous muscles*—bipennate muscles attached to both sides of the metacarpal bones, serving to abduct (spread) the fingers.
2. Three *palmar interosseous muscles*—unipennate muscles that arise from metacarpals II, IV, and V and adduct the fingers (draw them together).
3. Four *lumbrical muscles*—wormlike muscles that flex the metacarpophalangeal joints (proximal knuckles) but extend the interphalangeal joints (distal knuckles).

Dorsal Interosseous[4] Muscles (IN-tur-OSS-ee-us) (four muscles)

Abduct digits II–IV; flex metacarpophalangeal joints; extend interphalangeal joints

O: two heads on facing sides of adjacent metacarpals	I: proximal phalanges II–IV	N: ulnar n.

Palmar Interosseous Muscles (three muscles)

Adduct digits II, IV, and V; flex metacarpophalangeal joints; extend interphalangeal joints

O: metacarpals II, IV, and V	I: proximal phalanges II, IV, and V	N: ulnar n.

Lumbricals[5] (four muscles)

Flex metacarpophalangeal joints; extend interphalangeal joints

O: tendons of flexor digitorum profundus	I: proximal phalanges II–V	N: median and ulnar nn.

[4]*inter* = between + *osse* = bone
[5]*lumbric* = earthworm

(continued)

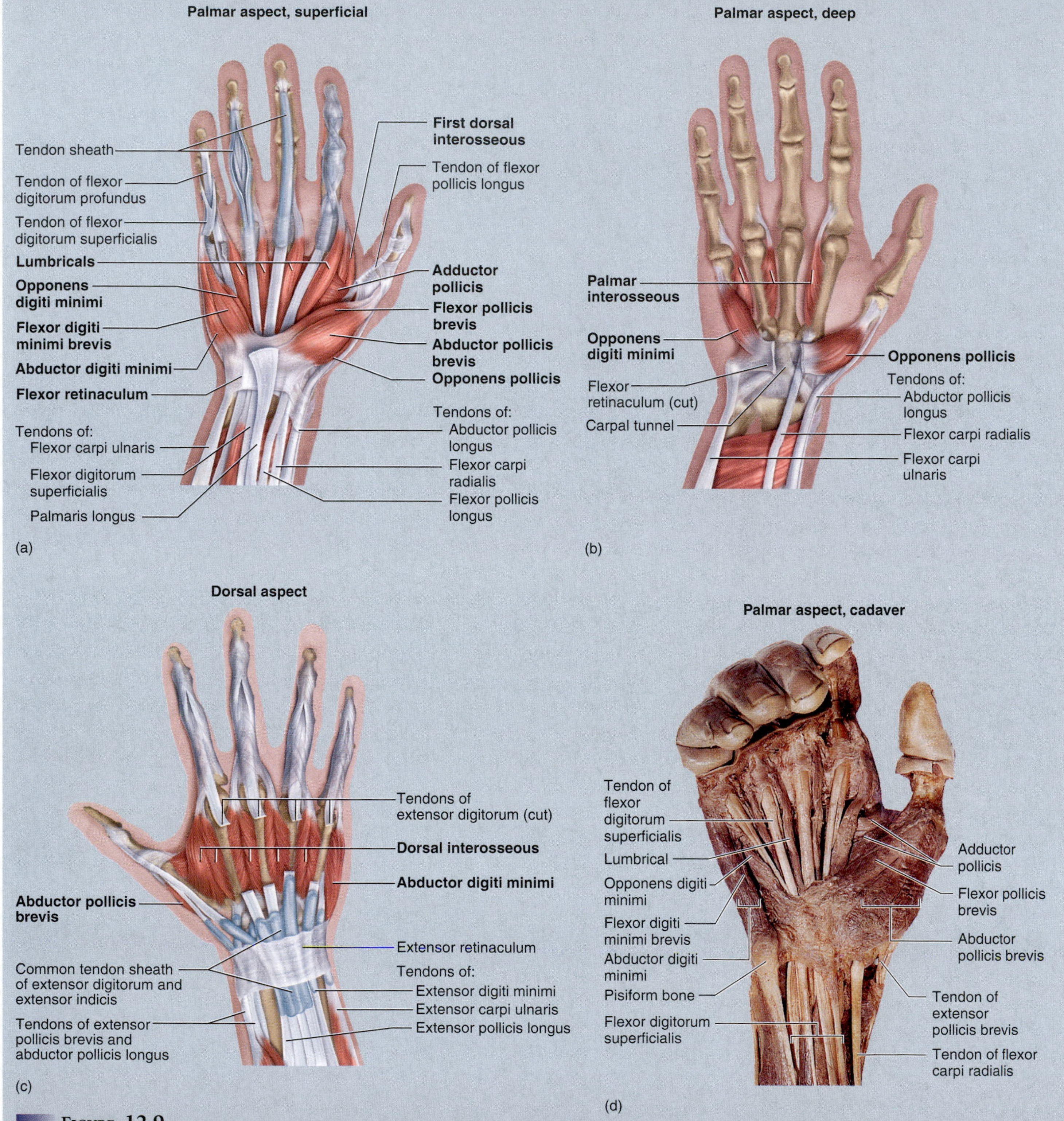

FIGURE 12.9

Intrinsic Muscles of the Hand. Muscles labeled in boldface are: (*a*) superficial muscles, anterior (palmar) view; (*b*) deep muscles, anterior view; (*c*) superficial muscles, posterior (dorsal) view. (*d*) Anterior (palmar) view of cadaver hand.

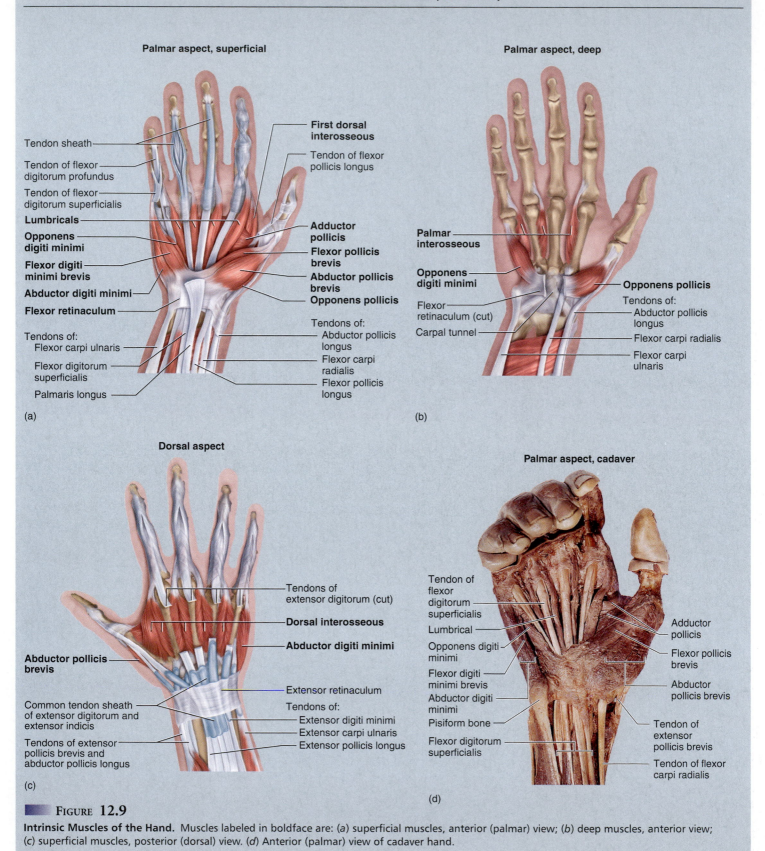

TABLE 12.8

Actions of the Wrist and Hand

Italics indicate prime movers; others are synergists. Parentheses indicate only a slight effect.

Wrist Flexion	Wrist Extension
Flexor carpi radialis	*Extensor digitorum*
Flexor carpi ulnaris	Extensor carpi radialis longus
Flexor digitorum superficialis	Extensor carpi radialis brevis
(Palmaris longus)	Extensor carpi ulnaris
(Flexor pollicis longus)	

Wrist Abduction	Wrist Adduction
Flexor carpi radialis	Flexor carpi ulnaris
Extensor carpi radialis longus	Extensor carpi ulnaris
Extensor carpi radialis brevis	
Abductor pollicis longus	

Finger Flexion	Finger Extension	Thumb Opposition
Flexor digitorum superficialis	Extensor pollicis longus	
Flexor digitorum profundus	Extensor pollicis brevis	Opponens pollicis
Flexor pollicis longus	Extensor digitorum	Opponens digiti minimi
	Extensor indicis	

Table 12.8 summarizes the muscles responsible for the major movements of the wrist and hand.

Before You Go On

Answer the following questions to test your understanding of the preceding section:

1. Name a muscle that inserts on the scapula and plays a significant role in each of the following actions: (*a*) pushing a stalled car, (*b*) paddling a canoe, (*c*) squaring the shoulders in military attention, (*d*) lifting the shoulder to carry a heavy box on it, and (*e*) lowering the shoulder to lift a suitcase.
2. Describe three contrasting actions of the deltoid muscle.
3. Name the four rotator cuff muscles and describe the scapular surfaces against which they lie.
4. Name the prime movers of elbow flexion and extension.
5. Identify three functions of the biceps brachii.
6. Name three extrinsic muscles and two intrinsic muscles that flex the phalanges.

MUSCLES ACTING ON THE HIP AND LOWER LIMB

Objectives

When you have completed this section, you should be able to

- name and locate the muscles that act on the hip, knee, ankle, and toe joints;
- relate the actions of these muscles to the joint movements described in chapter 9; and
- describe the origin, insertion, and innervation of each muscle.

The largest muscles are found in the lower limb. Unlike those of the upper limb, they are adapted less for precision than for the strength needed to stand, maintain balance, walk, and run. Several of them cross and act upon two or more joints, such as the hip and knee. To avoid confusion in this discussion, remember that in the anatomical sense the word *leg* refers only to that part of the limb between the knee and ankle. The term *foot* includes the tarsal region (ankle), metatarsal region, and toes. Tables 12.9 through 12.12 group the muscles of the lower limb into those that act on the femur and hip joint, those that act on the leg and knee joint, extrinsic (leg) muscles that act on the foot and ankle joint, and intrinsic (foot) muscles that act on the arches and toes.

TABLE 12.9

Muscles Acting on the Hip and Femur (figs. 12.10–12.12)

O = origin, I = insertion, N = innervation (n. = nerve, nn. = nerves)

Anterior Muscles of the Hip. Most muscles that act on the femur originate on the os coxae. The two principal anterior muscles are the *iliacus,* which fills most of the broad iliac fossa of the pelvis, and the *psoas major,* a thick, rounded muscle that originates mainly on the lumbar vertebrae. Collectively, they are called the **iliopsoas** (fig. 12.10). They converge on a single tendon that inserts on the femur and flexes the hip joint—for example, when you bend forward at the waist, swing the leg forward in walking, or raise the thigh in a marching stance.

Iliacus (ih-LY-uh-cus)

Flexes hip joint

O: iliac crest and fossa, sacral ala, anterior sacroiliac ligaments	I: lesser trochanter of femur, psoas major tendon	N: femoral n.

Psoas Major (SO-ass)

Flexes hip joint

O: vertebral bodies T12–L5	I: lesser trochanter of femur	N: lumbar plexus

(continued)

Table 12.9
Muscles Acting on the Hip and Femur *(continued)*

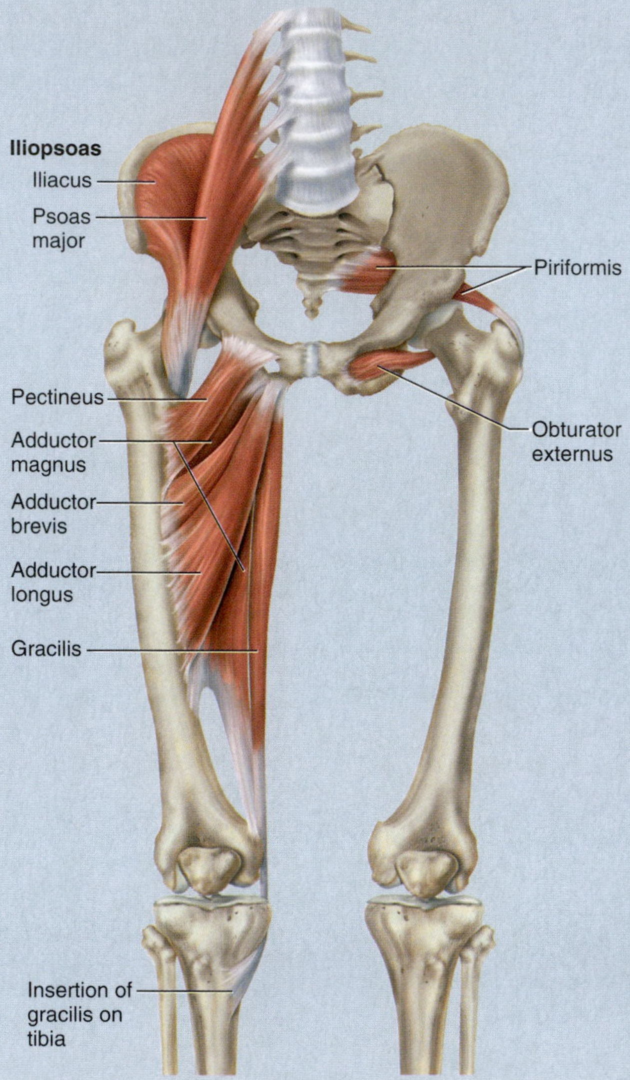
Iliopsoas
Iliacus
Psoas major
Pectineus
Adductor magnus
Adductor brevis
Adductor longus
Gracilis
Insertion of gracilis on tibia
Piriformis
Obturator externus

Figure 12.10
Muscles That Act on the Hip and Femur. Anterior view.

Lateral and Posterior Muscles of the Hip. On the lateral and posterior sides of the hip are the *tensor fasciae latae* and three gluteal muscles—the *gluteus maximus, gluteus medius,* and *gluteus minimus* (fig. 12.11). The gluteus maximus is the largest muscle of this group and forms most of the lean mass of the buttocks. It is an extensor of the hip joint that produces the backswing of the leg in walking and provides most of the lift when you climb stairs. It generates its maximum force when the thigh is flexed at a 45° angle to the trunk. This is the advantage in starting a foot race from a crouched position.

The **fascia lata**[6] is a fibrous sheath that encircles the thigh like a subcutaneous stocking and tightly binds its muscles. On the lateral surface, it combines with the tendons of the gluteus maximus and tensor fasciae latae to form the **iliotibial band,** which extends from the iliac crest to the lateral condyle of the tibia (figs. 12.12 and 12.14). The tensor fasciae latae tautens the iliotibial band and braces the knee, especially when we raise the opposite foot.

Deep fasciae divide the thigh into three compartments, each with its own nerve and blood supply: the *anterior (extensor) compartment, medial (adductor) compartment,* and *posterior (flexor) compartment.* Muscles of the anterior compartment function mainly as extensors of the knee, those of the medial compartment as adductors of the femur, and those of the posterior compartment as extensors of the hip and flexors of the knee.

[6]*fasc* = band + *lata* = broad

(continued)

TABLE 12.9

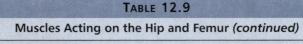

Muscles Acting on the Hip and Femur *(continued)*

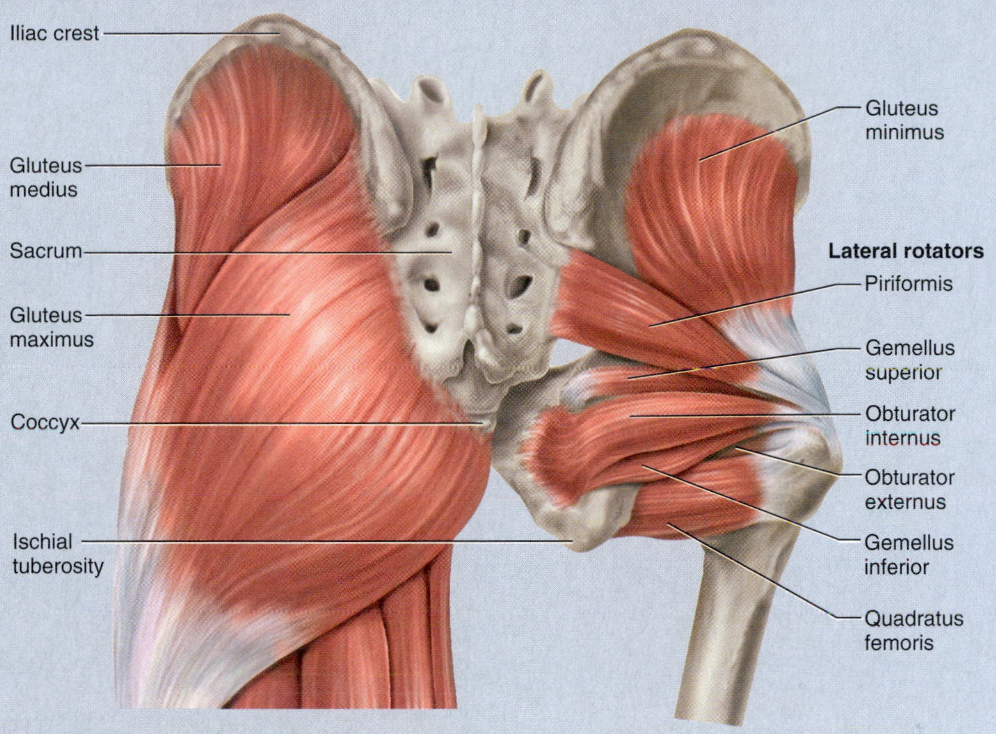

FIGURE 12.11
Gluteal Muscles. Superficial muscles are shown on the *left* and deep muscles on the *right*.

Tensor Fasciae Latae (TEN-sor FASH-ee-ee LAY-tee)

Flexes hip joint; abducts and medially rotates femur; tenses fascia lata and braces knee when opposite foot is lifted from ground

O: iliac crest and anterior superior spine	I: lateral condyle of tibia via iliotibial band	N: superior gluteal n.

Gluteus Maximus

Extends hip joint; abducts and laterally rotates femur; important in the backswing of the stride, climbing stairs, and rising from a sitting position

O: ilium, sacrum, coccyx	I: lateral condyle of tibia via iliotibial band; gluteal tuberosity of femur	N: inferior gluteal n.

Gluteus Medius and Gluteus Minimus

Abduct and medially rotate femur; maintain balance by shifting body weight during walking

O: ilium	I: greater trochanter of femur	N: superior gluteal n.

Lateral Rotators. The deep lateral rotators of the pelvic region (see fig. 12.11) rotate the femur laterally, as when you cross your legs to rest an ankle on your knee and your femur turns slightly on its longitudinal axis. Thus, they oppose medial rotation by the gluteus medius and minimus. Most of them also abduct or adduct the femur. The abductors are important in walking because when we lift one foot from the ground, they shift the body weight to other leg and prevent us from falling over.

Gemellus Superior (jeh-MEL-us) and Gemellus Inferior

Laterally rotate femur when hip is extended; abduct femur when hip is flexed

O: body of ischium	I: obturator internus tendon	N: L5 and S1

Obturator Externus (OB-too-RAY-tur)

Laterally rotates femur

O: anterior margin of obturator foramen	I: greater trochanter of femur	N: obturator n.

(continued)

Table 12.9

Muscles Acting on the Hip and Femur *(continued)*

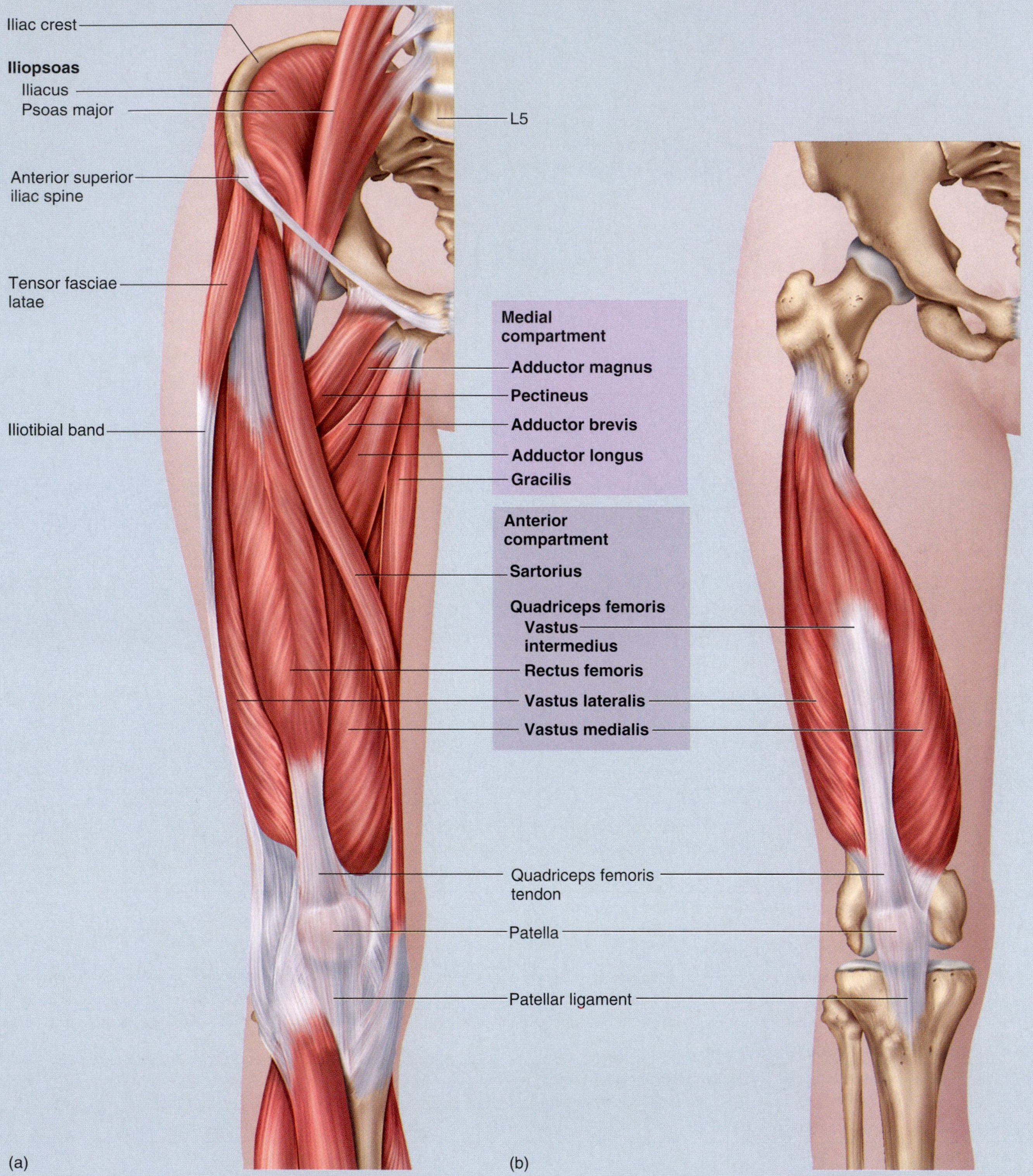

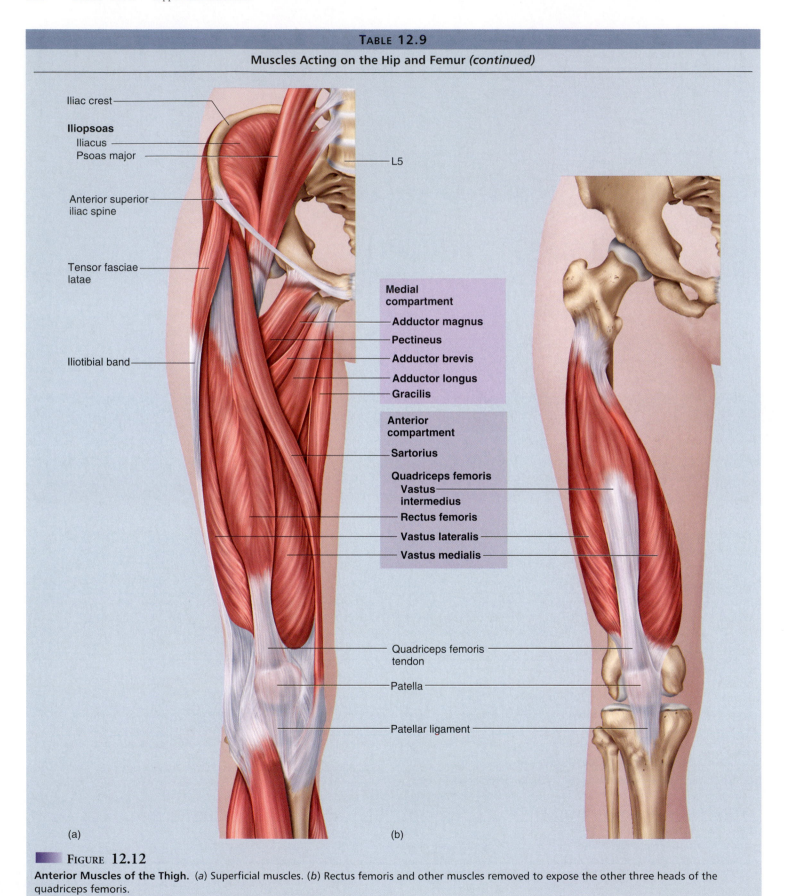

Iliac crest
Iliopsoas
Iliacus
Psoas major
L5
Anterior superior iliac spine
Tensor fasciae latae
Iliotibial band
Medial compartment
Adductor magnus
Pectineus
Adductor brevis
Adductor longus
Gracilis
Anterior compartment
Sartorius
Quadriceps femoris
Vastus intermedius
Rectus femoris
Vastus lateralis
Vastus medialis
Quadriceps femoris tendon
Patella
Patellar ligament
(a)
(b)

Figure 12.12

Anterior Muscles of the Thigh. (*a*) Superficial muscles. (*b*) Rectus femoris and other muscles removed to expose the other three heads of the quadriceps femoris.

(continued)

TABLE 12.9

Muscles Acting on the Hip and Femur *(continued)*

Obturator Internus		
Abducts and laterally rotates femur		
O: posterior margin of obturator foramen	I: greater trochanter of femur	N: L5 and S1
Piriformis (PIR-ih-FOR-miss)		
Abducts and laterally rotates femur		
O: anterolateral aspect of sacroiliac region	I: greater trochanter of femur	N: ventral rami of S1–S2
Quadratus Femoris (quad-RAY-tus FEM-oh-riss)		
Adducts and laterally rotates femur		
O: ischial tuberosity	I: intertrochanteric crest of femur	N: sacral plexus
Medial (adductor) Compartment of the Thigh. In the medial compartment are five muscles that act on the hip joint—the *adductor longus, adductor brevis, adductor magnus, gracilis,* and *pectineus* (see fig. 12.10). All of them adduct the thigh, but some cross both the hip and knee joints and have additional actions noted below.		
Adductor Longus and Adductor Brevis		
Adduct and laterally rotate femur		
O: pubis	I: linea aspera of femur	N: obturator n.
Adductor Magnus		
Anterior part adducts and laterally rotates femur and flexes hip joint; posterior part extends hip joint		
O: ischium and pubis	I: linea aspera of femur	N: obturator and tibial nn.
Gracilis[7] (GRASS-ih-lis)		
Adducts femur; flexes knee; medially rotates tibia		
O: pubis	I: medial aspect of proximal tibia	N: obturator n.
Pectineus[8] (pec-TIN-ee-us)		
Adducts and laterally rotates femur; flexes hip		
O: pubis	I: posterior aspect of proximal femur	N: femoral n.

[7] *gracil* = slender
[8] *pectin* = comb

INSIGHT 12.2 CLINICAL APPLICATION

INTRAMUSCULAR INJECTIONS

Muscles with thick bellies are commonly used for intramuscular (I.M.) drug injections. Since drugs injected into these muscles are absorbed into the bloodstream gradually, it is safe to administer relatively large doses (up to 5 mL) that could be dangerous or even fatal if injected directly into the bloodstream. Intramuscular injections also cause less tissue irritation than subcutaneous injections.

Knowledge of subsurface anatomy is necessary to avoid damaging nerves or accidentally injecting a drug into a blood vessel. Anatomical knowledge also enables a clinician to position a patient so that the muscle is relaxed, making the injection less painful.

Amounts up to 2 mL are commonly injected into the deltoid muscle about two finger widths below the acromion. A misplaced injection into the deltoid can injure the axillary nerve and cause atrophy of the muscle. Drug doses over 2 mL are commonly injected into the gluteus medius, in the superolateral quadrant of the gluteal area, at a safe distance from the sciatic nerve and major gluteal blood vessels. Injections are often given to infants and young children in the vastus lateralis of the thigh, because their deltoid and gluteal muscles are not well developed.

TABLE 12.10

Muscles Acting on the Knee and Leg (figs. 12.13–12.14)

O = origin, I = insertion, N = innervation (n. = nerve)

The following muscles form most of the mass of the thigh and produce their most obvious actions on the knee joint. Some of them, however, cross both the hip and knee joints and produce actions at both, moving the femur, tibia, and fibula.

Anterior (extensor) Compartment of the Thigh. The anterior compartment of the thigh contains the large *quadriceps femoris* muscle, the prime mover of knee extension and the most powerful muscle of the body (figs. 12.12 and 12.13). As the name implies, it has four heads—the *rectus femoris, vastus lateralis, vastus medialis,* and *vastus intermedius.* All four converge on a single **quadriceps (patellar) tendon,** which extends to the patella, then continues as the **patellar ligament** and inserts on the tibial tuberosity. (Remember that a tendon usually extends from muscle to bone, and a ligament from bone to bone.) The patellar ligament is struck with a rubber reflex hammer to test the knee-jerk reflex. The quadriceps extends the knee when you stand up, take a step, or kick a ball. It is very important in running because, together with the iliopsoas, it flexes the hip in each airborne phase of the leg's cycle of motion. The rectus femoris also flexes the hip in such actions as high kicks or simply in drawing the leg forward during a stride.

Crossing the quadriceps from the lateral side of the hip to the medial side of the knee is the narrow, straplike *sartorius,* the longest muscle of the body. It flexes the hip and knee joints and laterally rotates the thigh, as in crossing the legs. It is colloquially called the "tailor's muscle" after the cross-legged stance of a tailor supporting his work on the raised knee.

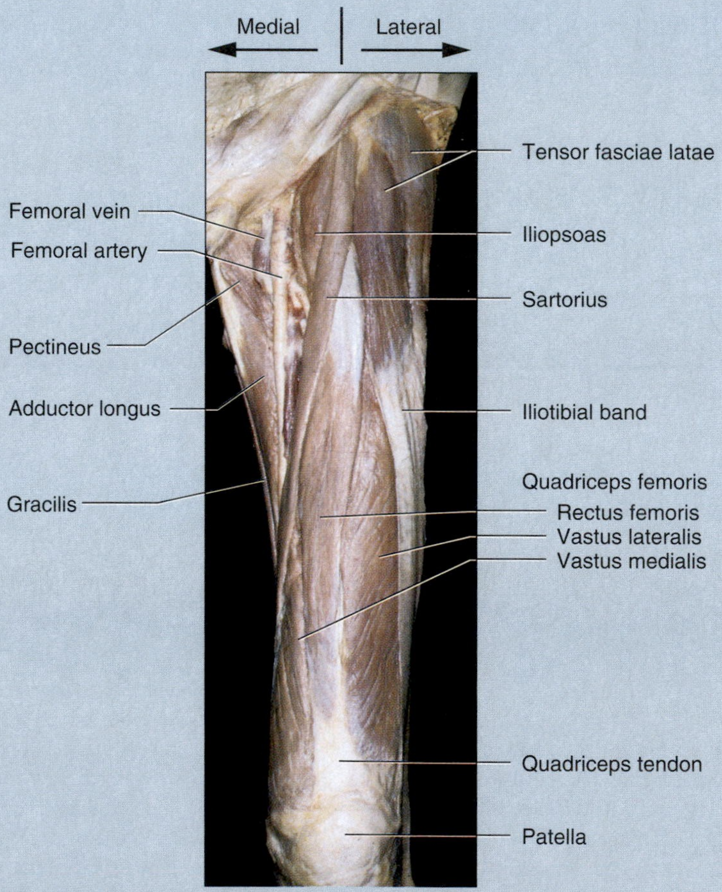
Medial
Lateral
Tensor fasciae latae
Femoral vein
Femoral artery
Iliopsoas
Sartorius
Pectineus
Adductor longus
Iliotibial band
Gracilis
Quadriceps femoris
Rectus femoris
Vastus lateralis
Vastus medialis
Quadriceps tendon
Patella

FIGURE 12.13
Anterior Superficial Thigh Muscles of the Cadaver. Left limb.

Quadriceps femoris (QUAD-rih-seps FEM-oh-riss)

Extends knee; rectus femoris also flexes hip

O: *rectus femoris*—anterior inferior spine of ilium; *vastus lateralis*—greater trochanter and linea aspera of femur; *vastus medialis*—linea aspera; *vastus intermedius*—anterior and lateral shaft of femur	I: tibial tuberosity	N: femoral n.

(continued)

TABLE 12.10

Muscles Acting on the Knee and Leg *(continued)*

Sartorius[9]

Flexes, abducts, and laterally rotates thigh at hip; flexes knee; used in crossing legs

O: on and below anterior superior spine of ilium I: medial aspect of tibial tuberosity N: femoral n.

Posterior (flexor) Compartment of the Thigh (hamstring group). The posterior compartment contains the *biceps femoris, semimembranosus,* and *semitendinosus* (fig. 12.14). These muscles are colloquially known as the "hamstrings" because their tendons at the knee of a hog are commonly used to hang a ham for curing. They flex the knee, and aided by the gluteus maximus, they extend the hip during walking and running. The semitendinosus is named for the fact that it is often bisected by a transverse tendinous band. The pit at the rear of the knee, called the *popliteal fossa,* is bordered by the biceps tendon on the lateral side and the tendons of the semimembranosus and semitendinosus on the medial side. When wolves attack large prey, they often attempt to sever the hamstring tendons, because this renders the prey helpless.

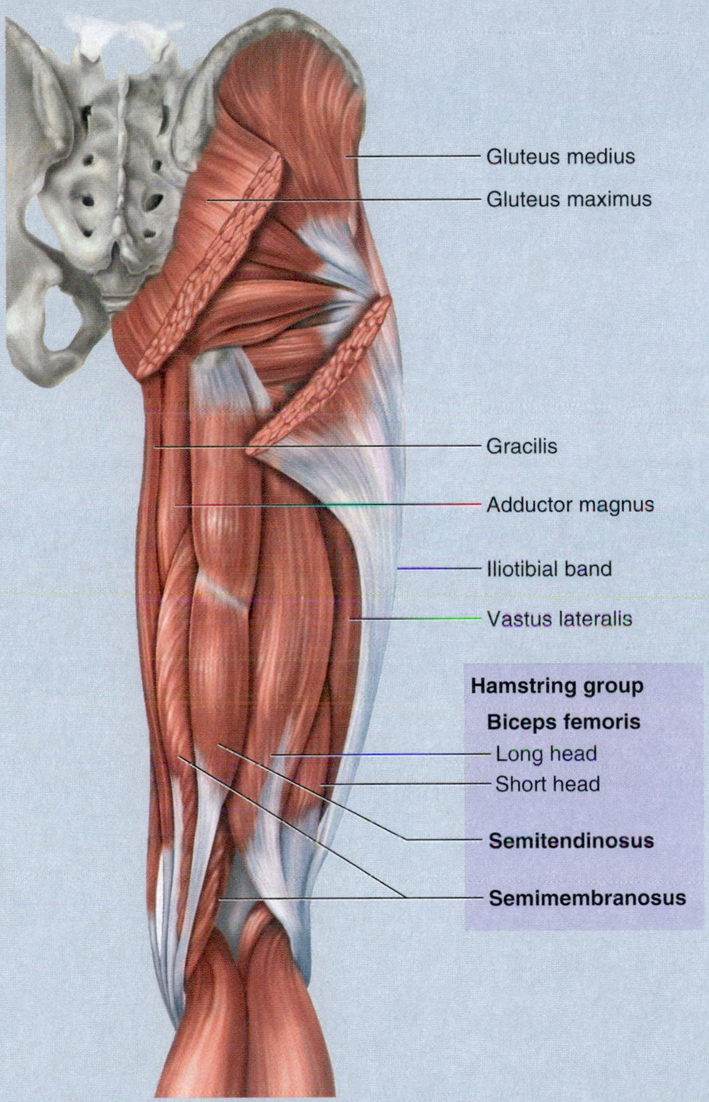

Gluteus medius
Gluteus maximus
Gracilis
Adductor magnus
Iliotibial band
Vastus lateralis
Hamstring group
Biceps femoris
Long head
Short head
Semitendinosus
Semimembranosus

FIGURE 12.14
Gluteal and Thigh Muscles. Posterior view. The gluteus maximus is cut to expose the origins of the hamstring muscles.

[9]*sartor* = tailor

(continued)

TABLE 12.10

Muscles Acting on the Knee and Leg *(continued)*

Biceps Femoris

Flexes knee; extends hip; laterally rotates leg

O: *long head*—ischial tuberosity; *short head*—lower half of linea aspera of femur	I: head of fibula	N: *long head*—tibial n.; *short head*—common fibular n.

Semimembranosus (SEM-ee-MEM-bran-OH-sis)

Flexes knee; extends hip; medially rotates tibia; tenses joint capsule of knee

O: ischial tuberosity	I: medial condyle of tibia, collateral ligament of knee	N: tibial n.

Semitendinosus

Flexes knee; extends hip; medially rotates tibia

O: ischial tuberosity	I: medial shaft of tibia near tibial tuberosity	N: tibial n.

Posterior Compartment of Leg. Most muscles in the posterior compartment of the leg act on the ankle and foot and are reviewed in table 12.11, but the *popliteus* acts on the knee.

Popliteus (pop-LIT-ee-us)

Unlocks knee to allow flexion; flexes knee; medially rotates tibia

O: lateral condyle of femur	I: posterior proximal tibia	N: tibial n.

INSIGHT 12.3 CLINICAL APPLICATION

Hamstring Injuries

Hamstring injuries are common among sprinters, soccer players, and other athletes who depend on quick extension of the knee to kick or jump forcefully. Rapid knee extension stretches the hamstrings and often tears the proximal tendons where they originate on the ischial tuberosity. These muscle strains are excruciatingly painful. Hamstring injuries often result from failure to warm up adequately before competition or practice.

TABLE 12.11

Muscles Acting on the Foot (figs. 12.15–12.18)

O = origin, I = insertion, N = innervation (n. = nerve)

The fleshy mass of the leg proper (below the knee) is formed by a group of **crural muscles,** which act on the foot. These muscles are divided into anterior, lateral, and posterior compartments.

Anterior Compartment of the Leg. Muscles of the anterior compartment dorsiflex the ankle and prevent the toes from scuffing the ground during walking. These are the *extensor digitorum longus* (extensor of toes II–V), *extensor hallucis longus* (extensor of the great toe), *fibularis (peroneus) tertius,* and *tibialis anterior.* Their tendons are held tightly against the ankle and kept from bowing by two *extensor retinacula* similar to the one at the wrist (fig. 12.15).

Extensor Digitorum Longus (DIDJ-ih-TOE-rum)

Extends toes II–V; dorsiflexes and everts foot

O: lateral condyle of tibia, shaft of fibula, interosseous membrane	I: middle and distal phalanges II–V	N: deep fibular n.

Extensor Hallucis[10] Longus (hal-OO-sis)

Extends hallux (great toe); dorsiflexes and inverts foot

O: medial aspect of fibula, interosseous membrane	I: distal phalanx I	N: deep fibular n.

Fibularis Tertius (FIB-you-LERR-iss TUR-she-us)

Dorsiflexes and everts foot; not always present

O: distal shaft of fibula and interosseous membrane	I: metatarsal V	N: deep fibular n.

Tibialis Anterior (TIB-ee-AY-lis)

Dorsiflexes and inverts foot

O: lateral tibia, interosseous membrane	I: medial cuneiform, metatarsal I	N: deep fibular n.

[10] *halluc* = great toe

(continued)

TABLE 12.11

Muscles Acting on the Foot *(continued)*

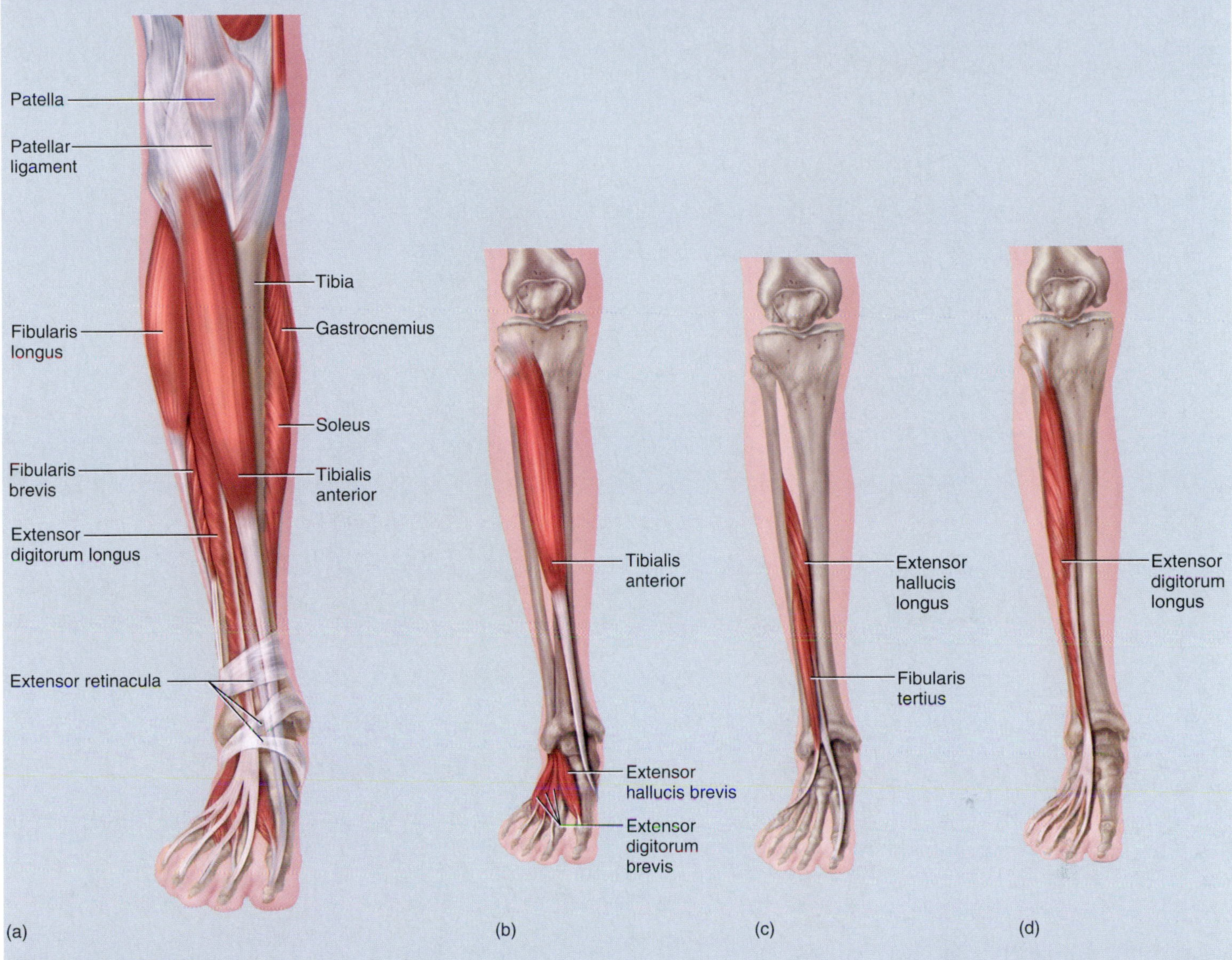

FIGURE 12.15

Anterior Muscles of the Leg. (*a*) A view showing some muscles of the anterior, lateral, and posterior compartments. (*b–d*) Individual muscles of the anterior compartment.

Posterior Compartment of the Leg—Superficial Group. The posterior compartment has superficial and deep muscle groups. The three muscles of the superficial group are plantar flexors—the *gastrocnemius, soleus,* and *plantaris* (fig. 12.16). The first two of these, collectively known as the **triceps surae,**[11] insert on the calcaneus by way of the **calcaneal (Achilles) tendon.** This is the strongest tendon of the body but is nevertheless a common site of sports injuries resulting from sudden stress. The plantaris, a weak synergist of the triceps surae, inserts medially on the calcaneus by a tendon of its own. It is a relatively unimportant muscle and is absent from many people. Surgeons often use plantaris tendon for tendon grafts needed in other parts of the body.

Gastrocnemius[12] (GAS-trock-NEE-me-us)

Flexes knee; plantar flexes foot

O: lateral condyle and popliteal surface of femur	I: calcaneus	N: tibial n.

Soleus[13] (SO-lee-us)

Plantar flexes foot

O: proximal third of tibia and fibula	I: calcaneus	N: tibial n.

[11]*sura* = calf of leg
[12]*gastro* = belly + *cnem* = leg
[13]*soleus* = sole, a flatfish

(continued)

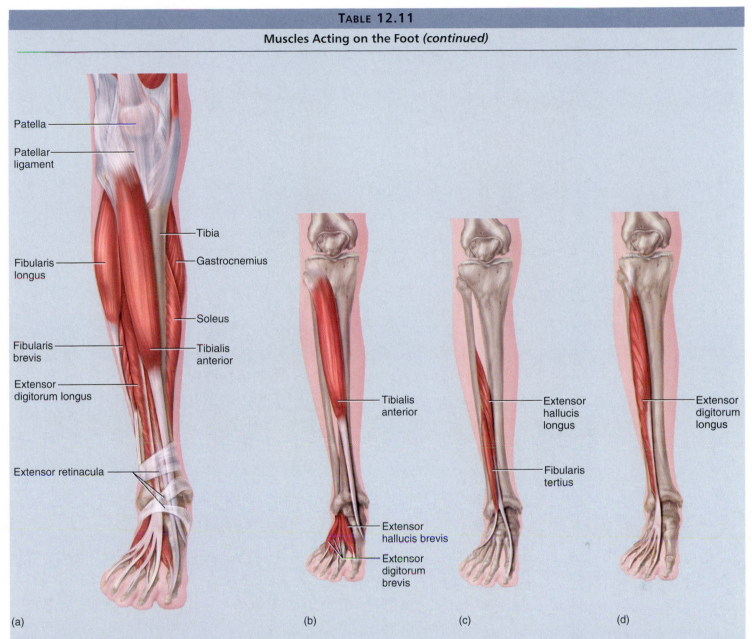

Table 12.11

Muscles Acting on the Foot *(continued)*

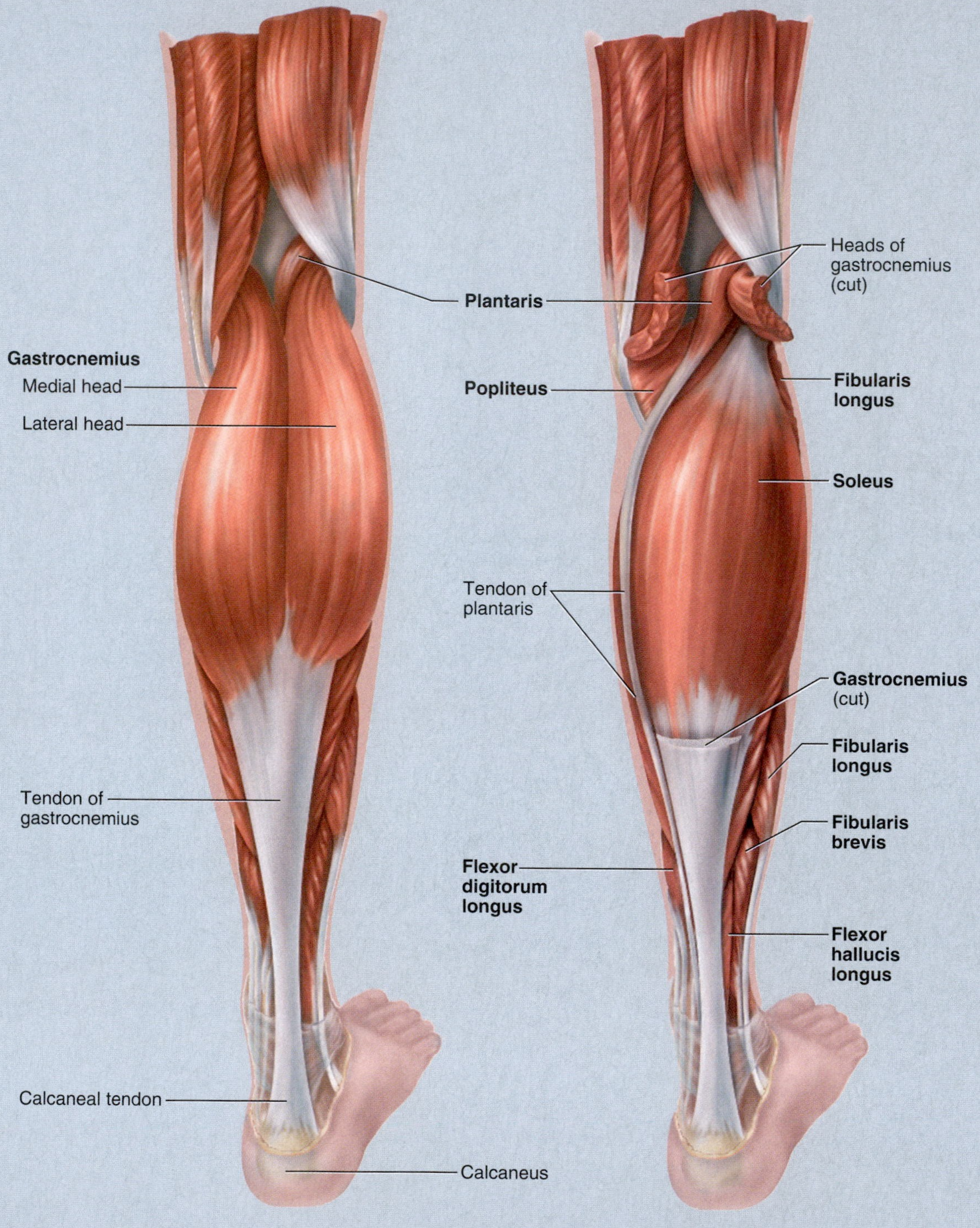
Gastrocnemius
Medial head
Lateral head
Tendon of gastrocnemius
Calcaneal tendon
Calcaneus
Plantaris
Popliteus
Tendon of plantaris
Flexor digitorum longus
Heads of gastrocnemius (cut)
Fibularis longus
Soleus
Gastrocnemius (cut)
Fibularis longus
Fibularis brevis
Flexor hallucis longus

Figure 12.16

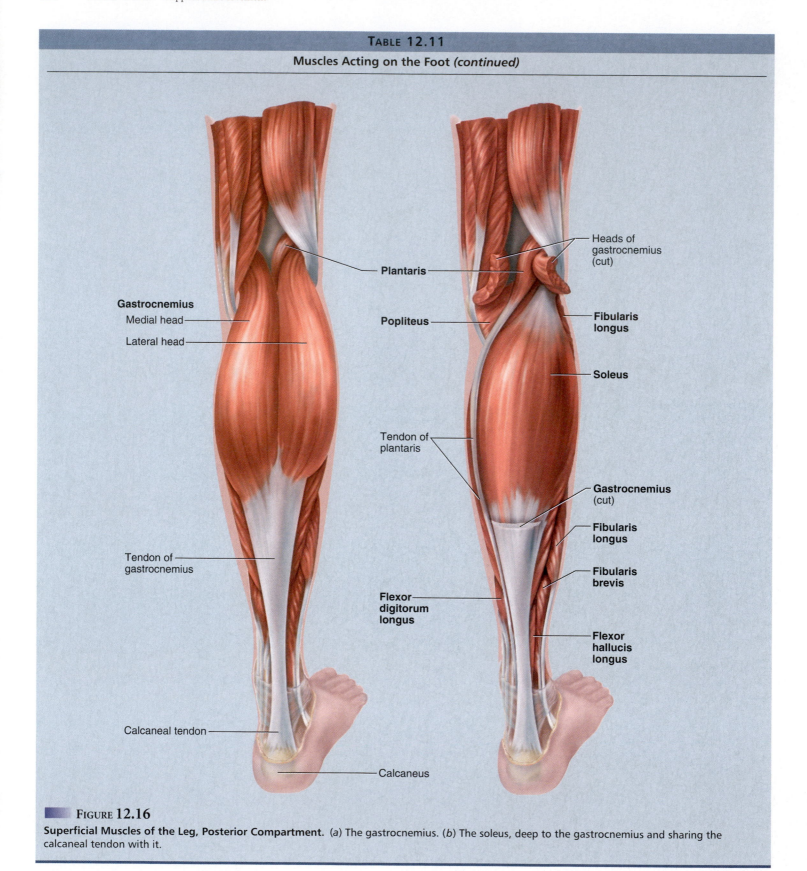

Superficial Muscles of the Leg, Posterior Compartment. (*a*) The gastrocnemius. (*b*) The soleus, deep to the gastrocnemius and sharing the calcaneal tendon with it.

(continued)

TABLE 12.11

Muscles Acting on the Foot *(continued)*

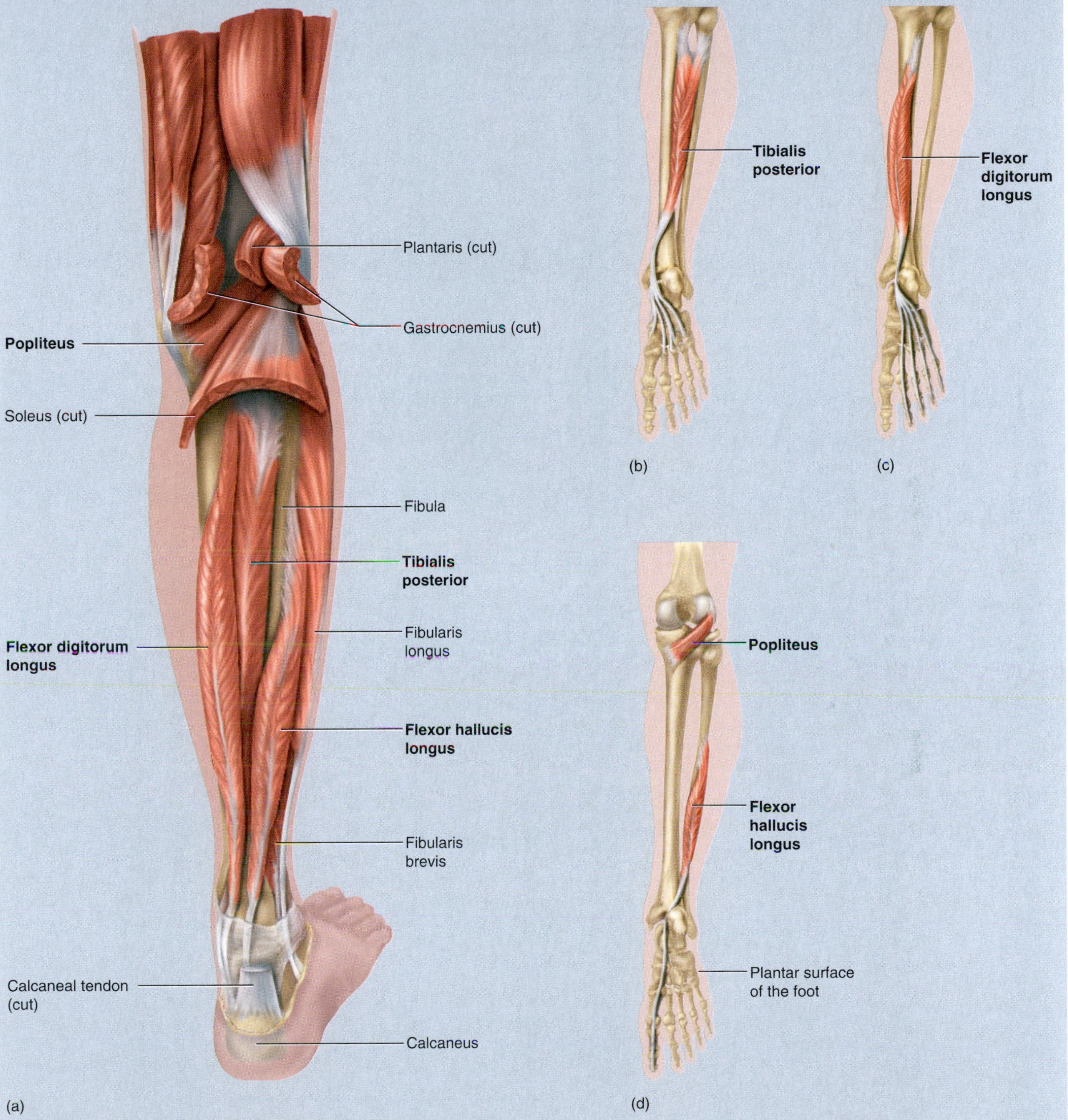

Plantaris (cut)
Gastrocnemius (cut)
Popliteus
Soleus (cut)
Fibula
Tibialis posterior
Fibularis longus
Flexor digitorum longus
Flexor hallucis longus
Fibularis brevis
Calcaneal tendon (cut)
Calcaneus
(a)
Tibialis posterior
(b)
Flexor digitorum longus
(c)
Popliteus
Flexor hallucis longus
Plantar surface of the foot
(d)

FIGURE 12.17

Deep Muscles of the Leg, Posterior and Lateral Compartments. (*a*) Muscles deep to the soleus. (*b–d*) Exposure of some individual deep muscles with the foot plantar flexed (sole facing viewer).

(continued)

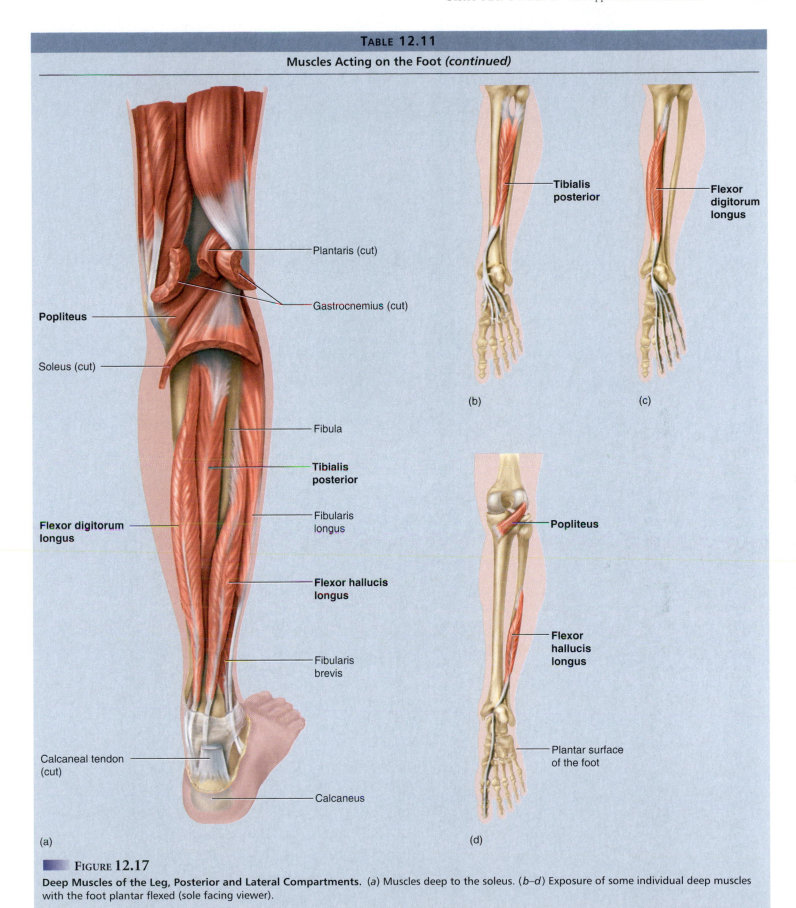

TABLE 12.11

Muscles Acting on the Foot *(continued)*

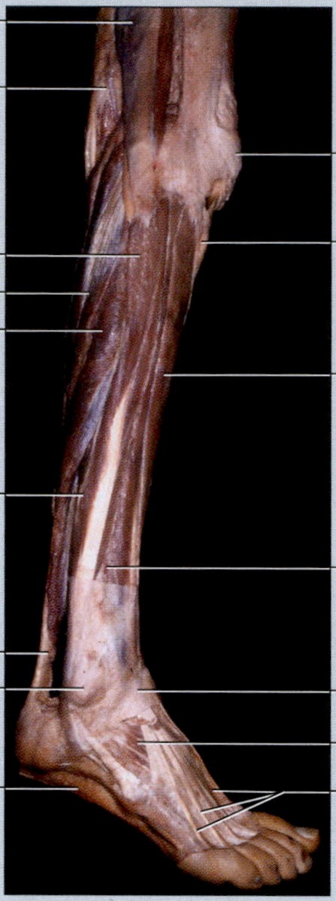

FIGURE 12.18

Superficial Muscles of the Leg of the Cadaver. Right lateral view.

Plantaris[14] (plan-TERR-is)

Flexes knee; plantar flexes foot.

O: lateral supracondylar line of femur	I: calcaneus	N: tibial n.

Posterior Compartment of the Leg—Deep Group. There are four muscles in the deep group (fig. 12.17). The *flexor digitorum longus, flexor hallucis longus,* and *tibialis posterior* are plantar flexors. The fourth muscle, the *popliteus,* is described in table 12.10 because it acts on the knee rather than on the foot.

Flexor Digitorum Longus

Flexes toes II–V; plantar flexes and inverts foot

O: midshaft of tibia	I: distal phalanges II–V	N: tibial n.

Flexor Hallucis Longus

Flexes hallux (great toe); plantar flexes and inverts foot

O: shaft of fibula	I: distal phalanx I	N: tibial n.

Tibialis Posterior

Plantar flexes and inverts foot

O: proximal half of tibia, fibula, interosseous membrane	I: navicular, cuneiforms, metatarsals II–IV	N: tibial n.

Lateral (fibular) Compartment of the Leg. The lateral compartment includes the *fibularis (peroneus) brevis* and *fibularis (peroneus) longus* (figs. 12.15*a*, 12.17, and 12.18). They plantar flex and evert the foot. Plantar flexion is important not only in standing on tiptoes but in providing lift and forward thrust each time you take a step.

Fibularis Brevis

Plantar flexes and everts foot

O: shaft of fibula	I: base of metatarsal V	N: superficial fibular n.

Fibularis Longus

Plantar flexes and everts foot

O: proximal half of fibula, lateral condyle of tibia	I: medial cuneiform, metatarsal I	N: superficial fibular n.

[14]*planta* = sole of foot

Muscles of the leg are tightly bound by deep fasciae which compress them and aid in the return of blood from the legs. The fasciae separate the crural muscles into the anterior, lateral, and posterior compartments noted in table 12.11, each compartment with its own nerve and blood supply (fig. 12.19). The tight binding of muscles furnished by these fasciae can cause a serious problem called *compartment syndrome* when one of the muscles is injured (see insight 12.4).

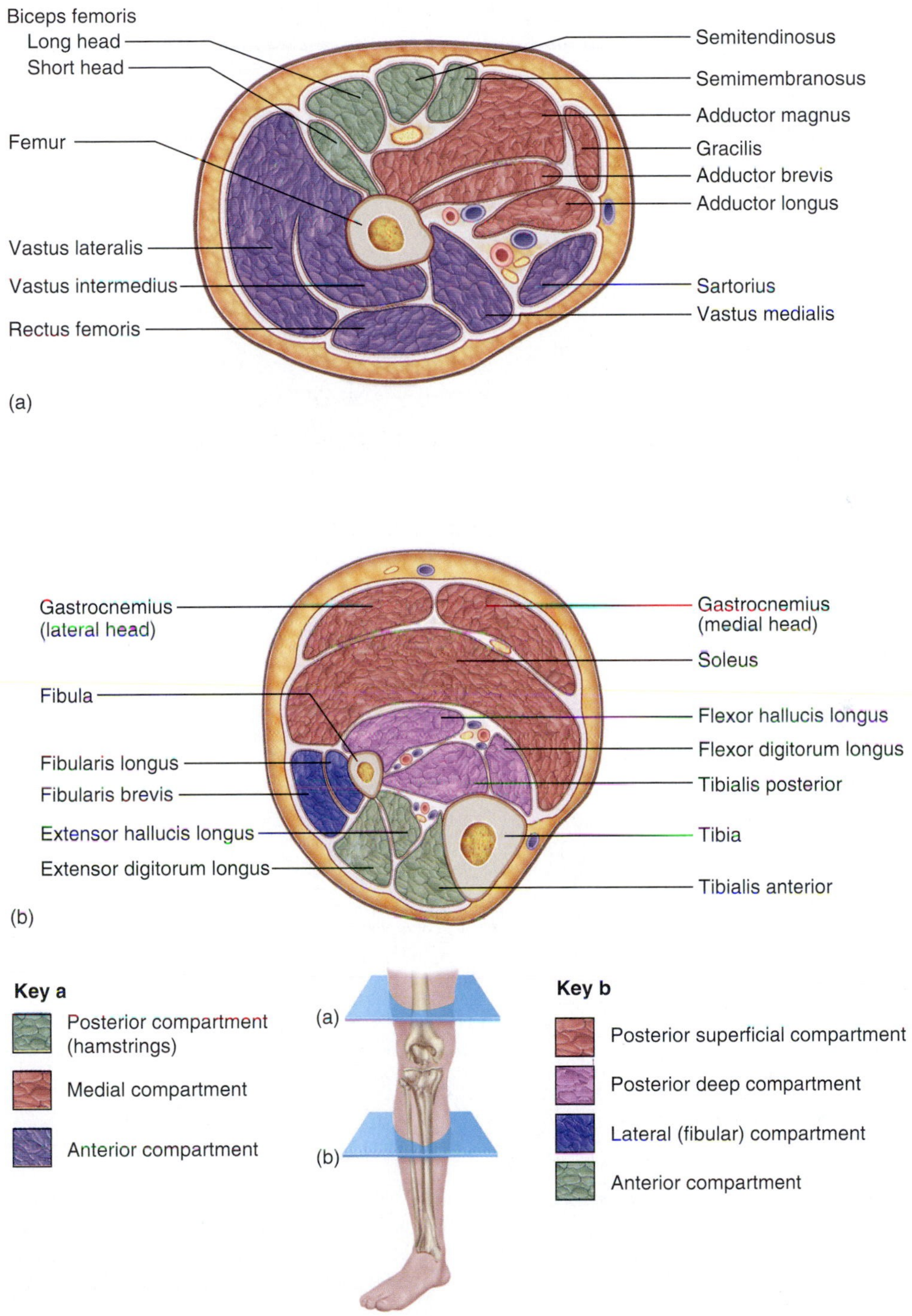

FIGURE 12.19

Serial Cross Sections Through the Lower Limb. Each section is taken at the correspondingly lettered level in the figure at the *bottom* and is pictured with the posterior muscle compartment facing the *top* of the page.

INSIGHT 12.4 CLINICAL APPLICATION

Compartment Syndrome

The fasciae of the upper and lower limb enclose the muscle compartments very snugly. If a blood vessel in a compartment is damaged by overuse or contusion, blood and tissue fluid accumulate in the compartment. The fasciae prevent the compartment from expanding to relieve the pressure. Mounting pressure on the muscles, nerves, and blood vessels triggers a sequence of degenerative events called *compartment syndrome.* Blood flow to the compartment is obstructed by pressure on its arteries. If *ischemia* (poor blood flow) persists for more than 2 to 4 hours, nerves begin to die. After 6 hours, muscle also dies. Nerves can regenerate after the pressure is relieved, but muscle necrosis is irreversible. The breakdown of muscle releases myoglobin into the blood. *Myoglobinuria,* the presence of myoglobin in the urine, gives the urine a dark color and is one of the key signs of compartment syndrome and some other degenerative muscle disorders. Compartment syndrome is treated by immobilizing and resting the limb and, if necessary, making an incision *(fasciotomy)* to relieve the compartment pressure.

Table 12.12

Intrinsic Muscles of the Foot (fig. 12.20)

O = origin, I = insertion, N = innervation (n. = nerve, nn. = nerves)

The intrinsic muscles of the foot support the arches and act on the toes in ways that aid locomotion. Several of them are similar in name and location to the intrinsic muscles of the hand. Only one of these muscles, the *extensor digitorum brevis,* is on the dorsal side of the foot.

Dorsal Aspect of the Foot

Extensor Digitorum Brevis[15] (DIDJ-ih-TOE-rum)

Extends toes

O: dorsal aspect of calcaneus	I: tendons of extensor digitorum longus	N: deep fibular n.

The other intrinsic muscles of the foot are ventral or lie between the metatarsals. They are grouped in four layers. Several of the muscles in the first three layers originate on a broad **plantar aponeurosis,** a fibrous sheet between the plantar skin and muscles. It diverges like a fan from the calcaneus to the bases of all five toes.

Ventral Layer 1 (most superficial). The most superficial layer of intrinsic muscles includes the stout *flexor digitorum brevis* medially, with four tendons that supply all the digits except the hallux. It is flanked by the *abductor digiti minimi* laterally and *abductor hallucis* medially; the tendons of these two muscles serve the little toe and great toe, respectively (fig. 12.20*a*).

Flexor Digitorum Brevis

Flexes toes II–V

O: calcaneus, plantar aponeurosis	I: middle phalanges II–V	N: medial plantar n.

Abductor Digiti Minimi[16]

Abducts and flexes little toe; supports lateral longitudinal arch

O: calcaneus, plantar aponeurosis	I: proximal phalanx V	N: lateral plantar n.

Abductor Hallucis[17] (hal-OO-sis)

Flexes hallux (great toe); supports medial longitudinal arch

O: calcaneus, plantar aponeurosis	I: proximal phalanx I	N: medial plantar n.

Ventral Layer 2. The second layer, deep to the first, consists of the thick medial *quadratus plantae,* which joins the tendons of the flexor digitorum longus, and the four *lumbrical muscles* located between the metatarsals (fig. 12.20*b*).

Quadratus Plantae (quad-RAY-tus PLAN-tee)

Flexes toes

O: calcaneus, plantar aponeurosis	I: tendons of flexor digitorum longus	N: lateral plantar n.

Lumbricals (four muscles)

Flex metatarsophalangeal joints; extend interphalangeal joints

O: tendons of flexor digitorum longus	I: extensor tendons to digits II–V	N: lateral and medial plantar nn.

Ventral Layer 3. The third layer includes the *adductor hallucis, flexor digiti minimi brevis,* and *flexor hallucis brevis* (fig. 12.20*c*). The adductor hallucis has an oblique head that crosses the foot and inserts at the base of the great toe, and a transverse head that passes across the bases of digits II–V and meets the long head at the base of the hallux.

(continued)

[15]"short extensor of the digits"
[16]"abductor of the little toe"
[17]"abductor of the hallux (great toe)"

TABLE 12.12

Intrinsic Muscles of the Foot *(continued)*

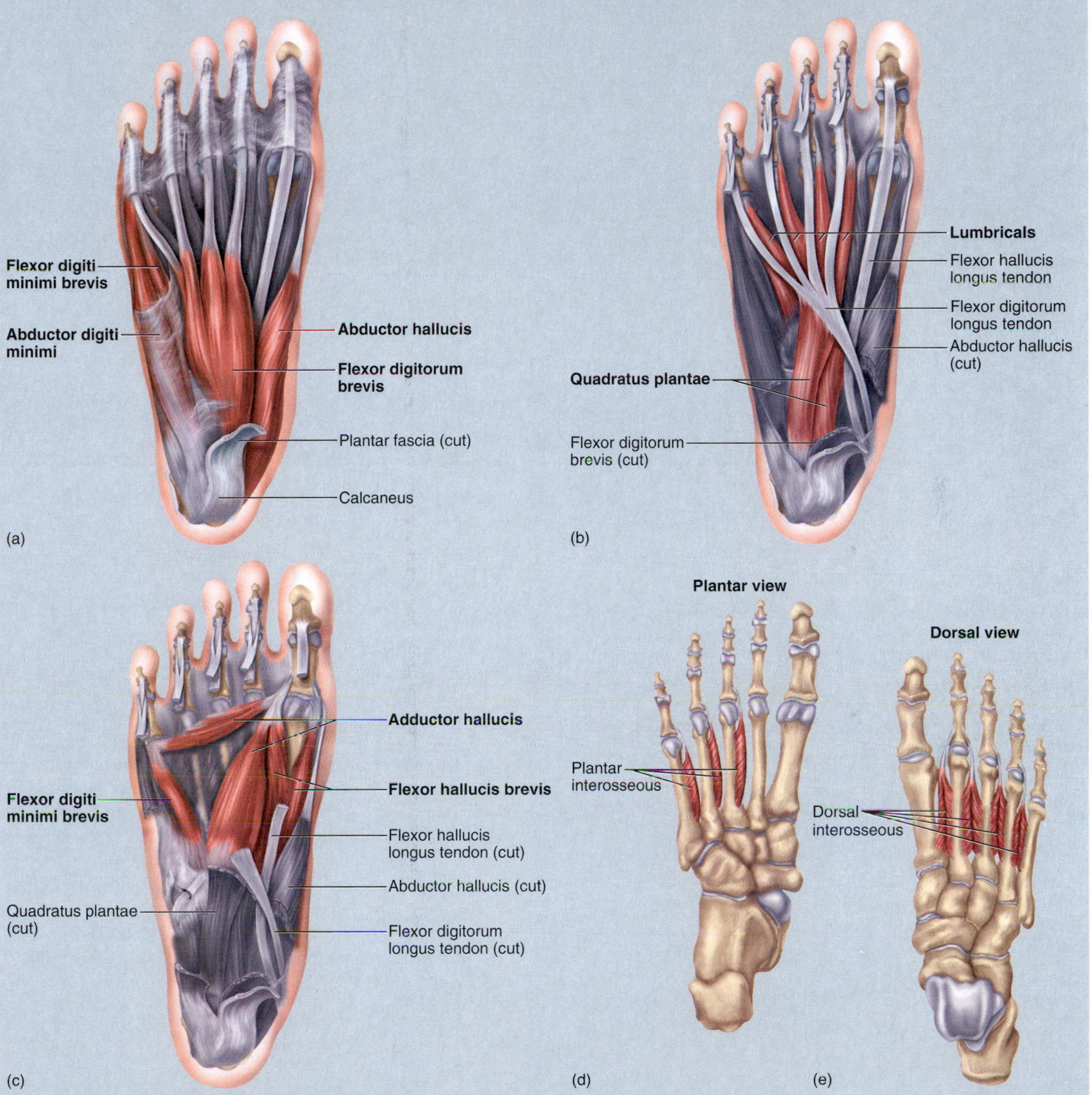

FIGURE 12.20

Intrinsic Muscles of the Foot. (*a–d*) First through fourth layers, respectively, in ventral (plantar) views. (*e*) Fourth layer, dorsal view.

(continued)

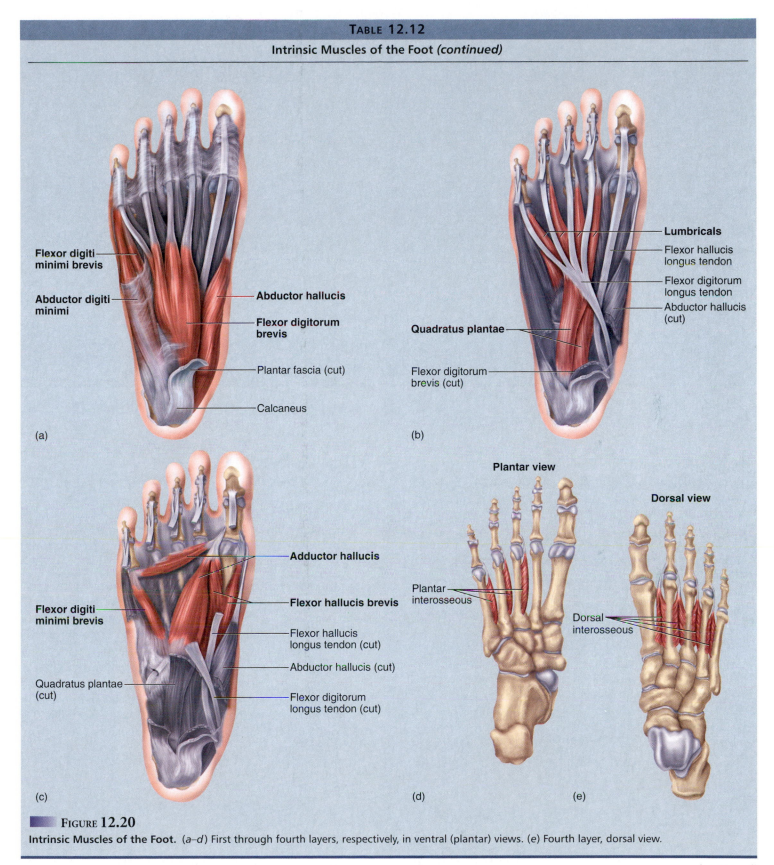

Table 12.12

Intrinsic Muscles of the Foot *(continued)*

Adductor Hallucis		
Adducts hallux		
O: metatarsals II–IV	I: proximal phalanx I	N: lateral plantar n.
Flexor Digiti Minimi Brevis		
Flexes little toe		
O: metatarsal V, plantar aponeurosis	I: proximal phalanx V	N: lateral plantar n.
Flexor Hallucis Brevis		
Flexes hallux		
O: cuboid, lateral cuneiform	I: proximal phalanx I	N: medial plantar n.
Ventral Layer 4 (deepest). The deepest layer consists of four *dorsal interosseous muscles* and three *plantar interosseous muscles* located between the metatarsals. Each dorsal interosseous muscle is bipennate and originates on two adjacent metatarsals. The plantar interosseous muscles are unipennate and originate on only one metatarsal each (fig. 12.20*d*, e).		
Dorsal Interosseous Muscles (four muscles)		
Abduct toes II–IV		
O: each with two heads arising from adjacent metatarsals	I: proximal phalanges II–IV	N: lateral plantar n.
Plantar Interosseous Muscles (three muscles)		
Adduct toes III–V		
O: medial aspect of metatarsals III–V	I: proximal phalanges III–V	N: lateral plantar n.

Think About It!

Not everyone has the same muscles. From the information provided in this chapter, identify two muscles that are lacking in some people.

Before You Go On

Answer the following questions to test your understanding of the preceding section:

7. In the middle of a stride, you have one foot on the ground and you are about to swing the other leg forward. What muscles produce the movements of that leg?
8. Name the muscles that cross both the hip and knee joints and produce actions at both.
9. List the major actions of the muscles of the anterior, medial, and posterior compartments of the thigh.
10. Describe the role of plantar flexion and dorsiflexion in walking. What muscles produce these actions?

Muscle Injuries

Objectives

When you have completed this section, you should be able to

- explain how to reduce the risk of muscle injuries; and
- define several types of muscle injuries often incurred in sports and recreation.

Although the muscular system suffers fewer diseases than most organ systems, it is particularly vulnerable to injuries resulting from sudden and intense stress placed on muscles and tendons. Each year, thousands of athletes from high school to professional level sustain some type of muscle injury, as do increasing numbers of people who have taken up running and other forms of physical conditioning. Overzealous exertion without proper preparation and warm-up is frequently the cause. Some of the most common athletic injuries are briefly described in table 12.13. (See table 10.5, p. 277, for more general disorders of the muscular system).

TABLE 12.13

Muscle Injuries

Baseball Finger	Tears in the extensor tendons of the fingers resulting from the impact of a baseball with the extended fingertip
Blocker's Arm	Ectopic ossification in the lateral margin of the forearm as a result of repeated impact with opposing players
Pitcher's Arm	Inflammation at the origin of the flexor carpi resulting from hard wrist flexion in releasing a baseball
Pulled Groin	Strain in the adductor muscles of the thigh; common in gymnasts and dancers who perform splits and high kicks
Pulled Hamstrings	Strained hamstring muscles or a partial tear in the tendinous origin, often with a hematoma (blood clot) in the fascia lata; frequently caused by repetitive kicking (as in football and soccer) or long, hard running
Rider's Bones	Ectopic ossification in the tendons of the thigh adductors; results from prolonged abduction of the thighs when riding horses
Shinsplints	General term for several kinds of injury with pain in the crural region—tendinitis of the tibialis posterior, inflammation of the tibial periosteum, and anterior compartment syndrome; May result from unaccustomed jogging, walk-a-thons, walking on snowshoes, or any vigorous activity of the legs after a period of inactivity
Tennis Elbow	Inflammation at the origin of the extensor carpi muscles on the lateral epicondyle of the humerus occurs when these muscles are repeatedly tensed during backhand strokes and then strained by sudden impact with the tennis ball. Any activity that requires rotary movements of the forearm and a firm grip of the hand (for example, using a screwdriver) can cause the symptoms of tennis elbow.
Tennis Leg	Partial tear in the lateral origin of the gastrocnemius; results from repeated strains put on the muscle while supporting the body weight on the toes

Disorders Described Elsewhere

Back injuries 306
Carpal tunnel syndrome 326
Compartment syndrome 342
Hamstring injuries 336
Hernias 306
Rotator cuff injury 319

Most athletic injuries can be prevented by proper conditioning. A person who suddenly takes up vigorous exercise may not have sufficient muscle and bone mass to withstand the stresses such exercise entails. These must be developed gradually. Stretching exercises keep ligaments and joint capsules supple and therefore reduce injuries. Warm-up exercises promote more efficient and less injurious musculoskeletal function in several ways. Most of all, moderation is important, as most injuries simply result from overuse of the muscles. "No pain, no gain" is a dangerous misconception.

Muscular injuries can be treated initially with "RICE": **r**est, **i**ce, **c**ompression, and **e**levation. Rest prevents further injury and allows repair processes to occur; ice reduces swelling; compression with an elastic bandage helps to prevent fluid accumulation and swelling; and elevation of an injured limb promotes drainage of blood from the affected area and limits further swelling. If these measures are not enough, anti-inflammatory drugs such as hydrocortisone and aspirin may be employed.

Before You Go On

Answer the following questions to test your understanding of the preceding section:

11. Explain why stretching exercises reduce the incidence of muscle injuries.
12. Explain the reason for each of the four treatments in the RICE approach to muscle injuries.

CHAPTER REVIEW

Review of Key Concepts

Muscles Acting on the Shoulder and Upper Limb (p. 314)

1. Muscles that act on the pectoral girdle originate on the axial skeleton and insert on the clavicle and scapula. They fall into an anterior group that includes the *pectoralis minor* and *serratus anterior,* and a posterior group that includes the superficial *trapezius* and three deeper muscles, the *levator scapulae, rhomboideus major,* and *rhomboideus minor* (table 12.1).
2. Muscles that act on the arm (humerus) originate mainly on the pectoral girdle and axial skeleton and cross the shoulder joint. These include the *pectoralis major, latissimus dorsi, deltoid, teres major, coracobrachialis,* and four *rotator cuff (SITS)* muscles: the *supraspinatus, infraspinatus, teres minor,* and *subscapularis* (table 12.2).
3. Muscles that act on the forearm and elbow are located in both the arm (the *biceps brachii, brachialis,* and *triceps brachii*) and the forearm (the *anconeus, brachioradialis, pronator teres, pronator quadratus,* and *supinator*) (table 12.4).
4. Most muscles whose origins and bellies are in the forearm have insertions in, and act upon, the wrist and hand (table 12.6). Several of them, however, have origins on the humerus and also cross the elbow joint. Therefore, they also contribute slightly to actions of the elbow. Deep fasciae divide the forearm muscles into *anterior* and *posterior compartments* and separate those of each compartment into superficial and deep layers.
5. Anterior compartment muscles are mainly flexors of the wrist and hand. Superficial muscles of the anterior compartment include the *palmaris longus* and *flexor carpi radialis,* which form the two most prominent tendons of the anterior wrist, and the *flexor carpi ulnaris* and *flexor digitorum superficialis.* The deep muscles of the anterior compartment include the *flexor pollicis longus* and *flexor digitorum profundus.*
6. Posterior compartment muscles are mainly extensors of the wrist and hand. Superficial muscles of the posterior compartment include the *extensor carpi radialis longus, extensor carpi radialis brevis, extensor carpi ulnaris, extensor digitorum,* and *extensor digiti minimi.* Deep muscles of the posterior compartment include the *extensor indicis, abductor pollicis longus, extensor pollicis longus,* and *extensor pollicis brevis.*
7. Most tendons of the forearm muscles pass under a *flexor retinaculum* on the anterior side of the wrist or an *extensor retinaculum* on the posterior side of the wrist. The space between the flexor retinaculum and carpal bones is called the *carpal tunnel.*
8. Intrinsic muscles of the hand assist the forearm muscles and make movements of the digits more precise. They are divided into a thenar, hypothenar, and midpalmar group (table 12.7).
9. The *thenar group* muscles form the thick fleshy mass at the base of the thumb and the web between the thumb and palm. They move the thumb. They include the *abductor pollicis brevis, adductor pollicis, flexor pollicis brevis,* and *opponens pollicis.*
10. The *hypothenar group* muscles form the fleshy *hypothenar eminence* at the base of the little finger, and are concerned with movements of that digit. They include the *abductor digiti minimi, flexor digiti minimi brevis,* and *opponens digiti minimi.*
11. The *midpalmar group* of muscles span the palm and include four *dorsal interosseous muscles,* three *palmar interosseous muscles,* and four *lumbrical muscles,* located between the metacarpal bones. They act on digits II through V.

Muscles Acting on the Hip and Lower Limb (p. 329)

1. Most muscles that act on the femur originate on the os coxae (table 12.9). The two major anterior muscles of this group are the *iliacus* and the *psoas major,* collectively called the *iliopsoas.*
2. Superficial muscles on the lateral and posterior sides of the hip include the *tensor fasciae latae, gluteus maximus, gluteus medius,* and *gluteus minimus.* The tendons of the first two of these muscles join the *fascia lata* to form the fibrous *iliotibial band* on the lateral aspect of the thigh.
3. Deep muscles on the lateral aspect of the hip, known as the *lateral rotators,* include the *gemellus superior, gemellus inferior, obturator externus, obturator internus, piriformis,* and *quadratus femoris.* The principal actions of these muscles are abduction and lateral rotation of the femur.
4. Deep fasciae divide the other thigh muscles into an *anterior (extensor) compartment, medial (adductor) compartment,* and *posterior (flexor) compartment.*
5. Muscles of the anterior compartment act mainly as extensors of the knee. These include the *sartorius* and the four heads of the *quadriceps femoris: rectus femoris, vastus lateralis, vastus medialis,* and *vastus intermedius.*
6. Muscles of the medial compartment act as adductors of the femur. These include the *adductor longus, adductor brevis, adductor magnus, gracilis,* and *pectineus* (table 12.9).
7. Muscles of the posterior compartment act as extensors of the hip and flexors of the knee. These are the *biceps femoris, semimembranosus,* and *semitendinosus,* known colloquially as the *hamstring muscles.*
8. Muscles of the leg are divided into anterior, posterior, and lateral compartments (table 12.11). Most of them act on the foot.
9. Anterior compartment muscles of the leg include the *extensor digitorum longus, extensor hallucis longus, fibularis tertius,* and *tibialis anterior.*
10. Superficial posterior compartment muscles include the *popliteus,* which acts on the knee; two muscles, the *gastrocnemius* and *soleus,* collectively also known as the *triceps surae* (these share the calcaneal tendon to the heel); and the *plantaris.*
11. Deep posterior compartment muscles include the *flexor digitorum longus, flexor hallucis longus,* and *tibialis posterior.*
12. Lateral compartment muscles include the *fibularis brevis* and *fibularis longus.*
13. Intrinsic muscles of the foot support the arches and act on the toes, and resemble intrinsic muscles of the hand. The *extensor digitorum brevis* is located dorsally. The others are ventral and are arranged in layers.

14. The layer 1 (most superficial) intrinsic muscles of the foot are the *flexor digitorum brevis, abductor digiti minimi,* and *abductor hallucis;* layer 2 comprises the *quadratus plantae* and four *lumbrical muscles;* layer 3 includes the *adductor hallucis, flexor digiti minimi brevis,* and *flexor hallucis brevis;* and layer 4 (deepest) includes four *dorsal interosseous muscles* and three *plantar interosseous muscles.*

Muscle Injuries (p. 344)

1. Athletic and recreational injuries to the appendicular muscles are especially common, and often result from overly zealous or vigorous exercise without proper conditioning or warm-up. Table 12.13 defines many well-known injuries of the appendicular muscles.

Testing Your Recall

1. Which of the following muscles could you most easily do without?
   a. flexor digitorum profundus
   b. trapezius
   c. palmaris longus
   d. triceps brachii
   e. tibialis anterior
2. Which of the following has the least in common with the other four?
   a. vastus intermedius
   b. vastus lateralis
   c. vastus medialis
   d. rectus femoris
   e. biceps femoris
3. The triceps surae is a muscle group composed of
   a. the flexor hallucis longus and brevis.
   b. the gastrocnemius and soleus.
   c. lateral, medial, and long heads.
   d. the biceps brachii and triceps brachii.
   e. the vastus lateralis, medialis, and intermedius.
4. The interosseous muscles lie between
   a. the ribs.
   b. the tibia and fibula.
   c. the radius and ulna.
   d. the metacarpal bones.
   e. the phalanges.
5. Which of these muscles does *not* contribute to the rotator cuff?
   a. the supraspinatus
   b. the infraspinatus
   c. the subscapularis
   d. the teres major
   e. the teres minor
6. Which of these actions is *not* performed by the trapezius?
   a. extension of the neck
   b. depression of the scapula
   c. elevation of the scapula
   d. rotation of the scapula
   e. adduction of the humerus
7. Both the hands and feet are acted upon by a muscle or muscles called
   a. the extensor digitorum.
   b. the abductor digiti minimi.
   c. the flexor digitorum profundus.
   d. the abductor hallucis.
   e. the flexor digitorum longus.
8. Which of the following muscles does *not* extend the hip joint?
   a. quadriceps femoris
   b. gluteus maximus
   c. biceps femoris
   d. semitendinosus
   e. semimembranosus
9. Both the gastrocnemius and _____ muscles insert on the heel by way of the calcaneal tendon.
   a. semimembranosus
   b. tibialis posterior
   c. tibialis anterior
   d. soleus
   e. plantaris
10. Which of these is *not* in the anterior compartment of the thigh?
    a. semimembranosus
    b. rectus femoris
    c. vastus intermedius
    d. vastus lateralis
    e. sartorius
11. The major superficial muscle of the shoulder, where injections are often given, is the _____.
12. If a muscle has the word *hallucis* in its name, it must cause movement of the _____.
13. Pronation of the forearm is achieved by two muscles, the pronator _____ just distal to the elbow and the pronator _____ near the wrist.
14. The three large muscles on the posterior side of the thigh are collectively known by the colloquial name of _____ muscles.
15. Connective tissue bands called _____ prevent flexor tendons from rising like bowstrings.
16. The web between your thumb and palm consists mainly of the _____ muscle.
17. The patella is embedded in the tendon of the _____ muscle.
18. The _____ muscle, named for its origin and insertion, originates on the coracoid process of the scapula, inserts on the humerus, and adducts the arm.
19. The most medial adductor muscle of the thigh is the long, slender _____.
20. Like the tendinous intersections of the rectus abdominis (chapter 11), a transverse tendinous band also subdivides the _____ muscle of one of the limbs.

Answers in the Appendix

True or False

Determine which five of the following statements are false, and briefly explain why.

1. All plantar flexors in the posterior compartment of the calf insert on the heel by way of the calcaneal tendon.
2. The trapezius can act as both a synergist and antagonist of the levator scapulae.
3. To push someone away from you, you would use the serratus anterior muscle more than the trapezius.
4. Both the extensor digitorum and extensor digiti minimi extend the little finger.
5. The interosseous muscles are fusiform.
6. The actions of the palmaris longus and plantaris muscles are weak and relatively dispensable.
7. The psoas major is an antagonist of the rectus femoris.
8. Rapid flexion of the knee often causes hamstring injuries.
9. Curling your toes employs the quadratus plantae muscle.
10. The tibialis posterior and tibialis anterior are synergists.

Answers in the Appendix

Testing Your Comprehension

1. Radical mastectomy, once a common treatment for breast cancer, involved removal of the pectoralis major along with the breast. What functional impairments would result from this? What synergists could a physical therapist train a patient to use to recover some lost function?
2. Table 12.6 describes a simple test for determining whether you have a palmaris longus muscle. Why do you think the other major tendon of the anterior wrist, the flexor carpi radialis tendon, does not stand out conspicuously in such a test?
3. Poorly conditioned, middle-aged people may suffer a rupture of the calcaneal tendon when the foot is suddenly dorsiflexed. Explain each the following signs of a ruptured calcaneal tendon: (*a*) a prominent lump typically appears in the calf; (*b*) the foot can be dorsiflexed farther than usual; and (*c*) the patient cannot plantar flex the foot very effectively.
4. Women who habitually wear high heels may suffer painful "high heel syndrome" when they go barefoot or wear flat shoes. What muscle(s) and tendon(s) are involved? Explain.
5. A student moving out of a dormitory kneels down, in correct fashion, to lift a heavy box of books. What prime movers are involved as he straightens his legs to lift the box?

Answers at the Online Learning Center

www.mhhe.com/saladinha1

Visit the Online Learning Center for practice tests, answer keys, and other learning aids for this chapter. Enhance your understanding of human anatomy with our interactive art labeling exercises, supplemental photo atlases, web links, puzzles, flashcards, and much more.

B

ATLAS B

Surface Anatomy

The Importance of External Anatomy *350*

The Head and Neck *351*

The Trunk *352*
- The Thorax and Abdomen 352
- The Back and Gluteal Region 353
- The The Pelvic Region 354
- The Axillary Region 355

The Upper Limb *356*
- The Lateral Aspect 356
- The Antebrachium (forearm) 356
- The Wrist and Hand 357

The Lower Limb *358*
- The Thigh and Knee 358
- The Leg and Foot 359
- Foot 362

Muscle Self-Test *364*

The Importance of External Anatomy

In the study of human anatomy, it is easy to become so preoccupied with internal structure that we forget the importance of what we can see and feel externally. Yet external anatomy and appearance are major concerns in giving a physical examination and in many aspects of patient care. A knowledge of the body's surface landmarks is essential to one's competence in physical therapy, cardiopulmonary resuscitation, surgery, making X rays and electrocardiograms, giving injections, drawing blood, listening to heart and respiratory sounds, measuring the pulse and blood pressure, and finding pressure points to stop arterial bleeding, among other procedures. A misguided attempt to perform some of these procedures while disregarding or misunderstanding external anatomy can be very harmful and even fatal to a patient.

Having just studied skeletal and muscular anatomy in the preceding chapters, this is an opportune time for you to study the body surface. Much of what we see there reflects the underlying structure of the superficial bones and muscles. A broad photographic overview of surface anatomy is given in atlas A (fig. A.5). In the following pages, we examine the body literally from head (fig. B.1) to toe (fig. B.14), studying its regions in more detail. To make the most profitable use of this atlas, refer back to the skeletal and muscular anatomy in chapters 7 to 12. Relate drawings of the clavicles in chapter 8 to the photograph in figure B.1, for example. Study the shape of the scapula in chapter 8 and see how much of it you can trace on the photographs in figure B.3. See if you can relate the tendons visible on the hand (fig. B.8) to the muscles of the forearm illustrated in chapter 12.

For learning surface anatomy, there is a resource available to you that is far more valuable than any laboratory model or textbook illustration—your own body. For the best understanding of human structure, compare the art and photographs in this book with your body or with structures visible on a study partner. In addition to bones and muscles, you can palpate a number of superficial arteries, veins, tendons, ligaments, and cartilages, among other structures. By palpating regions such as the shoulder, elbow, or ankle, you can develop a mental image of the subsurface structures better than you can obtain by looking at two-dimensional textbook images. And the more you can study with other people, the more you will appreciate the variations in human structure and be able to apply your knowledge to your future patients or clients, who will not look quite like any textbook diagram or photograph you have ever seen. Through comparisons of art, photography, and the living body, you will get a much deeper understanding of the body than if you were to study this atlas in isolation from the earlier chapters.

At the end of this atlas, you can test your knowledge of externally visible muscle anatomy. The two photographs in figure B.15 have 30 numbered muscles and a list of 26 names, some of which are shown more than once in the photographs and some of which are not shown at all. Identify the muscles to your best ability without looking back at the previous illustrations, and then check your answers in the appendix at the back of the book.

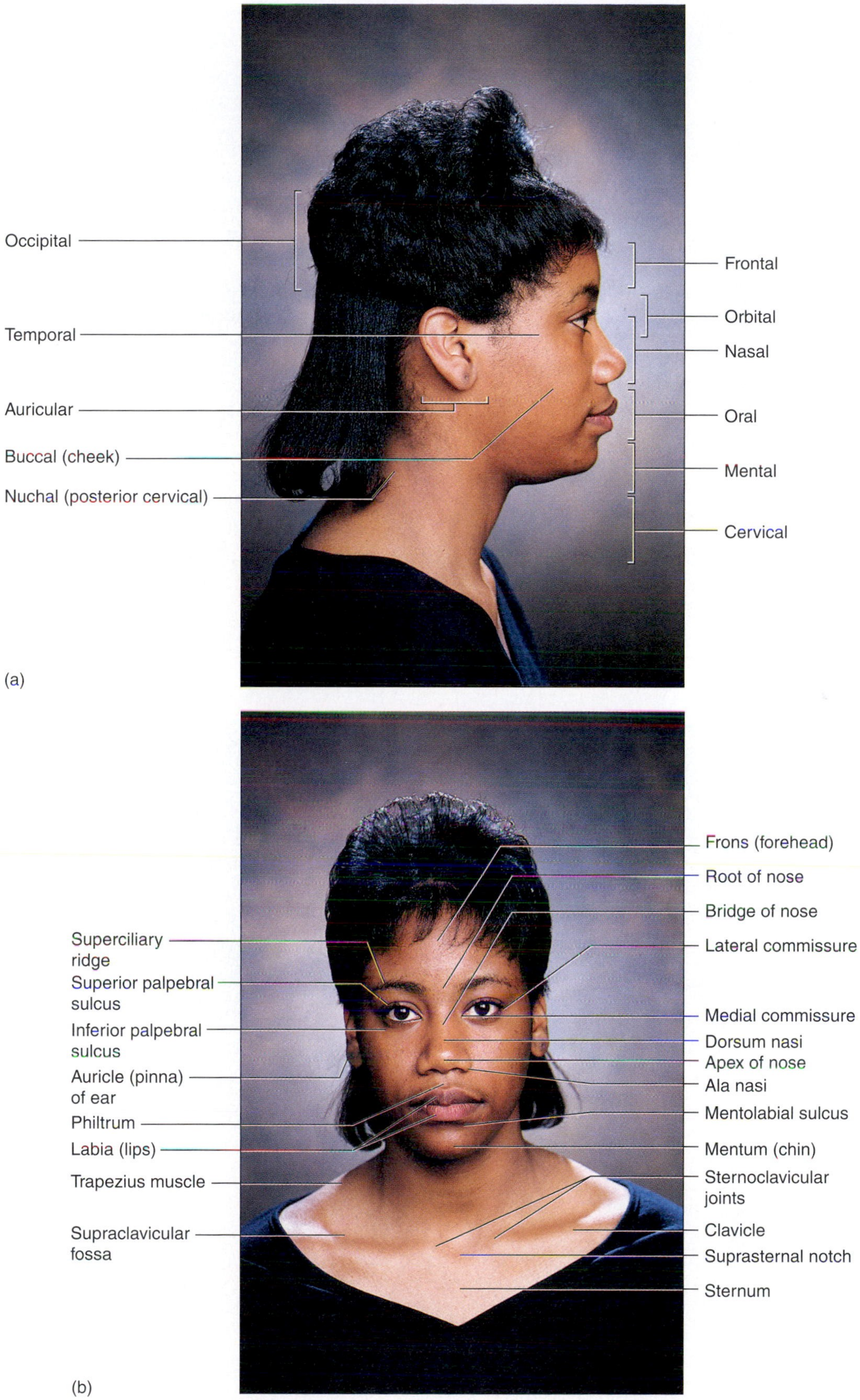
Occipital
Temporal
Auricular
Buccal (cheek)
Nuchal (posterior cervical)
Frontal
Orbital
Nasal
Oral
Mental
Cervical
(a)
Superciliary ridge
Superior palpebral sulcus
Inferior palpebral sulcus
Auricle (pinna) of ear
Philtrum
Labia (lips)
Trapezius muscle
Supraclavicular fossa
Frons (forehead)
Root of nose
Bridge of nose
Lateral commissure
Medial commissure
Dorsum nasi
Apex of nose
Ala nasi
Mentolabial sulcus
Mentum (chin)
Sternoclavicular joints
Clavicle
Suprasternal notch
Sternum
(b)

Figure B.1

The Head and Neck. (*a*) Anatomical regions of the head, lateral aspect. (*b*) Features of the facial region and upper thorax.

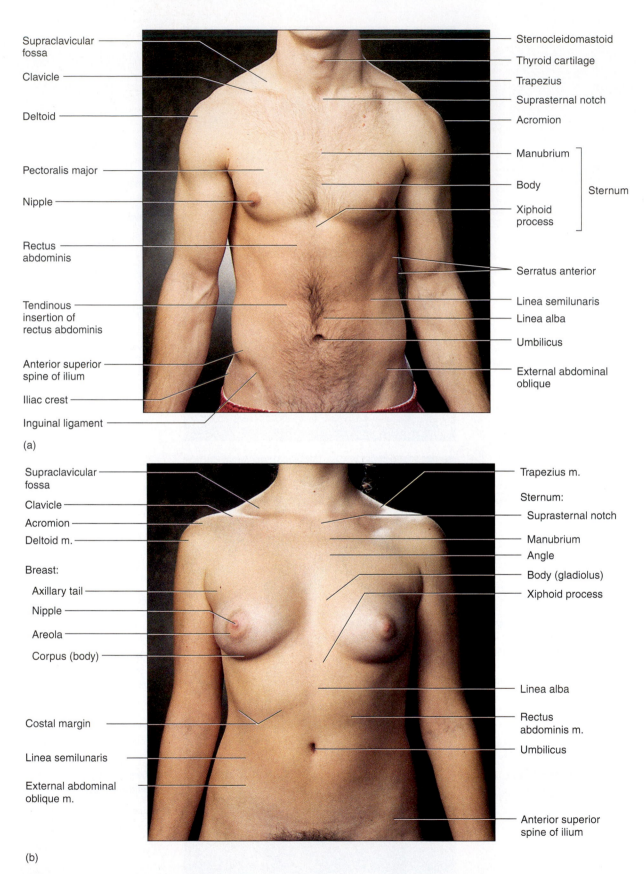

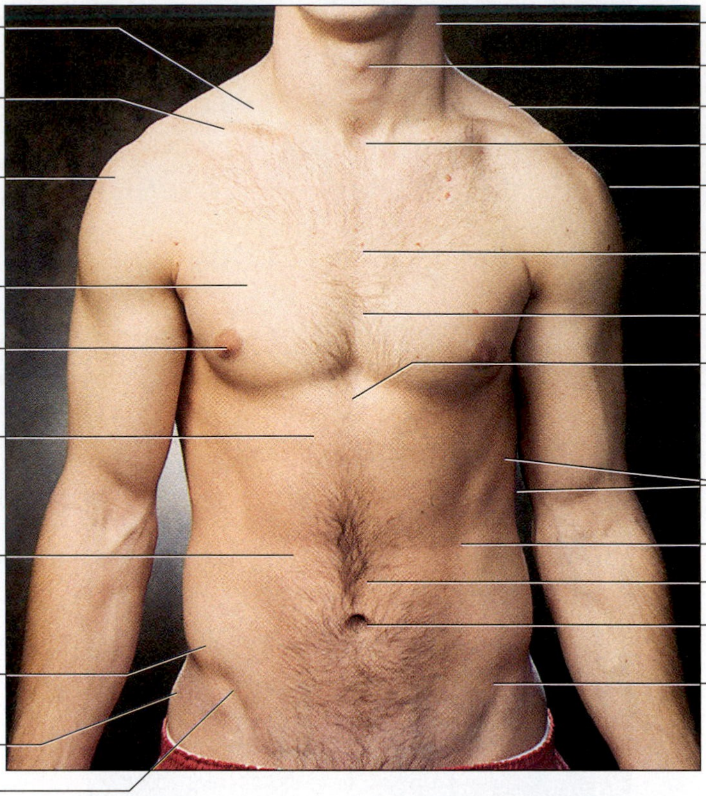
Supraclavicular fossa
Clavicle
Deltoid
Pectoralis major
Nipple
Rectus abdominis
Tendinous insertion of rectus abdominis
Anterior superior spine of ilium
Iliac crest
Inguinal ligament
Sternocleidomastoid
Thyroid cartilage
Trapezius
Suprasternal notch
Acromion
Manubrium
Body
Xiphoid process
Sternum
Serratus anterior
Linea semilunaris
Linea alba
Umbilicus
External abdominal oblique

(a)

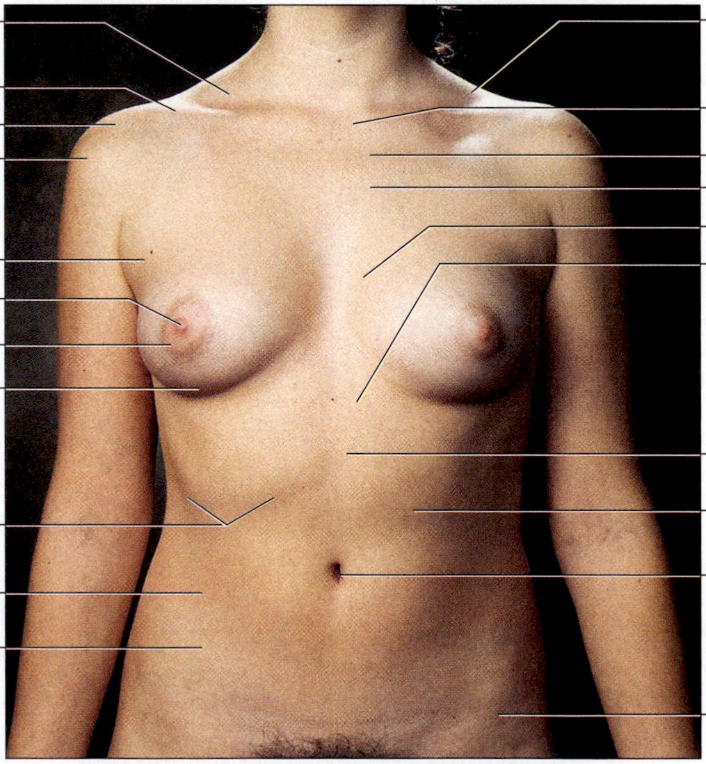
Supraclavicular fossa
Clavicle
Acromion
Deltoid m.
Breast:
Axillary tail
Nipple
Areola
Corpus (body)
Costal margin
Linea semilunaris
External abdominal oblique m.
Trapezius m.
Sternum:
Suprasternal notch
Manubrium
Angle
Body (gladiolus)
Xiphoid process
Linea alba
Rectus abdominis m.
Umbilicus
Anterior superior spine of ilium

(b)

FIGURE B.2

The Thorax and Abdomen, Ventral Aspect. (*a*) Male. (*b*) Female. Except for the breast, all of the features labeled are common to both sexes, though some are labeled only on the photograph that shows them best.

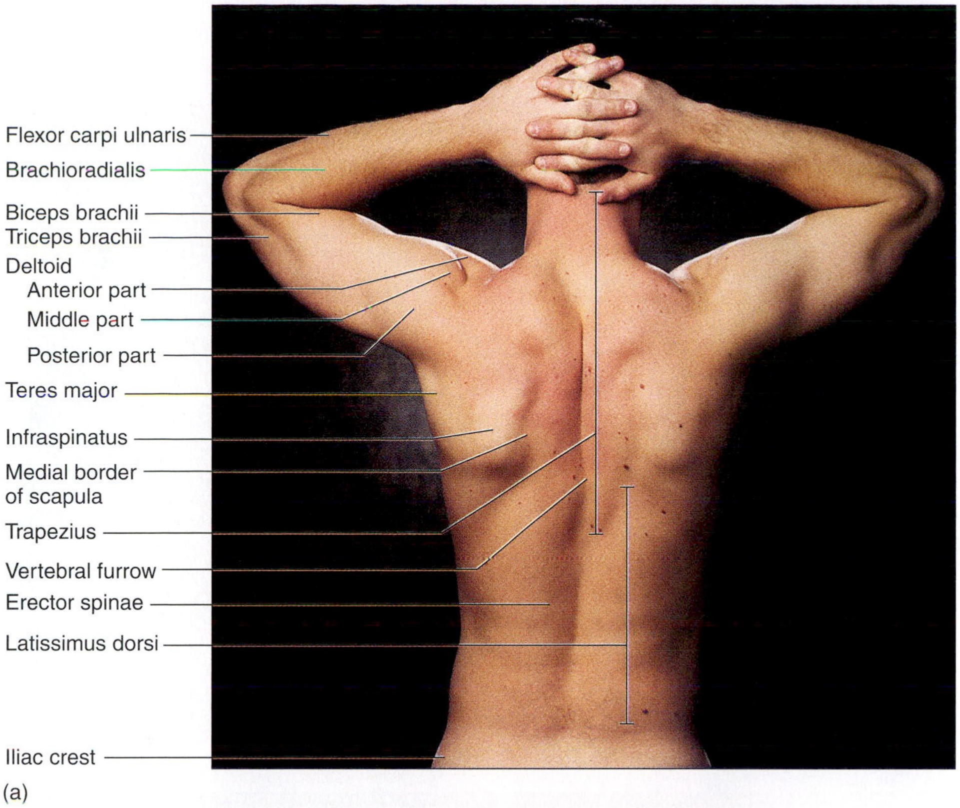
Flexor carpi ulnaris
Brachioradialis
Biceps brachii
Triceps brachii
Deltoid
Anterior part
Middle part
Posterior part
Teres major
Infraspinatus
Medial border of scapula
Trapezius
Vertebral furrow
Erector spinae
Latissimus dorsi
Iliac crest

(a)

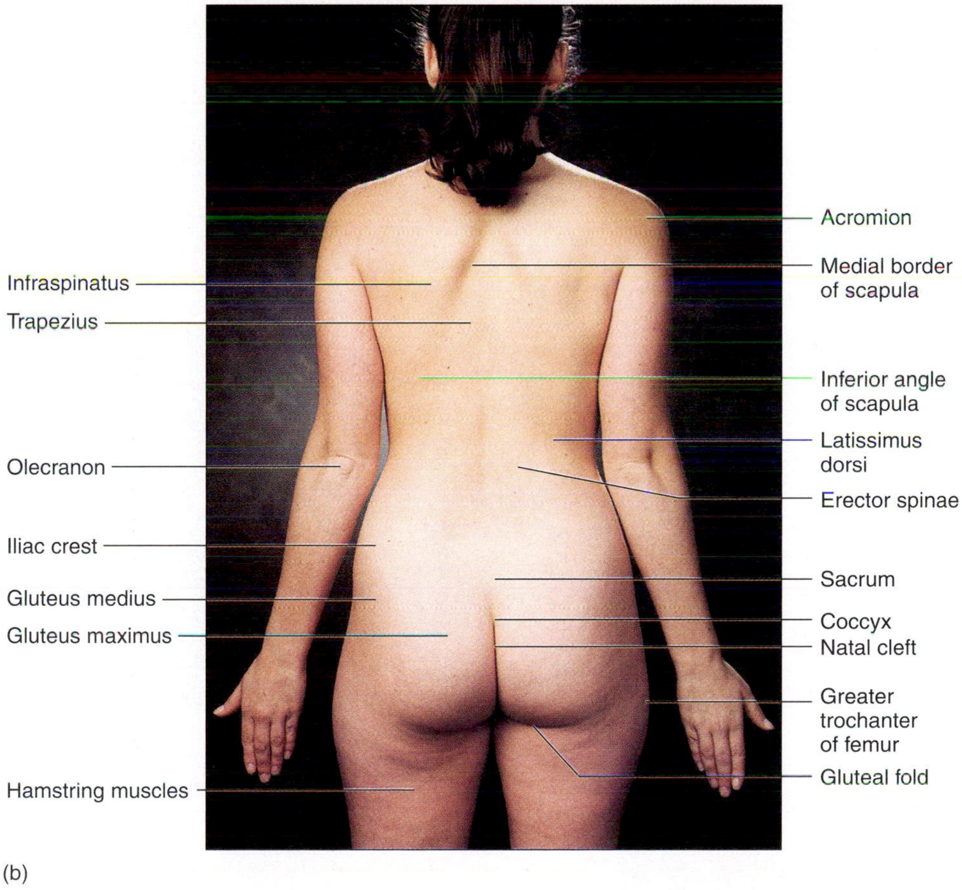
Infraspinatus
Trapezius
Olecranon
Iliac crest
Gluteus medius
Gluteus maximus
Hamstring muscles
Acromion
Medial border of scapula
Inferior angle of scapula
Latissimus dorsi
Erector spinae
Sacrum
Coccyx
Natal cleft
Greater trochanter of femur
Gluteal fold

(b)

FIGURE B.3

The Back and Gluteal Region. (*a*) Male. (*b*) Female. All of the features labeled are common to both sexes, though some are labeled only on the photograph that shows them best.

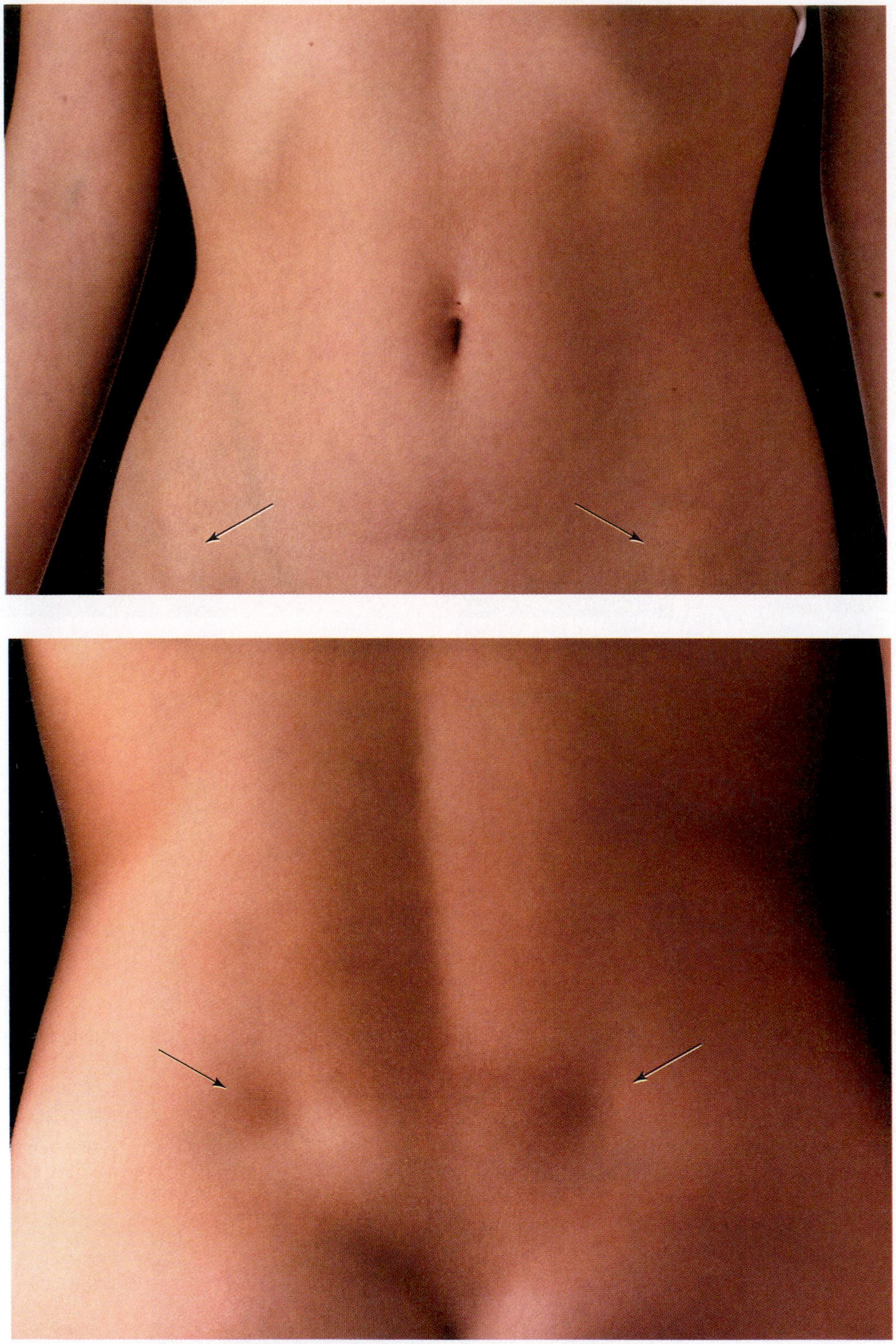

(a)

(b)

FIGURE B.4

The Pelvic Region. (*a*) The anterior superior spines of the ilium are marked by anterolateral protruberances *(arrows)*. (*b*) The posterior superior spines are marked in some people by dimples in the sacral region *(arrows)*.

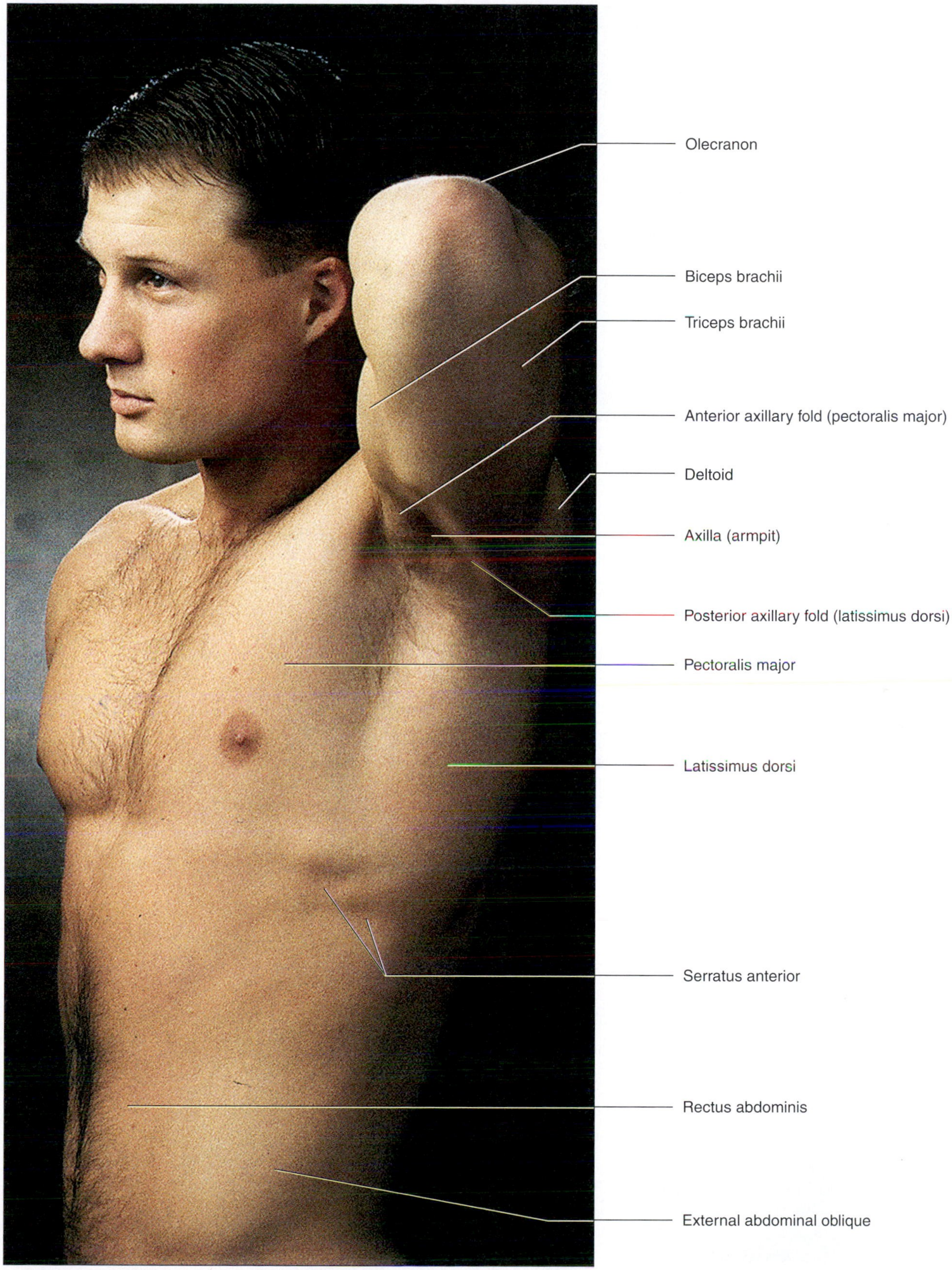

Olecranon
Biceps brachii
Triceps brachii
Anterior axillary fold (pectoralis major)
Deltoid
Axilla (armpit)
Posterior axillary fold (latissimus dorsi)
Pectoralis major
Latissimus dorsi
Serratus anterior
Rectus abdominis
External abdominal oblique

FIGURE B.5
The Axillary Region.

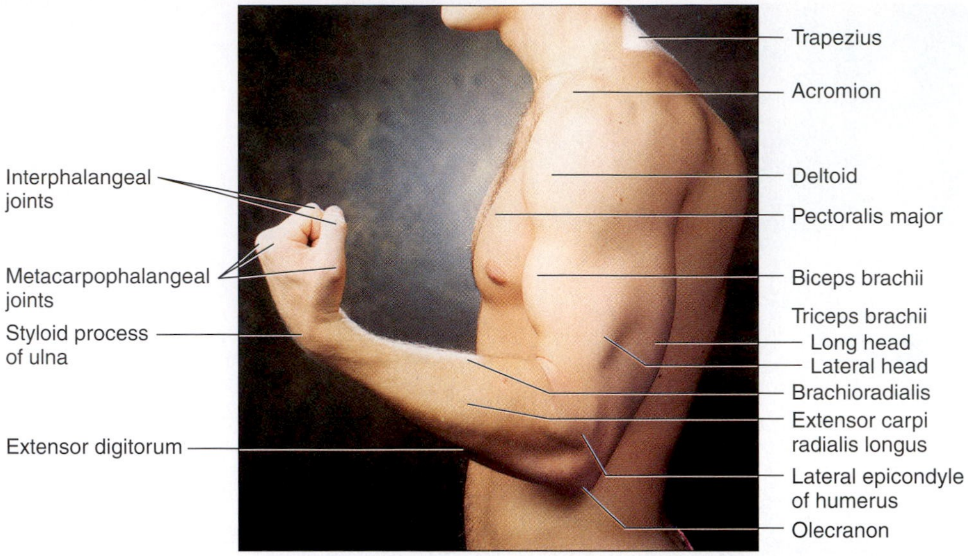
Trapezius
Acromion
Deltoid
Pectoralis major
Biceps brachii
Triceps brachii
Long head
Lateral head
Brachioradialis
Extensor carpi radialis longus
Lateral epicondyle of humerus
Olecranon
Interphalangeal joints
Metacarpophalangeal joints
Styloid process of ulna
Extensor digitorum

Figure B.6

The Upper Limb, Lateral aspect.

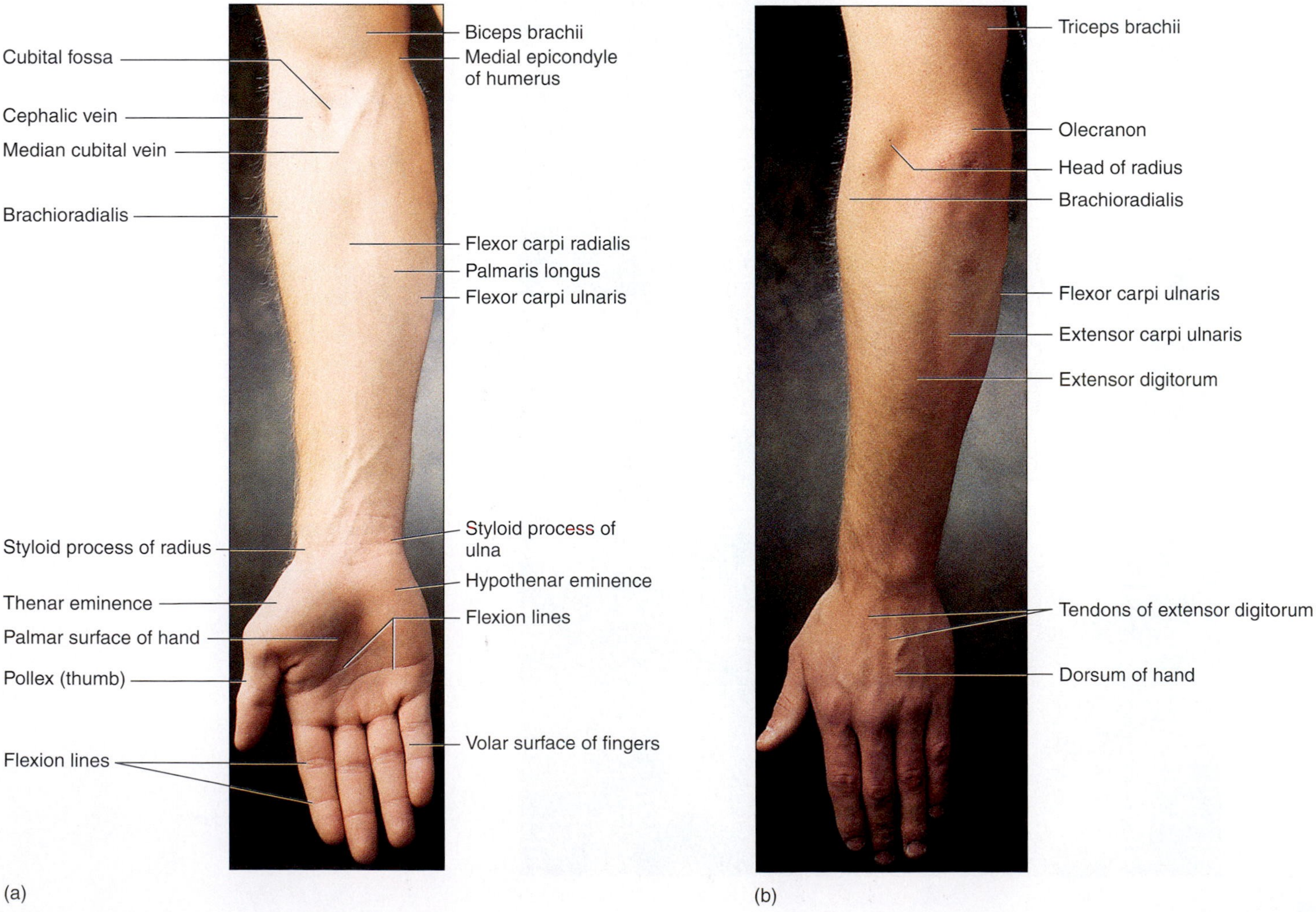
Cubital fossa
Cephalic vein
Median cubital vein
Brachioradialis
Styloid process of radius
Thenar eminence
Palmar surface of hand
Pollex (thumb)
Flexion lines
Biceps brachii
Medial epicondyle of humerus
Flexor carpi radialis
Palmaris longus
Flexor carpi ulnaris
Styloid process of ulna
Hypothenar eminence
Flexion lines
Volar surface of fingers
(a)
Triceps brachii
Olecranon
Head of radius
Brachioradialis
Flexor carpi ulnaris
Extensor carpi ulnaris
Extensor digitorum
Tendons of extensor digitorum
Dorsum of hand
(b)

Figure B.7

The Antebrachium (forearm). (*a*) Anterior (ventral) aspect. (*b*) Posterior (dorsal) aspect.

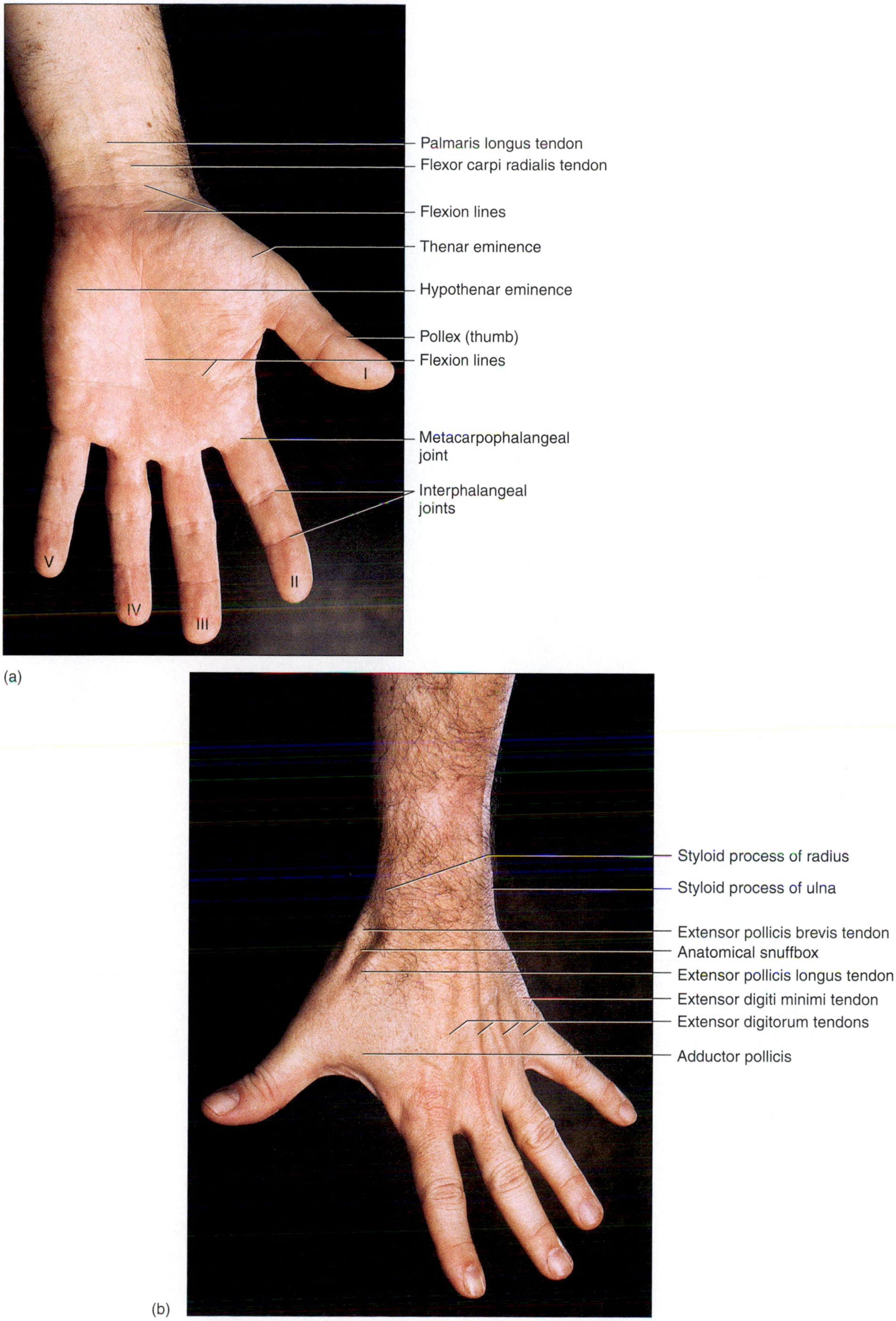

Palmaris longus tendon
Flexor carpi radialis tendon
Flexion lines
Thenar eminence
Hypothenar eminence
Pollex (thumb)
Flexion lines
I
Metacarpophalangeal joint
Interphalangeal joints
V
II
IV
III
(a)
Styloid process of radius
Styloid process of ulna
Extensor pollicis brevis tendon
Anatomical snuffbox
Extensor pollicis longus tendon
Extensor digiti minimi tendon
Extensor digitorum tendons
Adductor pollicis
(b)

Figure B.8

The Wrist and Hand. (*a*) Anterior (ventral) aspect. (*b*) Posterior (dorsal) aspect.

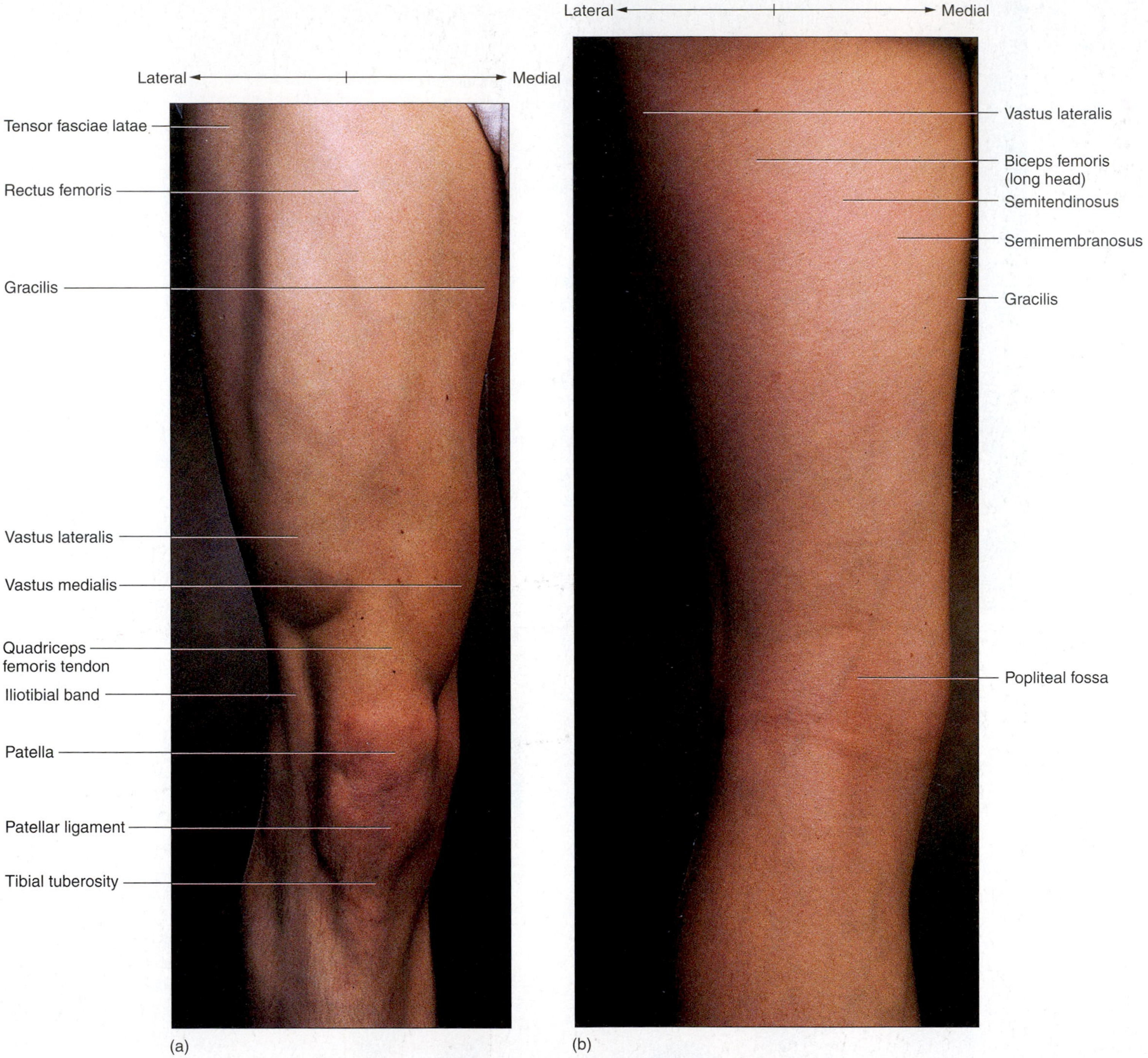

Lateral
Medial
Tensor fasciae latae
Rectus femoris
Gracilis
Vastus lateralis
Vastus medialis
Quadriceps femoris tendon
Iliotibial band
Patella
Patellar ligament
Tibial tuberosity
(a)
Lateral
Medial
Vastus lateralis
Biceps femoris (long head)
Semitendinosus
Semimembranosus
Gracilis
Popliteal fossa
(b)

FIGURE B.9

The Thigh and Knee. (*a*) Anterior (ventral) aspect. (*b*) Posterior (dorsal) aspect.

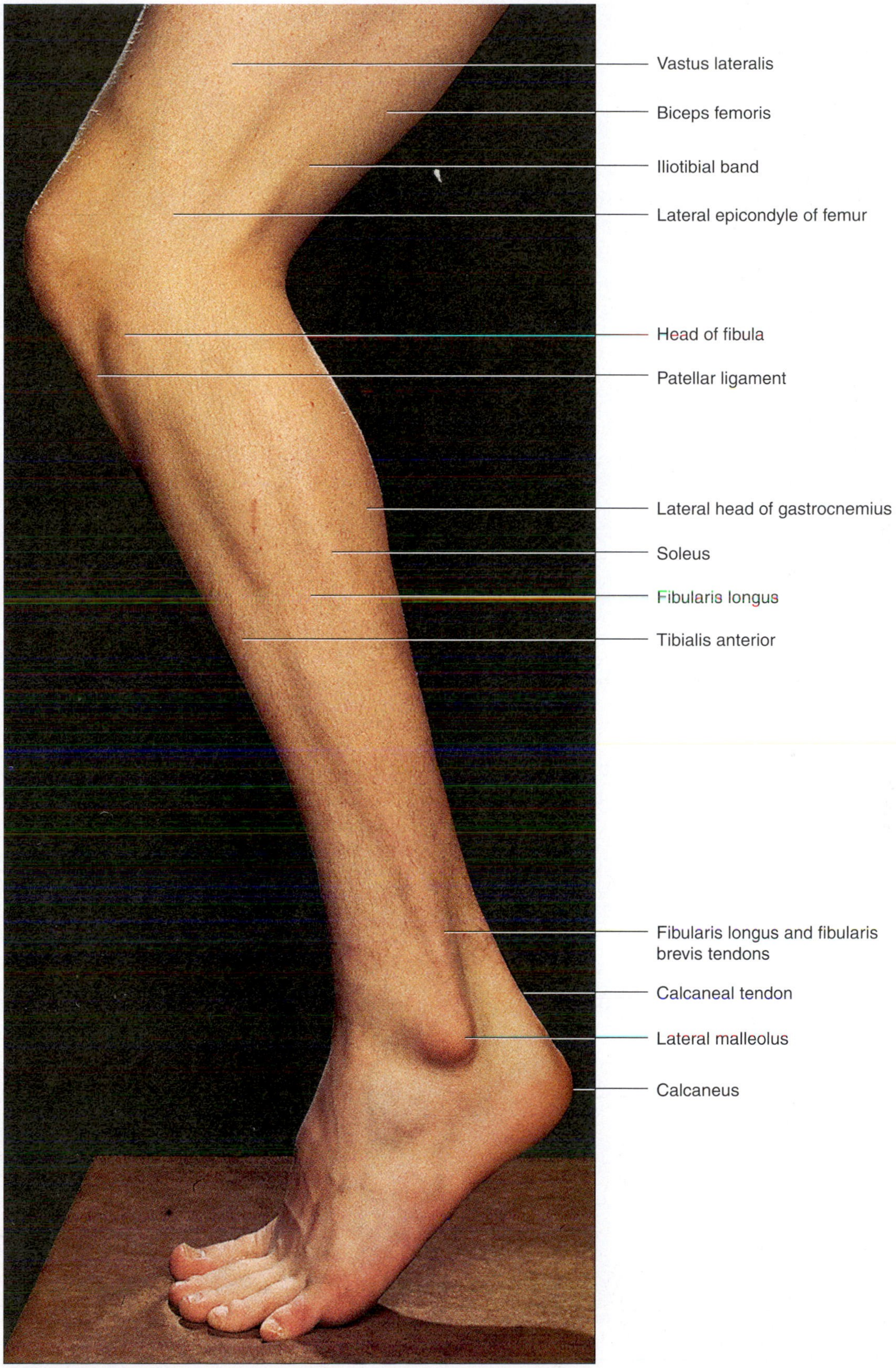

Vastus lateralis
Biceps femoris
Iliotibial band
Lateral epicondyle of femur
Head of fibula
Patellar ligament
Lateral head of gastrocnemius
Soleus
Fibularis longus
Tibialis anterior
Fibularis longus and fibularis brevis tendons
Calcaneal tendon
Lateral malleolus
Calcaneus

FIGURE B.10
The Leg and Foot, Lateral aspect.

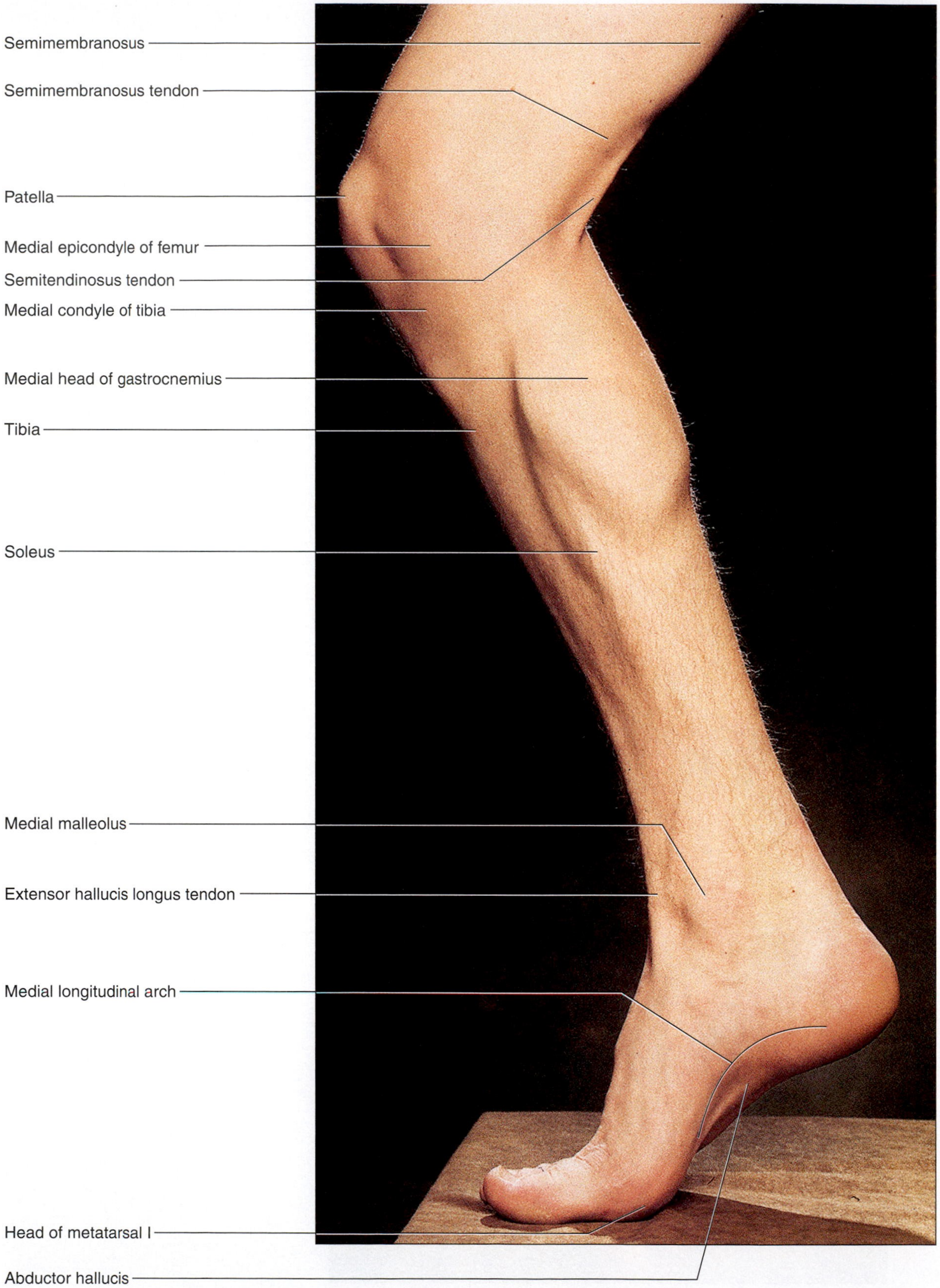
Semimembranosus
Semimembranosus tendon
Patella
Medial epicondyle of femur
Semitendinosus tendon
Medial condyle of tibia
Medial head of gastrocnemius
Tibia
Soleus
Medial malleolus
Extensor hallucis longus tendon
Medial longitudinal arch
Head of metatarsal I
Abductor hallucis

FIGURE B.11
The Leg and Foot, Medial Aspect.

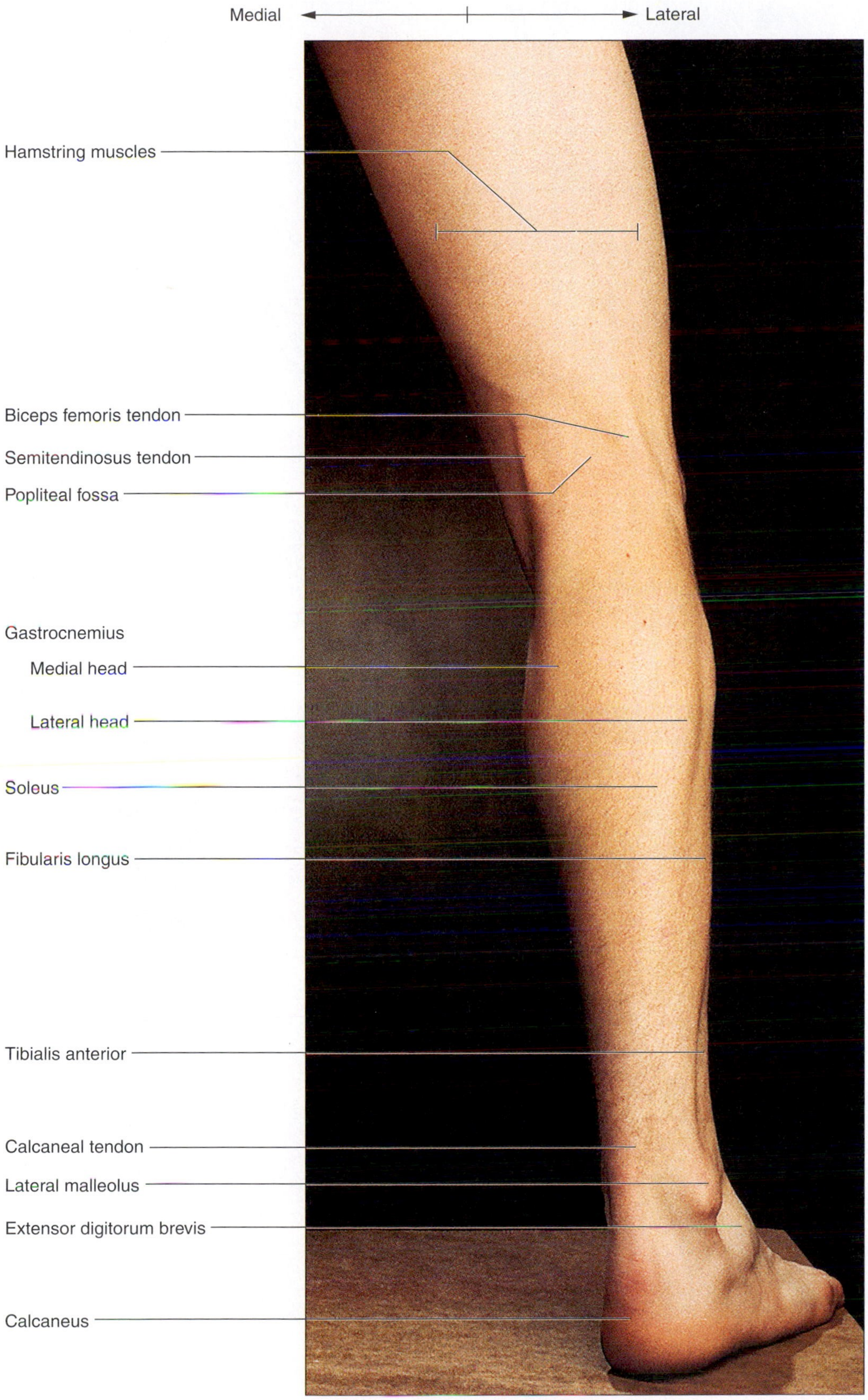
Medial
Lateral
Hamstring muscles
Biceps femoris tendon
Semitendinosus tendon
Popliteal fossa
Gastrocnemius
Medial head
Lateral head
Soleus
Fibularis longus
Tibialis anterior
Calcaneal tendon
Lateral malleolus
Extensor digitorum brevis
Calcaneus

FIGURE B.12
The Leg and Foot, Dorsal Aspect.

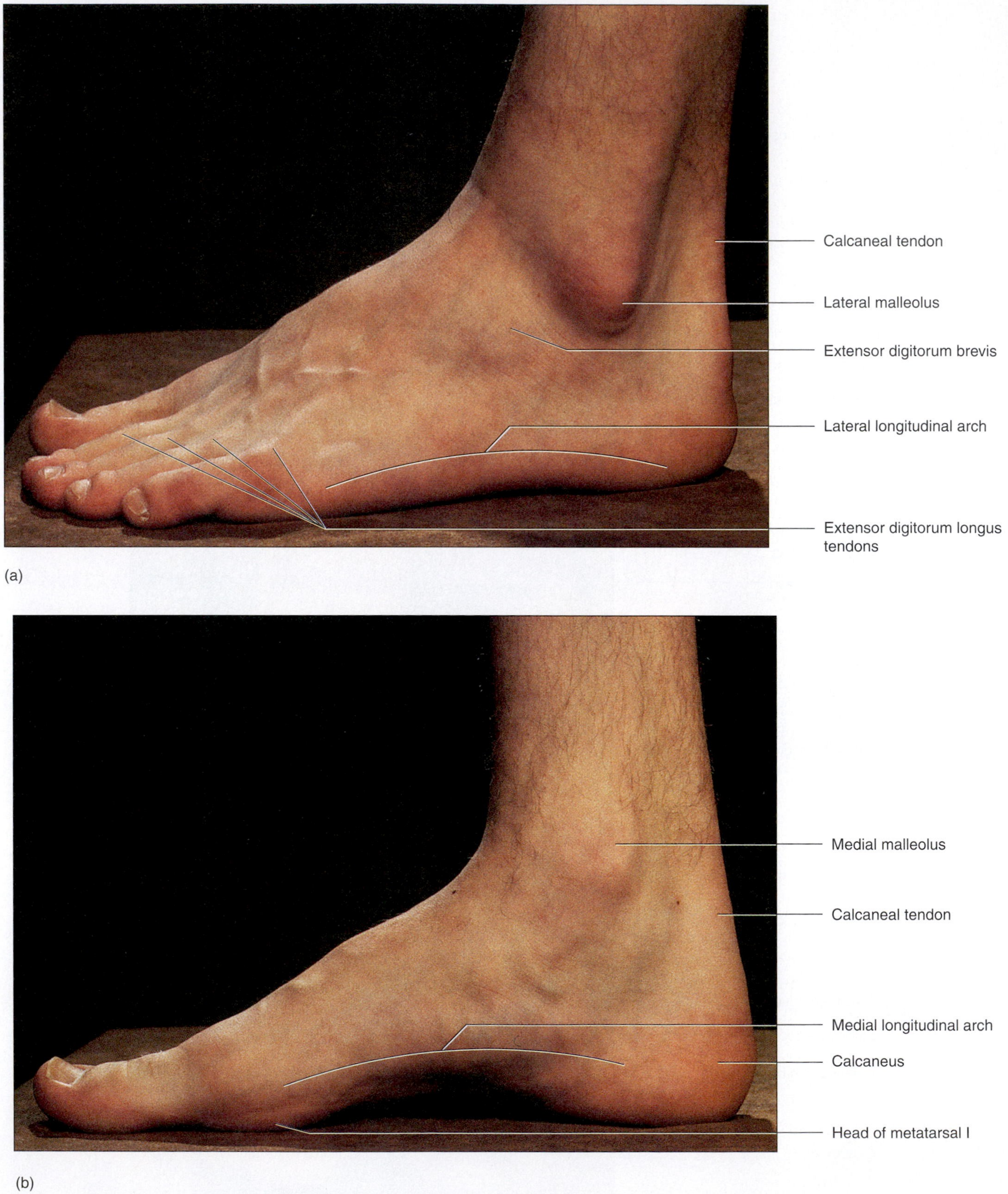
Calcaneal tendon
Lateral malleolus
Extensor digitorum brevis
Lateral longitudinal arch
Extensor digitorum longus tendons
(a)
Medial malleolus
Calcaneal tendon
Medial longitudinal arch
Calcaneus
Head of metatarsal I
(b)

FIGURE B.13

The Foot. (*a*) Lateral aspect. (*b*) Medial aspect.

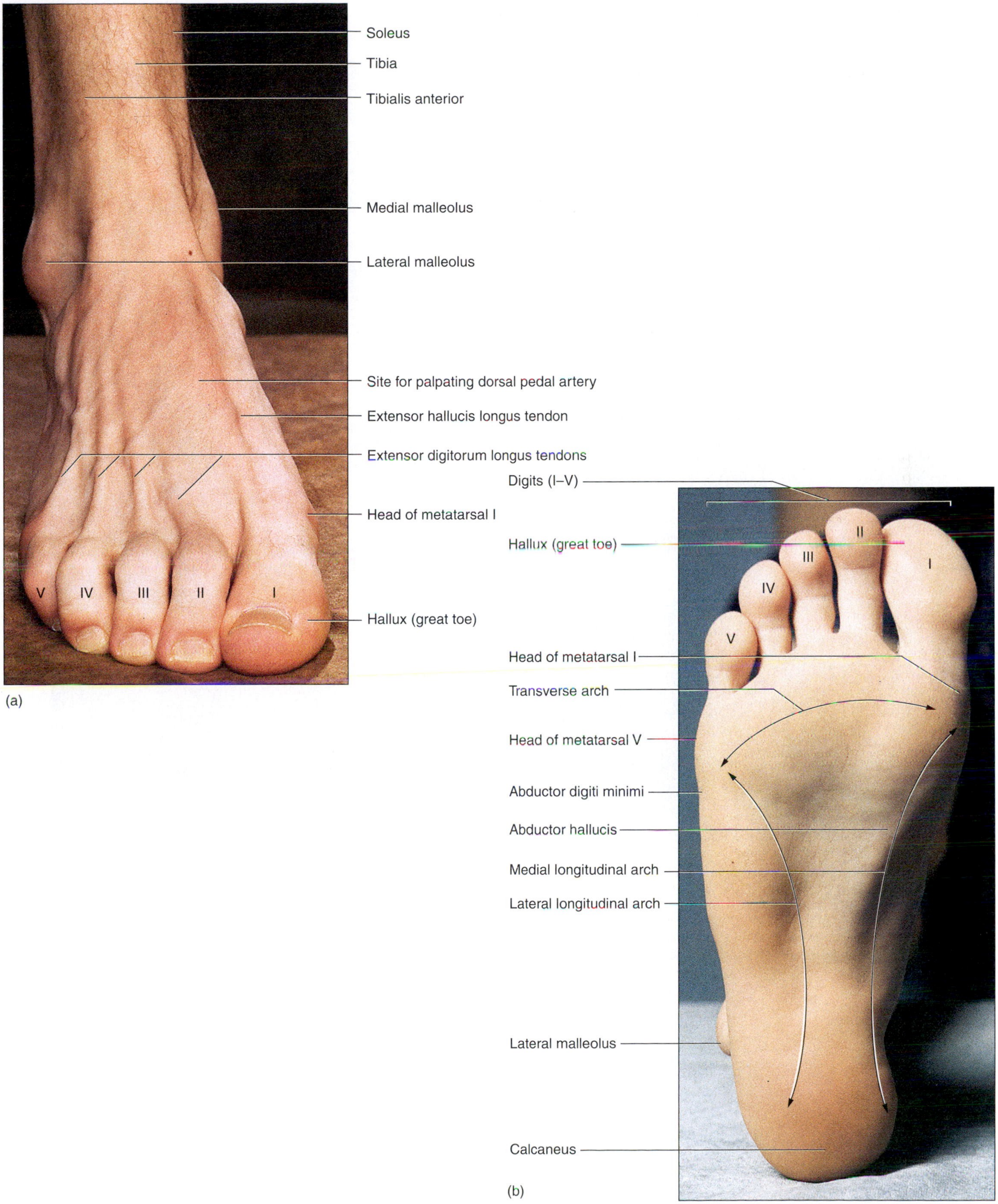
Soleus
Tibia
Tibialis anterior
Medial malleolus
Lateral malleolus
Site for palpating dorsal pedal artery
Extensor hallucis longus tendon
Extensor digitorum longus tendons
Head of metatarsal I
V
IV
III
II
I
Hallux (great toe)
(a)
Digits (I–V)
Hallux (great toe)
II
III
I
IV
V
Head of metatarsal I
Transverse arch
Head of metatarsal V
Abductor digiti minimi
Abductor hallucis
Medial longitudinal arch
Lateral longitudinal arch
Lateral malleolus
Calcaneus
(b)

Figure B.14

The Foot. (*a*) Dorsal aspect. (*b*) Plantar aspect.

FIGURE B.15

Muscle Self-Test. To test your knowledge of muscle anatomy, match the 30 labeled muscles on these photographs to the alphabetical list of muscles below. Answer as many as possible without referring back to the previous illustrations. Some of these names will be used more than once, since the same muscle may be shown from different perspectives, and some of these names will not be used at all. The answers are in the appendix.

a. biceps brachii
b. brachioradialis
c. deltoid
d. erector spinae
e. external oblique
f. flexor carpi ulnaris
g. gastrocnemius
h. gracilis
i. hamstrings
j. infraspinatus
k. latissimus dorsi
l. pectineus
m. pectoralis major
n. rectus abdominis
o. rectus femoris
p. serratus anterior
q. soleus
r. splenius capitis
s. sternocleidomastoid
t. subscapularis
u. teres major
v. tibialis anterior
w. transversus abdominis
x. trapezius
y. triceps brachii
z. vastus lateralis

13

CHAPTER THIRTEEN

Nervous Tissue

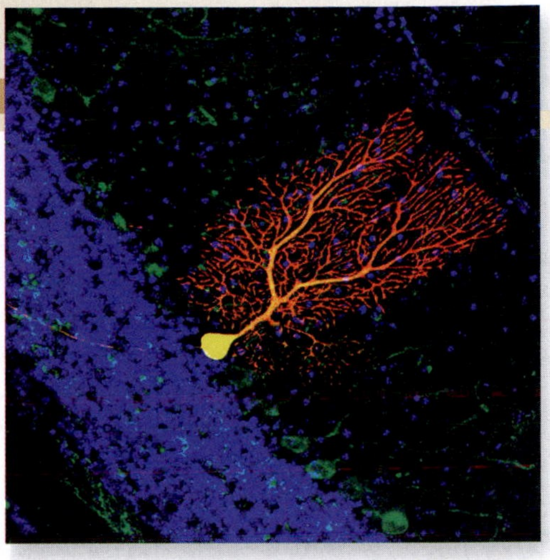

A Purkinje cell, a neuron from the cerebellum of the brain

CHAPTER OUTLINE

Overview of the Nervous System 366

Nerve Cells (Neurons) 367
- Universal Properties of Neurons 367
- Functional Classes of Neurons 367
- Structure of a Neuron 368
- Neuronal Variety 370

Supportive Cells (Neuroglia) 371
- Types of Neuroglia 371
- Myelin 371
- Unmyelinated Nerve Fibers 374
- Myelin and Signal Conduction 374

Synapses and Neural Circuits 374
- Synapses 374
- Neuronal Pools and Circuits 376

Developmental and Clinical Perspectives 378
- Development of the Nervous System 378
- Developmental Disorders of the Nervous System 380

Chapter Review 382

INSIGHTS

13.1 Clinical Application: Glial Cells and Brain Tumors 372

13.2 Clinical Application: Diseases of the Myelin Sheath 373

BRUSHING UP

To understand this chapter, you may find it helpful to review the following concepts:

- General structure of nerve cells, especially the soma, dendrites, and axon (p. 94)
- Early embryonic development (pp. 110–113)
- Introduction to synapses and neurotransmitters (p. 266)

If the body is to maintain homeostasis and function effectively, its trillions of cells must work together in a coordinated fashion. If each cell behaved without regard to what others were doing, the result would be physiological chaos and death. This is prevented by two communication systems—the nervous system (fig. 13.1), which is specialized for the rapid transmission of signals from cell to cell, and the endocrine system, which is specialized for sending chemical messengers, the hormones, through the blood. The most important aspect of both systems is that they detect changes in an organ and modify the activity of other organs. Thus, these systems functionally coordinate the organs of the body and play a central role in maintaining homeostasis.

The nervous system is the subject of chapters 13 to 17, and the endocrine system is discussed in chapter 18. This chapter is primarily concerned with individual nerve cells. The next four chapters are concerned with the organization of the nervous system at the organ level.

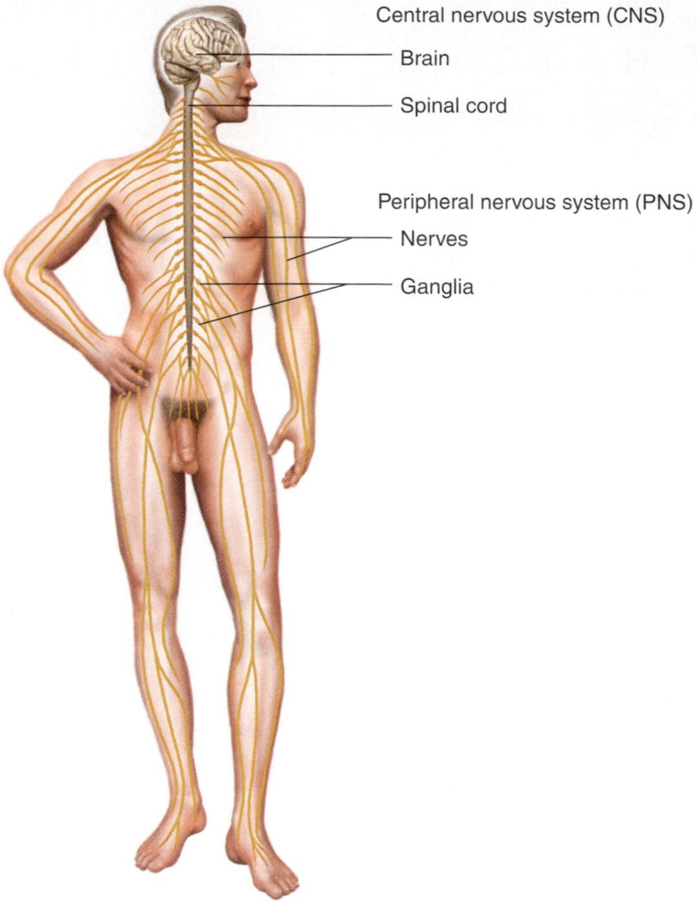

FIGURE 13.1
The Nervous System.

OVERVIEW OF THE NERVOUS SYSTEM

Objectives

When you have completed this section, you should be able to

- state the function of the nervous system;
- identify the major subdivisions of the nervous system;
- define *nerve* and *ganglion;* and
- define *receptor* and *effector.*

The fundamental purpose of the nervous system is (1) to receive information about changes in the body and its external environment; (2) to process this information and determine the appropriate response, if any; and (3) to issue commands to cells that carry out the response. To serve these purposes, the nervous system has two major anatomical subdivisions (fig. 13.2):

- The **central nervous system (CNS)** consists of the brain and spinal cord, which are enclosed and protected by the cranium and vertebral column. Most of the information processing by the nervous system occurs here.
- The **peripheral nervous system (PNS)** comprises the remainder of the nervous system and serves for input to and output from the CNS. It is composed of nerves and ganglia. A **nerve** is a bundle of nerve fibers wrapped in fibrous connective tissue. Nerves emerge from the CNS through foramina of the skull and vertebral column and carry signals to and from other organs of the body. A **ganglion**[1] (plural, *ganglia*) is a knotlike swelling in a nerve where the cell bodies of neurons are concentrated (see fig. 14.8).

The peripheral nervous system is functionally divided into *sensory* and *motor* divisions, and each of these has *somatic* and *visceral* subdivisions.

- The **sensory (afferent**[2]**) division** carries sensory signals by way of afferent nerve fibers from sensory **receptors** (cells and organs that detect stimuli) to the CNS.

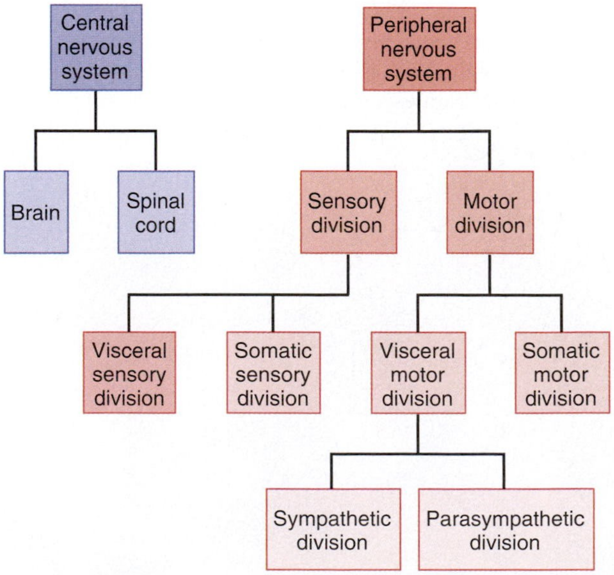

FIGURE 13.2
Subdivisions of the Nervous System.

[1]*gangli* = knot
[2]*af* = *ad* = toward + *fer* = to carry

- The **visceral sensory division** carries signals mainly from the viscera of the thoracic and abdominal cavities, such as the heart, lungs, stomach, and urinary bladder.
- The **somatic**[3] **sensory division** carries signals from receptors in the skin, muscles, bones, and joints.

- The **motor (efferent**[4]**) division** carries motor signals by way of *efferent nerve fibers* from the CNS to **effectors** (cells and organs that carry out the body's responses; mainly gland and muscle cells).
  - The **visceral motor division (autonomic**[5] **nervous system)** carries signals to glands, cardiac muscle, and smooth muscle. We usually have no voluntary control over these effectors, and this system operates at an unconscious level. The responses of this system and its effectors are *visceral reflexes.* The autonomic nervous system has two further divisions:
    - The **sympathetic division** tends to prepare the body for action, for example by accelerating the heartbeat and increasing respiratory airflow, but it inhibits digestion.
    - The **parasympathetic division** tends to adapt the body to a state of rest, reducing the heart rate and respiratory airflow, for example, but stimulating digestion.
  - The **somatic motor division** carries signals to the skeletal muscles. This output produces muscular contractions that are under voluntary control, as well as involuntary muscle contractions called *somatic reflexes.*

The foregoing terminology may give the impression that the body has several nervous systems—central, peripheral, sensory, motor, somatic, visceral, sympathetic, and parasympathetic. These are just terms of convenience, however. There is only one nervous system, and these subsystems are interconnected parts of the whole.

Before You Go On

Answer the following questions to test your understanding of the preceding section:

1. What is a receptor? Give two examples of effectors.
2. Distinguish between the central and peripheral nervous systems, and between the visceral and somatic divisions of the sensory and motor systems.
3. What is another name for the visceral motor nervous system? What are the two subdivisions of this system and how do they differ in their functions?

[3]*somat* = body + *ic* = pertaining to
[4]*ef* = *ex* = out, away + *fer* = to carry
[5]*auto* = self + *nom* = law, governance

Nerve Cells (Neurons)

Objectives

When you have completed this section, you should be able to

- describe the properties that neurons must have to carry out their function;
- identify and define three functional classes into which all neurons fall;
- describe the structure of a representative neuron;
- describe some variations in neuron structure;

Universal Properties of Neurons

The functional unit of the nervous system is the **nerve cell,** or **neuron;** neurons carry out the system's communicative role. These cells have three fundamental physiological properties that are necessary to this function:

1. **Excitability (irritability).** All cells possess excitability, the ability to respond to environmental changes called **stimuli.** Neurons have developed this property to the highest degree.
2. **Conductivity.** Neurons respond to stimuli by producing traveling electrical signals that quickly reach other cells at distant locations.
3. **Secretion.** When the electrical signal reaches the end of a nerve fiber, the neuron usually secretes a chemical called a *neurotransmitter* that crosses the gap and stimulates the next cell.

Functional Classes of Neurons

There are three general classes of neurons (fig. 13.3) corresponding to the three major aspects of nervous system function listed earlier:

1. **Sensory (afferent) neurons** are specialized to detect stimuli such as light, heat, pressure, and chemicals, and to transmit information about them to the CNS. These neurons can begin in almost any organ of the body but always end in the brain or spinal cord; the word *afferent* refers to signal conduction *toward* the CNS. Some sensory receptors, such as pain and smell receptors, are themselves neurons. In other cases, such as taste and hearing, the receptor is a separate cell that communicates directly with a sensory neuron.
2. **Interneurons**[6] **(association neurons)** lie entirely within the CNS. They receive signals from many other neurons and carry out the integrative function of the nervous system—that is, they process, store, and retrieve information and "make decisions" about how the body responds to stimuli. About 90% of human neurons are interneurons. The word

[6]*inter* = between

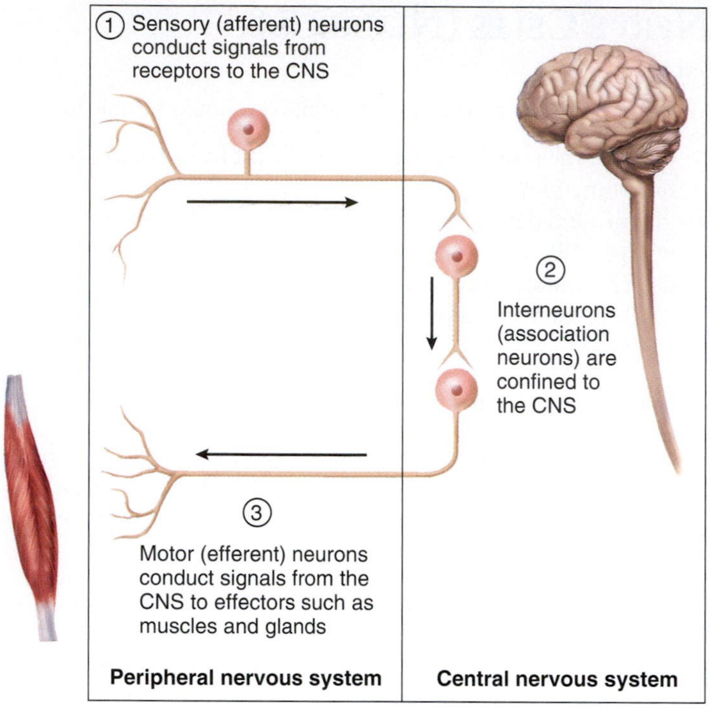
① Sensory (afferent) neurons conduct signals from receptors to the CNS
② Interneurons (association neurons) are confined to the CNS
③ Motor (efferent) neurons conduct signals from the CNS to effectors such as muscles and glands
Peripheral nervous system
Central nervous system

FIGURE 13.3
Functional Classes of Neurons. All neurons can be regarded as either sensory neurons, interneurons, or motor neurons.

interneuron refers to the fact that they lie *between,* and interconnect, the incoming sensory pathways and the outgoing motor pathways of the CNS.

3. **Motor (efferent) neurons** send signals predominantly to muscle and gland cells, the effectors that carry out the body's responses to stimuli. They are called *motor* neurons because most of them lead to muscle cells, and *efferent* neurons to signify the signal conduction *away from* the CNS.

Structure of a Neuron

There are several varieties of neurons, as we shall see, but a good starting point for discussing neuronal stucture is a motor neuron of the spinal cord (fig. 13.4). The control center of the neuron is its **soma,**[7] or **cell body.** It has a single, centrally located nucleus with a large nucleolus. The cytoplasm contains mitochondria, lysosomes, a Golgi complex, numerous inclusions, and an extensive rough endoplasmic reticulum and cytoskeleton. The cytoskeleton consists of a dense mesh of microtubules and **neurofibrils** (bundles of actin filaments) that compartmentalize the rough ER into dark-staining regions called *Nissl*[8] *bodies,* unique to neurons (fig. 13.4*c, d*). Nissl bodies are a helpful clue to identifying neurons in tissue sections with mixed cell types. Mature neurons lack centrioles and apparently undergo no further mitosis after adolescence, but they are unusually long-lived cells, capable of functioning for over a hundred years. Even in old age, however, there are unspecialized stem cells in the CNS that can divide and develop into new neurons.

The major cytoplasmic inclusions in a neuron are glycogen granules, lipid droplets, melanin, and a golden brown pigment called *lipofuscin*[9] (LIP-oh-FEW-sin)—an end product of lysosomal digestion of worn-out organelles and other products. Lipofuscin collects with age and pushes the nucleus to one side of the cell. Lipofuscin granules are also called "wear-and-tear granules" because they are most abundant in old neurons, but they are apparently harmless.

The soma of a neuron usually gives rise to a few thick processes that branch into a vast number of **dendrites**[10]—named for their striking resemblance to the bare branches of a tree in winter. The dendrites are the primary site for receiving signals from other neurons. Some neurons have only one dendrite and some have thousands. The more dendrites a neuron has, the more information it can receive from other cells and incorporate into its decision making. As tangled as the dendrites may seem, they provide exquisitely precise pathways for the reception and processing of neural information.

On one side of the soma is a mound called the **axon hillock,** from which the **axon (nerve fiber)** originates. An axon is specialized for rapid conduction of nerve signals to points remote from the soma. It is cylindrical and relatively unbranched for most of its length; however, it may give rise to a few branches called *axon collaterals* along the way, and most axons branch extensively at their distal end. Its cytoplasm is called the **axoplasm** and its membrane the **axolemma.**[11] A neuron never has more than one axon, and some neurons in the retina and brain have none.

Somas range from 5 to 135 μm in diameter, while axons range from 1 to 20 μm in diameter and from a few millimeters to more than a meter long. Such dimensions are more impressive when we scale them up to the size of familiar objects. If the soma of a spinal motor neuron were the size of a tennis ball, its dendrites would form a huge bushy mass that could fill a 30-seat classroom from floor to ceiling. Its axon would be up to a mile long but a little narrower than a garden hose. This is quite a point to ponder. The neuron must assemble molecules and organelles in its "tennis ball" soma and deliver them through its "mile-long garden hose" to the end of the axon. In a process called *axonal transport,* neurons employ *motor proteins* that can carry organelles and macromolecules as they crawl along the cytoskeleton of the nerve fiber to distant destinations in the cell.

At the distal end, axons usually have a **terminal arborization**[12]—an extensive complex of fine branches. Each branch ends in a **synaptic knob (terminal button).** As described in chapter 10, the synaptic knob is a little swelling that forms a junction (*synapse*[13]) with a muscle cell, gland cell, or another neuron. Synapses are described in detail later in this chapter.

[7]*soma* = body
[8]Franz Nissl (1860–1919), German neuropathologist
[9]*lipo* = fat, lipid + *fusc* = dusky, brown
[10]*dendr* = tree, branch + *ite* = little
[11]*axo* = axis, axon + *lemma* = husk, peel, sheath
[12]*arbor* = treelike
[13]*syn* = together + *aps* = to touch, join

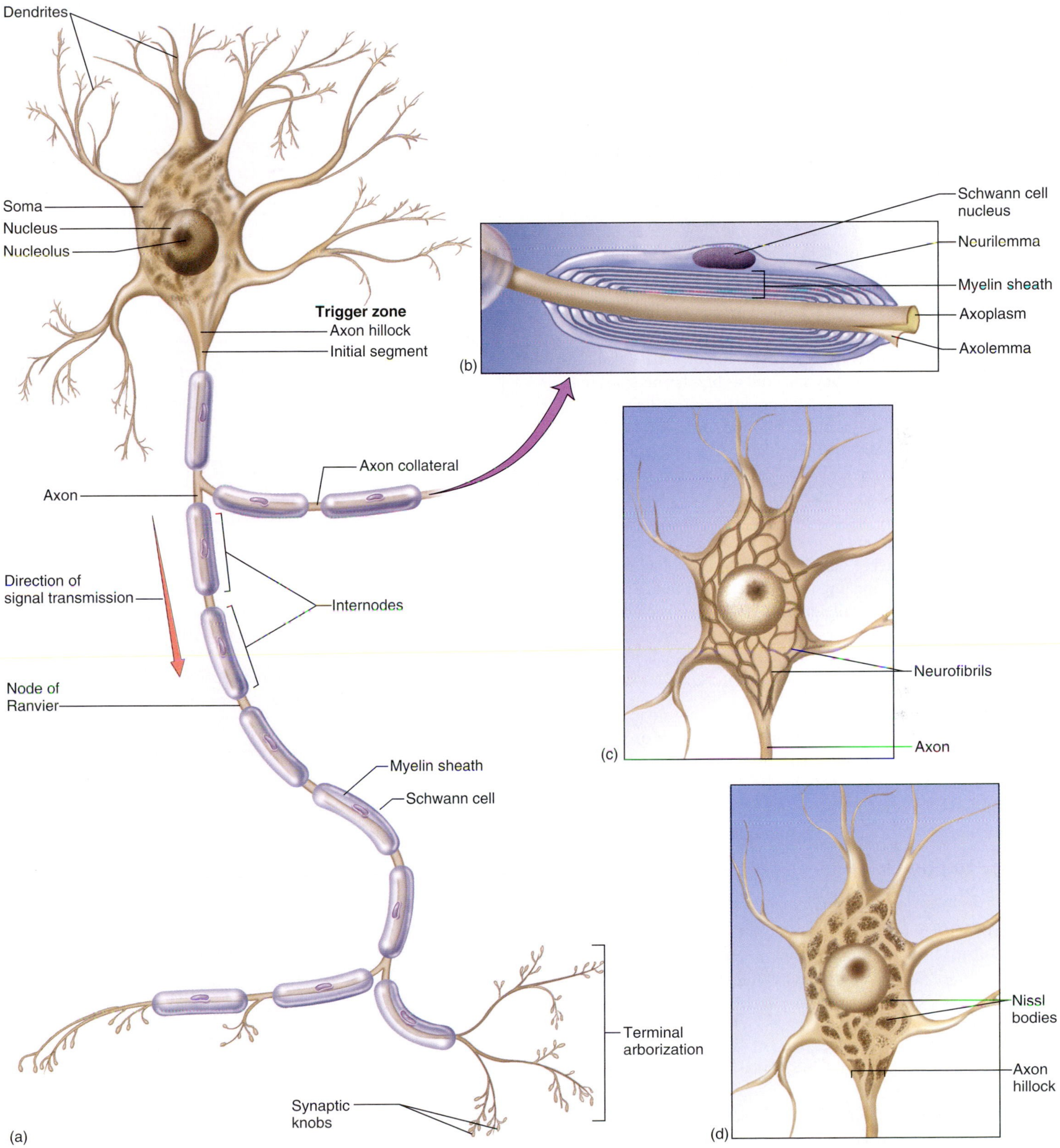

FIGURE 13.4

A Representative Neuron. (*a*) A multipolar neuron such as a spinal motor neuron. (*b*) Detail of myelin sheath. (*c*) Neurofibrils of the soma. (*d*) Nissl bodies, stained masses of rough ER separated by bundles of neurofibrils. The Schwann cells and myelin sheath are explained later in this chapter.

Neuronal Variety

Not all neurons fit every detail of the preceding description. Neurons are classified structurally according to the number of processes extending from the soma (fig. 13.5):

- **Multipolar neurons** are those, like the preceding, that have one axon and two or more (usually many) dendrites. This is the most common type of neuron and includes most neurons of the brain and spinal cord.
- **Bipolar neurons** have one axon and one dendrite. Examples include olfactory cells of the nasal cavity, some neurons of the retina, and sensory neurons of the inner ear.
- **Unipolar neurons** have only a single process leading away from the soma. They are represented by the neurons that carry sensory signals to the spinal cord. These neurons are also called *pseudounipolar* because they start out as bipolar neurons in the embryo, but their two processes fuse into one as the neuron matures. A short distance away from the soma, the process branches like a T, with a *peripheral fiber* carrying signals from the source of sensation and a *central fiber* continuing into the spinal cord. In most other neurons, a dendrite carries signals toward the soma and an axon carries them away. In unipolar neurons, however, there is one long fiber that bypasses the soma and carries nerve signals directly to the spinal cord. The dendrites are the branching receptive endings in the skin or other place of origin, while the rest of the fiber is considered to be the axon (defined in these neurons by the presence of myelin and the ability to generate action potentials—two concepts explained later in this chapter).
- **Anaxonic neurons** have multiple dendrites but no axon. They communicate through their dendrites and do not produce action potentials. Anaxonic neurons are found in the brain and retina. In the retina, they help in visual processes such as the perception of contrast.

Before You Go On

Answer the following questions to test your understanding of the preceding section:

4. Explain why neurons could not function without the properties of excitability, conductivity, and secretion.
5. Distinguish between sensory neurons, interneurons, and motor neurons.
6. Define each of the following and explain its importance to neuronal function: dendrites, soma, axon, synaptic knob, and synaptic vesicles.
7. Make a simple sketch of a multipolar, bipolar, unipolar, and anaxonic neuron and next to each sketch, state one place where such a neuron could be found.

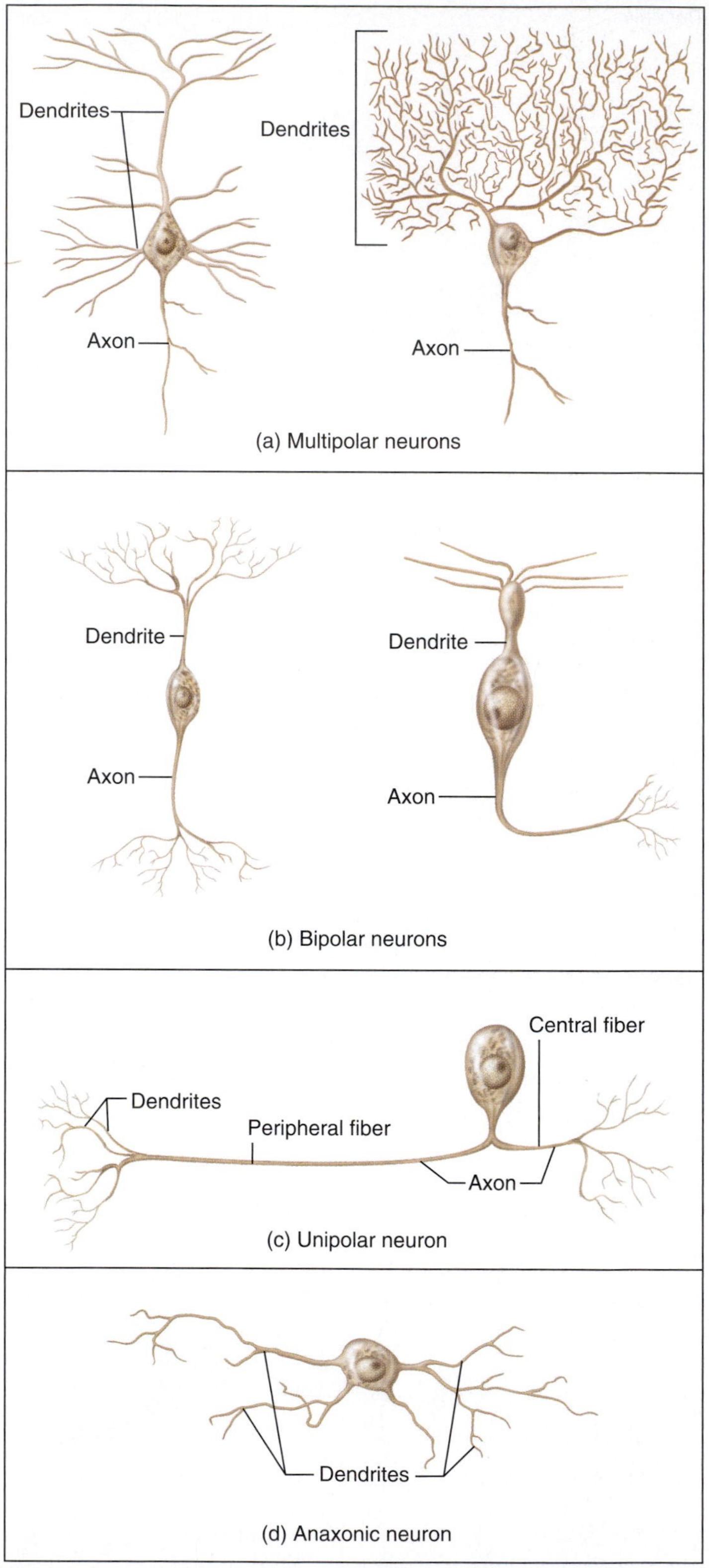

FIGURE 13.5

Variation in Neuronal Structure. (*a*) Two multipolar neurons of the brain—a pyramidal cell of the cerebral cortex *(left)* and a Purkinje cell of the cerebellum. (*b*) Two bipolar neurons—a bipolar cell of the retina *(left)* and an olfactory cell of the nose. (*c*) A unipolar neuron of the type that detects stimuli in the skin, muscles, and joints. (*d*) An anaxonic neuron of the retina.

Supportive Cells (Neuroglia)

Objectives

When you have completed this section, you should be able to

- name the cells that aid neuronal function and state their locations and functions;
- describe the myelin sheath that is formed around certain nerve fibers; and
- describe how the speed of nerve signal conduction varies with nerve fiber diameter and the presence or absence of myelin.

There are about a trillion (10^{12}) neurons in the nervous system—10 times as many neurons in one body as there are stars in our galaxy! Yet the neurons are outnumbered as much as 50 to 1 by supportive cells called **neuroglia** (noo-ROG-lee-uh), or **glial** (GLEE-ul) **cells.** Glial cells protect the neurons and aid their function. The word *glia,* which means "glue," implies one of their roles—they bind neurons together. In the fetus, glial cells form a scaffold that guides young migrating neurons to their destinations. Wherever a mature neuron is not in synaptic contact with another cell, it is covered with glial cells. This prevents neurons from contacting each other except at points specialized for signal transmission, and thus gives precision to their conduction pathways.

Types of Neuroglia

There are six major categories of neuroglia, each with a unique function (table 13.1). Four types occur in the central nervous system (fig. 13.6):

1. **Oligodendrocytes**[14] (OL-ih-go-DEN-dro-sites) somewhat resemble an octopus; they have a bulbous body with as many as 15 armlike processes. Each process reaches out to a nerve fiber and spirals around it like electrical tape wrapped repeatedly around a wire. This spiral wrapping, called the *myelin sheath,* insulates the nerve fiber from the extracellular fluid and speeds up signal conduction in the nerve fiber.
2. **Astrocytes**[15] are the most abundant and functionally diverse glia in the CNS and constitute over 90% of the tissue in some areas of the brain. They are many-branched and have a somewhat starlike shape. Astrocytes cover the entire brain surface and most nonsynaptic regions of the neurons in the gray matter of the CNS. They form a supportive framework for the nervous tissue. They issue numerous extensions, called *perivascular feet,* that contact the endothelial cells of the blood capillaries and stimulate them to form tight junctions. These junctions contribute to a *blood-brain barrier* that strictly controls which substances are able to get from the bloodstream into the brain tissue (see chapter 15). Astrocytes convert blood glucose to lactate and supply this to the neurons for nourishment. They secrete growth factors that promote neuron growth and synapse formation. They communicate electrically with neurons and may influence future synaptic signalling between them. Astrocytes also regulate the chemical composition of the tissue fluid—when neurons transmit signals, they release neurotransmitters and potassium ions; astrocytes absorb these substances and prevent them from accumulating in the tissue fluid. When neurons are damaged, astrocytes form hardened masses of scar tissue and fill space formerly occupied by neurons. This process is called *astrocytosis* or *sclerosis.*
3. **Ependymal**[16] (ep-EN-dih-mul) **cells** resemble a cuboidal epithelium lining the internal cavities of the brain and spinal cord. Unlike epithelial cells, however, they have no basement membrane and they exhibit rootlike processes that penetrate into the underlying nervous tissue. Ependymal cells produce *cerebrospinal fluid (CSF),* a clear liquid that bathes the CNS and fills its internal cavities. They have patches of cilia on their apical surfaces that help to circulate the CSF. Ependymal cells and CSF are considered in more detail in chapter 15.
4. **Microglia** are small macrophages that develop from white blood cells called monocytes. They wander through the CNS and phagocytize dead nervous tissue, microorganisms, and other foreign matter. They become concentrated in areas damaged by infection, trauma, or stroke. Pathologists look for clusters of microglia in histological sections of the brain as a clue to sites of injury.

The other two types of glial cells occur in the peripheral nervous system:

5. **Schwann**[17] (shwon) **cells** envelop nerve fibers of the PNS, forming a sleeve called the neurilemma around them. In most cases, a Schwann cell winds repeatedly around a nerve fiber and produces a myelin sheath between the neurilemma and nerve fiber. This is similar to the myelin sheath produced by oligodendrocytes in the CNS, but there are differences in the way myelin is produced, as described later. In addition to myelinating peripheral nerve fibers, Schwann cells assist in the regeneration of damaged fibers.
6. **Satellite cells** surround the neuron cell bodies in ganglia of the PNS. Little is known of their function.

Myelin

The **myelin** (MY-eh-lin) **sheath** is an insulating layer around a nerve fiber, somewhat like the rubber insulation on a wire. It is formed by oligodendrocytes in the central nervous system and Schwann cells in

[14]*oligo* = few + *dendro* = branches + *cyte* = cell
[15]*astro* = star + *cyte* = cell
[16]*ependyma* = upper garment
[17]Theodore Schwann (1810–82), German histologist

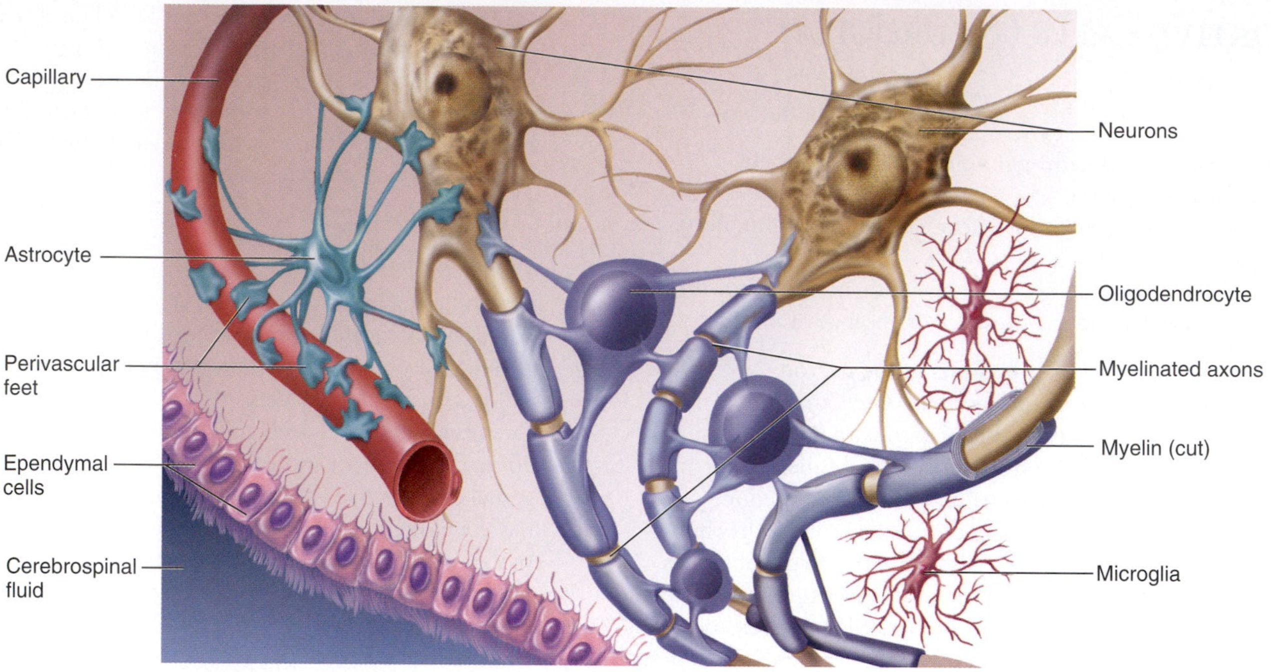
Capillary
Astrocyte
Perivascular feet
Ependymal cells
Cerebrospinal fluid
Neurons
Oligodendrocyte
Myelinated axons
Myelin (cut)
Microglia

FIGURE 13.6
Neuroglia of the Central Nervous System.

TABLE 13.1
Types of Glial Cells

Type	Location	Functions
Oligodendrocytes	CNS	Form myelin in brain and spinal cord
Astrocytes	CNS	Cover brain surface and nonsynaptic regions of neurons; stimulate formation of blood-brain barrier; remove neurotransmitters and K^+ from extracellular fluid (ECF) of brain and spinal cord; help to regulate composition of ECF; form supportive framework in CNS; form scar tissue to replace damaged nervous tissue
Ependymal cells	CNS	Line cavities of brain and spinal cord; secrete and circulate cerebrospinal fluid
Microglia	CNS	Phagocytize and destroy microorganisms, foreign matter, and dead nervous tissue
Schwann cells	PNS	Form neurilemma around all PNS nerve fibers and myelin around most of them; aid in regeneration of damaged nerve fibers
Satellite cells	PNS	Surround somas of neurons in the ganglia; function uncertain

the peripheral nervous system. Since it consists of the plasma membranes of these glial cells, its composition is like that of plasma membranes in general. It is about 20% protein and 80% lipid, the latter including phospholipids, glycolipids, and cholesterol.

The formation of myelin is called **myelination.** In the CNS, each oligodendrocyte reaches out to several nerve fibers in its immediate vicinity. Its armlike process spirals repeatedly around the nerve fiber, laying down many compact layers of its own membrane with almost no cytoplasm between the membranes (fig. 13.7*a*). The growing edge of the oligodendrocyte pushes into the space beneath the previous layer of myelin, so the myelin layers spiral inward toward the axon. These layers constitute the myelin sheath. A nerve fiber is much longer than the reach of a single oligodendrocyte, so it requires many oligodendrocytes to cover one nerve fiber.

In the PNS, a Schwann cell spirals around a single nerve fiber, putting down as many as a hundred layers of membrane (fig. 13.7*b*). Here, the innermost coil of myelin is the first to be deposited, and the myelin spirals outward as the Schwann cell grows, leaving its trailing edge behind. The outermost coil of the Schwann cell, external to the myelin sheath, is the **neurilemma**[19] (noor-ih-LEM-ah). Here, the bulging body of the Schwann cell contains its nucleus and most of its cytoplasm. To visualize this, imagine wrapping an almost-empty

INSIGHT 13.1 CLINICAL APPLICATION

Glial Cells and Brain Tumors

A tumor consists of a mass of rapidly dividing cells. Mature neurons, however, have little capacity for mitosis and seldom form tumors. Some brain tumors arise from the meninges (protective membranes of the CNS) or arise by metastasis from tumors elsewhere, such as malignant melanoma and colon cancer. Most adult brain tumors, however, are composed of glial cells, which are mitotically active throughout life. Such tumors, called *gliomas,*[18] grow rapidly and are highly malignant. Because of the blood-brain barrier, brain tumors usually do not yield to chemotherapy and must be treated with radiation or surgery.

[18]*glia* = glial cells + *oma* = tumor

[19]*neuri* = nerve + *lemma* = husk, peel, sheath

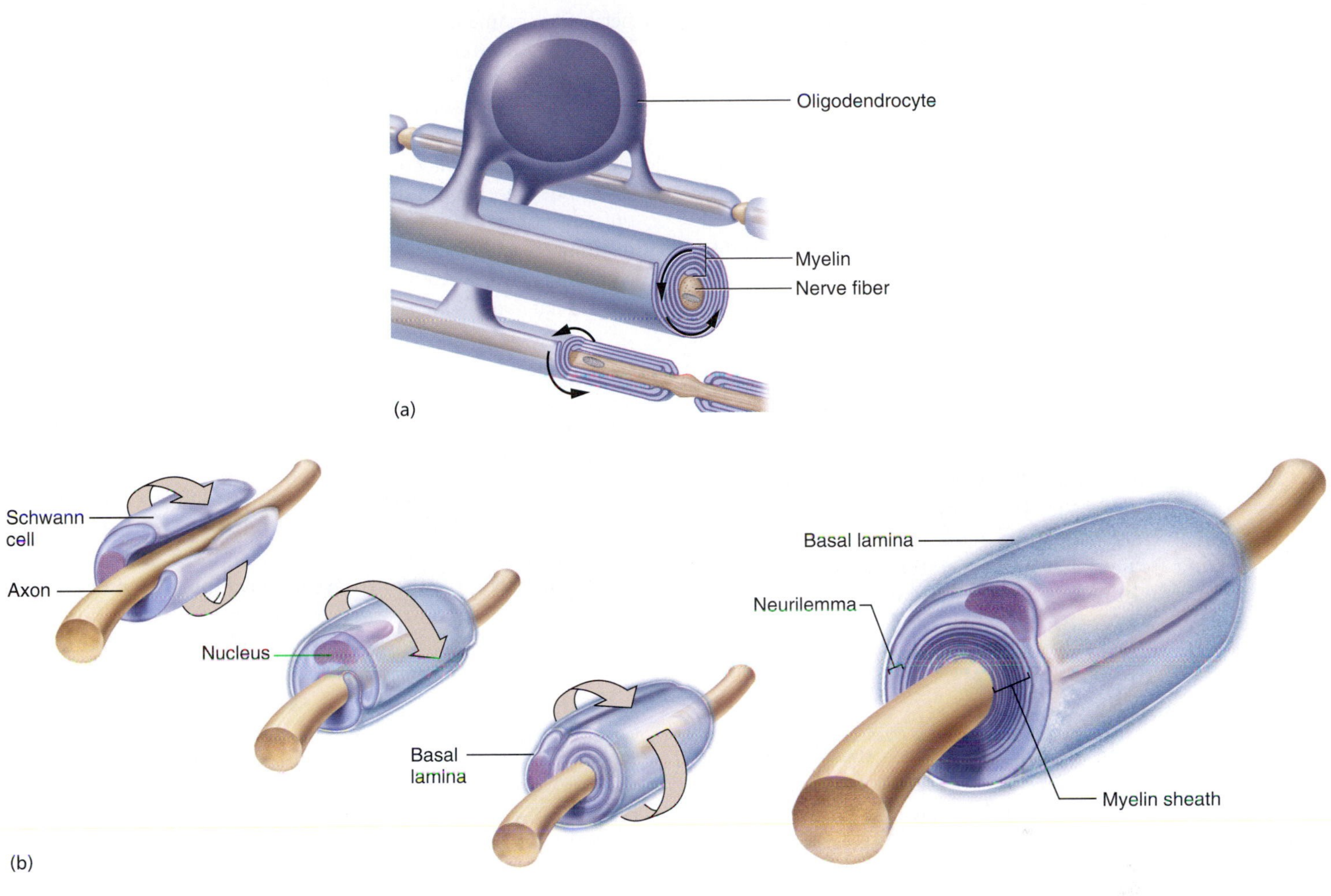

FIGURE 13.7

Formation of the Myelin Sheath. (*a*) In the central nervous system, each oligodendrocyte reaches out to multiple nerve fibers and myelinates them. The myelin sheath spirals inward toward the nerve fiber. (*b*) In the peripheral nervous system, a Schwann cell coils repeatedly around a single axon. The myelin sheath thus spirals outward away from the nerve fiber, and the myelin is covered by a neurilemma and basal lamina.

INSIGHT 13.2 CLINICAL APPLICATION

Diseases of the Myelin Sheath

Multiple sclerosis and Tay-Sachs disease are degenerative disorders of the myelin sheath. In *multiple sclerosis (MS)*, the oligodendrocytes and myelin sheaths of the CNS deteriorate and are replaced by hardened scar tissue, especially between the ages of 20 and 40. Nerve conduction is disrupted with effects that depend on what part of the CNS is involved—numbness, double vision, blindness, speech defects, neurosis, or tremors. Patients experience variable cycles of milder and worse symptoms until they eventually become bedridden. Most die from 7 to 32 years after the onset of the disease. The cause of MS remains uncertain; most hypotheses suggest that it results from an immune disorder triggered by a virus in genetically susceptible individuals. There is no cure.

Tay-Sachs[20] *disease* is a hereditary disorder seen mainly in infants of Eastern European Jewish ancestry. It results from the abnormal accumulation of a glycolipid called GM_2 (ganglioside) in the myelin sheath. GM_2 is normally decomposed by a lysosomal enzyme, but this enzyme is lacking from those who inherit the recessive Tay-Sachs gene from both parents. As GM_2 accumulates, it disrupts the conduction of nerve signals and the victim typically suffers blindness, loss of coordination, and dementia. Signs begin to appear before the child is a year old and most victims die by the age of three or four. Asymptomatic adult carriers can be identified by a blood test and advised by genetic counselors on the risk of their children having the disease.

[20]Warren Tay (1843–1927), English physician; Bernard Sachs (1858–1944), American neurologist

tube of toothpaste tightly around a pencil. The pencil represents the axon, and the spiral layers of toothpaste tube (with the toothpaste squeezed out) represent the myelin. The toothpaste would be forced to one end of the tube, which would form a bulge on the external surface of the wrapping, like the body of the Schwann cell.

Since each glial cell (Schwann cell or oligodendrocyte) myelinates only part of an axon, the myelin sheath is segmented. The gaps between the segments are called **nodes of Ranvier**[21] (RON-vee-AY), and the myelin-covered segments are called **internodes** (see fig. 12.4). The internodes are about 0.2 to 1.0 mm long in the PNS. The short section of nerve fiber between the axon hillock and the first glial cell is called the **initial segment.** Since the axon hillock and initial segment play an important role in initiating a nerve signal, they are collectively called the **trigger zone.**

In peripheral nerve fibers, there is a basal lamina external to the neurilemma, and then a thin sleeve of fibrous connective tissue called the *endoneurium.* The neurilemma and endoneurium are necessary for the repair (regeneration) of damaged nerve fibers. Nerve fibers of the CNS have no neurilemma or endoneurium and are incapable of regeneration, although being enclosed in the cranium and vertebral column, they are better protected from injury.

Unmyelinated Nerve Fibers

Not all nerve fibers are myelinated, but even the **unmyelinated fibers** of the PNS are associated with Schwann cells. In such cases, one Schwann cell harbors from 1 to 12 small nerve fibers in grooves in its surface. Some nerve fibers lie in shallow grooves, while others are enclosed in deep infoldings of the Schwann cell membrane (fig. 13.8). A basal lamina surrounds the entire Schwann cell along with its nerve fibers. Figure 13.9 contrasts myelinated and unmyelinated nerve fibers.

Myelin and Signal Conduction

The significance of a myelin sheath lies in its effect on the conduction of nerve signals. In an unmyelinated nerve fiber, the signal spreads by the diffusion of sodium and potassium ions through the plasma membrane at every point along the fiber. This ion movement creates a sudden voltage change called an **action potential** at each point on the membrane. Each action potential triggers another one just ahead of it, like a burning fuse igniting the unburnt fuse just ahead of it. The **nerve signal** consists of a wave of action potentials traveling down the axon.

In small unmyelinated fibers (2–4 μm in diameter), nerve signals travel at speeds of 0.5 to 2 m/sec. In myelinated fibers of the same size, signals travel 3 to 15 m/sec, and in large myelinated fibers (up to 20 μm in diameter) they travel as fast as 120 m/sec. The increased speed in myelinated fibers in general is due to the fact that ion movements through the membrane, a relatively slow process, occur only at the nodes of Ranvier. In the internodes, signals travel by a much faster process of ion diffusion along the length of the nerve fiber immediately under the plasma membrane. Since most of the fiber is covered with myelin internodes, the signal travels by the faster method most of the way.

One might wonder why all of our nerve fibers are not large, myelinated, and fast, but if this were so, our nervous system would be either impossibly bulky or limited to far fewer fibers. Slow unmyelinated fibers are quite sufficient for stimulating processes in which quick responses are not particularly important, such as the secretion of stomach acid or dilation of the pupil. Fast myelinated fibers are employed where speed is more important, as in motor commands to the skeletal muscles and sensory signals for vision and balance.

Before You Go On

Answer the following questions to test your understanding of the preceding section:

8. From memory, make your own table of the six kinds of glial cells and the functions of each. Which type has the most varied functions?
9. Summarize the major ways in which oligodendrocytes and Schwann cells differ in the way they produce a myelin sheath, and state where the glial cell body of each type is located relative to the myelin and nerve fiber.
10. Explain why damaged nerve fibers in the PNS can regenerate but damaged fibers in the CNS cannot.
11. Explain why nerve signals travel faster in myelinated fibers than in unmyelinated ones. This being the case, why aren't *all* nerve fibers in the body myelinated?

Synapses and Neural Circuits

Objectives

When you have completed this section, you should be able to

- describe the synaptic junctions between one neuron and another;
- describe the variety of interconnections that exist between two neurons; and
- describe four basic variations in the circuitry or "wiring patterns" of the nervous system.

No neuron functions in isolation from others; neurons work in groups of cells that are interconnected in patterns similar to the electrical circuits of radios and other electronic devices. In this section, we examine the connections between neurons and the functional circuits of neuronal groups.

Synapses

The meeting point between a neuron and any other cell is called a **synapse.** The other cell may be an epithelial, muscular, glandular, or other cell type, but in most cases, it is another neuron. Synapses make neural integration (information processing) possible; each synapse is a "decision-making" device that determines whether a second cell will respond to signals from the first. Without synapses, signals would simply be transmitted automatically from receptors to effectors, effectors would respond to every stimulus, and the

[21] L. A. Ranvier (1835–1922), French histologist and pathologist

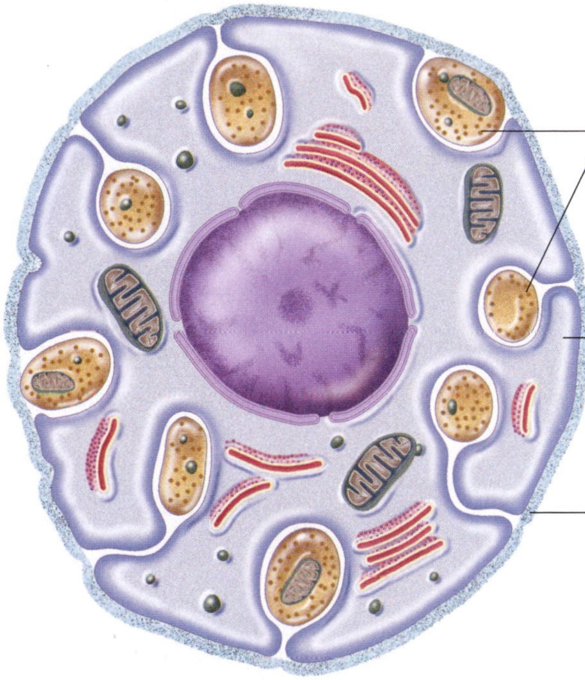

FIGURE 13.8

Unmyelinated Nerve Fibers. Multiple unmyelinated fibers are enclosed in channels in the surface of a single Schwann cell.

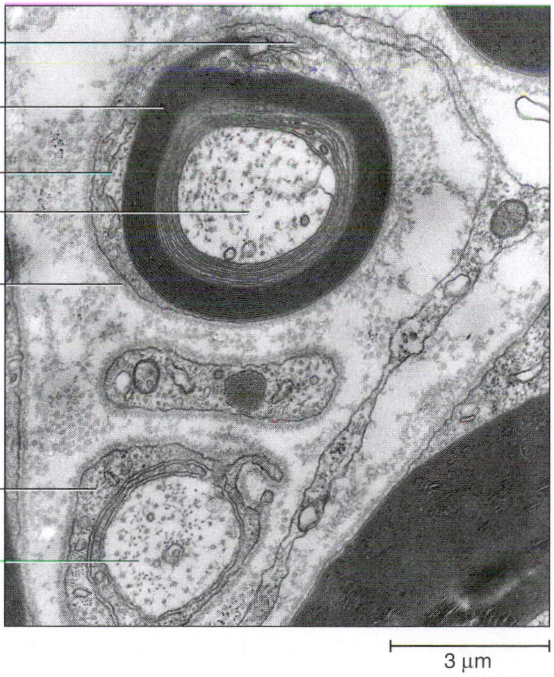

FIGURE 13.9

Myelinated and Unmyelinated Axons (TEM).

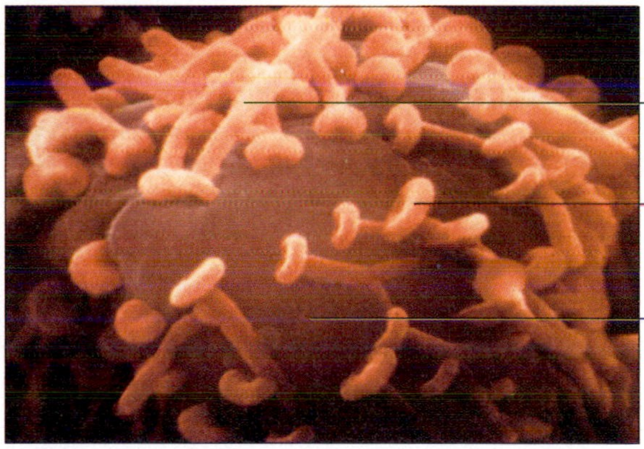

FIGURE 13.10

Synaptic Knobs on the Soma of a Neuron in a Marine Slug, Aplysia (SEM).

nervous system would be incapable of any decision making. But in reality, one neuron can have an enormous number of synapses and thus a great deal of information-processing capability (fig. 13.10). For example, a spinal motor neuron receives about 8,000 synaptic contacts from other neurons on its dendrites and another 2,000 on its soma. In part of the brain called the cerebellum, one neuron can have as many as 100,000 synapses. The cerebral cortex (the main information-processing tissue of the brain) is estimated to have 100 trillion (10^{14}) synapses. To get some impression of this number, imagine trying to count them. Even if you could count two synapses per second, night and day without a coffee break, and you were immortal, it would take you 1.6 million years.

A nerve signal arrives at a synapse by way of the **presynaptic neuron,** then continues on its way via the **postsynaptic neuron** (fig. 13.11*a*). When a presynaptic axon ends at the dendrite of a postsynaptic neuron, the two cells are said to form an **axodendritic synapse.** When the presynaptic axon terminates on the soma of the next cell, they form an **axosomatic synapse.** When it terminates on the axon of the next cell, they form an **axoaxonic synapse** (fig. 13.11*b*).

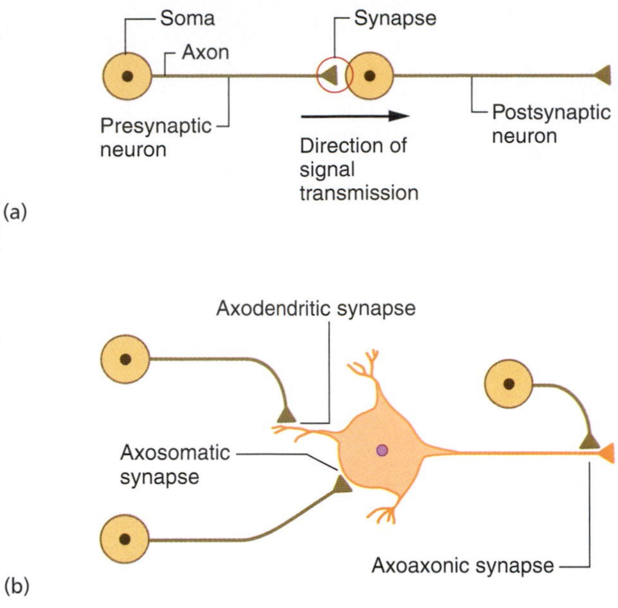

FIGURE 13.11

Synaptic Relationships Between Neurons. (*a*) Pre- and postsynaptic neurons. (*b*) Types of synapses defined by the site of contact on the postsynaptic neuron.

CHEMICAL SYNAPSES AND NEUROTRANSMITTERS

A **chemical synapse** is a junction at which the presynaptic neuron releases a neurotransmitter to stimulate the postsynaptic cell. The neuromuscular junction (NMJ) described in chapter 10 is an example of this. The NMJ and many other synapses employ *acetylcholine* as a neurotransmitter. Postsynaptic neurons of the sympathetic nervous system use *norepinephrine.* Some neurotransmitters are *excitatory* and tend to cause the postsynaptic cell to generate a nerve signal. Some widely used excitatory neurotransmitters in the central nervous system (CNS) are *glutamate* in the brain and *aspartate* in the spinal cord. Other neurotransmitters are *inhibitory* and suppress responses in the postsynaptic cell. The most widely used inhibitory neurotransmitters in the CNS are *gamma-aminobutyric acid (GABA)* in the brain and *glycine* in the spinal cord. Some other well-known neurotransmitters are *dopamine, serotonin, histamine,* and *beta-endorphin.* There are over 100 known neurotransmitters.

At a chemical synapse, a terminal branch of the presynaptic nerve fiber ends in a swelling, the **synaptic knob.** The knob is separated from the next cell by a 20- to 40-nm gap called the **synaptic cleft** (fig. 13.12). The knob contains membrane-bounded secretory vesicles called **synaptic vesicles,** which contain the neurotransmitter. Many of these vesicles are "docked" at release sites on the inside of the plasma membrane, ready to release their neurotransmitter on demand. Neurotransmitter release is achieved by exocytosis (see chapter 2). A reserve pool of synaptic vesicles is located a little farther away from the membrane, clustered near the release sites and tethered to the cytoskeleton by protein microfilaments. These vesicles stand by and "step forward" to dock on the membrane and release their neurotransmitter after the previously docked vesicles have expended their contents. Synaptic vesicles are found in a few cells other than neurons, such as the sensory cells of taste, hearing, and equilibrium. They release neurotransmitter to stimulate a nearby nerve cell.

A postsynaptic neuron does not show such conspicuous specializations. At this end, a neuron has no synaptic vesicles and cannot release neurotransmitters. Its membrane does, however, contain proteins that function as neurotransmitter receptors, and it may be folded to increase its receptor-laden surface area, and therefore its sensitivity to the neurotransmitter. A signal always travels in only one direction across a chemical synapse, from the presynaptic cell with synaptic vesicles to the postsynaptic cell with neurotransmitter receptors. This one-way transmission ensures the precise routing of nerve signals in the body.

Synaptic transmission begins when a nerve signal arrives at the ends of the presynaptic neuron and triggers the exocytosis of synaptic vesicles. Neurotransmitter is released into the synaptic cleft, diffuses across to the postsynaptic cell, and binds to receptors on that cell's membrane. Depending on the neurotransmitter and the type of receptor, this may either stimulate or inhibit the postsynaptic cell. The postsynaptic cell "decides" whether or not to initiate a new nerve signal based on the composite effects of excitatory and inhibitory input through the many synapses on its dendrites and soma.

ELECTRICAL SYNAPSES

Another type of synapse, called an **electrical synapse,** connects some neurons, neuroglia, and cardiac and single-unit smooth muscle cells. Here, adjacent cells are joined by gap junctions that allow ions to diffuse directly from one cell into the next. These junctions have the advantage of quick transmission because there is no delay for the release and binding of neurotransmitter. Their disadvantage, however, is that they cannot integrate information and make decisions.

Neuronal Pools and Circuits

Neurons function in ensembles called **neuronal pools.** One neuronal pool may consist of thousands to millions of interneurons concerned with a particular body function—one to control the rhythm of your breathing, one to move your limbs rhythmically as you walk, one to regulate your sense of hunger, and another to interpret smells, for example. The functioning of a neuronal pool hinges on the anatomical organization of its neurons, much like the functioning of a radio depends on the particular way its transistors, diodes, and capacitors are laid out. The interconnections between neurons are called **neuronal circuits.** A wide variety of neuronal functions result from the operation of four principal kinds of neuronal circuits (fig. 13.13):

1. In a **diverging circuit,** one nerve fiber branches and synapses with several postsynaptic cells. Each of those may synapse with several more, so input from just one neuron may produce output through dozens more. Such a circuit allows one motor neuron of the brain, for example, to ultimately cause thousands of muscle fibers to contract.

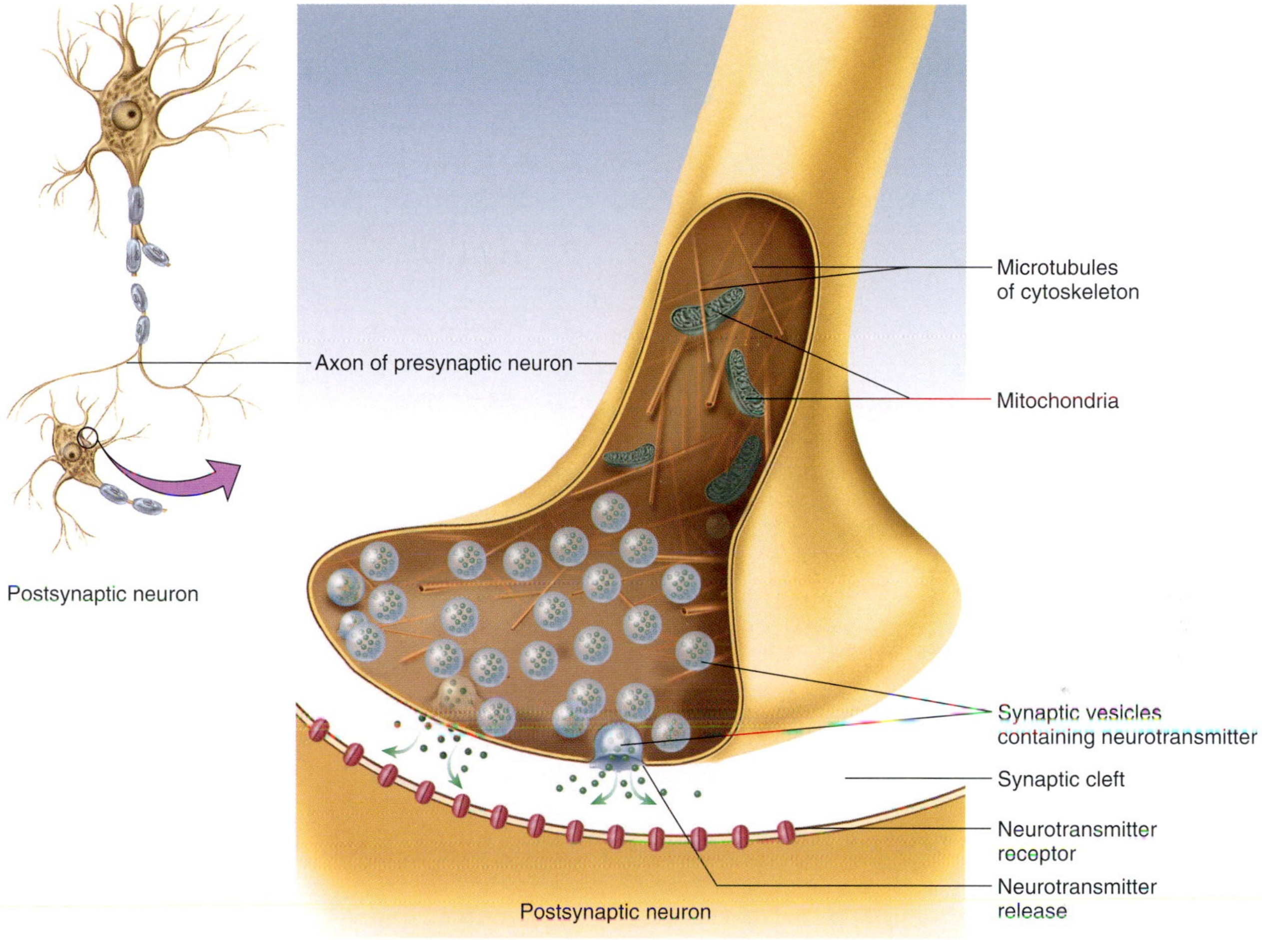

Axon of presynaptic neuron
Postsynaptic neuron
Microtubules of cytoskeleton
Mitochondria
Synaptic vesicles containing neurotransmitter
Synaptic cleft
Neurotransmitter receptor
Neurotransmitter release
Postsynaptic neuron

FIGURE 13.12
Structure of a Chemical Synapse.

2. A **converging circuit** is the opposite of a diverging circuit—input from many different sources is funneled to one neuron or neuronal pool. Through neuronal convergence, a respiratory center in the brainstem receives input from other parts of the brain, from receptors for blood chemistry in the arteries, and from stretch receptors in the lungs. The respiratory center can then produce an output that takes all of these factors into account and sets an appropriate pattern of breathing.
3. In a **reverberating circuit,** neurons stimulate each other in a linear sequence such as A → B → C → D, but neuron *C* sends an axon collateral back to *A.* As a result, every time *C* fires it not only stimulates output neuron *D,* but also restimulates *A* and starts the process over. Such a circuit produces a repetitive output that lasts until one or more neurons in the circuit fail to fire, or until an inhibitory signal from another source stops one of them from firing. A reverberating circuit sends repetitious signals to the diaphragm and intercostal muscles, for example, to make you inhale. When the circuit stops firing, you exhale; the next time it fires, you inhale again. Reverberating circuits may also be involved in short-term memory (for example, in the way a telephone number "echoes" in your memory from the time you look it up in the phone book until the time you dial it), and they may play a role in the uncontrolled "storms" of neuronal activity that occur in epilepsy.
4. In a **parallel after-discharge circuit,** an input neuron diverges to stimulate several chains of neurons. Each chain has a different number of synapses, but eventually they all reconverge on the same output neuron. Each synapse delays a nerve signal by about 0.5 millisecond, so the more synapses there are in a pathway, the longer it takes a nerve signal to get through that pathway to the output neuron. The output neuron, receiving signals from multiple pathways, may go on firing for some time after the input has ceased. Unlike a reverberating circuit, this type has no feedback loop. Once all the neurons in the circuit have fired, the output ceases. Continued firing after the stimulus stops is called *after-discharge.* It explains why you can stare at a lamp, then close your eyes and continue to see an image of it for a while. Such a circuit is also important to certain

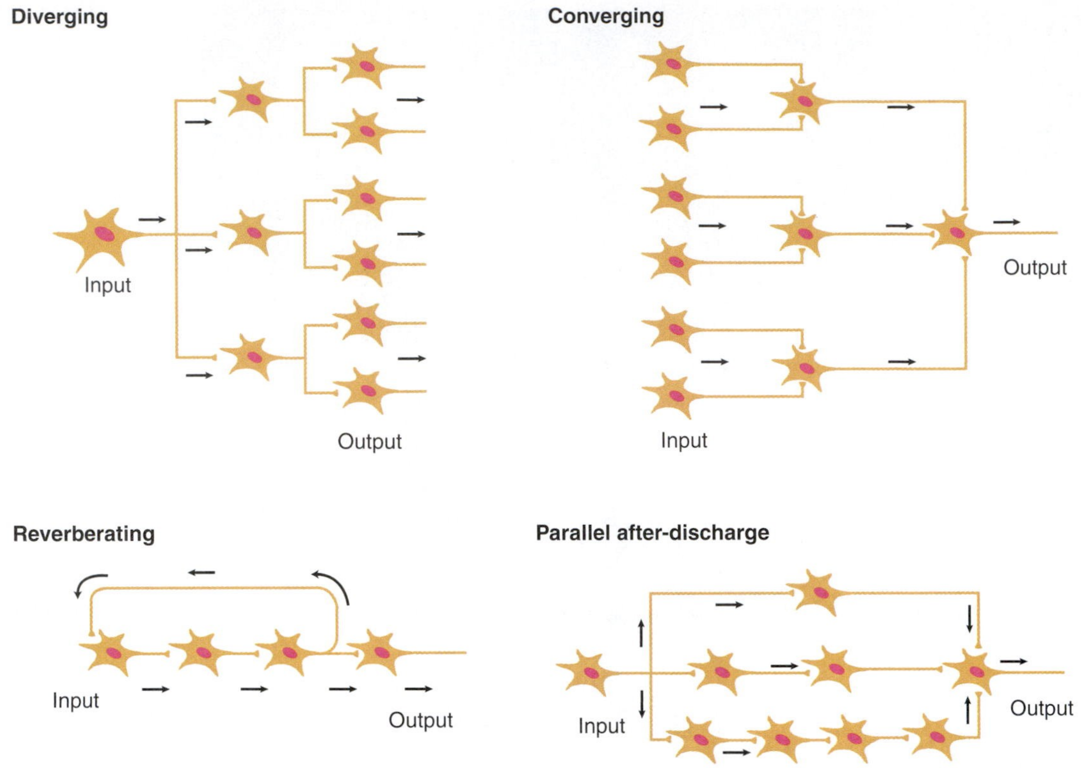
Diverging
Input
Output
Converging
Input
Output
Reverberating
Input
Output
Parallel after-discharge
Input
Output

FIGURE 13.13
Four Types of Neuronal Circuits. *Arrows* indicate the direction of signal transmission.

reflexes, for example when a brief pain produces a longer-lasting output to the limb muscles and causes you to draw back your hand or foot from danger. (See the discussion of *reflex arcs* in chapter 14)

Before You Go On

Answer the following questions to test your understanding of the preceding section:

12. At a given synapse, what features are present on the presynaptic neuron that are absent from the postsynaptic neuron?
13. In synaptic transmission, where does a neurotransmitter come from? How does it affect the postsynaptic neuron?
14. Name any four neurotransmitters and state some functional differences between them.
15. What is an electrical synapse? Where can electrical synapses be found? Identify an advantage and a disadvantage of an electrical synapse compared to a chemical synapse.
16. What is the difference between a neuronal pool and a neuronal circuit?
17. Name the four types of neuronal circuits and briefly describe the functional differences between them, or an advantage of each type for certain purposes.

Developmental and Clinical Perspectives

Objectives

When you have completed this section, you should be able to

- describe how the nervous system develops in an embryo; and
- describe a few birth defects that result from abnormalities of this developmental process.

Development of the Nervous System

Some aspects of nervous system development, or **neurulation,** were briefly discussed in chapter 4. Further understanding of this process will form a basis for understanding the brain and spinal cord anatomy presented in chapters 14 and 15.

The first embryonic trace of the central nervous system appears early in the third week of development. A dorsal streak called the *neuroectoderm* appears along the length of the embryo and thickens to form a **neural plate** (fig. 13.14). This is destined to give rise to all neurons and glial cells except microglia, which come from mesoderm. As development progresses, the neural plate sinks and the edges of it thicken, thus forming a **neural groove** with a raised **neural fold** along each side. The neural folds then fuse along the midline, somewhat like a closing zipper, beginning in the cervical

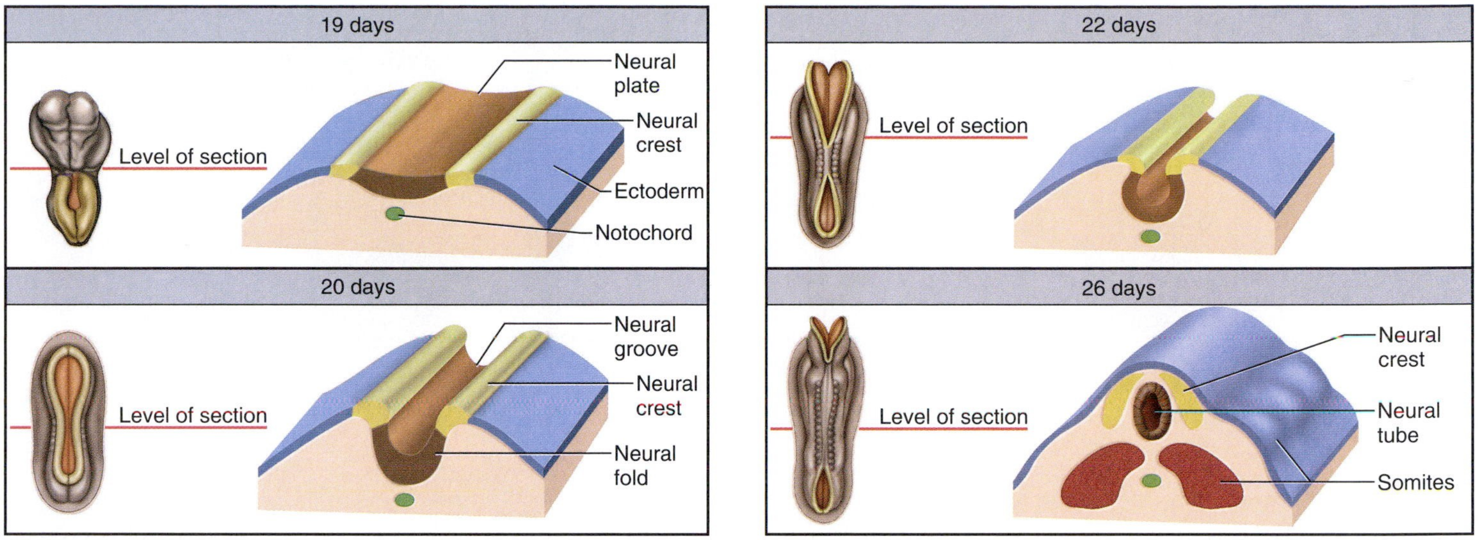

FIGURE 13.14

Formation of the Neural Tube. The *left-hand* figure in each case is a dorsal view of the embryo and the *right-hand* figure is a three-dimensional representation of the tissues at the indicated level on the respective embryo.

(neck) region of the neural groove and progressing rostrally (toward the head) and caudally (toward the tail). By 4 weeks, this process creates a hollow channel called the **neural tube.** For a time, the neural tube is open to the amniotic fluid at the rostral and caudal ends. These openings close at 25 and 27 days, respectively. The lumen of the neural tube becomes a fluid-filled space that later constitutes the *central canal* of the spinal cord and *ventricles* of the brain.

Following closure, the neural tube separates from the overlying ectoderm, sinks a little deeper, and grows lateral processes that later give rise to motor nerve fibers. Some ectodermal cells that originally lay along the margin of the neural groove separate from the rest and form a longitudinal column on each side called the **neural crest.** Some neural crest cells become sensory neurons, while others migrate to other locations and give rise to sympathetic neurons, ganglia, Schwann cells, and the *adrenal medulla,* a gland described in chapter 18.

By the fourth week, the neural tube exhibits three anterior dilations, or *primary vesicles,* called the **forebrain** (*prosencephalon*[22]) (PROSS-en-SEF-uh-lon), **midbrain** (*mesencephalon*[23]) (MEZ-en-SEF-uh-lon), and **hindbrain** (*rhombencephalon*[24]) (ROM-ben-SEF-uh-lon) (fig. 13.15). While these vesicles develop, the neural tube bends at the junction of the hindbrain and spinal cord to form the **cervical flexure,** and in the midbrain region to form the **cephalic flexure.**

By the fifth week, the neural tube undergoes further flexion and subdivides into five *secondary vesicles.* The forebrain divides into two of them, the **telencephalon**[25] (TEL-en-SEFF-uh-lon) and **diencephalon**[26] (DY-en-SEF-uh-lon); the midbrain remains undivided and retains the name **mesencephalon;** and the hindbrain divides into two vesicles, the **metencephalon**[27] (MET-en-SEF-uh-lon) and **myelencephalon**[28] (MY-el-en-SEF-uh-lon). The telencephalon has a pair of lateral outgrowths that later become the *cerebral hemispheres,* and the diencephalon exhibits a pair of small cuplike *optic vesicles* that become the retinas of the eyes. Figure 13.15*c* shows structures of the fully developed brain that arise from each of the secondary vesicles.

In week 14, Schwann cells and oligodendrocytes begin spiraling around the nerve fibers, laying down layers of myelin and giving the fibers a white appearance. Yet very little myelin is present in the brain at birth, and there is little visible distinction between the *gray matter* and *white matter* of the newborn brain. Myelination proceeds rapidly in infancy and it is this, far more than the multiplication or enlargement of neurons, that accounts for most postnatal brain growth. Myelination is not completed until late adolescence. Since myelin has such a high lipid content, dietary fat is important to early nervous system development. Well-meaning parents can do their children significant harm by giving them the sort of low-fat diets (skimmed milk, etc.) that may be beneficial to an adult.

In the third month of development, the spinal cord extends for the full length of the embryo. As the vertebrae develop (see chapter 7), *spinal nerves* arise from the cord and pass straight laterally to emerge between the vertebrae, through the intervertebral foramina. Subsequently, however, the vertebral column grows faster than the spinal cord. By birth, the cord ends in the vertebral canal of the third lumbar vertebra (L3), and by adulthood, it ends at the level of L1 to L2. As the vertebral column elongates, the spinal nerve roots elongate, so they still emerge between the same vertebrae but the lower vertebral canal is occupied by a bundle of nerve roots instead of spinal cord. The resulting adult anatomy is described in chapter 14.

[22]*pros* = before, in front + *encephal* = brain
[23]*mes* = middle
[24]*rhomb* = rhombus
[25]*tele* = end, remote
[26]*di* = through, between
[27]*met* = behind, beyond, distal to
[28]*myel* = spinal cord

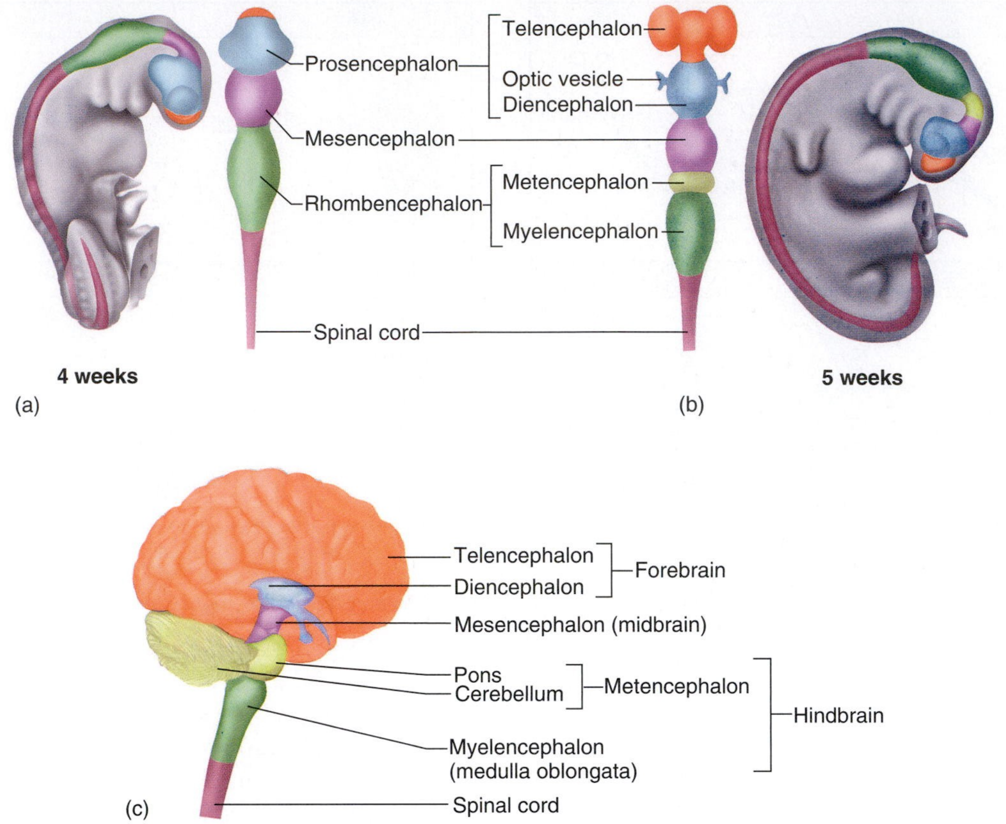

FIGURE 13.15

Primary and Secondary Vesicles of the Embryonic Brain. (*a*) The three primary vesicles at 4 weeks. (*b*) The secondary vesicles at 5 weeks. (*c*) The fully developed brain, color-coded to relate its structures to the secondary embryonic vesicles.

Developmental Disorders of the Nervous System

The central nervous system is subject to multiple aberrations in embryonic development. Approximately 1 out of 100 live-born infants exhibit major defects in brain development. Common among these are **neural tube defects (NTDs)** such as **spina bifida** (SPY-nuh BIF-ih-duh). Spina bifida occurs when one or more vertebrae fail to form a complete neural arch for enclosure of the spinal cord. It is especially common in the lumbosacral region. The mildest form, *spina bifida occulta,*[29] involves only one to a few vertebrae and causes no functional problems. Its only external sign is a dimple or patch of hairy pigmented skin on the lower back. *Spina bifida cystica*[30] is more serious (fig. 13.16). A sac protrudes from the spine and may contain parts of the spinal cord and nerve roots, meninges (membranes that enclose the CNS), and cerebrospinal fluid. In extreme cases, inferior spinal cord function is absent, causing paralysis of the lower limbs and urinary bladder and lack of bowel control. Bladder paralysis can lead to chronic urinary infections and renal failure. Pregnant women can significantly reduce the risk of spina bifida by taking supplemental folic acid (a B vitamin) during early pregnancy.

Other severe neural tube defects include microcephaly and anencephaly. In **microcephaly,**[31] the face is of normal size but the brain and calvaria are abnormally small. Microcephaly is accompanied by profound mental retardation. **Anencephaly**[32] results from failure of the rostral end of the neural tube to close. This leaves the brain exposed to the amniotic fluid. The brain tissue degenerates, most of the brain is absent at birth, and the head is relatively flat or truncated above the eyes. Such infants generally die within a few hours. Neural tube defects sometimes run in families but can also be caused by teratogens and nutritional deficiencies.

Other disorders of the nervous system are described in chapters 14 through 17.

[29] *bifid* = divided, forked + *occult* = hidden
[30] *cyst* = sac, bladder
[31] *micro* = small + *cephal* = head
[32] *an* = without + *encephal* = brain

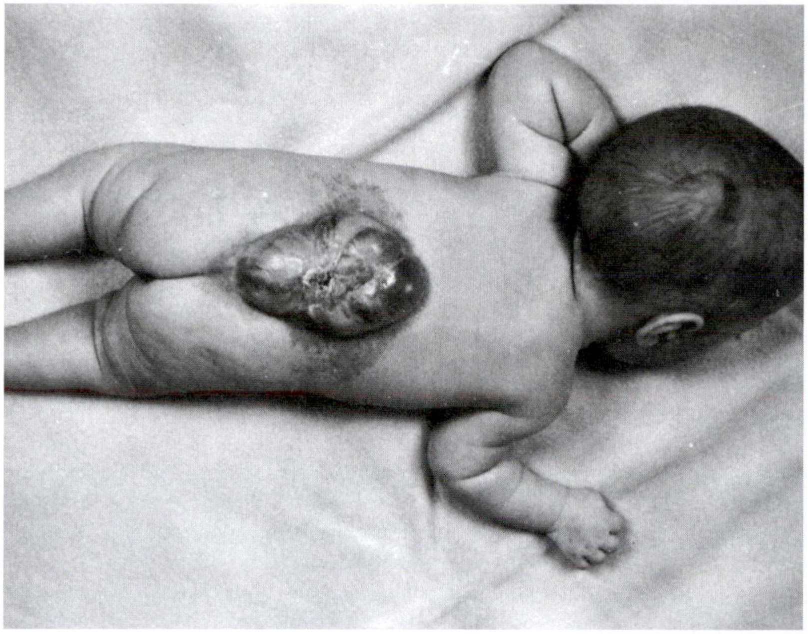

FIGURE 13.16
Spina Bifida Cystica.

Before You Go On

Answer the following questions to test your understanding of the preceding section:

18. How does the neural crest originate? What cells or tissues arise from it?
19. Where does closure of the neural tube begin? What are the last regions to close?
20. What single adult structure arises from all five of the secondary vesicles of the neural tube?
21. When does myelination begin? When does it end?

CHAPTER REVIEW

Review of Key Concepts

Overview of the Nervous System (p. 366)

1. The nervous and endocrine systems are the body's two principal mechanisms of internal communication and coordination. The nervous system is specialized for the rapid transmission of signals from cell to cell.
2. The two major divisions of the nervous system are the *central nervous system (CNS)* (brain and spinal cord) and the *peripheral nervous system (PNS)* (nerves and ganglia).
3. The PNS has *sensory (afferent)* and *motor (efferent)* divisions, and each of these, in turn, has subdivisions called the visceral and somatic divisions. The *visceral division* innervates organs of the body cavities such as the heart and stomach, and the *somatic division* innervates the skin, muscles, bones, and joints.
4. The visceral motor division is also called the *autonomic nervous system.* It innervates glands, cardiac muscle, and smooth muscle, and controls unconscious, involuntary visceral reflexes. It consists of *sympathetic* and *parasympathetic* divisions, which often have contrasting effects on the same target organs.

Nerve Cells (Neurons) (p. 367)

1. The functional unit of the nervous system is the *neuron.* Neurons are able to communicate only because they possess the properties of excitability, conductivity, and secretion.
2. There are three functional classes of neurons: *sensory (afferent) neurons,* which convey signals to the CNS; *interneurons,* contained entirely within the CNS; and *motor (efferent) neurons,* which convey signals away from the CNS to effectors such as muscle and gland cells.
3. A neuron consists of a *soma* (cell body); usually multiple *dendrites,* which receive signals and convey them to the soma; and a single *axon* (nerve fiber), which carries nerve signals away from the soma. The soma contains the nucleus and protein-synthesizing organelles of the cell. The axon arises from an *axon hillock.* This and the first segment of the axon are the *trigger zone,* where action potentials are generated. The axon branches into a *terminal arborization* at its distal end, and each branch ends in a dilated *synaptic knob.*
4. Neurons are described as *multipolar* (with an axon and two or more dendrites), *bipolar* (with an axon and one dendrite), *unipolar* (with only a single process arising from the soma), or *anaxonic* (with dendrites but no axon).

Supportive Cells (Neuroglia) (p. 371)

1. Most cells of the nervous system are not neurons but *neuroglia (glial cells).*
2. Four kinds of neuroglia occur in the CNS: *oligodendrocytes* (which produce myelin), *astrocytes* (with numerous roles in the supportive framework of the CNS, the blood-brain barrier, nourishment of neurons, homeostatic maintenance of the extracellular fluid, and repair of damaged CNS tissue); *ependymal cells* (which line the internal cavities of the CNS and produce cerebrospinal fluid); and *microglia* (macrophages of the CNS).
3. Two kinds of glial cells occur in the PNS: *Schwann cells* (which produce myelin) and *satellite cells* (of little-known function).
4. Myelin is an insulating sheath around certain nerve fibers. It consists of spiral layers of plasma membrane arising from oligodendrocytes in the CNS and Schwann cells in the PNS.
5. In the PNS, the outermost coil of the Schwann cell is called the *neurilemma.* It is covered with a basal lamina and then a thin connective tissue sheath, the *endoneurium.* Neurilemma and endoneurium are required for the regeneration of damaged nerve fibers. They are absent from the CNS, and damaged fibers there cannot regenerate.
6. One glial cell myelinates only a short segment of a nerve fiber. Therefore the myelin sheath around a nerve fiber is segmented, with long *internodes* separated by interruptions of the myelin sheath called *nodes of Ranvier.*
7. Schwann cells also envelop *unmyelinated neurons,* but enfold them in only one coil of plasma membrane and do not form myelin around them. Each Schwann cell can have several surface grooves, each accommodating one unmyelinated nerve fiber.
8. A nerve signal consists of a chain reaction of electrical changes called *action potentials.* Nerve signals travel relatively slowly (up to 2 m/sec) in unmyelinated fibers; faster in myelinated fibers of the same diameter; and fastest (up to 120 m/sec) in large myelinated fibers. Myelin speeds up signal conduction because the signal travels by a relatively rapid means in the internodes, which cover most of a nerve fiber.

Synapses and Neural Circuits (p. 374)

1. The point where a nerve fiber ends at a target cell (such as another neuron or a muscle or gland cell) is called a *synapse.* Synapses are the decision-making, information-processing points in the nervous system; the more synapses a neuron or neural circuit has, the more data the neuron or circuit can process.
2. With respect to the direction of signal transmission, the neuron before the synapse is called the *presynaptic neuron,* and the one after the synapse is the *postsynaptic neuron.*
3. A presynaptic neuron can terminate on the dendrites, soma, or axon of a postsynaptic neuron; such junctions are respectively called *axodendritic, axosomatic,* and *axoaxonic synapses.*
4. A *chemical synapse* is one at which the presynaptic neuron releases a chemical *neurotransmitter,* which diffuses across the *synaptic cleft* and binds to receptors on the postsynaptic cell.
5. Some familiar neurotransmitters are acetylcholine, norepinephrine, epinephrine, glutamate, aspartate, GABA, glycine, dopamine, serotonin, histamine, and beta-endorphin. There are many others.
6. Neurotransmitters are stored in *synaptic vesicles* of the presynaptic neuron. The arrival of a nerve signal stimulates the release of neurotransmitter by vesicle exocytosis.
7. Some cells are linked by *electrical synapses* (gap junctions)—cardiac and single-unit smooth muscle, and some neurons and neuroglia. Electrical synapses allow for very rapid signal transmission but no decision-making.
8. Neurons function in groups called *neuronal pools,* aggregations of neurons collectively dedicated to a certain purpose such as breathing or sensory perception. Within a pool, the neurons are connected along pathways called *neuronal circuits.*
9. There are four principal types of neuronal circuits: *diverging, converging, reverberating,* and *parallel after-discharge circuits.*

Developmental and Clinical Perspectives (p. 378)

1. The early stages of central nervous system development are a middorsal thickening of ectoderm called the *neural plate,* developing into a *neural groove* flanked by raised *neural folds,* and then developing into an enclosed *neural tube.*
2. A longitudinal column of ectodermal tissue separates from the neural groove on each side to become the *neural crest;* this gives rise to sensory and sympathetic neurons, neuroglia, and other cell types.
3. The neural tube develops anterior dilations that form three *primary vesicles* (*forebrain, midbrain,* and *hindbrain*), then undergoes flexion and subdivision of the forebrain and hindbrain, producing five *secondary vesicles.*
4. Myelination begins in the fourth month, but most brain myelination occurs after birth.
5. The spinal cord initially occupies the entire vertebral canal, but the vertebral column grows faster than the spinal cord and by adulthood, the spinal cord ends at the level of vertebrae S1 to S2.
6. *Neural tube defects (NTDs)* are deformities of the brain or spinal cord that result from failure of the neural tube to close or otherwise to develop normally. NTDs range from the relatively mild *spina bifida occulta* to the more serious *spina bifida cystica, microcephaly,* and *anencephaly.* NTDs can be genetic or caused by teratogens and nutritional deficiencies.

Testing Your Recall

1. The integrative functions of the nervous system are performed mainly by
   a. afferent neurons.
   b. efferent neurons.
   c. neuroglia.
   d. sensory neurons.
   e. interneurons.
2. Neurons arise from embryonic
   a. endoderm.
   b. epidermis.
   c. mesoderm.
   d. mesenchyme.
   e. ectoderm.
3. The soma of a mature neuron lacks
   a. a nucleus.
   b. endoplasmic reticulum.
   c. lipofuscin.
   d. centrioles.
   e. ribosomes.
4. The glial cells that destroy microorganisms in the CNS are
   a. microglia.
   b. satellite cells.
   c. ependymal cells.
   d. oligodendrocytes.
   e. astrocytes.
5. A _____ circuit produces a continuous stream of output even after the input has stopped.
   a. diverging
   b. converging
   c. presynaptic
   d. reverberating
   e. parallel after-discharge
6. Neurotransmitters are found in
   a. the cell bodies of neurons.
   b. the dendrites.
   c. the axon hillock.
   d. the synaptic knob.
   e. the postsynaptic plasma membrane.
7. Another name for the axon of a neuron is
   a. nerve fiber.
   b. neurofibril.
   c. neurilemma.
   d. axoplasm.
   e. endoneurium.
8. Nerves that control the motility of the stomach or rate of the heartbeat would belong to
   a. the central nervous system.
   b. the somatic sensory division.
   c. the somatic motor division.
   d. the visceral motor division.
   e. the visceral sensory division.
9. The glial cells that guide migrating neurons in the developing fetal brain are
   a. astrocytes.
   b. oligodendrocytes.
   c. satellite cells.
   d. ependymal cells.
   e. microglia.
10. Which of the following appears earlier than all the rest in prenatal development of the nervous system?
    a. the neural groove.
    b. the primary vesicles.
    c. the neural plate.
    d. the neural crest.
    e. the neural tube.
11. Neurons that convey information to the CNS are called sensory, or _____, neurons.
12. Motor effects that depend on repetitive output from a neuronal pool are most likely to use the _____ type of neuronal circuit.
13. Prenatal degeneration of the forebrain results in a birth defect called _____.
14. Neurons receive incoming signals by way of specialized processes called _____.
15. In the central nervous system, cells called _____ perform one of the same functions that Schwann cells do in the peripheral nervous system.
16. A/an _____ synapse is formed when a presynaptic neuron synapses with the cell body of a postsynaptic neuron.
17. All of the nervous system except the brain and spinal cord is called the _____.
18. Whether or not it forms myelin, a Schwann cell always forms a sleeve called the _____ around a peripheral nerve fiber. If myelin is present, it lies between the nerve fiber and this sleeve.
19. The _____ of a neuron consists of the axon hillock and the exposed part of the axon between the soma and the first segment of the myelin sheath.
20. At a given synapse, the _____ neuron has neurotransmitter receptors.

Answers in the Appendix

True or False

Determine which five of the following statements are false, and briefly explain why.

1. Neurons are incapable of mitosis.
2. Most neurons have more dendrites than axons.
3. Dendrites never contain synaptic vesicles.
4. Interneurons connect sense organs to the CNS.
5. Nerve signals travel faster in myelinated nerve fibers than in unmyelinated ones.
6. The myelin sheath covers the neurilemma of a nerve fiber.
7. Nodes of Ranvier are present only in myelinated fibers of the PNS.
8. The outermost tissue layer of a nerve is the perineurium.
9. Unipolar neurons cannot produce action potentials because they have no axon.
10. There are more glial cells than neurons in the nervous system.

Answers in the Appendix

Testing Your Comprehension

1. Suppose some hypothetical disease prevented the formation of astrocytes in the fetal brain. How would you expect this to affect brain development?
2. How would nervous system function be affected if both the presynaptic and postsynaptic neurons at every synapse had both synaptic vesicles and neurotransmitter receptors?
3. What unusual characteristic of neurons can be attributed to their lack of centrioles?
4. Of the three properties of neurons described on p. 367, which ones are also characteristic of skeletal muscle fibers (see chapter 10)? Explain why these properties are needed by both types of cells. Which is absent from skeletal muscle? Explain why neurons require this property but skeletal muscle does not.
5. What division of subdivision of the peripheral nervous system would control each of the following: constriction of the pupils in bright light; the movements of your hand as you write; the sensation of a stomach ache; blinking as a particle of dust is blown toward your eye; your awareness of the position of your hand as you touch your nose with your eyes closed. Briefly explain each answer.

Answers at the Online Learning Center

www.mhhe.com/saladinha1

Visit the Online Learning Center for practice tests, answer keys, and other learning aids for this chapter. Enhance your understanding of human anatomy with our interactive art labeling exercises, supplemental photo atlases, web links, puzzles, flashcards, and much more.

14

CHAPTER FOURTEEN

The Spinal Cord and Spinal Nerves

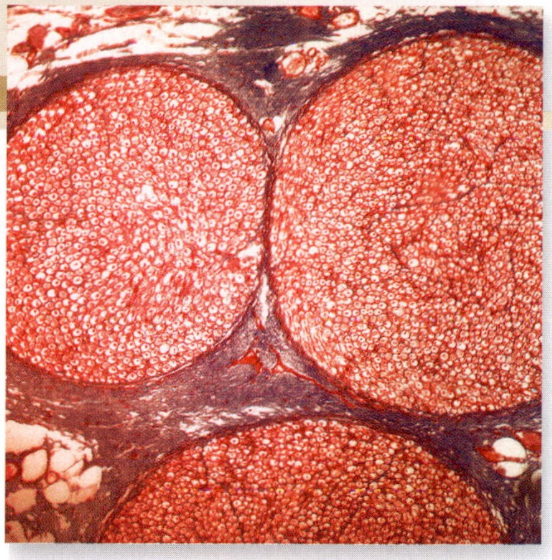

Cross section of a nerve showing parts of three fascicles

CHAPTER OUTLINE

The Spinal Cord 386
- Functions 386
- Surface Anatomy 386
- Meninges of the Spinal Cord 386
- Cross-Sectional Anatomy 387
- Spinal Tracts 389

The Spinal Nerves 393
- General Anatomy of Nerves and Ganglia 393
- Spinal Nerves 395
- Nerve Plexuses 397
- Cutaneous Innervation and Dermatomes 405

Somatic Reflexes 405

Clinical Perspectives 407

Chapter Review 409

INSIGHTS

14.1 Clinical Application: Spinal Taps 388

14.2 Clinical Application: Poliomyelitis and Amyotrophic Lateral Sclerosis 393

14.3 Clinical Application: Shingles 395

14.4 Clinical Application: Spinal Nerve Injuries 405

BRUSHING UP

To understand this chapter, you may find it helpful to review the following concepts:

- Divisions of the nervous system (p. 366)
- Embryonic development of the spinal cord (pp. 378–379)

We studied the nervous system at a cellular level in chapter 13. In these next four chapters, we move up the structural hierarchy to study the nervous system at the organ and system levels of organization. We begin with the *spinal cord*, an "information highway" between the brain and the trunk and limbs. It is about as thick as a finger, and extends through the vertebral canal as far as the first or second lumbar vertebra. At regular intervals, it gives off a pair of *spinal nerves*, which receive sensory input from the skin, muscles, bones, joints, and viscera, and which issue motor commands back to muscle and gland cells. The spinal cord is a component of the central nervous system and the spinal nerves are a component of the peripheral nervous system, but these central and peripheral components are so closely linked structurally and functionally that it is appropriate that we consider them together in this chapter. The brain and cranial nerves will be discussed in chapter 15.

THE SPINAL CORD

Objectives

When you have completed this section, you should be able to

- name the two types of tissue in the central nervous system and state their locations;
- describe the gross and microscopic anatomy of the spinal cord; and
- name the major conduction pathways of the spinal cord and state their functions.

Functions

The spinal cord serves three principal functions:

1. **Conduction.** The spinal cord contains bundles of nerve fibers that conduct information up and down the body, connecting different levels of the trunk with each other and with the brain. It enables sensory information to reach the brain, motor commands to reach the effectors, and input received at one level of the cord to affect output from another level.
2. **Locomotion.** Walking involves repetitive, coordinated contractions of several muscle groups in the limbs. Motor neurons in the brain initiate walking and determine its speed, distance, and direction, but the simple repetitive muscle contractions that put one foot in front of another, over and over, are coordinated by groups of neurons called **central pattern generators** in the cord. These neuronal circuits produce the sequence of outputs to the extensor and flexor muscles that cause alternating movements of the legs.
3. **Reflexes.** Reflexes are involuntary stereotyped responses to stimuli. They involve the brain, spinal cord, and peripheral nerves.

Surface Anatomy

The **spinal cord** (fig. 14.1) is a cylinder of nervous tissue that begins at the foramen magnum of the skull and passes through the vertebral canal as far as the inferior margin of the first lumbar vertebra (L1) or slightly beyond. In adults, it averages about 1.8 cm thick and 45 cm long. Thus, it occupies only the upper two-thirds of the vertebral canal; the lower one-third is described shortly. The cord gives rise to 31 pairs of spinal nerves. The first pair pass between the skull and vertebra C1, and the rest pass through the intervertebral foramina. Although the spinal cord is not visibly segmented, the part supplied by each pair of spinal nerves is called a *segment.* The cord exhibits longitudinal grooves on its ventral and dorsal sides—the *ventral median fissure* and *dorsal median sulcus,* respectively.

The spinal cord is divided into **cervical, thoracic, lumbar,** and **sacral regions.** It may seem odd that it has a sacral region when the cord itself ends well above the sacrum. These regions, however, are named for the level of the vertebral column from which the spinal nerves emerge, not for the vertebrae that contain the cord itself.

The cord widens at two points along its course: a **cervical enlargement** in the inferior cervical region, where it gives rise to nerves of the upper limbs; and a similar **lumbar enlargement** in the lumbosacral region, where it gives rise to nerves of the pelvic region and lower limbs. Inferior to the lumbar enlargement, the cord tapers to a point called the **medullary cone.** The lumbar enlargement and medullary cone give off a bundle of nerve roots that occupy the vertebral canal from L2 to S5. This bundle, named the **cauda equina**[1] (CAW-duh ee-KWY-nah) for its resemblance to a horse's tail, innervates the pelvic organs and lower limbs.

THINK ABOUT IT!

Spinal cord injuries commonly result from fractures of vertebrae C5 to C6, but never from fractures of L3 to L5. Explain both observations.

Meninges of the Spinal Cord

The spinal cord and brain are enclosed in three connective tissue membranes called **meninges** (meh-NIN-jeez)—singular, *meninx*[2] (MEN-inks). These membranes separate the soft tissue of the central nervous system from the bones of the vertebrae and skull. From superficial to deep, they are the dura mater, arachnoid mater, and pia mater.

The **dura mater**[3] (DOO-ruh MAH-tur) forms a loose-fitting sleeve called the **dural sheath** around the spinal cord. It is a tough collagenous membrane with a thickness and texture similar to a rubber kitchen glove. The space between the sheath and vertebral bone, called the **epidural space,** is occupied by blood vessels, adipose tissue, and loose connective tissue (fig. 14.2*a*). Anesthetics are sometimes introduced to this space to block pain signals during childbirth or surgery; this procedure is called *epidural anesthesia.*

[1]*cauda* = tail + *equin* = horse
[2]*menin* = membrane
[3]*dura* = tough + *mater* = mother, womb

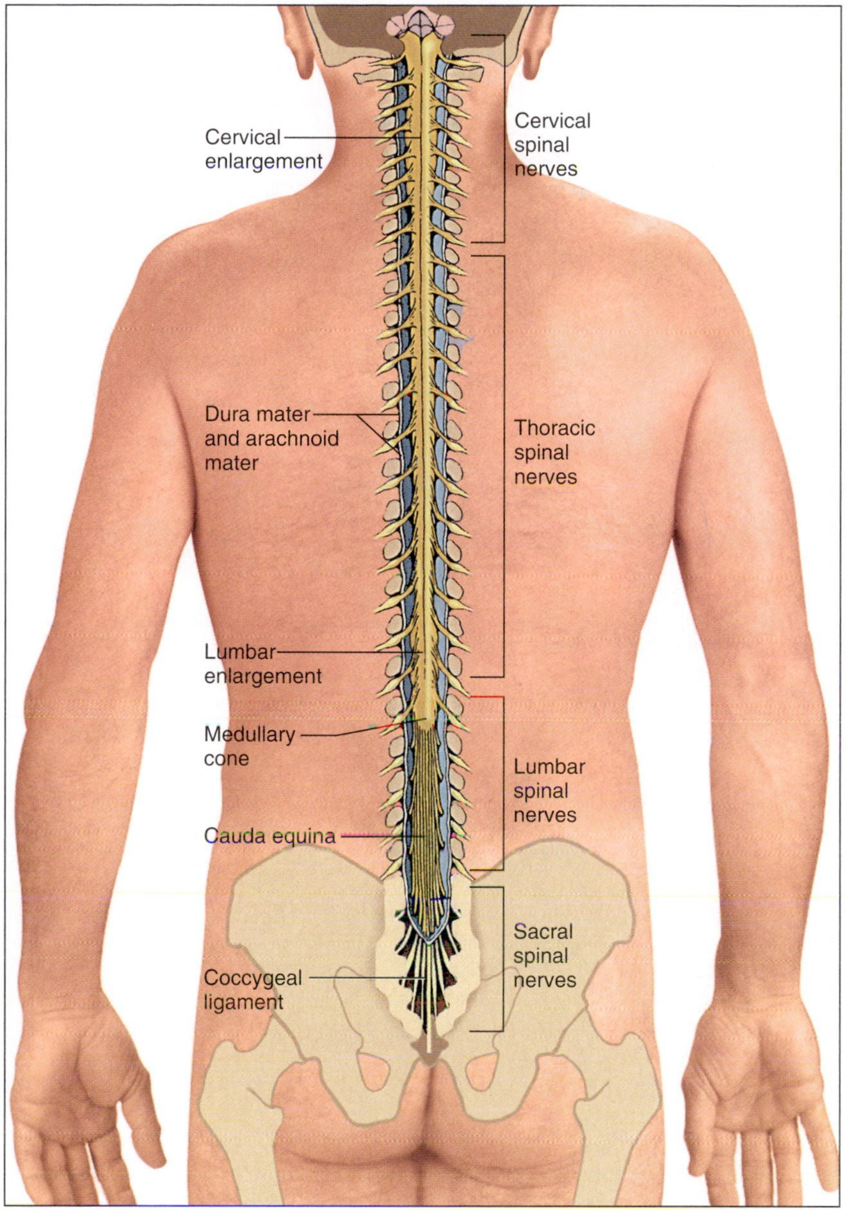

FIGURE 14.1
The Spinal Cord, Dorsal Aspect.

The **arachnoid**[4] (ah-RACK-noyd) **mater** consists of a simple squamous epithelium, the *arachnoid membrane,* adhering to the inside of the dura, and a loose mesh of collagenous and elastic fibers spanning the gap between the arachnoid membrane and the pia mater. This gap is called the **subarachnoid space.** Inferior to the medullary cone, the subarachnoid space is called the **lumbar cistern,** a space occupied by the cauda equina and cerebrospinal fluid (CSF), a clear liquid discussed in chapter 15.

The **pia**[5] (PEE-uh) **mater** is a delicate, translucent membrane that closely follows the contours of the spinal cord. It continues beyond the medullary cone as a fibrous strand, the *terminal filum,* forming part of the **coccygeal ligament** that anchors the cord to vertebra L2. At regular intervals along the cord, extensions of the pia called **denticulate ligaments** extend through the arachnoid to the dura, anchoring the cord and preventing side-to-side movements.

Cross-Sectional Anatomy

Figure 14.2*a* shows the relationship of the spinal cord to a vertebra and spinal nerve, and figure 14.2*b* shows the cord itself in more detail. The spinal cord, like the brain, consists of two kinds of nervous tissue called gray and white matter. **Gray matter** has a relatively dull color because it contains little myelin. It contains the somas, dendrites, and proximal parts of the axons of neurons. It is the site of synaptic contact between neurons, and therefore the site of all synaptic integration (information

[4]*arachn* = spider, spider web + *oid* = resembling
[5]*pia* = tender, soft

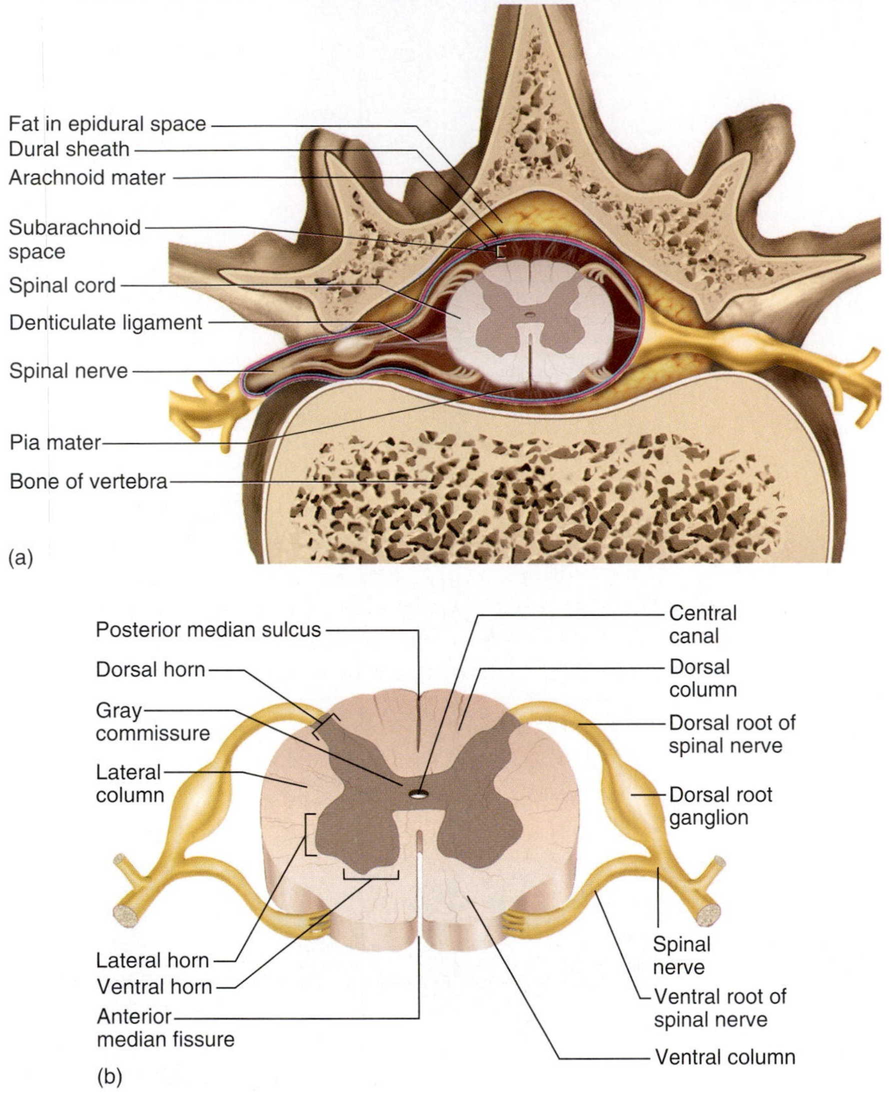
Fat in epidural space
Dural sheath
Arachnoid mater
Subarachnoid space
Spinal cord
Denticulate ligament
Spinal nerve
Pia mater
Bone of vertebra
(a)
Posterior median sulcus
Dorsal horn
Gray commissure
Lateral column
Lateral horn
Ventral horn
Anterior median fissure
(b)
Central canal
Dorsal column
Dorsal root of spinal nerve
Dorsal root ganglion
Spinal nerve
Ventral root of spinal nerve
Ventral column

FIGURE 14.2

Cross Section of the Thoracic Spinal Cord. (*a*) Relationship to the vertebra, meninges, and spinal nerve. (*b*) Anatomy of the spinal cord itself.

INSIGHT 14.1 CLINICAL APPLICATION

SPINAL TAPS

Several neurological diseases are diagnosed in part by examining cerebrospinal fluid for bacteria, blood, white blood cells, or abnormalities of chemical composition. CSF is obtained by a procedure called a *spinal tap* or *lumbar puncture*. The patient leans forward or lies on one side with the spine flexed, thus spreading the vertebral laminae and spinous processes apart. The skin over the lumbar vertebrae is anesthetized and a needle is inserted between the spinous processes of L3 and L4 (sometimes L4 and L5). This is the safest place to obtain CSF because the spinal cord does not extend this far and is not exposed to injury by the needle. At a depth of 4 to 6 cm, the needle punctures the dura mater and enters the lumbar cistern. CSF normally drips out at a rate of about 1 drop per second. A lumbar puncture is not performed if a patient has signs of high intracranial pressure, because the sudden release of pressure (causing CSF to jet from the puncture) can cause fatal herniation of the brainstem and cerebellum into the vertebral canal.

processing) in the central nervous system. **White matter** contains an abundance of myelinated axons, which give it a bright, pearly white appearance. It is composed of bundles of axons, called **tracts,** that carry signals from one part of the CNS to another. In silver-stained nervous tissue sections, gray matter tends to have a brown or golden color and white matter a lighter tan to yellow color.

GRAY MATTER

The spinal cord has a central core of gray matter that looks somewhat butterfly- or H-shaped in cross sections. The core consists mainly of two **dorsal (posterior) horns,** which extend toward the dorsolateral surfaces of the cord, and two thicker **ventral (anterior) horns,** which extend toward the ventrolateral surfaces. The right and left sides are connected by a **gray commissure.** In the middle of the commissure is the **central canal,** which is collapsed in most areas of the adult spinal cord, but in some places (and in young children) remains open, lined with ependymal cells, and filled with CSF. The canal is a remnant of the lumen of the embryonic neural tube (see p. 379).

As a spinal nerve approaches the cord, it branches into a *dorsal root* and *ventral root.* The dorsal root carries sensory nerve fibers, which enter the dorsal horn of the cord and sometimes synapse with an interneuron there. Such interneurons are especially numerous in the cervical and lumbar enlargements and are quite evident in histological sections at these levels. The ventral horns contain the large somas of the somatic motor neurons. Axons from these neurons exit by way of the ventral root of the spinal nerve and lead to the skeletal muscles. The spinal nerve roots are described more fully later in this chapter.

In the thoracic and lumbar regions, an additional **lateral horn** is visible on each side of the gray matter. It contains neurons of the sympathetic nervous system, which send their axons out of the cord by way of the ventral root along with the somatic efferent fibers.

WHITE MATTER

The white matter of the spinal cord surrounds the gray matter. It consists of bundles of axons that course up and down the cord and provides avenues of communication between different levels of the CNS. These bundles are arranged in three pairs called **columns,** or **funiculi**[6] (few-NIC-you-lie)—a **dorsal (posterior), lateral,** and **ventral (anterior) column** on each side. Each column consists of subdivisions called **tracts** or **fasciculi**[7] (fah-SIC-you-lye).

Spinal Tracts

Knowledge of the locations and functions of the spinal tracts is essential in diagnosing and managing spinal cord injuries. **Ascending tracts** carry sensory information up the cord and **descending tracts** conduct motor impulses down. All nerve fibers in a given tract have a similar origin, destination, and function.

Several of these tracts undergo **decussation**[8] (DEE-cuh-SAY-shun) as they pass up or down the brainstem and spinal cord—meaning that they cross over from the left side of the body to the right, or vice versa. As a result, the left side of the brain receives sensory information from the right side of the body and sends its motor commands to that side, while the right side of the brain senses and controls the left side of the body. A stroke that damages motor centers of the right side of the brain can thus cause paralysis of the left limbs and vice versa. When the origin and destination of a tract are on opposite sides of the body, we say they are **contralateral**[9] to each other. When a tract does not decussate, so the origin and destination of its fibers are on the same side of the body, we say they are **ipsilateral.**[10]

The major spinal cord tracts are summarized in table 14.1 and figure 14.3. Bear in mind that each tract is repeated on the right and left sides of the spinal cord.

[6]*funicul* = little rope, cord
[7]*fascicul* = little bundle
[8]*decuss* = to cross, form an X
[9]*contra* = opposite + *later* = side
[10]*ipsi* = the same + *later* = side

ASCENDING TRACTS

Ascending tracts carry sensory signals up the spinal cord. Sensory signals typically travel across three neurons from their origin in the receptors to their destination in the sensory areas of the brain: a **first-order neuron** that detects a stimulus and transmits a signal to the spinal cord or brainstem; a **second-order neuron** that continues as far as a "gateway" called the *thalamus* at the upper end of the brainstem; and a **third-order neuron** that carries the signal the rest of the way to the sensory region of the cerebral cortex. The axons of these neurons are called the first- through third-order nerve fibers. Deviations from the pathway described here will be noted for some of the sensory systems to follow.

The major ascending tracts are as follows. The names of most ascending tracts consist of the prefix *spino-* followed by a root denoting the destination of its fibers in the brain.

- The **gracile**[11] **fasciculus** (GRAS-el fah-SIC-you-lus) carries signals from the midthoracic and lower parts of the body. Below vertebra T6, it composes the entire dorsal column. At T6, it is joined by the cuneate fasciculus, discussed next. The gracile fasciculus consists of first-order nerve fibers that travel up the ipsilateral side of the spinal cord and terminate at the *gracile nucleus* in the medulla oblongata of the brainstem. These fibers carry signals for vibration, visceral pain, deep and discriminative touch (touch whose location one can precisely identify), and especially *proprioception*[12] from the lower limbs and lower trunk. (Proprioception is the nonvisual sense of the position and movements of the body.)
- The **cuneate**[13] (CUE-nee-ate) **fasciculus** (fig. 14.4*a*) joins the gracile fasciculus at the T6 level. It occupies the lateral portion of the dorsal column and forces the gracile fasciculus medially. It carries the same type of sensory signals, originating from level T6 and up (from the upper limb and chest). Its fibers end in the *cuneate nucleus* on the ipsilateral side of the medulla oblongata. In the medulla, second-order fibers of the gracile and cuneate systems decussate and form the **medial lemniscus**[14] (lem-NIS-cus), a tract of nerve fibers that leads the rest of the way up the brainstem to the thalamus. Third-order fibers go from the thalamus to the cerebral cortex. Because of decussation, the signals carried by the gracile and cuneate fasciculi ultimately go to the contralateral cerebral hemisphere.
- The **spinothalamic** (SPY-no-tha-LAM-ic) **tract** (fig. 14.4*b*) and some smaller tracts form the *anterolateral system,* which passes up the anterior and lateral columns of the spinal cord. The spinothalamic tract carries signals for pain, temperature, pressure, tickle, itch, and light or crude touch. Light touch is the sensation produced by stroking hairless skin with a feather or cotton wisp, without indenting the skin; crude touch is touch whose location one can only vaguely identify. In this pathway, first-order neurons end in the dorsal horn of the

[11]*gracil* = thin, slender
[12]*proprio* = one's own + *cept* = receive, sense
[13]*cune* = wedge
[14]*lemniscus* = ribbon

TABLE 14.1
Major Spinal Tracts

Tract	Column	Decussation	Functions
Ascending (sensory) Tracts			
Gracile fasciculus	Dorsal	In medulla	Limb and trunk position and movement, deep and discriminative touch, visceral pain, vibration, below level T6
Cuneate fasciculus	Dorsal	In medulla	Same as gracile fasciculus, from level T6 up
Spinothalamic	Lateral and ventral	In spinal cord	Light and crude touch, tickle, itch, temperature, pain, and pressure
Dorsal spinocerebellar	Lateral	None	Feedback from muscles (proprioception)
Ventral spinocerebellar	Lateral	In spinal cord	Same as dorsal spinocerebellar
Descending (motor) Tracts			
Lateral corticospinal	Lateral	In medulla	Fine control of limbs
Ventral corticospinal	Ventral	None	Fine control of limbs
Tectospinal	Lateral and ventral	In midbrain	Reflexive head-turning in response to visual and auditory stimuli
Lateral reticulospinal	Lateral	None	Balance and posture; regulation of awareness of pain
Medial reticulospinal	Ventral	None	Same as lateral reticulospinal
Vestibulospinal	Ventral	None	Balance and posture

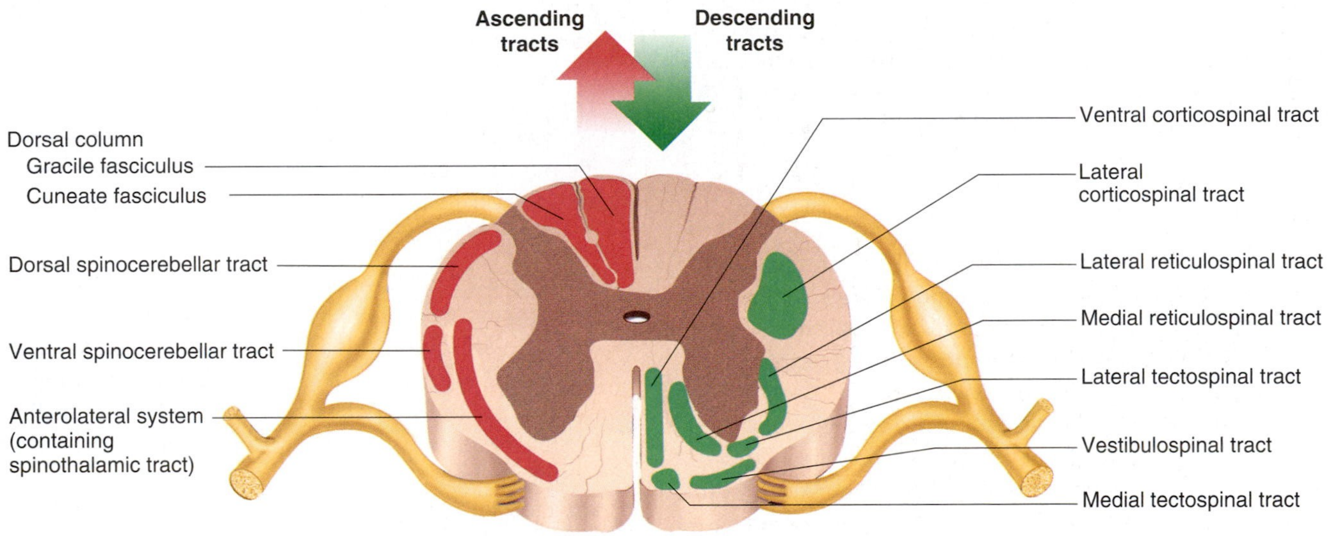

FIGURE 14.3
Tracts of the Spinal Cord. All of the illustrated tracts occur on both sides of the cord, but only the ascending sensory tracts are shown on the *left* (*red*), and only the descending motor tracts on the *right* (*green*).

spinal cord near the point of entry. Second-order neurons decussate to the opposite side of the spinal cord and there form the ascending spinothalamic tract. These fibers lead all the way to the thalamus. Third-order neurons continue from there to the cerebral cortex.

- The **dorsal** and **ventral spinocerebellar** (SPY-no-SERR-eh-BEL-ur) **tracts** travel through the lateral column and carry proprioceptive signals from the limbs and trunk to the cerebellum, a large motor control area at the rear of the brain. The first-order neurons of this system originate in the muscles and tendons and end in the dorsal horn of the spinal cord. Second-order neurons send their fibers up the spinocerebellar tracts and end in the cerebellum. Fibers of the dorsal tract travel up the ipsilateral side of the spinal cord. Those of the ventral tract cross over and travel up the contralateral side but then cross back in the brainstem to enter the ipsilateral cerebellum. Both tracts provide the cerebellum with feedback needed to coordinate muscle action, as discussed in chapter 15.

DESCENDING TRACTS

Descending tracts carry motor signals down the brainstem and spinal cord. A descending motor pathway typically involves two neurons called the upper and lower motor neuron. The **upper motor neuron** begins with a soma in the cerebral cortex or brainstem and has an axon that terminates on a **lower motor neuron** in the brainstem or spinal cord. The axon of the lower motor neuron then leads the rest of the way to the muscle or other target organ. The

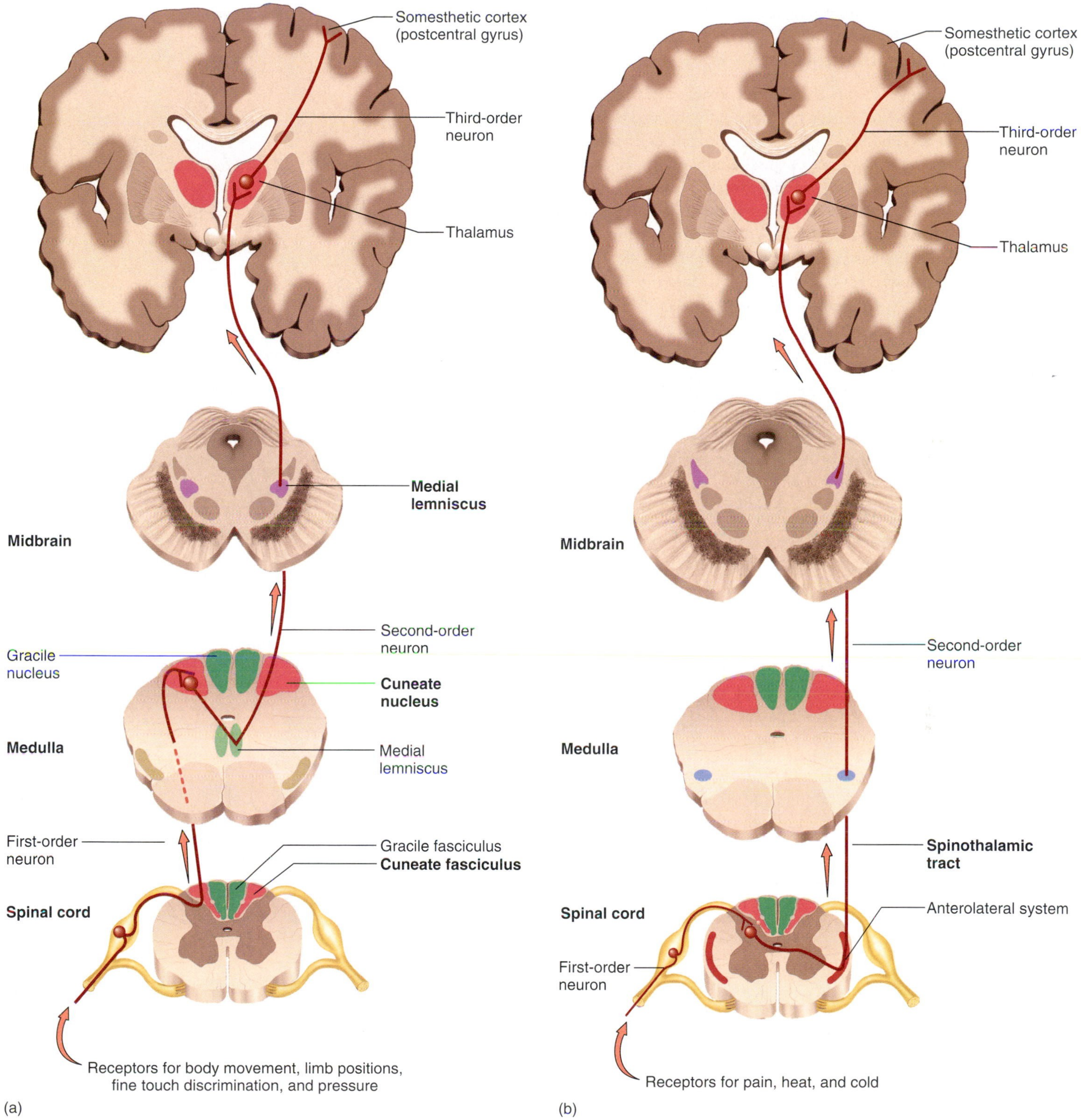
Somesthetic cortex (postcentral gyrus)
Third-order neuron
Thalamus
Medial lemniscus
Midbrain
Second-order neuron
Gracile nucleus
Cuneate nucleus
Medulla
Medial lemniscus
First-order neuron
Gracile fasciculus
Cuneate fasciculus
Spinal cord
Receptors for body movement, limb positions, fine touch discrimination, and pressure
(a)
Somesthetic cortex (postcentral gyrus)
Third-order neuron
Thalamus
Midbrain
Second-order neuron
Medulla
Spinothalamic tract
Spinal cord
Anterolateral system
First-order neuron
Receptors for pain, heat, and cold
(b)

FIGURE 14.4

Two Ascending Pathways of the CNS. (*a*) The cuneate fasciculus and medial lemniscus. (*b*) The spinothalamic tract. The spinal cord, medulla, and midbrain are shown in cross section and the cerebrum and thalamus (*top*) in frontal section. Nerve signals enter the spinal cord at the *bottom* of the figure and carry somatosensory (somesthetic) information up to the cerebral cortex.

names of most descending tracts consist of a word root denoting the point of origin in the brain, followed by the suffix *-spinal.* The major descending tracts are described here.

- The **corticospinal** (COR-tih-co-SPY-nul) **tracts** carry motor signals from the cerebral cortex for precise, finely coordinated limb movements. The fibers of this system form ridges called *pyramids* on the ventral surface of the medulla oblongata, so these tracts were once called *pyramidal tracts.* Most corticospinal fibers decussate in the lower medulla and form the **lateral corticospinal tract** on the contralateral side of the spinal cord. A few fibers remain uncrossed and form the **ventral corticospinal tract** on the ipsilateral side (fig. 14.5). Fibers of the ventral tract decussate lower in the spinal cord, however, so even they control contralateral muscles.
- The **tectospinal** (TEC-toe-SPY-nul) **tract** begins in a midbrain region called the *tectum* and crosses to the contralateral side of the brainstem. In the lower medulla, it branches into *lateral* and *medial tectospinal tracts* of the upper spinal cord. These are involved in reflex movements of the head, especially in response to visual and auditory stimuli.
- The **lateral** and **medial reticulospinal** (reh-TIC-you-lo-SPY-nul) **tracts** originate in the *reticular formation* of the brainstem. They control muscles of the upper and lower limbs, especially to maintain posture and balance. They also contain *descending analgesic pathways* that reduce the transmission of pain signals to the brain (see chapter 17).
- The **vestibulospinal** (vess-TIB-you-lo-SPY-nul) **tract** begins in a brainstem *vestibular nucleus* that receives impulses for balance from the inner ear. The tract passes down the ventral column of the spinal cord and controls muscles that maintain balance and posture.

Rubrospinal tracts are prominent in other mammals, where they aid in muscle coordination. Although often pictured in illustrations of human anatomy, they are almost nonexistent in humans and have little functional importance.

Before You Go On

Answer the following questions to test your understanding of the preceding section:

1. Name the four major regions and two enlargements of the spinal cord.
2. Describe the distal (inferior) end of the spinal cord and the contents of the vertebral canal from level L2 to S5.
3. Sketch a cross section of the spinal cord showing the dorsal and ventral horns. Where are the gray and white matter? Where are the columns and tracts?
4. Give an anatomical explanation as to why a stroke in the right cerebral hemisphere can paralyze the limbs on the left side of the body.

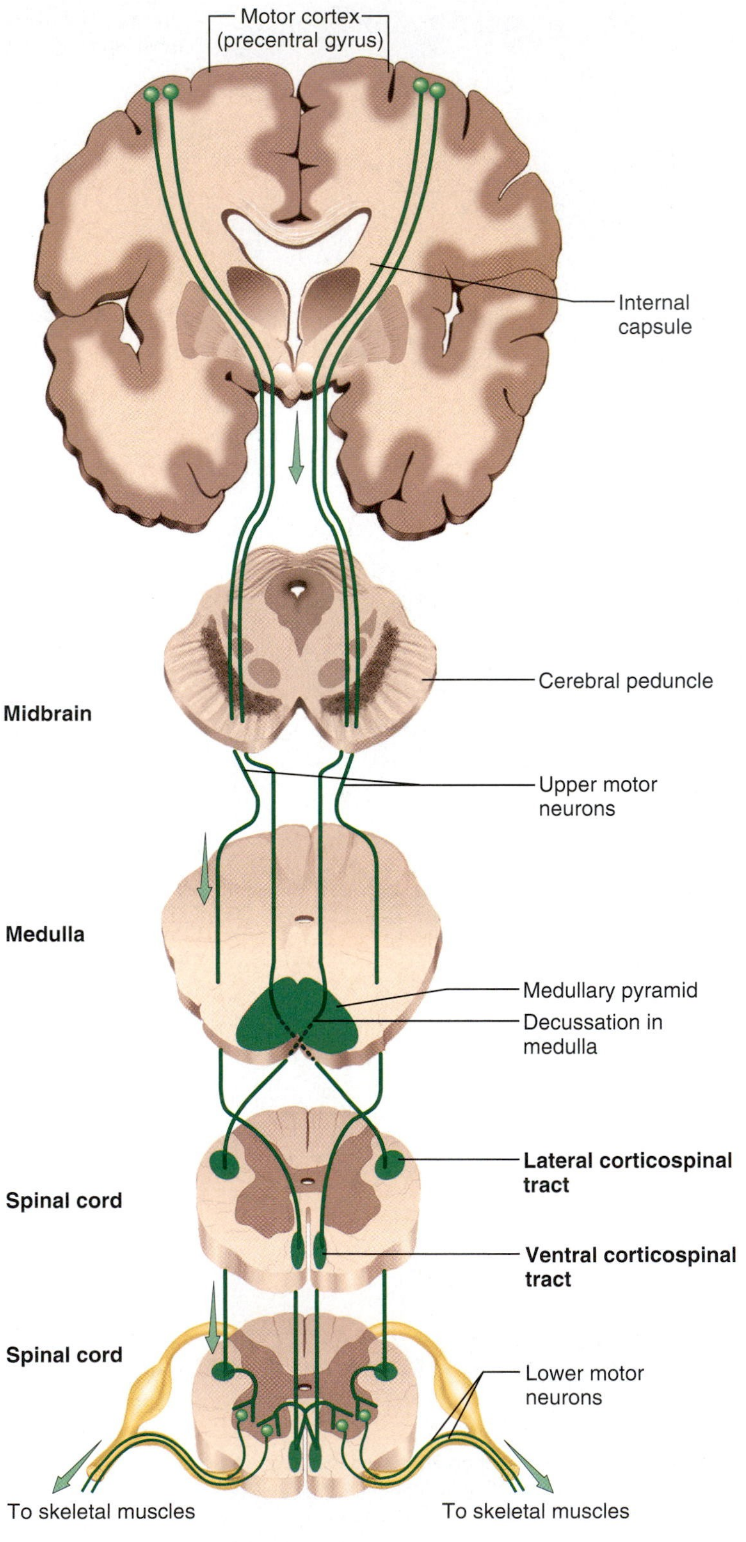

FIGURE 14.5

Two Descending Pathways of the CNS. The lateral and ventral corticospinal tracts, which carry signals for voluntary muscle contraction. Nerve signals originate in the cerebral cortex at the *top* of the figure and carry motor commands down the spinal cord.

INSIGHT 14.2 CLINICAL APPLICATION

Poliomyelitis and Amyotrophic Lateral Sclerosis

Poliomyelitis[15] and *amyotrophic lateral sclerosis*[16] (ALS) are two diseases that result from the destruction of motor neurons. In both diseases, the skeletal muscles atrophy from lack of innervation.

Poliomyelitis is caused by the poliovirus, which destroys motor neurons in the brainstem and ventral horn of the spinal cord. Signs of polio include muscle pain, weakness, and loss of some reflexes, followed by paralysis, muscular atrophy, and sometimes respiratory arrest. The virus spreads through water contaminated by feces. Historically, polio afflicted many children who contracted the virus from contaminated swimming pools. The polio vaccine has nearly eliminated new cases.

ALS is also known as Lou Gehrig[17] disease after the baseball player who had to retire from the sport because of it. It is marked not only by the degeneration of motor neurons and atrophy of the muscles, but also sclerosis (scarring) of the lateral regions of the spinal cord—hence its name. Most cases occur when astrocytes fail to reabsorb the neurotransmitter glutamate from the tissue fluid, allowing it to accumulate to a neurotoxic level. The early signs of ALS include muscular weakness and difficulty in speaking, swallowing, and using the hands. Sensory and intellectual functions remain unaffected, as evidenced by the accomplishments of astrophysicist and best-selling author Stephen Hawking (fig. 14.6), who was stricken with ALS while he was in college. Despite near-total paralysis, he remains highly productive and communicates with the aid of a speech synthesizer and computer. Tragically, many people are quick to assume that those who have lost most of their ability to communicate their ideas and feelings have no ideas and feelings to communicate. To a victim, this may be more unbearable than the loss of motor function itself.

[15]*polio* = gray matter + *myel* = spinal cord + *itis* = inflammation
[16]*a* = without + *myo* = muscle + *troph* = nourishment
[17]Lou Gehrig (1903–41), American baseball player

Figure 14.6
Stephen Hawking (1942–), Lucasian Professor of Mathematics at Cambridge University.

The Spinal Nerves

Objectives

When you have completed this section, you should be able to

- describe the attachment of a spinal nerve to the spinal cord;
- trace the branches of a spinal nerve distal to its attachment;
- name the five plexuses of spinal nerves and describe their general anatomy;
- name some major nerves that arise from each plexus; and
- explain the relationship of dermatomes to the spinal nerves.

General Anatomy of Nerves and Ganglia

The spinal cord communicates with the rest of the body by way of the spinal nerves. Before we discuss those specific nerves, however, it is necessary to be familiar with the structure of nerves and ganglia in general.

A **nerve** is a cordlike organ composed of numerous nerve fibers (axons) bound together by connective tissue (fig. 14.7). If we compare a nerve fiber to a wire carrying an electrical current in one direction, a nerve would be comparable to an electrical cable composed of thousands of wires carrying currents in opposite directions. A nerve contains anywhere from a few nerve fibers to more than a million. Nerves usually have a pearly white color and resemble frayed string as they divide into smaller and smaller branches.

Nerve fibers of the peripheral nervous system are ensheathed in Schwann cells, which form a neurilemma and often a myelin sheath around the axon (see chapter 13). External to the neurilemma, each fiber is surrounded by a basal lamina and then a thin sleeve of loose connective tissue called the **endoneurium.** In most nerves, the nerve fibers are gathered in bundles called **fascicles,** each wrapped in a sheath called the **perineurium.** The perineurium is composed of one to six layers of overlapping, squamous, epithelium-like cells. Several fascicles are then bundled together and wrapped in an outer

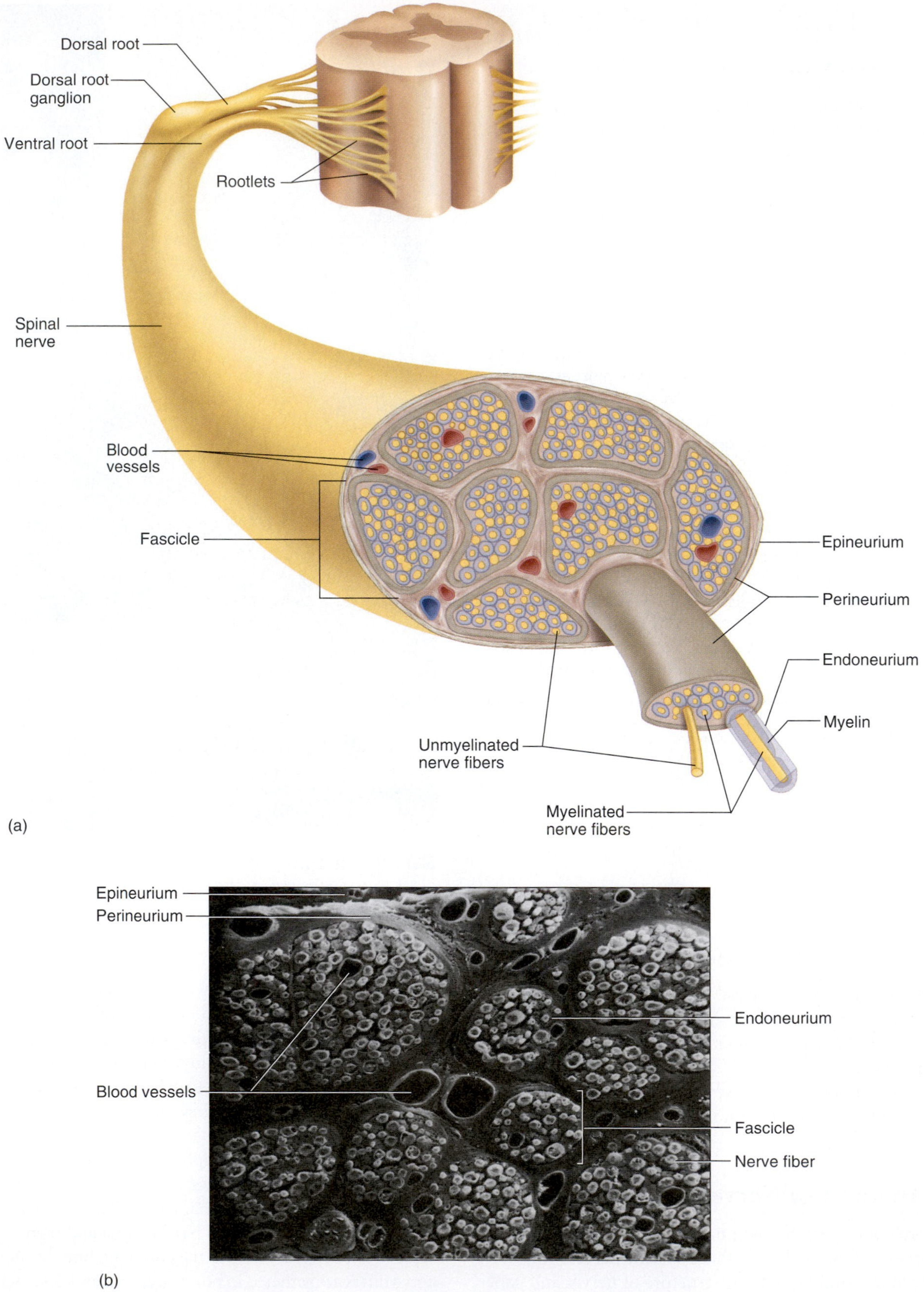

Dorsal root
Dorsal root ganglion
Ventral root
Rootlets
Spinal nerve
Blood vessels
Fascicle
Epineurium
Perineurium
Endoneurium
Myelin
Unmyelinated nerve fibers
Myelinated nerve fibers
(a)
Epineurium
Perineurium
Endoneurium
Blood vessels
Fascicle
Nerve fiber
(b)

FIGURE 14.7

Anatomy of a Nerve. (*a*) A spinal nerve and its association with the spinal cord. (*b*) Cross section of a nerve (SEM). Myelinated nerve fibers appear as white rings and unmyelinated fibers as solid gray. Credit for *b:* Richard E. Kessel and Randy H. Kardon, *Tissues and Organs: A Text-Atlas of Scanning Electron Microscopy,* 1979, W. H. Freeman and Company.

epineurium to compose the nerve as a whole. The epineurium consists of dense irregular fibrous connective tissue and protects the nerve from stretching and injury. Nerves have a high metabolic rate and need a plentiful blood supply. Blood vessels penetrate as far as the perineurium, and oxygen and nutrients diffuse through the extracellular fluid from there to the nerve fibers.

THINK ABOUT IT!

How does the structure of a nerve compare to that of a skeletal muscle? Which of the descriptive terms for nerves have similar counterparts in muscle histology?

Peripheral nerve fibers are of two kinds: *sensory (afferent) fibers* carry signals from sensory receptors to the CNS, and *motor (efferent) fibers* carry signals from the CNS to muscles and glands. Both sensory and motor fibers can also be described as *somatic* or *visceral* and as *general* or *special* depending on the organs they innervate (table 14.2).

A **mixed nerve** consists of both sensory and motor fibers and thus transmits signals in two directions, although any one fiber within the nerve transmits signals one way only. Most nerves are mixed. **Sensory nerves,** composed entirely of sensory axons, are less common; they include the olfactory and optic nerves discussed in chapter 15. Nerves that carry only motor fibers are called **motor nerves.** Many nerves often described as motor are actually mixed because they carry sensory signals of proprioception from the muscle back to the CNS.

If a nerve resembles a thread, a **ganglion**[18] resembles a knot in the thread. A ganglion is a cluster of cell bodies (somas) outside the CNS. It is enveloped in an epineurium continuous with that of the nerve. Among the somas are bundles of nerve fibers leading into and out of the ganglion. Figure 14.8 shows a type of ganglion associated with the spinal nerves.

TABLE 14.2
The Classification of Nerve Fibers

Class	Description
Afferent fibers	Carry sensory signals from receptors to the CNS
Efferent fibers	Carry motor signals from the CNS to effectors
Somatic fibers	Innervate skin, skeletal muscles, bones, and joints
Visceral fibers	Innervate blood vessels, glands, and viscera
General fibers	Innervate widespread organs such as muscles, skin, glands, viscera, and blood vessels
Special fibers	Innervate more localized organs in the head, including the eyes, ears, olfactory and taste receptors, and muscles of chewing, swallowing, and facial expression

Spinal Nerves

There are 31 pairs of **spinal nerves:** 8 cervical (C1–C8), 12 thoracic (T1–T12), 5 lumbar (L1–L5), 5 sacral (S1–S5), and 1 coccygeal (Co) (fig. 14.9). The first cervical nerve emerges between the skull and atlas, and the others emerge through intervertebral foramina, including the anterior and posterior foramina of the sacrum.

PROXIMAL BRANCHES

Each spinal nerve has two points of attachment to the spinal cord (fig. 14.10). Dorsally, a branch of the spinal nerve called the **dorsal root** divides into six to eight *nerve rootlets* that enter the spinal cord (fig. 14.11). A little distal to the rootlets is a swelling, the **dorsal root ganglion,** which contains the somas of unipolar afferent neurons. Ventrally, another row of six to eight rootlets leave the spinal cord and converge to form the **ventral root.** The dorsal and ventral roots merge, penetrate the dural sac, enter the intervertebral foramen, and there form the spinal nerve proper.

Spinal nerves are mixed nerves, with a two-way traffic of afferent (sensory) and efferent (motor) signals. Afferent signals approach the cord by way of the dorsal root and enter the dorsal horn of the gray matter. Efferent signals begin at the somas of motor neurons in the ventral horn and leave the spinal cord via the ventral root. Some viruses invade the central nervous system by way of these roots (see insight 14.3).

The dorsal and ventral roots are shortest in the cervical region and become longer inferiorly. The roots that arise from segments L2 to Co of the cord form the cauda equina.

INSIGHT 14.3 CLINICAL APPLICATION

SHINGLES

Chickenpox *(varicella)*, a common disease of early childhood, is caused by the *varicella-zoster* virus. It produces an itchy rash that usually clears up without complications. The virus, however, remains for life in the dorsal root ganglia. The immune system normally keeps it in check, but if the immune system is compromised, the virus can travel down the sensory nerves by axonal transport and cause *shingles (herpes zoster)*. This is characterized by a painful trail of skin discoloration and fluid-filled vesicles along the path of the nerve. These signs usually appear in the chest and waist, often on just one side of the body. Shingles usually occurs after the age of 50. While it can be very painful and may last 6 months or longer, it eventually heals spontaneously and requires no special treatment other than aspirin and steroidal ointment to relieve pain and inflammation.

[18] *gangli* = knot

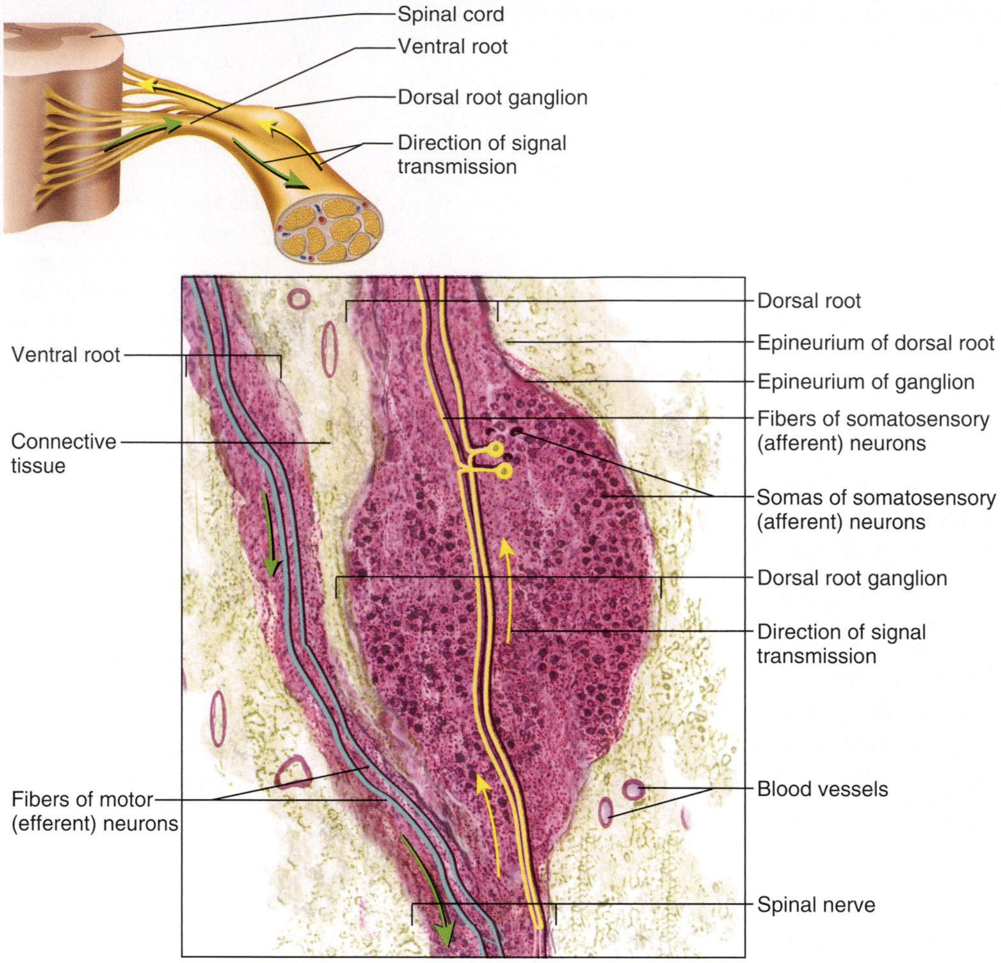

FIGURE 14.8

Anatomy of a Ganglion. The dorsal root ganglion contains the somas of unipolar sensory neurons conducting signals to the spinal cord. To the *left* of it is the ventral root of the spinal nerve, which conducts motor signals away from the spinal cord. (The ventral root is not part of the ganglion.)

DISTAL BRANCHES

Distal to the vertebrae, the branches of a spinal nerve are more complex (fig. 14.12). Immediately after emerging from the intervertebral foramen, the nerve divides into a **dorsal ramus,**[19] a **ventral ramus,** and a small **meningeal branch.** The meningeal branch (see fig. 14.10) reenters the vertebral canal and innervates the meninges, vertebrae, and spinal ligaments. The dorsal ramus innervates the muscles and joints in that region of the spine and the skin of the back. The ventral ramus innervates the ventral and lateral skin and muscles of the trunk and gives rise to nerves of the limbs.

[19]*ramus* = branch

THINK ABOUT IT!

Do you think the meningeal branch is sensory, motor, or mixed? Explain your reasoning.

The ventral ramus differs from one region of the trunk to another. In the thoracic region, it forms an **intercostal nerve** that travels along the inferior margin of a rib and innervates the skin and intercostal muscles (thus contributing to breathing), as well as the internal oblique, external oblique, and transversus abdominis muscles. All other ventral rami form the *nerve plexuses* described next.

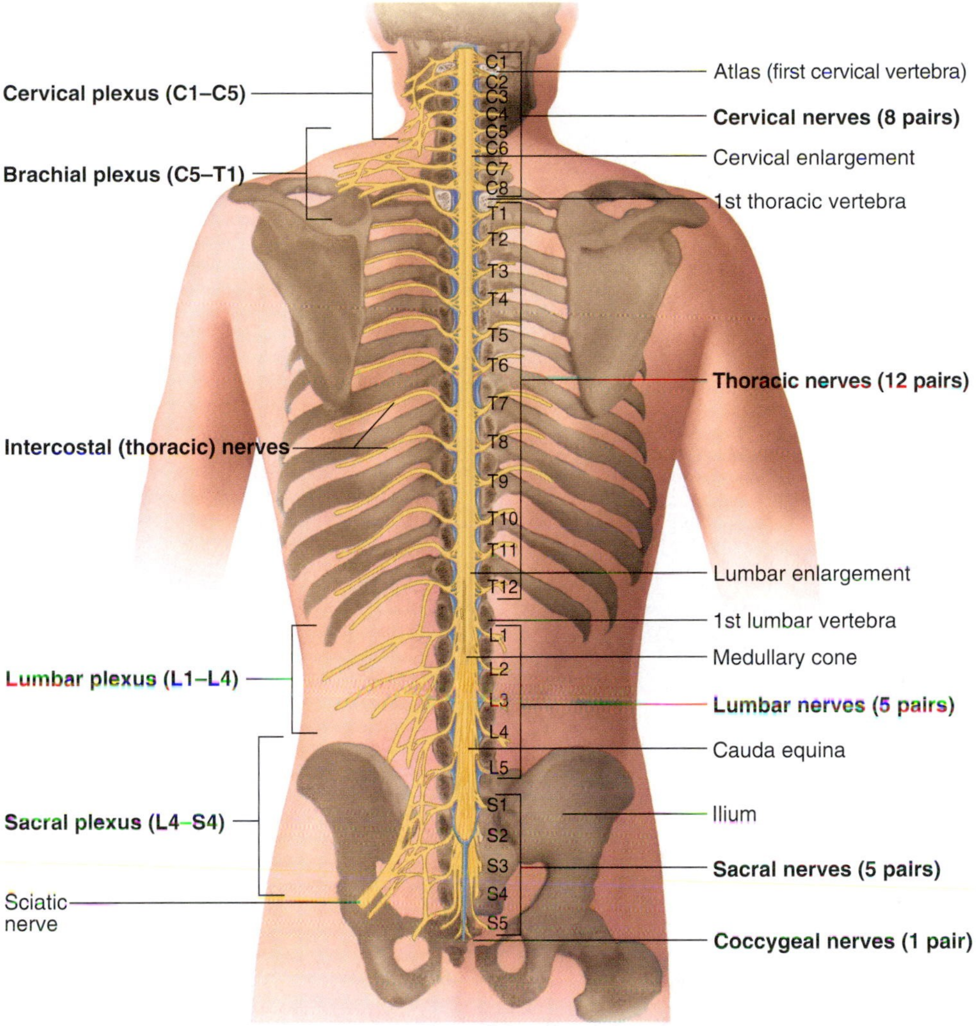

Cervical plexus (C1–C5)
Brachial plexus (C5–T1)
Intercostal (thoracic) nerves
Lumbar plexus (L1–L4)
Sacral plexus (L4–S4)
Sciatic nerve
C1
C2
C3
C4
C5
C6
C7
C8
T1
T2
T3
T4
T5
T6
T7
T8
T9
T10
T11
T12
L1
L2
L3
L4
L5
S1
S2
S3
S4
S5
Atlas (first cervical vertebra)
Cervical nerves (8 pairs)
Cervical enlargement
1st thoracic vertebra
Thoracic nerves (12 pairs)
Lumbar enlargement
1st lumbar vertebra
Medullary cone
Lumbar nerves (5 pairs)
Cauda equina
Ilium
Sacral nerves (5 pairs)
Coccygeal nerves (1 pair)

FIGURE 14.9
The Spinal Nerve Roots and Plexuses, Dorsal View.

Nerve Plexuses

Except in the thoracic region, the ventral rami branch and anastomose (merge) repeatedly to form five weblike nerve plexuses: the small **cervical plexus** deep in the neck, the **brachial plexus** near the shoulder, the **lumbar plexus** of the lower back, the **sacral plexus** immediately inferior to this, and finally the tiny **coccygeal plexus** adjacent to the lower sacrum and coccyx. A general view of these plexuses is shown in figure 14.9; they are illustrated and described in tables 14.3 through 14.6. The muscle actions controlled by these nerves are described in the muscle tables in chapter 10. The *somatosensory* function listed for many of these nerves means that they carry sensory signals from bones, joints, muscles, and the skin, in contrast to sensory input from the viscera or from special sense organs such as the eyes and ears. (See chapter 17 for different modes of sensory function.)

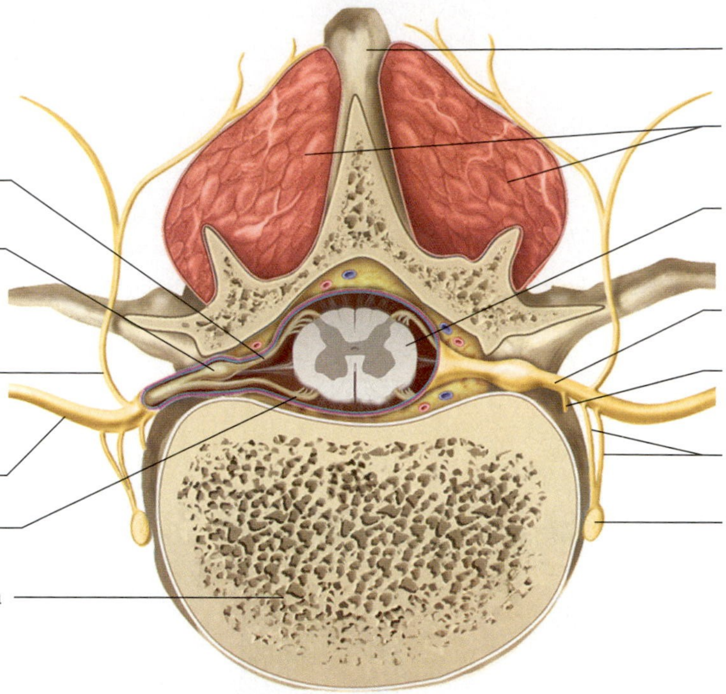

Posterior
Spine of vertebra
Deep muscles of back
Spinal cord
Spinal nerve
Meningeal branch
Communicating rami
Sympathetic ganglion
Dorsal root
Dorsal root ganglion
Dorsal ramus
Ventral ramus
Ventral root
Body of vertebra
Anterior

FIGURE 14.10
Branches of a Spinal Nerve in Relation to the Spinal Cord and Vertebra (cross section).

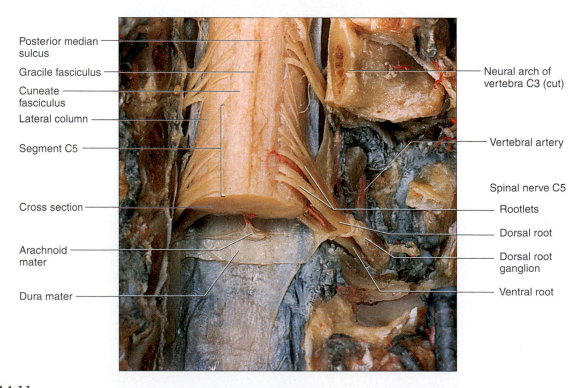

Posterior median sulcus
Gracile fasciculus
Cuneate fasciculus
Lateral column
Segment C5
Cross section
Arachnoid mater
Dura mater
Neural arch of vertebra C3 (cut)
Vertebral artery
Spinal nerve C5
Rootlets
Dorsal root
Dorsal root ganglion
Ventral root

FIGURE 14.11
The Point of Entry of Two Spinal Nerves into the Spinal Cord. Dorsal view with vertebrae cut away. Note that each dorsal root divides into several rootlets that enter the spinal cord. A segment of the spinal cord is the portion receiving all the rootlets of one spinal nerve.

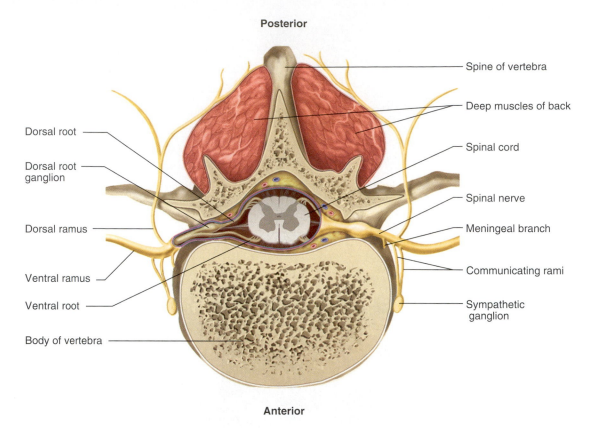

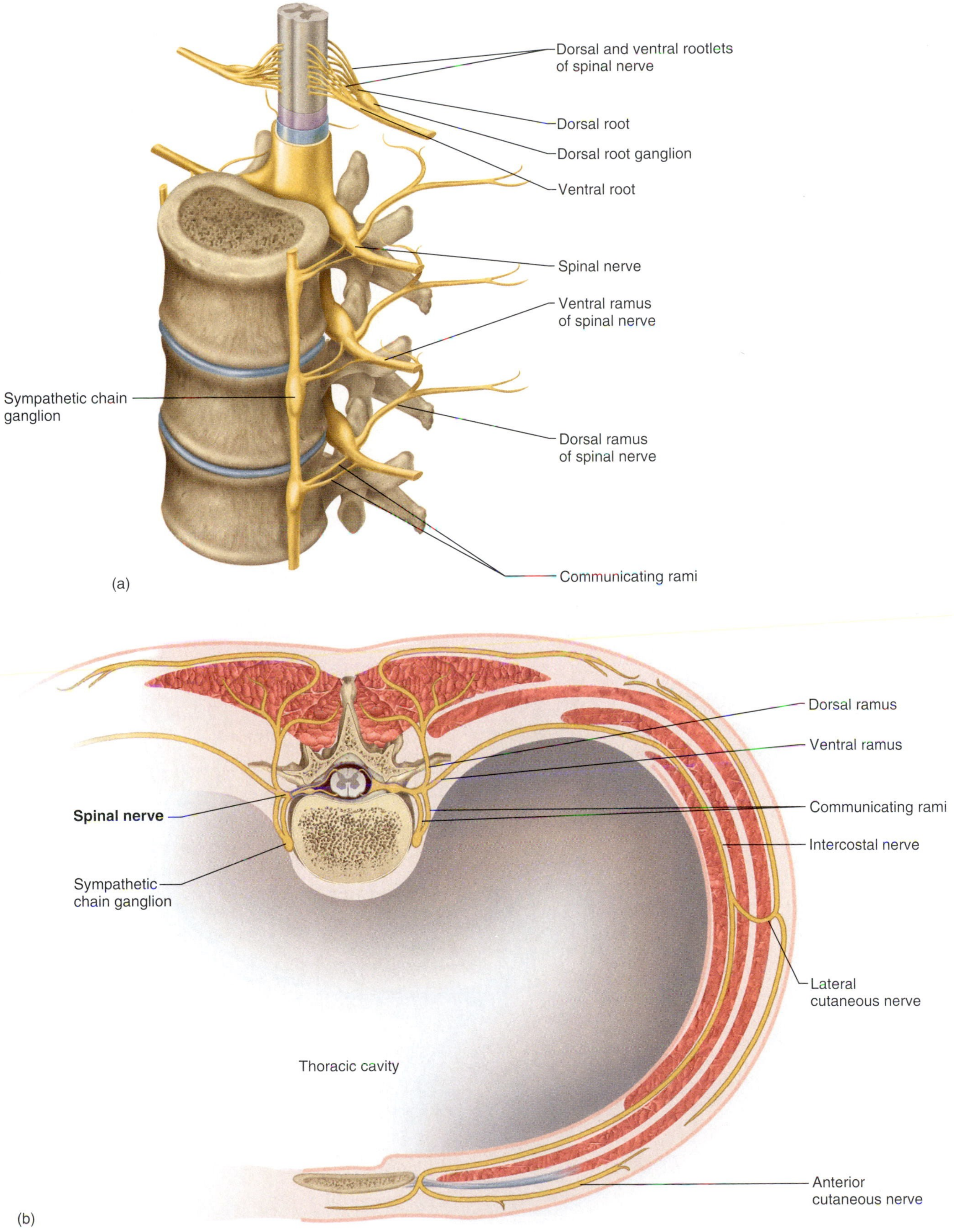
Dorsal and ventral rootlets of spinal nerve
Dorsal root
Dorsal root ganglion
Ventral root
Spinal nerve
Ventral ramus of spinal nerve
Sympathetic chain ganglion
Dorsal ramus of spinal nerve
Communicating rami
(a)
Dorsal ramus
Ventral ramus
Spinal nerve
Communicating rami
Intercostal nerve
Sympathetic chain ganglion
Lateral cutaneous nerve
Thoracic cavity
Anterior cutaneous nerve
(b)

FIGURE 14.12

Rami of the Spinal Nerves. (*a*) Anterolateral view of the spinal nerves and their subdivisions in relation to the spinal cord and vertebrae. (*b*) Cross section of the thorax showing innervation of muscles of the chest and back.

Table 14.3

The Cervical Plexus

The cervical plexus (fig. 14.13) receives fibers from the ventral rami of nerves C1 to C5 and gives rise to the nerves listed below, in order from superior to inferior. The most important of these are the *phrenic*[20] *nerves,* which travel down each side of the mediastinum, innervate the diaphragm, and play an essential role in breathing. In addition to the major nerves listed here, there are several motor branches that innervate the geniohyoid, thyrohyoid, scalene, levator scapulae, trapezius, and sternocleidomastoid muscles.

Lesser Occipital Nerve

Composition: Somatosensory

Innervation: Skin of lateral scalp and dorsal part of external ear

Great Auricular Nerve

Composition: Somatosensory

Innervation: Skin of and around external ear

Transverse Cervical Nerve

Composition: Somatosensory

Innervation: Skin of ventral and lateral neck

Ansa Cervicalis

Composition: Motor

Innervation: Omohyoid, sternohyoid, and sternothyroid muscles

Supraclavicular Nerve

Composition: Somatosensory

Innervation: Skin of lower ventral and lateral neck, shoulder, and ventral chest

Phrenic (FREN-ic) Nerve

Composition: Motor

Innervation: Diaphragm

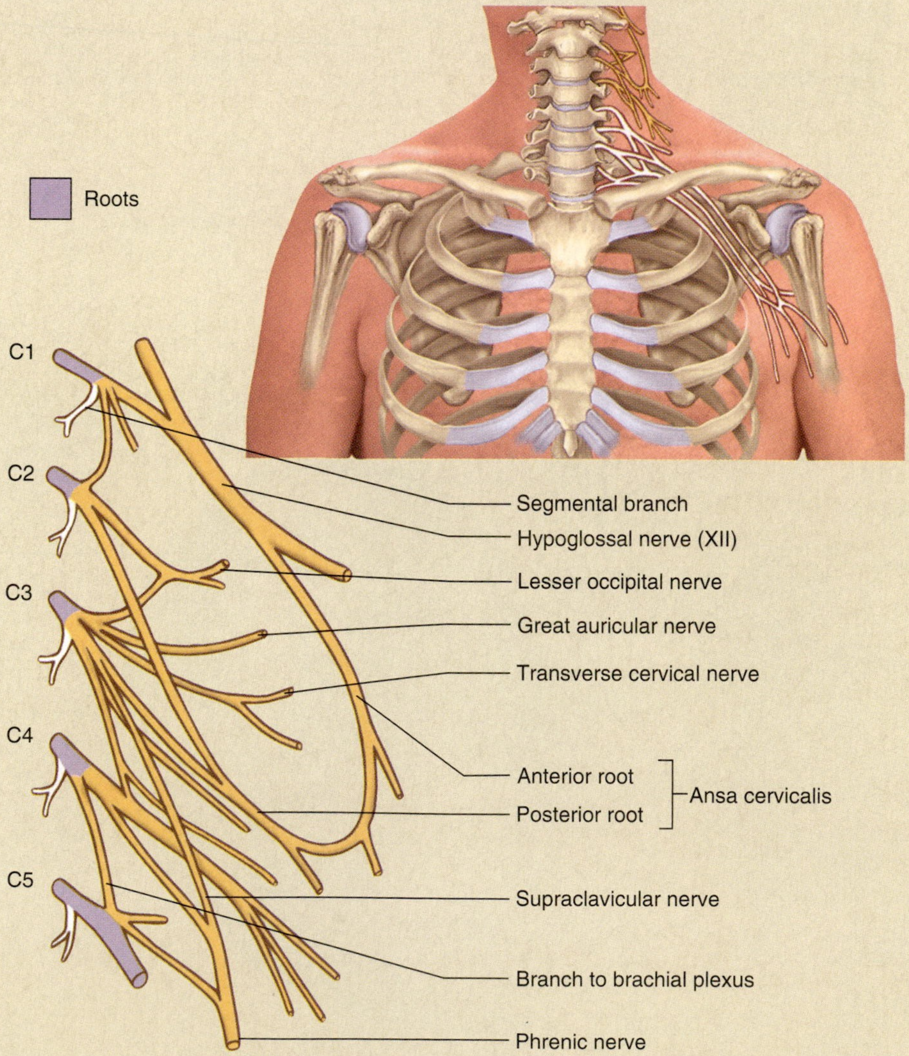
Roots
C1
C2
C3
C4
C5
Segmental branch
Hypoglossal nerve (XII)
Lesser occipital nerve
Great auricular nerve
Transverse cervical nerve
Anterior root
Posterior root
Ansa cervicalis
Supraclavicular nerve
Branch to brachial plexus
Phrenic nerve

Figure 14.13
The Cervical Plexus.

[20]*phren* = diaphragm

TABLE 14.4

The Brachial Plexus

The brachial plexus (figs. 14.14 and 14.15) is formed by the ventral rami of nerves C4 to T2. It passes over the first rib into the axilla and innervates the upper limb and some muscles of the neck and shoulder.

The subdivisions of this plexus are called *roots, trunks, divisions,* and *cords* (color-coded in figure 14.14). The five **roots** are the ventral rami of nerves C5 to T1, which provide most of the fibers to this plexus (C4 and T2 contribute partially). The five roots unite to form the **upper, middle,** and **lower trunks.** Each trunk divides into an **anterior** and **posterior division,** and finally the six divisions merge to form three large fiber bundles—the **posterior, medial,** and **lateral cords.** From these cords arise the following major nerves, which serve for cutaneous sensation, muscle contraction, and proprioception from the joints and muscles.

Axillary Nerve

Composition: Motor and somatosensory
Origin: Posterior cord of brachial plexus
Sensory innervation: Skin of lateral shoulder and arm; shoulder joint
Motor innervation: Deltoid and teres minor

Radial Nerve

Composition: Motor and somatosensory
Origin: Posterior cord of brachial plexus
Sensory innervation: Skin of posterior arm, forearm, and wrist; joints of elbow, wrist, and hand
Motor innervation: Muscles of posterior arm and forearm: triceps brachii, supinator, anconeus, brachioradialis, extensor carpi radialis brevis, extensor carpi radialis longus, and extensor carpi ulnaris

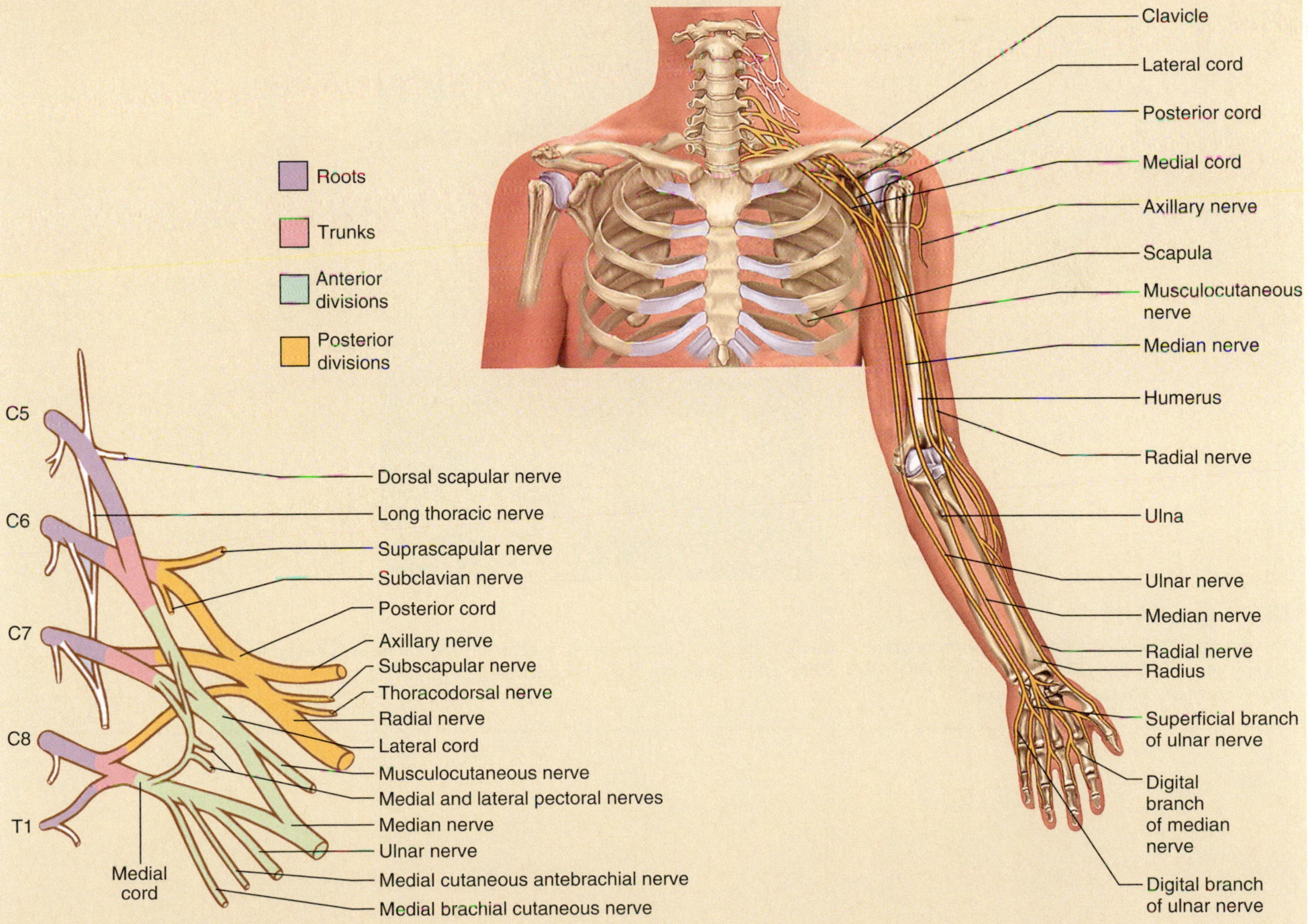

FIGURE 14.14
The Brachial Plexus.

(continued)

TABLE 14.4

The Brachial Plexus *(continued)*

Musculocutaneous Nerve

Composition: Motor and somatosensory
Origin: Lateral cord of brachial plexus
Sensory innervation: Skin of lateral forearm
Motor innervation: Muscles of anterior arm: coracobrachialis, biceps brachii, and brachialis

Median Nerve

Composition: Motor and somatosensory
Origin: Medial cord of brachial plexus
Sensory innervation: Skin of lateral two-thirds of hand, joints of hand
Motor innervation: Flexors of anterior forearm; thenar muscles; first and second lumbricals

Ulnar Nerve

Composition: Motor and somatosensory
Origin: Medial cord of brachial plexus
Sensory innervation: Skin of medial hand; joints of hand
Motor innervation: Flexor carpi ulnaris, flexor digitorum profundus, adductor pollicis, hypothenar muscles, interosseous muscles, and third and fourth lumbricals

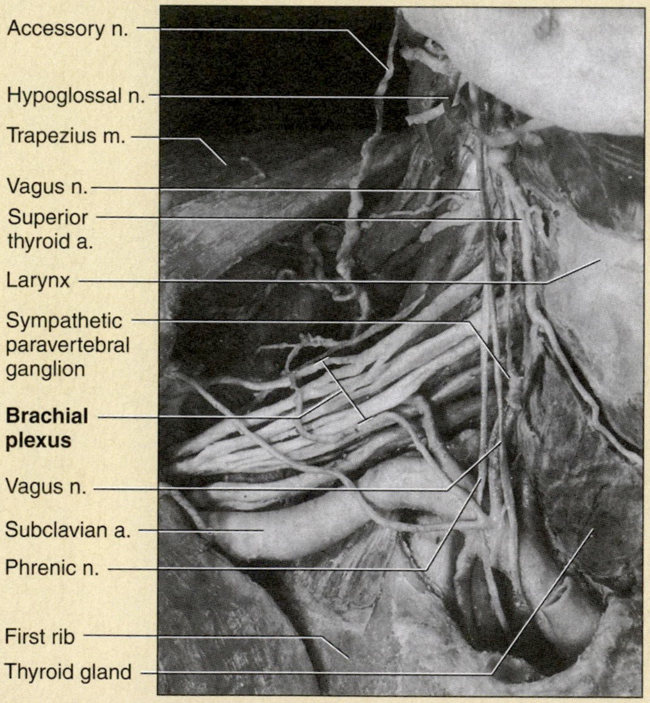
Accessory n.
Hypoglossal n.
Trapezius m.
Vagus n.
Superior thyroid a.
Larynx
Sympathetic paravertebral ganglion
Brachial plexus
Vagus n.
Subclavian a.
Phrenic n.
First rib
Thyroid gland

FIGURE 14.15
The Brachial Plexus of a Cadaver. Anterior view of the right shoulder, also showing three of the cranial nerves, the sympathetic trunk, and the phrenic nerve (a branch of the cervical plexus). Most of the other structures resembling nerves in this photograph are blood vessels.
(*a.* = artery; *m.* = muscle; *n.* = nerve.)

TABLE 14.5
The Lumbar Plexus

The lumbar plexus (fig. 14.16) is formed from the ventral rami of nerves L1 to L4 and some fibers from T12. With only five roots and two divisions, it is less complex than the brachial plexus. It gives rise to the following nerves.

Iliohypogastric Nerve

Composition: Motor and somatosensory

Sensory innervation: Skin of anterior abdominal wall

Motor innervation: Internal and external obliques and transversus abdominis

Ilioinguinal Nerve

Composition: Motor and somatosensory

Sensory innervation: Skin of upper medial thigh; male scrotum and root of penis; female labia majora

Motor innervation: Joins iliohypogastric nerve and innervates the same muscles

Genitofemoral Nerve

Composition: Somatosensory

Sensory innervation: Skin of middle anterior thigh; male scrotum and cremaster muscle; female labia majora

Lateral Femoral Cutaneous Nerve

Composition: Somatosensory

Sensory innervation: Skin of lateral thigh

Femoral Nerve

Composition: Motor and somatosensory

Sensory innervation: Skin of anterior and lateral thigh; medial leg and foot

Motor innervation: Anterior muscles of thigh and extensors of leg; iliacus, psoas major, pectineus, quadriceps femoris, and sartorius

Saphenous (sah-FEE-nus) Nerve

Composition: Somatosensory

Sensory innervation: Skin of medial leg and foot; knee joint

Obturator Nerve

Composition: Motor and somatosensory

Sensory innervation: Skin of superior medial thigh; hip and knee joints

Motor innervation: Adductor muscles of leg: external obturator, pectineus, adductor longus, adductor brevis, adductor magnus, and gracilis

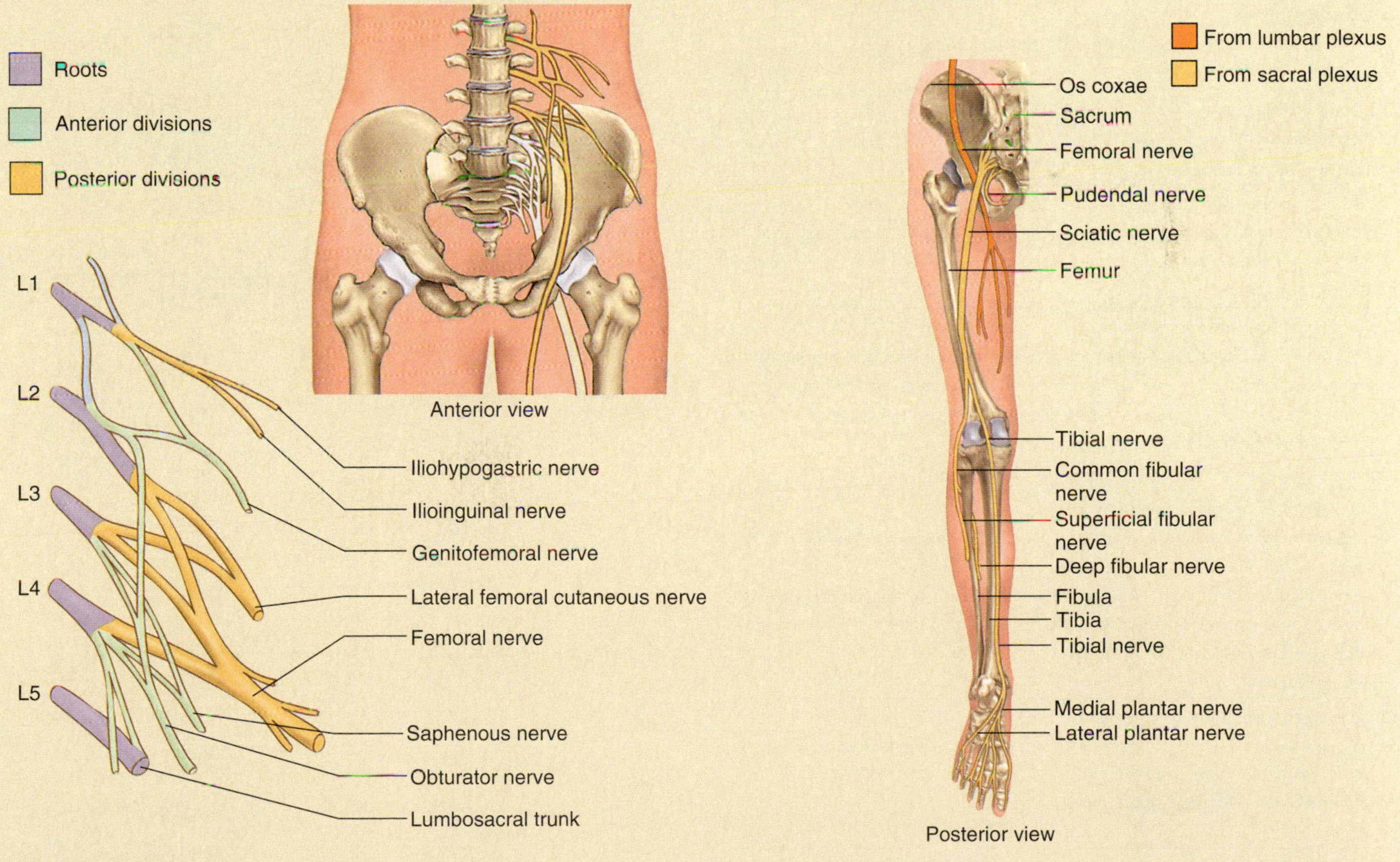

Roots
Anterior divisions
Posterior divisions
L1
L2
L3
L4
L5
Iliohypogastric nerve
Ilioinguinal nerve
Genitofemoral nerve
Lateral femoral cutaneous nerve
Femoral nerve
Saphenous nerve
Obturator nerve
Lumbosacral trunk
Anterior view
From lumbar plexus
From sacral plexus
Os coxae
Sacrum
Femoral nerve
Pudendal nerve
Sciatic nerve
Femur
Tibial nerve
Common fibular nerve
Superficial fibular nerve
Deep fibular nerve
Fibula
Tibia
Tibial nerve
Medial plantar nerve
Lateral plantar nerve
Posterior view

FIGURE 14.16
The Lumbar Plexus.

Table 14.6

The Sacral and Coccygeal Plexuses

The sacral plexus is formed from the ventral rami of nerves L4, L5, and S1 to S4. It has six roots and anterior and posterior divisions. Since it is connected to the lumbar plexus by fibers that run through the *lumbosacral trunk,* the two plexuses are sometimes referred to collectively as the *lumbosacral plexus.* The coccygeal plexus is a tiny plexus formed from the ventral rami of S4, S5, and Co (fig. 14.17).

The *tibial* and *common fibular nerves* listed in this table travel together through a connective tissue sheath; they are referred to collectively as the **sciatic** (sy-AT-ic) **nerve.** The sciatic nerve passes through the greater sciatic notch of the pelvis, extends for the length of the thigh, and ends at the popliteal fossa. Here, the tibial and common fibular nerves diverge and follow their separate paths into the leg. The sciatic nerve is a common focus of injury and pain.

Superior Gluteal Nerve

Composition: Motor
Motor innervation: Gluteus minimus, gluteus medius, and tensor fasciae latae

Inferior Gluteal Nerve

Composition: Motor
Motor innervation: Gluteus maximus

Nerve to Piriformis

Composition: Motor
Motor innervation: Piriformis

Nerve to Quadratus Femoris

Composition: Motor and somatosensory
Sensory innervation: Hip joint
Motor innervation: Quadratus femoris and gemellus inferior

Nerve to Internal Obturator

Composition: Motor
Motor innervation: Internal obturator and gemellus superior

Perforating Cutaneous Nerve

Composition: Somatosensory
Sensory innervation: Skin of posterior aspect of buttock

Posterior Cutaneous Nerve

Composition: Somatosensory
Sensory innervation: Skin of lower lateral buttock, anal region, upper posterior thigh, upper calf, scrotum, and labia majora

Tibial Nerve

Composition: Motor and somatosensory
Sensory innervation: Skin of posterior leg and sole of foot; knee and foot joints
Motor innervation: Semitendinosus, semimembranosus, long head of biceps femoris, gastrocnemius, soleus, flexor digitorum longus, flexor hallucis longus, tibialis posterior, popliteus, and intrinsic muscles of foot

Common Fibular (Peroneal) Nerve

Composition: Motor and somatosensory
Sensory innervation: Skin of anterior distal one-third of leg, dorsum of foot, and toes I and II; knee joint
Motor innervation: Short head of biceps femoris, fibularis tertius, fibularis brevis, fibularis longus, tibialis anterior, extensor hallucis longus, extensor digitorum longus, and extensor digitorum brevis

Pudendal Nerve

Composition: Motor and somatosensory
Sensory innervation: Skin of penis and scrotum of male; clitoris, labia majora and minora, and lower vagina of female
Motor innervation: Muscles of perineum

Coccygeal Nerve

Composition: Motor and somatosensory
Sensory innervation: Skin over coccyx
Motor innervation: Muscles of pelvic floor

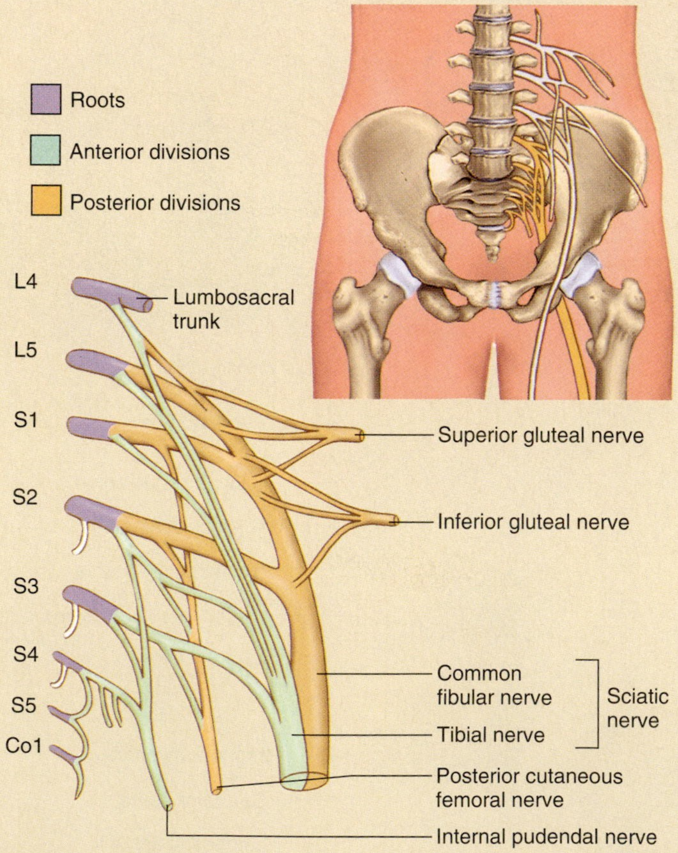

FIGURE 14.17
The Sacral and Coccygeal Plexuses.

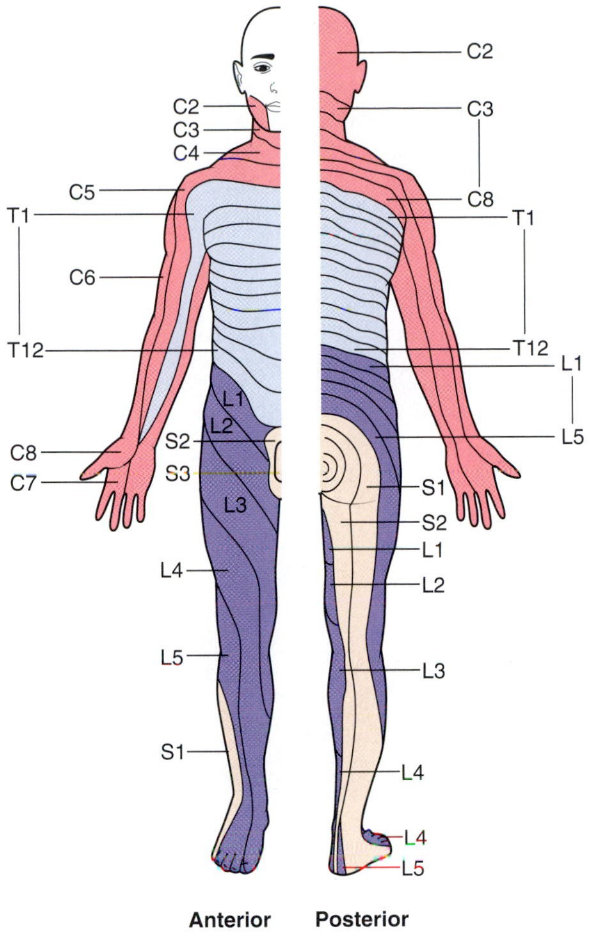

FIGURE 14.18

A Dermatome Map of the Body. Anterior and posterior views. Each zone of the skin is innervated by sensory branches of the spinal nerves indicated by the labels. Nerve C1 does not innervate the skin.

Cutaneous Innervation and Dermatomes

Each spinal nerve except C1 receives sensory input from a specific area of skin called a **dermatome,**[21] derived from the embryonic dermatomes described in chapter 4. A *dermatome map* (fig. 14.18) is a diagram of the cutaneous regions innervated by each spinal nerve. Such a map is very simplified, however, because the dermatomes overlap at their edges by as much as 50%. Therefore, severance of one sensory nerve root does not entirely deaden sensation from a dermatome. It is necessary to sever or anesthetize three successive spinal nerves to produce a total loss of sensation from one dermatome. Spinal nerve damage is assessed by testing the dermatomes with pinpricks and noting areas in which the patient has no sensation.

[21]*derma* = skin + *tome* = segment, part

Before You Go On

Answer the following questions to test your understanding of the preceding section:

5. What is meant by the dorsal and ventral roots of a spinal nerve? Which of these is sensory and which is motor?
6. Where are the somas of the dorsal root located? Where are the somas of the ventral root?
7. List the five plexuses of spinal nerves and state where each one is located.
8. State which plexus gives rise to each of the following nerves: axillary, ilioinguinal, obturator, phrenic, pudendal, radial, and sciatic.

INSIGHT 14.4 CLINICAL APPLICATION

SPINAL NERVE INJURIES

The radial and sciatic nerves are especially vulnerable to injury. The radial nerve, which passes through the axilla, may be compressed against the humerus by improperly adjusted crutches, causing *crutch paralysis.* A similar injury often resulted from the discredited practice of trying to correct a dislocated shoulder by putting a foot in a person's armpit and pulling on the arm. One consequence of radial nerve injury is *wrist drop*—the fingers, hand, and wrist are chronically flexed because the extensor muscles supplied by the radial nerve are paralyzed.

Because of its position and length, the sciatic nerve of the hip and thigh is the most vulnerable nerve in the body. Trauma to this nerve produces *sciatica,* a sharp pain that travels from the gluteal region along the posterior side of the thigh and leg as far as the ankle. Ninety percent of cases result from a herniated intervertebral disc or osteoarthritis of the lower spine, but sciatica can also be caused by pressure from a pregnant uterus, dislocation of the hip, injections in the wrong area of the buttock, or sitting for a long time on the edge of a hard chair. Men sometimes suffer sciatica because of the habit of sitting on a wallet carried in the hip pocket.

SOMATIC REFLEXES

Objectives

When you have completed this section, you should be able to

- define *reflex* and explain how reflexes differ from other motor actions;
- describe the general components of a typical reflex arc; and
- describe some common variations in reflex arcs.

Reflexes are quick, involuntary, stereotyped reactions of glands or muscles to stimulation. This definition sums up four important properties of a reflex:

1. Reflexes *require stimulation*—they are not spontaneous actions but responses to sensory input.
2. Reflexes are *quick*—they generally involve few if any interneurons and minimal synaptic delay.

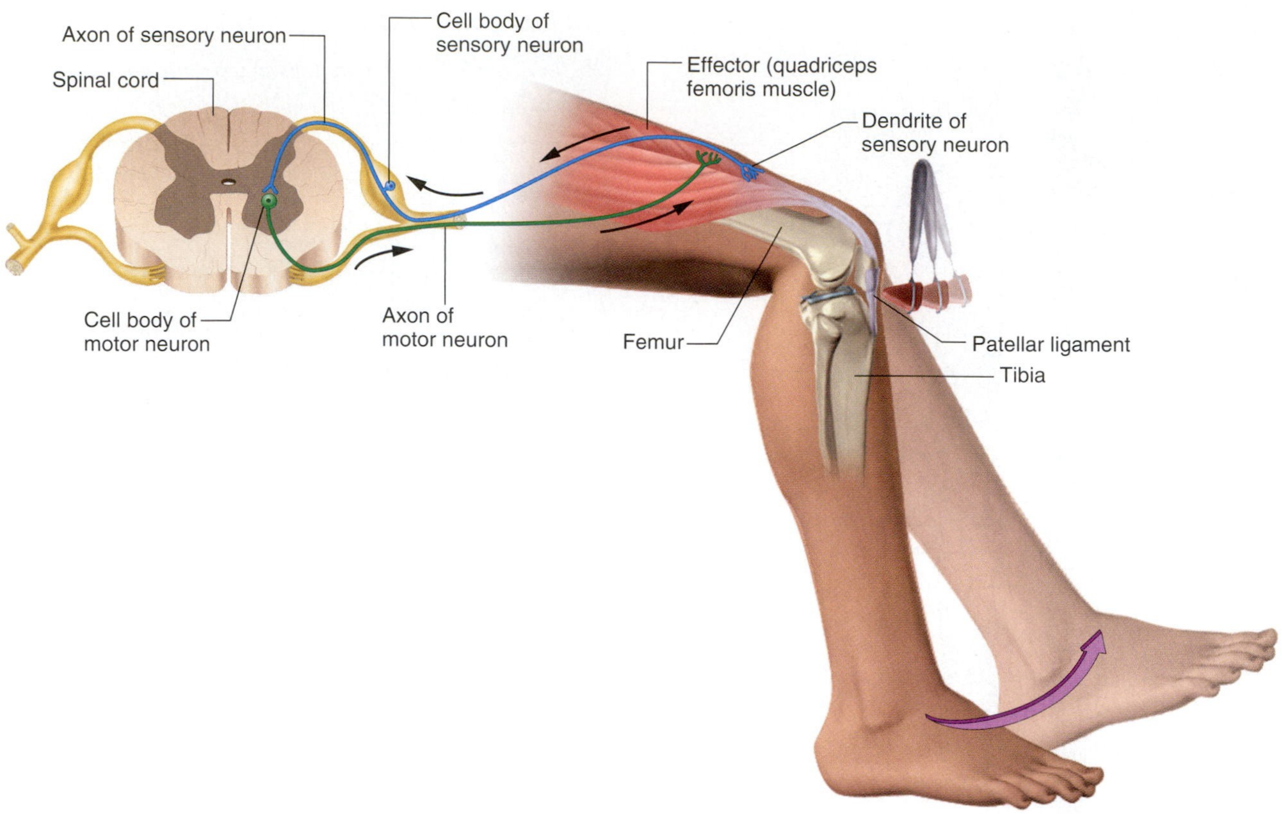

Figure 14.19

A Representative Reflex Arc. The monosynaptic reflex arc of the patellar tendon reflex.

3. Reflexes are *involuntary*—they occur without intent, often without our awareness, and they are difficult to suppress. Given an adequate stimulus, the response is essentially automatic. You may become conscious of the stimulus that evoked a reflex, and this awareness may enable you to correct or avoid a potentially dangerous situation, but awareness is not a part of the reflex itself. It may come after the reflex action has been completed, and some reflexes occur even if the spinal cord has been severed so that no stimuli reach the brain.
4. Reflexes are *stereotyped*—they occur in essentially the same way every time; the response is very predictable.

Visceral reflexes are responses of glands, cardiac muscle, and smooth muscle. They are controlled by the autonomic nervous system and discussed in chapter 16. **Somatic reflexes** are responses of skeletal muscles, such as the quick withdrawal of your hand from a hot stove or the lifting of your foot when you step on something sharp. They are controlled by the somatic nervous system. Somatic reflexes will be briefly discussed here from the anatomical standpoint. They have traditionally been called *spinal reflexes,* although some of them are mediated more by the brain than by the spinal cord.

A somatic reflex employs a rather simple neural pathway called a **reflex arc,** from a sensory nerve ending to the spinal cord or brainstem and back to a skeletal muscle. The components of a reflex arc are:

1. *Somatic receptors* in the skin, a muscle, or a tendon. These include simple nerve endings for heat and pain in the skin, specialized stretch receptors called *muscle spindles* embedded in the skeletal muscles, and other types (see chapter 17).
2. *Afferent nerve fibers,* which carry information from these receptors into the dorsal horn of the spinal cord.
3. An *integrating center,* a point of synaptic contact between neurons in the gray matter of the spinal cord or brainstem. In most reflex arcs, there are one or more interneurons in the integrating center. Synaptic events in the integrating center determine whether the efferent (output) neuron issues a signal to the muscle.
4. *Efferent nerve fibers,* which originate in the ventral horn of the spinal cord and carry motor impulses to the skeletal muscles.
5. *Skeletal muscles,* the somatic effectors that carry out the response.

In the simplest type of reflex arc, there is no interneuron. The afferent neuron synapses directly with an efferent neuron, so this kind of pathway is called a **monosynaptic reflex arc** (fig. 14.19). Synaptic delay is minimal, and the response is especially quick. Most reflex arcs, however, have one or more interneurons, and indeed often involve multineuronal circuits with many synapses. Such reflex arcs produce more prolonged muscular responses and, by way of diverging circuits, may stimulate multiple muscles at once.

TABLE 14.7

Types of Somatic Reflexes

Stretch Reflex	Increased muscle tension in response to stretch. Serves to maintain equilibrium and posture, stabilize joints, and make joint actions smoother and better coordinated. The knee-jerk reflex (patellar reflex, fig. 14.19) is a familiar monosynaptic spinal reflex.
Flexor Reflex	Contraction of flexor muscles resulting in withdrawal of a limb from an injurious stimulus, as in withdrawal from a burn or pinprick.
Crossed Extension Reflex	Contraction of extensor muscles in one limb when the flexor muscles of the opposite limb contract. Stiffens one leg, for example, when the opposite leg is lifted from the ground so that one does not fall over.
Golgi Tendon Reflex	Inhibition of muscle contraction when a tendon is excessively stretched, thus preventing tendon injuries.

THINK ABOUT IT!

There is actually a second synapse in a "monosynaptic" reflex arc. Identify its location.

A reflex like the one diagrammed in figure 14.19 is described as an **ipsilateral reflex** because the CNS input and output are on the same side of the body. Others such as the crossed extension reflex (table 14.7) are called **contralateral reflexes** because the sensory input enters the spinal cord on one side of the body and the motor output leaves from the opposite side. In an **intersegmental reflex,** the sensory signal enters the spinal cord at one level (segment) and the motor output leaves the cord from a higher or lower level. For example, if you step on something sharp and lift your foot from the ground, some motor output leaves the spinal cord higher up and goes to trunk muscles that flex your waist. This shifts your center of gravity over the leg still on the ground, preventing you from falling over.

Table 14.7 describes several types of somatic reflexes. These reflexes are controlled primarily by the cerebrum and cerebellum of the brain, but a weak response is mediated through the spinal cord and persists even if the spinal cord is severed from the brain. The spinal component can be more pronounced if the stimulus is sudden or intense, as in the clinical testing of the knee-jerk (patellar) reflex and other stretch reflexes.

Before You Go On

Answer the following questions to test your understanding of the preceding section:

9. Define *reflex.* Distinguish between somatic and visceral reflexes.
10. List and define the five components of a typical somatic reflex arc.
11. Describe a situation in which each of the following would be functionally relevant: an ipsilateral, a contralateral, and an intersegmental reflex arc.

CLINICAL PERSPECTIVES

Objectives

When you have completed this section, you should be able to

- describe some effects of spinal cord injuries; and
- define the types of paralysis and explain the basis for their differences.

Some developmental abnormalities of the spinal cord are described in chapter 13. In children and adults, the most significant disorder of the spinal cord is trauma. Each year in the United States, 10,000 to 12,000 people become paralyzed by spinal cord trauma, usually as a result of vertebral fractures. The group at greatest risk is males from 16 to 30 years old, because of their high-risk behaviors. Fifty-five percent of their injuries are from automobile and motorcycle accidents, 18% from sports, and 15% from gunshot and stab wounds. Elderly people are also at above-average risk because of falls, and in times of war, battlefield injuries account for many cases.

Complete *transection* (severance) of the spinal cord causes immediate loss of motor control at and below the level of the injury. Victims also lose all sensation from the level of injury and below, although some patients temporarily feel burning pain within one or two dermatomes of the level of the lesion.

THINK ABOUT IT!

Respiratory paralysis typically results from spinal cord transection above level C4, but not from injuries below that level. Explain.

In the early stage, victims exhibit a syndrome (a suite of signs and symptoms) called **spinal shock.** The muscles below the level of injury exhibit flaccid paralysis and an absence of reflexes because of the lack of stimulation from higher levels of the CNS. For 8 days to 8 weeks after the accident, the patient typically lacks bladder and bowel reflexes and thus retains urine and feces. Lacking sympathetic stimulation to the blood vessels, a patient may exhibit *neurogenic shock* in which the vessels dilate and blood pressure drops dangerously low. Spinal shock can last from a few days to 3 months, but typically lasts 7 to 20 days.

As spinal shock subsides, somatic reflexes begin to reappear, at first in the toes and progressing to the feet and legs. Autonomic reflexes also reappear. Contrary to the earlier urinary and fecal retention, a patient now has the opposite problem, incontinence, as the rectum and bladder empty reflexively in response to stretch. Both the somatic and autonomic nervous systems typically exhibit exaggerated reflexes, a state called *hyperreflexia* or the *mass reflex reaction.* Stimuli such as a full bladder or cutaneous touch can trigger an extreme cardiovascular reaction. The systolic blood pressure, normally about 120 mmHg, jumps to as high as 300 mmHg, sometimes causing a stroke. Pressure receptors in the major arteries sense this rise in blood pressure and activate a reflex that slows the heart, sometimes to a rate as low as 30 or 40 beats/minute *(bradycardia).*

Men at first lose the capacity for erection and ejaculation. They may recover these functions later and become capable of climaxing and fathering children, but still lack sexual sensation.

The most serious permanent effect of spinal cord trauma is paralysis. The flaccid paralysis of spinal shock later changes to spastic paralysis as reflexes are regained, but lack inhibitory control from the brain. Spastic paralysis typically starts with chronic flexion of the hips and knees *(flexor spasms)* and progresses to a state in which the limbs become straight and rigid *(extensor spasms)*. Three forms of muscle paralysis are **paraplegia,** a paralysis of both lower limbs resulting from spinal cord lesions at levels T1 to L1; **quadriplegia,** the paralysis of all four limbs resulting from lesions above level C5; and **hemiplegia,** paralysis of one side of the body, resulting not from spinal cord injuries but usually from a stroke or other brain lesion. Spinal cord lesions from C5 to C7 can produce a state of partial quadriplegia—total paralysis of the lower limbs and partial paralysis (*paresis,* or weakness) of the upper limbs.

Treatment of spinal cord injuries is an area of intense medical research today, with hopes for recovery of spinal functions stimulated by new insights on the physiological mechanisms of spinal cord tissue death and the potential for embryonic stem cells to regenerate damaged cord tissue.

Table 14.8 describes some injuries and other disorders of the spinal cord and spinal nerves.

Before You Go On

Answer the following questions to test your understanding of the preceding section:

12. Describe the signs of spinal shock.
13. Describe the difference between flaccid paralysis and spastic paralysis.
14. Distinguish between the causes of paraplegia, quadriplegia, and hemiplegia.

TABLE 14.8

Some Disorders of the Spinal Cord and Spinal Nerves

Guillain-Barré Syndrome	An acute demyelinating nerve disorder often triggered by viral infection, resulting in muscle weakness, elevated heart rate, unstable blood pressure, shortness of breath, and sometimes death from respiratory paralysis
Neuralgia	General term for nerve pain, often caused by pressure on spinal nerves from herniated intervertebral discs or other causes
Paresthesia	Abnormal sensations of prickling, burning, numbness, or tingling; a symptom of nerve trauma or other peripheral nerve disorders
Peripheral Neuropathy	Any loss of sensory or motor function due to nerve injury; also called *nerve palsy*
Rabies (hydrophobia)	A disease usually contracted from animal bites, involving viral infection that spreads via somatic motor nerve fibers to the CNS and then out of the CNS via autonomic nerve fibers, leading to seizures, coma, and death; invariably fatal if not treated before CNS symptoms appear
Spinal Meningitis	Inflammation of the spinal meninges due to viral, bacterial, or other infection

Disorders Described Elsewhere

Amyotrophic lateral sclerosis 393	Multiple sclerosis 373	Sciatica 405
Carpal tunnel syndrome 326	Poliomyelitis 393	Shingles 395
Crutch paralysis 405	Paraplegia 408	Spina bifida 380
Diabetic neuropathy 480	Quadriplegia 408	Spinal cord trauma 407
Leprosy 480		

CHAPTER REVIEW

Review of Key Concepts

The Spinal Cord (p. 386)

1. The spinal cord conducts signals up and down the body, contains *central pattern generators* that control locomotion, and mediates many reflexes.
2. The spinal cord occupies the vertebral canal from vertebrae C1 to L1. A bundle of nerve roots called the *cauda equina* occupies the canal from C2 to S5.
3. The cord is divided into cervical, thoracic, lumbar, and sacral regions, named for the levels of the vertebral column through which the spinal nerves emerge. The portion served by each spinal nerve is called a *segment* of the cord.
4. *Cervical* and *lumbar enlargements* are wide points in the cord marking the emergence of nerves that control the limbs.
5. The spinal cord is enclosed in three fibrous *meninges.* From superficial to deep, these are the *dura mater, arachnoid mater,* and *pia mater.* An *epidural space* exists between the dura mater and vertebral bone, and a *subarachnoid space* between the arachnoid and pia mater.
6. The pia mater issues periodic *denticulate ligaments* that anchor it to the dura, and continues inferiorly as a *coccygeal ligament* that anchors the cord to vertebra L2.
7. In cross section, the spinal cord exhibits a central H-shaped core of *gray matter* surrounded by white matter. The gray matter contains the somas, dendrites, and synapses while the white matter consists of nerve fibers (axons).
8. The *dorsal horn* of the gray matter receives afferent (sensory) nerve fibers from the dorsal root of the spinal nerve. The *ventral horn* contains the somas that give rise to the efferent (motor) nerve fibers of the ventral root. A *lateral horn* in the thoracic and lumbar regions contains somas of the sympathetic neurons.
9. The white matter is divided into *dorsal, lateral,* and *ventral columns* on each side of the cord. Each column consists of one or more *tracts,* or bundles of nerve fibers. The nerve fibers in a given tract are similar in origin, destination, and function.
10. *Ascending tracts* carry sensory information up the cord to the brain. Their names and functions are listed in table 14.1.
11. From receptor to cerebral cortex, sensory signals typically travel through three neurons (first- through third-order) and cross over *(decussate)* from one side of the body to the other in the spinal cord or brainstem. Thus, the right cerebral cortex receives sensory input from the left side of the body (from the neck down) and vice versa.
12. *Descending tracts* carry motor commands from the brain downward. Their names and functions are also listed in table 14.1.
13. Motor signals typically begin in an *upper motor neuron* in the cerebral cortex and travel to a *lower motor neuron* in the brainstem or spinal cord. The latter neuron's axon leaves the CNS in a cranial or spinal nerve leading to a muscle.

The Spinal Nerves (p. 393)

1. A nerve is a cordlike organ composed of nerve fibers (axons) and connective tissue.
2. Each nerve fiber is enclosed in its own fibrous sleeve called an *endoneurium.* Nerve fibers are bundled in groups called *fascicles* separated from each other by a *perineurium.* A fibrous *epineurium* covers the entire nerve.
3. Nerve fibers are classified as *afferent* or *efferent* depending on the direction of signal conduction, *somatic* or *visceral* depending on the types of organs they innervate, and *special* or *general* depending on the locations of the organs they innervate (table 14.2).
4. A *sensory nerve* is composed of afferent fibers only, a *motor nerve* of efferent fibers only, and a *mixed nerve* is composed of both. Most nerves are mixed.
5. A *ganglion* is a swelling along the course of a nerve containing the cell bodies of the peripheral neurons.
6. There are 31 pairs of *spinal nerves,* which enter and leave the spinal cord and emerge mainly through the intervertebral foramina. Within the vertebral canal, each branches into a *dorsal root* which carries sensory signals to the dorsal horn of the spinal cord, and a *ventral root* which receives motor signals from the ventral horn. The dorsal root has a swelling, the *dorsal root ganglion,* containing unipolar neurons of somatic sensory neurons.
7. Distal to the intervertebral foramen, each spinal nerve branches into a *dorsal ramus, ventral ramus,* and *meningeal branch.*
8. The ventral ramus gives rise to *intercostal nerves* in the thoracic region and *nerve plexuses* in all other regions. The nerve plexuses are weblike networks adjacent to the vertebral column: the *cervical, brachial, lumbar, sacral,* and *coccygeal plexus.* The nerves arising from each are described in tables 14.3 through 14.6.

Somatic Reflexes (p. 405)

1. A reflex is a quick, involuntary, stereotyped reaction of a gland or muscle to a stimulus.
2. *Visceral reflexes* are reactions of glands, cardiac muscle, and smooth muscle, controlled by the autonomic nervous system. *Somatic (spinal) reflexes* are responses of skeletal muscles, controlled by the somatic nervous system.
3. A somatic reflex employs a simple neural pathway called a *reflex arc,* in which signals travel from a somatic receptor through an afferent nerve fiber to the spinal cord or brainstem, an integrating center in the CNS, an efferent nerve fiber leaving the CNS, and finally to a skeletal muscle.
4. A *monosynaptic reflex arc* has no interneuron between the afferent and efferent neurons; thus it has minimal synaptic delay and especially quick responses. Most reflex arcs, however, are polysynaptic, involving one or more interneurons.
5. *Ipsilateral reflex arcs* have the sensory input and motor output on the same side of the CNS; *contralateral reflex arcs* have their output on the side opposite from the input; and *intersegmental reflex arcs* have their output at a different vertical level of the spinal cord than their input. Four specific classes of somatic reflexes are described in table 14.7.

Clinical Perspectives (p. 407)

1. Trauma is the most common disorder of the spinal cord, usually resulting from accidents.
2. Complete transection of the spinal cord immediately abolishes sensation and motor control in areas below the injury. *Spinal shock* typically lasts up to 20 days from the injury. Somatic and autonomic reflexes then begin to reappear, and may be exaggerated *(hyperreflexia).* Flaccid paralysis is typically replaced by spastic paralysis as reflex functions return. *Paraplegia* and *quadriplegia* are common consequences of spinal cord injury, while *hemiplegia* usually results from a brain lesion.
3. Other disorders of the spinal cord and spinal nerves are described in table 14.8.

Testing Your Recall

1. Below L2, the vertebral canal is occupied by a bundle of spinal nerve roots called
   a. the terminal filum.
   b. the descending tracts.
   c. the gracile fasciculus.
   d. the medullary cone.
   e. the cauda equina.
2. The brachial plexus gives rise to all of the following nerves *except*
   a. the axillary nerve.
   b. the radial nerve.
   c. the saphenous nerve.
   d. the median nerve.
   e. the ulnar nerve.
3. Between the dura mater and vertebral bone, one is most likely to find
   a. arachnoid mater.
   b. denticulate ligaments.
   c. cartilage.
   d. adipose tissue.
   e. spongy bone.
4. Which of these tracts carry motor signals destined for the postural muscles?
   a. the gracile fasciculus
   b. the cuneate fasciculus
   c. spinothalamic tracts
   d. vestibulospinal tracts
   e. tectospinal tracts
5. A patient has a gunshot wound that caused a bone fragment to nick the spinal cord. The patient now feels no pain or temperature sensations from that level of the body down. Most likely, the _____ was damaged.
   a. gracile fasciculus
   b. medial lemniscus
   c. tectospinal tract
   d. lateral corticospinal tract
   e. spinothalamic tract
6. Which of these is *not* a region of the spinal cord?
   a. cervical
   b. thoracic
   c. pelvic
   d. lumbar
   e. sacral
7. In the spinal cord, the somas of the lower motor neurons are found in
   a. the cauda equina.
   b. the dorsal horns.
   c. the ventral horns.
   d. the dorsal root ganglia.
   e. the fasciculi.
8. The outermost connective tissue wrapping of a nerve is called the
   a. epineurium.
   b. perineurium.
   c. endoneurium.
   d. arachnoid membrane.
   e. dura mater.
9. The intercostal nerves between the ribs arise from which spinal nerve plexus?
   a. cervical
   b. brachial
   c. lumbar
   d. sacral
   e. none of them
10. All somatic reflexes share all of the following properties except
    a. they are quick.
    b. they are monosynaptic.
    c. they require stimulation.
    d. they are involuntary.
    e. they are stereotyped.
11. Outside the CNS, the somas of neurons are clustered in swellings called _____.
12. Distal to the intervertebral foramen, a spinal nerve branches into a dorsal and ventral _____.
13. The cerebellum receives feedback from the muscles and joints by way of the _____ tracts of the spinal cord.
14. In the _____ reflex, contraction of flexor muscles in one limb is accompanied by the contraction of extensor muscles in the contralateral limb.
15. Modified muscle fibers serving primarily to detect stretch are called _____.
16. The _____ nerves arise from the cervical plexus and innervate the diaphragm.
17. The crossing of a nerve fiber or tract from the right side of the CNS to the left, or vice versa, is called _____.
18. The nonvisual awareness of the body's position and movements is called _____.
19. The _____ ganglion contains the somas of neurons that carry sensory signals to the spinal cord.
20. The sciatic nerve is a composite of two nerves, the _____ and _____.

Answers in the Appendix

True or False

Determine which five of the following statements are false, and briefly explain why.

1. The gracile fasciculus is a descending spinal tract.
2. At the inferior end, the adult spinal cord ends before the vertebral column does.
3. Each spinal cord segment has only one pair of spinal nerves.
4. Some spinal nerves are sensory and others are motor.
5. The dura mater adheres tightly to the bone of the vertebral canal.
6. The dorsal and ventral horns of the spinal cord are composed of gray matter.
7. The corticospinal tracts carry motor signals down the spinal cord.
8. The dermatomes are nonoverlapping regions of skin innervated by different spinal nerves.
9. Somatic reflexes are those that do not involve the brain.
10. The Golgi tendon reflex acts to inhibit muscle contraction.

Answers in the Appendix

Testing Your Comprehension

1. Jillian is thrown from a horse. She strikes the ground with her chin, causing severe hyperextension of the neck. Emergency medical technicians properly immobilize her neck and transport her to a hospital, but she dies 5 minutes after arrival. An autopsy shows multiple fractures of vertebrae C1, C6, and C7 and extensive damage to the spinal cord. Explain why she died rather than being left quadriplegic.
2. Wallace is the victim of a hunting accident. A bullet grazed his vertebral column and bone fragments severed the left half of his spinal cord at segments T8 through T10. Since the accident, Wallace has had a condition called *dissociated sensory loss,* in which he feels no sensations of deep touch or limb position on the *left* side of his body below the injury and no sensations of pain or heat on the *right* side. Explain what spinal tract(s) the injury has affected and why these sensory losses are on opposite sides of the body.
3. Anthony gets into a fight between rival gangs. As an attacker comes at him with a knife, he turns to flee, but stumbles. The attacker stabs him on the medial side of the right gluteal fold and Anthony collapses. He loses all use of his right limb, being unable to extend his hip, flex his knee, or move his foot. He never fully recovers these lost functions. Explain what nerve injury Anthony has most likely suffered.
4. Stand with your right shoulder, hip, and foot firmly against a wall. Raise your left foot from the floor without losing contact with the wall at any point. What happens? Why? What principle of this chapter does this demonstrate?
5. When a patient needs a tendon graft, surgeons sometimes use the tendon of the palmaris longus, a relatively dispensable muscle of the forearm. The median nerve lies nearby and looks very similar to this tendon. There have been cases where a surgeon mistakenly removed a section of this nerve instead of the tendon. What effects do you think such a mistake would have on the patient?

Answers at the Online Learning Center

www.mhhe.com/saladinha1

Visit the Online Learning Center for practice tests, answer keys, and other learning aids for this chapter. Enhance your understanding of human anatomy with our interactive art labeling exercises, supplemental photo atlases, web links, puzzles, flashcards, and much more.

15

CHAPTER FIFTEEN

The Brain and Cranial Nerves

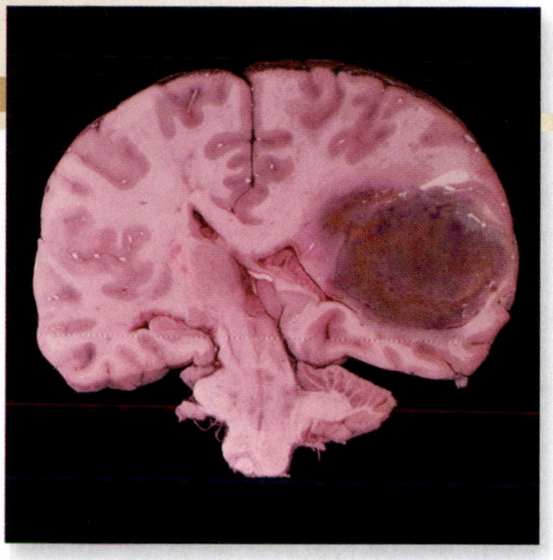

Frontal section of a brain with a large tumor (glioblastoma) in the left cerebral hemisphere

CHAPTER OUTLINE

Overview of the Brain 414
- Major Landmarks 414
- Gray and White Matter 414
- Meninges 414
- Ventricles and Cerebrospinal Fluid 417
- Blood Supply and the Brain Barrier System 418

The Hindbrain and Midbrain 421
- The Medulla Oblongata 421
- The Pons 421
- The Cerebellum 421
- The Midbrain 425
- The Reticular Formation 426

The Forebrain 427
- The Diencephalon 428
- The Cerebrum 428
- Higher Forebrain Functions 432

The Cranial Nerves 440
- Classification 440
- Nerve Pathways 440
- An Aid to Memory 440

Developmental and Clinical Perspectives 449
- The Aging Central Nervous System 449
- Two Neurodegenerative Diseases 449

Chapter Review 451

INSIGHTS

15.1 Clinical Application: Meningitis 417
15.2 Medical History: An Accidental Lobotomy 438
15.3 Clinical Application: Some Cranial Nerve Disorders 449

BRUSHING UP

To understand this chapter, you may find it helpful to review the following concepts:

- Anatomy of the cranium (pp. 175–183)
- Glial cells and their functions (p. 371)
- Embryonic development of the central nervous system (pp. 378–379)
- Meninges (p. 386)
- Gray and white matter (pp. 388–389)
- Tracts of the spinal cord (pp. 389–392)
- Structure of nerves and ganglia (pp. 393–395)

The mystique of the brain intrigues modern biologists and psychologists even as it did the philosophers of antiquity. Aristotle thought that the brain was only a radiator for cooling the blood, but generations earlier, Hippocrates had expressed a more accurate view of its functions. "Men ought to know," he said, "that from the brain, and from the brain only, arise our pleasures, joy, laughter and jests, as well as our sorrows, pains, griefs and tears. Through it, in particular, we think, see, hear, and distinguish the ugly from the beautiful, the bad from the good, the pleasant from the unpleasant." Brain function is so strongly associated with what it means to be alive and human that the cessation of brain activity is taken as a clinical criterion of death even when other organs of the body are still functioning.

The brain communicates with the rest of the body by two routes: the spinal cord, examined in chapter 14, and 12 pairs of *cranial nerves.* The cranial nerves arise from the base of the brain and emerge through the skull foramina. They lead to sensory and motor nerve endings mainly in the head-neck region. Because of the close anatomical and functional relationship between the brain and cranial nerves, they are considered together in this chapter.

Overview of the Brain

Objectives

When you have completed this section, you should be able to

- describe the major subdivisions and anatomical landmarks of the brain;
- state the locations of the gray and white matter of the brain;
- describe the meninges of the brain;
- describe a system of fluid-filled chambers within the brain;
- discuss the production, flow, and function of the cerebrospinal fluid in these chambers; and
- explain the significance of the brain barrier system.

Major Landmarks

Before we study the form and function of specific regions of the brain, it is necessary to have an overview of its major landmarks (figs. 15.1 and 15.2). These will provide important points of reference as we progress through a more detailed study.

The terms *rostral* and *caudal,* though used in some earlier chapters, are especially useful in describing brain anatomy. **Rostral**[1] means "toward the nose" and **caudal**[2] means "toward the tail." These are clear descriptions for rats and other mammals on which so much neuroanatomy has been done, but a little less obvious for humans. Our upright stance creates a situation in which the CNS rises vertically through the spinal cord and brainstem, then turns about 90° and continues toward the forehead. In human brain anatomy, one structure is considered rostral to another if it is closer to the forehead and caudal to another if it is closer to the spinal cord or rear of the head (fig. 15.3).

The average adult brain weighs about 1,600 g (3.5 lb) in men and 1,450 g (3.2 lb) in women. Its size is proportional to body size, not intelligence—the Neanderthal people had larger brains than modern humans.

The brain is divided into three major portions—the *cerebrum, cerebellum,* and *brainstem.* The **cerebrum** (SER-eh-brum or seh-REE-brum) consists of a pair of large *cerebral hemispheres* that dominate the brain and somewhat conceal the other structures. The **cerebellum**[3] (SER-eh-BEL-um) is the second-largest part of the brain. It lies inferior to the cerebrum in the posterior cranial fossa.

Authorities differ on how they define the **brainstem.** It is here considered to be all of the brain except the cerebrum and cerebellum. Its major components, from caudal to rostral, are the *medulla oblongata, pons, midbrain,* and *diencephalon.* The most common alternative definition includes only the first three of these. In a living person, the brainstem is oriented like a vertical stalk with the cerebrum perched on top of it like a mushroom cap. Postmortem changes give it a more oblique angle in the cadaver and consequently in many medical illustrations. The brainstem ends at the foramen magnum of the skull; the central nervous system continues below this as the spinal cord.

Gray and White Matter

The brain, like the spinal cord, is composed of gray and white matter. Gray matter—the site of the neuron cell bodies, dendrites, and synapses—forms a surface layer called the **cortex** over the cerebrum and cerebellum, and deeper masses called **nuclei** surrounded by white matter (see fig. 15.5*c*). The white matter thus lies deep to the cortical gray matter in most of the brain, opposite from the relationship of gray and white matter in the spinal cord. As in the spinal cord, the white matter is composed of **tracts,** or bundles of axons, which here connect one part of the brain to another. It gets its bright white color from myelin.

Meninges

Like the spinal cord, the brain is enveloped in connective tissue membranes, the meninges, which lie between the nervous tissue and bone. The meninges of the brain are basically the same as those of the spinal cord—dura mater, arachnoid mater, and pia mater—although there are some differences in the dura mater (fig. 15.4). In the cranial cavity, the dura consists of two layers—an outer *periosteal layer,* equivalent to the periosteum of the cranial bone, and an inner *meningeal layer.* Only the meningeal layer continues into the vertebral canal. The cranial dura mater lies closely against the cranial bone, with no intervening epidural space like the one

[1]*rostr* = nose
[2]*caud* = tail
[3]*cereb* = brain + *ellum* = little

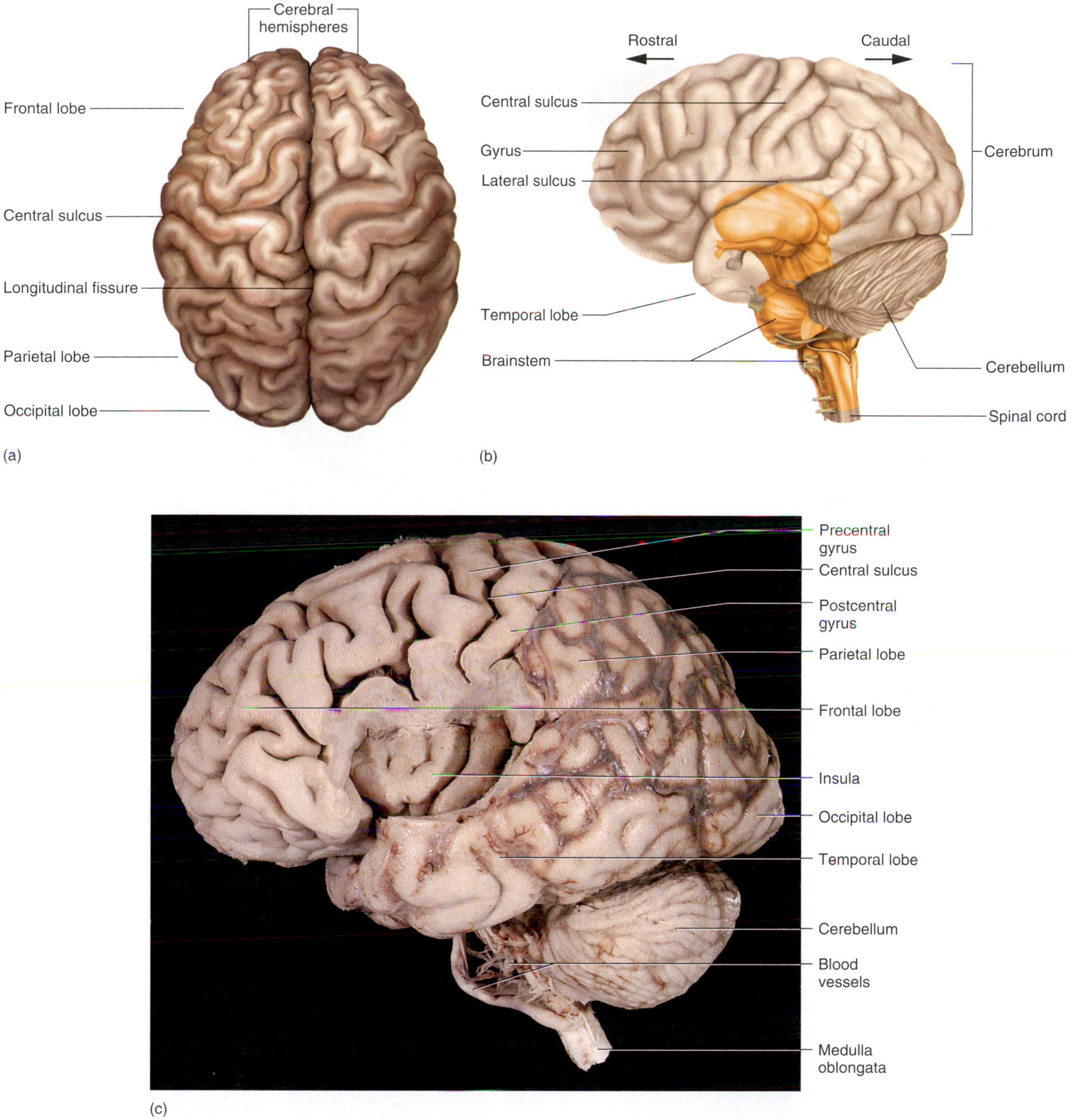
Cerebral hemispheres
Frontal lobe
Central sulcus
Longitudinal fissure
Parietal lobe
Occipital lobe
(a)
Rostral
Caudal
Central sulcus
Gyrus
Lateral sulcus
Cerebrum
Temporal lobe
Brainstem
Cerebellum
Spinal cord
(b)
Precentral gyrus
Central sulcus
Postcentral gyrus
Parietal lobe
Frontal lobe
Insula
Occipital lobe
Temporal lobe
Cerebellum
Blood vessels
Medulla oblongata
(c)

FIGURE 15.1

Surface Anatomy of the Brain. (*a*) Superior view of the cerebral hemispheres. (*b*) Left lateral view, with the brainstem in *orange.* The portion of the brainstem above the cerebellum is represented as showing through the cerebrum to convey its location. (*c*) The partially dissected brain of a cadaver. Part of the left hemisphere is cut away to expose the insula. The arachnoid mater is removed from the anterior (rostral) half of the brain to expose the gyri and sulci; the arachnoid with its blood vessels is seen on the posterior (caudal) half. Blood vessels of the brainstem are left in place.

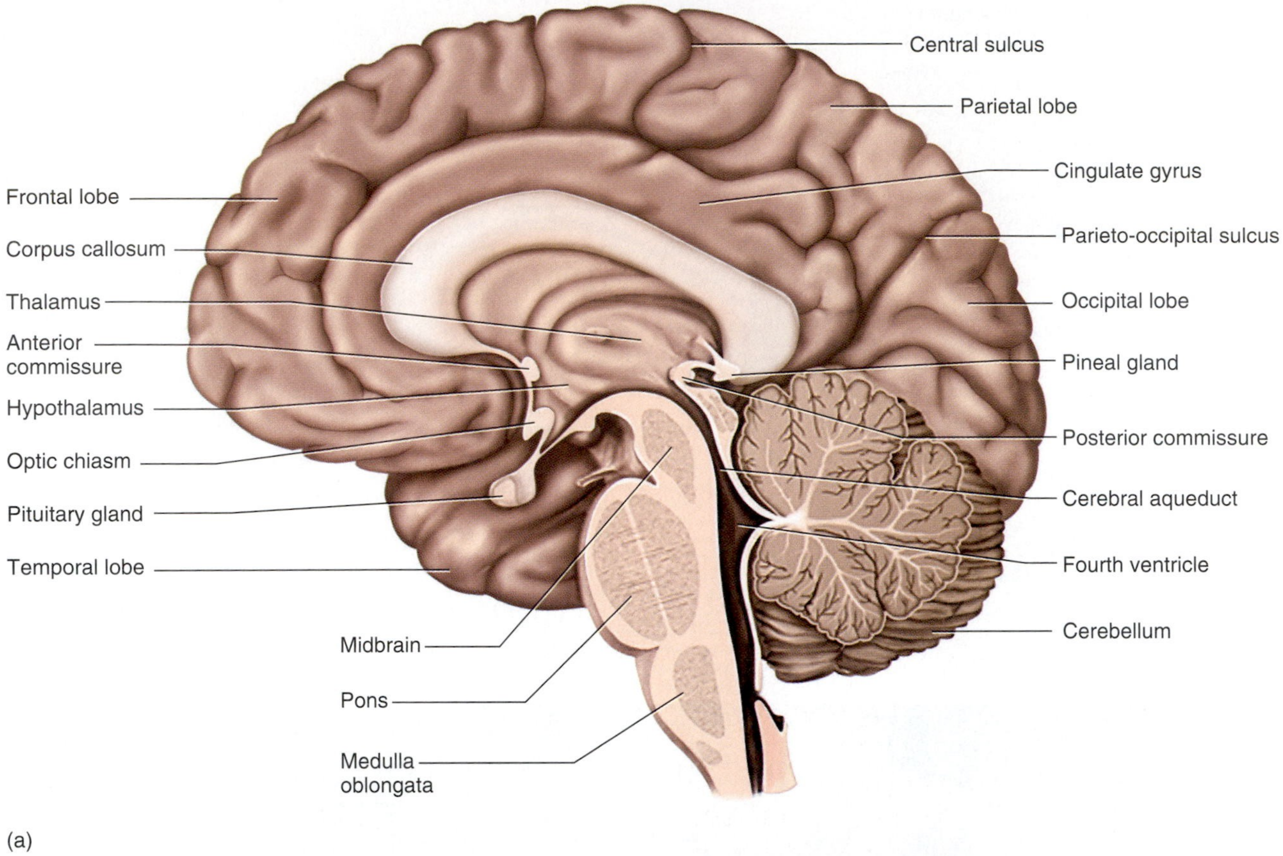
Central sulcus
Parietal lobe
Cingulate gyrus
Parieto-occipital sulcus
Occipital lobe
Pineal gland
Posterior commissure
Cerebral aqueduct
Fourth ventricle
Cerebellum
Frontal lobe
Corpus callosum
Thalamus
Anterior commissure
Hypothalamus
Optic chiasm
Pituitary gland
Temporal lobe
Midbrain
Pons
Medulla oblongata
(a)

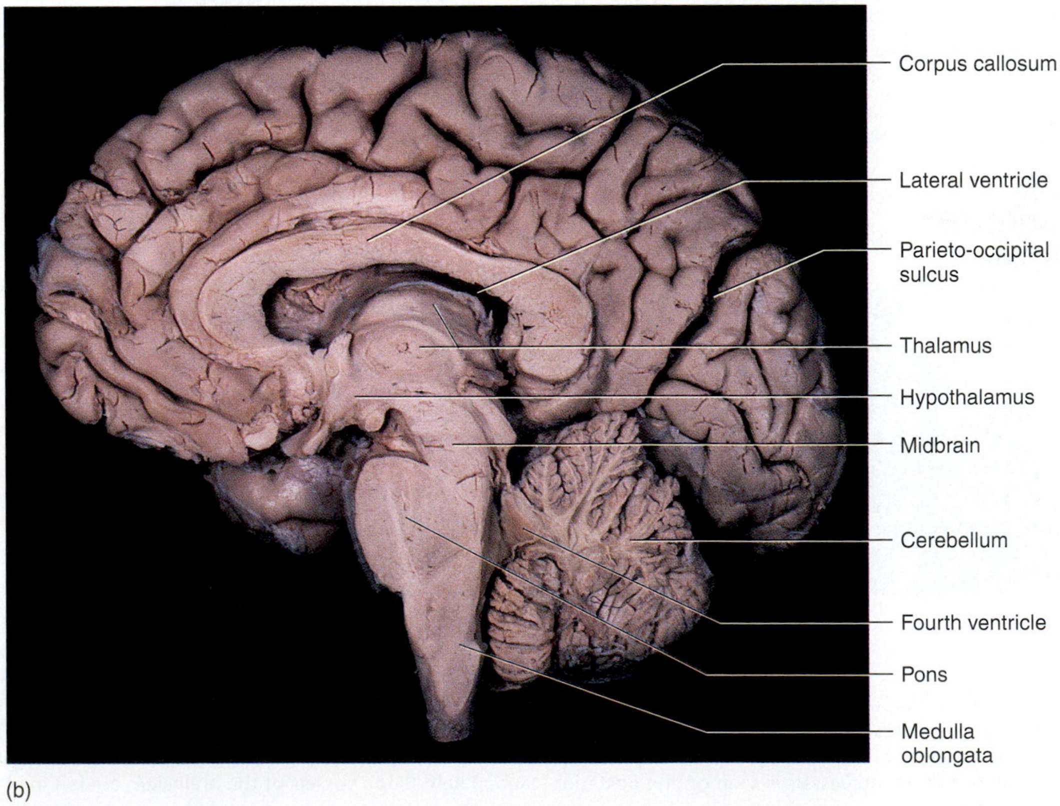
Corpus callosum
Lateral ventricle
Parieto-occipital sulcus
Thalamus
Hypothalamus
Midbrain
Cerebellum
Fourth ventricle
Pons
Medulla oblongata
(b)

FIGURE 15.2
Medial Aspect of the Brain. (*a*) Median section, left lateral view. (*b*) Median section of the cadaver brain.

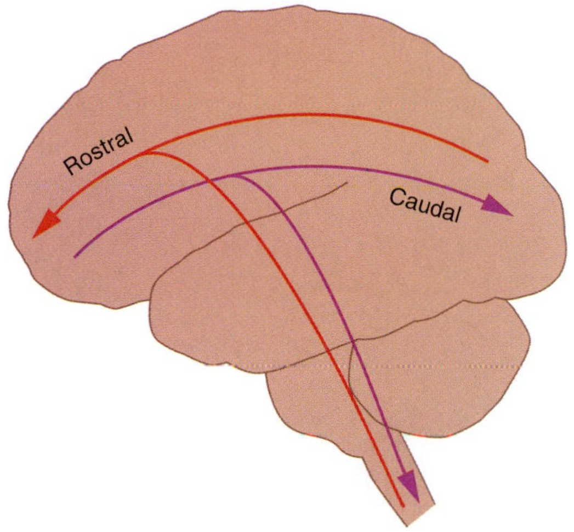

Figure 15.3

Directional Terms in CNS Anatomy. The *rostral* direction is from the rear of the head or from lower points in the brainstem or spinal cord toward the forehead. The *caudal* direction is from the forehead toward the rear of the head or toward lower points in the brainstem or spinal cord.

INSIGHT 15.1 CLINICAL APPLICATION

Meningitis

Meningitis—inflammation of the meninges—is one of the most serious diseases of infancy and childhood. It occurs especially between 3 months and 2 years of age. Meningitis is caused by a variety of bacteria and viruses that invade the CNS by way of the nose and throat, often following respiratory, throat, or ear infections. The pia mater and arachnoid are most often affected, and from here the infection can spread to the adjacent nervous tissue. In bacterial meningitis, the brain swells, the ventricles enlarge, and the brainstem may exhibit hemorrhages. Signs include a high fever, stiff neck, drowsiness, and intense headache and may progress to vomiting, loss of sensory and motor functions, and coma. Death can occur within hours of the onset. Infants and toddlers with a high fever should therefore receive immediate medical attention.

Meningitis is diagnosed partly by examining the cerebrospinal fluid (CSF) for bacteria and white blood cells. The CSF is obtained by making a *lumbar puncture (spinal tap)* between two lumbar vertebrae and drawing fluid from the subarachnoid space (see insight 14.1).

around the spinal cord. It is attached to the cranial bone only in limited places—around the foramen magnum, the sella turcica, the crista galli, and the sutural lines of the skull.

In some places, the two layers of dura are separated by **dural sinuses,** spaces that collect blood that has circulated through the brain. Two major dural sinuses are the **superior sagittal sinus,** found just under the cranium along the midsagittal line, and the **transverse sinus,** which runs horizontally from the rear of the head toward each ear. These sinuses meet like an inverted T at the back of the brain and ultimately empty into the internal jugular veins of the neck. The anatomy of the dural sinuses is detailed in chapter 21.

In certain places, the meningeal layer of the dura mater folds inward to separate major parts of the brain from each other: the *falx*[4] *cerebri* (falks SER-eh-bry) extends into the longitudinal fissure as a vertical wall between the right and left cerebral hemispheres, and is shaped like the curved blade of a sickle; the *tentorium*[5] (ten-TOE-ree-um) *cerebelli* stretches horizontally like a roof over the posterior cranial fossa and separates the cerebellum from the overlying cerebrum; and the vertical *falx cerebelli* partially separates the right and left halves of the cerebellum on the inferior side.

The arachnoid mater and pia mater are similar to those of the spinal cord. The arachnoid mater is a transparent membrane over the brain surface, visible in the caudal half of the cerebrum in figure 15.2*b*. A *subarachnoid space* separates the arachnoid from the pia, and in some places, a *subdural space* separates the dura from the arachnoid. The pia mater is a very thin, delicate membrane that closely follows all the contours of the brain surface, even dipping into the sulci. It is not generally visible without a microscope.

[4]*falx* = sickle
[5]*tentorium* = tent

Ventricles and Cerebrospinal Fluid

The brain has four internal chambers called **ventricles.** The largest are the two **lateral ventricles,** which form an arc in each cerebral hemisphere (fig. 15.5). Through a pore called the **interventricular foramen,** each lateral ventricle is connected to the **third ventricle,** a narrow medial space inferior to the corpus callosum. From here, a canal called the **cerebral aqueduct** passes down the core of the midbrain and leads to the **fourth ventricle,** a small triangular chamber between the pons and cerebellum (see fig. 15.2). Caudally, this space narrows and forms a **central canal** that extends through the medulla oblongata into the spinal cord.

On the floor or wall of each ventricle, there is a spongy mass of blood capillaries called a **choroid** (CO-royd) **plexus** (fig. 15.5*c*), named for its histological resemblance to the chorion of a fetus. Ependymal cells, a type of neuroglia, cover each choroid plexus and the entire interior surface of the ventricles and canals of the brain and spinal cord. The choroid plexuses produce some of the cerebrospinal fluid.

Cerebrospinal fluid (CSF) is a clear, colorless liquid that fills the ventricles and canals of the CNS and bathes its external surface. The brain produces about 500 mL of CSF per day, but the fluid is constantly reabsorbed at the same rate and only 100 to 160 mL are present at one time. About 40% of it is formed in the subarachnoid space external to the brain, 30% by the general ependymal lining of the brain ventricles, and 30% by the choroid plexuses. CSF forms partly by the filtration of blood plasma through the choroid plexuses and other capillaries of the brain. The ependymal cells chemically modify the filtrate as it passes through them into the ventricles and subarachnoid space.

The CSF is not a stationary fluid but continually flows through and around the CNS, driven partly by its own pressure and partly by rhythmic pulsations of the brain produced by each heartbeat. The CSF secreted in the lateral ventricles flows through

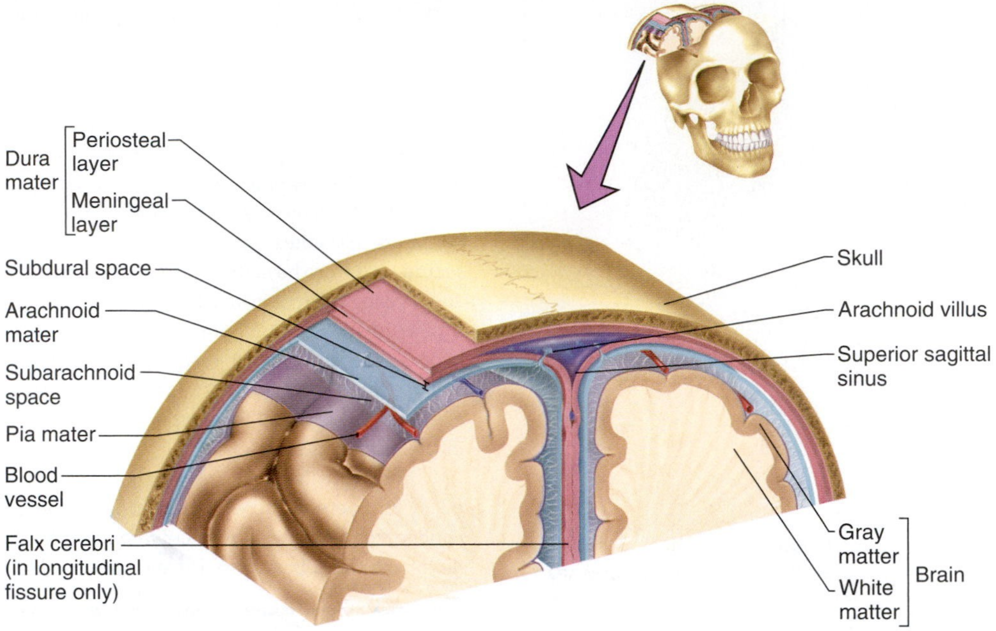
Dura mater
Periosteal layer
Meningeal layer
Subdural space
Arachnoid mater
Subarachnoid space
Pia mater
Blood vessel
Falx cerebri (in longitudinal fissure only)
Skull
Arachnoid villus
Superior sagittal sinus
Gray matter
White matter
Brain

FIGURE 15.4
The Meninges of the Brain. Frontal section of the head.

the interventricular foramina into the third ventricle (fig. 15.6), then down the cerebral aqueduct to the fourth ventricle. The third and fourth ventricles and their choroid plexuses add more CSF along the way. A small amount of CSF fills the central canal of the spinal cord, but ultimately, all of it escapes through three pores in the walls of the fourth ventricle—a *median aperture* and two *lateral apertures.* These lead into the subarachnoid space on the surface of the brain and spinal cord. On the brain surface, the CSF is absorbed from this space by **arachnoid villi,** cauliflower-like extensions of the arachnoid meninx that protrude through the dura mater into the superior sagittal sinus. CSF penetrates the walls of the arachnoid villi and mixes with the blood in the sinus.

Cerebrospinal fluid serves three purposes:

1. **Buoyancy.** Because the brain and CSF are very similar in density, the brain neither sinks nor floats in the CSF. It hangs from delicate specialized fibroblasts of the arachnoid meninx. A human brain removed from the body weighs about 1,500 g, but when suspended in CSF its effective weight is only about 50 g. By analogy, consider how much easier it is to lift another person when you are standing in a lake than it is on land. This buoyancy effect allows the brain to attain considerable size without being impaired by its own weight. If the brain rested heavily on the floor of the cranium, the pressure would kill the nervous tissue.
2. **Protection.** CSF also protects the brain from striking the cranium when the head is jolted. If the jolt is severe, however, the brain still may strike the inside of the cranium or suffer shearing injury from contact with the angular surfaces of the cranial floor. This is one of the common findings in child abuse (shaken child syndrome) and in head injuries (concussions) from auto accidents, boxing, and the like.
3. **Chemical stability.** The flow of CSF rinses metabolic wastes from the nervous tissue and homeostatically regulates its chemical environment. Slight changes in CSF composition can cause malfunctions of the nervous system. For example, a high glycine concentration disrupts the control of temperature and blood pressure, and a high pH causes dizziness and fainting.

THINK ABOUT IT!

What effect would you expect from a small brain tumor that blocked the left interventricular foramen?

Blood Supply and the Brain Barrier System

Although the brain constitutes only 2% of the adult body weight, it receives 15% of the blood (about 750 mL/min) and consumes 20% of the oxygen and glucose. But despite its critical importance to the brain, blood is also a source of agents such as bacterial toxins that can harm the brain tissue. Consequently, there is a brain barrier system that strictly regulates what substances get from the bloodstream into the tissue fluid of the brain.

One component of this system is the **blood-brain barrier (BBB),** which seals nearly all of the blood capillaries throughout the brain tissue. In the developing brain, astrocytes reach out and contact the capillaries with their perivascular feet. They do not fully surround the capillary, but stimulate the formation of tight junctions between the *endothelial cells* that line it. These junctions and the basement membrane around them constitute the BBB. Anything passing from the blood into the tissue fluid has to pass through the endothelial cells themselves, which are more selective than gaps between the cells can be.

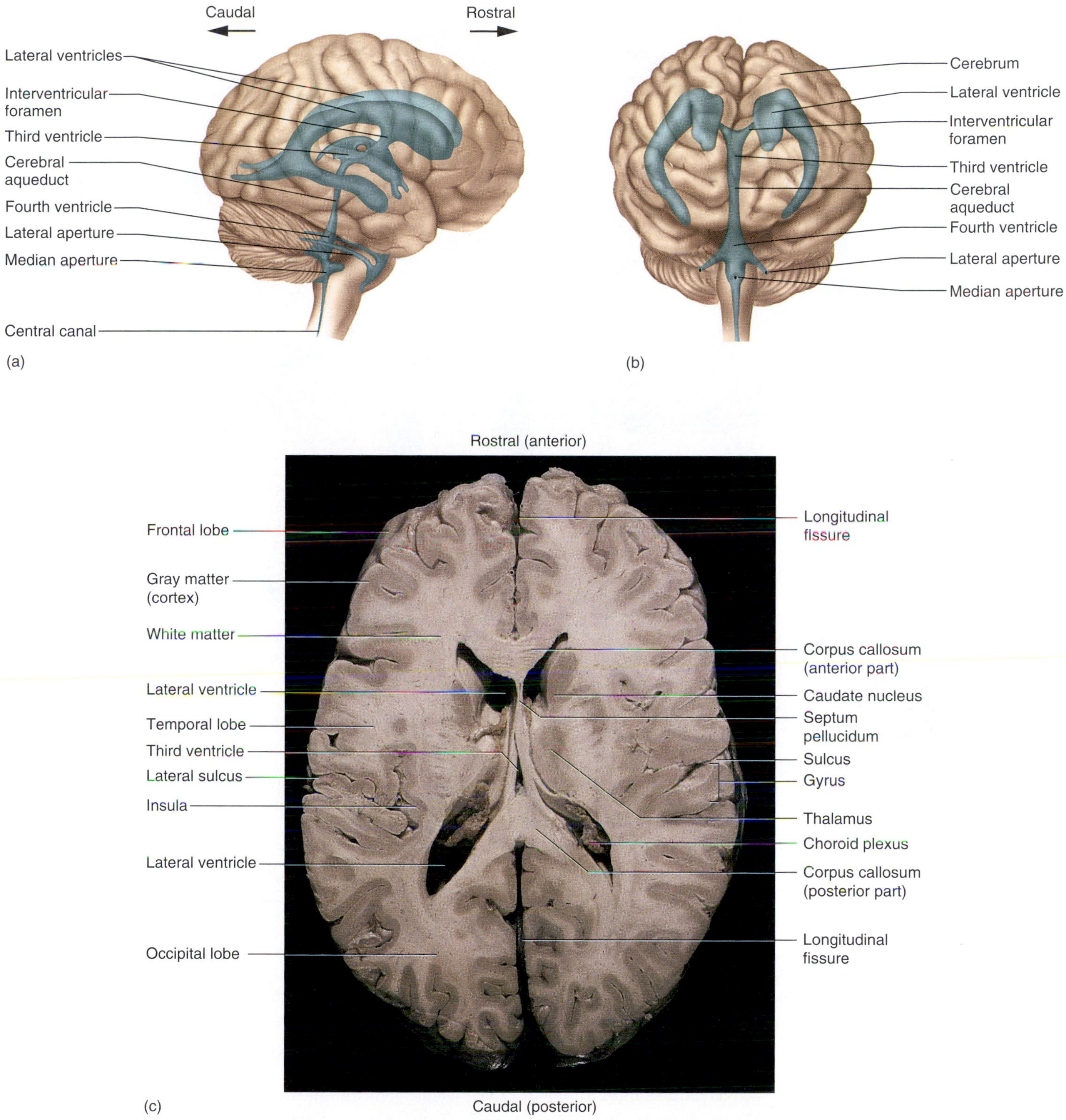
Caudal
Rostral
Lateral ventricles
Interventricular foramen
Third ventricle
Cerebral aqueduct
Fourth ventricle
Lateral aperture
Median aperture
Central canal
(a)
Cerebrum
Lateral ventricle
Interventricular foramen
Third ventricle
Cerebral aqueduct
Fourth ventricle
Lateral aperture
Median aperture
(b)
Rostral (anterior)
Frontal lobe
Gray matter (cortex)
White matter
Lateral ventricle
Temporal lobe
Third ventricle
Lateral sulcus
Insula
Lateral ventricle
Occipital lobe
Longitudinal fissure
Corpus callosum (anterior part)
Caudate nucleus
Septum pellucidum
Sulcus
Gyrus
Thalamus
Choroid plexus
Corpus callosum (posterior part)
Longitudinal fissure
(c)
Caudal (posterior)

FIGURE 15.5

Ventricles of the Brain. (*a*) Right lateral aspect. (*b*) Frontal aspect. (*c*) Horizontal section of the cerebrum, superior view, showing the lateral and third ventricles and other features of the cerebrum.

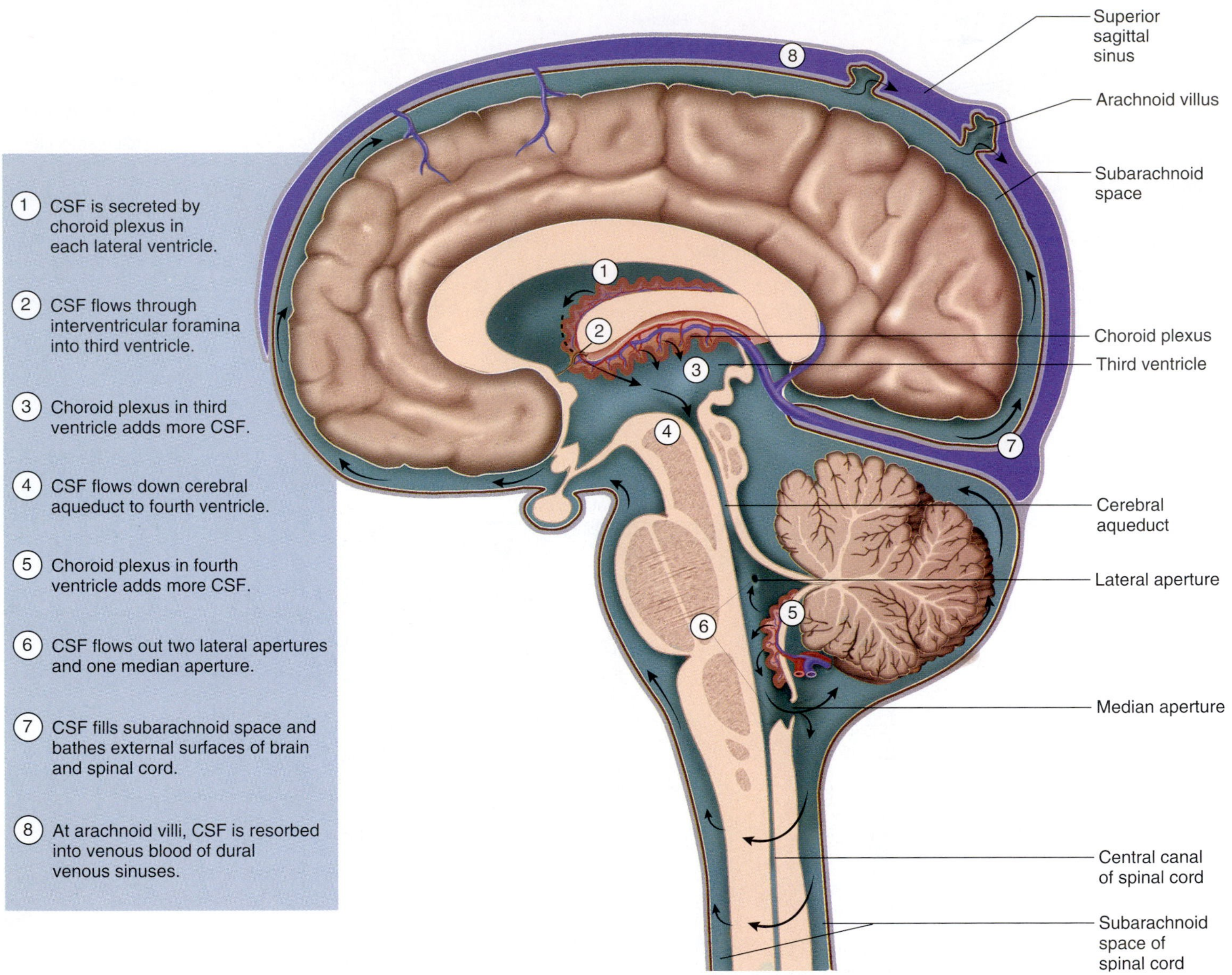
Superior sagittal sinus
Arachnoid villus
Subarachnoid space
Choroid plexus
Third ventricle
Cerebral aqueduct
Lateral aperture
Median aperture
Central canal of spinal cord
Subarachnoid space of spinal cord
1 CSF is secreted by choroid plexus in each lateral ventricle.
2 CSF flows through interventricular foramina into third ventricle.
3 Choroid plexus in third ventricle adds more CSF.
4 CSF flows down cerebral aqueduct to fourth ventricle.
5 Choroid plexus in fourth ventricle adds more CSF.
6 CSF flows out two lateral apertures and one median aperture.
7 CSF fills subarachnoid space and bathes external surfaces of brain and spinal cord.
8 At arachnoid villi, CSF is resorbed into venous blood of dural venous sinuses.

Figure 15.6
The Flow of Cerebrospinal Fluid.

At the choroid plexuses, there is a similar **blood-CSF barrier,** composed of ependymal cells joined by tight junctions. Tight junctions are absent from ependymal cells elsewhere, because it is important to allow exchanges between the brain tissue and CSF. That is, there is no brain-CSF barrier.

The brain barrier system (BBS) is highly permeable to water, glucose, and lipid-soluble substances such as oxygen and carbon dioxide, and to drugs such as alcohol, caffeine, nicotine, and anesthetics. While the BBS is an important protective device, it is an obstacle to the delivery of drugs such as antibiotics and cancer drugs, and thus complicates the treatment of brain diseases.

The BBB is absent from patches called **circumventricular organs (CVOs)** on the walls of the third and fourth ventricles. Here, the blood has direct access to the brain tissue, enabling the brain to monitor and respond to fluctuations in blood chemistry. Unfortunately, the CVOs also afford a route for the human immunodeficiency virus (HIV) to invade the brain.

Before You Go On

Answer the following questions to test your understanding of the preceding section:

1. List the three major parts of the brain and describe their locations.
2. Define *gyrus* and *sulcus.*
3. Name the parts of the brainstem from caudal to rostral.
4. Name the three meninges from superficial to deep.
5. Describe three functions of the cerebrospinal fluid.
6. Where does the CSF originate and what route does it take through and around the CNS?
7. Name the two components of the brain barrier system and explain the importance of this system.

The Hindbrain and Midbrain

Objectives

When you have completed this section, you should be able to

- list the components of the hindbrain and midbrain;
- describe the major features of their anatomy; and
- explain the functions of each hindbrain and midbrain region.

We will survey the functional anatomy of the brain in a caudal to rostral direction, beginning with the hindbrain and its relatively simple functions, and progressing to the forebrain, the seat of such complex functions as thought, memory, and emotion. This survey will be organized around the five secondary vesicles of the embryonic brain and their mature derivatives, as described in chapter 13.

The Medulla Oblongata

As noted earlier, the embryonic hindbrain differentiates into two subdivisions, the myelencephalon and metencephalon (see fig. 13.15). The myelencephalon then develops into just one structure, the **medulla oblongata** (meh-DULL-uh OB-long-GAH-ta).

The medulla begins at the foramen magnum of the skull and extends for 3 cm rostrally, ending at a groove marking the boundary between medulla and pons. The medulla contains all nerve fibers that travel between the brain and spinal cord; several nuclei concerned with basic physiological functions such as respiration and the heartbeat; and nuclei for the origin or termination of the last four (most caudal) pairs of cranial nerves.

The anterior surface of the medulla bears a pair of clublike ridges, the **pyramids.** Resembling side-by-side baseball bats, the pyramids are wider at the rostral end, taper caudally, and are separated by a longitudinal groove, the *anterior median fissure,* continuous with that of the spinal cord (fig. 15.7*a*). They contain the *corticospinal tracts,* composed of motor nerve fibers descending from the cerebral cortex on their way to the spinal cord. Near the caudal end of the pyramids, at a point called the *pyramidal decussation,* about 90% of these fibers cross over to the opposite side of the body. As a result, each side of the brain controls muscles on the contralateral side of the body below the neck.

Lateral to each pyramid is a bulge called the **olive.** It contains a wavy layer of gray matter, the **inferior olivary nucleus.** This nucleus receives information from many levels of the brain and spinal cord and relays it mainly to the cerebellum.

The gracile and cuneate fasciculi on the dorsal side of the spinal cord continue into the medulla and appear as a pair of dorsal ridges with the same names (fig. 15.7*b*). They contain sensory nerve fibers that terminate in the **gracile** and **cuneate nuclei.** Here these fibers synapse with second-order nerve fibers that continue up the brainstem to the thalamus, en route to the cerebrum and one's conscious awareness.

The medulla oblongata is the origin or termination of cranial nerves IX, X, part of XI, and XII. The nuclei of these nerves are columns of gray matter that extend through part of the medulla and, in the case of XI, into the spinal cord. The Latin names and individual functions of these nerves are provided in table 15.3. Collectively, their sensory functions include touch, pressure, temperature, taste, and pain. Their motor functions include chewing, swallowing, speech, respiration, cardiovascular control, gastrointestinal motility and secretion, and head, neck, and shoulder movements.

Several of the most basic physiological functions of the body are regulated by nuclei in the medulla. Among these nuclei are the **cardiac center,** which regulates the rate and force of the heartbeat; the **vasomotor center,** which regulates blood pressure and flow by dilating and constricting arteries; two **respiratory centers,** which regulate the rate and depth of breathing; and nuclei involved in speech, coughing, sneezing, salivation, swallowing, gagging, vomiting, and sweating.

The Pons

The metencephalon develops into two structures, the pons and cerebellum. Superficially, the **pons**[6] appears as a broad anterior bulge in the brainstem rostral to the medulla (fig. 15.7). It measures about 2.5 cm long from its caudal junction with the medulla to its rostral junction with the midbrain. Its white matter includes tracts that conduct signals from the cerebrum down to the cerebellum and medulla; tracts that carry sensory signals up to the thalamus; and tracts that cross the pons horizontally and connect the right and left hemispheres of the cerebellum.

Cranial nerves V, VI, VII, and VIII arise from the pons, with the last three emerging from the groove between the pons and medulla. Again, see table 15.3 for the individual names and functions of these nerves. Collectively, their sensory functions include hearing, equilibrium, taste, and facial sensations such as touch and pain. Their motor functions include eye movements, facial expressions, chewing, swallowing, and the secretion of saliva and tears. Other *pontine*[7] nuclei are concerned with sleep, respiration, and bladder control. The pons also is the source of most nerve fibers carrying signals from the brainstem to the cerebellum.

The Cerebellum

The **cerebellum,** the other derivative of the metencephalon, is the largest part of the hindbrain (fig. 15.8). It is not usually considered part of the brainstem, but we consider it here because of its developmental association with the pons. Its primary function is motor coordination. The cerebellum consists of right and left **cerebellar hemispheres** connected by a narrow wormlike bridge, the **vermis.**[8] It is connected to the brainstem by three pairs of stalks called the **cerebellar peduncles**[9] (peh-DUN-culs): (1) The *inferior peduncles* connect it to the medulla oblongata. Feedback on muscle performance ascends the spinal cord in the spinocerebellar tracts and enters the cerebellum by way of these peduncles. (2) The *middle peduncles* connect the cerebellum to the pons, and are the largest of the three. Signals from the motor regions of the cerebral cortex enter the cerebellum by this route. (3) The *superior peduncles* connect it to the midbrain and are the route of most output from the cerebellum.

[6]*pons* = bridge
[7]*pontine* = pertaining to the pons
[8]*verm* = worm
[9]*ped* = foot + *uncle* = little

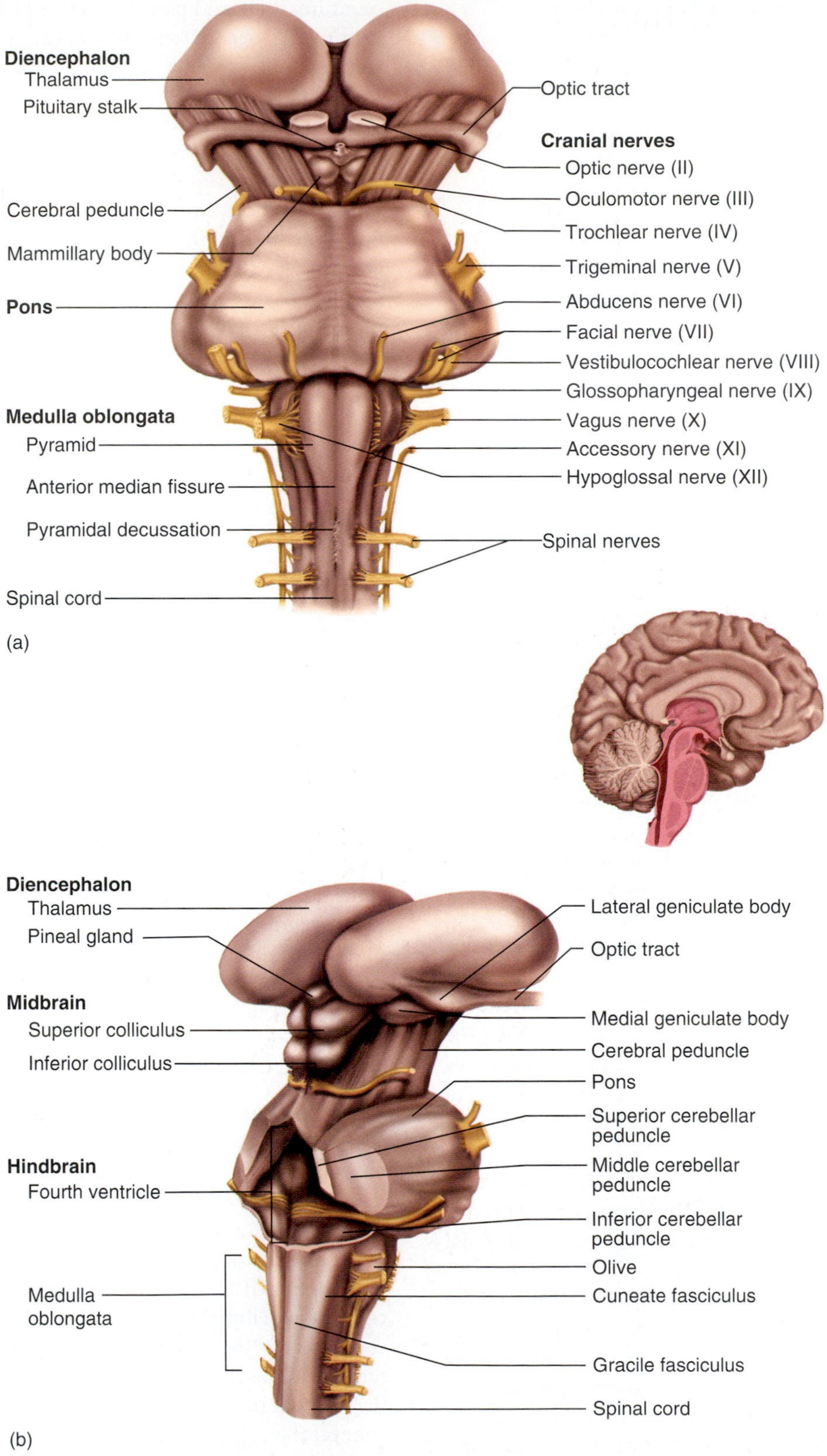
Diencephalon
Thalamus
Pituitary stalk
Cerebral peduncle
Mammillary body
Pons
Medulla oblongata
Pyramid
Anterior median fissure
Pyramidal decussation
Spinal cord
Optic tract
Cranial nerves
Optic nerve (II)
Oculomotor nerve (III)
Trochlear nerve (IV)
Trigeminal nerve (V)
Abducens nerve (VI)
Facial nerve (VII)
Vestibulocochlear nerve (VIII)
Glossopharyngeal nerve (IX)
Vagus nerve (X)
Accessory nerve (XI)
Hypoglossal nerve (XII)
Spinal nerves
(a)
Diencephalon
Thalamus
Pineal gland
Midbrain
Superior colliculus
Inferior colliculus
Hindbrain
Fourth ventricle
Medulla oblongata
Lateral geniculate body
Optic tract
Medial geniculate body
Cerebral peduncle
Pons
Superior cerebellar peduncle
Middle cerebellar peduncle
Inferior cerebellar peduncle
Olive
Cuneate fasciculus
Gracile fasciculus
Spinal cord
(b)

FIGURE 15.7

The Brainstem. (*a*) Anterior view. (*b*) Right dorsolateral view. The cerebellum has been cut off at the peduncles. Some authorities do not include the diencephalon in the brainstem.

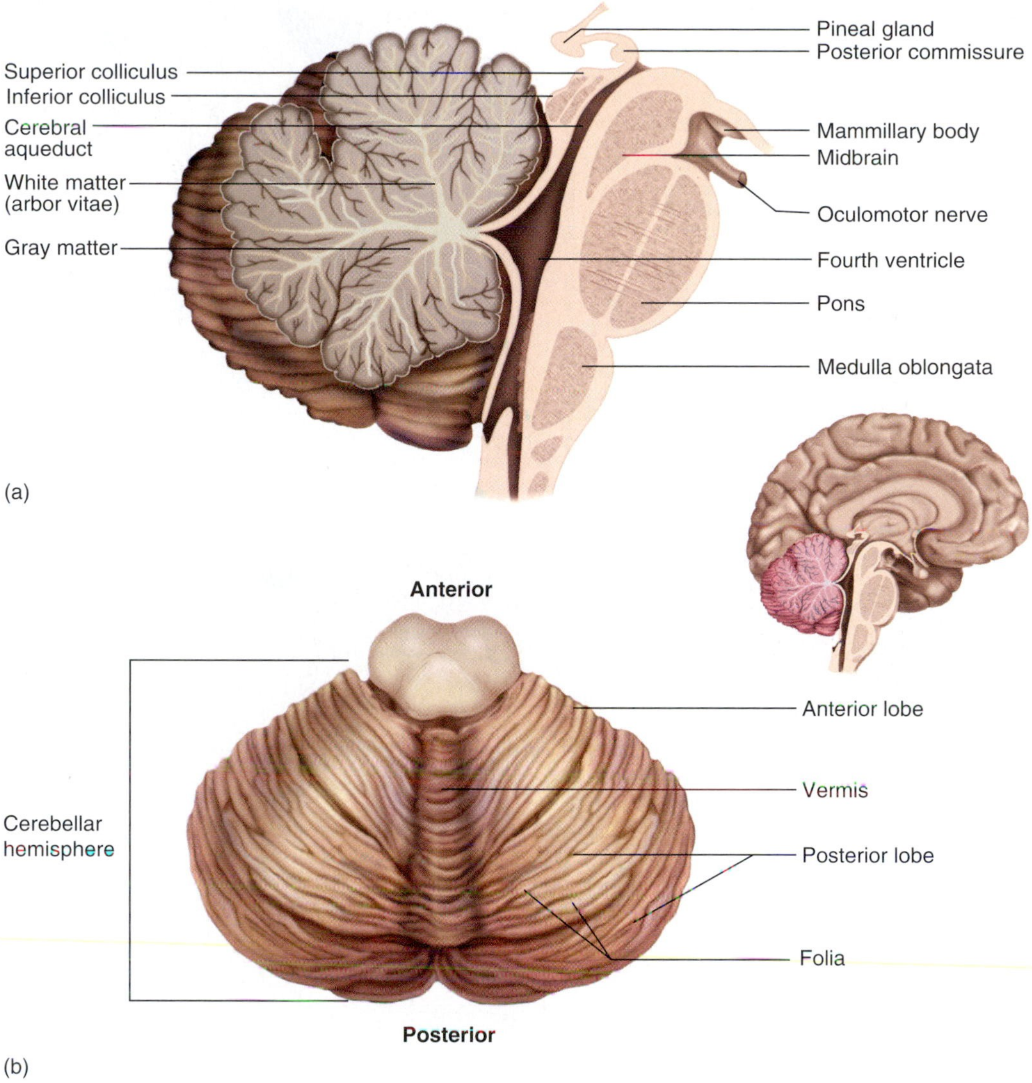
Pineal gland
Posterior commissure
Superior colliculus
Inferior colliculus
Cerebral aqueduct
Mammillary body
Midbrain
White matter (arbor vitae)
Oculomotor nerve
Gray matter
Fourth ventricle
Pons
Medulla oblongata
(a)
Anterior
Anterior lobe
Vermis
Cerebellar hemisphere
Posterior lobe
Folia
Posterior
(b)

FIGURE 15.8

The Cerebellum. (*a*) Median section, showing relationship to the brainstem. (*b*) Superior aspect.

Each cerebellar hemisphere exhibits slender, parallel folds called **folia**[10] (gyri) separated by shallow sulci. The cerebellum has a surface cortex of gray matter and a deeper layer of white matter. In a sagittal section, the white matter, called the **arbor vitae,**[11] exhibits a branching, fernlike pattern (fig. 15.8*a*). Each hemisphere has four **deep nuclei,** masses of gray matter embedded in the white matter. All input to the cerebellum goes to the cortex and all of its output comes from the deep nuclei.

The cerebellum contains about 100 billion neurons. The most distinctive of these are the **Purkinje**[12] (pur-KIN-jee) **cells**—unusually large, globose neurons with a tremendous profusion of dendrites, arranged in a single row in the cortex (see fig. 13.5*a*, p. 370 and the photograph on p. 365). Their axons travel to the deep nuclei, where they synapse on output neurons that issue fibers to the brainstem.

[10]*foli* = leaf
[11]"tree of life"
[12]Johannes E. von Purkinje (1787–1869), Bohemian anatomist

The cerebellum receives input from four principal sources (left side of figure 15.9): the cerebral cortex, the ear and eye, the reticular formation of the brainstem (described shortly), and the spinocerebellar tracts of the spinal cord. Collectively, these provide the cerebellum with information about the planned and actual movements of the body, enabling it to compare the plan with the performance. Output from the cerebellum (right half of figure 15.9) goes, for the most part, back to the cerebral cortex and to the skeletal muscles by way of the reticulospinal and vestibulospinal tracts of the spinal cord.

The cerebellum smooths muscle contractions, maintains muscle tone and posture, coordinates the motions of different joints with each other (such as the shoulder and elbow joints in pitching a baseball), coordinates eye and body movements, and serves in learning and storing motor skills. As the cerebrum issues motor commands to the muscles, it sends a "copy" through the middle peduncles to the cerebellum. As the muscles begin to carry out the command, feedback on their performance travels up the

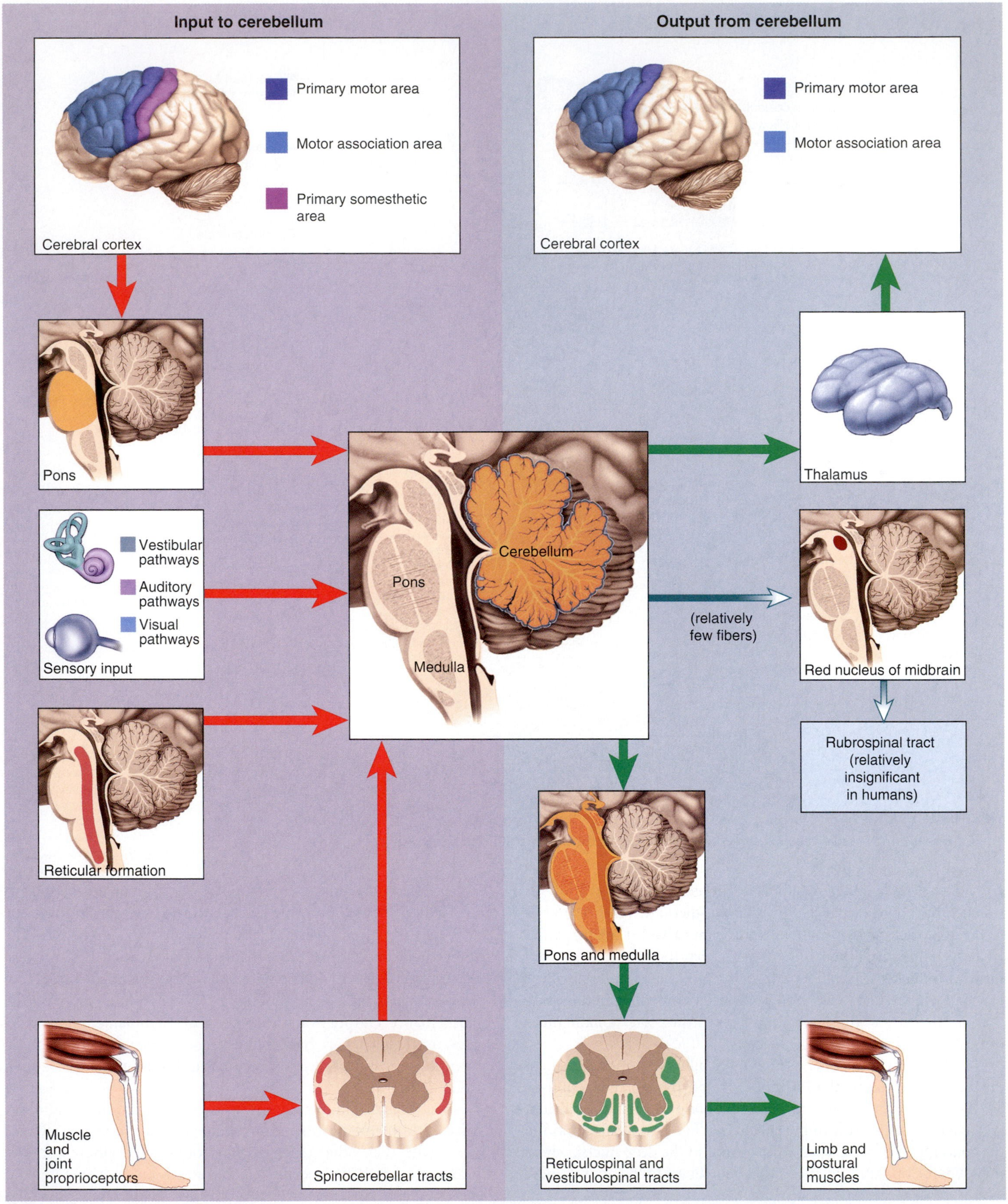

FIGURE 15.9

Motor Pathways Involving the Cerebellum. The cerebellum receives its input from the afferent pathways (*red*) on the *left* and sends its output through the efferent pathways (*green*) on the *right*.

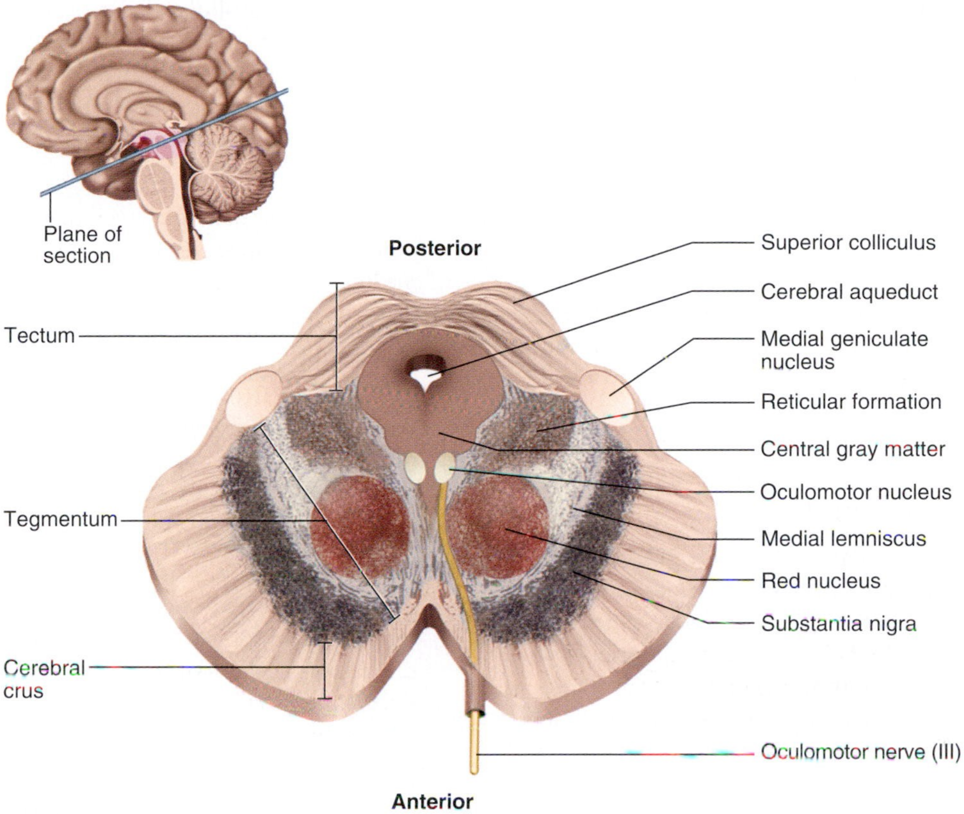

Plane of section
Posterior
Tectum
Tegmentum
Cerebral crus
Superior colliculus
Cerebral aqueduct
Medial geniculate nucleus
Reticular formation
Central gray matter
Oculomotor nucleus
Medial lemniscus
Red nucleus
Substantia nigra
Oculomotor nerve (III)
Anterior

FIGURE 15.10
Cross Section of the Midbrain.

spinocerebellar tracts and enters the cerebellum by way of the inferior peduncles. The Purkinje cells compare the performance to the original command, and the cerebellum issues a "report" to the cerebrum by way of the superior peduncles. The cerebrum then makes adjustments to improve muscle performance. Without this comparison of performance to command—for example, when the cerebellum has been injured—people have a clumsy, awkward gait *(ataxia)* and some tasks such as climbing a flight of stairs are virtually impossible. Recent evidence has suggested an ever broader and more surprising range of cerebellar functions, including roles in awareness, emotion, and judging the passage of time. Some additional functions of the cerebellum in cognition (thought or awareness) are described later in the chapter.

The Midbrain

The embryonic mesencephalon develops into just one mature brain structure, the midbrain—a short segment of the brainstem that connects the hindbrain and forebrain (see figs. 15.2 and 15.8). It contains the cerebral aqueduct and gives rise to two cranial nerves that control eye movements: cranial nerves III and IV (see table 15.3). Some major regions of the midbrain are (fig. 15.10):

- The **superior cerebellar peduncles,** described earlier.
- The **cerebral crura** (CROO-ra; singular, *crus*[13]), which anchor the cerebrum to the brainstem. Corticospinal and other tracts from the cerebrum descend through the cerebral crura on their way to lower levels of the brainstem and spinal cord. (The crura plus other structures listed here, the tegmentum and substantia nigra, are collectively called the *cerebral peduncles*).
- The **medial lemniscus,**[14] a continuation of the gracile and cuneate tracts of the spinal cord and brainstem.
- The **tectum,**[15] a rooflike region dorsal to the cerebral aqueduct. The tectum exhibits four bulges: a rostral pair called the **superior colliculi**[16] (col-LIC-you-lye) and a caudal pair called the **inferior colliculi.** The superior colliculi function in visual attention, such as visually tracking moving objects and reflexively turning the eyes and head in response to a visual stimulus, for example to

[13]*crus* = leg
[14]*lemn* = ribbon + *iscus* = little
[15]*tectum* = roof, cover
[16]*colli* = hill + *cul* = little

look at something that you catch sight of in your peripheral vision. The inferior colliculi receive and process auditory input from lower levels of the brainstem and relay it to other parts of the brain, especially the thalamus. They are sensitive to time delays between sounds heard by the two ears, and thus aid in locating the source of a sound in space, and they function in auditory reflexes such as turning the head toward a sound or the startle response to a loud noise. All four midbrain colliculi are collectively called the *corpora quadrigemina.*[17]

- The **tegmentum,**[18] the main mass of the midbrain, located ventral to the cerebral aqueduct. The tegmentum contains the **red nucleus,** named for the pink color that it gets from a high density of blood vessels. Fibers from the red nucleus form the *rubrospinal tract* in most mammals, but in humans its connections go mainly to and from the cerebellum, with which it collaborates in fine motor control.
- The **substantia nigra**[19] (sub-STAN-she-uh NY-gruh), a dark gray to black nucleus pigmented with melanin, located between the cerebral crura and tegmentum. This is a motor center that relays inhibitory signals to the thalamus and basal nuclei (both of which are discussed later). It improves motor performance by suppressing unwanted muscle contractions. Degeneration of the substantia nigra leads to the uncontrollable muscle tremors of Parkinson disease.
- The **central (periaqueductal) gray matter,** a large arrowhead-shaped region of gray matter surrounding the cerebral aqueduct. It collaborates with the *reticulospinal tracts* in controlling our awareness of pain (see chapter 17).

The Reticular Formation

The **reticular**[20] **formation** is a loosely organized web of gray matter that runs vertically through all levels of the brainstem and projects to many areas of the cerebrum (fig. 15.11). It occupies much of the space between the white fiber tracts and the more anatomically distinct brainstem nuclei. It consists of more than 100 small neural networks without well defined boundaries. The functions of these networks include the following:

- **Somatic motor control.** Some motor neurons of the cerebral cortex and cerebellum send their axons to reticular formation nuclei, which then give rise to the *reticulospinal tracts* of the spinal cord. These tracts adjust muscle tension to maintain tone, balance, and posture, especially during body movements. The reticular formation also relays signals from the eyes and ears to the cerebellum so the cerebellum can integrate visual, auditory, and vestibular (balance and motion) stimuli into its role in motor coordination. Other reticular formation motor nuclei include *gaze centers,* which enable the eyes to track and fixate on objects, and *central pattern generators*—neuronal pools that produce rhythmic signals to the muscles of breathing and swallowing.
- **Cardiovascular control.** The reticular formation includes the previously mentioned cardiac and vasomotor centers of the medulla oblongata.
- **Pain modulation.** The reticular formation is one route by which pain signals from the lower body reach the cerebral cortex. The reticular formation is also the origin of the *descending analgesic pathways* mentioned in chapter 14—nerve fibers that act in the spinal cord to block the transmission of pain signals to the brain.
- **Sleep and consciousness.** The reticular formation has projections to the cerebral cortex and thalamus that allow it some control over what sensory signals reach the cerebrum and come to our conscious attention. It plays a central role in states of consciousness such as alertness and sleep. Injury to the reticular formation can result in irreversible coma.
- **Habituation.** This is the process in which the brain learns to ignore repetitive, inconsequential stimuli while remaining sensitive to others. In a noisy city, for example, a person can sleep through traffic sounds but wake promptly to the sound of an alarm clock or a crying baby. The reticular formation screens out insignificant stimuli, preventing them from arousing cerebral centers of consciousness, while it permits important sensory signals to pass. Reticular formation nuclei that modulate activity of the cerebral cortex are called the *reticular activating system* or *extrathalamic cortical modulatory system*

Table 15.1 summarizes the hindbrain and midbrain functions discussed in the last several pages.

Before You Go On

Answer the following questions to test your understanding of the preceding section:

8. Name the visceral functions controlled by nuclei of the medulla.
9. Describe the anatomical and functional relationship of the pons to the cerebellum.
10. Describe the general functions of the cerebellum.
11. What are the functions of the corpora quadrigemina, substantia nigra, and central gray matter?
12. Describe the reticular formation and list several of its functions.

[17] *corpora* = bodies + *quadrigemina* = quadruplets
[18] *tegmen* = cover
[19] *substantia* = substance + *nigra* = black
[20] *ret* = network + *icul* = little

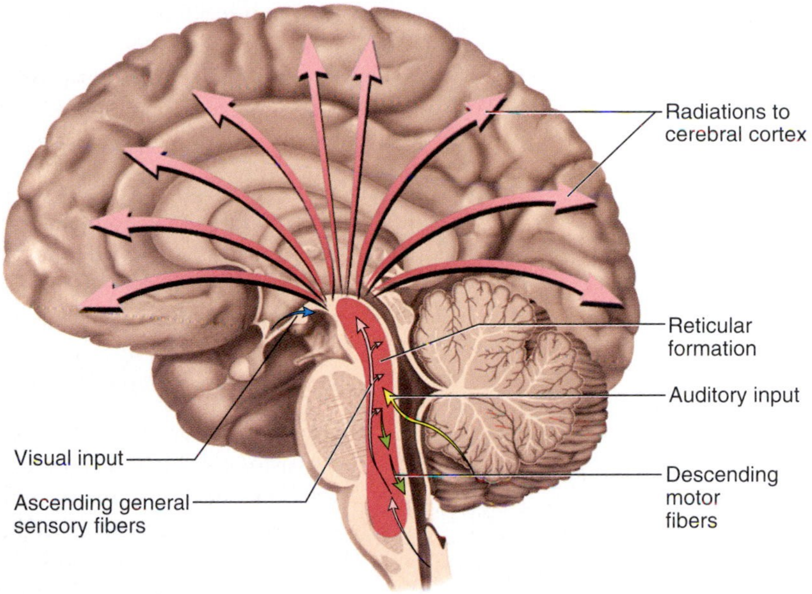

Radiations to cerebral cortex
Reticular formation
Auditory input
Visual input
Ascending general sensory fibers
Descending motor fibers

FIGURE 15.11

The Reticular Formation. The formation consists of over 100 neuronal pools scattered through the general brainstem region indicated in *red. Arrows* represent the breadth of its projections to and from the cerebral cortex and other CNS regions.

TABLE 15.1

Hindbrain and Midbrain Functions

Medulla Oblongata	Origin or termination of cranial nerves IX–XII. Sensory nuclei receive input from the taste buds, pharynx, and thoracic and abdominal viscera. Motor nuclei include the cardiac center (adjusts the rate and force of the heartbeat), vasomotor center (controls blood vessel diameter and blood pressure), two respiratory centers (control the rate and depth of breathing), and centers involved in speech, coughing, sneezing, salivation, swallowing, gagging, vomiting, sweating, gastrointestinal secretion, and movements of the tongue and head.
Pons	Sensory terminations and motor origins of cranial nerves V–VIII. Sensory nuclei receive input from the face, eye, oral and nasal cavities, sinuses, and meninges, concerned with pain, touch, temperature, taste, hearing, and equilibrium. Cranial nerve motor nuclei control chewing, swallowing, eye movements, middle- and inner-ear reflexes, facial expression, and secretion of tears and saliva. Other nuclei of pons relay signals from cerebrum to cerebellum (provides most of the input to the cerebellum) or function in sleep, respiration, bladder control, and posture.
Midbrain	Origin of cranial nerves III–IV (concerned with eye movements). Red nucleus is concerned with fine motor control. Substantia nigra relays inhibitory signals to thalamus and basal nuclei of forebrain. Central gray matter modulates awareness of pain. Superior colliculi concerned with visual attention and tracking movements of eyes, and visual reflexes such as shifting gaze to objects seen moving in peripheral vision. Inferior colliculi relay auditory signals to thalamus and mediate auditory reflexes such as the startle response to a loud noise.
Cerebellum	Muscular coordination, fine motor control, muscle tone, posture, equilibrium, judging passage of time; some involvement in emotion, processing tactile input, spatial perception, and language.
Reticular Formation	A network of over 100 nuclei extending throughout brainstem, including some nuclei described earlier in this table. Involved in somatic motor control, equilibrium, visual attention, breathing, swallowing, cardiovascular control, pain modulation, sleep, and consciousness.

THE FOREBRAIN

Objectives

When you have completed this section, you should be able to

- name the three major components of the diencephalon and describe their locations and functions;
- identify the five lobes of the cerebrum;
- describe the three types of tracts in the cerebral white matter;
- describe the distinctive cell types and histological arrangement of the cerebral cortex; and
- describe the locations and functions of the basal nuclei and limbic system.

The forebrain consists of the diencephalon and cerebrum. As noted earlier, some authorities treat the diencephalon as the most rostral part of the brainstem, while others exclude it from the brainstem.

The Diencephalon

The embryonic diencephalon has three major derivatives: the *thalamus, hypothalamus,* and *epithalamus.* These structures surround the third ventricle of the brain.

THE THALAMUS

About four-fifths of the diencephalon is **thalamus**[21] (figs. 15.7 and 15.12*a*), an oval mass of gray matter underlying each cerebral hemisphere (see the brain dissections in figs. 15.2*b* and 15.5*c*). The thalamus protrudes into the lateral ventricle, and medially protrudes into the third ventricle. In about 70% of people, a narrow *intermediate mass* connects the right and left thalami to each other.

The thalamus is the "gateway to the cerebral cortex." Nearly all sensory signals and other input to the cerebrum pass by way of nuclei in the thalamus, and the thalamus processes these signals and relays coded information to areas of the cerebral cortex specialized to interpret it. In return, nearly all regions of the cerebral cortex send signals back to the thalamic nuclei from which they receive their input. Thus there is a two-way traffic of neural signals between cerebral cortex and thalamus. The only sensory signals that can reach the cerebral cortex without passing through the thalamus are for olfaction (smell), but even olfactory pathways include branches that pass through the thalamus. Figure 15.12 illustrates the major groups of thalamic nuclei and some of their functions.

THE HYPOTHALAMUS

The **hypothalamus** (figs. 15.2 and 15.12*b*) forms part of the walls and floor of the third ventricle. It extends anteriorly to the *optic chiasm* (ky-AZ-um), an X-shaped crossing of the optic nerves, and posteriorly to a pair of humps called the *mammillary*[22] *bodies.* The mammillary bodies are composed of nuclei belonging to the hypothalamus and limbic system (part of the forebrain to be discussed later); they are the primary route of sensory input to the hypothalamus, and their output goes also to the thalamus and lower brainstem. The pituitary gland is attached to the hypothalamus by a stalk between the optic chiasm and mammillary bodies.

The hypothalamus is the major control center of the autonomic nervous system and endocrine system and plays an essential role in the homeostatic regulation of nearly all organs of the body. Its nuclei include centers concerned with a wide variety of visceral functions.

- **Hormone secretion.** The hypothalamus secretes hormones that control the anterior pituitary gland. Acting through the pituitary, it regulates growth, metabolism, reproduction, and stress responses. It also produces two hormones that are stored in the posterior pituitary gland—*oxytocin* concerned with labor contractions and lactation, and *antidiuretic hormone* concerned with water conservation—and it sends nerve signals to the posterior pituitary to stimulate release of these hormones at appropriate times.
- **Autonomic effects.** The hypothalamus is a major integrating center for the autonomic nervous system. It sends descending fibers to nuclei lower in the brainstem that influence heart rate, blood pressure, pupillary diameter, and gastrointestinal secretion and motility, among other functions.
- **Thermoregulation.** The *hypothalamic thermostat* is a nucleus that monitors blood temperature. When the temperature becomes too high or low, the thermostat signals other hypothalamic nuclei—the *heat-losing center* or *heat-producing center,* respectively—which control cutaneous vasodilation and vasoconstriction, sweating, shivering, and piloerection. These reactions usually return the body temperature to normal.
- **Food and water intake.** Neurons of the *hunger* and *satiety centers* monitor blood glucose and amino acid levels and produce sensations of hunger and satisfaction of the appetite. Hypothalamic neurons called *osmoreceptors* monitor the osmolarity of the blood and stimulate the hypothalamic *thirst center* when the body is dehydrated. Thus, our drives to eat and drink are under hypothalamic control.
- **Sleep and circadian rhythms.** The caudal part of the hypothalamus is part of the reticular formation. In this region are nuclei that regulate falling asleep and waking. Superior to the optic chiasm, the hypothalamus contains a *suprachiasmatic nucleus* that controls our *circadian rhythm* (24-hour cycle of activity).
- **Emotional responses.** The hypothalamus contains nuclei for a variety of emotional responses including anger, aggression, fear, pleasure, and contentment; and for sexual drive, copulation, and orgasm. The mammillary bodies provide a pathway by which emotional states can affect visceral function—for example, when anxiety accelerates the heart or upsets the stomach.
- **Memory.** In addition to their role in emotional circuits, the mammillary bodies lie in the pathway from the hippocampus to the thalamus. The hippocampus is a center for the creation of new memories—the cerebrum's "teacher"—so as intermediaries between the hippocampus and cerebral cortex, the mammillary bodies are essential for the acquisition of new memories.

THE EPITHALAMUS

The **epithalamus** consists mainly of the **pineal gland** (an endocrine gland discussed in chapter 18), the **habenula** (a relay from the limbic system to the midbrain), and a thin roof over the third ventricle.

The Cerebrum

The embryonic telencephalon becomes the cerebrum, the largest and most conspicuous part of the human brain. Your cerebrum enables you to turn these pages, read and comprehend the words, remember ideas, talk about them with your peers, and take an examination. It is

[21]*thalamus* = chamber, inner room
[22]*mammill* = nipple

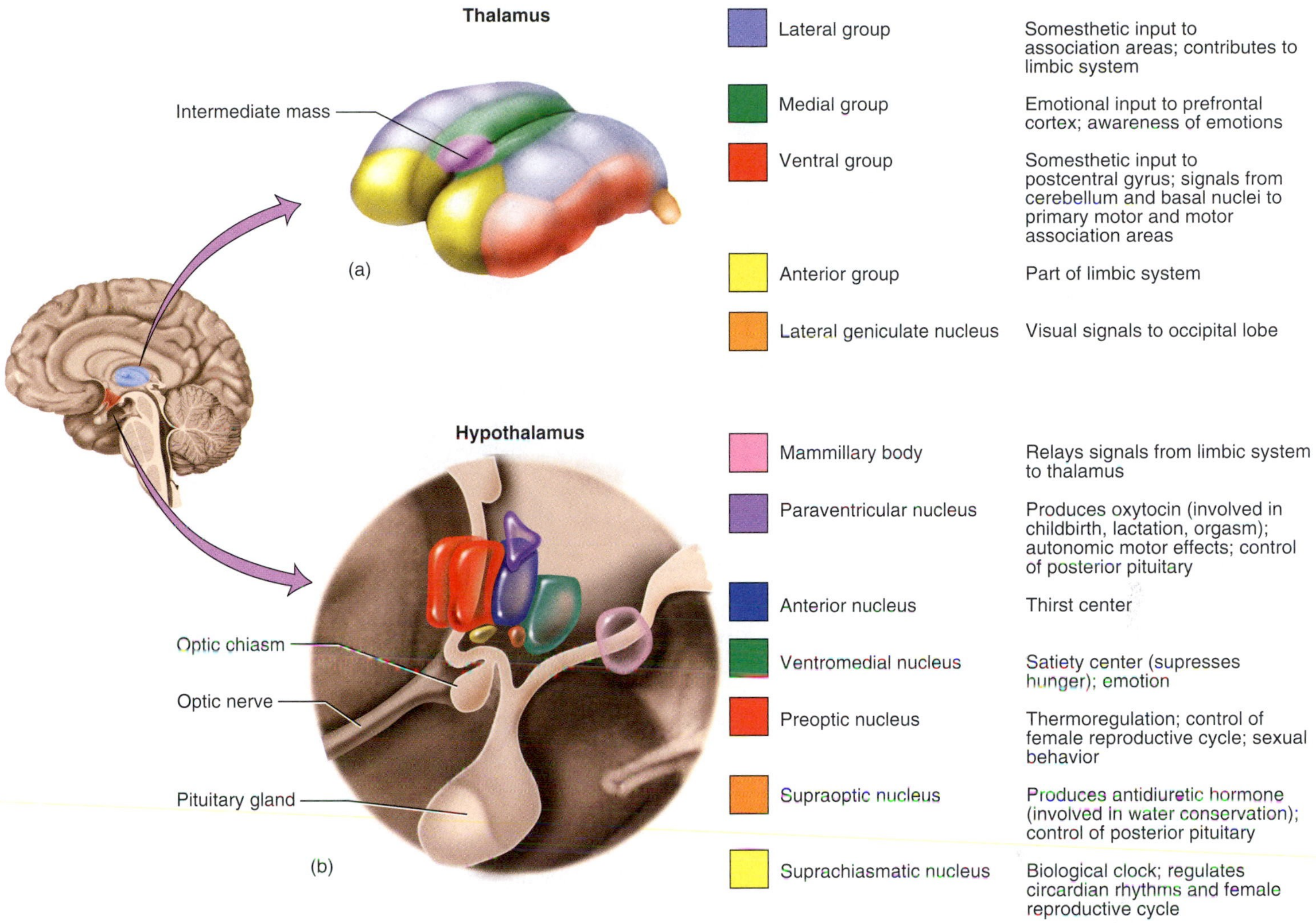
Thalamus
Intermediate mass
(a)
Lateral group
Somesthetic input to association areas; contributes to limbic system
Medial group
Emotional input to prefrontal cortex; awareness of emotions
Ventral group
Somesthetic input to postcentral gyrus; signals from cerebellum and basal nuclei to primary motor and motor association areas
Anterior group
Part of limbic system
Lateral geniculate nucleus
Visual signals to occipital lobe
Hypothalamus
Optic chiasm
Optic nerve
Pituitary gland
(b)
Mammillary body
Relays signals from limbic system to thalamus
Paraventricular nucleus
Produces oxytocin (involved in childbirth, lactation, orgasm); autonomic motor effects; control of posterior pituitary
Anterior nucleus
Thirst center
Ventromedial nucleus
Satiety center (supresses hunger); emotion
Preoptic nucleus
Thermoregulation; control of female reproductive cycle; sexual behavior
Supraoptic nucleus
Produces antidiuretic hormone (involved in water conservation); control of posterior pituitary
Suprachiasmatic nucleus
Biological clock; regulates circardian rhythms and female reproductive cycle

FIGURE 15.12

The Diencephalon. (*a*) Structure and nuclei of the thalamus. (*b*) Structure and nuclei of the hypothalamus. Only some of the nuclei of each are shown, and only some of their functions listed. These lists are by no means complete.

the seat of sensory perception, voluntary motor actions, memory, and mental processes such as thought, judgment, and imagination, which most distinguish humans from other animals. It is the most complex and challenging frontier of neurobiology.

GROSS ANATOMY

The cerebrum accounts for about 83% of the brain volume and so dwarfs and conceals the other structures that people often think of "cerebrum" and "brain" as synonymous. It consists of a pair of half-globes called the **cerebral hemispheres** (fig. 15.1*a*). Each hemisphere is marked by thick folds called **gyri**[23] (JY-rye; singular, *gyrus*) separated by shallow grooves called **sulci**[24] (SUL-sye; singular, *sulcus*). A very deep median groove, the **longitudinal fissure,** separates the right and left hemispheres from each other. At the bottom of this fissure, the hemispheres are connected by a thick C-shaped bundle of nerve fibers called the **corpus callosum**[25]—a prominent landmark for anatomical description (fig. 15.2).

The folding of the cerebral surface into gyri allows a greater amount of cortex to fit in the cranial cavity. The gyri give the cerebrum a surface area of about 2,500 cm^2, comparable to 4.5 pages of this book. If the cerebrum were smooth-surfaced, it would have only one-third as much area and proportionately less information-processing capability. This extensive folding is one of the greatest differences between the human brain and the relatively smooth-surfaced brains of most other mammals.

Some gyri have consistent and predictable anatomy, while others vary from brain to brain and from the right hemisphere to the left. Certain unusually prominent sulci divide each hemisphere into five anatomically and functionally distinct lobes, listed next.

[23]*gy* = turn, twist
[24]*sulc* = furrow, groove
[25]*corpus* = body + *call* = thick

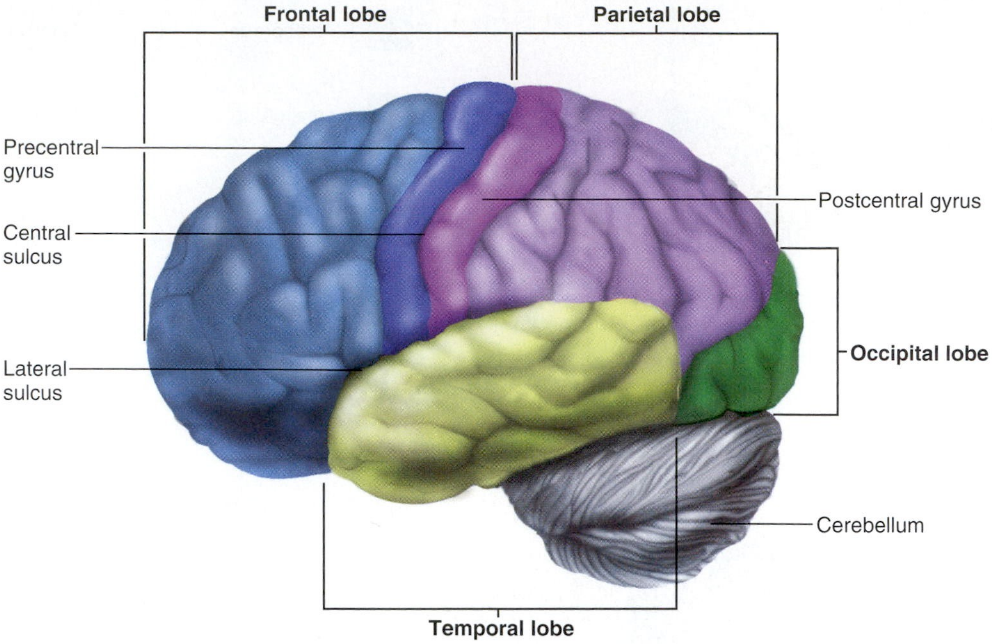

Frontal lobe
Parietal lobe
Precentral gyrus
Central sulcus
Lateral sulcus
Postcentral gyrus
Occipital lobe
Cerebellum
Temporal lobe

FIGURE 15.13
Lobes of the Cerebrum. The insula is not visible from the surface (see fig. 15.1*c*).

The first four of these are visible superficially and are named for the cranial bones overlying them (fig. 15.13); the fifth lobe is not visible from the surface.

1. The **frontal lobe** lies immediately behind the frontal bone, superior to the orbits. From the forehead, it extends caudally to a wavy vertical groove, the **central sulcus.** It is concerned with cognition (thought) and other "higher" mental processes, speech, and motor control.
2. The **parietal lobe** forms the uppermost part of the brain and underlies the parietal bone. Starting at the central sulcus, it extends caudally to the **parieto-occipital sulcus,** visible on the medial surface of each hemisphere (see fig. 15.2). It is the primary site for receiving and interpreting signals of the *general senses* described later in this chapter, as well as signals for taste, one of the *special senses.*
3. The **occipital lobe** is at the rear of the head, caudal to the parieto-occipital sulcus and underlying the occipital bone. It is the principal visual center of the brain.
4. The **temporal lobe** is a lateral, horizontal lobe deep to the temporal bone, separated from the parietal lobe above it by a deep **lateral sulcus.** Among its functions are hearing, smell, learning, memory, and some aspects of vision and emotion.
5. The **insula**[26] is a small mass of cortex deep to the lateral sulcus, made visible only by retracting or cutting away some of the overlying cerebrum (see figs. 15.1*c*, 15.5*c*, and 15.16). It is less understood than the other lobes but apparently plays roles in taste, hearing, and visceral sensation.

THE CEREBRAL WHITE MATTER

Most of the volume of the cerebrum is white matter. This is composed of glia and myelinated nerve fibers that transmit signals from one region of the cerebrum to another and between the cerebrum and lower brain centers. These fibers travel in bundles called tracts. There are three types of cerebral tracts (fig. 15.14):

1. **Projection tracts** extend vertically between higher and lower brain or spinal cord centers and carry information between the cerebrum and the rest of the body. The corticospinal tracts, for example, carry motor signals from the cerebrum to the brainstem and spinal cord. Other projection tracts carry signals upward to the cerebral cortex. Superior to the brainstem, such tracts form a dense band called the *internal capsule* between the thalamus and basal nuclei (described shortly), then radiate in a diverging, fanlike array (the *corona radiata*[27]) to specific areas of the cortex.
2. **Commissural tracts** cross from one cerebral hemisphere to the other through bridges called **commissures** (COM-ih-shurs). The great majority of commissural fibers pass through the corpus callosum. A few tracts pass through the much smaller **anterior** and **posterior commissures** (fig. 15.2*a*). Commissural tracts enable the two sides of the cerebrum to communicate with each other.
3. **Association tracts** connect different regions of the same hemisphere. *Long association fibers* connect different lobes to each other, whereas *short association fibers* connect different gyri within a single lobe. Among their roles,

[26]*insula* = island

[27]*corona* = crown + *radiata* = radiating

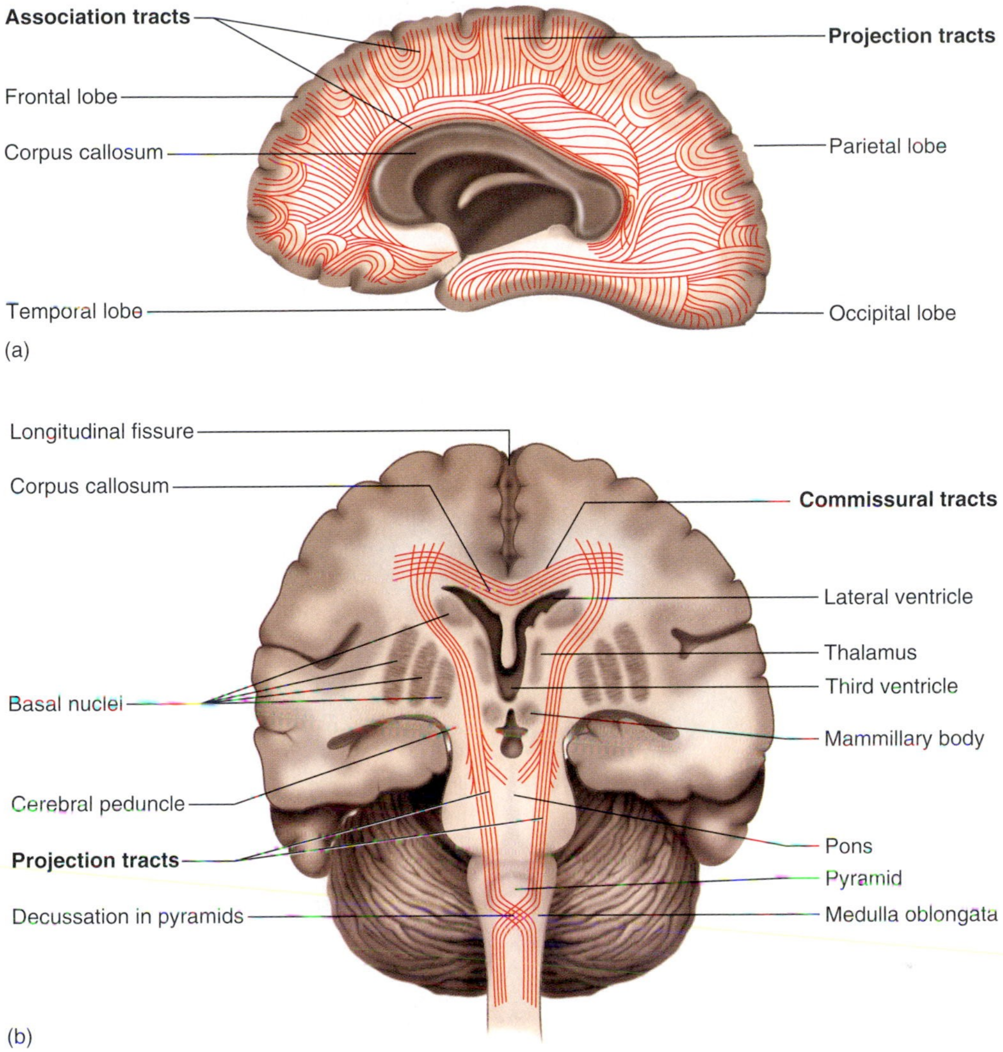
Association tracts
Frontal lobe
Corpus callosum
Temporal lobe
(a)
Projection tracts
Parietal lobe
Occipital lobe
Longitudinal fissure
Corpus callosum
Basal nuclei
Cerebral peduncle
Projection tracts
Decussation in pyramids
(b)
Commissural tracts
Lateral ventricle
Thalamus
Third ventricle
Mammillary body
Pons
Pyramid
Medulla oblongata

FIGURE 15.14
Tracts of Cerebral White Matter. (*a*) Left lateral aspect, showing association tracts. (*b*) Frontal section, showing commissural and projection tracts.

association tracts link perceptual and memory centers of the brain; for example, they enable you to smell a rose, name it, and picture what it looks like.

THE CEREBRAL CORTEX

Neural integration is carried out in the gray matter of the cerebrum, which is found in three places—the cerebral cortex, basal nuclei, and limbic system. The **cerebral cortex**[28] is a layer about 2 to 3 mm thick covering the surface of the hemispheres. It constitutes about 40% of the mass of the brain and contains 14 to 16 billion neurons. It possesses two principal types of neurons called stellate and pyramidal cells (fig. 15.15). **Stellate cells** have spheroidal somas with dendrites projecting for short distances in all directions. They are concerned largely with receiving sensory input and processing information on a local level. **Pyramidal cells** are tall and conical (triangular in tissue sections). Their apex points toward the brain surface and has a thick dendrite with many branches and small, knobby *dendritic spines.* The base gives rise to horizontally oriented dendrites and an axon that passes into the white matter. The axon also has collaterals that synapse with other neurons in the cortex or in deeper regions of the brain. Pyramidal cells are the output neurons of the cerebrum—they are the only cerebral neurons whose fibers leave the cortex and connect with other parts of the CNS.

About 90% of the human cerebral cortex is a six-layered tissue called **neocortex**[29] because of its relatively recent evolutionary origin. Although vertebrates have existed for about 600 million years, the neocortex did not develop significantly until about 60 million years ago, when there was a sharp increase in the diversity of mammals. It attained its highest development by far in the primates. The six layers of neocortex, numbered in figure 15.15, vary from one part of the cerebrum to another in relative thickness,

[28]*cortex* = bark, rind

[29]*neo* = new

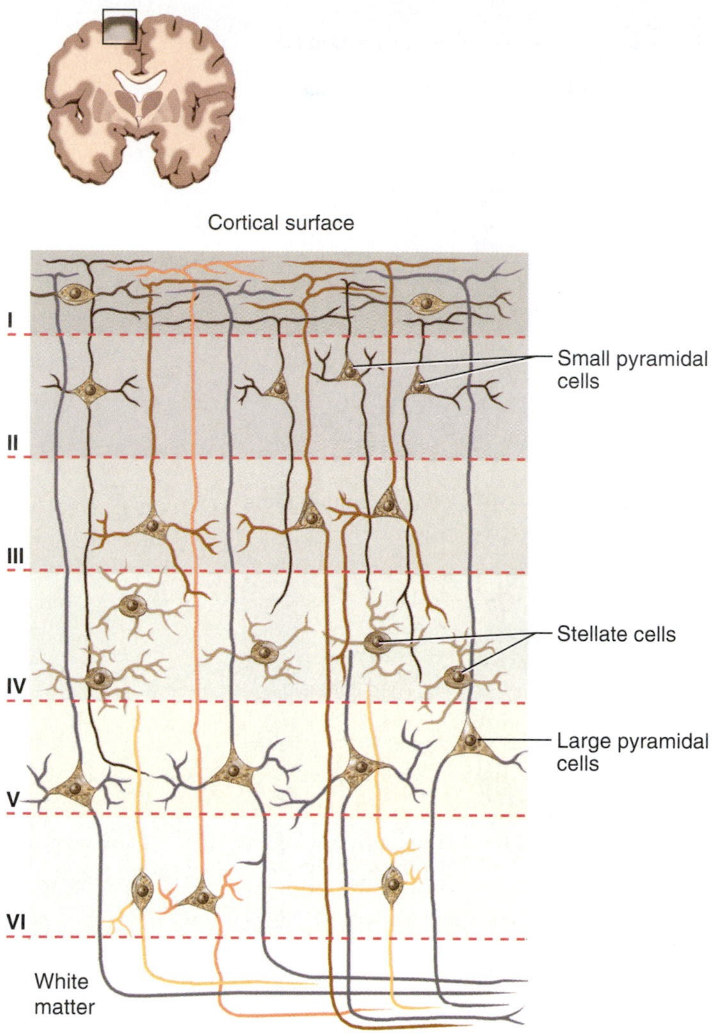
Cortical surface
I
II
III
IV
V
VI
White matter
Small pyramidal cells
Stellate cells
Large pyramidal cells

FIGURE 15.15
Histology of the Neocortex. Neurons are arranged in six layers.

cellular composition, synaptic connections, size of the neurons, and destination of their axons. Layer IV is thickest in sensory regions and layer V in motor regions, for example. All axons that leave the cortex and enter the white matter arise from layers III, V, and VI.

Some regions of cerebral cortex have fewer than six layers. The earliest type of cerebral cortex to appear in vertebrate evolution was a one- to five-layered tissue called *paleocortex* (PALE-ee-oh-cor-tex), limited in humans to part of the insula and certain areas of the temporal lobe concerned with smell. The next to evolve was a three-layered *archicortex* (AR-kee-cor-tex), found in the human hippocampus. The neocortex was the last to evolve.

THE BASAL NUCLEI

The **basal nuclei** are masses of cerebral gray matter buried deep in the white matter, lateral to the thalamus (fig. 15.16). They are often called *basal ganglia,* but the word *ganglion* is best restricted to clusters of neurons outside the CNS. Neuroanatomists disagree on how many brain centers to classify as basal nuclei, but agree on at least three: the **caudate**[30] **nucleus, putamen,**[31] and **globus pallidus.**[32] The putamen and globus pallidus are also collectively called the *lentiform*[33] *nucleus,* while the putamen and caudate nucleus are collectively called the *corpus striatum* because of their striped appearance. The basal nuclei are involved in motor control.

THE LIMBIC SYSTEM

The **limbic**[34] **system** is one of the brain's most important centers of emotion and learning. It was originally described in the 1850s as a ring of structures on the medial side of the cerebral hemisphere, encircling the corpus callosum and thalamus. Its most anatomically prominent components are the **cingulate**[35] (SING-you-let) **gyrus** that arches over the top of the corpus callosum in the frontal and parietal lobes, the **hippocampus**[36] in the medial temporal lobe (fig. 15.17), and the **amygdala**[37] (ah-MIG-da-luh) immediately rostral to the hippocampus, also in the temporal lobe. There are still differences of opinion on what structures to consider as parts of the limbic system, but these three are agreed upon; other components include the mammillary bodies and other hypothalamic nuclei, some thalamic nuclei, parts of the basal nuclei, and parts of the prefrontal cortex. Limbic system components are interconnected through a complex loop of fiber tracts allowing for somewhat circular patterns of feedback among its nuclei and cortical neurons. All of these structures are bilaterally paired; there is a limbic system in each cerebral hemisphere.

The limbic system was long thought to be associated with smell because of its close association with olfactory pathways, but beginning in the early 1900s and continuing even now, experiments have abundantly demonstrated more significant roles in emotion and memory. Most limbic system structures have centers for both gratification and aversion. Stimulation of a gratification center produces a sense of pleasure or reward; stimulation of an aversion center produces unpleasant sensations such as fear or sorrow. Gratification centers dominate some limbic structures, such as the *nucleus accumbens* (not illustrated), while aversion centers dominate others such as the amygdala. The roles of the amygdala in emotion and the hippocampus in memory are described in the next section, on higher forebrain functions.

Higher Forebrain Functions

We will here examine a number of "higher" functions of the forebrain—sensory awareness, motor control, language, emotion, thought, and memory. These processes call attention especially to the cerebral cortex, but are not limited to the cerebrum; they involve also the diencephalon and cerebellum. It is impossible in many cases to assign a specific function to a specific brain region. Functions of the brain do not have such easily defined boundaries as its anatomy. Some functions overlap anatomically, some cross anatomical boundaries from one region to another, and some functions such as consciousness and memory are widely distributed through the cerebrum.

[30] *caudate* = tailed, tail-like
[31] *putam* = pod, husk
[32] *glob* = globe, ball + *pall* = pale
[33] *lenti* = lens + *form* = shape
[34] *limbus* = border
[35] *cingul* = girdle
[36] *hippocampus* = sea horse, named for its shape
[37] *amygdala* = almond

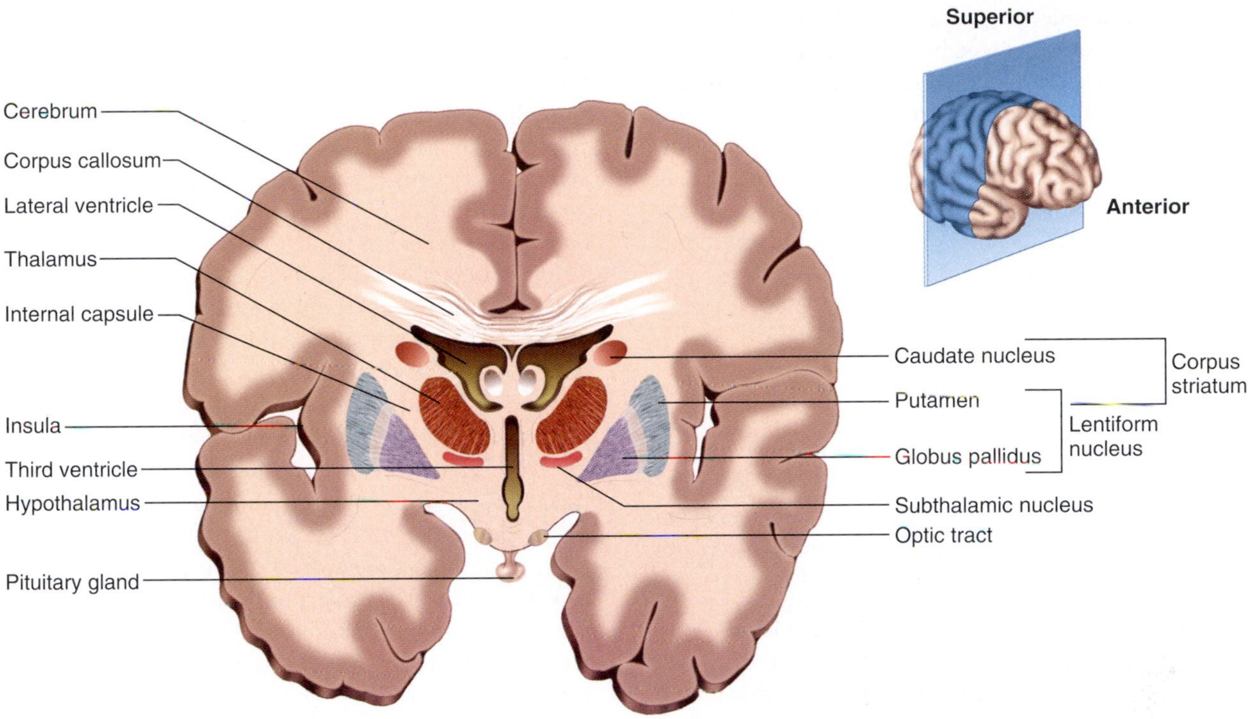
Superior
Anterior
Cerebrum
Corpus callosum
Lateral ventricle
Thalamus
Internal capsule
Insula
Third ventricle
Hypothalamus
Pituitary gland
Caudate nucleus
Putamen
Globus pallidus
Corpus striatum
Lentiform nucleus
Subthalamic nucleus
Optic tract

FIGURE 15.16
The Basal Nuclei. Frontal section of the brain.

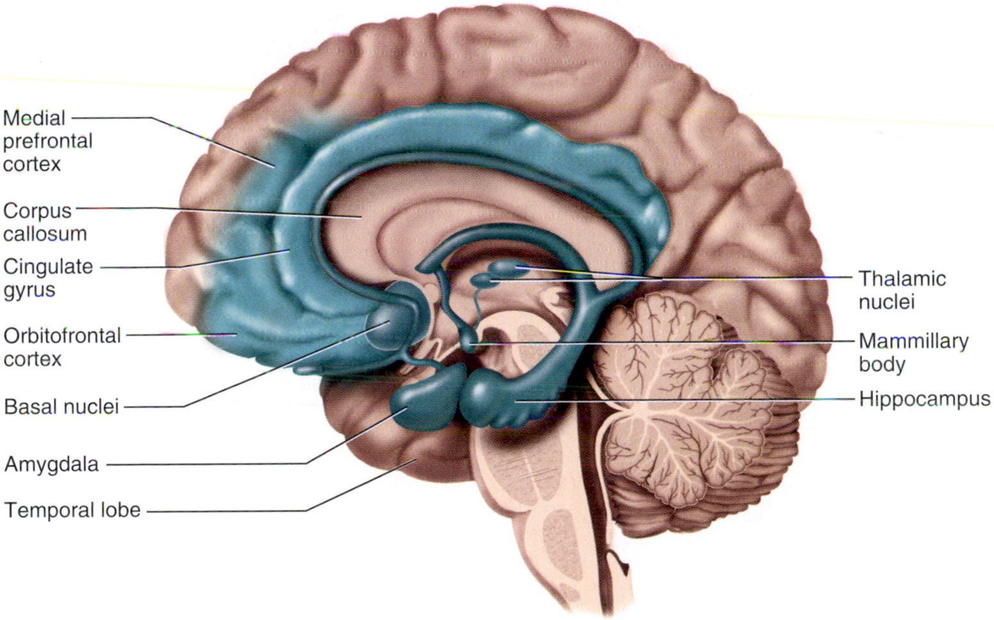
Medial prefrontal cortex
Corpus callosum
Cingulate gyrus
Orbitofrontal cortex
Basal nuclei
Amygdala
Temporal lobe
Thalamic nuclei
Mammillary body
Hippocampus

FIGURE 15.17
The Limbic System.

As a general principle for the functional discussion of the cerebrum, we distinguish between primary cortex and association cortex. **Primary cortex** consists of those regions that receive input directly from the sense organs or brainstem, or issue motor nerve fibers directly to the brainstem for distribution of the motor commands to cranial and spinal nerves. **Association cortex** consists of all regions other than the primary cortex, involved in integrative functions such as interpretation of sensory input, planning of motor output, cognitive (thought) processes, and the storage and retrieval of memories. About 75% of the mass of the cerebral cortex is association cortex. Typically an area of primary cortex has an association area immediately adjacent to it, concerned with the same general function. For example, the primary visual cortex, which receives input from the eyes, is bordered by visual association cortex, which interprets and makes cognitive sense of the visual stimuli. Some areas of association cortex are *multimodal*—instead of processing information from a single sensory source, they receive input from multiple senses and integrate this into our overall perception of our surroundings. The association cortex of the frontal lobe is a very important center of cognitive and emotional function, the **prefrontal cortex.**

SPECIAL SENSES

The *special senses* are the senses of taste, smell, hearing, equilibrium, and vision, mediated by relatively complex sense organs in the head. Signals from these sense organs are routed to areas of primary sensory cortex in widely separated regions of the cerebrum. The pathways taken by special sensory signals are detailed in chapter 17. Here we will only briefly identify the regions of cerebral cortex concerned with each of these senses (fig. 15.18):

- **Vision.** Visual signals are received by the **primary visual cortex** in the far posterior region of the occipital lobe. This is bordered anteriorly by the **visual association area,** which includes all the remainder of the occipital lobe and much of the inferior temporal lobe.
- **Hearing.** Auditory signals are received by the **primary auditory cortex** in the superior region of the temporal lobe and in the nearby insula. The **auditory association area** occupies areas of temporal lobe inferior to the primary auditory cortex and deep within the lateral sulcus.
- **Equilibrium.** Signals from the inner ear for equilibrium (balance and the sense of motion) project mainly to the cerebellum and several brainstem nuclei concerned with head and eye movements and visceral functions. Some fibers of this system, however, are routed through the thalamus to areas of association cortex in the roof of the lateral sulcus and near the lower end of the central sulcus. This is the seat of consciousness of our body movements and orientation in space.
- **Taste.** Gustatory signals are received by the **primary gustatory cortex** in the inferior end of the postcentral gyrus of the parietal lobe (discussed shortly) and an anterior region of the insula. See the following discussion for the location of the gustatory association cortex.
- **Smell.** Olfactory signals are the only sensory signals that can reach the cortex without going through the thalamus. The primary olfactory cortex lies in the medial surface of the temporal lobe and inferior surface of the frontal lobe. The **orbitofrontal cortex** (see fig. 15.17) is an association area that integrates gustatory, olfactory, and visual information to create a sense of the overall flavor and desirability (or rejection) of food.

GENERAL SENSES

The *general (somesthetic* or *somatosensory) senses* are widely distributed over the body and have relatively simple receptors (see chapter 17). They include such senses as touch, pressure, stretch, temperature, and pain. From the neck down, somesthetic signals travel up the spinal cord in the gracile and cuneate fasciculi and spinothalamic tracts. Somesthetic nerve fibers decussate in the spinal cord or medulla before reaching the thalamus (see table 14.1 and fig. 14.4). Consequently, these sensory signals ultimately arrive in the contralateral cerebral cortex—stimuli on the right side of the body are perceived in the left cerebral hemisphere and vice versa. The thalamus routes all somesthetic signals to one specific fold of the cerebrum, the **postcentral gyrus.** This gyrus lies immediately caudal to the central sulcus and forms the anterior border of the parietal lobe (fig. 15.19*a*). It rises from the lateral sulcus up to the crown of the head and then descends into the longitudinal fissure. The cortex of this gyrus is called the **primary somesthetic**[38] **cortex.**

This gyrus is like an upside-down sensory map of the contralateral side of the body, traditionally diagrammed as a *sensory homunculus*[39] (fig. 15.19*b*). As the diagram shows, receptors in the lower limb project to superior and medial parts of the gyrus, and receptors in the face project to the inferior and lateral parts. There is a point-for-point correspondence, or **somatotopy,**[40] between an area of the body and an area of the postcentral gyrus. The reason for the bizarre, distorted appearance of the homunculus is that the amount of cerebral tissue devoted to a given body region is proportional to how richly innervated and sensitive that region is, not to its size. Thus, the hands and face are represented by a much larger region of somesthetic cortex than the trunk is.

The **somesthetic association area** is found in the parietal lobe posterior to the postcentral gyrus and in the roof of the lateral sulcus (see fig. 15.18).

MOTOR CONTROL

The **primary motor cortex** is in the **precentral gyrus,** immediately anterior to the central sulcus (fig. 15.20). This gyrus forms the posterior margin of the frontal lobe. Like the primary somesthetic cortex, it exhibits a somatotopic, upside-down map of the contralateral side of the body, but its map represents muscle control rather than sensation. Thus, the primary motor cortex deep in the longitudinal fissure controls muscles of the foot and leg; as we progress up the gyrus toward the crown of the head, we find cortex that controls muscles of the hip and trunk; and descending the gyrus on the lateral surface of the brain, we find cortex in control of the upper limb, then face, and the tongue and pharynx. Like the sensory homunculus, the *motor homunculus* in figure 15.20 is distorted because the amount of cortex dedicated to a particular body region is proportional to the number of muscles and motor units in that region, not the size of the region. Thus, it requires more motor cortex to control a hand than it does to control the muscles of the trunk.

Those pyramidal cells of the precentral gyrus whose axons descend into the white matter are called **upper motor neurons.** They project caudally into the brainstem. About 19 million of these nerve fibers end in nuclei of the brainstem. The remaining 1 million form the corticospinal tracts, which constitute the pyramids of the

[38]*som* = body + *esthet* = sensation
[39]*hom* = man + *unculus* = little
[40]*somato* = body + *top* = place

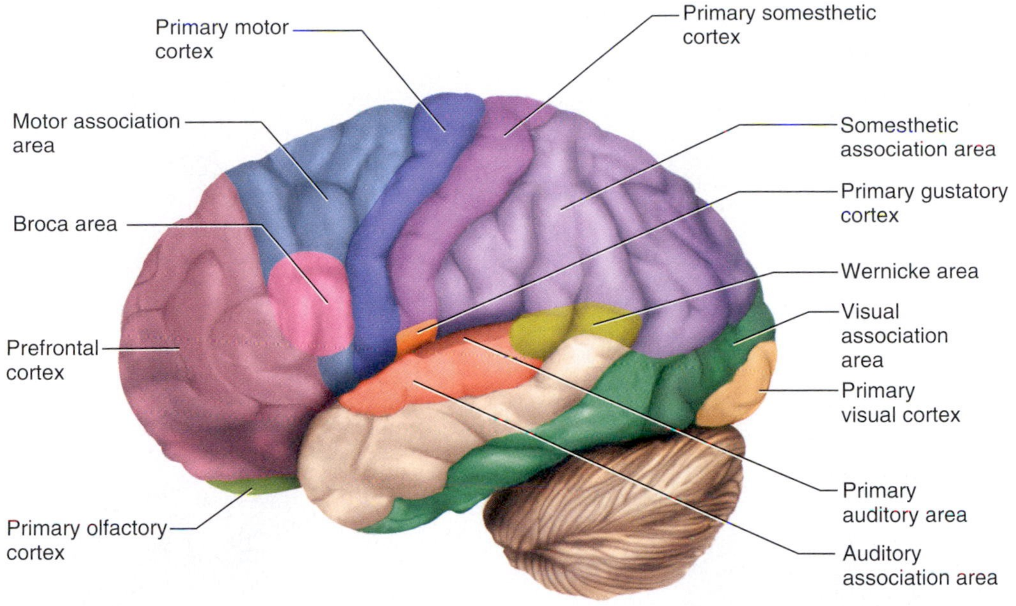

FIGURE 15.18

Some Functional Regions of the Cerebral Cortex. Left hemisphere.

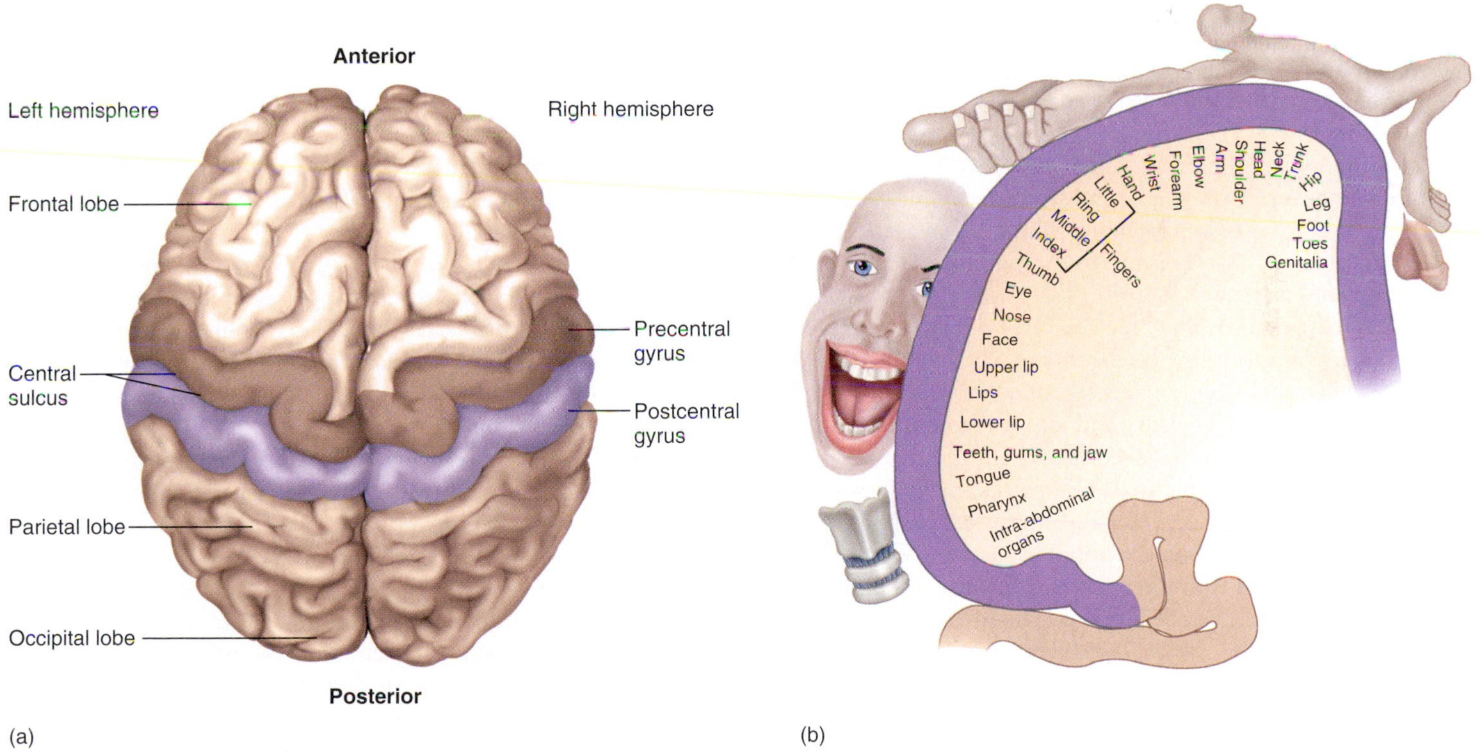

FIGURE 15.19

The Primary Somesthetic Area (postcentral gyrus). (*a*) Location, superior view. (*b*) Sensory homunculus, drawn so that body parts are in proportion to the amount of cortex dedicated to their sensation.

medulla oblongata and continue into the spinal cord. The upper motor neurons synapse in the brainstem and spinal cord with **lower motor neurons,** whose axons leave the CNS and directly innervate the muscles. About 90% of the corticospinal tract fibers decussate in the medulla oblongata and the rest do so in the spinal cord. Therefore the right motor cortex controls muscles on the left side of the body and vice versa. A stroke that kills motor cortex in one hemisphere thus produces paralysis on the contralateral side of the body.

The **motor association (premotor) area** is a large area of the cerebrum rostral to the precentral gyrus (see fig. 15.18). This is the

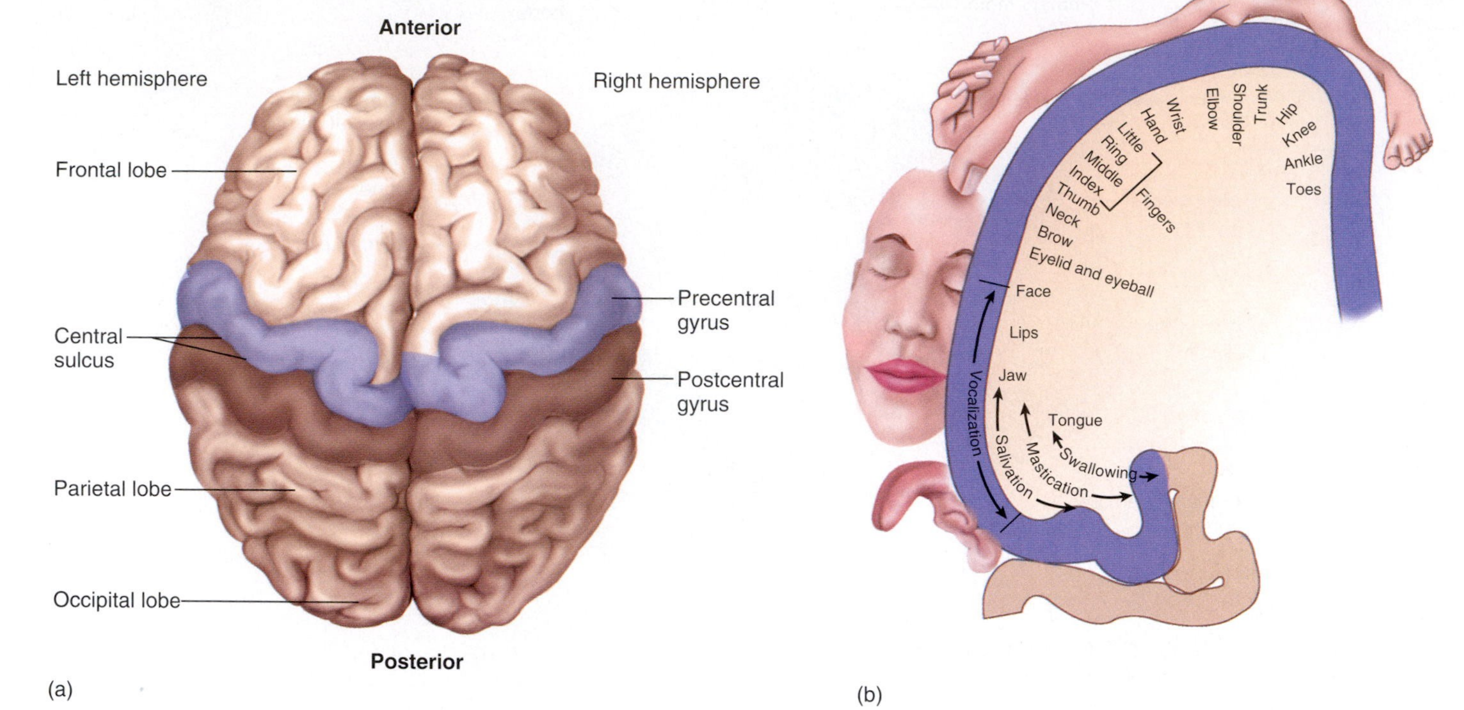
Anterior
Left hemisphere
Right hemisphere
Frontal lobe
Precentral gyrus
Central sulcus
Postcentral gyrus
Parietal lobe
Occipital lobe
Posterior
(a)
Hip
Knee
Ankle
Toes
Trunk
Shoulder
Elbow
Wrist
Hand
Little
Ring
Middle
Index
Thumb
Fingers
Neck
Brow
Eyelid and eyeball
Face
Lips
Jaw
Tongue
Swallowing
Mastication
Salivation
Vocalization
(b)

FIGURE 15.20

The Primary Motor Area (precentral gyrus). (*a*) Location, superior view. (*b*) Motor homunculus, drawn so that body parts are in proportion to the amount of primary motor cortex dedicated to their control.

seat of conscious planning of one's movements. Neurons here compile a program for the degree and sequence of muscle contractions needed to carry out an intended action such as dancing, typing, or speaking. They transmit this program caudally to the primary motor cortex, which executes the program and issues signals through the corticospinal tracts to carry out the motion.

The *basal nuclei* of the cerebrum are part of a feedback circuit involved in the planning and execution of movement. They receive input from many sensory and motor regions of the cortex, most importantly the prefrontal cortex. Their output goes by way of the thalamus back to the prefrontal cortex, motor association area, and precentral gyrus. The basal nuclei are responsible for controlling highly practiced behaviors such as tying your shoes or driving a car—skilled movements that you carry out with little thought. Recall that the cerebellum is also highly important in motor control. Its function was explained earlier in the chapter.

LANGUAGE

Language includes several abilities—reading, writing, speaking, and understanding spoken and printed words—assigned to different regions of cerebral cortex. The **Wernicke**[41] (WUR-nih-kee) **area** is responsible for the recognition of spoken and written language. It lies just posterior to the lateral sulcus, usually in the left hemisphere, at the "crossroads" between visual, auditory, and somesthetic areas of the cortex (see fig. 15.18). It is a sensory association area that receives input from all these neighboring regions of primary sensory cortex. The *angular gyrus,* part of the parietal lobe caudal and superior to the Wernicke area, is important in the ability to read and write.

When we intend to speak, the Wernicke area formulates phrases according to learned rules of grammar and transmits a plan of speech to the **Broca**[42] **area,** located in the inferior prefrontal cortex in the same hemisphere. PET scans show a rise in the metabolic activity of the Broca area as we prepare to speak. The Broca area generates a motor program for the muscles of the larynx, tongue, cheeks, and lips to produce speech. This program is then transmitted to the primary motor cortex, which executes it—that is, it issues commands to the lower motor neurons that supply the relevant muscles.

The emotional aspect of language is controlled by regions in the opposite hemisphere that mirror the Wernicke and Broca areas. Opposite the Broca area is the *affective language area.* Lesions to this area result in *aprosodia*—flat, emotionless speech. The cortex opposite the Wernicke area is concerned with recognizing the emotional content of another person's speech. Lesions here can result in such problems as the inability to understand a joke. Lesions in the language areas of the brain tend to produce a variety of language deficits called **aphasias**[43] (ah-FAY-zee-uhs). They may include a complete inability to speak; slow speech with difficulty choosing words; invention of words that only approximate the correct word; babbling, incomprehensible speech filled with invented words and illogical word order; inability to comprehend another person's

[41]Karl Wernicke (1848–1904), German neurologist

[42]Pierre Paul Broca (1824–80), French surgeon and anthropologist

[43]*a* = without + *phas* = speech

written or spoken words; or inability to name objects that a person sees. Since cranial nerves VII and XII (the facial and hypoglossal nerves) control several of the muscles of speech, speech deficits can also result from lesions to these nerves or their brainstem nuclei.

Think About It!

Mr. Thompson has had a stroke that destroyed his Wernicke area. Ms. Meyers has had a stroke that destroyed her Broca area. What differences in language deficits would you expect between these two patients?

EMOTION

Emotional feelings and memories are not exclusively cerebral functions, but rather result from an interaction between areas of the prefrontal cortex and the diencephalon. In the diencephalon, the hypothalamus and amygdala play especially important roles in emotion. Experiments by Swiss physiologist Walter Hess, leading to a 1949 Nobel Prize, showed that stimulation of various nuclei of the hypothalamus in cats could induce rage, attack, and other emotional responses. Nuclei involved in the senses of reward and punishment have been identified in the hypothalamus of cats, rats, monkeys, and other animals.

The *amygdala,* one of the most important centers of human emotion, is a major component of the limbic system described earlier. It receives processed information from the senses of vision, hearing, taste, smell, and general somesthetic and visceral senses. Thus, it is able to mediate emotional responses to such stimuli as a disgusting odor, a foul taste, a beautiful image, pleasant music, or a stomach ache. It is especially important in the sense of fear. Output from the amygdala goes in two directions of special interest: (1) Some output projects to the hypothalamus and lower brainstem, and thus influences the somatic and visceral motor systems. An emotional response to a sight or sound may, through these connections, make one's heart race, make the hair stand on end (piloerection), or induce vomiting. (2) Other output projects to areas of the prefrontal cortex that mediate conscious control and expression of the emotions, such as our ability to express love or control anger.

Many important aspects of personality depend on an intact, functional amygdala and hypothalamus. When specific regions of the amygdala or hypothalamus are destroyed or artificially stimulated, humans and other animals exhibit blunted or exaggerated expressions of anger, fear, aggression, self-defense, pleasure, pain, love, sexuality, and parental affection, as well as abnormalities in learning, memory, and motivation.

COGNITION

Cognition[44] is the range of mental processes by which we acquire and use knowledge—sensory perception, thought, reasoning, judgment, memory, imagination, and intuition. Cognitive abilities of various kinds are widely distributed through the association areas of the cerebral cortex. This is the most difficult area of brain research, and the most incompletely understood area of cerebral function. Much of what we know about cognitive functions of the brain has come from studies of patients with brain lesions—areas of tissue destruction resulting from cancer, stroke, and trauma. The many brain injuries incurred in World War I and II yielded a special abundance of insights into regional brain functions.

Attention to objects in the environment is based in the parietal lobe on the side opposite the Broca and Wernicke speech centers. Lesions here can produce *contralateral neglect syndrome,* in which a patient seems unaware of objects on one side of the body, ignores all words on the left side of a page of reading, or fails even to recognize, dress, and take care of the left side of his or her own body. Such patients are also unable to find their way around—say, to describe the route from home to work, or navigate within a familiar building.

The prefrontal cortex is concerned with many of our most distinctive abilities, such as abstract thought, foresight, judgment, responsibility, a sense of purpose, and a sense of socially appropriate behavior. Lesions here tend to render a person easily distracted from a task, irresponsible, exceedingly stubborn, unable to anticipate future events, and incapable of any ambition or planning for the future (insight 15.2).

Cognition is not limited to the cerebrum. The cerebellum has lately shown a surprising amount of involvement in cognitive function. Brain imaging techniques such as PET scans show increased cerebellar activity in connection with analyzing sensory input, telling time, solving spatial puzzles, and even performing language tasks. For example, if a person is given an noun such as *apple* and asked to think of a related verb such as *eat,* the cerebellum shows more activity than when the person is asked simply to repeat *apple.* Rubbing sandpaper over a person's fingers activates the cerebellum to some degree, but not as much as when a person is asked to rate the relative coarseness of two different sandpapers. The cerebellum is also involved in making short-term predictions about movement, such as predicting where a baseball will be in a second or two so that one can reach out and catch it.

MEMORY

Memory is one of the cognitive functions, but warrants special attention. There are two kinds of memory—**procedural memory,** the retention of motor skills such as how to tie one's shoes, play the violin, or type on a keyboard; and **declarative memory,** the retention of events and facts that one can put into words, such as names, dates, or facts important to an upcoming examination. At the cellular level, both forms of memory probably involve the same processes: the creation of new synaptic contacts and physiological changes that make synaptic transmission more efficient along certain pathways.

The limbic system has important roles in the establishment of memories. The amygdala creates emotional memories, such as the chilling fear of being stung when a wasp alights on the skin. The *hippocampus* (see fig. 15.17) is critical to the creation of long-term declarative memories. It does not store memories, but organizes sensory and cognitive experiences into a unified long-term memory. The hippocampus learns from sensory input while an experience is happening, but it has a short memory. Later, perhaps when one is sleeping, it plays this memory repeatedly to the cerebral cortex, which is a "slow

[44] *cognit* = to know, to learn

INSIGHT 15.2 MEDICAL HISTORY

An Accidental Lobotomy

Accidental but nonfatal destruction of parts of the brain has afforded many clues to the function of various regions. One of the most famous incidents occurred in 1848 to Phineas Gage, a laborer on a railroad construction project in Vermont. Gage was packing blasting powder into a hole with a 3 1/2 ft tamping iron when the powder prematurely exploded. The tamping rod was blown out of the hole and passed through Gage's maxilla, orbit, and the frontal lobe of his brain before emerging from his skull near the hairline and landing 50 ft away (fig. 15.21). Gage went into convulsions, but later sat up and conversed with his crewmates as they drove him to a physician in an oxcart. On arrival, he stepped out on his own and told the physician, "Doctor, here is business enough for you." His doctor, John Harlow, reported that he could insert his index finger all the way into Gage's wound. Yet 2 months later, Gage was walking around town carrying on his normal business.

He was not, however, the Phineas Gage people had known. Before the accident, he had been a competent, responsible, financially prudent man, well liked by his associates. In an 1868 publication on the incident, Harlow said that following the accident, Gage was "fitful, irreverent, indulging at times in the grossest profanity." He became irresponsible, lost his job, worked for a while as a circus sideshow attraction, and died a vagrant 12 years later.

A 1994 computer analysis of Gage's skull indicated that the brain injury was primarily to the ventromedial region of both frontal lobes. In Gage's time, scientists were reluctant to attribute social behavior and moral judgment to any region of the brain. These functions were strongly tied to issues of religion and ethics and were considered inaccessible to scientific analysis. Based partly on Phineas Gage and other brain injury patients like him, neuroscientists today recognize that planning, moral judgment, and emotional control are among the functions of the prefrontal cortex.

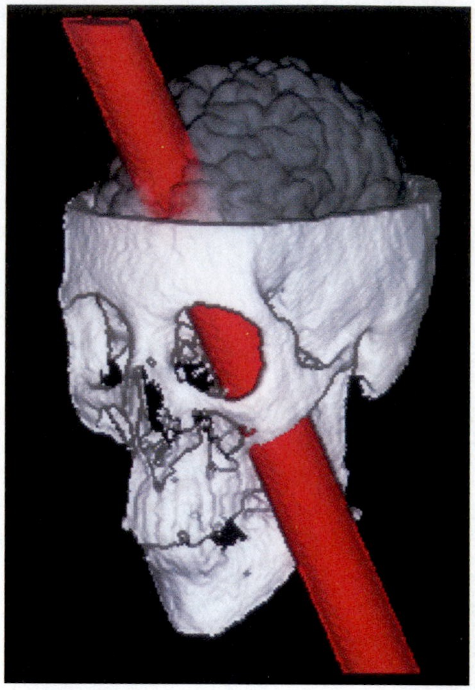

Figure 15.21
Phineas Gage's 1848 Accident. This is a computer-generated image made in 1994 to show the path taken by the tamping bar through Gage's skull and brain.

learner" but forms longer-lasting memories. This process of "teaching the cerebral cortex" until a long-term memory is established is called **memory consolidation.** Lesions of the hippocampus do not abolish old memories, but abolish the ability to form new ones.

Long-term memories are stored in various areas of cerebral cortex. The memory of language (vocabulary and grammatical rules) resides in the Wernicke area. Our memory of faces, voices, and familiar objects is stored in the superior temporal lobe. Memories of one's social role, appropriate behavior, goals, and plans are stored in the prefrontal cortex. Procedural memories are stored in the motor association area, basal nuclei, and cerebellum.

CEREBRAL LATERALIZATION

The two cerebral hemispheres look identical at a glance, but close examination reveals a number of differences. For example, in women the left temporal lobe is longer than the right. In left-handed people, the left frontal, parietal, and occipital lobes are usually wider than those on the right. The two hemispheres also differ in some of their functions (fig. 15.22). Neither hemisphere is "dominant," but each is specialized for certain tasks. This difference in function is called **cerebral lateralization.**

One hemisphere, usually the left, is called the *categorical hemisphere.* It is specialized for spoken and written language and for the sequential and analytical reasoning employed in such fields as science and mathematics. This hemisphere seems to break information into fragments and analyze it in a linear way. The other hemisphere, usually the right, is called the *representational hemisphere.* It perceives information in a more integrated, holistic way. It is the seat of imagination and insight, musical and artistic skill, perception of patterns and spatial relationships, and comparison of sights, sounds, smells, and tastes.

Table 15.2 summarizes the forebrain functions described in the last several pages.

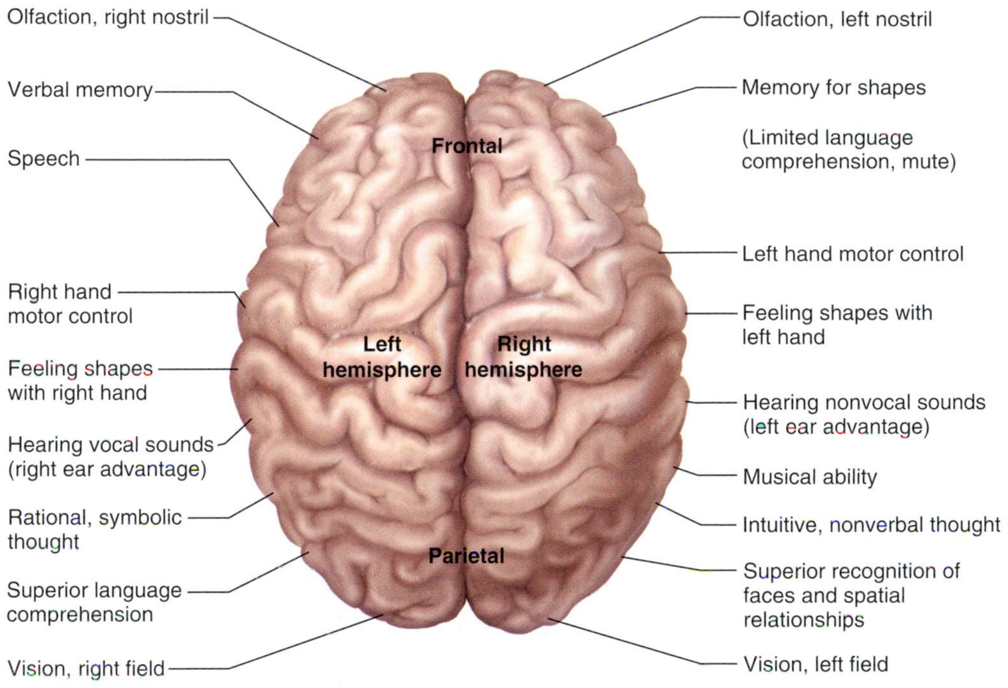
Olfaction, right nostril
Verbal memory
Speech
Right hand motor control
Feeling shapes with right hand
Hearing vocal sounds (right ear advantage)
Rational, symbolic thought
Superior language comprehension
Vision, right field
Frontal
Left hemisphere
Right hemisphere
Parietal
Olfaction, left nostril
Memory for shapes
(Limited language comprehension, mute)
Left hand motor control
Feeling shapes with left hand
Hearing nonvocal sounds (left ear advantage)
Musical ability
Intuitive, nonverbal thought
Superior recognition of faces and spatial relationships
Vision, left field

FIGURE 15.22
Lateralization of Cerebral Functions.

TABLE 15.2
Forebrain Functions

Diencephalon	
Thalamus	Sensory processing; relay of sensory and other signals to cerebrum; relay of cerebral output to other parts of brain
Hypothalamus	Hormone synthesis; control of pituitary secretion; autonomic responses affecting heart rate, blood pressure, pupillary diameter, digestive secretion and motility, and other visceral functions; thermoregulation; hunger and thirst; sleep and circadian rhythms; emotional responses; sexual function; memory
Epithalamus	Hormone secretion; relay of signals between midbrain and limbic system
Cerebral lobes	
Frontal lobe	Smell; motor aspects of speech; voluntary control of skeletal muscles; procedural memory; cognitive functions such as abstract thought, judgment, responsibility, foresight, ambition, planning, and ability to stay focused on a task
Parietal lobe	Somesthetic senses, taste, awareness of body movement and orientation, language recognition, nonmotor aspects of speech
Occipital lobe	Vision
Temporal lobe	Hearing, smell, interpreting visual information, learning, memory, emotion
Insula	Hearing, taste, visceral sensation
Basal nuclei	Motor control; procedural memory
Limbic system	Learning, emotion, gratification (pleasure) and aversion responses

Before You Go On

Answer the following questions to test your understanding of the preceding section:

13. What is the role of the thalamus in sensory function?
14. List at least six functions of the hypothalamus.
15. Name the five lobes of the cerebrum and describe their locations and boundaries.
16. Where are the basal nuclei located? What is their general function?
17. Where is the limbic system located? What component of it is involved in emotion? What component is involved in memory?
18. Describe the locations and functions of the somesthetic, visual, auditory, and frontal association areas.
19. Describe the somatotopy of the primary motor cortex and primary sensory cortex.
20. What are the roles of the Wernicke area, Broca area, and precentral gyrus in language?

The Cranial Nerves

Objectives

When you have completed this section, you should be able to

- list the 12 cranial nerves by name and number;
- identify where each cranial nerve originates and terminates; and
- state the functions of each cranial nerve.

To be functional, the brain must communicate with the rest of the body. Most of its input and output travels by way of the spinal cord, but it also communicates by way of the **cranial nerves,** which arise primarily from the base of the brain, exit the cranium through its foramina, and lead to muscles and sense organs primarily in the head and neck. There are 12 pairs of cranial nerves, numbered I to XII starting with the most rostral (fig. 15.23). Each nerve also has a descriptive name such as *optic nerve* and *vagus nerve.* The cranial nerves are illustrated and described in table 15.3.

Classification

Cranial nerves are traditionally classified as sensory (I, II, and VIII), motor (III, IV, VI, XI, and XII), or mixed (V, VII, IX, and X). In reality, only cranial nerves I and II (for smell and vision) are purely sensory, whereas all the rest contain both afferent and efferent fibers and are therefore mixed nerves. Those traditionally classified as motor not only stimulate muscle contractions but also contain afferent fibers of proprioception, which provide your brain with unconscious feedback for controlling muscle contraction, and which make you consciously aware of such things as the position of your tongue and orientation of your head. Cranial nerve VIII, concerned with hearing and equilibrium, is traditionally classified as sensory, but it has motor fibers that return signals to the inner ear and "tune" it to sharpen our sense of hearing. The nerves traditionally classified as mixed have sensory functions quite unrelated to their motor functions—for example, the facial nerve (VII) has a sensory role in taste and a motor role in controlling facial expressions.

In order to teach the traditional classification (which is relevant for such purposes as board examinations and comparison to other books), yet remind you that all but two of these nerves are mixed, table 15.3 describes many of the nerves as *predominantly* sensory or motor.

Nerve Pathways

The motor fibers of the cranial nerves begin in nuclei of the brainstem and lead to glands and muscles. The sensory fibers begin in receptors located mainly in the head and neck and lead mainly to the brainstem. Pathways for the special senses are described in chapter 17. Sensory fibers for proprioception begin in the muscles innervated by the motor fibers of the cranial nerves, but they often travel to the brain in a different nerve than the one which supplies the motor innervation.

Most cranial nerves carry fibers between the brainstem and ipsilateral receptors and effectors. Thus, a lesion to one side of the brainstem causes a sensory or motor deficit on the same side of the head. This contrasts with lesions to the motor and somesthetic cortex of the cerebrum, which, as we saw earlier, cause sensory and motor deficits on the contralateral side of the body. The exceptions are the optic nerve (cranial nerve I), where half the fibers decussate to the opposite side of the brain (see chapter 17), and the trochlear nerve (cranial nerve IV), in which all efferent fibers go to a muscle of the contralateral eye.

An Aid to Memory

Generations of biology and medical students have relied on mnemonic (memory-aiding) phrases and ditties, ranging from the sublimely silly to the unprintably ribald, to help them remember the cranial nerves and other anatomy. An old classic began, "On old Olympus' towering tops . . . ," with the first letter of each word matching the first letter of each cranial nerve (olfactory, optic, oculomotor, etc.). Some cranial nerves have changed names, however, since that passage was devised. One of the author's former students[†] devised a better mnemonic that can remind you of the first two to four letters of most cranial nerves:

Old	**ol**factory (I)	**fe**els	**fa**cial (VII)
Opie	**op**tic (II)	**ve**ry	**ve**stibulocochlear (VIII)
occasionally	**oc**ulomotor (III)	**glo**omy,	**glo**ssopharyngeal (IX)
tries	**tr**ochlear (IV)	**vagu**e,	**vagu**s (X)
trigonometry	**trig**eminal (V)	**a**nd	**a**ccessory (XI)
and	**a**bducens (VI)	**hypo**active.	**hypo**glossal (XII)

Another student's mnemonic, but using only the first letter of each nerve's name, is "Oh, once one takes the anatomy final, very good vacation ahead."[‡] The first two letters of *ahead* represent nerves XI and XII.

[†] Courtesy of Marti Haykin, M.D.

[‡] Courtesy of Sarah Veramay

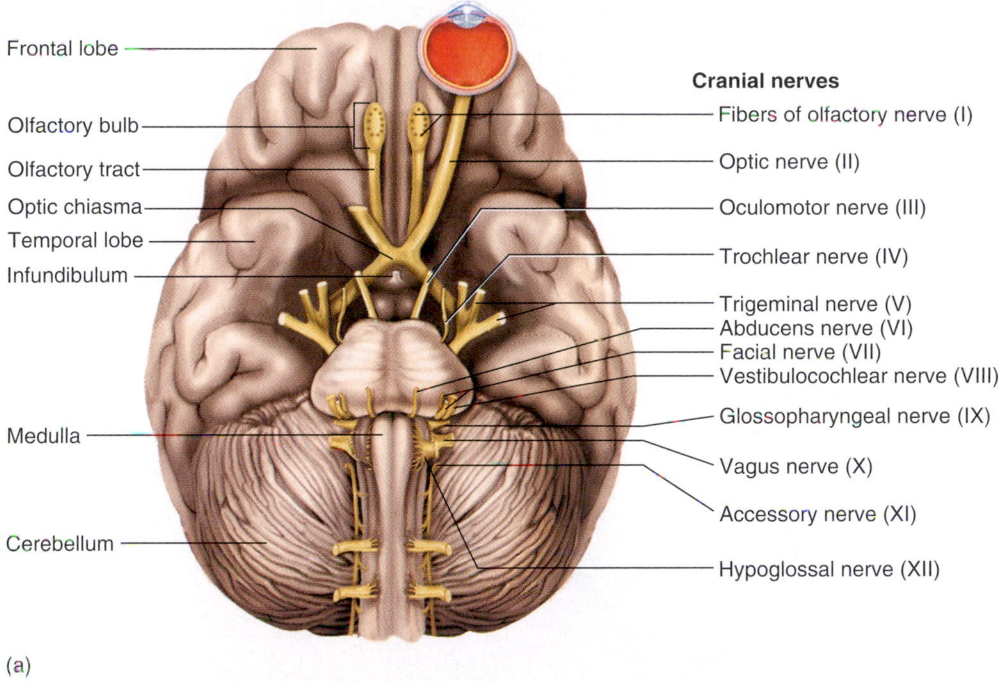
Frontal lobe
Olfactory bulb
Olfactory tract
Optic chiasma
Temporal lobe
Infundibulum
Medulla
Cerebellum
Cranial nerves
Fibers of olfactory nerve (I)
Optic nerve (II)
Oculomotor nerve (III)
Trochlear nerve (IV)
Trigeminal nerve (V)
Abducens nerve (VI)
Facial nerve (VII)
Vestibulocochlear nerve (VIII)
Glossopharyngeal nerve (IX)
Vagus nerve (X)
Accessory nerve (XI)
Hypoglossal nerve (XII)
(a)

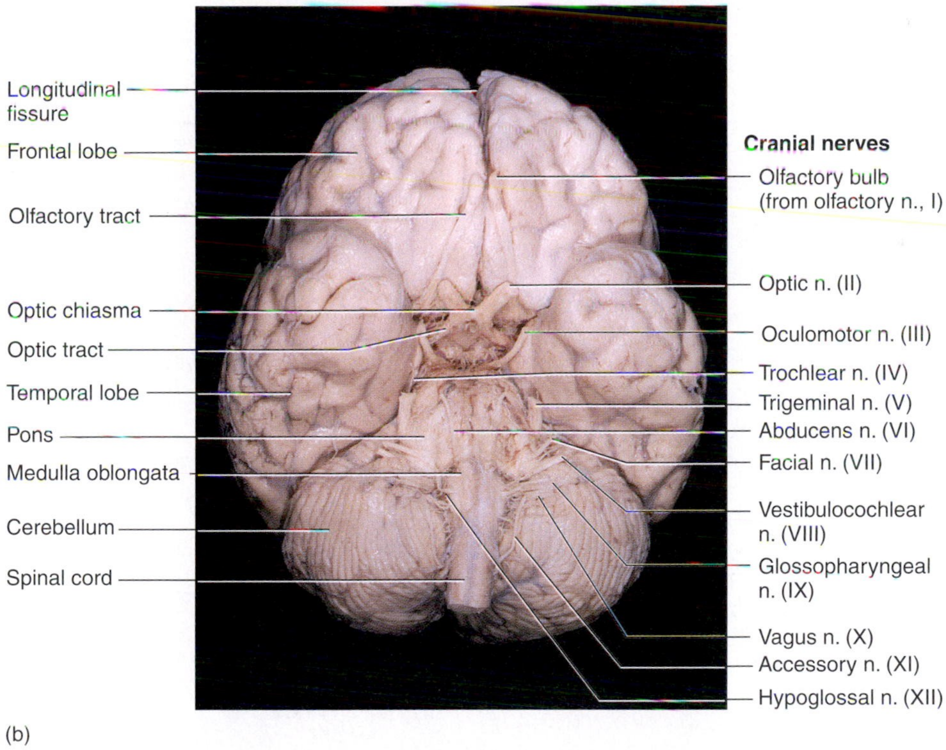
Longitudinal fissure
Frontal lobe
Olfactory tract
Optic chiasma
Optic tract
Temporal lobe
Pons
Medulla oblongata
Cerebellum
Spinal cord
Cranial nerves
Olfactory bulb (from olfactory n., I)
Optic n. (II)
Oculomotor n. (III)
Trochlear n. (IV)
Trigeminal n. (V)
Abducens n. (VI)
Facial n. (VII)
Vestibulocochlear n. (VIII)
Glossopharyngeal n. (IX)
Vagus n. (X)
Accessory n. (XI)
Hypoglossal n. (XII)
(b)

FIGURE 15.23

The Cranial Nerves. (*a*) Base of the brain, showing the 12 pairs of cranial nerves. (*b*) Cranial nerves on the cadaver brain.

Table 15.3
The Cranial Nerves

Origins of proprioceptive fibers are not tabulated; they are the muscles innervated by the motor fibers. Nerves listed as mixed or sensory are agreed by all authorities to be either mixed or purely sensory nerves. Nerves classified as *predominantly* motor or sensory are traditionally classified that way but contain some fibers of the other type.

I. Olfactory Nerve

Composition: Sensory
Function: Smell
Origin: Olfactory mucosa in nasal cavity
Termination: Olfactory bulbs beneath frontal lobe of brain

Cranial passage: Cribriform plate of ethmoid bone
Effects of damage: Impaired sense of smell
Clinical test: Determine whether subject can smell (not necessarily identify) aromatic substances such as coffee, vanilla, clove oil, or soap

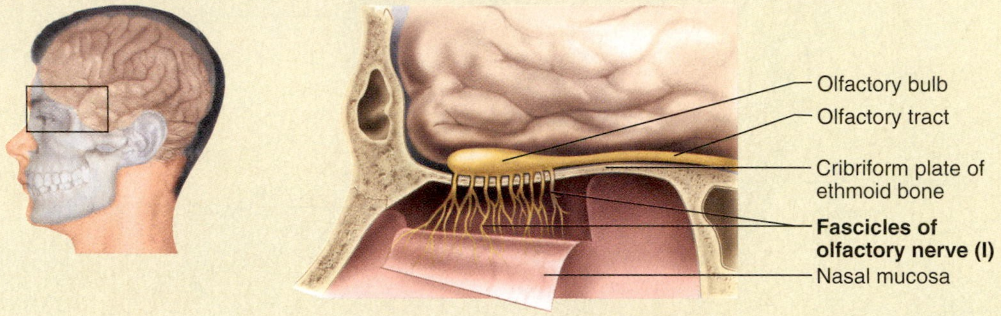
Olfactory bulb
Olfactory tract
Cribriform plate of ethmoid bone
Fascicles of olfactory nerve (I)
Nasal mucosa

FIGURE 15.24
The Olfactory Nerve.

II. Optic Nerve

Composition: Sensory
Function: Vision
Origin: Retina
Termination: Thalamus

Cranial passage: Optic foramen
Effects of damage: Blindness in part or all of the visual field
Clinical test: Inspect retina with ophthalmoscope; test peripheral vision and visual acuity

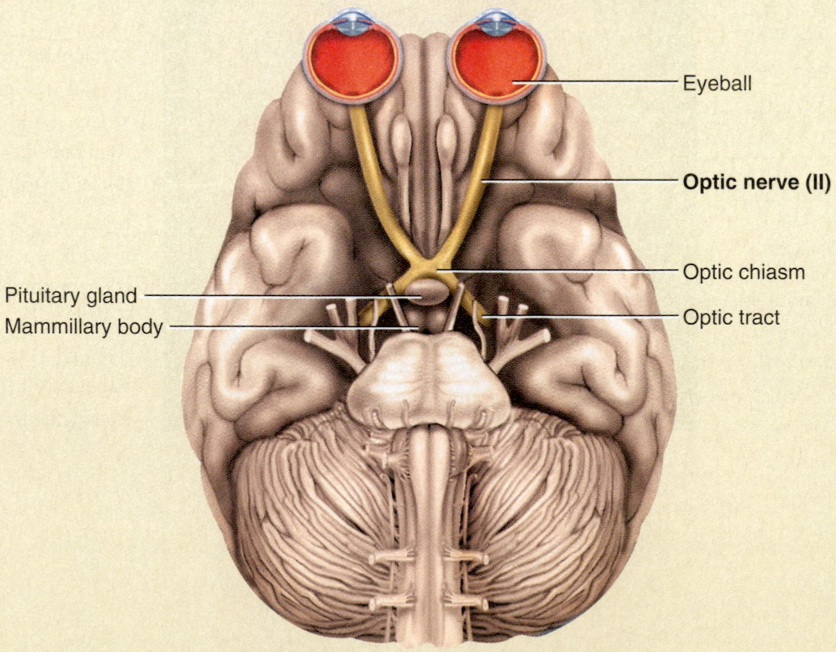
Eyeball
Optic nerve (II)
Optic chiasm
Optic tract
Pituitary gland
Mammillary body

FIGURE 15.25
The Optic Nerve.

(continued)

TABLE 15.3
The Cranial Nerves *(continued)*

III. Oculomotor (OC-you-lo-MO-tur) Nerve

Composition: Predominantly motor

Function: Eye movements, opening of eyelid, constriction of pupil, focusing, proprioception

Origin: Midbrain

Termination: Somatic fibers lead to levator palpebrae superioris; superior, medial, and inferior rectus; and inferior oblique muscles. Parasympathetic fibers enter eyeball and lead to constrictor of iris and ciliary muscle of lens.

Cranial passage: Superior orbital fissure

Effects of damage: Drooping eyelid, dilated pupil, inability to move eye in certain directions, tendency of eye to rotate laterally at rest, double vision, and difficulty focusing

Clinical test: Look for differences in size and shape of right and left pupil; test pupillary response to light; test ability to track moving objects

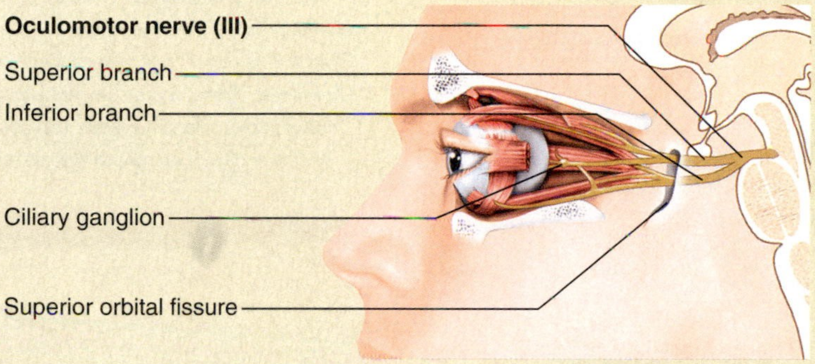
Oculomotor nerve (III)
Superior branch
Inferior branch
Ciliary ganglion
Superior orbital fissure

FIGURE 15.26
The Oculomotor Nerve.

IV. Trochlear (TROCK-lee-ur) Nerve

Composition: Predominantly motor

Function: Eye movements and proprioception

Origin: Midbrain

Termination: Superior oblique muscle of eye

Cranial passage: Superior orbital fissure

Effects of damage: Double vision and inability to rotate eye inferolaterally. Eye points superolaterally, and patient often tilts head toward affected side.

Clinical test: Test ability to rotate eye inferolaterally

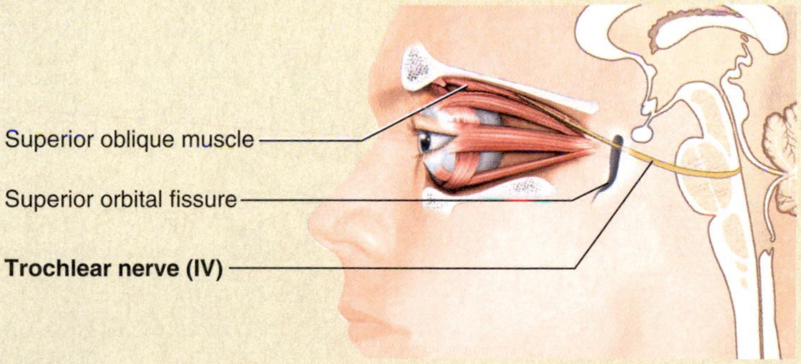
Superior oblique muscle
Superior orbital fissure
Trochlear nerve (IV)

FIGURE 15.27
The Trochlear Nerve.

(continued)

Table 15.3

The Cranial Nerves *(continued)*

V. Trigeminal[45] (tri-JEM-ih-nul) Nerve

Largest of the cranial nerves; consists of three divisions designated V_1 to V_3

V_1. Ophthalmic Division

Composition: Sensory
Function: Main sensory nerve of upper face (touch, temperature, pain)
Origin: Superior region of face as illustrated, surface of eyeball, tear gland, superior nasal mucosa, frontal and ethmoid sinuses
Termination: Pons
Cranial passage: Superior orbital fissure
Effects of damage: Loss of sensation
Clinical test: Test corneal reflex—blinking in response to light touch to eyeball

V_2. Maxillary Division

Composition: Sensory
Function: Same sensations as V_1 lower on face
Origin: Middle region of face as illustrated, nasal mucosa, maxillary sinus, palate, upper teeth and gums
Termination: Pons
Cranial passage: Foramen rotundum and infraorbital foramen
Effects of damage: Loss of sensation
Clinical test: Test sense of touch, pain, and temperature with light touch, pinpricks, and hot and cold objects

V_3. Mandibular Division

Composition: Mixed
Function: Same sensations as V_1–V_2 lower on face; mastication
Sensory origin: Inferior region of face as illustrated, anterior two-thirds of tongue (but not taste buds), lower teeth and gums, floor of mouth, dura mater
Sensory termination: Pons
Motor origin: Pons
Motor termination: Anterior belly of digastric, masseter, temporalis, mylohyoid, pterygoids, and tensor tympani of middle ear
Cranial passage: Foramen ovale
Effects of damage: Loss of sensation; impaired chewing
Clinical test: Assess motor functions by palpating masseter and temporalis muscles while subject clenches teeth; test ability of subject to move mandible from side to side and to open mouth against resistance

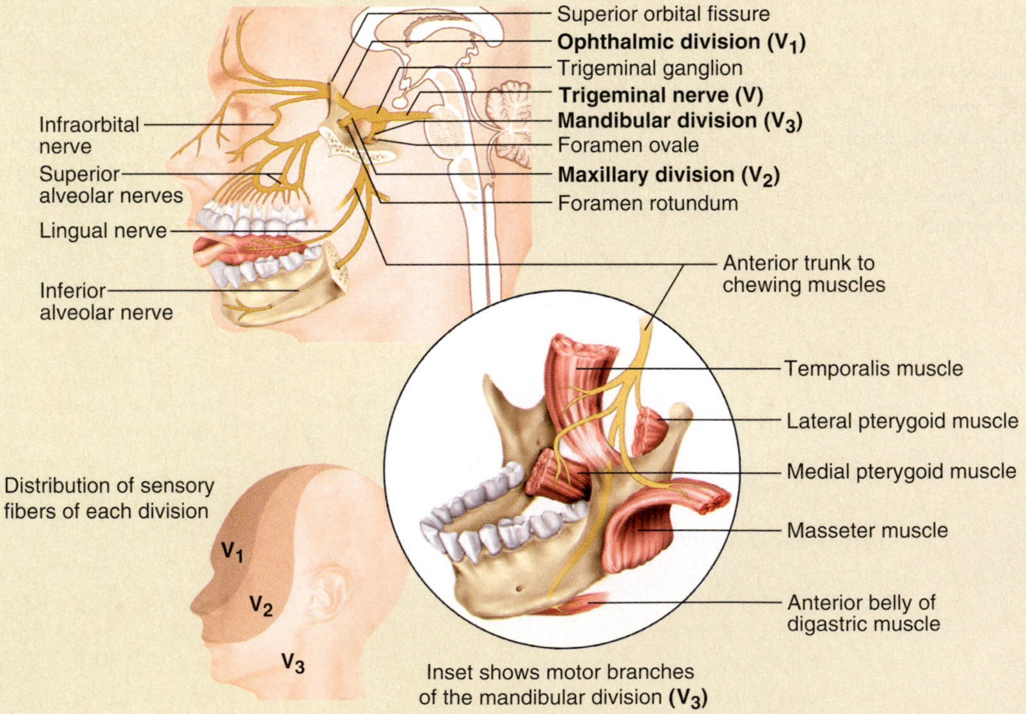
Superior orbital fissure
Ophthalmic division (V1)
Trigeminal ganglion
Trigeminal nerve (V)
Mandibular division (V3)
Foramen ovale
Maxillary division (V2)
Foramen rotundum
Infraorbital nerve
Superior alveolar nerves
Lingual nerve
Inferior alveolar nerve
Anterior trunk to chewing muscles
Temporalis muscle
Lateral pterygoid muscle
Medial pterygoid muscle
Masseter muscle
Anterior belly of digastric muscle
Distribution of sensory fibers of each division
V1
V2
V3
Inset shows motor branches of the mandibular division (V3)

Figure 15.28
The Trigeminal Nerve.

[45] *tri* = three + *gemin* = twin

(continued)

Table 15.3

The Cranial Nerves *(continued)*

VI. Abducens (ab-DOO-senz) Nerve

Composition: Predominantly motor
Function: Eye movements
Origin: Inferior pons
Termination: Lateral rectus muscle of eye
Cranial passage: Superior orbital fissure
Effects of damage: Inability to rotate eye laterally; at rest, eye rotates medially because of action of antagonistic muscles
Clinical test: Test lateral eye movements

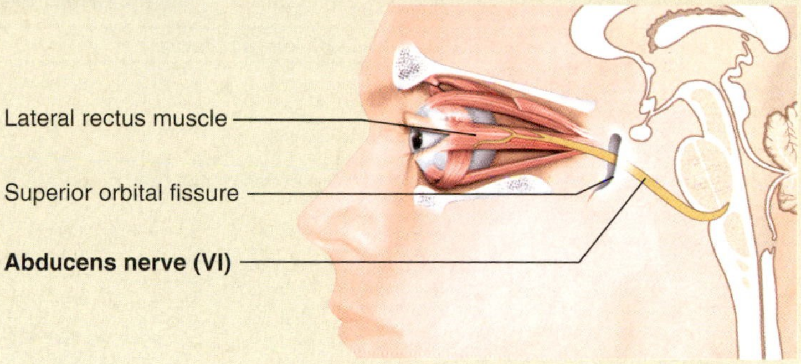

Lateral rectus muscle
Superior orbital fissure
Abducens nerve (VI)

Figure 15.29
The Abducens Nerve.

VII. Facial Nerve

Composition: Mixed
Function: Major motor nerve of facial expression; autonomic control of tear glands, nasal and palatine glands, submandibular and sublingual salivary glands; sense of taste
Sensory origin: Taste buds on anterior two-thirds of tongue
Sensory termination: Thalamus
Motor origin: Pons
Motor termination: Divides into *temporal, zygomatic, buccal, mandibular,* and *cervical branches.* Somatic motor fibers end on digastric muscle, stapedius muscle of middle ear, and muscles of facial expression; autonomic fibers end on submandibular and sublingual salivary glands
Cranial passage: Stylomastoid foramen
Effects of damage: Inability to control facial muscles; sagging resulting from loss of muscle tone; distorted sense of taste, especially for sweets
Clinical test: Test anterior two-thirds of tongue with substances such as sugar, salt, vinegar (sour), and quinine (bitter); test response of tear glands to ammonia fumes; test motor functions by asking subject to close eyes, smile, whistle, frown, raise eyebrows, etc.

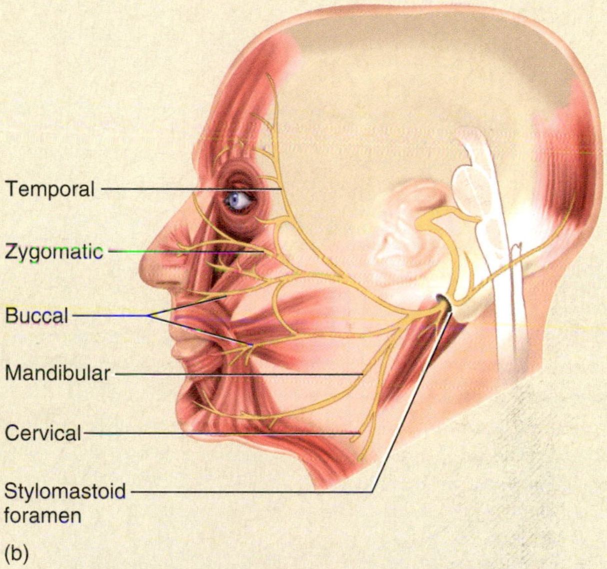

Temporal
Zygomatic
Buccal
Mandibular
Cervical
Stylomastoid foramen
(b)

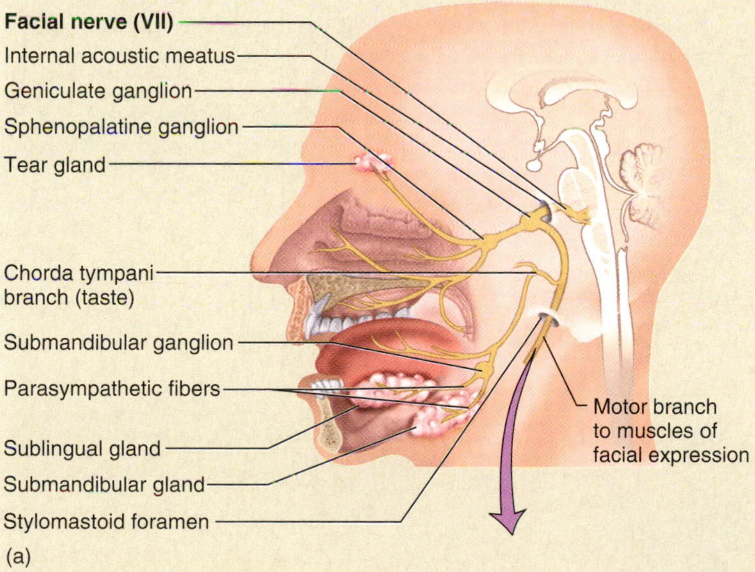

Facial nerve (VII)
Internal acoustic meatus
Geniculate ganglion
Sphenopalatine ganglion
Tear gland
Chorda tympani branch (taste)
Submandibular ganglion
Parasympathetic fibers
Sublingual gland
Submandibular gland
Stylomastoid foramen
Motor branch to muscles of facial expression
(a)

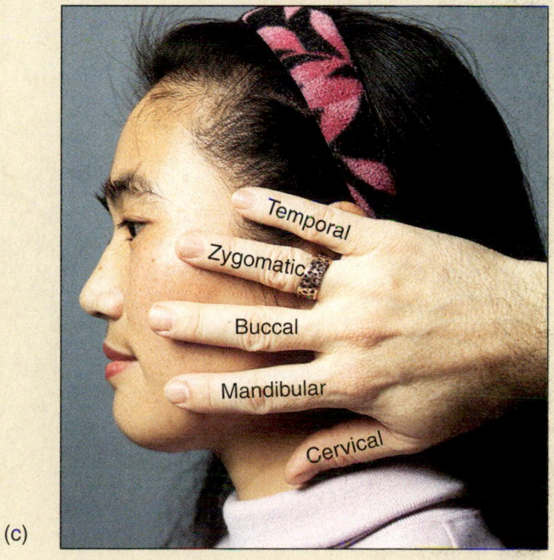

Temporal
Zygomatic
Buccal
Mandibular
Cervical

(c)

Figure 15.30
The Facial Nerve. (*a*) The facial nerve and associated organs. (*b*) The five major branches of the facial nerve. (*c*) A way to remember the distribution of the five major branches of the facial nerve.

(continued)

TABLE 15.3

The Cranial Nerves *(continued)*

VIII. Vestibulocochlear (vess-TIB-you-lo-COC-lee-ur) Nerve

Composition: Predominantly sensory

Function: Hearing and equilibrium

Sensory origin: Inner ear

Sensory termination: Fibers for equilibrium end at junction of pons and medulla; fibers for hearing end in medulla

Motor origin: Pons

Motor termination: Outer hair cells of cochlea of inner ear (see chapter 17)

Cranial passage: Internal acoustic meatus

Effects of damage: Nerve deafness, dizziness, nausea, loss of balance, and nystagmus (involuntary oscillation of the eyes from side to side)

Clinical test: Test hearing, balance, and ability to walk a straight line

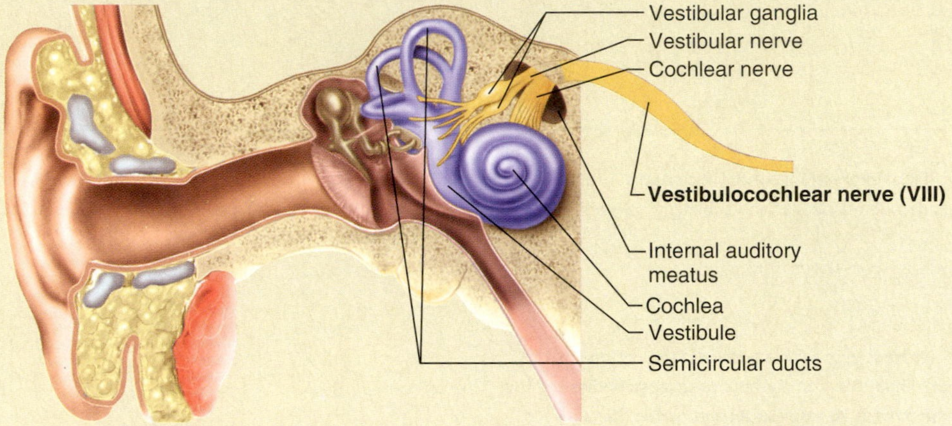

FIGURE 15.31
The Vestibulocochlear Nerve.

IX. Glossopharyngeal (GLOSS-oh-fah-RIN-jee-ul) Nerve

Composition: Mixed

Function: Swallowing, salivation, gagging; regulation of blood pressure and respiration; touch, pressure, taste, and pain sensations from tongue and pharynx; touch, pain, and temperature sensations from outer ear

Sensory origin: Pharynx, middle and outer ear, posterior one-third of tongue (including taste buds), internal carotid arteries

Sensory termination: Medulla oblongata

Motor origin: Medulla oblongata

Motor termination: Parotid salivary gland, glands of posterior tongue, stylopharyngeal muscle (which dilates the pharynx during swallowing)

Cranial passage: Jugular foramen

Effects of damage: Loss of bitter and sour taste; impaired swallowing

Clinical test: Test gag reflex, swallowing, and coughing; note speech impediments; test posterior one-third of tongue using bitter and sour substances

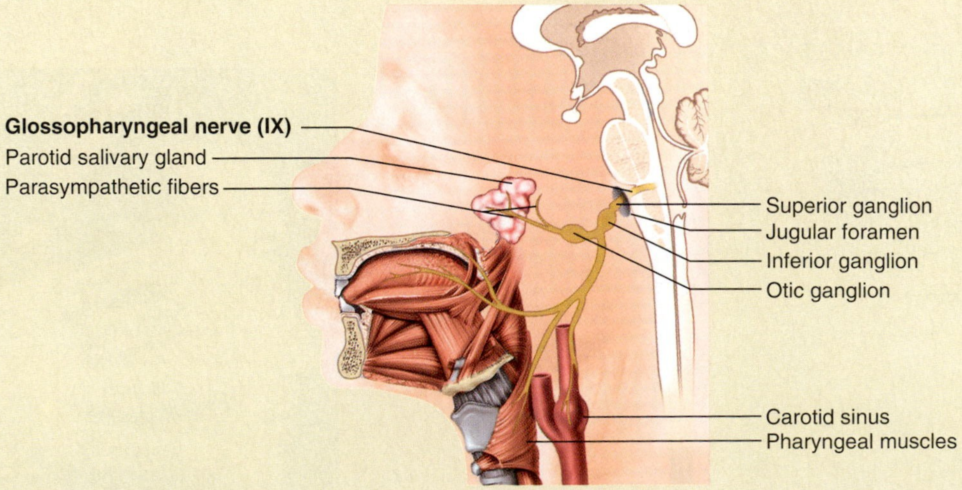

FIGURE 15.32
The Glossopharyngeal Nerve.

(continued)

TABLE 15.3

The Cranial Nerves *(continued)*

X. Vagus[46] (VAY-gus) Nerve

Composition: Mixed

Function: Swallowing; taste; speech; pulmonary, cardiovascular, and gastrointestinal regulation; sensations of hunger, fullness, and intestinal discomfort

Sensory origin: Thoracic and abdominal viscera, root of tongue, epiglottis, pharynx, larynx, outer ear, dura mater

Sensory termination: Medulla oblongata

Motor origin: Medulla oblongata

Motor termination: Tongue, palate, pharynx, larynx, thoracic and abdominal viscera

Cranial passage: Jugular foramen

Effects of damage: Hoarseness or loss of voice; impaired swallowing and gastrointestinal motility; fatal if both vagus nerves are damaged

Clinical test: Test with cranial nerve IX

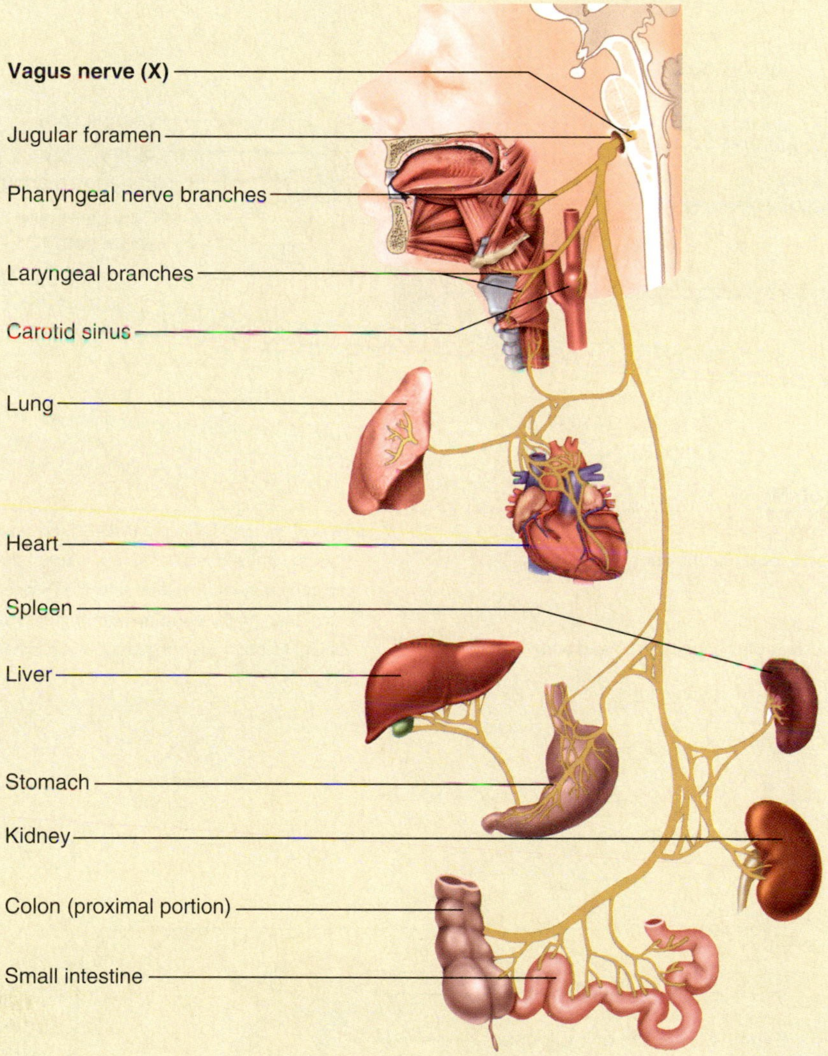
Vagus nerve (X)
Jugular foramen
Pharyngeal nerve branches
Laryngeal branches
Carotid sinus
Lung
Heart
Spleen
Liver
Stomach
Kidney
Colon (proximal portion)
Small intestine

FIGURE 15.33
The Vagus Nerve.

[46] *vag* = wandering

(continued)

Table 15.3

The Cranial Nerves *(continued)*

XI. Accessory (Spinal Accessory) Nerve

Composition: Predominantly motor

Function: Swallowing; head, neck, and shoulder movements

Origin: Medulla oblongata and segments C1 through C5 or C6 of spinal cord

Termination: Palate, pharynx, sternocleidomastoid and trapezius muscles

Cranial passage: Jugular foramen

Effects of damage: Impaired movement of head, neck, and shoulders; difficulty in shrugging shoulders on damaged side; paralysis of sternocleidomastoid, causing head to turn toward injured side

Clinical test: Test ability to rotate head and shrug shoulders against resistance

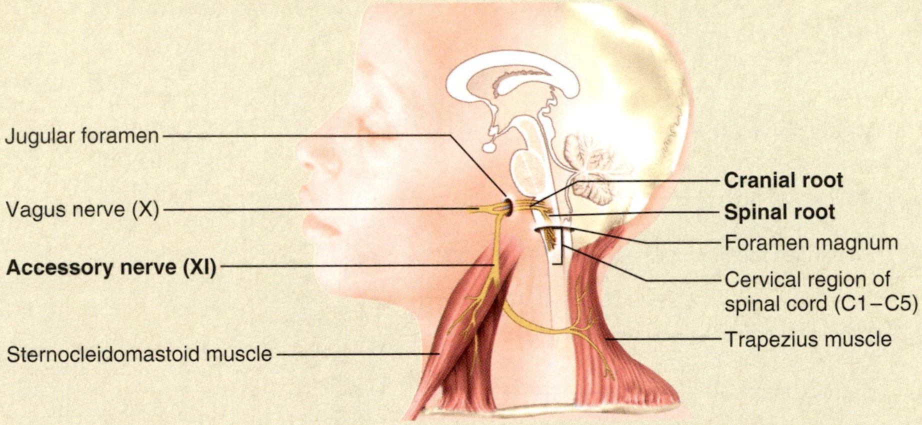

Figure 15.34
The Accessory Nerve.

XII. Hypoglossal (HY-po-GLOSS-ul) Nerve

Composition: Predominantly motor

Function: Tongue movements of speech, food manipulation, and swallowing

Origin: Medulla oblongata

Cranial passage: Hypoglossal canal

Termination: Intrinsic and extrinsic muscles of tongue, thyrohyoid and geniohyoid muscles

Effects of damage: Difficulty in speech and swallowing; inability to protrude tongue if both right and left nerves are injured; deviation toward injured side, and atrophy of tongue on that side, if only one nerve is damaged

Clinical test: Note deviations of tongue as subject protrudes and retracts it; test ability to protrude tongue against resistance

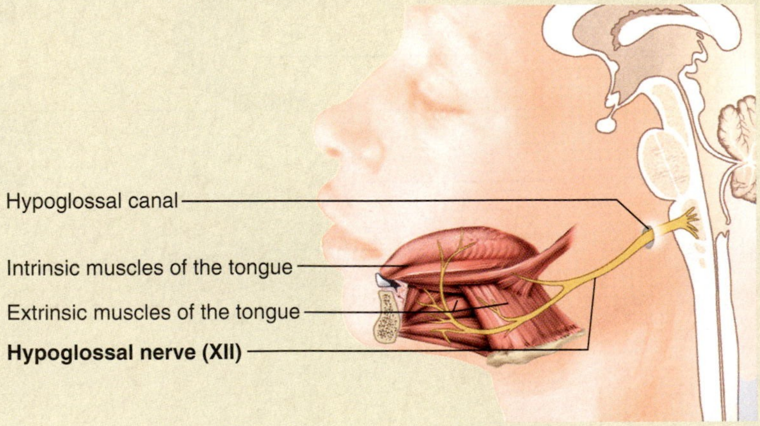

Figure 15.35
The Hypoglossal Nerve.

INSIGHT 15.3 CLINICAL APPLICATION

Some Cranial Nerve Disorders

Trigeminal neuralgia[47] (*tic douloureux*[48]) is a syndrome characterized by recurring episodes of intense stabbing pain in the trigeminal nerve. The cause is unknown; there is no visible change in the nerve. It usually occurs after the age of 50 and mostly in women. The pain lasts only a few seconds to a minute or two, but it strikes at unpredictable intervals and sometimes up to a hundred times a day. The pain usually occurs in a specific zone of the face, such as around the mouth and nose. It may be triggered by touch, drinking, tooth brushing, or washing the face. Analgesics (pain relievers) give only limited relief. Severe cases are treated by cutting the nerve, but this also deadens most other sensation in that side of the face.

Bell[49] *palsy* is a degenerative disorder of the facial nerve, probably due to a virus. It is characterized by paralysis of the facial muscles on one side with resulting distortion of the facial features, such as sagging of the mouth or lower eyelid. The paralysis may interfere with speech, prevent closure of the eye, and cause excessive tear secretion. There may also be a partial loss of the sense of taste. Bell palsy may appear abruptly, sometimes overnight, and often disappears spontaneously within 3 to 5 weeks.

[47] *neur* = nerve + *algia* = pain
[48] *douloureux* = painful
[49] Sir Charles Bell (1774–1842), Scottish physician

Before You Go On

Answer the following questions to test your understanding of the preceding section:

21. List the purely sensory cranial nerves by name and number, and state the function of each.
22. What is the only cranial nerve to extend beyond the head-neck region? In general terms, where does it lead?
23. If the oculomotor, trochlear, or abducens nerve were damaged the effect would be similar in all three cases. What would that effect be?
24. Which cranial nerve carries sensory signals from the largest area of the face?
25. Name two cranial nerves involved in the sense of taste and describe where their sensory fibers originate.

Developmental and Clinical Perspectives

Objectives

When you have completed this section, you should be able to

- describe some ways in which neuronal function and cerebral anatomy change in old age; and
- discuss Alzheimer and Parkinson diseases at the levels of neurotransmitter function and brain anatomy.

The Aging Central Nervous System

In chapter 13, we examined development of the nervous system at the beginning of life. As so many of us are regretfully aware, the nervous system also exhibits some marked changes at the other end of the life span. The nervous system reaches its peak development and efficiency around age 30. By age 75, the average brain weighs slightly less than half what it does at 30. The cerebral gyri are narrower, the sulci are wider, the cortex is thinner, and there is more space between the brain and meninges. The remaining neurons show signs that their metabolism is slowing down, such as less rough ER and Golgi complex. Old neurons accumulate lipofuscin pigment and begin to show *neurofibrillary tangles*—dense mats of cytoskeletal elements in their cytoplasm. In the extracellular material, *senile plaques* appear, especially in people with Down syndrome and Alzheimer disease. These are composed of cells and altered nerve fibers surrounding a core of *amyloid protein.*

Old neurons are also less efficient at signal conduction and transmission. The degeneration of myelin sheaths slows down conduction along the axon. The neurons have fewer synapses, and for multiple reasons, signals are not transmitted across the synapses as well as in the younger years: The neurons produce less neurotransmitter, they have fewer receptors, and the neuroglia around the synapses is more leaky and allows neurotransmitter to diffuse away. Target cells have fewer receptors for norepinephrine, and the sympathetic nervous system thus becomes less able to regulate such variables as body temperature and blood pressure.

Not all functions of the central nervous system are equally affected by aging. Language skills and long-term memory hold up better than motor coordination, intellectual function, and short-term memory. Elderly people are often better at remembering things in the distant past than remembering recent events.

Two Neurodegenerative Diseases

Like a machine with a great number of moving parts, the nervous system is highly subject to malfunctions. Neurological disorders fill many volumes of medical textbooks and can hardly be touched upon here. We have considered meningitis, one peculiar case of the effects of cerebral trauma, and two cranial nerve disorders in insights 15.1 through 15.3; several other neurological disorders are briefly described in table 15.4. We will close with a brief look at two of the most common brain dysfuntions, Alzheimer and Parkinson diseases. Both of these relate to neurotransmitter imbalances in the brain, and are considered to be *neurodegenerative diseases.* A basic understanding of these two diseases lends added clinical relevance to some areas of the brain studied in this chapter.

Alzheimer[50] **disease (AD)** affects about 11% of the U.S. population over the age of 65 and 47% by age 85. It accounts for nearly half of all nursing home admissions and is a leading cause of death among the elderly. AD may begin before the age of 50 with symptoms so slight and ambiguous that early diagnosis is difficult. One of its first symptoms is memory loss, especially for recent events. As the disease progresses, patients exhibit reduced attention span and may become disoriented and lost in previously familiar places. The AD patient may become moody, confused, paranoid, combative, or hallucinatory, and may eventually lose even the ability to read, write, talk, walk, and eat. Death ensues from pneumonia or other complications of confinement and immobility.

[50] Alois Alzheimer (1864–1915), German neurologist

Table 15.4

Some Disorders Associated with the Brain and Cranial Nerves

Cerebral Palsy	Muscular incoordination resulting from damage to the motor areas of the brain during fetal development, birth, or infancy; causes include prenatal rubella infection, drugs, or radiation exposure; oxygen deficiency during birth; and hydrocephalus
Concussion	Damage to the brain typically resulting from a blow, often with loss of consciousness, disturbances of vision or equilibrium, and short-term amnesia
Encephalitis	Inflammation of the brain, accompanied by fever, usually caused by mosquito-borne viruses or herpes simplex virus; causes neuronal degeneration and necrosis; can lead to delirium, seizures, and death
Epilepsy	Disorder causing sudden, massive discharge of neurons (seizures) resulting in motor convulsions, sensory and psychic disturbances, and often impaired consciousness; may result from birth trauma, tumors, infections, drug or alcohol abuse, or congenital brain malformation
Migraine Headache	Recurring headaches often accompanied by nausea, vomiting, dizziness, and aversion to light, often triggered by such factors as weather changes, stress, hunger, red wine, or noise; more common in women and sometimes running in families
Schizophrenia	A thought disorder involving delusions, hallucinations, inappropriate emotional responses to situations, incoherent speech, and withdrawal from society, resulting from hereditary or developmental abnormalities in neuronal networks

Disorders Described Elsewhere

Alzheimer disease 449	Cerebellar ataxia 425	Parkinson disease 450
Aphasia 436	Cranial nerve injuries 442–448	Poliomyelitis 393
Aprosodia 436	Hydrocephalus 124	Tay-Sachs disease 373
Bell palsy 449	Meningitis 417	Trigeminal neuralgia 449
Brain tumors 372	Multiple sclerosis 373	

Diagnosis of AD is confirmed on autopsy. There is atrophy of some of the gyri of the cerebral cortex and hippocampus. Neurofibrillary tangles and senile plaques are abundant (fig. 15.36). Cholinergic neurons are in reduced supply and the level of acetylcholine in affected areas of the brain is consequently low. Intense research efforts are currently geared toward identifying the cause of AD and developing treatment strategies. Researchers have identified three genes on chromosomes 1, 14, and 21 for various forms of early- and late-onset AD.

Parkinson[51] disease (PD), also called *paralysis agitans* or *parkinsonism,* is a progressive loss of motor function beginning in a person's 50s or 60s. It is due to degeneration of dopamine-releasing neurons in the substantia nigra. A gene has recently been identified for a hereditary form of PD, but most cases are nonhereditary and of little-known cause. Dopamine is an inhibitory neurotransmitter that normally prevents excessive activity in the basal nuclei. Degeneration of the dopamine-releasing neurons leads to hyperactivity of the basal nuclei, and therefore involuntary muscle contractions. These take such forms as shaking of the hands (tremor) and compulsive "pill-rolling" motions of the thumb and fingers. In addition, the facial muscles may become rigid and produce a staring, expressionless face with a slightly open mouth. The patient's range of motion diminishes. He or she takes smaller steps and develops a slow, shuffling gait with a forward-bent posture and a tendency to fall forward. Speech becomes slurred and handwriting becomes cramped and eventually illegible. Tasks such as buttoning clothes and preparing food become increasingly laborious.

Patients cannot be expected to recover from PD, but drugs and physical therapy can lessen the severity of its effects. A surgical

[51]James Parkinson (1755–1824), British physician

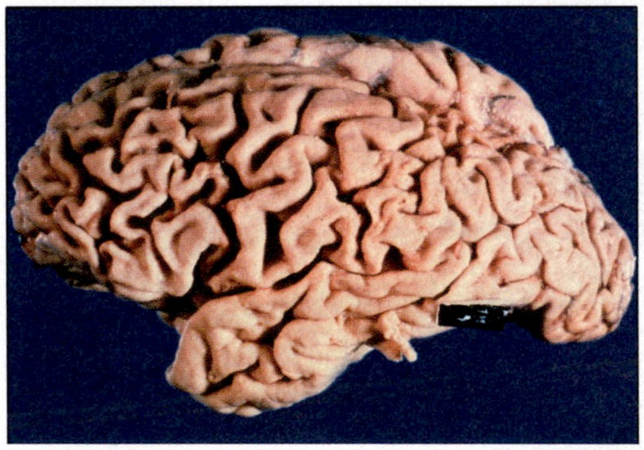

(a)

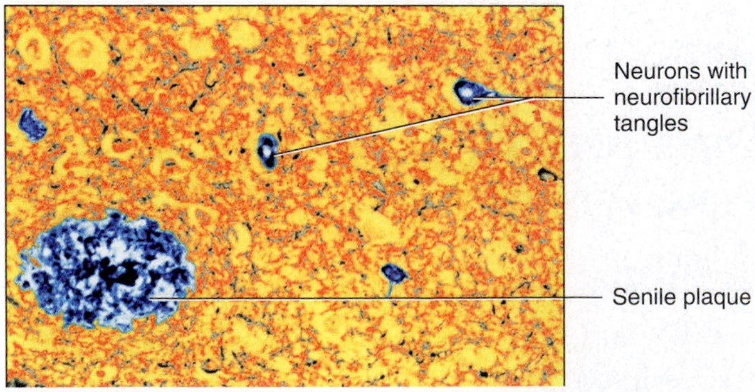
Neurons with neurofibrillary tangles
Senile plaque

(b)

Figure 15.36

Alzheimer Disease. (a) Brain of a person who died of AD. Note the shrunken gyri and wide sulci. (b) Cerebral tissue from a person with AD. Neurofibrillary tangles are present in the neurons, and a senile plaque is evident in the extracellular matrix.

technique called *pallidotomy* has been used since the 1940s to alleviate severe tremors. It involves the destruction of a small portion of the globus pallidus, one of the basal nuclei. Pallidotomy fell out of favor in the late 1960s when the drug L-DOPA came into common use. By the early 1990s, however, the limitations of L-DOPA had become apparent, while MRI- and CT-guided methods had improved surgical precision and reduced the risks of pallidotomy. The procedure has thus made a comeback. Other surgical treatments for parkinsonism target certain nuclei in and near the thalamus, relieving symptoms by ablating (destroying) small areas of tissue or implanting stimulating electrodes.

Before You Go On

Answer the following questions to test your understanding of the preceding section:

26. Describe two respects in which neurons function less efficiently in old age.
27. Describe some changes seen in the brain with aging.
28. Describe the neuroanatomical and behavioral changes seen in Alzheimer and Parkinson diseases.

CHAPTER REVIEW

Review of Key Concepts

Overview of the Brain (p. 414)

1. The adult brain weighs 1,450 to 1,600 g. It is divided into the *cerebrum, cerebellum,* and *brainstem.*
2. The brain is composed of two kinds of nervous tissue, *gray matter* and *white matter.* Gray matter constitutes the surface *cortex* and deeper *nuclei* of the cerebrum and cerebellum, and nuclei of the brainstem. White matter lies deep to the cortex and consists of *tracts* of myelinated nerve fibers.
3. The brain is surrounded by dura mater, arachnoid mater, and pia mater. The dura mater is divided into two layers, *periosteal* and *meningeal,* which in some places are separated by a blood-filled *dural sinus.* In some places, the dura folds inward to separate major brain regions. A *subdural space* separates some areas of dura from the arachnoid, and a *subarachnoid space* separates arachnoid from pia.
4. The brain has four internal, interconnected chambers: two *lateral ventricles* in the cerebral hemispheres, a *third ventricle* between the hemispheres, and a *fourth ventricle* between the pons and cerebellum.
5. The ventricles and canals of the CNS are lined with ependymal cells, and each ventricle contains a *choroid plexus* of blood capillaries.
6. These spaces are filled with cerebrospinal fluid (CSF), which is produced by the ependyma and choroid plexuses and in the subarachnoid space around the brain. The CSF of the ventricles flows from the lateral to the third and then fourth ventricle, out through foramina in the fourth, into the subarachnoid space around the brain and spinal cord, and finally returns to the blood by way of arachnoid villi.
7. CSF provides buoyancy, physical protection, and chemical stability for the CNS.
8. The brain has a high demand for glucose and oxygen and thus receives a copious blood supply.
9. The *blood-brain barrier* and *blood-CSF barrier* tightly regulate what substances can escape the blood and reach the nervous tissue.

The Hindbrain and Midbrain (p. 421)

1. The *medulla oblongata* is the most caudal part of the brain, just inside the foramen magnum. It conducts signals up and down the brainstem and between the brainstem and cerebellum, and contains nuclei involved in numerous visceral functions and some muscular control (table 15.1). Cranial nerves IX through XII arise from the medulla. Most descending motor fibers decussate in the medullary pyramids.
2. The *pons* is immediately rostral to the medulla. It conducts signals up and down the brainstem and between the brainstem and cerebellum. Cranial nerve V arises from the pons, and nerves VI through VIII arise between the pons and medulla. Physiological functions of the pontine nuclei are listed in table 15.1.
3. The *cerebellum* is the largest part of the hindbrain and receives most of its input by way of the pons. It has three pairs of *cerebellar peduncles* that attach it to the medulla, pons, and midbrain.
4. Histologically, the cerebellum exhibits a fernlike pattern of white matter called the *arbor vitae, deep nuclei* of gray matter embedded in the white matter, and unusually large neurons called *Purkinje cells.*
5. The cerebellum is concerned mainly with motor coordination, equilibrium, and memory of motor skills, but it also has emotional and cognitive functions (table 15.1).
6. The *midbrain* is rostral to the pons. It conducts signals up and down the brainstem and between the brainstem and cerebellum, and gives rise to cranial nerves III and IV. It includes important centers for vision, hearing, pain, and motor control (table 15.1).
7. The *reticular formation* is a loosely organized network of gray matter in the core of the brainstem, including over 100 small neural networks. Because of the number and variety of nuclei, it has wide-ranging functions in motor control, visceral function, pain modulation, sleep, and consciousness (table 15.1).

The Forebrain (p. 427)

1. The forebrain consists of the diencephalon and cerebrum.
2. The *diencephalon* is composed of the *thalamus, hypothalamus,* and *epithalamus.*
3. All sensory signals pass through the *thalamus,* which processes them and relays coded signals to the appropriate regions of the cerebral cortex; it is the "gateway to the cerebral cortex." It also relays signals from the cerebral cortex to other regions of the brain. Figure 15.12 summarizes the functions of its major nuclei.

4. The *hypothalamus* is inferior to the thalamus and forms the walls and floor of the third ventricle. It is a major homeostatic control center, and acts through the pituitary gland and autonomic nervous system to regulate many fundamental visceral functions (table 15.2). It is also involved in emotion and memory.
5. The *epithalamus* lies above the thalamus and includes the *pineal gland* (an endocrine gland) and *habenula* (a relay from limbic system to midbrain).
6. The cerebrum is the largest part of the brain. It is divided into two hemispheres separated by the longitudinal fissure. The hemispheres are prominently marked with gyri and sulci. The two hemispheres are connected chiefly through a large fiber tract, the *corpus callosum.*
7. Each hemisphere has five lobes: *frontal, parietal, occipital,* and *temporal lobes* and the *insula.* Their respective functions are summarized in table 15.2.
8. Nerve fibers of the cerebral white matter are bundled in tracts of three kinds: *projection tracts* that extend between higher and lower brain centers, *commissural tracts* that cross between the right and left cerebral hemispheres through the corpus callosum and the *anterior* and *posterior commissures;* and *association tracts* that connect different lobes and gyri within a single hemisphere.
9. The cerebral cortex is gray matter with two types of neurons: *stellate cells* and *pyramidal cells.* All output from the cortex travels by way of axons of the pyramidal cells. Most of the cortex is *neocortex,* in which there are six layers of nervous tissue. Evolutionarily older parts of the cerebrum have one- to five-layered *paleocortex* and *archicortex.*
10. The *basal nuclei* are masses of cerebral gray matter lateral to the thalamus, concerned with motor control. They include the *caudate nucleus, putamen,* and *globus pallidus.*
11. The *limbic system* is a loop of specialized structures on the medial border of each cerebral hemisphere. Its major components include the *cingulate gyrus, hippocampus,* and *amygdala.* Parts of the hypothalamus, thalamus, basal nuclei, and prefrontal cortex are also often regarded as belonging to the limbic system. Major functions of the limbic system include memory and emotion.
12. The cerebral cortex includes areas of *primary cortex* that either directly receive sensory input or provide the cerebral output to the muscular system, and much more extensive *association areas* that integrate sensory information, plan motor outputs, and are the seat of memory and other cognitive processes.
13. The *special senses* originate in relatively complex sense organs of the head and involve distinct regions of primary sensory cortex and association areas. Vision resides in the occipital lobe and inferior temporal lobe; hearing in the superior temporal lobe; equilibrium in the cerebellum and brainstem, but with centers of consciousness of body movements and position low in the parietal lobe; taste in the parietal lobe and insula; smell in the frontal and temporal lobes; and there is an association area in the frontal lobe for taste and smell combined.
14. The primary cortex for *somesthetic sensation* is in the postcentral gyrus of the parietal lobe, where there is a point-for-point correspondence *(somatotopy)* with specific regions on the contralateral side of the body. The somesthetic association area is a large region of parietal lobe caudal to this gyrus.
15. Motor control resides in the *motor association area* and *precentral gyrus* of the frontal lobe. The precentral gyrus shows a somatotopic correspondence with muscles on the contralateral side of the body. It contains the *upper motor neurons* whose axons project to *lower motor neurons* in the brainstem and spinal cord.
16. The basal nuclei and cerebellum play important roles in motor coordination and learned motor skills (procedural memory).
17. Language is coordinated largely by the Wernicke and Broca areas. Recognizing language and formulating what one will say or write occur in the Wernicke area; the Broca area compiles the motor program of speech; and commands to the muscles of speech originate in the precentral gyrus.
18. Emotional responses are controlled by the hypothalamus, amygdala, and prefrontal cortex.
19. Cognitive functions are widely distributed through the association cortex of all the cerebral lobes. The prefrontal cortex is the seat of many of our most distinctively human cognitive abilities such as social judgment and abstract thought.
20. The limbic system, especially the hippocampus, is an important site for the creation of new memories although not for memory storage; it essentially "teaches the cerebral cortex," which stores memories for the long term. The amygdala is important in creating emotional memories, such as associating fear with dangerous situations.
21. The brain exhibits *cerebral lateralization:* Some functions are coordinated mainly by the left hemisphere and others by the right. The *categorical hemisphere* (in most people, the left) is responsible for verbal and mathematical skills and logical, linear thinking. The *representational hemisphere* (usually the right) is a seat of imagination, insight, spatial perception, musical skill, and other "holistic" functions.
22. The forebrain regions and their functions are summarized in table 15.2.

The Cranial Nerves (p. 440)

1. Twelve pairs of *cranial nerves* arise from the floor of the brain, pass through foramina of the skull, and lead primarily to structures in the head and neck.
2. Cranial nerves (CN) I and II are purely sensory. All the rest are mixed, although the sensory components of some are only proprioceptive and aid in motor control, so they are often regarded as motor nerves (CN III, IV, VI, XI, and XII).
3. The functions and other characteristics of the cranial nerves are detailed in table 15.3.

Developmental and Clinical Perspectives (p. 449)

1. The brain exhibits substantial atrophy in old age, and neurons exhibit less efficient signal conduction and synaptic transmission.
2. Alzheimer disease (AD) is the most common neurodegenerative disease of old age, and a major cause of death in the elderly. It involves memory deficits, personality derangement, and a loss of motor and cognitive skills. At the structural level, AD shows neurofibrillary tangles and senile plaques in the cerebral tissue, a loss of cholinergic neurons, and a low level of acetylcholine in affected areas of the brain.
3. Parkinson disease (PD) results from degeneration of dopamine-releasing neurons of the midbrain substantia nigra, and is characterized by involuntary muscle contractions and progressive difficulty in motor tasks such as walking and speech. It is incurable, but medical and surgical treatments can reduce its severity and progression.
4. Several other neurological disorders are described in table 15.4.

Testing Your Recall

1. Which of these is caudal to the hypothalamus?
   a. the thalamus
   b. the optic chiasm
   c. the cerebral aqueduct
   d. the pituitary gland
   e. the corpus callosum
2. Hearing is associated mainly with
   a. the limbic system.
   b. the prefrontal cortex.
   c. the occipital lobe.
   d. the temporal lobe.
   e. the parietal lobe.
3. The blood-CSF barrier is formed by
   a. blood capillaries.
   b. endothelial cells.
   c. protoplasmic astrocytes.
   d. oligodendrocytes.
   e. ependymal cells.
4. The pyramids of the medulla oblongata contain
   a. descending corticospinal fibers.
   b. commissural fibers.
   c. ascending spinocerebellar fibers.
   d. fibers going to and from the cerebellum.
   e. ascending spinothalamic fibers.
5. Which of the following is *not* involved in vision?
   a. the temporal lobe
   b. the occipital lobe
   c. the midbrain tectum
   d. the trochlear nerve
   e. the vagus nerve
6. While studying in a noisy cafeteria, you get sleepy and doze off for a few minutes. You awaken with a start and realize that all the cafeteria sounds have just "come back." While you were dozing, this auditory input was blocked from reaching your auditory cortex by
   a. the temporal lobe.
   b. the thalamus.
   c. the reticular activating system.
   d. the medulla oblongata.
   e. the vestibulocochlear nerve.
7. Because of a brain lesion, a certain patient never feels full, but eats so excessively that she now weighs nearly 600 pounds. The lesion is most likely in her
   a. hypothalamus.
   b. amygdala.
   c. hippocampus.
   d. basal nuclei.
   e. pons.
8. The _____ is most closely associated with the cerebellum in embryonic development and remains its primary source of input fibers throughout life.
   a. telencephalon
   b. thalamus
   c. midbrain
   d. pons
   e. medulla
9. Damage to the _____ nerve could result in defects of eye movement.
   a. optic
   b. vagus
   c. trigeminal
   d. facial
   e. abducens
10. All of the following *except* the _____ nerve begin or end in the orbit.
    a. optic
    b. oculomotor
    c. trochlear
    d. abducens
    e. accessory
11. The right and left cerebral hemispheres are connected to each other by a thick C-shaped bundle of fibers called the _____.
12. The brain has four chambers called _____ filled with _____ fluid.
13. In a medial section, the cerebellar white matter exhibits a branching pattern called the _____.
14. Part of the limbic system involved in forming new memories is the _____.
15. Cerebrospinal fluid is secreted partly by a mass of blood capillaries called the _____ in each ventricle.
16. The primary motor area of the cerebrum is the _____ gyrus of the frontal lobe.
17. Your personality is determined mainly by which lobe of the cerebrum?
18. Areas of cerebral cortex that identify or interpret sensory information are called _____.
19. Linear, analytical, and verbal thinking occurs in the _____ hemisphere of the cerebrum, which is on the left in most people.
20. The motor pattern for speech is generated in an area of cortex called the _____ and then transmitted to the primary motor cortex to be carried out.

Answers in the Appendix

True or False

Determine which five of the following statements are false, and briefly explain why.

1. The two hemispheres of the cerebellum are separated by the longitudinal fissure.
2. Degeneration of the substantia nigra causes Alzheimer disease.
3. The midbrain is caudal to the thalamus.
4. The Broca area is ipsilateral to the Wernicke area.
5. Most of the cerebrospinal fluid is produced by the choroid plexuses.
6. Hearing is a function of the occipital lobe.
7. Respiration is controlled by nuclei in both the pons and medulla oblongata.
8. The trigeminal nerve carries sensory signals from a larger area of the face than the facial nerve does.
9. Unlike other cranial nerves, the vagus nerve extends far beyond the head-neck region.
10. The optic nerve controls movements of the eye.

Answers in the Appendix

Testing Your Comprehension

1. Which cranial nerve conveys pain signals to the brain in each of the following situations: (*a*) sand blows into your eye; (*b*) you bite the rear of your tongue; and (*c*) your stomach hurts from eating too much?
2. How would a lesion in the cerebellum and a lesion in the basal nuclei differ in their effects on skeletal muscle function?
3. Suppose that a neuroanatomist performed two experiments on an animal with the same basic spinal and brainstem structure as a human's: In experiment 1, he selectively transected (cut across) the pyramids on the ventral side of the medulla oblongata, and in experiment 2, he selectively transected the gracile and cuneate fasciculi on the dorsal side. How would the outcomes of the two experiments differ?
4. A person can survive destruction of an entire cerebral hemisphere but cannot survive destruction of the hypothalamus, which is a much smaller mass of brain tissue. Explain this difference and describe some ways that destruction of a cerebral hemisphere would affect one's quality of life.
5. What would be the most obvious effects of lesions that destroyed each of the following: (*a*) the hippocampus, (*b*) the amygdala, (*c*) the Broca area, (*d*) the occipital lobe, and (*e*) the hypoglossal nerve?

Answers at the Online Learning Center

www.mhhe.com/saladinha1

Visit the Online Learning Center for practice tests, answer keys, and other learning aids for this chapter. Enhance your understanding of human anatomy with our interactive art labeling exercises, supplemental photo atlases, web links, puzzles, flashcards, and much more.

16

CHAPTER SIXTEEN

The Autonomic Nervous System and Visceral Reflexes

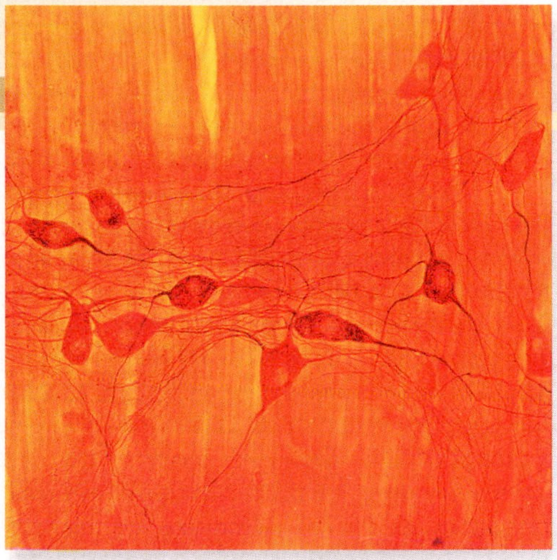

Autonomic neurons in the myenteric plexus of the digestive tract.

CHAPTER OUTLINE

General Properties of the Autonomic Nervous System 456
- General Actions 456
- Visceral Reflexes 456
- Divisions of the Autonomic Nervous System 457
- Neural Pathways 457

Anatomy of the Autonomic Nervous System 458
- The Sympathetic Division 458
- The Adrenal Glands 463
- The Parasympathetic Division 463
- The Enteric Nervous System 465

Autonomic Effects 465
- Neurotransmitters and Receptors 465
- Dual Innervation 468
- Central Control of Autonomic Function 468

Developmental and Clinical Perspectives 469
- Development and Aging of the Autonomic Nervous System 469
- Disorders of the Autonomic Nervous System 469

Chapter Review 471

INSIGHTS

16.1 Clinical Application: Biofeedback 456

16.2 Clinical Application: Drugs and the Autonomic Nervous System 466

BRUSHING UP

To understand this chapter, you may find it helpful to review the following concepts:

- Innervation of smooth muscle (p. 274)
- Neurotransmitters and receptors (p. 376)
- General anatomy of nerves and ganglia (p. 393)
- Branches of the spinal nerves (pp. 395–396)
- The limbic system and hypothalamus (pp. 428, 432)
- Cranial nerves, especially III, VII, IX, and X (pp. 443–447)

We have studied the somatic nervous system and somatic reflexes, and we now turn to the *autonomic nervous system (ANS)* and *visceral reflexes*—reflexes that regulate such primitive functions as blood pressure, heart rate, body temperature, digestion, energy metabolism, respiratory airflow, pupillary diameter, defecation, and urination. In short, the ANS quietly manages a multitude of unconscious processes responsible for the body's homeostasis.

Walter Cannon, who coined such expressions as *homeostasis* and the *fight or flight* reaction, dedicated his career to the physiology of the autonomic nervous system. Cannon found that an animal can live without a functional sympathetic nervous system, but it must be kept warm and free of stress; it cannot survive on its own or tolerate any strenuous exertion. The autonomic nervous system is more necessary to survival than many functions of the somatic nervous system; an absence of autonomic function is fatal because the body cannot maintain homeostasis. We are seldom aware of what our autonomic nervous system is doing, much less able to control it; indeed, it is difficult to consciously alter or suppress autonomic responses, and for this reason they are the basis for polygraph ("lie detector") tests. Nevertheless, for an understanding of bodily function and health care, we must be well aware of how this system works.

INSIGHT 16.1 CLINICAL APPLICATION

BIOFEEDBACK

Biofeedback is a technique in which an instrument produces auditory or visual signals in response to changes in a subject's blood pressure, heart rate, muscle tone, skin temperature, brain waves, or other physiological variables. It gives the subject awareness of changes that he or she would not ordinarily notice. Some people can be trained to control these variables in order to produce a certain tone or color of light from the apparatus. Eventually they can control them without the aid of the monitor. Biofeedback is not a quick, easy, infallible, or inexpensive cure for all ills, but it has been used successfully to treat hypertension, stress, and migraine headaches.

General Properties of the Autonomic Nervous System

Objectives

When you have completed this section, you should be able to

- explain how the autonomic and somatic nervous systems differ in form and function; and
- explain how the two divisions of the autonomic nervous system differ in general function.

General Actions

The **autonomic nervous system (ANS)** can be defined as a motor nervous system that controls glands, cardiac muscle, and smooth muscle. It is also called the **visceral motor system** to distinguish it from the somatic motor system, which controls the skeletal muscles. The primary target organs of the ANS are the viscera of the thoracic and abdominal cavities and some structures of the body wall, including cutaneous blood vessels, sweat glands, and piloerector muscles.

Autonomic literally means "self-governed."[1] The ANS usually carries out its actions involuntarily, without our conscious intent or awareness, in contrast to the voluntary nature of the somatic motor system. This voluntary-involuntary distinction is not, however, as clear-cut as it once seemed. Some skeletal muscle responses are quite involuntary, such as the somatic reflexes, and some skeletal muscles are difficult or impossible to control, such as the middle-ear muscles. On the other hand, therapeutic uses of biofeedback (see insight 16.1) show that some people can learn to voluntarily control such visceral functions as blood pressure.

Visceral effectors do not depend on the autonomic nervous system to function, but only to adjust (modulate) their activity to the body's changing needs. The heart, for example, goes on beating even if all autonomic nerves to it are severed, but the ANS modulates the heart rate in conditions of rest or exercise. If the somatic nerves to a skeletal muscle are severed, the muscle exhibits flaccid paralysis—it no longer contracts at all. But if the autonomic nerves to cardiac or smooth muscle are severed, the muscle exhibits exaggerated responses (*denervation hypersensitivity*).

Visceral Reflexes

The ANS is responsible for the body's **visceral reflexes**—unconscious, automatic, stereotyped responses to stimulation, much like the somatic reflexes discussed in chapter 14, but involving visceral receptors and effectors and somewhat slower responses. Some authorities regard the visceral afferent (sensory) pathways as part of the ANS, while most prefer to limit the term *ANS* to the efferent (motor) pathways. Regardless of this preference, however, autonomic activity involves a visceral reflex arc that includes receptors (nerve endings that detect stretch, tissue damage, blood chemicals, body temperature, and other internal stimuli), afferent neurons leading to the CNS, interneurons in the CNS, efferent neurons carrying motor signals away from the CNS, and finally effectors.

For example, high blood pressure activates a visceral *baroreflex*.[2] It stimulates stretch receptors called *baroreceptors* in the carotid arteries and aorta, and they transmit signals via the glossopharyngeal nerves to the medulla oblongata (fig. 16.1). The medulla integrates this input with other information and transmits efferent signals back to the heart by way of the vagus nerves. The vagus nerves slow down the heart and reduce blood pressure, thus completing a homeostatic negative feedback loop. A separate baroreflex arc accelerates the heart when blood pressure above the heart drops, as when we change from lying down to standing up and gravity draws blood away from the upper body.

[1] *auto* = self + *nom* = rule

[2] *baro* = pressure

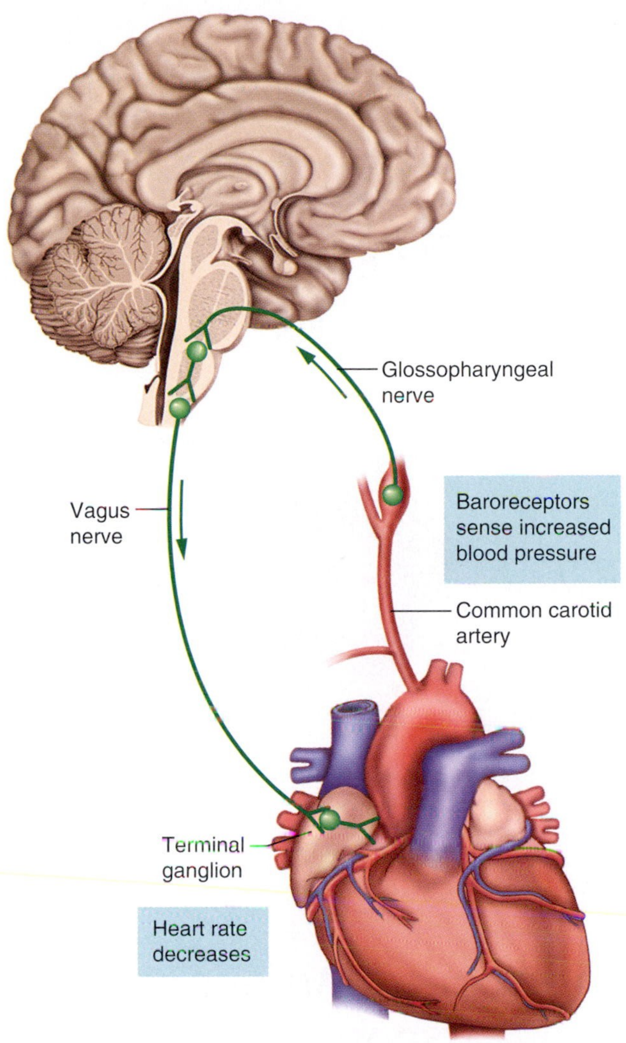
Glossopharyngeal nerve
Vagus nerve
Baroreceptors sense increased blood pressure
Common carotid artery
Terminal ganglion
Heart rate decreases

FIGURE 16.1

An Autonomic Reflex Arc in the Regulation of Blood Pressure. A rise in blood pressure is detected by baroreceptors in the carotid artery. The glossopharyngeal nerve transmits signals to the medulla oblongata, resulting in parasympathetic output from the vagus nerve that reduces the heart rate and lowers blood pressure.

Divisions of the Autonomic Nervous System

The ANS has two subsystems, the sympathetic and parasympathetic divisions. These divisions differ in anatomy and function, but they often innervate the same target organs and may have cooperative or contrasting effects on them. The **sympathetic division** adapts the body in many ways for physical activity—it increases alertness, heart rate, blood pressure, pulmonary airflow, blood glucose concentration, and blood flow to cardiac and skeletal muscle, but at the same time, it reduces blood flow to the skin and digestive tract. Cannon referred to extreme sympathetic responses as the "fight or flight" reaction because they come into play when an animal must attack, defend itself, or flee from danger. In our own lives, this reaction occurs in many situations involving arousal, competition, stress, danger, anger, or fear. Ordinarily, however, the sympathetic division has more subtle effects that we notice barely, if at all. The **parasympathetic division,** by comparison, has a calming effect on many body functions. It is associated with reduced energy expenditure and normal bodily maintenance, including such functions as digestion and waste elimination. This can be thought of as the "resting and digesting" state.

This does not mean that the body alternates between states where one system or the other is active. Normally both systems are active simultaneously. They exhibit a background rate of activity called **autonomic tone,** and the balance between *sympathetic tone* and *parasympathetic tone* shifts in accordance with the body's changing needs. Parasympathetic tone, for example, maintains smooth muscle tone in the intestines and holds the resting heart rate down to about 70 to 80 beats/minute. If the parasympathetic vagus nerves to the heart are cut, the heart beats at its own intrinsic rate of about 100 beats/min. Sympathetic tone keeps most blood vessels partially constricted and thus maintains blood pressure. A loss of sympathetic tone can cause such a rapid drop in blood pressure that a person goes into shock.

Neither division has universally excitatory or calming effects. The sympathetic division, for example, excites the heart but inhibits digestive and urinary functions, while the parasympathetic division has the opposite effects.

Neural Pathways

The ANS has components in both the central and peripheral nervous systems. It includes control nuclei in the hypothalamus and other regions of the brainstem, motor neurons in the spinal cord and peripheral ganglia, and nerve fibers that travel through the cranial and spinal nerves described in chapters 14 and 15.

The autonomic motor pathway to a target organ differs significantly from somatic motor pathways. In somatic pathways, a motor neuron in the brainstem or spinal cord issues a myelinated axon that reaches all the way to a skeletal muscle. In autonomic pathways, the signal must travel across two neurons to get to the target organ, and it must cross a synapse where these two neurons meet in an autonomic ganglion (fig. 16.2). The first neuron, called the **preganglionic neuron,** has a soma in the brainstem or spinal cord. Its axon terminates in a ganglion, where it synapses with a **postganglionic neuron** whose axon extends the rest of the way to the target cells. (Some call this cell the *ganglionic neuron* since its soma is in the ganglion and only its axon is truly postganglionic.) The axons of these neurons are called the *pre-* and *postganglionic fibers.*

Differences between the somatic and autonomic nervous systems are summarized in table 16.1.

Before You Go On

Answer the following questions to test your understanding of the preceding section:

1. How does the autonomic nervous system differ from the somatic motor system?
2. How does a visceral reflex resemble a somatic reflex? How does it differ?

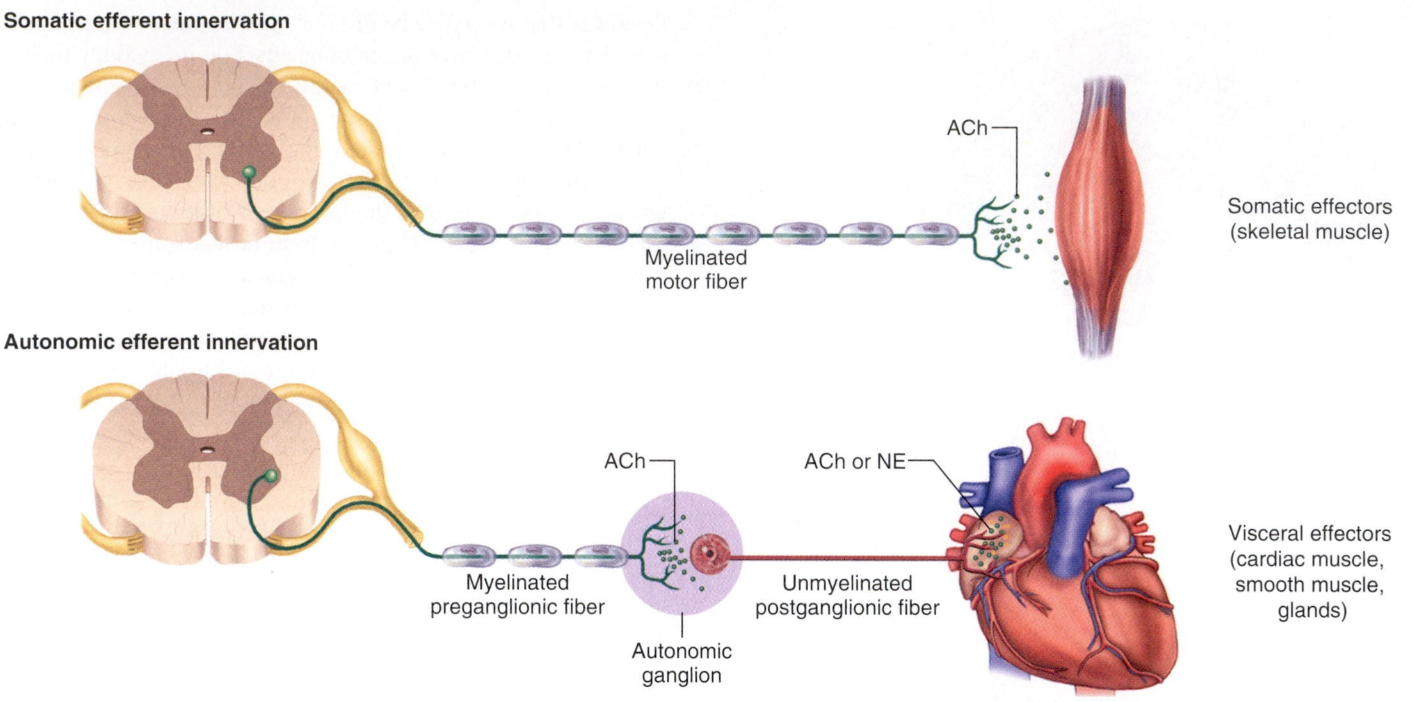
Somatic efferent innervation
ACh
Somatic effectors (skeletal muscle)
Myelinated motor fiber
Autonomic efferent innervation
ACh
ACh or NE
Visceral effectors (cardiac muscle, smooth muscle, glands)
Myelinated preganglionic fiber
Unmyelinated postganglionic fiber
Autonomic ganglion

FIGURE 16.2

Comparison of Somatic and Autonomic Efferent Pathways. The entire distance from CNS to effector is spanned by one neuron in the somatic system and two neurons in the autonomic system. Only acetylcholine (ACh) is employed as a neurotransmitter by the somatic neuron and the autonomic preganglionic neuron, but autonomic postganglionic neurons can employ either ACh or norepinephrine (NE).

3. What are the two divisions of the ANS? How do they functionally differ from each other?
4. Define *preganglionic* and *postganglionic neuron.* Why are these terms not used in describing the somatic motor system?

TABLE 16.1

Comparison of the Somatic and Autonomic Nervous Systems

Feature	Somatic	Autonomic
Effectors	Skeletal muscle	Glands, smooth muscle, cardiac muscle
Control	Usually voluntary	Usually involuntary
Efferent pathways	One nerve fiber from CNS to effector; no ganglia	Two nerve fibers from CNS to effector; synapse at a ganglion
Neurotransmitters	Acetylcholine (ACh)	ACh and norepinephrine (NE)
Effect on target cells	Always excitatory	Excitatory or inhibitory
Effect of denervation	Flaccid paralysis	Denervation hypersensitivity

ANATOMY OF THE AUTONOMIC NERVOUS SYSTEM

Objectives

When you have completed this section, you should be able to

- identify the anatomical components and nerve pathways of the sympathetic and parasympathetic divisions; and
- discuss the relationship of the adrenal glands to the sympathetic nervous system.

The Sympathetic Division

The sympathetic division is also called the *thoracolumbar division* because it arises from the thoracic and lumbar regions of the spinal cord. It has relatively short preganglionic and long postganglionic fibers. The preganglionic somas are in the lateral horns and nearby regions of the gray matter of the spinal cord. Their fibers exit by way of spinal nerves T1 to L2 and lead to the nearby **sympathetic chain** of ganglia (**paravertebral**[3] **ganglia**) along each side of the vertebral

[3]*para* = next to + *vertebr* = vertebral column

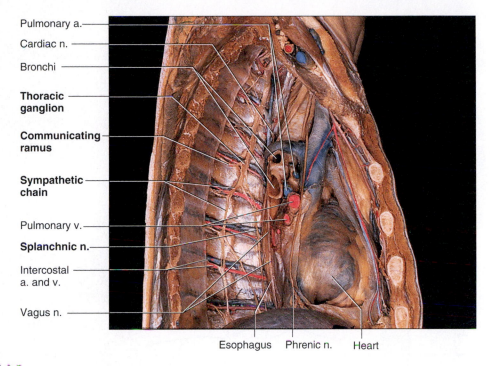

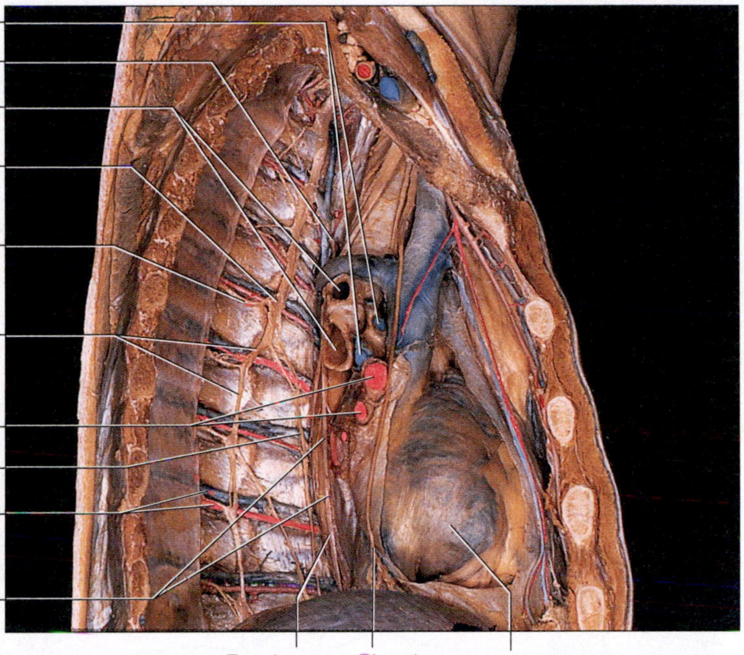
Pulmonary a.
Cardiac n.
Bronchi
Thoracic ganglion
Communicating ramus
Sympathetic chain
Pulmonary v.
Splanchnic n.
Intercostal a. and v.
Vagus n.
Esophagus
Phrenic n.
Heart

FIGURE 16.3
The Sympathetic Chain Ganglia. Right lateral view of the thoracic cavity. (*a.* = artery; *n.* = nerve; *v.* = vein)

column (figs. 16.3 and 16.4). Although these chains receive input from only the thoracolumbar region of the cord, they extend into higher and lower regions as well; some nerve fibers entering at levels T1 to L2 travel up or down to reach cervical, sacral, and coccygeal ganglia of the chain. The number of ganglia varies from person to person, but usually there are 3 cervical (*superior, middle,* and *inferior*), 11 thoracic, 4 lumbar, 4 sacral, and 1 coccygeal ganglion in each chain.

In the thoracolumbar region, each paravertebral ganglion is connected to a spinal nerve by two branches called *communicating rami* (fig. 16.5). The preganglionic fibers are small myelinated fibers that travel from the spinal nerve to the ganglion by way of the **white communicating ramus,**[4] which gets its color and name from the myelin. Unmyelinated postganglionic fibers leave the ganglion by various routes including a **gray communicating ramus,** named for its lack of myelin and duller color. These long fibers extend the rest of the way to the target organ.

THINK ABOUT IT!

Would autonomic postganglionic fibers have faster or slower conduction speeds than somatic motor fibers? Why? (See hints in chapter 13.)

[4]*ramus* = branch

After entering the sympathetic chain, preganglionic fibers may follow any of three courses:

- Some end in the ganglion that they enter and synapse immediately with a postganglionic neuron.
- Some travel up or down the chain and synapse in ganglia at other levels. It is these fibers that link the paravertebral ganglia into a chain. They are the only route by which ganglia at the cervical, sacral, and coccygeal levels receive input.
- Some pass through the chain without synapsing and continue as *splanchnic* (SPLANK-nic) *nerves,* to be considered shortly.

Nerve fibers leave the paravertebral ganglia by three routes: spinal, sympathetic, and splanchnic nerves. These are numbered in figure 16.5 to correspond to the following descriptions:

1. **The spinal nerve route.** Some postganglionic fibers exit by way of the gray ramus, return to the spinal nerve, and travel the rest of the way to the target organ. This is the route to most sweat glands, piloerector muscles, and blood vessels of the skin and skeletal muscles.
2. **The sympathetic nerve route.** Other postganglionic fibers leave the chain by way of **sympathetic nerves** that extend to the heart, lungs, esophagus, and thoracic blood vessels. These nerves form a plexus around each carotid artery and issue fibers from there to effectors in the head—including sweat, salivary, and nasal glands; piloerector muscles; blood vessels; and dilators of the iris. Some fibers from the superior and middle cervical ganglia form *cardiac nerves* to the heart.

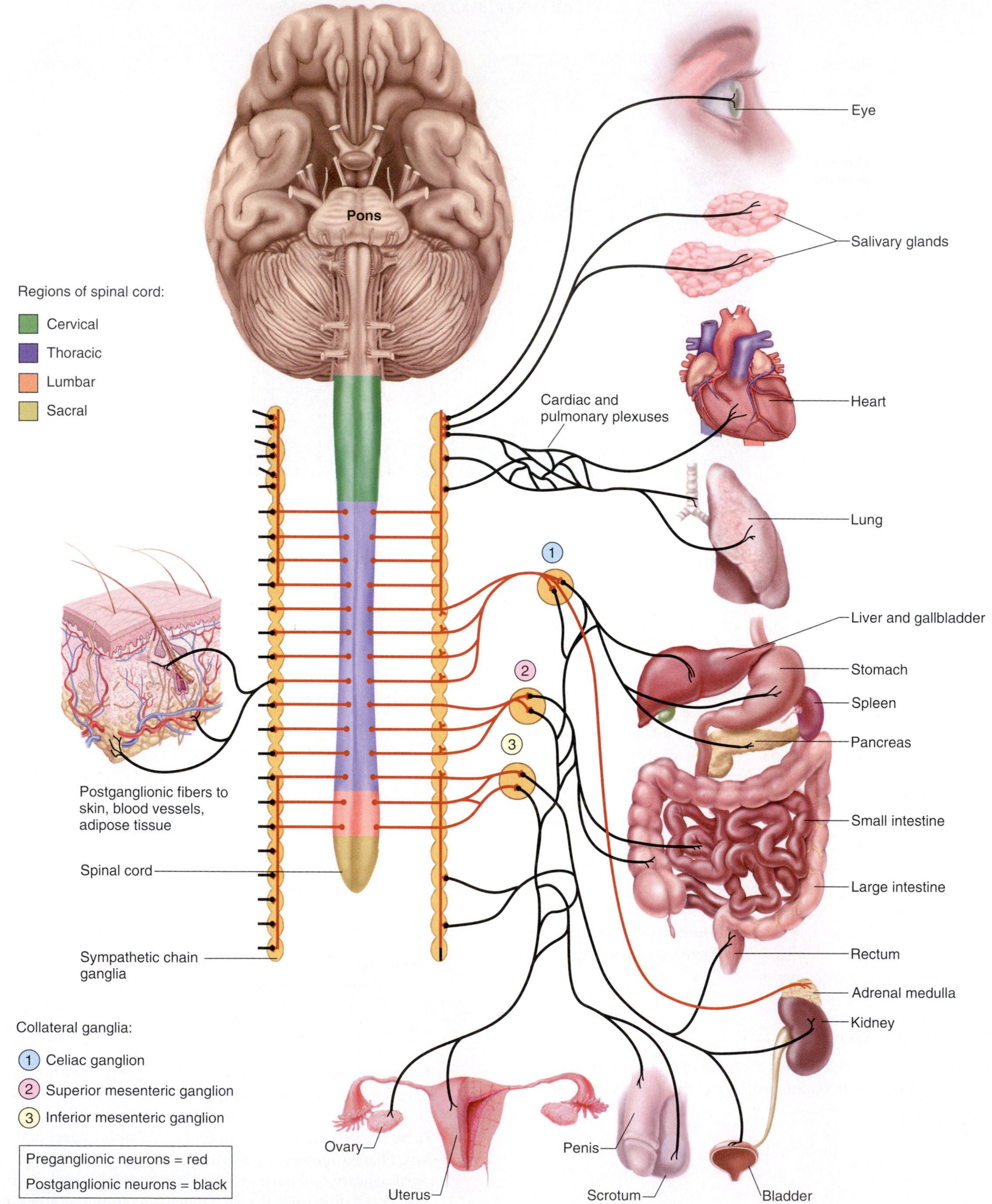
Pons
Eye
Salivary glands
Regions of spinal cord:
Cervical
Thoracic
Lumbar
Sacral
Heart
Cardiac and pulmonary plexuses
Lung
1
Liver and gallbladder
Stomach
Spleen
Pancreas
2
3
Postganglionic fibers to skin, blood vessels, adipose tissue
Small intestine
Spinal cord
Large intestine
Sympathetic chain ganglia
Rectum
Adrenal medulla
Kidney
Collateral ganglia:
1 Celiac ganglion
2 Superior mesenteric ganglion
3 Inferior mesenteric ganglion
Preganglionic neurons = red
Postganglionic neurons = black
Ovary
Uterus
Penis
Scrotum
Bladder

FIGURE 16.4
Sympathetic Pathways.

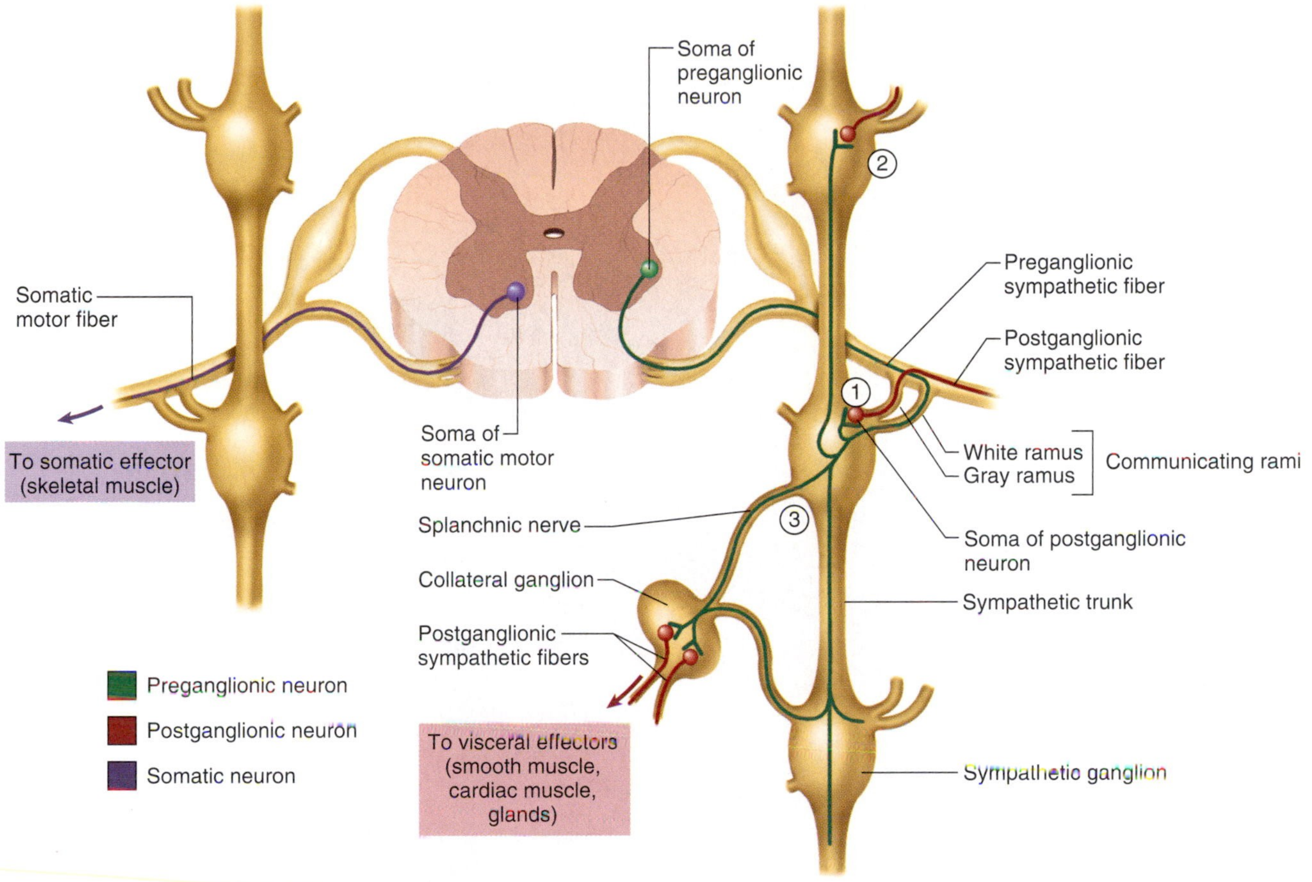

FIGURE 16.5

Sympathetic Pathways (*right*) Compared to Somatic Efferent Pathways (*left*). Sympathetic fibers can follow any of the three numbered routes: (*1*) the spinal nerve route, (*2*) the sympathetic nerve route, or (*3*) the splanchnic nerve route.

TABLE 16.2

Innervation To and From the Collateral Ganglia

Sympathetic Ganglia and Splanchnic Nerve →	Collateral Ganglion →	Postganglionic Target Organs
From thoracic ganglion 5 to 9 or 10 via greater splanchnic nerve	Celiac ganglion	Stomach, spleen, liver, small intestine, and kidneys
From thoracic ganglia 9 and 10 via lesser splanchnic nerve	Celiac and superior mesenteric ganglia	Small intestine and colon
From lumbar ganglia via lumbar splanchnic nerve	Inferior mesenteric ganglion	Rectum, urinary bladder, and reproductive organs

3. **The splanchnic[5] nerve route.** This route is formed by fibers that originate predominantly from spinal nerves T5 to T12 and pass through the ganglia without synapsing. Beyond the ganglia, they form the **greater, lesser,** and **least splanchnic nerves.** These nerves lead to the **collateral (prevertebral) ganglia,** which contribute to a network called the **abdominal aortic plexus** wrapped around the aorta (fig. 16.6). There are three major collateral ganglia in this plexus—one or more **celiac** (SEE-lee-ac) **ganglia,** the **superior mesenteric ganglion,** and the **inferior mesenteric ganglion**—located at points where arteries of the same names branch off the aorta. The postganglionic fibers accompany these arteries and their branches to the target organs. (The term *solar plexus* is regarded by some authorities as a collective designation for the celiac and superior mesenteric ganglia, and by others as a synonym for the celiac ganglion only. The term comes from the nerves radiating from the ganglion like rays of the sun.) Innervation to and from the three major collateral ganglia is summarized in table 16.2.

In summary, effectors in the muscles and body wall are innervated mainly by sympathetic fibers in the spinal nerves; effectors in the head and thoracic cavity by sympathetic nerves; and effectors in the abdominal cavity by splanchnic nerves.

There is no simple one-to-one relationship between preganglionic and postganglionic neurons in the sympathetic division.

[5] *splanchn* = viscera

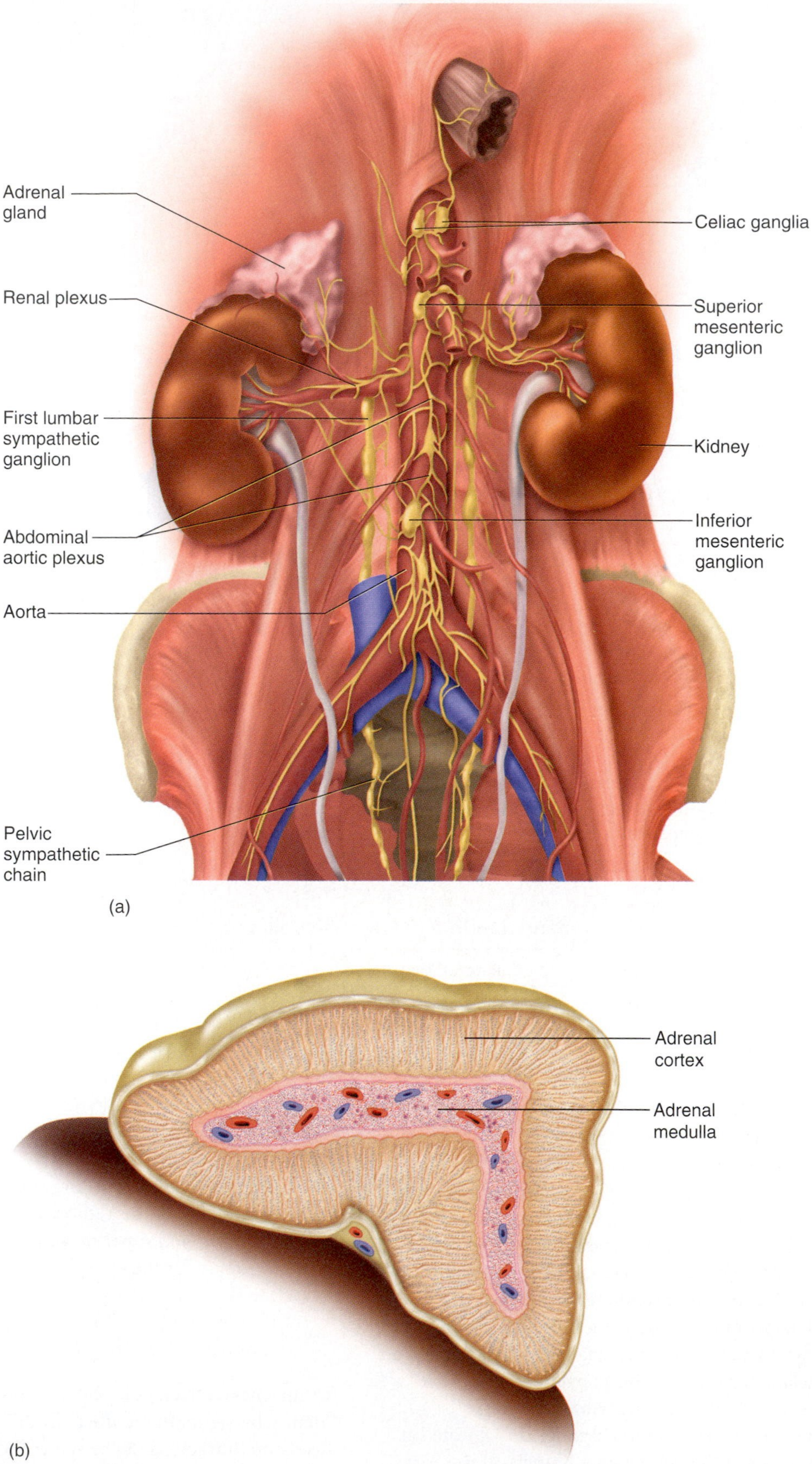

FIGURE 16.6

Abdominal Components of the Sympathetic Nervous System. (*a*) Collateral ganglia and the abdominal aortic plexus. (*b*) The adrenal gland, frontal section showing the sympathetic adrenal medulla.

For one thing, each postganglionic cell may receive synapses from multiple preganglionic cells, thus exhibiting the principle of *neuronal convergence* discussed in chapter 13. Furthermore, each preganglionic fiber branches and synapses with multiple postganglionic fibers, thus showing *neuronal divergence.* On average, each sympathetic preganglionic neuron synapses with about 17 postganglionic neurons. This means that when one preganglionic neuron fires, it can excite multiple postganglionic fibers leading to different target organs. The sympathetic division thus tends to have relatively widespread effects—as suggested by the name *sympathetic.*[6]

The Adrenal Glands

The paired **adrenal[7] glands** rest like hats on the superior pole of each kidney (fig. 16.6*a*). Each adrenal is actually two glands with different functions and embryonic origins. The outer rind, the **adrenal cortex,** secretes steroid hormones discussed in chapter 18. The inner core, the **adrenal medulla,** is a modified sympathetic ganglion (fig. 16.6*b*). It consists of modified postganglionic neurons without dendrites or axons. Sympathetic preganglionic fibers penetrate through the cortex and terminate on these cells. The sympathetic nervous system and adrenal medulla are so closely related in development and function that they are referred to collectively as the *sympathoadrenal system.* When stimulated, the adrenal medulla secretes a mixture of hormones into the bloodstream—about 85% epinephrine (adrenaline), 15% norepinephrine (noradrenaline), and a trace of dopamine.

The Parasympathetic Division

The parasympathetic division is also called the *craniosacral division* because it arises from the brain and sacral region of the spinal cord; its fibers travel in certain cranial and sacral nerves. Somas of the preganglionic neurons are located in the pons, medulla oblongata, and segments S2 to S4 of the spinal cord (fig. 16.7). They issue long preganglionic fibers which end in **terminal ganglia** in or near the target organ (see fig. 16.1). (If a terminal ganglion is embedded in the wall of a target organ, it is also called an *intramural*[8] *ganglion.*) Thus, the parasympathetic division has long preganglionic fibers, reaching almost all the way to the target cells, and short postganglionic fibers that cover the rest of the distance.

There is some neuronal divergence in the parasympathetic division, but much less than in the sympathetic. The parasympathetic division has a ratio of about two postganglionic fibers to every preganglionic. Furthermore, the preganglionic fiber reaches the target organ before even this slight divergence occurs. The parasympathetic division is therefore more selective than the sympathetic in its stimulation of target organs.

Parasympathetic fibers leave the brainstem by way of the following four cranial nerves. The first three supply all parasympathetic innervation to the head and the last one supplies viscera of the thoracic and abdominal cavities.

1. **Oculomotor nerve (III).** The oculomotor nerve carries parasympathetic fibers that control the lens and pupil of the eye. The preganglionic fibers enter the orbit and terminate in the *ciliary ganglion.* Postganglionic fibers enter the eyeball and innervate the *ciliary muscle,* which thickens the lens, and the *pupillary constrictor,* which narrows the pupil.
2. **Facial nerve (VII).** The facial nerve carries parasympathetic fibers that regulate the tear glands, salivary glands, and nasal glands. Soon after the facial nerve emerges from the pons, its parasympathetic fibers split away and form two smaller branches. The upper branch ends at the *sphenopalatine ganglion* near the junction of the maxilla and palatine bones. Postganglionic fibers then continue to the tear glands and glands of the nasal cavity, palate, and other areas of the oral cavity. The lower branch crosses the middle-ear cavity and ends at the *submandibular ganglion* near the angle of the mandible. Postganglionic fibers from here supply salivary glands in the floor of the mouth.
3. **Glossopharyngeal nerve (IX).** The glossopharyngeal nerve carries parasympathetic fibers concerned with salivation. The preganglionic fibers leave this nerve soon after its origin and form the *tympanic nerve,* which ends in the *otic*[9] *ganglion* near the foramen ovale. The postganglionic fibers then follow the trigeminal nerve to the *parotid salivary gland* just in front of the earlobe.
4. **Vagus nerve (X).** The vagus nerve carries about 90% of all parasympathetic preganglionic fibers. It travels down the neck and forms three networks in the mediastinum—the **cardiac plexus,** which supplies fibers to the heart; the **pulmonary plexus,** whose fibers accompany the bronchi and blood vessels into the lungs; and the **esophageal plexus,** whose fibers regulate swallowing.

   At the lower end of the esophagus, these plexuses give off anterior and posterior **vagal trunks,** each of which contains fibers from both the right and left vagus. These penetrate the diaphragm, enter the abdominal cavity, and contribute to the extensive *abdominal aortic plexus* mentioned earlier. As noted earlier, sympathetic fibers synapse here. The parasympathetic fibers, however, pass through the plexus without synapsing and lead to the liver, pancreas, stomach, small intestine, kidney, ureter, and proximal half of the colon.

The remaining parasympathetic fibers arise from levels S2 to S4 of the spinal cord. They travel a short distance in the ventral rami of the spinal nerves and then form **pelvic splanchnic nerves** that lead to the **inferior hypogastric (pelvic) plexus.** Some parasympathetic fibers synapse here, but most pass through this plexus and travel by way of **pelvic nerves** to the terminal ganglia in their target organs: the distal half of the large intestine, the rectum,

[6] *sym* = together + *path* = feeling
[7] *ad* = near + *ren* = kidney
[8] *intra* = within + *mur* = wall

[9] *ot* = ear + *ic* = pertaining to

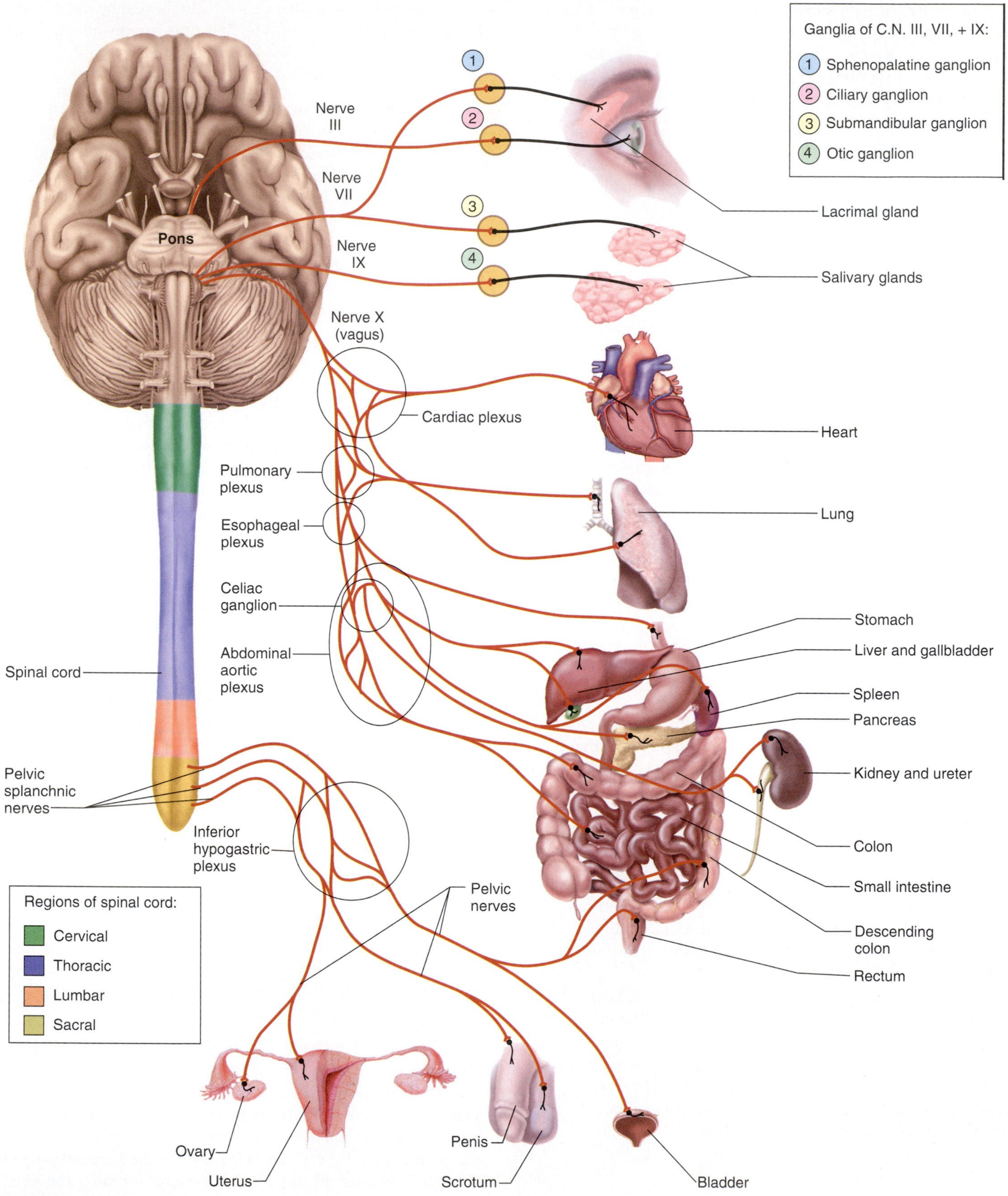
Ganglia of C.N. III, VII, + IX:
1 Sphenopalatine ganglion
2 Ciliary ganglion
3 Submandibular ganglion
4 Otic ganglion
Nerve III
Nerve VII
Nerve IX
Nerve X (vagus)
Pons
Lacrimal gland
Salivary glands
Cardiac plexus
Heart
Pulmonary plexus
Esophageal plexus
Lung
Celiac ganglion
Abdominal aortic plexus
Stomach
Liver and gallbladder
Spleen
Pancreas
Kidney and ureter
Spinal cord
Pelvic splanchnic nerves
Inferior hypogastric plexus
Colon
Small intestine
Pelvic nerves
Descending colon
Rectum
Regions of spinal cord:
Cervical
Thoracic
Lumbar
Sacral
Ovary
Uterus
Penis
Scrotum
Bladder

FIGURE 16.7
Parasympathetic Pathways.

TABLE 16.3

Comparison of the Sympathetic and Parasympathetic Divisions

Feature	Sympathetic	Parasympathetic
Origin in CNS	Thoracolumbar	Craniosacral
Location of ganglia	Paravertebral ganglia adjacent to spinal column and prevertebral ganglia anterior to it	Terminal ganglia near or within target organs
Fiber lengths	Short preganglionic Long postganglionic	Long preganglionic Short postganglionic
Neuronal divergence	Extensive (about 1:17)	Minimal (about 1:2)
Effects of system	Often widespread and general	More local and specific

urinary bladder, and reproductive organs. The parasympathetic system does not innervate body wall structures (sweat glands, piloerector muscles, or cutaneous blood vessels).

The sympathetic and parasympathetic divisions of the ANS are compared in table 16.3.

THINK ABOUT IT!

Would autonomic functions be affected if the ventral roots of the cervical spinal nerves were damaged? Why or why not?

The Enteric Nervous System

The digestive tract has a nervous system of its own called the **enteric**[10] **nervous system.** Unlike the ANS proper, it does not arise from the brainstem or spinal cord, but like the ANS, it innervates smooth muscle and glands. Thus, opinions differ on whether it should be considered part of the ANS. It consists of about 100 million neurons embedded in the wall of the digestive tract (see photograph on p. 455)—perhaps more neurons than there are in the spinal cord—and it has its own reflex arcs. The enteric nervous system regulates the motility of the esophagus, stomach, and intestines and the secretion of digestive enzymes and acid. To function normally, however, these digestive activities also require regulation by the sympathetic and parasympathetic systems. The enteric nervous system is discussed in more detail in chapter 24.

Before You Go On

Answer the following questions to test your understanding of the preceding section:

5. Explain why the sympathetic division is also called the thoracolumbar division even though its paravertebral ganglia extend all the way from the cervical to the sacral region.
6. Describe or diagram the structural relationships among the following: preganglionic fiber, postganglionic fiber, ventral ramus, gray ramus, white ramus, and paravertebral ganglion.
7. Explain in anatomical terms why the parasympathetic division affects target organs more selectively than the sympathetic division does.
8. Trace the pathway of a parasympathetic fiber of the vagus nerve from the medulla oblongata to the small intestine.

[10]*enter* = intestines + *ic* = pertaining to

AUTONOMIC EFFECTS

Objectives

When you have completed this section, you should be able to

- name the neurotransmitters employed by the ANS and define terms for neurons and synapses with different neurotransmitter and receptor types;
- in terms of neurotransmitters and receptors, explain why the two divisions of the ANS can have contrasting effects on the same organs;
- explain how the two divisions of the ANS interact when they both innervate the same organ; and
- describe how the central nervous system regulates the ANS.

Neurotransmitters and Receptors

As noted earlier, the divisions of the ANS often have contrasting effects on an organ. The sympathetic division accelerates the heartbeat and the parasympathetic division slows it down, for example. But each division also can have contrasting effects on different organs. For example, while the parasympathetic division inhibits the contraction of cardiac muscle, it stimulates the contraction of intestinal smooth muscle.

The key to understanding such seemingly contradictory effects lies in differences in the neurotransmitters released by autonomic neurons and the types of neurotransmitter receptors found on their target cells. Some nerve fibers of the ANS are **cholinergic,** meaning they secrete acetylcholine (ACh). Others are **adrenergic,** meaning they secrete norepinephrine (NE) (also called noradrenaline, the source of the term *adrenergic*). Cholinergic fibers include the preganglionic fibers of both divisions, the postganglionic fibers of the parasympathetic division, and a few sympathetic postganglionic fibers (those that innervate sweat glands and some blood vessels). Most sympathetic postganglionic fibers are adrenergic (fig. 16.8, table 16.4).

Both the sympathetic and parasympathetic divisions have excitatory effects on some target cells and inhibitory effects on others. The difference is due to the fact that different effector cells have different kinds of receptors for the foregoing neurotransmitters. The receptors for ACh and NE are called **cholinergic** and **adrenergic receptors,** respectively, and each of these has subclasses that lend further flexibility to autonomic function.

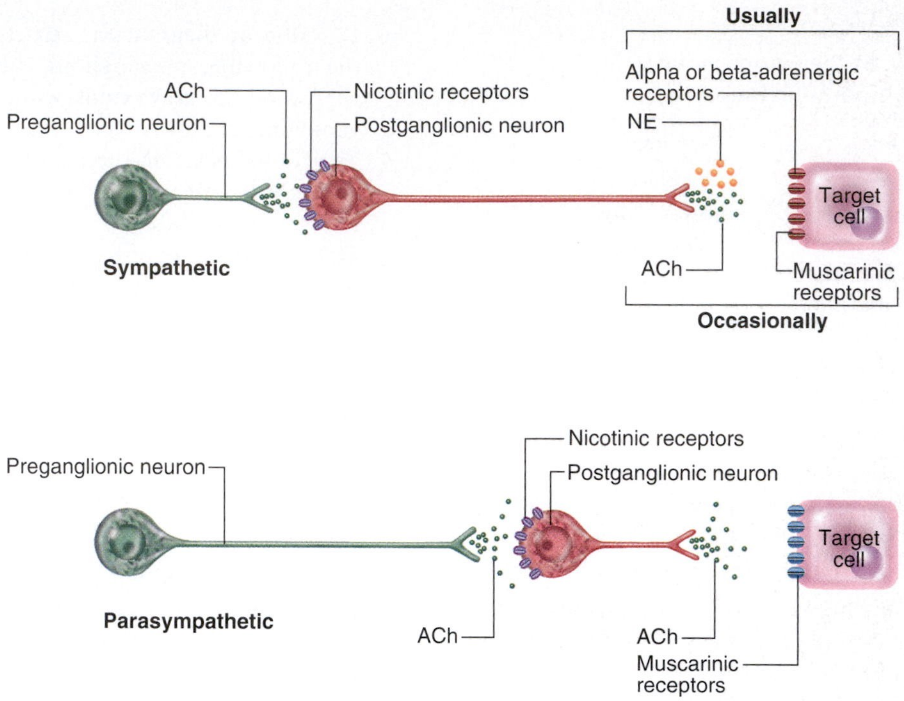

FIGURE 16.8

Neurotransmitters and Receptors of the Autonomic Nervous System. A given postganglionic fiber releases either ACh or NE, but not both. Both are shown in the *top* illustration only to emphasize that some sympathetic fibers are adrenergic and some are cholinergic.

INSIGHT 16.2 CLINICAL APPLICATION

DRUGS AND THE AUTONOMIC NERVOUS SYSTEM

The design of many drugs has been based on an understanding of autonomic neurotransmitters and receptor classes. *Sympathomimetics*[11] are drugs that enhance sympathetic action by stimulating adrenergic receptors or promoting norepinephrine release. For example, phenylephrine, found in such cold medicines as Chlor-Trimeton and Dimetapp, aids breathing by stimulating certain α-adrenergic receptors and thus dilating the bronchioles, constricting nasal blood vessels, and reducing swelling in the nasal mucosa. *Sympatholytics*[12] are drugs that suppress sympathetic action by inhibiting norepinephrine release or binding to adrenergic receptors without stimulating them. Propranolol, for example, is a β-*blocker.* It blocks the action of epinephrine and norepinephrine on the heart and blood vessels.

Parasympathomimetics enhance parasympathetic effects. Pilocarpine, for example, relieves glaucoma (excessive pressure within the eye) by dilating a vessel that drains fluid from the eye. *Parasympatholytics* inhibit ACh release or block its receptors. Atropine, for example, blocks muscarinic receptors and is sometimes used to dilate the pupils for eye examinations and to dry the mucous membranes of the respiratory tract before inhalation anesthesia. It is an extract of the deadly nightshade plant, *Atropa belladonna.*[13] Women of the Middle Ages used nightshade to dilate their pupils, which was regarded as a beauty enhancement.

The branch of medicine that deals with the effects of drugs on the nervous system—especially drugs that mimic, enhance, or inhibit the action of neurotransmitters—is called *neuropharmacology.*

[11] *mimet* = imitate, mimic
[12] *lyt* = break down, destroy
[13] *bella* = beautiful, fine + *donna* = woman

TABLE 16.4

Locations of Cholinergic and Adrenergic Fibers in the ANS

Division	Preganglionic Fibers	Postganglionic Fibers
Sympathetic	Always cholinergic	Mostly adrenergic; a few cholinergic
Parasympathetic	Always cholinergic	Always cholinergic

Acetylcholine binds to two classes of cholinergic receptors—**nicotinic** (NIC-oh-TIN-ic) and **muscarinic** (MUSS-cuh-RIN-ic) **receptors**—named for toxins (nicotine and muscarine) that were used to identify and distinguish them. ACh excites cells with nicotinic receptors, but can have either excitatory or inhibitory effects on cells with muscarinic receptors. Norepinephrine similarly binds to two broad classes of receptors called **α-adrenergic** and **β-adrenergic receptors.** It usually excites cells with α-adrenergic receptors and inhibits cells with β-adrenergic receptors, but there are exceptions to both, having to do with different subclasses of these receptor types.

Table 16.5 compares the effects of sympathetic and parasympathetic stimulation on many target organs and cells.

TABLE 16.5

Effects of the Sympathetic and Parasympathetic Nervous Systems

Target	Sympathetic Effect	Parasympathetic Effect
Eye		
Dilator of Pupil	Pupillary dilation	No effect
Constrictor of Pupil	No effect	Pupillary constriction
Ciliary Muscle and Lens	Relaxation for far vision	Contraction for near vision
Lacrimal (tear) Gland	None	Secretion
Integumentary System		
Merocrine Sweat Glands (cooling)	Secretion (muscarinic)	No effect
Apocrine Sweat Glands (scent)	Secretion	No effect
Piloerector Muscles	Hair erection	No effect
Adipose Tissue	Decreased fat breakdown	No effect
	Increased fat breakdown	No effect
Adrenal Medulla	Hormone secretion	No effect
Circulatory System		
Heart Rate and Force	Increased	Decreased
Deep Coronary Arteries	Vasodilation	Slight vasodilation
Blood Vessels of Most Viscera	Vasoconstriction	Vasodilation
Blood Vessels of Skeletal Muscles	Vasodilation	No effect
Blood Vessels of Skin	Vasoconstriction	Vasodilation, blushing
Platelets (blood clotting)	Increased clotting	No effect
Respiratory System		
Bronchi and Bronchioles	Bronchodilation	Bronchoconstriction
Mucous Glands	Decreased secretion Increased secretion	No effect
Urinary System		
Kidneys	Reduced urine output	No effect
Bladder Wall	No effect	Contraction
Internal Urinary Sphincter	Contraction, urine retention	Relaxation, urine release
Digestive System		
Salivary Glands	Thick mucous secretion	Thin serous secretion
Gastrointestinal Motility	Decreased	Increased
Gastrointestinal Secretion	Decreased	Increased
Liver	Glycogen breakdown	Glycogen synthesis
Pancreatic Enzyme Secretion	Decreased	Increased
Pancreatic Insulin Secretion	Decreased Increased	No effect
Reproductive System		
Penile or Clitoral Erection	No effect	Stimulation
Glandular Secretion	No effect	Stimulation
Orgasm, Smooth Muscle Roles	Stimulation	No effect
Uterus	Relaxation Labor contractions	No effect

THINK ABOUT IT!

Table 16.5 notes that the sympathetic nervous system has an adrenergic effect on blood platelets and promotes clotting. How can it do this, considering that platelets are drifting cell fragments in the bloodstream with no nerve fibers leading to them?

Dual Innervation

Most of the viscera receive nerve fibers from both the sympathetic and parasympathetic divisions and thus are said to have **dual innervation.** In such cases, the two divisions may have either *antagonistic* or *cooperative* effects on the same organ. Antagonistic effects oppose each other. Thus, the sympathetic division dilates the pupil and the parasympathetic division constricts it (fig. 16.9), among other examples already discussed and contrasted in table 16.5. Cooperative effects occur when the two divisions act on different effector cells in an organ to produce a unified overall effect. For example, the parasympathetic division stimulates the secretion of salivary enzymes and the sympathetic division stimulates the secretion of salivary mucus.

Dual innervation is not always necessary for the ANS to produce opposite effects on an organ. The adrenal medulla, piloerector muscles, sweat glands, and many blood vessels receive only sympathetic fibers. An example of control without dual innervation is the regulation of blood flow. The sympathetic fibers to a blood vessel have a baseline sympathetic tone which keeps the vessels in a state of partial constriction called *vasomotor tone* (fig. 16.10). An increase in sympathetic stimulation causes vasoconstriction by increasing smooth muscle contraction. A drop in sympathetic stimulation allows the smooth muscle to relax and the vessel to dilate.

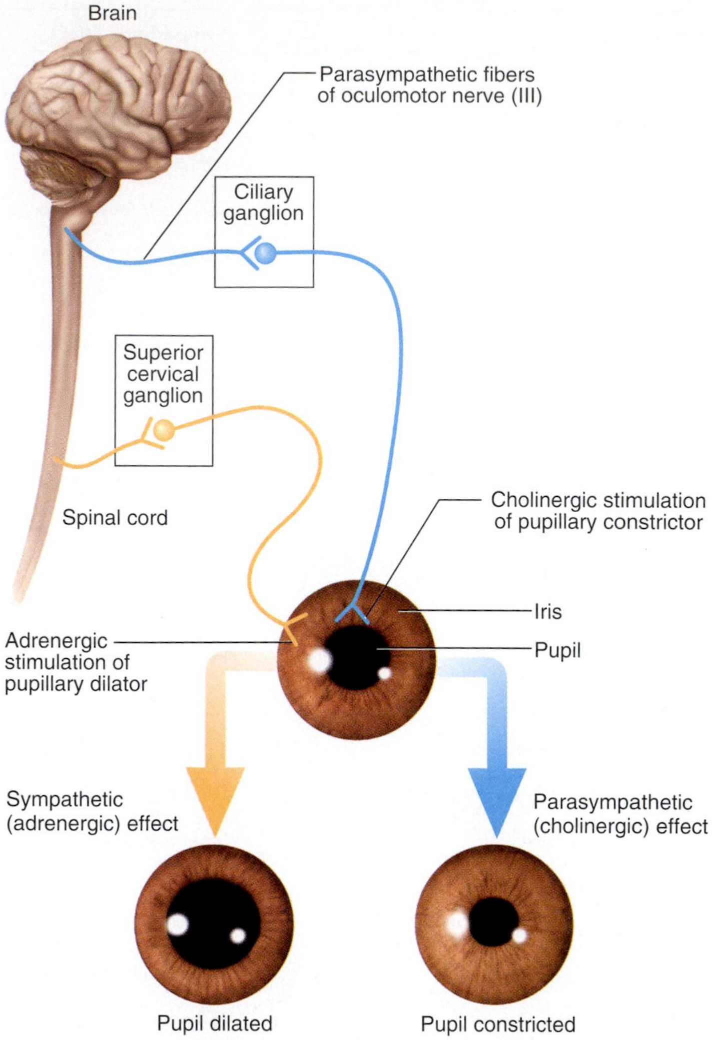

FIGURE 16.9

Dual Innervation of the Iris. Shows antagonistic effects of the sympathetic and parasympathetic divisions.

Central Control of Autonomic Function

In spite of its name, the ANS is not an independent nervous system. All of its output originates in the CNS, and it receives input from the cerebral cortex, hypothalamus, medulla oblongata, and somatic branch of the PNS.

Effects of the cerebral cortex on autonomic function are evident when anger raises the blood pressure, fear makes the heart race, thoughts of good food make the stomach rumble, and sexual thoughts or images increase blood flow to the genitals. The limbic system (p. 432) is involved in many emotional responses and has extensive connections with the hypothalamus, an important autonomic control center. Thus, the limbic system provides a pathway connecting sensory and mental experiences with the autonomic nervous system.

The hypothalamus contains many nuclei for primitive autonomic functions, including hunger, thirst, thermoregulation, and sexual response. Artificial stimulation of different regions of the hypothalamus can activate the fight or flight response typical of the sympathetic nervous system or have the calming effects typical of the parasympathetic. Output from the hypothalamus travels largely to nuclei in more caudal regions of the brainstem and from there to the cranial nerves and the sympathetic preganglionic neurons in the spinal cord.

The midbrain, pons, and medulla oblongata house the nuclei of cranial nerves that mediate several autonomic responses: the oculomotor nerve (pupillary constriction), facial nerve (lacrimal, nasal, palatine, and salivary gland secretion), glossopharyngeal nerve (salivation, blood pressure regulation), and vagus nerve (the chief parasympathetic supply to the thoracic and abdominal viscera).

The spinal cord also contains autonomic control nuclei. Such autonomic responses as the defecation and micturition (urination) reflexes are regulated here. Fortunately, the brain is able to inhibit these responses consciously, but when injuries sever the spinal cord from the brain, the autonomic spinal reflexes alone control the elimination of urine and feces.

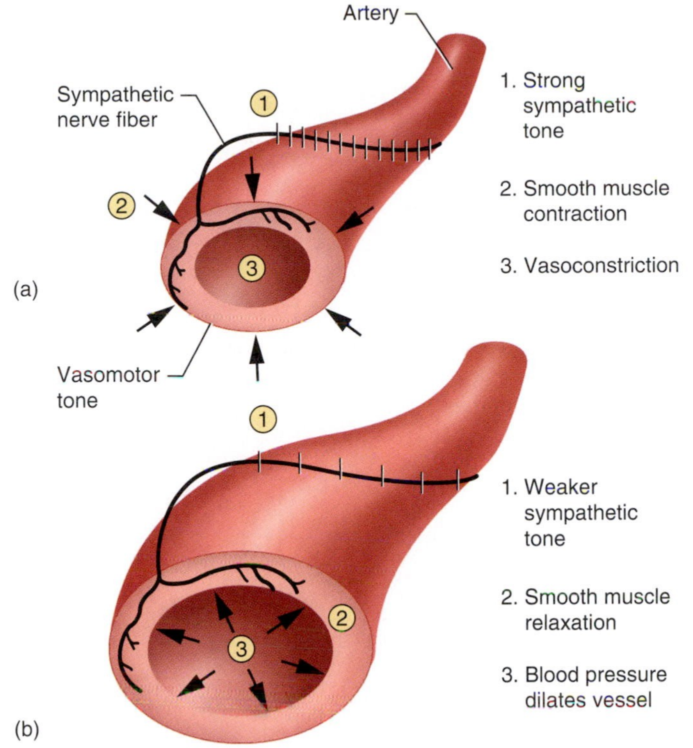
Artery
Sympathetic nerve fiber
1
1. Strong sympathetic tone
2
2. Smooth muscle contraction
3
3. Vasoconstriction
(a)
Vasomotor tone
1
1. Weaker sympathetic tone
2
2. Smooth muscle relaxation
3
3. Blood pressure dilates vessel
(b)

FIGURE 16.10
Sympathetic and Vasomotor Tone. (*a*) Vasoconstriction in response to a high rate of sympathetic nerve firing. (*b*) Vasodilation in response to a low rate of sympathetic nerve firing.

Before You Go On

Answer the following questions to test your understanding of the preceding section:

9. To what neurotransmitters do the terms *adrenergic* and *cholinergic* refer?
10. Why is a single autonomic neurotransmitter able to have opposite effects on different target cells?
11. What are the two ways in which the sympathetic and parasympathetic divisions can interact when they both innervate the same target organ? Give examples.
12. How can the sympathetic nervous system have contrasting effects in a target organ without dual innervation?
13. What system in the brain connects our conscious thoughts and feelings with the autonomic control centers of the hypothalamus?
14. List some autonomic responses that are controlled by nuclei in the hypothalamus.
15. What is the role of the midbrain, pons, and medulla in autonomic control?
16. Name some visceral reflexes controlled by the spinal cord.

Developmental and Clinical Perspectives

Objectives
When you have completed this section, you should be able to

- describe the embryonic origins of the autonomic neurons and ganglia;
- describe some consequences of aging of the autonomic nervous system; and
- describe a few disorders of autonomic function.

Development and Aging of the Autonomic Nervous System

Preganglionic neurons of the autonomic nervous system develop from the *neural tube* described in chapter 13; their somas remain embedded in the brainstem and spinal cord for life. Autonomic ganglia and postganglionic neurons, however, develop from the *neural crest* adjacent to the neural tube. During the fifth week of embryonic development, some neural crest cells migrate and assume positions alongside the vertebral bodies to become the sympathetic chain ganglia; others assume positions alongside the aorta to form the abdominal aortic plexus; and others migrate to the heart, lungs, digestive tract, and other viscera to form the terminal ganglia of the parasympathetic division.

The adrenal medulla arises from cells that separate from a nearby sympathetic ganglion, and thus ultimately comes from neural crest (ectodermal) cells. During its development, the medulla is surrounded by cells of mesodermal origin that produce the outer layer of the adrenal gland, the adrenal cortex (which is not part of the autonomic nervous system).

The efficiency of the ANS declines in old age, like that of the rest of the nervous system (see chapter 15). Target organs of the ANS have fewer neurotransmitter receptors in old age, and are thus less responsive to autonomic stimulation. As a result, elderly people may experience dry eyes and more eye infections; slower and less effective adaptation of the eye to changing light intensities, and poorer night vision; less efficient control of blood pressure; and reduced intestinal motility and increasing constipation. Because of reduced efficiency of the baroreflex described earlier in this chapter, some elderly people experience *orthostatic hypotension,* a drop in blood pressure when they stand up, sometimes causing dizziness, loss of balance, or fainting.

Disorders of the Autonomic Nervous System

Table 16.6 describes some dysfunctions of the autonomic nervous system.

TABLE 16.6

Some Disorders of the Autonomic Nervous System

Achalasia[14] *of the Cardia*	A defect in autonomic innervation of the esophagus, resulting in impaired swallowing, accompanied by failure of the lower esophageal sphincter to relax and allow food to pass into the stomach. (The region of the stomach at its junction with the esophagus is called the *cardia.*) Results in enormous dilation of the esophagus and inability to keep food down. Most common in young adults; cause remains poorly understood.
Horner[15] *Syndrome*	Chronic unilateral pupillary constriction, sagging of the eyelid, withdrawal of the eye into the orbit, flushing of the skin, and lack of facial perspiration. Results from lesions in the cervical ganglia, upper thoracic spinal cord, or brainstem that interrupt sympathetic innervation of the head.
Raynaud[16] *Disease*	Intermittent attacks of paleness, cyanosis, and pain in the fingers and toes, caused when cold or emotional stress triggers excessive vasoconstriction in the digits; most common in young women. In extreme cases, causes gangrene and may require amputation. Sometimes treated by severing sympathetic nerves to the affected regions.

Disorders Described Elsewhere

Autonomic effects of cranial nerve injuries 443, 447
Mass reflex reaction 407
Orthostatic hypotension 469

[14]*a* = without + *chalas* = relaxation
[15]Johann F. Horner (1831–86), Swiss ophthalmologist
[16]Maurice Raynaud (1834–81), French physician

Before You Go On

Answer the following questions to test your understanding of the preceding section:

17. How do the pre- and postganglionic neurons of the ANS differ in embryonic origin?
18. Briefly state how the intestines, eyes, and blood pressure are affected in old age by the declining efficiency of the autonomic nervous system.

CHAPTER REVIEW

Review of Key Concepts

General Properties of the Autonomic Nervous System (p. 456)

1. The autonomic nervous system (ANS) carries out many visceral reflexes that are crucial to homeostasis. It is a visceral motor system that acts on glands, cardiac muscle, and smooth muscle.
2. Functions of the ANS are largely, but not entirely, unconscious and involuntary.
3. Autonomic innervation is not necessary for smooth or cardiac muscle to contract, but the ANS does modulate their activity. It functions through *visceral reflex* arcs similar to those of the somatic reflexes except for the type of effector at the end of the arc.
4. The *sympathetic division* of the ANS prepares the body for physical activity and is especially active in stressful "fight or flight" situations.
5. The *parasympathetic division* has a calming effect on many body functions, but stimulates digestion; it is especially active in "resting and digesting" states.
6. Although the balance of activity may shift from one division to the other, both divisions are normally active simultaneously. Each maintains a background level of activity called *autonomic tone.*
7. The ANS is composed of nuclei in the brainstem, motor neurons in the spinal cord and ganglia, and nerve fibers in the cranial and spinal nerves.
8. Most autonomic efferent pathways, unlike somatic motor pathways, involve two neurons: a *preganglionic neuron* whose axon travels to a peripheral ganglion, and a *postganglionic neuron* whose axon leads the rest of the way to the target cells.

Anatomy of the Autonomic Nervous System (p. 458)

1. Sympathetic (thoracolumbar) preganglionic neurons arise from thoracic and lumbar segments of the spinal cord. They travel through spinal nerves T1 through L2 to a *sympathetic chain* of ganglia adjacent to the vertebral column.
2. The sympathetic chain extends above and below the thoracic and lumbar regions of the spinal cord; it usually has 3 cervical, 11 thoracic, 4 lumbar, 4 sacral, and 1 coccygeal ganglion.
3. Most preganglionic fibers synapse with postganglionic neurons in one of the ganglia of this chain, sometimes at a higher or lower level than the ganglion at which they enter. Some fibers pass through the chain without synapsing.
4. Preganglionic sympathetic fibers travel from the spinal nerve to the sympathetic ganglion by way of a white communicating ramus.
5. Postganglionic fibers may leave the ganglion through a gray communicating ramus that returns to the spinal nerve, or through *sympathetic nerves* that lead to target organs of the head and thorax. Other sympathetic fibers pass through the sympathetic ganglia without synapsing, and travel by way of *splanchnic nerves* to synapses in the ganglia of the *abdominal aortic plexus.* The *celiac, superior mesenteric,* and *inferior mesenteric ganglia* of this plexus then give off postganglionic fibers to the abdominopelvic viscera.
6. Sympathetic pathways show substantial neuronal divergence, with the average preganglionic neuron synapsing with 17 postganglionic neurons. Sympathetic stimulation therefore tends to have widespread effects on multiple target organs.
7. The *adrenal medulla* is a modified sympathetic ganglion composed of modified neurons. These cells secrete mainly epinephrine and norepinephrine into the blood when stimulated.
8. The parasympathetic division issues relatively long preganglionic fibers through cranial nerves III, VII, IX, and X, and spinal nerves S2 through S4, to their target organs.
9. Parasympathetic preganglionic fibers in cranial nerves III, VII, and IX terminate in the *ciliary ganglion* (III), *sphenopalatine* and *submandibular ganglia* (VII), and *otic ganglion* (IX) in the head; postganglionic fibers complete the route to such target organs as the eye, tear glands, salivary glands, and nasal glands.
10. The vagus nerve (X) carries about 90% of all parasympathetic preganglionic fibers, innervates viscera of the thoracic and abdominal cavities, and has the most extensive and complex pathway. It forms *cardiac, pulmonary,* and *esophageal plexuses* in the thoracic cavity, then penetrates the diaphragm as a pair of *vagal trunks* and contributes to the abdominal aortic plexus.
11. The parasympathetic fibers arising from the sacral spinal cord form *pelvic splanchnic nerves,* the *inferior hypogastric plexus,* and *pelvic nerves,* which supply the viscera of the pelvic cavity.
12. Parasympathetic preganglionic fibers end in *terminal ganglia* in or near the target organ. Relatively short postganglionic fibers complete the route to specific target cells.
13. The wall of the digestive tract contains an *enteric nervous system,* sometimes considered part of the ANS because it innervates smooth muscle and glands of the tract.

Autonomic Effects (p. 465)

1. The autonomic effects on a target cell depend on the neurotransmitter released and the type of receptors that the target cell has.
2. *Cholinergic* fibers secrete acetylcholine (ACh) and include all preganglionic fibers, all parasympathetic postganglionic fibers, and some sympathetic postganglionic fibers. Most sympathetic postganglionic fibers are *adrenergic* and secrete norepinephrine (NE).
3. ACh binds to two classes of cholinergic receptors called *nicotinic* and *muscarinic* receptors. The binding of ACh to a nicotinic receptor always excites a target cell, but binding to a muscarinic receptor can have excitatory effects on some cells and inhibitory effects on others, owing to different subclasses of muscarinic receptors.
4. NE binds to two major classes of receptors called α and β receptors. Binding to an α-adrenergic receptor is usually excitatory, and binding to a β-adrenergic receptor is usually inhibitory, but there are exceptions to both owing to subclasses of each receptor type.
5. Many organs receive *dual innervation* by both sympathetic and parasympathetic fibers. In such cases, the two divisions may have either antagonistic or cooperative effects on the organ.
6. The sympathetic division can have contrasting effects on an organ even without dual innervation, by increasing or decreasing the firing rate of the sympathetic neuron.

7. All autonomic output originates in the CNS and is subject to control by multiple levels of the CNS.
8. The hypothalamus is an especially important center of autonomic control, but the cerebral cortex, midbrain, pons, and medulla oblongata are also involved in autonomic responses.
9. Some autonomic reflexes such as defecation and micturition are regulated by nuclei in the spinal cord.

Developmental and Clinical Perspectives (p. 469)

1. Preganglionic neurons of the ANS develop from the neural tube. Postganglionic neurons, ganglia, and the adrenal medulla develop from the neural crest.
2. The ANS becomes less efficient with age, resulting in such conditions as dry eyes and eye infections; poorer adaptation to changing light intensities, and deficient night vision; inefficient control of blood pressure, sometimes resulting in orthostatic hypotension; and reduced intestinal motility.
3. Dysfunctions of the ANS include achalasia of the cardia, Horner syndrome, and Raynaud syndrome (table 16.6).

Testing Your Recall

1. The autonomic nervous system innervates all of these *except*
   a. cardiac muscle.
   b. skeletal muscle.
   c. smooth muscle.
   d. salivary glands.
   e. blood vessels.
2. Muscarinic receptors bind
   a. epinephrine.
   b. norepinephrine.
   c. acetylcholine.
   d. cholinesterase.
   e. neuropeptides.
3. All of the following cranial nerves except the _____ carry parasympathetic fibers.
   a. vagus
   b. facial
   c. oculomotor
   d. glossopharyngeal
   e. hypoglossal
4. Which of the following cranial nerves carries sympathetic fibers?
   a. oculomotor
   b. facial
   c. trigeminal
   d. vagus
   e. none of the cranial nerves
5. Which of these ganglia is *not* involved in the sympathetic division?
   a. intramural
   b. superior cervical
   c. paravertebral
   d. inferior mesenteric
   e. celiac
6. Epinephrine is secreted by
   a. sympathetic preganglionic fibers.
   b. sympathetic postganglionic fibers.
   c. parasympathetic preganglionic fibers.
   d. parasympathetic postganglionic fibers.
   e. the adrenal medulla.
7. The major autonomic control center within the CNS is
   a. the cerebral cortex.
   b. the limbic system.
   c. the midbrain.
   d. the hypothalamus.
   e. the sympathetic chain ganglia.
8. The gray communicating ramus contains
   a. visceral sensory fibers.
   b. parasympathetic motor fibers.
   c. sympathetic preganglionic fibers.
   d. sympathetic postganglionic fibers.
   e. somatic motor fibers.
9. The neural crest gives rise to all of the following *except*
   a. sympathetic chain ganglia.
   b. the celiac ganglion.
   c. parasympathetic preganglionic neurons.
   d. parasympathetic postganglionic neurons.
   e. the adrenal medulla.
10. Which of these does *not* result from sympathetic stimulation?
   a. dilation of the pupil
   b. acceleration of the heart
   c. digestive secretion
   d. enhanced blood clotting
   e. piloerection
11. Nerve fibers that secrete norepinephrine are called _____ fibers.
12. _____ is a state in which a target organ receives both sympathetic and parasympathetic fibers.
13. _____ is a state of continual background activity of the sympathetic and parasympathetic divisions.
14. Most parasympathetic preganglionic fibers are found in the _____ nerve.
15. The digestive tract has a semi-independent nervous system called the _____ nervous system.
16. The embryonic tissue that gives rise to all autonomic ganglia and postganglionic neurons, but not to any preganglionic neurons, is _____.
17. The adrenal medulla consists of modified postganglionic neurons of the _____ nervous system.
18. The sympathetic nervous system has short _____ and long _____ nerve fibers.
19. Orthostatic hypotension is the result of inefficiency of the _____ reflex.
20. Sympathetic stimulation of blood vessels maintains a state of partial vasoconstriction called _____.

Answers in the Appendix

True or False

Determine which five of the following statements are false, and briefly explain why.

1. The parasympathetic nervous system shuts down when the sympathetic nervous system is active, and vice versa.
2. Blood vessels of the skin receive no parasympathetic innervation.
3. Voluntary control of the ANS is not possible.
4. The sympathetic nervous system stimulates digestion.
5. Some sympathetic postganglionic fibers are cholinergic.
6. Urination and defecation cannot occur without signals from the brain to the bladder and rectum.
7. Some parasympathetic nerve fibers are adrenergic.
8. Parasympathetic effects are more localized and specific than sympathetic effects.
9. The parasympathetic division shows less neuronal divergence than the sympathetic division does.
10. The two divisions of the ANS have antagonistic effects on the iris.

Answers in the Appendix

Testing Your Comprehension

1. You are dicing raw onions while preparing dinner, and the vapor makes your eyes water. Describe the afferent and efferent pathways involved in this response.
2. Suppose you are walking alone at night when you hear a dog growling close behind you. Describe the ways your sympathetic nervous system would prepare you to deal with this situation.
3. Suppose that the cardiac nerves were destroyed. How would this affect the heart and the body's ability to react to a stressful situation?
4. What would be the advantage to a wolf in having its sympathetic nervous system stimulate the piloerector muscles? What happens in a human when the sympathetic system stimulates these muscles?
5. Pediatric literature has reported many cases of poisoning of children with Lomotil, an antidiarrheic medicine. Lomotil works by means of the morphine-like effects of its chief ingredient, diphenoxylate, but it also contains atropine. Considering the mode of action described for atropine in insight 16.2, why might it contribute to the antidiarrheic effect of Lomotil? In atropine poisoning, would you expect the pupils to be dilated or constricted? The skin to be moist or dry? The heart rate to be elevated or depressed? The bladder to retain urine or void uncontrollably? Explain each answer.

Answers at the Online Learning Center

www.mhhe.com/saladinha1

Visit the Online Learning Center for practice tests, answer keys, and other learning aids for this chapter. Enhance your understanding of human anatomy with our interactive art labeling exercises, supplemental photo atlases, web links, puzzles, flashcards, and much more.

17

CHAPTER SEVENTEEN

Sense Organs

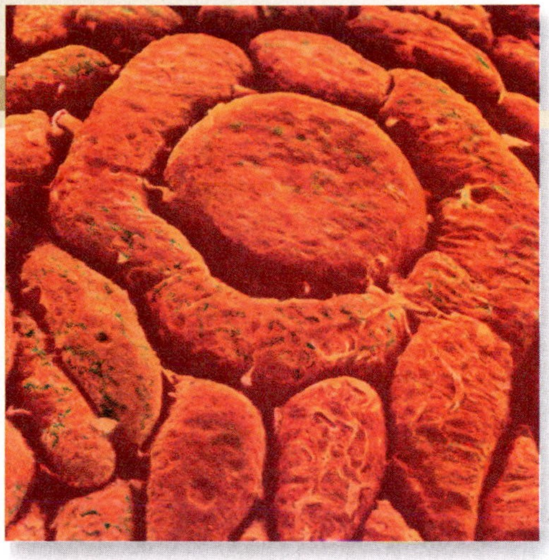

A vallate papilla of the tongue, where most taste buds are loacated (SEM).

CHAPTER OUTLINE

Receptor Types and the General Senses 476
- Classification of Receptors 476
- The General Senses 476
- The Receptive Field 478
- Somesthetic Projection Pathways 479
- Pain Pathways 479

The Chemical Senses 481
- Taste 481
- Smell 482

The Ear 484
- Anatomy of the Ear 485
- Auditory Function 487
- The Auditory Projection Pathway 489
- The Vestibular Apparatus 489
- Vestibular Projection Pathways 492

The Eye 494
- Accessory Structures of the Orbit 494
- Anatomy of the Eyeball 495
- Formation of an Image 499
- Structure and Function of the Retina 499
- The Visual Projection Pathway 502

Developmental and Clinical Perspectives 504
- Development of the Ear 504
- Development of the Eye 504
- Disorders of the Sense Organs 506

Chapter Review 508

INSIGHTS

17.1 Clinical Application: The Value of Pain 480
17.2 Clinical Application: Middle-Ear Infection 486
17.3 Clinical Application: Deafness 489
17.4 Clinical Application: Cataracts and Glaucoma 498

BRUSHING UP

To understand this chapter, you may find it helpful to review the following concepts:
- Converging circuits of neurons (p. 377)
- Spinal cord tracts (p. 389)
- Decussation (p. 389)
- Sensory areas of the cerebral cortex (p. 434)

Anyone who enjoys music, art, fine food, or a good conversation appreciates the human senses. Yet their importance extends beyond deriving pleasure from the environment. In the 1950s, behavioral scientists at Princeton University studied the methods used by Soviet Communists to extract confessions from political prisoners, including solitary confinement and sensory deprivation. Student volunteers were immobilized in dark soundproof rooms or suspended in dark chambers of water. In a short time, they experienced visual, auditory, and tactile hallucinations, incoherent thought patterns, deterioration of intellectual performance, and sometimes morbid fear or panic. Similar effects have been seen in burn patients who are immobilized and extensively bandaged (including the eyes) and thus suffer prolonged lack of sensory stimulation. Patients connected to life-support equipment and confined under oxygen tents sometimes become delirious. In short, sensory input is vital to the integrity of the personality and intellectual function.

Furthermore, much of the information communicated by the sense organs never comes to our conscious attention—blood pressure, body temperature, and muscle tension, for example. By monitoring such conditions, however, the sense organs initiate somatic and visceral reflexes that are indispensable to homeostasis and to our very survival in a ceaselessly changing and challenging environment.

Receptor Types and the General Senses

Objectives

When you have completed this section, you should be able to

- define *receptor* and *sense organ;*
- outline three ways of classifying receptors;
- define *general senses,* list several types, and describe their receptors;
- explain the meaning and relevance of a sensory neuron's receptive field;
- describe the pathways that the general senses take to the cerebral cortex; and
- describe the types of pain and its projection pathways.

A **receptor** is any structure specialized to detect a stimulus. Some receptors are simple nerve endings (sensory dendrites), whereas others are **sense organs**—nerve endings combined with connective, epithelial, or muscular tissues that enhance or moderate the response to a stimulus. Our eyes and ears are obvious examples of sense organs, but there are also innumerable microscopic sense organs in our skin, muscles, joints, and viscera.

Classification of Receptors

Receptors can be classified by multiple overlapping systems:

1. By **modality** (type of stimulus):
   - **Chemoreceptors** respond to chemicals, including odors, tastes, and composition of the body fluids.
   - **Thermoreceptors** respond to heat and cold.
   - **Nociceptors**[1] (NO-sih-SEP-turs) are pain receptors; they respond to tissue damage resulting from trauma (blows, cuts), ischemia (poor blood flow), or excessive stimulation by agents such as heat and chemicals.
   - **Mechanoreceptors** respond to physical forces on cells caused by touch, pressure, stretch, tension, or vibration. They include the organs of hearing and balance and many receptors of the skin, viscera, and joints.
   - **Photoreceptors,** the eyes, respond to light.
2. By the distribution of receptors in the body:
   - **General (somesthetic, somatosensory) senses** employ receptors that are widely distributed in the skin, muscles, tendons, joint capsules, and viscera. They detect touch, pressure, stretch, heat, cold, and pain, as well as many stimuli that we do not perceive consciously, such as blood pressure and blood chemistry.
   - **Special senses** are mediated by relatively complex sense organs of the head, innervated by the cranial nerves. They include vision, hearing, equilibrium, taste, and smell.
3. By the origins of the stimuli:
   - **Interoceptors** detect stimuli in the internal organs and produce feelings of visceral pain, nausea, stretch, and pressure.
   - **Proprioceptors** sense the position and movements of the body or its parts. They occur in muscles, tendons, and joint capsules.
   - **Exteroceptors** sense stimuli external to the body; they include the receptors for vision, hearing, taste, smell, and the cutaneous (skin) senses.

The General Senses

Receptors for the general senses are relatively simple in structure and physiology. They consist of one or a few sensory nerve fibers and usually a sparse amount of connective tissue. Depending on the presence or absence of connective tissue, they are classified as unencapsulated or encapsulated nerve endings (table 17.1). Nine types of simple receptors for the general senses are described here and illustrated in figure 17.1.

Unencapsulated nerve endings are sensory dendrites that lack a connective tissue wrapping. They include the following:

- **Free nerve endings.** These include *warm receptors,* which respond to rising temperature; *cold receptors,* which respond to falling temperature; and *nociceptors,* or pain receptors. They are bare dendrites with no special association with any specific accessory cells or tissues. They are most abundant in connective tissues and epithelia. They typically show profuse, fine branches that ramify through the connective tissue or between epithelial cells.

[1]*noci* = pain

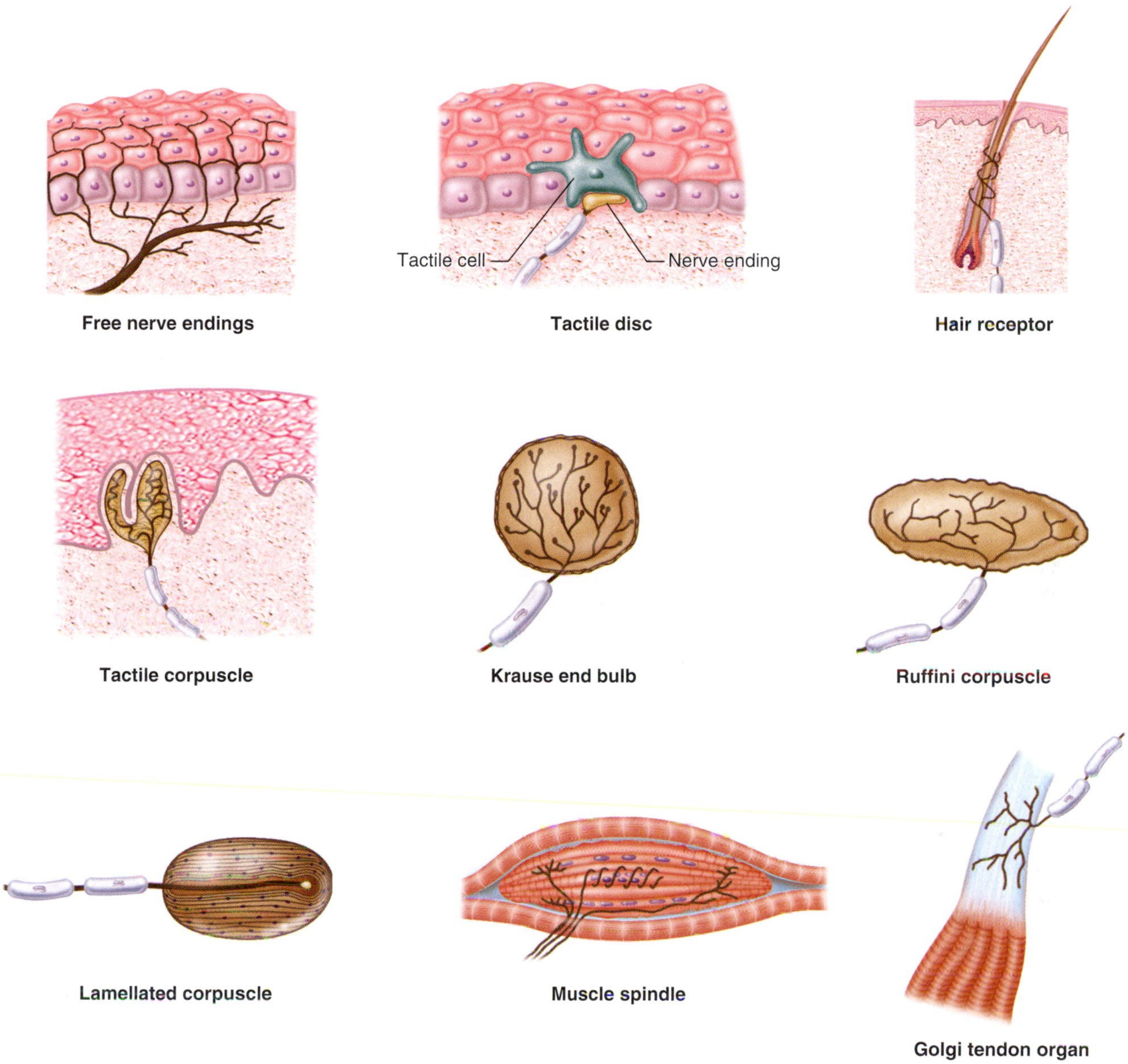
Tactile cell
Nerve ending
Free nerve endings
Tactile disc
Hair receptor
Tactile corpuscle
Krause end bulb
Ruffini corpuscle
Lamellated corpuscle
Muscle spindle
Golgi tendon organ

FIGURE 17.1
Receptors of the General (somesthetic) Senses.

- **Tactile (Merkel[2]) discs.** These are receptors for light touch and pressure on the skin. A tactile disc is a flattened nerve ending associated with a specialized *tactile (Merkel) cell* at the base of the epidermis.
- **Hair receptors (peritrichial[3] endings).** These are nerve fibers entwined around a hair follicle that monitor movements of the hair. Because they adapt quickly, we are not constantly annoyed by our clothing bending the body hairs. However, when an ant crawls across our skin, bending one hair after another, we are very aware of it.

Encapsulated nerve endings are dendrites wrapped in glial cells or connective tissue. Most of them are mechanoreceptors for touch, pressure, and stretch. The connective tissues around a sensory dendrite enhance the sensitivity or specificity of the receptor. They include the following:

- **Tactile (Meissner[4]) corpuscles.** These are receptors for light touch, texture, and low-frequency vibration. They occur in the dermal papillae of the skin, especially in sensitive hairless areas such as the fingertips, palms, eyelids, lips, nipples, and genitals. They are tall, ovoid to pear-shaped, and consist of two or three nerve fibers meandering upward through a mass of connective

[2]Friedrich S. Merkel (1845–1911), German anatomist and physiologist
[3]*peri* = around + *trich* = hair
[4]George Meissner (1829–1905), German histologist

TABLE 17.1

Receptors of the General Senses

Receptor Type	Locations	Modality
Unencapsulated Endings		
Free Nerve Endings	Widespread, especially in epithelia and connective tissues	Pain, heat, cold
Tactile Discs	Stratum basale of epidermis	Light touch, pressure
Hair Receptors	Around hair follicle	Light touch, movement of hairs
Encapsulated Nerve Endings		
Tactile (Meissner) Corpuscles	Dermal papillae of fingertips, palms, eyelids, lips, tongue, nipples, and genitals	Light touch, texture, low-frequency vibration
Krause End Bulbs	Mucous membranes	Similar to tactile corpuscles
Ruffini Corpuscles	Dermis, subcutaneous tissue, and joint capsules	Heavy continuous touch or pressure; joint movements
Lamellated (pacinian) Corpuscles	Dermis, joint capsules, breasts, genitals, and some viscera	Deep pressure, stretch, high-frequency vibration
Muscle Spindles	Skeletal muscles near tendon	Muscle stretch (proprioception)
Golgi Tendon Organs	Tendons	Tension on tendons (proprioception)

tissue. Tactile corpuscles enable one to tell the difference between silk and sandpaper, for example, through light strokes of your fingertips.

- **Krause[5] end bulbs.** These resemble tactile corpuscles in structure and function, but occur in mucous membranes rather than in the skin.
- **Ruffini[6] corpuscles.** These are receptors for constant heavy pressure and joint movements. They are flattened, elongated capsules containing a few nerve fibers, and are located in the dermis, subcutaneous tissue, and joint capsules.
- **Lamellated (pacinian[7]) corpuscles.** These are receptors for deep pressure, stretch, and high-frequency vibration. They consist of numerous concentric layers of Schwann cells surrounding a core of one to several sensory nerve fibers. They occur in the pancreas, mesenteries, some other viscera, and deep in the dermis—especially on the hands, feet, breasts, and genitals.
- **Muscle spindles.** These receptors detect stretch in a muscle and trigger a variety of skeletal muscle (somatic) reflexes. A muscle spindle has an elongated fibrous capsule, about 4 to 10 mm long, with a fusiform[8] shape (thick in the middle and tapered at the ends). It contains 3 to 12 modified muscle fibers called **intrafusal fibers,** which lack striations and the ability to contract except at the ends. Different types of sensory nerve fibers twine around the middle the intrafusal fibers or have flowerlike endings that contact the ends of the muscle fibers.
- **Golgi tendon organs.** These receptors detect stretch in a tendon and trigger a reflex that inhibits muscle contraction to avoid muscle or tendon injury. A tendon organ is about 1 mm long and consists of a tangle of knobby nerve endings squeezed into the spaces between the collagen fibers of the tendon.

The Receptive Field

The area monitored by a single sensory neuron is called its **receptive field.** Any information arriving at the CNS by way of that neuron is interpreted as coming from that sensory field, no matter where in the field the stimulus is applied. Furthermore, if two stimuli are simultaneously applied within the same field, the brain cannot perceive them as separate, because all its input is received through a single nerve fiber. If, for example, you touch someone on the arm at two points only 20 mm apart, he or she is likely to feel only one touch, because both points usually fall within the receptive field of a single tactile neuron (fig. 17.2*a*). A separation of 47 mm is needed here for two points of contact to fall in separate receptive fields and to be felt separately (fig. 17.2*b*). On the palm of the hand, a much more sensitive area, two stimuli need be only 13 mm apart to be felt separately. The minimum separation needed to ensure stimulation of two separate tactile neurons is called the *two-point touch threshold.*

Think About It!

Braille uses symbols composed of dots that are raised about 1 mm from the page surface and spaced about 2.5 mm apart, which a person scans with the fingertips. When a blind person reads Braille, do you think he or she employs neurons with large receptive fields or small ones? Explain.

[5]William J. F. Krause (1833–1910), German anatomist
[6]Angelo Ruffini (1864–1929), Italian anatomist
[7]Filippo Pacini (1812–83), Italian anatomist
[8]*fusi* = spindle + *form* = shaped

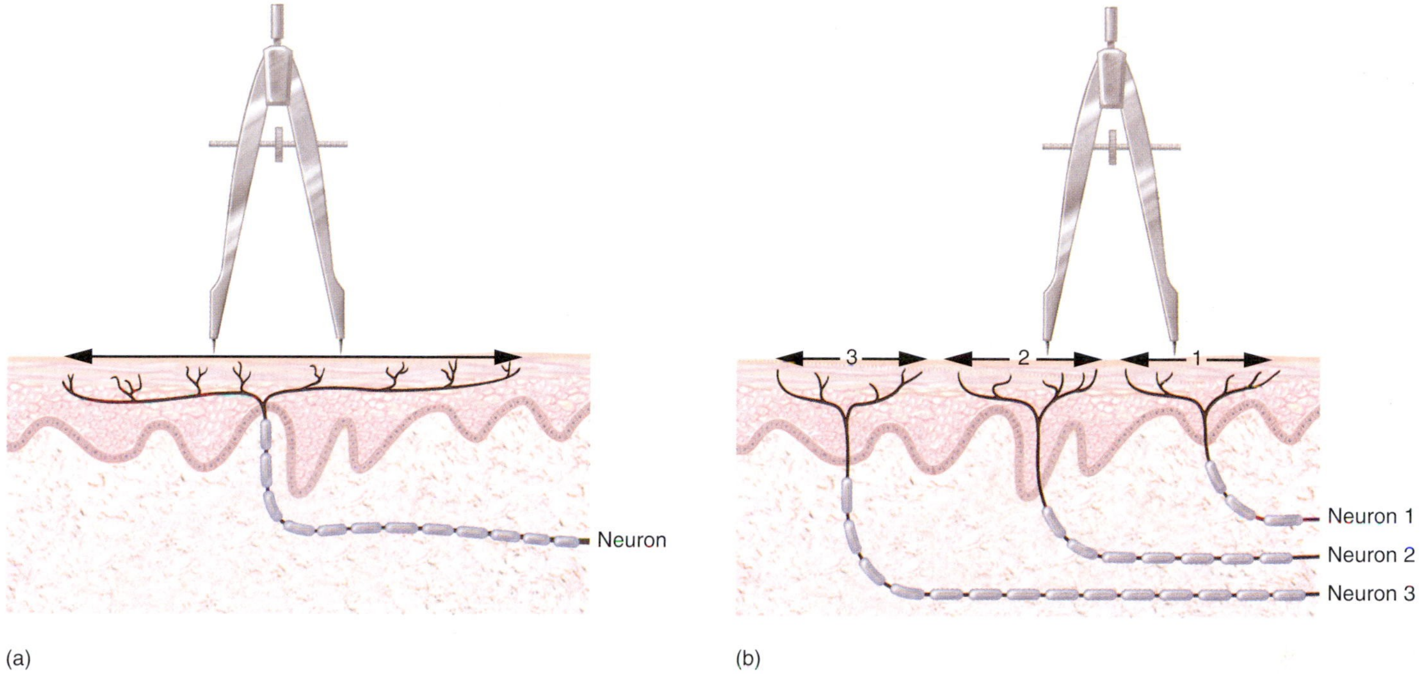
Neuron
3
2
1
Neuron 1
Neuron 2
Neuron 3
(a)
(b)

FIGURE 17.2

Receptive Fields of Sensory Neurons. (*a*) In areas of skin with relatively low sensitivity, neurons have large receptive fields. Two points of touch within the same receptive field stimulate only one neuron and are felt as a single touch. (*b*) In areas of greater sensitivity, there are more neurons in a given area and each has a smaller receptive field. Two points of contact separated by the same distance as in (*a*) may stimulate two different neurons and be felt as separate touches. *Black arrows* indicate width of receptive fields.

Somesthetic Projection Pathways

The pathways followed by sensory signals to their ultimate destinations in the CNS are called **projection pathways.** From the receptor to the final destination in the brain, most somesthetic signals travel by way of three neurons called the **first-, second-,** and **third-order neurons.** Their axons are called first- through third-order nerve fibers. The first-order (afferent) fibers for touch, pressure, and proprioception are large, myelinated, and fast; those for heat and cold are small, unmyelinated, and slow.

Somesthetic signals from the head, such as facial sensations, travel by way of several cranial nerves (especially V, the trigeminal nerve) to the pons and medulla oblongata. In the brainstem, the first-order fibers of these neurons synapse with second-order neurons that decussate and end in the contralateral thalamus. Third-order neurons then complete the route to the cerebrum. Proprioceptive signals are an exception, as the second-order fibers carry these signals to the cerebellum.

Below the head, the first-order fibers enter the dorsal horn of the spinal cord. Signals ascend the spinal cord in the spinothalamic and other pathways detailed in chapter 14 (see table 14.1 and figure 14.4). These pathways decussate either at or near the point of entry into the spinal cord, or in the brainstem, so the primary somesthetic cortex in each cerebral hemisphere receives signals from the contralateral side of the body.

Signals for proprioception below the head travel up the spinocerebellar tracts to the cerebellum. Signals from the thoracic and abdominal viscera travel to the medulla oblongata by way of sensory fibers in the vagus nerve (cranial nerve X).

Pain Pathways

Pain is a discomfort caused by tissue injury or noxious stimulation, and typically leading to evasive action. It makes us conscious of potentially injurious situations or actual tissue injuries, allowing us to avoid injury or, failing that, to favor an injured region so that it has a better chance to heal (see insight 17.1).

Pain is not merely an effect of overstimulation of somesthetic receptors. It has its own specialized receptors (nociceptors) and purpose. Nociceptors are especially dense in the skin and mucous membranes, and occur in virtually all organs, although not in the brain. In some brain surgery, the patient must remain conscious and able to talk with the surgeon; such patients need only a local scalp anesthetic.

There are two types of nociceptors corresponding to different pain sensations. Myelinated pain fibers conduct at speeds of 12 to 30 m/sec and produce the sensation of *fast (first) pain*—a feeling of sharp, localized, stabbing pain perceived at the time of injury. Unmyelinated pain fibers conduct at speeds of 0.5 to 2.0 m/sec and produce the *slow (second) pain* that follows—a longer-lasting, dull, diffuse feeling. Pain from the skin, muscles, and joints is called *somatic pain,* while pain from the viscera is called *visceral pain.* The latter often results from stretch, chemical irritants, or *ischemia* (poor blood flow), and it is often accompanied by nausea.

Pain signals from the face travel mainly by way of the trigeminal nerve to the pons, while somatic pain signals from the neck down travel by way of spinal nerve fibers to the dorsal horn of the spinal cord. These fibers synapse in the dorsal horn with second-order neurons that decussate and ascend the contralateral spinothalamic tract. The gracile fasciculus carries signals for visceral pain. By any of these

INSIGHT 17.1 CLINICAL APPLICATION

The Value of Pain

Although we generally regard pain as undesirable, we would be far worse off without it. Leprosy (Hansen disease) provides a good example of the protective function of pain. The infection of nerves by leprosy bacteria abolishes the sense of pain from affected areas. People fail to notice minor injuries such as scrapes and splinter wounds. Their neglect of the wounds leads to serious secondary infections that damage the bone and other deeper tissues. About 25% of untreated victims suffer crippling losses of fingers or toes as a result. Diabetes mellitus is also notorious for causing nerve damage *(diabetic neuropathy)* and loss of pain, contributing to lesions that often cost people their limbs.

pathways, pain signals arrive at the thalamus, where they are relayed to neurons that carry them to their final destination in the primary somesthetic cortex (postcentral gyrus) of the cerebrum (fig. 17.3). Pain signals also travel up the spinoreticular tract to the reticular formation and ultimately to the hypothalamus and limbic system. Pain signals arriving here activate visceral, emotional, and behavioral reactions to pain.

Pain in the viscera is often mistakenly thought to come from the skin or other superficial sites—for example, the pain of a heart attack is felt "radiating" along the left shoulder and medial side of the arm. This phenomenon, called **referred pain,** results from the convergence of neuronal pathways in the CNS. In the case of cardiac pain, for example, spinal cord segments T1 to T5 receive input from the heart as well as the chest and arm. Pain fibers from the heart and skin in this region converge on the same spinal interneurons, then follow the same pathway from there to the thalamus and cerebral cortex. The brain cannot distinguish which source the arriving signals are coming from. It acts as if it assumes that signals arriving by this path are most likely coming from the skin, since skin has more pain receptors than the heart and suffers injury more often. Knowledge of the origins of referred pain is essential to the skillful diagnosis of organ dysfunctions (fig. 17.4).

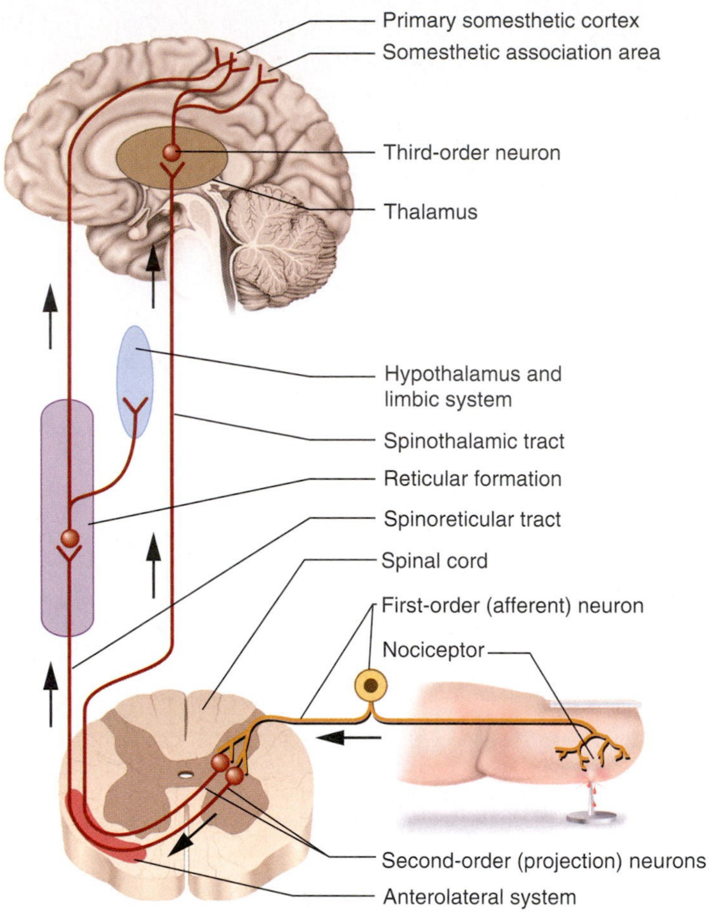

FIGURE 17.3

Projection Pathways for Pain. A first-order neuron conducts a pain signal to the dorsal horn of the spinal cord, a second-order neuron conducts it to the thalamus, and a third-order neuron conducts it to the cerebral cortex. Signals from the spinothalamic tract pass through the thalamus. Signals from the spinoreticular tract bypass the thalamus on the way to the sensory cortex.

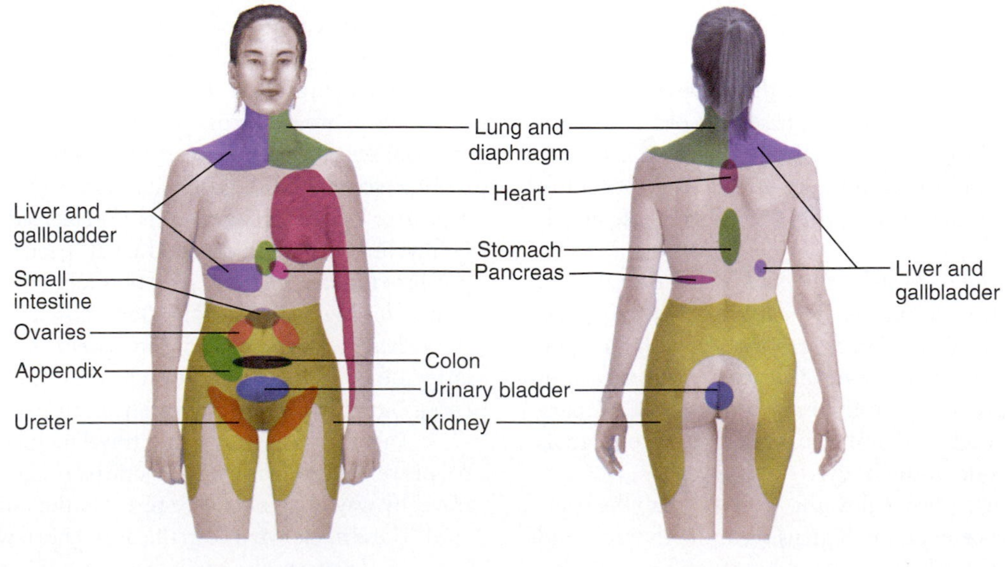

FIGURE 17.4

Referred Pain. Pain from the viscera is often felt in specific areas of the skin.

The reticular formation also issues nerve fibers back down the spinal cord, called **descending analgesic**[9] (pain-relieving) **fibers.** They travel down the reticulospinal tracts and synapse with the axons of the first-order pain neurons. Here, they secrete pain-relieving neurotransmitters called *enkephalins* and *dynorphins,* which inhibit the first-order pain neurons. Pain signals are thus stopped at the dorsal horn and do not reach the brain, so one feels less pain or none at all.

Before You Go On

Answer the following questions to test your understanding of the preceding section.

1. Distinguish between general and special senses.
2. Three schemes of receptor classification were presented in this section. In each scheme, how would you classify the receptors for a full bladder? How would you classify taste receptors?
3. What stimulus modalities are detected by free nerve endings?
4. Name any four encapsulated nerve endings and identify the stimulus modalities for which they are specialized.
5. Where do most second-order somesthetic neurons synapse with third-order neurons?
6. How do the spinothalamic tract and reticulospinal tract differ in their roles in the perception of pain?

The Chemical Senses

Objectives

When you have completed this section, you should be able to

- describe the anatomy of taste and smell receptors; and
- describe the projection pathways for these two senses.

Taste and smell are the chemical senses. In both cases, environmental chemicals bind to receptor cells and trigger nerve signals in certain cranial nerves. Other chemoreceptors, not discussed in this section, are located in the brain and blood vessels and monitor the chemistry of cerebrospinal fluid and blood.

Taste

Gustation (taste) results from the action of chemicals on the **taste buds.** There are about 4,000 taste buds distributed on the tongue, soft palate, pharynx, epiglottis, and inside the cheeks. The tongue, where the sense of taste is best developed, is marked by four types of surface projections called **lingual papillae** (fig. 17.5*a*):

1. **Filiform**[10] **papillae** are tiny spikes without taste buds. They are responsible for the rough feel of a cat's tongue and are important to many mammals for grooming the fur. They are the most abundant papillae on the human tongue (see photograph on p. 675), but they are small and play no gustatory role. They are, however, important to appreciation of the texture of food.
2. **Foliate**[11] **papillae** are also weakly developed in humans. They form parallel ridges on the sides of the tongue about two-thirds of the way back from the tip. Most of their taste buds degenerate by the age of 2 or 3 years.
3. **Fungiform**[12] (FUN-jih-form) **papillae** are shaped somewhat like mushrooms. Each one has about three taste buds, located mainly on its apex. These papillae are widely distributed but are especially concentrated at the tip and sides of the tongue.
4. **Vallate**[13] **(circumvallate) papillae** are large papillae arranged in a V at the rear of the tongue. Each is surrounded by a deep circular trench. There are only 7 to 12 vallate papillae, but they contain about half of all our taste buds—around 250 each, located on the wall of the papilla facing the trench (p. 475 and fig. 17.5*b*).

Regardless of location and sensory specialization, all taste buds look alike (fig. 17.5*c, d*). They are lemon-shaped groups of 40 to 60 cells of three kinds—*taste cells, supporting cells,* and *basal cells.* **Taste (gustatory) cells** are more or less banana-shaped and have a tuft of apical microvilli called **taste hairs,** which serve as receptor surfaces for taste molecules. The hairs project into a pit called a **taste pore** on the epithelial surface of the tongue. Taste cells are epithelial cells, not neurons; but at their bases, they synapse with sensory nerve fibers and have synaptic vesicles for the release of stimulatory neurotransmitters. A taste cell lives 7 to 10 days and is then replaced by mitosis and differentiation of basal stem cells. Supporting cells lie between the taste cells and have a similar shape, but no taste hairs.

There are five *primary taste* sensations: sweet, salty, sour, bitter, and umami. Umami, the most recently discovered primary taste, is a meaty taste stimulated by certain amino acids such as glutamate and aspartate. Pronounced "ooh-mommy," the word is Japanese and loosely means "delicious." All of the primary taste sensations can be detected throughout the tongue, but certain regions are more sensitive to one modality than to another. The tip of the tongue is most sensitive to sweets, the lateral margins to salty and sour, and the rear of the tongue (the vallate papillae) to bitter. Umami is not yet as well understood, but taste cells have been found to have umami receptors different from the receptors for any other taste. It was once popular to show "taste maps" of the tongue indicating where these modalities were localized, but sensory physiologists long ago discarded this concept because the tongue has no regional specializations of any significance to the brain's interpretation of modality.

The many flavors we perceive are not simply a mixture of these five primary tastes, but are also influenced by food texture, aroma, temperature, appearance, and one's state of mind, among other things. Food scientists refer to the texture of food as *mouthfeel.* Filiform and fungiform papillae of the tongue are innervated by the lingual nerve (a branch of the trigeminal) and are sensitive to texture. Many flavors depend on smell; without its aroma, cinnamon merely has a faintly sweet taste, and coffee and peppermint are bitter. Some flavors such as pepper are due to stimulation of free endings of the trigeminal nerve.

[9]*an* = without + *alges* = pain
[10]*fili* = thread + *form* = shaped
[11]*foli* = leaf + *ate* = like
[12]*fungi* = mushroom + *form* = shaped
[13]*vall* = wall + *ate* = like, possessing

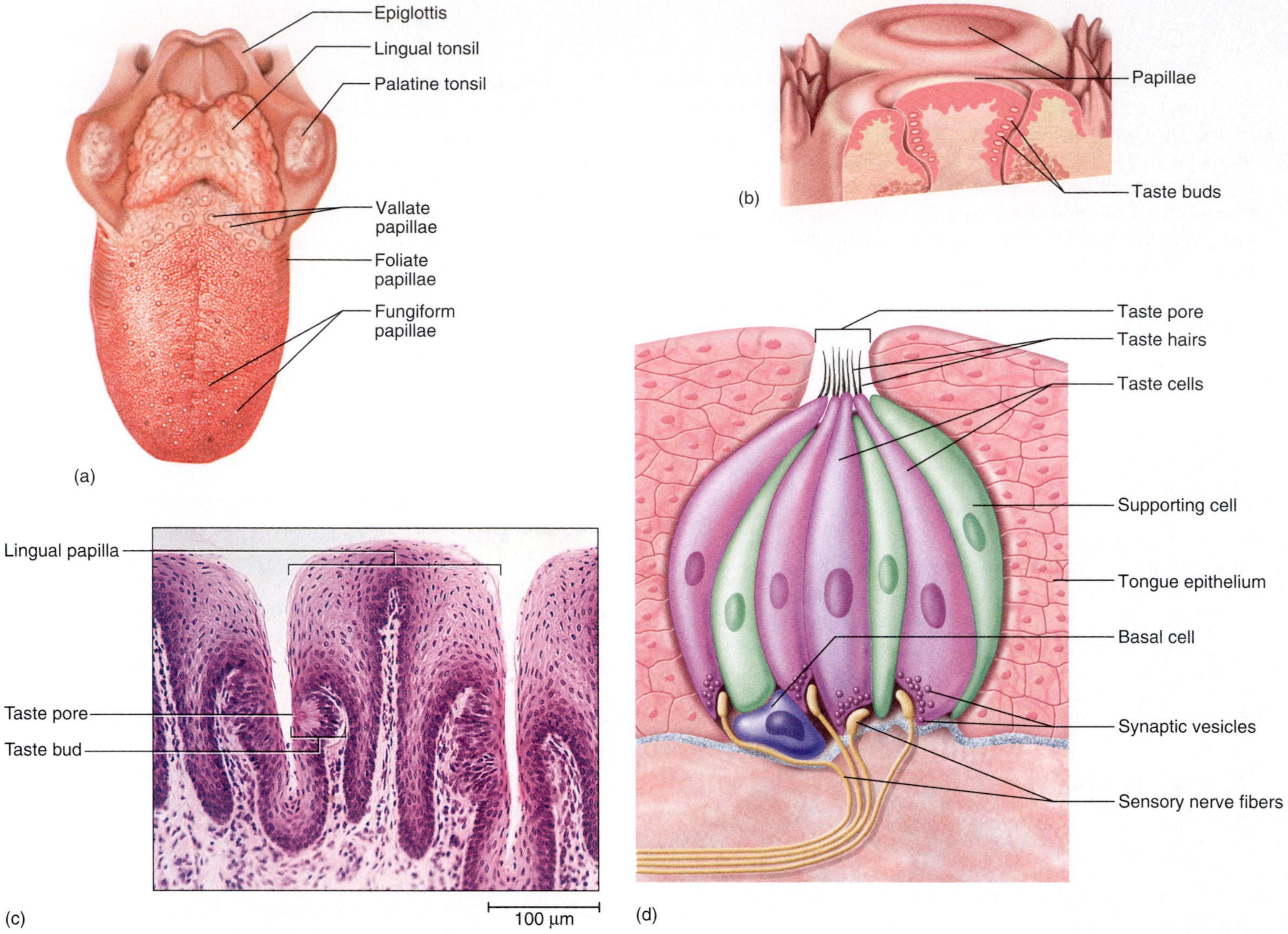

FIGURE 17.5

Taste Receptors. *(a)* Dorsal view of the tongue and locations of its papillae. *(b)* Detail of the vallate papillae. *(c)* Taste buds on the walls of two adjacent foliate papillae. *(d)* Structure of a tast bud.

Taste buds of the anterior two-thirds of the tongue stimulate the facial nerve (VII), those of the posterior one-third stimulate the glossopharyngeal nerve (IX), and those of the palate, pharynx, and epiglottis stimulate the vagus nerve (X). All taste fibers project to the *solitary nucleus* in the medulla oblongata. Second-order neurons from this nucleus relay the signals to two destinations: (1) nuclei in the hypothalamus and amygdala that activate autonomic reflexes such as salivation, gagging, and vomiting, and (2) the thalamus, which relays signals to three regions of cerebral cortex—the insula, postcentral gyrus, and roof of the lateral sulcus (fig. 17.6). Here we become conscious of the taste. Processed signals are further relayed to the orbitofrontal cortex (see fig. 15.17) where they converge with signals from the nose and eyes, and we form an overall impression of the flavor and palatability of food.

Smell

Olfaction (smell) resides in a patch of epithelium, the **olfactory mucosa,** on the roof of the nasal cavity (fig. 17.7). It covers about 5 cm^2 of the superior concha and nasal septum; the rest of the nasal cavity is lined by a nonsensory *respiratory mucosa.* This location places the olfactory cells close to the brain, but it is poorly ventilated; forcible sniffing is often needed to identify an odor or locate its source. Nevertheless, the sense of smell is highly sensitive. We can detect extremely low concentrations of odor molecules, and most people can distinguish 2,000 to 4,000 different odors; some can distinguish as many as 10,000. On average, women are more sensitive to odors than men are, and they are more sensitive to some odors near the time of ovulation than during other phases of the menstrual cycle.

The olfactory mucosa has 10 to 20 million **olfactory neurons** as well as epithelial supporting cells and basal cells. It has a yellowish tint due to lipofuscin in the supporting cells. Note that olfactory cells are neurons whereas taste cells are not. Olfactory cells are the only neurons in the body directly exposed to the external environment. Apparently this is hard on them, because they have a life span of only 60 days. Unlike most neurons, however, they are replaceable. The basal stem cells continually divide and differentiate into new olfactory cells.

An olfactory cell is shaped a little like a bowling pin. Its widest part, the soma, contains the nucleus. The neck and head of the cell are a modified dendrite with a swollen tip bearing 10 to 20 immobile

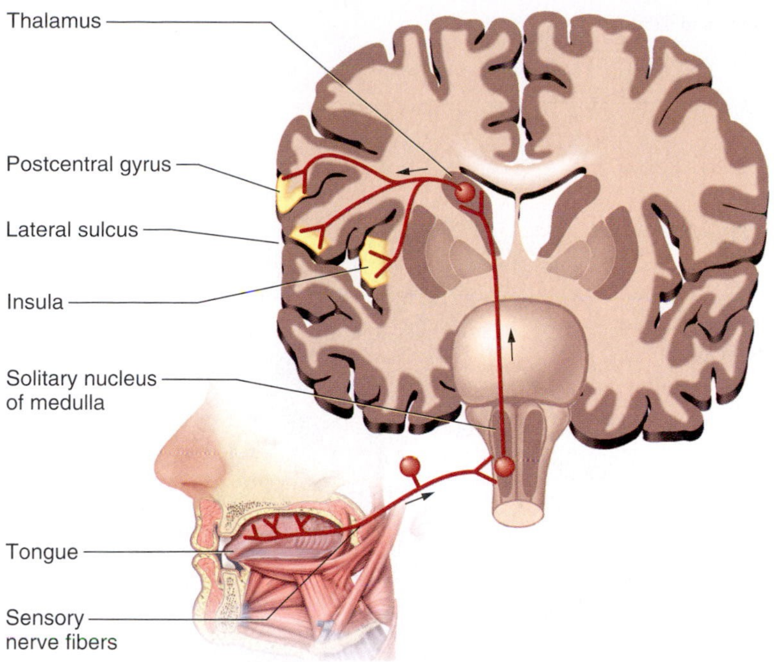
Thalamus
Postcentral gyrus
Lateral sulcus
Insula
Solitary nucleus of medulla
Tongue
Sensory nerve fibers

FIGURE 17.6

Gustatory Projection Pathways to the Cerebral Cortex. Other pathways not shown carry taste signals from the solitary nucleus to the hypothalamus and amygdala.

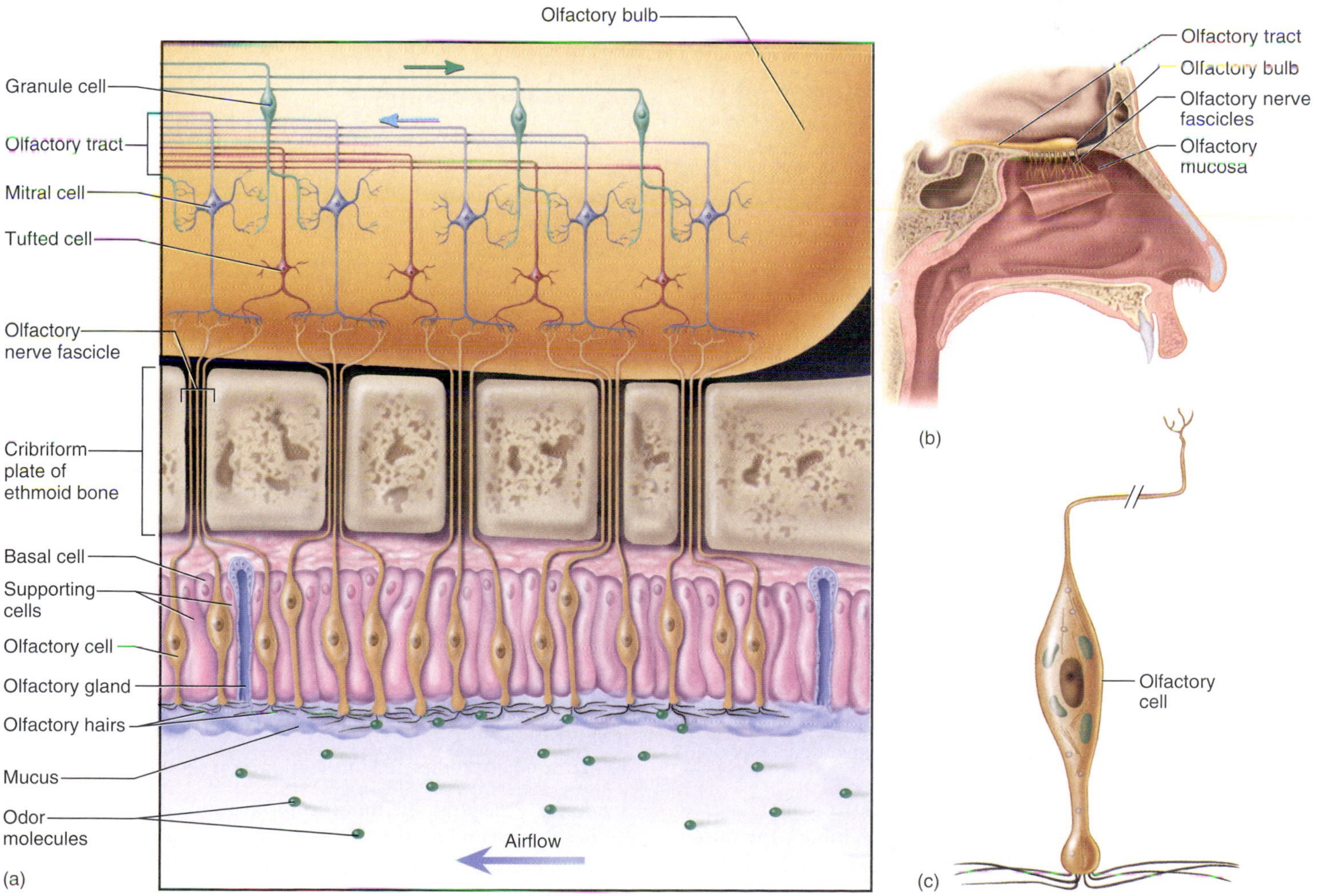
Olfactory bulb
Granule cell
Olfactory tract
Mitral cell
Tufted cell
Olfactory nerve fascicle
Cribriform plate of ethmoid bone
Basal cell
Supporting cells
Olfactory cell
Olfactory gland
Olfactory hairs
Mucus
Odor molecules
Airflow
(a)
Olfactory tract
Olfactory bulb
Olfactory nerve fascicles
Olfactory mucosa
(b)
Olfactory cell
(c)

FIGURE 17.7

Olfactory Receptors. (*a*) Neural pathways from the olfactory mucosa of the nasal cavity to the olfactory tract of the brain. (*b*) Location of the major structures in relation to the nasal and cranial cavities. (*c*) An olfactory cell.

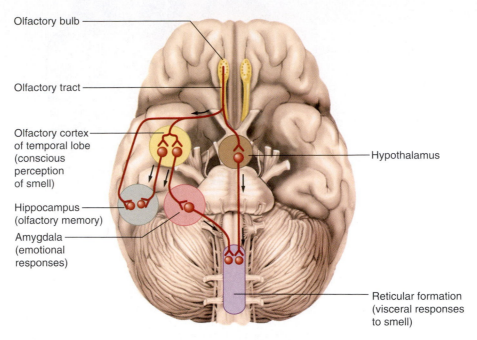

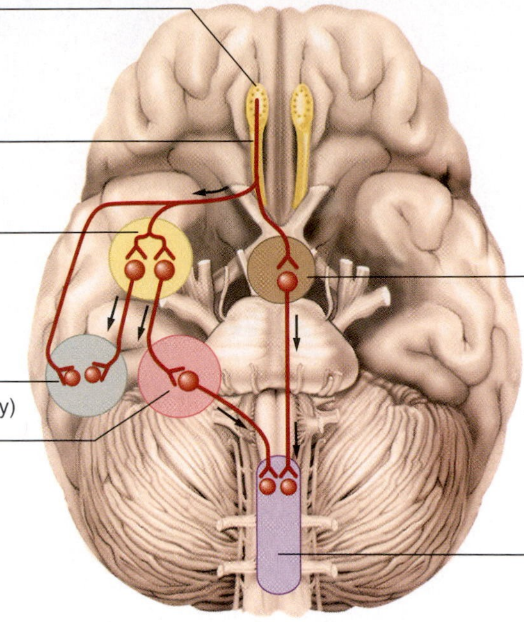
Olfactory bulb
Olfactory tract
Olfactory cortex of temporal lobe (conscious perception of smell)
Hippocampus (olfactory memory)
Amygdala (emotional responses)
Hypothalamus
Reticular formation (visceral responses to smell)

FIGURE 17.8
Olfactory Projection Pathways in the Brain.

cilia called **olfactory hairs.** These cilia bear the binding sites for odor molecules, and lie in a tangled mass embedded in a thin layer of mucus on the epithelial surface. The basal end of each cell tapers to become an axon. These axons collect into small fascicles that leave the nasal cavity through pores *(olfactory foramina)* in the cribriform plate of the ethmoid bone. Collectively, the fascicles are regarded as cranial nerve I (the olfactory nerve).

When olfactory fibers pass through the cribriform plate, they enter a pair of **olfactory bulbs** beneath the frontal lobes of the brain. In the bulbs, they synapse with neurons called *mitral cells* and *tufted cells* (fig. 17.7*b*), whose axons form bundles called the **olfactory tracts.** The tracts follow a complex pathway leading to the medial side of the temporal lobe (fig. 17.8). Olfactory input to the limbic system and hypothalamus can trigger emotional and reflex reactions. For example, we may react emotionally to the odor of certain foods, perfume, a hospital, or decaying flesh, or we may experience such reflexes as sneezing, coughing, salivating, vomiting, or gastric secretion.

There is an unusual aspect of this pathway: olfactory signals, unlike other sensory input, can reach the cerebral cortex before passing through the thalamus. Signals concerned with the conscious awareness of smell, however, do pass through the thalamus and from there to the orbitofrontal cortex mentioned earlier, where olfactory, gustatory, and visual stimuli are integrated (especially food-related stimuli).

The cerebral cortex also sends feedback to *granule cells* in the olfactory bulbs. The granule cells, in turn, inhibit the mitral cells. An effect of this is that odors can change in quality and significance under different conditions. Food may smell more appetizing when you are hungry, for example, than it does after you have just eaten.

THINK ABOUT IT!

Which taste sensations could be lost after damage to (1) the facial nerve and (2) the glossopharyngeal nerve? Why? A fracture of which cranial bone would most likely eliminate the sense of smell? Why?

Before You Go On

Answer the following questions to test your understanding of the preceding section.

7. What is the difference between a lingual papilla and a taste bud? Which is visible to the naked eye?
8. Which cranial nerves carry gustatory impulses to the brain?
9. What part of an olfactory cell bears the binding sites for odor molecules?
10. What region of the brain receives subconscious input from the olfactory cells? What region receives conscious input?

THE EAR

Objectives

When you have completed this section, you should be able to

- describe the gross and microscopic anatomy of the ear;
- briefly explain how the ear converts vibrations to nerve impulses and discriminates between sounds of different intensity and pitch;
- explain how the anatomy of the vestibular apparatus relates to our ability to interpret the body's position and movements; and

• describe the pathways taken by auditory and vestibular signals to the brain.

Hearing is a response to vibrating air molecules, and *equilibrium* is the sense of motion and balance. These senses reside in the inner ear, a maze of fluid-filled passages and sensory cells encased in the temporal bone.

Anatomy of the Ear

The ear has three sections called the *outer, middle,* and *inner ear.* The first two are concerned only with transmitting sound to the inner ear, where vibration is converted to nerve signals.

OUTER EAR

The **outer (external) ear** is essentially a funnel for conducting airborne vibrations to the eardrum. It begins with the fleshy **auricle (pinna)** on the side of the head, shaped and supported by elastic cartilage except for the earlobe. The auricle has a predictable arrangement of whorls and recesses that direct sound into the auditory canal (fig. 17.9).

The **auditory canal** is the passage through the temporal bone. Beginning at the external opening, the **external acoustic meatus,** it follows a slightly S-shaped course for about 3 cm to the eardrum (fig. 17.10). It is lined with skin and supported by fibrocartilage at its opening and by the temporal bone for the rest of its length. The canal has ceruminous and sebaceous glands whose secretions mix with dead skin cells and form **cerumen** (earwax). Cerumen normally dries up and falls from the canal, but sometimes it becomes impacted and interferes with hearing.

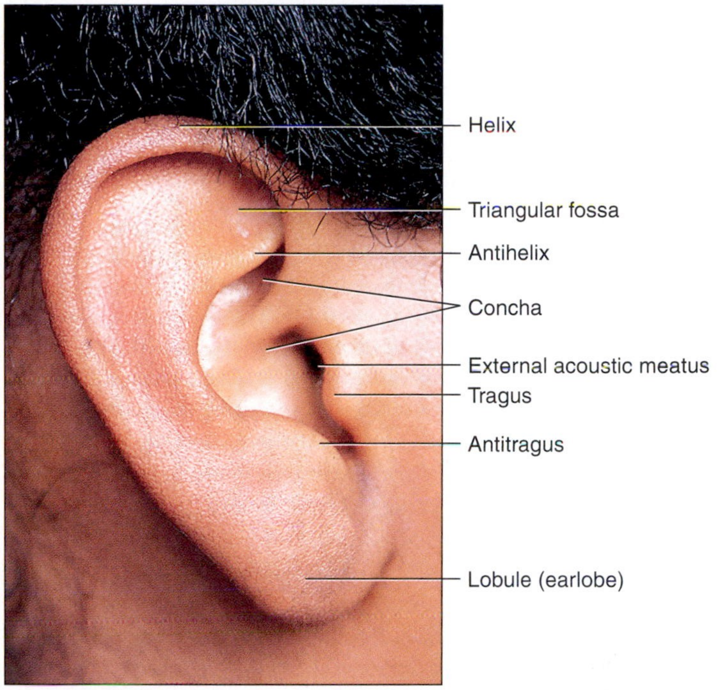

FIGURE 17.9
External Anatomy of the Ear.

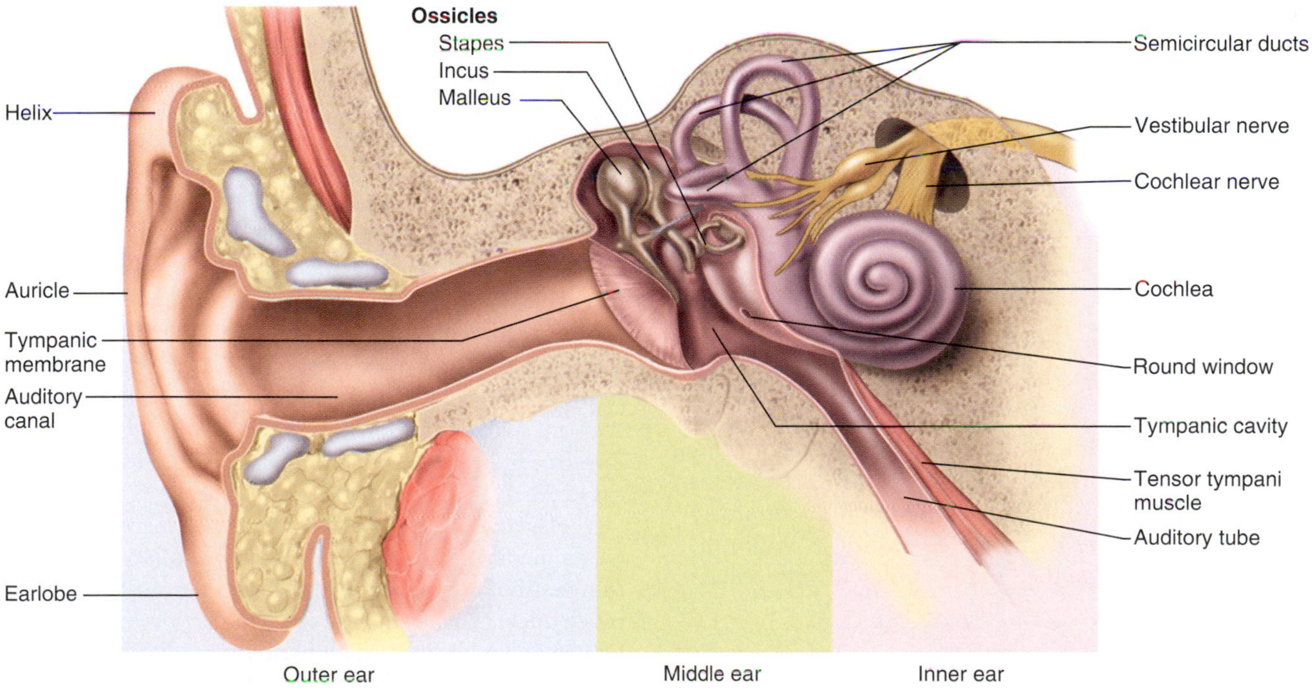

FIGURE 17.10
Internal Anatomy of the Ear.

MIDDLE EAR

The **middle ear** consists mainly of tiny bones and muscles housed in the **tympanic cavity** of the temporal bone. It begins with the eardrum, or **tympanic**[14] **membrane,** which closes the inner end of the auditory canal and separates it from the middle ear. The eardrum is about 1 cm in diameter and slightly concave on its outer surface. It is suspended in a ring-shaped groove in the temporal bone and vibrates freely in response to sound. It is innervated by sensory branches of the vagus and trigeminal nerves and is highly sensitive to pain.

Posteriorly, the tympanic cavity is continuous with the mastoidal air cells in the mastoid process. The cavity is filled with air that enters by way of the **auditory (eustachian**[15]**) tube,** a passage to the nasopharynx. (Be careful not to confuse *auditory tube* with *auditory canal.*) The auditory tube is normally flattened and closed, but swallowing or yawning opens it and allows air to enter or leave the tympanic cavity. This equalizes air pressure on both sides of the eardrum, allowing it to vibrate freely. Excessive pressure on one side or the other dampens the sense of hearing. Unfortunately, the auditory tube also allows throat infections to spread to the middle ear (see insight 17.2).

The tympanic cavity contains the three smallest bones and the two smallest skeletal muscles of the body. The bones, called the **auditory ossicles,**[16] span the 2 to 3 mm distance from the eardrum to the inner ear. Progressing inward, the first is the **malleus,**[17] which has an elongated *handle* attached to the inner surface of the eardrum; a *head,* which is suspended from the wall of the tympanic cavity; and a *short process,* which articulates with the next ossicle. The second bone, the **incus,**[18] articulates in turn with the **stapes**[19] (STAY-peez). The stapes has an arch and footplate that give it a shape like a stirrup. The *footplate,* shaped like the sole of a steam iron, is held by a ringlike ligament in an opening called the **oval window,** where the inner ear begins.

The muscles of the middle ear are the stapedius and tensor tympani. The **stapedius** (stay-PEE-dee-us) arises from the posterior wall of the cavity and inserts on the stapes. The **tensor tympani** (TEN-sor TIM-pan-eye) arises from the wall of the auditory tube, travels alongside it, and inserts on the malleus. In response to loud noises, these muscles contract and dampen the vibration of the ossicles, thus protecting the delicate sensory cells of the inner ear; this is called the *tympanic reflex.* The sensory cells can nevertheless be irreversibly damaged by sudden loud noises such as gunshots, and by sustained loud noise such as factory noise and loud music.

INNER EAR

The **inner (internal) ear** is housed in a maze of temporal bone passages called the **bony labyrinth,** which is lined by a system of fleshy tubes called the **membranous labyrinth** (fig. 17.11). Between the bony and membranous labyrinths is a cushion of fluid, similar to cerebrospinal fluid, called **perilymph** (PER-ih-limf). Within the membranous labyrinth is another fluid, similar to intracellular fluid, called **endolymph.**

[14]*tympan* = drum
[15]Bartholomeo Eustachio (1520–74), Italian anatomist
[16]*oss* = bone + *icle* = little
[17]*malleus* = hammer, mallet
[18]*incus* = anvil
[19]*stapes* = stirrup

INSIGHT 17.2 CLINICAL APPLICATION

Middle-Ear Infection

Otitis[20] *media* (middle-ear infection) is especially common in children because their auditory tubes are relatively short and horizontal. Upper respiratory infections spread easily from the throat to the tympanic cavity and mastoidal air cells. Fluid accumulates in the cavity and causes pressure, pain, and impaired hearing. If otitis media goes untreated, it may spread from the mastoidal air cells and cause meningitis, a potentially deadly infection of the meninges. Otitis media can also cause fusion of the middle-ear bones, preventing them from vibrating freely and thus causing hearing loss. It is sometimes necessary to drain fluid from the tympanic cavity by lancing the eardrum and inserting a tiny drainage tube—a procedure called *myringotomy.*[21] The tube, which is eventually discharged spontaneously by the ear, relieves the pressure and permits the infection to heal.

[20]*ot* = ear + *itis* = inflammation
[21]*myringo* = eardrum + *tomy* = cutting

The membranous labyrinth begins with a chamber called the **vestibule,** which contains organs of equilibrium to be discussed later. The organ of hearing is the **cochlea**[22] (COC-lee-uh), a coiled tube that arises from the anterior side of the vestibule. In other vertebrates, the cochlea is straight or slightly curved. In most mammals, however, it assumes the form of a snaillike spiral, allowing a longer cochlea to fit in a compact space. In humans, the spiral is about 9 mm wide at the base and 5 mm high, with the apex pointing anterolaterally. The cochlea winds for about 2.5 coils around an axis of spongy bone called the **modiolus**[23] (mo-DY-oh-lus). The modiolus is shaped like a screw; its threads form a spiral platform that supports the fleshy tube of the cochlea.

A vertical section cuts through the cochlea about five times (fig. 17.12*a*). A single cross section through it looks like figure 17.12*b.* It is important to realize that the structures seen in cross section actually have the form of spiral strips winding around the modiolus from base to apex.

The cochlea has three fluid-filled chambers separated by membranes. The superior chamber is called the **scala**[24] **vestibuli** (SCAY-la vess-TIB-you-lye) and the inferior one is the **scala tympani** (TIM-pan-eye). These are filled with perilymph and communicate with each other through a narrow channel called the *helicotrema* (HEL-ih-co-tree-muh) at the apex of the cochlea. The scala vestibuli begins near the oval window and spirals to the apex; from there, the scala tympani spirals back down to the base and ends at the **round window** (see fig. 17.11). The round window is covered by a membrane called the *secondary tympanic membrane.*

The middle chamber is a triangular space, the **cochlear duct (scala media).** It is separated from the scala vestibuli above by a thin **vestibular membrane,** and from the scala tympani below by a much thicker **basilar membrane.** Unlike those chambers, it is filled with endolymph rather than perilymph. Within the cochlear duct,

[22]*cochlea* = snail
[23]*modiolus* = hub
[24]*scala* = staircase

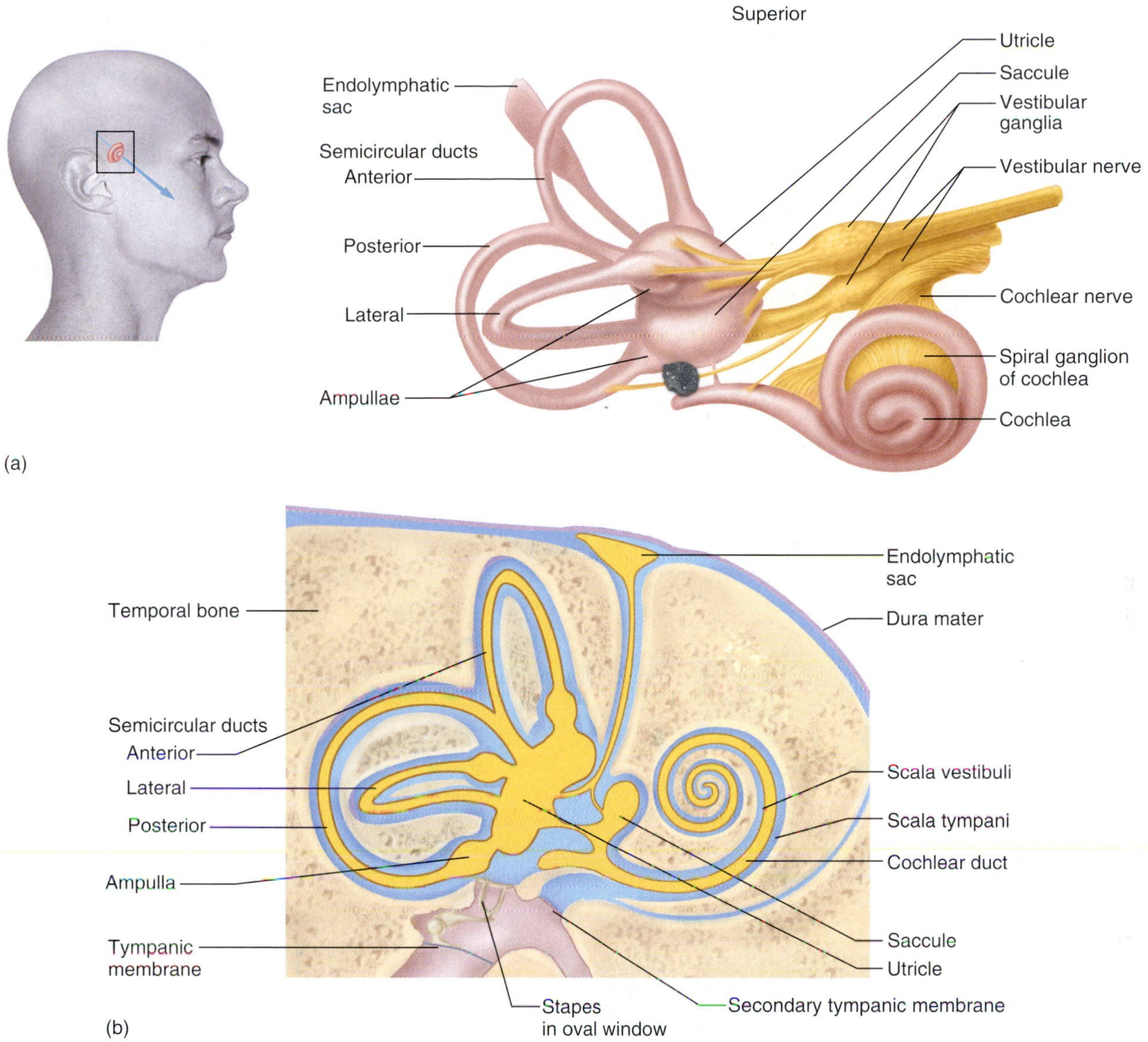
Superior
Utricle
Saccule
Vestibular ganglia
Vestibular nerve
Endolymphatic sac
Semicircular ducts
Anterior
Posterior
Lateral
Ampullae
Cochlear nerve
Spiral ganglion of cochlea
Cochlea
(a)
Endolymphatic sac
Dura mater
Temporal bone
Semicircular ducts
Anterior
Lateral
Posterior
Ampulla
Tympanic membrane
Scala vestibuli
Scala tympani
Cochlear duct
Saccule
Utricle
Stapes in oval window
Secondary tympanic membrane
(b)

FIGURE 17.11

Anatomy of the Inner Ear. (*a*) The membranous labyrinth. (*b*) Relationship of the perilymph (*blue*) and endolymph (*yellow*) to the labyrinth.

supported on the basilar membrane, is the **organ of Corti**[25] (COR-tee), a thick epithelium with associated structures (fig. 17.12*c*). This is the device that converts vibrations into nerve impulses, so we must pay particular attention to its structural details.

The organ of Corti has an epithelium composed of **hair cells** and **supporting cells.** Hair cells are named for the long, stiff microvilli called **stereocilia**[26] on their apical surfaces. (Stereocilia should not be confused with true cilia. They do not have an axoneme of microtubules as seen in cilia, and they do not move by themselves.) Resting on top of the stereocilia is a gelatinous **tectorial**[27] **membrane.**

The organ of Corti has four rows of hair cells spiraling along its length (fig. 17.13). About 3,500 of these, called **inner hair cells (IHCs),** are arranged in a row by themselves on the medial side of the basilar membrane (facing the modiolus). Each IHC has a cluster of 50 to 60 stereocilia, graded from short to tall. Another 20,000 **outer hair cells (OHCs)** are neatly arranged in three rows across from the inner hair cells. Each OHC has about 100 stereocilia arranged in a V, with their tips embedded in the tectorial membrane. All that we hear comes from the IHCs, which supply 90% to 95% of the sensory fibers of the cochlear nerve. The function of the OHCs is to adjust the response of the cochlea to different frequencies and enable the IHCs to work with greater precision. Hair cells are not neurons, but they synapse with nerve fibers at their base—the OHCs with both sensory and motor neurons and the IHCs with sensory neurons only.

Auditory Function

To make functional sense of this anatomy, we will examine the essential aspects of auditory function. Figure 17.14 is a mechanical model that presents some of the basic mechanisms in simple form. When sound waves vibrate the eardrum, the three auditory ossicles

[25]Alfonso Corti (1822–88), Italian anatomist
[26]*stereo* = solid
[27]*tect* = roof

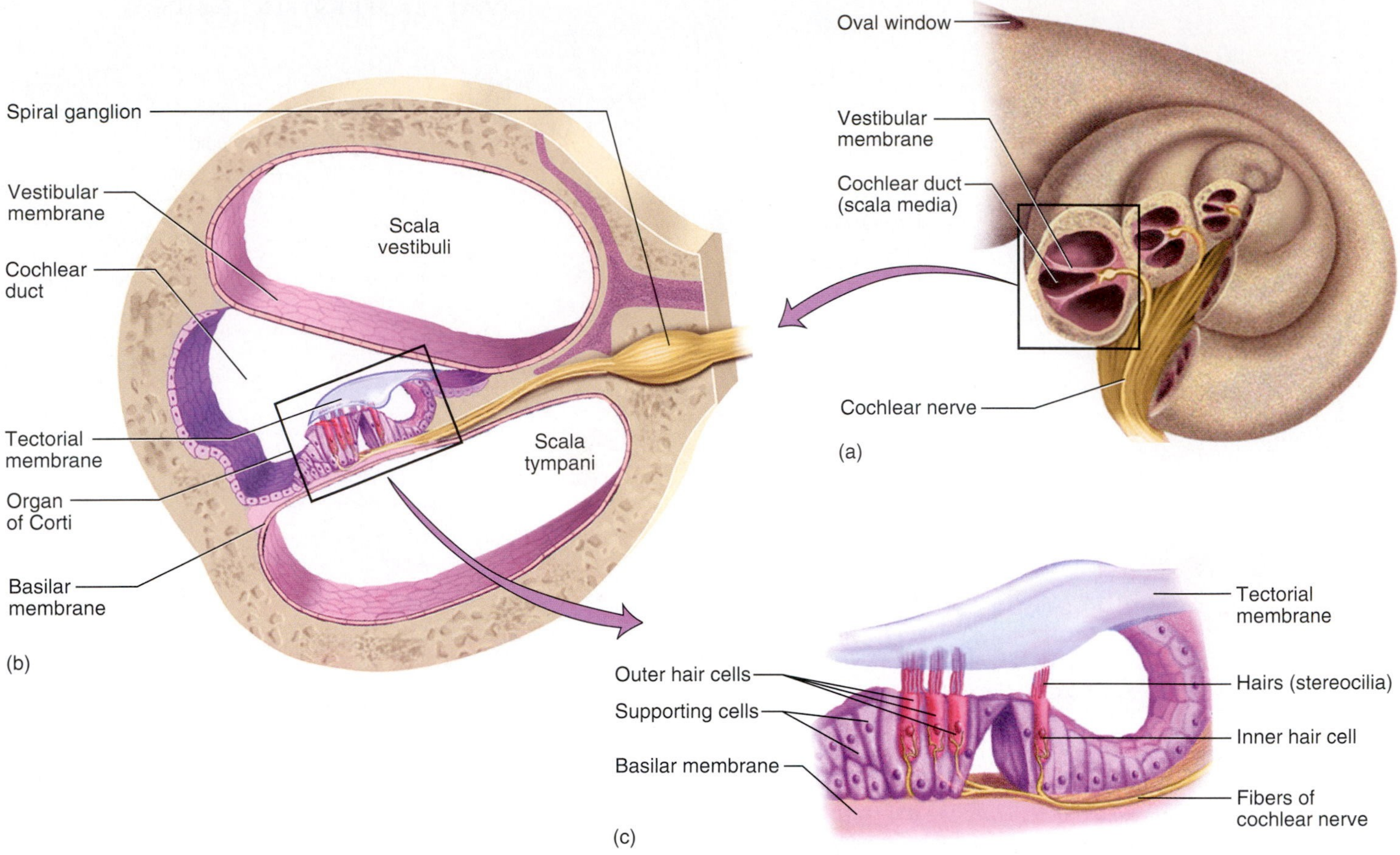

FIGURE 17.12

Anatomy of the Cochlea. (*a*) Vertical section. The apex of the cochlea faces downward and anterolaterally in anatomical position. (*b*) Detail of one section through the cochlea. (*c*) Detail of the organ of Corti.

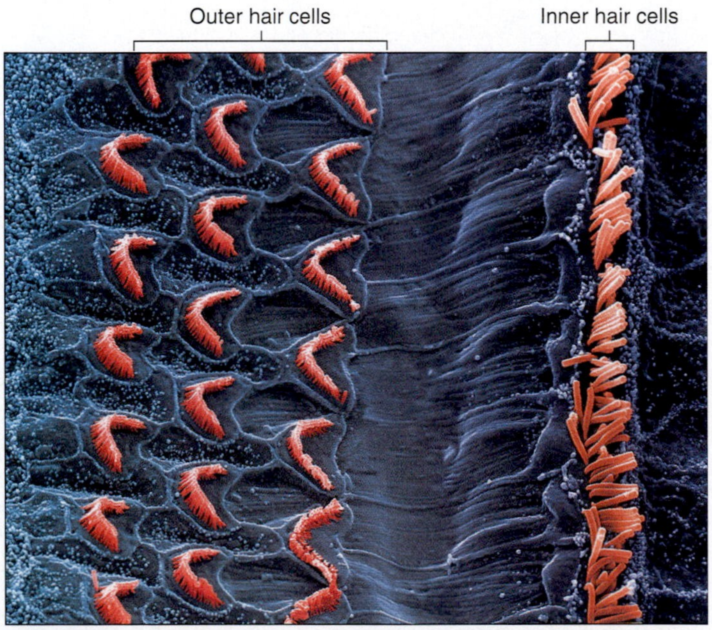

FIGURE 17.13

Surface of the Organ of Corti Showing Cochlear Hair Cells (SEM). On the *left* are the three rows of outer hair cells, which serve only to tune the cochlea. Each cell has a V-shaped row of stereocilia. On the *right* is the single row of inner hair cells, which generate the signals we hear.

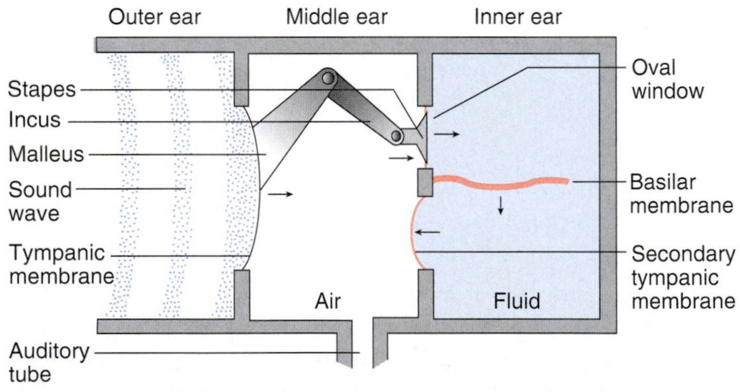

FIGURE 17.14

Mechanical Model of Auditory Function. Each inward movement of the tympanic membrane pushes inward on the auditory ossicles of the middle ear and fluid of the inner ear. This pushes down on the basilar membrane, and pressure is relieved by an outward bulge of the secondary tympanic membrane. Thus the basilar membrane vibrates up and down in synchrony with the vibrations of the tympanic membrane.

transfer these vibrations to the inner ear. The footplate of the stapes moves the fluid in the inner ear, and fluid movements cause the basilar membrane to vibrate up and down. As it does so, the hair cells on the basilar membrane are thrust up and down, while the tectorial membrane immediately above them remains relatively still, forcing the stereocilia on the hair cells to rock back and forth. The stereocilia have mechanically gated potassium channels at their tips; the rocking opens these gates and lets a burst of K^+ ions into the hair cell with each vibration. In response to the voltage change caused by this K^+ inflow, the hair cell releases a neurotransmitter from its base that stimulates the first-order sensory neuron. This neuron conducts signals through the cochlear nerve, as described shortly.

The cochlea must transmit signals that the brain can distinguish as differences in loudness and pitch. Loud sounds produce more vigorous vibrations of the organ of Corti over a broader area of the basilar membrane, and trigger a higher frequency of action potentials in the cochlear nerve fibers. High-frequency (high-pitched) sounds cause the free end of the basilar membrane, near the tip of the cochlea, to vibrate more than the attached, basal end, whereas low-frequency sounds cause the basal end to vibrate more. Thus, the brain can distinguish loudness and pitch from the number of hair cells responding, how frequently the cochlear nerve fibers are firing, and the relative intensity of signaling coming from different regions of the organ of Corti.

In order to tune the cochlea and sharpen its frequency discrimination, the brainstem sends motor signals back through the cochlear nerve to the outer hair cells (OHCs). The OHCs are anchored to the basilar membrane below and anchored to the tectorial membrane through their stereocilia above. They contract in response to signals from the brain, tugging on the basilar and tectorial membranes and thus suppressing the vibration of specific regions of the basilar membrane. This enhances the ability of the brain to tell one sound frequency from another—an ability that is important for distinguishing the words in someone else's speech, among other purposes.

The Auditory Projection Pathway

Winding around the modiolus is a **spiral ganglion** composed of somas of the bipolar sensory neurons of the cochlea (see fig. 17.11*a*). The dendrites of these neurons come from the bases of the hair cells, and their axons lead away to form the **cochlear nerve.** The cochlear nerve joins the *vestibular nerve,* discussed later, and the two together become the *vestibulocochlear nerve* (cranial nerve VIII).

Cochlear nerve fibers project to the *cochlear nucleus* on each side of the medulla oblongata. They synapse with second-order neurons that lead to the nearby *superior olivary nucleus* of the pons (fig. 17.15). The superior olivary nucleus has multiple connections and functions:

- It sends signals back to the cochlea by way of cranial nerve VIII, stimulating the outer hair cells for the purpose of cochlear tuning.
- It sends signals by way of cranial nerves V_3 and VII to the tensor tympani and stapedius muscles, respectively, which are responsible for the protective tympanic reflex.

INSIGHT 17.3 CLINICAL APPLICATION

DEAFNESS

Deafness means any hearing loss, from mild and temporary to complete and irreversible. *Conduction deafness* is a type that results from the impaired transmission of vibrations to the inner ear. It can result from a damaged eardrum, otitis media, or blockage of the auditory canal. Another cause is *otosclerosis,*[28] fusion of the auditory ossicles to each other or fusion of the stapes to the oval window, preventing the bones from vibrating freely. *Sensorineural (nerve) deafness* results from the death of hair cells or any of the nervous elements concerned with hearing. It is a common occupational disease of musicians and others who work in noisy environments. Deafness leads some people to develop delusions of being talked about, disparaged, or cheated. Beethoven said his deafness drove him nearly to suicide.

[28]*oto* = ear + *scler* = hardening + *osis* = process, condition

- It plays a role in *binaural*[29] *hearing*—comparing signals from the right and left ears to identify the direction from which a sound is coming.
- It issues fibers up the brainstem to the inferior colliculi of the midbrain.

The inferior colliculi aid in binaural hearing and issue fibers to the thalamus. In the thalamus, these in turn synapse with neurons that continue to the primary auditory cortex in the superior part of each temporal lobe. The temporal lobe is the site of conscious perception of sound; it completes the information processing essential to binaural hearing. The interpretation of sound in relation to memory—for example, the ability to recognize what a sound is—occurs in the auditory association area bordering the primary auditory cortex (see fig. 15.18).

Extensive connections exist between the right and left nuclei of hearing throughout the brainstem, allowing for comparison of the inputs from the right and left ears and the localization of sounds in space. Thus, unlike the somesthetic cortex, the auditory cortex on each side of the brain receives signals from both ears. Because of this extensive decussation, damage to the right or left auditory cortex does not cause a unilateral loss of hearing.

The Vestibular Apparatus

The original function of the ear in vertebrate history was not hearing, but **equilibrium**—coordination and balance. Only later did vertebrates evolve the cochlea, middle-ear structures, and auditory function of the ear. In humans, the receptors for equilibrium constitute the **vestibular apparatus,** which consists of three **semicircular ducts** and two chambers—an anterior **saccule**[30] (SAC-yule) and a posterior **utricle**[31] (YOU-trih-cul) (see fig.17.11).

The sense of equilibrium is divided into **static equilibrium,** the perception of the orientation of the head when the body is stationary, and **dynamic equilibrium,** the perception of motion or

[29]*bin* = two + *aur* = ears
[30]*saccule* = little sac
[31]*utricle* = little bag

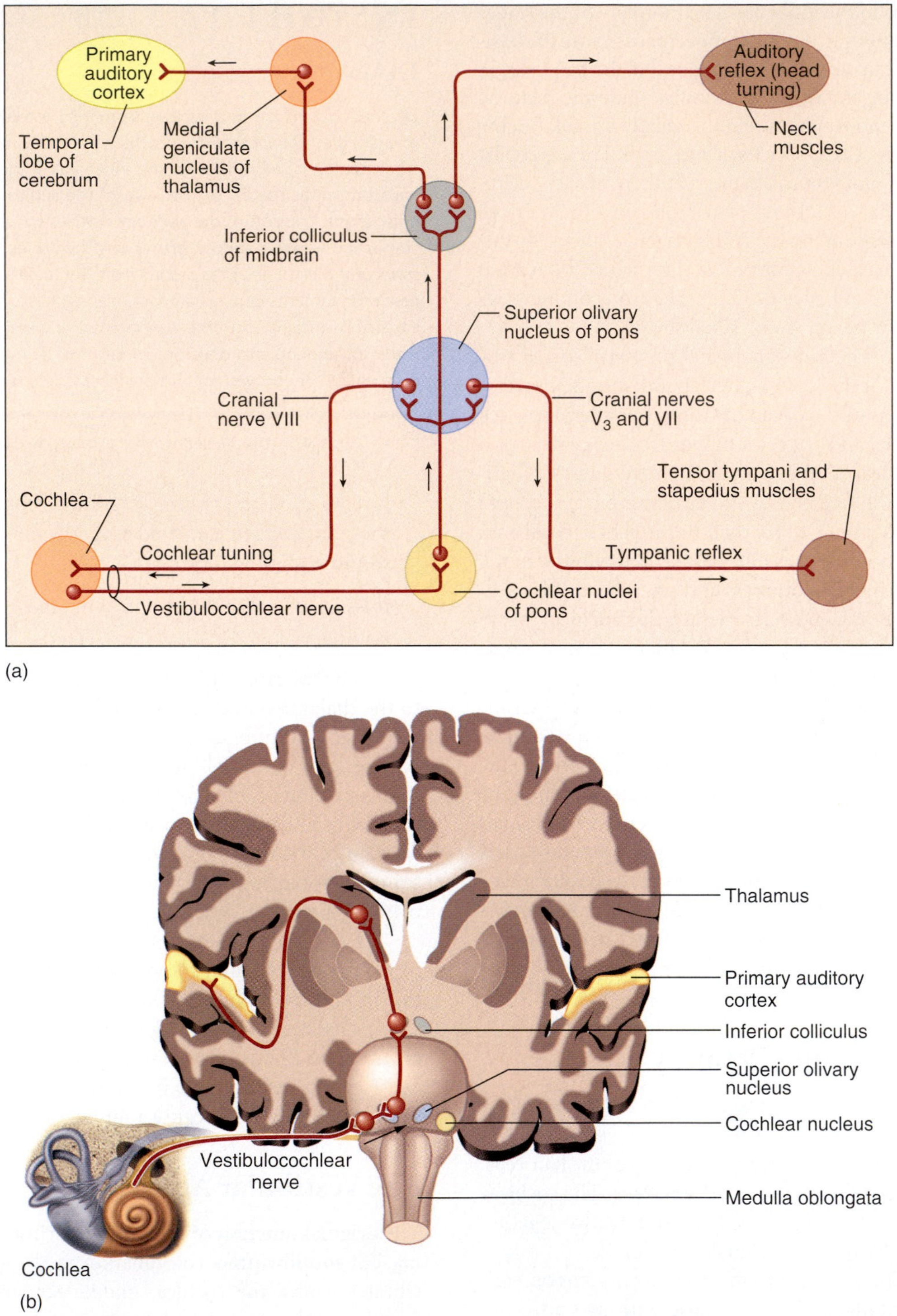

FIGURE 17.15

Auditory Pathways in the Brain. (*a*) Schematic. (*b*) Brainstem and frontal section of the cerebrum, showing the locations of auditory processing centers.

acceleration. Acceleration is divided into *linear acceleration,* a change in velocity in a straight line, as when riding in a car or elevator, and *angular acceleration,* a change in the rate of rotation. The saccule and utricle are responsible for static equilibrium and the sense of linear acceleration; the semicircular ducts detect only angular acceleration.

THE SACCULE AND UTRICLE

The saccule and utricle each have a 2-by-3 mm patch of hair cells and supporting cells called a *macula.*[32] The **macula sacculi** lies

[32]*macula* = spot

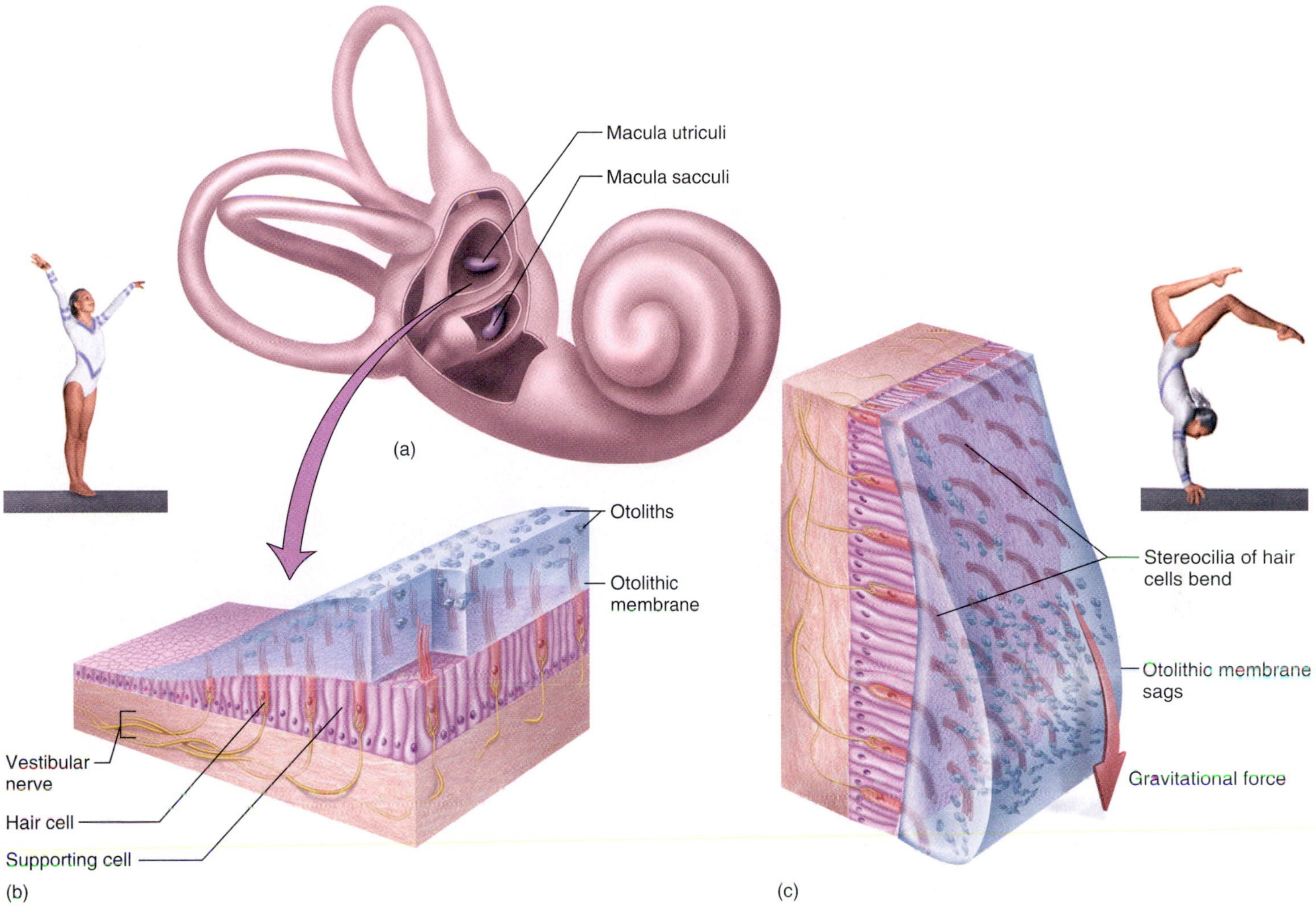
Macula utriculi
Macula sacculi
(a)
Otoliths
Otolithic membrane
Vestibular nerve
Hair cell
Supporting cell
(b)
Stereocilia of hair cells bend
Otolithic membrane sags
Gravitational force
(c)

FIGURE 17.16

The Saccule and Utricle. (*a*) Locations of the macula sacculi and macula utriculi. (*b*) Structure of a macula. (*c*) Action of the otolithic membrane on the hair cells when the head is tilted.

nearly vertically on the wall of the saccule, and the **macula utriculi** lies nearly horizontally on the floor of the utricle (fig. 17.16*a*). Each hair cell of a macula has 40 to 70 stereocilia and one motile true cilium, called a **kinocilium.**[33] The tips of the stereocilia and kinocilium are embedded in a gelatinous **otolithic membrane.** This membrane is weighted with granules called **otoliths,**[34] composed of calcium carbonate and protein (fig.17.16*b*). By adding to the density and inertia of the membrane, the otoliths enhance the sense of gravity and motion.

Figure 17.16*c* shows how the macula utriculi detects tilt of the head. With the head erect, the otolithic membrane bears directly down on the hair cells and stimulation is minimal. When the head is tilted, however, the weight of the membrane bends the stereocilia and stimulates the hair cells. Any orientation of the head causes a combination of stimulation to the utricles and saccules of the two ears, which enables the brain to sense head orientation by comparing these inputs to each other and to other input from the eyes and stretch receptors in the neck. The macula sacculi works similarly except that its vertical orientation makes it more responsive to up-and-down movements of the body, for example when you stand up or jump down from a height.

The inertia of the otolithic membranes is especially important in detecting linear acceleration. Suppose you are sitting in a car at a stoplight and then begin to move. The heavy otolithic membrane of the macula utriculi briefly lags behind the rest of the tissues, bends the stereocilia backward, and stimulates the cells. When you stop at the next intersection, the macula stops but the otolithic membrane keeps on going for a moment, bending the stereocilia forward. The hair cells convert this pattern of stimulation to nerve signals, and the brain is thus advised of changes in your linear velocity.

If you are standing in an elevator and it begins to move up, the otolithic membrane of the vertical macula sacculi lags behind briefly and pulls down on the hair cells. When the elevator stops, the otolithic membrane keeps on going for a moment and bends the hairs upward. The macula sacculi thus detects vertical acceleration.

[33]*kino* = moving
[34]*oto* = ear + *lith* = stone

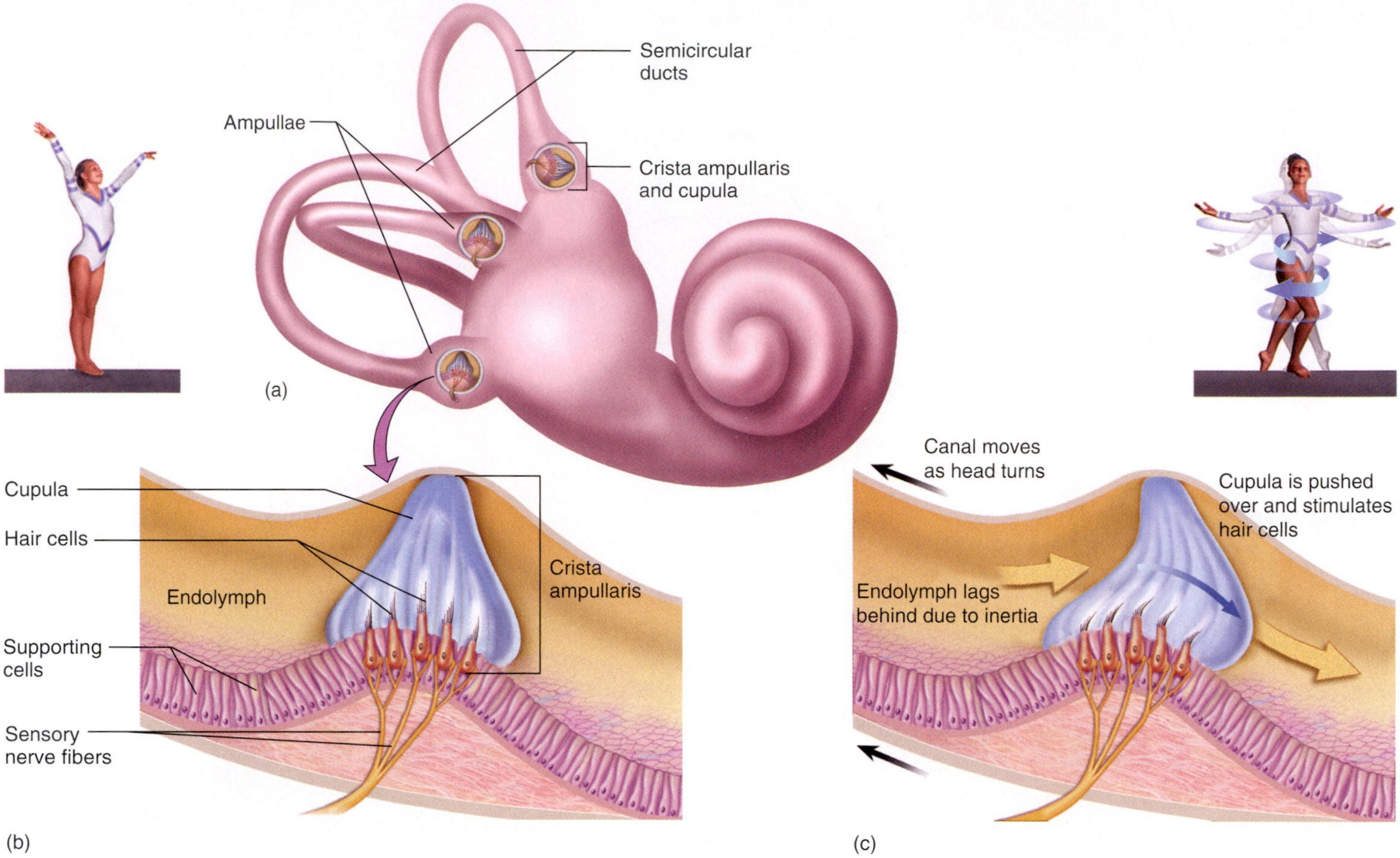

Semicircular ducts
Ampullae
Crista ampullaris and cupula
(a)
Cupula
Hair cells
Endolymph
Crista ampullaris
Supporting cells
Sensory nerve fibers
(b)
Canal moves as head turns
Cupula is pushed over and stimulates hair cells
Endolymph lags behind due to inertia
(c)

FIGURE 17.17

Structure and Function of the Semicircular Ducts. (*a*) Structure of the semicircular ducts, with each ampulla opened to show the crista ampullaris and cupula. (*b*) Detail of the crista ampullaris. (*c*) Action of the endolymph on the cupula and hair cells when the head is rotated.

THE SEMICIRCULAR DUCTS

Angular acceleration is detected by the three *semicircular ducts* (fig. 17.17), each housed in an osseous *semicircular canal* of the temporal bone. The *anterior* and *posterior semicircular ducts* are positioned vertically, at right angles to each other. The *lateral semicircular duct* is about 30° from horizontal. The orientation of the ducts causes a different duct to be stimulated by rotation of the head in different planes—turning it from side to side as in gesturing "no," nodding up and down as in gesturing "yes," or tilting it from side to side as in touching your ears to your shoulders.

The semicircular ducts are filled with endolymph. Each duct opens into the utricle and has a dilated sac at one end called an **ampulla.**[35] Within the ampulla is a mound of hair cells and supporting cells called the **crista ampullaris**[36]**.** The hair cells have stereocilia and a kinocilium embedded in the **cupula,**[37] a gelatinous membrane that extends from the crista to the roof of the ampulla. When the head turns the duct rotates, but the endolymph lags behind and pushes the cupula. This bends the stereocilia and stimulates the hair cells. After 25 to 30 seconds of continual rotation, however, the endolymph catches up with the movement of the duct, and stimulation of the hair cells ceases even though motion continues.

[35]*ampulla* = little jar
[36]*crista* = crest, ridge + *ampullaris* = of the ampulla
[37]*cupula* = little tub

THINK ABOUT IT!

*The semicircular ducts do not detect motion itself, but only acceleration—*a change in the rate *of motion. Explain why.*

Vestibular Projection Pathways

Hair cells of the macula sacculi, macula utriculi, and semicircular ducts synapse at their bases with sensory fibers of the **vestibular nerve.** This nerve joins the cochlear nerve, forming the vestibulo-

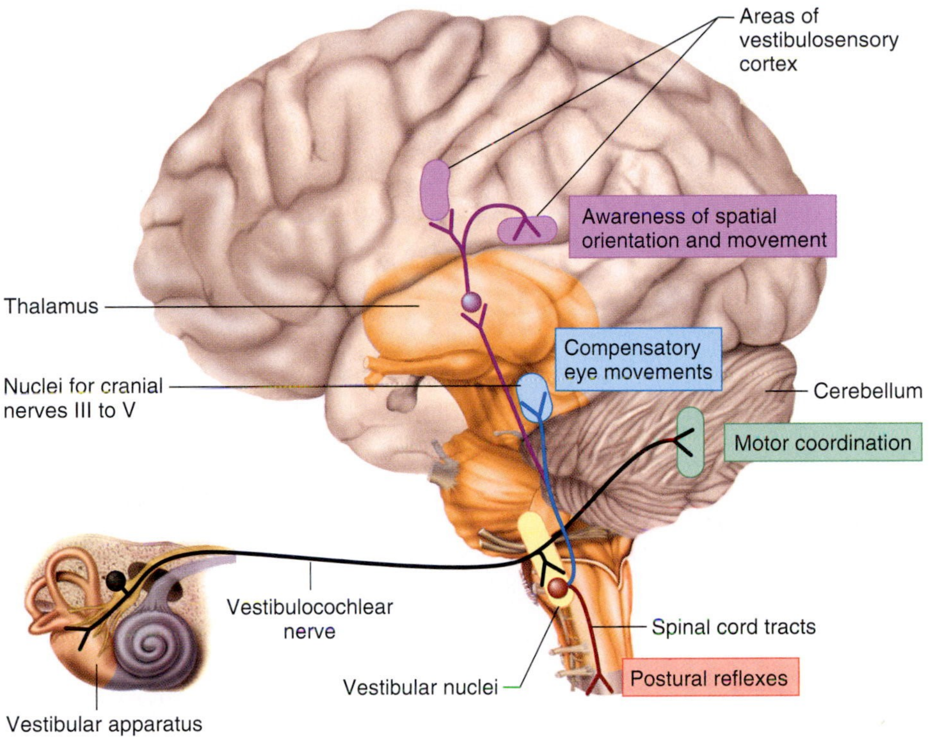

Areas of vestibulosensory cortex
Awareness of spatial orientation and movement
Thalamus
Compensatory eye movements
Nuclei for cranial nerves III to V
Cerebellum
Motor coordination
Vestibulocochlear nerve
Spinal cord tracts
Postural reflexes
Vestibular nuclei
Vestibular apparatus

FIGURE 17.18
Vestibular Projection Pathways in the Brain.

cochlear nerve (VIII). Some vestibular nerve fibers project directly to the cerebellum without synapsing in the brainstem, but most of them end in four pairs of **vestibular nuclei** in the pons and medulla. From here, second-order fibers project by complex pathways to three basic destinations (fig. 17.18):

1. Spinal cord tracts that produce postural reflexes of the skeletal muscles, enabling one to maintain balance when moving.
2. Nuclei of the oculomotor, trochlear, and abducens nerves (cranial nerves III, IV, and VI), which produce compensatory eye movements as the head moves. This enables one to fixate visually on a point in space while the head is moving. To observe this effect, hold this book in front of you at a comfortable reading distance and fixate on the middle of the page. Move the book left and right about once per second, keeping your eyes still, and you will be unable to read it. Now hold the book still and shake your head from side to side at the same rate. This time you will be able to read it because the reflex pathway compensates for your head movements and keeps your eyes fixated on the target.
3. Certain nuclei of the thalamus, which relay the signals by way of third-order neurons to two principal areas of cerebral cortex—one in the roof of the lateral sulcus and one at the lower end of the central sulcus. These areas of sensory cortex are responsible for our awareness of the body's movements and orientation in space.

Before You Go On

Answer the following questions to test your understanding of the preceding section.

11. What is the benefit of having three auditory ossicles and two muscles in the middle ear?
12. Explain how vibration of the tympanic membrane ultimately produces fluctuations of membrane voltage in a cochlear hair cell.
13. How does the brain recognize the difference between high C and middle C of a piano? Between a loud sound and a soft one?
14. How does the function of the semicircular ducts differ from the function of the saccule and utricle?
15. How is the sensory mechanism of the semicircular ducts similar to that of the saccule and utricle?

The Eye

Objectives

When you have completed this section, you should be able to

- describe the anatomy of the eye and its accessory structures;
- describe the histological structure of the retina and its receptor cells;
- explain why different types of receptor cells and neuronal circuits are required for day and night vision; and
- trace the visual projection pathways in the brain.

Accessory Structures of the Orbit

Before considering the eye itself, we will survey the accessory structures located in and around the orbit (figs. 17.19 and 17.20). These include the *eyebrows, eyelids, conjunctiva, lacrimal apparatus,* and *extrinsic eye muscles.*

- The **eyebrows** probably serve mainly to enhance facial expressions and nonverbal communication, but they may also protect the eyes from glare and help to keep perspiration from running into the eye.
- The **eyelids,** or **palpebrae** (pal-PEE-bree), close periodically to moisten the eye with tears, sweep debris from the surface, block foreign objects from the eye, and prevent visual stimuli from disturbing our sleep. The two eyelids are separated by the **palpebral fissure,** and the corners where they meet are called the **medial** and **lateral commissures (canthi).** The eyelid consists largely of the orbicularis oculi muscle covered with skin (fig. 17.20*a*). It also has a supportive connective tissue layer called the **tarsal plate.** Within this plate are 20 to 25 **tarsal glands** that open along the edge of the eyelid. They secrete an oil that coats the eye and reduces tear evaporation. The **eyelashes** are guard hairs that help to keep debris from the eye. Touching the eyelashes stimulates hair receptors and triggers the blink reflex.
- The **conjunctiva** (CON-junk-TY-vuh) is a transparent mucous membrane that covers the inner surface of the eyelid and anterior surface of the eyeball except the cornea. Its primary purpose is to secrete a thin mucous film that prevents the eyeball from drying. It is richly innervated and highly sensitive to pain. It is also very vascular, which is especially evident when the vessels are dilated and the eyes are "bloodshot." Because it is vascular and the cornea is not, the conjunctiva heals more quickly than the cornea when injured.
- The **lacrimal**[38] **apparatus** (fig. 17.20*b*) consists of the lacrimal (tear) gland and a series of ducts that drain the tears into the nasal cavity. The **lacrimal gland,** about the size and shape of an almond, is nestled in a shallow fossa of the frontal bone in the superolateral corner of the orbit. About 12 short ducts lead from the gland to the surface of the conjunctiva. Tears function to cleanse and lubricate the eye surface, deliver oxygen and nutrients to the conjunctiva, and prevent infection by means of a bactericidal enzyme, *lysozyme.* After washing across the conjunctiva, the tears collect at the **lacrimal caruncle**[39] (CAR-un-cul), the pink fleshy mass near the medial commissure of the eye. Near the caruncle, each eyelid has a tiny pore called a lacrimal **punctum,**[40] which collects the tears and conveys them through a short **lacrimal canal** into a **lacrimal sac.** From here, a **nasolacrimal duct** carries the tears to the inferior meatus of the nasal cavity—thus an abundance of tears from crying or "watery eyes" can result in a runny nose. Normally, the tears are swallowed. When you have a cold, the nasolacrimal ducts become swollen and obstructed, the tears cannot drain, and they may overflow from the brim of your eye.
- The **extrinsic eye muscles** are six muscles attached to the walls of the orbit and to the external surface of each eyeball. *Extrinsic* means arising from without, and distinguishes these from the *intrinsic* muscles inside the eyeball, to be considered later. The extrinsic muscles are responsible for movements of the eye (fig. 17.21). They include four *rectus* ("straight") muscles and two *oblique* muscles. The **superior, inferior, medial,** and **lateral rectus** originate on the posterior wall of the orbit and insert on the anterior region

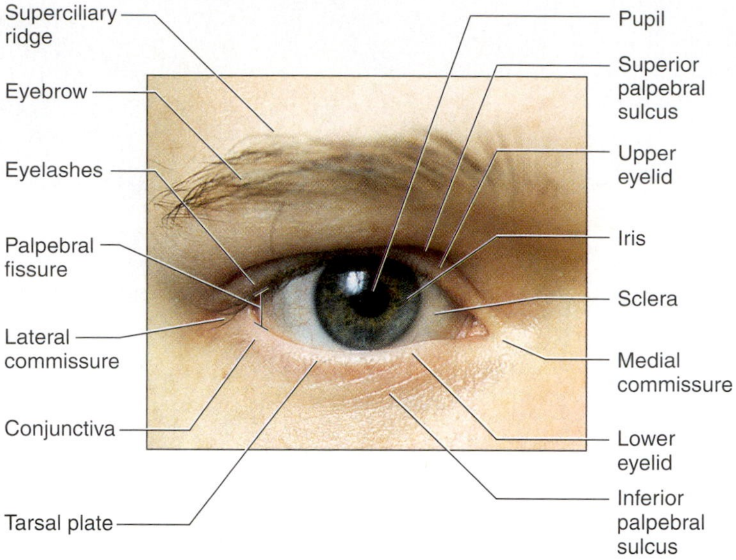

Figure 17.19

External Anatomy of the Orbital Region.

[38]*lacrim* = tear

[39]*car* = fleshy mass + *uncle* = little

[40]*punct* = point

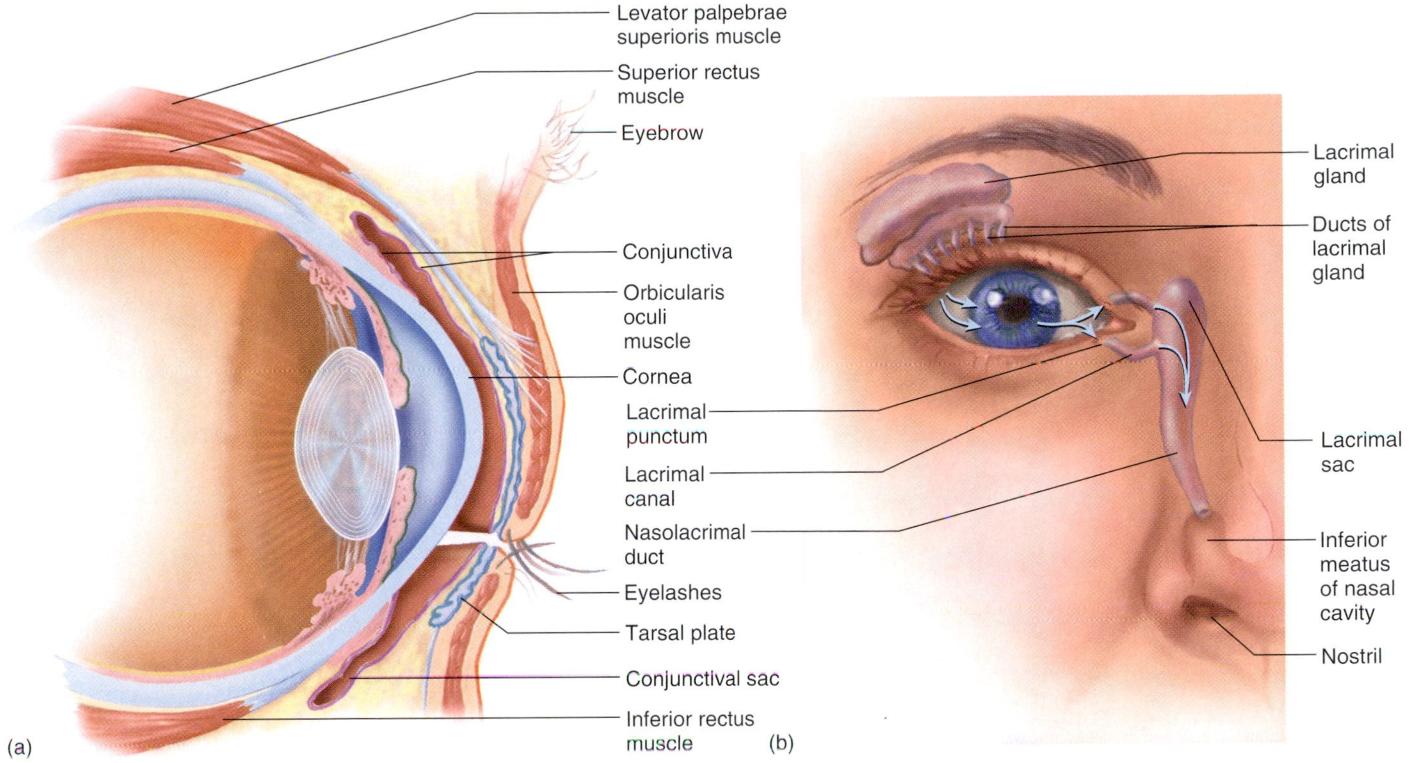
Levator palpebrae superioris muscle
Superior rectus muscle
Eyebrow
Conjunctiva
Orbicularis oculi muscle
Cornea
Lacrimal punctum
Lacrimal canal
Nasolacrimal duct
Eyelashes
Tarsal plate
Conjunctival sac
Inferior rectus muscle
(a)
(b)
Lacrimal gland
Ducts of lacrimal gland
Lacrimal sac
Inferior meatus of nasal cavity
Nostril

FIGURE 17.20
Accessory Structures of the Orbit. (*a*) Sagittal section of the eye and orbit. (*b*) The lacrimal apparatus.

of the eyeball, just beyond the visible "white of the eye." They move the eye up, down, medially, and laterally. The **superior oblique** travels along the medial wall of the orbit. Its tendon passes through a fibrocartilage loop, the **trochlea**[41] (TROCK-lee-uh), and inserts on the superolateral aspect of the eyeball. The **inferior oblique** extends from the medial wall of the orbit to the inferolateral aspect of the eye. To visualize the function of the oblique muscles, suppose you turn your eyes to the right. The superior oblique muscle slightly depresses your right eye, while the inferior oblique slightly elevates the left eye. The opposite occurs when you look to the left. This is the primary function of the oblique muscles, but they also rotate the eyes, turning the "twelve o'clock pole" of each eye slightly toward or away from the nose. The superior oblique muscle is innervated by the trochlear nerve (IV), the lateral rectus muscle by the abducens nerve (VI), and the rest of these muscles by the oculomotor nerve (III).

The eye is surrounded on the sides and back by **orbital fat.** It cushions the eye, gives it freedom of motion, and protects blood vessels and nerves as they pass through the rear of the orbit.

[41]*trochlea* = pulley

Anatomy of the Eyeball

The eyeball itself is a sphere about 24 mm in diameter (fig. 17.22) with three principal components: (1) three layers (tunics) that form the wall of the eyeball; (2) optical components that admit and focus light; and (3) neural components, the retina and optic nerve. The retina is not only a neural component but also part of the inner tunic. The cornea is part of the outer tunic as well as one of the optical components.

THE TUNICS

There are three tunics forming the wall of the eyeball:

- The outer **fibrous layer** *(tunica fibrosa)* is divided into two regions, the sclera and cornea. The **sclera**[42] (white of the eye) covers most of the eye surface and consists of dense collagenous connective tissue perforated by blood vessels and nerves. The **cornea** is the anterior transparent region of modified sclera that admits light into the eye.
- The middle **vascular layer** *(tunica vasculosa)* is also called the **uvea**[43] (YOU-vee-uh) because it resembles a peeled grape in

[42]*scler* = hard, tough
[43]*uvea* = grape

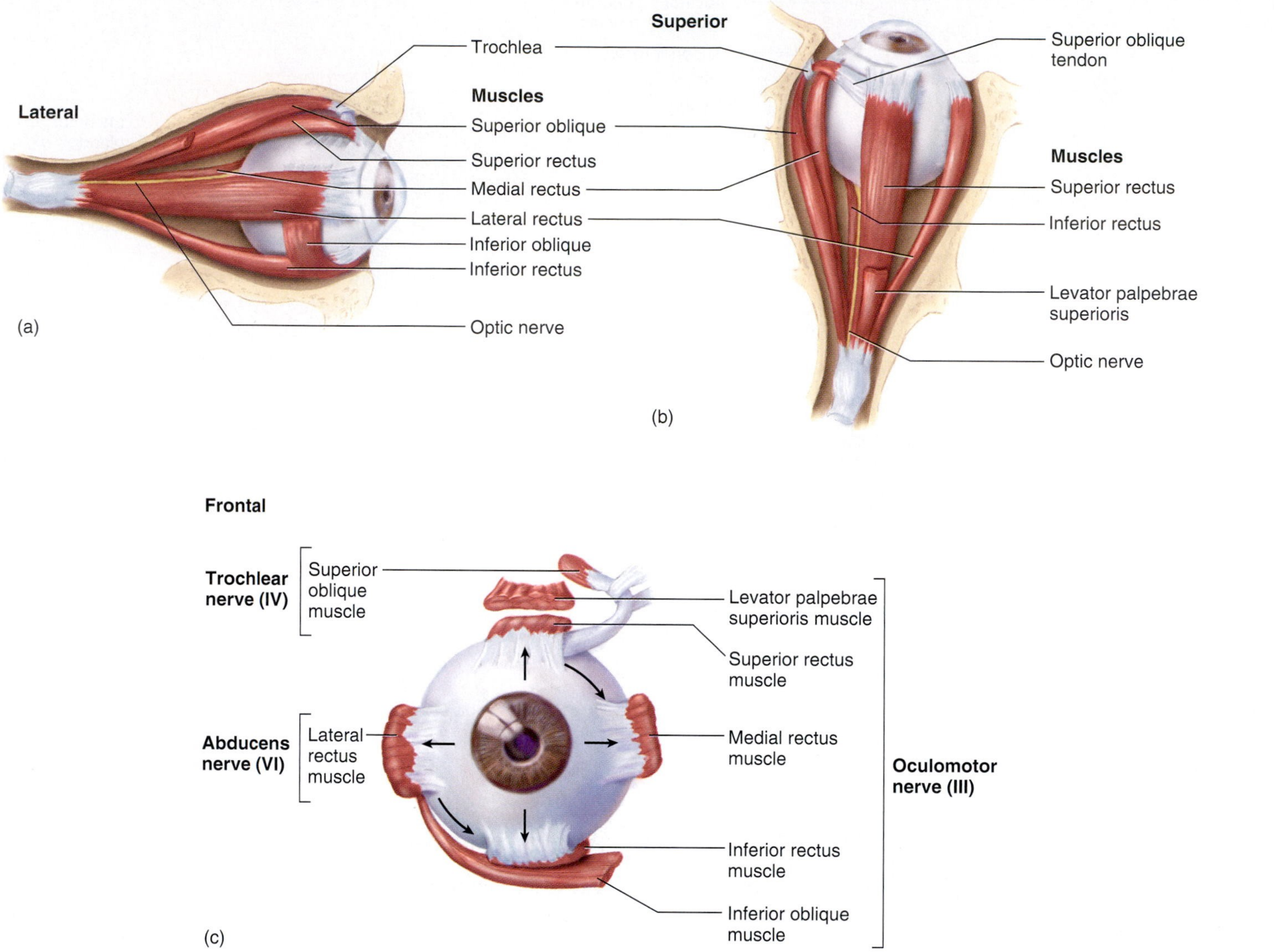

FIGURE 17.21

Extrinsic Muscles of the Eye. (*a*) Lateral view of the right eye. (*b*) Superior view of the right eye. (*c*) Innervation of the extrinsic muscles; *arrows* indicate the eye movement produced by each muscle.

fresh dissection. It consists of three regions—the choroid, ciliary body, and iris. The **choroid** (CO-royd) is a highly vascular, deeply pigmented layer of tissue behind the retina. It gets its name from a histological resemblance to the chorion of a fetus. The **ciliary body,** a thickened extension of the choroid, forms a muscular ring around the lens. It supports the iris and lens and secretes a fluid called the aqueous humor. The **iris** is an adjustable diaphragm that controls the diameter of the **pupil,** its central opening. The iris has two pigmented layers. One is a posterior *pigment epithelium* that blocks stray light from reaching the retina. The other is the *anterior border layer,* which contains pigmented cells called **chromatophores.**[44] High concentrations of melanin in the chromatophores give the iris a black, brown, or hazel color. If the melanin is scanty, light reflects from the posterior pigment epithelium and gives the iris a blue, green, or gray color.

The diameter of the pupil is controlled by two sets of contractile elements in the iris. The **pupillary constrictor** consists of concentric circles of smooth muscle cells around the pupil. The **pupillary dilator** consists of a spokelike arrangement of modified contractile epithelial cells called *myoepithelial cells.* When they contract, they dilate the pupil and admit up to five times as much light as a fully constricted pupil. The pupillary constrictor is innervated by parasympathetic nerve fibers and the pupillary dilator by sympathetic nerve fibers. The pupils constrict in response to increased light intensity and when we focus on nearby objects; they dilate in dimmer light and when we focus on more distant objects. Their response to light is called the *photopupillary reflex.*

[44]*chromato* = color + *phore* = bearer

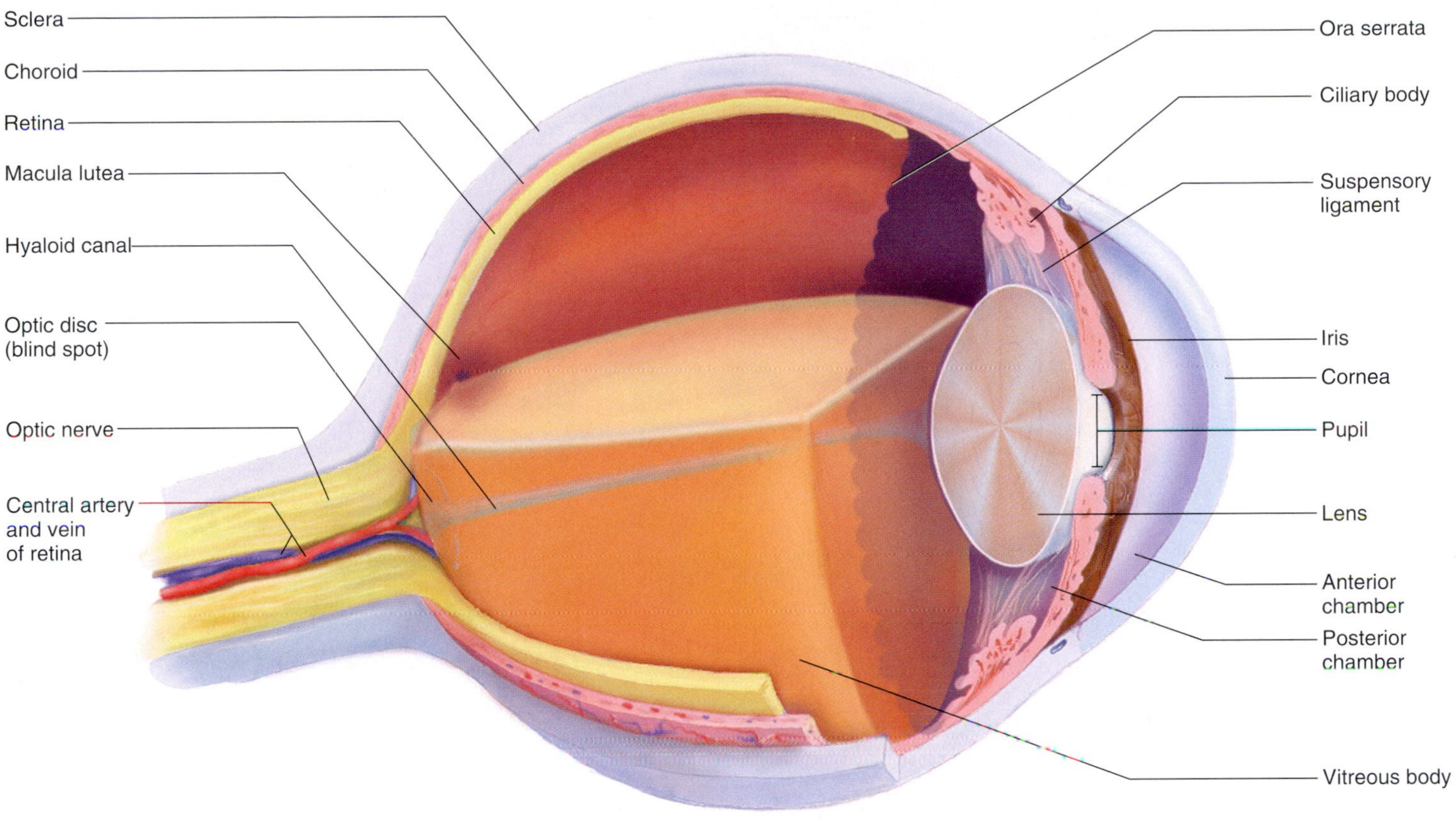

Sclera
Choroid
Retina
Macula lutea
Hyaloid canal
Optic disc (blind spot)
Optic nerve
Central artery and vein of retina
Ora serrata
Ciliary body
Suspensory ligament
Iris
Cornea
Pupil
Lens
Anterior chamber
Posterior chamber
Vitreous body

FIGURE 17.22
Anatomy of the Eye. The vitreous body has been omitted from the upper half to reveal structures behind it.

- The **inner layer** *(tunica interna)* consists of the *retina,* which internally lines the posterior two-thirds of the eyeball.

OPTICAL COMPONENTS

The optical components of the eye are transparent elements that admit light rays, bend (refract) them, and focus images on the retina. They include the *cornea, aqueous humor, lens,* and *vitreous body.* The cornea has been described already.

- The **aqueous humor** is a serous fluid secreted by the ciliary body into a space between the iris and lens called the **posterior chamber** (fig. 17.23). It flows through the pupil forward into the **anterior chamber,** a space between the cornea and iris. From here, it is reabsorbed by a ringlike vessel called the **scleral venous sinus** *(canal of Schlemm*[45]*).* Normally the rate of reabsorption balances the rate of secretion (see insight 17.4 for an important exception).
- The **lens** is composed of flattened, tightly compressed cells called **lens fibers.** It is suspended behind the pupil by a fibrous ring called the **suspensory ligament** (figs. 17.22 and 17.24), which attaches it to the ciliary body. Tension on the ligament somewhat flattens the lens, so it is about 9.0 mm in diameter and 3.6 mm thick at the middle. When the lens is removed from the eye and not under tension, it relaxes into a more spheroid shape and resembles a plastic bead.
- The **vitreous**[46] **body** *(vitreous humor)* is a transparent jelly that fills the large space behind the lens. An oblique channel through this body, called the *hyaloid canal,* is the remnant of a *hyaloid artery* present in the embryo (see fig. 17.22).

NEURAL COMPONENTS

The neural components of the eye are the retina and optic nerve. The retina is a thin transparent membrane attached at only two points—a scalloped anterior margin called the **ora serrata,** and the **optic disc,** where the optic nerve leaves the rear of the eye. The rest of the retina is held smoothly against the rear of the eyeball by the pressure of the vitreous body. It can detach (buckle away from the wall of the eyeball) because of blows to the head or insufficient pressure from the vitreous body. A *detached retina* may cause blurry

[45]Friedrich S. Schlemm (1795–1858), German anatomist

[46] *vitre* = glassy

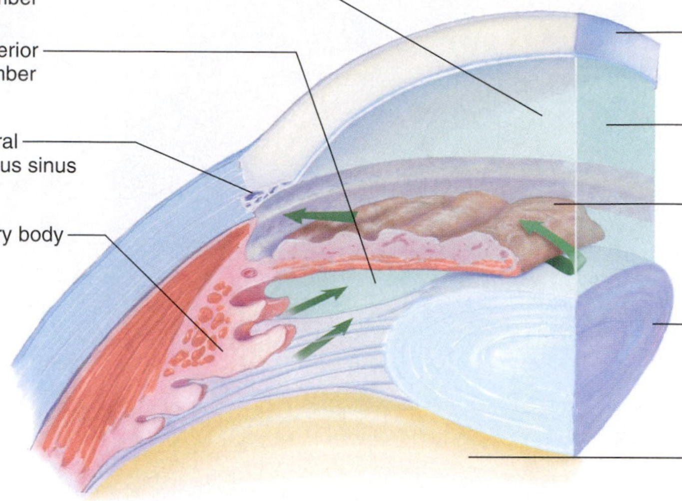

Anterior chamber
Posterior chamber
Scleral venous sinus
Ciliary body
Cornea
Aqueous humor
Iris
Lens
Vitreous body

FIGURE 17.23
Production and Reabsorption of Aqueous Humor.

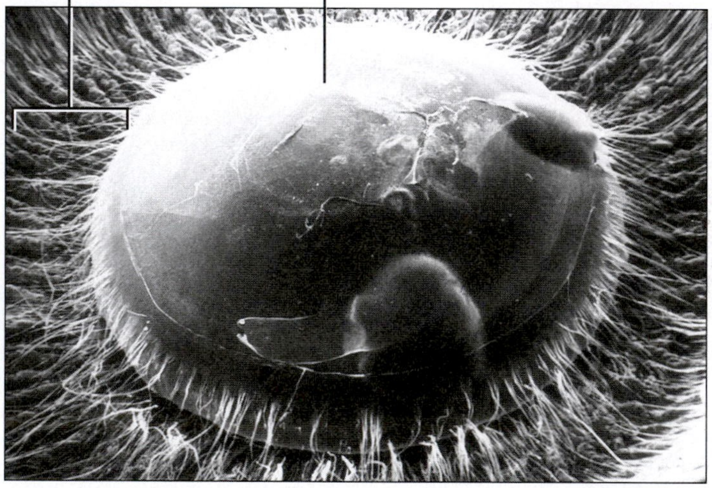

Suspensory ligament
Lens
2 mm

FIGURE 17.24
Lens of the Eye (SEM). Posterior view of the lens and the suspensory ligament that anchors it to the ciliary body.

areas in the field of vision. It can lead to blindness if the retina remains separated for too long from the choroid, on which it depends for oxygen, nutrition, and waste removal.

The inside rear of the eyeball, called the **fundus,** is routinely examined with an illuminating and magnifying instrument called an *ophthalmoscope.* Directly posterior to the center of the lens, on the visual axis of the eye, is a patch of cells called the **macula lutea,**[47] about 3 mm in diameter (fig. 17.25). In the center of the macula is a tiny pit, the **fovea**[48] **centralis,** which produces the most finely detailed images.

[47]*macula* = spot + *lutea* = yellow
[48]*fovea* = pit, depression

INSIGHT 17.4 CLINICAL APPLICATION

CATARACTS AND GLAUCOMA

The two most common causes of blindness are cataracts and glaucoma. A *cataract* is clouding of the lens. It occurs as the lens thickens with age, and it is a common complication of diabetes mellitus. It causes the vision to appear milky or as if one were looking from behind a waterfall.[49] Cataracts are also caused by heavy smoking and by the ultraviolet radiation in sunlight. They can be treated by replacing the natural lens with a plastic one. The implanted lens improves vision almost immediately, but glasses still may be needed for near vision.

Glaucoma is a state of elevated pressure within the eye that occurs when the scleral venous sinus is obstructed so aqueous humor is not reabsorbed as fast as it is secreted. Pressure in the anterior and posterior chambers drives the lens back and puts pressure on the vitreous body. The vitreous body presses the retina against the choroid and compresses the blood vessels that nourish the retina. Without a good blood supply, retinal cells die and the optic nerve may atrophy, producing blindness. Symptoms often go unnoticed until the damage is irreversible. In late stages, they include dimness of vision,[50] a narrowed visual field, and colored halos around artificial lights. Glaucoma can be halted with drugs or surgery, but lost vision cannot be restored. This disease can be detected at an early stage in the course of regular eye examinations. The field of vision is checked, the optic nerve is examined, and the intraocular pressure is measured with an instrument called a *tonometer.*

[49]*cataract* = waterfall
[50]*glauco* = grayness

The reason for this will be apparent later. About 3 mm medial to the macula lutea is the **optic disc.** Nerve fibers converge on this point from neurons throughout the retina and leave here in a bundle that constitute the optic nerve. Blood vessels travel through the optic nerve and enter and leave the eye at the optic disc. Eye examinations thus serve for more than evaluating the visual system; they allow for a direct, noninvasive examination of blood vessels for signs of hypertension, diabetes mellitus, atherosclerosis, and other vascular diseases.

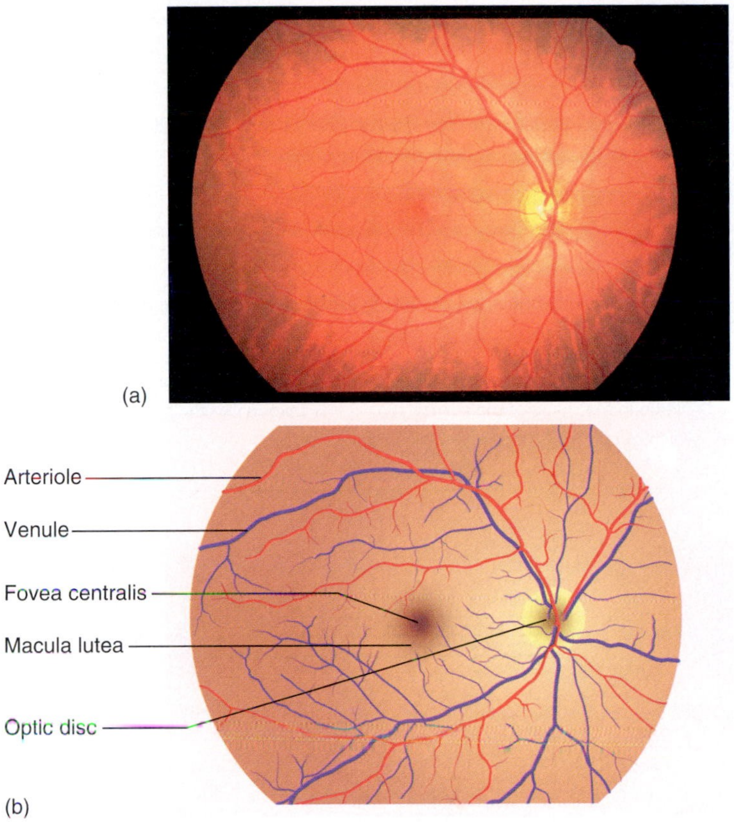
(a)
Arteriole
Venule
Fovea centralis
Macula lutea
Optic disc
(b)

FIGURE 17.25
Fundus of the Eye. (*a*) As seen with an ophthalmoscope. (*b*) Anatomical features of the fundus. Note the blood vessels diverging from the optic disc, where they enter the eye with the optic nerve. An eye examination also serves as a partial check on cardiovascular health.

The optic disc contains no receptor cells, however, so it produces a **blind spot** in the visual field of each eye. To see this effect, close your right eye and gaze straight ahead with your left; fixate on an object across the room. Now hold up a pencil about 30 cm (1 ft) from your face at eye level. Begin moving the pencil toward the left, but be sure you keep your gaze fixed on that point across the room. When the pencil is about 15° away from your line of vision, the end of it will disappear because its image falls on the blind spot of your left eye. The reason you do not normally notice a blind patch in your visual field is that the brain uses the image surrounding the blind spot to fill in that area with similar, essentially "imaginary" information.

Formation of an Image

The visual process begins when light rays enter the eye, become focused on the retina, and produce a tiny inverted image. The cornea refracts incoming light rays toward the optical axis of the eye, and the lens makes relatively slight adjustments to fine focus the image. When you focus on something more than 6 m (20 ft) away, the lens flattens to a thickness of about 3.6 mm at the center and refracts light less. When you focus on something closer than 6 m, the lens thickens to about 4.5 mm at the center and refracts light rays more strongly. These changes in the lens are called *accommodation.* Abnormalities in lens flexibility, the shape of the cornea, or the length of the eyeball result in various deficiencies of vision explained in table 17.2 and figure 17.26.

Structure and Function of the Retina

The conversion of light energy into action potentials occurs in the retina. Its cellular organization is shown in figure 17.27. The most posterior layer is the **pigment epithelium,** composed of darkly pigmented cuboidal cells whose basal processes interdigitate with receptor cells of the retina. The pigment here is not involved in nerve signal generation; rather, its purpose is to absorb light that is not absorbed first by the receptor cells, and to prevent it from degrading the visual image by reflecting back into the eye. It acts like the blackened inside of a camera to reduce stray light.

THINK ABOUT IT!

The vertebrate eye is often called a camera eye *for its many resemblances to the mechanisms of a camera. List as many comparisons as you can.*

The neural apparatus of the retina consists of three principal cell layers. Progressing from the rear of the eye forward, the major retinal cells are the *photoreceptors* (*mainly rods* and *cones*), *bipolar cells,* and *ganglion cells.*

1. **Photoreceptors.** The photoreceptors are all cells that absorb light and generate a chemical or electrical signal. There are three kinds of photoreceptors in the retina: rods, cones, and some of the ganglion cells. Only the rods and cones produce visual images; the ganglion cells are discussed shortly. Each rod or cone has an **outer segment** that points toward the wall of the eye and an **inner segment** facing the interior (fig. 17.28). The two segments are separated by a narrow constriction containing nine pairs of microtubules; the outer segment is actually a highly modified cilium specialized to absorb light. The inner segment contains mitochondria and other organelles. At its base, it gives rise to a cell body, which contains the nucleus, and to processes that synapse with retinal neurons in the next layer.

   In a **rod,** the outer segment is cylindrical and somewhat resembles a stack of coins in a paper roll—there is a plasma membrane around the outside and a neat stack of about 1,000 membranous discs inside. Each disc is densely studded with globular proteins—the visual pigment *rhodopsin.* The membranes hold these pigment molecules in a position that results in the most efficient light absorption. Rod cells are responsible for **night (scotopic[51]) vision** and cannot distinguish colors from each other. Even in ordinary indoor lighting, they are *saturated* (overstimulated) and nonfunctional.

[51] *scot* = dark + *op* = vision

TABLE 17.2

Common Defects of Image Formation

Myopia	Nearsightedness—a condition in which the eyeball is too long. Light rays come into focus before they reach the retina and begin to diverge again by the time they fall on it. Corrected with concave lenses, which cause light rays to diverge slightly before entering the eye.
Hyperopia	Farsightedness—a condition in which the eyeball is too short. The retina lies in front of the focal point of the lens, and the light rays have not yet come into focus when they reach the retina. Causes the greatest difficulty when viewing nearby objects. Corrected with convex lenses, which cause light rays to converge slightly before entering the eye.
Presbyopia	Reduced ability to accommodate for near vision with age because of declining flexibility of the lens. Results in difficulty in reading and doing close handwork. Corrected with reading glasses or bifocal lenses.
Astigmatism	Inability to simultaneously focus light rays that enter the eye on different planes. Focusing on vertical lines, such as the edge of a door, may cause horizontal lines, such as a tabletop, to go out of focus. Caused by a deviation in the shape of the cornea so that it is shaped like the back of a spoon rather than like part of a sphere. Corrected with "cylindrical" lenses, which refract light more in one plane than another.

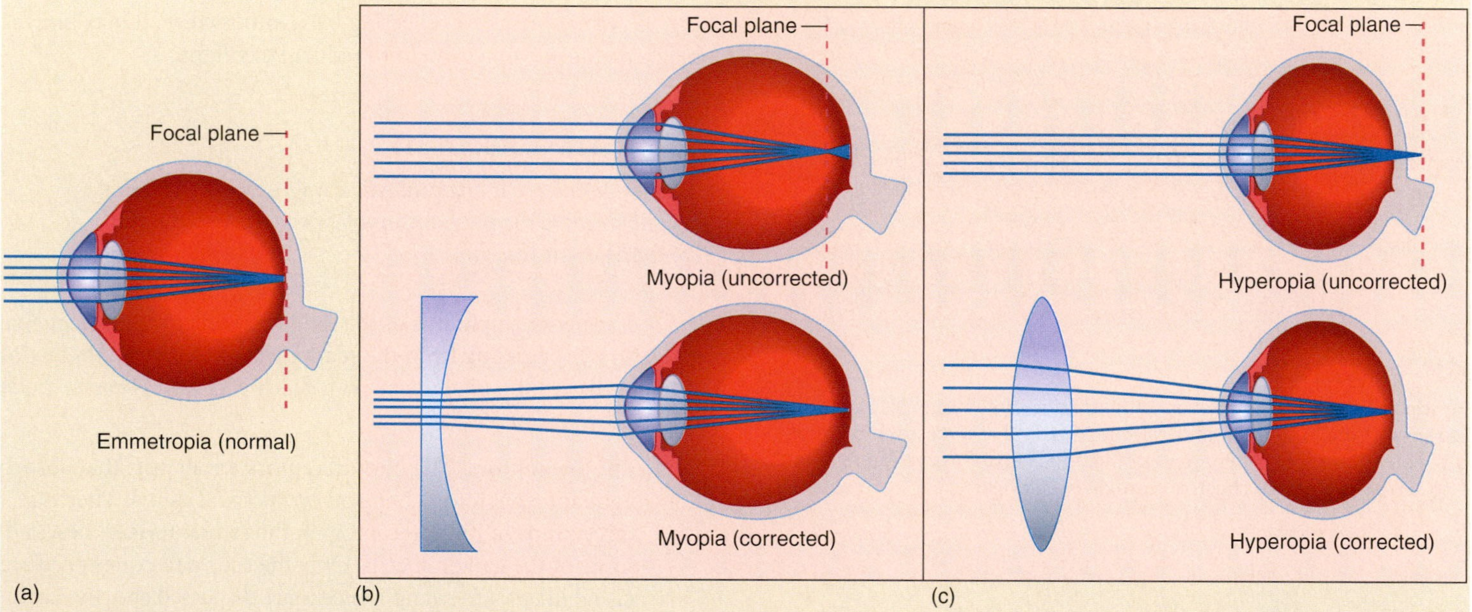

FIGURE 17.26

Two Common Visual Defects and the Effects of Corrective Lenses. (*a*) The normal emmetropic eye, with light rays converging on the retina. (*b*) Myopia (nearsightedness) and the corrective effect of a concave lens. In myopia, the eyeball is abnormally long, so light rays come to a focal point before they reach the retina. (*c*) Hyperopia (farsightedness) and the corrective effect of a convex lens. In hyperopia, the eyeball is abnormally short, so light rays have not yet reached their focal point at the time they fall on the retina.

A **cone** is similar except that the outer segment tapers to a point and the discs are not detached from the plasma membrane but are parallel infoldings of it. Cones begin to respond in light as dim as starlight and are the only receptor cells functional in daylight intensities; thus they are entirely responsible for our **day (photopic**[52]**) vision.** They are also responsible for **color vision,** because unlike rods, cones do not all carry the same visual pigment. Their pigments are called *photopsins.* Some cones have a photopsin that responds best at a wavelength of 420 nanometers (nm), a deep blue light, others respond best at 531 nm (green), and still others at 558 nm (orange-yellow). All colors we see are the result of various mixtures of input to the brain from these three types of cones.

2. **Bipolar cells.** Rods and cones synapse with the dendrites of bipolar neurons, the first-order neurons of the visual pathway. These, in turn, feed directly or indirectly into the ganglion cells described next (see fig. 17.27*b*).
3. **Ganglion cells.** These are the largest neurons of the retina, arranged in a single layer close to the vitreous body. They are the second-order neurons of the visual pathway. Most ganglion cells receive input from multiple bipolar cells. Their axons form the optic nerve. Some ganglion cells absorb light directly and transmit signals to brainstem nuclei that control pupillary diameter and the body's circadian rhythms. They do not contribute to visual images but detect only light intensity. Their sensory pigment is called *melanopsin.*

[52]*phot* = light + *op* = vision

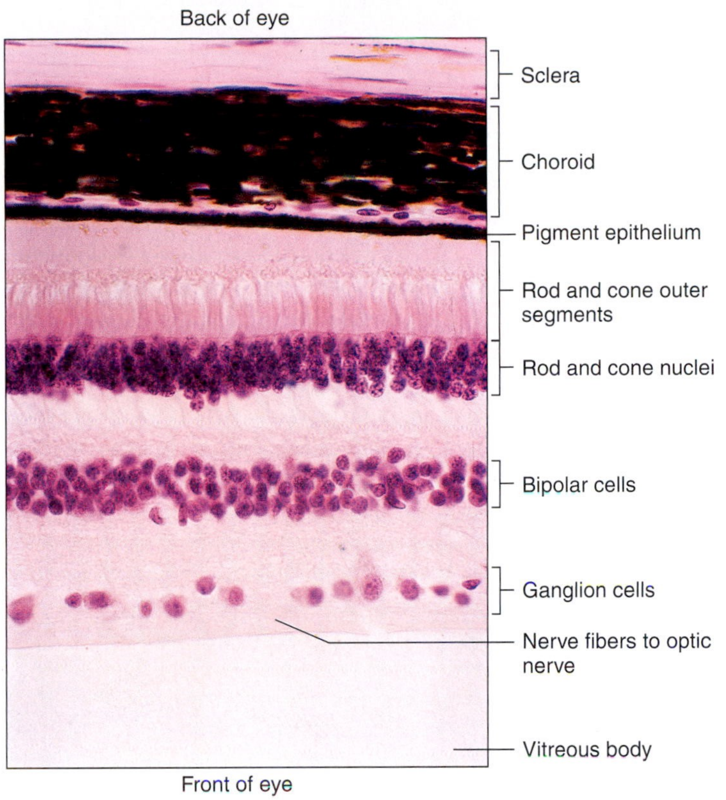
Back of eye
Sclera
Choroid
Pigment epithelium
Rod and cone outer segments
Rod and cone nuclei
Bipolar cells
Ganglion cells
Nerve fibers to optic nerve
Vitreous body
Front of eye

(a)

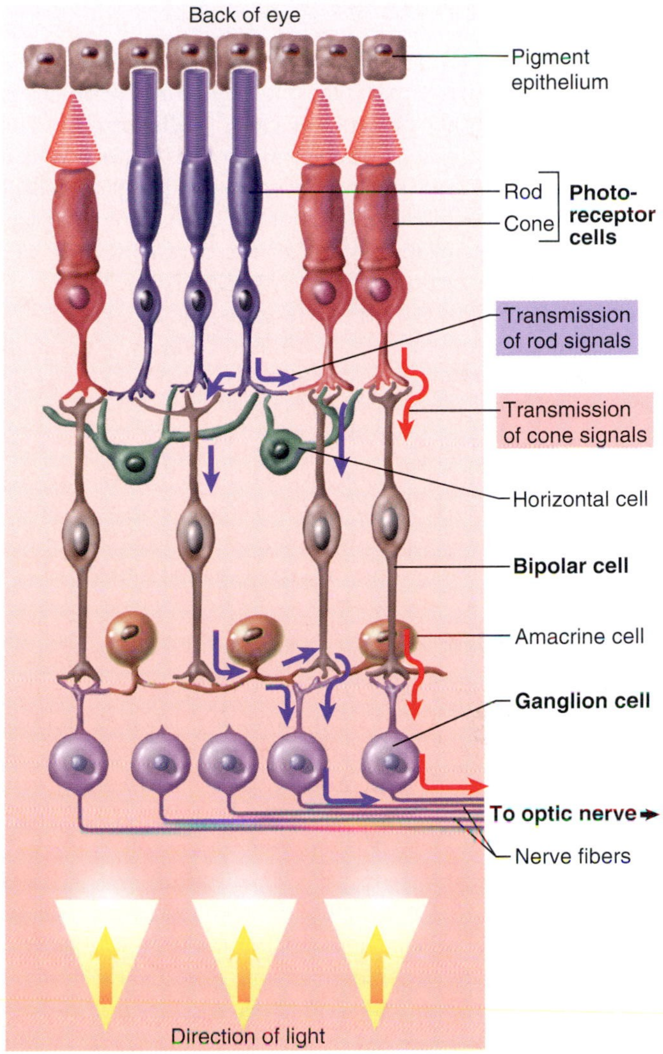
Back of eye
Pigment epithelium
Rod
Cone
Photo-receptor cells
Transmission of rod signals
Transmission of cone signals
Horizontal cell
Bipolar cell
Amacrine cell
Ganglion cell
To optic nerve
Nerve fibers
Direction of light

(b)

FIGURE 17.27

Histology of the Retina. (*a*) Photomicrograph. (*b*) Schematic of the layers and circuitry of the retinal cells.

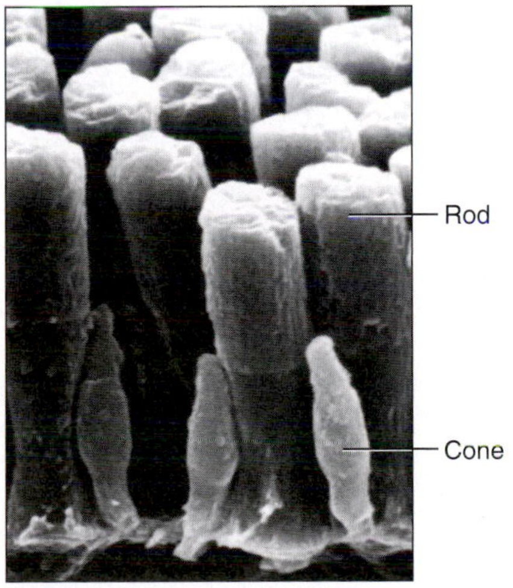
Rod
Cone

(a)

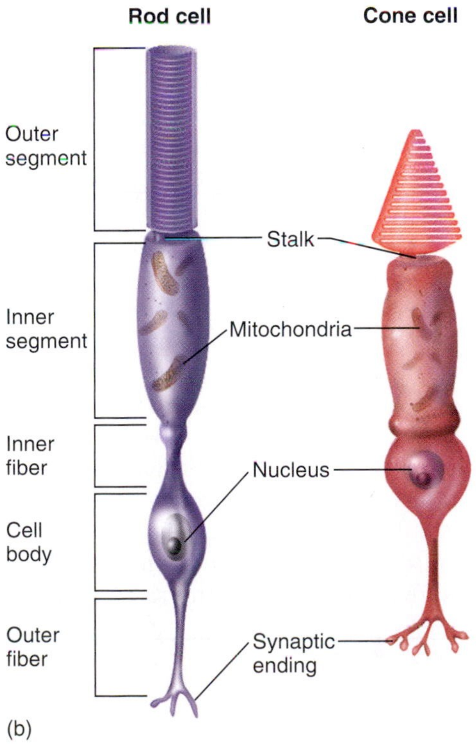
Rod cell
Cone cell
Outer segment
Inner segment
Inner fiber
Cell body
Outer fiber
Stalk
Mitochondria
Nucleus
Synaptic ending

(b)

FIGURE 17.28

Rod and Cone Cells. (*a*) Rods and cones of a salamander retina (SEM). The tall cylindrical cells are rods and the short, tapered cells in the foreground are cones. (*b*) Structure of rods and cones.

There are other retinal cells, but they do not form layers of their own. **Horizontal cells** and **amacrine**[53] **cells** form horizontal connections among rod, cone, and bipolar cells. Bipolar cells that carry rod signals do not synapse directly with ganglion cells, but only by way of amacrine cells. Horizontal and amacrine cells play diverse roles in enhancing the perception of contrast, the edges of objects, and changes in light intensity. In addition, much of the mass of the retina is composed of astrocytes and other types of glial cells.

There are approximately 130 million rods and 6.5 million cones in one retina, but only 1.2 million nerve fibers in the optic nerve. With a ratio of 114 receptor cells to 1 optic nerve fiber, it is obvious that there must be substantial *neuronal convergence* and information processing in the retina itself before signals are transmitted to the brain proper. Convergence begins where multiple rod or cone cells converge on one bipolar cell, and occurs again where multiple bipolar cells converge on a single ganglion cell.

THE DUPLICITY THEORY

You may wonder why we have two types of photoreceptor cells, the rods and cones. Why can't we simply have one type that would produce detailed color vision, both day and night? The answer to this lies largely in the concept of neuronal convergence (see chapter 13). The **duplicity theory** of vision holds that a single type of receptor cell cannot produce both high sensitivity and high resolution. It takes one type of cell and neuronal circuit, working at its maximum capacity, to provide sensitive night vision and a different type of receptor and circuit to provide high-resolution daytime vision.

The high sensitivity of rods in dim light stems partly from a cascade of chemical reactions that occurs when rhodopsin absorbs light. The cascade amplifies the effect of the light, so that a small stimulus produces a relatively large output from each rod. But in addition, up to 600 rods converge on each bipolar cell and multiple bipolar cells converge (via amacrine cells) on each ganglion cell (fig. 17.29*a*). Thus, weak stimulation of many rod cells can produce an additive effect on one bipolar cell, and several bipolar cells can collaborate to excite one ganglion cell. Thus, a ganglion cell can respond in dim light that only weakly stimulates an individual rod. A shortcoming of this system is that it cannot resolve finely detailed images. One ganglion cell receives input from all the rods in about 1 mm^2 of retina—its receptive field. What the brain perceives is therefore a coarse, grainy image similar to an overenlarged newspaper photograph.

Around the edges of the retina, receptor cells are especially large and widely spaced. If you fixate on the middle of this page, you will notice that you cannot read the words near the margins. Visual acuity decreases rapidly as the image falls away from the fovea centralis. Our peripheral vision is a low-resolution system that serves mainly to alert us to motion in the periphery and to stimulate us to look that way to identify what is there.

When you look directly at something, its image falls on the fovea, which is occupied by about 4,000 tiny cones and no rods. The other neurons of the fovea are displaced to the sides, like parted hair, so they do not interfere with light falling on the cones. The smallness of these cones is like the smallness of the dots in a high-quality photograph; it is partially responsible for the high-resolution images formed at the fovea. In addition, the cones here show no neuronal convergence. Each cone synapses with only one bipolar cell and each bipolar cell with only one ganglion cell. This gives each foveal cone a "private line" to the brain, and each ganglion cell of the fovea reports to the brain on a receptive field of just 2 μm^2 of retinal area (fig. 17.29*b*). Cones distant from the fovea exhibit some neuronal convergence but not nearly as much as rods do. The price of this lack of convergence at the fovea, however, is that cone cells cannot have additive effects on the ganglion cells, and the cone system therefore is less sensitive to light (requires brighter light to function).

THINK ABOUT IT!

If you look directly at a dim star in the night sky, it disappears, and if you look slightly away from it, it reappears. Why?

The Visual Projection Pathway

The first-order neurons in the visual pathway are the bipolar cells of the retina. They synapse with the second-order neurons, the retinal ganglion cells, whose axons are the fibers of the optic nerve (cranial nerve II). The optic nerves leave each orbit through the optic foramen, then converge to form an X, the **optic chiasm**[54] (ky-AZ-um), inferior to the hypothalamus and anterior to the pituitary. Beyond this, the fibers continue as a pair of **optic tracts** (see fig. 15.25). Within the chiasm, half the fibers of each optic nerve cross over to the opposite side of the brain (fig. 17.30). This is called **hemidecussation,**[55] since only half of the fibers decussate. As a result, objects in the left visual field, whose images fall on the right half of each retina (the medial half of the left eye and lateral half of the right eye), are perceived by the right cerebral hemisphere. Objects in the right visual field are perceived by the left hemisphere. Since the right brain controls motor responses on the left side of the body and vice versa, each side of the brain sees what is on the side of the body where it exerts motor control.

The optic tracts pass laterally around the hypothalamus, and most of their axons end in the **lateral geniculate**[56] (jeh-NIC-you-late) **nucleus** of the thalamus. Third-order neurons arise here and form the **optic radiation** of fibers in the white matter of the cerebrum. These fibers project to the primary visual cortex of the occipital lobe, where the conscious perception of an image occurs. A stroke that destroys occipital lobe tissue can cause blindness even if the eyes are fully functional. Association tracts connect the primary visual cortex to the visual association area just anterior to it, where this sensory input is interpreted.

[53]*a* = without + *macr* = long + *in* = fiber

[54]*chiasm* = cross, X

[55]*hemi* = half + *decuss* = to cross, form an X

[56]*geniculate* = bent like a knee

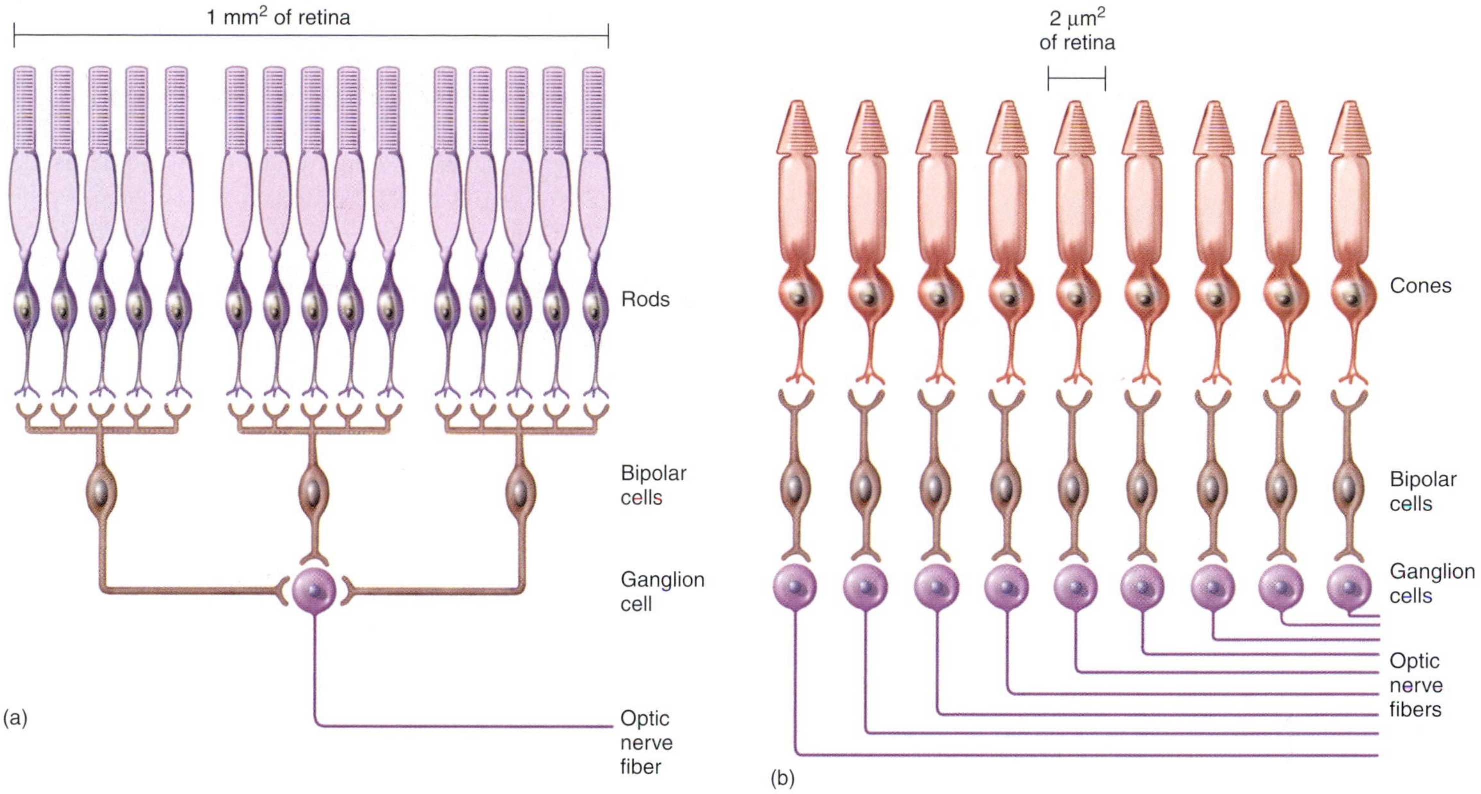

Figure 17.29

The Duplicity Theory of Vision. (*a*) In the scotopic (night vision) system, many rod cells converge on each bipolar cell, and many bipolar cells converge on each ganglion cell. This allows extensive spatial summation—many rods add up their effects to stimulate a ganglion cell even in dim light. However, it means that each ganglion cell (and its optic nerve fiber) represents a relatively large area of retina and produces a grainy image. (*b*) In the photopic (day vision) system, there is little neuronal convergence. In the fovea, represented here, each cone cell has a "private line" to the brain, so each optic nerve fiber represents a tiny area of retina, and vision is relatively sharp. However, the lack of convergence prevents spatial summation. Photopic vision does not function in dim light because weakly stimulated cone cells cannot collaborate to stimulate a ganglion cell.

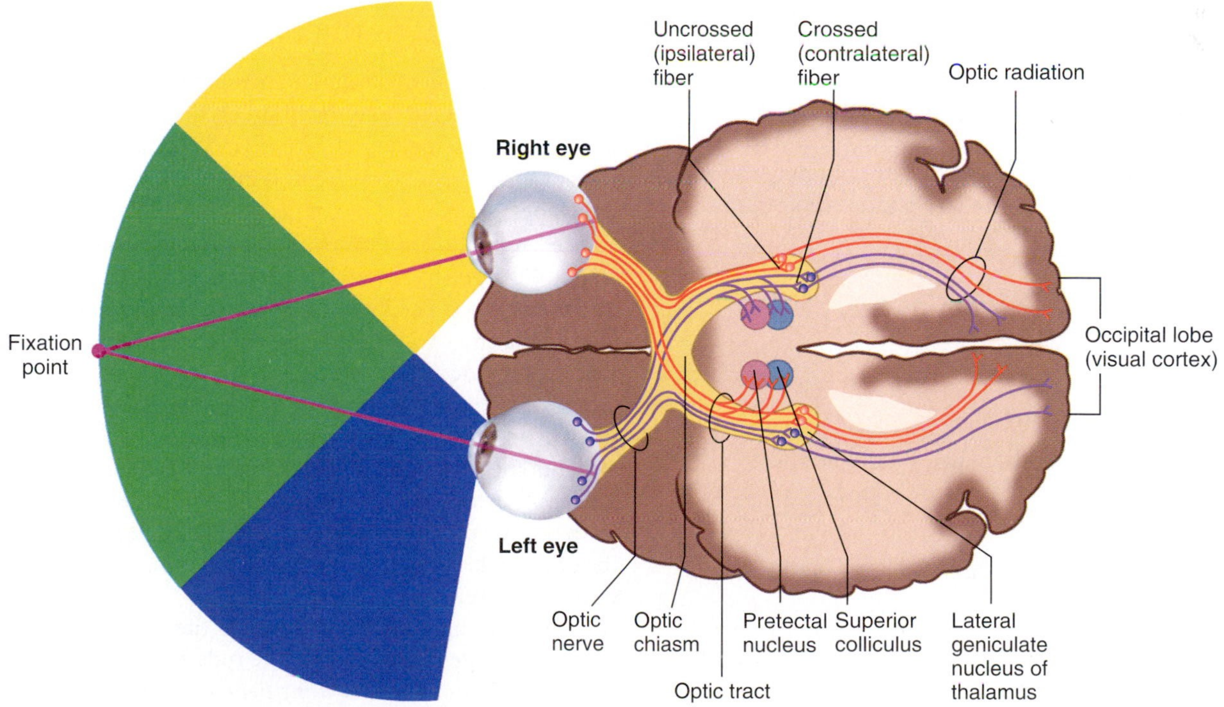

Figure 17.30

The Visual Projection Pathway. Note the hemidecussation at the optic chiasm and projection of fibers from both eyes to the primary visual cortex of each cerebral hemisphere.

Optic nerve fibers from the photosensitive ganglion cells take a different route; they project to the midbrain and terminate in the superior colliculi and adjacent *pretectal nuclei.* The superior colliculi control the visual reflexes of the extrinsic muscles, and the pretectal nuclei are involved in the photopupillary and accommodation reflexes. Thus, these retinal ganglion cells (about 1% to 2% of the total) are not engaged in producing the images we see, but only in providing input to evoke these somatic and autonomic reflexes of the eyes.

The processes of visual information processing in the brain are very complex and beyond the scope of this book. Some processing, such as contrast, brightness, motion, and stereoscopic vision, begins in the retina. The primary visual cortex in the occipital lobe is connected by association tracts to nearby visual association areas in the posterior part of the parietal lobe and inferior part of the temporal lobe. These association areas process visual data to extract information about the location, motion, color, shape, boundaries, and other qualities of the objects we observe. They also store visual memories and enable the brain to identify what we are seeing—for example, to recognize printed words or name the objects we see. What is yet to be learned about visual processing promises to have important implications for biology, medicine, psychology, and even philosophy.

Before You Go On

Answer the following questions to test your understanding of the preceding section.

16. List the refractile media of the eye and state the role of each one in the formation of an image.
17. List as many structural and functional differences between rods and cones as you can.
18. Trace the signal pathway from the point where a retinal cell absorbs light to the point where an optic nerve fiber synapses in the thalamus.
19. Discuss the duplicity theory of vision, summarizing the advantage of having two types of retinal photoreceptor cells.

Developmental and Clinical Perspectives

Objectives

When you have completed this section, you should be able to

- describe the prenatal development of the major features of the eye and ear; and
- describe some disorders of taste, vision, hearing, equilibrium, and somesthetic sensation.

Development of the Ear

The ear develops from the first pharyngeal pouch (described in chapter 4) and adjacent tissues of the first and second pharyngeal arches. The first section to develop is the inner ear. Its earliest trace is a thickening called the **otic placode**[57] **(otic disc),** which develops in the ectoderm near the hindbrain in week 3. The placode invaginates, forming an **otic pit** and then a fully enclosed **otic vesicle,** which detaches from the overlying ectoderm by the end of week 4 (fig. 17.31*a–c*). The otic vesicle differentiates into two chambers, the utricle and saccule. In week 5, the ventral tip of the saccule elongates and begins to form the cochlea, and soon after, three pouches grow from the utricle and begin to form the semicircular ducts (fig. 17.31*d–f*). These structures are still embedded in mesenchyme, which in week 9 chondrifies and forms a cartilaginous **otic capsule** enclosing the inner-ear structures. The capsule later ossifies to form the petrous part of the temporal bone and its bony labyrinth.

The middle-ear cavity and auditory tube arise by elongation of the first pharyngeal pouch, beginning in week 5. Mesenchyme of the first two pharyngeal arches gives rise to the three auditory ossicles and the two middle-ear muscles. These bones and muscles remain solidly embedded in mesenchyme until the last month of fetal development, at which time the mesenchyme degenerates and leaves the hollowed-out middle-ear cavity (fig. 17.31*g–i*).

At the same time as the middle ear begins to form (week 5), the facing edges of the first and second pharyngeal arches form three pairs of humps called *auricular hillocks.* These enlarge, fuse, and differentiate into the folds and whorls of the auricle by week 7 (fig. 17.31*j–l*). As this is happening, the first pharyngeal groove between these two arches begins to elongate, grow inward, and form the auditory canal. The eardrum arises from the wall between the auditory canal and tympanic cavity, and thus has an outer ectodermal layer and inner endodermal layer, with a thin layer of mesoderm between.

Development of the Eye

An early indication of the development of the eye is the **optic vesicle,** seen by day 24 as an outgrowth on each side of the diencephalon, continuous with the neural tube (fig. 17.32*a, b*). As the optic vesicle reaches the overlying ectoderm, it invaginates and forms a double-walled **optic cup,** while its connection to the diencephalon narrows and becomes a hollow **optic stalk.** The outer wall of the optic cup becomes the *pigment retina,* later becoming the pigment epithelium discussed earlier. The inner wall becomes the *neural retina.* Starting around the end of week 6, the neural retina produces waves of cells that migrate toward the vitreous body and arrange themselves into layers of receptor cells (rods and cones) and neurons. The narrow space between the pigment retina and neural retina becomes obliterated, but these two walls of the cup never fuse. This is why the retina is so easily detached later in life. By the eighth month, all cell layers of the retina are present. Nerve fibers grow from the ganglion cells into the optic stalk, occupying and eliminating its lumen as they become the optic nerve.

A seemingly peculiar aspect of the retinas of humans and other vertebrate animals is that the rods and cones face the back of the eye, *away from the incoming light.* This arrangement, called an *inverted retina,* is quite opposite the seemingly more sensible

[57]*ot* = ear + *ic* = pertaining to + *plac* = plate + *ode* = form, shape

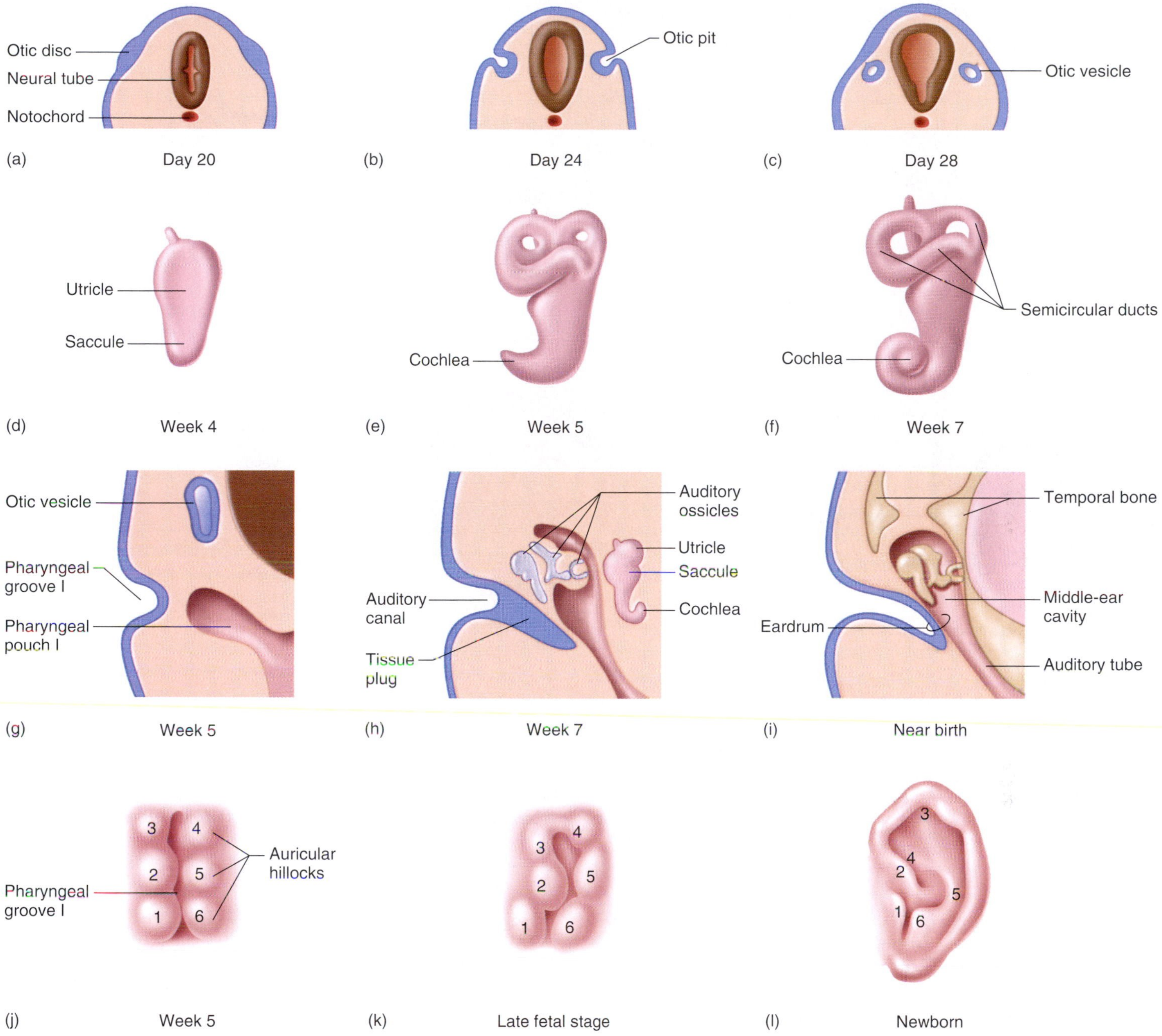

FIGURE 17.31

Development of the Ear. (*a–c*) Development of the otic vesicle from about 20 to 28 days in cross sections of the embryo. (*d–f*) Differentiation of the otic vesicle into the utricle, saccule, cochlea, and semicircular ducts of the membranous labyrinth, from week 4 to week 7. (*g–i*) Development of the middle ear, auditory canal, and auditory tube from about week 5 to the last month of gestation. (*j–l*) Development of the auricle from week 5 to newborn. Numbers *1* to *6* indicate regions of the newborn auricle that arise from each of the six auricular hillocks of the embryo.

arrangement of the octopus eye, which is like ours in many respects but has its receptor cells aimed toward the light. The reason for the inverted retina of humans is that the rods and cones are homologous to the ependymal cells of the neural tube—that is, they develop from the same embryonic origin, and thus face inward toward the lumen of the optic cup just like the mature ependymal cells that face inward toward the ventricles of the brain. The homology between these retinal cells and ependymal cells is also seen in their cilia. Ependymal cells have conventional motile cilia, whereas rod and cone outer segments are modified cilia.

As the optic cup contacts the ectoderm on the surface of the embryo, it induces ectodermal cells to thicken into a **lens placode** (fig. 17.32*c*). The placode invaginates, forms a **lens pit,** and by day 32, separates from the ectoderm and becomes a **lens vesicle** nestled

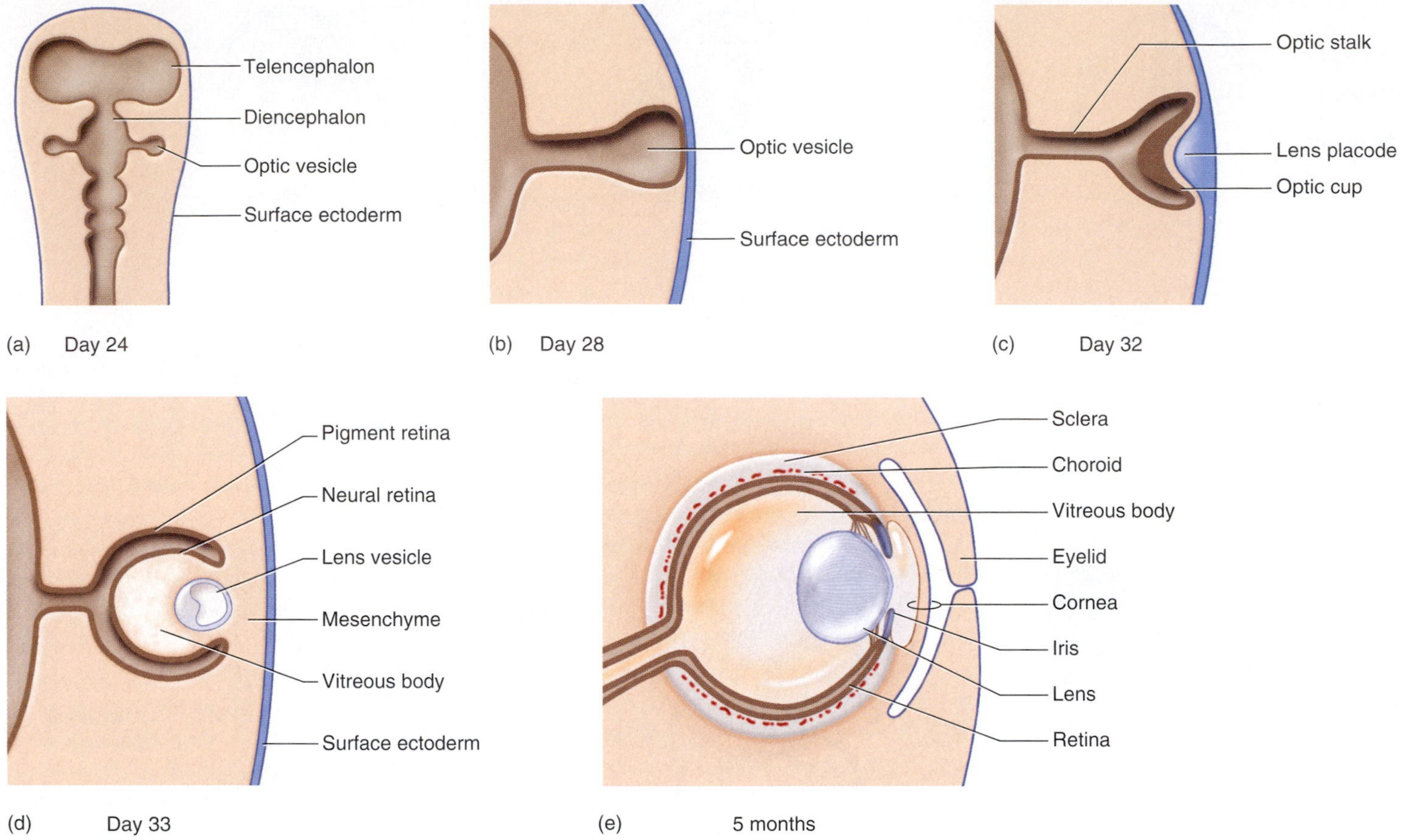
Telencephalon
Diencephalon
Optic vesicle
Surface ectoderm
(a) Day 24
Optic vesicle
Surface ectoderm
(b) Day 28
Optic stalk
Lens placode
Optic cup
(c) Day 32
Pigment retina
Neural retina
Lens vesicle
Mesenchyme
Vitreous body
Surface ectoderm
(d) Day 33
Sclera
Choroid
Vitreous body
Eyelid
Cornea
Iris
Lens
Retina
(e) 5 months

FIGURE 17.32

Development of the Eye. (*a*) Growth of the optic vesicles from the embryonic diencephalon. (*b*) Around day 28, the leading edge of the optic vesicle contacts the surface ectoderm. (*c*) The optic vesicle invaginates and becomes an optic cup, while inducing the ectoderm to form the lens placode. (*d*) The lens vesicle is now separated from the surface ectoderm and nestled in the optic cup, and a gelatinous vitreous body has been secreted between the lens vesicle and retina. (*e*) At 5 months, the lens vesicle is solidly filled with lens fibers; the sclera, choroid, iris, and cornea are partially formed; and the eyelids are just about to separate and reopen the eye.

within the optic cup (fig. 17.32*d*). (This is quite similar to the development of the otic vesicle of the ear.) The lens vesicle is hollow at first, but becomes filled in with lens fibers by day 33. The vitreous body develops from a gelatinous secretion that accumulates in the space between the lens vesicle and optic cup.

Mesenchyme grows to completely surround and enclose the optic cup, and differentiates into the extrinsic muscles and some other accessory structures of the orbit, and some components of the eyeball including the choroid and sclera. The choroid is homologous to the pia mater and arachnoid mater, and the sclera is homologous to the dura mater—that is, they arise from the same embryonic tissues as these meninges. The mesenchyme lateral to the optic cup (near the embryo surface) develops a split that becomes the anterior chamber of the eye. The cornea develops from both the lateral mesenchyme and the overlying ectoderm. The iris grows inward from the anterior margin of the optic cup around the end of month 3. The eyelids develop from folds of ectoderm with a mesenchymal center. The upper and lower eyelids approach each other and fuse, closing the eyes, at the end of month 3; they separate again, and the eyes open, between months 5 and 7 (fig. 17.32*e*).

Disorders of the Sense Organs

Several disorders of the senses have already been discussed in this chapter. These are listed in table 17.3 along with brief descriptions of five additional disorders of multiple sensory modalities.

Before You Go On

Answer the following questions to test your understanding of the preceding section.

20. Describe the contributions of the first pharyngeal pouch and the first two pharyngeal arches to the development of the ear.
21. From an embryological standpoint, explain why rod and cone outer segments face away from the incoming light.
22. How are paresthesia and tinnitus similar (table 17.3)?
23. How are anosmia and ageusia similar (table 17.3)?

TABLE 17.3

Some Sensory Disorders

Ageusia	Loss of the sense of one or more taste modalities, often due to damage to the hypoglossal nerve (loss of bitter taste) or facial nerve (loss of sweet, sour, and salty tastes)
Color Blindness	Inability to distinguish certain colors from each other, such as green and orange, due to a hereditary lack of one of the three types of cones. A sex-linked recessive trait that affects more men than women.
Ménière Disease	A disorder of proprioception in which one experiences episodes of vertigo (dizziness) often accompanied by nausea, tinnitus, and pressure in the ears. Usually accompanied by progressive hearing loss.
Paresthesia	Feelings of numbness, prickling, tingling, heat, or other sensations in the absence of stimulation; a symptom of nerve injuries and other neurological disorders.
Tinnitus	Perception of imaginary sounds such as whistling, buzzing, clicking, or ringing in the ear. May be temporary or permanent, intermittent or constant; typically associated with hearing loss in the high frequencies. May result from cochlear damage, aspirin or other drugs, ear infections, Ménière disease, or other causes.

Disorders Described Elsewhere

Anosmia 183	Detached retina 497	Leprosy 480
Astigmatism 500	Diabetic neuropathy 480	Middle-ear infection 486
Cataracts 498	Glaucoma 498	Myopia 500
Deafness 489	Hyperopia 500	Presbyopia 500

CHAPTER REVIEW

Review of Key Concepts

Receptor Types and the General Senses (p. 476)

1. Sensory *receptors* range from simple nerve endings to complex sense organs.
2. Receptors can also be classified by stimulus modality into *chemoreceptors, thermoreceptors, nociceptors, mechanoreceptors,* and *photoreceptors.*
3. The senses are also classified as *general (somesthetic) senses* and *special senses.* The general senses have receptors widely distributed over the body and include touch, pressure, stretch, temperature, and pain. *Special senses* have receptors in the head only and include vision, hearing, equilibrium, taste, and smell.
4. Receptors are classified according to the origins of their stimuli as *interoceptors, proprioceptors,* and *exteroceptors.*
5. Some general senses employ unencapsulated nerve endings, which are simple sensory dendrites with no connective tissue; these include *free nerve endings* for heat, cold, and pain; *tactile discs* for light touch and pressure on the skin; and *hair receptors,* which sense hair movements.
6. Encapsulated nerve endings are dendrites enclosed in glial or connective tissue cells. These include *tactile corpuscles, Krause end bulbs, Ruffini corpuscles, lamellated (pacinian) corpuscles, muscle spindles,* and *Golgi tendon organs.*
7. A sensory neuron receives stimuli within an area called its receptive field. Two tactile stimuli applied within the same receptive field stimulate a single neuron and cannot be distinguished from each other. The minimum spatial separation needed to fall within the receptive fields of two different neurons and be felt separately is the *two-point touch threshold.*
8. Sensory signals typically travel a three-neuron *projection pathway* from the receptor to the final destination of the sensory signal in the brain. The three neurons are called *first-, second-,* and *third-order neurons.*
9. Somesthetic signals from the head travel the trigeminal and other cranial nerves to the brainstem, and those below the head travel up the spinothalamic tract and other pathways. Most such signals pass through the thalamus en route to the cerebral cortex, but second-order proprioceptive fibers project to the cerebellum. Somesthetic pathways decussate in the spinal cord or medulla oblongata and project to the cerebral hemisphere contralateral to the origin of the stimulus.
10. Pain is a sensation that occurs when nociceptors detect tissue damage or potentially injurious situations.
11. *Fast pain* is a relatively quick, localized response mediated by myelinated nerve fibers; it may be followed by a less localized *slow pain* mediated by unmyelinated fibers.
12. *Somatic pain* arises from the skin, muscles, and joints. *Visceral pain* arises from the viscera and is often accompanied by nausea.
13. Pain from the face travels mainly by way of the trigeminal nerve to the pons. Somatic pain from the neck down travels by way of spinal nerves to the spinothalamic tract. Visceral pain signals travel up the gracile fasciculus. All of these pathways converge on the thalamus, which relays them to the primary somesthetic cortex in the postcentral gyrus of the cerebrum.
14. Pain signals also travel the spinoreticular tract to the reticular formation and from there to the hypothalamus and limbic system, producing visceral and emotional responses to pain.
15. *Referred pain* is the brain's misidentification of the location of pain resulting from convergence in sensory pathways.
16. The reticular formation issues *descending analgesic fibers* which can block pain signals in the dorsal horn of the spinal cord and prevent them from being transmitted to the brain; thus they have a pain-relieving effect.

The Chemical Senses (p. 481)

1. Taste *(gustation)* results from the action of chemicals on the *taste buds,* which are groups of sensory cells located on some of the *lingual papillae* and in the palate, pharynx, and epiglottis.
2. *Foliate, fungiform,* and *vallate papillae* have taste buds; *filiform papillae* lack taste buds but sense the texture of food. Foliate papillae carry few or no taste buds after early childhood; vallata papillae bear about half of all adult taste buds.
3. A taste bud is a lemon-shaped aggregation of *taste cells,* nonsensory *supporting cells,* and *basal cells.* Taste cells are not neurons but synapse with sensory dendrites at their bases. Basal cells are stem cells that replace expired taste cells.
4. The primary taste sensations are salty, sweet, sour, bitter, and umami. Flavor is a combined effect of these tastes and the texture, aroma, temperature, and appearance of food. Some flavors result from the stimulation of free endings of the trigeminal nerve.
5. Taste signals travel from the tongue through the facial and glossopharyngeal nerves, and from the palate, pharynx, and epiglottis through the vagus nerve. They travel to the *solitary nucleus* of the medulla oblongata and then by one route to the hypothalamus and amygdala, and by another route to the thalamus and cerebral cortex. The primary gustatory cortex is in the insula, lower postcentral gyrus, and roof of the lateral sulcus.
6. Smell *(olfaction)* results from the action of chemicals on *olfactory neurons* in the roof of the nasal cavity. Nerve fibers from the olfactory cells assemble into fascicles that collectively constitute cranial nerve I, pass through foramina of the cribriform plate, and end in the olfactory bulbs beneath the frontal lobes of the cerebrum.
7. Olfactory signals travel the *olfactory tracts* from the bulbs to the temporal lobes, limbic system, and hypothalamus, and follow another pathway via the thalamus to the *orbitofrontal cortex.* An association area in the orbitofrontal cortex integrates food-related stimuli from the olfactory, gustatory, and visual systems into an overall sense of flavor and palatability.
8. The cerebral cortex also sends inhibitory signals back to the olfactory bulbs and can modify the perception of odors according to varying conditions such as hunger and satiety.

The Ear (p. 484)

1. The ear is divided into three sections: the outer, middle, and inner ears. The *outer ear* consists of the *auricle* and *auditory canal.* The *middle ear* consists of the *tympanic membrane* and an air-filled *tympanic cavity* containing three bones (*malleus, incus,* and *stapes*) and two muscles (*tensor tympani* and *stapedius*). The inner ear consists of fluid-filled chambers and tubes (the *membranous labyrinth*) including the *vestibule, semicircular ducts,* and *cochlea.*
2. The most important part of the cochlea, the organ of hearing, is the spiral *organ of Corti,* which includes rows of sensory *hair cells*

supported on a movable *basilar membrane.* A row of 3,500 *inner hair cells* generates the signals we hear, and three rows of *outer hair cells* tune the cochlea to enhance its pitch discrimination.

3. Vibrations in the ear move the basilar membrane of the cochlea up and down. As the hair cells move up and down, their stereocilia bend against the relatively stationary tectorial membrane above them. This opens K^+ channels at the tip of each stereocilium, and the inflow of K^+ depolarizes the cell. This triggers neurotransmitter release, which initiates a nerve signal.
4. Loudness determines the amplitude of basilar membrane vibration and the firing frequency of the associated auditory neurons. Pitch determines which regions of the basilar membrane vibrate more than others, and which auditory nerve fibers respond most strongly.
5. The cochlear nerve joins the vestibular nerve to become the vestibulocochlear nerve (cranial nerve VIII). Cochlear nerve fibers project to the *cochlear nucleus* of the medulla oblongata and from there to the *superior olivary nucleus* of the pons. That nucleus issues output to the outer hair cells of the cochlea; to the muscles of the middle ear; and to the inferior colliculi of the midbrain; and it functions in binaural hearing. The inferior colliculi function in auditory reflexes and further aid in binaural hearing, and issue fibers to the thalamus, which relays signals to the primary auditory cortex of the temporal lobe.
6. The *vestibular apparatus* consists of inner-ear structures concerned with *static equilibrium,* the sense of the orientation of the head; *dynamic equilibrium* is the sense of motion or acceleration. Acceleration can be *linear* or *angular.*
7. The *saccule* and *utricle* are chambers in the vestibule of the inner ear, each with a *macula* containing sensory hair cells. The *macula sacculi* is nearly vertical and the *macula utriculi* is nearly horizontal.
8. The hair cell stereocilia are capped by a weighted gelatinous *otolithic membrane.* When pulled by gravity or linear acceleration of the body, these membranes stimulate the hair cells.
9. Any orientation of the head causes a combination of stimulation to the four maculae, sending signals to the brain that enable it to sense the orientation. Vertical acceleration also stimulates each macula sacculi, and horizontal acceleration stimulates each macula utriculi.
10. Each inner ear also has three *semicircular ducts* with a sensory patch of hair cells, the *crista ampullaris,* in each duct. The stereocilia of these hair cells are embedded in a gelatinous *cupula.*
11. Tilting or rotation of the head moves the ducts relative to the fluid (endolymph) within, causing the fluid to push the cupula and stimulate the hair cells. The brain detects angular acceleration of the head from the combined input from the six ducts.
12. Signals from the utricle, saccule, and semicircular ducts travel the *vestibular nerve,* which joins the cochlear nerve in cranial nerve VIII. Vestibular nerve fibers lead to the cerebellum and to *vestibular nuclei* in the pons and medulla. Second-order fibers from these nuclei project to the spinal cord for postural reflexes; to the nuclei of cranial nerves III, IV, and VI for eye movements; and to thalamic nuclei that relay signals to the sensory cortex in the lateral sulcus and at the lower end of the central sulcus, where awareness of the body's movements and orientation in space resides.

The Eye (p. 494)

1. Accessory structures of the orbit include the *eyebrows, eyelids, conjunctiva, lacrimal apparatus,* and *extrinsic eye muscles.*
2. The wall of the eyeball is composed of an outer *fibrous layer* composed of *sclera* and *cornea;* a middle *vascular layer* composed of *choroid, ciliary body,* and *iris;* and an *inner layer* composed of the *retina.*
3. The optical components of the eye admit and bend (refract) light rays and bring images to a focus on the retina. They include the *cornea, aqueous humor, lens,* and *vitreous body.*
4. The neural components of the eye absorb light and encode the stimulus in action potentials transmitted to the brain. They are the *retina* and *optic nerve.* The sharpest vision occurs in a region of retina called the *fovea centralis,* while the *optic disc,* where the optic nerve originates, is a blind spot with no receptor cells.
5. As light enters the eye, it is refracted mainly by the cornea, with the lens making slight adjustments in focus.
6. Light falling on the retina is absorbed by visual pigments in the *outer segments* of the *rod* and *cone* cells. Rods function at low light intensities (producing night, or *scotopic,* vision) but produce monochromatic images with poor resolution. Cones require higher light intensities (producing day, or *photopic,* vision) and produce color images with finer resolution.
7. Rods and cones synapse with *bipolar cells.* Bipolar cells, in turn, stimulate *ganglion cells.* Ganglion cells are the first cells in the pathway that generate action potentials; their axons form the optic nerve. Some ganglion cells contain their own sensory pigment and respond directly to light rather than to input from rod and cone pathways. They do not contribute to the visual image but to such nonvisual responses to light as the pupillary reflexes and circadian rhythms.
8. *Horizontal cells* and *amacrine cells* are additional types of retinal neurons; they are involved in the perception of contrast, edges, and changes in light intensity.
9. The *duplicity theory* explains that a single type of receptor cell cannot produce both high light sensitivity (like the rods) and high resolution (like the cones). The neuronal convergence responsible for the high light sensitivity of rod pathways reduces resolution, while the lack of convergence responsible for the high resolution of cones reduces light sensitivity.
10. Fibers of the optic nerves *hemidecussate* at the *optic chiasm,* so images in the left visual field project from both eyes to the right cerebral hemisphere, and images on the right project to the left hemisphere.
11. Beyond the optic chiasm, the optic nerve fibers continue as *optic tracts.* Most of these nerve fibers end in the *lateral geniculate nucleus* of the thalamus. Here they synapse with neurons whose fibers form the *optic radiation* leading to the primary visual cortex of the occipital lobe.
12. Optic nerve fibers from the photosensory ganglion cells lead to the superior colliculi and pretectal nuclei of the midbrain. These midbrain nuclei control visual reflexes of the extrinsic eye muscles, pupillary reflexes, and accommodation of the lens in near vision.

Developmental and Clinical Perspectives (p. 504)

1. The inner ear begins its embryonic development as an ectodermal thickening, the *otic placode,* which invaginates and eventually separates from the ectoderm as an *otic vesicle.* The vesicle differentiates into the saccule and utricle, which soon exhibit outgrowths that become the cochlea and semicircular ducts.
2. The middle-ear cavity and auditory tube develop from an outgrowth of the first pharyngeal pouch. Mesenchyme of the first two pharyngeal arches, flanking this pouch, gives rise to the auditory ossicles and middle-ear muscles.
3. The facing margins of the first two pharyngeal arches develop six *auricular hillocks* which enlarge, fuse, and become the folds of the auricle of the outer ear. The pharyngeal groove between these arches invaginates to become the auditory canal.

4. The eye begins its development as an outgrowth of the diencephalon called the *optic vesicle,* a hollow sac continuous with the neural tube. The optic vesicle invaginates to form an *optic cup* and its connection to the diencephalon constricts and becomes the *optic stalk.* The outer layer of the optic cup becomes the pigment epithelium of the retina and the inner layer gives rise to the layers of receptor cells and retinal neurons.
5. The optic cup induces the overlying ectoderm to thicken into a *lens placode,* which invaginates, separates from the ectoderm, and becomes a *lens vesicle,* nestled within the optic cup. The vesicle fills in with lens fibers.
6. The *primary vitreous body* appears as a gelatinous secretion between the optic cup and lens vesicle.
7. Mesenchyme grows around the lateral side of the optic vesicle to completely enclose the developing eye. A space opens within the lateral mesenchyme to become the anterior chamber of the eye, and the cornea develops from mesenchyme and the overlying ectoderm. The iris grows from the anterior margin of the optic cup, and the eyelids develop as folds of ectoderm with a thin mesenchymal center.
8. Some disorders of the sensory systems are described in table 17.3.

Testing Your Recall

1. Hot and cold stimuli are detected by
   a. free nerve endings.
   b. proprioceptors.
   c. Krause end bulbs.
   d. lamellated corpuscles.
   e. tactile corpuscles.
2. Sensory signals for all of the following except _____ must pass through the thalamus before they can reach the cerebral cortex.
   a. smell
   b. taste
   c. hearing
   d. equilibrium
   e. vision
3. The vallate papillae are more sensitive to _____ than to any of these other tastes.
   a. bitter
   b. sour
   c. sweet
   d. umami
   e. salty
4. The ear is somewhat protected from loud noises by
   a. the vestibule.
   b. the modiolus.
   c. the stapes.
   d. the stapedius.
   e. the superior rectus.
5. The sensory neurons that begin in the organ of Corti end in
   a. the cochlea.
   b. the cochlear nucleus.
   c. the superior olivary nucleus.
   d. the inferior colliculus.
   e. the temporal lobe.
6. The organ of Corti rests on
   a. the tympanic membrane.
   b. the secondary tympanic membrane.
   c. the tectorial membrane.
   d. the vestibular membrane.
   e. the basilar membrane.
7. The acceleration you feel when an elevator begins to rise is sensed by
   a. the anterior semicircular duct.
   b. the organ of Corti.
   c. the crista ampullaris.
   d. the macula sacculi.
   e. the macula utriculi.
8. The highest density of cone cells is found in
   a. the crista ampullaris.
   b. the optic disc.
   c. the fovea centralis.
   d. the chorion.
   e. the basilar membrane.
9. The dilated blood vessels seen in "bloodshot" eyes are vessels of
   a. the retina.
   b. the cornea.
   c. the conjunctiva.
   d. the sclera.
   e. the choroid.
10. A person would look crosseyed if _____ muscles contracted at once.
    a. both medial rectus
    b. both lateral rectus
    c. the right medial rectus and left lateral rectus
    d. both superior oblique
    e. the left superior oblique and right inferior oblique
11. The most finely detailed vision occurs when an image falls on a pit in the retina called the _____.
12. Fibers of the optic nerve come from the _____ cells of the retina.
13. A sensory nerve ending specialized to detect tissue injury and produce a sensation of pain is called a/an _____.
14. The gelatinous membranes of the macula sacculi and macula utriculi are weighted by calcium carbonate–protein granules called _____.
15. Three rows of _____ in the cochlea have V-shaped arrays of stereocilia and tune the frequency sensitivity of the cochlea.
16. The _____ is a tiny bone that vibrates in the oval window and thereby transfers sound vibrations to the inner ear.
17. The _____ of the midbrain receive auditory input and triggers the head-turning auditory reflex.
18. The apical microvilli of a gustatory cell are called _____.
19. Olfactory neurons synapse with mitral cells and tufted cells in the _____, which lies inferior to the frontal lobe.
20. In the phenomenon of _____, pain from the viscera is perceived as coming from an area of the skin.

Answers in the Appendix

True or False

Determine which five of the following statements are false, and briefly explain why.

1. Interoceptors belong to the general senses.
2. The sensory (afferent) neurons for touch end in the thalamus.
3. The right cerebral hemisphere perceives things we touch with our left hand.
4. Descending analgesic fibers prevent pain signals from reaching the spinal cord.
5. Sweets are tasted with the filiform papillae at the tip of the tongue.
6. Some chemoreceptors are exteroceptors and some are interoceptors.
7. Humans have more photoreceptor cells than taste cells.
8. Human neurons are never exposed to the external environment of the body.
9. The eardrum has no nerve fibers.
10. The vitreous body occupies the posterior chamber of the eye.

Answers in the Appendix

Testing Your Comprehension

1. The principle of neuronal convergence was explained in chapter 13. Discuss its relevance to referred pain and scotopic vision.
2. What type of cutaneous receptor enables you to feel an insect crawling through your hair? What type enables you to palpate a patient's pulse? What type enables a blind person to read braille?
3. Predict the consequences of a hypothetical disorder in which the eye begins to break down or reabsorb the gel of the vitreous body.
4. Suppose a virus were able to selectively invade and destroy the following nervous tissues. Predict the sensory consequences of each infection: (*a*) the spiral ganglion, (*b*) the vestibular nucleus, (*c*) the motor fibers of cranial nerve VIII, (*d*) the motor fibers of cranial nerve VII, (*e*) the dorsal horns of spinal cord segments L3 to S5.
5. Summarize the similarities and differences between olfactory cells and taste cells.

Answers at the Online Learning Center

www.mhhe.com/saladinha1

Visit the Online Learning Center for practice tests, answer keys, and other learning aids for this chapter. Enhance your understanding of human anatomy with our interactive art labeling exercises, supplemental photo atlases, web links, puzzles, flashcards, and much more.

18

CHAPTER EIGHTEEN

The Endocrine System

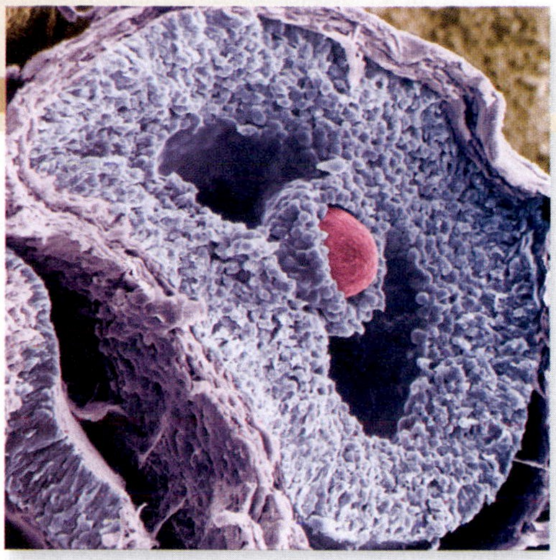

An ovarian follicle (SEM). Layered cells of the follicle wall secrete steroid hormones.

CHAPTER OUTLINE

Overview of the Endocrine System 514
- Hormone Chemistry and Action 514
- Comparison of the Nervous and Endocrine Systems 515

The Hypothalamus and Pituitary Gland 517
- Anatomy 518
- Hypothalamic Hormones 518
- Anterior Pituitary Hormones 518
- The Pars Intermedia 520
- Posterior Pituitary Hormones 520
- Actions of the Pituitary Hormones 522

Other Endocrine Glands 523
- The Pineal Gland 523
- The Thymus 523
- The Thyroid Gland 523
- The Parathyroid Glands 525
- The Adrenal Glands 526
- The Pancreatic Islets 527
- The Gonads 528
- Endocrine Cells in Other Organs 528

Developmental and Clinical Perspectives 532
- Prenatal Development of the Endocrine Glands 532
- The Aging Endocrine System 532
- Endocrine Disorders 532

Chapter Review 537

INSIGHTS

18.1 Clinical Application: Pituitary Trauma and Endocrine Dysfunction 522

18.2 Clinical Application: Pineal Tumors and Precocial Puberty 523

18.3 Clinical Application: A Fatal Effect of Postsurgical Hypoparathyroidism 526

BRUSHING UP

To understand this chapter, you may find it helpful to review the following concepts:

- Embryonic development of the pharyngeal pouches (p. 113)
- Embryonic development of the neural crest (p. 379)
- The hypothalamus (p. 428)

The body requires systems of internal communication if it is to function as an integrated homeostatic whole. The major communication systems are the nervous and endocrine systems, which communicate with neurotransmitters and hormones, respectively. You probably have at least some prior acquaintance with the endocrine system. Perhaps you have heard of the pituitary gland and thyroid gland; secretions such as growth hormone, estrogen, and insulin; and endocrine disorders such as dwarfism, goiter, and diabetes mellitus. This chapter is about **endocrinology,** the science of the endocrine system, its hormones, and the diagnosis and treatment of its dysfunctions.

Overview of the Endocrine System

Objectives

When you have completed this section, you should be able to

- define *hormone* and *endocrine system;*
- describe how endocrine glands differ from exocrine glands;
- describe the chemical classes of hormones;
- describe the general ways in which hormones affect their target cells; and
- compare and contrast the nervous and endocrine systems.

Hormones[1] are chemical messengers that are secreted into the bloodstream and stimulate physiological responses in distant organs. They are secreted by **endocrine glands** (fig. 18.1) and specialized cells found in many organs not usually thought of as glands, such as the brain, heart, and small intestine. The **endocrine system** consists of all of these hormone-producing cells and glands.

Exocrine glands secrete their products onto the surface of an epithelium, usually by way of a duct. Endocrine glands, by contrast, lack ducts and, instead, secrete their products into the bloodstream. Consequently, endocrine glands have an unusually high density of blood capillaries, and their capillaries are often of an especially porous type called *fenestrated capillaries* (see chapter 20), facilitating the entry of hormones into the bloodstream. Hormones used to be called the "internal secretions" of the body.[2] Hormones go everywhere the blood goes; they cannot be sent selectively to a particular organ. However, the only cells that respond to a hormone are those that have receptors for it. We call these the **target cells.**

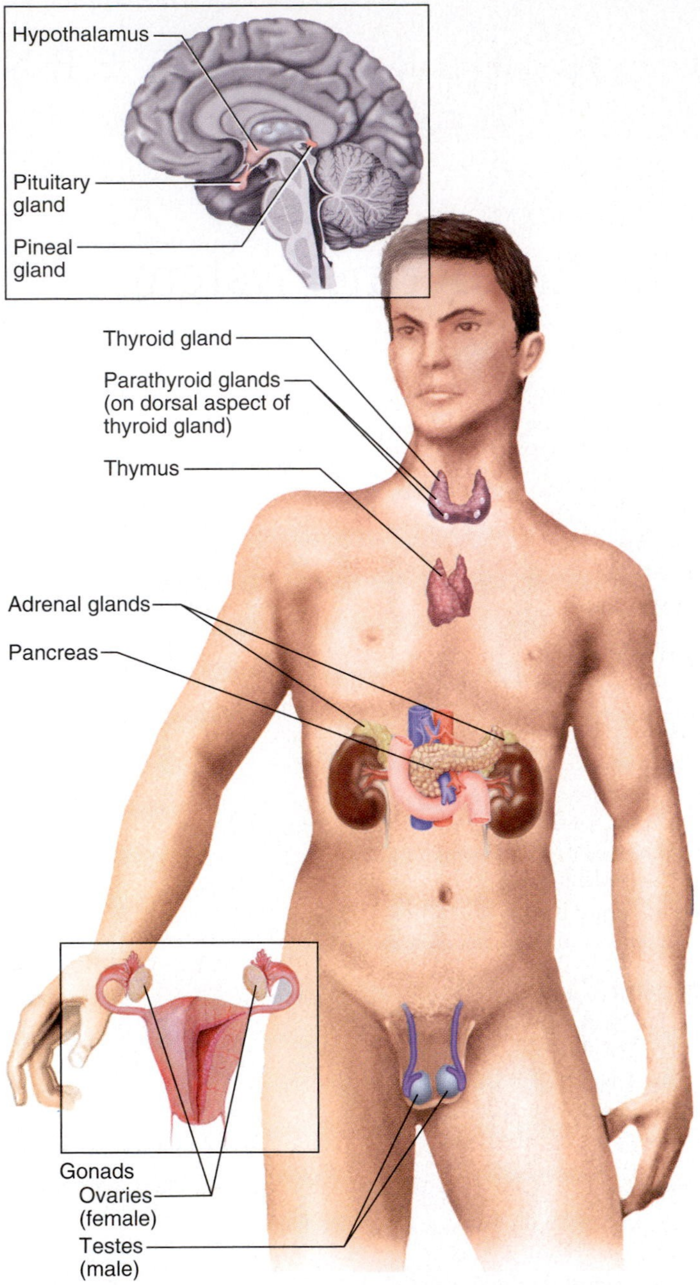

Figure 18.1

Major Organs of the Endocrine System. This system also includes gland cells in many organs not shown here.

Hormone Chemistry and Action

Hormones fall into three chemical classes: *steroids,* which are lipids synthesized from cholesterol; *monoamines,* which are small molecules, bearing an amino group, synthesized from the amino acids tyrosine and tryptophan; and *peptides,* which are chains of about 3 to 200 amino acids (table 18.1, fig. 18.2). The largest peptides (50 amino acids and longer) are proteins, and some of these have small carbohydrate chains bonded to them, making them *glycoproteins.*

Some hormones, especially monoamines and peptides, are hydrophilic—they mix freely with water and are therefore easily transported in the blood. Steroid hormones and thyroid hormone, however, are hydrophobic. They do not mix freely with water, and must bind to a *transport protein* in the blood plasma to be carried to their target cells. Transport proteins also temporarily protect the hormones from being broken down by enzymes in the blood plasma and liver, and from being filtered and excreted by the kidneys. Thus, they prolong the action of a hormone. Free hormone may be broken down or removed from the blood in a few minutes, whereas hormone bound to a transport protein may circulate from hours to weeks.

[1]*hormone,* = to excite, set in motion
[2]*endo* = internal + *crin* = to secrete

TABLE 18.1

Chemical Classification of Hormones

Steroids and Steroid Derivatives	
Aldosterone	Estrogens
Calcitriol	Progesterone
Cortisol	Testosterone
Corticosterone	
Monoamines	
Dopamine	Serotonin
Epinephrine	Thyroxine
Melatonin	Triiodothyronine
Norepinephrine	
Peptides	
Oligopeptides (3–10 amino acids)	
Angiotensin II	Oxytocin
Antidiuretic hormone	Thyrotropin-releasing hormone
Gonadotropin-releasing hormone	
Polypeptides (14–199 amino acids)	
Adrenocorticotropic hormone	Growth hormone–releasing hormone
Atrial natriuretic peptide	Insulin
Calcitonin	Pancreatic polypeptide
Corticotropin-releasing hormone	Parathyroid hormone
Glucagon	Prolactin
Growth hormone	Somatostatin
Glycoproteins (204–210 amino acids and a carbohydrate chain)	
Follicle-stimulating hormone	Luteinizing hormone
Human chorionic gonadotropin	Thyroid-stimulating hormone
Inhibin	

THINK ABOUT IT!

If equal amounts of estrogen or oxytocin were injected into a person, which hormone would persist longer in the bloodstream? Why?

When a hormone reaches a target cell, it can do one of two things depending on its chemistry. Steroids and thyroid hormone diffuse into the cell and bind to receptors in the cytoplasm or nucleus. Monoamines and peptides cannot enter the cell, but bind to receptors on its surface. In this case, the hormone acts as a *first messenger* to the target cell and triggers the formation of a *second messenger* inside the cell—a chemical such as *cyclic adenosine monophosphate (cAMP)* or *inositol triphosphate (IP_3)*. The second messenger then stimulates metabolic changes in the cell.

However a hormone stimulates a target cell, the target cell synthesizes new enzymes or activates or inhibits preexisting ones. This results in a metabolic effect on the target cell such as glycogen breakdown, muscle growth, or sperm development.

Comparison of the Nervous and Endocrine Systems

The nervous and endocrine systems both serve for internal communication, but they are not redundant; they complement rather than duplicate each other's function (table 18.2, fig. 18.3). One important difference between the systems is the speed with which they start and stop responding to a stimulus. The nervous system typically responds in just a few milliseconds, whereas it takes from several seconds to days for a hormone to act. When a stimulus ceases, the nervous system stops responding almost immediately. Hormonal effects, however, may last for several days or longer. On the other hand, under long-term stimulation, neurons adapt and their response declines. The endocrine system shows more persistent responses. For example, thyroid hormone level rises in cold weather and remains elevated as long as it remains cold. Finally, an efferent nerve fiber innervates only one organ and a limited number of cells within that organ; its effects, therefore, are precisely targeted and relatively specific. Some hormones, by contrast, have very widespread effects on the body, such as thyroid hormone and growth hormone.

These differences, however, should not blind us to the similarities and interactions between the two systems. Both communicate chemically, and several chemicals function as both neurotransmitters and hormones, including norepinephrine, dopamine, cholecystokinin, and thyrotropin-releasing hormone. Some hormones, such as oxytocin and some monoamines, are secreted by **neuroendocrine cells**—neurons that release their secretions into the blood. Some neurotransmitters and hormones produce identical effects on the same cells. For example, both norepinephrine and glucagon stimulate the liver to break down glycogen and release glucose. The nervous and endocrine systems continually regulate each other as they coordinate the activities of other organ systems. Some neurons trigger hormone secretion, and some hormones stimulate or inhibit neurons.

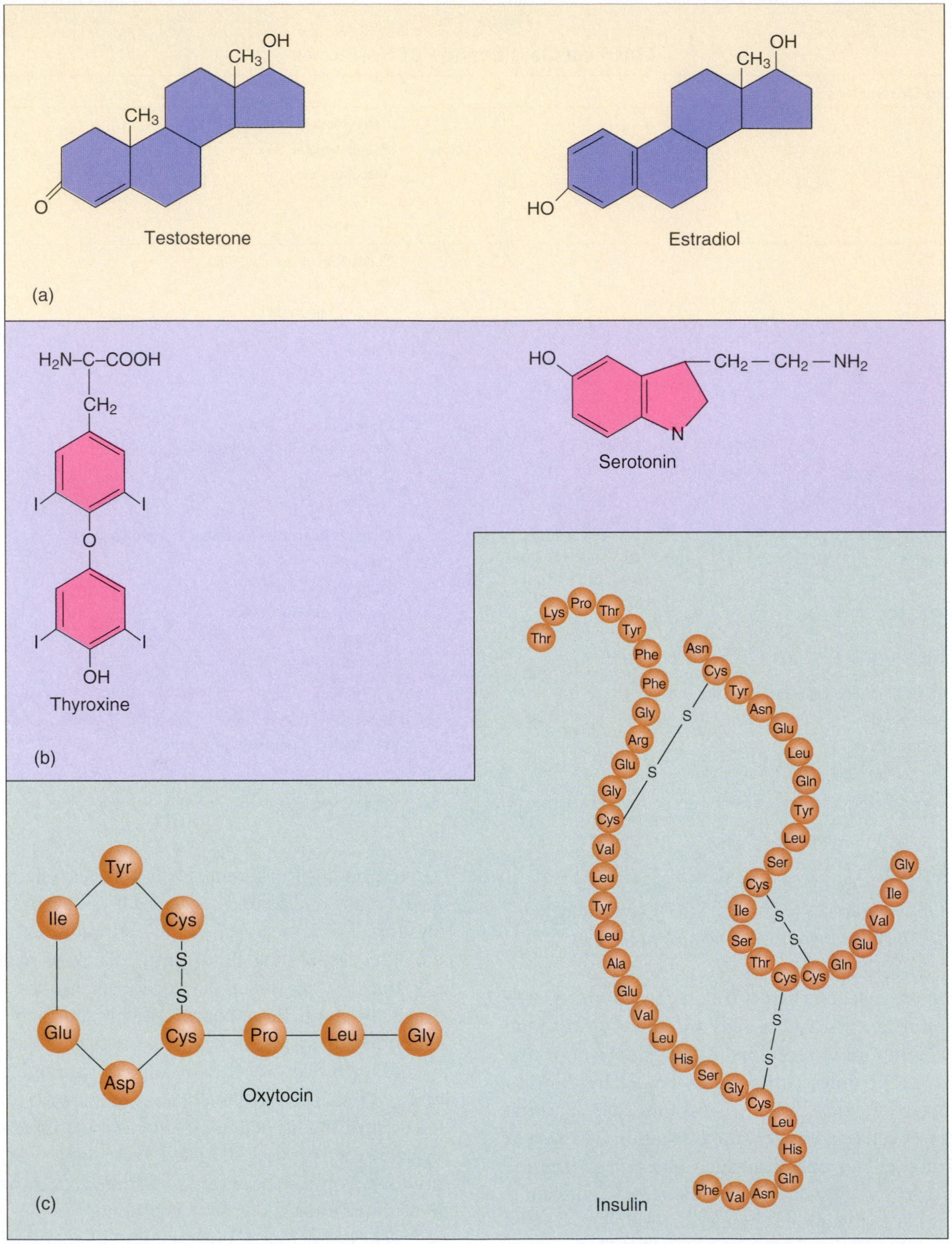

FIGURE 18.2

The Chemical Classes of Hormones. (*a*) Two steroid hormones, testosterone and estradiol (an estrogen). (*b*) Two monoamines, thyroxine and serotonin. Note the presence of an amino (–NH_2) group, which defines a monoamine. (*c*) Two peptide hormones, oxytocin and insulin. The three-letter labels are standard symbols for the various amino acids.

TABLE 18.2

Comparison of the Nervous and Endocrine Systems

Nervous System	Endocrine System
Communicates with electrical impulses and neurotransmitters	Communicates with hormones
Releases neurotransmitters at synapses at specific target cells	Releases hormones into bloodstream for general distribution throughout body
Has relatively local, specific effects on target organs	Sometimes has very general, widespread effects on many organs in body
Reacts quickly to stimuli, usually within 1 to 10 msec	Reacts more slowly to stimuli, often taking seconds to days
Stops quickly when stimulus stops	May continue responding long after stimulus stops
Adapts relatively quickly to continual stimulation	Adapts slowly; may continue responding for days to weeks of continual stimulation

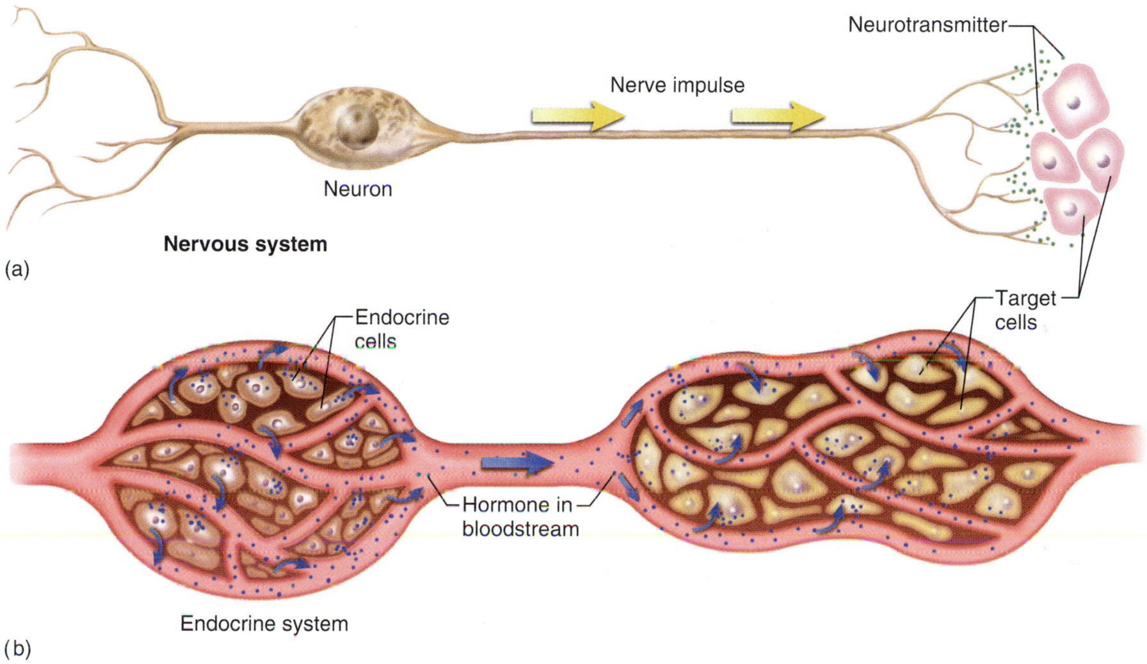

FIGURE 18.3

Communication by the Nervous and Endocrine Systems. (*a*) A neuron has a long fiber that delivers its neurotransmitter to the immediate vicinity of its target cells. (*b*) Endocrine cells secrete a hormone into the bloodstream. At a point often remote from its origin, the hormone leaves the bloodstream and enters or binds to its target cells.

Before You Go On

Answer the following questions to test your understanding of the preceding section.

1. Define the word *hormone* and distinguish a hormone from a neurotransmitter. Why is this an imperfect distinction?
2. Name two steroid hormones, two monoamine hormones, and two peptide hormones. Explain what distinguishes these three categories.
3. Describe where the receptors for hormones are located and which classes of hormones bind to receptors in the different locations.
4. List some similarities and differences between the endocrine and nervous systems.

THE HYPOTHALAMUS AND PITUITARY GLAND

Objectives

When you have completed this section, you should be able to

- describe the location and anatomy of the pituitary gland and its anatomical relationship with the hypothalamus; and
- list the hormones produced by the hypothalamus and pituitary gland and state the functions of those hormones.

There is no "master control center" that regulates the entire endocrine system, but the pituitary gland and a nearby region of the brain, the hypothalamus, have a more wide-ranging influence than any other endocrine gland. They are an appropriate place to start our survey.

Anatomy

The **hypothalamus,** shaped like a flattened funnel, forms the floor and walls of the third ventricle of the brain (see fig. 15.2). It regulates primitive functions of the body ranging from water balance to sex drive. Many of its functions are carried out by way of the pituitary gland.

The **pituitary gland (hypophysis**[3]**)** is housed in the sella turcica of the sphenoid bone. A sheet of dura mater covers the sella turcica and separates the pituitary from the brain, except for a stalk that perforates the dura and connects the pituitary to the hypothalamus. The pituitary is an ovoid gland about 1.3 cm in diameter, but grows about 50% larger in pregnancy. It is actually composed of two structures—the adenohypophysis and neurohypophysis—which arise independently in the embryo and have entirely separate functions.

The **adenohypophysis**[4] (AD-eh-no-hy-POFF-ih-sis) is the anterior three-quarters of the pituitary (figs. 18.4 and 18.5). It has two parts: a large **anterior lobe,** also called the *pars distalis* ("distal part") because it is most distal to the pituitary stalk, and the **pars tuberalis,** a small mass of cells adhering to the anterior side of the stalk. In the fetus there is also a **pars intermedia,** a strip of tissue between the anterior lobe and neurohypophysis. During subsequent development, its cells mingle with those of the anterior lobe; in adults, there is no longer a separate pars intermedia.

The anterior pituitary has no nervous connection to the hypothalamus, but is connected to it by a complex of blood vessels called the **hypophyseal portal system.** This system begins with a network of *primary capillaries* in the hypothalamus. They drain into *portal venules* (small veins) that travel down the pituitary stalk to a complex of *secondary capillaries* in the anterior pituitary. Hormones are secreted by the hypothalamus into the primary capillaries, travel to the secondary capillaries, and leave the bloodstream to stimulate cells of the anterior pituitary. These hypothalamic hormones either trigger or inhibit the release of pituitary hormones.

The **neurohypophysis** is the posterior one-quarter of the pituitary. It has three parts: the **posterior lobe** *(pars nervosa),* the **stalk** *(infundibulum)* that connects it to the hypothalamus, and the **median eminence,** an extension of the hypothalamic floor. The neurohypophysis is not a true gland but a mass of neuroglia and axons arising from certain hypothalamic neurons (fig. 18.5). The axons form a bundle called the **hypothalamo-hypophyseal tract,** which runs down through the stalk and ends in the posterior lobe. Hormones are synthesized by the neurons in the hypothalamus, transported down the axons, and stored in the posterior pituitary until a nerve signal triggers their release.

Hereafter, we can largely disregard all parts of the pituitary except the anterior and posterior lobes; they secrete all of the pituitary hormones we will consider.

Hypothalamic Hormones

The hypothalamus produces nine hormones of importance to this chapter. Seven of them travel through the portal system and regulate the activities of the anterior pituitary. These are listed in figure 18.4 and table 18.3. The *releasing hormones* stimulate the anterior pituitary to secrete its hormones, and the *inhibiting hormones* suppress pituitary secretion. In most cases, the name of the hypothalamic hormone indicates the pituitary hormone whose secretion it stimulates or inhibits. Gonadotropin-releasing hormone, however, controls two pituitary hormones called *gonadotropins:* follicle-stimulating hormone and luteinizing hormone. Prolactin-inhibiting hormone is the neurotransmitter dopamine, a monoamine. All of the other hypothalamic releasing and inhibiting hormones are peptides.

The other two hypothalamic hormones are **oxytocin (OT)** and **antidiuretic hormone (ADH).** These are oligopeptides composed of nine amino acids, only two of which differ between OT and ADH. Oxytocin is produced by a pair of hypothalamic neuron clusters called the **paraventricular**[5] **nuclei** because they lie in the walls of the third ventricle. Antidiuretic hormone is produced by another pair called the **supraoptic**[6] **nuclei** because they lie just above the optic chiasm. OT and ADH are transported through nerve fibers of the hypothalamo-hypophyseal tract and stored in the posterior pituitary. When the hypothalamic nuclei sense a need for one of these hormones, the neurons send nerve signals to the posterior pituitary to release the hormone into the blood.

Anterior Pituitary Hormones

The anterior lobe of the pituitary synthesizes and secretes six principal hormones: follicle-stimulating hormone (FSH), luteinizing hormone (LH), thyroid-stimulating hormone (TSH), adrenocorticotropic hormone (ACTH), prolactin (PRL), and growth hormone (GH) (table 18.4). They are all polypeptides or glycoproteins. The first four of these are **tropic,** or **trophic,**[7] hormones—pituitary hormones whose target organs are other endocrine glands. The first two are called **gonadotropins** because their target organs are the gonads.

The relationship between the pituitary, its tropic hormones, and their target endocrine glands is called an axis—a convenient term for referring to the way these endocrine glands influence each other. There are three such axes: the **pituitary-gonadal axis** involving FSH and LH, the **pituitary-thyroid axis** involving TSH, and the **pituitary-adrenal axis** involving ACTH (fig. 18.6). We can also include the hypothalamus in these axes, resulting in a more cumbersome terminology but acknowledging the role of the hypothalamic releasing and inhibiting hormones. For example, the *hypothalamo-pituitary-thyroid axis* includes the hypothalamus, pituitary, and thyroid gland, and the hormones thyrotropin-releasing hormone (TRH), thyroid-stimulating hormone (TSH), and thyroid hormone (TH).

---

[3]*hypo* = below + *physis* = growth
[4]*adeno* = gland
[5]*para* = next to + *ventricular* = pertaining to the ventricle
[6]*supra* = above
[7]*trop* = to turn, change; *troph* = to feed, nourish

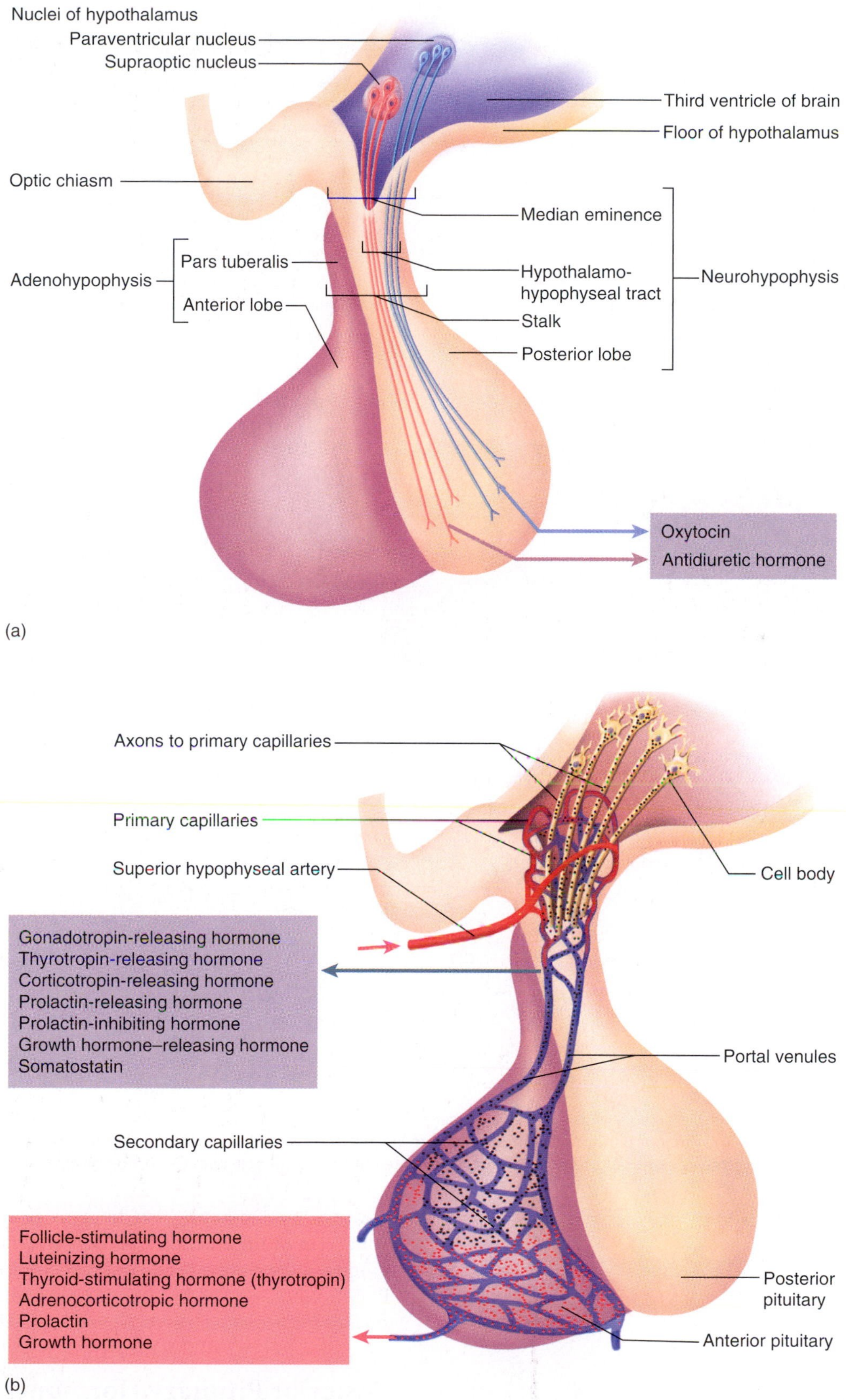

FIGURE 18.4

Gross Anatomy of the Pituitary Gland. (*a*) Major structures of the pituitary and hormones of the neurohypophysis. Note that these hormones are produced by two nuclei in the hypothalamus and later released from the posterior lobe of the pituitary. (*b*) The hypophyseal portal system. The hormones in the *violet box* are secreted by the hypothalamus and travel in the portal blood vessels to the anterior pituitary. The hormones in the *red box* are secreted by the anterior pituitary under the control of the hypothalamic releasers and inhibitors.

TABLE 18.3

Hypothalamic Releasing and Inhibiting Hormones That Regulate the Anterior Pituitary

Hormone	Principal Effects on Pituitary
TRH: Thyrotropin-releasing hormone	Promotes TSH and PRL secretion
CRH: Corticotropin-releasing hormone	Promotes ACTH secretion
GnRH: Gonadotropin-releasing hormone	Promotes FSH and LH secretion
PRH: Prolactin-releasing hormone	Promotes PRL secretion
PIH: Prolactin-inhibiting hormone (dopamine)	Inhibits PRL secretion
GHRH: Growth hormone–releasing hormone	Promotes GH secretion
GHIH: Growth hormone–inhibiting hormone (somatostatin)	Inhibits GH and TSH secretion

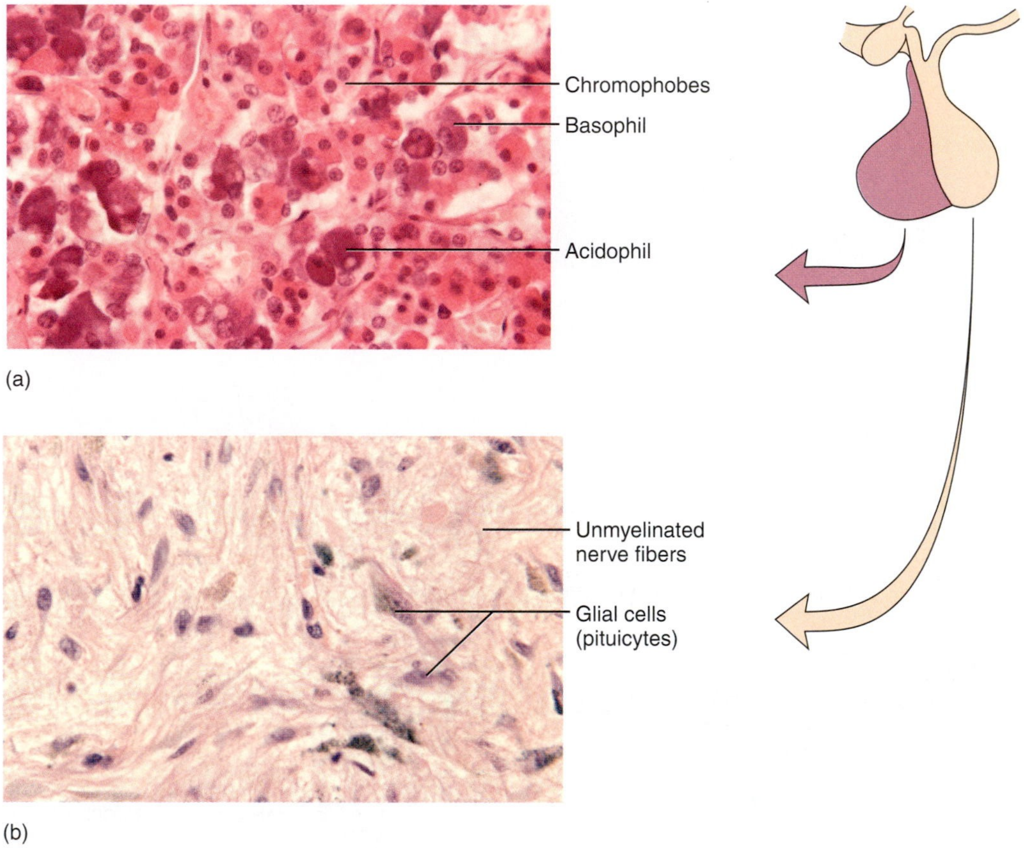

FIGURE 18.5
Histology of the Pituitary Gland. (*a*) The anterior pituitary. Chromophobes are inactive cells. Basophils include gonadotropes, thyrotropes, and corticotropes. Acidophils include somatotropes and lactotropes. These subtypes are not distinguishable with this histological stain. (*b*) The posterior pituitary.

THINK ABOUT IT!

Give a similarly complete description for what would be called the hypothalamo-pituitary-adrenal axis.

The Pars Intermedia

As mentioned earlier, the pars intermedia is absent from the adult human pituitary, but present in other animals and the human fetus. In other species, it secretes *melanocyte-stimulating hormone (MSH)*, which influences pigmentation of the skin, hair, or feathers. Until recently, it was thought to have a similar effect on the human skin, but evidence now indicates that humans have no circulating MSH. Some cells of the anterior lobe, derived from the pars intermedia, produce a large polypeptide called *pro-opiomelanocortin (POMC)*. POMC is not secreted, but is processed within the pituitary to yield smaller fragments such as ACTH and pain-inhibiting endorphins.

Posterior Pituitary Hormones

As we have already seen, the posterior lobe produces no hormones of its own. It stores and releases oxytocin and antidiuretic hormone, which are synthesized by the hypothalamus. Since they are released into the blood by the posterior pituitary, however, these are treated as pituitary hormones for convenience.

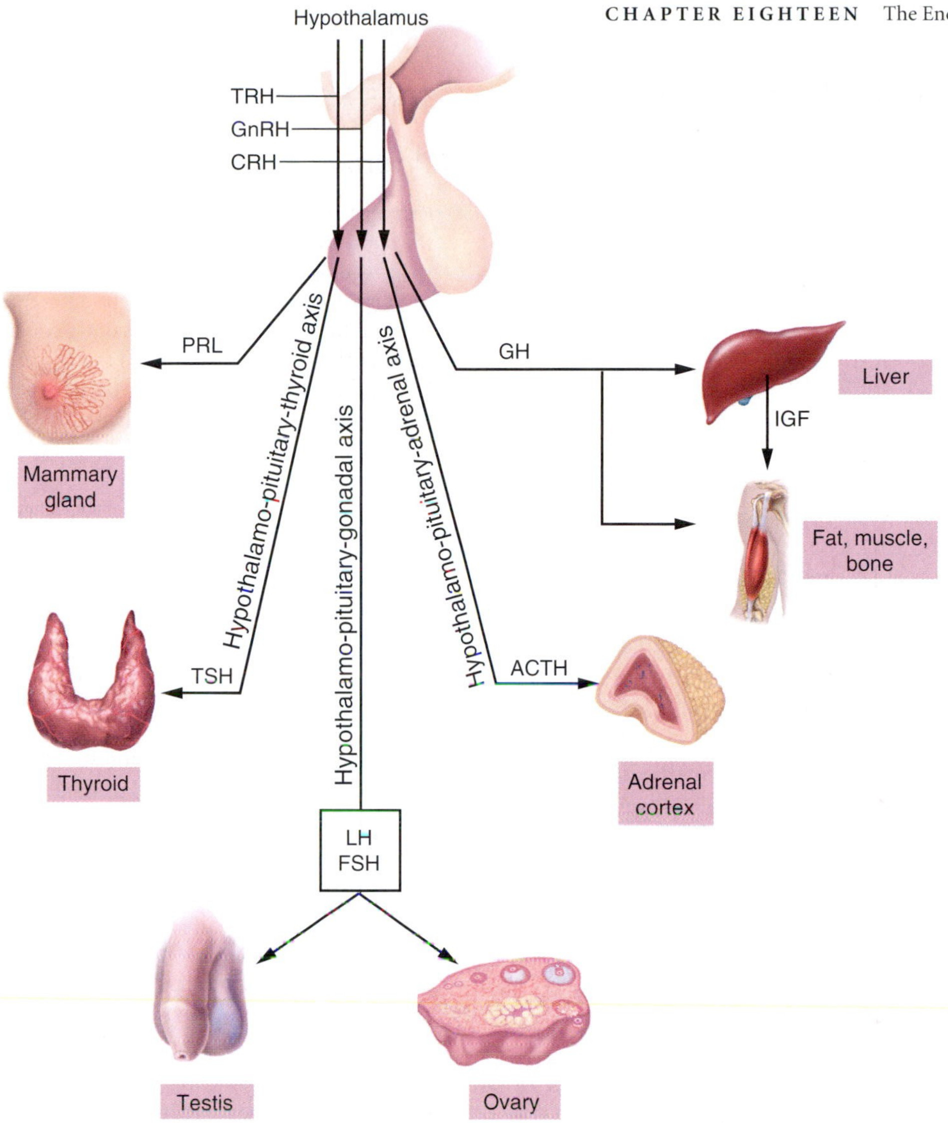

FIGURE 18.6

Hormones and Target Organs of the Anterior Pituitary Gland. The three axes physiologically link pituitary function to the function of other endocrine glands. Growth hormone acts both directly on target tissues such as fat, muscle, and bone, and indirectly through insulin-like growth factors (IGFs) secreted by the liver.

TABLE 18.4

Pituitary Hormones

Hormone	Target Organ	Principal Effects
Anterior Pituitary		
FSH: Follicle-stimulating hormone	Ovaries, testes	Female: growth of ovarian follicles and secretion of estrogen Male: sperm production
LH: Luteinizing hormone	Ovaries, testes	Female: ovulation, maintenance of corpus luteum Male: testosterone secretion
TSH: Thyroid-stimulating hormone	Thyroid gland	Growth of thyroid, secretion of thyroid hormone
ACTH: Adrenocorticotropic hormone	Adrenal cortex	Growth of adrenal cortex, secretion of glucocorticoids
PRL: Prolactin	Mammary glands, testes	Female: milk synthesis Male: increased LH sensitivity and testosterone secretion
GH: Growth hormone	Many organs	Widespread tissue growth, especially in cartilage, bone, muscle, and fat
Posterior Pituitary		
ADH: Antidiuretic hormone	Kidneys	Water retention
OT: Oxytocin	Uterus, mammary glands	Labor contractions, milk release; possibly involved in ejaculation, sperm transport, and sexual affection

Actions of the Pituitary Hormones

Following is a closer look at what these pituitary hormones do. Their actions are summarized in table 18.4.

ANTERIOR LOBE HORMONES

1. **Follicle-stimulating hormone (FSH).** FSH is one of the two gonadotropins. It is secreted by pituitary cells called *gonadotropes,* and its target organs are the ovaries and testes. In the ovaries, it stimulates the development of eggs and the bubblelike *follicles* that contain them, and the secretion of ovarian hormones. In the testes, it stimulates the production of sperm.
2. **Luteinizing hormone (LH).** LH is another gonadotropin secreted by the gonadotropes. In females, it stimulates *ovulation* (the release of an egg). LH is named for the fact that after ovulation, the remainder of a follicle is called the **corpus luteum** ("yellow body"). LH also stimulates the corpus luteum to secrete progesterone, a hormone important to pregnancy. In males, it stimulates the testes to secrete testosterone.
3. **Thyroid-stimulating hormone (TSH),** or **thyrotropin.** TSH is secreted by pituitary cells called *thyrotropes.* It stimulates growth of the thyroid gland and the secretion of thyroid hormone, whose effects are described later.
4. **Adrenocorticotropic hormone (ACTH),** or **corticotropin.** ACTH is secreted by pituitary cells called *corticotropes.* It is important in regulating the body's response to stress. It is named for its effect on the adrenal cortex, the outer layer of an endocrine gland near the kidney. ACTH stimulates the adrenal cortex to secrete hormones called *glucocorticoids,* which are important in glucose, fat, and protein metabolism.
5. **Prolactin**[8] **(PRL).** PRL is secreted by *lactotropes (mammotropes),* which increase greatly in size and number during pregnancy. They secrete PRL during pregnancy and for as long as a woman nurses, although the PRL has no effect until after she gives birth. Then, it stimulates the mammary glands to secrete milk. In males, PRL has a gonadotropic effect that makes the testes more sensitive to LH. Thus, it indirectly enhances their secretion of testosterone.
6. **Growth hormone (GH),** or **somatotropin.** GH is secreted by *somatotropes,* the most numerous cells in the anterior pituitary. The pituitary produces at least a thousand times as much GH as any other hormone. The general effect of GH is to promote mitosis and cellular differentiation, and thus to promote widespread tissue growth. Unlike the foregoing hormones, GH is not targeted to one or a few organs, but has widespread effects on the body, especially on cartilage, bone, muscle, and fat. GH not only stimulates these tissues directly, but also stimulates the liver and other tissues to secrete small polypeptides called **insulin-like growth factors (IGF-I** and **II),** or **somatomedins.**[9] IGFs then stimulate target cells in diverse other tissues (fig. 18.6). Most of these effects are caused by IGF-I, but IGF-II is important to fetal growth. GH and IGFs support tissue growth by mobilizing energy from fat, increasing levels of calcium and other electrolytes, and stimulating protein synthesis. Its most conspicuous effects occur during childhood and adolescence.

[8]*pro* = favoring + *lact* = milk

INSIGHT 18.1 CLINICAL APPLICATION

Pituitary Trauma and Endocrine Dysfunction

Perhaps surprisingly, the most frequently fractured bone of the skull is the sphenoid, which houses the pituitary gland. Sphenoid fractures can sever the pituitary stalk, including the hypothalamo-hypophyseal tract or portal system. Such injuries cut off communication from the brain to the pituitary and therefore disrupt pituitary functions that depend on hormonal or neural signals from the brain. Anterior lobe effects include loss of sexual functions (menstrual irregularity, sterility, and impotence due to the loss of gonadotropins) and inadequate thyroid and adrenal gland function due to TSH and ACTH hyposecretion. Growth hormone secretion also declines markedly, but this has no clinical effect on adults. For further consideration of the endocrine effects of a sphenoid bone fracture, see Testing Your Comprehension question 1 at the end of this chapter.

POSTERIOR LOBE HORMONES

1. **Antidiuretic**[10] **hormone (ADH).** ADH increases water retention by the kidneys, reduces urine volume, and helps prevent dehydration. It is also called *vasopressin* because it causes vasoconstriction at high concentrations. These concentrations are so unnaturally high for the human body, however, that this effect is of doubtful significance except in pathological states. ADH also functions as a brain neurotransmitter and is usually called vasopressin in the neuroscience literature.
2. **Oxytocin**[11] **(OT).** OT has two reproductive roles. In childbirth, it stimulates labor contractions, and in lactating mothers, it stimulates the flow of milk from the mammary gland acini to the nipple. In both sexes, OT surges during sexual arousal and orgasm. It may play a role in the propulsion of semen through the male reproductive tract, in uterine contractions that help transport sperm up the female reproductive tract, and in feelings of sexual satisfaction and emotional bonding.

[9]Acronym for *somato*tropin *med*iating prote*in*
[10]*anti* = against + *diuret* = to pass through, urinate
[11]*oxy* = sharp, quick + *toc* = childbirth

Before You Go On

Answer the following questions to test your understanding of the preceding section.

5. What are two good reasons for considering the pituitary to be two separate glands?
6. Describe two anatomical routes by which the hypothalamus sends signals to the pituitary gland.
7. Construct a three-column table. In the middle column, list the six hormones of the anterior pituitary gland. On the left, list the hypothalamic hormones that control these pituitary secretions. On the right, list the target organ and effect of each anterior pituitary hormone.
8. Name the two posterior lobe hormones, state where they are produced, and state their functions.

Other Endocrine Glands

Objectives

When you have completed this section, you should be able to

- describe the structure and location of the remaining endocrine glands;
- name the hormones that these endocrine glands produce and state their functions; and
- discuss hormones produced by endocrine cells in organs other than the classic endocrine glands.

The Pineal Gland

The **pineal**[12] **gland (epiphysis cerebri)** is a pine cone–shaped growth attached to the roof of the third ventricle, beneath the posterior end of the corpus callosum (see figs. 15.2 and 18.1). The philosopher René Descartes (1596–1650) thought it was the seat of the human soul. If so, children must have more soul than adults—a child's pineal gland is about 8 mm long and 5 mm wide, but after age seven it regresses rapidly and is no more than a tiny shrunken mass of fibrous tissue in the adult. Pineal secretion peaks between the ages of 1 and 5 years; by the end of puberty it declines 75%. Such shrinkage of an organ is called **involution.** Involution is accompanied by the appearance of granules of calcium phosphate and calcium carbonate called *pineal sand.* These stony granules are visible on X rays, enabling radiologists to determine the position of the gland. Displacement of the pineal from its normal location is evidence of a brain tumor or other structural abnormality.

We no longer look for the human soul in the pineal gland, but this little organ remains an intriguing mystery. It produces two monoamines, **serotonin** and **melatonin.** Melatonin secretion rises in the dark and occurs at a low level in daylight. Thus, its secretion fluctuates seasonally with changes in day length. In animals with seasonal breeding, the pineal regulates the gonads and the annual breeding cycle. Melatonin may suppress gonadotropin secretion; removal of the pineal from animals causes premature sexual maturation (see insight 18.2). The pineal may be involved in mood disorders in humans. Melatonin levels are elevated in seasonal affective disorder and premenstrual syndrome, but it is still uncertain whether the melatonin level or pineal dysfunction is the cause of these disorders.

INSIGHT 18.2 CLINICAL APPLICATION

Pineal Tumors and Precocial Puberty

Some physiologists believe that the pineal gland may play a role on the timing of puberty in humans. Tentative evidence of this comes from the effect of pineal tumors. Pineal tumors occur most commonly in childhood and sometimes cause puberty to begin prematurely, especially in boys. However, precocial puberty occurs only when the tumor also damages the hypothalamus, so it is difficult to say whether the effect is due specifically to pineal damage. Other effects of pineal tumors stem from the anatomical relationship of this gland to nearby brain structures (see Testing Your Comprehension question 2 at the end of this chapter).

The Thymus

The **thymus** plays a role in three systems: endocrine, lymphatic, and immune. It is a bilobed gland in the mediastinum superior to the heart, behind the sternal manubrium. In the fetus and infant, it is enormous in comparison to adjacent organs, sometimes protruding between the lungs nearly as far as the diaphragm, and extending upward into the neck (fig. 18.7*a*). It continues to grow until the age of 5 or 6 years, although not as fast as other thoracic organs, so its *relative* size decreases. After the age of 14, it undergoes rapid **involution** (shrinkage). It weighs barely 10 g in most adults, and in the elderly, it is a small fibrous and fatty remnant barely distinguishable from the surrounding mediastinal tissues (fig. 18.7*b*).

The thymus is a site of maturation for certain white blood cells that are critically important for immune defense (T lymphocytes, *T* for *thymus*). It secretes several hormones *(thymopoietin, thymosin,* and *thymulin)* that stimulate the development of other lymphatic organs and regulate the development and activity of the T lymphocytes. Its histology is discussed more fully in chapter 22 in relation to its immune function.

The Thyroid Gland

The **thyroid gland** is the largest endocrine gland in adults, weighing 20 to 25 g. It is composed of two lobes that lie adjacent to the trachea, immediately below the larynx. It is named for the nearby, shieldlike *thyroid*[13] *cartilage* of the larynx. Each lobe of the thyroid gland is bulbous at the inferior end and tapers superiorly. Near the inferior end, the two lobes are joined by a narrow bridge of tissue, the **isthmus,** which crosses the front of the trachea (fig. 18.8*a*). About 50% of people have an accessory *pyramidal lobe,* usually small, growing upward from the isthmus. Some people lack an isthmus, and some have thyroid tissue embedded in the root of the tongue, the thymus, or other places in the neck.

[12] *pineal* = pine cone

[13] *thyr* = shield + *oid* = like, resembling

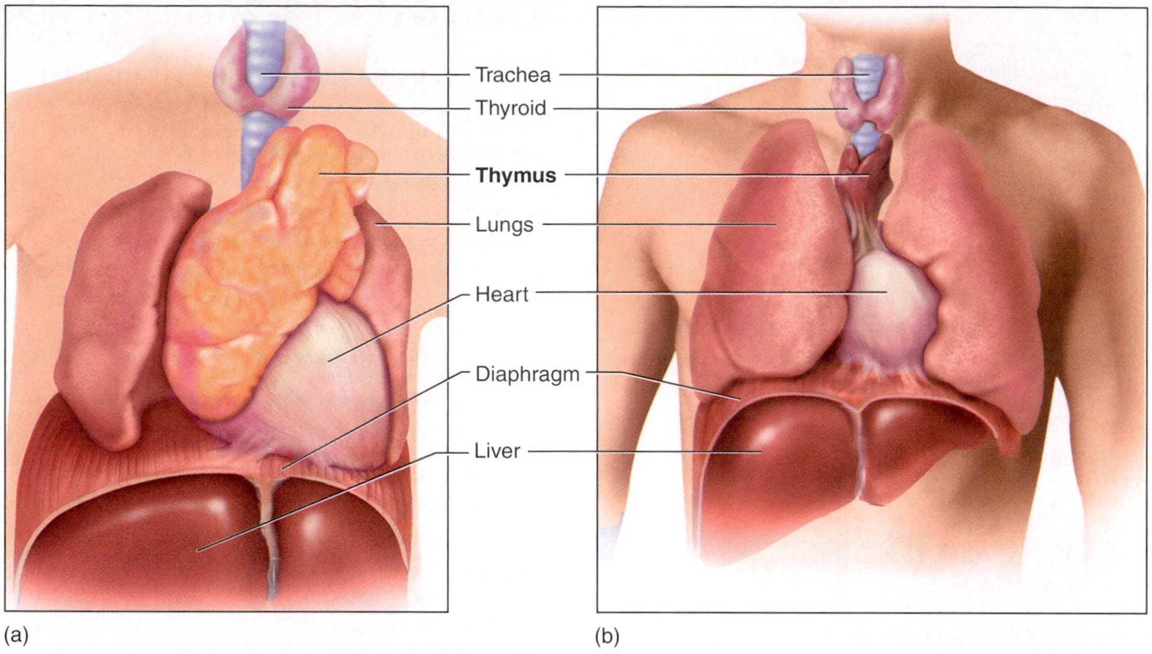

FIGURE 18.7
The Thymus. (*a*) Thymus of the neonate, showing its large size. (*b*) Atrophied thymus of the adult.

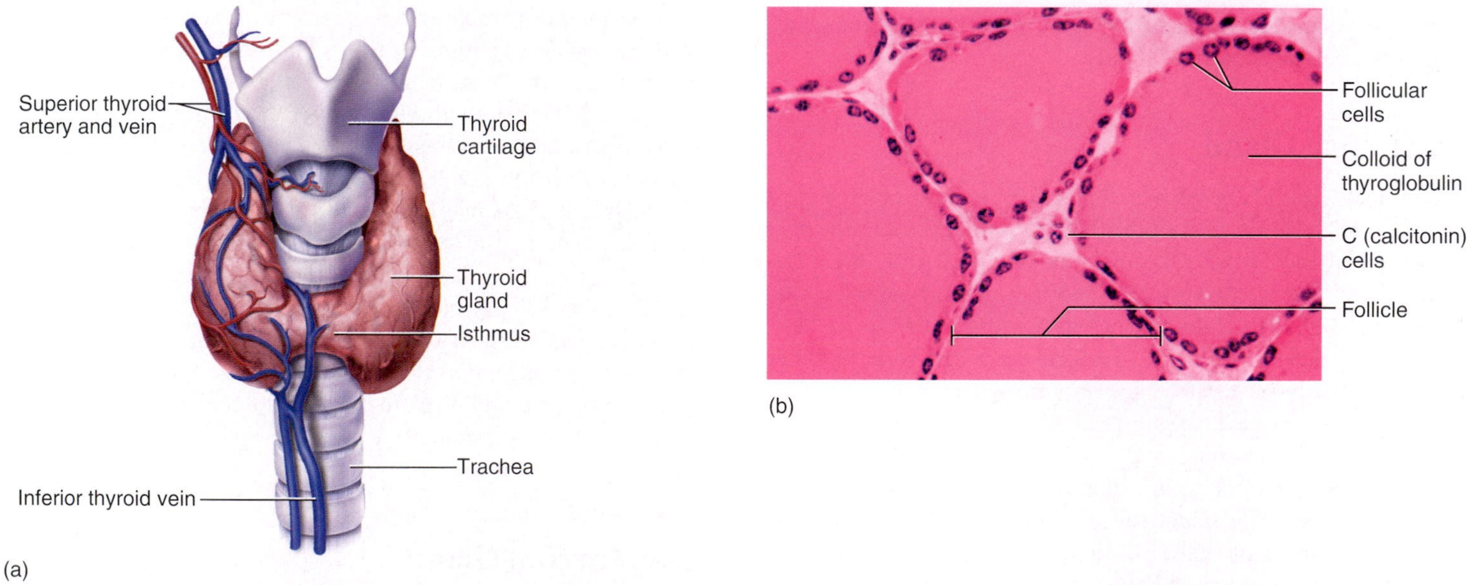

FIGURE 18.8
The Thyroid Gland. (*a*) Gross anatomy, anterior aspect. Major blood vessels are shown only on the anatomical right. (*b*) Histology.

The thyroid gland receives one of the body's highest rates of blood flow per gram of tissue. Its abundant blood vessels give the gland a dark reddish brown color. The thyroid is supplied by a pair of *superior thyroid arteries* that arise from the external carotid arteries of the neck, and a pair of *inferior thyroid arteries* that arise from the subclavian arteries beneath the clavicles. It is drained by two to three pairs of thyroid veins (*superior, middle,* and *inferior*), which drain into the internal jugular and brachiocephalic veins.

The main histological feature of the thyroid is that it is composed mostly of sacs called **thyroid follicles** (fig. 18.8*b*), lined by a simple cuboidal epithelium of **follicular cells** and filled with a protein-rich colloid. Follicular cells secrete two monoamine hormones—**thyroxine,** also known as T_4 or **tetraiodothyronine** (TET-ra-EYE-oh-doe-THY-ro-neen), and T_3, or **triiodothyronine** (try-EYE-oh-doe-THY-ro-neen). The expression **thyroid hormone (TH)** refers to T_3 and T_4 collectively.

TH synthesis begins when the follicular cells secrete a large protein called *thyroglobulin* into the follicle. Each thyroglobulin chain has 123 molecules of the amino acid tyrosine. A thyroid hormone molecule is made by linking two tyrosines together and adding three (T_3) to four (T_4) atoms of iodine to them (see fig. 18.2*b*). Upon stimulation by TSH, the follicular cells split TH from the thyroglobulin and release it into the blood.

Thyroid hormone increases the concentration and activity of mitochondrial enzymes that make ATP, and it stimulates the activity of Na^+-K^+ pumps. These effects increase the oxygen consumption and heat production of a cell, giving TH a **calorigenic**[14] **effect.** Cold weather stimulates the hypothalamus to release thyrotropin-releasing hormone (TRH); this stimulates the anterior pituitary to secrete thyroid-stimulating hormone (TSH); and TSH stimulates the thyroid to secrete TH. Consequently, the body consumes more calories and generates more heat, compensating for its cold-weather heat loss. T_3 also binds to ribosomes and nuclear receptors, and thus promotes protein synthesis, and it stimulates the pituitary gland to produce growth hormone. Thyroid hormone is especially important in the development of the nervous system of the fetus and child.

Between the thyroid follicles are clusters of less numerous **C (parafollicular) cells,** named for the peptide hormone **calcitonin.** The C cells secrete calcitonin when the blood calcium level rises above normal (a state called *hypercalcemia*). Calcitonin inhibits the bone-resorbing activity of osteoclasts. The usual balance between bone deposition and resorption is thus tipped in favor of deposition, and the blood calcium level falls as calcium is incorporated into the bones. This effect, however, is significant only in children and young animals. In healthy adults, calcitonin seems to have a negligible effect. No disease results from an excess or deficiency of calcitonin.

The Parathyroid Glands

The **parathyroid glands** are small ovoid glands in the neck measuring about 3 to 8 mm long and 2 to 5 mm wide. Usually there are four of them, but about 5% of people have more. They most often adhere to the posterior side of the thyroid gland in the approximate positions shown in figure 18.9*a*, but the parathyroids are highly variable in location and not always attached to the thyroid. They can occur as far superiorly as the hyoid bone and as far inferiorly as the aortic arch. They have a thin fibrous capsule separating them from the thyroid tissue (fig. 18.9*b*). They are supplied with blood and drained by the same vessels as the thyroid gland.

Hypocalcemia, a calcium deficiency, stimulates the **chief cells** of the parathyroids to secrete **parathyroid hormone (PTH).** PTH raises blood calcium levels by promoting intestinal calcium absorption, inhibiting urinary calcium excretion, and indirectly stimulating osteoclasts to resorb bone.

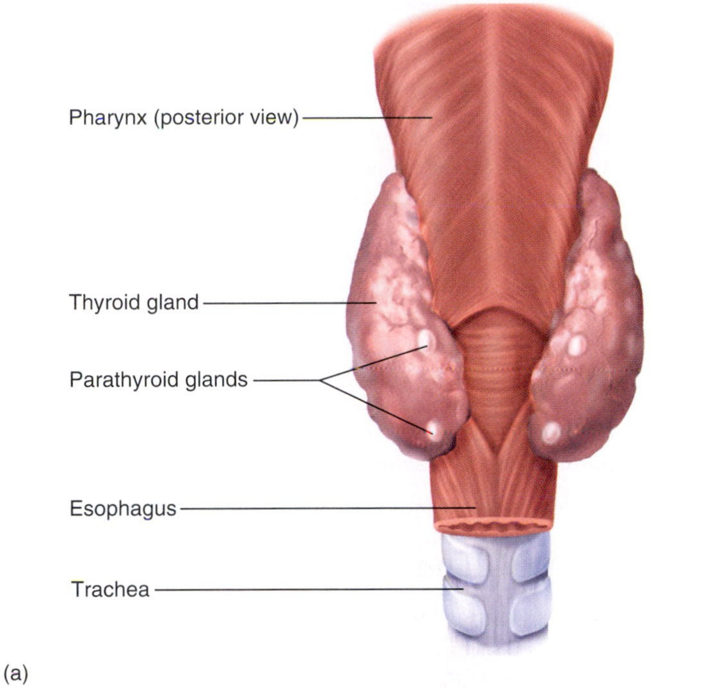
Pharynx (posterior view)
Thyroid gland
Parathyroid glands
Esophagus
Trachea
(a)

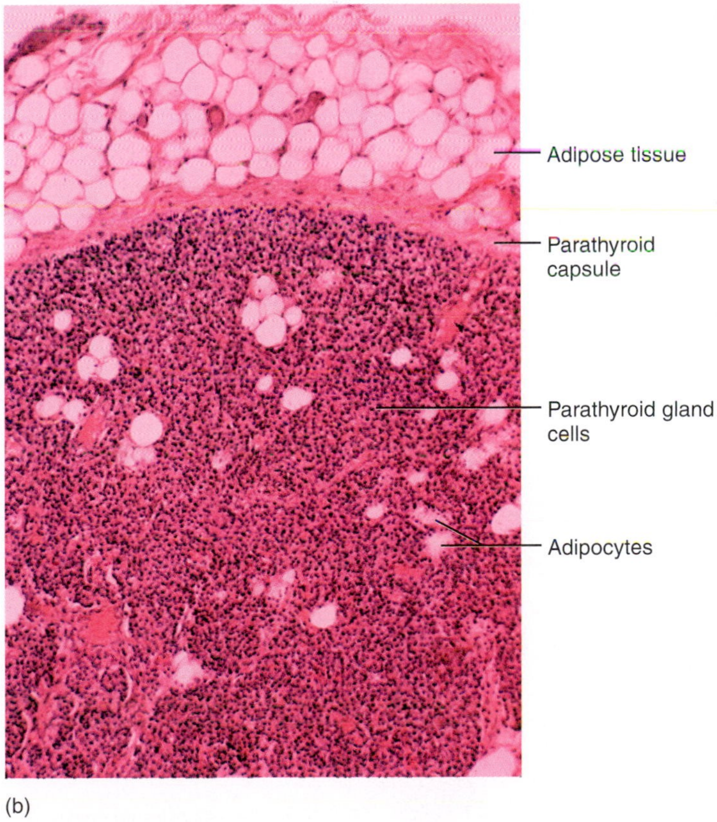
Adipose tissue
Parathyroid capsule
Parathyroid gland cells
Adipocytes
(b)

FIGURE 18.9

The Parathyroid Glands. (*a*) Gross anatomy. There are usually four parathyroid glands embedded in the posterior surface of the thyroid gland. (*b*) Histology.

[14]*calor* = heat + *genic* = producing

INSIGHT 18.3 CLINICAL APPLICATION

A Fatal Effect of Postsurgical Hypoparathyroidism

Thyroid cancer and other dysfunctions sometimes require the surgical removal of thyroid tissue. Because of their variable location and small size, the parathyroid glands are sometimes accidentally removed along with the thyroid tissue. Without hormone replacement therapy, the lack of parathyroid hormone leads to a rapid decline in blood calcium levels. The resulting hypocalcemia makes the skeletal muscles overly excitable and prone to exhibit spasmodic contractions called *hypocalcemic tetany.* One sign of this is spasmodic contraction of the hands and feet *(carpopedal spasm)*. A more serious effect is tetany of the laryngeal muscles, closing off the airway and causing suffocation. A patient can die in as little as 3 or 4 days without treatment. Because of this danger, surgeons usually try to leave the posterior part of the thyroid gland intact.

The Adrenal Glands

An **adrenal (suprarenal) gland** is attached to the superior to medial aspect of each kidney (fig. 18.10). The right adrenal gland is more or less triangular and rests on the superior pole of the kidney. The left adrenal gland is more crescent-shaped and extends from the medial indentation (hilum) of the kidney to its superior pole. In adults, the adrenal is about 5 cm wide at the base and 1 cm thick. It weighs about 7 to 10 g in the adult but about twice as much at birth.

The adrenal gland receives blood from three arteries: a *superior suprarenal artery* arising from the phrenic artery of the diaphragm; a *middle suprarenal artery* arising from the aorta; and an *inferior suprarenal artery* arising from the renal artery of the kidney. It is drained by a *suprarenal vein,* which empties from the right adrenal into the inferior vena cava and from the left adrenal into the renal vein.

Like the pituitary gland, the adrenal gland is formed by the merger of two fetal glands with different origins and functions. Its inner core, the **adrenal medulla,** is 10% to 20% of the gland. Surrounding it is a much thicker **adrenal cortex.** The adrenal medulla is functionally a part of the sympathetic nervous system, but behaves as an endocrine gland by releasing its secretions into the blood. Sympathetic preganglionic nerve fibers cross the cortex to get to the cells of the medulla.

THE ADRENAL MEDULLA

The adrenal medulla was discussed as part of the sympathoadrenal system in chapter 16. It is essentially a sympathetic ganglion consisting of *chromaffin cells,* named for their histological staining properties. These are modified neurons devoid of dendrites and axons. They are richly innervated by sympathetic fibers and respond to stimulation by secreting catecholamines (a subgroup of monoamines); the secretion is about three-quarters epinephrine, one-quarter norepinephrine, and a trace of dopamine. These hormones mimic the arousing effects of the sympathetic nervous system. Their secretion rises sharply in conditions of fear, pain, and other kinds of stress.

Think About It!

Chapter 13 described the classification of neurons according to the number of dendrites and axons issuing from the soma. In that classification scheme, how would you classify the adrenal chromaffin cells if we regard them as sympathetic neurons?

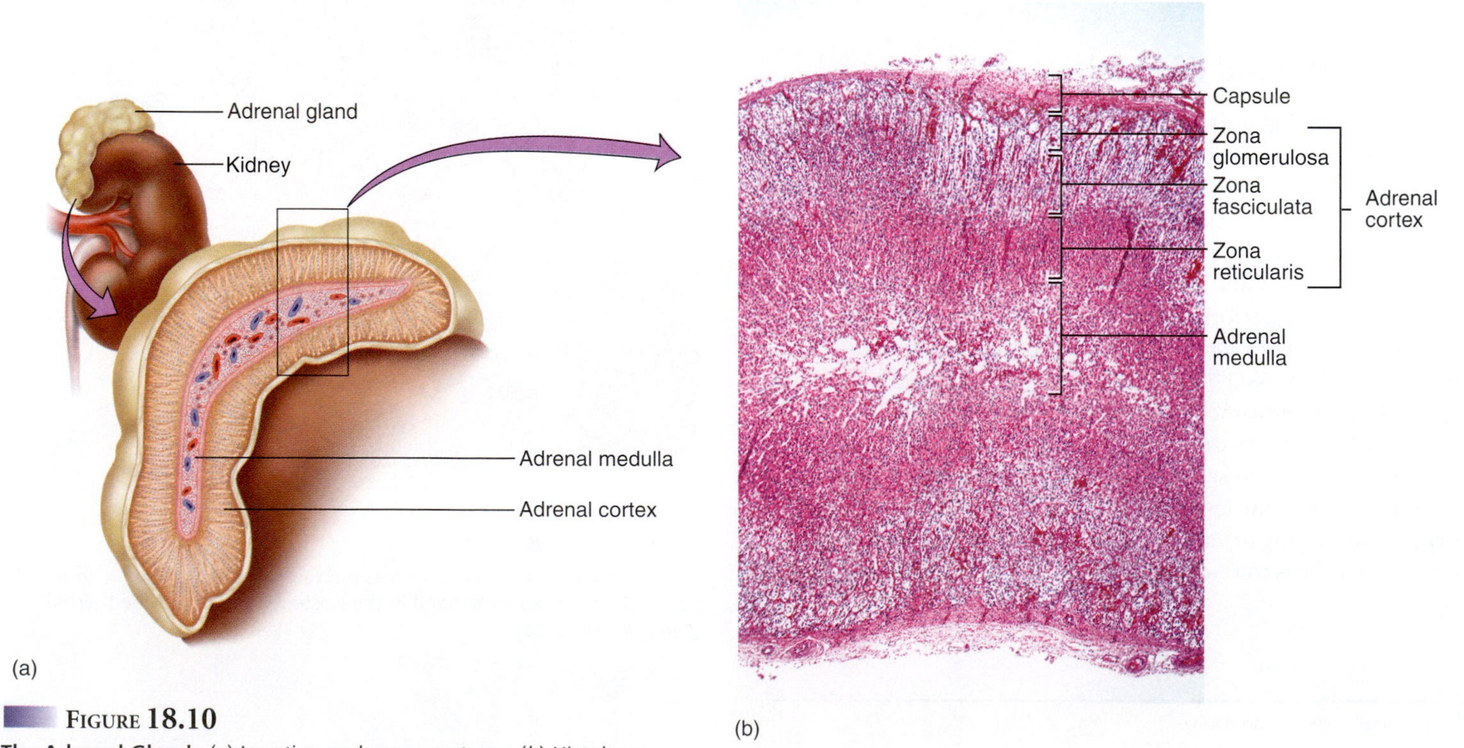

Figure 18.10
The Adrenal Gland. (*a*) Location and gross anatomy. (*b*) Histology.

THE ADRENAL CORTEX

The adrenal cortex synthesizes more than 25 steroid hormones known collectively as the **corticosteroids,** or **corticoids.** All of them are synthesized from cholesterol, and the adrenal cortex has a yellowish color owing to its high concentration of cholesterol and other lipids.

The adrenal cortex consists of three tissue layers (fig. 18.10*b*):

1. The **zona glomerulosa**[15] (glo-MER-you-LO-suh), the most superficial layer, is named for its round clusters of secretory cells.
2. The **zona fasciculata**[16] (fah-SIC-you-LAH-ta) is a thick middle layer constituting about three-quarters of the cortex. Here the cells are arranged into parallel cords perpendicular to the adrenal surface, separated by blood capillaries running parallel to the cords. The cells of the fasciculata are called **spongiocytes** because of a foamy appearance caused by their abundance of cytoplasmic lipid droplets.
3. The **zona reticularis**[17] (reh-TIC-you-LAR-iss) is a narrow layer adjacent to the adrenal medulla. Its cells form a branching network for which the zone is named.

The corticosteroids fall into three categories:

1. **Mineralocorticoids.** These steroids, secreted by the zona glomerulosa, control electrolyte balance. The principal mineralocorticoid is **aldosterone,** which acts on the kidneys to retain Na^+ in the body fluids and excrete K^+ in the urine. Since water is retained with the sodium by osmosis, aldosterone helps to maintain blood volume and pressure.
2. **Glucocorticoids.** These steroids are secreted mainly by the zona fasciculata in response to ACTH. They stimulate fat and protein catabolism, *gluconeogenesis* (the synthesis of glucose from amino acids and fats), and the release of fatty acids and glucose into the blood. They help the body adapt to stress and to repair damaged tissues. Nearly all glucocorticoid effects are caused by one member of the family, **cortisol (hydrocortisone). Corticosterone** is a less potent glucocorticoid. Recently it has been discovered that some cells of the adrenal medulla extend into the cortex. When stress activates the sympathoadrenal system, catecholamines from the medulla stimulate the cortex to secrete corticosterone.
3. **Sex steroids.** These are weak androgens and smaller amounts of estrogens secreted by the zona reticularis. **Androgens** are present in both sexes but best known for controlling many aspects of male development and reproductive physiology. The most potent androgen is **testosterone,** secreted by the testes, but the main androgen secreted by the adrenal cortex is **dehydroepiandrosterone (DHEA).** Although DHEA is much weaker than testosterone, tremendous amounts of it are produced by the large fetal adrenal glands. Androgens stimulate the libido (sex drive) and growth of pubic and axillary hair in both sexes.

[15]*glomerul* = little balls + *osa* = full of
[16]*fascicul* = little bundle + *ata* = possessing
[17]*reticul* = little network + *aris* = pertaining to

The Pancreatic Islets

The **pancreas** is an elongated, spongy gland located below and behind the stomach, mostly superficial to the peritoneum (fig. 18.11). It is approximately 15 cm long and 2.5 cm thick. Most of it is an exocrine digestive gland, but scattered among the exocrine acini are about 1 million endocrine cell clusters called the **pancreatic islets (islets of Langerhans**[18]**).** Although they are less than 2% of the mass of the pancreas, the islets secrete hormones of vital importance to metabolism—especially insulin and glucagon. The major effect of these hormones is to regulate *glycemia,* the concentration of glucose in the blood.

A pancreatic islet has from a few to 3,000 cells, belonging to five classes:

1. **Alpha (α) cells,** which secrete **glucagon.** Glucagon secretion rises between meals when the blood glucose level falls. In the liver, it stimulates glycogenolysis (glycogen breakdown) and gluconeogenesis (glucose synthesis), and the release of glucose into the blood. In adipose tissue, it stimulates fat catabolism and the release of free fatty acids. Glucagon is also secreted in response to rising amino acid levels in the blood after a high-protein meal. By promoting amino acid absorption, it provides cells with raw material for gluconeogenesis.
2. **Beta (β) cells,** which secrete **insulin.** Insulin secretion rises during and immediately after a meal in response to rising concentrations of blood glucose and amino acids. It stimulates cells to absorb these nutrients and store or metabolize them. Insulin promotes the synthesis of glycogen, fat, and protein and antagonizes the effects of glucagon. Brain, liver, and red blood cells absorb and use glucose without need of insulin stimulation.
3. **Delta (δ) cells,** which secrete **somatostatin** (growth hormone–inhibiting hormone). Somatostatin secretion rises shortly after a meal in response to the rising blood glucose and amino acid concentrations. It inhibits some digestive enzyme secretion and nutrient absorption, and also acts locally in the pancreas to moderate the other types of islet cells. In this local role, somatostatin acts as a **paracrine secretion**—a chemical messenger that diffuses to target cells a short distance away in the same organ, rather than traveling in the blood to more remote target organs.
4. **PP (F) cells,** which secrete **pancreatic polypeptide.** This hormone inhibits gallbladder contraction and the secretion of digestive enzymes by the pancreas.
5. **G cells,** which secrete **gastrin.** This hormone stimulates acid secretion, motility, and emptying of the stomach. The small intestine and the stomach itself also secrete gastrin.

The proportions of these pancreatic cells are about 20% alpha, 70% beta, 5% delta, and smaller numbers of PP and G cells.

[18]Paul Langerhans (1847–88), German anatomist

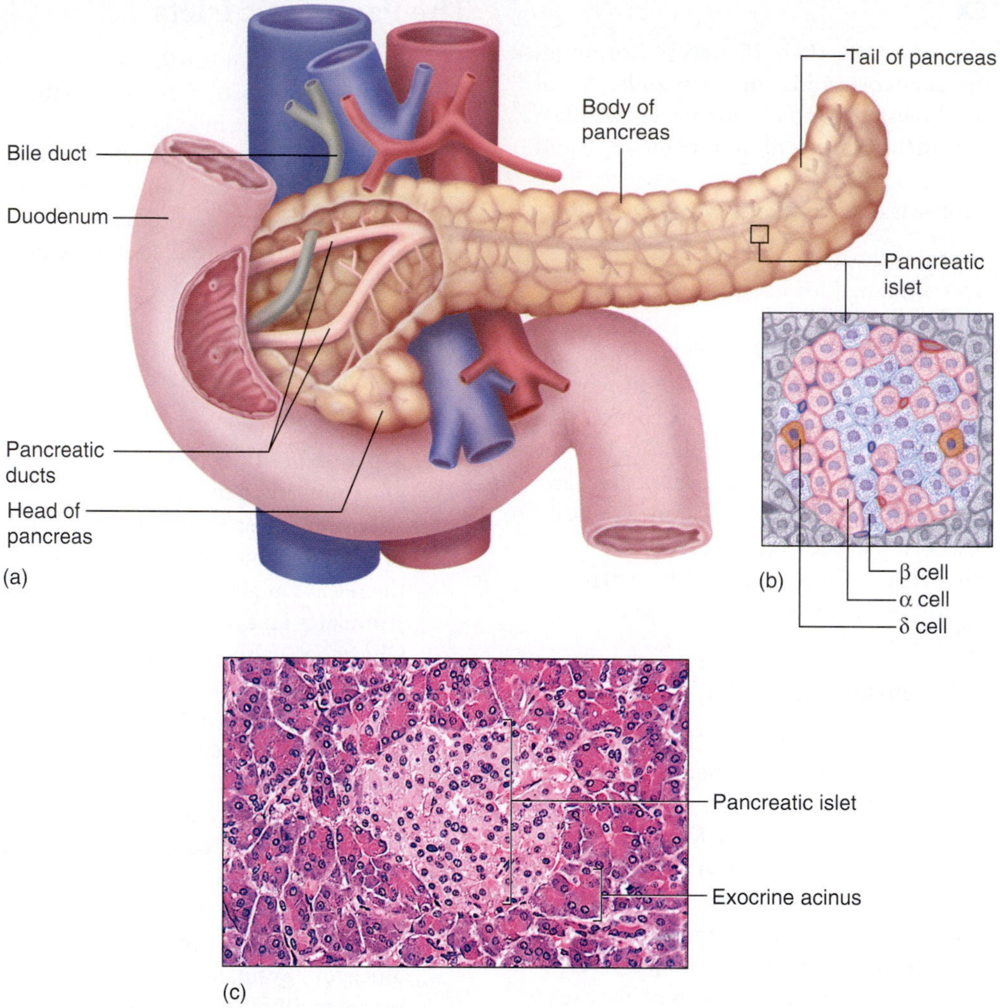
Tail of pancreas
Body of pancreas
Bile duct
Duodenum
Pancreatic islet
Pancreatic ducts
Head of pancreas
(a)
(b)
β cell
α cell
δ cell
Pancreatic islet
Exocrine acinus
(c)

FIGURE 18.11

The Pancreatic Islets. (*a*) Gross anatomy of the pancreas and relationship to the duodenum and other nearby organs. (*b*) Cell types of a pancreatic islet. PP and G cells are not shown; they are few in number and cannot be distinguished with ordinary histological staining. (*c*) Light micrograph of a pancreatic islet amid the darker exocrine acini.

The Gonads

Like the pancreas, the **gonads** are both endocrine and exocrine glands. Their exocrine products are eggs and sperm, and their endocrine products are the *gonadal hormones,* most of which are steroids.

Each follicle of the ovary is lined by a wall of **granulosa cells** (fig. 18.12*a*). They produce an estrogen called **estradiol** in the first half of the menstrual cycle. The corpus luteum that remains after ovulation secretes **progesterone** for the next 12 days or so, or for several weeks in the event of pregnancy. The functions of estradiol and progesterone are discussed in chapter 26. In brief, they contribute to the development of the reproductive system and feminine physique, regulate the menstrual cycle, sustain pregnancy, and prepare the mammary glands for lactation. The follicle and corpus luteum also secrete **inhibin,** which inhibits FSH secretion by the anterior pituitary.

The testis consists mainly of microscopic tubules that produce sperm, but nestled between the tubules are clusters of **interstitial cells** (**cells of Leydig**[19]) (fig. 18.12*b*). These endocrine cells produce testosterone and lesser amounts of weaker androgens and estrogen. Testosterone stimulates development of the male reproductive system in the fetus and adolescent, the development of the masculine physique in adolescence, and the sex drive. It sustains sperm production and the sexual instinct throughout adult life. **Sustentacular** (**Sertoli**[20]) **cells** of the testis secrete inhibin, which suppresses FSH secretion and thus homeostatically stabilizes the rate of sperm production.

Endocrine Cells in Other Organs

Several other organs have hormone-secreting cells:

- **The heart.** High blood pressure stretches the heart wall and stimulates muscle cells in the atria to secrete **atrial**

[19]Franz von Leydig (1821–1908), German histologist
[20]Enrico Sertoli (1824–1910), Italian histologist

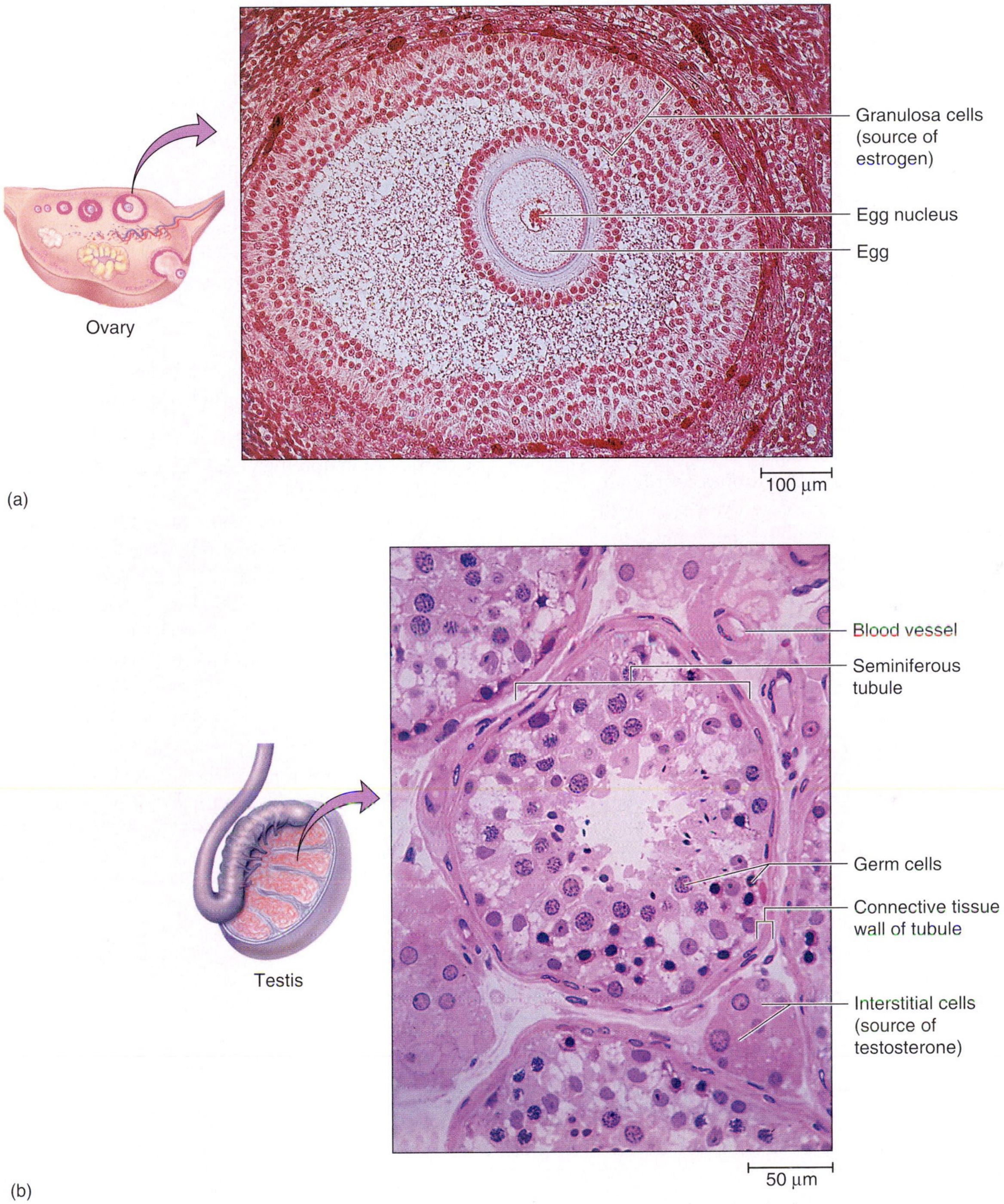
Ovary
Granulosa cells (source of estrogen)
Egg nucleus
Egg
100 µm
(a)
Testis
Blood vessel
Seminiferous tubule
Germ cells
Connective tissue wall of tubule
Interstitial cells (source of testosterone)
50 µm
(b)

FIGURE 18.12

The Gonads. (*a*) A follicle of the ovary. The granulosa cells lining the follicle secrete estrogen. (*b*) Histology of the testis. Sperm form from the germ cells in the seminiferous tubule, whereas the interstitial cells between tubules secrete testosterone.

natriuretic[21] **peptide (ANP).** ANP increases urine output and sodium excretion and interferes with the actions of angiotensin II, described shortly. Together, these effects lower the blood pressure.

- **The skin.** Keratinocytes of the epidermis carry out the first step in the synthesis of **calcitriol,** a process completed by the liver and kidneys. Calcitriol is a form of vitamin D that promotes calcium absorption by the small intestine, somewhat inhibits calcium loss in the urine, and thus makes more calcium available for bone deposition and other metabolic needs.
- **The liver.** The liver has four endocrine functions: (1) it secretes about 15% of the body's **erythropoietin** (eh-RITH-ro-POY-eh-tin) (EPO), a hormone that stimulates the bone

[21] *natri* = sodium + *uretic* = pertaining to urine

marrow to produce red blood cells; (2) it carries out the second step in calcitriol synthesis; (3) it secretes the insulin-like growth factors that mediate the action of growth hormone; and (4) it secretes a protein called *angiotensinogen,* the precursor of a hormone called **angiotensin II.** Enzymes in the kidneys and lungs carry out a two-step conversion of antiotensinogen into angiotensin II. Angiotensin II stimulates vasoconstriction and aldosterone secretion. Together, these effects raise blood pressure.

- **The kidneys.** The kidneys have three endocrine functions: (1) they secrete an enzyme, *renin,* that converts angiotensinogen into angiotensin I; (2) they secrete about 85% of the body's erythropoietin; and (3) they carry out the final stage of calcitriol synthesis.
- **The stomach and small intestine.** These have various *enteroendocrine cells,*[22] which secrete at least 10 different **enteric hormones.** In general, enteric hormones coordinate the different regions and glands of the digestive system with each other. For example, **gastrin** stimulates the stomach to secrete hydrochloric acid, and **cholecystokinin** stimulates the gallbladder to release bile.
- **The placenta.** This organ performs many functions in pregnancy, including fetal nutrition and waste removal. But it also secretes estrogen, progesterone, and other hormones that regulate pregnancy and stimulate development of the fetus and the mother's mammary glands.

[22]*entero* = intestine

Think About It!

Often, two hormones have opposite (antagonistic) effects on the same target organs. For example, oxytocin stimulates labor contractions and progesterone inhibits premature labor. Name some other examples of antagonistic effects among the hormones in this chapter.

You can see that the endocrine system is extensive. It includes numerous discrete glands as well as individual cells in the tissues of other organs. The endocrine organs and tissues other than the hypothalamus and pituitary are reviewed in table 18.5.

Before You Go On

Answer the following questions to test your understanding of the preceding section.

9. Name two endocrine glands that are larger in children than in adults. What are their functions?
10. What hormone increases the body's heat production in cold weather? What other functions does this hormone have?
11. Name a glucocorticoid, a mineralocorticoid, and a catecholamine secreted by the adrenal gland, and state the function of each.
12. What is the difference between a gonadal hormone and a gonadotropin?
13. What hormones are most important in regulating blood glucose concentration? What cells produce them? Where are these cells found?
14. Name one hormone produced by each of the following organs—the heart, liver, and placenta—and state the function of each hormone.

Table 18.5

Hormones from Sources Other than the Hypothalamus and Pituitary

Hormone	Target	Principal Effects
Pineal Gland		
Melatonin and serotonin	Brain	Influence mood; may regulate the timing of puberty
Thymus		
Thymopoietin, thymosin, thymulin	T lymphocytes	Stimulate T lymphocyte development and activity
Thyroid		
Triiodothyronine (T_3) and thyroxine (T_4)	Most tissues	Elevate metabolic rate and heat production; promote alertness, protein synthesis, fetal and childhood growth, and CNS development
Calcitonin	Bone	Promotes Ca^{2+} deposition and ossification; reduces blood Ca^{2+} level
Parathyroids		
Parathyroid hormone (PTH)	Bone, kidneys, small intestine	Increases blood Ca^{2+} level by stimulating bone resorption, calcitriol synthesis, and intestinal Ca^{2+} absorption, and reducing urinary Ca^{2+} excretion

TABLE 18.5

Hormones from Sources Other than the Hypothalamus and Pituitary *(continued)*

Hormone	Target	Principal Effects
Adrenal Medulla		
Epinephrine, norepinephrine, dopamine	Most tissues	Adaptive responses to arousal and stress
Adrenal Cortex		
Aldosterone	Kidney	Promotes Na^+ retention and K^+ excretion, maintains blood pressure and volume
Cortisol and corticosterone	Most tissues	Stimulate fat and protein catabolism, gluconeogenesis, stress resistance, and tissue repair; inhibit inflammation and immunity
Androgen (DHEA) and estrogen	Bone, muscle, integument, many other tissues	Growth of pubic and axillary hair, bone growth, sex drive, male prenatal development
Pancreatic Islets		
Glucagon	Primarily liver	Stimulates gluconeogenesis, glycogen and fat breakdown, release of glucose and fatty acids into circulation
Insulin	Most tissues	Stimulates glucose and amino acid uptake; lowers blood glucose level; promotes glycogen, fat, and protein synthesis
Somatostatin	Stomach, small intestine, pancreatic islet cells	Inhibits digestion and nutrient absorption, inhibits glucagon and insulin secretion
Pancreatic polypeptide	Pancreas, gallbladder	Inhibits release of bile and digestive enzymes
Gastrin	Stomach	Stimulates motility and acid secretion
Ovaries		
Estradiol	Many tissues	Stimulates female reproductive development and adolescent growth, regulates menstrual cycle and pregnancy, prepares mammary glands for lactation
Progesterone	Uterus, mammary glands	Regulates menstrual cycle and pregnancy, prepares mammary glands for lactation
Inhibin	Anterior pituitary	Inhibits FSH secretion
Testes		
Testosterone	Many tissues	Stimulates reproductive development, skeletomuscular growth, sperm production, and sex drive
Inhibin	Anterior pituitary	Inhibits FSH secretion
Heart		
Atrial natriuretic peptide	Kidney	Lowers blood volume and pressure by promoting Na^+ and water loss
Skin		
Vitamin D_3	—	First step in calcitriol synthesis (see kidneys)
Liver		
Calcidiol	—	Second step in calcitriol synthesis (see kidneys)
IGF-I and II	Many tissues	Mediate action of growth hormone
Erythropoietin	Red bone marrow	Promotes red blood cell production
Angiotensinogen	Blood vessels	Precursor of angiotensin II, a vasoconstrictor
Kidneys		
Calcitriol	Small intestine, kidneys	Promotes bone deposition by increasing calcium and phosphate absorption in small intestine and reducing their urinary loss
Erythropoietin	Red bone marrow	Promotes red blood cell production
Stomach and Small Intestine		
Enteric hormones	Stomach and intestines	Coordinate digestive motility and secretion
Placenta		
Estrogen, progesterone, and others	Many tissues of mother and fetus	Enhance effects of ovarian hormones on fetal development, maternal reproductive system, and preparation for lactation

Developmental and Clinical Perspectives

Objectives

When you have completed this section, you should be able to

- describe the embryonic development of each major endocrine gland;
- identify which endocrine glands or hormone levels change the most in old age, and state some of the consequences of these changes; and
- describe a few common disorders of the endocrine system, especially diabetes mellitus.

Prenatal Development of the Endocrine Glands

The endocrine glands, like other glands, develop mainly from embryonic epithelia, but lose their connection to the epithelial surface as they mature; hence the absence of ducts. All three embryonic germ layers—ectoderm, mesoderm, and endoderm—contribute to the endocrine system.

The pituitary has a dual origin (fig. 18.13*a-c*). The adenohypophysis begins with a pocket called the *hypophyseal pouch* that grows upward from the ectoderm of the pharynx. The pouch breaks away from the ectodermal surface and forms a hollow sac that continues to migrate upward. Meanwhile, growing down toward it, the neurohypophysis arises as an extension of the hypothalamus called the *neurohypophyseal bud.* The bud retains its connection to the brain throughout life as the pituitary stalk. The pouch and bud come to lie side by side and to be enclosed in the sella turcica of the sphenoid bone. They become so closely joined to each other that they look like a single gland.

The other endocrine gland associated with the brain, the pineal gland, develops from ependymal cells lining the third ventricle. A trace of the third ventricle persists as a canal in the stalk of the pineal gland.

The thyroid gland begins with an endodermal pouch *(thyroid diverticulum)* growing from the ventral floor of the pharynx slightly posterior to the hypophyseal pouch. It migrates posteriorly to its position slightly inferior to the future larynx (fig. 18.13 *a, d, e*).

In and near the neck, the thymus, parathyroid glands, and thyroid C cells arise from the *pharyngeal pouches* described in chapter 4. Cell masses break away from the third and fourth pouches and then split into dorsal and ventral cell groups. The dorsal groups form the parathyroid glands, which normally join the migrating thyroid and adhere to its dorsal side. The ventral groups migrate medially and merge with each other to become the thymus. Cells from the fifth pharyngeal pouch become the thyroid C cells; they mingle with the rest of the thyroid tissue as the future thyroid gland migrates toward the larynx (see fig. 4.8).

The adrenal gland, like the pituitary, has a dual origin (fig. 18.14). Recall from chapter 13 that ectodermal *neural crest cells* break away from the neural tube and give rise to sympathetic neurons and other cells. Some of the neural crest cells become the adrenal medulla, which is innervated by preganglionic sympathetic nerve fibers. These neural crest masses in turn stimulate cell multiplication in the overlying mesothelium, the serous lining of the embryonic peritoneal cavity. The mesothelium thickens and grows around the medulla, eventually completely enclosing it and becoming the adrenal cortex. The adrenal gland is not fully developed until the age of 3 years.

The Aging Endocrine System

The endocrine system tolerates aging better than most organ systems. Most hormones continue to be secreted at fairly stable levels even into old age. Target cell sensitivity declines, however, so some hormones have less effect. For example, in stress the anterior pituitary normally secretes ACTH, ACTH stimulates the adrenal cortex to secrete cortisol, and cortisol then signals the pituitary to diminish its ACTH output. In old age, however, the pituitary is less responsive to cortisol, so the ACTH level remains elevated and the stress response lasts longer than it would in a younger person. Diabetes mellitus is more common in old age, largely because target cells have fewer insulin receptors and elderly people have less muscle mass. Muscle is a major absorber of blood glucose and therefore important for stabilizing glycemia.

There are exceptions, of course, to the relative stability of endocrine function. The pineal gland and thymus undergo pronounced involution after puberty. In elderly people, they are shriveled vestiges of the glands of childhood, with most of the parenchyma replaced by fat or fibrous tissue. Thyroid hormone, growth hormone, and testosterone levels decline steadily after adolescence; an 80-year-old man secretes about 20% as much testosterone, for example, as he did at the age of 20. Declining GH secretion is one of the causes of muscle atrophy in old age. In women, the supply of ovarian follicles is exhausted at menopause and estrogen secretion thus drops abruptly.

Endocrine Disorders

Hormones are very potent chemicals—a little goes a long way toward producing major physiological changes. It is therefore necessary to tightly regulate their secretion and blood concentration. Slight variations in hormone concentration or target cell sensitivity often have pronounced effects on the body (fig. 18.15).

Inadequate hormone release is called **hyposecretion.** There are several potential causes of hyposecretion:

- A lesion can interfere with the ability of an endocrine gland to receive signals from another source (see insight 18.1).
- A tumor can destroy an endocrine gland (see precocial puberty, insight 18.2).
- Autoantibodies can destroy endocrine cells (as in insulin-dependent diabetes mellitus).
- A necessary nutrient may be lacking from the diet (as when a dietary iodine deficiency causes *hypothyroidism*).

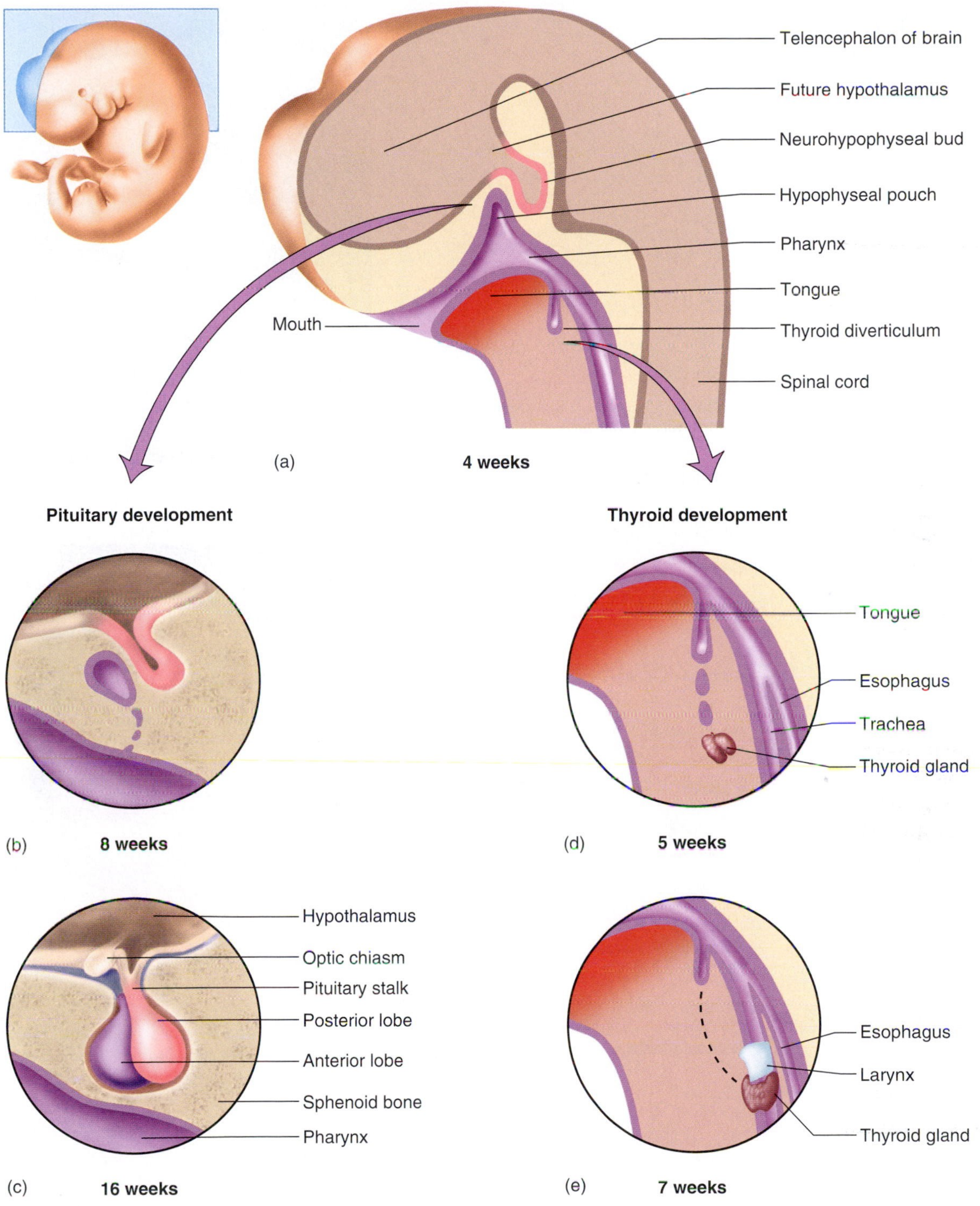

Telencephalon of brain
Future hypothalamus
Neurohypophyseal bud
Hypophyseal pouch
Pharynx
Tongue
Mouth
Thyroid diverticulum
Spinal cord
(a)
4 weeks
Pituitary development
Thyroid development
Tongue
Esophagus
Trachea
Thyroid gland
(b)
8 weeks
(d)
5 weeks
Hypothalamus
Optic chiasm
Pituitary stalk
Posterior lobe
Anterior lobe
Sphenoid bone
Pharynx
Esophagus
Larynx
Thyroid gland
(c)
16 weeks
(e)
7 weeks

FIGURE 18.13

Embryonic Development of the Pituitary Gland and Thyroid. (*a*) Sagittal section of a 4-week embryo showing the early buds (future glands) of the anterior pituitary, posterior pituitary, and thyroid glands. (*b*) At 8 weeks, the hypophyseal pouch separates from the pharynx. (*c*) At 16 weeks, the structure of the pituitary is essentially complete. (*d*) At 5 weeks, the thyroid gland descends through the neck and its connection to the tongue breaks down. (*e*) By 7 weeks, the thyroid has reached its final location on the trachea inferior to the larynx.

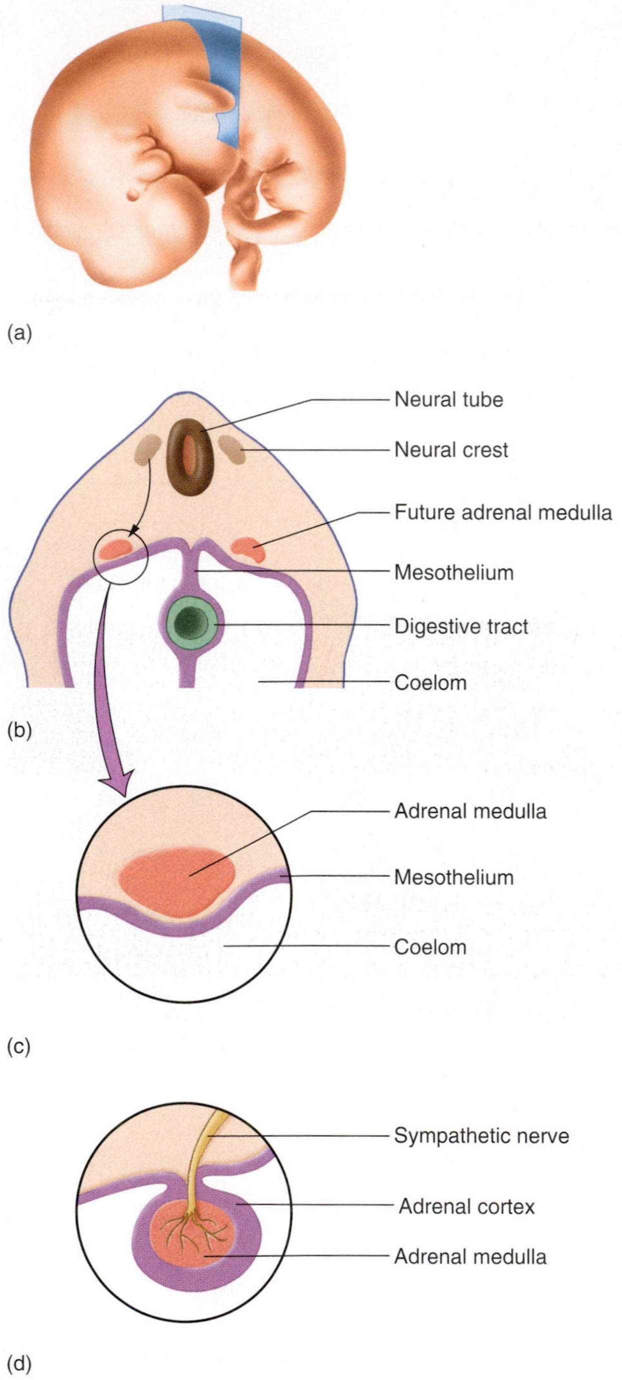
(a)
Neural tube
Neural crest
Future adrenal medulla
Mesothelium
Digestive tract
Coelom
(b)
Adrenal medulla
Mesothelium
Coelom
(c)
Sympathetic nerve
Adrenal cortex
Adrenal medulla
(d)

FIGURE 18.14
Embryonic Development of the Adrenal Gland. (*a*) A 4-week embryo with the plane of section seen in *b*. (*b*) The future adrenal medulla begins as a mass of cells that separate from the neural crest. (*c*) Growth of the adrenal medulla and bulging of the mesothelium into the body cavity (coelom). (*d*) The mesothelium thickens and encloses the adrenal medulla, giving rise to the adrenal cortex.

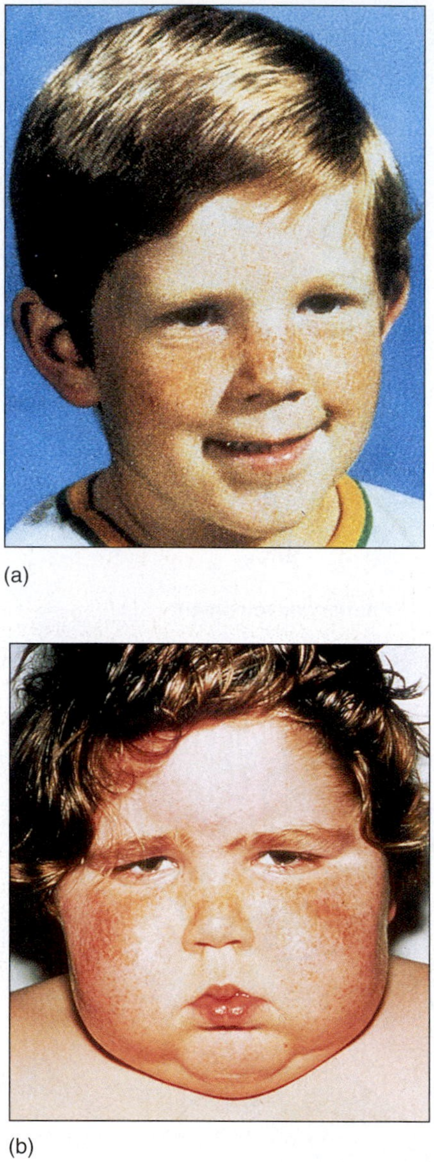
(a)
(b)

FIGURE 18.15
Cushing Syndrome. (*a*) Patient before the onset of the syndrome. (*b*) The same boy, only 4 months later, showing the "moon face" characteristic of Cushing syndrome (cortisol hypersecretion; see table 18.6).

Syndromes of hyposecretion can often be treated in part with *hormone replacement therapy (HRT),* in which hormone doses are administered orally, by injection, or by nasal spray.

A hormone excess is called **hypersecretion.** This can result from some tumors in which there is an overgrowth of functional endocrine cells. For example, a *pheochromocytoma,* a tumor of the adrenal medulla, secretes excess epinephrine and norepinephrine, causing hypertension, nervousness, indigestion, and an elevated metabolic rate. Some tumors in nonendocrine organs also secrete hormones. Certain lung tumors, for example, secrete ACTH and thus stimulate the adrenal gland to secrete excess cortisol. And while some autoantibodies cause endocrine hyposecretion, others cause hypersecretion. In *toxic goiter (Graves disease),* an autoantibody binds to the thyroid and mimics the effect of TSH, thus overstimulating the thyroid gland.

Aside from abnormal levels of hormone secretion, disorders can result from abnormalities in target cell sensitivity. In *androgen-insensitivity syndrome (AIS),* for example, the body lacks testosterone receptors. Even though testosterone is present, it has no effect. In males with AIS, the genitals develop female anatomy, and at puberty, the small amount of estrogen secreted by the testes causes males to develop feminine breasts and other female characteristics (fig. 18.16).

The most common endocrine disorder, indeed the most widespread of all metabolic diseases, is **diabetes mellitus** (**DM**) Ten percent of cases, called *type I* or *insulin-dependent diabetes mellitus (IDDM),* result from the destruction of pancreatic β cells. The stage for this is set when a person genetically susceptible to IDDM contracts a viral infection. Antibodies against the virus cross-react with the β cells and trigger their destruction. One can tolerate a remarkable degree of β cell loss, but when only 10% to 20% of them remain, insulin falls to a critically low level and the signs of diabetes begin to appear.

Ninety percent of diabetics have a form called *non-insulin-dependent (type II) diabetes mellitus (NIDDM),* which is a classic example of hormone insensitivity. In NIDDM, insulin levels are often normal or even elevated, but the insulin has little or no effect. Insulin receptors are either absent or defective.

In either case, insulin-dependent tissues such as muscle cannot absorb glucose and amino acids from the blood. This leads to a familiar suite of signs and symptoms:

- *Hyperglycemia,* elevated blood sugar. The blood glucose level remains elevated long after a meal because insulin-dependent tissues such as muscle cannot absorb it.
- *Glycosuria,* glucose in the urine. Blood glucose enters the kidney tubules faster than they can return it to the blood, so the excess appears in the urine.
- *Polyuria,* abnormally abundant urine.Water is osmotically retained in the urine with the glucose, so urine output increases dramatically.
- *Polydipsia,* insatiable thirst and compulsive water drinking. This results from the dehydration caused by polyuria.
- *Polyphagia,* insatiable hunger. This results from the inability of cells to absorb and use glucose.
- *Emaciation,* dramatic weight loss and muscular atrophy. Unable to absorb glucose, cells resort to breaking down protein and fat for energy. Muscle and adipose tissue thus waste away.
- *Ketonuria,* the presence of chemicals called ketones in the urine. These are by-products of the incomplete fat oxidation that occurs as cells unable to use glucose rapidly catabolize fats.

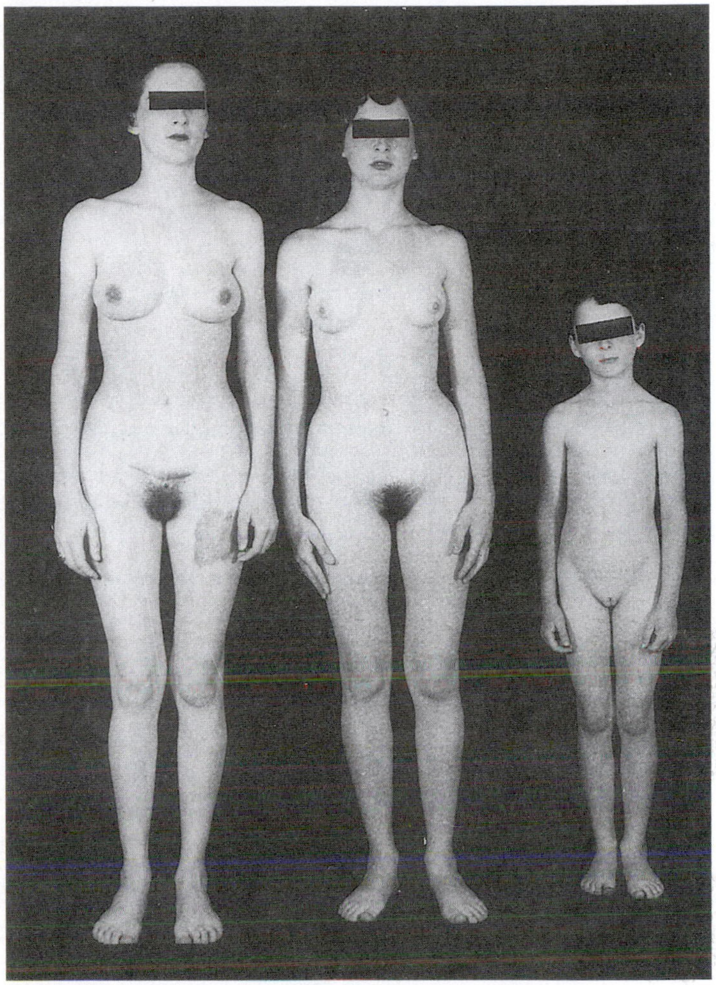

FIGURE 18.16
Androgen-Insensitivity Syndrome. Three siblings who are genetically male (XY) but have a hereditary lack of testosterone receptors. Testes are present and secrete testosterone, but for lack of receptors, the testosterone cannot exert an effect on development. The external features are feminine, but there are no ovaries, uterus, or vagina.

- *Atherosclerosis,* fatty degeneration of the arteries induced by chronic hyperglycemia. Atherosclerosis leads to blindness, kidney failure, and gangrene.
- *Diabetic neuropathy,* nerve degeneration caused by hyperglycemia. This results in a loss of sensation and contributes to gangrene of the extremities.
- *Ketoacidosis,* abnormally low blood pH caused by the acidic ketones. Ketoacidosis depresses nervous system function and thus leads to coma and death.

Type I diabetes mellitus is managed with dietary modification and insulin injections. Type II is managed mainly with exercise and weight loss, sometimes supplemented with oral medications to increase insulin sensitivity and with occasional insulin injections.

Some other endocrine disorders are briefly described in table 18.6.

Table 18.6

Disorders of the Endocrine System

Acromegaly	A result of adult growth hormone hypersecretion, resulting in thickening of the bones and soft tissues, especially noticeable in the face, hands, and feet
Addison Disease	Hyposecretion of adrenal glucocorticoids or mineralocorticoids, causing hypoglycemia, hypotension, weight loss, weakness, loss of stress resistance, darkening or bronzing (metallic discoloration) of the skin, and potentially fatal dehydration and electrolyte imbalances
Adrenogenital Syndrome	Hypersecretion of adrenal androgens. Prenatal hypersecretion can cause girls to be born with masculinized genitalia and to be misidentified as boys. In children, it often causes enlargement of the penis or clitoris and premature puberty. In women, it causes masculinizing effects such as increased body hair, beard growth, and deepening of the voice.
Congenital Hypothyroidism	Thyroid hyposecretion present from birth, resulting in stunted physical development, thickened facial features, low body temperature, lethargy, and irreversible brain damage in infancy
Cushing Syndrome	Cortisol hypersecretion resulting from overactivity of the adrenal cortex. Results in disruption of carbohydrate and protein metabolism, hyperglycemia, edema, loss of bone and muscle mass, and sometimes abnormal fat deposition in the face (fig. 18.15) or between the shoulders.
Endemic Goiter	Enlargement of the thyroid gland, combined with thyroid hormone hyposecretion, as a result of dietary iodine deficiency
Myxedema	A result of severe or prolonged adult hypothyroidism, characterized by low metabolic rate, sluggishness and sleepiness, weight gain, constipation, hypertension, dry skin and hair, abnormal sensitivity to cold, and tissue swelling
Pituitary Dwarfism	Abnormally short stature, with a normal proportion of limbs to trunk, resulting from growth hormone hyposecretion in childhood
Pituitary Gigantism	Abnormally tall stature, with a normal proportion of limbs to trunk, resulting from growth hormone hypersecretion in childhood

Disorders Described Elsewhere

Androgen-insensitivity syndrome 535	Pheochromocytoma 534	Pituitary trauma 522
Diabetes mellitus 535	Pineal tumors 523	Toxic goiter 535
Hypocalcemic tetany 526		

Before You Go On

Answer the following questions to test your understanding of the preceding section.

15. List some causes of hormone hyposecretion and name an endocrine disease that exemplifies each one.
16. List some causes and examples of hormone hypersecretion.
17. Describe two disorders in which hormone levels may be normal, but a person lacks receptors for the hormone and therefore does not respond to it.

CHAPTER REVIEW

Review of Key Concepts

Overview of the Endocrine System (p. 514)

1. *Hormones* are chemical messengers that are secreted into the bloodstream and stimulate distant *target cells.* The glands and cells that secrete hormones constitute the *endocrine system.*
2. Hormones fall into three chemical classes: steroids, monoamines, and peptides (table 18.1).
3. Monoamines and peptides are usually hydrophilic and mix freely with the blood plasma, but steroids are hydrophobic and must be carried by *transport proteins.* Transport proteins also prolong the action of a hormone by protecting it from rapid breakdown or excretion.
4. Steroids and thyroid hormone diffuse into their target cells and bind to receptors in the cytoplasm or nucleus. Other monoamines and peptides bind to receptors on the target cell surface and stimulate the production of a *second messenger* within the cell.
5. Hormones alter the target cell's metabolism by stimulating it to synthesize new enzymes or by activating or inhibiting enzymes that are already present.
6. The endocrine and nervous systems complement each other in serving internal communication in the body (table 18.2). The nervous system is generally quicker to respond to a stimulus but the endocrine system gives more prolonged responses. The two systems have many similarities and overlapping functions, however, and several chemical messengers serve as both hormones and neurotransmitters.

The Hypothalamus and Pituitary Gland (p. 517)

1. The hypothalamus and pituitary do not control all endocrine functions, but have more wide-reaching influences than any other endocrine gland.
2. The pituitary is attached to the hypothalamus by a stalk. It is composed of two structures with separate embryonic origins—the *adenohypophysis* and *neurohypophysis.*
3. The most significant part of the adenohypophysis is the *anterior lobe.* It is connected to the hypothalamus by a network of blood vessels, the *hypophyseal portal system,* through which the hypothalamus sends it chemical signals (releasing and inhibiting hormones).
4. The most significant part of the neurohypophysis is the *posterior lobe.* It is connected to the hypothalamus by a bundle of nerve fibers, the *hypothalamo-hypophyseal tract,* which travels through the pituitary stalk and through which the hypothalamus sends it nerve signals.
5. The hypothalamus produces seven releasing and inhibiting hormones (table 18.3) which determine when the anterior pituitary secretes its own hormones.
6. The hypothalamus also produces *oxytocin* and *antidiuretic hormone,* which are stored in the posterior pituitary and released on command from the brain.
7. The anterior pituitary secretes follicle-stimulating hormone (FSH), luteinizing hormone (LH), thyroid-stimulating hormone (TSH), adrenocorticotropic hormone (ACTH), prolactin (PRL), and growth hormone (GH).
8. Pituitary hormones whose target organs are other endocrine glands are called *tropic* hormones (the first four of these). Those targeted to the gonads (FSH and LH) are called *gonadotropins.* The relationship between the pituitary, a tropic hormone, and the target endocrine gland is called an *axis.*
9. The pars intermedia of the pituitary produces pro-opiomelanocortin (POMC), which breaks down into ACTH and endorphins. The pars intermedia does not directly secrete any circulating hormones in humans.
10. Actions of the pituitary hormones are summarized in table 18.4.

Other Endocrine Glands (p. 523)

1. The *pineal gland,* located above the third ventricle of the brain, produces serotonin and melatonin. It may be involved in mood and the timing of puberty.
2. The *thymus,* located above the heart in the mediastinum, secretes *thymic hormones* that stimulate the development of other lymphatic organs and regulate T lymphocyte activity.
3. The *thyroid gland,* located in the neck below the larynx, secretes *thyroid hormones* (T_3 and T_4) that affect the body's metabolic rate, protein synthesis, fetal growth, and nervous system development. It also secretes calcitonin, which lowers blood calcium concentration but has only negligible effects in adults.
4. The *parathyroid glands,* usually located on the dorsal side of the thyroid, secrete *parathyroid hormone,* which increases blood calcium concentration.
5. The *adrenal glands,* located at the superior end of each kidney, are composed of an *adrenal cortex* and *adrenal medulla* with different functions and embryonic origins. The adrenal medulla secretes mainly *epinephrine* and *norepinephrine (catecholamines),* which complement the effects of the sympathetic nervous system. The adrenal cortex secretes steroid hormones *(corticosteroids)* including *androgens (sex steroids), aldosterone* (a *mineralocorticoid*), and *cortisol* and *corticosterone (glucocorticoids).*
6. The *pancreas* is mainly an exocrine digestive gland, but the pancreatic islets secrete *glucagon, insulin, somatostatin, gastrin,* and *pancreatic polypeptide.* These hormones regulate digestion, nutrient absorption, and the metabolism of carbohydrates, amino acids, and lipids. Its most important secretions are insulin, which promotes glucose uptake by cells during and after a meal, and glucagon, which promotes glucose release by the liver between meals.
7. The *gonads* (ovaries and testes) contain endocrine cells that secrete *estrogens, progesterone, testosterone,* and *inhibin*—hormones with a variety of reproductive functions.
8. Several other organs have endocrine cells. The heart secretes *atrial natriuretic peptide;* the skin, liver, and kidneys collaborate to synthesize *calcitriol (vitamin D);* the liver also secretes *insulin-like growth factors, angiotensinogen,* and *erythropoietin;* the kidneys secrete erythropoietin and collaborate with the lungs to convert angiotensinogen to *angiotensin II;* the stomach and small intestine produce *gastrin, cholecystokinin,* and other *enteric hormones;* and the placenta produces estrogen, progesterone, and other hormones involved in pregnancy.
9. Hormones from sources other than the pituitary and hypothalamus are summarized in table 18.5.

Developmental and Clinical Perspectives (p. 532)

1. The anterior pituitary gland arises from a hypophyseal pouch that grows from the roof of the pharynx. The posterior pituitary

gland arises from a neurohypophyseal bud that grows downward from the floor of the hypothalamus. The two glands come to lie side by side, enclosed in the sella turcica.
2. The thyroid gland arises from a pouch that grows downward from the floor of the pharynx.
3. At 4 to 5 weeks, the embryo develops five pairs of pharyngeal pouches budding from the walls of the pharynx. Pharyngeal pouches III and IV each give rise to a dorsal and ventral cell mass. The four dorsal cell masses become the parathyroid glands, and the ventral masses fuse medially to become the thymus. Pouch V gives rise to the C cells of the thyroid gland.
4. The adrenal medulla develops from a group of ectodermal cells that break away from the neural crest. The adrenal cortex develops from mesothelial cells that surround the medulla.
5. The endocrine system as a whole shows less functional decline in old age than most other organ systems, although hormonal effects decline due to the presence of fewer hormone receptors.
6. The pineal gland and thymus atrophy markedly after puberty. Thyroid hormone, growth hormone, and testosterone levels decline steadily throughout middle and old age. Female sex steroids drop sharply after menopause.
7. Endocrine dysfunctions can result from hormone hyposecretion, hypersecretion, or insensitivity (receptor defects).
8. The most common endocrine disorder is diabetes mellitus (DM), resulting from insulin hyposecretion (type I or insulin-dependent DM) or insulin insensitivity (type II or non-insulin-dependent DM). Its classic complaints are polyuria, polydipsia, and polyphagia; its clinical signs include hyperglycemia, glycosuria, and ketonuria. Blindness, kidney failure, and gangrene are common consequences of DM, and death often follows from ketoacidosis and coma.

Testing Your Recall

1. Which of the following hormones is *not* synthesized by the brain?
   a. thyrotropin-releasing hormone
   b. antidiuretic hormone
   c. prolactin-releasing hormone
   d. follicle-stimulating hormone
   e. oxytocin
2. Which of the following hormones has the least in common with the others?
   a. adrenocorticotropic hormone
   b. follicle-stimulating hormone
   c. thyrotropin
   d. thyroxine
   e. prolactin
3. Which hormone would no longer be secreted if the hypothalamo-hypophyseal tract were destroyed?
   a. oxytocin
   b. follicle-stimulating hormone
   c. growth hormone
   d. adrenocorticotropic hormone
   e. corticosterone
4. Which of the following is *not* a hormone?
   a. prolactin
   b. thymosin
   c. renin
   d. atrial natriuretic peptide
   e. insulin-like growth factor
5. Which of the following do (does) *not* secrete steroid hormones?
   a. the placenta
   b. the ovaries
   c. the testes
   d. the pituitary gland
   e. the adrenal cortex
6. Which of these glands develops from the pharyngeal pouches?
   a. anterior pituitary
   b. posterior pituitary
   c. thyroid gland
   d. thymus
   e. adrenal gland
7. Which of these glands has more exocrine than endocrine tissue?
   a. the pancreas
   b. the adenohypophysis
   c. the thyroid gland
   d. the pineal gland
   e. the adrenal gland
8. ______ leads to increased osteoclast activity and thus elevates the blood calcium concentration.
   a. Parathyroid hormone
   b. Calcitonin
   c. Calcitriol
   d. Aldosterone
   e. ACTH
9. Which of these endocrine glands is most directly involved in immune function?
   a. the pancreas
   b. the thymus
   c. the adenohypophysis
   d. the adrenal glands
   e. the thyroid gland
10. Both the ______ are involved in the synthesis of calcitriol and erythropoietin.
    a. anterior and posterior pituitary
    b. thyroid gland and thymus
    c. liver and kidneys
    d. parathyroids and pancreatic islets
    e. epidermis and liver
11. The ______ develops from the hypophyseal pouch of the embryo.
12. Antidiuretic hormone is produced by a group of neurons in the hypothalamus called the ______.
13. Growth hormone hypersecretion in adulthood causes a disease called ______.
14. ______ is a hormone that increases urine output.
15. Adrenal steroids that regulate glucose metabolism are collectively called ______.
16. The hypophyseal portal system is a means for the brain to communicate with the ______.
17. In females, testosterone is secreted by the ______ gland.
18. In males, testosterone is secreted mainly by the ______ cells.
19. The thickest layer of the adrenal cortex is the ______.
20. The hormones secreted by the stomach and small intestine are collectively called ______.

Answer in the Appendix

True or False

Determine which five of the following statements are false, and briefly explain why.

1. Tumors can lead to either hormone hypersecretion or hyposecretion.
2. All hormones are secreted by endocrine glands.
3. If fatty plaques of atherosclerosis blocked the arteries of the hypophyseal portal system, it could cause the ovaries and testes to malfunction.
4. The pineal gland and thymus become larger as one ages.
5. The tissue at the center of the adrenal gland is called the zona reticularis.
6. Unlike neurotransmitters, hormones cannot be selectively delivered to just one particular target organ.
7. The adenohypophysis and thyroid gland are more similar to each other in their embryonic origin than the adenohypophysis and neurohypophysis are.
8. Oxytocin and antidiuretic hormone are secreted through a duct called the pituitary stalk or infundibulum.
9. Of the endocrine glands covered in this chapter, only the adrenal glands are paired. The rest are single.
10. Enlargement of the thyroid gland would produce a swelling in the neck.

Answer in the Appendix

Testing Your Comprehension

1. A young man is involved in a motorcycle accident that fractures his sphenoid bone and severs the pituitary stalk. Shortly thereafter, he begins to excrete enormous amounts of urine, up to 30 liters per day, and suffers intense thirst. His neurologist diagnoses his problem as *diabetes insipidus*. Explain how his head injury resulted in these effects on urinary function.
2. Examine the anatomical relationship between the pineal gland and nearby brain structures, and as necessary, review the functions of those brain structures in chapter 15. In light of this information, explain why a large pineal tumor might result in the following effects: (*a*) hydrocephalus; (*b*) a loss of photopupillary and accommodation reflexes of the eye; and (*c*) paralysis of some eye movements.
3. Renal failure puts a person at risk of anemia and hypocalcemia. To prevent this, renal dialysis patients are routinely given hormone replacement therapy. Explain the hormonal connection between renal failure and each of these conditions, and identify what hormones would be administered to correct or prevent them.
4. To which chemical class do the hormones of the adrenal cortex belong? To which class do the hormones of the pancreatic islets belong? In view of this, what major difference would you expect to see in the organelles of an adrenal spongiocyte and a pancreatic beta cell if you compared them with an electron microscope?
5. Review the difficulty of sharply distinguishing between nervous and endocrine function (p. 515) and then give two examples of this based on this chapter's survey of the endocrine glands.

Answers at the Online Learning Center

www.mhhe.com/saladinha1

Visit the Online Learning Center for practice tests, answer keys, and other learning aids for this chapter. Enhance your understanding of human anatomy with our interactive art labeling exercises, supplemental photo atlases, web links, puzzles, flashcards, and much more.

19

CHAPTER NINETEEN

The Circulatory System I—Blood

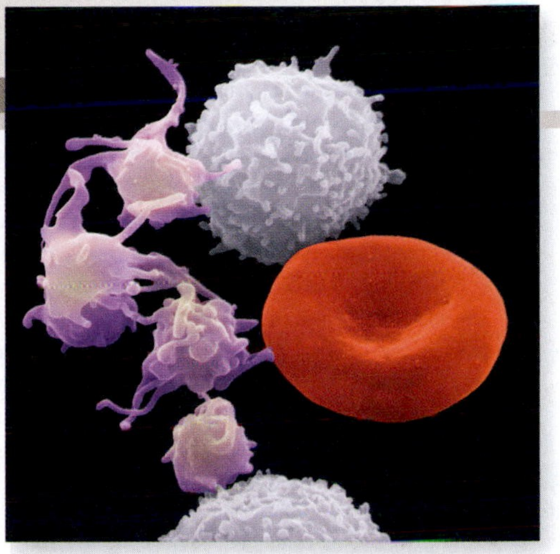
Blood cells and platelets (colorized SEM)

CHAPTER OUTLINE

Introduction 542
- Functions of the Circulatory System 542
- Components and General Properties of Blood 542
- Blood Plasma 543

Erythrocytes 545
- Form and Function 545
- Quantity of Erythrocytes 546
- Hemoglobin 546
- The Erythrocyte Life Cycle 546
- Blood Types 548

Leukocytes 548
- Form and Function 548
- Types of Leukocytes 548
- The Leukocyte Life Cycle 552

Platelets 552
- Form and Function 552
- Platelet Production 554
- Hemostasis 554

Clinical Perspectives 555
- Hematology in Old Age 555
- Disorders of the Blood 555

Chapter Review 558

INSIGHTS

19.1 Evolutionary Medicine: The Packaging of Hemoglobin 547
19.2 Clinical Application: The Complete Blood Count 552
19.3 Clinical Application: Sickle-cell Disease 556

BRUSHING UP

To understand this chapter, you may find it helpful to review the following concepts:
- Osmosis (p. 58)
- Blood as a connective tissue (p. 92)
- Red bone marrow (p. 158)
- Erythropoietin (p. 529)

PART FOUR

Blood has always had a special mystique. From time immemorial, people have seen blood flow from the body and with it, the life of the individual. People thus presumed that blood carried a mysterious "vital force," and Roman gladiators drank it to fortify themselves for battle. Even today, we become especially alarmed when we find ourselves bleeding, and the emotional impact of blood is enough to make many people faint at the sight of it. From ancient Egypt to nineteenth-century America, physicians drained "bad blood" from their patients to treat everything from gout to headaches, from menstrual cramps to mental illness. It was long thought that hereditary traits were transmitted through the blood, and people still use such unfounded expressions as "I have one-quarter Cherokee blood."

Scarcely anything meaningful was known about blood until blood cells were seen with the first microscopes. Even though blood is a uniquely accessible tissue, most of what we know about it dates only to the last 50 years. Recent developments in **hematology**[1]—the study of blood—have empowered us to save and improve the lives of countless people who would otherwise suffer or die.

Introduction

Objectives

When you have completed this section, you should be able to

- describe the functions and major components of the circulatory system;
- describe the components and physical properties of blood; and
- describe the composition of blood plasma.

Functions of the Circulatory System

The **circulatory system** consists of the heart, blood vessels, and blood. The term **cardiovascular system**[2] refers only to the heart and blood vessels, which are the subject of chapters 20 and 21.

The fundamental purpose of the circulatory system is to transport substances from place to place in the blood. Blood is the liquid medium in which these materials travel, blood vessels ensure the proper routing of blood to its destinations, and the heart is the pump that keeps the blood flowing.

More specifically, the functions of the circulatory system are as follows:

TRANSPORT

- The blood carries oxygen from the lungs to all of the body's tissues, while it picks up carbon dioxide from those tissues and carries it to the lungs to be removed from the body.
- It picks up nutrients from the digestive tract and delivers them to all of the body's tissues.
- It carries other metabolic wastes to the kidneys for removal.
- It carries hormones from endocrine cells to their target cells.
- It helps to regulate body temperature by carrying heat to the body surface for removal.

PROTECTION

- The blood plays several roles in inflammation, a mechanism for limiting the spread of infection.
- White blood cells destroy microorganisms and cancer cells.
- Antibodies and other blood proteins neutralize toxins and help to destroy pathogens.
- Platelets secrete factors that initiate blood clotting and other processes for minimizing blood loss.

REGULATION

- By absorbing or giving off fluid under different conditions, the blood capillaries help to stabilize fluid distribution in the body.
- By buffering acids and bases, blood proteins help to stabilize the pH of the extracellular fluids.

Considering the importance of efficiently transporting nutrients, wastes, hormones, and especially oxygen from place to place, it is easy to understand why an excessive loss of blood is quickly fatal, and why the circulatory system needs mechanisms for minimizing such losses.

Components and General Properties of Blood

All of the foregoing functions depend, of course, on the characteristics of the blood. Blood is a liquid connective tissue with two main components—the plasma and formed elements. **Plasma** is a clear extracellular matrix. It is no longer present on prepared slides of blood and would not be visible even if it were. **Formed elements,** by contrast, have a visible structure. They are cells and cell fragments: red blood cells, white blood cells, and platelets (fig. 19.1).

The formed elements are classified as follows:

Erythrocytes[3] (red blood cells, RBCs)
Platelets
Leukocytes[4] (white blood cells, WBCs)
  Granulocytes
    Neutrophils
    Eosinophils
    Basophils
  Agranulocytes
    Lymphocytes
    Monocytes

Thus, there are seven kinds of formed elements: the erythrocytes, platelets, and five kinds of leukocytes. The five leukocyte types are divided into two categories, the *granulocytes* and *agranulocytes,* on grounds explained later.

The ratio of formed elements to plasma can be seen by taking a sample of blood in a tube and spinning it for a few minutes in a centrifuge (fig. 19.2). Erythrocytes are the densest elements,

[1]*hem, hemato* = blood + *logy* = study of
[2]*cardio* = heart + *vas* = vessel
[3]*erythro* = red + *cyte* = cell
[4]*leuko* = white + *cyte* = cell

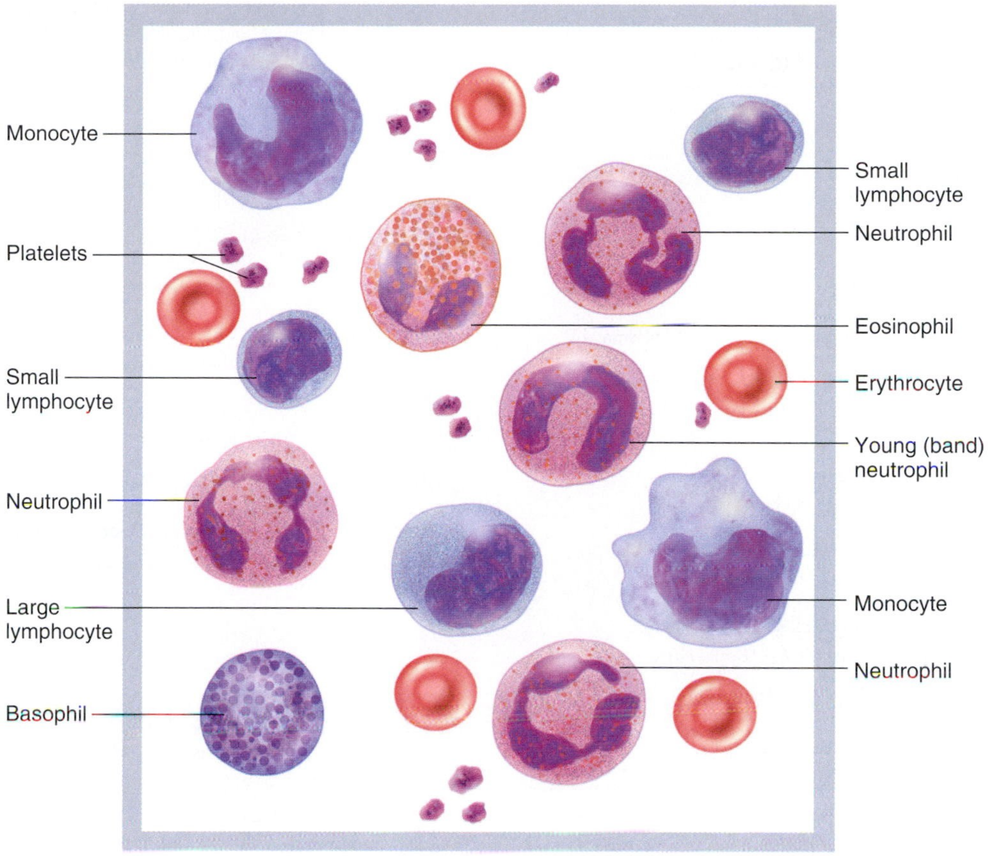

FIGURE 19.1
The Formed Elements of Blood.

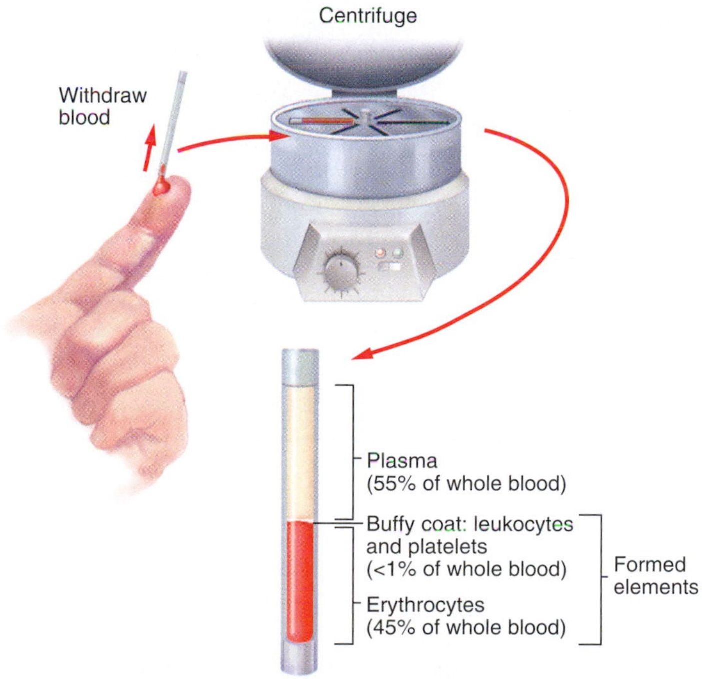

FIGURE 19.2
Separating the Plasma and Formed Elements of Blood. A small sample of blood is taken in a glass tube and spun in a centrifuge to separate the cells from the plasma. Erythrocytes are the densest components and settle to the bottom of the tube, platelets and WBCs are next, and plasma remains at the top of the tube.

settle to the bottom of the tube, and typically constitute about 45% of the total volume. Leukocytes and platelets settle into a narrow cream- or buff-colored layer called the *buffy coat* just above the erythrocytes; they total 1% or less of the blood volume. At the top of the tube is the plasma, which has a pale yellow color and accounts for slightly over half of the volume.

Some general properties of blood are listed in table 19.1. Some of the terms in that table are defined later in this chapter.

Blood Plasma

Even though blood plasma has no anatomy that we can study visually, we cannot ignore its importance as the matrix of this liquid connective tissue we call blood. Plasma is a complex mixture of water, proteins, nutrients, electrolytes, nitrogenous wastes, hormones, and gases (table 19.2). When the blood clots and the solids are removed, the remaining fluid is the blood **serum.** Serum is essentially identical to plasma except for the absence of the clotting protein fibrinogen.

Protein is the most abundant plasma solute by weight, totaling 6 to 9 g/dL. Plasma proteins play a variety of roles including clotting, defense, and transport of other solutes such as iron, copper, lipids, and hydrophobic hormones. There are three major categories of proteins: the albumins, globulins, and fibrinogen (table 19.3). Many other plasma proteins are indispensable to survival, but they account for less than 1% of the total.

TABLE 19.1

General Properties of Blood*

Mean Fraction of Body Weight	8%
Volume in Adult Body	Female: 4–5 L; male: 5–6 L
Volume/Body Weight	80–85 mL/kg
Mean Temperature	38°C (100.4°F)
pH	7.35–7.45
Viscosity (relative to water)	Whole blood: 4.5–5.5; plasma: 2.0
Osmolarity	280–296 mOsm/L
Mean Salinity (mainly NaCl)	0.9%
Hematocrit (packed cell volume)	Female: 37%–48% Male: 45%–52%
Hemoglobin	Female: 12–16 g/dL Male: 13–18 g/dL
Mean RBC Count	Female: 4.2–5.4 million/μL Male: 4.6–6.2 million/μL
Platelet Count	130,000–360,000/μL
Total WBC Count	5,000–10,000/μL

*Values vary slightly depending on the testing methods used.

THINK ABOUT IT!

Based on your body weight, estimate the volume (in liters) and weight (in kilograms) of your own blood.

Albumin is the smallest and most abundant plasma protein. It serves to transport various plasma solutes and buffer the pH of the blood plasma. It also makes a major contribution to two physical properties of blood: its *viscosity* (thickness, or resistance to flow) and *osmolarity* (the concentration of particles that cannot pass through the walls of the blood vessels). Through its effects on these two variables, changes in albumin concentration can significantly affect blood volume, pressure, and flow. **Globulins** are divided into three subclasses; from smallest to largest in molecular weight, they are the alpha (α), beta (β), and gamma (γ) globulins. Globulins play various roles in solute transport, clotting, and immunity. **Fibrinogen** is a soluble precursor of *fibrin,* a sticky protein that forms the framework of a blood clot. Some of the other plasma proteins are enzymes involved in the clotting process.

The liver produces as much as 4 g of plasma protein per hour, contributing all of the major proteins except γ globulins. The γ globulins come from *plasma cells*—connective tissue cells that are descended from white blood cells called *B lymphocytes.*

THINK ABOUT IT!

How could a disease such as liver cancer or hepatitis result in impaired blood clotting?

TABLE 19.2

Composition of Blood Plasma

Water	92% by weight
Proteins	Albumins, globulins, fibrinogen, other clotting factors, enzymes, and others
Nutrients	Glucose, amino acids, lactic acid, lipids (cholesterol, fatty acids, lipoproteins, triglycerides, and phospholipids), iron, trace elements, and vitamins
Electrolytes	Salts of sodium, potassium, magnesium, calcium, chloride, bicarbonate, phosphate, and sulfate
Nitrogenous Wastes	Urea, uric acid, creatinine, creatine, uric acid, bilirubin, and ammonia
Hormones	All hormones are transported in the blood.
Gases	Oxygen, carbon dioxide, and nitrogen

TABLE 19.3

Major Proteins of the Blood Plasma

Proteins	Functions
Albumin (60%)*	Major contributor to blood viscosity and osmolarity; transports lipids, hormones, calcium, and other solutes; buffers blood pH
Globulins (36%)*	
Alpha (α) Globulins	
Haptoglobulin	Transports hemoglobin released by dead erythrocytes
Ceruloplasmin	Transports copper
Prothrombin	Promotes blood clotting
Others	Transport lipids, fat-soluble vitamins, and hormones
Beta (β) Globulins	
Transferrin	Transports iron
Complement proteins	Aid in destruction of toxins and microorganisms
Others	Transport lipids
Gamma (γ) Globulins	Antibodies; combat pathogens
Fibrinogen (4%)*	Becomes fibrin, the major component of blood clots

*Percentage of the total plasma protein by weight

Before You Go On

Answer the following questions to test your understanding of the preceding section:

1. List some transport, protective, and regulatory functions of the blood.
2. What are the two principal components of the blood? Outline the classification of its formed elements.

3. What blood components make the greatest contributions to its viscosity and osmolarity? Why are viscosity and osmolarity important?
4. List the three major classes of plasma proteins. Which one is missing from blood serum?
5. What are the functions of blood albumin?
6. List some organic and inorganic components of plasma other than protein.

ERYTHROCYTES

Objectives

When you have completed this section, you should be able to

- describe the morphology and functions of erythrocytes (RBCs);
- explain some clinical measurements of RBC and hemoglobin quantities;
- describe the structure and function of hemoglobin;
- discuss the formation, life span, death, and disposal of RBCs; and
- describe the chemical and immunological basis and the clinical significance of blood types.

Erythrocytes, or **red blood cells (RBCs),** have two principal functions: (1) to pick up oxygen from the lungs and deliver it to tissues elsewhere, and (2) to pick up carbon dioxide from the tissues and unload it in the lungs. RBCs are the most abundant formed elements of the blood and therefore the most obvious things one sees upon its microscopic examination. They are also the most critical to survival; while a severe deficiency of leukocytes or platelets can be fatal within a few days, a severe deficiency of erythrocytes can be fatal within a few minutes. It is the deficiency of life-giving oxygen, carried by erythrocytes, that leads rapidly to death in cases of major trauma or hemorrhage.

Form and Function

An erythrocyte is a discoid cell with a thick rim and a thin sunken center. It is about 7.5 µm in diameter and 2.0 µm thick at the rim (fig. 19.3). While most cells, including white blood cells, have an abundance of organelles, RBCs lose nearly all organelles during their development and are thus remarkably devoid of internal structure. When viewed with the transmission electron microscope, the interior of an RBC appears uniformly gray (fig. 19.4). Lacking mitochondria, RBCs rely exclusively on anaerobic fermentation to produce ATP. The lack of aerobic respiration prevents them from consuming the oxygen that they must transport to other tissues. Lacking a nucleus and DNA, RBCs also are incapable of protein synthesis and mitosis.

The plasma membrane of a mature RBC has glycoproteins and glycolipids on the outer surface that determine a person's blood type. On its inner surface are two cytoskeletal proteins, *spectrin* and *actin,* that give the membrane resilience and durability. This is especially important when RBCs pass through small blood capillaries and sinusoids. Many of these passages are narrower than the diameter of an RBC, forcing the RBCs to stretch, bend, and fold as they squeeze through. When they enter larger vessels, RBCs spring back to their discoid shape, like an air-filled inner tube.

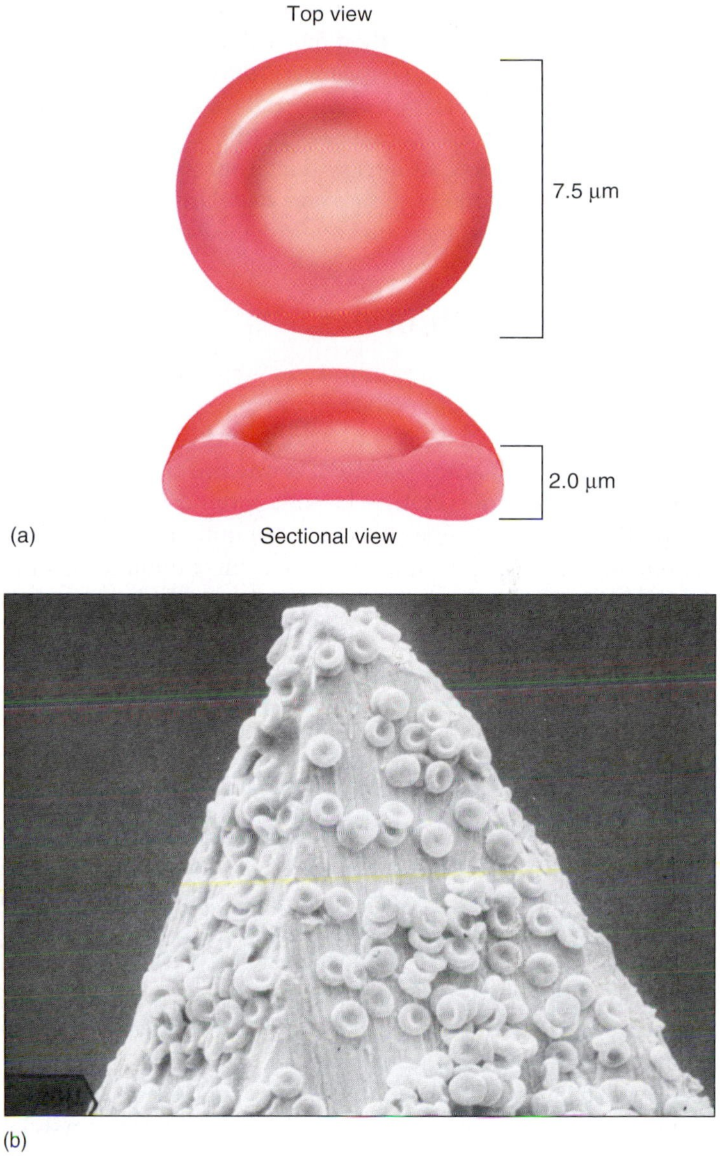
Top view
7.5 µm
2.0 µm
(a)
Sectional view
(b)

FIGURE 19.3
The Structure of Erythrocytes. (*a*) Dimensions and shape of an erythrocyte. (*b*) Erythrocytes on the tip of a hypodermic needle.

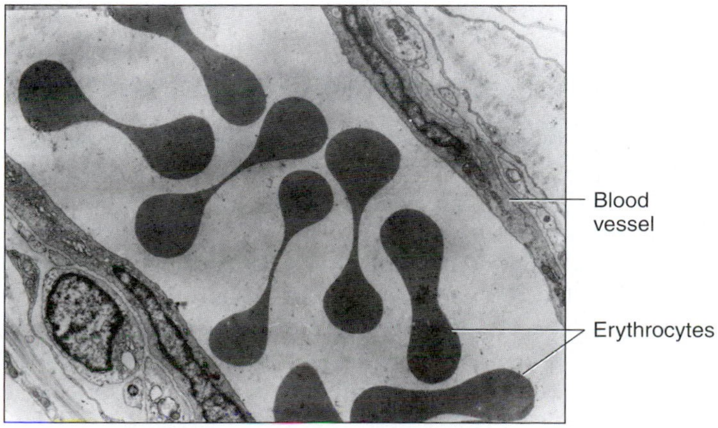
Blood vessel
Erythrocytes

FIGURE 19.4
Erythrocytes in a Blood Capillary (TEM). Note the lack of internal structure in the cells.

The cytoplasm of an RBC consists mainly of a 33% solution of hemoglobin (about 280 million molecules per cell). Hemoglobin is known especially for its oxygen-transport function, but it also aids in the transport of carbon dioxide and the buffering of blood pH. While the lack of a nucleus makes an RBC unable to repair itself, it has an overriding advantage: the biconcave shape gives the cell a much greater ratio of surface area to volume, which enables O_2 and CO_2 to diffuse quickly to and from the hemoglobin.

Quantity of Erythrocytes

The quantity of circulating erythrocytes is critically important to health, because it determines the amount of oxygen the blood can carry. Two of the most routine measurements in hematology are measures of RBC quantity: the **RBC count** and the **hematocrit,**[5] or **packed cell volume (PCV).** The RBC count is normally 4.6 to 6.2 million RBCs/μL in men and 4.2 to 5.4 million/μL in women. (A microliter, μL, is the same volume as a cubic millimeter, mm^3; RBC counts are also expressed as RBCs/mm^3.) The hematocrit is the percentage of blood volume composed of RBCs (see fig. 19.2). In men, it normally ranges between 42% and 52%; in women, between 37% and 48%.

Hemoglobin

The red color of blood is due to its **hemoglobin (Hb),** an iron-containing gas-transport protein normally found only in the RBCs. Hemoglobin consists of four protein chains called **globins** (fig. 19.5*a*). In adult hemoglobin, two of these, the *alpha (α) chains,* are 141 amino acids long, and the other two, the *beta (β) chains,* are 146 amino acids long. A nonprotein component called the **heme** group is bound to each protein chain (fig. 19.5*b*). At the center of each heme is a ferrous ion (Fe^{2+}), the binding site for oxygen. Having four heme groups, each hemoglobin molecule can transport up to 4 O_2. About 5% of the CO_2 in the bloodstream is also transported by hemoglobin, but this is bound to the globin component rather than to the heme, and a hemoglobin molecule can therefore transport both gases simultaneously. The **hemoglobin concentration** of whole blood is normally 13 to 18 g/dL in men and 12 to 16 g/dL in women.

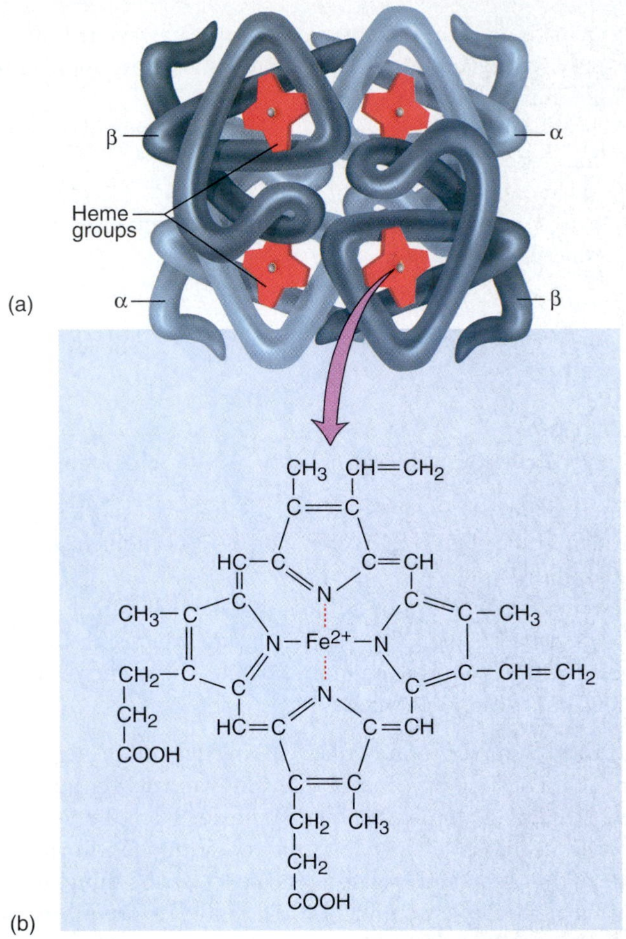

FIGURE 19.5
The Structure of Hemoglobin. (*a*) The hemoglobin molecule consists of two α proteins and two β proteins, each conjugated to a nonprotein heme group. (*b*) Structure of the heme group. Oxygen binds to Fe^{2+} at the center of the heme.

The Erythrocyte Life Cycle

The production of red blood cells is called **erythropoiesis**[6] (eh-RITH-ro-poy-EE-sis). It is one aspect of the more general process called **hemopoiesis** (HE-mo-poy-EE-sis), the production of all formed elements of the blood. A knowledge of hemopoiesis provides a foundation for understanding leukemia, anemia, and other blood disorders. We will survey some general aspects of hemopoiesis before examining erythropoiesis specifically.

The tissues that produce blood cells are called **hemopoietic tissues.** The first hemopoietic tissues of the human embryo form in the *yolk sac,* a membrane associated with all vertebrate embryos. In most vertebrates (fish, amphibians, reptiles, and birds), this sac encloses the egg yolk, transfers its nutrients to the growing embryo, and produces the forerunners of the first bood cells. Even animals that don't lay eggs, however, have a yolk sac that retains its hemopoietic function. (It is also the source of cells that later produce eggs and sperm.) Cell clusters called *blood islands* form here by the third week of human development. They produce primitive *stem cells* that migrate into the embryo proper and colonize the bone marrow, liver, spleen, and thymus. Here, the stem cells multiply and give rise to blood cells throughout fetal development. The liver stops producing blood cells around the time of birth. The spleen stops producing RBCs soon after, but it continues to produce lymphocytes for life.

From infancy onward, the red bone marrow produces all seven kinds of formed elements, while lymphocytes are produced not only there but also in the lymphatic tissues and organs—especially the thymus, tonsils, lymph nodes, spleen, and patches

---

[5]*hemato* = blood + *crit* = to separate
[6]*erythro* = red + *poiesis* = formation of

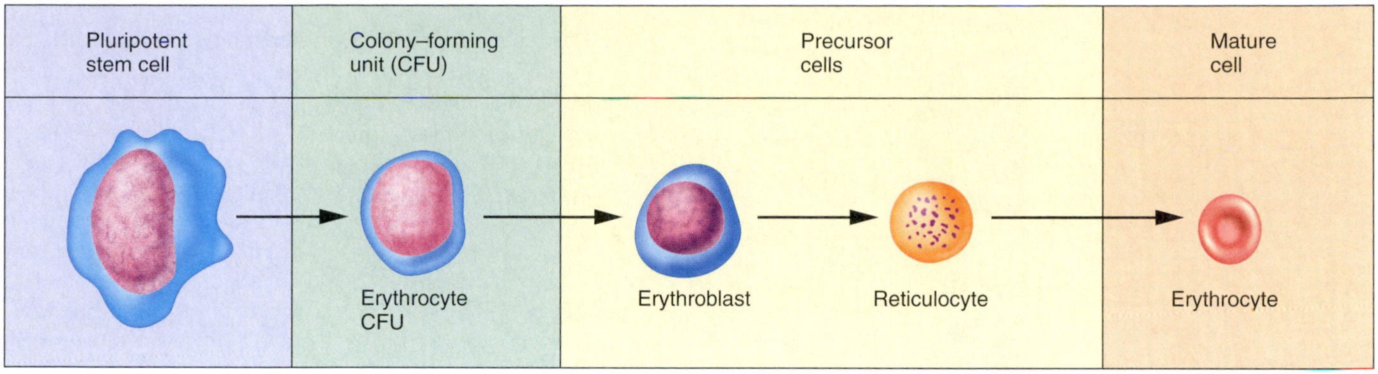
Pluripotent stem cell
Colony–forming unit (CFU)
Precursor cells
Mature cell
Erythrocyte CFU
Erythroblast
Reticulocyte
Erythrocyte

FIGURE 19.6
Erythropoiesis. Stages in the development of erythrocytes.

INSIGHT 19.1 EVOLUTIONARY MEDICINE

The Packaging of Hemoglobin

The gas-transport pigments of earthworms, snails, and many other animals are dissolved in the plasma rather than contained in blood cells. You might wonder why human hemoglobin must be contained in RBCs. The main reason is osmotic. The osmolarity of blood depends on the number of particles in solution. A "particle," for this purpose, can be a sodium ion, an albumin molecule, or a whole cell. If all the hemoglobin contained in the RBCs were free in the plasma, it would drastically increase blood osmolarity, since each RBC contains about 280 million molecules of hemoglobin. The circulatory system would osmotically absorb excess water and become enormously congested. Circulation would be severely impaired. The blood simply could not contain that much free hemoglobin and support life. On the other hand, if it contained a safe level of free hemoglobin, it could not transport enough oxygen to support the high metabolic demand of the human body. By having our hemoglobin packaged in RBCs, we are able to have much more of it and hence to have more efficient gas transport and more active metabolism.

Another reason for packaging the hemoglobin in RBCs is that some of the body's capillaries (*fenestrated capillaries* found in the kidneys and endocrine glands, for example) are permeable to proteins. Hemoglobin would leak out into the tissues if it were not contained in cells too big to pass through the capillary wall.

of lymphatic tissue in the mucous membranes. Blood formation in the bone marrow and lymphatic organs is called, respectively, **myeloid**[7] and **lymphoid hemopoiesis.**

All formed elements trace their origins to a common type of bone marrow stem cell, the **pluripotent stem cell (PPSC)** (formerly called a hemocytoblast[8]). PPSCs are so-named because they have the potential to develop into multiple mature cell types. They multiply at a relatively slow rate and thus maintain a small population in the bone marrow. Some of them go on to differentiate into a variety of more specialized cells called **colony-forming units (CFUs),** each type destined to produce one or another class of formed elements. Through a series of **precursor cells,** the CFUs divide and differentiate into mature formed elements.

Erythropoiesis itself begins when a PPSC becomes an *erythrocyte colony-forming unit (ECFU)* (fig. 19.6). The hormone *erythropoietin (EPO)* stimulates the ECFU to develop into a *proerythroblast,* followed by an *erythroblast.* Erythroblasts multiply and synthesize hemoglobin. When this task is completed, the nucleus shrivels and is discharged from the cell. The cell is now called a *reticulocyte,*[9] named for a temporary network composed of clusters of ribosomes *(polyribosomes).* Reticulocytes leave the bone marrow and enter the circulating blood. When the last of the polyribosomes disintegrate and disappear, the cell is a mature erythrocyte. Normally, about 0.5% to 1.5% of the circulating RBCs are reticulocytes, but this percentage rises when the body is making RBCs especially rapidly, as when compensating for blood loss.

The entire process of transformation from PPSC to mature RBC takes 3 to 5 days and involves four major developments—a reduction in cell size, an increase in cell number, the synthesis of hemoglobin, and the loss of the nucleus and other organelles. The process normally generates about 2.5 million RBCs per second, or 20 mL of packed RBCs per day.

The average RBC lives about 120 days after its release from the bone marrow. As it ages, its membrane proteins (especially spectrin) deteriorate and the membrane grows increasingly fragile. Without a nucleus or ribosomes, an RBC cannot synthesize new spectrin. Eventually, it ruptures as it tries to flex its way through narrow capillaries and sinusoids. The spleen has been called an "erythrocyte graveyard" because RBCs have an especially difficult time passing through its small channels. Here the old cells become trapped, broken up, and destroyed.

The 120-day life span of an RBC is often described as relatively short, owing to the fact that it lacks protein synthesis organelles and cannot repair itself. However, this is actually a relatively long life compared to other formed elements. Most leukocytes (which have nuclei) live less than a week and platelets live about 10 days. Only monocytes and lymphocytes outlive the RBCs.

[7] *myel* = bone marrow
[8] *hemo* = blood + *cyto* = cell + *blast* = precursor
[9] *reticulo* = little network + *cyte* = cell

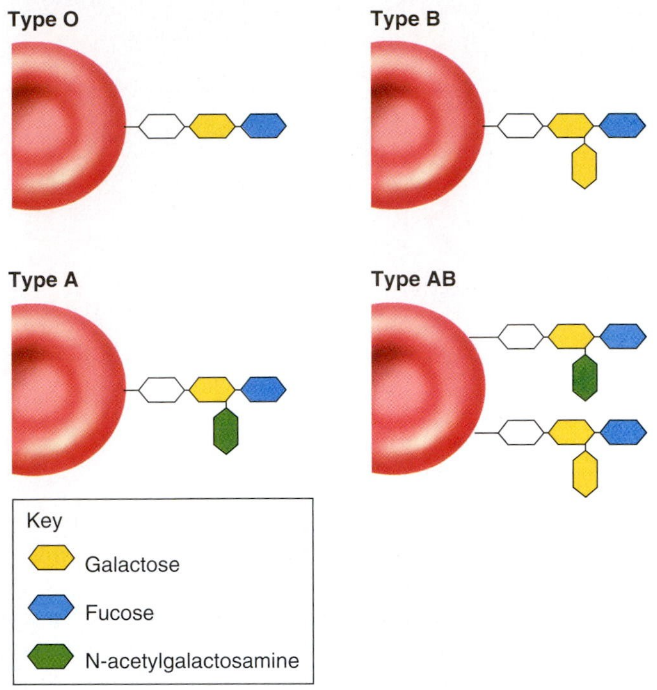
Type O
Type B
Type A
Type AB
Key
Galactose
Fucose
N-acetylgalactosamine

FIGURE 19.7
Chemical Basis of the ABO Blood Types. Blood types are determined by the chemistry of glycolipids of the RBC plasma membrane. Types A, B, AB, and O erythrocytes differ only in the terminal three to four carbohydrates of the glycolipid molecules. All of them end with galactose and fucose, but they differ in the presence or absence, and type, of additional sugar bonded to the galactose. The additional sugar is absent in type O; it is N-acetylgalactosamine in type A; it is another galactose in type B; and type AB cells have both the type A and B chains.

Blood Types

There are numerous genetically determined *blood groups* in the human population, each of which contains multiple **blood types.** The most familiar of these are the ABO group (with blood types A, B, AB, and O) and Rh group (with blood types Rh-positive and Rh-negative). These types differ with respect to the chemical composition of glycoproteins and glycolipids on the RBC surface (fig. 19.7). These molecules act as *antigens,* substances capable of evoking an immune reaction. The blood plasma contains *antibodies* that react against incompatible antigens on foreign RBCs. The RBC antigens and plasma antibodies thus determine the compatibility of donor and recipient blood in transfusions. In the event of an incompatibility in Rh type between the mother and fetus, they can also cause a severe anemia in the newborn infant.

Before You Go On

Answer the following questions to test your understanding of the preceding section:

7. What are the two main functions of RBCs?
8. Define *hematocrit* and *RBC count* and state some normal clinical values for each.
9. Describe the structure of a hemoglobin molecule. Explain where O_2 and CO_2 are carried on a hemoglobin molecule.
10. Name the stages in the production of an RBC, and state the differences between them.
11. Explain what plasma and RBC components are responsible for blood types, and why blood types are clinically important.

LEUKOCYTES

Objectives

When you have completed this section, you should be able to

- describe the appearance of the five kinds of leukocytes;
- explain the function of WBCs in general and the individual roles of each WBC type;
- describe the formation and life history of WBCs; and
- describe the production, death, and disposal of leukocytes.

Form and Function

Leukocytes, or **white blood cells (WBCs),** are the least abundant formed elements, totalling only 5,000 to 10,000 WBCs/μL. Yet we cannot live long without them, because they afford protection against infectious microorganisms and other pathogens. WBCs are easily recognized in stained blood films because they have conspicuous nuclei that stain from light violet to dark purple with the most common blood stains. They are much more abundant in the body than their low number in blood films would suggest, because they spend only a few hours in the bloodstream, then migrate through the walls of the capillaries and venules and spend the rest of their lives in the connective tissues. It's as if the bloodstream were merely the subway that the WBCs take to work; in blood films, we see only the ones on their way to work, not the WBCs already at work in the tissues.

Leukocytes differ from erythrocytes in that they retain their organelles throughout life; thus, when viewed with the transmission electron microscope, they show a complex internal structure (fig. 19.8). Among these organelles are the usual instruments of protein synthesis—the nucleus, rough endoplasmic reticulum, ribosomes, and Golgi complex—for leukocytes must synthesize proteins in order to carry out their functions. Some of these proteins are packaged into lysosomes and other organelles, which appear as conspicuous cytoplasmic granules that distinguish one WBC type from another.

Types of Leukocytes

As outlined at the beginning of this chapter, there are five kinds of leukocytes (table 19.4). They are distinguished from each other by their relative size and abundance, the size and shape of their nuclei, the presence or absence of cytoplasmic granules, the coarseness and staining properties of those granules, and most importantly by their functions. Three of the WBC types—the neutrophils, eosinophils, and basophils—are called **granulocytes** because their cytoplasm contains lysosomes and other membrane-bounded organelles that

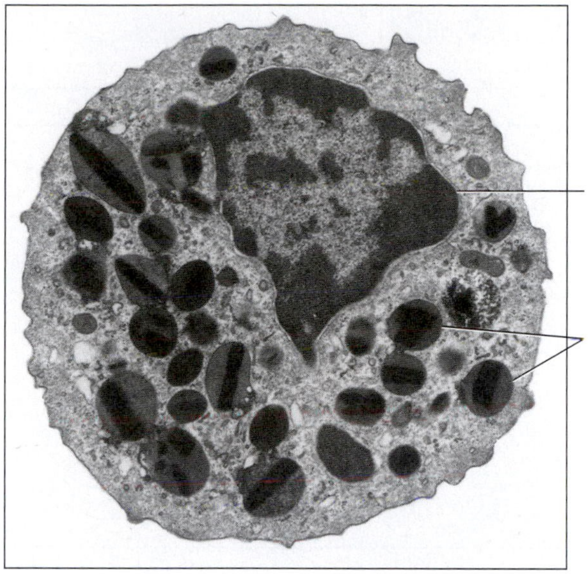

FIGURE 19.8
Structure of an Eosinophil. In contrast to an RBC, the WBC cytoplasm is crowded with organelles, including a nucleus.

appear as conspicuous colored granules in stained blood films. The other two WBC types—lymphocytes and monocytes—are called **agranulocytes** because cytoplasmic granules are scarce or absent.

GRANULOCYTES

- **Neutrophils** are the most abundant WBCs—generally about 4,150 cells/μL and constituting 60% to 70% of the circulating leukocytes. The nucleus is clearly visible and in a mature neutrophil, typically consists of three to five lobes connected by slender nuclear strands. These strands are sometimes so delicate that they are scarcely visible, and the neutrophil may seem as if it had multiple nuclei. Young neutrophils have an undivided nucleus shaped like a band or a knife puncture; thus they are called *band cells* or *stab cells.* Neutrophils are also called *polymorphonuclear leukocytes (PMNs)* because of their varied nuclear shapes.

  The cytoplasm contains fine reddish to violet granules, which contain lysozyme, peroxidase, and other antibiotic agents. Neutrophils are named for the way these granules take up histological stains at pH 7—some stain with acidic dyes and others with basic dyes, and the combined effect gives the cytoplasm a pale lilac color.

TABLE 19.4

The White Blood Cells (Leukocytes)

Neutr'ophils

Percent of WBCs	60%–70%
Mean count	4,150 cells/μL
Diameter	9–12 μm

*Appearance**
Nucleus usually with 3–5 lobes in S- or C-shaped array
Fine reddish to violet granules in cytoplasm

Differential Count
Increases in bacterial infections

Functions
Phagocytosis of bacteria
Release of antimicrobial chemicals

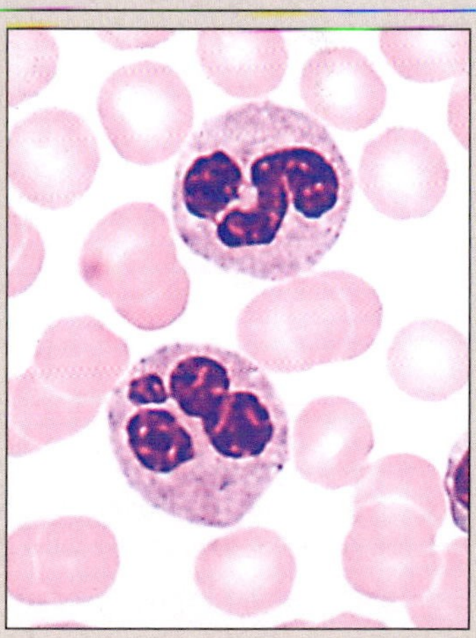

Neutrophils

Eosinophils

Percent of WBCs	2%–4%
Mean count	170 cells/μL
Diameter	10–14 μm

*Appearance**
Nucleus usually has two large lobes connected by thin strand
Large orange-pink granules in cytoplasm

Differential Count
Fluctuates greatly from day to night, seasonally, and with phase of menstrual cycle
Increases in parasitic infections, allergies, collagen diseases, and diseases of spleen and central nervous system

Functions
Phagocytosis of antigen-antibody complexes, allergens, and inflammatory chemicals
Release enzymes that weaken or destroy parasites such as worms

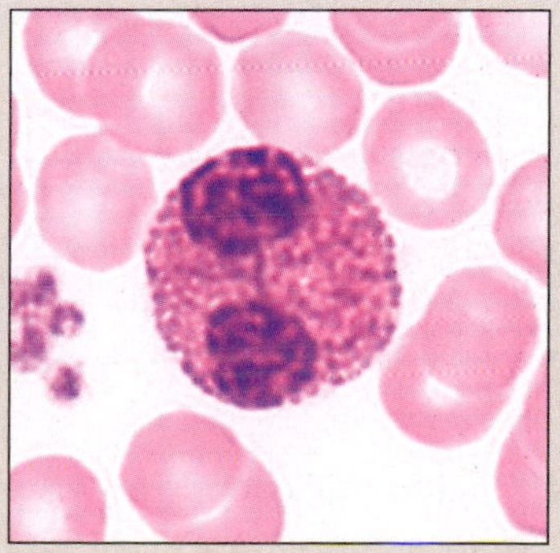

Eosinophil

(continued)

TABLE 19.4

The White Blood Cells (Leukocytes)

Basophils

Percent of WBCs	<0.5%–1%
Mean count	40 cells/μL
Diameter	8–10 μm

*Appearance**
Nucleus large and irregularly shaped, but typically obscured from view
Coarse, abundant, dark violet granules in cytoplasm

Differential Count
Relatively stable
Increases in chicken pox, sinusitis, diabetes mellitus, myxedema, and polycythemia

Functions
Secrete histamine (a vasodilator), which increases blood flow to a tissue
Secrete heparin (an anticoagulant), which promotes mobility of other WBCs by preventing clotting

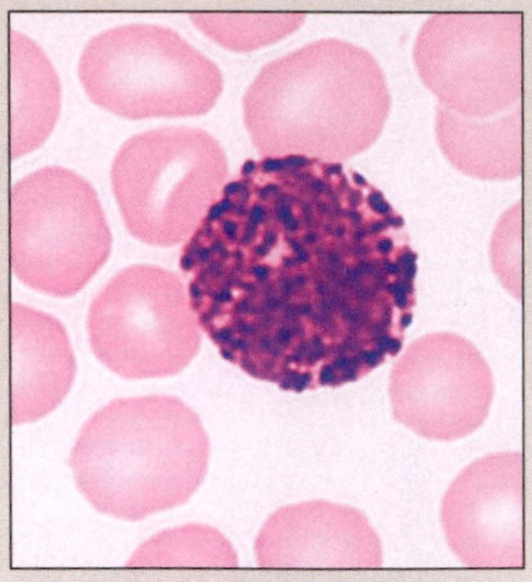
Basophil

Monocytes

Percent of WBCs	3%–8%
Mean count	460 cells/μL
Diameter	12–15 μm

*Appearance**
Nucleus ovoid, kidney-shaped, or horseshoe-shaped; light violet
Abundant cytoplasm with sparse, fine granules
Sometimes very large with stellate or polygonal shapes

Differential Count
Increases in viral infections and inflammation

Functions
Differentiate into macrophages (large phagocytic cells of the tissues)
Phagocytize pathogens, dead neutrophils, and debris of dead cells
"Present" antigens to activate other cells of immune system

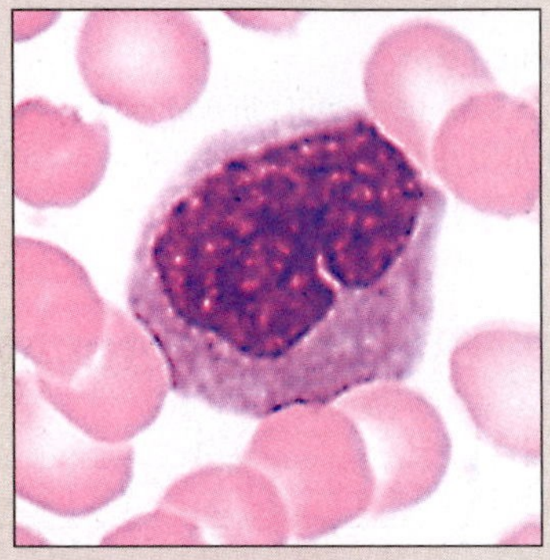
Monocyte

Lymphocytes

Percent of WBCs	25%–33%
Mean count	2,200 cells/μL
Diameter	
Small class	5–8 μm
Medium class	10–12 μm
Large class	14–17 μm

*Appearance**
Nucleus round, ovoid, or slightly dimpled on one side, of uniform dark violet color
In small lymphocytes, nucleus fills nearly all of the cell and leaves only a scanty rim of clear, light blue cytoplasm
In larger lymphocytes, cytoplasm is more abundant; large lymphocytes may be hard to differentiate from monocytes

Differential Count
Increases in diverse infections and immune responses

Functions
Several functional classes indistinguishable by light microscopy
Destroy cancer cells, cells infected with viruses, and foreign cells
"Present" antigens to activate other cells of immune system
Coordinate actions of other immune cells
Secrete antibodies
Serve in immune memory

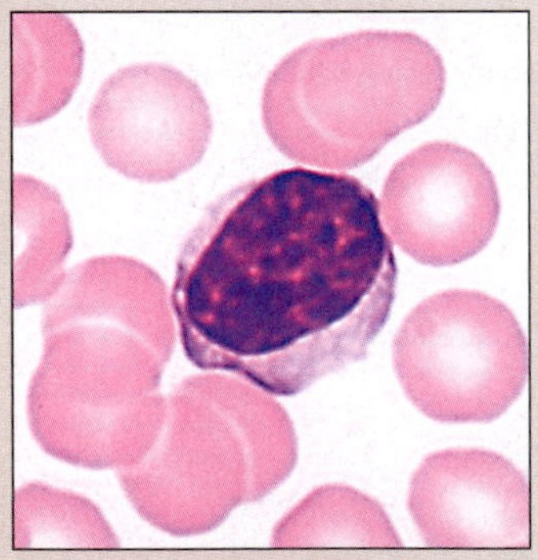
Lymphocyte

* Appearance pertains to blood films dyed with Wright's stain

The neutrophil count rises—a condition called *neutrophilia*—in response to bacterial infections. The primary task of the neutrophils is to destroy bacteria, which they achieve in two ways. One is to phagocytize the bacteria and digest them. The other is to release a potent mix of toxic chemicals, including hypochlorite (HClO) (the active agent in household bleach) and the superoxide anion ($O_2{\bullet}^-$), which reacts with hydrogen ions to produce hydrogen peroxide

(H_2O_2). Just as bleach and hydrogen peroxide are often used around the home as disinfectants, they are deadly to bacteria in the tissues. These chemicals form a **killing zone** around the neutrophil, lethal to the invaders but also to the neutrophil itself. Neutrophils are thus the body's suicidal guardians against infection.

- **Eosinophils** (EE-oh-SIN-oh-fills) are harder to find in a blood film, because they are only 2% to 4% of the WBC total, typically numbering about 170 cells/µL. The eosinophil count fluctuates greatly, however, from day to night, seasonally, and with the phase of the menstrual cycle. It rises *(eosinophilia)* in allergies, parasitic infections, collagen diseases, and diseases of the spleen and central nervous system. Although relatively scanty in the blood, eosinophils are abundant in the mucous membranes of the respiratory, digestive, and lower urinary tracts. The eosinophil nucleus usually has two large lobes connected by a thin strand, and the cytoplasm has an abundance of coarse rosy to orange-colored granules. Eosinophils secrete chemicals that weaken or destroy relatively large parasites such as hookworms and tapeworms, too big for any one WBC to phagocytize. Eosinophils also phagocytize and dispose of inflammatory chemicals, antigen-antibody complexes (masses of antigen molecules stuck together by antibodies), and allergens (foreign antigens that trigger allergies).
- **Basophils** are the rarest of the WBCs and, indeed, of all formed elements. They number about 40 cells/µL and constitute from less than 0.5% to about 1% of the WBC count. They can be recognized mainly by an abundance of very coarse, dark violet cytoplasmic granules. The nucleus is largely hidden from view by these granules, but is large, pale, and typically S- or U-shaped. Basophils secrete two chemicals which aid in the body's defense processes: (1) **histamine,** a vasodilator that widens the blood vessels, speeds the flow of blood to an injured tissue, and makes the blood vessels more permeable so that blood components such as neutrophils and clotting proteins can get into the connective tissues more quickly; and (2) **heparin,** an anticoagulant that inhibits blood clotting and thus promotes the mobility of other WBCs in the area.

AGRANULOCYTES

- **Monocytes** are the largest WBCs, often two or three times the diameter of an RBC. They number about 460 cells/µL and about 3% to 8% of the WBC count. The nucleus is large and clearly visible, often a relatively light violet, and typically ovoid, kidney-shaped, or horseshoe-shaped. The cytoplasm is abundant and contains sparse, fine granules. In prepared blood films, monocytes often assume sharply angular to spiky shapes (see fig. 19.1). The monocyte count rises in inflammation and viral infections. Monocytes go to work only after leaving the blood stream and transforming into large tissue cells called **macrophages**[10] (MAC-ro-fay-jez). Macrophages are highly phagocytic cells that consume up to 25% of their own volume per hour. They destroy dead or dying host and foreign cells, pathogenic chemicals and microorganisms, and other foreign matter. They also chop up or "process" foreign antigens and "display" fragments of them on the cell surface to alert the immune system to the presence of a pathogen. Thus, they and a few other cells are called *antigen-presenting cells (APCs).* There are several kinds of macrophages in the body descended from monocytes or from the same hemopoietic stem cells as monocytes. Macrophages of the loose connective tissues are generally called simply *macrophages* or sometimes *histiocytes.* Macrophages in some other localities have special names:
    - **dendritic cells** in the epidermis (see chapter 5) and the mucous membranes of the mouth, esophagus, and vagina;
    - **microglia,** a type of neuroglia in the central nervous system (see chapter 13);
    - **alveolar macrophages** in the pulmonary alveoli (see chapter 23); and
    - **hepatic macrophages (Kupffer cells)** in the liver sinusoids (see chapter 24).
- **Lymphocytes** include the smallest WBCs; at 5 to 17 µm in diameter, they range from smaller than RBCs to two and a half times as large, but those in circulating blood are generally at the small end of the range. They are second to neutrophils in abundance and are thus quickly spotted when you examine a blood film. They number about 2,200 cells/µL and are 25% to 33% of the WBC count. The lymphocyte nucleus is round, ovoid, or slightly dimpled on one side. It usually stains dark violet and fills nearly the entire cell, especially in small lymphocytes. The cytoplasm, which stains a clear light blue color, thus forms a narrow and often barely detectable rim around the nucleus, although it is more abundant in the larger lymphocytes. Small lymphocytes are sometimes difficult to distinguish from basophils, but most basophils are conspicuously grainy whereas the lymphocyte nucleus is uniform or merely mottled. Basophils also lack the rim of clear cytoplasm seen in most lymphocytes. Large lymphocytes are sometimes difficult to distinguish from monocytes.

The lymphocyte count increases in diverse infections and immune responses. Some of them function in nonspecific defense of the body against viruses and cancer, but most of them are involved in specific *immunity,* a defense in which the body recognizes a certain antigen it has encountered before and mounts such a quick response that a person notices little or no symptoms of illness. The various lymphocytes are not distinguishable by light microscopy, but differ in their functions. Certain T lymphocytes attack foreign cells directly, while B lymphocytes fight pathogens by differentiating into plasma cells and secreting antibodies.

---
[10]*macro* = big + *phage* = eater

INSIGHT 19.2 CLINICAL APPLICATION

The Complete Blood Count

One of the most common clinical procedures, in both routine physical examinations and the diagnosis of disease, is a *complete blood count (CBC)*. The CBC yields a highly informative profile of data on multiple blood values. Hematocrit and hemoglobin concentrations are measured from fresh whole blood. Living blood cells are microscopically examined for abnormal RBC or WBC size or structure. A CBC includes total RBCs, reticulocytes, WBCs, and platelets per microliter of blood, and a *differential WBC count*, a count of what percentage of the WBC count consists of each WBC type.

A differential WBC count used to require the microscopic examination of stained blood films, but most laboratories now use electronic instruments such as a Coulter counter, which forces blood cells through a narrow orifice and counts them as voltage pulses that vary with cell type and size. Such *electronic cell counters* give much faster and more accurate results. They can distinguish RBCs, platelets, and different types of WBCs from each other, measure hematocrit and hemoglobin concentration, and measure various *RBC indices* such as RBC size *(mean corpuscular volume, MCV)* and hemoglobin concentration per RBC *(mean corpuscular hemoglobin, MCH)*.

The wealth of information gained from a CBC is too vast to give more than a few examples here. An elevated reticulocyte count may indicate recovery from anemia, and too low a count may indicate a nutritional deficiency or unresponsiveness to EPO. Various forms of anemia are indicated by low RBC counts or abnormalities of RBC size, shape, and hemoglobin content. A high neutrophil count suggests bacterial infection and a high eosinophil count suggests an allergy or parasitic infection. If a CBC does not provide enough information or if it suggests other disorders, other tests may be done such as coagulation time and bone marrow biopsy.

The Leukocyte Life Cycle

Leukopoiesis (LOO-co-poy-EE-sis), production of white blood cells, begins with the same pluripotent stem cells (PPSCs) as erythropoiesis. Some PPSCs differentiate into the distinct types of colony-forming units, which then go on to produce the following cell lines (fig. 19.9):

1. *Myeloblasts,* which ultimately differentiate the three types of granulocytes (neutrophils, eosinophils, and basophils).
2. *Monoblasts,* which look identical to myeloblasts but lead ultimately to monocytes.
3. *Lymphoblasts,* which give rise to three types of lymphocytes (B lymphocytes, T lymphocytes, and natural killer cells).

Granulocytes and monocytes stay in the red bone marrow until they are needed; the marrow contains 10 to 20 times as many of these cells as the circulating blood does. Lymphocytes, by contrast, begin developing in the bone marrow but then migrate elsewhere. **B lymphocytes (B cells)** mature in the bone marrow and some remain there, while others disperse and colonize the lymph nodes, spleen, tonsils, and mucous membranes. To remember their site of maturation, it may help to think "B for bone marrow," although these cells were actually named for an organ in chickens (the *bursa of Fabricius*) where they were discovered. **T lymphocytes (T cells)** begin development in the bone marrow but then migrate to the thymus (a gland in the mediastinum just above the heart) and mature there. In this case, the *T* really does stand for *thymus.* Mature T lymphocytes disperse from the thymus and colonize the same organs as B lymphocytes. **Natural killer (NK) cells** develop in the bone marrow like B cells.

Circulating WBCs do not stay in the blood for very long. Granulocytes circulate for 4 to 8 hours and then migrate into the tissues, where they live for another 4 or 5 days. Monocytes travel in the blood for 10 or 20 hours, then migrate into the tissues and transform into a variety of macrophages, which can live as long as a few years. Lymphocytes, responsible for long-term immunity, have a life span ranging from weeks to decades.

When leukocytes die, they are generally phagocytized and digested by macrophages. Dead neutrophils, however, are responsible for the creamy color of pus, and are sometimes disposed of by the rupture of a blister onto the skin surface.

Before You Go On

Answer the following questions to test your understanding of the preceding section:

12. What is the purpose of WBCs in general?
13. Name the five kinds of WBCs and state what contributions each type makes to that general purpose.
14. Describe the key features that enable one to microscopically identify each WBC type.
15. What are macrophages? What class of WBCs do they arise from? Name some types of macrophages.

Platelets

Objectives

When you have completed this section, you should be able to

- describe the structure of blood platelets;
- explain the multiple roles played by platelets in hemostasis and blood vessel maintenance;
- describe platelet production and state their longevity; and
- describe the general processes of hemostasis.

Form and Function

Platelets are not cells but small fragments of marrow cells called megakaryocytes. Platelets are the second most abundant formed elements, after erythrocytes; a normal platelet count in blood from a fingerstick ranges from 130,000 to 400,000 platelets/μL (averaging about 250,000). The platelet count can vary greatly, however, under different physiological conditions and in blood samples taken from various places in the body. In spite of their numbers, platelets are so small (2 to 4 μm in diameter) that they contribute even less than the WBCs to the blood volume.

Platelets have a complex internal structure that includes lysosomes, mitochondria, microtubules and microfilaments, **granules** filled with platelet secretions, and a system of channels called the **open**

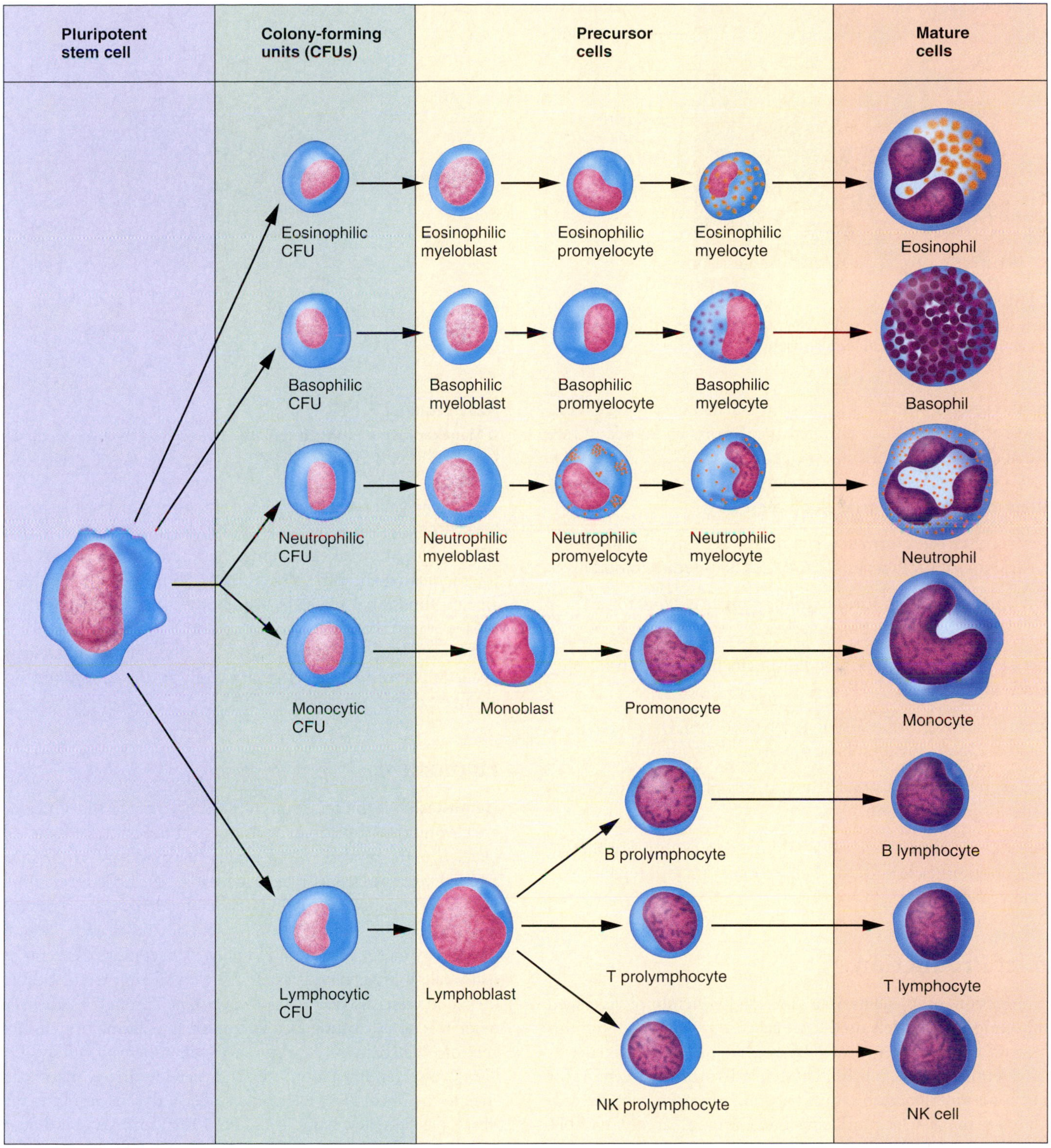
Pluripotent stem cell
Colony-forming units (CFUs)
Precursor cells
Mature cells
Eosinophilic CFU
Eosinophilic myeloblast
Eosinophilic promyelocyte
Eosinophilic myelocyte
Eosinophil
Basophilic CFU
Basophilic myeloblast
Basophilic promyelocyte
Basophilic myelocyte
Basophil
Neutrophilic CFU
Neutrophilic myeloblast
Neutrophilic promyelocyte
Neutrophilic myelocyte
Neutrophil
Monocytic CFU
Monoblast
Promonocyte
Monocyte
B prolymphocyte
B lymphocyte
Lymphocytic CFU
Lymphoblast
T prolymphocyte
T lymphocyte
NK prolymphocyte
NK cell

FIGURE 19.9
Leukopoiesis. Stages in the development of leukocytes.

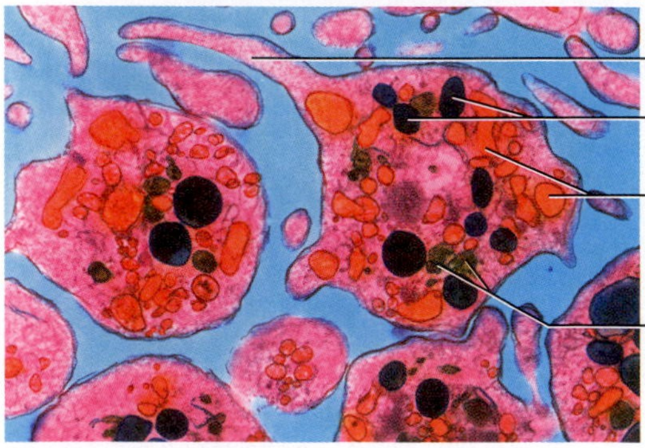
Pseudopod
Granules
Open canalicular system
Mitochondria

FIGURE 19.10
Platelets (TEM). Platelets have no nuclei but do possess other organelles such as endoplasmic reticulum and lysosomes.

canalicular system, which opens onto the platelet surface (fig. 19.10). They have no nucleus. When activated, they form pseudopods and are capable of ameboid movement.

Despite their small size, platelets have a greater variety of functions than any of the true blood cells:

- They secrete *vasoconstrictors,* chemicals that cause spasmodic constriction of broken vessels and thus help reduce blood loss.
- They stick together to form temporary *platelet plugs* to seal small breaks in injured blood vessels.
- They secrete *procoagulants,* or clotting factors, which promote blood clotting.
- They initiate the formation of a clot-dissolving enzyme that dissolves blood clots that have outlasted their usefulness.
- They secrete chemicals that attract neutrophils and monocytes to sites of inflammation.
- They secrete *growth factors* that stimulate mitosis in fibroblasts and smooth muscle and thus help to maintain and repair blood vessels.

Platelet Production

The production of platelets is a division of hemopoiesis called **thrombopoiesis.** (Platelets used to be called *thrombocytes,*[11] but this term is now reserved for nucleated true cells in other animals such as birds and reptiles.) Some pluripotent stem cells produce receptors for the hormone *thrombopoietin,* thus becoming *megakaryoblasts,* cells committed to the platelet-producing line. The megakaryoblast duplicates its DNA repeatedly without undergoing nuclear or cytoplasmic division. The result is a **megakaryocyte**[12] (meg-ah-CAR-ee-oh-site), a gigantic cell up to 150 μm in diameter, visible to the naked eye, with a huge multilobed nucleus and multiple sets of chromosomes (fig. 19.11). Most megakaryocytes live in the bone marrow, but some of them colonize the lungs and produce platelets there.

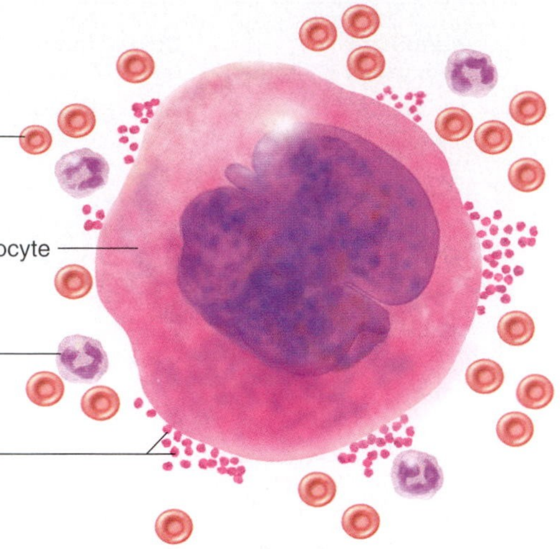
RBC
Megakaryocyte
WBC
Platelets

FIGURE 19.11
A Megakaryocyte Producing Platelets. Several red and white blood cells are shown for size comparison.

A megakaryocyte develops infoldings of the plasma membrane that divide its marginal cytoplasm into little compartments. The cytoplasm breaks up along these lines of weakness into tiny fragments that enter the bloodstream. Some of these fragments are already functional platelets, while others are larger particles that break up into platelets as they pass through the lungs. About 25% to 40% of the platelets are stored in the spleen and released as needed. The remainder circulate freely in the blood and live for about 10 days.

Hemostasis

Hemostasis[13] is the cessation of bleeding. The details of hemostasis are beyond the scope of an anatomy textbook, but the basic roles of platelets in the process will be briefly surveyed here. Upon injury to a blood vessel, platelets release *serotonin.* This chemical stimulates *vasoconstriction,* or narrowing of the blood vessel, to reduce blood loss. Platelets also adhere to the vessel wall and to each other, forming a sticky mass called a *platelet plug.* Platelet plugs can temporarily seal breaks in small blood vessels. Platelets and injured tissues around the blood vessel also release *clotting factors.* Through a series of enzymatic reactions, clotting factors convert the plasma protein fibrinogen into the sticky protein *fibrin.* Fibrin adheres to the wall of the blood vessel, and as blood cells and platelets arrive, many of them stick to the fibrin like insects in a spider web. The resulting mass of fibrin, platelets, and blood cells (fig. 19.12) forms a clot that ideally seals the break in the blood vessel long enough for the vessel to heal.

Once the leak is sealed and the crisis has passed, platelets secrete *platelet-derived growth factor (PDGF),* a substance that stimulates fibroblasts and smooth muscle to proliferate and replace the damaged tissue of the blood vessel. When tissue repair is completed and the blood clot is no longer needed, the clot must be disposed

[11]*thrombo* = clotting + *cyte* = cell
[12]*mega* = giant + *karyo* = nucleus + *cyte* = cell
[13]*hemo* = blood + *stasis* = stability

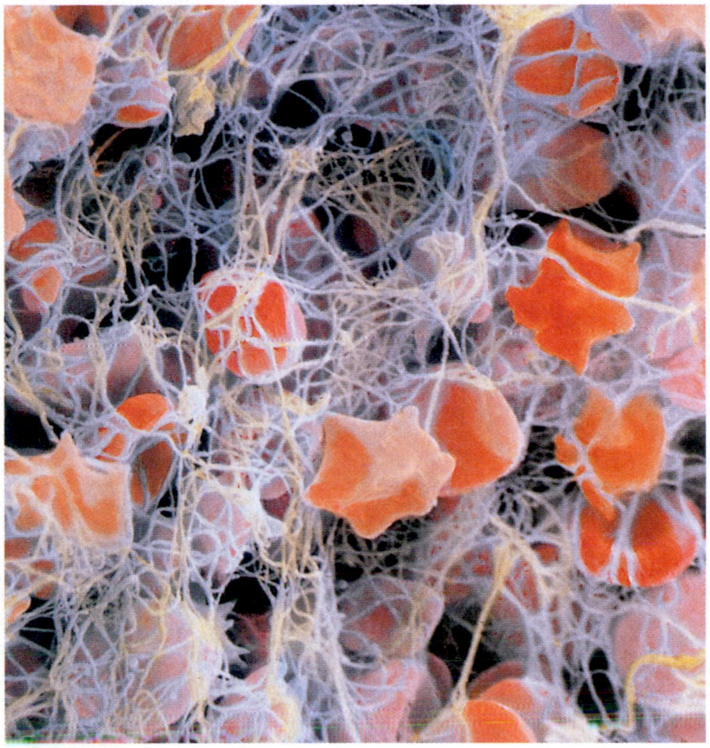

FIGURE 19.12

A Blood Clot (colorized SEM). Platelets are seen trapped in a sticky mesh of fibrin.

of. Platelets then secrete *factor XII,* a protein that initiates a series of reactions leading to the formation of a fibrin-digesting enzyme called *plasmin.* Plasmin dissolves the old blood clot.

Before You Go On

Answer the following questions to test your understanding of the preceding section:

16. List several functions of blood platelets.
17. How are blood platelets produced? How long do they live?
18. Briefly describe the stages in which platelets help to stop bleeding and repair a damaged blood vessel.

CLINICAL PERSPECTIVES

Objectives

When you have completed this section, you should be able to

- describe changes that occur in the blood in old age; and
- describe some common abnormalities of RBC, WBC, and platelet count and morphology, and the consequences of these abnormalities.

Hematology in Old Age

We have considered the embryonic origin and ongoing hemopoietic development of the blood in previous sections. At the other end of the life span, aging has multiple effects on the blood. Evidence suggests that the baseline rate of erythropoiesis does not change much with age; cell counts, hemoglobin concentration, and other variables are about the same among healthy people in their 70s as in the 30s. However, older people do not adapt well to stress on the hemopoietic system, perhaps because of the senescence of other organ systems. The stomach atrophies in old age, thus producing less of the *intrinsic factor* needed for the absorption of dietary vitamin B_{12}. A deficiency of vitamin B_{12} causes *pernicious anemia.* Anemia can also stem from atrophy of the kidneys in old age, since the kidneys secrete erythropoietin and erythropoietin is the principal stimulus for RBC production. There may also be a limit to how many times the hemopoietic stem cells can divide and continue producing new blood cells.

Anemia may also result from nutritional deficiencies, inadequate exercise, disease, and other causes. The factors that cause anemia in older people are very complicated and interrelated. It is almost impossible to determine whether aging alone causes anemia in the absence of other contributing factors such as poor exercise or nutritional habits.

Thrombosis also becomes increasingly problematic in old age. Plaques of atherosclerosis in the blood vessels can act as sites of blood clotting. The blood clots especially easily in the veins, where blood flow is slowest. About 25% of people over age 50 experience venous blockage by thrombosis, especially people who do not exercise regularly and people confined to a bed or wheelchair.

Disorders of the Blood

We conclude with a survey of some clinical aspects of hematology, especially disorders that affect the relative numbers of formed elements, and thus the appearance of stained blood films, or that alter the appearance of the individual formed elements. Some common nonstructural blood disorders are described in table 19.5.

ERYTHROCYTE DISORDERS

The two principal RBC disorders are **anemia**[14] (an RBC or hemoglobin deficiency) and **polycythemia**[15] (an RBC excess). The latter is also known as *erythrocytosis.*

There are three fundamental causes of anemia:

1. Depressed erythropoiesis or hemoglobin synthesis, so that erythropoiesis fails to keep pace with the normal death of RBCs. A lack of iron or certain vitamins in the diet can cause nutritional anemias (such as *iron deficiency anemia*). Radiation, viruses, and some poisons cause anemia by destroying bone marrow. The role of gastric and renal atrophy in anemia of the elderly was discussed earlier.
2. Excess RBC destruction, exceeding the rate of erythropoiesis. Anemias in this category are called *hemolytic anemia,* and may result from a variety of poisons, drug reactions, blood-destroying parasitic infections such as malaria, or sickle-cell disease.

[14]*an* = without + *emia* = blood
[15]*poly* = many + *cyt* = cells + *hemia* = blood condition

Table 19.5

Some Disorders of the Blood

Disseminated Intravascular Coagulation (DIC)	Widespread clotting within unbroken vessels, limited to one organ or occurring throughout the body. Usually triggered by septicemia but also occurs when blood circulation slows markedly (as in cardiac arrest). Marked by widespread hemorrhaging, congestion of the vessels with clotted blood, and tissue necrosis in blood-deprived organs.
Embolism	The presence of any abnormal object *(embolus)* traveling in the bloodstream, such as an air bubble *(air embolism)*, agglutinated RBCs or bacteria, or traveling blood clot *(thromboembolism)*. Presents a danger of blocking small blood vessels and shutting off the blood flow to vital tissues, thus causing stroke, heart failure, kidney failure, or pulmonary failure.
Hemophilia	Abnormally slow blood clotting as a result of the hereditary deficiency of a clotting factor, usually factor VIII (a liver product). Prolonged bleeding results in the painful pooling of clotted blood *(hematomas)* in such sites as the muscles and joints, or in fatal blood loss. Treatable with injections of the missing clotting factor.
Infectious Mononucleosis	Infection of B lymphocytes with Epstein-Barr virus. Usually transmitted by exchange of saliva, as in kissing; most common in adolescents and young adults. Causes fever, fatigue, sore throat, inflamed lymph nodes, and leukocytosis. Usually self-limiting and resolves within a few weeks.
Septicemia	Bacteria in the bloodstream, stemming from infection elsewhere in the body. Often causes fever, chills, and nausea, and may cause septic shock.
Thrombosis	Abnormal clotting in unbroken blood vessels, triggered by atherosclerosis and other defects, or by immobility of people confined to bed or a wheelchair. Stationary blood clots can cause stroke, heart failure, etc. (see *embolism* in this table), and clots can break free and cause *thromboembolism*.

Disorders Described Elsewhere

Anemia 555	Leukopenia 557	Thalassemia 556
Leukemia 557	Polycythemia 556	Thrombocytopenia 557
Leukocytosis 557	Sickle-cell disease 556	

3. Blood loss, resulting in *hemorrhagic anemia.* This can be a consequence of trauma such as gunshot, automobile, or battlefield injuries; hemophilia; bleeding ulcers; ruptured aneurysms; or heavy menstruation.

While anemia most obviously affects RBC count, it can also affect RBC morphology. *Thalassemia,* for example, is a hereditary blood disease among people of Mediterranean descent. It is characterized by deficient hemoglobin synthesis, and not only is the RBC count reduced but the existing RBCs are *microcytic* (abnormally small) and *hypochromic* (pale). Iron-deficiency anemia is characterized by these structural abnormalities as well as *poikilocytosis,*[16] in which RBCs assume teardrop, pencil, and other variable and abnormal shapes. Sickle-cell disease is another well-known hereditary anemia with abnormal RBC morphology (insight 19.3).

Polycythemia (POL-ee-sih-THEME-ee-uh), an excess RBC count, can result from cancer of the bone marrow *(primary polycythemia)* or from a multitude of other conditions *(secondary polycythemia)* such as an abnormally high oxygen demand (as in people who engage in over-zealous aerobic exercise) or low oxygen supply

[16]*poikilo* = variable + *cyt* = cell + *osis* = condition

INSIGHT 19.3 CLINICAL APPLICATION

Sickle-cell Disease

Sickle-cell disease is a hereditary hemoglobin defect occurring mostly among people of African descent; its symptoms occur in about 1.3% of American blacks, and about 8.3% are asymptomatic carriers with the potential to pass it to their children. The disease is caused by a defective gene that results in the substitution of valine for a glutamic acid in each β hemoglobin chain. The abnormal hemoglobin (HbS) turns to gel at low oxygen levels, as when blood passes through the oxygen-hungry skeletal muscles. The RBCs become elongated and pointed (sickle-shaped; fig. 19.13); they are sticky and thus agglutinate and block small vessels; and they are fragile and hemolyze, thus producing anemia. Blockage of the blood vessels produces severe pain and can lead to kidney or heart failure, stroke, or paralysis, among many other effects. Without treatment, a child with sickle-cell disease has little chance of living to age 2, and even with the best treatment, few victims live to an age of 50.

Sickle-cell disease exists because in the areas of Africa where it originated, millions of lives are lost to malaria. HbS is indigestible to the malaria parasite, so this gene confers a resistance to malaria. The lives saved by HbS in Africa far outnumber the deaths from sickle-cell disease, so natural selection favors the persistence of the gene rather than its elimination. But in North America, where malaria is not prevalent, the lost lives and suffering caused by the sickle-cell gene far outweigh any of its benefits.

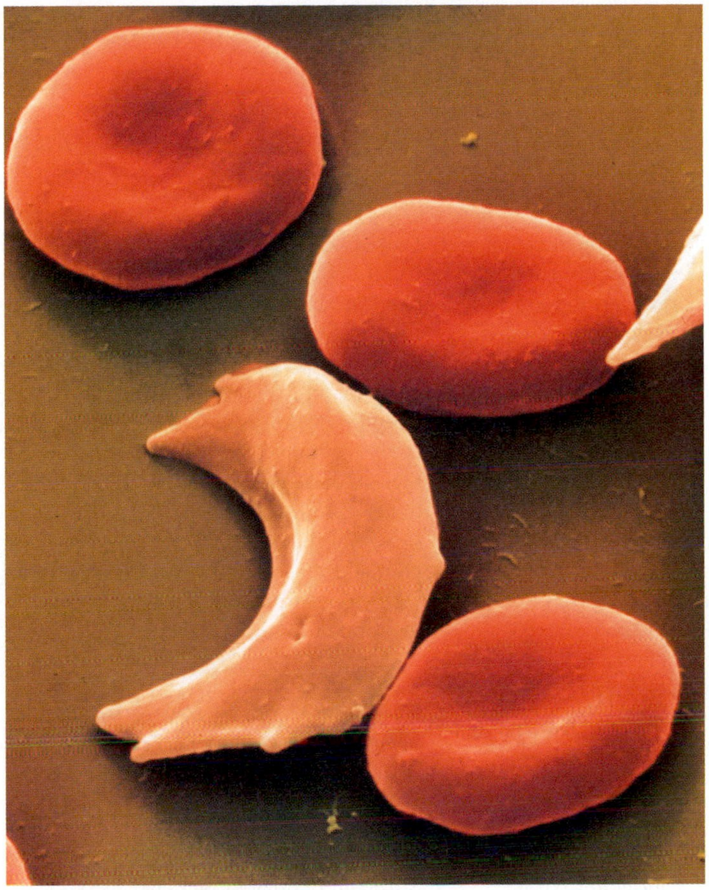

FIGURE 19.13
Sickle-cell Disease. The lower left RBC has become deformed into the pointed sickle shape diagnostic of this genetic disorder.

(as in people who live at high altitudes or suffer lung diseases such as emphysema). RBC counts can rise as high as 11 million RBCs/μL and hematocrit as high as 80%. The thick blood "sludges" in the vessels, tremendously increases blood pressure, and puts a dangerous strain on the cardiovascular system that can lead to heart failure or stroke.

LEUKOCYTE DISORDERS

A WBC deficiency is called **leukopenia,** and can result from heavy metal poisoning, radiation exposure, and infectious diseases such as measles, chicken pox, polio, and AIDS. Such a deficiency of disease-fighting WBCs leaves a person susceptible to *opportunistic infections,* infections that a normal person could fight off. An abnormally high WBC count is called **leukocytosis.** It usually results from an infection or allergy, but can also stem from such causes as emotional stress and dehydration (WBCs become more concentrated when water is lost from the bloodstream). **Leukemia** is a cancer of the hemopoietic tissues that results in a high number of circulating WBCs. It, too, makes a person vulnerable to opportunistic infection, because even though the WBC count is high, these are immature WBCs incapable of performing their normal defensive roles. Leukemia tends to lead to anemia and thrombocytopenia (see the next section) because stem cells are diverted into the rapid production of WBCs instead of RBC and platelet production.

PLATELET DISORDERS

Thrombocytopenia, a platelet count less than 100,000/μL, results from such causes as leukemia, radiation, or bone marrow poisoning. It results not only in impaired clotting when a vessel is injured, but also in increased bleeding because of a loss of the normal blood vessel maintenance function of platelets.

Some other common blood disorders are briefly described in table 19.5.

Before You Go On

Answer the following questions to test your understanding of the preceding section:

19. What are the terms for an excess and deficiency of RBCs? An excess and deficiency of WBCs?
20. What are the three basic categories of the causes of anemia?
21. In what way are leukocytosis and leukemia alike? What is the difference between them?
22. Describe some causes and effects of thrombocytopenia.

CHAPTER REVIEW

Review of Key Concepts

Introduction (p. 542)

1. The *circulatory system* is composed of the heart, blood vessels, and blood; *cardiovascular system* refers to the heart and blood vessels only.
2. The circulatory system serves to transport O_2, CO_2, nutrients, wastes, hormones, and heat; it provides protection against infection and other pathogens; it includes mechanisms for minimizing blood loss from broken vessels; and it helps to stabilize the distribution and pH of the body fluids.
3. The blood is about 55% plasma and 45% formed elements.
4. Formed elements are cells and cell fragments; they include erythrocytes, platelets, and five kinds of leukocytes.
5. Blood plasma is a mixture of water, proteins, nutrients, electrolytes, nitrogenous wastes, hormones, and gases. Protein is the most abundant solute by weight. The three major plasma proteins are albumin, globulins, and fibrinogen.
6. The liver produces all the plasma proteins except γ globulins (antibodies), which are produced by *plasma cells.*

Erythrocytes (p. 545)

1. Erythrocytes (RBCs) serve to transport O_2 and CO_2. They are discoid cells with a sunken center and no organelles, but they do have a cytoskeleton of *spectrin* and *actin* that reinforces the plasma membrane.
2. The most important component of the cytoplasm is *hemoglobin (Hb).* Hb transports nearly all of the O_2 and some of the CO_2 in the blood.
3. RBC quantities are expressed as *RBC count* and *hematocrit (packed cell volume).* Collectively, the two sexes normally have RBC counts of 4.2 to 6.2 million RBCs/μL and hematocrits of 37% to 52%. The average values for women are somewhat lower than those for men.
4. Hemoglobin consists of four proteins *(globins)*—two α and two β chains—each with a nonprotein *heme* group.
5. Oxygen binds to the Fe^{2+} at the center of each heme. About 5% of the CO_2 in the blood binds to the globin component.
6. The hemoglobin concentration of the blood is normally 12 to 18 g/dL, averaging less in women than in men.
7. The production of the formed elements of blood is called *hemopoiesis.* The prenatal hemopoietic tissues and organs are the yolk sac, bone marrow, liver, spleen, and thymus; after birth, hemopoiesis is limited to the bone marrow and lymphatic organs and tissues.
8. *Erythropoiesis,* the production of RBCs, progresses through the following stages: *pluripotent stem cell* → *proerythroblast* → *erythroblast* → *reticulocyte* → *erythrocyte.* It is stimulated by the hormone *erythropoietin.*
9. An RBC lives for about 120 days, grows increasingly fragile, and then breaks apart, especially in the spleen.
10. There are multiple *blood types* based on genetically determined antigens on the RBC surface. The most familiar and clinically relevant are types in the ABO and Rh groups.

Leukocytes (p. 548)

1. Leukocytes (WBCs) number 5,000 to 10,000 cells/μL. Leukocytes have nuclei and a full complement of cytoplasmic organelles.
2. WBCs play various roles in defending the body from pathogens.
3. Three types of WBCs—neutrophils, eosinophils, and basophils—are classified as *granulocytes* because of prominent granules (lysosomes and other membrane-bounded organelles) in the cytoplasm. The other two types—lymphocytes and monocytes—are classified as *agranulocytes* because such granules are scarce or absent. The identifying structural characteristics of the WBC types are shown in table 19.4.
4. *Neutrophils* act primarily to combat bacteria, which they destroy by phagocytosis and digestion and by producing a *killing zone* of toxic oxidizing agents.
5. *Eosinophils* secrete antiparasitic chemicals and phagocytize and dispose of allergens, inflammatory chemicals, and antigen-antibody complexes.
6. *Basophils* secrete the vasodilator histamine and the anticoagulant heparin, thus promoting increased blood flow to inflamed tissues.
7. *Monocytes* transform into a variety of *macrophages,* which phagocytize foreign and host cells and act as antigen-presenting cells.
8. *Lymphocytes* are of several kinds (chiefly T and B lymphocytes) involved in nonspecific defense and specific immune reactions.
9. Leukocyte production is called *leukopoiesis.* It follows multiple pathways from pluripotent stem cells to the different types of mature WBCs.
10. Granulocytes, monocytes, and B lymphocytes mature in the red bone marrow. T lymphocytes are produced there but mature in the thymus. Both B and T lymphocytes then colonize various lymphatic organs and tissues throughout the body.
11. WBC life spans range from a few days for granulocytes to a few years for monocytes (after transformation to macrophages) and decades for some lymphocytes.

Platelets (p. 552)

1. Platelets are not cells, but small, mobile, phagocytic fragments of cytoplasm from bone marrow cells called *megakaryocytes.* They have a variety of organelles but no nuclei.
2. Platelets function primarily to plug broken vessels and stop bleeding. They secrete vasoconstrictors, which cause spasmodic constriction of damaged vessels; form *platelet plugs,* which can temporarily seal a small break in a vessel; secrete clotting factors that lead to formation of blood clots; help to produce an enzyme that dissolves old blood clots; secrete chemicals that recruit WBCs to sites of inflammation; and secrete growth factors that promote repair and maintenance of blood vessels.
3. Platelet production, called *thrombopoiesis,* is stimulated by the hormone *thrombopoietin.* The stages in platelet production are *pluripotent stem cell* → *megakaryoblast* → *megakaryocyte* → *platelets.* Platelets are membrane-bounded fragments of cytoplasm that break away from the megakaryocyte surface.

Clinical Perspectives (p. 555)

1. Healthy elderly people have blood values (hematocrit, hemoglobin, etc.) comparable to those of healthy young people. However, the hematologic system in old age is less responsive to stress, partly because of lower levels of intrinsic factor and erythropoietin. Thrombosis is also increasingly common in old age.
2. *Anemia* is a deficiency of erythrocytes or hemoglobin. The various causes of anemia

relate to depressed erythropoiesis, excessive RBC destruction, or hemorrhage. Anemia is often characterized by RBCs that are *microcytic* (small), *hypochromic* (pale), or *poikilocytic* (variable in shape).

3. *Polycythemia* is an excessive RBC count. It results from bone marrow cancer or is secondary to causes such as zealous exercise, lung disease, or high-altitude living. It thickens the blood, stresses the heart, and may cause stroke or heart failure.
4. *Leukopenia* is a WBC deficiency. It can be caused by poisoning, radiation, or certain infections, and results in a reduction in immune defense and therefore vulnerability to *opportunistic infections.*
5. *Leukocytosis* is an elevated WBC count. It can result from infections, allergies, dehydration, stress, or cancer. In the last case, it is called *leukemia.* Leukemia is a cancer characterized by overproduction of WBCs that are functionally immature. It also presents a high risk of opportunistic infection. It leads as well to anemia and thrombocytopenia as stem cells are diverted to WBC production.
6. *Thrombocytopenia* is a platelet deficiency. It results in spontaneous bleeding and slow blood clotting.

Testing Your Recall

1 Antibodies belong to a class of plasma proteins called
   a. albumins.
   b. γ globulins.
   c. α globulins.
   d. procoagulants.
   e. agglutinins.

2. Serum is blood plasma minus its
   a. sodium ions.
   b. calcium ions.
   c. fibrinogen.
   d. albumin.
   e. cells.

3. The most abundant formed elements seen in most stained blood films are
   a. erythrocytes.
   b. neutrophils.
   c. lymphocytes.
   d. platelets.
   e. monocytes.

4. Heparin and histamine are secreted by
   a. plasma cells.
   b. basophils.
   c. B lymphocytes.
   d. platelets.
   e. neutrophils.

5. _____ have a finely granular cytoplasm and a nucleus typically divided into three to five lobes.
   a. Basophils
   b. Eosinophils
   c. Lymphocytes
   d. Monocytes
   e. Neutrophils

6. Platelets have all of the following functions *except*
   a. coagulation.
   b. plugging broken blood vessels.
   c. stimulating vasoconstriction.
   d. transporting oxygen.
   e. recruiting neutrophils.

7. Which of these is a granulocyte?
   a. a monocyte
   b. a lymphocyte
   c. a macrophage
   d. an eosinophil
   e. an erythrocyte

8. Allergies stimulate an increased _____ count.
   a. erythrocyte
   b. platelet
   c. eosinophil
   d. monocyte
   e. neutrophil

9. Which of the following leads to pernicious anemia?
   a. hypoxemia
   b. iron deficiency
   c. malaria
   d. lack of intrinsic factor
   e. lack of erythropoietin

10. Oxygen binds to the _____ of a hemoglobin molecule.
   a. valine
   b. Fe^{2+}
   c. globin
   d. spectrin
   e. β-chain

11. Production of all the formed elements of blood is called _____.
12. The percentage of blood volume composed of RBCs is called the _____.
13. Microglia and dendritic cells are two kinds of _____.
14. An excessively low WBC count is called _____.
15. The hereditary lack of factor VIII causes a disease called _____.
16. The overall cessation of bleeding, involving several mechanisms, is called _____.
17. _____ results from a mutation that changes one amino acid in each β chain of the hemoglobin molecule.
18. An excessively high RBC count is called _____.
19. Intrinsic factor enables the small intestine to absorb _____.
20. The kidney hormone _____ stimulates RBC production.

Answers in the Appendix

True or False

Determine which five of the following statements are false, and briefly explain why.

1. By volume, the blood usually contains more plasma than blood cells.
2. An increase in the albumin concentration of the blood tends to affect blood pressure.
3. Anemia is caused by a low oxygen concentration in the blood.
4. The most important WBCs in combating a bacterial infection are basophils.
5. Platelets and erythrocytes lack nuclei.
6. Lymphocytes are the most abundant WBCs in the blood.
7. Platelet count is often depressed in people with leukemia.
8. All formed elements of the blood come ultimately from pluripotent stem cells.
9. Since RBCs have no nuclei, they do not live as long as the granulocytes do.
10. Leukemia is a severe deficiency of white blood cells.

Answers in the Appendix

Testing Your Comprehension

1. Considering the quantity of hemoglobin in an erythrocyte and the oxygen-binding properties of hemoglobin, calculate how many molecules of oxygen one erythrocyte could carry.
2. A patient is found to be seriously dehydrated and to have an elevated RBC count. Does the RBC count necessarily indicate a disorder of erythropoiesis? Why or why not?
3. Patients suffering from renal failure are typically placed on hemodialysis and erythropoietin (EPO) replacement therapy. Explain the reason for giving EPO and predict what the consequences would be of not including this in the treatment regimen.
4. A leukemia patient exhibits minute hemorrhagic spots *(petechiae)* in her skin. Explain why leukemia could produce this effect.
5. Do you think platelets can synthesize proteins? Why or why not?

Answers at the Online Learning Center

www.mhhe.com/saladinha1

Visit the Online Learning Center for practice tests, answer keys, and other learning aids for this chapter. Enhance your understanding of human anatomy with our interactive art labeling exercises, supplemental photo atlases, web links, puzzles, flashcards, and much more.

20

CHAPTER TWENTY

The Circulatory System II—The Heart

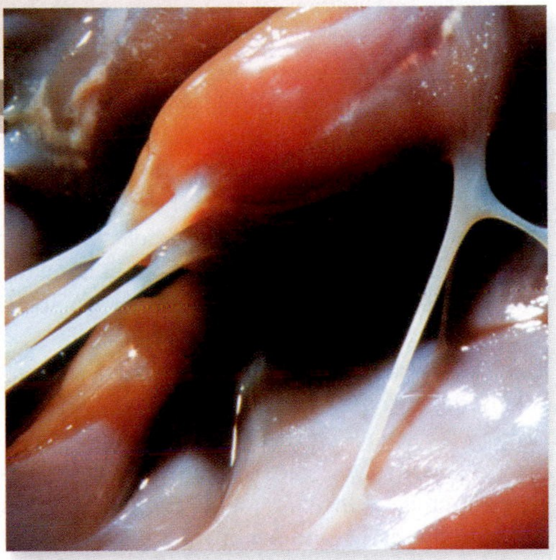

Interior of the human heart, showing a papillary muscle (top center), tendinous cords (white cords on the *left,*) and Purkinje fibers (branched fibers on the *right*)

CHAPTER OUTLINE

Overview of the Cardiovascular System 562
- The Pulmonary and Systemic Circuits 562
- Position, Size, and Shape of the Heart 563
- The Pericardium 563

Gross Anatomy of the Heart 565
- The Heart Wall 565
- The Chambers 566
- The Valves 567
- Blood Flow Through the Chambers 568

Blood Supply to the Cardiac Muscle 569
- Arterial Supply 569
- Venous Drainage 571

The Cardiac Conduction System and Cardiac Muscle 573
- The Conduction System 574
- Structure of Cardiac Muscle 574
- Nerve Supply to the Heart 576

Developmental and Clinical Perspectives 576
- Prenatal Development of the Heart 577
- Changes at Birth 578
- The Aging Heart 578
- Heart Disease 580

Chapter Review 581

INSIGHTS

20.1 Clinical Application: Coronary Artery Disease 572
20.2 Clinical Application: Heart Block 574
20.3 Clinical Application: Patent Ductus Arteriosus 579

BRUSHING UP

To understand this chapter, you may find it helpful to review the following concepts:
- Thoracic cavity anatomy (p. 29)
- Desmosomes and gap junctions (p. 62)
- Endothelium (p. 98)
- Ultrastructure of striated muscle (pp. 263, 273)
- Comparisons of cardiac and skeletal muscle (p. 273)

This chapter is on **cardiology,**[1] a science that embraces the study of the heart, the clinical evaluation of cardiac function and disorders, and treatment of cardiac diseases. We are more conscious of our heart than we are of most organs, and more wary of its failure. Speculation about the heart is at least as old as written history. Some ancient Chinese, Egyptian, Greek, and Roman scholars correctly surmised that the heart is a pump for filling the vessels with blood. Aristotle's views, however, were a step backward. Perhaps because the heart quickens its pace when we are emotionally aroused, and because grief causes "heartache," he regarded it primarily as the seat of emotion, as well as a source of heat to aid digestion. During the Middle Ages, Western medical schools clung dogmatically to the ideas of Aristotle. Perhaps the only significant advance came from Muslim medicine, when thirteenth-century physician Ibn an-Nafis described the role of the coronary blood vessels in nourishing the heart. The sixteenth-century dissections and anatomical charts of Vesalius, however, greatly improved knowledge of cardiovascular anatomy and set the stage for a more scientific study of the heart and treatment of its disorders.

In the early decades of the twentieth century, little could be recommended for heart disease other than bed rest. Then nitroglycerin was found to improve coronary circulation and relieve the pain resulting from physical exertion, digitalis proved effective for treating abnormal heart rhythms, and diuretics were first used to reduce hypertension. In the last few decades, such advances as coronary bypass surgery, replacement of diseased valves, clot-dissolving enzymes, heart transplants, artificial pacemakers, and artificial hearts have made cardiology one of the most dramatic and attention-getting fields of medicine.

Overview of the Cardiovascular System

Objectives

When you have completed this section, you should be able to

- define and distinguish between the pulmonary and systemic circuits;
- describe the general location, size, and shape of the heart; and
- describe the pericardial sac that encloses the heart.

The Pulmonary and Systemic Circuits

The **cardiovascular system** consists of the heart and the blood vessels that carry the blood to and from the body's organs. The system has two major divisions: a **pulmonary circuit,** which carries blood to the lungs for gas exchange and returns it to the heart, and a **systemic circuit,** which supplies blood to every organ of the body (fig. 20.1), including other parts of the lungs and the wall of the heart itself. The right side of the heart serves the pulmonary circuit. It receives blood that has circulated through the body, unloaded its oxygen and nutrients, and picked up a load of carbon dioxide and other wastes. It pumps this oxygen-poor blood into a large artery, the *pulmonary trunk,* which immediately divides into *right* and *left pulmonary arteries.* These transport blood to the air sacs *(alveoli)* of the lungs, where carbon dioxide is unloaded and oxygen is picked up. The oxygen-rich blood then flows by way of the *pulmonary veins* to the left side of the heart.

The left side serves the systemic circuit. Blood leaves it by way of another large artery, the *aorta.* The aorta takes a sharp inverted U-turn, the *aortic arch,* and passes downward, dorsal to the heart.

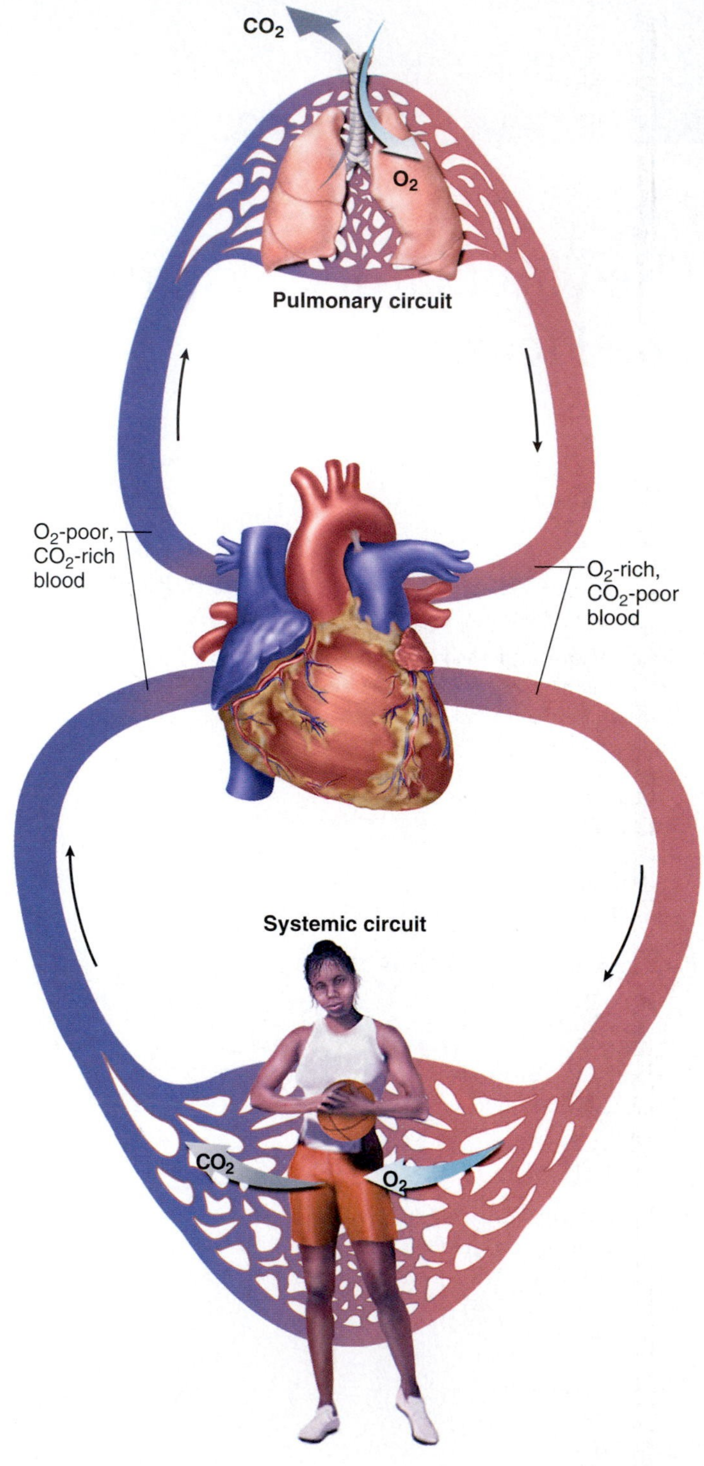

FIGURE 20.1
General Schematic of the Cardiovascular System.

[1]*cardio* = heart + *logy* = study

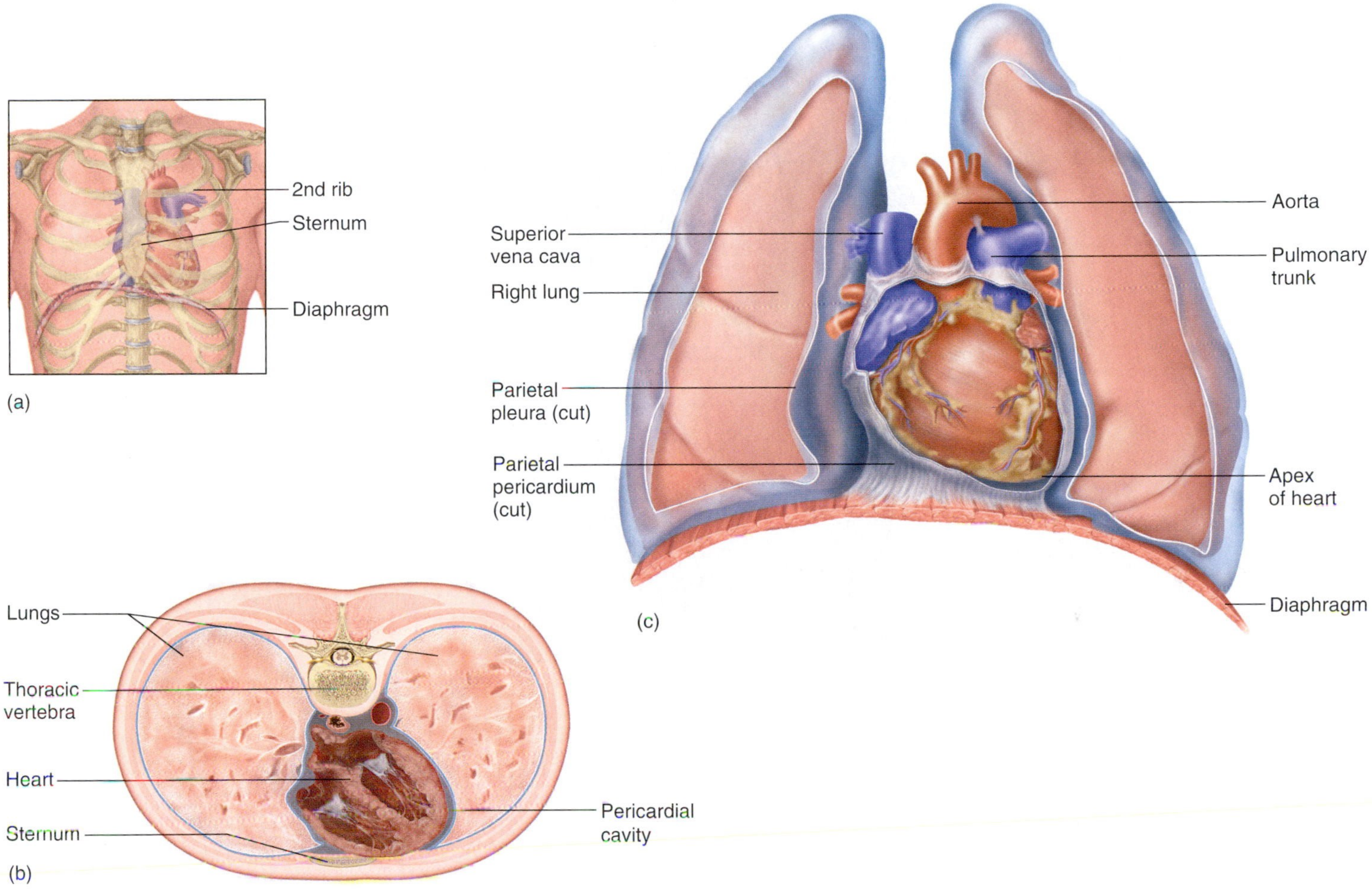
2nd rib
Sternum
Diaphragm
(a)
Superior vena cava
Right lung
Parietal pleura (cut)
Parietal pericardium (cut)
Aorta
Pulmonary trunk
Apex of heart
Diaphragm
(c)
Lungs
Thoracic vertebra
Heart
Sternum
Pericardial cavity
(b)

FIGURE 20.2

Position of the Heart in the Thoracic Cavity. (*a*) Relationship to the thoracic cage. (*b*) Cross section of the thorax at the level of the heart. (*c*) Frontal section of the thoracic cavity with the lungs slightly retracted and the pericardial sac opened.

The aortic arch gives off arteries that supply the head, neck, and upper limbs. The aorta then travels through the thoracic and abdominal cavities and issues smaller arteries to the other organs. After circulating through the body, the now-deoxygenated systemic blood returns to the right side of the heart mainly by way of two large veins, the *superior vena cava* (draining the head, neck, upper limbs, and thoracic organs) and *inferior vena cava* (draining the organs below the diaphragm). The major arteries and veins entering and leaving the heart are called the *great vessels* (*great arteries* and *veins*) because of their relatively large diameters.

Position, Size, and Shape of the Heart

The heart is located in the thoracic cavity in the mediastinum, between the lungs and deep to the sternum. From its superior to inferior midpoints, it is tilted toward the left, so about two-thirds of the heart lies to the left of the median plane (figs A.18–A.19 and 20.2). The broad superior portion of the heart, called the **base,** is the point of attachment for the great vessels described previously. The inferior end tapers to a blunt point, the **apex** of the heart, immediately above the diaphragm (fig. 20.3).

The adult heart is about 9 cm (3.5 in.) wide at the base, 13 cm (5 in.) from base to apex, and 6 cm (2.5 in.) from anterior to posterior at its thickest point—roughly the size of one's fist. It weighs about 300 g (10 oz).

The Pericardium

The heart is enclosed in a double-walled sac called the **pericardium.** The outer wall, called the **parietal pericardium (pericardial sac),** has a tough, superficial *fibrous layer* of dense irregular connective tissue and a deep, thin *serous layer.* The serous layer turns inward at the base of the heart and forms the **visceral pericardium (epicardium)** covering the heart surface (fig. 20.4). The pericardial sac is anchored by ligaments to the diaphragm below and the sternum anterior to it, and more loosely anchored by fibrous connective tissue to mediastinal tissue dorsal to the heart.

Between the parietal and visceral membranes is a space called the **pericardial cavity** (see figs. 20.2*b* and 20.4). It contains 5 to 30 mL of **pericardial fluid,** exuded by the serous pericardium. The pericardial fluid lubricates the membranes and allows the heart to beat almost without friction. In *pericarditis*—inflammation of the

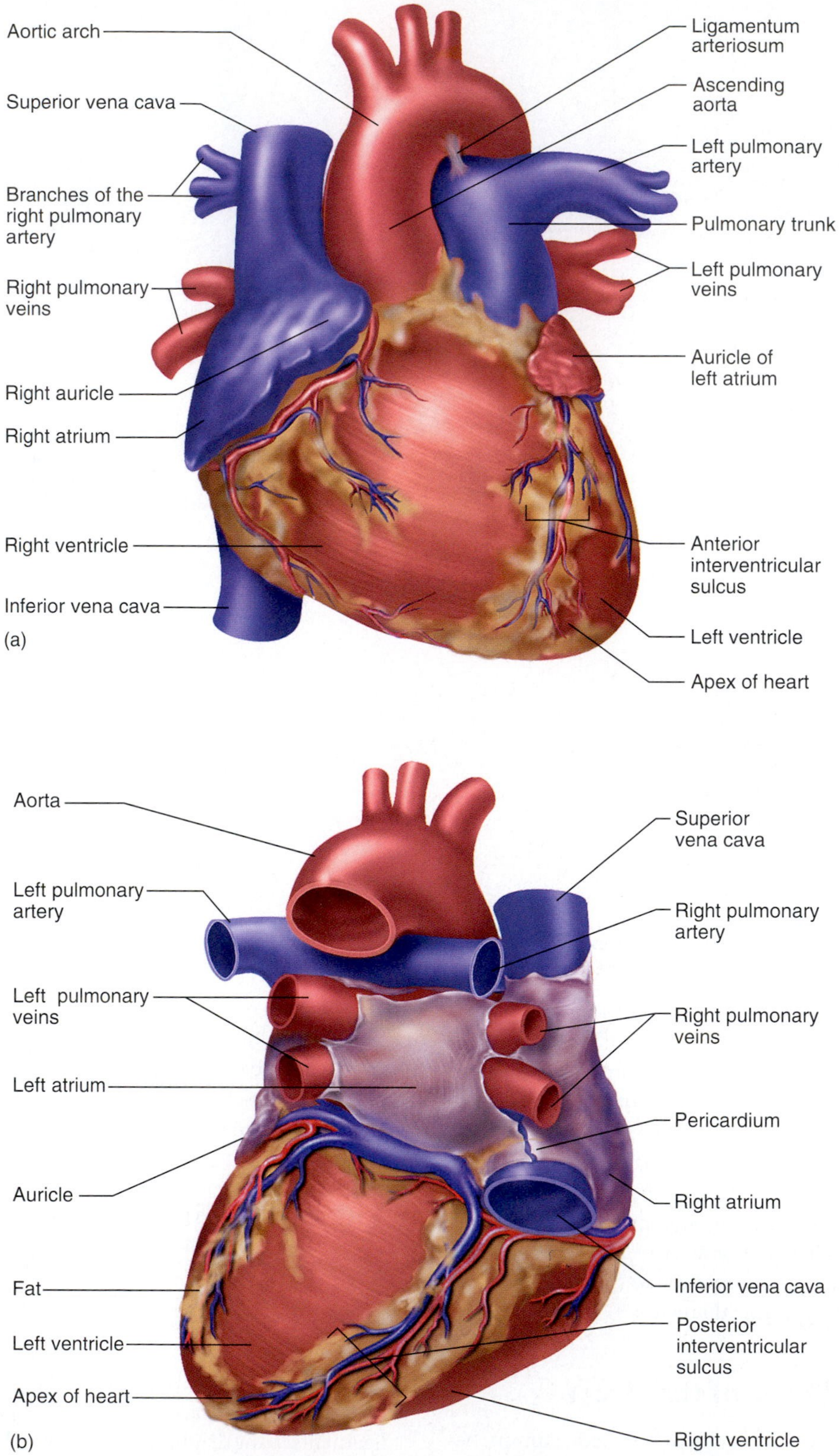
Aortic arch
Superior vena cava
Branches of the right pulmonary artery
Right pulmonary veins
Right auricle
Right atrium
Right ventricle
Inferior vena cava
(a)
Ligamentum arteriosum
Ascending aorta
Left pulmonary artery
Pulmonary trunk
Left pulmonary veins
Auricle of left atrium
Anterior interventricular sulcus
Left ventricle
Apex of heart
Aorta
Left pulmonary artery
Left pulmonary veins
Left atrium
Auricle
Fat
Left ventricle
Apex of heart
(b)
Superior vena cava
Right pulmonary artery
Right pulmonary veins
Pericardium
Right atrium
Inferior vena cava
Posterior interventricular sulcus
Right ventricle

FIGURE 20.3

External Anatomy of the Heart. (*a*) Anterior aspect. (*b*) Posterior aspect. Figures shown about 60% life size. The coronary blood vessels on the heart surface are identified in figure 20.11.

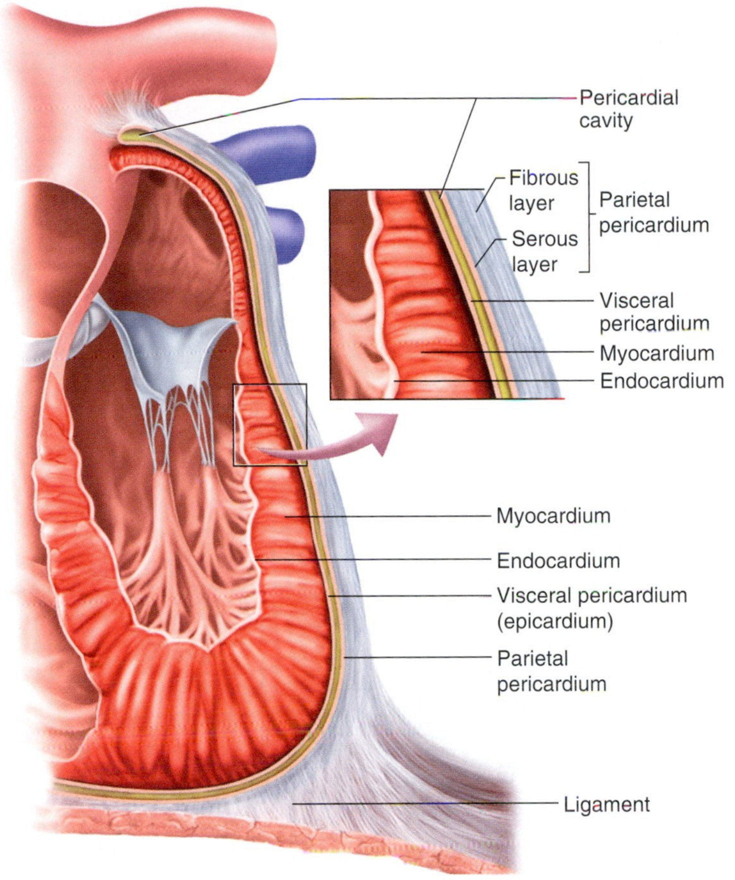

FIGURE 20.4

The Pericardium and Heart Wall. The inset shows the layers of the heart wall in relationship to the pericardium.

pericardium—the membranes may become dry and produce a painful *friction rub* with each heartbeat. In addition to reducing friction, the pericardium isolates the heart from other thoracic organs, allows the heart room to expand, yet resists excessive expansion. (See *cardiac tamponade* in table 20.1, p. 579).

Before You Go On

Answer the following questions to test your understanding of the preceding section:

1. Distinguish between the pulmonary and systemic circuits and state which part of the heart serves each one.
2. Make a two-color sketch of the pericardium; use one color for the fibrous pericardium and another for the serous pericardium and show their relationship to the heart wall and pericardial cavity.

Gross Anatomy of the Heart

Objectives

When you have completed this section, you should be able to

- describe the three layers of the heart wall;
- identify the four chambers of the heart;
- identify the surface features of the heart and correlate them with its internal four-chambered anatomy;
- identify the four valves of the heart; and
- trace the flow of blood through the four chambers of the heart and adjacent blood vessels.

The Heart Wall

The heart wall consists of three layers—a thin *epicardium* covering its external surface, a thick muscular *myocardium* in the middle, and a thin *endocardium* lining the interior of the chambers (fig. 20.4).

The **epicardium**[2] (visceral pericardium) is a serous membrane on the heart surface. It consists mainly of a simple squamous epithelium overlying a thin layer of areolar tissue. In some places, it also includes a thick layer of adipose tissue, whereas in other areas it is fat-free and translucent, so the muscle of the underlying myocardium shows through (fig. 20.5). The largest branches of the coronary blood vessels travel through the epicardium. A similar layer, the **endocardium,**[3] lines the interior of the heart chambers. It is a simple squamous endothelium overlying a thin areolar tissue layer; it has no adipose tissue. The endocardium covers the valve surfaces and is continuous with the endothelium of the blood vessels.

It is the layer between these two, the **myocardium,**[4] that is thickest by far and performs the work of the heart. Its thickness varies greatly from one heart chamber to another and is proportional to the workload on the individual chambers. Its muscle spirals around the heart (fig. 20.6), so when the ventricles contract, they exhibit a twisting or wringing motion. The microscopic structure of the cardiac muscle cells (*cardiac myocytes* or *cardiocytes*) is detailed later.

The heart also has a meshwork of collagenous and elastic fibers that make up the **fibrous skeleton.** This tissue is especially concentrated in the walls (septa) between the heart chambers, in *fibrous rings (annuli fibrosi)* around the openings of the heart valves, and in sheets of tissue that interconnect these rings. The fibrous skeleton has multiple functions: (1) It provides structural support for the heart, especially around the valves and the openings of the great vessels; it holds the valve orifices open and prevents them from being excessively stretched when blood surges through them. (2) It anchors the myocytes and gives them something to pull against. (3) As a nonconductor of electricity, it serves as electrical insulation between the atria and the ventricles, so the atria cannot stimulate the ventricles directly. This insulation is important to the timing and coordination of electrical and contractile activity. (4) Some authorities think (while others disagree) that elastic recoil of the fibrous skeleton may aid in refilling the heart with blood after each beat, like a hollow rubber ball that expands when you relax your grip.

Think About It!

Parts of the fibrous skeleton sometimes become calcified in old age. How would you expect this to affect cardiac function?

[2]*epi* = upon + *cardi* = heart
[3]*endo* = internal + *cardi* = heart
[4]*myo* = muscle + *cardi* = heart

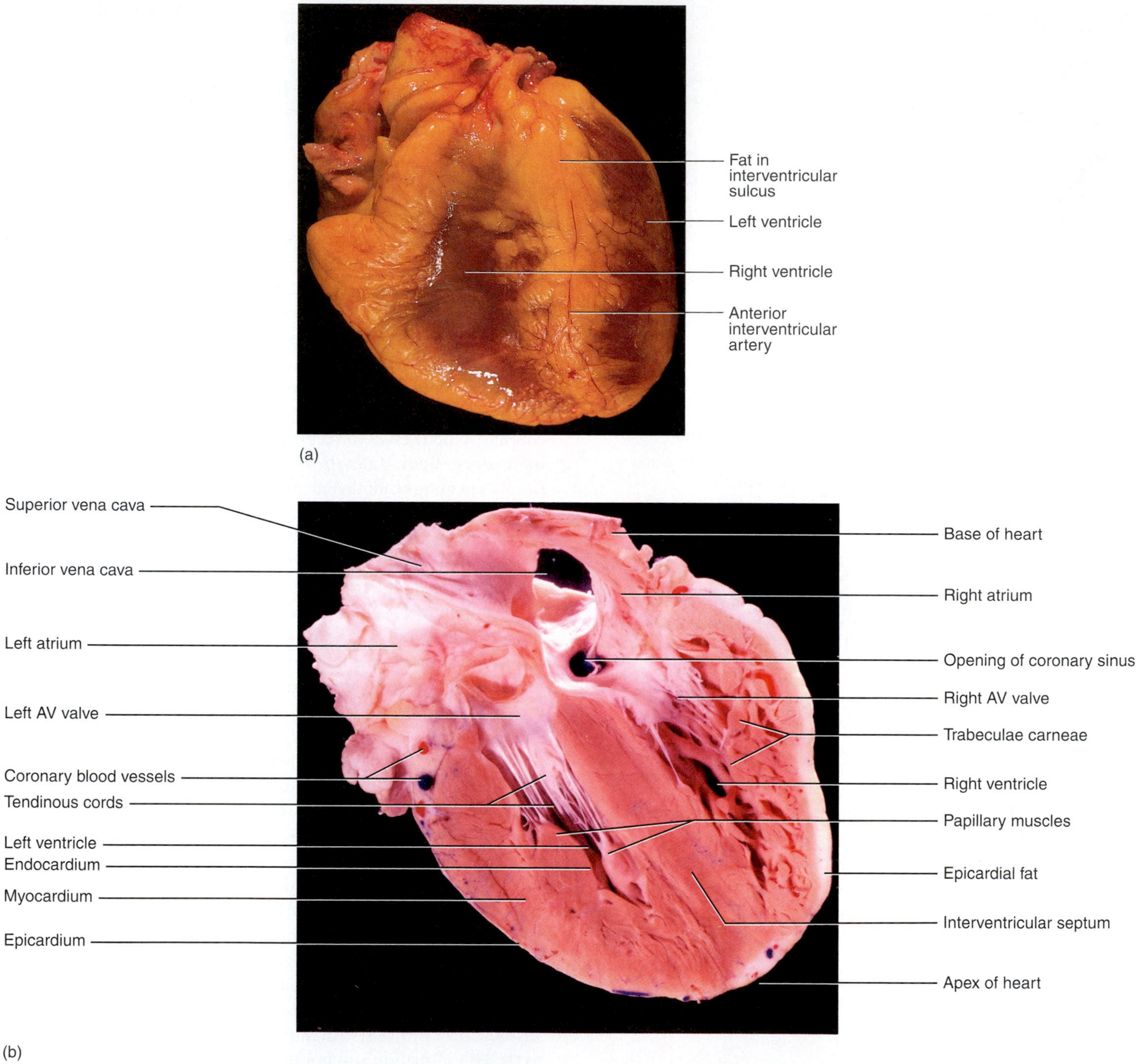
Fat in interventricular sulcus
Left ventricle
Right ventricle
Anterior interventricular artery
(a)
Superior vena cava
Inferior vena cava
Left atrium
Left AV valve
Coronary blood vessels
Tendinous cords
Left ventricle
Endocardium
Myocardium
Epicardium
Base of heart
Right atrium
Opening of coronary sinus
Right AV valve
Trabeculae carneae
Right ventricle
Papillary muscles
Epicardial fat
Interventricular septum
Apex of heart
(b)

Figure 20.5

The Human Heart. (*a*) Anterior aspect. (*b*) Internal anatomy. Figure *b* is photographed from a dorsal perspective, so note that the left ventricle is page-right in figure *a* but page-left in figure *b.* The left ventricle can be identified from its relatively thick myocardium.

The Chambers

The heart has four chambers, best seen in a frontal section (fig. 20.7). The two at the superior pole (base) of the heart are the **right** and **left atria** (AY-tree-uh; singular *atrium*[5]). They are thin-walled receiving chambers for blood returning to the heart by way of the great veins. Most of the mass of each atrium is on the posterior side of the heart, so only a small portion is visible from the anterior aspect. Here, each atrium has a small earlike extension called an *auricle*[6] that slightly increases its volume.

The two inferior heart chambers, the **right** and **left ventricles,**[7] are the pumps that eject blood into the arteries and keep it flowing around the body. The right ventricle constitutes most of the anterior aspect of the heart, while the left ventricle forms the apex and inferoposterior aspect.

[5]*atrium* = entryway

[6]*auricle* = little ear

[7]*ventr* = belly, lower part + *icle* = little

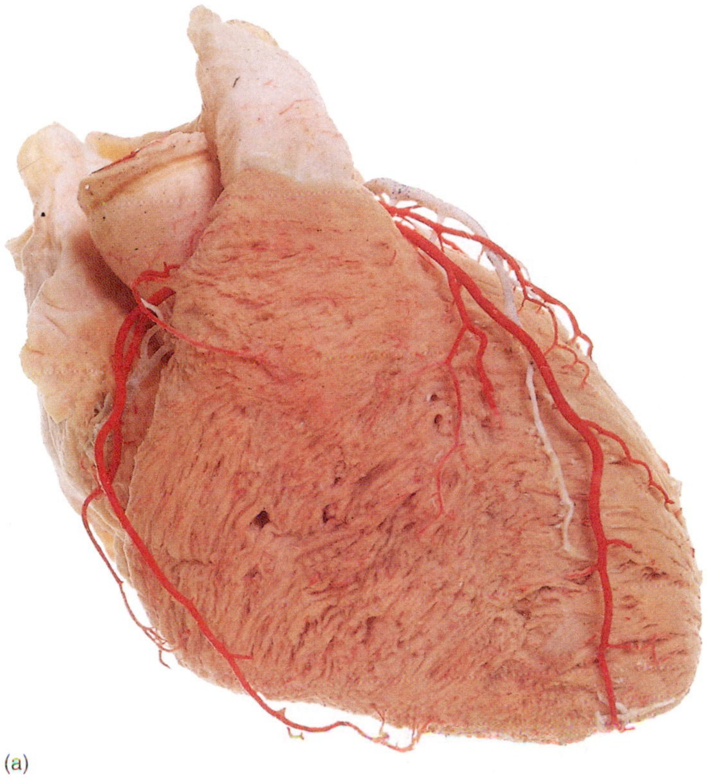

(a)

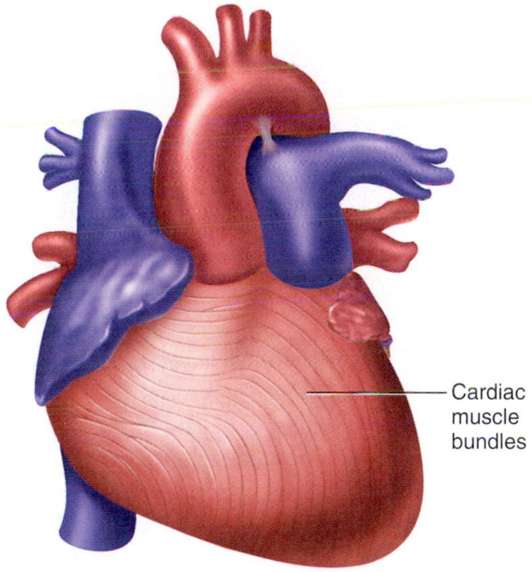
Cardiac
muscle
bundles

(b)

FIGURE 20.6

Twisted Orientation of Myocardial Muscle. (*a*) A heart with the epicardium stripped off to expose the spiraling bundles of myocardial muscle. (*b*) Orientation of muscle in the ventricles.

On the surface, the boundaries of the four chambers are marked by three sulci (grooves). The sulci are occupied largely by fat and coronary blood vessels (see fig. 20.5*a*). The **coronary**[8] **(atrioventricular) sulcus** encircles the heart near the base and separates the atria above from the ventricles below. It can be exposed by lifting the margins of the atria. The other two sulci extend obliquely down the heart from the coronary sulcus toward the apex, one on the front of the heart called the **anterior interventricular sulcus** and one on the back called the **posterior interventricular sulcus.** These sulci overlie an internal *interventricular septum* that divides the right ventricle from the left. The coronary sulcus and two interventricular sulci harbor the largest of the coronary blood vessels.

The atria exhibit thin flaccid walls corresponding to their light workload—all they do is pump blood into the ventricles immediately below. They are separated from each other by a wall called the **interatrial septum.** The right atrium and both auricles exhibit internal ridges of myocardium called **pectinate**[9] **muscles.** The **interventricular septum** is a much more muscular, vertical wall between the ventricles. The right ventricle pumps blood only to the lungs and back, so its wall is only moderately muscular. The wall of the left ventricle is two to four times as thick because it bears the greatest workload of all four chambers, pumping blood through the entire body. Both ventricles exhibit internal ridges called **trabeculae carneae**[10] (trah-BEC-you-lee CAR-nee-ee).

The Valves

To pump blood effectively, the heart needs valves that ensure a predominantly one-way flow. There is a valve between each atrium and its ventricle and another at the exit from each ventricle into its great artery (fig. 20.7), but there are no valves where the great veins empty into the atria. Each valve consists of two or three fibrous flaps of tissue called **cusps** or **leaflets,** covered with endothelium.

The **atrioventricular (AV) valves** regulate the openings between the atria and ventricles. The **right AV (tricuspid) valve** has three cusps and the **left AV (bicuspid) valve** has two (fig. 20.8). The left AV valve is also known as the **mitral** (MY-trul) **valve** after its resemblance to a miter, the headdress of a church bishop. Stringlike **tendinous cords** *(chordae tendineae),* reminiscent of the shroud lines of a parachute, connect the valve cusps to conical **papillary muscles** on the floor of the ventricle.

The **semilunar**[11] **valves** (pulmonary and aortic valves) regulate the flow of blood from the ventricles into the great arteries. The **pulmonary valve** controls the opening from the right ventricle into the pulmonary trunk, and the **aortic valve** controls the opening from the left ventricle into the aorta. Each has three cusps shaped somewhat like shirt pockets. There are no tendinous cords on the semilunar valves.

The opening and closing of heart valves is the result of pressure gradients between the upstream and downstream sides of the valve (fig. 20.9). When the ventricles are relaxed and their pressure is low, the AV valve cusps hang down limply and both AV valves are open. Blood flows freely from the atria into the ventricles even before the atria contract. When the ventricles have filled with blood and begin to contract, their internal pressure rises and blood surges against the AV valves. This pushes their cusps together, seals the openings, and prevents blood from flowing back into the atria. The papillary muscles contract with the rest of the ventricular myocardium and tug on

[8]*coron* = crown + *ary* = pertaining to

[9]*pectin* = comb + *ate* = like

[10]*trabecula* = little beam + *carne* = flesh, meat

[11]*semi* = half + *lunar* = like the moon

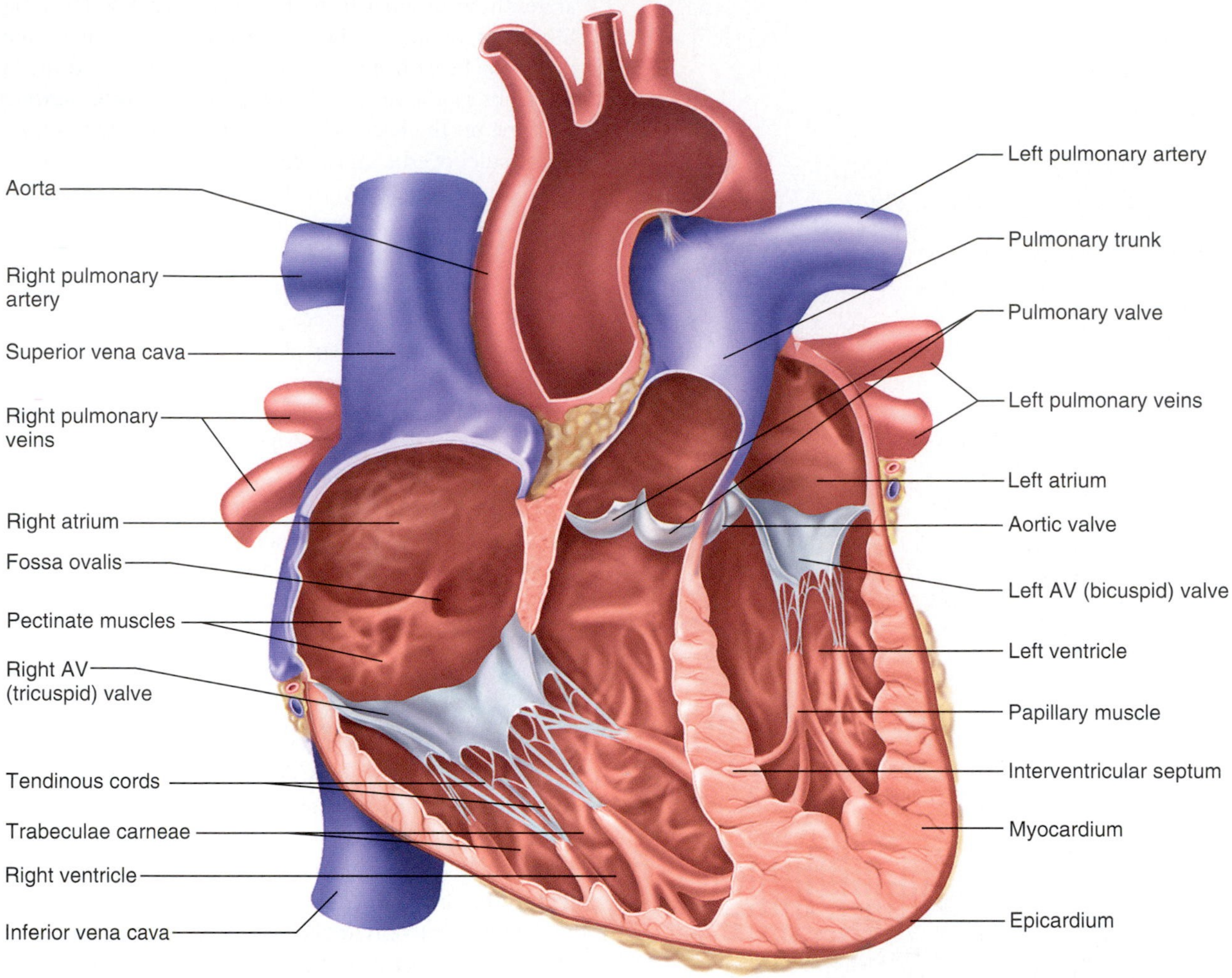
Aorta
Right pulmonary artery
Superior vena cava
Right pulmonary veins
Right atrium
Fossa ovalis
Pectinate muscles
Right AV (tricuspid) valve
Tendinous cords
Trabeculae carneae
Right ventricle
Inferior vena cava
Left pulmonary artery
Pulmonary trunk
Pulmonary valve
Left pulmonary veins
Left atrium
Aortic valve
Left AV (bicuspid) valve
Left ventricle
Papillary muscle
Interventricular septum
Myocardium
Epicardium

FIGURE 20.7
Internal Anatomy of the Heart (anterior view).

the tendinous cords, which prevents the valves from bulging excessively (prolapsing) into the atria or turning inside out like windblown umbrellas (see *mitral valve prolapse* in table 20.1, p. 579).

As the pressure rises in the contracting ventricles, it soon exceeds the pressure downstream from it in the pulmonary trunk and aorta. The ventricular blood then forces the pulmonary and aortic valves open and blood is ejected from the heart. Then as the ventricles relax again and their pressure falls below that in the arteries, arterial blood briefly flows backward and fills the pocketlike cusps of the semilunar valves. The three cusps meet in the middle of the orifice and seal it, thereby preventing blood from reentering the heart.

Blood Flow Through the Chambers

Until the sixteenth century, anatomists thought that blood flowed directly from the right ventricle to the left through invisible pores in the septum. This of course is not true. Blood in the right and left chambers of the heart is kept entirely separate. Figure 20.10 shows the pathway of the blood as it travels from the right atrium through the body and back to the starting point. The figure is numbered to correspond to the following description.

Blood returns to the heart through two large veins, the *superior vena cava* draining the head, neck, upper limbs, and thoracic cavity, and the *inferior vena cava* draining the abdominal cavity and lower limbs. Both of these veins enter the right atrium (1), approaching from above and below the heart, respectively. Blood in the right atrium flows through the right AV valve (2) into the right ventricle (3).

When the right ventricle contracts, the AV valve closes and blood is forced through the pulmonary valve (4) into the *pulmonary trunk* (5). This artery ascends from the front of the heart and branches into the right and left *pulmonary arteries* (6), which lead to the respective lungs. In the lungs, this blood unloads its carbon dioxide and picks up a load of oxygen.

The oxygen-enriched blood returns by way of several veins which converge to form four *pulmonary veins* (7) by the time they reach the heart. These four empty into the left atrium (8). Blood flows from there past the left AV valve (9) into the left ventricle (10). The left ventricle contracts at the same time as the right, and expels this blood through the aortic valve (11) into the *ascending aorta* (12). Blood in the aorta flows to every organ in the body (13), unloading some of its O_2, picking up CO_2 from the tissues, and returning to the heart via the venae cavae (14).

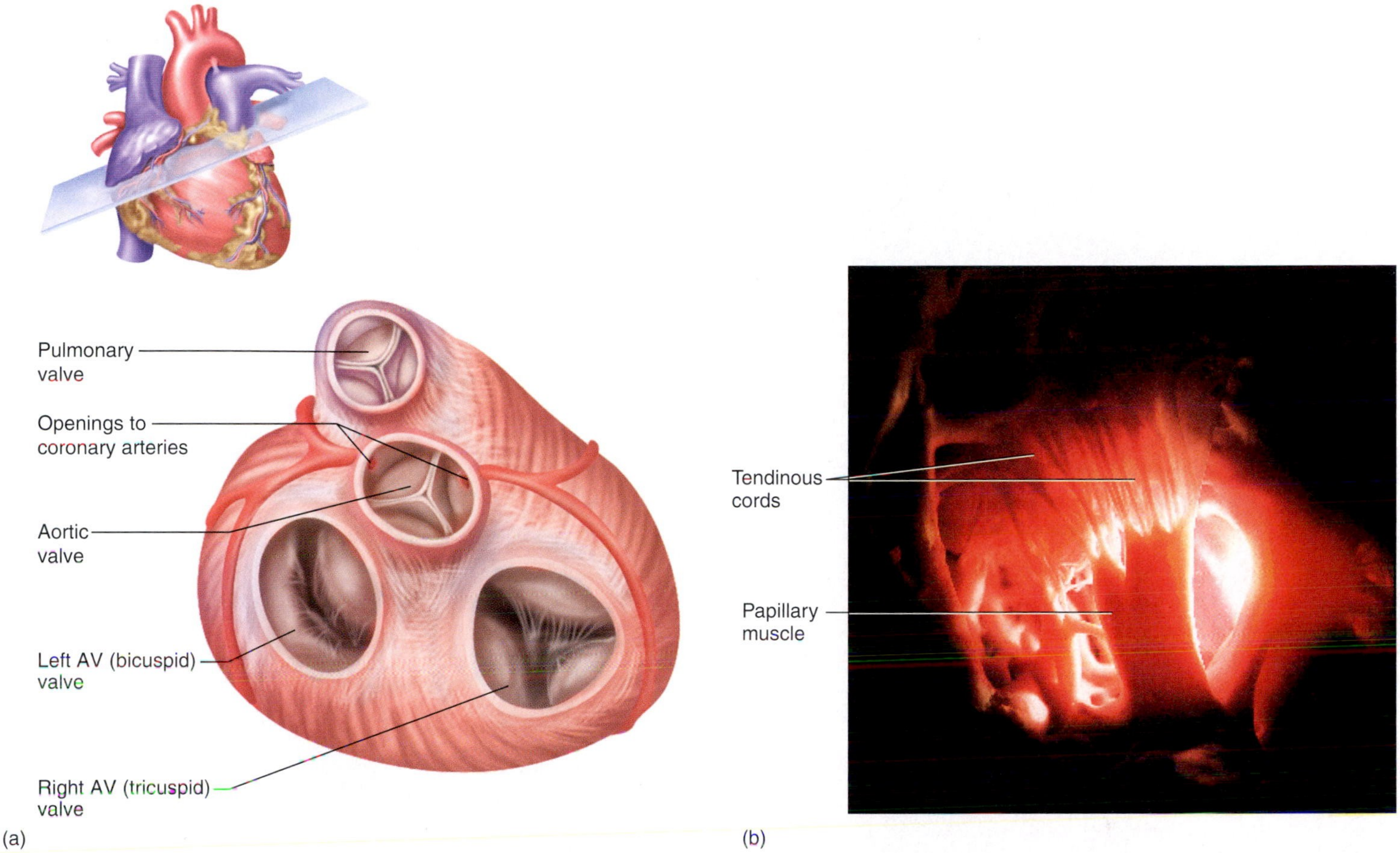
Pulmonary valve
Openings to coronary arteries
Aortic valve
Left AV (bicuspid) valve
Right AV (tricuspid) valve
(a)
Tendinous cords
Papillary muscle
(b)

FIGURE 20.8
The Heart Valves. (*a*) Superior view of the heart with the atria removed. (*b*) Papillary muscle and tedinous cords seen from within the right ventricle.

Before You Go On

Answer the following questions to test your understanding of the preceding section:

3. Name the three layers of the heart and describe their structural differences.
4. What are the functions of the fibrous skeleton?
5. Trace the flow of blood through the heart, naming each chamber and valve in order.

Blood Supply to the Cardiac Muscle

Objectives

When you have completed this section, you should be able to

- describe the arteries that nourish the myocardium and the veins that drain it; and
- define *myocardial infarction* and relate it to the coronary arteries.

If your heart lasts for 80 years and beats an average of 75 times a minute, it will beat more than 3 billion times and pump more than 200 million liters of blood. It is, in short, a remarkably hard-working organ, and understandably, it needs an abundant supply of oxygen and nutrients. These needs are not met to any appreciable extent by the blood in the heart chambers, because the diffusion of nutrients from there to the myocardium would be too slow. Instead, the myocardium has its own supply of arteries and capillaries that deliver blood to every muscle cell. The blood vessels of the heart wall constitute the **coronary circulation.**

At rest, the coronary blood vessels supply the myocardium with about 250 mL of blood per minute. Approximately 5% of the circulating blood goes to meet the metabolic needs of the heart, even though the heart is only 0.5% of the body's weight. It receives ten times its "fair share" to sustain its strenuous workload.

Arterial Supply

The coronary circulation is the most variable aspect of cardiac anatomy. The following description covers only the largest coronary blood vessels, and describes only the pattern seen in about 70% to 85% of persons.

Immediately after the aorta leaves the left ventricle, it gives off a right and left coronary artery. The orifices of these two arteries lie deep in the pockets formed by the aortic valve cusps (see

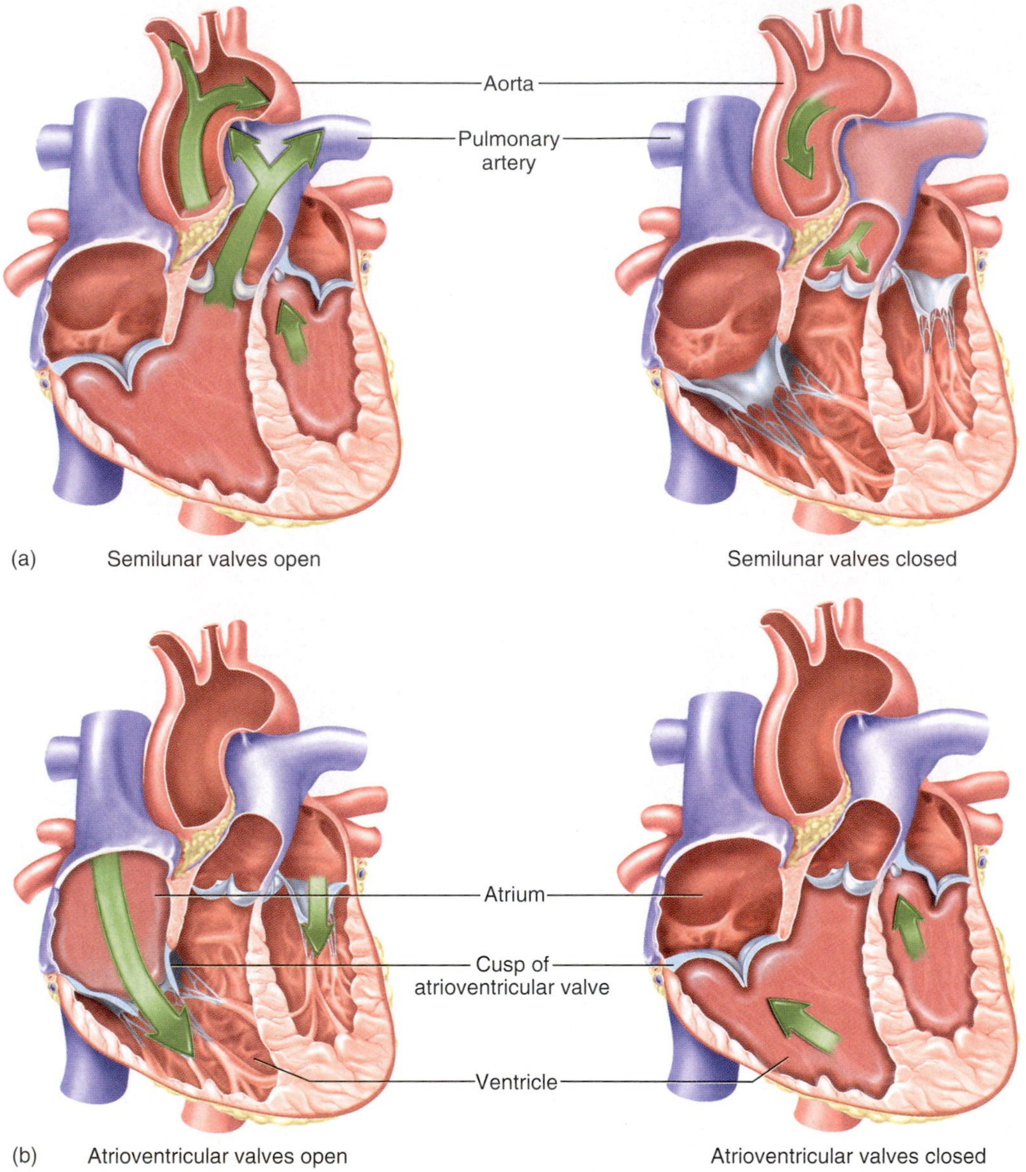

Aorta
Pulmonary artery
(a)
Semilunar valves open
Semilunar valves closed
Atrium
Cusp of atrioventricular valve
Ventricle
(b)
Atrioventricular valves open
Atrioventricular valves closed

FIGURE 20.9

Operation of the Heart Valves. (*a*) The semilunar valves. When the pressure in the ventricle is greater than the pressure in the artery, the valve is forced open and blood is ejected. When ventricular pressure is lower than arterial pressure, arterial blood holds the valve closed. (*b*) The atrioventricular valves. When atrial pressure is greater than ventricular pressure, the valve opens and blood flows through. When ventricular pressure rises above atrial pressure, the blood in the ventricle pushes the valve cusps closed.

fig. 20.8*a*). The **left coronary artery (LCA)** travels through the coronary sulcus under the left auricle and divides into two branches (fig. 20.11):

1. The **anterior interventricular branch** travels down the anterior interventricular sulcus to the apex, rounds the bend, and travels a short distance up the posterior side of the heart. There it anastomoses with (joins) the posterior interventricular branch described shortly. Clinically, it is also called the *left anterior descending (LAD) branch.* This artery supplies blood to both ventricles and the anterior two-thirds of the interventricular septum.
2. The **circumflex branch** continues around the left side of the heart in the coronary sulcus. It gives off a **left marginal branch** that passes down the left margin of the heart and furnishes blood to the left ventricle. The circumflex branch then ends on the posterior side of the heart. It supplies blood to the left atrium and posterior wall of the left ventricle.

The **right coronary artery (RCA)** supplies the right atrium and sinoatrial node (pacemaker), continues along the coronary sulcus under the right auricle, and gives off two branches of its own:

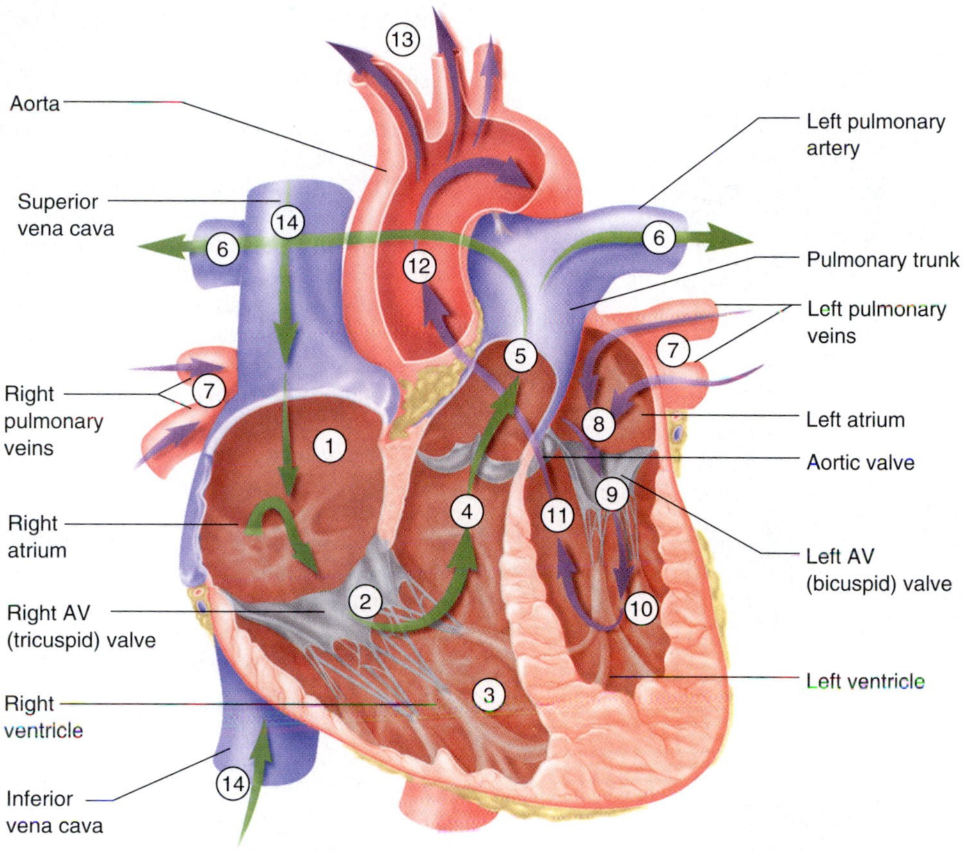
Aorta
Superior vena cava
Right pulmonary veins
Right atrium
Right AV (tricuspid) valve
Right ventricle
Inferior vena cava
Left pulmonary artery
Pulmonary trunk
Left pulmonary veins
Left atrium
Aortic valve
Left AV (bicuspid) valve
Left ventricle
1
2
3
4
5
6
7
8
9
10
11
12
13
14

FIGURE 20.10

The Pathway of Blood Flow Through the Heart. See text for explanation of numbers. The pathway from *5* to *7* is the pulmonary circuit, and the pathway from *12* to *14* is the systemic circuit.

1. The **right marginal branch** runs toward the apex of the heart and supplies the lateral aspect of the right atrium and ventricle.
2. The RCA continues around the right margin of the heart to the posterior side, sends a small branch to the atrioventricular node, then gives off a large **posterior interventricular branch.** This branch travels down the corresponding sulcus and supplies the posterior walls of both ventricles as well as the posterior portion of the interventricular septum. It ends by anastomosing with the circumflex and anterior interventricular branches of the LCA.

The energy demand of the cardiac muscle is so critical that an interruption of the blood supply to any part of the myocardium can cause necrosis within minutes. A fatty deposit or blood clot in a coronary artery can cause a **myocardial infarction**[12] **(MI),** the sudden death of a patch of tissue deprived of its blood flow (see insight 20.1). Some protection from MI is provided by the aforementioned *anastomoses* (ah-NASS-tih-MO-seez), points where two arteries come together and combine their blood flow to points farther downstream. Anastomoses provide an alternative route, called **collateral circulation,** that can supply the heart tissue with blood if the primary route becomes obstructed.

Most organs receive more arterial blood when the ventricles contract than when they relax, but the opposite is true in the coronary arteries. There are three reasons for this. (1) Contraction of the myocardium compresses the arteries and obstructs blood flow. (2) During *ventricular systole* (contraction of the ventricles), the aortic valve is forced open and the valve cusps cover the openings to the coronary arteries, blocking blood from flowing into them. (3) During *ventricular diastole* (relaxation), blood in the aorta briefly surges back toward the heart. It fills the aortic valve cusps and some of it flows into the coronary arteries, like sand filling a shirt pocket and flowing out through a hole in the bottom. In the coronary blood vessels, therefore, diastolic blood flow is greater than systolic blood flow.

Venous Drainage

Venous drainage refers to the route by which blood leaves an organ. After flowing through capillaries of the heart wall, about 20% of the coronary blood empties directly from multiple small *thebesian*[13] *veins* into the right ventricle. The other 80% returns to the right atrium by the following route (fig. 20.11):

[12]*infarct* = to stuff

[13]Adam Christian Thebesius (1686–1732), German physician

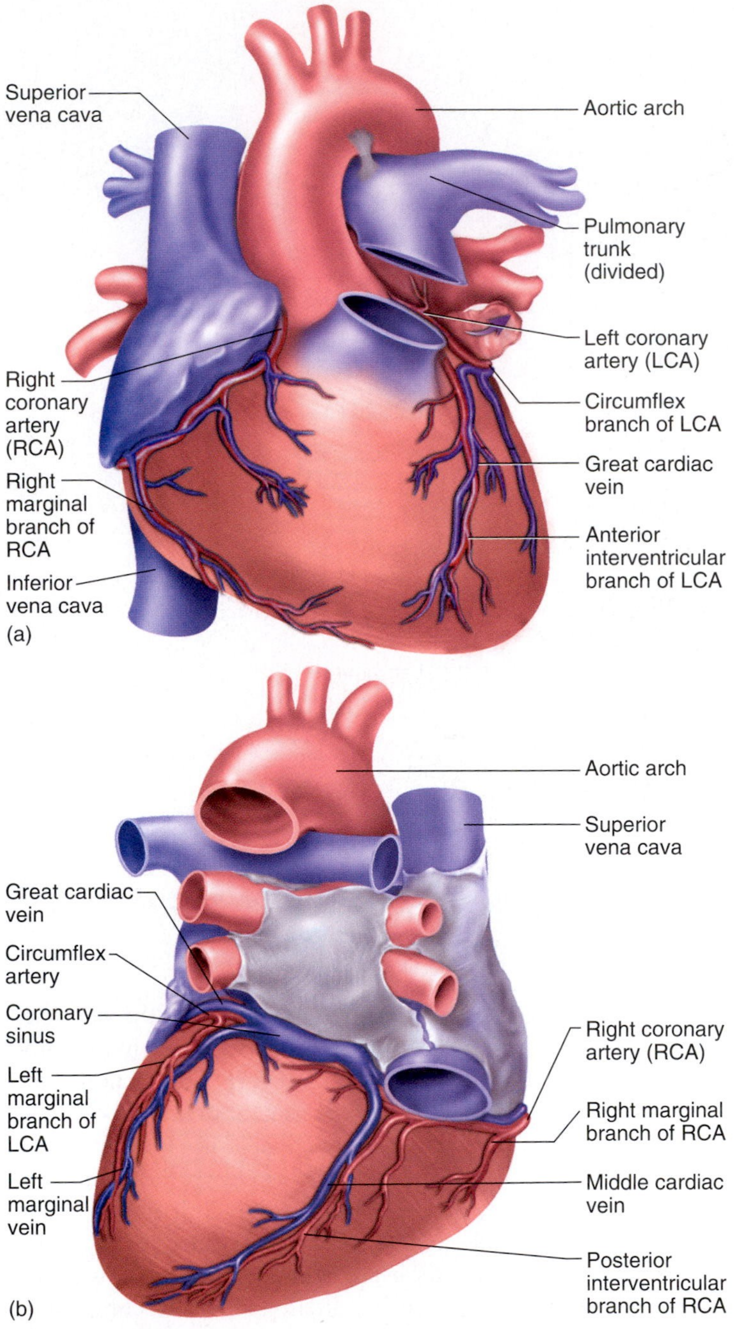

Superior vena cava
Aortic arch
Pulmonary trunk (divided)
Left coronary artery (LCA)
Right coronary artery (RCA)
Circumflex branch of LCA
Great cardiac vein
Right marginal branch of RCA
Anterior interventricular branch of LCA
Inferior vena cava
(a)
Aortic arch
Superior vena cava
Great cardiac vein
Circumflex artery
Coronary sinus
Right coronary artery (RCA)
Left marginal branch of LCA
Right marginal branch of RCA
Left marginal vein
Middle cardiac vein
Posterior interventricular branch of RCA
(b)

FIGURE 20.11

The Coronary Blood Vessels (*a*) Anterior aspect. (*b*) Posterior aspect.

- The **great cardiac vein** collects blood from the anterior aspect of the heart and travels alongside the anterior interventricular artery. It carries blood from the apex of the heart toward the coronary sulcus, then arcs around the left side of the heart and empties into the coronary sinus.

INSIGHT 20.1 CLINICAL APPLICATION

Coronary Artery Disease

Coronary artery disease (CAD) is a narrowing of the coronary arteries resulting in insufficient blood flow to maintain the myocardium. It is usually caused by *atherosclerosis,* a vascular disorder in which fatty deposits form in an arterial wall, causing arterial degeneration and obstructed blood flow. The atherosclerotic *plaque (atheroma)* is composed of lipids, smooth muscle, and scar tissue, and may progress to a calcified *complicated plaque,* causing the arterial walls to become rigid. *Myocardial infarction* (heart attack) can occur when the artery becomes so occluded that cardiac muscle begins to die from lack of oxygen. Partial obstruction of an artery can cause a temporary sense of heaviness and chest pain called *angina pectoris* when the artery constricts.

There are multiple ways in which an atheroma can lead to heart attack. The atheroma itself may block so much of the artery that blood flow is insufficient to support the cardiac muscle (fig. 20.12), especially during exercise when the metabolic need of the myocardium increases sharply. Platelets often adhere to atheromas and produce blood clots. If the vessel space (lumen) is already largely closed off by the atheroma, a blood clot may finish the job. Furthermore, a clot can break free from the atheroma and block a smaller coronary artery downstream.

- The **posterior interventricular** (**middle cardiac**) **vein,** found in the posterior sulcus, collects blood from the posterior aspect of the heart. It, too, carries blood from the apex upward and drains into the same sinus.
- The **left marginal vein** travels from a point near the apex of the heart up the left margin, and also empties into the coronary sinus.
- The **coronary sinus,** a large transverse vein in the coronary sulcus on the posterior side of the heart, collects blood from all three of the aforementioned veins as well as some smaller ones. It empties blood into the right atrium.

Before You Go On

Answer the following questions to test your understanding of the preceding section:

6. What are the three principal branches of the left coronary artery? Where are they located on the heart surface? What are the branches of the right coronary artery, and where are they located?
7. What is the medical significance of anastomoses in the coronary arterial system?
8. Why do the coronary arteries carry a greater blood flow during ventricular diastole than they do during ventricular systole?
9. What are the three major veins that empty into the coronary sinus?

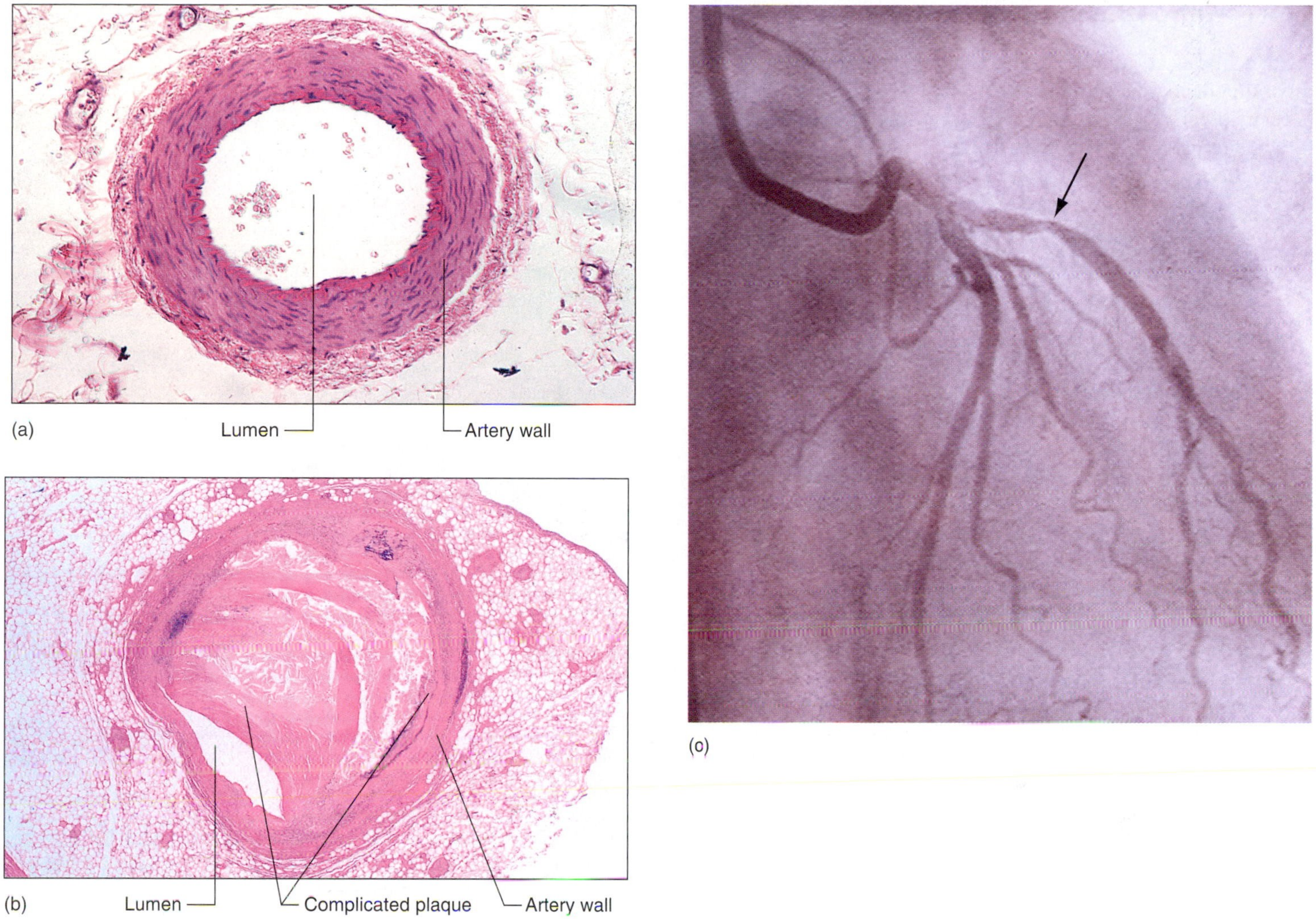
(a)
Lumen
Artery wall
(b)
Lumen
Complicated plaque
Artery wall
(c)

FIGURE 20.12

Coronary Atherosclerosis. (*a*) Cross section of a healthy artery. (*b*) Cross section of an artery with advanced atherosclerosis. Most of the original lumen is obstructed by a complicated plaque composed of calcified scar tissue. The small remaining space could easily be blocked by thrombosis, embolism, or vasoconstriction. (c) Coronary arteriogram showing 60% obstruction of the anterior interventricular artery *(arrow)*.

The Cardiac Conduction System and Cardiac Muscle

Objectives

When you have completed this section, you should be able to

- describe the heart's electrical conduction system.
- contrast the structure of cardiac and skeletal muscle; and
- describe the types and significance of intercellular junctions between cardiac muscle cells.

The most obvious physiological fact about the heart is its rhythmicity. It contracts at regular intervals, typically about 75 beats per minute (bpm) in a resting adult. Among invertebrates such as clams, crabs, and insects, each heartbeat is triggered by a pacemaker in the nervous system. The vertebrate heartbeat, however, is said to be *myogenic*[14] because the signal originates within the heart itself. Indeed, we can remove the heart from the body, keep it in aerated saline, and it will beat for hours. Cut the heart into little pieces, and each piece continues its own rhythmic pulsations. Thus it is obviously not dependent on the nervous system for its rhythm. The heart has its own pacemaker and electrical conduction system, and it is to this system that we now turn our attention.

[14]*myo* = muscle + *genic* = arising from

The Conduction System

Cardiac myocytes (muscle cells) are said to be **autorhythmic**[15] because they depolarize spontaneously at regular time intervals. Some of them lose the ability to contract and become specialized, instead, for generating action potentials. These cells constitute the **cardiac conduction system,** which controls the route and timing of electrical conduction to ensure that the four chambers are coordinated with each other. Electrical signals arise and travel through the cardiac conduction system in the following order (fig. 20.13):

1. The **sinoatrial (SA) node,** a patch of modified myocytes in the right atrium, just under the epicardium near the superior vena cava. This is the **pacemaker** that initiates each heartbeat and determines the heart rate. Signals from the SA node spread throughout the atria, as shown by the yellow arrows in figure 20.13.
2. The **atrioventricular (AV) node,** located near the right AV valve at the lower end of the interatrial septum. This node acts as an electrical gateway to the ventricles; the fibrous skeleton acts as an insulator to prevent currents from getting to the ventricles by any other route.
3. The **atrioventricular (AV) bundle** *(bundle of His*[16]*)*, a pathway by which signals leave the AV node. The AV bundle soon forks into **right** and **left bundle branches,** which enter the interventricular septum and descend toward the apex.
4. **Purkinje**[17] (pur-KIN-jee) **fibers,** nervelike processes that arise from the lower end of the bundle branches and turn upward to spread throughout the ventricular myocardium. Purkinje fibers distribute the electrical excitation to the myocytes of the ventricles. They form a more elaborate network in the left ventricle than in the right.

Although the SA node is the normal pacemaker of the heart, other autorhythmic sites in the heart can fire and stimulate cardiac contraction if the SA node fails to fire first. These other sites usually fire at slower rates than the SA node and therefore do not assume a pacemaker role unless the SA node is diseased. When another site does assume such a role, it is called an **ectopic focus.** The AV node is the most common ectopic focus, but other areas of myocardium can also take on this role. Ectopic foci usually fire too slowly to sustain life.

Think About It!

Some people have cords or bridges of myocardium that extend from atrium to ventricle, bypassing the AV node and other parts of the conduction system. How would you expect this to affect the cardiac rhythm?

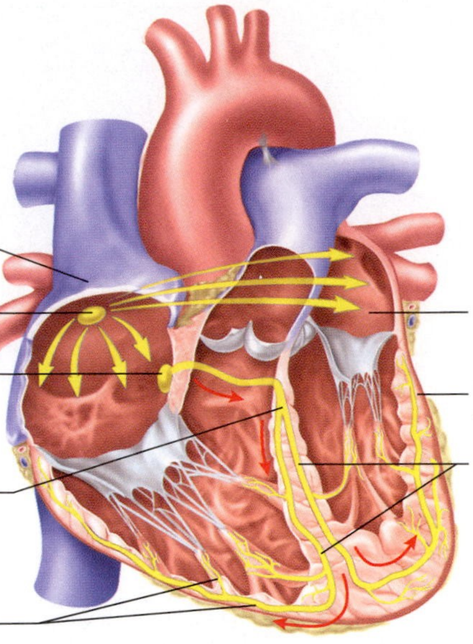
Right atrium
Sinoatrial node (pacemaker)
Atrioventricular node
Atrioventricular bundle
Purkinje fibers
Left atrium
Purkinje fibers
Bundle branches

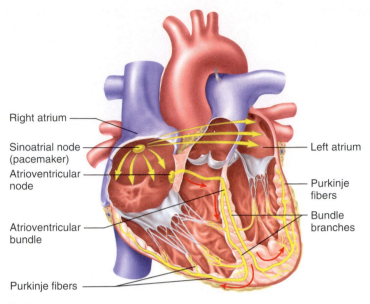

Figure 20.13
The Cardiac Conduction System. Electrical signals travel along the pathway indicated by the *arrows.*

INSIGHT 20.2 CLINICAL APPLICATION

Heart Block

Heart block is a condition in which electrical signals cannot travel normally through the cardiac conduction system because of disease and degeneration of the conduction system fibers. A *bundle branch block* exists when one or both of the atrioventricular bundle branches are diseased. *Total heart block* results from disease of the AV node. Heart block is one of the causes of cardiac arrhythmia, an irregularity in the heartbeat (such as *fibrillation*). In total heart block, signals from the SA node stop at the diseased AV node and cannot reach the ventricular myocardium. The ventricles then beat at their own intrinsic rhythm of about 20 to 40 beats/min, out of synchrony with the atria and at a rate too slow to sustain life.

Structure of Cardiac Muscle

The traveling electrical signal does not end with the Purkinje fibers, and Purkinje fibers do not reach every myocyte. Rather, the myocytes pass the signal from cell to cell. This is something that skeletal muscle cannot do, so to understand how the heartbeat is coordinated, one must understand microscopic anatomy of cardiac myocytes and how they differ from skeletal muscle.

Cardiac muscle is striated like skeletal muscle but otherwise differs from it in many structural and physiological ways. Cardiac myocytes, or *cardiocytes,* are relatively short, thick, branched cells, typically 50 to 100 μm long and 10 to 20 μm wide (fig. 20.14). They

[15]*auto* = self
[16]Wilhelm His, Jr. (1863–1934), German physiologist
[17]Johannes E. Purkinje (1787–1869), Bohemian physiologist

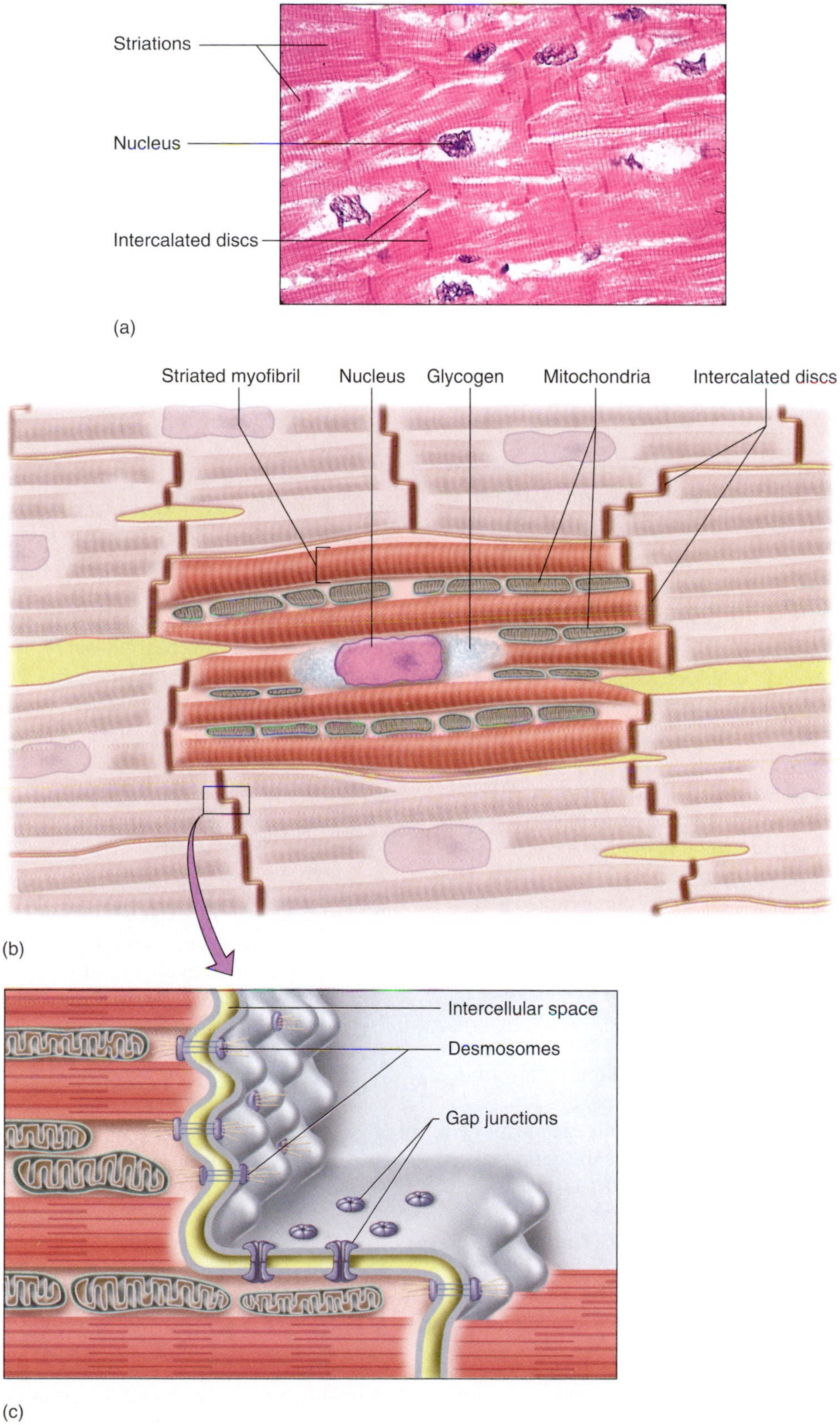

Striations
Nucleus
Intercalated discs
(a)
Striated myofibril
Nucleus
Glycogen
Mitochondria
Intercalated discs
(b)
Intercellular space
Desmosomes
Gap junctions
(c)

FIGURE 20.14

Cardiac Muscle. (*a*) Light micrograph. (*b*) Structure of a cardiac myocyte (center) and its relationship to adjacent myocytes. At each end, a myocyte is typically linked to two or more neighboring myocytes through the mechanical and electrical junctions of their intercalated discs. (c) Structure of an intercalated disc.

usually have only one, centrally placed nucleus. The sarcoplasmic reticulum (SR) is less developed than in skeletal muscle; it lacks terminal cisternae, although it does have footlike sacs associated with the T tubules. The T tubules are much larger than in skeletal muscle. During excitation of the cell, they admit supplemental calcium ions from the extracellular fluid to activate muscle contraction. Cardiocytes have especially large mitochondria, which make up about 25% of the cell volume, compared to skeletal muscle mitochondria which are much smaller and only 2% of the cell volume.

Think About It!

Why should mitochondria be larger and more abundant in cardiac muscle than in skeletal muscle?

The myocytes are joined end to end by thick connections called **intercalated** (in-TUR-ku-LAY-ted) **discs,** which appear as dark lines (thicker than the striations) in properly stained tissue sections. An intercalated disc is a complex steplike structure with three distinctive features not found in skeletal muscle:

1. **Interdigitating folds.** The plasma membrane at the end of the cell is folded somewhat like the inside of an egg carton. The folds of adjoining cells interlock with each other and increase the surface area of intercellular contact.
2. **Mechanical junctions.** The cells are tightly joined by two types of mechanical junctions—the fascia adherens and desmosomes. The *fascia adherens*[18] (FASH-ee-ah ad-HEER-enz) is the most extensive. It is a broad band in which the actin of the thin myofilaments is anchored to the plasma membrane, and via transmembrane proteins, one cell is linked to the next. The fascia adherens is interrupted here and there by *desmosomes.* Described in more detail in chapter 2, desmosomes are weldlike mechanical junctions between cells. In the contracting heart, they enable the myocytes to pull on each other without pulling apart.
3. **Electrical junctions.** The intercalated discs also contain *gap junctions,* which form channels that allow ions to flow from the cytoplasm of one cell directly into the next (again, see chapter 2 for details). These junctions enable each myocyte to electrically stimulate its neighbors. Thus the entire myocardium of the two atria behaves almost as if it were a single cell, as does the entire myocardium of the two ventricles. This unified action is essential for the effective pumping of a heart chamber.

Skeletal muscle contains satellite cells that can divide and replace dead muscle fibers to some extent. Cardiac muscle lacks satellite cells, however, so the repair of damaged cardiac muscle is almost entirely by fibrosis (scarring). A limited capacity for myocardial mitosis and regeneration was discovered in 2001.

[18]*fascia* = band + *adherens* = adhering

Nerve Supply to the Heart

Even though the heart has its own pacemaker, it does receive both sympathetic and parasympathetic nerves, which modify the heart rate and contraction strength. Sympathetic stimulation is able to raise the heart rate to as high as 230 beats/min, while parasympathetic stimulation can slow the heart rate to as low as 20 beats/min or even stop the heart for a few seconds.

The sympathetic pathway to the heart originates with neurons in the lower cervical to upper thoracic spinal cord. Efferent fibers from these neurons pass from the spinal cord to the sympathetic chain and travel up the chain to the three cervical ganglia. **Cardiac nerves** arise from the cervical ganglia (see fig. 16.4) and lead mainly to the ventricular myocardium, where they increase the force of contraction. Some fibers, however, innervate the atria. Sympathetic fibers to the coronary arteries dilate them and increase coronary blood flow during exercise.

The parasympathetic pathway to the heart is through the vagus nerves. The right vagus nerve innervates mainly the SA node and the left vagus nerve innervates mainly the AV node, although there is some cross-innervation from each nerve to both nodes. The ventricles receive little or no vagal stimulation. The vagus nerves slow the heartbeat. Without this influence, the SA node would produce an average resting heart rate of about 100 beats/min, but steady background firing of the vagus nerves *(vagal tone)* normally holds the resting rate down to about 70 to 80 beats/min.

Before You Go On

Answer the following questions to test your understanding of the preceding section:

10. Why is the human heart described as myogenic? Where is its pacemaker and what is it called?
11. List the components of the cardiac conduction system in the order traveled by signals from the pacemaker.
12. What organelle(s) are less developed in cardiac muscle than in skeletal muscle? What organelle(s) are more developed? What is the functional significance of these differences?
13. Name two types of cell junctions in the intercalated discs and explain their functional importance.
14. Identify the nerve supplies to the SA node, AV node, and ventricular myocardium.

Developmental and Clinical Perspectives

Objectives

When you have completed this section, you should be able to

- describe the embryonic development of the human heart;
- explain how and why the heart changes in old age; and
- define or briefly describe several of the most common heart diseases.

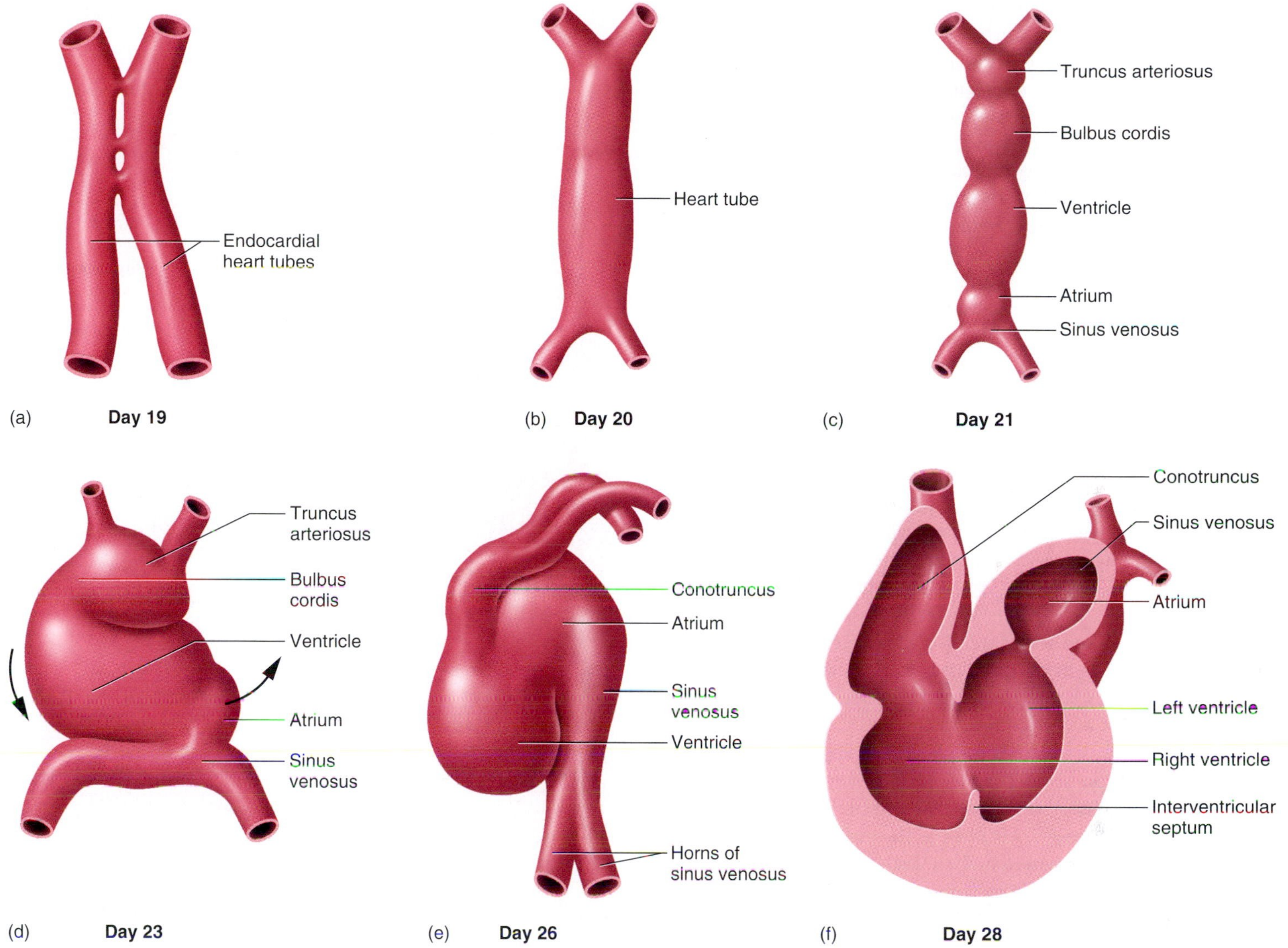
Endocardial heart tubes
(a) Day 19
Heart tube
(b) Day 20
Truncus arteriosus
Bulbus cordis
Ventricle
Atrium
Sinus venosus
(c) Day 21
Truncus arteriosus
Bulbus cordis
Ventricle
Atrium
Sinus venosus
(d) Day 23
Conotruncus
Atrium
Sinus venosus
Ventricle
Horns of sinus venosus
(e) Day 26
Conotruncus
Sinus venosus
Atrium
Left ventricle
Right ventricle
Interventricular septum
(f) Day 28

FIGURE 20.15

Embryonic Development of the Heart. (*a*) The endocardial heart tubes beginning to fuse at day 19. (*b*) Complete fusion by day 20, forming the heart tube. (*c*) Division of the heart tube into five dilated segments by day 21. The heart begins beating about a day later. (*d*) The heart begins looping around day 23, with the bulbus cordis migrating caudally *(left arrow)* and the atrium and sinus venosus migrating rostrally *(right arrow)*. Blood circulates throughout the embryo within a day of this stage. (*e*) Looping is nearly completed by day 26. (*f*) Frontal section of the heart at 28 days. As the interventricular septum develops, the conotruncus will divide longitudinally into the ascending aorta and pulmonary trunk, receiving blood from the left and right ventricles, respectively. The single atrium seen here divides into the right and left atria by day 33.

Prenatal Development of the Heart

The heart is one of the earliest organs to form and begin functioning in the embryo. The first traces of the heart appear in week 3; by 22 to 23 days (usually before the mother is aware that she is pregnant) the heart is already beating; by day 24, it circulates blood throughout the embryo.

In week 3, a region of mesoderm at the far anterior end of the embryo condenses into a pair of longitudinal cellular cords. By day 19, these become hollow, parallel **endocardial heart tubes** (fig. 20.15*a*). As the embryo grows and the head region folds, these tubes are pushed closer together, the tissues dividing them break down, and they fuse into a single **heart tube** (fig. 20.15*b*). As the tubes are fusing, surrounding mesoderm forms a primordial myocardium, responsible for the inception of the heartbeat just a few days later. The fetal heartbeat first becomes audible with a stethoscope in the fourth month of gestation.

With continued folding of the head region, the heart tube elongates and segments into five dilated spaces, some of them corresponding to future heart chambers. From rostral to caudal, these are the **truncus arteriosus,**[19] **bulbus cordis,**[20] **ventricle, atrium,** and **sinus venosus** (fig. 20.15*c*). Two of these, the ventricle and bulbus cordis, grow more rapidly than the others, causing the heart to loop

[19]*truncus* = trunk + *arteriosus* = arterial
[20]*bulbus* = bulb + *cordis* = of the heart

into a U and then an S shape similar to a fish heart (fig. 20.15*d–e*). In the course of this looping, the bulbus cordis shifts caudally, the ventricle shifts to the left, and the atrium and sinus venosus shift rostrally, as indicated by the arrows in the figure. During this looping, the heart bulges into the pericardial cavity. Looping is completed by day 28 and results in the forerunners of the adult atria and ventricles assuming their final relationship to each other (the future atria are now superior, or rostral, to the future ventricles). The primordial ventricle seen at day 21 becomes the left ventricle of the adult heart, and the inferior part of the bulbus cordis becomes the right ventricle. The superior part of the bulbus cordis and the truncus arteriosus are now collectively called the **conotruncus** (fig 20.15*e*). This passage soon gives rise to the aorta and pulmonary trunk.

The next phase of development is the partitioning of the heart tube into separate chambers (two atria and two ventricles) through the growth of the interatrial and interventricular septa. The interatrial septum begins to form at the end of week 4 and is well established by about 33 days, except for an opening between the atria called the *foramen ovale.* This foramen persists until after birth; its significance is discussed in the next section. The sinus venosus is initially a separate heart chamber, but it becomes extensively remodeled. Originally, it branches into a right and left horn at its inferior end. The right horn enlarges and receives systemic blood from the superior and inferior venae cavae. It eventually becomes part of the right atrium. The left horn shrinks. Parts of it become the coronary sinus, the sinoatrial node (pacemaker), and a portion of the atrioventricular node.

The interventricular septum begins to appear on the floor of the ventricle at the end of week 4 (fig. 20.15*f*) and increases in height as the two ventricles grow on either side of it. The septum is complete by the end of week 7. Meanwhile, the inner portion of the ventricular wall becomes honeycombed with cavities which give rise to the trabeculae carneae, papillary muscles, and tendinous cords.

Yet another septum forms during week 5 within the bulbus cordis and truncus arteriosus, dividing this outflow passage in two along its length. The separate halves of the passage become the ascending aorta and pulmonary trunk. The passage twists about 180° as the septum forms. The evidence of this twisting shows in the way the adult pulmonary trunk twists around the aorta. This twisting must be closely coordinated with the closure of the interventricular septum so that the right ventricle will open into the pulmonary trunk and the left ventricle will open into the aorta. Developmental irregularities at this stage are responsible for many cardiac birth defects.

Changes at Birth

There is little point to pumping all of the blood through the lungs of the fetus, because the fetal lungs are not yet inflated or functional. They receive enough blood to meet their metabolic and developmental needs, but most blood bypasses the pulmonary circuit by way of two anatomical shortcuts or *shunts* (fig. 20.16). One shunt is the **foramen ovale,** the opening through the interatrial septum. Some of the blood entering the right atrium passes through this opening directly into the left atrium and from there into the left ventricle and systemic circuit. The other shunt is a short vessel, the **ductus arteriosus,** from the left pulmonary artery to the aorta. Most of the blood that the right ventricle pumps into the pulmonary trunk takes this bypass directly into the aorta instead of following the usual path to the lungs.

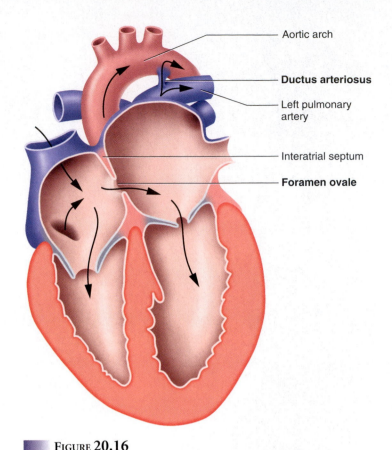

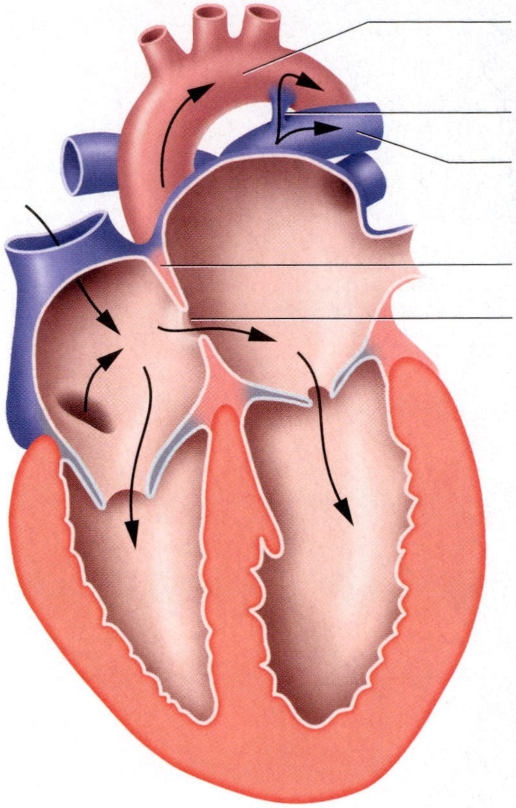

FIGURE 20.16

The Fetal Heart. Note the two shunts (foramen ovale and ductus arteriosus) that allow most blood to bypass the nonfunctional lungs.

At birth, the lungs inflate and their resistance to blood flow drops sharply. The sudden change in pressure gradients causes a flap of tissue to seal the foramen ovale. Blood in the right atrium is no longer able to flow directly into the left and bypass the lungs. In most people, the tissues grow together and permanently seal the foramen, leaving only a depression in the right atrial wall, the *fossa ovalis,* marking its former location. The foramen remains unsealed in about 15% of adults, but the tissue flap acts as a valve that prevents blood from passing through. The ductus arteriosus normally constricts around 10 to 15 hours after birth and becomes a permanently closed fibrous cord *(ligamentum arteriosum)* by the age of 2 to 3 weeks (but see insight 20.3).

The Aging Heart

The most noticeable effect of aging on the cardiovascular system is a stiffening of the arteries. While that itself is not a cardiac disease, it has important repercussions on the heart. Normally, when the ventricles eject blood, the arteries expand to accommodate the surge in pressure. When arteries are stiffened by age, or even calcified by arteriosclerosis, they cannot do so. They resist blood flow

INSIGHT 20.3 CLINICAL APPLICATION

Patent Ductus Arteriosus

Patent[21] *ductus arteriosus (PDA)* is the failure of the ductus arteriosus to close. For a short time after birth, PDA causes no problems; but as the lungs become better inflated and more functional, pulmonary blood pressure drops below aortic blood pressure. Blood may then begin to flow from the aortic arch back into the pulmonary circuit for an immediate second trip through the lungs. Since this blood soon returns to the left ventricle, it adds markedly to the left ventricular workload. The lungs sometimes respond to the persistent high blood flow with vascular changes that increase pulmonary resistance and stress the right ventricle as well.

The signs of PDA include poor weight gain in early childhood, frequent respiratory illnesses, *dyspnea* (difficulty breathing) on exertion, and *cardiomegaly* (enlargement of the heart). PDA is usually suspected at about 6 to 8 weeks of age because of a persistent "machinery-like" heart murmur; it is confirmed by X ray and other cardiac imaging methods.

Usually, the ductus arteriosus can be stimulated to close with a prostaglandin inhibitor, but if this fails, surgery is required. Surgery is ideally performed between 1.5 and 2.5 years of age because there is a rising risk of *infective endocarditis* (see table 20.1) if it is delayed. The usual procedure is to tie off the DA with several ligatures. This is a low-risk surgery with almost no mortality. However, other methods are available for blocking blood flow through the DA that are less invasive and give an easier recovery for the young patient.

[21]*patent* = open

more than younger arteries, and the heart has to work harder to overcome this resistance. Like any other muscle, when the heart works harder, it grows. The ventricles enlarge, especially the left ventricle, which has to work the hardest to overcome the most resistance. In ventricular hypertrophy, the heart wall and interventricular septum can become so thick that the space within the ventricle is severely diminished. Cardiac output sometimes declines to the point of heart failure.

Many other changes are seen in the aging heart: The valve annuli become more fibrous or even calcified, and the AV valves (especially the mitral valve) thicken and tend to prolapse. The interventricular septum often deviates to the left and interferes with the ejection of blood into the aorta. The fibrous skeleton becomes less elastic, so it has less capacity to rebound in diastole and aid in the filling of the heart. There is a loss of cells from the SA node and conduction system, so impulse conduction is less efficient and more irregular. Degeneration of the conduction system increases the risk of cardiac arrhythmia or heart block. Myocytes die off in the myocardium and the heart thus becomes weaker. Exercise tolerance is further diminished by decreasing sensitivity to sympathetic stimulation in the elderly heart.

Table 20.1

Common Cardiac Pathologies

Cardiac Tamponade	Compression of the heart by serous fluid or clotted blood in the pericardial cavity, rendering the heart unable to expand and fill completely during diastole, thus reducing systolic output
Cardiomyopathy	Any disease of the myocardium from causes other than valvular dysfunction or vascular diseases. Can cause atrophy or hypertrophy of the heart wall and interventricular septum, or dilation and failure of the heart.
Congestive Heart Failure (CHF)	Failure of either ventricle to pump as much blood as the other one, resulting in accumulation of blood and edema (congestion) in peripheral tissues. Left ventricular failure results in pulmonary congestion and right ventricular failure in systemic congestion (once called *dropsy*). Failure of one ventricle stresses the other and may lead to its subsequent failure.
Infective Endocarditis	Inflammation of the endocardium, usually due to bacterial infection with streptococci or staphylococci
Mitral Valve Prolapse (MVP)	A valvular defect in which one or both mitral valve cusps balloon into the atrium during ventricular contraction. Often hereditary, affecting 1 out of 40 people overall, and young women especially. Causes significant illness in only 3% of cases, including chest pain, fatigue, shortness of breath, and occasionally infective endocarditis, arrhythmia, or stroke.
Rheumatic Fever	Autoimmune disease triggered by a bacterial infection. Antibodies against streptococci or other bacteria attack tissues of the heart valves, causing scarring and constriction (stenosis) of the valves, especially the mitral valve. Regurgitation of blood through the incompetent valve causes turbulence heard as a *heart murmur.*
Septal Defects	Abnormal openings in the interatrial or interventricular septum, allowing blood to flow directly between right and left heart chambers. Results in pulmonary hypertension, difficulty breathing, and fatigue. Often fatal in childhood if not corrected.
Ventricular Fibrillation	Squirming, uncoordinated contractions of the ventricular myocardium with no effective ejection of blood. Often caused by myocardial infarction (MI); the usual cause of death in heart attack.

Disorders Described Elsewhere

Angina pectoris 572
Arrhythmia 574
Coronary artery disease 572
Heart block 574
Myocardial infarction 572
Patent ductus arteriosus 579
Pericarditis 563

Heart Disease

Heart disease is the leading cause of death in the United States (about 30% of deaths per annum, averaged across all age groups). The most common form of heart disease is coronary atherosclerosis, often leading to myocardial infarction. However, there are a multitude of other heart diseases. The principal categories of heart disease are congenital defects in cardiac anatomy, myocardial hypertrophy or degeneration, inflammation of the pericardium and heart wall, valvular defects, and cardiac tumors. Several examples are described in the insight sidebars in this chapter and in table 20.1.

Before You Go On

Answer the following questions to test your understanding of the preceding section:

15. When does the embryonic heart begin to beat? At what gestational age does it become audible?
16. What are the five primitive chambers that develop from the heart tube? What becomes of each as the heart continues to develop?
17. Describe the two routes (shunts) by which fetal blood bypasses the lungs. What happens to each one shortly after birth?
18. Why does the heart tend to enlarge in old age? Why does the risk of cardiac arrhythmia increase?

CHAPTER REVIEW

Review of Key Concepts

Overview of the Cardiovascular System (p. 562)

1. The cardiovascular system is divided into a pulmonary circuit served by the right heart and a systemic circuit served by the left.
2. The heart is located in the mediastinum between the lungs, with about two-thirds of it to the left of the median plane.
3. The heart is enclosed in a fibrous, two-layered *pericardium.* The space between the parietal and visceral pericardium is the pericardial cavity, and contains lubricating pericardial fluid.

Gross Anatomy of the Heart (p. 565)

1. The heart wall is composed of a thin outer *epicardium,* a thick muscular *myocardium,* and a thin inner *endocardium.*
2. The heart has a connective tissue *fibrous skeleton* which supports the myocardium and valves, anchors the myocytes, electrically insulates the ventricles from the atria, and may aid in ventricular filling by means of elastic recoil.
3. The two upper chambers of the heart are the *atria,* and serve to receive blood from the venae cavae and pulmonary veins. The two lower chambers are the *ventricles,* which eject blood into the pulmonary trunk and aorta. The ventricles are much more muscular than the atria. The atrioventricular and interventricular sulci on the heart surface mark the boundaries of these chambers.
4. The chambers are internally separated by an *interatrial septum* between the atria and *interventricular septum* between the ventricles.
5. The passages between the atria and ventricles are regulated by the atrioventricular valves (*tricuspid valve* on the right and *bicuspid,* or *mitral, valve* on the left). The cusps of these valves are connected by *tendinous cords* to *papillary muscles* on the floor of the ventricles.
6. The openings into the pulmonary trunk and aorta are regulated by the *semilunar* (*pulmonary* and *aortic*) *valves.* The opening and closing of the heart valves is caused by changes in the pressure difference on the two sides of a valve.
7. Systemic blood is received by the right atrium and flows into the right ventricle. The right ventricle pumps it into the pulmonary trunk, from which it flows to the lungs. Pulmonary blood returning from the lungs is received by the left atrium and flows into the left ventricle. The left ventricle pumps it into the aorta, the beginning of the systemic circulation.

Blood Supply to the Cardiac Muscle (p. 569)

1. The myocardium has a high workload and metabolic rate and needs an abundant oxygen and nutrient supply. It gets this not from the blood in its chambers but from a system of blood vessels called the *coronary circulation.*
2. The *left coronary artery* arises behind an aortic valve cusp near the beginning of the aorta, and gives rise mainly to the *anterior interventricular* and *circumflex branches.* The circumflex gives off a *left marginal branch.*
3. The *right coronary artery* gives off mainly a *right marginal branch* and *posterior interventricular branch.*
4. The myocardium is drained mainly by the *great cardiac, posterior interventricular,* and *left marginal veins,* all of which empty into the *coronary sinus.* The coronary sinus and several small *thebesian veins* empty into the right atrium.
5. Obstruction of a coronary artery deprives the downstream myocardium of a blood supply and may cause *myocardial infarction* (heart attack). The risk of this is reduced to some extent by several *anastomoses* in the coronary circulation.

The Cardiac Conduction System and Cardiac Muscle (p. 573)

1. The cardiac rhythm is set by its own internal pacemaker, the *sinoatrial (SA) node.* Electrical signals originating here spread through the atrial myocardium and then travel via the *atrioventricular (AV) node, AV bundle, bundle branches,* and *Purkinje fibers* to reach the ventricular myocytes.
2. If the SA node fails, other areas of myocardium called *ectopic foci* may take over the timing of the heartbeat, but usually at a rate too slow to sustain life.
3. Cardiac myocytes are striated muscle cells with a single nucleus, a poorly developed sarcoplasmic reticulum, very large T tubules, and large abundant mitochondria. They are joined end to end by *interdigitating folds* in their *intercalated discs.* The discs also include intercellular mechanical junctions (*fascia adherens* and *desmosomes*) and electrical (gap) junctions. The latter enable cardiocytes to electrically communicate directly with each other.
4. Although the heart can beat independently of the nervous system, it is innervated by the autonomic nervous system, which modifies the heart rate and contraction force. Sympathetic nerves supply mainly the ventricular myocardium and parasympathetic (vagus) nerves supply the SA and AV nodes. Sympathetic stimulation increases heart rate and contraction strength, and parasympathetic stimulation reduces the heart rate.

Developmental and Clinical Perspectives (p. 576)

1. Embryonic heart development begins when a pair of mesodermal *endothelial tubes* appear in the third week of gestation and fuse into a single median *heart tube.* The heart tube divides into five regions: the *truncus arteriosus, bulbus cordis, ventricle, atrium,* and *sinus venosus.* As the heart tube loops into U and S shapes around days 26 to 28, the truncus arteriosus and upper bulbus cordis divide longitudinally to become the pulmonary trunk and ascending aorta; the lower bulbus cordis becomes the right

ventricle; the single embryonic ventricle develops into the mature left ventricle; the single atrium divides into the right and left atria; and the sinus venosus contributes to the right atrium, coronary sinus, sinoatrial node, and atrioventricular node.
2. In the fetal circulation, most blood bypasses the lungs by way of the *foramen ovale* between the right and left atria and the *ductus arteriosus* between the left pulmonary artery and the aortic arch. These passages close soon after birth so that all blood from the right ventricle is forced to flow through the lungs.
3. In old age, the heart has to work harder against increasing arterial resistance. This tends to lead to ventricular hypertrophy. The aged heart also exhibits fibrosis or calcification of the valve annuli, valvular dysfunction, deviation of the interventricular septum, stiffening of the fibrous skeleton, loss of cells from both the myocardium and the cardiac conduction system, and reduced sensitivity to sympathetic stimulation. These changes lead to weakening of the myocardium, reduced cardiac output, less exercise tolerance, and increasing risk of arrhythmia or heart block. Heart failure is the leading cause of death in the United States.
4. Heart diseases are very diverse and include congenital defects in anatomy, hypertrophy and atrophy of the myocardium, inflammation of the pericardium and heart wall, valvular defects, and cardiac tumors.

Testing Your Recall

1. The cardiac conduction system includes all of the following *except*
   a. the SA node.
   b. the AV node.
   c. the bundle branches.
   d. the tendinous cords.
   e. the Purkinje fibers.
2. To get from the right atrium to the right ventricle, blood flows through the right AV, or _____, valve.
   a. pulmonary
   b. tricuspid
   c. bicuspid
   d. aortic
   e. mitral
3. There is/are _____ pulmonary vein(s) emptying into the right atrium of the heart.
   a. no
   b. one
   c. two
   d. four
   e. more than four
4. The coronary blood vessels are part of the _____ circuit of the circulatory system.
   a. cardiac
   b. pulmonary
   c. systematic
   d. systemic
   e. cardiovascular
5. The outermost layer of the heart wall is known as
   a. the pericardial sac.
   b. the epicardium.
   c. the visceral pericardium.
   d. both *a* and *c*.
   e. both *b* and *c*.
6. The thickest myocardium is found in _____ because it is this chamber that works the hardest to circulate blood.
   a. the right atrium
   b. the left atrium
   c. both atria
   d. the right ventricle
   e. the left ventricle
7. The ascending aorta and pulmonary trunk develop from the embryonic
   a. bulbus cordis only.
   b. truncus arteriosus only.
   c. horns of the sinus venosus.
   d. conotruncus.
   e. ventricle.
8. The _____ prevent the AV valves from flipping inside out during ventricular systole.
   a. tendinous cords
   b. pectinate muscles
   c. trabeculae carneae
   d. AV nodes
   e. cusps
9. Blood in the anterior interventricular branch of the left coronary artery flows into myocardial blood capillaries and next drains into
   a. the superior vena cava.
   b. the great cardiac vein.
   c. the left atrium.
   d. the middle cardiac vein.
   e. the coronary sinus.
10. Which of these is *not* characteristic of the heart in old age?
   a. ventricular enlargement
   b. thickening of the atrial walls
   c. a less elastic fibrous skeleton
   d. fewer cells in the conduction system
   e. less sensitivity to norepinephrine
11. The contraction of any heart chamber is called _____ and its relaxation is called _____.
12. The circulatory route from aorta to the venae cavae is the _____ circuit.
13. The circumflex branch of the left coronary artery travels in a groove called the _____.
14. The finest passages through which electrical signals pass before reaching the ventricular myocytes are called _____.
15. Electrical signals pass quickly from one cardiac myocyte to another through the _____ of the intercalated discs.
16. The abnormal bulging of a bicuspid valve cusp into the atrium is called _____.
17. The _____ nerves innervate the heart and tend to reduce the heart rate.
18. The death of cardiac tissue from lack of blood flow is commonly known as a heart attack, but clinically called _____.
19. Blood in the heart chambers is separated from the myocardium by a thin membrane called the _____.
20. The sinoatrial node develops from an embryonic heart tube chamber called the _____.

Answers in the Appendix

True or False

Determine which five of the following statements are false, and briefly explain why.

1. All blood that has circulated through the myocardium eventually flows into the coronary sinus and from there to the right atrium.
2. The aorta is the body's largest artery.
3. Normally, the only way electrical signals can get from the atria to the ventricles is to pass through the AV node and AV bundle.
4. The epicardium contains adipose tissue but the endocardium does not.
5. If all nerves from the central nervous system to the heart were severed, the heart would stop beating.
6. The thickest myocardium is normally found in the left ventricle.
7. Many of the cardiac veins have anastomoses that ensure that the myocardium will receive blood even if one of the veins becomes blocked.
8. During embryonic development, a ventricular septum grows and divides the single primordial ventricle into right and left ventricles.
9. Blood in the superior and inferior venae cavae flows through the semilunar valves as it enters the right atrium.
10. Cardiac myocytes transmit electrical signals to each other by way of their gap junctions.

Answers in the Appendix

Testing Your Comprehension

1. Mr. Jones, 78, dies of a massive myocardial infarction triggered by coronary thrombosis. Upon autopsy, necrotic myocardium is found in the lateral and posterior right ventricle and posterior interventricular septum. Based on the information in this chapter, where in the coronary circulation do you think the thrombosis occurred?
2. Becky, age 2, was born with a hole in her interventricular septum (*ventricular septal defect,* or *VSD*) Considering that the blood pressure in the left ventricle is significantly higher than blood pressure in the right ventricle, predict the effect of the VSD on Becky's pulmonary blood pressure, systemic blood pressure, and long-term changes in the ventricular walls.
3. Marcus is born with *transposition* of the great arteries, in which the aorta arises from the right ventricle and the pulmonary artery arises from the left. Assuming no other anatomical abnormalities, trace the flow of blood through the pulmonary and systemic routes in his case. Predict the consequences, if any, for the ability of Marcus's cardiovascular system to deliver oxygen to the systemic tissues. Do you think Marcus would require immediate surgical correction in early infancy; correction at the age of 2 or 3 years; or that it could be left alone and not seriously affect his life expectancy?
4. A man hiking in the woods is mistaken for a deer and shot by a bow hunter. The arrow pierces his right pulmonary cavity and causes the right lung to collapse, while his left lung remains normal. Predict how and why this could affect the position of his heart when viewed on X ray.
5. In dilated cardiomyopathy of the left ventricle, the ventricle can become enormously enlarged. Explain why this might lead to regurgitation of blood through the mitral valve (blood flowing from the ventricle back into the left atrium) during ventricular systole.

Answers at the Online Learning Center

www.mhhe.com/saladinha1

Visit the Online Learning Center for practice tests, answer keys, and other learning aids for this chapter. Enhance your understanding of human anatomy with our interactive art labeling exercises, supplemental photo atlases, web links, puzzles, flashcards, and much more.

21

CHAPTER TWENTY-ONE

The Circulatory System III—Blood Vessels

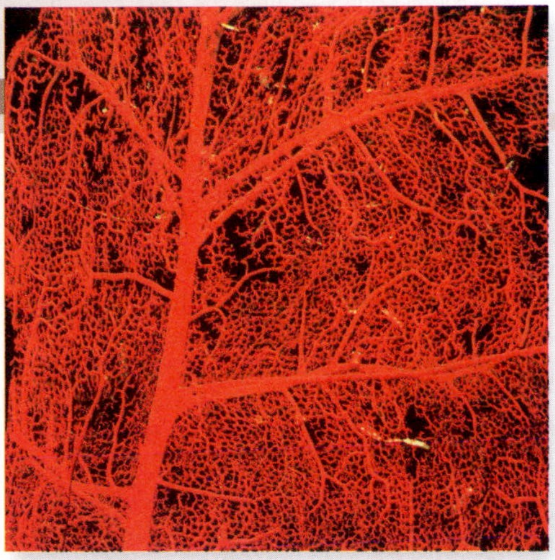

Blood capillary beds

CHAPTER OUTLINE

General Anatomy of the Blood Vessels 586
- The Vessel Wall 586
- Arteries 588
- Capillaries 589
- Veins 593
- Circulatory Routes 594

Anatomy of the Pulmonary Circuit 595

Anatomy of the Systemic Arteries 596
- The Aorta and Its Major Branches 598
- Arterial Supply to the Head And Neck 599
- Arterial Supply to the Upper Limb 601
- Arterial Supply to the Thorax 602
- Arterial Supply to the Abdomen 604
- Arterial Supply to the Pelvic Region and Lower Limb 607
- Arterial Pressure Points 609

Anatomy of the Systemic Veins 610
- Venous Drainage of the Head and Neck 611
- Venous Drainage of the Upper Limb 612
- The Azygos System 614
- Major Tributaries of the Inferior Vena Cava 615
- The Hepatic Portal System 616
- Venous Drainage of the Lower Limb and Pelvic Organs 618

Developmental and Clinical Perspectives 620
- Embryonic Development of the Blood Vessels 620
- Changes at Birth 622
- The Aging Vascular System 622
- Vascular Diseases 622

Chapter Review 625

INSIGHTS

21.1 Clinical Application: Aneurysm 588

21.2 Clinical Application: Varicose Veins 594

BRUSHING UP

To understand this chapter, you may find it helpful to review the following concepts:

- Primary germ layers of the embryo (p. 110)
- The pulmonary and systemic circuits (p. 562)
- The great vessels associated with the heart (p. 562–563)
- Cardiac systole and diastole (p. 571)

The route taken by the blood after it leaves the heart was a point of much confusion until the seventeenth century. Chinese emperor Huang Ti (2697–2597 B.C.E.) correctly believed that it flowed in a complete circuit around the body and back to the heart. But in the second century, Roman physician Claudius Galen argued that it flowed back and forth in the veins, like air in the bronchial tubes. He believed that the liver received food from the small intestine and converted it to blood, the heart pumped the blood through the veins to all other organs, and those organs consumed it.

Huang Ti was right, but the first experimental demonstration of this did not come until the seventeenth century. English physician William Harvey (1578–1657) studied the filling and emptying of the heart in snakes, tied off the vessels above and below the heart to observe the effects on cardiac filling and output, and measured cardiac output in a variety of living animals. He concluded that (1) the heart pumps more blood in half an hour than there is in the entire body, (2) not enough food is consumed to account for the continual production of so much blood, and (3) since the planets orbit the sun and (as he believed) the human body is modeled after the solar system, it follows that the blood orbits the body. So for a peculiar combination of experimental and superstitious reasons, Harvey argued that the blood must return to the heart rather than being consumed by the peripheral organs. He could not explain how, since the microscope had yet to be developed to the point that allowed Marcello Malpighi (1628–1694) and Antony van Leeuwenhoek (1632–1723) to discover the blood capillaries.

Harvey published his findings in 1628 in a short but elegant book entitled *Exercitio Anatomica de Motu Cordis et Sanguinis in Animalibus (Anatomical Studies on the Motion of the Heart and Blood in Animals).* This landmark in the history of biology and medicine was the first experimental study of animal physiology. But so entrenched were the ideas of Aristotle and Galen in the medical community, and so strange was the idea of doing experiments on living animals, that Harvey's contemporaries rejected his ideas. Indeed, some of them regarded him as a crackpot because his conclusion flew in the face of common sense—if the blood was continually recirculated and not consumed by the tissues, they reasoned, then what purpose could it possibly serve?

Harvey lived to a ripe old age, served as physician to the kings of England, and later did important work in embryology. His case is one of the most interesting in biomedical history, for it shows how empirical science overthrows old theories and spawns better ones, and how common sense and blind allegiance to authority can interfere with acceptance of the truth. But most importantly, Harvey's contributions represent the birth of experimental physiology.

General Anatomy of the Blood Vessels

Objectives

When you have completed this section, you should be able to

- describe the structure of a blood vessel;
- describe the different types of arteries, capillaries, and veins;
- trace the general route usually taken by the blood from the heart and back again; and
- describe some variations on this route.

There are three principal categories of blood vessels—arteries, veins, and capillaries. **Arteries** are the efferent vessels of the cardiovascular system—that is, vessels that carry blood away from the heart. **Veins** are the afferent vessels—vessels that carry blood back to the heart. **Capillaries** are microscopic, thin-walled vessels that connect the smallest arteries to the smallest veins. Aside from their general location and direction of blood flow, these three categories of vessels also differ in the histological structure of their walls.

The Vessel Wall

The walls of arteries and veins are composed of three layers called *tunics* (figs. 21.1 and 21.2):

1. The **tunica interna (tunica intima)** lines the inside of the vessel and is exposed to the blood. It consists of a simple squamous epithelium called the **endothelium,** overlying a basement membrane and a sparse layer of loose connective tissue. The endothelium acts as a selectively permeable barrier to materials entering or leaving the bloodstream; it secretes chemicals that stimulate the muscle of the vessel wall to contract or relax, thus narrowing or widening the vessel; and it normally repels blood cells and platelets so that they flow freely without sticking to the vessel wall. When the endothelium is damaged, however, platelets may adhere to it and form a blood clot; and when the tissue around a vessel is inflamed, the endothelial cells produce *cell-adhesion molecules* that induce leukocytes to adhere to the surface. This causes leukocytes to congregate in tissues where their defensive actions are needed.
2. The **tunica media,** the middle layer, is usually the thickest. It consists of smooth muscle, collagen, and in some cases, elastic tissue. The relative amounts of smooth muscle and elastic tissue vary greatly from one vessel to another and form a basis for classifying vessels as described in the next section. The principal functions of the tunica media are to strengthen the vessels and prevent the blood pressure from rupturing them, and to provide for *vasomotion,* changes in the diameter of a blood vessel. The widening of a vessel is called *vasodilation* and a narrowing is called *vasoconstriction.*
3. The **tunica externa (tunica adventitia**[1]**)** is the outermost layer. It consists of loose connective tissue that often merges with that of neighboring blood vessels, nerves, or other organs. It anchors the vessel and provides passage for small nerves, lymphatic vessels, and smaller blood vessels. Small vessels called the **vasa vasorum**[2] (VAY-za vay-SO-rum) supply blood to at least the outer half of the wall of a larger vessel. Tissues of the inner half of the wall are thought to be nourished by diffusion from blood in the lumen.

[1]*advent* = added to
[2]*vasa* = vessels + *vasorum* = of the vessels

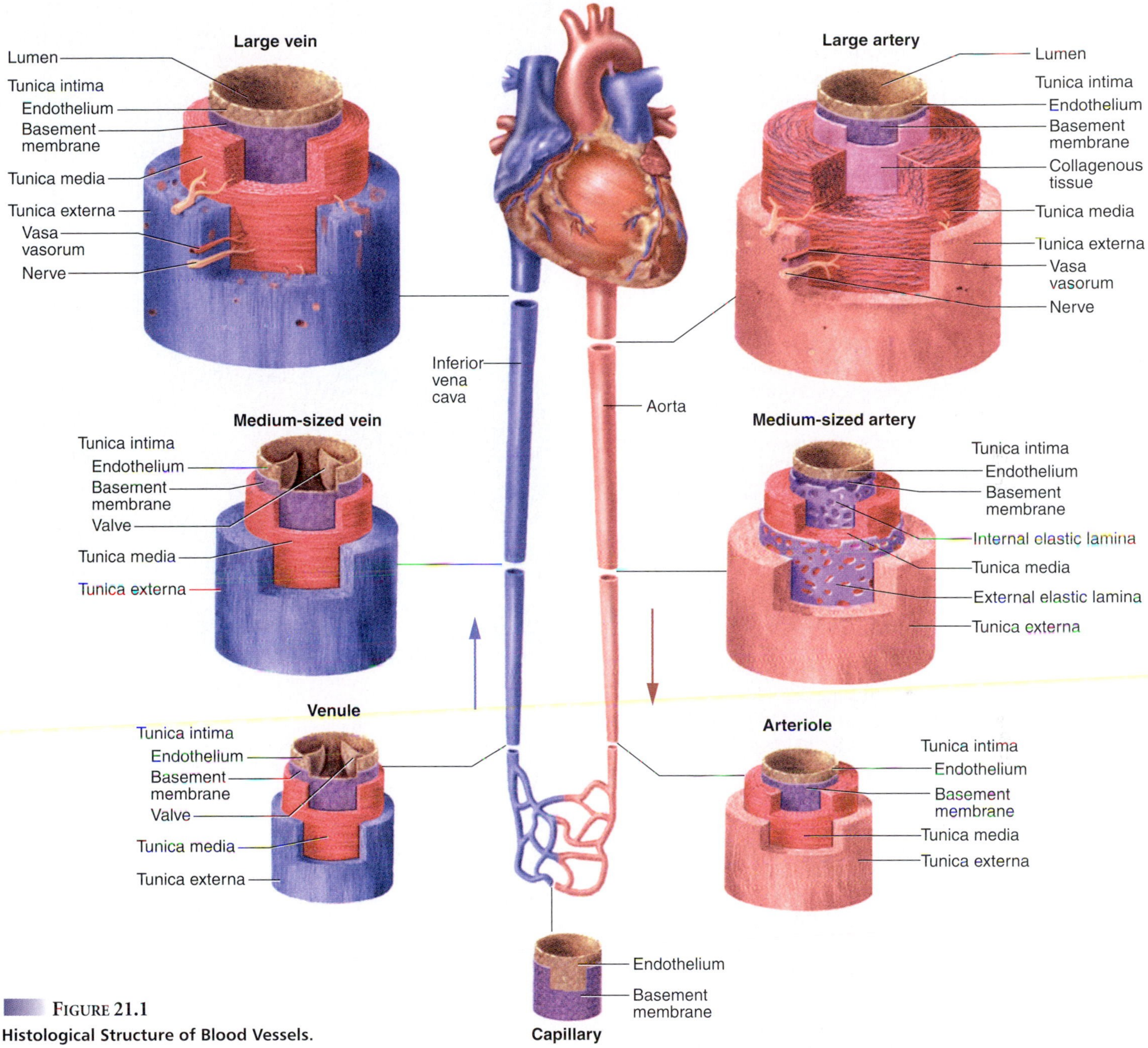
Large vein
Lumen
Tunica intima
Endothelium
Basement membrane
Tunica media
Tunica externa
Vasa vasorum
Nerve
Large artery
Lumen
Tunica intima
Endothelium
Basement membrane
Collagenous tissue
Tunica media
Tunica externa
Vasa vasorum
Nerve
Inferior vena cava
Aorta
Medium-sized vein
Tunica intima
Endothelium
Basement membrane
Valve
Tunica media
Tunica externa
Medium-sized artery
Tunica intima
Endothelium
Basement membrane
Internal elastic lamina
Tunica media
External elastic lamina
Tunica externa
Venule
Tunica intima
Endothelium
Basement membrane
Valve
Tunica media
Tunica externa
Arteriole
Tunica intima
Endothelium
Basement membrane
Tunica media
Tunica externa
Endothelium
Basement membrane
Capillary

FIGURE 21.1
Histological Structure of Blood Vessels.

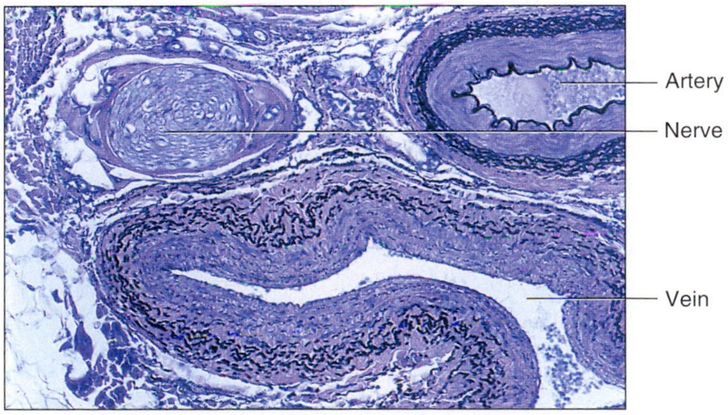
Artery
Nerve
Vein

FIGURE 21.2
A Neurovascular Bundle. A small artery, small vein, and nerve traveling together in a common sheath of connective tissue. The dark wavy line near the lumen of the artery is the internal elastic lamina.

Arteries

Arteries are sometimes called the *resistance vessels* of the cardiovascular system because they have a relatively strong, resilient tissue structure that resists the high blood pressure within. Each beat of the ventricles creates a surge of pressure in the arteries as blood is ejected into them. Arteries are constructed to withstand these pressure surges. They are more muscular than veins, so they retain their round shape even when empty, and they appear relatively circular in tissue sections. They are divided into three categories by size, but of course there is a smooth transition from one category to the next.

1. **Conducting** (**elastic** or **large**) **arteries** are the biggest arteries. The aorta, common carotid and subclavian arteries, pulmonary trunk, and common iliac arteries are examples of conducting arteries. There is a very thin layer of elastic fibers called the *internal elastic lamina* at the border between the intima and media, but it is sparse and not very visible microscopically. The tunica media consists of 40 to 70 layers of elastic sheets, perforated like slices of Swiss cheese, alternating with thin layers of smooth muscle, collagen, and elastic fibers. In histological sections, the view is dominated by this elastic tissue. There is a thin *external elastic lamina* at the border between the media and externa. The tunica externa is relatively thick and well supplied with vasa vasorum.

   Conducting arteries expand during ventricular systole to receive blood, and recoil during diastole. Their expansion takes some of the pressure off the blood so that smaller arteries downstream are subjected to less systolic stress. Their recoil between heartbeats prevents the blood pressure from dropping too low while the heart is relaxing and refilling. These effects lessen the fluctuations in blood pressure that would otherwise occur. Arteries stiffened by atherosclerosis cannot expand and recoil as freely. Consequently, the downstream vessels are subjected to greater stress and are more likely to develop aneurysms and rupture (see insight 21.1).
2. **Distributing** (**muscular** or **medium**) **arteries** are smaller branches that distribute blood to specific organs. You could compare a conducting artery to an interstate highway and distributing arteries to the exit ramps and state highways that serve specific towns. Most arteries that have specific anatomical names are in these first two size classes. The brachial, femoral, renal, and splenic arteries are examples of distributing arteries. Distributing arteries typically have up to 40 layers of smooth muscle constituting about three-quarters of the wall thickness. This smooth muscle is more conspicuous than the elastic tissue in histological specimens of distributing arteries. Both the internal and external elastic laminae, however, are thick and often histologically conspicuous in distributing arteries.

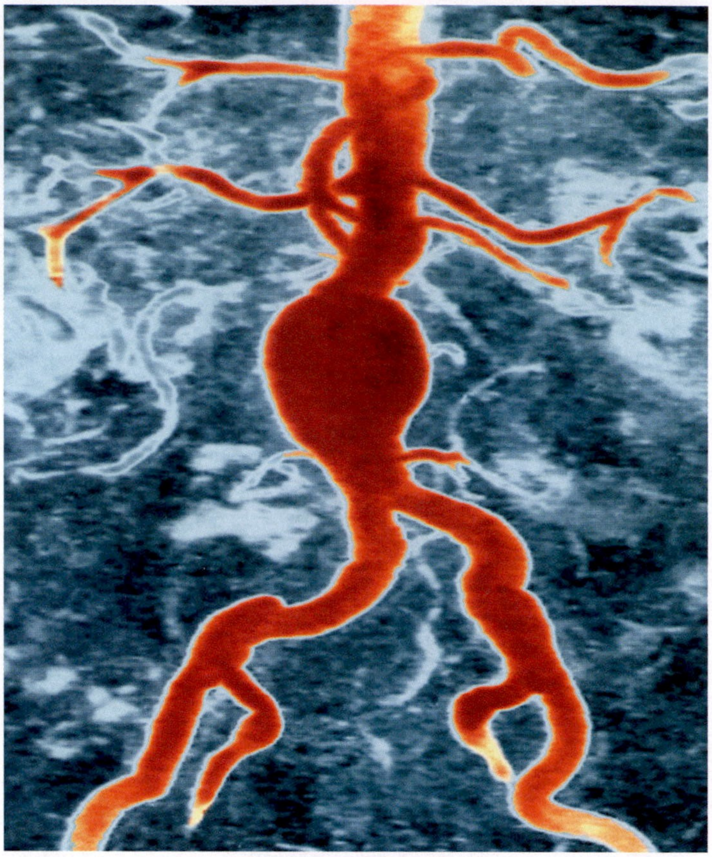

FIGURE 21.3
An Aortic Aneurysm. This is a magnetic resonance angiogram (MRA) of the abdominal aorta of a patient with hypertension, showing a prominent bulge (aneurysm) of the aorta *(red)* immediately superior to the common iliac arteries.

INSIGHT 21.1 CLINICAL APPLICATION

Aneurysm

An aneurysm is a weak point in an artery or in the heart wall. It forms a thin-walled, bulging sac that pulsates with each beat of the heart and may eventually rupture. In a *dissecting aneurysm,* blood pools between the tunics of an artery and separates them, usually because of degeneration of the tunica media. The most common sites of aneurysms are the abdominal aorta (fig. 21.3), renal arteries, and the arterial circle at the base of the brain. Even without hemorrhaging, aneurysms can cause pain or death by putting pressure on brain tissue, nerves, adjacent veins, pulmonary air passages, or the esophagus. Other consequences include neurological disorders, difficulty in breathing or swallowing, chronic cough, or congestion of the tissues with blood. Aneurysms sometimes result from congenital weakness of the blood vessels and sometimes from trauma or bacterial infections such as syphilis. The most common cause, however, is the combination of atherosclerosis and hypertension.

3. **Resistance (small) arteries** are usually too variable in number and location to be given individual names. They exhibit up to 25 layers of smooth muscle and relatively little elastic tissue. Their tunica media is thicker in proportion to the lumen than that of larger arteries. The smallest of these arteries, about 40 to 200 μm in diameter and with only one to three layers of smooth muscle, are the **arterioles.** Arterioles have very little tunica externa.

Metarterioles[3] are short vessels that link arterioles and capillaries. Instead of a continuous tunica media, they have individual muscle cells spaced a short distance apart, each forming a **precapillary sphincter** that encircles the entrance to one capillary. Constriction of these sphincters reduces or shuts off blood flow through their respective capillaries and diverts blood to tissues or organs elsewhere.

ARTERIAL SENSE ORGANS

Certain major arteries above the heart have sensory structures in their walls that monitor blood pressure and chemistry (fig. 21.4). They transmit information to the brainstem that is used to regulate the heartbeat, vasomotion, and respiration. The sensory receptors are of three kinds:

1. **Carotid sinuses.** These are *baroreceptors* (pressure sensors) that respond to changes in blood pressure. Ascending the neck on each side is a *common carotid artery* which branches near the angle of the mandible, forming the *internal carotid artery* to the brain and *external carotid artery* to the face. The carotid sinuses are located in the wall of the internal carotid artery just above the branch point. The carotid sinus has a relatively thin tunica media and an abundance of glossopharyngeal nerve fibers in the tunica externa. A rise in blood pressure easily stretches the thin media and stimulates these nerve fibers. The glossopharyngeal nerve then transmits signals to the vasomotor and cardiac centers of the brainstem, and the brainstem responds by lowering the heart rate and dilating the blood vessels, thereby lowering the blood pressure.
2. **Carotid bodies.** Also located near the branch of the common carotid arteries, these are oval receptors about 3 × 5 mm innervated by sensory fibers of the vagus and glossopharyngeal nerves. They are *chemoreceptors* that monitor changes in blood composition. They primarily transmit signals to the brainstem respiratory centers, which adjust breathing to stabilize the blood pH and its CO_2 and O_2 levels.
3. **Aortic bodies.** These are one to three chemoreceptors located in the aortic arch near the arteries to the head and arms. They are structurally similar to the carotid bodies and have the same function.

[3]*meta* = beyond, next in a series

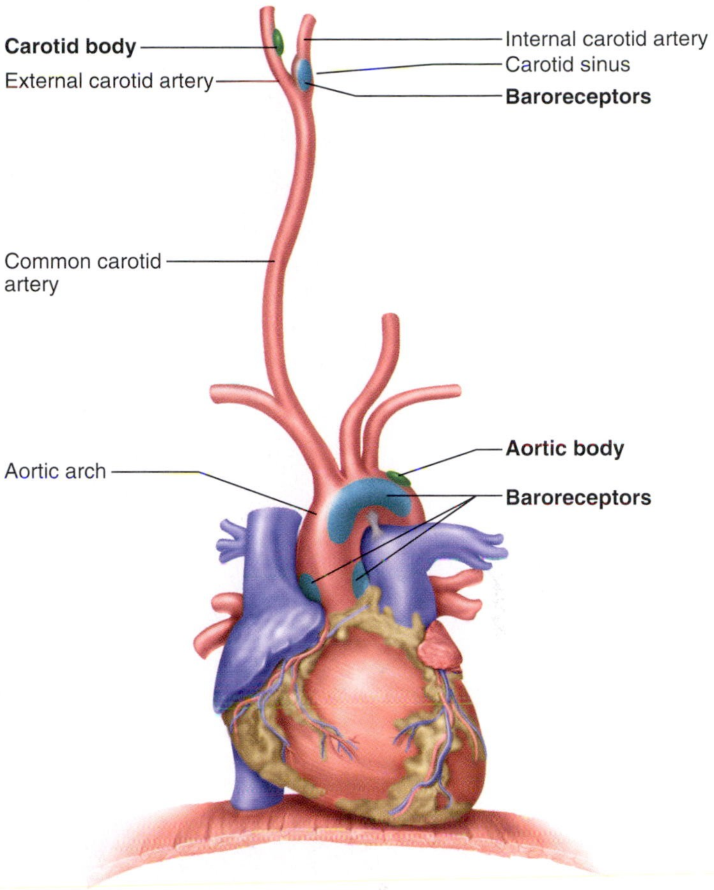

FIGURE 21.4
Baroreceptors and Chemoreceptors in the Arteries Superior to the Heart. The structures shown here are repeated in the left carotid arteries.

Capillaries

For the blood to serve any purpose, materials such as nutrients, wastes, and hormones must be able to pass between the blood and the tissue fluids, through the walls of the blood vessels. There are only two places in the circulation where this can occur—the capillaries and venules. Since capillaries greatly exceed venules in number and permeability, they are the more important of the two. Capillaries are sometimes called the *exchange vessels* of the cardiovascular system. We can think of them as the "business end" of the system, because all the rest of the circulatory system exists to serve the exchange processes that occur in the capillaries and smallest of the venules.

Blood capillaries (fig. 21.5) are composed of only an endothelium and basement membrane. Their walls are as thin as 0.2 to 0.4 μm. They average about 5 μm in diameter at the proximal end (where they receive arterial blood), they widen to about 9 μm at the distal end (where they empty into a small vein), and they often branch along the way. Since an erythrocyte is about 7 μm in diameter, erythrocytes often have to stretch into elongated shapes to squeeze through the smallest capillaries.

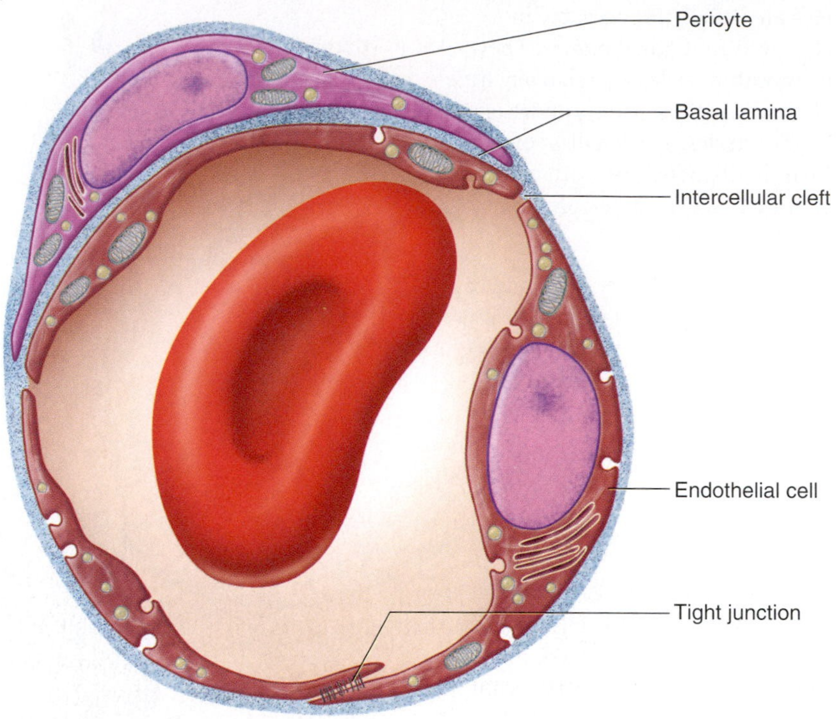
Pericyte
Basal lamina
Intercellular cleft
Endothelial cell
Tight junction

(a)

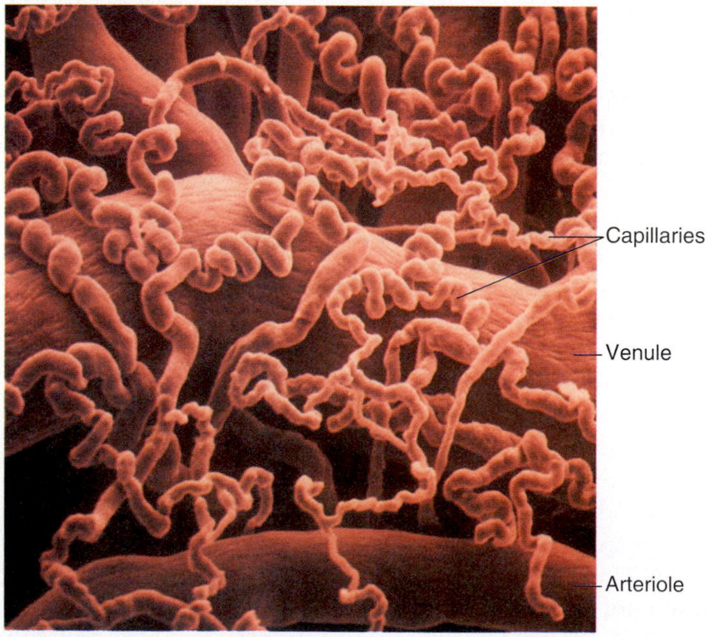
Capillaries
Venule
Arteriole

(b)

FIGURE 21.5
Structure of a Continuous Capillary. (*a*) A continuous capillary and pericyte. (*b*) Continuous capillaries of the eye. This is a cast prepared by injecting the vessels with a polymer, digesting away all tissue to leave a replica of the vessels, and photographing the cast through the SEM.

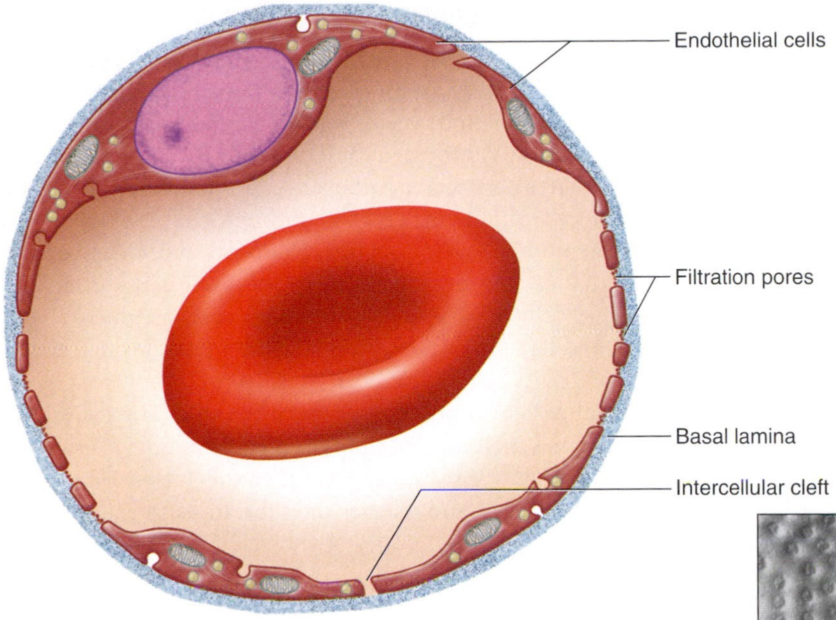

(a)

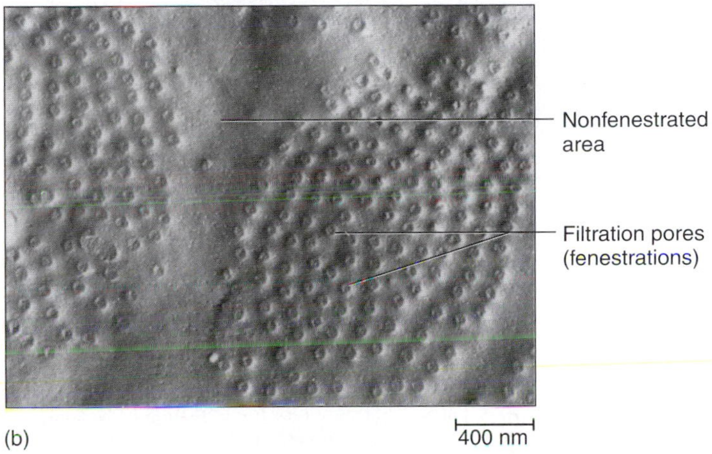

(b)

FIGURE 21.6

Structure of a Fenestrated Capillary. (*a*) Structure of the capillary wall. (*b*) Surface view of a fenstrated endothelial cell (SEM). The cell has patches of filtration pores (fenestrations) separated by nonfenestrated areas.

The number of capillaries has been estimated at a billion and their total surface area at 6,300 m^2. But a more important point is that scarcely any cell in the body is more than 60 to 80 μm (about four to six cell widths) away from the nearest capillary (see page 585). There are a few exceptions: Capillaries are scarce in tendons and ligaments and absent from cartilage, epithelia, and the cornea and lens of the eye.

TYPES OF CAPILLARIES

There are three types of capillaries, distinguished by the ease with which they allow substances to pass through their walls and by structural differences that account for their greater or lesser permeability.

1. **Continuous capillaries** (fig. 21.5) occur in most tissues, such as skeletal muscle. Their endothelial cells, held together by tight junctions, form an uninterrupted tube. A thin protein-carbohydrate layer, the **basal lamina,** surrounds the endothelium and separates it from the adjacent connective tissues. The endothelial cells are separated by narrow **intercellular clefts** about 4 nm wide. Small solutes such as glucose can pass through these clefts, but plasma proteins, other large molecules, and formed elements of the blood are held back. The continuous capillaries of the brain lack intercellular clefts and have more complete tight junctions that form the blood-brain barrier discussed in chapter 15.

   Some continuous capillaries exhibit cells called **pericytes** that lie external to the endothelium. Pericytes have elongated tendrils that wrap around the capillary. They contain the same contractile proteins as muscle, and it is thought that they contract and regulate blood flow through the capillaries. They also can differentiate into endothelial and smooth muscle cells and thus contribute to vessel growth and repair.
2. **Fenestrated capillaries** have endothelial cells riddled with holes called **filtration pores (fenestrations**[4]**)** (fig. 21.6). Filtration pores are about 20 to 100 nm in diameter, and are often spanned by a thin glycoprotein membrane. They allow for the rapid passage of small molecules but still retain most proteins and larger particles in the bloodstream. Fenestrated capillaries are important in organs that engage in rapid absorption or filtration—the kidneys, endocrine glands, small intestine, and choroid plexuses of the brain, for example.
3. **Sinusoids (discontinuous capillaries)** are irregular blood-filled spaces in the liver, bone marrow, spleen, and some other organs (fig. 21.7). They are twisted, tortuous passageways, typically 30 to 40 μm wide, that conform to the shape of the surrounding tissue. The endothelial cells are separated by wide gaps with no basal lamina, and the cells also frequently have especially large fenestrations through them. Even proteins and blood cells can pass through these pores; this is how albumin, clotting factors, and other proteins synthesized by the liver enter the blood, and how newly formed blood cells enter the circulation from the bone marrow and lymphatic organs. Sinusoids may contain macrophages or other specialized cells.

[4]*fenestra* = window

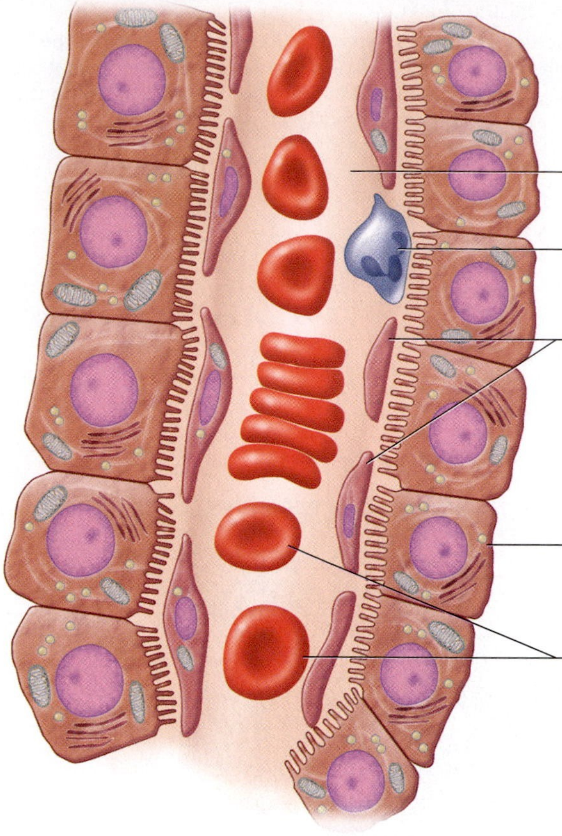
Sinusoid
Macrophage
Endothelial
cells
Liver cell
(hepatocyte)
Erythrocytes
in sinusoid

FIGURE 21.7

A Sinusoid of the Liver. Large gaps between the endothelial cells allow blood plasma to directly contact the liver cells, but retain blood cells in the lumen of the sinusoid.

CAPILLARY PERMEABILITY

The structure of a capillary wall is closely related to its permeability, the ease with which substances can pass through it from the blood to the tissue fluid or vice versa. There are three routes that materials can travel through a capillary wall (fig. 21.8): (1) the intercellular clefts between endothelial cells; (2) the filtration pores in fenestrated capillaries; and (3) through the endothelial cell plasma membrane and cytoplasm. Nonpolar substances such as O_2, CO_2, lipids, and thyroid hormone diffuse easily through the endothelial cells. Hydrophilic substances such as glucose, electrolytes, and large molecules such as insulin pass through the intercellular clefts and filtration slits, or cross through the endothelial cells by a process called *transcytosis:* The endothelial cell internalizes molecules or fluid droplets by endocytosis on one side of the capillary wall, transports the endocytotic vesicles to the other side of the cell, and releases the substances by exocytosis on that side.

CAPILLARY BEDS

Capillaries are organized into groups called **capillary beds**—usually 10 to 100 capillaries supplied by a single metarteriole (fig. 21.9; see also p. 585). Beyond the origins of the capillaries, the metarteriole continues as a **thoroughfare channel** leading directly to a venule. Capillaries empty into the distal end of the thoroughfare channel or directly into the venule.

When the precapillary sphincters are open, the capillaries are well perfused with blood and engage in exchanges with the tissue fluid. When the sphincters are closed, blood bypasses the capillaries, flows through the thoroughfare channel to a venule, and engages in relatively little fluid exchange. There is not enough blood

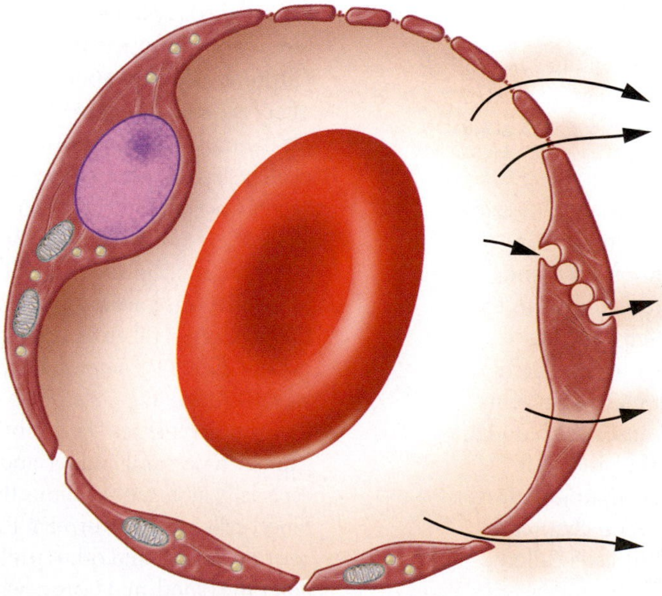
Filtration pores
Transcytosis
Diffusion through
endothelial cells
Intercellular clefts

FIGURE 21.8

Pathways of Capillary Fluid Exchange. Materials can move through the capillary wall through filtration pores (in fenestrated capillaries only), by transcytosis, by diffusion through the endothelial cells, or through intercellular clefts. Although this figure depicts material leaving the bloodstream, material can also enter the bloodstream by the same methods.

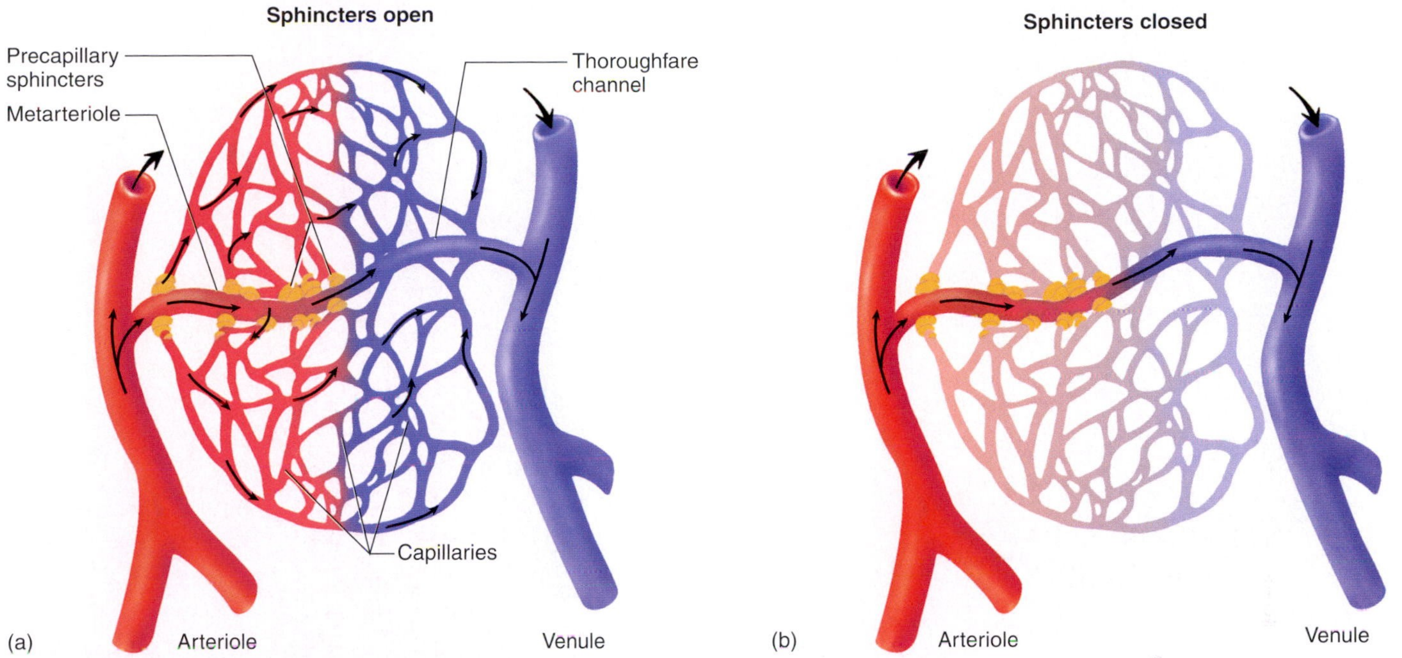
Sphincters open
Precapillary sphincters
Metarteriole
Thoroughfare channel
Capillaries
(a)
Arteriole
Venule
Sphincters closed
(b)
Arteriole
Venule

FIGURE 21.9

Perfusion of a Capillary Bed. (*a*) Precapillary sphincters dilated and capillaries well perfused. (*b*) Precapillary sphincters closed, with blood bypassing the capillaries.

in the body to fill the entire vascular system at once; consequently, about three-quarters of the body's capillaries are shut down at any given time. In the skeletal muscles, for example, about 90% of the capillaries have little to no blood flow during periods of rest. During exercise, they receive an abundant blood flow, while capillary beds elsewhere—for example, in the skin and intestines—shut down to compensate for this.

Veins

Veins are sometimes called the *capacitance vessels* of the cardiovascular system because they are relatively thin-walled and flaccid, and expand easily to accommodate an increased volume of blood; that is, they have a greater capacity for blood containment than arteries do. At rest, about 54% of the blood is found in the systemic veins as compared to only 11% in the systemic arteries (fig. 21.10). The reason that veins are so thin-walled and accommodating is that, being distant from the ventricles of the heart, they are subjected to relatively low blood pressure. In large arteries, blood pressure averages 90 to 100 mmHg (millimeters of mercury) and surges to 120 mmHg during systole, whereas in veins it averages about 10 mmHg. Furthermore, the blood flow in the veins is steady, rather than pulsating with the heartbeat like the flow in the arteries. Veins therefore do not require thick, pressure-resistant walls. Veins collapse when empty and thus have relatively flattened, irregular shapes in histological sections (see fig. 21.2).

As we trace blood flow in the arteries, we find it splitting off repeatedly into smaller and smaller *branches* of the arterial system. In the venous system, conversely, we find small veins merging to form larger and larger ones as they approach the heart. We refer to the smaller veins as *tributaries*, by analogy to the streams that converge and act as tributaries to rivers. In examining the types of veins, we will follow the direction of blood flow, working up from the smallest to the largest vessels.

1. **Postcapillary venules** are the smallest of the veins, beginning with diameters of about 15 to 20 μm. They receive blood from capillaries directly or by way of the distal ends of the thoroughfare channels. They have a tunica interna with only a few fibroblasts around it, and like capillaries, they are often surrounded by pericytes. Postcapillary venules are even more porous than capillaries; therefore venules also exchange fluid with the surrounding tissues. Most leukocytes emigrate from the bloodstream through the venule walls.
2. **Muscular venules** receive blood from the postcapillary venules. They are up to 1 mm in diameter. They have a tunica media of one or two layers of smooth muscle, and a thin tunica externa.
3. **Medium veins** range up to 10 mm in diameter. Most veins with individual names are in this category, such as the radial and ulnar veins of the forearm and the small and great saphenous veins of the lower limb. Medium veins have a tunica interna with an endothelium, basement membrane, loose connective tissue, and sometimes a thin internal elastic lamina. The tunica media is much thinner than it is in medium arteries; it exhibits bundles of smooth muscle, but not a continuous muscular layer as seen in arteries. The muscle is interrupted by regions of collagenous, reticular, and elastic tissue. The tunica externa is relatively thick.

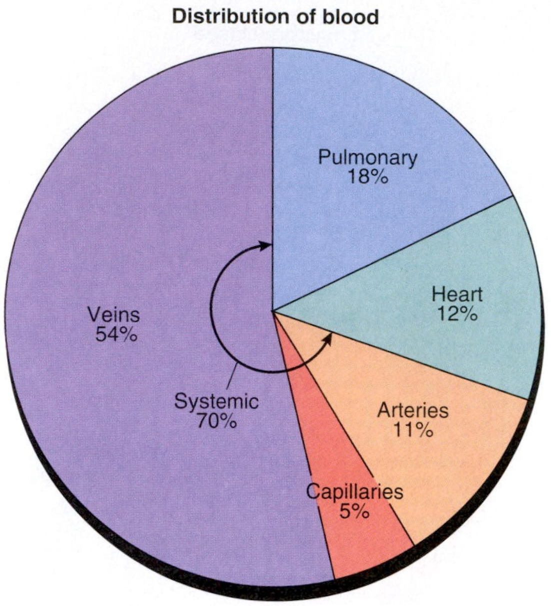

FIGURE 21.10
Typical Distribution of the Blood in a Resting Adult.

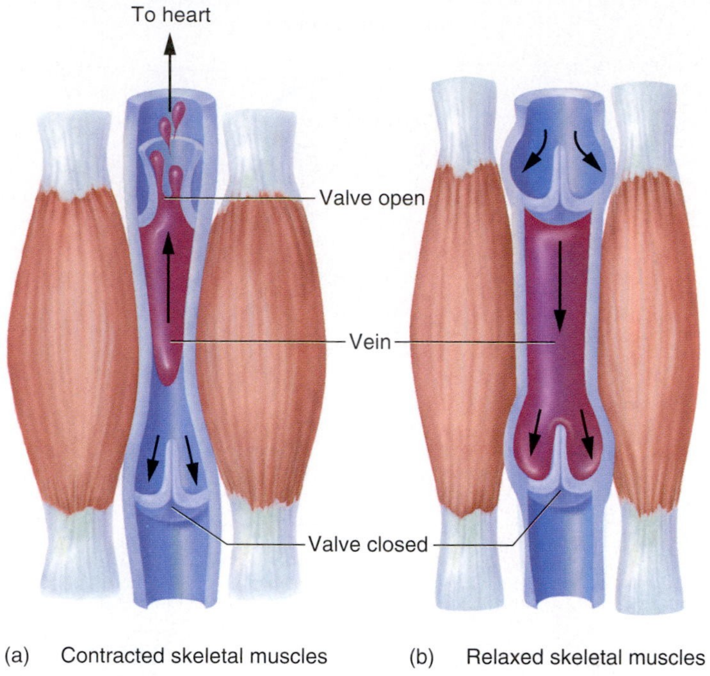

FIGURE 21.11
The Skeletal Muscle Pump. (*a*) Muscle contraction squeezes the deep veins and forces blood through the next valve in the direction of the heart. Valves below the point of compression prevent backflow. (*b*) When the muscles relax, blood flows back downward under the pull of gravity but can only flow as far as the nearest valve.

Many medium veins, especially in the limbs, exhibit infoldings of the tunica interna that meet in the middle of the lumen, forming **venous valves** directed toward the heart (fig. 21.11). The pressure in the veins is not high enough to push blood upward against the pull of gravity in a standing or sitting person. The upward flow of blood in these vessels depends partly on the massaging action of skeletal muscles and the ability of these valves to keep the blood from dropping down again when the muscles relax. When the muscles surrounding a vein contract, they force blood through these valves. The propulsion of venous blood by muscular massaging, aided by the venous valves, is a mechanism of blood flow called the *skeletal muscle pump.* Varicose veins result in part from the failure of the venous valves (see insight 21.2). Such valves are absent from very small and large veins, veins of the abdominal and thoracic cavities, and veins of the brain.

4. **Venous sinuses** are veins with especially thin walls, large lumens, and no smooth muscle. Examples include the coronary sinus of the heart and the dural sinuses of the brain. Unlike other veins, they are not capable of vasomotion.
5. **Large veins** have diameters greater than 10 mm. They have a relatively thin tunica media with only a moderate amount of smooth muscle; the tunica externa is the thickest layer and contains longitudinal bundles of smooth muscle. Some smooth muscle is found in all three layers of these veins. Large veins include the venae cavae, pulmonary veins, internal jugular veins, and renal veins.

INSIGHT 21.2 CLINICAL APPLICATION

VARICOSE VEINS

In people who stand for long periods, such as dentists and hairdressers, blood tends to pool in the lower limbs and stretch the veins. This is especially true of superficial veins, which are not surrounded by supportive tissue. Stretching pulls the cusps of the venous valves farther apart until the valves become incompetent to prevent the backflow of blood. As the veins become further distended, their walls grow weak and they develop into *varicose veins* with irregular dilations and twisted pathways. Obesity and pregnancy also promote development of varicose veins by putting pressure on large veins of the pelvic region and obstructing drainage from the legs. Varicose veins sometimes develop because of hereditary weakness of the valves. With less drainage of blood, tissues of the leg and foot may become edematous and painful. *Hemorrhoids* are varicose veins of the anal canal.

Circulatory Routes

The simplest and most common route of blood flow is heart → arteries → capillaries → veins → heart. Blood usually passes through only one network of capillaries from the time it leaves the heart until the time it returns (fig. 21.12*a*), but there are exceptions, notably portal systems and anastomoses.

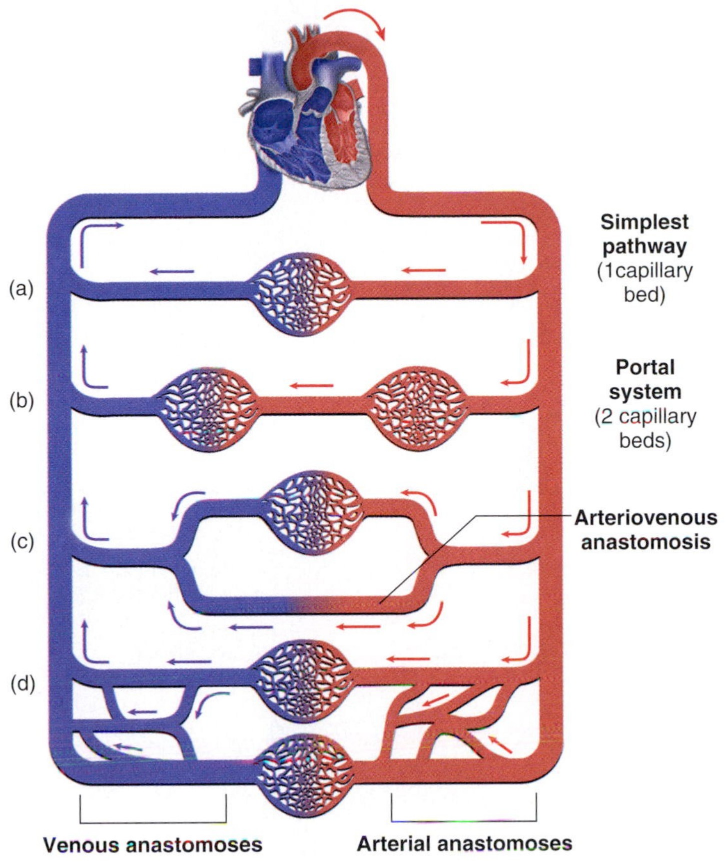

FIGURE 21.12
Variations in Circulatory Pathways.

In a **portal system** (fig. 21.12*b*), blood flows through two consecutive capillary networks before returning to the heart. Portal systems occur within the kidneys (chapter 25); connecting the hypothalamus and anterior pituitary (chapter 18); and connecting the intestines to the liver (table 21.11).

An **anastomosis** is a point where two veins or arteries merge with each other without intervening capillaries. In an **arteriovenous anastomosis (shunt),** blood flows from an artery directly into a vein (fig. 21.12*c*). Shunts occur in the fingers, palms, toes, and ears, where they reduce heat loss in cold weather by allowing warm blood to bypass these exposed surfaces. Unfortunately, this makes these poorly perfused areas more susceptible to frostbite. In an **arterial anastomosis,** two arteries merge, providing *collateral* (alternative) routes of blood supply to a tissue (fig. 21.12*d*). Those of the coronary circulation were mentioned in chapter 20. They are also common around joints where movement may temporarily compress an artery and obstruct one pathway. **Venous anastomoses,** where one vein empties directly into another, are the most common. They provide several alternative routes of drainage from an organ, so blockage of a vein is rarely as life-threatening as blockage of an artery. Several arterial and venous anastomoses are described later in this chapter.

Before You Go On

Answer the following questions to test your understanding of the preceding section:

1. Name the three tunics of a typical blood vessel and explain how they differ from each other.
2. Contrast the tunica media of a conducting artery, arteriole, and venule and explain how the histological differences are related to the functional differences between these vessels.
3. Describe the differences between a continuous capillary, a fenestrated capillary, and a sinusoid.
4. Describe the differences between a medium vein and a medium (muscular) artery. State the functional reasons for these differences.
5. Describe three routes by which substances can escape the bloodstream and pass through a capillary wall into the tissue fluid.

Anatomy of the Pulmonary Circuit

Objective

When you have completed this section, you should be able to

- trace the route of blood through the pulmonary circuit.

The remainder of this chapter centers on the names and pathways of the principal arteries and veins. The pulmonary circuit is described here, and the systemic arteries and veins in the two sections that follow.

The pulmonary circuit (fig. 21.13) begins with the **pulmonary trunk,** a large vessel that ascends diagonally from the right ventricle and branches into the right and left **pulmonary arteries.** Each pulmonary artery enters a medial indentation of the lung called the *hilum* and branches into one **lobar artery** for each lobe of the lung: three on the right and two on the left. These arteries lead ultimately to small basketlike capillary beds that surround the pulmonary alveoli (air sacs). This is where the blood unloads CO_2 and loads O_2. After leaving the alveolar capillaries, the pulmonary blood flows into venules and veins, ultimately leading to the **pulmonary veins** that exit the lung at the hilum. The left atrium of the heart receives two pulmonary veins on each side.

The purpose of the pulmonary circuit is to exchange CO_2 for O_2. It does not serve the metabolic needs of the lung tissue itself; there is a separate systemic supply to the lungs for that purpose, the *bronchial arteries* discussed later.

Before You Go On

Answer the following questions to test your understanding of the preceding section:

6. Trace the flow of an RBC from right ventricle to left atrium, naming the vessels along the way.
7. The lungs have two separate arterial supplies. Explain their functions.

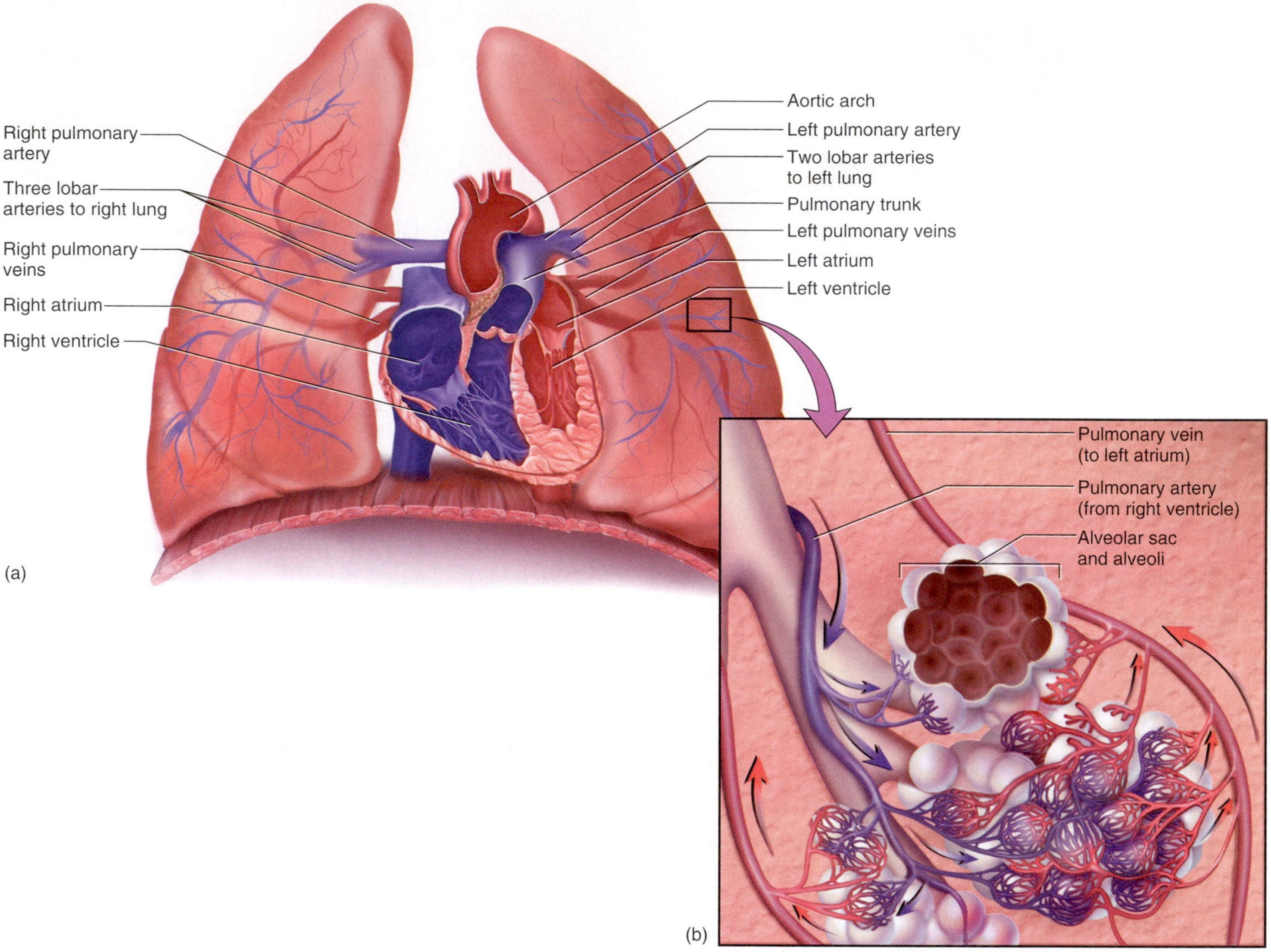
Right pulmonary artery
Three lobar arteries to right lung
Right pulmonary veins
Right atrium
Right ventricle
Aortic arch
Left pulmonary artery
Two lobar arteries to left lung
Pulmonary trunk
Left pulmonary veins
Left atrium
Left ventricle
(a)
Pulmonary vein (to left atrium)
Pulmonary artery (from right ventricle)
Alveolar sac and alveoli
(b)

FIGURE 21.13

The Pulmonary Circulation. (*a*) Gross anatomy. (*b*) Microscopic anatomy of the blood vessels that supply the pulmonary alveoli. All alveoli are surrounded by a basketlike mesh of capillaries; to show the alveoli, this drawing omits the capillaries from some of them.

Anatomy of the Systemic Arteries

Objectives

When you have completed this section, you should be able to

- identify the principal arteries of the systemic circuit; and
- trace the flow of blood from the heart to any major organ.

The systemic circuit supplies oxygen and nutrients to all the organs and removes their metabolic wastes. Part of it, the coronary circulation, was described in chapter 20. The other systemic arteries (fig. 21.14) are described in tables 21.1 through 21.6. The names of the blood vessels often describe their location by indicating the body region traversed (as in the *axillary* artery or *lumbar* artery); an adjacent bone (as in *radial* artery or *temporal* artery); or the organ supplied or drained by the vessel (as in *hepatic* artery or *renal* vein). There is a great deal of anatomical variation in the circulatory system from one person to another. The remainder of this chapter describes the most common pathways.

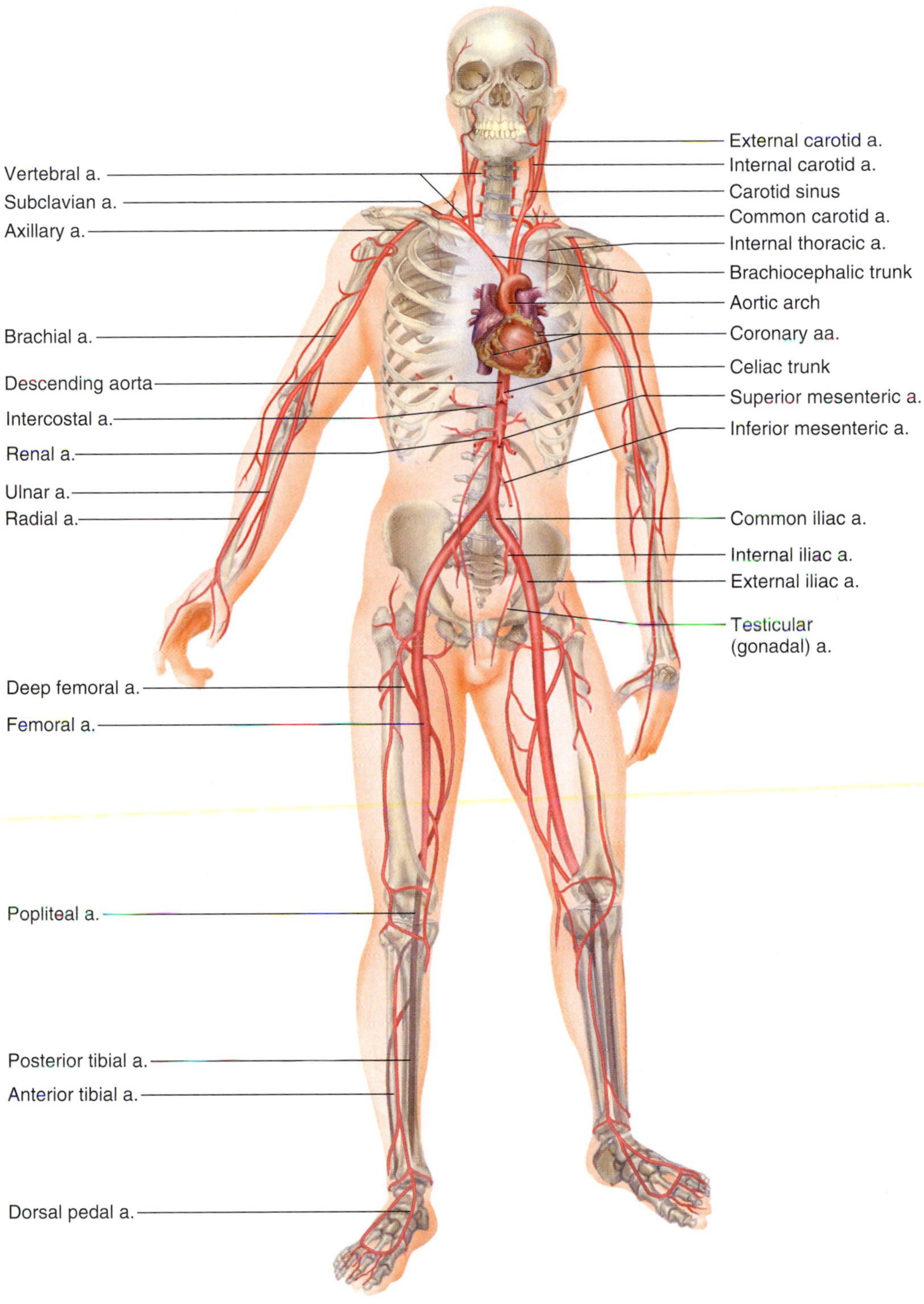
Vertebral a.
Subclavian a.
Axillary a.
Brachial a.
Descending aorta
Intercostal a.
Renal a.
Ulnar a.
Radial a.
Deep femoral a.
Femoral a.
Popliteal a.
Posterior tibial a.
Anterior tibial a.
Dorsal pedal a.
External carotid a.
Internal carotid a.
Carotid sinus
Common carotid a.
Internal thoracic a.
Brachiocephalic trunk
Aortic arch
Coronary aa.
Celiac trunk
Superior mesenteric a.
Inferior mesenteric a.
Common iliac a.
Internal iliac a.
External iliac a.
Testicular
(gonadal) a.

FIGURE 21.14
The Major Systemic Arteries. (*a.* = artery; *aa.* = arteries)

Table 21.1

The Aorta and Its Major Branches

All systemic arteries arise from the aorta, which has three principal regions (fig. 21.15):

1. The **ascending aorta** rises about 5 cm above the left ventricle. Its only branches are the coronary arteries, which arise behind two cusps of the aortic valve.
2. The **aortic arch** curves to the left like an inverted U superior to the heart. It gives off three major arteries in this order: the **brachiocephalic**[5] (BRAY-kee-oh-seh-FAL-ic) **trunk, left common carotid** (cah-ROT-id) **artery,** and **left subclavian**[6] (sub-CLAY-vee-un) **artery,** which are further traced in tables 21.2 and 21.3.
3. The **descending aorta** passes downward dorsal to the heart, at first to the left of the vertebral column and then anterior to it, through the thoracic and abdominal cavities. It is called the **thoracic aorta** above the diaphragm and the **abdominal aorta** below. It ends in the lower abdominal cavity by forking into the *right* and *left common iliac arteries,* which are further traced in table 21.6.

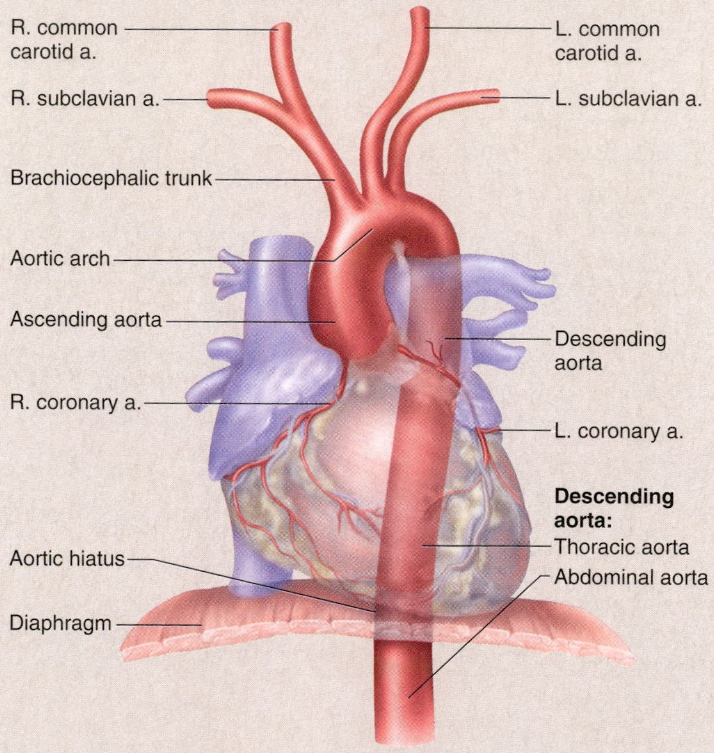
R. common
carotid a.
R. subclavian a.
Brachiocephalic trunk
Aortic arch
Ascending aorta
R. coronary a.
Aortic hiatus
Diaphragm
L. common
carotid a.
L. subclavian a.
Descending
aorta
L. coronary a.
Descending
aorta:
Thoracic aorta
Abdominal aorta

Figure 21.15

The Thoracic Aorta. The thoracic region includes the ascending aorta, aortic arch, and the part of the descending aorta that is superior to the diaphragm. (*R.* = right; *L.* = left; *a.* = artery)

[5]*brachio* = arm + *cephal* = head
[6]*sub* = below + *clavi* = clavicle, collarbone

TABLE 21.2

Arterial Supply to the Head and Neck

Origins of the Head-Neck Arteries

The head and neck receive blood from four pairs of arteries (fig. 21.16):

1. The **common carotid arteries.** The brachiocephalic trunk divides shortly after leaving the aortic arch, giving rise to the *right subclavian* and *right common carotid arteries.* The *left common carotid* artery arises directly from the aortic arch. The common carotids pass up the anterolateral aspect of the neck, alongside the trachea.
2. The **vertebral arteries** arise from the right and left subclavian arteries. Each travels up the neck through the transverse foramina of the cervical vertebrae and enters the cranial cavity through the foramen magnum.
3. The **thyrocervical**[7] **trunks** are tiny arteries that arise from the subclavian arteries lateral to the vertebral arteries; they supply the thyroid gland and some scapular muscles.
4. The **costocervical**[8] **trunks** (also illustrated in table 21.4) arise from the subclavian arteries a little farther laterally. They perfuse the deep neck muscles and some of the intercostal muscles of the superior rib cage.

Continuation of the Common Carotid Arteries

The common carotid arteries have the most extensive distribution of all the head-neck arteries. Near the laryngeal prominence (Adam's apple), each common carotid branches into an *external carotid artery* and an *internal carotid artery.*

1. The **external carotid artery** ascends the side of the head external to the cranium and supplies most external head structures except the orbits. The external carotid gives rise to the following arteries, in ascending order:
   a. the **superior thyroid artery** to the thyroid gland and larynx;
   b. the **lingual artery** to the tongue;
   c. the **facial artery** to the skin and muscles of the face;

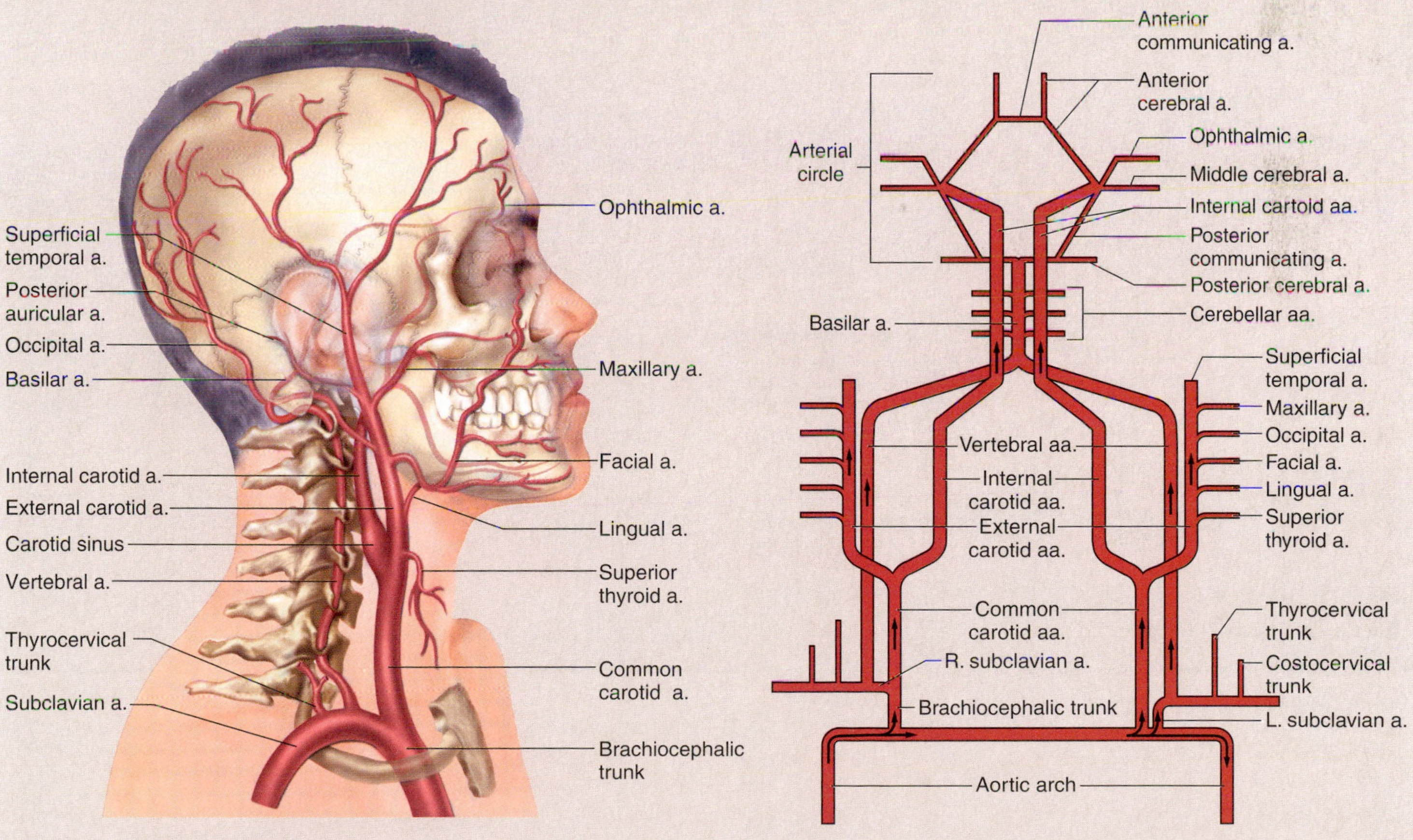

FIGURE 21.16
Arteries of the Head and Neck.

[7]*thyro* = thyroid gland + *cerv* = neck
[8]*costo* = rib

(continued)

Table 21.2

Arterial Supply to the Head and Neck *(continued)*

   d. the **occipital artery** to the posterior scalp;
   e. the **maxillary artery** to the teeth, maxilla, buccal cavity, and external ear; and
   f. the **superficial temporal artery** to the chewing muscles, nasal cavity, lateral aspect of the face, most of the scalp, and the dura mater surrounding the brain.
2. The **internal carotid artery** passes medial to the angle of the mandible and enters the cranial cavity through the carotid canal of the temporal bone. It supplies the orbits and about 80% of the cerebrum. Compressing the internal carotids near the mandible can therefore cause loss of consciousness. After entering the cranial cavity, each internal carotid artery gives rise to the following branches:
   a. the **ophthalmic artery** to the orbits, nose, and forehead;
   b. the **anterior cerebral artery** to the medial aspect of the cerebral hemisphere (see *arterial circle*); and
   c. the **middle cerebral artery,** which travels in the lateral sulcus of the cerebrum and supplies the lateral aspect of the temporal and parietal lobes.

Continuation of the Vertebral Arteries

The vertebral arteries give rise to small branches in the neck that supply the spinal cord and other neck structures, then enter the foramen magnum and merge to form a single **basilar artery** along the anterior aspect of the brainstem. Branches of the basilar artery supply the cerebellum, pons, and inner ear. At the pons-midbrain junction, the basilar artery divides and gives rise to the *arterial circle.*

The Arterial Circle

Blood supply to the brain is so critical that it is furnished by several arterial anastomoses, especially an array of arteries called the **arterial circle (circle of Willis[9]),** which surrounds the pituitary gland and optic chiasm. The arterial circle receives blood from the internal carotid and basilar arteries (fig. 21.17). Only 20% of people have a complete arterial circle. It consists of

1. two **posterior cerebral arteries,**
2. two **posterior communicating arteries,**
3. two **anterior cerebral arteries,** and
4. a single **anterior communicating artery.**

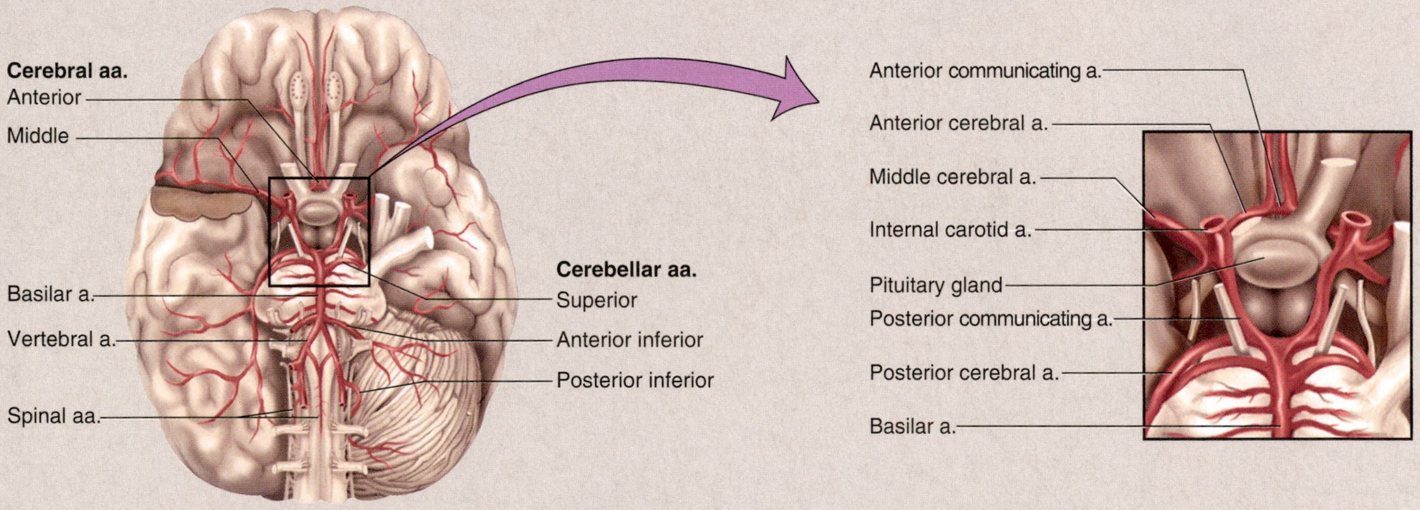
Cerebral aa.
Anterior
Middle
Basilar a.
Vertebral a.
Spinal aa.
Cerebellar aa.
Superior
Anterior inferior
Posterior inferior
Anterior communicating a.
Anterior cerebral a.
Middle cerebral a.
Internal carotid a.
Pituitary gland
Posterior communicating a.
Posterior cerebral a.
Basilar a.

Figure 21.17
The Cerebral Arterial Circle.

[9]Thomas Willis (1621–75), English anatomist

TABLE 21.3

Arterial Supply to the Upper Limb

The Shoulder and Arm (Brachium)

The origins of the subclavian arteries were described and illustrated in tables 21.1 and 21.2. We now trace these further to examine the blood supply to the upper limb (fig. 21.18). This begins with a large artery that changes name from *subclavian* to *axillary* to *brachial* along its course.

1. The **subclavian artery** travels between the clavicle and first rib. It gives off several small branches to the thoracic wall and viscera, considered later.
2. The **axillary artery** is the continuation of the subclavian artery through the axillary region. It also gives off small thoracic branches, discussed later, and then ends at the neck of the humerus. Here, it gives off the **circumflex humeral artery,** which encircles the humerus. This loop supplies blood to the shoulder joint and deltoid muscle.
3. The **brachial** (BRAY-kee-ul) **artery** is the continuation of the axillary artery beyond the circumflex. It travels down the medial side of the humerus and ends just distal to the elbow, supplying the anterior flexor muscles of the brachium along the way. It exhibits several anastomoses near the elbow, two of which are noted next. This is the most commonly used artery for routine blood pressure measurements.
4. The **deep brachial artery** arises from the proximal end of the brachial artery and supplies the triceps brachii muscle.
5. The **ulnar recurrent artery** arises about midway along the brachial artery and anastomoses distally with the ulnar artery. It supplies the elbow joint and the triceps brachii.
6. The **radial recurrent artery** leads from the deep brachial artery to the radial artery and supplies the elbow joint and forearm muscles.

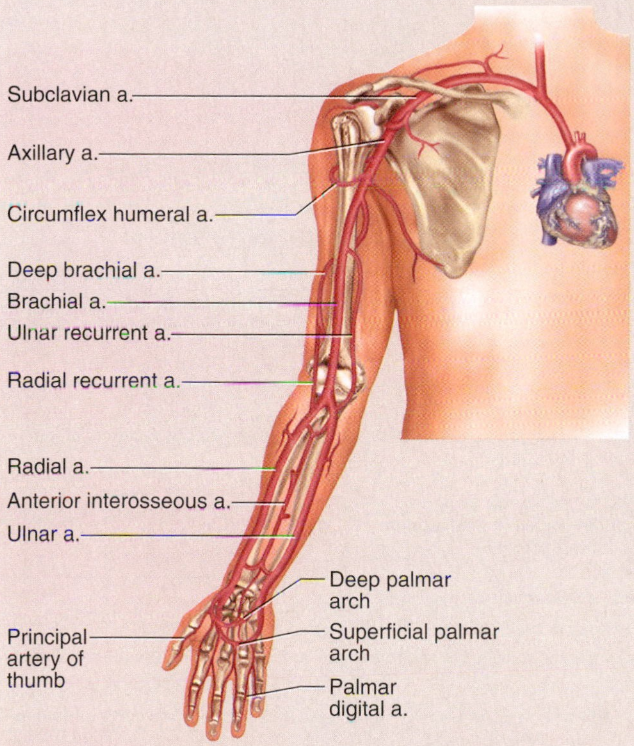

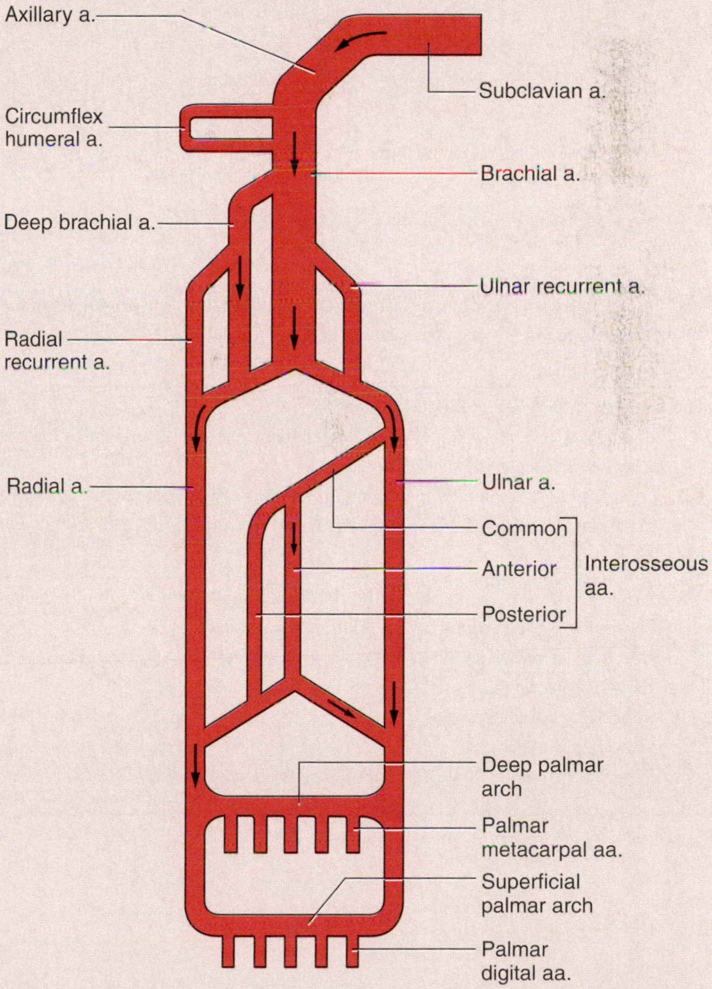

FIGURE 21.18
Arteries of the Upper Limb.

(continued)

Table 21.3

Arterial Supply to the Upper Limb *(continued)*

The Forearm (Antebrachium)

Just distal to the elbow, the brachial artery divides into the **radial artery** and **ulnar artery,** which travel alongside the radius and ulna, respectively. The most common place to take a pulse is at the radial artery, just proximal to the thumb. Near its origin, the radial artery receives the deep brachial artery. The ulnar artery gives rise, near its origin, to the **anterior** and **posterior interosseous**[10] **arteries,** which travel between the radius and ulna. Structures supplied by these three arteries are as follows:

1. Radial artery: lateral forearm muscles, wrist, thumb, and index finger.
2. Ulnar artery: medial forearm muscles, digits III to V, and medial aspect of index finger.
3. Interosseous arteries: deep flexors and extensors.

The Hand

At the wrist, the radial and ulnar arteries anastomose to form two *palmar arches:*

1. The **deep palmar arch** gives rise to the **metacarpal arteries** of the hand.
2. The **superficial palmar arch** gives rise to the **digital arteries** of the fingers.

[10] *inter* = between + *osse* = bones

Table 21.4

Arterial Supply to the Thorax

The thoracic aorta begins distal to the aortic arch and ends at the **aortic hiatus** (hy-AY-tus), a passage through the diaphragm. Along the way, it sends off numerous small branches to viscera and structures of the body wall (fig. 21.19).

Visceral Branches

These supply the viscera of the thoracic cavity:

1. **Bronchial arteries.** Two of these on the left and one on the right supply the visceral pleura, esophagus, and bronchi of the lungs. They are the systemic blood supply to the lungs mentioned earlier.
2. **Esophageal arteries.** Four or five of these supply the esophagus.
3. **Mediastinal arteries.** Many small mediastinal arteries (not illustrated) supply structures of the posterior mediastinum.

Parietal Branches

The following branches supply chiefly the muscles, bones, and skin of the chest wall; only the first is illustrated:

1. **Posterior intercostal arteries.** Nine pairs of these course around the posterior aspect of the rib cage between the ribs, then anastomose with the anterior intercostal arteries (see following). They supply the skin and subcutaneous tissue, mammary glands, spinal cord and meninges, and the pectoralis, intercostal, and some abdominal muscles.
2. **Subcostal arteries.** A pair of these arise from the aorta, inferior to the twelfth rib, and supply the posterior intercostal tissues, vertebrae, spinal cord, and deep muscles of the back.
3. **Superior phrenic**[11] (FREN-ic) **arteries.** These supply the posterior and superior aspects of the diaphragm.

[11] *phren* = diaphragm

(continued)

TABLE 21.4

Arterial Supply to the Thorax *(continued)*

The thoracic wall is also supplied by the following arteries. The first of these arises from the subclavian artery and the other three from the axillary artery:

1. The **internal thoracic (mammary) artery** supplies the breast and anterior thoracic wall and issues finer branches to the diaphragm and abdominal wall. Near its origin, it gives rise to the **pericardiophrenic artery,** which supplies the pericardium and diaphragm. As the internal thoracic artery descends alongside the sternum, it gives rise to **anterior intercostal arteries** that travel between the ribs and supply the ribs and intercostal muscles.
2. The **thoracoacromial**[12] (THOR-uh-co-uh-CRO-me-ul) **trunk** supplies the superior shoulder and pectoral regions.
3. The **lateral thoracic artery** supplies the lateral thoracic wall.
4. The **subscapular artery** supplies the scapula, latissimus dorsi, and posterior wall of the thorax.

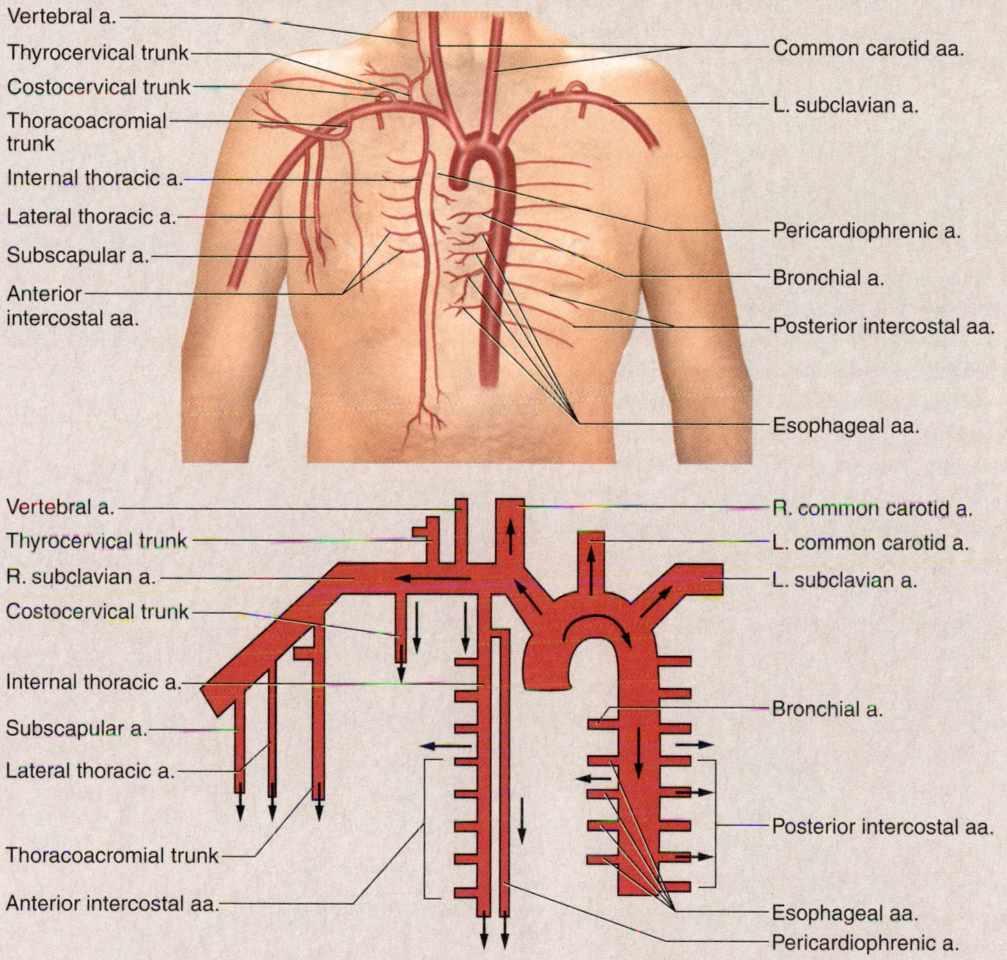

FIGURE 21.19
Arteries of the Thorax.

[12]*thoraco* = chest + *acr* = tip + *om* = shoulder

Table 21.5

Arterial Supply to the Abdomen

Major Branches of the Abdominal Aorta

After passing through the aortic hiatus, the aorta descends through the abdominal cavity. The abdominal aorta is retroperitoneal. It gives off arteries in the order listed here (fig. 21.20). Those indicated in the plural are paired right and left, and those indicated in the singular are single median arteries.

1. The **inferior phrenic arteries** supply the inferior surface of the diaphragm and issue a small **superior suprarenal artery** to each adrenal (suprarenal) gland.
2. The **celiac**[13] (SEE-lee-ac) **trunk** issues several branches to the upper abdominal viscera, further traced later in this table.
3. The **superior mesenteric artery** supplies the intestines (see mesenteric circulation later in this table).
4. The **middle suprarenal arteries** arise on either side of the superior mesenteric artery and supply the adrenal glands.
5. The **renal arteries** supply the kidneys and issue a small **inferior suprarenal artery** to each adrenal gland.
6. The **gonadal arteries** are long, narrow, winding arteries that descend from the midabdominal region to the female pelvic cavity or male scrotum. They are called the **ovarian arteries** in females and **testicular arteries** in males. The gonads begin their embryonic development near the kidneys. These arteries acquire their peculiar length and course as the gonads descend to the pelvic cavity during fetal development.
7. The **inferior mesenteric artery** supplies the distal end of the large intestine (see mesenteric circulation).
8. The **lumbar arteries** arise from the lower aorta in four pairs and supply the dorsal abdominal wall.
9. The **median sacral artery,** a tiny medial artery at the inferior end of the aorta, supplies the sacrum and coccyx.
10. The **common iliac arteries** arise as the aorta forks at its inferior end. They supply the lower abdominal wall, pelvic viscera (chiefly the urinary and reproductive organs), and lower limbs. They are further traced in table 21.6.

Branches of the Celiac Trunk

The celiac circulation to the upper abdominal viscera is perhaps the most complex route off the abdominal aorta. Because it has numerous anastomoses, the bloodstream does not follow a simple linear path but divides and rejoins itself at several points (fig. 21.21). As you study the following description, locate these branches in the figure and identify the points of anastomosis. The short, stubby celiac trunk is a median branch of the aorta. It immediately gives rise to three principal subdivisions—the *common hepatic, left gastric,* and *splenic arteries.*

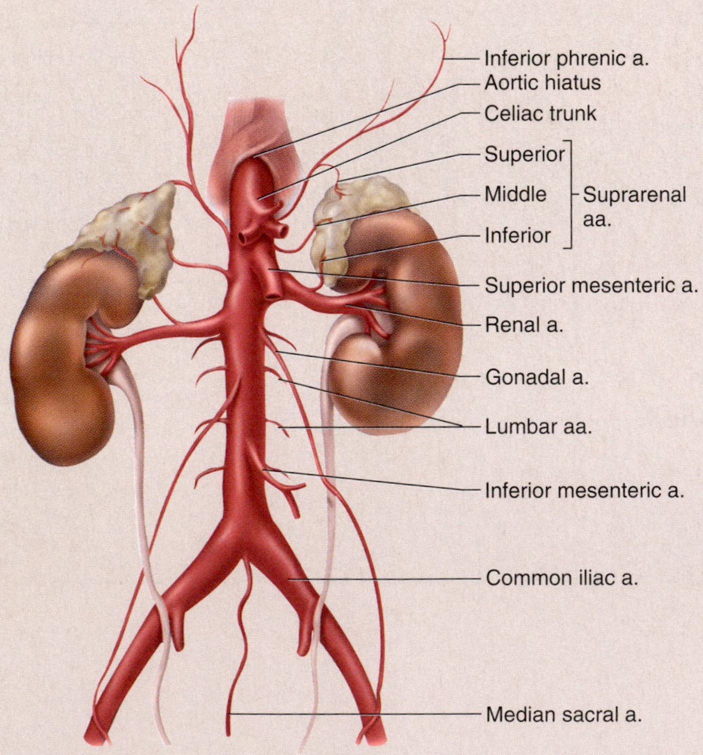

Inferior phrenic a.
Aortic hiatus
Celiac trunk
Superior
Middle
Inferior
Suprarenal aa.
Superior mesenteric a.
Renal a.
Gonadal a.
Lumbar aa.
Inferior mesenteric a.
Common iliac a.
Median sacral a.

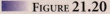

Figure 21.20
The Abdominal Aorta and Its Major Branches.

[13] *celi* = belly, abdomen

(continued)

TABLE 21.5

Arterial Supply to the Abdomen *(continued)*

I. The **common hepatic artery** issues two main branches:

a. the **gastroduodenal artery,** which supplies the stomach, anastomoses with the right gastroepiploic artery (see p 606), and then continues as the **inferior pancreaticoduodenal** (PAN-cree-AT-ih-co-dew-ODD-eh-nul) **artery,** which supplies the duodenum and pancreas before anastomosing with the superior mesenteric artery; and

b. the **proper hepatic artery,** which is the continuation of the common hepatic artery after it gives off the gastroduodenal artery. It enters the inferior surface of the liver and supplies the liver and gallbladder.

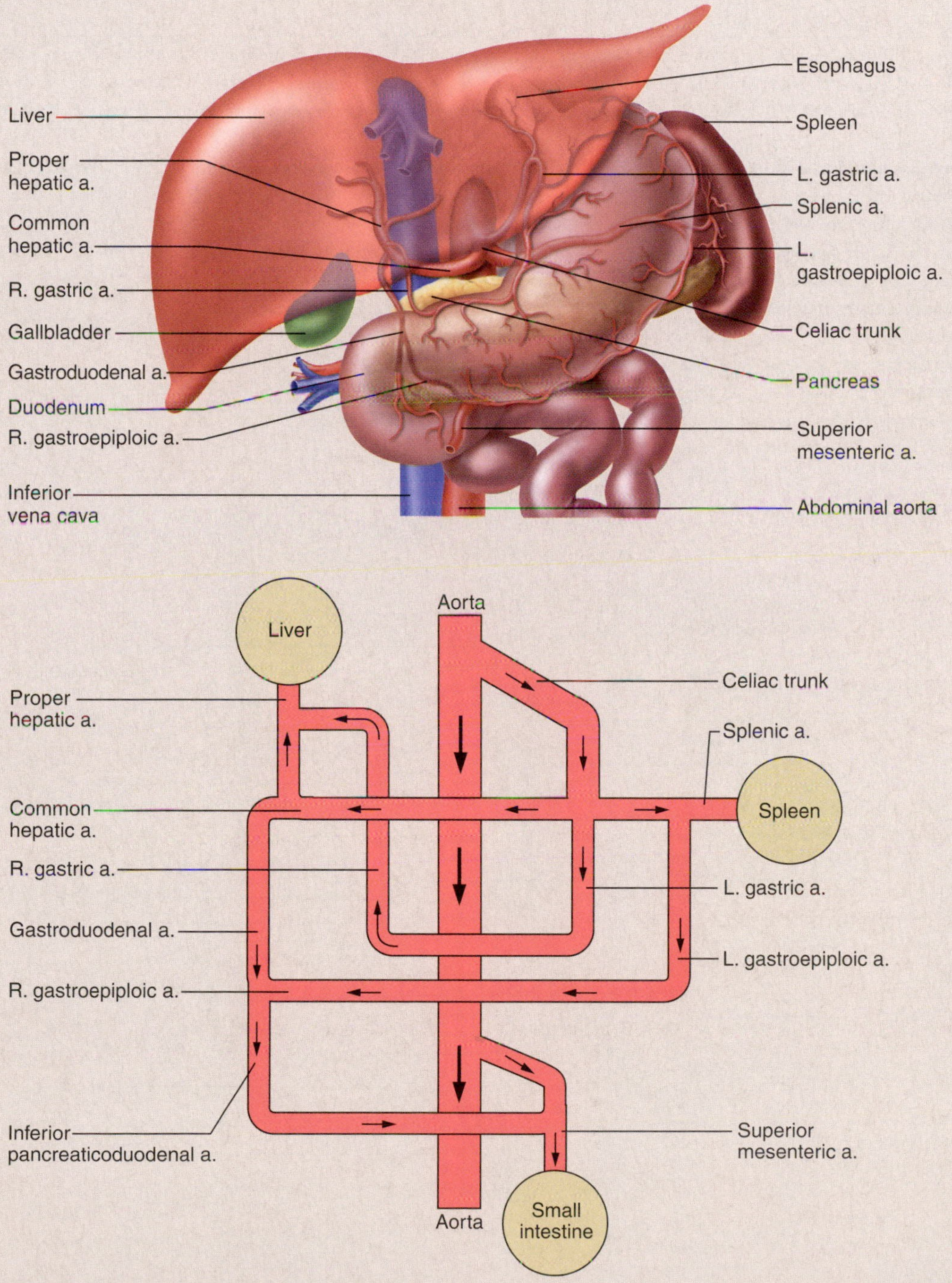

FIGURE 21.21
Branches of the Celiac Trunk.

(continued)

Table 21.5

Arterial Supply to the Abdomen *(continued)*

2. The **left gastric artery** supplies the stomach and lower esophagus, arcs around the *lesser curvature* of the stomach, becomes the **right gastric artery** (which supplies the stomach and duodenum), and then anastomoses with the hepatic artery.
3. The **splenic artery** supplies blood to the spleen, but gives off the following branches on its way there:
   a. the **pancreatic arteries** (not illustrated), which supply the pancreas; and
   b. the **left gastroepiploic**[14] (GAS-tro-EP-ih-PLO-ic) **artery,** which arcs around the *greater curvature* of the stomach, becomes the **right gastroepiploic artery,** and then anastomoses with the gastroduodenal artery. Along the way, it supplies blood to the stomach and *greater omentum* (a fatty membrane suspended from the greater curvature).

Mesenteric Circulation

The mesentery (see atlas A, p. 30) contains numerous mesenteric arteries, veins, and lymphatic vessels that perfuse and drain the intestines. The arterial supply issues from the *superior* and *inferior mesenteric arteries* (fig. 21.22); numerous anastomoses between these ensure collateral circulation and adequate perfusion of the intestinal tract even if one route becomes obstructed. The following branches of the **superior mesenteric artery** serve the small intestine and most of the large intestine, among other organs:

1. The **inferior pancreaticoduodenal artery,** already mentioned, is an anatomosis from the gastroduodenal to the superior mesenteric artery; it supplies the pancreas and duodenum.
2. The **intestinal arteries** supply nearly all of the small intestine (jejunum and ileum).
3. The **ileocolic** (ILL-ee-oh-CO-lic) **artery** supplies the ileum of the small intestine and the appendix, cecum, and ascending colon.
4. The **right colic artery** supplies the ascending colon.
5. The **middle colic artery** supplies the transverse colon.

Branches of the *inferior mesenteric artery* serve the distal part of the large intestine:

1. The **left colic artery** supplies the transverse and descending colon.
2. The **sigmoid arteries** supply the descending and sigmoid colon.
3. The **superior rectal artery** supplies the rectum.

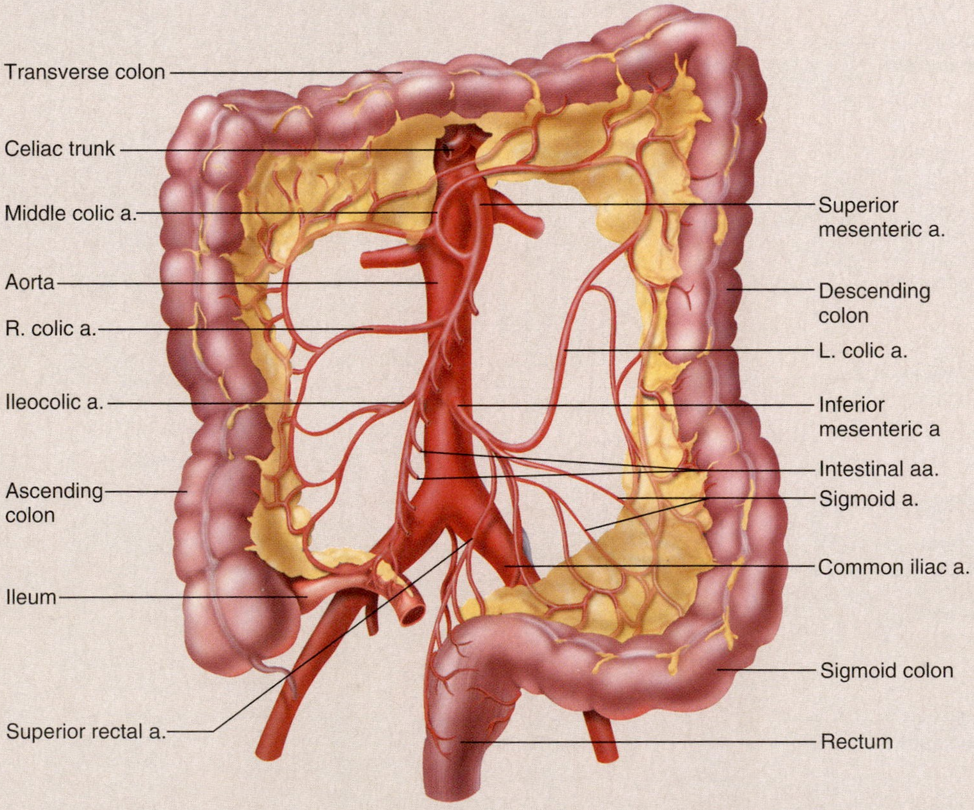

Figure 21.22
The Mesenteric Arteries.

[14]*gastro* = stomach + *epi* = upon, above + *ploic* = pertaining to the greater omentum

TABLE 21.6

Arterial Supply to the Pelvic Region and Lower Limb

The common iliac arteries arise from the aorta at the level of vertebra L4 and continue for about 5 cm. At the level of the sacroiliac joint, each divides into an internal and external iliac artery. The **internal iliac artery** supplies mainly the pelvic wall and viscera, and the **external iliac artery** supplies mainly the lower limb (figs. 21.23 and 21.24).

Branches of the Internal Iliac Artery

1. The **iliolumbar** and **lateral sacral arteries** supply the wall of the pelvic region.
2. The **middle rectal artery** supplies the rectum.
3. The **superior** and **inferior vesical**[15] **arteries** supply the urinary bladder.
4. The **uterine** and **vaginal arteries** supply the uterus and vagina.
5. The **superior** and **inferior gluteal arteries** supply the gluteal muscles.
6. The **obturator artery** supplies the adductor muscles of the medial thigh.
7. The **internal pudendal**[16] (pyu-DEN-dul) **artery** serves the perineum and external genitals; it supplies the blood for vascular engorgement during sexual arousal.

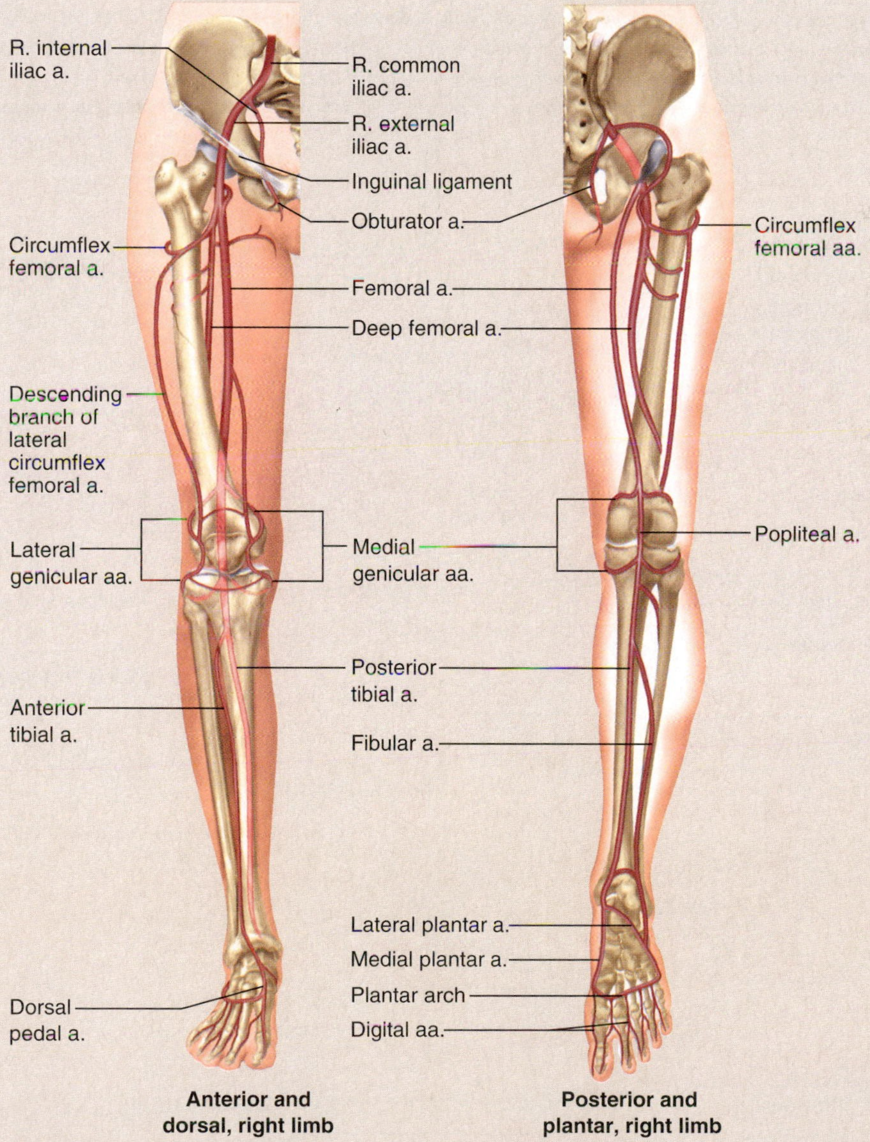

FIGURE 21.23
Arteries of the Lower Limb.

[15] *vesic* = bladder
[16] *pudend* = literally "shameful parts"; the external genitals

(continued)

TABLE 21.6

Arterial Supply to the Pelvic Region and Lower Limb *(continued)*

Branches of the External Iliac Artery

The external iliac artery sends branches to the skin and muscles of the abdominal wall and pelvic girdle. It then passes deep to the inguinal ligament and gives rise to branches that serve mainly the lower limbs.

1. The **femoral artery** passes through the femoral triangle of the upper medial thigh, where its pulse can be palpated. It gives off the following branches to supply the thigh region:
   a. the **deep femoral artery,** which supplies the hamstring muscles; and
   b. the **circumflex femoral arteries,** which encircle the neck of the femur and supply the femur and hamstring muscles.
2. The **popliteal artery** is a continuation of the femoral artery in the popliteal fossa at the rear of the knee. It produces anastomoses **(genicular arteries)** that supply the knee, then divides into the anterior and posterior tibial arteries.
3. The **anterior tibial artery** travels lateral to the tibia in the anterior compartment of the leg, where it supplies the extensor muscles. It gives rise to
   a. the **dorsal pedal artery,** which traverses the ankle and dorsum of the foot; and
   b. the **arcuate artery,** a continuation of the dorsal pedal artery that gives off the **metatarsal arteries** of the foot.
4. The **posterior tibial artery** travels through the posteromedial part of the leg and supplies the flexor muscles. It gives rise to
   a. the **fibular (peroneal) artery,** which arises from the proximal end of the posterior tibial artery and supplies the lateral peroneal muscles;
   b. the **lateral and medial plantar arteries,** which arise by bifurcation of the posterior tibial artery at the ankle and supply the plantar surface of the foot; and
   c. the **plantar arch,** an anastomosis from the lateral plantar artery to the dorsal pedal artery that gives rise to the digital arteries of the toes.

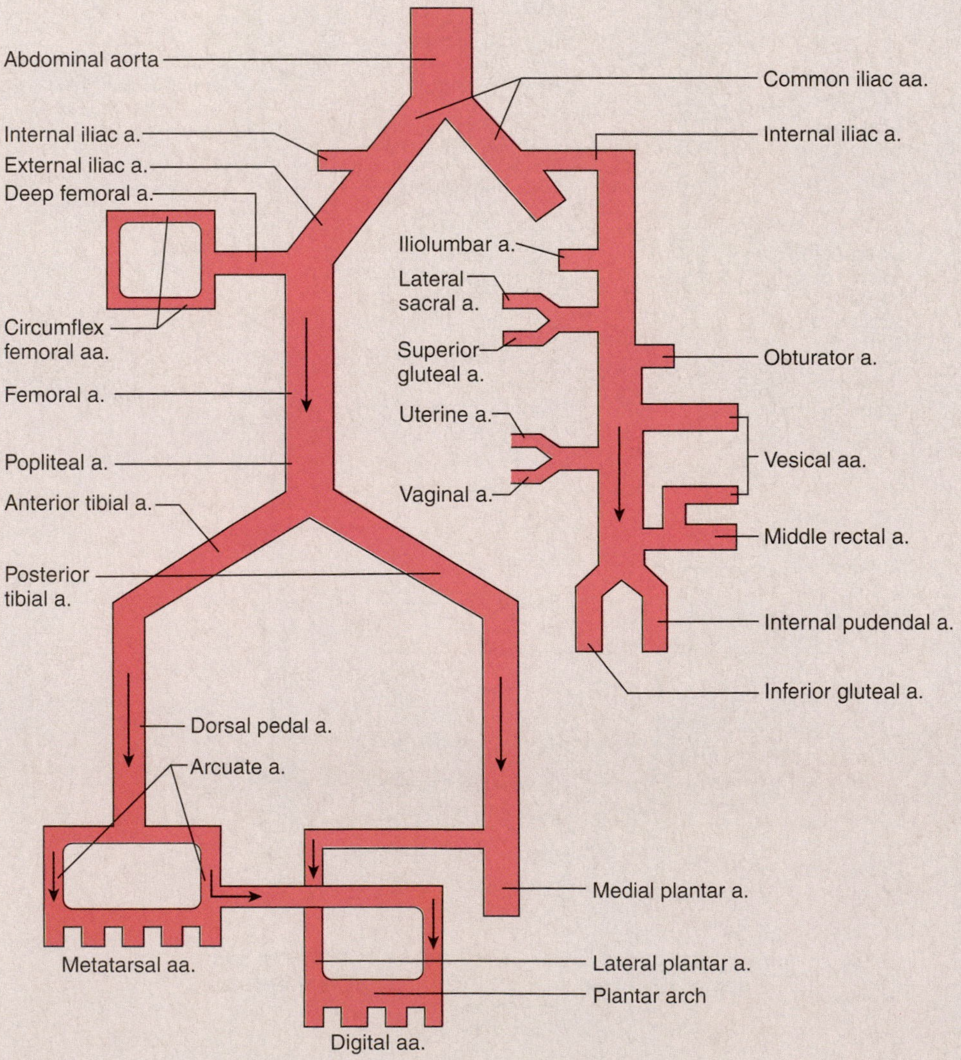

FIGURE 21.24
Arterial Flowchart of the Lower Limb.

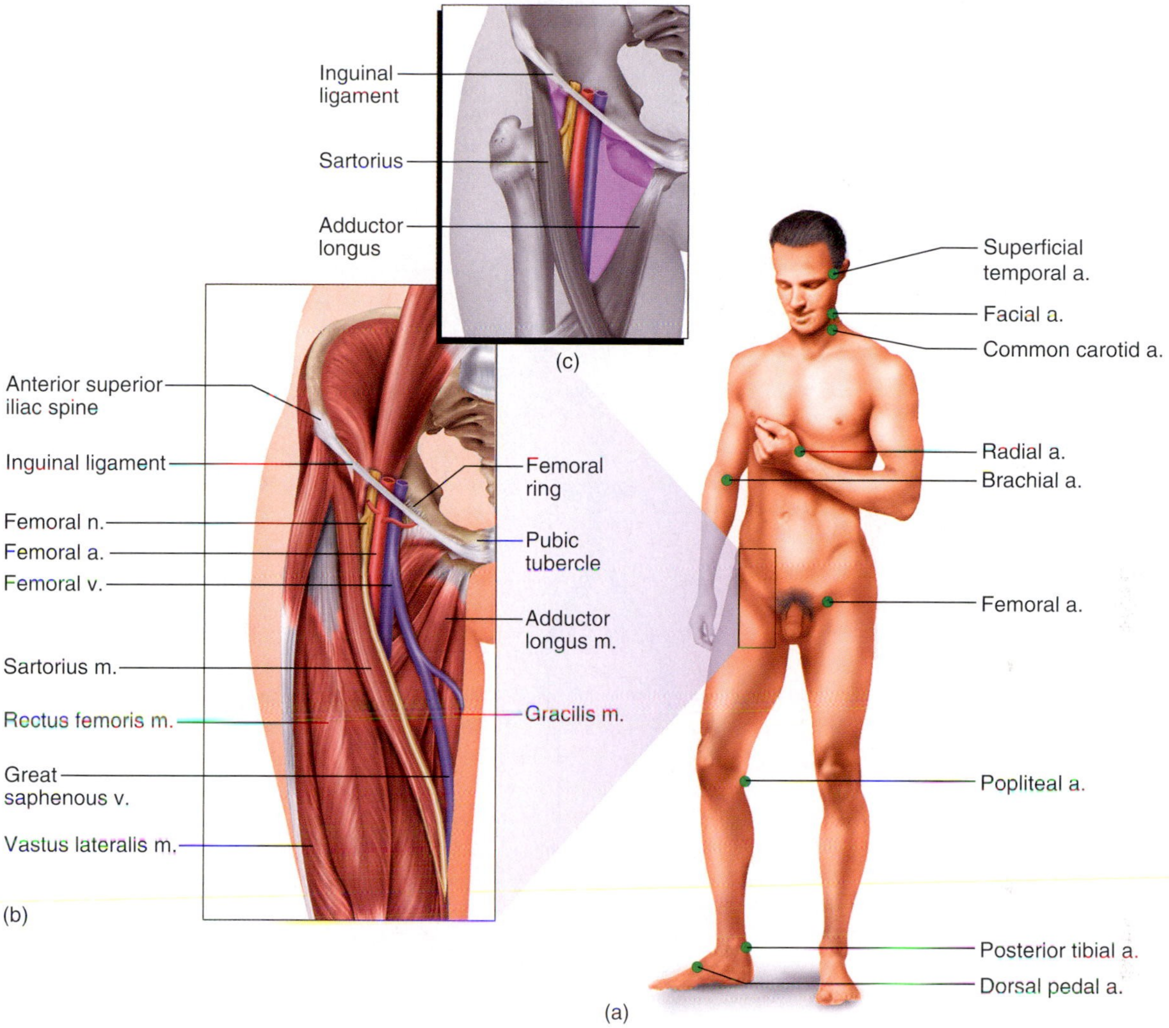

FIGURE 21.25

Arterial Pressure Points. (*a*) Areas where arteries lie close enough to the surface that a pulse can be palpated or pressure can be applied to reduce arterial bleeding. (*b*) Structures in the femoral triangle. (c) Boundaries of the femoral triangle.

Think About It!

There are certain similarities between the arteries of the hand and foot. What arteries of the wrist and hand are most comparable in arrangement and function to the arcuate artery and plantar arch of the foot?

Arterial Pressure Points

In some places, major arteries come close enough to the body surface to be palpated. These places can be used to take a pulse, and they can serve as emergency **pressure points** where firm pressure can be applied to temporarily reduce arterial bleeding (fig. 21.25*a*). One of these points is the **femoral triangle** of the upper medial thigh (fig. 21.25*b–c*). This is an important landmark for arterial supply, venous drainage, and innervation of the lower limb. Its boundaries are the sartorius muscle laterally, the inguinal ligament superiorly, and the adductor longus muscle medially. The femoral artery, vein, and nerve run close to the surface at this point.

Before You Go On

Answer the following questions to test your understanding of the preceding section:

8. Concisely contrast the destinations of the external and internal carotid arteries.
9. Briefly state the tissues that are supplied with blood by (a) the arterial circle, (b) the celiac trunk, (c) the superior mesenteric artery, and (d) the external iliac artery.
10. Trace the path of an RBC from the left ventricle to the metatarsal arteries. State two places along this path where you can palpate the arterial pulse.

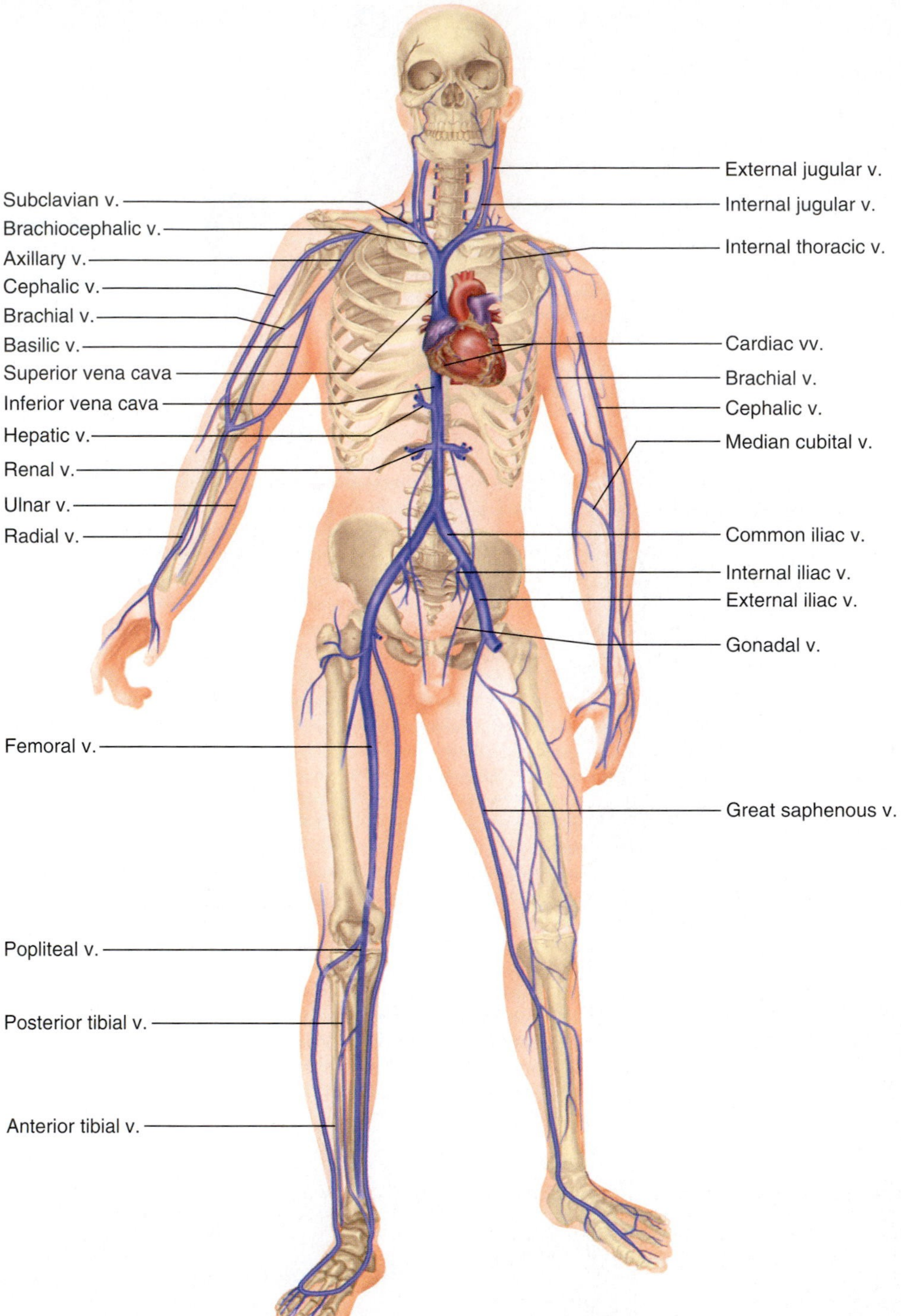
External jugular v.
Internal jugular v.
Internal thoracic v.
Subclavian v.
Brachiocephalic v.
Axillary v.
Cephalic v.
Brachial v.
Basilic v.
Superior vena cava
Inferior vena cava
Hepatic v.
Renal v.
Ulnar v.
Radial v.
Cardiac vv.
Brachial v.
Cephalic v.
Median cubital v.
Common iliac v.
Internal iliac v.
External iliac v.
Gonadal v.
Femoral v.
Great saphenous v.
Popliteal v.
Posterior tibial v.
Anterior tibial v.

FIGURE 21.26
The Major Systemic Veins. (*v.* = Vein; *vv.* = Veins)

ANATOMY OF THE SYSTEMIC VEINS

Objective

When you have completed this section, you should be able to

- identify the principal veins of the systemic circuit and trace the flow of blood from any major organ to the heart.

The principal veins of the systemic circuit (fig. 21.26) are detailed in tables 21.7 through 21.12. While arteries are usually deep and well protected, veins occur in both deep and superficial groups; you may be able to see quite a few of them in your arms and hands.

Deep veins run parallel to the arteries and often have similar names (*femoral artery* and *femoral vein,* for example); this is

(continued on p. 620)

Table 21.7

Venous Drainage of the Head and Neck

Most blood of the head and neck is drained by three pairs of veins—the *internal jugular, external jugular,* and *vertebral veins.* This table traces their origins and drainage and follows them to the formation of the *brachiocephalic veins* and *superior vena cava* (fig. 21.27).

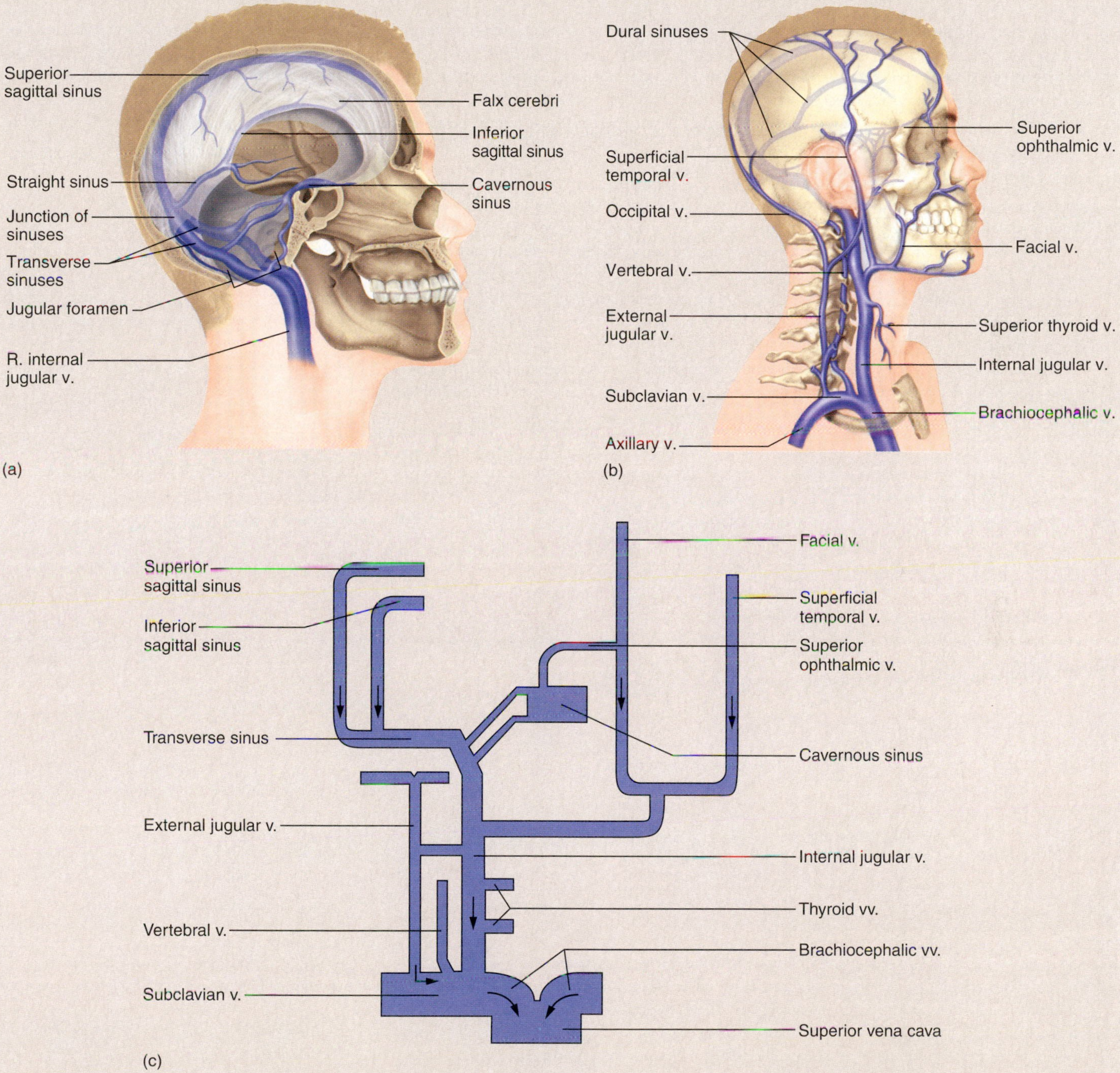

Figure 21.27

Veins of the Head and Neck. (*a*) Deep venous drainage. (*b*) Superficial venous drainage. (c) Flowchart of venous drainage.

(continued)

Table 21.7

Venous Drainage of the Head and Neck *(continued)*

Dural Sinuses

Large thin-walled veins called **dural sinuses** occur within the cranial cavity between layers of dura mater. They receive blood from the brain and face and empty into the internal jugular veins.

1. The **superior** and **inferior sagittal sinuses** are found in the falx cerebri between the cerebral hemispheres; they receive blood that has circulated through the brain.
2. The **cavernous sinuses** occur on each side of the body of the sphenoid bone; they receive blood from the **superior ophthalmic vein** draining the orbit, and the **facial vein** draining the nose and upper lip.
3. The **transverse (lateral) sinuses** encircle the inside of the occipital bone and lead to the jugular foramen on each side. They receive blood from the previously mentioned sinuses and empty into the internal jugular veins.

Major Veins of the Neck

Blood flows down the neck mainly through three veins on each side, all of which empty into the subclavian vein.

1. The **internal jugular**[17] (JUG-you-lur) **vein** courses down the neck, alongside the internal carotid artery, deep to the sternocleidomastoid muscle. It receives most of the blood from the brain, picks up blood from the **facial vein** and **superficial temporal vein** along the way, passes deep to the clavicle, and joins the subclavian vein. (Note that the facial vein empties into both the cavernous sinus and the internal jugular vein.)
2. The **external jugular vein** drains tributaries from the parotid gland, facial muscles, scalp, and other superficial structures. Some of this blood also follows venous anastomoses to the internal jugular vein. The external jugular vein courses down the side of the neck superficial to the sternocleidomastoid muscle and empties into the subclavian vein.
3. The **vertebral vein** travels with the vertebral artery in the transverse foramina of the cervical vertebrae. Although the companion artery leads to the brain, the vertebral vein does not come from there. It drains the cervical vertebrae, spinal cord, and some of the small deep muscles of the neck.

Drainage from Shoulder to Heart

From the shoulder region, blood takes the following path to the heart:

1. The **subclavian vein** drains the arm and travels inferior to the clavicle; receives the external jugular, vertebral, and internal jugular veins in that order; and ends where it receives the internal jugular.
2. The **brachiocephalic vein** is formed by union of the subclavian and internal jugular veins. It continues medially and receives tributaries draining the upper thoracic wall and breast.
3. The **superior vena cava** is formed by the union of the right and left brachiocephalic veins. It travels inferiorly for about 7.5 cm and empties into the right atrium. It drains all structures superior to the diaphragm except the pulmonary circuit and coronary circulation. It also receives considerable drainage from the abdominal cavity by way of the azygos system (see table 21.9).

[17] *jugul* = neck, throat

Table 21.8

Venous Drainage of the Upper Limb

Table 21.7 briefly noted the subclavian veins that drain each arm. This table begins distally in the forearm and traces venous drainage to the subclavian vein (fig. 21.28).

Deep Veins

1. The **digital veins** drain each finger into the **superficial palmar venous arch.**
2. The **metacarpal veins** parallel the metacarpal bones and drain blood from the hand into the **deep palmar venous arch.** Both the superficial and deep palmar venous arches are anastomoses between the next two veins, which are the major deep veins of the forearm.
3. The **radial vein** receives blood from the lateral side of both palmar arches and courses up the forearm alongside the radius.
4. The **ulnar vein** receives blood from the medial side of both palmar arches and courses up the forearm alongside the ulna.
5. The **brachial vein** is formed by the union of the radial and ulnar veins at the elbow; it courses up the brachium.
6. The **axillary vein** is formed at the axilla by the union of the brachial and basilic veins.
7. The **subclavian vein** is a continuation of the axillary vein into the shoulder inferior to the clavicle. The further course of the subclavian is explained in table 21.7.

(continued)

TABLE 21.8

Venous Drainage of the Upper Limb *(continued)*

Superficial Veins

These are easily seen through the skin of most people and are larger in diameter than the deep veins.

1. The **dorsal venous arch** (not illustrated) is a plexus of veins visible on the back of the hand; it empties into the major superficial veins of the forearm, the cephalic and basilic.
2. The **cephalic vein** arises from the lateral side of the dorsal venous arch, winds around the radius as it travels up the forearm, continues up the lateral aspect of the brachium to the shoulder, and joins the axillary vein there. Intravenous fluids are often administered through the distal end of this vein.
3. The **basilic**[18] (bah-SIL-ic) **vein** arises from the medial side of the dorsal venous arch, travels up the posterior aspect of the forearm, and continues into the brachium. About midway up the brachium it turns deeper and runs beside the brachial artery. At the axilla it joins the brachial vein, and the union of these two gives rise to the axillary vein.
4. The **median cubital vein** is a short anastomosis between the cephalic and basilic veins that obliquely crosses the cubital fossa (anterior bend of the elbow). It is clearly visible through the skin and is the most common site for drawing blood.
5. The **median antebrachial vein** originates near the base of the thumb, travels up the forearm between the radial and ulnar veins, and terminates at the elbow, emptying into the cephalic vein in some people and into the basilic vein in others.

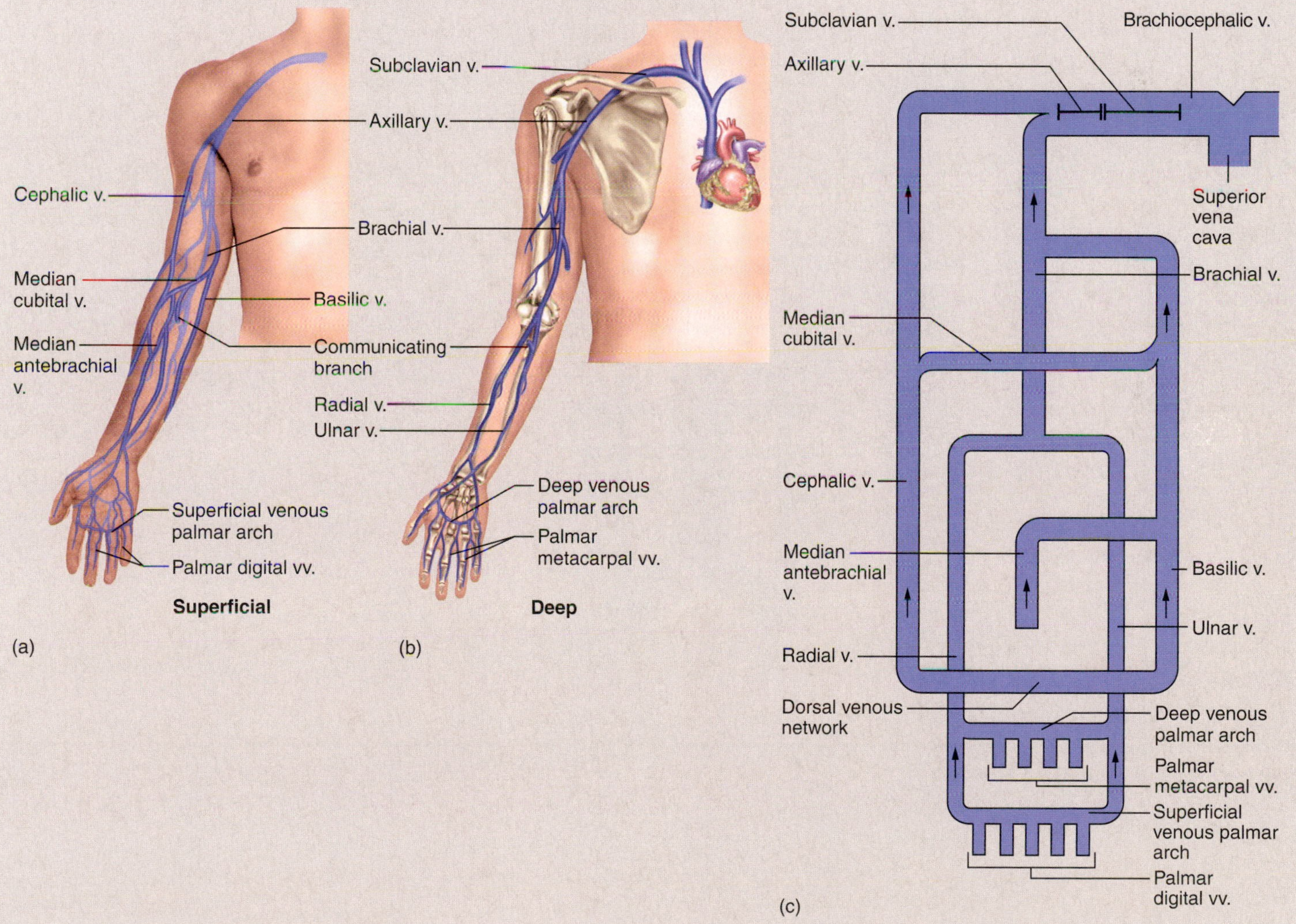

FIGURE 21.28

Veins of the Upper Limb. (*a*) Superficial venous drainage. (*b*) Deep venous drainage. (*c*) Flowchart of venous drainage.

[18] *basilic* = royal, prominent, important

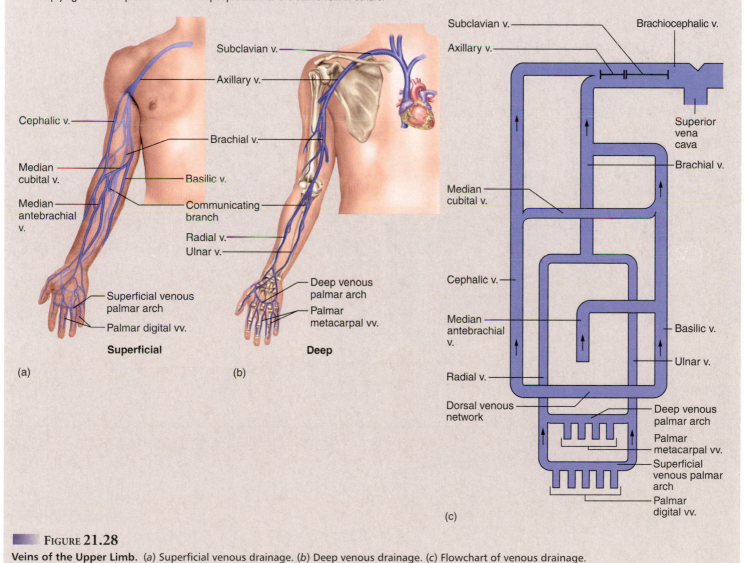

Table 21.9

The Azygos System

The superior vena cava receives extensive drainage from the thoracic and abdominal walls by way of the **azygos** (AZ-ih-goss) **system** (fig. 21.29).

Drainage of the Abdominal Wall

A pair of **ascending lumbar veins** receive blood from the common iliac veins below and a series of short horizontal **lumbar veins** that drain the abdominal wall. The ascending lumbar veins anastomose with the inferior vena cava beside them and ascend through the diaphragm into the thoracic cavity.

Drainage of the Thorax

Right side. After penetrating the diaphragm, the right ascending lumbar vein becomes the **azygos**[19] **vein** of the thorax. The azygos receives blood from the right **posterior intercostal veins,** which drain the chest muscles, and from the **esophageal, mediastinal, pericardial,** and **right bronchial veins.** It then empties into the superior vena cava at the level of vertebra T4.

Left side. The left ascending lumbar vein continues into the thorax as the **hemiazygos**[20] **vein.** The hemiazygos drains the ninth through eleventh posterior intercostal veins and some esophageal and mediastinal veins on the left. At midthorax, it crosses over to the right side and empties into the azygos vein.

The **accessory hemiazygos vein** is a superior extension of the hemiazygos. It drains the fourth through eighth posterior intercostal veins and the left bronchial vein. It also crosses to the right side and empties into the azygos vein.

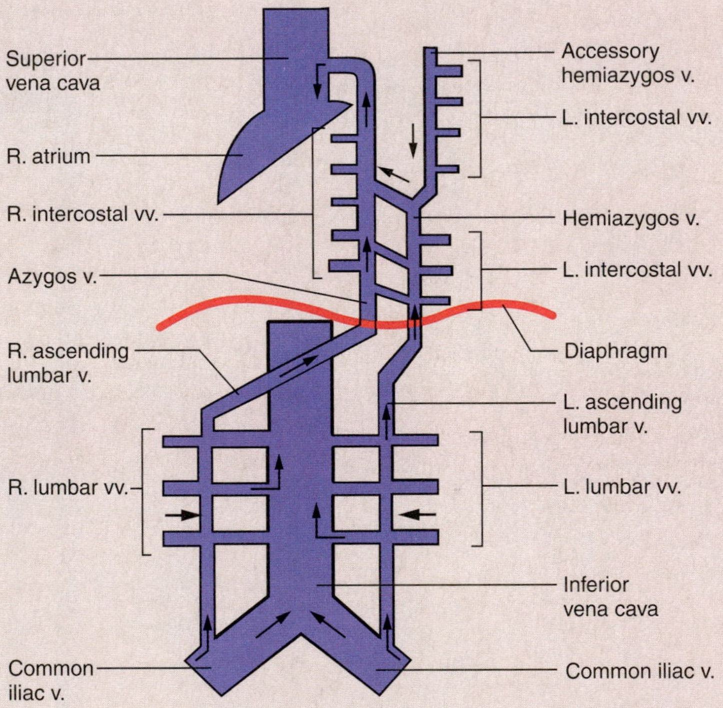

Figure 21.29
Flowchart of the Azygos System.

[19] unpaired; from *a* = without + *zygo* = union, mate
[20] *hemi* = half

TABLE 21.10

Major Tributaries of the Inferior Vena Cava

The **inferior vena cava (IVC)** is formed by the union of the right and left common iliac veins at the level of vertebra L5. It is retroperitoneal and lies immediately to the right of the aorta. Its diameter of 3.5 cm is the largest of any vessel in the body. As it ascends the abdominal cavity, the IVC picks up blood from numerous tributaries in the order listed here (fig. 21.30):

1. Some **lumbar veins** empty into the IVC as well as into the ascending lumbar veins described in table 21.9.
2. The **gonadal veins** (**ovarian veins** in the female and **testicular veins** in the male) drain the gonads. The right gonadal vein empties directly into the IVC, whereas the left gonadal vein empties into the left renal vein.
3. The **renal veins** drain the kidneys into the IVC. The left renal vein also receives blood from the left gonadal and left suprarenal veins.
4. The **suprarenal veins** drain the adrenal (suprarenal) glands. The right suprarenal empties directly into the IVC, and the left suprarenal empties into the renal vein.
5. The **hepatic veins** drain the liver, extending a short distance from its superior surface to the IVC.
6. The **inferior phrenic veins** drain the inferior aspect of the diaphragm.

After receiving these inputs, the IVC penetrates the diaphragm and enters the right atrium from below. It does not receive any thoracic drainage.

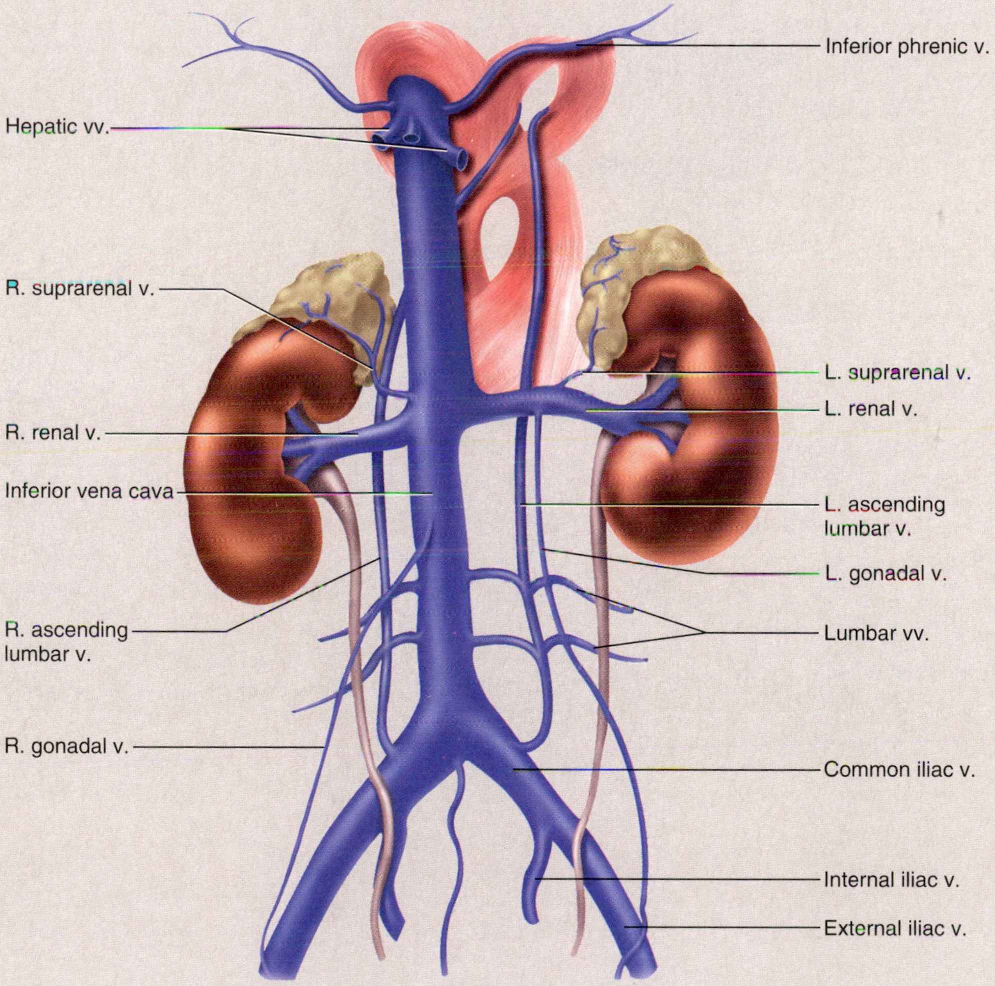

FIGURE 21.30
The Inferior Vena Cava and Its Tributaries.

Table 21.11

The Hepatic Portal System

The **hepatic portal system** connects capillaries of the intestines and other digestive organs to the **hepatic sinusoids** of the liver. The intestinal blood is richly laden with nutrients for a few hours following a meal. The hepatic portal system gives the liver "first claim" to these nutrients before the blood is distributed to the rest of the body. It also allows the blood to be cleansed of bacteria and toxins picked up from the intestines, an important function of the liver. The route from the intestines to the inferior vena cava follows (figs. 21.31 and 21.32):

1. The **inferior mesenteric vein** receives blood from the rectum and distal part of the large intestine. It converges in a fanlike array in the mesentery and empties into the splenic vein.
2. The **superior mesenteric vein** receives blood from the entire small intestine, ascending colon, transverse colon, and stomach. It, too, exhibits a fanlike arrangement in the mesentery and then joins the splenic vein to form the hepatic portal vein.
3. The **splenic vein** drains the spleen and travels across the abdominal cavity toward the liver. Along the way, it picks up the **pancreatic veins** from the pancreas and the inferior mesenteric vein. It changes name when it joins the superior mesenteric vein, as explained on the next page.

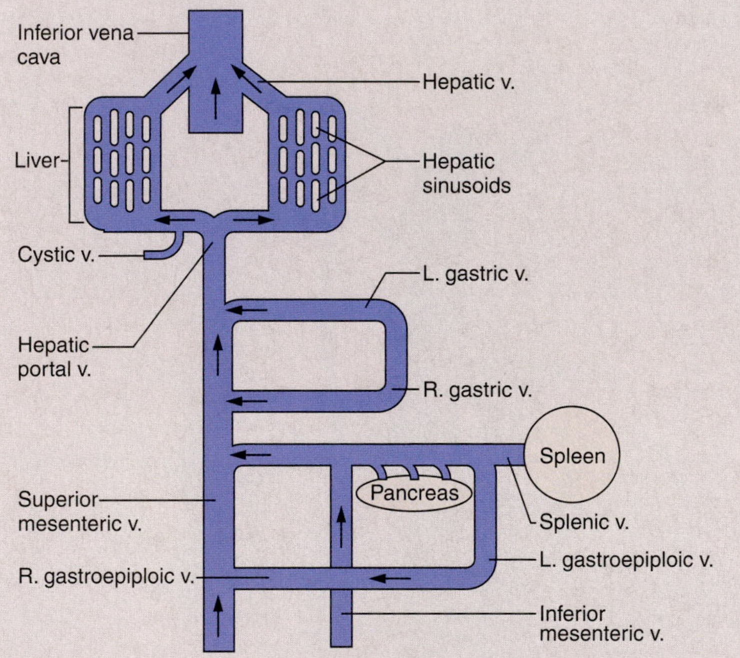

Figure 21.31
Flowchart of the Hepatic Portal System.

(continued)

TABLE 21.11

The Hepatic Portal System *(continued)*

4. The **hepatic portal vein** is formed by convergence of the splenic and superior mesenteric veins. It travels about 8 cm up and to the right and then enters the inferior surface of the liver. Near this point it receives the **cystic vein** from the gallbladder. In the liver, the hepatic portal vein ultimately leads to the innumerable microscopic hepatic sinusoids. Blood from the sinusoids empties into the hepatic veins described earlier. Circulation within the liver is described in more detail in chapter 24.
5. The left and right **gastric veins** form an arch along the lesser curvature of the stomach and empty into the hepatic portal vein.

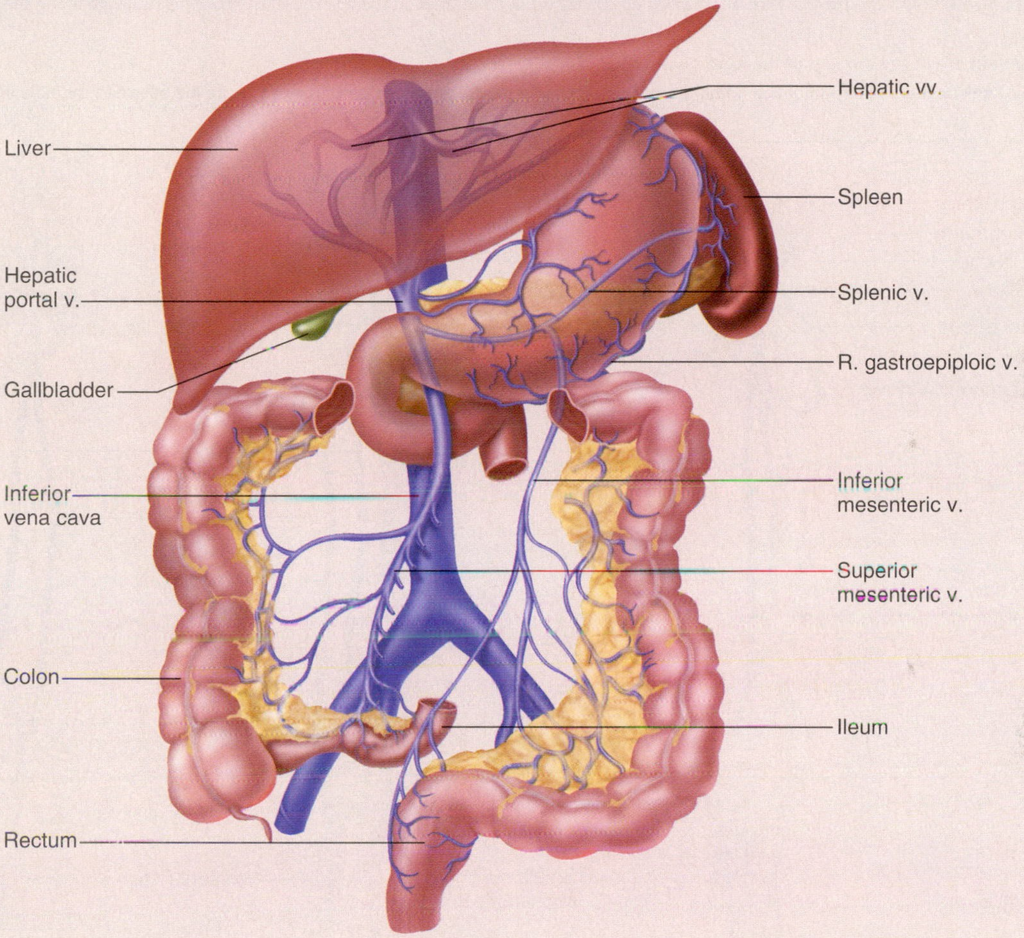

FIGURE 21.32
Anatomy of the Hepatic Portal System.

TABLE 21.12

Venous Drainage of the Lower Limb and Pelvic Organs

Drainage of the lower limb is described starting at the toes and following the flow of blood to the inferior vena cava (figs. 21.33 and 21.34). As in the upper limb, there are deep and superficial veins with anastomoses between them.

Deep Veins

1. The **plantar arch** drains the plantar aspect of the foot, receives blood from the **plantar digital veins** of the toes, and gives rise to the next vein.
2. The **posterior tibial vein** drains the plantar arch and passes up the leg embedded deep in the calf muscles, receiving drainage along the way from the **fibular (peroneal) vein.**
3. The **dorsal pedal vein** drains the dorsum of the foot.
4. The **anterior tibial vein** is a continuation of the dorsal pedal vein. It travels up the anterior compartment of the leg between the tibia and fibula.

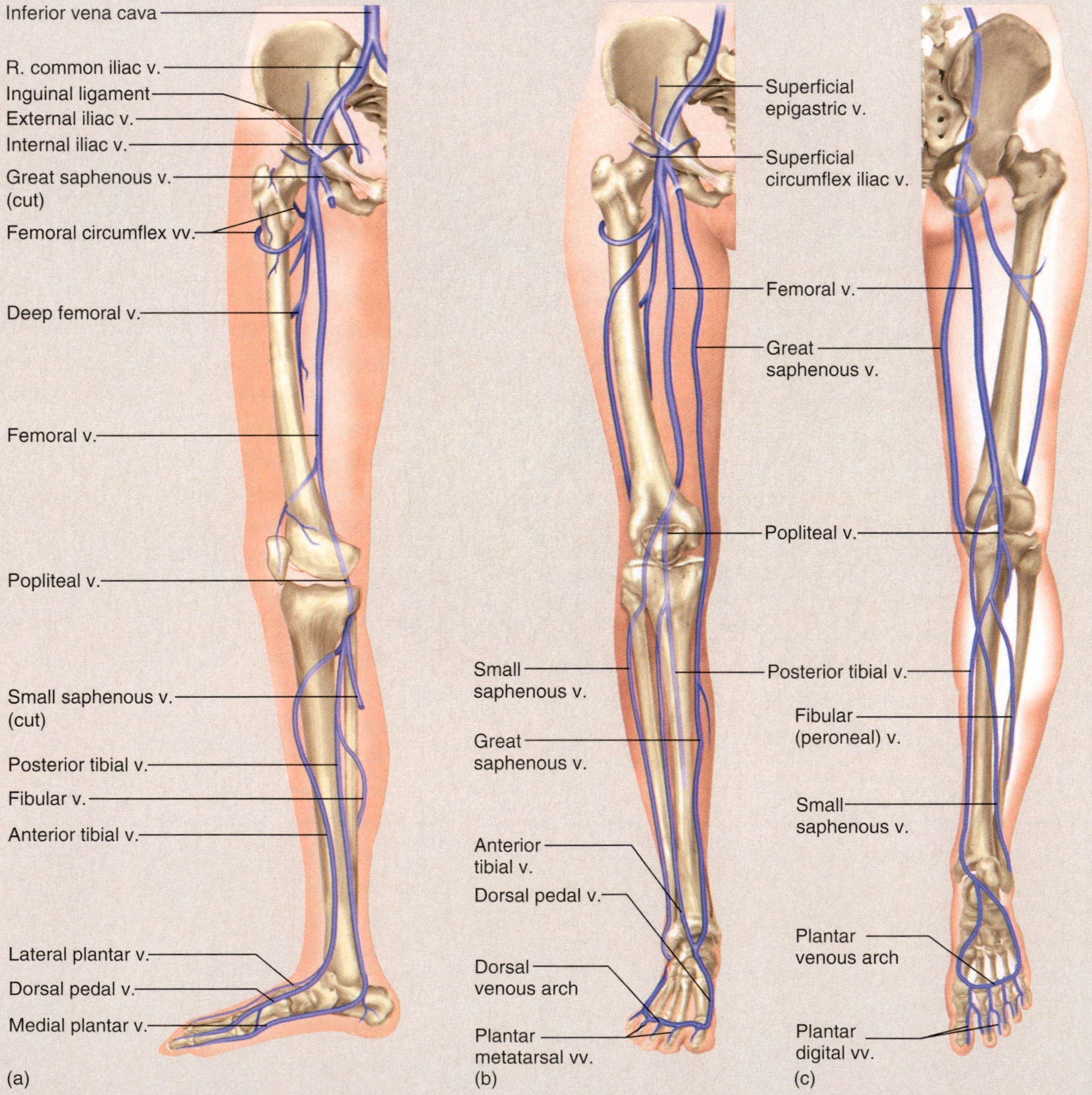

FIGURE 21.33

Veins of the Lower Limb. (*a*) Deep veins, anteromedial view of the right limb. (*b*) Anterior aspect of the right limb and dorsal aspect of the foot. (*c*) Posterior aspect of the right limb and plantar aspect of the foot.

(continued)

Table 21.12

Venous Drainage of the Lower Limb and Pelvic Organs *(continued)*

5. The **popliteal vein** is formed at the back of the knee by the union of the anterior and posterior tibial veins.
6. The **femoral vein** is a continuation of the popliteal vein into the thigh. It receives drainage from the deep thigh muscles and femur.
7. The **external iliac vein,** superior to the inguinal ligament, is formed by the union of the femoral vein and great saphenous vein (one of the superficial veins described in the next section).
8. The **internal iliac vein** follows the course of the internal iliac artery and its distribution. Its tributaries drain the gluteal muscles; the medial aspect of the thigh; the urinary bladder, rectum, prostate, and ductus deferens in the male; and the uterus and vagina in the female.
9. The **common iliac vein** is formed by the union of the external and internal iliac veins; it also receives blood from the ascending lumbar vein. The right and left common iliacs then unite to form the inferior vena cava.

Superficial Veins

1. The **dorsal venous arch** is visible through the skin on the dorsum of the foot. It has numerous anastomoses similar to the dorsal venous arch of the hand.
2. The **great saphenous**[21] (sah-FEE-nus) **vein,** the longest vein in the body, arises from the medial side of the dorsal venous arch. It traverses the medial aspect of the leg and thigh and terminates by emptying into the femoral vein, slightly inferior to the inguinal ligament. It is commonly used as a site for the long-term administration of intravenous fluids; it is a relatively accessible vein in infants and in patients in shock whose veins have collapsed. Portions of this vein are commonly excised and used as grafts in coronary bypass surgery.
3. The **small saphenous vein** arises from the lateral side of the dorsal venous arch, courses up the lateral aspect of the foot and through the calf muscles, and terminates at the knee by emptying into the popliteal vein. It has numerous anastomoses with the great saphenous vein. The great and small saphenous veins are among the most common sites of varicose veins.

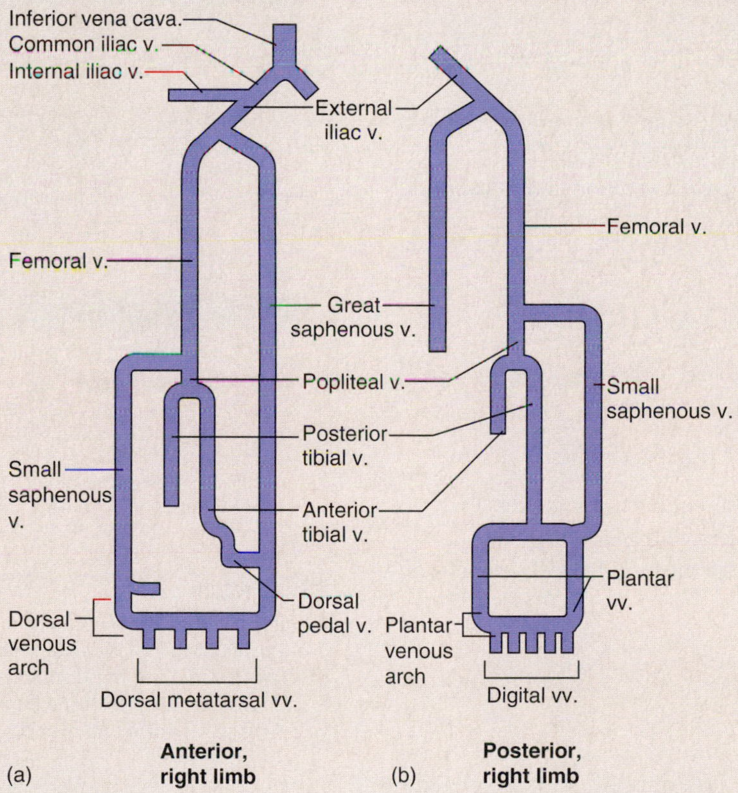

Figure 21.34

Flowchart of the Lower Limb Veins. (*a*) Anterior aspect of the right limb. (*b*) Posterior aspect of the same limb.

[21]*saphen* = standing

not true of the superficial veins, however. The deep veins are not described in as much detail as the arteries were, since it can usually be assumed that they drain the same structures as the corresponding arteries supply.

In general, we began the study of arteries with those lying close to the heart and progressed away. In the venous system, by contrast, we begin with those that are remote from the heart and follow the flow of blood as they join each other and approach the heart. Venous pathways have more anastomoses than arterial pathways, so the route of blood flow is often not as clear. Many anastomoses are omitted from figures 21.27 to 21.34 for clarity.

Before You Go On

Answer the following questions to test your understanding of the preceding section:

11. If you were dissecting a cadaver, where would you look for the internal and external jugular veins? What muscle would help you distinguish one from the other?
12. How do the vertebral veins differ from the vertebral arteries in their superior terminations?
13. By what route does blood from the abdominal wall reach the superior vena cava?
14. Trace one possible path of an RBC from the fingertips to the right atrium, naming the veins along the way.
15. State two ways in which the great saphenous vein has special clinical significance. Where is this vein located?

Developmental and Clinical Perspectives

Objectives

When you have completed this section, you should be able to

- describe the embryonic development of the blood vessels;
- explain how the circulatory system changes at birth; and
- describe the changes that occur in the blood vessels in old age.

Embryonic Development of the Blood Vessels

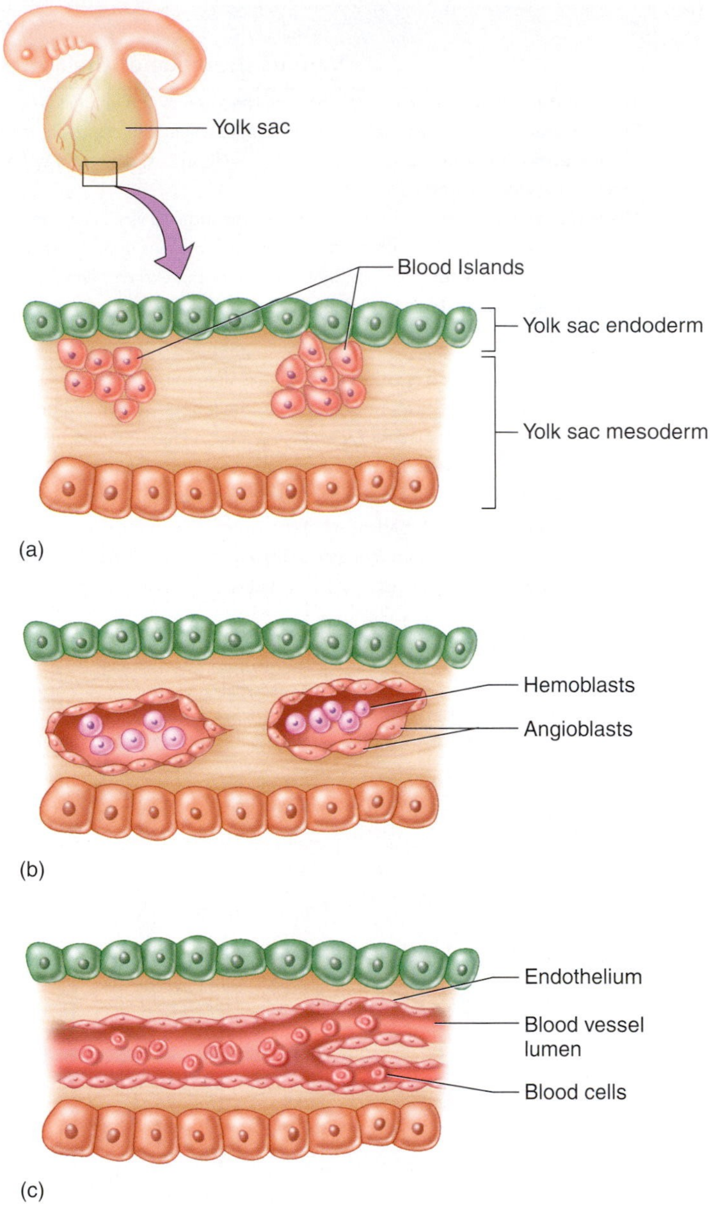

FIGURE 21.35

Development of Blood Vessels and Primitive Erythrocytes from Embryonic Blood Islands. (*a*) Early blood islands in the yolk sac. (*b*) Differentiation of mesenchymal cells into angioblasts and hemoblasts. (*c*) Merger of the lumens of the blood islands as a blood vessel begins to form and branch.

The development of blood vessels, both in the embryo and later in life, is called **angiogenesis.**[22] The first trace of embryonic angiogenesis appears at 13 to 15 days of gestation. At this time, the embryo is a three-layered disc of ectoderm, mesoderm, and endoderm, attached to the uterus by an *embryonic stalk* and associated with three membranes, the *chorion, amnion,* and *yolk sac* (see fig. 4.11). In the yolk sac, groups of mesenchymal cells differentiate into cell masses called *blood islands.* Spaces open in the middle of a blood island, and cells in these spaces differentiate into *hemoblasts,* the forerunners of the first blood cells. Cells of the margin become *angioblasts,* which give rise to the endothelium of the future blood vessel (fig. 21.35).

As blood islands proliferate and grow, they begin to connect with each other, and their internal spaces form the lumens of the blood vessels. By the end of week 3, the yolk sac is fully vascularized. In subsequent weeks, blood islands begin to appear in the liver, spleen, and bone marrow, and those of the yolk sac and other sites external to the embryo disappear. Also during this time, the heart tube forms and links up with the blood vessels. Around the end of

[22]*angio* = vessel + *genesis* = origin

week 3, the heart begins beating (see chapter 20), and a week later, a unidirectional blood flow is established. The blood vessels are not long uniform tubes at this time, but a network of irregularly shaped channels. Those channels that receive the greatest blood flow develop a tunica media and externa and become more tubular, thus becoming typical blood vessels. Those with a lesser flow either degenerate or remain composed of nothing but endothelium, thus becoming capillaries. Capillaries and the larger vessels sprout lateral branches, giving rise to the eventual anatomical circuitry of the mature cardiovascular system.

We will not examine the complex details of the embryonic development of all the major blood vessels, but primarily the major arteries and veins near the heart. Remember that the embryo forms five pairs of pharyngeal arches in weeks 4 to 5 (see chapter 4). As these develop, a pouch called the *aortic sac* appears at the rostral end of the heart. An artery arises from each side of the sac, loops through the first pharyngeal arch, and ends in the dorsal aorta on that side. The loop is *aortic arch I.* Five more pairs of **aortic arches** (II–VI) later form between the pharyngeal pouches (fig. 21.36*a*). This reflects a primitive vertebrate pattern seen in fish, where the six aortic arches supply blood to the gills, but it becomes highly modified in humans and other mammals. The six arches actually never appear all at once as shown in the figure. Arches I and II degenerate before the most caudal arches appear, and arch V never develops to any great extent in mammals. Arches III, IV, and VI, however, play major roles in human development (fig. 21.36*b*). Arch III gives rise to the common carotid artery and the proximal portion of the internal carotid artery; the external carotid artery buds from the common carotid. The common carotids are short at first, but elongate as the embryo grows and the heart moves caudally. Arch IV degenerates on the right, but on the left it produces the aortic arch. Arch VI gives rise to the pulmonary arteries.

Initially, the embryo has two *dorsal aortae* that pass side by side for the length of the body. Caudal to the pharyngeal arches, however, these soon fuse into a single **dorsal aorta,** the forerunner of the adult descending aorta (fig. 21.37). The dorsal aorta issues about 30 pairs of *intersegmental arteries,* which supply blood to the somites and their derivatives. Most of these degenerate, but the adult intercostal arteries, lumbar arteries, and common iliac arteries are remnants of some of the embryonic intersegmental arteries.

The principal veins associated with the heart (fig. 21.37) are the *anterior cardinal vein,* which drains the head region; the *posterior cardinal vein,* which drains the body caudal to the heart; the *vitelline veins* from the yolk sac; and the *umbilical veins* from the placenta. There are initially two umbilical veins, but only the left one persists until birth. The cardinal veins provide most of the venous drainage of the embryonic body. They meet at a *common cardinal vein* just before entering the heart. The future superior vena cava develops from the right anterior cardinal and common cardinal veins. The posterior cardinal veins largely degenerate, however, leaving only the common iliac veins and part of the azygos system as remnants. The inferior vena cava develops separately, not from the posterior cardinal veins.

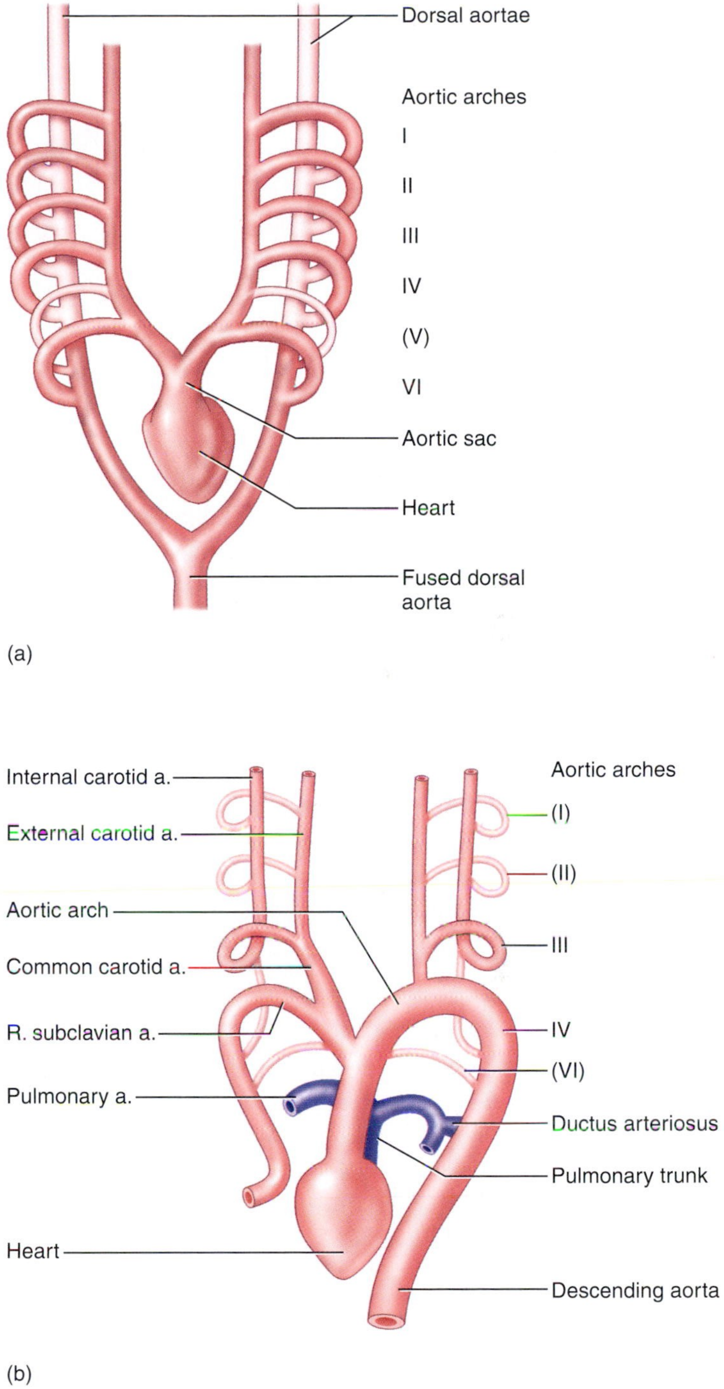

FIGURE 21.36

Development of Some Major Arteries from the Embryonic Aortic Arches. (*a*) The six aortic arches, dorsal view. This is a composite diagram representing developments from day 22 through day 29. In reality, arch I degenerates as arches III and IV form, and arch II degenerates as arch VI forms. Aortic arch V develops very little, and sometimes not at all, in humans. (*b*) Remodeled arterial system at about 8 weeks. Pale colors and arch numbers in parentheses indicate the former positions of aortic arches that no longer exist at this time, for comparison to figure *a*.

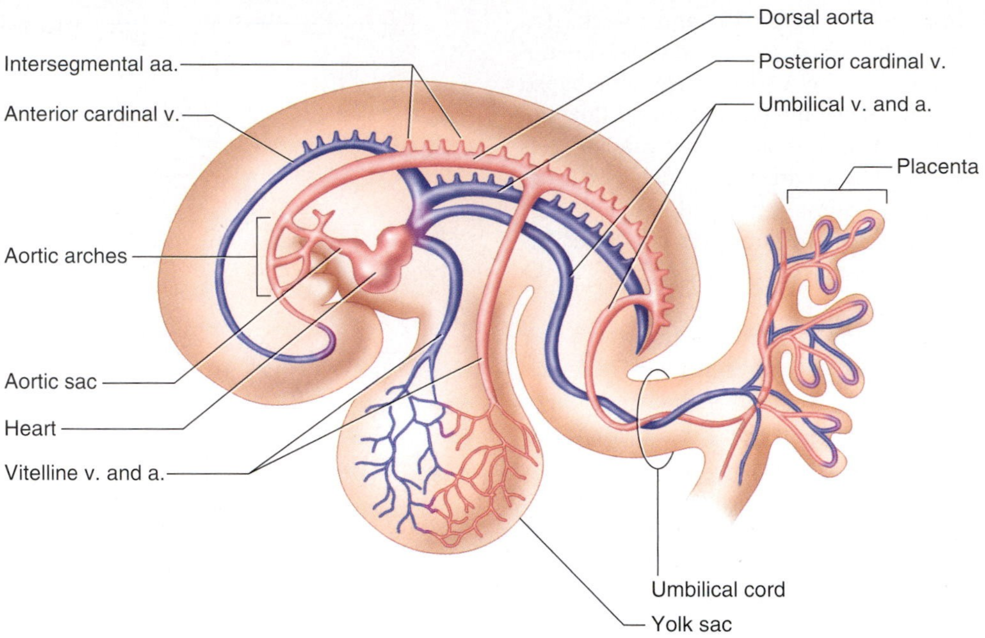
Dorsal aorta
Posterior cardinal v.
Umbilical v. and a.
Placenta
Intersegmental aa.
Anterior cardinal v.
Aortic arches
Aortic sac
Heart
Vitelline v. and a.
Umbilical cord
Yolk sac

FIGURE 21.37
Major Embryonic Blood Vessels at 26 Days. Left lateral view of the embryo, yolk sac, and part of the placenta.

Changes at Birth

As we saw in chapter 20, the fetus has certain *shunts* by which most blood bypasses the nonfunctional lungs: the *foramen ovale* and *ductus arteriosus.* After birth, when the lungs are functional, these shunts close, leaving a *fossa ovalis* in the interatrial septum and *ligamentum arteriosum* between the aortic arch and left pulmonary artery. Another fetal shunt called the *ductus venosus* bypasses the liver, which also is not very functional before birth. This is a vein; blood returning from the placenta enters the fetus through the umbilical vein and flows into the ductus venosus. The ductus venosus then empties into the inferior vena cava. After birth, the ductus venosus constricts and blood is forced to flow through the liver. The ductus venosus leaves a fibrous remnant, the *ligamentum venosum,* on the inferior surface of the liver.

The two umbilical arteries become the *superior vesical arteries* to the urinary bladder. The umbilical vein becomes a fibrous cord, the *round ligament,* which attaches the liver to the ventral body wall (fig. 21.38).

The Aging Vascular System

Atherosclerosis is the principal change seen in the blood vessels with advancing age. It is such a universal phenomenon that it is difficult to isolate and identify other age-related changes independent of atherosclerosis. However, even non-atherosclerotic vessels stiffen with age, owing to increasing deposition of collagen, cross-linking of collagen molecules (the same phenomenon that stiffens the skeletal joints, lens of the eye, and tissues elsewhere), and declining resilience of the elastic fibers. This arterial stiffening is less pronounced in elderly people who engage in routine vigorous exercise.

Another effect of aging is declining responsiveness of the baroreceptors, so vasomotor responses to changes in blood pressure are not as quick or efficient. Among some elderly people, the blunted response causes *orthostatic hypotension:* When one goes from a lying to a sitting or standing posture, blood is drawn away from the brain by gravity. Without a prompt corrective baroreflex, the drop in cerebral perfusion can cause dizziness or even fainting and falling, which in turn presents a risk of serious bone fractures.

Vascular Diseases

Atherosclerosis, the most common vascular disease, can lead to stroke, renal failure, or heart failure—most notoriously the last of these. Therefore, it was described in chapter 20 in the context of coronary artery disease. Table 21.13 describes a few other vascular diseases. Some of the hematologic and cardiac pathologies in the preceding chapters also include aspects of vascular pathology.

Before You Go On

Answer the following questions to test your understanding of the preceding section:

16. What do angioblasts develop from, and what do they develop into?
17. Describe, in humans, what becomes of the six pairs of aortic arches typical of vertebrate embryos.
18. Name two blood vessels that close off and become fibrous cords soon after birth.
19. Describe two changes that occur in the blood vessels in old age, other than specific vascular diseases.

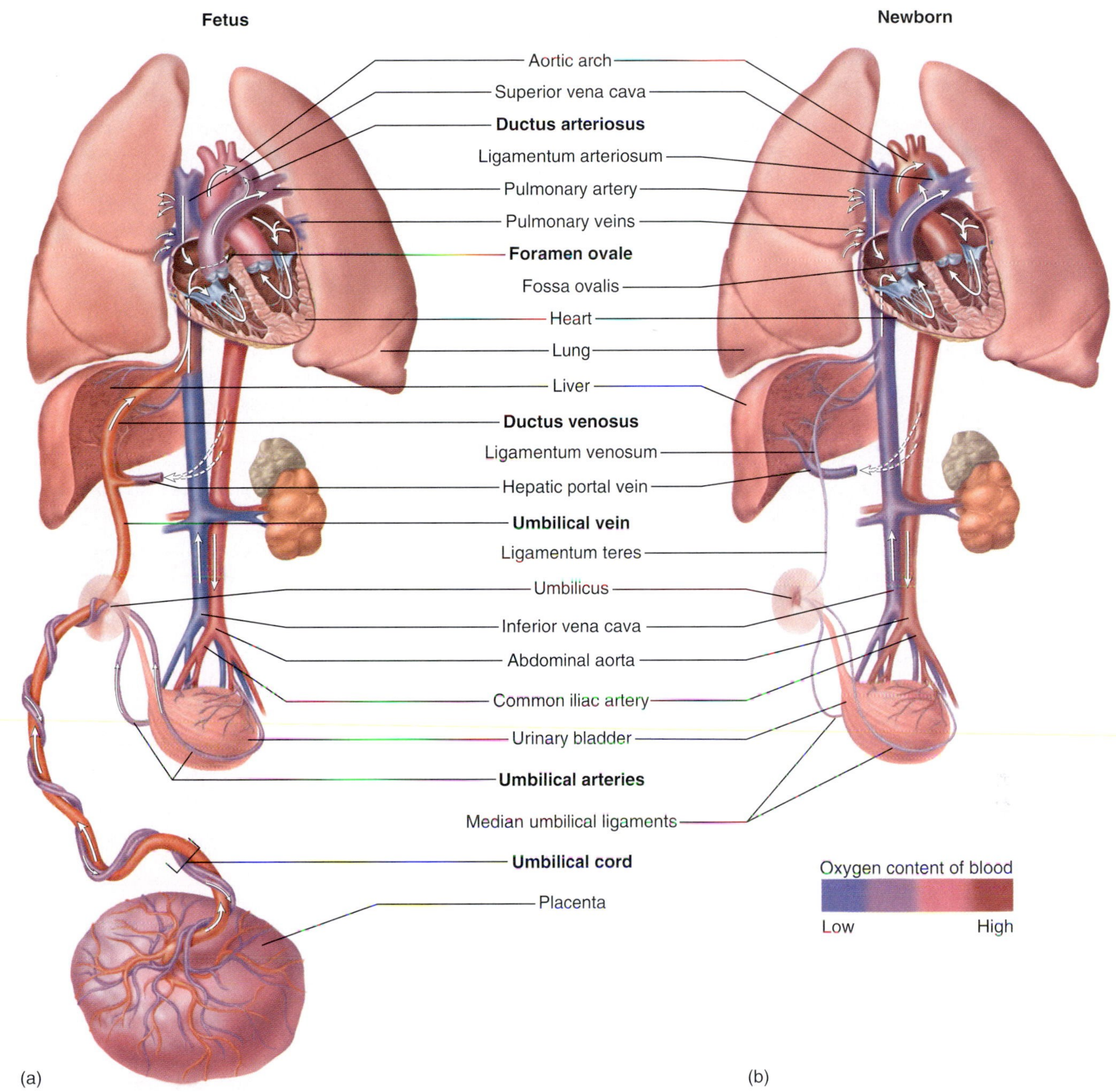
Fetus
Newborn
Aortic arch
Superior vena cava
Ductus arteriosus
Ligamentum arteriosum
Pulmonary artery
Pulmonary veins
Foramen ovale
Fossa ovalis
Heart
Lung
Liver
Ductus venosus
Ligamentum venosum
Hepatic portal vein
Umbilical vein
Ligamentum teres
Umbilicus
Inferior vena cava
Abdominal aorta
Common iliac artery
Urinary bladder
Umbilical arteries
Median umbilical ligaments
Umbilical cord
Placenta
Oxygen content of blood
Low
High
(a)
(b)

FIGURE 21.38

Some Circulatory Changes Occurring at Birth. (*a*) Circulatory system of the full-term fetus. (*b*) Circulatory system of the newborn.

Table 21.13

Some Vascular Pathologies

Hypertension	Abnormally high blood pressure. In a young adult, a BP up to 130/85 is considered normal, BP above 140/90 is considered hypertensive, and a BP between these ranges is borderline or "high normal." About 90% of cases of hypertension *(primary hypertension)* result from a poorly understood complex of hereditary, behavioral, and other factors. Risk factors include obesity, a sedentary life style, diet, smoking, sex, and race. *Secondary hypertension* (10% of cases) results from other identifiable disorders such as renal insufficiency, atherosclerosis, hyperthyroidism, and polycythemia. Treated with dietary modification, weight loss, and drugs such as beta-blockers (which reduce responsiveness of the blood vessels to sympathetic stimulation), calcium channel blockers (which relax the vascular smooth muscle), diuretics (which reduce blood volume), and ACE inhibitors (which inhibit synthesis of the vasoconstrictor angiotensin II).
Phlebitis	Inflammation of a vein, causing pain, tenderness, edema, and skin discoloration along its course. Often of unknown cause, but may follow surgery, childbirth, or infections.
Raynaud[23] *Disease*	Occasional spasmodic contractions of the digital arteries, causing pallor, numbness, and coldness of the fingers or toes. The digits may at first appear cyanotic, but then redden, with throbbing and paresthesia (tingling, burning, or itching sensations). Repeated and severe cases can lead to brittle nails and occasionally to gangrene and a necessity for amputation. Most common in young women and often triggered by emotional stress or brief exposure to cold.
Stroke (cerebrovascular accident)	The sudden death (infarction) of brain tissue occurring when cerebral atherosclerosis, thrombosis, or hemorrhage of a cerebral aneurysm cuts off blood flow to part of the brain. Effects range from unnoticeable to fatal, depending on the extent of tissue damage and function of the affected tissue. Blindness, paralysis, loss of speech, and loss of sensation are among the sublethal effects.
Vasculitis	Inflammation of any blood vessel (see also phlebitis in this table), usually caused by an immune response or infectious pathogen, but sometimes by radiation, trauma, or toxins. Produces a wide variety of symptoms including muscle and joint pain, fever, headache, myocardial ischemia, numbness, and blindness.

Disorders described elsewhere

Air embolism 556
Aneurysm 588
Atherosclerosis 572
Orthostatic hypotension 622
Patent ductus arteriosus 579
Varicose veins 594

[23] Maurice Raynaud (1834–81), French physician

CHAPTER REVIEW

Review of Key Concepts

General Anatomy of the Blood Vessels (p. 586)

1. Blood normally circulates from the heart through arteries, one bed of capillaries, veins, and back to the heart. In some places in the body, there are exceptions to this pattern called *portal systems* and *anastomoses.*
2. Arteries and veins have three layers: an outer *tunica externa* of loose connective tissue; a middle *tunica media* of smooth muscle and connective tissue; and an inner *tunica interna* consisting of an *endothelium* overlying a basement membrane and thin connective tissue layer. In large and medium arteries, there is an *internal elastic lamina* at the boundary between the interna and media, and *external elastic lamina* at the boundary between media and externa.
3. Arteries are the efferent blood vessels. Although there is a gradual transition from one type to another, they can be classified into three general types: large *conducting arteries* with an abundance of elastic connective tissue in the tunica media, adapted to withstanding blood pressure surges; medium *distributing arteries* with a more muscular tunica media; and smaller *resistance arteries* with a thinner wall of smooth muscle in the media. The smallest resistance arteries are *arterioles.*
4. Metarterioles link arterioles to blood capillaries. They have no continuous tunica media, but have a circular cuff of smooth muscle, the *precapillary sphincter,* at the beginning of each capillary.
5. Some of the great arteries above the heart have special sense organs in their walls: a pair of baroreceptors (blood pressure monitors) called *carotid sinuses,* and chemoreceptors (blood chemistry monitors) called *carotid bodies* and *aortic bodies.* These receptors communicate with the brainstem by way of the glossopharyngeal and vagus nerves and trigger corrective responses in heartbeat, vasomotion, and breathing to maintain a normal blood pressure, pH, and CO_2 and O_2 levels.
6. Blood capillaries have only an endothelium (no tunica media or externa) and are the main point at which materials leave the bloodstream for the tissues, or return from the tissue fluid to the bloodstream to be carried away. Some capillaries are smaller than an RBC in width, and few cells in the body are more than four to six cell-widths away from the nearest capillary.
7. Capillaries are arranged in groups called *capillary beds,* supplied by a metarteriole and drained by venules and thoroughfare channels.
8. There are three types of capillaries: *continuous capillaries, fenestrated capillaries,* and *sinusoids.*
9. Veins are the afferent blood vessels. The smallest veins are *venules.* The smallest venules are very thin-walled and are another point of fluid exchange with the tissues. Even large veins have thinner, less muscular walls than arteries of comparable size. *Venous sinuses* have large lumens, thin walls, and no muscle.
10. Veins are under relatively low blood pressure. Consequently they are thin-walled, they stretch and accommodate more blood than any other vessels, and in the limbs, they have valves to help produce a one-way flow of blood.

Anatomy of the Pulmonary Circuit (p. 595)

1. The pulmonary circuit begins with a *pulmonary trunk,* which arises from the right ventricle of the heart and branches into the right and left *pulmonary arteries* to the lungs. These divide into one *lobar artery* for each lobe of the respective lung. Finer branches lead to capillaries around the pulmonary alveoli, where gas exchange occurs.
2. Pulmonary blood empties into the left atrium of the heart by way of two pulmonary veins on each side.
3. The pulmonary circuit serves only for CO_2 unloading and O_2 loading. The lung tissue receives nourishment and waste removal by a separate set of vessels, the *bronchial arteries* of the systemic circuit.

Anatomy of the Systemic Arteries (p. 596)

1. The *ascending aorta* arises from the left ventricle and immediately gives off the two coronary arteries to the heart wall. It continues as the *aortic arch,* which gives off three large arteries to the neck, head, and upper limbs: the *brachiocephalic trunk, left common carotid artery,* and *left subclavian artery.* The brachiocephalic trunk quickly branches into the *right subclavian* and *right common carotid arteries.* Beyond the arch, the aorta turns downward and continues as the *descending aorta,* divided into thoracic and abdominal regions (table 21.1).
2. The common carotid arteries ascend beside the trachea and branch into an *external carotid artery* (supplying mainly head tissues external to the cranium) and *internal carotid artery* (supplying mainly the brain). The external carotid gives off branches to the thyroid gland, larynx, tongue, facial skin and muscles, teeth, buccal cavity, nasal cavity, external ear, scalp, and dura mater. The internal carotid gives off the ophthalmic artery and two cerebral arteries (table 21.2).
3. The subclavian arteries give off *vertebral arteries* to the brain and *thyrocervical* and *costocervical trunks* to the thyroid gland and various neck and thoracic muscles (table 21.2).
4. The internal carotid and vertebral arteries converge on an *arterial circle* at the base of the brain, surrounding the pituitary gland. Further blood flow to the cerebrum arises from this circle (table 21.2).
5. The subclavian artery continues as the *axillary* and then *brachial* artery supplying the arm. Just distal to the elbow, the brachial artery branches into the *radial* and *ulnar* arteries, which are the major blood supply to the forearm. These two arteries anastomose at the wrist to produce two *palmar arches,* which give rise to *metacarpal* and *digital* arteries to the hand and fingers (table 21.3).
6. The thoracic aorta gives off *bronchial, esophageal,* and *mediastinal* arteries to the thoracic viscera; *intercostal* and *subcostal* arteries to the skin, thoracic muscles, vertebrae, and other structures; and *superior phrenic* arteries to the diaphragm (table 21.4).
7. Other thoracic structures are supplied by small branches from the subclavian and axillary arteries (table 21.4).
8. The abdominal aorta gives off along its course: *inferior phrenic arteries* to the diaphragm; the *celiac trunk* to the upper abdominal viscera; a *superior mesenteric artery* to the intestines; *middle suprarenal arteries* to the adrenal glands; *renal arteries* to the kidneys; *gonadal arteries* to the ovaries or

testes; the *inferior mesenteric artery* to part of the large intestine; *lumbar arteries* to the dorsal abdominal wall; and a *median sacral artery* to the sacrum and coccyx. It then ends by branching into two *common iliac arteries* (table 21.5).

9. The celiac trunk supplies an extensive system of arteries with branches to the stomach, duodenum, pancreas, liver, gallbladder, spleen, and omenta (table 21.5).
10. The intestines are supplied by the *superior* and *inferior mesenteric arteries* (table 21.5).
11. The common iliac arteries continue for a short distance beyond the aorta and then branch into *internal* and *external iliac arteries.* The internal iliac issues branches to organs of the pelvic region including the pelvic wall, rectum, urinary bladder, reproductive organs, gluteal muscles, and some thigh muscles (table 21.6).
12. The external iliac artery supplies skin and muscles of the abdominal wall and pelvic region. It then continues as the *femoral artery* (supplying the thigh), *popliteal artery* (supplying the knee), *anterior* and *posterior tibial arteries* (supplying the leg), and more distal branches, including *metatarsal* and *digital arteries,* to the foot.
13. *Arterial pressure points* are points where arteries lie close enough to the body surface so that the pulse can be felt and pressure can be applied to stop bleeding.

Anatomy of the Systemic Veins (p. 610)

1. After flowing through the brain, blood collects in several *dural sinuses* between layers of dura mater. Blood then flows down the neck via the *internal jugular vein* to the *subclavian vein.* Blood from the more superficial tissues of the head and neck, the vertebrae, the spinal cord, and muscles of the region flows by way of the *external jugular vein* and *vertebral vein* to the subclavian (table 21.7).
2. When the subclavian and internal jugular veins converge, they form the *brachiocephalic vein.* After receiving small tributaries from other thoracic organs, the right and left brachiocephalics converge and form the *superior vena cava,* which empties into the right atrium from above (table 21.7).
3. Venous drainage of the upper limb begins with digital and metacarpal veins from the fingers and hand. These converge on *superficial* and *deep palmar venous arches* that drain into the *radial* and *ulnar veins* of the forearm (table 21.8).
4. The radial and ulnar veins converge at the elbow to form the *brachial vein,* which continues as the *axillary vein* and then subclavian vein (table 21.8).
5. The major superficial veins of the upper limb include the dorsal venous arch on the back of the hand, and the *cephalic* (lateral) and *basilic* (medial) *veins* which pass up the forearm and arm and lead to the axillary vein (table 21.8).
6. The *azygos system* is a network of thoracic veins that receive blood from the abdominal and thoracic walls and empty into the superior vena cava. The system receives *posterior intercostal, esophageal, mediastinal, pericardial,* and *bronchial veins* from thoracic structures and an *ascending lumbar vein* from the abdominal wall (table 21.9).
7. The *inferior vena cava* enters the right atrium from below. In its ascent through the abdominal cavity, it receives *lumbar, gonadal, renal, suprarenal, hepatic,* and *inferior phrenic veins,* in that order (table 21.10).
8. The *hepatic portal system* is a network of vessels connecting blood capillaries of the intestines to the sinusoids of the liver, carrying nutrients to the liver for processing. It receives blood from the *superior* and *inferior mesenteric veins* of the intestines, and the *splenic* and *pancreatic veins* from the spleen and pancreas. Union of the splenic and superior mesenteric veins forms the *hepatic portal vein.* This vein receives additional blood from the stomach and gallbladder, then enters the inferior surface of the liver. After circulating through the liver, the blood drains by way of a pair of *hepatic veins* into the inferior vena cava (table 21.11).
9. Drainage of the lower limb begins with *digital veins* of the toes. The *plantar arch* and *dorsal pedal vein* drain the plantar and dorsal sides of the foot. *Posterior* and *anterior tibial veins* travel up the leg and unite at the knee to form the *popliteal vein.* This continues into the thigh as the *femoral vein.*
10. The major superficial veins of the lower limb are the *small saphenous* and *great saphenous veins.* The small saphenous vein empties into the popliteal vein. The great saphenous vein joins the femoral vein, and their union forms the *external iliac vein.*
11. The *internal iliac vein* drains muscles of the thigh and gluteal regions and viscera of the pelvic region (rectum, urinary bladder, and reproductive organs). Union of the internal and external iliac veins then forms the short *common iliac vein,* and union of the right and left common iliacs forms the inferior vena cava (table 21.12).

Developmental and Clinical Perspectives (p. 620)

1. The first indications of developing blood vessels in humans are the *blood islands* that form from mesenchymal cells of the yolk sac. Cells in the middle of a blood island develop into *hemoblasts,* which give rise to blood cells, and cells on the periphery become *angioblasts,* which differentiate into the endothelium of the blood vessels. Convergence of blood islands gives rise to an irregular network of channels in the embryo that later become remodeled into tubular blood vessels.
2. Like other vertebrate embryos, the human embryo develops six pairs of *aortic arches* that connect the *aortic sac* of the heart with a pair of long *dorsal aortae.* Little becomes of arches I, II, and V, but arch III becomes the common carotid artery and part of the internal carotid; IV on the right becomes the aortic arch; and VI gives rise to the pulmonary arteries.
3. The two dorsal aortae fuse into a single dorsal aorta, which gives off numerous *intersegmental arteries.* Most intersegmental arteries degenerate, but some remain as the adult intercostal, lumbar, and common iliac arteries.
4. The body of the embryo is drained mainly by the *anterior* and *posterior cardinal veins.* Other major veins are the *vitelline* and *umbilical veins,* draining the yolk sac and placenta, respectively. The anterior cardinal vein eventually contributes to the superior vena cava, while the posterior cardinal vein degenerates except for the common iliac veins and part of the azygos system.
5. Shortly after birth, two vascular shunts close: the *ductus arteriosus* and *ductus venosus* (becoming fibrous cords called the *ligamentum arteriosum* and *ligamentum venosum,* respectively). This forces more blood to flow through the lungs and liver.
6. The aging vascular system exhibits stiffening of the vessels by deposition and cross-linking of collagen, and declining baroreflexes, resulting in less prompt adjustments to changes in posture and sometimes causing orthostatic hypotension.
7. Atherosclerosis is the most common disease of the aging blood vessels. Other disorders are described in table 21.13.

Testing Your Recall

1. Blood normally flows into a capillary bed from
   a. a distributing artery.
   b. a conducting artery.
   c. a metarteriole.
   d. a thoroughfare channel.
   e. a venule.
2. Plasma solutes enter the tissue fluid most easily from
   a. continuous capillaries.
   b. fenestrated capillaries.
   c. arteriovenous anastomoses.
   d. collateral vessels.
   e. venous anastomoses.
3. A blood vessel adapted to withstand a high pulse pressure would be expected to have
   a. an elastic tunica media.
   b. a thick tunica interna.
   c. one-way valves.
   d. a flexible endothelium.
   e. a rigid tunica media.
4. A circulatory pathway in which the blood flows through two capillary beds in series before it returns to the heart is called
   a. an arteriovenous anastomosis.
   b. an arterial anastomosis.
   c. a venous anastomosis.
   d. a venous return pathway.
   e. a portal system.
5. Intestinal blood flows into the liver by way of
   a. the superior mesenteric vein.
   b. the hepatic portal vein.
   c. the abdominal aorta.
   d. the hepatic sinusoids.
   e. the hepatic veins.
6. Blood islands first form in the embryonic
   a. spleen.
   b. yolk sac.
   c. placenta.
   d. liver.
   e. red bone marrow.
7. Blood flowing in all of the following arteries except _____ is destined to circulate through the brain before returning to the heart.
   a. the vertebral arteries
   b. the internal carotid arteries
   c. the basilar artery
   d. the superficial temporal artery
   e. the anterior communicating artery
8. All of the following blood vessels except _____ are located in the upper limb.
   a. the cephalic vein
   b. the small saphenous vein
   c. the brachial artery
   d. the circumflex humeral artery
   e. the metacarpal arteries
9. The adult aortic arch develops from the embryonic
   a. right aortic arch IV.
   b. left aortic arch IV.
   c. right aortic arch V.
   d. conus arteriosus.
   e. dorsal aorta.
10. To get from the posterior tibial vein to the femoral vein, blood flows through
   a. the anterior tibial vein.
   b. the popliteal vein.
   c. the internal iliac vein.
   d. the great saphenous vein.
   e. the basilic vein.
11. Filtration pores are characteristic of _____ capillaries.
12. The capillaries of skeletal muscles are of the structural type called _____.
13. The epithelium that lines the inside of a blood vessel is called _____.
14. The two _____ veins unite like an upside-down Y to form the inferior vena cava.
15. Carotid and aortic bodies are called _____ because they respond to changes in blood chemistry.
16. Movement across the capillary endothelium by the uptake and release of fluid droplets is called _____.
17. The two largest veins that empty into the right atrium are the _____ and _____.
18. The pressure sensors in the major arteries near the head are called _____.
19. Most of the blood supply to the brain comes from a ring of arterial anastomoses called _____.
20. The major superficial veins of the arm are the _____ on the medial side and _____ on the lateral side.

Answers in the appendix

True or False

Determine which five of the following statements are false, and briefly explain why.

1. The lungs receive both a pulmonary and a systemic blood supply.
2. The pancreas and spleen receive their blood supply mainly from the superior mesenteric artery.
3. Veins anastomose more than arteries do.
4. From the time blood leaves the heart to the time it returns, it always passes through only one capillary bed.
5. The superior vena cava begins where the two subclavian veins meet.
6. Erythrocytes and endothelial cells arise from the same embryonic stem cells.
7. The smooth muscle of the tunica media of a large vessel is nourished mainly by the diffusion of nutrients from blood in the vessel lumen.
8. Venous blood from the intestines flows through the liver before it flows through the heart.
9. In a few unusual cases, one or more arteries of the cerebral arterial circle are lacking.
10. Arteries to the ovaries and testes originate relatively high in the abdominal cavity, near the kidneys.

Answers in the appendix

Testing Your Comprehension

1. Suppose the posterior tibial vein were obstructed by thrombosis. Describe one or more alternative routes by which blood from the foot could get to the common iliac vein.
2. Why would a ruptured aneurysm of the basilar artery be more serious than a ruptured aneurysm of the anterior communicating artery?
3. What differences would you expect between a sample of blood taken from the superior mesenteric vein and a sample taken from a hepatic vein? Consider especially differences in nutrient levels and bacterial count, and look forward in the book if necessary for a preview of liver functions.
4. Why could a choke hold (a tight grip around the neck) cause a person to pass out? What arteries would be involved?
5. Why is it better to have baroreceptors in the carotid sinus rather than in some other location such as the abdominal aorta or common iliac arteries?

Answers at the Online Learning Center

www.mhhe.com/saladinha1

Visit the Online Learning Center for practice tests, answer keys, and other learning aids for this chapter. Enhance your understanding of human anatomy with our interactive art labeling exercises, supplemental photo atlases, web links, puzzles, flashcards, and much more.

22

CHAPTER TWENTY-TWO

The Lymphatic System and Immunity

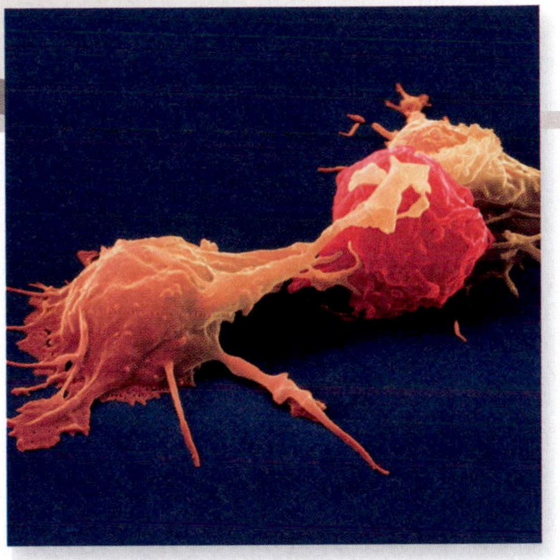

Natural killer cells *(orange)* attacking a human cancer cell *(red)*

CHAPTER OUTLINE

Lymph and Lymphatic Vessels 630
- Components and Functions of the Lymphatic System 630
- Lymph 630
- Lymphatic Vessels 631
- The Flow of Lymph 633

Lymphatic Cells, Tissues, and Organs 634
- Lymphatic Cells and Modes of Defense 634
- Lymphatic Tissues 636
- Overview of Lymphatic Organs 636
- Red Bone Marrow 636
- The Thymus 637
- Lymph Nodes 637
- Tonsils 640
- The Spleen 642

The Lymphatic System in Relation to Immunity 644
- B Cells and Humoral Immunity 644
- T Cells and Cellular Immunity 644

Developmental and Clinical Perspectives 646
- Embryonic Development 646
- The Aging Lymphatic-Immune System 646
- Lymphatic-Immune Disorders 646

Chapter Review 649

INSIGHTS

22.1 Clinical Application: Elephantiasis 631
22.2 Clinical Application: Lymph Nodes and Metastatic Cancer 639
22.3 Clinical Application: Splenectomy 643

BRUSHING UP

To understand this chapter, you may find it helpful to review the following concepts:

- General gland structure: capsule, septa, stroma, and parenchyma (pp. 96–97)
- Anatomy of the thymus (p. 523)
- Antigens and antibodies (p. 548)
- Leukocyte types, especially lymphocytes (pp. 548–551)

The **lymphatic system** (fig. 22.1) is a network of tissues, organs, and vessels that help to maintain the body's fluid balance, cleanse the body fluids of foreign matter, and provide immune cells for defense. Of all the body systems, it is perhaps the least familiar to most people. Yet without it, neither the circulatory system nor the immune system could function—circulation would shut down from fluid loss, and the body would be overrun by infection for lack of immunity. This chapter discusses the anatomy of the lymphatic system in relation to its roles in fluid recovery and immunity. The structure and function of the lymphatic system is so intimately tied to the immune system that this chapter will sometimes refer to them jointly as the *lymphatic-immune system*.

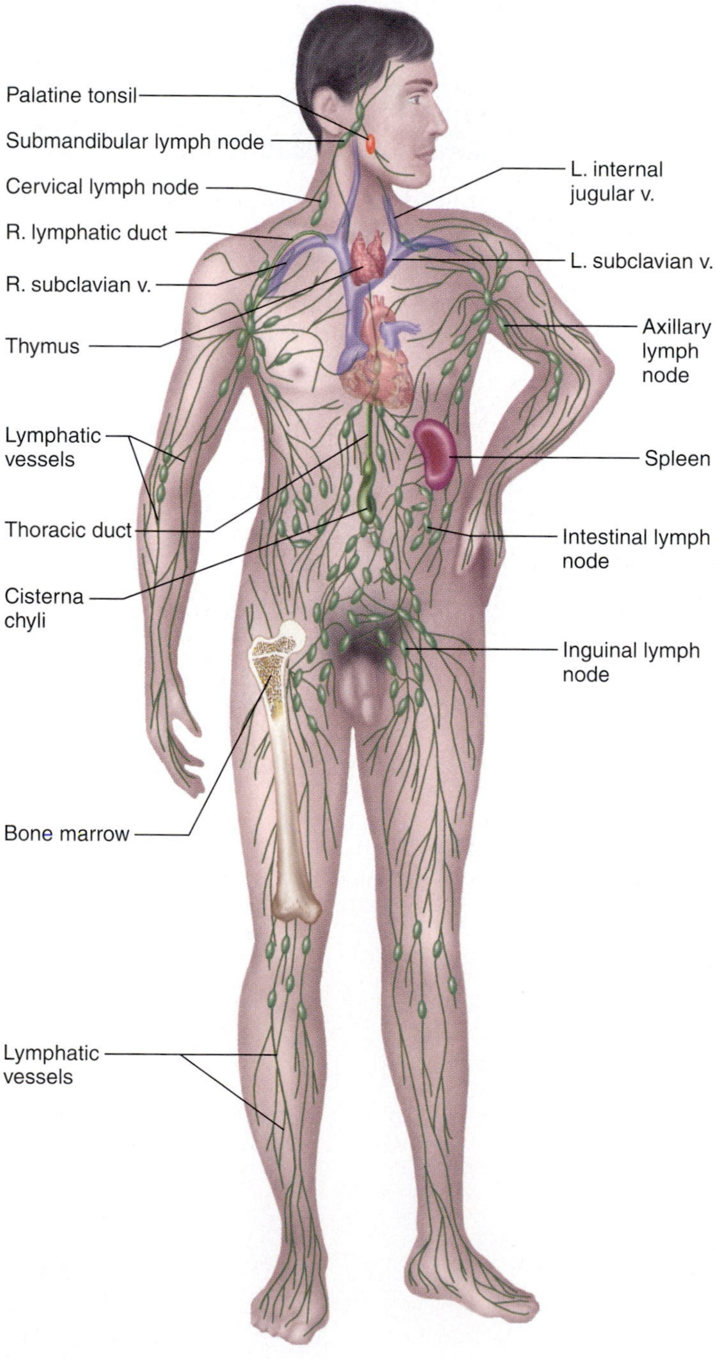
Palatine tonsil
Submandibular lymph node
Cervical lymph node
R. lymphatic duct
R. subclavian v.
Thymus
Lymphatic vessels
Thoracic duct
Cisterna chyli
Bone marrow
Lymphatic vessels
L. internal jugular v.
L. subclavian v.
Axillary lymph node
Spleen
Intestinal lymph node
Inguinal lymph node

FIGURE 22.1
The Lymphatic System.

Lymph and Lymphatic Vessels

Objectives

When you have completed this section, you should be able to

- list the functions and basic components of the lymphatic system;
- explain how lymph is formed;
- describe the route that lymph takes to get into the bloodstream; and
- explain what makes lymph flow through the lymphatic vessels.

Components and Functions of the Lymphatic System

The lymphatic system consists of the following components: (1) *lymph*, the fluid that the system collects from the interstitial spaces of the tissues and returns to the bloodstream; (2) *lymphatic vessels*, which transport the lymph; (3) *lymphatic tissue*, composed of aggregates of lymphocytes and macrophages that populate many organs of the body; and (4) *lymphatic organs*, in which these cells are especially concentrated and which are set off from surrounding organs by connective tissue capsules.

The functions of the lymphatic system are:

1. **Fluid recovery.** Fluid continually filters from our blood capillaries into the tissue spaces. The blood capillaries reabsorb about 85% of it, but the 15% that they do not absorb would amount, over the course of a day, to 2 to 4 L of water and one-quarter to one-half of the plasma protein. One would die of circulatory failure within hours if this water and protein were not returned to the bloodstream. One task of the lymphatic system is to reabsorb this excess and return it to the blood. Even partial interference with lymphatic drainage can lead to severe edema and sometimes even more grotesque consequences (see insight 22.1).
2. **Immunity.** As the lymphatic system recovers excess tissue fluid, it also picks up foreign cells and chemicals from the tissues. Some of these are **pathogens**—agents with the potential to cause disease. On its way back to the bloodstream, the fluid passes through lymph nodes, where immune cells stand guard against pathogens and activate protective immune responses.
3. **Lipid absorption.** In the small intestine, special lymphatic vessels called *lacteals* absorb dietary lipids that cannot be absorbed by the intestinal blood capillaries (see chapter 24).

Lymph

Lymph is usually a clear, colorless fluid, similar to blood plasma but low in protein. Its composition varies substantially from place to place. After a meal, for example, lymph draining from the small intestine has a milky appearance because of its high lipid content. This intestinal lymph is called *chyle*[1] (kile). Lymph leaving the

[1] *chyle* = juice

INSIGHT 22.1 CLINICAL APPLICATION

Elephantiasis

Any obstruction of the lymphatic vessels can block the return of fluid to the bloodstream and thus result in *edema*, the accumulation of excess tissue fluid. A particularly dramatic illustration of this is *elephantiasis* (fig. 22.2), a parasitic disease found in tropical climates worldwide, especially in Africa but also in India, southeast Asia, the Philippines, the Pacific Islands, and parts of South America (introduced by slave trading).

Elephantiasis is caused by mosquito-borne roundworms called *filariae* (fil-AIR-ee-ee), usually the species *Wuchereria bancrofti.* When an infected mosquito bites, tiny larvae escape from its proboscis, crawl into the bite wound, and enter the lymphatic vessels of the skin. They migrate to larger lymphatic vessels near the lymph nodes, where they mature into tightly coiled adults as large as 10 cm long and 0.3 cm wide. The worms cause intense inflammation of the lymphatic vessels and lymph nodes, especially in the lower half of the body. The vessels and nodes become swollen and painful, and infected males often suffer very painful edematous enlargement of the testes.

Infection with the worms, called *filariasis,* only occasionally leads to elephantiasis, but when it does, the effect can be horrible. The chronic blockage of lymph flow causes enormous enlargement and fibrosis of tissues upstream from the obstruction—notably the legs and arms, the scrotum of men *(lymph scrotum)*, and sometimes the vulva and breasts of women. The skin becomes fibrotic, thickened, and cracked, so it comes to resemble an elephant's hide—hence the name of the disease.

Thousands of American military personnel who served in the Pacific theater in World War II contracted filariasis. Alarmed by pictures of extreme cases, many of them feared having to carry their scrotum in a wheelbarrow. However, elephantiasis seldom develops in anyone whose first exposure to the filariae occurs in adulthood, and it requires many years of repetitive infection. Some servicemen had symptoms of filariasis for as long as 16 years after their return, but not one of them developed elephantiasis.

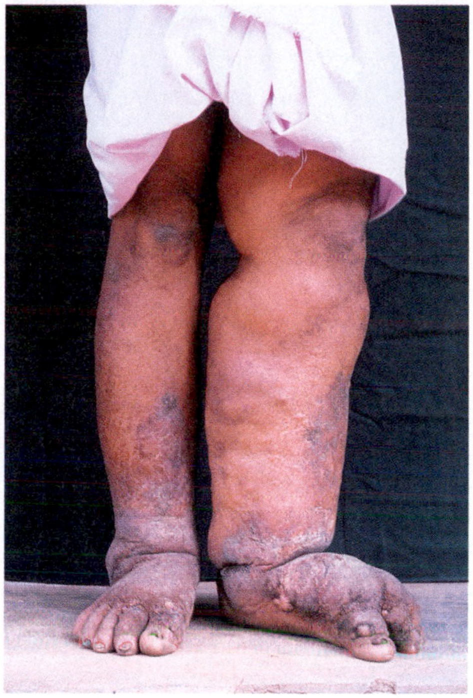

FIGURE 22.2
Elephantiasis of the Lower Limb.

lymph nodes contains a large number of lymphocytes—indeed, this is the main supply of lymphocytes to the bloodstream. Lymph can also contain macrophages, hormones, bacteria, viruses, cellular debris, and even traveling cancer cells.

Lymphatic Vessels

Lymph flows through a system of **lymphatic vessels (lymphatics)** similar to blood vessels. These begin with microscopic **lymphatic capillaries (terminal lymphatics),** which penetrate nearly every tissue of the body but are absent from the central nervous system, cartilage, cornea, bone, and bone marrow. They are closely associated with blood capillaries, but unlike them, they are closed at one end (fig. 22.3). A lymphatic capillary consists of a sac of thin endothelial cells that loosely overlap each other like the shingles of a roof. The cells are tethered to surrounding tissue by protein filaments that prevent the sac from collapsing.

Unlike the endothelial cells of blood capillaries, lymphatic endothelial cells are not joined by tight junctions, nor do they have a continuous basement membrane; indeed, the gaps between them are so large that bacteria, lymphocytes, and other cells and particles can enter along with the tissue fluid. Thus, the composition of lymph arriving at a lymph node is like a report on the state of the upstream tissues.

The overlapping edges of the endothelial cells act as valvelike flaps that can open and close. When tissue fluid pressure is high, it pushes the flaps inward (open) and fluid flows into the capillary. When pressure is higher in the lymphatic capillary than in the tissue fluid, the flaps are pressed outward (closed).

Think About It!

Contrast the structure of a lymphatic capillary with that of a continuous blood capillary. Explain why their structural difference is related to their functional difference.

The larger lymphatic vessels are similar to veins in their histology. They have a *tunica interna* with an endothelium and valves (fig. 22.4), a *tunica media* with elastic fibers and smooth muscle, and a thin outer *tunica externa.* Their walls are thinner and their valves are closer together than those of the veins.

As the lymphatic vessels converge along their path, they become larger and larger vessels with changing names. The route from the tissue fluid back to the blood stream is: lymphatic capillaries → collecting vessels → six lymphatic trunks → two collecting ducts → subclavian veins. Thus, there is a continual recycling of fluid from blood to tissue fluid to lymph and back to the blood (fig. 22.5).

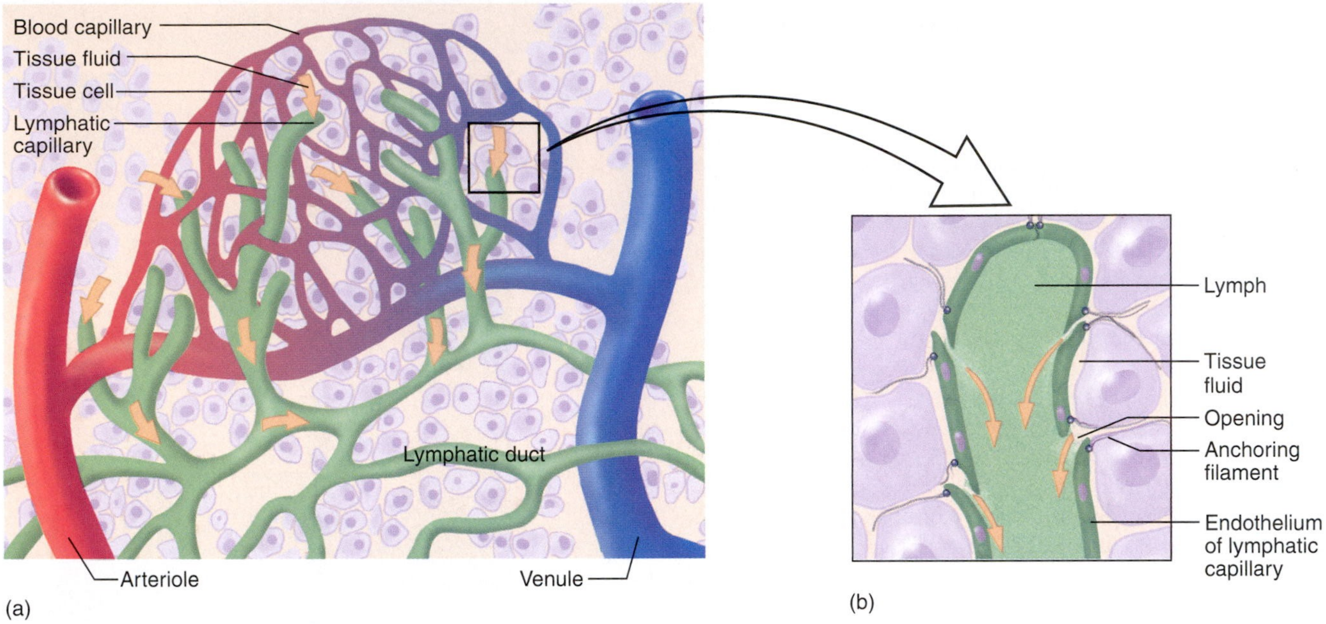

FIGURE 22.3

Lymphatic Capillaries. (*a*) Relationship of the lymphatic capillaries to a bed of blood capillaries. (*b*) Uptake of tissue fluid by a lymphatic capillary.

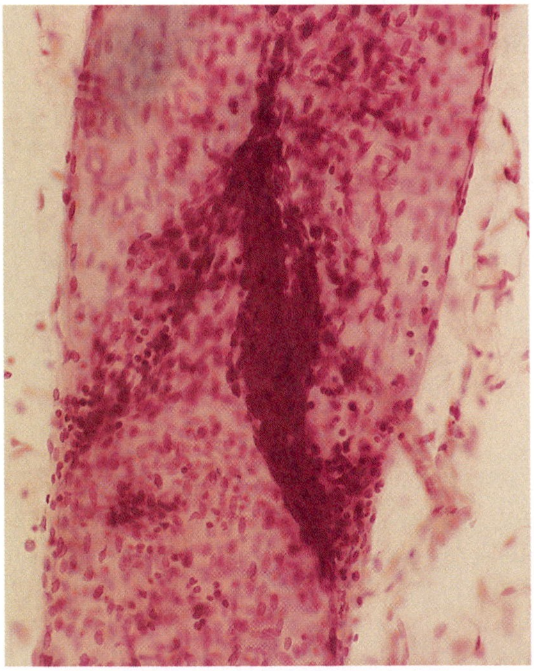

FIGURE 22.4

Valve in a Lymphatic Vessel.

The lymphatic capillaries converge to form **collecting vessels.** These often travel alongside veins and arteries and share a common connective tissue sheath with them. At irregular intervals, the collecting vessels empty into lymph nodes. The lymph trickles slowly through the node, where bacteria are phagocytized and immune cells monitor the fluid for foreign antigens. It leaves the other side of the node through another collecting vessel, traveling on and often encountering additional lymph nodes before it finally returns to the bloodstream.

Eventually, the collecting vessels converge to form larger **lymphatic trunks,** each of which drains a major portion of the body. There are six principal lymphatic trunks whose names indicate their locations and parts of the body they drain: the *jugular, subclavian, bronchomediastinal, intercostal, intestinal,* and *lumbar trunks.* The lumbar trunk drains not only the lumbar region but also the lower limbs.

The lymphatic trunks converge to form two **collecting ducts,** the largest of the lymphatic vessels (fig. 22.6):

1. The **right lymphatic duct** is formed by the convergence of the right jugular, subclavian, and bronchomediastinal trunks in the right thoracic cavity. It receives lymphatic drainage from the right upper limb and right side of the thorax and head, and empties into the right subclavian vein.
2. The **thoracic duct,** on the left, is larger and longer. It begins just below the diaphragm, anterior to the vertebral column at the level of the second lumbar vertebra. Here, the two lumbar trunks and the intestinal trunk join and form a prominent sac called the **cisterna chyli** (sis-TUR-nuh KY-lye), named for the large amount of chyle that it collects after a meal. The thoracic duct then passes through the diaphragm with the aorta and ascends the mediastinum, adjacent to the vertebral column. As it passes through the thorax, it receives additional lymph from the left bronchomediastinal, left subclavian, and left jugular trunks, then empties into the left subclavian vein.

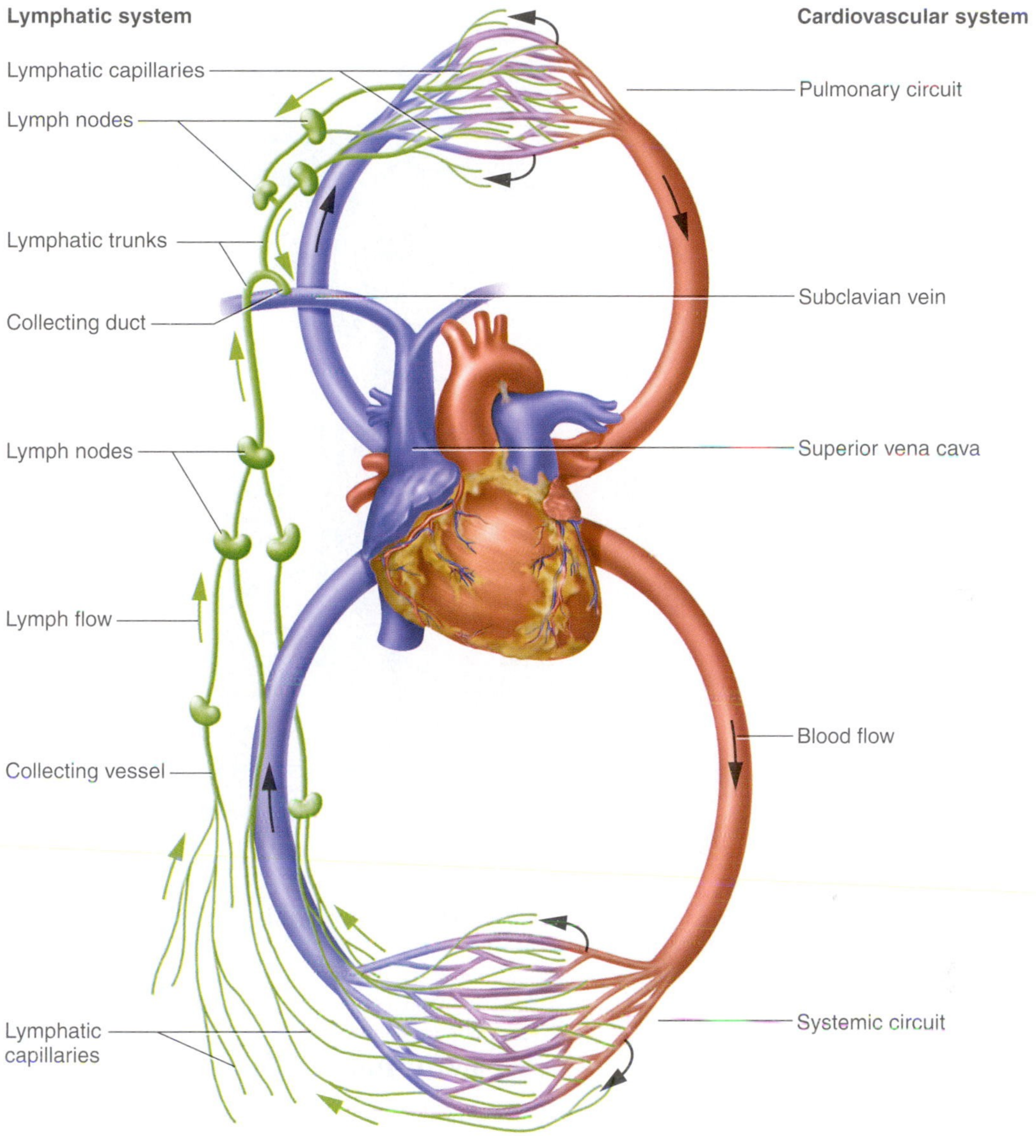
Lymphatic system
Cardiovascular system
Lymphatic capillaries
Pulmonary circuit
Lymph nodes
Lymphatic trunks
Subclavian vein
Collecting duct
Lymph nodes
Superior vena cava
Lymph flow
Blood flow
Collecting vessel
Lymphatic capillaries
Systemic circuit

FIGURE 22.5
Fluid Exchange Between the Circulatory and Lymphatic Systems. Blood capillaries lose fluid to the tissue spaces. The lymphatic system picks up excess tissue fluid and returns it to the bloodstream.

Collectively, this duct therefore drains all of the body below the diaphragm, and the left upper limb and left side of the head, neck, and thorax.

The Flow of Lymph

Lymph flows under forces similar to those that govern venous return, except that the lymphatic system has no pump like the heart, and it flows at even lower pressure and speed than venous blood. The primary mechanism of flow is rhythmic contractions of the lymphatic vessels themselves, which contract when the flowing lymph stretches them. The valves of lymphatic vessels, like those of veins, prevent the fluid from flowing backward. Lymph flow is also produced by skeletal muscles squeezing the lymphatic vessels, like the skeletal muscle pump that moves venous blood. Since lymphatic vessels are often wrapped with an artery in a common connective tissue sheath, arterial pulsation may also rhythmically squeeze the lymphatic vessels and contribute to lymph flow. A thoracic (respiratory) pump promotes the flow of lymph from the abdominal to the thoracic cavity as one inhales, just as it does in venous return. Finally, at the point where the collecting ducts empty into the subclavian veins, the rapidly flowing bloodstream draws the lymph into it.

THINK ABOUT IT!

Why does it make more functional sense for the collecting ducts to connect to the subclavian veins than it would for them to connect to the subclavian arteries?

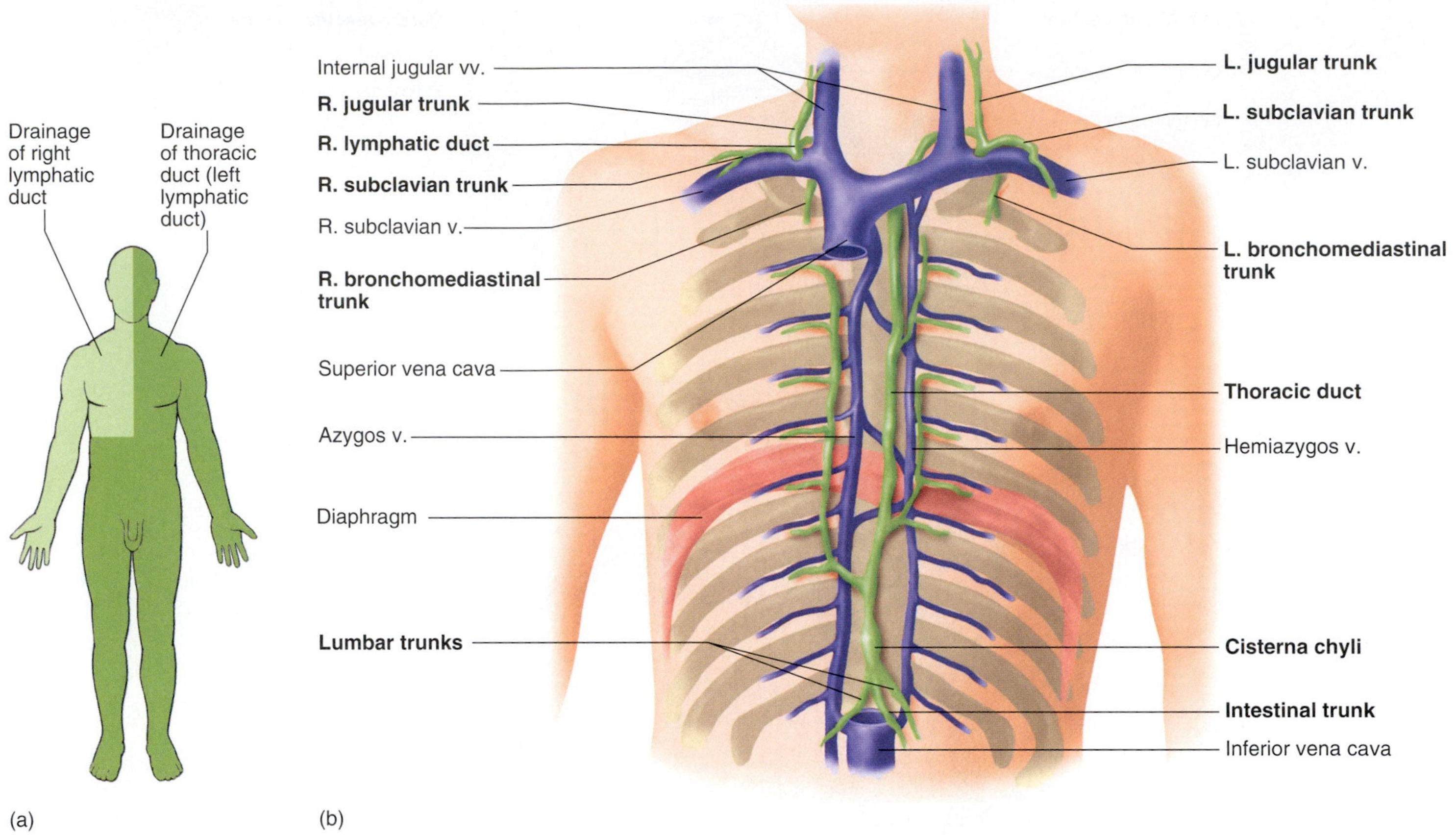
Drainage of right lymphatic duct
Drainage of thoracic duct (left lymphatic duct)
Internal jugular vv.
R. jugular trunk
R. lymphatic duct
R. subclavian trunk
R. subclavian v.
R. bronchomediastinal trunk
Superior vena cava
Azygos v.
Diaphragm
Lumbar trunks
L. jugular trunk
L. subclavian trunk
L. subclavian v.
L. bronchomediastinal trunk
Thoracic duct
Hemiazygos v.
Cisterna chyli
Intestinal trunk
Inferior vena cava
(a)
(b)

FIGURE 22.6

Lymphatics of the Thoracic Region. (*a*) Regions drained by the two collecting ducts. (*b*) Lymphatics of the thorax and upper abdomen, and their relationship to the subclavian veins, where the lymph returns to the bloodstream.

Before You Go On

Answer the following questions to test your understanding of the preceding section:

1. List the primary functions of the lymphatic system.
2. How does fluid get into the lymphatic system? What prevents it from draining back out?
3. Where does this fluid (lymph) go once it enters the lymphatic vessels? What makes it flow?

Lymphatic Cells, Tissues, and Organs

Objectives

When you have completed this section, you should be able to

- name the major types of cells in the lymphatic system and state their functions;
- describe the types of lymphatic tissue; and
- describe the anatomy and lymphatic-immune function of the red bone marrow, thymus, lymph nodes, tonsils, and spleen.

Lymphatic Cells and Modes of Defense

Another component of the lymphatic system is the lymphatic tissues, which range from loosely scattered cells in the mucous membranes of the digestive, respiratory, reproductive, and urinary tracts, to compact cell populations encapsulated in lymphatic organs. These tissues are composed of a variety of lymphocytes and other cells with various roles in defense and immunity.

LYMPHATIC CELL TYPES

The principal cell types of the lymphatic system are as follows:

1. **Natural killer (NK) cells.** These are large lymphocytes that attack and lyse bacteria, transplanted tissue cells, and *host cells* (cells of one's own body) that have either become infected with viruses or turned cancerous (see photo on p. 629). Their continual patrolling of the body "on the lookout" for abnormal cells is called *immunological surveillance,* and is one of the body's most important defenses against cancer.
2. **T lymphocytes (T cells).** These are so-named because they develop for a time in the thymus and later depend on

thymic hormones to regulate their activity. The *T* stands for *thymus-dependent.* There are several subclasses of T cells:

- *Cytotoxic T (T_C) cells* are the only T lymphocytes that directly attack and kill other cells. They are especially responsive to cells of transplanted tissues and organs, cancer cells, and host cells infected with viruses, bacteria, or intracellular parasites.
- *Helper T (T_H) cells* respond to antigens and activate various defense mechanisms, but do not carry out the attack themselves; instead, they help other immune cells respond to the threat. T_H cells play a central coordinating role in multiple forms of defense. They are also called T4, CD4, or CD4+ cells because of a surface glycoprotein called CD4.
- *Suppressor T (T_S) cells* moderate the action of T_C cells and prevent excessive immune responses. T_C and T_S cells are also called T8, CD8, or CD8+ cells for their surface glycoprotein, CD8. (*CD* stands for *cluster of differentiation,* a classification system for many cell-surface molecules.) There is some disagreement on whether T_S cells should continue to be recognized as a distinct class of T cells.
- *Memory T cells* provide long-lasting memory of an antigen. Upon reexposure, the immune system neutralizes the antigen so quickly that it causes no disease symptoms. This is what we mean by being immune to a disease.

3. **B lymphocytes (B cells).** These were originally named for an organ in chickens (the *bursa of Fabricius*[2]) in which they were first discovered; however, you may find it useful to think of *B* for *bone marrow,* the site where these cells mature. When activated, B cells differentiate into *plasma cells,* which secrete circulating antibodies, the protective immunoglobulins of the body fluids. Some B cells become memory B cells instead of plasma cells, functioning like memory T cells but in humoral (antibody-mediated) immunity instead of cellular immunity.

   In stained blood films, the different types of lymphocytes are not morphologically distinguishable, but they are about 80% T cells, 15% B cells, and 5% NK and stem cells.
4. **Macrophages.** These cells develop from monocytes that have emigrated from the bloodstream. Macrophages are very large, avidly phagocytic cells. They ingest and destroy tissue debris, dead neutrophils, bacteria, and other foreign matter (fig. 22.7). They also process foreign matter and transport antigenically active fragments of it *(antigenic determinants)* to the cell surface, where they "display" it to T_C and T_H cells. This stimulates the T cells to launch an immune response against the foreign invader. Macrophages, B lymphocytes, and reticular cells are collectively called **antigen-presenting cells (APCs)** because they display antigen fragments to other immune cells.

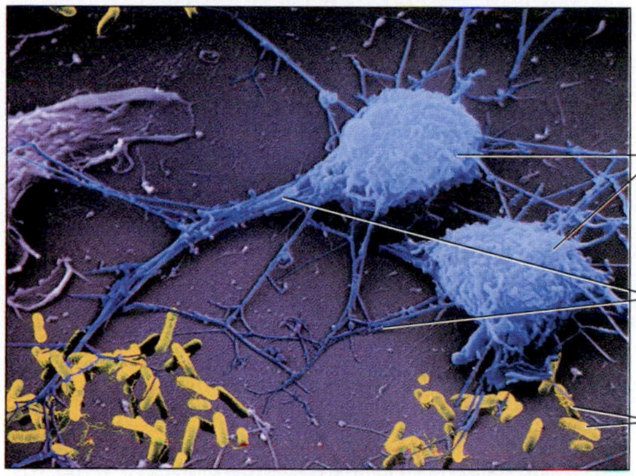

FIGURE 22.7
Macrophages Attacking Bacteria. Colorized SEM.

   The *macrophage (lymphoid-macrophage) system* includes all of the body's phagocytic cells except leukocytes. Some of these phagocytes are wandering cells that actively seek pathogens, while others are fixed in place and phagocytize only those pathogens that come to them—although they are strategically positioned for this to occur. Cells of the macrophage system include the macrophages of the loose connective tissue, *microglia* of the central nervous system, *alveolar macrophages* in the lungs, *hepatic macrophages* in the liver, and the *dendritic cells* discussed next. Alveolar and hepatic macrophages are described in chapters 23 and 24.
5. **Dendritic cells.** These are branched macrophages found in the epidermis, mucous membranes, and lymphatic organs. (In the skin, they are often called *Langerhans*[3] *cells.*) They engulf foreign matter by receptor-mediated endocytosis rather than phagocytosis, but they otherwise function like macrophages and are included in the macrophage system.
6. **Reticular cells.** These are branched stationary cells that contribute to the stroma of the lymphatic organs and act as APCs in the thymus (see fig. 22.10). (They should not be confused with reticular *fibers,* which are fine branched collagen fibers common in lymphatic organs.)

MODES OF DEFENSE

The types of defense provided by these cells can be classified into nonspecific defense and specific immunity. **Nonspecific defense** consists of mechanisms that are present from birth, defend the body against a broad spectrum of pathogens, and involve no mechanism for "remembering" any particular pathogen and reacting differently to it upon reexposure. The NK cells provide such indiscriminate defense, as do neutrophils and other leukocytes (except B and T cells). Other mechanisms of nonspecific defense include physical barriers to infection such as the skin and mucous membranes, antimicrobial proteins called *interferons* and the *complement system,* and processes such as inflammation and fever.

[2]Hieronymus Fabricius (Girolamo Fabrizzi) (1537–1619), Italian anatomist

[3]Theodor Langerhans (1839–1915), German pathologist

Specific immunity, by contrast, is a defense that arises upon exposure to a particular pathogen, remembers that pathogen as a result of the encounter, and reacts so quickly and efficiently to it the next time that the pathogen has little or no chance to produce illness. Immunity is specific to a particular pathogen, however. Immunity to chickenpox does not confer immunity to measles, for example; one must develop a separate immunity to each one. Thus, immunity has a specificity and memory that are lacking from nonspecific defense. Specific immunity is the function of the B and T lymphocytes.

Some of the aforementioned cells play roles in both nonspecific defense and specific immunity—notably the macrophages and helper T cells. Macrophages are fairly undiscriminating in what microbes they phagocytize and destroy, but they also act as the antigen-presenting cells that activate the lymphocytes of specific immunity. Helper T cells activate not only the B and T_C cells of specific immunity, but also help to mediate the inflammatory response of nonspecific defense.

Think About It!

Suppose a new virus emerged that selectively destroyed memory T and B cells. What would be the pathological effect of such a virus?

Lymphatic Tissues

Lymphatic (lymphoid) tissues are aggregations of lymphocytes in the connective tissues of mucous membranes and various organs. The simplest form is **diffuse lymphatic tissue,** in which the lymphocytes are scattered rather than densely clustered. It is particularly prevalent in body passages that are open to the exterior—the respiratory, digestive, urinary, and reproductive tracts—where it is called **mucosa-associated lymphatic tissue (MALT).** (In the respiratory and digestive tracts, it is sometimes called bronchus-associated and gut-associated lymphatic tissue, BALT and GALT, respectively).

In some places, lymphocytes and macrophages congregate in dense masses called **lymphatic nodules (follicles)** (fig. 22.8), which come and go as pathogens invade the tissues and the immune system answers the challenge. Abundant lymphatic nodules are, however, a relatively constant feature of the lymph nodes, tonsils, and appendix. In the ileum, the distal portion of the small intestine, they form clusters called **Peyer[4] patches.**

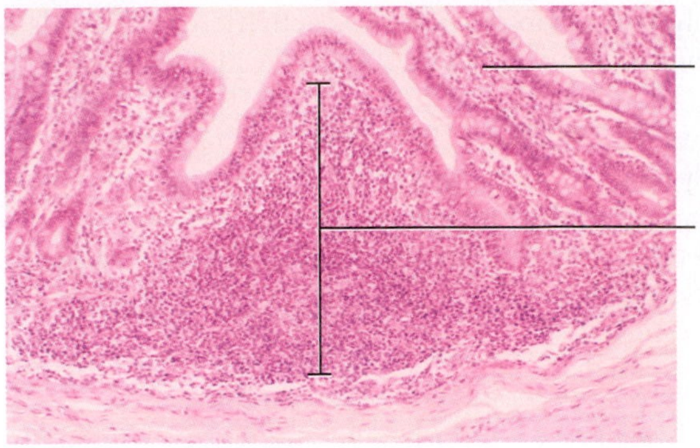

Figure 22.8
Lymphatic Nodule in the Mucous Membrance of the Small Intestine.

Overview of Lymphatic Organs

In contrast to the diffuse lymphatic tissue, **lymphatic (lymphoid) organs** have well-defined anatomical sites and at least partial connective tissue capsules that separate the lymphatic tissue from neighboring tissues. These organs include the thymus, red bone marrow, lymph nodes, tonsils, and spleen. The red bone marrow and thymus are regarded as *primary lymphatic organs* because they are the sites where T and B lymphocytes, respectively, become *immunocompetent*—that is, able to recognize and respond to antigens. The lymph nodes, tonsils, and spleen are called *secondary lymphatic organs* because they are populated with immunocompetent lymphocytes only after the cells have matured in the primary lymphatic organs.

Red Bone Marrow

As discussed in chapter 6, there are two kinds of bone marrow—red and yellow. Red bone marrow is involved in hemopoiesis (blood formation) and immunity; yellow bone marrow can be disregarded for our present purposes. In children, red bone marrow occupies the medullary spaces of nearly the entire skeleton. In adults, it is limited to parts of the axial skeleton and the proximal heads of the humerus and femur. Red bone marrow is an important supplier of lymphocytes to the immune system. Its role in the life history of lymphocytes is described later.

Red bone marrow is a soft, loosely organized, highly vascular material, separated from osseous tissue by the endosteum of the bone. It produces all classes of formed elements of the blood; its red color comes from the abundance of erythrocytes. Numerous small arteries enter *nutrient foramina* on the bone surface, penetrate the bone, and empty into large *sinusoids* (45 to 80 μm wide) in the marrow (fig. 22.9). The sinusoids drain into a *central longitudinal vein* that exits the bone via the same route that the arteries entered it. The sinusoids are lined by endothelial cells, like other blood vessels, and are surrounded by reticular cells and reticular fibers. The reticular cells secrete colony-stimulating factors that induce the formation of various leukocyte types. In the long bones of the limbs, aging reticular cells accumulate fat and transform into adipose cells, eventually replacing red bone marrow with yellow bone marrow.

[4]Johann Conrad Peyer (1653–1712), Swiss anatomist

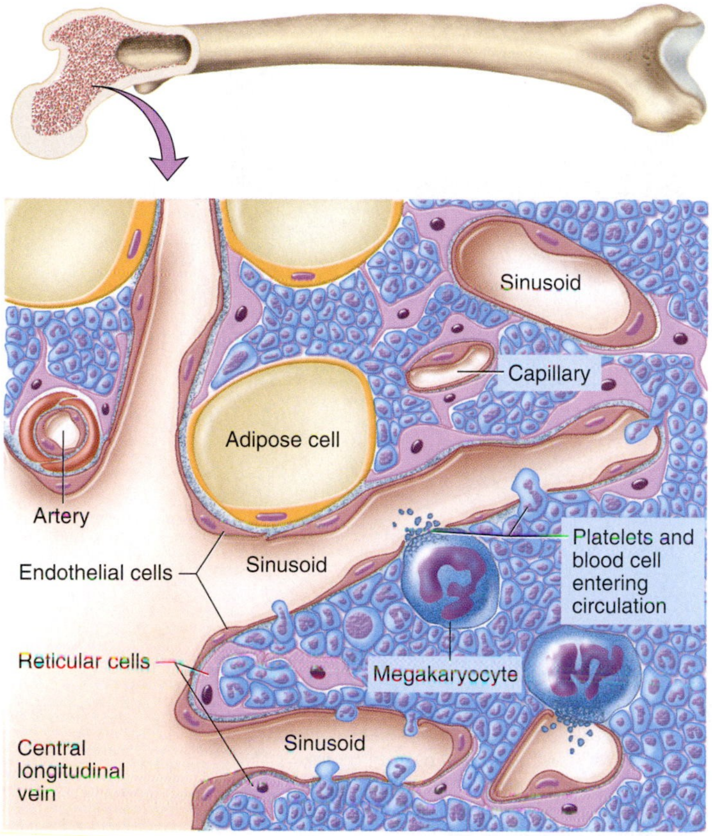

FIGURE 22.9

Histology of the Red Bone Marrow. The formed elements of blood squeeze through the endothelial cells into the sinuses, which converge on the central longitudinal vein at the lower left.

The spaces between the sinusoids are occupied by *islands* or *cords* of hemopoietic cells, composed of macrophages and blood cells in all stages of development. The macrophages destroy malformed blood cells and the nuclei discarded by developing erythrocytes. As blood cells mature, they push their way through the reticular and endothelial cells to enter the sinus and flow away in the bloodstream.

Think About It!

If we regard red bone marrow as a lymphatic organ and define lymphatic organs partly by the presence of a connective tissue capsule, what could we regard as the capsule of red bone marrow?

The Thymus

The gross anatomy of the **thymus** was described in chapter 18 (see fig. 18.7), since it is a member of both the endocrine and lymphatic systems. Here we will be concerned only with its microscopic anatomy and how this relates to its immune function.

The fibrous capsule of the thymus gives off trabeculae (septa) that divide the gland into several angular lobules. Each lobule has a dense, dark-staining *cortex* and a lighter *medulla* inhabited by T lymphocytes (fig. 22.10). **Reticular epithelial cells** seal off the cortex from the medulla and surround the blood vessels and lymphocyte clusters in the cortex. They thereby form a *blood-thymus barrier* that isolates developing lymphocytes from blood-borne antigens. In the medulla, the reticular epithelial cells form whorls called *thymic (Hassall[5]) corpuscles.* These have no known function but are useful for identifying the thymus histologically.

Besides forming the blood-thymus barrier, reticular epithelial cells secrete hormones called *thymosins, thymulin,* and *thymopoietin,* which promote the development and action of T cells. If the thymus is removed from newborn mammals, they waste away and never develop immunity. Other lymphatic organs also seem to depend on thymosins or T cells and develop poorly in thymectomized animals. The relationship of T cell maturation to thymic histology is discussed later in this chapter.

Lymph Nodes

Lymph nodes (fig. 22.11) are the most numerous lymphatic organs, numbering in the hundreds. Lymph nodes serve two functions: they cleanse the lymph and alert the immune system to pathogens. A lymph node is an elongated or bean-shaped structure, usually less than 3 cm long, often with an indentation called the *hilum* on one side (fig. 22.12). It is enclosed in a fibrous capsule with trabeculae that partially divide the interior of the node into compartments. Between the capsule and parenchyma is a narrow, relatively clear space called the *subcapsular sinus,* which contains reticular fibers, macrophages, and dendritic cells. Deep to this, the gland consists mainly of a stroma of reticular connective tissue (reticular fibers and reticular cells) and a parenchyma of lymphocytes and antigen-presenting cells.

The parenchyma is divided into an outer C-shaped **cortex** that encircles about four-fifths of the organ, and an inner **medulla** that extends to the surface at the hilum. The cortex consists mainly of ovoid to conical lymphatic nodules. When the lymph node is fighting a pathogen, these nodules acquire light-staining **germinal centers** where B cells multiply and differentiate into plasma cells. The medulla consists largely of a branching network of *medullary cords* composed of lymphocytes, plasma cells, macrophages, reticular cells, and reticular fibers. The cortex and medulla also contain lymph-filled sinuses continuous with the subcapsular sinus.

Several **afferent lymphatic vessels** lead into the node along its convex surface. Lymph flows from these vessels into the subcapsular

[5]Arthur H. Hassall (1817–94), British chemist and physician

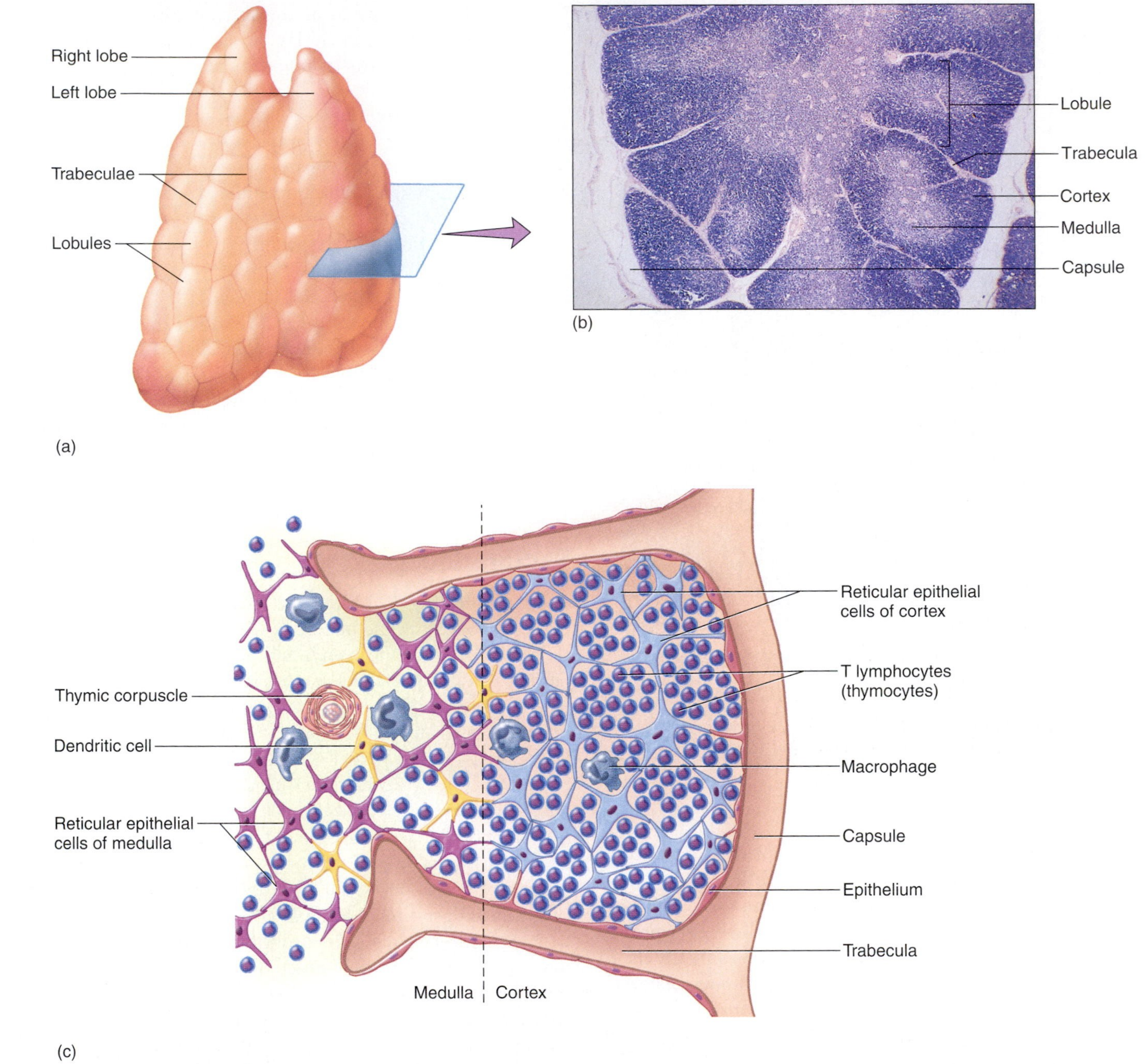

FIGURE 22.10

The Thymus. (*a*) Gross anatomy. (*b*) Histology. (*c*) Arrangement of the reticular epithelial cells to form the blood-thymus barrier.

sinus, percolates slowly through the sinuses of the cortex and medulla, and leaves the node through one to three **efferent lymphatic vessels** that emerge from the hilum. No other lymphatic organs have afferent lymphatic vessels; lymph nodes are the only organs that filter lymph as it flows along its course. The lymph node is a "bottleneck" that slows down lymph flow and allows time for cleansing it of foreign matter. The macrophages and reticular cells of the sinuses remove about 99% of the impurities before the lymph leaves the node. On its way to the bloodstream, lymph flows through one lymph node after another and thus becomes quite thoroughly cleansed of most impurities.

Blood vessels also penetrate the hilum of a lymph node. Arteries follow the medullary cords and give rise to capillary beds in the medulla and cortex. In the *deep cortex* near the junction with

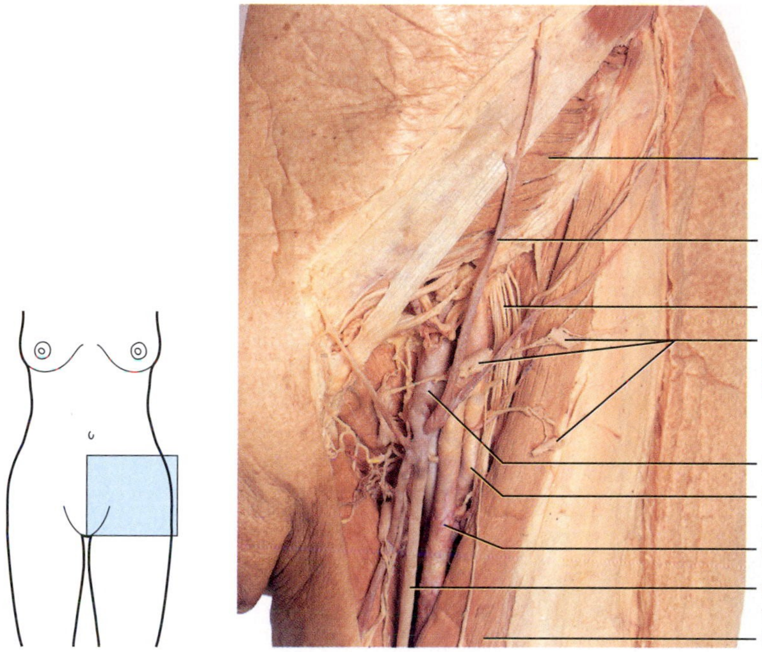

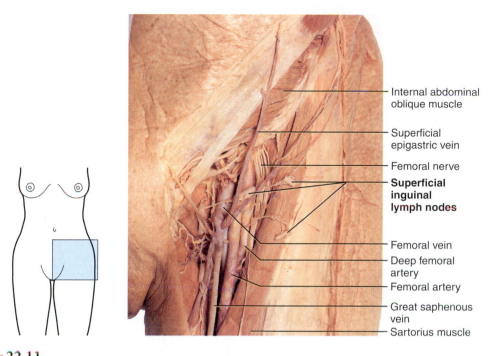

Internal abdominal oblique muscle
Superficial epigastric vein
Femoral nerve
Superficial inguinal lymph nodes
Femoral vein
Deep femoral artery
Femoral artery
Great saphenous vein
Sartorius muscle

FIGURE 22.11

Inguinal Lymph Nodes. Left inguinal region of a female cadaver.

the medulla, lymphocytes can emigrate from the bloodstream into the parenchyma of the node. Most lymphocytes in the deep cortex are T cells.

Lymph nodes are widespread but especially concentrated in the following locations:

- *Cervical lymph nodes* occur in deep and superficial groups in the neck, and monitor lymph coming from the head and neck.
- *Axillary lymph nodes* are concentrated in the armpit (axilla) and receive lymph from the upper limb and the female breast (fig. 22.13*a*).
- *Thoracic lymph nodes* occur in the thoracic cavity and receive lymph from the lungs, airway, and mediastinum.
- *Abdominal lymph nodes* monitor lymph from the urinary and reproductive systems.
- *Intestinal* and *mesenteric lymph nodes* monitor lymph from the digestive tract (fig. 22.13*b*).
- *Inguinal lymph nodes* occur in the groin (see fig. 22.11) and receive lymph from the entire lower limb.
- *Popliteal lymph nodes* occur at the back of the knee and receive lymph from the leg proper.

INSIGHT 22.2 CLINICAL APPLICATION

LYMPH NODES AND METASTATIC CANCER

Metastasis is a phenomenon in which cancerous cells break free of the original *primary tumor,* travel to other sites in the body, and establish new tumors. Because of the high permeability of lymphatic capillaries, metastasizing cancer cells easily enter them and travel in the lymph. They tend to lodge in the first lymph node they encounter and grow there, eventually destroying the node. Cancerous lymph nodes are swollen but relatively firm and usually painless. Cancer of a lymph node is called *lymphoma.*

Once a tumor is well established in one node, cells may emigrate from there and travel to the next. However, if the metastasis is detected early enough, cancer can sometimes be eradicated by removing not only the primary tumor, but also the nearest lymph nodes downstream from that point. For example, breast cancer is often treated with a combination of lumpectomy or mastectomy along with removal of the nearby axillary lymph nodes.

Physicians routinely palpate the superficial lymph nodes of the cervical, axillary, and inguinal regions for swelling *(lymphadenitis).* Lymph nodes are common sites of metastatic cancer (see insight 22.2).

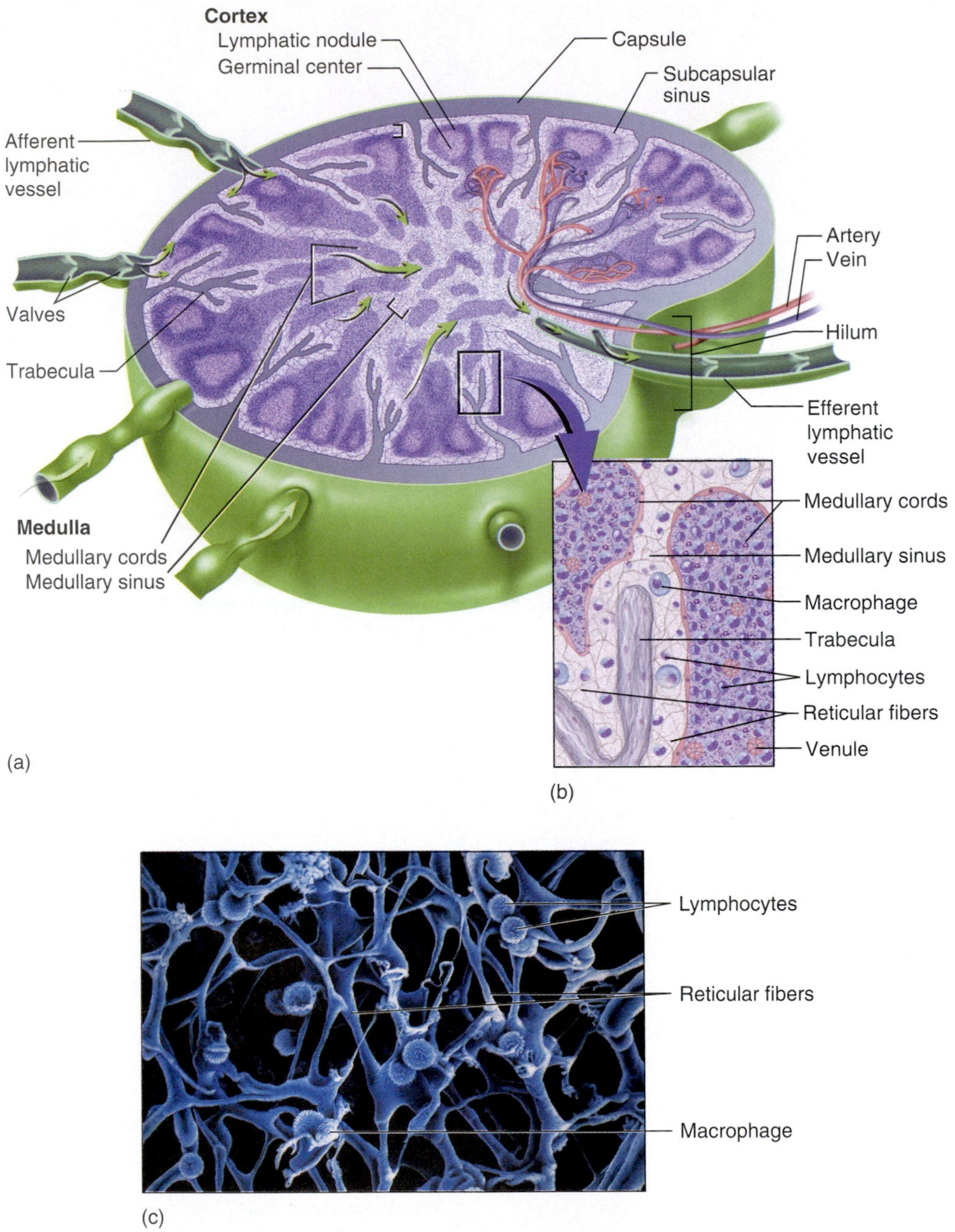
Cortex
Lymphatic nodule
Germinal center
Capsule
Subcapsular sinus
Afferent lymphatic vessel
Valves
Trabecula
Artery
Vein
Hilum
Efferent lymphatic vessel
Medulla
Medullary cords
Medullary sinus
(a)
Medullary cords
Medullary sinus
Macrophage
Trabecula
Lymphocytes
Reticular fibers
Venule
(b)
Lymphocytes
Reticular fibers
Macrophage
(c)

FIGURE 22.12

Anatomy of a Lymph Node. (*a*) Bisected lymph node showing pathway of lymph flow. (*b*) Detail of the boxed region in *a.* (*c*) Stroma and immune cells in a medullary sinus.

Tonsils

The **tonsils** are patches of lymphatic tissue located at the entrance to the pharynx, where they guard against ingested and inhaled pathogens. Each is covered by an epithelium and has deep pits called **tonsillar crypts** lined by lymphatic nodules (fig. 22.14). The crypts often contain food debris, dead leukocytes, bacteria, and antigenic chemicals. Below the crypts, the tonsils are partially separated from underlying connective tissue by an incomplete fibrous capsule.

There are three main sets of tonsils: (1) a single median **pharyngeal tonsil (adenoids)** on the wall of the pharynx just behind the nasal cavity, (2) a pair of **palatine tonsils** at the posterior margin of the oral cavity, and (3) numerous **lingual tonsils,** each with a single crypt, concentrated in a patch on each side of the root of the tongue (see fig. 24.5, p. 681).

The palatine tonsils are the largest and most often infected. *Tonsillitis* is an acute inflammation of the palatine tonsils, usually caused by a *Streptococcus* infection. Their surgical removal, called

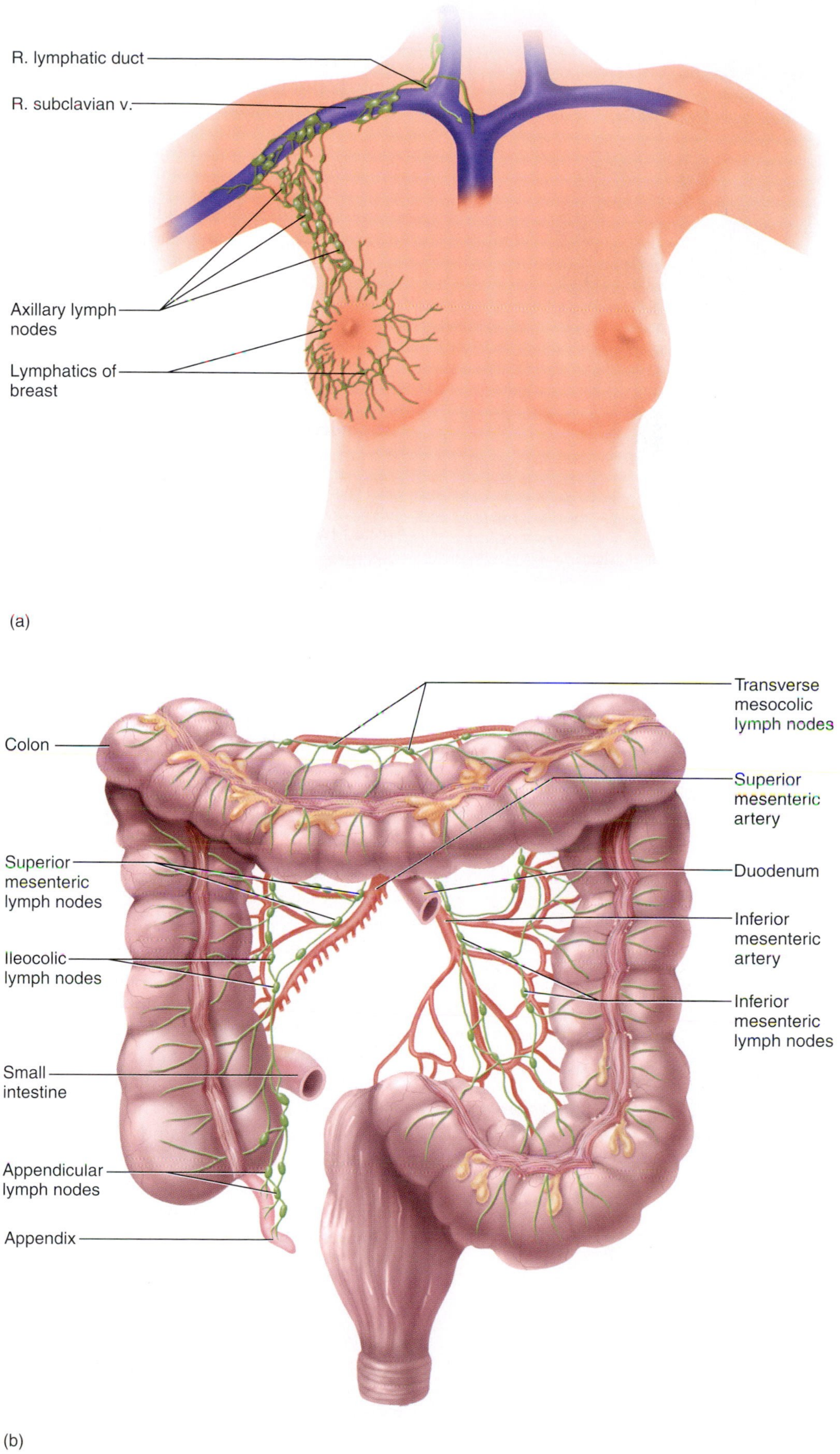

R. lymphatic duct
R. subclavian v.
Axillary lymph nodes
Lymphatics of breast
(a)
Colon
Transverse mesocolic lymph nodes
Superior mesenteric artery
Superior mesenteric lymph nodes
Duodenum
Inferior mesenteric artery
Ileocolic lymph nodes
Inferior mesenteric lymph nodes
Small intestine
Appendicular lymph nodes
Appendix
(b)

FIGURE 22.13

Some Areas of Lymph Node Concentration. (*a*) Axillary lymph nodes and lymphatics of the female breast. (*b*) Mesenteric lymph nodes.

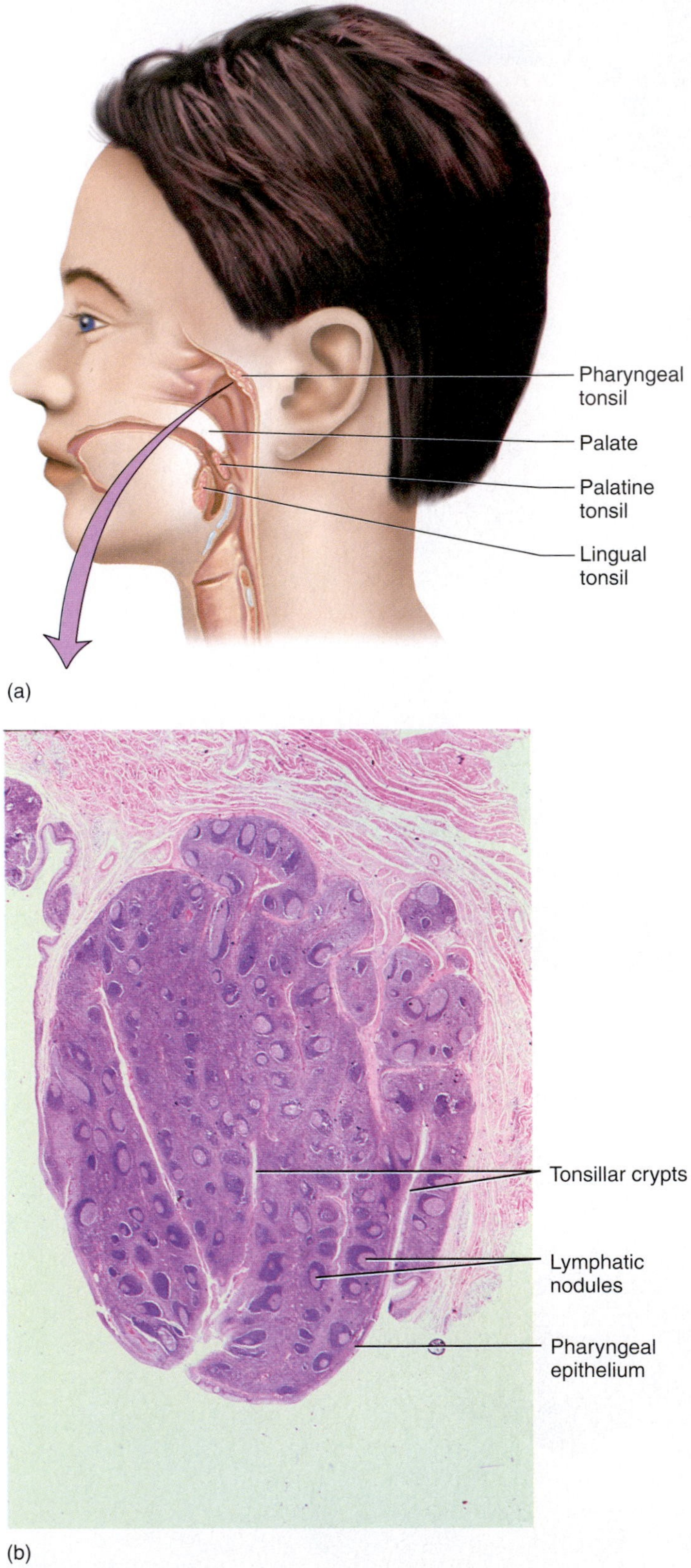
Pharyngeal tonsil
Palate
Palatine tonsil
Lingual tonsil
(a)
Tonsillar crypts
Lymphatic nodules
Pharyngeal epithelium
(b)

FIGURE 22.14

The Tonsils. (*a*) Locations of the tonsils. (*b*) Histology of the pharyngeal tonsil.

tonsillectomy, used to be one of the most common surgical procedures performed on children, but it is done less often today. Tonsillitis is now usually treated with antibiotics.

Think About It!

Which tonsil(s) is or are most likely to be affected by an inhaled pathogen?

The Spleen

The **spleen** is the body's largest lymphatic organ, measuring up to 12 cm long and weighing up to 160 g. It is located in the left hypochondriac region, just inferior to the diaphragm and dorsolateral to the stomach (fig. 22.15; see also fig. A.15, p. 38). It is protected by ribs 10 through 12. The spleen fits snugly between the diaphragm, stomach, and kidney and has indentations called the *gastric area* and *renal area* where it presses against these adjacent viscera. It has a medial hilum penetrated by the splenic artery, splenic vein, and lymphatic vessels.

The parenchyma exhibits two types of tissue named for their appearance in fresh specimens (not in stained sections): **red pulp,** which consists of sinuses gorged with concentrated erythrocytes, and **white pulp,** which consists of lymphocytes and macrophages aggregated like sleeves along small branches of the splenic artery. In tissue sections, white pulp appears as an ovoid mass of lymphocytes with an arteriole passing through it. However, it is important to bear in mind that the three-dimensional shape is not egglike but cylindrical.

These two tissue types reflect the multiple functions of the spleen. It produces blood cells in the fetus and may resume this role in adults in the event of extreme anemia. Lymphocytes and macrophages of the white pulp monitor the blood for foreign antigens, much like the lymph nodes do the lymph. The splenic blood capillaries are very permeable; they allow RBCs to leave the bloodstream, accumulate in the sinuses of the red pulp, and reenter the bloodstream later. The spleen is an "erythrocyte graveyard"—old, fragile RBCs rupture as they squeeze through the capillary walls into the sinuses. Macrophages phagocytize their remains, just as they dispose of blood-borne bacteria and other cellular debris. The spleen also helps to stabilize blood volume by transferring excess plasma from the bloodstream into the lymphatic system.

Think About It!

From an anatomical perspective, why are lymph nodes the only lymphatic organs that can filter the lymph?

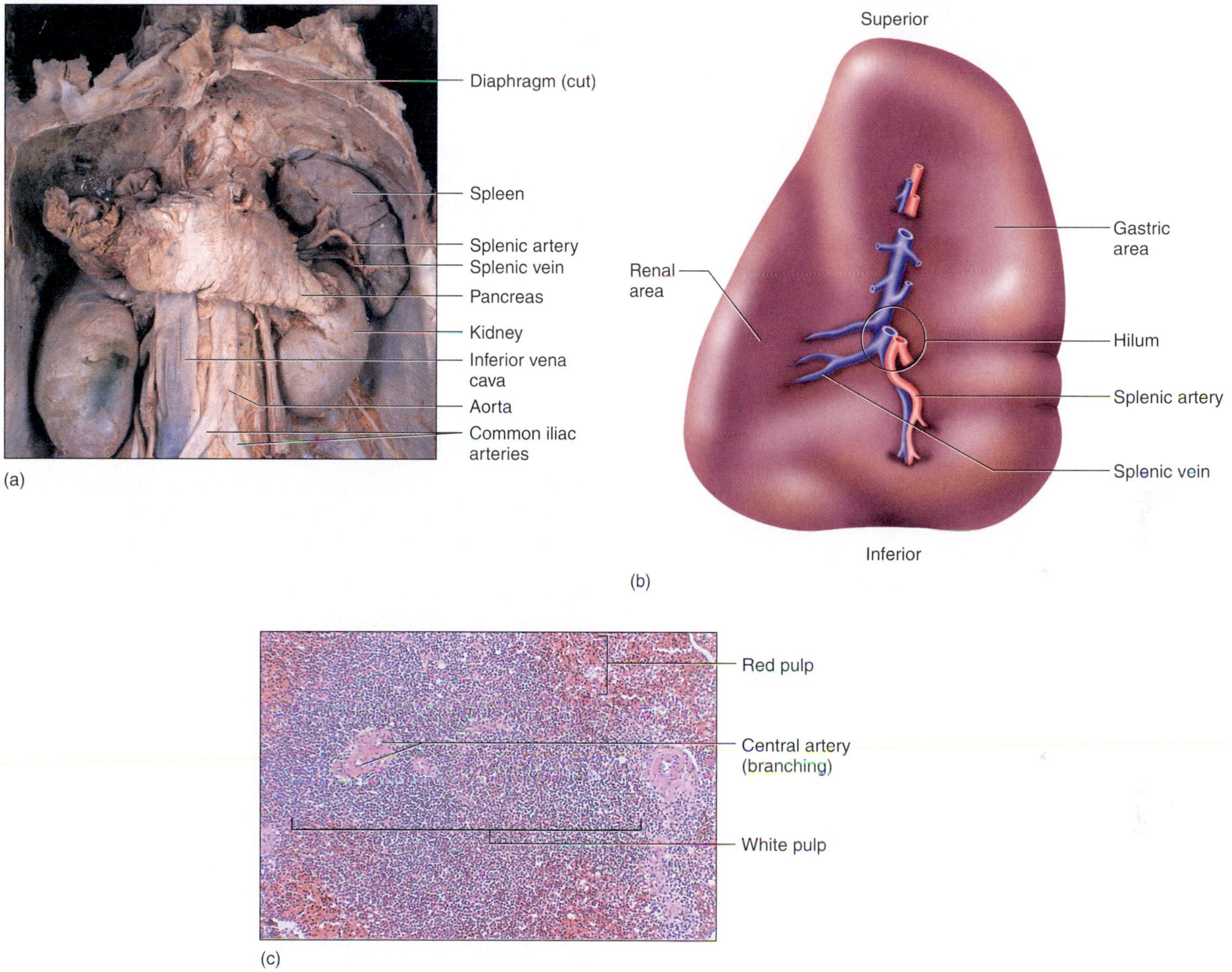

FIGURE 22.15

The Spleen. (*a*) Position of the spleen in the upper left quadrant of the abdominal cavity. (*b*) Gross anatomy of the medial surface. (*c*) Histology.

INSIGHT 22.3 CLINICAL APPLICATION

SPLENECTOMY

A ruptured spleen is one of the most common consequences of blows to the left thoracic or abdominal wall, as in sports injuries and automobile accidents. It is especially likely to rupture if the lower ribs are fractured, and sometimes it is nicked during abdominal surgery. The spleen is such a pulpy and vascular organ that it bleeds profusely, and its capsule is so thin and delicate that it is difficult to repair surgically. To prevent fatal hemorrhaging, it is often necessary to quickly remove the spleen and tie off the splenic artery. This procedure is called *splenectomy*.

The loss of splenic function, called *hyposplenism*, is usually not serious; its functions are adequately carried out by hepatic and bone marrow macrophages. However, it does leave a person somewhat more at risk of septicemia (bacteria in the blood) and pneumococcal infections. Therefore, if possible, surgeons try to leave some of the spleen in place; the spleen regenerates rapidly in such cases.

Some people have overactive spleens *(hypersplenism)*, in which excessive phagocytosis of the formed elements of blood can lead to anemia, leukopenia, or thrombocytopenia (see chapter 19). This can be another reason for performing a splenectomy.

Before You Go On

Answer the following questions to test your understanding of the preceding section:

4. What do T, B, and NK cells have in common? How do NK cells functionally differ from the other two? How do T and B cells functionally differ from each other?
5. What is the function of an antigen-presenting cell (APC)? Name three kinds of APCs.
6. What is a lymphatic nodule? Describe three places where lymphatic nodules can be found.
7. What are the two primary lymphatic organs? Why are they called that? Describe their collaborative relationship in producing the lymphocytes that populate other organs.
8. Describe the structural and functional differences between the cortex and medulla of a lymph node.
9. Name the three kinds of tonsils and state how they differ in number and location.
10. What are the two types of "pulp" in the spleen? What are their respective functions?
11. In what sense does the spleen serve the blood in the same way that the lymph nodes serve the lymph?

The Lymphatic System in Relation to Immunity

Objectives

When you have completed this section, you should be able to

- define *immune system* and explain its relationship to the lymphatic system;
- distinguish between humoral and cellular immunity; and
- describe the life histories and immune functions of B cells and T cells, and how these relate to the anatomy of the lymphatic organs.

The **immune system** is not an organ system, but rather a population of disease-fighting cells that reside in the mucous membranes, lymphatic organs, and other localities in the body. Although it does not have a specific anatomy distinct from what we have already studied in this chapter, a brief survey of immune function will enhance your understanding of the defensive role of the lymphatic system.

There are two forms of specific immunity called *humoral* and *cellular immunity.* **Humoral (antibody-mediated) immunity** is carried out by B lymphocytes and antibodies. It is called *humoral* because the antibodies circulate freely in the body fluids; *humors* is an archaic term for body fluids. **Cellular (cell-mediated) immunity** is carried out by cytotoxic T cells. We will not delve into the details of these defenses, but we will take a look at how the activities of B and T cells are related to the anatomy of the lymphatic organs. These lymphocytes are the most abundant cells of the lymphatic organs.

There are some things that B and T cells have in common, and we can address these before examining the differences in their life histories. Both types begin their development with *pluripotent stem cells* (PPSCs) in the red bone marrow. PPSCs divide and give rise to *lymphocyte colony-forming units,* which ultimately produce B and T lymphocytes (see fig. 19.9, p. 553). Before they can take part in immune reactions, both types of lymphocytes must develop antigen receptors on their surfaces, giving them **immunocompetence**—the ability to recognize, bind, and respond to an antigen. In addition, the body must get rid of lymphocytes that react against its own (host) antigens so that the immune system will not attack our own organs. The destruction or deactivation of self-reactive lymphocytes is called **negative selection.** Only about 2% of the lymphocytes survive this culling process. The developmental histories of B and T cells are contrasted in figure 22.16.

B Cells and Humoral Immunity

B cells achieve immunocompetence and go through negative selection in the red bone marrow. Many of the mature immunocompetent B cells remain there, while many more disperse and populate other sites such as the mucous membranes, spleen, and especially the cortical nodules of the lymph nodes, where they can sit and await the arrival of antigens in the incoming lymph.

When one of these B cells encounters an antigen, it internalizes and digests it, and presents fragments of the antigen to a helper T cell. The helper T cell secretes chemical *helper factors* that stimulate the B cell to divide still more. Most of its daughter cells differentiate into **plasma cells,** which are larger than B cells and have an abundance of rough endoplasmic reticulum (fig. 22.17)—as well they might, for plasma cells secrete antibodies at the astounding rate of up to 2,000 molecules per second for a life span of 4 or 5 days. Plasma cells develop mainly in the germinal centers of the nodules of the lymph nodes. About 10% of them remain there, while the rest emigrate from the lymph nodes and populate the bone marrow and other lymphatic organs and tissues. Their antibodies travel throughout the body in the blood and other fluids and react in various ways against antigens that they encounter.

Instead of becoming plasma cells, some of the B cells become memory cells. These live for months to years, and respond very quickly if they ever encounter the same antigen again. This provides long-lasting immunity to that pathogen.

T Cells and Cellular Immunity

T lymphocytes leave the bone marrow before reaching maturity. They migrate to the fetal thymus and colonize the cortex, where the blood-thymus barrier isolates them from premature exposure to blood-borne antigens. Reticular epithelial cells secrete thymic hormones that stimulate these T cells to develop antigen receptors, thus becoming immunocompetent. Following negative selection, the surviving T cells migrate into the medulla of the thymus, where they spend another 3 weeks. There is no blood-thymus barrier in the medulla, so T cells here can easily enter the blood and lymphatic vessels and disperse throughout the body. They colonize the same

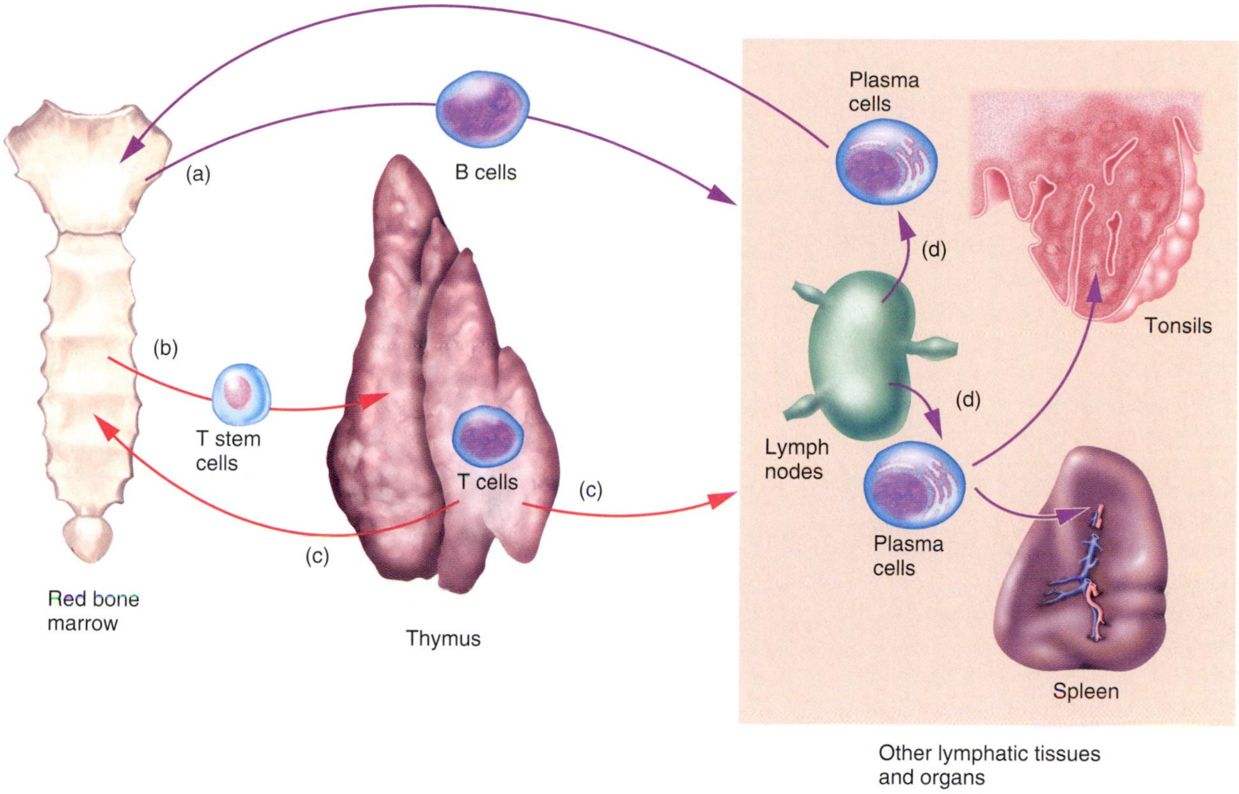

FIGURE 22.16

The Life History and Migrations of B and T Cells. Humoral immunity is represented by the *violet* pathways and cellular immunity by the *red*. (*a*) B cells achieve immunocompetence in the red bone marrow, and many emigrate to other lymphatic tissues and organs. (*b*) T stem cells emigrate from the bone marrow and attain immunocompetence in the thymus. (*c*) Immunocompetent T cells leave the thymus and colonize the bone marrow and other organs. (*d*) Plasma cells develop in the lymph nodes (among other sites) and emigrate to the bone marrow and other lymphatic organs, where they spend a few days secreting antibodies.

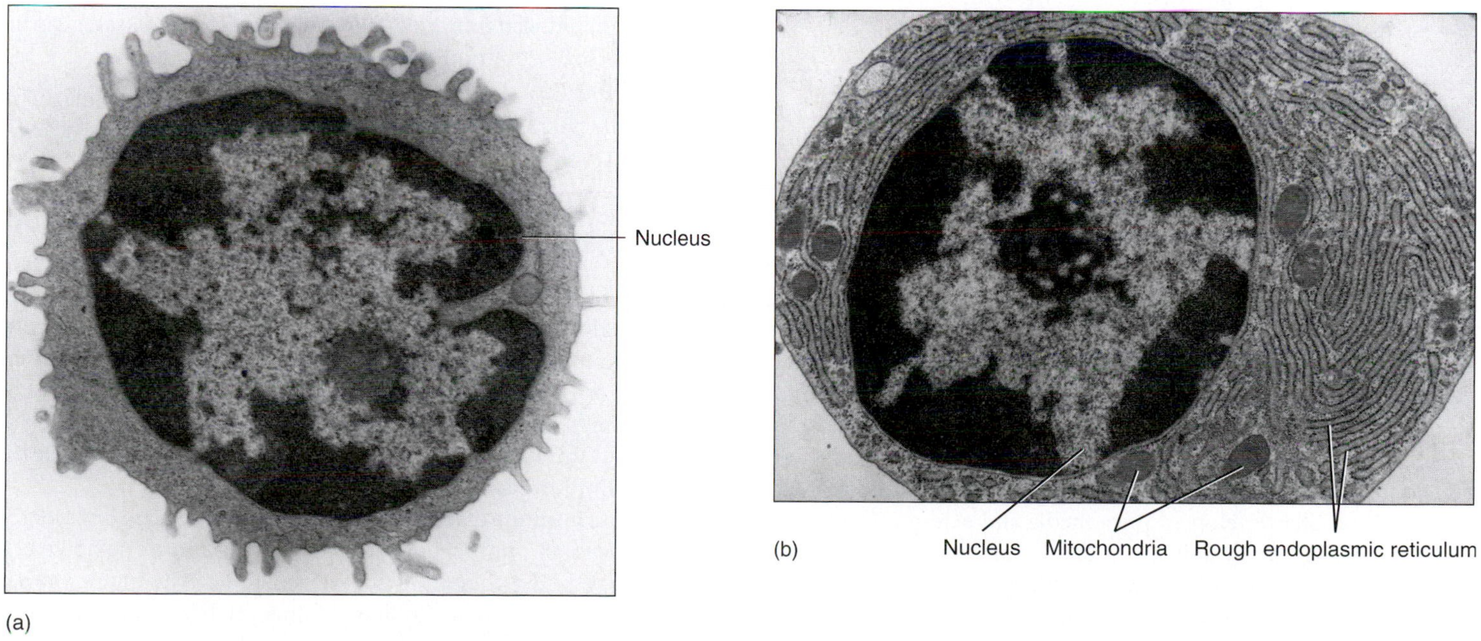

FIGURE 22.17

A B Cell and Plasma Cell. (*a*) A B cell, with a nucleus occupying almost the entire cell volume. (*b*) A plasma cell, showing the extreme proliferation of rough endoplasmic reticulum in keeping with its protein- (antibody-) synthesizing function.

sites as B cells do, including recolonizing the red bone marrow. They become especially concentrated in the deep cortex of the lymph nodes.

When cytotoxic (T_C) cells encounter an enemy cell, they attack it directly and destroy it by a *lethal hit* of toxic chemicals. This is why immunity carried out by T cells is called *cellular (cell-mediated) immunity.* As in humoral immunity, some T cells remain as memory T cells that live as long as a few decades and thus confer long-lasting protection.

Before You Go On

Answer the following questions to test your understanding of the preceding section:

12. What is the difference between humoral and cellular immunity?
13. Where do B cells acquire immunocompetence? Where do T cells do so?
14. What are the structural and functional differences between a B cell and a plasma cell?

Developmental and Clinical Perspectives

Objectives

When you have completed this section, you should be able to

- describe the embryonic origins of the lymphatic organs;
- describe changes in the lymphatic system that occur with old age; and
- describe some common disorders of the lymphatic-immune system.

Embryonic Development

Embryonic development of the thymus was described in chapter 18. Here we will examine the development of the lymphatic vessels, lymph nodes, and spleen.

Lymphatic vessels begin as endothelium-lined channels in the mesoderm called **lymph sacs.** Some of these originate by budding from the blood vessels and then detaching from them, while others originate as isolated mesodermal channels that fuse with each other and ultimately link up with the venous system. In a manner similar to the development of blood vessels (see chapter 19), lymph sacs proliferate, enlarge, and merge with each other to form larger and larger channels in the mesoderm (fig. 22.18*a*). Those with the greatest fluid flow later develop a tunica media and externa. The first to form are the *jugular lymph sacs* near the junction of the internal jugular and subclavian veins. By week 7, these sacs join the primitive veins and thus form the forerunners of the thoracic and right lymphatic ducts. The cisterna chyli arises from a *median lymph sac* that initially grows from the primitive vena cava and then breaks away from it. Smaller lymphatic vessels grow outward from the lymph sacs and follow blood vessels growing into the developing limbs.

Lymph nodes begin to develop as lymphocytes invade the lymph sacs and form cell clusters in the lumens. Blood vessels grow into these clusters, while a connective tissue capsule forms around them (fig. 22.18*b*).

The spleen develops from mesenchymal cells that invade the dorsal mesentery leading to the stomach. Thus it remains enveloped in this mesentery and permanently connected to the stomach by a *gastrosplenic ligament.* The spleen is poorly developed at birth. Invasion of the splenic tissue by immunocompetent lymphocytes stimulates its postnatal development.

The Aging Lymphatic-Immune System

The effects of old age on the lymphatic system are seen not so much in anatomical changes as in declining immune function. There are several reasons for the reduced immune responsiveness. The quantities of red bone marrow and lymphatic tissue decline, so there are fewer hemopoietic stem cells, leukocytes, and antigen-presenting cells. As the thymus involutes (see chapter 18), the level of thymic hormones declines. Perhaps because of this, an increasing percentage of lymphocytes fail to mature and achieve immunocompetence. There are fewer helper T cells, so both humoral and cellular immunity suffer from their absence. T_C cells are less responsive to antigens, and even antibody levels rise more slowly in response to infection. With fewer NK cells, immunological surveillance is weaker—one of multiple reasons that cancer becomes more common in old age. Paradoxically, while normal antibody responses are weaker in old age, the level of circulating *autoantibodies* rises. These are antibodies that fail to distinguish between host and foreign antigens, and therefore attack the body's own tissues, causing a variety of *autoimmune*[6] *diseases* such as rheumatoid arthritis.

With reduced immunity in old age, infectious diseases can be not only more common but also more serious. Epidemics of influenza (flu), for example, take a disproportionate toll of lives among the elderly. It becomes increasingly important in old age to be vaccinated against such acute seasonal diseases.

Lymphatic-Immune Disorders

It is a delicate balancing act for the body to discriminate between foreign and host antigens, ward off foreign pathogens, and mount immune responses that are not too weak, not too strong, and not misdirected. It comes as no surprise, therefore, that many things can go wrong. Most immune disorders can be classified into three categories: autoimmune diseases, hypersensitivity, and immunodeficiency.

Autoimmune diseases, as already mentioned, are diseases resulting from an immune attack misdirected against one's own tissues. Insulin-dependent diabetes mellitus, rheumatic fever, rheumatoid arthritis, and systemic lupus erythematosus are some examples.

[6]*auto* = self

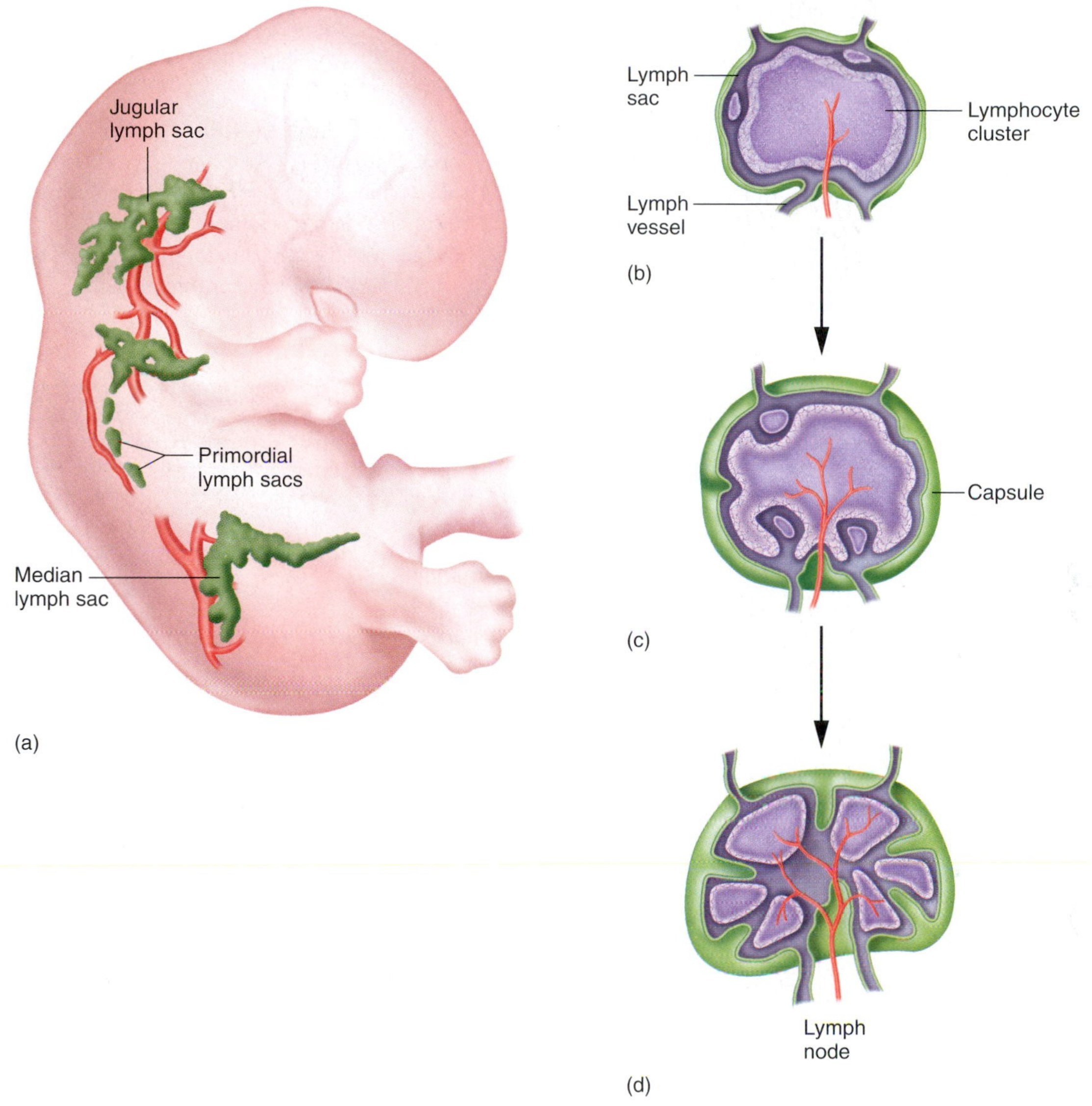

FIGURE 22.18

Embryonic Development of the Lymphatic Vessels and Lymph Nodes. (*a*) A 7-week embryo showing the right jugular lymph sac, which connects to the future subclavian vein; the primordial lymph sacs, which will merge to form the thoracic duct; and the median lymph sac, which will become the cisterna chyli. (*b–d*) Stages in the development of a lymph node. (*b*) Lymphocytes aggregate in a lymph sac and blood vessels grow into the cluster. (*c*) A fibrous capsule forms around the sac as blood vessels proliferate. (*d*) Trabeculae partially subdivide the interior as the lymph node takes shape.

Hypersensitivity is an exaggerated, harmful immune response to antigens. The most prevalent examples are *allergies*—excessive reactions to environmental antigens *(allergens)* that most people tolerate. Allergens are found in a broad range of substances such as bee and wasp venoms; toxins from poison ivy and other plants; mold; dust; pollen; animal dander; foods such as nuts, milk, eggs, and shellfish; cosmetics; latex; vaccines; and drugs such as penicillin, tetracycline, and insulin. In many cases, an allergen stimulates basophils and mast cells to release histamine and other chemicals that cause a broad range of symptoms: edema, congestion, watery eyes, runny nose, hives, cramps, diarrhea, vomiting, and sometimes catastrophic circulatory failure *(anaphylactic shock)*.

Immunodeficiency diseases are failures of the immune system to respond strongly enough to ward off disease. One of these is a congenital (inborn) condition—*severe combined immunodeficiency disease (SCID)*, in which an infant is born without a functional immune system and must live in a sterile enclosure to avoid fatal infections. The most notorious immunodeficiency disease, of course, is AIDS *(acquired immunodeficiency syndrome)*. Unlike SCID, this is not inborn but results from an infection with the human immunodeficiency virus (HIV), usually acquired by sexual intercourse or use of contaminated needles for drug injection. HIV targets especially the helper T (CD4) cells. When the T_H count drops from its normal level of 600 to 1,200 cells/μL of blood to less than 200 cells/μL, a person has "full-blown AIDS" and is highly susceptible to such opportunistic infections as *Toxoplasma* (a protozoan that infects brain tissue), *Pneumocystis* (a group of respiratory fungi), *Candida* (a fungus that grows in white patches on the oral mucosa), herpes simplex, cytomegalovirus, tuberculosis, and others. Opportunistic infection is the principal cause of death.

More specific to the lymphatic system, this chapter has already discussed filariasis and elephantiasis, lymph node cancer, tonsillitis, and ruptured spleen. A few more lymphatic disorders are briefly described in table 22.1.

Before You Go On

Answer the following questions to test your understanding of the preceding section:

15. How do lymph sacs form in the embryo? How do lymph nodes form?
16. Describe some reasons for the declining efficiency of the immune system in old age.
17. What are the three principal categories of immune system disorders? Give an example of each.

TABLE 22.1

Some Disorders of the Lymphatic System

Lymphadenitis[7] *(lim-FAD-en-EYE-tis)*	Inflammation of a lymph node in response to challenge from a foreign antigen; marked by swelling and tenderness.
Lymphadenopathy[8] *(lim-FAD-en-OP-a-thee)*	Collective term for all diseases of the lymph nodes
Lymphangitis[9] *(LIM-fan-JY-tis)*	Inflammation of a lymphatic vessel, as in filariasis (insight 22.1); marked by redness and pain along the course of the vessel.
Hodgkin[10] *Disease*	A lymph node malignancy, with early symptoms including enlarged painful lymph nodes, especially in the neck, and fever, anorexia, weight loss, night sweats, and severe itching. Diagnosis is confirmed by finding characteristic *Reed-Sternberg* cells in a lymph node biopsy. Often progresses to neighboring lymph nodes. Radiation and chemotherapy cure about three out of four patients.
Splenomegaly[11]	Enlargement of the spleen, sometimes without underlying disease but often indicating infections, autoimmune diseases, heart failure, cirrhosis, Hodgkin disease, and other cancers. The enlarged spleen may "hoard" erythrocytes, causing anemia, and it may become fragile and subject to rupture.
Non-Hodgkin Lymphoma	A lymphoma similar to Hodgkin disease, but more common, with more widespread distribution in the body (including axillary, inguinal, and femoral lymph nodes), and without Reed-Sternberg cells. Has a higher mortality rate than Hodgkin disease.

Disorders Described Elsewhere

AIDS 647	Filariasis 631
Allergy 647	Opportunistic infection 648
Autoimmune diseases 646	Ruptured spleen 643
Cancer of lymph nodes 639	Severe combined immunodeficiency disease 647
Elephantiasis 631	Tonsillitis 640

[7]*lymph* + *adeno* = gland + *itis* = inflammation
[8]*lymph* + *adeno* = gland + *pathy* = disease
[9]*lymph* + *ang* = vessel + *itis* = inflammation
[10]Thomas Hodgkin (1798–1866), British physician
[11]*megaly* = enlargement

CHAPTER REVIEW

Review of Key Concepts

Lymph and Lymphatic Vessels (p. 630)

1. The lymphatic system consists of lymph, lymphatic vessels, lymphatic tissues, and lymphatic organs.
2. The system serves to recover tissue fluid and maintain fluid balance; provide immune cells and monitor the body fluids for foreign matter; and transport dietary lipids from the small intestine to the blood.
3. Lymph is usually a colorless liquid similar to blood plasma, but is milky when absorbing digested lipids. It contains lymphocytes, macrophages, and hormones, and may contain metastasizing cancer cells, cellular debris, bacteria, and viruses.
4. Lymph originates in blind *lymphatic capillaries* that pick up tissue fluid throughout the body. The endothelial cells of lymphatic capillaries have large gaps between them that permit cells and other large particles to enter the lymph stream.
5. Lymphatic capillaries converge to form larger collecting vessels with a histology similar to that of blood vessels. Lymph nodes lie at irregular intervals along the collecting vessels and filter the lymph on its way back to the blood.
6. Collecting vessels converge to form six lymphatic trunks that drain specific regions of the body. The lymphatic trunks then converge to form two collecting ducts—the *right lymphatic duct* and *thoracic duct*—which empty lymph into the subclavian veins.
7. There is no heartlike pump to move the lymph; lymph flows under forces similar to those that drive venous return, and like some veins, lymphatic vessels have valves to ensure a one-way flow.

Lymphatic Cells, Tissues, and Organs (p. 634)

1. The cells of lymphatic tissue are natural killer (NK) cells, T lymphocytes, B lymphocytes, macrophages, dendritic cells, and reticular cells.
2. *NK cells* provide a nonspecific defense against bacteria, viruses, tissue transplants, and cancer. Their constant patrolling of the body for infected, cancerous, or defective cells is called *immunological surveillance.*
3. *T lymphocytes* are named for their development in the thymus. The types of T cells are *cytotoxic T (T_C) cells,* which directly attack and destroy enemy cells; *helper T (T_H) cells,* which stimulate other lymphocytes as well as inflammation; *suppressor T (T_S) cells,* which limit the immune response; and *memory T cells,* which provide lasting immunity after the initial exposure to an antigen.
4. *B lymphocytes* attain maturity in the bone marrow. Upon exposure to an antigen, some of them develop into *plasma cells,* which secrete protective antibodies, and others into *memory B cells.*
5. *Macrophages* are large, highly phagocytic cells that develop from monocytes. They engulf and destroy foreign matter and dead host cells, and act as *antigen-presenting cells (APCs)* to activate immune responses. There are several kinds of macrophages including microglia, dendritic cells, and alveolar and hepatic macrophages.
6. *Dendritic cells* are macrophages of the skin, mucous membranes, and lymphatic organs. They engulf foreign matter by receptor-mediated endocytosis rather than phagocytosis, but otherwise act like any other macrophage.
7. *Reticular cells* are branched stationary cells that make up part of the stroma and act as APCs in the lymphatic organs.
8. *Diffuse lymphatic tissue* is an aggregation of lymphatic cells in the walls of other organs, especially in the mucous membranes of the respiratory, digestive, urinary, and reproductive tracts—where it is called mucosa-associated lymphatic tissue (MALT). In some places, lymphocytes and macrophages form dense masses called *lymphatic nodules,* such as the *Peyer patches* of the ileum.
9. Lymphatic organs have well defined anatomical locations and have a fibrous capsule that at least partially separates them from adjacent organs and tissues. Two of these, the red bone marrow and thymus, are called *primary lymphatic organs* because lymphocytes mature here before colonizing the other sites. The other sites, called *secondary lymphatic organs,* are the lymph nodes, tonsils, and spleen.
10. Red bone marrow is a hemopoietic tissue and the point of origin of all immune cells of the lymphatic system. It consists of *islands* of hemopoietic tissue composed of macrophages and developing blood cells, separated by *sinusoids* that converge on a *central longitudinal vein.* Lymphocytes and other formed elements pass from the islands into the sinusoids and enter the blood stream.
11. The *thymus* is located in the mediastinum above the heart. It is divided into numerous polygonal lobules, each with a dense cortex and lighter medulla. Reticular epithelial cells separate the cortex from the medulla and surround the blood vessels, forming a blood-thymus barrier that isolates developing lymphocytes from blood-borne antigens. These cells also secrete thymic hormones that regulate T cell development and activity.
12. Lymph nodes are numerous small, bean-shaped organs that receive lymph through *afferent lymphatic vessels,* filter it, and pass it along via *efferent lymphatic vessels* that exit the hilum. They monitor the lymph for foreign antigens, remove impurities before it returns to the bloodstream, contribute lymphocytes to the lymph and blood, and mount immune responses to foreign antigens.
13. The parenchyma of a lymph node exhibits an outer *cortex* composed mainly of lymphatic nodules, and a deeper *medulla* with a network of *medullary cords.* B cells multiply and differentiate into plasma cells in the germinal centers of the nodules. T cells are very concentrated in the *deep cortex* next to the medulla.
14. Lymph nodes are widespread but especially concentrated in cervical, axillary, thoracic, abdominal, intestinal, mesenteric, inguinal, and popliteal groups.
15. The *tonsils* encircle the pharynx and include a medial *pharyngeal tonsil* in the nasopharynx, a pair of *palatine tonsils* at the rear of the oral cavity, and numerous *lingual tonsils* clustered in the root of the tongue. Their superficial surface is covered with epithelium and their deep surface with a fibrous partial capsule. They have deep pits called *tonsillar crypts* bordered by rows of lymphatic follicles.
16. The *spleen* lies in the left hypochondriac region between the diaphragm, stomach, and kidney. Its parenchyma is composed of *red pulp* containing concentrated RBCs and *white pulp* composed of lymphocytes and macrophages.

17. The spleen monitors the blood for foreign antigens, activates immune responses to them, disposes of old RBCs, and helps to regulate blood volume.

The Lymphatic System in Relation to Immunity (p. 644)

1. The immune system is a population of disease-fighting cells (mainly lymphocytes and macrophages) that inhabit the mucous membranes, connective tissues, and lymphatic organs among other sites.
2. The two forms of active immunity are humoral immunity, carried out by B cells and antibodies, and cellular immunity, carried out by cytotoxic T cells.
3. Both B and T cells arise from bone marrow pluripotent stem cells through *lymphocyte colony-forming units.* Before participating in immune reactions, they must become immunocompetent and survive the process of negative selection, which eliminates lymphocytes that would attack the body's own tissues.
4. B cells achieve immunocompetence and go through the selection process in the red bone marrow. Many of them emigrate to colonize the lymph nodes and other sites, where they stand guard against pathogens.
5. Upon encountering an antigen, a B cell displays fragments of it on its surface. T_H cells respond by secreting *helper factors,* which stimulate B cell multiplication. Most of the daughter B cells become antibody-secreting *plasma cells,* while some become the *memory B cells* that confer lasting immunity to that antigen.
6. T cells originate in the red bone marrow but achieve immunocompetence and undergo selection in the thymic cortex. Mature T cells then pass into the thymic medulla, emigrate from the thymus via blood and lymphatic vessels, and colonize the various lymphatic organs. T_C cells directly attack enemy cells and destroy them with a *lethal hit* of toxic chemicals. Some T cells remain as *memory T cells* that produce lasting immunity.

Developmental and Clinical Perspectives (p. 646)

1. Lymphatic ducts begin as *lymph sacs,* which are endothelium-lined channels in the mesoderm. Some of these arise by budding from the blood vessels and then detaching from them, and some as isolated mesodermal channels that fuse with each other and then (in the case of the two collecting ducts) join the venous system.
2. Embryonic lymph sacs are infiltrated with lymphocytes, which form dense clusters, become encapsulated, and develop into lymph nodes.
3. The spleen develops from a mass of mesenchymal cells in the dorsal mesentery to the stomach. Most of its development is postnatal.
4. In old age, the lymphatic system itself shows little anatomical change, but the quantity of lymphatic tissue and numbers of lymphocytes decline, and immune responses become less efficient. Elderly people are thus more susceptible to infectious diseases and cancer. Autoantibody levels rise in old age and cause such autoimmune diseases as rheumatoid arthritis.
5. There are three principal classes of immune system disorders: misdirected immune attacks (autoimmune diseases), hypersensitivity disorders such as allergy, and immunodeficiency diseases such as SCID and AIDS. AIDS, the most notorious immune disease, is caused by destruction of T_H cells by HIV, leaving a person susceptible to cancer and a variety of opportunistic infections.
6. Table 22.1 describes some disorders of the lymphatic system. Lymph node diseases in general are called *lymphadenopathy.* Inflammation of lymph nodes and lymphatic vessels, respectively, is called *lymphadenitis* and *lymphangitis.* Lymph node tumors are called *lymphomas.* Two categories of lymphoma are Hodgkin disease and non-Hodgkin lymphoma. *Splenomegaly* is enlargement of the spleen.

TESTING YOUR RECALL

1. The only lymphatic organ with both afferent and efferent lymphatic vessels is
   a. the spleen.
   b. a lymph node.
   c. a tonsil.
   d. a Peyer patch.
   e. the thymus.
2. Which of the following cells are involved in nonspecific defense but not in specific defense?
   a. helper T cells
   b. cytotoxic T cells
   c. natural killer cells
   d. B cells
   e. plasma cells
3. The lethal hit is used by _____ to kill enemy cells.
   a. neutrophils
   b. basophils
   c. mast cells
   d. NK cells
   e. cytotoxic T cells
4. Which of these is a macrophage?
   a. microglia
   b. a plasma cell
   c. a reticular cell
   d. a helper T cell
   e. a mast cell
5. Which of these lymphatic organs has a cortex and medulla: (I) spleen; (II) lymph node; (III) thymus; (IV) red bone marrow?
   a. II only
   b. III only
   c. II and III only
   d. III and IV only
   e. I, II, and III
6. What cells form the blood-thymus barrier?
   a. astrocytes
   b. Hassall corpuscles
   c. T cells
   d. dendritic cells
   e. reticular epithelial cells
7. Where do B cells attain immunocompetence?
   a. in the red bone marrow
   b. in the germinal centers of the lymph nodes
   c. in the thymic cortex
   d. in the thymic medulla
   e. in the splenic white pulp
8. If it were not for the process of negative selection, we would expect to see more
   a. allergies.
   b. lymphatic nodules in the MALT.
   c. antigen-presenting cells.
   d. immunodeficiency diseases.
   e. autoimmune diseases.
9. Lymph nodes tend to be especially concentrated in all of these sites *except*
   a. the cervical region.
   b. the popliteal region.
   c. the carpal region.
   d. the inguinal region.
   e. the mesenteries.

10. All lymph ultimately reenters the bloodstream at what point?
    a. the right atrium
    b. the common carotid arteries
    c. the internal iliac veins
    d. the subclavian veins
    e. the inferior vena cava
11. Any organism or substance capable of causing disease is called a/an _____.
12. The milky lymph, rich in fat, absorbed from the small intestine is called _____.
13. _____ is a condition in which lymph nodes are swollen and painful to the touch.
14. The two lymphatics that empty into the subclavian veins are the _____ on the right and the _____ on the left.
15. The latter duct in question 14 begins with a sac called the _____ below the diaphragm.
16. The only lymphocytes whose function lies entirely in nonspecific defense are the _____ cells.
17. Any cells that process antigens and display fragments of them to activate immune reactions are called _____.
18. The _____ is a lymphatic organ composed mainly of hemopoietic islands and sinusoids.
19. The ovoid masses of lymphocytes that line the tonsillar crypts are called _____.
20. Any disease in which antibodies attack one's own tissues is called a/an _____ disease.

Answers in the Appendix

True or False

Determine which five of the following statements are false, and briefly explain why.

1. B cells play roles in both nonspecific defense and specific immunity.
2. T lymphocytes undergo clonal deletion in the thymus.
3. Lymphatic capillaries are more permeable than blood capillaries.
4. T lymphocytes are involved only in cellular immunity.
5. The white pulp of the spleen gets its color mainly from lymphocytes and macrophages.
6. Obstruction of a major lymphatic vessel is likely to cause edema.
7. Lymph nodes are populated by B cells but not T cells.
8. Lymphatic nodules are permanent structures enclosed in fibrous capsules.
9. Tonsillectomy is regarded as the current treatment of choice for most cases of tonsillitis.
10. Most plasma cells form in the germinal centers of the lymph nodes.

Answers in the Appendix

Testing Your Comprehension

1. About 10% of people have one or more *accessory spleens*, typically about 1 cm in diameter and located near the hilum of the main spleen or embedded in the tail of the pancreas. If a surgeon is performing a splenectomy as a treatment for hypersplenism, why would it be important to search for and remove any accessory spleens? What might be the consequences of overlooking one of these?
2. In treating a woman for malignancy in the right breast, the surgeon removes some of her axillary lymph nodes. Following surgery, the patient experiences edema of her right arm. Explain why.
3. Explain why a detailed knowledge of the pathways of lymphatic drainage is important to the clinical management of cancer.
4. A burn research center uses mice for studies of skin grafting. To prevent graft rejection, the mice are thymectomized at birth. Even though B cells do not develop in the thymus, these mice show no humoral immune response and are very susceptible to infection. Explain why the removal of the thymus would improve the success of skin grafts but adversely affect humoral immunity.
5. Contrast the structure of a B cell with that of a plasma cell, and explain how their structural difference relates to their functional difference.

Answers at the Online Learning Center

www.mhhe.com/saladinha1

Visit the Online Learning Center for practice tests, answer keys, and other learning aids for this chapter. Enhance your understanding of human anatomy with our interactive art labeling exercises, supplemental photo atlases, web links, puzzles, flashcards, and much more.

23

CHAPTER TWENTY-THREE

The Respiratory System

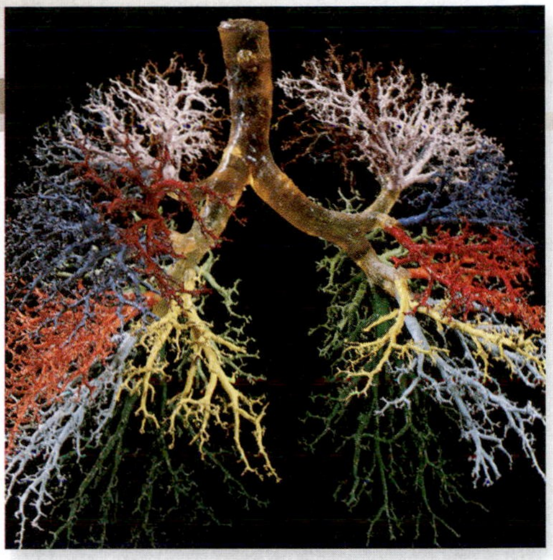

The bronchial trees, with each bronchopulmonary segment shown in a different color (resin cast)

CHAPTER OUTLINE

Overview of the Respiratory System 654

The Upper Respiratory Tract 655
- The Nose 655
- The Pharynx 655
- The Larynx 657

The Lower Respiratory Tract 659
- The Trachea and Bronchi 660
- The Lungs 660
- The Pleurae 665

Neuromuscular Aspects of Respiration 666
- Inspiration 666
- Expiration 666
- Respiratory Centers of the Brainstem 666

Developmental and Clinical Perspectives 668
- Prenatal and Neonatal Development 668
- The Aging Respiratory System 669
- Respiratory Pathology 669

Chapter Review 671

INSIGHTS

23.1 Clinical Application: Tracheostomy 661
23.2 Clinical Application: Pneumothorax and Atelectasis 666
23.3 Clinical Application: Ondine's Curse 668
23.4 Clinical Application: Premature Birth and Respiratory Distress Syndrome 669

BRUSHING UP

To understand this chapter, you may find it helpful to review the following concepts:

- Serous and mucous membranes (pp. 97–98)
- The maxilla, nasal bones, vomer, and ethmoid bone (pp. 183–185)
- The muscles of respiration (p. 299)
- Pulmonary blood circulation (p. 595)

The **respiratory system** is an organ system specialized primarily to provide oxygen to the blood and remove carbon dioxide from it. These functions are linked to the need for ATP; most ATP synthesis requires oxygen and generates carbon dioxide. The respiratory and cardiovascular systems collaborate to shuttle these gases between the tissues and the air. These two systems have a close spatial relationship in the thoracic cavity, and have such a close functional relationship that a disorder of the lungs can have direct and pronounced effects on the heart, and vice versa. The two systems are often considered jointly under the heading *cardiopulmonary system*.

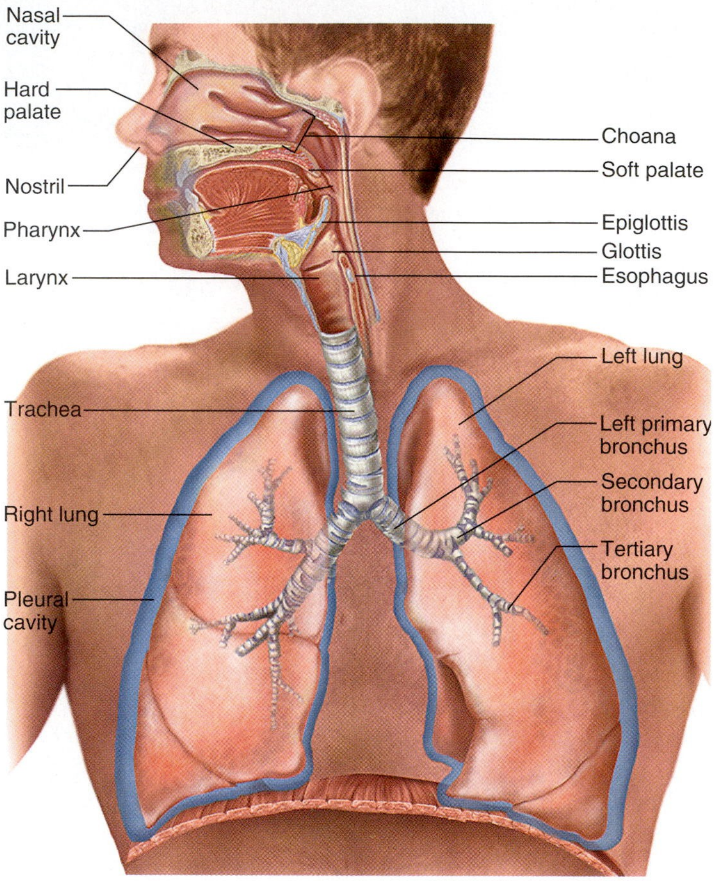

FIGURE 23.1
The Respiratory System.

Overview of the Respiratory System

Objectives

When you have completed this section, you should be able to

- state the functions of the respiratory system;
- name the principal organs of this system;
- distinguish between the conducting and respiratory divisions; and
- distinguish between the upper and lower respiratory tract.

The respiratory system has a broader range of functions than are commonly supposed:

1. It provides for oxygen and carbon dioxide exchange between the blood and air.
2. It serves for speech and other vocalizations (laughing, crying).
3. It provides the sense of smell, which is important in social interactions, food selection, and avoiding danger (such as a gas leak or spoiled food).
4. By eliminating CO_2, it helps to control the pH of the body fluids. Excess CO_2 reacts with water and releases hydrogen ions ($CO_2 + H_2O \rightarrow H_2CO_3 \rightarrow HCO_3^- + H^+$); therefore, if the respiratory system does not keep pace with the rate of CO_2 production, H^+ accumulates and the body fluids have an abnormally low pH *(acidosis)*.
5. The lungs carry out a step in the synthesis of a vasoconstrictor called *angiotensin II,* which helps to regulate blood pressure.
6. Breathing creates pressure gradients between the thorax and abdomen that promote the flow of lymph and venous blood.
7. Taking a deep breath and holding it while contracting the abdominal muscles (the *Valsalva*[1] *maneuver*) helps to expel abdominal contents during urination, defecation, and childbirth.

The principal organs of the respiratory system are the nose, pharynx, larynx, trachea, bronchi, and lungs (fig. 23.1). Within the lungs, air flows along a dead-end pathway consisting essentially of bronchi → bronchioles → alveoli (with some refinements to be introduced later). Incoming air stops in the *alveoli* (millions of thin-walled, microscopic air sacs), exchanges gases with the bloodstream across the alveolar wall, and then flows back out.

The **conducting division** of the respiratory system consists of those passages that serve only for airflow, essentially from the nostrils through the bronchioles. The **respiratory division** consists of the alveoli and other distal gas-exchange regions. The airway from the nose through the larynx is often called the **upper respiratory tract** (that is, the respiratory organs in the head and neck), and the regions from the trachea through the lungs compose the **lower respiratory tract** (the respiratory organs of the thorax). However, these are inexact terms and various authorities place the dividing line between the upper and lower tracts at different points.

Before You Go On

Answer the following questions to test your understanding of the preceding section:

1. What are some functions of the respiratory system other than supplying O_2 to the body and removing CO_2?
2. Which portions of the respiratory tract belong to the conducting division? What portions belong to the respiratory division? How do the two divisions differ functionally?
3. What is the distinction between the upper and lower respiratory tract?

[1]Antonio Maria Valsalva (1666–1723), Italian anatomist

The Upper Respiratory Tract

Objectives

When you have completed this section, you should be able to

- trace the flow of air from the nose through the larynx;
- describe the anatomy of these passages;
- relate the anatomy of any portion of the upper respiratory tract to its function; and
- describe the action of the vocal cords in speech.

The Nose

The **nose** has several functions: it warms, cleanses, and humidifies inhaled air; it detects odors in the airstream; and it serves as a resonating chamber that amplifies the voice. It extends from a pair of anterior openings called the **nostrils** or **anterior (external) nares** (NAIR-eze) to a pair of posterior openings called the **choanae**[2] (co-AH-nee), or **posterior (internal) nares** (fig. 23.2). The **nasal cavity** is divided into right and left halves called **nasal fossae** (FAW-see) by a wall of bone and hyaline cartilage, the **nasal septum.** The facial part of the nose is shaped by bone and hyaline cartilage as well. Its superior half is supported by a pair of small nasal bones medially and the maxillae laterally. The inferior half is supported by the **lateral** and **alar cartilages** (fig. 23.3). By palpating your own nose, you can easily find the boundary between the bone and cartilage. The flared portion at the lower end of the nose, called the **ala nasi**[3] (AIL-ah NAZE-eye), is shaped by the alar cartilages and dense connective tissue.

The nasal cavity begins with a small dilated chamber called the **vestibule** just inside the nostril, bordered by the ala nasi. This space is lined with stratified squamous epithelium like the facial skin, and has stiff **guard hairs,** or **vibrissae** (vy-BRISS-ee), that block insects and large airborne particles from the nose. The nasal septum has three components: the bony *vomer* forming the inferior part, the perpendicular plate of the *ethmoid bone* supporting the superior part, and a hyaline *septal cartilage* forming the anterior part. The roof of the nasal cavity is formed by the ethmoid and sphenoid bones, and the hard palate forms its floor. The palate separates the nasal cavity from the oral cavity and allows you to breathe while chewing food (see insight 7.2, p. 184). The nasal cavity receives drainage from the paranasal sinuses (see chapter 7) and the nasolacrimal ducts of the orbits (see chapter 17).

There is not much space in the nasal cavity. Most of it is occupied by three folds of tissue—the **superior, middle,** and **inferior nasal conchae**[4] (CON-kee)—that project from the lateral walls toward the septum (see fig. 23.2). Beneath each concha is a narrow air passage called a **meatus** (me-AY-tus). The narrowness of these passages and the turbulence caused by the conchae ensure that most air contacts the mucous membrane on its way through. As it does, most dust in the air sticks to the mucus, and the air picks up moisture and heat from the mucosa. The conchae thus enable the nose to cleanse, warm, and humidify the air more effectively than if the air had an unobstructed flow through a cavernous space.

Odors are detected by sensory cells in the **olfactory mucosa,** a small patch of epithelium that covers the roof of the nasal fossa and parts of the septum and superior concha. A ciliated pseudostratified **respiratory mucosa** not only lines the rest of the nasal cavity, but also extends deeply into the lungs. (In the lower reaches of the airway described later, it becomes a ciliated cuboidal epithelium.) It is a nonsensory epithelium with two principal types of cells: **goblet cells,** which secrete mucus, and **ciliated cells** which, in the nose, drive the mucus toward the posterior nares and into the pharynx so it can be swallowed and digested. The lamina propria also contains mucous glands, which supplement the mucus produced by the goblet cells. Pollen, dust, and other inhaled particles stick to the mucus, and lysozyme in the mucus destroys bacteria. The lamina propria is also well populated by lymphocytes that mount immune defenses against inhaled pathogens, and plasma cells that secrete antibodies into the tissue fluid.

The lamina propria contains large blood vessels that help to warm the air. The inferior concha has an especially extensive venous plexus called the **erectile tissue** *(swell body).* Every 30 to 60 minutes, the erectile tissue on one side becomes engorged with blood and restricts airflow through that fossa. Most air is then directed through the other nostril and fossa, allowing the engorged side time to recover from drying. Thus the preponderant flow of air shifts between the right and left nostrils once or twice each hour.

Nosebleed *(epistaxis)* usually results from trauma to the lower nasal septum, for example by nose-picking or blows to the face. In the absence of trauma, however, spontaneous nosebleeds often stem from the erectile tissue of the inferior nasal concha. This can be an early warning of hypertension.

The Pharynx

The **pharynx** (FAIR-inks) is a muscular funnel extending about 13 cm (5 in.) from the choanae to the larynx. It has three regions: the *nasopharynx, oropharynx,* and *laryngopharynx* (fig. 23.2*b*).

The **nasopharynx** lies posterior to the choanae and dorsal to the soft palate. It receives the auditory (eustachian) tubes from the middle ears and houses the pharyngeal tonsil. Inhaled air turns 90° downward as it passes through the nasopharynx. Relatively large particles (>10 μm) generally cannot make the turn because of their inertia. They collide with the posterior wall of the nasopharynx and stick to the mucosa near the tonsil, which is well positioned to respond to airborne pathogens.

The **oropharynx** is a space between the soft palate and root of the tongue. It extends inferiorly as far as the hyoid bone, and contains the palatine and lingual tonsils. Its anterior border is formed by the base of the tongue and the *fauces* (FAW-seez), the opening of the oral cavity into the pharynx.

The **laryngopharynx** (la-RIN-go-FAIR-inks) begins with the union of the nasopharynx and oropharynx at the level of the hyoid bone. It passes inferiorly and dorsal to the larynx and ends at the opening of the esophagus, at the level of the *cricoid cartilage* of the larynx (described next). The nasopharynx passes only air and is lined

[2]*choana* = funnel
[3]*ala* = wing + *nasi* = of the nose
[4]*concha* = seashell

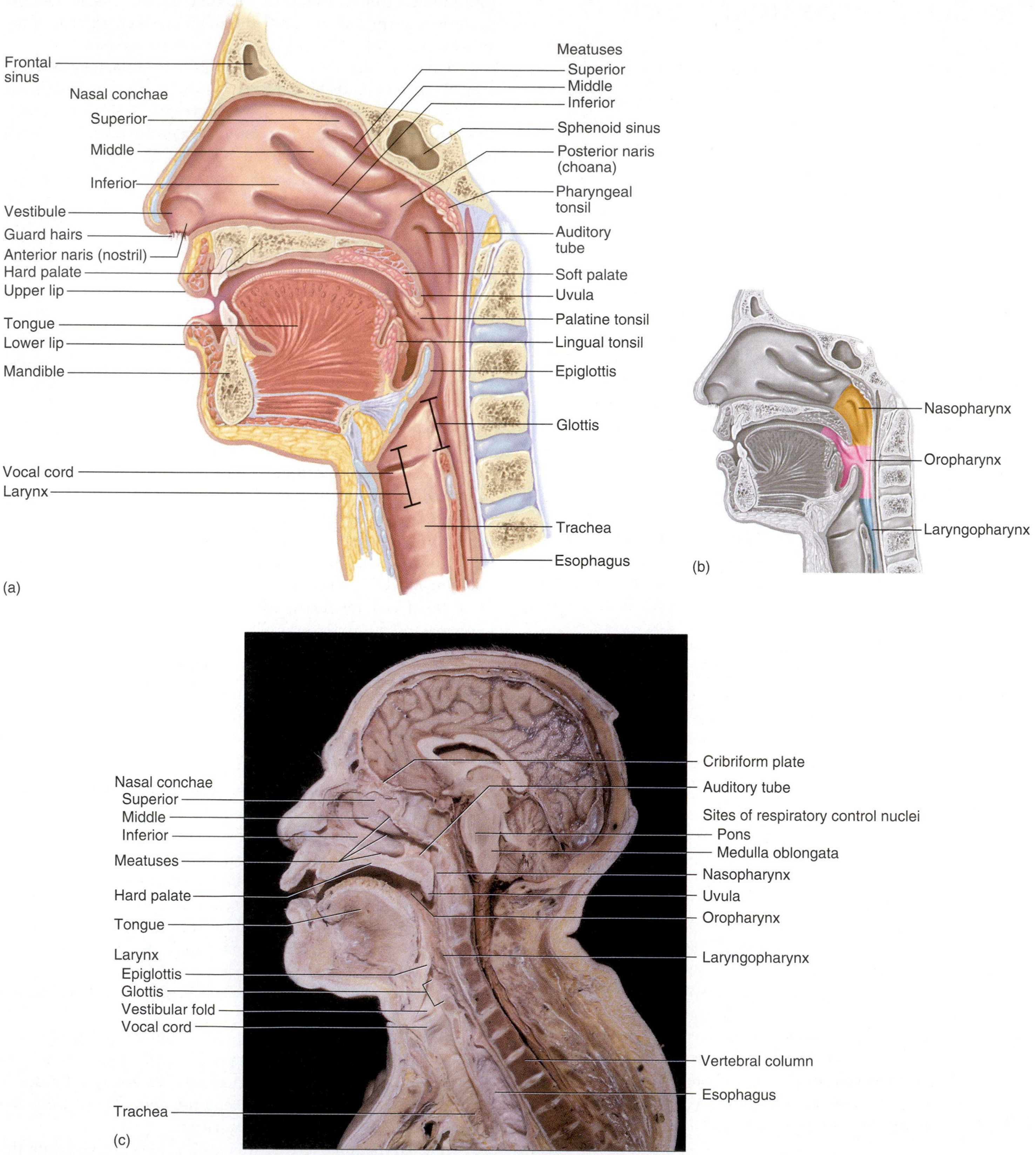
Frontal sinus
Nasal conchae
Superior
Middle
Inferior
Vestibule
Guard hairs
Anterior naris (nostril)
Hard palate
Upper lip
Tongue
Lower lip
Mandible
Vocal cord
Larynx
Meatuses
Superior
Middle
Inferior
Sphenoid sinus
Posterior naris (choana)
Pharyngeal tonsil
Auditory tube
Soft palate
Uvula
Palatine tonsil
Lingual tonsil
Epiglottis
Glottis
Trachea
Esophagus
(a)
Nasopharynx
Oropharynx
Laryngopharynx
(b)
Nasal conchae
Superior
Middle
Inferior
Meatuses
Hard palate
Tongue
Larynx
Epiglottis
Glottis
Vestibular fold
Vocal cord
Trachea
(c)
Cribriform plate
Auditory tube
Sites of respiratory control nuclei
Pons
Medulla oblongata
Nasopharynx
Uvula
Oropharynx
Laryngopharynx
Vertebral column
Esophagus

FIGURE 23.2

Anatomy of the Upper Respiratory Tract. (*a*) Internal anatomy of the nasal cavity, pharynx, and larynx. (*b*) Regions of the pharynx. (*c*) Median section of the head of a cadaver.

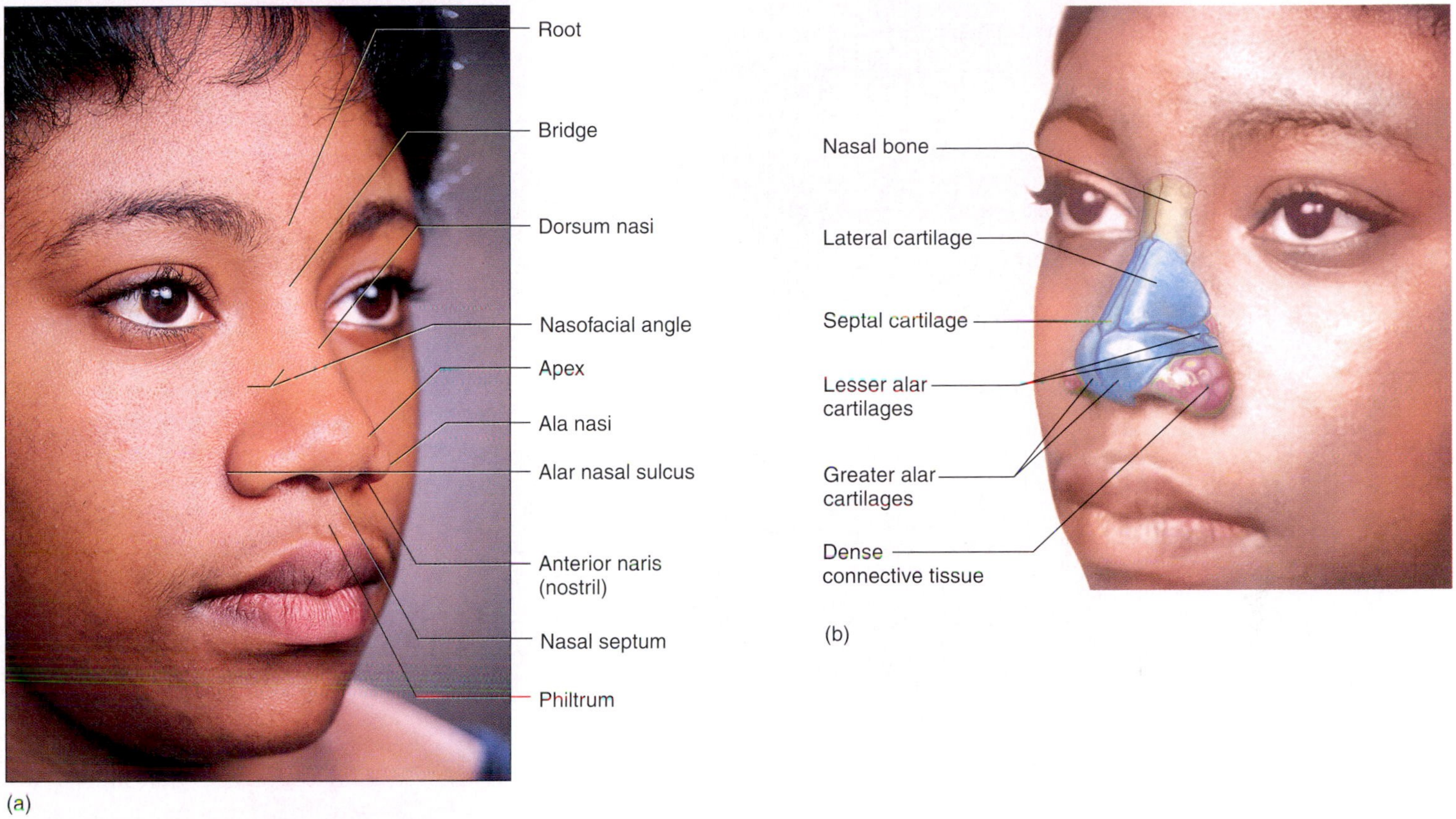
Root
Bridge
Dorsum nasi
Nasofacial angle
Apex
Ala nasi
Alar nasal sulcus
Anterior naris (nostril)
Nasal septum
Philtrum
(a)
Nasal bone
Lateral cartilage
Septal cartilage
Lesser alar cartilages
Greater alar cartilages
Dense connective tissue
(b)

FIGURE 23.3
Anatomy of the Nasal Region. (*a*) External anatomy. (*b*) Connective tissues that shape the nose.

by pseudostratified columnar epithelium, whereas the oropharynx and laryngopharynx pass air, food, and drink and are lined by stratified squamous epithelium.

The Larynx

The **larynx** (LAIR-inks), or "voicebox" (fig. 23.4), is a cartilaginous chamber about 4 cm (1.5 in.) long. Its primary function is to keep food and drink out of the airway, but it evolved the additional role of sound production *(phonation)* in many animals and achieved its highest vocal sophistication in humans.

The superior opening of the larynx, the **glottis,**[5] is guarded by a flap of tissue called the **epiglottis**[6] (fig. 23.5). At rest, the epiglottis usually stands almost vertically. During swallowing, however, *extrinsic muscles* of the larynx pull the larynx upward toward the epiglottis, the tongue pushes the epiglottis downward to meet it, and the epiglottis directs food and drink into the esophagus dorsal to the airway.

In infants, the larynx is relatively high in the throat and the epiglottis touches the soft palate. This creates a more or less continuous airway from the nasal cavity to the larynx and allows an infant to breathe continually while swallowing. The epiglottis deflects milk away from the airstream, like rain running off a tent while it remains dry inside. By age two, the root of the tongue becomes more muscular and forces the larynx to descend to a lower position. It then becomes impossible to breathe and swallow at the same time without choking.

The framework of the larynx consists of nine cartilages. The first three are solitary and relatively large. The most superior one, the **epiglottic cartilage,** is a spoon-shaped supportive plate in the epiglottis. The largest, the **thyroid**[7] **cartilage,** is named for its shieldlike shape. It broadly covers the anterior and lateral aspects of the larynx. The "Adam's apple" is an anterior peak of the thyroid cartilage called the *laryngeal prominence.* Testosterone stimulates the growth of this prominence, which is therefore larger in males than in females. Inferior to the thyroid cartilage is a ringlike **cricoid**[8] (CRY-coyd) **cartilage,** which connects the larynx to the trachea. The thyroid and cricoid cartilages essentially constitute the "box" of the "voicebox."

The remaining cartilages are smaller and occur in three pairs. Posterior to the thyroid cartilage are the two **arytenoid**[9] (AR-ih-TEE-noyd) **cartilages,** and attached to their upper ends are a pair of little horns, the **corniculate**[10] (cor-NICK-you-late) **cartilages.** The arytenoid and corniculate cartilages function in speech, as explained

[5]*glottis* = back of the tongue
[6]*epi* = above, upon
[7]*thyr* = shield + *oid* = resembling
[8]*crico* = ring + *oid* = resembling
[9]*aryten* = ladle + *oid* = resembling
[10]*corni* = horn + *cul* = little + *ate* = possessing

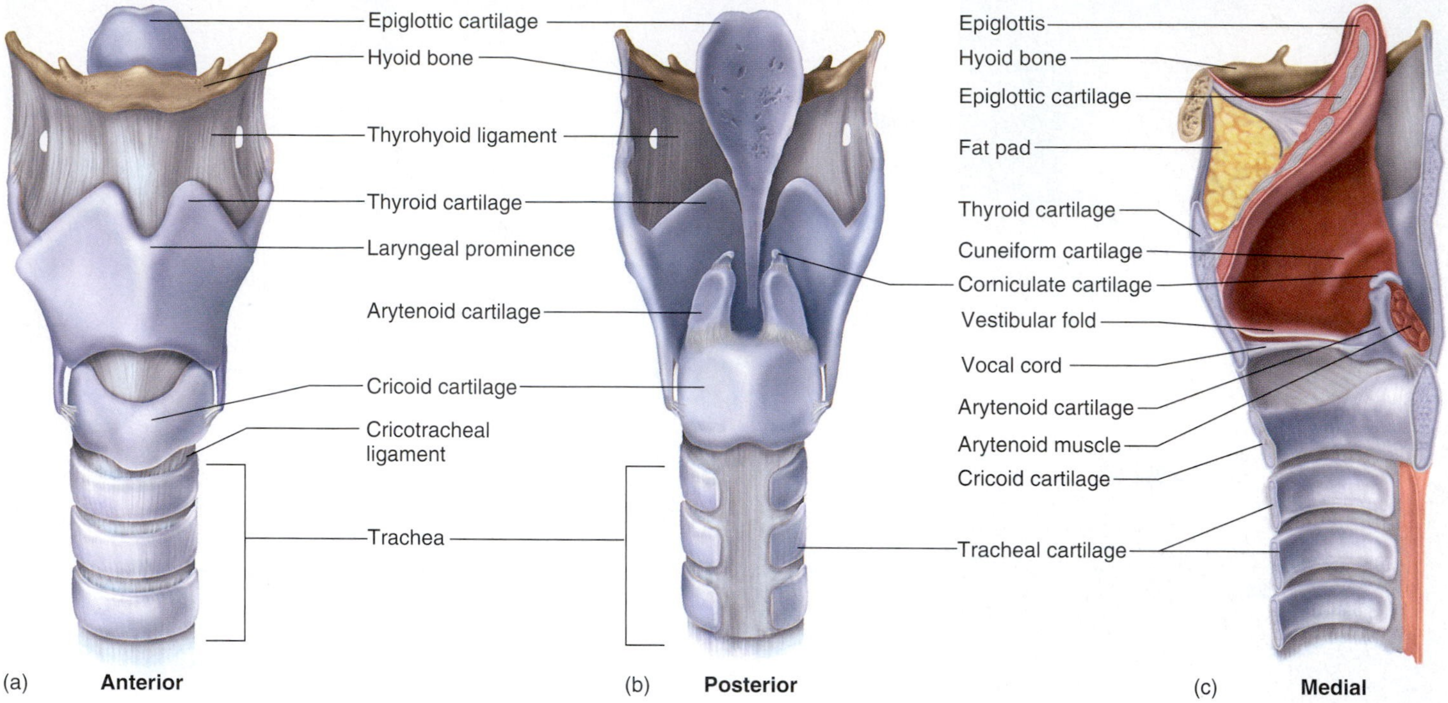

FIGURE 23.4

Anatomy of the Larynx. (*a*) Anterior aspect. (*b*) Posterior aspect. (*c*) Medial section, anterior aspect facing *left*.

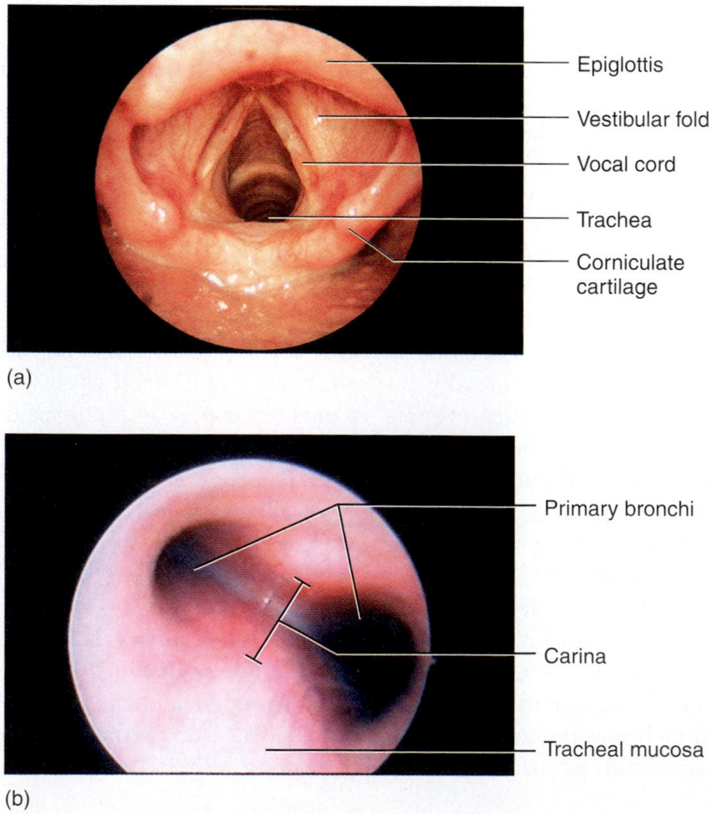

FIGURE 23.5

Endoscopic Views of the Respiratory Tract. (*a*) Superior view of the larynx, seen with a laryngoscope. (*b*) Lower end of the trachea, where it forks into the two primary bronchi, seen with a bronchoscope.

shortly. A pair of **cuneiform**[11] (cue-NEE-ih-form) **cartilages** support the soft tissues between the arytenoids and the epiglottis. The thyroid and cricoid cartilages and inferior part of the arytenoids are hyaline cartilage; the epiglottic, corniculate, and cuneiform cartilages and superior part of the arytenoids are elastic cartilage.

A group of fibrous ligaments bind the cartilages of the larynx together and to adjacent structures in the neck. Superiorly, a broad sheet called the **thyrohyoid ligament** joins the thyroid cartilage to the hyoid bone, and inferiorly, the **cricotracheal ligament** joins the cricoid cartilage to the trachea. These are collectively called the *extrinsic ligaments* because they link the larynx to other organs. The *intrinsic ligaments* are contained entirely within the larynx and link its nine cartilages to each other. Two pairs of intrinsic ligaments, the **vestibular** and **vocal ligaments,** extend from the thyroid cartilage anteriorly to the arytenoid cartilages posteriorly, and support the vestibular folds and vocal cords, described shortly.

The walls of the larynx are also quite muscular. The deep *intrinsic muscles* operate the vocal cords, and the superficial *extrinsic muscles* connect the larynx to the hyoid bone and elevate the larynx during swallowing. The extrinsic muscles, also called the *infrahyoid group,* are named and described in chapter 11 (table 11.3).

The interior wall of the larynx has two folds on each side that stretch from the thyroid cartilage in front to the arytenoid cartilages in back. The superior pair, called the **vestibular folds** (fig. 23.5*a*), play no role in speech but close the glottis during swallowing. They are supported by the aforementioned vestibular ligaments. The inferior pair, the **vocal cords (vocal folds),** produce sound when air

[11] *cune* = wedge + *form* = shape

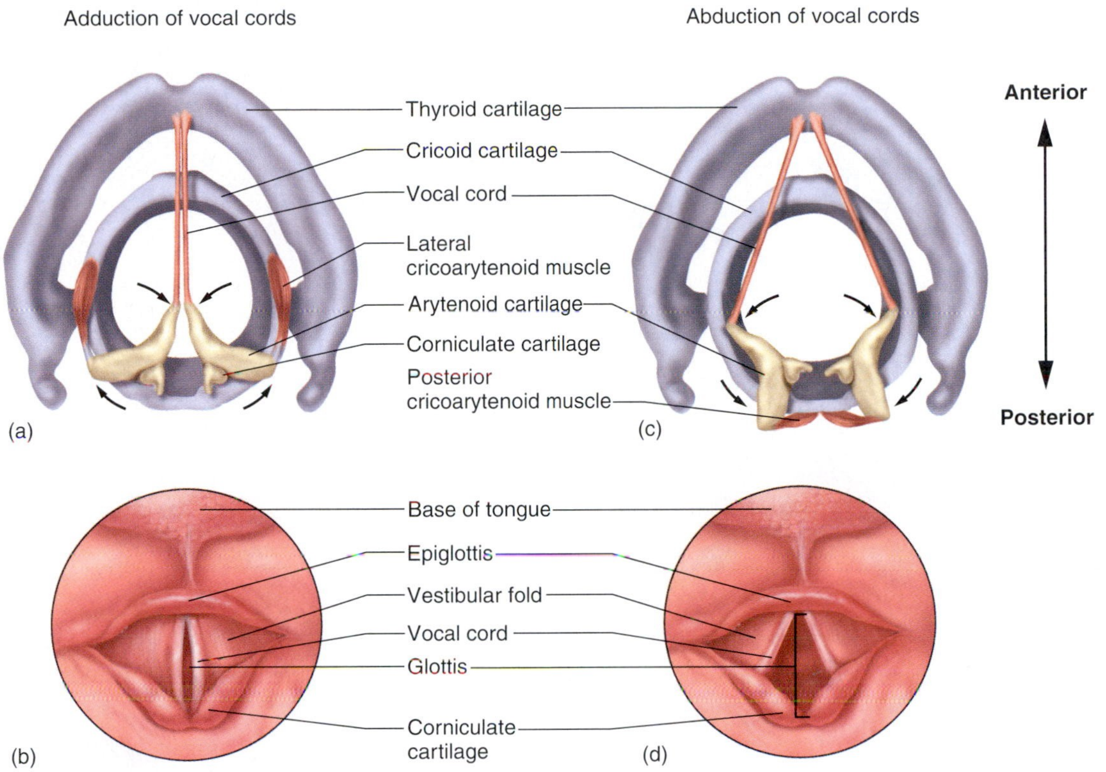

FIGURE 23.6

Action of Some of the Intrinsic Laryngeal Muscles on the Vocal Cords. (*a*) Adduction of the vocal cords by the lateral cricoarytenoid muscles. (*b*) Adducted vocal cords seen with the laryngoscope. (*c*) Abduction of the vocal cords by the posterior cricoarytenoid muscles. (*d*) Abducted vocal cords seen with the laryngoscope.

passes between them. They contain the vocal ligaments and are covered with stratified squamous epithelium, best suited to endure vibration and contact between the cords.

The intrinsic muscles control the vocal cords by pulling on the corniculate and arytenoid cartilages, causing the cartilages to pivot. Depending on their direction of rotation, the arytenoid cartilages abduct or adduct the vocal cords (fig. 23.6). Air forced between the adducted vocal cords vibrates them, producing a high-pitched sound when the cords are relatively taut and a lower-pitched sound when they are more relaxed. In adult males, the vocal cords are longer and thicker, vibrate more slowly, and produce lower-pitched sounds than in females. Loudness is determined by the force of the air passing between the vocal cords. Although the vocal cords alone produce sound, they do not produce intelligible speech. The crude sounds coming from the larynx are formed into words by actions of the pharynx, oral cavity, tongue, and lips.

Before You Go On

Answer the following questions to test your understanding of the preceding section:

4. Describe the histology of the mucous membrane of the nasal cavity and the functions of the cell types present.
5. Name the anterior and posterior openings that mark the beginning and end of the nasal cavity.
6. What are the right and left halves of the nasal cavity called? What are the three scroll-like folds on the wall of each nasal fossa called? What is their function?
7. Describe the roles of the intrinsic muscles, corniculate cartilages, and arytenoid cartilages in speech.

The Lower Respiratory Tract

Objectives

When you have completed this section, you should be able to

- trace the flow of air from the trachea to the pulmonary alveoli;
- describe the anatomy of these passages;
- relate the gross anatomy of any portion of the lower respiratory tract to its function;
- relate the microscopic anatomy of the pulmonary alveoli to their role in gas exchange; and
- describe the relationship of the pleurae to the lungs.

If you palpate your larynx, you will find the laryngeal prominence of the thyroid cartilage only slightly above the level where your clavicles articulate with the sternum. All the rest of the respiratory tract is in the thorax rather than the head and neck, and is thus called the lower respiratory tract. This portion extends from the trachea to the pulmonary alveoli.

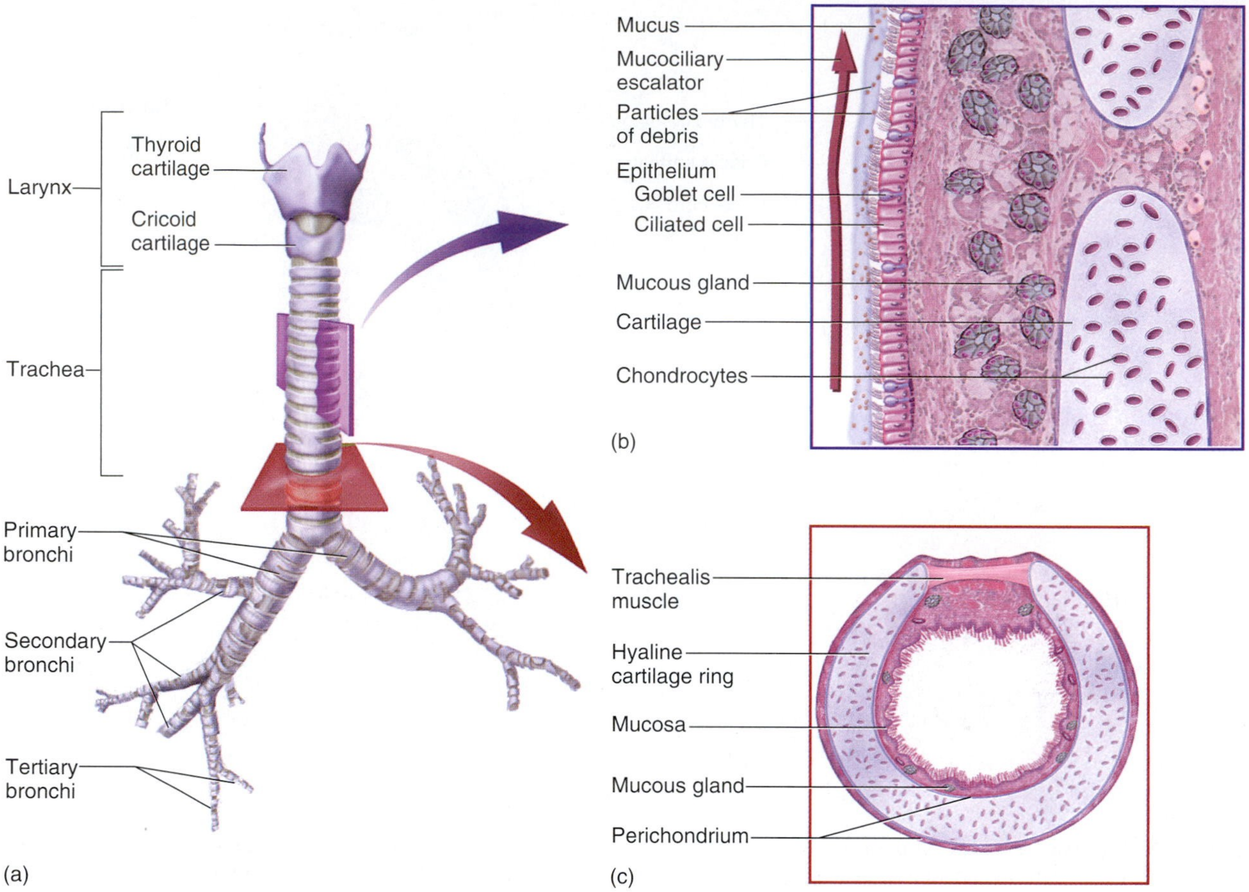

Larynx
Thyroid cartilage
Cricoid cartilage
Trachea
Primary bronchi
Secondary bronchi
Tertiary bronchi
(a)
Mucus
Mucociliary escalator
Particles of debris
Epithelium
Goblet cell
Ciliated cell
Mucous gland
Cartilage
Chondrocytes
(b)
Trachealis muscle
Hyaline cartilage ring
Mucosa
Mucous gland
Perichondrium
(c)

FIGURE 23.7

Anatomy of the Lower Respiratory Tract. (*a*) Anterior view. (*b*) Longitudinal section of the trachea showing the action of the mucociliary escalator. (*c*) Cross section of the trachea showing the C-shaped tracheal cartilage.

The Trachea and Bronchi

The **trachea** (TRAY-kee-uh), or "windpipe," is a tube about 12 cm (4.5 in.) long and 2.5 cm (1 in.) in diameter, lying anterior to the esophagus (fig. 23.7*a*). It is supported by 16 to 20 C-shaped rings of hyaline cartilage, some of which you can palpate between your larynx and sternum. The inner lining of the trachea is a pseudostratified columnar epithelium composed mainly of mucus-secreting goblet cells, ciliated cells, and short basal stem cells (figs. 23.7*b* and 23.8). The mucus traps inhaled particles, and the upward beating of the cilia drives the debris-laden mucus toward the pharynx, where it is swallowed. This mechanism of debris removal is called the **mucociliary escalator.**

The connective tissue beneath the tracheal epithelium contains lymphatic follicles, mucous and serous glands, and the tracheal cartilages. Like the wire spiral in a vacuum cleaner hose, the cartilage rings reinforce the trachea and keep it from collapsing when you inhale. The open part of the C faces dorsally and allows room for the esophagus to expand as swallowed food passes by. The gap is spanned by smooth muscle tissue called the **trachealis** (fig. 23.7*c*). Contraction of this muscle narrows or widens the trachea to adjust airflow. The outermost layer of the trachea, called the **adventitia,** is fibrous connective tissue that blends into the adventitia of other organs of the mediastinum.

At its inferior end, the trachea forks into right and left *primary bronchi.* The lowermost tracheal cartilage has an internal median ridge called the **carina**[12] (ca-RY-na) that directs the airflow to the right and left (see fig. 23.5*b*). The bronchi are further traced in the discussion of the *bronchial tree* of the lungs.

The Lungs

Each **lung** (fig. 23.9) is a somewhat conical organ with a broad, concave **base** resting on the diaphragm and a blunt peak called the **apex** projecting slightly above the clavicle. The broad **costal surface** is pressed against the rib cage, and the smaller concave **mediastinal surface** faces medially. The mediastinal surface exhibits a slit called the **hilum** through which the lung receives the primary bronchus, blood vessels, lymphatic vessels, and nerves. These structures constitute the **root** of the lung.

The lungs are crowded by adjacent viscera and therefore neither fill the entire rib cage, nor are they symmetrical. Inferior to the lungs and diaphragm, much of the space within the rib cage is occupied by the liver, spleen, and stomach (see fig. A.15, p. 38). The right lung is shorter than the left because the liver rises higher on

[12]*carina* = keel

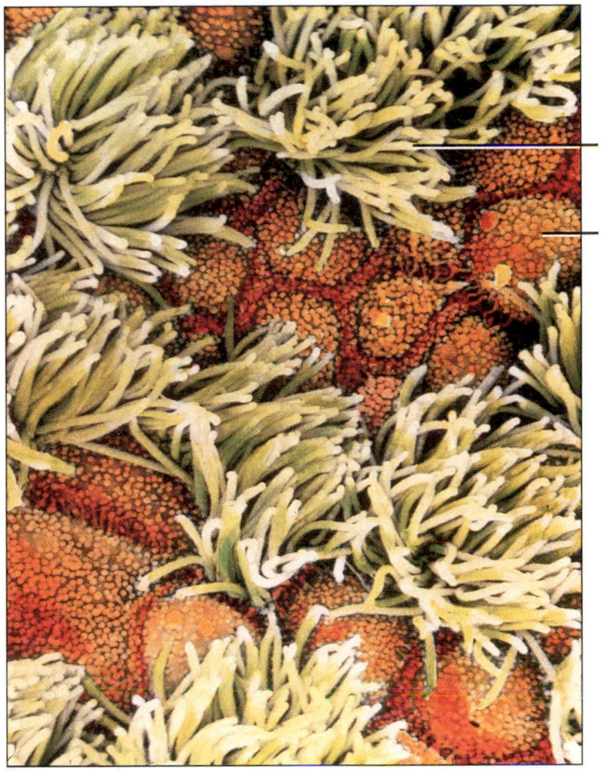

FIGURE 23.8
The Tracheal Epithelium Showing Ciliated Cells and Nonciliated Goblet Cells. The small bumps on the goblet cells are microvilli. (Colorized SEM.)

INSIGHT 23.1 CLINICAL APPLICATION

TRACHEOSTOMY

The functional importance of the nasal cavity becomes especially obvious when it is bypassed. If the upper airway is obstructed, it may be necessary to make a temporary opening in the trachea inferior to the larynx and insert a tube to allow airflow—a procedure called *tracheostomy.* This prevents asphyxiation, but the inhaled air bypasses the nasal cavity and thus is not humidified. If the opening is left for long, the mucous membranes of the respiratory tract dry out and become encrusted, interfering with the clearance of mucus from the tract and promoting infection. When a patient is on a ventilator and air is introduced directly into the trachea, the air must be filtered and humidified by the apparatus to prevent respiratory tract damage.

the right. The left lung, while taller, is narrower than the right because the heart tilts toward the left and occupies more space on this side of the mediastinum. On the medial surface, the left lung has an indentation called the **cardiac impression** where the heart presses against it. The right lung has three lobes—**superior, middle,** and **inferior**—separated by two fissures. The left lung has only a **superior** and **inferior lobe** and a single fissure.

THE BRONCHIAL TREE

The lung has a spongy parenchyma containing the **bronchial tree** (see photo on p. 653), a highly branched system of air tubes extending from the primary bronchus to about 65,000 *terminal bronchioles.* Two **primary bronchi** (BRON-kye) arise from the trachea near the level of the angle of the sternum. Each continues for 2 to 3 cm and enters the hilum of its respective lung. The right bronchus is slightly wider and more vertical than the left; consequently, *aspirated* (inhaled) foreign objects lodge in the right bronchus more often than in the left. Like the trachea, the primary bronchi have a ciliated pseudostratified epithelium and are supported by C-shaped hyaline cartilages. All divisions of the bronchial tree also have a substantial amount of elastic connective tissue, which is important in expelling air from the lungs.

Upon entering the hilum, the primary bronchus branches into one **secondary (lobar) bronchus** for each pulmonary lobe. Thus, there are two secondary bronchi in the left lung and three in the right. Each secondary bronchus divides into **tertiary (segmental) bronchi**—10 in the right lung and 8 in the left. The portion of the lung supplied by one tertiary bronchus is called a **bronchopulmonary segment.** Secondary and tertiary bronchi are supported by overlapping plates of cartilage, not rings.

Branches of the *pulmonary artery* closely follow the bronchial tree on their way to the alveoli. The bronchial tree itself is serviced by the *bronchial artery,* which arises from the aorta and carries systemic blood.

Bronchioles are continuations of the airway that lack supportive cartilage and are 1 mm or less in diameter. The portion of the lung ventilated by one bronchiole is called a **pulmonary lobule.** Bronchioles have a ciliated cuboidal epithelium and a well-developed layer of smooth muscle in their walls. Spasmodic contractions of this muscle at death cause the bronchioles to exhibit a wavy lumen in most histological sections.

Each bronchiole divides into 50 to 80 **terminal bronchioles,** the final branches of the conducting division. These measure 0.5 mm or less in diameter and have no mucous glands or goblet cells. They do have cilia, however, so that mucus draining into them from the higher passages can be driven back by the mucociliary escalator, thus preventing congestion of the terminal bronchioles and alveoli.

Each terminal bronchiole gives off two or more smaller **respiratory bronchioles,** which have alveoli budding from their walls. Respiratory bronchioles are the beginning of the respiratory division. Their walls have scanty smooth muscle, and the smallest of them are nonciliated. Each respiratory bronchiole divides into 2 to 10 elongated, thin-walled passages called **alveolar ducts,** which also have alveoli along their walls (fig. 23.10). The alveolar ducts and smaller divisions have nonciliated simple squamous epithelia. The ducts end in **alveolar sacs,** which are grapelike clusters of alveoli arrayed around a central space called the *atrium.* The distinction between an alveolar duct and atrium is their shape—an elongated duct, or an atrium with about equal length and width. It is sometimes a subjective judgment whether to regard a space as an alveolar duct or atrium.

Airflow to the alveoli varies with the diameter of the bronchioles. Contraction of the smooth muscle of the bronchioles causes a narrowing, or *bronchoconstriction,* that reduces airflow. Relaxation of the muscle causes *bronchodilation,* widening of the airway permitting increased airflow. The trachea and bronchi also constrict and dilate, but are more restricted by the cartilage in their walls.

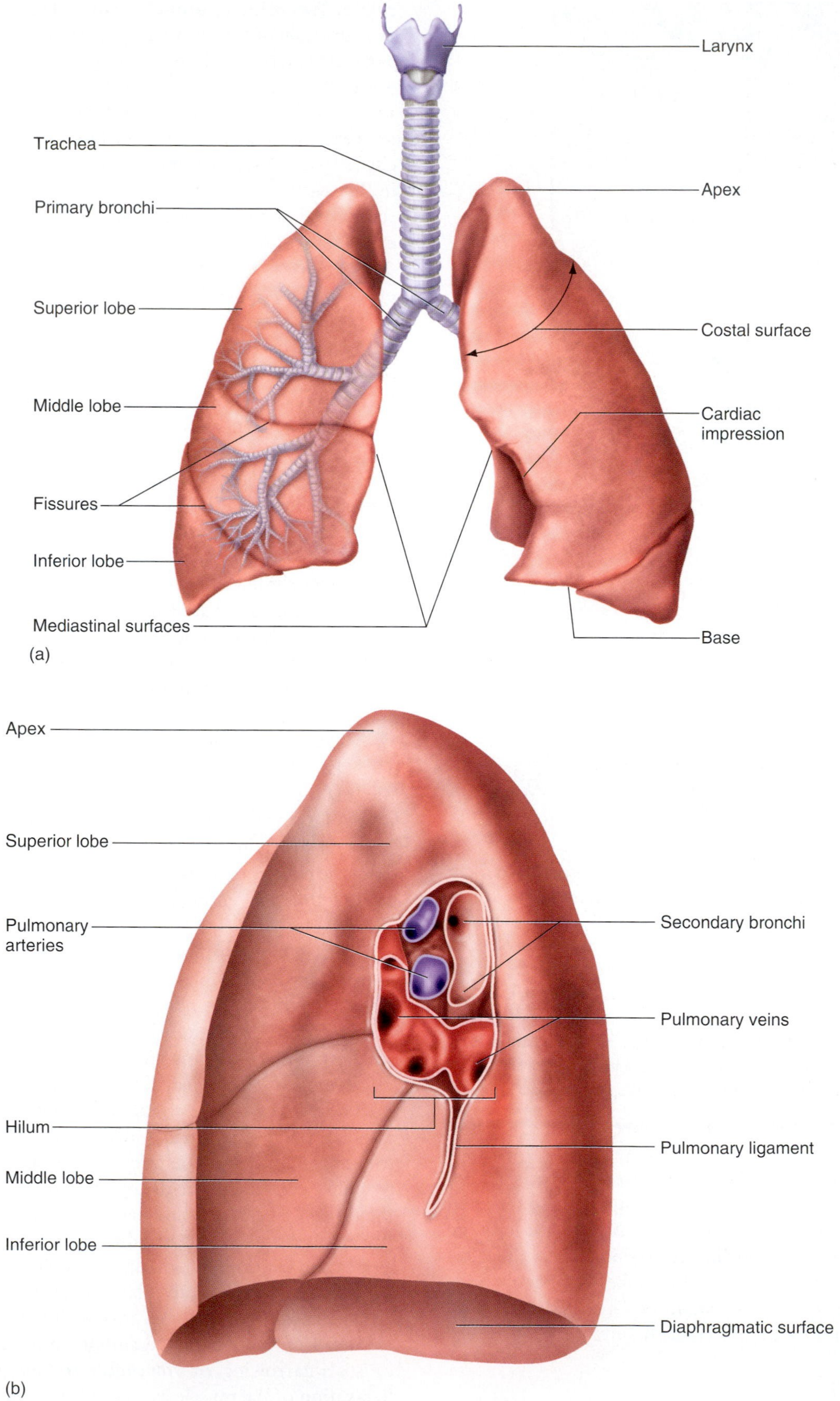
Larynx
Trachea
Apex
Primary bronchi
Superior lobe
Costal surface
Middle lobe
Cardiac impression
Fissures
Inferior lobe
Mediastinal surfaces
Base
(a)
Apex
Superior lobe
Pulmonary arteries
Secondary bronchi
Pulmonary veins
Hilum
Pulmonary ligament
Middle lobe
Inferior lobe
Diaphragmatic surface
(b)

FIGURE 23.9

Gross Anatomy of the Lungs. (*a*) Anterior view. (*b*) Mediastinal surface of the right lung.

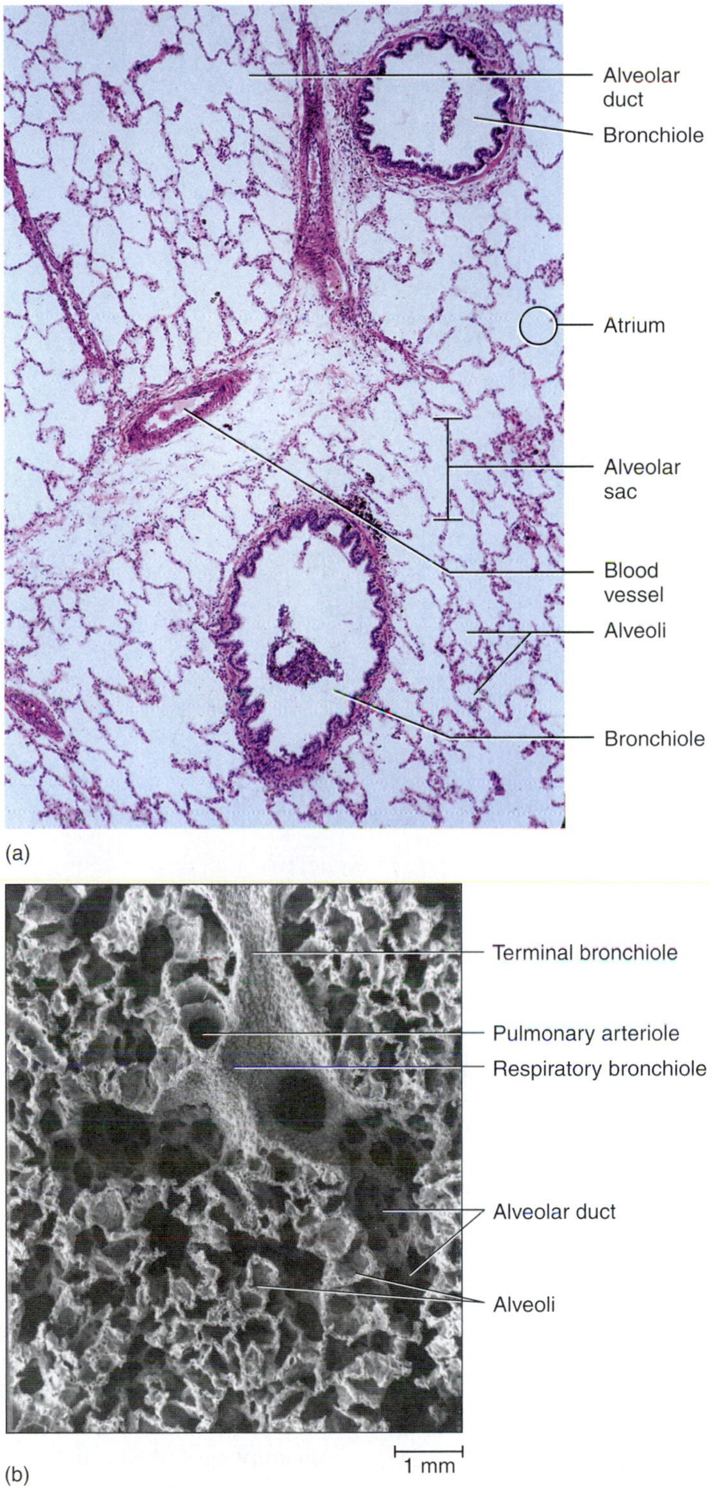
Alveolar duct
Bronchiole
Atrium
Alveolar sac
Blood vessel
Alveoli
Bronchiole
(a)
Terminal bronchiole
Pulmonary arteriole
Respiratory bronchiole
Alveolar duct
Alveoli
1 mm
(b)

FIGURE 23.10

Tissue of the Lung. (*a*) Light micrograph. (*b*) SEM. Note the spongy texture of the lung.

Air in the conducting division of the respiratory tract cannot exchange gases with the blood, and is therefore called *dead air.* The conducting division is also called the *anatomic dead space.* In a state of relaxation, the parasympathetic nervous system keeps the bronchioles partially constricted. This minimizes the dead space, so a greater percentage of the air inhaled goes to the alveoli, where it can oxygenate the blood. In exercise, the sympathetic nerves dilate the bronchioles. Even though this increases the dead space, it enables air to flow more easily and rapidly so the alveoli can be ventilated in proportion to the demands of exercise. The increased airflow more than compensates for the increased dead space.

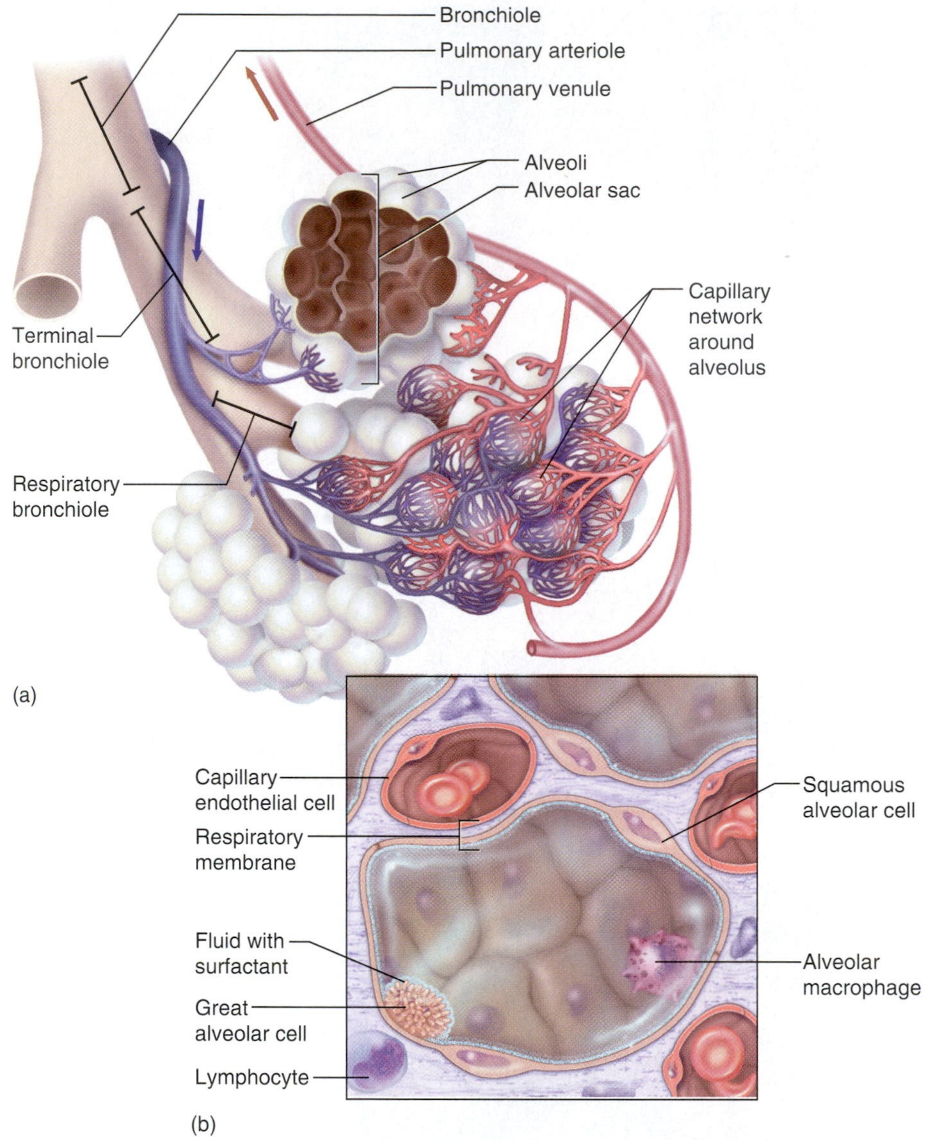
Bronchiole
Pulmonary arteriole
Pulmonary venule
Alveoli
Alveolar sac
Capillary network around alveolus
Terminal bronchiole
Respiratory bronchiole
(a)
Capillary endothelial cell
Respiratory membrane
Fluid with surfactant
Great alveolar cell
Lymphocyte
Squamous alveolar cell
Alveolar macrophage
(b)

FIGURE 23.11

Pulmonary Alveoli. (*a*) Clusters of alveoli and their blood supply. (*b*) Structure of an alveolus.

ALVEOLI

Each human lung is a spongy mass with about 150 million little sacs, the alveoli, which provide about 70 m^2 of surface for gas exchange. An **alveolus** (AL-vee-OH-lus) is a pouch about 0.2 to 0.5 mm in diameter (fig. 23.11). Thin, broad cells called **squamous (type I) alveolar cells** cover about 95% of the alveolar surface area. Their thinness allows for rapid gas diffusion between the alveolus and bloodstream. The other 5% is covered by round to cuboidal **great (type II) alveolar cells.** Even though they cover less surface area, great alveolar cells considerably outnumber the squamous alveolar cells. Great alveolar cells have two functions: (1) they repair the alveolar epithelium when the squamous alveolar cells are damaged, and (2) they secrete *pulmonary surfactant,* a mixture of phospholipids and protein that coats the alveoli and smallest bronchioles and prevents them from collapsing when one exhales. Without surfactant, the walls of a deflating alveolus would tend to cling together like sheets of wet paper, and it would be very difficult to reinflate them on the next inhalation (see insight 23.4).

The most numerous of all cells in the lung are **alveolar macrophages (dust cells),** which wander the lumens of the alveoli and the connective tissue between them. These cells keep the alveoli free of debris by phagocytizing dust particles that escape entrapment by mucus in the higher parts of the respiratory tract. In lungs that are infected or bleeding, the macrophages also phagocytize bacteria and loose blood cells. As many as 100 million alveolar macrophages perish each day as they ride up the mucociliary escalator to be swallowed and digested, thus ridding the lungs of their load of debris.

Each alveolus is surrounded by a basket of blood capillaries supplied by the pulmonary artery. The barrier between the alveolar air and blood, called the **respiratory membrane,** consists only of

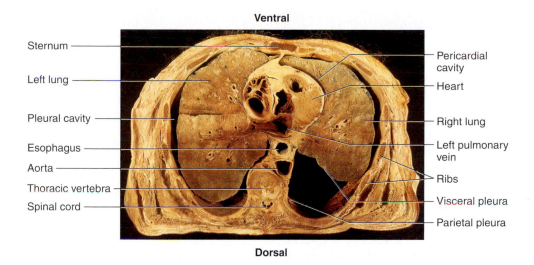

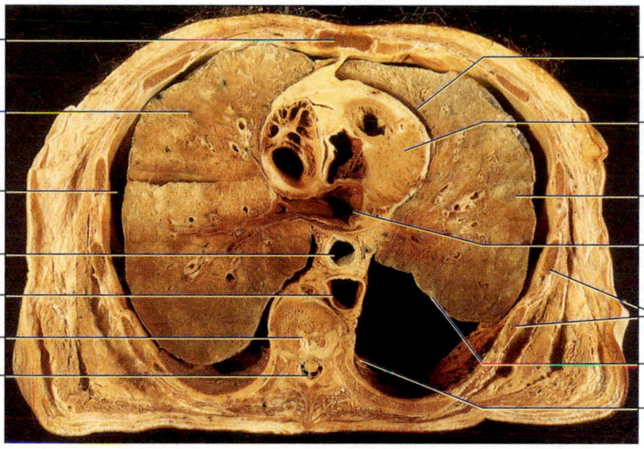
Ventral
Sternum
Left lung
Pleural cavity
Esophagus
Aorta
Thoracic vertebra
Spinal cord
Pericardial cavity
Heart
Right lung
Left pulmonary vein
Ribs
Visceral pleura
Parietal pleura
Dorsal

FIGURE 23.12

Lungs in Relation to the Pleurae. A cross section of the thoracic cavity. Postmortem retraction of the lungs makes the pleural cavities visible. Normally the pleurae adhere to each other and the cavity is only a potential space.

the squamous alveolar cell, the squamous endothelial cell of the capillary, and their shared basement membrane. These have a total thickness of only 0.5 μm, in contrast to the 7 μm diameter of the erythrocytes passing through the capillaries.

It is very important to prevent serous fluid from accumulating in the alveoli, because gases diffuse too slowly through liquid to sufficiently aerate the blood. Except for the film of moisture on the alveolar wall, the alveoli are kept dry by the absorption of excess liquid by the blood capillaries and abundant lymphatic capillaries of the lungs. The lungs have a more extensive lymphatic drainage than any other organ in the body.

The Pleurae

The surface of the lung is covered by a serous membrane, the **visceral pleura** (PLOOR-uh), which extends into the fissures. At the hilum, the visceral pleura turns back on itself and forms the **parietal pleura,** which adheres to the mediastinum, inner surface of the rib cage, and superior surface of the diaphragm (fig. 23.12). An extension of the parietal pleura, the *pulmonary ligament,* connects each lung to the diaphragm.

The space between the parietal and visceral pleurae is called the **pleural cavity.** The two membranes are normally separated only by a film of slippery **pleural fluid;** thus, the pleural cavity is only a *potential space,* meaning there is normally no room between the membranes. Under pathological conditions, however, this space can fill with air or liquid (see *pneumothorax* in insight 23.2).

The pleurae and pleural fluid have three functions:

1. **Reduction of friction.** Pleural fluid acts as a lubricant that enables the lungs to expand and contract with minimal friction. In some forms of *pleurisy,* the pleurae are dry and inflamed and each breath gives painful testimony to the function that the fluid should be serving.
2. **Creation of a pressure gradient.** During inspiration (inhalation), the rib cage expands and draws the parietal pleura outward along with it. The visceral pleura clings to the parietal pleura, and since the visceral pleura is the lung surface, its outward movement expands the lung. The air pressure within the lung thus drops below the atmospheric pressure outside the body, and outside air flows down its pressure gradient into the lung.
3. **Compartmentalization.** The pleurae, mediastinum, and pericardium compartmentalize the thoracic organs and prevent infections of one organ from spreading easily to neighboring organs.

THINK ABOUT IT!

In what ways do the structure and function of the pleurae resemble the structure and function of the pericardium?

Before You Go On

Answer the following questions to test your understanding of the preceding section:

8. A dust particle is inhaled and gets into an alveolus without being trapped along the way. Describe the path it takes, naming all air passages from nostrils to alveoli. What would happen to it after arrival in an alveolus?
9. Contrast the epithelium of the bronchioles with that of the alveoli and explain how the structural difference is related to their functional differences.
10. Describe the relationship of the parietal and visceral pleurae to the lungs and thoracic wall.

INSIGHT 23.2 CLINICAL APPLICATION

Pneumothorax and Atelectasis

If the thoracic wall is punctured—for example, by a knife wound or broken rib—inspiration can suck air through the wound into the pleural cavity, opening a real space between the pleurae. Air in the pleural cavity is called *pneumothorax.* When the visceral pleura (lung surface) no longer clings to the parietal pleura, the lung recoils from the thoracic wall and collapses. Partial or total collapse of a lung is called *atelectasis*[13] (AT-eh-LEC-ta-sis).

Atelectasis can also occur when a pulmonary alveolus ruptures through the lung surface and inhaled air passes through the lung into the pleural cavity, or when fluid accumulates in the pleural cavity. Yet another cause is obstruction of the airway by mucus or aspirated objects, or its compression by a tumor, swollen lymph node, or nearby aneurysm. The blood continues to draw air out of the alveoli, and when this is not replaced by a fresh inflow of air, the alveoli collapse. When one lung collapses, the positive pressure in that pleural cavity can shift the entire mediastinum (including the heart and major blood vessels) toward the other pleural cavity, compressing and partially collapsing that lung as well.

[13]*atel* = imperfect + *ectasis* = expansion

Neuromuscular Aspects of Respiration

Objectives

When you have completed this section, you should be able to

- describe the processes of inspiration and expiration and identify the muscles involved in each;
- describe the mechanisms of coughing and sneezing;
- identify the brainstem centers that control breathing, their efferent connections to the respiratory muscles, and the inputs they receive from other brain centers and the thorax.

The heartbeat and breathing are the two most conspicuously rhythmic processes in the body. In chapter 20, we didn't need to go beyond the heart to discover the basis for its rhythm; that organ has its own internal pacemaker and goes on beating even if all nerves to it are severed. The lungs, by contrast, contain no pacemaker; breathing depends on rhythmic stimuli from the brainstem. Why the difference? There are several reasons: (1) The lungs are passive organs; they do not ventilate themselves, but fill with air and then expel it as muscles of the thorax contract and relax. (2) These thoracic muscles are skeletal muscles, and as such, they cannot contract on their own; they require stimulation by the nervous system. (3) Breathing involves the coordinated action of multiple skeletal muscles and thus requires a central coordinating mechanism to ensure that they all work together. In this section, we will survey the skeletal muscles that cause breathing, the pacemakers of the central nervous system (CNS) that control the rhythm, and the nervous connections between the CNS and the respiratory muscles.

Inspiration

The **diaphragm** does most of the work of inspiration (inhalation). It is controlled by the right and left **phrenic nerves,** which arise from the cervical nerve plexus (see fig. 14.13, p. 400). The diaphragm is dome-shaped at rest, but when stimulated, it tenses and flattens somewhat, dropping about 1.5 cm in quiet respiration and as much as 7 cm in deep breathing.

The **external intercostal muscles** between the ribs are also of great importance to inspiration. They are innervated by the **intercostal nerves,** which arise from the ventral rami of the thoracic spinal nerves (see fig. 14.12*b*, p. 399). While the **scalenes** fix the first pair of ribs, the external intercostals lift the remaining ribs like bucket handles; the ribs swing up and out, increasing the dorsoventral dimension of the thoracic cavity. The combined action of the diaphragm and external intercostal muscles enlarges the thoracic cavity, creating a pressure gradient that draws air into the lungs. Deep inspiration is further aided by the **pectoralis minor, sternocleidomastoid,** and **erector spinae muscles.**

Think About It!

A patient has suffered damage to the left phrenic nerve, resulting in paralysis of the left side of the diaphragm but not the right. X rays show that during inspiration, the right side of the diaphragm descends as normal, but the left side rises. *Explain the unusual motion of the left diaphragm.*

Expiration

Normal expiration (exhalation) is an energy-saving passive process. It is achieved by the elasticity of the lungs and thoracic cage. The bronchial tree, the attachments of the ribs to the spine and sternum, and the tendons of the diaphragm and other respiratory muscles all have a degree of elasticity that causes them to spring back when muscular contraction ceases. As these structures recoil, the thoracic cage diminishes in size, the air pressure in the lungs rises above the atmospheric pressure, and the air flows out. The only muscular effort involved in normal expiration is a braking action. When inspiration ceases, the phrenic nerves continue to stimulate the diaphragm for a little while longer. This prevents the lungs from recoiling too suddenly, so it makes the transition from inspiration to expiration smoother.

Exhaling more completely or rapidly than usual—say, in blowing out the candles on a birthday cake—employs the internal intercostal and abdominal muscles. The **internal intercostal muscles** depress the ribs. The **internal** and **external abdominal obliques, transversus abdominis,** and **rectus abdominis** compress the abdominal organs, raise the intra-abdominal pressure, and push the viscera up against the diaphragm. This compresses the thoracic cavity and helps to expel air from the lungs. Abdominal control of expiration is important in singing, public speaking, coughing, and sneezing.

Respiratory Centers of the Brainstem

The rhythm of unconscious breathing is caused by nuclei in the reticular formation of the medulla oblongata and pons. In the medulla, these nuclei contain **inspiratory (I) neurons,** which fire during inspiration, and **expiratory (E) neurons,** which fire during forced expiration (but not during relaxed breathing). Fibers from these neurons travel down the CNS and synapse with lower motor neurons in the cervical to thoracic regions of the spinal cord. The lower motor neurons issue efferent nerve fibers via the phrenic and intercostal nerves to the respiratory muscles. No pacemaker neurons have been found that are analogous to the autorhythmic cells of the heart, and the exact mechanism for setting the rhythm of respiration remains unknown despite intensive research.

The medulla has two respiratory nuclei (fig. 23.13). One called the **inspiratory center,** or **dorsal respiratory group (DRG),** is composed primarily of I neurons, which feed into the motor neurons that innervate the diaphragm and external intercostal muscles. Firing of the I neurons stimulates the contraction of the muscles of inspiration. The other nucleus is the **expiratory center,** or **ventral respiratory group (VRG).** It is not involved in normal expiration, but only in forced or prolonged expiration.

The pons contains two respiratory nuclei that receive input from other brain centers and modulate the activity of the respiratory nuclei in the medulla. The lower pons has an **apneustic** (ap-NEW-stic) **center** whose function remains unclear; it seems to prolong inspiration. The upper pons has a **pneumotaxic** (NEW-mo-TAX-ic) **center,** which sends a continual stream of signals to the inspiratory center of the medulla. The pneumotaxic center regulates the duration and depth of each breath.

These respiratory centers receive input from several sources:

- Chemoreceptors respond to the pH and the CO_2 and O_2 concentrations of the blood and cerebrospinal fluid. There are two groups of chemoreceptors involved in respiration: (1) **Peripheral chemoreceptors** are located in the aortic bodies and carotid bodies of the major arteries near the heart (fig. 23.14). The aortic bodies communicate with the medulla by way of the vagus nerves and the carotid bodies by way of the glossopharyngeal nerves. (2) **Central chemoreceptors** are paired nuclei close to the surface of the medulla oblongata, ventral to the inspiratory center.

- Stretch receptors monitor inflation of the lungs. They are located in the bronchial tree and visceral pleura. Excessive inflation triggers a protective reflex that strongly inhibits inspiration.

- The vagus nerves transmit sensory signals from the respiratory mucosa to the inspiratory center. They are stimulated by irritants in the airway, such as smoke, dust, noxious fumes, or mucus, or by food or drink aspirated into the larynx or trachea. The vagal reflex arc produces the protective responses of bronchoconstriction, coughing, and sneezing.

- Higher brain centers also influence the respiratory nuclei, including the limbic system, hypothalamus, and cerebral cortex. This input allows for conscious control over breathing (as in holding one's breath) and for emotions to affect respiration—for example, in gasping, crying, and laughing, or when anxiety provokes a bout of hyperventilation (rapid breathing in excess of physiological need). Voluntary control over breathing originates in the motor cortex of the frontal lobe of the cerebrum, which sends signals down the corticospinal tracts to the respiratory neurons in the spinal cord, thus bypassing the brainstem respiratory centers.

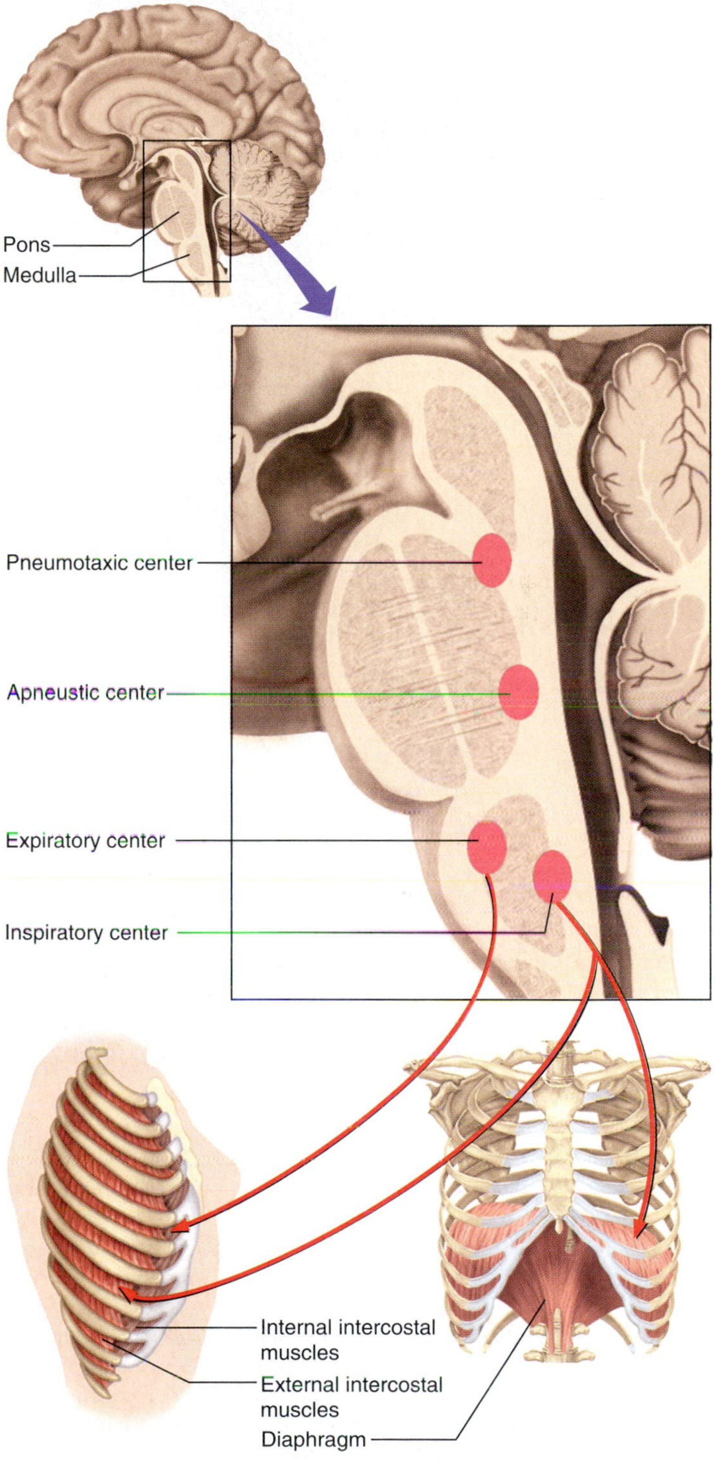
Pons
Medulla
Pneumotaxic center
Apneustic center
Expiratory center
Inspiratory center
Internal intercostal muscles
External intercostal muscles
Diaphragm

FIGURE 23.13
Respiratory Control Centers. Locations of the respiratory centers of the pons and medulla oblongata, and connections to the principal respiratory muscles.

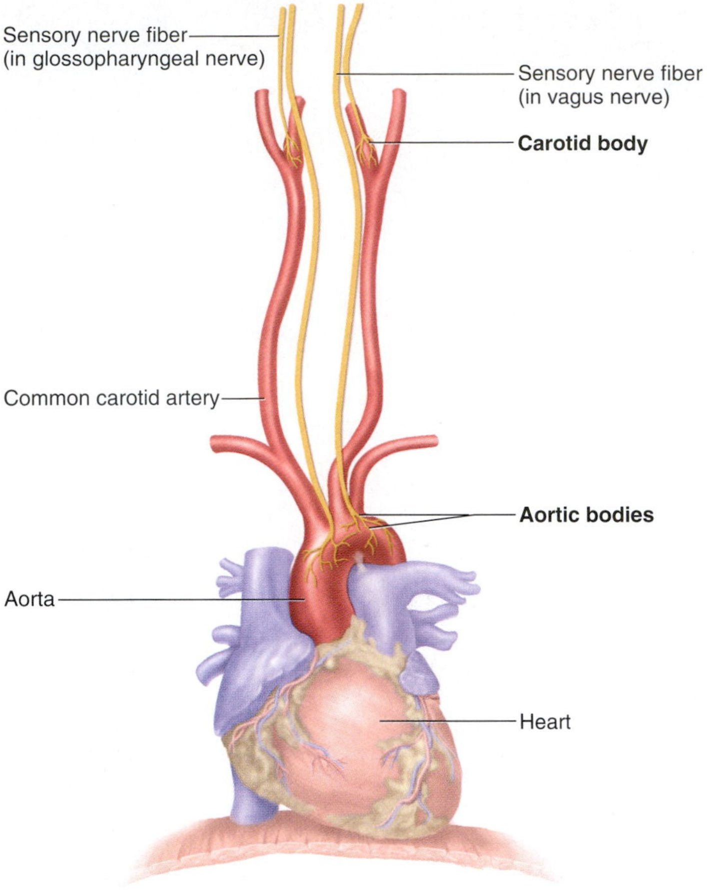

Sensory nerve fiber (in glossopharyngeal nerve)
Sensory nerve fiber (in vagus nerve)
Carotid body
Common carotid artery
Aortic bodies
Aorta
Heart

FIGURE 23.14
The Peripheral Chemoreceptors of Respiration.

INSIGHT 23.3 CLINICAL APPLICATION

Ondine's Curse

In German legend, there was a water nymph named Ondine who took a mortal lover. When he was unfaithful to her, the king of the nymphs put a curse on him that took away his automatic physiological functions. Consequently, he had to remember to take each breath, and he could not go to sleep or he would die of suffocation—which, as exhaustion overtook him, was indeed his fate.

Some people suffer a disorder called *Ondine's curse,* in which the automatic respiratory functions are disabled—usually as a result of brainstem damage from poliomyelitis or as an accident of spinal cord surgery. Victims of Ondine's curse must remember to take each breath and cannot go to sleep without the aid of a mechanical ventilator.

Before You Go On

Answer the following questions to test your understanding of the preceding section:

11. Explain why breathing is not controlled by a pacemaker in the lungs.
12. Name the major muscles involved in inspiration. What nerves innervate these muscles?
13. What is the normal mechanism of relaxed expiration? What muscles are involved in forced expiration?
14. What four brainstem nuclei regulate the respiratory rhythm? What role does each one play?
15. From what sources do the respiratory nuclei receive input that influences respiration?

Developmental and Clinical Perspectives

Objectives

When you have completed this section, you should be able to

- describe the embryonic and fetal development of the respiratory system;
- describe the changes that occur in the respiratory system in old age; and
- describe some common respiratory disorders.

Prenatal and Neonatal Development

The first embryonic trace of the respiratory system is a small pouch in the floor of the pharynx called the **pulmonary groove,** appearing at about 3.5 weeks. The groove grows down the mediastinum as an elongated tube, the future trachea, and branches into two **lung buds** by week 4 (fig. 23.15). The lung buds branch repeatedly and grow laterally and dorsally, occupying the space dorsal to the heart in the ventral body cavity. Repeated branching of the lung buds produces the bronchial tree, which is completed as far as the bronchioles by the end of month 6. For the remainder of gestation and after birth, the bronchioles bud off alveoli. The adult number of alveoli is attained around the age of 10 years.

By week 8, the lungs are isolated from the heart by the growth of the pericardium, and by week 9, the diaphragm forms and separates the lungs and pleurae from the abdominal cavity. At 28 weeks, the respiratory system is usually adequately developed to support independent life (see insight 23.4).

By 11 weeks, the fetus begins respiratory movements called **fetal breathing,** in which it rhythmically inhales and exhales amniotic fluid for as much as 8 hours per day. Fetal breathing stimulates lung development and conditions the respiratory muscles for life outside the womb. It ceases during labor. When the newborn infant begins breathing air, the fluid in the lungs is quickly absorbed by the pulmonary blood and lymphatic capillaries.

For the newborn infant, breathing is very laborious at first. The fetal lungs are collapsed and airless, and the neonate must take very strenuous first breaths to pop the alveoli open. Once they are fully inflated, the alveoli normally never collapse again. Even the pulmonary blood vessels are collapsed in the fetus, but as the infant takes its first breaths, the drop in thoracic pressure draws blood into the pulmonary circulation and expands the vessels. As

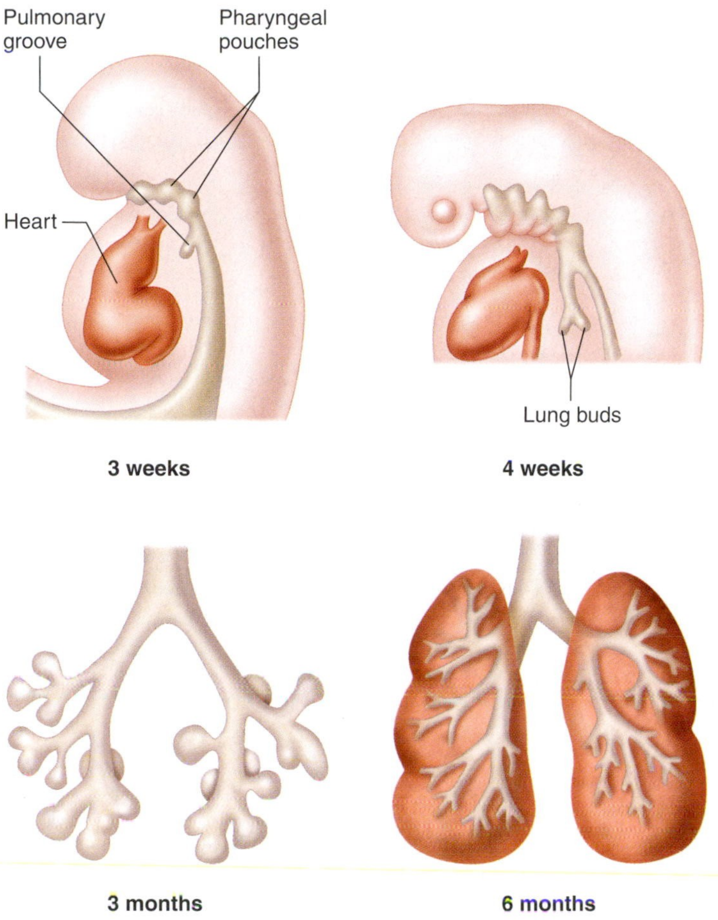
Pulmonary groove
Pharyngeal pouches
Heart
Lung buds
3 weeks
4 weeks
3 months
6 months

FIGURE 23.15
Embryonic Development of the Respiratory System.

pulmonary resistance drops, the foramen ovale and ductus arteriosus close (see chapter 20) and pulmonary blood flow increases to match the airflow.

Think About It!

In a certain criminal investigation, the pathologist performing an autopsy on an infant removes the lungs, places them in a pail of water, and concludes that the infant was live-born. What do you think the pathologist saw that led to this conclusion? What contrasting observation would suggest that an infant had been stillborn?

The Aging Respiratory System

Pulmonary ventilation declines steadily after the 20s and is one of several factors in a person's gradual loss of stamina. The costal cartilages and joints of the thoracic cage become less flexible, the lungs have less elastic tissue, and there are fewer alveoli in old age. There is a corresponding decline in the volume of air inhaled in each breath *(tidal volume),* the maximum amount of air a person can inhale *(vital capacity),* and the maximum speed of airflow *(forced expiratory volume).* The elderly are also less capable of clearing the lungs of irritants and pathogens, and therefore increasingly susceptible to respiratory infections. Pneumonia causes more deaths in old age than any other communicable disease and is often contracted in hospitals and nursing homes.

Chronic obstructive pulmonary diseases (see next section) are more common in old age since they represent the cumulative effects of a lifetime of degenerative change. They are among the leading causes of death in old age. Declining pulmonary function also contributes to cardiovascular disease and hypoxemia, and the latter is a factor in degenerative disorders of all other organ systems. Respiratory health is therefore a major concern in aging.

INSIGHT 23.4 CLINICAL APPLICATION

Premature Birth and Respiratory Distress Syndrome

Premature infants often suffer from *respiratory distress syndrome (RDS),* also called *hyaline membrane disease* (compare adult respiratory distress syndrome, table 23.1). They have not yet produced enough pulmonary surfactant to keep the alveoli open between inspirations. Consequently, the alveoli collapse during expiration, and a great effort is required to reinflate them. The infant becomes exhausted by the effort to breathe, and becomes progressively *cyanotic* (blue) because of the deficiency of oxygen in the blood *(hypoxemia).* Progressive destruction of the alveolar epithelium and capillary walls leads to leakage of plasma into the alveolar air spaces and connective tissue between the alveoli. The plasma coagulates, and the alveoli fill with stiff clear "membranes" of fibrin, fibrinogen, and cell debris. Eventually, the infant cannot inhale forcefully enough to inflate the alveoli again, and without treatment, death ensues from hypoxemia and carbon dioxide retention *(hypercapnia).*

RDS occurs in about 60% of infants born before 28 weeks of gestation, and 15% to 20% of those born between 32 and 36 weeks. It is the most common cause of neonatal death, with about 60,000 cases and 5,000 deaths per year in the United States. In addition to premature birth, some risk factors for RDS include maternal diabetes, oversedation of the mother during birth, aspiration of blood or amniotic fluid, and prenatal hypoxia caused by winding of the umbilical cord around the neck.

RDS can be treated with a ventilator that forces air into the lungs and keeps the alveoli inflated *(positive end-expiratory pressure, PEEP)* until the infant's lungs produce their own surfactant, and by giving a mist of surfactant from external sources such as calf lungs or genetically engineered bacteria. The infant may also be given oxygen therapy, but this is a limited and risky treatment because oxygen generates damaging free radicals that can cause blindness and severe bronchial problems. Oxygen toxicity can be minimized by a technique called *extracorporeal membrane oxygenation (ECMO),* which is similar to the heart-lung bypass procedure used in surgery. Blood flows from catheters in the baby's neck to a machine that oxygenates it, warms it, and returns it to the body.

Respiratory Pathology

Many of the respiratory disorders can be classified as *restrictive* or *obstructive disorders.* **Restrictive disorders** stiffen the lungs and reduce their *compliance* (ease of inflation) and vital capacity. An example is pulmonary fibrosis, in which much of the normal respiratory tissue of the lung is replaced by fibrous scar tissue. Fibrosis is an effect of such diseases as *tuberculosis* and the *black lung disease* of coal miners. **Obstructive disorders** narrow the airway

TABLE 23.1

Some Disorders of the Respiratory System

Acute Rhinitis	The common cold. Caused by many types of viruses that infect the upper respiratory tract. Symptoms include congestion, increased nasal secretion, sneezing, and dry cough. Transmitted especially by contact of contaminated hands with mucous membranes; not transmitted orally.
Adult Respiratory Distress Syndrome	Acute lung inflammation and alveolar injury stemming from trauma, infection, burns, aspiration of vomit, inhalation of noxious gases, drug overdoses, and other causes. Alveolar injury is accompanied by severe pulmonary edema and hemorrhage, followed by fibrosis that progressively destroys lung tissue. Fatal in about 40% of cases under age 60 and in 60% of cases over age 65.
Pneumonia	A lower respiratory infection caused by any of several viruses, fungi, or protozoans (most often the bacterium *Streptococcus pneumoniae*). Causes filling of alveoli with fluid and dead leukocytes, and thickening of the respiratory membrane, which interferes with gas exchange and causes hypoxemia. Especially dangerous to infants, the elderly, and people with compromised immune systems, such as AIDS and leukemia patients.
Sleep Apnea	Cessation of breathing for 10 seconds or longer during sleep; sometimes occurs hundreds of times per night, often accompanied by restlessness and snoring. Can result from altered function of CNS respiratory centers, airway obstruction, or both. Over time, may lead to daytime drowsiness, hypoxemia, polycythemia, pulmonary hypertension, congestive heart failure, and cardiac arrhythmia. Most common in obese people and in men.
Tuberculosis (TB)	Pulmonary infection with the bacterium *Mycobacterium tuberculosis,* which invades the lungs by way of air, blood, or lymph. Stimulates the lung to form fibrous nodules called tubercles around the bacteria. Progressive fibrosis compromises the elastic recoil and ventilation of the lungs. Especially common among impoverished and homeless people and becoming increasingly common among people with AIDS.

Disorders Described Elsewhere

Airway obstruction 670	Emphysema 670
Asthma 670	Lung cancer 670
Atelectasis 666	Neonatal respiratory distress syndrome 669
Black lung disease 669	Ondine's curse 668
Chronic bronchitis 670	Pneumothorax 666
Chronic obstructive pulmonary diseases 670	Pulmonary fibrosis 669

and interfere with airflow, so that expiration requires more effort and may be less complete than normal. Airway obstructions, bronchoconstriction, and tumors or aneurysms that compress the airways can cause obstructive disorders.

Among the leading causes of death in old age are the three **chronic obstructive pulmonary diseases (COPDs)**—chronic bronchitis, emphysema, and asthma. The first two of these are usually caused by cigarette smoking. **Asthma** is an allergic reaction to airborne antigens (allergens) that stimulate intense bronchoconstriction and airway inflammation, sometimes to the point of suffocation. **Chronic bronchitis** is characterized by airway congestion with *sputum,* a mixture of thick mucus and cellular debris, accompanied by chronic respiratory infection and bronchial inflammation. **Emphysema**[14] (EM-fih-SEE-muh) is characterized by a breakdown of alveolar walls, resulting in a reduction in the number of alveoli and reduced ability to oxygenate the blood. COPDs also lead to *cor pulmonale*—enlargement and potential failure of the right heart due to obstruction of the pulmonary circulation.

Lung cancer accounts for more deaths than any other form of cancer. The most important cause is cigarette smoking, distantly followed by air pollution. Lung cancer commonly follows or accompanies COPD. It begins with uncontrolled proliferation of cells of the surface epithelium or mucous glands of the bronchi. As the dividing epithelial cells invade the underlying tissues of the bronchial wall, the bronchus develops bleeding lesions. Dense masses of keratin and malignant cells appear in the lung parenchyma and replace functional respiratory tissue. Because of the extensive lymphatic drainage of the lungs, lung cancer quickly metastasizes to other organs—especially the pericardium, heart, bones, liver, lymph nodes, and brain. The chance of recovery is poor, with only 7% of patients surviving for 5 years after diagnosis.

Some other disorders of the respiratory system are briefly described in table 23.1.

Before You Go On

Answer the following questions to test your understanding of the preceding section:

16. When and where does the pulmonary groove appear? Concisely describe its further development.
17. What changes in the lungs occur at birth? What corresponding changes occur in the cardiovascular system?
18. Identify some reasons why a person's vital capacity declines in old age.
19. Name and compare two COPDs and describe some pathological effects that they have in common.
20. In what lung tissue does lung cancer originate? How does it kill?

[14]*emphys* = inflamed

CHAPTER REVIEW

REVIEW OF KEY CONCEPTS

Overview of the Respiratory System (p. 654)

1. The respiratory system enables the blood to exchange gases with the air; serves for vocalization; provides a sense of smell; regulates blood pH and blood pressure; and creates pressure gradients that aid in the flow of lymph and blood and in expelling the contents of some abdominal organs.
2. The *conducting division* of the respiratory system consists of the nose, pharynx, larynx, trachea, bronchi, and most bronchioles; it serves only for airflow.
3. The *respiratory division* consists of *respiratory bronchioles, alveoli,* and other distal gas-exchange regions of the lungs.
4. The *upper respiratory tract* consists of the respiratory organs of the head and neck, extending from the nose through the larynx. The *lower respiratory tract* consists of the respiratory organs of the thorax, including the trachea, bronchi, and lungs.

The Upper Respiratory Tract (p. 655)

1. The nose extends from the anterior nares (nostrils) to the posterior nares (choanae) and is internally divided by the nasal septum into right and left *nasal fossae.* The facial part of the nose is shaped by the maxillae, nasal bones, and the lateral and alar cartilages.
2. The nasal septum is composed of the bony perpendicular plate of the ethmoid bone above; another bone, the vomer, below; and the *septal cartilage* anteriorly. The roof of the nasal cavity is formed by parts of the ethmoid and sphenoid bones, and the floor by the hard palate.
3. Each fossa has three scroll-like *nasal conchae* covered with a ciliated mucous membrane. Air flows through narrow spaces called the meatuses between the conchae. The conchae warm, humidify, and cleanse the air flowing over them.
4. The nasal cavity is lined with a sensory *olfactory epithelium* high in each fossa, and with a ciliated pseudostratified *respiratory epithelium* throughout the rest of the cavity. The respiratory epithelium traps airborne particles in its mucus and propels the mucus to the pharynx to be swallowed.
5. Erectile tissues of the inferior nasal concha swell and shrink, usually in one fossa at a time, to periodically shift airflow from one fossa to the other and allow for recovery from drying.
6. The *pharynx* is a muscular passage divided into *nasopharynx, oropharynx,* and *laryngopharynx.*
7. The *larynx* is a cartilaginous chamber beginning superiorly at the *glottis* and ending about 4 cm below this at the trachea. It is supported by nine cartilages bound to each other by intrinsic ligaments, while two extrinsic ligaments attach the larynx to the hyoid bone above and the trachea below.
8. The larynx has a pair of superior *vestibular folds* that exclude food and drink from the airway, and a pair of inferior *vocal cords* that function in speech. *Extrinsic muscles* of the larynx help to close it during swallowing, and its *intrinsic muscles* operate the vocal cords during speech.

The Lower Respiratory Tract (p. 659)

1. The *trachea* is a 12-cm tube, supported by cartilaginous rings, that extends from the larynx above to the two primary bronchi below. The ciliated mucosa of the trachea acts as a *mucociliary escalator* to remove inhaled debris, stuck in the tracheal mucus, from the respiratory tract. The C-shaped cartilage rings hold the trachea open during inspiration. The gap on the dorsal side of the C allows for expansion of the esophagus during swallowing.
2. Each lung is a conical organ extending from the superior *apex* to the inferior, broad *base.* Its extensive *costal surface* lies against the rib cage and its indented *mediastinal surface* faces the heart. The mediastinal surface has a *hilum* through which it receives the primary bronchi, pulmonary blood vessels, nerves, and lymphatics.
3. The right lung is shorter but broader than the left and is divided by two deep fissures into superior, middle, and inferior lobes. The left lung is taller but narrower and has only a superior and inferior lobe, separated by a single fissure.
4. The bronchial tree is a branching system of air passages extending from one *primary bronchus* in each lung to *secondary bronchi* (2 in the left lung and 3 in the right), *tertiary bronchi* (8 in the left lung and 10 in the right), *bronchioles, terminal bronchioles,* and *respiratory bronchioles.* Each secondary bronchus supplies one lobe of the lung; each tertiary bronchus supplies one *bronchopulmonary segment.* The terminal bronchioles are the end of the conducting division; all branches beyond this have *alveoli* and belong to the respiratory division.
5. Respiratory bronchioles branch into 2 to 10 thin-walled *alveolar ducts.* Alveolar ducts end in grapelike clusters of alveoli called *alveolar sacs.*
6. *Bronchoconstriction* narrows the bronchioles and reduces airflow; *bronchodilation* widens them and increases airflow.
7. An alveolus is a thin-walled sac surrounded by a basket of blood capillaries. It is composed of squamous and great alveolar cells and contains alveolar macrophages, the last line of defense against inhaled debris.
8. Gases are exchanged through the thin *respiratory membrane* composed of the capillary endothelial cells, the squamous alveolar cells, and their shared basement membrane. The great alveolar cells secrete a *pulmonary surfactant* that prevents the alveoli from collapsing during expiration.
9. The surface of each lung is a serous membrane called the *visceral pleura.* It continues as the *parietal pleura,* which lines the inside of the rib cage. The space between the pleurae is the *pleural cavity,* and is lubricated with *pleural fluid.* The pleurae reduce friction during breathing, contribute to the pressure gradients that move air into and out of the lungs, and help compartmentalize the thoracic cavity.

Neuromuscular Aspects of Respiration (p. 666)

1. Breathing alternates between *inspiration* and *expiration.*
2. The lungs do not ventilate themselves. The respiratory rhythm is set by respiratory centers in the brainstem, and ventilation is achieved by the contractions of thoracic and abdominal skeletal muscles.
3. *Inspiration* is usually achieved by contractions of the diaphragm and external intercostal muscles, which are innervated by the phrenic and intercostal nerves, respectively. The scalenes aid in normal inspiration, while the pectoralis minor, sternocleidomastoid, and erector spinae muscles assist in deep inspiration. These muscles expand the thoracic cavity, lowering its internal air pressure below that of the surrounding atmosphere, causing air to flow down its pressure gradient into the lungs.

4. *Expiration* is caused mainly by elastic recoil of the lungs and thoracic cage when the inspiratory muscles relax. Deep expiration is assisted by the internal intercostal, abdominal oblique, transversus abdominis, and rectus abdominis muscles.
5. The medulla oblongata contains an *inspiratory center* composed mainly of *inspiratory (I) neurons.* When these neurons fire, signals are transmitted down the spinal cord to lower motor neurons which, in turn, stimulate contraction of the diaphragm and external intercostal muscles.
6. The medulla also has an *expiratory center* that is active only in deep expiration.
7. The pons contains an *apneustic center* that may serve to prolong inspiration, and a *pneumotaxic center* that regulates the duration and depth of each breath.
8. The brainstem respiratory centers receive input from *central* and *peripheral chemoreceptors* that monitor the gas concentrations and pH of the CSF and blood; stretch receptors that inhibit inspiration if the lungs expand excessively; sensory fibers in the respiratory mucosa that respond to irritants in the airway; and higher brain centers than allow for emotional and voluntary influences on breathing.

Developmental and Clinical Perspectives (p. 668)

1. The respiratory system begins development as a *pulmonary groove* that grows from the floor of the pharynx around 3.5 weeks of gestation. This groove grows into a tube that forks into two *lung buds,* then branches extensively to form the bronchial trees. By weeks 8 to 9, the pericardium and diaphragm isolate the lungs and pleural cavities from the heart and abdominal cavity.
2. The neonate's first breaths are very strenuous because of the need to inflate the alveoli. The pulmonary blood vessels also expand during these breaths, and the foramen ovale and ductus arteriosus gradually close to direct blood to the lungs.
3. With advancing age, the thoracic cage becomes less flexible, and the depth and rate of breathing decline. Irritants and pathogens cannot be cleared from the lungs as easily, so elderly people become more susceptible to respiratory infections. Pneumonia is a common cause of death in old age.
4. *Restrictive disorders* of the lung interfere with their inflation. Pulmonary fibrosis is an example. *Obstructive disorders* narrow the airway and interfere with the ease and speed of airflow. This can be caused by tumors, aneurysms, or bronchial congestion.
5. Chronic obstructive pulmonary diseases (COPDs) are leading causes of death in old age. The COPDs are *asthma, chronic bronchitis,* and *emphysema.* Asthma is an allergic disease involving intense inflammation and constriction of the bronchioles, sometimes severe enough to cause suffocation. The other COPDs are usually caused by tobacco smoke. Chronic bronchitis entails congestion of the airway with thick mucus, and susceptibility to respiratory infection. Emphysema entails destruction of pulmonary alveoli. The COPDs also frequently lead to right-sided heart failure.
6. Lung cancer also is usually caused by tobacco smoke. Tumors replace functional respiratory tissue, cause bleeding lesions of the lung, and metastasize quickly to adjacent thoracic organs.

Testing Your Recall

1. The nasal cavity is divided by the nasal septum into right and left
   a. nares.
   b. vestibules.
   c. fossae.
   d. choanae.
   e. conchae.
2. The intrinsic laryngeal muscles regulate speech by rotating
   a. the extrinsic laryngeal muscles.
   b. the thyroid cartilage.
   c. the arytenoid cartilages.
   d. the hyoid bone.
   e. the vocal cords.
3. The largest air passages that engage in gas exchange with the blood are
   a. the respiratory bronchioles.
   b. the terminal bronchioles.
   c. the primary bronchi.
   d. the alveolar ducts.
   e. the alveoli.
4. Respiratory arrest would most likely result from a tumor of the
   a. pons.
   b. midbrain.
   c. thalamus.
   d. cerebellum.
   e. medulla oblongata.
5. A deficiency of pulmonary surfactant is most likely to cause
   a. chronic obstructive pulmonary disease.
   b. atelectasis.
   c. pneumothorax.
   d. chronic bronchitis.
   e. asthma.
6. The source of pulmonary surfactant is
   a. the visceral pleura.
   b. tracheal glands.
   c. alveolar capillaries.
   d. squamous alveolar cells.
   e. great alveolar cells.
7. Which of the following are fewest in number but largest in diameter?
   a. alveoli
   b. terminal bronchioles
   c. alveolar ducts
   d. tertiary bronchi
   e. respiratory bronchioles
8. The rhythm of breathing is set by neurons in
   a. the medulla oblongata.
   b. the pons.
   c. the midbrain.
   d. the hypothalamus.
   e. the cerebral cortex.
9. Which of the following muscles aids in deep expiration?
   a. the scalenes
   b. the sternocleidomastoids
   c. the rectus abdominis
   d. the external intercostals
   e. the diaphragm
10. All regions of the respiratory division are characterized by the presence of
    a. mucous glands.
    b. ciliated cells.
    c. alveoli.
    d. cartilage rings.
    e. goblet cells.
11. The superior opening into the larynx is called the _____.
12. Within each lung, the airway forms a branching complex called the _____.
13. The flared areas of the nose lateral to the nostrils are shaped by _____ cartilages.
14. The three folds on the lateral walls of the nasal cavity are called _____.

15. _____ disorders reduce the speed of airflow through the airway.
16. Some inhaled air does not participate in gas exchange because it fills the _____ of the respiratory tract.
17. The largest cartilage of the larynx is the _____ cartilage.
18. Inspiration is caused by the firing of I neurons in the _____ of the medulla oblongata.
19. The primary bronchi and pulmonary blood vessels penetrate the lung at a medial slit called the _____.
20. The last line of defense against inhaled particles are phagocytic cells called _____.

Answers in the Appendix

True or False

Determine which five of the following statements are false, and briefly explain why.

1. The glottis is the opening from the larynx to the trachea.
2. The lungs contain more respiratory bronchioles than terminal bronchioles.
3. The lungs occupy the spaces between the parietal and visceral pleurae.
4. Expiration is normally caused by contraction of the internal intercostal muscles.
5. Atelectasis can result from causes other than pneumothorax.
6. Alveoli continue to be produced after birth.
7. Unlike bronchi, bronchioles have no cartilage.
8. Blood gases are monitored by the aortic and carotid sinuses.
9. The respiratory system begins its development by budding from the dorsal side of the esophagus.
10. Extrinsic ligaments link the larynx to adjacent nonlaryngeal structures.

Answers in the Appendix

Testing Your Comprehension

1. Discuss how the different functions of the conducting division and respiratory division relate to differences in their histology.
2. From the upper to the lower end of the trachea, the ratio of goblet cells to ciliated cells gradually changes. Would you expect the highest ratio of ciliated cells to goblet cells to be in the upper trachea or the lower trachea? Give a functional rationale for your answer.
3. The bronchioles are to the airway and air flow what the arterioles are to the circulatory system and blood flow. Explain or elaborate on this comparison.
4. Failure of the left ventricle of the heart causes blood to back up in the lungs, raising blood pressure in the alveolar capillaries. Explain why this could lead to hypoxemia and cyanosis.
5. An 83-year-old woman is admitted to the hospital, where a critical care nurse attempts to insert a nasoenteric tube ("stomach tube") for feeding. The patient begins to exhibit dyspnea, and a chest X ray reveals air in the right pleural cavity and a collapsed right lung. The patient dies 5 days later from respiratory complications. Name the conditions revealed by the X ray and explain how they could have resulted from the nurse's procedure.

Answers at the Online Learning Center

www.mhhe.com/saladinha1

Visit the Online Learning Center for practice tests, answer keys, and other learning aids for this chapter. Enhance your understanding of human anatomy with our interactive art labeling exercises, supplemental photo atlases, web links, puzzles, flashcards, and much more.

24

CHAPTER TWENTY-FOUR

The Digestive System

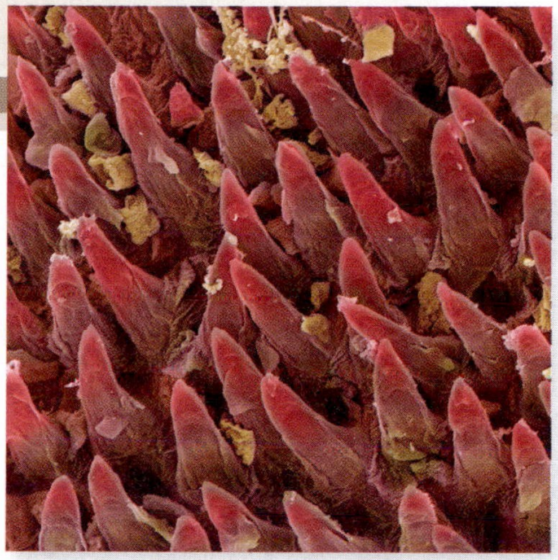

Filiform papillae of the human tongue

CHAPTER OUTLINE

Digestive Processes and General Anatomy 676
- Digestive System Functions 676
- General Anatomy 676
- Innervation 677
- Circulation 678
- Relationship to the Peritoneum 679

The Mouth Through Esophagus 680
- The Oral Cavity 680
- The Salivary Glands 683
- The Pharynx 684
- The Esophagus 684

The Stomach 685
- Gross Anatomy 685
- Microscopic Anatomy 686

The Small Intestine 689
- Gross Anatomy 690
- Microscopic Anatomy 690

The Large Intestine 692
- Gross Anatomy 694
- Microscopic Anatomy 694

Accessory Glands of Digestion 694
- The Liver 695
- The Gallbladder and Bile Passages 698
- The Pancreas 698

Developmental and Clinical Perspectives 699
- Prenatal Development 700
- The Aging Digestive System 701
- Digestive Disorders 701

Chapter Review 703

INSIGHTS

24.1 Clinical Application: Tooth and Gum Disease 682
24.2 Clinical Application: Gastroesophageal Reflux Disease 685
24.3 Clinical Application: Peptic Ulcer 689
24.4 Clinical Application: Gallstones 698

BRUSHING UP

To understand this chapter, you may find it helpful to review the following concepts:

- Brush borders and microvilli (p. 59)
- The embryonic disc and germ layers (p. 110)
- The autonomic nervous system (chapter 16)
- The celiac and mesenteric blood circulation (pp. 604–606)

Some of the most fundamental facts we know about digestion date to a grave accident that occurred in 1822. A Canadian fur trapper, Alexis St. Martin, was accidentally hit by a shotgun blast while standing outside a trading post on Mackinac Island, Michigan. An Army doctor summoned to the scene, William Beaumont, found part of St. Martin's lung protruding through the wound, and a hole in the stomach "large enough to receive my forefinger." Surprisingly, St. Martin survived, but the wound left a permanent opening (fistula) into his stomach, covered for the rest of his life by only a loose fold of tissue.

Beaumont saw this as an opportunity to learn something about digestion. Now disabled from hunting, St. Martin agreed to participate in Beamont's experiments in exchange for room and board. Working under crude frontier conditions with little idea of scientific methods, Beaumont nevertheless performed more than 200 experiments on St. Martin over a period of several years. He placed food into the stomach through the fistula and removed it hourly to observe the progress of digestion. He sent vials of gastric juice to chemists for analysis. He proved that digestion required hydrochloric acid and an unknown agent we now know to be the enzyme pepsin. St. Martin had a short temper and during his outbursts, Beamont observed that little digestion occurred; we now know this to be due to the inhibitory effect of the sympathetic nervous system on digestion.

In 1833, Beamont published a book on his results that laid a foundation for modern **gastroenterology,**[1] the scientific study and medical treatment of the digestive system. Many authorities continued to believe, for a time, that the stomach acted essentially as a grinding chamber, fermentation vat, or cooking pot. Some of them even attributed digestion to a supernatural spirit in the stomach. Beaumont was proven right, however, and his work had no equal until Russian physiologist Ivan Pavlov built upon it, receiving a Nobel Prize in 1904 for his studies of digestion.

Digestive Processes and General Anatomy

Objectives

When you have completed this section, you should be able to

- identify the functions and major processes of the digestive system;
- list the regions and accessory organs of the digestive system;
- identify the layers of the wall of the digestive tract;
- describe the enteric nervous system; and
- name the mesenteries and describe their relationship to the digestive system.

The digestive system is essentially a disassembly line—its primary purpose is to break nutrients down into forms that can be used by the body, and to absorb them so they can be distributed to the tissues. Most of what we eat cannot be used in the form found in the food. Nutrients must be broken down into smaller components, such as amino acids and monosaccharides, that are universal to all species. Consider what happens if you eat a piece of beef, for example. The myosin of beef differs very little from that of your own muscles, but the two are not identical, and even if they were, beef myosin could not be absorbed, transported in the blood, and incorporated into your muscles. Like any other dietary protein, it must be broken down into amino acids before it can be used. Since beef and human proteins are made of the same 20 amino acids, those of beef proteins might indeed become part of your own myosin but could equally well wind up in your insulin, fibrinogen, collagen, or any other protein.

Digestive System Functions

The **digestive system** is the organ system that processes food, extracts nutrients from it, and eliminates the residue. It does this in five stages:

1. **ingestion,** the selective intake of food;
2. **digestion,** the mechanical and chemical breakdown of food into a form usable by the body;
3. **absorption,** the uptake of nutrients into the blood and lymph;
4. **compaction,** absorbing water and consolidating the indigestible residue into feces; and finally
5. **defecation,** the elimination of feces.

General Anatomy

The digestive system has two anatomical subdivisions, the digestive tract and the accessory organs (fig. 24.1). The **digestive tract** is a muscular tube extending from mouth to anus, measuring about 9 m (30 ft) long in the cadaver. It is also known as the *alimentary*[2] *canal* or *gut.* It includes the oral cavity, pharynx, esophagus, stomach, small intestine, and large intestine. The stomach and intestines constitute the *gastrointestinal (GI) tract.* The **accessory organs** are the teeth, tongue, salivary glands, liver, gallbladder, and pancreas.

The digestive tract is open to the environment at both ends. Most of the material in it has not entered any body tissues and is considered to be external to the body until it is absorbed by epithelial cells of the alimentary canal. In the strict sense, defecated food residue was never in the body.

Most of the digestive tract follows the basic structural plan shown in figure 24.2, with a wall composed of the following tissue layers, in order from the inner to the outer surface:

- Mucosa
  - Epithelium
  - Lamina propria
  - Muscularis mucosae
- Submucosa
- Muscularis externa
  - Circular layer
  - Longitudinal layer
- Serosa
  - Areolar tissue
  - Mesothelium

[1]*gastro* = stomach + *entero* = intestine + *logy* = study of

[2]*aliment* = food

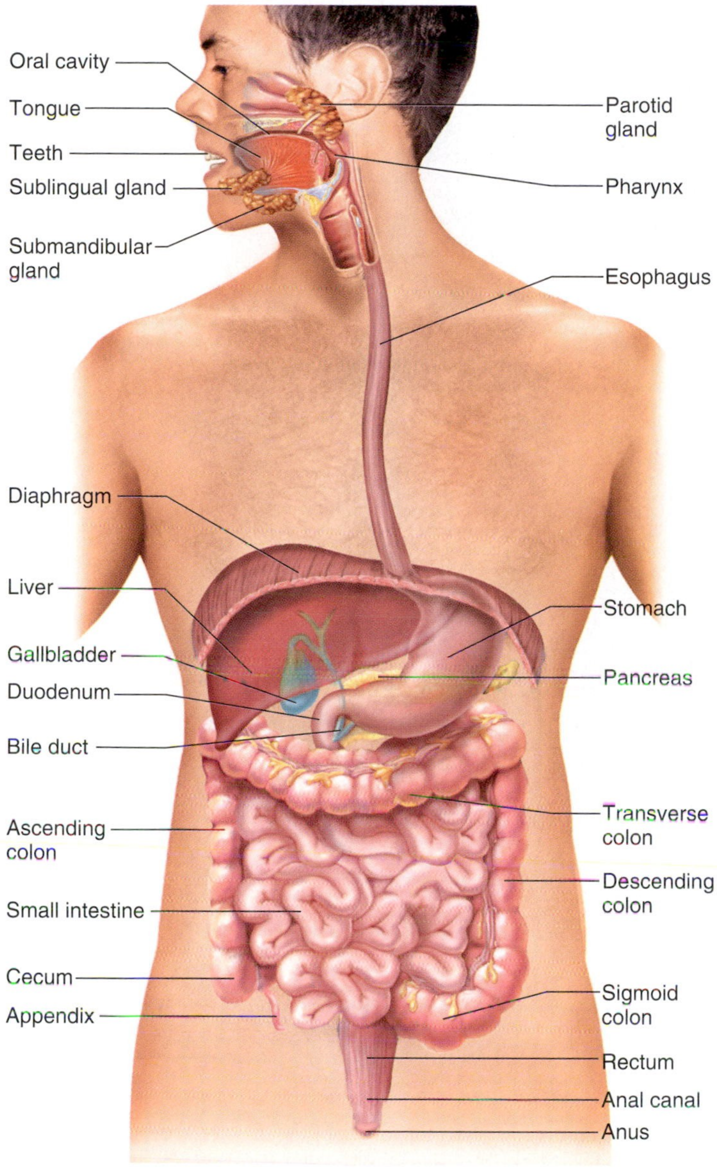

FIGURE 24.1
The Digestive System.

This basic pattern extends from the esophagus through the anal canal, but slight variations on the theme are found in different regions of the tract.

The **mucosa (mucous membrane),** lining the lumen, consists of an inner epithelium, a loose connective tissue layer called the **lamina propria,** and a thin layer of smooth muscle called the **muscularis mucosae** (MUSS-cue-LAIR-is mew-CO-see). The epithelium is simple columnar in most of the digestive tract, but the oral cavity, pharynx, esophagus, and anal canal differ. These upper and lower ends of the digestive tract are subject to more abrasion than the stomach and intestines, and thus have a nonkeratinized stratified squamous epithelium.The mucosa often exhibits an abundance of lymphocytes and lymphatic nodules—the *mucosa-associated lymphatic tissue (MALT)* (see chapter 22).

The **submucosa** is a thicker layer of loose connective tissue containing blood vessels, lymphatic vessels, a nerve plexus, and in some places, mucous glands. The MALT also extends into the submucosa in some parts of the GI tract.

The **muscularis externa** consists of usually two layers of smooth muscle near the outer surface. Cells of the inner layer encircle the tract while those of the outer layer run longitudinally. In some places, the circular layer is thickened to form valves (sphincters) that regulate the passage of material through the digestive tract.

The **serosa** is composed of a thin layer of areolar tissue topped by a simple squamous mesothelium. The serosa begins in the lower 3 to 4 cm of the esophagus and ends with the sigmoid colon. The oral cavity, pharynx, most of the esophagus, and the rectum are surrounded by a fibrous connective tissue layer called the **adventitia** (AD-ven-TISH-ah), which blends into the adjacent connective tissues of other organs.

Innervation

Tongue movements, mastication, and the initial actions of swallowing employ skeletal muscles innervated by somatic motor fibers in six of the cranial nerves (V, VII, and IX–XII) and in the ansa cervicalis; these muscles and their innervation are detailed in table 11.3. The salivary glands are innervated by sympathetic fibers from the superior cervical ganglion and parasympathetic fibers in cranial nerves VII and IX (see figs. 15.30 and 16.4).

From the lower esophagus to the anal canal, most of the muscle is smooth muscle (the external anal sphincter is the only exception), and therefore receives only autonomic innervation. Parasympathetic innervation dominates the digestive tract and comes mainly from the vagus nerves, which supply all of the tract from esophagus to transverse colon. The descending colon and rectum receive their parasympathetic innervation from pelvic nerves arising from the inferior hypogastric plexus (see fig. 16.7). The parasympathetic nervous system relaxes sphincter muscles and stimulates gastrointestinal motility and secretion. Thus, in general, it promotes digestion.

The sympathetic nervous system plays a lesser role, but in general it inhibits motility and secretion and keeps the GI sphincters contracted and closed. Thus, it inhibits digestion. Sympathetic efferent pathways travel through the celiac ganglion to the stomach, liver, and pancreas; through the superior mesenteric ganglion to the small intestine and most of the large intestine; and through the inferior mesenteric ganglion to the rectum (see fig. 16.4).

For all this autonomic innervation received from the CNS, the digestive tract can function independently even if these nerves are severed. This is because the esophagus, stomach, and intestines have their own extensive nervous network called the **enteric**[3] **nervous system.** The enteric nervous system is thought to have over 100 million neurons—more than the spinal cord. These include sensory neurons that monitor tension in the gut wall; sympathetic postganglionic nerve fibers; and parasympathetic ganglia and postganglionic fibers. These neurons are

[3]*enter* = intestine

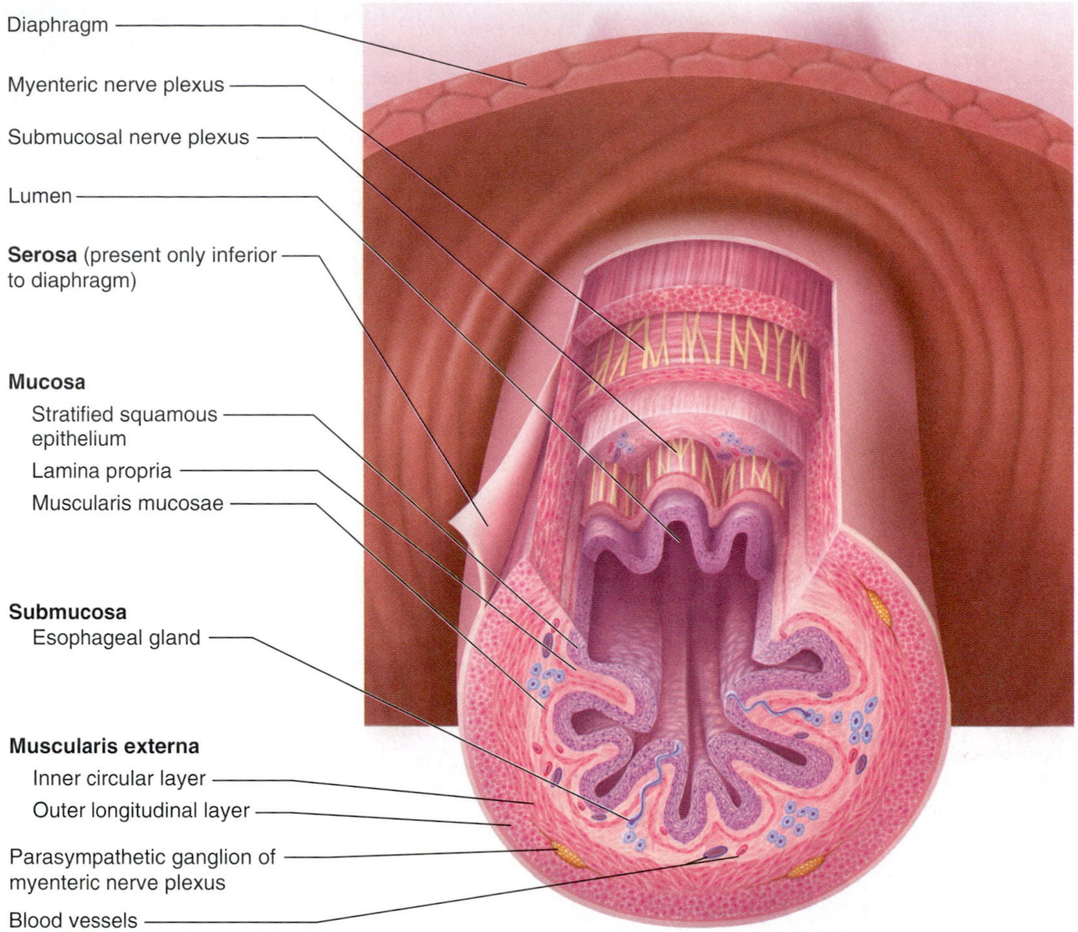
Diaphragm
Myenteric nerve plexus
Submucosal nerve plexus
Lumen
Serosa (present only inferior to diaphragm)
Mucosa
Stratified squamous epithelium
Lamina propria
Muscularis mucosae
Submucosa
Esophageal gland
Muscularis externa
Inner circular layer
Outer longitudinal layer
Parasympathetic ganglion of myenteric nerve plexus
Blood vessels

FIGURE 24.2

Tissue Layers of the Digestive Tract. A cross section of the esophagus just below the diaphragm, serving as an example of the most typical histological structure of the tract.

distributed in two networks: the **submucosal (Meissner[4]) plexus** in the submucosa and the **myenteric (Auerbach[5]) plexus** between the two layers of the muscularis externa. Parasympathetic preganglionic fibers terminate on the ganglia of the myenteric plexus. Postganglionic fibers arising in this plexus not only innervate the muscularis externa, but also pass through its inner circular layer and contribute to the submucosal plexus. The myenteric plexus controls peristalsis and other contractions of the muscularis externa, while the submucosal plexus controls movements of the muscularis mucosae and glandular secretion of the mucosa.

Sensory nerve fibers monitor stretching of the GI wall and chemical conditions in the lumen. These fibers carry signals to adjacent regions of the GI tract in **short (myenteric) reflex arcs** contained in the myenteric plexus, and to the central nervous system by way of **long (vagovagal) reflex arcs,** predominantly in the vagus nerves. These visceral reflex arcs enable different regions of the GI tract to regulate each other over both short and long distances.

Circulation

We will see near the end of this chapter that the embryonic digestive tract forms in three segments: the foregut, midgut, and hindgut. These segments are defined by their arterial blood supply (see table 21.5, pp. 604–606).

- The **foregut** includes the oral cavity, pharynx, esophagus, stomach, and the beginning of the duodenum (to the point where the bile duct empties into it). Above the diaphragm, the thoracic aorta gives off a series of **esophageal arteries** to the esophagus (see fig. 21.19). Below the diaphragm, the foregut components receive their blood from branches of the **celiac trunk** (see fig. 21.21).
- The **midgut** begins at the opening of the bile duct and includes the rest of the duodenum, the jejunum and ileum (the second and third portions of the small intestine), and the large

[4]Georg Meissner (1829–1905), German histologist
[5]Leopold Auerbach (1828–97), German anatomist

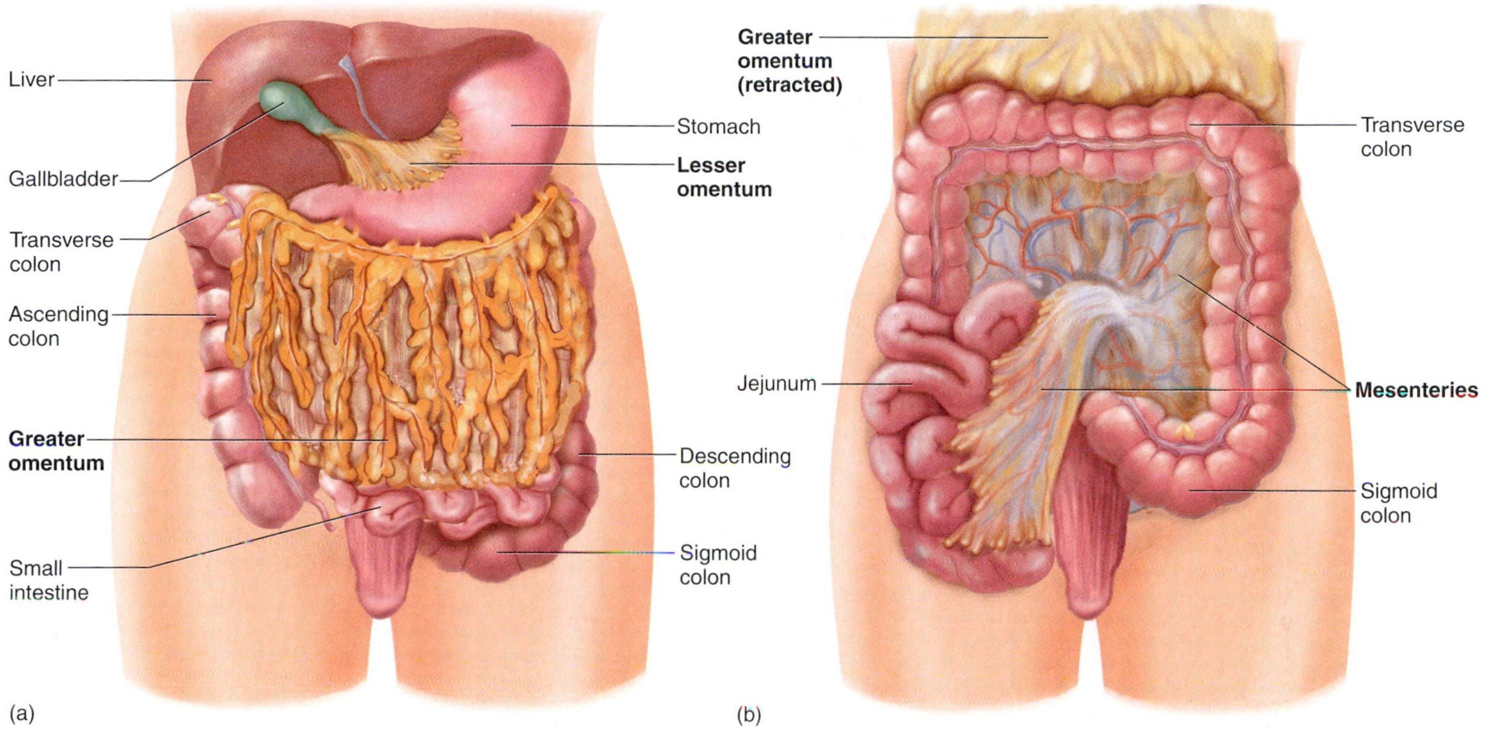
Liver
Gallbladder
Transverse colon
Ascending colon
Greater omentum
Small intestine
Stomach
Lesser omentum
Descending colon
Sigmoid colon
(a)
Greater omentum (retracted)
Jejunum
Transverse colon
Mesenteries
Sigmoid colon
(b)

FIGURE 24.3
Serous Membranes Associated with the Digestive Tract. (*a*) The greater and lesser omenta. (*b*) Greater omentum and small intestine retracted to show the mesenteries. These membranes contain the mesenteric arteries and veins.

intestine as far as the first two-thirds of the transverse colon. It receives blood from the **superior mesenteric artery** (see figs. 21.21 and 21.22).

- The **hindgut** includes the remainder of the large intestine, from the end of the transverse colon through the anal canal. It is supplied by branches of the **inferior mesenteric artery** (see fig. 21.22).

The most noteworthy general point about the venous drainage of the GI tract is that blood from the entire tract below the diaphragm ultimately drains into the **hepatic portal vein,** which enters the liver. The system of vessels connecting the lower digestive tract to the liver is the **hepatic portal system** (table 21.11, p. 616). It routes all blood from the stomach and intestines, as well as from some other abdominal viscera, through the liver before returning it to the general circulation. Like other portal systems, this one has two capillary networks in series. Capillaries in the small intestine receive digested nutrients, and capillaries in the liver (the *hepatic sinusoids* described later) deliver these nutrients to the liver cells. This gives the liver a chance to process most nutrients and cleanse the intestinal blood of bacteria.

Relationship to the Peritoneum

In processing food, the stomach and intestines undergo such strenuous contractions that they need freedom to move in the abdominal cavity. Thus, they are not tightly bound to the abdominal wall, but over most of its length, the tract is loosely suspended from it by connective tissue sheets called the **mesenteries** (see figs. A.9 and A.10 in atlas A). The mesenteries also hold the abdominal viscera in their proper relationship to each other and prevent the small intestine, especially, from becoming twisted and tangled by changes in body position and by the intestine's own contractions. Furthermore, the mesenteries provide passage for the blood vessels and nerves that supply the digestive tract, and contain many lymph nodes and lymphatic vessels.

Along the dorsal midline of the abdominal cavity, the parietal peritoneum turns inward and forms the **dorsal mesentery,** a translucent two-layered membrane extending to the digestive tract. Upon reaching an organ such as the stomach or small intestine, the two layers of the mesentery separate and pass around opposite sides of the organ, forming the serosa. In some places, the two layers come together again on the far side of that organ and continue as another sheet of tissue, the **ventral mesentery.** The ventral mesentery may hang freely in the abdominal cavity or attach to the ventral abdominal wall or other organs. The relationship between the dorsal and ventral mesenteries and the serosa are shown in figure A.9.

Along the right superior margin *(lesser curvature)* of the stomach, a ventral mesentery called the **lesser omentum** extends from the stomach to the liver (fig. 24.3). Another membrane, a fatty **greater omentum,** hangs from the left inferior margin *(greater curvature)* of the stomach and loosely covers the small intestine like an apron. At its inferior margin, the greater omentum turns back on itself and passes upward, thus forming a deep pouch between its deep and superficial layers. At its inner superior margin, it continues as a serosa enclosing the spleen and transverse colon, then continues still farther as the **mesocolon,** which anchors the transverse colon to the dorsal abdominal wall. The omenta have a loosely organized, lacy appearance due partly to many holes or gaps in the membranes and partly to an irregular distribution of adipose tissue. They adhere to perforations or inflamed areas of the stomach or intestines, contribute immune cells to the site, and isolate infections that might otherwise give rise to peritonitis.

When an organ is enclosed by mesentery (serosa) on both sides, it is considered to be within the peritoneal cavity, or **intraperitoneal.**[6] When an organ lies against the dorsal body wall and is covered by peritoneum on the ventral side only, it is said to be outside the peritoneal cavity, or **retroperitoneal.**[7] The duodenum, most of the pancreas, and parts of the large intestine are retroperitoneal. The stomach, liver, and other parts of the small and large intestines are intraperitoneal.

Before You Go On

Answer the following questions to test your understanding of the preceding section:

1. Name the principal regions of the digestive tract in order from mouth to anus.
2. What are the similarities and differences between the lamina propria and the submucosa?
3. What are the two components of the enteric nervous system? How do they differ in location and function?
4. What three major branches of the aorta supply the digestive tract? What is the relationship between these three arterial supplies and the embryonic development of the digestive tract?
5. Name the serous membranes that suspend the intestines from the dorsal body wall. Name the external layer of the intestines formed by an extension of this membrane.

The Mouth Through Esophagus

Objectives

When you have completed this section, you should be able to

- describe the teeth, tongue, and other organs of the oral cavity;
- state the names and locations of the salivary glands;
- describe the location and function of the pharyngeal constrictor muscles; and
- describe the gross anatomy and histology of the esophagus.

The Oral Cavity

The mouth is also known as the **oral cavity (buccal[8] cavity).** Its functions include ingestion (food intake), taste and other sensory responses to food, chewing, chemical digestion, swallowing, speech, and respiration. The mouth is enclosed by the cheeks, lips, palate, and tongue (fig. 24.4). Its anterior opening between the lips is the **oral orifice** and its posterior opening into the throat is the **fauces**[9] (FAW-seez). The oral cavity is lined with nonkeratinized stratified squamous epithelium.

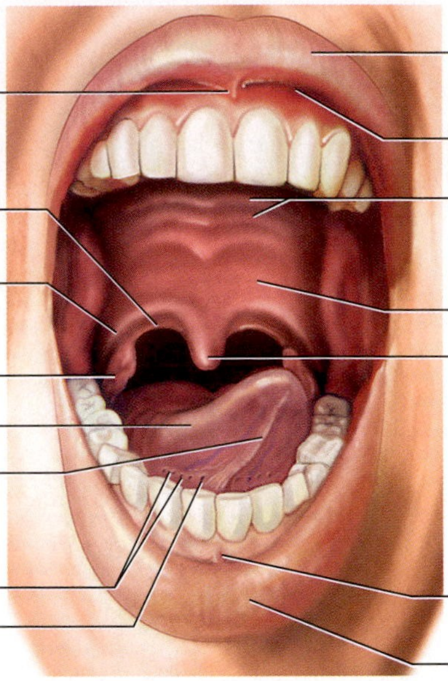

FIGURE 24.4
The Oral Cavity. For a photographic medial view see figure A.17 (atlas A).

THE CHEEKS AND LIPS

The cheeks and lips define the anterior and lateral limits of the mouth. They retain food and push it between the teeth for mastication, and are essential for articulate speech and for sucking and blowing actions, including suckling by infants. Their fleshiness is due mainly to subcutaneous fat, the buccinator muscles of the cheeks, and the orbicularis oris muscle of the lips. A median fold called the **labial frenulum**[10] attaches each lip to the gum, between the anterior incisors. The **vestibule** is the space between the cheeks or lips and the teeth—the space where you insert your toothbrush when brushing the outer surfaces of the teeth.

The lips are divided into three areas: (1) The *cutaneous area* is colored like the rest of the face and has hair follicles and sebaceous glands; on the upper lip, this is where a mustache grows. (2) The *red area (vermilion),* is the hairless region where the lips meet (where some people apply lipstick). It has unusually tall dermal papillae, which allow blood capillaries and nerve endings to come closer to the epidermal surface. Thus, this area is redder and more sensitive than the cutaneous area. (3) The *labial mucosa* is the inner surface of the lip, facing the gums and teeth.

THE TONGUE

The tongue (fig. 24.5), although muscular and bulky, is an agile and sensitive organ with several functions: it aids in food intake; its senses of taste, texture, and temperature are important in the

[6]*intra* = within
[7]*retro* = behind
[8]*bucca* = cheek
[9]*fauces* = throat
[10]*labi* = lip + *frenulum* = little bridle

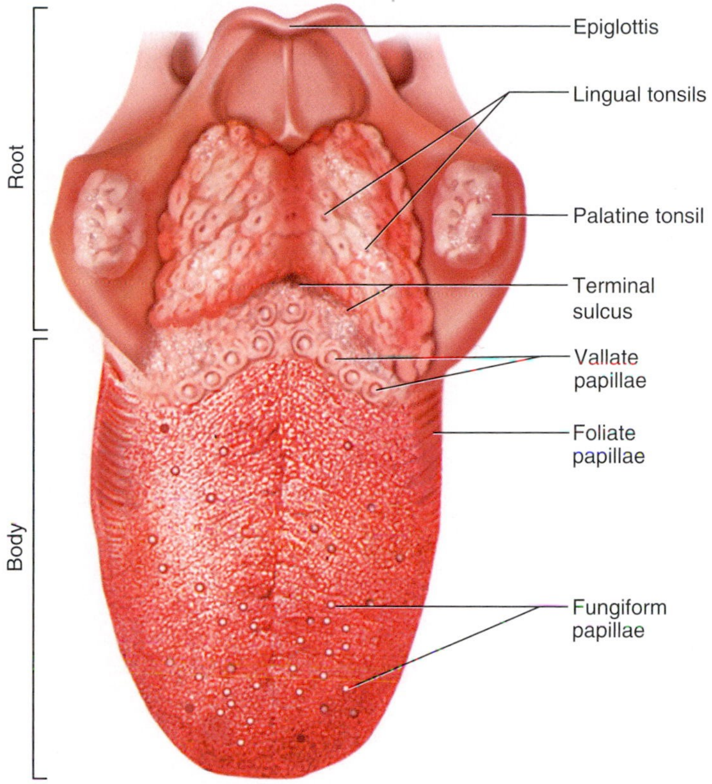
Root
Body
Epiglottis
Lingual tonsils
Palatine tonsil
Terminal sulcus
Vallate papillae
Foliate papillae
Fungiform papillae

FIGURE 24.5
The Tongue.

acceptance or rejection of food; it compresses and breaks up food; it maneuvers food between the teeth for mastication; it secretes mucus and enzymes; it compresses the chewed food into a soft mass, or *bolus,* that is easier to swallow; it initiates swallowing; and it is necessary for articulate speech. Its surface is covered with nonkeratinized stratified squamous epithelium and exhibits bumps and projections called **lingual papillae,** the site of the taste buds. The types of papillae and sense of taste are discussed in chapter 17.

The anterior two-thirds of the tongue, called the **body,** occupies the oral cavity, while the posterior one-third, the **root,** occupies the oropharynx. The boundary between the body and root is marked by a V-shaped row of *vallate papillae* and, behind these, a groove called the **terminal sulcus.** A ventral median fold called the **lingual frenulum** attaches the body of the tongue to the floor of the mouth. The root of the tongue contains the lingual tonsils. Amid the tongue muscles are serous and mucous **lingual glands,** which secrete a portion of the saliva.

The muscles of the tongue, which compose most of its mass, are described in chapter 11. The **intrinsic muscles,** contained entirely within the tongue, produce the relatively subtle tongue movements of speech. The **extrinsic muscles,** with origins elsewhere and insertions in the tongue, produce the stronger movements of food manipulation. The extrinsic muscles include the *genioglossus, hyoglossus, palatoglossus,* and *styloglossus* (see table 11.3).

THE PALATE

The palate, separating the oral cavity from the nasal cavity, makes it possible to breathe while chewing food. Its anterior portion, the **hard (bony) palate,** is supported by the palatine processes of the maxillae and by the smaller palatine bones. It has transverse *friction ridges (palatal rugae)* that aid the tongue in holding and manipulating food. Posterior to this is the **soft palate,** which has a more spongy texture and is composed mainly of skeletal muscle and glandular tissue, but no bone. It has a conical medial projection, the **uvula,**[11] visible at the rear of the oral cavity. The uvula helps to retain food in the oral cavity until one is ready to swallow.

A pair of muscular arches on each side of the oral cavity begin dorsally near the uvula and follow the wall of the cavity to its floor. The anterior one is the **palatoglossal arch** and the posterior one is the **palatopharyngeal arch.** The latter arch marks the beginning of the pharynx. The palatine tonsils are located on the wall between the arches.

THE TEETH

The teeth are collectively called the **dentition.** They serve to *masticate* the food, breaking it into smaller pieces that are easier to swallow and have more surface area exposed to the action of digestive enzymes. Adults normally have 16 teeth in the mandible and 16 in the maxilla. From the midline to the rear of each jaw, there are two incisors, a canine, two premolars, and up to three molars (fig. 24.6*a*). The **incisors** are chisel-like cutting teeth used to bite off a piece of food. The **canines** are more pointed and act to puncture and shred it. They serve as weapons in many mammals but became reduced in the course of human evolution until they now project barely above the other teeth. The **premolars** and **molars** have relatively broad surfaces adapted for crushing and grinding.

Each tooth is embedded in a socket called an **alveolus,** forming a joint called a *gomphosis* between the tooth and bone (fig. 24.7). The alveolus is lined by a **periodontal** (PERR-ee-oh-DON-tul) **ligament,** a modified periosteum whose collagen fibers penetrate into the bone on one side and into the tooth on the other. This anchors the tooth firmly in the alveolus, but allows for a slight amount of movement under the pressure of chewing. The gum, or **gingiva** (JIN-jih-vuh), covers the alveolar bone. Regions of a tooth are defined by their relationship to the gingiva: the **crown** is the portion above the gum, the **root** is the portion inserted into the alveolus below the gum, and the **neck** is the line where the crown, root, and gum meet. The space between the tooth and gum is the **gingival sulcus.** The hygiene of this sulcus is especially important to dental health (see insight 24.1).

Most of a tooth consists of hard yellowish tissue called **dentin,** covered with **enamel** in the crown and **cementum** in the root. Dentin and cementum are living connective tissues with cells or cell processes embedded in a calcified matrix. Cells of the cementum *(cementocytes)* are scattered more or less randomly and occupy tiny cavities similar to the lacunae of bone. Cells of the dentin *(odontoblasts)* line the pulp cavity and have slender

[11] *uvula* = little grape

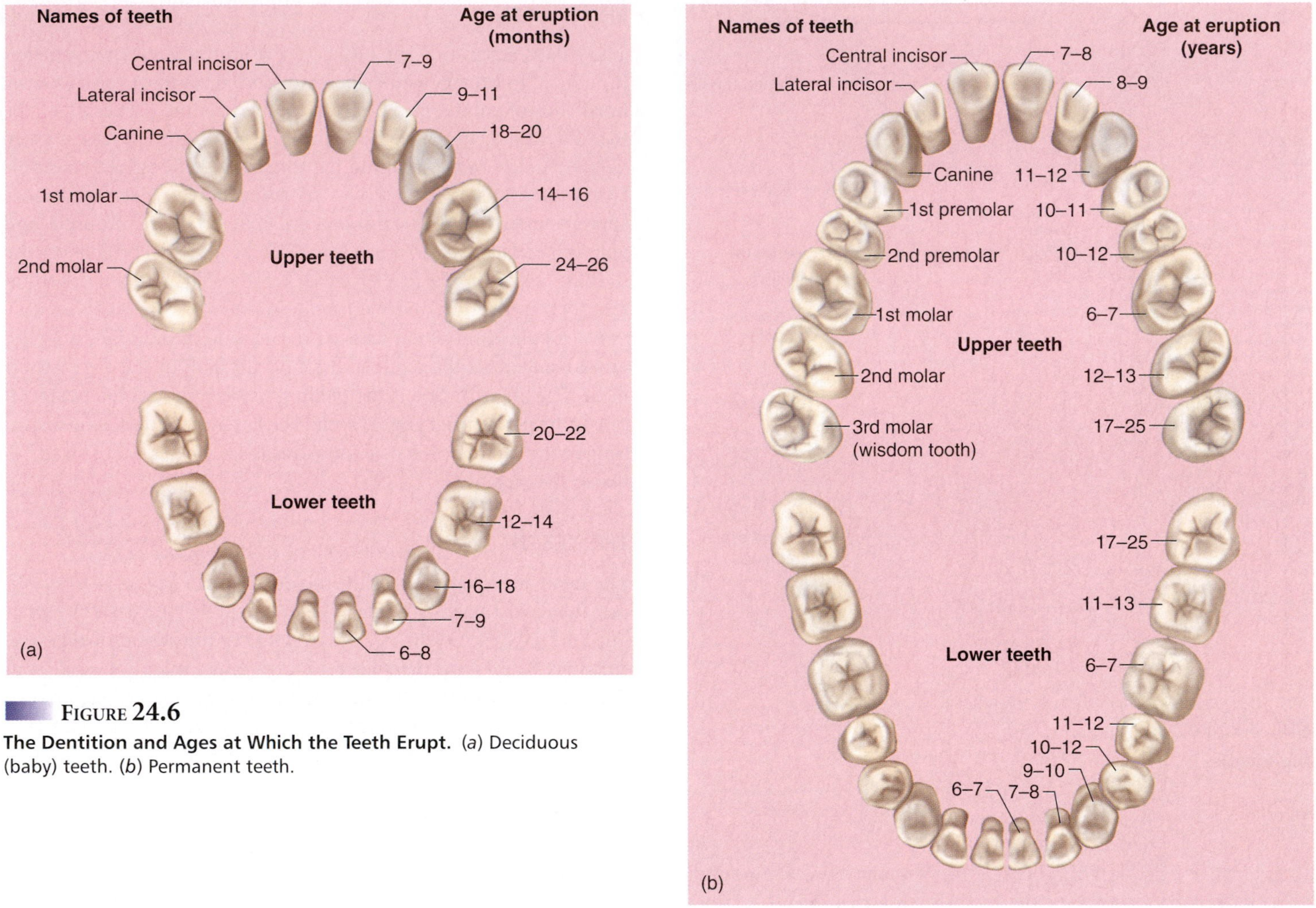

Names of teeth
Age at eruption (months)
Central incisor
7–9
Lateral incisor
9–11
Canine
18–20
1st molar
14–16
2nd molar
24–26
Upper teeth
20–22
Lower teeth
12–14
16–18
7–9
6–8
(a)
Names of teeth
Age at eruption (years)
Central incisor
7–8
Lateral incisor
8–9
Canine
11–12
1st premolar
10–11
2nd premolar
10–12
1st molar
6–7
Upper teeth
2nd molar
12–13
3rd molar (wisdom tooth)
17–25
17–25
11–13
Lower teeth
6–7
11–12
10–12
9–10
6–7
7–8
(b)

FIGURE 24.6

The Dentition and Ages at Which the Teeth Erupt. (*a*) Deciduous (baby) teeth. (*b*) Permanent teeth.

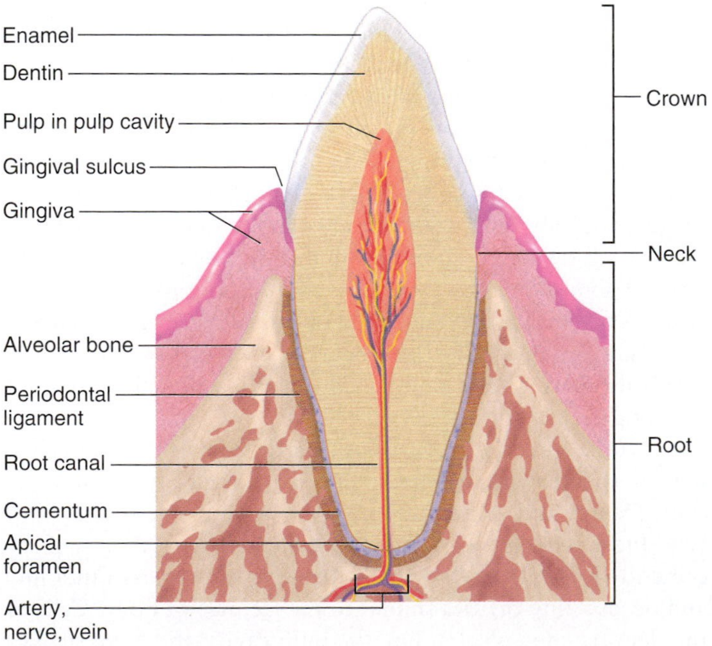

Enamel
Dentin
Pulp in pulp cavity
Gingival sulcus
Gingiva
Alveolar bone
Periodontal ligament
Root canal
Cementum
Apical foramen
Artery, nerve, vein
Crown
Neck
Root

FIGURE 24.7

Median Section of a Canine Tooth and Its Alveolus. Shows typical anatomy of a tooth and periodontal tissues.

INSIGHT 24.1 CLINICAL APPLICATION

TOOTH AND GUM DISEASE

Food leaves a sticky residue on the teeth called *plaque,* composed mainly of bacteria and sugars. If plaque is not thoroughly removed by brushing and flossing, bacteria accumulate, metabolize the sugars, and release lactic acid and other acids. These acids dissolve the minerals of enamel and dentin, and the bacteria enzymatically digest the collagen and other organic components. The eroded "cavities" of the tooth are known as *dental caries.*[12] If not repaired, caries may fully penetrate the dentin and spread to the pulp cavity. This requires either extraction of the tooth or *root canal therapy,* in which the pulp is removed and replaced with inert material.

When plaque calcifies on the tooth surface, it is called *calculus (tartar).* Calculus in the gingival sulcus wedges the tooth and gum apart and allows bacterial invasion of the sulcus. This leads to *gingivitis,* or gum inflammation. Nearly everyone has gingivitis at some time. In some cases, bacteria spread from the sulcus into the alveolar bone and begin to dissolve it, producing *periodontal disease.* About 86% of people over age 70 have periodontal disease and many suffer tooth loss as a result. This accounts for 80% to 90% of adult tooth loss.

[12]*caries* = rottenness

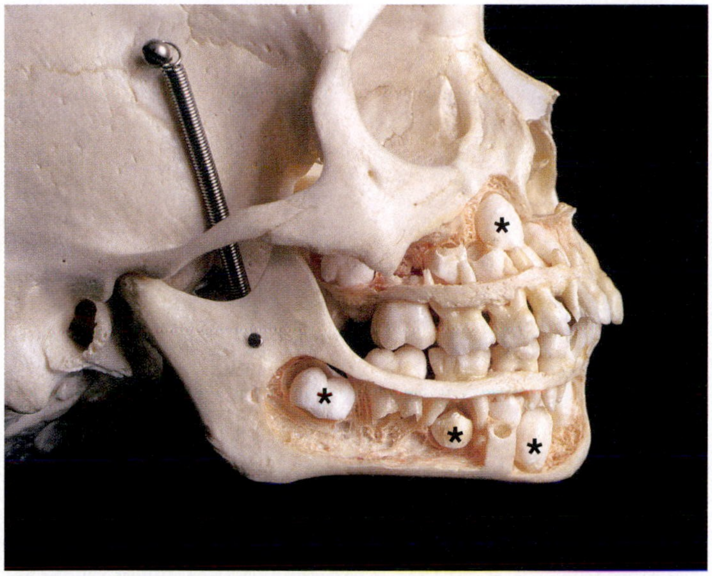

FIGURE 24.8
Permanent and Deciduous Teeth in a Child's Skull. This dissection shows erupted deciduous teeth and, below them and marked with asterisks, the permanent teeth waiting to erupt.

processes that travel through tiny parallel tubules in the dentin. Enamel is not a tissue but a noncellular secretion produced before the tooth erupts. Damaged dentin and cementum can regenerate, but damaged enamel cannot—it must be artificially repaired.

Internally, a tooth has a dilated **pulp cavity** in the crown and a narrow **root canal** in the root. These spaces are occupied by **pulp**—a mass of loose connective tissue, blood and lymphatic vessels, and nerves. These nerves and vessels enter the tooth through a pore, the **apical foramen,** at the inferior end of each root canal.

The surfaces at which the upper and lower teeth meet when the mouth closes are called the **occlusal surfaces.** The occlusal surface of a premolar has two rounded bumps called **cusps;** thus the premolars are also known as **bicuspids.** The molars have four to five cusps. Cusps of the upper and lower premolars and molars mesh when the jaws are closed and slide over each other as the jaw makes lateral chewing motions. This grinds and tears food more effectively than if the occlusal surfaces were flat.

Teeth develop beneath the gums and **erupt** (emerge) in predictable order. Twenty **deciduous teeth** (*milk teeth* or *baby teeth*) erupt from the ages of 6 to 30 months, beginning with the incisors (see fig. 24.6*a*). Between 6 and 25 years of age, these are replaced by 32 **permanent teeth.** As a permanent tooth grows below a deciduous tooth (fig. 24.8), the root of the deciduous tooth dissolves and leaves little more than the crown by the time it falls out. The third molars (wisdom teeth) erupt around ages 17 to 25, if at all. Over the course of human evolution, the face became flatter and the jaws shorter, leaving little room for the third molars. Thus, they often remain below the gum and become *impacted*—so crowded against neighboring teeth and bone that they cannot erupt.

The Salivary Glands

Saliva moistens the mouth, digests a small amount of starch and fat, cleanses the teeth, inhibits bacterial growth, dissolves molecules so they can stimulate the taste buds, and moistens food and binds particles together to aid in swallowing. It is a solution of 97.0% to 99.5% water and the following solutes:

- **salivary amylase,** an enzyme that begins starch digestion in the mouth;
- **lingual lipase,** an enzyme that is activated by stomach acid and digests fat after the food is swallowed;
- **mucus,** which binds and lubricates the food mass and aids in swallowing;
- **lysozyme,** an enzyme that kills bacteria;
- **immunoglobulin A (IgA),** an antibody that inhibits bacterial growth; and
- **electrolytes,** including sodium, potassium, chloride, phosphate, and bicarbonate ions.

There are two kinds of salivary glands, intrinsic and extrinsic. The **intrinsic salivary glands** are an indefinite number of small glands dispersed amid the other oral tissues. They include *lingual glands* in the tongue, *labial glands* on the inside of the lips, and *buccal glands* on the inside of the cheeks. They secrete saliva at a fairly constant rate whether we are eating or not, but in relatively small amounts.

The **extrinsic salivary glands** are three pairs of larger, more discrete organs located outside of the oral mucosa (fig. 24.9):

1. The **parotid**[13] **gland** is located just beneath the skin anterior to the earlobe. Its duct passes superficially over the masseter, pierces the buccinator, and opens into the mouth opposite the second upper molar tooth. *Mumps* is an inflammation and swelling of the parotid gland caused by a virus.
2. The **submandibular gland** is located halfway along the body of the mandible, medial to its margin, just deep to the mylohyoid muscle. Its duct empties into the mouth at a papilla on the side of the lingual frenulum, near the lower central incisors.
3. The **sublingual gland** is located in the floor of the mouth. It has multiple ducts that empty into the mouth posterior to the papilla of the submandibular duct.

These are all compound tubuloacinar glands with a treelike arrangement of branching ducts ending in acini (see chapter 3). Some acini have only mucous cells, some have only serous cells, and some have a mixture of both (fig. 24.10). Mucous cells secrete salivary mucus, and serous cells secrete a thinner fluid rich in amylase and electrolytes.

[13]*par* = next to + *ot* = ear

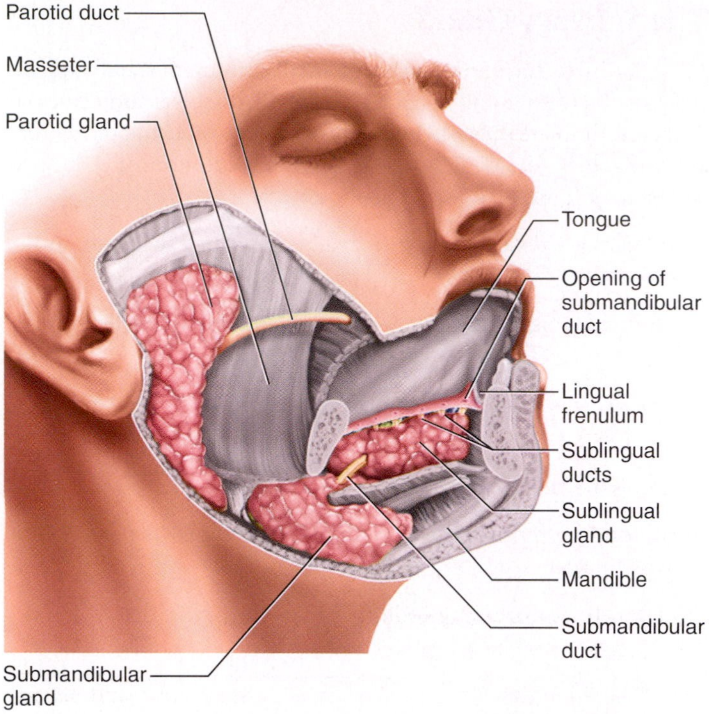

FIGURE 24.9

The Extrinsic Salivary Glands. Part of the mandible has been removed to expose the sublingual gland medial to it.

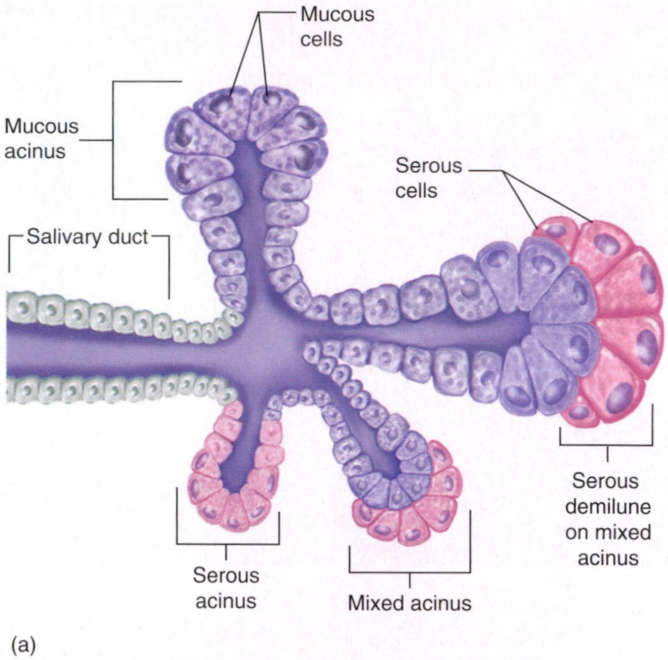

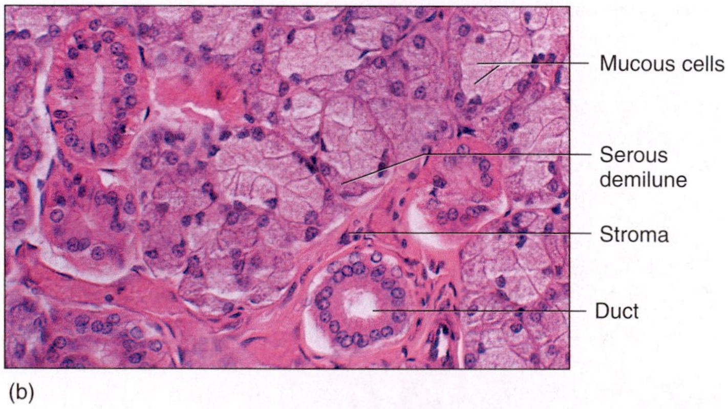

FIGURE 24.10

Microscopic Anatomy of the Salivary Glands. (*a*) Duct and acini of a generalized salivary gland with a mixture of mucous and serous cells. Serous cells often form crescent-shaped caps called *serous demilunes* over the ends of mucous acini. No one salivary gland shows all the features shown here. (*b*) Histology of the sublingual salivary gland.

Salivation is controlled by groups of neurons called **salivatory nuclei** in the medulla oblongata and pons. They receive signals from sensory receptors in the mouth as well as from higher brain centers that respond to the odor, sight, or thought of food. Efferent nerve pathways to the salivary glands were described earlier. Salivation is mostly under parasympathetic control, but the sympathetic nerves stimulate mucus secretion.

The Pharynx

The pharynx, as described in chapter 23, consists of three regions called the nasopharynx, oropharynx, and laryngopharynx (fig. 23.2*b*). The first is exclusively respiratory and is lined with pseudostratified columnar epithelium; the last two are shared by the respiratory and digestive tracts and are lined with nonkeratinized stratified squamous epithelium, an adaptation to withstanding abrasion by passing food.

The pharynx has a deep layer of longitudinally oriented skeletal muscle and a superficial layer of circular skeletal muscle. The circular muscle is divided into **superior, middle,** and **inferior pharyngeal constrictors,** which force food downward during swallowing. When one is not swallowing, the inferior constrictor remains contracted to exclude air from the esophagus. This constriction is regarded as the **upper esophageal sphincter,** although it is not an anatomical feature of the esophagus. It disappears at the time of death when the muscle relaxes. Thus it is regarded as a *physiological sphincter* rather than a constant anatomical structure.

The Esophagus

The **esophagus** is a straight muscular tube 25 to 30 cm long (see figs. 24.1 and 24.2), dorsal to the trachea. Its superior opening lies between vertebra C6 and the cricoid cartilage of the larynx. After passing downward through the mediastinum, the esophagus penetrates the diaphragm at an opening called the **esophageal hiatus,** continues another 3 to 4 cm, and meets the stomach at the level of vertebra T7. Its opening into the stomach is called the **cardiac orifice** (for its proximity to the heart). Food pauses briefly at this point before entering the stomach, because of a constriction called the **lower esophageal sphincter (LES).** The LES is also a physiological rather than an anatomical sphincter, and thus is not found in the cadaver. It is thought to be either a constriction of the diaphragm surrounding the esophageal hiatus, or muscle tone in the

INSIGHT 24.2 CLINICAL APPLICATION

Gastroesophageal Reflux Disease

It would seem that churning of the stomach would drive its contents back up the esophagus. Such backflow, or *gastroesophageal reflux,* is normally prevented, however, by the lower esophageal sphincter. Weakening of the LES leads to repetitive or chronic reflux, called *gastroesophageal reflux disease (GERD).* Stomach acid and sometimes bile acids and pancreatic enzymes regurgitate into the esophagus and irritate the mucosa. This causes the sensation of "heartburn," so-named for its location although it has nothing to do with the heart. GERD affects as much as 50% of the population in the United States, especially white males. Besides sex and race, risk factors include age (middle-aged or beyond), being overweight, and going to bed too soon after eating.

The heartburn sensation can often be managed with antacids and is commonly dismissed by physicians and patients as merely a nuisance. However, in a few cases, GERD can lead to more serious complications such as scarring and narrowing of the esophagus *(stricture),* erosion and inflammation of the esophageal wall *(erosive esophagitis),* a transformation (metaplasia) of esophageal epithelium to an intestinal-type columnar epithelium *(Barrett*[14] *esophagus),* and a form of esophageal cancer called *adenocarcinoma.* Although most people with Barrett esophagus and adenocarcinoma have a long-term history of GERD, only 5% to 15% of those with GERD progress to Barrett esophagus and fewer than 0.1% to adenocarcinoma.

[14]Norman R. Barrett (1903–79), British surgeon

smooth muscle of the esophagus. The LES prevents stomach contents from regurgitating into the esophagus, thus protecting the esophageal mucosa from the corrosive effect of the stomach acid (see insight 24.2).

The wall of the esophagus is organized into the tissue layers described earlier, with some regional specializations. The mucosa has a nonkeratinized stratified squamous epithelium. The submucosa contains **esophageal glands,** which secrete lubricating mucus into the lumen. When the esophagus is empty, the mucosa and submucosa are deeply folded into longitudinal ridges, giving the lumen a starlike shape in cross section.

The muscularis externa is composed of skeletal muscle in the upper one-third of the esophagus, a mixture of skeletal and smooth muscle in the middle one-third, and only smooth muscle in the lower one-third. This transition corresponds to a shift from voluntary to involuntary phases of swallowing as food passes down the esophagus.

Most of the esophagus is in the mediastinum. Here, it is covered with a connective tissue adventitia which merges into the adventitias of the trachea and thoracic aorta. The short segment below the diaphragm is covered by a serosa.

Swallowing, or **deglutition** (DEE-glu-TISH-un), is a complex action involving over 22 muscles in the mouth, pharynx, and esophagus, coordinated by the **swallowing center,** a nucleus in the medulla oblongata and pons. This center communicates with muscles of the pharynx and esophagus by way of the trigeminal, facial, glossopharyngeal, and hypoglossal nerves (cranial nerves V, VII, IX, and XII), and coordinates a complex series of muscle contractions to produce swallowing without choking.

Before You Go On

Answer the following questions to test your understanding of the preceding section:

6. List as many functions of the tongue as you can.
7. Name the four types of teeth in order from medial to lateral. How do they differ in function?
8. What is the difference in function and location between intrinsic and extrinsic salivary glands? Name the extrinsic salivary glands and describe their locations.
9. The upper and lower esophageal sphincters cannot be found in the cadaver. Why not? What forms these sphincters and what purpose does each one serve?

The Stomach

Objectives

When you have completed this section, you should be able to

- describe the gross and microscopic anatomy of the stomach;
- describe the nerve and blood supply to the stomach;
- state the function of each type of epithelial cell in the gastric mucosa; and
- explain how the stomach is protected from digesting itself.

The stomach is a muscular sac in the upper left abdominal cavity immediately inferior to the diaphragm (see fig. 24.1). It functions primarily as a food storage organ, with an internal volume of about 50 mL when empty and 1.0 to 1.5 L after a typical meal. When extremely full, it may hold up to 4 L and extend nearly as far as the pelvis. The stomach mechanically breaks up food particles, liquefies the food, and begins the chemical digestion of proteins and a small amount of fat. This produces a soupy or pasty mixture of semidigested food called **chyme**[15] (kime). Most digestion occurs after the chyme passes on to the small intestine.

Gross Anatomy

The stomach is somewhat J-shaped (fig. 24.11), relatively vertical in tall people and more horizontal in short people (see fig. 1.7*a*). The **lesser curvature** of the stomach extends the short distance (about 10 cm) from esophagus to duodenum along the medial to superior aspect, facing the liver, while the **greater curvature** extends the longer distance (about 40 cm) from esophagus to duodenum on the lateral to inferior aspect. As described earlier, the lesser omentum extends from the lesser curvature to the liver, and the greater omentum is suspended from the greater curvature and overhangs the intestines below.

The stomach is divided into four regions: (1) The **cardiac region (cardia)** is the small area within about 3 cm of the cardiac orifice. (2) The **fundic region (fundus)** is the dome-shaped portion superior to the esophageal attachment. (3) The **body (corpus)** makes up the greatest part of the stomach inferior to the cardiac orifice. (4) The **pyloric region** is a slightly narrower pouch at the inferior end; it is subdivided into a funnel-like **antrum**[16] and a narrower

[15]*chyme* = juice
[16]*antrum* = cavity

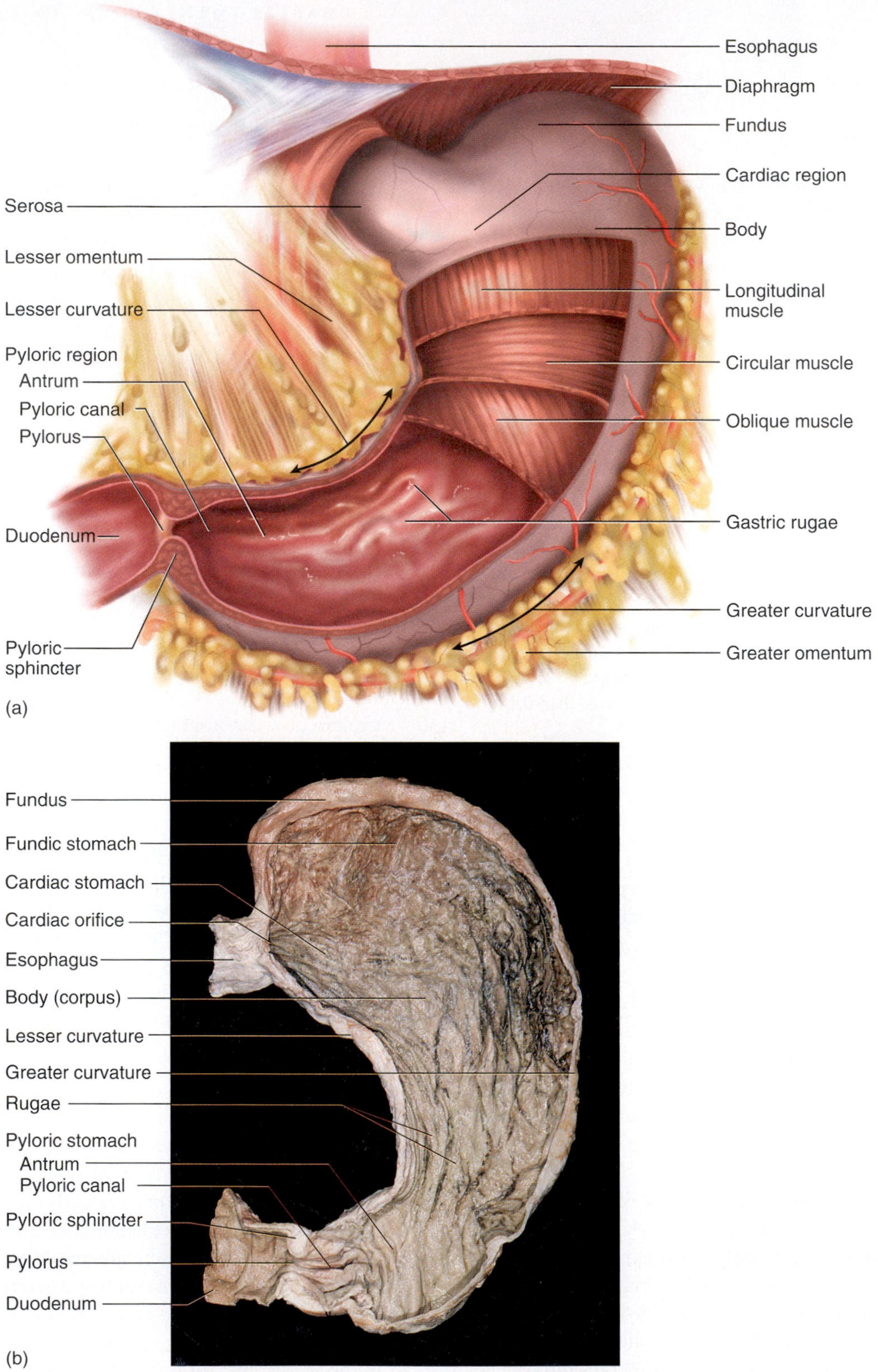

Esophagus
Diaphragm
Fundus
Cardiac region
Serosa
Body
Lesser omentum
Longitudinal muscle
Lesser curvature
Pyloric region
Antrum
Circular muscle
Pyloric canal
Oblique muscle
Pylorus
Gastric rugae
Duodenum
Greater curvature
Pyloric sphincter
Greater omentum
(a)
Fundus
Fundic stomach
Cardiac stomach
Cardiac orifice
Esophagus
Body (corpus)
Lesser curvature
Greater curvature
Rugae
Pyloric stomach
Antrum
Pyloric canal
Pyloric sphincter
Pylorus
Duodenum
(b)

FIGURE 24.11

The Stomach. (*a*) Gross anatomy. (*b*) Photograph of the internal surface.

pyloric canal. The latter terminates at the **pylorus,**[17] a narrow passage into the duodenum. The pylorus is surrounded by a thick ring of smooth muscle, the **pyloric (gastroduodenal) sphincter,** which regulates the passage of chyme into the duodenum.

[17]*pylorus* = gatekeeper

Microscopic Anatomy

The stomach wall has tissue layers similar to those of the esophagus, with some variations. The mucosa is covered with a simple columnar glandular epithelium (fig. 24.12). The apical regions of its surface cells are filled with mucin. After it is secreted, mucin swells with water and becomes mucus. When the stomach is full,

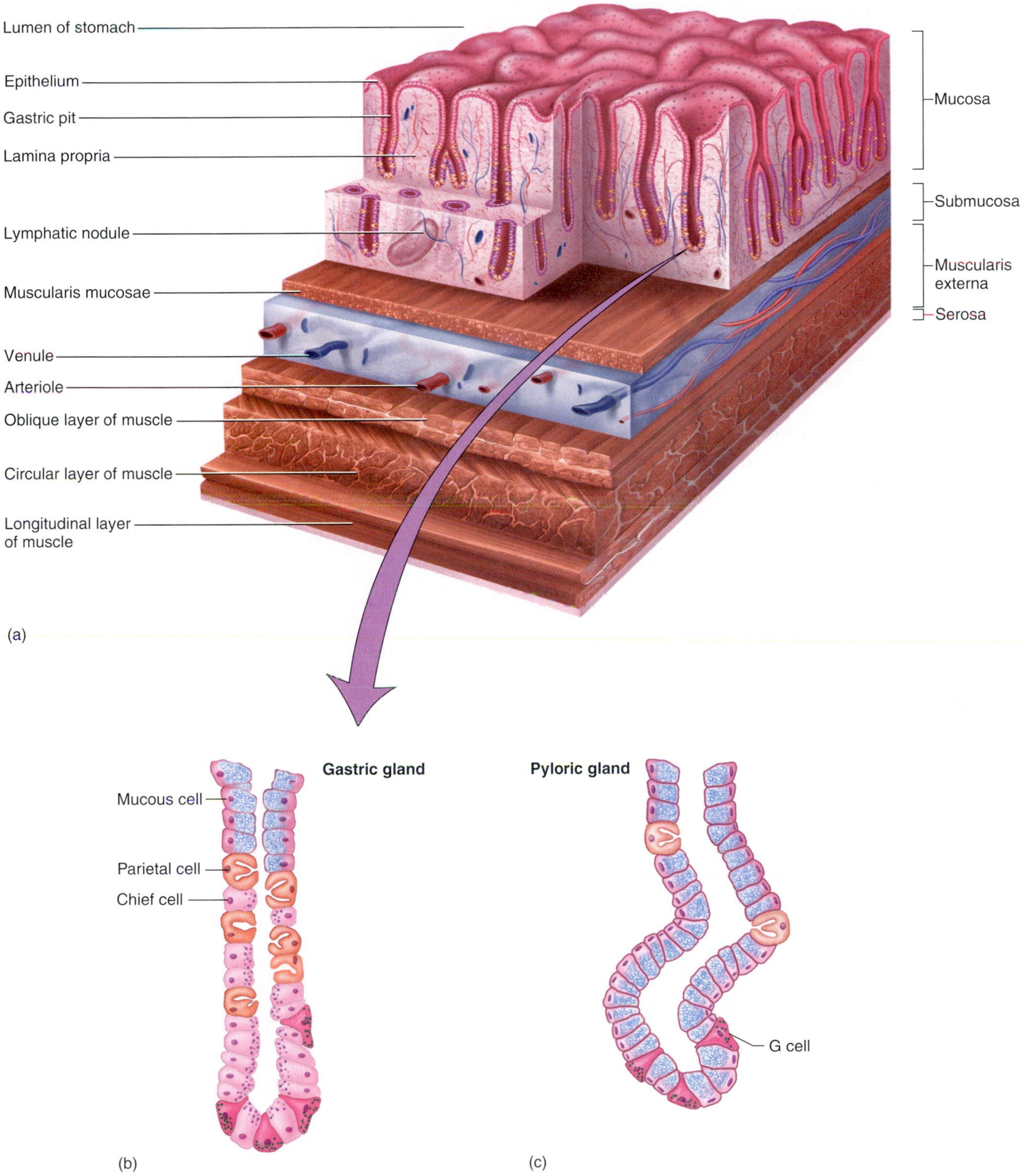

Lumen of stomach
Epithelium
Gastric pit
Lamina propria
Lymphatic nodule
Muscularis mucosae
Venule
Arteriole
Oblique layer of muscle
Circular layer of muscle
Longitudinal layer of muscle
Mucosa
Submucosa
Muscularis externa
Serosa
(a)
Gastric gland
Mucous cell
Parietal cell
Chief cell
Pyloric gland
G cell
(b)
(c)

FIGURE 24.12

Microscopic Anatomy of the Stomach Wall. (*a*) A block of tissue showing all layers from the mucosa *(top)* to the serosa *(bottom)*. (*b*) A gastric gland. (*c*) A pyloric gland.

the mucosa and submucosa are flat and smooth, but as it empties, these layers fold into longitudinal wrinkles called **gastric rugae**[18] (ROO-gee). The lamina propria is almost entirely occupied by tubular glands, to be described shortly. The muscularis externa has three layers, rather than two—an outer longitudinal, middle circular, and inner oblique layer (see fig. 24.11*a*).

Think About It!

Contrast the epithelium of the esophagus with that of the stomach. Why is each epithelial type best suited to the function of its respective organ?

The gastric mucosa is pocked with depressions called **gastric pits** lined with the same columnar epithelium as the mucosal surface. Two or three tubular glands open into the bottom of each gastric pit and span the rest of the lamina propria. The glands are simple wavy or coiled tubes of more or less uniform diameter, except for a constriction called the **neck** at the point where the gland opens into the pit. In the cardiac and pyloric regions they are called **cardiac glands** and **pyloric glands,** respectively. In the rest of the stomach, they are called **gastric glands** (fig. 24.12*b,c*). Collectively, the glands have the following cell types:

- **Mucous cells,** which secrete mucus, predominate in the cardiac and pyloric glands. In gastric glands, they are called *mucous neck cells* and are concentrated in the neck of the gland.
- **Regenerative (stem) cells,** found in the base of the pit and neck of the gland, divide rapidly and produce a continual supply of new cells. Newly generated cells migrate upward to the gastric surface as well as downward into the glands to replace cells that die and fall off into the lumen of the stomach.
- **Parietal cells,** found mostly in the upper half of the gland, secrete hydrochloric acid (HCl) and intrinsic factor. They are found mostly in the gastric glands, but a few occur in the pyloric glands.
- **Chief cells,** so-named because they are the most numerous, secrete chymosin (formerly called rennin) and lipase in infancy and pepsinogen throughout life. They dominate the lower half of the gastric glands but are absent from cardiac and pyloric glands.
- **Enteroendocrine cells,** concentrated especially in the lower end of a gland, secrete hormones and paracrine messengers that regulate digestion. They are found in all regions of the stomach, but are most numerous in the gastric and pyloric glands. There are at least eight different kinds in the stomach, each of which produces a different chemical messenger. **G cells,** for example, secrete a hormone called *gastrin* which stimulates the exocrine cells of the gastric glands to secrete acid and enzymes.

[18]*ruga* = fold, crease

Table 24.1

Major Secretions of the Gastric Glands

Secretory Cells	Secretion	Function
Mucous neck cells	Mucus	Protects mucosa from HCl and enzymes
Parietal cells	Hydrochloric acid	Activates pepsin and lingual lipase; helps liquefy food; reduces dietary iron to usable form (Fe^{2+}); destroys ingested pathogens
	Intrinsic factor	Enables small intestine to absorb vitamin B_{12}
Chief cells	Pepsinogen	Converted to pepsin, which digests protein
	Chymosin	Coagulates milk proteins in infant stomach; not secreted in adults
	Gastric lipase	Digests fats in infant stomach; not secreted in adults
Enteroendocrine cells	Gastrin	Stimulates gastric glands to secrete HCl and enzymes; stimulates intestinal motility; relaxes ileocecal valve
	Serotonin	Stimulates gastric motility
	Histamine	Stimulates HCl secretion
	Somatostatin	Inhibits gastric secretion and motility; delays emptying of stomach; inhibits secretion by pancreas; inhibits gallbladder contraction and bile secretion; reduces blood circulation and nutrient absorption in small intestine

In general, the cardiac and pyloric glands secrete mainly mucus; acid and enzyme secretion occur predominantly in the gastric glands; and hormones are secreted throughout the stomach. Table 24.1 describes the functions of the gastric gland secretions.

It may seem that the stomach would digest itself; we can, after all, digest tripe (animal stomachs) as readily as any other meat. The living stomach, however, is protected from self-digestion in three ways:

1. **Mucous coat.** A thick, highly alkaline mucus resists the action of acid and enzymes.
2. **Tight junctions.** The epithelial cells are joined by tight junctions that prevent gastric juice from seeping between them and digesting the connective tissue of the lamina propria or beyond.
3. **Epithelial cell replacement.** In spite of these other protections, the stomach's epithelial cells live only 3 to 6 days and are then sloughed off into the chyme and digested with the food. They are replaced just as rapidly, however, by the division of stem cells in the gastric pits.

The breakdown of these protective mechanisms can result in inflammation and peptic ulcer (see insight 24.3).

INSIGHT 24.3 CLINICAL APPLICATION

Peptic Ulcer

Inflammation of the stomach, called *gastritis*, can lead to a *peptic ulcer* as pepsin and hydrochloric acid erode the stomach wall (fig. 24.13). Peptic ulcers occur even more commonly in the duodenum and occasionally in the esophagus. If untreated, they can perforate the organ and cause fatal hemorrhaging or peritonitis. Most such fatalities occur in people over age 65.

There is no evidence to support the popular belief that peptic ulcers result from psychological stress. Hypersecretion of acid and pepsin is sometimes involved, but even normal secretion can cause ulceration if the mucosal defense is compromised by other causes. Many or most ulcers involve an acid-resistant bacterium, *Helicobacter pylori*, that invades the mucosa of the stomach and duodenum and opens the way to chemical damage to the tissue. Other risk factors include smoking and the use of aspirin and other nonsteroidal anti-inflammatory drugs (NSAIDs). NSAIDs suppress the synthesis of prostaglandins, which normally stimulate the secretion of protective mucus and acid-neutralizing bicarbonate. Aspirin itself is an acid that directly irritates the gastric mucosa.

Until recently, the most widely prescribed drug in the United States was Cimetidine (Tagamet), which was designed to treat peptic ulcers by reducing acid secretion. Histamine stimulates acid secretion by binding to sites on the parietal cells called *H_2 receptors*; Cimetidine, an *H_2 blocker*, prevents this binding. Lately, however, ulcers have been treated more successfully with antibiotics against *Helicobacter* combined with bismuth suspensions such as Pepto-Bismol. This is a much shorter and less expensive course of treatment and permanently cures about 90% of peptic ulcers, as compared with a cure rate of only 20% to 30% for H_2 blockers.

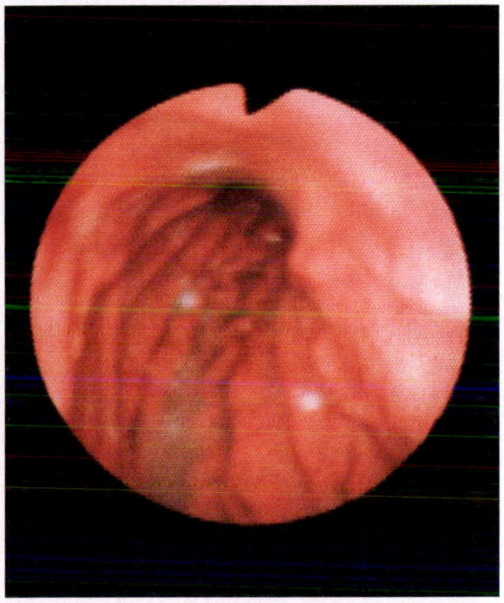

(a)

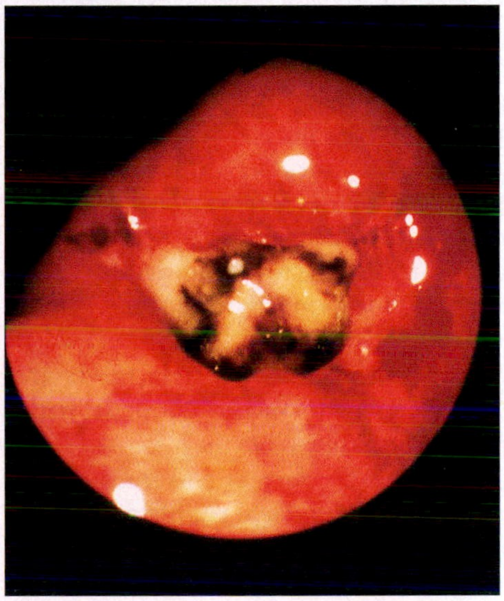

(b)

Figure 24.13

Endoscopic Views of the Gastroesophageal Junction. The esophagus can be seen opening into the cardiac stomach. (*a*) A view of the cardiac orifice from above, showing a healthy esophageal mucosa. The small white spots are reflections of light from the endoscope. (*b*) A bleeding peptic ulcer. A peptic ulcer typically has an oval shape and yellow-white color. Here the yellowish floor of the ulcer is partially obscured by black blood clots, and fresh blood is visible around the margin.

Before You Go On

Answer the following questions to test your understanding of the preceding section:

10. Distinguish between the cardiac region, fundic region, body, and pyloric region of the stomach.
11. Name the cell types in the gastric and pyloric glands and state what each one secretes.
12. Explain why the stomach does not digest itself.

The Small Intestine

Objectives

When you have completed this section, you should be able to

- describe the gross and microscopic anatomy of the small intestine; and
- describe the structural adaptations of the small intestine for digestion and nutrient absorption.

The stomach "spits" about 3 mL of chyme at a time into the small intestine. Nearly all chemical digestion and nutrient absorption occur here. To perform these roles efficiently, the small intestine must have a large surface area exposed to the chyme. This surface area is imparted to it by extensive folding of the mucosa, and by the great length of the small intestine. It measures about 2 m long in a living person, but in the cadaver, where there is no muscle tone, it is 6 to 7 m long. The expression *small* intestine refers not to its length but to its diameter—about 2.5 cm (1 in.).

Gross Anatomy

The small intestine is a coiled mass filling most of the abdominal cavity inferior to the stomach and liver. It is divided into three regions: the duodenum, jejunum, and ileum (fig. 24.14).

The **duodenum** (dew-ODD-eh-num or DEW-oh-DEE-num) constitutes the first 25 cm (10 in.). It begins at the pyloric valve, arcs around the head of the pancreas and passes to the left, and ends at a sharp bend called the **duodenojejunal flexure.** Its name refers to its length, about equal to the width of 12 fingers.[19] The first 2 cm of the duodenum is intraperitoneal, but the rest is retroperitoneal, along with the pancreas.

Internally, the duodenum exhibits transverse to spiral ridges, up to 10 mm high, called **circular folds** *(plicae circulares)* (see fig. 24.20). They cause the chyme to flow on a spiral path along the mucosa, slowing its progress, causing more contact with the mucosa, and promoting thorough mixing, digestion, and nutrient absorption.

Adjacent to the head of the pancreas, the duodenal wall has a prominent wrinkle called the **major duodenal papilla** where the bile and pancreatic ducts open into the intestine. This papilla marks the boundary between the foregut and midgut. In most people, there is a smaller **minor duodenal papilla** a little proximal to this, which receives an *accessory pancreatic duct.*

The duodenum receives and mixes the stomach contents, pancreatic juice, and bile. Stomach acid is neutralized here by bicarbonate in the pancreatic juice, fats are physically broken up (emulsified) by the bile, pepsin is inactivated by the rise in pH, and pancreatic enzymes take over the job of chemical digestion.

The **jejunum** (jeh-JOO-num) is the next 2.5 m (8 ft), or by definition, the first 40% of the small intestine beyond the duodenum. Its name refers to the fact that early anatomists typically found it to be empty.[20] The jejunum begins in the upper left quadrant of the abdomen but lies mostly within the umbilical region (see fig. A.6). The jejunum has large, tall, closely spaced circular folds. Most digestion and nutrient absorption occur here. Its wall is relatively thick and muscular, and it has an especially rich blood supply which gives it a relatively red color.

The **ileum**[21] forms the last 3.6 m (12 ft), or 60% of the post-duodenal small intestine. (The lengths given here are for the cadaver.) The ileum occupies mainly the hypogastric region and part of the pelvic cavity. Compared to the jejunum, its wall is thinner, less muscular, and less vascular, and it has a paler pink color. Its circular folds are smaller and more sparse, and are lacking from the distal end. On the side opposite from its mesenteric attachment, the ileum has prominent lymphatic nodules in clusters called **Peyer[22] patches,** which are readily visible to the naked eye and become progressively larger approaching the large intestine.

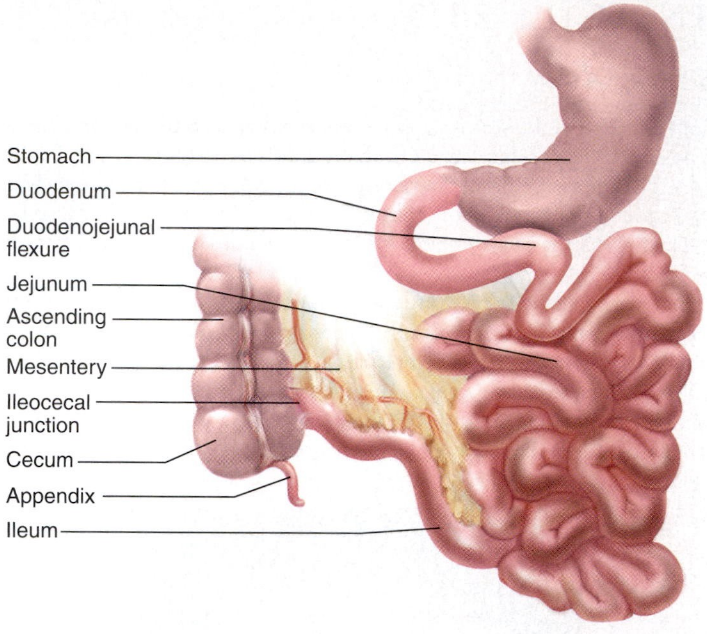

FIGURE 24.14
Gross Anatomy of the Small Intestine. The intestine is pulled aside to expose the mesentery and ileocecal junction.

The end of the small intestine is the **ileocecal** (ILL-ee-oh-SEE-cul) **junction,** where the ileum joins the *cecum* of the large intestine. The muscularis of the ileum is thickened at this point to form a sphincter, the **ileocecal** (ILL-ee-oh-SEE-cul) **valve,** which protrudes into the cecum and regulates the passage of food residue into the large intestine. Both the jejunum and ileum are intraperitoneal and thus covered with a serosa, which is continuous with the complex, folded mesentery that suspends the small intestine from the dorsal abdominal wall.

Microscopic Anatomy

As mentioned earlier, effective digestion and nutrient absorption requires that the small intestine have a large internal surface area. This is provided by its relatively great length and by three kinds of internal folds or projections: the circular folds, villi, and microvilli. If the mucosa of the small intestine were smooth, like the inside of a hose, it would have a surface area of about 0.3 to 0.5 m^2, but with these surface elaborations, its actual surface area is about 200 m^2—clearly a great advantage for nutrient absorption. The circular folds increase the surface area by a factor of 2 to 3; the villi by a factor of 10; and the microvilli by a factor of 20.

[19]*duoden* = 12
[20]*jejun* = empty, dry
[21]from *eilos* = twisted

[22]Johann K. Peyer (1653–1712), Swiss anatomist

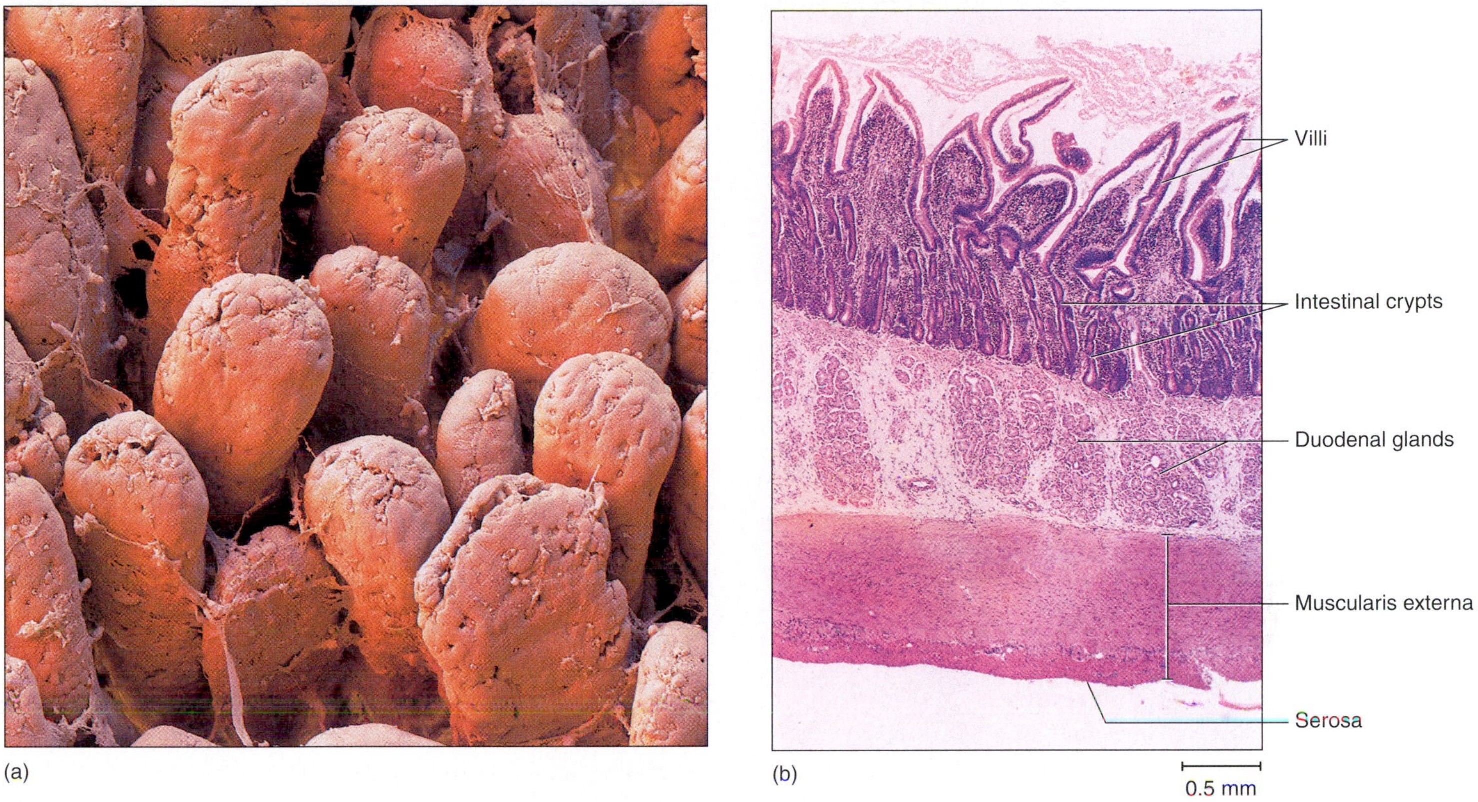
Villi
Intestinal crypts
Duodenal glands
Muscularis externa
Serosa
(a)
(b)
0.5 mm

FIGURE 24.15

Intestinal Villi. (*a*) Colorized SEM. (*b*) Histological section of the duodenum showing villi, intestinal crypts, and duodenal glands.

The largest of these elaborations, the circular folds, were described earlier. They occur from the duodenum to the middle of the ileum. They involve only the mucosa and submucosa; they are not visible on the external surface, which is smooth.

If the mucosa is examined more closely it appears fuzzy, like a terrycloth towel. This is due to the **villi** (VIL-eye; singular, *villus*), tongue- to finger-shaped projections about 0.5 to 1.0 mm high (fig. 24.15). The villi are largest in the duodenum and become progressively smaller in more distal regions of the small intestine. A villus is covered with two kinds of epithelial cells—columnar **absorptive cells** and mucus-secreting **goblet cells.** Like epithelial cells of the stomach, those of the small intestine are joined by tight junctions that prevent digestive enzymes from seeping between them.

The core of a villus is filled with areolar tissue of the lamina propria. Embedded in this tissue are an arteriole, a bed of blood capillaries, a venule, and a lymphatic capillary called a **lacteal** (LAC-tee-ul) (fig. 24.16). Blood capillaries of the villus absorb most nutrients, but the lacteal absorbs most dietary lipid. The reason for this difference is that when lipids pass through the intestinal absorptive cells, the Golgi complex packages them in protein- and phospholipid-coated droplets called *chylomicrons,* then releases them from the base of the epithelium into the core of the villus. Chylomicrons are too large (60 to 750 nm) to pass into the bloodstream through the blood capillary walls, but they can pass through the larger gaps between the cells of lymphatic capillaries and thus enter the lymph. The lymphatic system, of course, eventually delivers the chylomicrons to the bloodstream. The fatty lymph in the lacteal is called *chyle.* It has a milky appearance for which the lacteal is named.[23] The core of the villus also has a few smooth muscle cells that contract periodically. This enhances mixing of the chyme in the intestinal lumen and milks lymph down the lacteal to the larger lymphatic vessels of the submucosa.

Think About It!

Identify the exact place in the body where chylomicrons enter the blood. (Hint: see chapter 22.)

Each absorptive cell of a villus has a fuzzy brush border of microvilli about 1 μm high. The brush border increases the absorptive surface area of the small intestine and contains **brush border enzymes,** integral proteins of the plasma membrane. One of these, *enterokinase,* activates pancreatic enzymes. Others carry out some of the final stages in the enzymatic digestion of small carbohydrates and peptides. These enzymes are not released into the lumen; instead, the chyme must contact the brush border for digestion to occur. This process, called **contact digestion,** is one reason that thorough mixing of the chyme is so important.

On the floor of the small intestine, between the bases of the villi, there are numerous pores that open into tubular glands called

[23]*lact* = milk

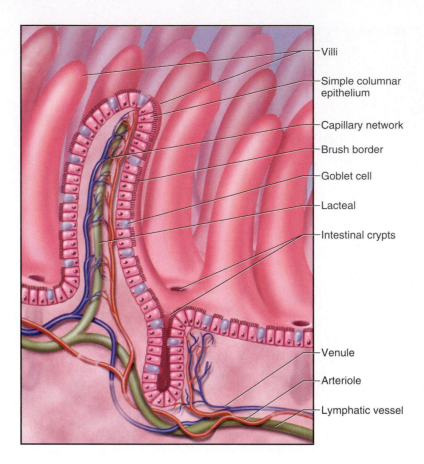

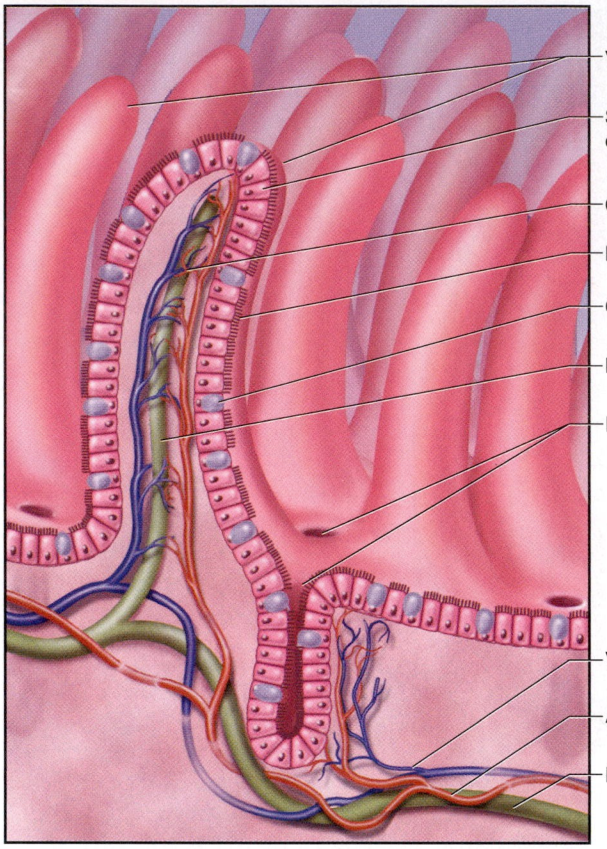
Villi
Simple columnar epithelium
Capillary network
Brush border
Goblet cell
Lacteal
Intestinal crypts
Venule
Arteriole
Lymphatic vessel

FIGURE 24.16
Structure of a Villus.

intestinal crypts (crypts of Lieberkühn;[24] LEE-ber-koohn). These crypts, similar to the gastric glands, extend as far as the muscularis mucosae. In the upper half they consist of absorptive and goblet cells like those of the villi. The lower half is dominated by dividing epithelial cells. In its life span of 3 to 6 days, an epithelial cell migrates up the crypt to the tip of the villus, where it is sloughed off and digested. Also seen deep in the crypts are enteroendocrine cells and **Paneth**[25] **cells.** Paneth cells secrete the antimicrobial enzyme *lysozyme,* which minimizes bacterial invasion of the mucosa.

The duodenum has prominent **duodenal (Brunner**[26]**) glands** in the submucosa. They secrete an abundance of alkaline mucus, which neutralizes stomach acid and shields the mucosa from its corrosive effects. Throughout the small intestine, the lamina propria and submucosa have a large population of lymphocytes that intercept pathogens before they can invade the bloodstream. In some places these are aggregated into conspicuous lymphatic nodules (see fig. 22.8), which become most conspicuous in the Peyer patches of the ileum.

The muscularis externa consists of a relatively thick inner circular layer and a thinner outer longitudinal layer. Ganglia of the myenteric nerve plexus occur between these layers.

[24]Johann M. Lieberkühn (1711–56), German anatomist
[25]Josef Paneth (1857–90), Austrian physician
[26]Johann C. Brunner (1653–1727), Swiss anatomist

Before You Go On

Answer the following questions to test your understanding of the preceding section:

13. Name the three regions of the small intestine and describe the distinctive features of each one.
14. Name the sphincters at the beginning and end of the small intestine.
15. What three structures increase the absorptive surface area of the small intestine?
16. Sketch a villus and label its epithelium, brush border, lamina propria, blood capillaries, and lacteal.

THE LARGE INTESTINE

Objectives

When you have completed this section, you should be able to

- describe the gross and microscopic anatomy of the large intestine; and
- contrast the mucosa of the colon with that of the small intestine.

The large intestine (fig. 24.17*a*) receives about 500 mL of indigestible food residue per day, reduces it to about 150 mL of feces by absorbing water and salts, and eliminates the feces by defecation.

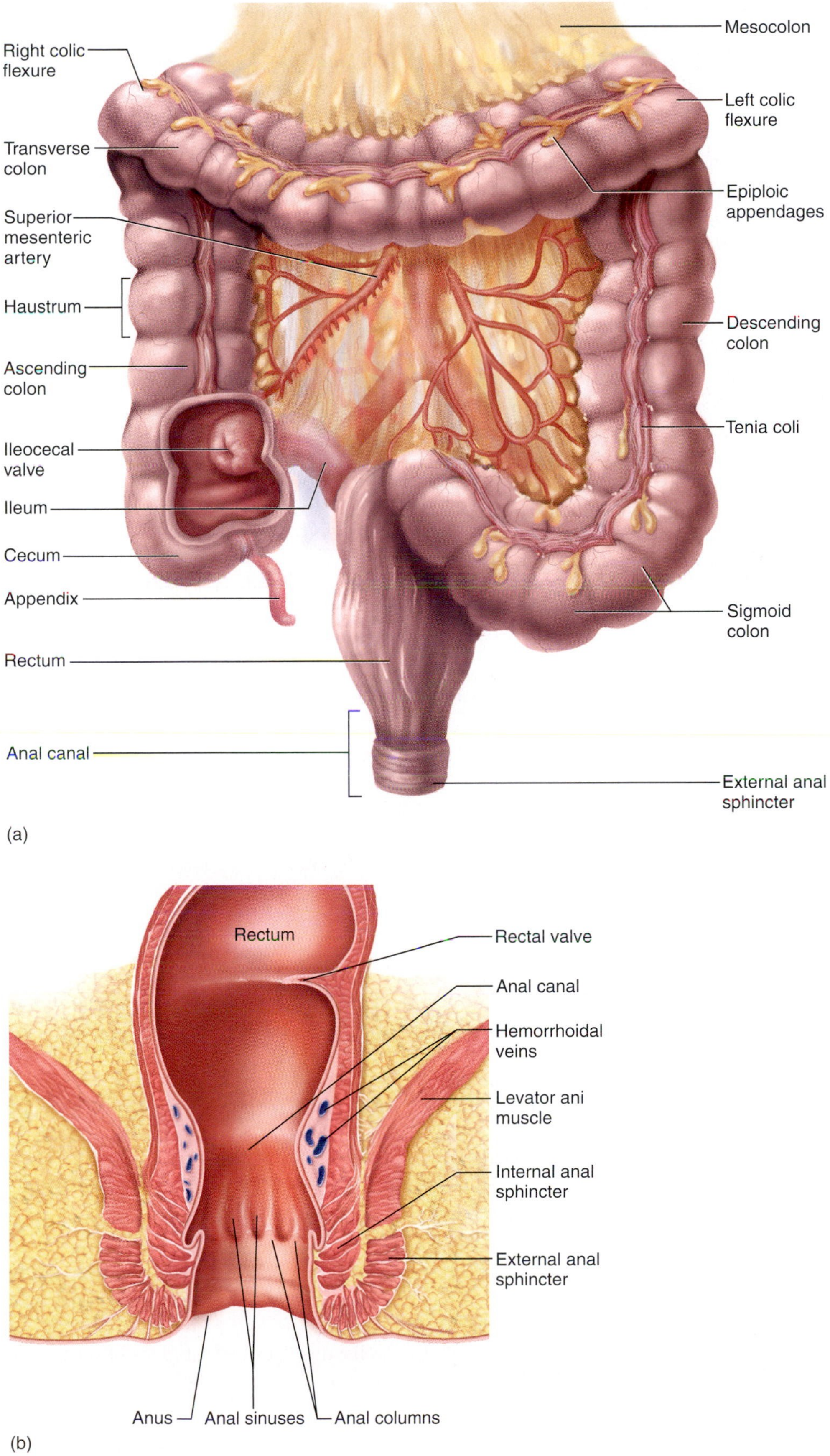
Mesocolon
Right colic flexure
Left colic flexure
Transverse colon
Epiploic appendages
Superior mesenteric artery
Haustrum
Descending colon
Ascending colon
Tenia coli
Ileocecal valve
Ileum
Cecum
Appendix
Sigmoid colon
Rectum
Anal canal
External anal sphincter
(a)
Rectum
Rectal valve
Anal canal
Hemorrhoidal veins
Levator ani muscle
Internal anal sphincter
External anal sphincter
Anus
Anal sinuses
Anal columns
(b)

FIGURE 24.17
The Large Intestine. (*a*) Gross anatomy. (*b*) Detail of the anal canal.

Gross Anatomy

The large intestine measures about 1.5 m (5 ft) long and 6.5 cm (2.5 in.) in diameter in the cadaver. It begins with the **cecum,**[27] a blind pouch in the lower right abdominal quadrant inferior to the ileocecal valve. Attached to the lower end of the cecum is the **appendix,** a blind tube 2 to 7 cm long. The mucosa of the appendix is densely populated with lymphocytes and is a significant source of immune cells. Herbivorous primates such as gorillas and orangutans have an enormous cecum, packed with bacteria which digest the plant fiber in their coarse diet. Humans, with more mixed and easily digested diet, have only the appendix as a vestige of the larger cecum.

The **colon** is that portion of the large intestine between the ileocecal junction and the rectum (not including the cecum, rectum, or anal canal). It is divided into the ascending, transverse, descending, and sigmoid regions. The **ascending colon** begins at the ileocecal valve and passes up the right side of the abdominal cavity. It makes a 90° turn at the **right colic (hepatic) flexure,** near the right lobe of the liver, and becomes the **transverse colon.** This passes horizontally across the upper abdominal cavity and turns 90° downward at the **left colic (splenic) flexure** near the spleen. Here it becomes the **descending colon,** which passes down the left side of the abdominal cavity. Ascending, transverse, and descending colons thus form a squarish, three-sided frame around the small intestine.

The pelvic cavity is narrower than the abdominal cavity, so at the pelvic inlet the colon turns medially and downward, forming a roughly S-shaped portion called the **sigmoid**[28] **colon.** (Visual examination of this region is performed with an instrument called a *sigmoidoscope.*) In the pelvic cavity, the large intestine continues as the **rectum,**[29] about 15 cm long. In spite of its name, the rectum is not straight but has three lateral curves as well as a dorsoventral curve. The rectal mucosa is smoother than that of the colon. It has three internal **transverse rectal folds** (**rectal valves**) that enable it to retain feces while passing gas. The large intestine contains about 7 to 10 L of gas per day, expelling about 500 mL/day anally as *flatus* and reabsorbing the rest through the colonic wall.

The final 3 cm of the large intestine is the **anal canal** (fig. 24.17*b*), which passes through the levator ani muscle of the pelvic floor and terminates at the anus. The mucosa of the anal canal forms longitudinal ridges called **anal columns** with depressions between them called **anal sinuses.** As feces pass through the canal, they press against the sinuses and cause them to exude extra mucus, which lubricates the canal during defecation. Large **hemorrhoidal veins** form superficial plexuses in the anal columns and around the orifice. Unlike veins in the limbs, they lack valves and are particularly subject to distension and venous pooling. *Hemorrhoids* are permanently distended veins that protrude into the anal canal or form bulges external to the anus.

The muscularis externa is unusual in that its longitudinal fibers do not encircle the colon but are divided into three ribbonlike strips called the **teniae coli** (TEE-nee-ee CO-lye). The muscle tone of the teniae coli contracts the colon lengthwise and causes its wall to bulge, forming pouches called **haustra**[30] (HAW-stra; singular, *haustrum*). In the rectum and anal canal, however, the longitudinal muscle forms a continuous sheet and haustra are absent. The anus is normally held shut by two muscular rings—an **internal anal sphincter** composed of smooth muscle of the muscularis externa, and an **external anal sphincter** composed of skeletal muscle of the pelvic diaphragm.

The ascending and descending colon are retroperitoneal, whereas the transverse and sigmoid colon are covered with serosa and anchored to the dorsal abdominal wall by the mesocolon. The serosa of these regions often has **epiploic**[31] **appendages,** clublike fatty pouches of peritoneum of unknown function.

Microscopic Anatomy

The mucosa of the large intestine has a simple columnar epithelium in all regions except the lower half of the anal canal, where it has a nonkeratinized stratified squamous epithelium. The latter provides more resistance to the abrasion caused by the passage of feces. There are no circular folds or villi in the large intestine, but there are intestinal crypts. They are deeper than in the small intestine and have a greater density of goblet cells; mucus is their only significant secretion. The lamina propria and submucosa have an abundance of lymphatic tissue, providing protection from the bacteria that densely populate the large intestine.

Before You Go On

Answer the following questions to test your understanding of the preceding section:

17. Name the regions of the large intestine in order from cecum to anus.
18. How does the mucosa of the large intestine differ from that of the small intestine? How does the muscularis externa differ?
19. How do the two anal sphincters differ in location, tissue composition, and function?

Accessory Glands of Digestion

Objectives

When you have completed this section, you should be able to

- describe the gross and microscopic anatomy of the liver, gallbladder, and bile duct system;
- describe the functions of the liver and bile;
- describe the gross and microscopic anatomy of the pancreas; and
- list the digestive secretions of the pancreas and their functions.

[27]*cec* = blind
[28]*sigm* = sigma or S + *oid* = resembling
[29]*rect* = straight
[30]*haustr* = to draw
[31]*epiploic* = pertaining to an omentum

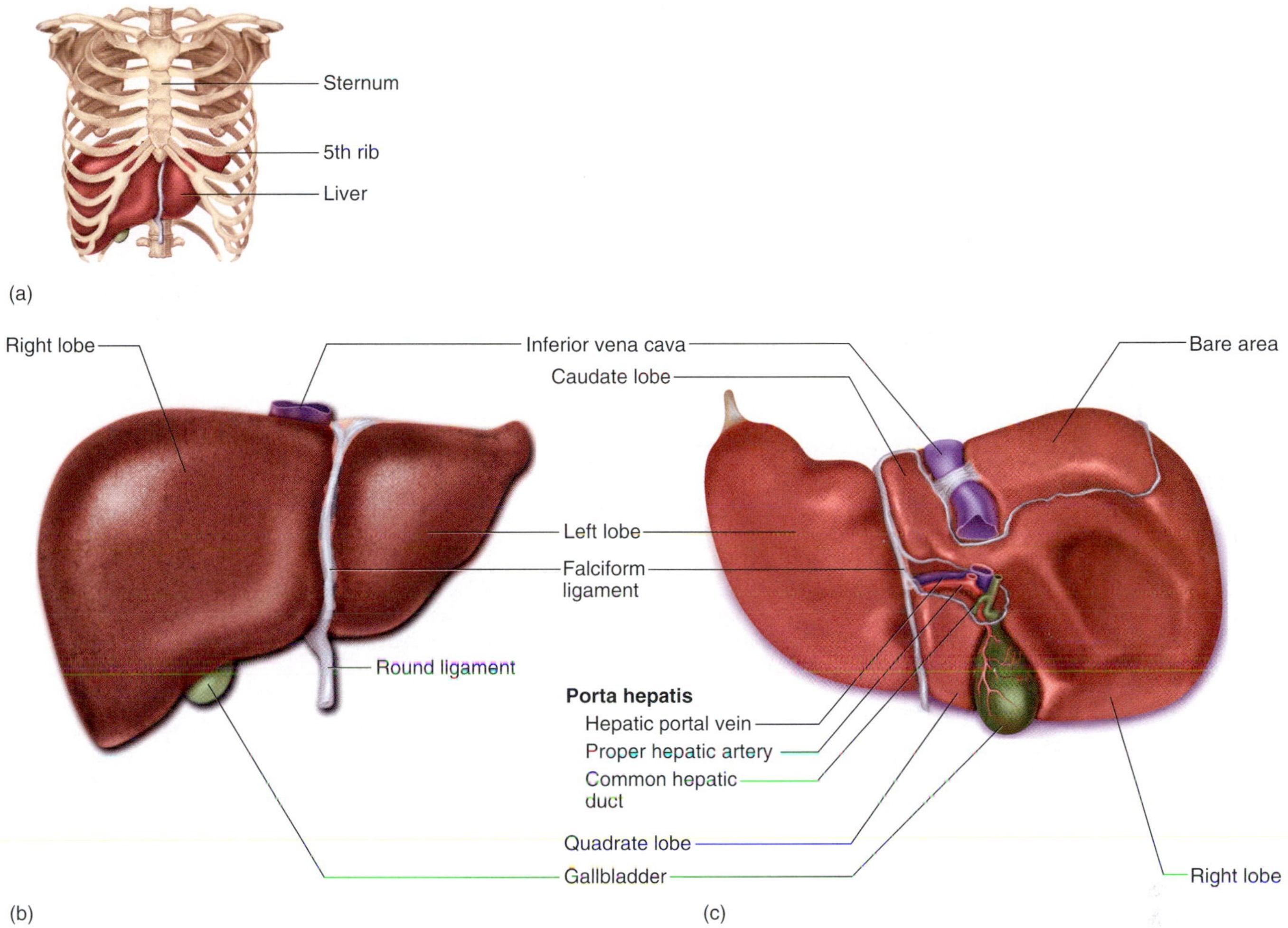

Sternum
5th rib
Liver
(a)
Right lobe
Inferior vena cava
Caudate lobe
Bare area
Left lobe
Falciform ligament
Round ligament
Porta hepatis
Hepatic portal vein
Proper hepatic artery
Common hepatic duct
Quadrate lobe
Gallbladder
Right lobe
(b)
(c)

FIGURE 24.18

Gross Anatomy of the Liver. (*a*) Relationship of the liver to the thoracic cage. (*b*) Anterior aspect, where only the left and right lobes can be seen. (*c*) Inferior aspect, showing the caudate and quadrate lobes and the porta hepatis.

The small intestine receives not only chyme from the stomach but also secretions from the liver and pancreas, which enter the digestive tract near the junction of the stomach and small intestine.

The Liver

The liver (fig. 24.18) is a reddish brown gland located immediately inferior to the diaphragm, filling most of the right hypochondriac and epigastric regions. It is the body's largest gland, weighing about 1.4 kg (3 lb). The liver has numerous functions (table 24.2). From their variety and importance, it is not hard to understand why such liver diseases as cirrhosis, hepatitis, and liver cancer are so serious and often fatal. Only one of these functions contributes to digestion—the secretion of bile. **Bile** is a yellow-green fluid containing minerals, cholesterol, neutral fats, phospholipids, bile pigments, and bile acids. The principal pigment is **bilirubin,** derived from the decomposition of hemoglobin. Bacteria of the large intestine metabolize bilirubin to **urobilinogen,** which is responsible for the brown color of feces. In the absence of bile secretion, the feces are grayish white and marked with streaks of undigested fat *(acholic feces)*. **Bile acids (bile salts)** are steroids synthesized from cholesterol. Bile acids and **lecithin,** a phospholipid, emulsify fat—breaking globules of dietary fat into smaller droplets with more surface area exposed to the action of pancreatic lipase. Emulsification greatly enhances the efficiency of fat digestion.

GROSS ANATOMY

The liver is enclosed in a fibrous capsule and, external to this, most of it is covered by serosa. The serosa is absent from the *bare area* where its superior surface is attached to the diaphragm.

The liver is superficially subdivided into the right, left, quadrate, and caudate lobes. From an anterior view, we see only the large **right lobe** and smaller **left lobe.** They are separated from each other by the **falciform**[32] **ligament,** a sheet of mesentery that attaches the liver to the anterior abdominal wall. Superiorly, the falciform ligament forks into right and left **coronary**[33] **ligaments,**

[32]*falci* = sickle + *form* = shape
[33]*coron* = crown + *ary* = like, resembling

TABLE 24.2

Functions of the Liver

Carbohydrate Metabolism
Converts dietary fructose and galactose to glucose. Stabilizes blood glucose concentration by storing excess glucose as glycogen *(glycogenesis)*, releasing glucose from glycogen when needed *(glycogenolysis)*, and synthesizing glucose from fats and amino acids *(gluconeogenesis)* when glucose demand exceeds glycogen reserves. Receives lactic acid generated by anaerobic fermentation in skeletal muscle and other tissues and converts it to pyruvic acid or glucose 6-phosphate for storage or energy-releasing metabolism.
Lipid Metabolism
Degrades chylomicron remnants. Carries out most of the body's fat synthesis *(lipogenesis)* and synthesizes cholesterol and phospholipids. Produces *very low-density lipoproteins* (VLDLs) to transport lipids to adipose tissue and other tissues for storage or use, and stores fat in its own cells. Carries out most fatty acid oxidation. Produces protein shells of *high-density lipoproteins (HDLs)*, which pick up excess cholesterol from other tissues and return it to the liver; liver excretes the excess cholesterol in bile.
Protein and Amino Acid Metabolism
Metabolizes amino acids; removes their $-NH_2$ and converts the resulting ammonia to *urea*, the major nitrogenous waste in the urine. Synthesizes some amino acids.
Synthesis of Plasma Proteins
Synthesizes nearly all the proteins of blood plasma, including albumin, α and β globulins, fibrinogen, prothrombin, and several other clotting factors. (Does not synthesize plasma enzymes, peptide hormones, or gamma globulins.)
Vitamin and Mineral Metabolism
Converts vitamin D_3 to calcidiol, a step in the synthesis of the hormone calcitriol; stores a 3- to 4-month supply of vitamin D. Stores a 10-month supply of vitamin A and enough vitamin B_{12} to last one to several years. Stores iron in ferritin and releases it as needed. Excretes excess calcium by way of the bile.
Digestion
Synthesizes bile acids and lecithin, which emulsify fat and promote its digestion.
Disposal of Drugs, Toxins, and Hormones
Detoxifies alcohol, antibiotics, and many other drugs. Metabolizes bilirubin from RBC breakdown and excretes it as bile pigments. Deactivates thyroxine and steroid hormones and excretes them or converts them to a form more easily excreted by the kidneys.
Phagocytosis
Hepatic macrophages cleanse the blood of bacteria and other foreign matter.

which suspend the liver from the diaphragm. The **round ligament** *(ligamentum teres)*, visible anteriorly at the inferior end of the falciform, is a fibrous remnant of the umbilical vein, which carries blood from the umbilical cord to the liver of a fetus.

From the inferior view, we also see a squarish anterior **quadrate lobe** next to the gallbladder and a tail-like **caudate**[34] **lobe** dorsal to that. An irregular opening between these lobes, the **porta hepatis,**[35] is a point of entry for the hepatic portal vein and hepatic artery and a point of exit for the bile passages. All of these blood vessels and bile passages travel in the lesser omentum. The gallbladder adheres to a depression on the inferior surface of the liver between the right and quadrate lobes. The posterior aspect of the liver has a deep groove (sulcus) occupied by the inferior vena cava.

MICROSCOPIC ANATOMY

The interior of the liver is filled with an enormous number of tiny cylinders called **hepatic lobules,** about 2 mm long by 0.7 mm in diameter. A lobule consists of a **central vein** passing down its core, surrounded by radiating sheets of cuboidal cells called **hepatocytes** (fig. 24.19). Imagine spreading a book wide open until its front and back covers touch. The pages of the book would fan out around the spine somewhat like the plates of hepatocytes fan out from the central vein of a liver lobule.

Each plate of hepatocytes is an epithelium one or two cells thick. The spaces between the plates are blood-filled channels called **hepatic sinusoids.** The sinusoids are lined by widely spaced endothelial cells that separate the hepatocytes from the blood cells. There is a space between the hepatocytes and endothelial cells, however, where the hepatocytes are in direct contact with the blood plasma and have abundant microvilli that enable them to rapidly absorb plasma-borne substances (see fig. 21.7). After a meal, as blood from the intestines circulates through the hepatic sinusoids, the hepatocytes rapidly remove glucose, amino acids, iron, vitamins, and other nutrients for metabolism or storage. They also remove and degrade hormones, toxins, bile pigments, and drugs. Conversely, they secrete albumin, lipoproteins, clotting factors, glucose, and other products into the blood. The sinusoids also contain phagocytic cells called **hepatic macrophages** (**Kupffer**[36] **cells**), which remove bacteria and debris from the blood.

[34]*caud* = tail

[35]*porta* = gateway, entrance + *hepatis* = of the liver

[36]Karl W. von Kupffer (1829–1902), German anatomist

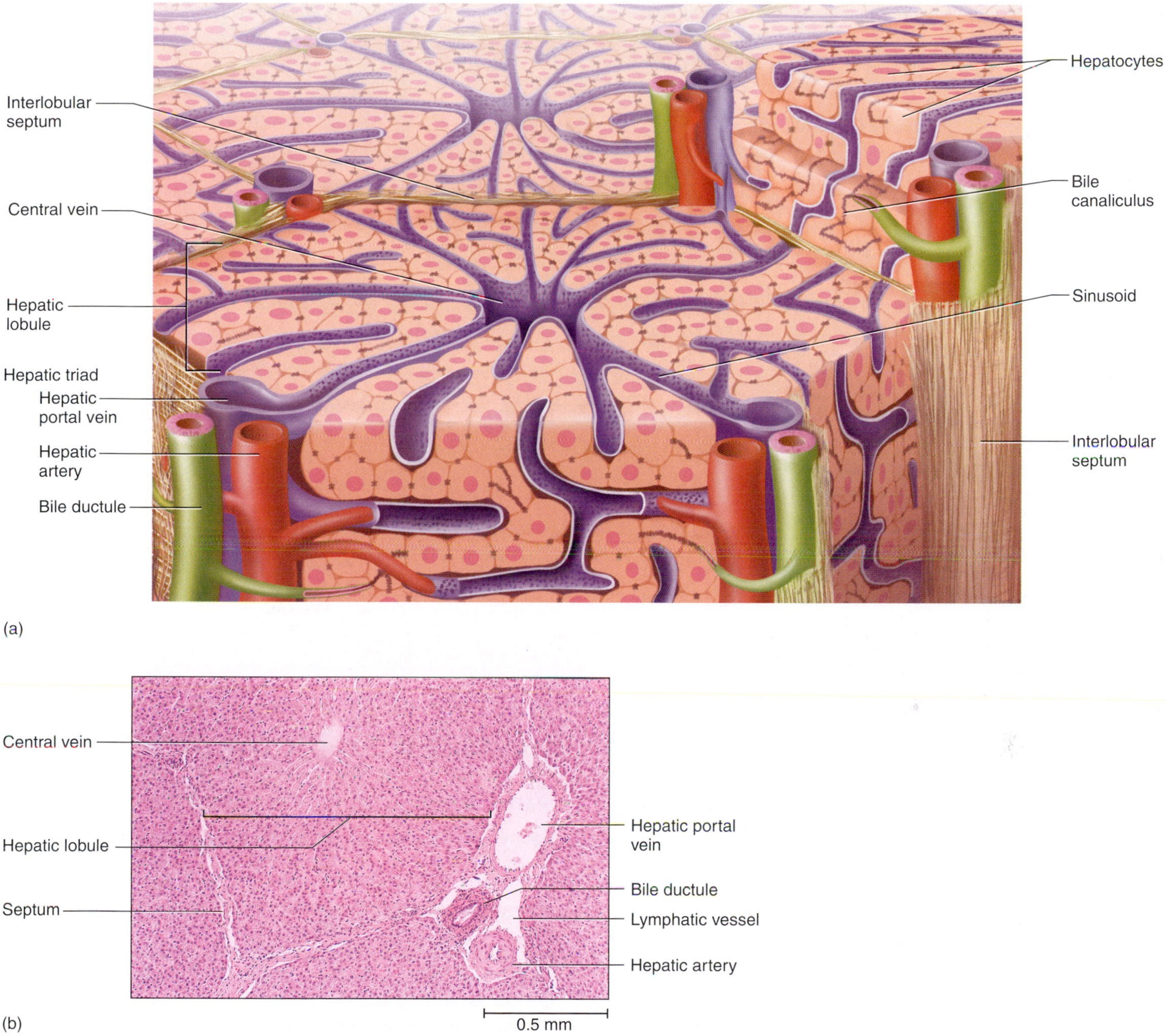
Interlobular septum
Central vein
Hepatic lobule
Hepatic triad
Hepatic portal vein
Hepatic artery
Bile ductule
Hepatocytes
Bile canaliculus
Sinusoid
Interlobular septum
(a)
Central vein
Hepatic lobule
Septum
Hepatic portal vein
Bile ductule
Lymphatic vessel
Hepatic artery
(b)
0.5 mm

FIGURE 24.19

Microscopic Anatomy of the Liver. (*a*) The hepatic lobules and their relationship to the blood vessels and bile tributaries. (*b*) Histological section of the liver.

The hepatocytes secrete bile into narrow channels, the **bile canaliculi,** between the plates of the lobules. Bile passes from there into the small **bile ductules** between lobules. These ductules lead ultimately to the **right** and **left hepatic ducts,** which exit the inferior surface of the liver at the porta hepatis.

The hepatic lobules are separated by a sparse connective tissue stroma. In cross sections, the stroma is especially visible in the triangular areas where three or more lobules meet. Here there is often a **hepatic triad** of two blood vessels and a bile ductule. The blood vessels are small branches of the hepatic artery and hepatic portal vein.

CIRCULATION

The liver receives blood from two sources: about 70% from the **hepatic portal vein** and 30% from the **hepatic artery.** The hepatic portal vein delivers blood received from the stomach, intestines, pancreas, and spleen, and carries nutrients from the intestines to the liver. The hepatic artery arises from the celiac trunk and delivers oxygen. The arterial and venous bloodstreams mix in the hepatic sinusoids. After processing by the hepatocytes, the blood collects in the central vein at the core of the lobule. Blood from the central veins ultimately converges in the right and left **hepatic veins,** which exit the superior surface of the liver and empty into the nearby inferior vena cava.

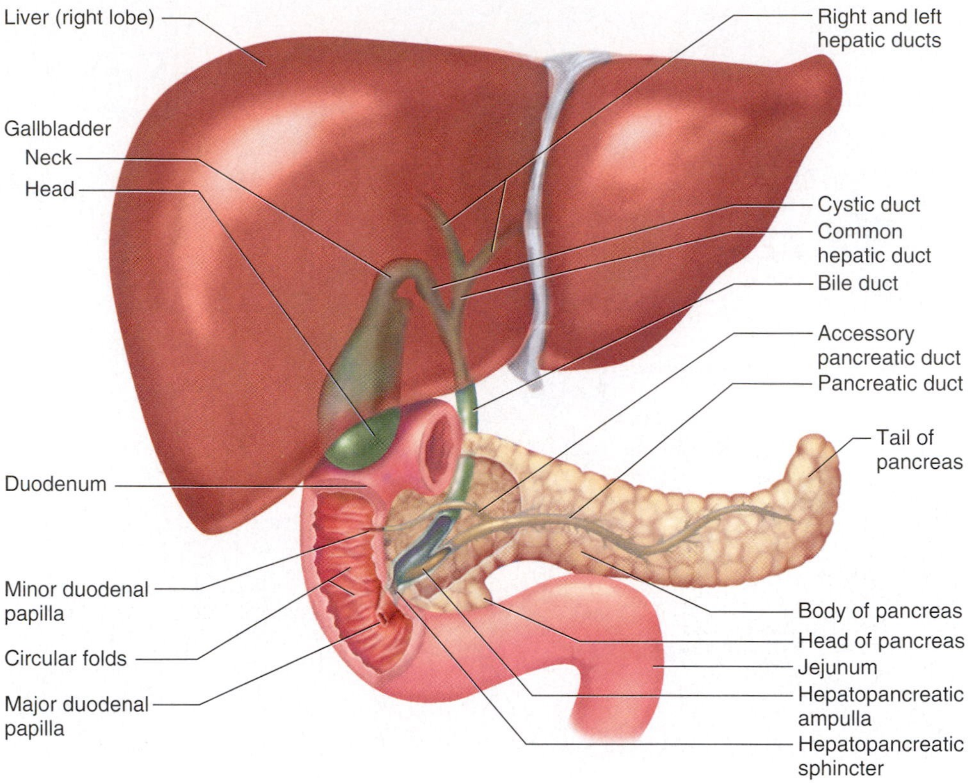

Liver (right lobe)
Gallbladder
Neck
Head
Duodenum
Minor duodenal papilla
Circular folds
Major duodenal papilla
Right and left hepatic ducts
Cystic duct
Common hepatic duct
Bile duct
Accessory pancreatic duct
Pancreatic duct
Tail of pancreas
Body of pancreas
Head of pancreas
Jejunum
Hepatopancreatic ampulla
Hepatopancreatic sphincter

FIGURE 24.20

Gross Anatomy of the Gallbladder, Pancreas, and Bile Passages.

The Gallbladder and Bile Passages

Since the only digestive role of the liver is bile secretion, we will further trace the flow of bile through organs associated with the liver. The most conspicuous of these organs is the **gallbladder,** a pear-shaped sac on the underside of the liver that serves to store and concentrate the bile (fig. 24.20). It is about 10 cm long and internally lined by a highly folded mucosa with a simple columnar epithelium. Its head *(fundus)* usually projects slightly beyond the inferior margin of the liver. Its neck *(cervix)* leads into the **cystic duct,** through which bile enters and leaves the gallbladder.

When the two hepatic ducts leave the porta hepatis, they converge almost immediately to form the **common hepatic duct.** This duct goes only a short distance before joining the cystic duct. Their union forms the **bile duct,** which descends through the lesser omentum to the duodenum. The bile duct and main pancreatic duct both approach the major duodenal papilla. Usually, just before emptying into the duodenum, the two ducts join each other and form an expanded chamber called the **hepatopancreatic ampulla.** A muscular **hepatopancreatic sphincter (sphincter of Oddi**[37]**)** regulates the release of bile and pancreatic juice from the ampulla into the duodenum.

The Pancreas

Most digestion is carried out by pancreatic enzymes. The **pancreas** (fig. 24.20) is a spongy digestive gland dorsal to the greater curvature of the stomach. It is about 15 cm long, and divided into a globose *head* encircled on the right by the duodenum, a midportion called the *body,* and a blunt, tapered *tail* on the left near the spleen. It has a very thin connective tissue capsule and a nodular surface. It is retroperitoneal; its ventral surface is covered by parietal peritoneum, whereas its dorsal surface contacts the aorta, left kidney, left adrenal gland, and other viscera on the dorsal body wall.

[37]Ruggero Oddi (1864–1913), Italian physician

INSIGHT 24.4 CLINICAL APPLICATION

GALLSTONES

Gallstones (biliary calculi) are hard masses in the gallbladder or bile ducts, usually composed of cholesterol, calcium carbonate, and bilirubin. *Cholelithiasis,* the formation of gallstones, is most common in obese women over the age of 40 and usually results from excess cholesterol. The gallbladder may contain several gallstones, some over 1 cm in diameter. Gallstones cause excruciating pain when they obstruct the bile ducts or when the gallbladder or bile ducts contract. When they block the flow of bile into the duodenum, they cause jaundice (yellowing of the skin due to bile pigment accumulation), poor fat digestion, and impaired absorption of fat-soluble vitamins. Once treated only by surgical removal, gallstones are now often treated with stone-dissolving drugs or by *lithotripsy,* the use of ultrasonic vibration to pulverize them without surgery. Reobstruction can be prevented by inserting a stent (tube) into the bile duct, which keeps it distended and allows gallstones to pass while they are still small.

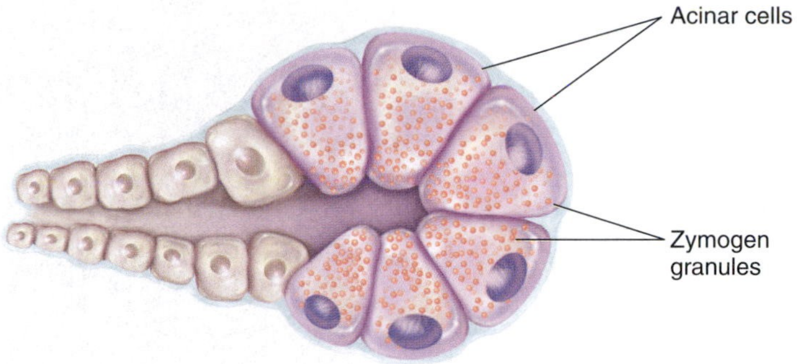
Acinar cells
Zymogen granules

(a)

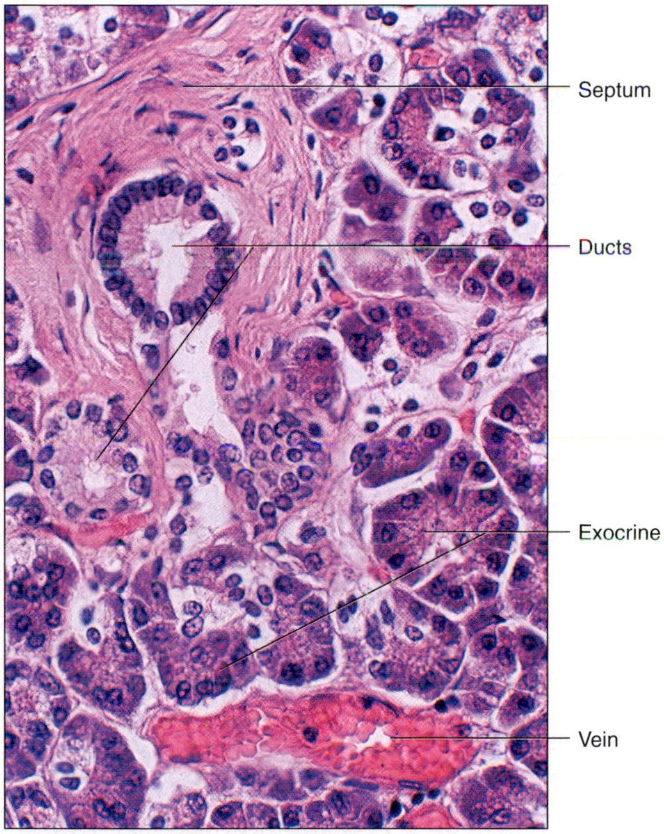
Septum
Ducts
Exocrine cells
Vein

(b)

FIGURE 24.21

Histology of the Pancreas. (*a*) An acinus. (*b*) Histological section of the exocrine tissue and some of the connective tissue stroma.

The pancreas is both an endocrine and exocrine gland. Its endocrine part is the pancreatic islets, which secrete the hormones insulin and glucagon (see chapter 18). Ninety-nine percent of the pancreas is exocrine tissue, which secretes enzymes and sodium bicarbonate. The exocrine pancreas is a compound tubuloacinar gland—that is, it has a system of branching ducts whose finest branches end in sacs of secretory cells, the acini. The cells of the acini exhibit a high density of rough ER and *zymogen granules,* which are vesicles filled with secretion (fig. 24.21). The smaller ducts converge on a main **pancreatic duct,** which runs lengthwise through the middle of the gland and joins the bile duct at the hepatopancreatic ampulla. Usually, there is a smaller **accessory pancreatic duct** that branches from the main pancreatic duct and opens independently into the duodenum at the minor duodenal papilla, proximal to the major papilla. The accessory duct bypasses the hepatopancreatic sphincter and allows pancreatic juice to be released into the duodenum even when bile is not.

TABLE 24.3

Exocrine Secretions of the Pancreas

Secretion	Function
Sodium bicarbonate	Neutralizes HCl
Zymogens	Converted to active digestive enzymes after secretion
Trypsinogen	Becomes trypsin, which digests protein
Chymotrypsinogen	Becomes chymotrypsin, which digests protein
Procarboxypeptidase	Becomes carboxypeptidase, which hydrolyzes the terminal amino acid from the carboxyl (–COOH) end of small peptides
Enzymes	
Pancreatic amylase	Digests starch
Pancreatic lipase	Digests fat
Ribonuclease	Digests RNA
Deoxyribonuclease	Digests DNA

The pancreas secretes 1,200 to 1,500 mL of **pancreatic juice** per day. This fluid is an alkaline mixture of water, sodium bicarbonate, other electrolytes, enzymes, and zymogens (table 24.3). Zymogens are inactive precursors of enzymes that are activated after they are secreted.

Before You Go On

Answer the following questions to test your understanding of the preceding section:

20. What does the liver contribute to digestion? List several of its nondigestive functions.
21. Describe the structure of a hepatic lobule and the blood flow through a lobule.
22. Describe the pathway that bile takes from a hepatocyte that secretes it to the point where it enters the duodenum.
23. Describe the pathway taken by pancreatic juice from a gland acinus to the duodenum.

DEVELOPMENTAL AND CLINICAL PERSPECTIVES

Objectives

When you have completed this section, you should be able to

- describe the prenatal development of the digestive tract, liver, and pancreas;
- describe the structural and functional changes in the digestive system in old age; and
- define and describe some common disorders of the digestive system.

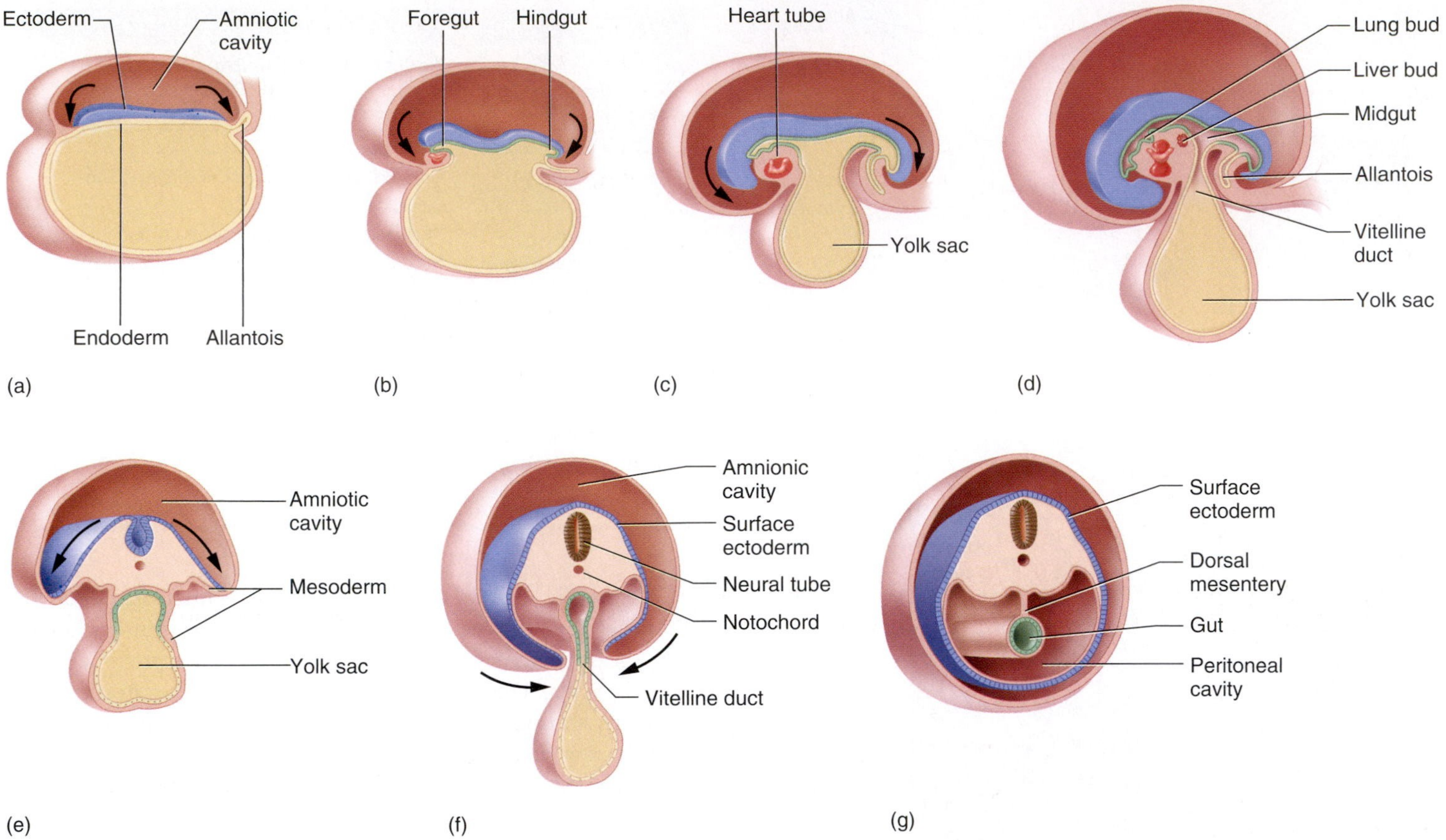
Ectoderm
Amniotic cavity
Endoderm
Allantois
(a)
Foregut
Hindgut
(b)
Heart tube
Yolk sac
(c)
Lung bud
Liver bud
Midgut
Allantois
Vitelline duct
Yolk sac
(d)
Amniotic cavity
Mesoderm
Yolk sac
(e)
Amnionic cavity
Surface ectoderm
Neural tube
Notochord
Vitelline duct
(f)
Surface ectoderm
Dorsal mesentery
Gut
Peritoneal cavity
(g)

FIGURE 24.22

Embryonic Development of the Digestive Tract. (*a–d*) Sagittal sections of the embryo from the end of week 2 (*a*) to the end of week 5 (*d*). Embryonic folding at the head and tail produce the foregut and hindgut. Folding of the lateral margins of the embryo encloses the midgut and reduces the opening into the yolk sac to a narrow vitelline duct. (*e–g*) Cross sections of the embryo, showing lateral folding, separation of the gut from the yolk sac, and formation of the dorsal mesentery.

Prenatal Development

The digestive system is one of the earliest organ systems to appear in the embryonic stage of development. Shortly after the three-layered embryonic disc is formed at 2 weeks, it elongates in the cephalocaudal (head-to-tail) direction. Endodermal pockets form at each end which become the foregut and hindgut (fig. 24.22*a,b*). Initially, there is a wide opening between the embryo and the yolk sac, but as the embryo continues to grow, the passage between them becomes constricted and a distinct tubular midgut appears. Temporarily, the midgut continues to communicate with the yolk sac through a narrow **vitelline duct** (fig. 24.22*c,d*). In week 4, the anterior end of the digestive tract breaks through to form the mouth, and 3 weeks later, the posterior end breaks through to form the anus.

Growth of the embryonic body segments (*somites;* see chapter 4) causes the lateral margins of the embryo to fold inward, changing the flat embryonic disc into a more cylindrical body and separating the embryonic body cavity from the yolk sac. By week 5, the gut is an elongated tube suspended from the body wall by the dorsal mesentery (fig. 24.22*e–g*). While the inner epithelial lining of the gut is endoderm, the tube is covered by a layer of mesoderm that gives rise to all other tissue layers of the digestive tract: the lamina propria, submucosa, muscle layers, and serosa.

At 4 weeks, the foregut exhibits a dilation that is the first sign of the future stomach (fig. 24.23). Further development of the digestive tract entails elongation, rotation, and differentiation of its regions into the esophagus, stomach, and small and large intestines. At 6 weeks, the body cavity is crowded by the rapidly growing liver, and the intestine is so long and crowded that a loop of it herniates into the umbilical cord. This loop normally withdraws back into the enlarged body cavity in week 10, but in some tragic cases it fails to do so, resulting in severely deformed infants with part of the bowel outside the body (*omphalocele*[38]).

The liver appears in the middle of week 3 as a **liver bud,** a pocketlike outgrowth of the endodermal tube at the junction of the foregut and midgut. Its connection to the gut narrows and

[38]*omphalo* = navel, umbilicus + *cele* = swelling, herniation

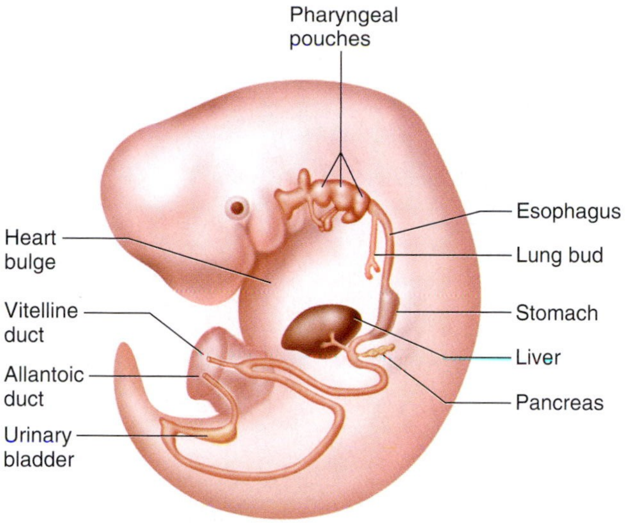
Pharyngeal pouches
Esophagus
Lung bud
Stomach
Liver
Pancreas
Heart bulge
Vitelline duct
Allantoic duct
Urinary bladder

FIGURE 24.23

Lateral View of the 5-week Embryo. The primordial stomach is present as a foregut dilation, and the liver and pancreatic buds are present. The loop of midgut approaching the allantoic duct will herniate into the umbilical cord within the next week.

becomes the bile duct. A small outgrowth from the ventral side of the bile duct becomes the gallbladder and cystic duct. By week 12, the liver secretes bile into the gut, so the gut contents become dark green. The liver produces blood cells throughout most of fetal development, but this function gradually subsides in the last 2 months.

The pancreas originates as two buds, a **dorsal** and **ventral pancreatic bud,** around week 4. The ventral bud eventually rotates dorsally and merges with the dorsal bud. Pancreatic islets appear in the third month and begin secreting insulin at 5 months.

At birth, the digestive tract contains dark, sticky feces called **meconium,** which is discharged in the first few bowel movements of the neonate.

The Aging Digestive System

Like most other organ systems, the digestive system shows significant degenerative change *(senescence)* in old age. Less saliva is secreted in old age, making food less flavorful, swallowing more difficult, and the teeth more prone to caries. Nearly half of people over age 65 wear dentures because they have lost their teeth to caries and periodontal disease. The stratified squamous epithelium of the oral cavity and esophagus is thinner and more vulnerable to abrasion.

The gastric mucosa atrophies and secretes less acid and intrinsic factor. Acid deficiency reduces the absorption of calcium, iron, zinc, and folic acid. The declining level of intrinsic factor reduces the absorption of vitamin B_{12}. Since this vitamin is needed for hemopoiesis, the deficiency can lead to a form of anemia called *pernicious anemia.*

Heartburn becomes more common in old age as the weakening lower esophageal sphincter fails to prevent reflux into the esophagus. The most common digestive complaint of older people is constipation, which results from the reduced muscle tone and weaker peristalsis of the colon. This seems to stem from a combination of factors: atrophy of the muscularis externa, reduced sensitivity to neurotransmitters, less fiber and water in the diet, and less exercise. The liver, gallbladder, and pancreas show only slightly reduced function in old age. Any drop in liver function, however, makes it harder to detoxify drugs and can contribute to overmedication.

Many older people reduce their food intake because of lower energy demand and appetite, because declining sensory functions make food less appealing, and because reduced mobility makes it more difficult to shop and prepare meals. However, they need fewer calories than younger people because they have lower basal metabolic rates and tend to be less physically active. Protein, vitamin, and mineral requirements remain essentially unchanged, although vitamin and mineral supplements may be needed to compensate for reduced food intake and intestinal absorption. Malnutrition is common among older people and is an important factor in anemia and reduced immunity.

Digestive Disorders

The digestive system is subject to a wide variety of disorders. Disorders of motility include difficulty swallowing *(dysphagia),* gastroesophageal reflux disease (GERD), and pyloric obstruction. Inflammatory diseases include esophagitis, gastritis, appendicitis, colitis, diverticulitis, pancreatitis, hepatitis, and cirrhosis. Cancer can occur in virtually every part of the digestive system: oral, esophageal, gastric, colon, rectal, hepatic, and pancreatic cancer. Colon and pancreatic cancer are among the leading causes of cancer deaths in the United States.

Digestive disorders can be manifested in a variety of signs and symptoms: anorexia (loss of appetite), vomiting, constipation, diarrhea, abdominal pain, or gastrointestinal bleeding. Many of these are nonspecific; they do not by themselves identify a particular digestive disorder. Gastrointestinal bleeding, for example, can result from *varices* (varicose veins) in the digestive tract wall, intestinal polyps, GI inflammation, hemorrhoids, peptic ulcers, parasitic infections, or cancer. Nausea is even less specific; it may result not only from nondigestive disorders but also from such causes as tumors in the abdomen or brainstem, trauma to the urogenital organs, or inner-ear dysfunction.

It was remarked earlier that the foregut, midgut, and hindgut are defined by differences in arterial blood supply. This embryonic division also extends to the nerve supply and to the perception of pain from the digestive tract. Gastrointestinal pain is often perceived as if it were coming from the abdominal wall (see *referred pain,* chapter 17). Pain arising from the foregut is referred to the epigastric region, midgut pain to the umbilical region, and hindgut pain to the hypogastric region.

Table 24.4 lists and describes some common digestive disorders.

TABLE 24.4

Some Digestive System Diseases

Acute Pancreatitis	Severe pancreatic inflammation perhaps caused by trauma leading to leakage of pancreatic enzymes into parenchyma, where they digest tissue and cause inflammation and hemorrhage.
Appendicitis	Inflammation of the appendix, with swelling, gangrene, pain, and a threat of perforation and peritonitis.
Ascites	Accumulation of serous fluid in the peritoneal cavity, often causing extreme distension of the abdomen. Most often caused by cirrhosis of the liver and frequently associated with alcoholism. The diseased liver "weeps" fluid into the abdomen.
Cirrhosis[39] *of the Liver*	An irreversible inflammatory disease of the liver often caused by alcoholism. Gives the liver a "cobbly" appearance and hard consistency due to fibrosis and nodular regeneration of damaged tissue. Obstruction of bile ducts causes jaundice, and obstruction of the circulation causes new vessels to grow and bypass the liver, leaving the liver subject to hypoxia, further necrosis, and failure.
Crohn[40] *Disease*	Inflammation of small and large intestines, similar to ulcerative colitis in symptoms and hereditary predisposition. Produces granular lesions and fibrosis of intestine; diarrhea; and lower abdominal pain.
Diverticulitis	Presence of inflamed herniations (outpocketings, diverticula) of the colon, associated especially with low-fiber diets. Diverticula may rupture, leading to peritonitis.
Dysphagia[41]	Difficulty swallowing. Can result from esophageal obstructions (tumors, constrictions) or impaired peristalsis (due to neuromuscular disorders).
Hiatal Hernia	Protrusion of part of the stomach into the thoracic cavity, where the negative thoracic pressure may cause it to balloon. Often causes gastroesophageal reflux (especially when a person is supine).
Ulcerative Colitis	Chronic inflammation resulting in ulceration of the large intestine, especially the sigmoid colon and rectum. Tends to be hereditary but exact causes are not well known.

Disorders Described Elsewhere

Constipation 701	Gingivitis 682	Mumps 683
Dental caries 682	Hemorrhoids 694	Omphalocele 700
Gallstones 698	Hepatitis 756	Peptic ulcer 689
Gastroesophageal reflux disease 685	Impacted molars 683	Periodontal disease 682

[39] *cirrho* = orange-yellow + *osis* = condition
[40] Burrill B. Crohn (1884–1983), American gastroenterologist
[41] *dys* = bad, difficult, abnormal + *phag* = eat, swallow

Before You Go On

Answer the following questions to test your understanding of the preceding section:

24. Explain why the foregut and hindgut appear in the embryo earlier than the midgut.
25. What accessory digestive gland arises as a single bud from the embryonic gut? What gland arises as a pair of buds that later merge?
26. Explain why dental caries, constipation, and heartburn become more common as the digestive system ages.
27. Explain why gastrointestinal bleeding and nausea provide only inconclusive evidence for the existence and location of a digestive system disorder.

CHAPTER REVIEW

Review of Key Concepts

Digestive Processes and General Anatomy (p. 676)

1. The digestive system processes food, extracts nutrients, and eliminates the residue. It does this in five stages: ingestion, digestion, absorption, compaction, and defecation.
2. The *digestive tract* is a tube consisting of the oral cavity, pharynx, esophagus, stomach, and small and large intestines. The *accessory organs* are the teeth, tongue, salivary glands, liver, gallbladder, and pancreas.
3. In most areas, the wall of the digestive tract consists of an inner *mucosa*, a *submucosa*, a *muscularis externa*, and an outer *serosa*. The mucosa is usually composed of an epithelium, *lamina propria*, and *muscularis mucosae*. In some areas a connective tissue *adventitia* replaces the serosa.
4. The upper digestive system is innervated by somatic motor fibers from several cranial nerves and the ansa cervicalis. Autonomic fibers innervate the salivary glands and most of the digestive tract from esophagus to rectum. In general, parasympathetic stimulation promotes digestion and sympathetic stimulation inhibits it.
5. The *enteric nervous system* regulates most digestive activity and consists of two nerve networks in the wall of the digestive tract: the *submucosal plexus* and the *myenteric plexus*.
6. The foregut receives blood from esophageal arteries above the diaphragm and the celiac trunk below. The midgut is supplied by the superior mesenteric artery and the hindgut by the inferior mesenteric artery. The abdominal organs of the digestive system are drained by the hepatic portal system.
7. Veins from all of the digestive tract below the diaphragm drain into the hepatic portal system, which routes blood to the liver. The liver thus has opportunity to extract nutrients from the intestinal blood before the blood flows to any other organ.
8. In the abdominal cavity, the *dorsal mesentery* suspends the digestive tract from the body wall, wraps around it to form the serosa, and in some places continues as a *ventral mesentery*. The ventral mesentery includes the *greater* and *lesser omenta*.
9. Digestive organs completely enclosed in a serosa are *intraperitoneal*. Organs that lie against the abdominal wall and are covered by a serosa only anteriorly are *retroperitoneal*.

The Mouth Through Esophagus (p. 680)

1. The *mouth (oral cavity)* serves for ingestion, sensory responses to food, mastication, chemical digestion, swallowing, speech, and respiration.
2. The mouth extends from the *oral orifice* anteriorly to the *fauces* posteriorly. Its anatomical elements include the lips, cheeks, tongue, hard and soft palates, teeth, and *intrinsic salivary glands*. It also receives saliva from three pairs of *extrinsic salivary glands: parotid, sublingual,* and *submandibular*.
3. The tongue functions in ingestion, the manipulation and physical breakdown of food, the sense of taste, mucus and enzyme secretion, and swallowing. It exhibits surface projections called *lingual papillae*, many of them with taste buds. *Intrinsic* and *extrinsic muscles* control tongue movements.
4. The bony *hard palate* separates the oral cavity from the nasal cavity. Posterior to this, the fleshy *soft palate* separates the oropharynx from the nasopharynx.
5. The adult teeth *(dentition)* include two incisors, one canine, two premolars, and up to three molars on each side of each jaw. A tooth is composed mainly of dentin, covered with cementum on the root and enamel on the crown. It encloses a *pulp cavity* and *root canal* occupied by blood vessels, nerves, and loose connective tissue.
6. Mastication breaks food into pieces small enough to be swallowed and exposes more food surface to the action of digestive enzymes, making digestion more efficient.
7. Saliva moistens the mouth, digests starch and fat, cleanses the teeth, inhibits bacterial growth, dissolves taste molecules, and binds food into a soft *bolus* to facilitate swallowing. It contains amylase, lipase, mucus, lysozyme, IgA, and electrolytes.
8. Saliva is produced by *intrinsic salivary glands* in the tongue, lips, and cheeks, and *extrinsic glands* (the *parotid, submandibular,* and *sublingual glands*) with ducts leading to the oral cavity.
9. Salivation is controlled by *salivatory nuclei* in the medulla oblongata and pons and occurs in response to the thought, odor, sight, taste, or oral feel of food.
10. The pharynx is a muscular funnel in the throat where the respiratory and digestive tracts meet. Its wall contains three sets of *pharyngeal constrictor* muscles that aid in swallowing.
11. The esophagus is a muscular tube from the pharynx to the *cardiac orifice* of the stomach. It is lined with a nonkeratinized stratified squamous epithelium, has a mixture of skeletal muscle (dominating the upper esophagus) and smooth muscle (dominating the lower), and is lubricated by mucous *esophageal glands* in the submucosa.
12. The upper end of the esophagus is normally held closed by the *upper esophageal sphincter*, a constriction maintained by muscle tone in the inferior pharyngeal constrictor. This sphincter prevents air from entering the esophagus during breathing.
13. The lower end of the esophagus is held closed by the *lower esophageal sphincter*, a constriction maintained by muscle tone around the esophageal hiatus of the diaphragm or in the esophageal smooth muscle. This sphincter protects the esophagus from regurgitation of stomach acid.
14. *Deglutition* (swallowing) requires the coordinated action of numerous muscles and is integrated by the *swallowing center* of the medulla oblongata and pons.

The Stomach (p. 685)

1. The stomach is primarily a food-storage organ, with a capacity of 4 L. It mechanically breaks up food, begins the chemical digestion of proteins and fat, and converts ingested food to soupy *chyme*.
2. The stomach extends from the *cardiac orifice* proximally to the *pylorus* distally. Its subdivisions are the *cardiac region, fundic region, body,* and *pyloric region*. The pylorus is regulated by the *pyloric sphincter*.
3. The gastric mucosa has a simple columnar epithelium of mucous cells and is marked by *gastric pits*. Two or three tubular glands open into the bottom of each pit. Most of the stomach has digestive *gastric glands*, whereas the cardiac and pyloric regions have mucous *cardiac glands* and *pyloric glands*, respectively. These glands contain mucous, regenerative, parietal, chief, and enteroendocrine cells.
4. The stomach is protected from self-digestion by its mucous coat, tight junctions between the epithelial cells, and a high rate of cellular replacement.

The Small Intestine (p. 689)

1. The small intestine carries out most digestion and nutrient absorption. Efficient and thorough digestion and absorption require a large surface area, which is provided by its great length and by its internal *circular folds, villi,* and *microvilli.*
2. The *duodenum* begins at the pylorus and extends for about 12 cm to the *duodenojejunal flexure.* It receives the stomach contents and secretions of the liver and pancreas. The submucosa contains mucous *duodenal glands.*
3. The *jejunum* extends for the next 2.5 m. Most digestion and nutrient absorption occur here. It has high, closely spaced circular folds and a relatively thick, muscular wall with a high density of blood vessels, giving it a reddish color.
4. The *ileum* is the final 3.6 m, ending at the *ileocecal junction* (gateway to the large intestine). It has a thinner wall and is paler in color; it has lower circular folds but more lymphocytes, with lymphatic nodules clustered in *Peyer patches.*
5. Villi are tongue- to finger-shaped projections of the mucosal surface, covered with a columnar epithelium of absorptive cells and goblet cells. The absorptive cells have a brush border of microvilli on their surface. The core of a villus contains blood capillaries, which absorb most nutrients, and a *lacteal* (lymphatic capillary), which absorbs lipids.
6. The brush border bears enzymes that carry out the terminal *contact digestion* of carbohydrates and peptides, and an enzyme, *enterokinase,* that activates the pancreatic enzyme trypsin.
7. Glandular *intestinal crypts* open onto the floor of the intestine between the villi. The crypt epithelium is composed of absorptive and goblet cells, stem cells, enteroendocrine cells, and antibacterial *Paneth cells.*

The Large Intestine (p. 692)

1. The large intestine receives the indigestible residue of food, absorbs water and salts, consolidates the residue into feces, and eliminates the feces by defecation.
2. The large intestine is about 1.5 m long and consists of the *cecum;* the *ascending, transverse, descending,* and *sigmoid colon; rectum;* and *anal canal.* The *colon* is the region from the ascending through sigmoid colon.
3. The longitudinal layer of the muscularis externa consists of three strips of muscle, the *teniae coli,* whose muscle tone folds the wall of the colon into pouches called *haustra.*
4. The anal canal has an involuntary *internal anal sphincter* of smooth muscle and a voluntary *external anal sphincter* of skeletal muscle.
5. The mucosa is mostly simple columnar epithelium except for the lower half of the anal canal, which is stratified squamous. The mucosa has intestinal crypts but lacks villi and circular folds. Mucus is the only substance secreted by the large intestine.

Accessory Glands of Digestion (p. 694)

1. The liver is the body's largest gland and has a broad range of metabolic functions (table 24.2). Its one digestive function is to secrete *bile acids* and *lecithin,* which emulsify dietary fats and facilitate their digestion by pancreatic lipase.
2. The liver is divided into four lobes: the right, left, caudate, and quadrate. An opening on the inferior surface, the *porta hepatis,* receives the hepatic portal vein and hepatic artery, and is the exit for the bile duct system.
3. The liver parenchyma is composed of microscopic, cylindrical *hepatic lobules.* Each lobule has sheets of *hepatocytes* (liver cells) that fan out around a *central vein.* Blood filters through narrow spaces called *hepatic sinusoids* between the sheets of hepatocytes.
4. The liver receives nutrient-rich intestinal blood from the hepatic portal vein and oxygen-rich arterial blood from the hepatic artery. The two bloodstreams mix in the sinusoids. The sinusoids converge on the central vein of the lobule. These ultimately lead to two *hepatic veins* that exit the superior surface of the liver and lead to the inferior vena cava.
5. Hepatocytes secrete bile into channels called the *bile canaliculi.* Bile exits the liver via the *right* and *left hepatic ducts.* These join to form the *common hepatic duct.* This, in turn, joins the *cystic duct* from the gallbladder to form the *bile duct.* The bile duct leads to the duodenum, usually joining the pancreatic duct just before emptying into the duodenum.
6. The gallbladder is a sac on the inferior surface of the liver that stores and concentrates the bile.
7. The pancreas produces the hormones insulin and glucagon, and about 1,200 to 1,500 mL of pancreatic juice per day. The pancreatic juice passes through a *pancreatic duct,* which joins the bile duct before emptying into the duodenum at the *major duodenal papilla.* Usually a smaller *accessory pancreatic duct* opens independently into the duodenum, at the *minor duodenal papilla* proximal to the major papilla.
8. The pancreatic secretions are summarized in table 24.3.

Developmental and Clinical Perspectives (p. 699)

1. As the embryonic disc begins to elongate at 2 weeks, the foregut and hindgut appear. The midregion becomes an enclosed midgut as the opening between the embryo and yolk sac constricts. The digestive mucosal epithelium forms from the embryonic endoderm, and the other layers of the GI tract from mesoderm.
2. The liver bud appears at 3.5 weeks, and the stomach and two pancreatic buds at 4 weeks. The GI tract continues its development by elongation, rotation, and differentiation of its regions. Between weeks 6 and 10, the body cavity is so crowded that a loop of intestine herniates into the umbilical cord.
3. In old age, declining salivation makes food less appealing and dental caries and periodontal disease more common. Gastric atrophy promotes poorer nutrient absorption. Heartburn (gastroesophageal reflux) is increasingly common, and reduced GI motility tends to cause constipation. Malnutrition in old age is common and has causes ranging from the difficulty of shopping and cooking to declining nutrient absorption by the intestinal mucosa.
4. The digestive system is subject to a wide range of disorders including abnormalities in motility, inflammatory diseases, and several kinds of cancer. Digestive disorders can be manifested in anorexia, vomiting, constipation, diarrhea, pain, or bleeding. GI pain is referred to different regions of the anterior abdominal wall correlated with the foregut, midgut, or hindgut origin of the pain.

Testing Your Recall

1. All of the following are retroperitoneal *except*
   a. the liver.
   b. the pancreas.
   c. the duodenum.
   d. the ascending colon.
   e. the descending colon.
2. The falciform ligament attaches the _____ to the abdominal wall.
   a. colon
   b. liver
   c. spleen
   d. pancreas
   e. stomach
3. A brush border is found on the
   a. goblet cells.
   b. intestinal absorptive cells.
   c. enteroendocrine cells.
   d. parietal cells.
   e. chief cells.
4. The yolk sac is connected to the embryonic _____ by way of the vitelline duct.
   a. liver bud
   b. stalk
   c. foregut
   d. midgut
   e. hindgut
5. Lacteals absorb dietary
   a. proteins.
   b. carbohydrates.
   c. enzymes.
   d. vitamins.
   e. lipids.
6. All of the following contribute to the absorptive surface area of the small intestine *except*
   a. its length.
   b. the brush border.
   c. haustra.
   d. circular folds.
   e. villi.
7. Which of the following is a periodontal tissue?
   a. gingiva
   b. enamel
   c. cementum
   d. pulp
   e. dentin
8. The _____ of the stomach most closely resemble the _____ of the small intestine.
   a. gastric pits, intestinal crypts
   b. pyloric glands, intestinal crypts
   c. rugae, Peyer patches
   d. parietal cells, goblet cells
   e. gastric glands, duodenal glands
9. Which of the following cells secrete digestive enzymes?
   a. chief cells
   b. mucous neck cells
   c. parietal cells
   d. goblet cells
   e. enteroendocrine cells
10. The tissue layer between the muscularis mucosae and muscularis externa of the digestive tract is
    a. the mucosa.
    b. the lamina propria.
    c. the submucosa.
    d. the serosa.
    e. the adventitia.
11. The alimentary canal has an extensive nervous network called the _____.
12. The passage of chyme from the stomach into the duodenum is controlled by a muscular ring called the _____.
13. The _____ salivary gland is named for its location near the ear.
14. A wave of contraction traveling along the esophagus or intestine is called _____.
15. Nervous stimulation of gastrointestinal activity is mediated mainly through the parasympathetic fibers of the _____ nerves.
16. Hydrochloric acid is secreted by _____ cells of the stomach.
17. Hepatic macrophages occur in blood-filled spaces of the liver called _____.
18. The superior opening into the stomach is called the _____.
19. The root of a tooth is covered with a calcified tissue called _____.
20. Protrusions of the tongue surface, some of which bear taste buds, are called _____.

Answers in the Appendix

True or False

Determine which five of the following statements are false, and briefly explain why.

1. The liver and pancreas are retroperitoneal.
2. A tooth is composed mostly of enamel.
3. Hepatocytes secrete bile into the hepatic sinusoids.
4. The small intestine is much shorter in a living person than it is after death.
5. Peristalsis is controlled by the myenteric nerve plexus.
6. Pepsinogen, trypsinogen, and procarboxypeptidase are enzymatically inactive zymogens.
7. The greater omentum suspends the stomach from the body wall.
8. Salivary acini can be composed of mucous cells, serous cells, or both.
9. In all parts of the digestive tract, the muscularis externa has two layers.
10. The external anal sphincter is under voluntary control; the internal sphincter is not.

Answers in the Appendix

Testing Your Comprehension

1. People who suffer from GERD (see insight 24.2) upon retiring often find their heartburn is worse when they lie on the right side than when they lie on the left. Give an anatomical explanation for this effect.
2. Cystic fibrosis (CF) is characterized by unusually thick, sticky mucus that obstructs the respiratory tract and pancreatic duct. Predict the effect of CF on digestion, nutrition, and growth in childhood.
3. Explain why the small intestine would function poorly if it had the same type of mucosal epithelium as the esophagus.
4. News reports of patients (especially children) in need of organ transplants often prompt people to call and offer to donate one of their organs, such as a kidney, to save the patient's life. If you worked in an organ donor program, what would you say to a well-meaning volunteer offering to donate a liver?
5. The hyposecretion of pancreatic bicarbonate and the hyposecretion of mucus by duodenal goblet cells could both contribute to the same pathological result. What would that result be, and why would it occur?

Answers at the Online Learning Center

www.mhhe.com/saladinha1

Visit the Online Learning Center for practice tests, answer keys, and other learning aids for this chapter. Enhance your understanding of human anatomy with our interactive art labeling exercises, supplemental photo atlases, web links, puzzles, flashcards, and much more.

CHAPTER TWENTY-FIVE

The Urinary System

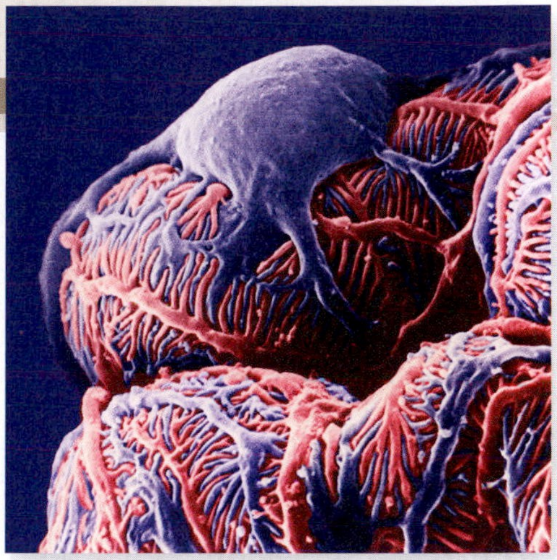

Two interlacing podocytes of the kidney

CHAPTER OUTLINE

Functions of the Urinary System 708

Anatomy of the Kidney 709
- Position and Associated Structures 709
- Gross Anatomy 709
- Circulation 710
- Innervation 710
- The Nephron 710

Anatomy of the Ureters, Urinary Bladder, and Urethra 717
- The Ureters 717
- The Urinary Bladder 718
- The Urethra 719

Developmental and Clinical Perspectives 719
- Prenatal Development 719
- The Aging Urinary System 720
- Urinary System Disorders 721

Chapter Review 723

INSIGHTS

25.1 Clinical Application: Blood and Protein in the Urine 714
25.2 Evolutionary Medicine: The Kidney and Life on Dry Land 716
25.3 Clinical Application: Kidney Stones 718
25.4 Clinical Application: Urinary Tract Infections 719
25.5 Clinical Application: Developmental Abnormalities of the Kidney 721

BRUSHING UP

To understand this chapter, you may find it helpful to review the following concepts:
- Transitional epithelium (p. 85)
- General exocrine gland architecture (pp. 96–97)
- Fenestrated capillaries (p. 591)

Metabolism produces a variety of waste products that cannot be allowed to accumulate in the body, for if they do, the body is quickly poisoned. The respiratory and digestive systems and the sweat glands eliminate some wastes, but the urinary system is the principal means of waste excretion. Its functions go far beyond that, however. It also collaborates with the endocrine, circulatory, and respiratory systems to regulate many aspects of homeostasis such as blood pressure, erythrocyte count, blood gases, and electrolyte and acid-base balance.

Anatomically, the urinary system is closely associated with the reproductive system. In many animals the eggs and sperm are emitted through the urinary tract, and the two systems share some aspects of evolutionary history, embryonic development, and adult anatomy. This is reflected in humans, where the systems develop together in the embryo and, in the male, the urethra continues to serve as a passage for both urine and sperm. Thus the urinary and reproductive systems are often collectively called the *urogenital (U-G)* or *genitourinary (G-U) system,* and *urologists* treat both urinary and reproductive disorders. Because of their anatomical and developmental relationship, we consider the urinary and reproductive systems in these last two chapters.

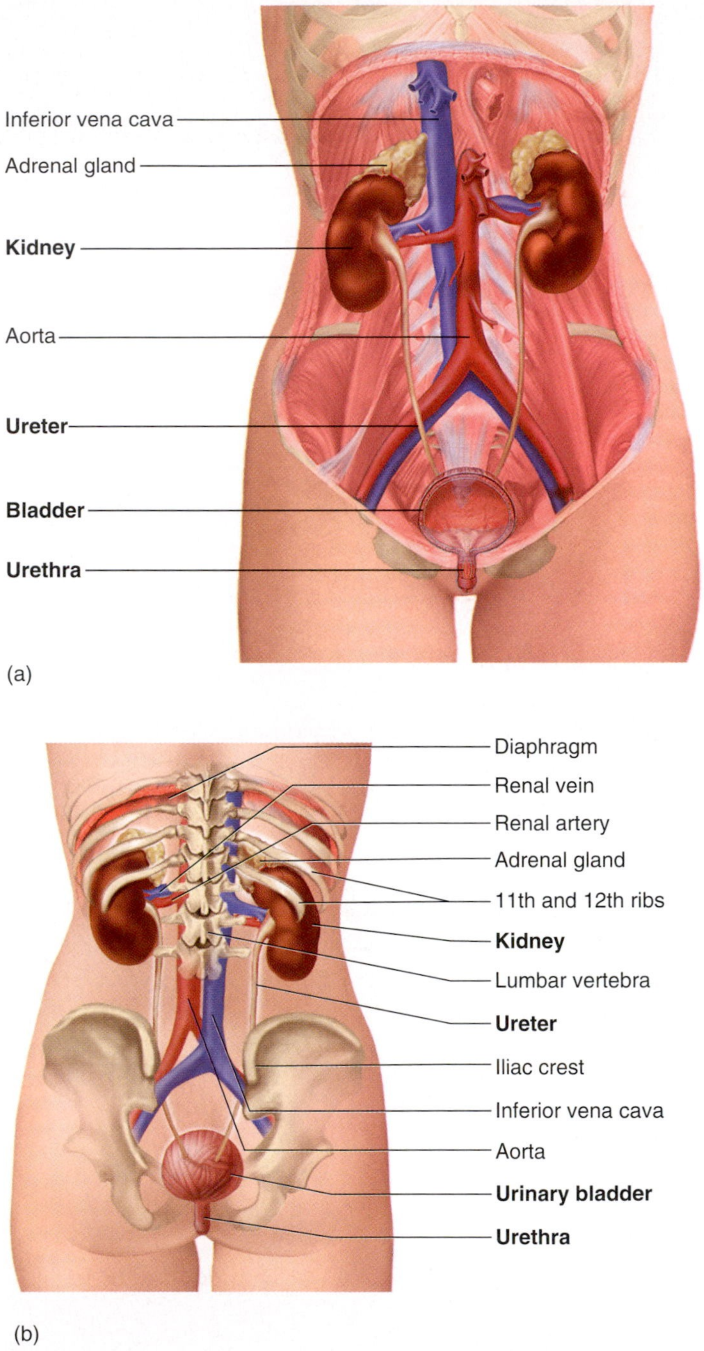

FIGURE 25.1
The Urinary System. (*a*) Ventral view, with peritoneum and other tissues removed to expose the retroperitoneal urinary organs. Organs of the urinary system are indicated in boldface. (*b*) Dorsal view, showing relationship of the urinary organs to skeletal landmarks.

Functions of the Urinary System

Objectives
When you have completed this section, you should be able to

- name and locate the organs of the urinary system;
- list several functions of the kidneys in addition to urine formation;
- define *excretion;* and
- identify the major nitrogenous waste excreted by the kidneys.

The **urinary system** consists of six organs: two **kidneys,** two **ureters,** the **urinary bladder,** and the **urethra** (fig. 25.1). Most of the focus of this chapter is on the kidneys.

The kidneys have more functions than are commonly realized:

- They filter blood plasma, separate wastes from the useful chemicals, and eliminate the wastes while returning the rest to the bloodstream.
- They regulate blood volume and pressure by eliminating or conserving water as necessary.
- They regulate the osmolarity of the body fluids by controlling the ratio of water to solutes retained in the body.
- They secrete the enzyme *renin,* which catalyzes a step in the synthesis of the hormone angiotensin II, a vasoconstrictor that raises blood pressure. Angiotensin II also stimulates the adrenal cortex to secrete aldosterone, the "salt-retaining hormone," which stimulates the kidneys to retain Na^+ and water.
- They secrete the hormone *erythropoietin,* which controls the red blood cell count and therefore the oxygen-carrying capacity of the blood.
- They carry out the final step in synthesizing the hormone *calcitriol* (vitamin D), and thereby contribute to calcium homeostasis.
- They collaborate with the lungs to regulate the CO_2 and acid-base balance of the body fluids.
- They detoxify free radicals and drugs.
- In times of starvation, they convert amino acids to glucose (a process called *gluconeogenesis*) and thus help to support blood glucose level.

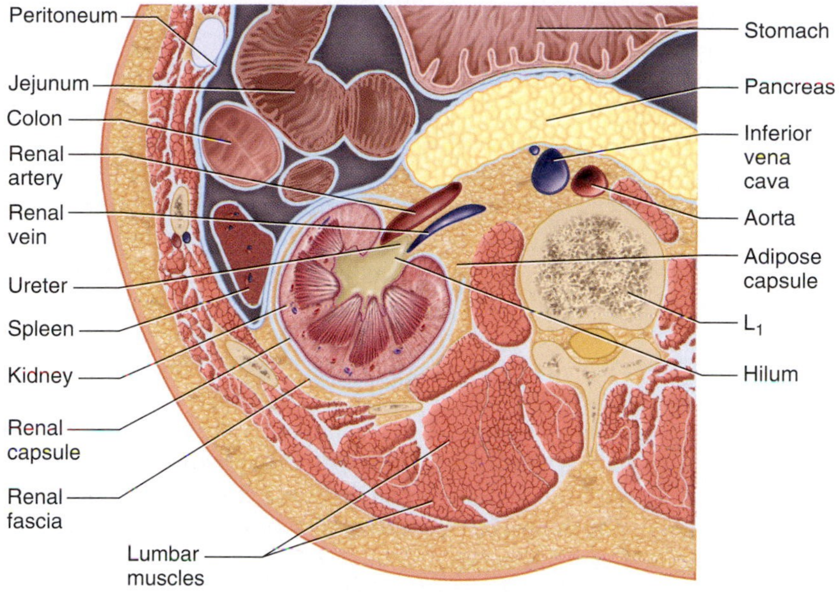
Peritoneum
Jejunum
Colon
Renal artery
Renal vein
Ureter
Spleen
Kidney
Renal capsule
Renal fascia
Lumbar muscles
Stomach
Pancreas
Inferior vena cava
Aorta
Adipose capsule
L_1
Hilum

FIGURE 25.2

Location of the Kidney. Cross section of the abdomen at the level of vertebra L1, showing the relationship of the kidney to the body wall and peritoneum.

Excretion, the most obvious function of the urinary system, is the process of extracting wastes from the body fluids and eliminating them, thus preventing metabolic poisoning of the body. Among other things, the kidneys excrete organic nitrogen-containing molecules called **nitrogenous wastes.** The most abundant of these is **urea,** a product of protein metabolism. If the kidneys do not function adequately, one develops a condition called **azotemia**[1] (AZ-oh-TEE-me-uh), in which the blood urea concentration (or *blood-urea nitrogen, BUN*) is abnormally high. In severe renal failure, azotemia progresses to **uremia** (you-REE-me-uh), a syndrome of diarrhea, vomiting, dyspnea (labored breathing), and cardiac arrhythmia. Convulsions, coma, and death can follow within a few days, underscoring the importance of adequate renal function.

Before You Go On

Answer the following questions to test your understanding of the preceding section:

1. State four functions of the kidneys other than forming urine.
2. What is the most abundant nitrogenous waste in the urine? What terms describe an abnormally high level of this waste in the blood, and poisoning by this waste?

Anatomy of the Kidney

Objectives

When you have completed this section, you should be able to

- describe the location and general appearance of the kidney, and its relationship to neighboring organs;
- identify the major external and internal features of the kidney;
- trace the flow of blood through the kidney;
- describe the nerve supply to the kidney;
- trace the flow of fluid through the renal tubules; and
- state the function of each segment of a renal tubule.

Position and Associated Structures

The kidneys lie against the dorsal abdominal wall at the level of vertebrae T12 to L3. Rib 12 crosses the approximate middle of the kidney. The right kidney is slightly lower than the left because of the space occupied by the large right lobe of the liver above it. The kidneys are retroperitoneal, along with the ureters, urinary bladder, renal artery and vein, and the adrenal[2] glands (fig. 25.2). The left adrenal gland rests on the superior pole of that kidney, while the right adrenal gland lies against the superomedial surface of its kidney. Their functions (see chapter 18) are not as directly related to the kidneys as their spatial relationship might suggest, although the kidneys and adrenals do influence each other.

Gross Anatomy

The kidney is a compound tubular gland containing about 1.2 million functional excretory units called **nephrons**[3] (NEF-rons). Each kidney weighs about 160 g and measures about 10 cm long, 5 cm wide, and 2.5 cm thick—about the size of a bar of bath soap. The lateral surface is convex while the medial surface is concave and has a slit, the **hilum,** where it receives the renal nerves, blood vessels, lymphatics, and ureter.

[1] *azot* = nitrogen + *emia* = blood condition

[2] *ad* = to, toward, near + *ren* = kidney + *al* = pertaining to
[3] *nephro* = kidney

The kidney is protected by three layers of connective tissue: (1) A fibrous **renal fascia,** immediately deep to the parietal peritoneum, binds the kidney and associated organs to the abdominal wall; (2) the **adipose capsule,** a layer of fat, cushions the kidney and holds it in place; and (3) the **renal capsule,** a fibrous sac, encloses the kidney like a cellophane wrapper anchored at the hilum, and protects it from trauma and infection. Collagen fibers extend from the renal capsule, through the fat, to the renal fascia. The renal fascia is fused with the peritoneum ventrally and with the deep fascia of the lumbar muscles dorsally. Thus the kidneys are suspended in place. Nevertheless, they drop about 3 cm when one goes from a supine to a standing position, and under some circumstances they become detached and drift even lower, with pathological results (see nephroptosis, or "floating kidney," in table 25.2 at the end of this chapter).

The renal parenchyma—the glandular tissue that forms the urine—appears C-shaped in frontal section. It encircles a medial space, the **renal sinus,** occupied by blood and lymphatic vessels, nerves, and urine-collecting structures. Adipose tissue fills the remaining space in the sinus and holds these structures in place.

The parenchyma is divided into two zones: an outer **renal cortex** about 1 cm thick and an inner **renal medulla** facing the sinus (fig. 25.3). Extensions of the cortex called **renal columns** project toward the sinus and divide the medulla into 6 to 10 **renal pyramids.** Each pyramid is conical, with a broad base facing the cortex and a blunt point called the **renal papilla** facing the sinus. One pyramid and the overlying cortex constitute one *lobe* of the kidney.

The papilla of each renal pyramid is nestled in a cup called a **minor calyx**[4] (CAY-lix), which collects its urine. Two or three minor calyces (CAY-lih-seez) converge to form a **major calyx,** and two or three major calyces converge in the sinus to form the funnel-like **renal pelvis.**[5] The ureter is a tubular continuation of the renal pelvis that drains the urine down to the urinary bladder.

Circulation

Although the kidneys account for only 0.4% of the body weight, they receive about 21% of the cardiac output (the *renal fraction*). This is a hint of how important the kidneys are in regulating blood volume and composition.

The larger divisions of the renal circulation are shown in figure 25.4. Each kidney is supplied by a **renal artery** arising from the aorta. Just before or after entering the hilum, the renal artery divides into a few **segmental arteries,** and each of these gives rise to a few **interlobar arteries.** An interlobar artery penetrates each renal column and travels between the pyramids toward the *corticomedullary junction,* the boundary between the cortex and medulla. Along the way, it branches again to form **arcuate arteries,** which make a sharp 90° bend and travel along the base of the pyramid. Each arcuate artery gives rise to several **interlobular arteries,** which pass upward into the cortex.

The finer branches of the renal circulation are shown in figure 25.5. As an interlobular artery ascends through the cortex, a series of **afferent arterioles** arise from it at nearly right angles like the limbs of a pine tree. Each afferent arteriole supplies one nephron. It leads to a spheroidal mass of capillaries called a **glomerulus**[6] (glo-MERR-you-lus), enclosed in a nephron structure called the *glomerular capsule,* to be discussed later. The glomerulus is drained by an **efferent arteriole.** The efferent arteriole usually leads to a plexus of **peritubular capillaries,** named for the fact that they form a network around the renal tubules. These capillaries pick up the water and solutes reabsorbed by the renal tubules.

From the peritubular capillaries, blood flows to **interlobular veins, arcuate veins, interlobar veins,** and the **renal vein,** in that order. These veins travel parallel to the arteries of the same names. (There are, however, no segmental veins corresponding to the segmental arteries.) The renal vein leaves the hilum and drains into the inferior vena cava.

The renal medulla receives only 1% to 2% of the total renal blood flow, supplied by a network of vessels called the **vasa recta.**[7] These arise from the nephrons in the deep cortex, closest to the medulla *(juxtamedullary nephrons).* Here, the efferent arterioles descend immediately into the medulla and give rise to the vasa recta instead of peritubular capillaries. The capillaries of the vasa recta lead into venules that ascend and empty into the arcuate and interlobular veins. Capillaries of the vasa recta are wedged into the tight spaces between the medullary parts of the renal tubule, and carry away water and solutes reabsorbed by those sections of the tubule. Figure 25.4*b* summarizes the route of renal blood flow.

THINK ABOUT IT!

Can you identify a portal system in the renal circulation?

Innervation

Renal nerves arise from the superior mesenteric ganglion (see p. 460) and enter the hilum of each kidney. They follow branches of the renal artery and innervate the afferent and efferent arterioles. These nerves consist mostly of sympathetic fibers that regulate the blood flow into and out of each nephron, and thus control the rate of filtration and urine formation. If the blood pressure falls, they also stimulate the secretion of renin, an enzyme that activates hormonal mechanisms for restoring the blood pressure.

The Nephron

A nephron (fig. 25.5) consists of two principal parts: a *renal corpuscle,* which filters the blood plasma, and a long *renal tubule,* which converts the filtrate to urine.

Before we embark on the microscopic anatomy of the nephron, it will be helpful to have a broad overview of the process of urine production. This knowledge will lend functional meaning to the structural details of the nephron. The kidney converts blood plasma to urine in three stages (fig. 25.6):

1. **Glomerular filtration** is the passage of fluid out of the bloodstream into the nephron, carrying not only wastes but

[4]*calyx* = cup
[5]*pelvis* = basin
[6]*glomer* = ball + *ulus* = little
[7]*vasa* = vessels + *recta* = straight

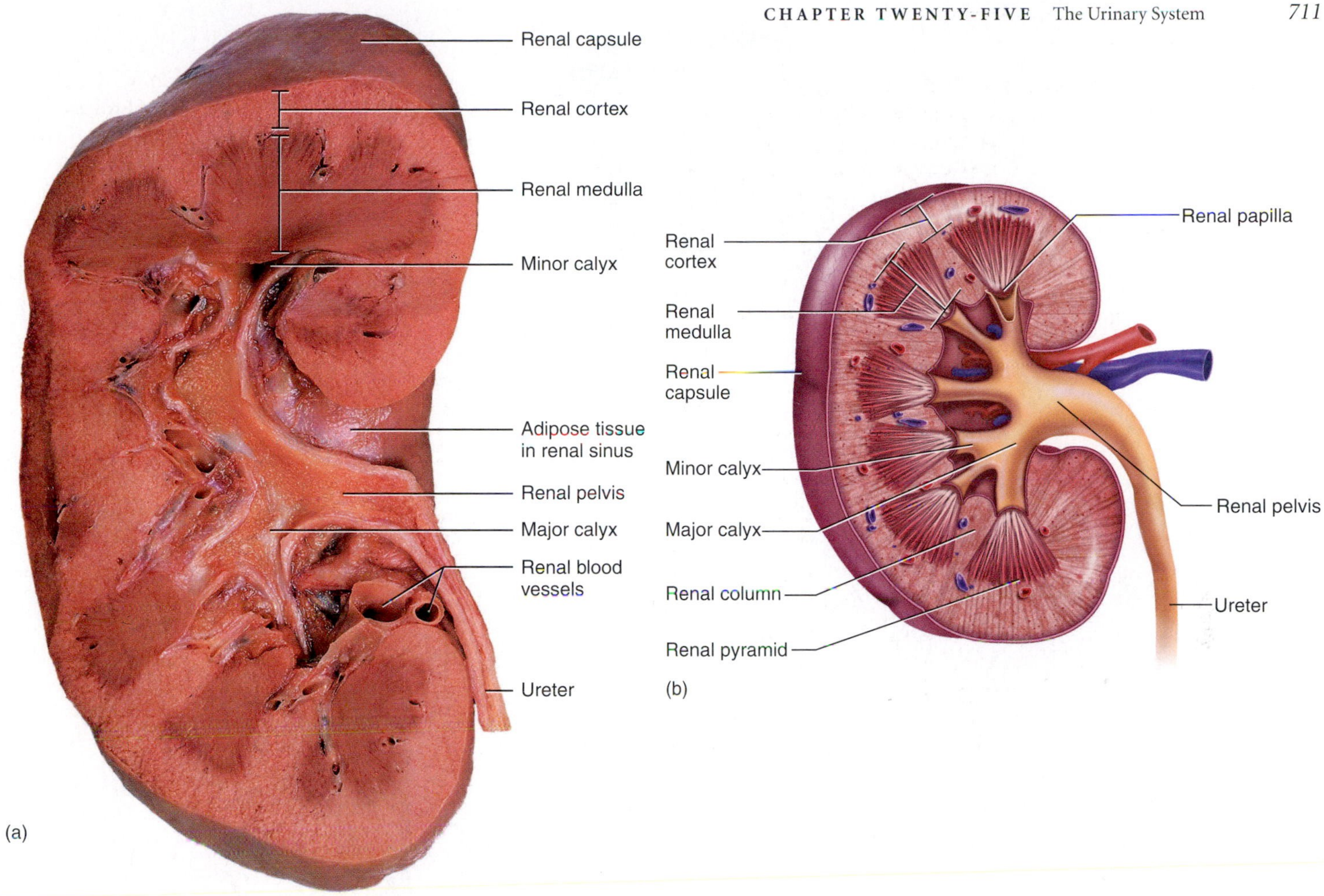
Renal capsule
Renal cortex
Renal medulla
Minor calyx
Adipose tissue in renal sinus
Renal pelvis
Major calyx
Renal blood vessels
Ureter
(a)
Renal papilla
Renal cortex
Renal medulla
Renal capsule
Minor calyx
Major calyx
Renal column
Renal pyramid
Renal pelvis
Ureter
(b)

FIGURE 25.3

Gross Anatomy of the Kidney. (*a*) Photograph of frontal section. (*b*) Major anatomical features.

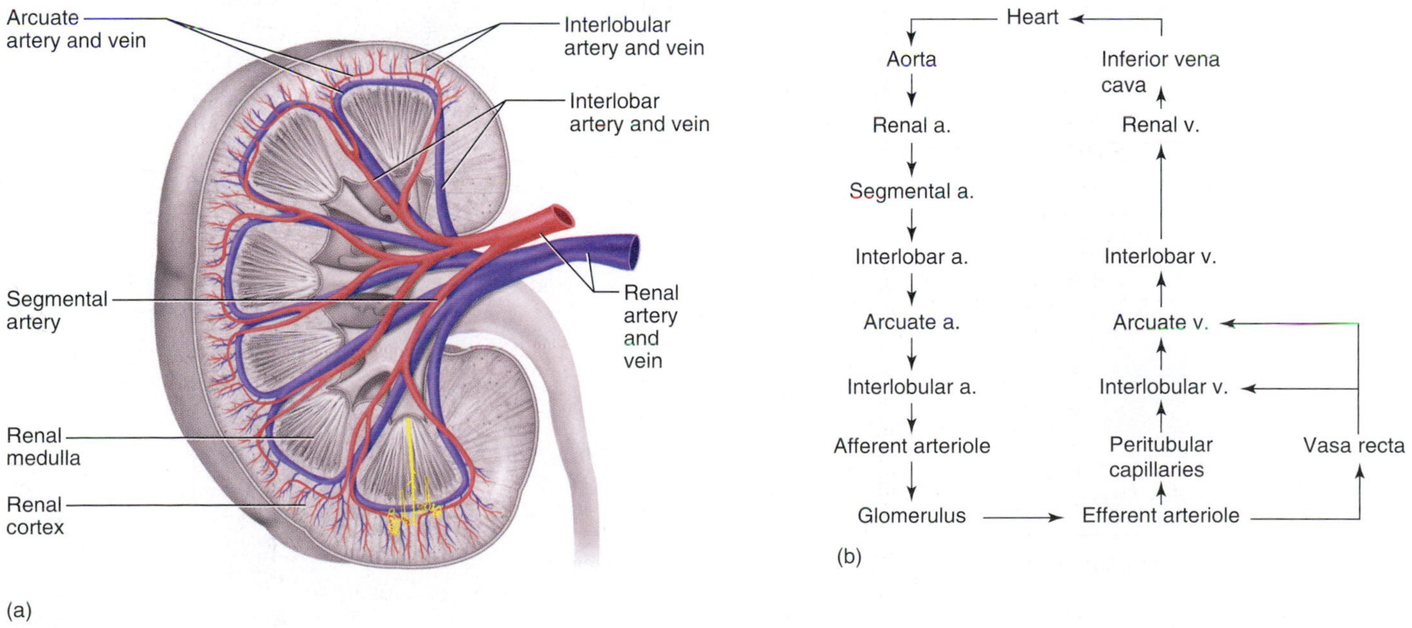
Arcuate artery and vein
Interlobular artery and vein
Interlobar artery and vein
Segmental artery
Renal artery and vein
Renal medulla
Renal cortex
(a)
Heart
Aorta
Renal a.
Segmental a.
Interlobar a.
Arcuate a.
Interlobular a.
Afferent arteriole
Glomerulus
Inferior vena cava
Renal v.
Interlobar v.
Arcuate v.
Interlobular v.
Peritubular capillaries
Efferent arteriole
Vasa recta
(b)

FIGURE 25.4

Renal Circulation. (*a*) The larger blood vessels of the kidney. (*b*) Flow chart of renal circulation. The pathway through the vasa recta (instead of peritubular capillaries) applies only to the juxtamedullary nephrons.

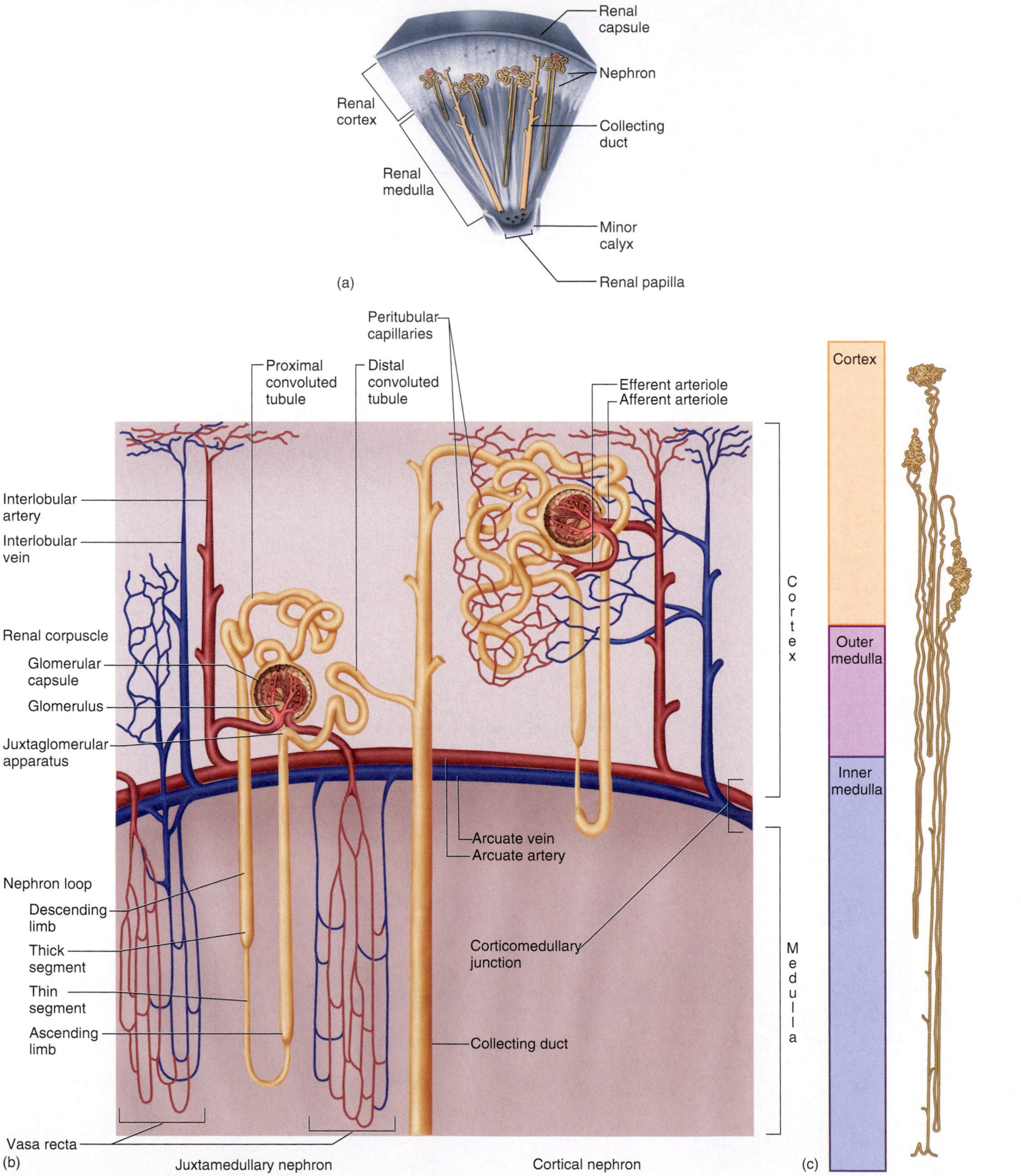

Renal capsule
Nephron
Renal cortex
Collecting duct
Renal medulla
Minor calyx
Renal papilla
(a)
Peritubular capillaries
Proximal convoluted tubule
Distal convoluted tubule
Efferent arteriole
Afferent arteriole
Interlobular artery
Interlobular vein
Renal corpuscle
Glomerular capsule
Glomerulus
Juxtaglomerular apparatus
Cortex
Arcuate vein
Arcuate artery
Nephron loop
Descending limb
Thick segment
Thin segment
Ascending limb
Corticomedullary junction
Medulla
Collecting duct
Vasa recta
(b)
Juxtamedullary nephron
Cortical nephron
(c)
Cortex
Outer medulla
Inner medulla

FIGURE 25.5

Structure of the Nephron. (*a*) Location of the nephrons in one wedge-shaped lobe of the kidney. (*b*) Structure of two nephrons. For clarity, vasa recta are shown only on the *left* and peritubular capillaries only on the *right,* and the vasa recta are shown to one side of the nephron loop, which travels through this capillary plexus. Note that juxtamedullary nephrons are closer to the corticomedullary junction and have longer nephron loops than cortical nephrons. Vasa recta come only from the nephrons closest to the medulla. (*c*) The true proportions of the nephron loops relative to the convoluted tubules.

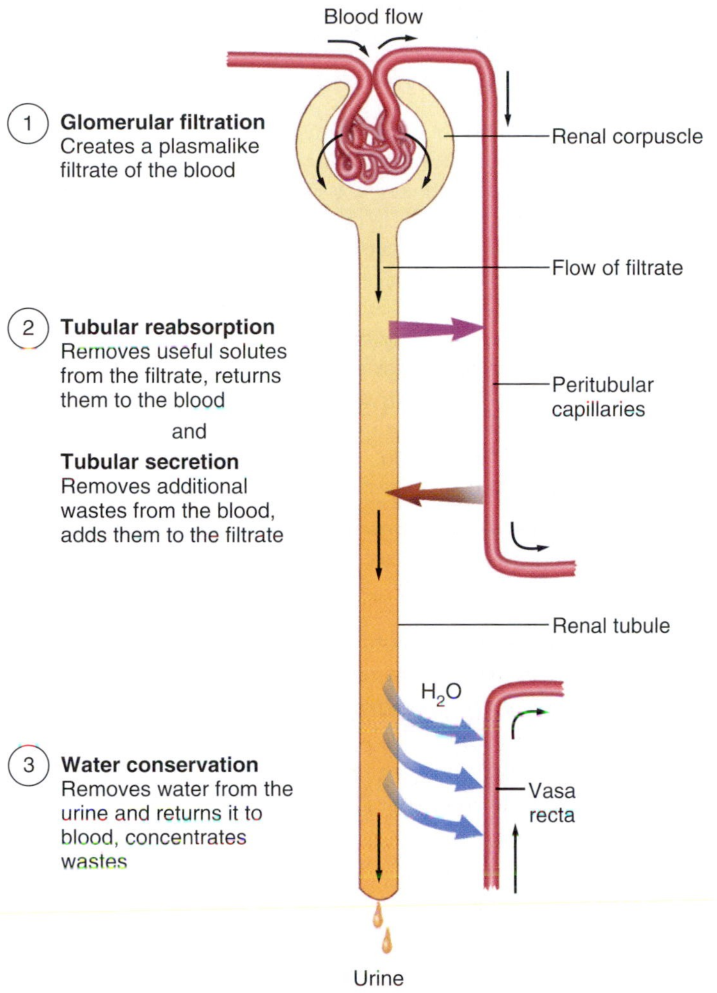

FIGURE 25.6
Basic Steps in the Formation of Urine.

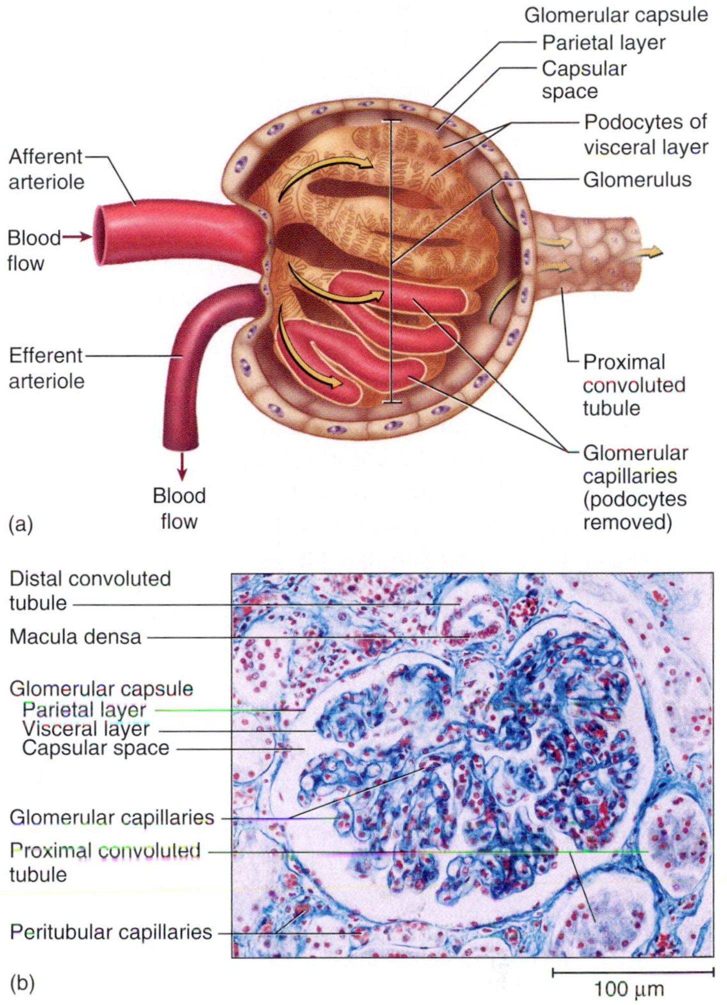

FIGURE 25.7
The Renal Corpuscle. (*a*) Anatomy of the corpuscle. (*b*) Light micrograph of the renal corpuscle and sections of the surrounding renal tubule.

also chemicals useful to the body. The fluid filtered from the blood is called *glomerular filtrate.* In contrast to the blood, it is free of cells and very low in protein. After it passes into the renal tubule, its composition is quickly modified by the following processes, and we call it *tubular fluid.*

2. **Tubular reabsorption** and **tubular secretion** are two simultaneous processes that alter the composition of the tubular fluid. Substances useful to the body, such as glucose, are reabsorbed from the tubular fluid and returned to the blood. Blood-borne substances such as hydrogen ions and some drugs, conversely, are extracted from the peritubular capillaries and secreted into the tubular fluid, thus becoming part of the urine.
3. **Water conservation** is achieved by reabsorbing variable amounts of water from the urine so that the body can eliminate metabolic wastes without losing excess water. Water reabsorption occurs in all parts of the renal tubule, but is the final change occurring in the urine as it passes through the collecting duct. The fluid is regarded as urine once it has entered this duct.

We can now examine the individual segments of the nephron, their contribution to the foregoing processes, and how their structures are adapted to their individual roles.

THE RENAL CORPUSCLE

The **renal corpuscle** (fig. 25.7) consists of the glomerulus and, enclosing it, a two-layered **glomerular (Bowman[8]) capsule.** The inner, or visceral, layer of the glomerular capsule consists of elaborate cells called *podocytes* wrapped around the capillaries. The parietal (outer) layer is a simple squamous epithelium. The two layers are separated by a urine-collecting **capsular space.** In tissue sections, the capsular space appears as an empty circular or C-shaped space around the glomerulus.

Opposite sides of the renal corpuscle are called the vascular pole and urinary pole. At the **vascular pole,** the afferent arteriole enters the capsule, bringing blood to the glomerulus, and the efferent arteriole exits the capsule and carries blood away. The afferent

[8]Sir William Bowman (1816–92), British physician

arteriole is conspicuously larger than the efferent arteriole. Thus, the glomerulus has a large inlet and small outlet. This gives its capillaries an unusually high blood pressure, which is the driving force of glomerular filtration. At the **urinary pole,** the parietal wall of the capsule turns away from the corpuscle and gives rise to the renal tubule. The simple squamous epithelium of the capsule becomes simple cuboidal in the renal tubule.

A **podocyte**[9] is shaped somewhat like an octopus, with a bulbous cell body and several thick arms (fig. 25.8). Each arm has numerous little extensions called **foot processes (pedicels**[10]**)** that wrap around the glomerular blood capillaries and interdigitate with each other, like wrapping your hands around a pipe and lacing your fingers together (see the photograph on p. 707). The foot processes have narrow **filtration slits** between them.

The job of the renal corpuscle is glomerular filtration: blood cells and plasma proteins are retained in the blood stream because they are too large to pass through the barriers described below. Water, however, freely passes through and carries along small solute particles such as urea, glucose, amino acids, and electrolytes. The high blood pressure in the glomerulus drives the water and small solutes out through the capillary walls, into the capsular space. Pressure in the capsular space drives the filtrate into the renal tubule and ultimately all the way to the calyces and renal pelvis.

Anything leaving the bloodstream must pass through a barrier called the **filtration membrane,** composed of three layers (fig. 25.8*c*):

1. **Capillary endothelium.** Glomerular capillaries have a fenestrated endothelium (see fig. 21.6, p. 591) honeycombed with large filtration pores about 70 to 90 nm in diameter. They are much more permeable than capillaries elsewhere, although the filtration pores are small enough to hold back blood cells.
2. **Basement membrane.** This is a layer of proteoglycan gel (a protein-carbohydrate complex) beneath the endothelial cells. For large molecules to pass through it is like trying to pass sand through a kitchen sponge. A few particles may penetrate its small spaces, but most are held back. On the basis of size alone, the basement membrane excludes any molecules larger than 8 nm. Some smaller molecules, however, are also held back by a negative electrical charge on the proteoglycans. Blood albumin is slightly less than 7 nm in diameter, but it is also negatively charged and thus repelled by the basement membrane. Therefore, the protein concentration is about 7% in the blood plasma but only 0.03% in the glomerular filtrate. The filtrate contains traces of albumin and smaller polypeptides, including some hormones.
3. **Filtration slits.** The slits between the pedicels are about 30 nm wide and are also negatively charged. This charge is a final barrier to large anions such as proteins.

Almost any molecule smaller than 3 nm passes freely through the filtration membrane. This includes water, electrolytes, glucose, fatty acids, amino acids, nitrogenous wastes, and vitamins. Such substances have about the same concentration in the filtrate as in the plasma. Some substances of low molecular weight are retained in the bloodstream because they are bound to plasma proteins that cannot get through the membrane. For example, most calcium, iron, and thyroid hormone in the blood are bound to plasma proteins that retard their filtration by the kidneys. The small fraction that is unbound, however, passes freely through the membrane and appears in the urine.

[9]*podo* = foot + *cyte* = cell
[10]*pedi* = foot + *cel* = little

INSIGHT 25.1 CLINICAL APPLICATION

Blood and Protein in the Urine

Urinalysis, one of the most routine procedures performed upon patient admission and in routine medical examinations, is an analysis of the physical and chemical properties of the urine. It includes tests for blood and protein, both of which are normally lacking from urine. Damage to the filtration membrane, however, can result in blood or protein in the urine, called *hematuria*[11] and *proteinuria (albuminuria),* respectively. These can be signs of kidney infections, trauma, and other kidney diseases (see table 25.1 at the end of the chapter). They can be temporary conditions of little concern, or they can be chronic and gravely serious. Long-distance runners and swimmers often show temporary proteinuria and hematuria. Strenuous exercise reduces perfusion of the kidneys as blood shifts to the muscles. With a reduced blood flow, the glomeruli deteriorate and leak protein and sometimes blood cells into the filtrate.

[11]*hemat* = blood + *uria* = urine condition

THE RENAL TUBULE

The **renal (uriniferous**[12]**) tubule** is a duct that leads away from the glomerular capsule and ends at the tip of a medullary pyramid. It is about 3 cm long and divided into four major regions: the *proximal convoluted tubule, nephron loop, distal convoluted tubule,* and *collecting duct* (see fig. 25.5). Only the first three of these are parts of an individual nephron; the collecting duct receives fluid from many nephrons. Each region of the renal tubule has unique physiological properties and roles in the production of urine.

The **proximal convoluted tubule (PCT)** arises from the glomerular capsule. It is the longest and most coiled of the four regions and thus dominates histological sections of renal cortex. The PCT has a simple cuboidal epithelium with prominent microvilli (a brush border), which attests to the great deal of absorption that occurs here. The microvilli give the epithelium a distinctively shaggy look in tissue sections.

The PCT carries out both tubular reabsorption and tubular secretion. It reabsorbs about 65% of the glomerular filtrate, and consumes about 6% of one's daily ATP expenditure in doing so. On the surface facing the tubular fluid, the epithelial cells have a variety of membrane transport proteins that carry solutes into the cells by active transport and facilitated diffusion. These solutes, and water, pass through the cell cytoplasm (the **transcellular**[13] **route**) and either diffuse out or are actively pumped out the basal and lateral cell surfaces, adjacent to peritubular blood capillaries

[12]*urin* = urine + *fer* = to carry
[13]*trans* = across

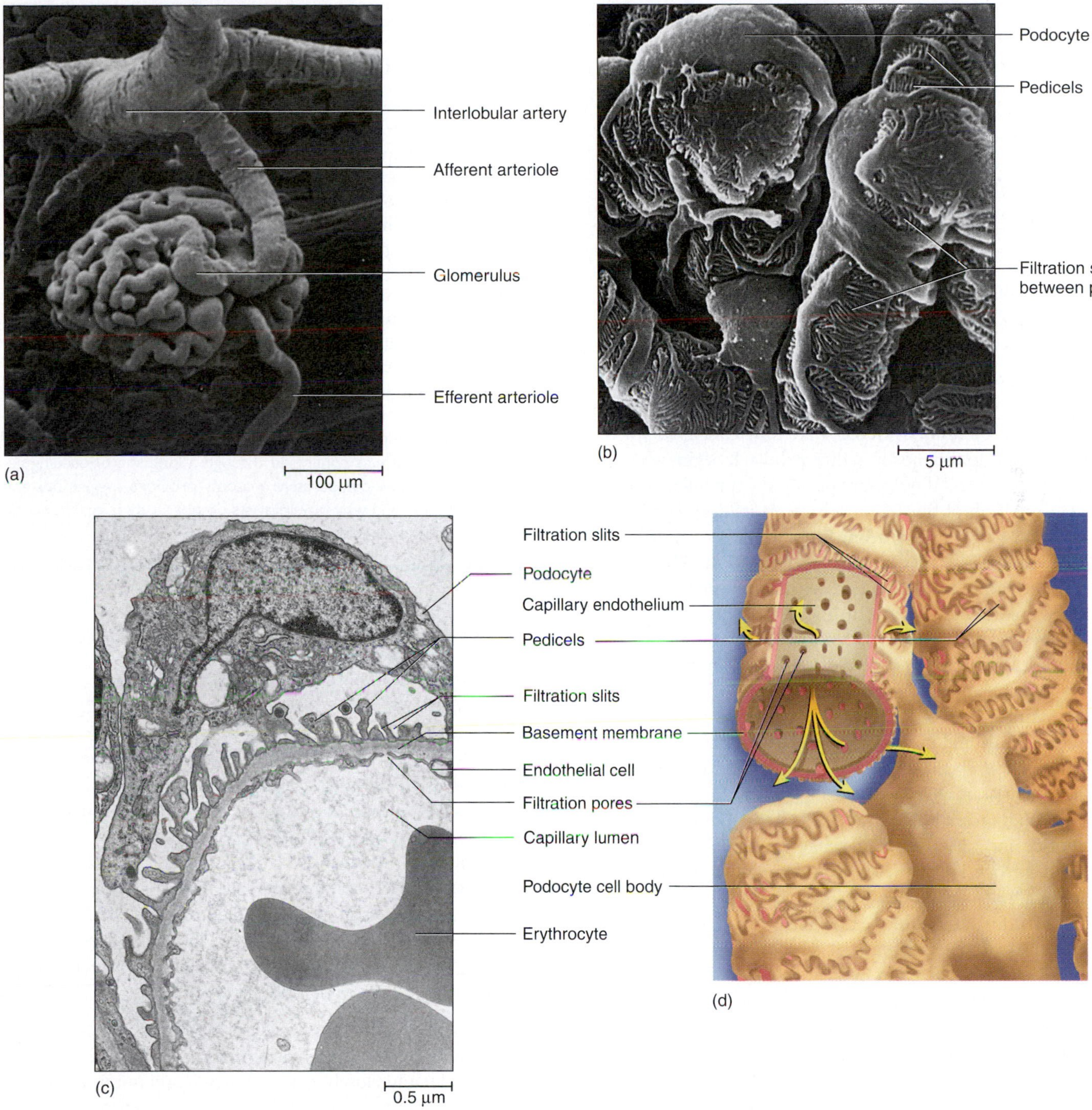
Interlobular artery
Afferent arteriole
Glomerulus
Efferent arteriole
(a)
100 μm
Podocyte cell body
Pedicels
Filtration slits between pedicels
(b)
5 μm
Filtration slits
Podocyte
Capillary endothelium
Pedicels
Filtration slits
Basement membrane
Endothelial cell
Filtration pores
Capillary lumen
Podocyte cell body
Erythrocyte
(c)
0.5 μm
(d)

FIGURE 25.8

Structure of the Glomerulus. (*a*) A resin cast of the glomerulus and nearby arteries (SEM). Note that the efferent arteriole is much narrower than the afferent arteriole, which causes blood pressure in the glomerulus to be unusually high. (*b*) Blood capillaries of the glomerulus closely wrapped in the spidery podocytes that form the visceral layer of the glomerular capsule (SEM). (c) A blood capillary and podocyte showing fenestrations and filtration slits (TEM). (*d*) The production of glomerular filtrate by the passage of fluid through the fenestrations and filtration slits. (*a*) From R. G. Kessel and R. H. Kardon, *Tissues and Organs: A Text-Atlas of Scanning Electron Microscopy* (W. H. Freeman, 1979).

waiting to receive them. Water and solutes also take a **paracellular**[14] **route** between the epithelial cells. Even though the cells are joined by tight junctions, these are quite leaky and allow a substantial amount of fluid to pass through.

[14]*para* = next to

Among the solutes reabsorbed by the PCT are sodium, potassium, magnesium, phosphate, chloride, bicarbonate, glucose, amino acids, lactate, protein, smaller peptides, urea, and uric acid. Water follows by osmosis.

By *tubular secretion,* the PCT extracts solutes from the peritubular capillaries and secretes them into the tubular fluid, so they can be passed in the urine. Secreted solutes include hydrogen and bicarbonate ions, ammonia, urea, uric acid, creatinine, bile acids, pollutants, and some drugs (aspirin, penicillin, and morphine, for example). Notice that urea and uric acid go both ways between the blood and tubular fluid, transported by both tubular reabsorption and tubular secretion. The kidneys do not completely cleanse the blood of these wastes; indeed they remove only about half of the urea, but this is sufficient to keep the blood urea concentration down to a safe level.

Think About It!

The proximal convoluted tubule exhibits some of the same structural adaptations as the small intestine, and for the same reason. Discuss what they have in common, and the reason for it.

The **nephron loop (loop of Henle**[15]**)** is a long U-shaped portion of the renal tubule found mostly in the medulla. It begins where the PCT straightens out and dips toward or into the medulla, forming the **descending limb** of the nephron loop. At its deep end, the loop turns 180° and forms the **ascending limb,** which returns to the cortex, traveling parallel and close to the descending limb. The loop is divided into thick and thin segments. The **thick segments** have a simple cuboidal epithelium. They form the initial part of the descending limb and part or all of the ascending limb. The cells here are heavily engaged in active transport of salts, so they have very high metabolic activity and are loaded with mitochondria. The **thin segment** has a simple squamous epithelium. It forms the lower part of the descending limb, and in some nephrons, it rounds the bend and continues partway up the ascending limb. The cells here have low metabolic activity but are very permeable to water.

The nephron loop reabsorbs about 25% of the sodium, potassium, and chloride and 15% of the water that was in the glomerular filtrate. Its primary function, however, is to maintain a gradient of salinity in the renal medulla. It does this by pumping Na^+, K^+, and Cl^- from the ascending limb into the medullary tissue fluid. At the corticomedullary junction, the tissue fluid is isotonic with the blood plasma (300 milliosmoles/liter), but deep in the medulla, it is four times as concentrated. The significance of this is explained later.

The nephron loops are not identical in all nephrons. Nephrons just beneath the renal capsule, close to the kidney surface, are called **cortical nephrons.** They have relatively short nephron loops that dip only slightly into the outer medulla before turning back (see fig. 25.5), or turn back even before leaving the cortex. Some cortical nephrons have no nephron loops at all. Nephrons close to the medulla are called **juxtamedullary**[16] **nephrons.** They have very long nephron loops that extend nearly to the apex of the renal pyramid. Only 15% of the nephrons are juxtamedullary, but these are almost solely responsible for maintaining the salinity gradient of the medulla.

The **distal convoluted tubule (DCT)** is a coiled part of the renal tubule located in the cortex beyond the nephron loop. It is shorter and less convoluted than the PCT, so fewer sections of it are seen in histological sections. It has a simple cuboidal epithelium with smooth-surfaced cells nearly devoid of microvilli. It absorbs variable amounts of sodium, calcium, chloride, and water, and secretes potassium and hydrogen into the tubular fluid. Unlike the PCT, which absorbs solutes and water at a constant rate, the DCT reabsorbs these at variable rates determined by the influence of two hormones: *aldosterone,* which regulates sodium and potassium excretion, and *parathyroid hormone,* which regulates calcium and phosphate excretion. The DCT is the end of a nephron.

The initial portion of the DCT, before it coils, contacts the afferent and efferent arterioles adjacent to the renal corpuscle (fig. 25.9), and the three structures form the **juxtaglomerular** (JUX-tuh-glo-MER-you-lur) **apparatus.** This is a device for monitoring the flow and composition of the urine and adjusting the performance of the nephron. Three specialized cell types are found here:

1. The **macula densa**[17] is a patch of epithelium in the DCT on the side facing the afferent arteriole. The epithelial cells here are slender and closely spaced. Much is still unknown about their physiology, but they apparently act as sensory cells that monitor the flow or composition of the tubular fluid and communicate with the cells described next.
2. **Juxtaglomerular (JG) cells** are enlarged smooth muscle cells found in the afferent arteriole and in smaller numbers in the efferent arteriole. When stimulated by the macula densa, they constrict and narrow the arterioles, especially the afferent. This reduces blood flow into the glomerulus, and thus reduces the glomerular filtration rate. JG cells also secrete renin, the enzyme mentioned earlier that triggers corrective changes in blood pressure.

INSIGHT 25.2 EVOLUTIONARY MEDICINE

The Kidney and Life on Dry Land

Physiologists first suspected that the nephron loop plays a role in water conservation because of their studies of a variety of animal species. Animals that must conserve water have longer, more numerous nephron loops than animals with little need to conserve it. Fish and amphibians lack nephron loops and produce urine that is isotonic to their blood plasma. Aquatic mammals such as beavers have short nephron loops and only slightly hypertonic urine.

But the kangaroo rat, a desert rodent, provides an instructive contrast. It lives on seeds and other dry foods and can live without drinking any water at all. The water produced by its aerobic respiration is enough to meet its needs because its kidneys are extraordinarily efficient at conserving it. They have extremely long nephron loops and produce urine that is 10 to 14 times as concentrated as their blood plasma (compared with about 4 times, at most, in humans).

Comparative studies thus suggested a hypothesis for the function of the nephron loop that was confirmed through a long line of ensuing research. This shows how comparative anatomy provides suggestions and insights into function and why physiologists do not study human function in isolation from other species.

[15]Friedrich G. J. Henle (1809–85), German anatomist
[16]*juxta* = next to
[17]*macula* = spot, patch + *densa* = dense

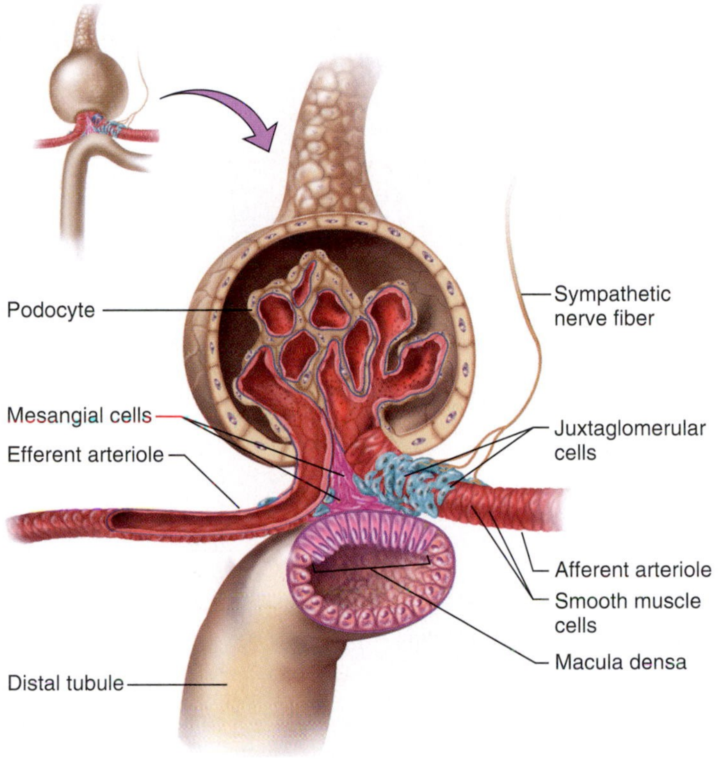
Podocyte
Mesangial cells
Efferent arteriole
Distal tubule
Sympathetic nerve fiber
Juxtaglomerular cells
Afferent arteriole
Smooth muscle cells
Macula densa

FIGURE 25.9
The Juxtaglomerular Apparatus.

3. **Mesangial**[18] (mez-AN-jee-ul) **cells** occupy the cleft between the afferent and efferent arterioles and the spaces between the capillaries of the glomerulus. They are connected to the macula densa and JG cells by gap junctions and perhaps mediate communication between those cells.

The **collecting duct** is a straight tubule that passes down into the medulla. It is part of the renal tubule but not part of the nephron; the nephron and collecting duct have separate embryonic origins. The cortical part of the collecting duct receives fluid from the DCTs of several nephrons. The duct then continues into the medulla, where the greater part of it lies. Near the renal papilla, several collecting ducts converge to form a larger, short stretch called the **papillary duct.** About 30 papillary ducts drain from each papilla into a minor calyx. Once the urine drains into a minor calyx, it undergoes no further change in composition or concentration.

The collecting duct is lined with a simple cuboidal epithelium with two types of cells—principal cells and intercalated ("in between") cells. **Intercalated cells** play a role in regulating the body's acid-base balance by secreting either H^+ or bicarbonate ions (HCO_3^-) into the urine. **Principal cells** reabsorb Na^+ and water and secrete K^+ into the urine. They represent the kidney's last chance to adjust the water content and thus the osmolarity of the urine. The principal cells also have water channels called *aquaporins* in their membranes. As the tubular fluid descends through the collecting duct, water passes by osmosis through these channels, out of the tubule and into the increasingly salty tissue fluid of the medulla. The salinity gradient created by the nephron loop makes this osmotic reabsorption of water possible. The reabsorbed water is carried away by the blood capillaries of the vasa recta.

The collecting duct is influenced by two hormones: *atrial natriuretic peptide,* which increases sodium excretion in the urine, and *antidiuretic hormone,* which promotes water retention and reduces urine volume.

To summarize, the flow of fluid from the point where the glomerular filtrate is formed to the point where urine leaves the kidney is: glomerular capsule → proximal convoluted tubule → nephron loop → distal convoluted tubule → collecting duct → papillary duct → minor calyx → major calyx → renal pelvis → ureter.

[18]*mes* = in the middle + *angi* = vessel

Before You Go On

Answer the following questions to test your understanding of the preceding section:

3. Arrange the following in order from the most numerous to the least numerous structures in a kidney: glomeruli, major calyces, minor calyces, interlobular arteries, interlobar arteries.
4. Trace the path taken by one red blood cell from the renal artery to the renal vein.
5. Concisely state the functions of the glomerulus, PCT, nephron loop, DCT, and collecting duct.
6. Describe the location and appearance of podocytes and explain their function.
7. Consider one molecule of urea in the urine. Trace the route that it took from the bloodstream to the point where it left the body.

Anatomy of the Ureters, Urinary Bladder, and Urethra

Objective

When you have completed this section, you should be able to

- describe the functional anatomy of the ureters, urinary bladder, and male and female urethra.

Urine is produced continually, but fortunately it does not drain continually from the body. Urination is episodic—occurring when we allow it. This is made possible by an apparatus for storing urine and by neural controls for its timely release.

The Ureters

The renal pelvis funnels urine into the ureter, a retroperitoneal, muscular tube that extends to the urinary bladder. The ureter is about 25 cm long and reaches a maximum diameter of about 1.7 cm near the bladder. The ureters pass dorsal to the bladder and enter it from below, penetrating obliquely through its muscular wall and opening onto its floor. A small flap of mucosa acts as a valve at the opening of each ureter into the bladder.

INSIGHT 25.3 CLINICAL APPLICATION

Kidney Stones

A *renal calculus*[19] (kidney stone) is a hard granule of calcium, phosphate, uric acid, and protein. Renal calculi form in the renal pelvis and are usually small enough to pass unnoticed in the urine flow. Some, however, grow as large as several centimeters and block the renal pelvis or ureter, which can lead to the destruction of nephrons as pressure builds in the kidney. A large, jagged calculus passing down the ureter stimulates strong contractions that can be excruciatingly painful. It can also damage the ureter and cause hematuria. Causes of renal calculi include hypercalcemia (excess calcium in the blood), dehydration, pH imbalances, frequent urinary tract infections, or an enlarged prostate gland causing urine retention. Calculi are sometimes treated with stone-dissolving drugs, but often they require surgical removal. A nonsurgical technique called *lithotripsy*[20] uses ultrasound to pulverize the calculi into fine granules easily passed in the urine.

[19] *calc* = calcium, stone + *ul* = little
[20] *litho* = stone + *tripsy* = crushing

The ureter has three layers: an adventitia, muscularis, and mucosa. The adventitia is a connective tissue layer that binds it to the surrounding tissues. It blends with the capsule of the kidney at the superior end and with the connective tissue of the bladder wall at the inferior end. The muscularis consists of two layers of smooth muscle over most of its length, but a third layer appears in the lower ureter. The inner muscular layer consists of longitudinal muscle cells; the cells in the next layer superficial to this have a circular arrangement; and the third and outermost layer in the lower ureter is again longitudinal. Peristaltic waves of contraction in the muscularis "milk" urine from the renal pelvis down to the bladder. The mucosa of the ureter has a transitional epithelium that begins in the minor calyces of the kidney; lines the major calyces, renal pelvis, and ureter; and continues into the urinary bladder. The lumen is very narrow and is easily obstructed by kidney stones (see insight 25.3).

The Urinary Bladder

The urinary bladder (fig. 25.10) is a muscular sac on the floor of the pelvic cavity, inferior to the peritoneum and dorsal to the pubic symphysis. It is covered by parietal peritoneum on its flattened superior surface and by a fibrous adventitia elsewhere. Its muscular layer, called the **detrusor**[21] (deh-TROO-zur) **muscle,** consists of three indistinctly separated layers of smooth muscle. The mucosa has a transitional epithelium. When the bladder is empty, this epithelium is five or six cells thick and the mucosa has conspicuous wrinkles called **rugae**[22] (ROO-gee). When the bladder fills, the stretching smooths out the rugae and the epithelium thins to about two or three cells thick. The bladder is a highly distensible organ, capable of holding up to 800 mL of urine.

[21] *de* = down + *trus* = push
[22] *ruga* = fold, wrinkle

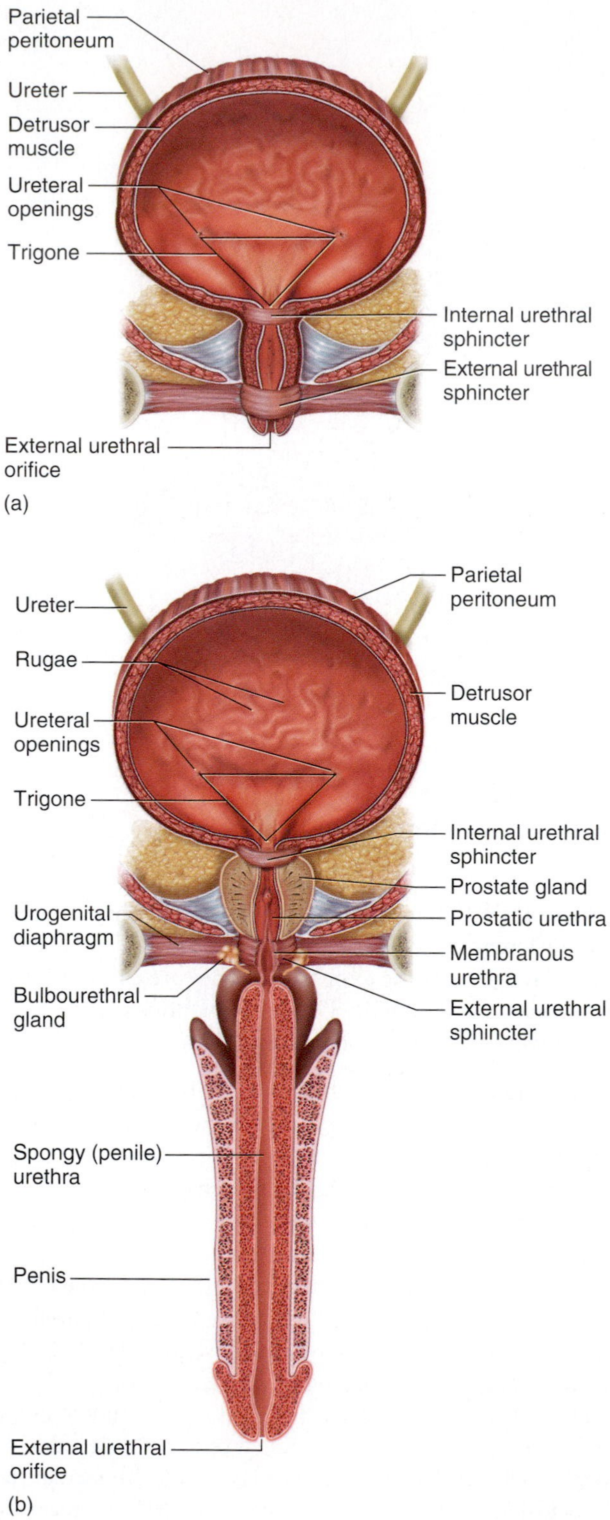

Figure 25.10
Anatomy of the Urinary Bladder and Urethra. (*a*) Female. (*b*) Male.

The openings of the two ureters and the urethra mark a smooth-surfaced triangular area called the **trigone**[23] on the bladder floor. This is a common site of bladder infection (see insight 25.4). For photographs of the relationship of the bladder and urethra to other pelvic organs in both sexes, see figure A.22 (p. 43).

INSIGHT 25.4 CLINICAL APPLICATION

Urinary Tract Infections

Infection of the urinary bladder is called *cystitis.*[24] It is especially common in females because bacteria such as *Escherichia coli* can travel easily from the perineum up the short urethra. Because of this risk, young girls should be taught never to wipe the anus in a forward direction. If cystitis is untreated, bacteria can spread up the ureters and cause *pyelitis,*[25] infection of the renal pelvis. If it reaches the renal cortex and nephrons, it is called *pyelonephritis.* Kidney infections can also result from invasion by bloodborne bacteria. Urine stagnation due to renal calculi or prostate enlargement increases the risk of infection.

[24]*cyst* = bladder + *itis* = inflammation
[25]*pyel* = pelvis

The Urethra

The urethra conveys urine out of the body. In the female, it is a tube 3 to 4 cm long bound to the anterior wall of the vagina by fibrous connective tissue (fig. 25.10*a*). Its opening, the **external urethral orifice,** lies between the vaginal orifice and clitoris. The male urethra (fig. 25.10*b*) is about 18 cm long and has three regions: (1) The **prostatic urethra** begins at the urinary bladder and passes for about 2.5 cm through the prostate gland. During orgasm, it receives semen from the reproductive glands. (2) The **membranous urethra** is a short (0.5 cm), thin-walled portion where the urethra passes through the muscular floor of the pelvic cavity. (3) The **spongy (penile) urethra** is about 15 cm long and passes through the penis to the external urethral orifice. It is named for the *corpus spongiosum,* an erectile tissue that surrounds the penile urethra (see chapter 26). The male urethra assumes an S-shape: it passes downward from the bladder, turns anteriorly as it enters the root of the penis, and then turns about 90° downward again as it enters the external, pendant part of the penis. The mucosa has a transitional epithelium near the bladder, a pseudostratified columnar epithelium for most of its length, and finally a stratified squamous epithelium near the external urethral orifice. There are mucous **urethral glands** in the wall of the penile urethra.

In both sexes, the detrusor muscle is thickened near the urethra to form an **internal urethral sphincter,** which compresses the urethra and retains urine in the bladder. Since this sphincter is composed of smooth muscle, it is under involuntary control. Where the urethra passes through the pelvic floor, it is encircled by an **external urethral sphincter** of skeletal muscle, which provides voluntary control over the voiding of urine.

[23]*tri* = three + *gon* = angle

Before You Go On

Answer the following questions to test your understanding of the preceding section:

8. Describe the anatomical relationship of the ureter to the renal pelvis and to the bladder wall.
9. Compare and contrast the structure and function of the internal and external urethral sphincters.
10. Contrast the structure of the bladder wall when the bladder is empty with when it is full.
11. Name and define the three segments of the male urethra.

Developmental and Clinical Perspectives

Objectives

When you have completed this section, you should be able to

- describe the embryonic development of the urinary system;
- describe the degenerative changes that occur in old age;
- describe the causes and effects of renal failure; and
- briefly define or describe several urinary system diseases.

Prenatal Development

Perhaps surprisingly, the embryonic urinary system develops two pairs of primitive, temporary kidneys before "settling down" and producing the permanent pair. The system develops as if replaying the evolutionary history of the vertebrate urinary system. Early in week 4, a rudimentary kidney called the *pronephros*[26] appears in the cervical region, resembling the kidneys of many fish and amphibian embryos and larvae. The pronephros disappears by the end of that week. As it degenerates, a second kidney, the *mesonephros,*[27] appears in the thoracic to lumbar region. The mesonephros functions in the embryos of all vertebrates, but is of minor importance in most mammals, where wastes are eliminated via the placenta. Most of the mesonephros disappears by the end of month 2, but its collecting duct, the *mesonephric duct,* remains and contributes importantly to the male reproductive tract (see chapter 26). This duct opens into an embryonic *cloaca,* a temporary rectumlike receiving chamber for the digestive, urinary, and reproductive systems. The final kidney, the *metanephros,*[28] appears in week 5 and thus overlaps the existence of the mesonephros.

The permanent urinary tract begins with a pouch called the **ureteric bud** growing from the lower end of each mesonephric duct. The closed, upper end of the bud dilates and branches to form

[26]*pro* = first + *nephros* = kidney
[27]*meso* = middle + *nephros* = kidney
[28]*meta* = beyond, next in a series + *nephros* = kidney

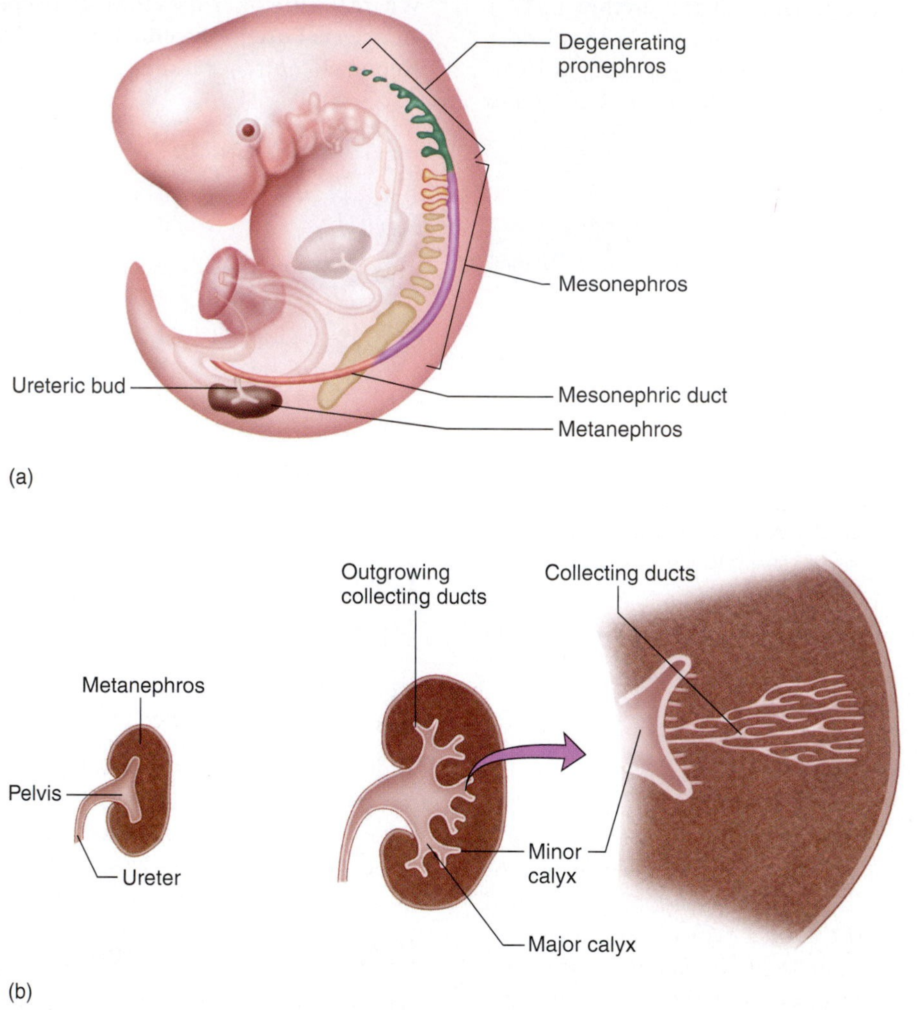

FIGURE 25.11

Embryonic Development of the Urinary Tract. (*a*) Relationship of the early ureteric bud and metanephros to the lower mesonephric duct. (*b*) Progression in the development of the ureter, renal pelvis, calyces, and collecting ducts, all of which arise from the ureteric bud.

the renal pelvis, then the major and minor calyces, and finally the collecting ducts (fig. 25.11). Each collecting duct has a cap of metanephric kidney tissue over its tip. The duct induces this cap to differentiate into an S-shaped tubule (fig. 25.12). Blood capillaries grow into one end of the tubule and form a glomerulus, as the tubule grows around it to form the double-walled glomerular capsule. The other end of the tubule breaks through to become continuous with the collecting duct. The tubule gradually lengthens and differentiates into the proximal convoluted tubule, nephron loop, and distal convoluted tubule. By the time of birth, each kidney will have formed over 1 million nephrons in this manner. No more nephrons form after birth, but the existing ones continue to grow. The kidney surface is lumpy at birth but smooths out because of nephron growth.

The kidneys originate in the pelvic region and later migrate superiorly—a movement called **ascent of the kidney.** Initially, the kidney is supplied by a pelvic branch of the aorta, but as it ascends, new arteries higher and higher on the aorta take over the job of supplying the kidney, while the lower arteries degenerate.

In weeks 4 to 7, the cloaca divides into a ventral *urogenital (U-G) sinus* and a dorsal *anal canal.* The superior part of the U-G sinus forms the urinary bladder and the inferior part forms the urethra. In infants and children, the urinary bladder is located in the abdomen. It begins to drop into the greater pelvis at about 6 years of age, but does not enter the lesser pelvis and become a true pelvic organ until after puberty. Early in its development, the bladder is connected to the allantois, an extraembryonic sac described in chapter 4. This connection eventually becomes a constricted passage, the *urachus* (yur-AY-kus), connecting the bladder to the umbilicus. In the adult, the urachus is reduced to a fibrous cord, the *median umbilical ligament.*

Urine production begins around week 12 of fetal development, but metabolic wastes are cleared by the placenta, not by the fetal urinary system. Fetal urine is continually recycled as the fetus voids it into the amniotic fluid, swallows it, and then excretes it again.

The Aging Urinary System

The kidneys exhibit a striking degree of atrophy in old age. From ages 25 to 85, the number of nephrons declines by 30% to 40%, and up to one-third of the remaining glomeruli become atherosclerotic,

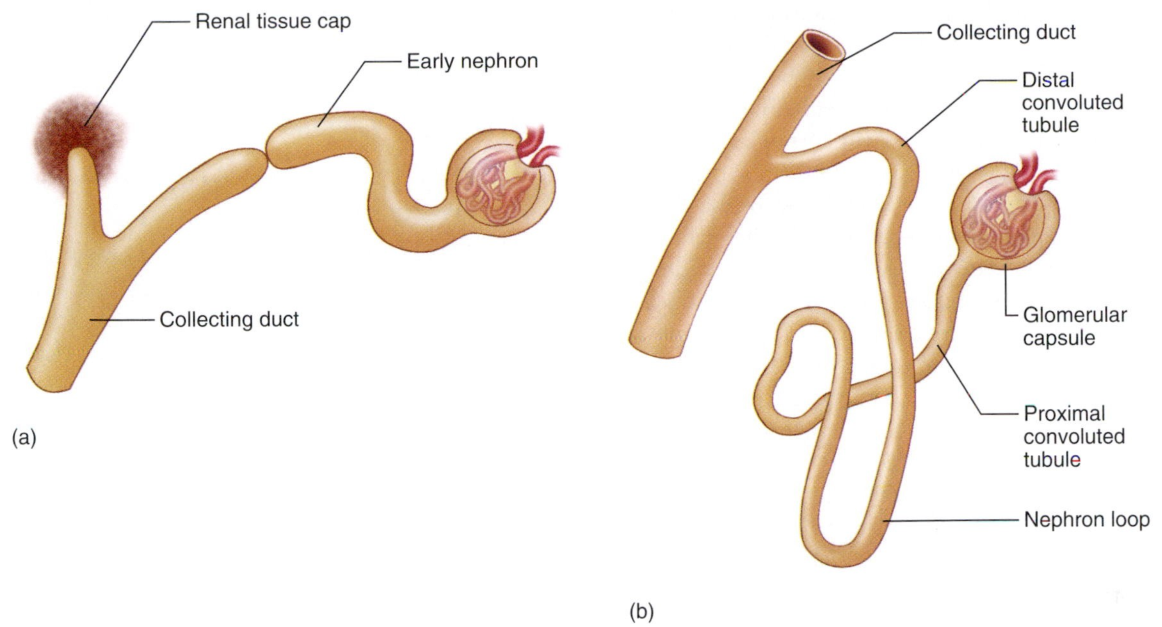

FIGURE 25.12

Embryonic Development of the Nephron. (*a*) The collecting duct induces mesoderm to differentiate into a renal (metanephric) tissue cap, as shown at the *far left.* This cap differentiates into an S-shaped tube that will become the nephron, as shown on the *right fork* of the duct (representing a later stage of development). (*b*) The renal tubule has begun to differentiate into proximal and distal convoluted tubules and nephron loop.

INSIGHT 25.5 CLINICAL APPLICATION

DEVELOPMENTAL ABNORMALITIES OF THE KIDNEY

Several anomalies can occur during the embryonic development of the kidneys (see fig. 1.7*c*). *Pelvic kidney* is a condition in which the kidney fails to ascend, and remains in the pelvic cavity for life. *Horseshoe kidney* is a single C-shaped kidney formed when the ascending kidneys are crowded together and merge into one. The C typically snags on the inferior mesenteric artery, preventing the horseshoe kidney from ascending any farther. Some people have two ureters arising from a single kidney, resulting from a splitting of the ureteric bud in early embryonic development. Usually the two ureters empty into the bladder, but in rare cases, one of them empties into the uterus, vagina, urethra, or elsewhere. This requires surgical correction so that urine does not dribble continually from the urethra or vagina. Some kidneys have an *accessory renal artery* resulting from failure of one of the early, temporary renal arteries to degenerate as the permanent renal artery forms. Most such irregularities cause no functional problems and usually go unnoticed, but they may be discovered in surgery, radiography, or cadaver dissection.

bloodless, and nonfunctional. The kidneys of a 90-year-old are 20% to 40% smaller than those of a 30-year-old and receive only half as much blood. They are proportionately less efficient at clearing wastes from the blood. Although baseline renal function is adequate even in old age, the kidneys have little reserve capacity; thus, other diseases can lead to surprisingly rapid renal failure. Drug doses often need to be reduced for the elderly because the kidneys cannot clear drugs from the blood as rapidly. Reduced renal function is a significant factor in overmedication of the aged.

Water balance becomes more precarious in old age because the kidneys are less responsive to antidiuretic hormone and because the sense of thirst is blunted. Even with free access to water, many elderly people do not drink enough to maintain normal blood osmolarity, and dehydration is common.

Voiding and bladder control become a problem for both men and women. About 80% of men over the age of 80 are affected by *benign prostatic hyperplasia,* a noncancerous growth of the prostate gland that compresses the urethra, reduces the force of the urine stream, and makes it harder to empty the bladder. Urine retention can cause pressure to back up in the kidneys, aggravating the failure of the nephrons. Older women become increasingly subject to *incontinence* (see table 25.2), especially if their history of pregnancy and childbearing has weakened the pelvic muscles and urethral sphincters. Senescence of the sympathetic nervous system and nervous disorders such as stroke and Alzheimer disease can also cause incontinence.

Urinary System Disorders

The most serious disorder of the urinary system is renal failure. *Acute renal failure* is an abrupt decline in kidney function, often caused by trauma or by a hemorrhage or thrombosis cutting off blood flow to the nephrons. *Chronic renal failure* is a long-term, progressive, irreversible loss of functional nephrons. It can result from such causes as prolonged or repetitive kidney infections, trauma, poisoning, atherosclerosis of the renal arteries (often in conjunction with diabetes mellitus), or an autoimmune disease called acute glomerulonephritis (table 25.1).

Nephrons can regenerate and restore kidney function after short-term injuries, and even when some nephrons are irreversibly destroyed, others can hypertrophy and compensate for their lost function. Indeed, a person can survive on as little as one-third of one kidney. When 75% of the nephrons are lost, however, the remaining ones cannot maintain homeostasis. The result is azotemia and acidosis, and if 90% of renal function is lost, uremia is likely. Loss of nephron function also leads to anemia, because erythrocyte production depends on the hormone *erythropoietin,* which is secreted mainly by the kidneys.

Renal insufficiency or failure must be treated either with a kidney transplant or with *hemodialysis.* The latter is a procedure in which, usually, arterial blood is pumped through a *dialysis machine.* In the machine, the blood passes through tubes of dialysis membrane immersed in dialysis fluid. Wastes and excess water diffuse from the blood into the fluid, which is discarded, and drugs can be added to the dialysis fluid to diffuse into the blood. Another method called *continuous ambulatory peritoneal dialysis (CAPD)* frees the patient from the dialysis machine and can be carried out at home. Dialysis fluid is introduced into the peritoneal cavity through a catheter, absorbs metabolic wastes, and then is drained from the body and discarded.

Some other urinary system disorders are briefly described in table 25.1.

Before You Go On

Answer the following questions to test your understanding of the preceding section:

12. Explain why the nephron has a different embryonic origin than the passages from collecting duct to urethra.
13. How does the number of functional nephrons change in old age? What are the implications of this change for homeostasis and medication dosages?
14. What are some causes of renal failure? Describe some clinical signs of renal failure.
15. Cover the right side of table 25.1 and define or describe the disorders on the left from memory.

TABLE 25.1

Some Disorders of the Urinary System

Acute Glomerulonephritis	An autoimmune inflammation of the glomeruli, often following a streptococcus infection. Results in destruction of glomeruli leading to hematuria, proteinuria, edema, reduced glomerular filtration, and hypertension. Can progress to chronic glomerulonephritis and renal failure, but most individuals recover from acute glomerulonephritis without lasting effect.
Hydronephrosis[29]	Increase in fluid pressure in the renal pelvis and calyces owing to obstruction of the ureter by kidney stones, nephroptosis, or other causes. Can progress to complete cessation of glomerular filtration and atrophy of nephrons.
Nephroptosis[30] *(NEFF-rop-TOE-sis)*	Slippage of the kidney to an abnormally low position *(floating kidney).* Occurs in people with too little body fat to hold the kidney in place, and in people who subject the kidneys to prolonged vibration, such as truck drivers, equestrians, and motorcyclists. Can twist or kink the ureter, which causes pain, obstructs urine flow, and potentially leads to hydronephrosis.
Nephrotic Syndrome	Excretion of large amounts of protein in the urine ($\geq$ 3.5 g/day) due to glomerular injury. Can result from trauma, drugs, infections, cancer, diabetes mellitus, lupus erythematosus, and other diseases. Loss of plasma protein leads to edema, ascites (pooling of fluid in the abdominal cavity), hypotension, and susceptibility to infection (because of immunoglobulin loss).
Urinary Incontinence	Inability to hold the urine; involuntary leakage from the bladder. Can result from incompetence of the urinary sphincters; bladder irritation; pressure on the bladder in pregnancy; an obstructed urinary outlet so that the bladder is constantly full and dribbles urine *(overflow incontinence)*; uncontrollable urination due to brief surges in bladder pressure, as in laughing or coughing *(stress incontinence)*; and neurological disorders such as spinal cord injuries.

Disorders Described Elsewhere

Azotemia 709	Proteinuria 714
Cystitis 719	Pyelitis 719
Hematuria 714	Pyelonephritis 719
Kidney stones 718	Uremia 709

[29]*hydro* = water + *nephr* = kidney + *osis* = medical condition
[30]*nephro* = kidney + *ptosis* = sagging, falling

CHAPTER REVIEW

Review of Key Concepts

Functions of the Urinary System (p. 708)

1. The urinary system consists of two kidneys, two ureters, the urinary bladder, and the urethra.
2. The kidneys filter blood plasma; separate wastes from useful chemicals; regulate blood volume and pressure; control blood osmolarity; secrete renin, erythropoietin, and calcitriol; regulate blood pH; detoxify free radicals and drugs; and carry out gluconeogenesis in times of starvation.
3. Excretion is the process of isolating wastes from the body fluids and eliminating them.
4. The kidneys excrete nitrogenous wastes, among other waste products. The most abundant nitrogenous waste in the urine is *urea.*
5. An elevated level of blood urea is called *azotemia,* and may progress to a serious syndrome called *uremia.*

Anatomy of the Kidney (p. 709)

1. The kidneys are located retroperitoneally against the superior dorsal abdominal wall. Each is provided with a renal artery and renal vein. The adrenal glands lie against the superior to medial aspects of the kidneys.
2. The kidney has a slit called the *hilum* on its concave side, where it receives renal nerves, blood and lymphatic vessels, and the ureter.
3. From superficial to deep, the kidney is enclosed by the *renal fascia, adipose capsule,* and *renal capsule.*
4. The renal parenchyma is a C-shaped tissue enclosing a space called the *renal sinus.* The parenchyma is divided into an outer *renal cortex* and inner *renal medulla.* The medulla is divided into 6 to 10 *renal pyramids.*
5. The apex, or *papilla,* of each pyramid projects into a receptacle called a *minor calyx,* which collects the urine from that lobe. Minor calyces converge to form *major calyces,* and these converge on the *renal pelvis,* where the ureter arises.
6. Each kidney contains about 1.2 million functional units called *nephrons.*
7. The renal artery branches and gives rise to *segmental arteries, interlobar arteries, arcuate arteries,* and then *interlobular arteries,* which penetrate into the cortex. For each nephron, an *afferent arteriole* arises from the interlobular artery and supplies the capillaries of the *glomerulus.* An *efferent arteriole* leaves the glomerulus and usually gives rise to a bed of *peritubular capillaries* around the renal tubules. Blood then flows through a series of *interlobular, arcuate,* and *interlobar veins,* before leaving the kidney by way of the *renal vein.*
8. The renal medulla is supplied by vessels called the *vasa recta,* which arise from the efferent arterioles of juxtamedullary nephrons and empty into arcuate and interlobular veins.
9. *Renal nerves* follow the renal artery and innervate the afferent and efferent arterioles. They provide sympathetic control over blood flow to the glomerulus, and thus regulate the rate of glomerular filtration and urine formation. They also stimulate renin secretion.
10. A nephron produces urine in three stages: *glomerular filtration, tubular reabsorption* and *secretion,* and *water conservation.*
11. A nephron begins with a double-walled *glomerular capsule* enclosing the glomerulus; the glomerulus and capsule constitute the *renal corpuscle.* The capsule consists of a parietal layer with a simple squamous epithelium and a visceral layer composed of *podocytes.* Podocytes have numerous *foot processes* that wrap around the glomerular capillaries.
12. To leave the blood and enter the glomerular capsule, water and solutes must pass through the fenestrations of the capillary endothelium, the basement membrane, and the *filtration slits* between the podocyte foot processes. The small sizes and negative charges on these barriers prevent blood cells and most protein from leaving the capillaries.
13. Filtrate collects in the *capsular space* between the capsule layers and then flows into the *renal tubule* leading away from the capsule.
14. The renal tubule consists of a highly coiled *proximal convoluted tubule (PCT),* a U-shaped *nephron loop,* a coiled *distal convoluted tubule (DCT),* and a *collecting duct.* The first three of these belong to a single nephron; the collecting duct receives fluid from many nephrons.
15. The proximal convoluted tubule (PCT) is the longest part of the renal tubule, and has a brush border of prominent microvilli. It reabsorbs about 65% of the glomerular filtrate and returns it to the peritubular capillaries. Materials pass through its epithelium by both *transcellular* and *paracellular routes.*
16. The PCT reabsorbs electrolytes, glucose, amino acids, lactate, protein and other peptides, amino acids, urea, uric acid, and water. It secretes urea, uric acid, creatinine, ammonia, bile acids, H^+, HCO_3^-, drugs, and other solutes into the tubular fluid.
17. The nephron loop reabsorbs Na^+, K^+, Cl^-, and water, and serves mainly to maintain a gradient of salinity in the tissue fluid of the renal medulla. In conjunction with the collecting duct, this makes it possible to excrete hypertonic urine, thus to eliminate wastes without excessive water loss.
18. The distal convoluted tubule (DCT) reabsorbs Na^+, Ca^{2+}, Cl^-, phosphate, and water, and secretes K^+ and H^+ at variable rates depending on the influence of aldosterone, atrial natriuretic peptide, and parathyroid hormone.
19. The *juxtaglomerular apparatus* is a junction of the DCT with the afferent and efferent arterioles near the glomerular capsule. It is a feedback device for monitoring the flow and composition of the tubular fluid and adjusting glomerular filtration.
20. The *collecting duct* receives filtrate from nephrons in the cortex and passes down through the medulla to the renal papilla. It secretes variable amounts of H^+, K^+, and HCO_3^- into the tubular fluid and reabsorbs Na^+. Most importantly, it reabsorbs variable amounts of water according to the influence of antidiuretic hormone, thus adjusting the osmolarity of the urine and regulating the body's water balance.

Anatomy of the Ureters, Urinary Bladder, and Urethra (p. 717)

1. The ureter is a muscular tube from the renal pelvis to the floor of the urinary bladder. It drives urine to the bladder by means of peristalsis.
2. The urinary bladder has a smooth muscle layer called the *detrusor muscle* with a thickened ring, the *internal urethral sphincter,* around the origin of the urethra. Most of its mucosa exhibits folds called rugae when the bladder is empty. The *trigone,* a smooth triangular area between the openings of the ureters and urethra, is the most common site of bladder infections.

3. The urethra is 3 to 4 cm long in the female, but in the male it is 18 cm long and divided into *prostatic, membranous,* and *spongy (penile)* segments. An *external urethral sphincter* of skeletal muscle encircles the urethra in both sexes where it passes through the pelvic floor.

Developmental and Clinical Perspectives (p. 719)

1. Temporary embryonic kidneys called the *pronephros* and *mesonephros* develop and degenerate before the permanent kidney, the *metanephros,* appears early in week 5.
2. At this time, a pouch called the *ureteric bud* arises from the lower end of the mesonephric duct. It elongates and branches, eventually giving rise to the ureter, renal pelvis and calyces, and collecting ducts.
3. As the ureteric bud penetrates into the mesonephric kidney tissue, it induces the formation of nephrons. These start as S-shaped tubules that eventually enfold glomerular capillaries at one end, connect with the collecting duct at the other end, and differentiate into PCT, nephron loop, and DCT.
4. The embryonic urinary tract initially opens into a cloaca that also receives the digestive and reproductive tracts, but in weeks 4 to 7 the cloaca divides into an anal canal and a urogenital sinus. The upper part of the urogenital sinus differentiates into the urinary bladder, and the lower end into the urethra.
5. In old age, the kidneys atrophy markedly and become less efficient at clearing wastes and drugs from the blood. Elderly women often experience urinary incontinence due to weakening of the pelvic muscles and urethral sphincters, whereas elderly men often experience urine retention and reduced flow during micturition because of pressure on the urethra from an enlarged prostate gland.
6. Renal failure can have numerous causes ranging from atherosclerosis to poisoning and trauma. It presents a danger of azotemia or uremia, and may require hemodialysis or a kidney transplant.

Testing Your Recall

1. Which of these is *not* a function of the kidneys?
   a. to secrete hormones
   b. to excrete nitrogenous wastes
   c. to store urine
   d. to control blood volume
   e. to control acid-base balance
2. The compact ball of capillaries in a nephron is called
   a. the nephron loop.
   b. the peritubular plexus.
   c. the renal corpuscle.
   d. the glomerulus.
   e. the vasa recta.
3. Which of these is *not* true of the position of the kidneys in the body?
   a. They are medial to the aorta.
   b. They are retroperitoneal.
   c. The right kidney is lower than the left.
   d. They are inferior to the liver and spleen.
   e. They lie partially within the rib cage.
4. Which of these lies closest to the renal cortex?
   a. the parietal peritoneum
   b. the renal fascia
   c. the renal capsule
   d. the adipose capsule
   e. the renal pelvis
5. The water that is reabsorbed by the collecting duct enters
   a. the nephron loop.
   b. the minor calyx.
   c. the ureter.
   d. the efferent arteriole.
   e. the vasa recta.
6. A glomerulus and glomerular capsule make up one
   a. renal capsule.
   b. renal corpuscle.
   c. kidney lobule.
   d. kidney lobe.
   e. nephron.
7. The kidney has more _____ than any of the other structures listed.
   a. arcuate arteries
   b. minor calyces
   c. medullary pyramids
   d. afferent arterioles
   e. collecting ducts
8. The _____ arises from the embryonic ureteric bud.
   a. nephron
   b. renal pelvis
   c. glomerulus
   d. urinary bladder
   e. proximal convoluted tubule
9. The _____ absorbs variable amounts of water depending on the level of antidiuretic hormone present.
   a. proximal convoluted tubule
   b. nephron loop
   c. distal convoluted tubule
   d. collecting duct
   e. urinary bladder
10. In cortical nephrons, blood of the efferent arteriole flows next into
    a. the peritubular capillaries.
    b. the arcuate artery.
    c. the arcuate vein.
    d. the vasa recta.
    e. the glomerulus.
11. The most abundant nitrogenous waste in the urine is _____.
12. The ureter, renal pelvis, calyces, and collecting duct arise from an embryonic pouch called the _____.
13. The openings of the two ureters and urethra form the boundaries of a smooth area called the _____ on the floor of the urinary bladder.
14. The _____ is a group of epithelial cells of the distal convoluted tubule that monitors the flow or composition of the tubular fluid.
15. To enter the capsular space, filtrate must pass between foot process of the _____, cells that form the visceral layer of the glomerular capsule.
16. What part of the nephron is characterized by a brush border and especially great length?
17. Epithelial cells in the _____ of the nephron loop have few mitochondria and low metabolic activity, but they are very permeable to water.
18. The smooth muscle of the bladder wall is called the _____.
19. Each renal pyramid drains into a separate cuplike urine receptacle called a _____.
20. Blood flows through the _____ arteries just before entering the interlobular arteries.

Answers in the Appendix

True or False

Determine which five of the following statements are false, and briefly explain why.

1. The ureters open through pores in the roof of the urinary bladder.
2. The kidneys secrete antidiuretic hormone to promote water retention and prevent dehydration.
3. The kidney has more distal convoluted tubules than collecting ducts.
4. Tight junctions prevent material from leaking between the epithelial cells of the renal tubule.
5. Many collecting ducts empty into each minor calyx.
6. The glomerulus is a complex of blood capillaries located in the capsular space of the glomerular capsule.
7. Each interlobular artery serves multiple nephrons.
8. Blood-borne solutes can become incorporated into the urine by either glomerular filtration or tubular secretion.
9. The kidneys are normally located in the pelvic cavity.
10. The kidneys develop in the pelvic region and ascend to a higher location in the abdominal cavity during fetal development.

Answers in the Appendix

Testing Your Comprehension

1. What function could the collecting duct *not* perform if there were no nephron loops? Why?
2. Why would a simple squamous epithelium function poorly as an inner lining of the urinary bladder?
3. In some infants, the urachus fails to close; it remains as an open passage *(urachal fistula)* from the urinary bladder to the umbilicus. What would you expect to be the most obvious sign of a urachal fistula?
4. Suppose the ureters entered the bladder from above instead of from below. Which disorder in table 25.1 would you expect to result from this? Explain why.
5. In what ways do the proximal and distal convoluted tubules differ in structure? How is this related to their functional difference?

Answers at the Online Learning Center

www.mhhe.com/saladinha1

Visit the Online Learning Center for practice tests, answer keys, and other learning aids for this chapter. Enhance your understanding of human anatomy with our interactive art labeling exercises, supplemental photo atlases, web links, puzzles, flashcards, and much more.

26

CHAPTER TWENTY-SIX

The Reproductive System

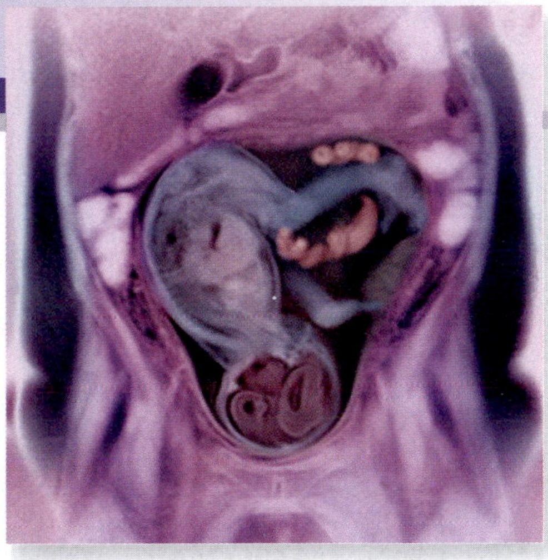

MRI scan of a fetus in the uterus at the 36th week of pregnancy

CHAPTER OUTLINE

Sexual Reproduction 728
- The Two Sexes 728
- Overview of the Reproductive System 728

Male Reproductive Anatomy 729
- The Testes 729
- The Scrotum 732
- Spermatogenesis and Sperm 732
- The Spermatic Ducts 736
- The Accessory Glands 737
- The Penis 737

Female Reproductive Anatomy 740
- The Ovaries 740
- Oogenesis and Ova 742
- The Uterine Tubes 743
- The Uterus 745
- The Vagina 747
- The External Genitalia 747
- The Breasts and Mammary Glands 749

Developmental and Clinical Perspectives 751
- Prenatal Development and Sexual Differentiation 751
- Puberty 753
- The Aging Reproductive System 754
- Reproductive Disorders of the Male 755
- Reproductive Disorders of the Female 755

Chapter Review 759

INSIGHTS

26.1 Clinical Application: Prostate Diseases 737

26.2 Clinical Application: Reproductive Effects of Pollution 739

26.3 Clinical Application: Cervical Cancer and Pap Smears 757

BRUSHING UP

To understand this chapter, you may find it helpful to review the following concepts:

- Chromosome structure (p. 69)
- Mitosis (p. 69)
- Muscles of the pelvic floor (p. 307)
- Hypothalamic releasing factors and pituitary gonadotropins (p. 518)
- Androgens (p. 527)
- Embryology of the mesonephric duct (p. 719)

PART FIVE

From all we have learned of the structure and function of the human body, it seems a wonder that it works at all! The fact is, however, that even with modern medicine we cannot keep it working forever. The body suffers degenerative changes as we age, and eventually we expire. Yet our genes live on in new containers—our offspring. This final chapter concerns the means of their production—the male and female reproductive systems.

Sexual Reproduction

Objectives

When you have completed this section, you should be able to

- define *sexual reproduction;*
- identify the most fundamental biological distinction between male and female; and
- define *primary sex organs, secondary sex organs,* and *secondary sex characteristics.*

The Two Sexes

The essence of sexual reproduction is that it is biparental—the offspring receive genes from two parents and therefore are not genetically identical to either one. To achieve this, the parents must produce **gametes**[1] (sex cells) that meet and combine their genes in a **zygote**[2] (fertilized egg). The gametes must have two properties for reproduction to be successful: motility so they can achieve contact, and nutrients for the developing embryo. A single cell cannot perform both of these roles optimally, because to contain ample nutrients means to be relatively large and heavy, and this is inconsistent with the need for motility. Therefore, these tasks are usually apportioned to two kinds of gametes. The small motile one—little more than DNA with a propeller—is the **sperm (spermatozoon),** while the large nutrient-laden one is the **egg (ovum).**

In any species, by definition, an individual that produces eggs is female and one that produces sperm is male. This criterion is not always that simple, as we see in certain abnormalities in sexual development. Genetically, however, any human with a Y sex chromosome is classified as male and anyone lacking a Y is classified as female. Normally, a male inherits an X from the mother and Y from the father, and his sex chromosomes are thus designated XY. A female inherits an X from each parent, and therefore has an XX chromosome pair.

In mammals, the female is also the parent that provides a sheltered internal environment for the development and prenatal nutrition of the embryo. For fertilization and development to occur in the female, the male must have a copulatory organ, the penis, for introducing his gametes into the female reproductive tract, and the female must have a copulatory organ, the vagina, for receiving the sperm. This is the most obvious difference between the sexes, but appearances can be deceiving (see fig. 18.16).

[1]*gam* = marriage, union
[2]*zygo* = yoke, union

Overview of the Reproductive System

The **male reproductive system** is an organ system that serves to produce sperm and introduce them into the female body. The **female reproductive system** is an organ system that serves to produce eggs, receive the sperm, provide for the union of these gametes, harbor the fetus, give birth, and nourish the offspring.

In both sexes, the reproductive system consists of primary and secondary sex organs, or **genitalia.** The **primary sex organs,** or **gonads,**[3] are the organs that produce gametes—testes of the male and ovaries of the female. The **secondary sex organs** are organs other than the gonads that are essential to reproduction. In the male, they constitute a system of ducts, glands, and the penis, concerned with the storage, survival, and conveyance of sperm. In the female, they include the uterine tubes, uterus, and vagina, concerned with uniting the sperm and egg and harboring the fetus.

According to location, the reproductive organs are classified as **internal** and **external genitalia** (table 26.1). Most of the internal genitalia are located in the pelvic cavity. The external genitalia are located in the perineum—the diamond-shaped region marked by the pubic symphysis anteriorly, the coccyx posteriorly, and the ischial tuberosities laterally (see figs. 11.17, 26.1, and 26.22). Most of them are externally visible, except for the accessory glands of the female perineum.

Secondary sex characteristics are features that develop in adolescence, further distinguish the sexes, and play a role in mate attraction. From the call of a bullfrog to the tail of a peacock, these are well known in the animal kingdom. In humans, the identification of secondary sex characteristics rests on somewhat subjective and culturally variable judgments of what is sexually attractive. Commonly considered among the secondary sex characteristics are the pubic, axillary, and male facial hair; apocrine scent glands

[3]*gon* = seed

Table 26.1
The Internal and External Genitalia

Internal Genitalia	External Genitalia
Male	
Testes (s. testis)	Penis
Epididymides (s. epididymis)	Scrotum
Ductus deferentes (s. ductus deferens)	
Seminal vesicles	
Prostate	
Bulbourethral glands	
Female	
Ovaries	Mons pubis
Uterine tube	Labia majora (s. labium majus)
Uterus	Labia minora (s. labium minus)
Vagina	Clitoris
	Vaginal orifice
	Vestibular bulbs
	Vestibular glands
	Paraurethral glands

associated with these patches of hair; differences in the texture and visibility of the hair on the limbs and trunk; the female breasts; differences in muscularity and the quantity and distribution of body fat; and differences in the pitch of the voice.

Before You Go On

Answer the following questions to test your understanding of the preceding section:

1. Define *gonad* and *gamete.* Explain the relationship between the terms.
2. Define *male, female, sperm,* and *egg.*

Male Reproductive Anatomy

Objectives

When you have completed this section, you should be able to

- describe the anatomy of the testes, scrotum, spermatic ducts, accessory glands, and penis;
- describe the stages in spermatogenesis, the production of sperm; and
- describe the structure of a sperm cell and the composition of semen.

Figures 26.1 and 26.2 are overviews of the male reproductive system from external and internal perspectives.

The Testes

The **testes (testicles)** are the male gonads—combined endocrine and exocrine glands that produce sex hormones and sperm. Each testis is oval and slightly flattened, about 4 cm long, 3 cm from anterior to posterior, and 2.5 cm from left to right (fig. 26.3). Its anterior and lateral surfaces are covered by the **tunica vaginalis,**[4] a saccular extension of the peritoneum. The testis itself has a white fibrous capsule called the **tunica albuginea**[5] (TOO-nih-ca AL-byu-JIN-ee-uh). Connective tissue septa extend from the capsule into the parenchyma of the testis, dividing it into 200 to 300 wedge-shaped lobules. Each lobule contains one to three **seminiferous**[6] (SEM-ih-NIF-er-us) **tubules**—slender ducts up to 1 m long in which the sperm are produced. Between the seminiferous tubules are clusters of **interstitial (Leydig**[7]**) cells,** the source of testosterone.

A seminiferous tubule has a narrow lumen lined by a thick **germinal epithelium** (fig. 26.4). The epithelium consists of several layers of **germ cells** in the process of becoming sperm, and a much smaller number of tall **sustentacular**[8] **(Sertoli**[9]**) cells,** which protect the germ cells and promote their development. The germ cells depend on the sustentacular cells for nutrients, waste removal, growth factors, and other needs. The sustentacular cells also secrete a hormone, *inhibin,* which regulates the rate of sperm production.

A sustentacular cell is shaped a little like a tree trunk whose roots spread out over the basement membrane, forming the boundary of the tubule, and whose thick trunk reaches to the tubule lumen. Tight junctions between adjacent sustentacular cells form a **blood-testis barrier (BTB),** which prevents antibodies and other large molecules in the blood and intercellular fluid from getting to the germ cells. This is important because the germ cells, being genetically different from other cells of the body, would

[4]*tunica* = coat + *vagina* = sheath
[5]*alb* = white
[6]*semin* = seed, sperm + *fer* = to carry
[7]Franz von Leydig (1821–1908), German anatomist
[8]*sustentacul* = support
[9]Enrico Sertoli (1842–1910), Italian histologist

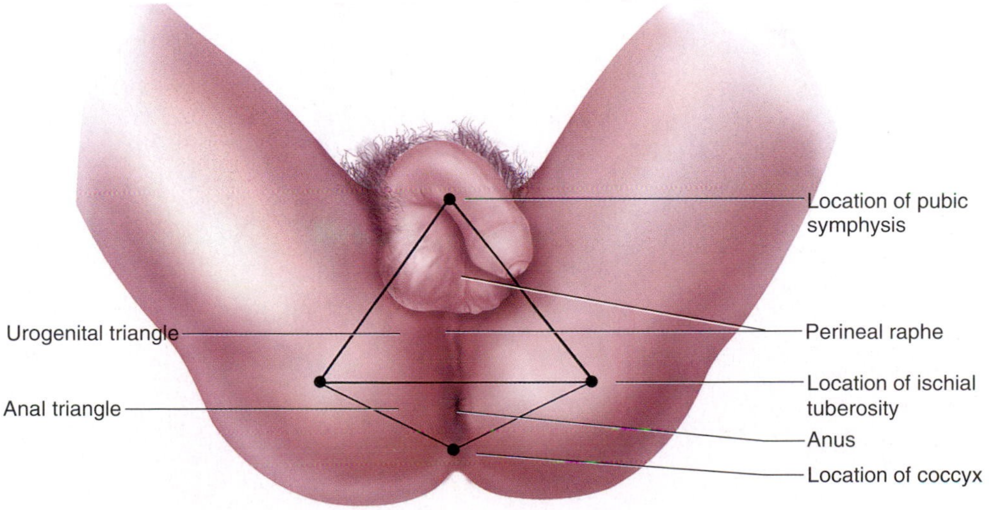
Location of pubic symphysis
Urogenital triangle
Perineal raphe
Location of ischial tuberosity
Anal triangle
Anus
Location of coccyx

Figure 26.1
The Male Perineum.

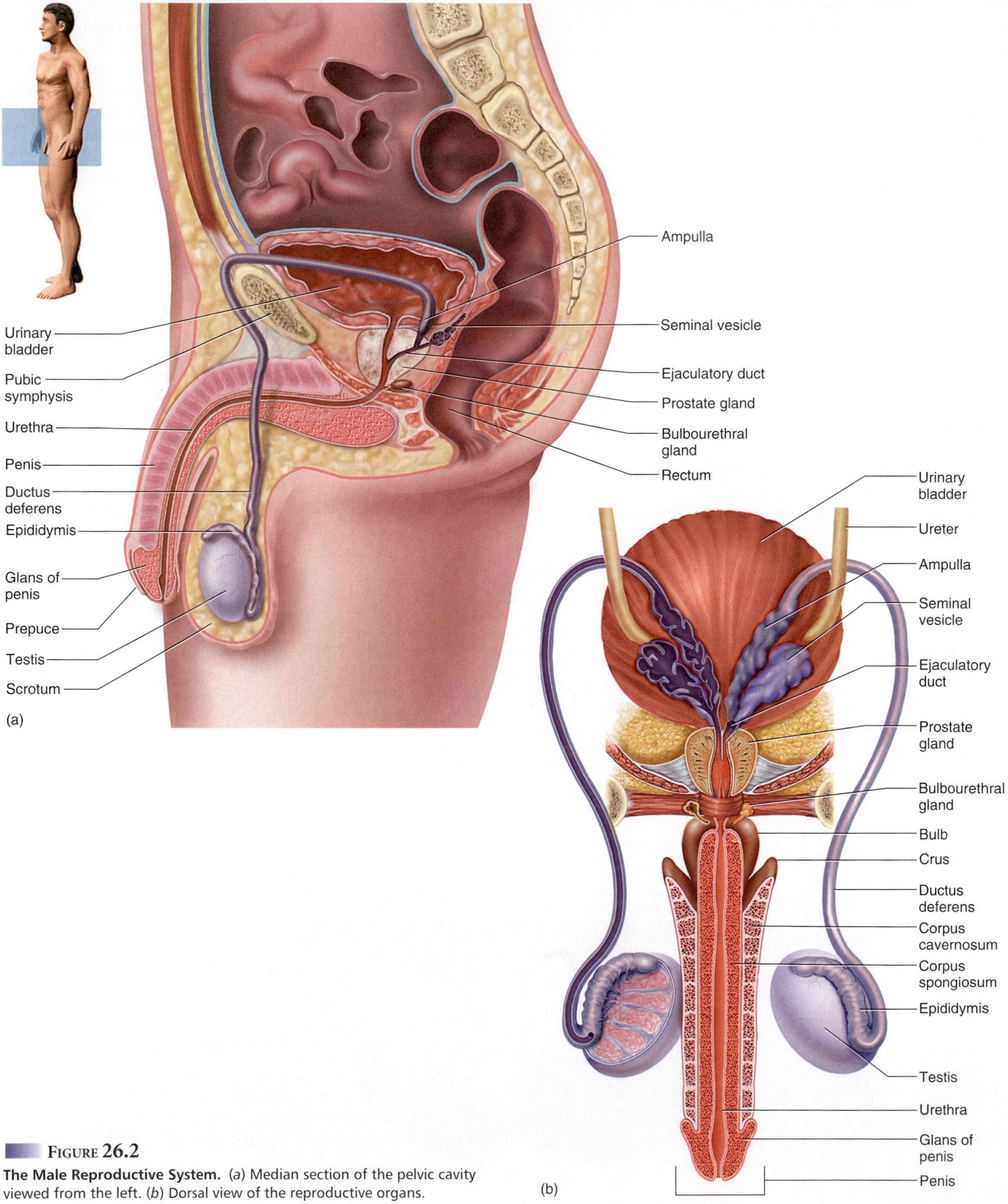

Ampulla
Urinary bladder
Seminal vesicle
Pubic symphysis
Ejaculatory duct
Prostate gland
Urethra
Bulbourethral gland
Penis
Rectum
Ductus deferens
Epididymis
Glans of penis
Prepuce
Testis
Scrotum
(a)
Urinary bladder
Ureter
Ampulla
Seminal vesicle
Ejaculatory duct
Prostate gland
Bulbourethral gland
Bulb
Crus
Ductus deferens
Corpus cavernosum
Corpus spongiosum
Epididymis
Testis
Urethra
Glans of penis
Penis
(b)

Figure 26.2

The Male Reproductive System. (*a*) Median section of the pelvic cavity viewed from the left. (*b*) Dorsal view of the reproductive organs.

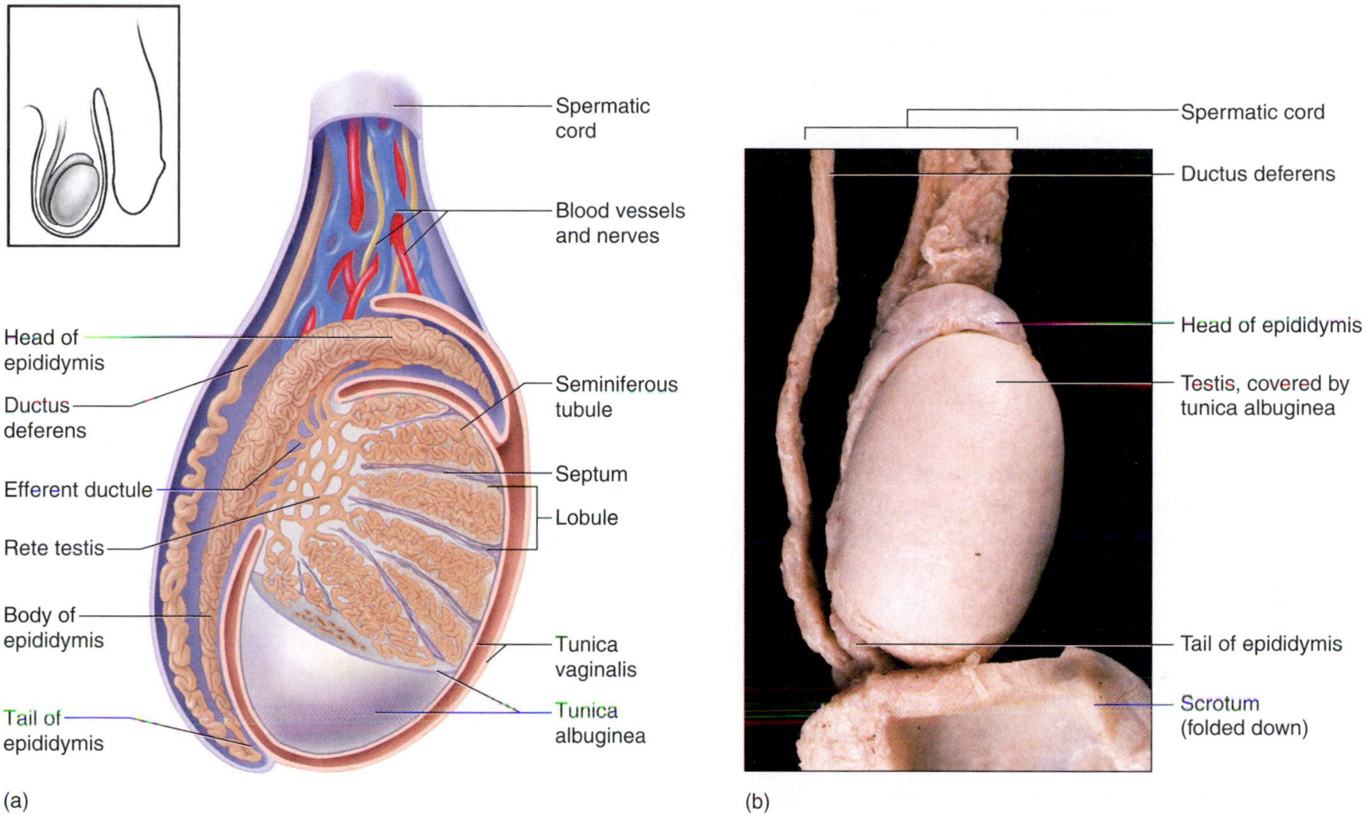
Spermatic cord
Blood vessels and nerves
Head of epididymis
Ductus deferens
Seminiferous tubule
Efferent ductule
Septum
Lobule
Rete testis
Body of epididymis
Tunica vaginalis
Tail of epididymis
Tunica albuginea
(a)
Spermatic cord
Ductus deferens
Head of epididymis
Testis, covered by tunica albuginea
Tail of epididymis
Scrotum (folded down)
(b)

FIGURE 26.3

The Testis and Associated Structures. (*a*) Anatomy of the testis, epididymis, and spermatic cord. (*b*) The testis and associated structures dissected free of the scrotum, shown life size.

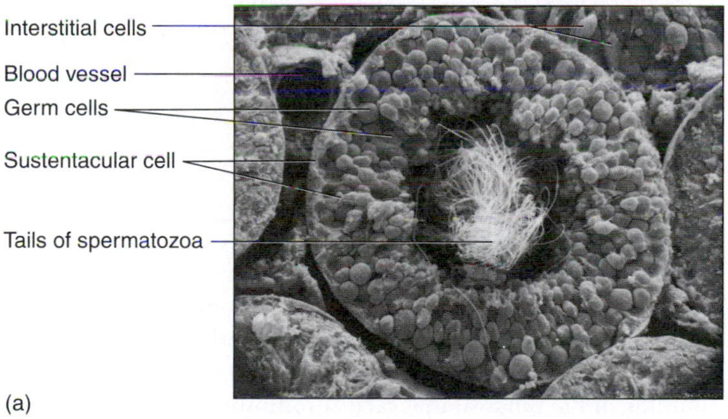
Interstitial cells
Blood vessel
Germ cells
Sustentacular cell
Tails of spermatozoa
(a)

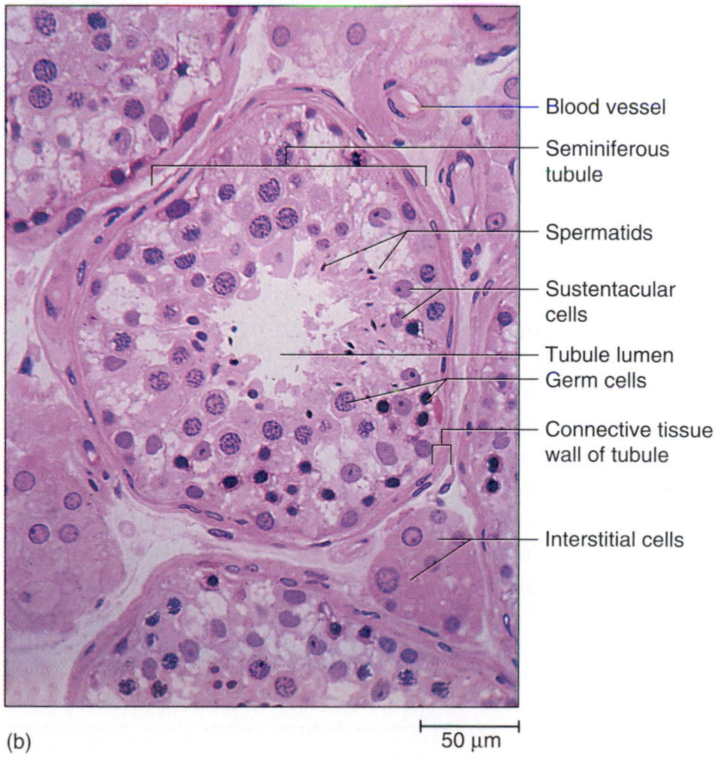
Blood vessel
Seminiferous tubule
Spermatids
Sustentacular cells
Tubule lumen
Germ cells
Connective tissue wall of tubule
Interstitial cells
50 μm
(b)

FIGURE 26.4

Histology of the Testis. (*a*) Scanning electron micrograph. (*b*) Light micrograph. Figure *b* is from a region of the tubule that did not have mature sperm at the time. Figure *a* from R. G. Kessel and R. H. Kardon, *Tissues and Organs: A Text-Atlas of Scanning Electron Microscopy,* W. H. Freeman, 1979.

otherwise be attacked by the immune system. Some cases of sterility occur when the BTB fails to form adequately in adolescence and the immune system produces autoantibodies against the germ cells.

Think About It!

Would you expect to find blood capillaries in the walls of the seminiferous tubules? Why or why not?

The seminiferous tubules lead into a network called the **rete**[10] (REE-tee) **testis,** embedded in the capsule on the posterior side. Sperm partially mature in the rete. They are moved along by the flow of fluid secreted by the sustentacular cells and by the cilia on some rete cells. Sperm do not swim while they are in the male.

Each testis is supplied by a **testicular artery,** which arises from the abdominal aorta just below the renal artery. This is a very long, slender artery that winds its way down the dorsal abdominal wall before passing through the inguinal canal into the scrotum. Its blood pressure is very low, and indeed this is one of the few arteries to have no pulse. Consequently, blood flow to the testes is quite meager and the testes receive a poor oxygen supply. The sperm appear to compensate by developing unusually large mitochondria, which may precondition them for survival in the hypoxic environment of the female reproductive tract.

Blood leaves the testis by way of a **testicular vein.** The right testicular vein drains into the inferior vena cava and the left one drains into the left renal vein. Lymphatic vessels also drain each testis and lead to the inguinal lymph nodes. **Testicular nerves** lead to the gonads from spinal cord segment T10. They are mixed sensory and motor nerves containing predominantly sympathetic but also some parasympathetic fibers.

The Scrotum

The testes are contained in a pouch, the **scrotum**[11] (fig. 26.5). The left testis is usually suspended lower than the right so the two are not compressed against each other between the thighs. The skin of the scrotum has sebaceous glands, sparse hair, rich sensory innervation, and somewhat darker pigmentation than skin elsewhere. The scrotum is divided into right and left compartments by an internal **median septum,** which protects each testis from infections of the other one. The location of the septum is externally marked by a seam called the **perineal raphe**[12] (RAY-fee), which also extends anteriorly along the ventral side of the penis and posteriorly as far as the margin of the anus (see fig. 26.1).

The **spermatic cord** is a cord of connective tissue that passes upward behind the testis, across the anterior side of the pubis, and into an opening called the *inguinal ring* in the muscles of the groin. From there, it travels about 4 cm through the *inguinal canal* and emerges into the pelvic cavity. It contains the *ductus deferens* (a sperm duct), blood and lymphatic vessels, and testicular nerves. The cord is easily palpated through the skin of the scrotum.

The original reason that a scrotum evolved is a subject of debate among reproductive biologists. For whatever reason human testes reside in the scrotum, however, they have adapted to this cooler environment and cannot produce sperm at the core body temperature of 37°C; they must be held at about 35°C. The scrotum has three mechanisms for regulating the temperature of the testes:

1. The **cremaster**[13] **muscle**—strips of the internal abdominal oblique muscle that enmesh the spermatic cord. When it is cold, the cremaster contracts and draws the testes closer to the body to keep them warm. When it is warm, the cremaster relaxes and the testes are suspended farther from the body.
2. The **dartos**[14] **muscle**—a subcutaneous layer of smooth muscle. It, too, contracts when cold, making the scrotum taut and wrinkled. This reaction holds the testes snugly against the warm body and reduces the surface area of the scrotum, thus reducing heat loss.
3. The **pampiniform**[15] **plexus**—an extensive network of veins from the testis that surround the testicular artery in the spermatic cord. As they pass through the inguinal canal, these veins converge to form the testicular vein. Without the pampiniform plexus, warm arterial blood would heat the testis and inhibit spermatogenesis. The pampiniform plexus, however, prevents this by acting as a *countercurrent heat exchanger.* The relatively cool blood (about 35°C) ascending through the plexus absorbs heat from the warmer blood descending through the testicular artery. By the time the arterial blood reaches the testis, it is 1.5° to 2.5°C cooler than the core body temperature.

Spermatogenesis and Sperm

Spermatogenesis is the process of sperm production. It occurs in the seminiferous tubules and involves three principal events: (1) remodeling of a relatively large germ cell into a small, mobile cell with a flagellum; (2) reduction of the chromosome number by one-half, so that when the sperm and egg combine we do not get a doubling of chromosome number in every generation; and (3) a shuffling of the genes so that each chromosome of the sperm carries new gene combinations that did not exist in the chromosomes received from one's parents. This ensures genetic variety in the offspring. The genetic recombination and reduction in chromosome number are achieved through a form of cell division called **meiosis,** which produces four daughter cells that subsequently differentiate into sperm (fig. 26.6).

[10]*rete* = network
[11]*scrotum* = bag
[12]*raphe* = seam
[13]*cremaster* = suspender
[14]*dartos* = skinned
[15]*pampin* = tendril + *form* = shape

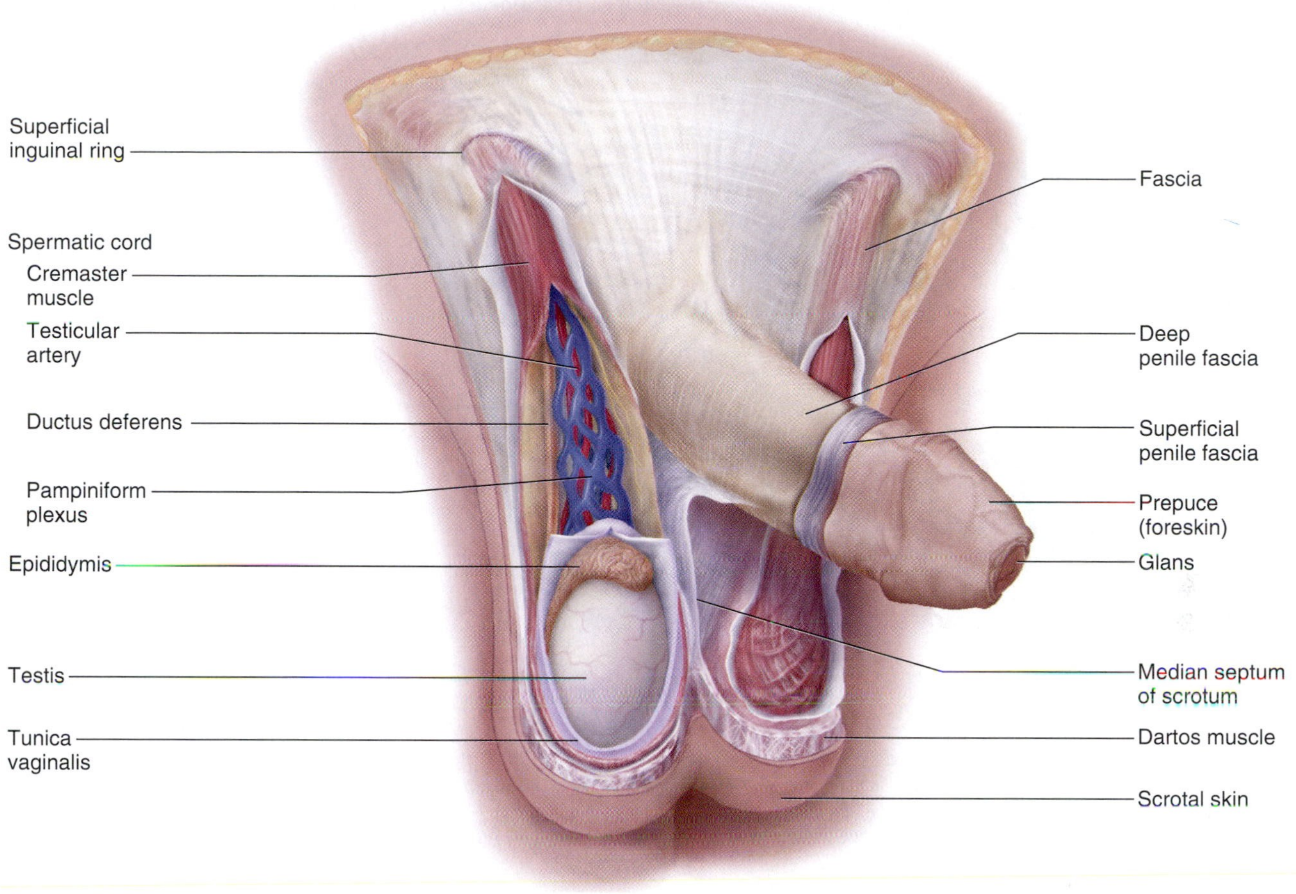

FIGURE 26.5
Anatomy of the Male Inguinal Region and External Genitalia.

The basic stages of spermatogenesis are as follows (fig. 26.7):

1. Early in prenatal development, **primordial germ cells** form in the yolk sac, migrate by ameboid motion into the embryo itself, and colonize the embryonic gonad. Here they become stem cells called **spermatogonia.** These cells, like nearly all other cells of the body, are **diploid**[16]—they have 46 chromosomes (23 pairs).
2. The spermatogonia remain dormant in childhood, but begin to divide mitotically at puberty. One daughter cell from each division remains near the tubule wall as a stem cell called the *type A spermatogonium.* The other one, called the *type B spermatogonium,* migrates slightly away from the wall on its way to becoming sperm.
3. The type B spermatogonium enlarges and becomes a **primary spermatocyte.** Since this cell is about to undergo meiosis and become genetically different from other cells of the body, it must be protected from the immune system. It migrates a little farther away from the tubule wall and the blood-testis barrier closes behind it, isolating it from the blood.
4. The primary spermatocyte undergoes *meiosis I,* a cell division that reduces the chromosome number by half. The daughter cells, called **secondary spermatocytes,** are therefore **haploid**[17]—they have 23 unpaired chromosomes.
5. The secondary spermatocytes undergo another division, *meiosis II,* and produce four daughter cells (two from each spermatocyte) called **spermatids.** Spermatids still have 23 chromosomes, but each chromosome is reduced from a double-stranded (two-chromatid) state to a single-stranded state.
6. A spermatid divides no further, but undergoes a process called **spermiogenesis,** in which it differentiates into a single spermatozoon (sperm). The fundamental changes in spermiogenesis are a loss of excess cytoplasm and the growth of a tail (flagellum), making the sperm a lightweight, mobile cell (fig. 26.8).

All stages from primary spermatocyte to spermatozoon are enfolded in tendrils of the sustentacular cells and bound to them by tight junctions and gap junctions. At the conclusion of spermiogenesis, the spermatozoon is released and washed down the tubule by fluid from the sustentacular cells. It takes about 74 days for a spermatogonium to become a mature spermatozoon. A young adult male produces about 300,000 sperm per minute, or 400 million per day.

[16]*diplo* = double, paired

[17]*haplo* = half

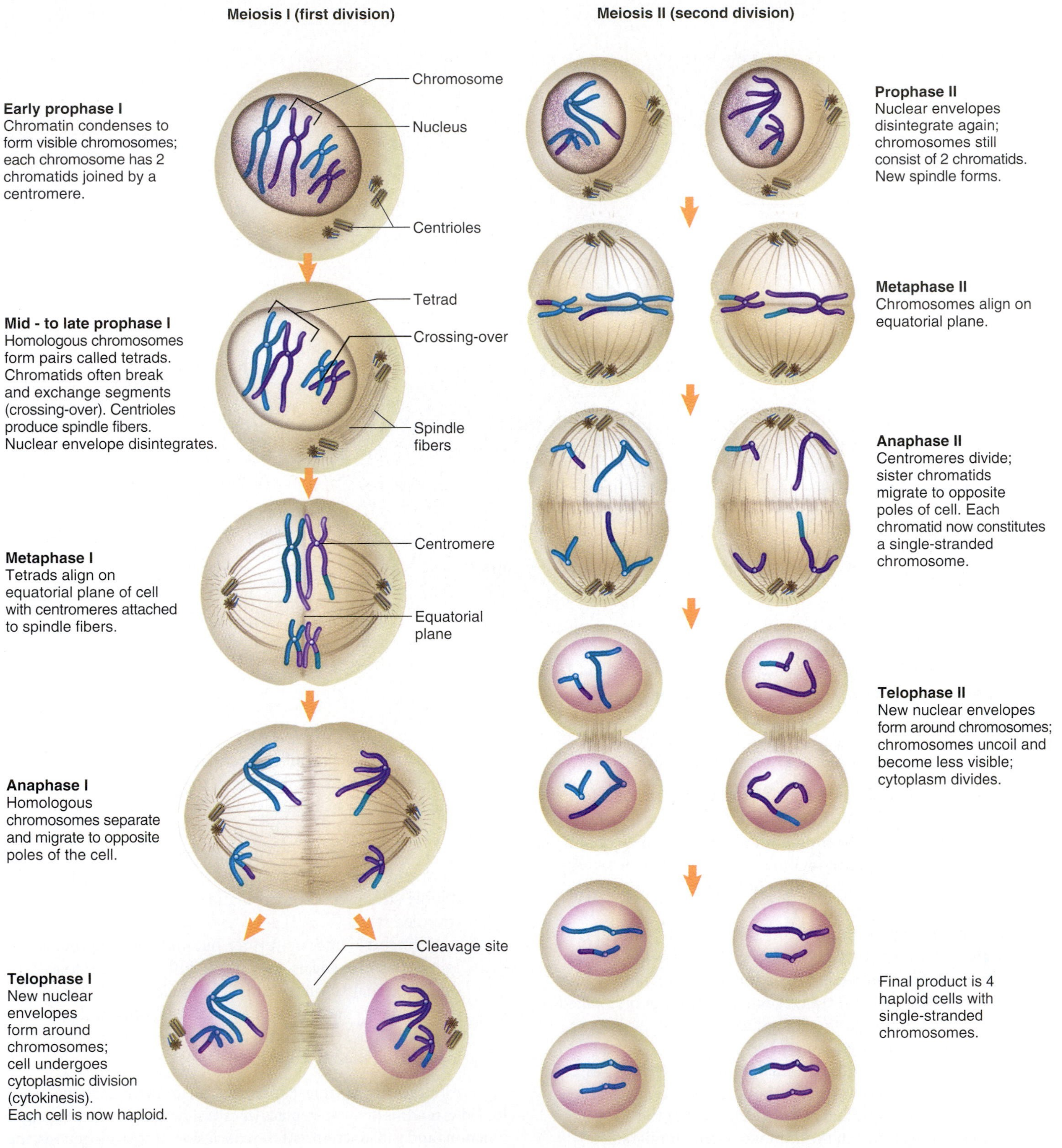

Meiosis I (first division)
Meiosis II (second division)
Chromosome
Nucleus
Centrioles
Early prophase I
Chromatin condenses to form visible chromosomes; each chromosome has 2 chromatids joined by a centromere.
Tetrad
Crossing-over
Spindle fibers
Mid - to late prophase I
Homologous chromosomes form pairs called tetrads. Chromatids often break and exchange segments (crossing-over). Centrioles produce spindle fibers. Nuclear envelope disintegrates.
Centromere
Equatorial plane
Metaphase I
Tetrads align on equatorial plane of cell with centromeres attached to spindle fibers.
Anaphase I
Homologous chromosomes separate and migrate to opposite poles of the cell.
Cleavage site
Telophase I
New nuclear envelopes form around chromosomes; cell undergoes cytoplasmic division (cytokinesis). Each cell is now haploid.
Prophase II
Nuclear envelopes disintegrate again; chromosomes still consist of 2 chromatids. New spindle forms.
Metaphase II
Chromosomes align on equatorial plane.
Anaphase II
Centromeres divide; sister chromatids migrate to opposite poles of cell. Each chromatid now constitutes a single-stranded chromosome.
Telophase II
New nuclear envelopes form around chromosomes; chromosomes uncoil and become less visible; cytoplasm divides.
Final product is 4 haploid cells with single-stranded chromosomes.

FIGURE 26.6
Meiosis. For simplicity, the cell is shown with only two pairs of homologous chromosomes. Human cells begin meiosis with 23 pairs.

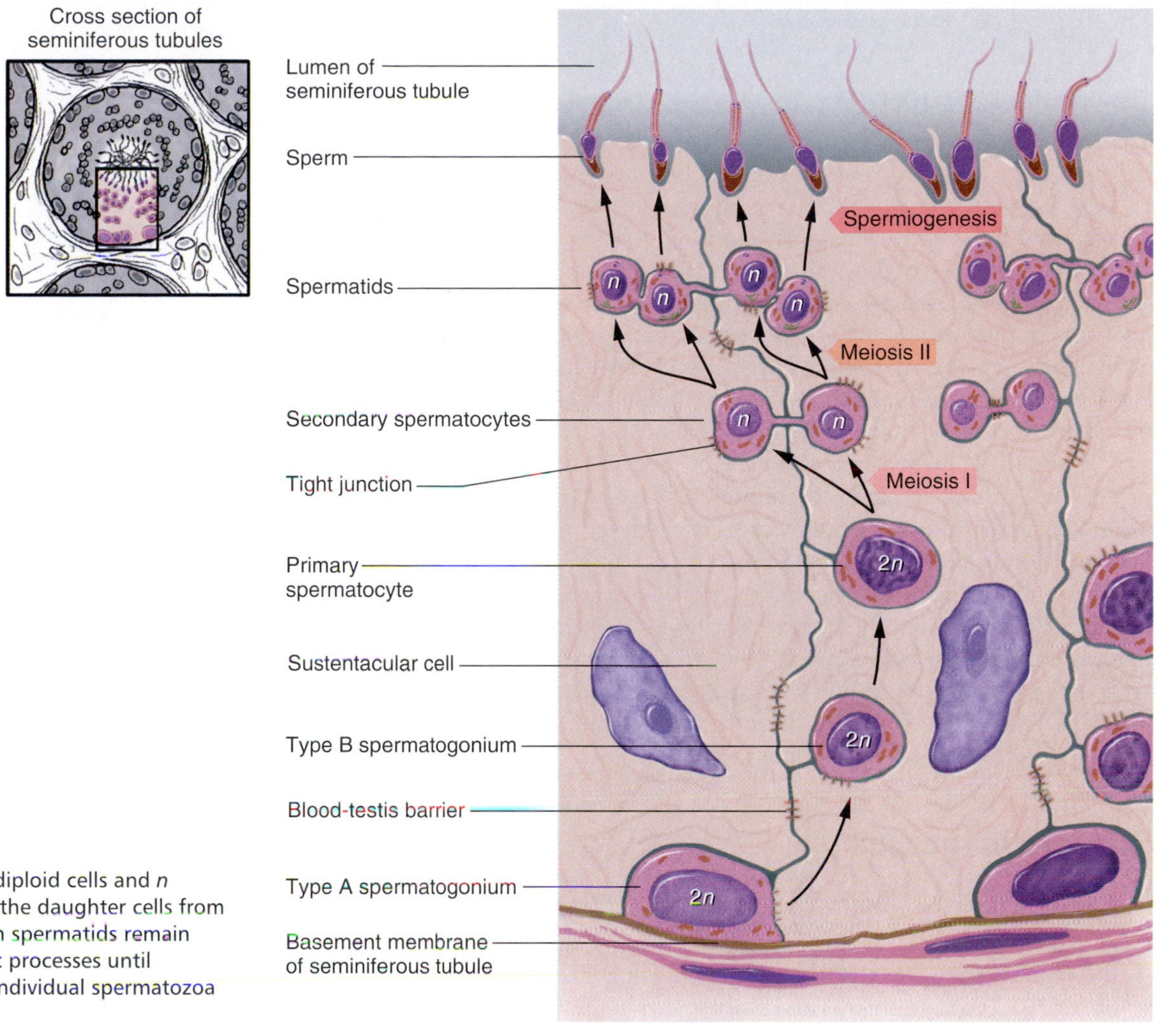

Cross section of seminiferous tubules
Lumen of seminiferous tubule
Sperm
Spermiogenesis
Spermatids
n
Meiosis II
Secondary spermatocytes
Tight junction
Meiosis I
Primary spermatocyte
2n
Sustentacular cell
Type B spermatogonium
Blood-testis barrier
Type A spermatogonium
Basement membrane of seminiferous tubule

FIGURE 26.7

Spermatogenesis. 2*n* indicates diploid cells and *n* indicates haploid cells. Note that the daughter cells from secondary spermatocytes through spermatids remain connected by slender cytoplasmic processes until spermiogenesis is complete and individual spermatozoa are released.

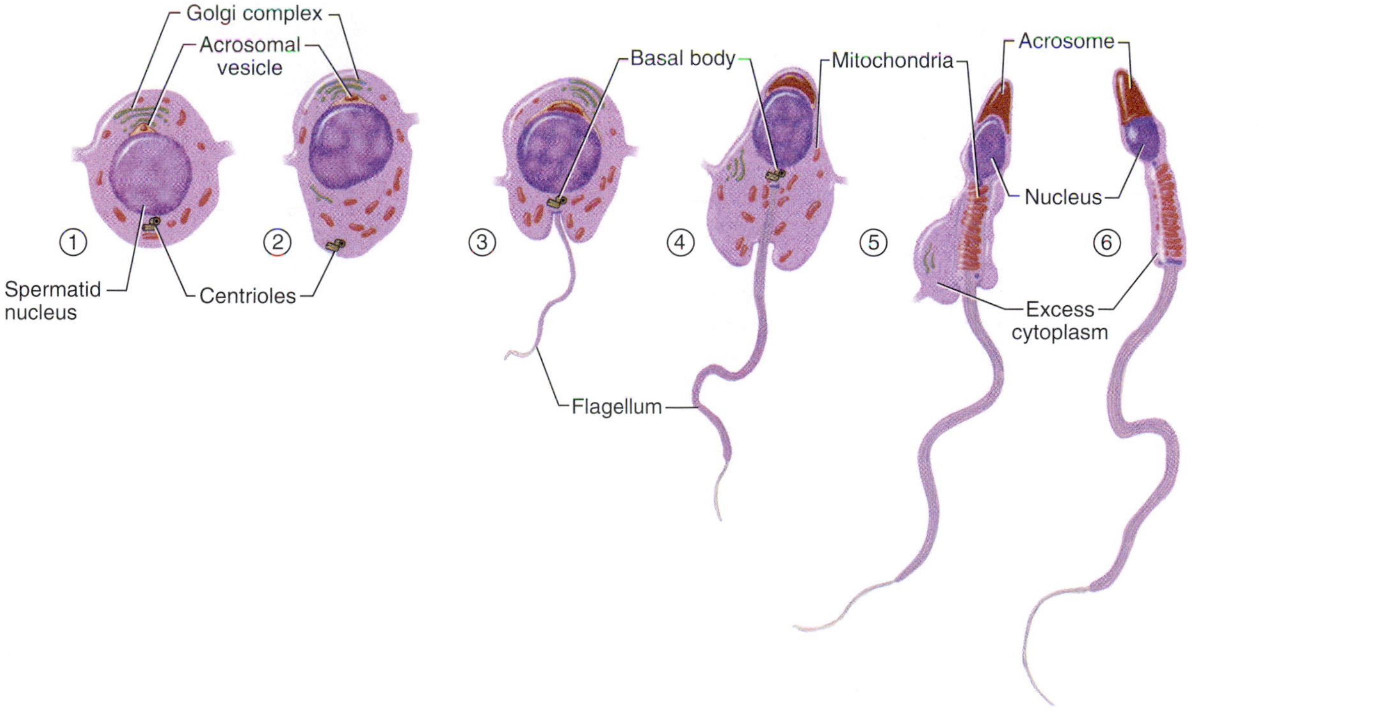

Golgi complex
Acrosomal vesicle
Basal body
Mitochondria
Acrosome
Nucleus
1
2
3
4
5
6
Spermatid nucleus
Centrioles
Excess cytoplasm
Flagellum

FIGURE 26.8

Spermiogenesis. In this process, the spermatids discard excess cytoplasm, grow tails, and become spermatozoa.

THE SPERMATOZOON

The spermatozoon has two parts: a pear-shaped head and a long tail (fig. 26.9). The **head,** about 4 to 5 μm long and 3 μm wide at its broadest part, contains three structures: a nucleus, acrosome, and flagellar basal body. The most important of these is the nucleus, which fills most of the head and contains a haploid set of condensed, genetically inactive chromosomes. The **acrosome**[18] is a lysosome in the form of a thin cap covering the apical half of the nucleus. It contains enzymes that are later used to penetrate the egg if the sperm is successful. The basal body of the tail flagellum is nestled in an indentation at the posterior end of the nucleus.

The **tail** is divided into three regions called the midpiece, principal piece, and endpiece. The **midpiece,** a cylinder about 5 to 9 μm long and half as wide as the head, is the thickest part. It contains numerous large mitochondria that spiral tightly around the axoneme of the flagellum. They produce the ATP needed for the beating of the tail when the sperm migrates up the female reproductive tract. The **principal piece,** 40 to 45 μm long, constitutes most of the tail and consists of the axoneme surrounded by a sheath of fibers. The **endpiece,** 4 to 5 μm long, consists of the axoneme only and is the narrowest part of the sperm.

The Spermatic Ducts

After leaving the testis, the sperm travel through a series of *spermatic ducts* to reach the urethra. These include the following (see fig. 26.2):

- **Efferent ductules.** About 12 small efferent ductules arise from the dorsal side of each testis and carry sperm to the epididymis. They have clusters of ciliated cells that help drive the sperm along.
- **Duct of the epididymis.** The **epididymis**[19] (EP-ih-DID-ih-miss; plural, *epididymides*) is a site of sperm maturation and storage. It adheres to the posterior side of the testis, measures about 7.5 cm long, and consists of a clublike *head* at the superior end, a long middle *body,* and a slender *tail* at its inferior end. It contains a single coiled duct, about 6 m (18 ft) long, embedded in connective tissue. This duct reabsorbs about 90% of the fluid secreted by the testis. Sperm are physiologically immature (incapable of fertilizing an egg) when they leave the testis, but mature as they travel through the head and body of the epididymis. In 20 days or so, they reach the tail. They are stored here and in the adjacent portion of the ductus deferens. Stored sperm remain fertile for 40 to 60 days, but if they become too old without being ejaculated, they disintegrate and the epididymis reabsorbs them.
- **Ductus (vas) deferens.** The duct of the epididymis straightens out at the tail, turns 180°, and becomes the ductus deferens. This is a muscular tube about 45 cm long and 2.5 mm in diameter. It passes upward from the scrotum, travels through the inguinal canal, and enters the pelvic cavity. There, it turns medially and approaches the urinary bladder. After passing

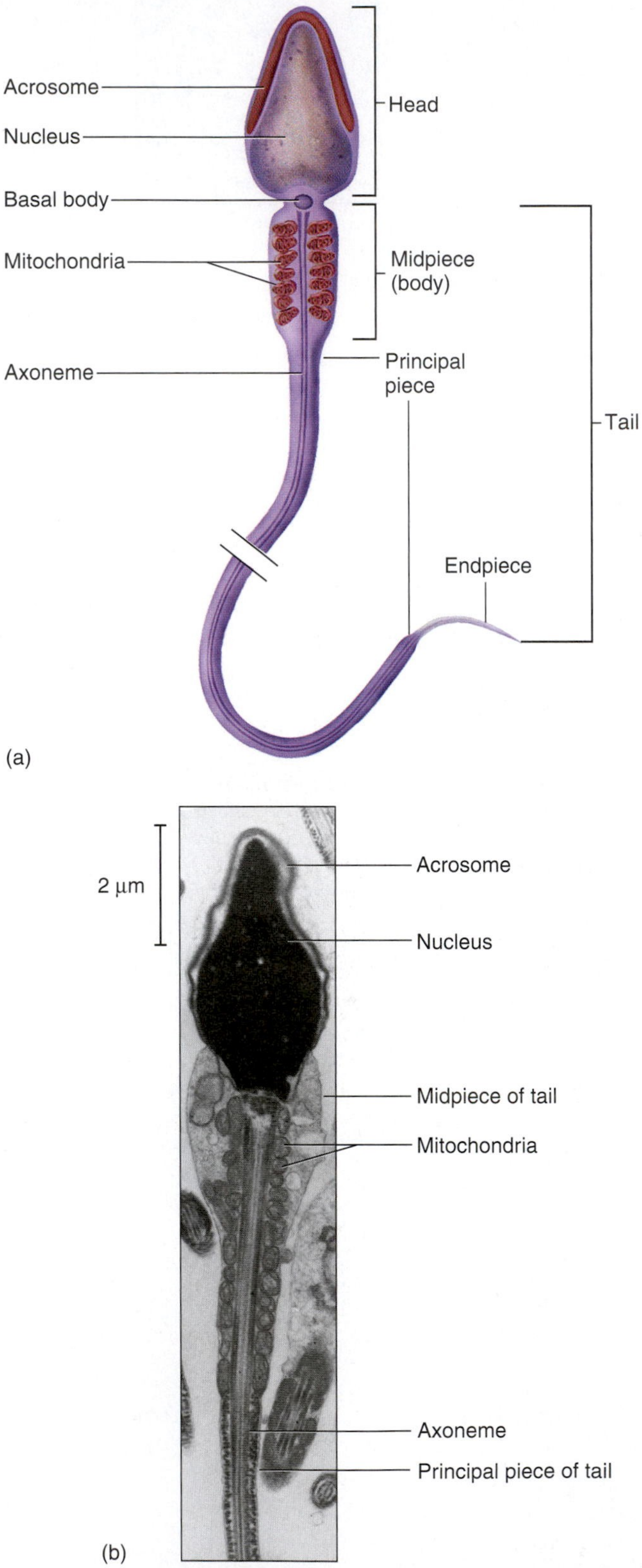

FIGURE 26.9
The Mature Spermatozoon. (*a*) Structure. (*b*) Head and part of the tail of a spermatozoon (TEM).

[18] *acro* = tip, peak + *some* = body
[19] *epi* = upon + *didym* = twins, testes

between the bladder and ureter, the duct turns downward behind the bladder and widens into a terminal **ampulla.** The ductus deferens ends by uniting with the duct of the seminal vesicle, a gland considered later. The duct has a very narrow lumen and a thick wall of smooth muscle well innervated by sympathetic nerve fibers.

- **Ejaculatory duct.** Where the ductus deferens and duct of the seminal vesicle meet, they form a short (2 cm) ejaculatory duct, which passes through the prostate gland and empties into the urethra. The ejaculatory duct is the last of the spermatic ducts.

The male urethra is shared by the reproductive and urinary systems. It is about 20 cm long and consists of three regions: the *prostatic, membranous,* and *penile urethra,* described in chapter 25 (see fig. 25.10*b*). Although it serves both urinary and reproductive roles, it cannot pass urine and semen simultaneously. During ejaculation, the internal urethral sphincter contracts to prevent the voiding of urine and exclude semen from the urinary bladder.

The Accessory Glands

There are three sets of *accessory glands* in the male reproductive system (see fig. 26.2*a*):

- **Seminal vesicles.** These are a pair of glands dorsal to the urinary bladder; one is associated with each ductus deferens. A seminal vesicle is about 5 cm long, or approximately the dimensions of one's little finger. It has a connective tissue capsule and underlying layer of smooth muscle. The secretory portion is a very convoluted duct with numerous branches that form a complex labyrinth. The duct empties into the ejaculatory duct. The yellowish secretion of the seminal vesicles constitutes about 60% of the semen.
- **Prostate**[20] (PROSS-tate) **gland.** This is a median structure that surrounds the urethra and ejaculatory duct immediately inferior to the urinary bladder. It measures about 3 cm in diameter. It is actually a composite of 30 to 50 compound tubuloacinar glands enclosed in a single fibrous capsule. These glands empty into the urethra through about 20 pores in the urethral wall. The stroma of the prostate consists of connective tissue and smooth muscle, like that of the seminal vesicles. The thin, milky secretion of the prostate contributes about 30% of the semen.
- **Bulbourethral (Cowper**[21]**) glands.** These are named for their position near a dilated bulb at the inner end of the penis and their association with the penile urethra. They are brownish, spherical glands about 1 cm in diameter. During sexual arousal, they produce a clear slippery fluid that lubricates the head of the penis in preparation for intercourse. Perhaps more importantly, though, the fluid protects the sperm by neutralizing the acidity of residual urine in the urethra.

[20]*pro* = before + *stat* = to stand; commonly misspelled and mispronounced "prostrate"
[21]William Cowper (1666–1709), British anatomist

INSIGHT 26.1 CLINICAL APPLICATION

Prostate Diseases

The prostate gland weighs about 20 g by age 20, remains at that weight until age 45 or so, and then begins to grow slowly again. By age 70, over 90% of men show some degree of *benign prostatic hyperplasia*—noncancerous enlargement of the gland. The major complication of this is that it compresses the urethra, slows the flow of urine, and sometimes promotes bladder and kidney infections.

Prostate cancer is the second most common cancer in men (after lung cancer); it affects about 9% of men over the age of 50. Prostate tumors tend to form near the periphery of the gland, where they do not obstruct urine flow; therefore, they often go unnoticed until they cause pain. Prostate cancer often metastasizes to nearby lymph nodes and then to the lungs and other organs. It is more common among American blacks than whites and is rare among the Japanese.

The position of the prostate immediately anterior to the rectum allows it to be palpated through the rectal wall to check for tumors. This procedure is called *digital rectal examination (DRE).* Prostate cancer can also be diagnosed from elevated levels of *prostate specific antigen (PSA)* and *acid phosphatase* (a prostatic enzyme) in the blood. Up to 80% of men with prostate cancer survive when it is detected and treated early, but only 10% to 50% survive if it spreads beyond the prostatic capsule.

SEMEN

Semen[22] **(seminal fluid)** is a complex mixture of sperm and glandular secretions (table 26.2). A typical ejaculation discharges 2 to 5 mL of semen. About 10% of it consists of sperm and fluids from the spermatic ducts, 30% is from the prostate, and 60% is from the seminal vesicles. The bulbourethral glands contribute a trace of fluid. The **sperm count** is normally 50 to 120 million sperm/mL. A count lower than 20 to 25 million/mL is usually associated with *infertility (sterility),* the inability to fertilize an egg (table 26.3).

In ejaculation, peristalsis of the ductus deferens drives sperm from the epididymis to the ampulla. Contraction of the ampulla then discharges the sperm into the prostatic urethra, where they mix with secretions of the accessory glands. Semen in the urethra triggers a reflex contraction of the *bulbospongiosus muscle,* which ensheaths the root of the penis. Rhythmic compression of the urethra by this muscle expels the semen from the penis. Most of the sperm are contained in the first 1 mL of ejaculate. Prostatic fluid follows, and seminal vesicle fluid comes last, flushing remaining sperm from the urethra.

The Penis

The **penis**[23] serves to deposit semen in the vagina. Half of it is an internal **root** and half is the externally visible **shaft** and **glans**[24] (see figs. 26.2 and 26.10). The external portion is about 8 to 10 cm (3–4 in.) long and 3 cm in diameter when flaccid (nonerect); the typical

[22]*semen* = seed
[23]*penis* = tail
[24]*glans* = acorn

Table 26.2

The Principal Constituents of Semen

From the Testes and Epididymis	
Sperm	Digest a path through the cervical mucus and through cells covering the ovum; fertilize the ovum
From the Prostate Gland	
Spermine	A base that neutralizes vaginal acidity to protect the sperm; gives the semen a pH of 7.35 to 7.50
Clotting Enzymes	Act on fibrinogen from the seminal vesicles, causing newly ejaculated semen to temporarily coagulate and become sticky
Fibrinolysin	Liquefies the coagulated semen 15 to 30 min after ejaculation, after semen has had time to be taken into the uterus; liberates the sperm from the coagulum
From the Seminal Vesicles	
Fibrinogen	Converted to fibrin by the prostatic clotting enzymes, causing coagulation of the semen so it adheres to the inner vagina and cervix until taken into the uterus
Fructose	Sugar that provides energy for sperm motility
Prostaglandins	Function obscure; possibly reduce viscosity of cervical mucus and promotes motility of the uterus and uterine tubes, thus promoting the spread of sperm in the female reproductive tract

dimensions of an erect penis are 13 to 18 cm (5–7 in.) long and 4 cm in diameter. The glans is the expanded head at the distal end of the penis with the external urinary meatus at its tip.

The skin is loosely attached to the shaft, allowing for expansion during erection. It continues over the glans as the **prepuce,** or foreskin, which is often removed by circumcision. A ventral fold of tissue called the *frenulum* attaches the skin to the proximal margin of the glans. The skin of the glans itself is thinner and firmly attached to the underlying erectile tissue. The glans and facing surface of the prepuce have sebaceous glands that produce a waxy secretion called **smegma.**[25]

The penis consists mainly of three cylindrical bodies called **erectile tissues,** which fill with blood during sexual arousal and account for its enlargement and erection. A single erectile body, the **corpus spongiosum,** passes along the ventral side of the penis and encloses the penile urethra. It expands at the distal end to fill the entire glans. The dorsal side of the penis, proximal to the glans, has a **corpus cavernosum** (plural, *corpora cavernosa*) on each side. Each corpus cavernosum is ensheathed in a tight fibrous sleeve called the **tunica albuginea,** and they are separated from each other by a **median septum.** (Note that the testes also have a tunica albuginea and the scrotum also has a median septum.)

All three cylinders of erectile tissue are spongy in appearance and contain numerous tiny blood sinuses called **lacunae.** The partitions between lacunae, called **trabeculae,** are composed of connective tissue and smooth **trabecular muscle.** In the flaccid penis, trabecular muscle tone collapses the lacunae, which appear as slits in the tissue.

At the body surface, the penis turns 90° dorsally and continues inward as the root. The corpus spongiosum terminates at the internal end of the root as a dilated **bulb,** which is ensheathed in the bulbospongiosus muscle and attached to the lower surface of the perineal membrane within the urogenital triangle (see p. 307). The corpora cavernosa diverge like the arms of a Y. Each arm, called a **crus** (cruss; plural, *crura*), attaches the penis to the pubic arch of the pelvis and to the perineal membrane on its respective side. Each crus is enveloped by an ischiocavernosus muscle.

The penis receives blood from a pair of **internal pudendal (penile) arteries,** which branch from the internal iliac arteries. As each artery enters the root of the penis, it divides in two. One branch, the **dorsal artery,** travels dorsally along the penis not far beneath the skin, supplying blood to the skin, fascia, and corpus spongiosum. The other branch, the **deep artery,** travels through the core of the corpus cavernosum and gives off smaller **helicine**[26] **arteries,** which penetrate the trabeculae and empty into the lacunae. There are numerous anastomoses between the dorsal and deep arteries, so neither of them is the exclusive source of blood to any one erectile tissue.

When the deep arteries dilate, the lacunae fill with blood and the erectile tissues swell. The tunica albuginea around the corpora cavernosa cannot expand, so pressure builds especially in these two erectile tissues and the penis becomes elongated and erect. The corpus spongiosum becomes less engorged. When the penis is flaccid, most of its blood supply comes from the dorsal arteries. A single **deep dorsal vein** drains blood from the penis. It runs between the two dorsal arteries beneath the deep fascia and empties into a plexus of prostatic veins.

The penis is richly innervated by sensory and motor nerve fibers. The glans has an abundance of tactile, pressure, and temperature receptors, especially on its proximal margin and frenulum.

[25]*smegma* = unguent, ointment, soap

[26]*helic* = coil, helix

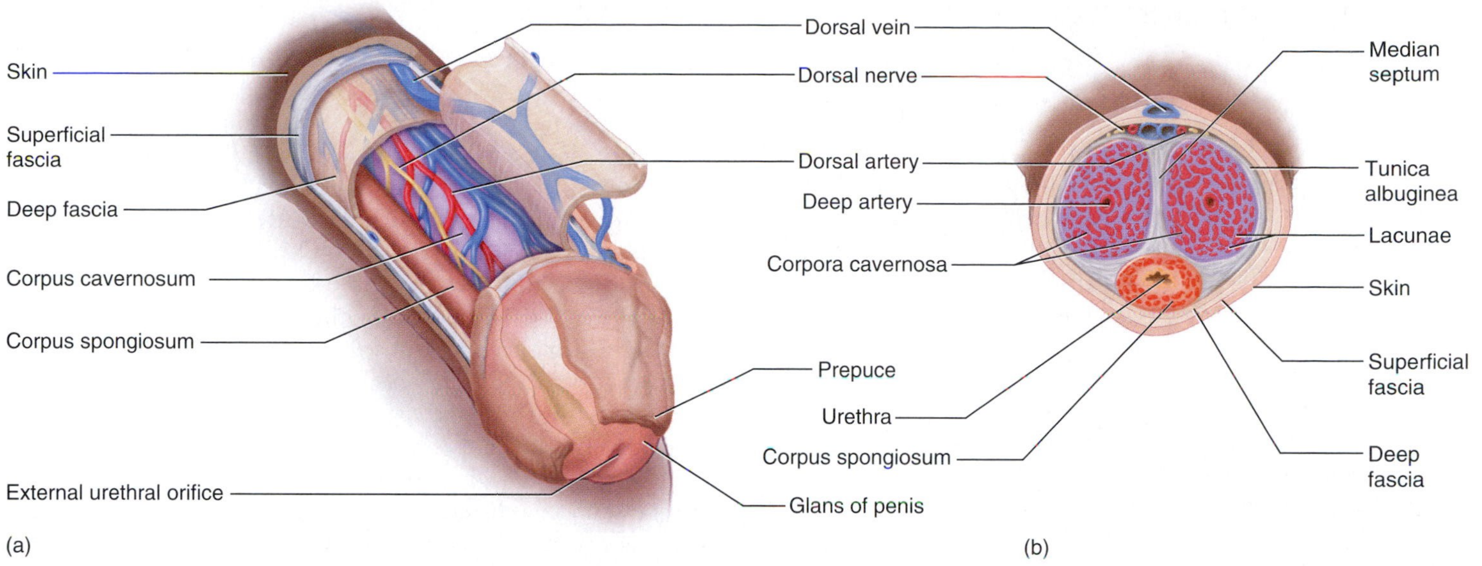

FIGURE 26.10
Anatomy of the Penis. (*a*) Dissection in lateral view. (*b*) Cross section.

They lead by way of a prominent **dorsal nerve** of the penis to the **pudendal nerve,** then via the sacral plexus to spinal cord segments S2 to S4. Sensory fibers of the shaft, scrotum, perineum, and elsewhere are also highly important to erotic stimulation.

INSIGHT 26.2 CLINICAL APPLICATION

REPRODUCTIVE EFFECTS OF POLLUTION

In recent decades, wildlife biologists have noticed increasing numbers of male birds, fish, and alligators with a variety of abnormalities in reproductive development. These deformities have been attributed to chemical pollutants called *endocrine disruptors* or *estrogen mimics.* Evidence is mounting that humans, too, are showing declining fertility and increasing anatomical abnormalities due to pollutants in water, meat, vegetables, and even breast milk and the uterine environment.

Over the last 50 years, there has been an alarming increase in the incidence of *cryptorchidism* (undescended testes) and *hypospadias* (a condition in which the urethra opens on the ventral side of the penis instead of at the tip). The rate of testicular cancer has more than tripled in that time. Data on 15,000 men from several countries also show a sharp drop in average sperm count—from 113 million/mL in 1940 to only 66 million/mL in 1990. Total sperm production decreased even more, because the average volume of semen per ejaculate dropped 19% over this period.

The pollutants implicated in this trend include a wide array of common herbicides, insecticides, industrial chemicals, and breakdown products of materials ranging from plastics to dishwashing detergents. Some authorities think these chemicals act by mimicking estrogens or blocking testosterone receptors. Other scientists, however, question the data and feel the issue may be overstated. While the debate continues, the U.S. Environmental Protection Agency is screening thousands of industrial chemicals for endocrine effects.

Both autonomic and somatic motor fibers carry signals from integrating centers in the spinal cord to the penis and other pelvic organs. Sympathetic nerve fibers arise from levels T12 to L2, pass through the hypogastric and pelvic nerve plexuses, and innervate the penile arteries, trabecular muscle, spermatic ducts, and accessory glands. They dilate the penile arteries and can induce erection even when the sacral region of the spinal cord is damaged. They also initiate erection in response to input from the special senses and to sexual thoughts.

Parasympathetic fibers extend from segments S2 to S4 of the spinal cord through the pudendal nerves to the arteries of the penis. They are involved in an autonomic reflex arc that causes erection in response to direct stimulation of the penis and other perineal organs.

Before You Go On

Answer the following questions to test your understanding of the preceding section:

3. Name the stages of spermatogenesis from spermatogonium to spermatozoon.
4. Name two types of cells in the testis other than the germ cells, and describe their locations and functions.
5. Describe the three major parts of a spermatozoon and state what organelles or cytoskeletal components are contained in each.
6. Name all the ducts that the sperm follow, in order, from the time they form in the testis to the time of ejaculation.
7. Describe the locations and functions of the seminal vesicles, prostate, and bulbourethral glands.
8. Name the erectile tissues of the penis, and describe their locations relative to each other.

Female Reproductive Anatomy

Objectives

When you have completed this section, you should be able to

- describe the structure of the ovary;
- describe the stages of oogenesis and how these relate to changes in histology of the ovarian follicles;
- trace the female reproductive tract and describe the gross anatomy and histology of each organ;
- describe changes in the uterine lining through the menstrual cycle;
- identify the ligaments that support the female reproductive organs;
- describe the blood supply to the female reproductive tract;
- identify the external genitalia of the female; and
- describe the structure of the nonlactating breast and lactating mammary gland.

Figure 26.11 shows the female reproductive tract. The principal reproductive organs of the female pelvic cavity are the ovaries, uterine tubes, uterus, and vagina, which will be described in that order.

The Ovaries

The female gonads are the **ovaries,**[27] which produce egg cells (ova) and sex hormones. The ovary is an almond-shaped organ nestled in a depression of the dorsal pelvic wall called the *ovarian fossa.* The ovary measures about 3 cm long, 1.5 cm wide, and 1 cm thick. Its capsule, like that of the testis, is called the **tunica albuginea.** The interior of the ovary is indistinctly divided into a central **medulla** and an outer **cortex** (fig. 26.12). The medulla is a zone of fibrous connective tissue occupied by the principal arteries and veins of the ovary. The cortex is the site of the **ovarian follicles,** each of which consists of one developing ovum surrounded by numerous small follicular cells. The ovary does not have a system of tubules like the testis; eggs are released one at a time by the bursting of the follicles *(ovulation).*

The ovaries and other internal genitalia are held in place by several connective tissue ligaments (fig. 26.13). The medial pole of the ovary is attached to the uterus by the *ovarian ligament* and its lateral pole is attached to the pelvic wall by the *suspensory ligament.*

[27]*ov* = egg + *ary* = place for

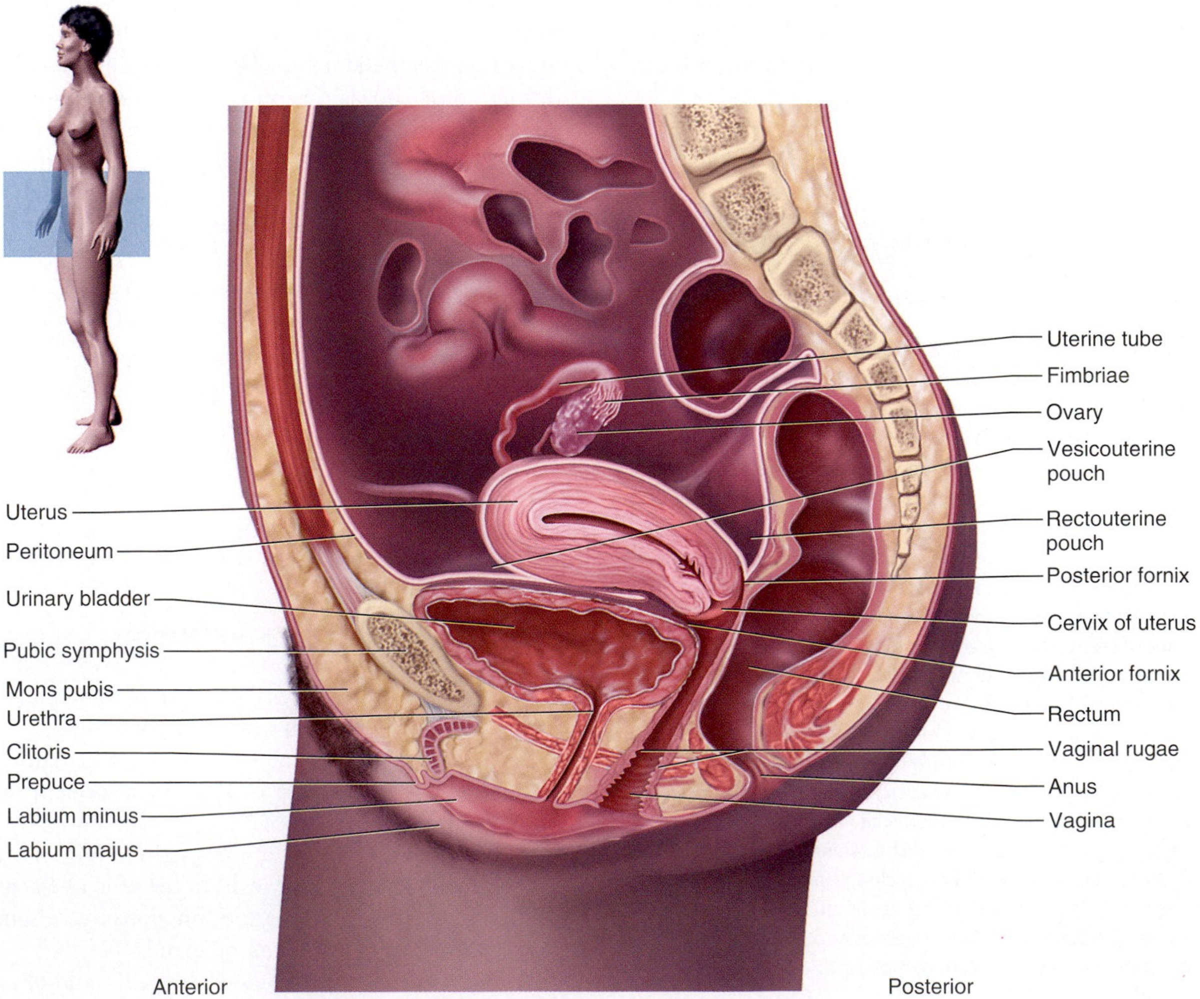
Uterine tube
Fimbriae
Ovary
Vesicouterine pouch
Rectouterine pouch
Posterior fornix
Cervix of uterus
Anterior fornix
Rectum
Vaginal rugae
Anus
Vagina
Uterus
Peritoneum
Urinary bladder
Pubic symphysis
Mons pubis
Urethra
Clitoris
Prepuce
Labium minus
Labium majus
Anterior
Posterior

Figure 26.11
The Female Reproductive System.

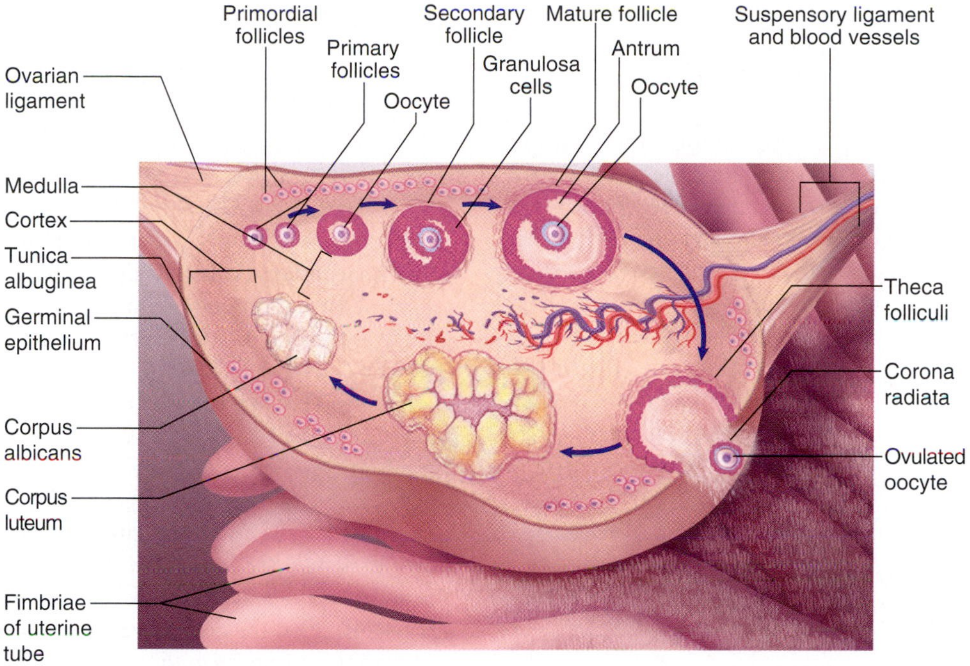

FIGURE 26.12

Structure of the Ovary and the Developmental Sequence of the Ovarian Follicles.

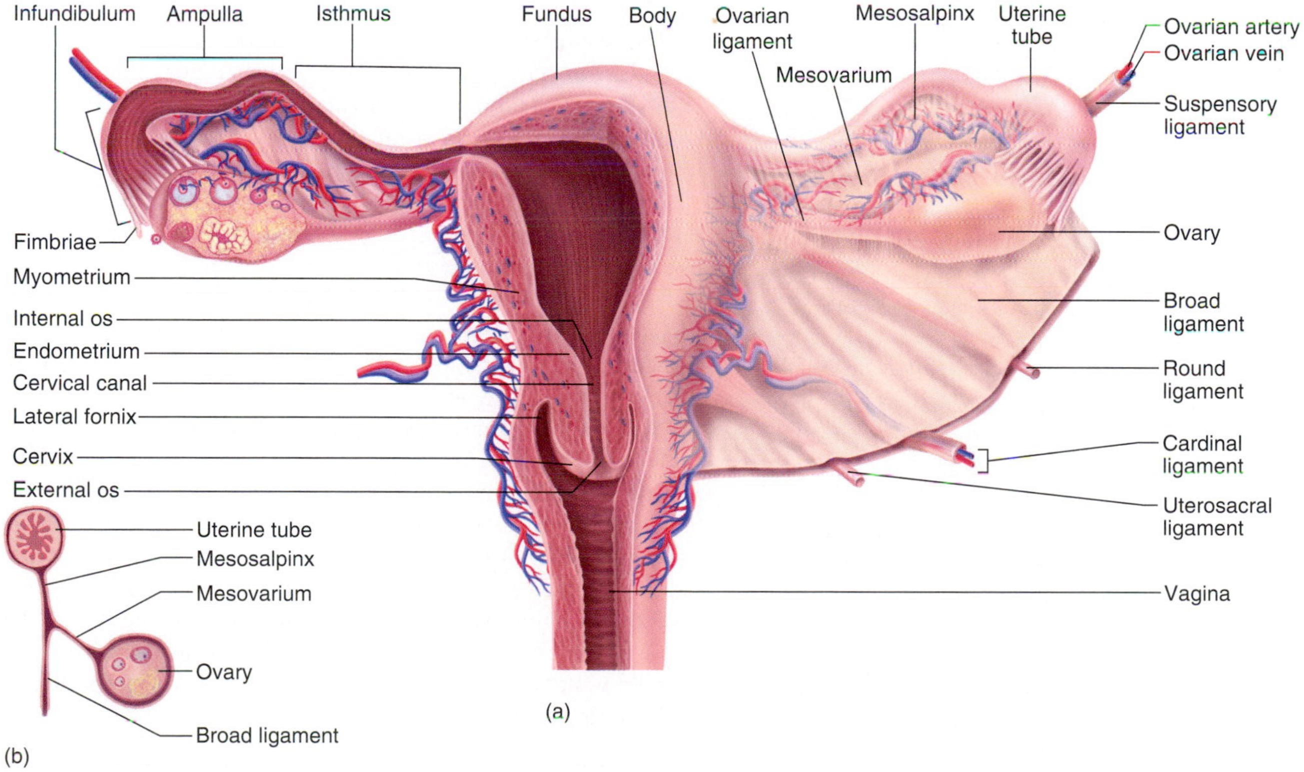

FIGURE 26.13

The Female Reproductive Tract and Supportive Ligaments. (*a*) Drawing of the reproductive tract, dorsal view. (*b*) Relationship of the ligaments to the uterine tube and ovary.

The anterior margin of the ovary is anchored by a peritoneal fold called the *mesovarium.*[28] This ligament extends to a sheet of peritoneum called the *broad ligament,* which flanks the uterus and encloses the uterine tube in its superior margin.

The ovary receives blood from two arteries: the **ovarian branch of the uterine artery,** which passes through the mesovarium and approaches the medial pole of the ovary, and the **ovarian artery,** which passes through the suspensory ligament and approaches the lateral pole. The ovarian artery is the female equivalent of the testicular artery described earlier, arising high on the aorta and traveling down to the gonad along the dorsal body wall. The ovarian and uterine arteries anastomose along the margin of the ovary and give off multiple small arteries that enter the ovary on that side. Ovarian veins and nerves also travel through the suspensory ligament.

Oogenesis and Ova

The production of female gametes is called **oogenesis**[29] (OH-oh-JEN-eh-sis) (fig. 26.14). Like spermatogenesis, it employs meiosis and produces haploid gametes. It differs from spermatogenesis in other respects, however: It is not a continual process, but occurs in a rhythm called the **ovarian cycle,** and for each original germ cell (oogonium), it produces only one functional gamete. The other daughter cells are tiny **polar bodies** that soon die.

The female primordial germ cells arise, like those of the male, from the yolk sac of the embryo. They colonize the gonadal ridges in the first 5 to 6 weeks and then differentiate into **oogonia** (OH-oh-GO-nee-uh). Oogonia multiply until the fifth month of gestation, reach 6 to 7 million in number, and then go into a state of arrested development until shortly before birth. At that time, some of them transform into **primary oocytes** and go as far as early meiosis I. Any stage from the primary oocyte to the time of fertilization can be called an egg, or **ovum.**

Most primary oocytes degenerate even before a girl is born; only 2 million remain at birth. Furthermore, most of these degenerate during childhood, and by the onset of puberty, only about 400,000 remain. This is the female's lifetime supply of germ cells. The degeneration of oocytes and follicles without maturation is called *atresia* (ah-TREE-zhee-uh).

During a woman's reproductive years, about 20 to 25 oocytes and follicles begin to develop each month. Normally just one of these reaches maturity and ovulates, and the rest degenerate. The stages of oogenesis are accompanied by pronounced changes in the follicle (fig. 26.15).

The primary oocyte is initially enclosed in a **primordial follicle,** composed of a single layer of squamous follicular cells applied tightly to the oocyte. About 3 days before the menstrual period begins, pituitary secretion of follicle-stimulating hormone (FSH) stimulates several primordial follicles to develop into **primary follicles.** The follicular cells thicken into a cuboidal epithelium, multiply, and become stratified. They are now called **granulosa cells.** The ovarian stroma adjacent to the follicle condenses into a fibrous capsule called the **theca**[30] **folliculi** (THEE-ca fol-IC-you-lye). The theca and granulosa cells collaborate to synthesize **estrogens.** Among other effects, estrogens stimulate regrowth of the uterine lining (endometrium) after menstruation.

Most primary follicles degenerate with no further development. In a few, however, the granulosa cells secrete pools of estrogen-rich **follicular fluid.** These pools grow and merge until they form a single fluid-filled cavity, the **antrum.** The follicle is now called a **secondary (antral) follicle.** This is the state of development when menstruation ends around day 5. (Day 1 of the cycle is regarded as the first day of menstruation.)

By day 10 or so, all but one of the developing follicles usually degenerate. That one enlarges to as much as 2.5 cm in diameter and bulges like a balloon from the surface of the ovary. This **mature (graafian**[31]**) follicle** (fig. 26.15*b*) is the one destined to ovulate. The oocyte in this follicle is held against the follicular wall by a mound of granulosa cells called the **cumulus oophorus**[32] (CUE-mew-lus oh-OFF-or-us). A clear space called the **zona pellucida**[33] separates the granulosa cells from the oocyte. The innermost layer of cumulus cells is called the **corona radiata.**[34] Microvilli from the corona cells and the oocyte span the zona pellucida.

The primary oocyte, suspended in prophase I of meiosis, now completes its division. It divides into a large **secondary oocyte** and a small *first polar body* (see fig. 26.14). Meiosis I reduces the chromosome number by half, so the secondary oocyte is haploid. It retains as much of the cytoplasm as possible, so that if it is fertilized, it can divide repeatedly and produce numerous daughter cells. Splitting each oocyte into four equal but small parts would run counter to this purpose. The first polar body is simply a way of discarding the other haploid set of chromosomes; it soon dies. The secondary oocyte begins meiosis II and then goes into developmental arrest again until after ovulation. If this egg is fertilized, it completes meiosis II and produces a *second polar body.* If not fertilized, it dies and never finishes meiosis.

Ovulation, the release of an oocyte, typically occurs on day 14, the midpoint of the average cycle. As ovulation approaches, the oocyte and cumulus oophorus become detached from the follicular wall and drift in the antrum. Ovulation takes only 2 or 3 minutes. A nipplelike **stigma** appears on the ovarian surface over the follicle. Follicular fluid seeps from the stigma for 1 or 2 minutes, and then the follicle ruptures. The remaining follicular fluid oozes out, carrying the oocyte and cumulus cells (fig. 26.16).

When the oocyte is expelled, the follicle collapses and bleeds into the antrum. As the clotted blood is slowly absorbed, granulosa and theca interna cells multiply and fill the antrum, and a dense bed of blood capillaries grows amid them. The ovulated follicle has now become a structure called the **corpus luteum,**[35] named for a yellow lipid that accumulates in the theca interna cells (see fig. 26.12). These cells are now called **lutein cells.** The corpus luteum secretes a large amount of progesterone, which stimulates the uterus to prepare for possible pregnancy.

---

[28] *mes* = middle + *ovari* = ovary
[29] *oo* = egg + *genesis* = production

[30] *theca* = box, case
[31] Reijnier de Graaf (1641–73), Dutch physiologist and histologist
[32] *cumulus* = little mound + *oo* = egg + *phor* = to carry
[33] *zona* = zone + *pellucid* = clear, transparent
[34] *corona* = crown + *radiata* = radiating
[35] *corpus* = body + *lute* = yellow

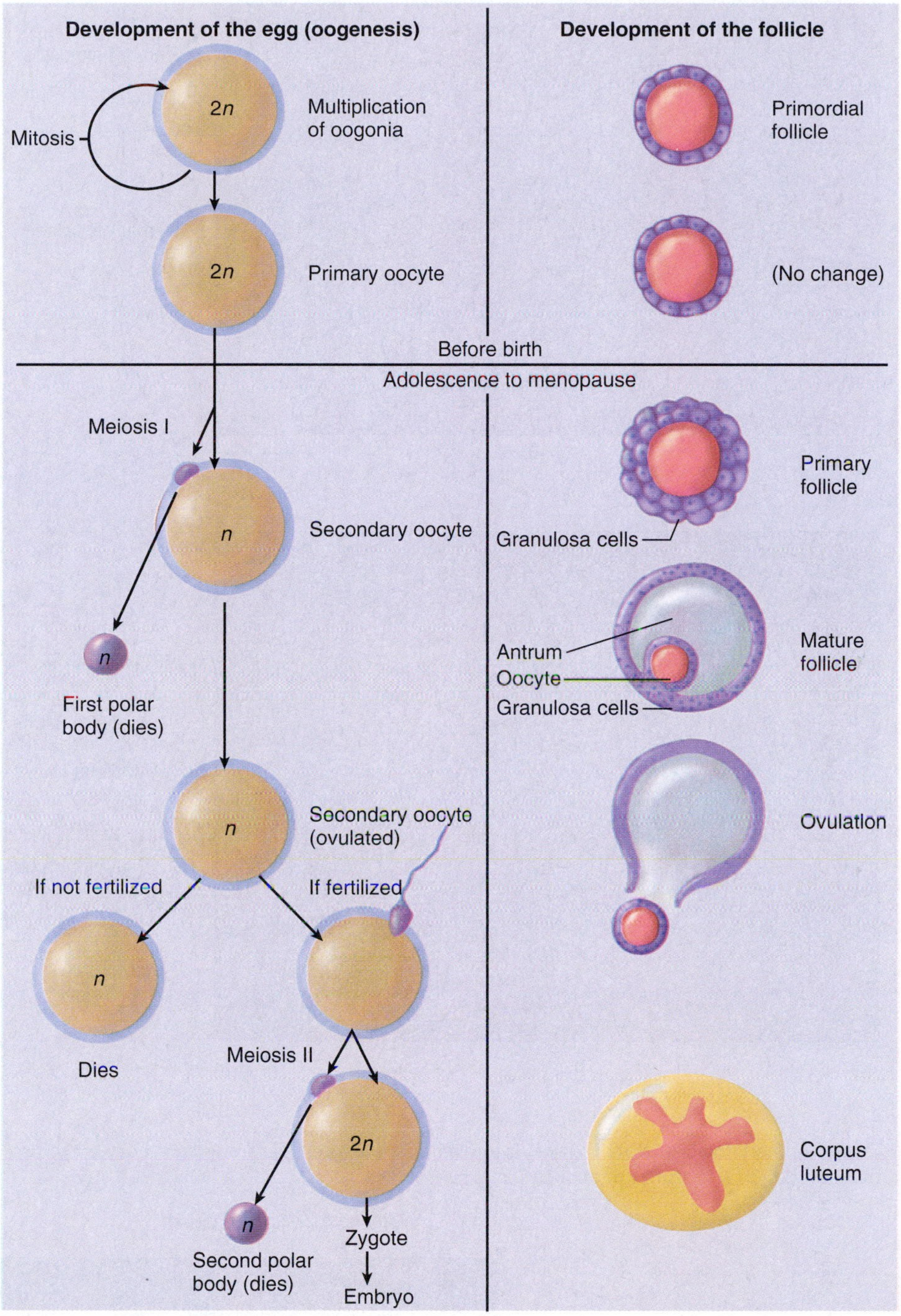

FIGURE 26.14
Oogenesis *(left)* and Corresponding Development of the Follicle *(right)*.

If pregnancy does not occur, the corpus luteum atrophies from days 24 to 26—a process called *involution.* As it does, the level of circulating progesterone declines, and this brings about menstruation. By day 26 or so, involution is complete and the corpus luteum becomes an inactive scar, the **corpus albicans.**[36] If pregnancy occurs, however, the corpus luteum remains active for about 3 months. Its progesterone is necessary to sustain the early pregnancy. Eventually the placenta takes over the role of progesterone secretion, among other functions, and the corpus luteum is no longer needed.

These events in the ovarian cycle are correlated with changes in uterine histology, which we will examine shortly.

The Uterine Tubes

An ovulated oocyte is received into the **uterine tube,** also called the **oviduct** or **fallopian**[37] **tube** (fig. 26.17). The tube is a ciliated canal about 10 cm long leading from the ovary to the uterus. At the

[36] *corpus* = body + *alb* = white

[37] Gabriele Fallopio (1528–62), Italian anatomist and physician

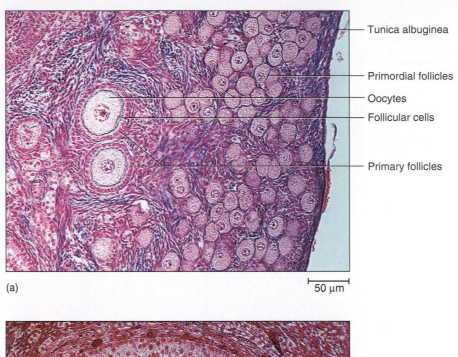

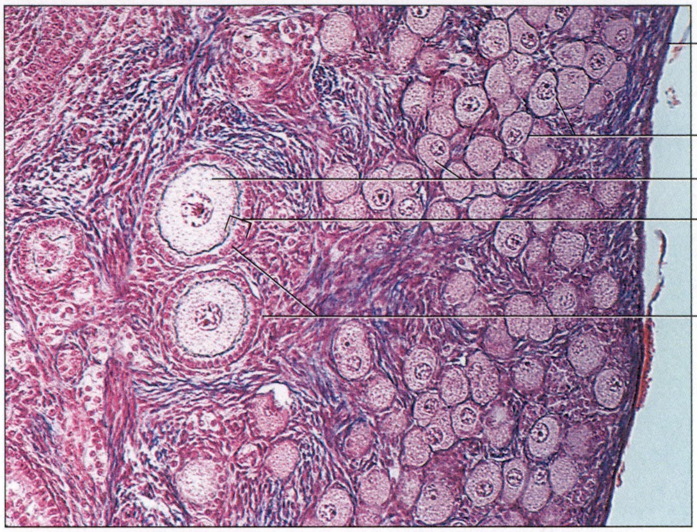
Tunica albuginea
Primordial follicles
Oocytes
Follicular cells
Primary follicles
(a)
50 µm

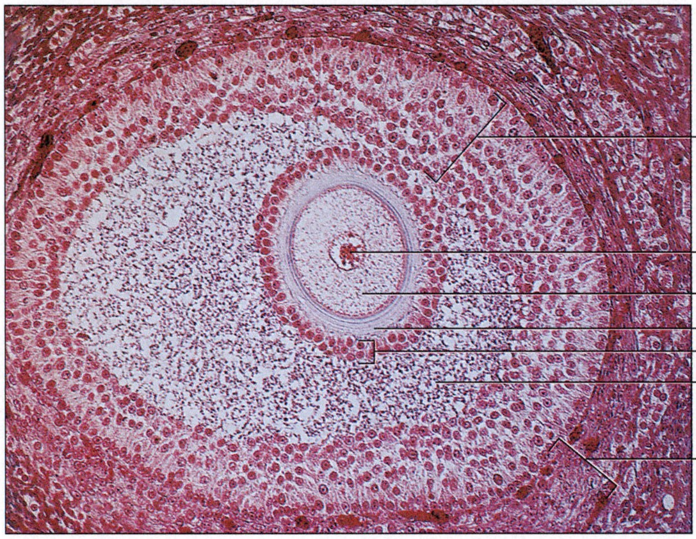
Granulosa cells
Egg nucleus
Egg
Zona pellucida
Cumulus oophorus
Antrum
Theca folliculi
(b)
100 µm

FIGURE 26.15

Ovarian Follicles. (*a*) Primordial and primary follicles. Note the very thin layer of squamous cells around the oocyte in a primordial follicle, and the single layer of cuboidal cells in a primary follicle. (*b*) A mature (graafian) follicle.

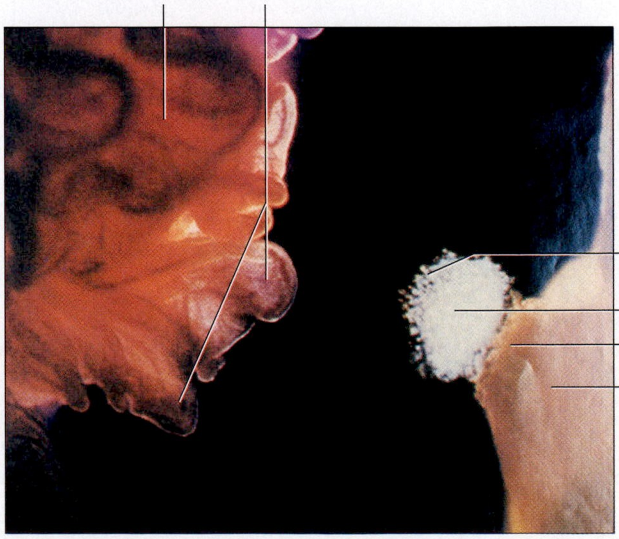
Uterine tube
Fimbriae
Corona radiata
Oocyte
Stigma
Ovary

FIGURE 26.16

Ovulation of a Graafian Follicle, Viewed by Endoscopy.

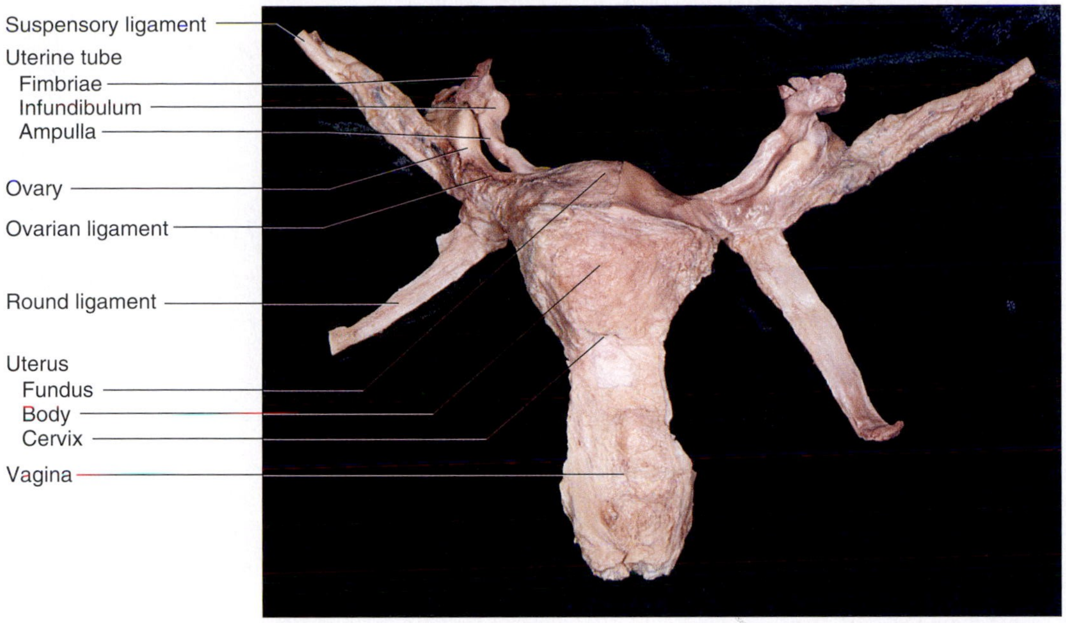
Suspensory ligament
Uterine tube
Fimbriae
Infundibulum
Ampulla
Ovary
Ovarian ligament
Round ligament
Uterus
Fundus
Body
Cervix
Vagina

FIGURE 26.17
Dissection of the Female Reproductive Tract. The ovaries and uterine tubes are turned upward for visibility.

distal (ovarian) end, it flares into a trumpet-shaped **infundibulum**[38] with feathery projections called **fimbriae**[39] (FIM-bree-ee); the middle and longest part of the tube is the **ampulla;** and the segment near the uterus is a narrower **isthmus.** The uterine tube is enclosed in the **mesosalpinx**[40] (MEZ-oh-SAL-pinks), which is the superior margin of the broad ligament.

The wall of the uterine tube is well endowed with smooth muscle. Its mucosa is extremely convoluted. It has an epithelium of ciliated cells and a smaller number of glandular **peg cells** (fig. 26.18). The cilia beat toward the uterus and, with the help of muscular contractions of the tube, convey the egg in that direction.

The Uterus

The **uterus**[41] is a thick muscular chamber that opens into the roof of the vagina and usually tilts forward over the urinary bladder (see figs. 26.11, 26.13, and 26.17). Its function is to harbor the fetus, provide it with a source of nutrition (the placenta, composed partially of uterine tissue), and expel the fetus at the end of gestation (pregnancy). It is somewhat pear-shaped, with a broad superior curvature called the **fundus,** a midportion called the **body (corpus),** and a narrow inferior end called the **cervix.** The uterus measures about 7 cm from cervix to fundus, 4 cm wide at its broadest point on the fundus, and 2.5 cm thick, but it is somewhat larger in women who have been pregnant.

FIGURE 26.18
Epithelial Lining of the Uterine Tube. Secretory cells are shown in *red* and *green,* and cilia of the ciliated cells in *yellow* (colorized SEM).

[38]*infundibulum* = funnel
[39]*fimbria* = fringe
[40]*meso* = mesentery + *salpin* = trumpet
[41]*uterus* = womb

The lumen of the uterus is roughly triangular, with its two upper corners opening into the uterine tubes. It communicates with the vagina by way of a narrow passage through the cervix called the **cervical canal.** The superior opening of this canal into the body of the uterus is the *internal os*[42] (oss) and its opening into the vagina is the *external os.* The canal contains **cervical glands** that secrete mucus, thought to prevent the spread of microorganisms from the vagina into the uterus. Near the time of ovulation, the mucus becomes thinner than usual and allows easier passage for sperm.

THE UTERINE WALL

The uterine wall consists of an external serosa called the *perimetrium,* a middle muscular layer called the *myometrium,* and an inner mucosa called the *endometrium.* The **perimetrium** is composed of simple squamous epithelium overlying a thin layer of areolar tissue. The **myometrium,**[43] about 1.25 cm thick in the nonpregnant uterus, constitutes most of the wall. It is composed mainly of bundles of smooth muscle that sweep downward from the fundus and spiral around the body of the uterus. The myometrium is less muscular and more fibrous near the cervix; the cervix itself is almost entirely collagenous. The muscle cells of the myometrium are about 40 μm long immediately after menstruation, twice this long at the middle of the menstrual cycle, and up to 10 times as long in pregnancy. The function of the myometrium is to produce the labor contractions that help to expel the fetus.

The inner lining of the uterus, or mucosa, is called the **endometrium.**[44] It has a simple columnar epithelium, compound tubular glands, and a stroma populated by leukocytes, macrophages, and other cells (fig. 26.19). The superficial half to two-thirds of it, called the **stratum functionalis,** is shed in each menstrual period. The deeper layer, called the **stratum basalis,** stays behind and regenerates a new functionalis in the next cycle. When pregnancy occurs, the endometrium is the site of attachment of the embryo and forms the maternal part of the placenta.

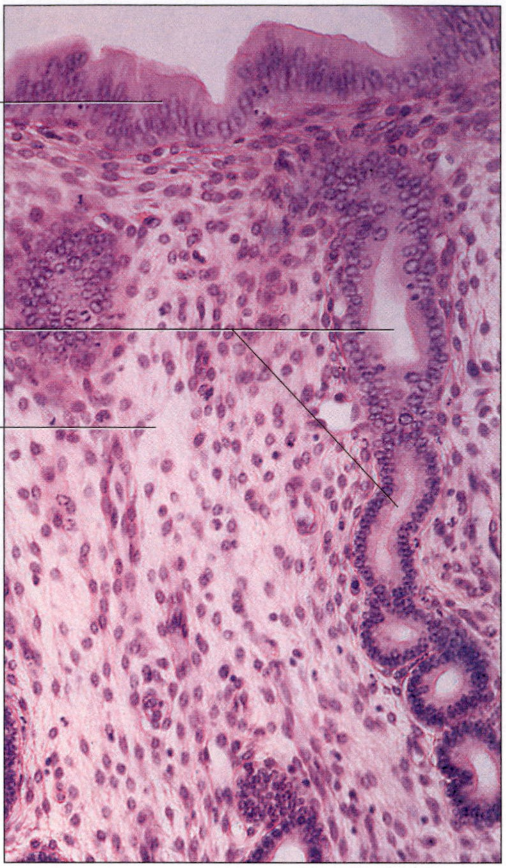
Surface epithelium
Endometrial gland
Lamina propria

FIGURE 26.19
Histology of the Endometrium.

LIGAMENTS

The uterus is supported by the muscular floor of the pelvic outlet and folds of peritoneum that form supportive ligaments around the organ, as they do for the ovary and uterine tube (see fig. 26.13). The **broad ligament** has two parts: the *mesosalpinx* mentioned earlier and the *mesometrium* on each side of the uterus. The cervix and superior part of the vagina are supported by **cardinal (lateral cervical) ligaments** extending to the pelvic wall. A pair of **uterosacral ligaments** attach the dorsal side of the uterus to the sacrum, and a pair of **round ligaments** arise from the ventral surface of the uterus, pass through the inguinal canals, and terminate in the labia majora.

As the peritoneum folds around the various pelvic organs, it creates several dead-end recesses and pouches. Two major ones are the **vesicouterine**[45] **pouch,** which forms the space between the uterus and urinary bladder, and **rectouterine pouch** between the uterus and rectum (see fig. 26.11).

BLOOD SUPPLY

The uterine blood supply is particularly important to the menstrual cycle and pregnancy. A **uterine artery** arises from each internal iliac artery and travels through the broad ligament to the uterus (fig. 26.20). It gives off several branches that penetrate into the myometrium and lead to **arcuate arteries.** Each arcuate artery travels in a circle around the uterus and anastomoses with the arcuate artery on the other side. Along its course, it gives rise to smaller arteries that penetrate the rest of the way through the myometrium, into the endometrium, and produce the **spiral arteries.** The spiral arteries coil between the endometrial glands toward the surface of the mucosa. They rhythmically constrict and dilate, making the mucosa alternately blanch and flush with blood.

THE MENSTRUAL (UTERINE) CYCLE

The **menstrual cycle** consists of a buildup of the endometrium followed by its breakdown and vaginal discharge, in parallel with the

[42]*os* = mouth
[43]*myo* = muscle + *metr* = uterus
[44]*endo* = inside + *metr* = uterus
[45]*vesico* = bladder

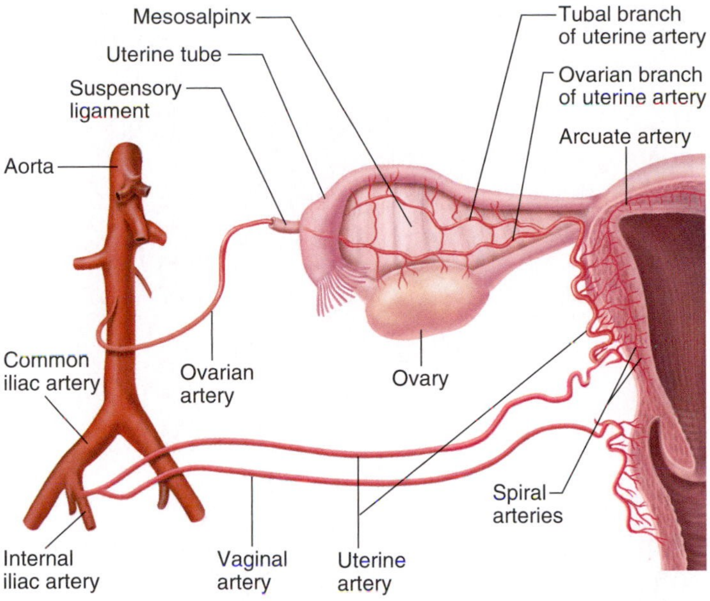

FIGURE 26.20
Blood Supply to the Female Reproductive Tract.

ovarian cycle. The menstrual cycle is divided into a *menstrual phase, proliferative phase, secretory phase,* and *premenstrual phase,* in that order. The menstrual phase averages 5 days long. The reason for menstruation is best understood after becoming acquainted with the buildup of endometrial tissue that precedes it, so we begin with the proliferative phase.

The **proliferative phase** is a time of rebuilding of endometrial tissue lost at the last menstruation. At the end of menstruation, around day 5, the endometrium is about 0.5 mm thick and consists only of the stratum basalis. The stratum functionalis is rebuilt by mitosis from day 6 to day 14, under the influence of estrogen from the growing ovarian follicles. By day 14, the endometrium is about 2 to 3 mm thick (fig. 26.21*a*).

The **secretory phase** is a period of further endometrial thickening, but results from secretion and fluid accumulation rather than mitosis. It extends from day 15 (after ovulation) to day 26 of a typical cycle, and is stimulated by progesterone from the corpus luteum. In this phase, the endometrial glands grow wider, longer, and more coiled. As a result of the coiling, a vertical section through the endometrium shows these glands with a sawtooth or zigzag appearance (fig. 26.21*b*). Endometrial cells and the uterine stroma accumulate glycogen during this phase. By the end of the secretory phase, the endometrium is about 5 to 6 mm thick—a soft, wet, nutritious bed available for embryonic development in the event of pregnancy.

The **premenstrual phase** is a period of endometrial degeneration occuring in the last two days or so of the menstrual cycle. When the corpus luteum involutes, the spiral arteries exhibit spasmodic contractions that cause endometrial ischemia (interrupted blood flow). The premenstrual phase is therefore also called the **ischemic** (iss-KEE-mic) **phase.** Ischemia leads to tissue necrosis. As the endometrial glands, stroma, and blood vessels degenerate, pools of blood accumulate in the stratum functionalis. Necrotic endometrium falls away from the uterine wall, mixes with blood and serous fluid in the lumen, and forms the **menstrual fluid** (fig. 26.21*c*).

The **menstrual phase (menses)** begins when enough menstrual fluid has accumulated in the uterus that it begins to be discharged vaginally. The first day of external discharge marks day 1 of a new menstrual cycle.

The Vagina

The **vagina,**[46] or birth canal, is a tube about 8 to 10 cm long that allows for the discharge of menstrual fluid, receipt of the penis and semen, and birth of a baby. The vaginal wall is thin but very distensible. It consists of an outer adventitia, a middle muscularis, and an inner mucosa. The vagina tilts dorsally between the urethra and rectum; the urethra is bound to its anterior wall. The vagina has no glands, but it is lubricated by the *transudation* ("vaginal sweating") of serous fluid through its walls and by mucus from the cervical glands above it. The vagina extends slightly beyond the cervix and forms blind-ended spaces called **fornices**[47] (FOR-nih-sees; singular, *fornix*) surrounding it (see fig. 26.11).

At its lower end, the vaginal mucosa folds inward and forms a membrane, the **hymen,** which stretches across the orifice. The hymen has one or more openings to allow menstrual fluid to pass through, but it usually must be ruptured to allow for intercourse. A little bleeding often accompanies the first act of intercourse; however, the hymen is commonly ruptured before then by tampons, medical examinations, or strenuous exercise. The lower end of the vagina also has transverse friction ridges, or **vaginal rugae,** which stimulate the penis and help induce ejaculation.

The vaginal epithelium is simple cuboidal in childhood, but the estrogens of puberty transform it into a stratified squamous epithelium. This is an example of *metaplasia,* the transformation of one tissue type to another. The epithelial cells are rich in glycogen. Bacteria ferment this to lactic acid, which produces a low vaginal pH (about 3.5–4.0) that inhibits the growth of pathogens. The mucosa also has antigen-presenting cells called **dendritic cells,** which are a route by which HIV invades the female body.

THINK ABOUT IT!

Why do you think the vaginal epithelium changes type at puberty? Of all types of epithelium it might become, why stratified squamous?

The External Genitalia

The external genitalia of the female occupy most of the perineum, and are collectively also known as the **vulva**[48] or **pudendum**[49] (fig. 26.22). The perineum has the same skeletal landmarks in the female as in the male.

[46] *vagina* = sheath
[47] *fornix* = arch, vault
[48] *vulva* = covering
[49] *pudend* = shameful

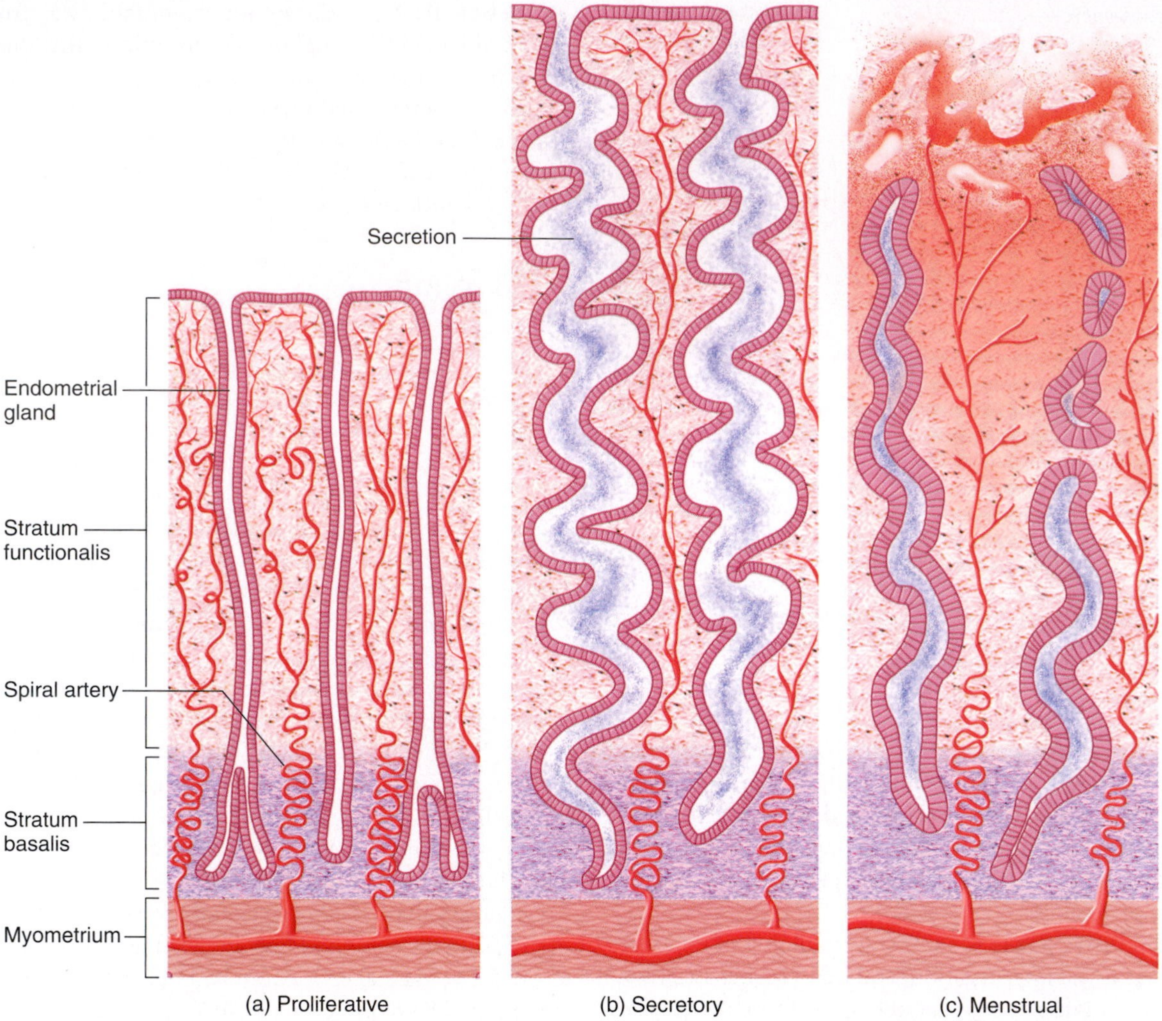

Secretion
Endometrial gland
Stratum functionalis
Spiral artery
Stratum basalis
Myometrium
(a) Proliferative
(b) Secretory
(c) Menstrual

Figure 26.21

Endometrial Changes Through the Menstrual Cycle. (*a*) Late proliferative phase. The endometrium is 2 to 3 mm thick and has relatively straight, narrow endometrial glands. Spiral arteries penetrate upward between the endometrial glands. (*b*) Secretory phase. The endometrium has thickened to 5 to 6 mm thick by accumulation of glycogen and mucus. The endometrial glands are much wider and more distinctly coiled, showing a zigzag or "sawtooth" appearance in histological sections. (*c*) Menstrual phase. Ischemic tissue has begun to die and fall away from the uterine wall, with bleeding from broken blood vessels and pooling of blood within the tissue and in the uterine cavity.

The **mons**[50] **pubis** (see fig. A.5) consists mainly of an anterior mound of adipose tissue overlying the pubic symphysis, covered with skin and bearing pubic hair. The **labia majora**[51] (singular, *labium majus*) are a pair of thick folds of skin and adipose tissue inferior to the mons, between the thighs; the slit between the labia majora is the *pudendal cleft.* Pubic hair grows on the lateral surfaces of the labia majora at puberty, but the medial surfaces remain hairless. Medial to the labia majora are the much thinner, entirely hairless **labia minora**[52] (singular, *labium minus*). The area enclosed by them, called the **vestibule,** contains the urinary and vaginal orifices. At the anterior margin of the vestibule, the labia minora meet and form a hoodlike **prepuce** over the clitoris.

The **clitoris** is structured much like a miniature penis but has no urinary role. Its function is entirely sensory, serving as the primary center of erotic stimulation. Unlike the penis, it is almost entirely internal, it has no corpus spongiosum, and it does not enclose the urethra. Essentially, it is a pair of corpora cavernosa enclosed in connective tissue. Its **glans** protrudes slightly from the prepuce. The **body (corpus)** passes internally, inferior to the pubic symphysis (see fig. 26.11). At its internal end, the corpora cavernosa diverge like a Y as a pair of **crura,** which, like those of the penis, attach the clitoris to each side of the pubic arch. The circulation and innervation of the clitoris are largely the same as for the penis.

Just deep to the labia majora, a pair of subcutaneous erectile tissues called the **vestibular bulbs** bracket the vagina like parentheses. They become congested with blood during sexual excitement and cause the vagina to tighten somewhat around the penis, enhancing sexual stimulation.

[50]*mons* = mountain
[51]*labi* = lip + *major* = larger, greater
[52]*minor* = smaller, lesser

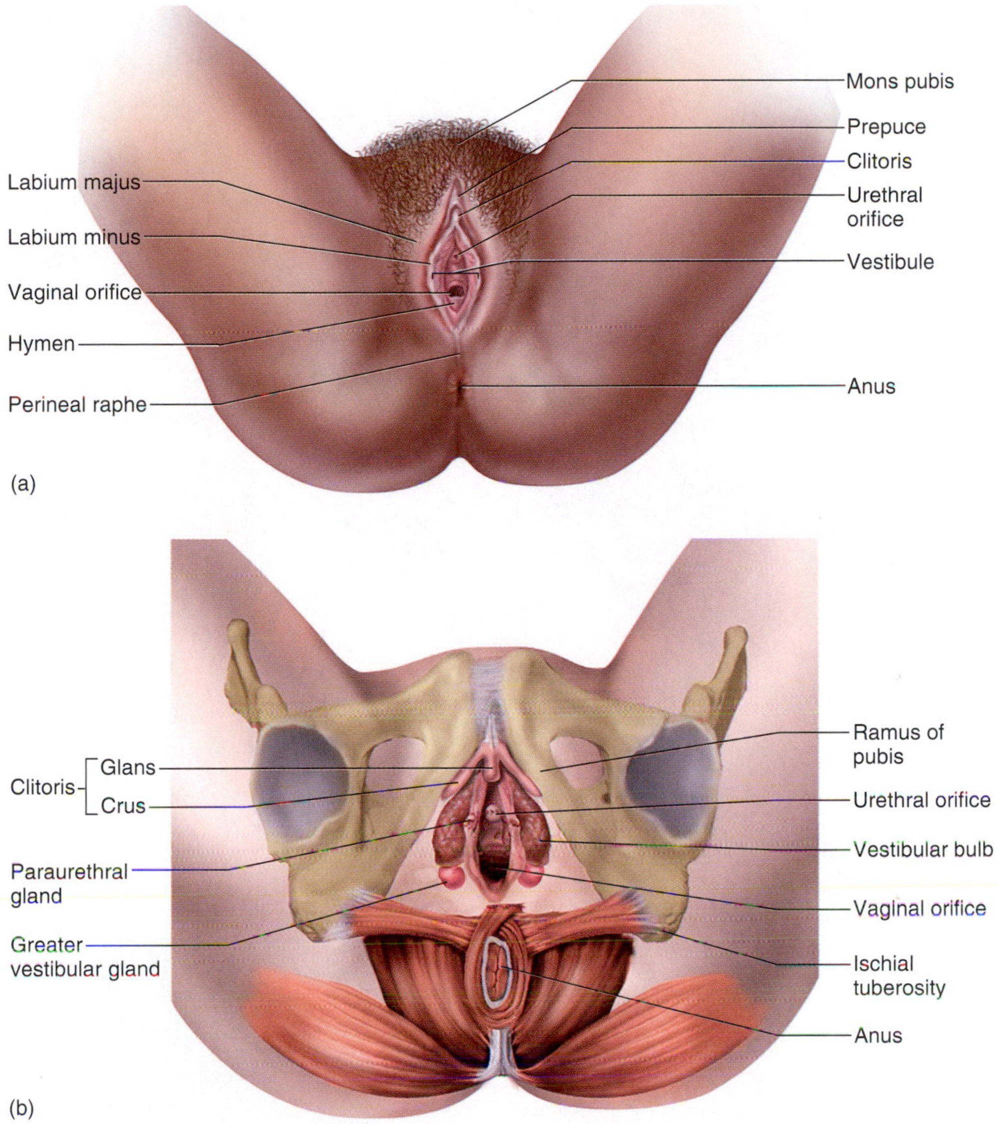
Mons pubis
Prepuce
Clitoris
Urethral orifice
Vestibule
Anus
Labium majus
Labium minus
Vaginal orifice
Hymen
Perineal raphe
(a)
Clitoris
Glans
Crus
Paraurethral gland
Greater vestibular gland
Ramus of pubis
Urethral orifice
Vestibular bulb
Vaginal orifice
Ischial tuberosity
Anus
(b)

FIGURE 26.22

The Female Perineum. (*a*) Surface anatomy. (*b*) Subcutaneous structures.

Next to the vagina are a pair of pea-sized **greater vestibular (Bartholin[53]) glands** with short ducts opening into the vestibule or lower vagina (fig. 26.22*b*). These are the counterpart to the bulbourethral glands of the male. They keep the vulva moist, and during sexual excitement they provide most of the lubrication for intercourse. The vestibule is also lubricated by a number of **lesser vestibular glands.** A pair of mucous **paraurethral (Skene[54]) glands,** homologous to the male prostate, open into the vestibule near the external urethral orifice. The paraurethral glands may eject fluid, sometimes abundantly, during orgasm ("female ejaculation"). This fluid is similar to the secretion of the prostate gland.

[53]Caspar Bartholin (1655–1738), Danish anatomist
[54]Alexander J. C. Skene (1838–1900), American gynecologist

The Breasts and Mammary Glands

The breast (fig. 26.23) is a mound of tissue overlying the pectoralis major. It develops at puberty and remains for life, but most of this time it contains very little mammary gland. The mammary gland develops within the breast during pregnancy, remains active in the lactating breast, and atrophies when a woman ceases to nurse.

The breast has two principal regions: the conical to pendulous **body,** with the nipple at its apex, and an extension toward the armpit called the **axillary tail.** Lymphatics of the axillary tail are especially important as a route of breast cancer metastasis.

The nipple is surrounded by a circular colored zone, the **areola.** Dermal blood capillaries and nerves come closer to the surface here than in the surrounding skin and make the areola more sensitive and more reddish in color. In pregnancy, the areola and nipple often darken, making them more visible to the indistinct vision of a nursing infant. Sensory nerve fibers of the areola are important in

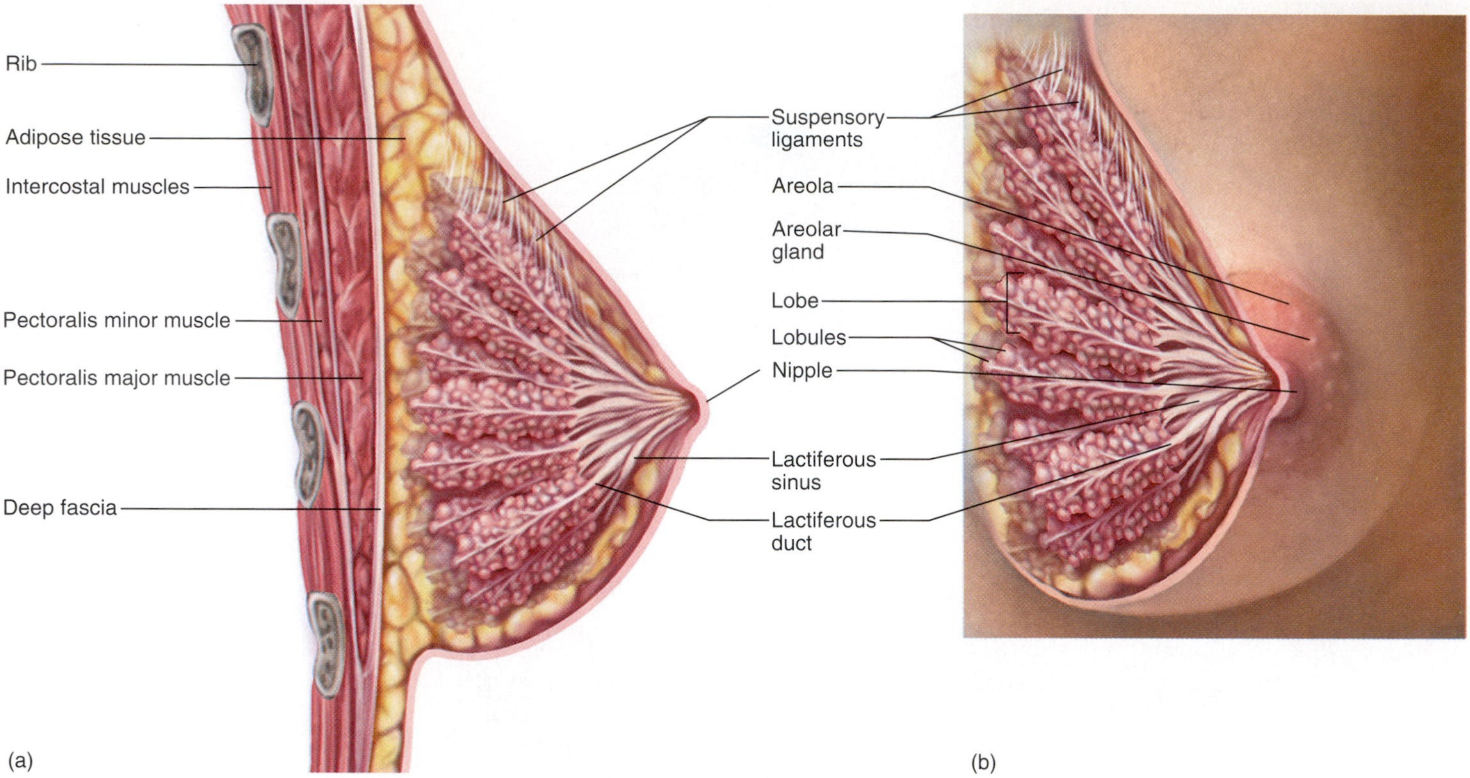
Rib
Adipose tissue
Intercostal muscles
Pectoralis minor muscle
Pectoralis major muscle
Deep fascia
(a)
Suspensory ligaments
Areola
Areolar gland
Lobe
Lobules
Nipple
Lactiferous sinus
Lactiferous duct
(b)

FIGURE 26.23

Anatomy of the Lactating Breast. (*a*) Sagittal section of breast. (*b*) Surface anatomy of the breast with cutaway view of the lobes of mammary gland; anterior view of left breast.

triggering a *milk ejection reflex* when an infant nurses. The areola has sparse hairs and **areolar glands,** visible as small bumps on the surface. These glands are intermediate between sweat glands and mammary glands in their degree of development. When a woman is nursing, the areola is protected from chapping and cracking by secretions of the areolar glands and sebacous glands of the areola. The dermis of the areola has smooth muscle fibers that contract in response to cold, touch, and sexual arousal, wrinkling the skin and erecting the nipple.

Internally, the nonlactating breast consists mostly of adipose and collagenous tissue (fig. 26.24). Breast size is determined by the amount of adipose tissue and has no relationship to the amount of milk the mammary gland can produce. **Suspensory ligaments** attach the breast to the dermis of the overlying skin and to the fascia of the pectoralis major. The nonlactating breast contains very little glandular tissue, but it does have a system of ducts branching through its connective tissue stroma and converging on the nipple.

When the mammary gland develops during pregnancy, it exhibits 15 to 20 lobes arranged radially around the nipple, separated from each other by fibrous stroma. Each lobe is drained by a **lactiferous**[55] **duct,** which dilates to form a **lactiferous sinus** opening onto the nipple. Internally, this duct branches repeatedly with the finest branches ending in secretory acini. The acini are organized into grapelike clusters (lobules) within each lobe of the breast.

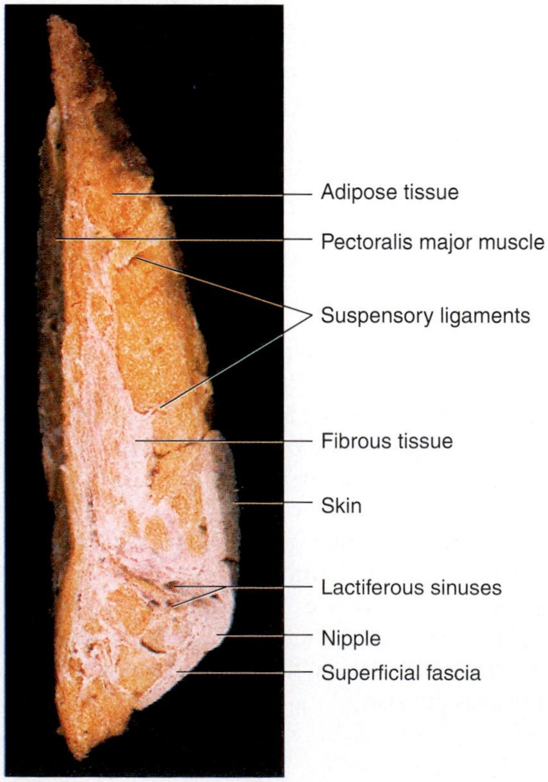
Adipose tissue
Pectoralis major muscle
Suspensory ligaments
Fibrous tissue
Skin
Lactiferous sinuses
Nipple
Superficial fascia

FIGURE 26.24

Sagittal Section of the Breast of a Cadaver.

[55]*lact* = milk + *fer* = to carry

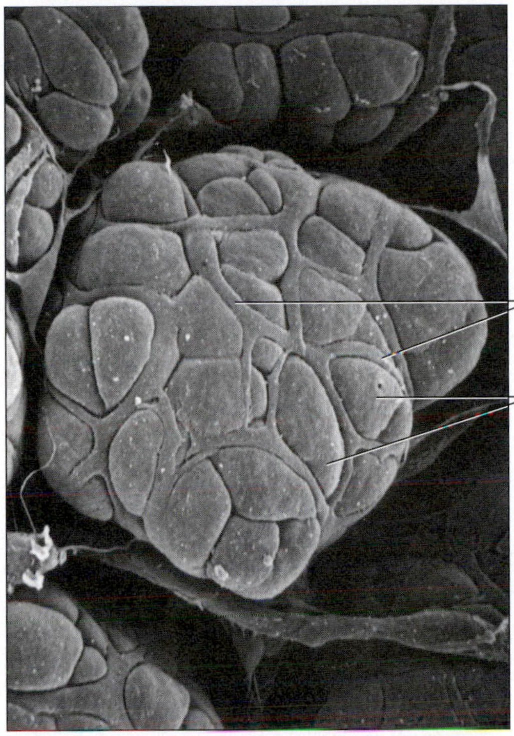

FIGURE 26.25
Acinus of a Mammary Gland. Myoepithelial cells can be seen forming a mesh around the secretory cells. The myoepithelial cells contract and force milk from the acinus into the duct.

Each acinus consists of pyramidal secretory cells arranged around a central lumen, and a network of contractile **myoepithelial cells** around the secretory cells (fig. 26.25). When a woman nurses, stimulation of the nipple induces the posterior lobe of the pituitary gland to secrete oxytocin. Oxytocin stimulates the myoepithelial cells to contract, squeezing milk from the acini into the lactiferous ducts.

Before You Go On

Answer the following questions to test your understanding of the preceding section:

9. What are the stages of oogenesis? Describe all the ways in which it differs from spermatogenesis.
10. Describe the changes that occur in the ovarian follicles and uterine endometrium over the course of the ovarian and menstrual cycle.
11. How does the structure of the uterine tube mucosa relate to its function?
12. Describe the external anatomy of the vulva.
13. Name the accessory glands of the vulva and state their locations and functions.
14. How does the structure of the nonlacting breast differ from that of the lactating breast? What is the difference between a breast and a mammary gland?

Developmental and Clinical Perspectives

Objectives

When you have completed this section, you should be able to

- explain how sexual differentiation is determined by the sex chromosomes and prenatal hormones;
- describe the embryonic development of the reproductive system;
- describe what aspects of anatomical development the male and female have in common, and how they become differentiated from each other;
- describe changes in reproductive function that occur in old age; and
- describe some reproductive disorders of each sex.

Prenatal Development and Sexual Differentiation

The male and female reproductive systems begin with embryonic sex organs that are anatomically "indifferent," yet genetically destined to differentiate into the genitalia of one sex or the other. Thus, the testes and ovaries develop from an initially indistinguishable *indifferent gonad,* the penis and clitoris develop from a single embryonic *phallus,* and the scrotum and labia majora develop from the embryonic *labioscrotal folds.* Organs that develop from the same embryonic precursor are said to be **homologous** to each other.

THE INTERNAL GENITALIA

The gonad appears at 5 to 6 weeks as a **gonadal ridge** near the mesonephros, the primitive kidney. Adjacent to each gonadal ridge are two ducts, the **mesonephric**[56] **(wolffian**[57]**) duct** described in chapter 25, and the **paramesonephric**[58] **(müllerian**[59]**) duct.** In males, the mesonephric ducts develop into parts of the reproductive tract and the paramesonephric ducts degenerate. In females, the opposite occurs (fig. 26.26).

Their differentiation of these ducts into the organs of one sex or the other is determined by an interaction between genes and hormones. If the zygote has sex chromosomes X and Y, it is normally destined to develop into a male; if it has two X chromosomes and no Y, it will develop into a female. Thus the sex of a child is determined at conception (fertilization), depending on whether the egg (always X) is fertilized by an X- or a Y-bearing sperm.

But why? The answer lies in the Y chromosome, where a gene called **SRY** (sex-determining region of the Y) codes for a protein called **testis-determining factor (TDF).** TDF then interacts with genes on some of the other chromosomes, including a gene on the

[56] *meso* = middle + *nephr* = kidney; named for a temporary embryonic kidney, the mesonephros
[57] Kaspar F. Wolff (1733–94), German anatomist
[58] *para* = next to
[59] Johannes P. Müller (1801–58), German physician

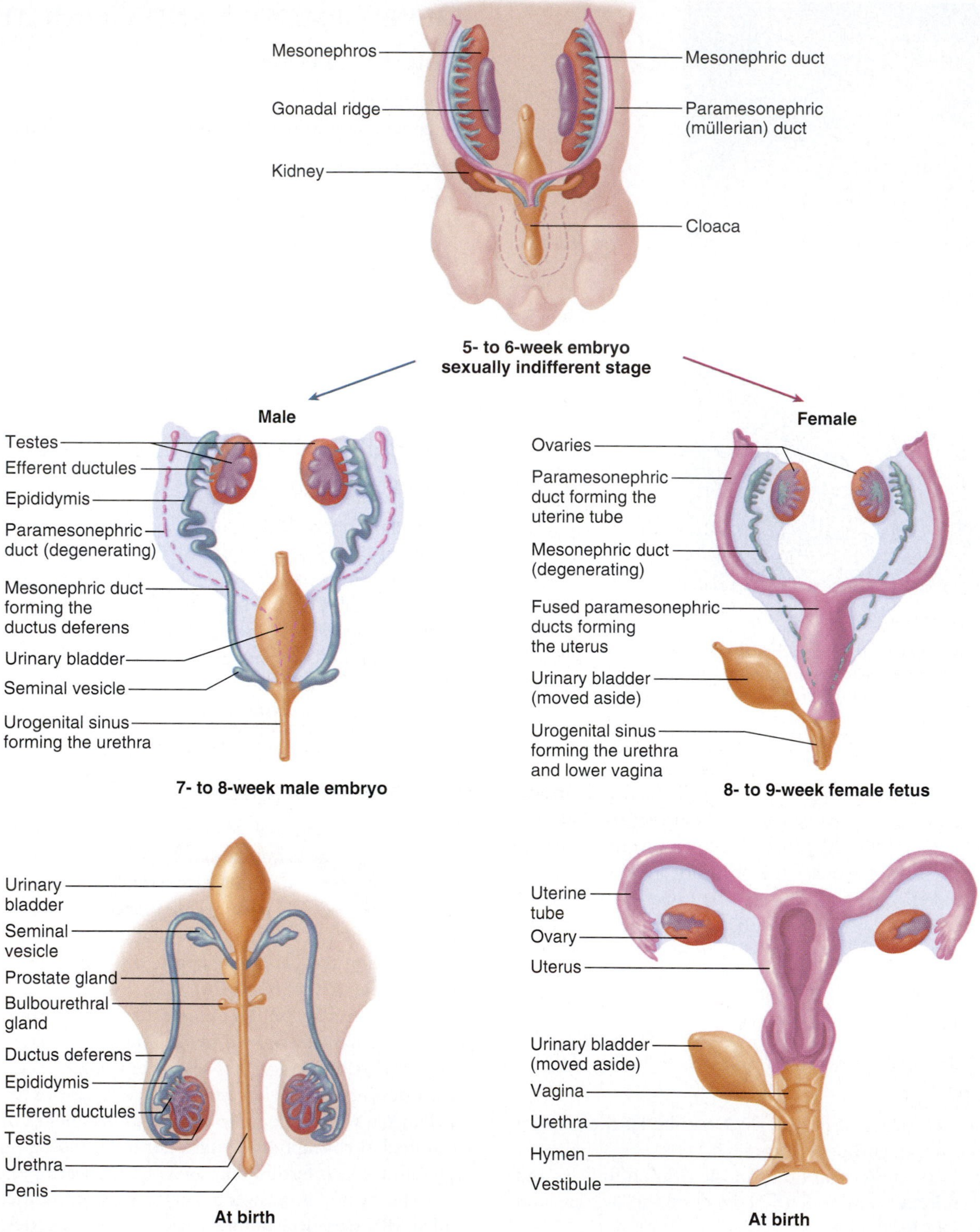
Mesonephros
Gonadal ridge
Kidney
Mesonephric duct
Paramesonephric (müllerian) duct
Cloaca
5- to 6-week embryo sexually indifferent stage
Male
Testes
Efferent ductules
Epididymis
Paramesonephric duct (degenerating)
Mesonephric duct forming the ductus deferens
Urinary bladder
Seminal vesicle
Urogenital sinus forming the urethra
7- to 8-week male embryo
Female
Ovaries
Paramesonephric duct forming the uterine tube
Mesonephric duct (degenerating)
Fused paramesonephric ducts forming the uterus
Urinary bladder (moved aside)
Urogenital sinus forming the urethra and lower vagina
8- to 9-week female fetus
Urinary bladder
Seminal vesicle
Prostate gland
Bulbourethral gland
Ductus deferens
Epididymis
Efferent ductules
Testis
Urethra
Penis
At birth
Uterine tube
Ovary
Uterus
Urinary bladder (moved aside)
Vagina
Urethra
Hymen
Vestibule
At birth

FIGURE 26.26

Embryonic Development of the Male and Female Reproductive Tracts. Note that the male tract develops from the mesonephric duct and the female tract from the paramesonephric duct, while the other duct in each sex degenerates.

X chromosome for androgen receptors and genes that initiate the development of male anatomy. By 8 to 9 weeks, the male gonadal ridge has become a rudimentary testis whose interstitial cells begin to secrete testosterone. Testosterone stimulates the mesonephric duct to develop into the system of male reproductive ducts. Sustentacular cells of the fetal testis secrete a hormone called **müllerian-inhibiting factor (MIF),** which causes atrophy of the paramesonephric duct. Even an adult male, however, retains a tiny

Y-shaped vestige of the paramesonephric ducts, like a vestigial uterus and uterine tubes, in the area of the prostatic urethra. It is named the *uterus masculinus.*

In a female fetus, the absence of testosterone causes the mesonephric ducts to degenerate, and in the absence of MIF, the paramesonephric ducts develop "by default" into a female reproductive tract. Each duct differentiates into one of the uterine tubes. At their inferior end, the ducts fuse to form the single uterus and the upper one-third of the vagina. The lower two-thirds of the vagina develops as an outgrowth of the urogenital sinus described in chapter 25.

It may seem as if androgens should induce the formation of a male reproductive tract and estrogens induce a female reproductive tract. However, estrogen levels are always high during pregnancy, so if this mechanism were the case, it would feminize all fetuses. Thus the development of a female results from the low level of androgens, not the presence of estrogens.

THE EXTERNAL GENITALIA

At 8 weeks, the external genitalia are represented by the following sexually undifferentiated structures (fig. 26.27):

- the **phallus,**[60] a small shaft of tissue with a swollen glans;
- **urogenital folds,** a pair of medial tissue folds slightly posterior to the phallus; and
- **labioscrotal folds,** a larger pair of tissue folds lateral to the urogenital folds.

These organs begin to show sexual differentiation by the end of week 9, and either male or female genitalia are distinctly identifiable by the end of week 12. In the female, the phallus, urogenital folds, and labioscrotal folds become the clitoris, labia minora, and labia majora, respectively. In the male, the phallus elongates to form the penis, the urogenital folds fuse to enclose the urethra within the penis, and the labioscrotal folds fuse to form the scrotum.

The homology of the male and female genitalia becomes strikingly evident in certain abnormalities of sexual development. In the presence of excess androgen, the clitoris may enlarge and the labioscrotal folds fuse, so closely resembling a penis and scrotum that a newborn girl can be misidentified as a boy. In other cases, the ovaries descend into the labia majora as if they were testes descending into a scrotum.

DESCENT OF THE GONADS

Both male and female gonads initially develop high in the abdominal cavity and migrate into the pelvic cavity (ovaries) or scrotum (testes). The most pronounced migration is the **descent of the testes** (fig. 26.28). In the embryo, a ligament called the **gubernaculum**[61] (GOO-bur-NACK-you-lum) extends from the gonad to the floor of the scrotum. It shortens as the fetus grows and guides the testis through a passageway in the groin called the **inguinal canal.**

The testes begin to descend in week 6 and usually enter the scrotum by week 28. They are accompanied by ever-elongating testicular arteries and veins and by lymphatic vessels, nerves, the future ductus deferens, and extensions of the internal abdominal oblique muscle, which become the cremaster muscle. About 3% of boys, however, are born with undescended testes, or *cryptorchidism* (table 26.3).

The ovaries also descend, but to a much lesser extent. A gubernaculum extends from the inferior pole of the ovary to the labioscrotal fold. The ovaries eventually lodge just inferior to the brim of the lesser pelvis. The inferior part of the gubernaculum becomes the round ligament of the uterus and the superior part becomes the ovarian ligament.

Puberty

Unlike any other organ system, the reproductive system remains dormant for several years after birth. Around age 10 to 12 in most boys and 8 to 10 in most girls, however, the hypothalamus begins to secrete gonadotropin-releasing hormone (GnRH) and the pituitary responds by secreting the two gonadotropins, follicle-stimulating hormone (FSH) and luteinizing hormone (LH). These hormones, in turn, stimulate the gonads to secrete estrogens, progesterone, and testosterone. Combined with surges in the secretion of growth hormone and other hormones, the adolescent body exhibits pronounced anatomical changes. **Puberty,**[62] the first few years of **adolescence,**[63] has begun.

In boys, the earliest sign of puberty is usually enlargement of the testes and scrotum, and in girls, it is breast development, called **thelarche**[64] (thee-LAR-kee). These changes are soon followed by **pubarche** (pyu-BAR-kee), the growth of pubic and axillary hair, sebaceous glands, and apocrine glands. In girls, the third principal event is **menarche**[65] (men-AR-kee), the first menstrual period. Menarche does not immediately signify fertility. A girl's first few menstrual cycles are typically *anovulatory* (no egg is ovulated). Most girls begin ovulating regularly about a year after they begin menstruating. The male counterpart to menarche is the onset of ejaculation. Puberty ends when an individual is fully capable of reproducing, while adolescence extends until a person reaches full height in the late teens to early twenties.

Puberty entails many other changes too numerous for the scope of this book. The internal and external genitalia enlarge. Changes in muscularity and fat deposition bring about some of the secondary sex characteristics of the two sexes. The male voice deepens as the larynx enlarges. Testosterone, estrogens, and growth hormone cause rapid elongation of the long bones, and thus the adolescent growth in stature. And to the great anxiety of the parents of adolescents, the anatomical readiness for reproduction is accompanied by psychological interest, the *libido,* elicited in both sexes by testosterone. (Testosterone is produced not only by the testes but also, in small amounts, by the ovaries and adrenal cortex.)

[60]*phallo* = penis
[61]*gubern* = rudder, to steer, guide

[62]*puber* = grown up
[63]*adolesc* = to grow up
[64]*thel* = breast, nipple + *arche* = beginning
[65]*men* = monthly

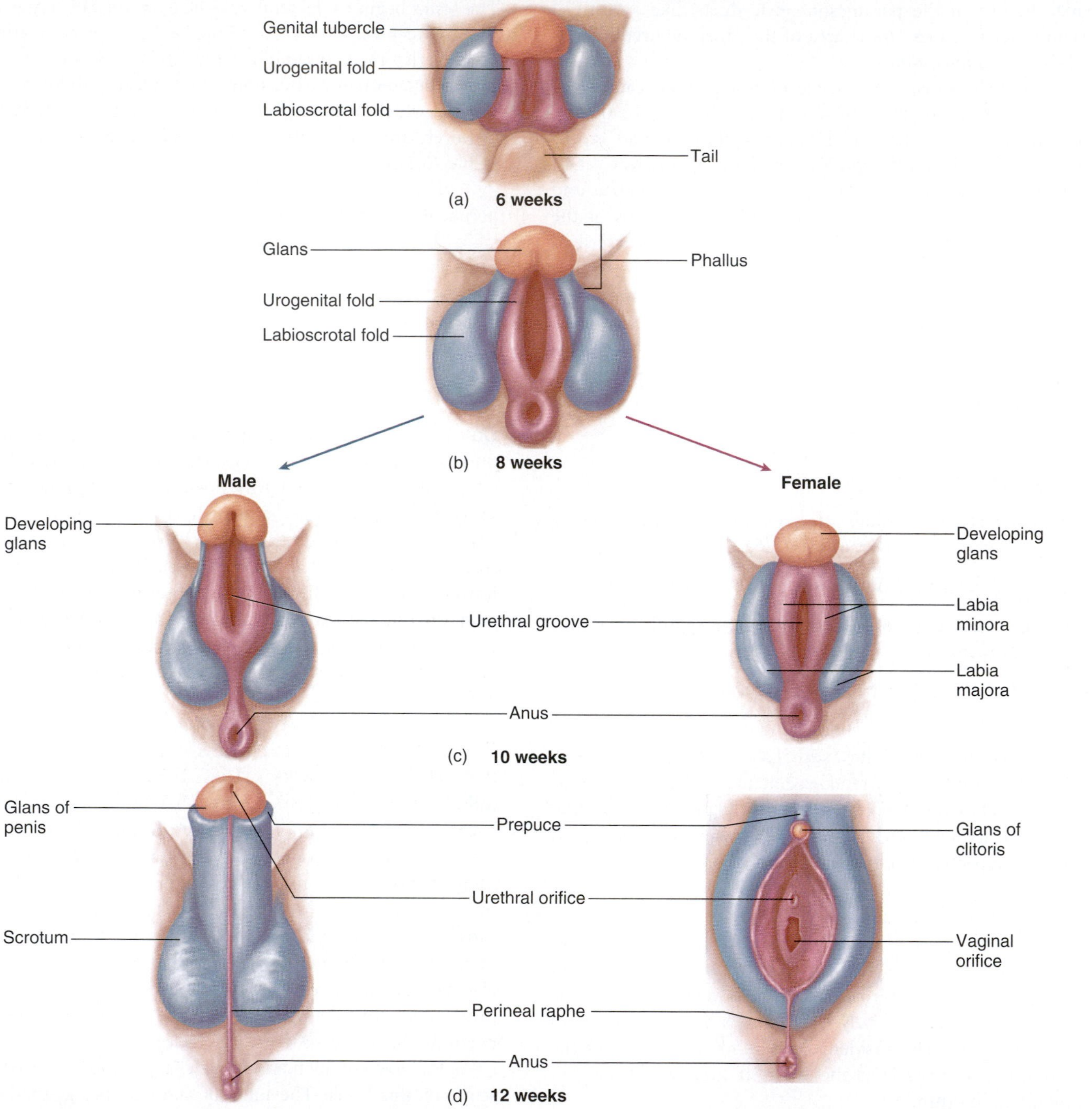

Genital tubercle
Urogenital fold
Labioscrotal fold
Tail
(a) 6 weeks
Glans
Phallus
Urogenital fold
Labioscrotal fold
(b) 8 weeks
Male
Female
Developing glans
Urethral groove
Developing glans
Labia minora
Labia majora
Anus
(c) 10 weeks
Glans of penis
Prepuce
Glans of clitoris
Urethral orifice
Scrotum
Vaginal orifice
Perineal raphe
Anus
(d) 12 weeks

FIGURE 26.27

Development of the External Genitalia. (*a*) By 6 weeks, the embryo has three primordial structures—the phallus, urogenital folds, and labioscrotal folds—that will become the male or female genitalia. (*b*) At 8 weeks these structures have grown, but the sexes are still indistinguishable. (*c*) Slight sexual differentiation is noticeable at 10 weeks. (*d*) The sexes are fully distinguishable by 12 weeks. Matching colors identify homologous structures of the male and female.

The Aging Reproductive System

Fertility and sexual function decline in and beyond middle age, owing to declining levels of testosterone and estrogen. Around ages 50 to 55, both men and women go through a period of physical and psychological change called **climacteric,** although (jokes about "male menopause" aside) only women experience **menopause,** the cessation of menses.

MALE CLIMACTERIC

In males, testosterone secretion peaks at about 7 mg/day at age 20 and then declines steadily to as little as one-fifth of this level by age 80. There is a corresponding decline in the number and secretory activity of the interstitial cells (the source of testosterone) and sustentacular cells (the source of inhibin). Along with the declining testosterone level, the sperm count and libido diminish.

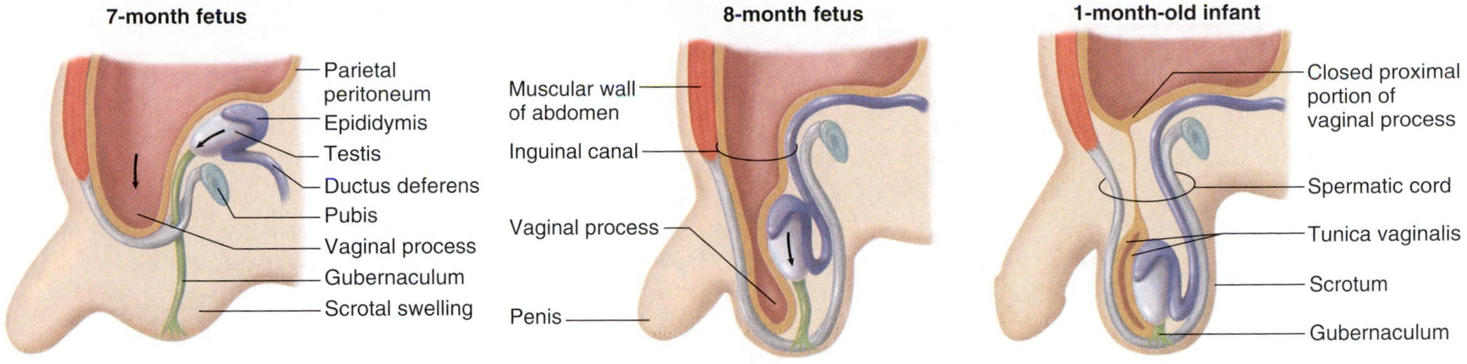
7-month fetus
8-month fetus
1-month-old infant
Parietal peritoneum
Epididymis
Testis
Ductus deferens
Pubis
Vaginal process
Gubernaculum
Scrotal swelling
Muscular wall of abdomen
Inguinal canal
Vaginal process
Penis
Closed proximal portion of vaginal process
Spermatic cord
Tunica vaginalis
Scrotum
Gubernaculum

FIGURE 26.28

Descent of the Testis. Note that the testis and spermatic ducts are retroperitoneal. An extension of the peritoneum called the *vaginal process* follows the testis through the inguinal canal and becomes the tunica vaginalis.

By age 65, sperm count is typically about one-third of what it was in a man's 20s. Nevertheless, men remain capable of fathering a child far into old age.

Male climacteric is brought on by falling levels of testosterone and inhibin. When the pituitary is less inhibited by these hormones, it secretes elevated levels of FSH and LH. In some cases, these gonadotropins cause mood changes, hot flashes, or even illusions of suffocation—symptoms similar to those that occur in perimenopausal women. Most men, however, pass through climacteric with little or no effect.

About 20% of men in their 60s and 50% of men in their 80s experience *erectile dysfunction (impotence),* the frequent inability to maintain a sufficient erection for intercourse (see table 26.3). Erectile dysfunction can also result from hypertension, atherosclerosis, medication, diabetes mellitus, and psychological causes. Over 90% of men with erectile dysfunction nevertheless remain able to ejaculate.

FEMALE CLIMACTERIC AND MENOPAUSE

Female climacteric is brought on by declining ovarian function. It generally begins when the ovaries are down to their last 1,000 eggs or so, and the follicles and ova that remain are less responsive to gonadotropins. Consequently, the follicles secrete less estrogen and progesterone. Without these steroids, the uterus, vagina, and breasts atrophy. Intercourse may become uncomfortable, and vaginal infections more common, as the vagina becomes thinner, less distensible, and drier. The skin becomes thinner, cholesterol levels rise (increasing the risk of cardiovascular disease), and bone mass declines (increasing the risk of osteoporosis). Blood vessels constrict and dilate in response to shifting hormone balances, and the sudden dilation of cutaneous arteries may cause *hot flashes*—a spreading sense of heat from the abdomen to the thorax, neck, and face. Hot flashes may occur several times a day, sometimes accompanied by headaches resulting from the sudden vasodilation of arteries in the head. In some people, the changing hormonal profile also causes mood changes.

Female climacteric is accompanied by **menopause,** the cessation of menstruation and end of fertility. Menopause usually occurs between the ages of 45 and 55. The average age has increased steadily in the last century and is now about 52. It is difficult to precisely establish the time of menopause because the menstrual periods can stop for several months and then begin again. Menopause is generally considered to have occurred when there has been no menstruation for a year or more.

Reproductive Disorders of the Male

Prostate cancer (insight 26.1) is the most common cancer of the male reproductive system, but not the only one. Men are also subject to testicular, penile, and breast cancer. Testicular cancer is the most common of these three and often strikes at a relatively young age compared to prostate cancer. Testicular self-examination combined with regular physical examinations are important preventive measures. Some additional facts about these cancers and other male reproductive disorders are given in table 26.3

Reproductive Disorders of the Female

The most important malignancies of the female reproductive system are breast and cervical cancer, although the increase in cigarette smoking has caused lung cancer to surpass both of these as a cause of female mortality in the United States. Breast cancer occurs in one out of every eight or nine American women. Breast tumors originate in cells of the mammary ducts and may metastasize to other organs by way of mammary and axillary lymphatics. Although some breast cancer is genetic, many nonhereditary risk factors are also known, including age, early menarche and late menopause, high alcohol or fat consumption, and smoking. Over 70% of cases of breast cancer, however, lack any identifiable risk factor. Early detection through regular breast X rays (mammograms) are currently regarded as the best protection. Breast self-examination (BSE) may also be helpful, but recent research has cast doubt on whether tumors are detected early enough by BSE to significantly reduce female mortality.

Uterine cancer is of two kinds, endometrial and cervical. Cervical cancer is a slow-growing neoplasia of the lower cervical canal and can be detected by microscopic examination of squamous cells from the cervix (a Pap smear; insight 26.3).

TABLE 26.3

Some Male Reproductive Disorders

Breast Cancer	Accounts for 0.2% of male cancers in the United States, usually seen after age 60 but sometimes in children and adolescents. About 175 females get breast cancer for every male who does so. Usually felt as a lump near the nipple, often with crusting and discharge from the nipple. Often quite advanced by the time of diagnosis, with poor prospects for recovery, because of denial and delay in seeking treatment.
Cryptorchidism[66] *(crip-TOR-ki-dizm)*	Failure of one or both testes to descend completely into the scrotum. Leads to infertility if not corrected, because undescended testes are too warm for spermatogenesis. In most cases, the testes descend spontaneously in the first year of infancy; otherwise, the condition can be corrected with hormone injections or surgery.
Hypospadias[67] *(HY-po-SPAY-dee-us)*	A congenital defect in which the urethra opens on the ventral side or base of the penis rather than at the tip; usually corrected surgically at about 1 year of age.
Infertility	Inability to fertilize an egg because of a low sperm count (lower than 20–25 million/mL), poor sperm motility, or a high percentage of deformed sperm (two heads, defective tails, etc.). May result from malnutrition, gonorrhea and other infections, toxins, or testosterone deficiency.
Penile Cancer	Accounts for 1% of male cancers in the United States; most common in black males aged 50 to 70 and of low income. Most often seen in men with nonretractable foreskins *(phimosis)* combined with poor penile hygiene; least common in men circumcised at birth.
Testicular Cancer	The most common solid tumor in men 15 to 34 years old, especially white males of middle to upper economic classes. Typically begins as a painless lump or enlargement of the testis. Highly curable if detected early. Men should routinely palpate the testes for normal size and smooth texture.
Varicocele (VAIR-ih-co-seal)	Abnormal dilation of veins of the spermatic cord, so that they resemble a "bag of worms." Occurs in 10% of males in the United States. Caused by absence or incompetence of venous valves. Reduces testicular blood flow and often causes infertility.
Disorders Described Elsewhere	
Androgen-insensitivity syndrome 535	Prostate cancer 737
Benign prostatic hyperplasia 737	

[66]*crypt* = hidden + *orchid* = testes
[67]*hypo* = below + *spad* = to draw off (the urine)

There are many other disorders of female reproductive function, too numerous to discuss here. Pregnancy adds its own risks to women's health. Table 26.4 describes some of the more common complications of pregnancy.

More common than any of the foregoing male or female reproductive disorders are the **sexually transmitted diseases (STDs),** caused by infectious microorganisms transmitted by intercourse, other sexual activity, and to infants during or before birth. Most of these are caused by viruses and bacteria. Currently the most serious viral STDs are **AIDS** (caused by the human immunodeficiency virus, HIV) and **hepatitis C.** HIV infects helper T lymphocytes among other cells, thus exerting a devastating effect on the immune system and leaving a person susceptible to certain forms of cancer and opportunistic infection (see chapter 22). The hepatitis C virus (HCV) is a common cause of liver failure and is the leading reason for liver transplants in the United States. Other viral STDs include genital herpes (usually caused by herpes simplex virus type 2, HSV-2) and genital warts (caused by 60 or more human papillomaviruses, HPVs). Some forms of HPV are associated with cervical, vaginal, penile, and anal cancer.

Bacterial STDs include gonorrhea and syphilis, caused by the bacteria *Neisseria gonorrhoeae* and *Treponema pallidum*, respectively. Cases of these two diseases are outnumbered, however, by chlamydia (caused by *Chlamydia trachomatis*), which affects 3 to 5 million people per year in the United States.

Before You Go On

Answer the following questions to test your understanding of the preceding section:

15. What are mesonephric and paramesonephric ducts? What factors determine which one develops and which one regresses in the fetus?
16. What male structures develop from the phallus and labioscrotal folds?
17. Define the *gubernaculum* and describe its function.
18. Which of the following occur in both men and women—thelarche, pubarche, climacteric, menopause, breast cancer, and cryptorchidism? Explain.

INSIGHT 26.3 CLINICAL APPLICATION

Cervical Cancer and Pap Smears

Cervical cancer is common among women from ages 30 to 50, especially those who smoke, who began sexual activity at an early age, and who have histories of frequent sexually transmitted diseases or cervical inflammation. It begins in the epithelial cells of the lower cervix, develops slowly, and remains a local, easily removed lesion for several years. If the cancerous cells spread to the subepithelial connective tissue, however, the cancer is said to be *invasive* and is much more dangerous, potentially requiring *hysterectomy*[68] (removal of the uterus).

The best protection against cervical cancer is early detection by means of a *Pap*[69] *smear*—a procedure in which loose cells are scraped from the cervix and vagina and microscopically examined. Figure 26.29 shows normal and cancerous Pap smears. The findings are rated on a five-point scale:

Class I—no abnormal cells seen
Class II—atypical cells suggestive of inflammation, infection, or irritation
Class III—nonmalignant but mildly abnormal cell growth *(dysplasia)*
Class IV—cells typical of localized cancer
Class V—cells typical of invasive cancer

An average woman is typically advised to have annual Pap smears for 3 years and may then have them less often at the discretion of her physician. Women with any of the risk factors listed may be advised to have more frequent examinations.

[68]*hyster* = uterus + *ectomy* = cutting out
[69]George N. Papanicolaou (1883–1962), Greek-American physician and cytologist

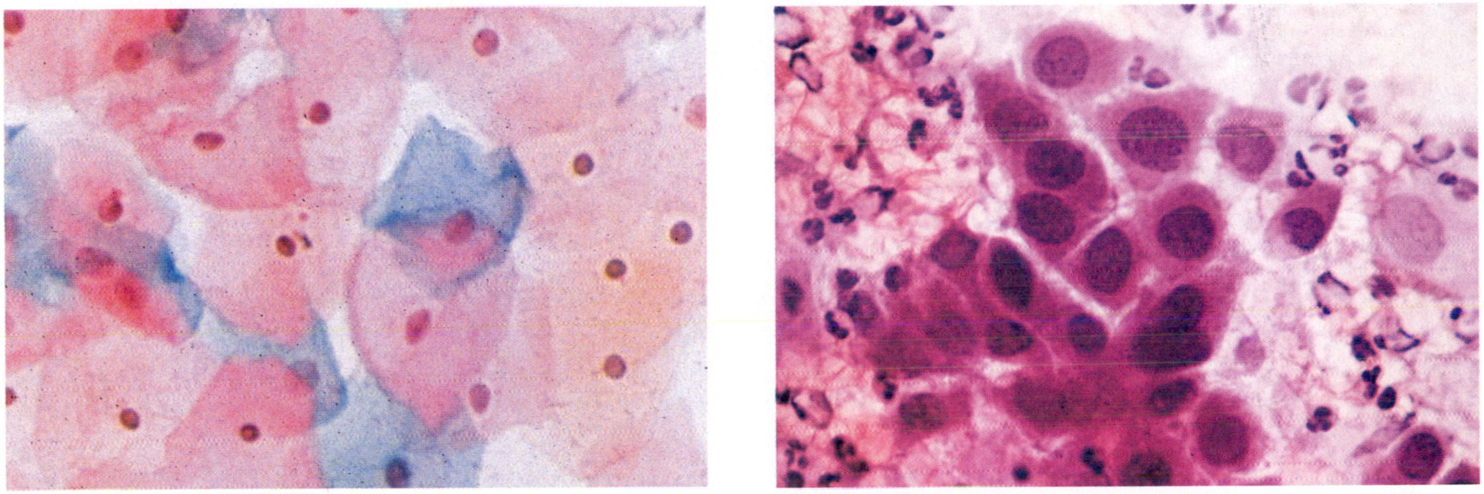

(a) (b)

Figure 26.29

Pap Smears. These are smears of squamous epithelial cells scraped from the cervix. (*a*) A normal Pap smear. (*b*) Pap smear from a patient with class V cervical cancer. Note the greatly enlarged nuclei in proportion to the volume of cytoplasm in these malignant cells.

Table 26.4

Some Female Reproductive Disorders

Disorders of Pregnancy	
Abruptio Placentae[70]	Premature separation of the placenta from the uterine wall, often associated with preeclampsia or cocaine use. May require birth by cesarian section.
Ectopic[71] *Pregnancy*	Implantation of the conceptus anywhere other than the uterus; usually starts in the uterine tube *(tubal pregnancy)* and may progress to *abdominal pregnancy* if the tube ruptures.
Gestational Diabetes	A form of diabetes mellitus that develops in about 1% to 3% of pregnant women, characterized by insulin insensitivity, hyperglycemia, glycosuria, and a risk of excessive fetal size and birth trauma. Glucose metabolism usually returns to normal after delivery of the infant, but 40% to 60% of women with gestational diabetes develop diabetes mellitus within 15 years after the pregnancy.
Hyperemesis Gravidarum[72]	Prolonged vomiting, dehydration, alkalosis, and weight loss in early pregnancy, often requiring hospitalization to stabilize fluid, electrolyte, and acid-base balance; sometimes associated with liver damage.
Placenta Previa[73]	Blockage of the cervical canal by the placenta, preventing birth of the infant before the placenta separates from the uterus. Requires birth by cesarian section.
Preeclampsia[74] *(toxemia of pregnancy)*	Rapid onset of hypertension and edema, with swelling especially of the face and hands; proteinuria and reduced glomerular filtration rate; increased blood clotting; sometimes with headaches, visual disturbances, and small cerebral infarctions. Seen in about 4% of pregnancies, especially in the third trimester of women pregnant for the first time. Can progress to *eclampsia,* with seizures and widespread vascular spasms sometimes fatal to the mother, fetus, or both. Eclampsia usually occurs shortly before or after childbirth.
Spontaneous Abortion	Occurs in 10% to 15% of pregnancies, usually because of fetal deformities or chromosomal abnormalities incompatible with survival, but may also result from maternal abnormalities, infectious disease, and drug abuse.
Other Reproductive Disorders	
Amenorrhea[75]	Absence of menstruation. Normal in pregnancy, lactation, early adolescence, and perimenopausal years, but can also result from gonadotropin hyposecretion, genetic disorders, CNS disorders, or excessively low body fat content.
Dysmenorrhea[76]	Painful menstruation in the absence of pelvic disease, caused by excessive endometrial prostaglandin secretion. Prostaglandins stimulate painful contractions of myometrium and uterine blood vessels. Usually begins around age 15 or 16 and affects up to 75% of women from 15 to 25 years old.
Endometriosis	Growth of endometrial tissue in any site other than the uterus, including the uterine tubes, ovaries, unirary bladder, vagina, pelvic cavity, small or large intestine, or even the lungs or pleural cavity. May cause dysmenorrhea, abnormal vaginal bleeding, and infertility.
Leiomyomas[77] *(uterine fibroids)*	Benign tumors of uterine smooth muscle. Usually small and asymptomatic, but may cause abnormal uterine bleeding and pain or heavy menstruation.
Pelvic Inflammatory Disease (PID)	Acute, painful inflammation due to infection of the uterus, uterine tubes, or ovaries, usually with the organisms of sexually transmitted diseases. Causes abdominopelvic pain, pain on urination, and irregular bleeding.
Disorders Described Elsewhere	
Breast cancer 755	Sexually transmitted diseases 756
Cervical cancer 757	

[70]*ab* = away + *rupt* = to tear + *placentae* = of the placenta
[71]*ec* = out of + *top* = place
[72]*hyper* = excessive + *emesis* = vomiting + *gravida* = pregnant woman
[73]*pre* = before + *via* = the way (obstructing the way)
[74]*ec* = forth + *lampsia* = shining
[75]*a* = without + *meno* = monthly + *rrhea* = flow
[76]*dys* = painful, abnormal + *meno* = monthly + *rrhea* = flow
[77]*leio* = smooth + *myo* = muscle + *oma* = tumor

CHAPTER REVIEW

Review of Key Concepts

Sexual Reproduction (p. 728)

1. Sexual reproduction is the production of offspring that combine genes from two parents.
2. Sexual reproduction entails the union of two *gametes* to form a *zygote* (fertilized egg). The gametes are a small motile *sperm* produced by the male and a large, immobile, nutrient-laden *egg (ovum)* produced by the female.
3. The sexes are genetically defined by their sex chromosomes. Normally, males have an X and a Y chromosome and females have two X chromosomes.
4. The *gonads* (testes and ovaries) are the *primary sex organs. Secondary sex organs* are other anatomical structures needed to produce offspring, such as the male glands, ducts, and penis and the female uterine tubes, uterus, and vagina. *Secondary sex characteristics* are features not essential to reproduction but which help to attract mates.

Male Reproductive Anatomy (p. 729)

1. The testis has a fibrous capsule, the *tunica albuginea.* Fibrous septa divide the interior of the testis into 250 to 300 compartments called *lobules.* Each lobule contains 1 to 3 sperm-producing *seminiferous tubules.* Testosterone-secreting *interstitial cells* lie in clusters between the tubules.
2. The epithelium of a seminiferous tubule consists of *germ cells* and *sustentacular cells.* The germ cells develop into sperm, while the sustentacular cells support and nourish them, form a *blood-testis barrier* between the germ cells and nearest blood supply, and secrete *inhibin,* which regulates the rate of sperm production.
3. Each testis is supplied by a long, slender *testicular artery* and drained by a *testicular vein,* and is supplied with *testicular nerves* and lymphatic vessels.
4. The *scrotum* contains the testes and the *spermatic cord*—a bundle of connective tissue, testicular blood vessels, and a sperm duct, the *ductus deferens.* The spermatic cord passes up the back of the scrotum and through the inguinal ring into the inguinal canal.
5. Sperm cannot develop at the core body temperature of 37°C. The testes are kept about 2°C cooler than this by three structures in the scrotum: the *cremaster muscle* of the spermatic cord, which relaxes when it is warm and contracts when it is cool, thus lowering or raising the scrotum and testes; the *dartos muscle* in the scrotal wall, which contracts and tautens the scrotum when it is cool; and the *pampiniform plexus* of blood vessels in the spermatic cord, which acts as a *countercurrent heat exchanger* to cool the blood on its way to the testis.
6. Sperm formation, or *spermatogenesis,* begins with primordial germ cells that migrate from the embryonic yolk sac into the gonad and become *spermatogonia.* Beginning at puberty, spermatogonia give rise to a series of cells called *primary spermatocytes, secondary spermatocytes, spermatids,* and finally *spermatozoa (sperm).*
7. The sperm consists of a head containing the nucleus and *acrosome,* and a tail composed of a mitochondria-stuffed *midpiece,* a long *principal piece,* and a short *endpiece.*
8. *Spermatic ducts* carry sperm from the testis to the urethra. They include *efferent ductules* leaving the testis; the *duct of the epididymis,* a highly coiled structure adhering to the posterior side of the testis; a muscular *ductus deferens* that travels through the spermatic cord into the pelvic cavity; and a short *ejaculatory duct* that carries sperm and seminal vesicle secretions the last 2 cm to the urethra. The urethra completes the path of the sperm to the outside of the body.
9. The male has three sets of accessory glands: a pair of *seminal vesicles* dorsal to the urinary bladder; a single *prostate gland* inferior to the bladder, enclosing the prostatic urethra; and a pair of small *bulbourethral glands* that secrete into the proximal end of the penile urethra. The seminal vesicles and prostate secrete most of the semen. The bulbourethral glands produce a small amount of clear slippery fluid that lubricates the urethra and neutralizes its pH.
10. *Semen* is a mixture of sperm (10% of the volume) and fluids from the prostate (30%) and seminal vesicles (60%). It contains fructose, fibrinogen, clotting enzymes, fibrinolysin, prostaglandins, and spermine (table 26.2).
11. The *penis* is divided into an internal *root* and an external *shaft* and *glans.* It is covered with loose skin that extends over the glans as the *prepuce,* or foreskin.
12. Internally, the penis consists mainly of three long *erectile tissues*—a pair of dorsal *corpora cavernosa,* which engorge with blood and produce most of the effect of erection, and a single ventral *corpus spongiosum,* which contains the urethra. All three tissues have blood sinuses called *lacunae* separated by *trabeculae* composed of connective tissue and smooth muscle *(trabecular muscle).*
13. At the proximal end of the penis, the corpus spongiosum dilates into a *bulb* that receives the urethra and ducts of the bulbourethral glands, and the corpora cavernosa diverge into a pair of *crura* that anchor the penis to the pubic arch and perineal membrane.
14. The penis is supplied by a pair of internal pudendal arteries. Each branches into a *dorsal artery,* which travels dorsally under the skin of the penis, and a *deep artery,* which travels through the corpus cavernosum and supplies blood to the lacunae. The dorsal arteries supply most of the blood when the penis is flaccid, and the deep arteries during erection.
15. Nerves of the penis converge on the *dorsal nerve,* which leads via the *internal pudendal nerve* to the sacral plexus and then the spinal cord. The penis receives sympathetic, parasympathetic, and somatic motor nerve fibers.

Female Reproductive Anatomy (p. 740)

1. The ovary has a central *medulla,* a surface *cortex,* and an outer fibrous capsule, the *tunica albuginea.*
2. Each egg develops in its own bubblelike *follicle.* Follicles are located primarily in the cortex.
3. The ovary is supported by a medial *ovarian ligament,* lateral *suspensory ligament,* and anterior *mesovarium.* It receives blood from a branch of the *uterine artery* medially and from the *ovarian artery* laterally.
4. *Oogenesis* is the production of eggs. Unlike spermatogenesis, it occurs in a monthly rhythm (the *ovarian cycle*) and usually produces only one gamete (egg) per month.
5. Oogenesis begins with primordial germ cells, which colonize the fetal gonad and become *oogonia.* These multiply until the fifth month of fetal development. Some of these develop into *primary oocytes* and begin meiosis I before birth. Most primary oocytes undergo *atresia* during childhood, leaving about 400,000 at puberty.

6. Each month, about 20 to 25 *primordial follicles* begin to develop. The single layer of squamous follicular cells around the oocyte thicken into cuboidal cells, then multiply and become stratified *granulosa cells.* The follicle is then a *primary follicle.* The granulosa cells secrete *follicular fluid,* which forms small pools amid the granulosa cells. These pools eventually coalesce to form a single cavity, the *antrum,* and the follicle is then called a *secondary (antral) follicle.* Normally only one of these secondary follicles becomes a fully *mature (graafian) follicle,* destined to ovulate.
7. At the secondary follicle stage, the primary oocyte completes meiosis I and becomes a *secondary oocyte.* This is the stage that ovulates. It never develops further unless it is fertilized.
8. After ovulation, the remnant of the follicle becomes a *corpus luteum,* which secretes progesterone. If pregnancy does not occur, the corpus luteum undergoes involution (degenerates) from days 24 to 26 of an average 28-day cycle. It persists for about 3 months if pregnancy does occur.
9. The *uterine (fallopian) tube* is a ciliated duct that extends from the ovary to the uterus. Its flared distal end, near the ovary, is called the *infundibulum* and has feathery projections called *fimbriae* to receive the ovulated egg. Its long midportion is the *ampulla,* and the short constricted zone near the uterus is the *isthmus.*
10. The uterus is a thick muscular chamber superior to the urinary bladder. It consists of an upper *fundus,* middle *corpus* (body), and lower *cervix* (neck), where it meets the vagina.
11. The uterine wall is three-layered: an outer serosa called the *perimetrium,* a thick muscular *myometrium,* and an inner mucosa called the *endometrium.* The endometrium contains numerous tubular glands and is divided into two layers—a thick superficial *stratum functionalis,* which is shed in each menstrual period, and a thinner basal *stratum basalis,* which is retained from cycle to cycle.
12. The uterus is supported by a pair of lateral winglike *broad ligaments* and cordlike *cardinal, uterosacral,* and *round ligaments.* It receives blood from a pair of *uterine arteries.*
13. The endometrium undergoes cyclic histological changes called the *menstrual cycle,* governed by the shifting hormonal secretions of the ovaries. The *proliferative phase* of the menstrual cycle consists of mitotic rebuilding of tissue lost in the previous menstrual period. The *secretory phase* consists of further thickening of the endometrium by accumulation of secretions, not by mitosis. The *premenstrual phase* is characterized by ischemia and necrosis of the endometrium. The *menstrual phase* begins when endometrial tissue and blood is first discharged from the vagina, and marks day 1 of a new cycle.
14. The *vagina* tilts dorsally between the urethra and rectum. It has no glands but is moistened by transudation of serous fluid through the vaginal wall and by mucus from glands in the cervical canal. The adult vagina is lined with a stratified squamous epithelium populated by antigen-presenting *dendritic cells.*
15. The *vulva* (*pudendum* or *external genitalia*) include the *mons pubis, labia majora* and *minora, clitoris,* vaginal orifice, accessory glands (*greater* and *lesser vestibular glands* and *paraurethral glands*) and erectile tissues *(vestibular bulbs).* The urethra also opens into the vulva.
16. The breast is internally divided into lobes, each with a *lactiferous duct* that conveys milk to the nipple. Outside of pregnancy or lactation, the breast contains only small traces of mammary gland. During pregnancy, the ducts grow and branch, and secretory acini develop at the ends of the smallest branches.

Developmental and Clinical Perspectives (p. 751)

1. Many organs of the male and female reproductive systems develop from the same embryonic organs. Organs with the same embryonic precursor are said to be *homologous* to each other—for example, the scrotum and labia majora are homologous because both develop from the labioscrotal folds.
2. Both sexes initially exhibit a pair of *gonadal ridges, mesonephric ducts,* and *paramesonephric ducts.*
3. In an XY (male) fetus, the *SRY gene* codes for a protein called *testis-determining factor,* which initiates the development of male genitalia. The gonadal ridge becomes a testis, which secretes testosterone and *müllerian-inhibiting factor.* These hormones cause the paramesonephric (müllerian) duct to degenerate, while the mesonephric duct develops into the male reproductive tract.
4. In a female fetus, lacking a Y chromosome, the gonads become ovaries, the mesonephric ducts degenerate, and the paramesonephric ducts develop into a female reproductive tract.
5. The external genitalia of both sexes begin as a *phallus,* a pair of *urogenital folds,* and a pair of *labioscrotal folds.* By week 12 of prenatal development, the phallus differentiates into the penis of a male or clitoris of a female; the urogenital folds enclose the urethra of the male or become the labia minora of a female; and the labioscrotal folds become the scrotum of a male or labia majora of a female.
6. During male development, a cord called the *gubernaculum* pulls the fetal testes through the inguinal canal into the scrotum; this is called *descent of the testes.* In the female fetus, the ovaries also descend to a point just below the brim of the lesser pelvis.
7. Puberty is initiated by the secretion of GnRH by the hypothalamus, inducing secretion of FSH and LH by the anterior pituitary. The visible changes at puberty result from many hormones including testosterone, estrogens, growth hormone, and others. Puberty is at an end when the individual attains fertility. Adolescence continues until full adult height is attained.
8. The earliest visible sign of male puberty is enlargement of the testes and scrotum; the completion of puberty is marked by the ejaculation of fertile sperm.
9. Female puberty is marked by *thelarche* (breast development), *pubarche* (growth of pubic and axillary hair, also occurring in males), and *menarche* (the onset of menstruation). Regular ovulation and fertility are attained about a year after menarche.
10. At midlife, both sexes go through a period of hormonal and physical change called *climacteric.* This is marked by a decline in testosterone or estrogen secretion, and a rise in the secretion of FSH and LH. In females, climacteric is accompanied by *menopause,* the cessation of ovarian function and fertility.
11. Among the important issues for male reproductive health are prostate, testicular, penile, and breast cancer. Other reproductive disorders are described in table 26.3.
12. The most important diseases of the female reproductive system are uterine and breast cancer. Pregnancy also carries a risk of varied complications described in table 26.4
13. The most common reproductive disorders in both sexes are sexually transmitted diseases. Most of these are caused by viruses (HIV, HCV, HSV, HPV) and bacteria (*Neisseria, Treponema,* and *Chlamydia*).

Testing Your Recall

1. The ductus deferens develops from the _____ of the embryo.
   a. mesonephric duct
   b. paramesonephric duct
   c. phallus
   d. labioscrotal folds
   e. urogenital folds
2. Descent of the testes is achieved by contraction of a cord called
   a. the gubernaculum.
   b. the spermatic cord.
   c. the ductus deferens.
   d. the pampiniform plexus.
   e. the rete testis.
3. Four spermatozoa arise from each
   a. primordial germ cell.
   b. type A spermatogonium.
   c. type B spermatogonium.
   d. secondary spermatocyte.
   e. spermatid.
4. Prior to ejaculation, sperm are stored primarily in
   a. the seminiferous tubules.
   b. the epididymis.
   c. the seminal vesicles.
   d. the bulb of the penis.
   e. the ejaculatory ducts.
5. The principal source of testosterone is (are)
   a. the seminiferous tubules.
   b. the sustentacular cells.
   c. the interstitial cells.
   d. the seminal vesicles.
   e. the prostate gland.
6. The fluid-filled central cavity of a mature ovarian follicle is
   a. the antrum.
   b. the zona pellucida.
   c. the theca folliculi.
   d. the granulosa.
   e. the stigma.
7. The tissue lost in menstruation is
   a. perimetrium.
   b. myometrium.
   c. stratum basalis.
   d. stratum functionalis.
   e. stratum corneum.
8. The male scrotum is homologous to the female
   a. ovaries.
   b. vagina.
   c. labia majora.
   d. vestibular bulbs.
   e. clitoris.
9. The narrowest part of the uterus is
   a. the fundus.
   b. the infundibulum.
   c. the body.
   d. the ampulla.
   e. the cervix.
10. The vesicouterine pouch is a space in the peritoneal cavity between the uterus and
    a. the fornices.
    b. the uterine tube.
    c. the sacrum.
    d. the urinary bladder.
    e. the rectum.
11. Under the influence of androgens, the embryonic _____ duct develops into the male reproductive tract.
12. Spermatozoa obtain energy for locomotion from _____ in the semen.
13. The _____, a network of veins in the spermatic cord, helps keep the testes cooler than the core body temperature.
14. Each egg cell develops in its own fluid-filled space called a/an _____.
15. The mucosa of the uterus is called the _____.
16. Over half of the semen consists of secretions from a pair of glands called the _____.
17. The blood-testis barrier is formed by tight junctions between the _____ cells.
18. The male organ(s) homologous to the female's paraurethral glands is (are) the _____.
19. A yellowish structure called the _____ secretes progesterone during the secretory phase of the menstrual cycle.
20. The funnel-like distal end of the uterine tube is called the _____ and has feathery processes called _____.

Answers in the Appendix

True or False

Determine which five of the following statements are false, and briefly explain why.

1. After ovulation, a follicle begins to move down the uterine tube to the uterus.
2. The uterine tubes develop from the embryonic mesonephric ducts.
3. Thelarche normally precedes menarche in female puberty.
4. The follicle that ovulates is called the primary follicle.
5. Sperm cannot develop at the core body temperature.
6. A high testosterone level makes a fetus develop a male reproductive system, and a high estrogen level makes it develop a female reproductive system.
7. Most ovarian follicles degenerate before a girl reaches puberty.
8. The pampiniform plexus serves to keep the testes warm.
9. Prior to ejaculation, sperm are stored mainly in the epididymis.
10. The thickest layer of the uterine wall is the myometrium.

Answers in the Appendix

Testing Your Comprehension

1. The most common method of male sterilization is vasectomy, in which the ductus (vas) deferens is tied, cut, or both. What is the equivalent method of female sterilization? Why are the difficulty and risks of that procedure greater than for a vasectomy?
2. *Uterus bicornis* (*bicorn* = two horns) is a rare condition in which a woman has two separate uteri, each opening by its own cervix into the vagina. What abnormal event of embryonic development do you think could account for this?
3. Suppose the corpus spongiosum became as engorged with blood as the corpora cavernosa do during erection of the penis. What problem would this create for sexual function? In light of this, why is it beneficial that the corpus spongiosum is not enclosed in a tunica albuginea?
4. What male structure(s) do you think is (are) homologous to the vestibular bulbs of the female? Explain your reasoning.
5. An oocyte lives for only 24 hrs after ovulation if it is not fertilized. The trip down the uterine tube, from infundibulum to uterus, takes about 72 hrs. In light of this, where do you think fertilization normally occurs?

Answers at the Online Learning Center

www.mhhe.com/saladinha1

Visit the Online Learning Center-for-practice tests, answer keys, and other learning aids for this chapter. Enhance your understanding of human anatomy with our interactive art labeling exercises, supplemental photo atlases, web links, puzzles, flashcards, and much more.

APPENDIX

Answers to Chapter Review Questions

Answers are provided here for the Testing Your Recall and True or False questions at the end of each chapter. Answers to Think About It and Testing Your Comprehension questions are in the Instructor's Manual, and answers to Testing Your Comprehension are also available to students at the Online Learning Center, www.mhhe.com/saladinha1.

Chapter 1

Testing Your Recall

1. a
2. b
3. c
4. a
5. e
6. c
7. a
8. d
9. c
10. b
11. dissection
12. Robert Hooke
13. metabolism
14. homeostasis
15. morphology
16. palpation
17. computed tomography
18. organ
19. stereoscopic
20. prehensile, opposable

True or False

(these items are false for the reasons given; all others are true)

3. Auscultation is listening to sounds made by the body.
4. Radiology is concerned with all methods of medical imaging.
6. Every cell contains many organelles.
7. Leeuwenhoek was a textile merchant by trade.
9. Stereoscopic vision was probably an adaptation to the arboreal habitat.

Atlas A

Testing Your Recall

1. d
2. c
3. e
4. d
5. d
6. a
7. a
8. d
9. b
10. e
11. supine
12. parietal
13. mediastinum
14. occipital, nuchal
15. hand, foot
16. meninges
17. retroperitoneal
18. medial
19. inferior
20. cubital, popliteal

True or False

(these items are false for the reasons given; all others are true)

4. The diaphragm is inferior to the lungs.
5. The esophagus is in the ventral body cavity.
6. The liver is in the right hypochondriac and epigastric regions.
9. The peritoneum lines the abdominal cavity and external surfaces of the stomach and intestines.
10. The sigmoid colon is in the lower left quadrant.

Chapter 2

Testing Your Recall

1. e
2. d
3. b
4. b
5. e
6. a
7. d
8. a
9. d
10. b
11. micrometers (μm)
12. receptor
13. gates
14. multipotent
15. scanning electron
16. squamous
17. mitochondrion, nuclear envelope
18. peroxisomes, smooth ER
19. cell-adhesion molecules
20. phagocytosis

True or False

(these items are false for the reasons given; all others are true)

2. The most important quality is resolution.
6. Movement down a gradient does not employ ATP.
7. Osmosis is a case of simple diffusion.
9. Desmosomes provide no channels from one cell to another.
10. The nucleolus is not an organelle.

Chapter 3

Testing Your Recall

1. a
2. b
3. c
4. e
5. d
6. a
7. b
8. e
9. b
10. b
11. apoptosis (programmed cell death)
12. mesothelium
13. lacunae
14. fibers
15. collagen
16. fibrocartilage
17. basement membrane
18. matrix
19. holocrine
20. simple

True or False

(these items are false for the reasons given; all others are true)

2. The noncellular components include ground substance and fibers.
5. The tongue epithelium is nonkeratinized.
6. Macrophages develop from monocytes.
8. Brown fat produces no ATP.
9. In metaplasia, one mature tissue type transforms into another.

Chapter 4

Testing Your Recall

1. b
2. b
3. d
4. c
5. a
6. e
7. c
8. a
9. e
10. d
11. teratogens
12. nondisjunction
13. neural groove or neural tube
14. implantation
15. chorionic villi
16. acrosome
17. uterine tube
18. somites
19. polyspermy
20. embryo

True or False

(these items are false for the reasons given; all others are true)

1. Sperm must undergo capacitation first.
2. Fertilization occurs in the uterine tube.
3. Several sperm must digest a path for the one that fertilizes the egg.
8. Oogenesis produces one large oocyte and small discarded polar bodies.
10. The energy comes from the midpiece mitochondria.

Chapter 5

Testing Your Recall

1. d
2. c
3. d
4. b
5. a
6. e
7. c
8. a
9. a
10. d
11. dermato-, cutane-
12. piloerector (arrector pili)
13. keratin, collagen
14. cyanosis
15. dermal papillae
16. earwax
17. sebaceous glands
18. cuticle
19. dermal papilla
20. second-degree

True or False

(these items are false for the reasons given; all others are true)

3. The dermis is mainly collagen.
4. Vitamin D synthesis begins in the keratinocytes and is completed in the liver and kidneys.
7. The hypodermis is not part of the skin.
8. People of all races have similar densities of melanocytes.
9. The genetic lack of melanin causes albinism.

Chapter 6

Testing Your Recall

1. e
2. a
3. d
4. c
5. b
6. c
7. d
8. e
9. b
10. d
11. hydroxyapatite
12. canaliculi
13. appositional
14. osteons
15. parathyroid
16. articular cartilage
17. osteoblasts
18. osteoporosis
19. metaphysis
20. intramembranous ossification

True or False

(these items are false for the reasons given; all others are true)

3. The most common bone disorder is osteoporosis.
4. The growth zone is the epiphyseal plate.
5. Osteoclasts develop from stem cells related to monocytes.
7. The protein of the bone matrix is collagen.
9. Only red bone marrow is hemopoietic.

Chapter 7

Testing Your Recall

1. b
2. e
3. a
4. d
5. a
6. b
7. a
8. a
9. c
10. c
11. fontanels
12. temporal
13. sutures
14. sphenoid
15. annulus fibrosus
16. dens
17. auricular
18. false, floating
19. costal cartilages
20. xiphoid process

True or False

(these items are false for the reasons given; all others are true)

1. The vertebral bodies are derived from the sclerotomes.
2. Adults have fewer bones than children do.
4. The zygomatic processes of the temporal bone and maxilla also contribute to the arch.
5. The dura mater lies loosely against most of the cranium.
10. Lumbar vertebrae have transverse processes, but not transverse costal facets.

Chapter 8

Testing Your Recall

1. a
2. e
3. c
4. b
5. a
6. d
7. c
8. b
9. e
10. b
11. pollex, hallux
12. scapula
13. 56
14. epicondyles
15. hamate
16. interpubic disc
17. crural
18. styloid
19. trochanters
20. medial longitudinal

True or False

(these items are false for the reasons given; all others are true)

2. Each hand and foot has 14 phalanges.
3. The upper limb is attached at the humeroscapular joint.
5. The arm contains only the humerus but the leg contains the tibia and fibula.
7. The most frequently broken bone is the clavicle.
10. That opening is the pelvic inlet.

Chapter 9

Testing Your Recall

1. c
2. b
3. a
4. e
5. c
6. c
7. c
8. d
9. b
10. b
11. synovial fluid
12. bursa
13. pivot
14. kinesiology
15. gomphosis
16. serrate
17. extension
18. range of motion
19. rheumatologist
20. menisci

True or False

(these items are false for the reasons given; all others are true)

1. Osteoarthritis is much more common than rheumatoid arthritis.
2. A doctor who treats arthritis is a rheumatologist.
3. Synovial joints are diarthroses.
5. This action hyperextends the shoulder; the elbow cannot be hyperextended.
9. Synovial fluid fills the bursae but is secreted by the synovial membrane of the joint capsule.

Chapter 10

Testing Your Recall

1. e
2. c
3. c
4. a
5. a
6. e
7. e
8. c
9. d
10. b
11. synaptic vesicles

12. motor end plate
13. terminal cisternae
14. myosin
15. superficial fascia (hypodermis)
16. myoglobin
17. Z discs
18. sphincter
19. myoblasts
20. peristalsis

True or False

(these items are false for the reasons given; all others are true)

4. The expression *slow oxidative* refers to skeletal muscle fibers.
5. One motor neuron can supply from a few to a thousand muscle fibers.
6. Calcium binds to troponin, not to myosin.
9. The blood vessels become wavy when the muscle shortens.
10. Muscle growth involves an increase in muscle fiber thickness, not in their number.

Chapter 11

Testing Your Recall

1. b
2. c
3. a
4. c
5. e
6. e
7. a
8. a
9. c
10. d
11. erector spinae
12. bulbospongiosus
13. levator palpebrae superioris
14. hypoglossal
15. digastric
16. urogenital triangle
17. linea alba
18. larynx
19. sternocleidomastoid
20. trapezius

True or False

(these items are false for the reasons given; all others are true)

1. The mastoid process is the insertion, not the origin.
5. Normal exhalation is passive and does not employ the internal intercostals.
6. The floor of the mouth is formed by the mylohyoid muscle.
7. The scalenes are deep to the trapezius.
10. Only cranial nerves III, V, VII, XI, and XII innervate head and neck muscles.

Chapter 12

Testing Your Recall

1. c
2. e
3. b
4. d
5. d
6. e
7. b
8. a
9. d
10. a
11. deltoid
12. great toe
13. teres, quadratus
14. hamstring
15. retinacula
16. adductor pollicis
17. quadriceps femoris
18. coracobrachialis
19. gracilis
20. semitendinosus

True or False

(these items are false for the reasons given; all others are true)

1. The plantaris muscle inserts on the foot by a tendon of its own.
5. The interosseous muscles are pennate.
7. The psoas major and rectus femoris are synergists in flexing the hip.
8. Hamstring injuries usually result from rapid extension of the knee, not flexion.
10. These muscles are on opposite sides of the tibia and act as antagonists.

Atlas B

Muscle Self-Test (fig. B.15)

1. f
2. b
3. k
4. p
5. h
6. z
7. o
8. x
9. c
10. a
11. y
12. m
13. n
14. e
15. g
16. v
17. f
18. c
19. y
20. x
21. k
22. d
23. s
24. b
25. a
26. u
27. j
28. i
29. g
30. q

Chapter 13

Testing Your Recall

1. e
2. e
3. a
4. a
5. e
6. d
7. a
8. d
9. a
10. c
11. afferent
12. reverberating
13. anencephaly
14. dendrites
15. oligodendrocytes
16. axosomatic
17. peripheral nervous system
18. neurilemma
19. trigger zone
20. postsynaptic

True or False

(these items are false for the reasons given; all others are true)

4. Sensory (afferent) neurons connect sense organs to the CNS.
6. The myelin sheath is deep to the neurilemma.
7. Nodes of Ranvier also exist in the CNS.
8. The outermost layer of a nerve is the epineurium.
9. Unipolar neurons have an axon and produce action potentials.

Chapter 14

Testing Your Recall

1. e
2. c
3. d
4. d
5. e
6. c
7. c
8. a
9. e
10. b
11. ganglia
12. rami
13. spinocerebellar
14. crossed extension
15. intrafusal fibers
16. phrenic
17. decussation
18. proprioception
19. dorsal root
20. tibial, common fibular

True or False

(these items are false for the reasons given; all others are true)

1. The gracile fasciculus is a sensory (ascending) tract.
4. All spinal nerves are mixed nerves.
5. The dura mater is set off from the vertebral bone by the epidural space.
8. Dermatomes overlap each other by up to 50%.
9. Many somatic reflexes involve the brain.

Chapter 15

Testing Your Recall

1. c
2. d
3. e
4. a
5. e
6. c
7. a
8. d
9. e
10. e
11. corpus callosum
12. ventricles, cerebrospinal
13. arbor vitae
14. hippocampus
15. choroid plexus
16. precentral
17. frontal
18. association cortex
19. categorical
20. Broca area

True or False
(these items are false for the reasons given; all others are true)
1. The longitudinal fissure separates the cerebral hemispheres, not cerebellar hemispheres.
2. Degeneration of the substantia nigra causes Parkinson disease.
5. The choroid plexuses produce only 30% of the CSF.
6. Hearing is a temporal lobe function.
10. The optic nerve carries visual signals, not motor signals.

Chapter 16
Testing Your Recall
1. b
2. c
3. e
4. e
5. a
6. e
7. d
8. d
9. c
10. c
11. adrenergic
12. dual innervation
13. autonomic tone
14. vagus
15. enteric
16. neural crest
17. sympathetic
18. preganglionic, postganglionic
19. baroreceptor
20. vasomotor tone

True or False
(these items are false for the reasons given; all others are true)
1. Normally both divisions are active simultaneously.
3. With biofeedback and other methods, some degree of voluntary control is possible.
4. It inhibits digestion.
6. These reflexes occur even without involvement of the brain, but are less controllable.
7. All parasympathetic fibers are cholinergic.

Chapter 17
Testing Your Recall
1. a
2. a
3. a
4. d
5. b
6. e
7. d
8. c
9. c
10. a
11. fovea centralis
12. ganglion
13. nociceptor
14. otoliths
15. outer hair cells
16. stapes
17. inferior colliculi
18. taste hairs
19. olfactory bulb
20. referred pain

True or False
(these items are false for the reasons given; all others are true)
2. Afferent touch fibers end in the spinal cord and medulla oblongata.
4. Pain signals are blocked after entering the spinal cord, before ascending to the brain.
5. The filiform papillae have no taste buds.
9. The eardrum has sensory fibers of the vagus and trigeminal nerves.
10. The posterior chamber, between the iris and lens, is filled with aqueous humor.

Chapter 18
Testing Your Recall
1. d
2. d
3. a
4. c
5. d
6. e
7. a
8. a
9. b
10. c
11. anterior pituitary
12. supraoptic nucleus
13. acromegaly
14. atrial natriuretic peptide
15. glucocorticoids
16. anterior pituitary
17. adrenal
18. interstitial
19. zona fasciculata
20. enteric hormones

True or False
(these items are false for the reasons given; all others are true)
2. The heart, brain, stomach, and kidneys secrete hormones but are not usually thought of as endocrine glands.
4. The pineal gland and thymus shrink with age.
5. The center of the adrenal gland is the adrenal medulla.
8. The pituitary stalk is not a duct.
9. There are two pairs of parathyroids and one pair of gonads.

Chapter 19
Testing Your Recall
1. b
2. c
3. a
4. b
5. e
6. d
7. d
8. c
9. d
10. b
11. hemopoiesis
12. hematocrit
13. macrophages
14. leukopenia
15. hemophilia
16. hemostasis
17. sickle-cell disease
18. polycythemia
19. vitamin B_{12}
20. erythropoietin

True or False
(these items are false for the reasons given; all others are true)
3. Anemia is a cause, not the result, of low blood oxygen content.
4. Neutrophils are the most actively antibacterial WBCs.
6. Neutrophils are the most abundant WBCs.
9. RBCs live longer than most WBCs.
10. WBC count is elevated in leukemia.

Chapter 20
Testing Your Recall
1. d
2. b
3. a
4. d
5. e
6. e
7. d
8. a
9. b
10. b
11. systole, diastole
12. systemic
13. coronary sulcus
14. Purkinje fibers
15. gap junctions
16. valvular prolapse
17. vagus
18. myocardial infarction
19. endocardium
20. sinus venosus

True or False
(these items are false for the reasons given; all others are true)
1. About 20% of the blood returns to the right atrium by way of thebesian veins.
5. The heart does not require nervous stimulation to beat.
7. It is anastomoses of the arteries, not veins, that serve this purpose.
8. The primordial ventricle develops into the left ventricle only.
9. There are no valves at the entrances to the right atrium.

Chapter 21
Testing Your Recall
1. c
2. b
3. a
4. e
5. b
6. b
7. d
8. b
9. b
10. b
11. fenestrated
12. continuous capillaries
13. endothelium
14. common iliac
15. chemoreceptors
16. transcytosis
17. superior and inferior venae cavae
18. carotid sinuses
19. cerebral arterial circle
20. basilic, cephalic

True or False

(these items are false for the reasons given; all others are true)

2. They receive blood from the celiac trunk.
4. Blood sometimes passes through portal systems (two capillary beds) or anastomoses (bypassing capillaries).
5. It is formed by the union of the two brachiocephalic veins.
7. The tunica media is nourished mainly by capillaries of the vasa vasorum.
9. One or more arteries of the circle are missing in 80% of people.

Chapter 22

Testing Your Recall

1. b
2. c
3. e
4. a
5. c
6. e
7. a
8. e
9. c
10. d
11. pathogen
12. chyle
13. lymphadenitis
14. right lymphatic duct, thoracic duct
15. cisterna chyli
16. natural killer
17. antigen-presenting cells
18. red bone marrow
19. lymphatic nodules
20. autoimmune

True or False

(these items are false for the reasons given; all others are true)

1. B cells are involved in specific immunity only.
4. Helper T cells also play a role in humoral immunity.
7. Both B and T cells populate lymph nodes.
8. Lymphatic nodules are temporary and have no capsules.
9. Tonsillectomy is now a much less common treatment than it used to be.

Chapter 23

Testing Your Recall

1. c
2. c
3. a
4. e
5. b
6. e
7. d
8. a
9. c
10. c
11. glottis
12. bronchial tree
13. alar
14. conchae
15. obstructive
16. dead space
17. thyroid
18. inspiratory center
19. hilum
20. alveolar macrophages

True or False

(these items are false for the reasons given; all others are true)

1. The glottis is the superior opening into the larynx.
3. The space between the parietal and visceral pleura contains only a thin film of pleural fluid.
4. Normal expiration is not produced by muscular contraction.
8. The aortic and carotid sinuses monitor blood pressure.
9. The respiratory system begins as a bud arising from the floor of the pharynx.

Chapter 24

Testing Your Recall

1. a
2. b
3. b
4. d
5. e
6. c
7. a
8. a
9. a
10. c
11. enteric nervous system
12. pyloric sphincter
13. parotid
14. peristalsis
15. vagus
16. parietal
17. hepatic sinusoids
18. cardiac orifice
19. cementum
20. lingual papillae

True or False

(these items are false for the reasons given; all others are true)

1. The pancreas is retroperitoneal but the liver is not.
2. A tooth is composed mostly of dentin.
3. Bile is secreted into bile canaliculi, not the sinusoids.
7. The greater omentum is not attached to the body wall.
9. The muscularis externa of the stomach has three layers.

Chapter 25

Testing Your Recall

1. c
2. d
3. a
4. c
5. e
6. b
7. d
8. b
9. d
10. a
11. urea
12. ureteric bud
13. trigone
14. macula densa
15. podocytes
16. proximal convoluted tubule
17. thin segment
18. detrusor
19. minor calyx
20. arcuate

True or False

(these items are false for the reasons given; all others are true)

1. The ureters open into the floor of the bladder.
2. ADH is secreted by the posterior lobe of the pituitary gland.
4. A substantial amount of fluid passes through the tight junctions.
6. The glomerulus is not located within the capsular space.
9. The kidneys are normally in the abdominal cavity.

Chapter 26

Testing Your Recall

1. a
2. a
3. c
4. b
5. c
6. a
7. d
8. c
9. e
10. d
11. mesonephric
12. fructose
13. pampiniform plexus
14. follicle
15. endometrium
16. seminal vesicles
17. sustentacular
18. prostate gland
19. corpus luteum
20. infundibulum, fimbriae

True or False

(these items are false for the reasons given; all others are true)

1. The follicle does not leave the ovary.
2. The uterine tubes develop from paramesonephric ducts.
4. The follicle that ovulates is the mature (graafian) follicle.
6. The female reproductive system develops as a result of a low androgen level.
8. The pampiniform plexus serves to keep the testes cool.

GLOSSARY

This glossary defines terms likely to be most useful to the reader of this particular book, especially terms that are reintroduced most often and cannot be defined again at every introduction. Terms are defined only in the sense that they are used in this book. Some have broader meanings, even within biology and medicine, that are beyond its scope. Pronunciation guides are provided for words whose pronunciations may not be obvious. These guides should be quite intuitive, but a key at the end of the glossary indicates how to pronounce letter sequences within the guides if help is needed. Figures are cited where they will help convey the meaning of a term. Figure references such as A.3 and B.7 refer to figures in atlas A (p. 23) and atlas B (p. 349).

A

abdominal cavity The body cavity between the diaphragm and pelvic brim. (fig. A.7)

abdominopelvic cavity Collective name for the abdominal and pelvic cavities, which constitute a continuous space between the diaphragm and pelvic floor. (fig. A.7)

abduction Movement of a body part away from the median plane, as in raising the right arm away from the body to point to the right. (fig. 9.8)

accessory organ A smaller organ associated with or embedded in another and performing a related function; for example, the hair, nails, and sweat glands are accessory organs of the skin.

acetylcholine (ACh) (ASS-eh-till-CO-leen) A neurotransmitter released by somatic motor nerve fibers, parasympathetic nerve fibers, and some other neurons, composed of choline and an acetyl group.

acetylcholinesterase (AChE) (ASS-eh-till-CO-lin-ESS-ter-ase) An enzyme found in synaptic clefts and on postsynaptic cells that breaks down acetylcholine and stops synaptic signal transmission.

acinar gland (ah-SEE-nur) A gland in which the secretory cells form a dilated sac or acinus. (fig. 3.29)

acinus (ASS-in-nus) A sac of secretory cells at the inner end of a gland duct. (fig. 3.28)

acromial region The apex of the shoulder.

actin A filamentous intracellular protein that provides cytoskeletal support and interacts with other proteins, especially myosin, to cause cellular movement; important in muscle contraction and membrane actions such as phagocytosis, ameboid movement, and cytokinesis. *See also* microfilament.

action The movement produced by the contraction of a muscle.

action potential A rapid voltage change in which a plasma membrane briefly reverses electrical polarity; has a self-propagating effect that produces a traveling wave of excitation in nerve and muscle cells.

acute Pertaining to a disorder with a sudden onset, severe effects, and brief duration. *Compare* chronic.

adaptation **1.** An evolutionary process leading to the establishment of species characteristics that favor survival and reproduction. **2.** Any characteristic of anatomy, physiology, or behavior that promotes survival and reproduction. **3.** A sensory process in which a receptor adjusts its sensitivity or response to the prevailing level of stimulation, as in *dark adaptation* of the eye.

adduction (ah-DUC-shun) Movement of a body part toward the median plane, such as bringing the feet together from a spread-legged position. (fig. 9.8)

adenohypophysis The anterior two-thirds of the pituitary gland, consisting of the anterior lobe and pars tuberalis; synthesizes and secretes gonadotropins, thyrotropin, adrenocorticotropin, growth hormone, and prolactin. (fig. 18.4)

adenosine triphosphate (ATP) A molecule composed of adenine, ribose, and three phosphate groups that functions as a universal energy-transfer molecule; briefly captures energy in its phosphate bonds and transfers it to other chemical reactions, yielding adenosine diphosphate and a free phosphate group upon hydrolysis.

adipocyte A fat cell.

adipose tissue A loose connective tissue composed predominantly of adipocytes; fat. (fig. 3.16)

adrenal gland (ah-DREE-nul) An endocrine gland on the superior pole of each kidney; consists of an outer adrenal cortex and inner adrenal medulla, with separate functions and embryonic origins. (fig. 18.10)

adult stem cell Any of several kinds of undifferentiated cells that populate the body's organs, where they multiply and differentiate to replace cells that are lost to damage or normal cellular turnover. Adult stem cells have more limited developmental potential than embryonic stem cells. *See also* embryonic stem cell.

adventitia (AD-ven-TISH-uh) Loose connective tissue forming the outermost sheath around organs such as a blood vessel or the esophagus.

afferent Carrying toward, as in *afferent neurons,* which carry signals toward the central nervous system, and *afferent arterioles,* which carry blood toward a tissue.

afferent neuron *See* sensory neuron.

aging Any changes in the body that occur with the passage of time, including growth, development, and senescence. *See also* senescence.

agonist *See* prime mover.

agranulocyte Either of the two leukocyte types (lymphocytes and monocytes) that lack prominent cytoplasmic granules. (fig. 19.1)

alveolus (AL-vee-OH-lus) **1.** A microscopic air sac of the lung. **2.** A gland acinus. **3.** A pit or socket in a bone, such as a tooth socket. **4.** Any small anatomical space.

Alzheimer disease (AD) (ALTS-hy-mur) A degenerative disease of the senescent brain, typically beginning with memory lapses and progressing to severe losses of mental and motor functions and ultimately death.

ameboid movement Ameba-like crawling of cells such as leukocytes by means of pseudopods.

amnion A transparent membrane that surrounds the developing fetus and contains the amniotic fluid; the "bag of waters" that breaks during labor. (fig. 4.15)

ampulla (AM-pyu-luh) A wide or saclike portion of a tubular organ such as a semicircular duct or uterine tube.

anastomosis (ah-NASS-tih-MO-sis) An anatomical convergence, the opposite of a branch; a point where two blood vessels merge and combine their bloodstreams or where two nerves or ducts converge. (fig. 21.12)

anatomical position A reference posture on which certain standardized anatomical terminology is based. A subject in anatomical position is standing with the feet flat on the floor, arms down to the sides, and the palms and eyes directed forward. (fig. A.1)

anatomy **1.** Structure of the body. **2.** The study of structure. *See also* morphology.

anemia A deficiency of erythrocytes or hemoglobin.

aneurysm (AN-you-riz-um) A weak, bulging point in the wall of a heart chamber or blood vessel that presents a threat of hemorrhage. (fig. 21.3)

angiogenesis The growth of new blood vessels, both prenatally and postnatally.

angiography The process of visualizing blood vessels by injecting them with a radiopaque substance and photographing them with X rays. (fig. 1.2)

antagonist **1.** A muscle that opposes the prime mover at a joint. **2.** Any agent, such as a hormone or drug, that opposes another.

antebrachium (AN-teh-BRAY-kee-um) The region from elbow to wrist; the forearm.

anterior Pertaining to the front (facial-abdominal aspect) of the body; ventral.

antibody A protein of the gamma globulin class that reacts with an antigen; found in the blood plasma, in other body fluids, and on the surfaces of certain leukocytes and their derivatives.

antigen (AN-tih-jen) Any large molecule capable of binding to an antibody and triggering an immune response.

antigen-presenting cell (APC) A cell that phagocytizes an antigen and displays fragments of it on its surface for recognition by other cells of the immune system; chiefly macrophages and B lymphocytes.

antrum A saccular or pouchlike space, such as at the inferior end of the stomach or in an ovarian follicle.

aorta A large artery that extends from the left ventricle to the lower abdominal cavity and gives rise to all other arteries of the systemic circulation. (fig. 21.15)

aortic arch **1.** In the embryo, any of six pairs of blood vessels that arise rostral to the heart and loop mainly through the pharyngeal arches; some of these later give rise to carotid and pulmonary arteries and the permanent aortic arch. (fig. 21.36) **2.** A segment of the adult aorta that arches over the heart like an inverted U; gives rise to the brachiocephalic trunk, left common carotid artery, and left subclavian artery; then continues dorsal to the heart as the descending aorta. (fig. 21.15)

apex The summit or a pointed part of an organ or body region such as the heart, lung, or shoulder.

apical surface The uppermost surface of an epithelial cell, opposite from the basement membrane, usually exposed to the lumen of an organ. (fig. 2.5)

apocrine Pertaining to certain sweat glands with large lumens and relatively thick, aromatic secretions, and to similar glands such as the mammary gland; formerly thought to form secretions by pinching off bits of apical cytoplasm. (fig. 5.10)

aponeurosis A broad, flat tendon that attaches a muscle to a bone or to other soft tissues in such locations as the abdominal wall and deep to the scalp.

apoptosis The normal death of cells that have completed their function, usually in a process involving self-destruction of the cell's DNA, shrinkage of the cell, and its phagocytosis by a macrophage; also called *programmed cell death. Compare* necrosis.

appendicular (AP-en-DIC-you-lur) Pertaining to the limbs and their supporting skeletal girdles. (fig. 7.1)

arcuate (AR-cue-et) Making a sharp L- or U-shaped bend (arc), or forming an arch, as in *arcuate arteries* of the kidneys and uterus.

areolar tissue (AIR-ee-OH-lur) A fibrous connective tissue with loosely organized, widely spaced fibers and cells and an abundance of fluid-filled space; found under nearly every epithelium, among other places. (fig. 3.14)

arrector pili *See* piloerector.

arteriole A small artery that empties into a metarteriole or capillary.

artery Any blood vessel that conducts blood away from the heart, or in the case of coronary arteries, away from the aorta and into the heart wall.

articular cartilage A thin layer of hyaline cartilage covering the articular surface of a bone at a synovial joint, serving to reduce friction and ease joint movement. (fig. 9.4)

articulation A skeletal joint; any point at which two bones meet; may or may not be movable.

aspect A particular view of the body or one of its structures, or a surface that faces in a particular direction, such as the anterior aspect.

association area A region of the cerebral cortex that does not directly receive sensory input or control skeletal muscles, but serves to interpret sensory information, to plan motor responses, and for memory and cognition.

atherosclerosis A degenerative disease of the blood vessels characterized by the presence of plaques on the vessel wall composed of lipid, smooth muscle, and macrophages; can lead to arterial occlusion, loss of arterial elasticity, hypertension, heart attack, kidney failure, and stroke.

ATP *See* adenosine triphosphate.

atrioventricular (AV) node (AY-tree-oh-ven-TRIC-you-lur) A group of autorhythmic cells in the interatrial septum of the heart that relays excitation from the atria to the ventricles.

atrioventricular (AV) valves The bicuspid (right) and tricuspid (left) valves between the atria and ventricles of the heart.

atrium **1.** Either of the two superior chambers of the heart, which receive systemic and pulmonary blood. **2.** The central space of an alveolar sac into which individual pulmonary alveoli open.

atrophy Shrinkage of a tissue due to age, disuse, or disease.

auditory ossicles Three small middle-ear bones that transfer vibrations from the tympanic membrane to the inner ear; the malleus, incus, and stapes.

Auerbach plexus *See* myenteric plexus.

auricle **1.** The portion of the ear external to the cranium; the pinna. **2.** An ear-shaped structure, such as the auricles of the heart.

autoantibody An antibody that fails to distinguish the body's own molecules from foreign molecules and thus attacks host tissues, causing autoimmune diseases.

autoimmune disease Any disease in which antibodies fail to distinguish between foreign and self-antigens and attack the body's own tissues; for example, systemic lupus erythematosus and rheumatic fever.

autolysis (aw-TAHL-ih-sis) Digestion of cells by their own internal enzymes.

autonomic nervous system (ANS) A motor division of the nervous system that innervates glands, smooth muscle, and cardiac muscle; consists of sympathetic and parasympathetic divisions and functions largely without voluntary control. *Compare* somatic nervous system.

autosome Any chromosome except the sex chromosomes. Genes on the autosomes are inherited without regard to the sex of the individual.

axial Pertaining to the head, neck, and trunk; the part of the body excluding the appendicular portion. (fig. 7.1)

axillary (ACK-sih-LERR-ee) Pertaining to the armpit.

axon A process of a neuron that conducts action potentials away from the soma; also called a *nerve fiber.* There is only one axon to a neuron, and it is usually much longer and much less branched than the dendrites. (fig. 13.4)

axoneme The core of a cilium or flagellum, usually composed of a "9 + 2" array of microtubules that provide support and motility. (fig. 2.13)

B

baroreceptor (BARE-oh-re-SEP-tur) Pressure sensor located in the heart, aortic arch, and carotid sinuses that triggers autonomic reflexes in response to fluctuations in blood pressure.

basal lamina A thin layer of collagen, proteoglycan, and glycoprotein that binds epithelial and other cells to adjacent connective tissue; forms part of the basement membrane of an epithelium, and surrounds some nonepithelial cells such as muscle fibers and Schwann cells. (fig. 13.9)

basal nuclei Masses of deep cerebral gray matter that play a role in the coordination of posture and movement and the performance of learned motor skills; also called *basal ganglia.* (fig. 15.16)

base The broadest part of a tapered organ such as the uterus, or the inferior aspect of an organ such as the brain.

basement membrane A thin layer of matter that underlies the deepest cells of an epithelium and binds them to the underlying connective tissue; consists of the basal lamina of the epithelial cells and fine reticular fibers of the connective tissue. (fig. 3.30)

basophil (BASE-oh-fill) A leukocyte with coarse cytoplasmic granules that produces heparin, histamine, and other chemicals involved in inflammation. (fig. 19.1)

belly The thick part of a skeletal muscle between its origin and insertion. (fig. 10.4)

bipedalism The habit of walking on two legs; a defining characteristic of the family Hominidae that underlies many skeletal and other characteristics of humans.

blastocyst A hollow spheroidal stage of the conceptus that implants in the uterine wall; consists of an inner cell mass, or embryoblast, enclosed in a saclike outer cell mass, or trophoblast. (fig. 4.3)

blood A liquid connective tissue composed of plasma, erythrocytes, platelets, and five kinds of leukocytes.

blood-brain barrier A barrier between the bloodstream and nervous tissue of the brain that is impermeable to many blood solutes and thus prevents them from affecting the brain tissue; formed by the tight junctions between capillary endothelial cells, the basement membrane of the endothelium, and the perivascular feet of astrocytes.

B lymphocyte A lymphocyte that functions as an antigen-presenting cell and, in humoral immunity, differentiates into an antibody-producing plasma cell; also called a *B cell.*

body **1.** The entire organism. **2.** Part of a cell, such as a neuron, containing the nucleus and most other organelles. **3.** The largest or principal part of an organ such as the stomach or uterus; also called the *corpus.*

bolus A mass of matter, especially food or feces traveling through the digestive tract.

bone **1.** A calcified connective tissue; also called *osseous tissue.* **2.** An organ of the skeleton composed of osseous tissue, fibrous connective tissue, marrow, cartilage, and other tissues.

Bowman capsule *See* glomerular capsule.

brachial (BRAY-kee-ul) Pertaining to the brachium.

brachium The region between the shoulder and elbow; the arm proper.

brainstem The stalklike lower portion of the brain, composed of all of the brain except the cerebrum and cerebellum. (Many authorities exclude the diencephalon and regard only the medulla oblongata, pons, and midbrain as the brainstem.) (fig. 15.7)

bronchiole (BRON-kee-ole) A pulmonary air passage that is usually 1 mm or less in diameter and lacks cartilage, but has relatively abundant smooth muscle, elastic tissue, and a simple cuboidal, usually ciliated epithelium.
bronchus (BRON-kus) A relatively large pulmonary air passage with supportive cartilage in the wall; any passage beginning with the primary bronchus at the fork in the trachea and ending with tertiary bronchi, from which air continues into the bronchioles.
brush border A fringe of microvilli on the apical surface of an epithelial cell, serving to enhance surface area and promote absorption. (fig. 2.12)
buccal Pertaining to the cheek.
bulb A dilated terminal part of an organ such as the penis or hair, or the olfactory bulb at the beginning of the olfactory tract.
bursa A sac filled with synovial fluid at a synovial joint, serving to facilitate muscle or joint action. (fig. 9.15)

C

calcaneal tendon (cal-CAY-nee-ul) A thick tendon at the heel that attaches the triceps surae muscles to the calcaneus; also called the *Achilles tendon.* (fig. 12.16)
calcification The hardening of a tissue due to the deposition of calcium salts; also called *mineralization.*
calculus A calcified mass, especially a renal calculus (kidney stone) or biliary calculus (gallstone).
calvaria (cal-VERR-ee-uh) The rounded bony dome that forms the roof of the cranium; the general portion of the skull superior to the eyes and ears; skullcap.
calyx (CAY-lix) (plural, *calices*) A cuplike structure, as in the kidneys. (fig. 25.3)
canal A tubular passage or tunnel such as the auditory, semicircular, or condylar canal.
canaliculus (CAN-uh-LIC-you-lus) A microscopic canal, as in osseous tissue. (fig. 6.4)
cancellous bone *See* spongy bone.
capillary (CAP-ih-LERR-ee) The narrowest type of vessel in the cardiovascular and lymphatic systems; engages in fluid exchanges with surrounding tissues.
capillary bed A network of blood capillaries that arise from a single metarteriole and converge on a thoroughfare channel or venule. (fig. 21.9)
capsule The fibrous covering of a structure such as the spleen or a synovial joint.
carbohydrate A hydrophilic organic compound composed of carbon and a 2:1 ratio of hydrogen to oxygen; includes sugars, starches, glycogen, and cellulose.
cardiac center A nucleus in the medulla oblongata that regulates autonomic reflexes for controlling the rate and strength of the heartbeat.
cardiac muscle Striated involuntary muscle of the heart. (fig. 20.14)
cardiopulmonary system Collective name for the heart and lungs, emphasizing their close spatial and physiological relationship.
cardiovascular system An organ system consisting of the heart and blood vessels, serving for the transport of blood. *Compare* circulatory system.
carotid body (ca-ROT-id) A small cellular mass near the branch in the common carotid artery, containing sensory cells that detect changes in the pH and the carbon dioxide and oxygen content of the blood. (fig. 21.4)
carotid sinus A dilation at the base of the internal carotid artery; contains baroreceptors, which monitor changes in blood pressure. (fig. 21.4)
carpal Pertaining to the wrist (carpus).
carrier **1.** A protein that transports solutes through a cell membrane. **2.** A person who does not exhibit a particular hereditary disorder, but who has the gene for it and may pass it to the next generation.
cartilage A connective tissue with a rubbery matrix, cells (chondrocytes) contained in lacunae, and no blood vessels; covers the joint surfaces of many bones and supports organs such as the ear and larynx.
caudal (CAW-dul) **1.** Pertaining to a tail or narrow tail-like part of an organ. **2.** Pertaining to the inferior part of the trunk of the body, where the tail of other animals arises. *Compare* cranial. **3.** Relatively distant from the forehead, especially in reference to structures of the brain and spinal cord; for example, the medulla oblongata is caudal to the pons. *Compare* rostral.
celiac Pertaining to the abdomen.
celiac trunk An arterial trunk that arises from the abdominal aorta near the diaphragm, and quickly branches to give off arteries that supply the stomach, spleen, pancreas, liver, and other viscera of the upper abdominal cavity. (fig. 21.21)
cell The smallest subdivision of a tissue considered to be alive; consists of a plasma membrane enclosing cytoplasm and, in most cases, a nucleus.
cell body The main part of a cell, especially a neuron, where the nucleus is located; also called the *soma.*
central Located relatively close to the median axis of the body, as in *central nervous system;* opposite of *peripheral.*
central canal **1.** A canal that passes through the core of an osteon in bone, and contains blood vessels and nerves; also called a *haversian canal* or *osteonic canal.* **2.** A canal that passes through the center of the spinal cord, containing cerebrospinal fluid.
central nervous system (CNS) The brain and spinal cord. *Compare* peripheral nervous system.
central pattern generator A nucleus of neurons in the CNS that generates a repetitive motor output, producing rhythmic muscle contractions for such purposes as walking and breathing.
centriole (SEN-tree-ole) An organelle composed of a short cylinder of nine triplets of microtubules, usually paired with another centriole perpendicular to it; origin of the mitotic spindle; identical to the basal body of a cilium or flagellum. (fig. 2.17)
cephalic (seh-FAL-ic) Pertaining to the head.
cerebellum (SERR-eh-BEL-um) A large portion of the brain dorsal to the brainstem and inferior to the cerebrum, responsible for equilibrium, motor coordination, some timekeeping functions, and learning of motor skills. (fig. 15.8)
cerebrospinal fluid (CSF) (SERR-eh-bro-SPY-nul, seh-REE-bro-SPY-nul) A liquid that fills the ventricles of the brain, the central canal of the spinal cord, and the space between the CNS and dura mater.
cerebrum (SERR-eh-brum, seh-REE-brum) The largest and most superior part of the brain, divided into two convoluted cerebral hemispheres separated by a deep longitudinal fissure.
cervical (SUR-vih-cul) Pertaining to the neck or any cervix.
cervix (SUR-vix) **1.** The neck. **2.** A narrow or necklike part of an organ such as the uterus and gallbladder. (fig. 26.11)
chemoreceptor An organ or cell specialized to detect chemicals, as in the carotid bodies and taste buds.
chief cells The majority type of cell in an organ or tissue such as the parathyroid glands or gastric glands.
choana (co-AN-ah) Opening of a nasal fossa into the pharynx; also called a *posterior naris.* (fig. 7.5)
chondrocyte (CON-dro-site) A cartilage cell; a former chondroblast that has become enclosed in a lacuna in the cartilage matrix. (fig. 6.8)
chordate Any animal, including humans, that has a notochord, pharyngeal pouches, a dorsal hollow nerve cord, and a tail extending beyond the anus, in at least some stage of prenatal development or postnatal life; any member of the phylum Chordata.
chorion (CO-ree-on) A fetal membrane external to the amnion; forms part of the placenta and has diverse functions including fetal nutrition, waste removal, and hormone secretion. (fig. 4.11)
chromatid (CRO-muh-tid) One of two genetically identical rodlike bodies of a metaphase chromosome, joined to its sister chromatid at the centromere. (fig. 2.21)
chromatin (CRO-muh-tin) Filamentous material in the interphase nucleus, composed of DNA and protein; all of the chromosomes collectively.
chromosome A strand of DNA and protein carrying the genetic material of a cell's nucleus, having a fine filamentous structure during interphase and a condensed rodlike structure during mitosis and meiosis. Normally there are 46 chromosomes in the nucleus of each cell except germ cells. (fig. 2.21)
chronic Pertaining to a disorder with a gradual onset, slow progression, and long duration. *Compare* acute.
chyme (kime) A slurry of partially digested food in the stomach and small intestine.
cilium (SIL-ee-um) A hairlike process, with an axoneme, projecting from the surface of many or most cells; usually immobile, solitary, and serving a sensory or unknown role; found in large numbers on the apical surfaces of some epithelial cells (as in the respiratory tract and uterine tube), where they are motile and serve to propel matter across the surface of the epithelium. (fig. 2.13)
circulatory system An organ system consisting of the heart, blood vessels, and blood. *Compare* cardiovascular system.
circumduction A joint movement in which one end of an appendage remains relatively stationary and the other end moves in a circle. (fig. 9.11)
cisterna (sis-TUR-nuh) A fluid-filled space or sac, such as the cisterna chyli of the lymphatic system and the cisternae of the endoplasmic reticulum and Golgi complex. (fig. 2.17)
coelom A body cavity bounded on all sides by mesoderm and lined with peritoneum. The embryonic coelom becomes the thoracic and abdominopelvic cavities.
collagen (COLL-uh-jen) The most abundant protein in the body, forming the fibers of many connective tissues in places such as the dermis, tendons, and bones.
colloid A chemical mixture of particles ranging 1 to 100 nanometers in size, dispersed through another medium (usually water); such particles are too large to pass through most selectively permeable membranes, and render colloids cloudy in appearance, but they are small enough to remain uniformly mixed in the medium and to be invisible to the light microscope. Examples

include the proteins of blood plasma and tissue gel, the contents of thyroid follicles, and the matrix of cartilage.

colony-forming unit (CFU) A bone marrow cell that differentiates from a pluripotent stem cell and gives rise to precursor cells, which in turn produce a specific class of formed elements. (fig. 19.9)

commissure (COM-ih-shur) **1.** A bundle of nerve fibers that crosses from one side of the brain or spinal cord to the other. **2.** A corner or angle at which the eyelids, lips, or genital labia meet; in the eye, also called the *canthus.* (fig. 17.19)

compact bone A form of osseous tissue found on bone surfaces and composed predominantly of osteons, with the tissue completely filled with mineralized matrix (other than lacunae and central canals) and leaving no room for bone marrow; also called *dense bone.* (fig. 6.4) *Compare* spongy bone.

computed tomography (CT) A method of medical imaging that uses X rays and a computer to create an image of a thin section of the body; the image is called a *CT scan.* (fig. 1.4)

conception The fertilization of an egg, producing a zygote; the beginning of pregnancy.

conceptus All products of conception, ranging from a fertilized egg to the full-term fetus with its fetal membranes, placenta, and umbilical cord. *Compare* embryo; fetus; preembryo.

condyle (CON-dile) An articular surface on a bone, usually in the form of a knob (as on the mandible), but relatively flat on the proximal end of the tibia. (fig. 9.19)

congenital Present at birth; for example, an anatomical defect, a syphilis infection, or a hereditary disease.

congenital anomaly The abnormal structure or position of an organ at birth, resulting from a defect in prenatal development; a birth defect.

connective tissue A tissue usually composed of more extracellular than cellular volume and usually with a substantial amount of extracellular fiber; forms supportive frameworks and capsules for organs, binds structures together, holds them in place, stores energy (as in adipose tissue), or transports materials (as in blood).

contralateral On opposite sides of the body, as in reflex arcs where the stimulus comes from one side of the body and a response is given by muscles on the other side. *Compare* ipsilateral.

convergent Coming together, as in a *convergent muscle* and a converging neuronal circuit.

cornified Having a heavy surface deposit of keratin, as in the stratum corneum of the epidermis.

corona A halo- or crownlike structure, such as the corona radiata or coronal suture of the skull.

coronal plane *See* frontal plane.

corona radiata **1.** An array of nerve tracts in the brain that arise mainly from the thalamus and fan out to different regions of the cerebral cortex. **2.** The first layer of cuboidal cells immediately external to the zona pellucida around an egg cell.

coronary **1.** Crownlike; encircling. **2.** Pertaining to the heart.

coronary artery Either of two branching arteries that arise from the aorta near the heart and supply blood to the heart wall.

coronary circulation A system of blood vessels that serve the wall of the heart. (fig. 20.11)

corpus **1.** A body of tissue, such as the corpus cavernosum of the penis. **2.** The principal part (body) of an organ such as the uterus or stomach, as opposed to smaller regions of an organ such as its head, tail, fundus, or cervix.

corpus callosum (COR-pus ca-LO-sum) A prominent C-shaped band of nerve tracts that connect the right and left cerebral hemispheres to each other, seen superior to the third ventricle in a median section of the brain. (fig. 15.2)

corpus luteum A yellowish cellular mass that forms in the ovary from a follicle that has ovulated; secretes progesterone, hormonally regulates the second half of the menstrual cycle, and is essential to sustaining the first 7 weeks of pregnancy.

cortex (plural, *cortices*) The outer layer of some organs such as the adrenal gland, cerebrum, lymph node, and ovary; usually covers or encloses tissue called the medulla.

corticospinal tract A bundle of nerve fibers that descend through the brainstem and spinal cord and carry motor signals from the cerebral cortex to the neurons that innervate the skeletal muscles of the limbs; important in the fine control of limb movements. (fig. 14.5)

costal (COSS-tul) Pertaining to the ribs.

costal cartilage A bladelike plate of hyaline cartilage that attaches the distal end of a rib to the sternum; collectively the costal cartilages constitute much of the anterior part of the thoracic cage.

coxal Pertaining to the hip.

cranial **1.** Pertaining to the cranium. **2.** In a position relatively close to the head or a direction toward the head. *Compare* caudal.

cranial nerve Any of 12 pairs of nerves connected to the base of the brain and passing through foramina of the cranium.

cranium That portion of the skull that encloses the cranial cavity and protects the brain; also called the *braincase.* Comprises the frontal, parietal, temporal, occipital, sphenoid, and ethmoid bones.

crest A narrow ridge, such as the neural crest or the crest of the ilium.

cricoid cartilage The most inferior cartilage of the larynx, connecting the larynx to the trachea.

crista A crestlike structure, such as the crista galli of the ethmoid bone, crista ampullaris of the inner ear, or the crista of a mitochondrion.

cross section (c.s.) A cut perpendicular to the long axis of the body or of an organ. (fig. 3.2)

crural (CROO-rul) Pertaining to the leg proper or to the crus of a organ. *See also* crus.

crus (cruss) (plural, *crura*) **1.** The region from the knee to the ankle; the leg proper. **2.** A leglike extension of an organ such as the penis and clitoris. (fig. 26.2) *See also* crural.

CT scan An image of the body made by computed tomography. (fig. 1.4)

cubital region The anterior region at the bend of the elbow.

cuboidal (cue-BOY-dul) A shape that is roughly like a cube or in which the height and width are about equal.

cuneiform (cue-NEE-ih-form) Wedge-shaped, as in *cuneiform cartilages* and *cuneiform bone.*

cusp **1.** One of the flaps of a valve of the heart, veins, and lymphatic vessels. **2.** A conical projection on the occlusal surface of a premolar or molar tooth.

cutaneous Pertaining to the skin; integumentary.

cuticle **1.** The outermost layer of a hair, consisting of a single layer of overlapping squamous cells. **2.** A layer of dead epidermal cells that cover the proximal end of a nail; also called the *eponychium.*

cytokinesis (SY-toe-kih-NEE-sis) Division of the cytoplasm of a cell into two cells following nuclear division.

cytology The study of cell structure and function.

cytoplasm The contents of a cell between the plasma membrane and the nuclear envelope, consisting of cytosol, organelles, inclusions, and the cytoskeleton.

cytoskeleton A system of protein microfilaments, intermediate filaments, and microtubules in a cell, serving for physical support, cellular movement, and the routing of molecules and organelles to their destinations within the cell. (fig. 2.15)

cytosol A clear, featureless, gelatinous colloid in which the organelles and other internal structures of a cell are embedded.

cytotoxic T cell A T lymphocyte that directly attacks and destroys infected body cells, cancerous cells, and the cells of transplanted tissues.

D

darwinian medicine *See* evolutionary medicine.

daughter cells Cells that arise from a parent cell by mitosis or meiosis.

decussation (DEE-cuh-SAY-shun) The crossing of nerve fibers from the right side of the central nervous system to the left or vice versa, especially in the spinal cord, medulla oblongata, and optic chiasm. *Compare* hemidecussation.

deep Relatively far from the body surface; opposite of *superficial.* For example, the bones are deep to the skeletal muscles.

dendrite Process of a neuron that receives information from other cells or from environmental stimuli and conducts signals to the soma. Dendrites are usually shorter, more branched, and more numerous than the axon and are incapable of producing action potentials. (fig. 13.4)

dendritic cell An antigen-presenting cell of the epidermis, vaginal mucosa, and some other epithelia.

denervation atrophy The shrinkage of skeletal muscle that occurs when the motor neuron dies or is severed from the muscle.

dense bone *See* compact bone.

dense connective tissue A connective tissue with a high density of fiber, relatively little ground substance, and scanty cells; seen in tendons and the dermis, for example. Classified as *regular* if the extracellular fibers are more or less parallel and *irregular* if the fibers travel in highly varied directions. (fig. 3.18)

depression **1.** A sunken place on the surface of a bone. **2.** A joint movement that lowers a body part, as in dropping the shoulders or opening the mouth. (fig. 9.9)

dermal papilla **1.** A bump or ridge of dermis that extends upward to interdigitate with the epidermis, creating a wavy boundary that resists stress and slippage of the epidermis. (fig. 5.4) **2.** A projection of the dermis into the bulb of a hair, supplying blood to the hair. (fig. 5.7)

dermatome **1.** In the embryo, a group of mesodermal cells that arise from a somite and gives rise to the dermis on one side of one segment of the body. *Compare* myotome; sclerotome. **2.** In the adult, a region of skin on the neck, trunk, or limbs that is innervated by one spinal nerve. (fig. 14.18)

dermis The deeper of the two layers of the skin, underlying the epidermis and composed of fibrous connective tissue.

desmosome (DEZ-mo-some) A patchlike intercellular junction that mechanically links two cells together. (fig. 2.14)

desquamation *See* exfoliation.

diaphragm A muscular partition that separates the thoracic cavity from the abdominal cavity and plays a major role in respiration.

diaphysis (dy-AFF-ih-sis) The shaft of a long bone. (fig. 6.2)

diarthrosis *See* synovial joint.

diencephalon (DY-en-SEFF-uh-lon) A portion of the brain between the midbrain and corpus callosum; composed of the thalamus, epithalamus, and hypothalamus. (fig. 15.12)

differentiation Development of a relatively unspecialized cell or tissue into one with a more specific structure and function.

digestive system The organ system specialized for the intake and chemical breakdown of food, absorption of nutrients, and discharge of the indigestible residue.

digit A finger or toe.

digital rays The first ridgelike traces of fingers or toes to appear in the embryonic hand plate and foot plate.

dilation (dy-LAY-shun) Widening of an organ or passageway such as a blood vessel or the pupil of the eye.

diploid (2*n*) Pertaining to a cell or organism with chromosomes in homologous pairs. All nucleated cells of the human body are diploid except for germ cells beyond the meiosis I stage of cell division.

distal Relatively distant from a point of origin or attachment; for example, the wrist is distal to the elbow. *Compare* proximal.

dizygotic (DZ) twins Two individuals who developed simultaneously in one uterus but originated from separate fertilized eggs and therefore are not genetically identical.

dorsal Toward the back (spinal) side of the body.

dorsal root A branch of a spinal nerve that enters the spinal cord on its dorsal side, composed of sensory fibers; also called the *posterior root.* (fig. 14.10)

dorsal root ganglion A swelling in the dorsal root of a spinal nerve, near the spinal cord, containing the somas of the afferent neurons of the nerve. (fig. 14.10)

dorsiflexion (DOR-sih-FLEC-shun) A movement of the ankle that reduces the joint angle and raises the toes. (fig. 9.13)

dorsum The dorsal surface of a body region, especially the back of the hand or top of the foot.

Down syndrome *See* trisomy-21.

duct An epithelium-lined, tubular passageway, such as a semicircular duct or a gland duct.

duodenum (DEW-oh-DEE-num, dew-ODD-eh-num) The first portion of the small intestine extending for about 25 cm from the pyloric valve of the stomach to a sharp bend called the duodenojejunal flexure; receives chyme from the stomach and secretions from the liver and pancreas.

dural sheath An extension of the dura mater into the vertebral canal, loosely enclosing the spinal cord.

dura mater The thickest and most superficial of the three meninges around the brain and spinal cord.

dynein (DINE-een) A motor protein involved in the beating of cilia and flagella and in the movement of molecules and organelles within cells.

dyspnea Labored breathing.

E

ectoderm The outermost of the three primary germ layers of an embryo; gives rise to the nervous system and epidermis.

ectopic (ec-TOP-ic) In an abnormal location; for example, ectopic pregnancy and ectopic pacemakers of the heart.

edema Accumulation of excess tissue fluid, resulting in swelling of a tissue.

effector A molecule, cell, or organ that carries out a response to a stimulus.

efferent (EFF-ur-ent) Carrying away or out, such as a blood vessel that carries blood away from a tissue or a nerve fiber that conducts signals away from the central nervous system.

efferent neuron *See* motor neuron.

elastic cartilage A form of cartilage with an abundance of elastic fibers in its matrix, lending flexibility and resilience to the cartilage; found in the epiglottis and ear pinna. (fig. 3.20)

elastic fiber A connective tissue fiber, composed of the protein elastin, that stretches under tension and returns to its original length when released; responsible for the resilience of organs such as the skin and lungs.

elasticity The tendency of a stretched structure to return to its original dimensions when tension is released.

elastin A fibrous protein with the ability to stretch and recoil, found in the skin, pulmonary airway, arteries, and elastic cartilage, among other locations.

elevation A joint movement that raises a body part, as in hunching the shoulders or closing the mouth. (fig. 9.9)

embryo In humans, a developing individual from the time the ectoderm, mesoderm, and endoderm have all formed at about 16 days, through the end of 8 weeks when all organ systems are represented; preceded by the *preembryo* and followed by the *fetus.* In other animals, any unborn stage of development beginning with the two-celled stage. *Compare* conceptus; fetus; preembryo.

embryogenesis A process of prenatal development that occurs during implantation of the blastocyst and gives rise to the three primary germ layers; embryogenesis ends with the existence of an embryo.

embryology The scientific study of prenatal development, from fertilization to birth.

embryonic disc A flat plate of cells in early embryonic development, composed of initially two and then three cell layers.

embryonic stage The stage of prenatal development from day 16 through the end of week 8. *See also* embryo; fetal stage.

embryonic stem cell An undifferentiated cell from a preembryo of up to 150 cells, capable of developing into any type of embryonic or adult cell.

encapsulated nerve ending Any sensory nerve ending that is surrounded by or associated with specialized connective tissues, which enhance its sensitivity or its mode of responding to stimulation.

endocardium A tissue layer that lines the inside of the heart, composed of a simple squamous epithelium overlying a thin layer of areolar tissue.

endochondral ossification A process of bone development in which the bone is preceded by a model of hyaline cartilage in roughly the shape of the bone to come, and the cartilage is then replaced by osseous tissue. *Compare* intramembranous ossification. (fig. 6.7)

endocrine gland (EN-doe-crin) A ductless gland that secretes hormones into the bloodstream; for example, the thyroid and adrenal glands. *Compare* exocrine gland.

endocrine system A system of internal chemical communication composed of all endocrine glands and the hormone-secreting cells found in other tissues and organs.

endocytosis Any process of vesicular transport of materials from the extracellular material into a cell; includes pinocytosis, receptor-mediated endocytosis, and phagocytosis. (fig. 2.11)

endoderm The innermost of the three primary germ layers of an embryo; gives rise to the mucosae of the digestive and respiratory tracts and to their associated glands.

endogenous (en-DODJ-eh-nus) Originating internally, such as the endogenous cholesterol synthesized in the body in contrast to the exogenous cholesterol coming from the diet. *Compare* exogenous.

endometrium (EN-doe-MEE-tree-um) The mucosa of the uterus; the site of implantation and source of menstrual discharge.

endoplasmic reticulum (ER) (EN-doe-PLAZ-mic reh-TIC-you-lum) An extensive system of interconnected cytoplasmic tubules or channels; classified as *rough ER* or *smooth ER* depending on the presence or absence of ribosomes on its membrane. (fig. 2.17)

endothelium (EN-doe-THEEL-ee-um) A simple squamous epithelium that lines the lumens of the blood vessels, heart, and lymphatic vessels.

enteric (en-TERR-ic) Pertaining to the small intestine, as in *enteric hormones.*

eosinophil (EE-oh-SIN-oh-fill) A leukocyte with a large, often bilobed nucleus and coarse cytoplasmic granules that stain with eosin; phagocytizes antigen-antibody complexes, allergens, and inflammatory chemicals and secretes enzymes that combat parasitic infections. (fig. 19.1)

epiblast The layer of cells in the early embryonic disc facing the amniotic cavity. These cells migrate during gastrulation to replace the hypoblast with endoderm, then to form the mesoderm, after which the remaining surface epiblast cells are called the ectoderm.

epicondyle A bony projection or ridge superior to a condyle, for example at the distal ends of the humerus and femur. (fig. 8.10)

epidermis A stratified squamous epithelium that constitutes the superficial layer of the skin, overlying the dermis. (fig. 5.2)

epigastric Pertaining to a medial region of the abdomen superior to the umbilical region, bordered inferiorly by the subcostal line and laterally by the midclavicular lines. (fig. A.6)

epiglottis A flap of tissue in the pharynx that covers the glottis during swallowing, deflecting swallowed matter away from the airway and into the esophagus.

epiphyseal plate (EP-ih-FIZZ-ee-ul) A plate of hyaline cartilage between the epiphysis and diaphysis of a long bone in a child or adolescent, serving as a growth zone for bone elongation. (fig. 6.7)

epiphysis (eh-PIFF-ih-sis) **1.** The head of a long bone. (fig. 6.2) **2.** The pineal gland (epiphysis cerebri).

epithelium A type of tissue consisting of one or more layers of closely adhering cells with little intercellular material and no blood vessels; forms the coverings and linings of many organs and the parenchyma of the glands.

equilibrium **1.** The sense of balance. **2.** A state in which opposing processes occur at comparable rates and balance each other so that there is little or no net change in the system, such as a chemical equilibrium.

erectile tissue A tissue that functions by swelling with blood, as in the penis, clitoris, and inferior concha of the nasal cavity.

erythrocyte (eh-RITH-ro-site) A red blood cell.

erythropoiesis (eh-RITH-ro-poy-EE-sis) The production of erythrocytes.

eversion A movement of the foot that turns the sole laterally. (fig. 9.13)

evolution A change in the genetic composition of a population over a period of time; the mechanism that produces adaptations in human form and function. *See also* adaptation.

evolutionary medicine A science that examines the evolutionary history of the human species for insights into human structure, function, and especially dysfunction; also called *darwinian medicine.*

excitability The ability of a cell to respond to a stimulus, especially the ability of nerve and muscle cells to produce membrane voltage changes in response to stimuli; irritability.

excretion The process of eliminating metabolic waste products from a cell or from the body. *Compare* secretion.

excursion A side-to-side movement of the mandible, as in chewing. (fig. 9.10)

exfoliation The shedding of squamous cells from the surface of a stratified squamous epithelium. Also called *desquamation.* Sampling and examination of these cells, such as a Pap smear, is called *exfoliate cytology.* (fig. 3.8)

exocrine gland (EC-so-crin) A gland that secretes its products into another organ or onto the body surface, usually by way of a duct; for example, salivary and gastric glands. *Compare* endocrine gland.

exocytosis A mode of vesicular transport in which a secretory vesicle of a cell fuses with the plasma membrane and releases its contents from the cell; a mode of glandular secretion and discharge of cellular wastes. (fig. 2.11)

exogenous (ec-SODJ-eh-nus) Originating externally, such as exogenous (dietary) cholesterol; extrinsic. *Compare* endogenous.

expiration **1.** Exhaling. **2.** Dying.

extension Movement of a joint that increases the angle between articulating bones (straightens the joint). *Compare* flexion. (fig. 9.7)

external acoustic meatus A canal in the temporal bone that conveys sound waves to the eardrum; also called the *external auditory meatus.*

exteroceptor A sensory receptor that responds to stimuli originating outside the body, such as the eye or ear. *Compare* interoceptor.

extracellular fluid (ECF) Any body fluid that is not contained in the cells; for example, blood, lymph, and tissue fluid.

extrinsic (ec-STRIN-sic) **1.** Originating externally, such as extrinsic blood-clotting factors; exogenous. **2.** Not fully contained within an organ but acting on it, such as the *extrinsic muscles* of the hand and eye. *Compare* intrinsic.

F

facet A smooth articular surface on a bone; may be flat, slightly concave, or slightly convex; for example, the articular facets of the vertebrae.

facilitated diffusion A process of solute transport through a cellular membrane, down its concentration gradient, with the aid of a carrier protein; the carrier does not consume ATP.

fallopian tube *See* uterine tube.

fascia (FASH-ee-uh) A layer of connective tissue between the muscles (deep fascia) or between the muscles and the skin (superficial fascia). (fig. 10.2)

fascicle (FASS-ih-cul) A bundle of muscle or nerve fibers ensheathed in connective tissue; multiple fascicles bound together constitute a muscle or nerve as a whole. Also called a *fasciculus.* (fig. 10.2)

fat **1.** A triglyceride molecule. **2.** Adipose tissue.

female In humans, any individual with no Y chromosome; normally, one possessing two X chromosomes in each somatic cell, and having reproductive organs that serve to produce eggs, receive sperm, provide sites of fertilization and prenatal development, expel the full-term fetus, and nourish the infant.

femoral Pertaining to the femur or the thigh.

femoral region The region between the hip and knee; the thigh.

femoral triangle A triangular region of the groin bounded by the sartorius muscle, adductor longus muscle, and inguinal ligament, and through which the femoral artery, femoral vein, and femoral nerve pass close to the body surface. (fig. 21.25)

fenestrated (FEN-eh-stray-ted) Perforated with holes or slits, as in certain blood capillaries and the elastic sheets of large arteries. (fig. 21.6)

fetal stage The period of prenatal development from the beginning of week 9 until birth; a period in which all organ systems are represented at the outset, and grow and differentiate until capable of supporting life outside the uterus. *See also* embryonic stage; fetus.

fetus In human development, an individual from the beginning of the ninth week when all of the organ systems are present, through the time of birth. *See also* conceptus; embryo.

fiber **1.** In muscular histology, a skeletal muscle cell *(muscle fiber).* **2.** In neurohistology, the axon of a neuron *(nerve fiber).* **3.** Any long threadlike structure, such as a Purkinje fiber of the heart, a collagen or elastic fiber of the connective tissues, or cellulose and other digestion-resistant dietary fiber.

fibrinogen A protein synthesized by the liver and present in blood plasma, semen, and other body fluids; precursor of the sticky protein fibrin, which forms the matrix of a clot.

fibroblast A connective tissue cell that produces collagen fibers and ground substance; the only type of cell in tendons and ligaments.

fibrocartilage A form of cartilage with coarse bundles of collagen fibers in the matrix, found in the intervertebral discs, joint menisci, pubic symphysis, and some tendon-bone junctions. (fig. 3.21)

fibrosis Replacement of damaged tissue with fibrous scar tissue rather than by the original tissue type; scarring. *Compare* regeneration.

fibrous connective tissue Any connective tissue with a preponderance of fiber, such as areolar, reticular, dense regular, and dense irregular connective tissues.

filament A fine threadlike structure such as the myofilaments of muscle and the microfilaments and intermediate filaments of the cytoskeleton.

filtration A process in which a fluid is physically forced through a membrane that allows water and some solutes to pass, and holds back larger particles; especially important in the emission of fluid from blood capillaries.

finger Any of the five digits of the hand, including the thumb.

first-order neuron An afferent (sensory) neuron that carries signals from a receptor to a second-order neuron in the spinal cord or brain. (fig. 14.4) *See also* second-order neuron; third-order neuron.

fissure **1.** A slit through a bone, such as the orbital fissure. **2.** A deep groove, such as the longitudinal fissure between the cerebral hemispheres.

fix **1.** To hold a structure in place, for example by fixator muscles that prevent unwanted joint movements. **2.** To preserve a tissue by means of a fixative.

fixative A chemical that prevents tissue decay, such as formalin or ethanol.

fixator A muscle that minimizes or prevents bone movement in certain joint actions, such as the rhomboideus major holding the scapula stationary while the biceps brachii flexes the elbow.

flagellum (fla-JEL-um) A long, motile, usually single hairlike extension of a cell; the tail of a sperm cell is the only functional flagellum in humans.

flat bone A bone with a platelike shape, such as the parietal bone or sternum.

flexion A joint movement that, in most cases, decreases the angle between two bones. (fig. 9.7) *Compare* extension.

flexor A muscle that flexes a joint.

fMRI *See* functional magnetic resonance imaging.

follicle (FOLL-ih-cul) A small space, such as a hair follicle, thyroid follicle, or ovarian follicle. (fig. 26.15) *See also* lymphatic nodule.

foramen (fo-RAY-men) A hole through a bone or other organ, in most cases providing passage for blood vessels and nerves.

foramen magnum The largest opening into the cranial cavity, at the point where the occipital bone articulates with the vertebral column; allows passage of the spinal cord and vertebral arteries into the cranial cavity.

foramen ovale **1.** An ovoid foramen in the sphenoid bone that allows for passage of the mandibular division of the trigeminal nerve. **2.** An opening in the fetal interatrial septum that allows blood to flow directly from the right atrium into the left atrium and bypass the pulmonary circulation.

forebrain The most rostral part of the brain, consisting of the cerebrum and diencephalon. (fig. 13.5)
foregut **1.** The most rostral part of the embryonic digestive tract; all of the tract rostral to the initial attachment of the yolk sac. (fig. 4.6) **2.** In adults, all of the digestive tract from the oral cavity to the major duodenal papilla, with a blood supply and innervation separate from those of the midgut and hindgut.
formed element An erythrocyte, leukocyte, or platelet; any normal component of blood or lymph that is a cell or cell fragment, as opposed to the extracellular fluid component.
fossa (FOSS-uh) A depression in an organ or tissue, such as the fossa ovalis of the heart or a cranial fossa of the skull.
fovea (FOE-vee-uh) A small pit, such as the fovea capitis of the femur or fovea centralis of the retina.
free nerve ending A bare sensory nerve ending, lacking associated connective tissue or specialized cells; includes receptors for heat, cold, and pain; also called an *unencapsulated nerve ending.*
frontal plane An anatomical plane that passes through the body or an organ from right to left and superior to inferior; also called a *coronal plane.* (fig. A.3)
functional magnetic resonance imaging (fMRI) A variation on MRI that enables the visualization of moment-to-moment changes in the metabolic activity of a tissue, rather than static images; used to study quickly changing patterns of brain activity, among other diagnostic and research purposes.
fundus The base or broadest part of certain organs such as the stomach and uterus.
funiculus (few-NICK-you-lus) Any of the three major divisions of the white matter of the spinal cord, composed of multiple fascicles, or tracts; also called a *column.* The three funiculi on each side of the cord are the dorsal, lateral, and ventral columns.
fusiform (FEW-zih-form) Spindle-shaped; elongated, thick in the middle, and tapered at both ends, such as the shape of a smooth muscle cell or a muscle spindle. (fig. 2.3)

G

gamete (GAM-eet) An egg or sperm cell.
gametogenesis (GAM-eh-toe-JEN-eh-sis) The production of eggs or sperm.
ganglion (GANG-glee-un) A cluster of nerve cell bodies in the peripheral nervous system, often resembling a knot in a string.
gangrene Tissue necrosis resulting from ischemia.
gap junction A junction between two cells consisting of a pore surrounded by a ring of proteins in the plasma membrane of each cell, allowing solutes to diffuse from the cytoplasm of one cell to the next; functions include cell-to-cell nutrient transfer in the developing embryo and electrical communication between cells of cardiac and smooth muscle. (fig. 2.14)
gastric Pertaining to the stomach.
gastrointestinal (GI) system The part of the digestive tract composed of the stomach and intestines.
gate A protein channel in a cellular membrane that can open or close in response to chemical, electrical, or mechanical stimuli, thus controlling when substances are allowed to pass through the membrane.
general senses Senses such as touch, heat, cold, pain, vibration, and pressure, mediated by relatively simple sense organs that are distributed throughout the body. *See also* somesthetic; special senses.
genitalia The pelvic reproductive organs including the *internal genitalia* in the pelvic cavity and *external genitalia* in the perineum; most of the external genitalia are externally visible, but some are subcutaneous, between the skin and the muscles of the pelvic floor.
genitourinary (G-U) system *See* urogenital (U-G) system.
germ cell A gamete or any precursor cell destined to become a gamete.
germ layer Any of three tissue layers of an embryo: the ectoderm, mesoderm, or endoderm.
gestation (jess-TAY-shun) Pregnancy.
gland Any organ specialized for secretion or excretion; in some cases a single cell, such as a goblet cell.
glial cell (GLEE-ul, GLY-ul) Any of the six types of supporting cells of the nervous system (oligodendrocytes, astrocytes, microglia, and ependyma in the CNS; Schwann cells and satellite cells in the PNS); constitute most of the bulk of the nervous system and perform various protective and supportive roles for the neurons. Also called *neuroglia.*
glomerular capsule (glo-MERR-you-lur) A double-walled capsule around each glomerulus of the kidney; receives glomerular filtrate and empties into the proximal convoluted tubule. Also called the *Bowman capsule.* (fig. 25.7)
glomerulus A spheroid mass of blood capillaries in the kidney that filters plasma and produces glomerular filtrate, which is further processed to form the urine. (fig. 25.7)
glucose A monosaccharide ($C_6H_{12}O_6$) also known as blood sugar; glycogen, starch, cellulose, and maltose are made entirely of glucose, and glucose constitutes half of a sucrose or lactose molecule. The isomer involved in human physiology is also called *dextrose.*
gluteal Pertaining to the buttocks.
glycocalyx (GLY-co-CAY-licks) A layer of carbohydrate covalently bonded to the phospholipid and protein molecules of a plasma membrane; forms a surface coat on all human cells. (fig. 2.12)
glycogen A glucose polymer synthesized by liver, muscle, uterine, and vaginal cells that serves as an energy-storage polysaccharide.
glycolipid A phospholipid molecule with carbohydrate covalently bonded to it, found in the plasma membranes of cells.
glycoprotein A protein-carbohydrate complex in which the protein is dominant; found in mucus and the glycocalyx of cells, for example.
glycosaminoglycan (GAG) (GLY-cose-am-ih-no-GLY-can) A polysaccharide composed of modified sugars with amino groups; the major component of a proteoglycan. GAGs are largely responsible for the viscous consistency of tissue gel and the stiffness of cartilage.
goblet cell A mucus-secreting gland cell, shaped somewhat like a wineglass, found in the epithelia of many mucous membranes. (fig. 3.7)
Golgi complex (GOAL-jee) An organelle composed of several parallel cisternae, somewhat like a stack of saucers, that modifies and packages newly synthesized proteins and synthesizes carbohydrates. (fig. 2.17)
Golgi vesicle A membrane-bounded vesicle pinched from the Golgi complex, containing its chemical product; may be retained in the cell as a lysosome or become a secretory vesicle that releases the product by exocytosis.
gonad The ovary or testis.
gonadal ridge The earliest trace of the embryonic gonad, a streak of tissue adjacent to the kidney, populated by the first germ cells arriving from the yolk sac at 5 to 6 weeks of gestation; also called a *genital ridge.*
granulocyte (GRAN-you-lo-site) Any of three types of leukocytes (neutrophils, eosinophils, or basophils) with prominent cytoplasmic granules. (fig. 19.1)
granulosa cells Cells that form a stratified cuboidal epithelium lining an ovarian follicle; source of steroid sex hormones. (fig. 26.15)
gray matter A zone or layer of tissue in the central nervous system where the neuron cell bodies, dendrites, and synapses are found; forms the core of the spinal cord, nuclei of the brainstem, basal nuclei of the cerebrum, cerebral cortex, and cerebellar cortex. (fig. 15.5)
great toe The large medial toe; also called the *hallux.*
great vessels The largest of the blood vessels attached directly to the heart; the superior and inferior venae cavae, pulmonary trunk, and aorta.
gross anatomy Bodily structure that can be observed without magnification.
ground substance The clear, featureless material in which the fibers and cells of a connective tissue are embedded; includes the liquid plasma of the blood, tissue gel of areolar tissue, and calcified tissue of bone.
guard hairs Coarse, stiff hairs that prevent insects, debris, or other foreign matter from entering the ear, nose, or eye; also called *vibrissae.*
gustatory Pertaining to the sense of taste.
gyrus (JY-rus) A wrinkle or fold in the cortex of the cerebrum or cerebellum. (fig. 15.1)

H

hair cell A sensory cell of the cochlea, semicircular ducts, utricle, and saccule, with a fringe of surface microvilli that respond to the relative motion of a gelatinous membrane at their tips; responsible for the senses of hearing and equilibrium. (fig. 17.13)
hair follicle An oblique epithelial pit in the skin that contains a hair and extends into the dermis or hypodermis.
hair receptor Free sensory nerve endings entwined around a hair follicle, responsive to movement of the hair.
hallux The great toe; the medial digit of the foot.
haploid *(n)* Having a single set of unpaired chromosomes. In humans, the only haploid cells are germ cells past the meiosis I stage of cell division, including the mature egg and sperm.
haversian canal *See* central canal.
head **1.** The uppermost part of the human body, above the neck. **2.** The expanded end of an organ such as a bone, the pancreas, or the epididymis.
helper T cell A type of lymphocyte that performs a central coordinating role in humoral and cellular immunity; target of the human immunodeficiency virus (HIV).
hematocrit (he-MAT-oh-crit) The percentage of blood volume that is composed of erythrocytes.
hematoma (HE-muh-TOE-muh) A mass of clotted blood in the tissues; forms a bruise when visible through the skin.
hemidecussation Crossing over of one half of the nerve fibers in a nerve or tract to the opposite side

of the central nervous system, especially at the optic chiasm. *Compare* decussation.

hemoglobin The red pigment of erythrocytes; binds and transports about 98.5% of the oxygen and 5% of the carbon dioxide carried in the blood.

hemopoiesis (HE-mo-poy-EE-sis) Production of any of the formed elements of blood.

hemopoietic tissue Any tissue in which hemopoiesis occurs, especially red bone marrow and lymphatic tissue.

hemostasis The cessation of bleeding by the mechanisms of vascular spasm, a platelet plug, and blood clotting.

hepatic Pertaining to the liver.

hepatic macrophage A macrophage found in the sinusoids of the liver; also called a *Kupffer cell.*

hepatic portal system A network of blood vessels that connect capillaries of the intestines to capillaries (sinusoids) of the liver, thus delivering newly absorbed nutrients directly to the liver.

hepatocyte Any of the cuboidal gland cells that constitute the parenchyma of the liver.

hiatus (hy-AY-tus) An opening or gap, such as the esophageal hiatus through the diaphragm.

hilum (HY-lum) A point on the surface of an organ where blood vessels, lymphatic vessels, or nerves enter and leave, usually marked by a depression and slit; the midpoint of the concave surface of any organ that is roughly bean-shaped, such as the lymph nodes, kidneys, and lungs. Also called the *hilus.* (fig. 23.9)

hindbrain The most caudal part of the brain, composed of the medulla oblongata, pons, and cerebellum. (fig. 13.5)

hindgut **1.** The most caudal part of the embryonic digestive tract; all of the tract caudal to the initial attachment of the yolk sac. (fig. 4.6) **2.** In adults, all of the digestive tract from the end of the transverse colon through the anal canal, with a blood supply and innervation separate from those of the foregut and midgut.

histological section A thin slice of tissue, usually mounted on a slide and artificially stained to make its microscopic structure more visible.

histology **1.** The microscopic structure of tissues and organs. **2.** The study of such structure.

holocrine gland An exocrine gland whose secretion is formed by the breakdown of entire gland cells; for example, a sebaceous gland.

homeostasis (HO-me-oh-STAY-sis) The tendency of a living body to maintain relatively stable internal conditions in spite of greater changes in its external environment.

homologous (ho-MOLL-oh-gus) **1.** Having the same embryonic or evolutionary origin but not necessarily the same function, such as the scrotum and labia majora. **2.** Pertaining to two chromosomes with identical structures and gene loci but not necessarily identical alleles; each member of the pair is inherited from a different parent.

hormone A chemical messenger that is secreted into the blood by an endocrine gland or isolated gland cell and triggers a physiological response in distant cells with receptors for it.

human Any species of primate classified in the family Hominidae, characterized by bipedal locomotion, relatively large brains, and usually articulate speech; currently represented only by *Homo sapiens* but including extinct species of *Homo* and *Australopithecus.*

hyaline cartilage (HY-uh-lin) A form of cartilage with a relatively clear matrix and fine collagen fibers but no conspicuous elastic fibers or collagen bundles as in other types of cartilage. (fig. 3.19)

hyaluronic acid (HY-uh-loo-RON-ic) A glycosaminoglycan that is particularly abundant in connective tissues, where it becomes hydrated and forms the tissue gel.

hyperextension A joint movement that increases the angle between two bones beyond 180°. (fig. 9.7)

hyperplasia (HY-pur-PLAY-zhuh) The growth of a tissue through cellular multiplication, not cellular enlargement. *Compare* hypertrophy.

hypertrophy (hy-PUR-tro-fee) The growth of a tissue through cellular enlargement, not cellular multiplication; for example, the growth of muscle under the influence of exercise. *Compare* hyperplasia.

hypoblast The layer of cells in the early embryonic disc facing away from the amniotic cavity; forms the yolk sac and is then replaced during gastrulation by migrating epiblast cells.

hypochondriac Pertaining to an area on each side of the abdomen superior to the subcostal line and lateral to the midclavicular line. (fig. A.6)

hypodermis (HY-po-DUR-miss) A layer of connective tissue deep to the skin; also called *superficial fascia, subcutaneous tissue,* or when it is predominantly adipose, *subcutaneous fat.*

hypogastric Pertaining to a medial area of the lower abdomen inferior to the intertubercular line and medial to (between) the midclavicular lines; also called the *pubic region.* (fig. A.6)

hypophyseal portal system A circulatory pathway that connects a capillary plexus in the hypothalamus to a capillary plexus in the anterior pituitary; carries hypothalamic releasing and inhibiting hormones to the anterior pituitary. (fig. 18.4)

hypophysis The pituitary gland.

hypothalamic thermostat A nucleus of neurons in the hypothalamus responsible for the homeostatic regulation of body temperature.

hypothalamo-hypophyseal tract A bundle of nerve fibers that begin in nuclei in the hypothalamus, travel through the pituitary stalk, and terminate in the posterior lobe of the pituitary gland. They deliver the hormones oxytocin and antidiuretic hormone to the posterior pituitary for storage, and signal the pituitary when to release them into the blood. (fig. 18.4)

hypothalamus (HY-po-THAL-uh-mus) The inferior portion of the diencephalon of the brain, forming the walls and floor of the third ventricle and giving rise to the posterior pituitary gland; controls many fundamental physiological functions such as appetite, thirst, and body temperature. (fig. 15.2)

hypothesis An informed conjecture that is capable of being tested and potentially falsified by experimentation or data collection.

hypoxemia A deficiency of oxygen in the blood.

hypoxia A deficiency of oxygen in any tissue; may lead to tissue necrosis.

I

immune system A population of cells, including leukocytes and macrophages, that occur in most organs of the body and protect against foreign organisms, some foreign chemicals, and cancerous or other aberrant host cells.

immunity The ability to ward off a specific infection or disease, usually as a result of prior exposure and the body's production of antibodies or lymphocytes against a pathogen.

implantation The process in which a conceptus attaches to the uterine endometrium and then becomes embedded in it.

inclusion Any visible object in the cytoplasm of a cell other than an organelle or cytoskeletal element; usually a foreign body or a stored cell product, such as a virus, dust particle, lipid droplet, glycogen granule, or pigment.

infarction **1.** The sudden death of tissue resulting from a loss of blood flow, often resulting from the occlusion of an artery; for example, cerebral infarction and myocardial infarction. **2.** A region of tissue that has died from lack of blood; also called an *infarct.*

inferior Lower than another structure or point of reference from the perspective of anatomical position; for example, the stomach is inferior to the diaphragm.

infundibulum (IN-fun-DIB-you-lum) Any funnel-shaped passage or structure, such as the distal portion of the uterine tube and the stalk that attaches the pituitary gland to the hypothalamus.

inguinal (IN-gwih-nul) Pertaining to the groin. (fig. A.6)

innervation (IN-ur-VAY-shun) The nerve supply to an organ.

insertion The point at which a muscle attaches to another tissue (usually a bone) and produces movement, opposite from its stationary origin. (fig. 10.4) *Compare* origin.

inspiration Inhaling.

integral protein A protein that extends through a plasma membrane and contacts both the extracellular and intracellular fluid; also called a *transmembrane protein.* (fig. 2.8)

integument The skin.

integumentary system An organ system consisting of the skin, cutaneous glands, hair, and nails.

interatrial septum The wall between the atria of the heart.

intercalated disc (in-TUR-kuh-LAY-ted) A complex of fascia adherens, gap junctions, and desmosomes that join two cardiac muscle cells end to end, microscopically visible as a dark line which helps to histologically distinguish this muscle type; functions as a mechanical and electrical link between cells. (fig. 20.14)

intercellular Between cells.

intercellular junction A point at which two cells are joined together; includes desmosomes, tight junctions, and gap junctions. (fig. 2.14)

intercostal (IN-tur-COSS-tul) Between the ribs, as in the *intercostal* muscles, arteries, veins, and nerves.

interdigitate To fit together like the fingers of the folded hands; for example, at the dermal-epidermal boundary, podocytes of the kidney, and intercalated discs of the heart.

interneuron (IN-tur-NEW-ron) A neuron that is contained entirely in the central nervous system and, in the path of signal conduction, lies anywhere between an afferent pathway and an efferent pathway.

interoceptor A sensory receptor that responds to stimuli originating within the body. *Compare* exteroceptor.

interosseous membrane (IN-tur-OSS-ee-us) A fibrous membrane that connects the radius to the ulna and the tibia to the fibula along most of the shaft of each bone. (fig. 8.4)

interstitial (IN-tur-STISH-ul) **1.** Pertaining to the extracellular spaces in a tissue. **2.** Located between other structures, as in the *interstitial cells* of the testis.

interstitial fluid Fluid in the interstitial spaces of a tissue, also called *tissue fluid.*
intervertebral disc A cartilaginous pad between the bodies of two adjacent vertebrae.
intracellular Contained within a cell.
intracellular fluid (ICF) The fluid contained in the cells; one of the major fluid compartments.
intramembranous ossification A process of bone development in which there is no cartilage precursor; rather, the bone develops directly from a sheet of condensed mesenchyme. (fig. 6.11) *Compare* endochondral ossification.
intraperitoneal Within the peritoneal cavity. *Compare* retroperitoneal.
intrinsic **1.** Arising from within, such as intrinsic blood-clotting factors; endogenous. **2.** Fully contained within an organ, such as the intrinsic muscles of the hand and eye. *Compare* extrinsic.
inversion Movement of the foot that turns the sole medially. (fig. 9.13)
involuntary Not under conscious control, as in the case of the autonomic nervous system and cardiac and smooth muscle contraction.
involution Shrinkage of a tissue or organ by autolysis, such as shrinkage of the thymus after childhood and of the uterus after pregnancy.
ipsilateral (IP-sih-LAT-ur-ul) On the same side of the body, as in reflex arcs in which a muscular response occurs on the same side of the body as the stimulus. *Compa*re contralateral.
ischemia A state in which the blood flow to a tissue is inadequate to meet its metabolic needs; may lead to tissue necrosis from hypoxia or waste accumulation.
isthmus A narrow zone of tissue connecting two larger masses, for example at the front of the thyroid gland and connecting the uterine tube ampulla to the uterus.

J

joint *See* articulation.

K

keratin A tough protein formed by keratinocytes that constitutes the hair, nails, and stratum corneum of the epidermis.
keratinized Covered with keratin, such as the epidermis.
keratinocyte A cell of the epidermis that synthesizes keratin, then dies; most cells of the epidermis are keratinocytes, with dead ones constituting the stratum corneum.
Kupffer cell *See* hepatic macrophage.
kyphosis An exaggerated thoracic spinal curvature, often resulting from osteoporosis; also called *widow's hump* or *dowager's hump.*

L

labium (LAY-bee-um) A lip, such as those of the mouth and the labia majora and minora of the vulva.
lacrimal Pertaining to the tears or tear glands.
lacteal A lymphatic capillary located in the core of an intestinal villus, serving to absorb dietary lipids.
lacuna (la-CUE-nuh) A small cavity or depression in a tissue such as bone, cartilage, and the erectile tissues.
lamella A little plate or sheet of tissue, such as a lamella of bone. (fig. 6.4)
lamellated corpuscle A bulbous sensory receptor with one or a few dendrites enclosed in onionlike layers of Schwann cells; found in the dermis, mesenteries, pancreas, and some other viscera, and responsive to deep pressure, stretch, and high-frequency vibration. Also called a *pacinian corpuscle.* (fig. 17.1)
lamina (LAM-ih-nuh) A thin layer, such as the lamina of a vertebra or the lamina propria of a mucous membrane. (fig. 7.25)
lamina propria (PRO-pree-uh) A thin layer of areolar tissue immediately deep to the epithelium of a mucous membrane. (fig. 3.30)
laryngopharynx (la-RIN-go-FAIR-inks) The portion of the pharynx formed by the union of the oropharynx and nasopharynx, beginning at the level of the hyoid bone and extending inferiorly to the opening of the esophagus. (fig. 23.2)
larynx (LAIR-inks) A cartilaginous chamber in the neck containing the vocal cords; the voicebox. (fig. 23.4)
lateral Away from the midline of an organ or median plane of the body; toward the side. *Compare* medial.
leg **1.** That part of the body between the knee and ankle; the crural region. **2.** A leglike extension of an organ. *See also* crus.
lesion A circumscribed zone of tissue injury, such as a skin abrasion or myocardial infarction.
leukocyte (LOO-co-site) Any nucleated blood cell; a neutrophil, eosinophil, basophil, lymphocyte, or monocyte. Also called a *white blood cell.* (fig. 19.1)
leukopoiesis The process of leukocyte development from hemopoietic stem cells.
libido The sex drive; a psychological desire for sex.
ligament A cord or band of tough collagenous tissue binding one organ to another, especially one bone to another, and serving to hold organs in place; for example, the cruciate ligaments of the knee and falciform ligament of the liver.
light microscope (LM) A microscope that produces images with visible light.
limb **1.** An appendage of the body arising from the shoulder or hip; *see also* lower limb; upper limb. **2.** An appendage or extension of another structure, such as the descending limb of the nephron loop.
limb bud An outgrowth of the embryo that develops into an upper or lower limb.
limbic system A ring of brain structures that encircle the corpus callosum and thalamus, including the cingulate gyrus, hippocampus, amygdala, and other structures; functions include learning and emotion. (fig. 15.17)
line **1.** Any long narrow mark. *See also* linea. **2.** An elongated, slightly raised ridge on a bone, such as the nuchal lines of the skull. (fig. 7.5)
linea (LIN-ee-uh) An anatomical line, such as the linea alba.
lingual (LING-wul) Pertaining to the tongue.
lipid A hydrophobic organic compound with a high ratio of hydrogen to oxygen; includes steroids, fatty acids, triglycerides (fats), phospholipids, and prostaglandins.
LM **1.** Light microscope. **2.** Light micrograph, a photograph made through the light microscope.
load **1.** To pick up oxygen or carbon dioxide for transport in the blood. **2.** The resistance acted upon by a muscle.
lobe **1.** A structural subdivision of an organ such as a gland, a lung, or the brain, bounded by a visible landmark such as a fissure or septum. **2.** The inferior, noncartilaginous, often pendant part of the ear pinna; the earlobe.
lobule (LOB-yool) A small subdivision of an organ or of a lobe of an organ, especially of a gland.
long bone A bone such as the femur or humerus that is markedly longer than wide and that generally serves as a lever.
longitudinal section (l.s.) A cut along the longest dimension of the body or of an organ. (fig. 3.2)
loose connective tissue Areolar, reticular, or adipose tissue; a connective tissue which, in the first two cases, has an abundance of ground substance and relatively widely spaced fibers and cells.
lower limb The appendage that arises from the hip, consisting of the thigh from hip to knee; the crural region from knee to ankle; the ankle; and the foot. Loosely called the leg, although that term properly refers only to the crural region.
lumbar Pertaining to the lower back and sides, between the thoracic cage and pelvis.
lumen The internal space of a hollow organ such as a blood vessel or the esophagus, or a space surrounded by cells as in a gland acinus.
lymph The fluid contained in lymphatic vessels and lymph nodes, produced by the absorption of tissue fluid.
lymphatic nodule A temporary, dense aggregation of lymphocytes in such places as mucous membranes and lymphatic organs; also called a *lymphatic follicle.* (fig. 22.8)
lymphatic system An organ system consisting of lymphatic vessels, lymph nodes, the tonsils, spleen, and thymus; functions include tissue fluid recovery and immunity.
lymph node A small organ found along the course of a lymphatic vessel; filters the lymph and contains lymphocytes and macrophages, which respond to antigens in the lymph. (fig. 22.12)
lymphocyte (LIM-foe-site) A relatively small leukocyte with numerous types and roles in nonspecific defense, humoral immunity, and cellular immunity. (fig. 19.1)
lysosome A membrane-bounded organelle containing a mixture of enzymes with a variety of intracellular and extracellular roles in digesting foreign matter, pathogens, and expired organelles. (fig. 2.5)
lysozyme An enzyme found in tears, milk, saliva, mucus, and other body fluids that destroys bacteria by digesting their cell walls; also called *muramidase.*

M

macrophage (MAC-ro-faje) Any cell of the body, other than a leukocyte, that is specialized for phagocytosis; usually derived from blood monocytes and often functioning as antigen-presenting cells.
macula (MAC-you-luh) A patch or spot, such as the macula lutea of the retina and macula sacculi of the inner ear.
magnetic resonance imaging (MRI) A method of producing a computerized image of the interior of the body using a strong magnetic field and radio waves. (fig. 1.5)
male In humans, any individual with a Y chromosome; normally, one possessing one X and one Y chromosome in each somatic cell, and having reproductive organs that serve to produce and deliver sperm.
mammary gland The milk-secreting gland that develops within the breast in pregnancy and lactation; only minimally developed in the breast of a nonpregnant or nonlactating woman.

mast cell A connective tissue cell, similar to a basophil, that secretes histamine, heparin, and other chemicals involved in inflammation; often concentrated along the course of a blood capillary.
matrix **1.** The extracellular material of a tissue. **2.** The substance or framework within which other structures are embedded, such as the fibrous matrix of a blood clot. **3.** A mass of epidermal cells from which a hair root or nail root develops. **4.** The fluid within a mitochondrion containing enzymes of the citric acid cycle.
meatus (me-AY-tus) An opening into a canal, such as an acoustic meatus.
mechanoreceptor A sensory nerve ending or organ specialized to detect mechanical stimuli such as touch, pressure, stretch, or vibration.
medial Toward the midline of an organ or median plane of the body. *Compare* lateral.
median plane The sagittal plane that divides the body or an organ into equal right and left halves; also called the *midsagittal plane.* (fig. A.3) *Compare* sagittal plane.
mediastinum (ME-dee-ah-STY-num) The thick median partition of the thoracic cavity that separates one pleural cavity from the other and contains the heart, great blood vessels, and thymus. (fig. A.7)
medical imaging Any of several noninvasive or minimally invasive methods for producing images of the interior of the body, including X rays, MRI, PET, CT, and sonography.
medulla (meh-DUE-luh, meh-DULL-uh) Tissue deep to the cortex of certain organs such as the adrenal glands, lymph nodes, hairs, and kidneys.
medulla oblongata (OB-long-GAH-ta) The most caudal part of the brainstem, immediately superior to the foramen magnum of the skull, connecting the spinal cord to the rest of the brain. (fig. 15.2)
meiosis (my-OH-sis) A form of cell division in which a diploid cell divides twice and produces four haploid daughter cells; occurs only in gametogenesis.
Meissner plexus *See* submucosal plexus.
melanin A brown or black pigment synthesized by melanocytes and some other cells; provides color to the skin, hair, eyes, and some other organs and tissues.
melanocyte A cell of the stratum basale of the epidermis that synthesizes melanin and transfers it to the keratinocytes.
meninges (meh-NIN-jeez) (singular, *meninx*) Three fibrous membranes between the central nervous system and surrounding bone: the dura mater, arachnoid mater, and pia mater. (fig. 15.4)
merocrine (MERR-oh-crin) Pertaining to gland cells that release their product by exocytosis; also called *eccrine.* (fig. 5.10)
mesenchyme (MEZ-en-kime) A gelatinous embryonic connective tissue derived from the mesoderm; differentiates into all permanent connective tissues and most muscle.
mesentery (MEZ-en-tare-ee) A serous membrane that binds the intestines together and suspends them from the abdominal wall; the visceral continuation of the peritoneum. (fig. 24.3)
mesocolon A dorsal mesentery that anchors parts of the colon to the abdominal wall. (fig. 24.17)
mesoderm (MEZ-oh-durm) The middle layer of the three primary germ layers of an embryo; gives rise to muscle and connective tissue.
mesonephric ducts A pair of embryonic ducts that form in association with the temporary mesonephric kidney; they degenerate in the female, while in the male they develop into parts of the reproductive tract. (fig. 26.26)
mesothelium (MEZ-oh-THEEL-ee-um) A simple squamous epithelium that covers the serous membranes.
metaphysis A growth zone at the junction between the diaphysis and epiphysis of a long bone, where cartilage is replaced by osseous tissue and the bone grows in length. (fig. 6.8)
metaplasia Transformation of one mature tissue type into another; for example, a change from pseudostratified columnar to stratified squamous epithelium in an overventilated nasal cavity.
metarteriole A short blood vessel that links an arteriole to a bed of blood capillaries, with no tunica media except for a smooth muscle precapillary sphincter at the opening to each capillary. (fig. 21.9)
metastasis (meh-TASS-tuh-sis) The spread of cancer cells from the original tumor to a new location, where they seed the development of a new tumor.
microfilament A thin filament of actin in the cytoskeleton of a cell, involved especially in the supportive core of a microvillus, the membrane skeleton just deep to the plasma membrane, and in muscle contraction. *See also* actin.
micrograph A photograph made with a microscope.
micrometer (μ**m**) One thousandth of a millimeter, or 10^{-6} meter; a convenient unit of length for expressing the sizes of cells.
microtubule An intracellular cylinder composed of the protein tubulin, forming centrioles, the axonemes of cilia and flagella, and part of the cytoskeleton.
microvillus An outgrowth of the plasma membrane that increases the surface area of a cell and functions in absorption and some sensory processes; distinguished from cilia and flagella by its smaller size and lack of an axoneme.
midbrain A short section of the brainstem between the pons and diencephalon. (fig. 15.2)
midgut **1.** The middle part of the embryonic digestive tract, located at the attachment of the yolk sac. (fig. 4.6) **2.** In adults, all of the digestive tract from the major duodenal papilla through the end of the transverse colon, with a blood supply and innervation separate from those of the foregut and hindgut.
midsagittal plane *See* median plane.
mineralization *See* calcification.
mitochondrion (MY-toe-CON-dree-un) An organelle specialized to synthesize ATP, enclosed in a double unit membrane with infoldings of the inner membrane called cristae.
mitosis A form of cell division in which a cell divides once and produces two genetically identical daughter cells; sometimes used to refer only to the division of the genetic material or nucleus and not to include cytokinesis, the subsequent division of the cytoplasm.
mixed nerve A nerve containing both afferent (sensory) and efferent (motor) nerve fibers.
monocyte A leukocyte specialized to migrate into the tissues and transform into a macrophage. (fig. 19.1)
monozygotic (MZ) twins Two individuals that develop from the same zygote and are therefore genetically identical.
morphology Anatomy, especially as interpreted from a functional perspective.
morula A preembryonic stage of development consisting of 16 or more identical-looking cells having a bumpy surface appearance reminiscent of a mulberry. The morula develops into a blastocyst and then implants on the uterine wall.
motor end plate A depression in a muscle fiber where it has synaptic contact with a nerve fiber and has a high density of neurotransmitter receptors. (fig. 10.12)
motor neuron A neuron that transmits signals from the central nervous system to any effector (muscle or gland cell); also called an *efferent neuron.* The axon of a motor neuron is an *efferent nerve fiber.*
motor protein Any protein that produces movements of a cell or its components owing to its ability to undergo quick repetitive changes in conformation and to bind reversibly to other molecules; for example, myosin, dynein, and kinesin.
motor unit One motor neuron and all the skeletal muscle fibers innervated by it.
MRI *See* magnetic resonance imaging.
mucosa (mew-CO-suh) A tissue layer that forms the inner lining of an anatomical tract that is open to the exterior (the respiratory, digestive, urinary, and reproductive tracts). Composed of epithelium, connective tissue (lamina propria), and often smooth muscle (muscularis mucosae). Also called a *mucous membrane.* (fig. 3.30)
mucosa-associated lymphatic tissue (MALT) Aggregations of lymphocytes, including lymphatic nodules, in the mucous membranes.
mucous gland A gland that secretes mucus, such as the glands of the large intestine and nasal cavity. *Compare* serous gland.
mucous membrane *See* mucosa.
mucus A viscous, slimy or sticky secretion produced by mucous cells and mucous membranes and consisting of a hydrated glycoprotein, mucin; serves to bind particles together, such as bits of masticated food, and to protect the mucous membranes from infection and abrasion.
multipotent Pertaining to a stem cell that is capable of differentiating into multiple, but not unlimited, adult cell types; for example, bone marrow colony-forming units that can produce multiple types of leukocytes.
muscle fiber One skeletal muscle cell. *Compare* myocyte.
muscularis externa The external muscular wall of certain viscera such as the esophagus and small intestine. (fig. 24.2)
muscularis mucosae (MUSS-cue-LERR-iss mew-CO-see) A layer of smooth muscle immediately deep to the lamina propria of a mucosa. (fig. 3.30)
muscular system An organ system composed of the skeletal muscles, specialized mainly for maintaining postural support and producing movements of the bones. Cardiac and smooth muscle are not regarded as part of the muscular system.
muscular tissue A tissue composed of elongated, electrically excitable cells specialized for contraction; the three types are skeletal, cardiac, and smooth muscle.
mutagen (MEW-tuh-jen) Any agent that causes a mutation, including viruses, chemicals, and ionizing radiation.
mutation Any change in the structure of a chromosome or a DNA molecule, often resulting in a change of organismal structure or function.
myelin (MY-eh-lin) A lipid sheath around a nerve fiber, formed from closely spaced spiral layers of the plasma membrane of an oligodendrocyte or Schwann cell. (fig. 13.9)

myelination The process in which an oligodendrocyte or Schwann cell deposits myelin around a nerve fiber.
myeloid tissue Bone marrow.
myenteric plexus A plexus of parasympathetic neurons located between the layers of the muscularis externa of the digestive tract; controls peristalsis. Also called the *Auerbach plexus.*
myocardium The middle, muscular layer of the heart.
myocyte A muscle cell, especially a cell of cardiac or smooth muscle. *Compare* muscle fiber.
myoepithelial cell An epithelial cell that has become specialized to contract like a muscle cell; important in dilation of the pupil and ejection of secretions from gland acini.
myofibril (MY-oh-FY-bril) A bundle of myofilaments forming an internal subdivision of a cardiac or skeletal muscle cell. (fig. 10.8)
myofilament A protein microfilament responsible for the contraction of a muscle cell, composed mainly of myosin or actin. (fig. 10.9)
myosin A motor protein that constitutes the thick myofilaments of muscle and has globular, mobile heads of ATPase that bind to actin molecules.
myotome A group of mesodermal cells that arise from a somite in the fourth week of development and give rise to body wall muscles in that region of the trunk of the body. *Compare* dermatome; sclerotome.

N

nasal concha One of three curved or scroll-like plates of bone and mucous membrane that extends from the lateral wall toward the septum in each nasal fossa; serves to warm, cleanse, and humidify inhaled air. (fig. 23.2)
nasal septum A wall of bone and cartilage that separates the right and left nasal fossae.
nasopharynx That region of the pharynx that lies caudal to the nasal choanae and dorsal or superior to the soft palate. (fig. 23.2)
natural killer (NK) cell A lymphocyte that attacks and destroys cancerous or infected cells of the body without requiring prior exposure or a specific immune response; one of the body's nonspecific defenses.
necrosis (neh-CRO-sis) Pathological tissue death due to such causes as infection, trauma, or hypoxia. *Compare* apoptosis.
neonate An infant up to 6 weeks old.
neoplasia (NEE-oh-PLAY-zee-uh) Abnormal growth of new tissue, such as a tumor, with no useful function.
nephron One of approximately 1.2 million blood-filtering, urine-producing units in each kidney; consists of a glomerulus, glomerular capsule, proximal convoluted tubule, nephron loop, and distal convoluted tubule. (fig. 25.5) *Compare* renal tubule.
nerve A cordlike organ of the peripheral nervous system composed of multiple nerve fibers ensheathed in connective tissue.
nerve fiber The axon of a single neuron.
nerve impulse A wave of self-propagating action potentials spreading along a nerve fiber; the nerve signal.
nervous system An organ system composed of the brain, spinal cord, nerves, and ganglia, specialized for rapid communication of information.
nervous tissue A tissue composed of neurons and neuroglia.
neural crest A mass of ectoderm that begins at the edges of the neural groove, then separates from the neural tube and gives rise primarily to nerves, ganglia, and the adrenal medulla. (fig. 13.14)
neural groove A longitudinal depression in the ectoderm of the embryo which closes up to form the neural tube, forerunner of the central nervous system.
neural tube A dorsal hollow tube in the embryo that develops into the central nervous system. (fig. 13.14)
neuroglia (noo-ROG-lee-uh) All cells of nervous tissue except neurons; cells that perform various supportive and protective roles for the neurons.
neurohypophysis The posterior one-third of the pituitary gland, consisting of the posterior lobe, a stalk that attaches the pituitary to the hypothalamus, and the median eminence of the hypothalamic floor; stores and secretes antidiuretic hormone and oxytocin. (fig. 18.4)
neuromuscular junction (NMJ) A synapse between a nerve fiber and a muscle cell. (fig. 10.12)
neuron (NOOR-on) A nerve cell; an electrically excitable cell specialized for producing and transmitting action potentials and secreting chemicals that stimulate adjacent cells.
neuronal circuit A group of interconnected neurons that conduct signals along defined pathways to produce a sustained, repetitive, convergent, or divergent output. (fig. 13.13)
neuronal pool (noor-OH-nul) A group of interconnected neurons of the central nervous system that perform a single collective function; for example, the vasomotor center of the brainstem and speech centers of the cerebral cortex.
neurotransmitter A chemical released at the distal end of an axon that stimulates an adjacent cell; for example, acetylcholine, norepinephrine, and serotonin.
neutrophil (NOO-tro-fill) A leukocyte, usually with a multilobed nucleus, that serves especially to destroy bacteria by means of phagocytosis, intracellular digestion, and secretion of bactericidal chemicals. (fig. 19.1)
nitrogenous waste Any nitrogen-containing substance produced as a metabolic waste and excreted in the urine; chiefly ammonia, urea, uric acid, and creatinine.
nociceptor (NO-sih-SEP-tur) A nerve ending specialized to detect tissue damage and produce a sensation of pain; pain receptor.
node of Ranvier A gap between adjacent segments of myelin in a myelinated nerve fiber; the point where action potentials are generated in a myelinated fiber.
nonkeratinized Pertaining to a stratified squamous epithelium that lacks a surface layer of dead compacted keratinocytes; found in the oral cavity, pharynx, esophagus, anal canal, and vagina.
notochord A middorsal supportive rod that develops in all chordate embryos, including humans; represented in the adult only by the nuclei of the intervertebral discs.
nuchal Pertaining to the back of the neck.
nuclear envelope (NEW-clee-ur) A pair of unit membranes enclosing the nucleus of a cell, with prominent pores allowing traffic of molecules between the nucleoplasm and cytoplasm. (fig. 2.17)
nuclear medicine Any use of radioisotopes to treat disease or form diagnostic images of the body.
nucleus (NEW-clee-us) **1.** A cell organelle containing DNA and surrounded by a double unit membrane. **2.** A mass of neurons (gray matter) surrounded by white matter of the brain, including the basal nuclei and brainstem nuclei. **3.** A central structure, such as the nucleus pulposus of an intervertebral disc or nucleus of an atom.
nucleus pulposus The gelatinous center of an intervertebral disc.

O

oblique section A cut through an elongated organ on a slant, between a longitudinal and a cross section. (fig. 3.2)
occlusion **1.** Meeting of the surfaces of the teeth when one bites. **2.** Obstruction of an anatomical passageway, such as blockage of an artery by a thrombus or atherosclerotic plaque.
olfactory Pertaining to the sense of smell.
omentum A ventral mesentery that extends from the stomach to the liver *(lesser omentum)* or is suspended from the greater curvature of the stomach and overhangs the intestines *(greater omentum).* (fig. 24.3)
oocyte (OH-oh-site) In the development of an egg cell, a haploid stage between meiosis I and fertilization.
oogenesis (OH-oh-JEN-eh-sis) The production of a fertilizable egg cell through a series of mitotic and meiotic cell divisions; female gametogenesis.
ophthalmic (off-THAL-mic) Pertaining to the eye or vision; optic.
opposition A movement of the thumb in which it approaches or touches any fingertip of the same hand. (fig. 9.12)
optic Pertaining to the eye or vision.
optic chiasm An X-shaped point at the base of the brain, immediately rostral to the hypothalamus, where the two optic nerves meet and continue as optic tracts.
oral cavity The space enclosed by the lips anteriorly, the cheeks laterally, and the fauces posteriorly; also called the *buccal cavity.*
orbit The eye socket of the skull.
organ Any anatomical structure that is composed of at least two different tissue types, has recognizable structural boundaries, and has a discrete function different from the structures around it. Many organs are microscopic and many organs contain smaller organs, such as the skin containing numerous microscopic sense organs.
organelle Any structure within a cell that carries out one of its metabolic roles, such as mitochondria, centrioles, endoplasmic reticulum, and the nucleus; an intracellular structure other than the cytoskeleton and inclusions.
organism Any living individual; the entire body of any living thing such as a bacterium, plant, or human.
organogenesis The prenatal developmental process in which embryonic germ layers differentiate into specific organs and organ systems; the process that converts an embryo to a fetus, occurring between day 16 and the end of week 8 of gestation.
organ system Any of 11 systems of interconnected or physiologically interrelated organs that perform one of the body's basic functions; for example, the digestive, urinary, and respiratory systems.
origin The relatively stationary attachment of a skeletal muscle. (fig. 10.4) *Compare* insertion.

oropharynx That part of the pharynx that is caudal to the fauces at the rear of the oral cavity, and anterior or inferior to the soft palate. (fig. 23.2)
os coxae (OSS COX-ee) (plural, *ossa coxae*) The hip bone, formed by the fusion of the ilium, ischium, and pubis of childhood. (fig. 8.7)
osmoreceptor (OZ-mo-re-SEP-tur) A neuron of the hypothalamus that responds to changes in the osmolarity of the extracellular fluid.
osmosis The diffusion of water through a selectively permeable membrane from the side with less concentrated solutes to the side with more concentrated solutes.
osseous (OSS-ee-us) Pertaining to bone.
ossification (OSS-ih-fih-CAY-shun) Bone formation; also called *osteogenesis. See also* endochondral ossification; intramembranous ossification.
osteoarthritis (OA) A chronic degenerative joint disease characterized by loss of articular cartilage, growth of bone spurs, and impaired movement; occurs to various degrees in almost all people with age.
osteoblast A bone-forming cell that arises from an osteogenic cell, deposits bone matrix, and eventually becomes an osteocyte.
osteoclast A macrophage of the bone surface that dissolves the matrix and returns minerals to the extracellular fluid.
osteocyte A mature bone cell formed when an osteoblast becomes surrounded by its own matrix and entrapped in a lacuna.
osteogenesis *See* ossification.
osteon A structural unit of compact bone consisting of a central canal surrounded by concentric cylindrical lamellae of matrix. (fig. 6.4)
osteoporosis (OSS-tee-oh-pore-OH-sis) A degenerative bone disease characterized by a loss of bone mass, increasing susceptibility to spontaneous fractures, and sometimes deformity of the vertebral column; causes include aging, estrogen hyposecretion, and insufficient resistance exercise.
ovary The female gonad; produces eggs, estrogen, and progesterone.
oviduct *See* uterine tube.
ovulation The release of a mature oocyte by the bursting of an ovarian follicle.
ovum Any stage of the female gamete from the conclusion of meiosis I until fertilization; a primary oocyte; an egg.

P

pacinian corpuscle *See* lamellated corpuscle.
palate A horizontal partition between the oral and nasal cavities.
palatine Pertaining to the palate, such as palatine bones and tonsils.
palmar region The anterior surface (palm) of the hand.
pancreas A gland of the upper abdominal cavity, near the stomach, that secretes digestive enzymes and sodium bicarbonate into the duodenum and secretes hormones into the blood.
pancreatic islets (PAN-cree-AT-ic EYE-lets) Small clusters of endocrine cells in the pancreas that secrete insulin, glucagon, somatostatin, and other intercellular messengers; also called *islets of Langerhans.* (fig. 18.11)
papilla (pa-PILL-uh) A conical or nipplelike structure, such as a lingual papilla of the tongue or the papilla of a hair bulb.
papillary (PAP-ih-lerr-ee) **1.** Pertaining to or shaped like a nipple, such as the papillary muscles of the heart. **2.** Having papillae, such as the papillary layer of the dermis.
paramesonephric ducts A pair of embryonic ducts that form beside the mesonephric ducts; they degenerate in the male, while in the female they form the uterine tubes, uterus, and part of the vagina. (fig. 26.26)
parasympathetic nervous system (PERR-uh-SIM-pa-THET-ic) A division of the autonomic nervous system that issues efferent fibers through the cranial and sacral nerves and exerts cholinergic effects on its target organs. (fig. 16.7)
parathyroid glands (PERR-uh-THY-royd) Small endocrine glands, usually four in number, adhering to the posterior side of the thyroid gland. (fig. 18.9)
parenchyma (pa-REN-kih-muh) The tissue that performs the main physiological functions of an organ, especially a gland, as opposed to the tissues (stroma) that mainly provide structural support.
parietal (pa-RY-eh-tul) **1.** Pertaining to a wall, as in the *parietal cells* of the gastric glands and *parietal bone* of the skull. **2.** The outer or more superficial layer of a two-layered membrane such as the pleura, pericardium, or glomerular capsule. (fig. A.8) *Compare* visceral.
pathogen Any disease-causing chemical or organism.
pectoral Pertaining to the chest.
pectoral girdle The circle of bones that connect the upper limb to the axial skeleton; composed of the two scapulae and the two clavicles.
pedal region The foot.
pedicel *See* pedicle.
pedicle (PED-ih-cul) A small footlike process, as in the vertebrae and the renal podocytes; also called a *pedicel.* (fig. 7.22)
pelvic cavity The space enclosed by the true (lesser) pelvis, containing the urinary bladder, rectum, and internal reproductive organs. (fig. A.7)
pelvic girdle A ring of three bones—the two hip bones (ossa coxae) and the sacrum—which attach the lower limbs to the axial skeleton. *See also* pelvis.
pelvis **1.** The basinlike cradle of bones that enclose the pelvic cavity and provide attachment for the lower limbs; includes the ossa coxae, sacrum, and coccyx. (fig. 8.6) **2.** A basinlike structure such as the renal pelvis of the kidney. (fig. 25.3)
perfusion Loosely, blood flow to any tissue or organ. More specifically, the volume of blood received by a given mass of tissue in a given unit of time, such as milliliters per gram per minute.
pericardial cavity A narrow space between the parietal and visceral layers of the pericardium, containing pericardial fluid.
pericardium A two-layered serous membrane that folds around the heart. Its visceral layer forms the heart surface *(epicardium)* while its parietal layer forms a fibrous *pericardial sac* around the heart. (fig. 20.4)
perichondrium (PERR-ih-CON-dree-um) A layer of fibrous connective tissue covering the surface of hyaline or elastic cartilage. (fig. 3.19)
perineum (PERR-ih-NEE-um) The region between the thighs bordered by the coccyx, pubic symphysis, and ischial tuberosities; contains the orifices of the urinary, reproductive, and digestive systems. (fig. 11.17)
periosteum (PERR-ee-OSS-tee-um) A layer of fibrous connective tissue covering the surface of a bone. (fig. 6.4)
peripheral Away from the center of the body or of an organ, as in *peripheral vision* and *peripheral blood vessels;* opposite of central.
peripheral nervous system (PNS) A subdivision of the nervous system composed of all nerves and ganglia; all of the nervous system except the central nervous system. *Compare* central nervous system.
peristalsis (PERR-ih-STAL-sis) A wave of constriction traveling along a tubular organ such as the esophagus or ureter, serving to propel its contents.
peritoneum (PERR-ih-toe-NEE-um) A serous membrane that lines the peritoneal cavity of the abdomen and covers the mesenteries and viscera.
perivascular (PERR-ih-VASS-cue-lur) Pertaining to the region surrounding a blood vessel.
peroxisome An organelle composed of a unit membrane enclosing a mixture of enzymes; serves to detoxify free radicals, alcohol, and other drugs, and to break down fatty acids; named for the hydrogen peroxide that it generates in the course of these activities.
PET *See* positron emission tomography.
phagocytosis (FAG-oh-sy-TOE-sis) A form of endocytosis in which a cell surrounds a foreign particle with pseudopods and engulfs it, enclosing it in a cytoplasmic vesicle called a phagosome. (fig. 2.11)
pharyngeal arch One of five pairs of bulbous swellings in the pharyngeal region of an embryo. (fig. 4.9)
pharyngeal pouch One of six pairs of outpocketings of the pharynx between and adjacent to the pharyngeal arches of an embryo; these form gill slits in fishes and amphibians but in humans give rise to such structures as the middle-ear cavities, palatine tonsils, thymus, parathyroid glands, and C cells of the thyroid gland. (fig. 4.8)
pharynx (FAIR-inks) A muscular passage in the throat at which the respiratory and digestive tracts cross. (fig. 23.2)
phospholipid A lipid composed of a hydrophilic head with a phosphate group and a nitrogenous group such as choline, and two hydrophobic fatty acid tails; especially important as the most numerous molecules of the plasma membrane and other unit membranes of a cell, but also involved in emulsification of dietary fat and as a component of pulmonary surfactant. (fig. 2.7)
photoreceptor Any cell or organ specialized to absorb light and generate a nerve signal; the eye and its rods, cones, and some of its ganglion cells.
phrenic (FREN-ic) **1.** Pertaining to the diaphragm, as in *phrenic nerve.* **2.** Pertaining to the mind, as in *schizophrenic.*
physiology **1.** The functional processes of the body. **2.** The study of such function.
piloerector A bundle of smooth muscle cells associated with a hair follicle, responsible for erection of the hair; also called *arrector pili.* (fig. 5.7)
pineal gland (PIN-ee-ul) A small conical endocrine gland arising from the roof of the third ventricle of the brain; produces melatonin and serotonin and may be involved in mood and timing the onset of puberty. (fig. 15.2)
pinocytosis A form of endocytosis in which the plasma membrane sinks in and internalizes a droplet of extracellular fluid in a pinocytotic vesicle. (fig. 2.11)
pituitary gland (pih-TOO-ih-terr-ee) An endocrine gland suspended from the

hypothalamus and housed in the sella turcica of the sphenoid bone; secretes numerous hormones, most of which regulate the activities of other glands. (fig. 18.4)

placenta (pla-SEN-tuh) A thick discoid organ on the wall of the pregnant uterus, composed of a combination of maternal and fetal tissues, serving multiple functions in pregnancy including gas, nutrient, and waste exchange between mother and fetus. (fig. 4.14)

plantar (PLAN-tur) Pertaining to the sole of the foot.

plantar flexion A movement of the ankle that points the toes downward, as in pressing on the gas pedal of a car or standing on tiptoes. (fig. 9.13)

plaque A small scale or plate of matter, such as dental plaque, the fatty plaques of atherosclerosis, and the amyloid plaques of Alzheimer disease.

plasma The noncellular portion of the blood.

plasma cell A connective tissue cell that differentiates from a B lymphocyte and secretes antibodies. (fig. 22.17)

plasma membrane The unit membrane that encloses a cell and controls the traffic of molecules into and out of it. (fig. 2.6)

platelet A formed element of the blood derived from the peripheral cytoplasm of a megakaryocyte, known especially for its roles in stopping bleeding but also serves in dissolving blood clots, stimulating inflammation, and promoting tissue growth. (fig. 19.10)

pleura (PLOOR-uh) A two-layered serous membrane that folds around the lung. Its visceral layer forms the lung surface while its parietal layer lines the inside of the rib cage. (fig. 23.12)

pleural cavity A narrow space between the parietal and visceral layers of the pleura, containing pleural fluid. (fig. 23.12)

plexus A network of blood vessels, lymphatic vessels, or nerves, such as a choroid plexus of the brain or the brachial plexus of nerves. (fig. 14.14)

pluripotent Pertaining to an embryonic stem cell from the morula that is capable of producing any type of embryonic or adult cell. More loosely, describing certain adult stem cells with an especially broad developmental potential, able to produce a wide variety of differentiated cell types. *See also* pluripotent stem cell.

pluripotent stem cell (PPSC) A stem cell of the bone marrow that can produce any of the formed elements of blood.

pollex The thumb.

pons The section of the brainstem between the midbrain and medulla oblongata.

popliteal (po-LIT-ee-ul) Pertaining to the posterior aspect of the knee.

portal system A circulatory pathway in which blood passes through two capillary beds in series in a single trip from the heart and back. (fig. 21.12)

positron emission tomography (PET) A method of producing a computerized image of the physiological state of a tissue using injected radioisotopes that emit positrons. (fig. 1.6)

posterior Near or pertaining to the back or spinal side of the body; dorsal.

postganglionic Pertaining to a neuron that transmits signals from a ganglion to a more distal target organ. (fig. 16.2)

postsynaptic Pertaining to a neuron or other cell that receives signals from the presynaptic neuron at a synapse. (fig. 13.11)

potential space An anatomical space that is usually obliterated by contact between two membranes but opens up if air, fluid, or other matter comes between them. Examples include the pleural cavity and the lumen of the uterus.

preembryo A developing human that has not yet formed ectoderm, mesoderm, and endoderm; when those germ layers have formed, the individual is regarded as an embryo. *Compare* conceptus; embryo.

preembryonic stage Any stage of prenatal development from fertilization through 16 days, when the primary germ layers exist and the embryonic stage begins.

preganglionic Pertaining to a neuron that transmits signals from the central nervous system to a ganglion. (fig. 16.2)

prepuce A fold of tissue over the glans of the penis or clitoris; the penile foreskin or clitoral hood.

presynaptic Pertaining to a neuron that transmits signals to a synapse. (fig. 13.11)

prime mover The muscle primarily responsible for a given joint action; agonist.

process An outgrowth of bone or other tissue, such as the mastoid process of the skull.

programmed cell death (PCD) *See* apoptosis.

projection pathway The route taken by nerve signals from their point of origin (such as a sense organ) to their point of termination (such as the primary sensory cortex). (fig. 17.15)

pronation A rotational movement of the forearm that turns the palm downward or dorsally. (fig. 9.12)

prone **1.** A forearm position in which the anterior surface faces dorsally or downward. **2.** Lying face down.

proprioception (PRO-pree-oh-SEP-shun) The nonvisual perception, usually subconscious, of the position and movements of the body, resulting from input from proprioceptors and the vestibular apparatus of the inner ear.

proprioceptor (PRO-pree-oh-SEP-tur) A sensory receptor of the muscles, tendons, and joint capsules that detects muscle contractions and joint movements.

prostate gland (PROSS-tate) A male reproductive gland that encircles the urethra immediately inferior to the bladder and contributes to the semen. (fig. 26.2)

protein A polypeptide of 50 amino acids or more.

proteoglycan A protein-carbohydrate complex in which the carbohydrate is dominant; forms a gel that binds cells and tissues together, fills the umbilical cord and eye, lubricates the joints, and forms the rubbery texture of cartilage. Formerly called *mucopolysaccharide.*

protraction Forward movement of a body part in the horizontal plane, such as moving the mandible forward in preparation to take a bite from an apple. (fig. 9.10)

protuberance A bony outgrowth or protruding part, such as the mental protuberance of the mandible.

proximal Relatively near a point of origin or attachment; for example, the shoulder is proximal to the elbow. *Compare* distal.

pseudopod (SOO-doe-pod) A temporary cytoplasmic extension of a cell used for locomotion (ameboid movement) and phagocytosis.

pseudostratified columnar epithelium An epithelium in which every cell contacts the basement membrane but not all of them reach the free surface, thus giving an appearance of stratification. (fig. 3.7)

pubic Concerning the region of the genitalia. *See also* hypogastric.

pudendum *See* vulva.

pulmonary Pertaining to the lungs.

pulmonary circuit A route of blood flow that supplies blood to the pulmonary alveoli for gas exchange and then returns it to the heart; all blood vessels between the right ventricle and the left atrium of the heart. (fig. 21.13)

R

radiography The use of X rays to form an image of the interior of the body. (fig. 1.2)

radiology The branch of medicine concerned with producing images of the interior of the body, using such methods as X rays, sonography, MRI, CT, and PET.

ramus (RAY-mus) An anatomical branch, as in a nerve or in the pubis.

receptive field An area of the environment or of an epithelial surface from which a given neuron receives sensory information. (fig. 17.2)

receptor **1.** A cell or organ specialized to detect a stimulus, such as a taste cell or the eye. **2.** A protein molecule that binds and responds to a chemical such as a hormone, neurotransmitter, or odor molecule.

receptor-mediated endocytosis A mode of vesicular transport in which cell surface receptors bind a specific molecule in the extracellular fluid, then cluster together to be internalized by the cell. (fig. 2.11)

rectus Straight; used in muscle names such as *rectus femoris* and *rectus abdominis.*

reflected Folded back or away from something, often to expose another structure in anatomical demonstrations. (fig. 11.13)

reflex A stereotyped, automatic, involuntary response to a stimulus; includes somatic reflexes, in which the effectors are skeletal muscles, and visceral (autonomic) reflexes, in which the effectors are usually visceral muscle, cardiac muscle, or glands.

reflex arc A simple neural pathway that mediates a reflex; involves a receptor, an afferent nerve fiber, sometimes one or more interneurons, an efferent nerve fiber, and an effector. (fig. 14.19)

regeneration Replacement of damaged tissue with new tissue of the original type. *Compare* fibrosis.

renal (REE-nul) Pertaining to the kidney.

renal tubule A urine-forming duct that converts glomerular filtrate to urine by processes of reabsorption and secretion of water and solutes. Consists of the proximal convoluted tubule, nephron loop, and distal convoluted tubule of an individual nephron, plus a collecting duct and papillary duct shared by multiple nephrons. (fig. 25.5) *Compare* nephron.

renin An enzyme produced by the kidney that converts angiotensinogen to angiotensin I, the first step in producing the vasoconstrictor angiotensin II.

reproductive system An organ system specialized for the production of offspring.

resistance **1.** Opposition to the flow of fluid, such as blood in a vessel or air in a bronchiole. **2.** Opposition to the movement of a joint; the load against which a muscle works. **3.** A nonspecific ability to ward off an infection or disease, as opposed to the pathogen-specific defense provided by immunity.

respiratory system An organ system specialized for the intake of air and exchange of gases with the blood, consisting of the lungs and the air passages from the nose to the bronchi.

reticular cell (reh-TIC-you-lur) A delicate, branching cell in the reticular connective tissue of the lymphatic organs.

reticular fiber A fine, branching collagen fiber coated with glycoprotein, found in the stroma of lymphatic organs and some other tissues and organs.

reticular tissue A connective tissue composed of reticular cells and reticular fibers, found in bone marrow, lymphatic organs, and in lesser amounts elsewhere. (fig. 3.15)

retraction Movement of a body part dorsally on the horizontal plane; for example, retracting the mandible to grind food between the molars. (fig. 9.10)

retroperitoneal Located between the peritoneum and body wall, rather than in the peritoneal cavity; descriptive of certain abdominal viscera such as the kidneys, ureters, and pancreas. (fig. A.9) *Compare* intraperitoneal.

ribosome A granule found free in the cytoplasm or attached to the rough endoplasmic reticulum and nuclear envelope, composed of ribosomal RNA and enzymes; specialized to read the nucleotide sequence of messenger RNA and assemble a corresponding sequence of amino acids to make a protein.

risk factor Any environmental factor or characteristic of an individual that increases one's chance of developing a particular disease; includes such intrinsic factors as age, sex, and race and such extrinsic factors as diet, smoking, and occupation.

root **1.** Part of an organ that is embedded in other tissue and therefore not externally visible, such as the root of a tooth, a hair, or the penis. *Compare* shaft. **2.** The proximal end of a spinal nerve, adjacent to the spinal cord.

rostral Relatively close to the forehead, especially in reference to structures of the brain and spinal cord; for example, the frontal lobe is rostral to the parietal lobe. *Compare* caudal.

rotation Movement of a body part such as the humerus or forearm around its longitudinal axis. (fig. 9.11)

rough endoplasmic reticulum Regions of endoplasmic reticulum characterized by flattened, parallel cisternae externally studded with ribosomes; involved in making proteins for export from the cell, among other functions. *See also* endoplasmic reticulum; smooth endoplasmic reticulum.

ruga (ROO-ga) **1.** An internal fold or wrinkle in the mucosa of a hollow organ such as the stomach and urinary bladder; typically present when the organ is empty and relaxed but not when the organ is full and stretched. **2.** Tissue ridges in such locations as the hard palate and vagina. (fig. 24.11)

S

sagittal plane (SADJ-ih-tul) Any plane that extends from ventral to dorsal and cephalic to caudal, and divides the body into right and left portions. (fig. A.3) *Compare* median plane.

sarcomere (SAR-co-meer) In skeletal and cardiac muscle, the portion of a myofibril from one Z disc to the next, constituting one contractile unit. (fig. 10.10)

sarcoplasmic reticulum (SR) The smooth endoplasmic reticulum of a muscle cell, serving as a calcium reservoir. (fig. 10.8)

satellite cell **1.** A type of glial cell found surrounding the somas of neurons in ganglia of the peripheral nervous system. **2.** Stem cells of skeletal muscle that can multiply in response to muscle injury and contribute to some extent to regeneration of muscle fibers.

scanning electron microscope (SEM) A microscope that uses an electron beam in place of light to form high-resolution, three-dimensional images of the surfaces of objects; capable of much higher magnifications than a light microscope. *Compare* transmission electron microscope.

Schwann cell A glial cell that forms the neurilemma around all peripheral nerve fibers and the myelin sheath around many of them; also encloses neuromuscular junctions. (fig. 13.9)

sclerosis (scleh-RO-sis) Hardening or stiffening of a tissue, usually with scar tissue, as in *multiple sclerosis* of the central nervous system and *atherosclerosis* of the blood vessels.

sclerotome A group of mesodermal cells that arise from a somite in the fourth week of development and give rise to a segment of the vertebral column. *Compare* dermatome; myotome.

sebaceous gland A holocrine gland that is usually associated with a hair follicle and produces an oily secretion, sebum. (fig. 5.10)

sebum (SEE-bum) An oily secretion of the sebaceous glands that keeps the skin and hair pliable.

secondary sex characteristic Any feature that develops at puberty, further distinguishes the sexes from each other, and is not required for reproduction but promotes attraction between the sexes; examples include the distribution of subcutaneous fat, pitch of the voice, female breasts, male facial hair, and apocrine scent glands.

secondary sex organ An organ other than the ovaries and testes that is essential to reproduction, such as the external genitalia, internal genital ducts, and accessory reproductive glands.

second-order neuron An interneuron that receives sensory signals from a first-order neuron and relays them to a more rostral destination in the central nervous system (usually the thalamus). (fig. 14.4) *See also* first-order neuron; third-order neuron.

secretion **1.** A chemical released by a cell to serve a physiological function, such as a hormone or digestive enzyme, as opposed to a waste product. **2.** The process of releasing such a chemical, usually by exocytosis. *Compare* excretion.

secretory vesicle An organelle that arises from the Golgi complex and carries a secretion to the cell surface to be released by exocytosis.

section *See* histological section.

selection pressure A force of nature that favors the reproduction of some individuals over others and thus drives the evolutionary process; includes climate, predators, diseases, competition, and food supply. Human anatomy and physiology reflect adaptations to selection pressures encountered in the evolutionary history of the species.

SEM **1.** Scanning electron microscope. **2.** Scanning electron micrograph, a photograph taken with the scanning electron microscope. *Compare* TEM.

semen The fluid ejaculated by a male, including spermatozoa and the secretions of the prostate gland and seminal vesicles.

semicircular duct A ring-shaped, fluid-filled tube of the inner ear that detects angular acceleration of the head; enclosed in a bony passage called the semicircular canal. There are three semicircular ducts in each ear. (fig. 17.11)

semilunar valve A valve that consists of crescent-shaped cusps, including the aortic and pulmonary valves of the heart and valves of the veins and lymphatic vessels. (fig. 20.7)

senescence Degenerative changes that occur with age. *See also* aging.

sense organ Any organ that is specialized to respond to stimuli and generate a meaningful pattern of nerve signals; may be microscopic and simple, such as a tactile corpuscle, or macroscopic and complex, such as the eye or ear; may respond to stimuli arising within the body or from external sources.

sensory neuron A nerve cell that responds to a stimulus and conducts signals to the central nervous system; also called an *afferent neuron.* The axon of a sensory neuron is an *afferent nerve fiber.*

septum An anatomical wall between two structures or spaces, such as the nasal septum or the interventricular septum of the heart.

serosa *See* serous membrane.

serous fluid (SEER-us) A watery fluid similar to blood serum, formed as a filtrate of the blood or tissue fluid or as a secretion of serous gland cells; moistens the serous membranes.

serous gland A gland that secretes a relatively nonviscous product, such as the pancreas or a tear gland. *Compare* mucous gland.

serous membrane A membrane such as the peritoneum, pleura, or pericardium that lines a body cavity or covers the external surfaces of the viscera; composed of a simple squamous mesothelium and a thin layer of areolar connective tissue.

Sertoli cell *See* sustentacular cell.

serum **1.** The fluid that remains after blood has clotted and the solids have been removed; essentially the same as blood plasma except for a lack of fibrinogen. Used as a vehicle for vaccines. **2.** Serous fluid.

sex chromosomes The X and Y chromosomes, which determine the sex of an individual.

shaft **1.** The midpart, or diaphysis, of a long bone. **2.** The external, cylindrical part of an organ such as a hair or the penis. *Compare* root.

short bone A bone that is not markedly longer than it is wide, such as the bones of the wrist and ankle.

sign An objective indication of disease that can be verified by any observor, such as cyanosis or a skin lesion. *Compare* symptom.

simple columnar epithelium An epithelium composed of a single layer of cells that are noticeably taller than they are wide. (fig. 3.6)

simple cuboidal epithelium An epithelium composed of a single layer of cells that are about equal in height and width; often, but not always, the cells appear squarish in tissue sections. (fig. 3.5)

simple diffusion Net movement of particles from a place of high concentration to a place of low concentration (down their concentration gradient), resulting from their own spontaneous motion; may or may not involve passage through a cell membrane or other membranes such as dialysis tubing.

simple squamous epithelium An epithelium composed of a single layer of thin, flat cells. (fig. 3.4)
sinoatrial (SA) node A mass of autorhythmic cells near the surface of the right atrium of the heart which serves as the pacemaker of the cardiac rhythm.
sinus **1.** An air-filled space in the cranium. (fig. 7.8) **2.** A modified, relatively dilated vein that lacks smooth muscle and is incapable of vasomotion, such as the dural sinuses of the cerebral circulation and coronary sinus of the heart. **3.** A small fluid-filled space in an organ such as the lymph nodes. **4.** Pertaining to the sinoatrial node of the heart, as in *sinus rhythm.*
sinusoid An irrregularly shaped, blood-filled space in a tissue, with wide gaps between the endothelial cells; found in the liver, bone marrow, spleen, and some other organs. (fig. 22.9)
skeletal muscle Striated voluntary muscle, almost all of which is attached to the bones. (fig. 10.1)
skeletal system An organ system consisting of the bones, ligaments, bone marrow, periosteum, articular cartilages, and other tissues associated with the bones.
smear A tissue prepared for microscopic study by wiping it across a slide, rather than by sectioning; for example, blood, bone marrow, spinal cord, and Pap smears.
smooth endoplasmic reticulum Regions of endoplasmic reticulum characterized by tubular, branching cisternae lacking ribosomes; involved in detoxification, steroid synthesis, and in muscle, storage of calcium ions. *See also* endoplasmic reticulum; rough endoplasmic reticulum.
smooth muscle Nonstriated involuntary muscle found in the walls of the blood vessels, many of the viscera, and other places. (fig. 3.27)
sodium-potassium pump An active transport protein which, in each cycle of activity, pumps three sodium ions out of a cell and two potassium ions into the cell, with the expenditure of one ATP.
soma *See* cell body.
somatic **1.** Pertaining to the body as a whole. **2.** Pertaining to the skin, bones, and skeletal muscles as opposed to the viscera. **3.** Pertaining to all cells other than germ cells.
somatic motor fiber A nerve fiber that innervates skeletal muscle and stimulates its contraction, as opposed to autonomic fibers.
somatic nervous system A division of the nervous system that includes afferent fibers mainly from the skin, muscles, and skeleton and efferent fibers to the skeletal muscles. *Compare* autonomic nervous system.
somatosensory *See* somesthetic.
somatotopy A point-for-point correspondence between the locations where stimuli arise and the locations in the brain or spinal cord to which the sensory signals project, thus producing in the CNS a sensory "map" of part of the body. (fig. 15.19)
somesthetic **1.** Pertaining to widely distributed *general senses* in the skin, muscles, tendons, joint capsules, and viscera, as opposed to the *special senses* found in the head only; also called *somatosensory.* **2.** Pertaining to the cerebral cortex of the postcentral gyrus, which receives input from such receptors. *See also* general senses; special senses.
somite One of the segmental blocks of embryonic mesoderm that begin to appear around day 20 and eventually number up to 44 pairs; a somite subdivides into three tissue masses—dermatome, myotome, and sclerotome, which give rise to certain aspects of the skin, muscles, and vertebrae. *See also* dermatome; myotome; sclerotome. (fig. 4.9)
sonography Production of an image of the interior of the body by means of ultrasound. (fig. 1.3)
special senses The senses of taste, smell, hearing, equilibrium, and vision, mediated by sense organs that are confined to the head and in most cases are relatively complex in structure. *See also* general senses.
sperm **1.** A spermatozoon. **2.** The fluid ejaculated by the male; semen. Contains spermatozoa and glandular secretions.
spermatogenesis (SPUR-ma-toe-JEN-eh-sis) The production of sperm cells through a series of mitotic and meiotic cell divisions; male gametogenesis.
spermatozoon (SPUR-ma-toe-ZOE-on) A sperm cell; the male gamete. (fig. 26.9)
sphincter (SFINK-tur) A ring of muscle that opens or closes an opening or passageway; found, for example, in the eyelids, around the urinary orifice, and at the junction of the stomach and duodenum. (fig. 24.11)
spinal column *See* vertebral column.
spinal cord The nerve cord that passes through the vertebral column and constitutes all of the central nervous system except the brain.
spinal nerve Any of the 31 pairs of nerves that arise from the spinal cord and pass through the intervertebral foramina. (fig. 14.10)
spindle **1.** An elongated structure that is thick in the middle and tapered at the ends (fusiform). **2.** A football-shaped complex of microtubules that guide the movement of chromosomes in mitosis and meiosis. (fig. 2.20) **3.** A stretch receptor in the skeletal muscles. (fig. 17.1)
spine **1.** The vertebral column. **2.** A pointed process or sharp ridge on a bone, such as the styloid process of the cranium and spine of the scapula. (fig. 8.2)
spinothalamic tract A bundle of nerve fibers that ascend the spinal cord and brainstem and carry signals to the thalamus for light touch, tickle, itch, heat, cold, pain, and pressure. (fig. 14.4)
splanchnic (SPLANK-nic) Pertaining to the digestive tract.
spongy bone A form of osseous tissue found in the interiors of flat, irregular, and short bones and the epiphyses of long bones, with a matrix that forms a porous network of plates and bars, enclosing connected channels filled with bone marrow; also called *cancellous bone. Compare* compact bone. (fig. 6.4)
squamous Flat, scalelike, as in the surface epithelial cells of the epidermis and serous membranes. (fig. 3.10)
stain A pigment applied to tissues to color and enhance the contrast between their nuclei, cytoplasm, extracellular material, and other tissue components.
stem cell Any undifferentiated cell that can divide and differentiate into more functionally specific cell types such as blood cells and germ cells.
stenosis The pathological constriction or narrowing of a tubular passageway or orifice of the body, such as the esophagus, uterine tube, or a valve orifice of the heart.
stereocilium An unusually long, sometimes branched microvillus lacking the axoneme and motility of a true cilium; serves such roles as absorption in the epididymis and sensory transduction in the inner ear.
sternal Pertaining to the breastbone or sternum, or the overlying region of the chest.
stimulus A chemical or physical agent in a cell's surroundings that is capable of creating a physiological response in the cell, especially agents detected by sensory cells, such as chemicals, light, and pressure.
strain The extent to which a bone or other structure is deformed when subjected to stress. *Compare* stress.
stratified **1.** Layered. **2.** A class of epithelia in which there are two or more cell layers, with some cells resting atop others rather than contacting the basement membrane.
stratified cuboidal epithelium An epithelium composed of two or more layers of cells in which the cells at the surface are about equal in height and width. (fig. 3.11)
stratified squamous epithelium An epithelium composed of two or more layers of cells in which the cells at the surface are flat and thin. (fig. 3.10)
stratum Any layer of tissue, such as the stratum corneum of the skin or stratum basalis of the uterus.
stratum corneum The surface layer of dead keratinocytes of the skin. (fig. 5.2)
stress **1.** A mechanical force applied to any part of the body; important in stimulating bone growth, for example. *Compare* strain. **2.** A condition in which any environmental influence disturbs the homeostatic equilibrium of the body and stimulates a physiological response, especially involving the increased secretion of hormones of the pituitary-adrenal axis.
striated muscle Muscular tissue in which the cells exhibit striations; skeletal and cardiac muscle. *See also* striations.
striations Alternating light and dark bands in skeletal and cardiac muscle produced by the pattern of overlapping myofilaments. (fig. 10.1)
stroma The connective tissue framework of a gland, lymphatic organ, or certain other viscera, as opposed to the tissue (parenchyma) that performs the physiological functions of the organ.
subcutaneous Beneath the skin.
submucosa A layer of loose connective tissue deep to the mucosa of an organ. (fig. 3.30)
submucosal plexus A plexus of parasympathetic neurons in the submucosa of the digestive tract, responsible for controlling glandular secretion by the mucosa and movements of the muscularis mucosae; also called the *Meissner plexus.*
sulcus A groove in the surface of an organ, as in the cerebrum, the heart, or a bone. (fig. 15.1)
superficial Relatively close to the surface; opposite of *deep.* For example, the ribs are superficial to the lungs.
superior Higher than another structure or point of reference from the perspective of anatomical position; for example, the lungs are superior to the diaphragm.
supination (SOO-pih-NAY-shun) A rotational movement of the forearm that turns the palm so that it faces upward or forward. (fig. 9.12)
supine **1.** A position of the forearm in which the palms face anteriorly or upward. **2.** A position in which the body is lying face up.
suprarenal Pertaining to the adrenal (suprarenal) glands, as in *suprarenal artery.*
surfactant A chemical that interferes with the formation of hydrogen bonds between water molecules, and thus reduces the cohesion of water; in the lung, a mixture of phospholipid

and protein that prevents the alveoli from collapsing during expiration.
sustentacular cell **1.** A cell in the wall of a seminiferous tubule of the testis that supports and protects the germ cells and secretes the hormone inhibin; also called a *Sertoli cell.* (fig. 26.7) **2.** In many epithelia, such as taste buds and olfactory mucosa, any cell that supports and spaces the primary functional cells of the tissue; also called a *supporting cell.* (fig. 17.5)
suture A line along which any two bones of the skull are immovably joined, such as the coronal suture between the frontal and parietal bones. (fig. 7.6)
sympathetic nervous system A division of the autonomic nervous system that issues efferent fibers through the thoracic and lumbar nerves and usually exerts adrenergic effects on its target organs; includes a chain of paravertebral ganglia adjacent to the vertebral column, and the adrenal medulla. (fig. 16.4)
symphysis (SIM-fih-sis) A joint in which two bones are held together by fibrocartilage; for example, between bodies of the vertebrae and between the right and left pubic bones. (fig. 8.6)
symptom A subjective indication of disease that can be felt by the person who is ill but not objectively observed by another person, such as nausea or headache. *Compare* sign.
synapse (SIN-aps) **1.** A junction at the end of an axon where it stimulates another cell. **2.** A gap junction between two cardiac or smooth muscle cells at which one cell electrically stimulates the other; called an *electrical synapse.* (fig. 10.12)
synaptic cleft A narrow space between the synaptic knob of an axon and the adjacent cell, across which a neurotransmitter diffuses. (fig. 10.12)
synaptic knob The swollen tip at the distal end of an axon; the site of synaptic vesicles and neurotransmitter release. (fig. 13.10)
synaptic vesicle A spheroid organelle in a synaptic knob; contains neurotransmitter. (fig. 10.12)
syndrome A group of signs and symptoms that occur together and characterize a particular disease.
synergist (SIN-ur-jist) A muscle that works with the agonist to contribute to the same overall action at a joint.
synovial fluid (sih-NO-vee-ul) A lubricating fluid similar to egg white in consistency, found in the synovial joint cavities and bursae.
synovial joint A point where two bones are separated by a narrow, encapsulated space filled with lubricating synovial fluid; most such joints are relatively mobile. Also called a *diarthrosis.* (fig. 9.4)
systemic Widespread or pertaining to the body as a whole, as in *systemic circulation.*
systemic circuit All blood vessels that convey blood from the left ventricle to all organs of the body and back to the right atrium of the heart; all of the cardiovascular system except the heart and pulmonary circuit. (fig. 20.1)

T

tactile Pertaining to the sense of touch.
tail **1.** A slender process at one end of an organ, such as the tail of the pancreas or epididymis. (fig. 24.20) **2.** In vertebrate animals, an appendage that extends beyond the anus and contains part of the vertebral column; in humans, limited to the embryo. (fig. 4.9)
target cell A cell acted upon by a nerve fiber or by a chemical messenger such as a hormone.
tarsal **1.** Pertaining to the ankle. **2.** Pertaining to the margin of the eyelid.
T cell A type of lymphocyte involved in nonspecific defense, humoral immunity, and cellular immunity; occurs in several forms including helper, cytotoxic, and suppressor T cells.
TEM **1.** Transmission electron microscope. **2.** Transmission electron micrograph, a photograph taken with the transmission electron microscope. *Compare* SEM.
temporal **1.** Pertaining to time, as in *temporal summation* in neurons. **2.** Pertaining to the side of the head, as in *temporal bone.*
tendinous cords Fibers that extend from the papillary muscles to the atrioventricular valve cusps in each ventricle of the heart, and serve to keep the valves from prolapsing during ventricular systole; also called *chordae tendineae.* (fig. 20.8)
tendon A collagenous band or cord associated with a muscle, usually attaching it to a bone and transferring muscular tension to it. *See also* aponeurosis.
teratogen Any agent capable of causing birth defects, including chemicals, infectious microorganisms, and radiation.
Terminologia Anatomica A code of standard anatomical terms developed by an international committee of anatomists, the Federative Committee on Anatomical Terminology, and published in 1998; provides a worldwide standard for naming human structures.
teres (TERR-eez) Round, cylindrical; used in the names of muscles and ligaments such as *teres major* and *ligamentum teres.*
testis The male gonad; produces spermatozoa and testosterone.
thalamus (THAL-uh-muss) The largest part of the diencephalon, located immediately inferior to the corpus callosum and bulging into each lateral ventricle; a point of synaptic relay of nearly all signals passing from lower levels of the CNS to the cerebrum. (fig. 15.12)
theory An explanatory statement, or set of statements, that concisely summarizes the state of knowledge on a phenomenon and provides direction for further study; for example, the fluid mosaic theory of the plasma membrane and the sliding filament theory of muscle contraction.
thermoreceptor A neuron specialized to respond to heat or cold, found in the skin and mucous membranes, for example.
third-order neuron An interneuron of the brain that receives sensory signals from a second-order neuron (often at the thalamus) and usually relays them to their final destination in the primary sensory cortex of the brain; in a few cases, a fourth-order neuron completes the pathway. (fig. 14.4) *See also* first-order neuron; second-order neuron.
thoracic Pertaining to the chest.
thorax A region of the trunk between the neck and the diaphragm; the chest.
thymus A lymphatic organ in the mediastinum superior to the heart; the site where T lymphocytes differentiate and become immunocompetent. (fig. 18.7)
thyroid cartilage A large shieldlike cartilage that encloses the larynx anteriorly and laterally and provides an anterior anchorage for the vocal cords and insertion for the infrahyoid muscles. (fig. 23.4)
thyroid gland An endocrine gland in the neck, partially encircling the trachea immediately inferior to the larynx. (fig. 18.8)
tight junction A zipperlike junction between epithelial cells that limits the passage of substances between them. (fig. 2.14)
tissue An aggregation of cells and extracellular materials, usually forming part of an organ and performing some discrete function for it; the four primary classes are epithelial, connective, muscular, and nervous tissue.
tissue gel The viscous colloid that forms the ground substance of many tissues; gets its consistency from hyaluronic acid or other glycosaminoglycans.
trabecula (tra-BEC-you-la) A thin plate or sheet of tissue, such as the calcified trabeculae of spongy bone or the fibrous trabeculae that subdivide a gland. (fig. 6.4)
trachea (TRAY-kee-uh) A cartilage-supported tube from the inferior end of the larynx to the origin of the primary bronchi; conveys air to and from the lungs; the "windpipe."
tract **1.** In the central nervous system, a bundle of nerve fibers with a similar origin, destination, and function, such as the corticospinal tracts of the spinal cord and commissural tracts of the cerebrum. **2.** A continuous anatomical pathway such as the digestive tract.
transitional epithelium A stratified epithelium of the urinary tract that is capable of changing thickness and number of cell layers from relaxed to stretched states. (fig. 3.12)
transmission electron microscope (TEM) A microscope that uses an electron beam in place of light to form high-resolution, two-dimensional images of ultrathin slices of cells or tissues; capable of extremely high magnification. *Compare* scanning electron microscope.
transverse section *See* cross section.
transverse (T) tubule A tubular extension of the plasma membrane of a muscle cell that conducts action potentials into the sarcoplasm and excites the sarcoplasmic reticulum. (fig. 10.8)
trauma Physical injury caused by external forces such as falls, gunshot wounds, motor vehicle accidents, or burns.
trisomy-21 The presence of three copies of chromosome 21 instead of the usual two; causes variable degrees of mental retardation, a shortened life expectancy, and structural anomalies of the face and hands. Also called *Down syndrome.*
trochanter Either of two massive processes serving for muscle attachment at the proximal end of the femur.
trunk **1.** That part of the body excluding the head, neck, and limbs. **2.** A major blood vessel, lymphatic vessel, or nerve that gives rise to smaller branches; for example, the pulmonary trunk and spinal nerve trunks. (fig. 14.14)
T tubule *See* transverse tubule.
tubercle A rounded process on a bone, such as the greater tubercle of the humerus.
tuberosity A rough area on a bone, such as the tibial or ischial tuberosity.
tubuloacinar gland A gland in which secretory cells are found in both the tubular and acinar portions. (fig. 3.29)
tunic A layer that encircles or encloses an organ, such as the tunics of a blood vessel or eyeball; also called a *tunica.* (fig. 21.1)
tympanic membrane The eardrum.

U

ultrastructure Structure at or near the molecular level, made visible by the transmission electron microscope.

umbilical (um-BIL-ih-cul) **1.** Pertaining to the cord that connects a fetus to the placenta. **2.** Pertaining to the navel (umbilicus).
undifferentiated Pertaining to a cell or tissue that has not yet attained a mature functional form; capable of differentiating into one or more specialized functional cells or tissues; for example, stem cells and embryonic tissues.
unencapsulated nerve ending *See* free nerve ending.
unipotent Pertaining to a stem cell that is capable of differentiating into only one type of mature cell, such as a spermatogonium able to produce only sperm, or an epidermal basal cell able to produce only keratinocytes.
unit membrane Any cellular membrane composed of a bilayer of phospholipids and embedded proteins. A single unit membrane forms the plasma membrane and encloses many organelles of a cell, whereas double unit membranes enclose the nucleus and mitochondria. (fig. 2.6)
unmyelinated Lacking a myelin sheath. (fig. 13.9)
upper limb The appendage that arises from the shoulder, consisting of the brachium from shoulder to elbow, the antebrachium from elbow to wrist, the wrist, and the hand; loosely called the *arm,* but that term properly refers only to the brachium.
urea The most abundant nitrogenous waste in urine, formed in the liver by a reaction between ammonia and carbon dioxide.
urethra The passage that conveys urine from the urinary bladder to the outside of the body; in males, it also conveys semen and acts as part of both the urinary and reproductive tracts.
urinary system An organ system specialized to filter the blood plasma, excrete waste products from it, and regulate the body's water, acid-base, and electrolyte balance.
urogenital (U-G) system Collective term for the reproductive and urinary tracts; also called the *genitourinary (G-U) system.*
uterine tube A duct that extends from the ovary to the uterus and conveys an egg or conceptus to the uterus; also called the *fallopian tube* or *oviduct.*

V

varicose vein A vein that has become permanently distended and convoluted due to a loss of competence of the venous valves; especially common in the lower limb, esophagus, and anal canal (where they are called *hemorrhoids*).
vas (vass) (plural, *vasa*) A vessel or duct.
vascular Possessing or pertaining to blood vessels.
vasoconstriction (VAY-zo-con-STRIC-shun) The narrowing of a blood vessel due to muscular constriction of its tunica media.
vasodilation (VAY-zo-dy-LAY-shun) The widening of a blood vessel due to relaxation of the muscle of its tunica media and the outward pressure of the blood exerted against the wall.
vasomotion Any constriction or dilation of a blood vessel.
vein Any blood vessel that carries blood toward either atrium of the heart.
ventral Pertaining to the front of the body, the regions of the chest and abdomen; anterior.
ventral root The branch of a spinal nerve that emerges from the anterior side of the spinal cord and carries efferent (motor) nerve fibers; also called the *anterior root.* (fig. 14.10)
ventricle A fluid-filled chamber of the brain or heart. (fig. 15.5)
venule (VEN-yool) The smallest type of vein, receiving drainage from capillaries.
vertebra (VUR-teh-bra) One of the bones of the vertebral column.
vertebral column (VUR-teh-brul) A dorsal series of usually 33 vertebrae; encloses the spinal cord, supports the skull and thoracic cage, and provides attachment for the limbs and postural muscles. Also called the *spine* or *spinal column.*
vesicle **1.** A fluid-filled tissue sac such as the seminal vesicle. **2.** A fluid-filled spheroidal organelle such as a synaptic or secretory vesicle.
vestibular apparatus Structures of the inner ear concerned with equilibrium and the perception of the movements and orientation of the head, including the semicircular ducts, utricle, and saccule.
vestibule An anatomical receiving chamber; for example, the vestibule between the teeth and cheek, the space immediately inside the nostril, and the chamber of the inner ear to which the cochlea and semicircular ducts are attached.
viscera (VISS-er-uh) (singular, *viscus*) The organs contained in the body cavities, such as the brain, heart, lungs, stomach, intestines, and kidneys.
visceral **1.** Pertaining to the viscera. **2.** The inner or deeper layer of a two-layered membrane such as the pleura, pericardium, or glomerular capsule. (fig. A.8) *Compare* parietal.
visceral muscle Single-unit smooth muscle found in the walls of blood vessels and the digestive, respiratory, urinary, and reproductive tracts.
volar Pertaining to the anterior surfaces of the fingers (the surfaces continuous with the palmar skin).
voluntary Under conscious control, as in skeletal muscle.
vulva The female external genitalia; the mons pubis, labia majora, and all superficial structures between the labia majora; also called the *pudendum.* (fig. 26.22)

W

white matter White myelinated nervous tissue deep to the cortex of the cerebrum and cerebellum and superficial to the gray matter of the spinal cord. (fig. 15.5)

X

X chromosome The larger of the two sex chromosomes; males have one X chromosome and females have two in each somatic cell.
xiphoid process (ZIFF-oyd, ZYE-foyd) A small pointed cartilaginous or bony process at the inferior end of the sternum. (fig. 7.27)
X ray **1.** A high-energy, penetrating electromagnetic ray with wavelengths in the range of 0.1 to 10 nm; used in diagnosis and therapy. **2.** A photograph made with X rays; radiograph.

Y

Y chromosome The smaller of the two sex chromosomes, found only in males and having little if any genetic function except development of the testis.
yolk sac An embryonic membrane that encloses the yolk in vertebrates that lay eggs and serves in humans as the origin of the first blood and germ cells. (fig. 4.6)

Z

zygomatic arch An arch of bone anterior to the ear, formed by the zygomatic processes of the temporal, frontal, and zygomatic bones; origin of the masseter muscle. (fig. 7.5)
zygote A single-celled, fertilized egg.

Key to pronunciation guides.

Pronounce letter sequences in the pronunciation guides as follows:

ah	as in father
al	as in pal
ay	as in day
bry	as in bribe
byu	as in bureau
c	as in calculus
cue	as in ridiculous
cuh	as in cousin
cul	as in bicycle
cus	as in custard
dew	as in dual
eez	as in ease
eh	as in feather
err	as in merry
fal	as in fallacy
few	as in fuse
ih	as in fit
iss	as in sister
lerr	as in lair
lur	as in learn
ma	as in man
mah	as in mama
me	as in meat
merr	as in merry
mew	as in music
muh	as in mother
na	as in corona
nerr	as in nary
new	as in news
nuh	as in nothing
odj	as in dodger
oe	as in go
oh	as in home
ol	as in alcohol
oll	as in doll
ose	as in gross
oss	as in floss
perr	as in pair
pew	as in pewter
ruh	as in rugby
serr	as in serration
sterr	as in stereo
sy	as in siren
terr	as in terrain
tirr	as in tyranny
thee	as in theme
uh	as in mother
ul	as in bicycle
verr	as in very
y	as in why
zh	as in measure
zy	as in enzyme

CREDITS

Photographs

Chapter 1

Opener: © SPL/Photo Researchers, Inc.; **1.1:** National Library of Medicine/Peter Arnold, Inc.; **1.2a:** © U.H.B. Trust/Tony Stone Images/Getty Images; **1.2b:** © Custom Medical Stock; **1.3top:** © Alexander Tsiaras/Photo Researchers, Inc.; **1.3 bottom:** © Scott Camazine/Sue Trainor/Photo Researchers, Inc.; **1.4:** © CNR/Phototake; **1.5:** © Monte S. Buchsbaum, Mt. Sinai School of Medicine, New York, NY; **1.6:** © Tony Stone Images/Getty Images; **1.9:** National Library of Medicine/Peter Arnold, Inc.; **1.10:** © SPL/Photo Researchers, Inc.; **1.11a:** Kathy Talaro/Visuals Unlimited; **1.11b:** Courtesy of the Armed Forces Institute of Pathology; **1.12:** © Dr. Landrum Shettles; **1.14:** © Tim Davis/Photo Researchers, Inc.

Atlas A

A.1, A3, A.5a-d: © The McGraw-Hill Companies, Inc. /Joe DeGrandis, photographer; **A.17:** © CNR/Phototake; **A.18:** © Tony Stone Images/Getty Images; **A.19, A.20, A.21:** © The McGraw-Hill Companies, Inc./Rebecca Gray, photographer/Don Kincaid, dissections; **A.22a:** © The McGraw-Hill Companies, Inc./Dennis Strete, photographer; **A.22b:** © The McGraw-Hill Companies, Inc./Rebecca Gray, photographer/Don Kincaid, dissections

Chapter 2

Opener: © Dr. Donald Fawcett & Dr. Porter/Visuals Unlimited; **2.1a,b:** From *Cell Ultrastructure* by William A. Jenson and Roderick B. Park. © 1967 by Wadsworth Publishing Co., Inc. Reprinted by permission of the publisher; **2.2a:** © Ed Reschke; **2.2b,c:** David Phillips/Photo Researchers, Inc.; **2.6a:** © Don Fawcett/Photo Researchers, Inc.; **2.12a:** Courtesy of Dr. S. Ito, Harvard Medical School; **2.12b:** Biophoto Associates/Photo Researchers, Inc.; **2.13a:** © Don Fawcett/Photo Researchers, Inc.; **2.13c:** Biophoto Associates/Photo Researchers, Inc.; **2.15a:** © K.G. Murti/Visuals Unlimited; **2.15b(inset):** Biology Media/Photo Researchers, Inc.; **2.15b:** © K.G. Murti/Visuals Unlimited; **2.20(all):** © Ed Reschke; **2.21c:** © Biophoto Associates/Science Source / Photo Researchers, Inc.

Chapter 3

Opener: © Dr. Andrejs Liepins/Photo Researchers, Inc.; **3.4a, 3.5a:** © The McGraw-Hill Companies, Inc./Dennis Strete, photographer; **3.6a:** © Ed Reschke; **3.7a:** © The McGraw-Hill Companies, Inc./Dennis Strete, photographer; **3.8:** From R.G. Kessel and R.H. Kardon, *Tissues and Organs: A Text-Atlas of Scanning Electron Microscopy* (W.H. Freeman, 1979); **3.9a, 3.10a:** © Ed Reschke; **3.11a, 3.12a:** © The McGraw-Hill Companies, Inc./Dennis Strete, photographer; **3.13:** © The McGraw-Hill Companies, Inc./Rebecca Gray, photographer/on Kincaid, dissections; **3.14a:** © imagingbody.com; **3.15a, 3.16a, 3.17a, 3.18a, 3.19a, 3.20a:** © The McGraw-Hill Companies, Inc./Dennis Strete, photographer; **3.21a:** © Ed Reschke/Peter Arnold, Inc.; **3.22a:** © The McGraw-Hill Companies, Inc./Dennis Strete, photographer; **3.23a, 3.24a, 3.25a, 3.26a:** © Ed Reschke; **3.27a:** © The McGraw-Hill Companies, Inc./Dennis Strete, photographer; **3.31:** © Ken Greer/Visuals Unlimited

Chapter 4

Opener: © Petit Format/Nestle/Photo Researchers, Inc.; **4.9:** © 2003 L.B. Shettles, MD; **4.10:** © Neil Hardin/Tony Stone Images/Getty Images; **4.14a,b:** Dr. Kurt Benirschke; **4.15:** © Martin Rotker/Phototake; **4.18a:** Photo Researchers, Inc.; **4.18b:** © Biophoto Associates/Photo Researchers, Inc.; **4.19:** Courtesy Dr. Almuth Lenz

Chapter 5

Opener: © SPL/Photo Researchers, Inc.; **5.2b:** © The McGraw-Hill Companies, Inc./Dennis Strete, photographer; **5.3(top):** © Tom McHugh/Photo Researchers, Inc.; **5.3(bottom):** © The McGraw-Hill Companies, Inc./Joe DeGrandis, photographer; **5.4a:** © The McGraw-Hill Companies, Inc./Dennis Strete, photographer; **5.4b&c:** From R.G. Kessel and R.H. Kardon, *Tissues and Organs: A Text-Atlas of Scanning Electron Microscopy* (W.H. Freeman, 1979.); **5.6a:** © Lester V. Bergman/CORBIS; **5.6b:** © Fred Hossler/Visuals Unlimited; **5.7b:** © CBS/Phototake; **5.7c:** © P.M. Motta/SPL/Custom Medical Stock Photos; **5.8a-d, 5.10** (all): © The McGraw-Hill Companies, Inc./Joe DeGrandis, photographer; **5.13a:** © NMSB/Custom Medical Stock Photos; **5.13b:** © Biophoto Associates/Photo Researchers, Inc.; **5.13c:** © James Stevenson/SPL/Photo Researchers, Inc.; **5.14a:** © SPL/Custom Medical Stock Photos; **5.14b&c:** © John Radcliffe/Photo Researchers, Inc.

Chapter 6

Opener: © Univ. Zurich/Photo Researchers, Inc.; **6.4a:** D.W. Fawcett/Visuals Unlimited; **6.4d:** Visuals Unlimited; **6.5:** © Robert Calentine/Visuals Unlimited; **6.8:** Victor Eroschenko; **6.9:** © Biophoto Associates/Photo Researchers, Inc.; **6.9:** Courtesy of Utah Valley Regional Medical Center, Department of Radiology; **6.11:** © Ken Saladin; **6.12:** © The McGraw-Hill Companies, Inc./Joe DeGrandis, photographer; **6.14a:** © Doug Sizemore/Visuals Unlimited; **6.14b:** © SIU/Visuals Unlimited; **6.16a:** © Michael Klein/Peter Arnold, Inc.; **6.16b:** © Dr. P. Marazzi/Photo Researchers, Inc.; **6.17:** Custom Medical Stock

Chapter 7

Opener: Mehau Kulyk/Science Photo Library/Photo Researchers, Inc.; **7.30:** © Biophoto Associates/Photo Researchers, Inc.; **7.34:** © The McGraw-Hill Companies, Inc./Bob Coyle, photographer

Chapter 8

Opener: © NHS Trust/Tony Stone Images/Getty Images; **8.16b:** © Walter Reiter/Phototake; **8.17a:** © SPL/Photo Researchers, Inc.; **8.17b:** Jim Stevenson/Science Photo Library/Photo Researchers, Inc.

Chapter 9

Opener: CNRI/Science Photo Library/Photo Researchers, Inc.; **9.7a-9.13d:** © The McGraw-Hill Companies, Inc./Tim Vacula, photographer; **9.15d, 9.18d, 9.20:** © The McGraw-Hill Companies, Inc./Rebecca Gray, photographer / Don Kincaid, dissections; **9.23:** From Vidic/Saurez, *Photographic Atlas of the Human Body,* Copyright © 1984 Mosby-Year Book, In., St. Louis, MO; **9.24:** © Richard Anderson; **9.25a:** © SIU/Visuals Unlimited; **9.25b:** © Ron Mensching/Phototake; **9.25c:** © SIU/Peter Arnold, Inc.; **9.25d:** © Mehau Kulyk/SPL/Photo Researchers, Inc.

Chapter 10

Opener: © Don W. Fawcett/Photo Researchers, Inc.; **10.1:** © Ed Reschke; **10.2c:** Victor Eroschenko; **10.10a:** Visuals Unlimited; **10.11:** Victor B. Eichler; **10.14:** From R.G. Kessel and R.H. Kardon, *Tissues and Organs: A Text-Atlas of Scanning Electron Microscopy* (W.H. Freeman, 1979); **10.17a:** © Ed Reschke; **10.17b:** © The McGraw-Hill Companies, Inc./Dennis Strete, photographer

Chapter 11

Opener: © Simon Fraser/Photo Researchers, Inc.; **11.2:** © The McGraw-Hill Companies, Inc./Rebecca Gray, photographer/Don Kincaid, dissections; **11.4(all), 11.8:** © The McGraw-Hill Companies, Inc./Joe DeGrandis, photographer; **11.9:** © The McGraw-Hill Companies, Inc./Rebecca Gray, photographer/Don Kincaid, dissections; **11.12:** © imagingbody.com; **11.16:** © The McGraw-Hill Companies, Inc./Rebecca Gray, photographer/ Don Kincaid, dissections

Chapter 12
Opener: Custom Medical Stock; **12.3a,b, 12.9, 12.13, 12.18:** © The McGraw-Hill Companies, Inc./Rebecca Gray, photographer/Don Kincaid, dissections

Atlas B
B.1ab, B.2ab, B.3ab: © The McGraw-Hill Companies, Inc./Joe DeGrandis, photographer **B.4ab:** © The McGraw-Hill Companies, Inc./Tim Vacula, photographer **B.5, B.6, B.7ab, B.8ab, B.9ab, B.10, B.11, B.12, B.13ab, B.14ab, B.15ab:** © The McGraw-Hill Companies, Inc./Joe DeGrandis, photographer

Chapter 13
Opener: © SPL/Photo Researchers, Inc.; **13.9:** © The McGraw-Hill Companies, Inc./Dr. Dennis Emery, Dept. of Zoology and Genetics, Iowa State University, photographer; **13.10:** © Omikron/Science Source/Photo Researchers, Inc.; **13.16:** Alfred I. DuPont Institute

Chapter 14
Opener: Visuals Unlimited; **14.6:** MMP / Campix; **14.7b:** From R.G. Kessel and R.H. Kardon, *Tissues and Organs: A Text-Atlas of Scanning Electron Microscopy* (W.H. Freeman, 1979); **14.11:** © From *A Stereoscopic Atlas of Anatomy of David L. Bassett.* Courtesy of Dr. Robert A. Chase, MD.; **14.15:** From Vidic/Saurez, *Photographic Atlas of the Human Body.* Copyright © 1984 Mosby-Year Book, Inc., St. Louis, MO

Chapter 15
Opener: © CNRI/Photo Researchers, Inc.; **15.1c:** © The McGraw-Hill Companies, Inc./Rebecca Gray, photographer/Don Kincaid, dissections; **15.2b:** © The McGraw-Hill Companies, Inc./Dennis Strete, photographer; **15.5c:** © The McGraw-Hill Companies, Inc./Rebecca Gray, photographer/Don Kincaid, dissections; **15.21:** from Damasio H., Grabowski, T., Frank R. Galaburda A.M., Damasio A.R.: *The Return of Phineas Gage; Clues about the brain from a famous patient. Science,* 264: 1102-1105, 1994. Department of Neurology and Image Analysis Facility, University of Iowa; **15.23b:** © The McGraw-Hill Companies, Inc./Rebecca Gray, photographer/Don Kincaid, dissections; **15.30c:** © The McGraw-Hill Companies, Inc./Joe DeGrandis, photographer; **15.36a:** Custom Medical Stock Photo, Inc.; **15.36b:** © Simon Fraser/Photo Researchers, Inc.

Chapter 16
Opener: © Innerspace Imaging/Photo Researchers, Inc.; **16.3:** © From *A Stereoscopic Atlas of Anatomy* by David L. Bassett. Courtesy of Dr. Robert A. Chase, MD.

Chapter 17
Opener: Photo Researchers, Inc.; **17.5c:** © Ed Reschke; **17.9:** © The McGraw-Hill Companies, Inc./Joe DeGrandis, photographer; **17.13:** Quest/Science Photo Library/Photo Researchers, Inc.; **17.24:** © Ralph C. Eagle, MD/Photo Researchers, Inc.; **17.25a:** © Lisa Klancher; **17.27a:** © The McGraw-Hill Companies, Inc./Joe DeGrandis, photographer; **17.28a:** Courtesy of Beckman Vision Center at UCSF School of Medicine/D. Copenhagen, S. Miltman, and M. Maglio

Chapter 18
Opener: © P. Bagavandoss/Photo Researchers, Inc.; **18.5a&b:** Visuals Unlimited; **18.8b:** © Robert Calentine/Visuals Unlimited; **18.9b:** © John Cunningham/Visuals Unlimited; **18.10b:** Visuals Unlimited; **18.11c:** © Ed Reschke; **18.12a:** © Manfred Kage/Peter Arnold, Inc.; **18.12b:** Ed Reschke; **18.15a,b:** From *Atlas of Pediatric Physical Diagnosis,* 3/e, by Zitelli & Davis, fig 9-17 1997. Mosby-Wolfe Europe Limited, London, UK; **18.16:** Courtesy Mihaly Bartalos, from M. Bartalos and T.A. Baramki, 1967 *Medical Cytogenetics,* Williams & Wilkins

Chapter 19
Opener: © Juergen Berger, Max-Planck Institute/Photo Researchers, Inc.; **19.3b:** © Richard J. Poole/Polaroid International Photomicrography Competition; **19.4:** © Don W. Fawcett/Visuals Unlimited; **19.8:** © Don W. Fawcett/Visuals Unlimited; **Table 19.4(all):** © Ed Reschke; **19.10:** NIBSC/Science Photo Library/Photo Researchers, Inc.; **19.12:** © P. Motta/SPL / Photo Researchers, Inc.; **19.13:** © Meckes/Ottawa / Photo Researchers, Inc.

Chapter 20
Opener: Friedler Michler/Science Photo Library/Photo Researchers, Inc.; **20.5a:** © The McGraw-Hill Companies, Inc.; **20.5b, 20.6a:** © imagingbody.com; **20.8b:** © The McGraw-Hill Companies, Inc.; **20.12a&b:** © Ed Reschke; **20.12c:** Custom Medical Stock Photo; **20.14a:** © Ed Reschke

Chapter 21
Opener: © Biophoto Associates/Photo Researchers Inc.; **21.2:** © The McGraw-Hill Companies, Inc./Dennis Strete, photographer; **21.3:** © Photo Researchers, Inc.; **21.5b:** © Wolf H. Fahrenbach/Visuals Unlimited; **21.6b:** Courtesy of S. McNutt

Chapter 22
Opener: © Eye of Science/Photo Researchers, Inc.; **22.2:** © SPL/Custom Medical Stock Photos; **22.4:** © The McGraw-Hill Companies, Inc./Dennis Strete, photographer; **22.7:** © Manfred Kage/Peter Arnold, Inc.; **22.8:** Custom Medical Stock Photo; **22.10b:** © The McGraw-Hill Companies, Inc./Dennis Strete, photographer; **22.11:** © imagingbody.com; **22.12c:** Photo Researchers, Inc.; **22.14b:** © Biophoto Associates/Photo Researchers, Inc.; **22.15a&c:** © The McGraw-Hill Companies, Inc./Dennis Strete, photographer; **22.17a,b:** Visuals Unlimited

Chapter 23
Opener: © imagingbody.com; **23.2c:** © The McGraw-Hill Companies, Inc./Rebecca Gray, photographer/Don Kincaid, dissections; **23.3a,b:** © The McGraw-Hill Companies, Inc./Joe DeGrandis, photographer; **23.5a:** Phototake; **23.5b:** © J. Siebert/Custom Medical Stock Photo; **23.8:** Custom Medical Stock; **23.10a:** Phototake; **23.10b:** © Fred Hossler/Visuals Unlimited; **23.12:** Visuals Unlimited

Chapter 24
Opener: © SPL/Photo Researchers, Inc.; **24.8:** © The McGraw-Hill Companies, Inc./Rebecca Gray, photographer/Don Kincaid, dissections; **24.10b:** © The McGraw-Hill Companies, Inc./Dennis Strete, photographer; **24.11b:** © The McGraw-Hill Companies, Inc./Rebecca Gray, photographer/Don Kincaid, dissections; **24.13a&b:** CNRI/SPL/ Photo Researchers, Inc.; **24.15a:** Meekes/Ottawa/Photo Researchers, Inc.; **24.15b, 24.19b2, 24.21b:** © The McGraw-Hill Companies, Inc./Dennis Strete, photographer

Chapter 25
Opener: © Dr. Donald Fawcett & Spinelli/Visuals Unlimited; **25.3a:** © imagingbody.com; **25.7b:** © Manfred Kage/Peter Arnold, Inc.; **25.8a:** From R.G. Kessel and R.H. Kardon, *Tissues and Organs: A Text-Atlas of Scanning Electron Microscopy* (W.H. Freeman, 1979); **25.8b:** Photo Researchers, Inc.; **25.8c:** Biological Photo Service

Chapter 26
Opener: © Simon Fraser/Photo Researchers, Inc.; **26.3b:** © The McGraw-Hill Companies, Inc./Dennis Strete, photographer; **26.4a:** From R.G. Kessel and R.H. Kardon, *Tissues and Organs: A Text-Atlas of Scanning Electron Microscopy* (W.H. Freeman, 1979); **26.4b:** © Ed Reschke; **26.9b:** Visuals Unlimited; **26.15a:** © Ed Reschke; **26.15b:** © Manfred Kage / Peter Arnold, Inc.; **26.16:** Dr. Landrum Shettles; **26.17:** © The McGraw-Hill Companies, Inc./Rebecca Gray, photographer/Don Kincaid, dissections; **26.18:** Photo Researchers, Inc.; **26.19:** © Ed Reschke; **26.24:** © The McGraw-Hill Companies, Inc./Dennis Strete, photographer; **26.25:** Courtesy Toshikazu Nagato, from *Cell and Tissue Research,* 209: 1-10, 3, pg. 6, Springer-Verlag, 1980; **26.29a:** © SPL/Photo Researchers, Inc.

INDEX

Following a page number, *f* indicates a figure, *t* indicates a table, and *i* indicates information in an Insight box. Specific arteries, veins, bones, muscles, nerves, and other such structures are indexed under the general term (see *Arteries, specific,* and similar entries).

A

Abbe, Ernst, 10
ABCD rule for malignant melanoma, 146
Abdominal cavity, 29*f*, 30, 42*f*, 306*i*
Abdominal region
arteries, 604*f*, 604*t*–606*t*, 605*f*, 606*f*
muscular anatomy, 300*f*, 300*t*–302*t*, 301*f*, 302*f*
surface anatomy, 26, 27*f*, 28*f*, 352*f*
veins, 615*f*, 615*t*–617*t*, 616*f*
Abdominopelvic cavity, 29, 29*f*, 43*f*
Abduction of joints, 235, 236*f*, 320*t*, 329*t*
Abortion, spontaneous, 120
Abruptio placentae, 118*i*, 758*t*
Accessory organs
of digestion, 676, 694–699
of integumentary system, 137–142
of reproductive system, 737
of synovial joints, 232
Accommodation of lens, 499, 504
Acetabular labrum, 242, 244*f*
Acetabulum, 214, 214*f*, 216*t*, 234, 242, 244*f*
Acetylcholine, 266–267, 269, 276, 376, 450, 465–466, 466*f*
Acetylcholine receptors, 267, 276, 277
Acetylcholinesterase, 267
Achalasia, 470*t*
Acid-base balance, 152
Acid mantle, 130
Acidosis, 654, 722
Acinus, 97, 97*f*, 684*f*, 699*f*, 751*f*
ACL injury, 247*i*
Acne, 145*t*
Acoustic meatuses, 176*f*, 177*f*, 180, 181, 181*f*, 485, 485*f*
Acquired immunodeficiency syndrome (AIDS), 58, 557, 647–648, 756
Acromegaly, 536*t*
Acromial region, 27*f*
Acromion, 209, 209*f*, 352*f*, 353*f*, 355*f*
Acronyms, 17
Acrosomal reaction, 107, 107*f*
Acrosome, 107, 107*f*, 735*f*, 736, 736*f*
Actin, 59, 60*f*, 263, 264, 265*f*, 266*t*, 269, 270*f*, 271*f*, 545
Action potential, 374
Active transport, 57*f*, 58
Acute glomerulonephritis, 721, 722*t*
Acute pancreatitis, 702*t*
Acute rhinitis, 670*t*
Adam's apple, 658
Adaptation, evolutionary, 14. *See also* Bipedalism
Addison disease, 136, 536*t*
Adduction of joints, 235, 236*f*, 320*t*, 329*t*
Adenohypophysis. *See* Endocrine system, pituitary gland
Adenoids, 640
Adenosine triphosphate (ATP), 68, 654
and muscle contraction, 269, 271*f*, 276
Adipocytes, 86, 89*f*
Adipose tissue, 86, 88, 89*f*, 89*t*, 120*t*
Adolescence, 753. *See also* Puberty
Adrenal gland. *See* Endocrine system, adrenal gland
Adrenergic neurons, 465, 466*t*
Adrenergic receptors, 465–466, 466*f*, 466*i*
Adrenocorticotropic hormone (ACTH), 518, 521*t*, 522, 532, 534–535
Adrenogenital syndrome, 536*t*
Adult respiratory distress syndrome, 670*t*
Adventitia
of blood vessels, 586, 587*f*
of digestive tract, 677, 685
of trachea, 660
of urinary tract, 718
of vagina, 747
Aerobic respiration, 272*t*, 273, 275, 545
Affective language area, 436
Afferent arteriole, 710, 711*f*, 712*f*, 713*f*, 715*f*, 716, 717*f*
Ageusia, 507*t*
Aggression, 437
Aging
of autonomic nervous system, 469
of central nervous system, 449
of circulatory system, 555, 578–579, 622
of digestive system, 701
of endocrine system, 532
of integumentary system, 145
of lymphatic and immune systems, 646
of muscular system, 275–276
of reproductive system, 754–755
of respiratory system, 669
of skeletal system, 163
of urinary system, 720–721
Agonist, 260
Agranulocytes, 549, 551
AIDS. *See* Acquired immunodeficiency syndrome
Airflow, 663, 670
Ala nasi, 655, 657*f*
Albinism, 136
Albumin, 544, 544*t*, 714
Albuminuria, 714*i*
Aldosterone, 527, 716
Alimentary canal. *See* Digestive tract
Allantoic duct, 85*f*, 115
Allantois, 112*f*, 115, 116*f*, 117*f*, 720
Allergy, 647
Allografts, 147*i*
Alveolar ducts and sacs, 661, 663*f*
Alveolar sac, 661, 663*f*
Alveolus
of a bone, 174*t*
dental, 184, 681, 682*f*
pulmonary, 596*f*, 661, 663*f*, 664–665, 664*f*, 668
Alzheimer disease, 449–450, 450*f*
Amacrine cells, 501*f*, 502
Amelia, 223
Amnion, 111*f*, 112*f*, 114, 116*f*, 117*f*, 118*f*, 119*f*
Amniotic cavity, 110, 111*f*, 112*f*, 116*f*, 700*f*
Amniotic fluid, 114, 116*f*, 144, 668, 720
Amphiarthroses, 230–231, 230*f*, 234*t*
Amphibians, 716*i*
Amphimixis, 107
Amygdala, 432, 433*f*, 437, 481, 484*f*
Amyotrophic lateral sclerosis (ALS), 393*f*, 393*i*
Anabolism, 12
Anaerobic fermentation, 271, 272*t*, 545
Anal canal, 43*f*, 693*f*, 694, 720
Anal triangle, 307*t*, 308*f*, 729*f*
Anaphase, 69, 69*f*, 70*f*, 734*f*
Anaphylactic shock, 647
Anastomoses, 571, 595, 595*f*
Anatomical planes, 24–25, 25*f*
Anatomical position, 24, 24*f*, 223*i*
Anatomical snuffbox, 324*t*, 357*f*
Anatomic dead space, 663
Anatomy
history, 8–11
subdisciplines, 2
terminology, 17–19
variation in, 5, 7*f*
Androgen-insensitivity syndrome, 535, 535*f*
Androgens, 527, 753. *See also* Testosterone
Anemia, 555–556, 556*i*, 643*i*
Anencephaly, 122, 124*t*, 380
Anesthesia, 186, 386
Aneuploidy, 121, 121*f*
Aneurysm, 588*f*, 588*i*, 670
Anger, 437, 468
Angioblasts, 620, 620*f*
Angiogenesis, 620–621, 620*f*
Angiography, 3, 4*f*
Angiotensin, 530, 654, 708
Angular gyrus, 436
Ankle, 218, 220, 246, 248*f*, 249*t*
surface anatomy, 359*f*, 360*f*, 361*f*, 362*f*
Ankylosis, 249
Anosmia, 183*i*
Ansa cervicalis, 400*f*, 400*t*, 677
Antagonistic muscles, 259*f*, 260
Antebrachial region, 26
blood vessels (*See* Limbs, upper)
muscular anatomy, 321*f*, 321*t*, 322*t*, 323*f*, 324*t*, 325*f*
skeletal anatomy, 209, 210*f*
surface anatomy, 27*f*, 356*f*
Anterior chamber of eye, 497, 497*f*, 498*f*, 498*i*
Anterior commissure, 416*f*, 430
Anterior superior spine of ilium, 39*f*, 213*f*, 214, 214*f*, 332*f*, 352*f*, 354*f*
Anterolateral system, 389, 480*f*
Antibodies, 548, 551, 644
Antidiuretic hormone (ADH), 428, 518, 520, 522, 717, 721
Antigen-presenting cells (APCs), 551, 635, 646
Antigens, 548
Anti-inflammatory drugs, 326*i*
Antrum of stomach, 685, 686*f*
Anus, 43*f*, 693*f*, 694, 749*f*
Aortic arch, adult, 564*f*, 598*f*, 598*t*
Aortic arches, embryonic, 621, 621*f*
Aortic bodies, 589, 589*f*, 668*f*
Aortic hiatus, 598*f*
Aortic sac, 621, 622*f*
Aortic valve, 567, 568*f*
Apes, 17*f*, 186, 211, 215, 216, 220
Aphasia, 436
Apneustic center, 667, 667*f*
Apocrine sweat glands, 131*f*, 138*f*, 140, 141*f*, 142, 142*t*, 144
Aponeuroses, 259
of external abdominal oblique, 301*f*, 302*f*
plantar, 342*t*
thoracolumbar fascia, 303*f*, 305*f*
Apoptosis, 67, 77*f*, 99, 100, 133, 221
Appendicitis, 702*t*
Appendicular region. *See* Limbs
Appendix, 7*f*, 677*f*, 693*f*, 694
Aprosodia, 436
Aqueous humor, 497, 498*f*
Arachnoid mater, 387, 388*f*, 398*f*, 414, 417, 418*f*, 506
Arachnoid villi, 418, 418*f*, 420*f*
Arbor vitae, 423, 423*f*
Arches
aortic, adult, 564*f*, 598*f*, 598*t*
aortic, embryonic, 621, 621*f*
of foot, 220, 221*f*
palatoglossal, 680*f*
palatopharyngeal, 680*f*
vertebral, 190, 190*f*

Archicortex, 432
Areola, 136, 352*f*, 749, 750*f*
Areolar tissue, 87–88, 89*f*, 89*t*
Aristotle, 8, 106, 562
Arm bud, 114, 114*f*
Arterial pressure points, 609, 609*f*
Arteries, specific
anterior cerebral, 599*f*, 600*f*, 600*t*
anterior communicating, 599*f*, 600*f*, 600*t*
anterior interventricular, 566*f*, 572*f*
anterior tibial, 597*f*, 607*f*, 608*f*, 608*t*
aorta, 38*f*, 41*f*, 42*f*, 564*f*, 568*f*, 571*f*, 597*f*, 598*f*, 598*t*, 604*f*, 604*t*, 665*f*
arcuate
of foot, 608*f*, 608*t*
of kidney, 710, 711*f*
of uterus, 746, 747*f*
axillary, 597*f*, 601*f*, 601*t*
basilar, 599*f*, 600*f*, 600*t*
brachial, 37*f*, 597*f*, 601*f*, 601*t*, 609*f*
brachiocephalic trunk, 39*f*, 597*f*, 598*f*, 598*t*, 599*f*
bronchial, 595, 602*t*, 603*f*, 661
carotid
common, 36*f*, 39*f*, 597*f*, 598*f*, 598*t*, 599*f*, 599*t*, 603*f*, 609*f*, 621, 621*f*, 668*f*
external, 597*f*, 599*f*, 599*t*, 621, 621*f*
internal, 597*f*, 599*f*, 600*f*, 600*t*, 621, 621*f*
celiac trunk, 597*f*, 604*f*, 604*t*, 605*f*, 606*f*, 678
cerebellar, 599*f*, 600*f*
cerebral arterial circle, 600*t*
circumflex
coronary, 570, 572*f*
femoral, 607*f*, 608*f*, 608*t*
humeral, 601*f*, 601*t*
colic, 606*f*, 606*t*
common hepatic, 605*f*, 605*t*
coronary, 569–571, 572*f*, 573*f*, 598*f*
costocervical trunk, 599*f*, 599*t*, 603*f*
deep
brachial, 601*f*, 601*t*
femoral, 597*f*, 607*f*, 608*f*, 608*t*, 639*f*
palmar arch, 601*f*, 602*t*
penile, 738
digital
of foot, 608*f*, 608*t*
of hand, 602*t*, 607*f*
dorsal
aortae, embryonic, 621, 621*f*
pedal, 363*f*, 597*f*, 607*f*, 608*f*, 608*t*, 609*f*
penile, 738
efferent arterioles of kidney, 710, 712*f*, 713*f*, 715*f*, 717*f*
esophageal, 602*t*, 603*f*, 678
facial, 599*f*, 599*t*, 609*f*
femoral, 36*f*, 334*f*, 597*f*, 607*f*, 608*f*, 608*t*, 609*f*, 639*f*
fibular, 607*f*, 608*t*
gastric, 605*f*, 606*t*
gastroduodenal, 605*f*, 605*t*
gastroepiploic, 605*f*, 606*t*
genicular, 608*t*
gluteal, 607*t*, 608*f*
gonadal, 597*f*, 604*f*, 604*t*, 732, 742
helicine, of penis, 738
hepatic, 697, 697*f*
ileocolic, 606*f*, 606*t*
iliac
common, 38*f*, 597*f*, 604*f*, 604*t*, 606*f*, 607*f*, 608*f*, 621
external, 597*f*, 607*f*, 607*t*–608*t*, 608*f*
internal, 597*f*, 607*f*, 607*t*, 608*f*
iliolumbar, 607*t*, 608*f*
inferior pancreaticoduodenal, 605*f*, 605*t*
intercostal, 597*f*, 603*f*, 603*t*, 621
interlobar, of kidney, 710, 711*f*
interlobular, of kidney, 710, 711*f*, 712*f*
internal
pudendal, 738
thoracic, 597*f*, 603*f*, 603*t*
interosseous, 601*f*, 602*t*
intersegmental, 621, 622*f*
intestinal, 606*f*, 606*t*
lateral
genicular, 607*f*
sacral, 607*t*, 608*f*
thoracic, 603*f*, 603*t*
lingual, 599*f*, 599*t*
lobar of lung, 595
lumbar, 604*f*, 604*t*, 621
maxillary, 599*f*, 600*t*
medial genicular, 607*f*
median sacral, 604*f*, 604*t*
mediastinal, 602*t*
mesenteric
inferior, 597*f*, 604*f*, 606*f*, 606*t*, 641*f*, 679
superior, 597*f*, 604*f*, 606*f*, 641*f*, 693*f*
metacarpal, 601*f*, 602*t*
metatarsal, 608*f*, 608*t*
middle
cerebral, 599*f*, 600*f*, 600*t*
rectal, 607*t*, 608*f*
obturator, 607*f*, 607*t*, 608*f*
occipital, 599*f*, 600*t*
ophthalmic, 182, 599*f*, 600*t*
ovarian, 604*f*, 604*t*, 741*f*, 742, 747*f*
palmar digital, 601*f*
pancreatic, 606*t*
penile, 738
pericardiophrenic, 603*f*, 603*t*
peritubular capillaries of kidney, 710
phrenic, 602*t*, 604*f*, 604*t*
plantar, 607*f*, 608*f*, 608*t*
plantar arch, 607*f*, 608*f*, 608*t*
popliteal, 597*f*, 607*f*, 608*f*, 608*t*, 609*f*
posterior
auricular, 599*f*
cerebral, 599*f*, 600*t*
communicating, 599*f*, 600*f*, 600*t*
interventricular, 572*f*
tibial, 597*f*, 607*f*, 608*f*, 608*t*, 609*f*
proper hepatic, 605*f*, 605*t*
pudendal, 607*t*
pulmonary, 564*f*, 568*f*, 571*f*, 595, 596*f*, 621*f*, 661
pulmonary lobar, 596*f*
pulmonary trunk, 564*f*, 568*f*, 571*f*, 595, 596*f*, 621*f*
radial, 597*f*, 601*f*, 602*t*, 609*f*
radial recurrent, 601*f*, 601*t*
renal, 597*f*, 604*f*, 604*t*, 708*f*, 710, 711*f*, 721*i*
segmental, of kidney, 710, 711*f*
sigmoid, 606*t*
spinal, 600*f*
spiral, of uterus, 746, 747, 748*f*
splenic, 605*f*, 606*t*, 643*f*
subclavian, 37*f*, 597*f*, 598*f*, 598*t*, 599*f*, 599*t*, 601*f*, 601*t*, 603*f*, 621*f*
subcostal, 602*t*
subscapular, 603*f*, 603*t*
superficial
palmar arch, 601*f*, 602*t*
temporal, 599*f*, 600*t*, 609*f*
superior
rectal, 606*f*, 606*t*
thyroid, 402*f*, 599*f*, 599*t*
vesical, 622
suprarenal, 604*f*, 604*t*
testicular, 604*f*, 604*t*, 732
thyrocervical trunk, 599*f*, 599*t*, 603*f*
ulnar, 597*f*, 601*f*, 602*t*
ulnar recurrent, 601*f*, 601*t*
umbilical, 116, 622, 622*f*, 623*f*
uterine, 607*t*, 608*f*, 742, 746, 747*f*
vaginal, 607*t*, 608*f*, 747*f*
vertebral, 597*f*, 599*f*, 599*t*–600*t*, 600*f*, 603*f*
vesical, 607*t*, 608*f*
vitelline, 622*f*
Arteries in general
definition, 586
histology, 88, 587*f*
of pulmonary circuit, 595, 596*f*
of systemic circuit, 596–608, 597*f*
types and structure, 587*f*, 588–589
Arterioles, 589, 590*f*, 593*f*
Arthritis, 227*f*, 247, 249, 250*f*, 646
Arthrology, 228
Arthroplasty, 250
Arthroscopy, 247*i*
Articular discs, 231, 239, 240*f*
Articulations. *See* Joints
Artificial joints, 250, 251*f*
Ascites, 702*t*
Ascorbic acid, 163
Aspartate, 376
Aspirin, 250, 326*i*, 689*i*
Association cortex, 433–434, 435*f*, 437
Association tracts, 430, 431*f*
Aster, 70*f*
Asthma, 670
Astigmatism, 500*f*, 500*t*
Astrocytes, 371, 372*f*, 372*t*, 418, 502
Astrocytosis, 371
Ataxia, 425
Atelectasis, 666*i*
Atheroma, 572*i*
Atherosclerosis, 535, 572*i*, 573*f*, 588*i*, 622, 721
Athletes, 272, 344, 714*i*
Atresia, 742
Atrial natriuretic peptide (ANP), 528–529, 717
Atria of heart, 564*f*, 566–568, 566*f*, 568*f*, 570, 570*f*, 571*f*, 572, 572*f*, 577–578
Atrioventricular (AV) bundle, 574, 574*f*
Atrioventricular (AV) node, 571, 574, 574*f*, 574*i*, 576
Atrioventricular (AV) valves, 568*f*, 569*f*, 570*f*, 571*f*
Atrophy, 99
of brain, 450
of digestive system, 701
of integumentary system, 145
of kidneys, 720–721
of muscular system, 269–270, 275–276
of reproductive system, 755
of skeletal system, 163
Atropine, 466*i*
Auditory association area, 435*f*, 489
Auditory canal, 485, 485*f*
Auditory (eustachian) tube, 485*f*, 486, 504, 656*f*
Auditory ossicles, 197, 485*f*, 486, 504, 505*f*
Auricle
of ear, 485*f*, 504
of heart, 564*f*, 566
Auricular region, 351*f*
Australopithecus, 14, 16, 16*t*, 17*f*
Autoantibodies, 249, 646, 732. *See also* Autoimmune diseases
Autografts, 147*i*
Autoimmune diseases, 63*i*, 249, 276, 646. *See also* Autoantibodies
Autonomic nervous system
and cardiac function, 273
compared to somatic, 458*t*
disorders, 470*t*
divisions, 457, 465*t*, 466*f*, 466*t*
embryonic development, 469
and hypothalamus, 428
in old age, 469
parasympathetic division, 463–464, 464*f*
physiology, 465–469, 466*f*, 466*t*, 467*t*, 468*f*, 469*f*
and smooth muscle, 274, 274*f*
sympathetic division, 458–463, 459*f*, 460*f*, 461*f*, 462*f*
and visceral reflex arcs, 456, 457*f*
Autonomic tone, 457
Autophagy, 67
Autorhythmicity, 274, 574
Avicenna, 9
Avulsion, 224*t*
Axial region, 26
muscular anatomy, 283–308
skeletal anatomy, 171–202
surface anatomy, 27*f*, 351*f*, 352*f*, 353*f*, 354*f*, 355*f*
Axillary folds, 316*t*, 355*f*
Axillary hair, 138
Axillary region, 27*f*, 355*f*
Axis, endocrine, 518, 521*f*
Axolemma, 368, 369*f*
Axon, 94, 94*f*, 368, 369*f*, 375*f*
Axonal transport, 368
Axoneme, 60, 61*f*, 65*f*
Axon hillock, 368, 369*f*, 374
Azotemia, 709, 722

B

Back. *See also* Vertebrae; Vertebral column
injuries, 306*i*
muscles, 303*f*, 303*t*–306*t*, 304*f*, 305*f*
surface anatomy, 27*f*, 353*f*
Balance, sense of, 392
Ball-and-socket joints, 215, 234*t*
Band cells, 543*f*, 549
Baroreceptors, 456, 457*f*, 589, 589*f*, 622
Baroreflex, 456, 457*f*
Barrett esophagus, 685*i*
Basal bodies, 60, 61*f*, 63, 68
Basal cell carcinoma, 146, 146*f*
Basal lamina, 267*f*, 373*f*, 374, 375*f*, 590*f*, 591*f*
Basal nuclei, 432, 433*f*, 436, 438, 439*t*
and Parkinson disease, 450–451
Baseball finger, 345*t*
Basement membrane, 52*f*, 80, 98*f*, 714, 715*f*
Basophils, 543*f*, 548, 550*f*, 550*t*, 551, 553*f*
B cells. *See* B lymphocytes
Beaumont, William, 676
Becquerel, Henri, 152*i*
Beethoven, Ludwig van, 489*i*

Belladonna, 466*i*
Bell palsy, 449*i*
Benign prostatic hyperplasia, 721, 737*i*
Beta-endorphin, 376
Bicuspid valve, 567, 568*f*, 569*f*, 571*f*
Bile, 695
Bile acids, 695
Bile canaliculi, 697, 697*f*
Bile duct, 698, 698*f*
Bile pigments, 696*t*
Bilirubin, 695, 696*t*
Biofeedback, 456*i*
Biomechanics, 228
Bipedalism
  evolution, 16
  limb adaptations, 211–212, 216, 218*f*, 220, 220*f*, 244, 246
  pelvic adaptations, 215
  skull adaptations, 186, 186*f*
  vertebral adaptations, 188–189, 189*f*
Bipolar neurons, 370, 370*f*
  cochlear, 489
  retinal, 500, 501*f*
Birth defects, 120–124, 124*t*
  in bone development, 87*i*, 88*i*, 124*t*, 163*f*, 163*i*, 380
  from chemical teratogens, 122, 124*f*
  in chromosome number, 121–122, 121*f*, 123*f*, 124*t*
  congenital hypothyroidism, 536*t*
  in heart, 579*i*
  in immune system, 647
  from infectious diseases, 122–123
  in limb development, 122, 124*f*, 124*t*, 223, 223*f*
  in muscular system, 276
  in neural tube, 124*t*, 380, 381*f*
  in organ position, 6*i*
  in reproductive system, 739*i*, 753, 756*t*
Birthmarks, 137*i*
Black lung disease, 669
Blastocyst, 109, 109*f*, 109*t*, 110*f*, 120
Blastomeres, 108
Blindness, 498, 498*i*, 502
Blind spot, 499
Blocker's arm, 345*t*
Blood, 541–557
  complete blood count, 552*i*
  as connective tissue, 92–93, 93*f*, 93*t*
  disorders, 555–557, 556*i*, 556*t*, 557*f*
  distribution in body, 593, 594*f*
  erythrocytes, 545–548, 545*f*
  general properties, 542–543, 544*t*
  leukocytes, 548–552, 549*f*, 549*t*–550*t*, 550*f*, 553*f*
  plasma, 92, 542–544, 543*f*, 544*t*
  platelets, 93, 93*f*, 541*f*, 542, 543*f*, 544*t*, 552–555, 554*f*, 555*f*
  in urine, 714*i*
  volume, 544*t*, 642
Blood-brain barrier, 371, 418
Blood clots, 554–555, 555*f*
Blood-CSF barrier, 420
Blood islands, 546, 620, 620*f*
Blood plasma. *See* Blood, plasma
Blood pressure
  autonomic regulation, 456
  in old age, 469
  in spinal cord trauma, 407
Blood-testis barrier, 729, 732, 733, 735*f*
Blood-thymus barrier, 637
Blood types, 548, 548*f*
Blood-urea nitrogen, 709
Blood vessels. *See also* Arteries; Capillaries; Veins
  aging, 622
  disorders, 622, 624*t*
  embryonic development, 620–621, 620*f*, 621*f*
  innervation, 468, 469*f*
  structure, 586–594, 587*f*
  visual examination, 498
B lymphocytes, 544, 552, 553*f*, 635, 637, 644, 645*f*
Body cavities, 26, 29–30, 29*f*, 29*t*, 31*f*
Body fat, 135, 135*f*, 166
Bone loss, 163. *See also* Osteoporosis
Bone marrow
  distribution, 158, 159*f*
  hemopoiesis, 546–547, 547*f*, 552, 553*f*, 554, 554*f*
  histology, 636–637, 637*f*
  lymphocyte development, 644–646, 645*f*
  red and yellow, 43*f*, 158, 159*f*, 636
Bones, specific
  atlas (vertebra C1), 188*f*, 190, 191*f*
  axis (vertebra C2), 188*f*, 190–191, 191*f*
  calcaneus, 173*f*, 219*f*, 220, 221*f*, 222, 248*f*, 359*f*, 363*f*
  capitate, 210, 212*f*
  clavicle, 173*f*, 194*f*, 208, 208*f*, 208*i*, 221, 351*f*
  coccyx, 13, 193*f*, 194, 216*t*, 353*f*, 729*f*
  conchae, nasal, 183, 183*f*, 184*f*, 185, 655, 656*f*
  cuboid, 219*f*, 220, 221*f*, 222
  cuneiforms, 219*f*, 220, 221*f*, 222
  ethmoid, 176*f*, 177*f*, 178*f*, 183, 183*f*, 183*i*, 185*f*, 655
  femur, 173*f*, 215–216, 217*f*, 218*f*, 218*i*, 221, 244*f*
  fibula, 173*f*, 218, 219*f*, 248*f*, 359*f*
  frontal, 173*f*, 175*f*, 178*f*, 179, 185*f*, 198*f*
  hamate, 210, 212*f*
  humerus, 173*f*, 209–210, 210*f*, 221, 242*f*
  hyoid, 186, 186*f*, 197, 292*f*, 294*f*, 658*f*
  ilium, 39*f*, 213*f*, 214, 216*f*, 216*t*, 221
  incus, 186, 485*f*, 486
  inferior nasal concha, 184*f*, 185
  innominate (*See* Bones, specific, os coxae)
  ischium, 213*f*, 214–215, 216*f*, 221
  lacrimal, 175*f*, 176*f*, 184*f*, 185, 185*f*
  lunate, 210, 212*f*
  malleus, 186, 485*f*, 486
  mandible, 173*f*, 175*f*, 176*f*, 185–186, 185*f*, 198*f*
    movements of, 235, 236*f*, 237, 239
    synostosis of, 228
  maxilla, 173*f*, 175*f*, 176*f*, 184, 184*f*, 228
  metacarpals, 211, 212*f*, 221
  metatarsals, 173*f*, 219*f*, 220, 221
  nasal, 175*f*, 176*f*, 184*f*, 185
  navicular
    of foot, 219*f*, 220, 221*f*
    of wrist (scaphoid bone), 210, 212*f*
  occipital, 173*f*, 176*f*, 177*f*, 178*f*, 181–182, 198*f*
  os coxae, 172, 173*f*, 213–215, 213*f*, 216*f*, 222
  palatine, 176*f*, 177*f*, 184, 184*f*, 185*f*
  parietal, 173*f*, 175*f*, 176*f*, 177*f*, 178*f*, 180, 198*f*
  patella, 172, 173*f*, 215, 216, 217*f*, 222, 243, 245*f*, 246*f*, 358*f*
  phalanges
    of foot, 219*f*, 220, 221*f*
    of hand, 211, 212*f*
  pisiform, 210, 212*f*, 328*f*
  pubis, 213*f*, 215, 216*f*, 221
  radius, 173*f*, 210, 211*f*, 221, 242*f*, 243*i*, 356*f*, 357*f*
  ribs, 41*f*, 173*f*, 194*f*, 195, 195*f*, 196*f*, 196*t*
    embryonic development, 200, 200*f*
  sacrum, 43*f*, 193–194, 193*f*, 213, 216*t*, 353*f*
  scaphoid, 210, 212*f*
  scapula, 173*f*, 194*f*, 208–209, 209*f*, 221, 315*f*, 353*f*
  sphenoid, 175*f*, 176*f*, 177*f*, 182–183, 182*f*, 184*f*, 198*f*, 522*i*
  stapes, 186, 485*f*, 486
  sternum, 36*f*, 41*f*, 173*f*, 194–195, 194*f*, 200, 351*f*, 352*f*
  talus, 219*f*, 220, 221*f*
  temporal, 173*f*, 176*f*, 177*f*, 180, 181*f*, 198*f*
  tibia, 173*f*, 217, 219*f*, 221, 246*f*, 248*f*
  trapezium, 210, 212*f*
  trapezoid, 210, 212*f*
  triquetral (triquetrum), 210, 212*f*
  turbinate (*See* Nasal conchae)
  ulna, 173*f*, 210, 211*f*, 221, 242*f*
  vomer, 175*f*, 176*f*, 177*f*, 185, 655
  zygomatic, 173*f*, 175*f*, 176*f*, 177*f*, 185
Bones in general
  disorders, 163–167, 163*f*, 163*i*, 167*t* (*See also* Fractures)
  general anatomy, 153–154, 153*f*, 154*f*, 174, 174*f*
  growth and remodeling, 162–163
  number and inventory, 172, 172*t*
  sesamoid, 172
  shapes, 153, 153*f*
  sutural, 172, 180
Bone spurs, 247
Bone tissue, 92*t*
  cells, 155, 156*f*
  compact, 92, 92*f*, 153, 156–158, 157*f*
  embryonic development, 159–162, 160*f*, 161*f*, 162*f*
  histology, 156*i*, 157*f*
  loss, 163
  matrix, 155
  membranous, 196
  spongy, 92, 151*f*, 153, 154*f*, 157*f*, 158, 158*f*, 161
    in osteoporosis, 166, 166*f*
  stress on, 158, 158*f*
Bony joints. *See* Synostoses
Brachial region, 26
  blood vessels (*See* Limbs, upper)
  muscles, 317*f*, 318*f*, 320*t*–321*t*
  skeletal anatomy, 209–210, 210*f*
  surface anatomy, 27*f*, 355*f*
Bradycardia, 407
Brain, 413–439. *See also* Brainstem; Cerebellum; Cerebrum
  base, 441*f*
  circulation, 418, 599*f*, 600*f*, 600*t*, 611*f*, 611*t*–612*t*
  embryonic development, 379, 380*f*
  general anatomy, 414, 415*f*, 416*f*
  meninges, 414, 417, 418*f*
  PET scan, 6*f*
  ventricles and cerebrospinal fluid, 417–418, 419*f*, 420*f*
  volume and weight, 16*t*, 414, 449
Brain barrier system, 418, 420
Brainstem, 40*f*, 414, 415*f*, 421, 422*f*, 425–426, 427, 427*t*, 428
Brain trauma, 438*f*, 438*i*
Brain tumors, 372*i*
Breasts, 352*f*, 749–751, 750*f*, 751*f*, 753
  cancer of, 71*i*, 639*i*, 755, 756*t*
  lymphatic system of, 639, 639*i*, 641*f*
  self-examination of (BSE), 755
Breathing
  difficulty in, 299*i*
  neural control of, 666–667, 668*f*, 668*i*
Brittle bone disease, 88*i*
Broca area, 435*f*, 436
Bronchi, 38*f*, 654*f*, 658*f*, 660*f*, 661, 662*f*
Bronchial tree, 653*f*, 661
Bronchioles, 661, 663*f*
Bronchoconstriction and bronchodilation, 661
Bronchopulmonary segments, 661
Bronchus-associated lymphatic tissue (BALT), 636
Bronzing, 136
Bruise, 137
Brush border, 52*f*, 59, 60*f*, 82*f*, 691, 692*f*
Buccal cavity, 178*f*, 680–683, 680*f*
Buccal region, 351*f*
Bulbourethral gland, 730*f*, 737, 752*f*
Bulbus cordis, 577–578, 577*f*
Bundle branch block, 574*i*
Burns, 146–147, 147*f*, 147*i*
Bursae, 232, 232*f*
  of elbow, 241, 242*f*
  of humeroscapular joint, 241*f*
  of knee, 245*f*, 246
  of shoulder, 241
Bursitis, 232
Byers, Eben, 152*i*

C

Cadaveric anatomy (photographs)
  abdominal cavity, 42*f*, 643*f*
  abdominal musculature, 301*f*
  back musculature, 297*f*, 305*f*, 318*f*
  foot, 340*f*
  hand, 87*f*, 327*f*
  head and neck, 40*f*, 287*f*, 297*f*, 656*f*
  heart, 566*f*, 567*f*
  inguinal region, 27*f*, 639*f*
  joints, 157*f*, 241*f*, 244*f*, 246*f*, 248*f*
  kidney, 711*f*
  lower limb musculature, 334*f*, 340*f*
  nervous system, 398*f*, 402*f*, 413*f*, 415*f*, 416*f*, 441*f*, 450*f*, 459*f*
  pelvic cavity, 43*f*
  reproductive system, female, 43*f*, 745*f*, 750*f*
  reproductive system, male, 43*f*, 731*f*
  stomach, 686*f*
  thoracic cavity, 41*f*, 665*f*
  thoracic musculature, 318*f*
Calcaneal region, 27*f*, 362*f*
Calcitonin, 163, 166, 525
Calcitriol, 163, 529–530, 530*t*, 708
Calcium
  in blood, 163
  in bone matrix, 155, 162, 166
  dietary, 165*i*, 166–167
  and muscle contraction, 269, 270*f*, 271*f*, 273–274
Calculus
  biliary (gallstones), 698*i*
  dental, 682*i*
  renal (kidney stones), 718*i*
Calluses, 133, 136

Calvaria, 178
Calyces, renal, 710, 711f
Canal
- alimentary, 676, 677f
- anal, 44f, 693f, 694
- auditory, 180, 485, 485f
- carotid, 177f, 181
- central, of spinal cord, 379, 388, 388f, 417–418, 420f
- central (osteonic), 156, 157f, 158
- cervical, 740f, 741f, 746
- condylar, 177f, 182
- hypoglossal, 176f, 182
- inguinal, 736
- perforating, of bone, 157f, 158
- root, of tooth, 682f, 683
- sacral, 193f, 194
- semicircular, 492
- vertebral, 190

Canaliculi, 155, 157f
Cancer, 71i
- aging and, 646
- anal, 756
- bone, 152i
- breast, female, 639i, 755
- breast, male, 755, 756t
- cervical, 755, 756, 757f, 757i
- of digestive system, 701
- endometrial, 755
- esophageal, 685i
- leukemia, 557
- lung, 670, 755
- metastasis, 631, 639
- penile, 755, 756, 756t
- prostate, 737i, 755
- skin, 146, 146f
- testicular, 755, 756t
- uterine, 755
- vaginal, 756

Cancer cells, 77f, 629f, 757f
Candida, 648
Canine teeth, 681, 682f
Cannon, Walter, 456, 457
Capacitation of sperm, 106–107
Capillaries, blood, 585f, 586
- continuous, 590f, 591
- discontinuous, 591 (*See also* Sinusoids)
- discovery, 586
- fenestrated, 514, 591, 591f
- number, 591
- peritubular, 710, 711f, 712f
- permeability, 592, 592f
- structure, 587f, 589–591

Capillaries, lymphatic, 630, 631, 632f, 633f
Capillary beds, 585f, 592–593, 593f, 632f
Capillary fluid exchange, 592, 592f
Capitulum of humerus, 210, 210f
Carcinogens, 71i
Cardiac arrhythmia, 579
Cardiac center, 421
Cardiac conduction system, 573–574, 574f
Cardiac muscle. *See* Muscular tissue, cardiac
Cardiac tamponade, 579t
Cardiocytes, 273–274, 273f, 574–576, 575f
Cardiology, 562
Cardiomegaly, 579i
Cardiomyopathy, 579t
Cardiopulmonary system, 654
Cardiovascular system, 542, 561–628. *See also* Circulatory system
Caries, 682i, 701
Carina, 658f, 660
Carotene, 136
Carotid bodies, 589, 589f, 668f
Carotid sinuses, 446f, 447f, 457f, 589

Carpal bones, 173f, 210, 212f, 222. *See also* Bones, specific
Carpal region, 26, 27f, 232f
- skeletal anatomy, 209, 210, 212f, 326f
- surface anatomy, 26, 356f, 357f

Carpal tunnel, 324, 326f
Carpal tunnel syndrome, 326i
Carpopedal spasm, 526i
Carriers in plasma membrane, 55, 56f, 57f, 58
Cartilage in general, 90–92, 91t, 161
- articular, 154, 161, 231, 231f, 232i, 247, 249
- elastic, 91f, 91t
- in endochondral ossification, 159–161, 160f
- fibrocartilage, 91f, 91t
- hyaline, 91f, 91t, 159–161, 230

Cartilages, specific
- arytenoid, 657, 658f, 659, 659f
- bronchial, 661, 661f, 662f
- corniculate, 657, 658f, 659, 659f
- costal, 36f, 194f, 195, 230, 230f
- cricoid, 655, 657, 658f
- cuneiform, 658, 658f
- epiglottic, 657, 658f
- nasal, 655, 657f
- thyroid, 352f, 657, 658f, 659f, 660f
- tracheal, 658f, 660, 660f

Cartilaginous joints. *See* Amphiarthroses
Catabolism, 12
Cataracts, 498i
Catecholamines, 526
Categorical hemisphere, 438
Cauda equina, 386, 387f
Caudal direction, 414, 415f, 417f
Cavities
- buccal, 175, 178f, 680–683, 680f
- cranial, 5f, 29, 29f, 175, 176f, 177f, 178, 178f
- dental, 682i, 701
- ear, middle and inner, 175, 181, 485f, 486
- major body, 26, 29–30, 29f, 29t, 31f
- nasal, 175, 178f, 655, 656f
- orbital, 175f, 178f, 494–495
- peritoneal, 30, 31f, 42f, 111, 700f

C cells, thyroid, 113f, 114, 524f, 525, 532
CD4 and CD8 cells, 635, 647
Cecum, 693f, 694
Cell-adhesion molecules, 56, 56f, 586
Cell cycle, 68–69, 69f
Cell-identity markers, 55, 56f
Cell membrane. *See* Plasma membrane
Cells, 47–72. *See also specific types*
- cytoskeleton, 63, 64f, 65f
- general structure, 50–51, 52f
- intercellular junctions between, 60, 62–63, 62f
- life cycle, 68–72, 69f, 70f
- membrane transport, 57–58, 57f, 58f
- organelles, 63–68, 66f, 67f
- plasma membrane, 52–56, 54f, 56f
- shapes, 49–50, 50f
- size, 50, 51f, 53t
- surface outgrowths, 58–60, 60f, 61f
- surfaces of, 50, 52–56, 52f, 80

Cell theory, 10, 48, 48t
Cementocytes, 681
Cementum, 681, 682f, 683
Central gray matter of midbrain, 425f, 426
Central nervous system (CNS), 366, 366f. *See also* Brain; Spinal cord
- embryonic development, 378–379, 379f, 380f

Central pattern generators, 426
Central sulcus, 415f, 416f, 430, 430f
Centrioles, 52f, 63, 65f, 66f, 68, 70f
Centromere, 70f, 71f
Centrosome, 52f, 63, 68
Cephalic region, 27f
Cerebellum
- anatomy, 421, 423, 423f
- functions, 423–425, 424f, 427t, 437, 438, 493, 493f
- number of synapses, 375
- position, 40f, 414, 415f, 416f

Cerebral aqueduct, 416f, 417, 419f, 420f, 423f, 425, 425f
Cerebral arterial circle, 599f, 600f, 600t
Cerebral lateralization, 438, 439f
Cerebral palsy, 450t
Cerebrospinal fluid
- clinical examination, 388i, 417i
- functions, 418
- sources and flow, 371, 417–418, 420f
- of spinal cord, 388, 388i, 418, 420f

Cerebrovascular accident, 624t
Cerebrum, 40f, 414, 428–439. *See also* Cortex, cerebral
- atrophy, 449–450, 450f
- crura, 425, 425f
- embryonic development, 379, 380f
- functions, 430, 432–439, 438i, 439f, 468, 667
- gray and white matter, 419f, 430–431
- gross anatomy, 415f, 416f, 417f, 429–431
- hemispheres, 415f, 429, 438
- insula, 415f, 439t
- lobes, 430, 430f, 439t (*See also individual lobes*)
- peduncles, 422f
- spinal tracts, relationship to, 389, 391f, 392f

Cerumen, 142, 485
Ceruminous glands, 142, 142t
Cervical region. *See* Neck
Cervix, uterine, 43f, 740f, 741f, 745, 746
Channel proteins, 54f, 55, 56f
Charley horse, 277t
Charnley, Sir John, 250
Cheeks, 680
Chemoreceptors, 476, 481
- arterial, 589, 589f, 667

Chickenpox, 395i, 557
Chief cells, gastric, 687f, 688, 688t
Child abuse, 241
Childbirth, pelvic adaptations for, 215
Chimpanzee, 16f, 132f, 186f, 189, 189f, 215, 218f, 220f
Chlamydia trachomatis, 756
Choanae, nasal, 177f, 183, 654f, 655, 656f
Cholecystokinin (CCK), 515, 530
Cholelithiasis, 698i
Cholesterol, 53, 106
Cholinergic neurons, 450, 465, 466t
Cholinergic receptors, 465–466
Cholinesterase inhibitors, 267i
Chondrification, 196, 221
Chondroblasts, 90
Chondrocranium, 196
Chondrocytes, 90, 91f, 159
Chordae tendineae. *See* Tendinous cords
Chordata, 12f, 13, 114
Chorion, 111f, 115, 116f, 117f
Choroid of eye, 496, 497f, 498, 506
Choroid plexus, 417, 419f, 420, 420f
Chromaffin cells, 526
Chromatids, 69, 70f, 71f

Chromatin, 65, 70f
Chromatophore, 496
Chromosomes, 65, 70f, 71f, 121–122, 121f, 123f
- abnormal numbers, 121–122, 121f, 123f, 732

Chronic bronchitis, 670
Chronic obstructive pulmonary disease (COPD), 669–670
Chyle, 630
Chylomicrons, 691
Chyme, 685, 689
Chymosin, 688, 688t
Cilia, 59–60, 61f, 82f
- of nasal cavity, 655
- of retinal cells, 505
- of trachea, 61f, 660, 660f, 661f
- of uterine tube, 743, 745, 745f

Ciliary body of eye, 496, 497f, 498f
Cimetidine, 689i
Cingulate gyrus, 416f, 432, 433f
Circadian rhythms, 428, 500
Circle of Willis. *See* Cerebral arterial circle
Circular muscles, 259, 259f. *See also* Sphincters
Circulatory system, 33f, 541–628
- blood, 541–559
- blood vessels, 585–628
- definition, 542
- embryonic development, 577–578, 578f, 620–621, 620f, 621f, 623f
- fetal, 623f
- heart, 561–583

Circumduction of joints, 237, 237f
Circumventricular organs, 420
Cirrhosis, 702t
Cisterna chyli, 630f, 632, 634f
Cisternae, 65–67, 66f
Cleavage, embryonic, 108, 109f
Cleft lip and palate, 120, 122, 124t, 201t
Climacteric, 754–755
Clitoris, 740f, 748, 749f, 753, 754f
Clostridium tetani, 267i
Clotting factors, 554
Clubfoot (talipes), 124t, 223, 223f
Cochlea, 446f, 485f, 486–489, 487f, 488f
Coelom, 111
Cognition, 430, 437
Cold, common, 670t
Cold medicine, 466i
Cold receptors, 476, 478t
Collagen, 86–87, 87f, 89f, 90f, 91, 91f, 134f
- of bone, 156, 157f
- diseases, 87i, 93i

Collecting ducts
- lymphatic, 632–633, 633f
- renal, 712f, 714, 717

Colliculi of midbrain, 422f, 423f, 425, 425f, 426, 489, 490f, 503f, 504
Colon, 42f, 43f, 641f, 677f, 693f, 694
- mesenteries of, 679–680, 679f, 694

Colony-forming units (CFUs), 547, 553f
Color blindness, 507t
Color vision, 15, 500
Columns of spinal cord, 388f, 389, 390f
Commissural tracts, 430, 431f
Commissures
- cerebral, 416f, 430
- of eyelids, 494, 494f

Communicating rami, 459, 461f
Compartment syndrome, 342i
Complement system, 635
Complete blood count, 552i
Compliance, pulmonary, 669
Composite materials, 156i

Computed tomography (CT), 3
of brain, 5*f*
of muscle, 275–276, 283*f*
in Parkinson disease, 451
of skeleton, 171*f*
Concentration gradient, 57
Conceptus, 108
Concussion, 418, 450*t*
Conduction speed in nerve fibers, 374
Conductivity of cells, 257, 367
Condyles, 174*f*, 174*t*
femoral, 215, 217*f*
mandibular, 185*f*, 186
occipital, 177*f*, 181
tibial, 217, 219*f*
Condyloid joints, 234, 234*t*
Cones, retinal, 499–502, 501*f*, 503*f*
embryonic development, 504–505
Congenital anomalies. *See* Birth defects
Congestive heart failure, 579*t*
Conjunctiva, 494, 494*f*, 495*f*
Connectin, 263
Connective tissues, 79*t*, 86–93, 89*t*, 90*t*, 91*t*, 92*t*, 93*t*
Connexon, 62, 62*f*
Conotruncus, 577*f*, 578
Consciousness, 426
Constipation, 469, 701
Contact digestion, 691
Continuous ambulatory peritoneal dialysis (CAPD), 722
Contracture, 277*t*
Contralateral neglect syndrome, 437
Convergent muscles, 258, 259*f*
Convoluted tubules of kidney, 712*f*, 714–716
Coracoid process, 209, 209*f*
Cornea, 495, 495*f*, 497*f*
Corns, 133, 136
Coronal plane, 25, 25*f*
Corona radiata
cerebral, 430
ovarian, 742, 744*f*
Coronary artery disease, 572*i*, 573*f*
Coronary circulation, 569–572, 572*f*
Corpora quadrigemina. *See* Colliculi of midbrain
Cor pulmonale, 670
Corpus albicans, 741*f*, 743
Corpus callosum, 416*f*, 419*f*, 429, 430, 431*f*, 433*f*
Corpus cavernosum
clitoral, 748
penile, 43*f*, 730*f*, 738
Corpus luteum, 522, 741*f*, 742, 747
Corpus spongiosum, 43*f*, 719, 730*f*, 738, 739*f*
Corpus striatum, 432, 433*f*
Cortex
cerebellar, 414, 423, 423*f*
cerebral, 414, 419*f*, 431–432, 432*f* (*See also* Cerebrum)
association, 433–434, 435*f*
auditory, 434, 435*f*, 489, 490*f*
gustatory, 434, 435*f*, 482, 483*f*
motor, 434–436, 435*f*, 436*f*, 438
number of synapses, 375
olfactory, 434, 435*f*, 484, 484*f*
prefrontal, 433*f*, 434, 438*i*
primary, 433–434
somesthetic, 434, 435*f*, 480*f*
visual, 434, 435*f*, 502, 503*f*, 504
hair, 138*f*, 139, 139*f*
ovarian, 740
renal, 710, 711*f*
Cortical reaction, 107*f*
Corticoids, 527
Corticospinal tracts, 390*f*, 390*t*, 392, 392*f*, 421, 430, 434, 667
Corticosteroids, 527
Corticosterone, 527
Corticotropes, 522
Corticotropin. *See* Adrenocorticotropic hormone
Cortisol, 527, 532, 535
Costal margin, 194, 194*f*, 352*f*
Coughing, 421, 484, 666, 667
Coxal region, 27*f*, 242–243, 244*f*
Cranial bones, 175–183, 196. *See also* Bones, specific
Cranial fossae, 178, 180*f*
Cranial nerves, 440–448, 441*f*, 442*t*–448*t*, 448*f*. *See also* Nerves, specific
classification, 440
memorization aid, 440
Cranial region, 27*f*
Craniosynostosis, 201*t*
Cranium, 13, 40*f*, 175, 178
Crepitus, 247
Crest of a bone, 174*f*, 174*t*
Cribriform plate, 176*f*, 177*f*, 183, 183*f*, 442*f*, 483*f*, 656*f*
Cricoid cartilage, 657, 658*f*, 659*f*
Cri du chat, 124*t*
Crista ampullaris, 492, 492*f*
Crista galli, 176*f*, 177*f*, 183, 183*f*, 184*f*
Crohn disease, 702*t*
Crossed extension reflex, 407*t*
Cross sections, 78, 79*f*, 80*f*
Cruciate ligaments, 244, 245*f*, 247*f*, 247*i*
Crural region. *See* Leg
Crus (crura)
of cerebrum, 425
of clitoris, 748, 749*f*
of lower limb, 26
of penis, 730*f*, 738
Crush syndrome, 277*t*
Crutch paralysis, 405*i*
Cryosurgery, 146
Cryptorchidism, 739*i*, 753, 756*t*
Crypts of Lieberkühn, 692
CT scan. *See* Computed tomography
Cubital region, 27*f*, 356*f*
Cumulus oophorus, 742, 744*f*
Cuneate fasciculus, 389, 390*t*, 391*f*, 398*f*, 421, 422*f*, 434
Curare, 267*i*
Curie, Marie and Pierre, 152*i*
Curvatures, spinal, 188, 189*f*, 200*f*
Cushing syndrome, 534*f*, 536*t*
Cuticle
of hair, 138*f*, 139, 139*f*
of nail, 141*t*
Cyanosis, 136, 669*i*
Cyclic adenosine monophosphate (cAMP), 515
Cystic duct, 698, 698*f*
Cystitis, 719*i*
Cytokinesis, 70*f*, 71
Cytology, 48
Cytomegalovirus, 122, 648
Cytoplasm, 51, 63–68
Cytoskeleton, 54*f*, 56*f*, 63, 64*f*, 376
Cytosol, 51
Cytotoxic T cells, 635, 644, 646

D

Dance, 256
Dander, 133
Dartos muscle, 733*f*
Darwin, Charles, 14
Day vision, 500, 502
Dead air, 663
Deafness, 489*i*
Decidual cells, 116
Decubitus ulcer, 99
Decussation, 389, 422*f*, 431*f*, 434, 479, 489
Defecation, 468
Dehydration, 721
Dehydroepiandrosterone (DHEA), 527
Delayed onset muscle soreness (DOMS), 277*t*
Dendrites, 94, 94*f*, 368, 369*f*, 431
Dendritic cells, 131*f*, 132, 551, 635, 638*f*, 747
Dendritic spines, 431
Denervation atrophy, 270, 276
Denervation hypersensitivity, 456
Dense bodies, 274
Dentin, 681, 682*f*, 683
Dentition. *See* Teeth
Depression of a joint, 235, 236*f*
Dermal papillae
of hair, 138, 138*f*, 143
of skin, 131*f*, 133, 134*f*, 143, 477
Dermatitis, 145*t*
Dermatology, 130
Dermatomes, 114, 405, 405*f*
Dermis, 90*f*, 130, 131*f*, 133–135, 134*f*, 135*t*, 136*f*
embryonic development, 143, 143*f*
Descartes, René, 523
Descending analgesic pathways, 392, 426, 481
Desmosomes, 62, 62*f*, 63*i*, 576, 578*f*
Detoxification, 65, 152
Detrusor muscle, 718, 718*f*
Dextrocardia, 6*i*
Diabetes, gestational, 758*t*
Diabetes mellitus, 167, 480*i*, 535, 646
Diabetic neuropathy, 480*i*, 535
Diaphragm. *See* Muscles, specific, diaphragm
Diaphysis, 153, 154*f*, 230
Diarthroses. *See* Synovial joints
Diencephalon, 379, 380*f*, 422*f*, 427–428, 439*t*
and development of eye, 504, 506*f*
Differential WBC count, 552*i*
Differentiation, 12, 99
Diffusion, 57–58, 57*f*
Digestive system, 34*f*, 675–706, 677*f*
accessory glands (*See* Liver; Pancreas; Salivary glands)
disorders, 682*i*, 685*i*, 689*f*, 689*i*, 698*i*, 701, 702*t*
embryonic development, 111, 112*f*, 120*t*, 700–701, 700*f*, 701*f*
enteric hormones, 530, 531*t*, 688, 688*t*
gastric anatomy, 685–688, 686*f*, 687*f*
intestinal anatomy, 689–694, 690*f*, 691*f*, 692*f*, 693*f*
oral anatomy, 680–683, 680*f*, 681*f*, 682*f*, 683*f*, 684*f*
pharyngeal-esophageal anatomy, 678*f*, 684–685
relationship to mesenteries, 679–680, 679*f*
Digestive tract, 676. *See also* Digestive system
Digital rays, 120*t*, 221
Digital rectal examination, 737*i*
Digits. *See* Fingers; Toes
Diploe, 154, 154*f*, 177*f*, 180
Diploid cells, 106, 733
Directional terms, 25, 26*t*, 414, 415*f*, 417*f*
Disjunction, 121*f*
Dislocation, 224
elbow, 243*f*, 243*i*
hip, 242
jaw, 240
shoulder, 240–241
vertebrae, 202
Dissection, 3, 8–9
Disseminated intravascular coagulation, 556*t*
Distal convoluted tubule, 713*f*, 716, 717*f*
Diverticulitis, 702*t*
DNA, mitochondrial (mtDNA), 68*i*
L-DOPA, 451
Dopamine, 376, 450, 463, 515, 518, 526
Dorsal horn, 388, 388*f*, 479, 480
Dorsal respiratory group, 667, 667*f*
Dorsal root, 394*f*, 395, 396*f*, 398*f*
Dorsal root ganglion, 388*f*, 394*f*, 395, 396*f*, 398*f*
Dorsiflexion, 238, 238*f*, 246
Dorsum
of foot, 27*f*, 363*f*
of hand, 356*f*, 357*f*
Double-jointedness, 239
Down syndrome, 122, 123*f*, 449
Drug actions, 466*i*
Dual innervation, 468, 468*f*
Ducts
allantoic, 115, 701*f*
bile, 690, 698, 698*f*, 700
collecting, of kidney, 712*f*, 714, 717
common hepatic, 698, 698*f*
cystic, 698, 698*f*, 700
ductus arteriosus, 578, 578*f*, 579*i*, 621*f*, 622, 623*f*, 669
ductus deferens, 37*f*, 730*f*, 731*f*, 732, 736–737, 752*f*
ductus venosus, 622, 623*f*
efferent ductules of testis, 713*f*, 736, 752*f*
ejaculatory, 730*f*, 737
of epididymis, 731*f*, 736
gland, 96–97, 97*f*
hepatic, 697, 698*f*
lactiferous, 750, 750*f*
mesonephric, 719, 720*f*, 751–753, 752*f*
nasolacrimal, 494
pancreatic, 690, 698*f*, 699
papillary, of kidney, 717
paramesonephric, 751–753, 752*f*
right lymphatic, 632
semicircular, 485*f*, 487*f*, 489, 492, 492*f*
structure of, 97*f*, 98*f*
vitelline, 112*f*, 115, 700, 700*f*, 701*f*
Duodenojejunal flexure, 690, 690*f*
Duodenum, 42*f*, 686*f*, 690, 690*f*, 692
Duplicity theory of vision, 502, 503*f*
Dural sheath, 386, 388*f*
Dura mater, 178, 386, 398*f*, 414, 417, 418, 418*f*, 506
Dust cells, 664, 664*f*
Dwarfism, 163*f*, 163*i*
Dynein, 60, 61*f*, 65*f*
Dynorphins, 481
Dysphagia, 701, 702*t*
Dyspnea, 299*i*, 579*i*
Dystrophin, 276

E

Ear
anatomy, 485–487, 485*f*, 489–492
embryonic development, 504, 505*f*
and equilibrium, 489–493
and hearing, 487–489
Eardrum. *See* Tympanic membrane
Echocardiography, 3
Eclampsia, 758*t*
Ectoderm, 110, 111*f*, 112*f*, 113*f*, 113*t*, 378
Ectopic focus, 574
Ectopic pregnancy, 758*t*
Eczema, 145*t*
Edema, 630, 631*i*
Edward syndrome, 122
Efferent neurons and nerve fibers, 367, 368, 368*f*, 395, 396*f*, 406
somatic *versus* autonomic, 457, 458*f*, 458*t*
Egg (ovum), 107*f*, 728, 744*f*
Ehlers-Danlos syndrome, 87*i*
Ejaculation
female, 749
male, 408, 737
Ejaculatory duct, 730*f*
Elastic fibers, 87, 89*f*
Elastic filaments in muscle, 263, 266*t*
Elasticity, 257
Elastin, 87, 88*i*
Elbow. *See* Joints, specific
Electrodesiccation, 146
Electrolyte balance, 152
Eleidin, 133
Elephantiasis, 631*f*, 631*i*
Elevation of a joint, 235, 236*f*
Embolism, 556*t*
Embryo, 110–114. *See also* Embryonic stage
Embryoblast, 109, 110*f*
Embryogenesis, 108, 110
Embryology, history of, 106
Embryonic membranes, 114–115, 116*f*, 117*f*
Embryonic stage, 12*f*, 108, 109*t*, 110–118, 115*f*. *See also specific organs and systems*
embryonic disc, 110, 111*f*
embryonic folding, 111, 112*f*
embryonic membranes, 114–116, 116*f*
gastrulation leading to, 110, 111*f*
neurulation in, 113, 113*f*
nutrition in, 116–118, 116*f*, 117*f*
pharyngeal pouches and arches, 113–114, 113*f*, 114*f*
Embryonic stalk, 111*f*, 112*f*
Emmetropia, 500*f*
Emotion
cerebellum and, 425
hypothalamus and, 428, 437
language and, 436
limbic system and, 432, 437
temporal lobe and, 430
Emphysema, 670
Enamel, 681, 682*f*, 683
Encephalitis, 450*t*
Endemic goiter, 536*t*
Endocarditis, 579*i*, 579*t*
Endocardium, 565, 568*f*
Endocrine disruptors, 739*i*
Endocrine system, 33*f*, 513–539, 514*f*
adrenal gland, 38*f*, 462*f*, 526–527, 526*f*
cortex, 463, 526–527, 531*t*, 532
disorders, 536*t*
embryonic development, 532, 534*f*
medulla, 379, 463, 468, 526, 531*t*, 532, 534
aging of, 532
compared to nervous system, 516, 517*f*, 517*t*
disorders of, 532, 534–535, 534*f*, 535*f*, 536*t*
embryonic development, 532, 533*f*, 534*f*
gonads, endocrine role of, 528, 531*t*, 729, 752–755
hormone action, 515–516
hormone chemistry, 514, 515*f*, 516*t*
hypothalamus and, 517–518, 519*f*, 520*t*
pancreatic islets, 527, 528*f*, 531*t*
parathyroid glands, 113*f*, 114, 525, 525*f*, 526*i*, 530*t*
pineal gland, 416*f*, 422*f*, 423*f*, 428, 523, 523*i*, 530*t*, 532
pituitary gland
disorders, 536*t*
embryonic development, 532, 533*f*
histology, 520*f*
hormones, 518, 519*f*, 520, 521*f*, 521*t*, 522
location and anatomy, 416*f*, 518, 519*f*
relationship to hypothalamus, 433*f*, 518, 519*f*, 520*t*
trauma to, 522*i*
thymus
embryonic development, 113*f*, 114
functions, 523, 530*t*, 546
hemopoietic role, 546
histology, 637, 638*f*
hormones, 523, 530*t*
involution, 523, 532, 646
location and anatomy, 523, 524*f*, 630*f*, 638*f*
lymphocyte development in, 635, 644, 645*f*
thyroid gland
disorders, 535, 536*t*
embryonic development, 113*f*, 532, 533*f*
hormones, 515*f*, 524–525, 530*t*
location and anatomy, 514*f*, 523–524, 524*f*
Endocrinology, 514
Endocytosis, 58
Endoderm, 110, 111*f*, 112*f*, 113*t*
Endolymph, 486, 487*f*, 492
Endolymphatic sac, 487*f*
Endometrium, 110*f*, 741*f*, 746, 746*f*, 747, 748*f*
Endomysium, 256*f*, 257, 258*f*, 268, 273*t*
Endoneurium, 374, 393, 394*f*
Endoplasmic reticulum. *See also* Sarcoplasmic reticulum
rough, 52*f*, 65, 66*f*, 67*f*, 644, 645*f*
smooth, 52*f*, 65, 66*f*
Endoscopic views
gastroesophageal junction, 689*f*
larynx and vocal cords, 658*f*
Endosteum, 154, 154*f*
Endothelium, 98, 586, 587*f*
in blood-brain barrier, 418
of blood capillaries, 589, 590*f*, 591*f*, 592*f*
of bone marrow, 636, 637*f*
embryonic development, 620, 620*f*
glomerular, 714, 715*f*
lymphatic, 631, 632*f*
of pulmonary alveolus, 664*f*, 665
Endothermy, 13
Endurance, muscular, 271–272, 276, 277
Enkephalins, 481
Enteric hormones, 530, 531*t*, 688, 688*t*
Enteric nervous system, 455*f*, 465, 677–678, 678*f*
Enteroendocrine cells, 530, 688, 688*t*, 692
Enterokinase, 691
Enzymes of cell surface, 55
Eosinophils, 543*f*, 548, 549*f*, 549*t*, 551, 553*f*
Ependymal cells, 371, 372*f*, 372*t*, 417, 420, 505
and pineal embryology, 532
and retinal embryology, 505
Epiblast, 110, 117*f*
Epicardium, 563, 565, 568*f*
Epicondyles, 174*f*, 174*t*
of femur, 215, 359*f*
of humerus, 210, 355*f*
Epidermal ridges, 133
Epidermis, 130–133, 131*f*
embryonic development, 143, 143*f*
layers of, 131*f*, 132–133, 135*t*
Epididymis, 37*f*, 43*f*, 730*f*, 731*f*, 733*f*, 752*f*
Epidural anesthesia, 386
Epidural space, 386, 388*f*
Epigastric region, 28*f*
Epiglottis, 40*f*, 654*f*, 656*f*, 658*f*, 681*f*
Epilepsy, 450*t*
Epimysium, 257, 258*f*, 273*t*
Epinephrine, 463, 526
Epineurium, 394*f*, 395, 396*f*
Epiphyseal line, 154, 154*f*
Epiphyseal plate, 153, 154, 160*f*, 161, 161*f*, 222
Epiphysis, 153, 154*f*, 161*f*, 230
Epistaxis, 655
Epithalamus, 428, 439*t*
Epithelial tissue, 79*t*, 80–83
simple, 80–81, 81*f*, 81*t*–82*t*, 82*f*
stratified, 83, 84*f*, 84*t*–85*t*, 85*f*
Eponychium, 141*f*
Eponyms, 17
Equilibrium, 421, 485, 489–493, 490*f*, 491*f*
cerebellum and, 434
cerebral cortex for, 434
projection pathways for, 492–493, 493*f*
Erectile dysfunction, 755
Erectile tissues
nasal, 655
penile, 43*f*, 730*f*, 738, 739*f*
pudendal, 748, 749*f*
Erection, 408
Erosive esophagitis, 685*i*
Erythema, 136
Erythrocyte count, 544*t*
Erythrocytes, 49*f*, 92–93, 93*f*, 541*f*, 545–548, 545*f*
clinical assessment of, 556*i*
evolution of, 547*i*
in sickle-cell disease, 556*i*, 557*f*
Erythrocytosis, 555
Erythropoiesis, 546, 547*f*, 555
Erythropoietin (EPO), 529–530, 547, 708, 722
Eschar, 146
Escherichia coli, 719*i*
Esophagus, 38*f*, 40*f*, 41*f*, 654*f*, 656*f*, 665*f*, 677*f*, 684–685, 686*f*
Estradiol, 516*f*, 528
Estrogen, 163, 166, 516*f*, 528, 742, 747, 753, 754
Estrogen mimics, 739*i*
Ethmoid air cells, 183, 183*f*
Eumelanin, 135, 139, 139*f*
Eustachian (auditory) tube, 485*f*, 486
Eversion of the foot, 238–239, 238*f*
Evolution
evidence of, 15*i*
of lower limb, 16, 220
of mitochondria, 68*i*
of morning sickness, 114*i*
primate, 16*f*
of skull, 186, 186*f*
of upper limb, 15, 211–212
Evolutionary medicine, 16
Excitability, 12, 94, 257, 367
Excitable tissues, 94
Excretion, 709
Excursion of the mandible, 236*f*, 237
Exercise, 232*i*, 260*i*, 270
Exfoliation, 83, 83*f*, 133
Exocrine glands, 96–97, 97*f*, 98*f*
Exocytosis, 52*f*, 58, 59*f*
of synaptic vesicles, 267, 376
Expiratory neurons, 666–667
Extension of joints, 235, 322*t*
of fingers, 329*t*
of shoulder, 235*f*, 320*t*
of wrist, 235*f*, 329*t*
Extensor retinaculum
of ankle, 336*t*, 337*f*, 340*f*
of wrist, 324, 328*f*
Extensors, 259*f*
Exteroceptors, 476
Extracellular fluid (ECF), 51, 78
Extracorporeal membrane oxygenation (ECMO), 669*i*
Extrathalamic cortical modulatory system, 426
Extrinsic muscles, 260
Eye, 494–504
anatomy, 351*f*, 494*f*, 495–499, 495*f*, 497*f*
embryonic development, 504–506, 506*f*
extrinsic muscles of, 494–495, 496*f*, 504
intrinsic muscles of, 494
movements, 425, 495, 496*f*
Eyebrows and eyelashes, 138, 494, 494*f*, 495*f*
Eyelids, 494, 506

F

Facet of a bone, 174*t*
Facial bones, 184–186, 197. *See also* Bones, specific
embryonic development, 197
Facial expression, 130, 132*f*, 291*f*
muscles of, 288*t*–289*t*, 290*f*
Facial region, 27*f*, 175*f*, 178*f*, 290*f*, 351*f*
Fallen arches, 220
Fallopian tube. *See* Uterine tube
Falx cerebelli, 417
Falx cerebri, 417, 418*f*
Fascia
adherens, 576
deep and superficial, 257, 258*f*
lata, 330*t*
penile, 733*f*, 739*f*
plantar, 343*f*
thoracolumbar, 303*f*, 305*f*
Fascicles (fasciculi)
of muscle, 257–259, 258*f*, 259*f*
of nerve, 385*f*, 393, 394*f*
of spinal cord, 389
Fast glycolytic muscle fibers, 271–272, 272*f*, 272*t*
Fat. *See* Adipose tissue

Fatigue, muscular, 270–272, 273, 275, 276
Fauces, 680
Fear, 437
Feces, 692, 695
Female reproductive system. *See* Reproductive system
Femoral region, 27*f*
- blood vessels of (*See* Limbs, lower)
- muscular anatomy of, 330*f*, 332*f*, 334*f*, 335*f*
- skeletal anatomy of, 215–216, 217*f*, 218*f*
- surface anatomy of, 27*f*, 358*f*

Femoral triangle, 609, 609*f*
Fenestrated capillaries, 547*i*, 591, 591*f*, 592, 715*f*
Fertilization, 107*f*, 109*f*
Fetal alcohol syndrome, 122
Fetal breathing, 668
Fetal position, 120*t*, 727*f*
Fetal stage, 105*f*, 108, 109*t*, 118–119, 119*f*, 120*t*
Fever, 635
Fibrin, 554–555, 555*f*
Fibrinogen, 544, 544*t*, 554, 696*t*, 738*t*
Fibroblasts, 86, 88, 89*f*, 90*f*
Fibrocartilage, 91*f*, 91*t*, 92, 214, 231, 240, 242
Fibronectin, 80
Fibrosis, 100, 273, 576
Fibrous joints. *See* Synarthroses
Fibrous skeleton, cardiac, 565, 574
Fight or flight reaction, 457, 468
Filariasis, 631*i*
Filtration, 57, 57*f*, 592*f*
Filtration pores and slits, 591, 591*f*, 714, 715*f*
Fimbriae of uterine tube, 741*f*, 744*f*, 745
Fingerprints, 134, 137
Fingers, 26
- embryonic development, 115*f*, 120*t*, 221, 223, 223*f*
- muscles acting on, 322*t*, 323*f*, 324*t*, 325*f*, 327*t*, 328*f*, 329*t*
- skeletal anatomy of, 211, 212*f*
- surface anatomy of, 357*f*

First-order neuron, 389, 391*f*, 479
Fish, 716*i*
Fissures
- of bones, 174*t*, 182*f*, 183, 184, 185*f*
- of brain, 429

Fixatives, histological, 78
Fixators, 260
Flagellum, 60
Flat feet, 220, 224*t*
Flexion lines and creases, 137, 356*f*, 357*f*
Flexion of joints, 234–235, 235*f*, 322*t*
- fingers, 329*t*
- shoulder, 235*f*, 320*t*
- wrist, 235*f*, 329*t*

Flexor reflex, 407*t*
Flexor retinaculum, 232*f*, 324, 326*f*, 328*f*
Flexors, 259*f*
Fluid-mosaic model, 53, 54*f*
Folia of cerebellum, 423, 423*f*
Folic acid, 380
Follicles
- hair, 131*f*, 137–139, 138*f*, 141*f*, 143, 144*f*
- ovarian, 513*f*, 529*f*, 740, 742, 743*f*, 744*f*
- thyroid, 524, 524*f*

Follicle-stimulating hormone (FSH), 516*t*, 518, 519*f*, 522, 753, 755
Follicular fluid, 742
Fontanels, 198, 198*f*
Foot
- adaptations for bipedalism, 220, 220*f*
- arches of, 220, 221*f*, 360*f*, 362*f*
- congenital deformity, 223, 223*f*
- embryonic development, 114*f*, 115*f*, 120*t*, 221–222
- intrinsic muscles of, 342*t*–344*t*, 343*f*
- movements of, 238, 238*f*
- muscles acting on, 336*t*–337*t*, 337*f*, 338*f*, 339*f*, 340*f*, 340*t*, 341*f*
- skeleton of, 215, 218–220, 219*f*, 220*f*
- surface anatomy of, 362*f*, 363*f*

Foot plate, 114*f*, 115*f*, 221
Foot processes (pedicels) of podocyte, 714, 715*f*
Foramina, 174*t*, 175, 179*t*
- carotid canal, 177*f*, 181
- condylar canal, 177*f*, 182
- cribriform (olfactory), 177*f*, 183, 184*f*, 484
- foramen lacerum, 177*f*, 183
- foramen magnum, 40*f*, 177*f*, 181, 186, 186*f*
- foramen ovale
  - of heart, 578, 578*f*, 622, 623*f*, 669
  - of skull, 177*f*, 182*f*, 183
- foramen rotundum, 177*f*, 182*f*, 183
- foramen spinosum, 177*f*, 182*f*, 183
- greater palatine, 177*f*, 184
- hypoglossal canal, 176*f*, 182
- incisive, 177*f*, 184, 184*f*
- infraorbital, 175*f*, 176*f*, 184, 185*f*
- interventricular, 417, 419*f*
- intervertebral, 190, 191*f*
- jugular, 176*f*, 177*f*, 181, 182
- mandibular, 185*f*, 186
- mastoid, 177*f*, 181
- mental, 175*f*, 176*f*, 185*f*, 186
- nutrient, 154, 154*f*, 158
- obturator, 214, 214*f*, 215, 216*t*
- optic, 177*f*, 182, 182*f*, 185*f*
- orbital fissures, 182*f*, 183, 184, 185*f*
- parietal, 178*f*, 180
- sacral, 193–194, 193*f*
- stylomastoid, 177*f*, 181
- supraorbital, 179, 185*f*
- transverse, 191*f*, 192, 192*f*
- vertebral, 190, 190*f*, 200*f*
- zygomaticofacial, 176*f*, 185

Forced expiratory volume, 669
Forearm
- muscles acting on, 320*t*–322*t*
- muscles in, 322*t*, 323*f*, 324*t*, 325*f*
- pronation and supination of, 24, 24*f*, 238, 238*f*, 321*f*
- skeleton of, 210, 211*f*

Forebrain, 379, 380*f*, 427
Foregut, 112*f*, 678, 700, 700*f*
Formed elements of blood, 92–93, 542, 543*f*. *See also* Erythrocytes; Leukocytes; Platelets
Fornices, 740*f*, 741*f*, 747
Fossae, 174*f*, 174*t*
- cranial, 178, 180*f*
- cubital, 356*f*
- humeral, 210, 210*f*
- hypophyseal, 183
- nasal, 183, 184*f*, 655, 656*t*
- ovalis, 622, 623*f*
- ovarian, 740
- scapular, 208, 209*f*
- supraclavicular, 351*f*, 352*f*

Fovea centralis, 498, 499*f*
Foveae of bones, 174*f*, 174*t*, 215
Fractures, 164–166, 165*f*, 224
- avulsion, 224*t*
- clavicular, 208
- Colles, 224*t*
- epiphyseal, 224*t*
- femoral, 218*f*
- healing of, 166*f*
- Pott, 224*t*
- pubic, 215
- skull, 201–202, 201*f*
- types, 164*f*, 165*t*
- vertebral, 202, 202*f*

Freckles, 136, 137
Free nerve endings, 477*f*, 478*t*
Frenulum
- labial, 680, 680*f*
- lingual, 680*f*, 681
- penile, 738

Friction ridges, 134, 137
Friction rub, 565
Frontal lobe, 415*f*, 416*f*, 419*f*, 430, 430*f*
Frontal plane, 25, 25*f*
Frontal region, 351*f*
Functional morphology, 2
Funny bone, 210
Fusiform muscles, 258, 259*f*

G

Gage, Phineas, 438*f*, 438*i*
Gagging, 421, 481
Galea aponeurotica, 288*t*, 290*f*
Galen, Claudius, 9, 586
Gallbladder, 42*f*, 679*f*, 695*f*, 698, 698*f*, 700–701
Gallstones, 698*i*
Gametes, 106, 728
Gametogenesis, 106
Gamma-aminobutyric acid (GABA), 376
Ganglia
- autonomic, 457, 458*f*
- celiac, 460*f*, 462*f*, 464*f*, 677
- cervical, 576
- ciliary, 443*f*, 463, 464*f*
- collateral, 461, 461*f*, 461*t*
- dorsal root, 395, 396*f*
- embryonic development, 379
- geniculate, 445*f*
- histology, 395, 396*f*
- inferior mesenteric, 460*f*, 461, 462*f*, 677
- intramural, 463, 678*f*
- otic, 446*f*, 463, 464*f*
- parasympathetic terminal, 463, 469
- paravertebral (*See* Ganglia, sympathetic chain)
- prevertebral, 461, 461*f*
- sphenopalatine, 445*f*, 463, 464*f*
- submandibular, 445*f*, 463, 464*f*
- superior cervical, 677
- superior mesenteric, 460*f*, 461, 462*f*, 677
- sympathetic
  - chain, 398*f*, 399*f*, 402*f*, 458–459, 459*f*, 460*f*, 461*f*, 469
  - collateral, 461, 461*f*
- thoracic, 459*f*
- trigeminal, 444*f*
- vestibular, 446*f*

Ganglion cells, retinal, 499–500, 501*f*, 503*f*, 504
Gangrene, 99, 535
Gap junctions, 62, 62*f*, 273, 273*t*, 376, 576, 578*f*
Gap phases, 68–69, 69*f*
Gastric glands, 687*f*, 688*t*
Gastric lipase, 688*t*
Gastrin, 527, 530, 688, 688*t*
Gastritis, 689*i*
Gastroenterology, 676
Gastroesophageal reflux disease (GERD), 685*i*, 701
Gastrointestinal (GI) tract. *See* Digestive tract
Gastrulation, 110, 111*f*
Gates in plasma membrane, 55, 56*f*
Gaze centers, 426
G cells, 527, 687*f*, 688
Genital herpes, 756
Genitalia, 27*f*
- embryonic development, 120*t*, 751–753, 752*f*, 754*f*
- female, 43*f*, 740–749, 740*f*, 741*f*, 745*f*, 749*f*
- internal and external, 728, 728*t*
- male, 43*f*, 729–739, 729*f*, 730*f*, 731*f*, 733*f*, 739*f*

Genital tubercle, 754*f*
Genital warts, 756
Genitourinary (G-U) system, 708
Germ cells, 729, 731*f*, 733, 742
Germ layers, 108, 110, 111*f*, 113*t*
Gestation, 108
Gibbons, 211
Gingivae. *See* Gums
Gingivitis, 682*i*
Glabella, 175*f*
Glands. *See also* Endocrine system
- apocrine, 97, 140, 141*f*, 142
- areolar, 750, 750*f*
- Bartholin (greater vestibular), 749, 749*f*
- buccal, 683
- bulbourethral (Cowper), 730*f*, 737
- cardiac, 688
- ceruminous, 142, 485
- cervical, 746
- Cowper (bulbourethral), 730*f*, 737
- cutaneous, 140–142, 141*f*, 142*t*
- cytogenic, 97
- duodenal, 692
- endocrine *versus* exocrine, 96
- endometrial, 746, 746*f*, 747, 748*f*
- esophageal, 685
- exocrine structure, 96–97, 97*f*
- gastric, 687*f*, 688
- greater vestibular (Bartholin), 749, 749*f*
- holocrine, 97, 141*f*, 142
- labial, 683
- lacrimal, 494, 495*f*
- lingual, 681, 683
- liver, 695–697, 695*f*, 697*f*, 698*f*
- mammary, 142, 749–751, 750*f*, 751*f*
- merocrine, 97, 140, 141*f*
- mixed, 97, 684*f*
- mucous, 97
- pancreas, 698–699, 698*f*, 699*f*
- paraurethral (Skene), 749, 749*f*
- parotid, 683, 684*f*
- prostate, 43*f*, 718*f*, 719, 730*f*, 737, 737*i*, 738*t*, 752*f*
- pyloric, 687*f*, 688
- salivary, 683–684, 684*f*
- sebaceous, 141*f*, 142, 485
- seminal vesicles, 43*f*, 730*f*, 737, 738*t*, 752*f*
- serous, 98

Glands. *See also* Endocrine system—*Cont.*
Skene (paraurethral), 749, 749*f*
sublingual, 683, 684*f*
submandibular, 683, 684*f*
tarsal, 494
unicellular, 96
urethral, 719
vestibular, of vulva, 749, 749*f*
Glans
of clitoris, 748, 749*f*, 754*f*
of penis, 43*f*, 730*f*, 733*f*, 737–738
Glaucoma, 466*i*, 498*i*
Glenoid cavity, 209, 209*f*, 241*f*
Glenoid labrum, 240, 241*f*
Glial cells, 94, 94*f*, 371, 372*f*, 372*t*, 378, 430, 449
Gliding joints, 233, 233*f*, 234*t*
Glioblastoma (glioma), 372*i*, 413*f*
Globulins, 544, 544*t*, 696*t*
Globus pallidus, 433*f*, 451
Glomerular capsule, 712*f*, 713–714, 713*f*, 720
Glomerular filtration, 710, 713–714
Glomerulus, 710, 711*f*, 712*f*, 713*f*, 715*f*, 720
Glottis, 654*f*, 656*f*, 657
Glucagon, 515, 527
Glucocorticoids, 527
Gluconeogenesis, 696*t*, 708
Glutamate, 376
Gluteal fold, 353*f*
Gluteal region, 27*f*, 331*f*, 331*t*, 333*i*
Glycine, 267*i*, 376
Glycocalyx, 58–59, 60*f*
Glycogen, 263, 273, 276
Glycogenesis, 696*t*
Glycogenolysis, 696*t*
Glycolipids, 53, 54*f*, 58, 83
Glycoproteins, 54, 54*f*, 58, 66, 80, 86, 87, 88*i*, 514
Glycosaminoglycans (GAGs), 87, 91, 163
Glycosuria, 535
Goblet cells, 61*f*, 81, 82*f*, 98*f*, 655, 661, 661*f*, 691, 692*f*, 694
Golgi complex, 52*f*, 66, 66*f*, 67*f*, 691
Golgi tendon organ, 477*f*, 478, 478*t*
Golgi tendon reflex, 407*t*
Golgi vesicles, 52*f*, 66*f*, 67, 67*f*
Gomphoses, 229, 229*f*, 234*t*, 681
Gonadal ridge, 751–752, 752*f*
Gonadotropes, 522
Gonadotropin-releasing hormone (GnRH), 516*t*, 518, 519*f*, 520*t*, 753
Gonadotropins, 518, 522
Gonads, 728. *See also* Endocrine system; Ovaries; Testes
descent, 753, 755*f*
embryonic development, 751–753, 752*f*
Gonorrhea, 122, 756
Gorilla, 215
Gracile fasciculus, 389, 390*f*, 390*t*, 398*f*, 421, 422*f*, 434, 479
Granule cells, 483*f*, 484
Granulocytes, 548–551, 549*f*, 549*t*–550*t*, 550*f*
Granulosa cells, 107, 528, 741*f*, 742, 744*f*
Graves disease, 535
Gravity, sense of, 491
Gray matter, 379, 414
of cerebellum, 423, 423*f*
of cerebrum, 419*f*, 431–432
of spinal cord, 387–389
Greater pelvis, 213*f*, 214, 216*t*
Greater sciatic notch, 214, 214*f*, 216*t*
Greater vestibular glands, 749, 749*f*
Great toe, 219*f*, 220, 220*f*, 221*f*, 363*f*
Ground substance, 78, 87
Growth factors, 162, 554
Growth hormone (GH), 163, 516*t*, 518, 519*f*, 521*t*, 522, 522*i*, 532, 753
Growth hormone-inhibiting hormone (GHIH). *See* Somatostatin
Guard hairs, 494, 655, 656*f*
Gubernaculum, 753, 755*f*
Guillain-Barré syndrome, 408*t*
Gum disease, 682*i*
Gums, 681, 682*f*, 682*i*
Gustation. *See* Taste
Gut, embryonic, 111, 112*f*, 700–701, 700*f*, 701*f*
Gut-associated lymphatic tissue (GALT), 636, 636*f*
Gynecomastia, 122
Gyri, 415*f*, 419*f*, 429

H

Habenula, 428
Habituation, 426
Hair, 129*f*, 131*f*, 137–140, 138*f*, 139*f*, 140*t*, 143, 144*f*
Hair cells, cochlear, 487–489, 488*f*, 491*f*
Hair follicle. *See* Follicles, hair
Hair receptors, 131*f*, 138*f*, 139, 477, 477*f*, 478*t*
Hallux. *See* Great toe
Hamstring injuries, 336, 345*t*. *See also* Muscle groups, hamstrings
Hamulus, 210, 212*f*
Hand, 15, 209
congenital deformity, 223, 223*f*
embryonic development, 114*f*, 115*f*, 120*t*, 221–222, 223
intrinsic muscles, 327*t*–328*f*
muscles acting on, 322*t*–324*t*, 323*f*, 325*f*, 329*t*
skeleton, 211, 212*f*
surface anatomy, 357*f*
tendon sheaths and bursae, 232*f*
X rays, 161*f*, 207*f*, 227*f*
Hand plate, 114*f*, 115*f*, 221
Haploid cells, 106, 733, 734*f*, 735*f*, 742, 743*f*
Harvey, William, 586
Hawking, Stephen, 393*i*, 394*f*
H_2 blockers, 689*i*
Head
arteries, 599*f*, 599*t*–600*t*, 600*f*
of a bone (*See* Epiphysis)
median section, 40*f*, 656*f*
muscles, 287–297, 287*f*, 288*t*–289*t*, 290*f*, 291*f*, 292*f*, 292*t*–295*t*, 293*f*, 296*t*–297*t*
veins, 611*f*, 611*t*–612*t*
Hearing, 421, 426, 430, 485, 487–489, 488*f*, 490*f*
Heart, 41*f*, 563–580, 665*f*. *See also* Muscular tissue, cardiac
autonomic innervation, 457*f*
chambers, 566–568, 566*f*, 568*f*
blood flow through, 568, 571*f*, 578*f*
changes at birth, 578
conduction system, 573–574, 574*f*
coronary circulation, 569–572, 572*f*
diseases, 579*t*, 580
embryonic development, 112*f*, 120*t*, 275, 577–578, 577*f*, 578*f*
endocrine role, 528, 531*t*
innervation, 460*f*, 464*f*, 576
in old age, 578–579
pacemaker, 570, 574, 574*f*
position, 563, 563*f*
size and shape, 563
surface anatomy, 564*f*, 565*f*, 566*f*
valves, 567–568, 568*f*, 569*f*
wall, 565, 565*f*
Heart block, 574*i*
Heart bulge, 114, 114*f*
Heartburn, 685*i*, 701
Heart rate, 576
Heart tubes, 112*f*, 275, 577, 577*f*
Heat production and loss, 256, 428, 525
Heatstroke, 145
Heel spurs, 224*t*
Helicobacter pylori, 689*i*
Helper T cells, 635, 636, 646, 647, 756
Hemangioma, 137*i*
Hematocrit, 543*f*, 544*t*, 546
Hematology, 542
Hematoma, 137
Hematuria, 714*i*
Hemidecussation, 502, 503*f*
Hemidesmosomes, 62
Hemiplegia, 408
Hemoblasts, 620, 620*f*
Hemodialysis, 722
Hemoglobin, 136, 544*t*, 546, 546*f*, 547*i*, 556*i*
Hemophilia, 556*t*
Hemopoiesis, 152, 546–547, 701. *See also* Erythropoiesis; Leukopoiesis; Thrombopoiesis
Hemopoietic cells, 547, 547*f*, 553*f*, 554*f*, 637
Hemopoietic tissues, 158, 546
Hemorrhoidal veins, 693*f*
Hemorrhoids, 594*i*, 694
Hemostasis, 554
Heparan sulfate, 80
Heparin, 551
Hepatic lobules, 695
Hepatic portal system, 616*f*, 616*t*–617*t*, 617*f*, 679
Hepatic sinusoids, 591, 592*f*, 617*t*, 679, 695, 697, 697*f*
Hepatitis, 58, 756
Hepatocytes, 592*f*, 695, 697*f*
Hernia, 88*i*, 306*i*, 702*t*
Herniated disc, 202, 405*i*
Herophilus, 8
Herpes simplex, 122, 648, 756
Herpes zoster, 395*i*
Hess, Walter, 437
Heterodonty, 13. *See also* Teeth
Hiatal hernia, 702*t*
Hierarchy of structure, 6–8, 8*f*
Hindbrain, 379, 380*f*, 421–425, 422*f*
Hindgut, 112*f*, 679, 700, 700*f*
Hinge joints, 233, 234*t*
Hip
artificial, 250, 251*f*
fractures, 218*f*, 218*i*
joint, 242–243, 249*t*
muscles acting on, 329, 330*f*, 330*t*–333*t*, 331*f*, 332*f*
skeletal anatomy, 213–215, 213*f*, 214*f*, 215*f*
Hippocampus, 428, 432, 433*f*, 437–438, 484*f*
Hippocrates, 71*i*
Histamine, 376, 551, 688*t*
Histology, 2, 78–100
Histopathology, 2
Hives, 647
Hodgkin disease, 648*t*
Homeostasis, 12, 456, 476, 708
Homo, 14, 16, 16*t*, 17*f*
Homology, developmental, 506, 751
Homunculus
motor, 434, 436*f*
sensory, 434, 435*f*
Hooke, Robert, 10, 48
Horizontal cells, retinal, 501*f*, 502
Hormone replacement therapy (HRT), 534
Hormones, 514–530, 515*f*, 516*f*
action, 515–516
adrenal, 463, 526–527, 531*t*
cardiac, 528–529, 531*t*
chemistry, 514, 515*f*, 516*t*
enteric, 530, 531*t*, 688, 688*t*
gonadal, 162, 528, 531*t*, 729, 742–743
hepatic, 529–530, 531*t*
hypothalamic, 428, 429*f*, 518, 519*f*, 520*t*
monoamine, 514, 515–516, 515*f*
pancreatic, 527, 531*t*
parathyroid, 163, 525, 530*t*, 716
peptide, 514, 515–516, 515*f*
pineal, 523, 530*t*
pituitary, 162, 518–522, 519*f*, 521*t*
placental, 530, 531*t*
renal, 530, 531*t*, 708
steroid, 162, 514, 515, 515*f*
tables of, 516*t*, 520*t*, 521*t*, 530*t*–531*t*
thymic, 523, 530*t*, 637, 644
thyroid, 162, 515*f*, 524–525, 530*t*
Horner syndrome, 470*t*
Horseshoe kidney, 7*f*, 721*i*
Hot flashes, 755
Huang Ti, 586
Human, definition of, 12–13
Human immunodeficiency virus (HIV), 122, 420, 647, 747
Human papillomavirus (HPV), 756
Hunger, 428
Hyaline cartilage, 91*f*, 91*t*, 159–161, 160*f*, 230
Hyaline membrane disease, 669*i*
Hyaluronic acid, 231
Hydrocephalus, 124*t*
Hydrochloric acid, 688, 688*t*
Hydrocortisone, 527
Hydrogen peroxide, 550–551
Hydronephrosis, 722*t*
Hydroxyapatite, 155
Hyman, Flo, 88*i*
Hymen, 747
Hypercalcemia, 525
Hypercapnia, 669*i*
Hyperemesis gravidarum, 114*i*, 758*t*
Hyperextension of joints, 235, 235*f*
Hyperglycemia, 535
Hyperopia, 500*f*, 500*t*
Hyperplasia, 99, 274
Hyperreflexia, 407
Hypersecretion of hormones, 534–535
Hypersensitivity, immune, 647
Hypersplenism, 643*i*
Hypertension, 588*i*, 624*t*
Hypertrophy, 99, 160*f*, 270, 579
Hyperventilation, 667
Hypoblast, 110, 110*f*, 117*f*
Hypocalcemic tetany, 526*i*
Hypochlorite, 550
Hypochondriac region, 26, 28*f*
Hypodermis, 130, 131*f*, 135, 135*t*
Hypogastric region, 26, 28*f*
Hyponychium, 141*f*

Hypoparathyroidism, 526*i*
Hypophyseal portal system, 518, 519*f*, 522*i*
Hypophyseal pouch, 532, 533*f*
Hyposecretion of hormones, 532, 534
Hypospadias, 739*i*, 756*t*
Hyposplenism, 643*i*
Hypothalamo-hypophyseal tract, 518, 519*f*, 522*i*
Hypothalamus, 416*f*, 428, 433*f*, 439*t*, 484*f*
  and autonomic function, 468
  and endocrine system, 518, 519*f*
  hormones of, 428, 429*f*, 518, 519*f*, 520*t*
  nuclei, 428, 429*f*
  and olfaction, 484
  in respiratory control, 667
  and taste, 481
Hypothenar eminence, 357*f*
Hypothermia, 145
Hypoxemia, 669*i*
Hysterectomy, 757*i*

I

Ibn-Sina, Abu Ali (Avicenna), 9
Ileocecal junction and valve, 690, 690*f*
Ileum, 690, 690*f*, 693*f*
Iliac crest, 39*f*, 214, 214*f*, 353*f*
Iliotibial band, 286*f*, 330*t*, 332*f*, 334*f*, 335*f*, 340*f*, 358*f*, 359*f*
Ilium, 35*f*, 214, 214*f*
Immune system, 630, 644–651
  cells, 634–635
  disorders, 646–648
  relationship to lymphatic system, 630
Immunity, 551
  forms, 644, 646
  nonspecific defense compared to, 635–636
Immunocompetence, 636, 644, 645*f*
Immunodeficiency diseases, 647–648
Immunological surveillance, 634, 646
Implantation, 108, 109*f*, 110, 110*f*
Incisors, 681, 682*f*
Inclusions, 68
Incontinence, urinary, 721, 722*t*
Infant, premature, 669*i*
Infarction, 99
Infection
  congenital, 122
  opportunistic, 557
  protection from, 130, 132 (*See also* Immunity)
Infertility, 737, 756*t*
Inflammation, 635
Inguinal canal, 306*i*, 732, 753, 755*f*
Inguinal region, 26, 27*f*, 28*f*, 330*f*, 332*f*, 609*f*
Inhibin, 528, 729, 755
Injections, 256, 350
Inositol triphosphate (IP_3), 515
Inspiratory center, 666–667, 667*f*
Insula, 415*f*, 419*f*, 430, 433*f*, 434, 481, 483*f*
Insulin, 516*f*, 527, 535
Insulin-like growth factor (IGF), 522
Integument, 130
Integumentary system, 32*f*, 129–150
  aging, 145
  cutaneous glands, 140–142, 141*f*, 142*t*
  diseases, 145–147, 145*t*
  embryonic development, 143, 143*f*, 144*f*
  hair, 137–140, 138*f*, 139*f*
  nails, 140, 141*f*
  sensory receptors, 476–478, 478*t*, 479–480, 479*f*
  skin, 130–137, 131*f*, 134*f*, 136*f*
Interatrial septum, 567, 578*f*
Intercalated cells, 717
Intercalated discs, 95*f*, 96, 273, 275, 576, 578*f*
Intercellular clefts, 591, 592*f*
Intercellular junctions, 52*f*, 60, 62–63, 62*f*
Intercellular space, 52*f*, 54*f*
Interferons, 635
Intermediate filaments, 63, 64*f*
Intermediate muscle fibers, 271–272
Internal capsule of cerebrum, 430
Interneurons, 367–368, 368*f*, 406
Internodes, 369*f*, 374
Interoceptors, 476
Interosseous membrane, 210, 230, 248*f*
Interpubic disc, 213*f*, 214, 230*f*, 231
Interscapular region, 27*f*
Interstitial cells of testis, 528, 529*f*, 729, 731*f*
Interstitial fluid. *See* Tissue fluid
Interventricular foramen, 418, 420*f*
Interventricular septum, 567, 577*f*, 578
Intervertebral discs, 40*f*, 43*f*, 188, 191*f*
  anatomy, 190, 190*f*
  embryonic development, 198, 199*f*
  herniated (ruptured), 202*f*, 306*i*
  histology, 91*f*
  as symphyses, 230*f*, 231
Intestinal crypts, 691*f*, 692, 692*f*
Intestinal trunk, lymphatic, 634*f*
Intestines, 42*f*
  large, 36*f*, 37*f*, 42*f*, 43*f*, 677*f*, 692–694, 693*f*
  small, 37*f*, 38*f*, 42*f*, 43*f*, 677*f*, 689–692, 690*f*, 691*f*, 692*f*
Intracellular fluid (ICF), 51
Intramuscular injections, 333*i*
Intraperitoneal organs, 30, 680
Intrinsic factor, 555, 688, 688*t*, 701
Intrinsic muscles, 260
Inversion of the foot, 238, 238*f*
Inverted retina, 504–505
Involuntary muscle, 256
Involution
  of corpus luteum, 743
  of pineal gland, 523
  of thymus, 523
Iris, 468*f*, 494*f*, 496, 506
Ischemia, 747
Ischial tuberosity, 214–215, 214*f*, 728, 729*f*, 749*f*
Islets of Langerhans, 527
Itch, sense of, 389

J

Jaundice, 136
Jaws
  mandible, 173*f*, 175*f*, 176*f*, 185–186, 185*f*, 198*f*
  maxilla, 173*f*, 175*f*, 176*f*, 184, 184*f*, 228
  movements of, 235, 236*f*, 237, 239
  muscles of, 290*f*, 293*f*, 293*t*, 295*t*
  temporomandibular joint, 186, 231, 234, 239, 240*f*, 249*t*, 262
Jejunum, 690, 690*f*
Joint capsule, 231, 231*f*, 239
Joint prostheses, 250, 251*f*
Joints, specific
  acromioclavicular, 194*f*, 208, 231
  ankle, 246, 248*f*, 249*t*
  atlantoaxial, 191, 191*f*, 233
  atlanto-occipital, 191, 261
  costovertebral, 200, 230, 230*f*
  coxal, 242–243, 244*f*, 249*t*
  elbow, 233, 241, 242, 242*f*, 243*f*, 243*i*, 249*t*, 262
  glenohumeral, 239, 405*i*
  hip, 242–243, 244*f*, 249*t*
  humeroradial, 241, 242*f*
  humeroscapular (*See* Joints, specific, shoulder)
  humeroulnar, 241, 242*f*
  interphalangeal, 231*f*, 233, 355*f*
  knee, 233, 243–246, 245*f*, 246*f*, 247*f*, 249*t*
  metacarpophalangeal, 234, 355*f*
  patellofemoral, 243
  radiocarpal, 234
  radioulnar, distal, 231, 233
  sacroiliac, 194, 213–214
  shoulder, 238, 239, 241*f*, 249*t*, 405*i*
  sternoclavicular, 194*f*, 208, 231, 233, 234, 351*f*
  talocrural, 246, 248*f*
  temporomandibular (TMJ), 186, 231, 234, 239, 240*f*, 249*t*, 262
  tibiofemoral, 243–246, 245*f*, 247*f*
  trapeziometacarpal, 233–234, 233*f*
Joints in general, 227–251
  artificial, 250, 251*f*
  ball-and-socket, 233*f*, 234
  cartilaginous, 230–231, 230*f*
  classification of, 228–231, 234*t*
  condyloid, 233*f*, 234
  dislocation of, 202, 224, 240–242, 243*f*, 243*i*, 250*t*
  fibrous, 228–230, 229*f*
  gliding, 233, 233*f*
  hinge, 233, 233*f*
  movements of, 234–239, 235*f*, 236*f*, 237*f*, 238*f*
  pivot, 233, 233*f*
  saddle, 233–234, 233*f*
  synovial, 231–246, 231*f*, 233*f*
Jugular trunk, lymphatic, 634*f*
Juxtaglomerular apparatus, 716–717, 717*f*

K

Kangaroo rat, 716*i*
Keloids, 100*f*, 100*i*
Kennedy, John F., 136
Keratin, 63, 83, 130, 132, 137
Keratinocytes, 131*f*, 132, 133, 135, 143
Keratohyalin granules, 133
Ketoacidosis, 535
Ketonuria, 535
Kidneys, 708–717
  aging, 720–721
  anatomy
    gross, 708*f*, 709–710, 711*f*
    microscopic, 710, 712*f*, 713–717, 713*f*, 715*f*, 717*f*
    variations in, 6*i*, 7*f*, 721*i*
  ascent, 720
  circulation, 710, 711*f*, 712*f*
  diseases, 721–722, 722*t*
  embryonic, 719, 720*f*
  functions, 530, 531*t*, 708
  horseshoe, 7*f*, 721*i*
  pelvic, 6*i*, 7*f*, 721*i*
  position, 38*f*, 42*f*, 709, 709*f*
Kidney stones, 718*i*
Killer (cytotoxic) T cells, 77*f*, 635
Killing zone, 551
Kinesiology, 228, 234, 256
Kinetochore, 70*f*
Kinocilia, 491
Klinefelter syndrome, 122
Knee
  adaptations for bipedalism, 216, 218*f*
  injuries, 247*f*, 247*i*
  locking, 216, 244, 246
  muscles acting on, 334*t*–336*t*
  skeletal anatomy, 243–246, 245*f*, 246*f*, 249*t*
Knee reflex, 387*f*
Krause end bulb, 477*f*, 478, 478*t*
Kupffer cells, 551, 592*f*, 635, 695, 696*t*
Kyphosis, 166, 189*f*, 189*i*

L

Labia
  genital, 43*f*, 136, 740*f*, 748, 749*f*, 753, 754*f*
  oral, 351*f*, 680, 680*f*
Labioscrotal folds, 751, 754*f*
Labrum
  acetabular, 242, 244*f*
  glenoid, 240, 241*f*
Labyrinths of ear, 485*f*, 486, 487*f*, 491*f*, 492*f*
Lacrimal apparatus, 185, 445*f*, 494, 495*f*
Lacteal, 630, 691, 692*f*
Lactic acid, 271
Lactotropes, 522
Lacunae
  of bone, 92, 92*f*, 155
  of cartilage, 89, 91*f*
  of penis, 738, 739*f*
Lamellae of bone, 92, 92*f*, 156, 157*f*, 158
Lamellated corpuscle, 131*f*, 477*f*, 478, 478*t*
Lamina propria, 97, 98*f*, 678*f*
Laminectomy, 202
Laminin, 80
Langerhans cells. *See* Dendritic cells
Language, 436–438, 449
Lanugo, 120*t*, 138, 144
Large intestine. *See* Intestines, large
Laryngopharynx, 655, 656*f*
Larynx, 40*f*, 656*f*, 657–659, 658*f*
Lateral abdominal region, 26, 28*f*
Lateral aperture of brain, 418, 419*f*, 420*f*
Lateral horn, 388*f*, 389
Lateral sulcus, 419*f*, 430, 430*f*
Learning, 430
Lecithin, 55*f*, 695
Leeuwenhoek, Antony van, 10, 48, 586
Leg, 215
  blood vessels (*See* Limbs, lower)
  muscular anatomy, 336*t*–340*t*, 337*f*, 338*f*, 339*f*, 340*f*, 341*f*
  skeletal anatomy, 217–218
  surface anatomy, 27*f*, 359*f*, 360*f*, 361*f*
Leg bud, 114, 114*f*
Lens, 497, 497*f*, 498*f*
  corrective, 500*f*
  embryonic development, 505–506
Leprosy, 480*i*
Lesser pelvis, 213*f*, 214
Leukemia, 557
Leukocyte count, 544*t*
Leukocytes, 86, 88, 93, 541*f*, 548–552, 549*f*, 549*t*–550*t*
Leukocytosis, 557
Leukopenia, 557, 643*i*
Leukopoiesis, 552, 553*f*
Levers, 260–262, 260*f*, 262*f*

Leydig cells, 528, 529*f*, 729, 731*f*
Libido, 753
Life, definition of, 11–12
Lifting, injuries from, 306*i*
Ligaments, specific
broad, of uterus, 741*f*, 742, 746
cardinal, of uterus, 741*f*, 746
coccygeal, 387, 387*f*
coronary, of liver, 695
cricotracheal, 658, 658*f*
cruciate, 239, 245*f*, 246*f*, 247*f*, 247*i*
denticulate, 387, 388*f*
falciform, 695, 695*f*
inguinal, 35*f*, 300*t*–301*t*, 609*f*
ligamentum arteriosum, 564*f*, 578, 622, 623*f*
ligamentum teres, 623*f*, 695*f*, 696
ligamentum venosum, 622, 623*f*
median umbilical, 720
mesometrium, 746
mesosalpinx, 745, 746
mesovarium, 742
nuchal, 182, 191–192
ovarian, 740, 741*f*, 745*f*
patellar, 216, 332*f*, 337*f*, 340*f*, 358*f*
periodontal, 229, 681, 682*f*
pulmonary, 665
round
of liver, 622, 623*f*, 695*f*, 696
of uterus, 741*f*, 745*f*, 746
stylohyoid, 186
suspensory
of breast, 750, 750*f*
of eye, 497, 498*f*
of ovary, 740, 741*f*, 742, 745*f*, 747*f*
thyrohyoid, 658, 658*f*
uterosacral, 741*f*, 746
vestibular, 658
vocal, 658
Ligaments in general, 232. *See also* Ligaments, specific
of ankle, 246, 248*f*
composition, 88
of elbow, 241, 242*f*, 243*f*
of hip, 243, 244*f*
of knee, 243–244, 245*f*, 247*f*
sensory nerve endings in, 239
of shoulder, 240, 241*f*, 246
of temporomandibular joint, 239, 240*f*
Light microscope (LM), 48, 49*f*
Limb buds, 12*f*, 114*f*, 120*t*, 221
Limbic system
anatomy, 432, 433*f*
and autonomic function, 468
and aversion reactions, 432, 439*t*
and emotion, 432, 439*t*, 467
and memory, 432, 437, 439*t*
and olfaction, 484, 484*f*
and pain, 480, 480*f*
and pleasure, 432, 439*t*
and respiration, 667
Limbs
lower, 26
arteries, 607*f*, 607*t*–608*t*, 608*f*
joints, 242–246, 244*f*, 245*f*, 246*f*, 247*f*, 247*i*, 248*f*, 249*t*
muscular anatomy, 329–344, 329*t*–333*t*, 330*f*, 330*t*, 331, 332*f*, 334*f*, 334*t*–336*t*, 335*f*, 336*t*–340*t*, 337*f*, 338*f*, 339*f*, 340*f*, 341*f*, 342*t*–344*t*, 343*f*
skeletal anatomy, 215–220, 217*f*, 218*f*, 219*f*, 220*f*, 221*f*, 222*t*
surface anatomy, 27*f*, 358*f*, 359*f*, 360*f*, 361*f*, 362*f*, 363*f*
veins, 618*f*, 618*t*–619*t*, 619*f*
upper, 26
arteries, 601*f*, 601*t*–602*t*
joints, 233*f*, 240–242, 241*f*, 242*f*, 243*f*, 243*i*, 249*t*
muscular anatomy, 314–329, 317*f*, 318*f*, 320*t*–321*t*, 321*f*, 322*t*–324*t*, 323*f*, 324, 325*f*, 327*t*, 328*f*
skeletal anatomy, 209–212, 210*f*, 211*f*, 212*t*
surface anatomy, 27*f*, 356*f*, 357*f*
veins, 612*f*, 612*t*–613*t*
Lincoln, Abraham, 88*i*
Linea alba, 300*f*, 300*t*–301*t*, 301*f*, 352*f*
Linea semilunaris, 352*f*
Lines
on bones, 174*f*, 180, 182
nuchal, 177*f*, 186*f*
temporal, 176*f*
Lingual papillae, 481, 482*f*, 675*f*, 681, 681*f*
Lipase, 688
Lipid bilayer, 54*f*
Lipofuscin, 368, 449, 482
Lipogenesis, 696*t*
Lipoproteins, 696*t*
Lips, 351*f*, 680, 680*f*
Lithotripsy, 698*i*, 718*i*
Liver, 695–698
diseases, 136, 695, 702*t*
embryonic development, 622, 700–701, 701*f*
functions, 529–530, 531*t*, 546, 695, 696*t*
gross anatomy, 695–696, 695*f*, 698*f*
histology, 696–697, 697*f*
position, 31*f*, 36*f*, 42*f*, 660, 677*f*
Liver bud, 112*f*, 700, 701*f*
Lobes
atrioventricular, 574, 574*f*
of cerebrum, 430, 430*f*
of exocrine glands, 97, 97*f*
of kidney, 710, 712*f*
of lungs, 661, 662*f*
of pituitary gland, 518, 519*f*
sinoatrial, 574, 574*f*
of thyroid gland, 523
Lobules
of ear, 485, 485*f*
of exocrine glands, 97, 97*f*
hepatic, 696, 697*f*
pulmonary, 661
testicular, 729, 731*f*
thymic, 637, 638*f*
Longitudinal fissure, 415*f*, 419*f*, 429
Longitudinal sections, 79, 80*f*
Loop of Henle, 716
Lordosis, 189*f*, 189*i*
Lou Gehrig disease, 393*i*
Love, 437
Lower extremity. *See* Limbs, lower
Lower motor neurons, 392*f*, 435, 667
Lumbar cistern, 387
Lumbar puncture, 388*i*, 417*i*
Lumbar region, 26, 27*f*, 28*f*
muscles, 303*f*, 303*t*–306*t*, 304*f*, 305*f*, 306*i*
surface anatomy, 27*f*, 353*f*, 354*f*
vertebrae, 188*f*, 189*f*, 191*f*, 192*f*, 193
Lumbar trunks, lymphatic, 634*f*
Lung buds, 112*f*, 668, 669*f*
Lungs, 37*f*, 38*f*, 41*f*, 42*f*, 622, 654*f*, 660–664
embryonic development, 668–669, 669*f*
gross anatomy, 660–661, 662*f*, 665*f*
histology, 661, 663*f*, 664–665, 664*f*
position, 37*f*, 38*f*, 41*f*, 42*f*, 654*f*, 660
Lutein cells, 742
Luteinizing hormone (LH), 518, 522, 753, 755
Luxation, 250*t*
Lymph, 630–631, 632*f*, 633
Lymphadenitis, 639, 648*t*
Lymphadenopathy, 648*t*
Lymphangitis, 648*t*
Lymphatic nodules, 636, 636*f*, 640*f*, 642*f*
Lymphatic organs, 630, 636–643
Lymphatic system, 33*f*, 629–644, 630*f*, 648*t*. *See also* Lymph nodes
embryonic development, 646, 647*f*
lymph, 630–631
lymphatic cells and tissues, 630, 634–636, 646
lymphatic vessels, 630–633, 630*f*, 631–634, 632*f*, 633*f*, 634*f*, 637–638, 641*f*
red bone marrow, 636–637, 637*f*
in relation to immunity, 644–646
spleen, 642, 643*f*, 643*i*
thymus, 637, 638*f*
tonsils, 640, 642, 642*f*
Lymph nodes, 89*f*, 633*f*, 637–639, 640*f*
appendicular, 641*f*
axillary, 630*f*, 641*f*
B cells in, 644
cervical, 630*f*
embryonic development, 646, 647*f*
hemopoietic role, 546
ileocolic, 641*f*
inguinal, 630*f*, 639*f*
intestinal, 630*f*
locations, 639
mesenteric, 641*f*
mesocolic, 641*f*
submandibular, 630*f*
Lymphoblasts, 552
Lymphocyte colony-forming units, 644
Lymphocytes, 86, 93*f*, 543*f*, 550*f*, 550*t*, 551–552, 553*f*
in AIDS, 647–648
immune function, 634–636, 644, 646
in intestinal mucosa, 636, 636*f*, 692, 694
in lymph nodes, 637–639, 640*f*
in respiratory mucosa, 655
source of, 630–631
in spleen, 642
in thymus, 637, 638*f*
types, 634–635
Lymphoma, 639*i*
Lymph sacs, 646, 647*f*
Lymph scrotum, 631*i*
Lysosomes, 52*f*, 65, 66*f*, 67
Lysozyme, 683, 692

M

Macrophages
alveolar, 551, 635, 664, 664*f*
hepatic, 551, 592*f*, 635, 695, 696*t*
of lymph nodes, 638, 640*f*
phagocytosis by, 58, 635*f*
role in apoptosis, 99
role in immune system, 635, 636, 638*f*
of spleen, 642
types, 371, 551
Macula densa, 713*f*, 716
Macula lutea, 497*f*, 498, 499*f*
Macula sacculi, 490–491, 491*f*
Macula utriculi, 491, 491*f*
Magnetic resonance imaging (MRI), 4, 5*f*, 313*f*, 451, 727*f*
Magnification and resolution, 49, 49*f*
Malaria, 555, 556*i*
Male reproductive system. *See* Reproductive system
Malignant melanoma, 146, 146*f*
Malleoli, lateral and medial, 217, 219*f*, 246, 248*f*, 359*f*, 363*f*
Malocclusion, 239*i*
Malpighi, Marcello, 586
Mammals, 13, 716*i*
Mammary glands, 35*f*, 142, 142*t*, 749–751, 750*f*, 751*f*
Mammillary bodies, 422*f*, 423*f*, 428, 429*f*, 431*f*, 432, 433*f*
Mammogram, 755
Mammotropes, 522
Manus. *See* Hand
Marfan syndrome, 88*i*
Mass reflex reaction, 407
Mast cells, 86
Mastoid process, 176*f*, 180, 181*f*, 486
Matrix of a tissue, 78, 89, 91, 91*f*
Measles, 557
Meatus, 174*t*
acoustic, 176*f*, 177*f*, 179*t*, 180, 181, 181*f*, 485, 485*f*
nasal, 655, 656*f*
Mechanical advantage, 261, 261*f*
Mechanoreceptors, 476
Meconium, 120*t*, 700
Medial lemniscus, 389, 391*f*, 425, 425*f*
Medial prefrontal cortex, 433*f*
Median aperture of brain, 418, 419*f*, 420*f*
Median plane, 25, 25*f*
Mediastinum, 29, 29*f*, 665
Medical terminology, 17–19, 18*i*, 19*t*
Medication in old age, 721
Medulla
adrenal, 462*f*, 463, 468, 526
embryonic development, 379, 532, 534*f*
hair, 138*f*, 139, 139*f*
ovarian, 740, 741*f*
renal, 710, 711*f*
Medulla oblongata
and autonomic nervous system, 463, 468
in control of digestive system, 684, 685
decussation in, 389, 390*t*, 391*f*, 392, 392*f*, 421
and equilibrium, 493
functional overview, 421, 427*t*
and hearing, 489
location and anatomy, 416*f*, 421, 422*f*
pyramids, 392, 421, 422*f*
respiratory centers, 666–667, 667*f*
and taste, 481
Medullary cavity, 154*f*, 160*f*
Megakaryocyte, 552, 554, 554*f*, 637*f*
Meiosis, 69, 106, 732–733, 734*f*
Melanin
in hair, 139, 139*f*
in iris, 496
in midbrain, 426
in skin, 132, 135–136, 136*f*
Melanocytes, 131*f*, 132, 135
Melanocyte-stimulating hormone (MSH), 520
Melanopsin, 500
Melatonin, 523
Membrane-coating vesicles, 133
Membranes
basement, 52*f*, 80, 98*f*, 714, 715*f*

of body cavities, 29–30, 29*t*, 31*f*
embryonic, 114–115, 116*f*
interosseous, 210, 230, 248*f*
meninges, 29, 178, 414, 417, 418*f*
mesenteries, 30, 31*f*, 42*f*, 43*f*, 112*f*, 679, 679*f*, 690*f*
mucous, 97–98, 98*f*, 676, 678*f*
pericardium, 29, 30*f*, 36*f*, 563–564, 563*f*, 565*f*
peritoneum, 30, 31*f*, 42*f*, 679
plasma, 51, 52–56, 54*f*, 56*f*
pleurae, 29, 30*f*, 563*f*, 665, 665*f*, 666*i*
serous, 29–30, 31*f*, 98, 676, 678*f*, 686*f*
synovial, 99, 231, 231*f*
tympanic, 485*f*, 486, 487*f*
unit, 53, 54*f*
Membrane skeleton, 63
Membrane transport, 57–58
Memory, 430, 437–438
hypothalamus and, 428
limbic system and, 432
in old age, 449
Memory B and T cells, 635, 644, 646
Menarche, 753
Ménière disease, 507*t*
Meninges, 29, 178, 414, 417, 418*f*
Meningitis, 417*i*, 486*i*
Menisci, 231, 243–244, 245*f*, 246*f*, 247*f*, 247*i*
Menopause, 166, 754–755, 755
Menstrual cycle, 482, 746–747, 748*f*
Mental cleft, 288*t*
Mental protuberance, 175*f*, 185*f*, 186
Mental region, 290*f*, 351*f*
Mental symphysis, 197
Merkel discs, 477, 477*f*
Merocrine sweat glands, 131*f*, 140, 141*f*, 142, 142*t*, 144, 144*f*
Meromelia, 124*f*, 124*t*, 223
Mesangial cells, 717, 717*f*
Mesencephalon, 379, 380*f*, 425
Mesenchyme, 99, 110, 161*f*
Mesenteries, 30, 31*f*, 42*f*, 43*f*, 112*f*, 679, 679*f*, 690*f*
Mesocolon, 30, 679, 693*f*
Mesoderm, 110, 111*f*, 112*f*, 113*t*, 275
Mesonephric duct, 719, 720*f*, 751–753, 752*f*
Mesonephros, 719, 752*f*
Mesosalpinx, 740*f*, 741*f*, 747*f*
Mesothelium, 98, 677
Mesovarium, 740*f*, 741*f*, 742
Metabolism, 12
Metanephros, 719
Metaphase, 69, 69*f*, 70*f*, 734*f*
Metaphysis, 159, 160*f*
Metaplasia, 99, 747
Metarterioles, 589, 593*f*
Metastasis, 71*i*, 639*i*
Metencephalon, 379, 380*f*, 421
Microcephaly, 380
Microfilaments, 52*f*, 60*f*, 63, 64*f*
Microglia, 371, 372*f*, 372*t*, 378, 551, 635
Microscopes, 10, 11*f*, 48–49, 53*t*
Microscopic anatomy, 78
Microtubules, 52*f*, 60, 61*f*, 63, 64*f*, 66*f*
Microvilli, 52*f*, 59, 60*f*, 61*f*, 690, 714
Micturition. *See* Urination
Midbrain
anatomy, 416*f*, 422*f*, 425–426, 425*f*
embryonic development, 379, 380*f*
functions, 427*t*, 468, 489, 504
nerve tracts, 391*f*, 392, 392*f*
Middle ear, 114, 485*f*, 486, 486*i*, 504, 505*f*
Midgut, 112*f*, 678–679, 700, 700*f*
Migraine headache, 450*t*
Mineralocorticoids, 527
Mitochondria, 47*f*, 52*f*, 66*f*, 68
of cardiac muscle, 273, 576
evolution of, 68*i*
of skeletal muscle, 263, 264*f*, 271, 273, 276
of sperm, 107, 736, 736*f*
Mitochondrial DNA (mtDNA), 68, 68*i*, 107
Mitosis, 63, 69–71, 69*f*
Mitral cells, 483*f*, 484
Mitral (left AV) valve, 567, 568*f*, 579, 579*t*
Mixed nerves, 440
Molars, 681, 682*f*
Moles, 136, 137, 146
Monkeys, 220
Monoamines as hormones, 514, 515–516, 516*t*
Monoblasts, 552, 553*f*
Monocytes, 93*f*, 543*f*, 550*f*, 550*t*, 551, 553*f*, 635
Mononucleosis, 556*t*
Mons pubis, 27*f*, 35*f*, 740*f*, 748, 749*f*
Morning sickness, 114*i*, 758*t*
Morula, 108–109, 109*f*, 109*t*
Motion, sense of, 489–493
Motor association area, 435–436, 435*f*
Motor control, 386, 390, 392, 421–425, 424*f*, 430, 432, 434–436, 436*f*
Motor division of nervous system, 367
Motor end plate, 266, 267*f*
Motor homunculus, 434, 436*f*
Motor nerves, 440
Motor neurons, 255*f*, 266*f*, 268*f*, 276, 368, 396*f*
Motor proteins, 60, 368
Motor unit, 267–268, 268*f*
Mouthfeel, 481
MRI. *See* Magnetic resonance imaging
Mucociliary escalator, 660, 660*f*
Mucosa. *See* Mucous membranes
Mucosa-associated lymphatic tissue (MALT), 636, 636*f*, 677
Mucous cells, 684*f*, 688. *See also* Goblet cells
Mucous membranes, 31*f*, 97, 98*f*, 676, 678*f*
leukocytes in, 551, 644
sensory receptors of, 478, 479
Mucous neck cells, 688*t*
Mucus, 97, 686
Müllerian-inhibiting factor, 752
Multiple sclerosis, 373*i*
Muscarinic receptors, 466, 466*f*
Muscle-bound condition, 260*i*
Muscle compartments
of leg, 336*t*–337*t*, 340*t*, 341*f*
of thigh, 334*t*–336*t*, 341*f*
of upper limb, 322*t*, 324*t*, 325*f*
Muscle contraction, 269, 270*f*–271*f*
Muscle fibers, 256*f*
embryonic development, 275, 276*f*
fast glycolytic, 271–272, 272*f*, 272*t*
innervation of, 266–268, 266*f*, 267*f*, 268*f*
slow oxidative, 271–272, 272*f*, 272*t*
ultrastructure of, 263–266, 264*f*, 265*f*, 266*t*
Muscle groups, 259*f*, 260. *See also* Muscles, specific; Muscles in general
of abdomen, 283*f*, 300*f*, 300*t*–301*t*, 301*f*
acting on foot, 336*t*–337*t*, 340*t*
acting on forearm, 320*t*–321*t*, 321*f*, 322*t*
acting on head, 296*t*–297*t*
acting on hip and femur, 329*t*–331*t*, 333*t*
acting on humerus, 316*t*, 319*t*
acting on knee and leg, 334*t*–336*t*
acting on scapula, 314*t*–315*t*, 315*f*
acting on wrist and hand, 322*t*, 324*t*
of back, 303*f*, 303*t*–306*t*, 305*f*
of chewing and swallowing, 292*t*–295*t*, 293*f*
of facial expression, 287*f*, 288*t*–289*t*, 290*f*, 291*f*
of foot, intrinsic, 342*t*, 343*f*, 344*t*
of forearm, 321*f*, 323*f*
gluteal, 335*f*
hamstrings, 306*i*, 335*f*, 335*t*–336*t*, 341*f*, 360*f*
of hand, intrinsic, 327*t*, 328*f*, 329*t*
of hip, 313*f*, 330*f*
hypothenar group, 327*t*
infrahyoid, 295*t*, 658
lateral rotators, 331*f*, 331*t*
of leg, 337*f*
midpalmar group, 327*t*
of nuchal region, 286*f*, 297*f*
of pelvic floor, 307*t*, 308*f*
of respiration, 298*f*, 299*t*
of rotator cuff, 240, 319*f*, 319*t*
of speech, 288*t*–289*t*
thenar group, 327*t*
of thigh, 332*f*
triceps surae group, 337*t*
Muscle injuries, 344–345, 345*t*
Muscle relaxants, 267*i*
Muscle relaxation, 269, 271*f*
Muscles, specific. *See also* Muscle groups; Muscles in general
abdominal oblique, internal and external, 35*f*, 36*f*, 285*f*, 286*f*, 300*f*, 300*t*–301*t*, 301*f*, 302*f*, 303*f*, 304*f*, 318*f*, 355*f*, 666
abductor digiti minimi, 327*t*, 328*f*, 340*f*, 342*t*, 343*f*, 363*f*
abductor hallucis, 342*t*, 343*f*, 363*f*
abductor pollicis brevis, 327*t*, 328*f*
abductor pollicis longus, 323*f*, 324*t*
adductor brevis, 38*f*, 39*f*, 330*f*, 332*f*, 333*t*, 341*f*
adductor hallucis, 343*f*, 344*t*, 360*f*
adductor longus, 35*f*, 37*f*, 38*f*, 39*f*, 285*f*, 330*f*, 332*f*, 333*t*, 334*f*, 341*f*, 609*f*
adductor magnus, 37*f*, 39*f*, 286*f*, 330*f*, 332*f*, 333*t*, 335*f*, 341*f*
adductor pollicis, 327*t*, 328*f*, 357*f*
anconeus, 321*t*, 323*f*, 325*f*
auricularis, 15*i*
biceps brachii, 35*f*, 240, 241*f*, 259*f*, 271, 285*f*, 317*f*, 318*f*, 320*t*, 321*f*, 325*f*, 353*f*, 355*f*
biceps femoris, 286*f*, 335*f*, 335*t*–336*t*, 340*f*, 341*f*, 358*f*, 359*f*, 360*f*
brachialis, 317*f*, 320*t*, 325*f*
brachioradialis, 285*f*, 317*f*, 321*t*, 323*f*, 325*f*, 353*f*, 355*f*, 356*f*
buccinator, 289*t*, 290*f*, 293*f*
bulbospongiosus, 307*t*, 308*f*, 737
ciliary, 463
coccygeus, 307*t*, 308*f*
coracobrachialis, 36*f*, 37*f*, 241*f*, 316*t*, 317*f*, 325*f*
corrugator supercilii, 288*t*, 290*f*, 291*f*
cremaster, 732, 733*f*
cricoarytenoid, 659*f*
dartos, 732
deep transverse perineus, 307*t*, 308*f*
deltoid, 35*f*, 241*f*, 285*f*, 286*f*, 303*f*, 316*t*, 317*f*, 318*f*, 325*f*, 352*f*, 353*f*, 355*f*
depressor anguli oris, 287*f*, 289*t*, 290*f*, 291*f*
depressor labii inferioris, 287*f*, 289*t*, 290*f*, 291*f*
detrusor, 718
diaphragm, 29*f*, 30*f*, 31*f*, 36*f*, 39*f*, 42*f*, 298*f*, 299*i*, 299*t*, 306*i*, 563*f*, 666
digastric, 292*f*, 294*f*, 294*t*–295*t*
erector spinae, 42*f*, 286*f*, 303*f*, 304*f*, 305*f*, 306*i*, 353*f*, 666
esophageal sphincters, 684
extensor carpi radialis brevis and longus, 323*f*, 324*t*, 325*f*, 355*f*
extensor carpi ulnaris, 323*f*, 324*t*, 325*f*, 356*f*
extensor digiti minimi, 323*f*, 324*t*, 325*f*, 357*f*
extensor digitorum, 323*f*, 324*t*, 325*f*, 356*f*, 357*f*
extensor digitorum brevis, 337*f*, 340*f*, 342*t*, 361*f*, 362*f*
extensor digitorum longus, 285*f*, 336*t*, 337*f*, 340*f*, 341*f*, 363*f*
extensor hallucis brevis and longus, 336*t*, 337*f*, 341*f*, 360*f*, 363*f*
extensor indicis, 323*f*, 324*t*
extensor pollicis brevis and longus, 323*f*, 324*t*, 357*f*
external anal sphincter, 308*f*
external intercostal, 36*f*, 39*f*, 298*f*, 302*f*, 305*f*
external urethral sphincter, 308*f*
fibularis brevis, longus, and tertius, 285*f*, 336*t*, 337*f*, 338*f*, 340*f*, 340*t*, 341*f*, 359*f*, 361*f*
flexor carpi radialis, 285*f*, 322*t*, 323*f*, 325*f*, 356*f*, 357*f*
flexor carpi ulnaris, 322*t*, 323*f*, 325*f*, 353*f*, 356*f*
flexor digiti minimi brevis
of foot, 343*f*, 344*t*
of hand, 327*t*, 328*f*
flexor digitorum brevis, 342*t*, 343*f*
flexor digitorum longus, 338*f*, 339*f*, 340*t*, 341*f*
flexor digitorum profundus and superficialis, 322*t*, 323*f*, 324*t*, 325*f*, 326*f*
flexor hallucis brevis, 343*f*, 344*t*
flexor hallucis longus, 338*f*, 339*f*, 340*t*, 341*f*
flexor pollicis brevis, 327*t*, 328*f*
flexor pollicis longus, 322*t*, 323*f*, 325*f*
frontalis, 285*f*, 287*f*, 288*t*, 290*f*, 291*f*
gastrocnemius, 271, 272, 285*f*, 286*f*, 337*f*, 337*t*, 338*f*, 340*f*, 341*f*, 359*f*, 360*f*, 361*f*
gemellus superior and inferior, 331*f*, 331*t*
genioglossus, 292*f*, 292*t*, 681
geniohyoid, 292*f*, 295*t*
gluteus maximus, 215, 286*f*, 303*f*, 306*i*, 308*f*, 330*t*–331*t*, 331*f*, 335*f*, 353*f*
gluteus medius, 39*f*, 286*f*, 303*f*, 330*t*–331*t*, 331*f*, 335*f*, 353*f*
gluteus minimus, 330*t*–331*t*, 331*f*
gracilis, 35*f*, 37*f*, 38*f*, 39*f*, 285*f*, 286*f*, 330*f*, 332*f*, 333*t*, 334*f*, 335*f*, 341*f*, 358*f*
hyoglossus, 292*f*, 292*t*, 294*f*, 681

Muscles, Specific—*Cont.*
iliacus, 39*f*, 329*t*, 330*f*, 332*f*
iliocostalis, 304*f*, 304*t*, 305*f*
iliopsoas, 329*t*, 334*f*
infraspinatus, 240, 286*f*, 303*f*, 317*f*, 318*f*, 319*f*, 319*t*, 353*f*
intercostals, 36*f*, 39*f*, 298*f*, 299*i*, 299*t*, 302*f*, 666, 667*f*
interosseous
of foot, 343*f*, 344*t*
of hand, 327*t*, 328*f*
intrinsic, of tongue, 292*f*
ischiocavernosus, 307*t*, 308*f*, 738
latissimus dorsi, 286*f*, 303*f*, 305*f*, 316*t*, 317*f*, 318*f*, 353*f*, 355*f*
levator anguli oris, 289*t*, 290*f*, 291*f*
levator ani, 307*t*, 308*f*
levator labii superioris, 287*f*, 289*t*, 290*f*, 291*f*
levator palpebrae superioris, 288*t*, 291*f*, 495*f*
levator scapulae, 286*f*, 290*f*, 294*f*, 303*f*, 314*t*, 315*f*
longissimus, 297*f*, 304*f*, 304*t*, 305*f*
lumbricals
of foot, 342*t*, 343*f*
of hand, 327*t*, 328*f*
masseter, 186, 285*f*, 287*f*, 290*f*, 293*f*, 293*t*, 444*f*
mentalis, 288*t*–289*t*, 290*f*, 291*f*
multifidus, 304*f*, 306*t*
mylohyoid, 292*f*, 294*f*, 294*t*–295*t*
nasalis, 287*f*, 288*t*, 290*f*, 291*f*
oblique, of eye, 494–495, 496, 496*f*
obturator, 330*f*, 331*f*, 331*t*, 333*t*
occipitalis, 286*f*, 288*t*, 290*f*
occipitofrontalis, 288*t*
omohyoid, 36*f*, 290*f*, 294*f*, 295*t*
opponens digiti minimi, 327*t*, 328*f*
opponens pollicis, 327*t*, 328*f*
orbicularis oculi, 285*f*, 287*f*, 288*t*, 290*f*, 291*f*, 495*f*
orbicularis oris, 285*f*, 287*f*, 288*t*–289*t*, 290*f*, 291*f*, 293*f*
palatoglossus, 292*f*, 292*t*, 681
palmaris longus, 322*t*, 323*f*, 325*f*, 356*f*, 357*f*
pectineus, 37*f*, 38*f*, 330*f*, 332*f*, 333*t*, 334*f*
pectoralis major, 35*f*, 241*f*, 285*f*, 299*i*, 302*f*, 316*t*, 317*f*, 318*f*, 325*f*, 352*f*, 355*f*
pectoralis minor, 285*f*, 302*f*, 314*t*, 315*f*, 666
pharyngeal constrictors, 290*f*, 292*f*, 294*f*, 295*t*, 684
piriformis, 308*f*, 330*f*, 331*f*, 333*t*
plantaris, 338*f*, 340*t*
platysma, 35*f*, 285*f*, 287*f*, 289*t*, 290*f*, 291*f*
popliteus, 246, 336*t*, 338*f*, 339*f*
procerus, 287*f*, 288*t*, 290*f*, 291*f*
pronator quadratus and teres, 321*f*, 321*t*, 323*f*, 325*f*
psoas major, 39*f*, 329*t*, 330*f*, 332*f*
pterygoids, 293*f*, 293*t*, 444*f*
pupillary constrictor, 463, 496
quadratus femoris, 331*f*, 333*t*
quadratus lumborum, 39*f*, 304*f*, 306*t*
quadratus plantae, 342*t*, 343*f*
quadriceps femoris, 216, 243, 245*f*, 272, 332*f*, 334*f*, 358*f*
rectus femoris, 35*f*, 37*f*, 38*f*, 285*f*, 332*f*, 334*f*, 334*t*, 341*f*, 358*f*
vastus intermedius, 38*f*, 332*f*, 334*t*, 341*f*
vastus lateralis, 35*f*, 38*f*, 285*f*, 332*f*, 334*f*, 334*t*, 335*f*, 341*f*, 358*f*, 359*f*
vastus medialis, 38*f*, 285*f*, 332*f*, 334*f*, 334*t*, 341*f*, 358*f*
rectus, of eye, 445*f*, 494, 495*f*, 496*f*
rectus abdominis, 42*f*, 285*f*, 300*f*, 300*t*, 301*f*, 352*f*, 355*f*, 666
rectus femoris (*See* Muscles, specific, quadriceps femoris)
rhomboideus major and minor, 286*f*, 303*f*, 315*f*, 315*t*, 318*f*
risorius, 289*t*, 290*f*, 291*f*
sartorius, 35*f*, 38*f*, 285*f*, 332*f*, 334*f*, 335*t*, 341*f*, 609*f*, 639*f*
scalenes, 294*f*, 296*t*, 666
semimembranosus, 243, 286*f*, 335*f*, 335*t*–336*t*, 341*f*, 358*f*, 360*f*
semispinalis, 286*f*, 297*f*, 297*t*, 303*f*, 304*f*, 306*t*
semitendinosus, 286*f*, 335*f*, 335*t*–336*t*, 341*f*, 358*f*, 360*f*, 361*f*
serratus anterior, 285*f*, 286*f*, 299*i*, 302*f*, 303*f*, 314*t*, 315*f*, 318*f*, 352*f*, 355*f*
serratus posterior, 286*f*, 303*f*, 304*f*, 305*t*
soleus, 271, 272, 285*f*, 286*f*, 337*f*, 337*t*, 338*f*, 340*f*, 341*f*, 359*f*, 360*f*, 361*f*, 363*f*
spinalis, 304*f*, 305*f*, 305*t*
splenius, 286*f*, 294*f*, 297*t*, 303*f*, 304*f*
stapedius, 486, 489
sternocleidomastoid, 285*f*, 286*f*, 287*f*, 290*f*, 294*f*, 296*t*, 297*f*, 303*f*, 352*f*, 666
sternohyoid, 290*f*, 294*f*, 295*t*
sternothyroid, 290*f*, 294*f*, 295*t*
styloglossus, 292*f*, 292*t*, 681
stylohyoid, 186, 186*f*, 292*f*, 294*f*, 295*t*
subclavian, 39*f*, 612*t*
subscapularis, 36*f*, 240, 317*f*, 319*f*, 319*t*
superficial transverse perineus, 307*t*, 308*f*
supinator, 321*f*, 321*t*, 323*f*, 325*f*
suprahyoid group, 294*t*–295*t*
supraspinatus, 240, 286*f*, 303*f*, 317*f*, 319*f*, 319*t*
temporalis, 180, 290*f*, 293*f*, 293*t*, 444*f*
tensor fasciae latae, 35*f*, 37*f*, 38*f*, 285*f*, 330*t*–331*t*, 332*f*, 334*f*, 358*f*
tensor tympani, 485*f*, 486, 489
teres major, 286*f*, 303*f*, 316*t*, 317*f*, 318*f*, 325*f*, 353*f*
teres minor, 240, 286*f*, 303*f*, 317*f*, 319*f*, 319*t*
thyrohyoid, 290*f*, 294*f*, 295*t*
tibialis anterior, 285*f*, 336*t*, 337*f*, 340*f*, 341*f*, 359*f*, 361*f*, 363*f*
tibialis posterior, 339*f*, 340*t*, 341*f*
trachealis, 660, 660*f*
transversus abdominis, 36*f*, 285*f*, 300*f*, 301*f*, 301*t*, 666
trapezius, 35*f*, 285*f*, 286*f*, 294*f*, 297*f*, 297*t*, 303*f*, 305*f*, 314*t*, 315*f*, 351*f*, 352*f*, 353*f*, 355*f*
triceps brachii, 259*f*, 317*f*, 318*f*, 320*t*, 325*f*, 353*f*, 355*f*
vastus (*See* Muscles, specific, quadriceps femoris)
zygomaticus, 285*f*, 287*f*, 289*t*, 290*f*, 291*f*
Muscles in general
action, 260
atrophy, 270
belly, 259, 259*f*
blood supply, 268, 268*f*
connective tissues of, 257, 258*f*, 259
growth, 269–270
innervation, 266–268, 266*f*, 267*f*, 268*f*, 284
methods of studying, 287, 350
naming, 284, 284*t*
origin and insertion, 259, 259*f*
shapes and fascicle arrangement, 258–259, 259*f*
Muscle spindle, 406, 477*f*, 478, 478*t*
Muscle tone, 423
Muscle tremor, 450
Muscular dystrophy, 276
Muscularis externa, 274*f*, 676, 677, 678*f*, 685, 691*f*, 694
Muscularis mucosae, 97, 98*f*, 274*f*, 676–677, 678*f*, 687*f*
Muscular system, 32*f*, 255–348, 285*f*–286*f*. *See also* Muscles, specific
aging, 275–276
appendicular, 313–345
association with joints, 240, 243, 246
axial, 283–308
diseases, 276–277, 277*t*
injuries, 345*t*
Muscular tissue, 79*t*, 94–96
cardiac
embryonic development, 275
histology, 95–96, 95*f*, 95*t*, 273, 273*f*, 574–576, 575*f*
physiological properties, 256, 273–274, 273*t*, 574–576
skeletal
embryonic development, 275, 276*f*
histology, 95–96, 95*f*, 95*t*, 256
physiological properties, 257, 273*t*
smooth
embryonic development, 275
histology, 95–96, 95*f*, 95*t*, 256, 273*f*, 274*f*
physiological properties, 273*t*, 274–275
visceral, 96, 274*f*, 677
voluntary and involuntary, 95–96, 256, 276
Mutagens, 71*i*, 121
Mutation, 121
Myasthenia gravis, 276
Myelencephalon, 379, 380*f*, 421
Myelin, 371–374, 373*f*, 379, 388, 430
Myelinated nerve fibers, 369*f*, 372*f*, 373*f*, 375*f*, 394*f*
Myelin sheath, 369*f*, 371–374, 372*f*, 373*i*, 375*f*
Myeloblasts, 552
Myenteric plexus, 455*f*, 678, 678*f*, 692
Myoblasts, 263, 275, 276*f*
Myocardial infarction (MI), 571, 572*i*, 580
Myocardium, 565, 567*f*, 568*f*
Myocytes, 96, 273, 273*f*, 274, 274*f*
Myoepithelial cells, 140, 141*f*, 496, 751, 751*f*
Myofibrils, 263, 264*f*, 266*t*, 270, 276
Myofilaments, 263, 265*f*, 266*t*, 270*f*, 271*f*, 274, 275
Myoglobin, 263, 271, 273, 276
Myoglobinuria, 342*i*
Myology, 256
Myometrium, 741*f*, 746
Myopathy, 276
Myopia, 500*f*, 500*t*
Myosin, 263, 264, 265*f*, 266*t*, 269, 270*f*, 271*f*
Myotomes, 114, 275
Myotubes, 275, 276*f*
Myringotomy, 486*i*
Myxedema, 536*t*

N

an-Nafis, Ibn, 9, 562
Nails, 14, 120*t*, 140, 141*f*, 143
Nares, 655, 656*f*, 657*f*. *See also* Nasal choanae
Nasal cavity, 40*f*, 178*f*, 654*f*, 656*f*
Nasal choanae, 177*f*, 183, 654*f*, 655, 656*f*
Nasal conchae, 178*f*, 183, 183*f*, 184*f*, 185, 655
Nasal fossae, 183, 184*f*, 655, 656*f*
Nasal region, 351*f*, 657*f*
Nasal septum, 176*f*, 183, 185, 655, 657*f*
Nasopharynx, 655, 656*f*
Natal cleft, 353*f*
Natural killer (NK) cells, 552, 553*f*, 629*f*, 634, 635, 646
Natural selection, 14
Nausea, 701
Neck
arteries, 599*f*, 599*t*–600*t*
muscles, 294*f*, 294*t*–295*t*, 296*t*–297*t*, 297*f*
surface anatomy, 27*f*, 351*f*
triangles, 296*f*, 296*t*
veins, 611*f*, 611*t*–612*t*
vertebrae, 190–192, 191*f*, 192*f*
Necrosis, 99
Negative feedback, 456
Negative selection, 644
Neisseria gonorrhoeae, 756
Neocortex, 431, 432*f*
Neonate
congenital deformities (*See* Birth defects)
face, 198
immature state, 215
respiration, 668
size, 118–119
skull, 198, 198*f*
spinal curvature, 188, 199, 200*f*
Neoplasia, 99
Nephron loop, 712*f*, 716
Nephrons, 709–717, 712*f*, 721–722, 721*f*
Nephroptosis, 722*t*
Nephrotic syndrome, 722*t*
Nerve endings, free and encapsulated, 476–478, 477*f*, 478*t*
Nerve fibers, 94, 368
classification, 395, 395*t*
myelinated, 369*f*, 371–374, 372*f*, 373*f*, 375*f*
in reflex arcs, 406, 406*f*
unmyelinated, 374, 375*f*
Nerves, specific. *See also* Nerves in general
abducens (cranial nerve VI), 422*f*, 441*f*, 445*f*, 445*t*, 493, 495, 496*f*
accessory (cranial nerve XI), 402*f*, 422*f*, 441*f*, 448*f*, 448*t*, 677
alveolar, 444*f*
ansa cervicalis, 400*f*, 400*t*, 677
anterior cutaneous, 399*f*
axillary, 401*f*, 401*t*
cardiac, 459, 576

coccygeal, 404*t*
cochlear, 446*f*, 485*f*, 487*f*, 488*f*, 489
dorsal, of penis, 739
dorsal scapular, 401*f*
facial (cranial nerve VII), 422*f*, 441*f*, 445*f*, 445*t*, 449*i*, 463, 464*f*, 468, 481, 489, 677, 685
femoral, 36*f*, 403*f*, 403*t*, 609*f*, 639*f*
fibular, 403*f*, 404*f*, 404*t*
genitofemoral, 403*f*, 403*t*
glossopharyngeal (cranial nerve IX), 422*f*, 441*f*, 446*f*, 446*t*, 456, 457*f*, 463, 464*f*, 468, 481, 668*f*, 677, 685
gluteal, inferior and superior, 404*f*, 404*t*
great auricular, 400*f*, 400*t*
hypoglossal (cranial nerve XII), 400*f*, 402*f*, 422*f*, 441*f*, 448*f*, 448*t*, 677
iliohypogastric, 403*f*, 403*t*
ilioinguinal, 403*f*, 403*t*
infraorbital, 444*f*
intercostal, 396, 397*f*, 399*f*, 666, 667
internal obturator, nerve to, 404*t*
internal pudendal, 404*f*
lateral cutaneous, 399*f*
lateral femoral cutaneous, 403*f*, 403*t*
lesser occipital, 400*f*, 400*t*
lingual, 444*f*, 481
long thoracic, 401*f*
lumbosacral trunk, 403*f*, 404*f*
medial brachial cutaneous, 401*f*
medial cutaneous antebrachial, 401*f*
median, 326*f*, 326*i*, 401*f*, 402*t*
meningeal, 396, 398*f*
musculocutaneous, 401*f*, 402*t*
obturator, 403*f*, 403*t*
oculomotor (cranial nerve III), 422*f*, 423*f*, 425*f*, 441*f*, 443*f*, 443*t*, 463, 464*f*, 468, 493, 495, 496*f*
olfactory (cranial nerve I), 183*i*, 441*f*, 442*f*, 442*t*, 484
optic (cranial nerve II), 182, 422*f*, 440, 441*f*, 442*f*, 442*t*, 496*f*, 497*f*, 498, 501*f*, 502, 503*f*, 504
pectoral, 401*f*
pelvic, 463, 464*f*, 677
pelvic splanchnic, 463, 464*f*
perforating cutaneous, 404*t*
phrenic, 400*f*, 400*t*, 402*f*, 666, 667
piriformis, nerve to, 404*t*
plantar, 403*f*
posterior cutaneous, 404*t*
posterior cutaneous femoral, 404*f*
pudendal, 403*f*, 404*t*, 739
quadratus femoris, nerve to, 404*t*
radial, 401*f*, 401*t*, 405*i*
renal, 710
saphenous, 403*f*, 403*t*
sciatic, 403*f*, 404*f*, 404*t*, 405*i*
splanchnic, 459, 459*f*, 461, 461*f*, 461*t*
subclavian, 401*f*, 402*f*
subscapular, 401*f*
supraclavicular, 400*f*, 400*t*
suprascapular, 401*f*
testicular, 732
thoracodorsal, 401*f*
tibial, 403*f*, 404*f*, 404*t*
transverse cervical, 400*f*, 400*t*
trigeminal (cranial nerve V), 422*f*, 441*f*, 444*f*, 444*t*, 449*i*, 463, 479, 481, 486, 489, 677, 685
trochlear (cranial nerve IV), 422*f*, 440, 441*f*, 443*f*, 443*t*, 493, 495, 496*f*
tympanic, 463
ulnar, 401*f*, 402*t*
vagus (cranial nerve X), 402*f*, 422*f*, 441*f*, 447*f*, 447*t*, 456, 457*f*, 459*f*, 463, 464*f*, 468, 479, 481, 486, 576, 667, 668*f*, 677, 678, 685
vestibular, 446*f*, 485*f*, 487*f*
vestibulocochlear (cranial nerve VIII), 422*f*, 441*f*, 446*f*, 446*t*, 489, 490*f*, 492–493, 493*f*
Nerve signals, 374, 449
Nerves in general, 366, 366*f*
cervical, 397*f*
classification, 395, 440
coccygeal, 397*f*
cords, 401*f*, 401*t*
cranial, 287, 440, 441*f*, 442*t*–448*t*, 448*i*
divisions, 401*f*, 401*t*
histology, 385*f*, 393–395, 394*f*, 587*f*
lumbar, 397*f*
plexuses, 37*f*, 287, 397
brachial, 37*f*, 401*f*, 401*t*–402*t*, 402*f*
cervical, 400*f*, 400*t*
coccygeal, 404*f*, 404*t*
lumbar, 403*f*, 403*t*
sacral, 404*f*, 404*t*
rami, 287, 398*f*, 399*f*
roots, 401*f*, 401*t*, 403*f*, 404*f*
sacral, 397*f*
spinal, 284, 393–405
thoracic, 397*f*
trunks, 401*f*, 401*t*
Nervous system, 32*f*, 365–512, 366*f*. *See also* Autonomic nervous system; Brain; Nerves, specific; Nerves in general; Nervous tissue; Sense organs; Spinal cord
aging, 449
central (*See* Brain; Spinal cord)
compared to endocrine system, 515, 517*f*, 517*t*
development, 378–380, 379*f*, 380*f*
diseases, 380, 449–451, 450*t*
divisions, 366–367, 366*f*
peripheral (*See* Cranial nerves; Ganglia; Spinal nerves)
Nervous tissue, 79*t*, 94, 94*t*, 365–378
Neural crest, 379, 379*f*, 469, 532, 534*f*
Neural folds, 113*f*, 378
Neuralgia, 408*t*
Neural groove, 112*f*, 113*f*, 378, 379*f*
Neural plate, 378, 379*f*
Neural tube, 112*f*, 113, 113*f*, 379, 379*f*, 388, 469
defects, 120, 380, 381*f*
Neurilemma, 372, 373*f*, 374, 375*f*
Neurocranium, 196
Neuroectoderm, 378
Neuroendocrine cells, 515
Neurofibrillary tangles, 449–450, 450*f*
Neurofibrils, 368, 369*f*
Neurogenic shock, 407
Neuroglia. *See* Glial cells
Neurohypophyseal bud, 532
Neurohypophysis. *See* Endocrine system, pituitary gland
Neuromuscular junction, 255*f*, 266–267, 266*f*, 267*f*, 376
Neuronal circuits, 376–378, 378*f*
Neuronal convergence and divergence, 463, 502
Neuronal pools, 376
Neurons, 94
afferent (sensory), 367, 368*f*
anaxonic, 370, 370*f*
association (interneurons), 367–368, 368*f*
bipolar, 370, 370*f*, 489, 500, 501*f*, 502
classification, 367–368, 368*f*, 370, 370*f*
efferent (motor), 368, 368*f*
general structure, 94, 94*f*, 368, 369*f*
multipolar, 370, 370*f*
properties of, 367
unipolar, 370, 370*f*
upper and lower motor, 390, 392*f*, 434–435
Neuropharmacology, 466*i*
Neurotransmitters, 274, 367, 376, 449, 465, 466, 466*f*
receptors for, 449, 465–466, 469
Neurulation, 113, 113*f*, 378
Neutrophils, 67, 86, 93*f*, 543*f*, 548–551, 549*f*, 549*t*, 552, 553*f*
Nevus, 137
Nicotinic receptors, 466, 466*f*
Night vision, 499, 502
Nipples, 136, 142*i*, 749–750, 750*f*
Nissl bodies, 368, 369*f*
Nitrogenous wastes, 709
Nociceptors, 476, 479, 480*f*
Nodes of Ranvier, 369*f*, 374
Nomina Anatomica, 17
Nondisjunction, 121, 121*f*
Non-Hodgkin lymphoma, 648*t*
Nonspecific defense, 635
Norepinephrine, 376, 463, 465–466, 466*f*, 515, 526
Nose, 351*f*, 655, 656*f*, 657*f*
Nosebleed, 655
Notochord, 13, 111*f*, 198, 199*f*
Nuchal lines, 177*f*, 297*f*, 304*f*
Nuchal region, 27*f*, 351*f*
Nuclear envelope, 54*f*, 65, 66*f*, 70*f*
Nuclear medicine, 5
Nuclei of brain
anterior, of hypothalamus, 429*f*
basal, 432, 433*f*, 436
caudate, 419*f*, 432, 433*f*
cochlear, 489, 490*f*
cuneate, 391*f*, 421
deep cerebellar, 423
globus pallidus, 432, 433*f*
gracile, 421
inferior olivary, 421
lateral geniculate, 422*f*, 429*f*, 502, 503*f*
lentiform, 432, 433*f*
mammillary bodies, 422*f*, 423*f*, 428, 429*f*, 432
medial geniculate, 422*f*, 425*f*, 490*f*
nucleus accumbens, 432
oculomotor, 425*f*
paraventricular, 429*f*, 518
preoptic, 429*f*
pretectal, 503*f*, 504
putamen, 432, 433*f*
red, 424*f*, 425*f*, 426
salivatory, 684
solitary, 481, 483*f*
subthalamic, 433*f*
superior olivary, 432, 490*f*
suprachiasmatic, 428, 429*f*
supraoptic, 429*f*, 518
ventromedial, 429*f*
vestibular, 392, 493, 493*f*
Nucleolus, 52*f*, 65, 70*f*
Nucleoplasm, 51
Nucleus of a cell, 49*f*, 51, 52*f*, 54*f*, 65, 66*f*
Nutrition, prenatal, 116, 116*f*

O

Oblique sections, 78, 80*f*
Occipital condyles, 177*f*, 181
Occipital lobe, 415*f*, 416*f*, 419*f*, 430, 430*f*
Occipital protuberance, external, 177*f*
Occipital region, 351*f*
Occupational therapy, 256
Odontoblasts, 681
Olecranon, 210, 211*f*, 242*f*, 323*f*, 353*f*, 355*f*
Olfaction. *See* Smell
Olfactory bulbs, 442*f*, 483*f*, 484, 484*f*
Olfactory cortex, 484, 484*f*
Olfactory mucosa, 482, 483*f*, 655
Olfactory neurons, 482–484, 483*f*
Olfactory pathways, 428, 483*f*–484*f*, 484
Olfactory tracts, 483*f*, 484, 484*f*
Oligodendrocytes, 371–372, 372*f*, 372*t*, 373*f*, 374, 379
Olive, 421, 422*f*
Omentum, 30, 31*f*, 36*f*, 679, 679*f*, 686*f*
Omphalocele, 700
Oncogenes, 71*i*
Ondine's curse, 668*i*
Oocytes, 741*f*, 742, 743*f*, 744*f*. *See also* Egg
Oogenesis, 106, 742, 743*f*
Oogonia, 742, 743*f*
Opportunistic infection, 648
Opposition of the thumb, 238, 238*f*, 329*t*
Optic chiasm, 416*f*, 428, 441*f*, 442*f*, 502, 503*f*
Optic cup, 504, 505, 506*f*
Optic disc, 497, 497*f*, 498, 499, 499*f*
Optic radiation, 502, 503*f*
Optic tracts, 422*f*, 441*f*, 442*f*, 502, 503*f*
Optic vesicle, 379, 504, 506*f*
Oral cavity, 40*f*, 677*f*, 680–683, 680*f*
Oral region, 290*f*, 291*f*, 351*f*, 680, 680*f*
Orangutan, 211
Orbit, 175, 178*f*, 185*f*
Orbital region, 351*f*, 494–495, 494*f*, 495*f*
Orbitofrontal cortex, 433*f*, 434, 481, 484
Organ, definition of, 6
Organelles, 53*t*, 63–68
Organ of Corti, 488*f*
Organogenesis, 111
Organ systems, 6, 30–31, 32*f*–34*f*
Orgasm, 428
Oropharynx, 655, 656*f*
Orthopedics, 167, 239
Orthostatic hypotension, 469, 622
Osmoreceptors, 428
Osmosis, 58
Osseous tissue. *See* Bone tissue
Ossification
endochondral, 159–161, 160*f*, 196, 221
intramembranous, 161*f*, 162*f*, 221
Osteitis deformans, 167*t*
Osteoarthritis, 232*i*, 247, 249, 405*i*
Osteoblasts, 155, 156*f*, 159, 160*f*, 161–163, 162*f*
Osteoclasts, 155, 156*f*, 161, 163
Osteoclast-stimulating factor, 163
Osteocytes, 92, 155, 156*f*, 158, 161–162, 162*f*
Osteogenesis. *See* Ossification
Osteogenesis imperfecta, 87*i*, 88*i*
Osteogenic cells, 155, 156*f*, 161, 162
Osteoid tissue, 159, 162*f*
Osteology, 152

Osteomalacia, 167*t*
Osteons, 92, 92*f*, 156, 157*f*
Osteopenia, 163
Osteoporosis, 165–167, 166*f*, 167*f*, 218*i*, 755
Osteosarcoma, 152*i*
Otic placode, 504
Otitis media, 486*i*
Otoliths, 491, 491*f*
Ovarian cycle, 742
Ovaries, 38*f*, 109*f*, 740–743, 740*f*
  blood supply, 747*f*
  descent, 753
  embryonic development, 752*f*
  endocrine role, 528, 531*t*
  follicles, 85*f*, 740, 741*f*, 742–743, 743*f*, 744*f*
Ovulation, 740, 741*f*, 742, 743*f*, 744*f*
Oxalate, effect on bone healing, 165*i*
Oxytocin, 428, 516*f*, 518, 520, 522, 751

P

Pacemakers, 666, 667
Pacinian corpuscle. *See* Lamellated corpuscle
Packed cell volume (PCV), 546
Pain, 434, 437, 479, 480*i*
  and cranial nerves, 421, 449*i*
  modulation, 392, 426, 481
  projection pathways, 479–480, 480*f*
  receptors, 476, 478*t*, 479
  referred, 480, 480*f*, 701
  spinal tracts, 389, 390*t*, 391*f*
  visceral, 479, 480, 701
Palate, 40*f*, 177*f*, 184, 184*i*, 654*f*, 656*f*, 680*f*, 681
Palatine tonsil, 113*f*
Palatoglossal and palatopharyngeal arches, 680*f*, 681
Paleocortex, 432
Pallidotomy, 451
Pallor, 136
Palmar region, 27*f*, 356*f*, 357*f*
Palpebrae, 494, 494*f*, 495*f*
Pampiniform plexus, 732, 733*f*
Pancreas, 38*f*, 42*f*
  anatomy, 528*f*, 698, 698*f*
  embryonic development, 700–701, 701*f*
  histology, 528*f*, 699, 699*f*
  islets, 527, 528*f*, 531*t*
  secretions
    endocrine, 527, 531*t*
    exocrine, 699, 699*t*
Paneth cells, 692
Papillae
  dermal, 131*f*, 133–134, 134*f*, 138–139, 138*f*
  duodenal, 690, 698*f*
  lingual, 475*f*, 481, 482*f*, 675*f*, 681, 681*f*
Papillary muscles, 561*f*, 566*f*, 567, 568*f*, 569*f*, 578
Pap smear, 83, 757*f*, 757*i*
Paracrine secretions, 527
Parafollicular (C) cells, 524*f*, 525, 532
Parallel muscles, 258, 259*f*
Paralysis, 267*i*, 408
Paramesonephric duct, 751, 752*f*, 753
Paraplegia, 408
Parasites, defenses against, 551
Parasympathetic nervous system, 367, 457, 463, 465*t*
  efferent pathways, 463, 464*f*
  innervation of digestive tract, 677
  innervation of heart, 576
  innervation of penis, 739
  respiratory effects, 663
Parasympathetic tone, 457
Parasympatholytics, 466*i*
Parasympathomimetics, 466*i*
Parathyroid glands. *See* Endocrine system, parathyroid glands
Parathyroid hormone (PTH), 163, 525, 530*t*, 716
Paraurethral glands, 749, 749*f*
Parenchyma, 97
Paresis, 408
Paresthesia, 408*t*, 507*t*
Parietal cells, 687*f*, 688, 688*t*
Parietal lobe, 415*f*, 416*f*, 430, 430*f*
Parieto-occipital sulcus, 416*f*, 430
Parkinson disease, 426, 450–451
Pars intermedia, 518, 520
Patau syndrome, 122
Patellar region, 27*f*, 245*f*, 246*f*
Pathogens, 630
Pavlov, Ivan, 676
Pectinate muscles, 567, 568*f*
Pectoral girdle, 208, 212*t*
Pectoral region, 27*f*, 352*f*, 355*f*
Pedal region, 27*f*. *See also* Foot
Pedicels of glomerulus, 714, 715*f*
Peduncles
  cerebellar, 421, 422*f*, 423, 425
  cerebral, 422*f*, 425
Peg cells, 745, 745*f*
Pelvic cavity, 29*f*, 30, 43*f*
Pelvic diaphragm, 307*t*, 308*f*
Pelvic girdle, 173*f*, 213, 222*t*
  male *versus* female, 215, 216*f*, 216*t*
Pelvic inlet and outlet, 213*f*, 214
Pelvic kidney, 7*f*, 721*i*
Pelvis
  bony, 39*f*, 213*f*, 214
  renal, 710, 711*f*, 717, 719–720, 720*f*
Pemphigus vulgaris, 63*i*
Penis, 27*f*, 36*f*, 37*f*, 43*f*
  anatomy, 718*f*, 730*f*, 733*f*, 737–739, 739*f*
  cancer, 756*t*
  embryonic development, 753, 754*f*
Pennate muscles, 258–259, 259*f*
Pepsinogen, 688*t*
Peptic ulcer, 689*i*
Peptides as hormones, 514, 515
Pepto-Bismol, 689*i*
Perforating canals, 158
Perforating fibers, 154
Pericardial cavity, 29*f*, 30, 30*f*, 41*f*, 563, 563*f*, 578, 665*f*
Pericardial fluid, 30, 563
Pericarditis, 563, 565
Pericardium, 29, 30*f*, 36*f*, 563–564, 563*f*, 565*f*
Perichondrium, 91*f*, 92, 159
Pericytes, 590*f*, 591
Periderm, 143, 143*f*, 144
Perilymph, 486, 487*f*
Perimetrium, 746
Perimysium, 257, 258*f*, 273*t*
Perineal body, 307*t*, 308*f*
Perineal raphe, 308*f*, 729*f*, 732, 749*f*, 754*f*
Perineum, 27*f*, 307*t*, 308*f*
  female, 747, 749*f*
  male, 729*f*
Perineurium, 393, 394*f*
Periodontal disease, 682*i*, 701
Periosteum, 92, 154, 154*f*, 161, 162, 162*f*
Peripheral nervous system (PNS). *See* Nervous system
Peripheral neuropathy, 408*t*
Peristalsis, 275
  of digestive tract, 678, 701
  of ureter, 718
Peritoneal cavity, 30, 31*f*, 42*f*, 111, 700*f*
Peritoneal fluid, 30
Peritoneum, 30, 31*f*, 42*f*, 679
Peritrichial endings. *See* Hair receptors
Peritubular capillaries, 710, 712*f*, 713*f*
Perivascular feet, 371, 418
Pernicious anemia, 701
Peroxisomes, 68
Pes. *See* Foot
Pes planis, 220
Peyer patches, 636, 690, 692
Phagocytosis, 58, 59*f*
Phagosome, 58, 59*f*
Phallus, 751, 753, 754*f*
Pharyngeal arches, 13, 13*f*, 113, 113*f*, 114*f*, 196, 621
Pharyngeal pouches, 13, 113–114, 113*f*, 504, 505*f*, 532
Pharynx, 40*f*, 654*f*, 655, 656*f*, 657, 684
Phenylephrine, 466*i*
Pheochromocytoma, 534
Pheomelanin, 135, 139, 139*f*
Pheromones, 142
Philtrum, 351*f*, 657*f*
Phlebitis, 624*t*
Phospholipids, 53, 54*f*, 55*f*, 56*f*, 664, 695
Photoaging, 145
Photopsins, 500
Photopupillary reflex, 468*f*, 496, 504
Photoreceptors, 476, 499–500
Physical therapy, 234, 250, 256, 350
Physiology, comparative, 716*i*
Pia mater, 387, 388*f*, 414, 417, 418*f*, 506
Pigment epithelium, retinal, 499, 501*f*
Pilocarpine, 466*i*
Piloerector muscle, 15*i*, 131*f*, 138*f*, 139, 275, 468
Pineal gland. *See* Endocrine system
Pinocytosis, 58, 59*f*
Pitcher's arm, 345*t*
Pituitary dwarfism and gigantism, 536*t*
Pituitary gland. *See* Endocrine system, pineal gland
Pivot joints, 234*t*
Placenta, 105*f*, 118*f*, 119*f*
  circulation, 116, 117*f*, 623*f*
  development and anatomy, 115, 116, 117*f*
  endocrine role, 118*t*, 530, 531*t*
  excretory role, 118, 118*t*
  nutritive role, 116, 116*f*, 118*t*
Placenta previa, 118*i*, 758*t*
Planes
  anatomical, 24–25, 25*f*
  histological, 78, 79*f*
Plantar fascia, 343*f*
Plantar flexion, 238, 238*f*, 246
Plantar skin, 84*f*
Plantar surface, 27*f*
Plaque, atherosclerotic, 572*i*
Plasma, blood, 92, 542, 543–544, 543*f*, 544*t*
Plasma cells, 86, 544, 551, 635, 644, 645*f*
Plasma membrane, 51, 52–56, 54*f*, 56*f*
Plasmin, 554–555
Platelet count, 544*t*, 552
Platelet plug, 554
Platelets, 93, 93*f*, 541*f*, 543*f*, 552, 554–555, 554*f*, 555*f*
Pleasure, 432, 437, 439*t*
Pleurae, 29, 30*f*, 563*f*, 665, 665*f*, 666*i*
Pleural cavity, 29, 29*f*, 30*f*, 38*f*, 41*f*, 665, 665*f*
Pleural fluid, 29, 665
Pleurisy, 665
Plexus
  abdominal aortic, 461, 461*f*, 463, 464*f*, 469
  Auerbach (*See* Plexus, myenteric)
  brachial, 397, 397*f*, 401*f*, 401*t*, 402*f*, 402*t*
  cardiac, 463, 464*f*
  cervical, 397, 397*f*, 400*f*, 400*t*
  choroid, 417, 419*f*, 420, 420*f*
  coccygeal, 397, 404*f*, 404*t*
  esophageal, 463, 464*f*
  inferior hypogastric, 463, 464*f*, 677
  lumbar, 397, 397*f*, 403*f*, 403*t*
  Meissner (*See* Plexus, submucosal)
  myenteric, 455*f*, 678, 678*f*, 692
  pampiniform, 732
  pulmonary, 463, 464*f*
  sacral, 397, 397*f*, 403*f*, 404*t*
  solar, 461
  submucosal, 678, 678*f*
Plural suffixes, 19*t*
Pluripotent stem cells (PPSCs), 547, 547*f*, 552, 553*f*, 554
Pneumocystis, 648
Pneumonia, 166, 449, 669, 670*t*
Pneumotaxic center, 667, 667*f*
Pneumothorax, 666*i*
Podocytes, 707*f*, 713–714, 713*f*, 715*f*
Poikilocytosis, 556
Poisoning, 557
Polar bodies, 109*f*, 742, 743*f*
Poliomyelitis, 58, 393*i*, 557
Pollex, 357*f*. *See* Thumb
Pollutants and sexual abnormalities, 739*i*
Polycythemia, 555, 556
Polydactyly, 223, 223*f*
Polydipsia, 535
Polymastia, 142*i*
Polymorphonuclear leukocytes (PMNs), 549
Polyphagia, 535
Polyribosomes, 547
Polyspermy, 107
Polythelia, 142*i*
Polyuria, 535
Pons
  anatomy, 421, 422*f*
  and autonomic nervous system, 463, 468
  cranial nerves of, 421, 444*t*–446*t*, 463
  and digestion, 684, 685
  and equilibrium, 493
  functions, 421, 427*t*
  and hearing, 489
  location, 416*f*, 421, 423*f*
  and respiration, 667, 667*f*
Popliteal fossa, 335*t*, 358*f*
Popliteal region, 27*f*, 244
Porta hepatis, 695*f*, 696
Portal systems, 595, 595*f*
Positive end-expiratory pressure (PEEP), 669*i*
Positron emission tomography (PET), 5, 6*f*, 436, 437
Postcentral gyrus, 415*f*, 430*f*, 434, 435*f*, 481
Posterior chamber of eye, 497, 497*f*, 498*f*
Posterior commissure, 416*f*, 423*f*, 430
Posterior superior spine of ilium, 214, 214*f*, 354*f*
Postganglionic neuron, 457, 458*f*, 460*f*, 461*f*, 463, 465

Postsynaptic neuron, 375–376, 376*f*
Posture, 392, 423
Potential space, 665
Precapillary sphincters, 589, 592, 593*f*
Precentral gyrus, 415*f*, 430*f*, 434, 436*f*
Preeclampsia, 758*t*
Preembryonic stage, 108–110, 109*t*
Prefrontal cortex, 432, 433–434, 433*f*, 435*f*, 437, 438*i*
Preganglionic neuron, 457, 458*f*, 460*f*, 461*f*, 463, 465
Pregnancy. *See also* Gestation
  disorders, 111*i*, 114*i*, 118*i*, 758*t*
  ectopic, 111*i*, 122
  effects, 405*i*
Prehensile hand, 15, 15*f*. *See also* Thumb, opposable
Premenstrual syndrome (PMS), 523
Premolars, 681, 682*f*
Prenatal development, 106–124, 109*t*. *See also* Birth defects; *individual organ systems*
Prepuce
  of clitoris, 43*f*, 740*f*, 748, 749*f*
  of penis, 730*f*, 733*f*, 738
Presbyopia, 500*f*, 500*t*
Pressure, sense of, 389, 434, 477–478, 478*t*
Presynaptic neuron, 375–376, 376*f*
Primary cortex, 433–434
  auditory, 434, 435*f*, 489, 490*f*
  gustatory, 434, 435*f*, 482
  motor, 434, 435*f*
  olfactory, 434, 435*f*, 484
  regions, 435*f*
  somesthetic, 434, 435*f*, 479, 480, 480*f*
  visual, 434, 435*f*, 503*f*, 504
Primates, 13–14, 17*f*
Prime mover, 260, 320*t*
Primitive streak, 110, 111*f*
Primordial germ cells, 733
Principal cells, renal, 717
Procoagulants, 554
Progesterone, 528, 743, 747, 753
Programmed cell death. *See* Apoptosis
Projection pathways
  auditory, 489, 490*f*
  gustatory, 482, 483*f*
  olfactory, 484, 484*f*
  somesthetic, 389–390, 391*f*, 479–480, 480*f*
  vestibular, 492–493, 493*f*
  visual, 502–504, 503*f*
Projection tracts, 430, 431*f*
Prolactin-inhibiting hormone, 518, 519*f*, 520*t*
Prolactin (PRL), 518, 519*f*, 521*t*, 522
Pronation, 24, 24*f*, 238, 238*f*, 321*f*, 322*t*
Pronephros, 719, 720*f*
Pronuclei, 107, 109*f*
Pro-opiomelanocortin (POMC), 520
Propanolol, 466*i*
Prophase, 69, 69*f*, 70*f*, 734*f*
Proprioception, 389, 390*t*, 440, 479
Proprioceptors, 239, 424*f*, 476, 478, 478*t*
Prosencephalon, 379, 380*f*
Prostate gland. *See* Glands, prostate
Prostate specific antigen, 737*i*
Proteins
  of blood plasma, 543–544, 544*t*
  of connective tissue, 86–87
  of plasma membrane, 54–56, 54*f*, 56*f*
  synthesis, 65–67, 67*f*
  in urine, 714*i*
Proteoglycans, 87, 714
Protraction of joints, 235, 236*f*
Protuberance of a bone, 174*t*
Proximal convoluted tubule, 713*f*, 714–716
Pseudopods, 58, 59*f*, 635*f*
Psoriasis, 145*t*
Pterygoid plates, 177*f*, 293*f*
Pterygoid processes, 182*f*
Pubarche, 753
Puberty, 523*i*, 753
Pubic arch, 216*t*
Pubic hair, 138
Pubic symphysis, 43*f*, 214, 216*t*, 230*f*, 231, 729*f*, 730*f*, 740*f*
Pudendum (vulva), 747–749, 749*f*
Pulled elbow, 243*f*, 243*i*
Pulled groin, 345*t*
Pulled hamstrings, 345*t*
Pulmonary circuit, 562, 562*f*, 595, 596*f*
Pulmonary groove, 668
Pulmonary surfactant, 664
Pulmonary valve, 567, 568*f*, 571*f*
Pulmonary ventilation, 666–667
Pulp cavity, 682*f*, 683
Punishment, sense of, 437
Pupil, 468, 496, 497*f*, 500
Purkinje cells, 365*f*, 423, 425
Purkinje fibers, 561*f*, 574, 574*f*
Pus, 552
Putamen, 432, 433*f*
Pyelitis, 719*i*
Pyelonephritis, 719*i*
Pyloric glands, 687*f*, 688
Pyloric sphincter, 42*f*, 686, 686*f*
Pyramidal cells, 431, 432*f*, 434
Pyramids
  medullary, 392, 421, 422*f*, 431*f*, 434–435
  renal, 710, 711*f*

Q

Quadrants, abdominal, 26, 28*f*
Quadriplegia, 408
Quickening, 120*t*, 275

R

Rabies, 408*t*
Radioactivity, discovery of, 152*i*
Radiography, 3, 4*f*
Radiology, 3
Radioulnar joint, 231, 242, 242*f*
Radium, 152*i*
Rage, 437
Rami (nerve)
  communicating, 398*f*, 399*f*, 459, 459*f*, 461*f*
  dorsal and ventral, 396, 398*f*, 399*f*
Range of motion, 239, 260*i*
Raynaud disease, 470*t*, 624*t*
RBC count, 544*t*, 546, 556
Receptive field, 478, 479*f*, 502
Receptor-mediated endocytosis, 58, 59*f*
Receptors
  cell surface, 54–55, 56*f*
  for neurotransmitters, 267, 465–466, 466*f*
  sensory, 366, 406, 434, 476–478, 477*f*, 478*t* (*See also* Sense organs)
Rectum, 43*f*, 693*f*, 694
Rectus sheath, 35*f*, 300*f*, 300*t*, 301*f*
Red blood cells (RBCs). *See* Erythrocytes
Red muscles, 272
Referred pain, 480, 480*f*
Reflexes, 405–407
  somatic (spinal), 367, 406–407, 406*f*, 407*t*
  visceral (autonomic), 367, 406, 456, 457*f*
Regeneration, 100
  of muscular tissue, 273*t*, 274
  in nervous system, 374
Renal corpuscle, 710, 713–714, 713*f*
Renal failure, 721
Renal tubule, 710, 712*f*, 713*f*, 714–717
Renin, 530, 708, 716
Reposition of the thumb, 238, 238*f*
Representational hemisphere, 438
Reproductive system, 34*f*, 727–762. *See also specific organs*
  aging, 754–755
  disorders, 755–756, 756*t*, 757*i*, 758*t*
  embryonic development, 751–753, 752*f*, 755*f*
  female, 728*t*, 740–751, 740*f*, 741*f*, 745*f*, 749*f*
  male, 728*t*, 729–739, 729*f*, 730*f*, 733*f*
  at puberty, 753
Resolution
  of eye, 502, 503*f*
  of microscopes, 48, 49, 49*f*
Respiratory centers, 421, 666–667, 667*f*
Respiratory distress syndrome, 669*i*
Respiratory failure, 277
Respiratory system, 33*f*, 653–674, 654*f*. *See also* Breathing
  aging, 669
  anatomy, 653–665
  disorders, 661*i*, 666*i*, 668*i*, 669–670, 669*i*, 670*t*
  embryonic development, 668, 669*f*, 669*i*
  mucosa, 61*f*, 482, 655, 660*f*, 661*f*
Reticular activating system, 426
Reticular cells, 635, 636, 637*f*, 638
Reticular epithelial cells, 637, 638*f*
Reticular fibers, 87, 635, 640*f*
Reticular formation, 392, 424*f*, 425*f*, 426, 427*f*, 427*t*, 480*f*, 481, 484*f*, 666
Reticular tissue, 88, 89*f*, 89*t*
Reticulocyte, 547
Reticulospinal tract, 390*f*, 390*t*, 392, 423, 424*f*, 426
Retina, 497–502, 497*f*, 504
  embryonic development, 506*f*
  histology, 501*f*
  ophthalmoscopic view, 499*f*
Retraction of joints, 235, 236*f*
Retroperitoneal organs, 30, 31*f*, 680, 709, 709*f*
Reward, sense of, 437
Rhabdomyoma, 277*t*
Rhabdomyosarcoma, 277*t*
Rheumatic fever, 579*t*, 646
Rheumatoid arthritis (RA), 227*f*, 249, 250*f*, 646
Rheumatoid factor, 249
Rheumatology, 247
Rhodopsin, 499, 502
Rhombencephalon, 379, 380*f*
Ribosomes, 52*f*, 65, 67*f*
RICE treatment, 345
Rickets, 155, 167*t*
Rider's bones, 345*t*
Ringworm, 145*t*
Rods, retinal, 499, 501*f*, 502, 503*f*, 504–505
Root canal, 682*f*, 682*i*, 683
Rosacea, 145, 145*t*
Rostral direction, 414, 415*f*, 417*f*
Rotation of a joint, 237, 237*f*, 320*t*
Rotator cuff, 240, 241, 319*f*, 319*t*
Rubella, 122
Rubrospinal tract, 392, 424*f*, 426
Ruffini corpuscle, 477*f*, 478, 478*t*
Rugae
  of stomach, 686*f*
  of urinary bladder, 718, 718*f*
  of vagina, 740*f*

S

Saccule, 487*f*, 489–491, 492*f*
Sacral region, 27*f*, 193–194, 193*f*, 331*f*, 354*f*
Saddle joint, 233–234, 233*f*, 234*t*
Sagittal plane, 25, 25*f*
St. Martin, Alexis, 676
Salivary glands, 677*f*, 683–684, 684*f*
  innervation, 445*f*, 445*t*–446*t*, 446*f*, 460*f*, 464*f*, 467*t*, 677
Salivation, 421, 467, 468, 482, 484, 683–684
Sarcolemma, 263, 264*f*
Sarcomere, 265, 265*f*, 266*t*, 276
Sarcoplasmic reticulum
  in cardiac muscle, 273, 273*t*, 576
  role in muscle contraction, 269, 270*f*
  in skeletal muscle, 263, 264*f*, 266*t*, 273*t*
  in smooth muscle, 273*t*, 274
Satellite cells
  of muscle, 263, 273, 275, 576
  of nervous system, 371, 372*t*
Satiety, 428
Scalp, 138
Scanning electron microscope (SEM), 48–49, 49*f*
Scapular region
  muscular anatomy, 303*f*, 305*f*, 314*t*–315*t*, 315*f*, 317*f*, 318*f*, 319*f*
  skeletal anatomy, 208–209, 209*f*
  surface anatomy, 27*f*, 353*f*
Scar tissue, 100, 273, 371
Schizophrenia, 6*f*, 450*t*
Schleiden, Matthias, 10
Schwann, Theodor, 10
Schwann cells, 369*f*, 371, 372, 372*t*, 373*f*, 374, 375*f*, 379, 478
Sciatica, 405*i*
Sclera, 494*f*, 495, 497*f*, 506
Sclerosis, 371
Sclerotomes, 114, 198, 199*f*
Scoliosis, 189*f*, 189*i*
Scrotum, 27*f*, 36*f*, 37*f*, 43*f*, 730*f*, 732, 733*f*
  embryonic development, 753, 754*f*
Scurvy, 87*i*
Seasonal affective disorder (SAD), 523
Sebaceous glands, 131*f*, 138*f*, 141*f*, 142, 142*t*, 144, 144*f*
Sebum, 142
Second messengers, 515
Second-order neuron, 389, 390, 391*f*, 479, 480*f*
Secretory vesicles, 52*f*, 58, 67, 67*f*
Sections, histological, 78, 79*f*
Selection pressure, 14
Sella turcica, 176*f*, 177*f*, 182*f*, 183, 184*f*, 518
Semen, 719, 737, 738*t*, 739*i*
Semicircular canals, 492
Semicircular ducts, 485*f*, 487*f*, 489, 492, 492*f*
Semilunar valves, 567, 568*f*, 569*f*, 570*f*, 571*f*
Seminal vesicles. *See* Glands, seminal vesicles
Seminiferous tubules, 529*f*, 729, 731*f*, 732

Senile plaque, 449–450, 450*f*
Sense organs, 475–511
- classification, 476
- disorders, 500*f*, 500*t*, 506, 507*t*
- embryonic development, 504–506, 505*f*, 506*f*
- for equilibrium, 489–493, 491*f*, 492*f*
- for general (somesthetic) senses, 476–481, 477*f*, 478*t*
- for hearing, 484–489, 485*f*, 487*f*, 488*f*
- for smell, 482–484, 483*f*
- for taste, 481–482, 482*f*
- for vision, 494–504, 494*f*, 495*f*, 496*f*, 497*f*, 498*f*, 499*f*, 501*f*, 503*f*

Sensory deprivation, 476
Sensory division of nervous system, 366
Sensory homunculus, 434, 435*f*
Sensory nerves, 395, 440
Septal defects of heart, 579*t*
Septicemia, 556*t*
Septum
- of an exocrine gland, 97, 97*f*
- interatrial, 567, 578, 578*f*
- interventricular, 567, 577*f*, 578
- nasal, 176*f*, 183, 185, 655

Septum pellucidum, 419*f*
Series-elastic components, 257
Serosa (serous membrane), 29, 30, 31*f*, 98, 676, 677, 678*f*, 686*f*
Serotonin, 376, 516*f*, 523, 554, 688*t*
Serous fluid, 665
Serous membrane. *See* Serosa
Sertoli (sustentacular) cells, 528, 729, 731*f*, 735*f*
Serum, blood, 543
Severe combined immunodeficiency disease (SCID), 647
Sex characteristics, secondary, 728
Sex organs, primary and secondary, 728
Sex steroids. *See* Estrogen; Testosterone
Sexual dimorphism, pelvic, 215, 216*f*, 216*t*
Sexual drive, 428
Sexuality, brain centers for, 437
Sexually transmitted diseases (STDs), 756
Sexual reproduction, 728
Shaken child syndrome, 418
Shapes of cells, 50, 50*f*
Shingles, 395*i*
Shinsplints, 345*t*
Shoulder joint
- anatomy, 240–241, 241*f*
- muscles acting on, 314*t*–315*t*, 315*f*, 316*t*–319*t*, 317*f*, 318*f*, 319*f*, 320*t*

Shunts, circulatory, 595, 595*f*, 622
Sickle-cell disease, 555, 556*i*, 557*f*
Sinoatrial (SA) node, 274, 570, 574, 574*f*, 576, 579
Sinuses
- carotid, 589*f*, 599*f*
- cavernous, 611*f*
- coronary, 572, 572*f*
- dural, 417
- ethmoid, 175, 178*f*, 183*f*
- frontal, 40*f*, 175, 176*f*, 178*f*, 180, 184*f*, 656*f*
- inferior sagittal, 611*f*
- maxillary, 175, 178*f*
- paranasal, 175
- scleral venous, 497
- sphenoid, 175, 176*f*, 178*f*, 184*f*, 656*f*
- splenic, 642
- straight, 611*f*
- subcapsular, of lymph node, 637, 640*f*
- superior sagittal, 417, 418, 418*f*, 420*f*, 611*f*
- transverse, 417, 611*f*

Sinusoids, 591, 637*f*
- of bone marrow, 636, 637*f*
- hepatic, 591, 592*f*, 617*t*, 679, 695, 697, 697*f*

Sinus venosus, 577–578
Situs inversus, 6*i*
Size limits of cells, 50, 51*f*
Skeletal muscle. *See* Muscular tissue, skeletal
Skeletal muscle pump, 594, 594*f*, 633
Skeletal system, 32*f*, 151–254, 173*f*
- disorders
  - amelia and meromelia, 223
  - arthritis, 247, 249–250, 250*f*
  - cleft palate, 201*t*
  - clubfoot (talipes), 223, 223*f*
  - craniosynostosis, 201*t*
  - dislocations, 250*t*
  - dwarfism, 163*f*, 163*i*
  - elbow dislocation, 243*f*, 243*i*
  - fractures, 164–165, 164*f*, 165*f*, 165*i*, 165*t*, 201–202, 201*f*, 202*f*, 208*i*, 218*f*, 218*i*, 224*t*
  - gout, 250*t*
  - heel spurs, 224*t*
  - herniated discs, 202, 202*f*
  - jaw dislocation, 240
  - knee injuries, 247*f*, 247*i*
  - osteitis deformans, 167*t*
  - osteomalacia, 167*t*
  - osteoporosis, 166–167, 166*f*
  - Paget disease, 167*t*
  - polydactyly and syndactyly, 223, 223*f*
  - rheumatism, 246–247
  - rickets, 155, 167*t*
  - shoulder dislocation, 240–241
  - spinal stenosis, 201*t*
  - spondylosis, 201*t*
  - strain, 250*t*
  - subluxation, 250*t*
  - synovitis, 250*t*
  - TMJ syndrome, 239*i*
- embryonic development, 196–200, 198*f*, 199*f*, 200*f*, 221–222 (*See also* Ossification)
- general anatomy of bones, 153–154, 153*f*, 154*f*, 174, 174*f*, 174*t*
- joints, 227–254
- lower limb, 215–220, 242–246
- osseous tissue, 151–170
- pectoral girdle, 208–209
- pelvic girdle, 213–215
- skull, 175–187
- thoracic cage, 194–196
- upper limb, 209–212, 240–242
- vertebral column, 188–194

Skin, 130–137. *See also* Integumentary system
- aging, 145
- cancer, 146, 146*f*
- disorders, 145, 145*t*
- embryonic development, 143, 143*f*
- endocrine role, 529, 531*t*
- glands, 140–142, 141*f*, 144–145, 144*f*
- grafts, 147*i*
- incisions, 134*i*
- layers, 130, 131*f*, 132–134, 135*t*
- sensory receptors, 476–478, 477*f*, 478*t*
- structure, 130–135, 131*f*, 134*f*, 137

Skull, 175–186, 175*f*, 176*f*, 177*f*, 178*f*, 180*f*, 187*t*
- bones associated with, 186, 186*f*
- cranial bones, 175–183, 180*f*, 181*f*, 182*f*, 183*f*, 184*f*
- facial bones, 184–186, 185*f*
- fetal, 198*f*

Sleep, 421, 426, 428
Sleep apnea, 670*t*
Sliding filament theory, 269
Slow oxidative muscle fibers, 271–272, 272*f*, 272*t*
Small intestine, 42*f*, 530, 677*f*, 689–692, 690*f*
Smears, histological, 78
Smegma, 738
Smell, 430, 432, 482–484, 483*f*, 484*f*
Smoking, 167, 498*i*, 670, 689*i*, 755, 757*i*
Smooth muscle. *See* Muscular tissue, smooth
Smooth muscle tone, 275
Sneezing, 421, 484, 666, 667
Social behavior, brain and, 438, 438*i*
Sodium-potassium pump, 58
Somatomedin, 522
Somatosensory neurons, 396*f*
Somatosensory senses. *See* Somesthetic (general) senses
Somatostatin, 516*t*, 519*f*, 520*t*, 527, 688*t*
Somatotopy, 434
Somatotropes, 522
Somatotropin. *See* Growth hormone
Somesthetic association area, 434, 435*f*
Somesthetic (general) senses, 397, 430, 434, 476–481, 477*f*, 478*t*, 479*f*, 480*f*
- association cortex for, 434, 435*f*
- primary cerebral cortex for, 434, 435*f*

Somites, 114, 114*f*, 275
Sonography, 3, 4*f*, 6*i*, 118*i*
Special senses, 430, 433–434, 435*f*, 476. *See also* Sense organs
Specific immunity, 636
Spectrin, 545
Speech, 430, 658–659, 666
- cerebral areas for, 435*f*, 436–437, 439*f*
- medullary nuclei for, 421
- and Parkinson disease, 450

Sperm, 107*f*, 728, 733, 735*f*, 736, 736*f*
Spermatic cord, 731*f*, 732, 733*f*, 755*f*
Spermatids, 733, 735*f*
Spermatocytes, 733, 735*f*
Spermatogenesis, 106, 732–733, 735*f*
Spermatogonia, 733, 735*f*
Sperm count, 737, 739*i*
Spermiogenesis, 733, 735*f*
Sphincters, 95, 256, 259
- anal, 307*t*, 693*f*, 694
- esophageal, 684, 685*i*
- hepatopancreatic, 698, 698*f*, 699
- ileocecal, 690, 693*f*
- of Oddi, 698, 698*f*, 699
- precapillary, 589, 592, 593*f*
- pyloric, 686, 686*f*
- urethral, 307*t*, 718*f*, 719

Spina bifida, 380, 381*f*
Spinach, effect on bone healing, 165*i*
Spinal cord, 40*f*, 41*f*, 42*f*, 385–393, 398*f*
- cross-sectional anatomy, 387–389, 388*f*
- dimensions and extent, 379, 386
- diseases, 393*i*, 408*t*
- functions, 386
- gross anatomy, 386, 387*f*
- histology, 94*f*, 387–389
- injury, 202, 202*f*, 407–408
- meninges, 386–387
- tracts, 389–392, 390*f*, 390*t*, 391*f*, 392*f*

Spinal curvatures, 189*f*, 199, 200*f*
Spinal meningitis, 408*t*
Spinal nerves, 379, 387*f*, 389, 393–405, 394*f*, 396*f*, 397*f*, 399*f*
- dermatomes and, 405, 405*f*
- disorders, 395*i*, 405*i*, 408*t*
- plexuses, 397, 400*f*, 400*t*–404*t*, 401*f*, 402*f*, 403*f*, 404*f*
- relationship to spinal cord, 388*f*, 395, 398*f*

Spinal shock, 407
Spinal stenosis, 201*t*
Spinal tap, 388*i*, 417*i*
Spinal tracts. *See* Spinal cord, tracts
Spindle fibers, mitotic, 69, 70*f*
Spinocerebellar tract, 390, 390*f*, 390*t*, 424*f*, 425, 479
Spinoreticular tract, 480*f*
Spinothalamic tract, 389, 390*f*, 390*t*, 391*f*, 434, 479, 480*f*
Spiral ganglion, 488*f*, 489
Spleen, 38*f*, 642–643, 643*f*, 643*i*
- B cells in, 644
- embryonic development, 646
- erythrocyte disposal, 547
- hemopoietic role, 546
- platelets in, 554

Splenectomy, 643*i*
Splenomegaly, 648*t*
Spondylosis, 201*t*
Spongiocytes, 527
Spontaneous abortion, 758*t*
Sports medicine, 256
Sprains, 239, 246
Spreads, histological, 78
Squamous cell carcinoma, 146, 146*f*
SRY gene, 751
Stab cells, 549
Stains, histological, 78
Stellate cells, 431, 432*f*
Stem cells, 72, 688
- epidermal, 132
- pluripotent (PPSCs), 547, 547*f*, 552, 553*f*, 554, 644
- of yolk sac, 546, 733, 742

Stereocilia, 487, 489, 491
Stereoscopic vision, 15, 16*f*
Sterility, 737
Sternal bars, 200
Sternal region, 27*f*
Steroids
- as hormones, 514, 515, 515*f*, 516*t*
- in treatment of arthritis, 249–250

Stimulus modality, 476
Stomach, 42*f*, 530, 677*f*, 679*f*, 685–689
- anatomical variation, 7*f*
- gross anatomy, 685–686, 686*f*
- histology, 686–688, 687*f*
- secretions, 688, 688*t*

Stone-dissolving drugs, 698*i*, 718*i*
Strain of a joint, 250*t*
Streptococcus, 640
Stretch
- receptors, 478, 478*t*, 667
- sense of, 478

Stretch marks, 135
Stretch reflex, 406*f*, 407*t*
Striae, 135
Striations, 95, 95*f*, 256, 256*f*, 264–265, 265*f*, 266*t*, 273
Stride, 215, 220
Stroke, 624*t*
Stroma, 97
Styloid process
- of fibula, 218, 219*f*

of radius, 210, 211*f*, 356*f*, 357*f*
of temporal bone, 176*f*, 180, 181*f*
of ulna, 210, 211*f*, 356*f*, 357*f*
Subarachnoid space, 387, 388*f*, 417, 418*f*, 420*f*
Subclavian trunk, lymphatic, 634*f*
Subcutaneous tissue, 131*f*, 135, 135*f*
Subdural space, 417, 418*f*
Subluxation, 250*t*
Submucosa, 676, 677, 678*f*, 687*f*
Submucosal plexus, 678, 678*f*
Substantia nigra, 425*f*, 426, 450
Suffocation, 526*i*
Sulci
of bones, 174*t*
cerebellar, 423, 423*f*
cerebral, 419*f*, 429, 430, 430*f*
of heart, 564*f*, 567
mentolabial, 351*f*
Sunburn, 146
Sunscreens, 146
Suntan, 135
Superciliary ridge, 351*f*
Superficial perineal space, 307*t*, 308*f*
Superoxide anion, 550
Supination, 24, 24*f*, 238, 238*f*, 321*f*, 322*t*
Suppressor T cells, 635
Supraorbital margin, 175*f*, 179
Surface area of cells, 50, 51*f*
Surfactant, pulmonary, 664, 664*f*, 669*i*
Sustentacular (Sertoli) cells, 528, 729, 731*f*, 732, 733, 735*f*
Sutural bone, 172, 178*f*
Sutures, 175
classification, 228, 229*f*, 234*t*
coronal, 176*f*, 178*f*, 179, 180, 198*f*, 228
lambdoid, 176*f*, 177*f*, 178*f*, 180, 198*f*, 228
metopic, 197
sagittal, 178*f*, 180, 198*f*, 228
squamous, 176*f*, 180, 198*f*, 228
Swallowing, 421, 685
Swallowing center, 685
Sweat glands, 97, 140–142, 141*f*, 142*t*, 468
Sweating, 421
Sweat pores, 131*f*, 140
Sympathetic chain, 399*f*, 458–459, 459*f*, 460*f*, 461*f*
Sympathetic nervous system, 367, 457, 458–463
and adrenal gland, 526
efferent pathways, 458–461, 460*f*, 461*f*
embryonic development, 379
and heart, 576
innervation of penis, 739
in old age, 276
respiratory effects, 663
Sympathetic tone, 457, 468
Sympathoadrenal system, 463
Sympatholytics, 466*i*
Sympathomimetics, 466*i*
Symphyses, 230*f*, 231, 234*t*
Synapses, 266, 368, 374–376, 375*f*
electrical, 376
neuromuscular, 266–267, 266*f*, 267*f*, 268*f*
neurotransmitters, 376
signal transmission, 276, 376, 449
structure, 376, 377*f*
types, 375, 376*f*
Synaptic cleft, 266, 267*f*, 376, 377*f*
Synaptic knob, 266, 267*f*, 368, 375*f*, 376, 377*f*
Synaptic vesicles, 266, 267*f*, 376, 377*f*
Synarthroses, 228–230, 229*f*, 234*t*
Synchondroses, 230, 230*f*, 234*t*
Syndactyly, 223
Syndesmoses, 229*f*, 230, 234*t*
Synergists (muscles), 320*t*
Synostoses, 228, 234*t*
Synovial fluid, 231, 231*f*, 249
Synovial joints, 231–251, 249*t*
artificial, 250, 251*f*
classification, 233–234, 233*f*, 234*t*
coxal, 242–243, 244*f*
diseases, 247, 249–250, 250*f*, 250*t*
of elbow, 241–242, 242*f*, 243*f*, 243*i*
exercise, 232*i*
general anatomy, 231–232, 231*f*, 232*f*
humeroscapular, 240–241, 241*f*
knee, 243–246, 245*f*, 246*f*, 247*f*, 247*i*
movements, 234–239, 235*f*, 236*f*, 237*f*, 238*f*
range of motion, 239
temporomandibular, 239–240, 240*f*
Synovitis, 250*t*
Syphilis, 122, 588*i*, 756
Systemic circuit, 562, 562*f*
Systemic lupus erythematosus (SLE), 646

T

Tactile cells, 131*f*, 132
Tactile corpuscles, 131*f*, 477–478, 477*f*, 478*t*
Tactile discs, 477, 477*f*, 478*t*
Tagamet, 689*i*
Tail, embryonic, 13*f*, 114*f*
Talipes, 124*t*, 223, 223*f*
Target cells, 514, 515
Tarsal bones, 218–220, 219*f*, 221*f*, 222. *See also specific bones*
Tarsal plate, 494, 494*f*, 495*f*
Tarsal region, 26, 27*f*, 215, 359*f*, 360*f*, 361*f*, 362*f*, 363*f*
Taste, 421, 430, 481–482, 482*f*, 483*f*
Taste buds, 481, 482*f*
Tay-Sachs disease, 373*i*
T cells. *See* T lymphocytes
Tectospinal tract, 390*f*, 390*t*, 392
Tectum, 392, 425, 425*f*
Teeth, 13, 229, 681–683, 682*f*, 683*f*
Tegmentum, 425*f*, 426
Telencephalon, 379, 380*f*
Telophase, 69, 69*f*, 70*f*, 734*f*
Temperature. *See also* Thermoregulation
receptors, 130, 476, 478*t*, 738
sense, 389, 434, 476
and spinothalamic tract, 389, 390*t*
of testes, 732
Temporal lobe, 415*f*, 416*f*, 419*f*, 430, 430*f*, 435*f*, 484, 489
Temporal region, 176*f*, 287*f*, 290*f*, 351*f*
Temporomandibular joint (TMJ), 231, 239–240, 240*f*
TMJ syndrome, 239*i*
Tendinitis, 232
Tendinous cords, 561*f*, 566*f*, 567–568, 568*f*, 569*f*, 578
Tendons, 87*f*, 88, 232, 257, 258*f*
of biceps brachii, 240, 317*f*
calcaneal (Achilles), 220, 246, 286*f*, 338*f*, 361*f*, 362*f*
of foot, 343*f*
of forearm muscles, 323*f*, 326*f*, 328*f*
histology, 90*f*
of quadriceps femoris, 216, 332*f*, 334*f*, 358*f*
sensory nerve endings, 239
Tendon sheaths, 232, 232*f*
Tennis elbow, 345*t*
Tennis leg, 345*t*
Tension lines, 134*i*
Tentorium cerebelli, 417
Teratogens, 122, 223
Teratology, 120
Terminal cisternae, 263, 264*f*, 266*t*, 269
Terminal ganglia, parasympathetic, 463
Terminal hair, 138
Terminal web, 52*f*, 59, 63
Terminologia Anatomica, 17
Testes, 27*f*, 37*f*, 43*f*
anatomy, 528, 529*f*, 729–732, 730*f*, 731*f*
descent, 120*t*, 753, 755*f*
diseases, 755, 756*t*
embryonic development, 752–753, 752*f*
endocrine role, 528, 531*t*
sperm production, 732–733, 735*f*
Testis-determining factor, 751
Testosterone, 161, 516*f*, 527, 752, 753, 754, 755
and aging, 532
sources, 729
Tetanus (lockjaw), 267*i*
Tetraiodothyronine (T_4, thyroxine), 515*f*, 524
Texture, sense of, 477
Thalamus
anatomy, 422*f*, 428
and cerebellum, 424*f*
functions, 428, 429*f*, 439*t*
location, 416*f*, 419*f*, 422*f*, 428, 433*f*
nuclei, 429*f*
sensory pathways through, 389, 391*f*, 434
equilibrium, 493, 493*f*
hearing, 489, 490*f*
smell, 434, 484
somesthetic senses, 479–480, 480*f*
taste, 482, 483*f*
vision, 502, 503*f*
Thalassemia, 556
Thalidomide, 122, 124*f*, 223
Theca folliculi, 742, 744*f*
Thelarche, 753
Thenar eminence, 327*t*, 328*f*, 356*f*, 357*f*
Thermoreceptors, 130, 476, 478*t*, 738
Thermoregulation, 130, 145, 428
Thick filaments. *See* Myofilaments
Thigh. *See* Femoral region
Thin filaments. *See* Myofilaments
Third-order neuron, 389, 390, 391*f*, 479
Thirst, 428, 721
Thoracic cage, 171*f*, 194–195, 194*f*, 197*t*
Thoracic cavity, 29, 29*f*, 41*f*
Thoracic duct, lymphatic, 630*f*, 632, 634*f*
Thoracic region
arteries, 598*f*, 598*t*, 602*t*–603*t*, 603*f*
muscular anatomy, 302*f*, 303*f*, 304*f*, 305*f*
skeletal anatomy, 194–195, 194*f*
surface anatomy, 352*f*, 353*f*
veins, 614*f*, 614*t*
Thoracolumbar fascia, 303*f*, 305*f*
Thoroughfare channel, 592, 593*f*
Thrombocytopenia, 557, 643*i*
Thrombopoiesis, 554, 554*f*
Thrombopoietin, 554
Thrombosis, 555, 556*t*
Thumb, 356*f*, 357*f*
movements, 238, 238*f*, 329*t*
muscles acting on, 323*f*, 324*t*, 325*f*, 326*f*, 327*t*, 328*f*, 329*t*
opposable, 14, 15, 15*f*, 233
phalanges, 211, 212*f*
saddle joint, 233, 233*f*
Thymus. *See* Endocrine system, thymus
Thyroglobulin, 525
Thyroid cartilage, 657, 658, 658*f*, 659*f*
Thyroid diverticulum, 532, 533*f*
Thyroid gland. *See* Endocrine system, thyroid gland
Thyroid hormone, 162, 515*f*, 524–525, 530*t*
Thyroid-stimulating hormone (TSH), 516*t*, 518, 519*f*, 522, 535
Thyrotropes, 522
Thyrotropin, 516*t*, 518, 522
Thyrotropin-releasing hormone (TRH), 516*t*, 519*f*
Thyroxine, 515*f*, 524–525
Tibial tuberosity, 217, 219*f*, 358*f*
Tic douloureux, 449*i*
Tickle, sense of, 389, 390*t*
Tidal volume, 669
Tight junctions, 62, 62*f*
in blood-brain barrier, 418
in blood capillaries, 590*f*, 591
in blood-testis barrier, 729
in kidney, 715
in stomach, 688
Time, sense of, 425
Tinnitus, 239*i*, 507*t*
Tissue, 6–7, 78, 79*t*
Tissue fluid, 51, 78, 632*f*
Titin, 263, 266*t*
T lymphocytes, 552, 553*f*, 634–635, 638*f*, 645*f*
TMJ syndrome, 239*i*
Toe-off, 220
Toes
embryonic development, 115*f*, 120*t*, 221
muscles acting on, 339*f*, 340*f*, 340*t*, 341*f*, 342*t*, 343*f*, 344*t*
skeletal anatomy of, 26, 219*f*, 220
surface anatomy of, 363*f*
Tongue
anatomy, 40*f*, 482*f*, 680–681, 681*f*
innervation, 448*f*
muscles, 292*f*, 292*t*
Tonsillectomy, 642
Tonsillitis, 640
Tonsils, 114, 546, 640, 642
lingual, 642*f*, 656*f*, 681*f*
palatine, 642*f*, 656*f*, 680*f*, 681*f*
pharyngeal, 642*f*, 656*f*
Tooth decay, 682*i*
Total hip replacement, 250
Touch, sense of
cerebral cortex and, 434
cranial nerves and, 421
receptors, 477–478, 477*f*, 478*t*
spinal tracts and, 389, 390*t*, 391*f*
Toxemia of pregnancy, 758*t*
Toxic goiter, 535
Toxins, neuromuscular, 267*i*
Toxoplasma, 123, 648
Trabeculae
of bone, 151*f*, 154*f*, 157*f*, 161, 161*f*
of penis, 738
Trabeculae carneae, 566*f*, 567, 568*f*, 578
Trachea
embryonic development, 668, 669*f*
epithelium, 61*f*, 660*f*, 661*f*

Trachea—*Cont.*
location and anatomy, 654*f*, 656*f*, 658*f*, 660, 660*f*
Tracheostomy, 661*i*
Traction, 165, 242
Tracts
of brain, 414, 430–431, 431*f*
of spinal cord, 389–392, 390*f*, 390*t*, 391*f*, 392*f*
Transcytosis, 592, 592*f*
Transmission electron microscope (TEM), 48, 49*f*
Transport proteins
of blood plasma, 514, 543–544, 544*t*
of plasma membrane, 55–56*f*, 57*f*, 58
Transport vesicles, 66, 67*f*
Transverse plane, 25, 25*f*
Transverse (T) tubules, 263, 264*f*, 266*t*, 269, 273*t*, 274, 576
Treponema pallidum, 756
Triangles
anal, 307*t*, 308*f*, 729*f*
cervical, 296*f*, 296*t*
femoral, 609, 609*f*
urogenital, 307*t*, 308*f*, 729*f*
Tricuspid valve, 567, 568*f*, 569*f*, 571*f*
Trigeminal neuralgia, 449*i*
Trigger zone, 374
Trigone of bladder, 718*f*, 719
Triiodothyronine (T_3), 524
Trimesters of pregnancy, 108
Triplo-X syndrome, 121, 121*f*
Trisomy-21. *See* Down syndrome
Trochanters, 174*f*, 174*t*, 215, 217*f*, 353*f*
Trochlea
of eye, 495, 496*f*
of humerus, 210, 210*f*, 241
Trophoblast, 109, 110*f*, 117*f*
Trophoblastic nutrition, 116, 116*f*
Tropomyosin, 263, 264, 265*f*, 266*t*, 269, 270*f*, 271*f*
Troponin, 263, 264, 265*f*, 266*t*, 269, 270*f*, 271*f*
Truncus arteriosus, 577, 577*f*, 578
Trunk of body
muscular anatomy, 298–308
skeletal anatomy, 188–195
surface anatomy, 26, 27*f*, 28*f*, 352*f*, 353*f*, 355*f*
Trunks
arterial
brachiocephalic, 39*f*, 597*f*, 598*f*, 598*t*, 599*f*
celiac, 597*f*, 604*f*, 604*t*, 605*f*, 606*f*, 678
costocervical, 599*f*
pulmonary, 564*f*
thoracoacromial, 603*f*, 603*t*
lymphatic, 632, 633*f*, 634*f*
nerve, 401*f*, 401*t*
lumbosacral, 403*f*, 404*f*
vagal, 463
Tryptophan, 514
Tubercle of a bone, 174*t*
Tuberculosis, 648, 669, 670*t*
Tuberosity, 174*t*
calcaneal, 219*f*
deltoid, 210, 210*f*
ischial, 214, 214*f*
radial, 210, 211*f*
tibial, 217, 219*f*
ulnar, 211*f*
Tubular reabsorption, 713
Tubular secretion, 713, 715–716
Tubulin, 63, 65*f*
Tufted cells, 483*f*, 484
Tumors, 71*i*
Tumor-suppressor genes, 71*i*
Turner syndrome, 121*f*, 122
Two-point touch threshold, 478, 479*f*
Tympanic cavity, 113*f*, 485*f*, 486
Tympanic membrane, 485*f*, 486, 487*f*
Tympanic reflex, 486, 489
Tyrosine, 136, 514, 525

U

Ulcer, peptic, 689*f*, 689*i*
Ulcerative colitis, 702*t*
Ultrasound, 3, 718*i*
Ultrastructure, 48
of a muscle fiber, 263–266, 264*f*, 265*f*, 266*t*
Ultraviolet radiation, 135, 145, 146, 498*i*
Umami, 481
Umbilical cord, 105*f*, 116, 117*f*, 118*f*, 119*f*, 622*f*, 623*f*
Umbilical hernia, 306*i*
Umbilical region, 27*f*, 28*f*, 301*f*, 302*f*
Unit membrane, 53, 54*f*
Unmyelinated nerve fibers, 374, 394*f*
Upper extremity. *See* Limbs, upper
Upper motor neurons, 392*f*, 434–435
Urachus, 720
Urea, 709
Uremia, 709
Ureter, 38*f*, 708*f*, 711*f*, 717–718, 718*f*
Ureteric bud, 719–720, 720*f*
Urethra, 708, 708*f*
female, 43*f*, 718*f*, 719, 740*f*, 749*f*
fetal, 720
male, 43*f*, 718*f*, 719, 730*f*, 738, 739*f*
Urinary bladder, 38*f*, 43*f*, 708*f*, 718–720, 718*f*, 730*f*, 740*f*
control, 421
Urinary system, 707–725. *See also* Kidneys; Ureter; Urethra; Urinary bladder
aging, 720–721
disorders, 718*i*, 719*i*, 721–722, 722*t*
embryonic development, 719–720, 720*f*, 721*i*, 752*f*
Urination, 468
Urine, 713, 713*f*, 720
Urobilinogen, 695
Urogenital diaphragm, 307*t*, 308*f*
Urogenital fold, 753, 754*f*
Urogenital sinus, 720, 752*f*
Urogenital system, 708
Urogenital triangle, 308*f*, 729*f*
Uterine (endometrial) glands, 746, 746*f*, 747, 748*f*
Uterine milk, 116
Uterine tube, 38*f*, 106, 740*f*, 741*f*, 743–745, 745*f*
embryonic development of, 752*f*, 753
Uterus, 38*f*, 43*f*, 116*f*, 740*f*, 745–747
embryonic development of, 752*f*, 753
gross anatomy, 741*f*, 745–746, 745*f*
in menstrual cycle, 746–747, 748*f*
Uterus masculinus, 753
Utricle, 487*f*, 489–491, 492*f*
Uvea, 495–496
Uvula, 656*f*, 680*f*

V

Vagal tone, 576
Vagal trunks, 463
Vagina, 43*f*, 740*f*, 741*f*, 745*f*, 747, 749*f*, 752*f*, 753
epithelium, 83*f*, 84*f*, 747
Valsalva maneuver, 654
Valves. *See also* Sphincters
cardiac, 566*f*, 567–568, 569*f*
ileocecal, 690, 693*f*
lymphatic, 631, 632*f*, 640*f*
rectal, 694
venous, 594, 594*f*
Varicella, 395*i*
Varicocele, 756*t*
Varicose veins, 594*i*
Varicosities, autonomic, 274, 274*f*
Vasa recta, 710, 711*f*, 712*f*
Vasa vasorum, 586
Vasculitis, 624*t*
Vasomotion
autonomic control, 468, 469*f*
vasoconstriction, 130, 468, 469*f*, 554, 586
vasodilation, 130, 468, 469*f*, 586
Vasomotor center, 421
Vasomotor tone, 468
Vasopressin, 522. *See also* Antidiuretic hormone
Veins, specific
accessory hemiazygos, 614*f*, 614*t*
anterior tibial, 610*f*, 618*f*, 618*t*, 619*f*
arcuate, of kidney, 710, 711*f*
ascending lumbar, 614*f*, 614*t*, 615*f*
axillary, 610*f*, 611*f*, 612*t*, 613*f*
azygos, 614*f*, 614*t*, 621
basilic, 610*f*, 613*f*, 613*t*
brachial, 37*f*, 610*f*, 612*t*, 613*f*
brachiocephalic, 37*f*, 610*f*, 611*f*, 612*t*
bronchial, 614*t*
cardiac, 610*f*
cardinal, 621, 622*f*
cavernous sinuses, 611*f*, 612*t*
cephalic, 35*f*, 37*f*, 356*f*, 610*f*, 613*f*, 613*t*
cystic, 616*f*, 617*t*
deep dorsal, of penis, 738
deep femoral, 618*f*
deep palmar venous arch, 612*t*, 613*f*
digital
of foot, 618*f*, 618*t*, 619*f*
of hand, 612*t*, 613*f*
dorsal metatarsal, 619*f*
dorsal pedal, 618*f*, 618*t*, 619*f*
dorsal venous arch, 613*t*, 618*f*, 619*f*, 619*t*
dural sinuses, 611*f*, 612*t*
esophageal, 614*t*
external jugular, 36*f*, 611*f*, 612*t*
facial, 611*f*, 612*t*
femoral, 35*f*, 36*f*, 334*f*, 609*f*, 610*f*, 618*f*, 619*f*, 619*t*, 639*f*
femoral circumflex, 618*f*
fibular, 618*f*, 618*t*
gastric, 616*f*, 617*t*
gastroepiploic, 616*f*, 617*f*
gonadal, 610*f*, 615*f*, 615*t*
great cardiac, 572, 572*f*
great saphenous, 35*f*, 618*f*, 619*f*, 619*t*, 639*f*
hemiazygos, 614*f*, 614*t*
hemorrhoidal, 694
hepatic, 610*f*, 615*f*, 615*t*, 616*f*, 617*f*, 697
hepatic portal, 616*f*, 617*f*, 617*t*, 679, 697, 697*f*
iliac
common, 610*f*, 614*f*, 615*f*, 618*f*, 619*f*, 619*t*, 621
external, 610*f*, 615*f*, 618*f*, 619*f*, 619*t*
internal, 615*f*, 618*f*, 619*f*, 619*t*
inferior mesenteric, 616*f*, 616*t*, 617*f*
inferior phrenic, 615*f*, 615*t*
inferior sagittal sinus, 611*f*, 612*t*
inferior vena cava, 38*f*, 42*f*, 564*f*, 568*f*, 571*f*, 610*f*, 614*f*, 615*f*, 615*t*, 617*f*, 621
intercostal, 614*f*, 614*t*
interlobar, of kidney, 710, 711*f*
interlobular, of kidney, 710, 711*f*, 712*f*
internal thoracic, 610*f*
jugular, external and internal, 36*f*, 41*f*, 610*f*, 611*f*, 611*t*–612*t*
lumbar, 614*f*, 614*t*, 615*f*, 615*t*
median antebrachial, 613*f*, 613*t*
median cubital, 610*f*, 613*f*, 613*t*
mediastinal, 614*t*
metacarpal, 612*t*, 613*f*
middle cardiac, 572, 572*f*
occipital, 611*f*
ovarian, 615*f*, 615*t*, 741*f*
palmar digital, 612*t*, 613*f*
palmar metacarpal, 612*t*, 613*f*
pancreatic, 616*t*
pericardial, 614*t*
plantar arch, 618*t*
plantar digital, 618*f*, 618*t*
plantar metatarsal, 618*f*
plantar venous arch, 618*f*, 618*t*, 619*f*
popliteal, 610*f*, 618*f*, 619*t*
posterior tibial, 610*f*, 618*f*, 618*t*, 619*f*
pulmonary, 564*f*, 568*f*, 571*f*, 595, 596*f*
radial, 610*f*, 612*t*, 613*f*
renal, 610*f*, 615*f*, 615*t*, 708*f*, 710, 711*f*
saphenous, great and small, 610*f*, 618*f*, 619*t*
splenic, 616*f*, 616*t*, 617*f*, 643*f*
subclavian, 37*f*, 41*f*, 610*f*, 611*f*, 612*t*, 613*f*, 633*f*
lymph drainage into, 632–633
superficial circumflex iliac, 618*f*
superficial epigastric, 618*f*
superficial temporal, 611*f*, 612*t*
superficial venous palmar arch, 612*t*, 613*f*
superior mesenteric, 616*f*, 616*t*, 617*f*
superior ophthalmic, 611*f*, 612*t*
superior sagittal sinus, 611*f*, 612*t*
superior thyroid, 611*f*
superior vena cava, 37*f*, 564*f*, 568*f*, 571*f*, 610*f*, 612*t*, 614*f*, 633*f*
suprarenal, 615*f*, 615*t*
testicular, 615*f*, 615*t*, 732
thebesian, 571
transverse sinus, 611*f*, 612*t*
ulnar, 610*f*, 612*t*, 613*f*
umbilical, 116, 622, 622*f*, 623*f*
vasa recta, of kidney, 710, 711*f*, 712*f*
vertebral, 611*f*, 612*t*
vitelline, 622*f*
Veins in general
definition, 586
of pulmonary circuit, 595, 596*f*

of systemic circuit, 610, 610*f*, 611*t*–619*t*, 620
types and structure, 587*f*, 593–594, 594*f*
Vellus, 138, 144
Venous return, 594, 594*f*
Venous sinuses, 594
Venous valves, 594, 594*f*
Ventral horn, 388, 388*f*
Ventral respiratory group, 667, 667*f*
Ventral root, 394*f*, 395, 396*f*, 398*f*
Ventricles
of brain, 379, 416*f*, 417–418, 419*f*, 420*f*, 423*f*
of heart, 566, 566*f*, 568*f*, 577–578
Ventricular fibrillation, 579*t*
Venules, 590*f*, 593, 593*f*
Vermis, 421, 423*f*
Vernix caseosa, 120*t*, 144
Vertebrae, 43*f*, 188–194
cervical, 188*f*, 189*f*, 190–192, 191*f*, 192*f*
coccygeal, 193*f*, 194
embryonic development, 198–199, 199*f*, 200*f*
general anatomy, 190*f*, 191*f*
lumbar, 188*f*, 191*f*, 192*f*, 193
in osteoporosis, 166*f*, 167*f*
sacral, 193–194, 193*f*
thoracic, 188*f*, 192–193, 192*f*, 196*f*
Vertebral canal, 29, 29*f*, 190
Vertebral column, 40*f*, 173*f*, 188–194, 188*f*, 189*f*, 197*t*. *See also* Vertebrae
curvatures, 188–189, 189*i*, 199, 200*f*
embryonic development, 198–199, 199*f*
growth, 379
Vertebral furrow, 353*f*
Vertebral region, 27*f*, 188, 303*f*, 304*f*, 305*f*
Vertebrata, 13
Vertex position, 120*t*, 727*f*
Vertigo, 239*i*
Vesalius, Andreas, 9–10, 10*f*, 562
Vesicular transport, 58, 59*f*
Vestibular bulbs, 748, 749*f*
Vestibular folds, 656*f*, 658, 658*f*
Vestibular system, 489–493, 491*f*, 492*f*
Vestibule
of ear, 446*f*, 486
of mouth, 680, 680*f*
of nose, 655, 656*f*
of vulva, 748, 749*f*
Vestibulosensory cortex, 493*f*
Vestibulospinal tract, 390*f*, 390*t*, 392, 423, 424*f*
Vestigial organs, 15*i*
Vibration, sense of, 389, 477, 478
Vibrissae, 655
Villi, 690–691, 691*f*, 692*f*
Viscera, 26
Visceral motor system, 456
Visceral muscle, 274. *See also* Muscular tissue, smooth
Visceral pain, 479–480
Visceral reflexes, 367, 406
Visceral senses, 430
Viscerocranium, 196
Vision, 425, 430, 434
defects, 500*t*
mechanism, 499–502
projection pathways, 503*f*
Visual association area, 435*f*
Vital capacity, 669
Vitamin A, 161
Vitamin B_{12}, 555, 701
Vitamin C, 87*i*, 161, 167
Vitamin D, 130, 161, 529, 696*t*, 708
Vitelline duct, 112*f*, 115, 700, 700*f*, 701*f*
Vitreous body, 497, 497*f*, 504
Vocal cords, 40*f*, 656*f*, 658–659, 658*f*, 659*f*
Voice box. *See* Larynx
Volar surface, 356*f*, 357*f*
Voluntary muscle, 256
Vomiting, 421, 437, 481, 484
Vulva, 747–749, 749*f*

W

Warm receptors, 476
Warm-up exercises, 345
Warts, 145*t*
Wastes, metabolic, 708
Water barrier, epidermal, 133
Water conservation, 716*i*
Wernicke area, 435*f*, 436, 438
Whiplash, 202, 202*f*
White blood cells (WBCs). *See* Leukocytes
White matter, 379
of brain, 414, 419*f*, 423*f*, 430
of spinal cord, 388–389, 388*f*
White muscles, 272
Windpipe. *See* Trachea
Wrist, 233–234, 233*f*
carpal tunnel, 326, 326*f*, 326*i*
muscles acting on, 322*t*–323*t*, 323*f*, 329*t*
retinacula, 324, 326*f*
skeletal anatomy, 210, 212*f*
surface anatomy, 356*f*, 357*f*
tendon sheaths, 232*f*
Wrist drop, 405*i*

X

X chromosome, 121–122, 121*f*, 728, 751–752
Xenografts, 147*i*
Xiphoid process, 194*f*, 195
X rays, 3, 4*f*, 122, 161*f*, 207*f*, 227*f*

Y

Y chromosome, 121–122, 728, 751
Yellow elastic tissue, 88
Yolk sac
angiogenesis in, 620, 620*f*, 622*f*
and digestive system embryology, 700, 700*f*
embryonic development, 111*f*, 112*f*, 115, 116*f*, 117*f*
germ cell development, 733, 742
hemopoietic role, 546, 620

Z

Zeiss, Carl, 10
Zona pellucida, 107, 107*f*, 742, 744*f*
Zygomatic arch, 177*f*, 180, 290*f*
Zygomatic process, 180, 181*f*
Zygote, 108, 109*t*, 728
Zymogens, 699

Lexicon of Biomedical Word Elements

a- no, not, without (atom, agranulocyte)
ab- away (abducens, abduction)
acetabulo- small cup (acetabulum)
acro- tip, extremity, peak (acromion, acromegaly, acrosome)
ad- to, toward, near (adsorption, adrenal)
adeno- gland (lymphadenitis, adenohypophysis, adenoids)
aero- air, oxygen (aerobic, anaerobe, aerophagy)
af- toward (afferent)
ag- together (agglutination)
-al pertaining to (parietal, pharyngeal, temporal)
ala- wing (ala nasi)
albi- white (albicans, linea alba, albino)
algi- pain (analgesic, myalgia)
aliment- nourishment (alimentary, hyperalimentation)
allo- other, different (allele, allosteric)
amphi- both, either (amphiphilic, amphiarthrosis)
an- without (anaerobic, anemic)
ana- **1.** up, build up (anabolic, anaphylaxis). **2.** apart (anaphase, anatomy). **3.** back (anastomosis)
andro- male (androgen)
angi- vessel (angiogram, angioplasty, hemangioma)
ante- before, in front (antebrachium)
antero- forward (anterior, anterograde)
anti- against (antidiuretic, antibody, antagonist)
apo- from, off, away, above (apocrine, aponeurosis)
arbor- tree (arboreal, arborization)
artic- **1.** joint (articulation). **2.** speech (articulate)
-ary pertaining to (axillary, coronary)
-ase enzyme (polymerase, kinase, amylase)
ast-, astro- star (aster, astrocyte)
-ata, -ate **1.** possessing (Chordata, corniculate). **2.** plural of *-a* (stomata, carcinomata)
athero- fat (atheroma, atherosclerosis)
atrio- entryway (atrium, atrioventricular)
auri- ear (auricle, auricular)
auto- self (autolysis, autoimmune)
axi- axis, straight line (axial, axoneme, axon)
baro- pressure (baroreceptor, hyperbaric)
bene- good, well (benign, beneficial)
bi- two (bipedal, biceps, bifid)
bili- bile (biliary, bilirubin)
bio- life, living (biology, biopsy, microbial)
blasto- precursor, bud, producer (fibroblast, osteoblast, blastomere)
brachi- arm (brachium, brachialis, antebrachium)
brady- slow (bradycardia, bradypnea)
bucco- cheek (buccal, buccinator)
burso- purse (bursa, bursitis)
calc- calcium, stone (calcified, calcaneus, hypocalcemia)
callo- thick (callus, callosum)
calori- heat (calorie, calorimetry, calorigenic)
calv-, calvari- bald, skull (calvaria)
calyx cup, vessel, chalice (glycocalyx, renal calyx)
capito- head (capitis, capitate, capitulum)
capni- smoke, carbon dioxide (hypocapnia)
carcino- cancer (carcinogen, carcinoma)
cardi- heart (cardiac, cardiology, pericardium)
carot- **1.** carrot (carotene). **2.** stupor (carotid)
carpo- wrist, seize (carpus, metacarpal)
case- cheese (caseosa, casein)
cata- down, break down (catabolism)
cauda- tail (cauda equina, caudate nucleus)
-cel little (pedicel)
celi- belly, abdomen (celiac)
centri- center, middle (centromere, centriole)
cephalo- head (cephalic, encephalitis)
cervi- neck, narrow part (cervix, cervical)
chiasm- cross, X (optic chiasm)
choano- funnel (choana)
chole- bile (cholecystokinin, cholelithotripsy)
chondro- **1.** grain (mitochondria). **2.** cartilage, gristle (chondrocyte, perichondrium)
chromo- color (dichromat, chromatin, cytochrome)
chrono- time (chronotropic, chronic)
cili- eyelash (cilium, supraciliary)
circ- about, around (circadian, circumduction)
cis- cut (incision, incisor)
cisterna- reservoir (cisterna chyli)
clast- break down, destroy (osteoclast)
clavi- hammer, club (clavicle, supraclavicular)
-cle little (tubercle, corpuscle)
cleido- clavicle (sternocleidomastoid)
cnemo- lower leg (gastrocnemius)
co- together (coenzyme, cotransport)
collo- **1.** hill (colliculus). **2.** glue (colloid, collagen)
contra- opposite (contralateral)
corni- horn (cornified, corniculate, cornu)
corono- crown (coronary, corona, coronal)
corpo- body (corpus luteum, corpora quadrigemina)
corti- bark, rind (cortex, cortical)
costa- rib (intercostal, subcostal)
coxa- hip (os coxae, coxal)
crani- helmet (cranium, epicranius)
cribri- sieve, strainer (cribriform, area cribrosa)
crino- separate, secrete (holocrine, endocrinology)
crista- crest (crista ampullaris, mitochondrial crista)
crito- to separate (hematocrit)
cruci- cross (cruciate ligament)
-cule, -culus small (canaliculus, trabecula, auricular)
cune- wedge (cuneiform, cuneatus)
cutane-, cuti- skin (subcutaneous, cuticle)
cysto- bladder (cystitis, cholecystectomy)
cyto- cell (cytology, cytokinesis, monocyte)
de- down (defecate, deglutition, dehydration)
demi- half (demifacet, demilune)
den-, denti- tooth (dentition, dens, dental)
dendro- tree, branch (dendrite, oligodendrocyte)
derma-, dermato- skin (ectoderm, dermatology, hypodermic)
desmo- band, bond, ligament (desmosome, syndesmosis)
dia- **1.** across, through, separate (diaphragm, dialysis). **2.** day (circadian)
dis- **1.** apart (dissect, dissociate). **2.** opposite, absence (disinfect, disability)
diure- pass through, urinate (diuretic, diuresis)
dorsi- back (dorsal, dorsum, latissimus dorsi)
duc- to carry (duct, adduction, abducens)
dys- bad, abnormal, painful (dyspnea, dystrophy)
e- out (ejaculate, eversion)
-eal pertaining to (hypophyseal, arboreal)
ec-, ecto- outside, out of, external (ectopic, ectoderm, splenectomy)
ef- out of (efferent, effusion)
-el, -elle small (fontanel, organelle, micelle)
electro- electricity (electrocardiogram, electrolyte)
em- in, within (embolism, embedded)
emesi-, emeti- vomiting (emetic, hyperemesis)
-emia blood condition (anemia, hypoxemia, hypovolemia)
en- in, into (enzyme, parenchyma)
encephalo- brain (encephalitis, telencephalon)
enchymo- poured in (mesenchyme, parenchyma)
endo- within, into, internal (endocrine, endocytosis)
entero- gut, intestine (mesentery, myenteric)
epi- upon, above (epidermis, epiphysis, epididymis)
ergo- work, energy, action (allergy, adrenergic)
eryth-, erythro- red (erythema, erythrocyte)
esthesio- sensation, feeling (anesthesia, somesthetic)
eu- good, true, normal, easy (eukaryote, eupnea, aneuploidy)
exo- out (exopeptidase, exocytosis, exocrine)
facili- easy (facilitated)
fasci- band, bundle (fascia, fascicle)
fenestr- window (fenestrated, fenestra vestibuli)
fer- to carry (efferent, uriniferous)
ferri- iron (ferritin, transferrin)
fibro- fiber (fibroblast, fibrosis)
fili- thread (myofilament, filiform)
flagello- whip (flagellum)
foli- leaf (folic acid, folia)
-form shape (cuneiform, fusiform)
fove- pit, depression (fovea)
funiculo- little rope, cord (funiculus)
fusi- **1.** spindle (fusiform). **2.** pour out (perfusion)
gamo- marriage, union (monogamy, gamete)
gastro- belly, stomach (digastric, gastrointestinal)
-gen, -genic, -genesis producing, giving rise to (pathogen, carcinogenic, glycogenesis)
genio- chin (geniohyoid, genioglossus)
germi- **1.** sprout, bud (germinal, germinativum). **2.** microbe (germicide)
gero- old age (progeria, geriatrics, gerontology)
gesto- **1.** to bear, carry (ingest). **2.** pregnancy (gestation, progesterone)
glia- glue (neuroglia, microglia)
globu- ball, sphere (globulin, hemoglobin)
glom- ball (glomerulus)
glosso- tongue (hypoglossal, glossopharyngeal)
glyco- sugar (glycogen, glycolysis, hypoglycemia)
gono- **1.** angle, corner (trigone). **2.** seed, sex cell, generation (gonad, oogonium, gonorrhea)
gradi- walk, step (retrograde, gradient)
-gram recording of (sonogram, electrocardiogram)
-graph recording instrument (sonograph, electrocardiograph)
-graphy recording process (sonography, radiography)
gravi- severe, heavy (gravid, myasthenia gravis)
gyro- turn, twist (gyrus)
hallu- great toe (hallux, hallucis)
hemi- half (hemidesmosome, hemisphere, hemiazygos)
-hemia blood condition (polycythemia)
hemo- blood (hemophilia, hemoglobin, hematology)
hetero- different, other, various (heterotrophic, heterozygous)
histo- tissue, web (histology, histone)
holo- whole, entire (holistic, holocrine)
homeo- constant, unchanging, uniform (homeostasis, homeothermic)
homo- same, alike (homologous, homozygous)
hyalo- clear, glassy (hyaline, hyaluronic acid)
hydro- water (dehydration, hydrolysis, hydrophobic)
hyper- above, above normal, excessive (hyperkalemia, hypertonic)
hypo- below, below normal, deficient (hypogastric, hyponatremia, hypophysis)
-ia condition (anemia, hypocalcemia, osteomalacia)
-ic pertaining to (isotonic, hemolytic, antigenic)
-icle, -icul small (ossicle, canaliculus, reticular)
ilia- flank, loin (ilium, iliac)
-illa, -illus little (bacillus)
-in protein (trypsin, fibrin, globulin)
infra- below (infraspinous, infrared)
ino- fiber (inotropic, inositol)
insulo- island (insula, insulin)
inter- between (intercellular, intercalated, intervertebral)
intra- within (intracellular, intraocular)
iono- ion (ionotropic, cationic)
ischi- to hold back (ischium, ischemia)
-ism **1.** process, state, condition (metabolism, rheumatism). **2.** doctrine, belief, theory (holism, reductionism, naturalism)
iso- same, equal (isometric, isotonic, isomer)
-issimus most, greatest (latissimus, longissimus)
-ite little (dendrite, somite)
-itis inflammation (dermatitis, gingivitis)